The night sky in MARCH

Astronomy
The Evolving Universe

Astronomy

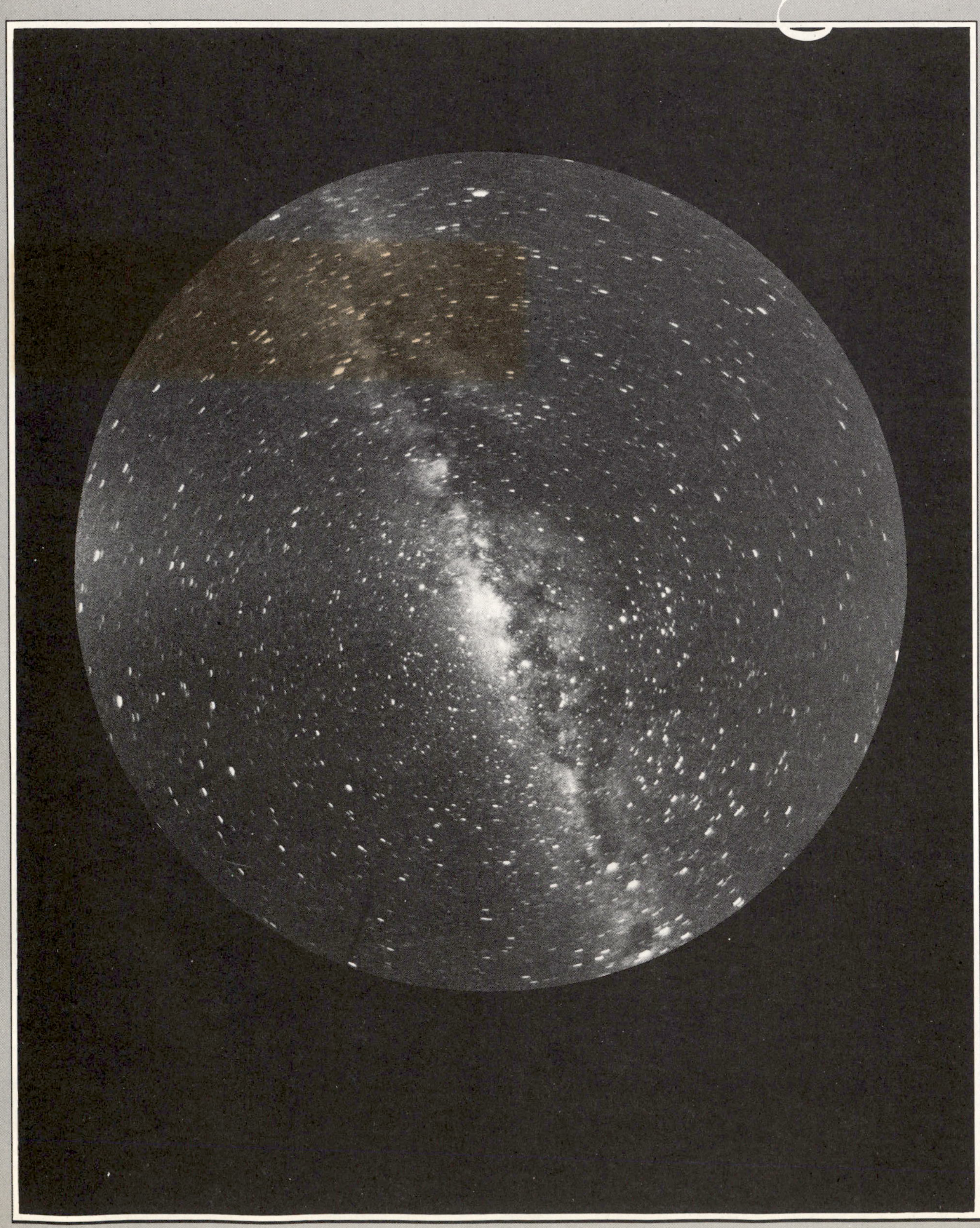

The Evolving Universe

Fourth Edition

Michael Zeilik
The University of New Mexico

PRESENTED TO ERINEO A. POLLACK
BY ALADDIN TRAVEL AGENCY
COMPLIMENTS OF:
DR. MICHAEL ZEILIK, APRIL 1986

Harper & Row, Publishers, New York
Cambridge, Philadelphia, San Francisco,
London, Mexico City, São Paulo,
Singapore, Sydney

To K^{2}L,

who keeps my feet on the ground when my head is in the stars

Sponsoring Editor: *Heidi Udell*
Project Editor: *Brigitte Pelner*
Text and Cover Design: *Gayle Jaeger*
Text Art: *J & R Services, Inc.*
Production Manager: *Jeanie Berke*
Compositor: *House of Equations Inc.*
Printer and Binder: *The Murray Printing Company*

Photo credits

Part 1: Coal Sack and Southern Cross / Part 2: NASA / Part 3: Kitt Peak National Observatory / Part 4: NASA / Chapter 1: Tom Gondola / Chapter 2:O. Gingerich / Chapter 3: Uraniborg, Tycho's Research Observatory / Chapter 4: Yerkes Observatory / Chapter 5: Hale Observatory / Chapter 6: Michael Zeilik, Haystack Observatory / Chapter 7: Mount Wilson and Palomar Observatories / Chapter 8: NASA / Chapter 9: Mount Wilson and Palomar Observatories / Chapter 10: Hale Observatory / Chapter 11: NASA / Chapter 12: Hale Observatory / Chapter 13: Institute for Astronomy, Haleakala Observatory / Chapter 14: F. Wright, Smithsonian Astrophysical Observatory / Chapter 15: Lick Observatory / Chapter 16: Mount Wilson and Palomar Observatories / Chapter 17: Steward Observatory / Chapter 18: Kitt Peak National Observatory / Chapter 19: Lick Observatory / Chapter 20: Kitt Peak National Observatory / Chapter 21: Dr. Harold E. Edgerton, Massachusetts Institute of Technology / Chapter 22: E. Barghoorn, The Biological Laboratories, Harvard University

The lines of poetry at the beginning of Chapter 19 are from *Prologue* by Edward Field. Reprinted by permission of Grove Press, Inc. Copyright © 1963 by Edward Field.

Figures 8.2, 8.3, and 8.4 are adapted from figures in *The Earth Sciences*, Second Edition, by A. N. Strahler, Harper & Row, 1971.

The song line at the beginning of Chapter 15 is from *Stardust*, lyrics by Mitchell Parish, music by Hoagy Carmichael, copyright © by Mills Music, Inc., used with permission of Mills Music.

About the cover

An eruption of Pele, the first volcano discovered on Io, moon of Jupiter. The plume from the eruption is visible above the limb of the moon. It rises to a height of around 300 km. Flows from past eruptions are visible around the complex of hills that are the source of the eruption. The intense volcanic activity on Io—at least eight active volcanoes have been seen on the surface—causes rapid changes in the surface features. Voyager 1 took the photos that were combined in this mosaic, which was made by Alfred S. McEwen of the U. S. Geological Survey. Photo courtesy of the Jet Propulsion Laboratory.

Library of Congress Cataloging in Publication Data

Zeilik, Michael.
Astronomy, the evolving universe.

Includes bibliographies and index.
1. Astronomy. I. Title.
QB43.2.Z44 1985 520 84-10911
ISBN 0-06-047374-6

84 85 86 87 9 8 7 6 5 4 3 2 1

Contents

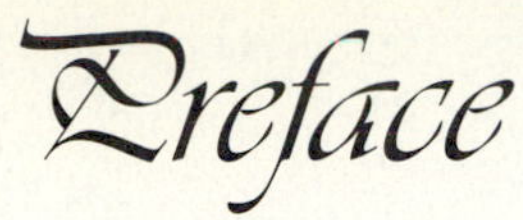

Preface

The civilized man has . . . a fine Geneva watch, but he has lost the skill to tell time by the sun. A Greenwich nautical almanac he has, and so being sure of the information when he wants it, the man in the street does not know a star in the sky. The solstice he does not deserve; the equinox he knows as little; and the whole bright calendar of the year is without a dial in his mind.

RALPH WALDO EMERSON, Self-Reliance

I have spent the past year studying the prehistoric astronomy in the Southwest. That task has taken me through foot-deep mud on the road to Chaco Canyon at the winter solstice, only to confront a bleak week of gray clouds and no sun. As I have made my monthly trips there, the truth of Emerson's statement has struck home, even with its quaint (to us) wording. Astronomy really plays no essential part in our daily lives. Yet for the people of the pueblos in the Southwest and their ancestors, astronomy formed the very core of their lives, both spiritual and practical. For them, the calendar of the sun and moon illuminated their lives.

Not so today, except for the lucky few, like myself, who can call themselves "astronomers" and can delight to the bright image of the cosmos in their minds. When I turn from my work with prehistory to Capilla Peak Observatory, the one I direct in the mountains of New Mexico, I'm really doing the same mental activity, though it might not seem so. Astronomers are the ultimate archaeologists: They shift through the space and time of the universe to discover its past and divine its future. In working on prehistoric astronomy, I also probe a past, albeit a more recent one than the one I see through my telescope. What do I hope to find? Some hints about the cosmic vision of the people of the past, who also puzzled over the skies above.

Our curiosity in the past few years has wrought changes in our cosmic vision. This fourth edition reflects those transformations in two main areas: the solar system and the distant galaxies. I have incorporated these new ideas in the fugal structure of this book. This structure divides the book into four main parts. Each part focuses on one key subtheme of cosmic evolution. And—like the cosmos—each part connects and relates to the others. So you can approach the parts in any order. Compared to earlier editions, I have made some key structural changes in Parts 3 and 4.

The four parts are as follows:

Part 1: Changing conceptions of the cosmos / This part concentrates on the evolution of cosmological thought from the nonscientific ideas of the Babylonians to the mind-boggling visions of modern astronomy. It starts with the simplest observations you can make from the earth and ends with the farthest reaches of the visible universe. This part also introduces the idea of *scientific models*, the conceptual core of scientific thought. Scientific models are used in the context of our changing ideas about the cosmos. The development and evolution of scientific models continues as a major subtheme throughout this book.

Part 2: The planets: Past and present / Flyby spacecraft and gangly landers—along with the manned exploration of the moon—have provoked dramatic revisions in our picture of the planets in the solar system. This part focuses on the physical properties of the planets to infer their origin and evolution. It first takes a comparative look at what we now know about the planets, especially our moon and the other planets like the earth—Mercury, Venus, and Mars. These planets show clear evidence of different degrees of evolution, with the earth the most evolved. The others—Jupiter, Saturn, Uranus, Neptune, and Pluto—have, in contrast, changed little since their birth. And Jupiter and Saturn are surrounded by systems of moons that are worlds unto themselves—mostly ice and rock, some scarred by violence in the past. I then turn from evolution to origin—the birth of the solar system from an interstellar cloud of gas and dust. This model implies that many other stars have planetary systems and that, perhaps, these other worlds resemble the local planets in general ways.

Part 3: The universe of the stars / The sun and the planets swing around in a vast island of stars called the Milky Way Galaxy. The sun is the nearest star to the earth in the Galaxy, so the part first investigates the sun as a model star. From our knowledge of the sun, we generalize about the natures of the stars in the Galaxy—some hundreds of billions of them. These stars are other suns, and they evolve: from birth in interstellar clouds of gas and dust, to ordinary lives as fusion reactions, to violent deaths, finally becoming bizarre corpses. This part deals with the complete span of lives of stars, an aging that our sun will follow, too.

Part 4: Galaxies and cosmic evolution / The universe is certainly a universe of galaxies, in which most of the luminous material resides. The part examines first ordinary galaxies and then hyperactive ones. We are just beginning to find that the layout of galaxies in three dimensions reveals a curious structure to the matter in the universe: long chains separated by pancakelike voids. This cosmic design in the layout of the galaxies was imprinted on the universe in the awesome explosion in which it began. That explosion linked the smallest and the largest things in the universe; it also left echoes that we have discovered. From that explosion 15 billion years ago, cosmic evolution shaped us to where we are now—intelligent beings on the earth who are curious about the possible existence of creatures elsewhere. And it forces us to ponder the possible future of the human race—the evolution of an advanced, technological civilization. I point out that in the cosmic view we will be forced to leave the earth—perhaps to travel to the stars.

I have designed this four-part structure so that you can investigate each part more or less independent of the others. Many cross-references, especially to basic physical and astronomical ideas, have been included to help you to read the parts intelligently in an order different from the one presented, if your instructor chooses such a sequence. The focus sections can also help in linking together different parts of the text. They are cross-referenced throughout the book, so they are easy to refer to, either forward or backward. I have set the focus sections off from the main flow of the text because they deal in more detail (occasionally mathematically) with physical and astronomical concepts.

NEW TO THIS EDITION

I have made structural changes in Parts 3 and 4 to give more unity to the material on stars and galaxies. Chapter 13 still deals with our sun as a model star; Chapter 14 then uses this concept immediately to examine the stars as suns. Then I develop the continuity of the evolution of stars, from birth to death, in Chapters 15 through 17. Chapter 18 covers the nearest conglomerate of stars—the Milky Way Galaxy, using the properties and evolution of stars to understand it. Then I turn to other galaxies, the universe, and the history of the universe in Chapters 19 through 21. Finally, I bring together all the themes of cosmic evolution in Chapter 22 to present our place in the cosmos from an astronomical perspective.

I have made many changes to make the presentation more concrete and to emphasize evolutionary ideas. For example, the earth is the local planet that has evolved the most since its birth, so it serves as a model for the evolution of other bodies like it—the moon, Mercury, Venus, and Mars. The sun serves as a model star, and the Milky Way as an ordinary galaxy.

Other changes have been motivated by the constant updating of astronomical knowledge and by my attempts to make matters clearer to you.

In Part 1, basic astronomical observations are presented more concretely; Chapter 2 simplifies Ptolemy's cosmology; Chapter 3 has a revision in the order of the presentation of the Copernican model; Chapter 4 ties together Newton's ideas more directly to those of Copernicus; Chapter 5 deals with spectra more concretely and limits the discussion to the sun and laboratory spectra (discussion of other stars has been moved to Chapter 14); and Chapter 7 has a simplified and shorter section on cosmology.

In Part 2, the material on the planets has been significantly updated; Chapter 10 includes the results of American and Soviet Venus probes; Chapter 11 presents the Voyager 1 and 2 flyby results on Jupiter, Saturn, and the moons of these mighty worlds; in Chapter 12, I have simplified, shortened, and made more concrete the nebular model.

In Part 3, Chapter 13 includes new ideas about solar activity; Chapter 14 pulls together material about the physical properties of stars, including stellar activity; Chapter 5 integrates starbirth and the interstellar medium; Chapter 16 has a shorter discussion of main-sequence stellar evolution; Chapter 17 includes an update on SS 433.

In Part 4, Chapter 18 revises the size, mass, and rotation curve of the Galaxy and includes new observations on the galactic center; Chapter 19 contains expanded material on the controversy over the value of the Hubble constant; Chapter 20 updates observations and models for activity in the nuclei of galaxies and the double quasar; Chapter 21 includes new material on the relationship of subatomic particles to the universe and the new model for the inflationary universe. The appendixes and glossary have been revised as a consequence of these changes in the text and other new information.

I have striven to clarify and simplify the writing by making it more direct and concrete. I have also tried to make the language as nonsexist as possible. To these ends, I often address the reader directly as "you" and make use of the first person to speak directly to you. Whenever I use *I* in the text, you know that I'm expressing my personal opinion.

Instructor's Resource Manual

Richard Reif, the director of the planetarium of the Albuquerque Public Schools System, and I put together the *Instructor's Resource Manual.* We tried to provide teachers with specific hints for presenting the material in each chapter plus references, lists of audio-visual materials, and suggested answers to the book's study questions. We have included sample test questions that are keyed to the learning objectives of each chapter. We also provide suggestions for using the parts of the book in a different sequence.

I thank the many people who have helped me to make this a better book for learning. These include Tom Harrison, John Gaustad, Linda Kelsey, Jack Burns, David King, Marc Price, Paul Heckert, Bob Anderson, Chuck Long, Owen Gingerich, Bernie McNamara, Alan Marscher, Tom Balonek, and Alan Peterson. Of course, any errors in the book are my responsibility.

If you have any comments on the book (especially if you find mistakes!), send them to me at the Department of Physics and Astronomy, The University of New Mexico, Albuquerque, New Mexico 87131.

Michael Zeilik

Note: If you have picked up this book because you're curious about astronomy, you may be interested in taking a college-level course for credit by mail. Write to Independent Study, Continuing Education, The University of New Mexico, 805 Yale N.E., Albuquerque, NM 87131. Ask for information about Astronomy 101C.

Astronomy
The Evolving Universe

part 1

I have wondered what the first stargazers felt as the heavens reeled above them. What magic did the dance of the heavens weave for them? How did they picture the cosmos and their place within it?

My work as an astronomer has given me hints to answers to these questions. On my first observing trip, I traveled to Mt. Hopkins Observatory in Arizona. Through sunset and dusk, I worked anxiously to set up my equipment. When all was ready, I walked out of the dome to take a curious look at the sky.

I was stunned. To the south, Scorpius balanced on its coiled tail. Faint constellations I'd never seen before patterned the sky in an unfamiliar quilt of lights. Mars shone like a red beacon in the sea of stars. I sensed then how ancient people watched the sky and wondered.

The problem of the design of the cosmos and our place in it has intrigued people for centuries. The picture of the universe painted by a culture displays its religious and philosophical beliefs. Yet astronomical observations set the outlines of the cosmic scheme. The evolution from fascinated stargazing to a design of the universe occurred in the past and still happens now.

Part 1 presents the evolution of conceptions of the cosmos from ancient musings to modern speculations. It starts with the basic astronomical observations, then investigates how astronomy prompted ideas about the universe and how well the cosmic schemes explained the known astronomy. I confine the tale to Western cultures and highlight crucial episodes in the evolution of astronomy. This perspective will give insight into the development of physical ideas that relate to cosmological ones and into the birth of scientific thought—an evolution of human ideas that marks a major theme in the grand scheme of cosmic evolution.

Changing conceptions of the cosmos

1/From chaos to cosmos

> ***The stars I know and recognize and even call by name. They are my names, of course; I don't know what others call the stars. Perhaps I should ask the priest. Perhaps the stars are God's to name, not ours to treat like pets . . .***
>
> ***Old woman quoted by Robert Coles in*** **The Old Ones of New Mexico**

Learning objectives

After studying this chapter, you should be able to:

_1._ Describe the daily motions of the sun, moon, stars, and planets relative to the horizon.
_2._ Describe the seasonal motion of the sun—at sunrise, noon, and sunset—relative to the horizon.
_3._ Describe the motions of the sun and the moon, as seen from the earth, relative to the stars of the zodiac.
_4._ Describe the motions of the planets, as seen from the earth, relative to the stars of the zodiac, with special attention to their retrograde motion.
_5._ Tell what astronomical events or cycles set the following time intervals: day, month, and year.
_6._ Describe the astronomical conditions necessary for the occurrence of a total solar and total lunar eclipse.
_7._ Describe the phases of the moon in relation to the moon's position in the sky relative to the sun.
_8._ Define the ecliptic and tell how to find its position in the sky.
_9._ Argue, on the basis of naked-eye observations, an order of the sun, moon, and planets from the earth.
_10._ Identify at least one astronomical achievement of a prehistoric culture.

Central question

What astronomical objects can you see without a telescope, and how do they change with time?

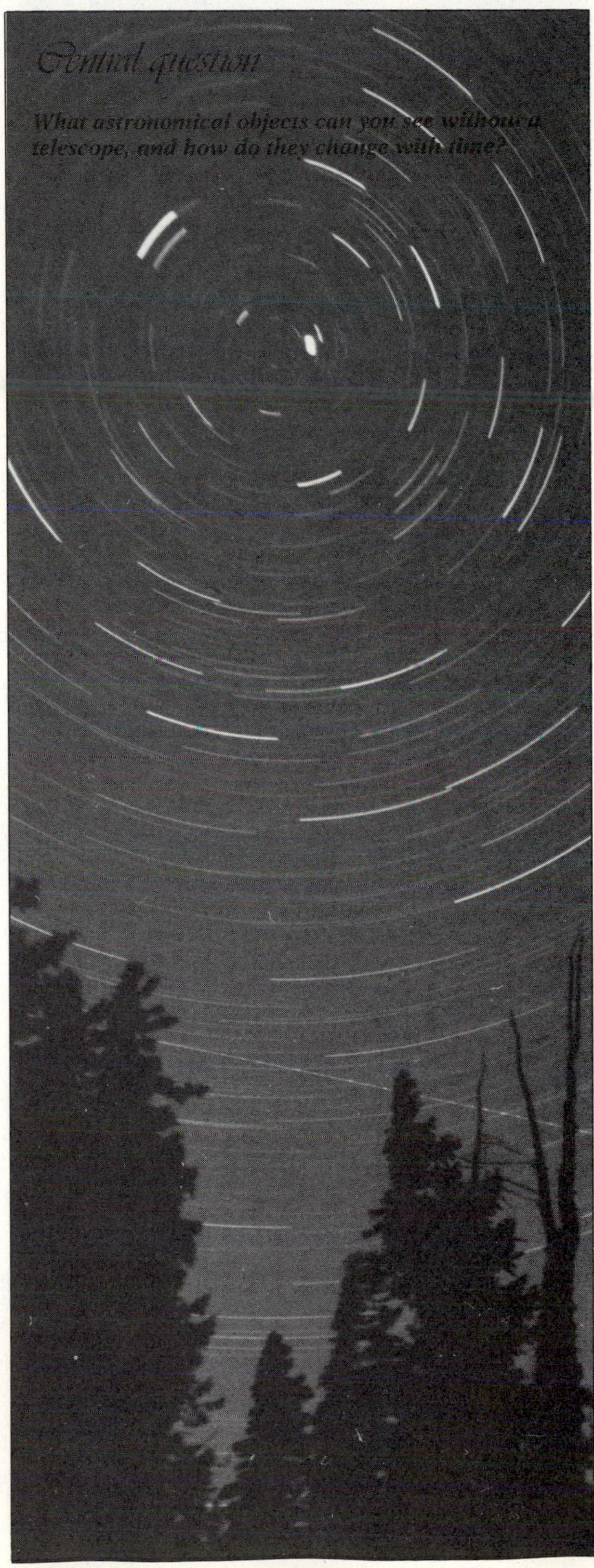

1.1
A time exposure of the stars in the constellation Orion and others nearby. The three bright stars making a diagonal line pointing to the top right form Orion's belt. The very bright star to the right of center is Rigel. This photo shows many faint stars invisible to the eye. See the constellation endpapers for winter to find Orion. (Courtesy Mt. Wilson Observatory, Carnegie Institution of Washington)

Do the heavens have an order? Can you make sense of the events in the sky? Early skywatchers, unhampered by light pollution, felt overwhelmed by what they saw. Their wonder probably gave way to a desire to find some order in the apparent chaos of the heavens.

This quest naturally drove them to concepts of space and time. The arching heavens far removed from the earth displayed cycles of celestial motions that served as cosmic clocks. Such cycles bring order to our ideas about the heavens. Early astronomers found an order in the heavens, a structure in space and a sequence in time. An orderly cosmos emerged from the initial chaos.

This chapter deals with the observations you can make of the sky without optical equipment—naked-eye observations, the same as made by early astronomers. From these you can sense the regular cycles of motions in the heavens. This chapter will not attempt to explain these motions. (That explanation will come in Chapters 2, 3, and 4.) But long-term observations do establish the periods of celestial cycles with amazing accuracy. The recognition of these rhythms marked a crucial step in the development of astronomy and early concepts of the cosmos.

1.1 The visible sky

Have you ever looked carefully and curiously at the night sky far away from a city? If so, you probably were amazed by the sparkle of stars set against the deep velvet of the sky. At first glance, you may have found no order in the uncountable number of stars, and no way to judge their distances, except to say that they are far away.

Constellations / If you take the time to study the stars for a while, you'll find that they fall into patterns, designs imposed by your mind (Fig. 1.1). Ancient skywatchers perceived stellar patterns and passed them on; such patterns are called *constellations*. Early constellations marked a convenient group of stars, with ill-defined boundaries, that typically outlined a mythological or realistic figure (Fig. 1.2). The oldest known constellations originated about 3000 B.C. in the Tigris-Euphrates valley of Mesopotamia. The constellations used today are handed down from the Greeks and Romans. (The endpapers of this book show the prominent modern constellations.)

If you observe nightly, you'll see that the shapes of the constellations don't change. In fact,

1.2
The mythological figure of Orion, the Hunter, from the star atlas* Uranometria, *published in 1603 by Johann Bayer. Albrecht Dürer made the engraving.

if you watched them for your whole life, you wouldn't notice any change. The stars appear to hold fixed positions relative to each other.

Angular measurement / How can you measure how far apart stars appear in the sky? You need a convenient sighting device, such as the extent of your hand held at arm's length. You can then measure the angle between one star and another; this angle is the *angular separation* or distance between them. Angular measurement is based on counting by 60: A circle is divided into 360 degrees (°), each degree into 60 minutes (′) of arc, and each minute into 60 seconds (″) of arc. (An *arc* is any part of the circumference of a circle.)

At arm's length, your fist covers about 10° of sky; each fingertip, about 1° (Fig 1.3). So you can use your hand to measure the angular size of the sun and moon (both about ½°, or half a fingertip) and the separation of the pointer stars in the Big Dipper (about 5°, or half a fist; Fig. 1.4).

The sun and the moon cover a certain visible size in the sky, called their *angular diameter*. These objects appear as disks, and their angular diameter is the angular distance between opposite edges. Angular diameter (and angular distance) depends on the distance to the objects involved, which you may not know. Here's how: The angular diameter of an object depends on its actual diameter and how far away it is. For example, imagine that someone stands 10 meters (m) away from you, and you measure their angular diameter. Now suppose that person moves 10 times farther away, to a distance of 100 m. Then their angular diameter will be only one-tenth as much. The same idea applies to the angular distance between two objects.

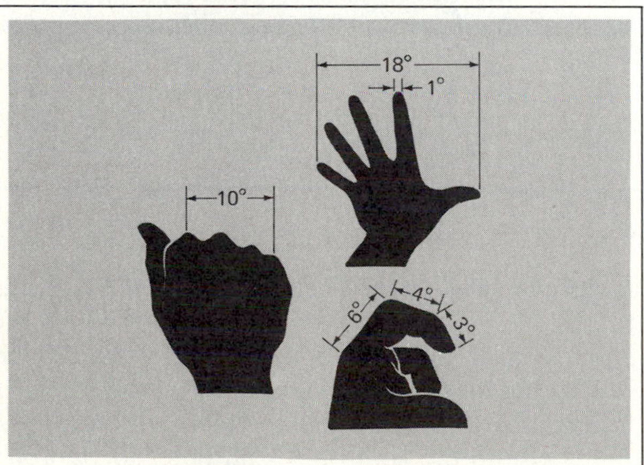

1.3
Angular measurements made with a hand extended at arm's length. The angles are approximate and typical for an average adult.

1.4
Measuring angles on the sky with a hand at arm's length. The angle between the two pointer stars in the Big Dipper is about 5°, or half a fist.

The celestial sphere / Try to imagine the sky that forms the backdrop to the stars as a dome to which they are attached. Then also picture another half to this dome, the half below your horizon. You then have reinvented the *celestial sphere* (Fig. 1.5), a concept first devised by the Greeks. The celestial sphere appears to be centered on the earth, and the stars seem to be fixed to it. This picture of the cosmos, typical of older ideas, is *geocentric*: centered on the earth.

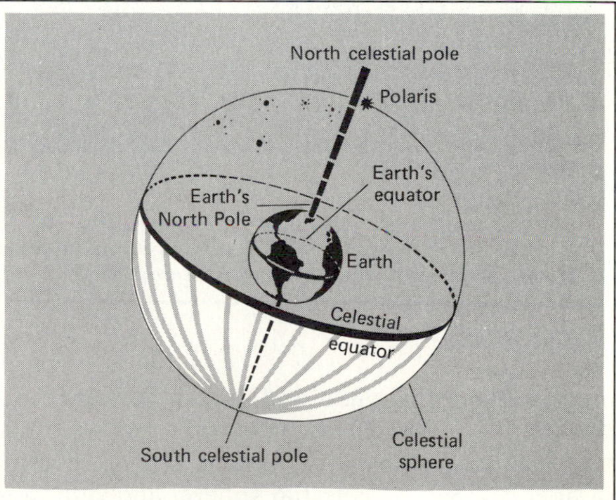

1.5
A simplified geocentric model of the earth and sky. The round earth rests in the center of a larger sphere of stars called the* celestial sphere. *The celestial poles lie on points directly above the earth's poles, and the celestial equator is a circle on the celestial sphere right above the earth's equator.

1.6
The motion of the stars relative to the horizon. This photo is a time exposure of a few hours; it shows the **east-to-west** *motion of the stars. (Courtesy Lick Observatory)*

Stay out one night and watch the stars from dusk to dawn. You'll notice that they move relative to your horizon—rising in the east, slowly traveling in arcs against the celestial sphere, and setting in the west (Fig. 1.6). If you live in the Northern Hemisphere and look north, you'll find that some stars never dive below your horizon. Instead, they trace complete circles above it (Fig. 1.7). These are the *circumpolar stars*. As they swing around, they draw concentric circles like the rings of a bull's-eye (Fig. 1.7). The center of these rings marks the *celestial pole*, the point about which the stars seem to pivot. A modestly bright star called *Polaris* lies close to the north celestial pole. Polaris is now the *north pole star*. (No bright star now falls close to the south celestial pole, so we don't have a south pole star.)

If you acquire a regular stargazing habit, you can discover that the constellations change with the seasons. Here's how: In winter, at midnight,

1.7

The motion of the stars around the north celestial pole. Time exposure showing apparent motion of the stars **counterclockwise** ***(east to west) about the north pole star, Polaris. The small arc traced by Polaris, near the center, indicates that it is not exactly at the north celestial pole. Amateur astronomers are gathered under the night sky; the bright blurs at the bottom result from the movement of their flashlights. (Courtesy T. Gondola)***

you can see Orion due south. Look due south the following nights also at midnight. Orion creeps slowly to the west, toward the sun. In summer, you can't see Orion at all because it's next to the sun and up during the day. In winter, a year later, Orion again lies due south at midnight. A constellation takes one year to return to its initial place in the sky relative to the sun.

Warning: Don't confuse this gradual yearly change relative to the sun with the much faster daily motion relative to your horizon (from east to west).

In summary: Over a human lifetime the stars do not move noticeably with respect to each other. They do move, daily, from east to west, with respect to the horizon and also yearly with respect to the sun.

1.2 *The motions of the sun*

The sun dominates all objects visible in the sky. Its daily motion—rising in the east, tracing an arched path in the sky, and setting in the west—sets the most basic time cycle in our world: day and night. Midway between sunrise and sunset, the sun ascends to its highest point relative to the horizon. This event defines *noon*. The interval from one noon to the next sets the length of the *solar day*.

Motions relative to the horizon / Place a short stick vertically in a flat place on the ground. You then have an instrument to study the sun's daily and seasonal motion in the sky relative to the horizon. (This device is called a *gnomon*; it is still used in sundials to make shadows.) Examine the shadow cast by a gnomon. The tip of the shadow marks the end of a line that connects the shadow's tip, the top of the stick, and the sun (Fig. 1.8). Note that the shadow points opposite the sun in the sky and that the length of the shadow tells the height

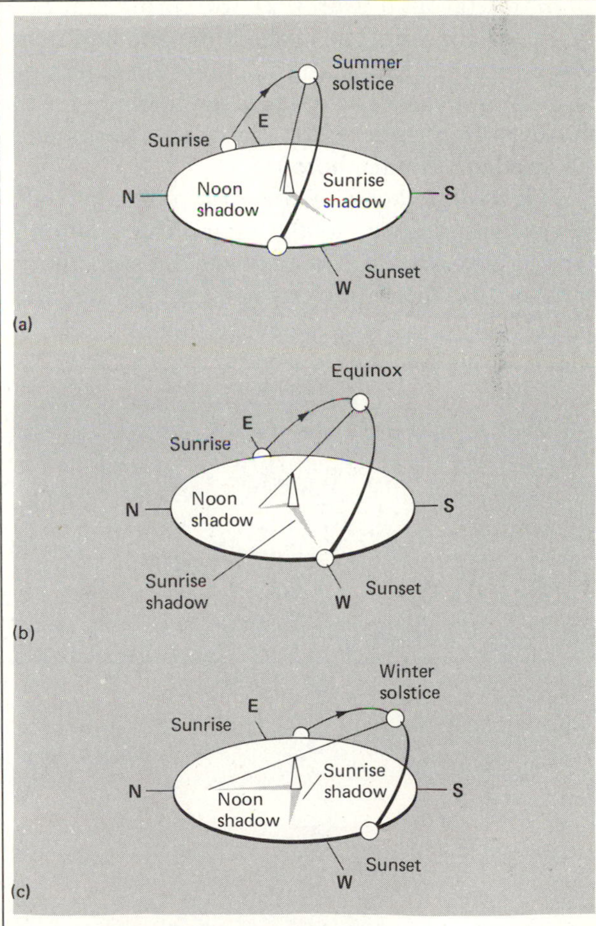

1.8

Gnomon shadows and the position of the sun in the sky. The higher the sun relative to the horizon, the shorter the shadow. The shortest noon shadow of the year occurs on the day of the summer solstice **(a)*****, the longest on the day of the winter solstice*** **(c)*****. At the spring and fall equinoxes*** **(b)*****, the length of the noon shadow falls between the extremes for winter and summer.***

of the sun relative to the horizon. When the sun hangs lower in the sky, the shadow is longer. At noon the shadow has its shortest length for that day. Also at noon in the northern latitudes, the shadow points due north—so the sun is due south.

Observe a gnomon's shadow throughout a year (Fig. 1.8). You'll find that the height of the sun in the sky at noon varies with the season. During the summer, the shadow falls the shortest at noon on the *summer solstice* (around June 21), the day with the greatest number of daylight hours. At the summer solstice, the noon sun hits its highest point in the sky for the year. In winter at noon, the shadow stretches longest on the *winter solstice* (around December 21), the day with the fewest daylight hours. The noon sun has dropped to its lowest point in the sky for the year. On the first day of spring and fall, the gnomon casts a shadow with a length between its summer minimum and winter maximum. These times are called the *equinoxes* (around March 21 for spring and September 21 for fall). The cycle of the gnomon's shadow defines the second basic unit of time: the year of seasons. (*Note*: The dates given here are for the Northern Hemisphere; the seasons are reversed in the Southern Hemisphere.)

Motion relative to the stars / The sun also moves with respect to the stars. This motion is hard to observe, for you can't see the stars during the day. Try this: Pick out a bright constellation visible just above your western horizon after sunset (Fig. 1.9a). Look again at the same time about a week later; the constellation will be closer to the horizon and so to the sun (Fig. 1.9b). Relative to the stars, the sun appears to move to the east. In one year the sun returns to the same position relative to the stars, so it moves 360° in a year, or about 1° a day.

How fast an object covers a certain angular distance is called it *angular speed.* Relative to the stars, the sun moves at an angular speed of 1° per day. Imagine that you recorded the sun's position among the stars for a year. If you drew an imaginary line through these points, you'd trace out a complete circle around the sky: It is called the *ecliptic* (Fig 1.10). The 12 constellations through which the sun moves define the *zodiac.*

Warning: Don't confuse this eastward motion of the sun with respect to the stars in a year with the much faster westward motion of the sun relative to the horizon in a day!

The sun's position along the ecliptic is labeled with reference to the constellations of the zodiac (Table 1.1). For instance, to say that the sun is "in Taurus" specifies a place along the ecliptic. The zodiac probably arose from a desire to mark the sun's position with respect to the stars.

The sun's location in the zodiac also roughly indicates the time of year. Twice yearly, in spring and fall, the equinoxes occur. Two other key times

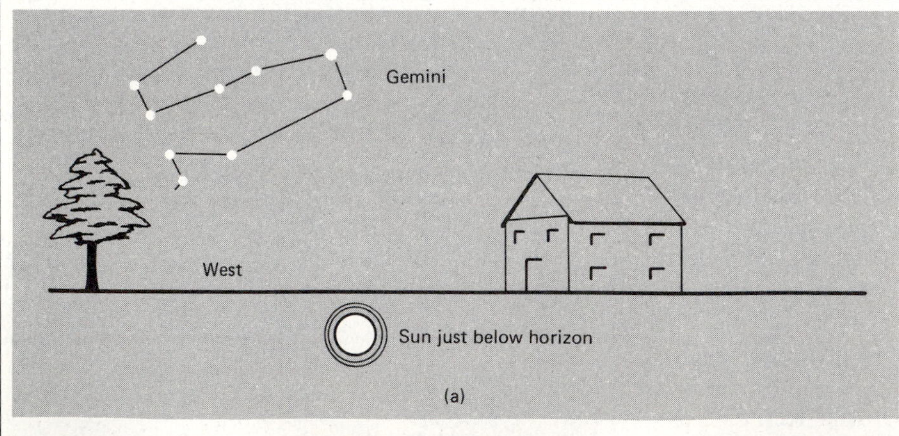

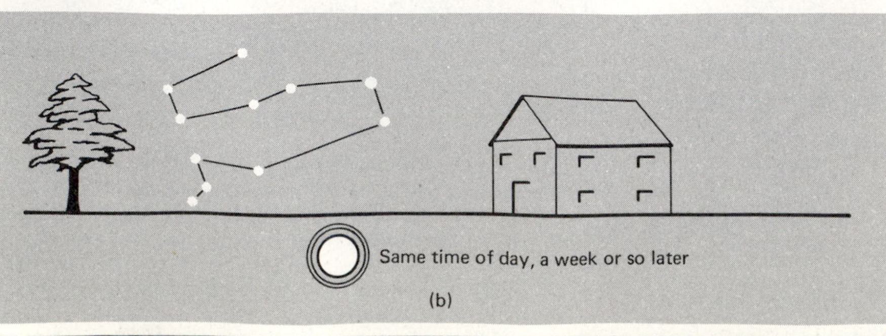

1.9

The motion of the sun relative to the stars. Find a constellation (such as Gemini, shown here) near the western horizon at sunset* (a). *About a week later, look for the same stars just after sunset* (b). *They will appear* closer *to the western horizon. Eventually, they will set before the sun goes down. So the stars seem to move to the* west *with respect to the sun. Or you can think of the sun as moving* east *with respect to the stars.

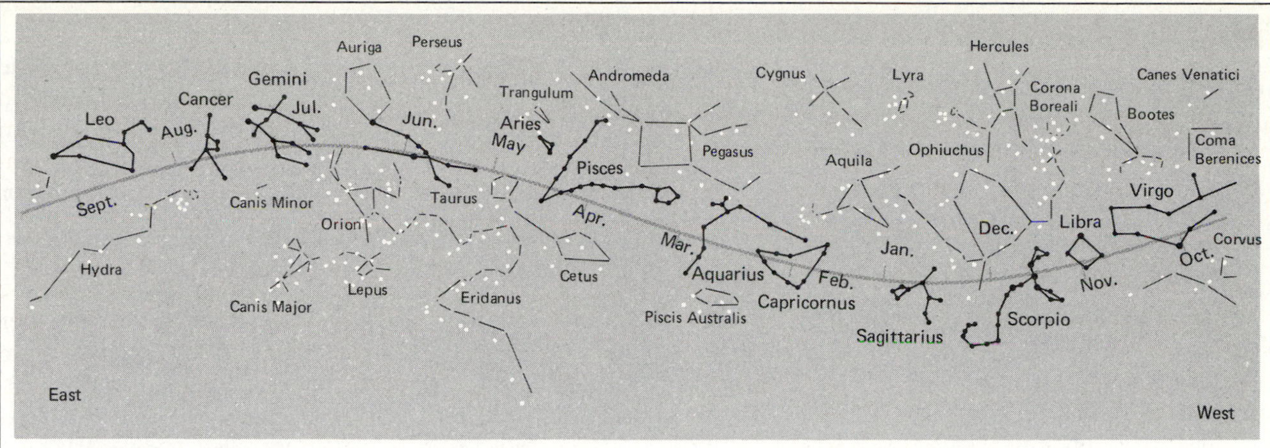

1.10
The ecliptic and zodiacal stars. With no atmosphere, you could see the sun's changing position with respect to background stars. The sun's path is called the* ecliptic. *This diagram indicates the sun's position by date along the ecliptic and shows the major constellations. The zodiacal stars are indicated by darker lines. Note that the sun moves* west to east *among the stars.

are the summer and winter solstices. Nowadays the sun lies in Aries in the spring, Gemini in the summer, Virgo in the fall, and Sagittarius in the winter (Fig. 1.10). These seasonal locations change slowly with time; for the Babylonians, for example, the sun was in Taurus in the spring (Focus 1.1).

***TABLE 1.1** The zodiacal constellations*

Constellation	*Astronomical symbol*	*Mesopotamian-Euphratean identity*
Aries, the Ram	♈	Ram, Messenger
Taurus, the Bull	♉	Bull in Front
Gemini, the Twins	♊	Great Twins
Cancer, the Crab	♋	Workman of the River Bed
Leo, the Lion	♌	Lion
Virgo, the Virgin	♍	Proclaimer of the Rain
Libra, the Balance	♎	Life-Maker of Heaven
Scorpius, the Scorpion	♏	Scorpion of Heaven
Sagittarius, the Archer	♐	Star of the Bow
Capricornus, the Goat	♑	Goat-Fish
Aquarius, the Water Bearer	♒	Urn
Pisces, the Fish	♓	Cord-Place Joining the Fish

Warning: It is standard practice to say that the sun is "in" a constellation. But the stars that form the constellation are far from the sun and not necessarily close together in space; they appear close together to us because they're roughly in the same direction along our line of sight. The statement "the sun is in Pisces" means that the sun is in the same direction in the sky as the pattern of stars we call Pisces. In fact, the stars are actually millions of times farther from the earth than the sun.

To sum up: As seen from the earth, the sun displays three motions: (1) daily, from east to west with respect to the horizon; (2) annually, from west to east through the zodiac; and (3) seasonally, higher (summer) and lower (winter) with respect to the horizon at noon.

1.3 The motions of the moon

If you watch the moon during a warm summer's night, you can spot two of its celestial motions. First, like the sun and the stars, the moon rises in the east and sets in the west. Second, the moon also journeys eastward against the backdrop of the zodiacal stars. Here's how to observe this eastward motion (Fig. 1.11): Wait until the moon appears close to a bright star. On the same night, observe the moon and star again a few hours later. The moon will have moved to the east with

Focus 11

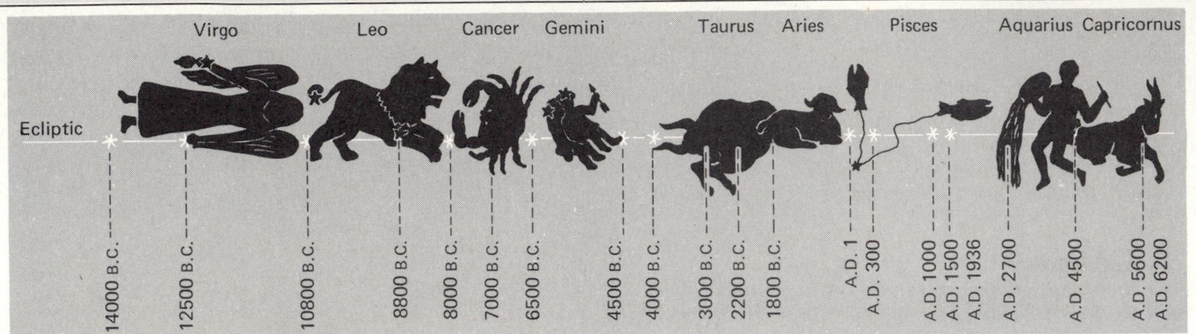

F.1
Precession and the change of position of the vernal equinox with respect to the zodiacal stars.

PRECESSION OF THE EQUINOXES

In 3000 B.C. the sun appeared in Taurus at the vernal equinox. Today you see the sun in Pisces at the start of spring. In the passage of 5000 years, the position of the vernal equinox has moved to the west out of Taurus, through Aries, and into Pisces. That is, in 5000 years it has moved through two constellations. So to circuit the whole zodiac will take six times as long, or about $6 \times 5000 = 30{,}000$ years. (A more precise calculation gives 25,780 years.) This slow westward drift of the equinoxes with respect to the stars is called the **precession of the equinoxes.**

The precession of the equinoxes changes the zodiacal location of the sun at the equinoxes and solstices—its most dramatic effect (Fig. F.1). Precession results in another, less obvious but important effect: The celestial poles move in the sky, so the north pole star changes (Fig. F.2). The north celestial pole now lies near the star Polaris. About 3000 years ago the north celestial pole was near the star Thuban (in Draco). About 12,000 years from now precession will carry the north celestial pole near to the bright star Vega in Lyra.

Precession is hard to observe without a telescope because it takes place so slowly. But if a culture kept astronomical records for a few centuries, its astronomers could notice the shift in the equinoxes and solstices with respect to rising stars. If so, it would prove the serious attention paid to celestial motions by early skywatchers.

Note on the focus sections / *The material set off in the focus sections would break up the flow of the main ideas if placed in the text. That material includes: (1) physical ideas developed in detail, (2) astronomical ideas presented in more depth, (3) curious sidelights, and (4) speculations. Some focus sections deal with the material at a deeper math level than the text. These are not meant to scare you; rather, they are here for readers who are curious about and can understand the material at such a level.*

Your instructor may assign specific focus sections for reading. Otherwise, check out each focus as you come to it and read it carefully if it interests you.

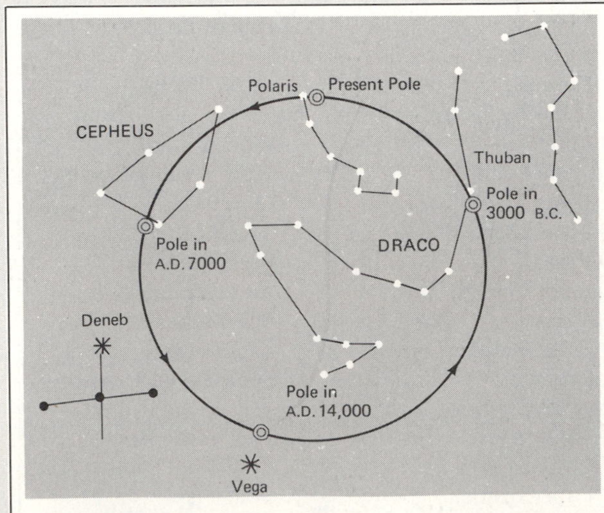

F.2
Precession and the change of position of the north celestial pole with respect to the stars. The pole's motion completes a circle in the sky in about 26,000 years. About 12,000 years from now, the pole will be near the bright star Vega.

(a)

(b)

1.11

The motion of the moon with respect to the planet Venus. The time span between the two photos is one hour. East is at the top and west is at the bottom. (Courtesy D. Hoff)

respect to the star (both will have moved westward in the sky with respect to the horizon). If you measure the moon's rate of motion, you'll find an angular speed of about ½° per hour. At this rate, the moon circuits the zodiac in about 27 days. (Although the moon's path does not fall right on the ecliptic, it lies close to it; so the moon stays within the zodiac.)

Watch the moon for a few nights; you'll note that the amount of its surface that is illuminated—its *phase*—follows a regular sequence. When the moon rises at sunset, its face is completely illuminated—a full moon. About 14½ days later, the moon is new and not visible in the sky. A few days later, the moon reappears at sunset partially illuminated—a crescent moon. A complete cycle of phases—say, from one full moon to the next—takes about 29½ days. It defines a third basic unit of time: the month of phases.

The different phases of the moon relate to specific alignments of the sun and the moon in the sky. At new moon, the angular separation of the sun and moon is small—less than a few degrees. At first quarter, the moon lies 90° east of the sun (Fig. 1.12a). At full, the moon is 180° from the sun (Fig. 1.12b); at last quarter, it is 90° west of the sun (Fig. 1.12c). First and last quarters refer to the position of the moon in the sky—one-quarter of a full circle away from the sun—not to the amount of illumination; the moon at quarter phase looks half full.

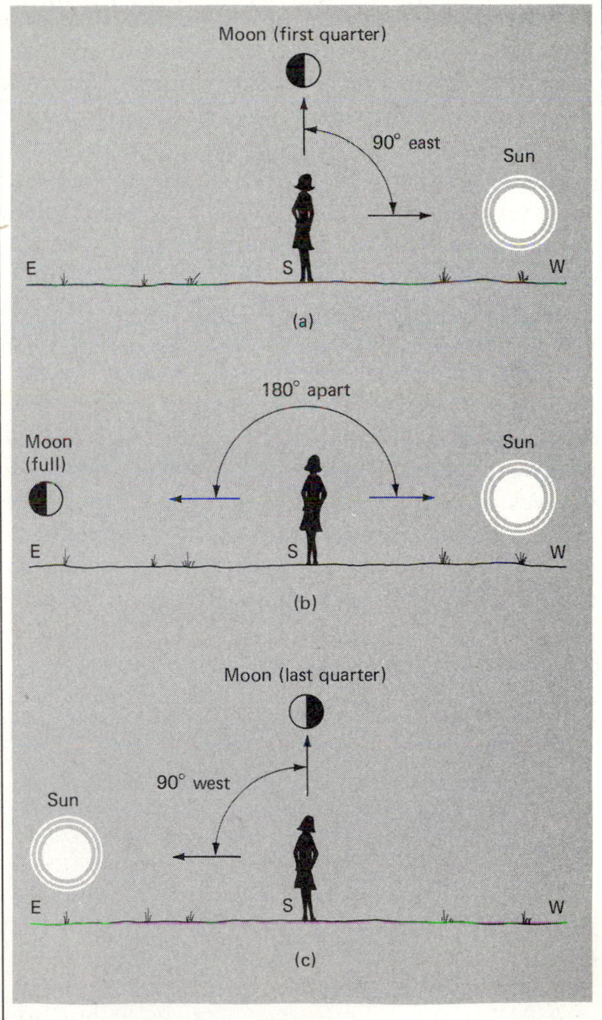

1.12

The orientation of the sun and moon in the sky for different phases of the moon. At first quarter* (a)*, the moon is 90° east of the sun (due south as the sun sets). At full* (b)*, it is 180° away (in opposition, so the full moon rises as the sun sets). At last quarter* (c)*, the moon is 90° west of the sun (due south as the sun rises). The view here is facing south.

To summarize: Daily, the moon rises in the east and sets in the west with respect to the horizon. Like the sun, the moon also moves eastward with respect to the stars, but it completes a circuit in about a month, whereas the sun takes one year.

1.4 The motions of the planets

If you observe the sky often, you can quickly pick out objects that don't belong to the familiar patterns of the stars. Five of these wander in regular ways through the stars; these are the planets Mercury, Venus, Mars, Jupiter, and Saturn. (Uranus, Neptune, and Pluto cannot be seen without a telescope.) Viewed by eye, the planets look pretty much like stars, though they tend not to twinkle as much as stars do. At times, some planets are brighter than the brightest stars, and all planets vary in brightness. It is their motions, however, that really separate the planets from the stars.

Retrograde motion / The planets display a peculiar motion that sets them apart from the other objects in the sky. An example: Suppose you observe Mars every night for a few months near the time of year when Mars appears brightest in the sky. At first Mars moves slowly eastward (with respect to the stars) through the zodiac close to the ecliptic (Fig. 1.13). In this respect, it moves like the sun and the moon. But later, Mars falters in its eastward motion with respect to the stars and stops for a short time. Next, for about three months, Mars moves westward—opposite its normal motion. After that, its westward motion slows down and stops. Then Mars resumes its normal eastward course. The planet's backward motion to the west is called *retrograde motion*. In the middle of its retrograde motion, Mars shines its brightest.

All planets loop along or near the ecliptic in retrograde motion—but generally not at the same time or for the same duration. For instance, Mars takes about 83 days to go through its retrograde motion; Saturn takes 139 days (Table 1.2). Ancient astronomers were puzzled by retrograde motion. The sun and the moon never move retrograde, and, in general, the planets travel eastward along or near the ecliptic.

TABLE 1.2 Fundamental observations of the visible planets

Planet	*Typical duration of retrograde motion (days)*	*Period around ecliptic (years)*
Mercury	34	1
Venus	43	1
Mars	83	2
Jupiter	118	12
Saturn	139	30

Note The lengths of retrograde motion (from start of westward displacement to renewal of eastward displacement) may vary a little from the above values from retrograde to retrograde.

Along with these motions, the planets move daily from east to west with respect to the horizon. So the planets display three motions: (1) daily, rising in the east and setting in the west; (2) eastward, in general, along the ecliptic; and (3) occasionally westward in retrograde loops. Note that a planet's motion with respect to the stars occurs much more slowly than its daily motion with respect to the horizon.

One more point about retrograde motion: The alignment of the sun and planet in the sky at the time of retrograde motion separates the visible planets into two groups. Mercury and Venus make up one group. These two planets never stray very

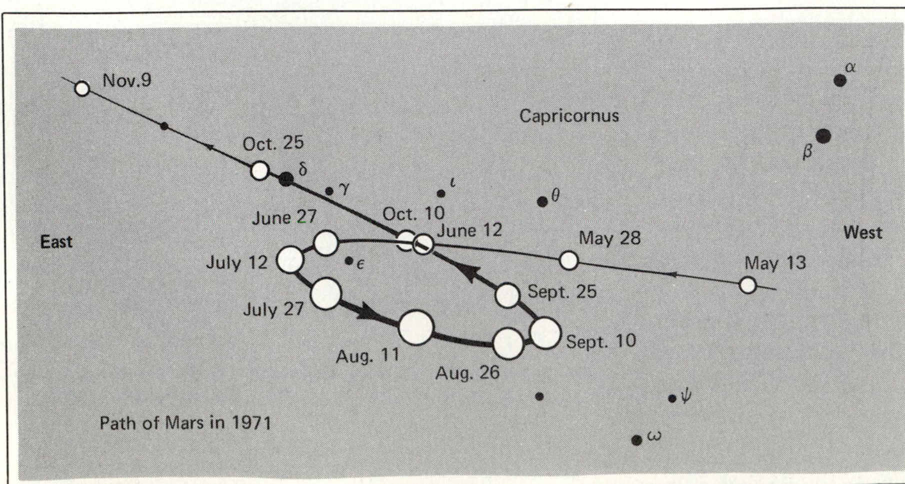

1.13 ***The retrograde motion of Mars, relative to the stars, during its August 1971 opposition. The stars of the constellation Capricorn are labeled by Greek letters. Note that from July 12 until September 10, Mars moved from east to west; this was the time of its retrograde motion. The size of the circles indicates how the brightness of Mars varied; the larger the circle, the brighter the planet appeared.***

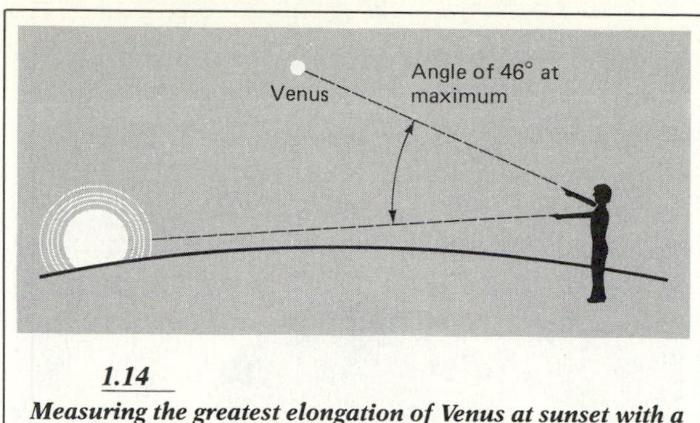

1.14

Measuring the greatest elongation of Venus at sunset with a fist. At this time the angle between Venus and the sun is about 46°. The same observation for Mercury gives a maximum angle that averages 23°.

far along the ecliptic from the sun; they act as if they were tied to it by a leash. Because they stick close to the sun, they are only visible as morning and evening "stars." That means that they hang over the western horizon after sunset and appear above the eastern horizon before sunrise. (But keep in mind that they are not stars.) You can use your fist to measure the maximum angular separation of Mercury and Venus from the sun. For Mercury, the average maximum separation is 23° (about 2½ fists); for Venus, about 46° (about 4½ fists). When either planet is at its greatest angular separation from the sun, it is at *maximum elongation* (Fig. 1.14). (Mercury's maximum elongation varies substantially; it can be as large as 28°. The figure given here is its *average value*.)

Mercury and Venus begin their retrograde motions after they have swung farthest east of the sun as evening stars. They then move westward, pass the sun, and reappear as morning stars west of the sun (Fig. 1.15). When Mercury or Venus is close together with the sun in the sky, the planet and the sun are in an alignment called *conjunction*. (Whenever two celestial objects appear to come close together in the sky, they are in conjunction.)

The second group of planets comprises Mars, Jupiter, and Saturn. In contrast to Mercury and Venus, these planets move freely along the ecliptic with respect to the sun. And they retrograde when they stand in *opposition* to the sun: opposite the sun in the sky. At opposition, the sun and the planet are separated by 180° (just as the moon at full). Then the planet rises as the sun sets. If you point one arm to Mars at opposition and the other at the sun, your arms would be 180° apart (they'd make a straight line). When at opposition, a planet crosses the middle of its retrograde loop and shines its brightest.

To sum up: Mercury and Venus can never be in opposition to the sun; they retrograde after passing their greatest eastern elongation and continue until reaching greatest western elongation. Mars, Jupiter, and Saturn can be in opposition or conjunction, but they retrograde only at opposition.

Relative distances of the planets / The time each planet spends in retrograde motion and that each takes to circle the ecliptic once provides a

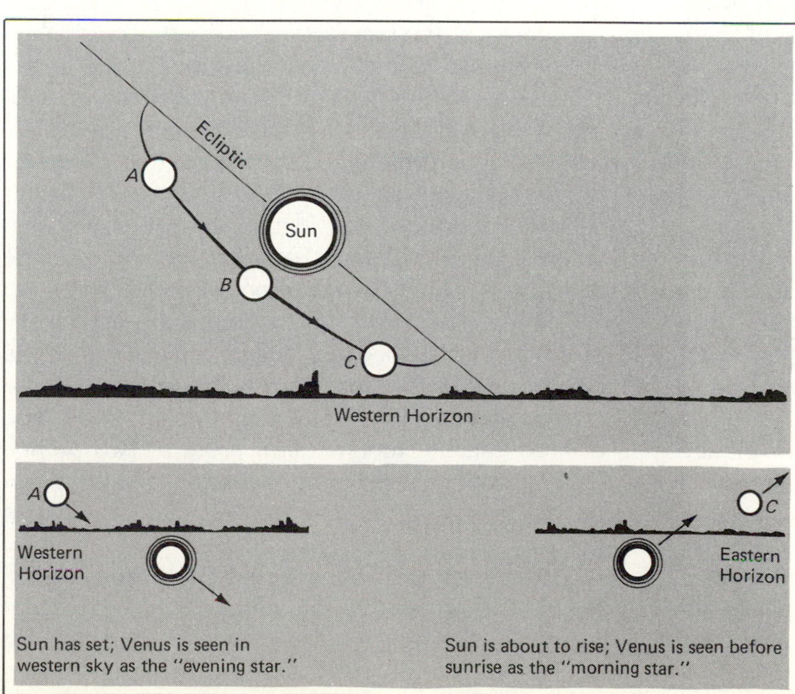

1.15

Retrograde motion of Venus. The upper part of the diagram shows how Venus moves (from A to C, east to west) with respect to the sun and ecliptic during retrograde motion. Venus goes through retrograde motion when it switches from being an evening "star" (A) to a morning "star" (C)—that is, from being on the east side of the sun to the west side.

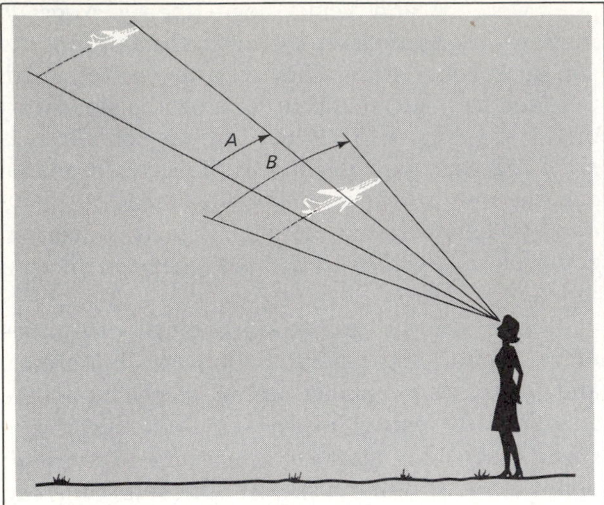

1.16
Judging relative distances from angular speeds. In the same time, plane A covers a smaller angle than plane B. So plane B must be the closer, if both planes fly at the same speed.

1.17
A total eclipse of the sun. As viewed from the earth, the moon covers the sun's visible disk, so the sun's outer atmosphere is visible. (Courtesy Yerkes Observatory)

clue to the relative distances of the planets from the earth. Assume that the planets move at the same speed. One that appears to move faster must be closer than another that moves at a slower angular rate. So the slowest-moving planet must be most distant from the earth; the swiftest, the nearest.

Here's an analogy (Fig. 1.16): Suppose you are watching the lights of two airplanes at night and wish to estimate their relative distances. Assume that both planes fly at the same speed. The one that appears to move faster must be the closer of the two. Apply the same argument to the moon, sun, and planets: The fastest (the moon) is closest to the earth; the slowest (Saturn) is the farthest. (It turns out that the speeds of the planets around the sun are not the same, but the argument still works out because the differences are not very great.)

Greek astronomers during the third century B.C. applied this reasoning and fixed the order from the earth as the moon, Mercury, Venus, the sun, Mars, Jupiter, and Saturn. Simple observations led to an ordering of the cosmos centered on the earth, one that was *geocentric*.

But the retrograde motion of the planets posed a sticky problem to those who wanted a geocentric cosmos. The explanation of this contrary motion eluded astronomers for centuries. The evolution of our ideas about the motions of the planets, especially the puzzle of retrograde motion, marks the central theme of the next three chapters.

1.5 *Eclipses of the sun and moon*

An eclipse of the sun (a *solar eclipse*) occurs when the moon passes in front of the sun (Fig. 1.17). It's a remarkable fact that, although the moon is actually smaller in diameter than the sun, it's closer by just the right amount so that the angular diameter of the sun and moon is almost the same—about ½°. So the moon can just cover the sun's disk when it passes directly between the sun and the earth, as it may do at new moon (Fig. 1.18).

Why don't eclipses happen each and every month? Mainly because the moon's path in the sky relative to the stars does *not* coincide with the ecliptic but is tilted at an angle of some 5°. The ecliptic and the moon's path cross at two points. Only at or near these points will the sun and moon be close enough to overlap so that an eclipse occurs.

During a total solar eclipse, the moon's shadow is only about 300 kilometers (km) wide. Only people in this narrow band as the shadow's tip sweeps across the earth can see the eclipse as a

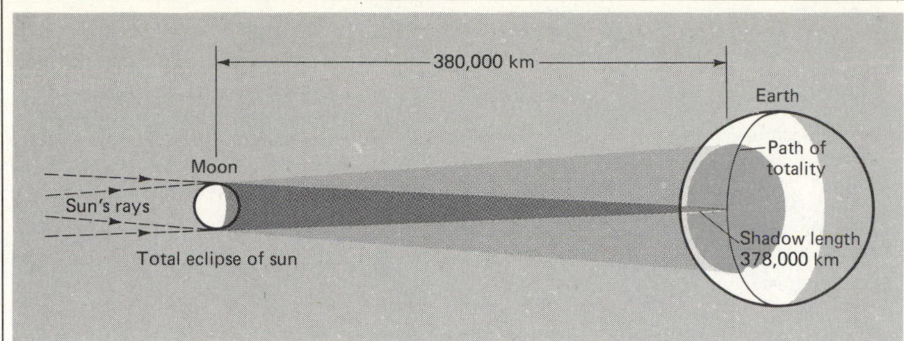

1.18
Alignment of the moon, sun, and earth for a total solar eclipse. The moon must be new and on or very close to the ecliptic. The length of the moon's central shadow is usually long enough to hit the earth; a total eclipse is seen in the path where the central shadow moves along the earth.

total one. Those just outside the central band see a partial eclipse, where the moon does not completely cover the sun.

An eclipse of the moon (a *lunar eclipse*) occurs when the moon passes directly through the shadow cast by the earth (Fig. 1.19). Then the sun's illumination is cut off from the moon (Fig. 1.20). A total eclipse of the moon takes place only when the moon is full—when the earth lies directly between the sun and the moon. Also, the moon must be close to the ecliptic; otherwise it will miss the earth's shadow.

Eclipses occur less frequently than the other celestial events described here. Their spectacular

1.19
Time sequence of some stages of a total lunar eclipse (July 1982). From left to right, these separate shots (made on the same negative) show the full moon emerging from the earth's shadow. (Courtesy Brian Walski)

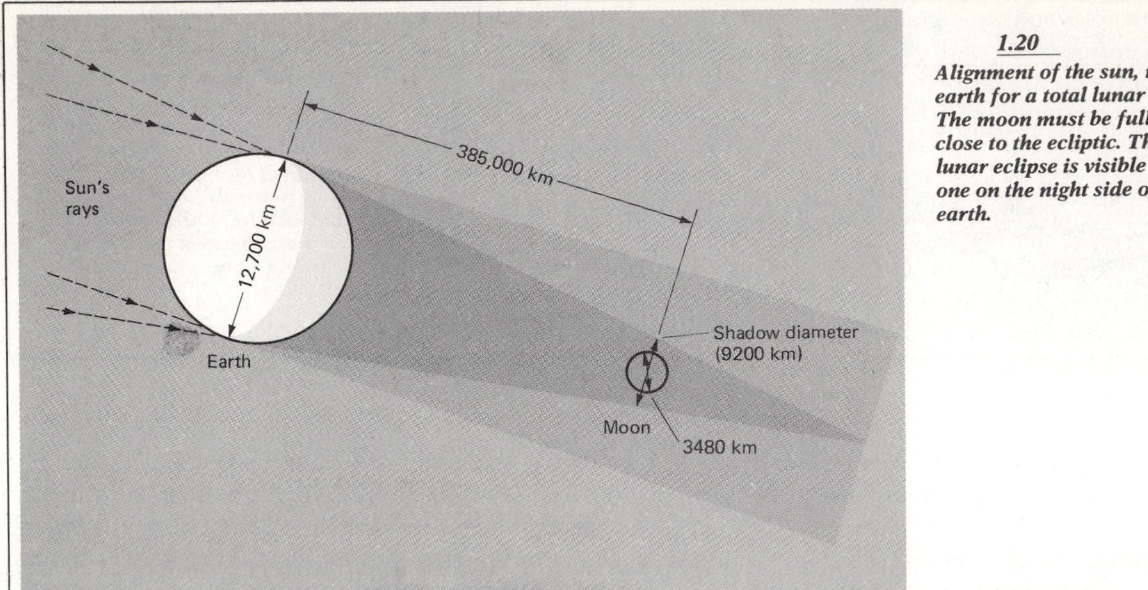

1.20
Alignment of the sun, moon, and earth for a total lunar eclipse. The moon must be full and on or close to the ecliptic. The total lunar eclipse is visible to everyone on the night side of the earth.

nature motivated people to study them. The Greeks, for one, found that eclipses were predictable from the motions of the sun and the moon. They also noted that solar eclipses prove that the moon must be closer than the sun, because the moon blots out the sun. So the sun must be larger in size than the moon.

Now you can see how the ecliptic got its name: Only when the moon lies on or close to the *ecliptic* can eclipses occur.

1.6 *Prehistoric astronomy*

How much astronomy did ancient peoples know? Prehistory has left few records. And astronomical knowledge was probably passed on by word of mouth in the form of symbolic stories. Fossil records of such oral traditions do exist: The Polynesian islanders, for example, knew how to navigate from Tahiti to Hawaii by starwatching. And in the pueblos of the Southwest, skywatchers today still keep their seasonal watch. Ancient peoples likely had much more astronomical knowledge than we usually credit them with. Some of this knowledge was incorporated in their art and architecture; only recently have we read the stories in the ruins.

Sunwatching of the Pueblos / After their conquest of Mexico, the Spanish turned northward on a fruitless search for the fantastic Seven Cities of Cibola, said to be made of gold. They found none. But they did encounter the adobe villages, which they called *pueblos*, of the native peoples who had lived in them at least a thousand years prior to the arrival of the Spanish. Many pueblos disappeared in historic times (from A.D. 1540 onward); those that survived are the cultural connection to the people called the *Anasazi*, who occupied a vast area in the Southwest, centered on the Four Corners. The Hopi Pueblos (in Arizona) and Zuni (in New Mexico) provide the best clues to the past because they are cultural descendants of the Anasazi (although we don't know from which specific Anasazi sites).

At Hopi and Zuni, sunwatching plays a central role in the agricultural and ceremonial life. The seasonal cycle of the sun sets the ritual calendar and determines the times of specific crop plantings. Sunwatching carries a practical weight as well as a religious one. The observing is invested in a religious officer, usually called the Sun Priest. He watches daily from a special spot within the pueblo or not far outside of it. The Sun Priest carefully observes sunrise (or sunset) relative to the horizon features. He knows from past experience what horizon points mark the summer and winter solstice and the times to plant crops. These he announces within the Pueblo, usually ahead of time so that ritual preparations can be carried out. Along with horizon features, the Zunis use windows and portholes in the Pueblo that allow sunlight to hit special plates or markings on the walls at significant times of the year. So light and shadow, along with horizon features, made up the basis of the Puebloan sunwatching.

One critical aspect of historic Pueblo astronomy is that the Sun Priest must *anticipate* ceremonials. The Pueblo culture requires a preparation time for a ceremony so that the participants will be in the proper frame of mind to make it effective. For example, the Sun Priest at Hopi announces the winter solstice on December 10–11, and he starts the sun watching on December 2–6.

Anticipation relates to an astronomical problem: prediction of the solstices. Because the sun does not move noticeably at that time, naked-eye observation cannot tell which day is, in fact, the solstice. The best way, astronomically, to determine the solstice is to observe the sun before the solstice, while it is still moving observable amounts along the horizon. Using just the eye and horizon features, I estimate that the minimum detectable solar motion occurs a week before the solstice. Watching for two weeks (or more) before makes the observations even more reliable. The Sun Priest can then do a day count to predict on which day the solstice will occur. Of course, this requires a few years of sunwatching to establish the anticipation times for various horizon markers. But once established, the horizon calendar remains fixed, and the knowledge can be handed down to others.

Anasazi sunwatching / Around A.D. 1000, the Anasazi prospered and built community houses up to five stories high, containing hundreds of rooms. I will discuss two possible solar observatories in such buildings: one at Chaco Canyon, the other at Hovenweep.

Chaco Canyon, in northwestern New Mexico, grew to be a center of Anasazi culture. By A.D. 1130, eight large villages were located within 15 km of the central canyon. One of the largest villages is called *Pueblo Bonito*. A D-shaped apartment house of over 800 rooms, Pueblo Bonito hugs the north wall of the canyon. Within it are a number of *corner* doorways and windows (Fig. 1.21a), which are rather unusual in Anasazi architecture: Pueblo Bonito contains over half of the known examples. The archaeologist Jonathan Reyman noted that two of the windows in rooms in the southeast part of the ruin have a clear view of the winter solstice sunrise—if, in fact, an outer wall did not obstruct the view. On the morning of the winter solstice, the rising sun's light streams through these windows to strike the opposite walls (Fig. 1.21b), confirming the solstice.

Both windows could also be used for anticipatory observations. They are fairly large, with an angular view of a few degrees from inside the room if you stand at the opposite wall. The winter solstice sun rises just about in the middle of the opening. You can move about 2 m south and still see sky until the cross jambs cut it off. About seven weeks before the winter solstice, the rising sun's light will first enter the rooms in the cross-

(a)

(b)

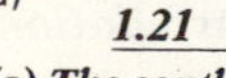

1.21

(a) ***The southeastern corner of the ruins of Pueblo Bonito in Chaco Canyon just after dawn on the winter solstice, 1981. Corner windows (arrows) in the second story align with the winter solstice sunrise.*** **(b)** ***Sunlight from the window on the left in*** **(a)** ***hitting the corner of the interior room where a student photographs the view out the window. (Photos by M. Zeilik)***

jamb alignment to throw a narrow beam within. As the sun moves southward, the beam widens and moves northward a total span of over 1.5 m. That gives an average motion of about 3 cm per day—easy to use to predict the winter solstice by markers on the wall and a day count.

One practical point: Low clouds often linger on the horizon in the winter and block out the sun's appearance. But it takes only one anticipatory observation and a day count from it for the Sun Priest to be able to predict the solstice to within a day or two.

The Hovenweep ruins bridge Colorado and Utah. Built at about the time that the last of the great villages were constructed at Chaco, Hovenweep contains many towerlike structures located on rocky outcrops. Most of the buildings have small ports built into them. The portals are typically 10 cm in diameter and made with considerable attention to their view: A large fraction are angled in their walls. In three of the ruins, we find rectangular rooms that are late additions to each structure. This is inferred from the different masonry style and the lack of integration with the rest of the structure. Ray Williamson has proposed that portals in these add-on rooms were used for solar observations, with alignments for the solstices and equinoxes. From an astronomical view, the portals define narrow angular views on the horizon and permit narrow shafts of light into the rooms.

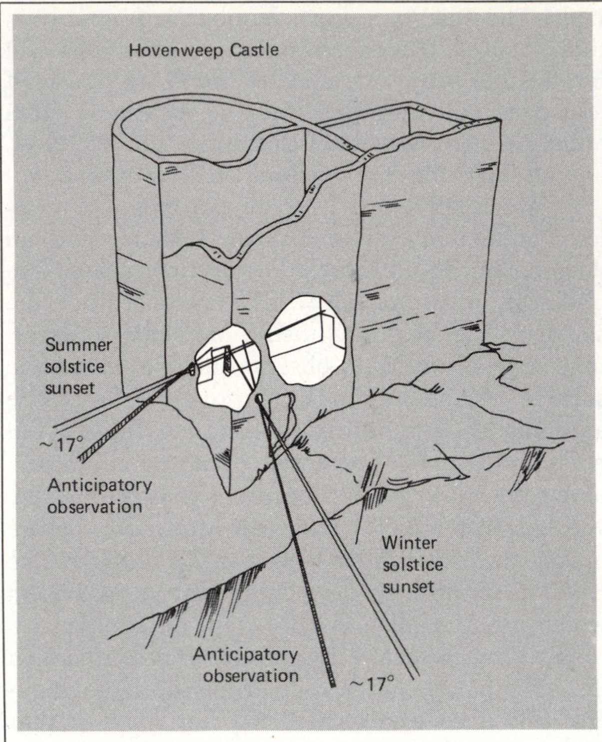

1.22
The solstice sunset ports at Hovenweep Castle. Both permit an angular span of about 17° on the horizon, so that sunlight first enters them about two months prior to each solstice.

In *Hovenweep Castle*, the possible observing room is situated on the south end of a D-shaped tower. Two portals are in the west side; one opens to the northwest for the summer solstice sunset, the other to the southwest for the winter solstice sunset (Fig. 1.22). The summer solstice sunset falls on the north side of the lintel of the east door. About 67 days before the summer solstice, the sunlight first enters the room, the beam striking the west side of the lintel of the north door. As the sun moves north, the sunset point tracks 2.4 m from one lintel to the other, a linear distance that corresponds to an angle of 17.3° on the horizon.

A similar pattern works for the winter solstice portal. At the winter solstice sunset, the light strikes the east side of the north doorway's lintel. The beam first appears 70 cm to the right of this position, or an angle of about 17°. So the sunset beam enters the room some 70 days before the winter solstice.

If markers were placed on the inside north wall, the tracks of the sunlight would make it quite easy to anticipate both solstices. The same technique could also set the planting cycle, which at Hopi runs from the middle of April to the summer solstice. (The climate at Hopi and Hovenweep is pretty similar.) In fact, sunlight first enters the summer solstice port on April 15—just the right date to begin planting key crops.

These examples show that astronomy infused the lives of the Anasazi and that simple observations reveal the basic cycles of the heavens.

KEY CONCEPTS

1. ***Celestial bodies participate in cyclic motions; some are short-term and others long-term in duration (Table 1.3).***
2. ***Relative to the horizon, the stars rise in the east and set in the west every day; the stars make up fixed patterns, called constellations, that do not change over a human lifetime; different constellations are visible in the night sky at different seasons.***
3. ***Relative to the horizon, the sun rises in the east and sets in the west daily; its highest point in the sky defines noon; the height of***

TABLE 1.3 *A summary of major celestial motions visible without optical aid*

Object	*Motion* Daily	Long-Term
Sun	E to W in about 12 hours from sunrise to sunset. Day length varies from season to season.	W to E along ecliptic 1° per day. Height of sun in sky at noon maximum in summer, minimum in winter.
Moon	E to W in about 12 hours, 25 minutes from moonrise to moonset. Rises about 50 minutes later each day.	W to E within 5° of ecliptic. Takes 27.3 days to travel 360° relative to the stars. Phases repeat in cycles of 29.5 days.
Planets	E to W in about 12 hours from rising to setting.	W to E within 7° of ecliptic. Time around ecliptic varies, shortest for Mercury and longest for Saturn. Retrograde motion from E to W at a time specific to each planet.
Stars	E to W in about 12 hours from starrise to starset. Rise about 4 minutes earlier each day. Circumpolar constellations never set; their motion centers on the celestial pole.	Relative to the sun, a constellation returns to the same position in one year. In fixed positions with respect to each other. The position of the celestial pole changes slowly, returning to its initial position in about 26,000 years.

the noon sun varies with season—it is highest at the summer solstice, lowest at the winter solstice, and midway at the equinoxes; relative to the stars, the sun moves eastward along a path called the ecliptic that cuts through the constellations of the zodiac; the sun completes one circuit of the zodiac in a year.

4 ***Relative to the horizon, the moon rises in the east and sets in the west; relative to the stars, the moon moves eastward (completing one circuit in about a month); as the moon moves around the sky relative to the sun, it goes through a cycle of phases, which depends on the angle between the sun and the moon.***

5 ***Relative to the horizon, the planets rise in the east and set in the west daily; generally, the planets move on or close to the ecliptic eastward with respect to the stars; occasionally, the planets loop westward relative to the stars in retrograde motion; of the five planets visible without a telescope, Mars, Jupiter, and Saturn retrograde only when in opposition to the sun, Mercury and Venus when in conjunction with the sun and moving from evening "star" to morning "star"; also, Mercury and Venus never move far from the sun.***

6 ***A lunar eclipse occurs when the full moon passes through the earth's shadow; a solar eclipse occurs when the moon passes between the sun and the earth; solar eclipses show that the moon is closer to the earth than the sun is.***

7 ***Prehistoric people paid close attention to the motions in the sky, especially the seasonal motion of the sun (necessary for agriculture), and left behind records of their knowledge in their art and architecture.***

STUDY EXERCISES

1 Tell how you can find the ecliptic in the sky. Tell how you can find the zodiac. (*Objectives 4 and 8*)

2 Draw a schematic diagram of the retrograde motion of a planet. Be sure to indicate east and west clearly. Alternative exercise: Plot a planet's positions during retrograde on a star map using an ephemeris, which gives the positions. (*Objective 4*)

3 What celestial bodies *never* show retrograde motion? (*Objective 3*)

4 Into what two groups can the planets be divided on the basis of their retrograde motion? (*Objective 4*)

5 When Mars is at opposition, at about what time will it rise? Set? (*Objective 4*)

6 For what two reasons did ancient astronomers believe (correctly) that the moon was closer to the earth than the sun? (*Objectives 6 and 9*)

7 (*a*) You go outside at about 9 P.M. and face south. The moon is up and off to your right, near the horizon. Is is rising or setting? What is its phase?
(*b*) The next night you go out again at 9 P.M. Where is the moon? Is it higher, lower, or not up at all? Did it move east or west with respect to the stars? Has its phase changed? If so, how? (*Objectives 3, 5, and 7*)

8 Describe the changing position of the rising sun on the eastern horizon throughout a year, with special emphasis on the solstices and equinoxes. (*Objective 2*)

9 What phase must the moon be in for a solar eclipse? A lunar eclipse? (*Objective 6*)

10 On what observational basis can you argue that Mars must be closer to the earth than Saturn? (*Objective 9*)

11 You are a novice Sun Priest and must establish your own day count to anticipate the winter solstice. How can you do it? (*Objectives 2 and 10*)

BEYOND THIS BOOK . . .

Astronomy and *Sky and Telescope* magazines have monthly star maps and planetary locations to help you observe stars and planets.

Hamlet's Mill (Gambit, Boston, 1969) by G. de Santillana and H. von Dechend analyzes the oral myths of preliterate people for astronomical knowledge.

In *The Roots of Civilization* (McGraw-Hill, New York, 1972), A. Marshack argues that the need to keep track of time, especially the month, led to the development of symbolic notation and language.

For a look at Stonehenge and other possible prehistoric observatories, read *Beyond Stonehenge* (Harper & Row, New York, 1973) by G. Hawkins. An excellent summary of sites in Great Britain and Europe is *Sun, Moon, and Standing Stones* (Oxford University Press, New York, 1978) by J. E. Wood.

For information on Chaco Canyon, see "Astronomy, Architecture, and Adaptation at Pueblo Bonito" by J. E. Reyman in *Science*, September 10, 1976, p. 957; and "The Sunwatchers of Chaco Canyon" by M. Zeilik in *Griffith Observer*, June 1983. A fine general reference is *Native American Astronomy* (University of Texas Press, Austin, 1977), edited by A. Aveni.

2/The birth of cosmological models

Marduk bade the moon come forth;
entrusted night to her,
Made her a creature of the dark, to measure time;
and every month, unfailingly, adorned her with
a crown.

ENUMA ELISH: translated by Thorkild Jacobson

Learning objectives

After studying this chapter, you should be able to:

1. Describe the essential aspects of a scientific model and evaluate cosmological models in the context of scientific model-making.

2. Cite one important astronomical achievement of the Babylonians.

3. Describe the physical basis of Aristotle's cosmological model and how it influenced his picture of the cosmos.

4. Describe what geometric devices Aristotle used to explain basic astronomical observations in the context of his model.

5. State the assumptions and physical basis for Ptolemy's cosmological model.

6. Sketch the basic Ptolemaic model for the motions of Venus, the sun, and Mars, and show how, using Venus and Mars as specific examples, the epicycle and deferent explained retrograde motion.

7. Indicate what geometric devices Ptolemy used to explain the main celestial motions and any variations in the major cycles.

8. Evaluate the assets of the Ptolemaic model that led to its wide, long-term acceptance.

9. Compare and contrast a mythical and scientific cosmological model with specific examples from Babylonian and Greek astronomy.

10. Use at least one specific case to show how geometric and aethestic concepts influenced Greek ideas about the cosmos.

What is a scientific model, and how did early cosmological models explain and predict astronomical observations?

How have people pictured the cosmos? Most ancient cultures viewed it as finite and geocentric, closed off by a shell of stars. Older cosmologies generally paid little attention to the details of the celestial motions, even if the cycles were carefully observed. Order in the cosmos tended to be explained in terms of religious myths.

The Greeks first attempted to take cosmological ideas beyond those explained by myths. Grappling with the vexing problem of planetary motion, Greek thinkers fleshed out the bare cosmological picture with geometric devices to account for the celestial cycles. These schemes marked the first earnest models: mental images that exhibited features like those observed in nature. Greek efforts culminated in the geocentric model of Claudius Ptolemy. So well did Ptolemy succeed that no one challenged his model for over 1400 years.

This chapter examines early cosmologies to see how closely early ideas related to actual observations. You'll see how these cultures tried to make sense of what they saw in the sky and observe the contrast between those cosmological ideas grounded in myths and those developed as scientific models. This evolution of thought resulted in the first comprehensive model of the cosmos, one that eventually forced the birth of the models used today.

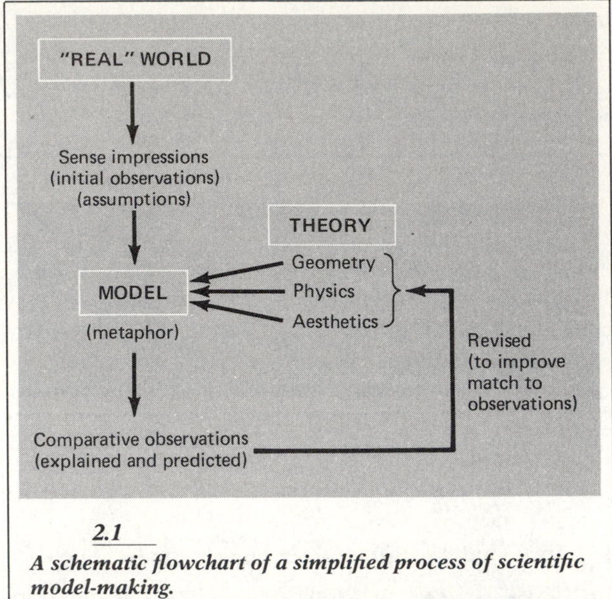

2.1
A schematic flowchart of a simplified process of scientific model-making.

2.1 Scientific models

Chapter 1 described naked-eye astronomical observations you can make. As you read, you probably felt the urge to place these observations into a grand design, a model for the operation of the heavens. (Notice that I presented the observations without any model for heavenly motions.)

People in general share your desire to make sense of what they see in the world. This natural drive gave birth to *scientific models*, conceptual plans designed to explain the workings of nature. Scientific models lie at the heart of the workings of science.

But this mental framework did not always underpin people's ideas about the world. The concept of a scientific model sprang from older ideas in a gradual evolution of thought. Ancient models tend to be mythical rather than scientific. As myths, they marked serious first attempts to understand the world. Astronomical observations naturally drove people to create models of the cosmos: mythical at first, later scientific.

What makes up a scientific model? It is a mental picture—based on geometric ideas, physical concepts, aesthetic notions, and basic assumptions—that tries to explain by analogy what is seen in nature. A model strives to come to grips with a seemingly chaotic world by casting it in familiar terms in a metaphorical way.

Some basic elements enter into a scientific model (Fig. 2.1). First, our sense impressions of the world, observations, provide the raw information and curiosity. Then the human mind tries to interpret this input by its intuition of geometry, physics, and aesthetics. The choices here are filtered by whatever assumptions, usually implicit, are held to apply to the situation. Geometry outlines a visual framework for the model. Physics deals with the motions and interactions of various parts of the model. And aesthetic ideas—gut judgments of what seems beautiful—select the simplest, most pleasing models from those that the fertile mind imagines. Finally comes the crucial test: How well do the features of the model correspond to observations? If the correspondence is good (within the errors of observations), the model is confirmed as workable. If not, various aspects of the model are changed to get a better fit.

A scientific model has two key features: (1) It explains what is seen, and (2) it predicts accurately what is seen. A model's predictions of future or past events must relate directly to observations and must do so with sufficient accuracy to be convincing. Although all good scientific models contain both explanation and prediction, one may overshadow the other.

The power of prediction prompts the drive to confirm a model by looking at how well predictions fit observations. The search for confirmation can make or break a model, depending on its success. This endless search ensures that all scientific models are tentative, for they contain the seeds of their own destruction. Scientific models must change when new or more accurate observations become available. As they evolve, we acquire a deeper understanding of the physical world.

2.2
Babylonian skywatching: The seeds of a science

About 1600 B.C. the Babylonians compiled the first star catalogs and began records of planetary motions. By 800 B.C. Babylonian astronomers could fix planetary locations with respect to the stars (of the zodiac) and kept records, on clay tablets, of these positions. Their early observations included Venus, Jupiter, and Mars; their records spanned several centuries.

Babylonian astronomy / Why did the Babylonians become such careful observers? In part from state support for the calendar and for astrology. These needs generated techniques to predict planetary positions and to record long-term observations. The records enabled the Babylonians to find the basic cycles of celestial motions (see Table 1.3).

The Babylonians went beyond these basic cycles. Their records helped them discover the *variations* in them. For instance, the angular size of a planet's retrograde loop and the duration of this motion changes from one retrograde to the next (Fig. 2.2). The Babylonian astronomers had tables of the major cycles and their variations on clay tablets; these were used to predict future positions and retrograde motions. This procedure did not require an explanation of the cycles, just a knowledge of their existence over a long period. The Babylonians had the power to predict but not to explain the cycles in the sky.

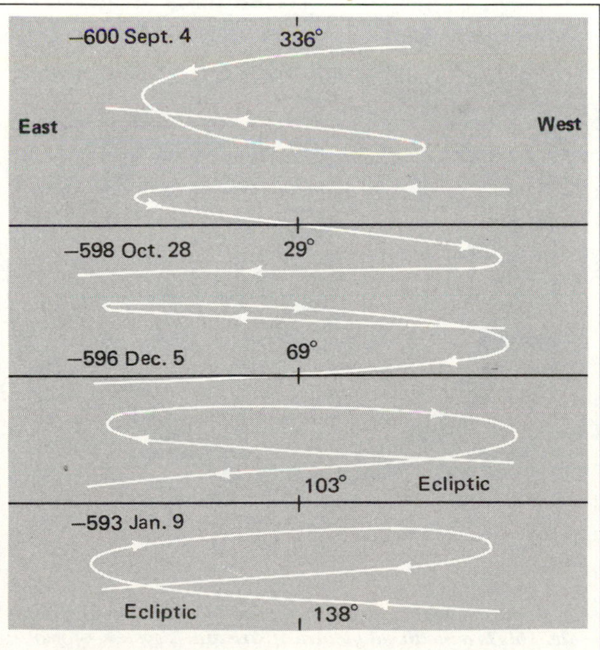

2.2
Retrograde motions of Mars in Babylonian times. This figure shows Mars at different positions along the ecliptic (indicated in degrees) for a few different times of retrograde motion. Note how the shape and size (and so the duration) of the retrograde loop vary. The motions shown here are relative to the stars; east is to the left, west to the right. (Courtesy O. Gingerich)

The Babylonians paid special attention to observations of the moon and kept month-by-month lunar observations. Their focus on the moon resulted in detailed observations of solar and lunar eclipses, one dating from 240 B.C. These records enabled the Babylonian astronomers to predict eclipses, at least in a rough way.

Babylonian cosmology / The Babylonian astronomers also served as priests. This dual role fostered the continuity of astronomical knowledge. But it also divorced Babylonian cosmology, the grand picture of the universe, from the cycles of astronomy. In the cosmic picture, the gods created, ordered, and controlled the world. These divine functions were explained in religious myths. The Babylonian tale of genesis, the *Enuma Elish* (literally, "when above"), dealt with the forming of the world from initial chaos by the god Marduk, who fashioned the stars from an unordered swirl of primeval waters.

Enuma Elish glosses over the details of observational astronomy. Tables of astronomical cycles paled before this grand scheme. Their cyclic function was strictly for predictions: No geometric framework lay behind them. The Babylonian astronomers could predict, but they did not explain celestial motions in terms of physical causes, a concept central to a scientific model.

2.3
Greek models of the cosmos

The Greeks paid only passing heed to the careful observational work of the Babylonians. Yet the Greek philosophers accomplished what the Babylonian astronomers never dreamed of: They devised geometrical, physical models of the cosmos. The Babylonians made sense of the world through myths. The Greeks had myths, too. But their curiosity drove them to develop models in order to grasp reality. A look at Greek astronomy and cosmology shows how scientific model-making was born in Western culture.

The music of the spheres / The Pythagorean cosmos (Fig. 2.3) first incorporated some aspects of a scientific model. In it, the earth had a spherical shape for reasons of symmetry and observa-

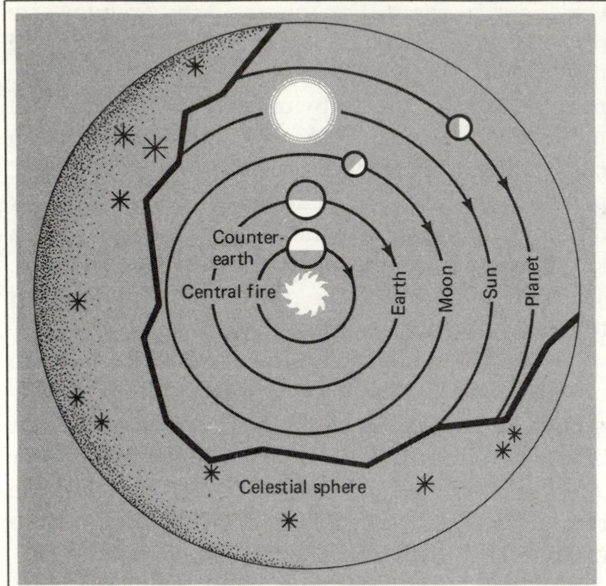

2.3

An artist's representation of the Pythagorean model of the cosmos. From symmetry, the Pythagoreans argued that the shapes of the celestial bodies and the entire cosmos must be spherical. Note that the earth is* not *in the center; rather, it revolves around a central fire (not the sun) once in 24 hours. (Adapted from a figure by R. E. Ridley)

tions (the shape of the shadow on the moon during a lunar eclipse, Fig. 1.19). A spherical shell bounded the cosmos and held the stars; other smaller spheres carried the planets around. The stellar sphere, driving all the other spheres with it, rotated daily east to west. At the same time, the planetary spheres rotated slowly at different rates from west to east. Their periods matched the time it takes each planet to circuit the zodiac (see Table 1.2).

This early model contained forceful aesthetic and geometric elements but lacked physical ideas. And its correspondence to observations was crude. For example, the model failed to account for retrograde motions. The Pythagorean model relied most strongly on the notions of harmony and symmetry. These aesthetic ideas play an essential role in scientific models.

The ideas of the philosopher Plato (428–348 B.C.) derived from Pythagorean concepts. Plato saw the perfection of the cosmos in the form of a sphere. In keeping with this shape, he assumed that all motions of the heavenly bodies were at a uniform rate around circles. Plato succinctly stated the problem of the planets: The goal of an astronomical model was to "save the appearances": to devise a model that explained the observed motions. This goal preoccupied astronomers for centuries.

A geometric model / Aristotle (384–322 B.C.), the most famous of Plato's pupils, devised a complex geometric model based on the idea of uniform, circular motion. To account for the motions of the planets, especially their retrograde motion, Aristotle's model (Fig. 2.4) had a total of 56 spheres centered on the earth. Even then, as Aristotle admitted, the model did not describe the celestial ballet very well. But this model did incorporate for the first time physical ideas of motion. By doing so, Aristotle made his model more scientific than earlier efforts.

What were these physical concepts? Aristotle viewed the cosmos as divided into two distinct realms: a region of change near the earth and an eternal region in the heavens. The realm of change contained bodies made of four basic elements: earth, air, fire, and water. Each element had its own natural motion toward its place of rest in the cosmos: the earth to the center, the fire to the greatest heights below the moon, the air below fire, and water between the earth and the air. In contrast, the heavens were made of an immutable, transparent substance.

These two realms had different versions of *natural motion*: motion without forces. In the heavens, the celestial spheres rotated naturally.

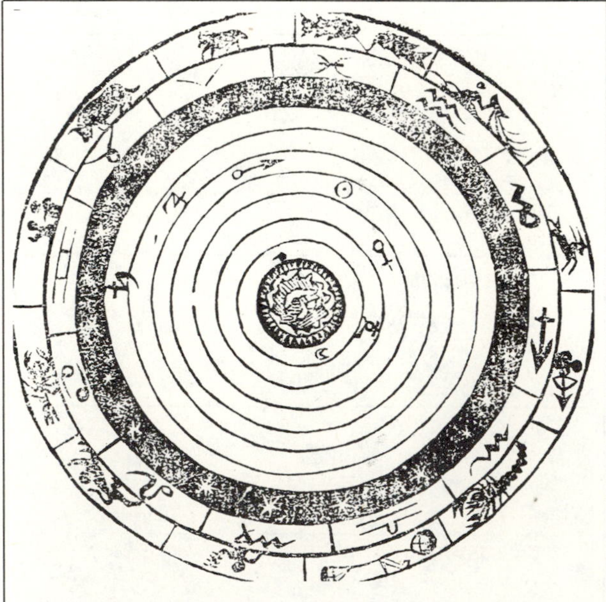

2.4

A geocentric model of the cosmos in the tradition of Aristotle. This is a medieval picture of Aristotle's system of geocentric spheres. The earth lies in the center; above it are the natural realms of water, air, and fire below the sphere of the moon. Beyond the moon are the heavenly spheres to which the planets and the stars are attached. Only one sphere for each planet is shown here; the actual scheme used a number of spheres to account for all the motions of each planet. (Courtesy Yerkes Observatory)

Focus 2.1

HELIOCENTRIC PARALLAX IN A FINITE COSMOS

Stellar parallax is an apparent shift in the positions of a star because of the motion of the observer. In a heliocentric model, stellar parallax arises from the earth's revolution around the sun; it is called heliocentric parallax. The details of heliocentric parallax differ, depending on whether the stars in space are confined to a thin shell (as in the Greek picture) or spread more or less throughout space (as in modern concepts).

Consider the situation if the stars are stuck in a thin shell that closes off the cosmos, such as in the model of Aristarchus (Fig. F.3). Pick out two stars close together on the celestial sphere (A and B in Fig. F.3). Observe them when they are due south at midnight (position 1 in Fig. F.3); they will appear some angular distance α apart. Just after sunset three months later, observe the stars again (position 2 in Fig. F.3); they will appear closer together (angle β is less than angle α), partially because you're now seeing them at an angle rather than face on. Observe them again six months later (position 3 in Fig. F.3); their angular separation is again β.

Imagine viewing these stars over a six-month cycle: from positions 3, 1, and 2. You'd see the stars close together (3, angle β), farther apart (1, angle α), and then closer together again (2, angle β). This cyclical shift in angular position (from β to α to β) is heliocentric stellar parallax. Note that the size of the earth's path compared to the size of the shell of stars determines the size of the shift: the smaller the ratio, the smaller the shift. Of course, if the earth were in the center of the stellar shell and did not move, no shift would occur.

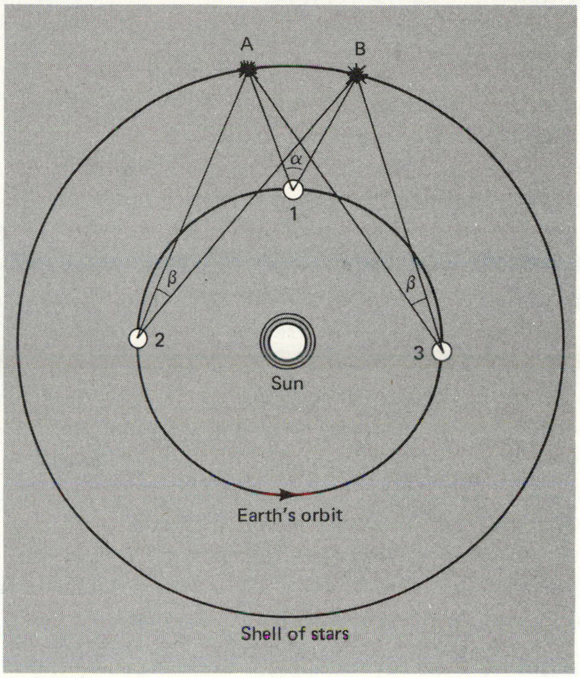

F.3
Parallax in a finite, heliocentric cosmos.

Greek astronomers did not observe this parallax. So the heliocentric model was inconsistent with observations and was in part rejected on this basis. We now know that heliocentric parallax is too small to detect without a telescope, because the stars are very far away.

So no forces were needed to move the planets around the earth. In the terrestrial realm, earth, air, fire, and water each had its natural motion. For example, the natural motion of earthy material was toward the center of the cosmos. But here forced motions could also occur. For instance, to keep a bicycle moving, a person must keep peddling it. Once this force stops, the bike rolls a bit but will eventually stop. Such motions, Aristotle reasoned, required a *force*, a push or a pull, to keep them going.

These physical ideas shaped Aristotle's image of the cosmos. He argued that the earth must be stationary and in the center of the universe. How? First, the natural motion of earthly material to seek the center of the cosmos explained the location of the earth there. Second, he reasoned that if the earth moved, bodies thrown upward would not drop back to their point of departure. Yet he noted that heavy objects thrown upward do return to their starting place. So the earth did not move. (Note how physical ideas of natural and forced motion influenced his model of the cosmos. Even now, the meaning of motion profoundly affects cosmological models.)

A model gains credibility when it explains

Focus 2.2

SURVEYING THE EARTH

Around 200 B.C. the Greek astronomer Eratosthenes, believing the earth to be round, measured its circumference. Here's how: Eratosthenes worked at the library in Alexandria. While on vacation in Syene (Aswan), Egypt, at the summer solstice, he noted that sunlight fell directly down a well at noon (Fig. F.4), which indicated that the sun was directly overhead. At noon on the same date the following year, he observed in Alexandria (located directly north of Syene) that a gnomon shadow indicated the sun was about 7° south of being directly overhead. Because the earth's circumference totals 360°, he concluded that the distance from Syene to Alexandria must be 7/360 of the circumference (Fig. F.5).

To find the length of the circumference, Eratosthenes needed the distance from Alexandria to Syene. Now Herodotus had recorded that the trip by camel between these two cities took about 50 days. The average camel then traveled about 100 stadia per day, so the trip covered 5000 stadia. (The stadium, *plural* stadia, *was an ancient unit of length about ⅙ km long; its exact value varied in the ancient world.) Eratosthenes calculated that the earth's circumference was 360/7 × 5000 = 250,000 stadia. Take the length of a stadium as ⅙ km, then the earth's circumference comes out to 42,000 km—surprisingly close to the modern value of 40,030 km. (This coincidence may be chance. Though Eratosthenes measured the angular separation of Syene and Alexandria very accurately, we do not know how well he knew the distance between the cities, nor do we know the exact length of the stadium in kilometers. In any case, he got the ratio right.) Divide the circumference by 2π and you have the radius: about 6700 km.*

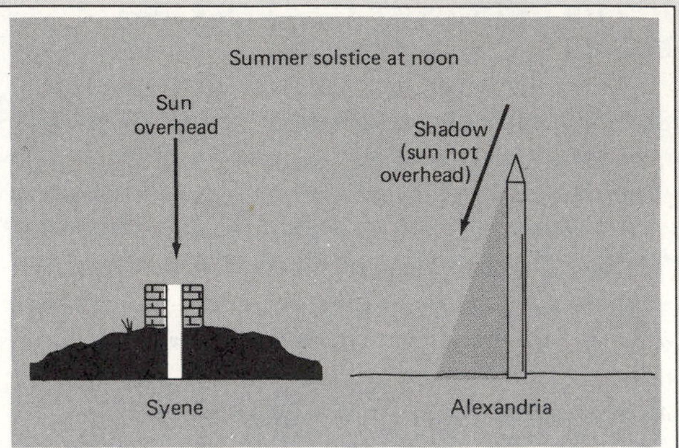

F.4
Eratosthenes's solar observations at noon at Alexandria and Syene (Aswan). He noted that at Syene on the summer solstice, the noon sun's light came down from directly overhead. In another year he saw that in Alexandria on the summer solstice, the noon sun's light came down at an angle of 7° from the vertical.

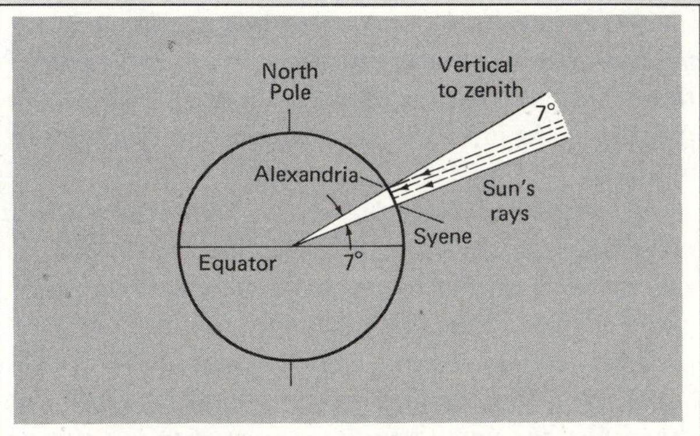

F.5
The geometric basis for calculating the earth's circumference. If the sun is so far away from the earth that its incoming rays are parallel, only a curved earth could simply account for the difference in the noon angle at the two locations.

observations in a natural way. In this respect, Aristotle's model was successful, for it explained two important observations: the lack of a stellar parallax and the spherical shape of the earth. Aristotle noted that if the earth moved around the sun, the stars must display an annual shift in their positions, called *parallax* (Focus 2.1). No one observed this change, so Aristotle concluded that the earth did not move around the sun, a conclusion that reinforced his geocentric view. (The motion does occur, but it is too small to detect with the naked eye.) Aristotle followed Pythagorean ideas and added observations of his own (such as the curve of the earth's shadow on the moon during a lunar eclipse) to conclude that the earth was a sphere. This led to a measurement of the earth's size by Aristotle, who cited a diameter of 5100 km, and a later Greek, Eratosthenes, who found a diameter of 13,400 km (Focus 2.2).

A contrary view: A sun-centered model / After the time of Alexander the Great (356–323 B.C.), the Greek scientific tradition centered upon the great library at Alexandria. Here worked astronomers such as Eratosthenes, who measured the earth's size, and Aristarchus, who proposed a sun-centered (*heliocentric*) model of the cosmos.

Aristarchus lived in the third century B.C. His heliocentric model had the earth rotating on its axis once a day to explain the daily motion of the sky. The earth also moved around the sun in one year; this explained the annual motion of the sun through the zodiac.

Unfortunately, when the library at Alexandria was burned, all the major writings of Aristarchus were destroyed. We know of his ideas from comments by others and from one fragment of a work. In it, Aristarchus worked out the earth-sun distance relative to the earth-moon distance and inferred that the sun was a body much larger than the earth. But we have no evidence that Aristarchus worked out planetary motions using his heliocentric scheme.

The heliocentric model was attacked on two fronts: (1) It contradicted Aristotle's physics in stating that the earth moved, and (2) it required a stellar parallax that was not observed (Focus 2.1).

2.5
Claudius Ptolemy (with the goddess Astronomy). "we shall only report what was rigorously proved by the ancients . . ." (Courtesy O. Gingerich)

For these reasons, and also because of the influence of Aristotle's ideas, Aristarchus's model languished.

Expanding the geometric model / Another distinguished astronomer lived and worked at Rhodes from 160 to 127 B.C. He was Hipparchus, who saw the importance of organizing the observations of the Babylonians in order to develop a better model of the cosmos.

To the basic picture of Aristotle, Hipparchus added the geometric devices of *eccentrics, epicycles*, and *deferents* to explain aspects of planetary motions that had previously been shrugged aside. Each device accounted for observed features of these motions: The epicycle and deferent together explained the retrograde motions; the eccentric, the nonuniform motion of the planets and sun through the zodiac (Focus 2.3). So Hipparchus used geometry to save the appearances and stuck to the assumption of uniform, circular motions.

2.4 Claudius Ptolemy: A complete geocentric model

Two-and-a-half centuries after Hipparchus, Claudius Ptolemy (Fig. 2.5) worked in Alexandria and so had access to records in the library. He molded the astronomical tradition into a comprehensive model that would endure for centuries. We know little of Ptolemy's life, not even the dates of his birth and death. His observations indicate that he worked around A.D. 125. Ptolemy's most influential astronomical work was the *Almagest*, the first professional astronomy textbook. Many of the geometric devices that Ptolemy used in the Almagest did not originate with him, but he was the first to design a complete system that accurately predicted planetary motions, with errors of usually not more than 5° and often less. The careful application of geometry described the celestial motions.

A geometric, geocentric model / At the start of the Almagest, Ptolemy pays his respects to Aristotle and then defines the problem: to use geometry to describe astronomical observations, especially the major cycles and their variations that had been known since Babylonian times. Note that Ptolemy is following Plato's dictum to "save the appearances."

Ptolemy assumes that the earth is spherical, in the center of the cosmos, has no motions, and is much smaller in size than the outer sphere of stars. In addition, he assumes that uniform motion around the centers of circles is the aesthetically pleasing motion for the celestial spheres car-

Focus 2.3

THE GEOMETRIC DEVICES OF HIPPARCHUS

Hipparchus used three geometric devices, each accounting for some aspect of planetary motion: the eccentric, the deferent, and the epicycle.

Following Greek tradition, Hipparchus demanded that the planets move at a uniform speed on circular paths. Yet he knew from observations that the planets' motions are not uniform but vary: The average angular speed of a planet is faster in one region of the zodiac and slower in the opposite region. Hipparchus explained this variation with an eccentric (Fig. F.6): The earth was displaced from the center of the planet's motion. Then the planet appeared to go faster through the zodiac when it was closer to the earth and slower when farther away. But its actual motion along the circle remained uniform.

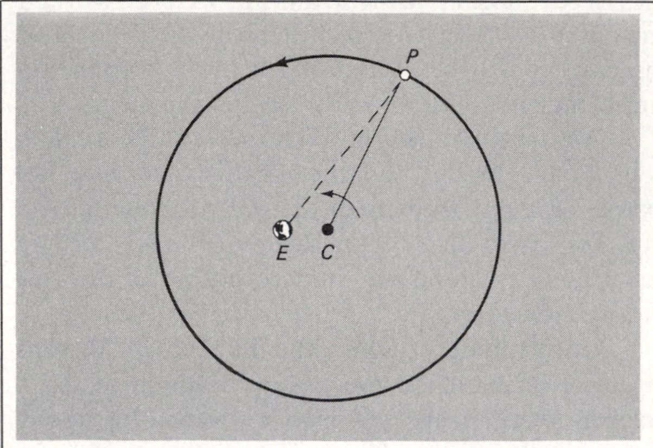

F.6

(a) The eccentric: *A planet (P) revolves with uniform circular motion about the center (C) of its path. The earth (E) is displaced from the circle's center, so the planet's motion as seen from the earth is not uniform.* **(b) The effect of the eccentric for modeling a planet's motion:** *As seen from the circle's center, the planet moves through 30° angles in equal amounts of time. But as seen from the earth, the planet covers different angles in the same times.*

But the eccentric did not explain retrograde motion. For that, Hipparchus used the epicycle and deferent (Fig. F.7). The deferent was a large circle, sometimes centered on the earth, sometimes offset (if so, the deferent acted as an eccentric). The center of the epicycle, usually a smaller circle, moved around the circumference of the deferent. The planet was fixed to the epicycle, so its motion was the combination of the circular motions of it around the epicyle and the epicycle's motion around the deferent. As the planet moves on the part of the epicycle interior to the deferent, it moves opposite the deferent's motion and so imitates the backward swing of retrograde (Fig. F.8).

rying the planets. So Ptolemy's model has a physical basis: that proposed by Aristotle.

In practice—and this is a key point—Ptolemy ends up violating this precept. Why? Recall that there are two main variations in the motions of the planets: (1) retrograde motion and (2) the nonuniform motion through the zodiac (ignoring the effect of retrograde motion). Also, retrograde motions show variations in size, shape, and duration (Fig. 2.2). How can uniform circular motion account for these variations?

Think of the problem of the motion of the planets as a model-building puzzle that needs a different device for each aspect. The two main aspects are: (1) eastward motion through the zodiac—with variations, and (2) retrograde motion—also with variations! Like Hipparchus, Ptolemy modeled the first with a planet moving around the earth at uniform speed on a circle. But he offset the earth from the center of the circle—the eccentric (Focus 2.3)—to account for variations. For the second, again like Hipparchus, he used smaller circles (epicycles) moving on the larger ones (deferents). Then, for variations in the retrograde motions, Ptolemy invented a new geometric device: the *equant*.

You do not need to understand the nitty-gritty details of the equant to appreciate Ptolemy's contribution to the evolution of cosmological models. The main point is that the equant required that celestial motions with it *no longer be uniform around the centers of circles*. The equant

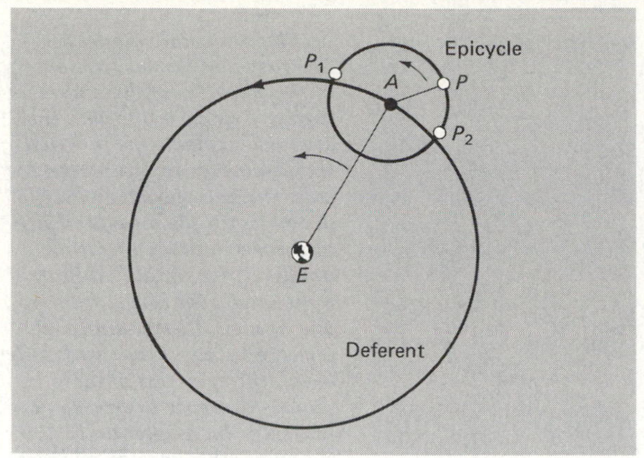

F.7

The epicycle and deferent for retrograde motion. A planet (P) is attached to a small circle (epicycle) *whose center rides on a larger circle* (deferent). *The earth (E) lies in the center of the deferent. The radius of the epicycle turns in the same direction as the radius of the deferent. So when the planet moves on the inside of the deferent (from P_1 to P_2), it moves* opposite *its normal motion with respect to the stars; this is its retrograde motion.*

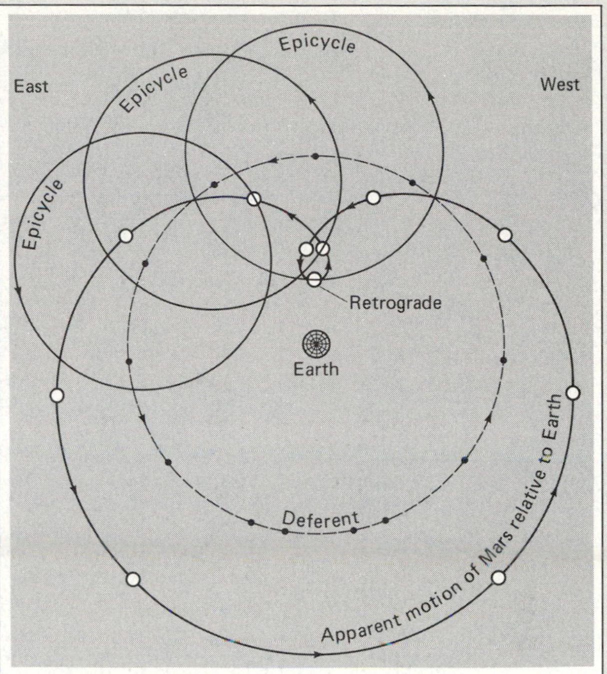

F.8

The motion of Mars, as seen from the earth, produced by an epicycle and deferent. Note that Mars is closest to the earth in the middle of its retrograde loop, so it appears at its brightest then. The period of the deferent is the average time it takes Mars to circuit the ecliptic; the period of the epicycle is the time between retrograde motions.

The motion of the deferent represented the planet's general motion west to east through the zodiac.

The periods of the epicycle and deferent are set from observations. They differ for the planets interior to and exterior to the sun. For the planets beyond the sun, the average time it takes a planet to return to the same place in the zodiac is the period of the deferent (Table 1.2). Mars, for example, has a deferent period of a little less than 2 years. The epicycle's period is just the average time between retrograde motions (with respect to the sun). With respect to the stars, though, the epicycle turns once in a year. For planets inside the sun, the period of the epicycle is the average time between retrograde motions. But the period of the deferent is one year. Note that although the treatments of the planets differ, they are symmetrical.

was a nonphysical, totally geometric device that broke with the assumption that planetary motion had to be uniform along circles.

Why did Ptolemy violate this aesthetic ideal? Probably because he demanded that his model fit observations reasonably well. With the equant, he could match the model predictions better to the observations, especially the annoying variations that the planets had in their overall smooth cycles. This correspondence apparently struck him as more important than sticking with the traditional precept of uniform, circular motion.

When completed, Ptolemy's model looked fairly simple (Fig. 2.6). But it treated Mercury and Venus differently from the rest of the planets. Recall (Section 1.4) that Mercury stays within about 23° of the sun, and Venus within 46°. To explain this fact, Ptolemy made Mercury's epicycle smaller than Venus's and demanded that the centers of these epicycles always lie on the line connecting the earth and the sun. So Mercury and Venus were constrained to stay near the sun (Fig. 2.6a).

In contrast, Mars, Jupiter, and Saturn can be anywhere along the zodiac relative to the sun. So the centers of their epicycles could be anywhere on the perimeter of the deferents. To ensure that these planets retrograde only at opposition, Ptolemy set the radii of their epicycles parallel to the earth-sun radius (Fig. 2.6b).

A key point: In Ptolemy's model the heavenly spheres are actual, physical spheres

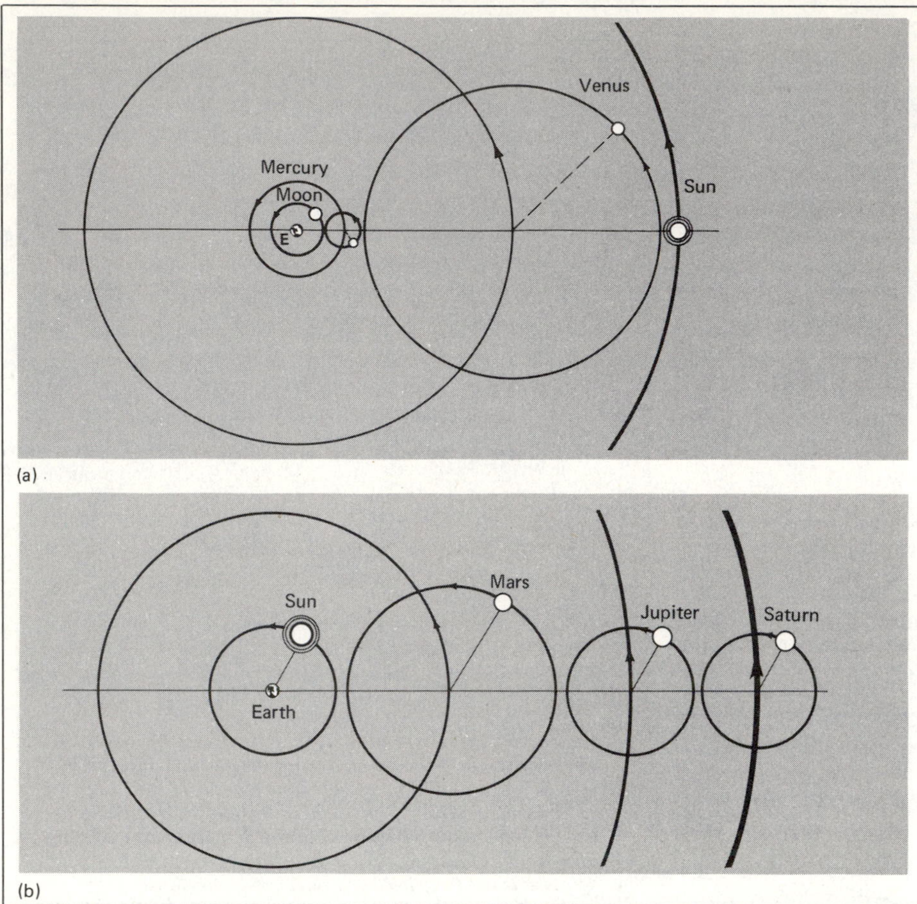

2.6
(a) *The Ptolemaic model for Mercury and Venus. The centers of the epicycles of these planets are fixed along a line with the sun; this explains the fact that they are never very far from the sun. The size of each epicycle accounts for the measured size of the maximum elongation angles.* (b) *A simplified Ptolemaic model for Mars, Jupiter, and Saturn. To account for the observed sizes of the retrograde loops, the epicycles of these planets decrease in size, so that Mars has the largest and Saturn the smallest. The radii of these epicycles must align with the earth-sun radius for the retrogrades to occur at opposition. The arrangement shown here emphasizes that point; in general, these planets can lie anywhere on their deferents.*

made of quintessence. The natural motion of these spheres—rotation—drives all the heavenly motions. No force is required.

Although Ptolemy's model (Fig. 2.7) lost acceptance centuries ago, you should not condemn it as a misguided step. This model remained in use for some 1400 years because it worked—it predicted planetary positions to the accuracy demanded by astronomers of the time (a few degrees). And it agreed with Greek physics and aesthetics. It survived because no other comprehensive model was advanced that competed effectively with it.

The size of the cosmos / Ptolemy's cosmos was finite; how large was it? Ptolemy gives the distances to the sun and the moon in terms of earth radii, as worked out by Aristotle and Hipparchus. To go farther out, he assumes that no space is wasted between the heavenly spheres. Then, with the earth-moon distance set at 59 earth radii, the distances out to Saturn can be laid out with the celestial spheres nesting tightly together. The sphere of the stars then lay 20,000 earth radii distant. It was a small cosmos—only about the distance we know today of the earth from the sun! But it did mark a reasonable try to establish the scale of the cosmos. Not until the twentieth century did astronomers finally break away from a small-sized universe.

KEY CONCEPTS

1 ***Scientific models are the core of the scientific process; models are designed to explain and predict what is observed; models contain geometric, physical, and aesthetic elements; scientific models are never complete or final—they evolve continually.***

2 ***Babylonian astronomers were excellent observers of celestial cycles but never developed scientific models for the cycles they observed; instead, they used mythical models.***

3 ***The Greeks made the first scientific models based on aesthetic notions and a desire to match observations; these included the concepts of harmony, symmetry, simplicity, and the use of geometry to "save the appearances."***

4 ***Aristotle developed a complete cosmological model based on physical ideas; different ideas of motion applied to the terrestrial and celestial realms; no forces were needed to turn the celestial spheres; the model was geo-***

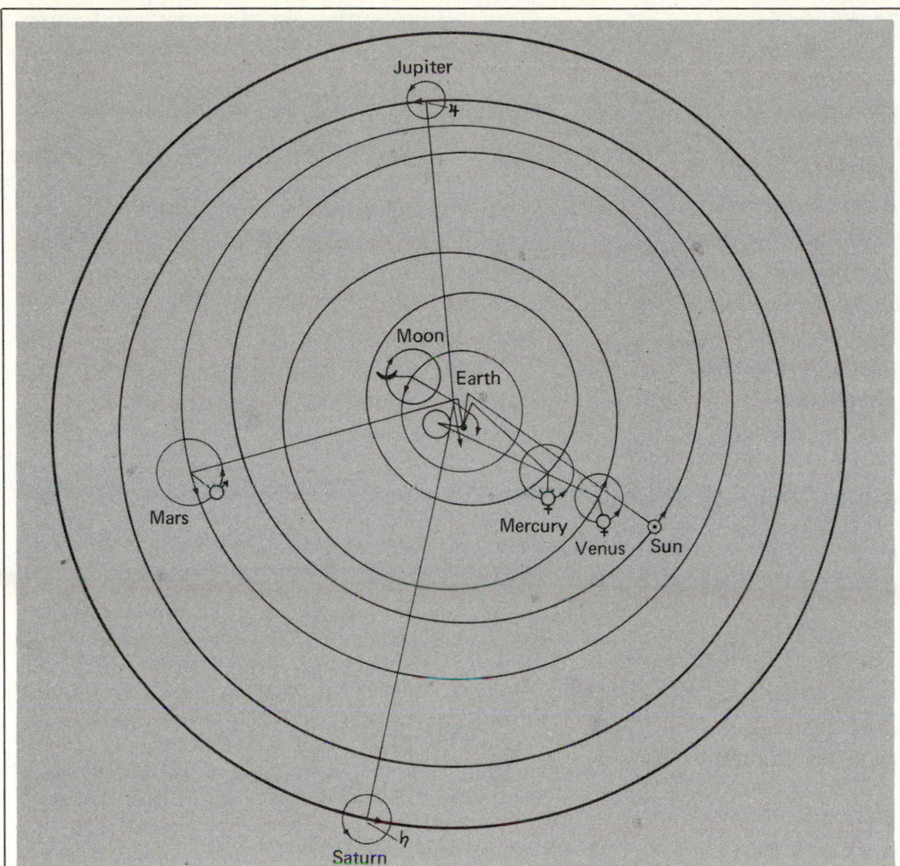

2.7
A view of the complete Ptolemaic model. Note that the centers of the deferents are offset from the earth; these have been made into eccentrics. This figure does* not *show the proper relative sizes of the epicycles or their correct alignments; these are given in Figure 2.6. (Adapted from a diagram by W. Stahlman)

TABLE 2.1 A summary of Ptolemy's model

Observation	*Explanation*
Motion of entire sky E to W in about 24 hours.	Daily motion E to W of the sphere of stars, carrying all other spheres with it.
Moon	
Monthly motion W to E compared to stars.	W to E motion of the moon's sphere in a month.
Sun	
Motion yearly W to E along ecliptic. Nonuniform rate along ecliptic.	Motion of sun's sphere W to E in a year. Eccentric—earth displaced from center of sun's circle.
Planets	
General motion W to E through zodiac.	Motion of deferent W to E; period set by observation of period of planet to go around ecliptic.
Retrograde motion.	Motion of epicycle in same direction as deferent; its period is the time between retrograde motions.
Variations in speed through zodiac, in retrograde motions.	Eccentrics, equants.
Mercury, Venus	
Greatest elongations of 23° and 46°; appear fixed to sun.	Size of deferents set by those angles; centers of deferents on earth-sun line.
Mars, Jupiter, Saturn	
Retrograde at opposition, when brightest.	Radii of epicycles aligned with earth-sun radius.

Notes 1. Motions of heavenly spheres were natural motions and did not require a force.
2. Use of equants *violated* the precept of uniform, circular motion for all heavenly bodies.
3. Overall accuracy: usually 5° or less, occasionally larger.

centric; it did not predict the motions very accurately.

5 ***Aristarchus proposed a heliocentric model, which never gained headway because it violated Aristotle's physics and predicted a stellar parallax, which was not observed.***

6 ***Claudius Ptolemy built up a comprehensive cosmological model (Table 2.1) based on the physics of Aristotle and the geometric ideas of the Greeks; for it he invented the equant, which violated the precept that celestial motions had to be uniform and circular; his model predicted planetary positions to acceptable accuracy (5° or less in general, usually about 1° or better).***

STUDY EXERCISES

1 What was one important astronomical achievement of the Babylonians? (*Objective 2*)

2 Contrast Babylonian and Greek astronomy in terms of their observational achievements. (*Objectives 1, 9, and 10*)

3 How did Aristotle's model explain
(a) the apparent lack of motion by the earth,
(b) the daily motion of the stars, and
(c) the annual motion of the sun through the zodiac? (*Objective 4*)

4 How did Aristotle argue against a heliocentric model? (*Objectives 3 and 4*)

5 In the Ptolemaic model, how did observations set the period of the epicycle? The deferent? (*Objective 6*)

6 How did Ptolemy treat Mercury and Venus differently from the other planets? (*Objective 6*)

7 How did Ptolemy violate his own precept of uniform, circular motion? (*Objectives 5, 6, and 7*)

8 What observation was Ptolemy trying to explain with each of the following geometric devices:
(a) epicycle,
(b) eccentric,
(c) equant? (*Objectives 6 and 7*)

9 State one strength of Ptolemy's model. (*Objective 8*)

BEYOND THIS BOOK . . .

A classic work in the astronomy of the Babylonians is *The Exact Sciences in Antiquity* (Dover, New York, 1969) by O. Neugebauer. Another one with a broader view is *A History of Science Through the Golden Age of Greece* (Harvard University Press, Cambridge, Mass., 1952) by G. Sarton.

You can find an English translation of the Almagest in volume 16 of *Great Books of the Western World* (Encyclopaedia Britannica, Chicago, 1952).

Was Ptolemy a fraud? Did he fabricate observations to conform with his model? Read *The Crime of Claudius Ptolemy* (Johns Hopkins University Press, Baltimore, Md., 1977) by R. Newton for one view. For a contrary view, read "Was Ptolemy a Fraud? " by O. Gingerich in *Quarterly Journal of the Royal Astronomical Society*, vol. 21 (1980), p. 253. Newton replies in the same journal, vol. 21 (1980), p. 388.

For more on the work of Aristarchus, read "From Aristarchus to Copernicus" by O. Gingerich in *Sky and Telescope*, November 1983, p. 410.

3/The new cosmic order

> *"He was rather weird," was Hans Castorp's view. "Some of the things he said were very queer: it sounded as if he meant to say that the sun revolves around the earth."*
>
> THOMAS MANN: *The Magic Mountain*

Learning objectives

After studying this chapter, you should be able to:

1. List the assumptions and arguments that Copernicus used to support his model and refute the Ptolemaic one.
2. Explain Copernicus's dislike of the equant and how it influenced the development of his heliocentric model.
3. Use a simple diagram to explain retrograde motion in the Copernican model.
4. Outline how in the Copernican model geometry and simple observations result in the relative distances of the planets from the sun.
5. Evaluate the strengths and weaknesses of the Copernican model compared to the Ptolemaic one.
6. Describe how Copernican ideas influenced the work of Kepler.
7. Argue that Kepler, not Copernicus, deserves the title of the *first astrophysicist.*
8. Describe the important geometric properties of ellipses.
9. Compare and contrast the Copernican model and the Keplerian one in terms of physics, simplicity, geometry, and prediction.
10. State Kepler's three laws of planetary motion and apply them to appropiate astronomical situations.

Central question

How did Copernicus's sun-centered model explain the motion of the planets and ignite a revolution in cosmological thought?

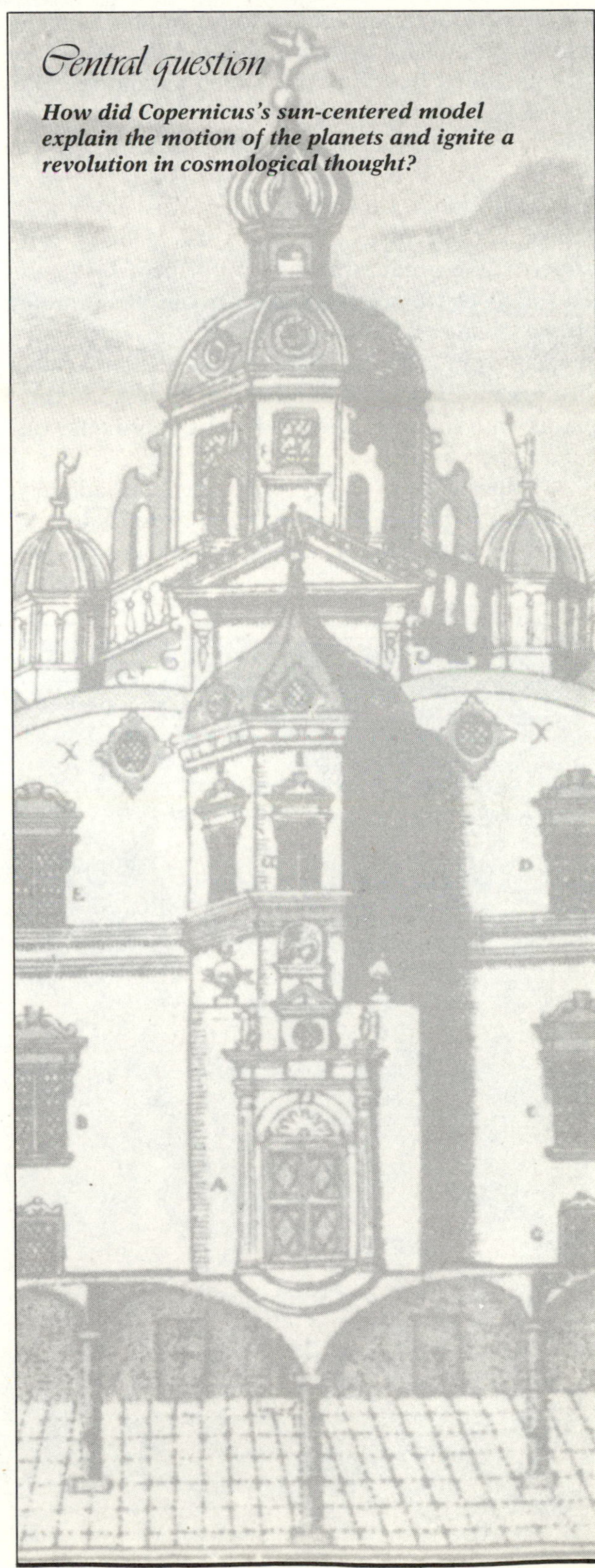

Astronomy languished with the decline of Greek civilization in the first few centuries after Ptolemy. Most of the Greek astronomical works were unknown in Europe until the twelfth century, when Arabic manuscripts of Greek thought were translated into Latin. In the thirteenth century Alfonso the Great of Spain sponsored the publication of the Alfonsine Tables, handy lists of planetary positions based on the Ptolemaic model. The essential Ptolemaic model worked well enough; practicing astronomers had no reason to discard it.

One medieval astronomer was unhappy: Nicolaus Copernicus. In the year of his death, Copernicus's great work was published: *De Revolutionibus Orbium Coelestium* (On the Revolutions of the Heavenly Spheres). In it, the earth was shaken from a static place at the center of the cosmos into a path around the sun—just another planet with the rest. The model of Copernicus was *heliocentric*: sun-centered.

Copernicus's view did not immediately wrench the minds of astronomers from their geocentric notions. Some, like the Danish observer Tycho Brahe, considered Copernican claims but finally rejected them. Others, like Johannes Kepler, were struck by essential harmonies in Copernicus's model. This appeal motivated Kepler to find a physical basis for it. The revolution in astronomy after Copernicus marked the first major shift in our concept of the earth's place in the cosmos: from geocentric to heliocentric. This revolution also injected a new concept into the models: that of a physical force working in the heavens.

3.1 Copernicus the conservative

In the sixteenth and seventeenth centuries a new model of the cosmos was born. Nicolaus Copernicus (1473–1543) initiated this revolution (Fig. 3.1) by developing a new model of the cosmos when the old one seemed adequate.

The heliocentric concept / How did the sun come to play a central role in Copernicus's mind? While in Italy to study medicine and law, Copernicus became familiar with the works of Aristotle, Pythagoras, and Plato. An offshoot of Plato's

3.1 ***Nicolaus Copernicus. "In the center rests the sun." (Painting by Maxim Kopf, courtesy Harvard College Observatory)***

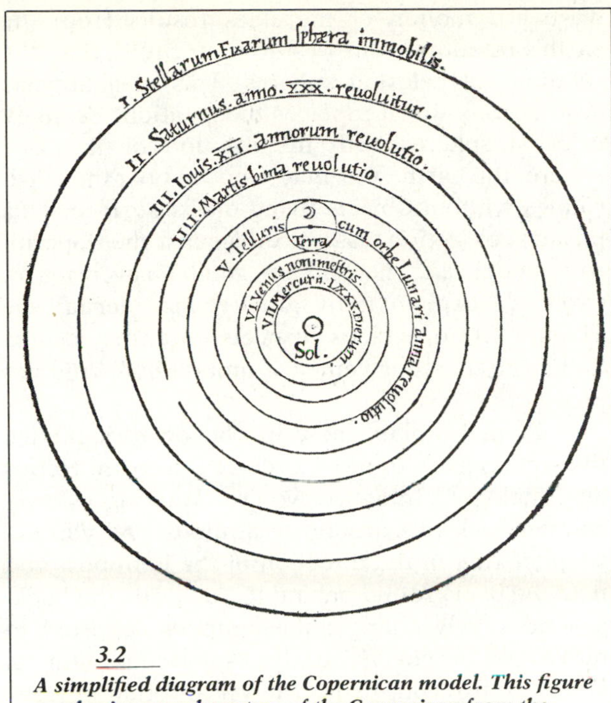

3.2

A simplified diagram of the Copernican model. This figure emphasizes one departure of the Copernican from the Ptolemaic model: The sun is placed in the center of the cosmos with the planets, including the earth, revolving around it. (Courtesy Yerkes Observatory)

philosophy—Neoplatonism—asserted that the sun is the source of the godhead and of all knowledge. These ideas singled out the sun as a body quite different from the planets and perhaps encouraged Copernicus to ponder a heliocentric model.

About 1529 Copernicus wrote a summary of his new model and circulated it to some friends. In it, he presented a general outline of his heliocentric model (Fig. 3.2): The sun replaced the earth at the center of the cosmos, and the earth revolved around the sun yearly and also rotated on its axis daily. (These ideas had been first outlined by Aristarchus; see Section 2.3.)

Copernicus still assumed that all celestial motions must be uniform, circular motions. He viewed Ptolemy's equant as a violation of the ideal and so attacked the Ptolemaic model as "not sufficiently pleasing to the mind." His new model would reinstate uniform circular motion to its proper aesthetic status.

Just what compelled Copernicus to offer a new model has always puzzled me. The basic Ptolemaic model worked well enough, by the demands of the day, in its planetary predictions. Contrary to some tales, the Ptolemaic model did not require the continual addition of circles to match observations and so did not become a monstrosity in medieval times. Copernicus would take 20 years to develop his model; yet his predictions came out no better than those based on the Ptolemaic model.

However, Copernicus asserted that his model surpassed the Ptolemaic one on aesthetic grounds. He focused on the notions of harmony and order. What were these aesthetically pleasing aspects of his model?

***The plan of* De Revolutionibus** / Long after Copernicus had privately distributed his ideas, his work became known to two astronomers at the University of Wittenberg: Georg Rheticus and Erasmus Reinhold. Fascinated by the new model, Rheticus visited the aging Copernicus, who showed him a final manuscript copy of *De Revolutionibus*. Rheticus begged Copernicus to allow its publication. (Not until Copernicus's lifetime had printed texts, rather than manuscript copies, become common.) Copernicus agreed. After much trouble with the printer, the book was published in April 1543. Meanwhile, Copernicus had suffered a stroke and was confined to his bed. Although the book was delivered to him before he died in June, he probably did not read it.

Well that he didn't, for the final version of his life's work had been given a new title! The original one was simply *De Revolutionibus* (On the Revolutions); the published title sported two more words: *De Revolutionibus Orbium Coelestium* (On the Revolutions of the Heavenly Spheres). In addition, an unsigned preface had been added. It stated that the work contained a new hypothesis, a heliocentric cosmos, to compute planetary positions. But it was not to be taken as a statement about the real world.

For some years this statement was attributed to Copernicus, with the implication that he did not believe in the reality of his model. But Johannes Kepler (see Section 3.4) discovered that the preface had been written by Andrew Osiander, a clergyman who oversaw the completion of the book's publication. Osiander had needed the disclaimer to elicit Protestant approval for publication of a book written by a Catholic. He may have also changed the title to emphasize the motions of the heavens rather than those of the earth.

De Revolutionibus took after the Almagest in outline and basic intention: to explain planetary motions using only uniform circular motions—the traditional precept of the Greeks. Copernicus thought that although Ptolemy's model matched observations, it clashed with this ideal because it used equants. An equant requires that a planet no longer move uniformly around the center of its deferent circle. Copernicus was greatly offended by the equant; it struck him as the major objection to the Ptolemaic model. In a basically conservative mood, he wished to devise a model faithful to uniform circular motion and so eliminate the equant. To do this, Copernicus had to discover a natural explanation for retrograde motion and a new harmony for the celestial spheres, which related the planets' distances to their periods around the sun.

3.2 The heliocentric model of Copernicus

In the introduction to *De Revolutionibus*, Copernicus, like Ptolemy, lays out his assumptions. First, he requires that the planets move in circular paths around the sun at uniform speeds. Second, the closer the planet is to the sun, the greater its speed. For instance, because Mercury is closer to the sun than the earth, its speed around the sun is faster than the earth's. Except for using the sun as the center of motion, these are identical to Ptolemy's assumptions. For the rest, Copernicus treads a different ground; here's a short summary of his ideas:

1. All the heavenly spheres revolve around the sun, with the sun at the center of the cosmos.
2. The distance from the earth to the sphere of stars is much greater than the distance from the earth to the sun.
3. The daily motion of the heavens relative to the horizon results from the earth's motion about its axis.
4. The apparent motion of the sun relative to the stars results from the annual revolution of the earth around the sun.
5. The planets' retrograde motions occur from the motion of the earth relative to the other planets.

Let me expand on these points to emphasize key differences between the Copernican and Ptolemaic models.

In point 1, Copernicus asserts that the cosmos is heliocentric rather than geocentric. He concludes so for philosophical and aesthetic reasons. But as the Greeks realized for Aristarchus's model, a heliocentric scheme demands a stellar parallax (see Focus 2.1) that had never been observed. Point 2 addresses this issue: If the stars are very far away, the parallax would be so small as not to be detectable by naked-eye observations. (Later telescopes would show that Copernicus's intuition about the vast distances to stars was correct.)

In point 3, Copernicus declares that the daily westward motion of the skies results from the earth's rotation (from west to east) rather than the rotations of celestial spheres. This idea appeals aesthetically, for it replaces the rotations of many celestial spheres with the rotation of just one sphere, the earth. You may wonder how the earth rotates without objects flying off its surface. This physical objection was voiced against the Copernican model, and he had no good answer for it. (Now we explain it by gravity and inertia; see Chapter 4.) Copernicus suspected that the cosmos had more than one center of motion but could not prove it.

Point 4 explains how, in a heliocentric model, the sun appears to move around the earth. Here's an analogy: Imagine you're walking slowly counterclockwise around a lamppost. As you do, keep looking in the direction of the lamppost and note the background behind it. You'll see the background slowly change; the lamppost appears to move counterclockwise (the same direction as you're going) with respect to background objects. Now imagine that the lamppost is the sun, you are the earth revolving around the sun, and the background is the stars of the zodiac (Fig. 3.3). From the earth you see the sun in a constellation, say Leo. As the earth revolves counterclockwise, the stars behind the sun change. After one month the sun appears in Virgo, one constellation to the east. The sun seems to have moved, relative to the stars, counterclockwise. Actually, the earth has moved; it's the line of sight from the earth to the sun that has changed.

Point 5 makes retrograde motion a natural result of the planets' revolutions. Retrograde motion arises from the planets chasing one another around, the faster (inner) planets regularly passing the slower (outer) ones. An analogy: When you pass a car on a highway, that slower car appears to move backward with respect to background objects. Similarly, as the earth speeds past a planet, that planet seems to move westward (backward) against the backdrop of stars.

Retrograde motion explained naturally / In the Ptolemaic model, epicycles moving on deferents generated the retrograde motions. By showing that retrograde motion arose naturally in his model, Copernicus eliminated five Ptolemaic epicycles. In this way, he made a simpler model by noting that the relative motions of the earth and planets produces retrograde motions. When the earth passes any of the outer ones or when the earth is passed by the inner ones, retrograde motion occurs. It is the passing that is the key.

Take Jupiter as an example (Fig. 3.4). Its retrograde motion takes place at opposition. In a

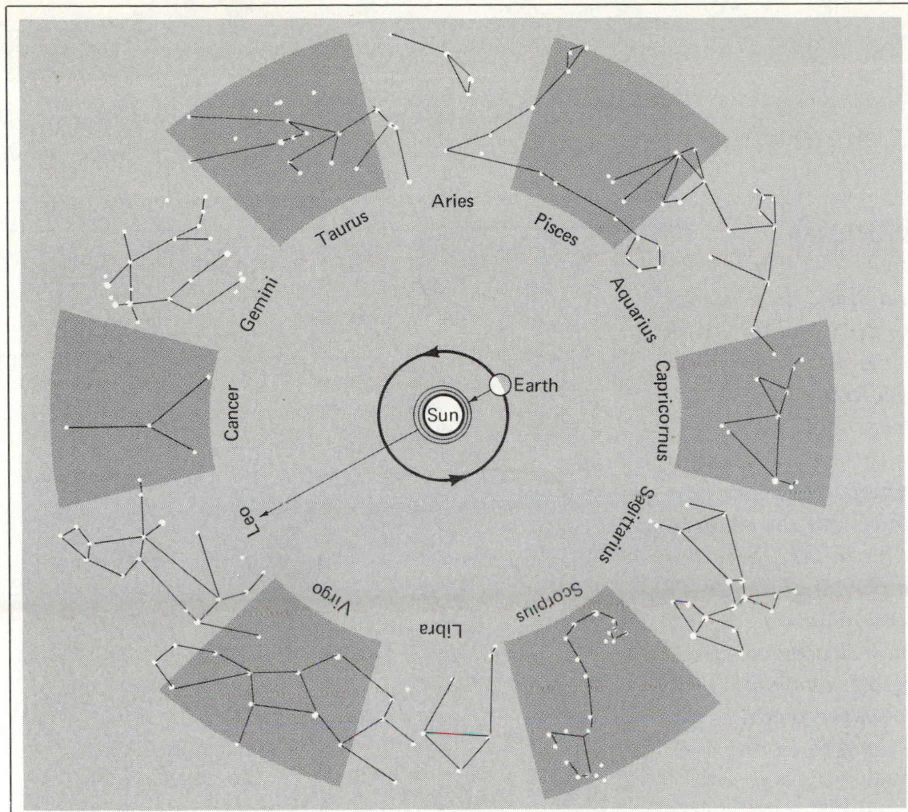

3.3

The sun's apparent motion through the zodiac in a heliocentric model. As the earth travels around the sun (counterclockwise; west to east), the line of sight to the sun and toward the background stars moves in the same direction (west to east). For example, start with the sun in Leo. A month later the earth will have moved eastward enough so that Virgo lies behind the sun. The sun seems to move eastward through the zodiacal stars at the rate of about one constellation a month.

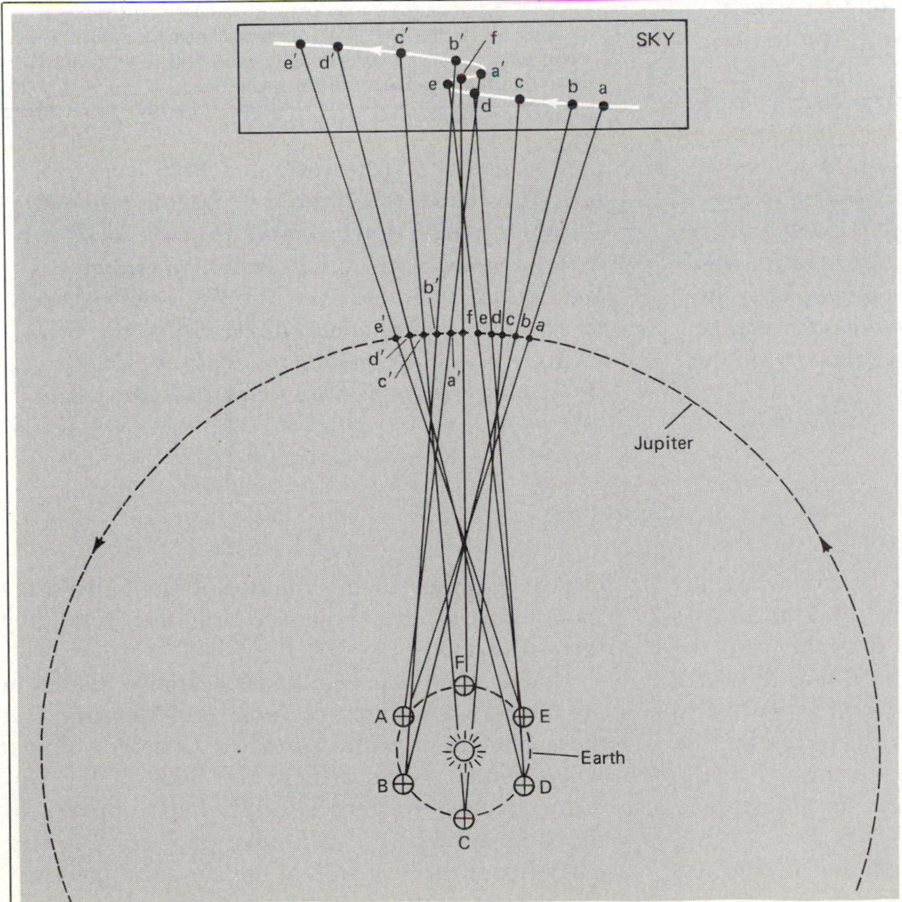

3.4

Retrograde motion of Jupiter in the Copernican model. As the earth comes around the same side of the sun as Jupiter, it is moving faster along its path and so passes Jupiter (point F). During this passing interval, Jupiter appears to move backward (to the west) with respect to the stars. Note that Jupiter appears in the middle of its retrograde motion just as the earth passes it, so Jupiter is in opposition to the sun as viewed from the earth. Also, it is closest to the earth, so it appears its brightest in the sky. The same basic diagram applies to Mars and Saturn; they undergo retrograde when the earth passes them. The illusion of retrograde motion results from the passing situation.

Focus 3.1

SIDEREAL AND SYNODIC PERIODS IN THE COPERNICAN MODEL

From the earth you can watch a planet in a particular alignment with the sun—opposition, for example. The time interval for that planet to return to the same alignment with respect to the sun is called the **synodic period**. *In contrast, consider the time it takes for a planet to return to a specific location in the zodiac. That's the planet's* **sidereal period**, *the amount of time it takes for the planet to go once around the sky relative to the stars. For example, Jupiter has a synodic period of about 400 days and a sidereal one of almost 12 years.*

Note that the synodic period is a geocentric property, while the sidereal period is a heliocentric one. From the earth, you can observe synodic periods directly. How can you transform these into sidereal ones? Follow Copernicus: Assume that the planets move in circles and that the closer a planet's path is to the sun, the faster it moves. So its sidereal period is shorter. You then have a chase situation, where an inner planet always catches up to an outer one.

Imagine that two adjacent planets start out lined up (Fig. F.9), with the outer planet in opposition as seen from the interior one. How long will it be until they are aligned again? Note that the inner planet will make one circuit around the sun and then catch up to the outer one to re-form the alignment. In other words, the inner planet laps the outer one. So as seen from the inner planet, the synodic period will be longer than its sidereal one.

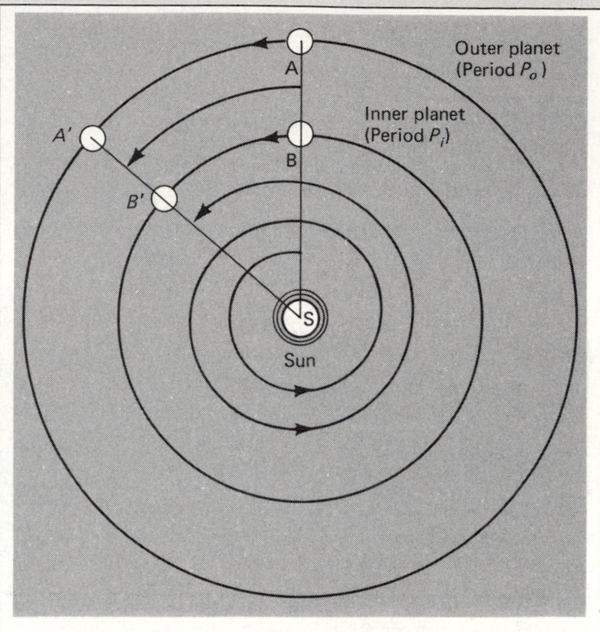

F.9
Synodic and sidereal periods in the Copernican model. The inner planet moves faster along its path and so must have a shorter sidereal period than the outer one.

The inner planet moves at $360/P_i$ degrees per day. (P_i is the sidereal period of the inner planet.) The outer planet, moving at the slower rate of $360/P_o$ degrees per day, lags behind the inner planet more and more each day. (P_o is the sidereal period of the outer planet.) At the end of one day, the inner planet has gained an angle of $(360/P_i - 360/P_o)$ degrees on the outer one. This day's gain

heliocentric model, opposition occurs when the earth is in line between the planet and the sun. As the earth approaches, the line of sight from the earth to Jupiter moves eastward. But as the earth passes Jupiter, the line of sight swings westward relative to the stars. As the earth moves on, the line of sight eventually moves eastward again. Jupiter undergoes the illusion of retrograde motion as the earth passes it.

Note (Fig. 3.4) that the earth and Jupiter are closest together in the middle of the retrograde motion. So Jupiter should be brightest in the sky then—and it is!

The same chase-and-pass scenario results in the retrograde motions of Venus and Mercury. But these planets pass the earth by. Consider Venus. When Venus moves around the back of the sun (from greatest western elongation), it appears to move eastward. But as Venus catches up to the earth (from the east side of the sun) and passes it,

is equal to 360/S degrees, where S is the synodic period of either planet seen from the other. In algebraic terms,

$$\frac{360^\circ}{S} = \frac{360^\circ}{P_i} - \frac{360^\circ}{P_o}$$

or, by dividing by 360_o,

$$\frac{1}{S} = \frac{1}{P_i} - \frac{1}{P_o}$$

Take the earth as the inner planet and Mars the outer one. Since the earth's sidereal period is one year, $P_i = 1$ ***and***

$$\frac{1}{S} = 1 - \frac{1}{P_{Mars}}$$

where P_{Mars}, ***the sidereal period of Mars, and S, the synodic period of the earth and Mars, are in years.***

Now consider the earth as the outer planet and Venus as the inner one. Then $P_o = 1$ ***is the earth's sidereal period, and***

$$\frac{1}{S} = \frac{1}{P_{Venus}} - 1$$

where all the quantities are in years. The synodic period of Venus is 585 days, or 585/365 = 1.6 years. Using the equation above,

$$\frac{1}{P_{Venus}} = 1 + \frac{1}{1.6}$$

$$\frac{1}{P_{Venus}} = 1.62$$

$$P_{Venus} = 0.62 \text{ year}$$

This transformation from synodic to sidereal periods played a key role in the Copernican model because it places the planets in a natural order from the sun, from the shortest (Mercury) to the longest (Saturn) periods.

TABLE 3.1 ***Copernicus's relative distances of the planets***

Planet	*Copernicus's value (AU)*	*Modern value (AU)*
Mercury	0.38	0.387
Venus	0.72	0.723
Earth	1.00	1.00
Mars	1.52	1.52
Jupiter	5.22	5.20
Saturn	9.17	9.54

Note AU is the abbreviation for *astronomical unit,* the average earth-sun distance.

you observe Venus to move westward with respect to the stars (toward the west side of the sun). So for the two inner planets, retrograde arises from relative motion, but the role of the earth is reversed. Note that the passing of any two planets produces the illusion of retrograde motion of one as seen from the other.

Planetary distances / Copernicus achieved another basic goal: the establishment of the order and distances of the planets from the sun from their observed synodic periods. (The *synodic period* is the time required for a planet in some alignment with the sun as seen from the earth—such as opposition—to move through the sky and return to that same alignment again.) From the synodic periods Copernicus calculated the *sidereal periods*: the periods of revolution of the planets with respect to the stars, as seen from the sun (Focus 3.1).

Copernicus found that the planetary order from the sun fell into a natural sequence when based on sidereal periods: Mercury, with the shortest period, was closest to the sun; Saturn, with the longest, was the farthest away. He felt a harmony in this sequence, which later scientists such as Kepler and Galileo considered an essential elegance of the heliocentric model.

Copernicus also calculated the distances of the planets from the sun relative to the earth-sun distance (which is now called the astronomical unit, abbreviated AU). His figures for the distances came from direct observations and simple geometric arguments (Table 3.1 and Focus 3.2). Copernicus was very pleased with this result; here he found a harmony of the motions of the planets and their distances from the sun.

Problems with the heliocentric model / Copernicus eliminated epicycles to explain retrograde motion and at the same time expelled the equant. Yet the heliocentric model in practice did not predict planetary positions any better than the Ptolemaic model. In fact, the Copernican model was not even simpler, judged on the basis of a count of circles. Copernicus had eliminated five, but he had to add many smaller ones to account for the variations in planetary motions, which Ptolemy had accounted for with the equant. His

Focus 3.2

RELATIVE DISTANCES OF THE PLANETS IN THE COPERNICAN MODEL

The planets' distances from the sun, relative to the earth-sun distance (called one astronomical unit, or AU) can be found directly from observations in the Copernican model. The method differs, however, for planets interior to the earth and for those exterior to it. It assumes circular orbits.

For Mercury and Venus, the inner planets, the method rests on the observed maximum elongation angle of the planet from the sun (see Section 1.4). This angle averages about 23° for Mercury and 46° for Venus. Here's the procedure for Venus.

In a heliocentric model, maximum elongation for Venus occurs when the line of sight from the earth to Venus hits tangent to Venus's orbit (Fig. F.10). So the line from the sun to the line of sight makes a right angle. Draw a triangle with one side the sun-Venus distance (SV), one side the sun-earth distance (SE, equal to 1 AU), and the third the earth-Venus line (EV). Note that it is a right triangle.

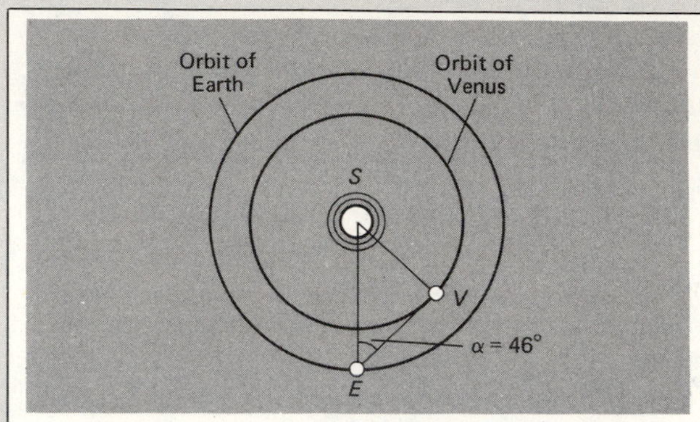

F.10
The distance of Venus from the sun. As seen from the earth at its time of greatest elongation, Venus makes an angle (α) of about 46° from the sun. The line of sight from the earth then touches tangent to the orbit of Venus. So the radius of Venus's orbit drawn to this point makes a right triangle with one side the sun-Venus distance (SV) and the other the earth-sun distance (ES).

complete model (Fig. 3.5) actually came out somewhat more complicated than Ptolemy's original model.

Copernicus also sidestepped the fact that his model violated Aristotelian physics. He did not offer new physical ideas to support his model. Those still needed development. Copernicus had created a new geometric model, not a physical one.

Warning: Don't fall into the trap of thinking that the Copernican model included a force between the sun and the planets. It did not. Planetary motions arose from the natural motions of the celestial spheres—uniform rotation. As in the Ptolemaic model, this was natural motion without forces.

The impact of the heliocentric model / Although Copernicus's claims clashed with traditional ones, some astronomers in the middle of the sixteenth century tested the new setup for the cosmos; they knew that the venerable Alfonsine Tables contained some inaccurate positions.

Foremost among those willing to try the new model was Erasmus Reinhold, the professor at Wittenberg who had encouraged Rheticus to visit Copernicus. Reinhold made up planetary position tables based on the new model; these were adopted by almanac makers as the Prutenic Tables. Yet Reinhold did not swing completely to the Copernican model; he saw it as a geometric rather than a physical design.

As noted earlier, the Copernican model did not predict planetary positions any better than the Ptolemaic one. Why not? Because Copernicus held to uniform circular motions, so his model was geometrically similar to Ptolemy's (with the exception of the placement of the earth and sun). What astronomers would switch if the new model offered no advantages over the old?

Then why did Copernicus bother to develop his model at all? Certainly, he was motivated by aesthetic reasons: He saw a new harmony in the "design of the universe and fixed symmetry in its parts" that was "pleasing to the mind." The "fixed

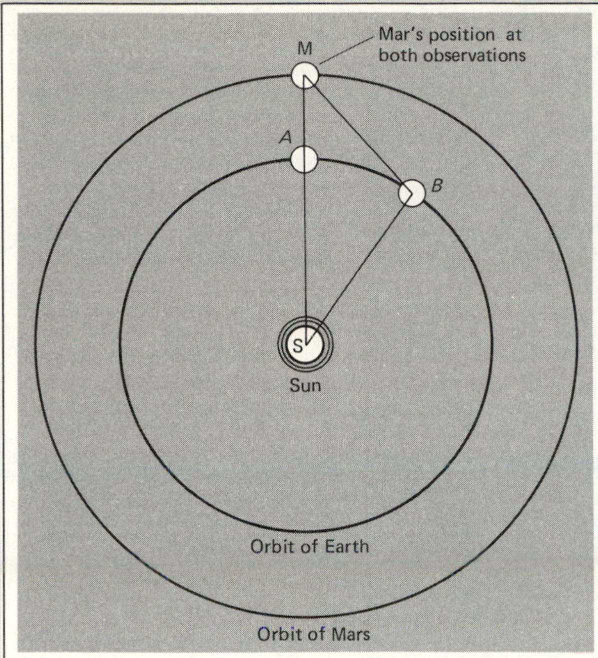

F.11

The distance of Mars from the sun. Start with the earth at A and Mars at M at opposition. Wait one sidereal period of Mars. Then Mars returns to the same place (M) in its orbit (as seen from the sun), but the earth has gone around almost twice (to B). The line of sight from the earth to Mars crosses the Martian orbit at the same spot that the earlier one did (from A). Call the earth-sun distance one unit. Then the distance from the sun to Mars is known to this scale because all the angles are known.

The elongation angle, α, is observed. Let SE be one unit long. Then draw a triangle to scale with α, and Venus's distance from the sun (SV) will be 0.72 the earth's (SE). And since SE is 1 AU, Venus is 0.72 AU from the sun. Similarly, use 23° as the maximum elongation angle for Mercury to find its distance as 0.39 AU. Note that just one observation establishes the distances of the inner planets in a heliocentric model.

A similar method applies to the outer planets. Take Mars as an example. Start with some alignment of the earth and Mars (A in Fig. F.11). Wait one sidereal period of Mars; it returns to the same position in its orbit relative to the stars as seen from the sun (see Focus 3.1). In contrast, the earth—moving faster—has gone around more than once (B in Fig. F.11). Triangulate the position of Mars from these two observations from the earth by extending the lines of sight for both. Where the lines intersect is the position of Mars in its orbit (M in Fig. F.11).

Suppose you have made an accurate scale drawing of these observations. Calling the earth-sun distance 1 AU, you can also measure the sun-Mars distance in AU. Or you can solve the triangle MSB for side EM using trigonometry, because you know all the angles. A similar technique works for the other outer planets. These distances (Table 3.1) are fixed by the planets' periods and the geometry of the heliocentric model.

symmetry" referred to the fact that in the Copernican model the spacing of the planets is fixed by observations; in the Ptolemaic model, each planet's circles can be treated independently of the others. Overall, Copernicus's model had better unity than the Ptolemaic one, so it was simpler.

3.3
Tycho Brahe: First master of astronomical measurement

Tycho Brahe (1546–1601) rejected the heliocentric model on both physical and observational grounds. He proposed an alternative model that mixed the classic geocentric model with the Copernican one. His work is infused with the attitudes typical of a professional astronomer at the time just after the publication of *De Revolutionibus*; it emphasizes that the reality of the Copernican model escaped most astronomers of the day, who saw the heliocentric model as merely a geometric device. Copernicus invited this stance because his model violated Aristotelian physics and offered no alternative physical basis.

Tycho Brahe (Fig. 3.6) was born in Denmark into a family of noble standing. While he was a child, his uncle—who had no children—stole him from his parents, supported the boy, and decided that his adopted son should become a lawyer. Young Tycho was sent off to study law, but secretly he worked on astronomy, spent his allowance on astronomy books, and sneaked out at night to make observations.

In November 1572 a new star, called then a *nova*, burst into view in the constellation of Cassiopeia. (Today we know that a nova is a star normally too faint to be seen that suddenly increases many times in brightness. Such an apparently new star is called a nova from the Latin stella nova, "new star." In the case of the 1572 nova, the event marked an extreme outburst now called a *supernova*.) At its brightest, the nova could be seen during the day.

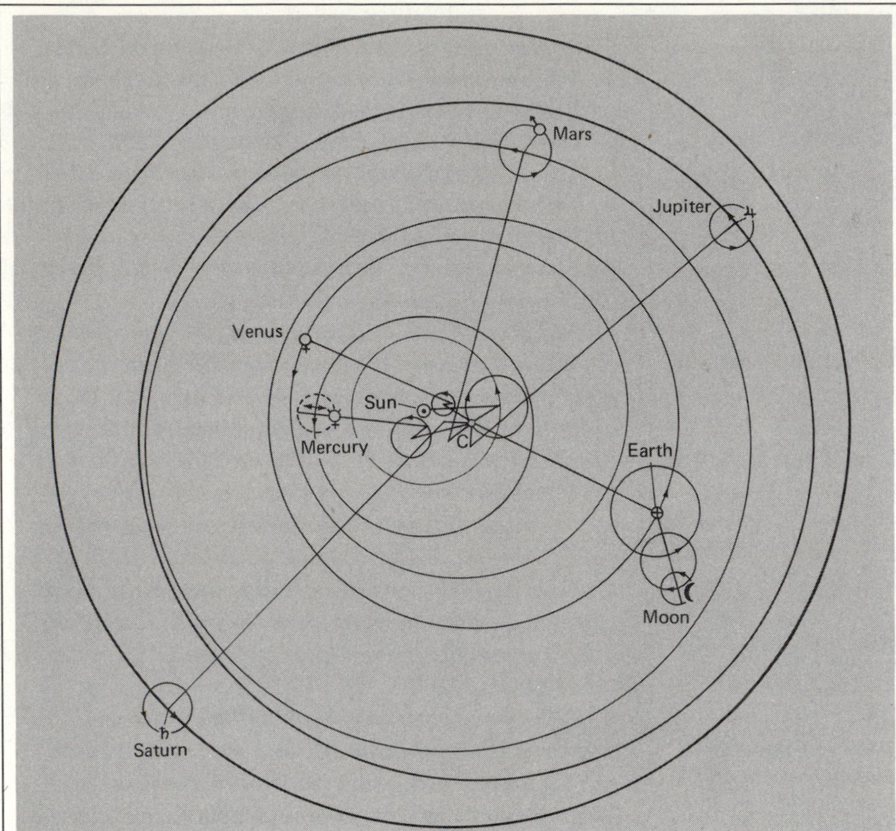

3.5

Details of the Copernican model. Observations forced Copernicus to add eccentrics and small epicycles (called* epicyclets*) to his model to account for irregularities in the planets' motions. The overall result resembled the complete Ptolemaic model. Compare to Fig. 2.7. (Adapted from a diagram by W. Stahlman)

3.6

A portrait of Tycho Brahe, showing the part of his nose that was cut off in a fencing duel and replaced with a silver piece.

Tycho observed this new star for over 2 years and collected observations from all over Europe of the star's position in the sky. From these and his own data, Tycho showed that the nova was in the same place in Cassiopeia from all places it was observed. That is, the nova did not have an observable angular shift from different places on the earth. So it was at a great distance from the earth; Tycho concluded that is must be in the sphere of stars. This fact contradicted the Aristotelian doctrine that the heavens must be immutable.

Tycho's work on the new star of 1572 catapulted him to fame as an astronomer. King Frederick II of Denmark then made him an offer he could not refuse: his own observatory on the island of Hveen. With royal funds—estimated by Tycho to be more than a ton of gold—Tycho built on Hveen the first modern astronomical observatory (Fig. 3.7): Uraniborg, Castle of the Heavens. Here he worked in grand style: He had the finest observing equipment (designed by himself) and a bevy of assistants. His instruments had an accuracy of 1′, as well as can be done with the naked eye. At Uraniborg, Tycho began the first continuous observational program. So planets were

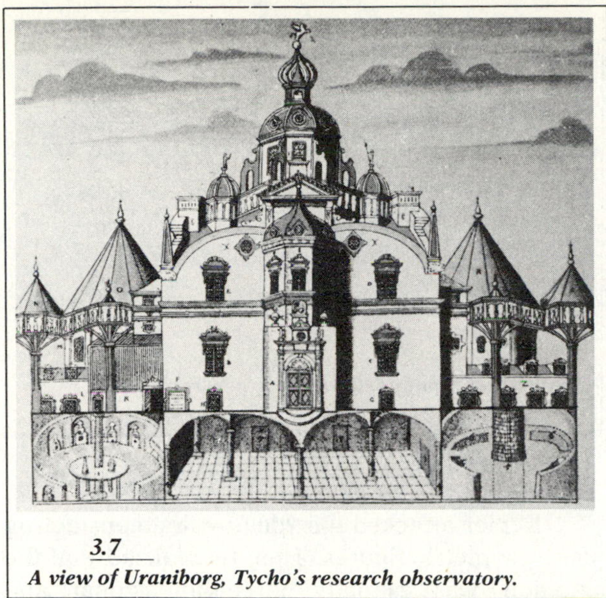

3.7
A view of Uraniborg, Tycho's research observatory.

observed not only at times of importance, such as during retrogrades, but also at times in between. No one had done this before. Tycho eventually amassed over 20 years (1576–1597) of records of precise planetary positions.

Tycho concluded that the nova of 1572 made the classic geocentric model untenable, so he devised his own. He had read about the Copernican model, but he could not accept the earth revolving around the sun. He thought that "the earth, that hulking, lazy body" was "unfit for a motion as quick as that of the ethereal torches." So Tycho made a model in which the moon and sun revolve around the earth, but all the planets revolve around the sun (Fig. 3.8). This was geometrically the same as the Copernican model—only the center of motion was changed.

Also, Tycho had an observational objection to a heliocentric model. He tried to measure a heliocentric parallax (see Focus 2.1) but failed to detect it. The lack of a detectable parallax fortified Tycho's belief that the Copernican model was invalid. (We now know that the largest heliocentric parallaxes are almost 100 times smaller than Tycho could detect.)

When Frederick II died, Tycho fell from royal favor because of his despotic ways. So he moved with his equipment to Prague to work for Emperor Rudolph of Bohemia. Here he took on the young Johannes Kepler as an assistant. When Tycho died in 1601, an era in astronomy ended. A new one was born when Kepler created a new astronomy based on Tycho's observations.

DE COMETA ANNI 1577. 189

NOVA MVNDANI SYSTEMATIS HYPOTYPOSIS ab Authore nuper adinuenta, qua tum vetus illa Ptolemaica redundantia & inconcinnitas, tum etiam recens Copernicana in motu Terræ Physica absurditas, excluduntur, omniaq; Apparentiis Cœlestibus aptissime correspondent.

A A 3 Pleniorem

3.8
A simplified version of Tycho's model for the cosmos, a geocentric modification of the Copernican. The sun moves around the earth, and all the planets circle around the sun. (Photo courtesy O. Gingerich; reprinted by permission of Houghton Library, Harvard University)

3.4 Johannes Kepler and the cosmic harmonies

Despite its making the earth a planet like the others, the Copernican model was conservative in its intent: to ensure the use of uniform, circular motions denied by the equant in the Ptolemaic model. This Copernicus did, but his model violated Aristotelian physics. Ptolemy's model, because it kept the earth in the center of the cosmos, was based on traditional physical ideas. So the heliocentric model lacked a physical basis.

Johannes Kepler (1571–1630) forged the new ideas about planetary motion that form the foundation of modern concepts about them (Fig. 3.9). Kepler shaped the Copernican model into a truly *physical* one in which the sun determines the planetary orbits by some force. Kepler also simplified the model and perfected its usefulness for the prediction of planetary positions.

The harmonies of the spheres / Kepler was born on December 27, 1571, in the small town of Weil der Stadt in southwestern Germany. In 1589 Kepler entered the University of Tübingen, where he was an outstanding student. In his last year at

3.9
Johannes Kepler. "Astronomy has two ends, to save the appearances and to contemplate the true form of the edifice of the world." (Courtesy O. Gingerich)

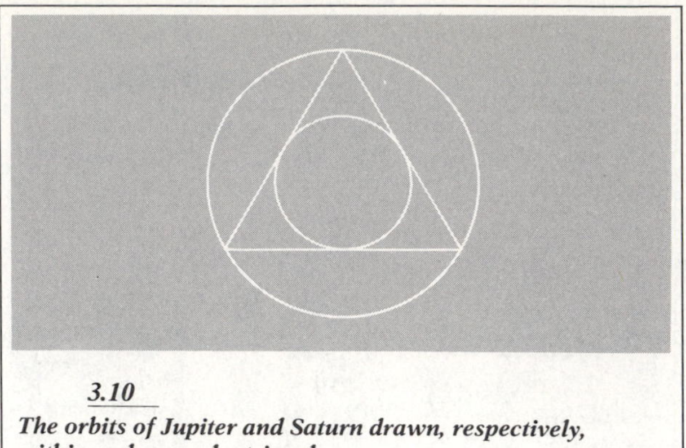

3.10
The orbits of Jupiter and Saturn drawn, respectively, within and around a triangle.

Tübingen, Kepler was selected to replace a teacher of mathematics at a Protestant high school in Graz, Austria. Here he made his first key astronomical discoveries.

One day while teaching, Kepler was struck accidentally by the most compelling insight of his life. He was telling his class about conjunctions of Jupiter and Saturn, which occur at regular intervals. He saw that the ratio of the radii of two circles—one drawn around a triangle and one drawn inside it (Fig. 3.10)—was about 2 to 1. That was almost the same as the ratio of the distances of Saturn and Jupiter from the sun in the Copernican model, 1.83 to 1. Kepler found this result exciting, for Saturn was the first planet in from the sphere of stars, and the triangle was the simplest plane figure. He felt he had found a geometric design for the solar system.

While at Tübingen, Kepler had learned about the Copernican model. He was pleased by the harmonies he found in the model, which he adopted in his thinking. One such harmony appears in the spacing of the planets and their periods around the sun (Focuses 3.1 and 3.2). Copernicus had not explained this relationship. Kepler thought he could: Perhaps a geometric layout determined the planets' spacings.

Kepler attacked this idea by first considering plane geometric figures. Then he realized that the situation was actually three-dimensional and required solid figures. From classical geometry Kepler knew that only five regular solids existed, those with faces having the same kind of regular plane figures. The known planets numbered six, separated by five spaces. So Kepler felt that these five regular solids might establish the spacing of the planetary spheres. He worked out a nesting scheme of solids and spheres (Fig. 3.11) that gives distances from the sun within a few percent of

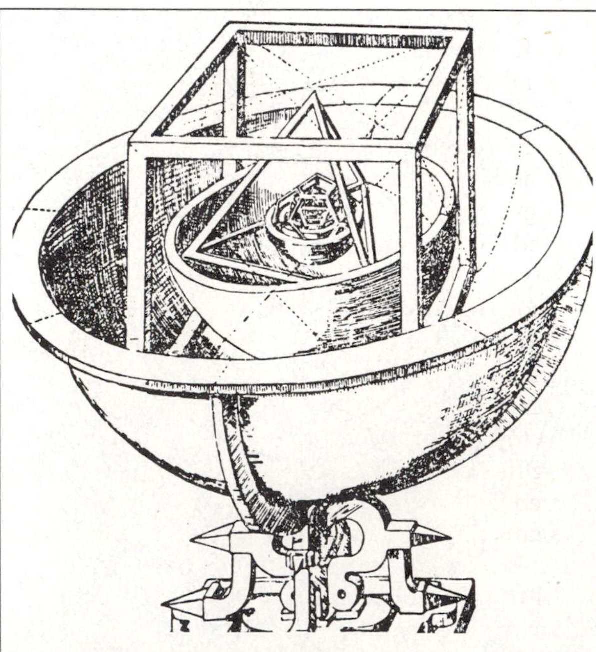

3.11
Kepler's model for the spacing of the planets. During Kepler's life, six planets were known, so five spaces existed between them. Kepler knew that only five regular solids exist. By arranging these solids between the spheres holding the planets, Kepler thought he explained the "edifice of the world." (Courtesy O. Gingerich)

those given by Copernicus. Kepler's intuition later proved wrong–the discovery of any new planet shoots it down–but it energized his later study of planetary motions. This work led to radical results.

Excitedly, Kepler wrote to his former astronomy teacher, Michael Maestlin, about this idea and another: that the sun must have some power or force to propel the planets in their orbits. This view seemed reasonable because Mercury, for example, moves faster in its orbit than Venus. A force from the sun would push Mercury harder than Venus, because Mercury is closer. Kepler failed to work this idea out at the time, but it was firmly implanted in his mind.

In 1594 Kepler published the results of his work in the *Mysterium Cosmographicum* (The Cosmic Mystery), which announced the details of his nesting scheme and declared his adherence to the Copernican model–the first such treatise by a professional astronomer. His intuition that the sun physically directs planetary motions set the stage for the next development of astronomy in terms of physical laws.

3.5 Kepler's new astronomy

Kepler was not yet satisfied with his geometric model; he needed better observational data to calculate the thicknesses of the shells. To obtain them, Kepler turned to Tycho Brahe. Now Tycho had read the *Mysterium Cosmographicum* but did not like its mystical approach to astronomy. So when he received Kepler's request for observations, he did not honor it. But he did invite Kepler to Prague to discuss the matter personally. Kepler traveled there to meet with him.

Tycho and Kepler met on February 4, 1600, an encounter fateful for them and the evolution of astronomy. Their personalities clashed so strongly that they could not work on astronomy together.

Later, Tycho was at a party with a baron and drank too much. Because it was impolite to leave during the affair, Tycho did not relieve himself and suffered a prostrate infection, which resulted in his death. In the days just before he died, Tycho urged Kepler to support Tycho's cosmological model from his observations. After Tycho's death, Kepler was promoted to Tycho's position and allowed access to his unpublished records. However, Tycho's heirs believed them to have financial value and so retained censorship rights. After some negotiations, Kepler attacked the problem of the orbit of Mars.

The battle with Mars / Kepler first tried to model Tycho's observations with a heliocentric scheme that had a circular orbit and an eccentric and an equant. With this, he could fit Tycho's data to within 2′ along the ecliptic. But his predictions above and below the ecliptic were wide of the mark. This discrepancy struck Kepler as important; he realized that he must treat the orbit in three dimensions. He then moved the center of the earth's orbit and fitted positions above and below the ecliptic correctly, but not along it, where he was 8′ off. Kepler was dissatisfied, for he knew that Tycho's observations were accurate to 1′. He demanded that his model work as least as well as the observations, the first time anyone did so seriously in astronomy.

Frustrated, Kepler pondered the idea that planetary motions were driven by the sun. What was that force? Influenced by William Gilbert's book *De Magnete*, Kepler imagined the sun as a fountain of magnetic force that directed the motion of the planets. Such an idea required that a planet move faster in its orbit when it was closer to the sun.

Back to Mars. Kepler applied this magnetic force idea to find it worked if the orbit of Mars were elliptical (Focus 3.3) rather than circular. He wrote, "With reasoning derived from physical principles agreeing with experience, there is no figure left for the orbit of the planet except a perfect ellipse." Kepler had found one of his famous three laws of planetary motion, which rested on a physical explanation of Tycho's observations.

Kepler's laws of planetary motion / Today, Kepler's fame rests mostly on his three laws. And rightly so, for they broke with the tradition of uniform motion on circular orbits. Yet these laws were small fragments of his wider vision for harmonies in the physics of celestial motions. Let's look briefly at each law in modern terminology. The dates given are those of publication, not of discovery.

Law 1. ***Law of Ellipses*** (1609). The orbit of each planet is an ellipse, with the sun at one focus (Fig. 3.12). The other focus is located in space. Note that the distance from the sun to the planet varies as it moves along its elliptical orbit.

Law 2. ***Law of Equal Areas*** (1609). A line drawn from a planet to the sun sweeps out equal areas in equal times. This law notes that the orbital velocities are nonuniform but vary in a regular way: The farther a planet is from the sun, the more slowly it moves (Fig. 3.13).

Law 3. ***Harmonic Law*** (1618). The square of the orbital period of a planet is directly proportional to the cube of its average dis-

Focus 3.3

PROPERTIES OF ELLIPSES

Ellipses are a lot like circles, so their basic properties are fairly easy to grasp. You can make an ellipse by taking a loop of string and holding it down to a board with two tacks. Keeping the string taut with a pencil held against it, draw a curve around the tacks. That curve is an ellipse. The two tacks mark the two* foci *of the ellipse. Each point on the ellipse has the property that the sum of its distances from the two foci (F_1 to P and F_2 to P in Fig. F.12) is the same.

The line through the foci to both sides of the ellipse (R_a to R_p) is called the* major axis. *Half this length is the* semimajor axis, *usually designated by* a. *So the major axis has length 2a. The distance from the center of the ellipse to a focus is designated* c, *so the distance between the two foci is 2c.

How much an ellipse differs from a circle (how "squashed" it looks) is defined as an ellipse's* eccentricity. *Imagine that you took the two tacks at the foci and moved them closer together. The ellipse would become more circular. In fact, when the two tacks coincide, the ellipse becomes a circle. Its eccentricity then is zero. Moving the two tacks farther apart increases the eccentricity until the tacks reach the limits of the string. The eccentricity is then 1. Note that the eccentricity increases as the distance between the foci does. To define it exactly, the eccentricity, e, is

$$e = \frac{c}{a}$$

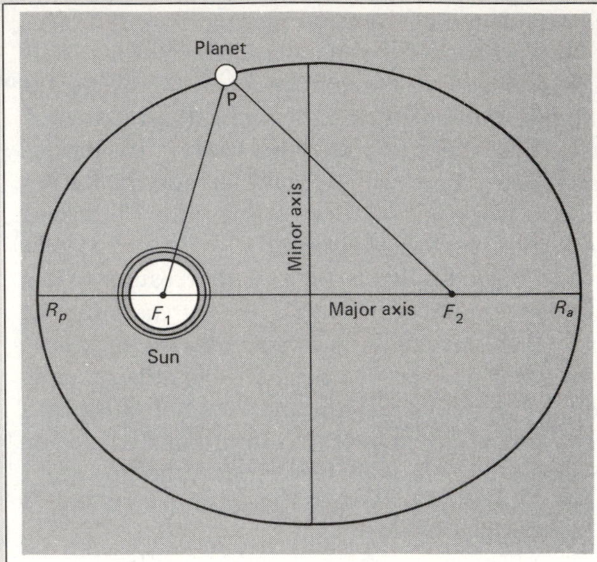

F.12
An ellipse. The sum of the distances of any point on the ellipse from the two foci (F_1 and F_2) is a constant.

If the sun is at one focus (F_1) of a planet's orbit, then R_p, the closest point to the sun, is the* perihelion *point in the orbit. The perihelion distance $R_p F_1$ is $a - c = a - ae = a(1 - e)$. R_a, the farthest point from the sun, is the* aphelion *point. The aphelion distance $R_a F_1$ is $a + c = a + ae = a(1 + e)$. Note that a is the average distance from one of the foci of a point moving on the ellipse. So for planetary orbits, a is the average distance of a planet from the sun.

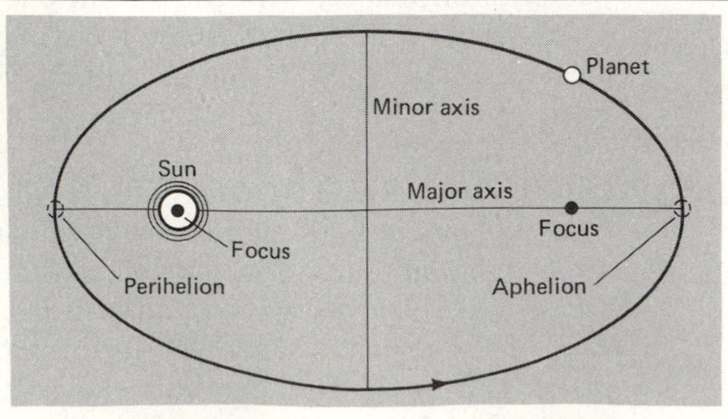

3.12
Kepler's first law. The shape of the planetary orbits is elliptical (greatly exaggerated here), with the sun located at one focus of the ellipse.

tance from the sun (Fig. 3.14). This law points out that the planets with larger orbits move more slowly around the sun, a fact implying that the sun-planet force decreases with distance.

Kepler discovered law 2 before law 1, when he was grappling with the orbit of Mars. Before finding Law 2, Kepler had concluded that the planes of all the orbits pass through the sun. This discovery set the stage for law 1.

Law 3 can be written algebraically as

$$P^2 = ka^3$$

where P is the orbital period, a the average distance from the sun, and k a constant. If P is in

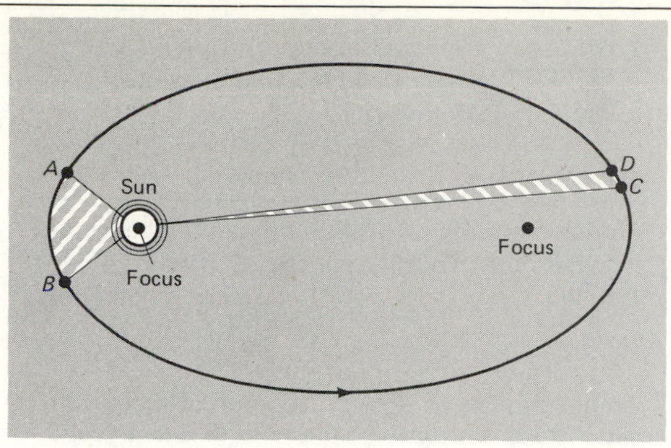

3.13
Kepler's second law. Consider two equal periods of time, one when the planet is closest to the sun (AB) and one when it is farther away (CD). At AB, the planet-sun distance is shortest, and the planet moves fastest in its orbit. At CD, the planet moves more slowly because it is farther from the sun. In both cases, the areas (shaded areas AB to the sun and CD to the sun) covered by the line drawn from the planet to the sun are equal.

years and a in AUs, then k equals 1. Law 3 can be used to find–for any body orbiting the sun (even spacecraft!)–the average distance from the period or the period from the average distance.

What do these laws mean? Law 1 points out that the shape of the orbits is elliptical, so a planet's distance from the sun varies. (Note that the sun does not lie in the center of the ellipse but at one focus.) Law 2 notes that as a planet's distance varies, so does its orbital speed: The closer a planet is to the sun, the faster it goes. You get the sense that the sun pulls a planet toward it; then it whips around the sun, and the sun slows it down as it moves away–a force from the sun acts on it. Law 3 states that the farther a planet's orbit is from the sun, the slower its average orbital motion will be. Both laws 2 and 3 imply that a force between the sun and the planets weakens with greater distances. Hidden in law 3 is an exact description of how the sun-planet force decreases with more distance, but Kepler was not able to figure it out. (Later, Issac Newton did; see Section 4.4.)

The new astronomy / As announced in his *Astronomia Nova* in 1609, Kepler had developed an astronomy based not only on geometry but also on physics–a force between the sun and the planets. Elliptical orbits satisfied both his intention and the observations of Tycho Brahe. Using ellipses, Kepler calculated the Rudolphine Tables (1627), which supplanted both the Prutenic Tables based on the Copernican model and the Alfonsine Tables based on the Ptolemaic one. The Rudolphine Tables completely revised the previously shoddy science of astronomy. Earlier, a few degrees of error between predictions and observations was considered good enough. Kepler's calculations were some 10 times more accurate than those based on either the Copernican or Ptolemaic model (Fig. 3.15). The reason: Both essentially tried to use uniform, circular motion for the planets when, in fact, the motions were neither circular nor uniform.

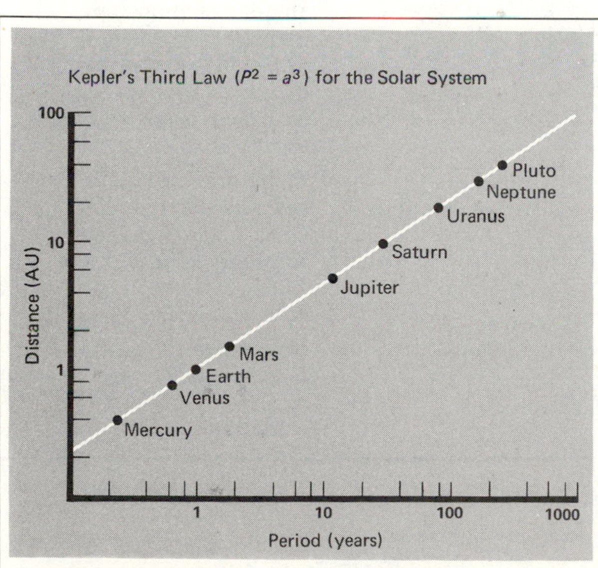

3.14
Kepler's third law. A planet's orbital period (once around the sun) is related to its average distance from the sun. This diagram shows for each (now known) planet a plot of its orbital period (in years) squared to its distance (in AU) cubed. The line is the values expected from the modern formulation of Kepler's third law.

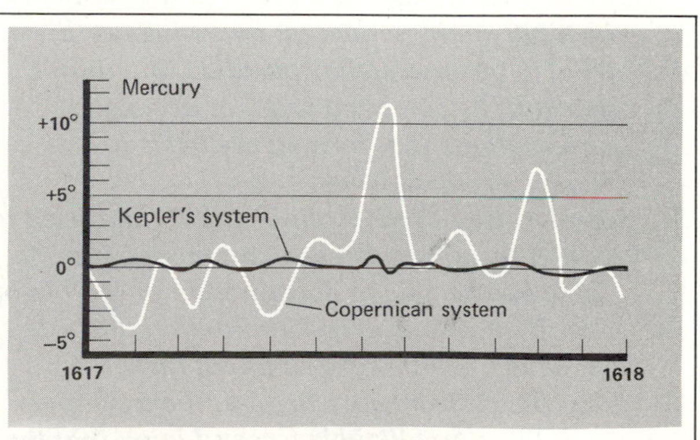

3.15
The accuracy of predictions of planetary positions. This diagram shows the difference between the predicted positions and the actual positions of Mercury in Kepler's model (heliocentric, elliptical orbits, nonuniform speeds) and the Copernican one (heliocentric, circular orbits, uniform speeds). In the Copernican model, the error is as large as 11°. For Kepler's scheme, the maximum error is 10 times smaller. (Courtesy O. Gingerich and B. Welther)

Kepler broke the ancient spell of perfect circles and uniform motion that had mesmerized astronomers for centuries. He rewove the fabric of the heavens into a pattern that proclaimed the end of an ancient era and the birth of astronomy as a physical science.

TABLE 3.2 ***Comparision of the Ptolemaic (P), Copernican (C), and Keplerian Models (K)***

Observed "fact"	*Explanation*
Motion of entire heavens daily from E to W	***P:*** Motion of all heavenly spheres E to W ***C:*** Reflection of rotation of earth W to E ***K:*** Same as ***C***
Annual motion of sun W to E through zodiac	***P:*** Rotation of sun's sphere W to E in a year ***C:*** Reflection of annual revolution of earth about sun ***K:*** Same as ***C***
Nonuniform motion of sun through zodiac	***P:*** Orbit of sun eccentric, speed uniform ***C:*** Orbit of earth eccentric, speed uniform ***K:*** Orbit of planet elliptical, speed nonuniform
Retrograde motions of the planets	***P:*** Epicycles and deferents ***C:*** Relative motions of planets, including earth, around sun ***K:*** Same as ***C***
Variations in retrograde motions	***P:*** Equant, eccentrics ***C:*** Epicyclets ***K:*** Nonuniform orbital motion; tilt of planetary orbits with respect to earth's
Distances of planets	***P:*** Arbitrary as long as angular relationships correct ***C:*** Relative distances set by observations ***K:*** Force relates distances to periods
"Cause" of planetary motions	***P:*** Natural motion of celestial spheres; *no force* ***C:*** Same as ***P:*** *no force* ***K:*** Magnetic *force* from sun
Accuracy of predictions	***P:*** Typically 5° or less; sometimes 10° ***C:*** Same as ***P*** ***K:*** Generally about 10′; sometimes as large as 1°

KEY CONCEPTS

1 ***Copernicus proposed his heliocentric model mainly for aesthetic reasons; he disliked the equant in the Ptolemaic model, which violated the traditional precept that heavenly motions be uniform about circles.***

2 ***The Copernican model (Table 3.2) views the daily motion of the sky as arising from the earth's rotation, the sun's eastward motion along the ecliptic as a reflection of the earth's revolution around the sun, and retrograde motion as an illusion when one planet passes another; the order of the planets is set by their periods of revolution about the sun; the planets' distances are set by observations and simple geometry.***

3 ***Copernicus's model did not predict planetary positions any better than the Ptolemaic model, nor was it simpler; it violated the accepted physics of the day (that of Aristotle); it required a stellar parallax that was not observed.***

4 ***Tycho Brahe typified the reaction of practicing astronomers toward the Copernican model: He thought it an interesting geometric idea but could not accept a moving earth; he tried to observe stellar parallax, but didn't detect it; he compiled years of accurate observations that formed the basis of Kepler's crucial work.***

5 ***Kepler used Tycho's observations and tried to match them by a Copernican model; his failure to be able to do so within the errors of the observations drove him to find that planetary orbits were elliptical (first law), that the planets move around them nonuniformly but predictably (second law), that the farther a planet is from the sun, the longer it takes to orbit (third law), and that the sun exerts a force that keeps the planets in their orbits.***

6 ***Kepler's model predicted planetary positions much more accurately than either the Ptolemaic or Copernican model because he broke with the tradition of uniform, circular motion for celestial objects; he was driven to this conclusion by his faith in the accuracy of Tycho's observations.***

STUDY EXERCISES

1 What was Copernicus's primary objection to the Ptolemaic model? (*Objectives 1 and 2*)

2 How did Copernicus account for retrograde motion in his model? (*Objective 3*)

3 If you were on Mars, when would you see Jupiter retrograde? The earth? (*Objective 3*)

4 What is the difference between a planet's synodic and sidereal periods? How can sidereal periods be found from synodic ones? (*Objective 4*)

5 In your opinion, what was the major advantage, if any, of the Copernican model over the Ptolemaic one? (*Objectives 1 and 5*)

6 What force keeps the planets moving around the sun in the Copernican cosmos? (*Objectives 6 and 9*)

7 Give at least two differences between the models of Copernicus and Kepler. (*Objective 9*)

8 What two advantages did Kepler's model have over that of Copernicus? (*Objective 9*)

9 According to Kepler's second law, a planet travels slowest at what point in its orbit? (*Objectives 8 and 10*)

BEYOND THIS BOOK . . .

For an excellent analysis of the work of Copernicus and the framework of thought in which his ideas developed, read *The Copernican Revolution* (Vintage Books, New York, 1959) by T. Kuhn.

For a modern variety of views about Copernicus and the impact of his work, look at *The Nature of Scientific Discovery* (Smithsonian Institution Press, Washington, D.C.), edited by O. Gingerich.

A good autobiography of Tycho Brahe is *Tycho Brahe: A Picture of Scientific Life and Work in the Sixteenth Century* (Dover, New York, 1963) by J. L. E. Dreyer.

A fine biography of Kepler is *Kepler* (Collier, New York, 1962) by M. Casper, translated by C. Hellman. A technical look at Kepler's work is in *An Account of the Astronomical Discoveries of Kepler* (University of Wisconsin Press, Madison, 1963) by R. Small.

You can find a controversial view of Copernicus, Tycho, Kepler, and Galileo in *The Sleepwalkers* (Grosset & Dunlap, New York, 1963) by A. Koestler; especially part 3, "The Timid Canon," and part 4, "The Watershed."

Volume 16 of *Great Books of the Western World* (Encyclopaedia Britannica, Chicago, 1952) contains a translation of *De Revolutionibus* and excerpts from Kepler's works.

4 / The clockwork universe

> ***I am much occupied with the investigation of physical causes. My aim in this is to show that the celestial machine is to be likened not to a divine organism, but rather a clockwork . . .***
>
> ***JOHANNES KEPLER***

Learning objectives

After studying this chapter, you should be able to:

1. **Describe Galileo's important telescopic discoveries and their impact on the controversy between the Copernican and Ptolemaic models.**
2. **Indicate Galileo's purpose in developing a new science of terrestrial motions and contrast Galileo's ideas about motion to those of Aristotle.**
3. **Contrast Galileo's astronomy and cosmology to those of Copernicus and Kepler.**
4. **Describe the difference between speed and velocity and between accelerated and unaccelerated motion; give examples of each.**
5. **Cite Newton's three laws of motion and describe each in simple, concrete examples.**
6. **Contrast Newton's concept of natural motion to that of Aristotle.**
7. **With the aid of a graph, describe Newton's law of gravitation in simple physical terms.**
8. **Outline how the moon-apple test supports Newton's law of gravitation.**
9. **Contrast Newton's astronomy and cosmology with that of Copernicus and Kepler.**
10. **Use Newton's physical ideas to support the Copernican model.**

Central question

How did Newton's laws of motion and gravitation unify the cosmos physically?

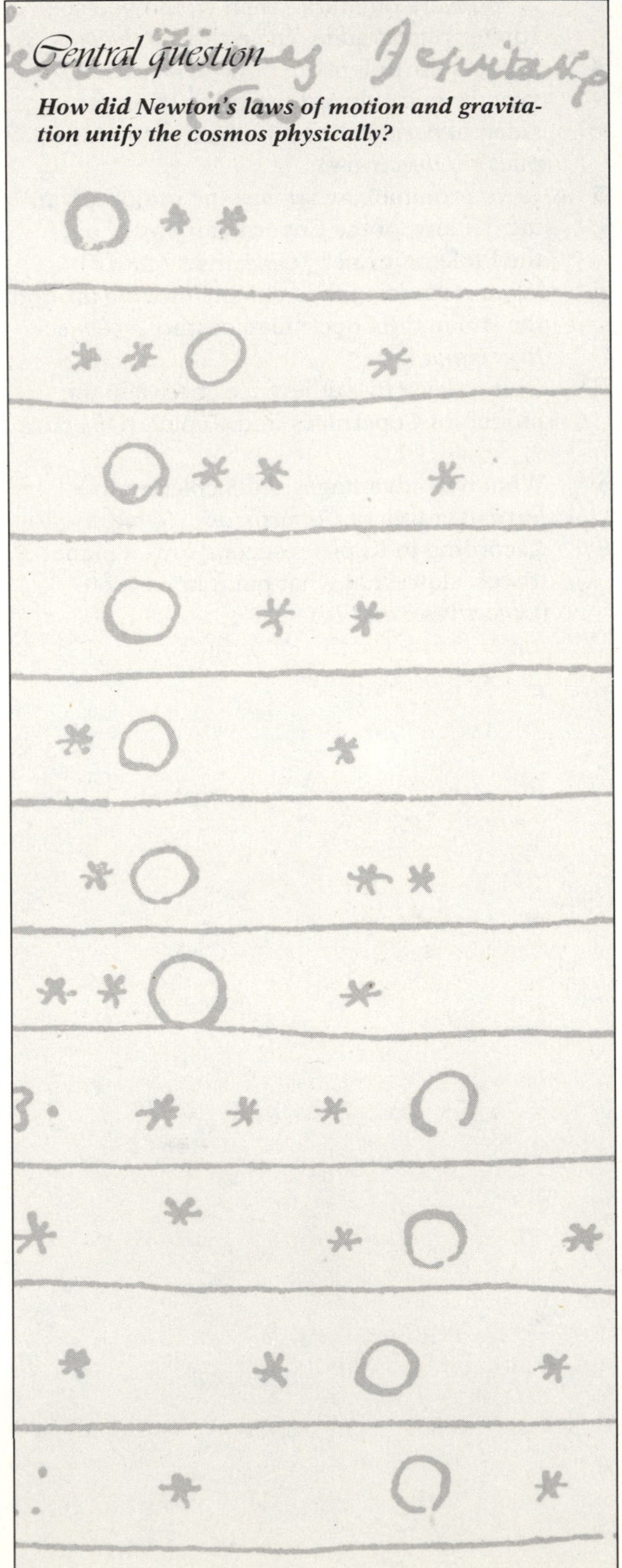

What makes the world go 'round? For people it may be love. For the planets, according to Isaac Newton, it's gravitation. By linking the cosmos with gravity, Newton achieved the goal that had eluded Kepler: a model of the cosmos run by a single force.

Newton arrived at this goal after crucial groundwork had been laid by Copernicus and Kepler. Copernicus made the inspired move of the sun to the center of the universe. To do so, he violated the physics of the day and offered no substitute; this gap was an embarrassment for the Copernican model. Kepler sought to establish a clockwork cosmos driven by a single force. His magnetic force finally failed. But his demand for a force from the sun driving the planets in their orbits directly influenced Newton's vision.

Newton erased the old separation between the earth and the heavens. He linked gravity—terrestrial physics—to the orbital motion of the planets—celestial physics. The links he forged were his laws of motion and gravitation, which sprang from a new concept of natural motion.

With Newton's ideas, the universe took on a new appearance. No more the closed, finite universe of Ptolemy, Copernicus, and tradition. In Newton's grand vision the universe grew to an infinite expanse, driven by a single force: gravitation.

4.1
Galileo Galilei. "By denying scientific principles, one may maintain any paradox." (Courtesy Yerkes Observatory)

4.1 *Galileo: Advocate of the heliocentric model*

The Italian scientist Galileo Galilei (1565–1642) desired to establish celestial physics on a firm experimental and mathematical basis. In contrast to Kepler, Galileo (Fig. 4.1) concerned himself almost exclusively with terrestrial motions, especially those of falling bodies. He felt that to establish the laws of terrestrial motions would do more to cement the structure of the Copernican cosmos than any observations, including those made with a telescope. He failed to achieve his goal, but his discoveries guided the later work of Newton.

The magical telescope / Galileo used his telescope to bolster the Copernican model. He did not invent optical lenses nor their use in a telescope. Rather, he promoted his astronomical ideas by utilizing the novelty and shock value of telescopic observations. The report of his observations in *Siderius Nuncius* (The Starry Messenger) was widely circulated and brought him public fame.

By Galileo's time, glass lenses had been known for about 300 years. Their date and place of origin are not clear, but they were used by eyeglass makers to correct defects of vision. The opticians of the day had no physical understanding of how their glasses functioned and had adopted a purely experimental approach to their construction.

In 1609 a messenger returning to Venice brought Galileo the news that a Dutchman had constructed a spyglass that made distant objects appear to be nearby. Although Galileo had little experience with optics, he set to work immediately in his workshop to duplicate the instrument (Fig. 4.2). Sparing neither labor nor expense, he ultimately succeeded in constructing an optical device that made objects appear 30 times closer than when viewed with the naked eye (see Chapter 6). He put this marvelous tube to astronomical use. Within a few weeks in 1609 and 1610 he made a series of astronomical discoveries that marked a new era in astronomy. Although Galileo was not the first person to build a telescope, he first recognized that the telescope increased our power to perceive reality.

Galileo first turned his telescope to the moon and saw that the moon's surface was not smooth and spherical, as Aristotelian ideas required for perfect heavenly bodies, but rough, with chains of mountains and valleys and many craters (Fig. 4.3). He determined the height of a lunar mountain from the length of its shadow: approximately 6 km was the astonishing (and correct) result.

(a)

(b)

4.2

(a) ***The lens from Galileo's largest telescope. It was accidently broken by him and mounted in this ivory frame in 1677.*** **(b)** ***Two telescopes made by Galileo. The upper one magnifies 14 times, the lower 20 times. (Both courtesy M. L. Righini Bionelli, Florence Museum of the History of Science)***

Galileo next peered at the stars. His instrument's power fragmented the faint band of the Milky Way into innumerable stars, more than could be seen individually with the unaided eye. This observation refuted the Aristotelian idea that the sky contained only a certain number of stars and that this known number could not change.

Then Galileo hit on what he considered to be his most important discovery: four new "planets" that no one had seen before. What he found were not actually new planets but rather the four brightest moons of Jupiter. (A total of 16 moons is now known; see Chapter 11.) To his amazement, Galileo found that these four bodies revolved around Jupiter (Fig. 4.4). From continuous nightly observations, he estimated the orbital periods of the Jovian satellites. Here he found another argument against tradition, for Jupiter and its satellites resembled a miniature solar system. This fact required a second center of revolution in the cosmos, a notion that contradicted the Aristotelian doctrine that only the center of the universe could be the center of revolution—and that was the earth.

4.3

Gaileo's drawings of the moon as seen through his telescope. Note the craters and mountains. (Courtesy Yerkes Observatory)

Galileo may not have been the first person to report on moons of Jupiter. An old Chinese record notes that Kan Te, an astronomer in China in the 4th century B.C., made many observations of Jupiter. In one of his books, he states that Jupiter looked like it had "a small reddish star attached to it." This record may point to the brightest of Jupiter's moons, which can, under ideal circumstances, be visible to the unaided eye. If so, Kan Te made note of this moon some 2000 years before its discovery by Galileo with a telescope!

The Starry Messenger / Galileo wrote up these observations in *Siderius Nuncius* (The Starry Messenger), which was published on March 12, 1610. His telescopic discoveries, especially the news of Jupiter's moons, grabbed the public imagination. The book sold as fast as it could be printed; it brought fame to Galileo and a demand for the telescopes produced in his private workshop. However, he was still cautious in print; although he used his observations to demolish the Aristotelian cosmos, he did not openly advocate the Copernican scheme or support it with his observations.

Critics quickly scorned Galileo's work by suggesting that he was seeing atmospheric phenomena or flaws in the lenses. Although such arguments seem strange to us today, they

4.4

Telescopic observations of the moons of Jupiter. The large circle represents Jupiter; the "stars" are the moons. Their changing positions next to the planet indicated that they were revolving around Jupiter and were not just background stars. (Courtesy Yerkes Observatory)

appeared reasonable enough to seventeenth-century people. The art of lens making was not far advanced, so many lenses were flawed by ghost images. (Galileo's telescopes produced images that were far inferior to those in a cheap modern telescope.) His opponents said that although Galileo may have honestly reported what he had seen, his observations might have no direct connection with reality.

Galileo, however, was convinced, by the regularity of the telescopic phenomena, that he was viewing reality and not illusion. In 1611 Kepler, who was well regarded in scientific circles, backed up Galileo with observations of his own and also developed a theory of optics to support the validity of telescopic observations.

Goaded by the opposition to what he considered indisputable facts, Galileo continued to scan the skies for new marvels. By projecting the image of the sun onto a piece of paper, he observed sunspots (Fig. 4.5). As defects on the supposedly perfect sun, the sunspots dealt another blow to the tenets of Aristotelian cosmology. In 1613 Galileo published his results in *The Letters on Sunspots* and made his first direct, printed declaration of his belief in the Copernican model.

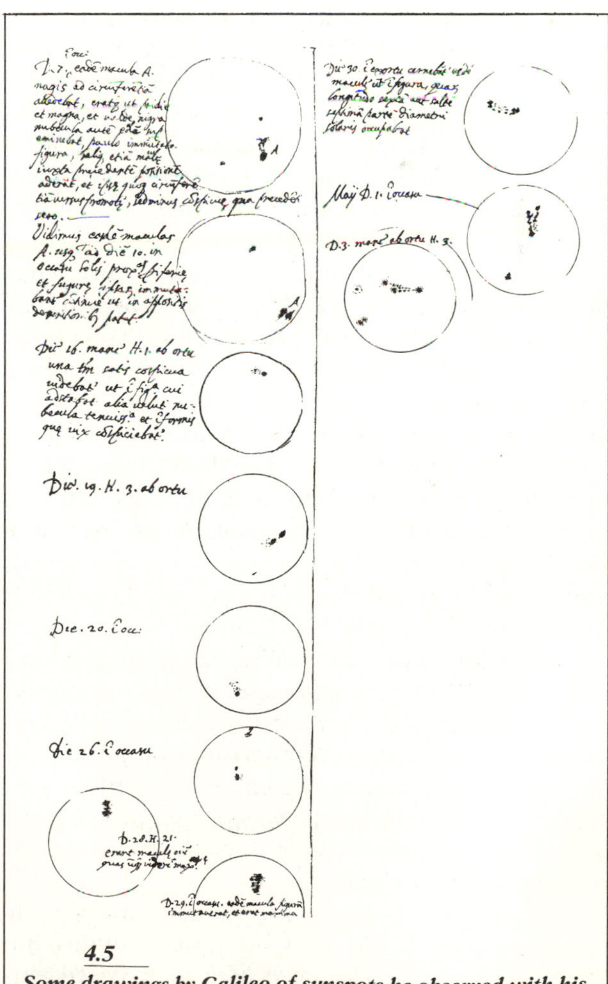

4.5

Some drawings by Galileo of sunspots he observed with his telescope. (Courtesy Yerkes Observatory)

4.2 *Galileo and a new physics of motion*

Despite his observational triumphs, Galileo still felt that the Copernican system lacked the anchor of a physical understanding of motion such as Aristotle had provided for his model. Kepler had made an important step by his mathematical description of *celestial* motions. Although Galileo ignored most of Kepler's work, he took the next step by devising a mathematical description of *terrestrial* motions, particularly those of bodies falling under the influence of gravity. Galileo aimed to justify the Copernican scheme by a systematic, experimental search for physical ideas of motion. He unraveled the details of motion produced by applied forces; this study is called *mechanics*.

To found the new science of mechanics, Galileo needed an important tool: a description of the motion of falling bodies, that is, the motion of masses under the influence of gravity. But before I explain Galileo's contribution, let me describe motion more precisely.

Acceleration, velocity, and speed / When you step on a car's accelerator, it does just that: accelerates. If you start out at rest, the car goes faster and faster, as you can tell from the speedometer. Of if you are cruising on the highway, you pass a slower car in front of you by stepping on the accelerator. In both instances your velocity changes; that's the meaning of *acceleration:* any change in velocity.

But what is velocity? It's not simply what you call speed, which tells you how fast you're going. Suppose, for example, that you travel from Albuquerque to Santa Fe, New Mexico; you go from one place to another at some *speed* in some *direction*. Velocity involves direction as well as speed. For example, if you drive from Albuquerque to Santa Fe, a distance of about 100 km, in one hour, your speed is about 100 km per hour (km/h). When you drive back at the same speed, your velocity is different because you're heading in a *different direction*.

Imagine now that you're riding on the outside horse of a merry-go-round. It turns around at a constant speed. Are you accelerating? Yes, because your velocity is constantly changing *direction* as you turn around the circle of the merry-go-round. Even though your speed does not change, your direction does. So your velocity is changing, and you are accelerating.

Of course, you are also accelerating if your speed changes while your direction remains the same. And you accelerate if *both* your speed and your direction change.

Note that speed and velocity have different physical meanings. Speed is the average rate of travel of something and is measured in distance per unit of time—miles per hour or meters per second, for instance. Velocity is speed with something added: direction. Acceleration is the rate at which *velocity* (not speed!) changes. It is measured in distance per unit of time per unit of time, such as meters per second per second (m/sec/sec).

Natural motion revisited / One of the nagging problems of Aristotelian physics was that of *inertia*, the tendency of a body forced into motion to retain that motion or of a body at rest to remain at rest. Aristotle had divided motions into two categories: forced and natural (see Section 2.3). An example of Aristotelian natural motion is the fall of a rock to the earth—a result of a supposedly natural tendency of earthly material to seek the central point of the cosmos, and it required no force. But the throwing of a rock, a motion at right angles to the natural downward motion, required a continuously acting force to keep up the unnatural, forced motion. However, as you know, this statement is not strictly true, for after you throw an object, its motion will continue for a while even after the force is removed.

Galileo inverted Aristotle's ideas of natural and forced motion. He concluded that the downward motion of objects resulted from an attractive force: gravity. In addition, he viewed the horizontal motion of objects flying through the air as due to their inertia. This inertial motion, he argued, was a natural one and would continue if no forces, such as air resistance, acted.

To take this important step, Galileo had to arrive at a description of inertia. He explained his ideas as follows (Fig. 4.6): If a perfectly smooth ball were placed on a hard, flat surface that sloped, the ball would roll down the slope forever if the surface were infinitely long. The greater the slope, the faster the ball would roll. By the same argument, a ball traveling up a slope would eventually stop. If, however, the surface had no slope and the ball were placed on the level with no horizontal velocity, the ball would remain at rest. A ball on a level surface, if pushed, would continue to move straight ahead at a constant velocity forever, since it is not slowed down by an ascent or speeded up by a descent.

With this concept of inertia, Galileo dealt with the problem of falling bodies. He did not view this motion as natural (as had Aristotle) but as *motion due to a force*: gravity. From a combination of experiments and intuition, Galileo concluded that such motion took place at a constant

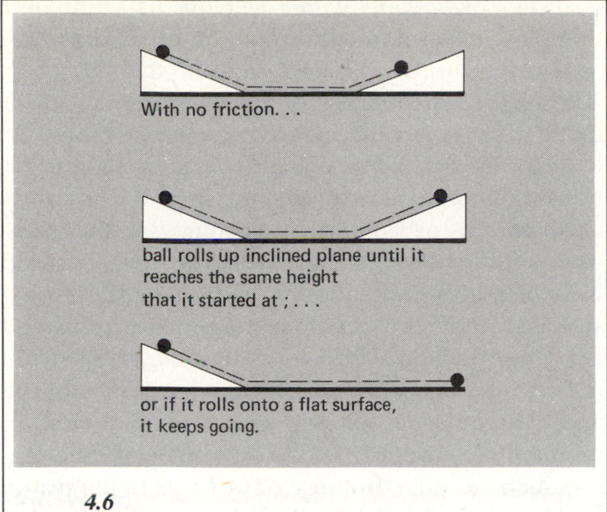

With no friction. . .
ball rolls up inclined plane until it
reaches the same height
that it started at ; . . .
or if it rolls onto a flat surface,
it keeps going.

4.6
Galileo's concept of inertia (natural motion). The trick is to ignore friction in this thought experiment of rolling a ball down an inclined plane.

acceleration, with the object's velocity changing at a constant rate as it fell. (The acceleration due to gravity at the earth's surface has a value of 9.8 m/sec/sec and is usually denoted by *g*. This means that for every second of fall an object gains 9.8 m/sec of speed.) He also found that *all* falling masses (without air resistance) have the *same* acceleration. When dropped, they reach the same velocity after the same time; they also fall the same distance in the same time. This conclusion—that all masses fall with the same acceleration near the earth—directly contradicted the traditional teaching that a massive body fell faster than a less massive one and so moved a greater distance in a given time.

Note that in the case of constant acceleration, such as falling bodies, the velocity continually increases as time passes. So a dropped object falls with greater and greater speed as it gets closer to the ground.

Legend has it that Galileo climbed to the top of the Leaning Tower of Pisa and dropped two different objects, which, to the astonishment of the skeptical professors gathered around the tower's base, hit the ground at the same time. As far as historians can determine, Galileo did *not* actually attempt this experiment, although a friend of his may have. It would have been a risky demonstration, for, more likely than not, the masses would not have struck the ground simultaneously because of air friction. The rumored Pisa experiment may not be the only one that Galileo did not do, even though he reported the results in his writings. Some experiments were really mental exercises, but Galileo intuitively reached the correct results more often than not. And he did perform many experiments to contradict Aristotle's concepts about motion. Galileo's views on motion became a pivot on which modern physics turned.

Galileo's cosmology / In his *Dialogue on the Two Chief World Systems*, published in 1632, Gaileo compares the traditional model to the Copernican one. In a drawing (Fig. 4.7), he shows the planetary system as essentially a pure Copernican scheme with no evidence of the ideas of Kepler, such as elliptical orbits. In addition, Galileo makes no attempt to apply his terrestrial mechanics to the motions of the planets.

This ignorance of physics and ellipses provides evidence that Galileo paid little attention to the use of the Copernican system in predicting the planetary positions. What struck Galileo was the Copernican cosmology, the order of the universe, rather than the detailed astronomy of planetary motions. He apparently felt, with the same conviction as Copernicus, that the harmony of the planetary order was established from observation. Of course, Galileo relied heavily on his telescopic observations, along with the general arguments of Copernicus (see Section 3.1).

Galileo placed the fixed stars in a thick shell far beyond the planets. This stellar shell remained from Greek ideas. However, his spokesman in the

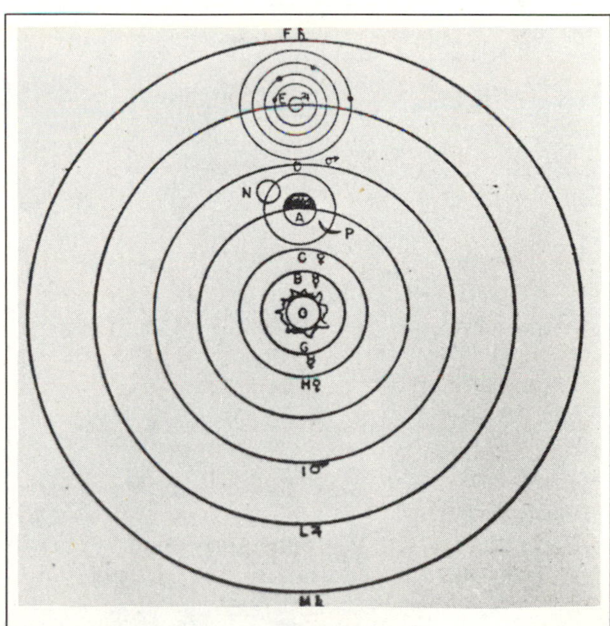

4.7
A drawing of Galileo's heliocentric model. Note that four moons orbit Jupiter. Also, the scheme lacks a sphere of stars just beyond Saturn; Galileo left open the possibility that the stars extend throughout space. All the orbits are circular, despite the discovery by Kepler that they are actually elliptical.

Dialogue keeps open the possibility that the universe is not spherical and finite but open and infinite, with the stars sprinkled "through the immense abyss of the universe." Here we find hints of the idea of the cosmos as an infinite universe rather than a closed space, a view that would be supported by Newton.

4.3 Newton: A physical model of the cosmos

Despite his interest in and insight into terrestrial motions, Galileo did not worry about the details of celestial motions. He saw his task as uncovering the principles of motion in general. Clearly, he believed that the model of Copernicus was on the right track, as was Kepler's idea of a physical force from the sun. But until the general concepts of motion were understood, Galileo thought it premature to consider physical explanations of planetary motion that would provide the physical foundation for the Copernican model.

Sir Isaac Newton (1642–1727) emerged as the genius destined to fuse the terrestrial and celestial realms and so to end the long-standing separation initiated by the Greeks (Fig. 4.8). The publication in 1687 of Newton's *Principia*, containing his new physics of motion and the concept of gravitation, resulted in a truly physical view of the universe.

4.8
Sir Isaac Newton. "I have laid down the principles of philosophy; principles not philosophical but mathematical . . ." (Courtesy Yerkes Observatory)

The prodigious young Newton / In the small English village of Woolsthorpe, on Christmas Day in 1642, Newton's widowed mother gave birth to a sickly child. The fragile baby was so small at birth that it is said he could have fit into a quart mug!

At 18 Newton enrolled in Trinity College at Cambridge University. He first intended to study mathematics as applied to astrology, but a meeting with Professor Isaac Barrow, who sensed Newton's abilities, encouraged him to study physics. In 1665 the bubonic plague overwhelmed England, and the University shut down. Newton returned to his home and mother at Woolsthorpe and, in quiet isolation, made discoveries in mathematics, optics, and the science of mechanics. As he wrote, "In those days I was in the prime of my invention, and minded mathematics and philosophy more than at any other time since."

This fertile period in Woolsthorpe generated the legend of the falling apple. As is common in a creative flash of genius, Newton—whether he in fact saw an apple fall or not—linked two seemingly unrelated phenomena: the fall of an apple and the orbit of the moon. He was puzzled by the fall of objects, such as apples, and wondered about the nature of the force that attracts masses, such as the moon, to the earth's center.

On his return to Trinity, Newton showed his work to Barrow, who soon resigned his position so that Newton could be elected to it. Because of his interest in optics, Newton came to invent the *reflecting telescope* (see Chapter 6), which uses a mirror as the primary light gatherer. His design was communicated to the Royal Society of London; after he constructed a small reflector for them, he was elected a Fellow of the Society. His election was not a completely happy one, for his work on light, which was also communicated to the Society, brought bitter controversy. Newton resolved never to publish his ideas again.

***The magnificent* Principia** / Newton broke his vow of nonpublication about 10 years later, when Edmund Halley (1656–1742) requested his advice on the problem of elliptical orbits described by Kepler's laws. Halley (Fig. 4.9) queried Newton on the nature of the force between the sun and the planets required to produce such orbits and was surprised to hear that Newton had solved the problem in exact detail. Absent-minded as ever, Newton had misplaced the solution and could not find it at the moment; he promised Halley that he would send it along later.

Recognizing the importance of Newton's discovery, Halley cajoled his introverted friend to publish the studies and promised to oversee and

4.9
Edmund Halley, who used Newton's laws to compute the orbit of the comet that bears his name. (Courtesy Yerkes Observatory)

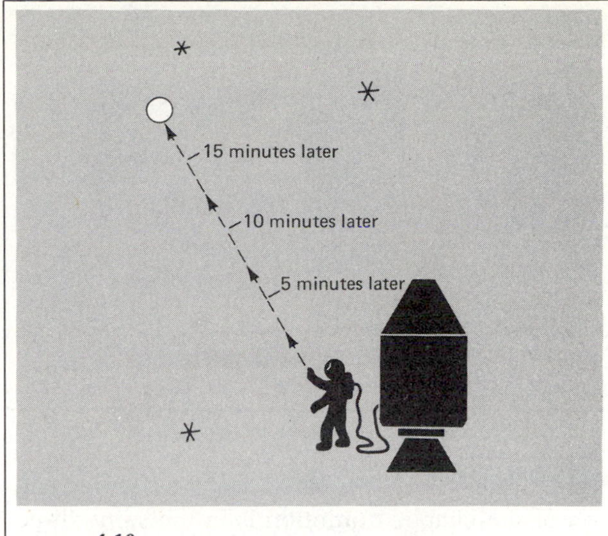

4.10
Newton's first law. An object thrown by an astronaut moves at a constant speed along a straight line (if no friction slows it down).

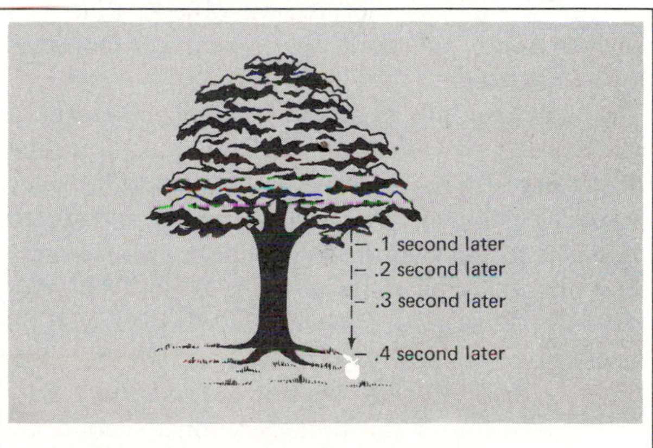

4.11
Newton's second law. A net force applied to an object changes its speed, direction of motion, or both in the direction of the applied force. Here, gravity pulls on an apple and accelerates it to the ground. As the apple accelerates, it falls faster so that it covers a greater distance in the same time (0.1 sec).

finance their publication. Newton, stimulated intellectually, labored for 2 years to complete his *Philosophiae Naturalis Principia Mathematica* (The Mathematical Principles of Natural Philosophy). Published in 1687, the *Principia* presented the solution to the vexing problem of planetary motions.

At the beginning of the *Principia*, Newton states his intention:

For the whole burden of philosophy seems to consist of this—from the phenomena of motions to investigate the forces of nature and then from these forces to demonstrate all other phenomena.

Newton's laws of motion / Newton defines mass, velocity, and acceleration. He then expounds his three Axioms, or Laws of Motion. In modern terms, these famous laws are:

Law 1. **The Inertial Law.** A body at rest or in motion at a constant velocity along a straight line remains in that state of rest or motion unless acted upon by a net outside force (Fig. 4.10).

Law 2. **The Force Law.** The change in a body's velocity due to an applied net force is in the same direction as the force and proportional to it but is inversely proportional to the body's mass (Fig. 4.11).

Law 3. **The Reaction Law.** For every applied force, a force of equal size but opposite direction arises (Fig. 4.12).

Newton's first law takes a logical step beyond Galileo's concept of inertia by postulating that constant, uniform motion is the *natural* state of moving mass anywhere in the universe. The first law gives you a way to judge whether or not a net force is acting on an object: You are told to look for a change either in an object's speed or in the direction of its motion, or in both speed and direction.

The second law extends the recognition from a force to the recognition of its results; the direc-

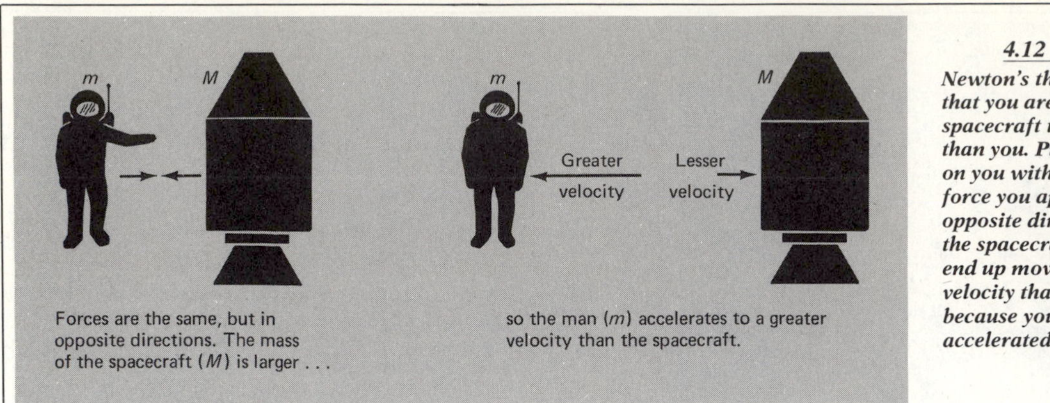
m
M
m
M
Greater
velocity
Lesser
velocity
Forces are the same, but in opposite directions. The mass of the spacecraft (M) is larger . . .
so the man (m) accelerates to a greater velocity than the spacecraft.

4.12
Newton's third law. Imagine that you are in space next to a spacecraft that has a larger mass than you. Push it; it pushes back on you with a force equal to the force you apply to it, but in the opposite direction. So you and the spacecraft move apart. You end up moving with a greater velocity than the spacecraft because you have less mass accelerated by the same force.

tion of the change in motion is in the same direction as the applied force. Also, the amount of acceleration, the change in the object's velocity, depends directly on the size of the force. So exerting a force (which you can think of simply as a push or a pull) accelerates an object; that is, it slows it down, speeds it up, or changes the direction of its motion.

For example, suppose you are floating in space next to a bunch of small objects. You push one. It accelerates and moves away, and it travels in the direction in which you pushed it. You can measure its acceleration, its change in velocity. Now push another mass with the same force. You measure its acceleration and find that it is half the acceleration of the first. You applied the same force to both objects, so the second must have twice the mass of the first. Newton's second law provides a means to measure mass along with the consequences of the application of forces to masses.

In algebraic form, Newton's second law is $F = ma$, where F is the net applied force, m the object's mass, and a the acceleration resulting from the force. Note that, like velocities, forces have directions, and so do accelerations. An object will accelerate in the *same* direction as an applied force.

The third law recognizes that forces are interactions and must act in simultaneous pairs. If you were at rest in space and pushed against a massive spacecraft, the ship would react to your applied force with an equal but oppositely directed force, pushing you away, as described by Newton's third law. Now the forces applied to you and the spacecraft are the same, but the resulting accelerations are different. According to Newton's second law, the acceleration is greater for you than for the spacecraft, because your mass is less. As a result, you would move away from the spacecraft quickly, while it would hardly budge. Also, you and the spacecraft would be moving in *opposite* directions.

4.4 Newton and gravitation

From these ideas about force and motion, Newton attacked the problem of the planets by devising the law of gravitation. Two questions needed to be answered: (1) In what *direction* does the force of gravity act, and (2) what is the *amount* of the force? The first question involves a recognition of the general nature of the force, and the second involves a recognition of the physical properties that determine the force's strength. To answer these questions, Newton built on Kepler's foundation: Newton's procedure combined his laws of motion with Kepler's planetary laws to arrive at a law of universal gravitation.

Newton demonstrated that the type of force that causes the elliptical orbits of Kepler's first law is a *central force*, one directed to the center of the motion. Also, he showed that planets moving in orbits under the influence of a central force follow Kepler's second law: the law of areas. Finally, Newton showed from the geometric properties of ellipses that the force may be described by an inverse-square law and then derived Kepler's third law. In this manner he ensured that his procedure fell in line with planetary laws as they were known in his time.

How do the moon and the famous apple enter into this scheme? Newton recognized that gravity causes the apple's fall. Might it not be that the earth's gravity, pulling on the moon, also keeps the moon in its orbit? For simplicity, assume that the moon's orbit is circular. Now, the direction of the moon's orbital motion changes constantly: The moon stays on a curved path

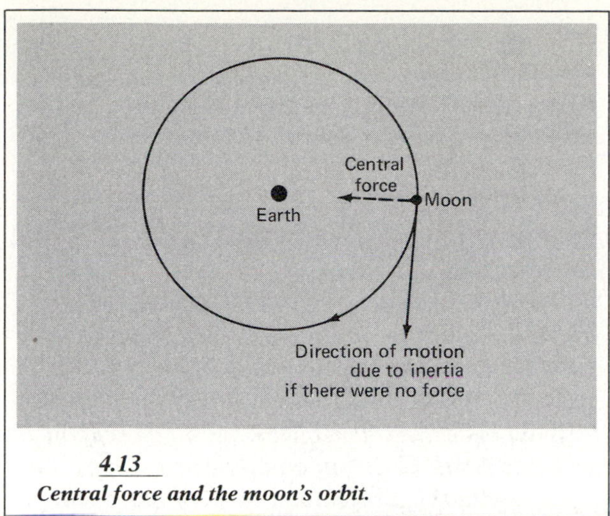

4.13
Central force and the moon's orbit.

rather than moving along a straight line away from the earth (Fig. 4.13). Newton's first law tells us that acting on the moon is a force that, according to the second law, results in an acceleration toward the earth's center; such a centrally directed force is called a *centripetal force*. (*Centripetal* means "directed toward the center.") The resulting acceleration is called *centripetal acceleration*.

Let's try to get a feel for centripetal force and acceleration. You experience it when you turn a corner in a car. Obviously, the faster you go around the corner, the greater the acceleration. But the size of the turn also affects the acceleration. For instance, if you make a long, gentle curve—even at 90 km/hr—you exert very little force on the steering wheel to change the car's direction. But if you take a sharp turn at 90 km/hr, you must exert a much greater force. So centripetal force depends on both the speed of an object around a circle and the size of the circle.

What causes the centripetal force that holds the moon in its orbit? Newton generalized from the apple to the moon: "I began to think of gravity extending to the orb of the moon." In this statement he made the creative leap toward a new insight that *every body in the universe attracts every other body with a gravitational force*. This statement became the first *universal* physical law.

In modern algebraic form, Newton's law of gravitation is

$$F = \frac{Gm_1m_2}{R^2}$$

where F is the gravitational attraction between two spherical bodies, m_1 and m_2, whose centers are separated by a distance R. The symbol G is a constant, a number whose value is assumed not to vary with time and location in the universe. The value for G in the mks (meter-kilogram-second) system is 6.67×10^{-11}. (See Appendix A for units.)

What does Newton's law of gravitation mean? First, that all masses in the universe attract all other masses. (This force can only attract; it does not repel.) Second, if you consider just two masses for a moment, the amount of the gravitational force depends directly on the amount of material *each* mass has. So if you doubled the mass of one and kept the distance between the two the same, the force would also double. (Note that the kind of material that makes up the masses does not matter.) Third, masses at greater separations have *less* gravitational force than those closer together, and this drop-off of force with distance happens in a special way—as the inverse square of the distance. Consider, for example, two masses 1 m apart. A certain amount of gravitational force attracts one to the other. Now move the masses so that they are 2 m apart. The force is less. How much less? By ½ squared, or one-fourth as much as when the masses were 1 m apart (Fig. 4.14).

Newton used the moon and the apple to test the validity of this law. He knew that the earth's gravity at its surface (1 earth radius from the

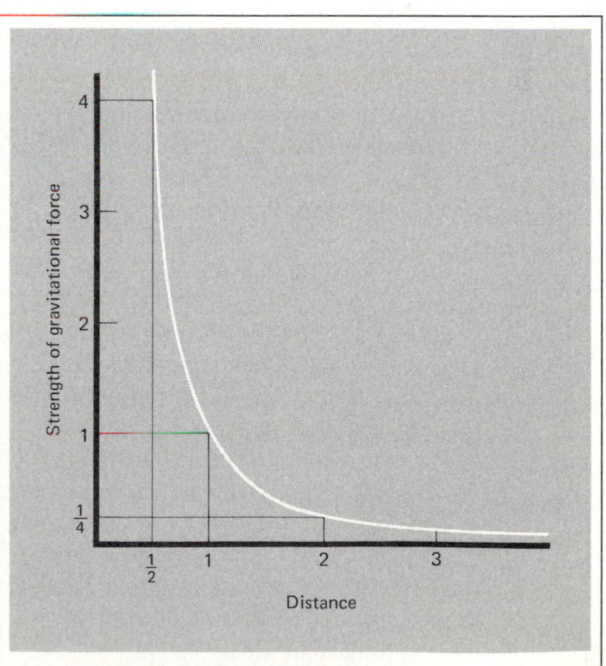

4.14
An inverse-square law. This graph shows how the strength of the gravitational force changes with the distance between masses. At a separation of 1 unit of distance, the force has 1 unit of strength. When the masses are twice as far apart, a distance of 2, the force has only one-fourth the original strength.

Focus 4.1

NEWTON, THE APPLE, AND THE MOON

How did Newton connect the apple's fall to the moon? He knew that the distance to the moon was approximately 60 earth radii. He also surmised that if the earth's influence extended to the moon, it grew weaker as the distance increased. How much did the force weaken? Newton guessed that it weakened as the inverse square of the distance (Fig. F.13). An apple dropping from a tree is 1 earth radius from the center of the earth. So, comparing the accelerations of the moon and the apple, Newton predicted that the moon's acceleration must be (1 earth radius)^2/(60 earth radii)^2 less, or 1/3600 as strong. The acceleration due to gravity at the earth's surface is 9.8 m/sec/sec. For the moon, then, the predicted acceleration is (9.8 m/sec/sec)/3600, or about 2.7 $\times$ 10^{-3} m/sec/sec. This acceleration of the moon toward the earth is a centripetal acceleration, for the necessary force, gravity, pulls the moon towards the earth's center.

Note that Newton did not need to know the mass of the apple or of the moon because their accelerations do not depend on their masses. Galileo (Section 4.1) had reached this same conclusion from his experiments. Newton made crucial use of this result in the moon-apple test.

How did this predicted value compare with the actual rate of the moon's fall? Newton had found that for a circular orbit, the centripetal acceleration has a value of

$$a = \frac{V^2}{R}$$

where a is the centripetal acceleration, R the radius of the orbit, and V the orbital velocity. For the moon, R equals 3.84 $\times$ 10^8 m. Its velocity is the distance it travels in one orbit divided by the period for one orbit, or

$$V = \frac{2\pi\,(3.84 \times 10^8\ \text{m})}{(27.3\ \text{days})}$$

One day contains 24 $\times$ 60 $\times$ 60, or 8.64 $\times$ 10^4 secs, so the moon's orbital velocity is

$$V = \frac{2.41 \times 10^9\ \text{m}}{(27.3\ \text{days})(8.64 \times 10^4\ \text{sec/day})} = \frac{2.41 \times 10^9\ \text{m}}{2.36 \times 10^6\ \text{sec}} = 1.02 \times 10^3\ \text{m/sec}$$

Then the moon's orbital acceleration is

$$a = \frac{(1.02 \times 10^3\ \text{m/sec})^2}{3.84 \times 10^4\ \text{m}} = 2.71 \times 10^{-3}\ \text{m/sec/sec}$$

This result comes close to the prediction made from an inverse-square law.

Newton did not have the modern values used in these calculations for the period and size of the moon's orbit and the acceleration due to gravity. He chose a value of the moon's distance that made his results compare closely. Newton did feel assured that his approach was correct, even though he fudged the figures slightly.

At the moon's distance the acceleration is
$\frac{1}{60^2} = \frac{1}{3600}$
less, if Newton's law of gravitation is correct.
60 earth radii
At the earth's surface the apple falls with a uniform acceleration of about 9.8 m/sec/sec
One earth radius
Center of earth

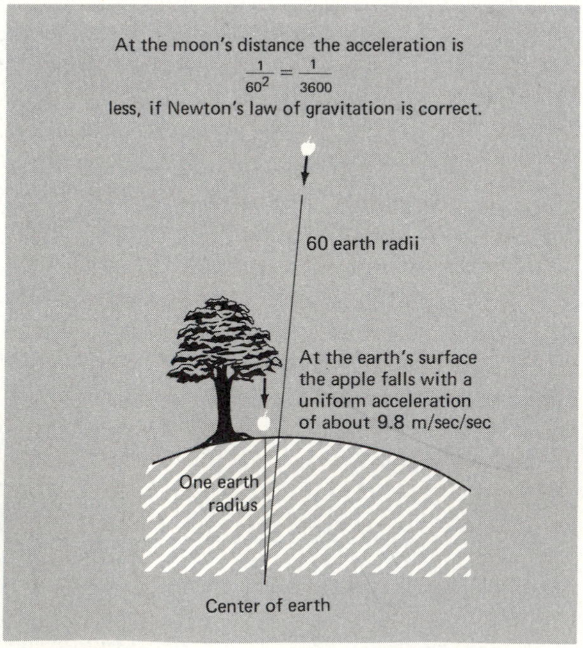

F.13

The moon and the apple. The apple is 1 earth radius from the earth's center and falls with an acceleration of 9.8 m/sec/sec. If an inverse-square law for gravity is correct, then the moon accelerates at a rate 1/3600 less than the apple.

center) caused the apple to fall. The earth-moon distance is about 60 earth radii, so if an inverse-square law correctly describes gravitational forces, the acceleration of the moon toward the earth must be $1/60^2$ or 1/3600 as much as the acceleration of the apple. (See Focus 4.1 for details.) Newton then compared his predicted centripetal acceleration with the centripetal acceleration derived from observations of the moon's orbit. As he put it, the predicted and the observed accelerations were "pretty nearly" the same. Newton concluded that the cause of the moon's centripetal acceleration is the same as that of the apple's: the earth's gravity. This force, extended out to the distance of the moon, kept the moon in its orbit.

You should now be able to understand Galileo's law of free-fall from Newton's standpoint. Imagine that you're going to repeat Galileo's rumored experiment at the Leaning Tower of Pisa by dropping a cannonball and a tennis ball from its uppermost story. The earth exerts a much greater gravitational force on the cannonball than on the tennis ball because of the cannonball's much greater mass. However, when the two are dropped, they fall side by side and land at the same time (if you neglect air resistance). So gravity's effect has been the same on both objects; more precisely, the *acceleration* of each is the same, because they speed up at the same rate. Although the forces are different, the accelerations turn out to be the same. Why? Because the extra force on the cannonball is exactly offset by its greater inertia to changes in motion.

4.5 Cosmic consequences of universal laws

The gravitational force of the earth causes the centripetal acceleration of the moon. This was Newton's central discovery: The earth's gravity keeps the moon swinging around. Newton's vision resulted in a new understanding of the planetary orbits. The most obvious: The sun's gravity locks the planets in their elliptical orbits. Newton had found the physical interaction between the sun and the planets first sought by Kepler. He derived a new form of Kepler's third law that included the gravitational constant and the masses of the interacting bodies (Focus 4.2). Related to basic physical laws, the revised third law became a potent tool in determining the masses of the planets (if the planet has a least one satellite) and of the sun—quantities never known before! Newton answered in detail the ancient question of how the planets moved. And he answered it precisely: His predictions of planetary positions were far more accurate than previous ones. Newton's ideas provided the physical support sorely needed for the Copernican model.

The earth's rotation / One objection to the Copernican model was that objects not tied down to the earth should fly off because of its rotation. Newton noted that gravity holds them down. Another objection was that dropped objects should land behind their starting position, because the turning earth leaves them behind. Newtonian physics explained that objects on the earth had inertia. When falling, they did not lose their foreward inertial motion but continued to move with the ground beneath them. So they landed at their starting points. A rotating earth did not leave behind unattached objects because of their inertia.

Here's how: When you throw an object straight up, it moves in the direction of the earth's rotation while it is in the air with the same speed it had while in your hand. So the object aloft keeps up with the turning earth, because no force acts on it to change its eastward velocity. When it returns to the ground, it then falls right back into your hand.

The earth's revolution and the sun's mass / You can use the earth's orbit to find the sun's mass. The earth-sun distance, a, is 1.50×10^{11} m. The earth's period, P, is 365.25 days. Then from Newton's second law and the law of gravitation, you can work out the mass of the sun (Focus 4.2).

The knowledge of these masses bolstered the Copernican model in the framework of Newtonian physics. Newton's laws show that the sun has roughly 3.3×10^5 the mass of the earth. Newton's third law of motion requires equal gravitational forces between the sun and the earth—the force of the sun on the earth equals that of the earth on the sun. However, the second law demands that the earth's acceleration be much greater than the sun's (in fact, 3.3×10^5 times greater!). This happens only if the motion of the sun is small, so that it has a small change in motion (acceleration) in one year. So the earth orbits the sun rather than the other way around.

Gravity and orbits / To add to its achievements, Newton's physics correctly described the orbits of comets. Through antiquity and the Middle Ages, astronomers believed that comets were objects confined to the earth's atmosphere. Tycho showed observationally that comets were not atmospheric phenomena. Newton and Halley decisively demonstrated that comets orbit the sun in accordance with the law of gravitation (Fig.

Focus 4.2

THE MASS OF THE SUN

By using his laws of motion and gravitation, Newton reworked Kepler's third law so that it had the form

$$P^2 = \frac{4\pi^2}{G(m_E + M_S)} a^3$$

Compare to Kepler's third law in the form

$$P^2 = ka^3$$

with $k = 4\pi^2/G(m_E + M_S)$. Note that the constant relates to the masses of the two bodies involved.

You can use the earth's orbit to find the sun's mass. The earth-sun distance, a, is 1.50×10^{11} m. The earth's period, P, is 365.25 days, or 3.16×10^7 secs. Because the mass of the sun is much larger than that of the earth, we can approximate $M_S + m_E$ by M_S. Then Newton's revision of Kepler's third law gives

$$M_S = \frac{39.5}{6.67 \times 10^{-11}} \frac{(1.50 \times 10^{11})^3}{(3.16 \times 10^7)^2}$$

$$= 1.99 \times 10^{30}\ \text{kg}$$

Note that you need to know the value for G to do this calculation.

If you don't know G, you can still use the third law to find relative masses. Use the earth again, but this time you also need the moon. The moon's distance is 3.84×10^8 m, or roughly 2.6×10^{-3} AU, and the moon's period is 7.5×10^{-2} yr. For the earth and moon,

$$m_E + m_M = \frac{4\pi^2}{G} \frac{a_{EM}^3}{P_{EM}^2}$$

and for the sun and earth,

$$M_S + m_E = \frac{4\pi^2}{G} \frac{a_{ES}^3}{P_{ES}^2}$$

If we don't know G, divide this equation by the other one to get

$$\frac{(M_S + m_E)}{(m_E + m_M)} = \frac{\dfrac{4\pi^2}{G} \dfrac{a_{ES}^3}{P_{ES}^2}}{\dfrac{4\pi^2}{G} \dfrac{a_{EM}^3}{P_{EM}^2}}$$

$$= \left[\frac{a_{ES}}{a_{EM}}\right]^3 \left[\frac{P_{EM}}{P_{ES}}\right]^2$$

As before, approximate $M_S + m_E$ by M_S, and $m_E + m_M$ by m_E. Then

$$\frac{M_S}{m_E} = \left\{\frac{1\ \text{AU}}{2.6 \times 10^{-3}\ \text{AU}}\right\}^3 \left\{\frac{7.5 \times 10^{-2}\ \text{year}}{1\ \text{year}}\right\}^2$$

$$= (5.7 \times 10^7)(5.6 \times 10^{-3})$$

$$= 3.3 \times 10^5$$

You have found the sun's mass relative to the earth's mass.

4.15). In fact, Halley correctly predicted the return of the comet that bears his name, but he did not live to see it return. (More on comets in Chapter 12.)

Newton's ideas also led to the discovery of Neptune in 1846, long after the first publication of the *Principia*. Neptune was the first planet to be found by its gravitational effects on another. Newton's laws predicted its existence *before* it was observed.

The discovery of Neptune rested on observed irregularities in the orbit of Uranus (discovered with a telescope in 1781). Astronomers had noted small discrepancies between the observed positions of Uranus and those predicted from Newton's laws. Such irregularities occur because all planets attract one another. Jupiter's tug, for instance, influences the orbit of Uranus so that it differs from that expected if Uranus and the sun were the only attracting bodies. The undiscovered Neptune revealed itself when Uranus's motion deviated from the path predicted from the effects of the known planets.

Applying Newton's laws to explain the deviations as due to a planet beyond Uranus, the Englishman John C. Adams in 1845 estimated

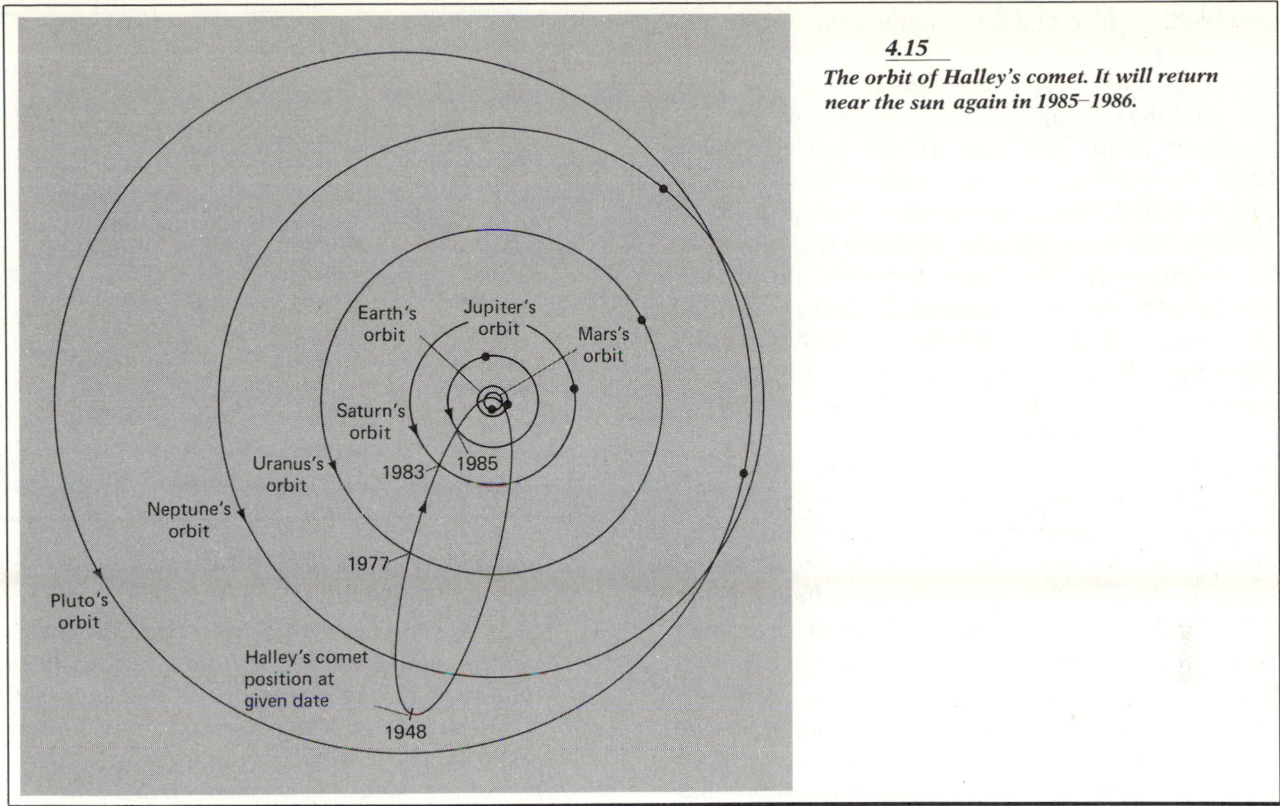

4.15
The orbit of Halley's comet. It will return near the sun again in 1985–1986.

where the unknown planet should be in the sky. He communicated his result to the Astronomer Royal, Sir George Airy, hoping that a search for the new planet would be begun immediately. But Airy did not believe the result, partly because he was one of those who thought that the gravitational force deviated from an inverse square law at large distances.

At almost the same time, Urbain J. J. Leverrier in Paris made similar calculations and transmitted them to Johann Galle in Berlin, who found the planet on September 23, 1846. Newton's laws triumphed. In the twentieth century, discrepancies in Neptune's orbit stimulated the search for other planets. The search resulted in the discovery of Pluto (see Chapter 11), even though we now know that Pluto actually has too small a mass to affect Neptune.

An equally dramatic discovery, which extended the validity of Newton's laws beyond the bounds of the solar system, was the observation of binary star systems in the late eighteenth century. (A binary star system consists of two stars, held together by their mutual gravity, orbiting each other.) William Herschel (1768–1822) and his sister, Caroline Herschel (1750–1848), observed many pairs of stars over a long time, looking for heliocentric parallax. Unexpectedly, the Herschels observed the orbital motions of some pairs. These observations showed that some pairs of stars in the sky lay close together not simply because they both happened to be in the same direction in the sky as seen from the earth but because they were actually linked by gravitational forces.

Calculations of the orbital periods of binary star systems and the separation between the stars confirmed that their motions followed Kepler's laws. These laws can be derived from Newton's laws of motion and gravitation. Since binary star systems obey Kepler's laws, they indirectly confirm Newton's laws. (See Chapter 15 for more details on binary star systems.)

Triumph after triumph hammered Newtonian ideas into people's minds. In terms of a scientific model, Newton's synthesis was complete. Its predictive success was astounding. Time after time, the numbers, not just the general ideas, came out right. Planetary positions could be predicted precisely, and also the motions of planetary satellites. And Newton's ideas were *universal*. No longer were separate laws needed for falling bodies, the moon, and binary stars. One simple force binds them all: gravity!

Newton's cosmology / The physical ideas that Newton found in the universe resulted in a new conception of the cosmos. For the Greeks, Ptolemy, and Copernicus, the cosmos—enclosed by the sphere of fixed stars—appeared finite and bounded. Galileo had left open the possibility that the universe might be infinite but did not strongly push his case. Newton argued that his laws required that the universe be infinite in extent. His argument went like this: If the universe were finite, gravitation would eventually pull all matter to the center. As a result, only one large mass would exist. However, we see other masses—stars for example. In an infinite universe, the matter would be pulled into an infinite number of small condensations. Newton believed that this picture was more like the real world, so he concluded that the universe was infinite.

Not all things in heaven and earth sat happily in Newton's infinite universe. The innermost planet, Mercury, posed an annoying problem: The axis of its orbit rotated in space. This motion could not be completely explained by the attraction of existing planets. Some astronomers thought that the excess rotation was caused by a planet between Mercury and the sun. Although observations of this hypothetical body—called Vulcan—have been reported, they have so far proved to be mistaken. The supposed Vulcan has never been seen, even when modern observation techniques have attempted to catch its swift flight. (The problem of Mercury's orbit was later solved by Einstein; see Chapter 7).

Newton was also sorely disturbed by the mutual forces among the planets, which he thought must eventually lead to the disintegration of the solar system. To avoid this awful event, Newton envisioned the hand of God occasionally descending to reset the clockwork mechanisms of planetary motions, like a conscientious artisan making adjustments. The order of the mechanical universe—ordained by the Divine Being—was maintained by His intervention, and the expanses of Newtonian space were benevolently watched by the distant God.

TABLE 4.1 Evolution of cosmological models

	Model			
	Ptolemy	*Copernicus*	*Kepler*	*Newton*
Planetary motions	Uniform, circular	Uniform, circular	Nonuniform elliptical	Nonuniform elliptical
Force on planets?	No, natural motion	No, natural motion	Yes, magnetic	Yes, gravitation
Kind of cosmos	Geocentric, finite	Heliocentric, finite	Heliocentric, finite	Centerless, infinite

KEY CONCEPTS

1 ***Galileo used his telescopic observations to promote the Copernican model (which he believed in for aesthetic reasons), even though none gave direct proof of its validity; he recognized that the Copernican model lacked a physical basis and sought one by analyzing the motions of falling objects near the earth; he concluded that falling objects had accelerated motion and so were subject to a force: gravity.***

2 ***With his laws of motion and gravitation, Newton abolished the physical distinction of motions in the terrestrial and celestial realms; it resulted in a physically unified cosmos (Table 4.1).***

3 ***Newton's laws of motion described natural motion and forced motion; he viewed natural motion as that of a body at rest or moving at a constant speed along a straight line; forces result in accelerations (changes in velocities); from these ideas, Newton concluded that planetary orbits resulted from a centrally directed force; that force is gravity; he used the moon's orbit to check that the force of gravity changes with the inverse square of the distance.***

4 ***Newton's physics justified the Copernican model, as revised by Kepler, with the correct force, gravity; the earth's rotation and revolution were naturally explained; Newton's work also resulted in very accurate predictions of planetary motions; the final result was simple, pleasing, and accurate.***

5 ***The discovery of binary stars whose motions follow Kepler's laws verified the validity of Newton's law of gravitation beyond the solar system.***

6 ***Newton's cosmology demanded that the universe be infinite in extent; otherwise gravity would bring all the material together, which was not the actual case.***

STUDY EXERCISES

1 Use one of Galileo's telescopic discoveries to support the Copernican model and refute the traditional (Aristotelian-Ptolemaic) one. (*Objective 1*)

2 What important discovery of Kepler's was ignored by Galileo? (*Objective 3*)

3 Describe Galileo's concept of inertia and contrast it to the Aristotelian one. (*Objective 2*)

4 You have two balls of the same size and shape. One is lead, the other wood. You drop them together. What happens? Explain. (*Objectives 2, 5, and 6*)

5 Imagine that you're out in space and push away from you an object with the same mass as you. What happens? Explain. (*Objective 5*)

6 Describe two ways in which Newton's model of the cosmos differed from that of Copernicus. (*Objective 9*)

7 Use Newtonian physics to argue that the earth rotates on its axis and revolves around the sun, answering the main physical objections to the Copernican model. (*Objective 10*)

8 Describe one way in which Newton and Kepler had similar ideas about the cosmos and one way in which their ideas differed. (*Objective 9*)

9 Give a simple example, different from those in the text, of each of Newton's laws of motion. (*Objective 5*)

10 Imagine you hold a ball in your hand with your arm out to your side. You walk along quickly toward a target on the floor. You release the ball just as it reaches a point above the target. Use Newton's concept of inertia to predict where the ball lands: behind, on, or in front of the target. (*Objectives 5 and 6*)

BEYOND THIS BOOK . . .

For a fictionalized look into Galileo the man, read *Galileo* by B. Brecht (better yet, see a production of the play) or *The Star-Gazer* (Putnam, New York, 1939) by Z. de Harsanyi, translated by P. Tabor.

For a view of Newton's work, nothing beats the *Principia: Mott's Translation Revised* (University of California Press, Berkeley, 1966), translated by F. Cajori.

You can get to know Galileo by reading *Dialogue Concerning the Two World Systems* (University of California Press, Berkeley, 1967), translated by S. Drake. *Discoveries and Opinions of Galileo*, also by Drake, contains a translation of *The Starry Messenger*.

For insight into how Galileo actually arrived at the correct analysis of the motion of falling bodies, read "Galileo's Discovery of the Law of Free Fall" by Stillman Drake in *Scientific American*, May 1973.

A short bibliography of Newton is "Isaac Newton" by I. B. Cohen in *Scientific American*, December 1955.

For details about how Neptune was found, look at *The Discovery of Neptune* by M. Grosser (Harvard University Press, Cambridge, Mass., 1962).

5/ The birth of astrophysics

> *We understand the possibility of determining their [celestial bodies'] shapes, their distances, their sizes and motion, whereas never, by any means, will we be able to study their chemical composition.*
>
> *AUGUST COMTE: Cours de Philosophie Positive (1835)*

Learning objectives

After studying this chapter, you should be able to:

_1._ Describe the difference in the appearance of the three basic types of spectra (continuous, absorption, emission) as seen through a spectroscope.
_2._ Explain how an understanding of spectra made it possible for astronomers to determine the chemical compositions of and physical conditions in stars.
_3._ Use Kirchhoff's rules to relate the three basic types of spectra to the physical conditions of their production.
_4._ Describe how wavelength, frequency, and speed of light are related.
_5._ Describe the relationship between energy and wavelength for light.
_6._ Briefly describe the electromagnetic spectrum with examples from each major region.
_7._ Use an energy-level diagram to explain in general how atoms absorb and emit light.
_8._ Use the energy-level diagram of a hydrogen atom to explain how the Balmer series is produced, both as emission and absorption lines.
_9._ Explain in simple physical terms how absorption lines occur in spectra.

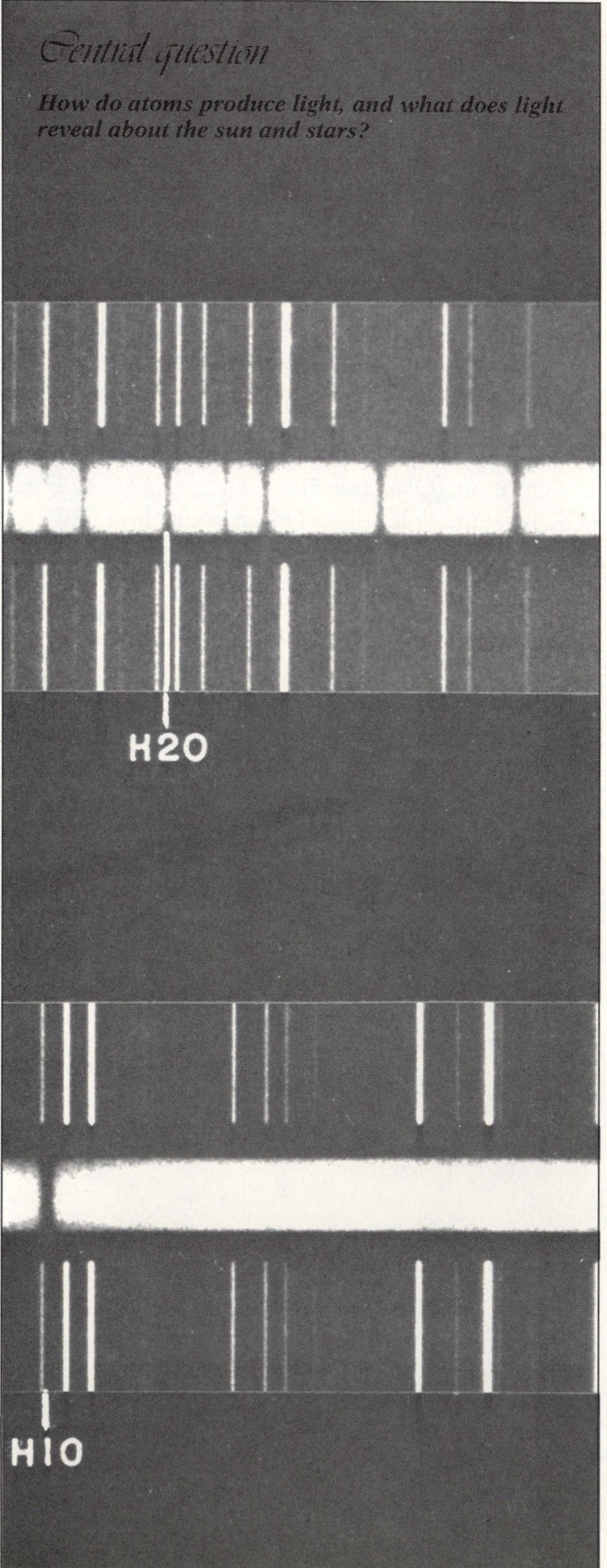

Central question

How do atoms produce light, and what does light reveal about the sun and stars?

Comte, a French philosopher of the nineteenth century, arrived at his pessimistic view for hard, practical reasons. Astronomers had for centuries charted the positions of the stars and planets. Their attention focused on the question of how the planets moved, a problem solved by Isaac Newton (see Chapter 4). In contrast, the stars played a passive role, a mere backdrop to the complex activity of the planets.

But how to find the chemical compositions and physical makeup of the stars and planets? Here Comte saw no hope. Astronomers had no means to bring a piece of a star or planet into the lab to examine firsthand. All they had in hand was the light funneled into their telescopes.

The birth of astrophysics at the end of the nineteenth century gave astronomers physical insight from light. Experimental attempts to understand light unexpectedly led to a study of atoms. The atoms in stars (and the sun) emit starlight (and sunlight), so an analysis of light reveals information about the physical conditions and chemical compositions of stars. Astronomers could finally penetrate the environments of the distant stars and the sun, and so proved Comte wrong.

Astrophysics transformed our conception of the cosmos. No more were the stars regarded as simple points of light. Now they appeared as other suns made of the same stuff as found in our sun and the earth.

5.1 *Sunlight and spectroscopy*

Matter makes up the sun—matter in the form of atoms. Atoms can give off and absorb light; the sunlight we see originates from atoms. The structure of atoms can be investigated by analyzing the light they emit and absorb. We can also investigate the physical environment of atoms by analyzing light. To understand more of the sun (and stars), you need to understand a little about how atoms and light interact.

Atoms and matter / In the eighteenth and nineteenth centuries, chemists discovered that substances can be divided into two classes: chemical *elements* and chemical *compounds*. *Elements* cannot be broken by chemical reactions into simpler substances. (See Appendix G for a periodic table of the elements.) They are the most basic substances, such as hydrogen (H), helium (He), carbon (C), oxygen (O), and so on. Although 92 elements occur in nature (and 14 more have been created in the laboratory), many are rare; only a few dozen are the most common. Most substances you encounter are not elements but *compounds*, substances made of two or more elements. For example, water (H_2O) is a chemical compound composed of the elements hydrogen and oxygen.

A *molecule* is the smallest unit of a compound that still has the chemical properties of the compound. However, a compound can be broken down into elements, so a unit of matter smaller than a molecule must exist. An *atom* is the smallest unit of an element that displays the chemical properties of the element. A compound is created when atoms join together to form molecules of the compound.

Physicists developed a useful model of the atom in the twentieth century. The modern concept of an atom pictures a tiny, dense *nucleus* surrounded by rapidly moving *electrons*. The study of electricity revealed that matter has two opposite charges, positive and negative. Like charges repell each other and opposite charges attract. *Electrons*, which carry a negative charge, are low-mass particles. *Protons* and *neutrons*, which are about 2000 times as massive as electrons, make up an atom's nucleus. The protons are positively charged, and the neutrons have no electrical charge. The nucleus of an atom, because of the protons it contains, is positively charged and attracts the negatively charged electrons. This attractive force binds the electrons to the nucleus and holds the atom together. (A strong *nuclear force* binds the protons and neutrons together in the nucleus, despite the mutual repulsion of the positively charged protons.) The electrons whiz around the nucleus in orbits, but their distance is very great in terms of the size of the nucleus. Most of an atom is empty space!

Elements differ from each other because their atoms differ. The nucleus holds the most important difference; the atoms of different elements contain different numbers of protons. The nucleus of a hydrogen atom, for example, contains one proton; helium, two; and carbon, six. Nuclei that contain the same number of protons but *different* number of neutrons are called *isotopes* of the element. For example, heavy hydrogen, sometimes called deuterium, has one proton and one neutron, whereas ordinary hydrogen has one proton and no neutrons (Fig. 5.1). They are the same element, but the atoms of heavy hydrogen each have more mass than the atoms of the ordinary form, because of the extra neutron.

Most elements contain approximately the same number of protons and neutrons. For example, atoms of the most common isotope of carbon contain six protons and six neutrons.

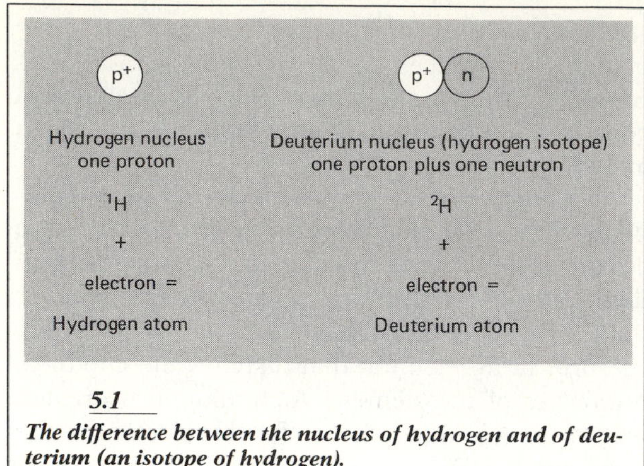

5.1
The difference between the nucleus of hydrogen and of deuterium (an isotope of hydrogen).

A normal atom has the same number of electrons orbiting the nucleus as it has protons in the nucleus, so it carries no net electric charge. If the number of electrons is less than or greater than the number of protons, the particle is called an *ion*.

Simple spectroscopy / Breaking up light into its component colors to detect how atoms and light interact is called *spectroscopy*. You've probably done a little spectroscopy without knowing it. I'm sure you've seen a rainbow. Next time you see one, look carefully to see that the colors run continuously from red to blue. What's happening? Raindrops are dispersing sunlight into a continuous band of colors.

You can use a prism to do the same at home or in a lab (Fig. 5.2). Use a slotted piece of cardboard to pass a beam of sunlight through a prism. Let the light coming out of the prism fall on a white sheet. You'll find the sunlight spread into an array of colors, with red light bent the least and violet the most. This sequence of colors is called a *spectrum*. A spectrum with no breaks in it is called a *continuous spectrum* (Color Plate 1).

Suppose you put another slotted piece of cardboard (or slit) in front of your prism and let just a *single color* pass through. You'd then see a bright line of that color, not a complete, continuous spectrum. This line is called a *spectral line*. White light produces a continuous spectrum with all colors running smoothly into one another and no colors missing. You can think of white light as a mixture of all colors of equal intensity.

Take your prism and slit and let sunlight pass through them; then magnify the spectrum with a telescope. You'll see a continuous spectrum filled with dark lines. There are some colors where light is missing. Something has removed, or absorbed, light of these colors. These dark lines are called *absorption lines*, and a spectrum with them is called an *absorption spectrum*. When light is emitted only at certain colors, as in this example, we speak of *emission lines* or an *emission spectrum*.

Note that straight lines appear in a spectrum because the slit admitting the light to the prism is a straight-line source of light. Each line is an image of the slit in the light of one color. You would also see a line spectrum if you looked at a straight hot wire. But if the slit, or other source of light, were curved in an S shape, the lines in the spectrum would also be S-shaped.

5.2 *Analyzing sunlight*

To unravel the message of sunlight, let's first do a few experiments in the lab with a *spectroscope* (Fig. 5.3), an instrument for observing fine details in a spectrum. A simple spectroscope consists of a

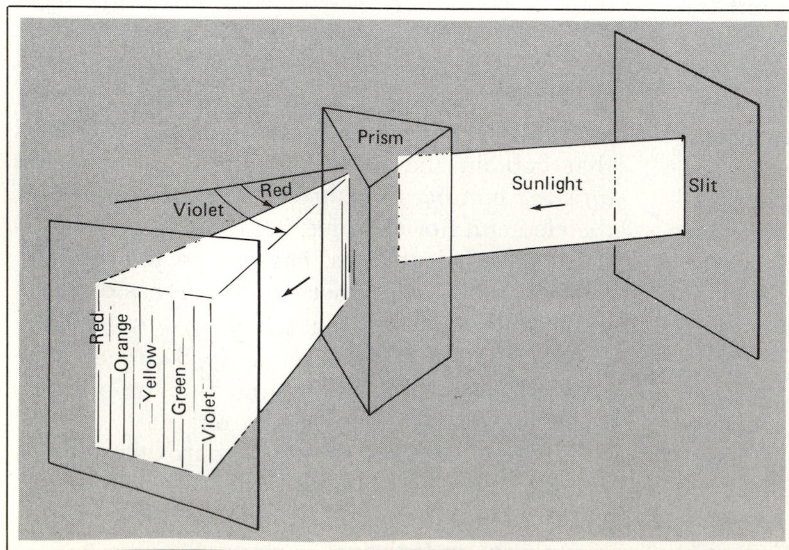

5.2
An experiment to display the sun's spectrum. The prism splits sunlight into the band of colors of the rainbow.

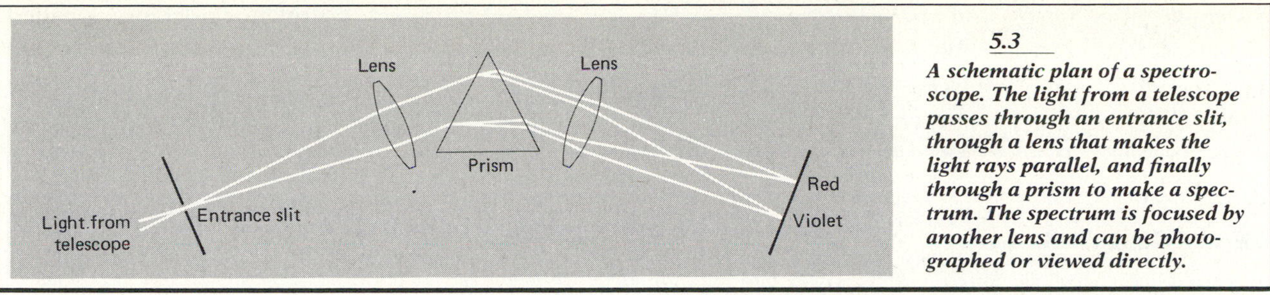

5.3

A schematic plan of a spectroscope. The light from a telescope passes through an entrance slit, through a lens that makes the light rays parallel, and finally through a prism to make a spectrum. The spectrum is focused by another lens and can be photographed or viewed directly.

lens behind a telescope, which feeds light from a slit into the prism. Another telescope allows viewing of the spectrum produced by the prism.

Consider putting sodium (in the form of table salt) in a hot, colorless flame. Visually, you'd see the flame turn yellow. Through the spectroscope, you'd see a series of bright lines, the brightest, a pair, in the yellow region (see Color Plate 1). With the spectroscope you can find that different chemical elements give off different patterns of bright lines. *Each chemical element displays a unique arrangement of bright lines.* A particular spectral pattern demonstrates the presence of a particular element. For example, the brightest lines of neon gas lie in the red region of the spectrum, so a pure neon advertising sign appears red. Hot mercury gas has the strongest bright lines in the yellow and green regions. These lines combined result in the greenish color of mercury street lamps. In contrast, sodium street lamps appear a ghastly yellow because the brightest lines from a hot sodium gas fall in the yellow region of the color spectrum.

To emit light, a hot gas must be giving of energy (Focus 5.1). To remain hot, an outside source of energy (such as the electric current in a street lamp) must be supplied. (Energy is conserved!) Some of this energy radiates away as light at the specific colors of the bright lines. The atoms in the gas produce the observed bright lines. Different gases composed of different atoms produce different patterns of spectral lines. So the bright-line spectrum of a hot gas is characteristic of the atoms in the gas. And we can tell by the pattern of the lines which elements the gas contains.

If you look at sunlight with a spectroscope, you see dark lines against a bright, continuous background of color (Color Plate 1 and Fig. 5.4). A pair of dark lines in the yellow region of the spectrum is particularly prominent. Astronomers call this pair the *D-lines*. Could these dark lines be related to the bright lines of sodium observed when salt is put in a flame?

To test this idea, pass light from a glowing solid (which gives off a continuous spectrum) through the sodium flame. A pair of dark lines appears in the spectrum at exactly the same position in the yellow regions as the pair in the sun's spectrum. So it must be sodium making these dark lines, *removing* yellow light from the solar spectrum. Now pass sunlight through a sodium flame into the spectroscope. What do you expect? That the added light of the sodium flame makes the lines less dark? Wrong! The pair of lines becomes even darker! The sodium in the flame absorbs even more of the sun's yellow light.

From such experiments, Gustav Kirchhoff (1824–1887) formulated empirical rules of *spectroscopic analysis*, the determination of the physi-

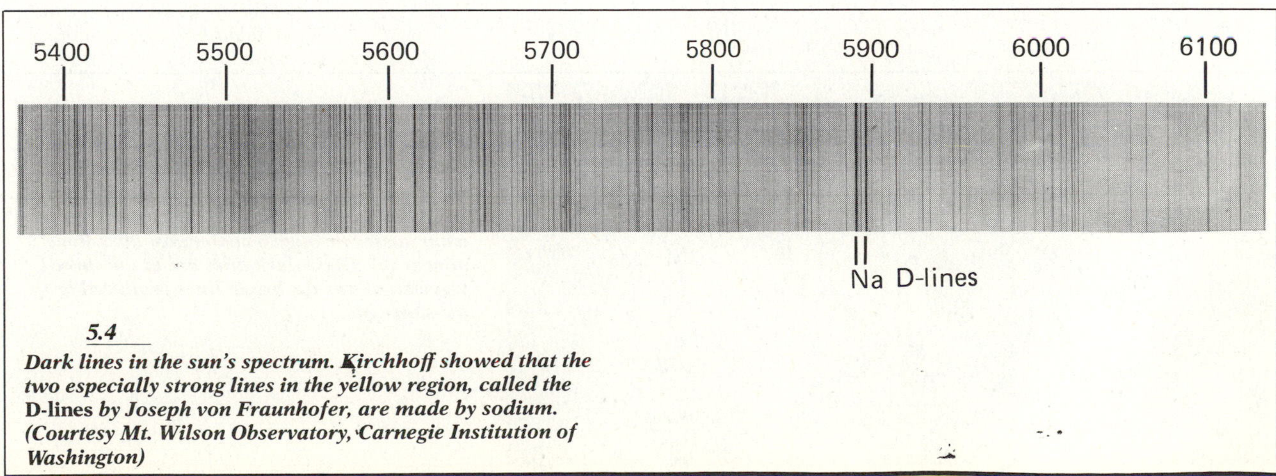

5.4

Dark lines in the sun's spectrum. Kirchhoff showed that the two especially strong lines in the yellow region, called the D-lines by Joseph von Fraunhofer, are made by sodium. (Courtesy Mt. Wilson Observatory, Carnegie Institution of Washington)

Focus 5.1

ENERGY AND WORK

Energy makes work possible and allows events to happen in the universe. The concept of energy has fundamental importance because, although energy comes in many forms, the total amount of energy that an object has can be described and assigned a number. If you have a bunch of objects, the sum of their individual total energies is the total energy of the group. The fact that the total energy of a group of objects, isolated from the rest of the world, has a constant value is known as the law of conservation of energy.

Energy exists in many different forms. Three common ones are (1) kinetic *energy, that due to motion; (2)* potential *energy, that due to position under an applied force; and (3)* radiative *energy, that which is carried by light.*

Kinetic energy can be understood in terms of work. For example, to stop a moving object requires an obvious effort, and the more massive the object or the faster its motion, the greater the effort needed to stop it.

You must distinguish between the kinetic energy of a large object, such as a baseball, and the average kinetic energy of the particles that make it up. A baseball's kinetic energy allows it to go somewhere but that doesn't affect the microscopic motions of its atoms, which move small distances at random within the ball. Their motion makes up the thermal energy of the ball, which is usually called heat. *Heat is a measure of the total kinetic energy of a large collection of particles. When you heat a gas, the average value of the random speeds of the molecules that make it up increases.*

Temperature measures the average kinetic energy of particles in, say, a gas. That, in turn, depends on the average speeds of the particles. The temperature scale that indicates this most directly is called the Kelvin *scale; at 0 Kelvin, all microscopic motions have ceased. (See Appendix A.)*

Keep these different concepts clear: Energy is the ability to do work; heat is the total random energy of motion of particles in an object; and temperature is a measure of the average speeds of the random motions of these particles.

Kinetic energy due to motion can be expressed in a simple equation:

$$KE = \frac{1}{2} mv^2$$

where m is the mass of an object and v its velocity. Note that the kinetic energy depends on the square *of an object's velocity. So of two cars, one traveling twice as fast as the other, the faster car has* four *times the kinetic energy and will be four times harder to stop. In other words, with locked brakes, the faster car skids four times the distance of the slower.*

Potential energy *is the stored energy of position. This possibility of motion for an object characterizes potential energy. An obvious example is the earth's gravitational attraction. All masses feel the earth's pull, although some, such as a ball held in your hand, do not immediately respond to the force. When you drop the ball, however, it gains kinetic energy, and the higher the point from which you drop it, the greater the kinetic energy it attains by the end of its fall. The*

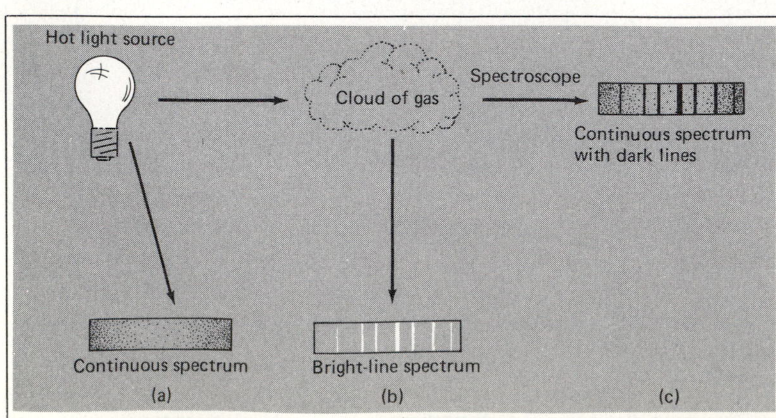

5.5

Kirchhoff's rules of spectral analysis. A hot, opaque solid or gas produces a continuous spectrum (a). *A hot, transparent gas makes an emission spectrum* (b). *If a continuous spectrum passes through a cooler gas, dark lines appear* (c). *These dark lines are at the same wavelengths as the bright lines produced by the same gas.*

position above the earth's surface relates to the total amount of potential energy the object has. Energy transforms from one form to another (potential to kinetic), yet the total energy of the ball remains constant, for as it loses potential energy, it gains kinetic energy.

The energy-level diagram of an atom describes a situation analogous to that of objects pulled by gravity. Instead of gravitational force, however, we have the force of electrical attraction between the positively charged nucleus and the negatively charged electrons. When an electron falls from one energy level to a lower one, it loses potential energy, which must manifest itself in some form, in this case a photon (radiative energy). To reverse the process requires the electron to gain energy, again produced from the absorption of a photon with exactly the energy required to raise the electron to the next level.

The amount of energy carried by light depends on its frequency (or wavelength). The fundamental relation is

$$E = hf$$

where E is the energy in joules, h a constant (called* Planck's constant*) with a value of 6.63×10^{-34} J-sec, and f the frequency in cycles per second. Note that light with twice the frequency carries twice the radiative energy; three times the frequency, three times the energy; and so on.

Now back to the concept of the conservation of energy. Experiments (such as the dropping of a ball) have shown that energy is naturally transformed from one form to another but that as long as all the energy is carefully accounted for, the total amount remains the same—it is conserved. One way to state the conservation of energy is:* Energy cannot be created or destroyed; it may be transformed but the total does not change. *(This principle will be modified somewhat because of Einstein's discovery that matter is actually a form of energy; see Section 7.2.)

cal state and the composition of an unknown mixture of elements. Briefly put, Kirchhoff's rules are (Fig. 5.5):

1 A hot and opaque solid, liquid, or highly compressed gas emits a continuous spectrum.

2 A hot, transparent gas produces a spectrum of bright lines (emission lines). The number and colors of these lines depend on which elements are present in the gas.

3 If a continuous spectrum passes through a gas at a lower temperature, the cooler gas causes the appearance of dark lines (absorption lines). Their colors and their number depend on the elements in the cool gas.

Essentially, the first rule says that an opaque, hot material produces a continuous spectrum. Recall that atoms usually emit and absorb at discrete wavelengths. However, when atoms are jammed together so densely that they become opaque to light, a continuous spectrum appears. For instance, the sun is so hot that it is gaseous, but it is still dense and opaque enough to produce a continuous spectrum.

A solar spectrum shows dark lines against a continuous background. What causes these dark lines? By Kirchhoff's third rule you expect that there must be a cooler and less dense gas between the visible surface of the sun and us. There are two places where this gas can be, in the atmosphere of the earth or the atmosphere of the sun. Although a few lines in the solar spectrum are produced by gases on earth (water vapor in particular), most lines are produced in the solar atmosphere. This atmospheric layer absorbs light from the continuous spectrum passing through it to produce the dark lines. The positions of the lines in the spectrum, their colors, tell which elements are present. The solar composition determined from the dark lines relates only to the region of the sun that produces the absorption spectrum (its atmosphere).

One key point: Dark lines do not mean the *complete* absence of light at those colors; rather, the lines appear dark in contrast to the continuous spectrum. That is the reason the sodium D-lines got darker when sunlight passed through the sodium flame.

With these ideas in mind, you should be able to understand what happens when sunlight passes through the sodium vapor to make darker D-lines. The sodium in the flame is *cooler* than the sun's visible surface, so the sodium *absorbs* the light in the D-lines. This experiment proves that the missing colors in the continuous solar spectrum are due to absorption of those colors by the atoms in the cool gas of the solar atmosphere.

Whether the atoms in the gas emit or absorb depends on the physical conditions in the gas. Emission requires high temperatures in a transparent gas; absorption occurs when a continuous spectrum passes through a cooler transparent gas. In either case, the pattern of emission lines is the same as the pattern of absorption lines for a gas with the same chemical composition.

Here's the fundamental point concerning emission and absorption lines: *The lines absorbed by a gas from a continuous spectrum are the same lines emitted by the gas when energy is put into it.*

5.3 Spectra and atoms

How can the physical conditions in stars be discovered from their spectra? The spectral code was cracked in a great revolution of physics in the twentieth century: the quantum theory and its explanation of the nature of the atom and of light. But to understand the power of spectroscopy, you must first know something about light.

Light and electromagnetic radiation / Today, we view light physically as having both particle and wave properties. In both cases, light carries energy—radiant energy. Let me emphasize here some of the wavelike features of light.

You probably have some experience with waves. Imagine that you're at a beach with waves arriving at regular intervals. You time them with your watch. The number of waves that arrive in a given time is their *frequency*. For instance, one wave hitting the shore every minute is a frequency of one per minute. The distance between the crests (peaks) of these waves is their *wavelength*. For example, if the crests are 10 m apart, the wavelength is 10 m. The wave *velocity* indicates how fast and in what direction the waves are traveling. Here the waves move 10 m in 1 minute, so their velocity is 10 m/min.

Be aware that waves do *not* involve the mass motion of material over long distances. If you're on the Pacific shore, for instance, the water in the waves pounding the sand is *not* the same water as that involved when the wave started at Hawaii. A wave carries energy of motion from one place to another, but it does not transport material.

To sum up: Waves have three fundamental properties (Fig. 5.6): wavelength, frequency, and velocity. The *wavelength* is the distance between two successive crests of a wave, the *frequency* is the number of waves that pass you each second, and the *velocity* is the distance covered in one second by a crest traveling in a certain direction. The units of velocity are length per time. When you multiply frequency (number per time) and wavelength (length), the result equals wave velocity (length per time):

frequency × wavelength = velocity

This fundamental rule applies to all kinds of waves. It tells you that for waves traveling through some material, a change in frequency requires a change in the wavelength if the wave velocity remains the same. For light, the velocity is the same for all wavelengths. The speed of light is 299,793 km/sec in a vacuum. It is usually designated by a lowercase *c* and is more easily recalled by its approximate value, 300,000 km/sec (3×10^8 m/sec). So for light waves,

$$f\lambda = c$$

where λ is the wavelength and f the frequency. Note that because c is constant for all kinds of light, different wavelengths must have different frequencies.

Light is one type of wave produced whenever electric charges are accelerated. Such energy is termed *electromagnetic radiation*. The range of all different wavelengths of electromagnetic radiation is called the *electromagnetic spectrum* (Fig. 5.7).

Although the wavelength of electromagnetic radiation is sometimes measured in meters or centimeters, scientists often use special wavelength units for different regions of the electromagnetic spectrum. For example, visible light is usually measured in *angstroms* (abbreviated Å), where 1 Å is 10^{-10} m. In the infrared region, the unit of length commonly used is the *micrometer* (abbreviated μm), with 1 μm equal to 10^{-6} m.

Sometimes frequencies are used instead of wavelengths. For example, in the radio region, astronomers commonly talk in terms of *frequency* rather than wavelength. The unit of frequency is the *hertz* (abbreviated *Hz*), which is one cycle (or vibration) per second. When you see a wave go by, from peak to peak, it has gone through one cycle. Even in the radio region, many cycles go by in a second. For example, the AM band (your car radio) covers the range from 540 to 1650 kHz (*kHz* is the abbreviation for *kilohertz*, 1000 Hz). The FM band ranges from 88 to 108 megahertz; 1 *megahertz* (*MHz*) is 1 million hertz. Radio astronomers work at such high frequencies that they use the unit of a *gigahertz* (*GHz*), 1 billion hertz. Police radars used to trap speeders typically operate at a frequency of about 10 GHz.

Atoms, light, and radiation / Although a wave model for light was championed during the nineteenth century, it did not explain how atoms

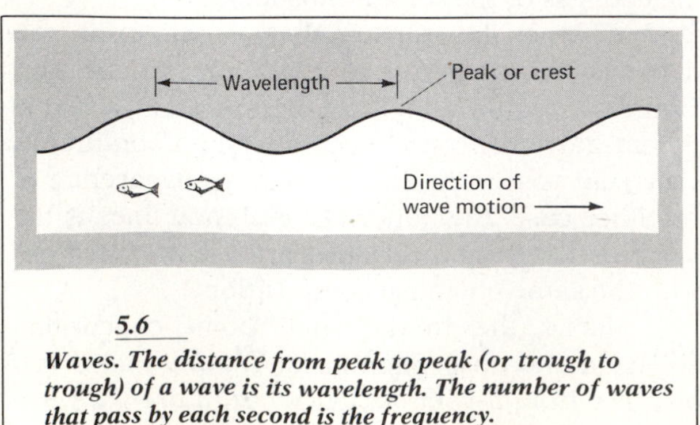

5.6
Waves. The distance from peak to peak (or trough to trough) of a wave is its wavelength. The number of waves that pass by each second is the frequency.

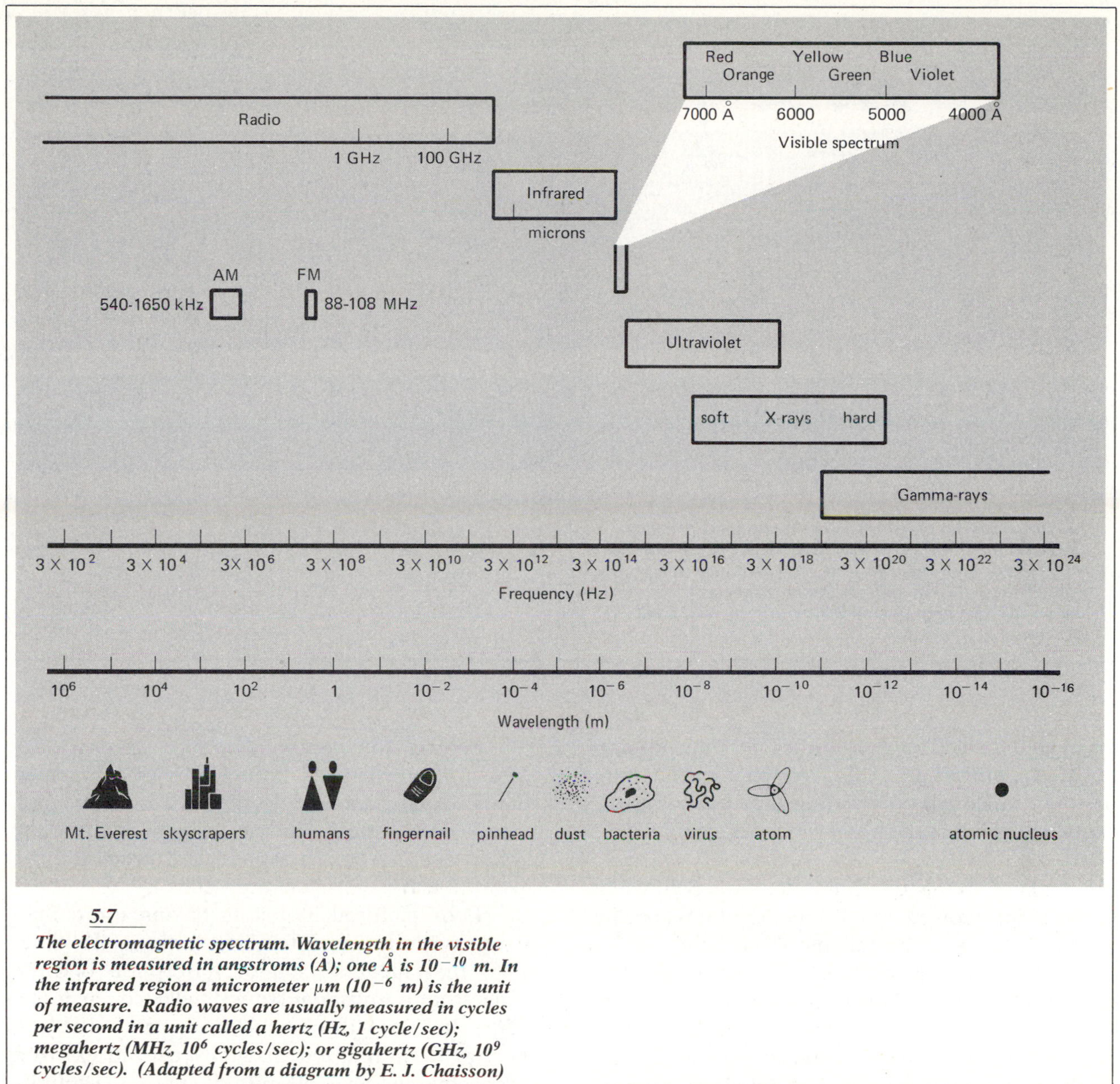

5.7

The electromagnetic spectrum. Wavelength in the visible region is measured in angstroms (Å); one Å is 10^{-10} m. In the infrared region a micrometer μm (10^{-6} m) is the unit of measure. Radio waves are usually measured in cycles per second in a unit called a hertz (Hz, 1 cycle/sec); megahertz (MHz, 10^6 cycles/sec); or gigahertz (GHz, 10^9 cycles/sec). (Adapted from a diagram by E. J. Chaisson)

produced waves. Physicists were puzzled by spectral lines. Why should stars emit light at only certain wavelengths and not others? The pattern of lines of hydrogen appeared in a simple series in the visible region of the spectrum. This series of hydrogen lines was named the *Balmer series* (Fig. 5.8), after the Swiss mathematician Johannes J. Balmer (1825–1898) who in 1885 devised an empirical formula that described the regular sequence of lines. Because the Balmer series followed a patterned sequence, the inference was that light emission and absorption related to a simple structure of atoms. But what was that structure?

In 1911 the British physicist Ernest Rutherford (1871–1937) proposed a model for the atom in which all the positive charge is concentrated in a tiny nucleus, with a surrounding negatively charged cloud of electrons. Rutherford imagined that the mutual attraction of unlike charges holds the atom together. If an atom loses one or more of the electrons, it becomes an ion. Physicists had noticed that an element's spectrum changes when it is ionized. For example, when hydrogen is ionized, it no longer produces the Balmer series of lines. The physicists inferred that the arrangement of electrons somehow determines its light-emitting properties.

In 1900 Max Planck (1858–1947) announced a revolutionary idea about atoms and light. Perplexed by experimental work on spectra, Planck

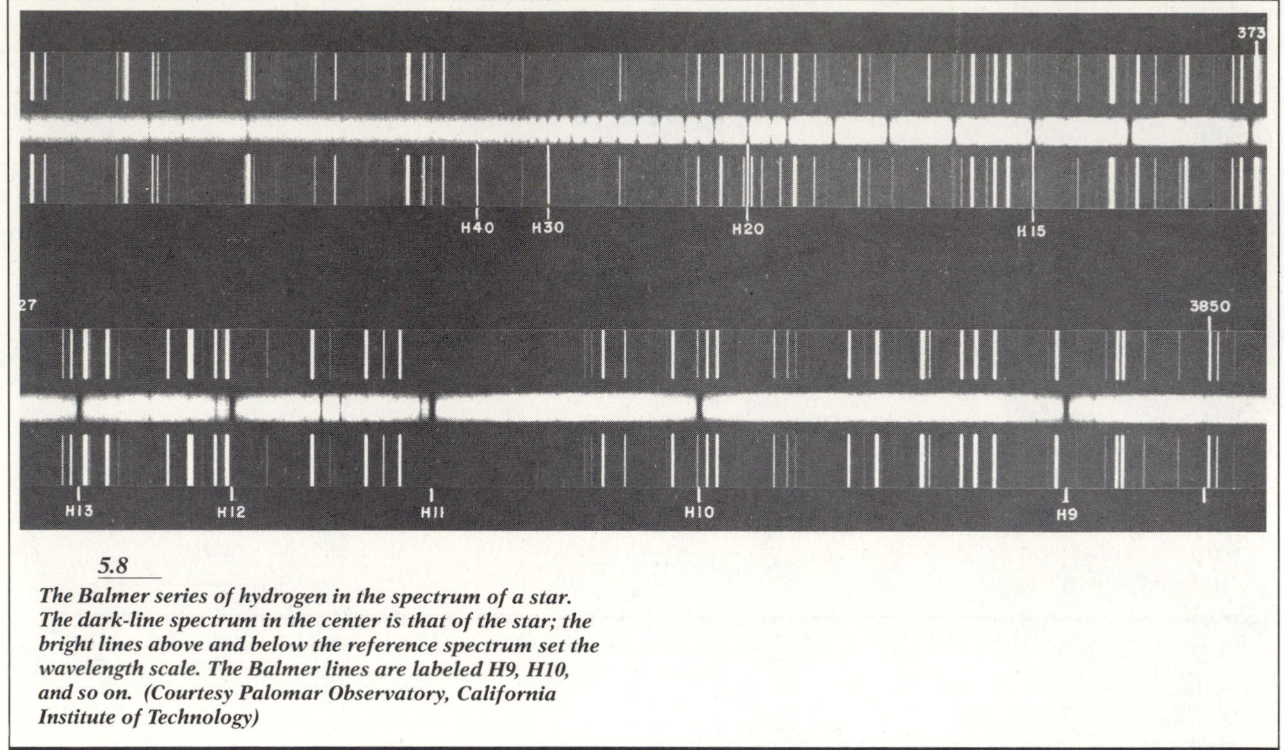

5.8

The Balmer series of hydrogen in the spectrum of a star. The dark-line spectrum in the center is that of the star; the bright lines above and below the reference spectrum set the wavelength scale. The Balmer lines are labeled H9, H10, and so on. (Courtesy Palomar Observatory, California Institute of Technology)

developed a startling new model for light: Radiating matter emits light in discrete chunks of energy that he called *quanta*. From this basis Planck explained some of the observed properties of radiation.

Taking up this quantum idea, Albert Einstein (1879–1955) showed that each quantum, usually called a *photon* when talking about light, carries an amount of energy that relates directly to its frequency. Light of high frequency (short wavelength) transports more energy than light of lower energy. For example, an X-ray photon carries much more energy than a visible-light photon. This relationship between energy and frequency is a direct one: A photon with *twice* the frequency carries *double* the radiative energy. Similarly, a photon with one-half the frequency transports half as much energy.

With this idea in mind, you should now realize that the sun's continuous spectrum is, in fact, an energy spectrum, from lower (red) to higher (blue) energies. You have seen that atoms emit and absorb certain colors at specific wavelengths. For each wavelength you can imagine a photon carrying off a certain amount of energy. Atoms must absorb and emit energy in the form of whole photons. This quantum model emphasizes that in the emission or absorption of light, the atom loses or gains energy in discrete amounts, the energies of individual photons.

Solving the puzzle of atomic spectra / The Danish scientist Niels Bohr (1885–1951) meshed Planck's quantum and Rutherford's atomic pictures. The resulting scheme, known as the *Bohr model* of the atom, explained and predicted the absorption and emission of photons.

Bohr pictured atoms as having more than one possible arrangement of electrons. He imagined that the emission or absorption of light arises from a *transition* between electron arrangements. In line with Planck's theory and observed spectra, Bohr realized that only certain electron arrangements are permitted, not an infinite number or arbitrary ones. Electron transitions can occur only between the special arrangements. You can visualize this atomic model by using concepts of energy (Focus 5.1). For a given nuclear charge (which is positive and varies from element to element), electrons have available a large number of *energy levels* (Fig. 5.9).

Consider the hydrogen atom: It has one proton for a nucleus and one electron attached by the electric force between it and the proton. (Other atoms have more protons and electrons but are held together in the same way.) The electron has a certain total energy; the essence of quantum theory is that electrons remain in stable states of specific energies. For example, the electron in the lowest energy level (called the *ground state*) securely orbits the nucleus. If enough energy is added to the atom—either by collision with another particle or by absorption of a photon with sufficient energy—the electron jumps up one or more energy levels. The atom is then *excited*. This condition does not last long; the electron drops to a lower level in about 10^{-8} second. However, for the electron to descend to a lower energy level requires that it lose some energy. The electron does so by emitting a photon with an energy equal to the amount it needs to lose. This energy relates directly to the photon's frequency.

If the electron gains enough energy, it flies away from the nucleus. The atom is then *ionized*. The loss of an electron changes the energy arrangements available and so changes the atom's spectrum also.

As an analogy, imagine moving a bowling ball up and down the stairs. Each step is an incremental change in the ball's energy, and only full-step changes are permitted. If we try to add less than one step's worth of energy to the ball, it remains at its initial level. If we add exactly one step of energy, the ball moves up one step. When the ball loses energy, it descends in steps until it hits the floor at the bottom of the staircase; this is the ground level, the state of lowest energy. The key point is this: Electron transitions can occur only between certain energy levels, and so an atom can produce only a particular pattern of lines.

Apply the energy-level concept to hydrogen. Every upward step requires the absorption of energy and every downward one the emission of energy. The greater the energy difference needed for a transition, the higher the frequency (and hence the shorter the wavelength of the photon produced from that transition. All transitions to and from the lowest energy level (level 1) involve large energy changes, so they correspond to wavelengths in the ultraviolet range of the spectrum. Although this set of lines, called the *Lyman series*, cannot be seen by the eye, it can be detected by photography. The set of transitions down to and up from the second energy level (level 2) is the Balmer series (Fig. 5.10); it lies in the visible region of the spectrum. The hydrogen absorption lines in the sun's spectrum are those in the Balmer series.

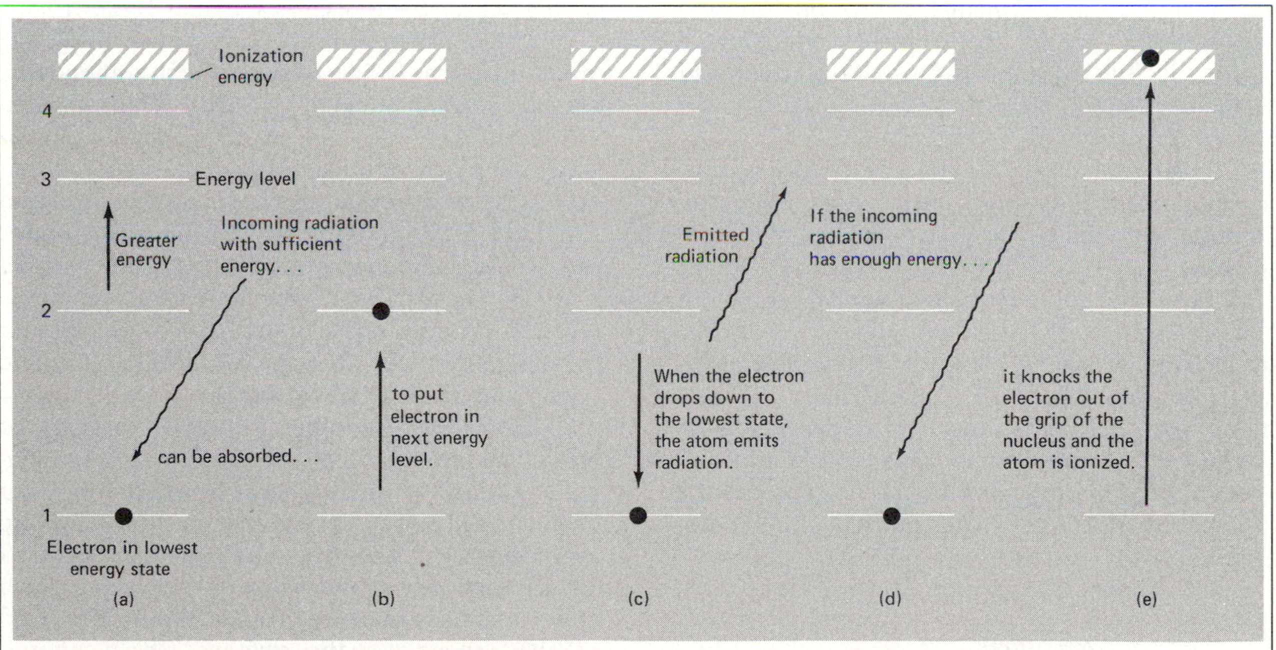

5.9

***A schematic energy-level diagram for an atom (the energy levels are not to scale). The electron is usually in the lowest energy level* (a). *The electron can gain energy by absorbing photons of certain energies, such as the difference in energy between levels 1 and 2. Absorbing such a photon, the electron jumps up to the second level* (b); *the atom is excited. When the electron drops back to level 1* (c), *it gives off a photon with the same energy as that of the one absorbed. The electron can gain so much energy that it is no longer bound to the nucleus* (d). *The atom is then ionized* (e).**

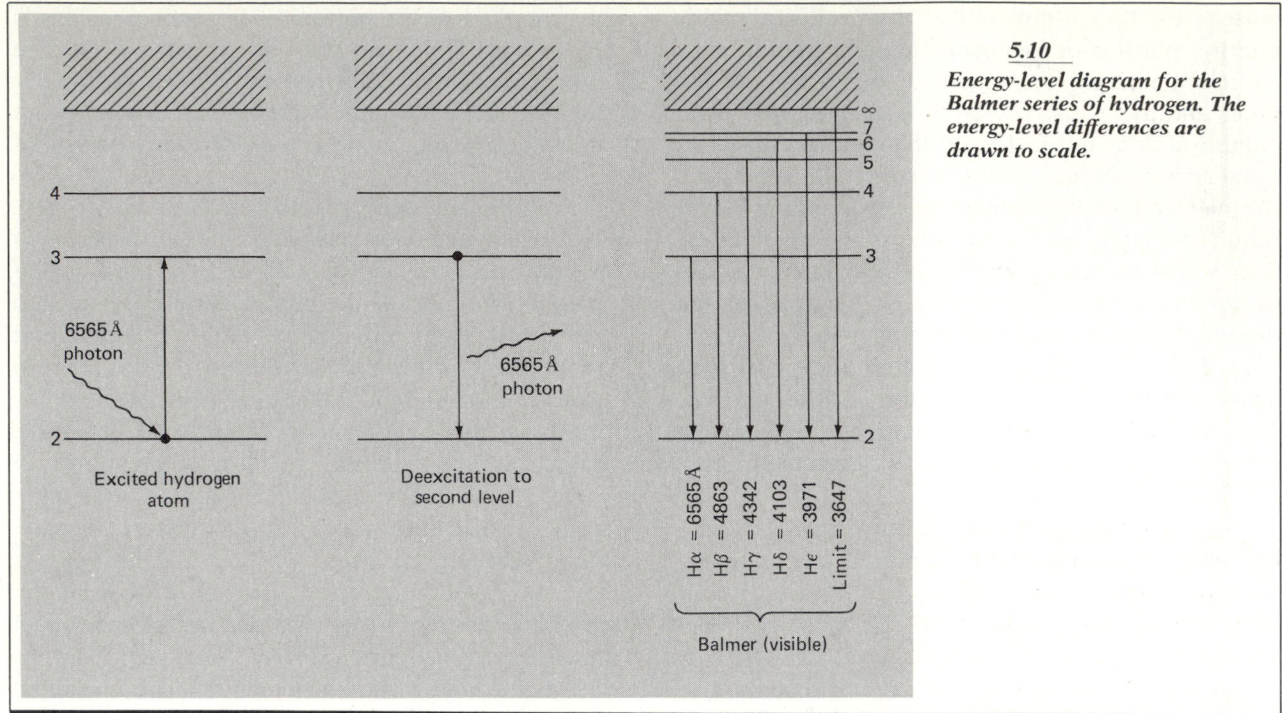

5.10
Energy-level diagram for the Balmer series of hydrogen. The energy-level differences are drawn to scale.

Other atoms / I have concentrated so far on the hydrogen atom because it has the simplest characteristics of all the elements and plays a dominant role in the cosmos. Also, the Bohr model correctly describes the emission and absorption of light by hydrogen atoms.

The simple Bohr quantum model needs drastic modification to work with other elements. Despite complications, one essential point remains: Each atom and each ion (even of the same atom, for more than one electron can be lost from bound states) has its own unique set of electron energy levels. So each has its unique energy-level diagram. Because spectral lines are produced by electronic transitions between energy levels, each element or ion has a unique set of spectral lines. Hence, the study of spectra reveals key information about the internal structure of atoms.

5.4 *Spectra from atoms*

To recap: Electrons have only certain stable energies in atoms. Add sufficient energy, and an electron moves to a higher energy level. Shortly it drops down to a lower energy level and emits a photon. The bigger the downward jump, the more energy the photon has. Any jump, however, produces or absorbs a photon with a specific energy and wavelength. This discrete jumping of electrons is how atoms absorb and emit photons, producing absorption and emission lines (Fig. 5.11).

These electron transitions between energy levels in which the electrons are tied to the atoms are called *bound-bound* transitions. When an atom is ionized and loses an electron from a bound state, the electron undergoes a *bound-free* transition. It is freed from a particular energy level and can have any energy over a wide, continuous range. When an ion captures an electron, we have a *free-bound* transition. Bound-bound transitions result is discrete lines. Bound-free or free-bound transitions produce absorption or emission over a continuous range of wavelengths.

Atoms have another important process of producing or absorbing radiation. Imagine an ionized gas, say of hydrogen. The electrons have been ripped away from the protons, leaving hydrogen ions, The free electrons and protons attract each other, and, since the electrons have much less mass than the protons, their paths bend as they speed by the protons. If a photon encounters an electron near a proton, the electron may jump suddenly to a path of higher energy. During this absorption, the electron stays free from the proton. So we call it *free-free absorption*. Similarly, an electron skirting around a proton may lose energy and emit a photon. That is *free-free emission*. Because an electron has a continuous range of energies over which it can change,

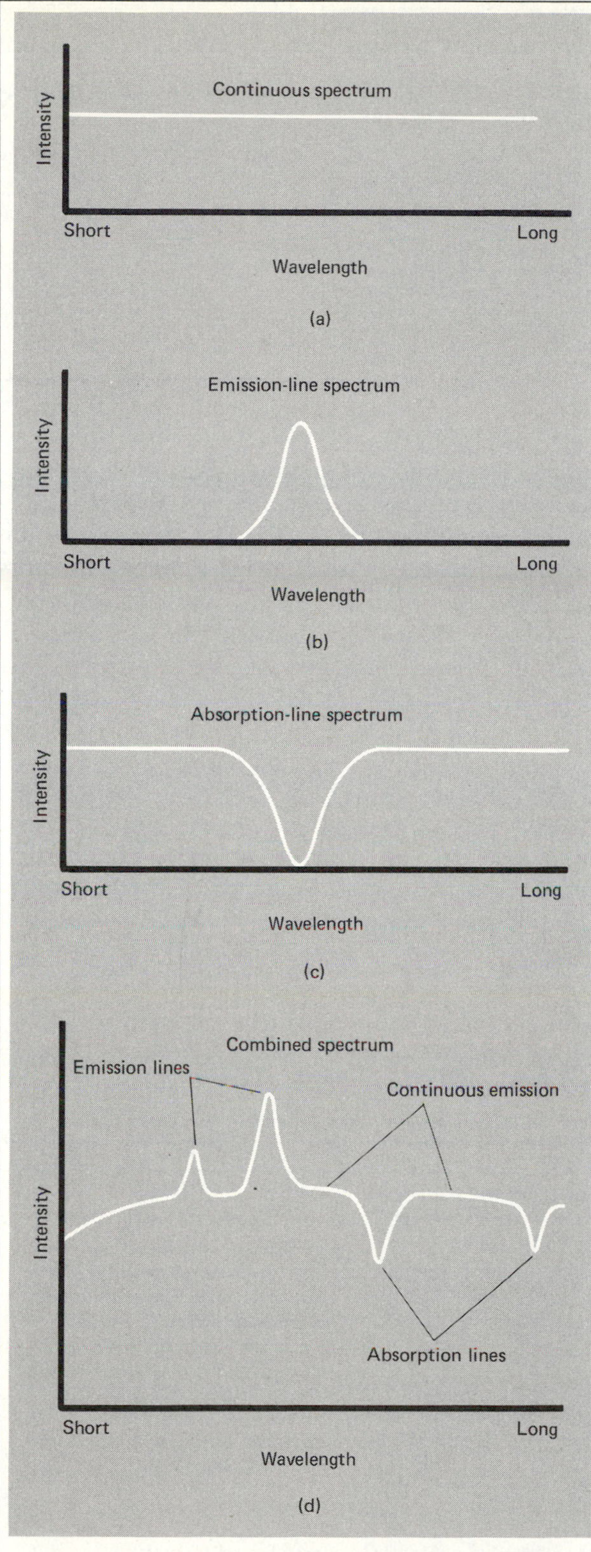

5.11

Schematic representations of the intensity profiles of the main type of spectra:* (a) *continuous,* (b) *emission-line,* (c) *absorption-line, and* (d) *a combination of continuous, emission, and absorption. Note that intensity is plotted versus wavelength, so emission lines rise above the level of the continous spectrum and absorption lines fall below it.

free-free transitions result in continuous (rather than discrete) emission and absorption.

What can excite atoms for emission or absorption? Imagine a box filled with a gas. The atoms of the gas collide with each other (and the sides of the box). Heat up the gas. The atoms then move faster and knock into each other harder. Collisions can bump electrons into higher energy levels; the harder the collision, the higher the level. This process is called *collisional excitation.* Some of the atom's energy of motion, called kinetic energy (Focus 5.1), transfers to an electron of another atom. Photons can also excite atoms when absorbed, but only certain photons, namely those with energies corresponding to the *difference* in energies of two energy levels of the atom. This process is called *photon excitation.* Whether excited by collisions or by photons, the excited atom usually radiates quickly.

Now return to the sodium vapor experiment with the quantum model in mind. When sodium chloride is placed in a flame, individual atoms of sodium are released from the salt. Some atoms collide with others and are excited. As the electrons return to lower levels, they emit photons of yellow light at a wavelength of 5893 Å (Color Plate 1).

Let's follow a photon on its way from the sun to the spectroscope through the sodium vapor cloud. If a photon encounters an unexcited sodium atom, the sodium atom *absorbs* the photon. The gas contains enough sodium atoms to absorb almost all the photons that try to pass through. The sodium atoms don't absorb practically any other wavelength in the visual region, so other photons pass right through the gas, preserving the continuous spectrum. There is a dark line at 5893 Å because these photons were absorbed by the sodium gas.

What happens to the sodium atom that has been excited by the photon absorption? An excited atom quickly emits a new 5893-Å photon. The original 5893-Å photon was headed directly for the spectroscope slit before it was absorbed by a sodium atom. The brand-new 5893-Å photon can be re-emitted in any direction (Fig. 5.12). Very rarely will it be emitted in the same direction in which the original photon was traveling. So only very few of these new photons enter the spectroscope slit. That's how the sodium line in the sun's spectrum becomes darker when passed through the sodium vapor. Note that if you observe the sodium vapor without the sun behind it, you would see only emission lines from the re-emitted 5893-Å photons.

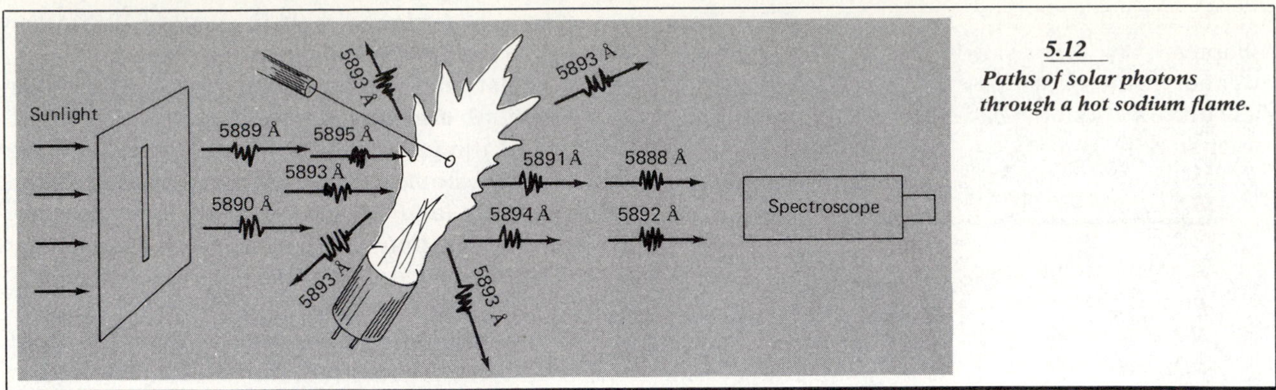

5.12
Paths of solar photons through a hot sodium flame.

That's the basic idea of how atoms in transparent gases produce line spectra. It applies to the sun (Fig. 5.13) and also to stars, because their spectra resemble the spectrum of the sun. By matching the spectral lines of stars to those made in labs from known elements (Fig. 5.14), astronomers inferred that stars contained almost all elements found on the earth. But one important difference emerged: In contrast to the earth, stars and the sun contained mostly hydrogen. (Details

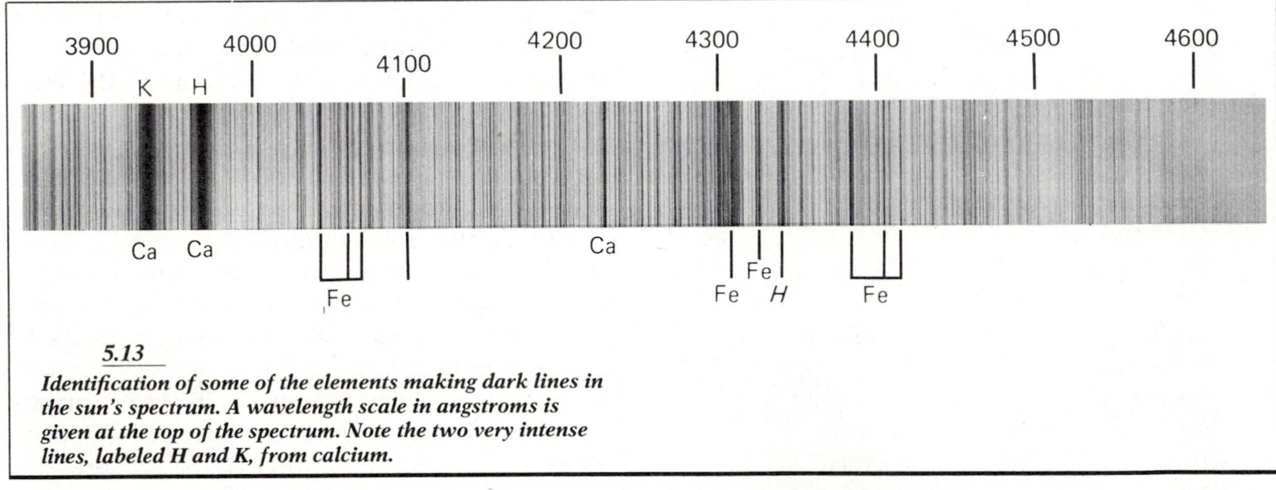

5.13
Identification of some of the elements making dark lines in the sun's spectrum. A wavelength scale in angstroms is given at the top of the spectrum. Note the two very intense lines, labeled H and K, from calcium.

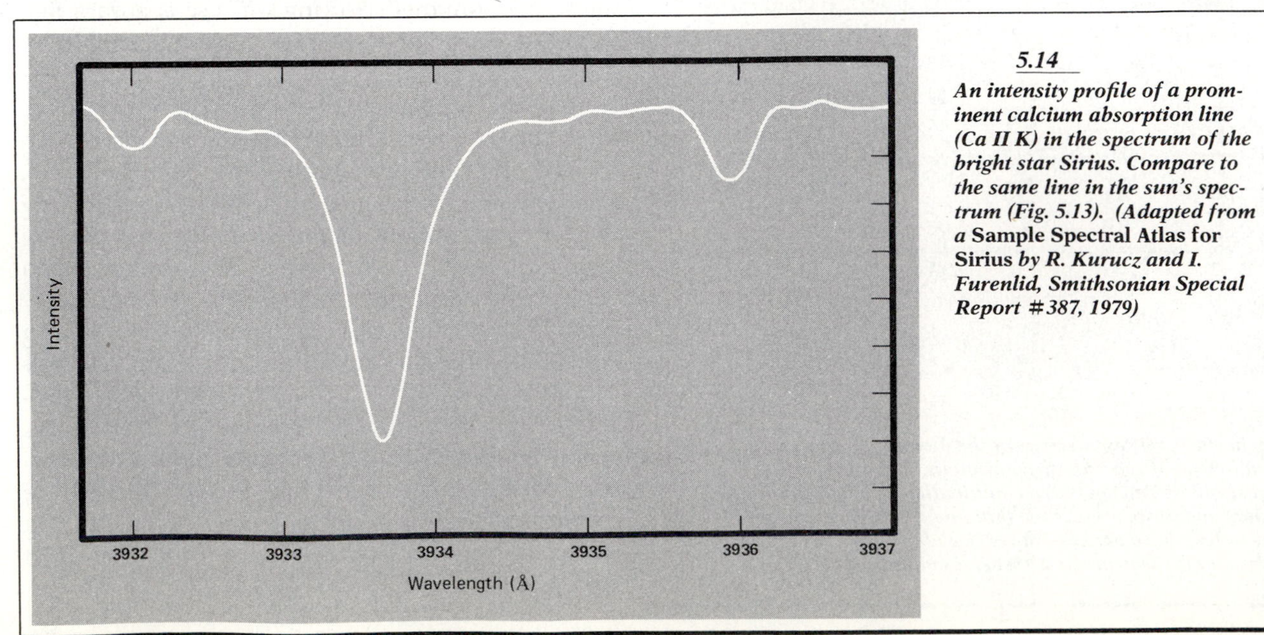

5.14
An intensity profile of a prominent calcium absorption line (Ca II K) in the spectrum of the bright star Sirius. Compare to the same line in the sun's spectrum (Fig. 5.13). (Adapted from a Sample Spectral Atlas for Sirius _by R. Kurucz and I. Furenlid, Smithsonian Special Report #387, 1979)_

will be found in Chapter 13 for the sun and in Chapter 14 for other stars.) The similarity of spectra and chemical composition gave astronomers a new insight about the nature of the stars: Stars are other suns.

KEY CONCEPTS

1 ***A spectroscope spreads light into its component colors, which is called a spectrum; three general types of spectra exist (Table 5.1): (a) continuous, with a smooth spread of colors; (b) emission, which shows discrete, bright lines; and (c) absorption, in which dark lines appear in a continuous spectrum.***

2 ***The sun (and most other stars) have absorption-line spectra; by matching the dark lines in a star's spectrum with those produced by elements in the lab, the chemical composition can be inferred; this works because each element has a unique set of spectral lines.***

3 ***Kirchhoff's rules of spectral analysis describe the physical conditions under which each type of spectrum is produced (Table 5.1): (a) continuous spectrum from a hot opaque gas, liquid, or solid; (b) emission spectrum from a hot transparent gas; (c) absorption line spectrum from a continuous spectrum passing through a cooler, transparent material (usually a gas).***

4 ***Light has wave properties; red light has longer wavelengths than blue light; light also carries energy; the shorter the wavelength, the greater the energy.***

5 ***In an energy-level model for the hydrogen atom, Balmer lines arise from an electron starting out in the second level and jumping to any higher level (for absorption lines) or dropping from a higher level to the second level (for emission lines).***

6 ***The energy-level model for the emission and absorption of light by atoms views the electrons as playing a critical role; electrons can only inhabit stable regions called energy levels; electrons can move up levels only if they gain energy; when they drop down levels, they give off energy (usually as photons); the greater the difference between levels, the greater the energy in the emitted photons.***

7 ***Spectroscopic analysis reveals that stars have physical environments and chemical compositions basically like the sun.***

TABLE 5.1 Basic types of spectra

Spectrum type	*Appearance*	*Physical conditions*
Emission	Distinct bright lines against a dark background	Hot, transparent gas
Continuous	Smooth blend of all colors	Hot, opaque gas, liquid, or solid
Absorption	Distinct dark lines against a bright background of colors	Light from a hot, opaque gas, liquid, or solid passing through a cooler, transparent material

STUDY EXERCISES

1 You throw ordinary table salt, which contains sodium, into the flame of a Bunsen burner. What kind of spectrum do you see when you look at the vaporized salt with a spectroscope? (*Objective 1*)

2 How do we know that sodium exists in the sun? (*Objectives 1, 2, and 3*)

3 The spectra of most stars look like the spectrum of the sun: absorption spectra. How can particular elements be identified in these spectra? (*Objective 2*)

4 Suppose you had a box of hydrogen ions. What kind of line spectrum would they produce? (*Objective 9*)

5 The Balmer lines in the sun's spectrum are absorption lines. Do electrons jump up or fall down energy levels to produce them? (*Objective 8*)

6 Arrange the following kinds of electromagnetic radiation in order from the least to the most energetic: X-rays, radio, ultraviolet, infrared. (*Objective 5*)

7 When you examine the spectrum of the moon, it resembles that of the sun. Explain by Kirchhoff's rules. (*Objective 3*)

8 What happens to a hydrogen atom when it absorbs light with enough energy to knock off its electron? (*Objective 9*)

BEYOND THIS BOOK . . .

For a more technical but very readable description of spectroscopy and stars, read *Atoms, Stars, and Nebulae* (Harvard University Press, Cambridge, Mass., 1971) by L. Aller.

The Nature of Light and Color in the Open Air (Dover, New York, 1954) by M. Minnaert investigates the nature of light through natural phenomena you may have seen. (Also relevant reading for Chapter 6.)

6/ Telescopes and our insight to the cosmos

All astronomical research must in the end be reducible to a visual observation . . .

AUGUST COMTE: Cours de Philosophie Positive

Learning objectives

After studying this chapter, you should be able to:

1. **Describe the impact of observations on scientific models.**
2. **Outline the main functions of a telescope (light-gathering, resolving, and magnifying power) and relate each to specific optical characteristics of a telescope's design.**
3. **Compare and contrast a telescope's light-gathering, resolving, and magnifying power.**
4. **Compare and contrast reflecting and refracting telescopes; include a sketch of the optical layout of each in your comparison.**
5. **Compare a radio telescope to an optical telescope in terms of functions, design, and use.**
6. **Cite a key drawback of a radio telescope compared with an optical telescope and describe how radio astronomers cope with this problem.**
7. **Describe the usefulness of a radio interferometer compared to a single-dish radio telescope.**
8. **Describe what is meant by the term *invisible astronomy*.**
9. **Contrast an infrared telescope to an optical telescope in terms of functions, design, and use.**
10. **Discuss at least two advantages a space telescope has over a ground-based telescope.**

Central question

How do astronomers collect the light from celestial objects?

This chapter turns from astronomical concepts (the focus of previous chapters) to astronomical tools. Telescopes (and other equipment, such as spectroscopes) spring from advances in technology. The technological development of astronomical tools affects not only *what* we observe but also *how* we observe. It expands, deepens, and sharpens our perceptions of the cosmos. What and how we observe act as a prelude to and confirmation of our models of astronomical objects.

In this century, technical advances have become so great that astronomers rarely labor alone. They now usually work in concert with colleagues whom they may never see during the course of a research project. Thc tcchnology has grown too complex for any one observer to master, and the astronomer now often works with engineers and technicians. This chapter will look at some of the developments providing new ways of observing the universe and the influence these have had on our astronomical models.

Observations accelerate the evolution of astronomical ideas. By them, the effectiveness of models are judged. From this judgment some models are discarded, new ones are proposed, and a few are finally adopted. Not all new observations have dramatic effects. In many instances, the change of view brought on by the change of vision is slow and subtle. Slowly or swiftly, new observations compel new conceptions of the cosmos.

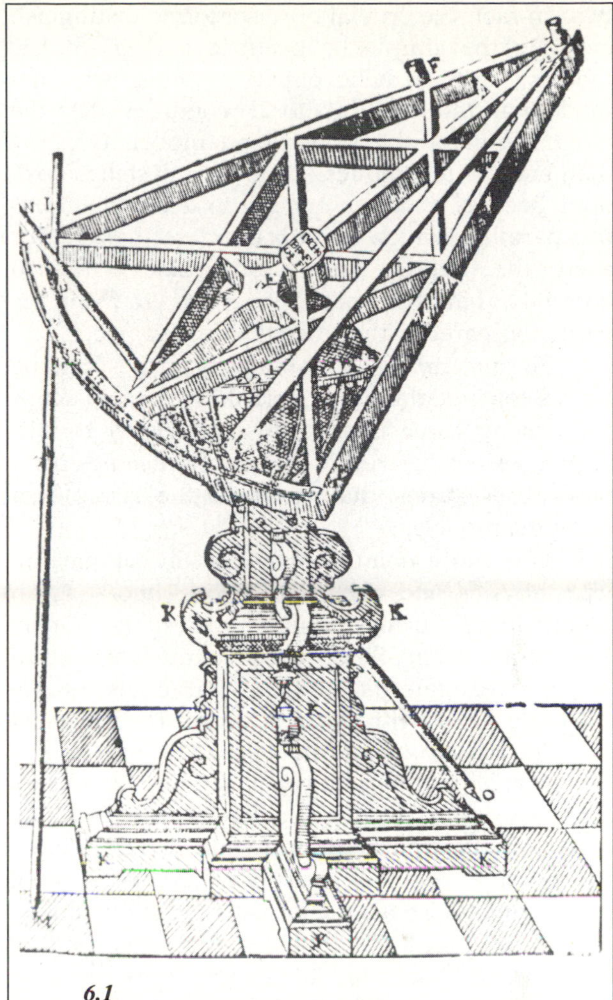

6.1
One of Tycho Brahe's exquisite instruments, an astronomical sextant used at Hveen.

6.1 *Observations and models*

How do observations and models interact? The process usually goes like this (see Section 2.1): Models spring from observations, whether straightforward or subtle. Basic astronomical observations naturally drove people to create models of the cosmos. The Ptolemaic model marked the first detailed attempt at a scientific explanation of these observations (see Section 2.4). This model had two key functions: (1) It used geometric, physical, and aesthetic ideas to *explain* what was seen, and (2) it also *predicted* planetary positions. A key aspect of the prediction is that it must have definite numbers attached to it. How well a model corresponds to actual observations becomes crucial to its acceptability.

The issue of how well is complex and revolves around the techniques of observation, as well as some gut judgment of how good is good enough. For example, astronomers knew for centuries that Ptolemaic predictions were frequently several degrees off from the actual planetary positions. This discrepancy did not bother astronomers at the time, for a few degrees was considered good enough, and their observations did not aim at any greater accuracy.

Not until Tycho Brahe (Section 3.3) did nontelescopic observations reach the limits imposed by the human eye (Fig. 6.1). Tycho's technical achievement compelled Kepler (Section 3.5) to take a discrepancy of only 8′ seriously. The failure of the Copernican model to fit the data with *circular* orbits resulted in Kepler's devising a model with *elliptical* orbits. Their success formed a critical link in the evolutionary chain to Newton's idea of gravitation and a model of a cosmos tied by an invisible force (Section 4.5).

Prior to Tycho's observations and the astronomical use of the telescope by Galileo, both the Ptolemaic and Copernican models explained the motions of the planets equally well. And both made predictions just about as badly. Because both models were equally confirmed observationally, astronomers had to rely on aesthetic and philosophical beliefs to make a choice between the two. In fact, the crucial observation to distinguish between the simple heliocentric and geocentric models, that of heliocentric parallax, was not made until the 1830s, about 2¼ centuries after the introduction of the Copernican model! Only by then had the techniques of measuring stellar positions become accurate enough to detect heliocentric parallax, which amounts to less than 1″ for even the nearest star. Copernicus guessed correctly that the stars were very far from the earth compared with the earth-sun distance.

To sum up: Observations form the building blocks for model-making, and they can also act as the driving force that causes models to be discarded. Their destructive effect often comes from new observations that don't fit the scheme of accepted models.

One more point: Astronomy, in comparison with physics and chemistry, works more as an *observational* science than an *experimental* one. We cannot physically bring a star into a terrestrial lab to investigate its physical characteristics. We can only work with what is given—for the most part, the light from celestial objects.

You should not, however, get the impression that astronomy totally lacks experimentation. Astronomers experiment in two basic ways. First, we can make different kinds of observations of the same objects. That's the importance of technological innovation, for it provides new tools for new experiments. Second, lab experiments such as those of spectroscopy can inform the analysis of observations. Third, we can play with theoretical models in light of whatever observations we have. Theoretical model-making may also be tied to technology; for example, electronic computers make it possible to manipulate quickly very detailed models of astronomical objects. And of course we make use of experiments in physics and chemistry that provide basic data on properties of matter and radiation.

6.2 Visible astronomy: Optical telescopes

As extensions of the human eye, telescopes amplified the power of detection without at first extending the spectral range of our vision. Today we can sense much more than the visual part of the electromagnetic spectrum (see Section 5.3), as you will find in the next section. This section restricts itself to optical telescopes, those that manipulate light detectable by the eye. Before I deal with telescopes, let me discuss a little about *optics*: how light is controlled. You need to understand some basic optics to understand how telescopes work.

Refraction, reflection, and images / When traveling through space or a uniform medium, light moves in a straight line. Although light has characteristics of both waves and particles, geometric optics pictures it as particles moving along straight-line paths, which are called *light rays*. Using lenses, mirrors, and prisms, we can change the direction of light rays or even break up white light into its component colors (light of different wavelengths). How light rays are affected by bouncing off or passing through materials is the essence of optics.

When light crosses the boundary from one transparent material to another (from air to glass, for example), its direction generally changes (Fig. 6.2). This bending of light rays is termed *refraction*.

Refraction occurs because light travels at different speeds in different media. Consider what happens as a consequence when a beam of light, a bunch of rays, hits glass from the air at some angle (Fig. 6.2). The first part of the beam to strike the glass enters it and slows down. The rest of the beam continues to move at a faster speed and so gains on the light already in the glass. This catch-up results in the front of the beam turning toward the glass.

Here's an analogy: Imagine a line of a marching band turning a corner. To do this and keep a straight line, the people on the inside march more slowly than those on the outside. The line turns around some angle and remains straight.

One key point: The amount of refraction depends on the wavelength of light, with blue light bent more than yellow, and yellow more than red. The shorter the wavelength, the greater the amount of refraction.

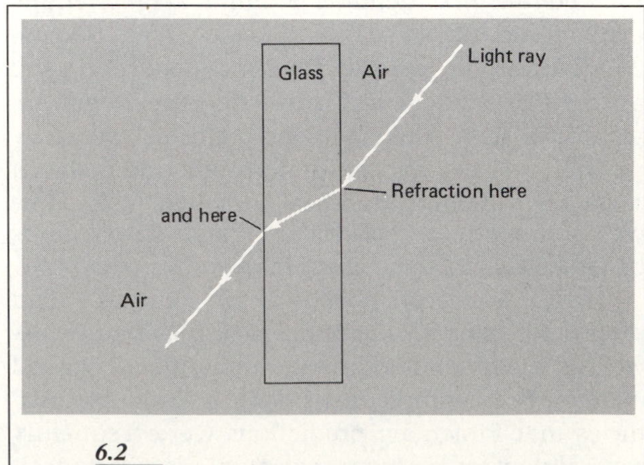

6.2
The refraction of light. When light rays cross the boundary between two different materials, their paths are bent.

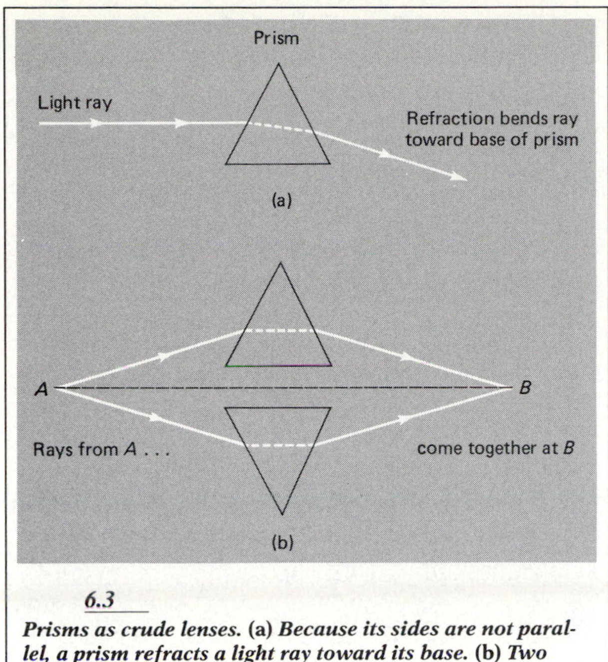

6.3

Prisms as crude lenses.* (a) *Because its sides are not parallel, a prism refracts a light ray toward its base.* (b) *Two prisms placed base to base refract rays from* A *to form at* B *in a crude image.

You have all seen yourselves in a mirror, so you know that smooth surfaces return light by bouncing back the rays. This process is *reflection*. A light ray bounces off a polished surface the same way a ball bounces off a smooth wall: The ball rebounds at the same angle at which it hits. Reflection does not depend on the wavelength of the light; red and blue light are reflected the same way at the same angle.

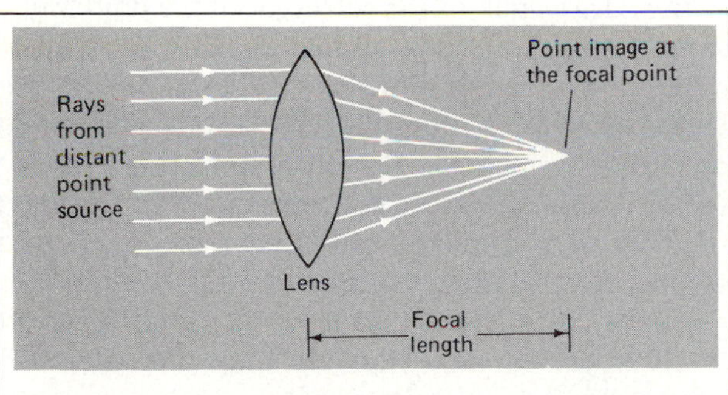

6.4

A lens forming a point image from a point source. A lens acts like base-to-base prisms (Fig. 6.3). Because its surface is smoothly curved, it forms a sharp focus at the focal point. The distance from the lens to the focal point is the focal length, for a distant source.

The point of optics is to make *images* by refraction and reflection. An image occurs when light rays are gathered together in the same relative alignment as when they left an object. You can recognize an image of an object as a visual representation of the object itself.

How can refraction form an image? Suppose that light travels through a glass prism (Fig. 6.3a). Because the sides of a prism are not parallel, a light ray does not come out along its original path but is bent toward the prism's base. Now place two prisms base to base (Fig. 6.3b). Then two light rays from a point source converge to a point. However, rays entering the prisms at different angles converge at different points. To get all rays to come to the same point requires a smoothly curved surface. Such a piece of glass is a *lens* (Fig. 6.4).

A lens brings rays from a point source to a point image at its focus. From an object of finite size, the lens makes an image of the object by focusing rays from each point of the source onto a separate point in the image. This image is generally (but not always) smaller than the object and upside down (Fig. 6.5). For objects at large distances, the distance from the lens to the image is approximately the same for all objects. This distance is termed the *focal length*.

How can a mirror make an image? You know that an image from a flat mirror is undistorted (but reversed right to left). An irregularly curved mirror, such as one in a fun house, creates a distorted image. A smoothly curved mirror—whose surface, for instance, follows the curve of a parabola—brings all the light to a focus (Fig. 6.6).

Telescopes / Basically, a telescope gathers up light and allows you to examine an image at a focus. To make a telescope, you need a lens or mirror, called the *objective*, to bring light to a focus. A lens called an *eyepiece* at the focus allows visual examination of the image, or a camera can be placed at the focus to photograph the image.

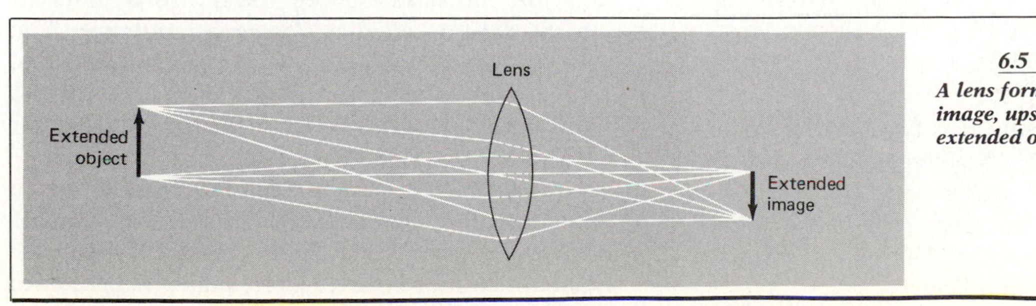

6.5

A lens forms an extended image, upside down, of an extended object.

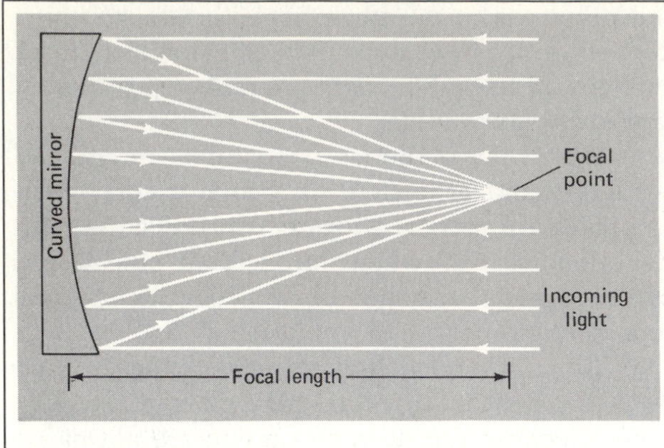

6.6

Making an image by reflection. A smoothly curved mirror reflects incoming light to a focus. Astronomical mirrors have their reflecting material on their front surfaces and follow a parabolic curve to make a clean focus.

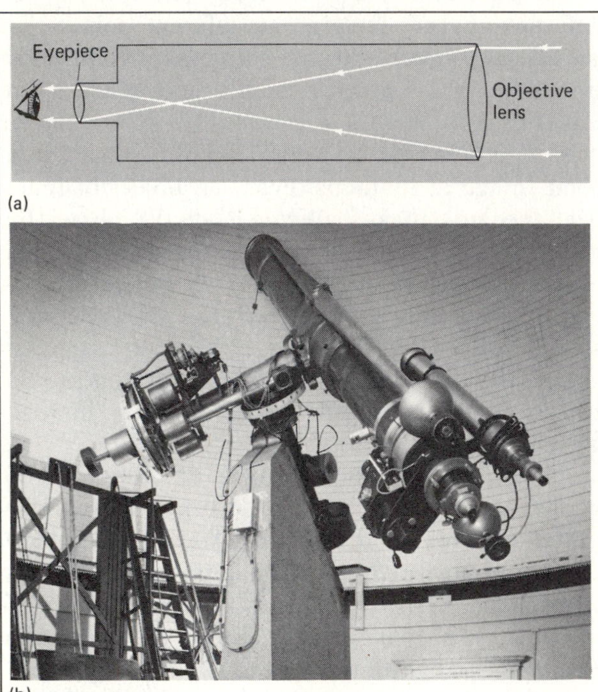

6.7

The refracting telescope. (a) *The design of a simple refracting telescope. An objective lens gathers the incoming light and brings it to a focus. An eyepiece allows viewing of a magnified image.* (b) *The Great Refractor of Harvard College Observatory. Built in the middle of the nineteenth century, this refracting telescope has a 15-inch objective lens. In its day, it was one of the largest telescopes in the world. (Courtesy Harvard College Observatory)*

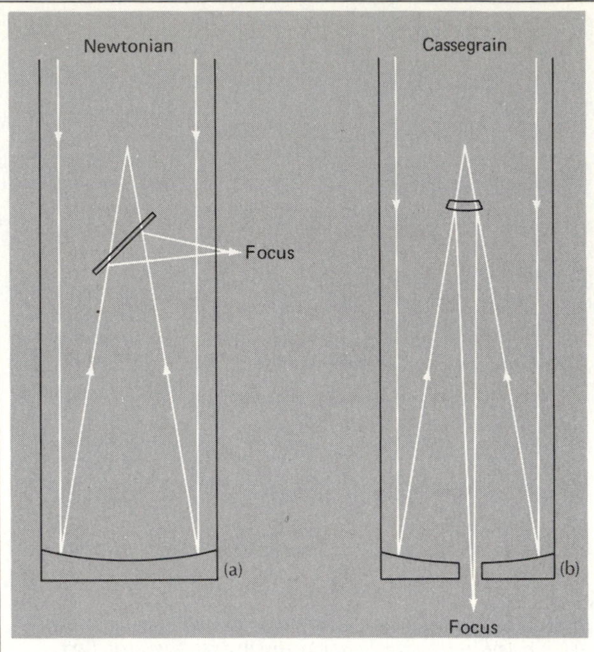

6.8

The design of reflecting telescopes. Newton placed a small, flat mirror at an angle to reflect the light out to one side, where an eyepiece is positioned (a). *This is called a Newtonian reflector. A common modern design uses a small convex mirror to reflect the light back down through a hole in the objective mirror for viewing* (b). *This is called a Cassegrain reflector.*

There are basically two types of telescopes, distinguished by their objectives: *Refracting telescopes* (or *refractors*) use a lens (Fig. 6.7), and *reflecting telescopes* (or *reflectors*) use a mirror. Galileo's telescope (Section 4.1) was a refractor, Newton's (Section 4.3) a reflector.

Newton designed the reflector because of a critical drawback of the refractor of his day. Since simple lenses act essentially as prisms, light of different colors has different focuses. So images viewed through a refractor have color halos (when one color is in focus the other colors are out of focus). Newton noted that reflection, in contrast to refraction, does not make light break up into colors. So the image formed by a reflector does not have color halos. (Later lenses were designed to eliminate most of the refractor's color problems.)

How to view the image formed by a reflector? Newton put a small mirror tilted at 45° to the path of light to direct the focus out the side of the telescope tube (Fig. 6.8), where an eyepiece is placed. Such a design is called a *Newtonian reflector*. Most telescopes today use a different optical design called a *Cassegrain reflector* (Fig. 6.8). All large telescopes built now are reflectors. In the very largest, the observer can get at the focus directly; he or she sits in a cage at the center of the telescope (Fig. 6.9).

Reflectors versus refractors / In addition to affecting all colors alike, reflectors have other advantages over refractors. First, they can be made much more cheaply, for, since the light does

6.9
An observer in the prime-focus cage of the Hale 5-m telescope on Palomar mountain. (Courtesy Palomar Observatory, California Institute of Technology)

not actually pass through the mirror, the glass can be of lower quality (in fact, since it only serves to support a reflecting surface, the mirror need not be glass at all). Second, reflectors can be made much larger, for they can be supported from the back and sides; a refractor can be supported only from its edges. On the other hand, all reflectors must use some kind of secondary mirror to bring the light to an accessible focus, which results in distortion of the light and images that are not as sharp as those produced by a refractor. So when large light-gathering power is desired, a reflector is the choice, but for precise positional measurements and observation of fine detail, refractors are still very useful.

Functions of a telescope / Whether a reflector or a refractor, a telescope has one primary function: to gather light. A telescope is basically a light bucket, collecting photons. How much light a telescope can collect depends on the *area* of its objective, which is proportional to the square of the diameter. So a mirror with twice the diameter of another can gather four times as much light. That's the reason astronomers want large telescopes—mirrors that have plenty of light-gathering power.

Light-gathering power has no absolute standard; it's a relative measure. For example, the diameter of your eye in the dark is about 0.5 cm. A telescope with a 50-cm objective (100 times bigger) would have a light-gathering power 10,000 times (100^2) greater than your eye. Similarly, the 5-m Hale telescope at Mount Palomar outdoes the 4-m Mayall telescope at the Kitt Peak Observatory by more than 50 percent.

The next most important function of a telescope, after gathering light, is to separate, or *resolve*, objects that are close together in the sky. This ability is called *resolving power*. It is usually expressed in terms of the minimum angle between two points that can be clearly separated. It depends directly on the diameter of the objective and inversely to the wavelength of the light. For the same wavelength, the resolving power depends directly on the objective's diameter. So a mirror *twice* the size of another has *double* the resolving power; that is, it can resolve objects half as far apart.

For example, a 10-cm telescope has a resolving power of 1.4″ at visual wavelengths. This means that if the telescope is aimed at two stars that are more than 1.4″ apart, you will see two separate star images. If the angular separation of the stars is less than 1.4″, you will see a single, elongated image.

The *theoretical* resolving power of a 500-cm telescope would be 0.02″. But ground-based telescopes never attain such performance. The resolving power of a big telescope is limited not by the telescope's optics but by the earth's atmosphere. You have probably noticed that stars twinkle. The twinkling comes from turbulence in the air that makes the atmosphere act like a huge, nonuniform lens. The motion of blobs of air, like the shimmering above a hot road, distorts and blurs images seen through a telescope. Even on the best of nights, the 5-m Hale telescope does not resolve better than a 10-cm telescope. At Kitt Peak, for example, it is a rare night when star images are smaller than 2″.

The limit that the earth's atmosphere sets on the resolving power makes a strong case for placing a large telescope in space. There a telescope's resolving power would be limited by the optics, not the atmosphere.

Finally, the least important of a telescope's function is its *magnifying power*, the apparent increase in the size of an object compared with visual observation. Magnifying power depends on the focal length of the objective and the focal length of the eyepiece. Changing the eyepiece on a telescope changes its magnifying power as follows: The *shorter* the focal length of the eyepiece, the *greater* the magnifying power. For example, if you put in an eyepiece with *half* the focal length of a previous one, you *double* the magnifying power.

An eyepiece with a shorter focal length would give a higher magnifying power, as high as you might like. But there is little point in using a magnifying power any greater than necessary to see clearly the smallest detail in the image, which is determined by the resolving power. Extremely high magnification merely makes the fuzziness larger.

Warning: Don't develop the impression the astronomers observe with their eyes. They do not. Some light-sensing device, called a *detector*, is usually placed at the focus. The detector may be a photographic plate (light-sensitive materials on glass rather than film); that's how many of the photos in this book were made. Or it may be an electronic detector similar, for example, to a television camera. These days astronomers rarely, if ever, use their eyes directly for observations.

To sum up: The three principal functions of a telescope are (1) to gather light, (2) to resolve fine detail, and (3) to magnify the image. Of these, I cannot overemphasize the importance of light-gathering power. Most astronomical objects are extremely faint. Without a telescope you can see about 6000 stars, but even a small 15-cm telescope allows you to see some *half a million* stars. That is the real power of a telescope—to enable us to see objects that we would otherwise not know existed.

Focus 6.1

UNDERSTANDING CONTOUR MAPS

Many of the figures in this book are* contour maps, *pictures of how the intensity of some kind of radiation (radio, visible, infrared, etc.) varies over some region of the sky. Such maps show a lot of wavy, connected lines labeled by numbers. What do they mean?

Here's an analogy you may be familiar with: a weather map (Fig. F.14). This is a map of atmospheric pressure across the United States, with high (H) and low (L) pressure systems indicated. What do the contours here tell you?

First, consider how a pressure contour is drawn. The weather stations around the United States report their local pressures. Each is put on the map. Then a contour line is drawn that connects all stations giving the same reading, providing that it does not cross a station of higher or lower reading. Then another contour (of higher or

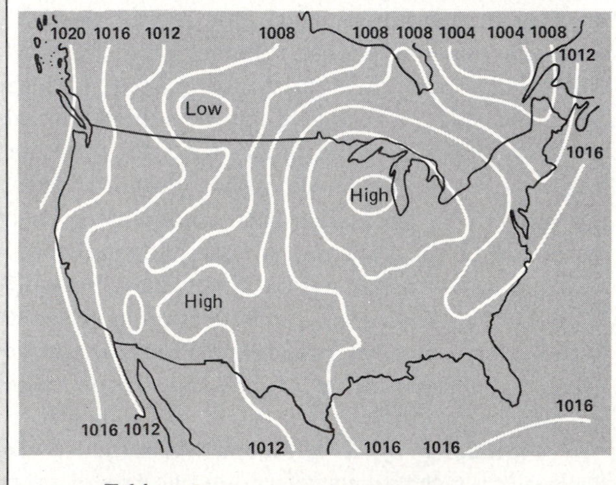

F.14
A weather map showing surface pressure over the United States. The contour lines connect regions with the same pressure readings. A "High" marks a peak in the surface pressure, a "Low" a local minimum.

6.3
Invisible astronomy

Your eye senses only a tiny sliver of the electromagnetic spectrum (see Section 5.3). When you have your teeth X-rayed at the dentist, the X-ray machine does not glow brightly when it's on. But the film placed in your mouth senses the X-rays and gives an internal picture of your teeth. When you stand next to an almost-dead fire, the coals look black. But your skin senses heat—infrared radiation—from the coals.

These examples focus on the critical aspect of invisible astronomy: the need for equipment that can detect the radiation you'd like to observe but can't see. The development of detectors is a technological enterprise, often with no direct drive from astronomy. (For instance, some sensitive infrared detectors basically evolved for military purposes.)

Invisible astronomy also has a less obvious aspect: whether or not the radiation from space can get to the earth's surface. Our atmosphere effectively absorbs large blocks of the electromagnetic spectrum, especially ultraviolet, X-rays, some infrared, and short-wavelength (millimeter) radio waves. What produces the absorption? In the case of infrared radiation, it is primarily absorbed by the water vapor in the atmosphere, which is found concentrated in the lower portions, below 20 km. The ultraviolet and X-ray radiation is primarily absorbed in the ionosphere, at an altitude of 100 km, well above the levels that can be reached by balloons and airplanes. The radiation

lower pressure) is drawn in, and so on. Notice that contour lines cannot cross. Why? Because if they did, it would mean that the same place has two different pressures, and that's impossible!

Second, note that there are places where the pressures hit a maximum (high) or a minimum (low). At the center of each is a last contour surrounding the region of highest or lowest pressure. The center of a high, for example, you can imagine as the peak of a pressure mountain. The contours around the peak tell you how the pressure falls from the peak. If the contour lines are close together, the pressure drops quickly over a short distance (the falloff is steep). If the contour lines are spread out, the pressure drops off slowly; you have a kind of ridge.

The same analysis applies to an intensity contour map (Fig. F.15). Instead of pressure, the map shows the intensity of radiation; in this case, radio waves of 2-cm wavelength from a cloud of ionized hydrogen (called an H II region). Note that there is one main peak where the radio intensity hits a maximum. To the right and below it lies a secondary peak. Between the peaks, the intensity does not fall off as quickly as it does moving away from either one. Each contour line is a set amount of intensity higher or lower than adjacent ones.

To help you see this better, the same intensity map is plotted in a three-dimensional way (Fig. F.16). The height of the contours here indicates the intensity. Notice that the H II region looks like a mountain! The peaks and steep sides stand out. This kind of image should pop into your mind when you see an ordinary contour map.

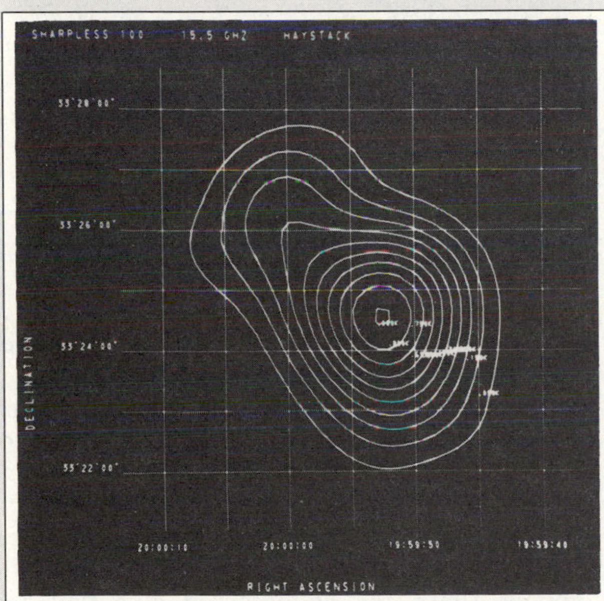

F.15
An intensity contour map of radio emission from a region of hot, ionized gas. This cloud is called Sharpless 100; its position in the sky is indicated by the coordinate grid, one axis labeled "right ascension," the other "declination." Each contour line is 0.1 unit lower than the one interior to it. (Observations by M. Zeilik at the Haystack Observatory)

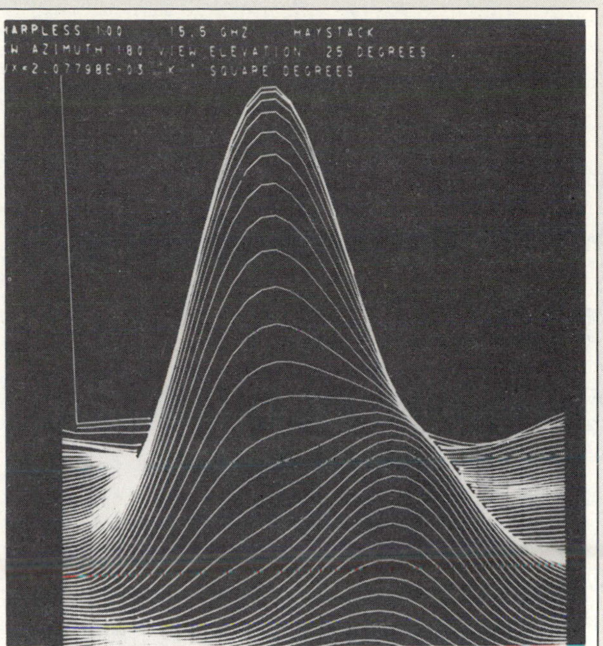

F.16
The observations in Figure F.15 plotted so that the height of a contour line represents its relative intensity. The view is from the top of Fig. F.15 looking down at an angle of 25°. (Map made by M. Zeilik at Haystack Observatory)

that is absorbed may have journeyed through space for millions or billions of years, only to be snuffed out in the last 0.001 second of its trip, never making it to the earth's surface.

The obvious way to get around atmospheric absorption is to go above it. This is space astronomy. (In this book, space astronomy includes rockets, balloons, and airplanes, as well as satellites and spacecraft.) So invisible astronomy has two natural divisions: that which can be done from the ground and that which must be accomplished in space.

6.10
Karl Jansky with the radio antenna in Holmdel, New Jersey, that he used to discover radio waves from space. (Courtesy Bell Labs)

6.11
One antenna of the Very Large Array (VLA) telescope in New Mexico. It has a diameter of 25 m. The surface of the dish acts like a mirror to reflect radio waves to a focus. (Photo by M. Zeilik)

Ground-based radio / Radio astronomy was born in 1930 when Karl Jansky undertook a study for the Bell Telephone Company of sources of static affecting transoceanic radiotelephone communications (Fig. 6.10). Jansky identified one source of noise as a celestial object: the Milky Way in Sagittarius. Jansky's discovery was published in 1932 but had little impact on the astronomers of the day.

However, an American radio engineer, Grote Reber, read Jansky's work and decided to search for cosmic radio static in his spare time. By the 1940s, Reber had made detailed maps of the radio sky. These maps of the radio intensity in different parts of the sky are one kind of *contour map* (Focus 6.1). He sensed that a new astronomy was in the making and took an astrophysics course at the University of Chicago to learn more about astronomy and discuss his discoveries with astronomers—only a few of whom were impressed.

World War II forced technical developments in radio and radar work. Accidentally, John S. Hey in Britain discovered that the sun emitted strong radio waves. After the war, Hey continued his astronomical pursuits at radio wavelengths. So did other groups in Britain, the Netherlands, and Australia. Radio astronomy was reborn as a technological fallout from research by scientists forced to deal with the practical problems of war.

A common type of radio telescope, a radio dish (Fig. 6.11), functions like a reflecting telescope. Essentially, it's a radio-wave bucket with a detector (a radio receiver) at the focus of the dish, which reflects and concentrates radio waves in much the same way a mirror does in a reflecting telescope (Fig. 6.12).

Radio astronomers cannot see radio sources as an optical astronomer can. The radio receiver that detects the incoming radio waves translates the signal into a voltage that can be measured and recorded. A computer then generates a map of the radio intensity over a region of the sky (Focus 6.1).

Our atmosphere allows some millimeter, centimeter, and longer wavelengths to reach the ground. These can be observed both day and night. Radio telescopes can even observe on cloudy days at the longer wavelengths. Because large radio dishes are easier to construct than large mirrors, radio telescopes are typically much larger than optical telescopes, and more sensitive because they can catch more radiation. These are a few of the advantages of radio compared to optical astronomy.

But radio telescopes have one major drawback: poor resolving power. How so? Resolving power depends on both the size of the objective and the wavelength of the gathered light. Radio waves are much longer than visible light, typically 100,000 times as long. So if an optical and a radio telescope had the same diameter, the radio telescope would have 100,000 times *less* resolving

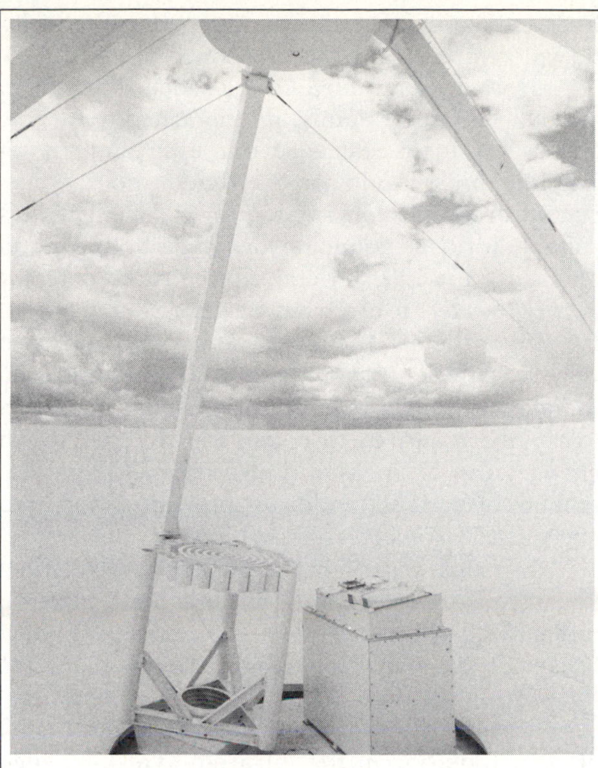

6.12
Surface of a radio dish, the antenna of Fig. 6.11. The metal surface reflects radio waves to a small convex dish (at top center) that reflects the radio signals down through a hole in the main dish (at left side of bottom center platform) to a radio receiver below. (Photo by M. Zeilik)

power. For example, for a radio telescope to have the same resolving power as the 5-m Hale telescope, it would have to have a diameter 100,000 times as great, about 500 km! Obviously, a single dish of this size cannot be built on earth.

There's a method for making a small radio telescope function as if it were a large one. Imagine two radio telescopes placed, say, 10 km apart. By synchronizing the signals received by both, they can be made to act like a single dish with a diameter of 10 km, but only for a strip across the sky. (Thet's because they act like two small pieces at the opposite ends of a large dish.) To get good resolving power in a small, more or less circular region of the sky requires an array of coordinated radio telescopes. The most advanced such telescope, called the Very Large Array (VLA), is in New Mexico (Fig. 6.13). Completed in late 1980, the VLA has a resolving power at centimeter wavelengths equivalent to that of a moderate-size optical telescope. Such devices are called *radio interferometers*.

How much resolving power does a radio interferometer have? Basically, it depends on the separation of the antennas and the wavelength observed. If the antennas are, say, 1 km apart and operate at a wavelength of 1 cm, the resolving power is about 2″, almost as good as optical telescopes and much better than either single dish could achieve.

6.13
The central section of the "Y" of the VLA. The antennas are spread out and electronically linked together to achieve high resolving power. (Courtesy NRAO)

How good can the resolution get? In the technique known as very long baseline interferometry (VLBI), the signals received by very distant antennas (even located on different continents) are recorded on magnetic tape and combined later in a computer. The maximum baseline extends to the diameter of the earth, 12,000 km, so, for this example, a resolving power of 2×10^{-4} arcsec is possible. In the future, radio telescopes in space, separated by larger baselines, will give even better resolution. Recently, the United States announced the development of the Very Large Baseline Array (VLBA) to be centered in New Mexico. Resembling the VLA in function, the VLBA will incorporate VLBI techniques in a coordinated array that will span the United States. The VLA will provide the central core of antennas.

Ground-based infrared / Carbon dioxide and water in the earth's atmosphere absorb much incoming infrared radiation. The infrared astronomer can observe at only a few restricted wavelength ranges: 2–25 μm, 30–40 μm, and 350–450 μm (remember, 1 μm equals 10^{-6} m). Such observations are best made from high sites in dry climates, where the atmosphere contains little water vapor above the telescope.

How does an infrared telescope differ from an optical one? Mainly in the detector at the telescope's focus. Because our eyes and photographic film sense infrared radiation poorly, special infrared detectors are required; sensitive ones suitable for astronomical work have been around only since the 1960s. A common infrared detector is a *bolometer*, a tiny chip of germanium (about the size of the head of a very small nail) cooled to very low temperatures, about 2 K. When infrared radiation strikes a bolometer, it heats up, and its resistance to an electric current changes. Such changes can be measured electronically, and the amount of variation indicates how much infrared energy the bolometer is absorbing.

Infrared observing has at least two distinct advantages over optical observing. First, infrared radiation is hindered less by interstellar dust (Chapter 15) than visible light, so you can see through dust clouds more readily. Second, cool celestial objects (3000 K and cooler) give off most of their radiation in the infrared. Typically, such

(a)

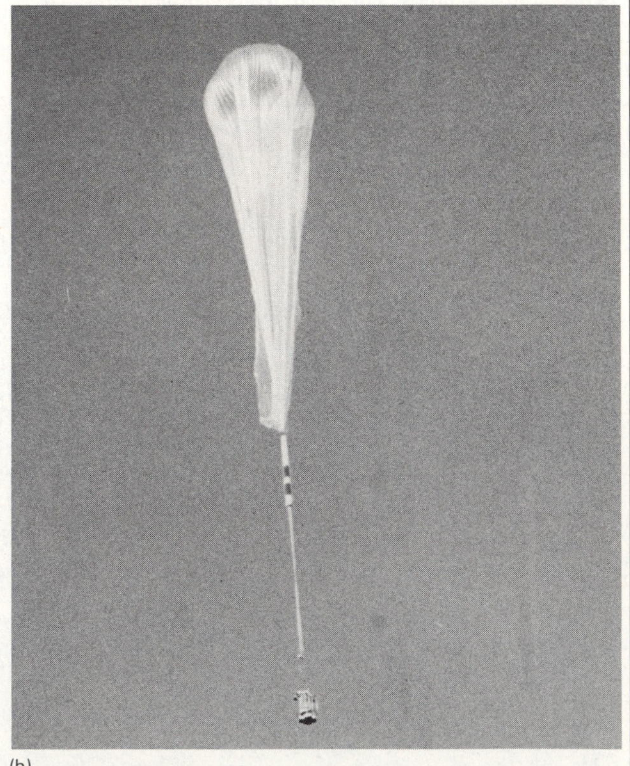

(b)

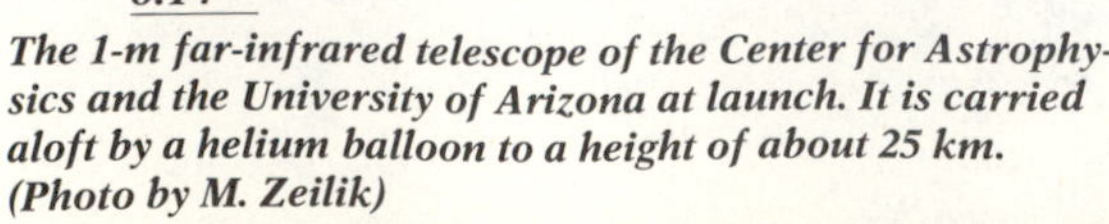

6.14

The 1-m far-infrared telescope of the Center for Astrophysics and the University of Arizona at launch. It is carried aloft by a helium balloon to a height of about 25 km. (Photo by M. Zeilik)

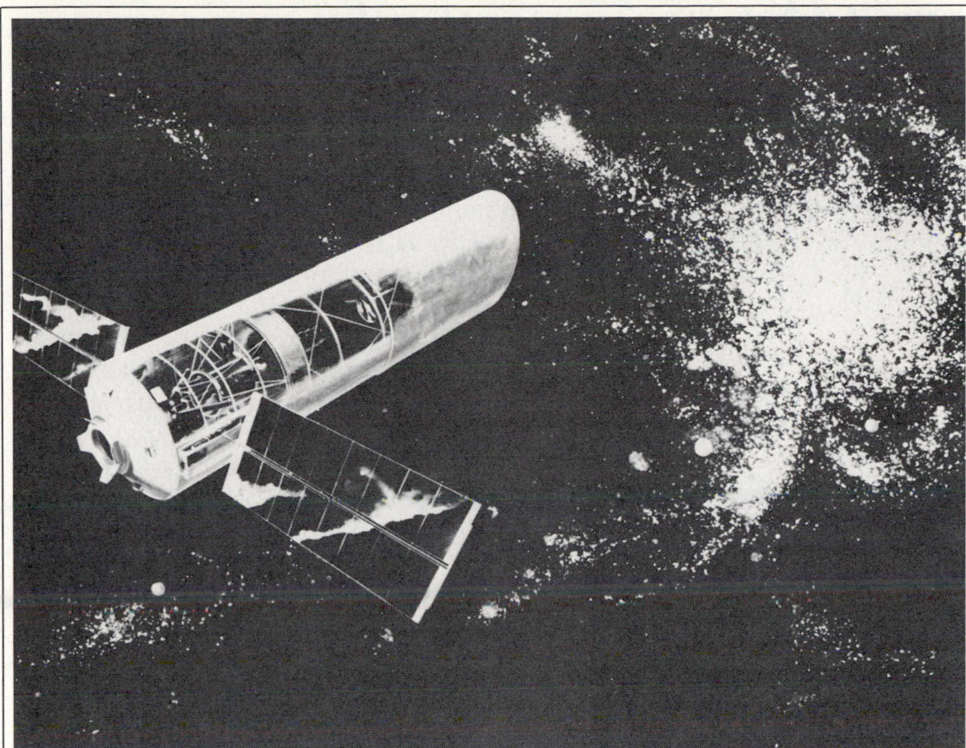

6.15
Artist's conception of the optical space telescope in earth orbit. This telescope will have a 2.4-m mirror and be placed 400 km above the earth by the shuttle. (Courtesy NASA)

cool objects cannot be seen in visible light but can be detected in the infrared. So infrared astronomy brings the cold universe into view.

Another practical advantage is that much infrared observing can be done during the day, when crowded telescopes are not wanted by optical astronomers. This is because very little sunlight at infrared wavelengths is scattered by air molecules, leaving the infrared sky as dark during the day as at night.

Space astronomy / What about the parts of the infrared spectrum that do not penetrate to the ground? And ultraviolet light and X-rays? Such radiation can only be detected above the earth's atmosphere from airplanes, balloons, rockets, satellites, or spacecraft.

For example, most of the far infrared (wavelength longer than about 40 μm) does not make it to the ground. But at altitudes of 15 to 20 km or so, very little of the earth's atmosphere remains. Far-infrared observations can be made at these altitudes from airplanes or balloon-borne telescopes (Fig. 6.14). The equipment is simply a reflecting telescope equipped with a bolometer.

For ultraviolet astronomy, methods of light gathering and detection remain similar to optical astronomy. Photographic plates and some television tubes respond well to ultraviolet light, so detection presents no serious problem. Because glass absorbs ultraviolet light, refracting telescopes cannot be used, but reflectors work perfectly well. All ultraviolet astronomy must be done from space, for the absorbing layer of the atmosphere is higher than balloons or aircraft can reach.

For the high-energy realm of X-ray and gamma-ray astronomy, gathering and focusing the light require special techniques, with the result that these telescopes have little resemblance to optical ones. X-rays pass through most ordinary matter and so are almost impossible to reflect. In recent years, X-ray telescopes have used the fact that X-rays can be reflected from certain surfaces if they strike at very small angles, almost parallel to the reflecting surface. Such reflections produce crude but reasonable images.

Gamma rays are so energetic that to focus them is almost an impossible task. Gamma-ray telescopes don't yet produce images as such, but they do indicate from what general direction in the sky gamma rays originate.

Finally, even optical astronomy in the near future will involve large space telescopes (Fig. 6.15). The main advantage here (besides the lack of the influence of weather) is that a large telescope can be used at its theoretical limit of resolving power rather than as limited by the atmosphere. To be launched by the space shuttle, the

U.S. Space Telescope will have a 2.4-m mirror and be placed in an orbit some 400 km above the earth. Its primary goal is observations of faint objects with high resolution, especially at ultraviolet wavelengths. It will also take wide-angle photos and measure quickly light intensity in the ultraviolet. Astronomers will operate the telescope remotely from the ground by a computer radio link.

KEY CONCEPTS

1 ***Astronomy is primarily an observational science; telescopes expand the astronomers' observing power by revealing faint objects, expanding the range of the electromagnetic spectrum that we can perceive, and increasing our ability to see fine detail; these abilities influence the evolution of astronomical models.***

2 ***Optical telescopes use lenses (refraction) or mirrors (reflection) to gather light and bring it to a focus; usually a photographic plate or light sensor is placed at the focus to detect the light; telescopes that use only lenses are called refracting telescopes, and those that use a mirror as the objective are called reflecting telescopes.***

3 ***Regardless of the type of telescope, it has three basic functions: (a) to gather light, (b) to resolve details, and (c) to magnify the image; the light-gathering power of a telescope depends on the area of its objective; resolving power depends on the diameter of the object and the wavelength of light being observed (as well as on the turbulence in the earth's atmosphere); magnifying power depends on the focal length of the objective and of the eyepiece.***

4 ***Invisible astronomy uses light outside of the visible region of the spectrum, such as radio, infrared, ultraviolet, and X-rays; many of these wavelengths are blocked by the earth's atmosphere, so telescopes must go above it.***

5 ***Radio telescopes are limited in resolution because radio wavelengths are so long (compared to visible light); special electronic techniques allow separated radio telescopes to function as a single one; this technique has been used in the design of the VLA in New Mexico, and the new VLBA, to be centered in New Mexico.***

6 ***Space telescopes have opened up our view of the ultraviolet and X-ray heavens; they are not hindered by bad weather, the atmosphere, or its lack of transparency to certain kinds of electromagnetic radiation.***

STUDY EXERCISES

1 How did Galileo's observations support the Copernican model and refute the traditional one? (*Objective 1*)

2 What telescopic observations were critical in the confirmation of the Copernican model? (*Objective 1*)

3 What is the most important function of a telescope? (*Objective 3*)

4 How do radio telescopes differ from optical ones? How are they similar? What are their advantages and disadvantages? (*Objectives 4 and 5*)

5 Suppose you're on a TV quiz show and shown optical, infrared, and radio telescopes with the same size objective. You're then asked to list them in order of *increasing* resolving power. What is the correct order? (*Objectives 3, 4, and 5*)

6 Imagine that you are going before a congressional hearing to justify the expense of putting a large telescope in space. What arguments would you use to pursuade the skeptical committee? (*Objective 10*)

7 Describe one advantage of an infrared telescope over an optical one. (*Objective 9*)

8 Why do X-ray telescopes have to be put above the earth's atmosphere? (*Objectives 8 and 10*)

BEYOND THIS BOOK . . .

For an intriguing analysis of the development of radio astronomy in Great Britain, read *Astronomy Transformed* by David Edge and Michael Mulkay (Wiley, New York, 1976). For a more American view on radio astronomy, try *The Invisible Universe* by Gerrit Verschuur (Springer-Verlag, New York, 1974).

The Evolution of Radio Astronomy by J. S. Hey (Science History Publications, New York, 1973) traces the development of techniques and ideas.

Colin Ronan in *Invisible Astronomy* (Lippincott, Philadelphia, 1972) makes a good case for how new techniques in astronomy lead to new views of the universe.

In *Scientific American* you can read about "Infrared Astronomy" by G. Murray and J. Westphal, August 1965; "Intercontinental Radio Astronomy" by K. Kellerman, February 1972; "Ultraviolet Astronomy" by L. Goldberg, June 1969; "X-Ray Astronomy" by H. Friedman, June 1964; "The X-Ray Sky" by H. Schnopper and J. Delvaille, July 1972; "Gamma-Ray Astronomy" by W. Krauschaar and G. Clark, May 1962; "Radio Astronomy by Very-Long-Baseline Interferometry" by A. C. S. Readhead, June 1982; and "The Space Telescope" by J. N. Bahcall and L. Spitzer, Jr., July 1982.

Sky and Telescope and *Astronomy* magazines contain ads and information about telescopes and accessories for them.

7/Einstein and the evolving universe

> ***What is inconceivable about the universe is that it is at all conceivable.***
>
> *ALBERT EINSTEIN*

Learning objectives

After studying this chapter, you should be able to:

1. State the principle of equivalence and illustrate it with a concrete example.
2. Show how the principle of equivalence leads to the local cancellation of gravitational forces.
3. Compare and contrast Aristotle's, Newton's, and Einstein's concepts of natural motion for bodies falling near the earth and for the motions of heavenly bodies.
4. Describe what is meant by the term *spacetime*.
5. Argue that concepts of natural motion must be coupled to a notion of the geometry of spacetime.
6. Sketch the Hubble law in graphical form and indicate what observations are needed to make it.
7. Describe the Hubble law for the line-of-sight velocities and distances of galaxies and, from it, find the value of the Hubble constant.
8. Interpret the Hubble law as a consequence of the uniform expansion of the universe.
9. Outline how to use the Hubble law to estimate the age of the universe and to determine its geometry: hyperbolic, spherical (closed), or flat.
10. Use the concept of escape velocity to relate the future of the universe to its geometry.

Central question

How did Einstein's ideas about gravitation lead to an evolving model of the cosmos?

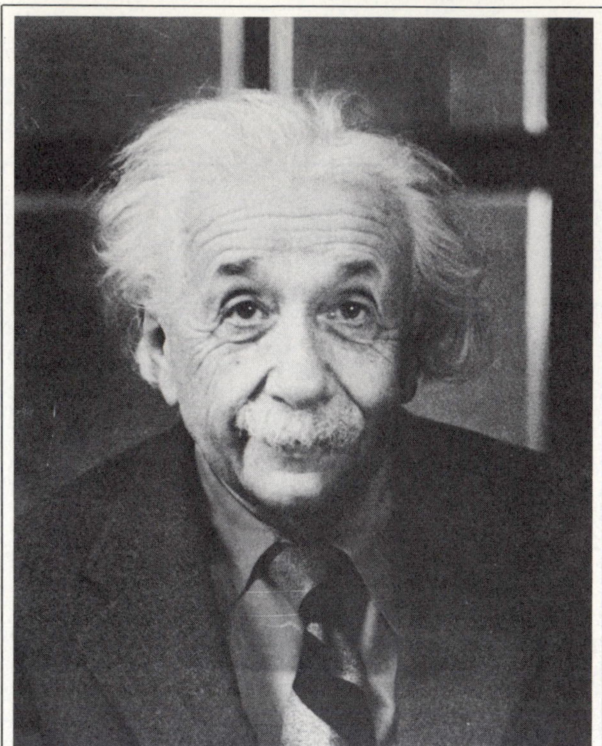

7.1
Albert Einstein. "What is inconceivable about the universe is that it is at all conceivable." (Courtesy Jewish Historical Society, Brandeis University)

The preceding chapters highlighted important changes in people's ideas of the universe. Two influential schemes, Aristotle's (Chapter 2) and Newton's (Chapter 4), rested on physical laws. However, Aristotle and Newton differed dramatically in their concepts of natural motion. These different foundations resulted in distinct pictures of the nature of the universe. Newton's cosmos was infinite, yet tied together by a universal physical law; Aristotle's cosmos was finite, with different laws for the terrestrial and celestial realms. Both models had one aspect in common: The universe was static.

Newton's grand model gained the authority of success after success. His infinite space, established firmly by his laws and intertwined with gravitation, seemed invincible.

Albert Einstein (Fig. 7.1) greatly admired Newton's synthesis. But he wondered if the last word had been said about the clockwork universe. Newton had failed to discuss the nature of gravitational force, although he did describe its effects. In addition to pondering gravity, Einstein was puzzled by the nature of light. From these two seemingly separate realms—gravity and light—Einstein synthesized a new view of the universe in his theory of relativity. One surprising result: In Einstein's vision, *gravity is no longer a force.*

In Einstein's theory, space, time, mass, and natural motion take on new and crucial relationships from which a new conception of the cosmos emerges—one in which the universe itself *evolves.* This chapter looks again at the concepts underlying Newtonian physics to find the themes of relativity and the consequences of relativistic ideas for the cosmos.

7.1 *Natural motion reexamined*

You live in a Newtonian world. By this statement we mean that you see the world the way Newton did. If an object falls or moves, you assume that a force is acting on it. You are unconsciously applying Newton's laws to what you observe. You know a force is acting because you see its effect, a change in motion: *acceleration.* You feel that the force is real because you can see the acceleration that results from it.

Take the force of gravity as an example: How do you know it's a force? Drop a stone. It falls, and its velocity continually increases as it falls. The stone exhibits a constant acceleration. What has changed? The stone's *natural motion,* which is, according to Newton, to be at rest or moving at a *constant speed along a straight line* (see Section 4.3). This was Newton's definition of natural motion. In contrast, Aristotle (Section 2.3) believed that a falling stone displayed the natural motion of earthy material, so no force was acting.

Here's the point: Newton's definition of natural motion leads directly to his concept of force, particularly gravitational force. Newton saw natural motion as ordinary, inconsequential, and so needing no explanation. To him, accelerated motion demanded explanation. After all, without accelerated motion, apples would not fall from trees, the moon would not orbit the earth, and the earth would not revolve around the sun.

To cope with accelerations, Newton had to deal with forces. Forces result in accelerations, changes in velocity. This emphasis on *change* of velocity rather than velocity itself is critical. All velocities are relative, so you can always find a frame of reference in which any given velocity disappears. For example, when you are stopped in your car and about to enter a highway, other cars go by you at about 90 km/h. You accelerate to 90 km/h, too. Then you've matched the velocities of the other cars so that compared with yours, they have zero velocity. The original 90-km/h difference has vanished. In any situation. an

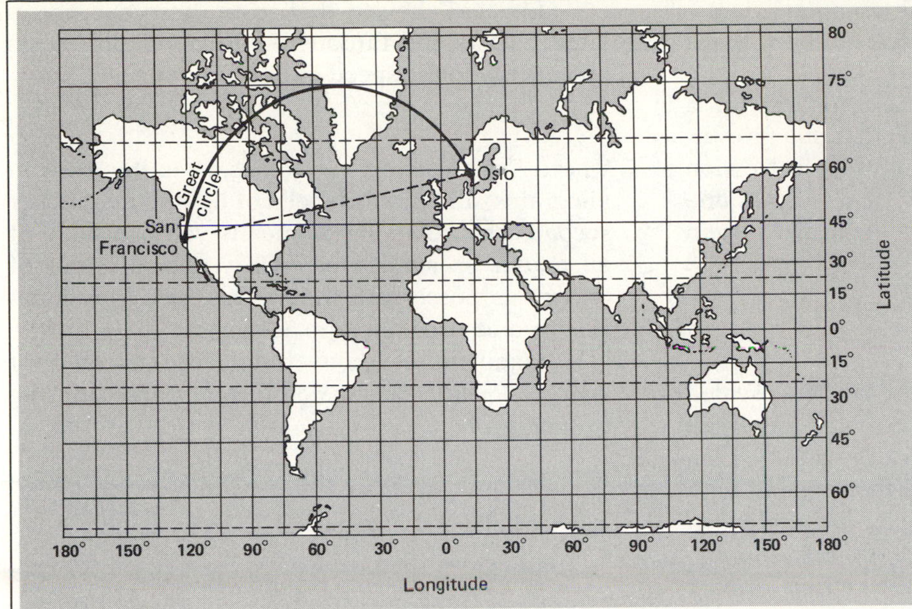

7.2
Straight lines and geometry on the earth. This Mercator map artificially makes the round earth appear flat. Straight lines on such a map are not the shortest distances between two points on the earth. That path is a great circle line, which appears curved on this flat map.

observer can match velocities with any other observer. Relative velocities can always be made to disappear. If this were not the case, astronauts wouldn't be able to dock their spacecraft!

Now imagine that forces related directly to velocities rather than to accelerations. Then, by matching velocities, you could always make the effects of a force vanish. No effects, no force! To be able to make forces disappear in such an arbitrary fashion repelled Newton. He believed that forces were real, that they could not be made to disappear arbitrarily. With forces connected to accelerations (and not to velocities), Newton seemed to have nailed down their absolute existence.

However, Newton had made an assumption that you may not have noticed. Recall in his definition of natural motion that he talks of velocity along a *straight line*. What's a straight line? The shortest distance between two points. You probably have an intuitive picture of a straight line drawn on some flat surface (like a blackboard). You then have the same kind of straight line in mind that Newton did: the straight line Euclid described in his geometry. That's the geometric assumption behind Newton's definition of natural motion: *straight line* means a Euclidean straight line. Newton really didn't have any choice, for in his time Euclidean geometry was thought to be the only possible geometry.

Can you imagine straight lines that aren't Euclidean? Consider a curved surface rather than a flat one: the earth's surface. A straight line, the shortest distance between two points, on the earth's surface differs from that on a flat surface. Look at a flat (Mercator) map of the world (Fig. 7.2). You might think that traveling on a straight line of the same latitude is the shortest distance between two points of the same latitude. But it's not. Airplanes, for instance, do not travel along latitude lines to go the shortest distance. Instead they travel part of a *great circle*, a circle on the earth's surface whose center is the earth's center.

You can make your own straight lines on the earth with a string. Stretch the string tightly against a globe with ends at the two places you want to connect (Fig. 7.3). Note that this path

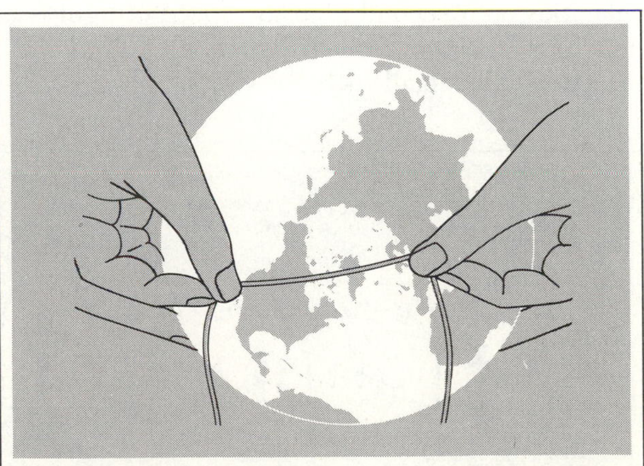

7.3
Straight lines on a sphere. You can check that great circle distances on a globe are the shortest distances (and so straight lines in a spherical geometry) by stretching a string along the globe's surface to connect two points.

curves relative to latitude lines! On a flat, Mercator map, such a line does not look straight. But it is—on a *curved* surface.

The point here is that the geometry of a curved surface is not the same as that of a flat surface. In his definition of natural motion, Newton assumed that our universe had the same geometric properties as a flat surface. Einstein challenged this assumption.

7.2 *The rise of relativity*

At the start of this century, Albert Einstein created a new vision of the cosmos. This picture emerged in two stages: the special theory of relativity, then the general theory. The special theory dealt with the laws of physics as seen by *unaccelerated* observers; those who experience uniform motion. The general theory goes beyond the special theory to cope with the nature of gravitation and so deals with *accelerated* observers.

Throughout the nineteenth century, scientists had puzzled over the nature of light. Einstein was displeased by the theories they had developed. To satisfy his aesthetic qualms, he was forced to reexamine Newton's concept of space and time. From this reexamination arose Einstein's special theory of relativity.

The special theory of relativity / Einstein's special theory can be summed up in one sentence: *The laws of physics are the same for all observers experiencing uniform motion.* I won't treat the special theory in any detail but will describe a few major consequences.

Einstein based the special theory on two fundamental postulates: first, that the laws of physics hold true for all observers traveling at constant velocities, and second, that the speed of light is a universal constant, which, when measured, has the same value to all such observers.

Einstein showed that from these two postulates follows the famous formula relating mass and energy to the speed of light:

$$E = mc^2$$

where E is the energy in joules, m is the mass in kilograms, and c is the speed of light in meters per second. This relationship means you can think of all matter as a form of energy and all kinds of energy as possessing mass. This mass produces gravity and also shows up as inertia.

Small masses convert to large amounts of energy. For instance, to completely transform 1 kg to energy:

$$\begin{aligned} E &= mc^2 \\ &= (1\ \text{kg})(3 \times 10^8\ \text{m/sec})^2 \\ &= 9 \times 10^{16}\ \text{J} \end{aligned}$$

which is enough energy to keep a 100-watt light bulb lit for about *30 million years*!

Another consequence of the special theory is that no mass can travel faster than light. More accurately, an object cannot be accelerated from a speed less than that of light to a speed greater than that of light. Why not? Einstein's theory shows that as you accelerate an object closer and closer to the speed of light, it takes a greater and greater force to accelerate it more. To get the object to go at the speed of light requires an *infinite* force—and that's impossible!

You may have heard about hypothetical particles that are believed to travel only faster than light. These particles, called *tachyons*, are a speculation at the moment; they have not been observed. Such particles do not contradict the special theory, because they are *always* traveling faster than light.

Spacetime / The special theory resulted in the unification of the concepts of space and time.

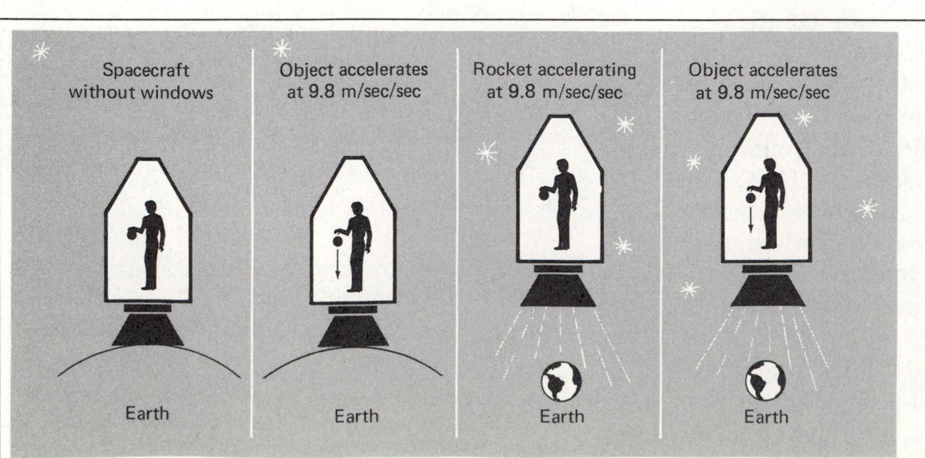

7.4 ***The principle of equivalence.*** *Imagine that you are in a small spacecraft with no windows. Drop a mass and measure its acceleration. If the spacecraft is on the earth, the acceleration will be 9.8 m/sec/sec. If the ship is in space far away from the earth and its engine is accelerating it at 9.8 m/sec/sec, the same experiment will yield the same results. You can't tell from it whether you're on the earth or in an accelerating ship.*

Scientists had looked at these as separate aspects of the universe (similar to the ancient astronomical separation of earth and sky). Newton, for instance, dealt with space and time as totally unrelated. In Einstein's view, all events in the universe involve space and time together: *spacetime*, of four dimensions (three of space and one of time).

Don't let the concept of spacetime throw you. You experience spacetime during your entire life. For example, when you arrange to meet a friend, you have to specify not only the place, but also the time. Otherwise you wouldn't get together!

Einstein made special note of what seems so obvious: When anything happens in the world, it takes place in *both* space and time. He called such a happening an *event*. You mark an event by noting where it took place (space) and when it took place (time). These events make up points in spacetime.

Here's another example of the intimate connection of space and time. As far as you're concerned, the events you see now are happening all at the same time. But are they really? Remember, light takes a finite time to reach you to let you see what's going on. So you can never see events happening at the same instant if they are spread out in space. You see more distant objects as they were in the past, compared to nearer ones.

Look up at the moon in the sky. You see it as it was about a second ago. When you look at the sun, your view is 8 minutes old. The stars at night give you a deeper look into the past. Sirius, as you see it now, shines as it did 8 years ago. As you look out into space, you look back in time. You now receive light emitted from different objects at many different times. In this sense, a telescope acts like a one-way time machine—it peers only into the past. Astronomers scan the history of the cosmos.

The general theory of relativity / From 1905 to 1915 Einstein tackled the problem of widening the scope of the special theory to include *accelerated* observers. This investigation led to the *general theory of relativity*, which dealt with gravitation.

Einstein rethought gravitation by questioning Newton's concepts of mass and inertia. He saw that Newton defined mass by two different operations: in the second law and in the law of gravitation. Suppose you apply a known force to an object and measure its acceleration. Using the force and acceleration, Newton's second law (Section 4.3) gives you the object's mass. Mass determined this way is called *inertial mass*. Now take the same object and weigh it. Weight is a force, the amount of gravitational force with which the earth (in this case) attracts the object. Mass measured this way is called *gravitational mass*.

Newton believed (and did not make the distinction) that an object's inertial mass and gravitational mass were the same. He knew this from Galileo's experiments with falling bodies (Section 4.1) as well as from his own careful experiments. These results demonstrated that, near the earth, *all masses fall with the same acceleration*.

Experimentally, this equality of gravitational and inertial mass holds true to a very high degree of accuracy: No difference has ever been detected to the limits of sensitivity with which tests can be done. The best experiment to date found the inertial and gravitational masses of gold and platinum to be the same to within one part in 10^{12}.

Einstein felt that the equality of gravitational and inertial mass was no accident. He saw it as a fundamental fact about the universe and gave it a special place in the general theory as the *principle of equivalence: You cannot distinguish accelerations due to gravitation from accelerations due to other kinds of forces*. Or, stated differently, why is the acceleration of gravity *independent* of the mass of an object?

Here is Einstein's own example of the principle of equivalence (Fig. 7.4): Imagine that you are on the earth in a spacecraft with no windows. If you drop objects in the spacecraft and measure their accelerations, you would find that they all fall with the same acceleration, 9.8 m/sec/sec. Now suppose that without your knowledge, you and the spacecraft were placed out in space and constantly accelerated at 9.8 m/sec/sec. Repeat your experiments dropping other objects. They will accelerate at 9.8 m/sec/sec. Where are you? You'd probably conclude that you were still on the earth—unless you could look out of the spacecraft. You cannot, by your experiments, distinguish between the effects due to a gravitational force and those due to the force of a rocket engine.

These examples of the principle of equivalence seem simple enough. Yet the principle has profound consequences: You now have a way to cancel out gravity locally. Put yourself in an elevator in a tall building. Let the elevator free-fall (Fig. 7.5). You find yourself weightless; gravity has vanished! Instantly transport yourself in space, far away from any large masses. Your condition is the same—you are weightless without gravity.

Note that by free-falling, an observer can make this gravity disappear. How strange! Newton saw gravity as mysterious, even though he was astute enough to describe its effects. To Einstein, the falling of objects was not mysterious at all.

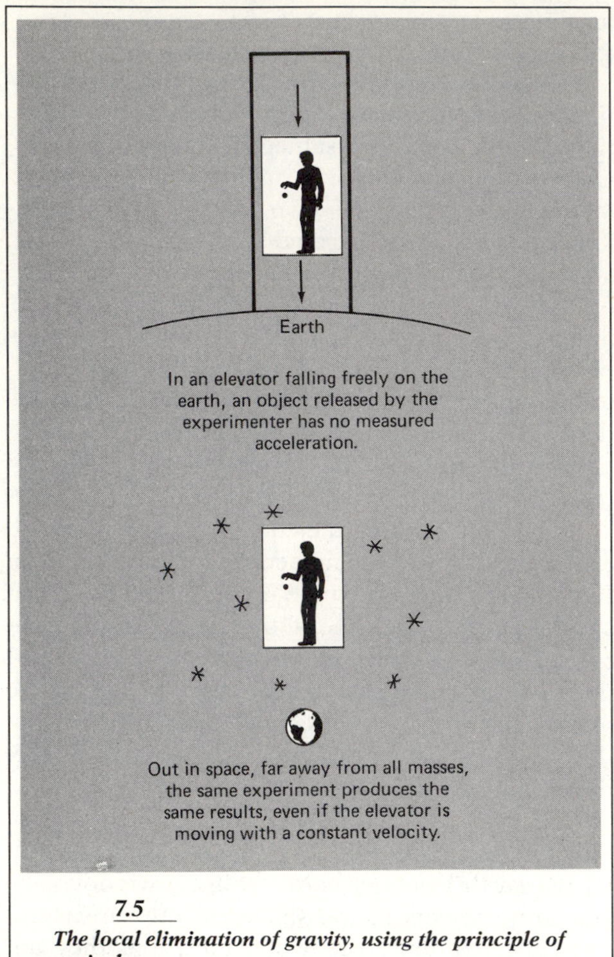

Earth
In an elevator falling freely on the earth, an object released by the experimenter has no measured acceleration.
Out in space, far away from all masses, the same experiment produces the same results, even if the elevator is moving with a constant velocity.

7.5
The local elimination of gravity, using the principle of equivalence.

The objects are simply following their *natural motion in spacetime*. In Einstein's view, gravity is not a force.

This statement would have made no sense to Newton. Because you live in a Newtonian world, you envision a force acting on a falling object. But you can make the effects of the force—acceleration—disappear quite simply: Free-fall with the object. Then the object's acceleration with respect to you is zero. No acceleration, no force! You are weightless.

Here's a rule of thumb in relativity: *When you are weightless, you are following your natural motion in spacetime*. Having weight is unnatural.

Don't confuse mass and weight! You are familiar enough with space travel to know that if you are far enough away from any large mass and if your spaceship is not accelerating, you will be *weightless*. What does that mean? Simply that if you place a scale beneath you, it would read zero. Now, when you stand on a scale on the earth, what are you measuring? In Newton's terms you are reading the amount of *gravitational force* exerted on you by the earth's mass. So *weight is a force*.

Mass relates to the inertial properties of matter; it is not a force. But forces can tell you how much mass you have. For example, to get around in a spaceship, even if you are weightless, you have to put out an effort. Suppose you want to go from one side of the ship to the other. The easiest way is to push yourself off a wall with a small amount of force. You'll then drift to the other side, and to stop yourself you'll need to push against the opposite wall with the same amount of force. If you measure the amount of force and your acceleration, you can use Newton's second law to calculate your mass—your *inertial* mass. No matter where you are—on the earth, or the moon, or in space—your inertial mass is always the same.

Einstein's view of natural motion matches Aristotle's (see Section 2.3) much more closely than Newton's. Drop an object. Aristotle says, "That's natural motion." Newton declares, "That's forced motion due to gravity." Einstein states, "That's natural motion in curved spacetime with no force." Aristotle saw the motion in the heavens as natural motion, Newton as motion influenced by the force of gravity. Einstein pictured such motions in free-fall as natural motion without forces.

But, you might say, the moon is also in free-fall, moving under the influence of gravity. If the moon's motion around the earth is its natural motion, why is the orbit curved rather than straight? To answer this question requires a look at geometry and its relationship to physics.

7.3 *The geometry of spacetime*

How can geometry and physics relate? Newton assumed that the geometry of the universe was Euclidean. Einstein makes no such assumption. He puts this question up to *experimental* confirmation:

> *Geometry is . . . evidently a natural science; we may in fact regard it as the most ancient branch of physics. Its affirmations rest on induction from experience. . . . Whether the . . . geometry of the universe is Euclidean or not has a clear meaning, and its answer can only be furnished by experience. . . . I attach special importance to this view of geometry . . . for without it I should have been* unable to formulate the theory of relativity.

What is Einstein's meaning here?

Geometry and physics / Geometry, developed by the Greeks into mathematics, derived from the practical surveying techniques of the Egyptians. So geometry, as we understand it, was developed from experience. For years people held that Euclidean geometry—and *only* Euclidean geometry—applied throughout space to physical measurements. Newton believed this assumption;

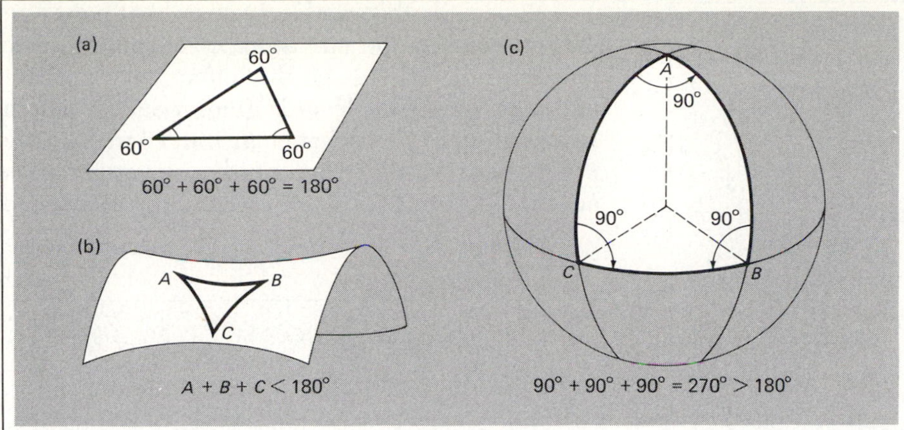

7.6
Properties of different geometries.* (a) *In a flat geometry, the sum of the angles of a triangle always equals 180°.* (b) *In a hyperbolic (open) geometry, the sum of the angles is always less than 180°.* (c) *In a spherical geometry, the sum of the angles is always greater than 180°.

he had no choice, for no other geometry had been devised by his time.

Because of Euclid's parallel-line postulate (essentially, that two parallel lines when extended to infinity remain the same distance apart and will never meet), Euclidean geometry is flat. Its flatness results in the Pythagorean theorem for right triangles and the statement that the sum of the angles of any triangle equals 180° (Fig. 7.6a).

In 1829 the Russian mathematician Nikolai I. Lobachevski (1793–1856) pointed out that Euclid's parallel-line postulate was not unique. Instead, Lobachevski proposed a new postulate that allowed parallel lines to diverge but still resulted in a self-consistent geometry. In two dimensions his geometry has properties similar to the surface of a saddle (Fig. 7.6b), on which all parallel lines diverge when extended; the geometry is curved. This geometry is sometimes termed *hyperbolic geometry*. When a triangle is drawn on a hyperbolic surface, the sum of its angles is less than 180°. Hyperbolic geometry is infinite, because extended parallel lines never meet. Both hyperbolic and flat geometries are for this reason called *open* geometries.

In 1854 Georg F. B. Riemann (1826–1866) devised yet another self-consistent geometry having a different parallel postulate. Riemann allowed parallel lines to converge when extended. The surface of a sphere has the resulting geometry (Fig. 7.6c), where the sum of a triangle's angles is *greater than* 180°. Such a geometry is sometimes called *spherical geometry* or *closed geometry*. Consider, as an analogy, the earth's surface with its lines of longitude and latitude. Two lines of longitude, both perpendicular to the equator (and so parallel to each other), intersect at the poles.

Both these non-Euclidean geometries may be characterized by their curvature, which does not typify a flat, Euclidean geometry. The hyperbolic geometry has a negative curvature because it bends away from itself. It extends infinitely far. Spherical geometry, in contrast, curves in on itself. Because of its positive curvature, spherical geometry is finite but unbounded: It has a definite size, but no edge. Consider again the earth's surface. You can travel around the earth's surface as many times as you like and in any direction you want without ever discovering a boundary. Yet the surface of the earth has a definite area. (Careful—I'm using two-dimensional examples here because they are easy to visualize, but the physical world exists in the four dimensions of spacetime. That's harder to picture, but the idea is the same.)

Whereas Newton had only one geometry at his disposal, by Einstein's time, three general categories of geometry—hyperbolic, spherical, and flat—were available. Which one was the appropriate choice to apply to the physical world? And how does this choice relate to gravity? Let's look at geometry and physics in a local region and then for the entire universe.

Local geometry and gravity / Imagine that you live in two dimensions on the earth's surface. You have no concept of a third dimension and no experience of it. You cannot conceive of an "up" that is off the surface.

You and a friend do an experiment. You both stand on the equator some distance apart (Fig. 7.7). You both walk away from the equator on paths that are at right angles to the equator. And you both believe that the geometry of the world you live in is flat.

As you walk—being very careful to keep on lines at right angles to the equator—something strange happens. You are moving closer together! Yet by all the precepts of the Euclidean geometry

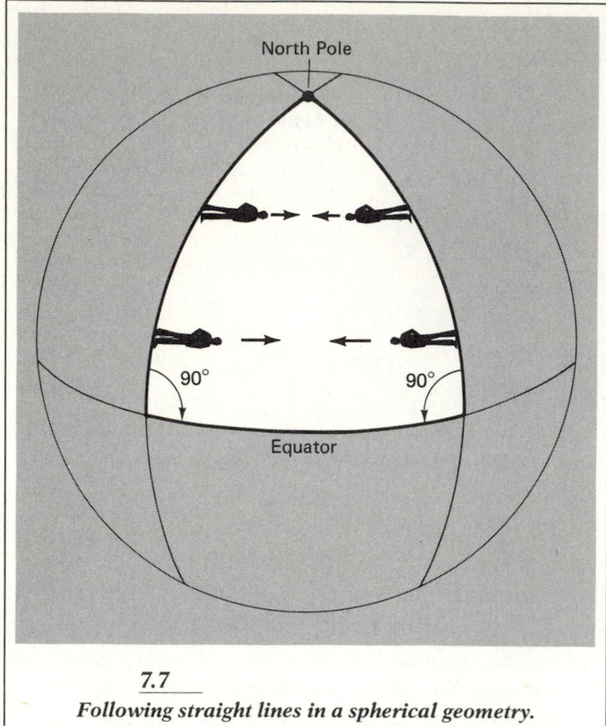

7.7
Following straight lines in a spherical geometry.

you learned in school, you should be traveling on parallel lines and so stay the same distance apart. What's happening?

You might still stick with the belief that your world is flat. Then you could explain your coming together by saying that there is a strange force of attraction bringing you and your friend together. You might even call this force "gravity" and see it as mysterious. Then you would be thinking like Newton.

Or you could say to your friend, "Maybe our assumption about this world's being flat is wrong. Maybe it's actually curved. We are moving on straight lines at right angles to another line. These lines should be parallel. Yet we move closer together. That's not what we should expect if the world is flat. So our experiment is telling us that our world is not flat but curved. Then our moving together has a natural explanation: It's our world informing us it's curved." No need for a mysterious force! An experiment has revealed the local geometry of the world in the region where you do the experiment.

The curvature of spacetime / How is spacetime curved? In Einstein's general theory of relativity, the distribution of mass (and energy) determines the geometry of spacetime. A massive object produces a curvature of nearby spacetime. And that curvature shows itself by accelerated motion, which Newton would say was caused by gravitational forces.

Einstein's theory also predicts that light rays will be bent by the curvature of spacetime near massive objects. To see this effect, let's go back into our Einstein elevator (Fig. 7.8a), which is accelerating by its rocket engine at 9.8 m/sec/sec. Put a laser in the front of the elevator to fire across its width at a target level on the other side.

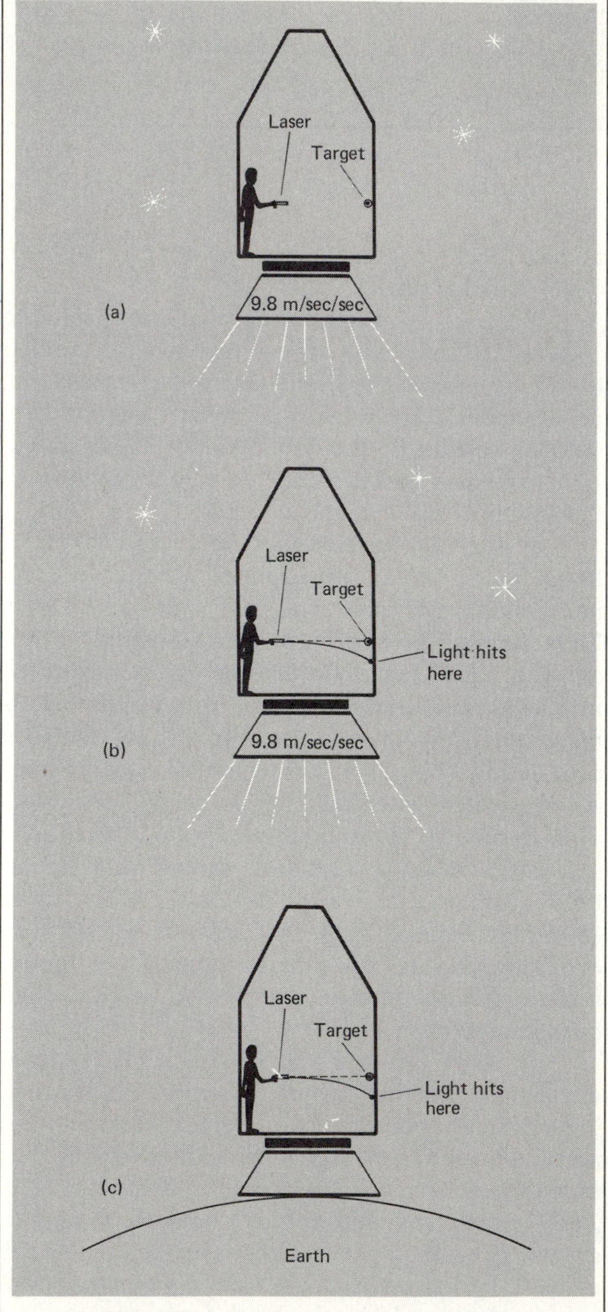

7.8
The bending of light in a gravitational field. In an accelerating spacecraft (a), *a laser beam shot at a target on the opposite wall will hit below the target* (b). *By the principle of equivalence, the same will happen if a spacecraft is at rest on the earth* (c).

Will the light beam hit the target? No, it will strike below it (Fig. 7.8b). Can you see the reason? As the light travels from one side of the elevator to the other, the elevator continues to accelerate "upward." The light path appears curved to an observer inside the elevator. Now, by the principle of equivalence, if you moved the elevator to the earth's surface, you shouldn't be able to tell the difference by an experiment inside the elevator. The light path will also bend, and by the same amount as before (Fig. 7.8c). So light paths should be deflected in all regions of curved spacetime, near all massive objects.

You should notice that the amount of bending in the elevator example above will be extremely small. The reason is simple: Light travels fast and crosses the width of the elevator in an extremely short time. Suppose, for instance, that the elevator were 3 m wide. Then the light transit time is a mere 10^{-8} sec! During this time, the elevator moves only 5×10^{-16} m. So testing this prediction of general relativity on the earth would be a difficult job.

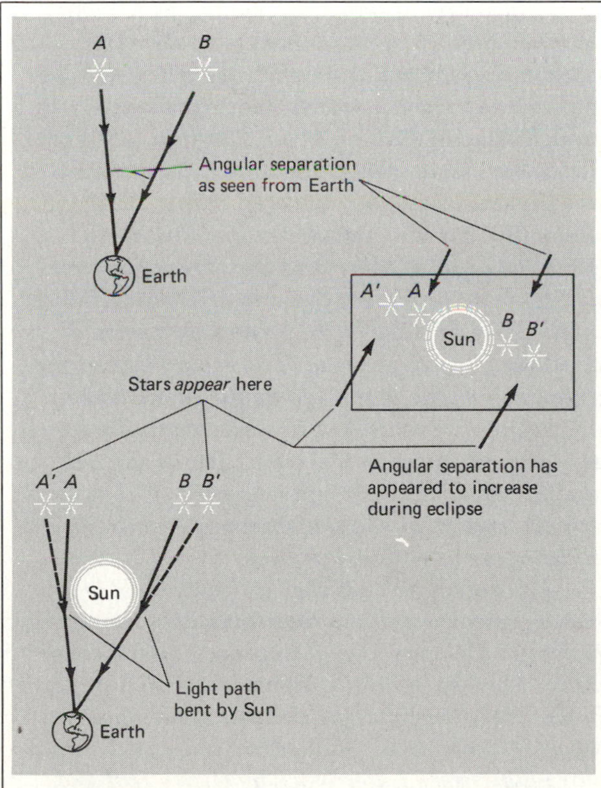

7.9
The effect of the sun on the apparent angular separation of two stars. Consider two stars (A and B) that have a measured angular separation when the sun is not nearby. During an eclipse, when the sun lies between the two stars, they appear to be farther apart due to the bending of the light paths.

Where is the most warped region of spacetime in the solar system? Where the most mass is located: at the sun. Relativity predicts that the sun's mass deflects light rays away from Euclidean, straight-line paths. How to test this? Take a picture of the stars and sun during a total solar eclipse. Later, when the sun is not in this part of the sky, take a picture of the stars. Compare the angular separation of two stars close to but on opposite sides of the sun. With the sun present, the angular separation of the stars is *greater* than when the sun is not there (Fig. 7.9). Is this shift visible? Yes. Such observations have been made through the toil of many people over many years. They come out very close to Einstein's prediction from general relativity of 1.75″. (That's about 1/1000 the angular diameter of the sun.)

Have I been fair to Newton? If he had known that light has a mass equivalent ($E = mc^2$), he could also have calculated, using his law of gravitation, a deflection from the sun's gravity. But the result is only one-half Einstein's value. Einstein wins in this experimental test! (It's not the only one; others are described in Focus 7.1).

What's happening? In Einstein's view, light pursues a straight line in spacetime. But the geometry in which it travels is curved by the sun's mass. You can picture the sun's mass as creating a warp in the geometry of spacetime (Fig. 7.10).

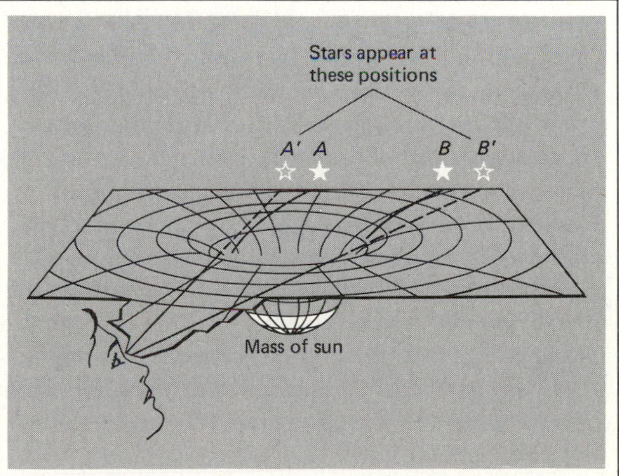

7.10
Einstein's explanation for the change in two stars' angular separation during an eclipse. The mass of the sun curves the spacetime around it; the closer to the sun, the greater the curvature. The light from the stars follows this curved spacetime when the sun lies between them as seen from the earth (A'B'). When the sun is not there, the light paths travel through a flatter region of spacetime, so the stars appear closer together (AB). (Adapted from a diagram by C. Misner, K. Thorne, and J. Wheeler)

Focus 7.1

EXPERIMENTAL TESTS OF GENERAL RELATIVITY

Einstein's model for gravitation would be no more than a fascinating idea if it did not make numerical predictions that could be tested experimentally, predictions of effects unknown or unexplained by Newton's model for gravitation. Let's look at two essential solar system tests here: the deflection of light in curved spacetime and the precession of Mercury's orbit.

Deflection of light / *The sun's mass strongly distorts spacetime in the solar system. Even light paths are bent, and this bending is greatest for light that just skirts the edge of the sun. So a solar eclipse is a natural event during which to try to measure the deflection of light from stars. General relativity predicts that the deflection should amount to 1.75″. These are hard observations to make, for precise photographs must be made of the sun and stars during eclipse and also of the stars without the sun in the vicinity.*

Results have ranged from 1.43″ to 2.7″. The prediction is 1.75″. Errors are hard to estimate and may be as large as 20 percent. The scatter in the observed values reflects the difficulty of doing this observation.

Modern technology permits a similar experiment to be done more precisely by radio astronomers. Radio and light are essentially the same, so these deflections should also occur for radio signals from distant objects. Some quasars (see Chapter 20) are intense radio sources, and each October the sun eclipses a quasar named 3C 279. By monitoring the position of 3C 279 relative to a nearby quasar, radio astronomers, using a technique called interferometry (Chapter 6), can accurately (to about 0.1″) measure the angular separation between the two quasars. This experiment has been done a number of times, and the observed separation has been within 10 percent of the value predicted by general relativity.

Precession of Mercury's orbit / *Astronomers have known for a long time that the major axis of Mercury's orbit does not remain fixed in space with respect to the stars (Fig. F.17). The axis rotates around, or precesses, in the plane of the orbit. Part of this shifting arises from the gravitational attraction of the other planets on Mercury. But when this effect and others are taken into account, there remains a residual shift of 41″ per earth century (which means the orbit turns through an extra 360° in about 3 million years).*

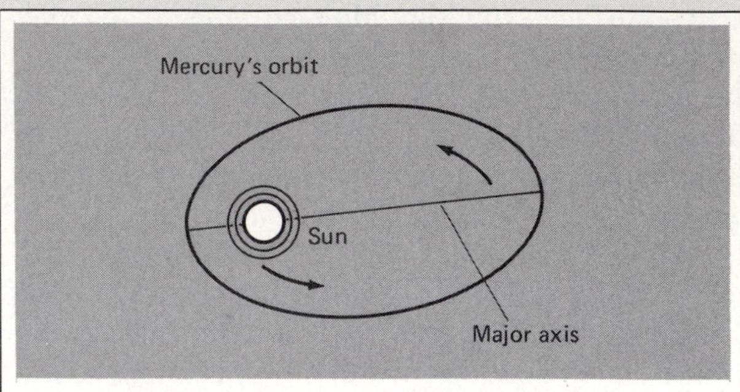

F.17
Precession of Mercury's orbit. The major axis of the orbit rotates in space with respect to the stars.

*What causes this precession? Newtonians have dreamt of an undiscovered planet, called Vulcan (no, not Mr. Spock's planet), orbiting within the orbit of Mercury. But no such planet has ever been observed. General relativity predicts a precession because of the strong curvature of spacetime close to the sun. (Precession due to curvature also happens for Venus, the earth, Mars, and the other planets, but it is much smaller for these planets than for Mercury.) The predicted value for Mercury is 43″ per century—*if *the sun is perfectly spherical. So the observed and predicted results agree to within 4 percent.*

The hooker here is the "if." If the sun is oblate, the non-spherical distribution of mass will also cause Mercury's orbit to precess. Because the sun is so bright, it's quite difficult to find out if it is oblate. Early experiments done at Princeton by Robert Dicke and M. Goldenberg indicated a mild solar oblateness. Later work by Henry Hill and colleagues, however, found no measurable oblateness. So it appears that general relativity accurately accounts for the precession of Mercury's orbit.

Light crossing the warp appears to us to take a curved path, although it is actually traveling on the shortest distance between two points through the nonflat region of spacetime.

You can imagine the situation as analogous to that of a golf ball moving along a poor putting green filled with depressions. Between depressions, the surface is flat and the ball moves along a straight line. When it crosses a depression, the ball follows a path that bends, compared to the path on the flat surface. The amount of bending depends on the depth of the depression and the speed of the ball. In Einstein's general theory of relativity, mass and energy create local warps in spacetime. The amount and density of mass-energy determine how steep the depression is. As an object moves through a warped region of spacetime, it follows a straight-line, natural-motion path. In warped spacetime, this path appears curved to us. But no force acts on the object. This is essentially how Einstein understood the nature of the gravitational force locally. Experimental tests (Focus 7.1) indicate that this understanding is correct.

Einstein considered the orbits of the planets in the same way. All the planets move on straight-line paths in spacetime. We view these paths as elliptical orbits in three dimensions. The paths appear curved to us because spacetime is warped, not flat. The amount of warping decreases as the distance from the sun increases. That's why the earth, for example, follows an orbit more strongly curved than that of Jupiter.

To sum up the difference in Newton's and Einstein's models for gravity: Newton concentrates on *forces*, Einstein on *courses*. Newton sees gravity as a force acting instantaneously between all matter, a force whose strength depends on the mass of the attracted object and the distance. Einstein focuses on the paths of objects in spacetime; such paths, for free-falling objects, don't depend on the mass of the object. (The demonstration is Galileo's experiment: Two objects of different mass fall with the same acceleration.) So the paths that free-falling objects take, their natural motions, are guided by the local geometry of spacetime.

Warning and notice: Einstein's geometric ideas about gravity and spacetime will *not* be used, in general, in the rest of this book. For most purposes, Newton's model of gravitation works perfectly well, and the chapters to come will use Newton's laws of motion and gravitation. So you do not need to master Einstein's ideas to the same depth and competency as you do Newton's. Einstein's general theory crops up only with the universe as a whole (see Chapter 21) and with black holes (see Chapter 17).

7.4 Geometry and the universe

So far I've described Einstein's picture of gravity in a small region of spacetime, such as near the sun. But his general theory applies also to the universe. With it, the geometry of the cosmos relates to the dynamics of the whole universe. Keep in mind that three basic geometries are possible: hyperbolic, spherical, and flat. Remember also that Einstein does not state which geometry must apply; he leaves this open to experimental confirmation. How to find out?

Imagine again living in two dimensions on the surface of a sphere. Start at any point and walk in a straight line away from it. Eventually you'll return to your starting point because a sphere has so much curvature that it comes back on itself. Suppose the universe has the same geometric properties (in spacetime) as a sphere's surface. If we sent out light signals, they would eventually return to us. Why? Just as for a two-dimensional surface, our four-dimensional spacetime, if curved enough, will close back on itself. (*Warning:* If the universe is closed, it does *not* have an edge, because it has no boundary. Consider again a sphere's surface: You can go around it many times and never find an edge.) We don't know for sure that our universe *is* closed—that depends on whether it contains enough matter and energy to curve it sufficiently. But if it does contain enough, then it will be finite and unbounded, just as the surface of a sphere is finite and unbounded.

Obviously this method of finding out if the universe is closed is impractical—the universe is large and the speed of light finite. Let's turn this idea around. The density of mass (and energy) determines the curvature of spacetime. If this density has a certain critical value (or greater), then the universe curves back on itself and is closed. Einstein's general theory gives this critical density: It is roughly 5×10^{-27} kg/m^3. (That's about one hydrogen atom for every cubic meter of space.)

In 1917 Einstein constructed a model of the universe with a closed geometry. How different from Newton's model, which had to be infinite (see Section 4.5)! In a geometric sense, Einstein's model was akin to the finite, closed picture of Aristotle (Section 2.3). In fact, Einstein's model was also static.

Our conception of the cosmos seems to have come full circle. But it did so with a difference. A few years after Einstein proposed his model, astronomers discovered it was *wrong*. They found that the universe is not static but *expanding*!

A quick tour of the universe / Before I investigate general relativity's impact on cosmological ideas, let's look at the content and scale of the universe. (The rest of this book will describe the contents of the astronomical universe in detail.)

Because the universe is a big place, we need a long measuring stick for it. I'll use what we consider the most natural one: light travel time. From science fiction, you're probably familiar with the term *light year* (abbreviated *ly*), the distance light travels in a year. That amounts to about 9.5×10^{15} m, or 63,000 AU. Other useful light units are a light second (3×10^8 m), light minute (1.8×10^{10} m), light hour (1.1×10^{12} m), light day (2.4×10^{13} m), and a light month (7.8×10^{14} m).

The nearest star, Alpha Centauri, lies some 4 light years away. The stars near the sun and all those you can see in the night sky (Fig. 7.11a) make up a disk of stars called the *Milky Way Galaxy*. The gravity of its contents—mostly stars, some 10^{12} of them—holds the Galaxy gracefully together. The Galaxy has a diameter of approximately 100,000 ly. The nearest galaxy that resembles our Milky Way Galaxy is the Andromeda galaxy (Fig. 7.11b)—so named because it appears in the constellation Andromeda as seen from the earth.

The Andromeda galaxy and the Milky Way Galaxy are parts of a local group of galaxies, bound by gravity. This neighborhood set of galaxies has a diameter of some 3 million ly. In this century, astronomers have found that the universe contains many groups and clusters of galaxies (Fig. 7.11c). The largest clusters are tens of millions of light years in diameter. Clusters of galaxies are spaced out, on the average, by hundreds of millions of light years. We see those galaxies in all directions, and with our biggest telescopes we can detect them at distances of several *billion* light years.

A key concept: Light travels at a fast but finite speed. Cosmic distances are huge. So when we look out into space, we are looking back in time. The farther out we peer, the deeper into the past we see. The light we receive now from clusters of galaxies left them many millions to billions of years ago.

Astronomers have found that these clusters of galaxies are not stationary. They are all moving away from each other. This motion tells us that the universe is expanding.

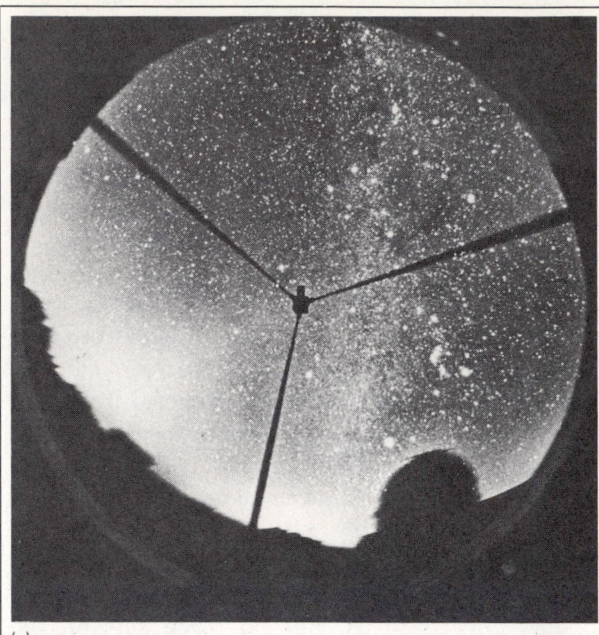

(a)

7.11

(a) *The Milky Way of stars as seen from the earth. (Courtesy Yerkes Observatory)* (b) *The Andromeda galaxy, the closest spiral galaxy to the Milky Way Galaxy. (Courtesy Kitt Peak National Observatory)* (c) *A cluster of galaxies in the constellation Hercules. (Courtesy Palomar Observatory, California Institute of Technology)*

The expanding universe / To understand how astronomers discovered the expanding universe, you'll need to know just a little about galaxies (more in Chapter 19). Here's the main point: Galaxies are made (mostly) of stars, bound together by gravity. Galaxies are spread throughout the visible universe, millions of light years apart—each one a visible marker of a distant region of space and time. In this century, astronomers found that these cosmic markers are moving apart, which implies that the universe is expanding.

How so? Observations with large telescopes provide two facts about galaxies: (1) their distances, and (2) their velocities relative to us. The velocities measured are those only along our line of sight: These are called *radial velocities*. Remarkably, the distances and radial velocities of galaxies are tied together—the surprising evidence that the universe is expanding. (Chapter 19 deals in detail with how distances and radial velocities of galaxies are measured.)

Hubble's law / Beginning in 1912, Vesto M. Slipher (1875–1969) began a project of measuring the radial velocities of galaxies. By 1928 he had observations of over 40 galaxies, and a trend emerged: Most galaxies were apparently moving away from our Galaxy, the Milky Way Galaxy.

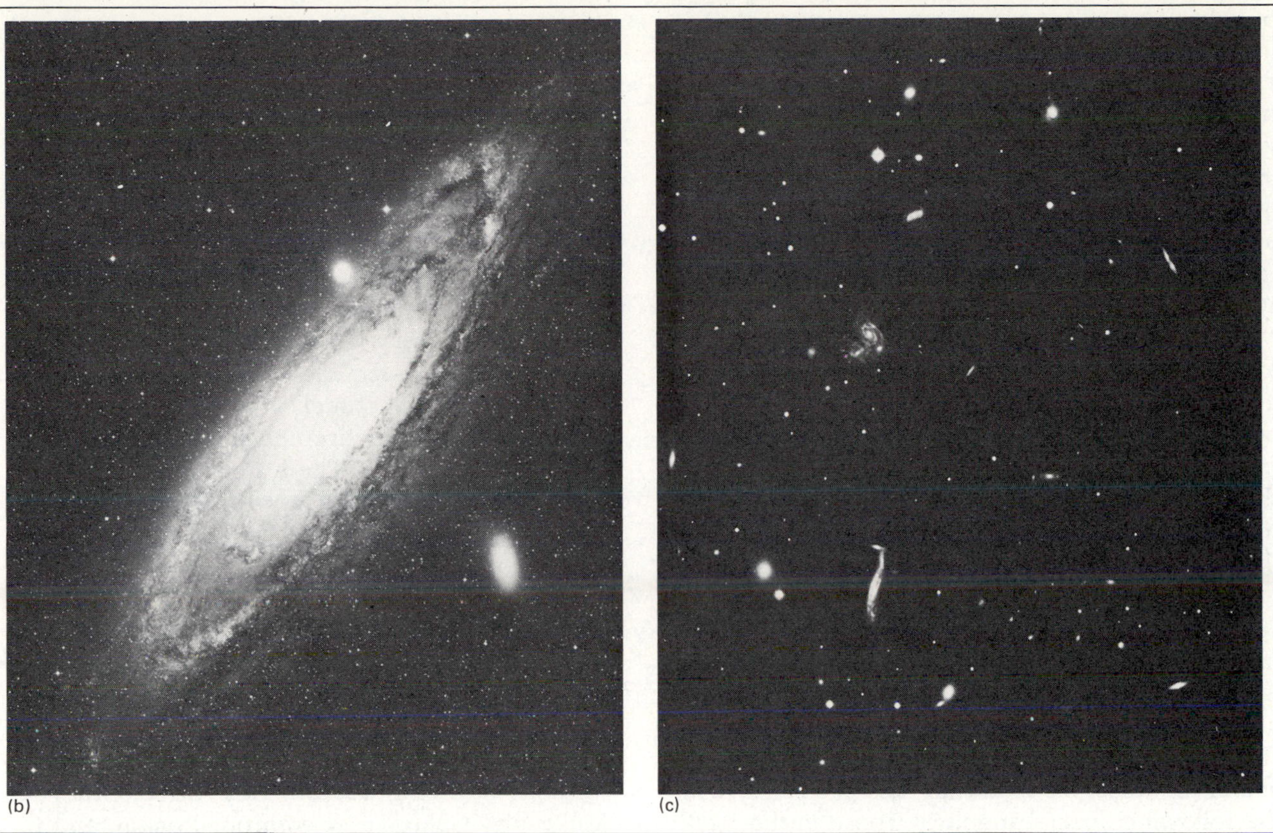

At about the same time, Edwin P. Hubble (1899–1953) had determined the distances to some galaxies and had noted an unexpected direct relationship between radial velocity and distance: For every million light years farther out, the galaxies had 170 km/sec more velocity. In later collaborative work, Hubble and Milton L. Humason (1891–1972), using the 100-inch (2.5-m) telescope, added more data to support the trend. This relationship, now known as *Hubble's law*, states that the distance to a galaxy and its recessional velocity are directly related (Fig. 7.12).

The number connecting the distance and velocity is called *Hubble's constant* and is usually indicated as *H*. It's the slope of the line on a radial velocity-distance plot such as Fig. 7.12. The steeper the slope, the larger the value of *H*.

Hubble and Humason thought the *H* was 170 km/sec/Mly. (When discussing the Hubble constant, I will use the abbreviation *Mly* for millions

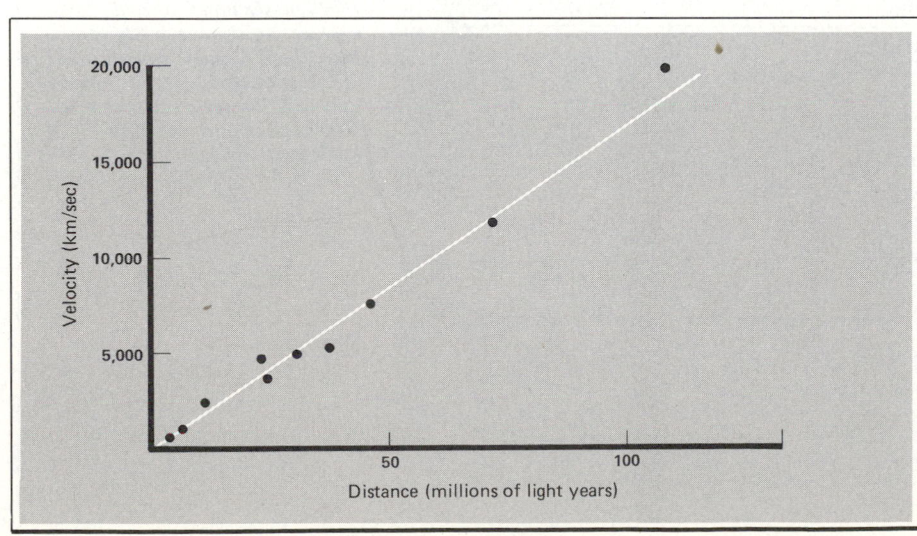

7.12

Hubble and Humason's distance-velocity relationship, now called Hubble's law. Plotted here are the recessional velocities (in kilometers per second) of galaxies against their distances (in millions of light years). The straight line indicates the trend of the data points.

of light years.) Today, astronomers believe that H lies between 15 and 27 km/sec/Mly. (The exact value is disputed. Chapter 19 deals with the controversy.) This book uses 20 km/sec/Mly. This value means that for every million light years we look away from our Galaxy, the radial velocity of other galaxies is 20 km/sec higher. For example, a galaxy at 5 Mly travels at 5×20, or 100 km/sec radial velocity, whereas one at 10 Mly moves at 10×20, or 200 km/sec.

The way I stated Hubble's law (the farther away a galaxy is from us, the faster it is moving away from us) seems to imply that some mysterious force accelerates galaxies as they move away from our own. Try turning Hubble's law around; things that are moving fast are far away. That makes pretty good sense. Since all the galaxies in the universe have been moving away from each other for the same amount of time, the ones that started out moving fastest relative to us will be the farthest away—and that's exactly what Hubble's law says.

Here's an analogy: Picture an astronomy class in a large lecture hall. Suppose that at the end of the class, the instructor lets people leave in the following special manner: Those in the back row go out with a higher velocity than those in front, and the speed with which people leave depends directly on how far away they are from the instructor. So if the people in the back row are 10 times as far away as the people in the front, they leave with 10 times the speed. Now suppose the instructor tells everyone to walk directly away from the lecture hall for some period of time (say 10 min) and then stop. If the instructor at the end of that time runs around the campus looking for the class, he or she would find the back-row people 10 times farther away from the lecture hall than those in the front row. This is an example of uniform expansion obeying Hubble's law.

The discovery of this trend, that more distant galaxies move faster, caused a radical revision of Einstein's static model for the universe. It appeared that the entire universe was expanding. As Hubble quipped, "The history of astronomy is the history of receding horizons." And Einstein later admitted that his 1917 static model was the "biggest mistake of my life."

The meaning of Hubble's law / Hubble and Humason were careful to term the galaxies' rush away as "apparent." Taken at face value, Hubble's law leaves us fixed in the center of the universe. Having been thrust away from the center by Copernicus, astronomers felt somewhat uncomfortable at being repositioned there. Was our Milky Way Galaxy now enthroned again, and the universe centered on it?

A simple argument demonstrates that our Galaxy does not really have a privileged status. From our viewpoint, the rest of the universe

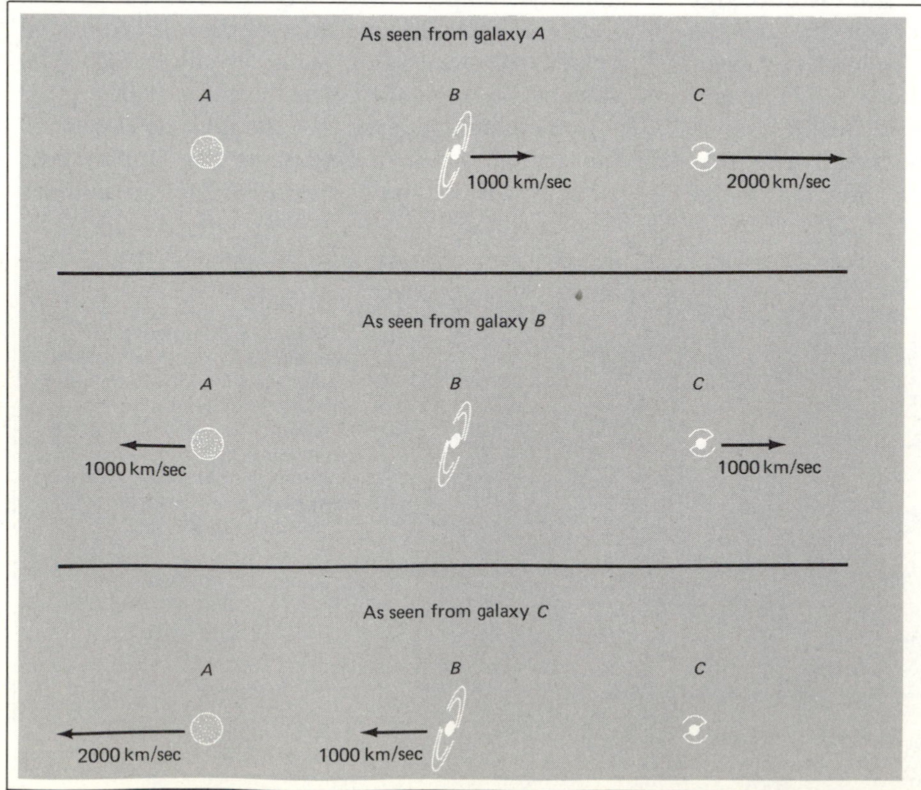

7.13

*Uniform expansion in two dimensions. Imagine three galaxies (**A**, **B**, and **C**) equal distances apart and arranged in a line. Picture yourself on galaxy **A**. If these galaxies expand uniformly, then **B** appears to be receding at 1000 km/sec and **C** at 2000 km/sec. Now move to **B**. From there, both **A** and **C** appear to be receding at 1000 km/sec. From galaxy **C**, **B** appears to recede at 1000 km/sec and **A** at 2000 km/sec.*

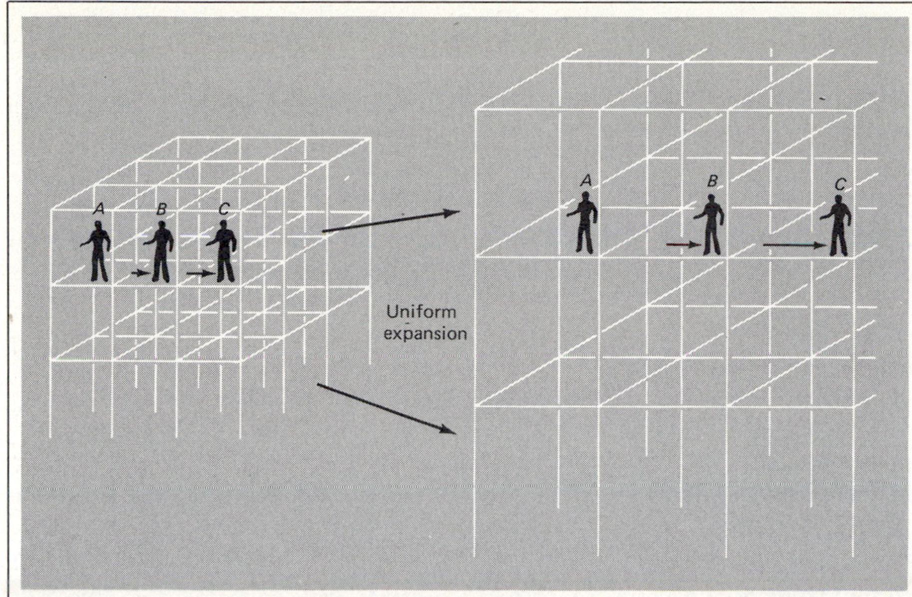

7.14
Uniform expansion in three dimensions. Imagine three people (A, B, and C) lined up at three intersections of a huge jungle gym. If it expands uniformly, then each person at each intersection sees the others recede, and each infers the same distance-velocity relationship.

appears to be retreating from our Galaxy. But if the expansion is *uniform*, then the view from any other galaxy will be the same (Fig. 7.13). Transported to another galaxy with the usual bundle of tools, an astronomer would plot the same Hubble's law as from our Galaxy. Another galaxy appears, to those in it, as the "center" of the expansion, so no privileged position actually exists.

Here's an analogy: Imagine a jungle gym with a person located at each intersection of the bars (Fig. 7.14). Suppose the bars expanded at the same constant rate. Each observer would see other observers moving away. How fast they moved would depend on their distance. One twice as far away as another would move twice as fast. (There are twice as many bars in between, each of which is expanding.) Since the expansion is uniform, all the observers would describe the expansion in the same way, and each would think that he or she was in the center. If everyone plotted a graph of distance versus velocity, all the graphs would look the same and resemble Hubble's law for galaxies. And the values for H obtained this way would all be the same.

This analysis implies that the expansion of the universe that we see now must have begun a finite time in the past. Having a value of Hubble's constant of roughly 20 km/sec/Mly, you can estimate when the expansion began (Focus 7.2): 15 billion years ago. (Note that this value assumes that the expansion rate has been constant all the time.) Astronomers call this beginning of the universe's expansion the Big Bang.

Warning: When first confronted by the expansion of the universe, people commonly have the misconception that the universe expands "into empty space." That's wrong; don't think of it! It is *space itself* that is expanding. And it does not expand into "something" or even "nothing"; the expression "expand into" is meaningless in this context.

7.5 *Relativity and the cosmos*

With the discovery of the expanding universe, cosmologists began devising models that would account for the expansion. These models must be consistent with Einstein's general theory of relativity and Hubble's expansion law. In them, Hubble's constant and the geometry of spacetime (whether flat, hyperbolic, or spherical) are physically interrelated. Let me show you how, using the concept of *escape velocity* (Focus 7.3).

Consider throwing a ball off the earth's surface. If the ball's velocity is less than the escape velocity, the ball slows down and falls back, regardless of how large or small its mass is. When thrown with a velocity greater than the escape velocity, the ball slows down a bit while it is still influenced strongly by the earth's gravity and then coasts outward as the earth's gravity affects it less and less. If it doesn't run into anything, the ball will travel out to infinity and never return to the earth. That's the meaning of escape velocity from any body: the minimum speed needed so any object moving off of it will not be pulled back by gravity.

Now consider the universe and the galaxies within it. Any one galaxy in the universe is analogous to the ball. The galaxies were once all "thrown away" from one another, for we now see

Focus 7.2

HUBBLE'S CONSTANT AND THE AGE OF THE UNIVERSE

You can infer from the measured value of H the time since the expansion of the universe began. To do this simply, you need to make the assumption that the expansion rate is uniform. (Warning: This assumption does not apply to the real universe. For any realistic model, velocities must decrease with time. However, this assumption allows a simple calculation of the oldest possible age of the universe from the present value of H.)

At a constant velocity, the distance traveled, d, is equal to the velocity, V, multiplied by the time, t. In algebraic form,

$$d = Vt$$

Conversely, the travel time is

$$t = \frac{d}{V}$$

For example, if you travel in a car from Boston to New York, a distance of 400 km at a constant speed of 80 km/h, the trip takes 5 hours (400/80 = 5).

Hubble's law written as an equation (rather than presented on a graph) is

$$V = Hd$$

Here V is the radial velocity of a galaxy, d the distance to the galaxy, and H Hubble's constant. If H has the units kilometers per second per million light years and d is given in millions of light years, then V comes out in kilometers per second. Now compare Hubble's law written as

$$\frac{1}{H} = \frac{d}{V}$$

to the trip formula

$$t = \frac{d}{V}$$

We see that the travel time equals 1/H. If H is 20 km/sec/Mly, then 1/H equals 1.5 × 10¹⁰ yr. To get this result we have to put in the conversion factors to get everything in the same units:

$$\begin{aligned} t = \frac{1}{H} &= \frac{1}{20\ \text{km/sec/Mly}} \\ &= \frac{1}{20\ \text{km/sec/Mly}} \times 10^6\ \text{ly/Mly} \times 9.5 \times 10^{13}\ \text{km/ly} \\ &= 4.5 \times 10^{17}\ \text{sec} \\ &= \frac{4.5 \times 10^{17}\ \text{sec}}{3 \times 10^7\ \text{sec/yr}} \\ &= 1.5 \times 10^{10}\ \text{yr} \end{aligned}$$

So the expansion of the universe started about 1.5 × 10¹⁰ years ago!

This conclusion is valid only if our assumption of uniform expansion is correct. However, even if the expansion were a bit faster in the past than it is now, the value of 1.5 × 10¹⁰ years is still in the right vicinity of the actual time since the expansion began.

an expansion. Consider a galaxy at some distance from us. Newton showed that the net effect of all matter spread within some space acted as if this total mass were concentrated at the center (as long as the matter is distributed uniformly). If there is enough mass within that space, the escape velocity will be larger than the expansion velocity. Note that "enough mass within that space" means a high enough density. (The meaning of *density* here is how closely galaxies are packed together; the closer they are, the higher the density.) So if the density is large enough, the galaxies will not have escape velocity; the expansion velocities will decrease with time and eventually reverse as gravity herds all the galaxies together. If the average density is too low, the galaxies will have more than escape velocity; gravity will never bring the galaxies all together, and the expansion will continue indefinitely.

Hyperbolic, flat, and spherical geometries correspond physically to the cases of greater than, equal to, and less than escape velocity. From Einstein's general relativity and the observed value of H, we can calculate the density required for the cosmos to be flat. This *calculated* density can then be compared with the *observed* cosmological density (which must include energy as well, since $E = mc^2$). If the observed cosmological density is greater than the critical density, the universe is spherical and closed. If less, the universe is hyperbolic and open. If they are exactly equal, the universe is flat. If H equals 20 km/sec/Mly, the critical density is 5×10^{-27} kg/m^3. So if we can observe the average density of matter and energy in the universe, we then have an *experimental* basis for finding out the appropriate geometry of spacetime! Observations of Hubble's constant combined with general relativity provide a way to find out if the universe is open or closed.

Making such an observation is not easy, however. First, astronomers can see only out to about 10 billion light years. Second, we can see only those masses that radiate or absorb radiation at detectable wavelengths in a manner intense enough to register on the instruments we have devised. Third, we need some method to determine the mass of observed objects. Note that our methods of observation cannot detect invisible objects (such as black holes). A density derived from estimating the mass in galaxies, about 4×10^{-28} kg/m^3, neglects the contribution of such undetected objects and all other unnoticed masses. We get this value by estimating the masses of all objects we can observe within a large volume of space, then dividing the total mass by the volume to get the density. The observed density amounts to about 12 times less than is needed to close the universe, but it falls tantalizingly near the necessary value. For cosmologists who prefer a closed universe, the close-but-not-quite results have spurred a search for the "missing matter" needed to close the universe. (The matter is not missing at all if the universe has an open geometry.)

Geometry and Hubble's constant / Whether or not Hubble's constant has the same value at all locations and times also depends on the geometry of the universe. Perhaps the expansion rate has been faster or slower in the past than it is now. As we peer deep into space, the value of H changes, and the Hubble diagram (Fig. 7.15) is not a straight line but bends up (if the expansion has been faster in the past) or down (if the expansion has been slower in the past). If the Hubble line curves up enough, the universe is closed and spherical; if it doesn't, the universe is open.

Examine a Hubble diagram (Fig. 7.15). Those galaxies that are the farthest away (in the upper right-hand part) are also those whose light comes to us from the farthest time in the past. In

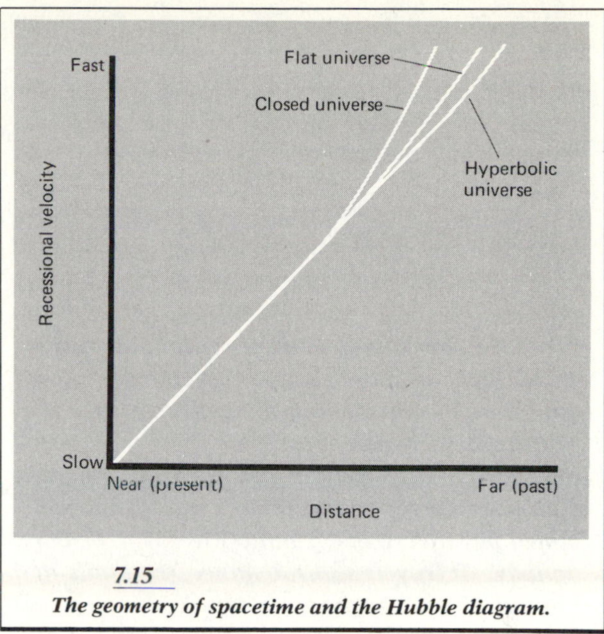

7.15
The geometry of spacetime and the Hubble diagram.

the past, the expansion rate (and so Hubble's constant) must have been much larger than it is now. The faraway galaxies that we see now mark the expansion rate in the past; if it were much larger, the slope of the distance-velocity must have been greater than it is now. (Remember, the slope of the graph is H.) Hubble's constant now is given by nearby galaxies (the lower left-hand part of the Hubble diagram). So you expect the Hubble plot to bend upward for the distant galaxies compared to the nearby ones. (So you see that H isn't really a constant; it's constant only in the sense that the ratio of velocity to distance is the same for all galaxies at any particular time; at different times that ratio is different.)

What's the future for the universe in the three cases? In each, there is some moment when the expansion begins. In a closed universe, the rate of expansion slows down and eventually stops; then the universe contracts. In a hyperbolic universe, the rate of expansion doesn't slow down as rapidly, and it never stops. Even after infinite time, the galaxies are still moving apart at a finite velocity. In the borderline case of a flat universe, the expansion slows down just enough so that it comes to a stop after infinite time.

The future of the universe / We seem to have two possible cosmic destinies: In one, the expansion grinds on like the wheel of karma forever, and the universe gradually thins out. In the other, the expansion slows down, stops, and reverses, and the universe collapses. During the time of diminishing size, the galaxies rush together into a dense conglomeration with (theoretically) zero

Focus 7.3

ESCAPE VELOCITY

From Newton's laws of motion and gravitation arises the concept of **escape velocity**, *the minimum velocity an object needs to escape the gravitational bonds of another.*

Imagine, as Newton did, a giant cannon placed on the top of a very high mountain and aimed parallel to the ground (Fig. F.18). Fire a cannonball. It travels some distance, then falls to the ground. If you use more powder in the cannon, the ball travels farther along the earth before it hits the ground. If you put a large enough charge in the cannon, the ball goes completely around the earth in a **circular** *orbit, returning to the cannon. The inertial motion of the ball just compensates for the falling due to gravity.*

Now dump in a larger charge. With a starting velocity greater than that needed for a circular orbit, the cannonball travels around the earth in an **elliptical** *orbit. Increasing the starting velocity makes the orbit more elliptical. Eventually the orbit becomes so elliptical that the semimajor axis is infinitely long, and it would take the ball an infinite time to return. So it never returns. This velocity is called the* **escape velocity**. *(Note from Kepler's second law that when the ball is an infinite distance away, its speed relative to the object it left is zero.) Any velocity larger than the escape velocity produces the same effect: The ball leaves the gravitational grip of the earth, never to return.*

What determines the escape velocity from an object? Consider the earth. Its escape velocity is about 11 km/sec. Suppose that you kept the earth at its present radius but increased its mass. Its escape velocity would increase. Now suppose that you kept its mass the same but decreased its radius. The escape velocity would again increase. Basically, an object's mass and radius determine its escape velocity. Greater mass results in a

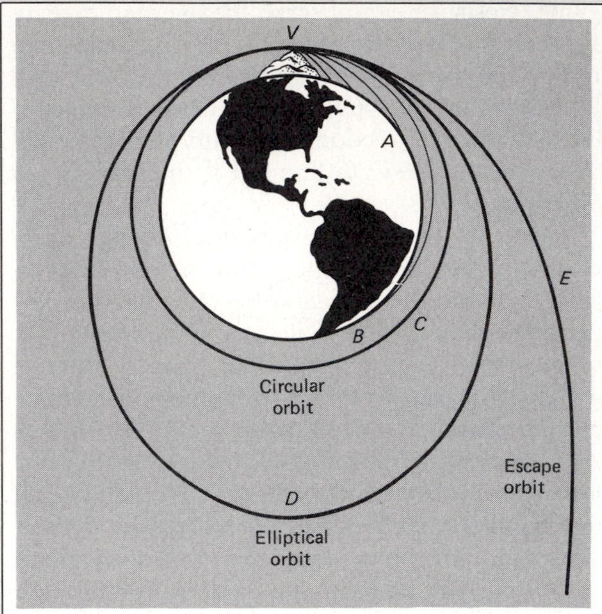

F.18
Launching of earth satellites, according to Newton. From a high mountain (V), cannonballs fired at low speeds hit the earth (A and B); a certain minimum speed puts the ball in a circular orbit (C); higher speeds result in elliptical orbits (D); and above some speed, the ball will escape from the earth (E).

radius. The universe would then crush everything into high-energy light—and many weird particles (Chapter 21).

Which fate does the Hubble diagram display for us? Unfortunately, the observations have not yet settled the issue. Here's one problem: As far out as galaxies can be observed, the observational errors are as large as the effect we are looking for. So deviations from a straight-line Hubble plot are hard to see (Fig. 7.15).

Results to date have been a mixed bag, some supporting an open, others a closed universe. Jerome Kristian, Allan Sandage, and James Westphal have extended the Hubble diagram out to red shifts of 75 percent of the speed of light (Fig. 7.16). At face value, the observations indicate the universe is closed. But if galaxies evolve rapidly with time, the results are consistent with a flat model. These astronomers state that "the case is not yet settled."

greater escape velocity, greater radius in a smaller one. Note that both greater mass and smaller radius mean a higher density. So for two objects of the same mass, the one with the higher density has the higher escape velocity (because it must have a smaller radius). Similarly, for two objects of the same radius, the one with the larger density has the higher escape velocity (because it must have a higher mass). So higher density generally means higher escape velocity.

Algebraically, the escape velocity formula is

$$V_e = \left[\frac{2GM}{R}\right]^{1/2}$$

where V_e is the escape velocity, M the mass of the object you want to escape from, and R its radius. (I've assumed the mass is spherical.) Note that the mass of the escaping object does not *enter into the calculation of escape velocity.*

Let's calculate, as an example, the earth's escape velocity. The mass of the earth is 6.0×10^{24} kg, its radius is 6.4×10^6 m, and $G = 6.7 \times 10^{-11}$ in mks units. So

$$V_e = \left[\frac{2 \times 6.7 \times 10^{-11} \times 6.0 \times 10^{24}}{6.4 \times 10^6}\right]^{1/2}$$

$$= (1.25 \times 10^8)^{1/2}$$

$$= 1.1 \times 10^4 \text{ m/sec}$$

$$= 11 \text{ km/sec}$$

Now consider throwing a ball off the earth's surface. If the ball's velocity is less than the escape velocity, the ball slows down and falls back, regardless of its mass. When thrown with a velocity greater than the escape velocity, the ball slows down a bit while it is still influenced strongly by the earth's gravity and then coasts out toward infinity.

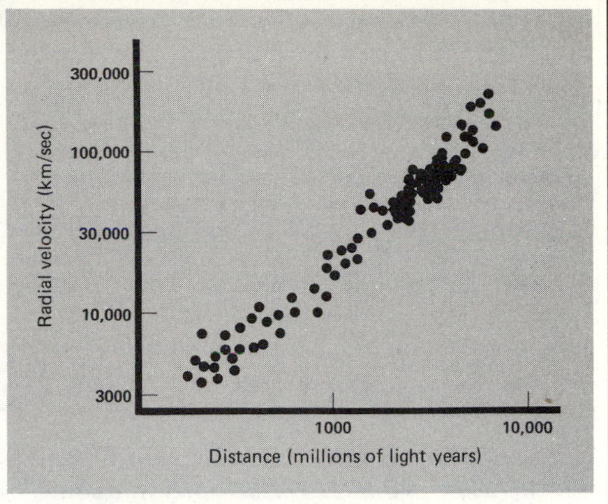

7.16

A Hubble diagram for galaxies with radial velocities as large as 75 percent of the speed of light. (From J. Kristian, A. Sandage, and J. Westphal, Astrophysical Journal, *vol. 221, p. 383, copyright © 1978 by the American Astronomical Society)*

KEY CONCEPTS

1 *Newton's concept of natural motion (and so of forces) rested on the assumption that the geometry of the universe was Euclidean (flat); Einstein, in developing his theory of gravitation—the general theory of relativity—challenged this assumption.*

2 *Before developing the general theory, Einstein worked out the special theory of relativity, which relates to the laws of physics as seen by observers moving at constant velocities (no acceleration); consequences of the special theory are the unification of space and time (in four-dimensional spacetime) and the intimate relationship between matter and energy.*

3 *The general theory deals with gravitation; it is based on the* principle of equivalence, *which states that accelerations due to gravity cannot be distinguished from accelerations from other kinds of forces; the principle of equivalence implies that in free-fall, when weightless, gravity disappears; free-falling is then viewed by Einstein as natural motion; gravity appears when the spacetime in which free-fall takes place is curved rather than flat.*

4 *Einstein's concept of gravity as the curvature of spacetime has been tested and found to predict accurately effects in the solar system, such as the bending of starlight near the sun; mass and energy cause the warping of spacetime.*

5 *Applied to the cosmos, the general theory allows it to have any one of three general geometric properties: hyperbolic, spherical (closed), and flat (Table 7.1); which one actually applies can be determined by observations.*

6 *Astronomers have found that the galaxies are moving away from each other in a systematic way: The farther away a galaxy is, the faster it recedes; this relationship is called* Hubble's law; *it implies that the cosmos is expanding; the rate of expansion is de-*

TABLE 7.1 *Geometry of the universe and observations*

	Hyperbolic	*Spherical*	*Flat*
Extent in space	Infinite	Finite	Infinite
Extent in time	Infinite	Finite	Infinite
Bounded or unbounded	Unbounded	Bounded	Unbounded
Average density (H = 20 km/sec/Mly)*	Less than 5×10^{-27} kg/m^3	More than 5×10^{-27} kg/m^3	Equals 5×10^{-27} kg/m^3
Hubble plot	Bends slightly upward	Bends strongly upward	Bends upward
Age to date (H = 20 km/sec/Mly)	More than 15 billion years	Less than 15 billion years	Equals 15 billion years
Future	Expansion forever	Expansion stops, collapse	Expansion forever

* Observed average density is about 10^{-28} kg/m^3

scribed by* Hubble's constant, *which has a value between 15 and 27 km/sec/Mly; the value is the same measured in different directions in space, which implies that the expansion is uniform; the beginning of this expansion was a cosmic explosion called the Big Bang.

7 ***If the geometry of the cosmos is hyperbolic or flat, the universe will expand forever; if closed, the expansion stops at some time in the future and then the universe contracts; to find out which is the case, compare the critical density (derived from Hubble's constant and general relativity) to the actual average density of the cosmos; if the actual is greater than the critical, the universe is closed; if less than the critical, it is open.***

8 ***The geometry of the cosmos can also be inferred from the amount of bending of the distance-velocity graph for very distant galaxies; most such tests imply that the cosmos is open; the issue is still widely debated.***

9 ***The geometry of the universe determines its future; the actual density of the universe relative to the critical density (determined from Hubble's constant and general relativity) indicates the overall geometry; if open or flat, the universe will expand forever, whereas a closed universe will eventually stop expanding and then contract.***

STUDY EXERCISES

1 Suppose you hear on the radio that astronomers have discovered a galaxy 500 Mly away. Use Hubble's law to estimate how fast the galaxy is receding. (*Objectives 6 and 7*)

2 For *H* equal to 20 km/sec/Mly, at what distance would a galaxy's recessional velocity equal 150,000 km/sec, one-half the speed of light? (*Objectives 6 and 7*)

3 Describe standing on the earth in Newton's terms and in Einstein's. (*Objectives 1, 2, and 3*)

4 Describe orbiting the earth in Newton's terms and contrast this to Einstein's description. (*Objectives 1, 2, and 3*)

5 In Einstein's model, how is the geometry of the universe related to its average density? (*Objective 9*)

6 Suppose you were an astronomer in the Andromeda galaxy with the same tools as an earthbound astronomer. You measure the recessional velocities from galaxies and your distance to them. What result would you expect? (*Objectives 6 and 7*)

7 Compare physically the case of the universe expanding and then contracting (a closed geometry) to the situation of firing a rocket off the earth with less than escape velocity. (*Objective 10*)

8 In what sense is a telescope a time machine? (*Objective 4*)

BEYOND THIS BOOK . . .

For a little help on the basics of Einstein's ideas, try *Relativity* (Crown, New York, 1961) by A. Einstein, *The Meaning of Relativity* (Princeton University Press, Princeton, N. J., 1956) also by Einstein, *Relativity and Common Sense* (Doubleday, Garden City, New York, 1964) by H. Bondi, and *Albert Einstein: Creator and Rebel* (Viking, N. Y., 1972) by B. Hoffman in collaboration with H. Dukas. The best book I've found on the nonscientific side of Einstein is R. Clark's *Einstein: The Life and Times* (World, New York, 1971).

Having trouble in picturing four dimensions? For a satirical analogy in fewer dimensions, read *Flatland* (Dover, New York, 1952) by E. A. Abbott.

For a witty exposition of Einstein's ideas with an emphasis on time, read *Space and Time in the Modern Universe* by P. C. W. Davies (Cambridge University Press, Cambridge, 1977)

The Red Limit (Bantam Books, New York, 1977) by T. Ferris gives good insights into the development of modern cosmology.

A good recent exposition of relativity and cosmology is *The Big Bang* by Joseph Silk (Freeman, San Francisco, 1980).

Epilogue part 1

I have guided you on a long trip though space and time. I hope that you have found it as fascinating as I do. I am amazed at how cosmological ideas grow from simple observations of the sky to working models of the universe.

What distinguishes contemporary concepts of the cosmos from those of the past? Basically, the demand that conceptual models both explain and predict what is observed. Such was not the case in some ancient cosmologies, such as the Babylonian. Prediction took place, but no explanations were advanced in terms of geometric or physical ideas. The Greeks originated the concept of model-making to try to understand the workings of the world. Ptolemy's geocentric picture represents the first crack at a complete scheme intended to explain and predict planetary motions. But Ptolemy did not believe that his model corresponded to the real world, that the planets really moved on epicycles and deferents.

Starting with the work of Copernicus, astronomers began to devise models that corresponded closely to reality, as given by observations. Kepler believed that a force from the sun actually pushed the planets in their orbits; he made the first try at a physical unification of the cosmos. Newton achieved that goal with gravitation and his laws of motion. But Newton would talk only about the effects of gravitation, not its cause. Einstein took the deeper step in the general theory of relativity. He envisioned gravitation as a manifestation of the curved geometry of spacetime, rather than a force. In so doing, he rediscovered the importance of geometry in the physical world—an attitude first taken by the Greeks. So Einstein's ideas were truly radical and transformed markedly our model of the cosmos.

That model relies on the work of modern astronomers with large telescopes. We have discovered that the universe contains innumerable galaxies, each holding billions of stars. These galaxies are moving away from us and each other: The universe is expanding. By applying Einstein's theory of relativity to the expanding universe, we can consider its past and ponder its future.

Scientific models form the core of contemporary scientific thought and will by used throughout this book. When you run across such a model, be sure that you can state (1) the aesthetic, geometric, and physical bases of the model, (2) the assumptions behind it and their reasonableness, (3) the key observations the model tries to explain and how well it does so, and (4) how to make predictions from the model.

Finally, remember that no scientific model is ever final. It is always subject to revision or complete replacement. This fact underlies the evolution of models, as you've seen in this part with our conceptions of the cosmos.

part 2

As a teenager, I built my own telescopes and used them first to observe the moon and the planets. I was especially fascinated by the detailed views of these other worlds. Later, the space programs of the United States and the Soviet Union conveyed to earth their amazing results. I watched in awe as the TV showed the pitted moon from a spacecraft plunging into it. And I recall myself and fellow students, beer in hand, fixed to the fuzzy images of Neil Armstrong stepping gently onto the lunar surface. Flyby and robot-lander missions, though, easily outstripped these first moments of men in space. Their fine photos provided a new vision of our planetary neighbors.

Part 2 focuses on the members of the solar system. I start off with the earth, the planet we know best. Our world provides the model and method to examine the other earthlike planets, including our companion, the moon. Then the focus shifts to the alien worlds, the Jupiterlike planets. These bodies will be compared in an evolutionary context, aided by the recent information from space missions. Then a brief look at the debris that floats among the planets: comets, asteroids, and meteoroids. Finally, this part closes with current models of the formation of the solar system.

Everything we see now in the solar system has evolved since its birth. By delving into the present physical properties of all, we can infer what they might have been like in the past and what their futures might bring. Our familiarity with the earth colors our picture of the evolution of the other planets, a scenario outlined by basic physical and chemical ideas. The goal is to fit the facts together to infer how the whole solar system has evolved.

The modern model for the solar system's origin and evolution implies that planets may form as a natural result of starbirth. So many other solar systems may exist in the Milky Way Galaxy—worlds on which other creatures may also wonder about the puzzle of the cosmos around them.

The planets: Past and present

8/The earth: An evolving planet

We travel together, passengers on a frail spaceship, dependent on its vulnerable reserve of air and soil; all committed for our safety to its security and peace; preserved from annihilation only by the work and love we give our fragile craft.

ADLAI STEVENSON

Central question

What are the basic physical features of the earth, and how have they changed since our planet's formation?

Learning objectives

After studying this chapter, you should be able to:

1. **Describe one method for determining the earth's mass and density.**
2. **Sketch the interior structure of the earth, indicating the composition of each general region and argue that the earth's interior structure implies that it must have been molten at some time in the past.**
3. **Argue simply that the earth's core probably has a metallic composition.**
4. **State the estimated age of the earth and explain the method by which this age is inferred.**
5. **Describe the overall structure of the earth's magnetic field and present a possible model for its source.**
6. **Describe at least two ways in which the earth's atmosphere affects astronomical observations and two ways it affects the earth's surface environment.**
7. **Explain how the earth's atmosphere acts like an insulating blanket that keeps the earth's surface relatively warm.**
8. **Outline a possible model for the evolution of the earth's crust and interior.**
9. **Outline a possible model for the evolution of the earth's oceans.**
10. **Outline a possible model for the evolution of the earth's atmosphere.**
11. **Summarize the processes that affect the evolution of the earth's atmosphere, crust, and interior.**

How small the earth is! Really a tiny planet, whirling around one ordinary star. And no longer the center of the universe, as people believed for over 4000 years. But that change in cosmic position does not mean we should value the earth any less. For the earth is our delicate ship, protecting us on our dark passage through space (Fig. 8.1).

As our home, the earth is a perfect planet. It has just the right range of temperatures, just the necessary atmospheric composition, just the ideal amount of water to foster living creatures such as we. Yet this planet was not always such a comfortable abode. Its past environment was much different than it is now. Our earth has changed tremendously since its birth, and we are a part of that change. One of the primary scientific goals of examining the earth in fine detail is to infer the history of the earth and also mull over its future and our impact on its future.

This chapter looks at the physical makeup of the earth, the only planet we know for certain to be inhabited. We live here, so we know this planet in more detail than any other. Our present understanding of the earth indicates that it is a highly evolved planet; it has changed dramatically in its physical structure since its formation. We will use our home planet as the basis of comparison for understanding the makeup and evolution of the other earthlike planets.

8.1
The earth from space, taken from Apollo 11. (Courtesy NASA)

8.1 The mass and density of the solid earth

How to find out the earth's mass? Here's one way: Recall (Section 4.2) that Galileo found that bodies at the earth's surface have the same acceleration due to gravity, *g*. Newton's law of gravitation (Section 4.3) relates this acceleration to the earth's mass and radius and to the gravitational constant, *G*. So if we know *G*, the earth's radius, and *g*, we can figure out the earth's mass from Newton's law. The mass comes out to about 6×10^{24} kg.

Knowing both the earth's mass and its volume (from its radius), we can compute its average density. *Density* is the amount of mass contained in a certain volume; if you divide the earth's mass by its volume, you obtain a density of 5500 kg/m^3, or 5.5 times the density of water (which is 1000 kg/m^3). This average density indicates that the earth, in bulk, consists of a combination of rocky and metallic materials. (Most rocks have a density between 2000 and 3000 kg/m^3; iron has a density of 8000 kg/m^3.)

Be clear about the concept of density! It is a measure of how well matter is packed into a given space. Consider two suitcases, both the same size. Imagine filling one with small rocks and the other with peanuts, both packed as tightly as possible. Close and lift the suitcases. Which one has more mass? Right, the one filled with rocks—it has more mass in the same amount of space (volume). So the suitcase filled with rocks is denser than the one jammed with peanuts. Note that different materials in general have different densities. Iron is more dense than water; wood, less so. This fact can be tested by placing both in water. Wood floats, and iron sinks.

Rocks near the earth's surface average 2400 kg/m^3, about half the average density of the earth. This difference implies that the core of the earth must be denser than the average. Present estimates indicate that the earth's core is perhaps 12,000 kg/m^3. This high density implies that the core contains dense materials, such as iron (density, 7800 kg/m^3). The weight of the overlying layers compresses the core, causing the material there to have a higher than normal density.

8.2 The earth's interior

The interior of the earth falls into three distinct layers: the core, the mantle, and the crust (Fig. 8.2). The *core* makes up the central zone and extends more than halfway to the surface. It consists mainly of iron in an alloy form with a small amount of sulfur or oxygen. Because of the melting properties of this alloy, the inner core is solid but the outer core is molten.

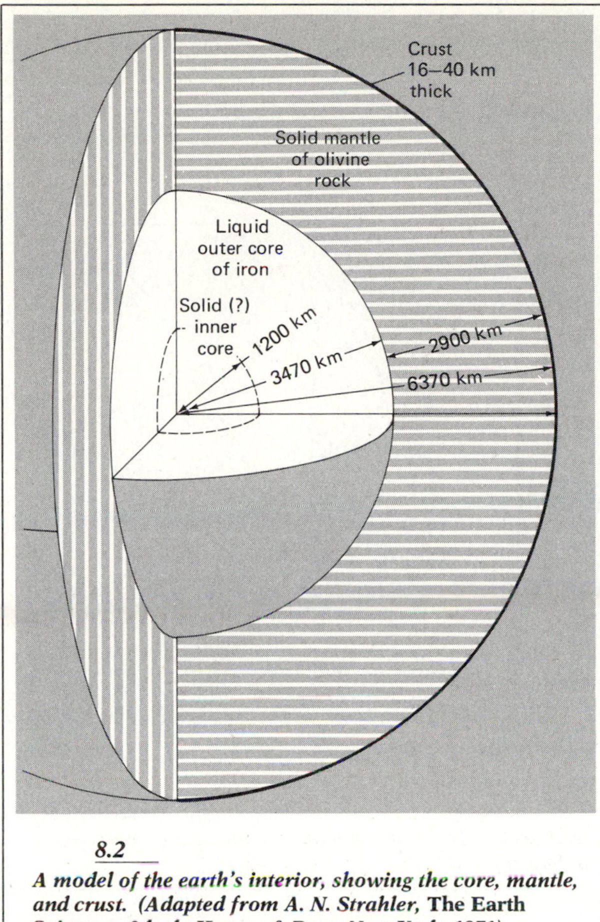

8.2 *A model of the earth's interior, showing the core, mantle, and crust. (Adapted from A. N. Strahler,* **The Earth Sciences,** *2d ed., Harper & Row, New York, 1971)*

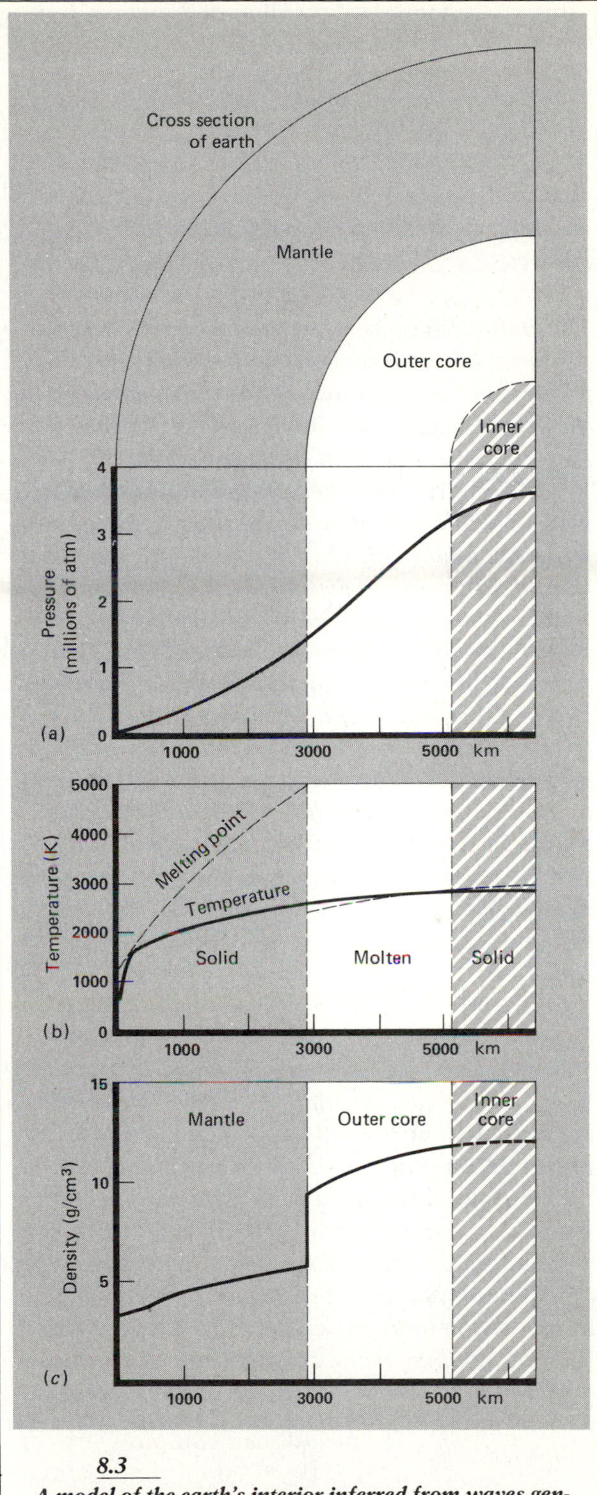

8.3 *A model of the earth's interior inferred from waves generated by earthquakes. The sharp increase in density* **(c)** *marks the transition between the mantle and the core. (Adapted from A. N. Strahler,* **The Earth Sciences,** *2d ed., Harper & Row, New York, 1971)*

Above the core extends the *mantle,* roughly 2900 km thick. The mantle material is rock made of iron and magnesium combined with silicon and oxygen (a silicate mineral called *olivine*). The mantle is plastic. Under slow, steady pressure such material flows like a liquid, but sudden changes in pressure make it snap and fragment like glass.

Encasing the mantle is the *crust,* the solid surface layer, which varies in depth from 16 to 40 km. Most of the crustal material consists of rocks that have solidified from molten lava (so they are called *igneous* rock). These rocks are *basalt,* a combination of oxygen, silicon, aluminum, magnesium, and iron. They comprise the ocean basins and the subcontinental sections of the crust. The continental masses are mostly *granite,* made of oxygen, silicon, aluminum, sodium, and potassium. Because the granite has a lower density than the basalt, the continental plates float on the basalt. Also, because the mantle is denser than the basalt and granite, the entire crust floats on the mantle.

You may wonder how we have any knowledge of the earth's interior, since we can't see into it. Geologists infer the structure by using the shock waves from earthquakes. These waves travel though the interior and are affected by it. So the waves convey clues to the interior's physical state (Fig. 8.3). Other clues come from the proper-

Focus 8.1

RADIOACTIVITY AND THE DATING OF ROCKS

The radioactive dating technique works because of the breakdown of the nuclei of radioactive elements, such as uranium. When these nuclei decay, they break apart into simpler nuclei. As they do, they release energy that heats the rock.

Given just one atom, you cannot estimate when it will decay because the process is random, but given a large number of atoms, you can determine a gross rate of disintegration. (An analogous random process is the popping of popcorn. It is impossible for you to predict which kernel will pop next, but you can estimate when the entire batch will be finished.) Half a piece of uranium-238 (^{238}U) decays to lead in 4.5 billion years, half again in the next 4.5 billion years, and so on. The length of time required for half the material to disintegrate is called the **half-life** *of the element (Fig. F.19). So you can calculate the amount of original uranium left at any time, even though the decay time for any one uranium atom cannot be specified.*

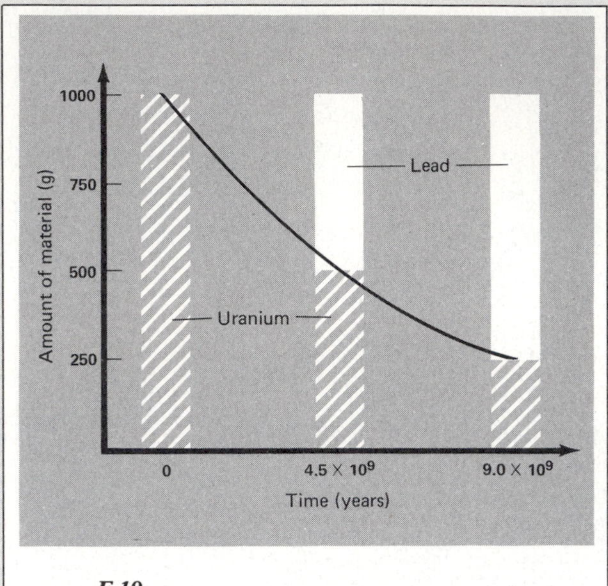

F.19
Radioactive decay of uranium-238, illustrating the concept of half-life.

By reversing this idea, you can estimate the age of a rock. Given a rock sample containing ^{238}U and lead and knowing the half-life of the uranium, we can calculate the age of the sample by noting how long it would take to form the observed amount of lead by radioactive decay.

Uranium is not the only element that can be used in radioactive dating: rubidium (^{87}Rb), which decays to strontium (^{87}Sr), with a half-life of 47 billion years; and potassium (^{40}K), which decays to the inert gas argon (^{40}Ar), with a half-life of 1.3 billion years, can also serve as radioactive clocks. Whatever elements are used, the derived age is the time elapsed since the rocks last solidified.

For example, suppose a rock sample now contains equal numbers of potassium-40 and argon-40 atoms. If there were no argon atoms in the rock originally, they must all have come from decay of potassium-40. Exactly half have decayed (and half remain), so the rock must be one half-life old, 1.3 billion years. How old is a rock that contains seven times as many argon-40 atoms as potassium-40 atoms? If all the argon came from decay of potassium, the remaining potassium-40 is one-eighth of the original. So three half-life periods must have elapsed ($1/8 = 1/2 \times 1/2 \times 1/2$), and the rock must be $3 \times 1.3 = 3.9$ billion years old. Note that the age inferred this way is the time since the rock last **solidified**. *If a rock forms, melts, then resolidifies, the age will be less, since some of the argon gas escapes during melting.*

ties of rocky materials. For other earthlike planets, we can get a firm idea of their interiors only if we also have information from shock waves going through them.

To sum up: The earth's interior is *differentiated.* This means that it consists of layers with the least dense materials at the surface and the most dense at the center. Basically, the interior consists of two zones: one iron rich (the core), the other silicate rich (the mantle and crust).

How did the earth get this way? Present theories (see Chapter 12) see the earth as being formed such that its material was well mixed up. Imagine the interior then heating up enough so that it was mostly molten (or at least fluid). Then the denser materials would settle at the core, and the less dense materials (silicates) would form a froth on top.

What heated the interior? One source is the heat generated by radioactive decay (Focus 8.1). In the past, the earth had more radioactive material than now—about six times as much. The heating from the decay of so much radioactive material could melt the interior enough for the material to flow and separate into regions of different density.

8.3 *The earth's age*

Geologists estimate the earth's age at 4.6 billion years from radioactive dating (Focus 8.1), although the oldest known rocks on the earth's surface are not this old. The most ancient known rock pieces, found in western Australia, have been dated at 4.2 billion years. The oldest known whole rocks, from Western Greenland, are 4.0 billion years old. (The most ancient rocks at other locations are all about 3.8 billion years old.) These rocks are *igneous rocks*, which means that they solidified from one-time molten material. Geologists estimate that about ½ billion years elapsed for the crust to melt and then cool to form the first rocks. So the earth's age is that of the oldest rocks plus the time to form them.

This estimate falls close to that for meteorite material (4.55 billion years) and lunar material (4.6 billion years), determined by the same radioactive dating techniques. The near-coincidence of these ages implies that the solar system formed in one single event about 4.6 billion years ago (see Chapter 12).

The exact times given here are still being debated, but most scientists agree that the earth is 4.5 to 5 billion years old, with 4.6 billion as the best estimate.

8.4 *The earth's magnetic field*

You can visualize the earth's magnetic properties by imagining a giant bar magnet located in the core (Focus 8.2). The magnetic field protrudes from the *south magnetic pole* in the Southern Hemisphere and returns to the *north magnetic pole* in the Nothern Hemisphere. The magnetic axis, which connects the magnetic poles, tilts about 20° from the spin axis and does not pass through the earth's center (Fig. 8.4). The part of this magnetic field that is parallel to the earth's surface orients a compass needle so that it points to the north and south magnetic poles.

The earth's magnetic field changes with time in both direction and intensity. We have evidence that the magnetic poles have reversed polarity at least nine times in the past 3.5 million years and probably many more times in earlier ages (see Section 8.6).

The source of the earth's magnetic field is its metallic core, which is both liquid (in part) and conducting. It acts like a giant dynamo and electromagnet, generating electricity and creating a magnetic field. The earth's rotation supposedly helps to stir the currents in the core that whirl

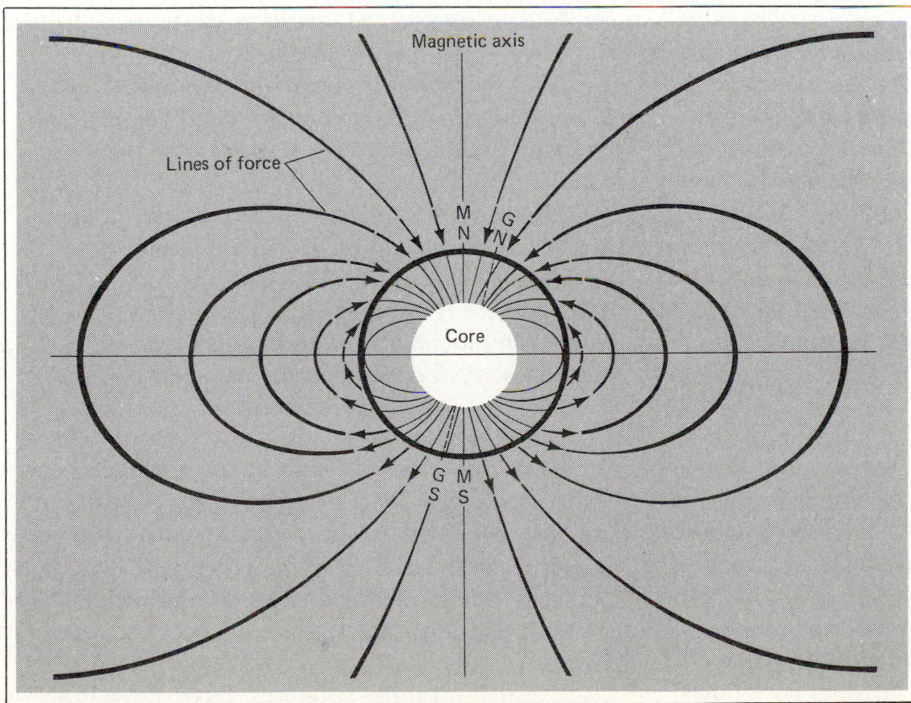

8.4
A model of the earth's magnetic field. Note that the magnetic axis (MN–MS) is not aligned with the spin axis (GN–GS) but is tilted about 20°. (Adapted from A. N. Strahler,* The Earth Sciences, *2d ed., Harper & Row, New York, 1971)

Focus 8.2

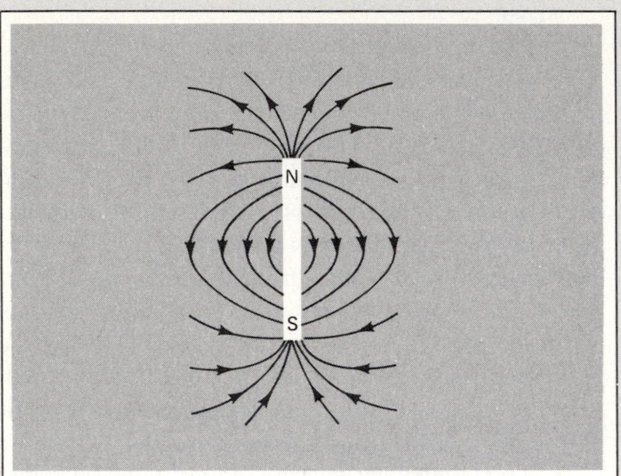

F.20
The magnetic field of a small bar magnet. Its field arises from the flow of electrons around the nuclei of iron atoms.

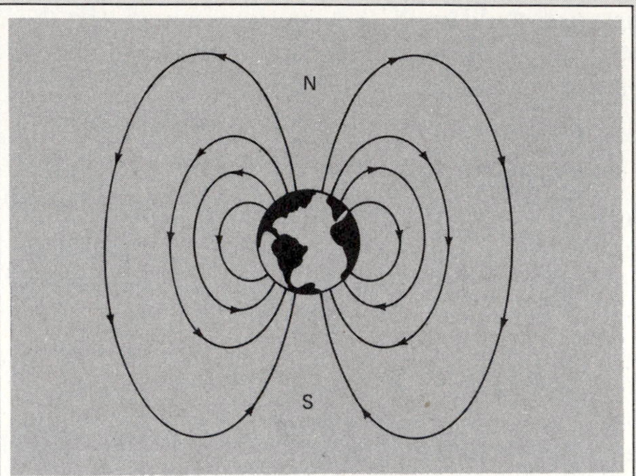

F.21
Overall magnetic field of the earth. Note that the shape of the field resembles that of a bar magnet.

MAGNETIC FIELDS AND FORCE

A field is another scientific model; it is a way of describing space that is somehow modified by the presence of matter. Magnetic fields are regions of space modified by electrical charges.

All magnetic fields come from electrical charges in motion. A bar magnet (Fig. F.20) has no outward sign of motion, but the circulation of electrons around iron nuclei sets up its magnetic field. You have probably seen an electromagnet in operation. An electric current, a flow of charged particles, passing through a loop of wire creates a magnetic field. The earth's magnetic field is probably generated by the internal circulation of charged particles. The magnetic fields around the earth and the sun have two poles, and so are called **dipole** *fields; this characteristic allows us to think of these fields as arising from a giant bar magnet buried in the earth (Fig. F.21), even though they really come from circulating electric currents.*

To help visualize a magnetic field, take a very small compass and move it around a magnet. The changing direction of the compass needle shows the direction of the magnetic lines of force. We usually draw the lines of force so that the spacing of the lines indicates the relative strength of the magnetic field: The closer the spacing, the stronger the field. Note that the farther away you are from the magnet, the weaker the field. In fact, far from a dipole, the field intensity drops off as $1/R^3$, an inverse-cube law. So a magnetic field decreases in intensity with distance from the source more rapidly than gravity.

Astronomers generally use a **gauss** *(abbreviated G) as the unit to measure magnetic-field strength. The strength of the earth's magnetic field at its surface is about ½ G.*

Charged particles and magnetic fields interact in such a way that the particles find it

around to generate current. This *dynamo model* for the earth's magnetic field, if correct, implies that any planet that exhibits a strong magnetic field must have a substantial fluid, conducting core and must rotate rapidly.

Warning: The dynamo model has yet to be worked out in detail. For instance, little agreement has developed to date on how the fluid core flows, what drives these motions, and how these flows generate the complex field at the surface. Recent work indicates that the flow may be driven by gravitational energy liberated as dense materials migrate to the center of the core and less dense materials flow outward. It also appears that the solid inner core affects the flow pattern in crucial ways.

Earth-orbiting satellites have detected two regions encircling the earth that contain a large

difficult to cross the field lines. If charged particles move parallel to the field lines, they feel no force. But if some part of their motion is perpendicular to the field lines, they will experience a force at right angles to the field lines and their motion. This force can result in spiraling paths along the field lines. The direction of the spiral twist depends on whether the particle is positively or negatively charged (Fig. F.22). So charged particles and magnetic fields are linked together by their interactions.

This linking is important for understanding what happens to a magnetic field that is immersed in an ionized gas (such as the sun). If the ionized gas moves, it carries the magnetic-field lines with it. For example, if the gas is moving turbulently, it tangles and jumbles up the direction of the magnetic-field lines.

To sum up: ***Moving charged particles produce magnetic fields. Magnetic fields, in turn, affect the motions of charged particles. The linking of magnetic fields and charged particles acts as a cosmic agitator to stir up matter and electromagnetic fields.***

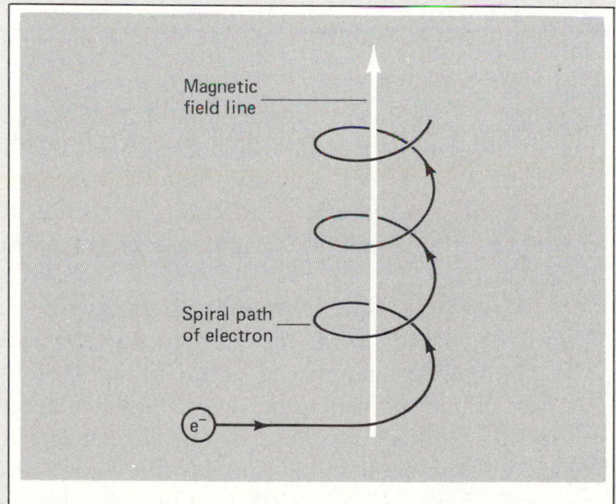

F.22
The path of an electron in a magnetic field. When an electron moves across a field line, a force acts on it and deflects it to a spiral path along the field line.

number of protons and electrons. These two doughnut-shaped belts of energetic particles trapped by the terrestrial magnetic fields are called the *Van Allen radiation belts* (Fig. 8.5), after their discoverer, James A. Van Allen. The sun blows off the particles that are trapped in the Van Allen belts.

The Van Allen belts mark one aspect of the interaction of the earth's magnetic field with charged particles that stream from the sun. In fact, the earth's magnetic field affects the flow of charged particles for many tens of earth radii out into space. This region is called the earth's *magnetosphere* (Fig. 8.6). As the particles from the sun run into the earth's field, they are forced along the field lines (Focus 8.2). Like the blunt bow of a boat in water, the earth's field deflects particles around it and leaves a wake in the direction opposite the sun. Most charged particles flow around the earth, but a few are caught in the field to make the Van Allen belts. When the particles dump out of the belts into the earth's atmosphere, they generate the *aurora*, a shimmering crown that makes visible the presence of the earth's magnetic field.

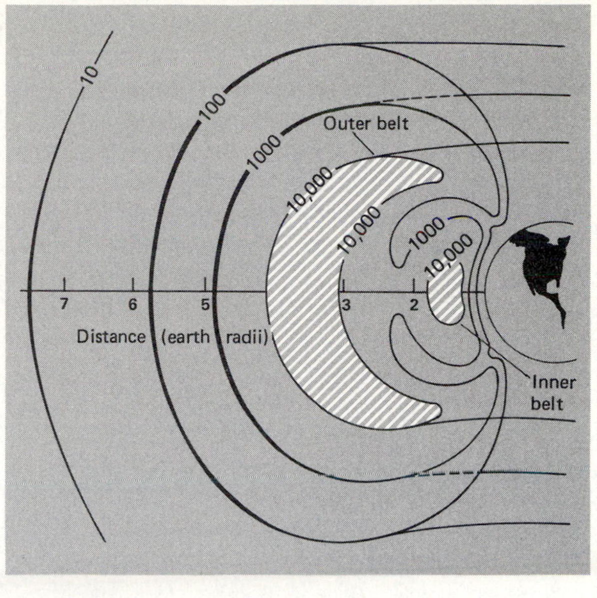

8.5
A model of the Van Allen radiation belts, which encircle the earth. The numbers on the contour lines indicate the average density of particles in a cubic centimeter. (Based on NASA data)

8.5
The blanket of the atmosphere

Although thin compared to the bulk of the solid earth, our atmosphere is a blessing. It provides oxygen for breathing, shields out the cancer-causing ultraviolet radiation of the sun, furnishes a thermal blanket to keep the surface warm, and spreads the heat around the earth. Understanding the earth's atmosphere provides information useful for the study of other planetary atmospheres in the solar system.

Relative to the total number of atoms and molecules available, our atmosphere contains approximately 79 percent molecular nitrogen (N_2), 20 percent molecular oxygen (O_2), 1 percent argon (Ar), 0.03 percent carbon dioxide (CO_2), and traces

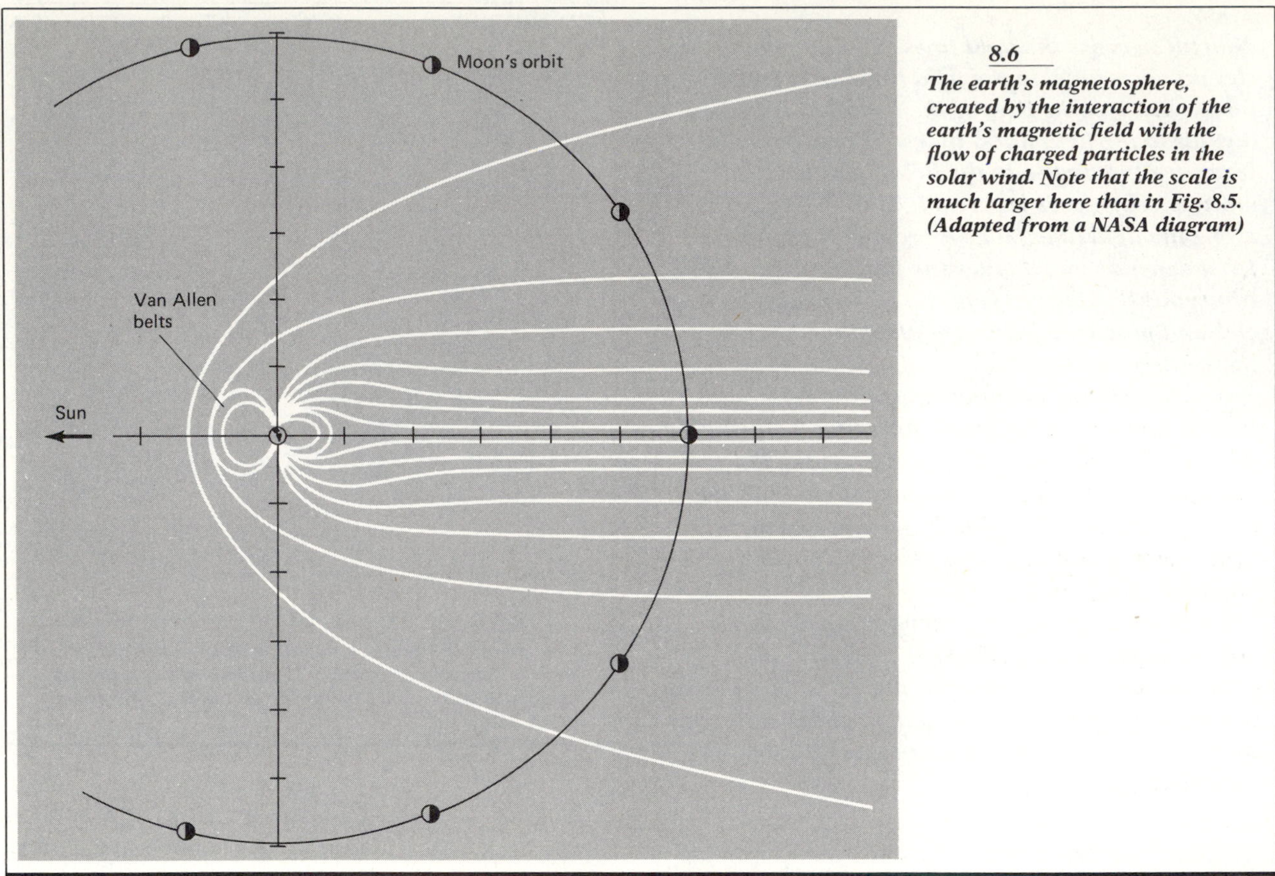

8.6
The earth's magnetosphere, created by the interaction of the earth's magnetic field with the flow of charged particles in the solar wind. Note that the scale is much larger here than in Fig. 8.5. (Adapted from a NASA diagram)

of other elements. Water vapor (H_2O) is present in variable amounts, sometimes as much as 4 percent near the surface.

The weight of the upper atmospheric layers makes the lower portion denser than the upper, just as a sandwich at the bottom of a pile is squashed by the weight of the sandwiches above it. The gas pressure at sea level on the earth's surface, where the entire atmosphere is piled above it, is called *one atmosphere* (1 atm) of pressure. (The *pressure* of a gas is the force it exerts on an area of surface.) The atmospheric pressure and density decrease with height, rapidly at first, then more slowly.

Seeing / Even on the clearest nights, atmospheric turbulence makes telescopic images flicker and so imposes the fundamental limitation on a telescope's effective resolving power (see Chapter 6). The extent to which the atmospheric turbulence affects the image is termed *seeing*. When the seeing is very good, stellar images are sharp, steady pinpoints, about 1″ in diameter. At times of bad seeing, the images waver like candle flames in a gentle breeze.

Extinction and reddening / The atmosphere absorbs and scatters some of the light that penetrates it. This reduction of light is called *atmospheric extinction*. The closer an object appears to the horizon, the greater the atmospheric thickness through which the object's light must pass, and so the dimmer it becomes. Because of atmospheric extinction, the rising full moon has about half the brightness of the same moon overhead.

Why is the sky blue? Air molecules scatter blue light (mostly from the sun) more than red. The atmosphere depletes a beam of light of its shorter (bluer) wavelengths, which are scattered uniformly through the sky. In any direction you look, you see blue light, and so the entire sky appears blue (Fig. 8.7). Light of longer wavelengths reaches you directly along the line of sight. The sinking sun appears a burning red because its radiation passes through a lot of atmosphere before reaching you. Along this path, most of the blue light scatters out, leaving mostly red light to strike your eye.

Albedo / All solid bodies shine by reflected sunlight, and so does the earth when viewed from space. A celestial body's reflecting ability is called its *albedo*, the ratio of the light reflected to the incoming light. If an object reflected all the light that struck its surface, its albedo would be 1. (If it absorbed it all, the albedo would be 0.0.) The clouds in the earth's atmosphere help to reflect visible light, and about 35 percent of the incident light reflects back into space; the earth's albedo equals 0.35. The atmosphere and surface absorb the other 65 percent.

The greenhouse effect / Of the incoming sunlight not reflected, 15 percent is absorbed in the lower atmosphere, and the remaining 50 percent of the original strikes the ground and heats it and the air in contact with the surface. However, if this direct solar radiation were the only source of heat, the temperature at the ground would be a frigid 253 K (or −20°C). Water would always be frozen! The average temperature at the surface is actually much higher, about 293 K (+20°C).

How does this heating happen? Visible light from the sun gets through to the surface and heats the earth. The earth in turn emits infrared radiation. If this infrared radiation simply escaped into space, the earth would be too cold for life. But this radiation doesn't escape completely. The infrared is absorbed by the earth's atmosphere (mostly by water vapor and carbon dioxide). The atmosphere heats up by absorbing this radiation. Some goes off into space (about 8 percent); the rest radiates back to the ground and heats it. Both direct sunlight and infrared radiation from the atmosphere heat the earth's surface. The atmosphere acts like a blanket, insulating the ground from space and so helping to warm the earth.

If you have ever visited a high region with an arid climate, such as New Mexico, you know what a dramatic effect water vapor in the atmosphere has on the ground temperature. Water vapor absorbs infrared radiation, so the more humid the atmosphere, the more opaque it is to infrared and the better it insulates. For instance, in Albuquerque on a clear winter day, the high temperature can typically reach 15°C and, if the night is also clear, drop to −10°C at night. But if it's cloudy at night, the low may be only 0°C or so. Why the difference? At night the ground radiates away, in the infrared, the energy it absorbed during the day. On a cloudy, high-humidity night, the additional water vapor in the air traps the outgoing infrared radiation from the ground more than on a clear night. So the air temperature stays higher because not as much heat escapes to space.

This warming of the ground by the atmospheric trapping of infrared radiation is often called the *greenhouse effect* by analogy to one process that keeps a greenhouse warm. Glass is transparent to visible light but opaque to infrared. So sunlight enters the greenhouse and warms the interior, which emits infrared. This heat can't radiate through the glass, so it stays to help warm up the interior.

To sum up: Any planetary atmosphere that is more or less transparent to sunlight but opaque to infrared will act to keep the planet's surface warmer than if the planet had no atmosphere.

Note: The greenhouse effect is probably misnamed. Experiments have shown that the absorption of infrared radiation is not the main process that keeps a greenhouse warm. The air inside, heated by contact with the hot inside of the

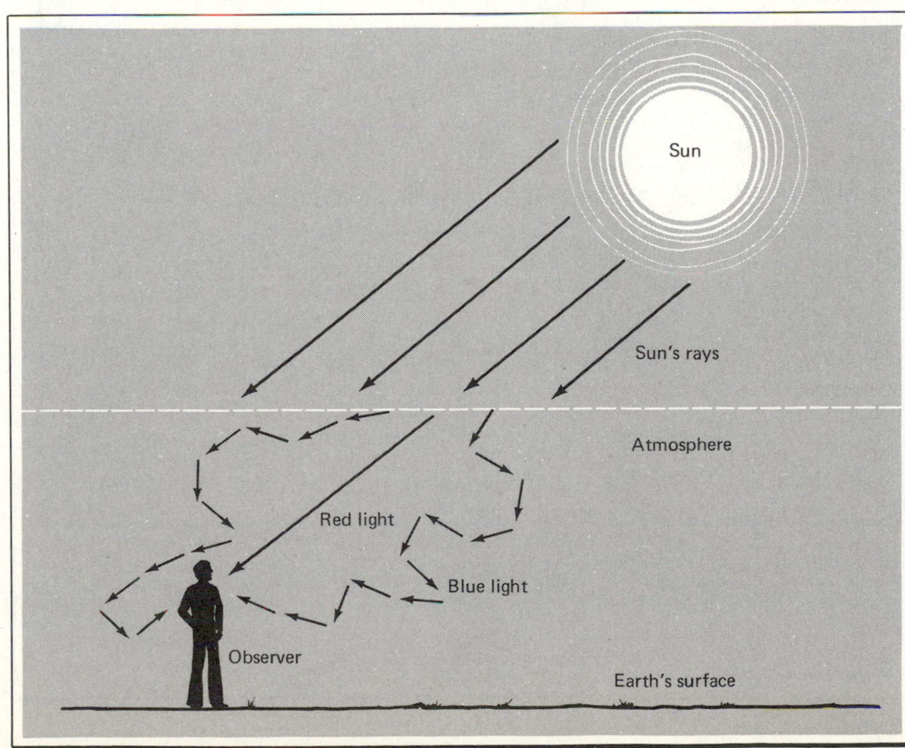

8.7

The blue sky. Air molecules let red light pass through relatively unhindered, but blue light is scattered in all directions. Looking toward the sun, you see red light directly. Since blue light scatters all around the air, in every direction you see blue light, hence a blue sky.

greenhouse, cannot escape. This inhibition of convective cooling occurs with any roof. However, the expression "greenhouse effect" is so ingrained in the astronomers' vocabulary, I'll continue to use it in this book.

Ozone layer / About 20 to 30 km up, the oxygen in the atmosphere strongly absorbs ultraviolet light from the sun. Ozone results from the combination of three oxygen atoms to form an ozone molecule O_3, and an *ozone layer* forms.

The ozone layer blocks out ultraviolet radiation from space. Some life forms, such as human beings, have developed on the earth in an environment sheltered from the ultraviolet light; they are so susceptible to this fairly lethal radiation that even a mild exposure results in a painful case of sunburn. A large dose can kill. Even relatively mild doses over a long time promote skin cancer in fair-skinned people. For example, the Anglo residents of New Mexico, which has clear skies and a high elevation, have one of the highest per capita skin cancer rates in the world.

8.6 The restless earth: Evolution of the crust

The unfolding of the earth's history seems rather uneventful. A fall of rain, a burst of wind, the transport of grains in a stream—these erosive processes serve to flatten the earth's surface relentlessly and to wash materials into the sea. Gently but inevitably the mountains are leveled to plains.

If so, why does the earth still have continents and ocean basins between the continents? The earth's surface divides roughly into two levels: the continents and the ocean basins, with an average height difference of about 5 km. Erosion and water transport of materials in a relatively short time (a few million years) should generally erase the difference in height between the oceanic and continental plains. Given the great age of the earth (see Section 8.3), the fact that these levels still stand apart implies that somehow the mountain

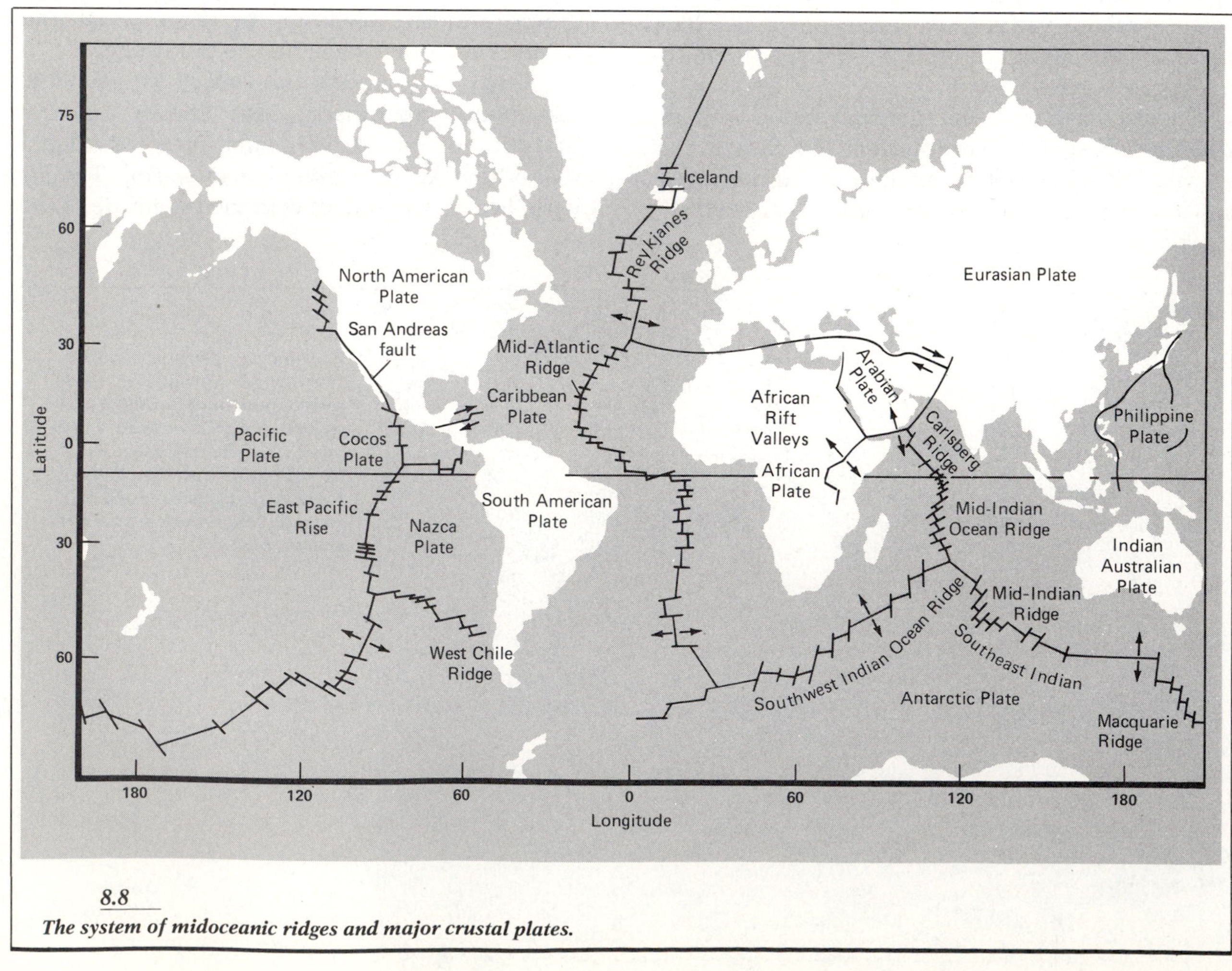

8.8
The system of midoceanic ridges and major crustal plates.

heights and ocean depths are regularly replenished.

Earthquakes and mountains / In the nineteenth century, geologists noted that zones of active volcanoes and frequent earthquakes are concentrated along the chains of young mountain ranges and submarine ridges. They argued that earthquakes and volcanic activity must be associated with mountain and island building. Modern research on the locations of earthquakes backs up this idea.

Another clue to the understanding of this activity shows up in the ocean basins, especially in the Atlantic. Modern sonar measurements have revealed a *midoceanic ridge*, an almost continuous submarine mountain chain that twists some 64,000 km through the ocean basins (Fig. 8.8). The midoceanic ridge indicates that crustal evolution takes place in the ocean basins.

Continental drift / These facts fell together as a coherent picture during the 1960s with the revival of the model of *continental drift*, the idea that the present continents were at one time a unified landmass that fragmented and drifted apart. In 1910 Alfred L. Wegener (1880–1930) of Germany suggested that displacements of the earth's crust could shift the positions of the continents. Wegener pointed to a number of remarkable geological connections, such as similar rock formations and fossils, between the lands on opposite sides of the Atlantic. In Wegener's picture, the continents were originally joined in one vast land area, which broke up about 200 million years ago (Fig. 8.9). Today's evidence points to two primordial landmasses, one in the Southern Hemisphere, the other in the Northern. These may have broken from a single, supercontinent. Note that the face of the earth looked quite different only 100 million years ago, when the present continental masses were crowded together. (Prior to the breakup 200 million years ago, the continents moved around in unknown configurations.)

Evidence from the magnetic characteristics of the ocean floors near ridges supports the continental-drift model. If the continents do move apart, the sea-floor between them must be spreading. Oceanographic cruises across the Atlantic have found that the sea-floor material contains remnants of ancient magnetism. When lava solidifies to form igneous rock, the iron minerals in the rock align with the earth's magnetic field. The directions and reversals of the direction of the earth's magnetic field in the past are preserved in the rock. They have a startling pattern. On both sides of the mid-Atlantic ridge, the reversal patterns appear identical—each side is a mirror reflection of the other. How did this come about? If the sea floor spreads, it needs a continuous supply of new material to add additional area. Lava flowing out pushes older material aside in both directions. When the lava solidifies, the rock on either side freezes in the magnetic field alignment of the time. So the sea-floor rocks act as a magnetic tape (a very slow-moving one!) that preserves the record of the past changes in the earth's magnetic field.

The alignment of magnetic-field reversals indicates that new material emerges from a rift in the center of the ridge and gives the rate of expansion of the sea-floor. The movement is about 3 cm per year at its fastest speed across the mid-Atlantic ridge. This rate amounts to enough to push apart the Old World and the New in 400 million years.

Plate tectonics / What accounts for the renewal of the ocean plains? It appears to be new material oozing out from the earth's interior. The separate continental plates float like large rafts on the basaltic basin material, which forms another plate. Where one continental plate crashes into another, the impact raises up mountains. In some regions, one plate may force another to fold under and descend into the mantle (Fig. 8.10). The plate's descent eliminates surface material essentially at the same rate it is created, so the earth's radius does not expand to accommodate the swelling plates. Because the plates' creation and destruction zones make natural fault areas, earthquakes and volcanoes predominate along the lines of plate collision.

This contemporary model of the earth's crustal activity and evolution—sea-floor spreading and the creation and destruction of the crust—is called *plate tectonics*, the dynamics of the crust's continental and oceanic plates.

What moves these plates? That's not completely clear. One popular model pictures the upper part of the mantle as divided into large whirls (Fig. 8.11) of flowing rock in the upper 700 km. The mantle's plasticity allows a slow flow upward, horizontally, and downward. At the region of horizontal flow, friction between the plate and the mantle drags the plate along with the mantle's flow. The upswelling magma supplies new materials to the plate. The energy source for such convection comes from heating by radioactive decay or from internal circulation persisting from the time when the earth's core formed.

Although many details are uncertain, the main point is clear: The earth's crust has evolved since its formation and is changing right now. When investigating other terrestrial planets, you

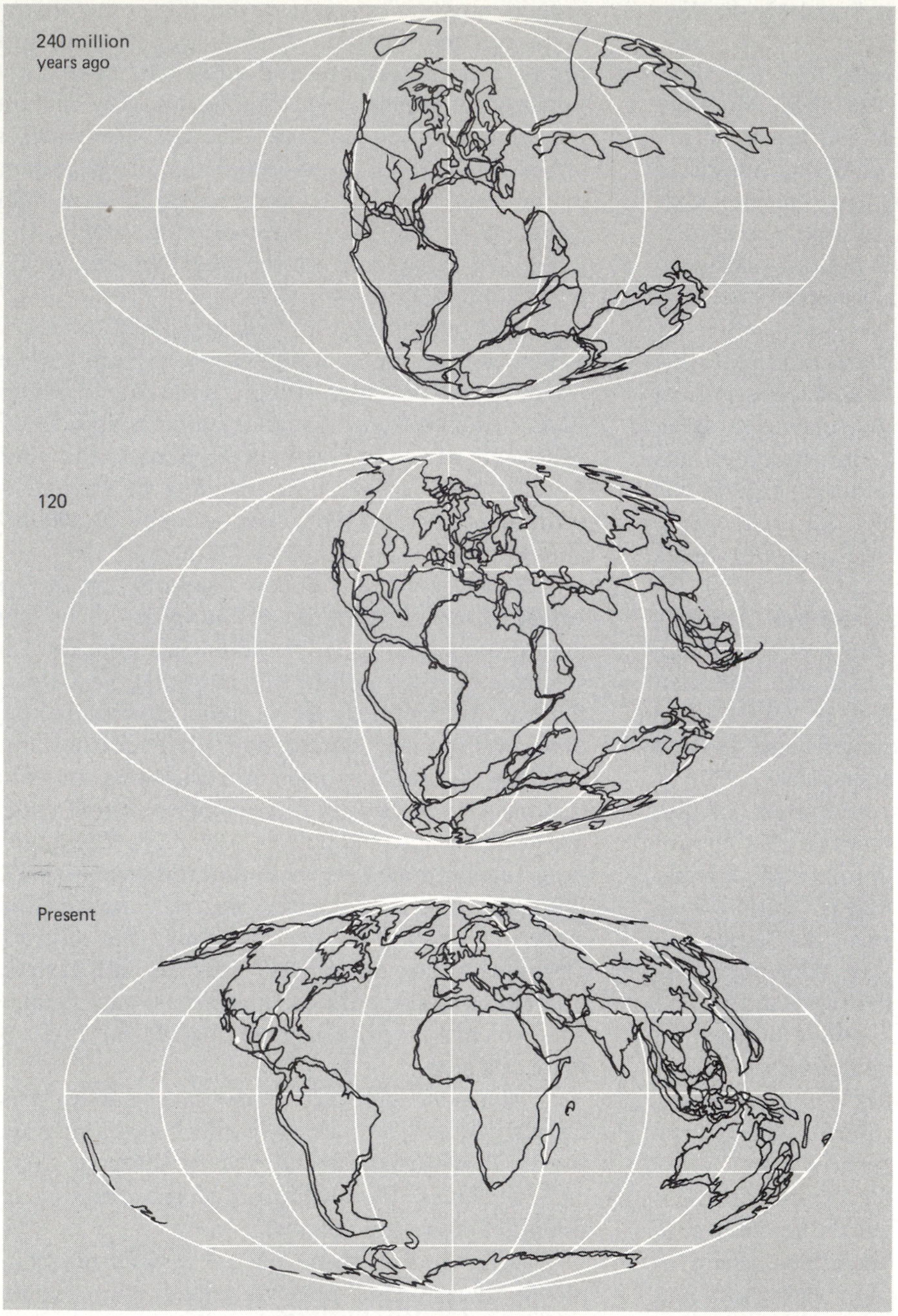

240 million years ago

120

Present

8.9

Computer-generated models of continental drift from 240 million years ago to the present. The positioning is based on magnetic evidence. (Adapted from maps prepared by A. M. Ziegler and C. S. Scotese of the University of Chicago's Paleographic Atlas

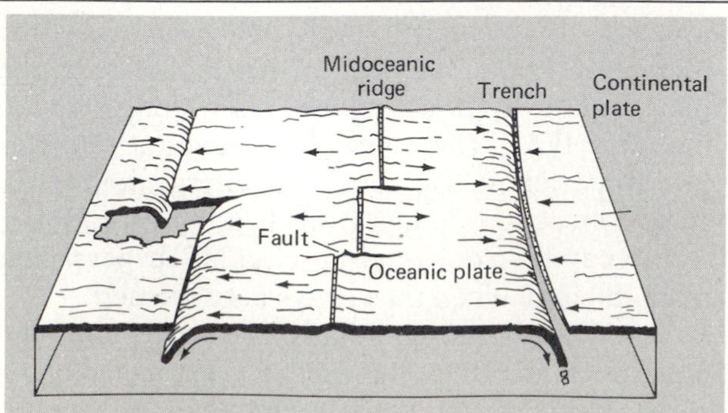

8.10

Interactions of oceanic and continental plates. The oceanic plates gain material from the outflow at oceanic ridges. As these plates expand, they crash into continental plates, provoking mountain building and earthquakes. (Adapted from A. N. Strahler,* The Earth Sciences, *2d ed., Harper & Row, New York, 1971)

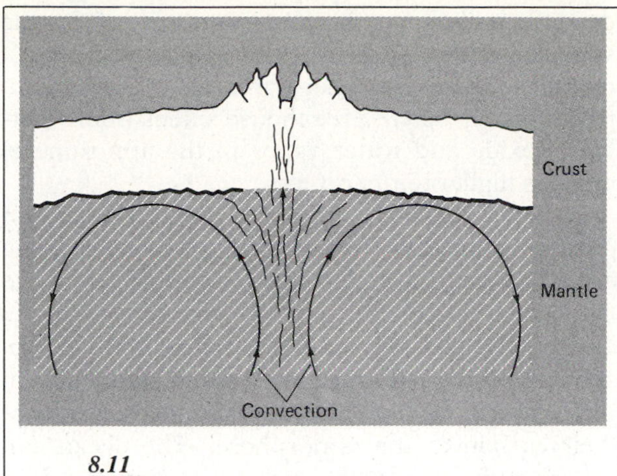

8.11

A convection model for the moving of crustal plates. Large convection currents in the plastic mantle carry along the plates that float on it.

should look for evidence of their crustal evolution. Note that the earth's ocean basins, because they have been recently formed, are the youngest parts of the earth's crust, while the center of continents are the oldest.

8.7 Evolution of the atmosphere and oceans

No other planet in the solar system has the earth's combination of an extensive atmosphere plus oceans of water. The earth's favorable environment for life results from the fact that the atmosphere and oceans transfer solar heat around the globe. This fine thermostat has been working for hundreds of millions of years. However, the atmosphere and oceans have changed in time, influenced by and influencing the earth's biological evolution. The future of life on the earth depends critically on the future of its fluid system.

Origin and development of the oceans / By the contemporary model of the formation of the planets of the solar system (see Chapter 12), the earth had no oceans when it formed 4.6 billion years ago. The primeval surface may have been fairly hot, about a few thousand Kelvins—certainly too hot for water to be liquid! When the surface had cooled to about 373 K (100°C), so that water could condense, perhaps one or two continents existed on the surface, and the rest of the surface comprised the initial ocean basin. This large tub contained very little water then, only a few percent of the present volume.

The rest of the oceanic water came from the earth's interior. When magma breaks through the crust, it carries a variety of gases, such as carbon dioxide, and also a large amount of water vapor. The steam arises from water trapped in the solid earth when the planet formed—water that has never before seen the light of day. The present rate of water production gassing out from the interior (a rate of about 10^{11} kg per year), if it has been constant for 4.5 billion years, accounts for the amount of water in the oceans today. That volume increases slowly. In fact, in the past, when the earth was hotter, the outgassing may have been much faster.

The earth's oceans now cover 71 percent of its surface with an average depth of 4 km. Together they contain some 10^{21} kg of material, mostly water. These figures have been roughly the same over the past billion years. But the configuration of the oceans must have been very different in the past because of plate tectonics.

The evolution of the atmosphere / The outgassing from the earth's interior that created the oceans also influenced the development of the earth's atmosphere. In fact, the earth's atmosphere evolved from the chemical interplay of the solid and fluid earth. Let's look at its interaction with the oceans. These interactions must be understood to work out the evolution of the atmosphere. We do *not* yet clearly understand all the details. So any discussion of the evolution of the earth's atmosphere must be taken as a working model, not the definitive one.

Here is one such reasonable model of atmospheric evolution. If our ideas of planetary formation are correct, the earth's first atmosphere did not resemble the present one at all. It may have contained mostly hydrogen and helium. But these gases, because they are so light, escaped from the earth into space. Our second atmosphere arose mostly from outgassings from the solid earth. Active volcanoes, for instance, now spew out carbon dioxide, sulfur dioxide, hydrogen, nitrogen, water, methane, and ammonia—gases that were trapped in the earth when it formed. (In addition, some gases, such as helium and argon, come from the decay of radioactive materials.) The outgassed materials (which are still entering the atmosphere) interact with the oceans, surface materials, and biomass in complex ways. For example, carbon dioxide is now added to the atmosphere by volcanoes, organic decay, and the combustion of fossil fuels. Carbon dioxide is taken out by plants and is being dissolved in the oceans, where much of it eventually ends up in rocks. The balance, however, has changed with time, so the atmospheric composition has evolved.

Michael H. Hart has made a computer simulation of how this evolution might have gone (Fig.

8.12). His models indicate that the earth's second atmosphere started out with a large amount of carbon dioxide. (The water vapor had quickly rained down to end up in oceans.) The carbon dioxide soon ended up in the oceans and rocks, so 3 billion years ago the atmosphere consisted mostly of methane and other hydrogen-carbon compounds.

At this time the atmosphere contained little free oxygen, so the earth had no ozone layer. Ultraviolet light readily penetrated and broke up methane, ammonia, and water. The hydrogen from these molecules fled into space. Some of the oxygen freed from the water combined with some of the methane and gradually eliminated the carbon. The rest of the oxygen eventually created an ozone layer, cutting out ultraviolet from interacting with most of the atmosphere. Nitrogen became the dominant constituent of the atmosphere.

The high abundance of atmospheric oxygen was produced (and is now maintained) by biological activity. Geologic evidence indicates that the transformation to an oxygen-rich atmosphere began roughly 2 billion years ago, when plant activity and photosynthesis bloomed (see Chapter 22). The increase in oxygen was probably a gradual, continuous one. About 1 billion years ago the atmosphere may have contained only 10 percent of the present amount of free oxygen. A large increase occurred about 600 million years ago. The oxygen content suddenly increased to present levels, along with a sudden proliferation of life.

The evolution of the earth's surface temperature / How hot it gets at the earth's surface depends on how much energy it receives from the sun and how effective the greenhouse effect operates. Less solar energy results in lower temperatures. A better greenhouse effect (more carbon dioxide and water vapor in the atmosphere) delivers higher temperatures.

People have disturbed the natural carbon dioxide balance by extracting fossil fuels from the earth, burning them for energy, and so adding to the carbon dioxide in the atmosphere. In addition, our destruction of forests has eliminated a substantial portion of the green plants that take in atmospheric carbon dioxide and added some of their carbon to the atmosphere. The ocean can absorb only a part of the excess.

Our activities have a net result of increasing the percentage of carbon dioxide in the earth's atmosphere. This increase is big enough to have been observed; observations at Mauna Loa Observatory in Hawaii indicate an increase of 5 percent over the past 30 years. Although there is still much controversy over what the impact of this increase will be, it may result in a temperature increase of about 2°C by 2020 A.D., if the overall water vapor and cloudiness do not change. This could have serious effects on climate and atmospheric circulation.

8.8 *The earth's evolution: An overview*

The earth is the most evolved of the planets. It's an active world now (powered by internal heat) and shows no sign of quitting. Geologic evidence implies it's been active since birth—4.6 billion

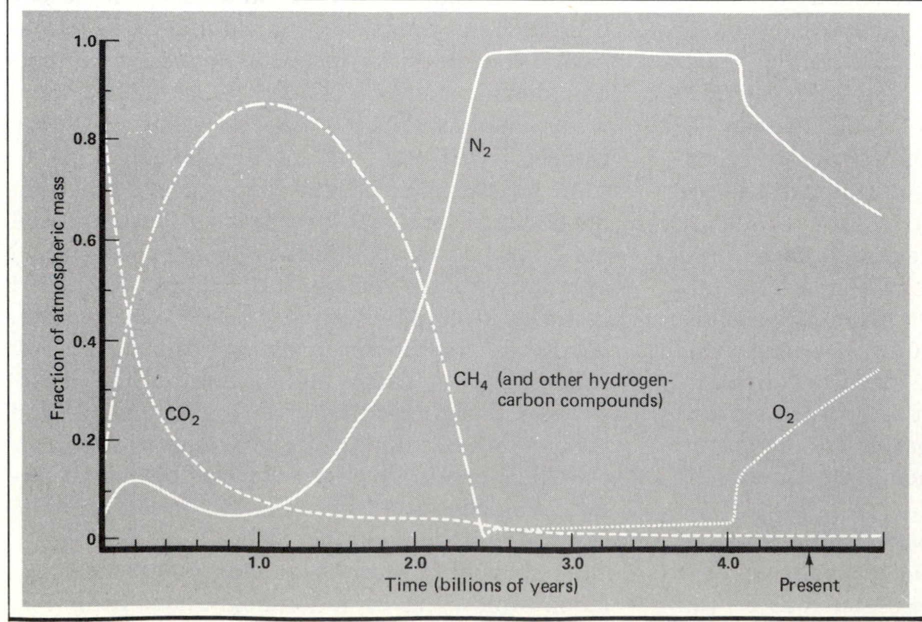

8.12
A model for the evolution of the earth's atmosphere. This graph shows, for a span of 5 billion years, the fraction of the earth's atmosphere in the form of carbon dioxide, molecular nitrogen, molecular oxygen, and methane. (Based on theoretical calculations by M. Hart)

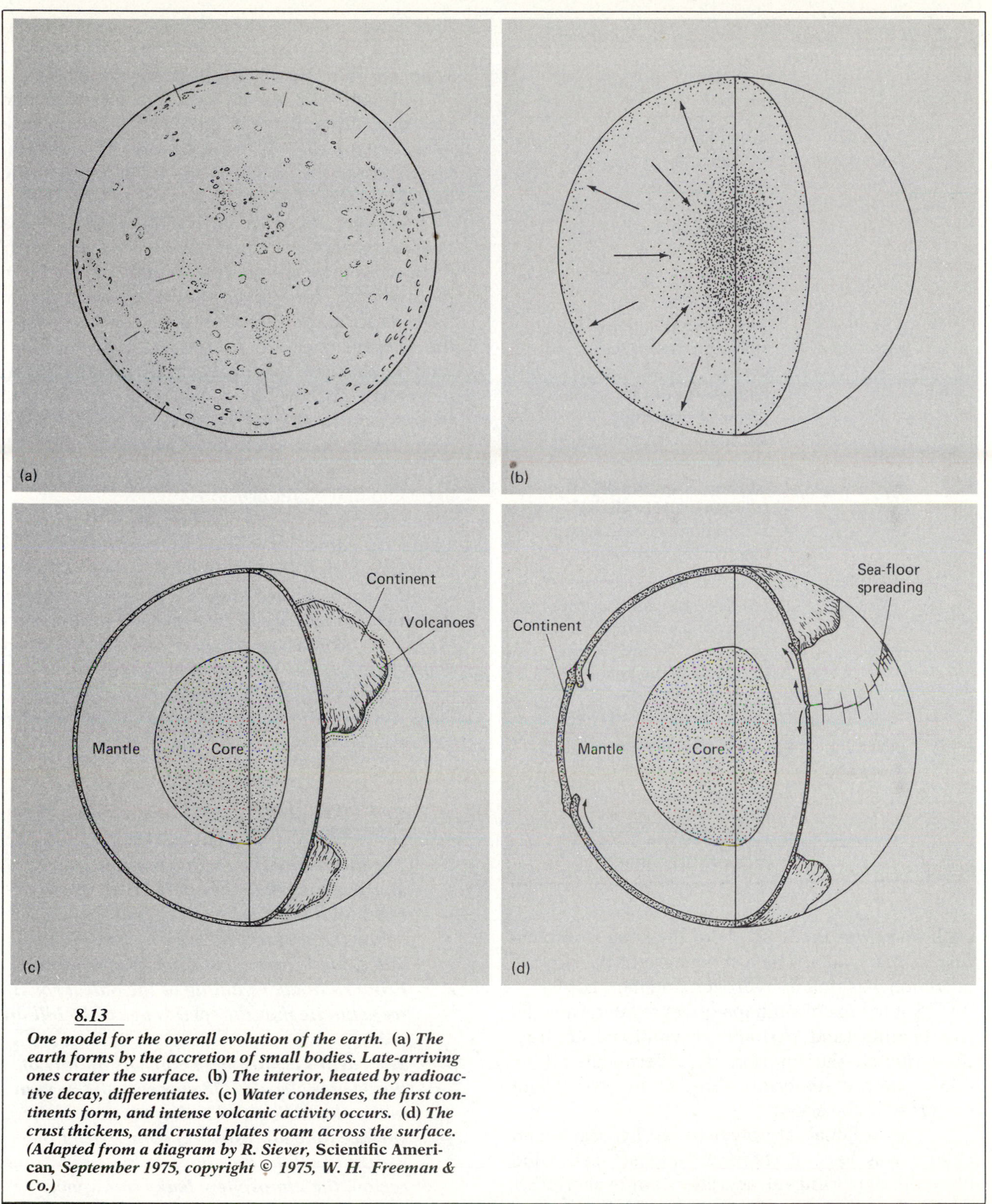

8.13

One model for the overall evolution of the earth.* (a) *The earth forms by the accretion of small bodies. Late-arriving ones crater the surface.* (b) *The interior, heated by radioactive decay, differentiates.* (c) *Water condenses, the first continents form, and intense volcanic activity occurs.* (d) *The crust thickens, and crustal plates roam across the surface. (Adapted from a diagram by R. Siever,* Scientific American, *September 1975, copyright © 1975, W. H. Freeman & Co.)

years. No other planet has been restless for so long!

Let's step back from the close-up presented of our planet and scan its overall evolution (Fig. 8.13 and Table 8.1). It falls into four large-scale stages.

First, there was the accretion of the earth from smaller bodies in the cloud of gas and dust that eventually formed the sun and planets (Fig. 8.13a). The buildup probably took place quickly, in only a few million years. It left a pockmarked planet of more or less uniform composition (since

TABLE 8.1 A possible order for the earth's evolution

Stage	Events
I. 4.6 billion years ago	1. Formation of earth by accretion from the solar nebula 2. Rapid internal heating from accretion and radioactive decay 3. Blowing off of primeval atmosphere
II. 4.5 billion years ago	1. First differentiation of interior, formation of crust and core 2. Outgassing of carbon dioxide, water, methane, ammonia, etc., to make second atmosphere 3. Infall of large objects to fracture crust in places 4. Formation of ocean basins; begin filling with water
III. 3.7 billion years ago	1. First tectonic movements, sea-floor spreading, formation of volcanoes and mountains 2. Cooling and thickening of crust 3. Ocean basins mostly filled
IV. 600 million years ago	1. Continuation of processes of stage III at slower rate 2. Enlargement of ocean basins 3. Formation and subsequent breakup of continent(s)

each body was made of about the same combination of materials). The atmosphere at the time of accretion was rich in hydrogen and inert gases.

Some tens of millions of years later, radioactive heating (and perhaps gravitational contraction) melted the interior. It differentiated (Fig. 8.13b). Dense materials sank to the core, light ones rose to the crust.

The original atmosphere of hydrogen and helium was lost. Perhaps an intense solar wind blew off the primeval accretion atmosphere. Or the low-mass gases simply escaped into space. This atmosphere was replaced by one containing much water, methane, ammonia, sulfur dioxide, and carbon dioxide. How? By volcanic activity caused by interior heating.

About 4.5 billion years ago, the earth's surface cooled enough for rain to fall and the oceans to form (Fig. 8.13c). About a billion years later, the first continents appeared. Plate tectonics began; mountains grew, only to fall to the weathering of wind and rain. Slowly the atmosphere evolved.

Roughly 2.2 billion years ago, crustal cooling thickened the crust enough to allow plate activity as we see it today. At the end of this stage 600 million years ago (Fig. 8.13d), the earth looked much as it does today.

To sum up: The earth's atmosphere, crust, and interior have changed greatly since the earth's formation. The interior's evolution is driven by internal heat: some from radioactive decay, some from its formation by accretion. The heat caused the internal layering. As the heat flows outward (and finally out to space), it drives the crustal movement and evolution. In the distant past, the crust was also shaped by impacts of solid bodies from space; few such impacts occur now. The atmosphere also transforms the crust by weathering: wind and water erosion.

Our planet is old, 4.6 billion years in age, a span of time almost incomprehensible to creatures as limited in life span as we. Although old, the earth is not static. The planet evolves, dynamically but slowly. The earth's structure and evolution, which we know in more detail and depth than for any other planet, serves as a model for the investigation of similar bodies within the solar system, those that are called the *terrestrial planets*.

KEY CONCEPTS

1 ***The interior of the earth can be probed by analyzing earthquakes; it is divided into zones that are denser the deeper they lie below the surface; the core is the densest part and probably consists of nickel and iron.***

2 ***From radioactive dating of the oldest rocks, we estimate that the earth's age is 4.6 billion years.***

3 ***The earth's atmosphere consists mainly of nitrogen molecules (78 percent) and oxygen molecules (21 percent); the pressure at the surface (one atmosphere) comes from the weight of all the air above it; in its outermost region, the atmosphere leaks into space.***

4 ***The atmosphere affects the light that enters it from space; ultraviolet is absorbed, forming the ozone layer; blue light is scattered (making the sky blue); and most of the visible light makes it to the surface (heating the ground).***

5 ***Carbon dioxide and water vapor trap the infrared emitted by the ground and so keep the earth warmer than if the infrared***

escaped into space; this warming is called the greenhouse effect.

6 ***The earth's crust evolves by sea-floor spreading and the motions of crustal plates, driven (perhaps) by convective flows below the surface; the ocean basins are the youngest parts of the earth's crust; the continents are older in comparison.***

7 ***The water in the oceans and gases in the atmosphere came from the release of materials from the earth's crust; both interact with each other and have changed because of the development of life on our planet.***

8 ***The evolution of the atmosphere is influenced by outgassing from the crust, escape of gases into space, biological activity, and the radiation from the sun.***

9 ***The interior and crust of the earth have been greatly modified since its formation 4.6 billion years ago.***

STUDY EXERCISES

1 Explain how you could determine the earth's mass by jumping off a building. (Explain it; don't do it!) (*Objective 1*)

2 Contrast the composition of the earth's core to that of its crust. (*Objectives 2 and 3*)

3 Discuss uncertainties in the statement, "The earth's age is 4.6 billion years." (*Objective 4*)

4 Make a simple argument to demonstrate that the earth's core must be denser than its crust. (*Objective 3*)

5 Describe two effects that the earth's atmosphere has on sunlight passing though it. (*Objective 6*)

6 Suppose the amount of water vapor in the atmosphere suddenly increased by a large amount. What would happen to the earth's surface temperature? (*Objective 7*)

7 How can volcanoes affect the evolution of the earth's oceans and atmosphere? (*Objectives 8 and 10*)

8 What was the composition of the earth's first atmosphere? What happened to it? What was the composition of the earth's second atmosphere? Where did it come from? What happened to it? (*Objective 10*)

9 Give two ways in which the oceans affect the atmosphere. (*Objective 9*)

10 Where did the oceans' water come from? (*Objective 9*)

BEYOND THIS BOOK . . .

R. Seiver compares the earth with the other planets in the solar system in "The Earth" in *Scientific American*, September 1975.

You can find a technical, comprehensive view about our atmosphere in *The Evolution of the Atmosphere* (Macmillian, New York, 1977) by J. C. G. Walker.

Chapter 13 of *Planetary Geology* (Prentice-Hall, Englewood Cliffs, N. J., 1975) by N. M. Short contains comparative information about the earth's evolution.

An excellent article on the fate and role of carbon dioxide in the atmosphere is "The Carbon Dioxide Question" by G. Woodwell in *Scientific American*, January 1978, p. 34.

For more on the dynamo model, see "The Source of the Earth's Magnetic Field" by C. Carrigan and D. Gubbins, *Scientific American*, February 1979, p. 118.

Read the details in "Plate Tectonics" by J. F. Dewey, *Scientific American*, November 1972.

The September 1983 issue of *Scientific American* is devoted to the earth; the lead article, "The Dynamic Earth" by R. Siever, presents a good summary of current ideas.

9/The moon and Mercury: Dead worlds

Everyone is a moon, and has a dark side which he never shows to anybody.

MARK TWAIN: Pudd'nhead Wilson's New Calendar

Learning objectives

After studying this chapter, you should be able to:

_1._ Compare the moon and Mercury in size, mass, and bulk density.
_2._ Describe the moon's major surface features and indicate a possible formation process for each.
_3._ Describe Mercury's major surface features and indicate a possible formation process for each.
_4._ Compare and contrast the surface environments (temperature, atmosphere, surface features) of the moon and Mercury to each other and to the earth.
_5._ Sketch the structure of the lunar interior and compare it to the earth's.
_6._ Sketch a model of Mercury's interior and contrast it to the interior of the earth and the moon.
_7._ Compare and contrast the magnetic fields (or lack thereof) of the moon, Mercury, and the earth.
_8._ Outline a possible history for the moon's evolution in light of Apollo results, and present evidence for each of the major stages.
_9._ Compare and contrast the evolution of the moon and Mercury to each other and to the earth.
_10._ Compare and contrast two models of the moon's origin, using Apollo results to support or refute the models.
_11._ Describe how cratering of planetary surfaces can be used to infer the relative ages of the surfaces.

Central question

What processes have driven the short evolution of the small, airless worlds of the moon and Mercury?

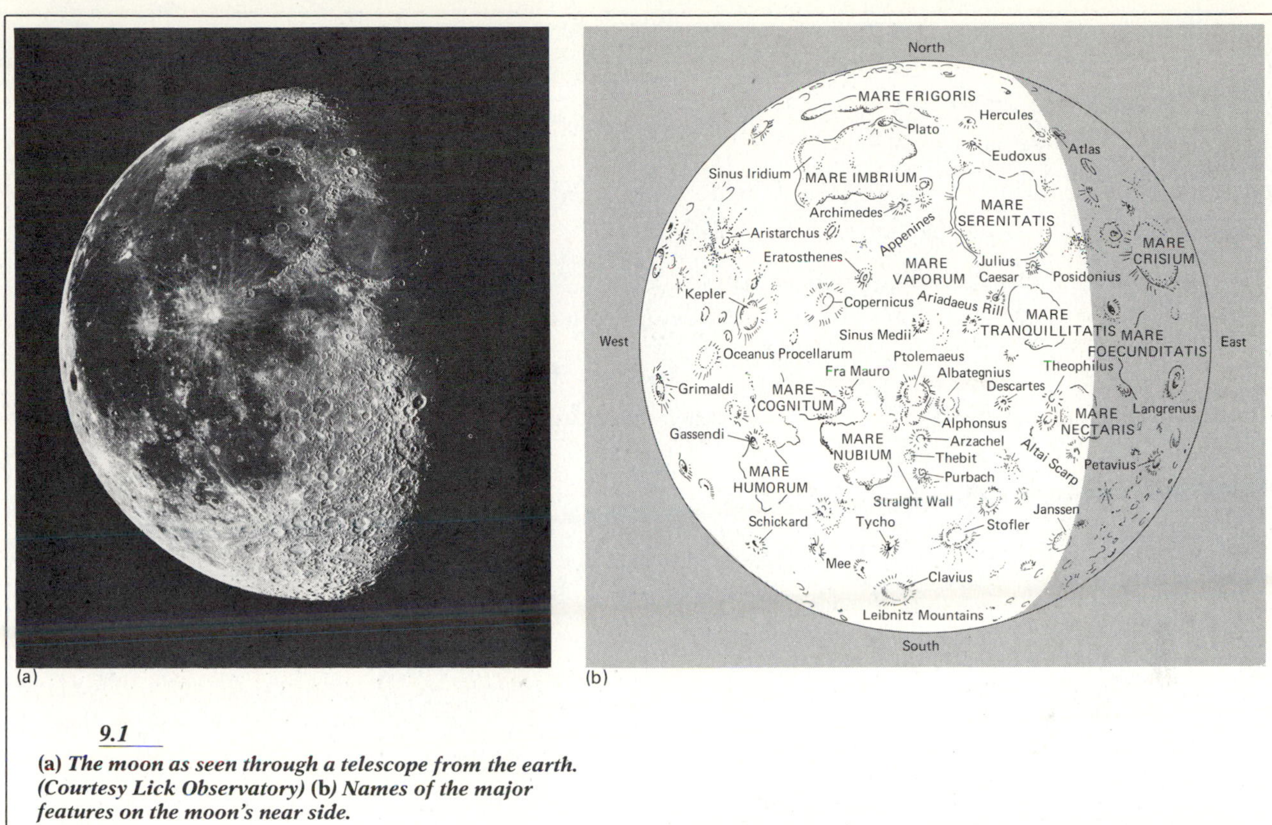

9.1
(a) ***The moon as seen through a telescope from the earth. (Courtesy Lick Observatory)*** **(b)** ***Names of the major features on the moon's near side.***

Through a small telescope, the moon strikes you as a stark world, tantalizingly close (Fig. 9.1). Fascination with this neighbor bred tales of traveling to the moon. The stories range from bird-borne excursions to the cannon-powered voyage described by Jules Verne. His astronauts, after their launch from Florida, circled the moon and returned home by plunging into the sea. Verne had the right idea! NASA carried out his vision, with the addition of a lunar landing, in July 1969. The event fulfilled many earlier dreams: the first visit to an alien world.

Another body in the solar system has a similar face: Mercury (Fig. 9.2). Like the moon, Mercury is a small, airless world pockmarked with many craters and scoured by intense sunlight.

Both the moon and Mercury are now dead worlds. Their interiors are now cooler than the earth; no heat drives the motions of crustal plates. No mountains rise; no volcanoes fume. Without atmospheres, no wind or water wears down their landscapes. Their heyday of activity has passed.

This chapter compares these tiny worlds to each other and to the earth to provide an insight into their evolution. The emphasis will fall on our moon because of the Apollo mission returns, which gives solid clues to reconstruct the moon's history and to infer that of Mercury in comparison.

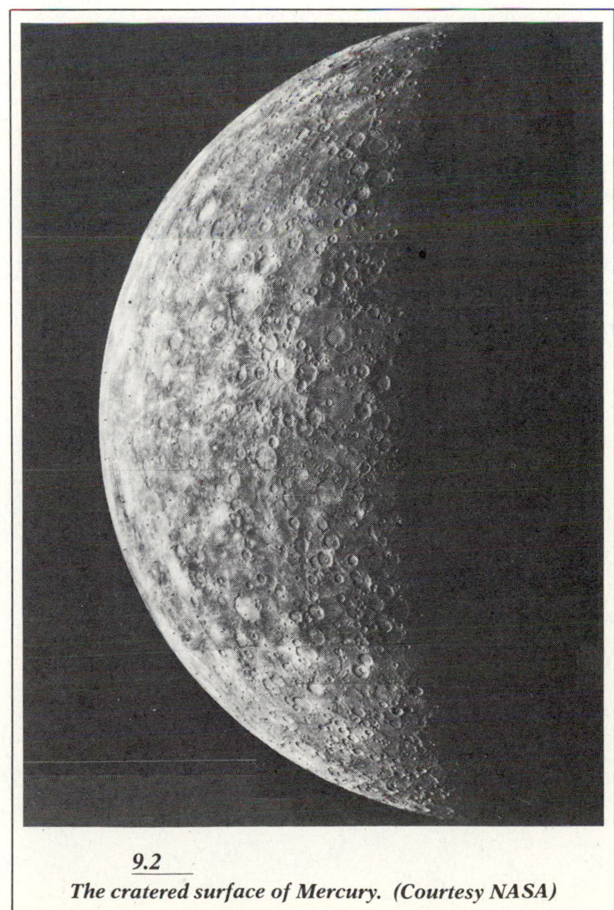

9.2
The cratered surface of Mercury. (Courtesy NASA)

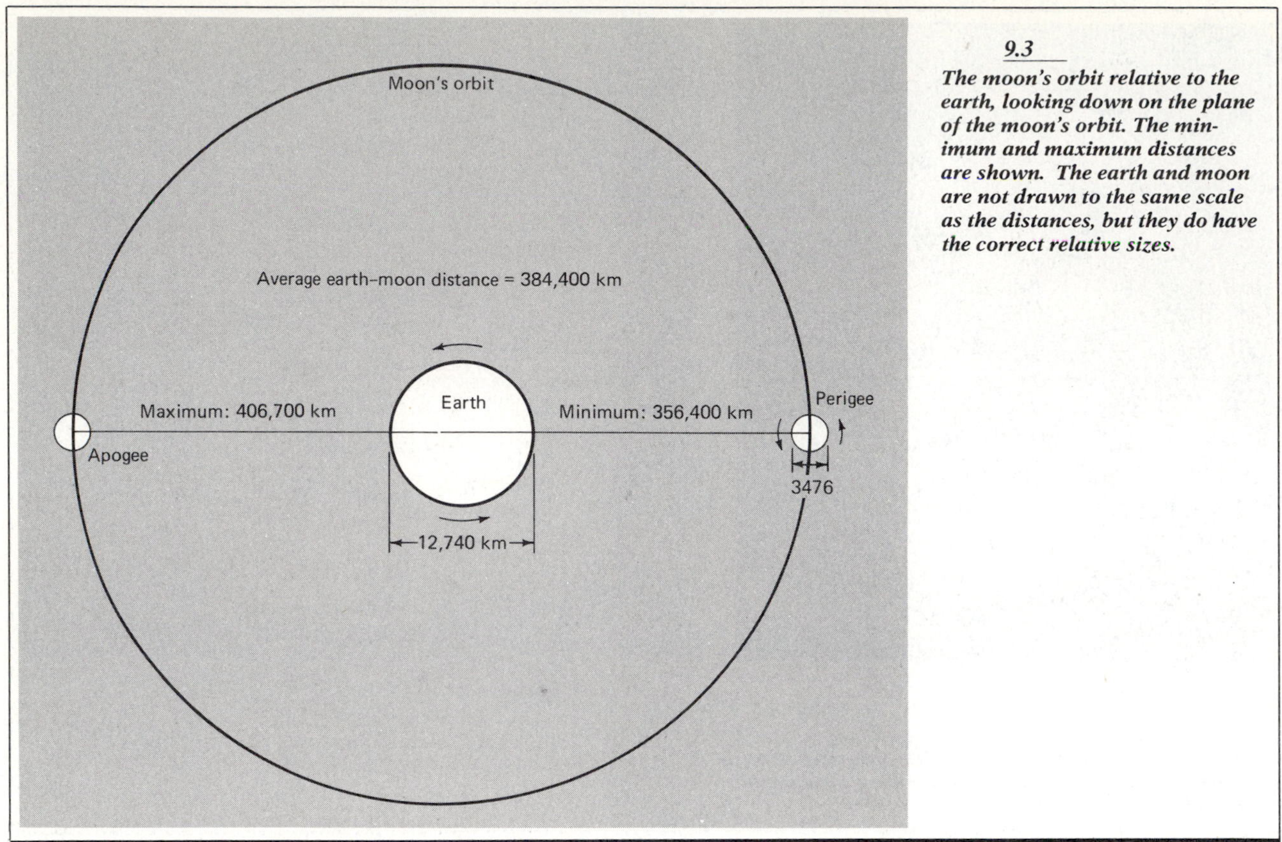

9.3
The moon's orbit relative to the earth, looking down on the plane of the moon's orbit. The minimum and maximum distances are shown. The earth and moon are not drawn to the same scale as the distances, but they do have the correct relative sizes.

9.1 *The moon's orbit, rotation, size, and mass*

The moon circles the earth in an elliptical orbit at an average distance of 30 earth diameters (Fig. 9.3), or about 384,400 km. In 1969 the Apollo 11 astronauts deposited special reflectors on the moon's surface. Later Apollo missions put down other such reflectors. These devices reflect back to the earth laser light bounced off the moon from the earth. Timing the round trip measures the earth-moon distance to within a mere 8 centimeters!

History of the moon's orbit / The orbit of the moon has not always been as we see it now. Gravity ties the earth and moon together. Tides are one consequence of this coupling. Tidal friction slows down the earth's rotation rate at about 2×10^{-3} second in a day per century. This decrease results in an increase of the earth-moon

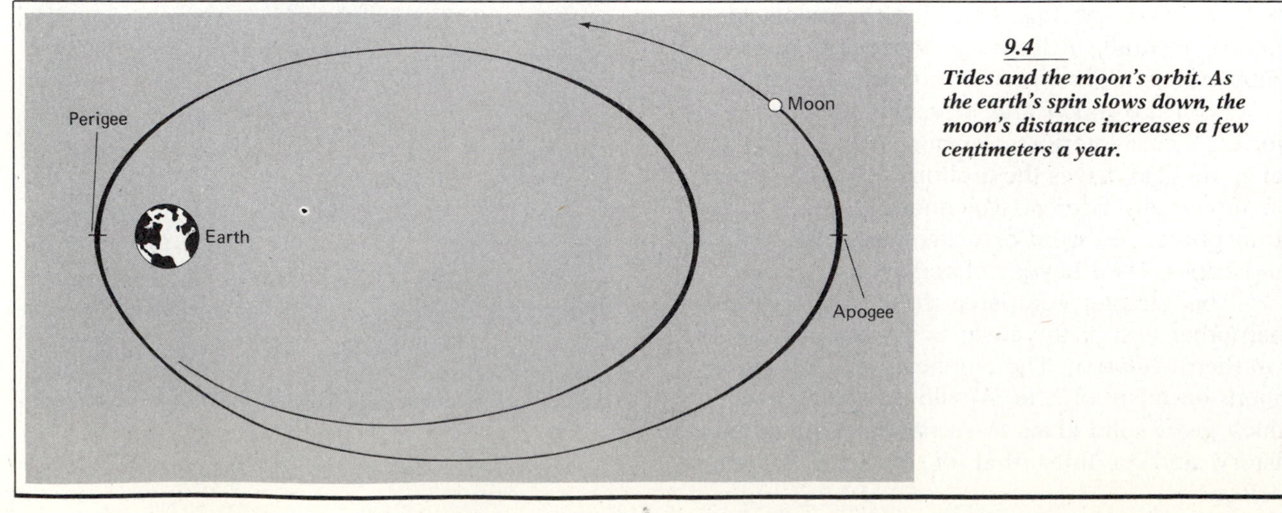

9.4
Tides and the moon's orbit. As the earth's spin slows down, the moon's distance increases a few centimeters a year.

distance (Fig. 9.4 and Focus 9.1). In the future the moon will be so far from the earth that the length of the month (longer than now) will equal the length of the day (also longer). That future day the day will be 55 present (24-hr) days long.

The moon must once have been much closer to the earth in the past. The closest approach took place about 1.2 billion years ago. At that time, the month was about 6.5 hours long, the day 5 hours long, and the moon only 18,000 km from the earth. The moon would have looked like an enormous hot-air balloon in the sky! It would have covered 11°, which is 22 times its present angular diameter.

The one-faced moon / If you have looked at the moon, even occasionally, you have grown used to the sight of its same face toward the earth. The moon must rotate on its axis to keep the same side facing the earth (Fig. 9.5). In fact, the moon rotates on its axis with a period of 27.3 days—the same as its period of revolution about the earth with respect to the stars. Until satellites orbited the moon, we had no direct view of the back side, the side turned away from the earth.

Lunar orbiters photographed the moon's once-mysterious far side (Fig. 9.6), which looks quite different from the near side (Fig. 9.1). It is almost completely cratered, with little of the dark-colored, smoother areas that cover so much of the near side.

If you stood on the moon's near side, you would see the earth suspended, never rising or setting, against the stars. With the earth always in sight, astronauts on the near side can communicate directly here by radio. By contrast, any astronaut on the far side would never see the earth and would have to rely on a lunar orbiter to relay radio signals to earth.

The lunar day / The moon keeps the same face to the earth, but not to the sun. It rotates once with respect to the sun in 29.5 days. So the lunar day is 29.5 earth days long. This also equals the time between corresponding phases of the moon. Suppose you were standing at the center of the surface of the moon at full moon. Where would the sun be? Directly overhead! When is the next time the sun is directly overhead? At the next full moon!

The moon's size, mass, and density / If you know the distance from the earth to the moon, you can find its physical diameter from its angular size. The result is 3476 km, about one-fourth the earth's diameter. If the earth were the size of your head, the moon would be about the size of a tennis ball. On the same scale, the diameter of its orbit would be about 12 m.

As you'd expect, the moon's mass is much less than the earth's: It is 1/81.3 times the mass of the earth, or 7.3×10^{22} kg. From the mass and radius, the average density comes out to be 3300 kg/m^3, about the same density as the rocks in the earth's mantle.

9.2 *The moon's surface environment*

Because the moon has a smaller mass and radius than the earth, its surface gravity is less. How

To distant stars
Radio antenna
E
(a)
(b)

9.5
The moon's rotation. Imagine a radio antenna on the moon.* (a) *If the moon did not rotate with respect to the stars, the antenna would not always point at the earth.* (b) *Because it does rotate at a rate equal to its revolution, an antenna always points to the earth.

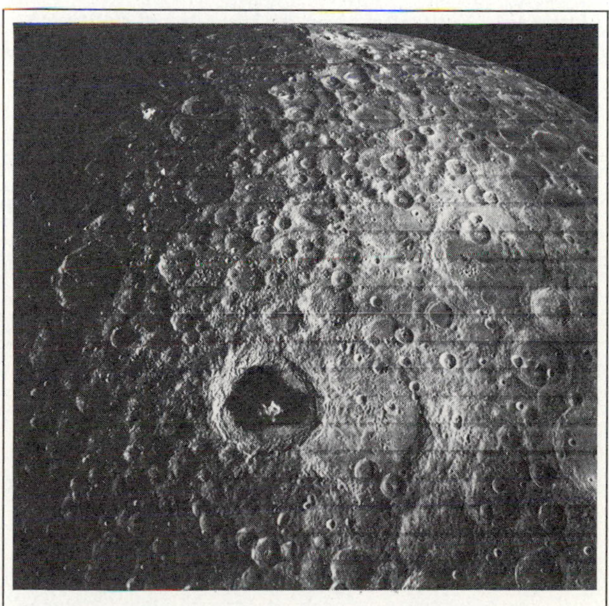

9.6
A close-up of the moon's far side. The smallest craters are about 1 km across. (Courtesy NASA)

Focus 9.1

TIDES AND THE MOON'S ORBIT

The main question about tides is this: Why are they associated with the moon so that typically two high tides and two low tides occur each day? I'll give you a general explanation of tidal dynamics using Newton's law of gravitation (see Section 4.4), but I warn you that it does not solve all the detailed problems about tides.

Imagine that the earth's surface is level and is covered completely with a layer of water. Consider for a moment the moon's gravitational attraction at three points lined up with the moon (Fig. F.23). Recall that the force of gravity decreases as the inverse square of the distance between masses. So the moon's gravitational force must be greater at A than at B and greater at B than at C. The greater the force acting on the same mass, the greater its acceleration. So a mass at A has a greater acceleration than a mass at C, and a mass at B has a greater acceleration than a mass at C. The **difference** *between these accelerations is crucial. Because of the difference, the water at A bulges ahead of the earth (point B), and the water at C lags behind the earth and forms a bulge on the side of the earth opposite the moon. So we have two high tides; one on the side of the earth toward the moon and one on the opposite side. At any point in the earth's oceans, these two high tides take place about a half day apart.*

Not all places have such a tidal rhythm. At some locations, for instance, water sloshes around the ocean basins, responding to the tidal forces. This can result, as in the Gulf of Mexico, in only one high tide a day.

Note that it is the **difference** *in gravitational forces that accounts for the tides; this difference is what is referred to as a* **tidal force**. *As an analogy, consider two masses at opposite ends of a spring. Pull more strongly on one end than the other. The spring will stretch, and the two masses move apart.*

Tidal forces affect the earth-moon system. The total angular momentum (Focus 9.2) of this system consists of two parts: the spin angular momentum (of the earth and the moon rotating about their axes) and the orbital angular momentum (of the moon revolving about the earth). The conservation of angular momentum says that the sum of all these must remain constant. Because the moon rotates slowly and its mass is small, the moon's spin adds very little to the total and can be ignored. So we need to consider only the spin of the earth and the revolution of the moon.

Because of tidal friction, the earth's rotation is slowing down, so its spin angular momentum decreases. For the system's total angular momentum to remain constant, the moon's orbital angular momentum must increase. This can happen if the moon moves away from the earth.

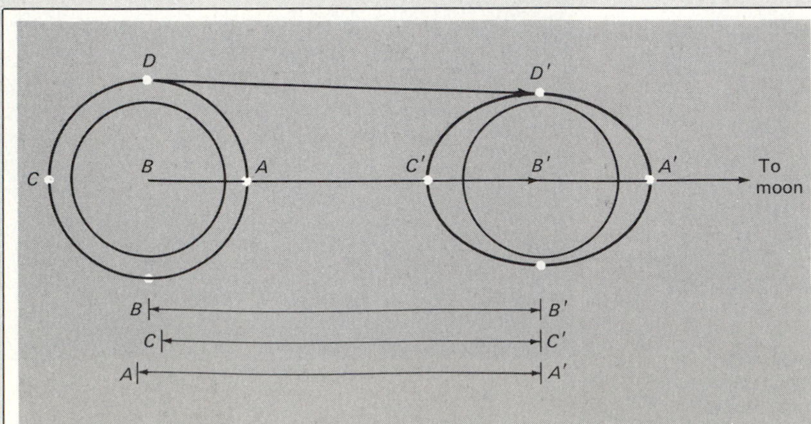

F.23

Tides. Imagine a smooth earth covered with water. Consider the water at **A**, **C**, *and* **D** *pulled by the moon's gravity. How far the water falls in some time depends on its distance from the moon; water at* **A** *falls farther than that at* **C**. *Water at* **D** *and the earth's center (***B***) fall the same distance (because the acceleration on them is the same). So as the earth falls toward the moon (from* **B** *to* **B′***), water on one side falls a bit more (from* **A** *to* **A′***) and on the other side a bit less (from* **C** *to* **C′***). Water flows from* **D** *to supply the bulges; the depth at* **D′** *is less than at* **A′** *or* **C′**.

much less? A planet's surface gravity is simply the acceleration of any object at its surface due to the planet's gravitational pull. The moon has a mass 1/81.3 that of the earth and a radius 0.273 that of the earth, so the moon's surface gravity is one-sixth the earth's. Objects weigh one-sixth as much on the moon as they do on the earth.

The moon has essentially no atmosphere. If you watch the moon move in front of a star, you will see the star suddenly vanish without warning! Even a thin atmosphere on the moon would dim the star gradually before it disappeared behind the edge.

Why no atmosphere? Gravity holds down any atmosphere of a celestial body. An atmosphere consists of a gas of molecules and atoms, moving at various velocities. The temperature of the gas is a measure of the average velocity of the particles in it. Typically, gas particles collide frequently. But in the thin upper layers of an atmosphere, far fewer collisions occur. If a gas particle has escape velocity here, it speeds off into outer space. For the moon, the escape velocity is 2.4 km/sec. At typical lunar temperatures, low-mass gas particles at the surface have escape velocity.

Most gases have escaped from the moon's gravitational grasp since its formation. Some material from the solar wind travels near the moon and stays briefly. But the surface density must be extremely low. The exhaust from the Apollo 11 landing dumped more gases into the atmosphere than had previously existed there! But these gases will not stay around very long either: For example, oxygen dumped at the moon's surface escapes in about 100 years.

The earth's atmosphere acts like an insulating blanket (see Section 8.5). During the night it retains much of the heat received in the previous day. Lacking such atmospheric insulation, the moon experiences a greater temperature range during a lunar "day" (which lasts a month of earth time).

We can estimate an airless planet's surface temperature from its energy budget. Suppose that only sunlight heats the planet's surface (Fig. 9.7). The surface absorbs some of the incoming sunlight and reflects some back into space. The moon reflects 7 percent of the light that hits it and absorbs the remaining 93 percent. (Even though a full moon seems to shine brightly in the sky, the moon's surface is really quite black.) The absorbed sunlight heats the surface; it radiates mostly infrared. The balance between the incoming sunlight and outgoing infrared determines the surface temperature of the sunlit side. At night, solar heating stops, and the infrared—because no atmosphere traps it—radiates away into space. Since the lunar night is so long (about 15 earth days), the surface has time to cool, and the temperature plummets to 125 K. The large noon-to-midnight temperature difference, some 250 K, occurs because the moon rotates slowly and has no atmosphere.

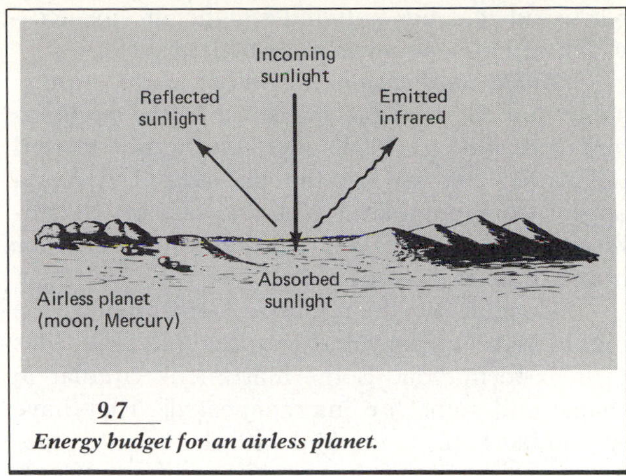

9.7
Energy budget for an airless planet.

Without a significant atmosphere, the moon has no shield from the lethal X-rays and ultraviolet radiation from the sun or from the small, solid particles coming in from space. The moon's cratered surface presents a fierce, unfriendly place for people. The Apollo astronauts, imprisoned in their bulky life-support systems, found the moon bleakly beautiful but uninviting.

9.3
The moon's surface: Pre-Apollo

Until the Apollo astronauts walked on the lunar surface and sampled it, the study of the moon was confined to viewing it from the earth, from orbiting satellites, or from lunar landers. This section briefly surveys the moon's surface as a preview to a discussion of the results of the Apollo landings.

Galileo first studied the moon carefully in 1609 (see Section 4.1). His pioneer work sparked an explosion of careful observations. Galileo named the dark areas *maria* (Latin for "seas"; the singular form is *mare*.) I will use maria rather than seas, because the English word implies water more than the Latin one.

Viewed through a small telescope or binoculars, the moon reveals a fascinating terrain. Lunar craters give the moon a pockmarked face. Mountains and craters irregularly rim the moon's edge. Some mountains stand alone in the maria, while others link in long ranges. Bright rays flower from some craters. The moon impresses you as a rough, old world that has suffered significant violence—violence that has carved its splotched surface.

Maria and basins / Photos from orbiters show that most of the maria lie in the northern half of the moon's hemisphere on the side toward the earth; few are on the far side. The maria appear dark compared with the rest of the surface. Their dark irregular stretches form the face of the "man in the moon."

Along with their darker appearance, the maria have other general features that give clues to their formation. Some maria look circular in shape, and some are interconnected. They have smooth surfaces compared with the brighter, cratered regions. Also, the maria have lower elevations, by about 3 km, than the rest of the surface. So the maria are called the lunar *lowlands*; the other areas, the *highlands*. The lowlands cover some 20 percent of the moon's total surface; the highlands, over 70 percent.

The vast flat extents of the maria look like solidified lava flows. If you look carefully at the regions around the maria, you can find craters flooded in by the dark material from the maria (Fig. 9.8). This indicates that the lava flow that formed the maria occurred *after* the formation of the lunar crust and certain craters. The lunar maria fill up large, shallow *basins* on the moon's surface. For example, Mare Imbrium basin (Fig. 9.9), on the moon's near side, is roughly circular in outline and has a diameter of about 1200 km. The moon's most striking basin is Mare Orientale (Fig. 9.10). Note how the concentric rings of mountains make it look like a bull's-eye! The outer rim of

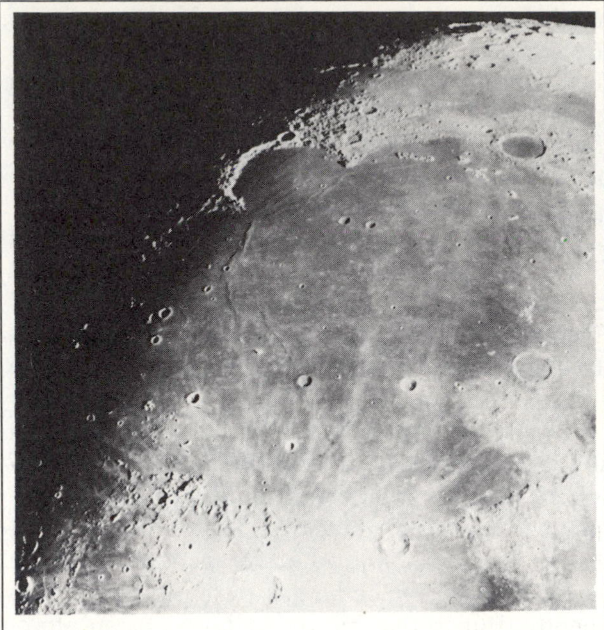

9.9

Mare Imbrium, in the moon's northern hemisphere on the near side. (Courtesy Yerkes Observatory)

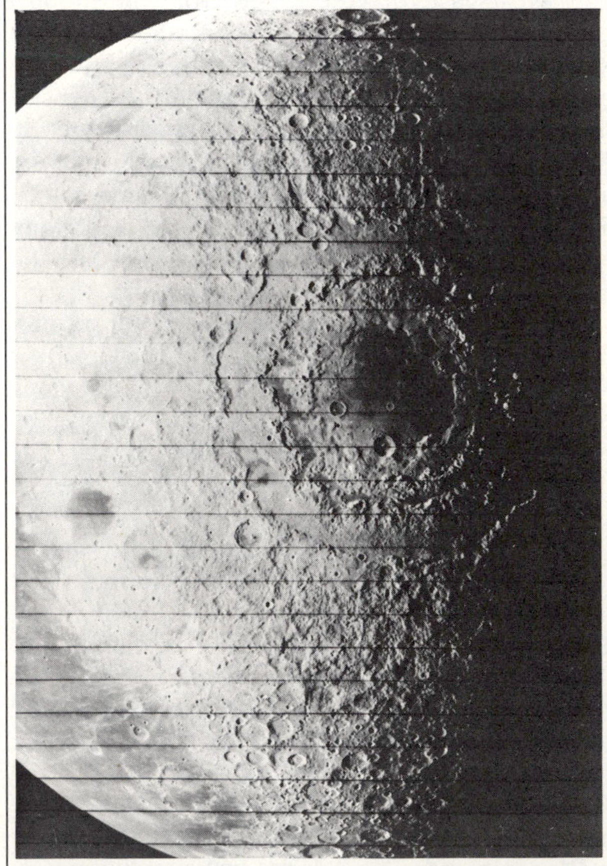

9.10

Mare Orientale, which the Cordillera Mountains ring like a bulls-eye. (Courtesy NASA)

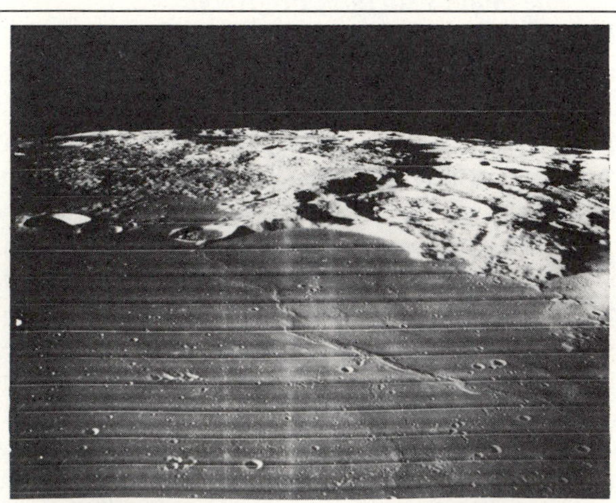

9.8

The surface of a mare (the Ocean of Storms). Note how much smoother, darker, and lower in height the mare is compared with the background highlands. The baylike area near the center (where the mare meets the highlands) is a filled-in crater (arrow). (Courtesy NASA)

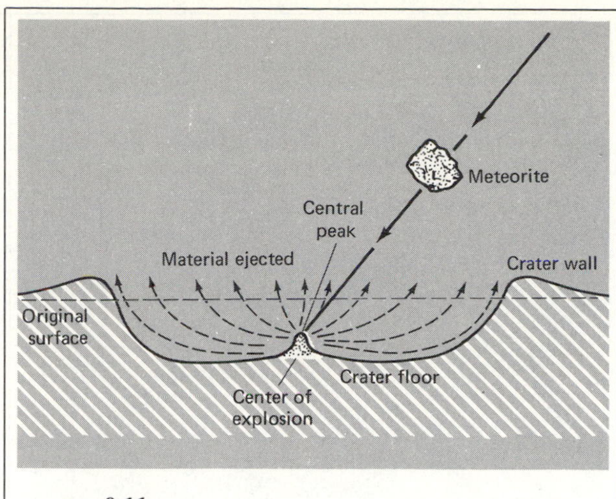

9.11
Formation of a crater by impact of a solid body. Ejected material can form rays and smaller craters surrounding the main impact.

mountains (the Cordillera Mountains) has a diameter of about 970 km and rises to a height of 7 km. Surrounding this basin for about 1000 km out lies a blanket of lighter material covering the older lunar surface. Mare Imbrium would look like Mare Orientale if most of its filling material were removed.

Planetary scientists have found concentrations of mass beneath most of the maria. These are called *mascons*. One mascon, for example, under Mare Imbrium, has a mass of 1.6×10^{19} kg and an average density of 3700 kg/m^3. (Note that this density is *greater* than the average density of the moon.) If this mascon were spread out to cover California, it would be 12 km thick! The fact that almost all maria have associated mascons implies that some common process produced large amounts of dense material under the maria.

Craters */ Craters* (from the Greek word meaning "cup" or "bowl") litter the moon almost everywhere. They range widely in size, from many smaller than a coin to five with diameters greater than 200 km. (That's about the size of Connecticut.) Craters are generally round. The heights of the rims of lunar craters are small compared with the diameters, and the floors are depressed compared with the surrounding landscape.

Most lunar craters were formed by objects from space that slammed into the moon and punched out the craters (Fig. 9.11). An impact produces a crater with undulating slopes, usually covered by debris, with a rippled terrain around it (Fig. 9.12). If the debris and crater wall were put back into an impact crater, they would fill it up. Material blasted out by an impact falls in streaks, leaving a ray-like pattern (Fig. 9.13). Large chunks of thrown-out material can create small secondary craters around impact craters.

What projectiles made these impacts? Small, solid bodies now orbit the sun; they are

9.12
The crater Aristarchus, about 60 km wide, with its ray system. Note the ripples in the surface just outside the crater; these formed from the shock of impact. (Courtesy NASA)

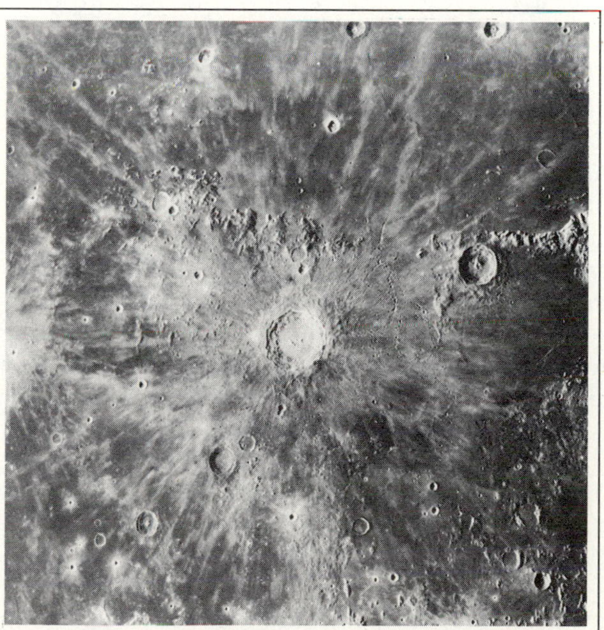

9.13
The crater Copernicus (center) with its ray system. (Courtesy Mt. Wilson Observatory, Carnegie Institute of Washington)

9.14
The moon's surface. It consists of fine particles, called soil, and larger rock fragments. Note how well it holds the imprints of the tires of the rover. (Courtesy NASA)

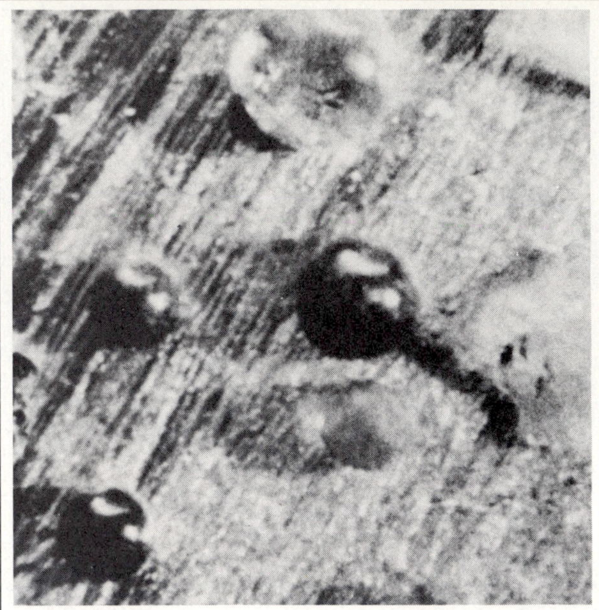

9.15
Typical small glass spheres, found in the lunar soil. (Courtesy NASA)

called *meteorites* when they strike the earth. If they pass close enough to the moon, they will collide with it. But we have seen no new, large craters form recently. So the era of heavy-impact cratering took place in the past. The Apollo missions hoped to find out when.

9.4 The moon close up: Apollo

The Apollo program obtained about 400 kg of lunar material from six different sites. Scientists have examined only about 10 percent in detail. Already these samples and other experiments in the Apollo program have provided the first deep understanding of a planet other than the earth. We can now sketch out many physical details of the moon and outline its history.

The lunar surface / The Apollo samples reveal the physical and chemical nature of the lunar surface. The very top layer (1 to 20 m deep) is a porous, somewhat adhesive layer of debris (Fig 9.14). It consists of fine particles (called lunar *soil*) and larger rock fragments. The soil samples contain a large amount of mostly round pieces of glass (Fig. 9.15), which make the surface slippery.

The moon rocks mostly fall into three categories: (1) dark, fine-grained rocks (Fig. 9.16a) similar to terrestrial basalts (magnesium-iron silicates) called *mare basalts*, (2) light-colored igneous rocks (Fig. 9.16b) with visible grains called *anorthosites* (aluminum/calcium silicates)—by far the most common rock on the surface, and (3) *breccias*, rock and mineral fragments cemented together (Fig 9.16c).

What do these characteristics imply? First, since the moon rocks are igneous rocks, they formed from the solidification of lava. The rate at which lava cools determines the grain size of igneous rocks: Fast cooling results in small grains, slow cooling in large ones. So the dark rocks (found in the lowlands) cooled faster than the light-colored ones (found in the highlands). In addition, the light-colored rocks are less dense than the dark ones. This density difference arises from a compositional difference, which indicates that the mare basalts formed from partial melting and slow cooling of silicates at a relatively low temperature (about 1300 K) inside the moon, after some of the heavier elements had settled toward the interior. In contrast, the anorthosites were hotter, cooled more rapidly, and formed from the less dense materials, probably on the surface.

Second, we can infer how the breccias formed. Imagine newly made igneous rocks on the moon's surface. Bodies from space pound into these rocks, fragmenting and heating them. The heat cements some fragments together to make breccias. The loose material left over makes up the lunar soil.

In a few key ways, moon rocks differ from the earth's igneous rocks. First, they contain more titanium, uranium, iron, and magnesium. Second, compared with earth rocks or meteorites, they are depleted of elements that would condense at rela-

(a)

(b)

(c)

9.16

(a) ***A typical mare basalt. The holes were air bubbles in the lava. The pit in the center resulted from an impact by a small meteorite.*** (b) ***A rock containing anorthosite (white area). This is a sample from Apollo 15 known as the "Genesis Rock."*** (c) ***A lunar breccia. The dark areas are melted rock. Large pieces of other rock embedded in it are visible. (All courtesy NASA)***

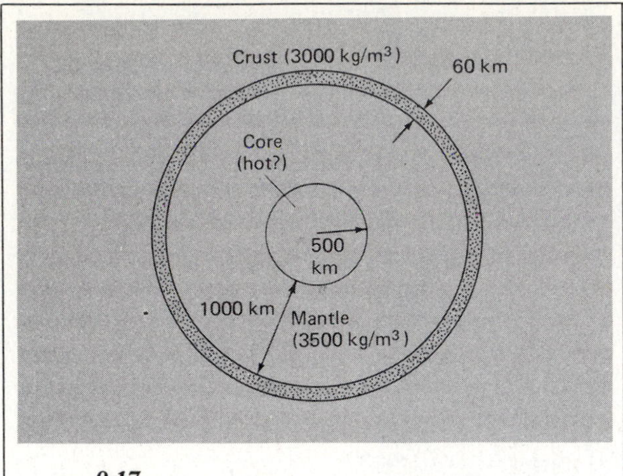

9.17

A model of the moon's interior inferred from Apollo data. The core may be hot enough to be molten, about 1500 K.

tively low temperatures (1300 K). These elements are called *volatiles* and include sodium, potassium, copper, argon, and chlorine. Third, the moon rocks contain *no* water. Earth rocks always have some water locked up in their minerals. The moon rocks turned out to be bone dry.

In one critical way, on a nuclear level, the moon resembles the earth. The isotopic composition of oxygen (relative abundances of oxygen-16, 17, and 18) in the lunar samples is the same as that for the earth. Studies of the oxygen isotopic compositions of meteorites show a distinct variation among samples. Some meteorites are thought to be primitive materials from the formation of the solar system. Indirect evidence suggests that the variation in oxygen isotopes corresponds to condensation in different parts of the solar system. The identity for the earth and moon shows that these bodies formed in the same general region of the solar system.

Finally, compare the bulk densities of the moon and earth: 3300 kg/m^3 versus 5500 kg/m^3. This difference implies that the moon contains less metals and probably does not have a large metallic core, as the earth does.

Ages of lunar samples / Moon rocks are dated by the same radioactive-decay techniques used to date earth rocks (see Focus 8.1). One caution: These methods give the time since the rock last solidified. If a rock was heated enough to melt it since its original formation, radioactive dating gives the age since that reheating.

Only a few rocks and some fragments from the lunar soil have ages as great as 4.6 billion years. The light-colored rocks from the highlands generally are the oldest: 4 billion years. The dark-colored rocks from the maria are the youngest, some only 3.2 billion years old and only a few as old as 3.8 billion years.

These ages imply that the moon formed a little more than 4.6 billion years ago. After formation, the present highlands solidified, about 4 billion years ago. Then the lava flows that made the maria took place some 3 billion years ago.

The moon's interior / The Apollo missions probed the moon's interior with the same shock-wave method used to look into the earth. They found the moon to be an inactive world (com-

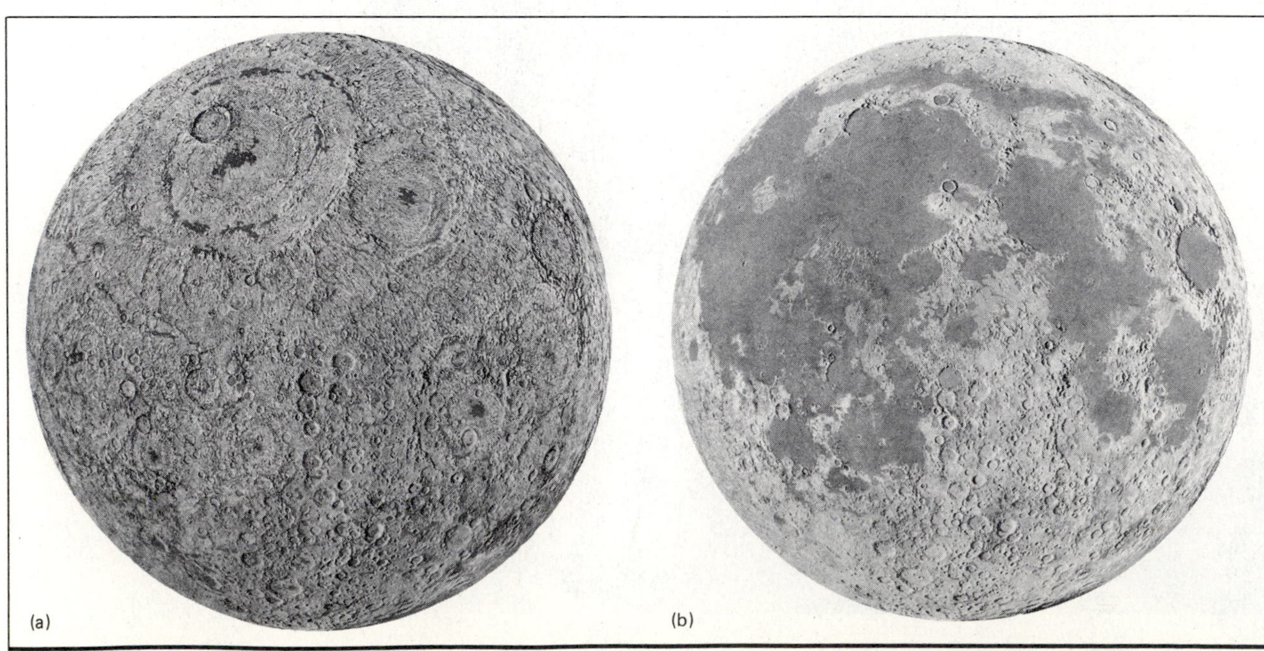

pared with the earth). Few moonquakes occur, and those few take place about 800 km below the surface. Such shocks release only about 100 to 100,000 J, barely a tremble by our standards. If you stood directly over the strongest moonquake so far recorded, you would not even feel the ground shake. Geologically the moon is a quiet world.

This low activity indicates that the moon is cold and solid down to a depth of about 1000 km (Fig. 9.17). This region makes up the moon's mantle, which very likely consists of silicates only a little more dense than the surface. Encasing the mantle is the crust, which is layered. On the surface lie the rocks and soil sampled in the Apollo missions.

We don't know for certain if the moon has a well-defined core. If it does, the core probably makes up the inner 500 km of the moon. It may be hot, about 1500 K, and molten in whole or in part. Evidence for a hot core comes from the Apollo measurement of heat flowing up through the lunar surface. The heat flows at a rate three times as great as that for the earth. So some part of the moon's interior must be relatively hot. We know that the mantle is solid and cool, so the core must be the source of the heat outflow.

However, the core cannot be liquid like the earth's. The moon's magnetic field is very weak, only 10^{-4} as strong as that of the earth. If the dynamo theory for the origin of a planet's magnetic field is correct (see Section 8.4), the moon's core cannot be completely molten or composed mainly of metals. (Also, the moon's density is too low for it to have a substantial metallic core.) On the other hand, some surface rock samples are magnetized much more than you would expect from such a weak magnetic field. Recall (Section 8.6) that iron minerals in an igneous rock preserve the magnetic field present at the time of solidification. So in the past the moon's magnetic field must have been stronger than it is now.

Conjectured lunar history / From the Apollo results we can concoct a scenario of the moon's history since it formed. The inferred sequence of events (Fig. 9.18 and Table 9.1) relies heavily on the dating of the lunar rocks.

About 4.6 billion years ago, the moon formed by the gathering together of chunks of material. These pieces continued to plunge into the moon after most of its mass gathered. During the first 200 million years after formation, these projectiles from space bombarded the surface and heated it enough so that it melted. Less dense materials floated to the surface of the melted shell; volatile materials were lost to space. The crust began to solidify from this melted shell about 4.4 billion years ago. From 4.1 to 4.4 billion years ago, the crust slowly cooled as the bombardment from space tapered off. The debris from this later bombardment made many of the craters now found in the highland areas.

Below the surface, the moon's material remained molten. About 4 billion years ago, a few huge chunks smashed the crust to produce basins that later became maria. For example, the Mare Orientale basin formed some 4 billion years ago when an object about 25 km across smashed into

(c)

9.18

A model for the moon's evolution. (a) *3.9 to 4.0 billion years ago. The surface has solidified and is cratered by the infall of solid objects; most of the surface is saturated with craters.* (b) *3.0 to 3.2 billion years ago. Fractures in the surface allow magma from the interior to flow out and fill the lowland basins with material darker than the highlands.* (c) *Now. Many of the large rayed craters blossomed on the surface after the formation of the maria. Parts of the maria have grown lighter in color from the material blown out by crater formation. (Paintings by D. Davis and D. Wilhelm, U.S. Geological Survey)*

TABLE 9.1 Evolution of the moon: A contemporary model

Event(s)	*Time (billion years ago)*	*Process(es)*
Formation	4.6	Accretion of small chunks of material
Melted shell	4.1–4.6	Infall of material and/or radioactive decay melts outer layer; volatile elements lost
Cratered highlands	4.1–4.4	Crust solidifies while debris still falls in to crater it
Large basins	3.9–4.1	Less debris hits surface, but a few large pieces smash crust to produce basins; basalts flow out from magma below solid crust
Maria	3.0–3.9	Flooding of basins by magma produced by radioactive decay
Quiet crust	3.0–present	Surface bombarded by small particles to pulverize and erode it.

SOURCE "Evolution of the Moon: The 1974 Model" by H. H. Schmitt, *Space Science Reviews*, 1975, vol. 18, p. 259.

the moon. Only later did the basins fill in. As the crust lost its original heat, short-lived radioactive elements (which decay rapidly) reheated sections of it. From 3.0 to 3.9 billion years ago, lava from the radioactive reheating punctured the thin crust beneath the basins, flowing into them to make the maria.

For the past 3 billion years, the crust has been inactive. However, small particles from space have incessantly plowed into the surface since it solidified. These sand-sized grains scoured the surface, smoothed it down, and pulverized it. Continued bombardment by larger bodies churned the fragmented surface. Impacts melted the soil, which swiftly cooled to form breccias and glass spheres. The moon's surface today resembles a heavily bombarded battlefield—constantly fragmented, stirred up, and melted.

9.5 The origin of the moon

The implications of the Apollo cargo so far have clarified a few aspects of the moon's origin, but none of the rival theories has completely claimed supremacy in explaining the moon's origin. These models fall into three broad categories:

1. the fission model
2. the binary-accretion model
3. the capture model

The *fission model*, the earliest of these probable ideas, was developed in 1879. It requires that the earth be spinning more rapidly in the past than now. A faster rotation rate would have created a greater equatorial bulge than now exists. If the earth were molten, its equatorial speed may have become so great that friction and gravitational attraction would no longer have been able to hold the bulge to the earth. A chunk of mantle detached. It then spiraled out from the earth, cooled, and formed the moon (Fig. 9.19).

The fact that the average density of the moon and the average density of the earth's mantle rock (3300 kg/m^3) are roughly the same lends support to the fission model. In fact, versions of the model viewed the protomoon as splitting off from the Pacific Ocean basin, and so it would consist of mantle material.

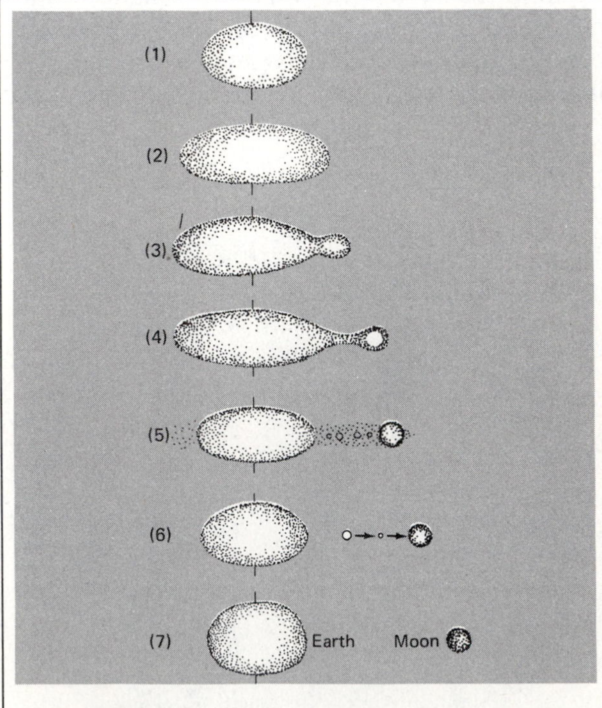

9.19
Formation of the moon by fission from the earth. Debris from the break-off rained down on the moon to crater it.

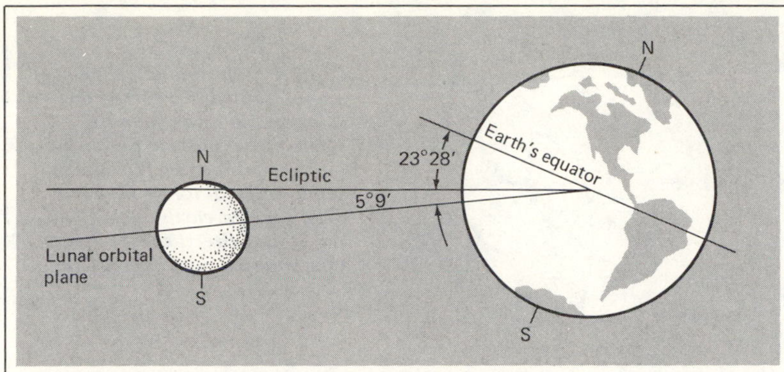

9.20
The tilt of the moon's orbit with respect to the earth's equator, which shows that it does not lie in the earth's equatorial plane.

The fission model was widely accepted for about 30 years. Then it ran into major difficulties. First, the angular momentum of the earth-moon system must be conserved (Focus 9.2). If the moon were joined to the earth now, the combined mass would spin much faster than the earth does now, about four times faster (once every 8 hours). But that rate is *not* fast enough to separate the lunar mass from the earth. Second, such a separation would take place at the earth's equator, where its rotation is the fastest. If angular momentum is conserved, the orbital plane of the broken-off mass must remain in the earth's equatorial plane. In fact, the moon's orbit is tilted with respect to the earth's equator (Fig. 9.20); it wobbles back and forth a bit, but the smallest angle it makes is 18.5°. Third, the evolution of crustal plates requires that the shape of the ocean basins change constantly and rapidly. So existing basin features cannot be used to argue that the moon left the earth in the distant past. Fourth, the present rate of tidal friction (Focus 9.1) would indicate that the moon separated from the earth only 1.2 billion years ago, but the earth was certainly solidified long before that.

Besides these problems, the Apollo results dealt the fission model a serious blow: The lunar basalts differ critically in chemical composition and water content from the terrestrial basalts that line the ocean basins. So the fission model, though once intriguing, is not a promising idea now.

The *binary-accretion model* views the moon as created out of the same cloud of material as the earth, rather than as born directly from our planet (Fig. 9.21). Dust particles grew from a gradual condensation of gas. These eventually accreted into the young moon and earth. The moon formed so close to the earth that earth's gravity held the moon in a close orbit. The leftover bits and pieces from the formation of the two planets fell into the moon and heated its primitive surface.

The composition of lunar minerals may discredit this idea. The moon and the earth are somewhat different in composition. And the moon has no water. In addition, the moon has no dense iron core. If the embryonic environment and materials of the two bodies were the same, how did the earth accumulate iron and the moon not? On the other hand, the oxygen isotopic composition of the earth and the moon are the same; this fact supports the condensation-accretion models.

The *capture model* (Fig. 9.22) was developed around 1955. It pictures the moon formed in some part of the solar system far from the earth. By chance, the moon traveled close enough to the

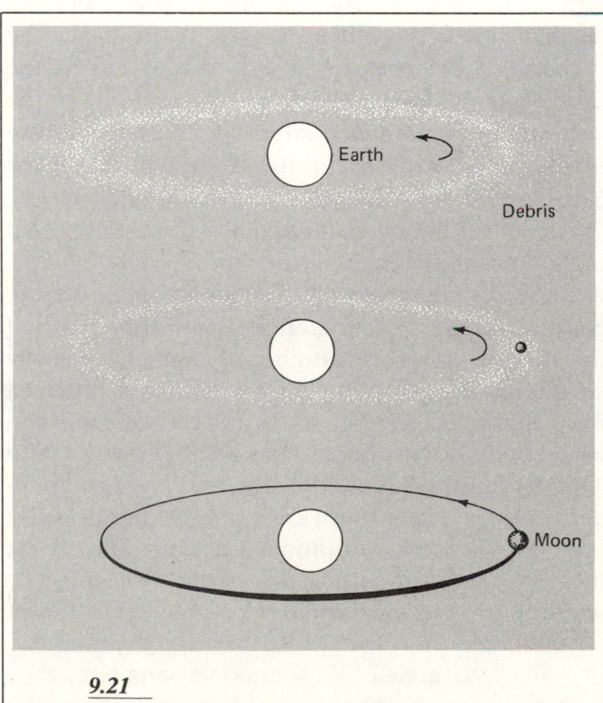

9.21
Formation of the moon by the condensation of debris, left over from the earth's formation.

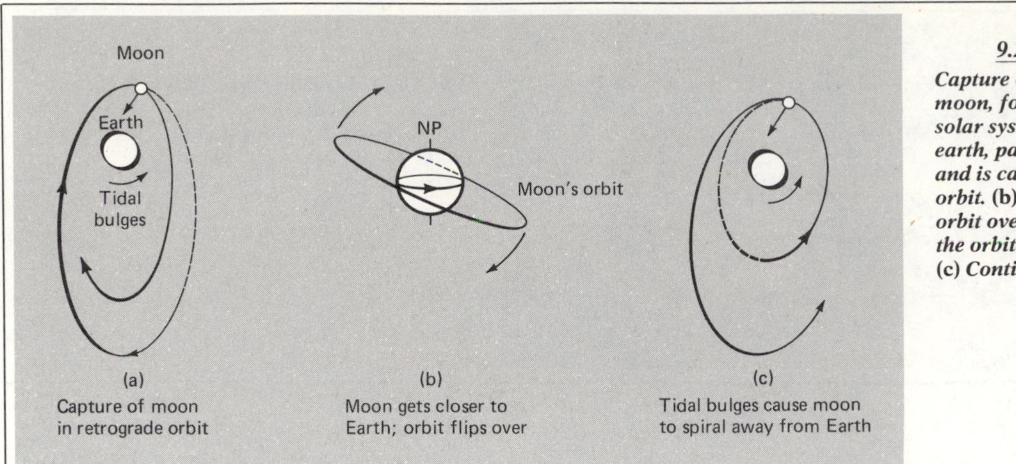
Moon
Earth
Tidal
bulges
NP
Moon's orbit
(a)
Capture of moon
in retrograde orbit
(b)
Moon gets closer to
Earth; orbit flips over
(c)
Tidal bulges cause moon
to spiral away from Earth

9.22

Capture of the moon.* (a) *The moon, formed in a part of the solar system away from the earth, passes close to the earth and is captured in a retrograde orbit.* (b) *Tidal forces flip the orbit over the earth's poles, and the orbital direction is reversed.* (c) *Continual tidal interactions

earth to be captured gravitationally. The earth caught the moon in a highly eccentric orbit in which the moon orbited the earth *opposite* the direction of the earth's rotation. The earth's tidal bulge acted to slow the moon down at its closest approach, so its orbit decreased in size (Focus 9.1). The tidal bulge also acted to change the inclination of the moon's orbit, tipping it over toward the direction in which the earth rotates. Eventually the orbit did get flipped, and now the tidal bulge works to speed up the moon on its closest approach and so make the orbit larger, as happens now at a rate of about 3 cm a year.

The capture model agrees in part with the inferred dynamic history of the earth-moon system. If the moon is now moving away from the earth, it must have been closer in the past. The capture model requires a closer-in moon at the time of its capture. Theoretical calculations of the earth-moon dynamics further back up the capture model. These show that if the moon were captured 5 billion years ago, it would have come closest to the earth 1.2 billion years ago and then moved out to its present distance.

Some other pieces of evidence support the capture model. For instance, if we suppose that the moon formed in another region of the early solar system, its early environment was different from that of the earth. So the differences between earth and moon—the moon's lack of water and iron, its lower average density, and its high abundance of uranium and rare-earth elements—arise from the different formation sites. However, if the moon were formed too far from the earth, it would not have the same isotopic composition of oxygen.

To date, none of these models wins out as the best explanation for the origin of the moon. The results from the Apollo mission have complicated the situation with a wave of details, still to be put into a more general framework.

9.6 *Interlude: Cratering*

This chapter and the next two show you that every terrestrial planet and nearly every satellite in the solar system are scarred by craters. That cratering happened throughout the solar system provides clues to processes that formed the terrestrial planets. And cratering serves as an easily applied tool to infer the relative ages of planetary surfaces.

The infall of massive objects on a planet's surface creates impact craters. The great abundance of craters on the planets and satellites implies that a storm of ancient impacts blasted them long ago. Since that time, any erosion and crustal evolution have modified the original pattern and changed the planets' surfaces. In addition, if we can estimate the rate at which cratering occurred, the number of craters gives a guide to the age of the surface on which they are found.

Let me outline a few major points about cratering. On any planet's surface, small craters greatly outnumber the larger ones. The size of a crater relates to the kinetic energy of the projectile that formed it, and so to the projectile's mass and size. The largest of these must have been rocks greater than 100 km in diameter. These objects strike planetary surfaces with speeds of 10 or more kilometers per second. They bang into the surface with enormous amounts of energy: some 10^{23} J for the mass that formed the crater Copernicus on the moon (Fig. 9.13). For comparison, the Mount St. Helens eruption in May 1980 blew up with an energy of 10^{17} J, an explosive yield equivalent to 35 megatons of TNT. So large craters involve explosions a *million* times greater!

Focus 9.2

ANGULAR MOMENTUM

The concept of momentum rests on that of the inertia of matter. To get an idea of the meaning of momentum, consider the following. A bicycle and a truck are coming at you at the same speed. Which would be easier for you to stop? The bicycle, because it has less mass. Now imagine two bicycles coming at you, one with twice the speed of the other. Which is easier to stop? The one moving slower. These examples show you that momentum depends on both the mass and the velocity of the object involved. In fact, momentum is defined as the product of an object's mass and velocity:

$$\text{momentum} = \text{mass} \times \text{velocity}$$

If we let p *be the momentum,* m *the mass, and* V *the velocity, then*

$$p = mV$$

Note that velocity has a direction, so momentum does too, the same as that of the velocity. You can think about momentum like this: Once you get a mass moving, you have to put out an effort to stop it.

Consider a spinning object, such as the earth rotating about its axis. What keeps it spinning? Its inertia about its spin axis. The faster it spins and the more mass spinning, the harder it is to stop. This spinning momentum is called **angular momentum.**

You can think of angular momentum as the tendency for bodies, because of inertia, to keep spinning (rotating) or orbiting (revolving). Angular momentum of a body is determined by its mass (but now the distribution of the mass about the center of motion complicates the picture), the velocity (around the center of motion), and the radius (the distance of the mass from the center of motion). For a single particle moving in a circle,

$$\text{angular momentum} = \text{mass} \times \text{circular velocity} \times \text{radius}$$

Let L *be the angular momentum,* m *the mass,* V *the circular velocity, and* r *the radius. Then*

$$L = mVr$$

For a body such as the earth, which is made up of many particles moving at different velocities at different distances from the axis of rotation, we must add up the angular momentum of all the particles.

The key aspect of angular momentum is that it is **conserved.** *If no twisting forces act on an object, its angular momentum remains the same. You can test this idea by performing a simple experiment (Fig. F.24). Tie a ball to the end of a string and whirl it around your head at a constant speed. Now, with your free hand, grab the string at half the distance you started with. The ball will move with double its circular speed. This is an example of the conservation of angular momentum. Note that if the distance from the center of spin decreases, the rate of spin increases.*

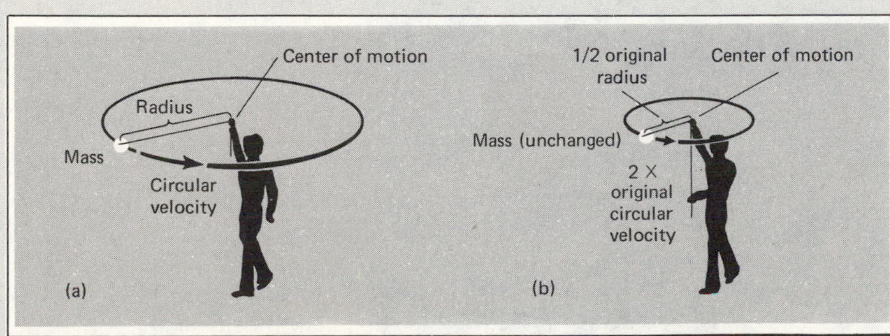

F.24
An illustration of the concept of the conservation of angular momentum.

Upon impact, the projectile's energy of motion converts into an explosive force below the ground at a depth only a few times the diameter of the infalling mass. A shock wave from the impact spreads through the rock, deforming it and throwing it outward. The volume blown out is much larger than that of the projectile, so even small objects result in big holes. Because the explosion occurs below the ground, even projectiles hitting at oblique angles leave round craters.

Even on an airless world, craters, once formed, can be wiped out by a number of processes: (1) later impacts, (2) materials thrown up from younger, nearby craters, and (3) lava flows. Large craters are less likely to be obliterated by these processes than small ones, so generally the largest craters on a surface are the oldest. Overall, the oldest part of the surface will have the highest density of craters.

The analysis of cratering leads to one key conclusion: The terrestrial planets have undergone different amounts and kinds of surface evolution. We can rank the planets and moons according to the relative amounts of modifications of their surfaces.

Our moon stands in stark evolutionary contrast to the earth. The oldest regions of the lunar highlands have not changed much since their formation. The analysis of cratering here combined with radioactive dating of Apollo samples tells us that a little over 4 billion years ago the cratering rate was 100 to 1000 times greater than now and that this intense influx tapered off, reaching the present rate about 3 billion years ago.

Now a look at another heavily cratered, little evolved world: Mercury.

9.7 Mercury: Orbital and physical characteristics

The Mariner 10 mission revealed that Mercury's surface resembles that of the moon in many ways and opened up a direct comparison of these inactive, desolate bodies. From this comparison, we can glean some ideas about what processes drive the evolution of planets without atmospheres or active interiors.

Mercury's orbit / Mercury speeds around the sun once every 88 days in a very eccentric orbit. When closest to the sun (at *perihelion*), its distance from the sun is 46 million km; at its greatest distance (at *aphelion*), it lies 70 million km out. This difference means that sunlight at perihelion falls on the planet 2.3 times more intensely than it does at aphelion.

Mercury's rotation / Mercury's small angular size and poor visibility from the earth render the study of its surface difficult (Fig. 9.23). But persistent observation reveals faint, dark, apparently permanent markings. In 1877 Giovanni Schiaparelli (1855–1910) constructed the first map of the Mercurian surface. Since the position of the surface features apparently did not change, he concluded that the rotation rate of Mercury equaled its revolution rate, 88 days. Schiaparelli thought Mercury always presented one side to the sun.

This long-accepted view changed radically in 1965, when radio astronomers Gordon H. Pettengill and Rolf B. Dyce, using the mammoth radio telescope at Arecibo, Puerto Rico, succeeded in measuring the rotational rate of Mercury using radar. They discovered that Mercury rotates in

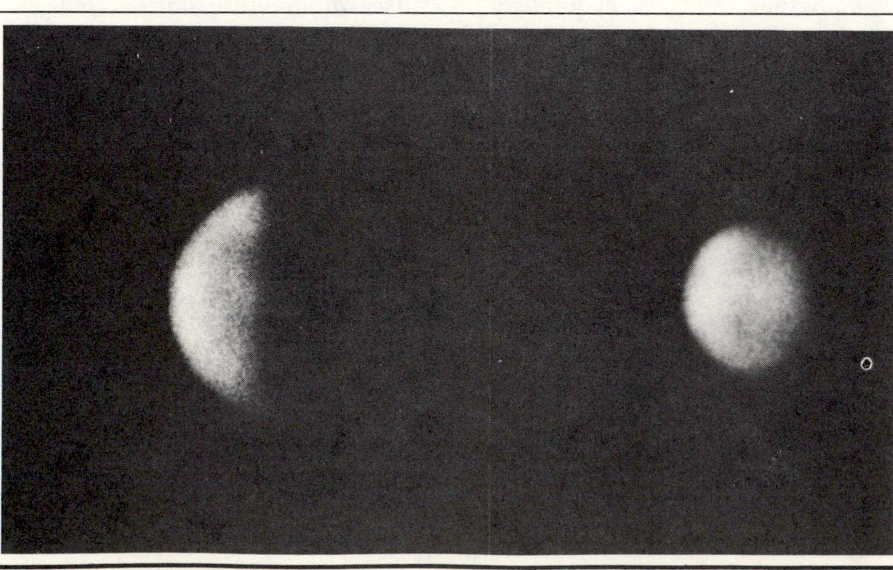

9.23
Two views of Mercury taken by earth-based telescopes. Note the vague, dusky features. (Photo by C. Knuckles, New Mexico State Observatory)

about 59 days. This new result astounded many astronomers, for all maps drawn since Schiaparelli's time had supported the 88-day rotation period.

So Mercury's rotation period is two-thirds its orbital period. How long, then, is the solar day on Mercury? It's the time from noon to noon. That's about 176 terrestrial days, just twice the length of Mercury's year.

Mercury's size, mass, and density / Knowing Mercury's orbit, we can find its physical size from its angular diameter. Mercury turns out to be a tiny world, only 4880 km in diameter. That's only about 40 percent larger than our moon.

Mercury has no known natural satellite. So it's hard to determine its mass accurately. During the Mariner 10 mission, the spacecraft sped past Mercury three times. Its acceleration from Mercury's gravity was accurately measured, so we now have a good value for Mercury's mass: 0.055 that of the earth, or about 3.3×10^{23} kg.

For such a small planet, Mercury has a fairly large mass. That means its bulk density is high: 5420 kg/m^3, essentially the same as the earth's. Comparing Mercury's interior with that of the earth, we expect a large iron alloy core (Fig. 9.24) and a rocky mantle. Although we haven't yet measured earthquakes to confirm this interior model in detail, it's probably correct by analogy to the earth.

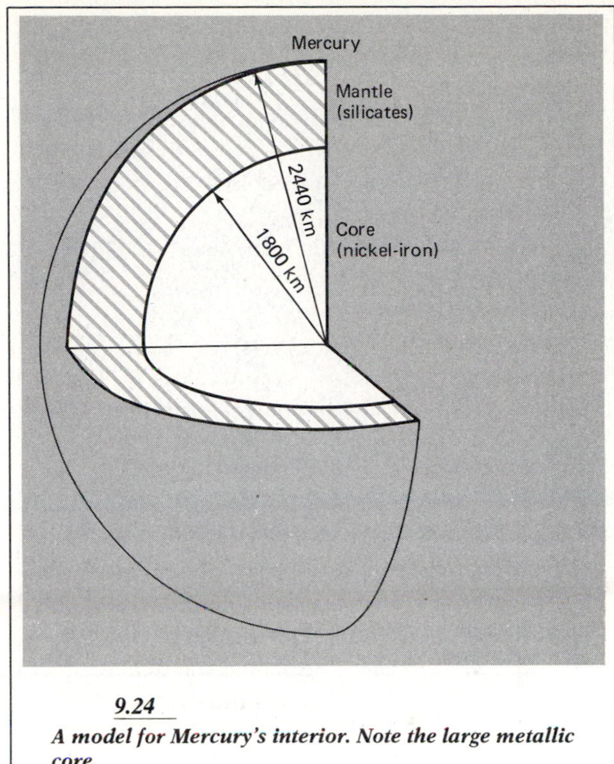

9.24
A model for Mercury's interior. Note the large metallic core.

9.8
Mercury's surface environment

Suppose that you stood on Mercury's equator at perihelion at noon. You would need to be made of sturdy stuff, because the surface temperature would be about 700 K! The surface temperature drops to 425 K at sunset and reaches about 100 K at midnight. This range of temperatures is the widest known in the solar system.

The long, hot solar day and low escape velocity (4.2 km/sec) make it unlikely that Mercury has much of an atmosphere. Gas molecules, even the most massive ones, would easily be heated to escape velocity. So any atmosphere could not be expected to last long. The Mariner 10 space probe had on board an ultraviolet spectrometer to search for an atmosphere. This device detected an atmosphere (of sorts) of helium and hydrogen. But the surface atmospheric pressure was very small, a little less than 10^{-15} that of the earth.

No atmosphere means no insulation from space. That's why the range of noon-to-midnight temperatures on Mercury is so severe. Let me contrast them to the moon's. Night on the moon and on Mercury are essentially the same: Without an insulating blanket, infrared radiation from the sunless side escapes directly into space during the long night. So both bodies have about the same midnight temperatures: about 100 K. What about at noon? The surface temperature then depends on the surface albedo and distance from the sun. The moon and Mercury have about the same albedo. But Mercury is 2½ times closer to the sun than the moon. So you'd expect that it would be hotter at noon, and it is.

With the day so hot and the escape velocity so low, why does Mercury have any atmosphere at all? One answer is the solar wind. The influx of material from the solar wind, which contains 10 percent helium, could possibly replenish the loss. Another source, at least for helium, is decay of radioactive elements in Mercury's interior.

9.9
The surface of Mercury close up

TV cameras aboard Mariner 10 scanned Mercury's surface to increase our resolution of its details by 5000 times. And what a view: a surface like our moon! There are differences, though: fewer large craters 20 to 50 km in size; no mountains; many shallow, scalloped cliffs, called *scarps* (Fig. 9.25), reaching lengths of hundreds of kilometers and rising to 1 km heights; fewer basins and large lava flows; and more relatively uncratered plains amid the heavily cratered regions. These

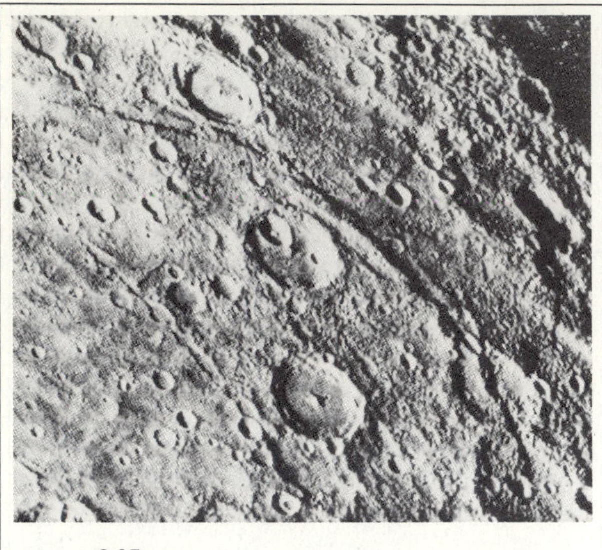

9.25
A scarp on Mercury's surface more than 300 km long. It extends diagonally from upper left to lower right in this photo. (Courtesy NASA)

differences are important, yet the general similarity jumps out at you, especially when compared to the far side of the moon. Mercury's highlands are riddled with craters like the moon's bleak highlands (Fig. 9.26). Light-colored rays spring from some of the craters, an indication that these were formed by violent impacts during Mercury's stormy past. Some craters are over 200 km in size, comparable to the biggest lunar craters.

And what of large maria basins, such as found on the moon's near side? Mariner 10 found only one: the Caloris ("hot") Basin (Fig. 9.27). Since the Caloris Basin sat on the sunrise line, only about half of it was photographed. It probably has an overall diameter of some 1300 km. The basin is bounded by rings of mountains about 2 km high. In size and structure, the Caloris Basin resembles the moon's Mare Orientale.

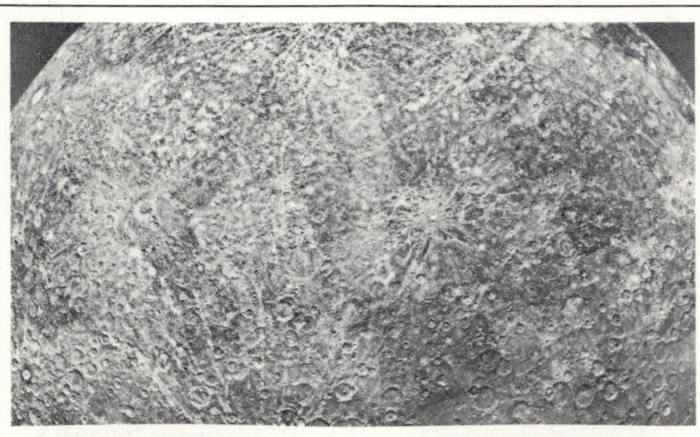

9.26
Impact craters in Mercury's south polar region. The craters with the brightest rays are the youngest; the largest ones are some 200 km in diameter. (Courtesy NASA)

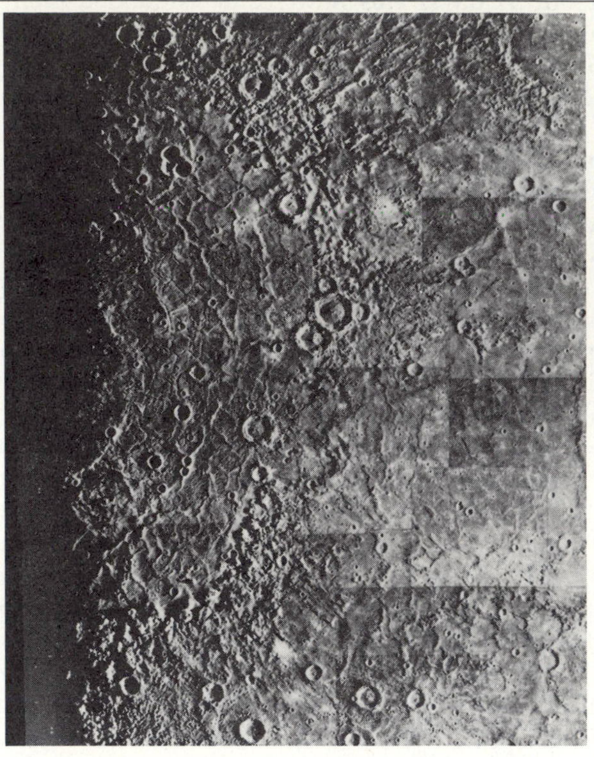

9.27
The Caloris Basin. Only a part shows in the right half of this photo; the rest is in shadow. The basin is about 1300 km across and rimmed by mountains 2 km high. (Courtesy NASA)

The Caloris Basin has a crinkled floor, perhaps fractures from rapid cooling of lava. Also visible are older craters flooded by the lava outpouring from the Caloris impact. The Caloris impact may have been so strong that it sent shock waves through to the other side of the planet, disrupting the surface there. On the side opposite Caloris lies a jumbled region, unique on Mercury's surface. Hills and ridges cut across craters and the intercrater plains. This region of weird terrain may be evidence of the cross-planet disruption powered by the Caloris impact.

All these similarities do not mean that the moon and Mercury are identical. Their surfaces differ in at least three ways: (1) Mercury's surface has scarps hundreds of kilometers long, (2) even the most heavily cratered regions are not completely saturated with craters, and (3) Mercury has fewer small craters compared to the number expected if the craters had formed at the same rate as on the moon. From the last two points, we can infer either that the cratering material had a different range of sizes for Mercury than for the

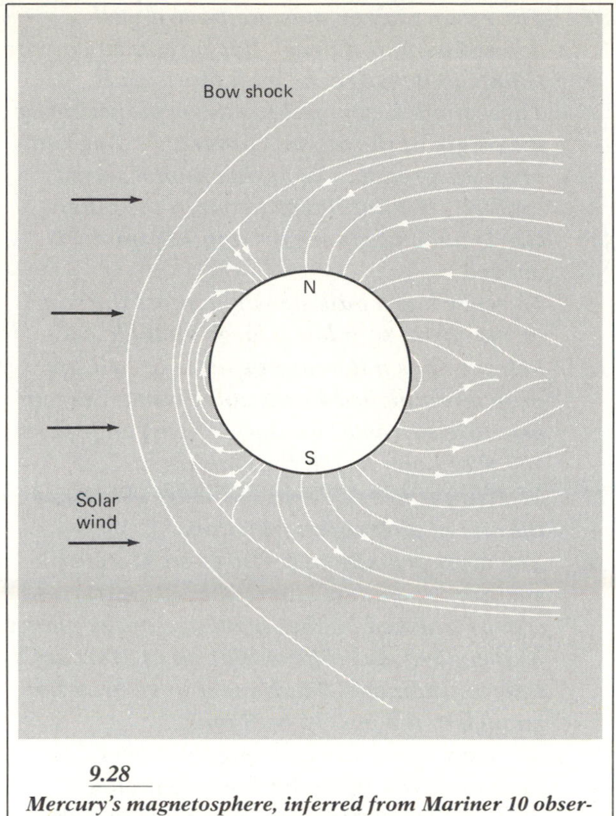

9.28
Mercury's magnetosphere, inferred from Mariner 10 observations. (Adapted from a NASA diagram)

moon or that some process modified the surface after the cratering.

Mercury's scarps vary in length from 20 km to over 500 km and have heights from a few hundred meters to 1 km. Individual scarps often travel over different types of terrain. If the regions photographed by Mariner 10 represent Mercury's overall surface, the characteristics of the scarps imply that Mercury's radius has shrunk some 1 to 3 km. How? Probably from cooling of its core, its crust, or both.

9.10 *The mystery of Mercury's magnetic field*

Mariner 10 detected a weak planetary magnetic field, about 1/100 times the strength of the earth's magnetic field at its surface. Small as this sounds, it's sufficient to carve out a magnetosphere in the solar wind (Fig. 9.28). Here the magnetic field deflects the charged particles (mostly protons) of the solar wind around the planet.

Mercury's field appears to be a dipole, more or less aligned with its spin axis. So, in general, Mercury's magnetic field is similar to the earth's, only weaker. And that's the problem. Mercury must, like the earth, have a metallic core. But it's presumed to be relatively cold and solid now because a small planet loses heat quickly. Recall (from Section 8.4) that the earth's magnetic field supposedly arises from swirling motions in its hot, liquid metallic core. The churning is thought to be driven by the earth's spinning. Because Mercury rotates much more slowly than the earth and is expected to have a cool core, no one really expected it to have a planetwide magnetic field.

What's the explanation? No one knows for sure. It may be that the dynamo model is incorrect; after all, Venus, which has a much larger and presumably hotter core than Mercury, has no detectable magnetic field. (But Venus rotates even more slowly.) Perhaps the field is left over from an older time. Maybe it originates from the planet's interaction with the solar wind. Or perhaps small planets with cool metallic cores produce magnetic fields by a mechanism not yet imagined.

9.11 *The evolution of the moon and Mercury compared*

What forces have shaped the surfaces of Mercury and the moon? The evidence is clear: impact cratering. Both the moon and Mercury lack substantial atmospheres, so weathering does not erode the surfaces. Both are tiny worlds with relatively cool interiors compared to the earth's interior. So neither has much (if any) volcanic activity now, and neither has undergone the continual surface evolution as the earth experiences from the shifting of crustal plates (see Section 8.6).

The lack of atmospheres and the short period of crustal evolution relate to the small masses of Mercury and the moon. Their surface gravities are so low that most gases have escape velocities, and atmospheres are not retained. The small masses also imply that internal heating from radioactive decay would be less than that for the earth, and the flow of heat outward would be so fast that both bodies would cool off quickly. The earth is hot at its interior, and the outward flow of heat sets up currents in the plastic mantle; these power the evolution of the earth's crust. Both Mercury and the moon lack this combination of a hot interior and a plastic mantle.

Without surface erosion and crustal evolution, the moon and Mercury retain the evidence of their early years. The similarities of surface features suggest similar histories. For example, the Caloris Basin on Mercury resembles maria on the moon. The lunar maria were probably formed by the impacts of large bodies, which created large basins; the inflow of lava from the interior filled them to make the maria. Such processes most likely also formed the Caloris Basin.

Using lunar analogies, we can set up a working model for the evolution of Mercury. (But remember that we have a lot less information about Mercury than the moon, so the details of the model are less certain.) Mercury probably went through the following general stages: (1) heating of the surface (by impacts or radioactive decay) and formation of a solid crust, (2) heavy cratering, (3) formation of impact basins, (4) filling in of basins by volcanism, and (5) low-intensity cratering. We can't date this sequence as we can for the moon because we do not have rock samples from Mercury's surface for radioactive dating. But a lunar comparison suggests that the intense sculpting of Mercury's surface took place about 4 billion years ago, not long after the planet formed.

The moon and Mercury are dead worlds, their dramatic evolution ended. Compared with the earth, they are fossil planets rather than living ones: They have passed through only the first stage of earthlike planetary evolution.

KEY CONCEPTS

1 ***The moon and Mercury are the two smallest terrestrial planets; their bulk densities indicate that they are made of mostly rocks (moon) and metals (Mercury).***

2 ***The most prominent surface features of the moon are craters, formed by the impacts of solid bodies from space some 4 billion years ago.***

3 ***The moon has essentially no atmosphere, so the surface temperature ranges widely, from 370 K at noon to 125 K at midnight.***

4 ***The tidal interactions of the moon and the earth keep the same side of the moon toward the earth, raise the ocean tides, and cause the moon to move away from the earth at a few centimeters a year.***

5 ***The main regions on the moon are the highlands (covered with low-density, light-colored rocks), and the lowlands (covered with darker, higher-density rocks); the Apollo samples, dated by using radioactive decay, show that the highlands are older than the lowlands; the maria are lava flows that filled in huge basins.***

6 ***Moon rocks differ from earth rocks as follows: (a) They contain more titanium, uranium, iron, and magnesium; (b) they have less volatiles; (c) they contain minute amounts of carbon compounds; and (d) they contain no water; they are similar in their density and isotopic abundances of oxygen.***

7 ***The moon may or may not have a well-defined core; if it does, that core must consist simply of dense rocks with few metals.***

8 ***Three models contend for the explanation of the origin of the moon: capture, fission, and binary accretion; the fission model seems unlikely, but the choice between the other two is not clear; perhaps another model is needed.***

9 ***Mercury has a bulk density almost the same as the earth, so it has a large metallic core.***

10 ***Mercury has a thin atmosphere of helium, probably supplied by the solar wind; the temperature extremes on the surface range from 700 K at noon to 100 K at midnight.***

11 ***Mercury rotates so that the same side faces the sun at alternate perihelions.***

12 ***The most prominent features on Mercury's surface are craters (formed by impacts) and scarps (formed by the shrinkage of the planet as it cooled); the differentiation of Mercury's interior indicates that it must have been hot enough in the past to be liquid.***

13 ***Mercury has a much stronger magnetic field than expected from the dynamo model.***

14 ***The surface of Mercury resembles that of the moon, so the processes that shaped it and the timing of the sequence of important events probably follow those of the moon.***

STUDY EXERCISES

1 Suppose you stepped out of TWA flight 101 onto the moon's surface. You look slowly around, at both the ground and the sky. How does what you see differ from what you would see on the earth? (*Objectives 2 and 4*)

2 Suppose you stepped out of TWA flight 102 onto the surface of Mercury. How would the scene differ from that on the earth? (*Objectives 3 and 4*)

3 Argue from a comparison of average density that the moon cannot have a metallic core like the earth's; that Mercury should have a metallic core. (*Objectives 1 and 6*)

4 How were most of the craters on the moon formed? On Mercury? Back up your statement with specific evidence. (*Objectives 2 and 3*)

5 What specific evidence do we have that the moon's lowland regions (maria) formed *after* the highlands? (*Objective 8*)

6 You are writing a grant proposal to NASA to do research on the origin of the moon. Describe the theory you plan to support in the best light possible. (*Objective 10*)

7 Compare the characteristics of the Orientale Basin on the moon to the Caloris Basin on Mercury. (*Objectives 2, 3, and 4*)

8 In *one* sentence, describe how the surfaces of the moon and Mercury *differ*. (*Objectives 2, 3, and 4*)

9 Neither the moon nor Mercury has a substantial atmosphere. Why not? (*Objective 4*)

10 In *one* sentence, describe the difference between the interiors of the moon and Mercury. (*Objectives 5, 6, and 7*)

BEYOND THIS BOOK . . .

The Apollo missions have produced a flood of data on the moon. You can find a short summary in "The Moon" by J. A. Wood in *Scientific American*, September 1975, p. 92.

For data about the moon as a planet, see *Planetary Geology* (Prentice-Hall, Englewood Cliffs, N.J., 1975) by N. M. Short. Another book with the same approach is *Geology of the Moon* (Princeton University Press, Princeton, N.J., 1970) by T. A. Mutch.

For a personal story about Apollo 11, read *Carrying the Fire* (Farrar, Straus & Giroux, New York, 1974) by M. Collins.

Voyages to the Moon (Macmillan, New York, 1960) by M. H. Nicolson is an amusing contrast to modern voyages to the moon.

C. Chapman writes about the evolution of Mercury in *The Inner Planets* (Scribner, New York, 1977).

For a pictorial tour of Mercury, look at *The Atlas of Mercury* (Crown, New York, 1977) by C. Cross and P. Moore.

B. Murray discusses the results of Mariner 10 in "Mercury," *Scientific American*, September 1975, p. 58.

10/Venus and Mars: Evolved worlds

Venus invited me in the evening whispers
Unto a fragrant field with roses crowned . . .

JOHN CLEVELAND

Learning objectives

After studying this chapter, you should be able to:

1. **Compare Venus and Mars in size, mass, and bulk density to each other and to the earth.**
2. **Compare and contrast the surface environments (temperature, atmosphere, general surface features) of Venus and Mars to each other and to the earth.**
3. **Compare the compositions of the atmospheres of Venus and Mars to the earth's.**
4. **Sketch a model for the interiors of Venus and Mars and compare them to the earth's interior.**
5. **List the major surface features of geologic importance on Venus and Mars and compare them to those on the earth.**
6. **Contrast the planetary magnetic fields of Venus and Mars to the earth's.**
7. **Evaluate your chances of survival on Venus and Mars.**
8. **Discuss the role of cratering in the shaping of the surfaces of Venus and Mars.**
9. **Discuss the implications of the volcanoes of Mars for the evolution of the Martian surface and atmosphere.**
10. **Compare and contrast the inferred histories of Venus, Mars, and the earth; justify an ordering of these planets from the least to the most evolved.**

Central question

What forces have shaped the evolution of Venus and Mars, planets with hot interiors and substantial atmospheres?

10.1
Venus, photographed in blue light at its crescent phase. (Courtesy Palomar Observatory, California Institute of Technology)

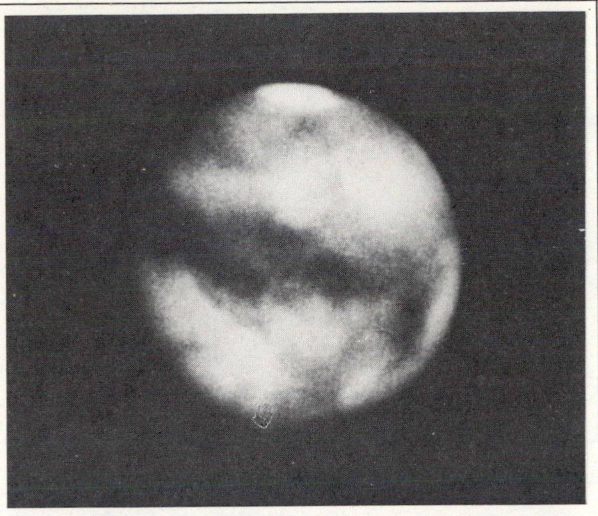

10.2
Mars as seen from the earth. Note the white polar cap at top and the fuzzy dark regions. (Courtesy Lick Observatory)

Venus, the brilliant light of love (Fig. 10.1); Mars, the red sign of war (Fig. 10.2). These two worlds come the closest to the earth in space. And these planets resemble the earth more closely than any others. They are truly terrestrial planets.

Yet in many ways the similarities are superficial; the differences go much deeper. Venus is blistering hot at its surface—hotter even than Mercury at noon! A thick, dense atmosphere of carbon dioxide presses heavily on the barren ground. On Mars, the atmosphere also consists of carbon dioxide. But it's thin, offering no protection from the incoming solar ultraviolet and no hindrance to the outgoing infrared. Mars is mostly a cold desert, where the water ceased to flow tens of millions of years ago.

How did Venus and Mars end up so different from the earth? What forces shaped their evolution? This chapter deals with the comparative evolution of these earthlike worlds.

10.1 Venus: Orbital and physical characteristics

Viewed from the earth, Venus can outshine every celestial body except the sun and the moon. If you know where to look, you can even spot Venus during the day! Venus's brilliance comes from its unbroken swirl of clouds, which reflects 77 percent of the incoming sunlight back into space (Fig. 10.1). This cloud cover completely frustrates any attempt to view its surface features through optical telescopes.

But we have pierced Venus's cloudy veil to uncover the surface environment. Radar beams, bouncing off the surface, paint its terrain. Spacecraft have sped by the planet, orbited around it, and plunged to its forbidding surface. What a difference from the earth! Surface temperatures hit 750 K. The atmosphere presses at an unrelenting 100 atm, and it contains mostly carbon dioxide. The surface resembles a cross between Mars and the moon, very little like the earth.

Venus has gained the reputation of being the earth's twin sister. In terms of size (the diameter of Venus is 12,104 km; earth, 12,756 km), average density (Venus, 5200 kg/m^3; earth, 5500 kg/m^3), and mass (Venus, 0.82; earth, 1.00), the sisterhood is appropriate. In most other ways, however, Venus has an environment that is tremendously different from the earth's—indeed, a veritable hell!

Revolution and rotation / Second planet from the sun, at an average distance of 0.72 AU, Venus completes one orbit in only 225 days. Venus's orbit is the most circular of all the planets', so Venus's distance from the sun varies by only 1.5 million km (1.4 percent).

Thwarted by the lack of a surface view because of the unyielding cloud cover, astronomers before 1961 had no real idea of Venus's rotation rate. Some proposed 24 hours (arguing that Venus is the earth's twin), and others decided that the rotation period must equal the revolution period, 225 days. (These people were misled by the

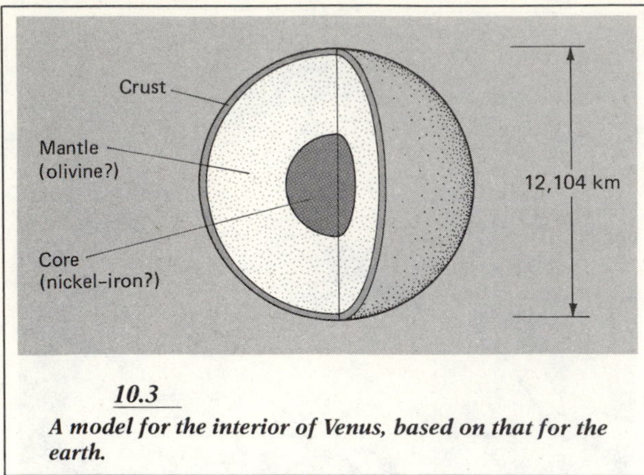

10.3

A model for the interior of Venus, based on that for the earth.

incorrect rotation rate agreed upon for Mercury.) Radar astronomers have now found that Venus rotates once every 243.01 days, retrograde. Retrograde rotation means that it spins from east to west, rather than west to east as does the earth. So on Venus, the sun (if you could see it through the clouds) rises in the west and sets in the east.

This long retrograde rotation results in a solar day on Venus of 117 terrestrial days. But because of the clouds, an inhabitant of Venus could never see the sun's disk directly. Like an overcast day on earth, it would be light during the day, but the sun's intensity on Venus is only about 1/100 that at the earth's surface.

Size, mass, and density / We can measure the earth-Venus distance directly by radar. Then, from the planet's angular diameter, we can calculate its physical diameter. Venus turns out to have a diameter of 12,104 km, only 5 percent smaller than the earth.

Like Mercury, Venus has no known natural moon, so we can accurately find the mass of Venus only when a spacecraft passes or orbits it. If in orbit, we simply use Kepler's third law. During a flyby, we measure the acceleration of the spacecraft (from the Doppler shift of its radio signals) and use Newton's law of gravitation. Venus's mass comes out to be 0.815 times that of the earth, or 4.86×10^{24} kg.

With the mass and size in hand, we can find Venus's bulk density: 5200 kg/m^3, almost the same as that of the earth. We guess that the interior of Venus should closely resemble the earth's interior (Fig. 10.3): a rocky crust (which the Venus landers confirmed), a mantle, and a metallic core. Because Venus has a lower density than the earth, we imagine that it has a somewhat smaller core.

Magnetic field / But this model has a severe problem. A metallic core, liquid in part, implies by comparison with the earth that Venus should have a planetary magnetic field. Because Venus rotates 243 times more slowly than the earth, we expect its internal dynamo to be weaker and the magnetic field to be less intense than the earth's, but still there. No probe to date has detected any magnetic field. If one exists, it must be at least 10,000 times weaker than the earth's magnetic field! That's much weaker than expected from a simple dynamo model. What a magnetic mess—Mercury has a planetary field, and Venus does not! Mars has a barely detectable magnetic field (see Section 10.5). Perhaps the dynamo model is not the correct one to apply to all planets.

One other possibility: We know that the earth's magnetic field periodically reverses its polarity (see Section 8.6). In the middle of a rever-

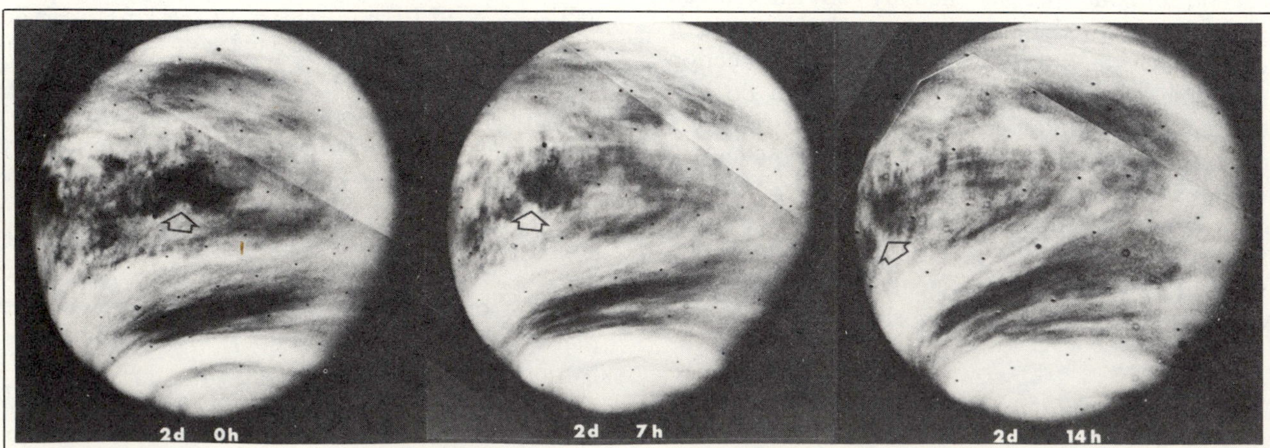

10.4

Venus, photographed in ultraviolet light by Mariner 10. The cloud patterns show the flow structure in the upper atmosphere. The clouds circle the planet once in about 4 days; shown here are 14 hr of that circulation. (Courtesy NASA)

sal, the magnetic field is essentially zero. Venus may be midway through a reversal and so has a very weak field right now.

10.2 *The atmosphere of Venus*

The atmosphere of Venus differs remarkably from ours—a key clue to understanding the planet's evolution. Substantial atmospheres directly affect changes in the surface of a terrestrial planet.

Clouds / Recent efforts have gathered much new information about the yellow-white clouds of Venus (Color Plate 3). The cloud tops, which you see in a telescope, reach about 65 km above the surface. (The highest clouds on the earth go up to about 16 km.) Ultraviolet photography by Mariner 10 revealed that these cloud tops flow with the upper atmosphere (Fig. 10.4), in patterns similar to jet streams of the earth. Ringing the equator, the clouds whiz around at roughly 300 km/h—fast enough to orbit the planet in only four days (Fig. 10.4). In addition to this planetwide circulation, winds also blow from the equator to the poles in large cyclones 100 to 500 km in diameter. They culminate in two giant cloud vortices that cap the polar regions.

What drives such fierce winds in the upper atmosphere of Venus? It's generally believed that solar heating does the trick, as on the earth, but details have yet to be worked out.

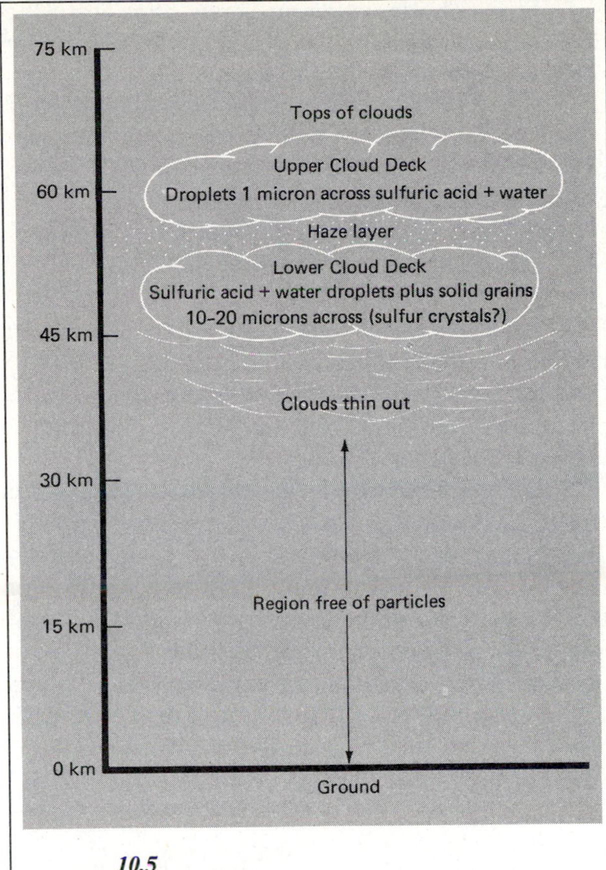

10.5
The structure of the clouds in the atmosphere of Venus. (Based on NASA data)

The Pioneer Venus probes found that clouds float in two broad layers (Fig. 10.5). The upper cloud deck tops at roughly 65 km and has a thickness of some 5 km. Below the upper deck sits a thin haze layer. Below that, at a 50-km height, is the lower cloud deck, by far the densest layer. The cloud particles here are both liquid and solid. Below 50 km, the clouds gradually thin out; below 33 km, and down to the surface, the atmosphere is clear of any particles.

In the densest part of the clouds, the visibility is less than 1 km, dangerous to fly through but still not a pea-soup fog. The clouds of Venus aren't particularly murky, just very thick.

That's the structure of the clouds. But what are they made of? The best proposal to date, made by Andrew Young, sees the upper level clouds as concentrated solutions of sulfuric acid. Models of the clouds designed to match their observed infrared spectrum imply that they contain a solution of 90 percent sulfuric acid mixed with water.

Although the atmosphere and clouds of Venus do contain some water vapor, it doesn't amount to very much compared with the total amount of water on the earth. If all the earth's water (in the atmosphere and oceans) were spread in a uniform layer over its surface, that sheet of liquid would be 3 km thick. All the water in the atmosphere of Venus (none exists on the surface because it's so hot) would amount to a layer only 30 cm thick. So our earth has about 10,000 times as much water as Venus does.

Atmosphere and surface temperature / Since 1932 we have known that Venus's atmosphere contains carbon dioxide, but we did not know how much. Interplanetary probes launched by both the United States and the Soviet Union indicate that the atmosphere contains about 96 percent carbon dioxide, 3 percent nitrogen, some argon, and traces of water vapor (0.1 to 0.4 percent), oxygen, hydrogen chloride, hydrogen fluoride, hydrogen sulfide, sulfur dioxide, helium, and carbon monoxide.

The Venera descents found the surface pressure to be 95 atm and the sunlit surface temperature about 740 K. This high temperature probably results from the effective trapping of surface heat (see Section 8.5), because carbon dioxide absorbs infrared radiation well. An extreme greenhouse effect keeps Venus very hot.

Atmospheric winds on Venus blow from the day to the night sides and from equatorial to polar regions. The wind flow carries heat. Along with the very effective atmospheric insulation, this helps to keep the temperatures fairly constant over Venus's surface. They vary about 10 K or less. So it doesn't cool off much at night.

10.3
The surface of Venus

How to investigate Venus's hidden surface? In two ways: by landing probes on the surface to examine it close up, and by bouncing radar (which penetrates the clouds) off the surface and analyzing the reflections.

Close-up views of the surface / The Soviet Union has sent four spacecraft (Venera 9, 10, 13, and 14) to the planet's surface. These landers worked only a short time after landing, because of the extreme pressure and temperature. They sent back close-up photos of the surface. The pictures from Venera 9 and 10 show slabby rocks about 60 cm long and 20 cm wide. A few rocks have small holes that look like they were once filled with gas; this implies a volcanic origin. Some rocks show jagged edges, indicating little erosion, and others show blunted, rounded edges, an indication of much erosion. The rocks rest on loose, coarse-grained dirt. What kind of rocks are these? Measurements by the landers of gamma-rays emitted by the radioactive potassium, uranium, and thorium in the rocks indicate that at one lander site they are basaltic, like those lining the earth's ocean basins (see Section 8.2), but at the other site they are granitic, like the earth's mountains. Both are igneous rocks formed from lava.

The photos from Veneras 13 and 14 showed quite different views (Fig. 10.6). Those from Venera 13 revealed a rocky plain with clustered outcroppings; the rocks were basaltic, similar to those on earth in continental rift zone. In contrast, those from Venera 14 showed even layers of broken, rocky plates that were basically volcanic, like the basalts in the earth's ocean floor near midoceanic ridges. So in both places (separated by 900 km), the rocks appear to be relatively young.

Radar mapping / The Venera pictures show a worm's-eye view of the surface. But what is the

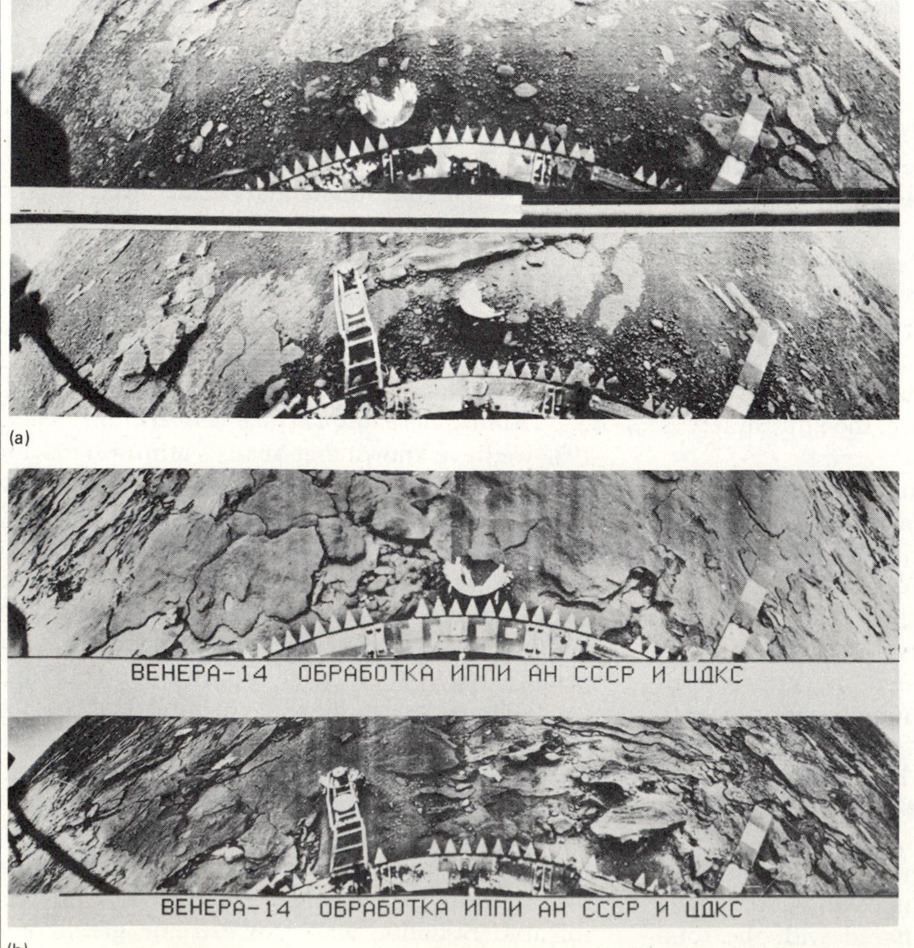

10.6
(a) *Venera 13 photo of the surface. The horizon is visible in the upper corners and the lander at bottom center. Note the slabs of rock, patches of dark soil, and pebbles close to the edge of the lander.* (b) *Venera 14 view of Venus. The flat expanses of rock are unbroken by any soil patches, and fewer pebbles are visible.*

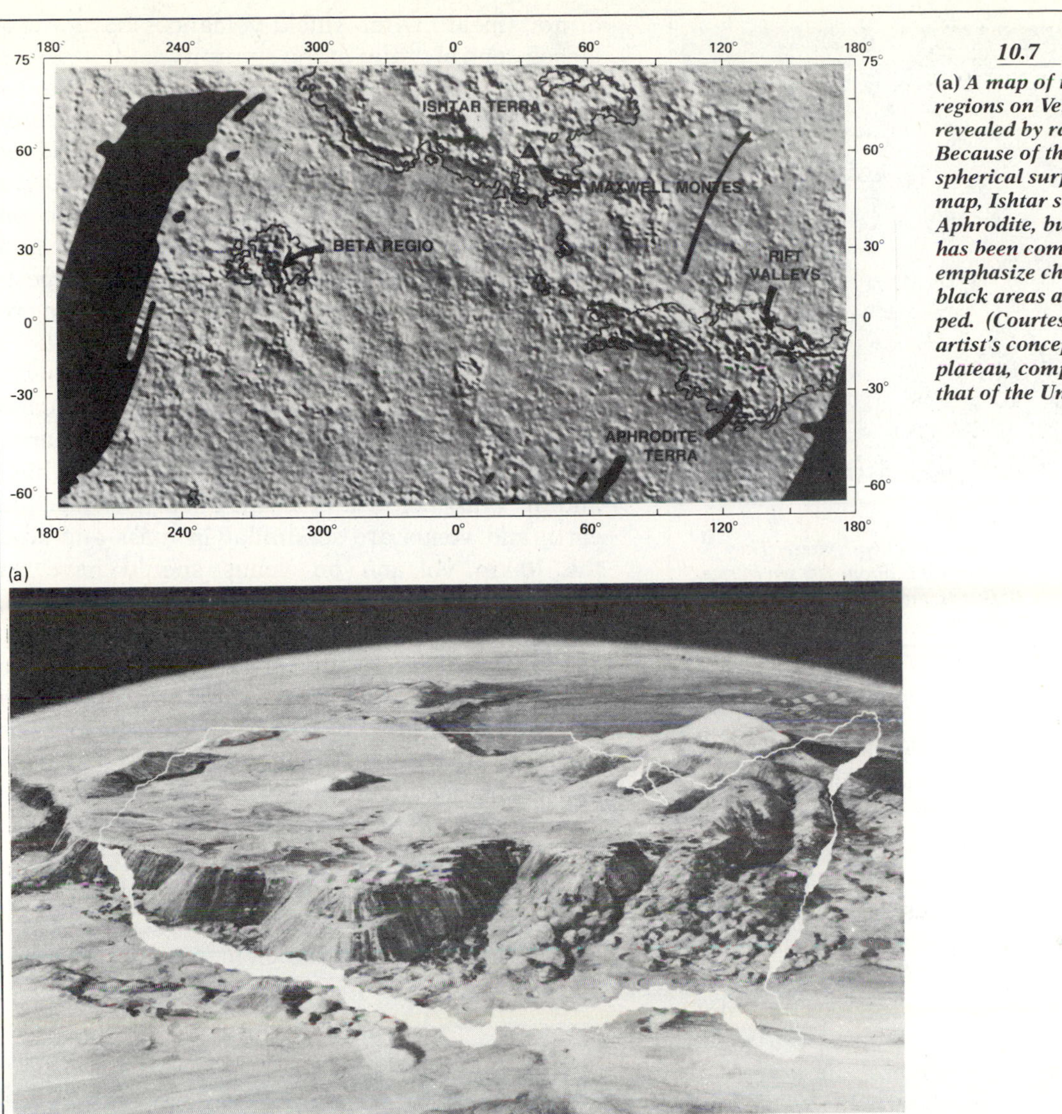

10.7
(a) *A map of the highland regions on Venus's surface, as revealed by radar mapping. Because of the distortion of a spherical surface drawn as a flat map, Ishtar seems larger than Aphrodite, but it isn't. The map has been computer processed to emphasize changes in elevation; black areas are those not mapped. (Courtesy NASA)* (b) *An artist's conception of the Ishtar plateau, comparing its size to that of the United States.*

more general lay of the land? We now have a pretty good overall idea from ground-based radar maps and radar mapping by the Pioneer Venus orbiter. These maps reveal a varied terrain: mountains, high plateaus, canyons, volcanoes, ridges, and impact craters. Overall, Venus looks fairly flat. Elevation differences are small, only 2 to 3 km, with the exception of a few highland regions. Here the land reaches up to 10 to 12 km, compared with 4-km highland-lowland differences on the moon and Mercury, 25-km differences on Mars, and a 20-km difference on the earth (Mt. Everest is 9 km high, the Japanese trench 11 km deep).

The northern and southern halves of the mapped face of Venus differ remarkably. The northern region is mountainous, with uncratered *upland plateaus*. The southern part consists of relatively flat *cratered terrain*.

Upland plateaus / The great northern plateau is called *Ishtar Terra*. Ishtar is huge (Fig. 10.7)—some 1000 by 1500 km. (That's larger than the biggest upland plateau on the earth, the Himalayan Plateau.) Three mountain ranges border Ishtar on the west, north, and east. The eastern range, called Maxwell Montes (Fig. 10.7), contains the highest elevations on Venus seen to date: some 12 km. Radar shows a rough terrain, as expected from a lava flow. And some radar views indicate a volcanic cone near Maxwell's center. The northern mountain range rises about 3 km above Ishtar; the western one reaches only 2 km above the plateau.

These three mountain ranges may have folded and risen from moving plates in Venus's crust, making them similar to mountains built from plate tectonics on the earth.

Cratered terrain / The southern half of

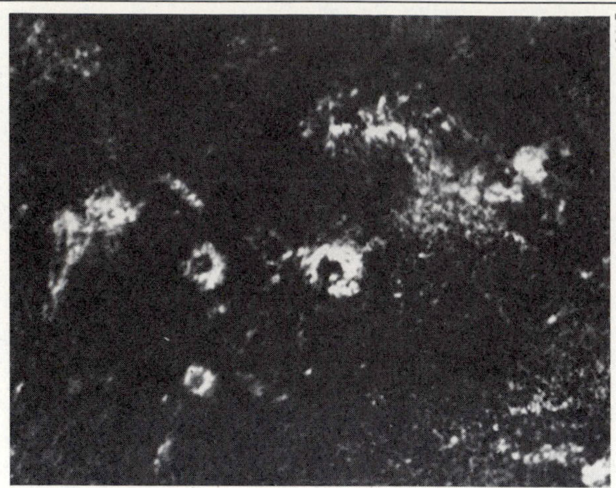

10.8
Ring-shaped features in the southern hemisphere. They may be impact craters. The largest is about 100 km in size. (Courtesy D. Campbell, B. Burns, and V. Boriakoff; observations at Arecibo Observatory)

Venus's face consists of low, rolling plains apparently punctuated by craters (Fig. 10.8), both large craters (up to 800 km in diameter) and smaller ones, less than 1 km in size. These are probably impact craters. (Craters smaller than this size do not form on Venus because the dense atmosphere completely vaporizes incoming debris before it reaches the ground.) The craters of Venus resemble those on the moon, Mercury, and Mars: They have high peaks in their centers, a direct clue to their impact origin. In general, the craters are shallow from erosion (probably by wind), only about 500 m deep.

One Venera probe landed in the midst of the southern terrain. Its instruments indicated that the rock was more granitic than basaltic—one clue that at least some of the surface here is old. Later lava flows could have poured out basaltic rock. Also, the fact that we still see craters here also implies that the surface must be old; otherwise volcanoes and mountain building would have obliterated all the craters.

One high region reaches up from the cratered terrain: Aphrodite. Aphrodite is almost 3000 km across, about half the size of Africa. It also has a rough surface that hits 7 to 10 km in height.

Volcanoes / The *Beta Regio,* which contains at least two separate volcanoes, appears to be an enormous volcanic complex that formed from a great north-south fracture zone (Fig. 10.9). The volcanoes here have gentle slopes; they are called *shield volcanoes.* (Instead of a sharply uplifted cone, shield volcanoes are relatively flat, like an armor shield.) Often shield volcanoes have a collapsed central crater at the summit.

One volcano in the Beta Regio has a diameter of 820 km, a height of 5 km, and a summit crater 60 by 90 km. In contrast, the island of Hawaii on the earth (a volcanic island) is 200 km across and 9 km high. (The largest volcano on Mars, Olympus Mons, is 550 km in diameter and 20 km high.) So Venus may have the largest-diameter volcanoes in the solar system! There are hints that some of the volcanoes may be active.

If the size and mass of a planet are known, geologists are able to relate the height of a volcano to the thickness of the crust below it. On earth, the Hawaiian volcano Mauna Loa is 9 km high and the crust below it is 57 km thick. Because the earth and Venus are so similar in mass and size, the 10-km volcano on Venus should have an underlying crust of about the same thickness, some 60 to 65 km.

Rift valleys / Venus also sports enormous canyons. On the side of Venus that faces the earth at inferior conjunctions, a huge canyon extends for more than 1300 km. It is about 150 km wide and 2 km deep. On the other side of Venus, an even larger canyon scars the landscape: It is 5 km deep, 320 km wide, and at least 1400 km long. (Its length has not yet been completely mapped.)

The canyons of Venus appear to be *rift valleys,* which form along fault zones rather than as a

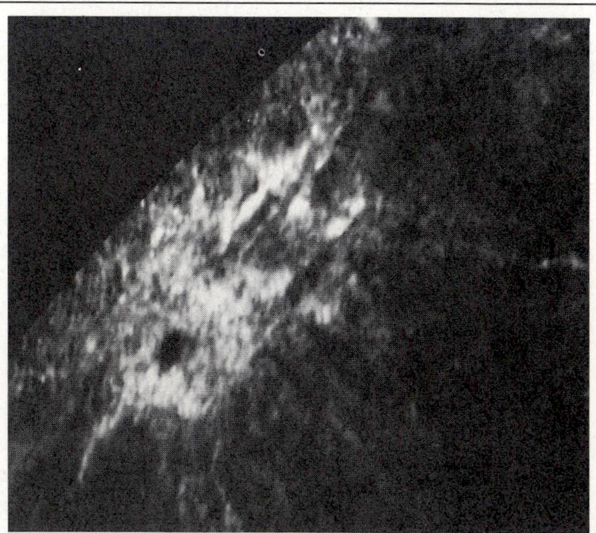

10.9
A radar map of the Beta Regio, which contains a large volcanic peak called Theia Mons. The central dark region is the volcano's crater, some 5 km high, and the streak radiating outward may be lava flows. (Courtesy D. Campbell, B. Burns, and V. Boriakoff; observations at Arecibo Observatory)

result of water erosion. (A large rift valley splits New Mexico; the Rio Grande flows down its length.) Rift valleys appear on the earth where the crust is spreading apart.

To sum up: The surface of Venus has mountains, plateaus, volcanoes, impact craters, lava flows, and rift valleys. For all this variety, the surface is remarkably flat: Only 18 percent of the mapped surface extends above 7 km, 11 percent above 10 km. In contrast, about 30 percent of the earth's surface reaches above 10 km.

Venus does not appear to have lunar-type basins, lowlands filled by lava flows. As indicated by the presence of craters, the lowlands of Venus must be older than the rest of the crust. The highlands are more evolved. That's just the opposite of the earth, where the lowlands (the ocean basins) make up the youngest part of the crust. The earth's continents are older. On Venus, the highland masses are younger than the lowlands.

10.4 The evolution of Venus

Overall, the surface of Venus differs from that of the earth. How did the differences develop, even though Venus has about the same size, density, mass, interior composition, and structure? Probably from the lack of liquid surface water. On earth, the oceans contain a large amount of the carbon dioxide. On Venus, the carbon dioxide cannot be trapped in this way. It stays in the atmosphere, keeping up the severe greenhouse effect.

The crust of Venus has experienced some crustal-plate movement. The evidence: rift valleys, where the plates have pulled apart, and mountain plateaus, where plates have collided. But the widespread cratered terrain indicates that plate movement has not been a planetwide process, as it has been on the earth. (Also, our observations to date do not quite have the resolution to show all the expected evidence for plate tectonics.) Finally, without water, large-scale erosion cannot carve landforms (such as river valleys) commonly found on the earth.

What do these observations imply for the geologic history of Venus? Right now we can really only speculate, using the earth as a guide. The early history of Venus (earlier than 4 billion years) must have followed the earth's early history, because both planets have similar mass, density, and size. The later history more closely resembled that of Mars.

We infer that Venus formed about 4.6 billion years ago with the other terrestrial planets (see Chapter 12). As happened to the earth (see Section 8.8), Venus's interior differentiated from internal heating. During the first 500 million years, a crust—part basalt and part granite—formed and solidified. About 3 to 4 billion years ago, large masses bombarded the surface, and fractured the crust. Volcanoes erupted. Bombardment by smaller bodies from space cratered the surface. About 3 billion years ago, intensive bombardment ended, and erosion has somewhat altered the ancient surface of Venus.

Since then, plate movement helped to push up some highland regions. Rift valleys appeared. Huge volcanoes vented through cracks in the surface; their cones formed the shield volcanoes of today. Venus seems to have evolved in a sequence similar to that of the earth, but more slowly and not as much in the later stages.

10.5 Mars: General characteristics

In this century Mars appeared to be the most likely home of extraterrestrial life in the solar system, as portrayed in the fantasies of H. G. Wells and other science fiction writers.

But the Viking landers have dampened this optimistic attitude: Mars now seems barren of life as we know it (more in Chapter 22). Viking and other spacecraft have described a new Mars (Fig. 10.10): a planet with plentiful, ancient craters, giant canyons, and huge volcanoes.

Martian orbit and day / Mars orbits the sun at an average distance of 1.52 AU. But Mars's distance from the sun varies considerably (by about 9 percent) because its orbit is fairly eccentric. So the distance between Mars and the earth at opposition, when both planets lie on the same side of the sun, ranges from less than 56 million km to more than 101 million km.

Surface markings visible through telescopes make the Martian rotation rate easy to measure. In 1659 the Dutch physicist Christian Huygens (1629–1695) observed the rotation rate to be close to 24 hours. Modern measurements place the value at 24 hours, 37 minutes; a day on Mars lasts only a bit longer than on the earth.

Size, mass, and density / Mars has a diameter of 6787 km. That's only 53 percent the earth's size. So Mars is about half the earth's size, but larger than Mercury (by about 40 percent). Mars has moons, so we can use Newton's form of Kepler's third law (see Focus 4.2) to find its mass from their orbits: It is only 6.4×10^{23} kg, about 11 percent of the earth's mass.

With the mass and size in hand, we can easily work out Mars's density: 3900 kg/m^3, only a

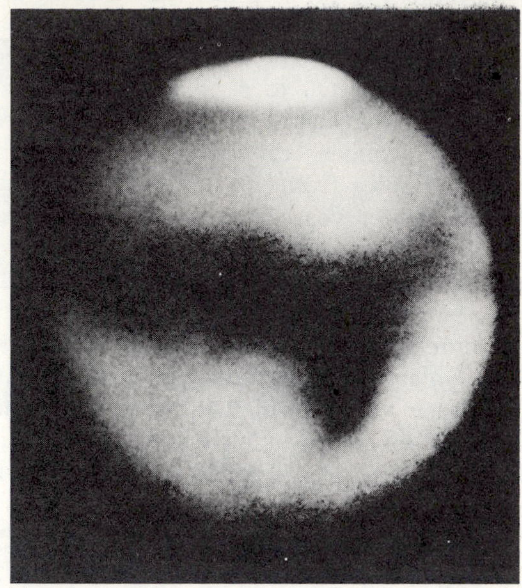
(a)

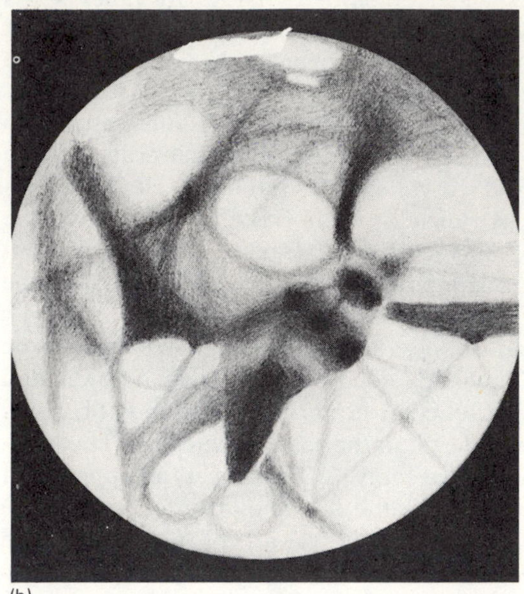
(b)

10.10

(a) Mars, a drawing made in 1926 of the surface features seen through a telescope. (Courtesy Lick Observatory) (b) Surface markings on Mars that are visible from the earth. A white polar cap tops the planet. (Courtesy Lick Observatory) (c) A mosaic of Mars compiled from Mariner 9 photos. The north pole ice cap is at top. Just below the center is the giant volcano Olympus Mons. (Courtesy NASA)

bit higher than the moon's (3300 kg/m^3) and much less than the earth's (5500 kg/m^3).

This comparatively low density implies that Mars's interior (Fig. 10.11) must be different from the earth's. In particular, its core must be smaller and probably consists of a mixture of iron and iron sulfide, which has a lower density than the materials in the earth's core. The Martian mantle probably has the same density as the earth's. The exact composition of the mantle is not known. Many different models have been developed. One, by John Lewis, has a mantle with olivine (an iron-magnesium silicate), iron oxide, and some water (0.3 percent).

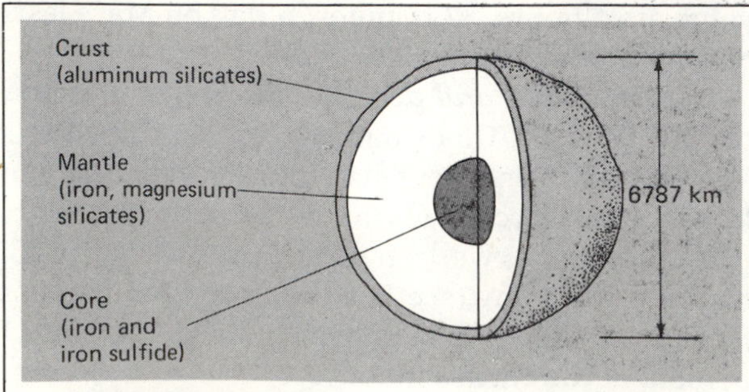

10.11

One model of the Martian interior. (Based on calculations by J. S. Lewis)

Magnetic field / Mars has an extremely weak planetwide magnetic field, only 1/500 the strength of the earth's. That small a value presents a puzzle if the dynamo model correctly describes the origin of planetary magnetic fields. Mars rotates as fast as the earth. Though the core is smaller, it should contain a substantial amount of iron or nickel-iron. We have no direct evidence that the core is liquid, but the evidence for past volcanic activity implies a hot mantle and so a hot, probably liquid core. So Mars should have a moderately strong field, but it does not.

Perhaps, as for Venus, we are viewing Mars in the middle of a magnetic-field reversal. But it is unlikely that we would catch both planets in the process of changing polarity. Again, the dynamo model is called into question by the weak field of another earthlike planet.

10.6 The Martian atmosphere and surface temperature

Astronomers have known for a long time that Mars has an atmosphere. Most believed that the Martian atmosphere was like the earth's but thinner. Well, thin it is, but nothing like the earth's! The Viking landers found average surface pressures of roughly 1/200 the earth's surface pressure. (You'd have to travel 40 km up in the

(c)

earth's atmosphere before the pressure falls that low.) This thin atmosphere consists of 95 percent carbon dioxide, 0.1 to 0.4 percent molecular oxygen, 2 to 3 percent molecular nitrogen, and about 1 to 2 percent argon, very similar in relative composition to the atmosphere of Venus.

The Viking orbiters measured the water vapor in the atmosphere and found the greatest amounts in the high northern latitudes. Peak concentrations were about 0.01 mm of precipitable water, which means that if all the water above that location rained to the surface, it would form a layer only 0.01 mm thick. On earth, the atmospheric water vapor is typically several centimeters of precipitable water, and, of course, the oceans are several kilometers thick. Mars is a very dry planet compared with the earth.

It cannot rain on Mars today because of the low surface pressure. Only in the deepest canyons, where the atmospheric pressure is higher, could water be liquid on the surface. However, it is common to have water ice on Mars, either on the surface or in the clouds (Fig. 10.12). Some evidence suggests water in a permanent frost layer beneath the surface.

Although the atmosphere contains mostly carbon dioxide, its low density does not provide much of a greenhouse effect against temperature extremes. At the Martian equator, when Mars is closest to the sun, the difference between noon and midnight amounts to almost 100 K. The summer tropical high of 310 K (37°C) is exceptional. For a period of two Martian months, the surface temperature remains below the freezing point of water both day and night.

At the Viking 1 site, 23° north latitude, the air temperature near the ground ranges from −85°C to −29°C. At the Viking 2 site, which is farther north, the temperature falls even lower, and water condenses on the surface. The layer of water ice that coats the rocks and soil is extremely thin, less than a millimeter. The frost remains for about 100 days.

10.7 The Martian surface viewed from the earth

Because an astronomer must peer through two atmospheres, the view of Mars is usually poor in clarity. A small telescope shows the main surface features: the white polar caps, light reddish-orange regions, and dark areas. Spacecraft visits improved dramatically the quality of our vision.

The sands of Mars / The definite surface features on Mars are dark, apparently greenish-gray areas, in contrast to the reddish-orange color of the rest of the surface. Some early observers thought that these features were oceans. This idea was discarded when the surface pressure was

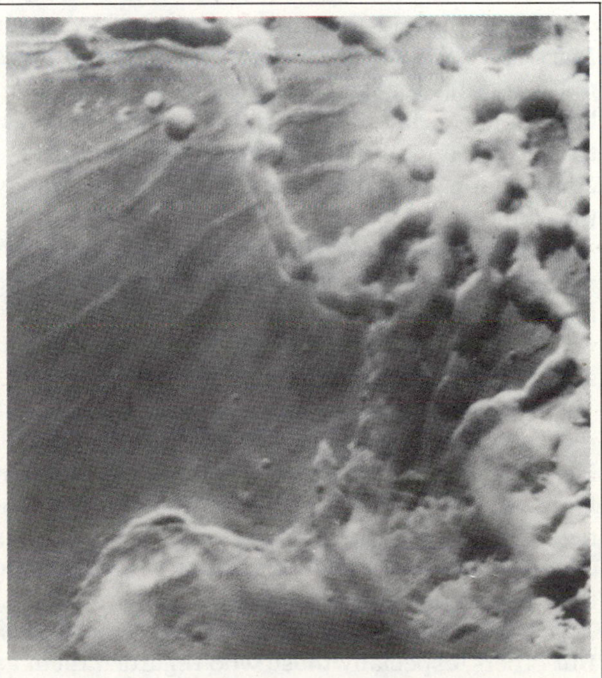

10.12
Canyons on Mars filled with fog in the morning. (Courtesy NASA)

found to be too low for liquid water. Other observers contended that the dark regions were green and so indicated vegetation. This greenness turns out to be an illusion caused by the contrast of the light and dark areas; the dark regions are not really green. They are actually red, just not as red as the light regions.

The light orange and yellow-brown regions make up almost 70 percent of the Martian surface. They give Mars its striking reddish appearance. In 1934 the American astronomer Rupert Wildt suggested that these areas contain ferric oxide—rusted iron! Iron oxides come in many forms on the earth; all are characteristically brown, yellow, and orange. Infrared observations have added support to the idea that the surface contains substantial rusted iron combined with water—perhaps as much as 1 percent of the surface is water bound up with minerals.

The Viking landers' measurements indicated a surface composition of about 19 percent ferric oxide (Fe_2O_3). In addition, they measured about 44 percent silica (SiO_2), which leads to the conclusion that silicate minerals make up a major part of the surface. The Martian surface is covered with rusty sand (Color Plate 2).

Global dust storms / The red sand, some of which is much finer than that on the earth's beaches, is blown up by fierce winds, greater than 100 km/h, to create planetwide dust storms. They occur most violently when Mars is closest to the sun. Then the dust clouds, whipped up to heights of 50 km, shroud the entire planet. They cover Mars in a yellow haze for about a month. It takes many months for the fine dust to completely settle back to the surface. These global storms sandblast the surface and mix it up so much that the surface composition over the planet becomes essentially the same.

We now know that this wind-driven dust causes most of the changes in Martian surface features seen in the past. For example, the windstorms blow dust into dunes or deposit it in streaks around mountains and craters.

Canals and polar caps / In 1877 Schiaparelli recorded Martian surface features in great detail. He charted a number of dark, almost straight features, which he called *canali*, Italian for "channels." The word was translated into English as canals, however, which implied to some people that they were artificial structures.

Some observers could not see any canals. But others, especially those who regarded them as natural waterways, continued to find more. These so-called canals ignited the curiosity of the American astronomer Percival Lowell (1855–1915). To pursue his interest in Mars, Lowell (Fig. 10.13) in 1894 founded an observatory near Flagstaff, Arizona, to take advantage of the excellent observing conditions there. Shortly afterward, he published Martian maps showing a mosaic of over 500 canals (Fig. 10.14). In a series of popular books, Lowell argued that the canals were artificial waterways, constructed by Martians to carry water from the polar caps to irrigate arid regions for farming.

10.13
Percival Lowell. "The solidarity of the Martian land system points to an efficient government . . ." (Courtesy Yerkes Observatory)

Lowell believed that the polar caps were water ice and that the dark regions were areas of vegetation that displayed seasonal growth, prompted by water from the polar caps. The polar caps are indeed largest in winter and smallest in summer.

We know now that the polar caps do consist mostly of water ice, especially the residual cap left in the summer, which ranges in thickness from year to year from 1 m to 1 km (Fig. 10.15). The outer reaches of the caps, prominent in winter, consist of carbon dioxide ice, which condenses at a lower temperature than water ice. (At Martian surface pressures, water ice condenses at about 190 K, carbon dioxide ice at 150 K). Lowell was right: Water does exist on Mars—but it does not flow freely on the surface because the temperature is too cold and the pressure too low. (If all the water in the polar caps could cover the surface as a liquid, it would form a layer only about 10 m deep.)

But Lowell was wrong about the vegetation and the canals. Some astronomers now believe that windblown dust deposits might have created

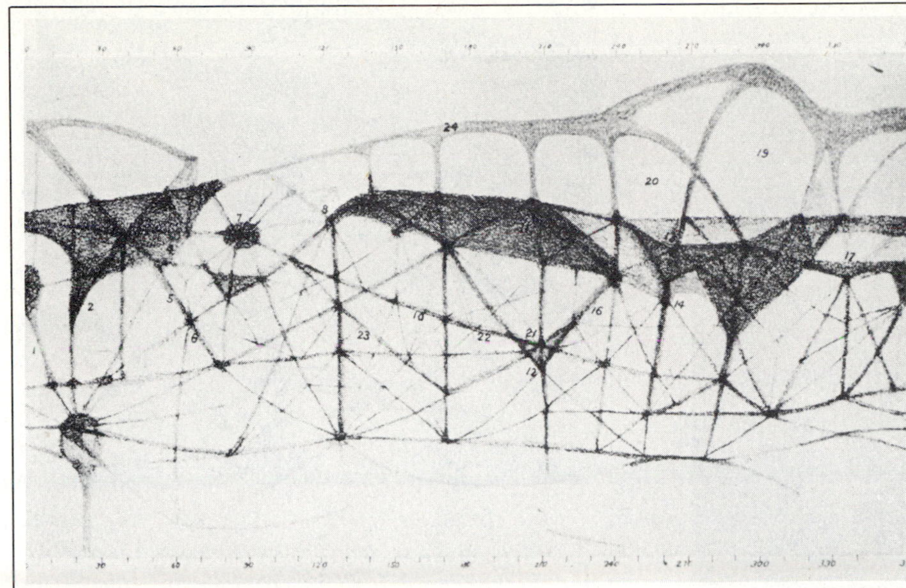

10.14
Surface features on Mars, including some "canals," drawn by Percival Lowell in 1896–1897. (Courtesy Lowell Observatory)

temporary features that were seen as the largest and fuzziest of the canals. The smallest ones were likely an optical illusion guided by wishful thinking. A comparison of Lowell's canal maps with orbiter photos indicates that only one real feature (part of Valles Marineris) corresponds to any of the so-called canals.

10.8 The Martian surface

Martians never had a chance to invade the earth—they don't exist. But we've invaded Mars—not to conquer, but to learn. We have found out that Mars was once an active world but now is a calm, cold desert.

General surface features / In 1969 Mariners 6 and 7 reinforced the view (first developed from the Mariner 4 pictures of 1965) that the surface of Mars resembles that of the moon. They photographed abundant Martian craters visible even under the thinner regions of the polar caps (Fig. 10.16). The Martian craters did look like impact craters but tended to be shallower than lunar ones. Their flat floors and low rims indicated that they had been strongly eroded, an important clue to conditions in the past.

Was Mars, like the moon, a dead world? No, for the photographs taken by Mariner 9 also showed spectacular features of a geologically active planet. The extensive photographic survey showed that the two Martian hemispheres have different topological characteristics: The southern hemisphere is relatively flat, older, and heavily cratered; the northern hemisphere is younger, with extensive lava flows, collapsed depressions, and huge volcanoes. Near the equator, separating

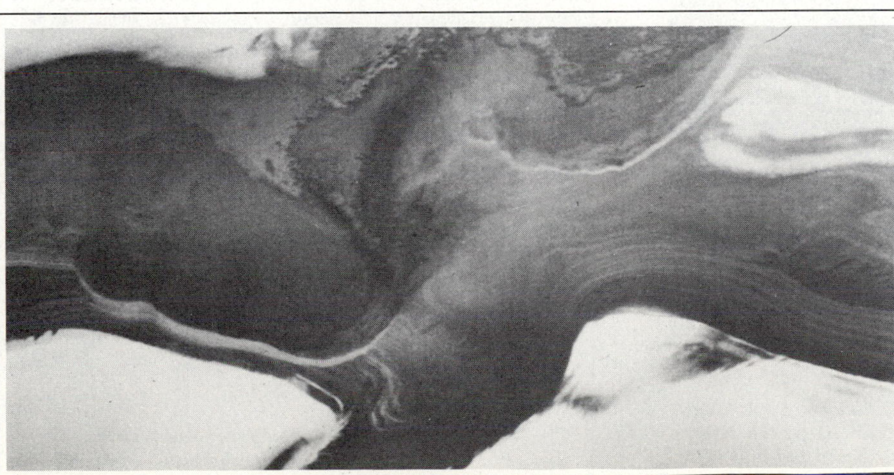

10.15
The residual north polar cap in the summer; note the layered appearance. The ice is water ice. Note the layered terrain beneath the ice. (Courtesy NASA)

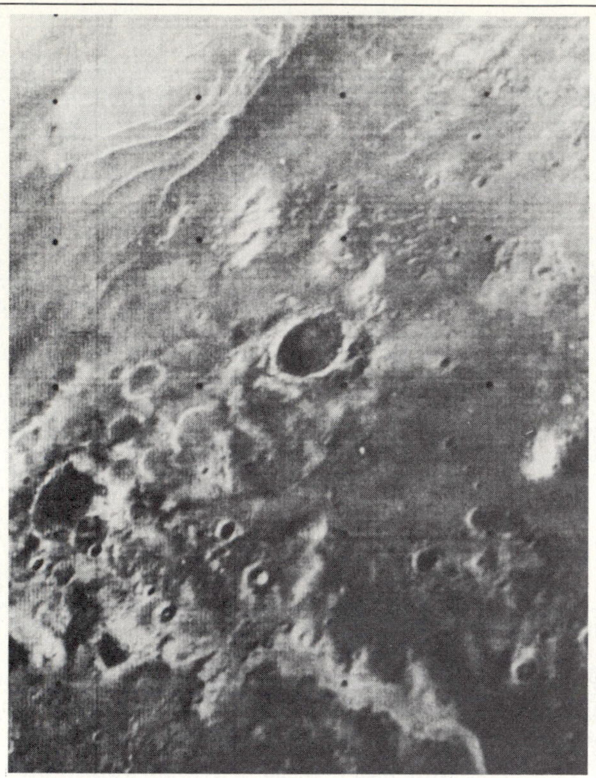

10.16
The south polar cap of Mars, photographed by Mariner 7. Craters are plainly visible through the ice, an indication that the layer is no more than a few meters thick. (Courtesy NASA)

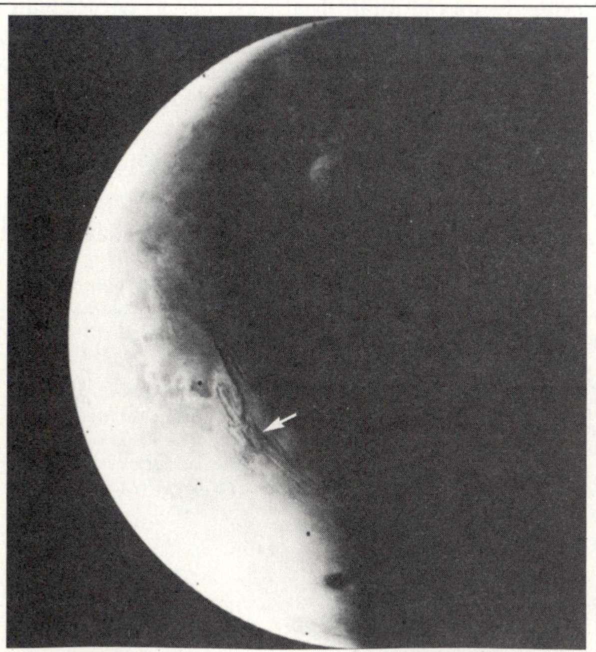

10.17
A wide-angle view of Valles Marineris taken by the Viking 1 orbiter. The canyon (arrow) is about 5 km deep at its west end, shallower to the east. It runs parallel to and just south of the equator. (Courtesy NASA)

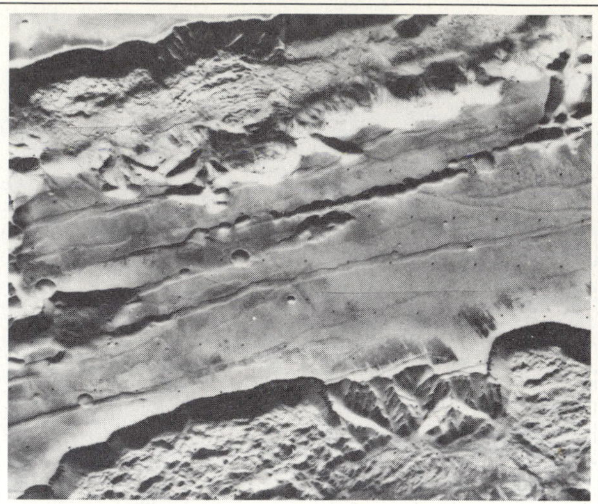

10.18
A close-up of a section of Valles Marineris. The two canyons visible are about 60 km wide and 1 km deep. (Courtesy NASA)

the two distinct hemispheres, lies a huge canyon, called Valles Marineris (Fig. 10.17). This chasm is 5000 km long (about the length of the United States) and some 500 km wide in places (Fig. 10.18).

The Viking landers touched down on Mars in 1969. Seen close up, the Martian surface is bleak and dry. Large rock boulders are strewn about, amid gravel, sand, and silt. The boulders are basaltic. Some contain small holes (Fig. 10.19) from which gas has apparently escaped; the holes make the rock look spongy. On earth, such basalts originate in frothy, gas-filled lava; the Martian rocks probably had a similar origin.

Both landers (Fig. 10.20) uncovered indirect evidence for once-flowing Martian surface water.

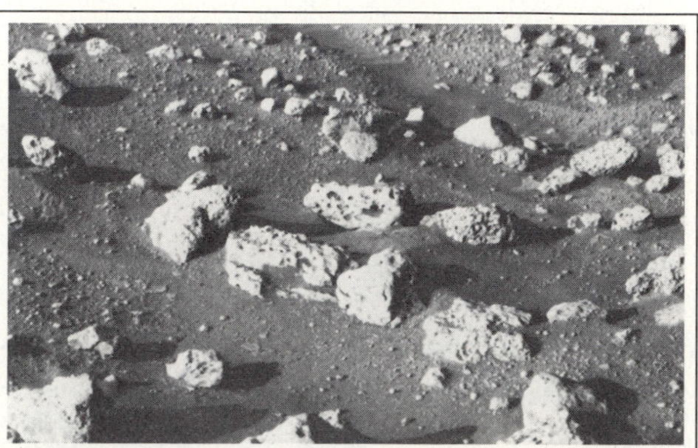

10.19
Close-up view of the Martian surface near Viking 2. The rocks are 10 to 25 cm in size. Note that most contain holes. (Courtesy NASA)

10.20
A wide-angle view of the Viking 2 surroundings. Note that many of the rocks have small holes in them, an indication of volcanic origin. (Courtesy NASA)

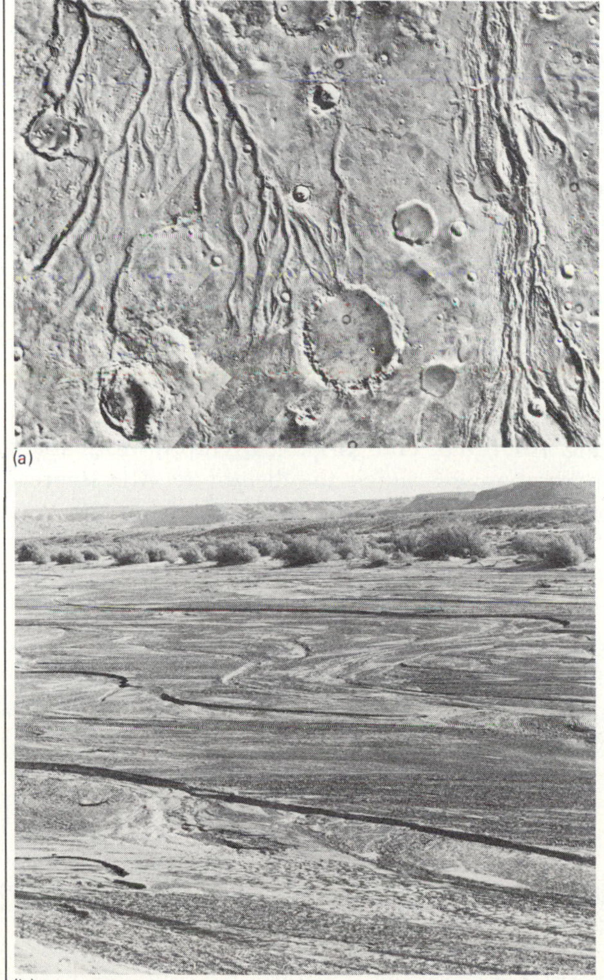

10.21
(a) *Large arroyos on Mars. The terrain slopes about 3 km. Note how some arroyos cut through craters, an indication that they formed after the craters. (Courtesy NASA)* **(b)** *An arroyo in New Mexico. Note the wavy patterns in the sand. (Photo by M. Zeilik)*

The region around Viking 1 seems to be a flood plain where water sorted the smaller rocks into gravel, sand, and silt. The ground there also resembles the hardened soil of the earth's deserts. Such soil forms when underground water percolates upward and evaporates at the surface. Upon evaporating, the water leaves behind minerals that harden the soil. Mineral analyses by Viking 1 and 2 indicated that the soil does contain such evaporated minerals, such as epsom salts.

Arroyos / The Mariner 9 mission discovered and the Viking orbiters confirmed a number of sinuous channels that appear to have been cut in the surface by running water (Fig. 10.21a). The largest ones have lengths up to 1500 km and widths as great as 100 km. (These channels are not the canals seen by Lowell and others; they are too small to be visible from the earth.) These channels resemble the arroyos commonly found in the Southwest of the United States. An *arroyo* is a channel in which water flows only occasionally (Fig. 10.21b).

What makes us think that the Martian channels were actually cut by flowing water? The evidence is pretty strong: (1) The flow direction is downhill; (2) the flow patterns meander (compare the channels in Fig. 10.21a with the channel in Fig. 10.21b); (3) tributary structures indicate where several flows merged to form a larger one; and (4) sandbars are cut by smaller flow channels, as is commonly found in arroyos on the earth.

The presence of Martian arroyos requires extensive running water for at least a short period of time. Since Mars does not have liquid surface water now, conditions for it must have occurred in

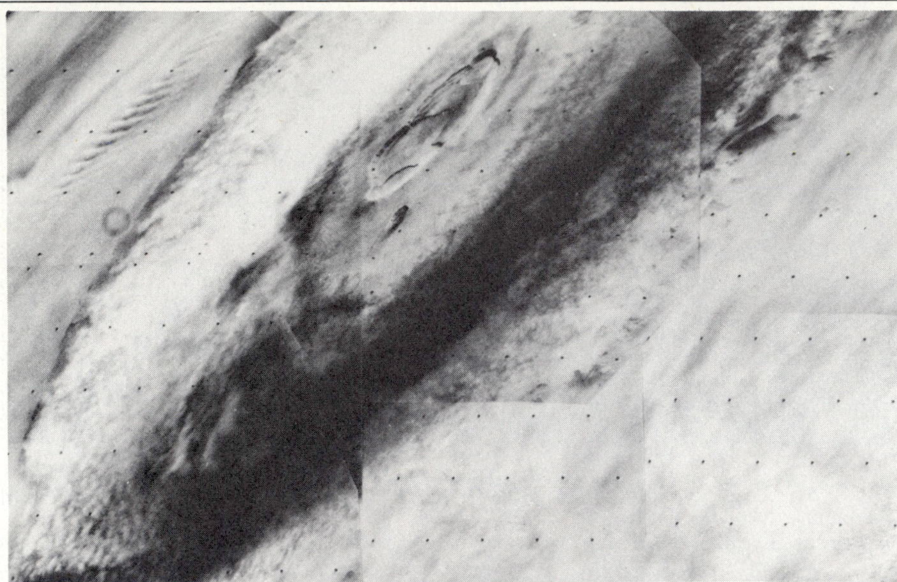

10.22
Olympus Mons wreathed by clouds. (Courtesy NASA)

the past and would have required a warmer climate and a denser atmosphere (Section 22.6).

Volcanoes / By far the most awesome Martian features are the shield volcanoes clustered on and near the Tharsis ridge. The largest is Olympus Mons, some 550 to 600 km across at its base (Fig. 10.22). The cone's surface shows a wavy texture that is the result of lava flows (Fig. 10.23). The cone reaches 25 km above the surrounding plain, and its base would span the bases of the islands of Hawaii, which are made of several volcanoes. If put down in California, the volcano would cover the territory from San Franciso to Los Angeles (Fig. 10.24). For comparison: Olympus Mons soars more than 2½ times the height of Mt. Everest above sea level! Indirect evidence from the lava

10.23
The top of Olympus Mons. (a) *A wide-angle view shows the crater at the top of the cone, one side of the volcano, and the surrounding plateau.* (b) *A closer view of the area in the white box in* (a) *shows the surface texture, which indicates a flow of material from the volcano's top. (Courtesy NASA)*

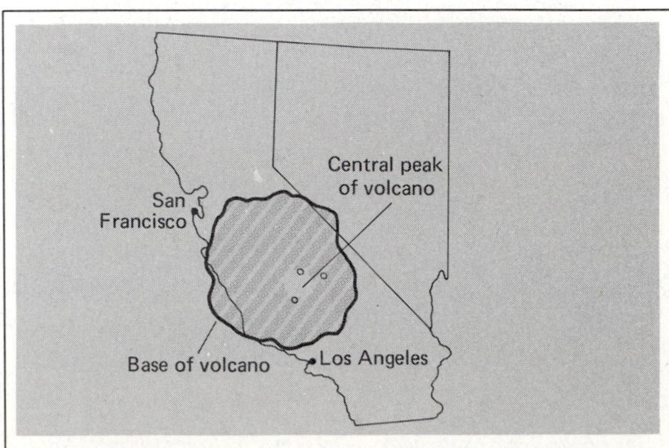

10.24
The size of Olympus Mons compared with the state of California.

flows implies that the volcanoes are about a billion years old.

The huge mass of Olympus Mons requires that the Martian crust beneath it be thicker than the crust beneath smaller such volcanoes on the earth. Geologists estimate the thickness of the Martian crust to be some 120 to 130 km, about twice that of the earth's.

Olympus Mons crowns a string of volcanoes situated on the Tharsis ridge. Occasionally, thin ice clouds have been seen decorating the tops of the volcanoes there. These clouds might result from erratic spurts of outgassing. On earth, volcanic activity spews forth gases (including water vapor) from the earth's interior. Such outgassing by the giant Martian volcanoes in the past may have contributed significantly to the Martian atmosphere. The clouds may also result from the cooling of air that is being blown up the sides of the volcanoes.

The Tharsis ridge is a hallmark of Mars's northern hemisphere, which differs so dramatically from the southern one. The ridge rises about 10 km above the average surface height for the planet and contains numerous volcanic structures. Very few impact craters are visible. In contrast, the southern hemisphere is basically a desert pockmarked by old, eroded craters. The geologic inference from this difference is that about 3 billion years ago in the northern hemisphere, a huge mass of lava oozed out from under the surface, creating the volcanic plains and the volcanoes over a long period of time. This flow wiped out the older deserts and craters in this region. Other flows have taken place in this region since the first.

The cratered southern hemisphere / The southern hemisphere of Mars has a cratered terrain (Fig. 10.25) that resembles the ancient highlands of the moon or the intercrater plains of Mercury. The landscape contains impact craters that range in size from huge, lava-filled basins down to some only a few meters across. The Martian craters come in the same varieties as lunar and Mercurian ones, some with central peaks that mark their impact origin.

In general, the Martian craters are shallower than the ones on the moon and Mercury. Many are filled with windblown dust. Wind scours the craters and piles the dust in dunes within the craters' bowls.

The Martian craters also do not usually have the rims of ejected material common to craters on the moon and Mercury. Instead, many Martian craters have bulges that protrude from their rims. We think that the impact melted frozen ice in the Martian soil; this melted water quickly turned the ground into mud that flowed in bulges away from the crater.

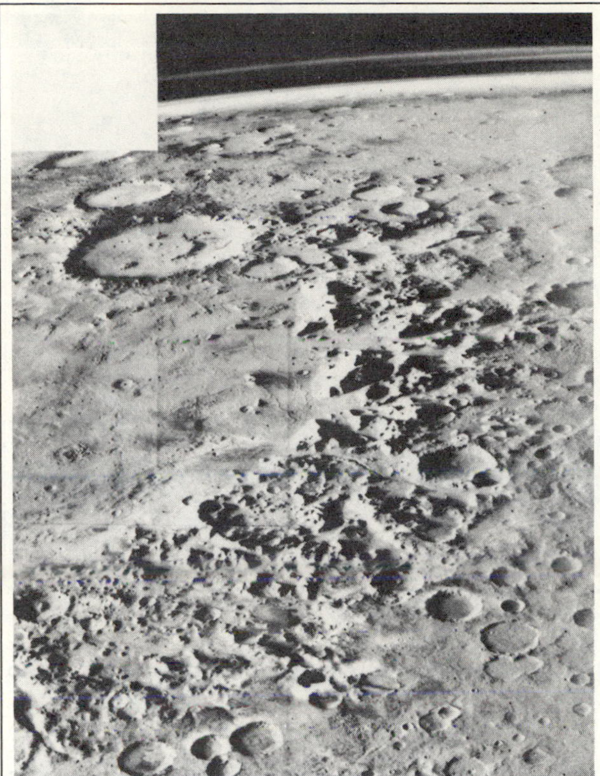

10.25
Argyre Planitia in the Martian southern hemisphere. Argyre is a large impact basin (left of center). Many relatively uneroded craters are on the surface. (Courtesy NASA)

An evolutionary history of Mars / Putting this mass of new information together to infer the Martian past is a tough task. Many uncertainties crop up, especially with respect to the sequence of events. Here is a preliminary working model for the geologic evolution of Mars.

First, after the formation of Mars by accretion, impact craters covered the surface. Shortly afterward, the planet differentiated (as did the earth; see Section 8.8) to form a crust, a mantle, and a core. Regions of thicker crust rose to higher elevations. In the second phase, thin regions of the crust fractured, and the Tharsis ridge uplifted, cracking the surface around it. During this time, a primitive atmosphere, more dense and warmer than at present, held large amounts of water vapor from the volcanic outgassing. Rainfall may have eroded the surface in furrows and then percolated to a depth of a few kilometers. Decreasing temperatures formed ice at shallow depths. When heated (perhaps by volcanic activity), this ice melted, leading to the formation of collapse and

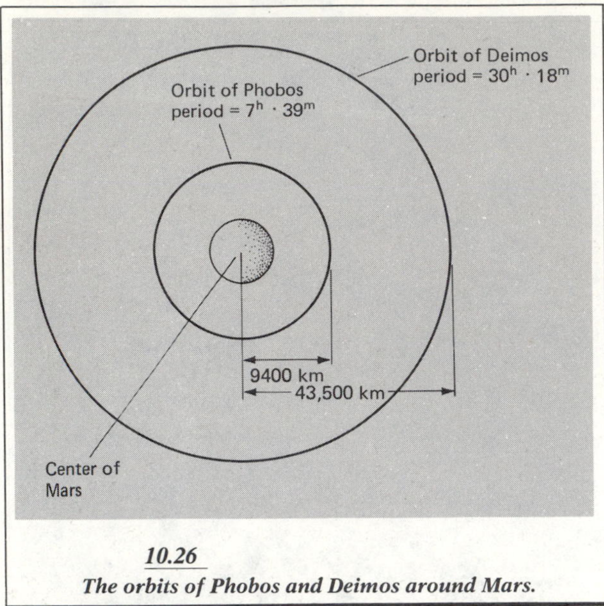

10.26
The orbits of Phobos and Deimos around Mars.

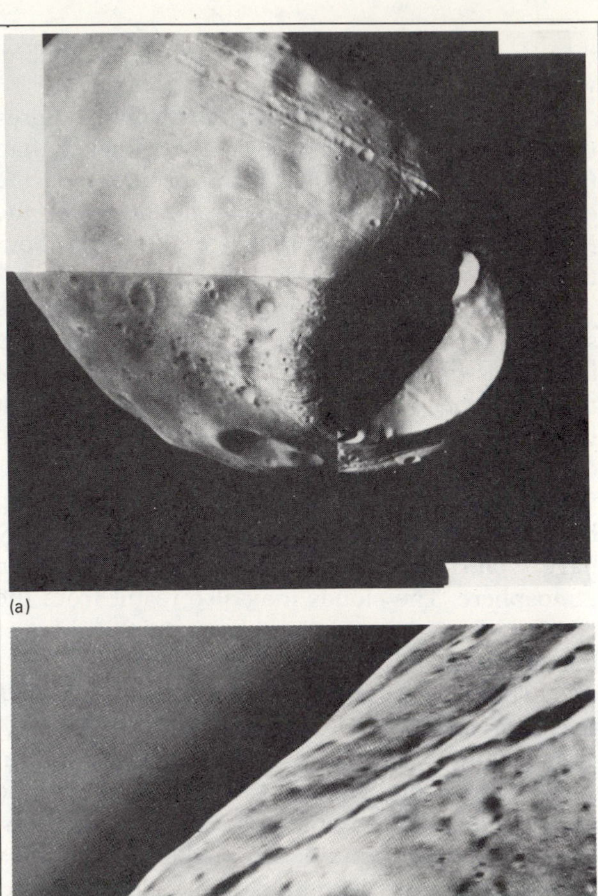

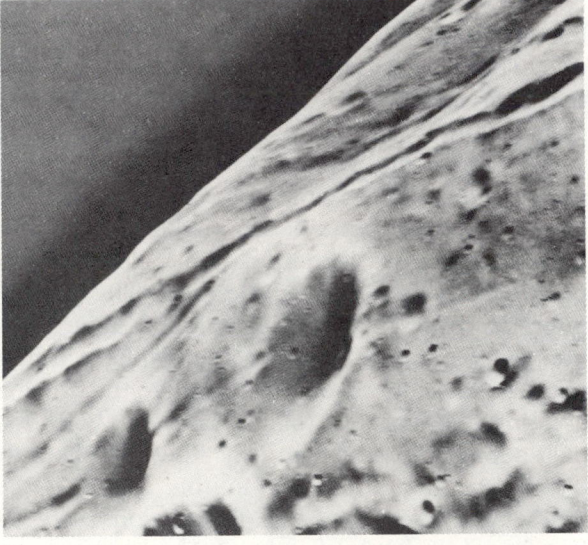

10.27
(a) *Phobos photographed from 612 km by the Viking 1 orbiter. This side of Phobos is the one that faces Mars. The largest crater is Stickney, 10 km in size.* **(b)** *Phobos from 120 km. The smallest craters visible are 10 m in size. The surface resembles that of the lunar highlands. (Both courtesy NASA)*

flow features. Planetwide water erosion carved the surface.

In the next phase, extensive volcanic activity occurred, especially in the northern hemisphere. The Tharsis region continued to uplift, generating more faults. Valles Marineris formed at this time. Finally, recent volcanism—most of it concentrated on the Tharsis ridge—broke the surface and spewed out great flows of lava.

Since that last time of great eruptions, it is mainly wind erosion that has sculpted the Martian surface. A few small impact craters have probably formed from time to time. Recently, strange, small craft from the earth have touched down on the surface, finding a cold, windy, dusty, and rock-strewn desert.

10.9 The moons of Mars

Two moons encircle the planet Mars; appropriately, they are named Phobos and Deimos ("Fear" and "Panic"), after the companions of the god Mars. Asaph Hall (1829–1907) at the United States Naval Observatory discovered the two moons in 1877.

Deimos and Phobos both lie close to Mars and orbit the planet rapidly (Fig. 10.26). Deimos, the outer moon, circles Mars in 30⅓ hours; Phobos, the inner moon, takes a mere 7⅓ hours. In fact, Phobos is one of two moons (the other is Jupiter's innermost satellite) that orbit their parent planet faster than that body spins. So while Deimos rises in the east and sets in the west, as our moon does, Phobos rises in the west and sets in the east! Like the earth's moon, the Martian moons keep the same face to the planet.

Spacecraft observations have found that Deimos and Phobos have the same general shape: an ellipsoid with three axes. Phobos, the larger, has axes about 27, 21, and 19 km long; Deimos's axes are only 15, 12, and 11 km. Photographs also show that Phobos (Fig. 10.27) and Deimos (Fig. 10.28) have cratered surfaces. The sizes and numbers of these craters indicate that the surfaces of these satellites are at least 2 billion years old.

The surfaces of these satellites are very dark. They reflect only 2 percent of they light that strikes them—much less than our moon, which

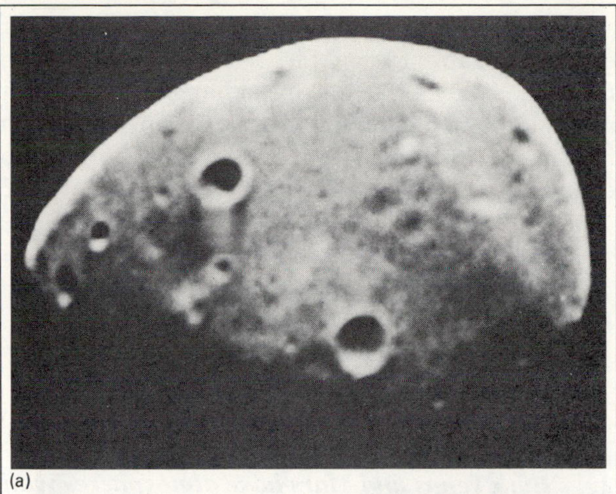

(a)

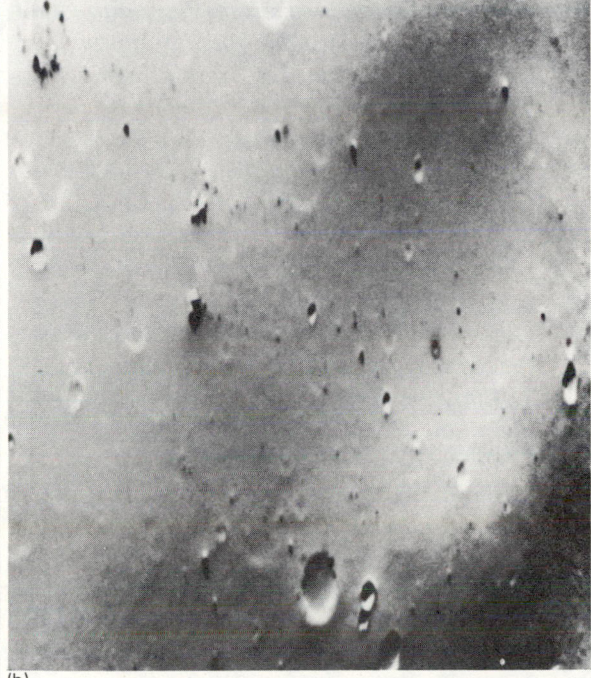

(b)

10.28
(a) *Deimos. The largest crater is 1.3 km in size. The illuminated part of the moon is about 12 by 8 km.* **(b)** *A close-up of Deimos. This photo shows a region 1.2 by 1.5 km and features as small as 3 m. The boulders are about the size of a house. (Both courtesy NASA)*

reflects about 7 percent. They rank among the blackest known objects in the solar system.

10.10 Venus, Earth, and Mars: An evolutionary comparison

Neighbors in space, Venus, the earth, and Mars share similar internal characteristics (as indicated by their densities), and all possess atmospheres. The atmospheres of Venus and Mars have about 95 percent carbon dioxide, with traces of other compounds; the earth's atmosphere contains only about 0.03 percent carbon dioxide. This difference in atmospheric constitution is as striking as the earth's unique treasure of oceans of water compared with the arid wastes of Mars and Venus. Why such profound differences among such otherwise similar planets? The answer probably lies in the different evolutionary rates and processes on the planets, which relate to their masses and distances from the sun. I'll present here a working model of these evolutionary differences.

The water question is a central one. When these planets formed (see Chapter 12), Venus contained the lowest percentage of water, earth somewhat more, and Mars the most. Now, water may be trapped underground, but geologic activity such as volcanoes releases it. Present conditions prohibit liquid water on the Martian surface, but it does exist as ice on the surface and may be underground. On Venus, it has always been so hot that volcanic-released water never flowed on the surface. So all three planets probably outgassed considerble quantities of water during their history.

Water can, however, be removed from the atmosphere if it is split into its components, hydrogen and oxygen. On Mars, the temperature in the upper atmosphere is high enough and the escape velocity low enough so that hydrogen escapes into space. The active oxygen then combines chemically to form some stable compound, such as ferric oxide.

What breaks up the water molecules? Ultraviolet radiation from the sun has enough energy to break the oxygen-hydrogen bonds; this process is called *photodissociation*. Mariner 6 and 7 found a substantial hydrogen cloud around Mars. (Earth also has such a cloud.) Because such a cloud could not last long, it must be supplied with hydrogen from photodissociation. Venus, because it is closer to the sun and hence subject to stronger ultraviolet radiation, should have suffered an even greater rate of photodissociation and consequently a greater loss of hydrogen. So the earth has lost its water less quickly than Mars, both because of a higher surface gravity and the protective ozone layer. Venus lost it more quickly than the earth because of the higher flux of ultraviolet radiation.

Now consider the carbon dioxide. The differences among the planets are more apparent than real if the carbon dioxide in the earth's crust and oceans is added to the total amount. When immersed in water, silicates in rocks react with carbon dioxide to form carbonates. Also, shells of sea creatures are primarily carbonate, from the carbon dioxide dissolved in the oceans. Added together, the crustal, oceanic, and atmospheric carbon dioxide total for the earth is about equal to the total in the atmosphere of Venus.

TABLE 10.1 Main stages in the evolution of terrestrial planets

1	Formation by accretion, heating of crust and interior, crust formation
2	Crust solidification, intense impacts, cratering of surface
3	Basin formation and flooding, lowlands formation
4	Low-intensity impacts, atmosphere formation by outgassing
5	Volcanoes, crustal movement, continent formation

Why did the carbon dioxide on Venus not end up in the crust rather than the atmosphere? The reason: lack of water because Venus is so close to the sun. About 4 billion years ago, the surface temperature was high enough to vaporize all the primeval water. Outgassing from the interior added to the carbon dioxide, further raising the temperature. So the surface temperature on Venus has always been high enough to boil water. Without liquid water, not much carbon dioxide can be deposited in the crust as carbonates. Furthermore, the rate of deposition is slower at higher temperatures, so fewer carbonates would have formed on Venus, even if the liquid water were available.

In summary, the evolution of terrestrial planets involves change to their interiors, surfaces, and atmospheres. These changes are driven mostly by the planet's internal heat: The more massive the planet, the more internal heat it generates (from radioactive decay) and the longer it retains this heat. So the greater it evolves.

Surfaces are modified by several processes: impacts of interplanetary debris, outflow of internal heat, volcanism, erosion by wind and water, and crustal movements (if the mantle is hot). The atmospheres change from interaction with sunlight, escape into space, degassing (from a hot interior), and life (if it exists).

Earthlike worlds have five general stages of evolution (Table 10.1). Mars has evolved the least, through stage 4. Venus has changed somewhat more, into the beginning of stage 5. Only the earth has made it through stage 5 and continues to evolve.

KEY CONCEPTS

1 ***Venus and Mars, the terrestrial planets most like the earth (Table 10.2), have bulk densities close to the earth and so have somewhat similar interior structures (Venus more so than Mars).***

2 ***Both Venus and Mars have atmospheres with essentially the same composition: almost all carbon dioxide. But the Martian atmosphere is very thin compared to that of Venus.***

3 ***The surface temperature of Venus is very high (700 K) compared to the earth (290 K); in contrast, Mars is quite cold (225 K). Venus is so hot because of a supergreenhouse effect from its atmosphere.***

4 ***The surface of Venus has mountains, volcanoes, high plateaus, rolling plains, large valleys, and impact craters; most of the surface is very flat with no large basins; the highland regions are the youngest, and the plains the oldest.***

5 ***Venus, because it has almost the same mass as the earth, went through a similar evolutionary sequence; Mars, with less mass, did not evolve as much.***

6 ***Mars is a desert world, with much less water than the earth (or Venus); most of the water is probably ice in the cores of the polar caps and under the surface.***

7 ***The surface of Mars has giant volcanoes (now extinct), arroyos cut by water (in the past), a huge valley, and impact craters; the northern hemisphere is higher and younger than the southern.***

8 ***The arroyos and eroded craters indicate that in the past Mars must have had an atmosphere extensive enough to permit liquid***

TABLE 10.2 Comparison of the terrestrial planets

	Diameter (earth = 1)	*Mass (earth = 1)*	*Density (water = 1)*	*Surface pressure (earth = 1)*	*Atmosphere's main constituents*
Mercury	0.38	0.055	5.4	10^{-15}	Helium, hydrogen
Venus	0.95	0.82	5.2	100	Carbon dioxide
Earth	1.0	1.0	5.5	1.0	Nitrogen, oxygen
Mars	0.53	0.11	3.9	0.005	Carbon dioxide

water to exist on the surface (now the surface pressure is too low and the temperature too cold).

9 ***Terrestrial planets go through similar evolutionary sequences, with mass mostly determining how far a planet will evolve (distance from the sun plays a secondary role); the less mass, the faster a planet loses internal heat and its atmosphere and the less it evolves.***

STUDY EXERCISES

1 How is the Martian surface similar to that of Venus? (*Objective 2*)

2 In what one respect is Venus most like the earth? Most different from the earth? (*Objectives 1, 2, and 3*)

3 How is Mars most like the earth? Most different from the earth? (*Objectives 1, 2, and 3*)

4 Suppose you are kidnapped by an evil alien creature who threatens to drop you on Venus or Mars with a limited amount of supplies. Which planet would you prefer, and what are the reasons for your choice? (*Objective 7*)

5 Under what conditions could the Martian arroyos have formed? (*Objectives 2, 5, and 9*)

6 Using your knowledge of terrestrial volcanoes, outline how the Martian volcanoes may have affected the evolution of the Martian atmosphere. (*Objective 9*)

7 In what major respects are the atmospheres of Mars and Venus similar? Different? (*Objectives 2 and 3*)

8 Compare the contents of the Martian polar caps in summer and in winter. (*Objective 2*)

9 What one surface feature can be used most effectively to compare the relative ages of the earth, Mars, and Venus? (*Objectives 5 and 10*)

BEYOND THIS BOOK . . .

In the September 1975 issue of *Scientific American*, you can find "Venus" by A. and L. Young and "Mars" by J. B. Pollack.

For a more detailed comparison of the terrestrial planets, read *Earthlike Planets* (Freeman, San Francisco, 1981) by B. Murray, M. Malin, and R. Greeley.

Perhaps Mariner 9's most important discovery was the giant volcanoes of Mars. For more about them, see "The Volcanoes of Mars" by M. Carr, *Scientific American*, Janurary 1976.

For details on the radar mapping of Venus, see "The Surface of Venus" by G. Pettengill, D. Campbell, and H. Masursky, *Scientific American*, August 1980.

To investigate the atmosphere of Venus, read "The Atmosphere of Venus" by G. Schubert and C. Convey, *Scientific American*, July 1981.

The New Solar System (Sky Publishing, Cambridge, Mass., 1983), edited by R. Beatty, B. O'Leary, and A. Chaikin, has nice summaries of recent explorations of the terrestrial planets.

Pictorial Guide to the Planets (Harper & Row, New York, 1981) by J. H. Jackson and J. H. Baumert has some excellent photos.

11/ The Jovian planets: Primitive worlds

MEPHISTOPHELES: Nor are the Names of Saturn . . . or Jupiter feign'd, but are erring stars.

CHRISTOPHER MARLOW:
The Tragical History of Doctor Faustus

Central question

How do the Jovian planets differ from the terrestrial ones, and what do these differences imply about different evolutions?

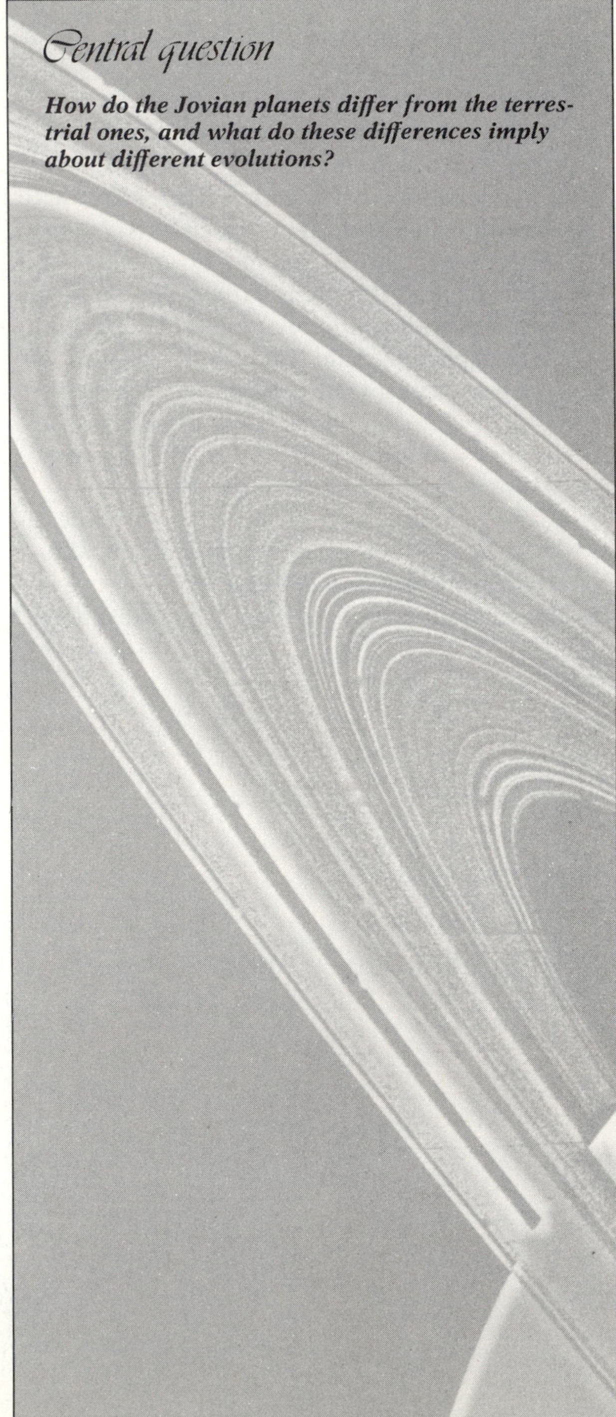

Learning objectives

After studying this chapter, you should be able to:

1. **Compare and contrast the Jovian planets as a group to the terrestrial planets, emphasizing the greatest differences.**
2. **Contrast the Jovian planets to one another in terms of their (a) relative sizes, (b) relative masses, (c) bulk densities, (d) atmospheric compositions, (e) internal structures, and (f) unique features.**
3. **Compare and contrast the interior of Jupiter to that of the earth.**
4. **Compare the rings of Saturn with those of Uranus and Jupiter in terms of size, shape, and possible composition.**
5. **Present the unique characteristics of Pluto that make it neither a Jovian nor a terrestrial planet.**
6. **Compare and contrast the general characteristics and surface features of the Galilean satellites of Jupiter: Io, Europa, Ganymede, and Callisto.**
7. **Compare and contrast the Galilean satellites to the earth's moon and to Pluto.**
8. **Compare Jupiter's and Saturn's magnetic field and magnetosphere to those of the earth.**
9. **Contrast Saturn's largest moon, Titan, with the other moons of Saturn, the Galilean moons of Jupiter, and the earth's moon.**
10. **Argue that, compared with the terrestrial planets, the Jovian planets have *not* evolved much since their formation.**

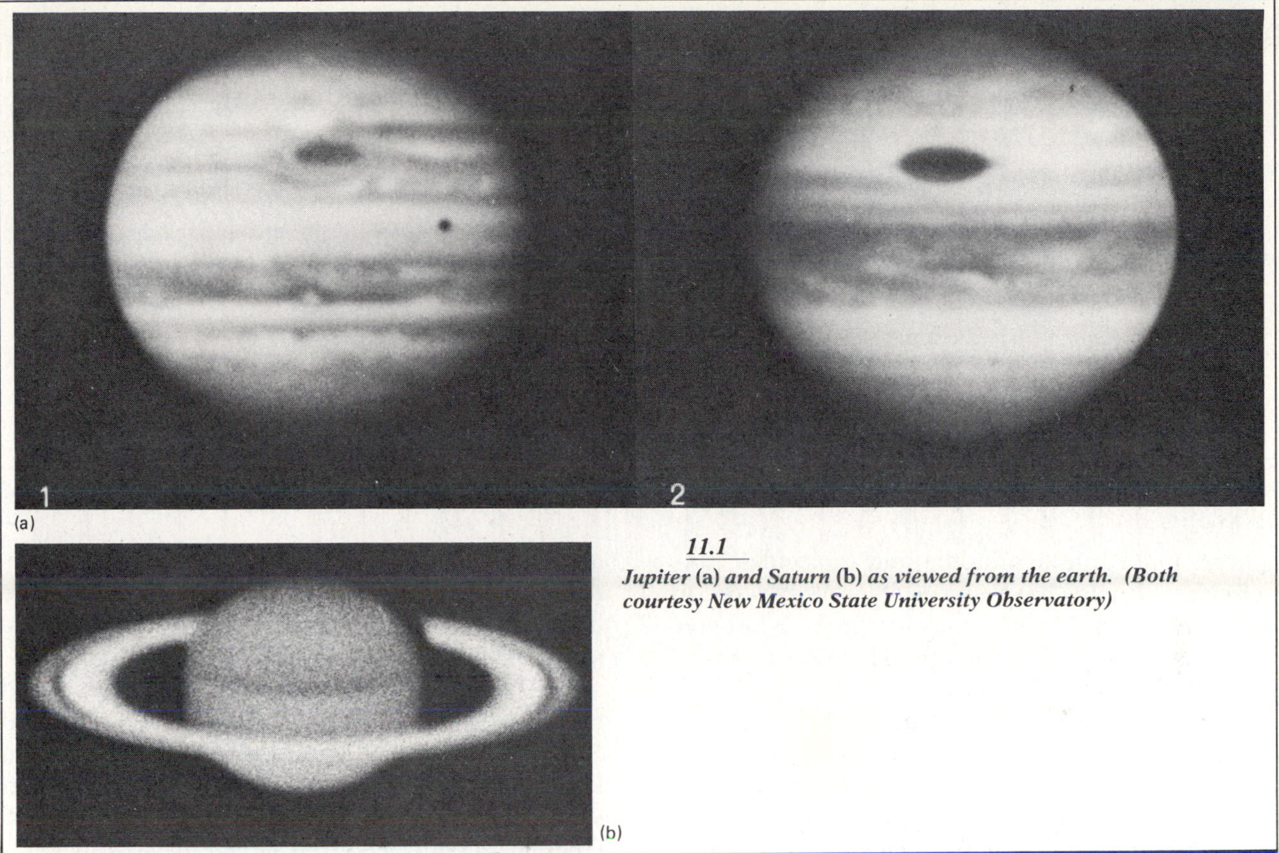

11.1
Jupiter (a) *and Saturn* (b) *as viewed from the earth. (Both courtesy New Mexico State University Observatory)*

When an amateur astronomer finishes constructing a telescope, he or she usually first turns it to the moon, and then to Jupiter and Saturn. And no wonder! These are impressive, awesome planets (Fig. 11.1)—two giant worlds languidly circling the sun. Banded Jupiter drags along its coterie of satellites. Saturn is set in its rings like a prize gem of space. You can barely grasp the weird environments of these giant worlds; planets 4 to 11 times the diameter of the earth, mostly gases and liquids, and without breathable atmospheres.

This chapter investigates the features of the Jovian planets that set them apart from the terrestrial ones. The major differences between the Jovian and terrestrial planets furnish additional clues about the evolution of the planets and the formation of the solar system. The key point is this: The Jovian planets are primitive worlds, looking today very much the same as when they were formed. These giant worlds are little evolved compared with the earth.

11.1 *Jupiter: Lord of the heavens*

Jupiter ranks as the largest and most massive body in the solar system (after the sun). Jupiter's total mass is about 2½ times that of all the other planets put together (and 318 times the earth's). Eleven earths placed edge to edge would stretch across Jupiter's visible disk—truly a giant planet!

Physical characteristics / Just as Jupiter's mass is the largest among the planets, so is its diameter, almost 140,000 km. Yet for all this size, Jupiter's material is less concentrated than the earth's, for Jupiter's density is only 1330 kg/m^3.

All the Jovian planets have low densities compared with the terrestrial planets. These low densities are a key difference between the two classes and imply that the Jovian planets are made of quite different stuff. The terrestrial planets are basically rocks and metals, made of elements such as iron, aluminum, oxygen, and silicon. Jupiter, in contrast, is made mostly of hydrogen and helium.

Almost all of this hydrogen and helium came together when Jupiter formed, and the giant planet has lost little, if any, since then. Why? Jupiter's huge mass means that its escape velocity (see Focus 7.3) is substantial, about 57 km/sec. Remote from the sun, Jupiter's upper atmosphere is cold, only about 130 K, and hydrogen molecules move at about 1 km/sec. So even hydrogen molecules do not have escape velocity. If the hydrogen cannot escape from the upper atmo-

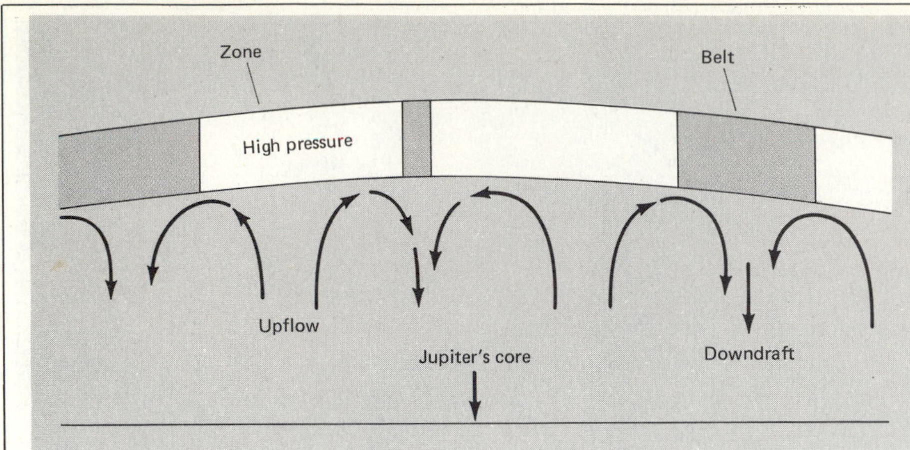

11.2

The circulation in Jupiter's upper atmosphere. Rising air creates high pressure regions (called* zones*), and the downflow makes low pressure areas (called* belts*). The zones generally appear lighter in color than the belts because they are higher up in the atmosphere.

sphere, more massive atoms and molecules cannot either. Jupiter has retained its atmosphere for eons and will hold it for eons to come. What you see now is basically the atmosphere and mass with which Jupiter was born.

General atmospheric features / The visible disk of Jupiter (Fig. 11.1) is *not* the planet's surface but its upper atmosphere, which shows alternating strips of light and dark regions that run parallel to the equator. The light regions, called *zones*, have lower temperatures than the dark regions, called *belts*. So the zones are higher up than the belts. These differences in temperature imply that the zones flag the tops of rising regions of high pressure, and the belts must be the descending areas of low pressure (Fig. 11.2). This atmospheric flow transports heat out to space from the planet's interior.

Markings in the clouds allow a measurement of Jupiter's rate of rotation. The rate varies with latitude: Jupiter spins in 9 hours and 50 minutes at its equator and 9 hours 55 minutes at its poles. This rapid rotation and Jupiter's large radius produce an equatorial velocity in excess of 43,000 km/h.

Such an enormous rotation speed drives the circulation in Jupiter's atmosphere. It causes the permanent high-pressure zones and low-pressure belts, which might otherwise, as on earth, be somewhat localized, to stretch out completely around the planet. High-speed winds (jet streams) speed along at the boundaries between the belts and zones, creating atmospheric disturbances.

The Voyager missions zoomed in on these complex streams and swirls of Jupiter's upper cloud layer. These stunning photos showed the turbulent atmospheric flow (Fig. 11.3). Earth-based telescopes have revealed complex changes in the belts and zones. Occasionally, dark blue, red, brown, and white ovals appear against the banded background (Fig. 11.3). These small oval spots last as long as a year or two.

The Red Spot / The most permanent and famous atmospheric disturbance is the Great Red Spot (Fig. 11.4), first observed by the Englishman Robert Hooke in 1630. The Red Spot changes in size; it is some 14,000 km wide and 30,000 to 40,000 km long—it could easily swallow the earth!

What is the Red Spot? The Pioneer flybys found that the Red Spot was a few degrees cooler than the surrounding zone and poked about 8 km above it, so it is a rising region of high pressure. In ground-based observations, astronomers at New

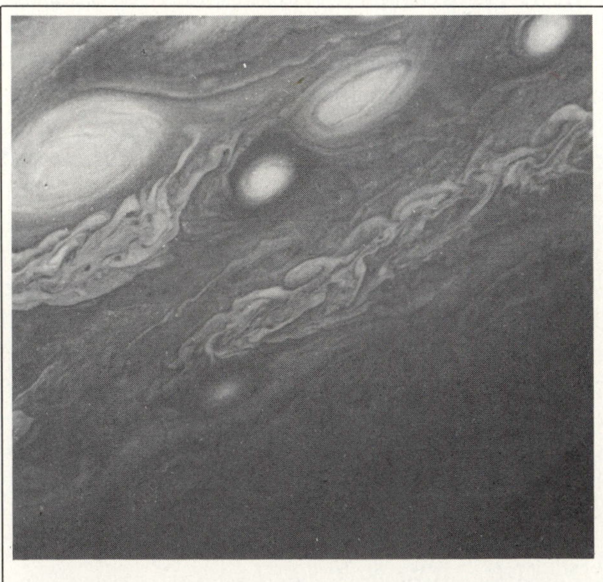

11.3

A close-up view of turbulent patterns in the atmospheric flow. Note the large white spots. (Courtesy NASA)

11.4
A close-up of the Red Spot and the turbulent region near it. The smallest details visible are about 30 km across. (Courtesy NASA)

Mexico State University noted that the Red Spot rotates counterclockwise like a vortex, just as expected from a high-pressure zone in Jupiter's southern hemisphere. The Red Spot turns like a huge wheel pushed by the surrounding atmospheric flow; in turn, the Red Spot deflects nearby clouds and forces them around it. Behind the Red Spot runs a region of turbulent flow (Fig. 11.5) caused by the atmosphere flowing past it.

Why is the Red Spot red? (Many other spots on Jupiter are white; see Color Plate 7.) We don't know for certain. It is likely that chemical reactions, driven by the heat flowing up from below, make colored compounds. These reactions may

11.5
The Great Red Spot and the region of turbulent flow to the west (left) of it. (Courtesy NASA)

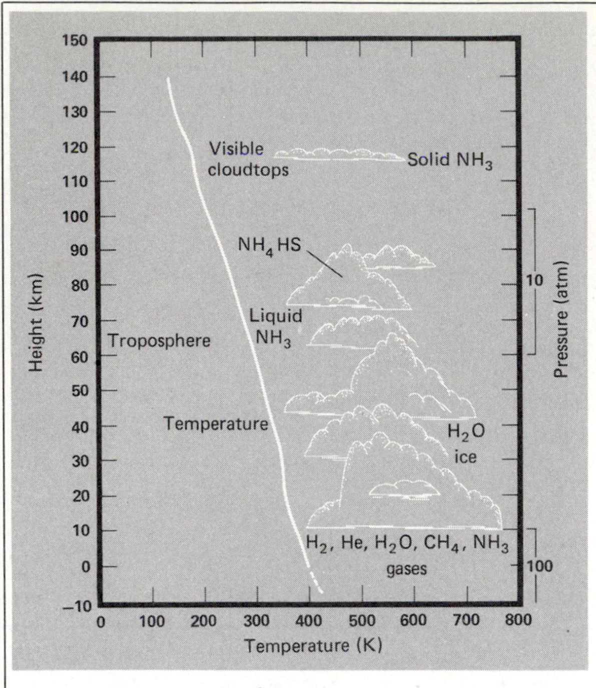

11.6

One model for the structure of Jupiter's atmosphere. Note that the temperature increases downward and hits room temperature (290 K) at about 60 to 70 km into the troposphere. (Adapted from a diagram by W. Hartmann; based on calculations by J. S. Lewis)

involve the molecule phosphine (chemical formula, PH_3), which creates a reddish color in chemical reactions powered by light.

Atmospheric composition / Infrared spectroscopy reveals the atmospheric composition above the clouds. In 1934 methane (CH_4) was discovered, the first molecule definitely identified. Later ammonia (NH_3), molecular hydrogen and helium were found. Other abundant molecules are: water, carbon monoxide, ethane (C_2H_6), and phosphine. A few molecules containing deuterium (heavy hydrogen) are also known to be present.

An analysis of Voyager spectroscopic data implies that Jupiter's upper atmosphere contains by mass about 79 percent hydrogen, 20 percent helium, and 1 percent all other elements, essentially the same composition as the sun. Most of this material is in the form of molecules.

The visible clouds at the tops of the zones are most likely ammonia ice crystals. Below them, according to theoretical models (Fig. 11.6), lies a layer of ammoniumhydrosulfide (NH_4HS) clouds. Below these float liquid ammonia and water-ice clouds. This model describes three separate cloud layers that make up the upper atmosphere: The tops are ammonia, below that are ammonia and hydrogen sulfide, and the bottom layer is water ice. The entire atmosphere may be 1000 km thick. In fact, there is no distinct boundary between atmosphere and interior. The atmosphere just gets denser and hotter farther in, gradually merging into the liquid interior.

A model of the interior / We can only infer Jupiter's internal structure from physical models that include two key pieces of information: (1) Jupiter's low density and atmospheric composition imply a solar mix of material throughout, and (2) Jupiter radiates into space more energy than it receives from the sun (about twice as much), so it must be *hot* inside. The internal heat is probably

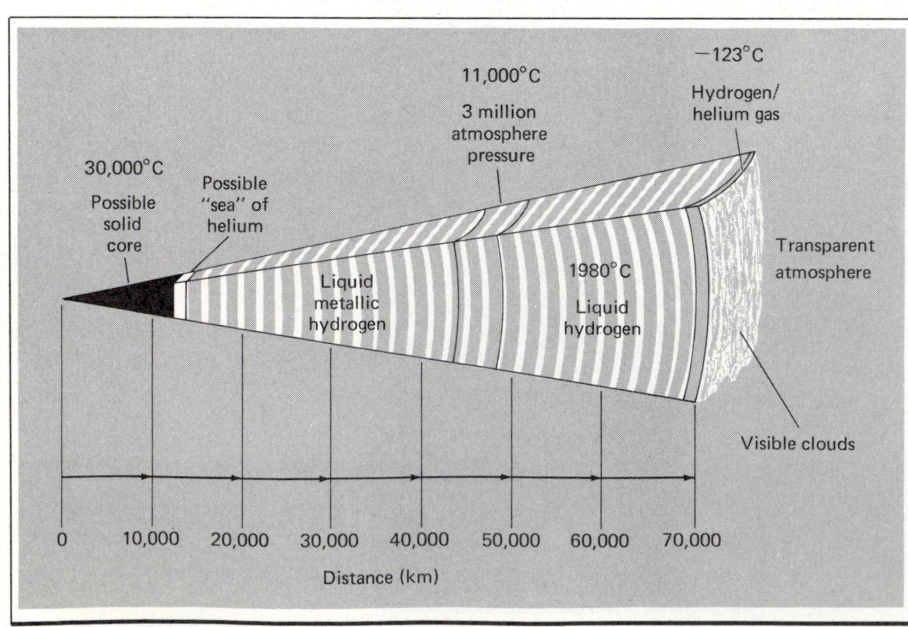

11.7

A model for Jupiter's interior structure. Below the atmosphere is a thick layer of liquid molecular hydrogen. The core is a dense material, perhaps silicates, that may be molten.

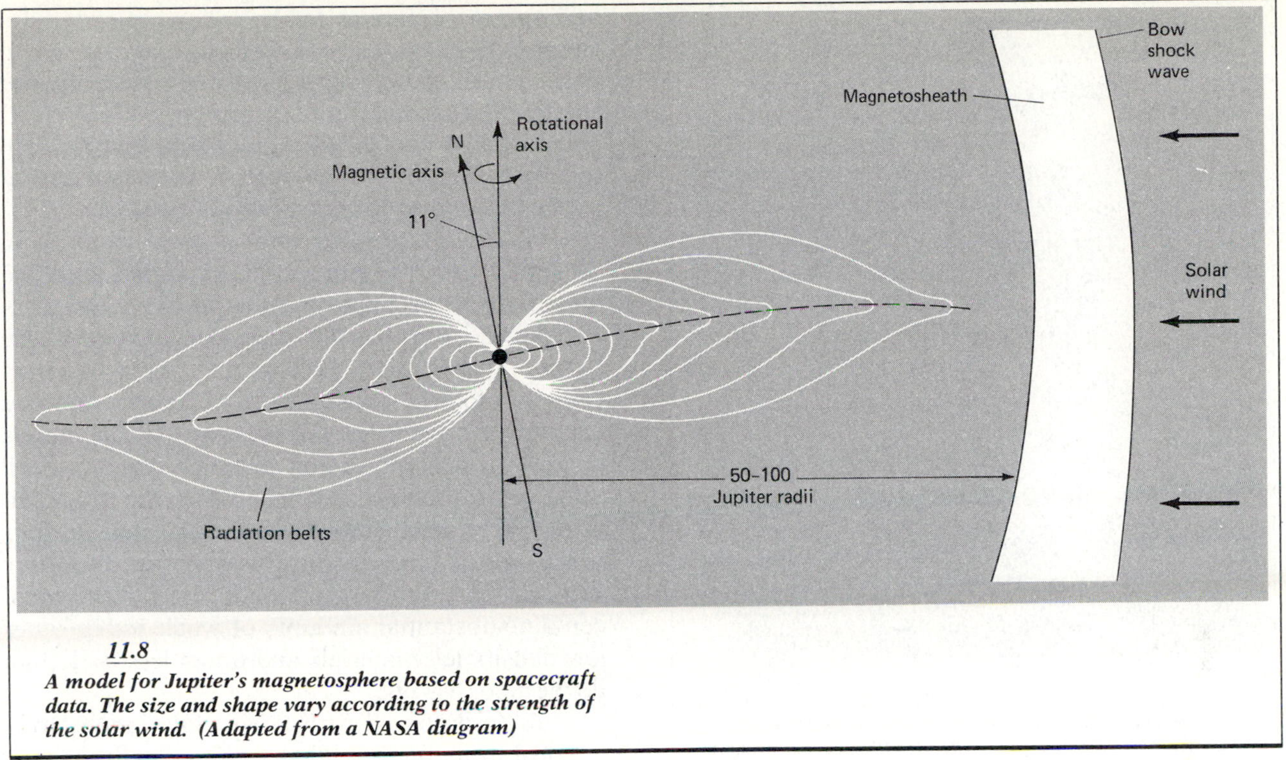

11.8
A model for Jupiter's magnetosphere based on spacecraft data. The size and shape vary according to the strength of the solar wind. (Adapted from a NASA diagram)

left over from Jupiter's formation. (Massive planets lose heat very slowly.)

Recent models of Jupiter, assuming a solar composition and internal heat, come up with the following picture (Fig. 11.7): The atmosphere covers the planet like a thin skin and consists mostly of molecular hydrogen. Going into the planet, the density, temperature, and pressure increase, so the hydrogen exists in a liquid state. At a pressure of about 2 to 3 million atmospheres, the hydrogen is squeezed so tightly that the molecules are separated into protons and electrons that move around freely and can conduct electricity. This state is called *metallic hydrogen*, and although it has never been observed in a lab on the earth, its existence is predicted from atomic physics. This strange state continues to within about 14,000 km of the planet's center. Here, perhaps, if Jupiter does have a solar composition, lies a core of heavy elements.

Most of Jupiter is hydrogen, and most of that hydrogen is liquid—quite a contrast to the earth's interior (see Section 8.2) and that of the other terrestrial planets. The core temperature may be about 10 times hotter than the earth's core—as high as 40,000 K. It is the flow of the heat outward from the core that drives the circulation to produce the beautiful banded atmosphere.

Magnetic field / Jupiter has a magnetic field at least 10 times as strong as the earth's. As particles from the sun plow into this intense field, it creates an enormous shock wave that enshrouds the planet (Fig. 11.8). In addition, the magnetic field traps energetic charged particles from the sun in belts similar to the Van Allen belts around the earth.

How does Jupiter generate such an intense magnetic field? Recall that a dynamo model (see Section 8.4) of the earth's magnetic field pictures currents in the liquid metal core generating the magnetic field like an electromagnet. Jupiter certainly does not have a metal core. But for a large part of Jupiter's interior, liquid metallic hydrogen can conduct electric current. So the conditions for a dynamo to operate are available: a fluid able to conduct electricity, filled with convective currents driven by heat and rapid rotation. A dynamo effect in the metallic hydrogen zone could produce Jupiter's magnetic field.

Radio emissions / Jupiter gives off intermittent bursts of radio noise of unexpectedly high intensity. At wavelengths of tens of meters, sharp bursts lasting about a second have been detected by radio telescopes. A strong burst generates a million times more energy than an intense lightning bolt on earth!

That these radio bursts are indeed generated by superbolts of lightning is supported by Voyager pictures that showed bright regions on the planet's night side (Fig. 11.9). You would expect the violent

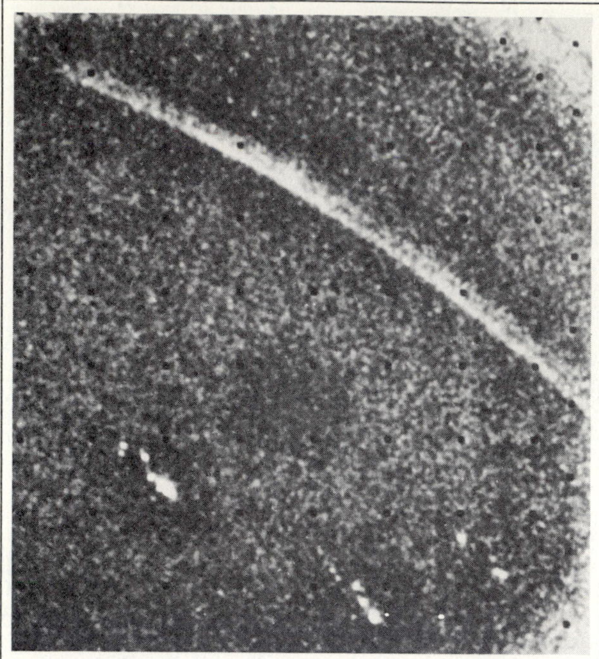

11.9

Jupiter's night side. The bright streak at the top is an aurora over the north pole. The bright spots below it are lightning flashes. (Courtesy NASA)

updrafts and turbulence in Jupiter's atmosphere to be able to generate lightning, in much the same way as thunderstorms do on the earth.

Auroras / Voyager pictures of Jupiter's night side showed polar *auroras* (Fig. 11.9) for the first time. We presume that these auroras happen for the same reason as on the earth: excitation of the upper atmosphere by energetic charged particles pouring in along the north and south magnetic poles. These particles flow from the sun and are trapped by Jupiter's magnetic field.

11.2 The many moons of Jupiter

Jupiter possesses an entourage of at least 16 moons (see Appendix B). The brightest and largest were first discovered by Galileo. Their orbits lie within 3° of Jupiter's equatorial plane, close to our line of sight. These huge moons orbit within 2 million km of Jupiter in the following order: Io, Europa, Ganymede, and Callisto. As our moon does to the earth, each keeps one face toward Jupiter. The Galilean moons are large (Table 11.1) compared to our moon: Ganymede and Callisto are both larger than Mercury. Io is somewhat larger than our moon, and Europa smaller.

Each Galilean moon is a world of its own, different from the others. These differences are hinted at by the bulk densities of the moons: Io, 3500 kg/m^3; Europa, 3000 kg/m^3; Ganymede, 2000 kg/m^3; and Callisto, 1800 kg/m^3. This list is in order of increasing distance from Jupiter, and you can see the pattern: The density of the moons decreases with increasing distance from Jupiter. Such density differences show that the compositions of Io and Europa resemble that of our moon—mostly rock, with perhaps a little icy material. In contrast, Ganymede and Callisto must contain substantial amounts of water ice or other low-density icy materials and much less rock than do the inner moons.

Io / Compared with terrestrial planets, Io is a world in its own right: three-fourths the size of Mercury, a giant satellite for a giant planet. Remarkably, Io has a thin atmosphere. (Only two other satellites, Saturn's Titan and Neptune's Triton, are known to have an atmosphere.) At the surface, the atmospheric pressure is about 10^{-10} atm, composed mainly of sulfur dioxide.

Io's atmosphere has a peculiar property: It gives off a bright, continuous glow of emission from sodium atoms. This sodium glow surrounds Io like a yellow halo out to a distance of about 30,000 km (Fig. 11.10). In fact, the sodium cloud extends about 200,000 km along Io's orbit, forming a partial ring of gas around Jupiter. Gases of potassium and sulfur have also been observed surrounding Io.

What produces Io's sodium cloud? Volcanic eruptions, at least in part. Io has at least 11 active volcanoes! In fact, it is volcanically one of the most active places in the solar system; erupting volcanoes and fuming lava lakes cover its surface (Color Plate 8). This activity implies that the interior must be hot.

TABLE 11.1 *Properties of the Galilean satellites*

	Diameter (km)	*Average distance from Jupiter (10^6 km)*	*Orbital period (days)*	*Bulk density (water = 1)*	*Mass (moon = 1)*
Io	3638	0.42	1.77	3.53	1.21
Europa	3126	0.67	3.55	3.03	0.66
Ganymede	5270	1.07	7.16	1.93	2.03
Callisto	4848	1.88	16.69	1.79	1.45

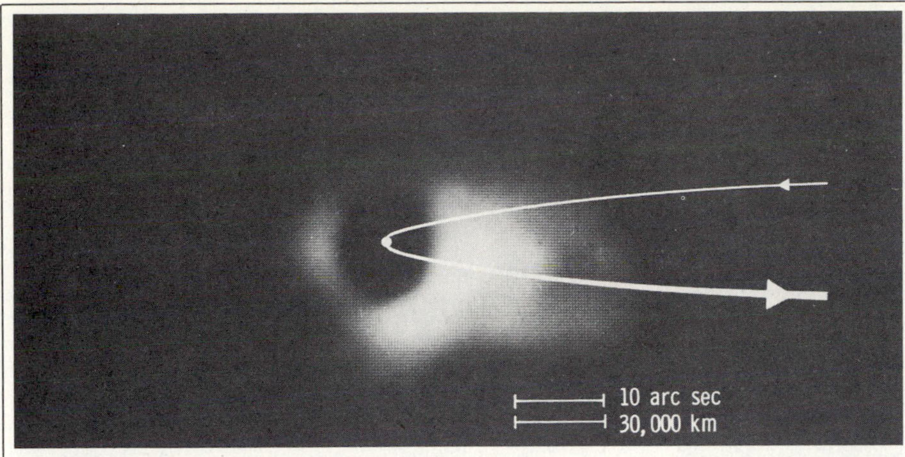

11.10
The sodium cloud around Io, as seen from the earth. The size and position of Io have been drawn in. (Courtesy B. Goldberg and the Jet Propulsion Lab)

Io's volcanoes eject plumes of gas and dust to heights of 250 km at velocities of up to 1000 m/sec. (In contrast, the earth's large volcanoes spit out material at about 50 m/sec.) On a nearly airless body like Io, the volcanic gas and dust crest like a fountain plume in several minutes and then spread and fall in a dome shape (Fig. 11.11). Io's escape velocity (2.5 km/sec) is greater than the speed with which the volcanic dust and gases erupt; thus there is little direct loss of the material to space.

Io's volcanoes have a different shape from those commonly found on the earth, Venus, and Mars. Few appear as cones or shields. They resemble collapsed volcanic craters. Lava simply pours out of a crater vent and spreads outward for hundreds of kilometers (Fig. 11.12). So dark-colored lava lakes surround many of Io's volcanoes. One of the largest is about the size of the island of Hawaii! The temperatures in these lava lakes are about 330 K.

Why is Io's interior so hot? The other Galilean moons force Io into an eccentric orbit, so its distance from Jupiter changes significantly. These distance variations cause large and variable tidal forces (see Focus 9.1) to act on Io. Its interior

11.11
Voyager 1 photo of a gigantic eruption on Io. The plume covered an area equal to the size of Alaska and rose more than 300 km above the surface. At the center of the photo is the source of the eruption: the complex of hills with a central valley. (Mosiac produced by A. S. McEwen, U. S. Geological Survey; photo courtesy Jet Propulsion Lab and NASA)

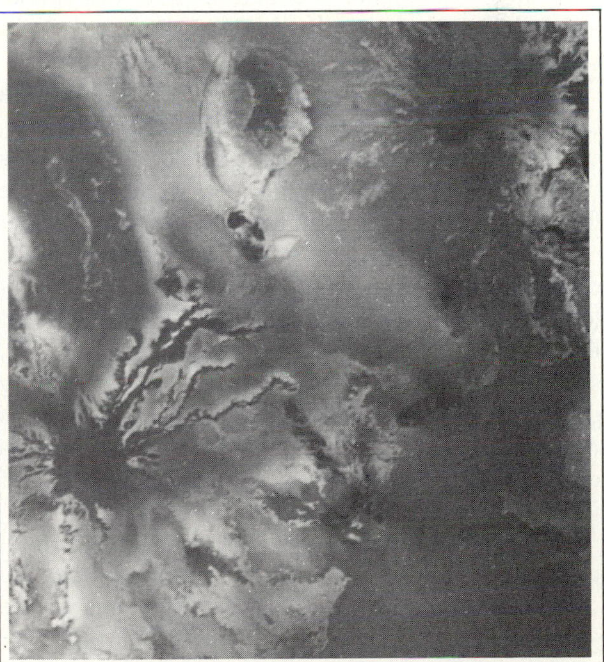

11.12
A close-up view of Io. The dark spot with the irregular radiating pattern (at the left) is a volcanic vent with lava flows. (Courtesy NASA)

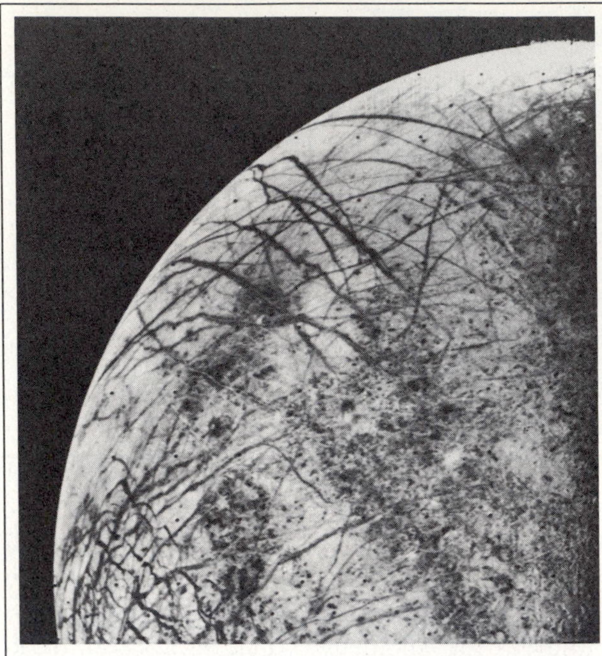

11.13
A view of Europa showing features as small as 4 km. Note the dark cracks along the surface. (Courtesy NASA)

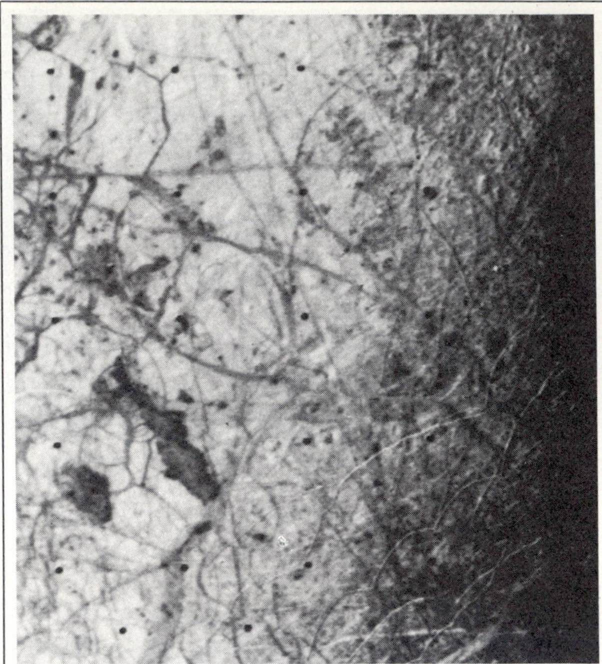

11.14
A close-up view of Europa. The photo covers an area of about 600 by 800 km along the day-night line, which highlights the surface features. The bright ridges are about 5 to 10 km wide and 100 km long. The dark bands are 20 to 40 km wide and thousands of kilometers long. (Courtesy NASA)

heats up from the continuous pushing and pulling from the tidal forces.

Because volcanic activity continually alters Io's surface, it must be very young. No impact craters appear on Io; volcanic flows have covered them up. Io's surface seems to be the youngest in the solar system, probably less than 1 million years old.

Europa / The surface features of Europa consist of bright areas of water ice among darker, orange-brown areas. Europa's surface is crisscrossed by stripes and bands that may be filled fractures in the moon's icy crust (Fig. 11.13).

The most impressive features on Europa are dark markings that crisscross its face, making it look like a cracked eggshell (Fig. 11.14). Some of these cracks extend for thousands of kilometers, split to widths of 50 to 200 km, but reaching depths of only 100 m or so. Europa's surface is really incredibly smooth. Compared to its size, its dark markings are no deeper than the thickness of ink drawn on a ping-pong ball.

Europa's surface is almost devoid of impact craters. So Europa's surface cannot be a primitive one; it must have evolved since its formation. The crust must have been warm and soft sometime after formation to wipe out evidence of the early, intense bombardment.

Europa's cracked surface indicates that its solid, icy crust is thin and its interior hot and primarily molten. How did Europa get this way? One tentative model proposes that its crust may long ago have been a slush kept partially melted by a hot interior. As Europa cooled, its crust turned to smooth, glassy ice that later cracked.

Ganymede / Largest moon of Jupiter, Ganymede ranks overall as the second largest moon in the solar system. (Only Titan of Saturn is bigger.) Its surface looks vaguely like our moon's, with dark, marialike regions. Yet it also has huge fault lines along its surface, like Europa.

Ganymede has two basic types of terrain (Fig. 11.15): *cratered* and *grooved*. Craters up to 150 km in size densely mark the cratered terrain. Their abundance indicates that the cratered terrain is old, some 4 billion years. Compared with those on the moon and Mercury, the craters are shallow for their size, and some have convex rather than concave floors. The craters of Ganymede also differ from those of the moon and Mercury in that they have central pits rather than central peaks. Many craters on Ganymede have very bright rays extending from them (Fig. 11.15), attesting to their formation by impacts on an icy surface.

The grooved terrain separates the cratered terrain into polygon-shaped segments. It consists of a mosaic of light ridges and darker grooves where the ground has slid, sheared, and torn apart. Long cracks, where the ground has moved

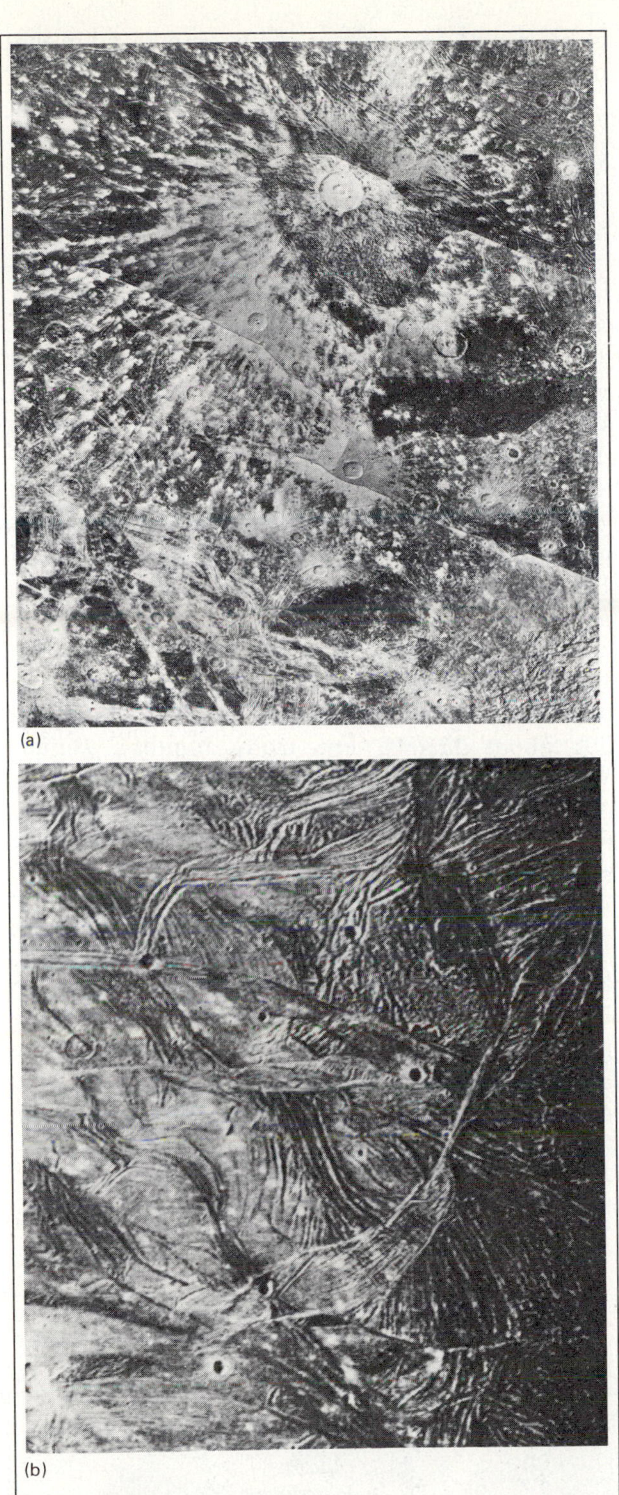

11.15

(a) Impact craters with bright rays on the surface of Ganymede. The large crater at the upper center is about 150 km in size. (b) Grooved and ridged terrain on Ganymede's surface. (Both courtesy NASA)

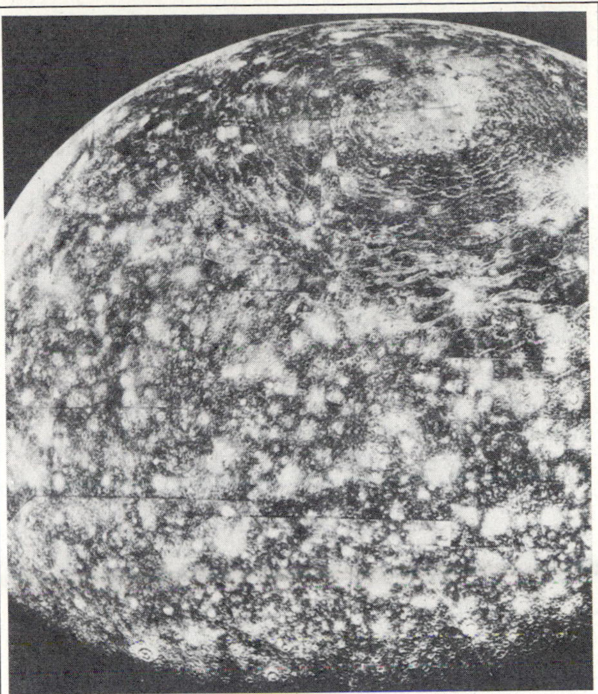

11.16

The giant ringed basin on Callisto (upper right). The bright spot at the basin's center is about 600 km across; the outer rings, 2600 km. Note the lack of ridges or mountains. (Courtesy NASA)

sideways for hundreds of kilometers, also cover the surface.

There are no large mountainous regions or large basins on Ganymede; in fact, nowhere on the satellite is there any relief greater than about 1 km. In some regions there are small bright patches with just a hint of surrounding walls, which look as if a crater has sunk into a soft surface. These features suggest that the crust of Ganymede is somewhat plastic. Crater rims and mountains slowly sink back into the surface; crater floors gradually fill in. This plastic flow probably occurs because of the large amount of water ice in Ganymede's crust.

Ganymede's bulk density is low, only about 2000 kg/m^3, so it must contain about half water and half rock. Occasional stresses on the water-rock crust have created the fracture patterns. Some ridges and grooves overlay others, an indication that many episodes of crustal deformation have happened in the past.

Studies of the earth's moon indicate that the era of intense cratering ended some 3 to 4 billion years ago. So Ganymede's surface is 3 to 4 billion years old; that is, Ganymede has been geologically inactive for over 3 billion years.

Callisto / Farthest out of the Galilean moons, Callisto (Fig. 11.16) has a surface that most resembles our moon and Mercury. It is riddled with craters of a wide range of sizes. Some have bright ice rays; others are filled with ice. Callisto's craters are shallow—several hundred meters deep or less. Why? Because the surface is a mixture of ice and rock; the surface slowly flows, flattening out the ups and downs of the land.

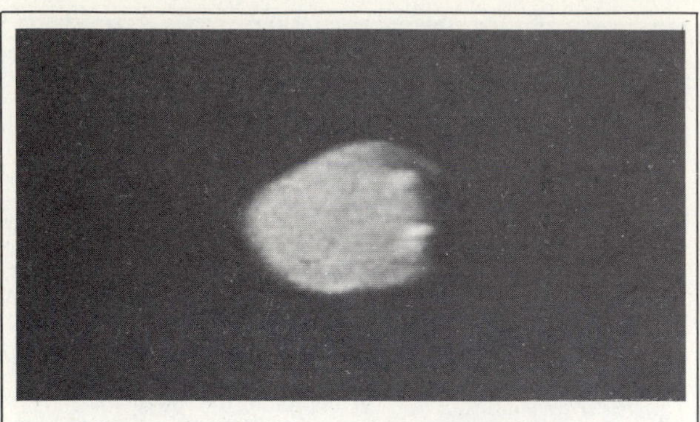

11.17
Amalthea, Jupiter's innermost moon. The indentations at top and bottom may be craters. (Courtesy NASA)

Callisto has one huge and beautiful multiringed feature (Fig. 11.16): Its central floor is 600 km in diameter; 20 to 30 mountainous rings that have diameters of up to 3000 km surround it like a bull's-eye. The rings look like a series of frozen waves. They might have been formed in a stupendous collision that melted subsurface ice, causing the water to spread in waves that quickly froze in the 95 K surface temperature. The ripple marks are preserved as rings: frozen blast waves.

The central floor of this ringed feature has fewer craters than the rest of the terrain. This difference indicates that the impact forming the rings occurred after much of the cratering, probably 4 billion years ago.

Asteroidal moons / Jupiter's other moons are asteroidlike bodies, and we expect that they are indeed captured asteroids. (This is likely, for, after all, Jupiter lies just outside the asteroid belt.) There are two groups of four, one group at a distance of about 12 million km that orbits counterclockwise and another at about 23 million km that orbits clockwise.

We have observed one moon, Amalthea, closely. Only 181,000 km out from Jupiter, it whizzes around once every 12 hours. It is elongated, 270 km by 155 km (Fig. 11.17); the surface is cratered and has a dark red color. This moon's irregular shape, small size, and dark, cratered surface imply its asteroidlike character (more about asteroids in Chapter 12).

The rings of Jupiter / Jupiter actually has millions of moons—tiny ones that make up its ring system, discovered by Voyager 1. The rings are so thin (less than 30 km thick) that they are essentially transparent. They are most visible when viewed edge on; then the particles scatter light well. To do so, the particles must be small, about 10 μm in size. We do not yet know what they are made of.

Dramatic pictures of the back-lit rings (Fig. 11.18) show that the rings have a definite structure. The outer, brightest part is 800 km wide and lies about 128,500 km from Jupiter's center. Within it is a broader ring some 6000 km wide. And within that ring lies a faint sheet of material that extends from 119,000 km out from Jupiter's center down to the cloud tops.

11.3 Saturn: Jewel of the solar system

Saturn bears a marked resemblance to Jupiter, but its ring system outranks in splendor that of the larger planet. Saturn has a slightly smaller size and less mass than Jupiter. It has the lowest density of any of the planets—only 680 kg/m^3, even less than that of water.

The atmospheric structure of Saturn resembles that of Jupiter. It also has belts running

11.18
Back-lit view of Jupiter's rings. Note the bright, sharp edge and diffuse inner region. (Courtesy NASA)

11.19
Jet streams and turbulence in the upper atmosphere of Saturn. (Courtesy NASA)

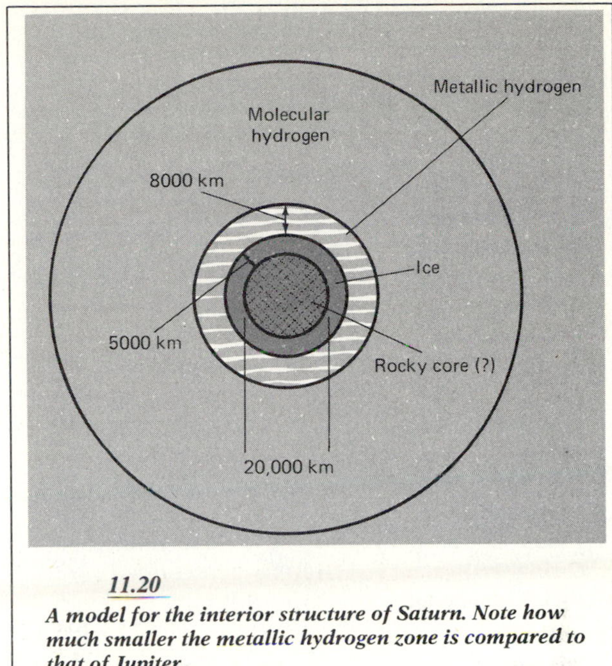

11.20
A model for the interior structure of Saturn. Note how much smaller the metallic hydrogen zone is compared to that of Jupiter.

parallel to the equator. Disturbances in the belts rarely occur (only 10 spots have been observed to date from the earth), compared with their frequency on Jupiter. Voyager 1 discovered a reddish spot, much smaller than the Great Red Spot on Jupiter; clouds only a few hundred kilometers across were detected at high latitudes.

The atmosphere of Saturn probably has much the same composition as that of Jupiter. So far methane, ammonia, ethane, phosphine, and molecular hydrogen have been found as the most abundant. The ammonia is less than that on Jupiter; probably just as much exists, but at the lower temperature of Saturn (93 K), it has frozen and fallen out of the upper atmosphere. Infrared spectroscopy has detected abundant molecular hydrogen, and a substantial percentage of helium is expected. Spacecraft observations give abundances (by mass) of 73 percent hydrogen and 26 percent helium, about the same as for Jupiter (and for the sun).

Saturn's clouds appear far less colorful than those of Jupiter—mostly a faint yellow and orange. Because of the low temperatures on Saturn compared to Jupiter, the clouds lie lower in the atmosphere, and a high-altitude haze subdues our view. However, the Voyager photos showed much of the same complexity of cloud patterns seen on Jupiter, with wind speeds much higher, up to 500 m/sec near the equator (Fig. 11.19).

Saturn's interior (Fig. 11.20) probably reflects Jupiter's composition; theoretical estimates are about 74 percent hydrogen, 24 percent helium, and 2 percent heavier elements. Again, this composition is roughly the same as that of the sun, but with more heavy elements. Saturn may have a small, rocky core some 20,000 km in diameter and a mass of about 20 earth masses. Other models have the metallic hydrogen region extending right to the center.

Saturn resembles Jupiter in two other important respects: (1) Infrared observations show that Saturn emits more energy, as infrared radiation, than it receives from the sun. The excess amount is about twice the energy Saturn receives from the sun. As with Jupiter, this excess heat may be left over from the planet's formation. (2) Radio and spacecraft observations of Saturn imply that it, too, has a magnetic field—but only 1/20 as strong as that of Jupiter (but that's still 40 times stronger than that of the earth). The magnetic axis aligns with Saturn's rotation axis, and the field pattern is much more regular than for Jupiter. The magnetic field is probably produced by a dynamo effect in the liquid metallic hydrogen zone of Saturn, in the same way it is presumably produced in Jupiter. The magnetic field creates Van Allen–like belts around Saturn, which trap charged particles from the sun.

TABLE 11.2 Saturnian satellites and rings

Object	Radius (km)*	Density (kg/m^3)	Distance from Saturn (km)
Cloud tops			60,330
D-ring inner edge			67,000
C-ring inner edge			74,400
B-ring inner edge			91,900
B-ring outer edge			117,400
A-ring inner edge			121,900
Encke division			133,400
A-ring outer edge			136,600
Atlas	10×20	?	137,670
1980 S27	$70 \times 50 \times 40$	?	139,350
F-ring			140,300
1980 S26	$55 \times 45 \times 35$	?	141,700
Epimetheus	$70 \times 60 \times 50$	?	151,420
Janus	$110 \times 100 \times 80$	?	151,470
G-ring			170,000
E-ring inner edge			180,000
Mimas	196 ± 3	1190 ± 50	185,540
Enceladus	250 ± 10	1200 ± 400	238,040
Telesto	$17 \times 14 \times 13$	?	294,670
Calypso	$17 \times 11 \times 11$	?	294,670
Tethys	530 ± 10	1210 ± 160	294,670
Dione	560 ± 5	1430 ± 60	377,420
1980 S6	$18 \times 16 \times 115$	?	377,420
E-ring outer edge			480,000
Rhea	765 ± 5	1330 ± 90	527,040
Titan	$2{,}575 \pm 2$	1880 ± 10	1,221,860
Hyperion	$205 \times 130 \times 110$	?	1,481,100
Iapetus	730 ± 10	1160 ± 90	3,561,300
Phoebe	110 ± 10	?	12,954,000

* Satellites with two or three dimensions listed are not spherical.

11.4 The moons and rings of Saturn

Saturn's band of moons totals at least 17 (Table 11.2). With two exceptions (Phoebe and Iapetus), all the moons stick close to Saturn's equatorial plane. Masses for some of the moons were determined from their gravitational attraction on spacecraft. The densities range from 1000 kg/m^3 for Tethys to 1400 kg/m^3 for Dione, similar to the densities of the outer Galilean moons of Jupiter.

To keep the general properties in mind, consider the moons of Saturn to fall into three groups: Titan by itself; the 6 large icy moons (Mimas, Enceladus, Tethys, Dione, Rhea, and Iapetus, in order outward from Saturn); and the 10 small moons (Phoebe, Hyperion, and the rest). Overall, their densities are less than 2000 kg/m^3, which implies that the moons are mostly ice (60 to 70 percent) with some rock (30 to 40 percent). In contrast to the Galilean moons, there is no trend of densities with distance from Saturn. Like Jupiter's moons, all except one (Phoebe) keep the same face toward Saturn.

Most of the moons are cratered. Some cratered terrain has been modified on the larger moons, which implies internal heating to melt parts of the icy surfaces. In contrast, the small moons, which are also cratered, show no changes—they still have their original surfaces. In fact, they may be pieces of an originally larger body.

With these basics in mind, let's look at the moons in some detail.

Titan / Titan, the largest moon, has a mass of 1.37×10^{23} kg and a diameter of 5150 km. Its density is 1900 kg/m^3, which implies a 50-50 composition of ice and rock.

Titan was the first moon found to have an atmosphere. The ultraviolet spectrometer on Voyager showed that it consists mostly of molecular nitrogen (99 percent), with about 1 percent methane and perhaps a trace of argon. Several

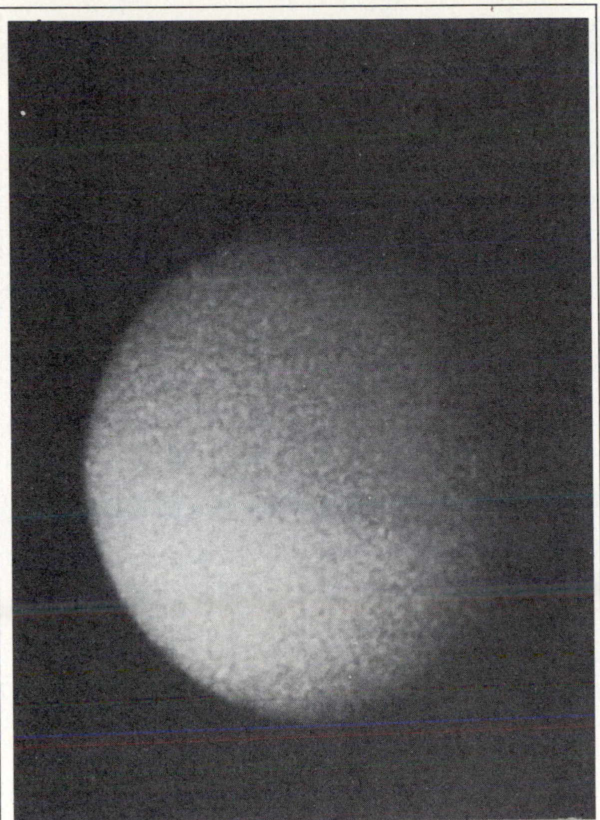

11.21
The clouds of Titan. Note that they appear darker in the northern hemisphere (top) than in the southern. (Courtesy NASA)

11.22
The surface of Dione. Note the many impact craters. The irregular valleys are old surface faults eroded by impacts. (Courtesy NASA)

hydrocarbons have also been detected, including ethane, acetylene, and ethylene. The atmosphere's surface pressure is about 1.5 atm. The surface temperature is roughly 94 K.

Close-up photos (Fig. 11.21) showed a stratospheric layur of orange smog as well as a blue color along Titan's edge. This coloration indicates that the atmosphere varies in composition. No surface features were seen. The surface was completely obscured from Voyager's view also, but the pressure and temperature data, along with the spectroscopic detections of nitrogen and hydrocarbons, have led to models of a surface covered with a frigid ocean of ethane, methane, and nitrogen up to 1 km deep, beneath which may reside a layer of acetylene.

Other moons / Saturn's four largest moons, after Titan, are Iapetus, Rhea, Dione, and Tethys, with diameters ranging from 1160 km to 1530 km. They appear heavily cratered (Fig. 11.22). In a few cases, wispy white streaks form rayed patterns around impact craters, but mostly they do not. They are probably deposits of frozen ice, but whether from material emanating from the interior or from debris deposited by colliding bodies is unknown.

Iapetus (Fig. 11.23a) has the most extremes of surface cover. The hemisphere leading in its orbit is only 1/15 as bright as that following (like the difference between a blackboard and a field of snow). The leading surface seems covered with dark debris picked up during its journey around the planet.

Only Enceladus does not have a surface thick with craters (Fig. 11.23b). That is a sure sign of recent modification of the surface. A hot interior can melt the icy surface; one photo showed a possible volcanic plume, which would clearly indicate a hot interior now.

The rest of the moons are all small bodies, 300 km or less in diameter. The largest is Hyperion, which has a strange shape (Fig. 11.24), like that of a thick hamburger. It has a cratered surface. The other moons are also cratered but much smaller. We presume that all these bodies are basically ice, as are the larger moons.

The ring system / In 1659 Huygens observed that Saturn "is surrounded by a thin, flat ring" that does not touch the body of the planet. Further observations by Cassini uncovered a gap in Huygens's single ring; this gap is known as *Cassini's division*. The rings lie tipped about 26° to the orbital plane, and because of their tilt, they change their appearance as viewed from the earth during the course of Saturn's revolution about the sun.

The near disappearance of the edge-on rings indicate that they are very thin, no more than 2 to

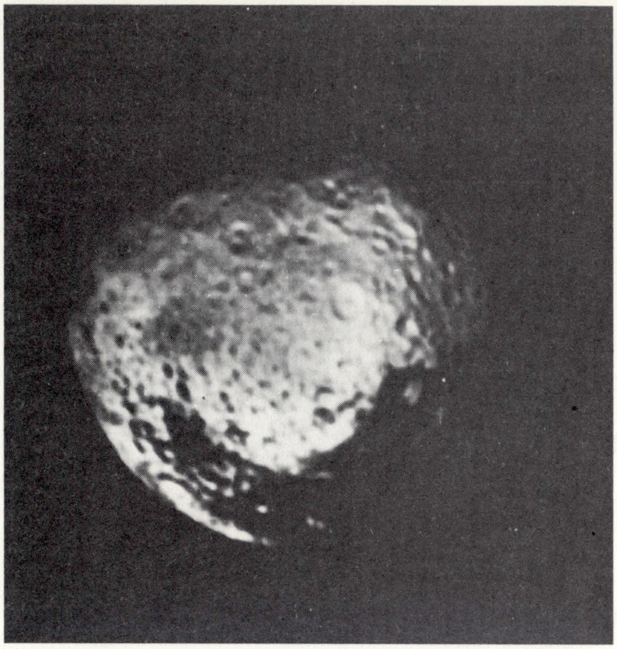

(a)

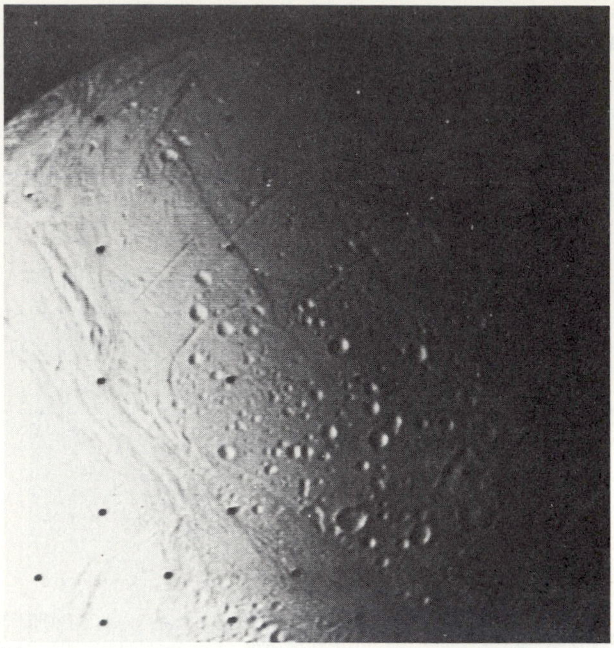

(b)

11.23

(a) *Iapetus, showing details as small as 20 km in size. Note the impact craters. The dark region at bottom covers the icy crust of the hemisphere that faces in the direction in which Iapetus orbits.* **(b)** *A close-up view of Enceladus. The grooves and linear features imply that the crust was deformed by internal heat. The largest crater visible is about 35 km in diameter. (Both courtesy NASA)*

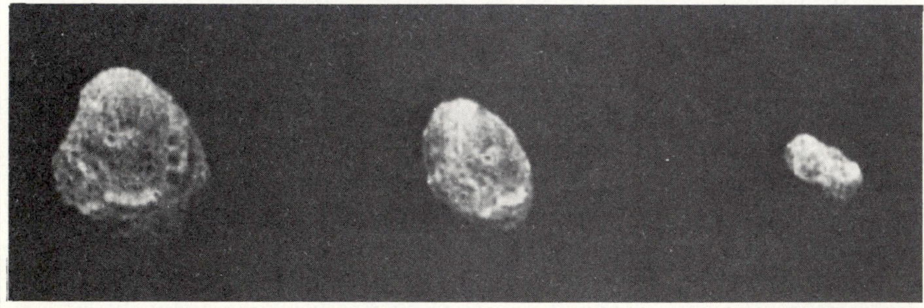

11.24

Three views of Hyperion, showing its unusual shape. Note the impact craters. (Courtesy NASA)

5 km thick. Although thin, the rings are wide; the three main rings visible from the earth reach from 71,000 km to 140,000 km from Saturn's center (Fig. 11.25).

The Voyager photographs revealed spectacular detail in the ring system. Although the A-ring is relatively smooth, the B- and C-rings break up into numerous small ringlets (Fig. 11.26), like grooves on a phonograph record. Many hundreds, perhaps a thousand, light and dark ringlets surround the planet, with widths as small as 2 km, the best resolution of the Voyager cameras. Some (in the C-ring) appear elliptical in shape, rather than circular. Even the Cassini division, apparently empty as seen from earth, was found to be filled with at least 20 ringlets.

Dark, spokelike features occur in the B-ring (Fig. 11.27). Typically, the spokes are about 10,000 km long and 1000 km wide. They consist of very small particles, much smaller than the average particle in the rings. Because the inner particles orbit faster than the outer ones, the spokes last only a few hours. We are not sure how they form.

Pioneer 11 discovered a new ring out beyond the previously known ones. Called the F-ring, it lies 3500 km outside the edge of the rings visible from the earth. The F-ring appears to be some 320 km wide and a mere 3 to 4 km thick. Voyager 1 photos resolved this ring into a complex system of knots and a braided structure of at least three strands. Voyager 2 photos nine months later showed that the braiding had disappeared!

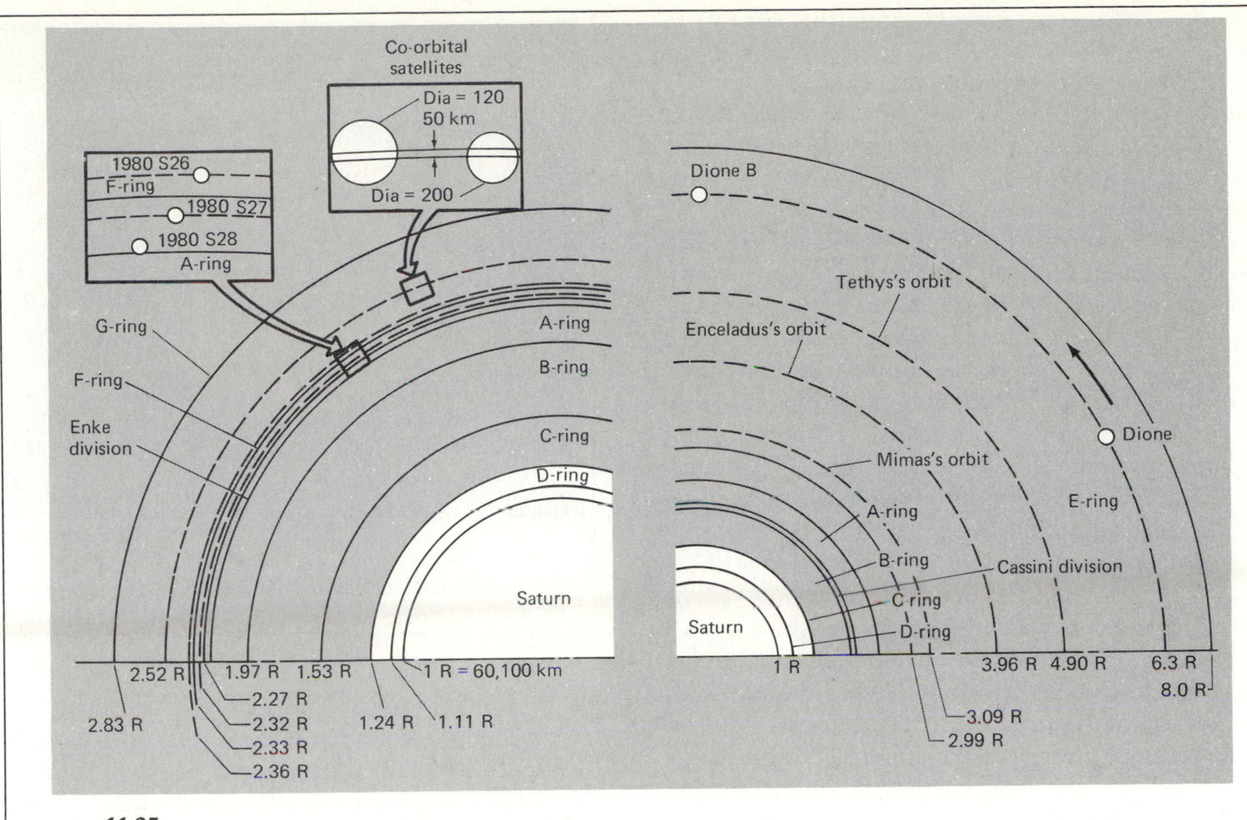

11.25

A schematic diagram of Saturn's ring system and the orbits of some of its moons. The view is from above Saturn's north pole. Note that the scales in the two halves differ. Distances are given in units of the planet's radius. (Adapted from a NASA diagram)

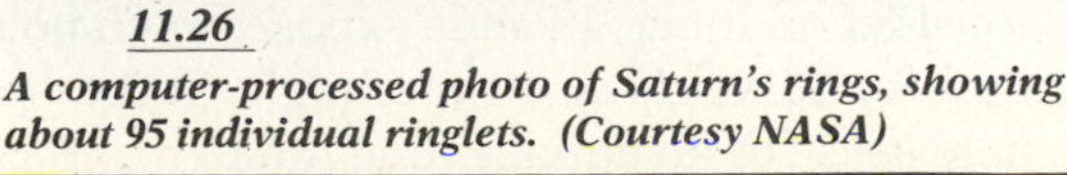

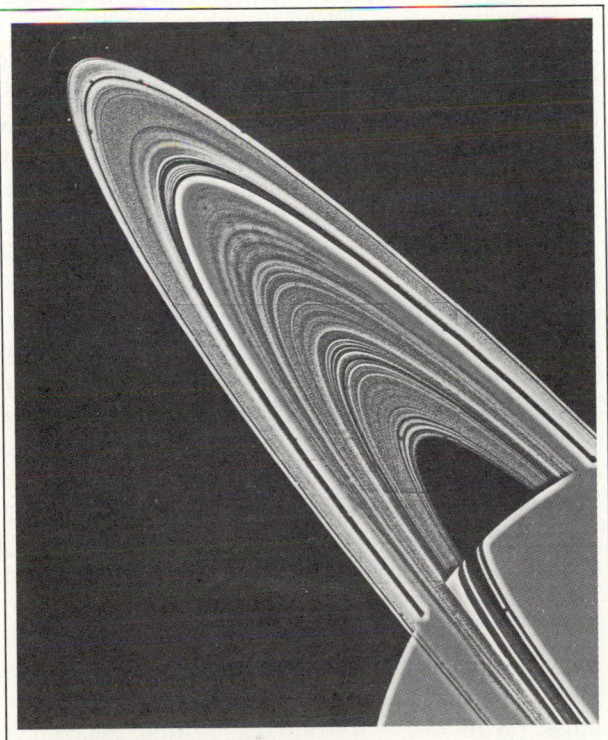

11.26

A computer-processed photo of Saturn's rings, showing about 95 individual ringlets. (Courtesy NASA)

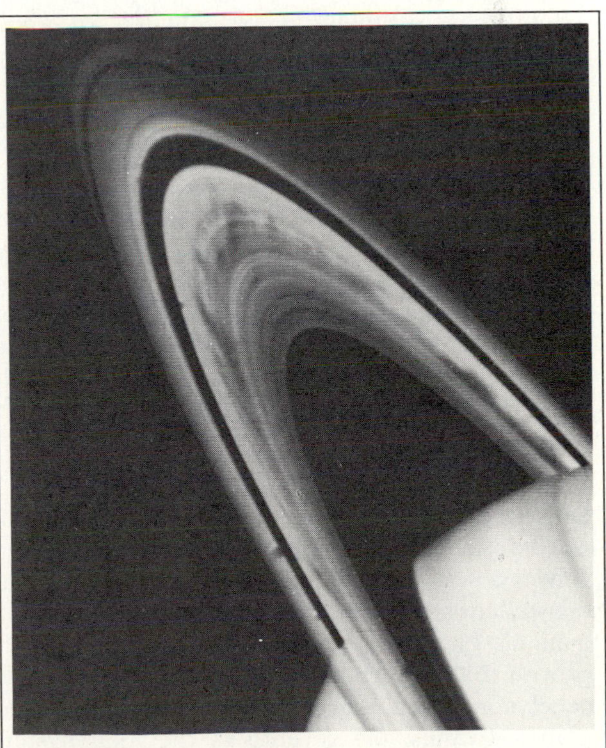

11.27

Dark spokes in the B-ring. The markings are about 12,000 km long. Note their diffuseness. (Courtesy NASA)

Another such very narrow ring (the G-ring) is 10,000 km farther out.

Two other extremely faint rings are known. The E-ring extends out beyond the F-ring to at least 6.5 Saturn radii (400,000 km). A ring inside the C-ring, called the D-ring, extends at least halfway to the surface of Saturn.

The rotation rate of the rings varies. The velocities range from 16 km/sec at the outer boundary of the A-ring to 20 km/sec at the inner boundary of the B-ring. The measured velocities agree with those expected from Kepler's third law for individual masses placed at the ring distances from Saturn; this agreement indicates that separate particles comprise the rings. (If the rings were solid, the velocity of the outer edge would be higher than that of the inner edge.)

Infrared observations of Saturn's rings show that they are made of particles of water ice or rocky particles coated with water ice. The ice does not evaporate because the surface temperature of the particles is only about 70 K, and the particles are occasionally screened from direct sunlight by each other and by Saturn's sphere. Observations of radio signals from Voyager reflected and transmitted by the rings indicate that the particles are about 1 m in diameter, but a range of sizes, from centimeters to tens of meters, probably exists.

Although deceptively solid in appearance and covering a large area of space, the rings have a total mass estimated to be only 10^{16} kg, about 10^{-6} the mass of our moon, and a mere 10^{-10} the mass of Saturn.

How did the rings form? Saturn's great mass creates strong tidal gravitational forces (see Focus 9.1). If a moon had approached close enough, it would have experienced tidal forces strong enough to tear it into pieces. Such may have been the fate of a former Saturnian moon whose demise scattered the particles that then spread to form the rings. Another possibility is that the tidal forces may have prevented close-in particles from ever forming into a moon.

11.5 Uranus: The first new world

On March 13, 1781, the then unknown amateur astronomer William Herschel (1738–1822) perceived a star "visibly larger than the rest" in the constellation Gemini and "suspected it to be a comet." Later observations in March and April proved that the orbit was not like a comet's. Herschel concluded that he had discovered a new planet—Uranus, the seventh in the solar system and the first to be discovered with a telescope. It was named Uranus, after the mythological father of Saturn.

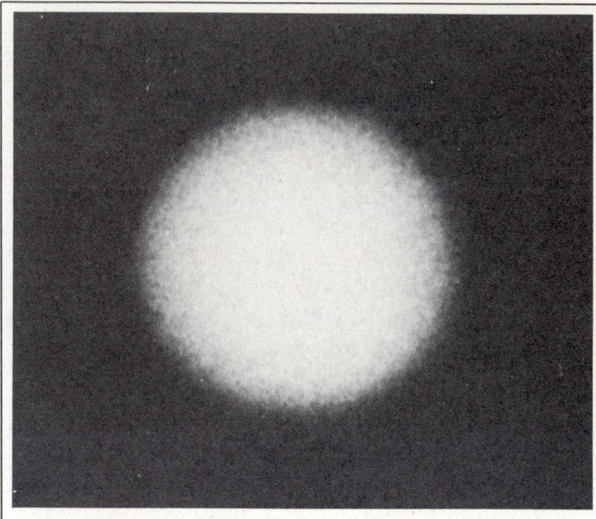

11.28
Uranus. Note the lack of any obvious banded structure in the upper atmosphere. (Courtesy NASA)

At an average distance of 19.2 AU, it takes Uranus (Fig. 11.28) 84 terrestrial years to journey around the sun. Far from the sun, the upper atmosphere must be very cold. Infrared observations put the temperature at 58 K. In such a deep freeze, all the ammonia has frozen out of the atmosphere and cannot be detected spectroscopically. Methane and molecular hydrogen do appear in the spectrum. Helium may also have been detected, but this result has not been confirmed.

Viewed through a telescope, Uranus has a distinctive pale green color. This greenish color comes from sunlight that penetrates deep into the planet's atmosphere; some red light is absorbed in the atmosphere, and much of the green is reflected back into space.

The low bulk density of Uranus, 1600 kg/m^3, implies that it contains mostly lightweight elements (Fig. 11.29). Uranus is thought to consist of roughly 15 percent hydrogen and helium, 60 percent icy materials (water, methane, and ammonia) and 25 percent earthy materials (silicates and iron).

For many years, astronomers believed that Uranus rotated in about 10.8 hours. Recent observations have been in conflict: 15, 15.6, 23, and 24 hours. The exact value is still being argued about, but I judge that it is about 15 hours.

The funny thing about Uranus's rotation is the direction of its rotational axis in space: It lies almost in the plane of the earth's orbit! Uranus spins on its side. Journeying around the sun in this lopsided manner, Uranus exposes each pole to sunlight for 21 years at a time; night at the opposite pole lasts for an equally long time.

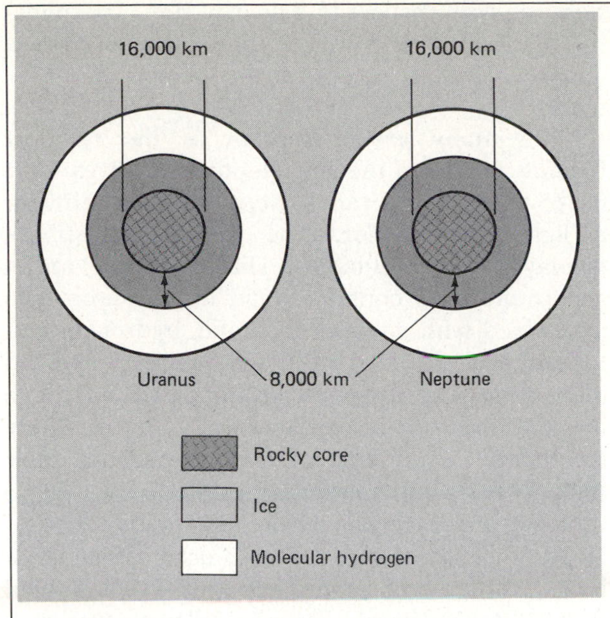

11.29
Models for the interiors of Uranus and Neptune. Both planets have the same mass and bulk composition, so their interiors are basically the same.

Five moons are so far known for Uranus (Fig. 11.30). They all move in the planet's equatorial plane and revolve in the same direction that the planet rotates (see Table B.9 in Appendix B). Because the moons lie in the same plane as Uranus's equator, their orbits as seen from the earth are alternately edge on, then fully open, every 42 years; in 1966 they appeared on edge, but in 1987 they will appear as circles.

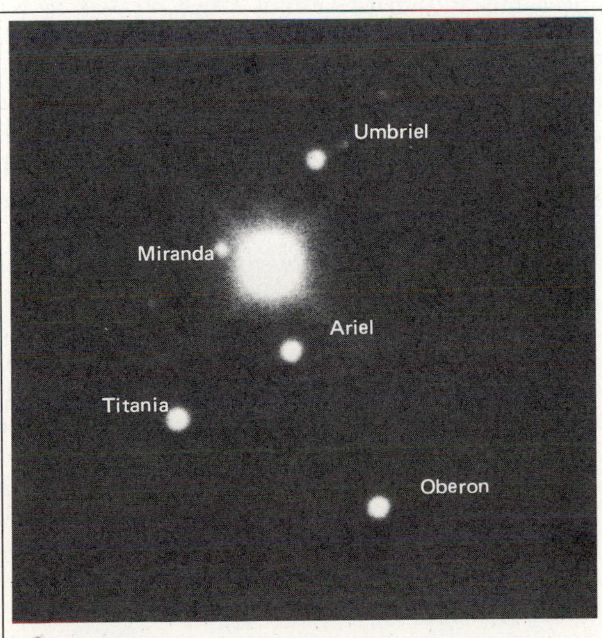

11.30
Uranus and its five moons. (Courtesy W. Liller; photo taken with the 4-m telescope at Cerro Tololo Interamerican Observatory)

The five moons are Miranda, Ariel, Umbriel, Titania, and Oberon. Miranda is the smallest (less than 200 km in diameter) and the closest to Uranus. The others range in diameter from 1110 km (Umbriel) to 1630 km (Oberon). Their surfaces appear to be made of a dirty ice, very much like the surface of Saturn's Hyperion. Recently, the masses of these moons have been estimated; they range from 0.22 to 0.81 our moon's mass. So we now know that the bulk densities range from 1300 to 2700 kg/m^3, which implies that these are bodies made of rock and ice.

Uranus has rings, at least nine in all, discovered accidentally in 1977 by three groups of astronomers—from Cornell, Lowell, and Perth observatories and the Indian Institute of Astrophysics. They watched Uranus pass in front of a faint star and were surprised to see the star momentarily dimmed a few times before Uranus covered it. Rings blocking out the starlight caused the dimmings. Later observations of Uranus passing in front of another star in 1978 showed a total of nine rings.

The observations to date paint a picture of a ring system (Fig. 11.31) dramatically different from that of Saturn. The nine rings circle the

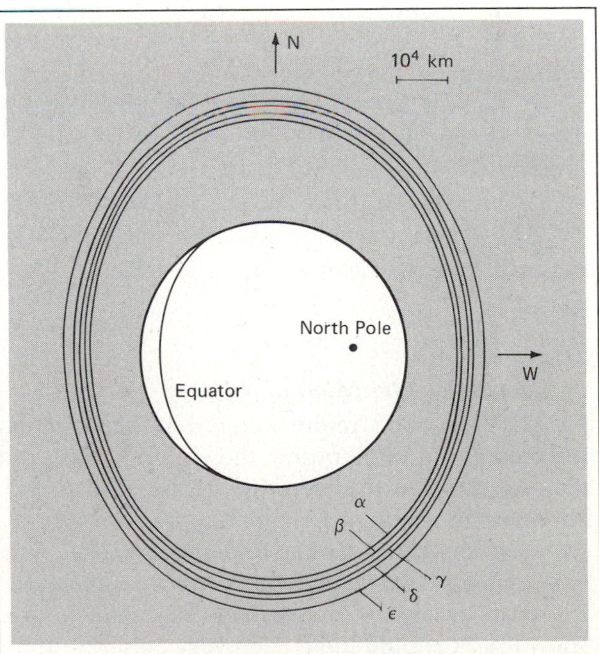

11.31
The five major rings of Uranus. Rings 4, 5, and 6 lie just inside the α ring; the η ring is between β and γ. (Adapted from a diagram by P. Nicholson, S. Persson, K. Matthews, P. Goldreich, and G. Neugebauer,* Astronomical Journal, *vol. 83, p. 1240, copyright © 1978 by the American Astronomical Society)

TABLE 11.3 The rings of Uranus

Ring	Semimajor axis (km)	Width (km)
6	41,860	5
5	42,270	5
4	42,600	5
α	44,750	9
β	45,700	16
η	47,210	5
γ	47,660	7
δ	48,330	5
ϵ	51,180	20–70

SOURCE J. L. Elliot, R. G. French, J. A. Frogel, J. H. Elias, D. Mink, and W. Liller, *Astronomical Journal*, 1981, vol. 86, p. 444.

planet in roughly three groups (Table 11.3): rings 6, 5, and 4 at about 42,500 km; α and β at 45,000 km; η, γ, and δ at 48,000 km; and the unique ϵ ring at 51,000 km from the center of Uranus. The smallest rings have widths of only about 5 km.

You may wonder why astronomers did not detect the rings of Uranus until now. (After all, we've known of Saturn's rings for about 300 years.) For one reason, the rings of Uranus are not very wide. More important, the material that makes up Uranus's rings is extremely black. Photos of Uranus indicate that the ring material can reflect no more than 5 percent of the light that strikes it. In contrast, because they are covered with water ice, the particles in the rings of Saturn reflect more than 80 percent of the light that hits them. So the particles in the rings of Uranus are probably bare of ice and likely made of dark carbon materials. We have no good idea yet of how large the rings' particles are. Observations so far indicate that they cannot be larger than 5 km in size.

11.6 Neptune: Guardian of the deep

Frankly, we don't really know very much about the cold world of Neptune, the eighth planet from the sun. It was discovered in 1846 because of its effects on the orbit of Uranus (see Section 4.5).

Evidence has come to light indicating that Neptune was first seen by none other than Galileo, 234 years earlier. Calculations of Neptune's orbit show that it should have been very close to Jupiter in the sky in January 1613. Galileo's journals have entries showing that he observed an object in the vicinity of Jupiter near Neptune's predicted position on December 27, 1612, and again on January 28, 1613, when Galileo detected a small motion of Neptune with respect to a nearby star. Inexplicably, Galileo never followed up on this discovery, and so failed to recognize the object as a new planet.

In many ways, Neptune is the twin of Uranus. Far from the sun, Neptune revolves once in 165 years. Like Uranus, Neptune shows off with a light green color. The upper atmosphere displays faint cloud bands. This cold (56 K) atmosphere probably contains water ice and ammonia ice, mixed with gaseous methane, hydrogen, and helium. One difference: Ethane has been detected in Neptune's atmosphere but not in Uranus'. The internal structure of Neptune (Fig. 11.29) probably resembles closely that of Uranus because their bulk densities and masses are similar.

We are uncertain about the rotation rate of Neptune. The value in common use has been 15.8 hours. Since 1977 others reported have ranged from 18 to 22 hours. For now, all we can say is that the rotation period is between 16 and 22 hours.

But all is not exactly the same with Uranus and Neptune. Infrared observations show that Neptune's surface temperature is about 56 K, compared with the expected value of 44 K if Neptune were heated only by the sun. So Neptune, unlike Uranus, has internal heat. It gives off 2.8 times as much energy as it receives from the sun, again most likely left over from its formation.

Does this mean that Neptune was formed with more internal heat than Uranus? No. Notice that the temperature of Neptune is now just about the same as that of Uranus (56 K versus 58 K). Theoretical models show that however they started, both planets should cool to about this temperature in 4.5 billion years. Uranus is close enough to the sun that solar heat can now keep it at this temperature. Neptune, farther away, will cool further before it comes into balance with sunlight.

Neptune may have weather, in the sense of changing meteorological conditions (Fig. 11.32). Infrared observations indicate that Neptune's atmospheric reflectivity, from 1 to 4 μm increased substantially over a one-year period. One explanation: the formation of an extensive cloud cover that then partially dissipated.

Neptune does have two known moons, Triton and Nereid. The orbit of the latter is extremely eccentric, and its distance from Neptune ranges from about 1 million to 10 million km. The larger satellite, Triton, revolves with a period of about 5 days in a retrograde (east-to-west) orbit that is inclined 20° to the plane of Neptune's equator. Triton has a diameter of between 3600 and 5200

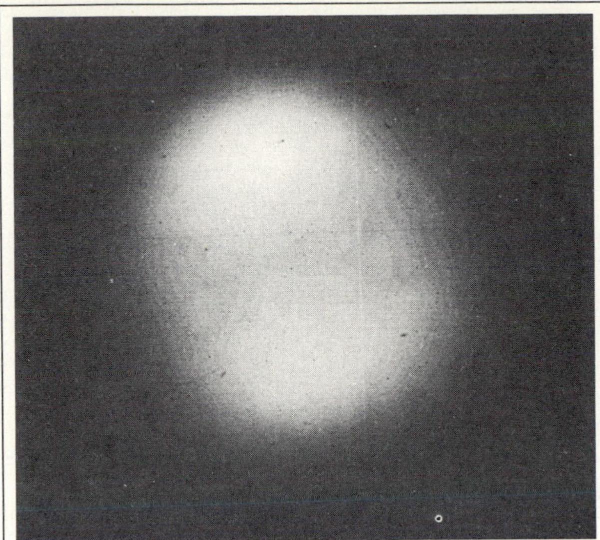

11.32
Neptune photographed in infrared light. The brighter regions are areas where methane ice clouds float in the atmosphere of methane gas. (Photo by H. Reitsema, B. Smith, and S. Larson, courtesy Lunar and Planetary Laboratory, University of Arizona)

km, which makes it one of the largest satellites in the solar system.

Triton now joins the short list of moons with an atmosphere. It contains gaseous methane. The pressure due to methane at Triton's surface is roughly 10^{-7} atm. Other observations indicate that the surface is mostly rocky, with a few patches of frozen methane. These patches are probably concentrated on the moon's night side. Nitrogen may also make up a part of the atmosphere, so there's a good chance that expanses of liquid nitrogen cover parts of the surface.

Neptune may have rings. When Neptune passed in front of a star in 1968, dips in the star's brightness occurred. If because of a ring, its radius is 28,600 km and its width 4300 km. Unfortunately, observations in 1981 did *not* indicate any rings, so their existence is still uncertain and doubtful.

11.7
Pluto: Guardian of the dark

Early in the twentieth century, Percival Lowell became fascinated with the problem of a planet beyond Neptune and initiated a search program at Lowell Observatory. After Lowell died in 1916, the search for Planet X was terminated until after the completion of a new telescope in 1929. It recorded over a million stars per photo in regions of the Milky Way.

Clyde W. Tombaugh worked at the new search, which started on April 1, 1929. Astronomers had assumed that Planet X would be similar to Neptune because of irregularities in Neptune's orbit (a parallel to the way Neptune was discovered; see Section 4.5). So they had searched for a visible disk of a new planet. Instead of searching for a disk, Tombaugh looked for Planet X's motion relative to the background stars.

The photographic search was trying and tedious, but on February 18, 1930, Tombaugh noted two images on different photographs, in the area near a star in Gemini, that had shifted slightly (Fig. 11.33). The shift was such that the object had to be a body orbiting the sun. The detection was quickly confirmed as a new planet, and the discovery was announced on March 13, 1930—Lowell's birthday. The name Pluto was officially accepted by Lowell Observatory.

I put Pluto in this chapter mostly because of its position just outside the orbits of the Jovian planets. Physically, Pluto is much smaller than the Jovian planets, but it may have a similar density. In fact, it is a lot like the large icy moons of Jupiter and Saturn.

Pluto's average distance from the sun is 39.44 AU. Since it has a highly eccentric orbit, it

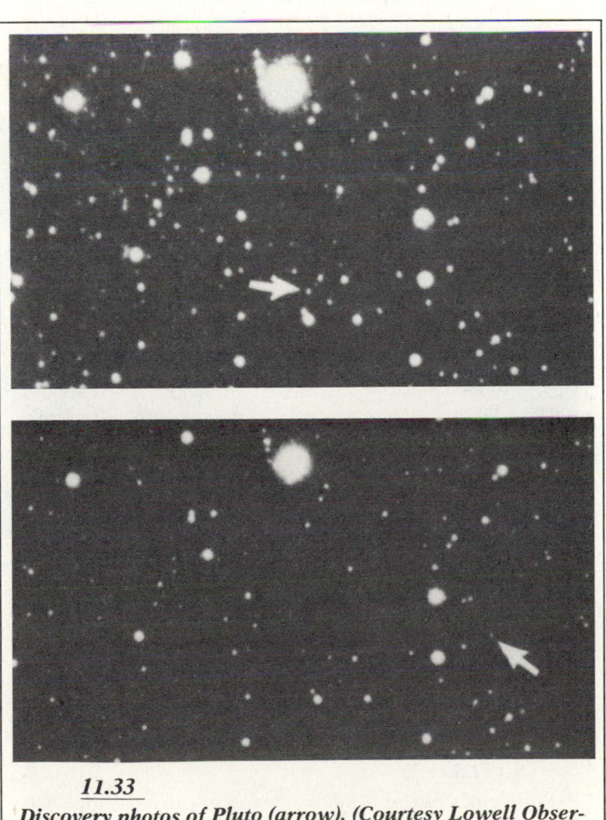

11.33
Discovery photos of Pluto (arrow). (Courtesy Lowell Observatory)

ranges from 29.7 AU to 49.3 AU from the sun, so it is never closer to the earth than 28.7 AU at closest approach (opposition). Because of its great distance and small diameter, Pluto strains the resolving power of large telescopes.

Recent attempts to measure Pluto's diameter have also been frustrating. One set of data indicates that Pluto is less than 6800 km in diameter; different observations with the Hale 5-m telescope indicate a diameter of 3000 to 3600 km.

Infrared observations show that methane ice coats Pluto's surface. The methane ice there means that the surface is bitter cold, no more than 40 K. This discovery also leads to a means of estimating Pluto's size by using how bright Pluto appears to us. Its brightness depends on its distance from the earth, its diameter, and what fraction of the surface is covered with ice. If the surface is completely covered by ice, then the diameter is less than 3000 km—smaller than our moon!

Observations of Pluto's brightness have uncovered a slight increase about every 6.4 days; this cyclic variation is the only evidence of rotation. The 6.4-day period is generally accepted as Pluto's rotation period.

In June 1978, James Christy of the U.S. Naval Observatory in Flagstaff, Arizona (Lowell's old haunt), noticed what appeared to be a bump on Pluto's image in a photo (Fig. 11.34). Checking older photos, Christy found seven showing the same bump, always oriented approximately north-south. He proposed that the bump was the faint image of Pluto's moon partially merged with the image of the planet. Christy named this moon Charon, after the mythological boatman who ferried souls across the river Styx to Pluto for judgment.

TABLE 11.4 ***Properties of the Pluto-Charon system***

Separation			17,500 km
Period of revolution			6.39 days
Pluto:	Mass (earth = 1)		0.0018
	Diameter		3000 km
	Density		800 kg/m^3
	Rotation period		6.39 days
Charon:	Mass (earth = 1)		0.0002
	Diameter		1300 km
	Density		800 kg/m^3
	Rotation period		6.39 days
Pluto-Charon mass ratio			10

Charon has a diameter of at least 1300 km, about half the size of Pluto (Table 11.4). The observations of Charon imply a revolution period of 6.4 days (the same as Pluto's rotation period) and a distance of 17,500 km from Pluto (Fig. 11.35). Now we can use Charon to find Pluto's mass by Kepler's third law. The result: Pluto has a mass about 0.002 that of the earth.

What a lightweight planet! Previous estimates of Pluto's mass, from its supposed gravitational influence on Neptune, came to 0.1 that of the earth, about 50 times higher than the new result. With Pluto's mass and diameter in hand, we can figure out its density. The result is 500 to 800 kg/m^3, which implies that Pluto consists mainly of ices and other frozen gases.

The idea that Pluto may be a small, icy planet, like a Jovian moon, fits with a speculation

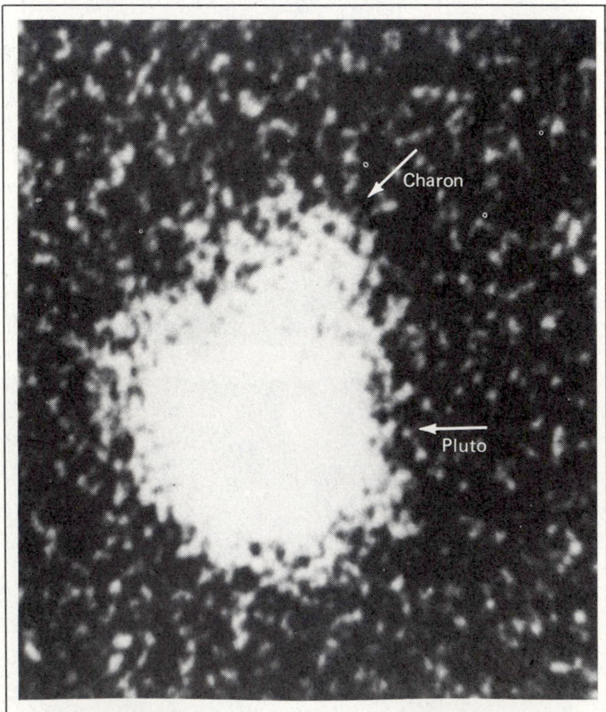

11.34
Pluto and Charon. The two bodies are so close together that their images merge. Charon (arrow) appears as the slight bump on Pluto. (Courtesy J. Christy)

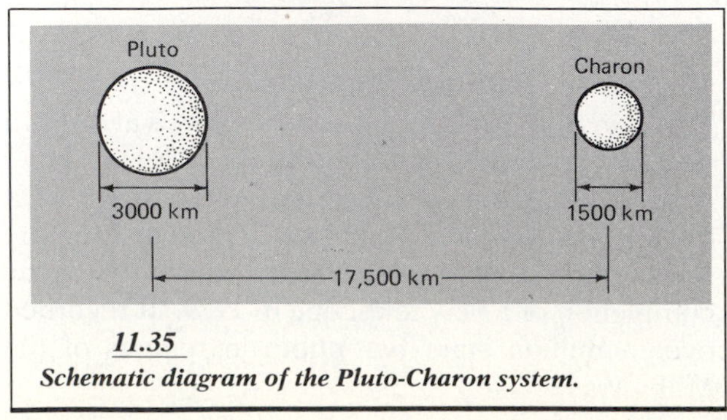

11.35
Schematic diagram of the Pluto-Charon system.

that Pluto is actually an escaped moon of Neptune. Why? The orbit of Pluto is so eccentric that it sometimes comes within Neptune's orbit, which means that at times Pluto is actually closer to the sun than Neptune. Recall that Neptune's Triton is a large satellite with an unusual retrograde orbit. If Pluto were originally a moon of Neptune along with Triton, a close encounter between the two might have caused Triton to reverse its orbital motion and also thrown Pluto free of Neptune. However, there are many details that this model fails to account for. So most astronomers consider it only a speculation.

By the way, Pluto is now within Neptune's orbit. On January 21, 1979, Pluto edged closer to the sun than Neptune. It will orbit closer to the sun than Neptune until March 1999. However, because of the high inclination of its orbit, Pluto is actually well above Neptune's orbital plane. Besides, Neptune is now about 60° around its orbit from Pluto, so there is no danger of a collision!

The search at Lowell Observatory for other planets beyond Neptune ended in 1945 without yielding further positive results. A planet with the same characteristics as Pluto, placed at a greater distance, has little chance for discovery. However, one like Uranus or Neptune might be just visible if it were closer than 80 AU. Some astronomers have postulated a new Planet X orbiting retrograde beyond Pluto. If this planet exists, it has not been observed to date. Tombaugh's search, which continued for 13 years, would have uncovered a planet like Neptune as far as 100 AU from the sun.

KEY CONCEPTS

1 ***As a group, the Jovian planets (Table 11.5) differ most from the terrestrial ones on the basis of lower bulk density, greater diameters, greater masses, and chemical composition.***

2 ***The Jovian planets largely consist of hydrogen and helium, with common molecules of methane, hydrogen, ammonia, and water.***

3 ***The large masses and low temperatures of the Jovian planets imply that their atmospheres have changed little since their time of formation.***

4 ***Jupiter, Saturn, and Neptune emit more heat into space than the energy from incoming sunlight, so their interiors must be hotter than if the sun were the sole source of their heat.***

5 ***Jupiter and Saturn have strong planetary magnetic fields, which implies that they have liquid, conducting cores—but their compositions rule out metal cores.***

6 ***Jupiter's largest moons decrease in density outward from the planet, an indication that at some time in the past Jupiter was luminous enough to affect them; these Galilean moons also have undergone different amounts of crustal evolution, as indicated by the extent of their impact cratering—Io the most and Callisto the least.***

7 ***Io is the most volcanically active body in the solar system, so it must have a hot interior.***

8 ***The rings of Jupiter are thin and consist of small particles; those of Saturn are wide, consist of larger, icy particles, and contain many small ringlets; those of Uranus are thin and narrow and consist of very dark, probably small particles.***

9 ***Except for Titan, Saturn's largest moons are basically ice and evolved some since the time of their formation, as indicated by their modified craters.***

10 ***The interior structures of Uranus and Neptune are essentially the same.***

11 ***Pluto is an icy world, with a methane-coated surface; its moon, Charon, is about half Pluto's size and 1/10 its mass.***

STUDY EXERCISES

1 In what significant respect is Jupiter most different from the other Jovian planets? (*Objective 1*)

TABLE 11.5 A comparison of the Jovian planets

	Diameter (earth = 1)	*Mass (earth = 1)*	*Bulk density (water = 1)*	*Atmosphere's main constituents*
Jupiter	11.0	318	1.3	Hydrogen, helium
Saturn	9.5	95	0.7	Hydrogen, helium
Uranus	4.1	15	1.0	Hydrogen, helium, methane
Neptune	3.9	17	1.7	Hydrogen, helium, methane
Pluto	0.2	0.002	0.8	Methane

2 Suppose you flew very close by Jupiter. What outstanding features would you see in the atmosphere? (*Objectives 2, 4, and 6*)

3 How do we know the bulk density of Pluto? (*Objectives 2 and 5*)

4 How do we know that the rings of Saturn are thin? (*Objective 4*)

5 What fact makes it relatively easy to find the masses of the Jovian planets? (*Objective 2*)

6 In one word, state the greatest difference between the Jovian and terrestrial planets. (*Objective 1*)

7 In two sentences, compare the rings of Saturn to those of Jupiter and to those of Uranus. (*Objective 5*)

8 How is Jupiter's magnetic field similar to the earth's? How is it different? Answer the same questions for Saturn. (*Objective 8*)

9 What features does Pluto have in common with the Galilean moons? (*Objective 7*)

10 In one short sentence, describe the interior compositions of the moons of the Jovian planets. (*Objectives 6, 7, and 9*)

BEYOND THIS BOOK . . .

Read "Jupiter" by J. H. Wolfe and "The Outer Planets" by D. M. Hunten in the September 1975 issue of *Scientific American.*

Contrast D. Cruikshank and D. Morrison's "The Galilean Satellites of Jupiter" in *Scientific American,* May 1976, with "The Galilean Moons of Jupiter" by L. Soderblom in the same magazine, January 1980.

The January 1980 issue of *National Geographic* contains an excellent article by R. Gore on Voyager observations of Jupiter and its satellites.

NASA has two excellent books on the results of the Voyager missions: *Voyage to Jupiter* (SP-439, 1980) by D. Morrison and J. Samz and *Voyages to Saturn* (SP-451, 1982) by D. Morrison.

For more details on the moons of Saturn, see "The Moons of Saturn" by L. Soderblom and T. Johnson in *Scientific American,* January 1982.

For a firsthand account of the discovery of Pluto, read *Out of the Darkness* (Stackpole, Harrisburg, Pa., 1980) by C. Tombaugh and P. Moore.

"Rings in the Solar System" by J. Pollack and J. Cuzzi in *Scientific American,* November 1981, compares the ring systems of Jupiter, Saturn, and Uranus.

12/ The origin and evolution of the solar system

But indeed the whole story of Comets and Planets and the Production of the World, is founded upon such poor and trifling Grounds, that I have often wondered how ingenious Men could spend all pains in making such fancies hang together.

CHRISTIAN HUYGENS

Learning objectives

After studying this chapter, you should be able to:

1. Identify at least two dynamic and two chemical properties of the solar system that any model of origin must try to explain.
2. Describe the general physical properties of comets, asteroids, meteoroids, and meteorites.
3. Specify what clues asteroids, comets, and meteorites provide about the formation of the solar system.
4. Explain what is meant by the *angular momentum problem* for nebular models.
5. Describe one possible way out of the angular momentum problem in modern nebular models.
6. Describe briefly the chemical condensation sequence, using one Jovian and one terrestrial planet to illustrate its use.
7. Describe the role of accretion in the formation of the planets.
8. Contrast the formation of Jupiter and Saturn to that of the terrestrial planets.
9. Sketch a modern scenario for the formation of the solar system, and evaluate how well it explains the known chemical and dynamic properties listed for objective 1.

Central question

What physical processes resulted in the formation of the planets from a uniform cloud of gas and dust?

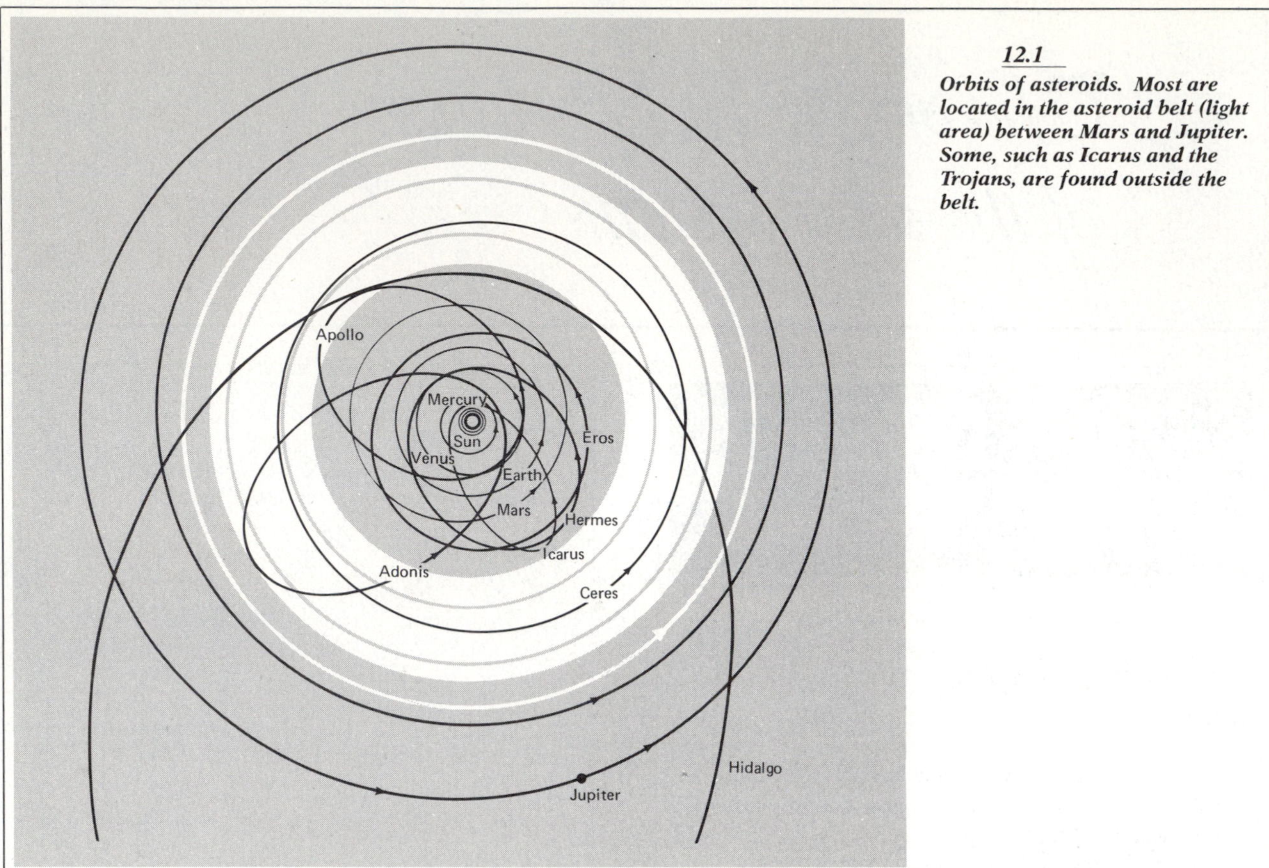

12.1
Orbits of asteroids. Most are located in the asteroid belt (light area) between Mars and Jupiter. Some, such as Icarus and the Trojans, are found outside the belt.

We have completed a grand tour of the solar system. By now you are probably wondering: How did the solar system originate? Such a question reflects people's concern with creation. You have seen (in Part 1) that most cultures have a story about the origin of the world. These myths tell of the world's development from some formless state, from chaos to cosmos.

Because it is still unresolved, the puzzle of the origin of the solar system still arouses astronomers' curiosity. Many models have been proposed. None has been completely successful. One scientific justification of the space program rested on finding possible information to support or refute theoretical ideas.

To unravel the puzzle of the solar system's genesis requires more than astronomy. It demands the interplay of astronomy with physics, chemistry, and geology. It refuses to be answered simply. This chapter investigates in some detail the contemporary approach to the question. In this picture, the formation of the solar system arises as a natural result of the formation of the sun from an interstellar cloud of gas and dust. The general outlines of this process reasonably explain the general features of the solar system. Many details and puzzles remain to be resolved; no one model yet fits them all together. In fact, the origin of the solar system has been a puzzle challenging astronomers for centuries!

12.1 *Debris between the planets*

Don't get the impression that the space between the planets is empty. It's not: Comets, meteroids, and asteroids are found there. These bodies, along with gas and dust, make up the interplanetary debris. Scooped up, the total mass amounts to less than the mass of the moon. Yet, as archaeologists have discovered, a city's trash heap holds valuable clues to a city's history, even though it contains much less mass than the city. Likewise, the solar system's debris is a fruitful hunting ground for clues to its past.

Asteroids: Minor planets / Most asteroids orbit the sun in a belt between Jupiter and Mars, at an average distance of 2.8 AU (Fig. 12.1). A few asteroids deviate widely from this value. Icarus actually skirts closer to the sun at perihelion than Mercury does. On June 15, 1968, Icarus careened past the earth with a closest approach of 6.4 million km, just 20 times farther away than the moon.

What is an asteroid? It is an irregular, rocky hunk, small both in size and in mass compared to

TABLE 12.1 The largest asteroids

Name	Diameter (km)
Ceres	1000
Pallas	608
Vesta	538
Hygcia	450
Euphrosyne	370
Interamnia	350
Davida	323
Cybele	309
Europa	289
Patientia	276

SOURCE D. Morrison, "Asteroid Sizes and Albedos," *Icarus*, 1977, vol. 31, pp. 185–220.

a planet. Ceres, the largest known asteroid, has a diameter of only about 1000 km (Table 12.1). (That's about one-third the size of the moon.)

Most asteroids fluctuate in brightness. Such variations indicate that they may have irregular surfaces or shapes (or both). From the observed brightness variations of Eros, for example, astronomers estimate it to have a bricklike shape, 6 km by 22 km. If Deimos and Phobos, the moons of Mars (see Section 10.9), are captured asteroids, their rough surfaces pitted with craters are what we expect other asteroids to look like.

The albedos of asteroids indicate that they fall into two compositional classes. Some are relatively bright (albedos of about 15 percent) and others are much darker (albedos of 2 to 5 percent), an indication that they contain a substantial percentage of dark compounds such as carbon. Those in the lighter class are dubbed *S-type* asteroids; the darker ones have been christened *C-type*. The S-type, in addition to having higher albedos, also show spectral features indicative of silicate materials. A third class, called *M-type*, has characteristics suggestive of metallic substances. They have albedos of about 10 percent. Only 5 percent of asteroids belong to this class.

Based on albedos, compositions in the asteroid belt vary with distance from the sun. Near the orbit of Mars, almost all asteroids have S-type characteristics. Farther out, we find fewer high-albedo ones and more dark ones. At the outer edge of the belt, 3 AU from the sun, some 80 percent of the asteroids are C-types. An explanation for this distribution is found in a current model for the formation of the solar system (see Section 12.7).

Comets: Snowballs in space / You probably associate comets with long, graceful tails. In fact, not all comets exhibit tails, even at perihelion, and when far from the sun, no comet has a visible tail (Fig. 12.2). When first sighted telescopically, a comet typically appears as a small, hazy dot. This bright head of the comet is called the *coma*. Sometimes the coma contains a small, starlike point called the *nucleus* (Fig. 12.3). We know that cometary nuclei are not larger than 1 km across,

12.2
Comet Cunnigham (1940). A comet far from the sun shows little or no tail. (By permission of Harvard College Observatory)

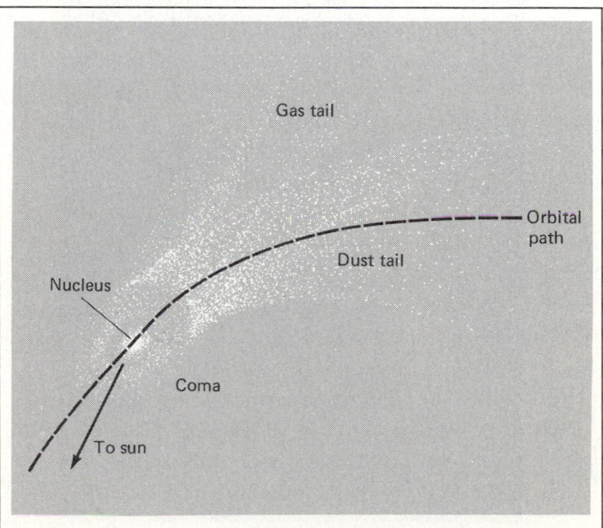

12.3
The main parts of a comet. The diffuse coma encloses a bright, starlike nucleus. The gas tail streams out opposite the sun. The dust tail follows the orbital path of the nucleus.

12.4
Comet Kohoutek (January 14, 1974). Note the kinks in the gas tail. (Courtesy Mt. Wilson Observatory, Carnegie Institute of Washington)

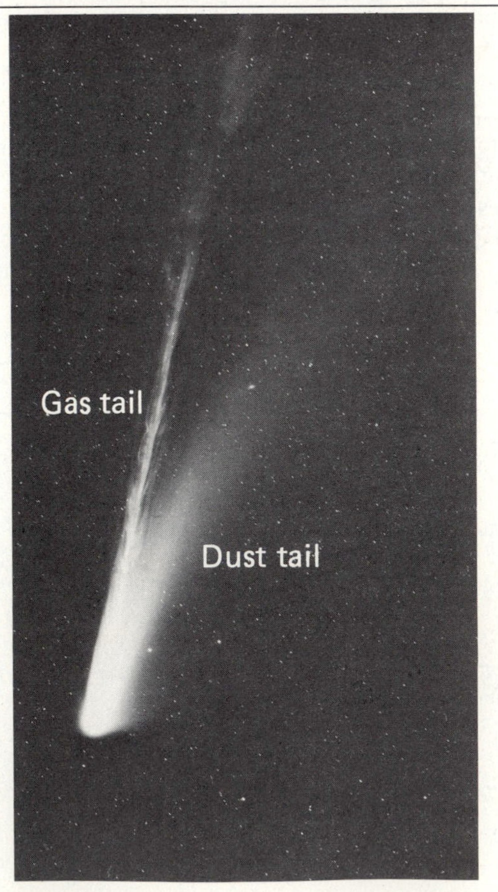

12.5
Comet Mrkos (1957), showing both a gas tail and a dust tail. (Courtesy Mt. Wilson Observatory, Carnegie Institution of Washington)

because none has ever been viewed as more than a point of light. As a comet heads toward perihelion, it grows brighter and sprouts a tail. A comet's tail may stretch for millions of kilometers and always points away from the sun (Fig. 12.4).

Comets may have two types of tails: gas and dust (Fig. 12.5). The physical differences between the two show up in their spectra. The spectrum of the gas tail has emission lines. The other tail does not show emission lines, but rather a spectrum of sunlight, reflected from dust expelled out of the coma. The pressure from sunlight detaches dust from the coma to form a tail.

The gas tail's spectrum shows carbon monoxide, carbon dioxide, molecular nitrogen, and free radicals of ammonia and methane. Puffs of ionized gas sometimes shoot through the tail. Gas tails point away from the sun because they are blown by the solar wind, which consists of ions carrying magnetic fields at high speeds through interplanetary space.

At great distances from the sun, the coma also shows a reflected solar spectrum, so the heads must contain solid particles that reflect sunlight. At about 1 AU from the sun, the head exhibits molecular emission bands of carbon, cyanogen (CN), molecular oxygen, and hydroxyl (OH). As the comet speeds nearer to the sun, emission lines of silicon, calcium, sodium, potassium, and nickel appear.

Infrared observations confirm that comets contain considerable amounts of dust, especially

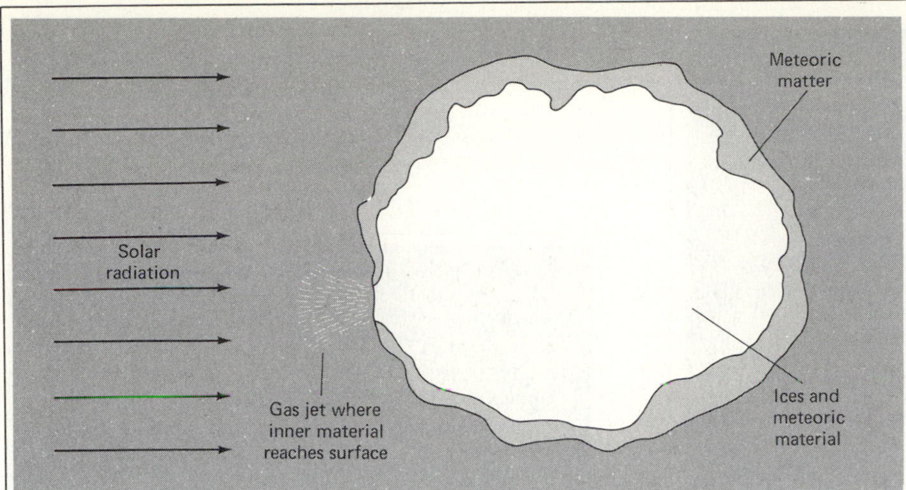

12.6
The dirty-snowball cometary model. The nucleus is a pudding of ices and rocky material encrusted by a thin, rocky shell. Sunlight heats the nucleus and vaporizes the ices. Gas streams out to make the coma and tail.

when close to the sun. Infrared spectra of some comets display evidence of silicate dust.

For all their stunning length against the sky, comets have very small masses. Astronomers can estimate comet masses only roughly because they are so small that they do not affect the orbits of other bodies. Halley's comet (Focus 12.1), one of the largest, has an estimated mass of only about 10^{16} kg (10^{-8} that of the earth) and loses about 10^{11} kg during each perihelion passage. With so little mass, a comet achieves its spectacular display only by spreading itself very thin.

The mass expelled from a comet, mostly in the form of gas, flies off into space. The nucleus supplies this material, but what is the nature of the nucleus? In 1950 astronomer Fred. L. Whipple developed the *dirty-snowball cometary model* (Fig. 12.6). It pictures comet nuclei as compact, solid bodies made of frozen gases (ices) of water, ammonia, and methane, embedded with rocky material. When a comet nears the sun, the icy material vaporizes. This released material enlarges the coma and creates the tail. As the ice evaporates, a thin coating of rocky material remains to form a solid, but fragile, crust on the nucleus.

As the comet rounds the sun, the semisolid nucleus can usually withstand the solar heat. Comet Ikeya-Seki, the great sun-grazing comet of 1965, passed within 470,000 km of the solar surface and survived. But some comets passing close to the sun are not so lucky. Comet West (1976) split into at least four pieces after its perihelion passage (Fig. 12.7); a few comets have actually collided with the sun (Fig. 12.8).

Meteors and meteorites / A *meteor* is the flash of light associated with the entry of a solid particle into the earth's atmosphere (Fig. 12.9). As it plunges through the air, the particle is burnt up by friction, and it may leave a bright trail, a glowing column of light, behind it. Before the particle meets its fiery doom in the upper air, it is called a *meteoroid*: a solid object traveling through interplanetary space. Of course, other objects (comets,

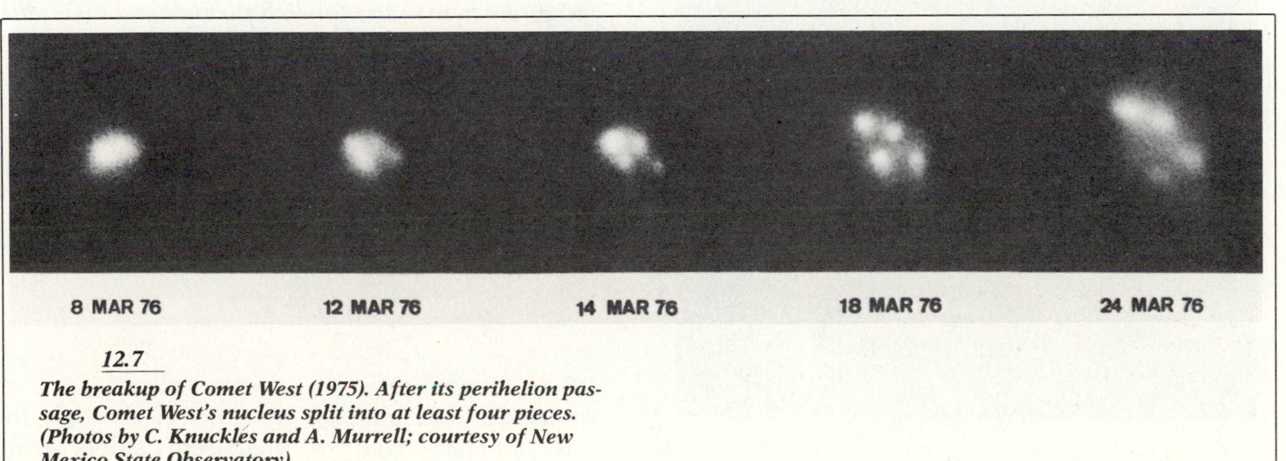

12.7
The breakup of Comet West (1975). After its perihelion passage, Comet West's nucleus split into at least four pieces. (Photos by C. Knuckles and A. Murrell; courtesy of New Mexico State Observatory)

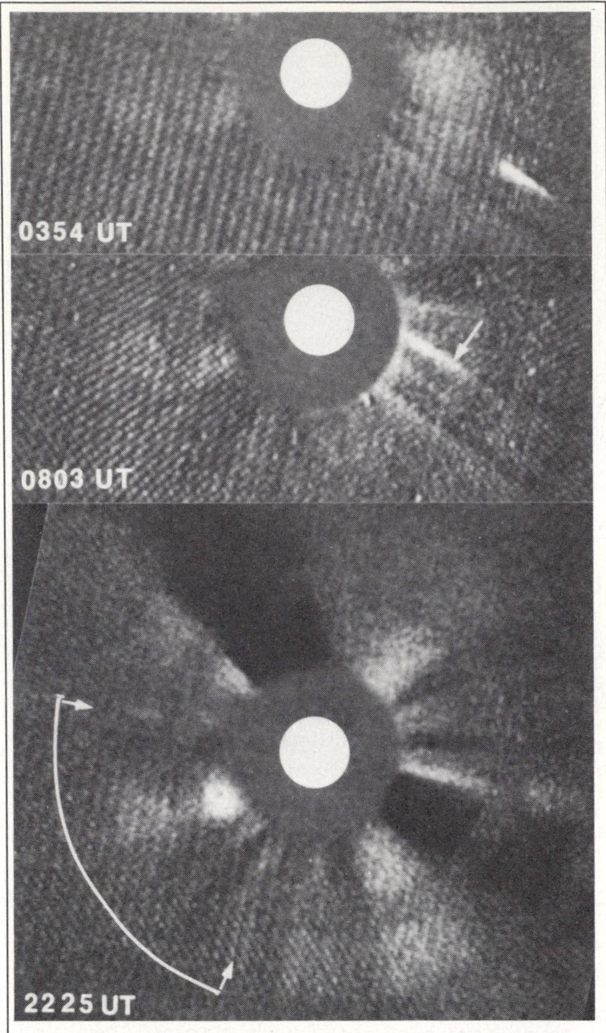

12.8
A sequence of photos taken in July 1981 showing a comet (right) inbound to the sun (white disk in center). (Courtesy N. R. Sheeley, Jr., U.S. Naval Observatory)

12.9
An unusually bright meteor called a fireball. (Courtesy F. Whipple)

Focus 12.1

F.25
The head of Halley's comet in 1910. (Courtesy Mt. Wilson Observatory, Carnegie Institution of Washington)

HALLEY'S COMET

When Edward the Confessor died in 1066, he left no direct heir to the English throne, and the nobles chose Harold II as their king. In the same year, a bright comet streaked across the sky—the commonly accepted sign of a ruler's death and misfortunes to follow. Meanwhile, in France, William of Normandy cleverly interpreted the comet as a sign portending his victory. With this psychological support for his men, he sailed to the British Isles and conquered the dispirited Saxon armies near

asteroids, and planets) also travel through the interplanetary void; a meteoroid differs from those chiefly in its small size, no more than a few meters in diameter, usually much less. If a

Hastings. By the end of the year, William the Conqueror was crowned king. The comet of 1066 was later to be known as Halley's comet.

Cometary orbits remained unknown until Edmund Halley (1656–1742), the Astronomer Royal and a friend of Newton, calculated the orbits of the comets of 1531, 1607, and 1682 by a method devised by Newton. Halley found, to his amazement, that the orbits were almost identical. Noting that the comets appeared at intervals of approximately 75 or 76 years, Halley concluded that these several comets were in fact one. He predicted it would return around 1758. Halley was right. His comet was sighted on Christmas night in 1758 and named in Halley's honor. Halley's comet (Fig. F.25). was the first comet to be recognized as a permanent member of the solar system, with an elliptical, periodic orbit.

Halley's comet is the granddaddy of all comets. Its passage near the sun has been recorded at least 29 times, as far back as 239 B.C., possibly even 466 B.C. In 1910 some people incorrectly believed the world might come to an end when the earth passed through the comet's tail. (A comet's tail is much too thin to have any effect on the earth.) Halley's comet moves in an extremely elongated ellipse. Having passed aphelion beyond Neptune's orbit in 1948, it will return to the earth's neighborhood in 1985–1986. It reaches perihelion on February 9, 1986, at a distance of only 0.59 AU. Before perihelion it will pass closest to the earth, at a distance of 0.62 AU, on November 27, 1985. After rounding the sun, the comet will again pass by the earth, closing to 0.42 AU on April 11, 1986.

Unfortunately, this pass of Halley's comet will not be a good one for observers in the mid-northern latitudes. Before perihelion, when it will be visible in the evening sky (Fig. F.26), the comet will be faint—certainly dimmer than Mars or Jupiter at opposition. After perihelion it will be brighter but low in the morning sky and so difficult to observe (especially from an urban area). Better views will occur south of the equator.

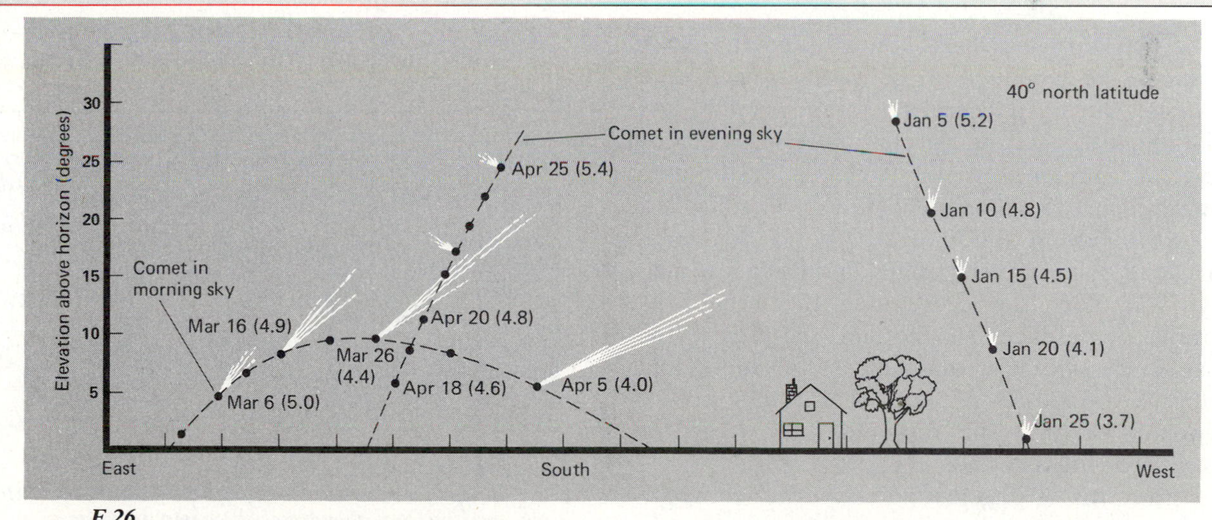

F.26
Observing conditions for Halley's comet in 1986 for 40° north latitude. Approximate visual magnitudes are given in parentheses next to the dates. (Adapted from a NASA diagram)

meteoroid survives its plunge through the atmosphere and strikes the earth's surface, the body is then called a *meteorite*.

Most meteoroids are fragile, delicate particles (Fig. 12.10) that crumble quickly in their contact with the air. A meteoroid is typically a flimsy dust speck whose demise is its only remarkable aspect.

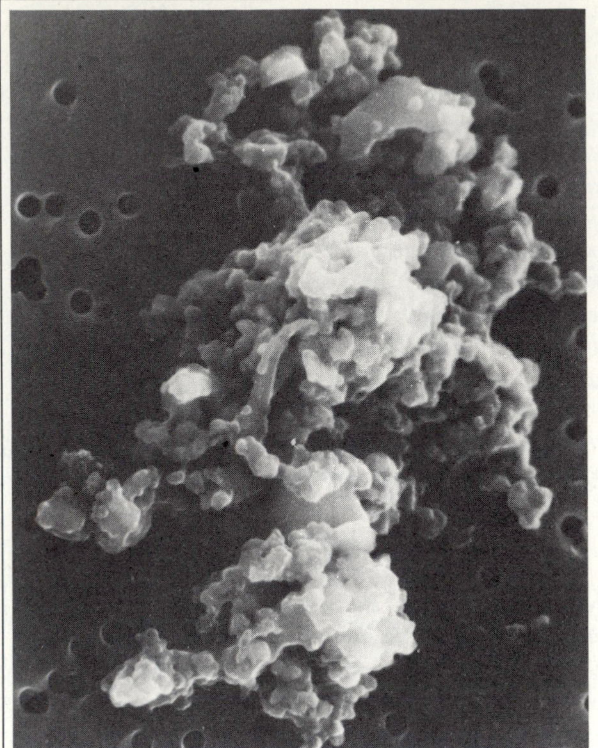

12.10
A particle of interplanetary dust; it is about 10 μm in size. Note its fragile structure. (Courtesy D. Brownlee, University of Washington)

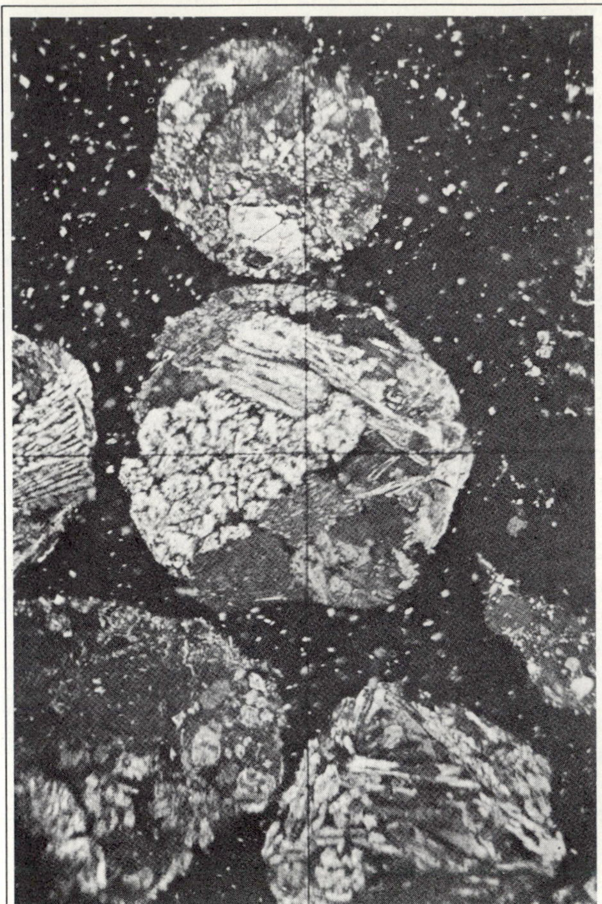

12.11
Chondrules, round silicate spheres found in chondrites. (Courtesy F. Whipple and the Smithsonian Astrophysical Observatory)

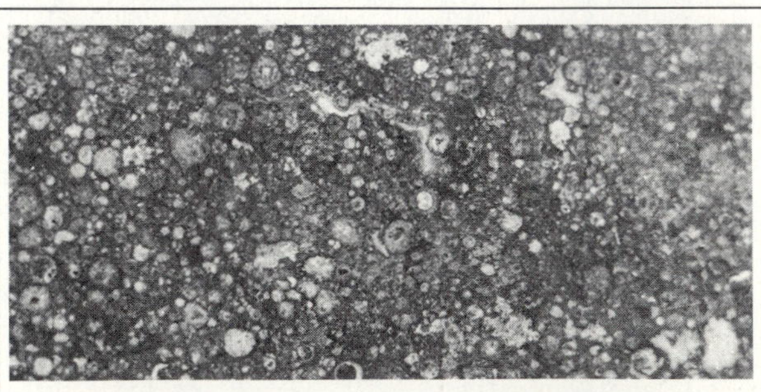

12.12
A close-up of a carbonaceous chondrite from Allende, Mexico, showing some chondrules. (Courtesy Institute of Meteoritics, University of New Mexico)

The dirty-snowball comet model readily explains the connection between comets and meteors. During a comet's passage by the sun, solar heating causes a continual loss of icy material from the cometary nucleus. So the dust and solid particles interspersed in the ices flake off and scatter in an untidy array around the comet. This solid debris is very fragile and has a low density. The older the comet and the greater its number of passages by the sun, the greater the loss of icy material and release of meteoroid material. About 99 percent of meteors are of cometary origin. (The remainder are probably associated with asteroids.)

In terms of physical and chemical composition, meteorites fall into three broad classifications: irons, stones, and stony-irons. The *irons*, which are generally about 90 percent iron and 9 percent nickel, with a trace of other elements, are the most commonly found. They are easy to identify because of their high density and obviously melted appearance. The *stones* are composed of light silicate materials similar to the earth's crustal rocks. When examined under a microscope, many stones are seen to contain silicate spheres, called *chondrules*, embedded in a smooth matrix (Fig. 12.11). These stones are known as *chondrites*. *Stony-iron* meteorites represent a crossbreed between the irons and the stones and commonly exhibit small stone pieces set in iron.

One of the most curious kinds of chondrites is the *carbonaceous chondrites* (Fig. 12.12). The chondrules in these meteorites are embedded in material that contains a large amount of carbon compared with other stony chondrites—typically about 1 to 4 percent carbon by mass. Their carbon content gives these meteorites a dark appearance.

Carbonaceous chondrites also contain significant fractions of water (ranging from about 3 to 20 percent) and volatile materials. In addition, the relative abundances of condensable elements in carbonaceous chondrites more closely resemble those found in the sun's photosphere than those in the crust of the earth. That is, if some gas were extracted from the sun and cooled, the condensed elements would be chemically quite different from the earth but very similar to carbonaceous chondrites. This similarity suggests that carbonaceous chondrites formed out of the same primordial material as the sun and have suffered no major bulk heating or changes since that time.

Most meteorites have too great a density to come from comet-related meteoroids (but most meteors do). They most resemble the inferred physical characteristics of asteroids. In cases where orbits of meteorites have been determined, they are like those of asteroids rather than of comets. Collisions between asteroids fragment them, and these pieces may be one source of meteorites.

What about the origin of iron meteorites? An important clue comes from etching with acid a polished surface of an iron meteorite. Large crystalline patterns, called *Widmanstätten figures* become visible (Fig. 12.13). Terrestrial iron does not show such patterns when etched. A nickel-iron mixture, when cooled slowly under low pressures from a melting temperature of about 1600 K, forms large crystals. The key point here is that the cooling must be very gradual (about 1 K every million years). But metals conduct heat well, and in the cold of space, a molten mass of nickel and iron would cool rapidly and not form large crystals. So how could nickel-iron meteorites grow Widmanstätten figures? They need protection from the cold. So it's likely that nickel-iron meteorite material solidified inside small bodies, termed *parent meteorite bodies*. To allow a cooling of only 1 K every million years, these must have been at least 100 km in diameter.

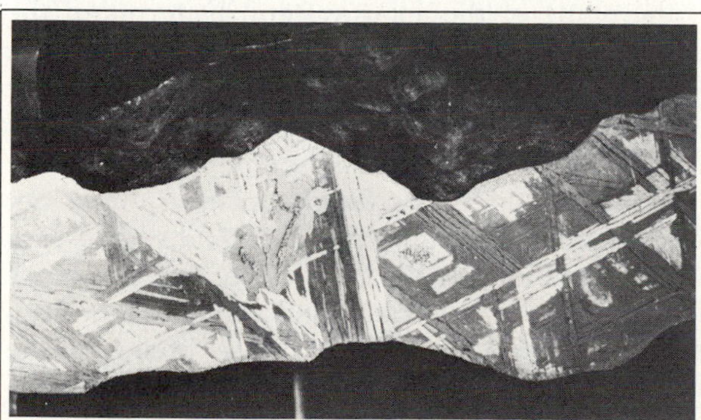

12.13
Widmanstätten figures in a nickel-iron meteorite from Glorietta, New Mexico. (Courtesy Institute of Meteoritics, University of New Mexico)

How did such bodies form? Probably with the formation of the solar system. Parent meteorite bodies are envisioned to have been only a few hundred kilometers across. Once formed, they could be heated by the radioactive decay of short-lived isotopes. When heated to melting, a parent meteorite body differentiates—the densest material falls to the center, and the least dense froths to the surface. So the object ends up with a core of metals and a cover of rocky material, which cools to form a crust. This insulates the molten metals and allows them to cool slowly and form Widmanstätten figures. Much later, the parent meteorite bodies collide and fragment. Pieces from the outer crust make stony meteorites, pieces from farther down become stony-iron meteorites, and the core produces the iron meteorites.

Are any parent meteorite bodies around now? Possibly yes—as asteroids. Recall that there are three main types of asteroids, the dark C-type, containing much carbon; the lighter S-type, composed of silicate materials; and the intermediate M-type, with metallic characteristics. These are probably related to the carbonaceous chondrites, the stony meteorites, and the irons, respectively.

You should have noticed that the view of meteorites outlined here implies that their parent bodies were among the first solid objects to form in the solar system. So the ages of meteorites should provide a direct indication of the age of the solar system. Meteorites can be dated by using radioactive-decay techniques (see Focus 8.1); such methods give ages very close to 4.6 billion years. Meteorites provide astronomers with a direct, reliable estimate of when the solar system formed.

12.2 Pieces and puzzles

The dynamic and chemical properties of the solar system impose crucial limitations on any model of its formation. These features serve as broad templates for shaping more specific questions.

Chemically, the solar system falls into three broad categories of material: solar, terrestrial, and icy (Table 12.2). Each group is distinguished primarily by its melting point. The solar and icy materials together are sometimes called *volatiles*. These are generally gaseous under the conditions expected during the solar system's formation. The bodies of the solar system are composed of various combinations of the three groups (Table 12.3).

TABLE 12.2 Generalized classes of solar system materials

Class	*Elements*	*Melting point*
Terrestrial	Silicon, magnesium, aluminum, iron, etc., plus oxygen in chemical composition	2000 K
Icy	Carbon, nitrogen, oxygen, plus hydrogen in chemical composition	273 K
Solar	Hydrogen, helium, neon, argon, etc.	14 K

TABLE 12.3 General compositon of the major bodies in the solar system

	Materials in composition (%)		
Bodies	*Terrestrial*	*Icy*	*Solar*
Terrestrial planets	70	30	0
Asteroids	70	30	0
Jupiter	1	10	90
Saturn	1	30	70
Uranus	10	80	10
Neptune	20	70	10
Comets	15	85	0

TABLE 12.4 Distribution of mass and angular momentum in the solar system

Object	*Mass (% of total)*	*Angular momentum (% of total)*
Sun	99.86	0.5
Jovian planets	0.132	99.0
Terrestrial planets	0.003	0.2
Asteroids	0.00003 (?)	0.1 (?)
Comets	0.0000003 (?)	0.2 (?)

Though one class of materials may dominate, any solar system body contains some of each group. For example, the sun contains terrestrial materials, but as gases (because the sun is hotter than 2000 K), not solids. In fact, in terms of total mass, the sun contains more terrestrial materials than all the terrestrial planets put together. The sun likewise contains the icy group as gases.

The solar system also displays a regular structure in terms of its dynamic properties, those that relate to its motions. Viewed from above the sun's north pole, the solar system shows the following regularities:

1 The planets revolve counterclockwise around the sun; the sun rotates in the same direction.
2 The major planets, except Mercury and Pluto, have orbital planes that are only slightly inclined with the plane of the ecliptic; that is, the orbits are *coplanar*.
3 With the exceptions of Mercury and Pluto, the planets move in orbits that are very nearly circular.
4 With the exceptions of Venus and Uranus, the planets rotate counterclockwise, in the same direction as their orbital motion.
5 The planets' orbital distances from the sun follow a regular spacing; roughly, each planet lies twice as far out as the previous one.
6 Most satellites revolve in the same direction as their parent planets rotate and lie close to their planets' equatorial planes.
7 Some satellites' orbital distances follow a regular spacing rule.
8 The planets together contain more angular momentum than the sun (Table 12.4).
9 Long-period comets have orbits that come in from all directions and angles, in contrast to the coplanar orbits of the planets, satellites, asteroids, and short-period comets.
10 Three of the Jovian planets are known to have rings.

A successful model must explain as many of these dynamic and chemical properties as possible. A good model is not simply the one that accounts for the greatest number of the listed characteristics; it must explain them in some internally consistent, simple fashion (see Section 2.1). (That is what is meant by a model's being "aesthetically pleasing.")

Note that the sun contains most of the mass of the solar system (Table 12.4). The rest lies close to the plane of the solar system. In terms of the layout of mass, the solar system is really very thin. If it were the size of an average pancake, the solar system would be only 1 cm thick, about the thickness of that average pancake.

Finally, a successful model must deal not only with the dynamic and chemical regularities of the system but also with the interplanetary debris: comets, asteroids, and meteoroids. Contemporary models consider these bodies to be important relics of the solar system's early history.

Most models today are variations of *nebular* models, in which the sun condenses from an interstellar cloud of gas and dust that also forms a disk, a solar *nebula*, out of which the planets condense. The nebular models interpret the existence of the solar system as a natural consequence of the sun's formation and, perhaps, of any star's formation. If nebular models are correct, planetary systems may be very common in our Galaxy and in other galaxies.

The rest of this chapter examines general nebular models to assess their strengths and weaknesses.

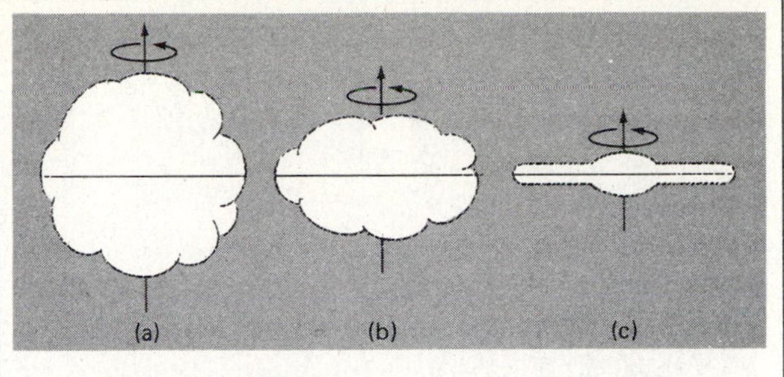

12.14
The contraction of the solar nebula. When it starts* (a)*, the cloud is slowly spinning. To conserve angular momentum, the materials falls in parallel to the spin axis* (b) *to finally form a disk* (c) *some 60 AU in diameter and 1 AU thick.

12.3 Basics of nebular models

The essential feature of nebular models is that the sun and then the planets form from a cloud of interstellar material. The sun's formation takes place in the center of a flattened cloud. The planets grow from the disk of the cloud. That's the basic picture and the nub of the problem: how to get the planets, in their present orbits, out of an originally diffuse cloud. Let's look at some clues.

We know that the solar system is now basically flat, with the sun in the center. Examining the rings of the Jovian planets, we see that they are flat and consist of small particles orbiting their parent planet. You can imagine that if somehow the particles could be stuck together, you'd end up with another moon. So the problem has two basic parts: (1) how to make a flat solar system, and (2) how to get the planets to grow out of the cloud.

To tackle the first part, we need to consider a basic physical idea: the conservation of angular momentum (see Focus 9.2). The basic point is this: Once a body has started spinning, it will keep on spinning as long as no outside influence affects it. The amount of angular momentum depends on how much mass the body has and how much it is spread out. If, by itself, the body changes size—for instance, if it contracts gravitationally—it will naturally spin faster to keep its angular momentum the same.

Probably the most familiar example of the conservation of angular momentum is the spin-up of an ice skater. She goes into the spin with her arms outstretched. As she pulls her arms in, her spin speeds up. But if she brought her head down to her chest, nothing would happen. Why not? Because angular momentum affects only the motion *around* a spin axis, not along it.

Consider now a large spherical cloud of particles, slowly spinning. Imagine that it pulls itself inward by its gravity. What happens (Fig. 12.14)? It will spin faster and collapse down along the spin axis to make a flat disk with a fat center. That's just what we need to form the solar system. As a natural result, we get the planets orbits aligned in a thin disk, and the sun rotates in the same direction as the planets revolve.

With this neat solution to key dynamic features of the solar system comes a serious objection: the present distribution of angular momentum (Table 12.4). Although the sun holds 99 percent of the system's mass, it contains less than 1 percent of the angular momentum. The outer planets have the most, 99 percent of the total. If all the planets with their present angular momenta were dumped into the sun, it would spin once every few hours rather than once a month. If the cloud's collapse goes as I've described, the sun forms from the central part. So it should be spinning very rapidly. Actually, the sun spins 400 times more slowly than this rate. The angular momentum is there, but not in the right place! To adopt a nebular model requires a process to account for the present distribution of angular momentum, a process still not well worked out.

One idea has been worked on in some detail. It involves the interaction of magnetic fields and charged particles to rearrange the distribution of angular momentum. The basic solution requires that the spin of the central part of the nebula be decreased and transferred to the outer regions.

Charged particles and magnetic fields interact so that the particles spiral along the magnetic lines of force (see Focus 8.2). Such interactions provide a way to transfer spin from the young sun (in the center of the nebula) to the outer parts of the nebula. As the sun forms, it heats up the interior regions of the nebula. Here the gas is ionized, so charged particles are abundant. The magnetic-field lines trap these particles. As the sun rotates, it carries its magnetic-field lines with it; these drag along the charged particles (Fig. 12.15), which in turn unite with and drag along the rest of the gas and dust. So the magnetic field spins around the material in the nebula near the

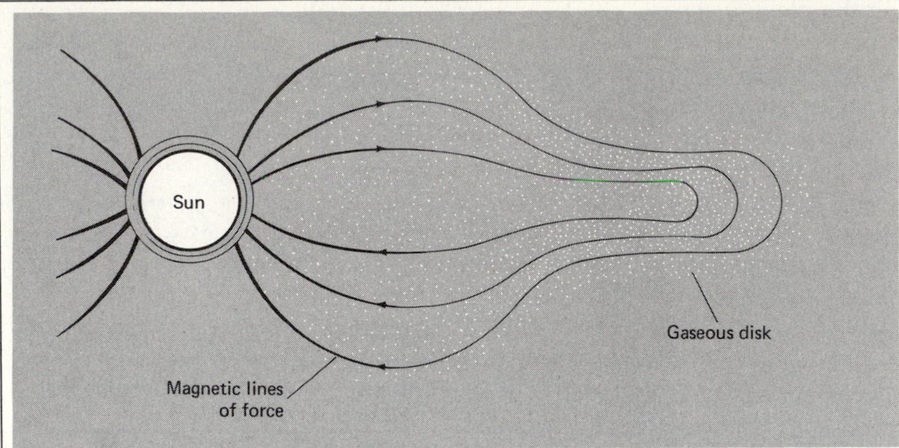

12.15
Side view of a possible primeval solar magnetic field. The field lines extend outward from the sun into the ionized gas of the solar nebula.

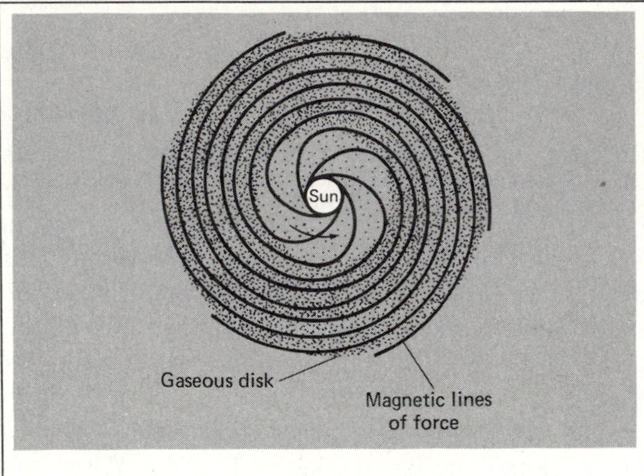

12.16
Top view of a possible primeval solar magnetic field. One end of the field lines rotates with the sun; the other end is dragged by the ionized material in the disk.

sun. At the same time, the mass of the nebula resists the rotation. This drag on the magnetic-field lines stretches them into a spiral shape (Fig. 12.16). The magnetic-field links the material in the nebula to the sun's rotation. So the nebular material gains rotation (and angular momentum) and in the process causes a drag on the sun's rotation, which slows it down. Note that this magnetic braking works *only* if the gas is ionized.

Whatever process transferred the angular momentum, the transfer must have taken place *before* large solid objects formed in the nebula. The transfer mechanism just described works effectively only on gases.

Finally, one other basic process occurs: heating from gravitational contraction. Whenever a mass pulls itself in by its own gravity, it gets hotter. The trick involves the transformation of energy from one kind (gravitational potential energy) into other kinds (heat and light).

Here's how. Consider a ball held above the earth's surface (Fig. 12.17). Its velocity is zero and it has no kinetic energy. But it does have potential energy, as you can tell when you drop it. The ball falls, and as it does, its velocity increases—it accelerates. So its kinetic energy increases the more it falls.

Instead of a ball, consider a cloud of gas containing a large number of particles (Fig. 12.18). Think of each as a small ball. Imagine that the cloud contracts gravitationally, from the combined attraction of all the particles. As it contracts, the particles gain velocity. Though at the start the velocities are directed inward, collisions will soon distribute them in random directions, with only a slow net motion inward. (Think of the collisions as a form of friction that slows down the gravitational collapse.) So the average kinetic energy of the particles increases. Temperature measures the average kinetic energy of the particles involved, so the temperature of the gas increases.

Meanwhile, the density also increases. So the particles collide more often, from the combined effect of the velocity and density increase. The collisions excite some atoms. These emit photons. So the net result of the gravitational contraction is

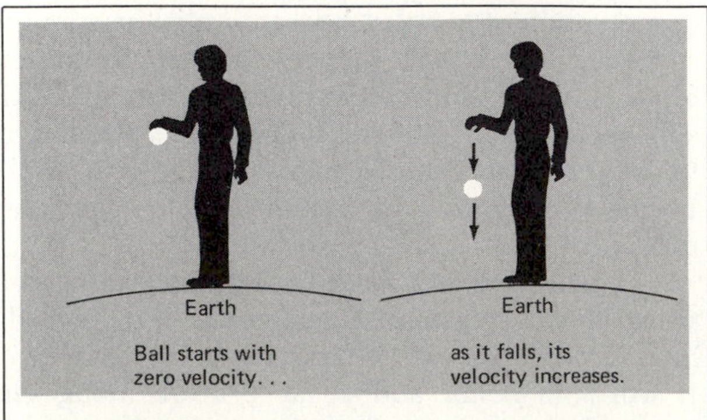

12.17
Conversion of gravitational energy into kinetic energy by the fall of a mass.

that some of the initial gravitational energy converts to heat (higher temperature) and some ends up as photons. In fact, one-half goes into raising the temperature and one-half into light. The key point for this chapter is: *as the cloud contracts, it gets hotter*.

With this in mind, let's turn to the next basic problem: how to form the planets.

12.4 The formation of the planets

A successful nebular model must account in some detail for four important stages in the solar system's evolution: (1) the formation of the nebula out of which the planets and sun originate, (2) the formation of the original planetary bodies, (3) the subsequent evolution of the planets, and (4) the dissipation of leftover gas and dust. Modern nebular models (there are more than one!) give tentative explanations for these stages, but many details are incomplete. No one model to date is really satisfactory.

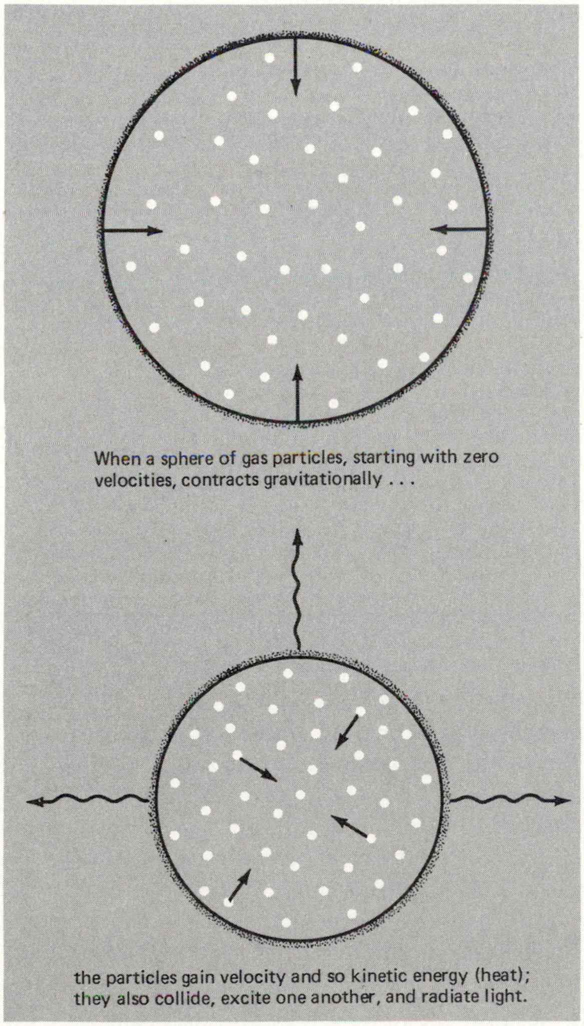

How does the growth of the planets happen? There are three main methods: (1) gravitational collapse, (2) accretion, and (3) condensation. *Gravitational collapse* works only if regions in the nebula have enough mass so that they contract by their own gravity to form a planet. *Accretion* occurs when small particles collide and stick together to form larger masses that eventually grow into planets. (An example: As snowflakes fall through the air, they can collide and stick to form clusters of snowflakes.) *Condensation* involves the growth of small particles by the sticking together of atoms and molecules. (An example: Water molecules combine in clouds to form raindrops.)

Two scenarios help to explain the formation of the planets: (1) The primeval nebula was turbulent, and violent motions helped the first steps in planetary formation. (2) The formation of the planets was a multistep process; first small bodies formed, and then larger ones that eventually evolved into the planets.

Turbulence promotes the accretion of grains in the nebula because it makes grains collide more frequently. Turbulent vortices occur in rapidly moving fluids and gases. For example, you have probably seen that the flow of smoke out of a smokestack breaks up into numerous swirls. The motions of gases in the nebula can cause a breakup into a pattern of turbulent cells. At the boundaries between cells, particles would collide and accrete, forming larger ones. These objects, from a few kilometers to a few hundred kilometers in size, are called *planetesimals*.

How do the planets grow? The model needs at least two stages for planetary growth. In the first, turbulence promotes the growth of planetesimals. Second, the planetesimals collide and accrete to make planet-sized masses (Fig. 12.19), called *protoplanets*. These evolved into the planets of today.

How to get from small dust grains to large protoplanets? Grains collide and accrete to form larger, perhaps pebble-sized, objects. These quickly fall into the plane of the nebula. These pebbles than accumulate into planetesimals by gravitational attraction. Whatever materials happen to be available at a certain distance from the center of the nebula make up a planetesimal. So the planetesimals reflect the compositions of local material.

Once the planetesimals have formed, they might gather into larger bodies, perhaps almost as large as the moon. Somehow these objects finally end up in a few protoplanets. Here gravity would

When a sphere of gas particles, starting with zero velocities, contracts gravitationally . . .

the particles gain velocity and so kinetic energy (heat); they also collide, excite one another, and radiate light.

12.18
Gravitational contraction converts gravitational energy into heat and light.

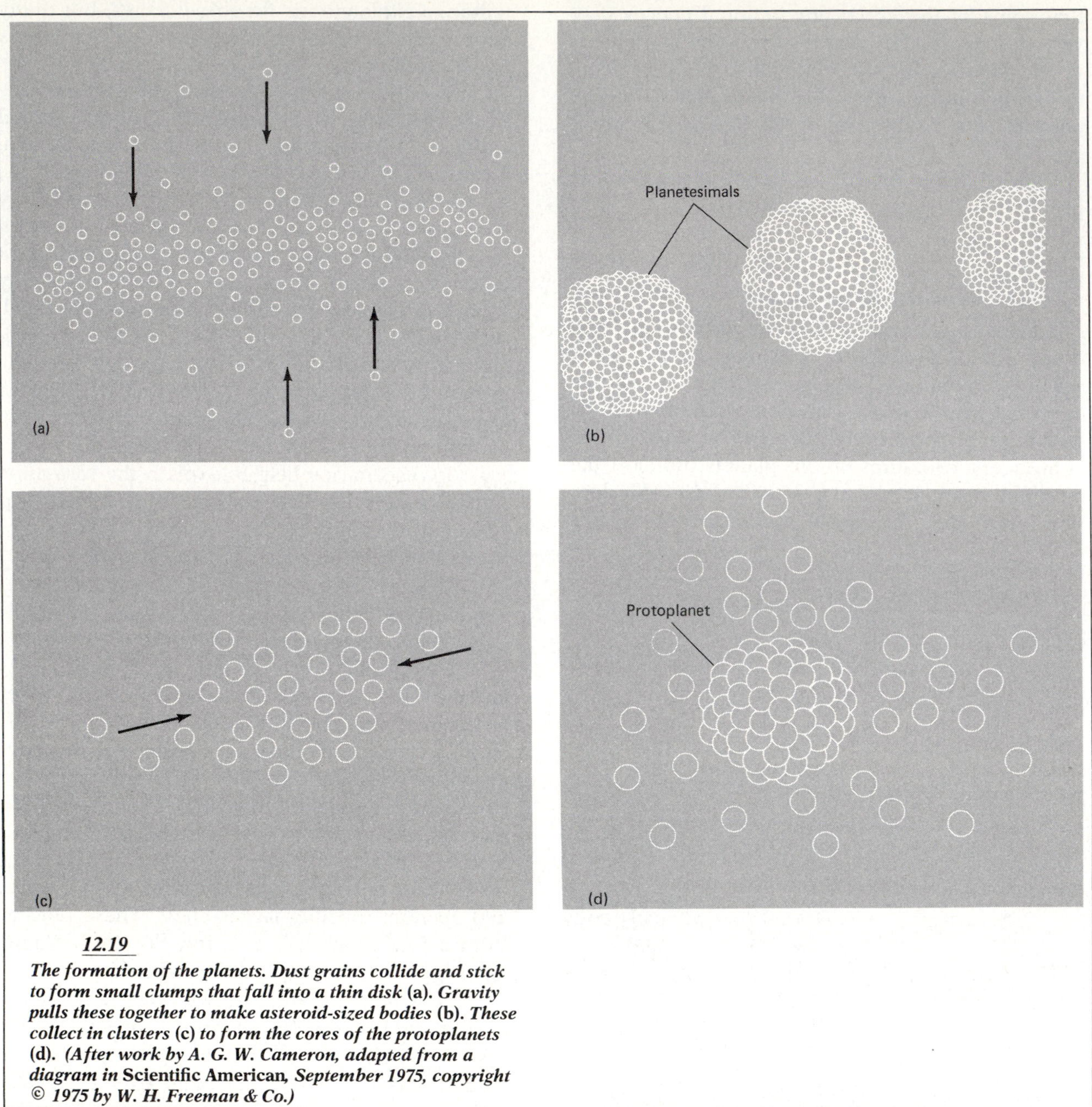

12.19
The formation of the planets. Dust grains collide and stick to form small clumps that fall into a thin disk (a). *Gravity pulls these together to make asteroid-sized bodies* (b). *These collect in clusters* (c) *to form the cores of the protoplanets* (d). *(After work by A. G. W. Cameron, adapted from a diagram in* Scientific American, *September 1975, copyright © 1975 by W. H. Freeman & Co.)*

help. Once the planetesimals have gathered (in a few tens of thousands of years) into several somewhat larger bodies (500 km in size), their masses would be enough to help pull in other smaller masses from a distance. So a growing planet will sweep clear a zone of the nebula to feed its mass. For the terrestrial planets to grow to their present sizes, calculations indicate an aggregation time of roughly 100 million years.

12.5 Chemistry and origin

How did the protoplanets acquire differences in chemical composition? Recent research has developed the concept of a *condensation sequence*. The basic idea is this: The nebula's center must have been hot, a few thousand degrees Kelvin. Here solid grains, even iron compounds and silicates, could not condense. Elsewhere, what materials would condense as new grains depended on the temperature (Fig. 12.20). Just below 2000 K, grains made of terrestrial materials would condense (Table 12.2); below 273 K, grains could form of both terrestrial and icy materials.

Recent research has reached even more specific conclusions about the sequence in which compounds could condense from a heated nebula (Table 12.5). At different temperatures, the gases available and the solids present react chemically to produce a variety of compounds. Here's the key

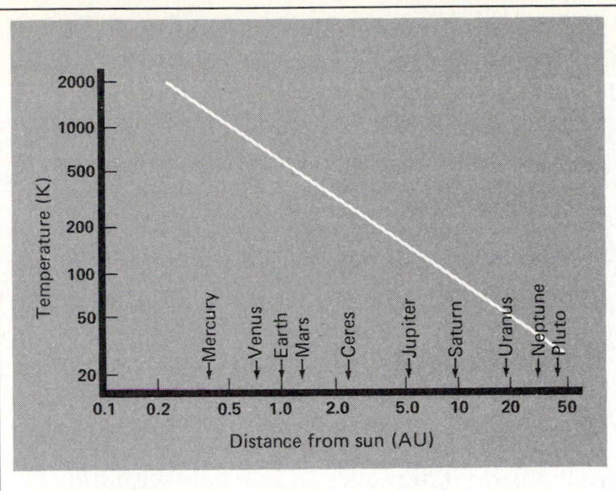

12.20
One model for how the temperature in the solar nebula might have decreased with increasing distance from the sun. At this stage, the major constituents of the protoplanets are condensing from the solar nebula. (Based on calculations by J. S. Lewis)

TABLE 12.5 Equilibrium condensation sequence in the solar nebula

Temperature (K)	Reactions
1600	Condensation of metals combined with oxygen (oxides)
1300	Condensation of iron-nickel
1200	Condensation of silicates
1000	Reaction of sodium with silicates and aluminum oxide to form common minerals such as feldspar
450–1200	Combination of iron and oxygen to form iron oxide, becoming olivine
680	Combination of iron with sulfur to form triolite (FeS)
425–550	Combination of water vapor with minerals such as olivine
175	Condensation of water ice
150	Reaction of ammonia (NH_3) with water ice
120	Reaction of methane (CH_4) with water ice
20–65	Condensation of inert elements such as argon and neon
1	Condensation of helium

result of the condensation sequence: The densities and compositions of the planets can be well explained with the condensation sequence *if* the temperature in the nebula drops *rapidly* from the center outward. Then, at different distances from the sun, different temperatures allowed different chemical compounds to condense and form grains that eventually made up the protoplanets. If a material could not condense because the temperature was too high, it would not end up in the protoplanet.

For instance, the terrestrial planets lack much of the icy and gaseous materials common in the Jovian planets because at their close distances to the sun, the temperatures were too high for the condensation of icy and gaseous materials. In contrast, the Jovian planets pretty much reflect the original composition of the nebula.

The condensation sequence also predicts chemical differences among the terrestrial and Jovian planets. Let's take Mercury, the earth, and Mars for comparison. At Mercury's distance, the temperature would be low enough for nickel-iron and silicates of magnesium to condense along with refractory oxides (including uranium, potassium, and thorium). The condensates of the radioactive elements provide one source of heating; accretion provides another. These heating sources drive the differentiation of the protoplanet. The resulting protoplanet would have a large nickel-iron core and a mantle of such silicates; we believe that Mercury does have such an interior structure (see Section 9.7). At the earth's distance, the temperature is considerably lower, about 600 K. Here, silicates of both iron and magnesium could condense, along with iron oxide and iron sulfide. Water would be bound up with minerals, perhaps as much as 5 percent of the total. So the model predicts a planet with a smaller nickel-iron core than Mercury's, containing sulfur (in the form of iron sulfide) and a larger mantle, rich in silicates and oxides of iron and magnesium. The earth does appear to have such an interior (see Section 8.2). What about Mars? Here the temperature was lower still. The resulting core would consist mostly of iron sulfide, not nickel-iron, and the mantle would be olivine, rich in iron oxide and water. We don't know the interior of Mars very well, but the prediction from the condensation sequence is compatible with our general inferences about the Martian interior (see Section 10.5).

Note that it is how *low* the temperature falls that determines the chemical reactions to produce condensates. For example, at the earth's position in the nebula, all the reactions in Table 12.5 down to 600 K can take place. The reactions at temperatures lower than 600 K do not happen.

In general, the condensation sequence requires a certain minimum temperature to be reached to account for the known chemical composition of the planets. Roughly, these temperatures are 1400 K for Mercury, 900 K for Venus, 600 K for the earth, 400 K for Mars, and 200 K for Jupiter.

When the central regions of the nebula finally form the sun, solar energy evaporates any icy material in the nebula and around the inner protoplanets or planetesimals. Intense solar radiation pressure and a strong solar wind (a solar gale!) push leftover gases out of the solar system.

Any leftover planetesimals bombard the planetary surfaces, leaving remnant craters on the terrestrial planets. This bombardment heats the surfaces of the planets. Radioactive decay heats the interiors and melts them. Dense elements, such as iron and nickel, sink down to form a core. Less dense materials, such as silicates, float to form surface froth that cools to become the crust. So the planets become differentiated—the first step in their evolution.

12.6 Jupiter and Saturn: A different story?

Jupiter and Saturn may have formed in a different way from the planetesimal accretion model just described. They may have condensed gravitationally from single large blobs of material in the nebula, rather than by accretion of planetesimals. The internal heat they gain comes from the conversion of gravitational potential energy into heat during gravitational contraction.

Theoretical models of the evolution of Jupiter start from a hot beginning, after the proto-Jupiter had formed. The proto-Jupiter is 16 times Jupiter's present size and has a central temperature of 16,000 K, a surface temperature of about 1000 K, and a luminosity of almost 1/100 the sun's present luminosity. The planet contracts; in only a million years, it shrinks to just about twice its present size, and its luminosity drops to about 10^{-5} the sun's luminosity. The shrinking then slows down because the planet's interior is liquid, and liquids are difficult to compress. In the next 4.5 billion years, Jupiter contracts to its present size, and its central temperature drops as it loses some of its heat to space. (Models for the early evolution of Saturn resemble that for Jupiter.)

The early high luminosity of Jupiter may explain, in the context of a condensation sequence, why the Galilean satellites decrease in density going outward from Jupiter (see Section 11.2). This density decrease implies, for example, that Callisto contains proportionally more icy materials than Io. At Io's closer distance, less ice condensed than at Callisto's distance. If Jupiter were hot at the time of the satellites' formation, the inner ones would not have accreted as much icy materials as the outer ones. So the Galilean moons may mimic the condensation and accretion of the terrestrial planets.

In fact, the evolution of Jupiter and Saturn may follow that of the solar system generally. Their rings may be unaccreted material.

12.7 Asteroids, meteorites, and evolution

Contemporary research on asteroids implies that they are planetesimals that just didn't get together to make a planet, rather than the popular idea that they are remnants of a planet that exploded.

Recall the composition variation across the asteroid belt. On the inner edge, it contains mostly S-type asteroids; at the outer edge, mostly C-type. These albedo differences fit in nicely with the condensation sequence if the C-types contain more carbon than the S-types. At the inside of the belt, the temperatures were low enough for silicates to condense but too high for carbon-bearing materials to do so. Farther out, both types of materials condensed to end up in planetesimals.

Why didn't they form a planet? Probably because of the gravitational influence of the proto-Jupiter. Recent theoretical calculations indicate that the proto-Jupiter formed quickly—in a time as short as perhaps a few months. Its formation happened fast simply because a lot of material gathered together, which drew in more material, and so on. Once the proto-Jupiter had formed, it would tug on the planetesimals just within its orbit. Meanwhile, the sun would pull these planetesimals toward it. In this tug-of-war, the orbits of the planetesimals changed from circular to elliptical. Some crashed into others, shattering them into smaller pieces. Some of these pieces caromed into the inner part of the solar system and eventually rained onto the surfaces of Mercury, Venus, the moon, the earth, and Mars, forming craters—we can see some of these bombarded regions today. The remainder of the broken planetesimals stayed mainly in the region about 3 AU from the sun; these remnants are today's asteroids.

How do meteorites fit into the nebular model? The characteristics of chondritic meteorites support the condensation picture (Fig. 12.21). Their chemical composition (similar to the sun) and unmixed structure suggest that they are the original condensed material of the nebula. The turbulence in the nebula may have created shock waves that swiftly melted the grains. After its passage, the drops cooled and solidified to form chondrules. The glassy spheres that resulted accumulated in planetesimals. Chondrules suggest that the bulk of their condensation took place at temperatures around 600 K. (According to the condensation sequence, this range produces materials

12.21
Chondrules. These are each a few millimeters in size. They are among the most primitive materials from the formation of the solar system. (Courtesy F. Whipple and the Smithsonian Astrophysical Observatory)

like those that make up the earth.) About a million years after formation, radioactive decay reheated some planetesimals, melting them to some extent and allowing them to differentiate into iron cores and stony mantles. The planetesimals that were not gathered into a protoplanet possibly became the parent meteorite bodies. These bodies collided and fragmented to make the pieces that we pick up as meteorites after they fall to the earth.

What happened to other planetesimals? Some may have collided at high speeds with others and disintegrated into small pieces. A few may have passed close enough to a protoplanet to be captured in an orbit as a satellite. Others may have experienced near misses. Their orbits might have changed enough to throw them out of the solar system. Near Jupiter, planetesimals would be mostly icy materials; these may have formed comets.

12.8 Evaluation of nebular models

How well does this amalgam of contemporary models match up with the chemical and dynamic properties of the solar system?

On the chemical side, it does pretty well. The condensation sequence tied into the nebular model explains why the planets fall into two compositional classes (terrestrial and Jovian) and why planets in the same class differ somewhat in chemical composition.

On the dynamic side, the model has some successes and a few failures. Let's compare the model's results to the features listed in Section 12.2.

1. The planets' revolution and sun's rotation: Well explained as the original rotation of the nebula.
2. Coplanar orbits: Well explained by the conservation of angular momentum applied to the nebula's rotation.
3. Circular orbits: Explained fairly well by the interactions and sweeping up of planetesimals.
4. Planets' rotation: Expected from the spin of the nebula but not clearly worked out.
5. Planets' spacings: Fairly well explained by the sweeping out of zones of planetesimals by the protoplanets.
6. Satellite systems: Well explained if formed as miniature solar systems.
7. Satellite spacings: Works out if the satellites formed by sweeping up material.
8. Angular momentum distribution: Decent attempts but no detailed solution.
9. Comets: Weakly explained as icy planetesimals that did not get caught by Jupiter, Saturn, Uranus, or Neptune but were thrown out by them.
10. Planetary ring systems: Explained vaguely, like the asteroids, as unaccumulated debris.

KEY CONCEPTS

1. *Any model for the origin of the solar system must account for its general chemical and dynamic properties in a unified way.*
2. *The key dynamic properties include: The system is flat; most of the mass is in the center (the sun); most of the angular momentum is in the planets; the planets' orbital directions are all the same; and the planets are regularly spaced.*
3. *The key chemical properties include the general division of the terrestrial and Jovian planets, the differences among planets in each group, and the compositions of asteroids, meteoroids, and comets.*
4. *Interplanetary debris provides important clues about the origin of the solar system, especially from the physical properties of comets (dirty snowballs), asteroids (rocks and metals), and meteorites (rocks and metals).*
5. *The crystalline patterns found in iron meteorites indicate that their material cooled slowly from a molten state, which implies that they formed in larger bodies.*
6. *A spinning cloud of interstellar material will flatten out (from the conservation of angular momentum) and heat up as it contracts gravitationally.*
7. *The condensation sequence works if the temperature in the solar nebula decreases rapidly from center to edge; which materials vaporize and condense depends on how high and then how low the temperature gets at a certain distance from the sun.*
8. *Magnetic fields transfer the spin of the sun to the rest of the solar nebula (before the planets form) to account for the distribution of angular momentum now.*
9. *The general process of planet formation may have been: condensation of grains; accretion of grains into planetesimals; clumping of planetesimals into protoplanets; evolution of the protoplanets into the bodies we see today.*
10. *Jupiter and Saturn may have formed by the gravitational contraction of large blobs of material rather than by planetesimal accretion.*
11. *Comets and asteroids are very likely leftover planetesimals.*

STUDY EXERCISES

1. Suppose you hopped in your spaceship on Saturday night and flew to an asteroid. What would you see? (*Objective 2*)
2. Then you speed off to a comet. What would you see? How would its appearance differ from the asteroid you just visited? (*Objective 2*)
3. After many perihelion passages of the sun, what happens to a comet in the dirty-snowball model? (*Objective 2*)
4. How can you tell an iron meteorite from a piece of terrestrial iron and nickel? (*Objective 2*)
5. What is the major weakness of the nebular model with respect to the dynamic properties of the solar system? (*Objectives 4 and 5*)
6. What one planet fits least well with the general chemical and dynamic properties of the solar system? (*Objectives 1 and 9*)
7. Use the chemical condensation sequence to explain the general chemical differences between the earth and Jupiter. (*Objective 6*)
8. How is it that the solar system is flat rather than spherical? (*Objectives 4 and 9*)

BEYOND THIS BOOK . . .

Up-to-date nontechnical material on the origin of the solar system is hard to come by. I suggest that you try, after reading this chapter, "The Origin and Evolution of the Solar System" by A. G. W. Cameron in *Scientific American*, September 1975. An explanation of the condensation sequence is "The Chemistry of the Solar System" by J. Lewis in *Scientific American*, March 1974.

W. Hartmann has a relatively nontechnical review of "The Planet-Forming State: Toward a Modern Theory" in *Protostars and Planets* (University of Arizona Press, Tucson, 1978), edited by T. Gehrels. Articles in part 5 are also pertinent but more technical; some of their introductions are good.

J. Lewis has a good summary in "Putting It All Together" in *The New Solar System* (Sky Publishing Co., Cambridge, Mass., 1983).

Epilogue part 2

This part has shown you the new richness in our knowledge of the solar system. But don't let this wealth of information detour you from the main point: What do the present properties of the solar system reveal about its origin and evolution?

By now you realize that the models we have developed to answer this question strike a very tentative chord. But we do have a pretty good idea of the general themes. The formation of the solar system naturally accompanied the birth of the sun from a cloud of interstellar materials. The closeness in age of the sun, moon, earth, and meteorites implies that the process of formation lasted a short time compared to their age of 4.5 billion years—at most a few hundreds of millions of years. The fast formation contrasts with the slow evolution afterward.

To glean the conditions in the solar system's early years, we must look at the fossil objects in it: the interplanetary debris of comets, asteroids, and meteoroids. These objects have evolved little (compared to the planets) and so give us the most direct clues to the past.

That past involves a nebular model for the solar system's formation. The start of the process requires the gravitational collapse of a huge interstellar cloud. What prompts such a collapse? We don't really know yet. But we do have evidence for the existence of enormous clouds between the stars. We can also see very young stars that have been born from such clouds. And that starbirth may well involve the formation of planets. So the fact that stars are common implies that planets are common, too. Stars and planets are connected in starbirth; that is one of the key themes of this book.

Since their birth, the planets have evolved, some more than others. The Jovian ones have changed little since their formation—they have essentially retained their youthful appearences. In contrast, the terrestrial planets have reshaped their interiors and surfaces and transformed their atmospheres. And the earth, a restless planet, has changed the most—its structure now differs greatly from its protoplanet state.

What a circus of planets orbits the sun! Each planet has its unique aspects. Mercury skirts closest to the sun, its marred surface seared by solar radiation. Venus hides a hellish surface behind a veil of clouds. The earth sports a surface covered mostly with water. Mars is rugged, with gigantic volcanoes and huge polar caps of water and carbon dioxide ice. Jupiter gives off more energy than it receives from the sun. Saturn spins rapidly in a setting of delicate rings. Uranus rotates on its side. Neptune sometimes orbits beyond Pluto. And distant Pluto has a surface frosted with methane ice.

I can't help wondering how many other planets circle other stars in the sky, especially when I observe on a winter's night. How many monsters such as Jupiter? How many jeweled Saturns? How many barren places like Mercury? And how many like the earth, where perhaps my counterpart also dreams of inhabited worlds?

part 3

The universe evolves—the whole universe and all its parts. That's what I mean by *cosmic evolution*. The universe has a history, and its parts do, too. Unraveling all these stories brings us insight into the grand scheme of things and our place within it: the cosmic connections that link us all.

Stars form the nexus of cosmic evolution. They make up the crucial links in the chain of cosmic connections. With their births, planets form. With their light, life survives. With their deaths, new elements are forged. Without the lives and deaths of stars, the flow of cosmic evolution would cease.

All this from what appear to be simply points of light in the sky—except for one star, the sun. The realization that stars are actually other suns marked a crucial transformation in astronomers' picture of the cosmos. Because we orbit a mere 8.3 light minutes from it, the sun is exposed to our close and careful view—a vision impossible for us to achieve with the stars. That view permits us to make the solar-stellar connection, using the sun as our basic model for stars.

Chapter 13 presents the sun in all its glory. Then Chapter 14 focuses on the stars as we understand them as suns. With the basic physical properties revealed, Chapter 15 turns to the birth of stars from the matter between stars. Once stars are born, they evolve through a life cycle that we can basically understand, even though we have not seen any one star do so. Finally, stars die, often in violent ways, and leave behind bizarre remains to mark their deaths. Chapter 17 tells the stories of their ends—and how these may connect to new beginnings.

The universe of stars

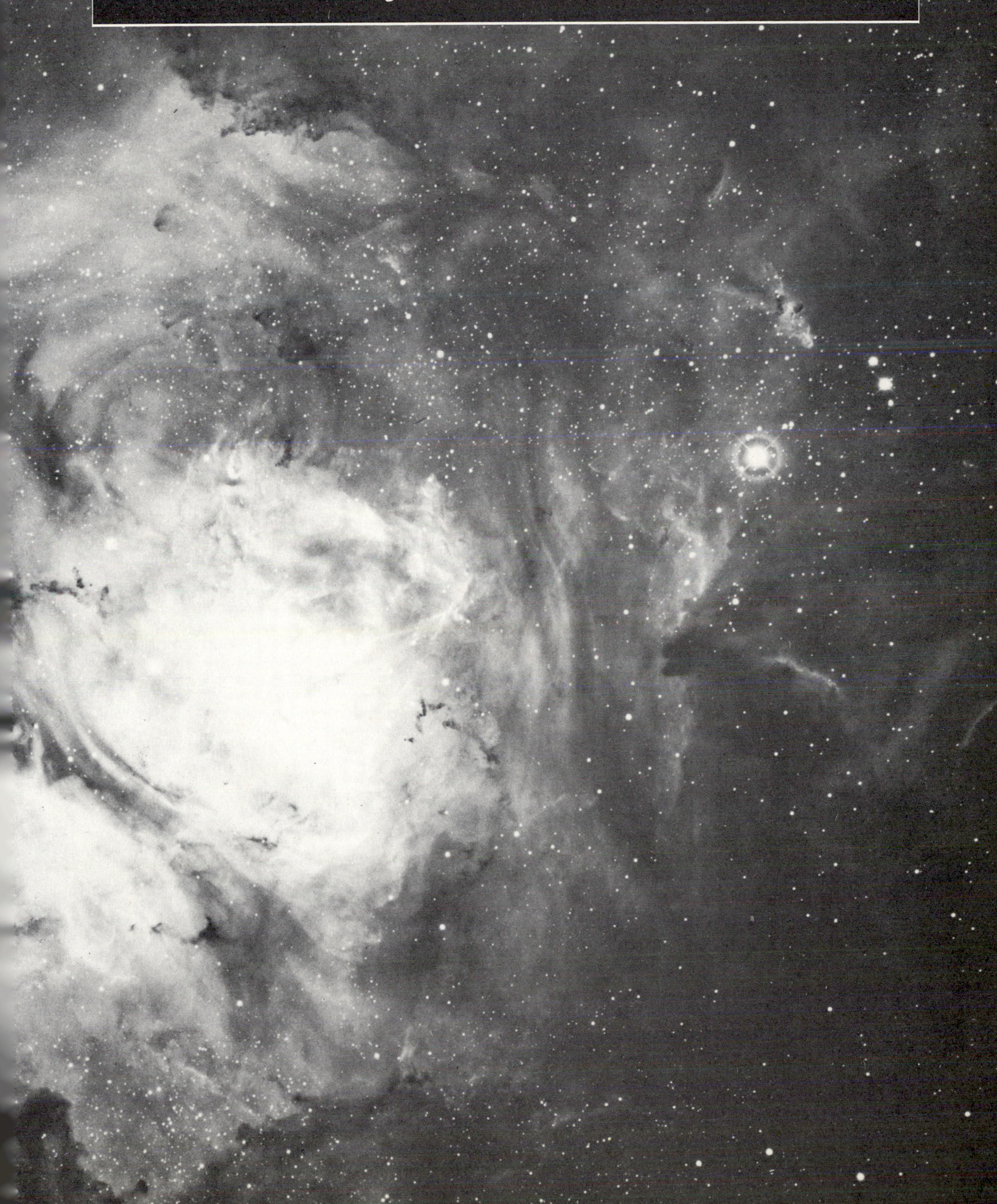

13/Our sun: Local star

Now this day,
My sun father,
Now that you have come out standing to your sacred place,
That from which we draw the water of life . . .

Zuni prayer at sunrise, translated by Ruth Bunzel

Learning objectives

After studying this chapter, you should be able to:

1. **Outline a contemporary method to find the distance to the sun.**
2. **Describe methods for finding out the sun's mass, size, density, luminosity, and surface temperature.**
3. **Define a blackbody, describe the characteristics of blackbody radiation, and apply these in astronomical situations (such as to the sun).**
4. **Describe the physical meaning of the term *opacity* and how opacity affects the flow of radiation through the sun.**
5. **Describe the appearance of the sun's spectrum and outline what atomic processes produce this spectrum.**
6. **List the sun's two most abundant elements and describe how these and others have been found.**
7. **In one sentence, explain the source of the sun's energy.**
8. **State the specific thermonuclear reactions that are supposed to produce the sun's energy and describe the conditions for them to take place.**
9. **Discuss the results and consequences of the solar neutrino experiment.**
10. **Trace the flow of energy from the sun's core to the earth and describe how the features of the quiet sun (photosphere, chromosphere, corona, and solar wind) result from this energy flow.**
11. **Select one feature of the active sun, describe its observed characteristics, and explain the main features in terms of energy flow and magnetic fields.**

Central question

How does the sun produce its life-giving energy, and how do its energy production and flow affect the sun's physical characteristics?

Sunlight gives life. All creatures on the earth are children of the sun. The sun now warms the earth and drives the weather. In the past, it raised the vegetation that produced our present fossil fuels. When these give out, we will likely turn to the sun directly for our usable energy.

What is the sun? Basically, a hot, huge ball of gas with fiery fusion reactions in its core. There, deep in the heart of sun, the energy bound up in matter is unleashed. Slowly, over millions of years, that energy (as light) erratically flows to the sun's surface. Free of the sun's material, it flies out to space. And 8.3 min later, a very small part of that light strikes the earth, and miracles occur.

For astronomers, the sun serves as our local link to the stars. Our sun has the same basic structure as the other stars in the sky, which we can see directly only as pinpoints of light. How to find out the physical characteristics of these distant lights? By using our sun as a guide. The sun serves as the local laboratory for testing our ideas about stars in general; our understanding of stars hinges on that of our sun.

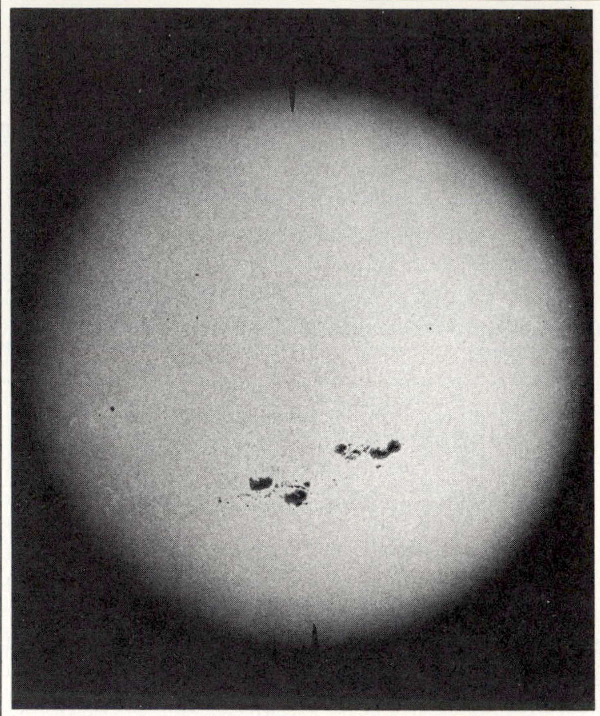

13.1
The sun's photosphere. Note that the photosphere has a granular look and fades out at the edge. The dark regions are sunspots. The dark spikes at top and bottom mark the sun's poles. (Courtesy Mt. Wilson Observatory, Carnegie Institution of Washington)

13.1
A solar physical checkup

How large is the sun (Fig. 13.1)? How massive is it? Before we can tackle such questions, we need an essential fact: How far is the earth from the sun? (You will find throughout this book that a critical question—perhaps *the* critical question in all astronomy— is finding the distance to celestial objects.) An analysis of the sun's light provides other facts without having to know the distance. Let's see what we know more or less directly about the sun.

How far? / The earth's orbit about the sun is elliptical. The semimajor axis of its orbit, which is the average earth-sun distance, has a length called one *astronomical unit* (*AU*). So the earth is 1 AU from the sun. How large is the AU in basic physical units, such as kilometers?

Recall that simple angular measurements establish the scale of the solar system in terms of the AU (Section 3.2) and that Kepler's laws (Section 3.5) completely describe the orbital motions of the planets, with AUs as the basic unit. So at any time you can draw up a scale map of the solar system with the planets' positions all neatly laid out in AUs.

To work out the distance scale in kilometers requires that only one piece, known in AUs, be measured in kilometers. An analogy: Suppose you're given a map of your region of the United States with no distance scale in kilometers but with all locations laid out correctly relative to one another. Hop in your car and drive between two points on the map while keeping close track of the distance in kilometers between these two points, which are separated on the map by so many centimeters. You've found the distance scale for the map.

To do the same for the solar system, we use the accurately known speed of light. Radar signals, which travel at light speed, are bounced off Venus, and the time between transmission and reception is accurately measured and converted into kilometers. Kepler's laws give the earth-Venus distance at the measurement time in AUs. So we know a fraction of an AU in kilometers, hence the AU in kilometers: 1.496×10^8 km. (Remember it by rounding off to 150 million km.)

Why bother bouncing radar off a planet? Why not bounce it off the sun and measure the AU directly? Mainly because the sun does not reflect radar signals very well.

How big? / Viewed from the earth, the average angular size of the sun is 32′. (That's about ½°, or half the tip of your finger at arm's length.) Because we know the AU in kilometers, we get the sun's actual size from its angular size. The sun's diameter is roughly 1.4 million km, or 109 times

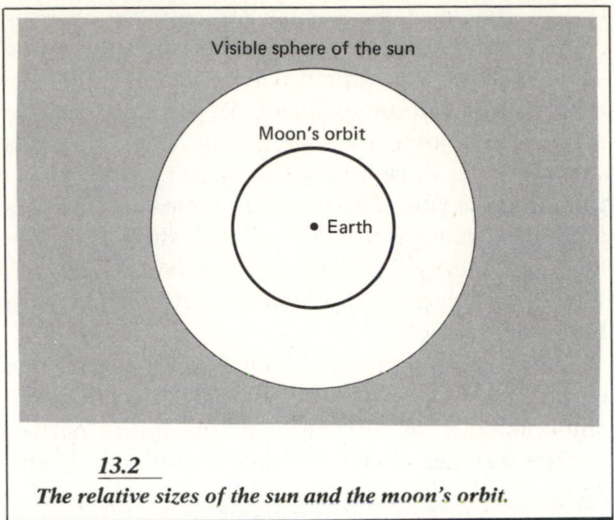

13.2
The relative sizes of the sun and the moon's orbit.

the earth's diameter (Fig. 13.2). Imagine the earth the size of a dime. The sun would be about 2 m in size and 200 m from the coin-sized earth.

How massive? / To find the sun's mass, you need the earth-sun distance, the earth's average orbital velocity, and Newton's second law of motion and gravitation (Section 4.5). Or you can use Newton's form of Kepler's third law (Focus 4.2). The sun's mass comes out to be about 2×10^{30} kg, or more than 300,000 times the earth's mass.

How dense? / The sun is a big, massive object. Yet its density (the mass divided by the volume) is low: 1400 kg/m^3, only 40 percent denser than water! This low average density, along with its hotness, implies that the sun is a gas.

Warning: Don't confuse density with mass or size alone. For example, the sun is much more massive and larger than the earth, yet it has a much lower average density. Why? The sun is a gas, while the earth is solid rocks and metals.

13.2 Ordinary gases

To know what happens inside the sun (and other stars), you must have some idea of how gases behave because stars are huge balls of gas. Let me describe a simple model for ordinary gases. (You will come across extraordinary gases in Chapter 17.)

A gas consists of particles—atoms, molecules, or ions, or perhaps all of these. Simplify the situation in a real gas by imagining all the particles to be the same and like hard spheres. Picture these spheres trapped in a small box. Once set into motion, the spheres keep moving, bounding off the walls of the box and colliding with each other (Fig. 13.3).

These collisions push any one sphere at times faster or slower, but over time the sphere has a definite average speed. Also from collisions, all the spheres in the box will have the same average speed. We use this average speed of particles in a gas as a measure of its *temperature*. If all the particles were motionless, a gas's temperature would be zero. (This book uses the Kelvin, or absolute, temperature scale, which is the same as the centigrade or Celsius scale but measured from absolute zero. On this scale, the freezing point of water is 273.15 Kelvins, abbreviated K. No degree mark is used in designating Kelvin temperatures.) At room temperature, about 300 K, the average speed in a gas of hydrogen particles is about 3 km/sec. Higher temperatures mean higher average speeds of the particles.

Imagine putting a partition in the gas container. The spheres ram and bounce off both sides of this partition; each collision exerts a small force on it. The combined force of all collisions is the *pressure* of the gas. This book uses the pressure of the atmosphere at sea level as the basic unit of pressure; it's called *one atmosphere* (atm).

Imagine that you increased the temperature of the gas. The average speed of the particles increases, so they collide with one another and into the partition more frequently and with greater force. The pressure increases. How, in relation to the increase in temperature? For ordinary gases, the increase is a direct one. So if you double the temperature (in Kelvins), the pressure doubles.

What happens to the pressure if you increase the number of particles in the box? (You've done this if you've pumped up a bicycle tube, for then

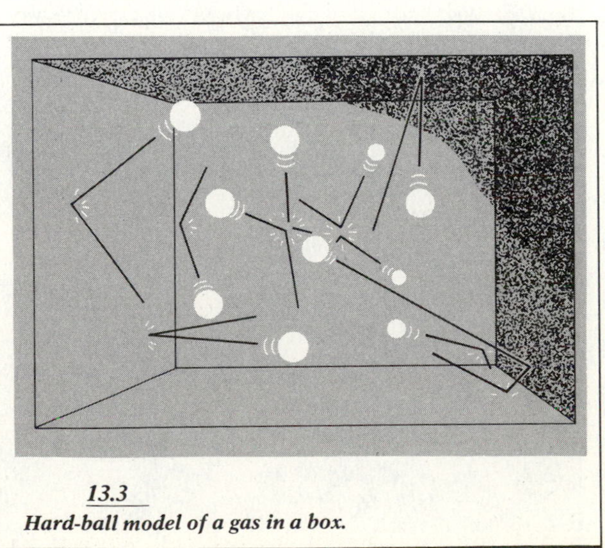

13.3
Hard-ball model of a gas in a box.

you forced more air particles into it.) Adding more particles to the same space increases the density of particles; on the average, each cubic meter contains more particles. For a gas, the number of particles in a unit volume is called the *number density*. Suppose you increased the number four times without changing the temperature. Each cubic meter now contains four times as many particles. So four times as many collisions occur, on the average, against the partition, and each collision has the same average force as before. The pressure increases; it is four times greater. Increasing the number density directly increases the pressure of an ordinary gas.

This hard-sphere model shows that gas pressure depends directly on *both* the number density and the temperature. With these basics about simple gases in mind, let's turn to that huge hot ball of gas that makes up the sun.

13.3 The sun's continuous spectrum

Most of what we know about the sun and other stars comes by way of light. Decoding the message of sunlight and starlight takes up much of the time, energy, and ingenuity of astronomers. Let's look at how the message is read.

Solar luminosity / The sun's *luminosity* is its total output in radiative energy each second. By the time the light reaches the earth (only 8.3 min after it leaves the sun's surface), it has spread out over a large region of space—in fact, over a sphere whose radius is the earth-sun distance. So you cannot directly measure this radiation, the sun's luminosity, all at once.

How to find the sun's luminosity? Place a special detector in a satellite orbiting the earth. (Why? Because the earth's atmosphere absorbs some sunlight.) Point the detector directly at the sun. Measure the radiant energy absorbed by the detector: It amounts to 1370 W for each square meter of the detector's area (Fig. 13.4). So on a surface 1 AU in radius (an imaginary sphere surrounding the sun), every square meter catches this amount of energy. Totaled up over the entire surface, the energy amounts to roughly 3.8×10^{26} W. That's enough power to light 3.8×10^{24} 100-W light bulbs all at once!

The earth intercepts only a small part of this energy, only about 10^{-10} of the total, but that still amounts to a lot. In one year, the entire earth's surface catches about 10^{18} kilowatt-hours (kWh). At the rate of 5 cents per kWh, the annual solar energy would cost 5×10^{16} dollars!

Geologic evidence indicates that the sun's luminosity has not varied more than a few tens of percent over the past 3.5 billion years. Satellite measurements show that it does not vary now more than a few tenths of a percent (mostly because of sunspots). Long-term observations indicate a variation of only 0.25 percent in the past 50 years.

Astronomers use a special name for the amount of energy passing though one square meter each second: *flux*. The sun's flux, 1370 W/m^2, serves as the standard of comparison for the flux from other celestial objects.

The sun's surface temperature / The sun's color (yellowish white) provides a clue to its surface temperature. We can assign a temperature to the sun's surface by examining its continuous spectrum.

To see how, consider heating a piece of metal in a very hot flame. At first the metal emits a dull

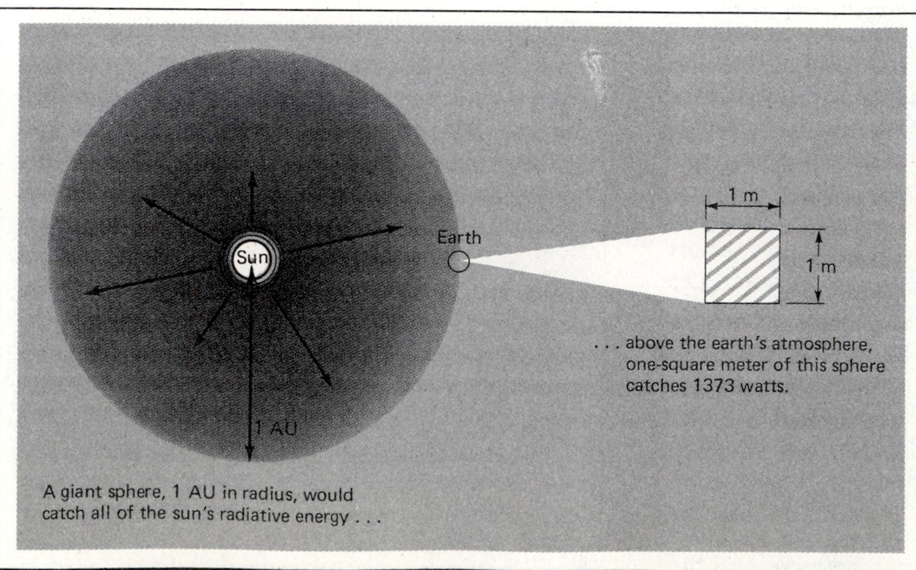
Sun
Earth
1 m
1 m
1 AU
. . . above the earth's atmosphere, one-square meter of this sphere catches 1373 watts.
A giant sphere, 1 AU in radius, would catch all of the sun's radiative energy . . .

13.4 ***Measuring the sun's luminosity from the earth. An imaginary sphere with a radius of 1 AU would capture all of the sun's luminosity, so 1 square meter catches only a small fraction, 1370 W, of it.***

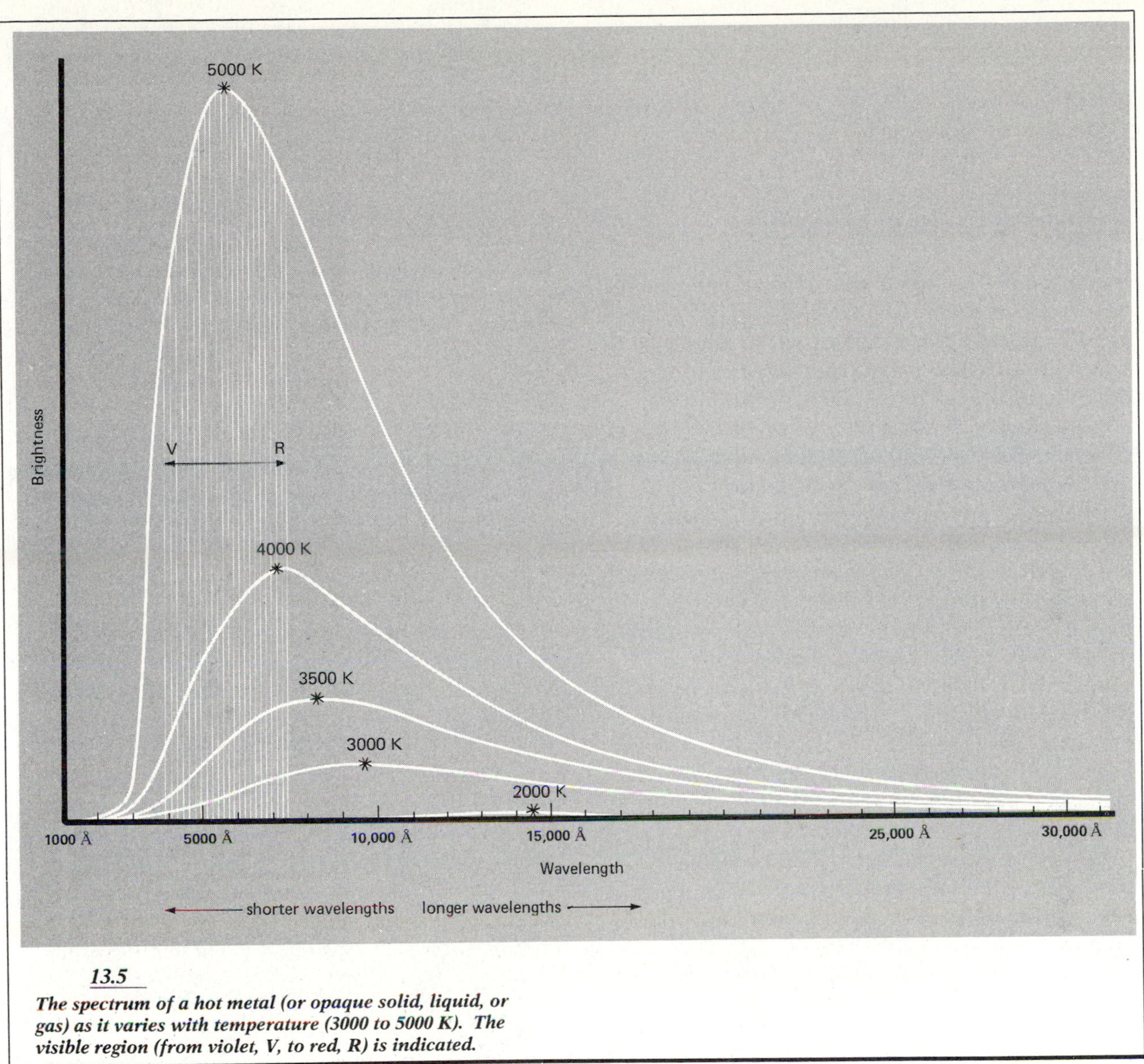

13.5

The spectrum of a hot metal (or opaque solid, liquid, or gas) as it varies with temperature (3000 to 5000 K). The visible region (from violet, V, to red, R) is indicated.

red light. Then, as it gets hotter, it glows more brightly reddish, then orange, yellowish, yellow-green, white, and, finally, bluish white. The overall color change in the visible region of the spectrum relates to the temperature of the metal. So a metal's continuous spectrum changes with temperature in a special way.

How do color and continuous spectra relate? Let's look at the metal's emission in some detail (Fig. 13.5). Measure the brightness at a range of wavelengths (from ultraviolet to infrared) for the metal at different temperatures. Then plot the metal's spectrum. Note three features in these spectra: (1) The emission *peaks* at some wavelength, (2) the peak shifts to *shorter* wavelengths as the metal gets hotter, and (3) the metal emits *more* intensely at *all* wavelengths at *higher* temperatures.

Your eye, because it responds only to the visible region of the spectrum (violet to red), sees only a small portion of the complete spectrum. (For example, your eye doesn't sense any infrared.) Yet for the temperatures considered here, your eye does reasonably discern the shift in the balance of colors at the peak emission.

An object whose spectrum has this characteristic shape (Fig. 13.5) and the variation with temperature just described is called a *blackbody radiator*, or simply a *blackbody* (Focus 13.1). The radiative output and spectrum of a blackbody depend only on its temperature and not on other properties (such as composition). Perfect blackbodies do not actually exist, but the sun (and other stars) emit radiation somewhat like an ideal blackbody. The wavelength at which a star's continuous spectrum peaks relates to its temperature: The

Focus 13.1

RADIATION FROM BLACKBODIES

A blackbody is an object that absorbs completely any radiation that falls on it. The radiation from a blackbody has a characteristic shape that depends only on the temperature of the body and not on anything else, such as its composition.

A blackbody has a number of special characteristics. First, a blackbody with a temperature greater than absolute zero emits **some** *wavelengths. Second, a hotter blackbody emits* **more** *energy at* **all** *wavelengths than does a cooler one. Third, a hotter blackbody emits a* **greater proportion** *of its radiation at* **shorter** *wavelengths than a cooler one. And fourth, the amount of radiation emitted by the surface of a blackbody depends on the* **fourth power** *of its temperature. So if one blackbody is twice as hot as another of the same size, it emits $2^4 = 2 \times 2 \times 2 \times 2$, or 16 times as much energy in total. The hotter one also emits 16 times as much energy from each square meter of surface as the cooler one.*

Two of the properties of blackbodies are described by simple equations. One, the amount of energy emitted for every square meter of a blackbody at temperature T Kelvins is

$$E = 5.67 \times 10^{-8}\, T^4 \text{ W}$$

This relation is called the **Stefan-Boltzmann law.** *For example, every square meter of the sun's photosphere (5800 K) emits about 6.3×10^7 W. Explicitly,*

$$\begin{aligned} E &= 5.67 \times 10^{-8}\,(5780)^4 \\ &= 5.67 \times 10^{-8}\,(1.12 \times 10^{15}) \\ &= 6.33 \times 10^7 \text{ W} \end{aligned}$$

Second, the wavelength at which the energy output from a blackbody peaks is

$$\lambda_{max} = \frac{2.9 \times 10^{-3}}{T}$$

where T is the temperature in Kelvins and λ_{max} is the wavelength, in meters, at which the peak output occurs. This is called **Wien's law.** *For the sun the peak for a temperature of 5800 K is about 4.9×10^{-7} m (0.49μm, or 4900 Å).*

Note that both these equations can be used to infer the sun's surface temperature. To use the first, you need to find out how much energy each square meter of the surface puts out. For the second, you have to determine at what wavelength the sun's spectrum peaks. These methods of determining temperature can be applied to other stars.

higher the temperature, the *shorter* the peak wavelength.

A common question about blackbodies is why they are called *black* when they give off light? Blackbodies are so-named from their light-absorbing abilities: They absorb light at any wavelength completely and reflect none. (If you beam light at the sun, none would reflect back.) When a blackbody absorbs radiative energy, it heats up and emits at all wavelengths, even though the peak of emission may not be visible to our eyes. Physically, a good absorber, when heated, is a good radiator. It's very difficult to make a perfect blackbody. But the sun and other stars emit radiation almost like a blackbody does, so in most cases we can apply this concept to their continuous radiation. Note that any opaque and hot solid, liquid, or gas will produce a blackbody spectrum.

Let's examine the sun's continuous spectrum (Fig. 13.6) as measured at the earth's surface. Note that when the spectrum is measured at sea level, the earth's atmosphere absorbs parts of it—especially the ultraviolet (absorbed by ozone) and the infrared (absorbed by water vapor and carbon dioxide). But very little of the visible light is absorbed. You see that the spectrum peaks at about 0.5 μm. Without the atmospheric absorption (Fig. 13.6), the sun's spectrum follows a more or less continuous curve with this one peak in the yellow-green part of the spectrum. To produce this peak requires a surface temperature of about 5800 K.

Opacity / If the sun's continuous spectrum has a blackbody shape, the emitting region must also be a good absorber of light. What does the absorbing?

Recall (Section 5.3) that atoms can absorb light by three basic processes: bound-bound, bound-free, and free-free transitions. Bound-bound transitions produce only lines; but bound-

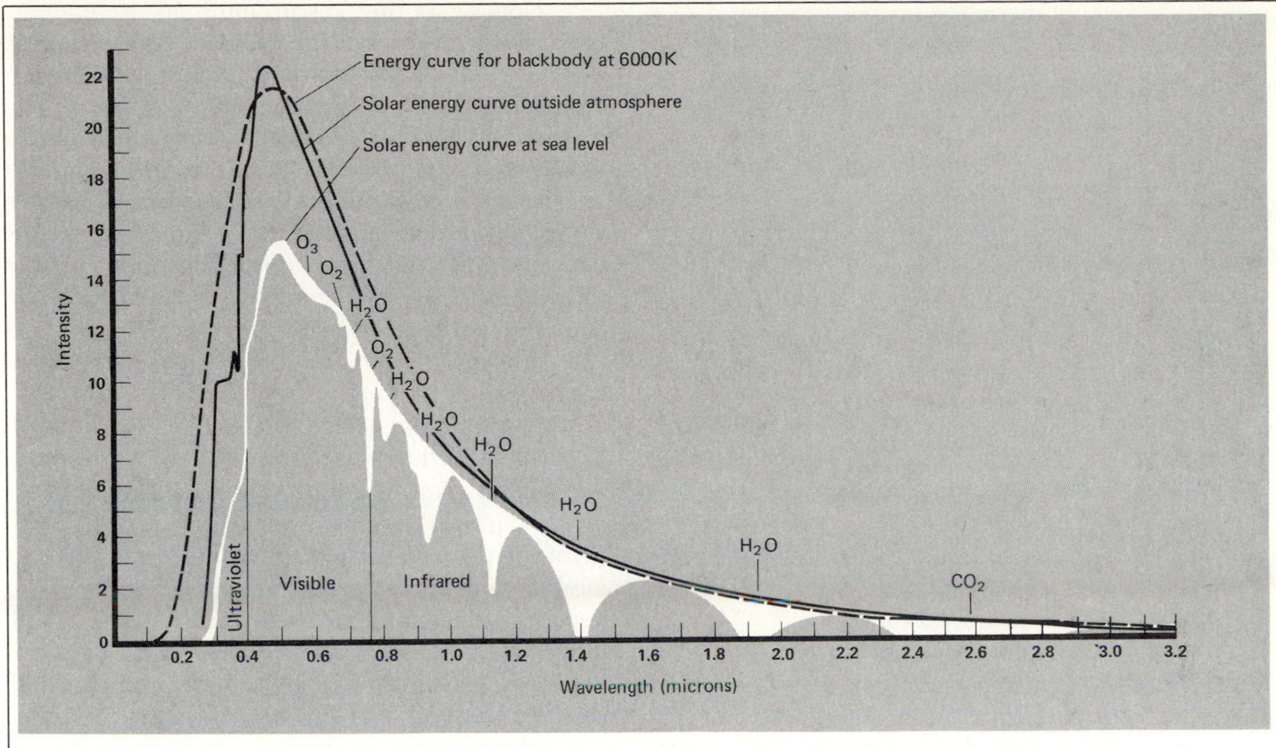

13.6

The absorption of the sun's radiation by the earth's atmosphere. The sun's radiation measured above the earth's atmosphere is indicated by the solid line. (Note how close this curve comes to a blackbody at 6000 K.) The shaded curve shows the actual measured spectrum at sea level. Indicated is whether the carbon dioxide (CO_2), water vapor (H_2O), oxygen (O_2), or ozone (O_3) is responsible for the absorption.

free and free-free transitions can contribute to a continuous spectrum. What transitions make the sun a good absorber?

To see this point, you first need to understand the concept of *opacity*. On a clear day, you can see to far distances (many kilometers) through the air. The air is transparent to light; its opacity is low. On a very foggy day, you can't see far at all (only a few meters in San Francisco!). The air is opaque to light.

What makes a gas opaque? Interactions of light with atoms and electrons. When a photon is absorbed, it no longer exists and so can't carry energy any farther. The photon's energy is not destroyed; it has just been transferred to an electron (in a bound-bound, bound-free, or free-free transition). When the electron loses the energy, it emits a photon. But—and this is the key point—that photon can be emitted in *any* direction, including back in the direction from which it originated. It heads off and moves only a short distance before it is absorbed. When another photon is re-emitted, it probably zips off in a different direction. So photons in an opaque gas travel very short distances before they are absorbed. In a gas of lower opacity, they travel greater distances. When a gas is opaque, photons walk slowly through it; when transparent, the photons fly straight through it.

So the opacity of a gas relates to how far photons can travel, on the average, between absorptions. The opacity depends on how much of the absorber is there (its density) and how effectively it absorbs. The culprit for the sun's visible opacity is a very strange type of hydrogen ion: a *negative hydrogen ion*. At a low enough temperature (roughly 6000 K), a hydrogen atom can acquire a second electron and still be stable. Infrared and visible light easily remove the extra electron. These free electrons then take part in free-free transitions. This process absorbs so well that the opacity in the visible and infrared is very high. If we define the visible surface of the sun as the layer where the gases become visibly transparent, that layer is only 100 to 200 km thick and has a temperature of 5800 K. This region essentially defines the sun's *photosphere*. It marks the place where photons can fly rather than walk in the sun's atmosphere.

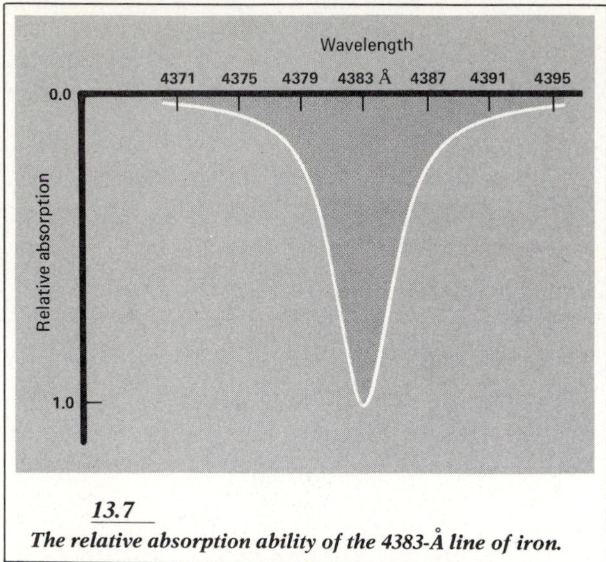

13.7
The relative absorption ability of the 4383-Å line of iron.

13.4
The solar absorption-line spectrum

Recall (Section 5.3) that a spectroscope shows the solar spectrum to consist of a continuous spectrum crossed with dark lines. How is this spectrum produced?

To see it, I'll use the concept of opacity again, this time for bound-bound transitions. Consider the transition that produces the 4383-Å line of iron (Fig. 13.7). We find when we measure the absorption or emission of this line that it is not infinitely narrow. It has a finite width, centered on 4383 Å, where the iron atom absorbs very well. At somewhat shorter or longer wavelengths, say 4379 Å or 4387 Å, the iron atom can still absorb light, but not as well. In other words, a gas containing iron has a much higher opacity for 4383-Å photons (at the line's center) than for those with wavelengths a few angstroms shorter or longer.

In the photosphere, the opacity at the centers of absorption lines traps photons. They travel very short distances, and so have little chance of escaping into space. The density of the photosphere drops rapidly from bottom to top. So the opacity decreases as fewer atoms per cubic meter are available to do the absorbing. Eventually it falls enough for photons, even at the wavelengths of line centers, to escape.

As a result, the absorption lines form at *different levels* in the photosphere. At their centers, where the opacity is high, the light emerges from higher up in the photosphere, where the gas is cooler and emits less intensely. Off the line centers, where the gas is more transparent, the light emerges from lower, hotter layers, which emit more intensely. So the lines are brighter (less dark) away from their centers.

Note again that absorption lines are *not* perfectly black and they do contain *some* energy. They are only dark relative to continuous emission at neighboring wavelengths.

In the photosphere, the temperature rises quickly as you go down into it: a few thousand Kelvins in a few hundred kilometers. It is this sharp temperature change that results in the sun's dark-line spectrum. In effect, you can consider the atmosphere as over a hot surface, just the situation needed, according to Kirchhoff's third rule (Section 5.2), to produce an absorption line spectrum.

Astronomers have analyzed more than 20,000 lines in the solar spectrum to find the chemical composition of the atmosphere. The intensity of the dark lines from a particular chemical element relates to how much of that element is found in the atmosphere. The line's intensity also depends on the temperature and density in the photosphere. Iron produces most of the absorption lines; other strong lines come from hydrogen, calcium, and sodium (look back at Fig. 5.13).

Identifying particular elements simply requires matching their "fingerprint" patterns. But it's a harder task to find out an element's actual abundance. To determine the abundance, you need to know how atoms of a particular element absorb and emit light. Then you must make a model of the sun's photosphere. You can construct this model from theoretical calculations and basic physical concepts; it consists of a list of temperatures, densities, and pressures at different depths in the photosphere. Then you add some amount of the element, and you can calculate how intense its absorption lines must be. You try to match the calculated intensity to the observed intensity by playing with different values of the abundance. A match between observed and calculated intensities gives the correct abundance.

Warning: This procedure gives you only the abundance in the *photosphere*. It does not tell you directly the abundance in the *interior*.

Most of the atmosphere's mass is hydrogen (74 percent) and helium (25 percent). All other elements (loosely called *metals* or *heavy elements* by astronomers) make up a mere 1 percent. Note that these abundances are relative to the total *mass*.

If an element's absorption lines don't show up in the visible spectrum, does it mean that it does not exist in the sun? Not necessarily. Perhaps so little of the element exists that it does not produce detectable absorption lines. Or the strongest absorption lines may be in a region of the sun's spectrum unobservable through the earth's atmosphere. Another possibility is that the temperature, pressure, and density of the sun's atmosphere may inhibit the formation of the element's spectral lines. If the conditions are too hot or too cool, the lines do not form.

Consider helium. If you look very carefully at the sun's visible spectrum, you won't find any absorption lines of helium. Why not? In the photosphere, the temperature is too low to excite helium to levels that can absorb photons of visible wavelength. Above the photosphere, the temperature rises to 40,000 K. Here the temperature is high enough to excite helium atoms, and during an eclipse you can find bright lines from helium. In fact, helium (from the Greek word *helios*, "sun") was discovered in the sun's spectrum before it was found on earth. However, too few excited helium atoms exist in this hot region to produce emission lines strong enough to be seen directly in contrast to the continuous spectrum.

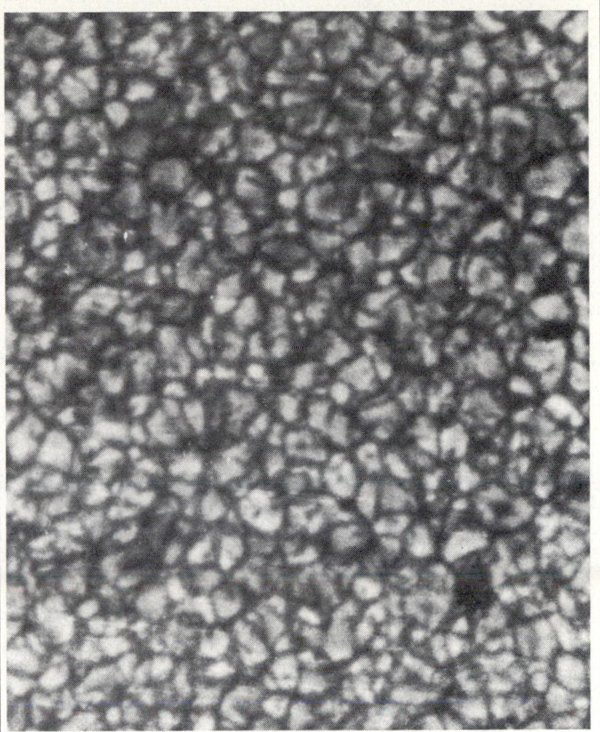

13.8
The granular structure of the photosphere of the sun. An individual granule is about 1500 km across. (Courtesy Pic du Midi Observatory, France)

13.5
The quiet sun

The term *quiet sun* refers to the seemingly placid day-to-day face of the sun. I use this expression to describe solar phenomena that are characteristic of large regions for long periods of time, such as the steady flow of energy out of the sun. This energy flow, from core to surface and beyond the atmosphere, controls the environment of the quiet sun.

The quiet sun that we see directly consists of the sun's outer layers, together known as the atmosphere. Because the sun's atmosphere is a gas, it does not have a distinct layer with sharp boundaries. But we have been able to discover that the atmosphere has three substantially different zones: the photosphere, the chromosphere, and the corona.

Photosphere / The sun's photosphere has a bubbly look, like the surface of a pot of boiling oatmeal (Fig. 13.8). Each bubble has an irregular shape about 2000 km across (half the size of the United States!) and lasts for about 10 min. This phenomenon is called *photospheric granulation*. You can visualize the photospheric granulation as the top layer of a seething zone where hot blobs of gas spurt to the surface, radiate energy, cool, and flow downward. Just below the seething photosphere is a region of convection (Focus 13.2). Here the outward flow of energy heats the gas and makes it rise. The base of the convection zone is thought to lie at about 0.8 solar radius.

Chromosphere / Just before and after totality in a solar eclipse, a bright, pink flash appears above the edge of the photosphere This is the chromosphere, the solar atmosphere just above the photosphere. The pink color comes from the emission of H-alpha, the first line of the Balmer series, in the red region of the spectrum.

A spectroscope directed at the chromosphere during its fleeting appearance shows a bright-line spectrum (Fig. 13.9). The temperature, density, and pressure in the chromosphere determine the intensities of its emission lines. So the line intensities provide clues to the physical conditions there.

The chromosphere begins a few hundred kilometers above the photosphere and extends only about 2000 km higher, where it merges into the corona. It is about 1000 times less dense than the photosphere but, surprisingly, gets much hotter. The temperature rises from 4300 K to above 400,000 K in the 2500 km of the chromosphere above the photosphere. This rise to high temperatures produces the emission lines from this region.

You may be wondering why we don't see emission lines from the chromosphere when we look through it down to the photosphere. The answer is that the chromosphere has such a low density that it is transparent to the light passing through it. The photospheric spectrum (continuous and absorption line) makes it through the chromosphere, which adds only a little emission.

Focus 13.2

ENERGY TRANSPORT: CONDUCTION, CONVECTION, RADIATION

Thermonuclear fires blaze in the sun's center, creating energy. Eventually that energy makes it to the earth. How does the energy get transported from deep in the sun to us?

In general, energy is carried by the processes of **conduction, convection,** *or* **radiation.** *If you've ever relaxed in front of a roaring fire, you've experienced all of these. You are directly warmed by the fire's heat, infrared radiation that travels directly from the fire to you to be absorbed by your skin. That's energy transported by radiation. Much of the energy from the fire, however, is wasted up the chimney. The fire heats the air just around it. Air is gas, and the air expands and becomes less dense. Cooler, denser air flows in and pushes the hotter air up and out of the house, where it cools off. This transport of energy by mass motions of a gas (or liquid) is convection. Finally, you may have by mistake left the poker in the fire. If you grab the handle without thinking, you yell upon finding it very hot. It got that way from conduction: The poker's electrons that were actually in the fire were heated and so moved around at high speeds. They banged into their neighbors, agitating them. These collided with their neighbors, and so on throughout the poker from one end to the other. The kinetic energy (temperature) was transformed by direct collisions across the poker.*

In ordinary stars like the sun, radiation or convection can carry energy along. (Conduction does not play an important role because the sun's material is a gas and not very dense. However, conduction does become important in extremely dense stars where matter is in a different state than ordinarily.) Which process is at work? That depends on the local conditions in the gas. The general rule is this: The transport process that can work most efficiently is the one that does operate. For most of the sun's interior—about 80 percent of the radius—energy flows most efficiently by radiation. Only in the outer 20 percent of the radius does convection come into play. We see the top of this convective zone as the turmoil of the photosphere (Fig. F.27). The hot gas bubbling up here radiates; these photons leap out to space.

Note that for a large part of the sun's radius, energy is carried by radiation. Why does the situation change? How well radiation transports energy through a material depends on a property

The chromosphere is constantly pierced by large spears of gas called *spicules* (Fig. 13.10). These jets of gas can spurt up to heights of 10,000 km and fade away again in several minutes. They have diameters of about 1000 km. Their temperatures are about 8000 K. Because of the fountains of spicules, the chromosphere is not a uniform layer. The chromosphere ends—jaggedly—at the tops of the spicules.

Why is the chromosphere generally hotter than the photosphere? If photons heated it, you'd expect that it would be cooler than the

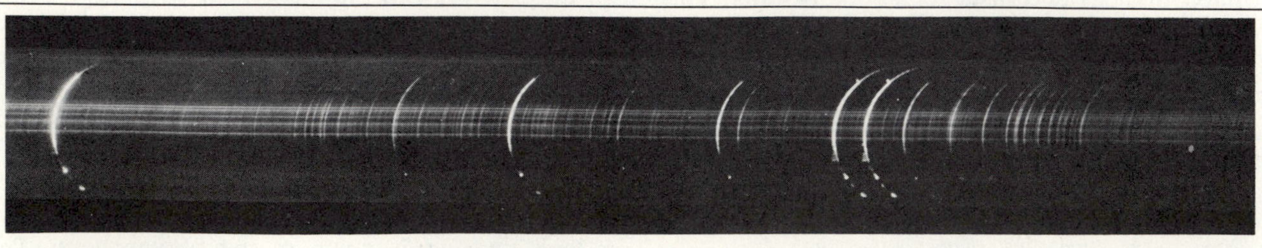

13.9
The emission spectrum of the chromosphere, visible just before total eclipse. The bright spectral lines are curved because the curved sliver of the chromosphere acts like a curved slit. (Courtesy Mt. Wilson Obervatory, Carnegie Institution of Washington)

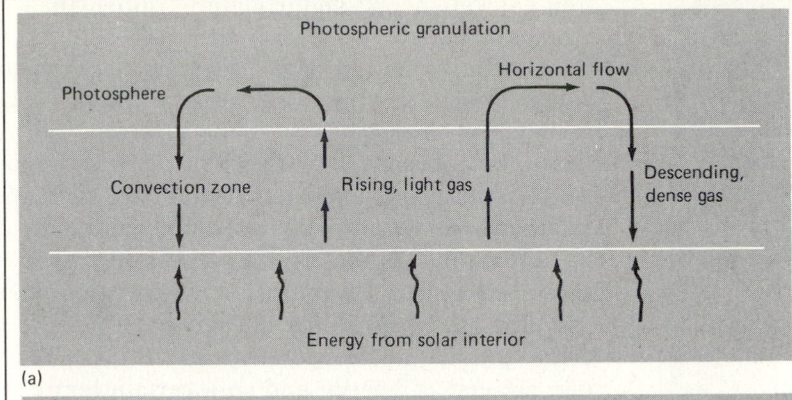

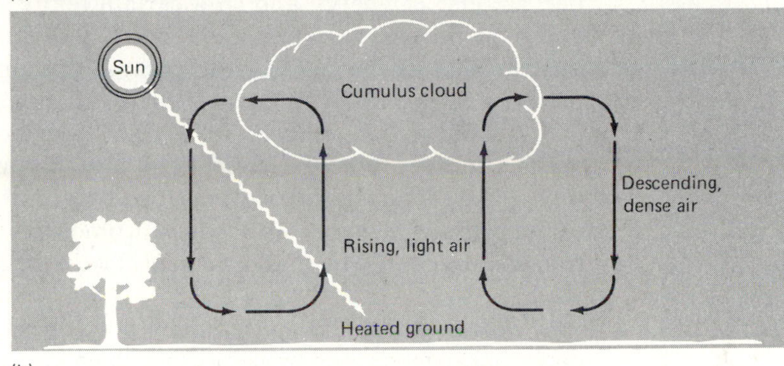

F.27

Flow of gas in the convective zone of the sun **(a)**. *Hotter, less dense gas rises to the top of the photosphere. There it radiates energy, cools off, and descends. A similar process forms cumulus clouds in the earth's atmosphere* **(b)**.

of that material known as its **opacity**. *Basically, opacity measures how efficiently a material absorbs photons. In the sun's core, photons travel about 1 mm between absorptions. The gas here is fairly opaque, but radiation is still more efficient than convection for transporting the energy. As photons journey toward the surface, they run into a region where the temperature is low enough so that hydrogen atoms form. These suddenly make the gas extremely opaque, so much that convection can transport energy more efficiently, and a convection zone forms.*

13.10

Jets of hot gas along the edge of the sun. These jets, called spicules, transport material and energy from the photosphere to the chromosphere. (Courtesy Mt. Wilson Observatory, Carnegie Institution of Washington)

photosphere—certainly not hotter. Some other energy source must do the heating. The secret lies with energy bound up in the magnetic fields.

You know that magnetic fields can store energy if you've ever ripped two magnets apart: You have to put out an effort—work—to do so. The trick, then, lies in a way to have the magnetic energy move upward from the photosphere. Magnets can transport energy by moving material or by generating electric currents that move through conducting materials. Currents flowing upward from the photosphere to the corona can leave some energy in the chromosphere. I'll tackle the evidence for that in Section 13.7.

Corona / You can see the sun's splendid corona directly during a total solar eclipse (Fig. 13.11 and Color Plate 14). Although as bright as the full moon, the corona is normally obscured by the sunlight scattered in the earth's atmosphere. During a total eclipse, when the photosphere is

13.11
The corona during a total solar eclipse, Mexico, March 7, 1970. (Courtesy NASA)

blocked out, the sky becomes dark enough for the corona to be visible. Stars and the bright planets can be seen next to and behind it.

Spectroscopes reveal that the corona emits bright lines, which were a mystery for many years because their patterns did not match any known elements. In 1940 the Swedish scientist Bengt Edlén demonstrated that highly ionized atoms of iron, nickel, neon, and calcium, rather than a strange element, produced the emission lines. (Recall that when an atom loses an electron, its energy levels change.) Because it takes large amounts of energy to rip many electrons off an atom, the corona must be very hot. For example, to strip iron of 16 of its normal 26 electrons requires a temperature greater than 2 million K.

What makes the corona so hot? Space telescopes have shown that most of the corona's emission lines show up in the ultraviolet. These lines come from highly ionized elements. It takes a certain amount of energy and so a certain temperature to form ions of various elements. These lines indicate that at the base of the corona, the temperature rises rapidly, roughly 500,000 K in just a few hundred kilometers in a thin transition region between the chromosphere and corona (Fig. 13.12). This sharp rise may result from the same physical process as in the chromosphere: transport of energy by magnetic fields (see Section 13.7).

Solar wind / The corona does not end suddenly but extends into space to distances greater than those indicated by its visual appearance. The corona moves; it flows out from the sun. The reason for the outflow of material is that, because of its high temperature, the coronal gas exerts a large outward pressure—a greater force than the inward pull of gravity. As a result, the gas from the corona streams away from the sun. Closer to the sun, the expansion is slow because the sun's gravity is stronger. With increasing distance, the flow speeds up because the gravity decreases. The gas stream becomes the *solar wind*, so named because of its high velocity.

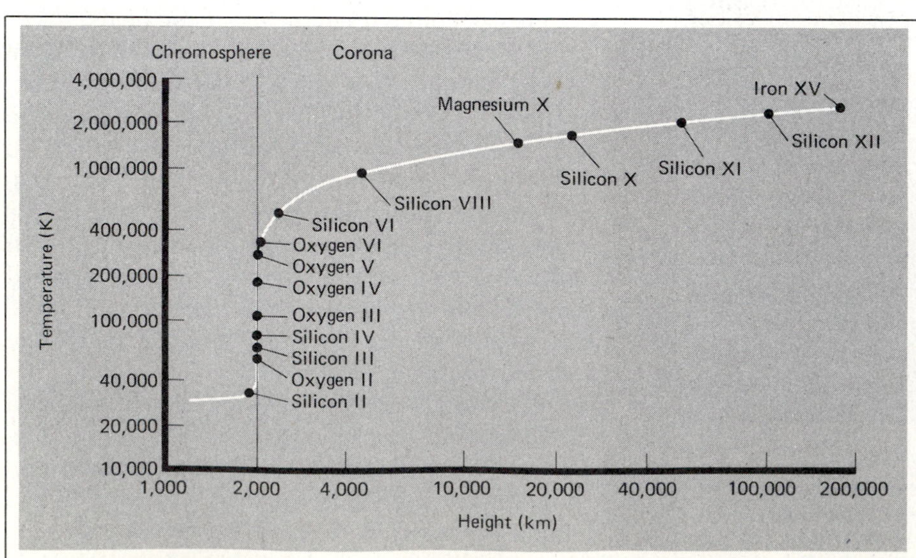

13.12
Ions used to flag temperatures in the corona. Since the corona is very hot, atoms in it are highly ionized. For a given element, it takes higher temperatures to ionize it more. So how much a certain element is ionized indicates the temperature at which it is found. Stages of ionization are marked by a Roman numeral next to the element's name; the numeral II, for example, denotes that the atom has lost one electron. Note that the corona's temperature increases rapidly above the chromosphere and then levels out to about 2 million K. (Data from Skylab)

At the earth's orbit, the solar wind whips by at roughly 700 km/sec. The particles in the solar wind (mostly protons and electrons) travel from the sun to the earth in about five days. (In contrast, light takes 8.3 min.) The earth swims through the solar spray and catches some of the particles in its magnetic field.

13.6 Energy from the solar interior

The interior of the sun contains the bulk of the sun's mass and the furnace that generates its energy. All the features of the quiet sun, from photosphere to solar wind, result from the energy flow out from the interior. How does the sun generate so much energy?

Energy sources / During the nineteenth century, when the earth was found to be very old (billions of years in age), the source of the sun's power became an embarrassingly difficult puzzle to explain. The crux of the problem is not the rate of energy production but its longevity. The sun has been at roughly the same luminosity for at least 3.5 billion years, according to present geologic evidence. Energy from ordinary chemical reactions, such as burning, could not provide the amount necessary for so long. If the sun were composed entirely of oxygen and coal, it would have burned to a dark cinder in a few tens of thousands of years. The sun could not be a coal-burning furnace; it wouldn't last long enough.

In the middle of the nineteenth century, Hermann von Helmholtz (1821–1890) and William Thompson, Lord Kelvin (1842–1907), proposed that the sun shone because it was releasing gravitational energy by shrinking. That is, gravitational contraction converted gravitational potential energy to radiative energy (see Section 12.3). Recall that half goes into thermal energy and the other half ends up as radiative energy.

Because of the sun's substantial mass, a contraction rate of only 40 meters a year would liberate the required energy. Gravitational energy stored in the sun would last for about 50 million years, far longer than the earth's age as determined in the nineteenth century. But when later geologic investigations expanded the earth's age to billions of years, the sun's resources of gravitational energy no longer sufficed to account for the sun's shining.

Albert Einstein provided the key idea about the sun's energy at the beginning of this century. In the special theory of relativity, mass and energy are related by the equation

$$E = mc^2$$

where E is the energy (in joules) released in the conversion of mass m (in kilograms), and c is the speed of light (in m/sec). Because c^2 is a large number (about 9×10^{16}), a minute mass stores enormous energy. For example, the conversion of 1000 kg of matter into energy unleashes about 10^{13} kWh (9×10^{19} J), roughly equal to the total energy consumption of the United States in one year!

Nuclear transformations / How can matter be changed into energy? Two operations in nature unleash the energy frozen in the nucleus of an atom: *nuclear fission* and *nuclear fusion*. In the process of nuclear fission, a nucleus of a heavy element (such as uranium or plutonium) splits into two lighter nuclei. The mass of the remnants adds up to less than that of the original nucleus. The deficit in mass is released as energy. In nuclear fusion, the nuclei of lighter elements are fused together to create a heavier nucleus. However, the product has less mass than the original particles that have been put in. The missing mass has been converted to energy.

The sun produces energy by fusion. Hydrogen is the most abundant element and also has the smallest nuclear charge (one proton). The fusion of hydrogen nuclei results in the production of helium. To make a helium nucleus from hydrogen nuclei releases 4.2×10^{-12} J of energy. That's a minuscule amount; it would raise the temperature of one gram of water only 10^{-13} K. Many hydrogen nuclei must react each second to supply the sun's luminosity.

Two sets of fusion reactions are possible to transform hydrogen to helium: the *proton-proton chain* (*PP chain* for short) and the *carbon-nitrogen-oxygen cycle* (*CNO cycle*). The CNO cycle contributes a minor amount to the energy of the sun, but it acts as a key source in more massive stars (see Chapter 16).

Let's look at the steps in the PP chain. A collision between two protons starts it off (Fig 13.13); if these nuclei collide with enough energy (a temperature of at least 8 million K), the protons stick together. A heavy hydrogen nucleus (^2H) forms, consisting of a proton and a neutron. The other positive charge breaks away as a positron. (A *positron* is the antiparticle of the electron and carries a positive instead of a negative charge.) Almost lost in the shuffle is a neutral, supposedly massless particle called a *neutrino*, which zips away at light speed. The dense solar interior offers no barrier to the neutrino's escape, and, in about 2 sec, the neutrino breaks into space, carrying away some energy.

In the solar interior, the positron quickly collides with an electron, and the two antiparticles

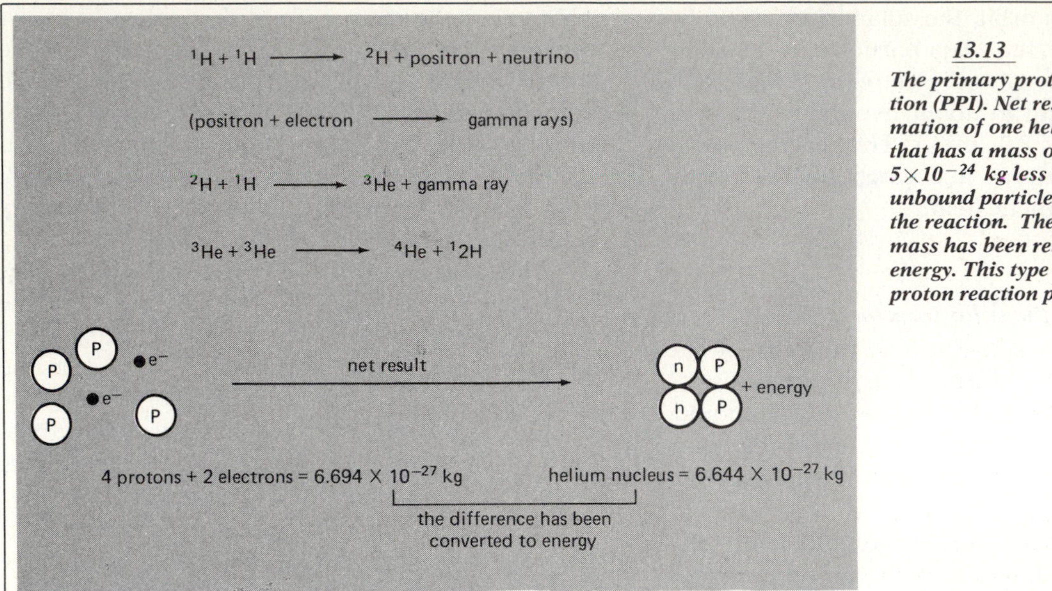

13.13
The primary proton-proton reaction (PPI). Net result is the formation of one helium nucleus that has a mass of about 5×10^{-24} kg less than the unbound particles that go into the reaction. The "missing" mass has been released as energy. This type of proton-proton reaction powers the sun.

annihilate to form two gamma-rays. Meanwhile, the heavy hydrogen crashes into another proton and forms light helium (^{3}He) and a gamma-ray. This chain occurs again. Another light helium nucleus is created; it collides with the one previously formed. In this final reaction of the PP chain, the usual result is ordinary helium (^{4}He) plus two protons and another gamma-ray.

Keep in mind that very high temperatures (at least 8 million K) and high densities are needed for protons to collide with enough energy to fuse and to collide frequently enough to generate as much energy as comes out of the sun. This requirement restricts the energy production to the sun's *core*, about the inner 25 percent of its radius. Only here does the temperature hit at least 8 million K.

In summary, the input for the PP chain is six protons and two electrons, and the usual output is one helium nucleus, two protons, two neutrinos, and miscellaneous gamma rays. Note that the net result of this nuclear cooking is the creation of helium and energy from four protons and two electrons. Each completed PP chain unlocks about 4.2×10^{-12} J. To account for the sun's luminosity, 1.4×10^{17} kg of matter must be converted to energy each year.

The neutrinos fly off into space with about 2 percent of the energy, but the gamma-rays, which carry off the bulk of the energy, find the sun's interior opaque (Fig. 13.14). The photons bounce along random paths as they are absorbed and reemitted. Slowly, the photons walk out toward the sun's surface, from regions of higher to lower temperature. As a consequence of this temperature decline, the average photon's energy declines. The original gamma-rays are degraded, by interaction with the sun's material, into lower-energy photons. In a few million years the photons break out of the photosphere in the form of less energetic, visible radiation.

The sun is a fusion furnace, forging helium from hydrogen in its core. Slowly, the core's helium abundance increases, and its hydrogen abundance decreases. The energy you see now was produced in the core many millions of years ago.

How long can the sun survive at this rate? The sun has enormous hydrogen supplies. However, only the hydrogen in and close to the core

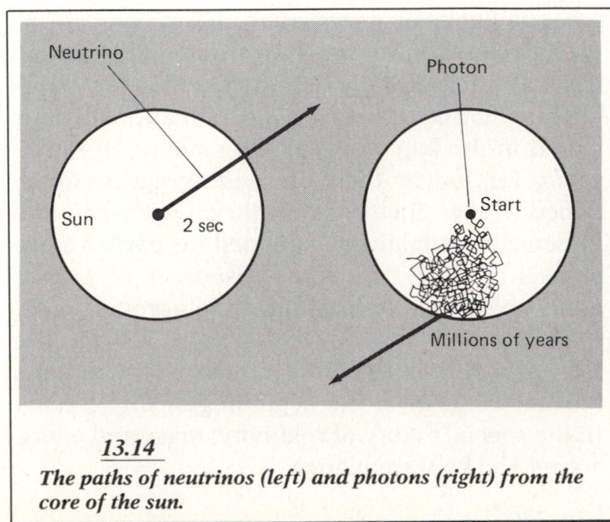

13.14
The paths of neutrinos (left) and photons (right) from the core of the sun.

can burn. This amounts to about 10 percent of the total. Also, the PP chain transforms only 0.7 percent of the mass into radiant energy. Even with these restrictions, the sun's fusion energy can last about 10 billion years. Fusion solves the energy problem for the sun.

Solar neutrino experiment / How to probe the sun's interior when we can't see below its surface? We have some idea of the interior conditions because astrophysics can make theoretical models of the sun. We do this by applying simple physical laws to a ball of hot gas, mostly hydrogen, at whose center proton-proton reactions take place. Electronic computers calculate those models, which match the observed characteristics of the sun. These models show, for example, how the sun's temperature increases from edge to center. Or how much the chemical composition in the core differs from that of the surface. Why should it differ? Because in the core, the hydrogen is fused to helium; at the surface, no such conversion takes place.

These models of the sun's interior have been constructed as carefully as possible. However, an ongoing experiment casts some doubt on them. Remember the neutrinos produced in the PP chain? These fly directly out of the sun's core and in about 8.3 min reach the earth. So if we could detect these neutrinos, we could see into the sun's interior.

13.15
The solar neutrino telescope. (Courtesy R. Davis and Brookhaven National Laboratory)

Raymond Davis and his colleagues have developed a strange telescope to catch the solar neutrinos (Fig 13.15). This telescope consists of about 378,000 liters (L) of a chlorine compound placed in a huge tank located about 1.6 km below the earth's surface in Lead, South Dakota. How does this telescope work? A chlorine atom can absorb a neutrino and convert to argon. By a very delicate procedure, the argon gas can be flushed out of the tank and the amount measured, so the number of neutrinos captured during a given time span can be known.

Nuclear theory and models of the solar interior predict how many argon atoms should be produced from the incorporation of neutrinos each day. We can compare the prediction with the actual experimental results.

The results? To date the detected solar neutrinos amount to only *one-third* the number predicted from standard models of the sun's interior and our current knowledge of the nuclear reactions involved! To put the result another way, the experimental result implies that PP reactions take place now at one-third the rate needed to account for the sun's present luminosity.

What's wrong? First, the experimental equipment may contain unknown problems. The chlorine capture experiment is the only one to have run for a long time. Its uniqueness demands results from another, independent experiment. Second, if the results are accurate, they are telling us that the standard solar model is incorrect or that something happens to the neutrinos on the way to the earth; that is, the experiment provides new information on the properties of neutrinos. Perhaps neutrinos have mass. Or perhaps solar models made with much lower abundances of heavy elements produce far fewer neutrinos—but they contradict our general picture of the manufacture of elements in stars. Solar models can be rigged to agree with the experiment, but then they contradict other astrophysical data or the earth's climatic history.

Among all proposals, no really satisfactory explanation exists. When the solution is found, we will know a little bit more than we do now about the sun's interior, and also about the interior of other stars.

13.7 *The active sun*

In contrast to the easygoing aspects of the quiet sun, events of the active sun are localized, short-lived phenomena on or near the solar surface.

The most important of these are sunspots, prominences, and flares. They are all associated with the locally intense solar magnetic field. In general, these areas of solar activity are called *active regions*.

One other solar property contributes to active-sun phenomena: the sun's nonuniform rotation. Because the sun is a gas, it spins fastest at the equator (one rotation every 25 days) than at the poles (one rotation every 31 days). So as you travel from a pole toward the equator on the sun, the photosphere's velocity increases; this effect is called *differential rotation*. It can distort solar magnetic-field lines and drive the development of active regions.

Magnetic fields play such a prominent role in solar activity that you can think of the sun as a mildly varying magnetic star.

Sunspots / If you view the sun through even a small, properly filtered telescope, you can see sunspots as dark blotches on the solar disk (Fig. 13.16). But they are not completely black. With a temperature of about 4200 K, a sunspot is relatively cooler than the photosphere and so appears dark in contrast. In fact, a sunspot is almost four times fainter than the photosphere. But if the sun could be whisked away, leaving its sunspots behind, you would see them shine with a bright red glow against the dark of space.

Sunspots have a strong tendency to form in groups. They are born in an active region where photospheric granules separate, and a tiny spot appears between them as a dark pore. Such pores have magnetic-field strengths of roughly 2500 gauss (G). Usually more soon become visible and coalesce over a period of several hours to form a sunspot. A large, single sunspot group may contain 100 individual spots and persist for two or more solar rotations.

Sunspot cycle / In 1843 Heinrich Schwabe, a German amateur astronomer, noted that the sunspot numbers vary periodically. About 11 years pass between sunspot *maxima* (times of greatest numbers of sunspots) or sunspot *minima* (times of least numbers of sunspots). Later the British astronomer Walter Maunder (1851–1928) discovered not only that sunspot numbers vary during a cycle but also that the sunspots' positions change. Sunspots usually appear only in the zone between the solar equator and 35° north or south latitude. At the start of a sunspot cycle, a few spots emerge at high latitudes. As the cycle progresses, the sunspot zone migrates toward the equator. As the survivors of one cycle expire near the equator (about 5° north or south latitude), new spots from the next cycle form at the higher latitudes.

In 1908 George Ellery Hale (1868–1938) detected intense magnetic fields associated with sunspots. The strongest magnetic sunspots have field strengths over 4000 G. (That's about 8000 times stronger than the earth's field at its surface.) Hale also noticed that sunspot groups contain spots of opposite magnetic polarity. For example, the east spot of a twin group might have north polarity, the west spot south. If this arrangement holds at a given time in the northern hemisphere of the sun, the situation at the same time in the southern hemisphere is reversed (Fig. 13.17). Later Hale discovered that the magnetic polarities for the west and east spots have a 22-year cycle, just double the 11-year cycle of sunspot numbers.

Recent investigations, especially by solar physicist John Eddy, indicate that little historical evidence exists to show an 11-year cycle in sunspot activity before 1700. Sunspots were discovered by Galileo with his new telescope in about 1613, but, as first noted by Maunder, hardly

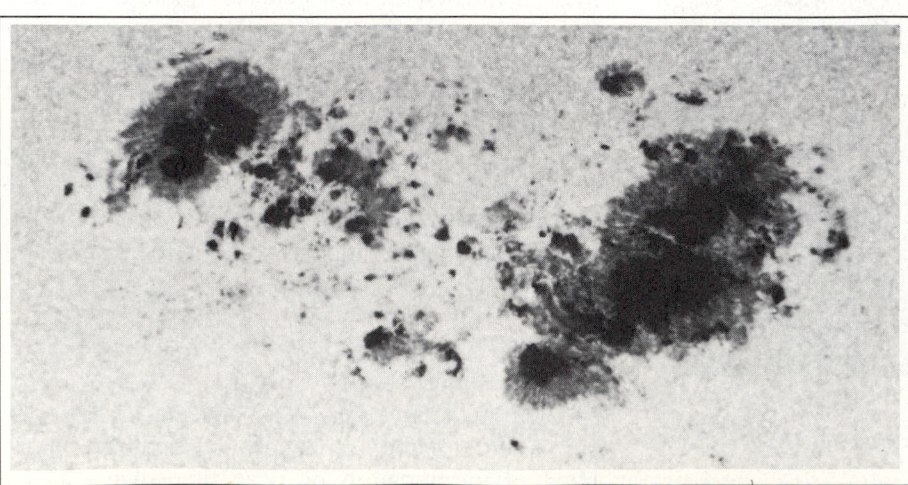

13.16

A large sunspot group, containing hundreds of sunspots. (Courtesy Mt. Wilson Observatory, Carnegie Institution of Washington)

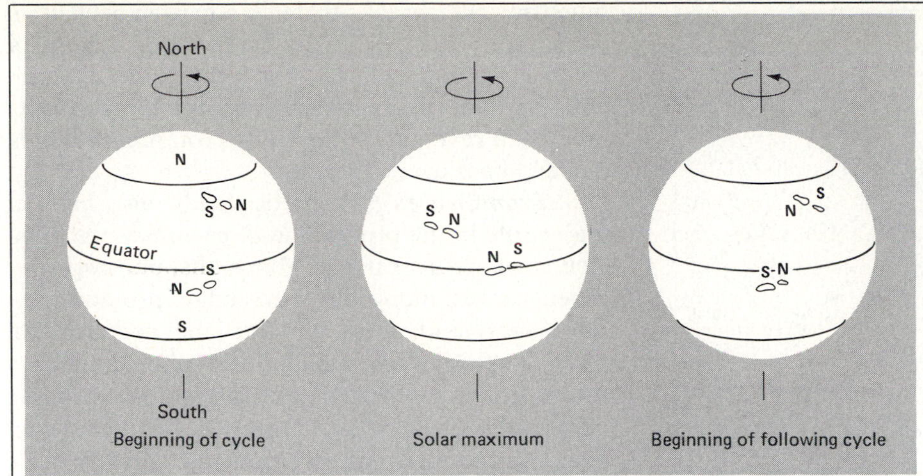

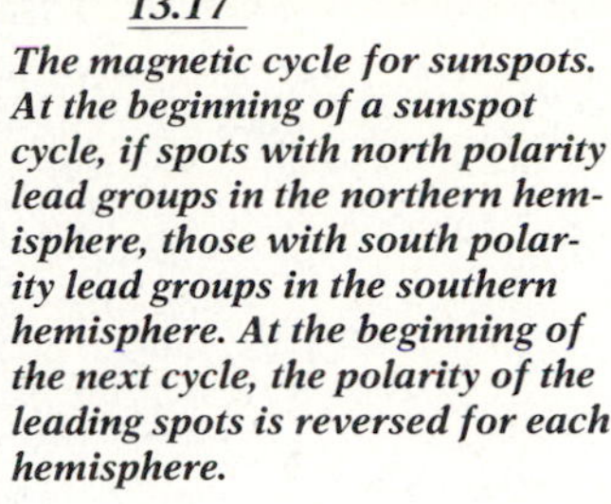
13.17

The magnetic cycle for sunspots. At the beginning of a sunspot cycle, if spots with north polarity lead groups in the northern hemisphere, those with south polarity lead groups in the southern hemisphere. At the beginning of the next cycle, the polarity of the leading spots is reversed for each hemisphere.

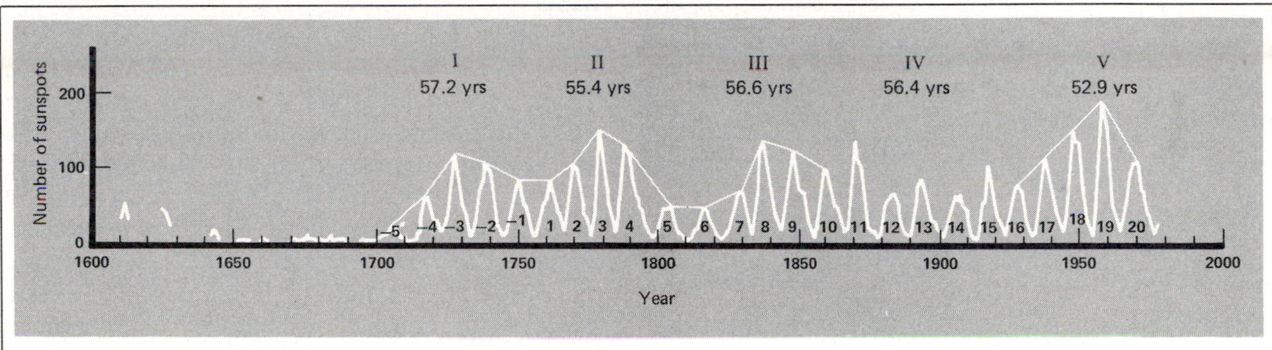

13.18

Sunspot cycles from 1600 to the 1970s. Plotted here are the number of observed sunspots during a year. The peaks (sunspot maxima) come roughly 11 years apart. Note how few sunspots were seen prior to 1700, compared to modern times. (Adapted from a diagram by H. Yosihmura, **Astrophysical Journal,** *vol. 227 (1979), p. 1047, copyright © 1979 by the American Astronomical Society)*

any sunspots were seen in the 60-year period from 1645 to 1705 (Fig. 13.18). The relative consistency of the cycle in modern times may be a phase of changes that take place over longer times.

Physical nature of sunspots / Sunspot magnetic fields are generated by enormous electric currents, much as the field of an electromagnet is produced. The hurricane of currents and fields in turn creates sunspots. Exactly how this formation happens is unclear. One possibility is that the sunspots' magnetic fields suppress the hot gas rising from the convective zone. The hot gas runs into a magnetic thicket and has trouble breaking through the surface in the sunspot's center (Fig. 13.19). A convective downflow may draw away some of the hot gas, removing heat from the region. As a result, the sunspot is cooler and darker than the surrounding photosphere.

Here is one possible model for the origin of sunspots. Because magnetic-field lines are dragged by the motions of the ionized solar gases,

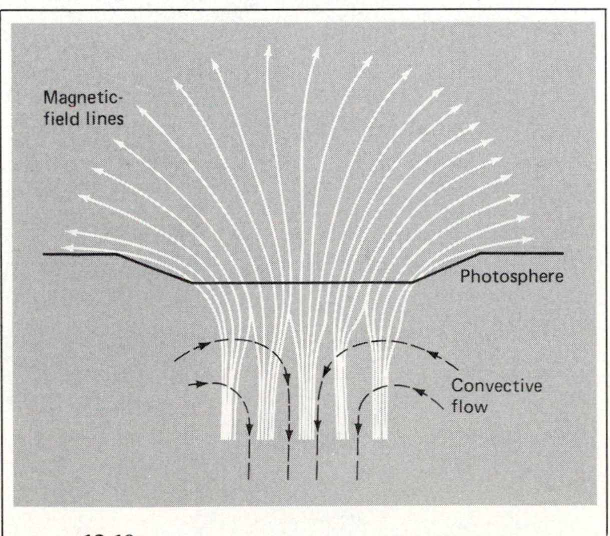

13.19

One model of the distribution of magnetic field lines in a sunspot. (Adapted from a figure by E. N. Parker, **Astrophysical Journal,** *vol. 230 (1979), p. 905, copyright © 1979 by the American Astronomical Society)*

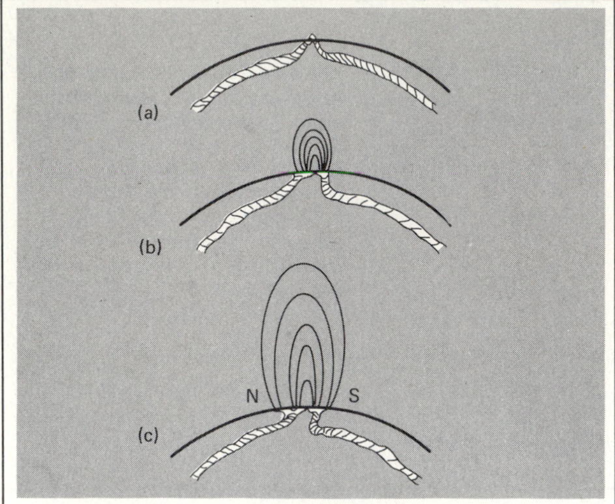

13.20

***A possible model for sunspots. The sun's nonuniform rotation may wind the magnetic field into tight ropes* (a). *Kinks develop in these regions and burst through the surface* (b). *Sunspots develop in pairs of opposite polarity* (c).**

the nonuniform rotation winds up the lines of force (Fig. 13.20a). Turbulence twists the trapped field lines into a ropelike structure (Fig. 13.20b). As the tubes are squeezed together by turbulence, kinks appear. A critical point is eventually reached when the kinked parts of magnetic field lines burst through the solar surface (Fig. 13.20c), much as water breaks through a garden hose at its weak spot. One side of the broken kink exhibits a north magnetic polarity, the other side a south polarity. The magnetic-tube model for sunspots predicts a reversal of polarity as rotation unwinds the field lines every 11 years.

Prominences / Magnetic fields also play a major role in the production of *prominences*, huge clouds of hydrogen gas above the photosphere. When viewed along the sun's edge, prominences often loop and surge up into the corona (Fig. 13.21). When seen against the photosphere, a prominence resembles a dark snake winding across the solar disk. Because they are cool clouds of coronal gas, prominences absorb some of the photospheric light and appear dark in contrast.

Prominences are almost always found near a sunspot group in an active region. Frequently, a young prominence may disappear, only to reappear a few days later relatively unchanged. This observation suggests that the basic underlying structure of the prominence is the magnetic field, made visible when gas is at the right temperature and density to radiate visible light.

Flares / Sunspots are floating islands of electromagnetic storms. In analogy to terrestrial thunderstorms, these solar storms generate short-lived, violent discharges of energy called *solar flares*. These energetic bursts appear with sun-

13.21

A prominence rising over 150,000 km above the photosphere. (Courtesy Mt. Wilson Observatory, Carnegie Institution of Washington)

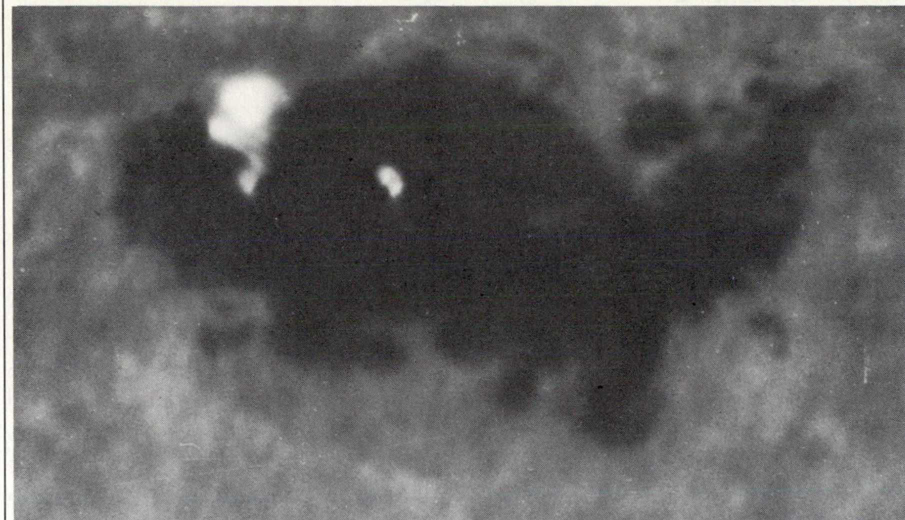

13.22

A small optical flare near a sunspot group. (Courtesy D. Neidig and Sacramento Peak Observatory, Association of Universities for Research in Astronomy, Inc.)

spots and sometimes bridge the gap between two close spots. Near large sunspots, about 100 small flares occur each day (Fig. 13.22). The elapsed time between the birth of a flare and its rise to intensity is only a few minute, even for a large flare, and the decay time is about an hour. Emitting myriad forms of energy—X-rays, ultraviolet and visible radiation, high-speed protons, and electrons—a large flare blows off about 10^{25} J, the equivalent of the energy released by a bomb of 2 billion megatons (Fig. 13.23)!

Because large active regions are the most

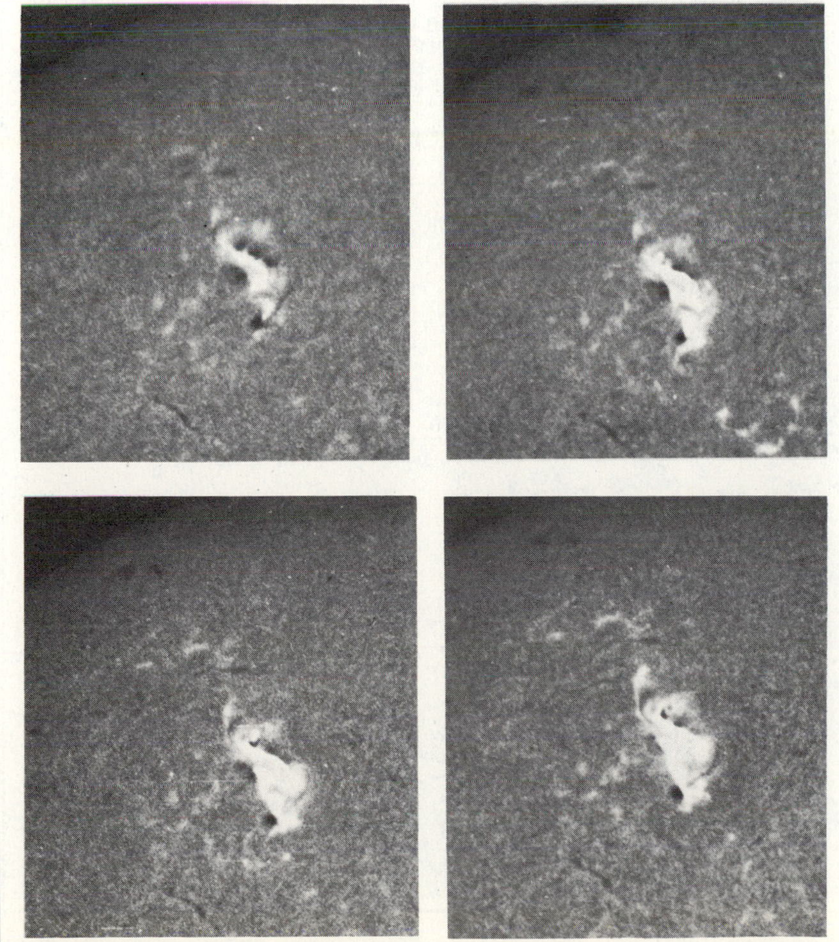

13.23

Outburst of a small solar flare (white region) taken in H-alpha light. The time span was about 45 min. (Courtesy A. Maxwell and Harvard College Observatory)

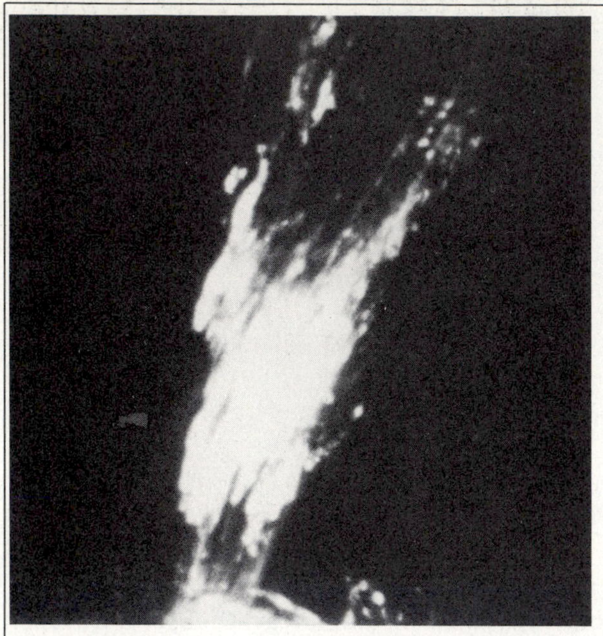

13.24
Spray of material from a solar flare, taken in H-alpha light. (Courtesy M. McCabe, Haleakala Observatory, University of Hawaii)

frequent locations for large solar flares, astronomers think that the concentration of energy arises from local twists and kinks in the magnetic-field loops. The energy accumulates until its release is triggered; how is not known yet. The flare starts in the corona and strikes down into the chromosphere in the form of high-speed electrons.

Flares blast energetic particles, mostly protons and electrons, into space (Fig. 13.24). A majority of the flare's energy, however, escapes as shortwave (X-ray and ultraviolet) radiation. Arriving at the earth about 8.3 minutes later, the flare's radiation rips through the upper atmosphere and tears electrons from neutral atoms. The increase in ionization disrupts shortwave long-distance radio communication. A few days later, the lagging protons and electrons approach the earth, but they are usually trapped by the earth's magnetic field.

Occasionally the earth's magnetic reservoirs overflow with charged particles, particularly at times of sunspot maxima when flare activity is at its peak. As the particles spill into the earth's upper atmosphere, the swift electrons bump into

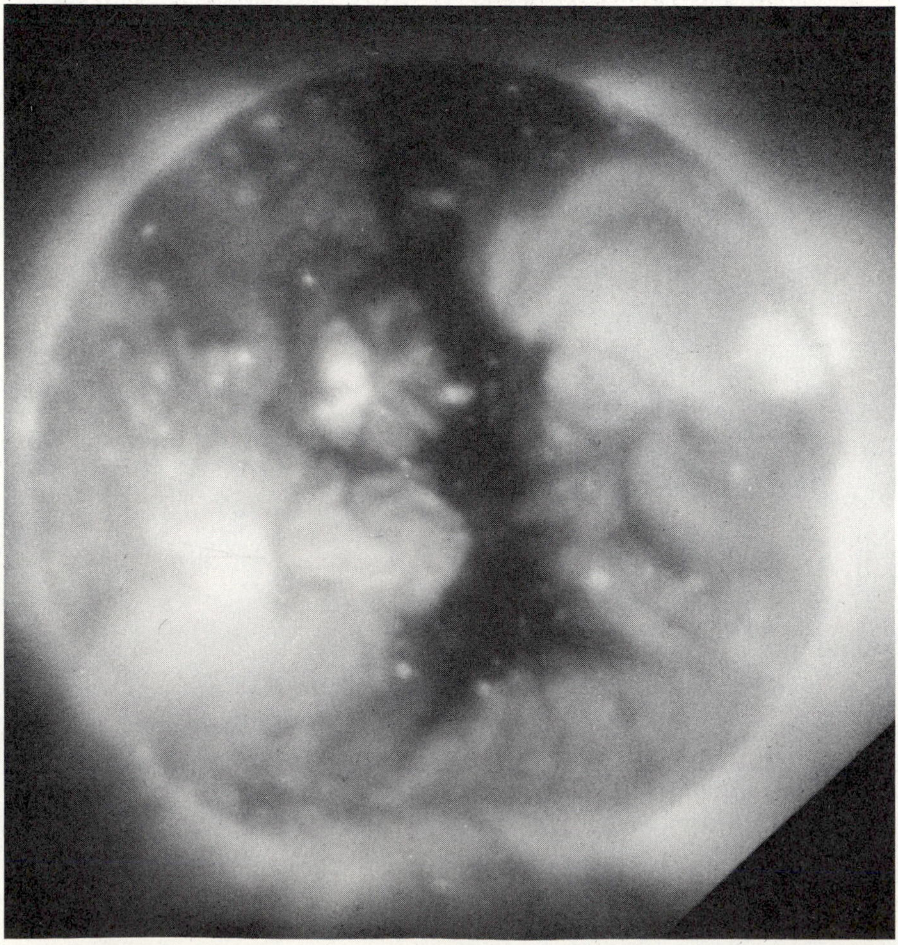

13.25
The sun seen in X-ray light. Very hot regions (1 million K or so) appear bright in this picture, so most of these regions lie in the sun's corona. The looped appearance of streamers in the corona arises from strong magnetic fields. The dark region running down the middle of the disk is a coronal hole. (Courtesy G. Vaiana and Harvard College Observatory)

atmospheric atoms and excite them. When the atoms deexcite, they emit visible radiation, causing a faint glow in the sky, often changing rapidly atmospheric atoms and excite them. When the atoms deexcite, they emit visible radiation, causing a faint glow in the sky, often changing rapidly in shape and color; this is called the *aurora*.

Coronal loops and holes / Because the coronal gas is so hot, it emits low-energy X-rays and shows up in X-ray photos of the sun (Fig. 13.25). These pictures show that the coronal gas has an irregular distribution above and around the sun. The large loop structures indicate where the ionized gas flows along magnetic fields that arch high above the sun's surface and return to it. The hot gas is trapped in these magnetic loops (see Focus 8.2). Solar physicists now view the corona as consisting primarily of such loops.

Note also that some regions of the corona appear dark, especially at the top pole and down the middle part of the sun. Here the coronal gas must be much less dense and less hot than usual; these regions are called *coronal holes*. The coronal holes at the poles do not appear to change very much, but those above other regions seem somehow related to solar activity.

What makes a coronal hole? Solar astronomers believe that coronal holes mark areas where fields from the sun continue outward into space rather than flow back to the sun in loops. So the coronal gas, not tied down in these regions, can flow away from the sun out of the coronal holes; this flow makes the solar wind (Fig. 13.26).

The coronal gas does *not* follow the differential rotation of the photosphere. Rather, it rotates at the same angular speed at all latitudes

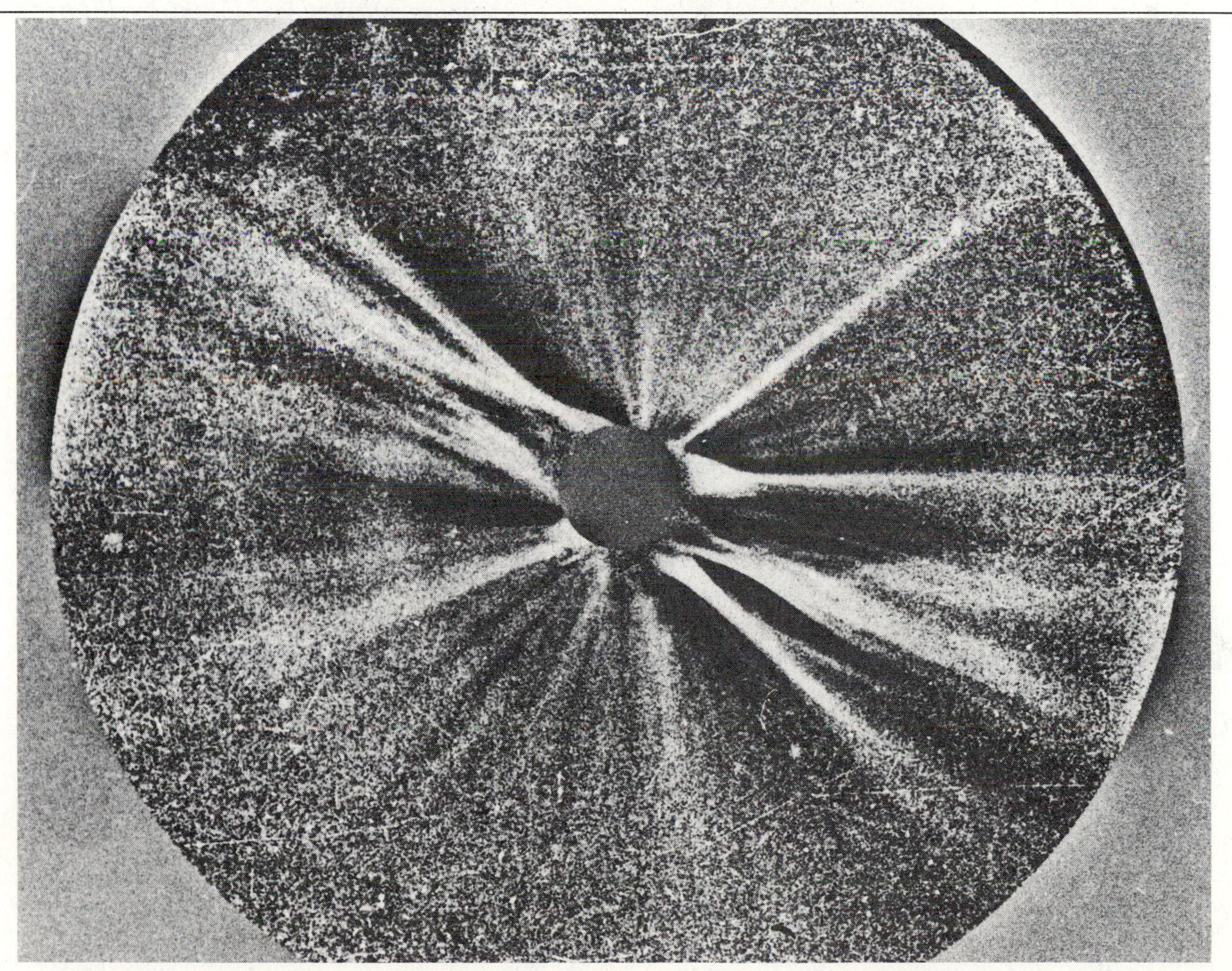

13.26

Streamers in the sun's corona (June 1973). The streamers are visible out to a distance of about 12 solar radii. They show the magnetic field configuration and solar wind flow out beyond the region near the photosphere. (Courtesy C. Keller and the Los Alamos National Laboratory)

(as the earth does). This fact implies that the bottoms of the magnetic loops are anchored deep below the photosphere, perhaps at the very bottom of the convective zone, where such fields might be generated by a solar dynamo.

Now back to why the corona (and chromosphere) is so hot. The magnetic loops, rising from the photosphere up to 400,000 km into the corona, play the key role, although we don't yet know how in detail. But here is a model: The magnetic field in a loop is twisted by photospheric motions at its base. If the twisting happens slowly, it generates electric fields that then heat the coronal gas. It doesn't take much energy to do so, because the coronal gas is so thin that it holds little heat.

KEY CONCEPTS

1 ***The astronomical unit (AU), the average earth-sun distance, is about 150 million km; it's found by bouncing radar signals off of Venus.***

2 ***With the distance known, the sun's size is found from its angular size, its mass from the earth's orbital period and Newton's law of gravitation, and its low density from its volume and mass.***

3 ***To find the sun's luminosity (total radiative energy output per second) requires a knowledge of the earth-sun distance in kilometers and a measurement of the sun's flux at the earth.***

4 ***The sun's surface temperature can be found from its color or the wavelength at which its continuous spectrum hits a peak intensity.***

5 ***The sun's surface—its photosphere—is where the gases become transparent to light; the continuous spectrum forms in this region; the absorption lines form in higher, cooler layers.***

6 ***The sun's absorption lines indicate what elements are in the photosphere; the sun contains by mass about 74 percent hydrogen, 25 percent helium, and 1 percent all other elements (generally called by astronomers*** heavy metals***).***

7 ***The photosphere has a bubbly structure, indicating that a zone of convection exists beneath it; this convection carries energy to the surface to be radiated into space; the chromosphere above the photosphere is hotter and visible by its emission lines during an eclipse; the corona is even hotter than the chromosphere, as indicated by emission lines of highly ionized atoms.***

8 ***The solar wind is the sun's corona evaporating into space.***

9 ***The sun's energy comes from fusion reactions; in them, four protons are fused into a helium nucleus with the release of energy; only in the sun's core is the temperature high enough for these fusion reactions to occur; the sun has enough hydrogen to fuel these reactions for billions of years; gradually, the sun's core increases in its percentage of helium as its hydrogen decreases.***

10 ***The solar neutrino telescope has found far fewer neutrinos than predicted by models of the sun and nuclear reactions.***

11 ***The sun undergoes cycles of activity in which active regions—where sunspots, flares, and prominences tend to be found—become more common and then less so; this cycle lasts 22 years, as indicated by the polarity of sunspots.***

12 ***Active regions relate to the development of strong (thousands of gauss) local magnetic fields.***

13 ***Solar flares emit electromagnetic radiation (especially X-ray, ultraviolet, and radio) and high-speed particles, which cause auroras when they reach the earth.***

14 ***The solar wind leaves the sun through coronal holes, cool regions in the corona where the sun's magnetic field streams into space.***

15 ***The sun (Table 13.1) serves as the model on which we base our understanding of other stars.***

TABLE 13.1 Properties of the sun

Property	*Value*
Distance from earth	1 AU = 1.496×10^{11} m
Radius	6.966×10^{8} m
Mass	1.991×10^{30} kg
Luminosity	3.86×10^{26} W
Photosphere's temperature	5780 K
Bulk density	1410 kg/m^3
Age	4.5–5.0 billion years
Photosphere's composition	74% hydrogen, 25% helium, 1% other elements

STUDY EXERCISES

1 For what reason do astronomers bounce radar signals off of *Venus* to find the distance to the *sun*? (*Objective 1*)

2 What *measurements* must be made to find out the sun's luminosity? Which of these is the *least* accurate? (*Objective 2*)

3 Suppose you examined sunlight with a spectroscope. Describe, in general, what you would see. (*Objective 5*)

4 Explain the appearance of the sun's spectrum by simple atomic processes. (*Objective 5*)

5 For what reason do you *not* see helium absorption lines in the sun's visible spectrum, even though helium is the sun's second most abundant element? (*Objective 5*)

6 Do you expect the chemical composition of the sun's core to differ from that of its photosphere? Why or why not? (*Objectives 6 and 8*)

7 Describe how to estimate the sun's surface temperature from its continuous spectrum. (*Objective 2*)

8 Use the concept of opacity to explain why it takes photons millions of years to walk out of the sun's core to the surface. (*Objective 4*)

9 In one short sentence, describe the source of the sun's energy. (*Objectives 7 and 8*)

BEYOND THIS BOOK . . .

E. N. Parker paints a contemporary picture of the nearest star in "The Sun," *Scientific American*, September 1975.

Some aspects of the active sun are treated in part 5 of *The New Astronomy and Space Science Reader* (Freeman, San Francisco, 1977), edited by J. C. Brandt and S. P. Maran.

For more details on the sun's outer atmosphere, read "The Solar Corona" by J. Pasachoff, *Scientific American*, October 1973, p. 68 and "The Active Solar Corona" by R. Wolfson, February 1983, p. 104.

Up-to-date books are *The Sun, Our Star* (Harvard University Press, Cambridge, Mass., 1982) by R. Noyes and *Our Turbulent Sun* (Prentice-Hall, Englewood Cliffs, N.J., 1982) by K. Frazier.

A nice review of the sun's activity cycle is in "The Solar Cycle" by C. Newkirk, Jr. and K. Frazier in *Physics Today*, April 1982, p. 25.

14 / The stars as suns

I have made the stars!
I have made the stars!
Above the earth I threw them.
All things above I made
And placed them to illumine.

A Pima Indian creation song by Earth Doctor

Learning objectives

After studying this chapter, you should be able to:

1. **Outline the methods astronomers use to find the following physical properties of stars: (a) surface temperature, (b) chemical composition, (c) size (radius or diameter), (d) mass, (e) luminosity, and (f) density.**
2. **Describe the relationship between a star's color and its surface temperature.**
3. **Describe the relationship of a star's luminosity, surface (effective) temperature, and size, assuming it radiates like a blackbody.**
4. **Explain the difference between flux and luminosity and relate flux, luminosity, and distance.**
5. **Describe what is meant by the *inverse-square law for light*.**
6. **Show by a simple diagram the relationship between a star's distance and its parallax.**
7. **Sketch a Hertzsprung-Russell diagram for stars, indicating the positions of the sun, main sequence, giants, supergiants, and white dwarfs.**
8. **Use the Hertzsprung-Russell diagram to infer the relative luminosities, surface temperatures, and sizes of stars represented on it.**
9. **Outline the steps by which astronomers determine the masses of stars.**
10. **Sketch and make use of the mass-luminosity relation for main-sequence stars; from it, argue that stars have finite lifetimes.**
11. **Give two observational indications that stars have activity cycles like the sun's.**

Central question

How do astronomers determine the physical properties of stars?

Many times I have peered intently at the stars set in the deep velvet of the sky. As an astronomer, I know what the stars are: other suns, enormous fusion reactors in space. That knowledge does not make the view any less fascinating to me. In fact, my image of space becomes even richer as I contemplate the stars as suns.

That view is recent, developed in full in this century. For many years, astronomers really didn't pay much attention to the stars. And the vast distances of stars from the sun reinforced their aloofness from human understanding.

With the development of spectroscopes and atomic theory (Chapter 5), we finally acquired the crucial tools to analyze starlight and compare the stars to the sun. By finding distances to nearby stars, we were able to infer basic physical properties that we then applied to more distant stars. These revelations reinforced the connection between the sun and the stars, *the* essential link to the rest of the cosmos. The stars are suns; the sun is a star.

14.1 Some messages of starlight

Go out on a clear January night to view the constellations. (See the constellation chart for January in this book's endpapers.) Face south. Orion (Fig. 14.1) immediately catches your eye. Two stars in Orion shine the brightest: Betelgeuse,

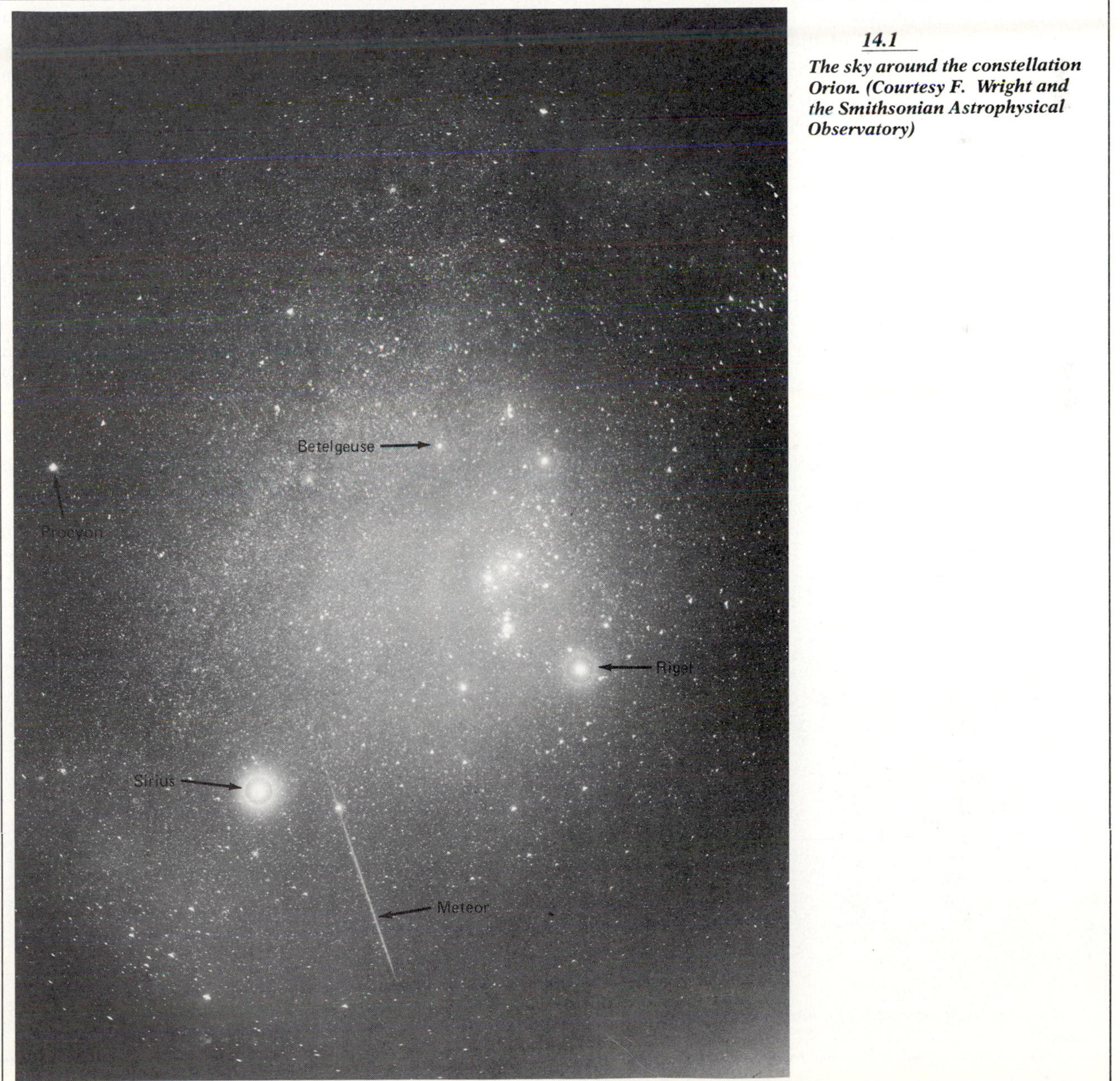

14.1 *The sky around the constellation Orion. (Courtesy F. Wright and the Smithsonian Astrophysical Observatory)*

which looks reddish, and Rigel, blazing bluish white. To the south and east of Orion lies Canis Major, the Great Dog, Orion's hunting companion. You can easily notice Sirius, the jewel of Canis Major, as the brightest star in the sky.

You are now faced with the astronomers' dilemma: What can we know about these distant stars? How do these stars compare with our sun? Are they larger? More luminous? Hotter? How to find out these physical properties? In what ways do these stars resemble the sun? How do they differ?

Starlight carries information about the physical properties of stars. By deciphering the messages of starlight and comparing it to those of the sun, we can infer the nature of stars. These physical properties include (1) chemical composition, (2) surface temperature, (3) radius, (4) luminosity, and (5) mass. For some of these properties, we do not need to know the distance to the stars; for others, we do. In all cases, we must observe, measure, and analyze the light from the stars as was done for the sun.

Brightness and flux / If you have even casually looked at the night sky, you know that stars do not appear to have the same brightness. You make this judgment with your eyes, which work essentially as a refracting telescope (see Section 6.2): The lens of your eye focuses the light onto your retina, which detects the light by transforming the radiative energy into electrochemical impulses that travel to your brain. How bright a star appears to you depends on how much energy, each second, strikes your retina. The more energy received, the greater brightness you perceive.

A telescope works the same way as your eye (Fig. 14.2). Its objective gathers light to a focus, and a detector there senses the radiative energy. How bright a star appears in a telescope depends on how much energy, each second, arrives at the earth from the star and how large the objective of the telescope is. Just to determine the brightness of the star, then, we need to get rid of the dependence on aperture size. We do this trick by considering only the energy crossing each square meter of the telescope's objective.

Flux is the name given to this quantity: the amount of energy striking each unit of area in a second. Since watts (see Appendix A) are a measure of energy per second, flux can have units such as watts per square meter. For example, the flux at the earth from the sun is 1370 watts per square meter. Compared to the sun, even the brightest star, Sirius, sends a small flux to the earth—a mere 10^{-7} watt per square meter!

Note that brightness actually means flux. *Brightness* is a more common word but lacks the specific physical meaning of *flux*. (Astronomers have a peculiar way of measuring flux; see Focus 14.1.)

Flux and luminosity / Remember how to work out the sun's luminosity (Section 13.1)? Measure the sun's flux, find the earth-sun distance in kilometers, construct an imaginary sphere with a 1 AU radius, and total up all the energy hitting

14.2

A comparison of the optics of a human eye and of a telescope.

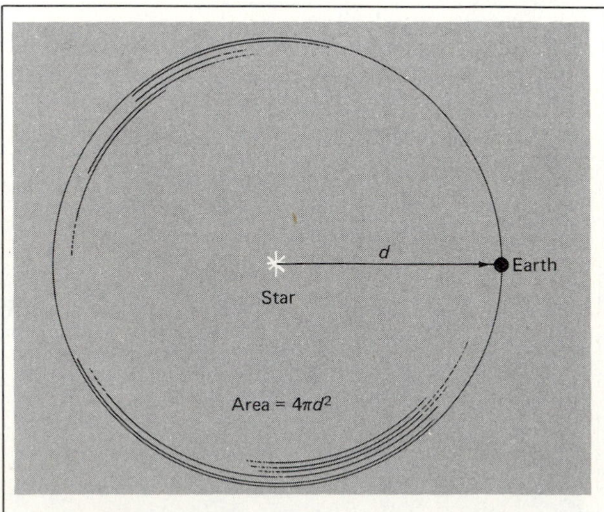

14.3

Determining the luminosity of a star from its distance and the flux at the earth.

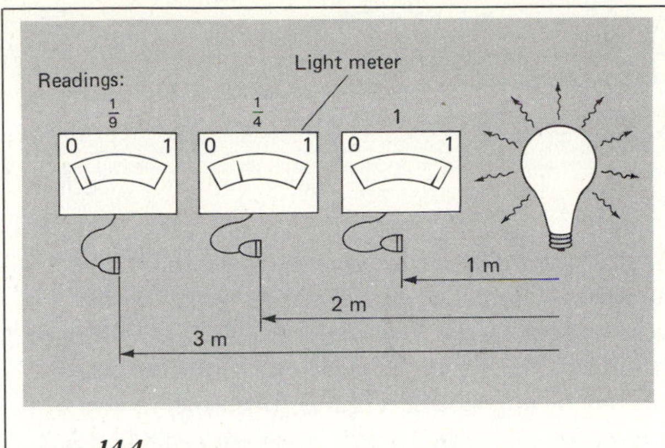

14.4
Light intensity and its change with distance measured with a light meter.

that sphere. We can find a star's luminosity by the same procedure (Fig. 14.3). Measure the star's flux at the earth. Then, knowing its distance, find the area of a sphere surrounding the star, and add up the flux over the total area.

Simple enough—but we need to know the distance to the star, and that may not be easy to come by (see Section 14.2). Why worry about a star's luminosity? Because it tells you how much radiative energy the star emits each second, and that's fundamental to understanding stars as suns.

Warning: To get a star's total luminosity requires that we measure its flux over the *complete* range of the electromagnetic spectrum. That might not be possible. First, our detector may work over only a limited range. Second, the earth's atmosphere may absorb some of the energy. Third, matter in interstellar space may also do some absorbing. If the flux is measured over a limited range of the spectrum, say the visual, astronomers report this fact by calling it the *visual flux*. Then the luminosity calculated from it is only the *visual luminosity* of the star. Fortunately, most stars emit most of their light in the visual part of the spectrum, so their visual luminosity is almost their total luminosity. Exceptions are very hot and very cool stars.

The inverse-square law for light / You may have noted that the distance plays a key role in relating flux to luminosity. This occurs because the brightness of a light source relates in a very specific way to your distance from it.

Imagine the following experiment (which you can do if you have the equipment). Put a bare light bulb (any wattage) in a socket and turn it on. Take a light meter—a device used to measure the intensity of light, commonly used with automatic cameras—and place it 1 m from the bulb (Fig. 14.4). Note the reading on the light meter, and call that one unit. Now move the light meter 2 m from the bulb; its reading will be one-fourth that at the 1-m position. Move the light meter to a distance of 3 m; the reading will now be one-ninth that at 1 m. Note that the light intensity decreases as the *inverse square* of the distance. (What would the reading be at 4 m? Right—1/16 that at 1 m).

Why does this inverse-square relation happen? Imagine a flashbulb placed in the center of two transparent spherical shells, one having twice the radius of the second (Fig. 14.5). Fire the flashbulb so that it emits a pulse of light. As the light moves away from the bulb, it expands in a sphere in all directions. But the total amount of energy in

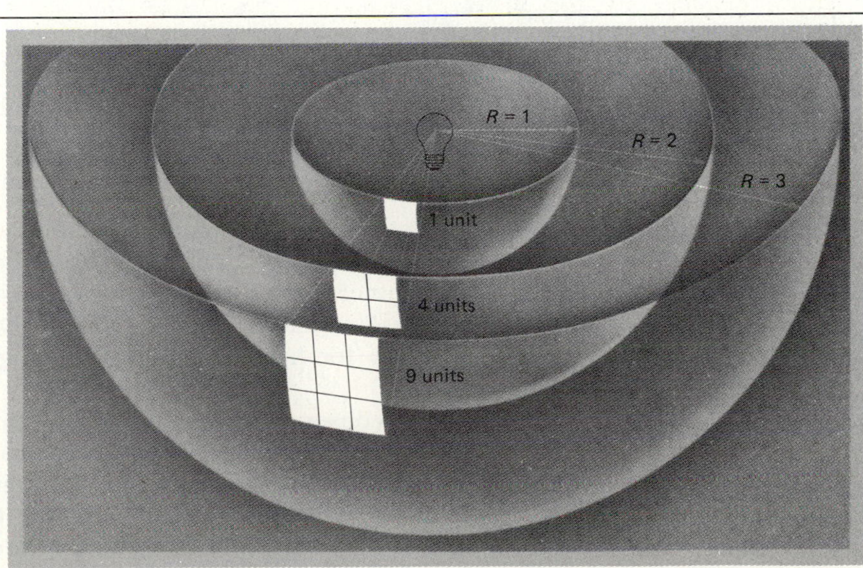

14.5
The geometry of the inverse-square law for light intensity. (Adapted from a figure by J. Pasachoff,* Contemporary Astronomy, *Saunders, Philadelphia, 1981)

Focus 14.1

FLUX AND MAGNITUDE

For reasons of history and convention, astronomers generally use a quirky way to talk about stellar brightness. It's called **apparent magnitude.** *Flux and apparent magnitude are two different methods used to describe the same property. I will* **not** *use magnitudes in this book because I think they are confusing to the novice, and they are not necessary to a basic understanding of the material. However, you may come across the concept elsewhere, so I'll do a little with it here.*

The magnitude scale on which stars are rated has evolved from a convention first established by Hipparchus (160–127 B.C.) and now become traditional. In his catalog of stars, he classified their apparent magnitude by rating the brightest star he could see as magnitude 1 and the faintest as magnitude 6. As this system evolved, some stars were found to be brighter than magnitude 1; for example, Vega is magnitude 0, and Sirius is magnitude −1.4. The first peculiarity to note about the magnitude scale is that the **larger** *the magnitude of a star (the more positive), the* **fainter** *it is, but the smaller the magnitude (the more negative), the brighter it is. That is, a star of magnitude 6.5 is fainter than one of magnitude 4.2, and one of magnitude −1.3 is fainter than one of magnitude −2.1; but a star of magnitude 12.5 is brighter than one of magnitude 17.9, and one of magnitude −1.8 is brighter than one of magnitude −0.9. Confusing? Perhaps it helps to think of magnitude as measuring the amount of* **faintness;** *larger numbers mean fainter stars (Fig. F.28).*

On Hipparchus's scale, stars of first magnitude were about 100 times brighter than stars of sixth magnitude. The modern system therefore **defines** *a difference of five magnitudes as corresponding to a brightness ratio of 100. A difference of one magnitude then amounts to a brightness ratio of 2.512. This strange number pops up because $2.512 \times 2.512 \times 2.512 \times 2.512 \times 2.512 = 2.512^5 = 100$, another way of stating that a difference of five magnitudes equals a ratio of 100 in brightness. Table F.1 will help you keep straight the magnitude differences and brightness*

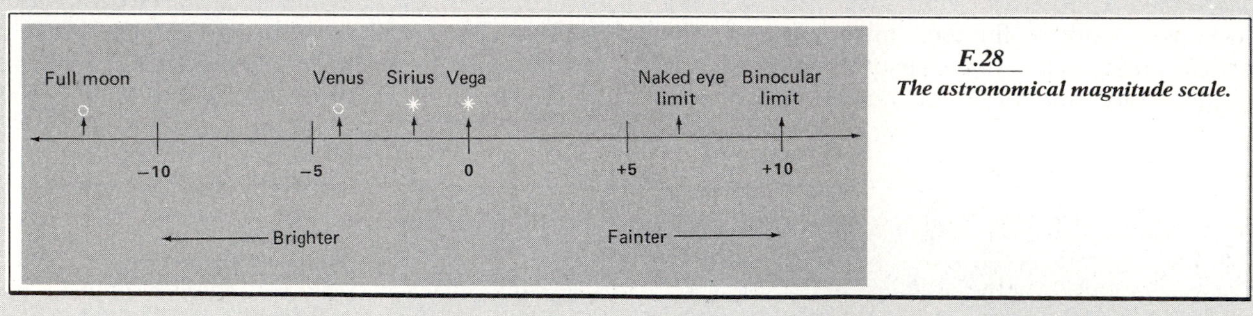

F.28
The astronomical magnitude scale.

the pulse remains the same, no matter how large the sphere. As the light passes through the first sphere, it covers a certain area. As it goes through the second, it covers a larger area. How much larger? The area of a sphere is directly related to the radius squared; so the larger sphere has four times the area of the smaller one, and the light spreads out four times as much.

Now pick any square meter of surface for both spheres. Because of the radiation's dilution over a larger area, the small patch you select on the larger sphere has only one-fourth as much light striking it as the patch on the smaller sphere. If the ratio of radii were increased to 3, the decrease in brightness would be by 9; if increased to 4, the decrease would be by 16. This is the *inverse-square law* for light (or any electromagnetic radiation).

Suppose you observe that the flux of one star is 100 times that of another. If you assume that both stars have the same luminosity, how do their distances compare? The brighter one must be 10 times as close as the fainter one, according to the inverse-square law.

TABLE F.1 *Conversion of magnitude differences to brightness ratios*

A magnitude difference of:	*Equals a brightness ratio of:*
0.0	1.0
0.2	1.2
1.0	2.5
1.5	4.0
2.0	6.3
2.5	10.0
3.0	15.6
4.0	40.0
5.0	100.0
7.5	1000.0
10.0	10,000.0

ratios. Note that differences in apparent magnitude provide a way of comparing fluxes if you convert to brightness ratios.

Astronomers also talk about luminosities in the form of magnitudes by a system called **absolute magnitude.** *Imagine that you could place all stars at the same distance from the earth. Then differing distances would not play a role in how bright stars appeared. Any differences in magnitudes among them would arise only from differences in luminosities.*

Astronomers do set stars, in an imaginary way, at a standard distance in order to compare their luminosities. However, we do not use the earth-sun distance as the standard distance; rather, we use a distance of 10 parsecs (pc), which is 32.6 light years.

Imagine that you could transport the stars in the sky, including the sun, to 10 pc from the earth. Stars that are in reality closer than this distance would appear fainter; those that are farther would get brighter, as expected from the inverse-square law. How bright a star would appear at 10 pc from us is termed its **absolute magnitude.**

Compare the absolute magnitudes of these stars: the sun, 4.8; Sirius, 1.4; Polaris, −4.6. Note that Polaris is actually the most luminous, followed by Sirius, with the sun last.

Once we know a star's absolute magnitude, we can compute its luminosity by comparing its absolute magnitude to that of the sun. The trick here is to convert magnitude differences into brightness ratios (Table F.1). Let's compare the sun with Sirius. In the visual range, the sun's absolute magnitude is 4.83, that of Sirius 1.41. The difference is roughly 3.4 magnitudes, so the brightness ratio is about $(2.512)^{3.4} \simeq 22$*. So at visual wavelengths, Sirius has roughly 22 times the sun's luminosity.*

Warning: *This comparison does* **not** *give the luminosity ratio over* **all** *wavelengths, because it compares just visual magnitudes. The magnitude of the sun (or any star) including light at all wavelengths is called its* **bolometric magnitude.** *The absolute bolometric magnitude of the sun is 4.75; that of Sirius 1.07. So the magnitude difference is 3.68 and the brightness ratio 26. Including all wavelengths, Sirius has 26 times the sun's luminosity.*

This example should show you that apparent magnitude, absolute magnitude, and distance are related to one another. If you know a star's distance and apparent magnitude, you can work out its absolute magnitude. In fact, if you know any two of the three properties, you can find the third. Algebraically, the relationship is

$$m - M = 5 \log d - 5$$

where m *is the apparent magnitude,* M *the absolute magnitude, and* d *the distance in parsecs.*

14.2 Stellar distances

The key to finding a star's luminosity from its flux is its distance. For nearby stars, we have a direct method to find distances: triangulation, similar to that done by surveyors on the earth. Stellar triangulation is called *heliocentric* or *trigonometric parallax*.

You actually observe parallaxes all the time, although you are usually unaware of it. Hold your hand out at arm's length with one finger extended. Now alternately open and close each eye. Your finger will appear to jump back and forth relative to the distant background. This angular shift in your finger's position is the parallax of your finger. Now if you measure the amount of shift (in angular measure) your finger appears to have and also the distance between your eyes, you can calculate how far your finger is from your head.

Imagine that your finger is a nearby star, the background more distant stars, and your eyes the sighting positions of the earth in orbit around the sun separated by a time of six months. From these two positions in the earth's orbit (separated by 2

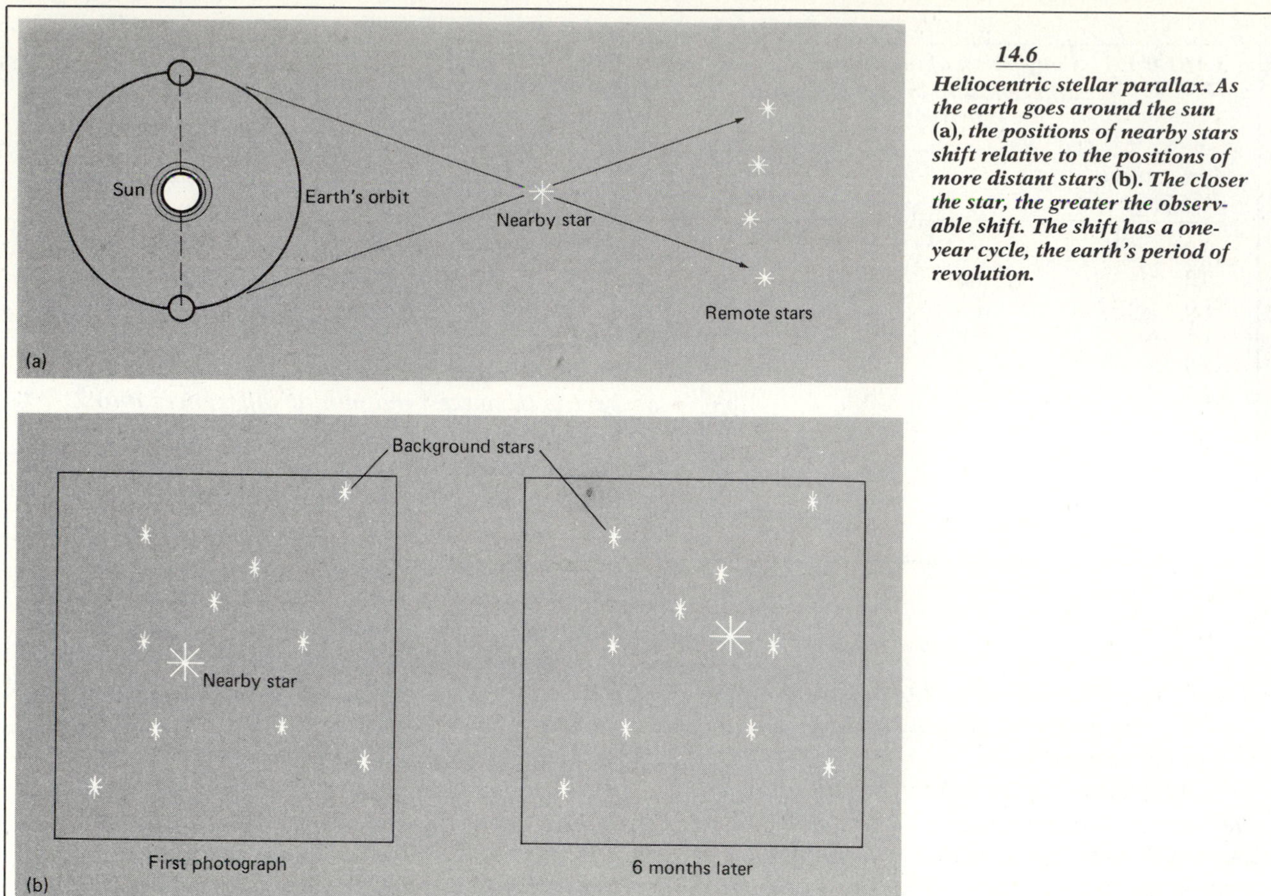

14.6
Heliocentric stellar parallax. As the earth goes around the sun* (a), *the positions of nearby stars shift relative to the positions of more distant stars* (b). *The closer the star, the greater the observable shift. The shift has a one-year cycle, the earth's period of revolution.

AU), the nearby star appears to shift in position (its parallax) compared with the more distant stars (Fig. 14.6). Measure the angular size of that shift. Half of this angular shift is the parallax of the star. Because you know the diameter of the earth's orbit, you can calculate the distance to the star. Note that the farther away a star is, the smaller its parallax will be. (To see this mathematically, see Focus 14.2.)

The parallax of Sirius is 0.38″. That implies a distance of 8.6 ly. The distance to Sirius is about 546,000 times the distance to the sun (546,000 AU). From this information, we can calculate Sirius's luminosity from its flux. This method relates to that for finding the sun's luminosity: We measure the flux from the star at the earth's distance from the star. With the star's distance known, the inverse-square law gives us the rate at which we would receive energy from the star, *if* it were 1 AU from us. We compare that figure to how much we do receive from the sun. From the comparison, we know how luminous the star is compared with the sun. For example, if you imagined we moved Sirius up to the sun, it would appear 23 times brighter than the sun. So Sirius is 23 times more luminous than the sun.

Heliocentric parallax works accurately only for close stars. Note that the parallax of Sirius, a very close star, is less than 1″, about 1/2000 the angular diameter of the moon! That's the size of a U.S. quarter at a distance of a little more than 5 km. People who measure parallaxes have refined their techniques to an accuracy of about 0.001″, so they can get accurate distances out to about 300 ly.

Many stars lie more distant than 300 ly. How are their distances measured? By ingenious, indirect methods. One, called spectroscopic parallaxes, comes later in this chapter.

14.3 *Stellar colors, temperatures, and sizes*

Let's return to the stars in the winter sky. I've dealt with one observable property: their fluxes. Another property you can observe is color. Betelgeuse looks flaming reddish, Rigel appears bluish white, Sirius looks white, and Capella shines yel-

Focus 14.2

HELIOCENTRIC (TRIGONOMETRIC) STELLAR PARALLAX

Parallax occurs when you view a relatively close object from each of two ends of a baseline. As the earth moves from one place in its orbit to another, the nearby star's position seems to change relative to the more distant stars. The maximum shift occurs when you view the star six months apart, so you are sighting from opposite sides of the earth's orbit. The angular shift (actually half the shift) is the star's parallax. *Measure the amount of that shift. Then, because you know the diameter of the earth's orbit, you can calculate the distance to the star.*

Now suppose you travel out into space and look back at the earth-sun separation (1 AU). You keep going until the angular size of the AU is 1″. Suppose you station a star at this point and hurry back to the earth. You observe this star for one year and measure its shift. One-half of the total shift for one year would equal 1″. We say the star is 1 parsec *from the sun, that is, the distance at which the* parallax is one second of arc, *a distance roughly equal to 3.26 light years. Suppose the star were twice as far away; it would have half the parallax, ½″. If at half the distance, its parallax would double and be 2″. Note this simple inverse relationship of parallax and distance. Algebraically,*

$$d = \frac{1}{p}$$

where d *is the distance in parsecs and* p *is the parallax in seconds of arc.*

An example: The star 40 Eridani has a parallax of 0.2″. How far is the star from the sun? Since the star's parallax is one-fifth that for a star 1 parsec (pc) away, 40 Eridani must be 5 pc distance. Explicitly,

$$d = \frac{1}{0.2} = 5 \text{ pc}$$

Here's a simple geometric explanation for the parallax formula. Travel out in space to some star. Then draw an imaginary circle (Fig. F.29) centered on the star and through the sun (S) so that the circle's radius is d. *Note that the parallax angle,* p *(in degrees), is some fraction of the total circle (360°). Also,* R, *the earth-sun distance (1 AU), corresponds closely to an arc on the circumference of the circle; its fraction of the circumference is the same as the fraction* p *is of 360°. So*

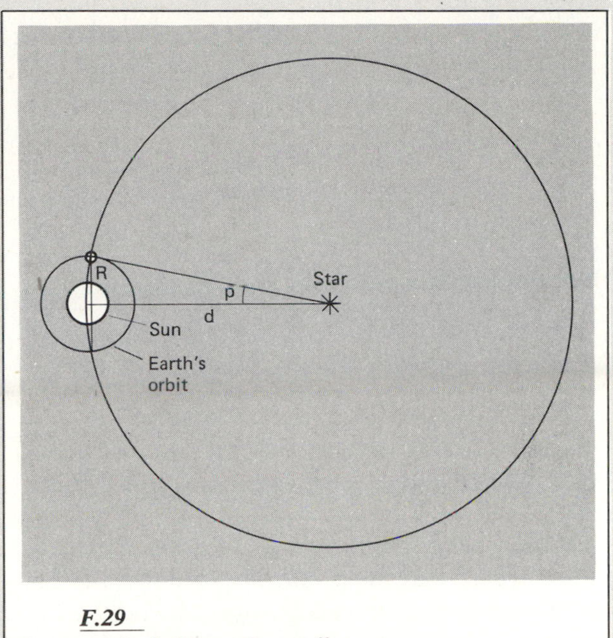

F.29
Geometry for heliocentric parallax.

$$\frac{R}{2\pi d} = \frac{p}{360^\circ}$$

and

$$d = \frac{360^\circ R}{2\pi p}$$

Convert 360° to seconds of arc:

$$d = \frac{360 \times 60 \times 60}{2\pi} \frac{R}{p} = 206{,}265 \frac{R}{p}$$

Now R *is 1 AU; define 1 pc as 206,265 AU. Then*

$$d\,(\text{pc}) = \frac{1}{p(\text{arcsec})}$$

or

$$d\,(\text{ly}) = \frac{3.26}{p(\text{arcsec})}$$

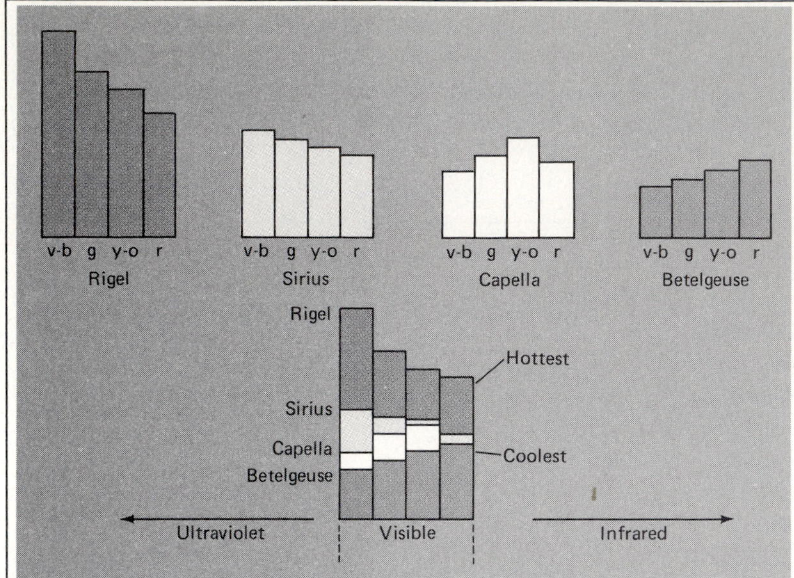

14.7
The relative intensities of the stars Rigel, Sirius, Capella, and Betelgeuse compared in the color bands violet-blue (v-b), green (g), yellow-orange (y-o), and red (r).

lowish white. What can we learn from these facts?

Color and temperature / These different colors suggest that these stars have different surface temperatures. (Recall Section 13.3 pertaining to the sun's surface temperature.) Rigel is the hottest of the four (Fig. 14.7); it is so hot that it emits more blue light than any of the other visible colors. In contrast, Betelgeuse is the coolest, for it radiates mostly red light. The colors of stars relate to their surface temperatures.

A word of caution: When we look at starlight with our eyes, we see only the visible part of its entire spectrum. Stars, in fact, radiate very much like blackbodies (see Focus 13.1), so their continuous spectra, from shortest to longest wavelengths, have the characteristic blackbody shape, called a *Planck curve*. The visible range covers but a small part of a blackbody's spectrum.

One way to measure a star's temperature is to find its peak in its Planck curve (Fig. 14.8). The hotter the star, the shorter the wavelength of the peak (Focus 13.1). Though simple, this technique has two drawbacks. First, we need to measure a wide range of the spectrum to find the peak. Second, at the ground, the earth's atmosphere may absorb the radiation at the wavelength of the peak. For example, the hottest stars have spectra that peak in the ultraviolet, which doesn't make it through the atmosphere to a ground-based telescope.

But the temperature of a blackbody can also be worked out by measuring the relative brightness at any two wavelengths—that's the meaning of color. A reddish star emits more than only red light, but the relative amount of the longer (red) wavelengths is more than that at the shorter (blue) wavelengths.

Temperature and radius / One reason I've concentrated on color is that the stellar temperatures obtained from the colors relate to the sizes of stars, if the stars radiate like blackbodies.

How can this estimate be done? In the constellation Scorpio, you can find the bright reddish star Antares (see the June star map in the endpapers). If you view Antares through a telescope, you'll find it has a faint bluish-white companion. Antares is a binary star (more on this in Section 14.6); it and the bluish-white companion revolve around each other, bound by gravity. The red star is called Antares A, the bluish-white companion Antares B. Antares A has a surface temperature of roughly 3000 K; its companion, about 15,000 K.

Antares A sends us about 40 times more flux than Antares B. Now both stars lie at the same distance from the earth, so the difference in flux cannot arise from different distances; it must come from differences in the radiative output at the stars' surfaces. So we must conclude that Antares A is 40 times more luminous than Antares B. But how can a cool star be so much more luminous than a hotter one?

Recall (Focus 13.1) that the amount of energy emitted *by each unit of area* of a blackbody's surface depends only on its temperature; in fact, on the fourth power of the temperature. For example, Antares B is about five times hotter than Antares A, so each square meter of

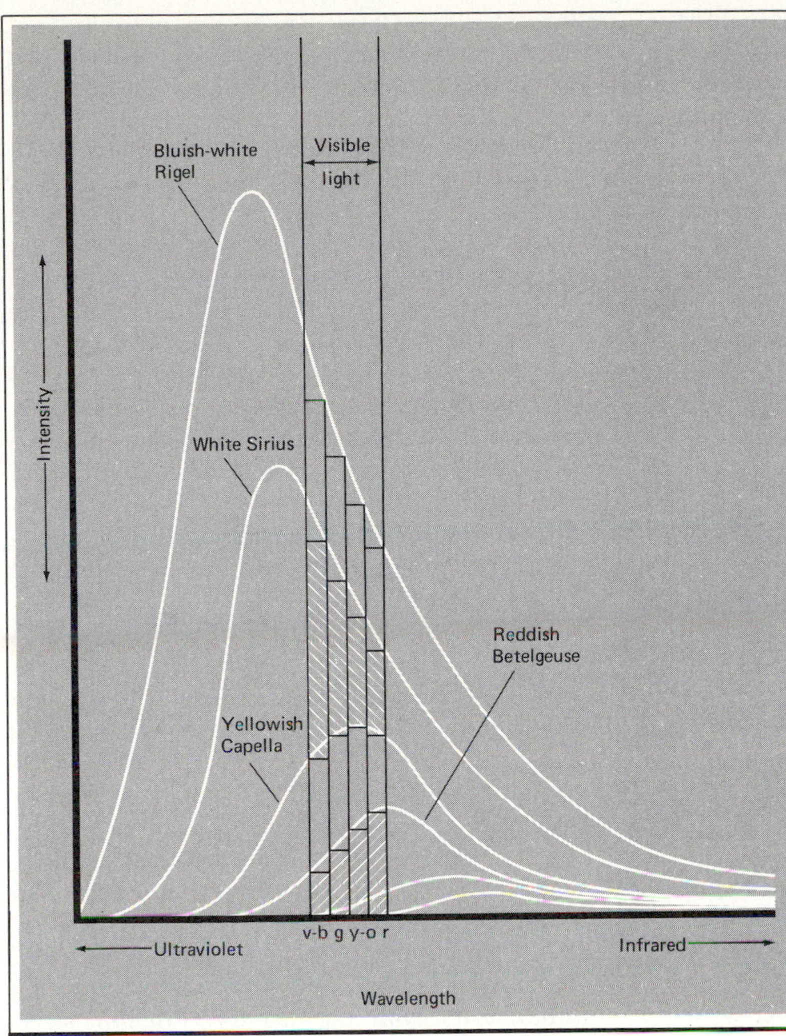

14.8

The continuous spectra of stars with different surface temperatures (those in Fig. 14.7). Their spectra have a distinctive shape known as a **Planck curve.** *Note that the visible range is only a small part of the total spectrum.*

Antares B emits 5^4, or 625 times the energy of Antares A each second (Fig. 14.9). But the number of square meters on the surface of each star is not the same. If Antares A had 625 times the surface area of Antares B, both stars would have the same luminosity. Because Antares A has 40 times the luminosity of Antares B, it must have 40 × 625, or 25,000 times the surface area.

Here's the main point: A star's luminosity is related to its surface temperature (which determines how much energy each square meter emits) *and* its surface area (which determines the total number of square meters doing the emitting).

Now you have a way to infer the radius of a star from its luminosity and surface temperature. As an example, take the star Capilla. Its surface temperature is 5200 K, almost the same as the sun's. But its luminosity is 130 times that of the sun. If Capilla radiates like a blackbody, its size is about 14 times larger in radius than our sun.

Direct measurement of diameters / If we could measure a star's angular diameter and its distance, we could calculate its actual diameter directly. The problem with this procedure is that stars are so far away, their disks have very small angular diameters—on the order of a few mil-

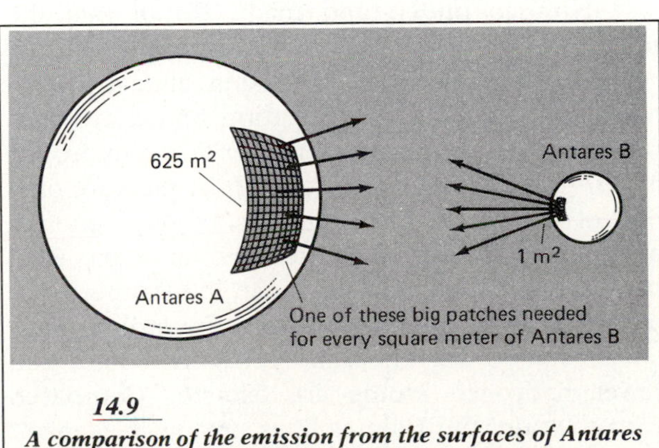

14.9

A comparison of the emission from the surfaces of Antares A and B, which have different surface temperatures.

liarcseconds (one milliarcsecond is 10^{-3} arcsecond), far too small to resolve with earth-based telescopes. (Stellar images rarely have angular diameters less than about one arcsecond, even during excellent seeing conditions.)

One method of obtaining the high angular resolution required involves *lunar occultations.* The moon occults a star when it passes in front of the star as viewed from the earth. Like a solar eclipse, an occultation is visible only across a narrow path on the earth. So an observatory at a fixed location can see only a small number of all occultations that occur. Occultation observers make very high speed measurements of a star's light as the moon occults it. Because we know well the moon's angular speed in the sky, we can find out a star's angular diameter from accurate timing of the occultation.

One star measured this way is Aldebaran, the red eye of Taurus the Bull. The results average about 21 milliarcseconds. Aldebaran is 67.8 ly from us. Its radius is 3.27×10^7 km, or about 50 times larger than the sun's radius.

14.4
Spectral classification of stars

The spectra of most stars resemble the spectrum of the sun. From the sun's spectrum, astronomers infer its photospheric composition. The same procedure can be applied to stars. But there's one trick: A star's spectrum is affected by its surface temperature as well as by its composition.

We now know that most stars have pretty much the same composition, but their spectra are *not* all alike. Consider, for example, the Balmer lines of hydrogen (Fig. 14.10). Stars cooler than the sun have spectra with weaker Balmer lines. Stars somewhat hotter than the sun show stronger Balmer lines. But stars *much* hotter than the sun again have spectra with weak Balmer lines.

How to understand this? Recall that the Balmer series arises from transitions between the second energy level of hydrogen and any level above it, upward for absorption and downward for emission. To absorb a Balmer line, a hydrogen atom must be excited to level 2. In the sun, only one of every 10^8 atoms is excited; not enough energetic collisions occur to kick up many electrons from the ground level. So the Balmer lines in the sun's spectrum are not very strong. In a cooler star, fewer energetic collisions occur, and fewer hydrogen atoms are excited. Compared with the sun, the Balmer lines are much weaker. In a hotter star with more energetic collisions, more atoms are excited, about one out of every million. So the Balmer lines are more intense (darker) in such a star. But if the star is very hot, collisions are so violent that many electrons are knocked out of the atom entirely, leaving hydrogen *ions* (protons) behind. So there are fewer excited atoms, the Balmer lines are weak (less dark).

The key point about the Balmer lines of hydrogen is that they are produced by atoms with an electron in the second energy level; that is, the atom must be *already* excited. If for some reason there aren't very many hydrogen atoms excited to the second level in a star, the Balmer lines in that star will be weak. This can happen if the star has a very high temperature, so that virtually all of its hydrogen is ionized, or if the star has a relatively low temperature, so that even though there is much neutral hydrogen, there are very few excited atoms in the second level.

At the turn of this century, workers at Harvard College Observatory classified stellar spectra by using absorption lines, especially hydrogen Balmer lines. Much of this work was done by Annie Jump Cannon (1863–1941; Fig. 14.11), who single-handedly classified the spectra of over 250,000 stars! The original classification scheme was set up strictly on the basis of the strength of various lines (Fig. 14.12), well before there was any understanding of the effects produced by different temperatures (to be explained shortly). The Balmer lines played an important role in this scheme: Stars with the strongest Balmer lines were called class A, those with slightly weaker lines class B, and so on. Some classes were later dropped because they contained too few stars or only very peculiar ones, and the order was rearranged to one of decreasing temperature, once the explanation of the line darknesses in terms of temperature was understood.

The sequence of stellar spectra in the Harvard classification now runs *O-B-A-F-G-K-M*. (The standard mnemonic for the sequence is: Oh, *B*e *A* *F*ine *G*irl (or *G*uy), *K*iss *M*e!) Almost all stellar spectra fit into this sequence. The *O*-stars have spectra with weak Balmer lines of hydrogen and lines of ionized helium. *A*-stars have the strongest Balmer lines. In *F*-stars the Balmer lines fade and many other lines appear, mostly of metals. The sequence from *O* to *M*, looking at the continuous spectra from the stars, is also a color sequence. *O*-stars appear bluish white, *G*-stars yellowish, and *M*-stars reddish. Astronomers further divide each class into subclasses, labeled from 0 to 9. For example: *G*0, *G*1, *G*2, *G*3, *G*4, *G*5, *G*6, *G*7, *G*8, *G*9; each subclass is distinguished by slightly different intensities of specific absorption lines.

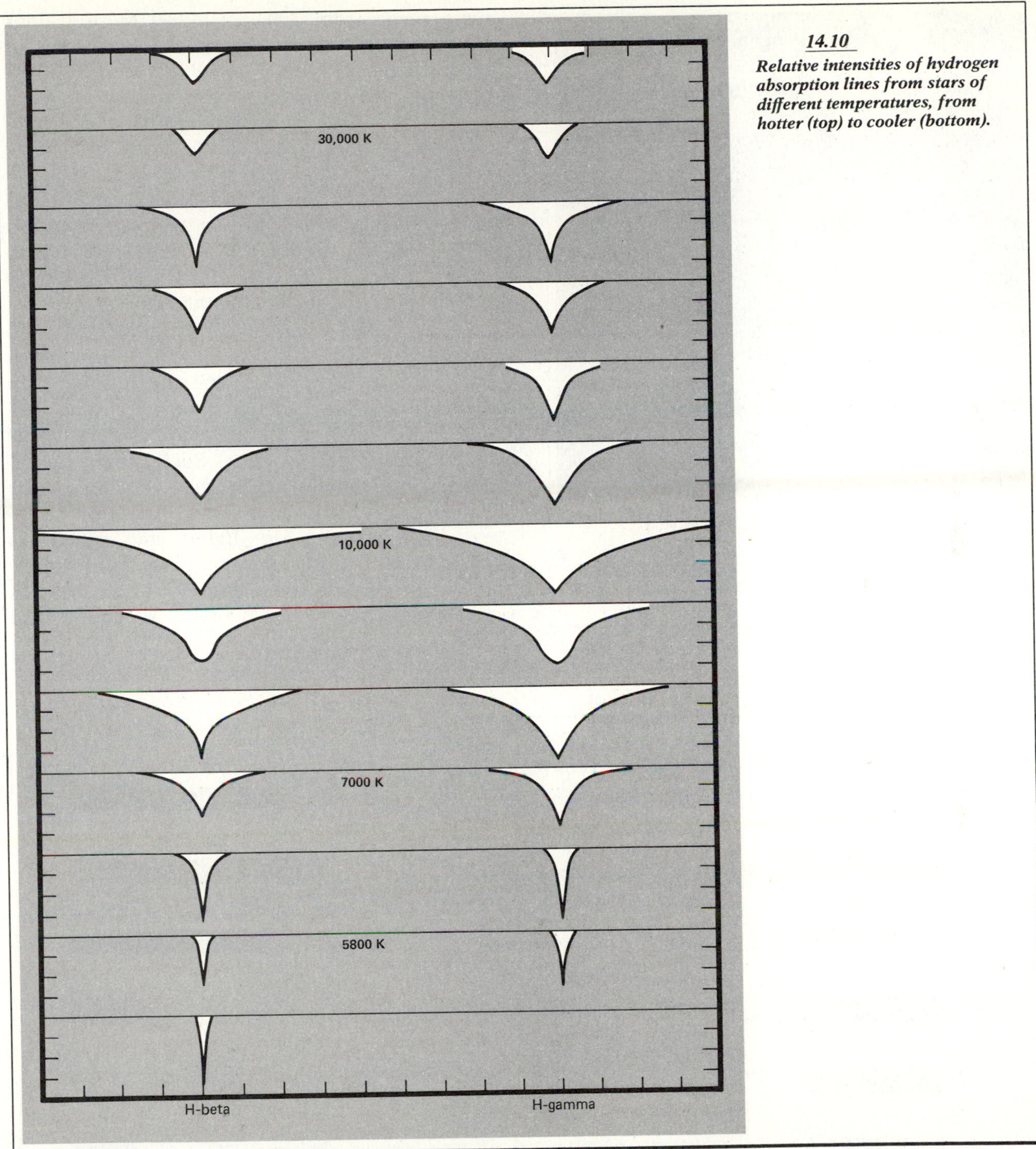

14.10
Relative intensities of hydrogen absorption lines from stars of different temperatures, from hotter (top) to cooler (bottom).

The strengths of the Balmer lines give a clue that the differences in stellar spectra reflect primarily differences in *temperature* and not in the abundance of elements. These temperature differences lead to different degrees of ionization and excitation of the atoms in the star. How many atoms are excited and how many are ionized determine the strength of the atom's spectral lines.

Consider first *O*-stars (Fig. 14.12). These have the hottest surface temperatures, 30,000 K and higher. At such high temperatures, atoms collide violently. The energies in such collisions can rip off the electrons from hydrogen atoms so that most of the hydrogen is ionized. Very few neutral atoms remain to absorb at wavelengths corresponding to the Balmer series. Because the hydrogen is mostly ionized in *O*-stars, the Balmer lines are weak.

Now turn to *A*-stars, whose surface tempera-

14.11

Annie Jump Cannon (left) and Henrietta Swan Leavitt (right), two workers at Harvard College Observatory who did fundamental work in astrophysics. (Courtesy Harvard College Observatory)

tures range from 8000 to 11,000 K. Collisions between atoms occur less violently, and most of the hydrogen is neutral. Recall that to absorb Balmer lines, the hydrogen atom must have its electron in the *second* energy level. Although the collisions are not strong enough to ionize the hydrogen, they do possess the energy to *excite* the electrons out of the lowest energy level. In *A*-stars, many hydrogen atoms are excited by collisions, so their electrons are in the second level. These excited atoms readily absorb light at the Balmer wavelengths, and the lines appear strong.

In *K*-stars, surface temperatures are still lower, roughly 4000 K. Very few hydrogen atoms are ionized. In addition, the impacts between atoms do not have enough energy to excite very many. Most hydrogen atoms have their electrons in the lowest energy state. Such atoms cannot absorb at Balmer wavelengths. As a result, the Balmer lines disappear almost completely.

The variation in Balmer-line intensities arises from collisions that excite and ionize atoms. How much the collisions ionize or excite depends on temperature. So each spectral type

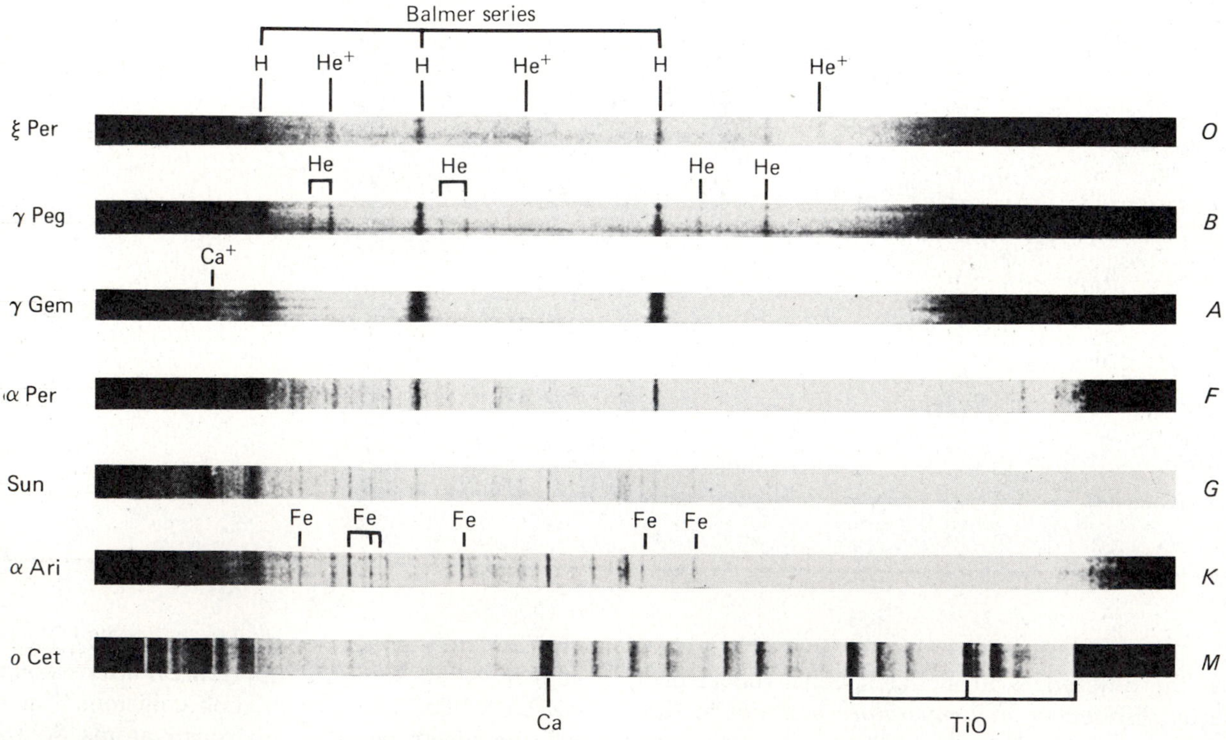

14.12

The basic spectral classification scheme from O (hottest) to M (coolest). The sequence is based on the intensities of the hydrogen Balmer lines and those of metals. Note that A-stars have the strongest hydrogen lines; stars above and below type A have weaker hydrogen lines. (Courtesy Yerkes Observatory)

TABLE 14.1 *Features of the stellar spectral classes*

Spectral class	*Color*	*Approximate temperature (K)*	*Principal features*	*Examples*
O	Bluish white	30,000–50,000	Relatively few absorption lines. Lines of ionized helium and other lines of highly ionized atoms. Hydrogen lines appear only weakly.	Naos
B	Bluish white	11,000–25,000	Lines of neutral helium. Hydrogen lines more pronounced than in *O*-type stars.	Rigel, Spica
A	Bluish white	7,500–11,000	Strong lines of hydrogen. Also lines of singly ionized magnesium, silicon, iron, titanium, calcium, and others. Lines of some neutral metals show weakly.	Sirius, Vega
F	Bluish white to white	6,000–7,500	Hydrogen lines are weaker than in *A*-type stars but are still conspicuous. Lines of singly ionized metals are present, as are lines of other neutral metals.	Canopus, Procyon
G	White to yellowish white	5,000–6,000	Lines of ionized calcium are the most conspicuous spectral features. Many lines of ionized and neutral metals are present. Hydrogen lines are weaker even than in *F*-stars.	Sun, Capella
K	Yellowish orange	3,500–5,000	Lines of neutral metals predominate.	Arcturus, Aldebaran
M	Reddish	3,500	Strong lines of neutral metals and molecules.	Betelgeuse, Antares

corresponds to a restricted range of surface temperatures. These are listed in Table 14.1.

Other lines from other elements can be analyzed in a fashion similar to that for the Balmer lines (Table 14.1). You can see that the observation of stellar spectra coupled with an understanding of the atom gives astronomers information about the physical conditions in the atmosphere of a star. An analysis of spectral lines based on atomic theory also provides information about the abundance of elements in stars in the same way as for the sun. Astronomers have found few surprises: Just like the sun, stars consist mostly of hydrogen and helium. (In Chapter 16, you'll see that stars do differ somewhat in composition relative to their abundance of heavy elements, but hydrogen and helium still make up the bulk of their mass.)

14.5 The Hertzsprung-Russell diagram

Let's return to the main stars in the winter sky. Table 14.2 summarizes the astronomical and physical properties of these stars. Examine this table for a minute. Can you find any pattern in it? Note the wide range of luminosities and sizes and the smaller range of temperatures. It's not obvious how these relate to the internal anatomy of the stars. Are they all like the sun? Or much different? Again we face the astronomer's dilemma: how to find out vital information from points of light in the sky. The solution lies in spectroscopy.

Let's turn the table into a picture: a temperature-luminosity diagram (Fig. 14.13) based on the spectral types of the stars. Such diagrams were set up early this century by Ejnar Hertzsprung (1873–1972) and Henry N. Russell (1879–1957). In their honor, such plots are called *Hertzsprung-Russell diagrams* (commonly abbreviated to *H-R diagrams*).

Examine Fig. 14.13. Do you see any patterns yet? Notice Sirius B in a corner by itself and Betelgeuse alone in the upper right-hand region. Regulus, Sirius A, Procyon, the sun, and Epsilon Eridani fall along a sloping line close to each other.

Let's now inspect a different H-R diagram, one for the nearest stars, all those within 20 ly of the sun (Fig. 14.14). Notice that most of the stars are less luminous and cooler than the sun. (In fact,

TABLE 14.2 Properties of key stars in the winter sky

Star	Distance (ly)	Spectral class	Temperature (K)	Radius (solar radii)	Luminosity (solar units)
Epsilon Orionis	1600(?)	*B*0	24,800	37.0	4.7×10^5
Rigel	880(?)	*B*8	11,550	74.0	8.9×10^4
Regulus	79.2	*B*7	12,210	3.63	2.7×10^2
Sirius A	8.63	*A*0	9,970	1.67	2.5×10^1
Procyon	11.4	*F*5	6,510	2.07	7.0
Sol	1.6×10^{-5}	*G*2	5,780	1.0	1.0
Capella*	41.1	*G*5, *G*0	5,200	14.1	1.3×10^2
Epsilon Eridani	10.8	*K*2	5,000 (est)	0.7	2.8×10^{-1}
Aldebaran	67.8	*K*5	3,780	61.0	6.9×10^2
Betelgeuse	650(?)	*M*2	3,250	1100.0	1.2×10^5
Sirus B	8.63	White dwarf	32,000	8×10^{-3}	5.8×10^{-2}

SOURCE Adapted from H. L. Shipman, *The Restless Universe*, (Houghton Mifflin, Boston, 1978), used with permission.
Note The distances of Epsilon Orionis, Rigel, and Betelgeuse are estimated, so the radii and luminosities of these three stars are approximate.
* Capella is a double star. The temperature, radius, and luminosity are those of the brighter and cooler component.

we can see them at all only because they are so close to us.) The star Alpha Centauri A has almost the same luminosity and temperature as the sun. This star is the sun's twin and is also the star nearest to us. (Alpha Centauri A is one member of a triple-star system with Alpha Centauri B and Proxima Centauri, and in this sense is unlike the sun.) Finally, note that the stars' properties clearly do not fall in a random scatter. Rather, there is a trend; if you draw a line through the points from luminous, hot, Sirius to the coolest, faintest star in the lower right-hand corner, you have identified the *main sequence*. Most nearby stars fall on the narrow strip of the main sequence in the H-R diagram. Note the few stars in the lower left-hand corner, Sirius B included. These stars have very high surface temperatures but low luminosities, so they must be very small. These peculiar stars are called *white dwarf stars* or *white dwarfs*.

Consider another H-R diagram (Fig. 14.15), one for the brightest stars you can see in the sky. Compare it with the previous plot. What a

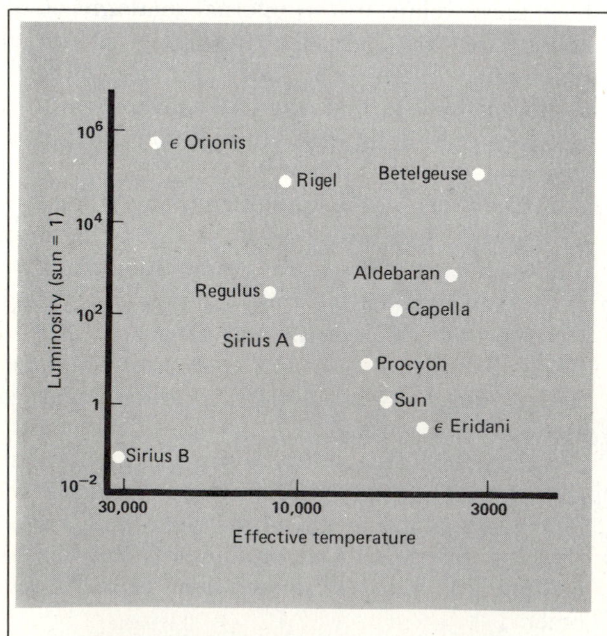

14.13
A temperature-luminosity diagram for important stars in the winter sky. (Adapted from a figure by H. Shipman,* The Restless Universe, *Houghton Mifflin, Boston, 1977)

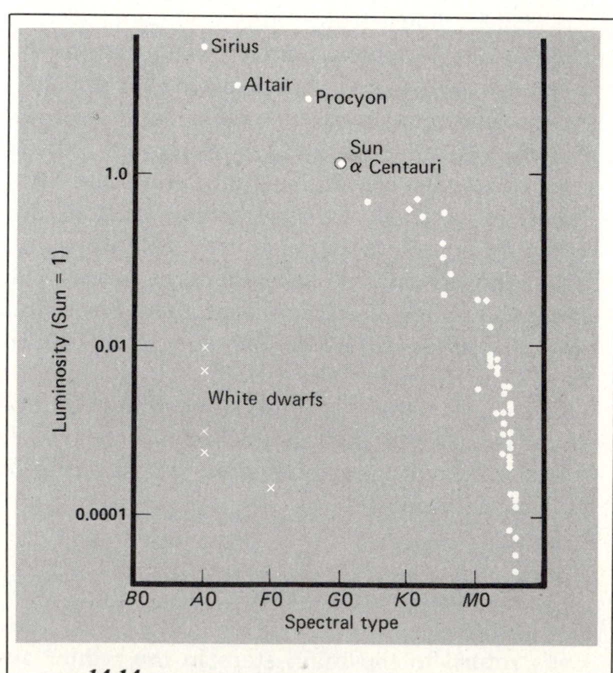

14.14
A Hertzsprung-Russell (luminosity-temperature) diagram for the stars nearest to the sun.

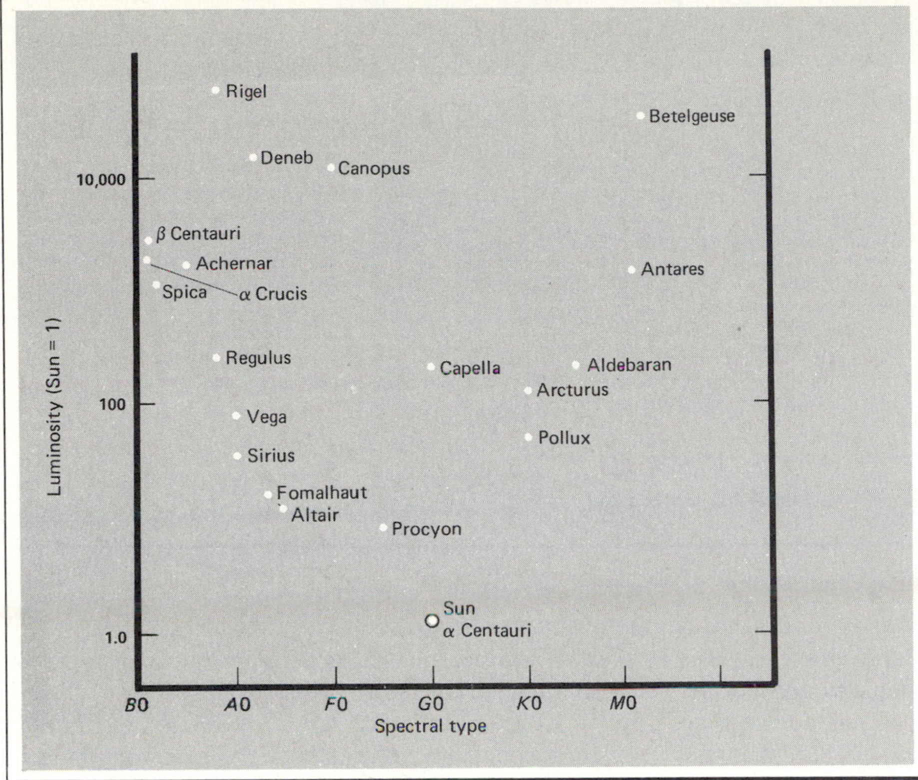

14.15
A Hertzsprung-Russell (luminosity-temperature) diagram for the brightest stars in the sky. Note that the sun and Alpha Centauri have almost the same luminosity and surface temperature, so they overlap on this diagram.

difference! Almost all of these stars have a much higher luminosity than the sun. And many of them are also much hotter (spectral class *O*- and *B*-stars). The main sequence no longer appears so obvious.

Note that only a handful of stars show up in both diagrams. One of these is Altair, the brightest star in the constellation Aquila (see the September starchart in the endpapers). It has a parallax of 0.19″ so its distance is 17.2 ly, and its luminosity is 11 solar luminosities. Observations hint that Altair may have a faint, low-mass companion. As you will find out, Sirius, Procyon, and Alpha Centauri all have companions—suggestive evidence that stars hotter than the sun may all have companion stars.

What are the physical differences among these stars? Take the star Betelgeuse, whose properties put it in the upper right-hand corner of the H-R diagram. Here is a star whose surface is much cooler than the sun's. So if Betelgeuse were the same size as the sun, it would be much less luminous. But Betelgeuse has a luminosity 120,000 times that of the sun. To be so much cooler and more luminous than the sun, Betelgeuse must be much larger (see Section 14.3), about 1200 times the size of the sun—a star so big that it could swallow up Mars if it were placed in the center of our solar system! Astronomers call Betelgeuse a *supergiant* star.

Here is the reason that the H-R diagram for the nearest stars (Fig. 14.14) differs from that for the brightest stars (Fig. 14.15): The first diagram contains ordinary stars with sizes like that of the sun; no giants are among the nearest stars, for they are very rare. The sun is a main-sequence star. Most of the stars in the sun's vicinity are also main-sequence stars, of spectral class *M*. So they are cool and not very luminous. The second diagram contains many giant and supergiant stars, still visible among the brighter stars because of their high luminosity, even though they are scattered widely through space.

Now piece these two diagrams together and add more stars (Fig. 14.16 and Color Plate 15). Most stars fall on the gentle curve of the main sequence. A scattering of stars cuts across the tip of the diagram; these are the very luminous supergiants. A group of relatively cool but luminous stars extends off the main sequence, these are the giants. Finally, note the white dwarf stars in the lower left-hand corner of the diagram.

Is the sun a typical star? / You might have heard that the sun is a typical or average star. In some sense this is true, for the sun is neither the largest nor the smallest kind of star; it ranks in the

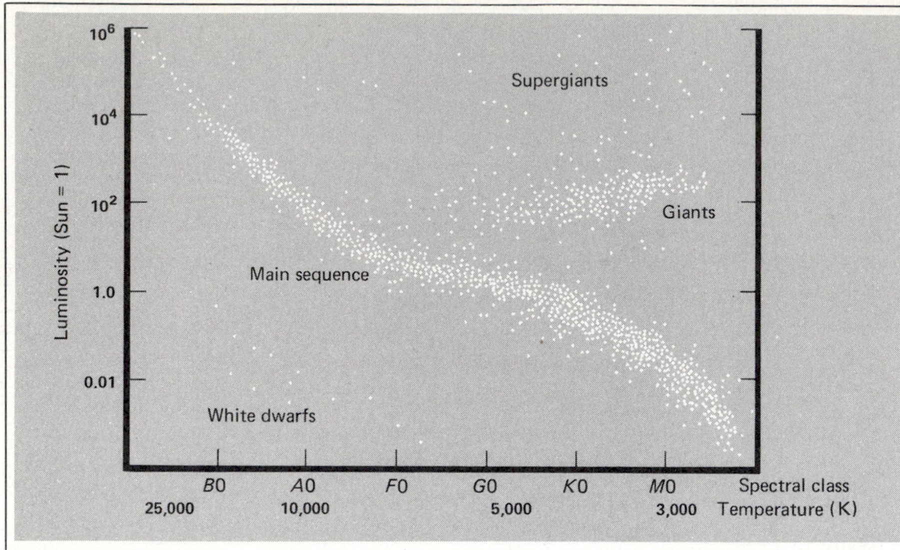

14.16

A schematic Hertzsprung-Russell diagram for a large number of stars. Each star is represented by a point that denotes its luminosity and spectral class (surface temperature). The stars fall into four general regions. Most lie along the main sequence. Many others fall into the giant region of luminous yellow to red stars. A small number make up the very luminous supergiants and the low-luminosity white dwarfs.

middle in both mass and luminosity, and it has a composition like most other stars. However, it is not true that most stars in the Galaxy are like the sun. Among the nearby stars, 44 are cooler than the sun and 9 as hot as or hotter than the sun. There are 49 whose luminosity is lower than the sun and 4 with the same or higher luminosity. Among the stars in the solar neighborhood, *G*-stars like the sun are relatively rare. The most common kind of star in our immediate vicinity, and in fact in the Galaxy in general, is a main-sequence *M*-star, a cool, red star of very low luminosity, on the lower end of the main sequence.

Spectroscopic parallaxes / Once we have an H-R diagram for many stars, we can use it to infer approximate distances to stars. How? First, find out the spectral type of a star. Suppose it's an *M*-star. Then look at the H-R diagram to find the luminosity. Then measure the flux and find the distance from the luminosity, flux, and the inverse-square law for light (Section 14.2).

But a problem arises: *M*-stars have a range of luminosities, from 1.6×10^{-5} the sun's luminosity for main-sequence *M*-stars to about 10^5 solar luminosities for supergiant ones. How do we decide what luminosity an *M*-star has?

Fortunately, we have a way to tell from a star's spectrum. Recall that the strengths of Balmer lines relate to a star's temperature. To absorb Balmer lines, the hydrogen atoms must be excited by collisions—collisions with sufficient energy to raise the electrons one energy level. How energetic the collisions are depends on the temperature in a star's atmosphere. But for gases at the same temperature, the *rate* of collisions depends on the density of the gases. So in a denser gas, collisions are more frequent than in a less dense gas at the same temperature (and so having the same energies in the collisions). Giant stars, because they are so huge, typically have lower-density atmospheres than main-sequence stars. The more frequent collisions in main-sequence stars make certain absorption lines in their spectra appear broader than the same lines in spectra of giant or supergiant stars (Fig. 14.17). So a star's size, and hence its luminosity, is given indirectly by the

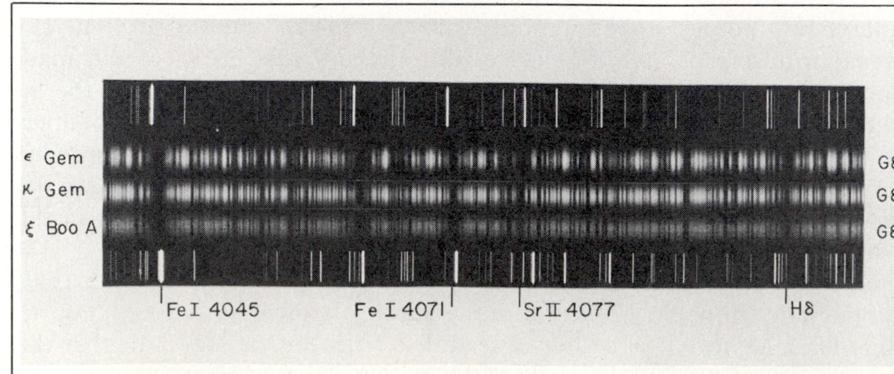

14.17

Luminosity class and spectra. Here are the spectra of three stars all with the same surface temperature (spectral class G8). The top spectrum is for a supergiant star, the middle for a giant, and the bottom for a dwarf. The different intensities of identified dark lines permit astronomers to infer the luminosity class of the stars. (Courtesy Lick Observatory)

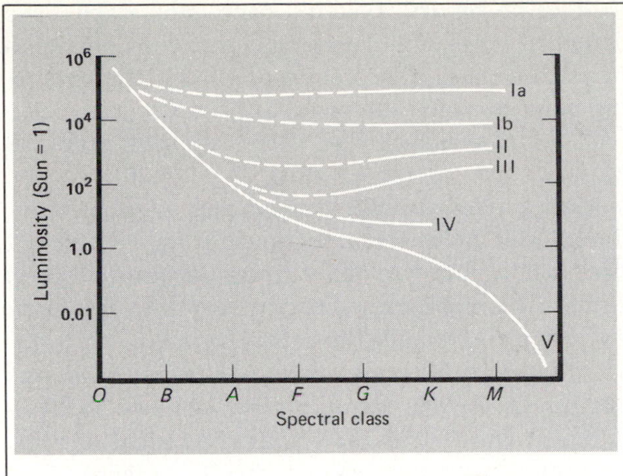

14.18

Luminosity classes of stars on the H-R diagram. The classes run from I, the largest supergiants, to V, stars like the sun. Classes Ia and Ib are supergiants; class II, luminous giants; class III, normal giants; class IV, subgiants; and class V, main-sequence stars.

widths of certain absorption lines when comparing the spectra of stars of the same spectral type.

Such an analysis reveals that stars fall into *luminosity classes*. The recognized luminosity classes (Fig. 14.18) are: Ia, most luminous supergiants; Ib, less luminous supergiants; II, luminous giants; III, normal giants; IV, subgiants, and V, main-sequence stars. The sun falls into luminosity class V.

So a star's spectrum allows it to be classified by spectral type *and* luminosity class. For a given spectral type, you can estimate the luminosity (Fig. 14.19) within a range of probable error (the width of a luminosity class on the H-R diagram). If you know a star's luminosity and its flux, you can calculate its distance. The procedure for working out distances from spectra is called *spectroscopic parallaxes*.

In summary: To make an H-R diagram, you need to find stars close enough to the sun to measure their distances reliably by parallax. Then calculate their luminosities from their distances and fluxes. Next, take spectra of the stars to find out their spectral class. From the spectra you determine how hot the stars are. Then plot the luminosity and spectral type. The result is a calibrated H-R diagram—calibrated in the sense that you have the *luminosities* of the stars plotted against their *surface temperatures*. The H-R diagram graphically summarizes some of the important physical properties of stars. It also serves as a visual sorting tool to bring to light different classes of stars. And it can be used as a tool for obtaining (by spectroscopic parallaxes) the distances of other stars.

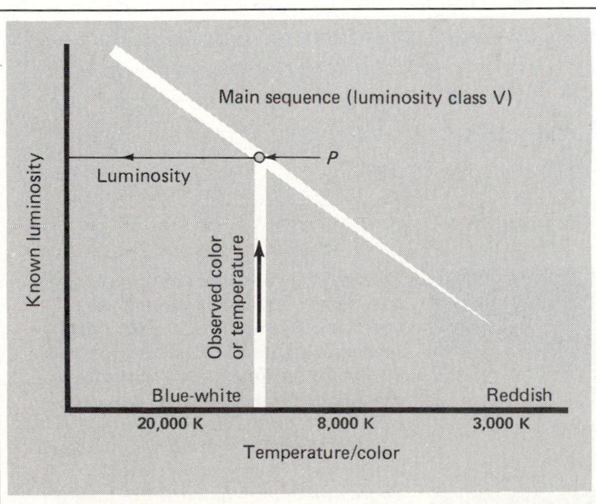

14.19

Using spectral class and luminosity class (in this case V, main sequence) to infer the luminosity and so the distance to a star, if its flux is known.

14.6 Weighing and sizing stars: Binary systems

You have seen how stars differ in properties such as luminosity and size. What about mass? Some stars are larger than the sun, some smaller. But sizes do not tell us directly if a star is more or less massive than the sun. How to find a star's mass?

Binary stars / We have no direct way of knowing the mass of an isolated star. To find masses, we need to examine the gravitational effects of one star on another. Recall how we find the sun's mass (Section 13.1): We look at the acceleration of the earth as it orbits the sun. Similarly, we use the accelerations of two stars orbiting one another to find their masses. Two stars bound by their mutual gravity are called *binary stars*.

If both stars in a binary system are visible, we can trace out their orbital motion by observing them over a long time, which gives us the angular size of the orbit and the orbital period. But that's not enough to get the masses! First, we need to find the distance to the binary system so that we can convert their angular separation into a physical one. Second, it is likely that the plane of a star's orbit is tilted from a direct face-on view; this orbital tilt needs to be accounted for. Then we have enough information to find the *sum* of the masses from Newton's revised form of Kepler's third law (see Focus 4.2). To find the *individual* masses, we must have one more piece of information: how far each star is from the center of mass of the system.

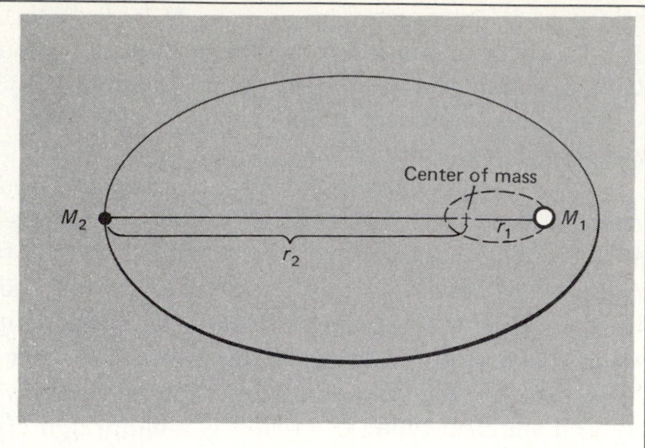

14.20
*Center of mass in a binary star system. Both stars (**M**$_1$ and **M**$_2$) move in elliptical orbits (r_1 and r_2) around the center of mass.*

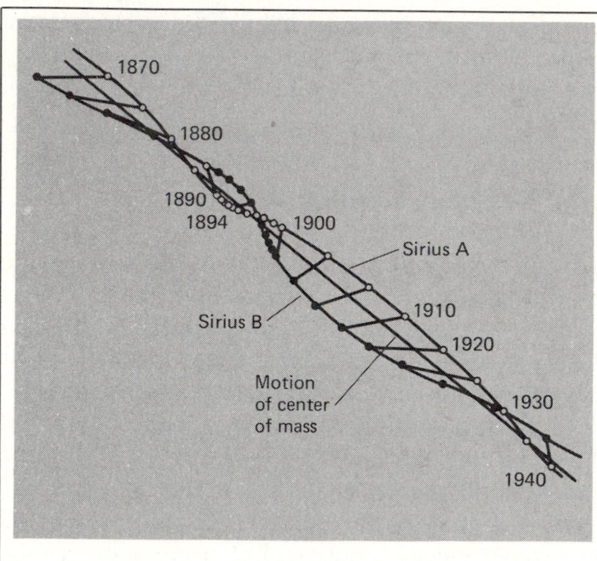

14.21
Motions of the binary star system Sirius A and its companion, Sirius B. This figure shows two motions: that of Sirius A and its companion about their center of mass, and that of the center of mass of the two stars relative to background stars as the center of mass moves through space. So the two stars make a corkscrew motion against the sky.

In a binary system, each star orbits the center of mass at a distance inversely proportional to its own mass. (The *center of mass* is the balancing point between the stars, as if they were on two ends of a seesaw.) The more massive star lies closer to the center of mass (Fig. 14.20). As the system travels through space, and so across our line of sight, the center of mass traces a straight line, while the two stars spiral around it (Fig. 14.21). This corkscrew motion identifies the stars as binary and locates the center of mass.

Types of binary systems / All binary systems are physically the same, but we observe them in different ways. This fact of astronomical life has prompted astronomers to divide binaries into three general classes: visual, spectroscopic, and eclipsing.

In a *visual binary*, a telescope clearly shows both stars. With enough observations, we can trace the orbital path of the fainter star (called the *secondary*) around the brighter one (called the *primary*). Because a supposed pair may be only an accidental line-of-sight juxtaposition of two physically separate stars, we must sometimes wait many years to confirm that the two stars really are bound by gravity. As you'd expect from Kepler's third law, binary stars with large separations will have large periods.

Suppose two stars are so close together that we cannot resolve them with a telescope and that their orbital periods are so short that the stars move quickly in their orbits. We can identify this binary by looking for two sets of lines in the spectrum (one from each star; Fig. 14.22) and measuring the Doppler shifts (Focus 14.3) produced by the orbital motion. This is a *spectroscopic binary*.

The Doppler shift / The *Doppler shift* results from the wave properties of light (see Focus 14.3 for details). Here's a brief description of the Doppler shift: Suppose you have a light source that emits one wavelength. You can measure the

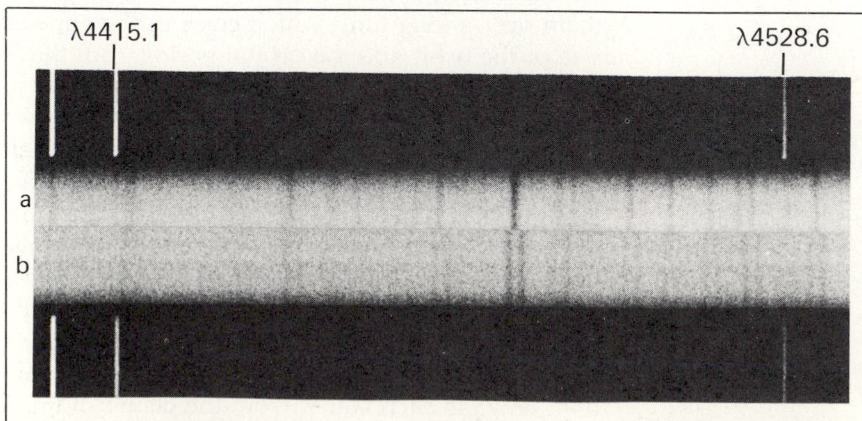

14.22
Spectra of a spectroscopic binary system (Mizar). The two spectra were taken at different times. In the upper spectrum (a), *only single dark lines are visible. In the lower one* (b), *the lines are double because the secondary has come out from behind the primary with a different radial velocity from the primary star. The bright-line spectra above and below the dark-line spectra serve to make the reference wavelengths of the spectral lines. (Courtesy Palomar Observatory, California Institute of Technology)*

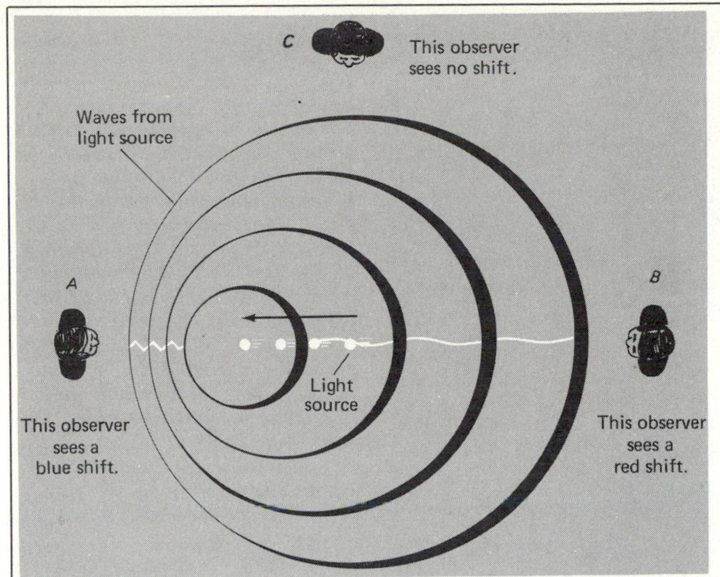
C
This observer
sees no shift.
Waves from
light source
A
B
Light
source
This observer
sees a
blue shift.
This observer
sees a
red shift.

14.23
The Doppler shift for a source moving to the left and emitting a single wavelength of light. The observer at **A** *sees a blue shift, the one at* **B** *a red shift, and the one at* **C** *no shift.*

wavelength of emission with a spectroscope. Now imagine the light source moving *away* from you at a constant velocity (Fig. 14.23). When you measure the wavelength again, it will be *longer* than you found previously. The emission has shifted toward the red end of the spectrum; it is *red shifted.* Now suppose the source moves toward you. You will find the wavelength to be shorter; it has shifted toward the blue end of the spectrum. The light is *blue shifted.* So you see a blue shift if you and the source approach and a red shift if you and it draw apart. The amount of the shift depends on the relative velocity of you and the source along your line of sight to the source. The greater the velocity, the greater the shift.

Warning: For speeds much less than light, it's only the part of the velocity *along the line of sight* that contributes to the Doppler shift. An object may, for example, move at some angle relative to your line of sight. Only the part of its velocity directly toward or away from you results in a Doppler shift. This line-of-sight part is called the *radial velocity.*

Let's see how to use the Doppler shift. Imagine the more massive star to be stationary, with the secondary revolving about it. As the secondary recedes from the earth, you see its spectral lines red-shifted compared with those of the primary; as the secondary approaches, you see its lines blue-shifted (Fig. 14.24). At the intermediate points, when the secondary travels across the line of sight, you see no shift. If the two stars do not differ greatly in brightness, both spectra can be observed, especially the cycle of shifts of the secondary with respect to the primary. (The smaller-mass star will have the higher velocity, because it is farther from the center of mass.) Sometimes the spectrum of the secondary is too faint to be seen. We then use the Doppler shift in the primary's spectrum alone.

These relative-wavelength shifts can be turned directly into relative-velocity shifts by using the Doppler effect (see Focus 14.3). We then use the velocities and the period to get the circumference of the orbit. Then we work out the radius of the orbit and so the separation of the two stars. So a spectroscopic binary gives us direct information on the system's orbit.

Note that in this method, because we can measure the actual velocity of the stars in kilometers per second, we get the actual radius of the orbit, in kilometers, not just the angular radius. So, unlike the case for a visual binary, we do not need to know the distance of a spectroscopic binary to determine the stars' masses.

However, the difficulty with this method is that we cannot get the stars' masses unless we know the *tilt* of the orbit with respect to our line of sight. Why? Because the Doppler shift gives only the velocity along the line of sight, not the total velocity. Generally, we don't know that tilt. But in a few cases, the orbits are tilted so that one star passes in front of the other, producing an eclipse. That's an *eclipsing binary system.*

Algol, the "demon star" in the constellation Perseus, is the prototype of eclipsing binaries. It has a period of 2.87 days and, in mideclipse, plummets sharply in brightness (Fig. 14.25)—easy to see by eye. (You can, with a little practice, observe Algol's light variation. *Sky and Telescope* provides

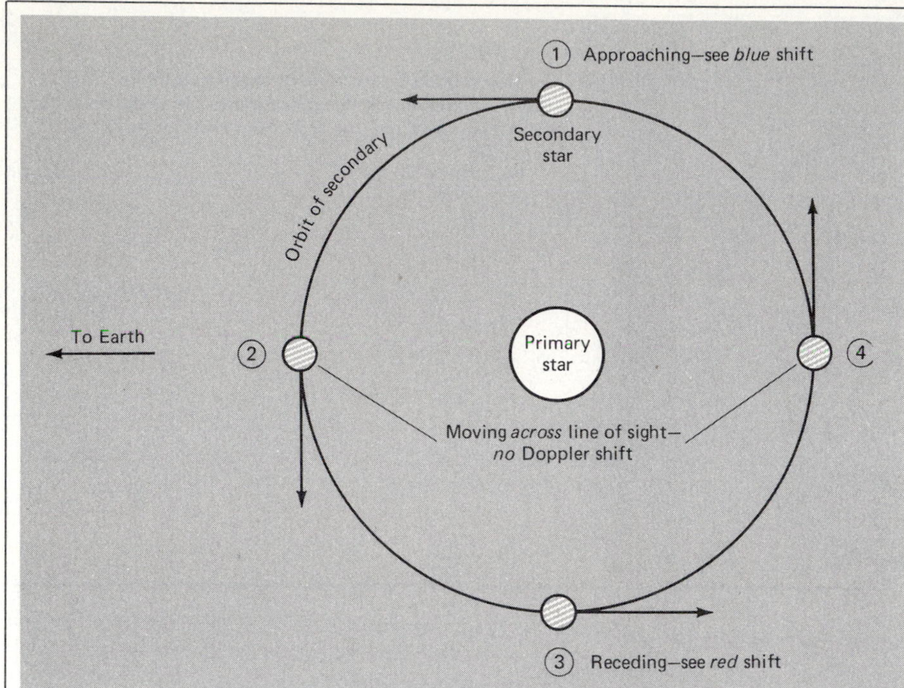

14.24
The cycle of Doppler shifts from a spectroscopic binary system. For simplicity, assume that the primary star is fixed (not revolving around the center of mass). When the secondary is at position 1, it is moving toward the earth, so its spectral lines appear blue shifted. At 2, the secondary moves across our line of sight, so we see no shift. The same is true at 4. At position 3, the secondary moves away from us, so we see a red shift. This means that from 1 to 2, we see a decreasing blue shift; from 2 to 3, an increasing red shift; from 3 to 4, a decreasing red shift; and so on.

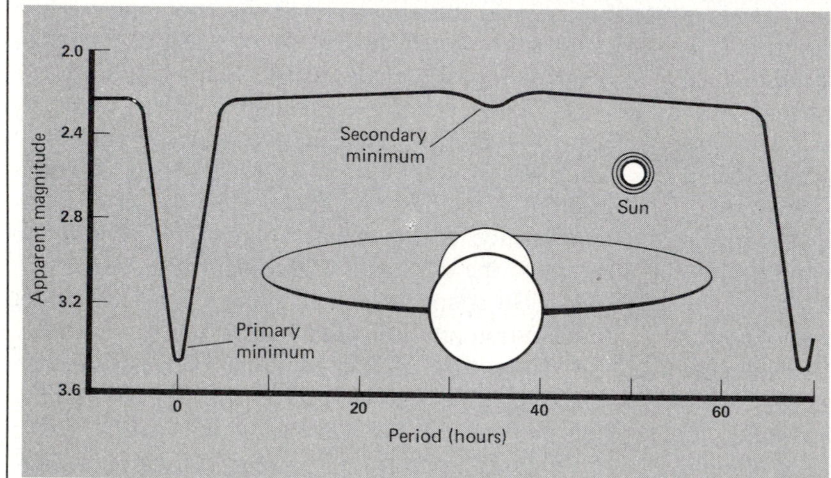

14.25
Algol, an eclipsing binary system. The orbits of the two main stars lie almost edge on as seen from the earth, so eclipses can occur. Algol A, the primary, is a B-star with a diameter about three times that of the sun. Its dimmer companion, Algol B, is cooler; it is a G-star with a surface temperature similar to the sun's. The eclipse of Algol A by its cool companion produces the largest dip (primary minimum) in the light curve, a plot of how the observed brightness changes with time. When the companion circles around Algol A, the loss of light is less, so a smaller dip (secondary minimum) occurs in the light curve.

the dates and times of eclipses.) Algol A, the brighter star (260 solar luminosities) is a *B*8 V (surface temperature, 13,000 K); its companion, Algol B, is a fainter (5 solar luminosities) *G*-star (surface temperature, 4600 K). Recent spectroscopic observations of Algol find that Algol B orbits at 201 km/sec and Algol A at 44 km/sec, if the orbit is inclined 82°. (A 90° inclination would put both orbits directly in our line of sight.) From the velocities and the period, Algol A orbits 1.71 million km from the center of mass, Algol B at 7.9 million km. So Algol A has a mass 4.6 times that of Algol B. The total separation is 9.6 million km. From Kepler's third law, the total mass of the system is 4.51 solar masses. Combining this with the spectroscopic mass ratio, we find that Algol A is 3.7 solar masses and Algol B 0.81 solar mass.

Eclipsing binaries give us directly one other property of the stars: their diameters. Let's see how. When Algol B passes in front of Algol A (primary eclipse, when the cooler star passes in front of the hotter one), the duration of the eclipse depends on the diameter of B relative to A and the relative orbital speeds. When Algol A swings in front of Algol B (secondary eclipse, when the hotter star crosses in front of the cooler one), the duration depends on the diameter of A relative to B and the relative orbital speeds. So we can find the radius of each star: 3.08 solar radii for Algol A and 3.23 solar radii for Algol B.

Sirius: A binary system / Finding stellar masses from binary stars is so important that I will do another example in detail for you. Sirius is a binary star; the main star, called Sirius A, is the one you see in the sky. Its companion, Sirius B, is much fainter, and you need a moderately large telescope to see it. The orbital motion of Sirius A and B is plotted in Fig. 14.26. From it, you can infer that the orbital period is close to 50 years (1880 to 1930). The distance to the stars is 8.64 ly; from this and Fig. 14.26, we find that the separation (semimajor axis) is about 20 AU. Use Kepler's third law with this information to find that Sirius A has about twice the mass of the sun, while Sirius B has about the sun's mass.

A more careful analysis shows that Sirius A has a mass of 2.143 solar masses, a radius of 1.678 solar radii, and an effective temperature of 9970 K. In contrast, Sirius B has a mass of 1.053 solar masses, a radius of 0.0073 solar radius, and an effective temperature of 29,500 K. Sirius B turns out to be a very strange kind of star called a white dwarf (see Section 17.1).

The mass-luminosity relation for the main sequence / Binary systems provide us with the only direct measure of stellar masses. In most cases, the luminosities of the two stars can also be determined. When the luminosities are plotted against the stars' masses, the points fall into a definite pattern (Fig. 14.27). For main-sequence stars, the mass determines the luminosity, and the resulting correlation is called the *mass-luminosity relation* (sometimes abbreviated *M-L relation*). Basically, the mass-luminosity relation shows that a star's luminosity is *roughly* proportional to the third power of its mass. For example, a star with a mass 10 times that of the sun has about 10^3 times the sun's luminosity. Main-sequence stars follow the mass-luminosity relation fairly well; hence, *the upward swing in luminosity of the main sequence from* M- *to* O-*stars reflects an increase in the stars' masses* (Fig. 14.28). So main-sequence *O*-stars are more massive than main-sequence *M*-stars. Astrophysicists had predicted the mass-luminosity relation theoretically, and its confirmation came from investigation of binary stars. (In Chapter 16 you will see the importance of the mass-luminosity relation for stellar evolution.)

From the mass-luminosity diagram you see that the masses of other stars do not differ widely from the sun's mass. According to theoretical considerations, stars with masses greater than about 100 solar masses are unstable, and bodies with masses less than roughly 0.1 solar mass cannot become hot enough to start nuclear reactions and become stars. The predicted narrow range of stellar masses is borne out by the H-R diagram and the mass-luminosity relation.

Note that the white dwarf stars plotted in Fig. 14.27 do *not* fall along the same line, as do main-sequence stars. This fact hints that white dwarf stars are different beasts. Perhaps their interior structures are much different? Let's check.

Stellar densities / From the mass-luminosity relation you can see that the masses of stars do not vary over a very wide range. Yet their sizes do. So stellar densities must vary widely.

The sun's average density is 1400 kg/m^3, 40 percent more than that of water. Let's compare the sun to Sirius B, a white dwarf. The mass of Sirius B is 1.05 solar masses, and its radius is 0.0073 solar radius. Amazing! Sirius B is about *4 million* times denser than water. This fact tells you that Sirius B—and other white dwarfs—cannot be ordinary stars and cannot have an internal structure like the sun. That's the reason they do not fall along the main sequence in the H-R diagram or along the M-L relation with other stars.

Stellar lifetimes / The existence of the mass-luminosity law immediately provides a way of comparing stellar lifetimes. The argument goes like this: The total amount of energy available to a star from the conversion of hydrogen to helium is proportional to its mass. The rate at which it loses energy is given by its luminosity. So the time over which a star can radiate before its energy is used up is proportional to the mass divided by the luminosity (the energy store divided by the rate at

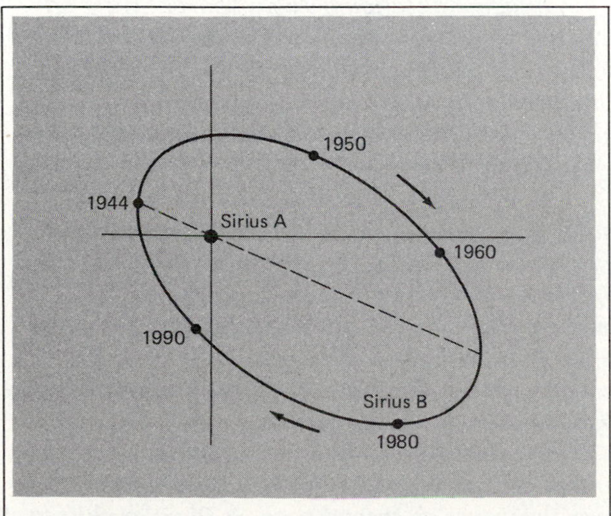

14.26
Orbit of Sirius B relative to Sirius A. Note that Sirius B was closest to Sirius A in 1944. The orbital period is 50 years.

Focus 14.3

THE DOPPLER SHIFT

The Doppler shift allows astronomers to find the line-of-sight velocities of luminous objects without having to know their distance. In a sense, the Doppler shift provides us with a cosmic speedometer.

The Doppler shift is named after Christian J. Doppler (1803–1853), who first noticed the effect in sound waves. Later the French physicist Armand Hippolyte Fizeau (1819–1896) applied the Doppler shift to light waves and recognized its importance in astronomical applications. In honor of these two men, some scientists like to call it the **Doppler-Fizeau shift.** *Most people call it simply the* **Doppler shift,** *and that is the name used in this book.*

The Doppler shift occurs with all kinds of waves—light, sound, and even water waves. Here's an example with water waves. Imagine you are out fishing in a small motorboat (Fig. F.30). You have been sitting in one spot for a while with little luck, and the rhythm of the waves, generated by a gentle wind, has lulled you almost to sleep. You decide to move for better fishing. First you go into the wind (the wave source). You notice that you bob up and down **more frequently** *than when you were at rest; the wavelength appears to have gotten* **shorter.** *Just for fun, you drive the boat with the wind. You discover that your bobbing is less frequent; the waves seem to you to have a* **longer** *wavelength. The explanation is simple: When you moved in the direction of the waves, they had to catch up with you; when you went into them, you went to meet them.*

That's the Doppler shift: When you are moving **toward** *a wave source, the waves appear more frequent and shorter in wavelength; in contrast, when you move* **away** *from a wave source, the waves appear less frequent and the wavelength longer. It's only the* **relative velocity** along the line of sight—*called the* **radial velocity**—*that causes the Doppler shift (for speeds much slower than that of light). Since velocities are relative, it makes no difference if you're moving, if the source is moving, or both.*

Let's look now at light waves, which you can't see directly. You can, however, see colors of light, which relate directly to wavelength. Keep in mind that red light has a longer wavelength than blue.

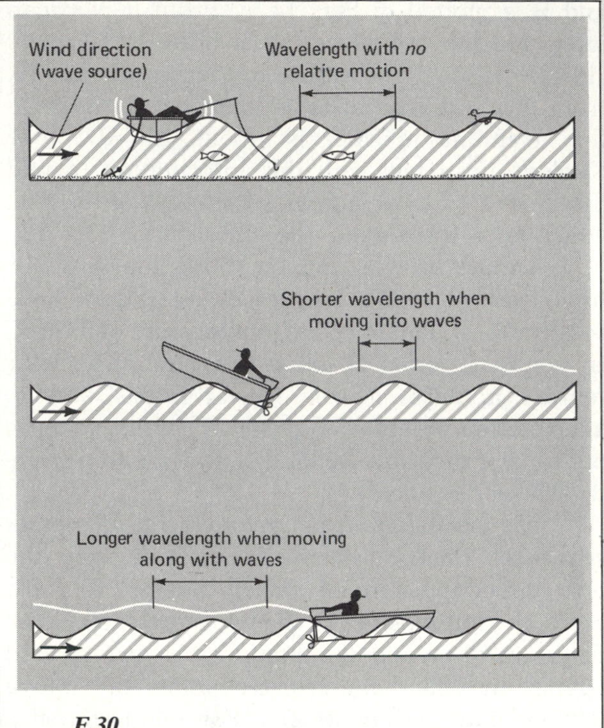

F.30
Wavelength and relative motion for water waves.

Imagine a stationary light source giving off just one particular wavelength every second (Fig. F.31). Each wave travels outward with velocity c, the speed of light. When the source is not moving, all the waves are concentric and separated by the wavelength of emission.

Now imagine that the source moves from point S_1 to point S_2 in one second, and so on. At each point (S_1, S_2, S_3, S_4) it emits a wave (1, 2, 3, 4) that travels out at c. In the direction of its motion, the source catches up a bit with the wave it has just emitted, so for an observer at A, the wavelength appears shorter. This observer sees the distance between the waves as compressed and so observes a Doppler shift to the short-wavelength end of the spectrum. In contrast, an observer at B sees the waves as spread apart, a Doppler shift to the long-wavelength end of the spectrum. An observer at C, at right angles to the direction of the source's motion, measures no change in the

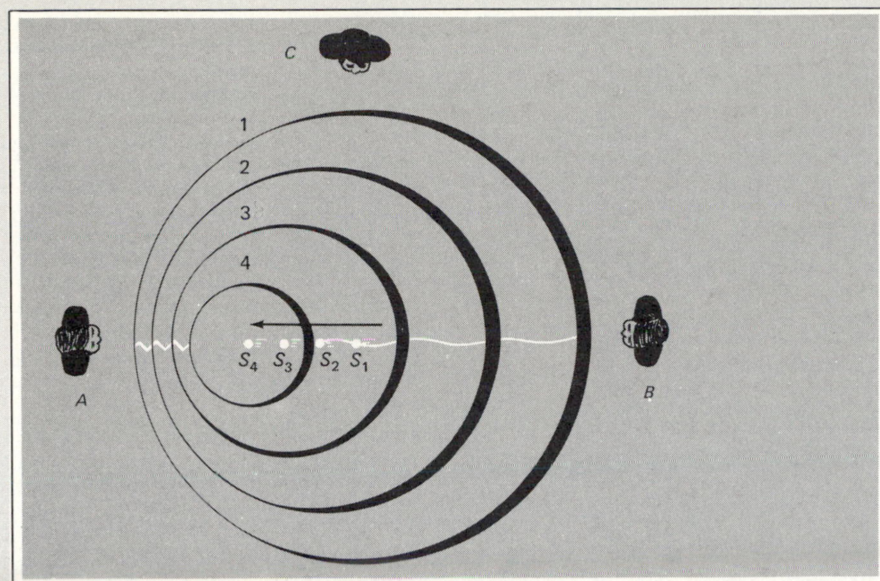

F.31
The Doppler shift in detail.

wavelength and consequently no Doppler shift. **Note that only the velocity along the line of sight contributes to the Doppler shift.** *It is also important to remember that the Doppler shift does* not *depend on the* distance *between the observer and the source, only on their relative radial* velocities.

The astronomer uses the Doppler shift to determine the radial velocities of celestial objects (such as stars) relative to the earth by using spectral lines to measure wavelength shifts. The astronomer takes a spectrogram of the object and at the same time superimposes a comparison spectrum from a local laboratory source (Fig. F.32). The comparison source is at rest with respect to the telescope and provides the normal (zero relative velocity) placement of the lines with respect to which the astronomer measures the shift. With the measured shift and the value of c, *the astronomer calculates the relative velocity between the source and earth by the expression*

$$V_r = (\frac{\Delta\lambda}{\lambda_0})$$

where V_r is the relative radial velocity, λ_0 the rest wavelength of the observed line, $\Delta\lambda$ the observed shift in the line ($\Delta\lambda = \lambda_{observed} - \lambda_0$), and c *the speed of light.*

Here's an example: A strong dark line from absorption by calcium has a rest wavelength of 3933 Å. Suppose you measure this line in the spectrum of a moving object and find it shifted to 3972 Å. Is the object moving away from or toward you? Away, since the wavelength is longer. How fast?

$$\Delta\lambda = 3972 - 3933 = 39 \text{ Å}$$

$$V_r = \frac{39}{3933}(3 \times 10^8 \text{ m/sec})$$

$$= 3 \times 10^4 \text{ m/sec}$$

$$= 30 \text{ km/sec}$$

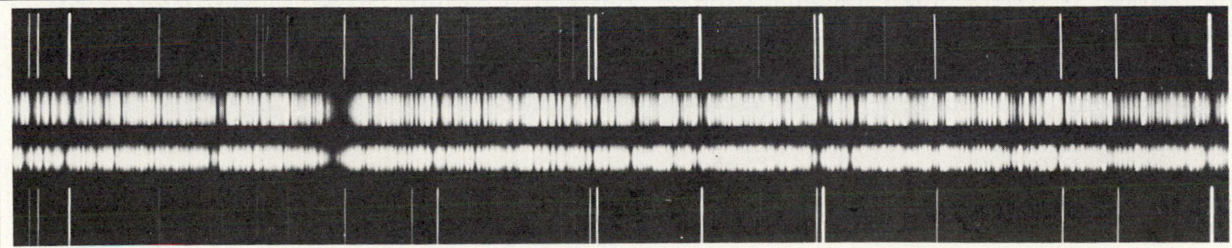

F.32
The Doppler shift in the spectrum of the star Arcturus. The two spectra were taken six months apart; above and below them are the bright-line spectra made at the telescope to provide the wavelength standard. The shift results from the orbital motion of the earth and the motion of Arcturus in space. (Courtesy Palomar Observatory, California Institute of Technology)

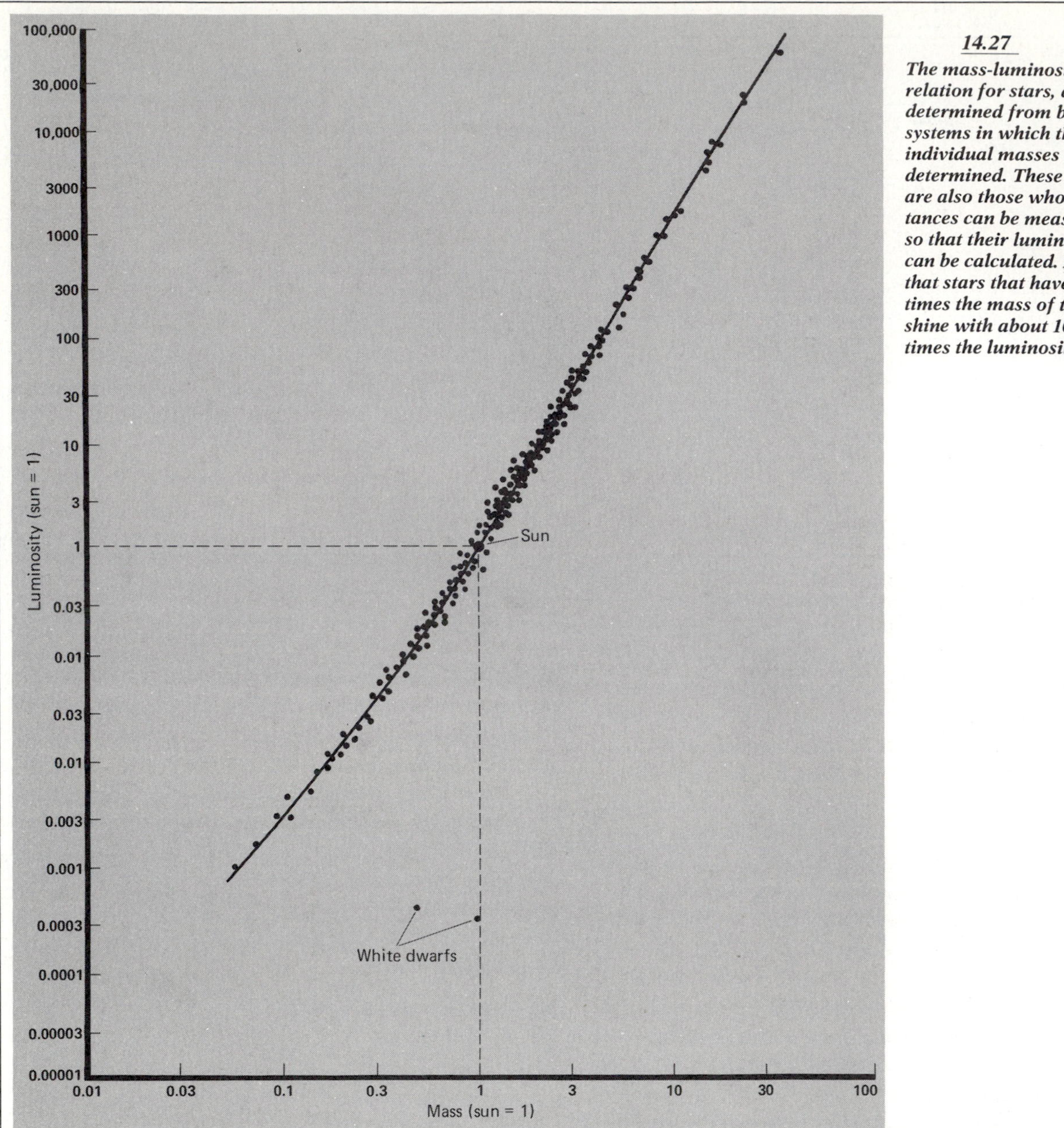

14.27
The mass-luminosity relation for stars, as determined from binary systems in which the individual masses can be determined. These stars are also those whose distances can be measured, so that their luminosities can be calculated. Note that stars that have 10 times the mass of the sun shine with about 1000 times the luminosity.

which the store is used up). Because the luminosity is proportional to the cube of the mass, the greater a star's mass, the shorter its lifetime (relative to the sun). For example, a star of 30 solar masses has a lifetime of a few million years, in contrast to the sun's 10 billion years.

14.7
Stellar activity

So far I've described the gross properties of the stars—those that are typical of the quiet sun. But if stars are suns, do they also show the aspects of the active sun (see Section 13.7), such as sunspots and flares? Remember that these are connected to active regions, which result from strong local magnetic fields. Let's see what recent observations reveal about the solar-stellar connection and magnetic-activity cycles.

Coronas / Loops of magnetic fields form the structure of the corona and probably serve to heat it to the measured temperature of about 1 million K (Fig. 14.29). We know that the sun's magnetic activity cycle influences the shape of the corona because pictures taken during eclipses show different structures at different times in the cycle.

Because stars are far away, we cannot see the delicate flower of their coronas (if they have them). How to observe them? Let's use the sun as a guide: The hot gas in the corona shows up in X-

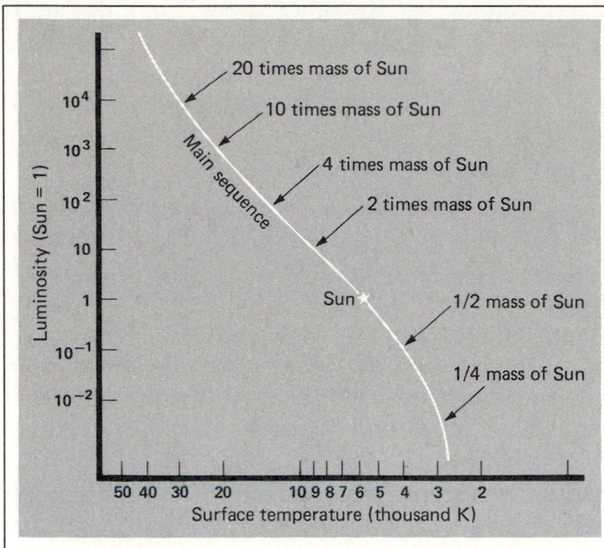

14.28
The approximate mass for stars on the main sequence of the Hertzsprung-Russell diagram. Note that the stars with the highest surface temperatures are the most massive; these are* O- *and* B-*stars.

ray photos as bright regions; in contrast, the photosphere and chromosphere appear dark. So X-ray observations of stars can open their coronas to our view—if stellar coronas resemble the sun's in general ways.

The Einstein X-Ray Observatory has examined a number of nearby stars for X-ray emission and has found more than expected. From these observations, we find that stars of *all* spectral classes have X-ray emission and, by deduction, coronas. The quiet sun seems to be one of the weaker X-ray emitters, below 10^{20} W. In contrast,

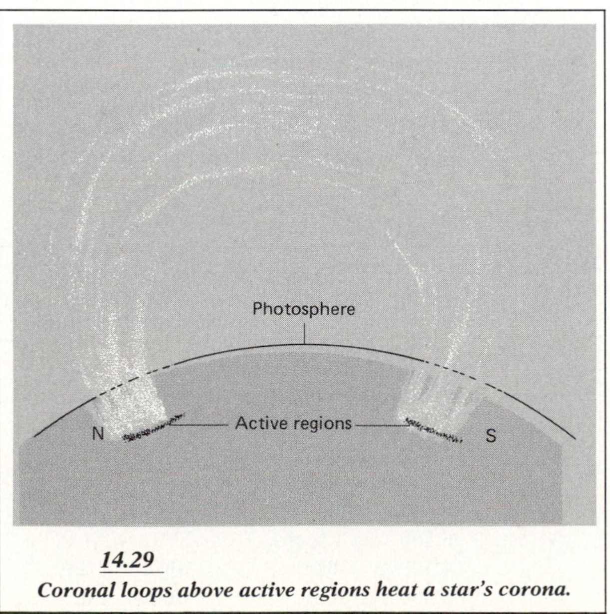

14.29
Coronal loops above active regions heat a star's corona.

the active sun puts out some 10^{22} W. Other *G*-stars emit upward of 10^{24} W—about 100 times more than the active sun. The implication is that such stars have more extensive and hotter coronas (temperatures up to 10 million K), connected with extensive active regions.

The bottom line of these X-ray observations is this: Coronas are common in other stars, and so must be magnetic loops within the coronas.

Starspots / During times of sunspot maxima, sunspots cover some 0.1 percent of the sun's total surface. If you observed the sun as a star from light-years away, you'd need extremely sensitive equipment to note any variation in the sun's light as it rotated and sunspots came into and passed out of view. So you might expect that it would be useless to try to search for starspots by the same kind of observations of changes in light output as stars rotate.

Luckily, this solar analogy proves wrong in scale but right in concept. Some stars, even *G*-stars like the sun, turn out to be hyperactive, with enormous concentrations of starspots compared to the sun—starspots covering such a large fraction of the surface that we can actually see them indirectly!

One type of such hyperactive stars are called *RS Canum Venaticorum* stars (*RS CVn* for short). The RS CVn stars are binary systems with orbital periods of about a week. One star is spectral type *G*, like the sun; the other is cooler, typically spectral type *K*. Both stars have pretty much the same mass as the sun.

Sounds pretty dull—after all, these are basically solar-type stars, and at a distance the sun seems awfully placid. The RS CVn stars, though, turn out to be curious beasts. If you observe their visual light (Fig. 14.30), it exhibits distinct variations in the shape of a fairly regular wave. How does this happen? The simplest explanation is that at least one of the stars in the system is heavily covered with starspots—a few tens of percent of the surface. These spots, for reasons not yet understood, concentrate in one hemisphere; the other is comparatively clean of spots. As the star rotates, we alternatively view the spotted and clean sides. When few spots face us, we see the most light from the star. When the spotted hemisphere turns our way, we see a decrease in the amount of light.

So the RS CVn stars have starspots—lots of them clustered in enormous active regions. Do these go through a cycle such as the sun's? Yes! Observations over many years indicate that the starspots vary in cycles that last some tens of years, similar to the solar activity cycle. So the

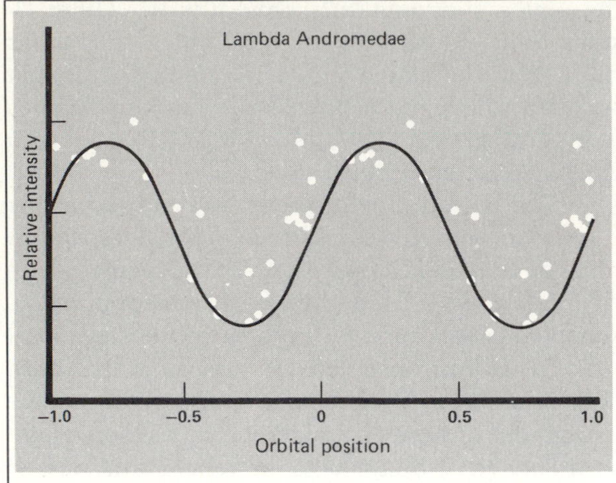

14.30
Variation of the intensity of light from an RS CVn binary system over one orbital revolution. Where the light intensity falls, the spotted side of the active star faces the earth. (Based on observations at Capilla Peak Observatory, University of New Mexico)

magnetic fields on these stars, though much stronger, evolve over times that are typical of the sun's magnetic-field changes.

Stellar flares / Now that I have you intrigued by the RS CVn stars, let me use them as one more example of stellar activity. In the sun, we know that flares occur in active regions. The same appears to be the case for the RS CVn stars, but the flares burst forth with much more energy than in the sun.

Here are two examples—one radio, one optical—of hyperactive flaring. In early 1978, one RS CVn system exploded with an enormous radio flare—millions of times stronger than the most energetic solar radio flare ever recorded. The flare lasted for about a day. During that time, the energy emitted in just the radio amounted to 10 times as much energy as the sun puts out at all wavelengths in one second!

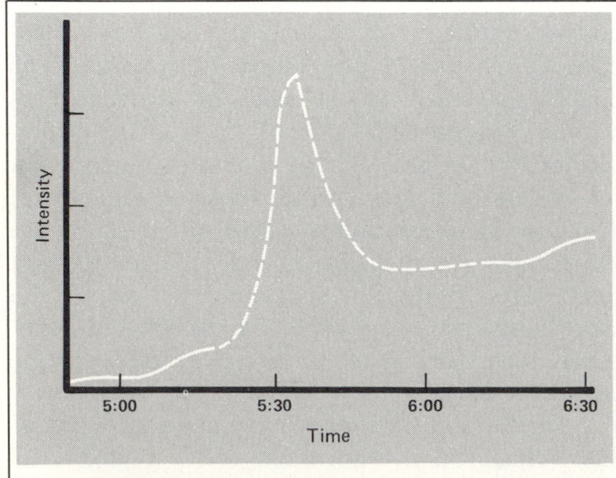

14.31
Intensity profile of a flare from an RS CVn system measured in ultraviolet light. (Based on observations at Capilla Peak Observatory, University of New Mexico)

Another RS CVn system banged out an optical flare in early 1982 (Fig. 14.31). The flare lasted for about 30 min (a duration common to solar flares) and gave off a total of some 10^{27} W, or a thousand times more than a similar solar flare.

The point of all this is not to dazzle you but simply to demonstrate observationally that other stars have active-sun characteristics, hyped up to vast energies. Strong magnetic fields, probably twisted and tangled, must lie at the roots of this activity. So stars are suns, in both their quiet and active guises.

KEY CONCEPTS

1. ***Stars are other suns, more or less like ours (Table 14.3), with similar physical properties; we can infer these by analyzing their light and measuring (or guessing) their distances; the most important physical properties are mass, luminosity, surface temperature, and chemical composition.***
2. ***The distances to stars can be estimated roughly from their brightness; generally, closer stars are brighter than stars that are farther away.***
3. ***Astronomers measure the brightness of stars as seen in the sky as their flux, how much energy from a star reaches the earth each second over a given area (such as a square meter).***
4. ***Light intensity changes with distance; the intensity goes as the inverse-square of the distance; this law applies to all kinds of electromagnetic waves.***

***TABLE 14.3** Average properties of main-sequence stars*

Spectral class	*Surface temperature (K)*	*Mass (solar masses)*	*Luminosity (solar luminosities)*	*Radius (solar radii)*
O	40,000	40.0	500,000.0	20.0
B	15,000	7.0	800.0	4.0
A	8,200	2.0	20.0	2.0
F	6,600	1.3	2.5	1.2
G	5,800	1.0	1.0	1.0
K	4,300	0.78	0.16	0.7
M	3,300	0.21	0.008	0.3

5 ***The distances of only the closest stars (within 300 ly) can be measured directly by a triangulation technique called*** **heliocentric parallax*****; the closer the star, the larger its parallax.***

6 ***If a star's distance is known, we can find its luminosity from its flux.***

7 ***Assuming that stars radiate like blackbodies, their colors indicate their surface temperatures; we can infer their sizes from their surface temperatures and radii.***

8 ***The Hertzsprung-Russell diagram is a graph of the surface temperatures and luminosities of stars; on it, stars fall into distinct groups: main sequence, giants, supergiants, and white dwarfs; once an H-R diagram has been made for a large number of stars, we can use it to estimate distances by spectroscopic parallaxes.***

9 ***Binary stars provide the only means of finding directly the masses of stars; for main-sequence stars, we find that the masses and luminosities are related (the more massive stars are more luminous); we infer that more massive stars have shorter lives than less massive ones.***

10 ***We are beginning to recognize that stars have activity, both erratic and cyclic, similar to that of the active sun; these include both starspots and flares that are driven by magnetic fields.***

STUDY EXERCISES

1 In the winter sky, you see stars with the following colors: Capella, yellowish; Betelgeuse, reddish; and Sirius, bluish. List these stars in order of *increasing* surface temperature. Estimate the surface temperature of Betelgeuse and of Sirius. Do they differ much? (*Objectives 1 and 2*)

2 Consider the following stars: *M*I, *G*III, and *A*V. Which star is the largest? Which is the most luminous? (*Objectives 3 and 10*)

3 Can you think of what considerations limit the accuracy of heliocentric parallax measurements? (*Objective 9*)

4 Refer to Fig. 14.15 to answer the following questions:
(*a*) Capella and the sun have roughly the same surface temperature. Which star is larger?
(*b*) Regulus and Capella have about the same luminosity. Which star is larger?
(*c*) Vega and Sirius have about the same surface temperature. Which star is more luminous?
(*d*) Which star would appear redder, Vega or Pollux? (*Objectives 10 and 11*)

5 Since we can't see the disks of other stars directly, how do we know that some of them have coronas like the sun? (*Objective 11*)

BEYOND THIS BOOK . . .

Atoms, Stars, and Nebulae (Harvard University Press, Cambridge, Mass., 1971) by L. Aller has more details about spectra and the physical properties of stars.

For original papers on spectra and stars, see sections 5 and 8 in *Source Book in Astronomy* (Harvard University Press, Cambridge, Mass., 1960), edited by H. Shapley.

Otto Struve and Velta Zebergs trace the development of ideas about stellar spectra in chapters 10 and 11 of *Astronomy in the 20th Century* (Macmillan, New York, 1962).

For an update on the Hertzsprung-Russell diagram, see "The H-R Diagram as an Astronomical Tool," *Sky and Telescope*, May 1978, p. 395, and *The H-R Diagram* (D. Reidel, Dordrecht, Netherlands, 1978) edited by A. G. D. Philip and D. S. Hayes.

For an incisive view about the physical properties of stars, read *Stars and Clusters* (Harvard University Press, Cambridge, Mass., 1979) by C. Payne-Gaposchkin.

For an introduction to RS CVn stars, read "The Strange RS Canum Venaticorum Binary Stars" by M. Zeilik, D. Hall, P. Feldman, and F. Walter, *Sky and Telescope*, February, 1979, p. 132.

PLATE 1

Major types of spectra. This figure has short wavelengths (blue) to the left with wavelength increasing to the right (to the red). At top (1) is a continuous spectrum such as you would observe from a glowing solid. Below it (2) is the absorption-line spectrum of the sun. Here only the most prominent dark lines are indicated by the element that produces them. Below the sun's spectrum are the bright-line spectra of selected elements: sodium (Na), hydrogen (H), calcium (Ca), mercury (Hg), and neon (Ne). Note how the bright lines of sodium, calcium, and hydrogen line up with the dark lines of the same elements in the sun's spectrum. (From General College Chemistry, fifth edition, by C. Keenen, J. Wood, and D. Kleinfelter, Harper & Row, 1976, reproduced with permission)

PLATE 2

A view of the surface of Mars, taken by the Viking 1 lander. Note that most of the rocks have small holes in them, an indication of a volcanic origin. Also note that the sky is NOT dark blue in color but a pink—the color comes from surface dust blown high up into the atmosphere by strong surface winds. (Courtesy NASA)

PLATE 3

Venus. Taken by the Pioneer Venus mission, this photo shows the circulation of the winds in the upper cloud layers of Venus's atmosphere. Note the yellowish color of the clouds—a color that is in accord with the idea that the clouds are made mostly of sulfuric acid. (Courtesy NASA)

PLATE 4

Comet West. The nucleus of this comet broke up into at least four pieces after it had passed the sun. (Photo taken with an 8″ Celestron Schmidt Camera, courtesy of Celestron International)

PLATE 5

A gigantic prominence erupting from the sun. This huge arch of hot gas spans more than 588,000 km across the solar surface. Photo taken from Skylab. (Courtesy of NASA)

PLATE 6

A close-up of the solar chromosphere, showing a small loop prominence. The colors here are not real, but are computer-generated to show the emission from lines in the ultraviolet. Red indicates emission from magnesium (9-times ionized), green from oxygen (5-times ionized), and blue from hydrogen (Lyman-alpha). The red region marks the corona above the (blue) chromophere. (Skylab photos Courtesy of R. Levine, Harvard College Observatory)

PLATE 7

Jupiter, as seen by Voyager 1 from a distance of 20 million km. The two moons visible are Io to the left, in front of the Red Spot, and Europa to the right. (Courtesy NASA)

PLATE 8

Volcanic eruption on Io, a Voyager 1 photo taken in March 1979. The eruption's plume stands out against the dark of space at the upper part of the picture. Computer processing has enhanced the brightness of the plume but preserved its greenish-white color. The plume reaches more than 200 km above Io's surface. (Courtesy NASA)

PLATE 9

Saturn, photographed by Voyager 1. Note the planet's shadow cast on the rings. (Courtesy NASA)

PLATE 10

Messier 17 (M 17), the Omega Nebula. This bright nebula is a cloud of hot, glowing gas surrounding a cluster of young, massive stars. Because the gas is largely hydrogen, it emits most of its light in the H-alpha line, which lies in the red region of the visible spectrum. So the nebula (and others like it) appears red. Note how the light is abruptly cut off at the top of the picture; here lie two fragments of a giant molecular cloud. The dust in the cloud cuts out starlight from stars behind them. Messier 17 formed out of one fragment of this much larger cloud. (Copyright by the Association of Universities for Research in Astronomy, Inc., Kitt Peak Observatory, reproduced with permission)

PLATE 11

Messier 20 (M 20), the Triffid Nebula in the constellation Sagittarius. Note that part of the emission is red (lower region) and part blue (upper region). The red light is from glowing hydrogen gas, heated by hot, young massive stars; the blue is starlight scattered by dust within the gas. The dark lanes mark regions where concentrations of dust block out light from the nebula. (Copyright by the Association of Universities for Research in Astronomy, Kitt Peak National Observatory, reproduced with permission)

PLATE 12

Messier 51 (M 51), a spiral galaxy, sometimes called the "Whirlpool Nebula." This photo, taken by James Wray, is designed to bring out the true colors in the different regions of the galaxy. Note how much redder the nucleus appears compared with the spiral arms. The blue regions in the arms are the locations of recent massive star formation. At the end of one spiral arm, at the bottom of the picture, lies a companion galaxy; note the blots of dust within it. (Courtesy J. Wray, copyright McDonald Observatory, the University of Texas)

PLATE 13

A VLA map of radio jets in the galaxy 3C388, a typical double radio source. Extending from the nucleus (red dot in center) are two huge lobes of radio emission, with sizes of about 300,000 ly. Note in the bottom lobe a clear indication of a jet pointing out from the nucleus to the strongest part of the lobe. This VLA radio map has been processed by computer so that the different colors reflect different levels of the intensity of radio emission: dark blue the weakest, light blue the next stronger level, light yellow the next, yellow next, and red the strongest. (Courtesy J. O. Burns; observations with the VLA by J. O. Burns and W. A. Christiansen)

PLATE 14

A total eclipse of the sun, February 26, 1979, taken from an aircraft at 40,000 ft. Note how far out the corona extends and the streams and knots visible within it. (Courtesy William H. Regan and Maxwell T. Sanford, Los Alamos National Laboratory)

PLATE 15

A color representation of the Hertzsprung-Russell diagram. The horizontal axis is temperature, the vertical one luminosity. Solid lines show where the luminosity classes fall: Ia supergiants at the very top, Ib supergiants below, III giants below them, and finally V main-sequence stars. The sizes of the stars are shown with the correct relative sizes but are not all to the same scale. The colors are true to those perceived by the eye.

PLATE 16

Voyager 2 photo of the wind flows in the northern hemisphere's upper atmosphere of Saturn. The colors have been computer-enhanced to bring out fine details. The greatest wind speeds here are about 100 m/sec. (Courtesy NASA)

PLATE 17

An all-sky map of the X-ray background at low energies. These energies are easily absorbed by neutral hydrogen gas in interstellar space, so the emission visible is from nearby, hot (one million Kelvins) gas. (Courtesy W. T. Sanders, University of Wisconsin at Madison)

PLATE 18

The Crab Nebula. Note the red filaments of the expanding gas. (Copyright by the California Institute of Technology and the Carnegie Institute of Washington, reproduced with permission)

PLATE 19

IRAS map in the infrared of the center of the Galaxy. In this computer-processed picture, the infrared emission is represented by the yellow color. (Courtesy NASA)

PLATE 20

IRAS map in the infrared of the Andromeda galaxy. This computer-processed picture uses red to show the strongest infrared emission, yellow next, and blue for the weakest. Note the strong emission from the nucleus and from starbirth regions in the spiral arms.

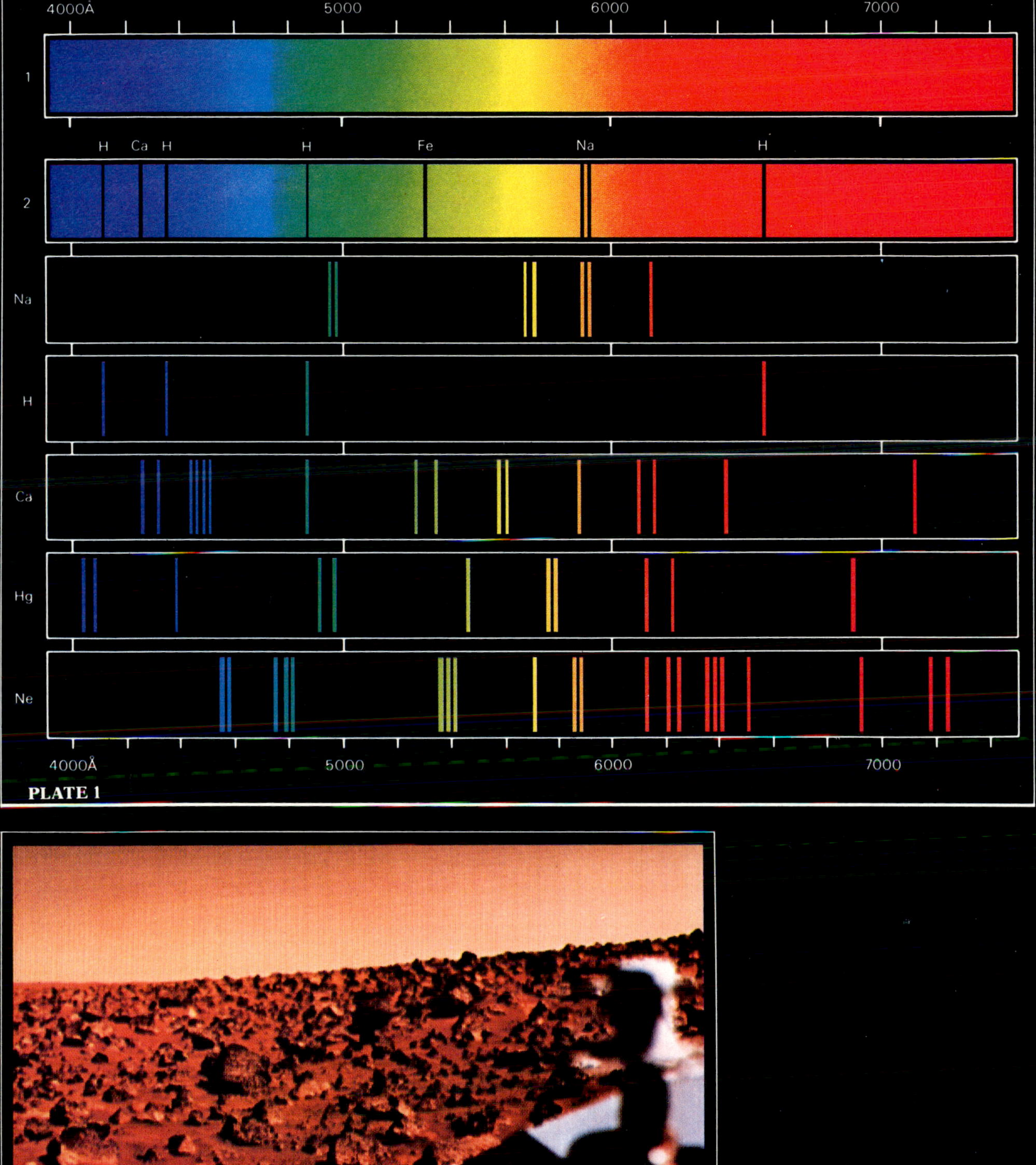

PLATE 1

PLATE 2

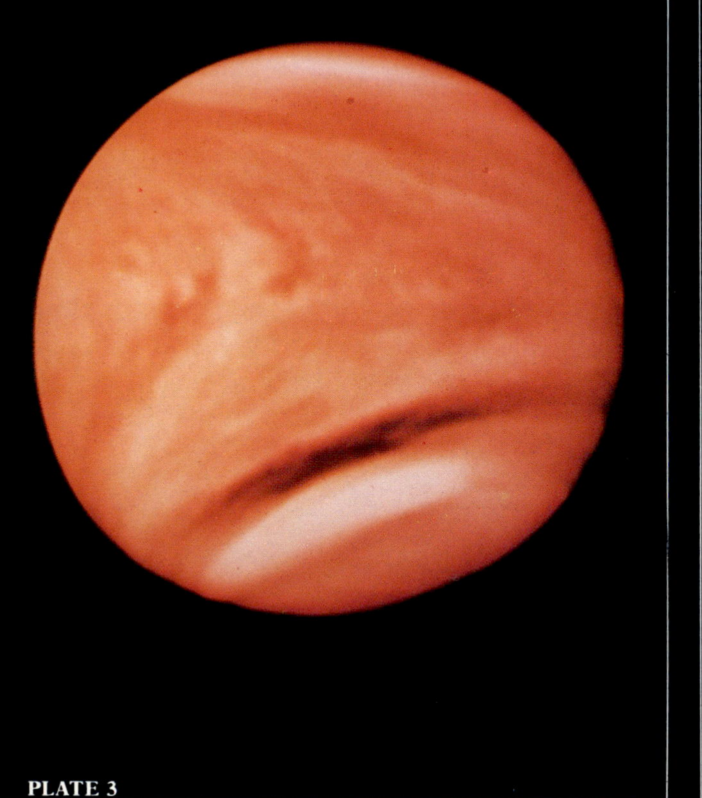

PLATE 3

PLATE 4

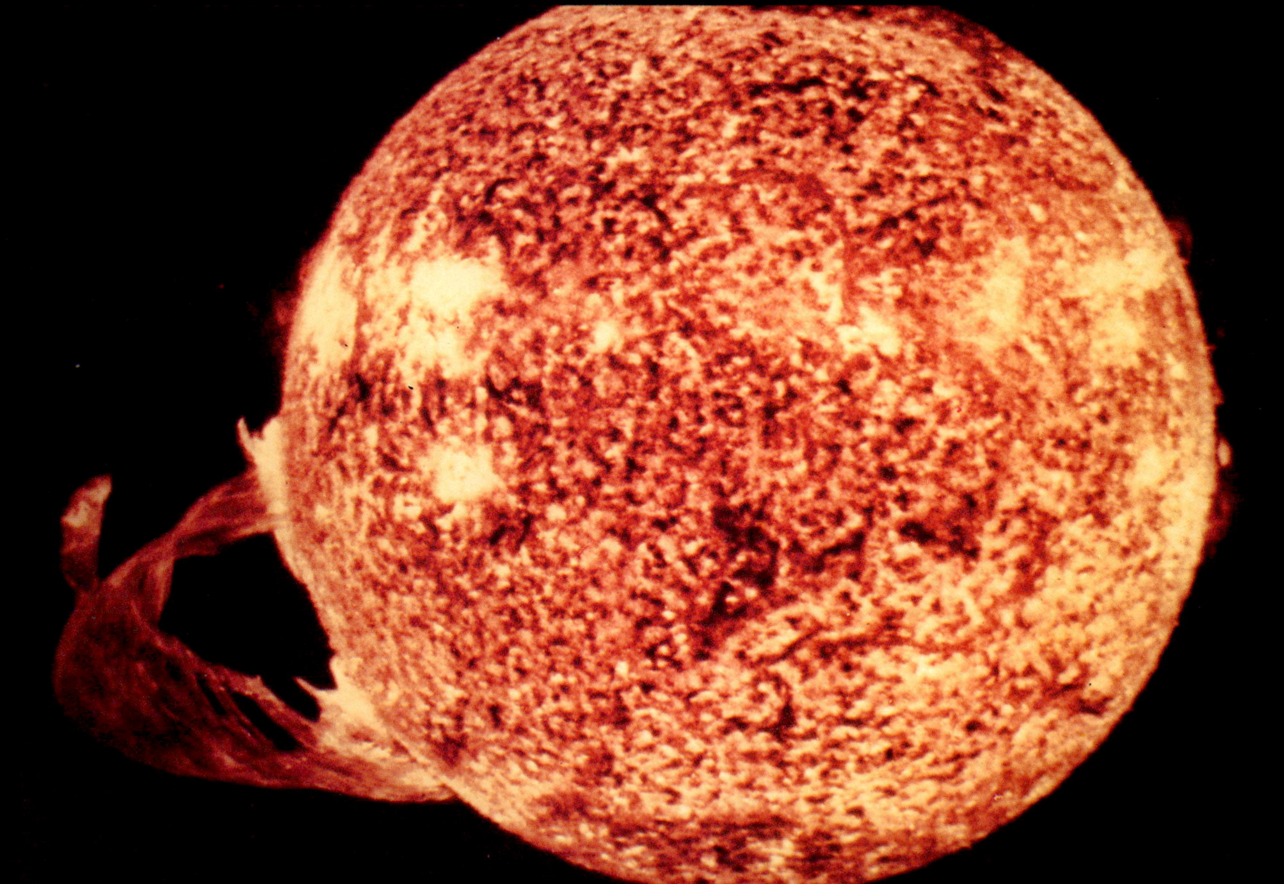

PLATE 5

PLATE 6

PLATE 7

PLATE 8

PLATE 9

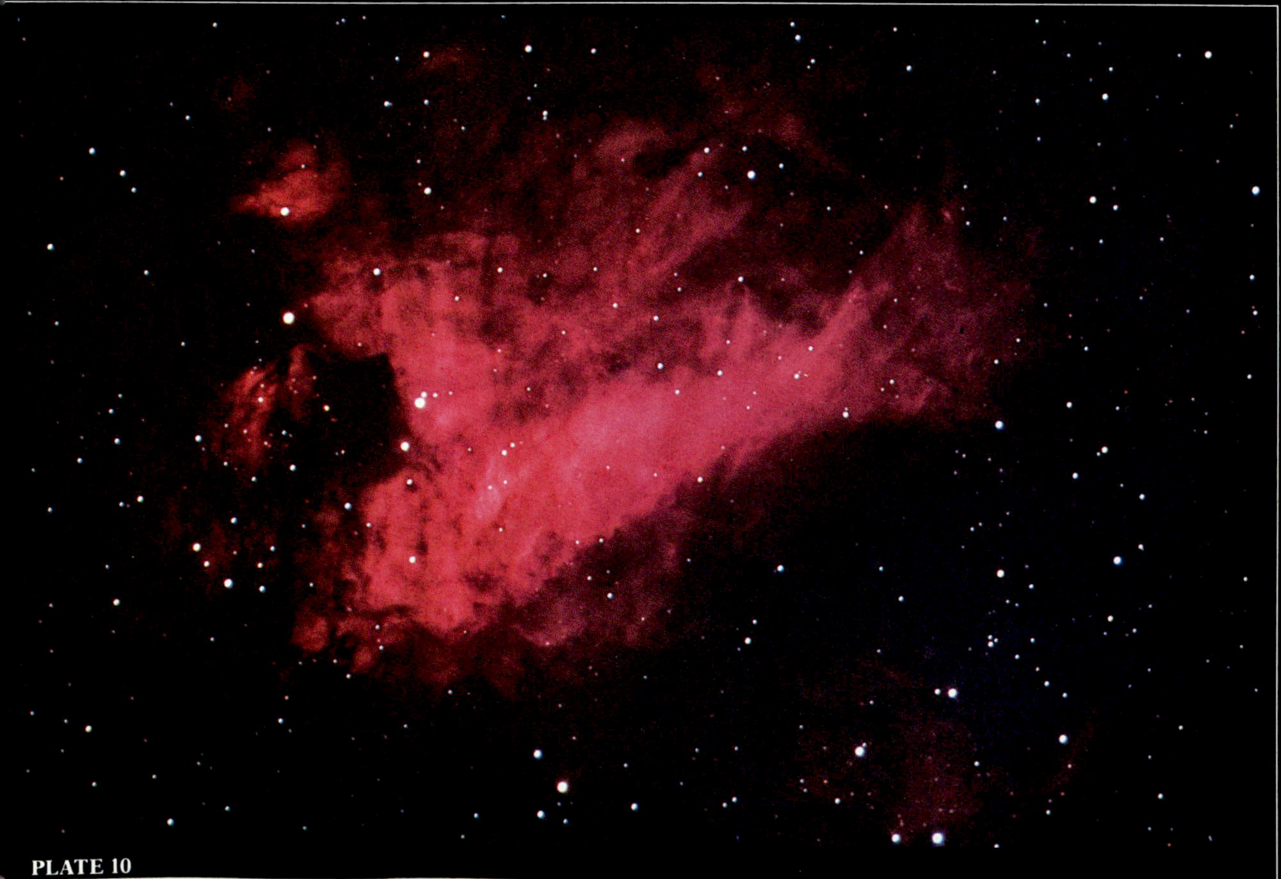

PLATE 10

PLATE 11

PLATE 12

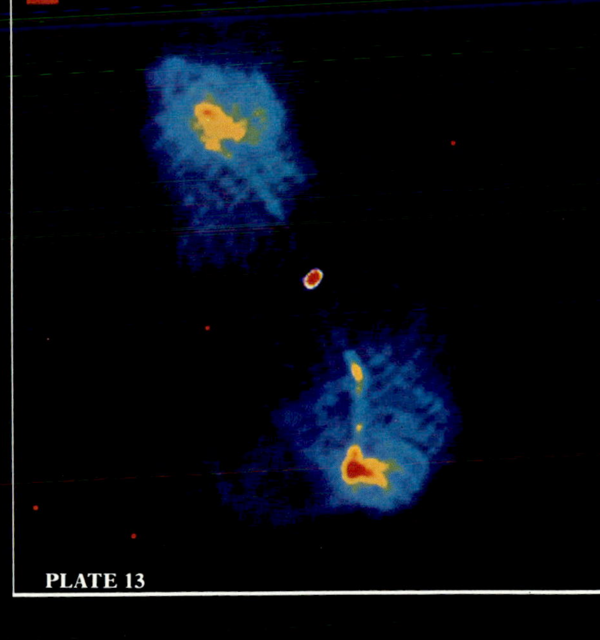

PLATE 13

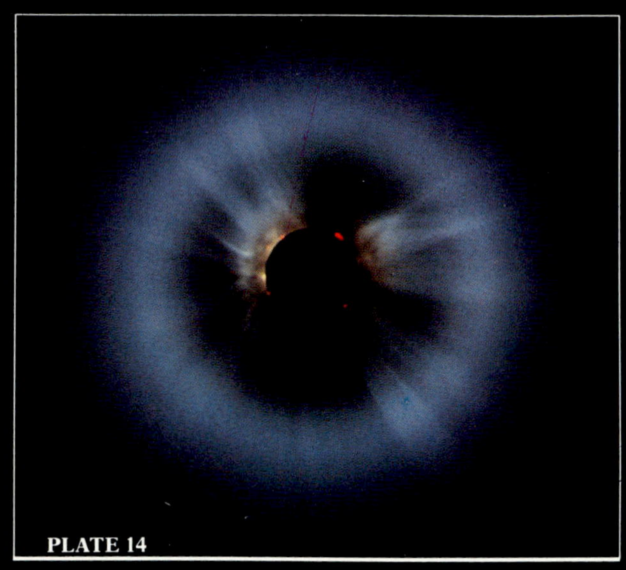

PLATE 14

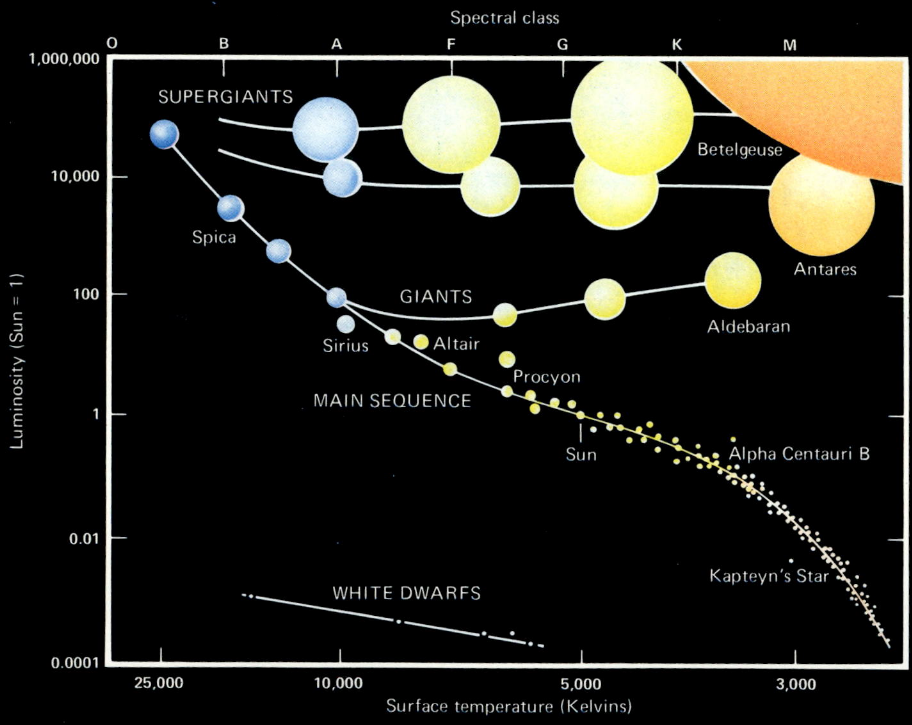

Spectral class

O B A F G K M

1,000,000

SUPERGIANTS

Betelgeuse

10,000

Spica

Antares

100

GIANTS

Aldebaran

Sirius

Altair

Procyon

MAIN SEQUENCE

1

Sun

Alpha Centauri B

0.01

Kapteyn's Star

WHITE DWARFS

0.0001

Luminosity (Sun = 1)

25,000 10,000 5,000 3,000

Surface temperature (Kelvins)

PLATE 15

PLATE 16

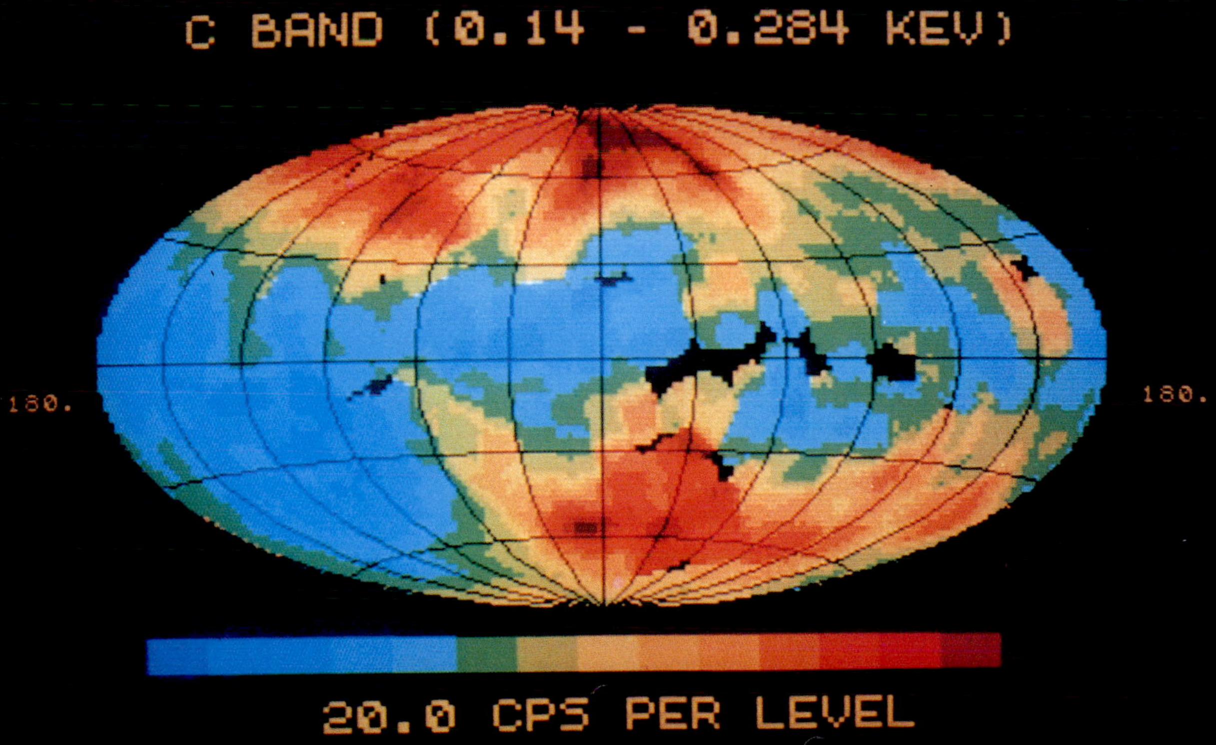

C BAND (0.14 - 0.284 KEV)

180.

180.

20.0 CPS PER LEVEL

PLATE 17

PLATE 18

PLATE 19

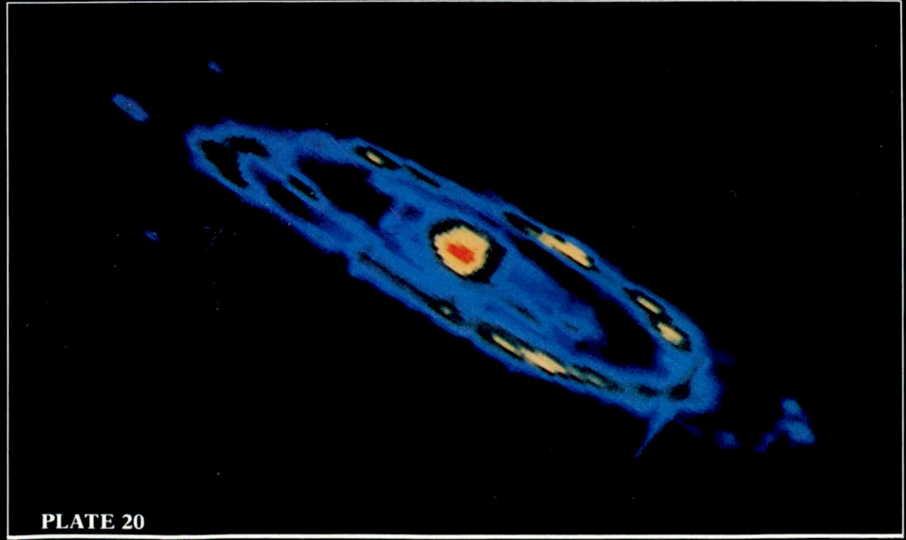
PLATE 20

15/Starbirth and interstellar Matter

Love is now the stardust of yesterday . . .
From "Stardust," lyrics by Mitchell Parrish

Learning objectives

After studying this chapter, you should be able to:

1. **Present observational evidence for the existence of gas and dust between the stars.**
2. **Compare and contrast the different forms in which the interstellar gas is found and tell how each form is observed.**
3. **Describe the effects of interstellar dust on starlight.**
4. **Describe possible physical properties of interstellar dust.**
5. **Indicate how interstellar molecules and dust might be formed.**
6. **Describe the basic physical ideas of gravitational collapse.**
7. **Sketch a scenario for the formation of massive stars from molecular clouds and indicate what observations support this model.**
8. **Outline possible processes for the birth of stars like the sun.**
9. **Sketch a model for the formation of massive stars and contrast it to that for solar-mass stars.**
10. **Argue that starbirth must be occurring now in our Galaxy, with a focus on infrared and radio observations.**

Central question

How are stars born out of the material between the stars?

How are stars born? You have found out that stars have finite lives—long by human standards, but limited nevertheless. How long a star lives depends on its mass. A star like the sun will survive for some 10 billion years. More massive stars live scant millions of years. The fact that we observe so many stars now means not only that starbirth occurred in the past but also that it must be going on now. Otherwise we would see fewer stars, especially the massive ones that die off relatively quickly.

Where are the wombs of starbirth? In clouds of gas and dust between the stars. Contrary to first impressions, interstellar space is *not* empty. It contains, thinly spread out and in clumps, both gas and dust. Hydrogen makes up most of the gas, which outnumbers the dust enormously. The interstellar gas is not simply made up of neutral atoms: Some is ionized and some is bound up in molecules. A large fraction of the interstellar gas is locked up in short-lived clouds. From these interstellar clouds, stars are born.

The interstellar medium marks the place where all the action is in the Galaxy. Here, stars form in vast stellar nurseries.

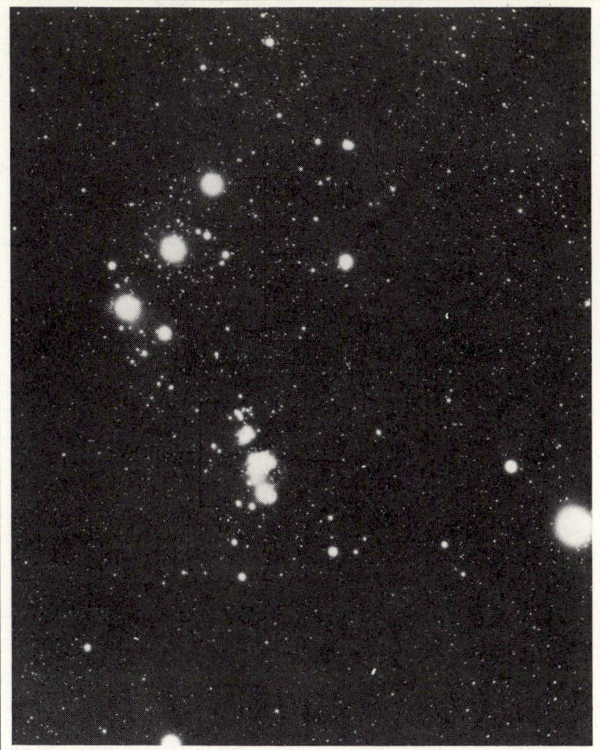

15.1
The belt and sword region of Orion (compare to Fig. 1.1). The Orion Nebula lies in the middle of the sword. (Courtesy Yerkes Observatory)

15.1
The interstellar medium: Gas

What do we know, in general, about the interstellar gas? First, it is made mostly of hydrogen. Second, it tends to clump in clouds. Third, a hot, dilute gas exists between the clouds. Fourth, the gas in different locations contains neutral atoms, ionized atoms, free electrons, and molecules. Fifth, on the average, the interstellar gas is very tenuous. The distance between interstellar atoms is roughly 100 million times larger than the size of the atoms themselves. If two people were separated by a proportional distance relative to their size, they would be about 100 million m apart, about the distance between the earth and moon! Sparse as it is, the interstellar gas occasionally clumps and forms stars.

Bright nebulas / On a winter night, you can easily spot the constellation Orion (see Fig. 1.1 and the January star chart in the endpapers). Dangling from Orion's belt is a short sword; if you look closely, the middle star appears fuzzy. A small telescope pointed at this fuzzy patch shows a diffuse, convoluted cloud surrounding a small cluster of stars (Fig. 15.1). This bright cloud is called the *Orion Nebula*. (*Nebula* is the Latin word for "cloud," and the use of *nebula* is an astronomical holdover from the last century.) Only 1500 ly from the earth, the Orion Nebula, roughly 20 ly in diameter, is typical of *diffuse* or *bright nebulas* (Fig. 15.2).

15.2
A close-up of the Orion Nebula, a typical bright, diffuse nebula. (Courtesy Lick Observatory)

At the end of the nineteenth century, spectroscopic analysis demonstrated that these bright nebulas did consist of hot gas. How? They showed spectra of emission lines. Recall from Kirchhoff's rules for spectra (Section 5.2) that an emission-line spectrum indicates a hot, diffuse gas. The Orion Nebula has lines of hydrogen, helium, and oxygen predominating in its bright-line spectrum.

Nebulas such as Orion do not shine by their own light; they absorb the energy from hot *O*- or *B*-stars located in or near them. The essential physical process is this: The gas absorbs high-energy ultraviolet photons given off by the central star (or stars) and gives off photons in emission lines at lower energies. For a hydrogen gas, the visible light comes out mainly in the red region of the spectrum as the Balmer line at 6563 Å. So emission nebulas glow predominantly with a red light (Color Plate 10). (For details of the emission process, see Focus 15.1.)

The star (or stars) in a bright nebula is quite hot, about 30,000 K, and it emits many photons with enough energy to ionize hydrogen. Most of these photons are absorbed by the gas surrounding the star so that, out to a considerable distance (a few tens of light years), the gas is almost totally ionized. This zone of ionized hydrogen is called an *H II region* (Fig. 15.3). (*H I* stands for neutral hydrogen and *H II* for ionized hydrogen. In general, a neutral atom is labeled I, an atom with one electron removed is labeled II, with two electrons removed, III, and so on.) Astronomers use the terms *H II region, bright nebula,* and *emission nebula* interchangeably.

Optical astronomers don't have a monopoly on viewing bright nebulas: Radio astronomers can see them in a special way. H II regions are visible by their continuous radio emission, produced by a free-free process when electrons speed past protons. As the opposite charges attract, the electrons are bent from their straight-line paths. The electrons are accelerated; their energy changes, and they emit electromagnetic radiation, mostly at millimeter and centimeter wavelengths. Since many different electrons undergo different energy changes at the same time, a continuous spectrum rather than an emission line, results. This continu-

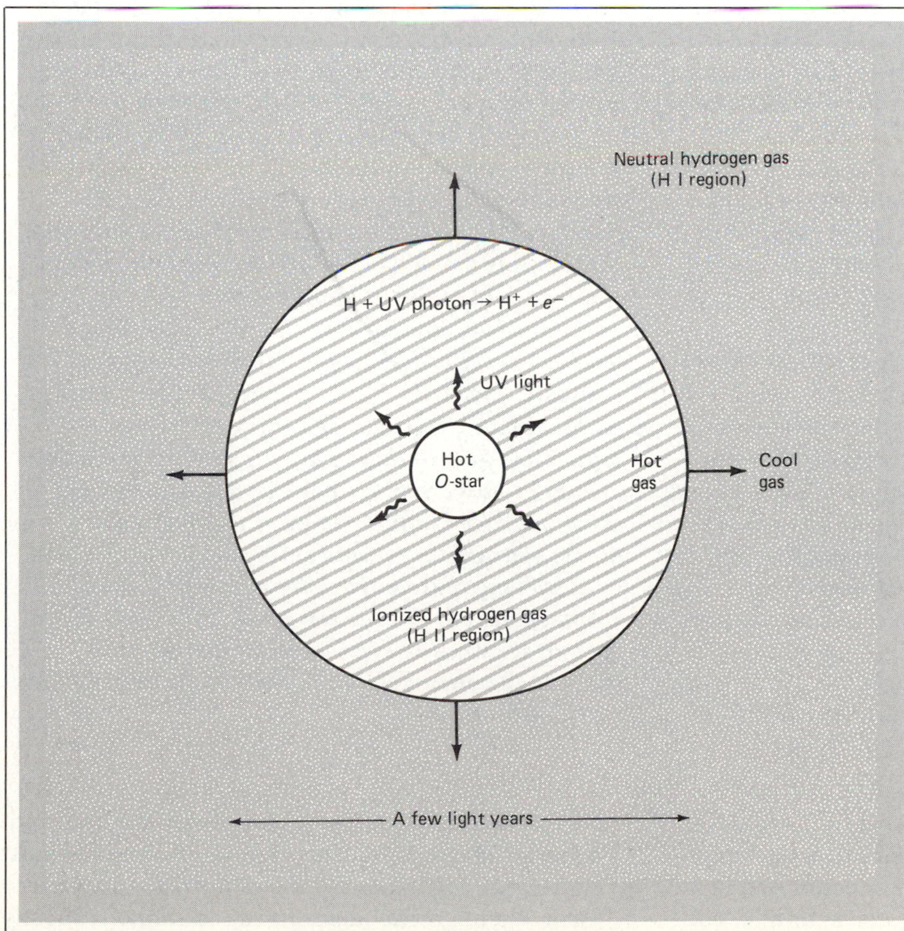

15.3

A schematic diagram of an H II region. A hot O-star (or stars) emits ultraviolet light that can ionize hydrogen for a few light years around. The absorption of the ultraviolet heats up the ionized gas to a temperature of about 10,000 K. The hot, ionized gas expands into the cooler, neutral gas surrounding it.

Focus 15.1

EMISSION NEBULAS

Let me describe in more detail the process by which a bright nebula emits light. Virtually all the atoms in an H II region are ionized. Most of these atoms are hydrogen. When a naked hydrogen nucleus (a proton) captures an electron (a process called **recombination***), the reformed atom gives off light. The details of this process are complicated by all the possible energy levels of the atom (see Section 5.3).*

To simplify matters, consider the lowest energy levels in hydrogen. Imagine a hydrogen atom with an electron in the lowest level. The atom absorbs a high-energy ultraviolet photon (wavelength 912 Å or shorter). This additional energy kicks the electron out of the nucleus's grip entirely and ionizes the atom. The electron zips around by itself with any one of a large possible range of energies. Ions and free electrons flying around are the typical state of the gas in an H II region. The ions travel around until they encounter an electron with which to combine. This does not happen often because the gas in the nebula is sparse.

Suppose the ion does pick up an electron. (Remember that they have opposite charges and so attract each other.) The electron does not necessarily end up in level 1. But if it does, it emits an ultraviolet photon that can ionize another atom. Quite quickly, another ultraviolet photon zaps the atom and ionizes it again.

Imagine that the electron is caught in one of the higher levels, say level 3 (Fig. F.33a). As it is caught, it emits a low-energy photon. Since the free electrons may have any energy, the photons emitted when the electrons recombine with an ion can also have any energy, as long as it is greater than the energy difference between the final level (3 in this case) and the fully ionized state. Hence this free-bound transition results in a continuous spectrum of emission.

The electron does not stay in level 3 for long, because it seeks the lowest energy state (level 1). It can get there by dropping through the levels one by one or by skipping over some, just as you could go down stairs one at a time or by skipping a few. However, as the electron drops down to the lowest level, it must emit a photon with each drop.

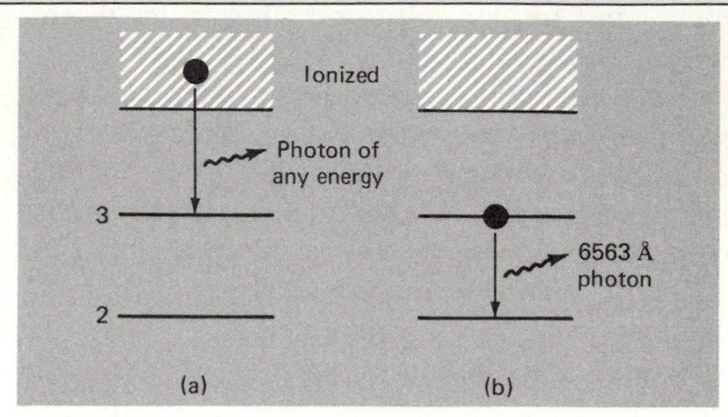

F.33
Emission of photons by hydrogen. (a) *When a hydrogen ion captures an electron, it can end up in any energy level below the starting level, so a photon of any energy can be emitted.* (b) *If it falls between bound levels, it can emit photons with certain energies, such as the H-alpha (6563-Å) photon emitted when the electron drops from level 3 to level 2.*

Suppose the atom, having caught the electron in level 3, then drops to level 2. It emits a Balmer photon with a wavelength of 6562.8 Å; the emission line that results is called the **H-alpha line** *(Fig. F.33b). It is the most common emission line from a nebula and makes it appear red.*

You may wonder why the next drop, from level 2 to level 1, does not produce the brightest emission line. This drop produces an ultraviolet photon at a wavelength of 1216 Å, the **Lyman-alpha line***. Few of these photons make it out of the nebula because they are constantly absorbed by other hydrogen atoms. Most of the hydrogen atoms are in their ground state and so are capable of absorbing these ultraviolet photons of the Lyman series of lines. Even if these photons did get out, we would never see them on the earth because the ultraviolet light does not penetrate the earth's atmosphere.*

ous radio emission can be mapped by a radio telescope (Fig. 15.4). From such maps, astronomers can infer how much ionized gas is contained in an H II region (the Orion Nebula, for example, contains about 300 solar masses).

Interstellar atoms / Investigations of bright nebulas show that they are composed almost entirely of hydrogen. So the neutral gas ionized to form them must also have contained mostly hydrogen, as a neutral atom (designated *H I*).

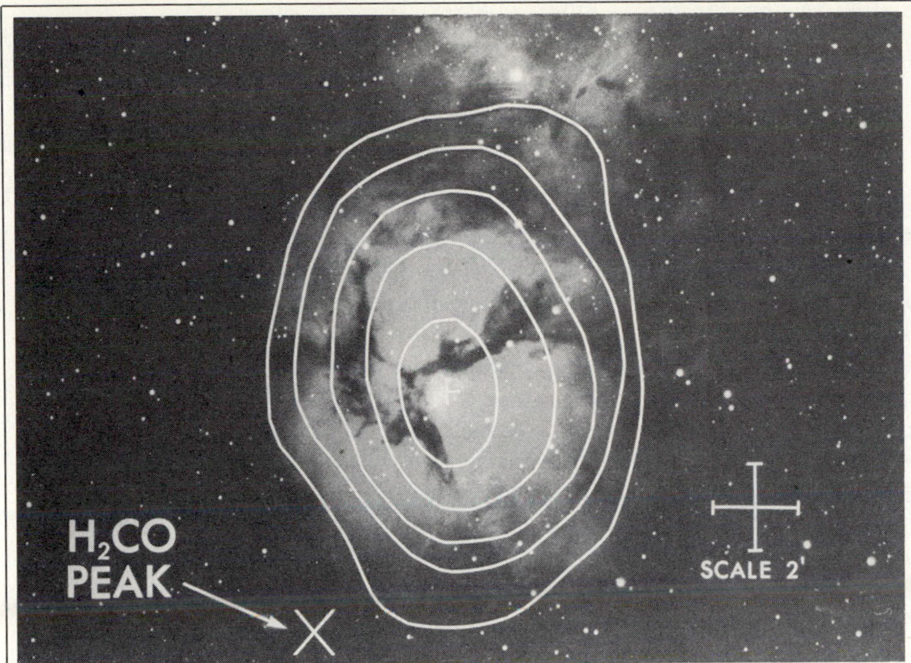

15.4

An intensity contour map of the free-free radio emission from an H II region overlayed on an optical photo. The radio observations were made at Haystack Observatory at a frequency of 8 GHz. Noted nearby is the peak of emission from the surrounding molecular cloud. (Courtesy E. Chaisson)

Astronomers surmised that hydrogen atoms populated interstellar space, but they did not observe the H I gas until 1951 with radio telescopes at a wavelength of 21 cm (Focus 15.2).

Surveys at 21 cm find that most neutral hydrogen is concentrated in the plane of the Milky Way. (The *Milky Way* is a faint, broad band of light in the sky; as Galileo discovered, it consists, when viewed with an optical telescope, of a multitude of faint stars, all belonging to our Galaxy. When I refer to our Galaxy, as opposed to all others, I call it *the Galaxy*. See Chapter 18 for more on the Galaxy.) On the average, the hydrogen atoms have a temperature of 70 K and a density of 3×10^5 in a cubic meter. So in a volume of space equivalent to the volume of your body, you'd find only 3×10^4 hydrogen atoms, whereas in fact your body contains some 10^{27} atoms.

As you might expect, there are also other gases in interstellar space. Even before atomic hydrogen was observed with radio telescopes, optical observations had revealed the presence of several other kinds of atoms. Superimposed on the spectra of some stars, astronomers found sharp, dark lines of elements such as sodium (Fig. 15.5). These narrow absorption lines are produced when starlight passes through cool regions of the interstellar gas.

Recall Kirchhoff's rules (Section 5.2): An absorption spectrum results when light from a continuous source passes through a cooler gas. But how do we know that these lines are formed by the interstellar gas rather than in the atmosphere of the star? For one thing, the lines are very narrow, an indication that the gas is cool. But the strongest evidence came when these lines were found in the spectra of some binary stars. As the stars move around each other, all the stellar absorption lines shift back and forth in

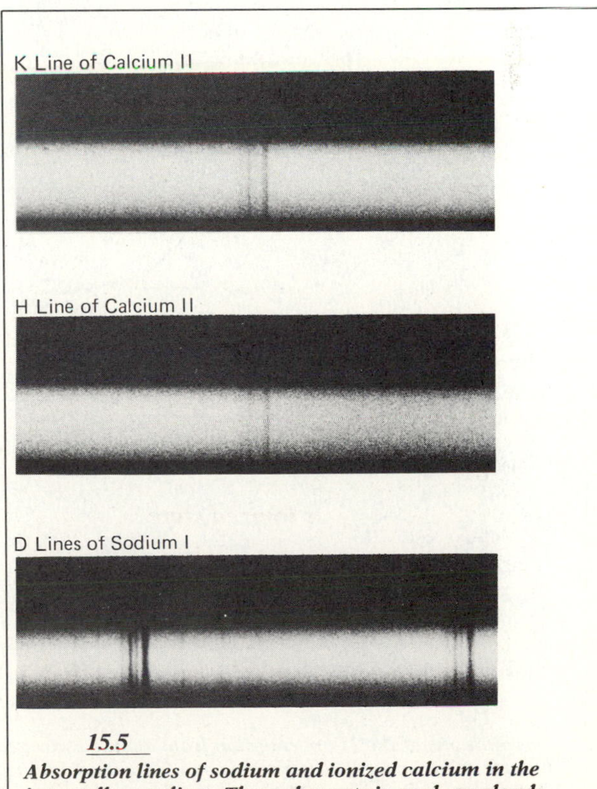

15.5

Absorption lines of sodium and ionized calcium in the interstellar medium. These elements in cool gas clouds absorb some of the starlight that passes through them. (Courtesy Palomar Observatory, California Institute of Technology)

Focus 15.2

21-cm EMISSION FROM HYDROGEN

Recall that the hydrogen atom has one proton in the nucleus and one electron in orbit around it. Both the proton and the electron have angular momentum. You can imagine them spinning like miniature tops. According to the rules of quantum physics, the electron and the proton can be oriented in the atom so that the two spins either align or oppose each other. If the spins oppose, the total energy of the atom is just a bit less than if the spins align. As usual, the atom prefers to be in the lower energy state. Suppose the spins are aligned; eventually the proton flips over and emits a low-energy photon—energy that corresponds to a wavelength of 21.11 cm (Fig. F.34).

How are the protons and electrons in hydrogen atoms aligned in the first place? It can happen by collisions with electrons and other atoms. The gas in interstellar space is very sparse, and collisions between two atoms occur only once every few million years. On the other hand, once the spins in a hydrogen atom are aligned, about 10 million years on the average must pass before the proton flips and the atom drops to its lowest energy state. So several collisions occur before the atom radiates. Some will misalign the spins again. But the final result is that an equilibrium is established in which three times as many atoms have spins aligned as have spins opposed. Every so often, once in 10 million years on the average, an aligned atom flips over spontaneously, without a collision, and emits a 21-cm photon. This is a rare event for any one atom. But because so many hydrogen atoms exist in interstellar space, enough are emitting 21-cm radiation at any give time that the interstellar gas radiates strongly at this wavelength and can be detected with radio telescopes (Fig. F.35).

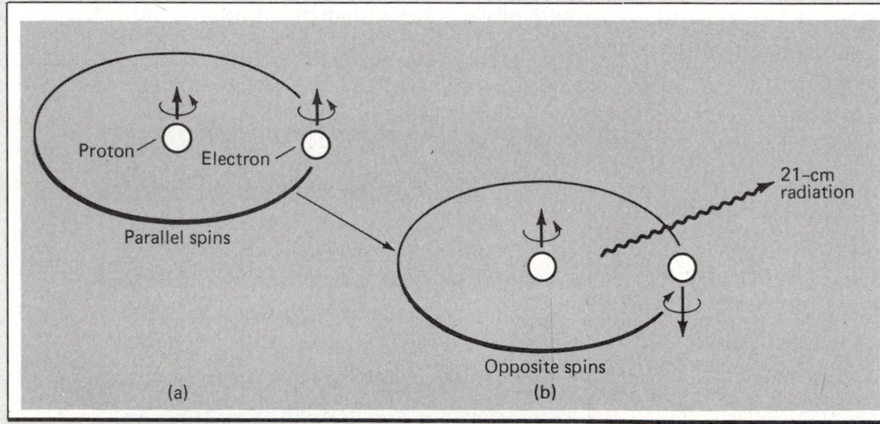

F.34
21-cm emission from hydrogen atoms. Collisions can line up the spins of the proton and electron so that they are parallel. After some time, if the atom does not hit another, the electron flips so that the spins are opposed, and the atom emits a 21.1-cm photon.

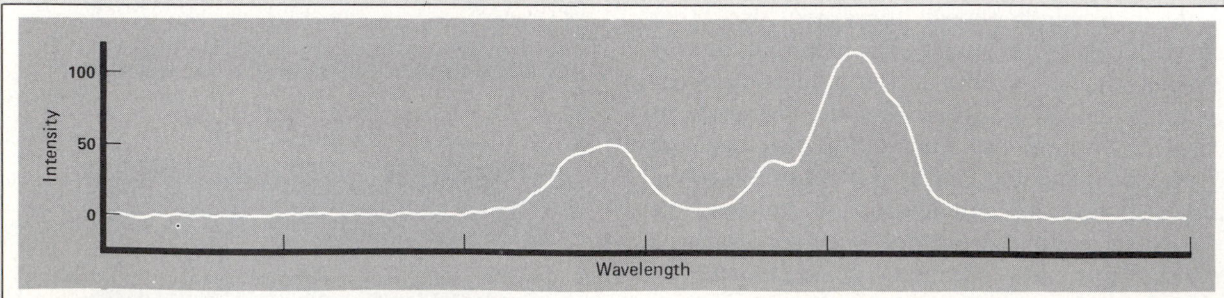

F.35
Observations of the 21-cm emission from atomic hydrogen in interstellar space. (Observations by W. Burton with the Dwingeloo radio telescope)

wavelength, due to the Doppler shift caused by their changing velocities. But the interstellar lines remain steady at a constant wavelength, indicating that the gas producing them has nothing to do with the stars themselves. These kinds of optical observations uncovered atoms of sodium, potassium, ionized calcium (Fig. 15.5), and iron in the interstellar medium.

These interstellar absorption lines were the first direct indication of the existence of a pervasive interstellar medium. They led to the idea that cool regions in the medium exist in the form of small clouds.

You may be wondering why hydrogen—by far the most abundant atom in the interstellar gas—wasn't first found optically. Because interstellar space is so empty and the temperature of the gas is so low, most hydrogen atoms remain in their lowest energy state, from which they can absorb only ultraviolet light. This cannot penetrate the earth's atmosphere, so these absorption lines cannot be observed with ground-based telescopes. With the advent of earth-orbiting ultraviolet telescopes, we have finally observed cold hydrogen optically.

Such ultraviolet observations have complemented the radio picture of the interstellar gas. They show that the neutral hydrogen gas has a very patchy distribution in clouds with diameters from tenths to tens of light years. The average density of neutral hydrogen is somewhat less than 1 million atoms per cubic meter. However, the observations also show directions in space where the neutral hydrogen density is ten times less. It doesn't seem likely that these regions have no gas at all, but rather that the gas is in a different form. For example, if the hydrogen is hot and ionized, it will no longer have any ultraviolet absorption spectrum and so will not be detectable by ultraviolet observations. Here is evidence that the same interstellar gas is (at least in part) ionized, although not dense enough to emit observable bright lines.

Intercloud gas / The spaces between interstellar clouds also contains gas. And, as you'd expect, it must consist mostly of hydrogen. Is it ionized or neutral? Well, to make life complex, it seems to be both! Observations at 21 cm indicate a thin, neutral gas. Other radio observations point to an even thinner, hotter (about 10,000 K) ionized gas also between the clouds.

Ultraviolet observations provide direct evidence for a *very* hot gas permeating the intercloud regions. The observations show absorption lines of oxygen stripped of five electrons (Fig. 15.6). (The symbol for five-times-ionized oxygen is *O VI*.) To

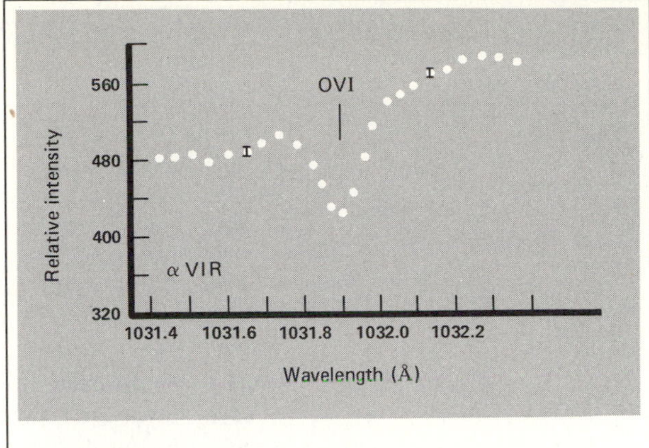

15.6
Interstellar absorption lines from O VI (five-times ionized oxygen) looking toward the star Alpha Virginis. (From observations with the Copernicus satellite by D. York,* Astrophysical Journal (Letters)*, vol. 193 (1974), p. L127, copyright © 1974 by the American Astronomical Society)

rip so many electrons from an oxygen atom requires a very high temperature—about a million K. Because this hot gas has about the same temperature as the sun's corona, it is called the *coronal interstellar gas*. Evidence for it comes also from the X-rays it emits (Color Plate 17). It must occupy a large fraction of the volume of interstellar space (perhaps 90 percent), because the O VI absorption lines, and the X-rays, have about the same intensity in all directions we look from the earth.

Interstellar molecules / So much for hot stuff. What about cold material? You might expect it to be in the form of simple molecules because atoms tend to combine into molecules at cold temperatures. Optical astronomers made the first discoveries of these. But the optical part of the spectrum is not the most fruitful region to search. A molecule consists of atoms linked together in particular arrangements by electron bonds. It can have different energy states if the atoms vibrate or the molecule spins in various ways. As with changes in electronic states in an atom, when a molecule changes its vibrational or rotational state, it can emit or absorb a photon. For changes in vibrational states, the photons are infrared ones; for rotational states, radio ones. In the cold regions where molecules can exist in interstellar space, occasional collisions between molecules (or perhaps with atoms) kick the molecules and get them spinning. Eventually, the molecules emit radio photons that can be observed as a radio emission line, generally at millimeter wavelengths.

The search for molecules began in earnest in the 1960s. More than 60 molecules have been found so far. Table 15.1 lists some of the key ones.

TABLE 15.1 Some interstellar molecules observed to date

Complexity	*Inorganic molecules*		*Organic molecules*	
Diatomic	H_2	Hydrogen	CH	Methylidyne radical
	HD	Deuterized hydrogen	CH^+	Methylidyne ion
	OH	Hydroxyl radical	CN	Cyanogen radical
	SiO	Silicon monoxide	CO	Carbon monoxide
	NS	Nitrogen sulfide		
	SO	Sulfur monoxide		
Triatomic	H_2O	Water	CCH	Ethynyl radical
	HDO	Heavy water	HCN	Hydrogen cyanide
	N_2H^+	Protonated nitrogen	DCN	Deuterium cyanide
	H_2S	Hydrogen sulfide	DNC	Deuterium isocyanide
	SO_2	Sulfur dioxide	HCO^+	Formyl ion
4-atomic	NH_3	Ammonia	H_2CO	Formaldehyde
			HNCO	Isocyanic acid
			H_2CS	Thioformaldehyde
			HC_2H	Acetylene
			C_3N	Cyanoethynyl radical
5-atomic			H_2CNH	Methanimine
			H_2NCN	Cyanamide
			HCOOH	Formic acid
			HC_3N	Cyanoacetylene
			H_2C_2O	Detene
6-atomic			CH_3OH	Methyl alcohol
			CH_3CN	Methyl cyanide
			$HCONH_2$	Formamide
7-atomic			CH_3NH_2	Methylamine
			CH_3C_2H	Methylacetylene
			$HCOCH_3$	Acetaldehyde
			H_2CCHCN	Vinyl cyanide
			HC_5N	Cyanodiacetylene
8-atomic			$HCOOCH_3$	Methyl formate
			CH_3C_3N	Methyl cyanoacetylene
9-atomic			$(CH_3)_2O$	Dimethyl ether
			CH_3CH_2OH	Ethyl alcohol
			HC_7N	Cyanotriacetylene
			CH_3CH_2CN	Ethyl cyanide

Note that carbon monoxide (CO) is one of the most common molecules in space. Other amusing ones include water and ethyl alcohol, which gives the kick to beer and wine. Note that many of the molecules are *organic*; that is, compounds in which carbon plays a central role. The most abundant atoms in these molecules—carbon, hydrogen, nitrogen, and oxygen—are also the most abundant in living creatures on the earth. (More on this in Chapter 22.)

By far the most abundant molecule is—you guessed it—molecular hydrogen, simply because hydrogen is the most abundant element in the universe. But even though radio telescopes can easily observe atomic and ionized hydrogen, they cannot detect molecular hydrogen, because it does not emit or absorb at radio wavelengths. The hydrogen molecule does absorb and emit ultraviolet and infrared wavelengths, however. Infrared emission lines of molecular hydrogen have been observed from heated interstellar clouds, and ultraviolet absorption lines from cool clouds have been detected in the spectra of hot stars (Fig. 15.7).

Molecular clouds / Although some interstellar molecules, such as carbon monoxide, pop up in almost every direction in the sky, most are concentrated in dark, dense, cold conglomerates called *molecular clouds*. These clouds often lie near H II regions. One of the closest molecular clouds sits behind the Orion Nebula, as we view it from earth. The molecular cloud here consists of two parts: a large, low-density cloud (inferred from carbon monoxide emission) surrounding a

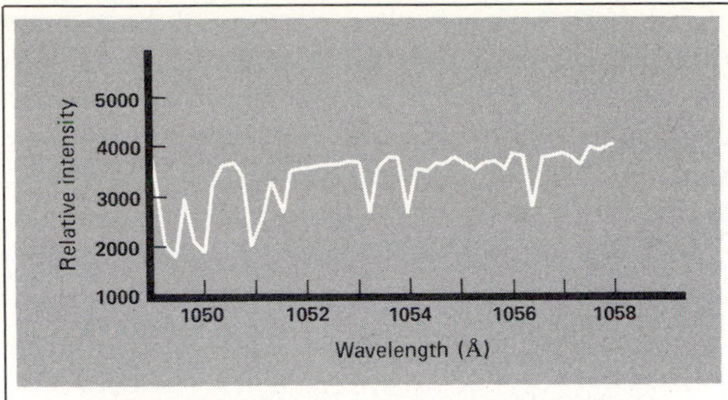

15.7
Interstellar absorption lines from molecular hydrogen, observed by the Copernicus satellite. (From J. Rogerson, L. Spitzer, J. Drake, K. Dressler, E. Jenkins, D. Morton, and D. York,* Astrophysical Journal (Letters), *vol. 181 (1973), p. L97, copyright © 1973 by the American Astronomical Society)

dense, small core. The low-density cloud has an enormous extent (Fig. 15.8): it is at least 30 ly across, has a peak density of 10^9 hydrogen molecules per cubic meter, and contains at least 10,000 solar masses of material. The core (Fig. 15.9) is only 0.5 ly in size, has a peak density of 10^{11} hydrogen molecules per cubic meter, and a mass of only 5 solar masses.

The Orion region presents an excellent example of a *giant molecular cloud.* Observations so far indicate that the bulk of the material of the interstellar medium is bound up in complexes of giant molecular clouds. These immense globs of molecules, held together by gravity, have the following typical properties:

1 They consist mostly of molecular hydrogen; many other molecules are present, but these make up only a small fraction of the mass.

2 The cloud complexes have average densities of a few hundred million molecules per cubic meter; the individual clouds are slightly denser, with a few billion molecules every cubic meter.

3 They have sizes of a few tens of light years.

4 The total masses of the complexes range from 10,000 to 10 million solar masses; 100,000 solar masses is typical. Masses of individual clouds are about 1000 solar masses.

The cores of these clouds are unusual places compared with the average interstellar medium. Here the temperatures are a frigid 10 K, and the densities get as high as 10^{12} molecules per cubic meter. That's an immense concentration by interstellar standards, yet it is only 10^{-13} times the density of molecules in the air at the earth's surface. Giant molecular clouds are so huge, though, that they contain an enormous number of molecules in total.

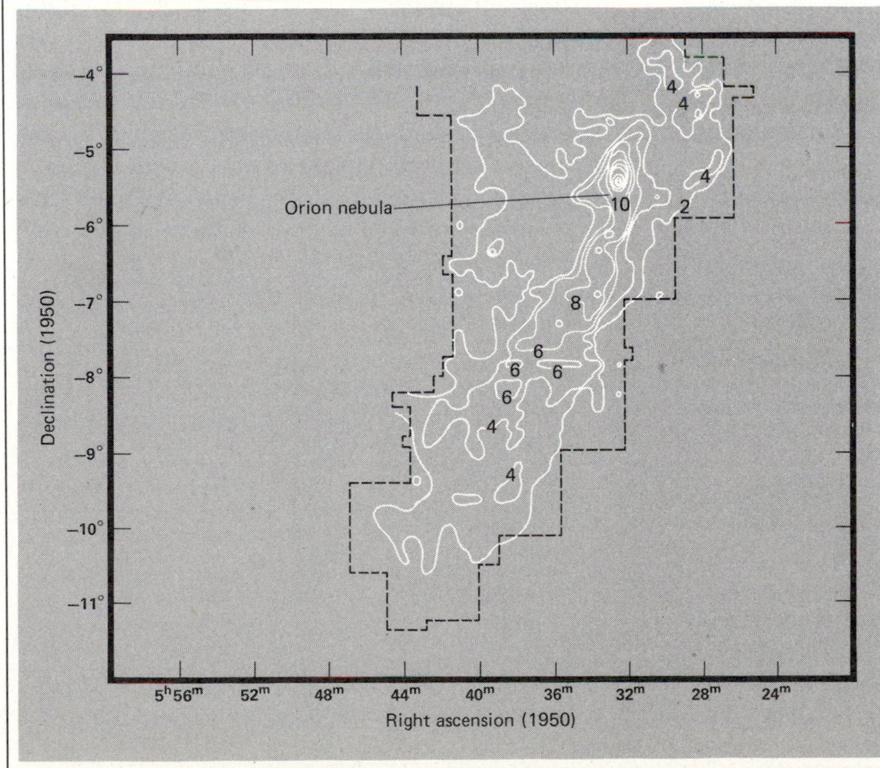

15.8
A carbon monoxide intensity map of the giant molecular cloud associated with the Orion Nebula. The hottest, densest part of the cloud lies just above the contour marked 10; the Orion Nebula lies just below and to the right of this peak. This map was made by the Columbia University millimeter-wave antenna. (Adapted from a figure by M. Kutner, K. Tucker, G. Chin, and P. Thaddeus,* Astrophysical Journal, *vol. 215 (1977), p. 521, copyright © 1977 by the Ameri-

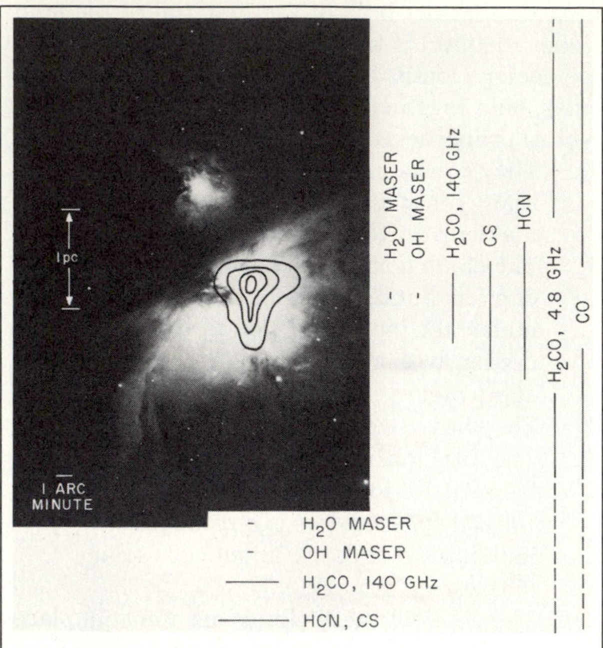

15.9

A map of the central part of the giant molecular cloud that lies behind the Orion Nebula. The dark contour lines show the dense core of cloud; the line labeled "1 pc" gives the scale (1 pc = 3.26 ly). The bars on the sides of the photo mark the extent of the core when mapped in the emission from carbon monoxide (CO), formaldehyde (H_2CO), and hydrogen cyanide (HCN). Very small regions of water and hydroxyl emission come from the "H_2O maser" and "OH maser" regions. (Figure courtesy E. Chaisson; photo courtesy Kitt Peak National Observatory)

There is one important property of the location of these clouds in space: Giant H II regions, which surround young, massive stars, are always found near molecular cloud complexes. This proximity suggests that giant molecular clouds play a key role in the process of star formation.

The parts of the interstellar gas: Summary / By now you may be wondering how all the forms of the interstellar gas fall together into a coherent picture of the interstellar medium. I have been involved with research on the interstellar medium for a number of years and have reached one firm conclusion: Our models for the interstellar medium have changed rapidly and will continue to do so.

Let's sum up what we know generally of the interstellar gas (Table 15.2):

1 *H II regions*. Zones of glowing, ionized hydrogen surrounding young, hot stars (spectral types *O* and *B*); contain a minor amount of the interstellar gas, perhaps a total of 10 million solar masses in the Galaxy.

2 *H I regions*. Clouds of cool, neutral hydrogen roughly 20 ly in diameter and each containing about 50 solar masses of material; total mass in the Galaxy may be 3 billion solar masses.

3 *Intercloud medium*. Two parts: One is a hot, ionized, but thin hydrogen gas between the clouds, some in the form of coronal gas; another part is cooler and neutral.

4 *Molecular clouds*. Small to giant in size, containing mostly molecular hydrogen; total mass of a few billion solar masses.

15.2
The interstellar medium: Dust

Dust also occupies interstellar space. There's not much out there; on the average, there's one dust particle in every million cubic meters—that's roughly a cube with sides one football field in size! But the dust amounts to about one percent of the total mass of interstellar matter, and it can cut out light from distant objects or from those shrouded in dense clouds. Piercing the dust veil has been an important goal of radio and infrared astronomers. That breakthrough has been critical in revealing the process of starbirth.

TABLE 15.2 Major parts of the interstellar medium

Component	*Indicator*	*Temperature (K)*	*Fraction by mass (%)*
Molecular clouds	CO	10–50	40
H I clouds	21 cm	50–100	40
Intercloud gas	21 cm	7000–10,000	20
H II regions	H-alpha Continuous radio	10^4	(little)
Coronal intercloud gas	O VI	10^6	0.1

SOURCE D. D. Clayton, "The Cloudy State of Interstellar Matter," in *Protostars and Planets*, ed. T. Gehrels (University of Arizona Press, Tucson, 1978).
Note The fractions by mass are very approximate and should only be taken as a guide.

15.10
The Horsehead Nebula, a dark nebula in Orion. (Courtesy Palomar Observatory, California Institute of Technology)

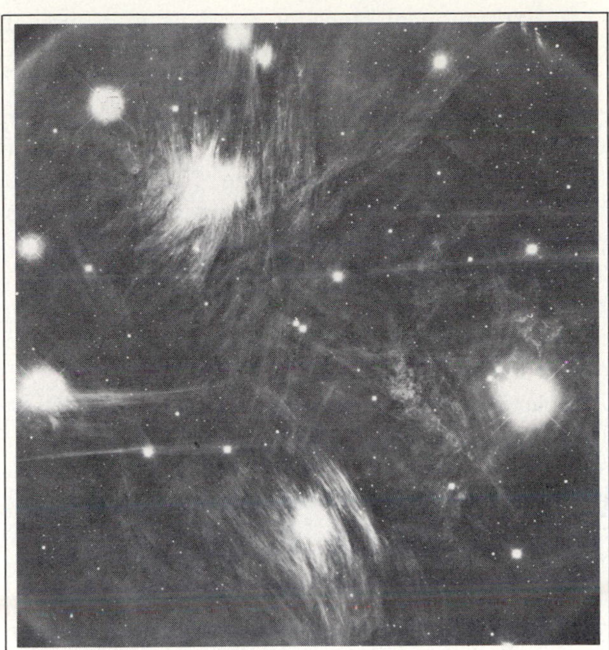

15.12
The Pleiades star cluster (Messier 45) immersed in a bright reflection nebula. (Courtesy Kitt Peak National Observatory)

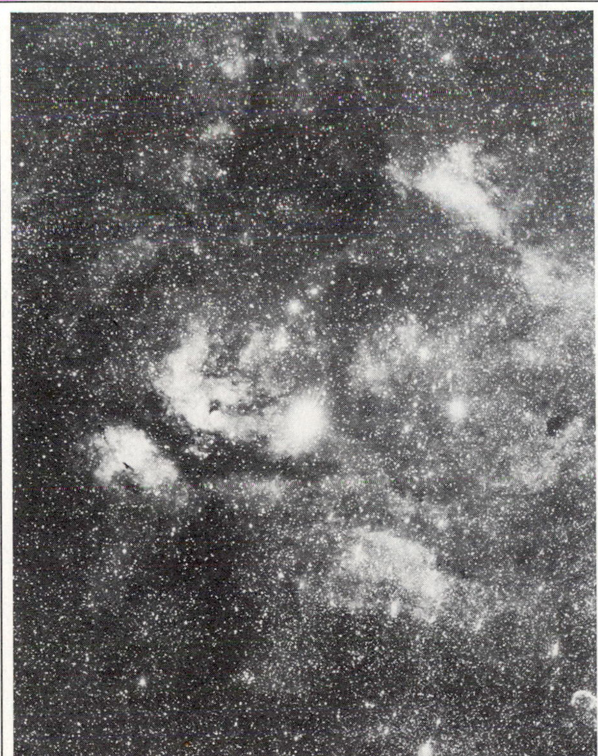

15.11
A region of the Milky Way in Cygnus. The dark areas (note especially the small globules) are regions of light absorption by interstellar dust. (Photo by B. Nelson, courtesy of Hopkins Observatory, Williams College)

Cosmic dust / Direct observations hint at dust between the stars. Dark nebulas, such as the famed Horsehead Nebula in Orion (Fig. 15.10), display dramatic cutoffs of light due to dust. The dark rifts and lanes in the Milky Way (Fig. 15.11), once thought to be due to the lack of stars, are actually regions heavily obscured by dust.

Some bright nebulas are not H II regions but clouds of dust illuminated by nearby stars. One of the best examples of such a nebula is that around the Pleiades (Fig. 15.12). The spectrum of this nebula does not exhibit the bright lines characteristic of an H II region. It shows simply the absorption-line spectrum of the Pleiades' stars—light reflected by dust. Bright nebulas that arise from the reflection of starlight by dust are called *reflection nebulas* (Color Plate 11).

Generally, interstellar dust makes itself known in two ways: (1) *extinction*, the dimming of starlight, and (2) *reddening*, the scattering of the blue wavelengths more than the longer wavelengths. Let's look at each of these.

Imagine starlight traveling though a dust cloud. The particles can absorb some of this light as it comes through. The dust particles can also scatter the starlight, so it goes off in a different direction from the original one. In either case, less light exits from the dust cloud than enters it. Astronomers call this dimming of starlight *extinction*.

Extinction of starlight happens in a special way; blue light is more strongly affected than red. So red light penetrates the dust cloud more readily than blue. When you observe a star through the dust cloud, more of its red light reaches your eye than does blue. The star appears redder than it actually is: astronomers call this process *reddening*. What about the blue light? The part that is scattered, and not absorbed, bounces around the dust cloud until it finally exits. So the cloud, a reflection nebula, appears blue (Color Plate 11).

Except for obvious dark nebulas, extinction is difficult to measure. It is much easier to measure reddening, since this is a color effect, and to estimate the quantity of the dust from the amount of reddening. From the H-R diagram we know that a certain spectral classification of a star corresponds to a certain color. As long as a spectrum of a star can be obtained, its spectral class can be determined (from the strength of absorption lines), even if its light is reddened. We measure the star's color compared with that expected for its spectral class. The difference in color, the reddening, tells how much dust lies along the line of sight to the star.

Dust and infrared observations / Interstellar dust blocks out visible light, which makes optical astronomers a bit unhappy, for it limits their view of distant stars and galaxies. Infrared astronomers feel rather happier, for infrared radiation penetrates dust. In fact, some infrared radiation from space comes from the dust itself. So infrared astronomers can both *see* dust and *see through* dust!

How does dust emit infrared? Basically, dust grains act roughly like (very small) blackbody radiators. If the grain has a temperature of around 100 K, its emission will roughly peak most strongly in the infrared.

The Orion Nebula marks a region studded with strong infrared sources. Optically, the core of the nebula (Fig. 15.13) is the densest part: hot gas (mostly hydrogen) ionized and excited by the *O*-stars there. The brightest area forms a trapezoid figure (easily seen in a small telescope) called the *Trapezium*. Now let's look at an infrared map (Fig. 15.14) of the core region. Quite a difference! The infrared emission does *not* peak around the Trapezium, but to the north and west.

What are we seeing here? Probably the infrared emission from cool dust (about 70 K) located somewhere at or near the center of the molecular cloud—dust heated by something capable of putting out 70,000 solar luminosities. The visible Orion Nebula, illuminated and sustained by the Trapezium stars, lies in front of the molecular cloud like a hot bubble (Fig. 15.15). Enough dust

15.13

A photo in red light of the core of the Orion Nebula. The small group of stars in the center is the Trapezium cluster (arrow). (Courtesy Lick Observatory)

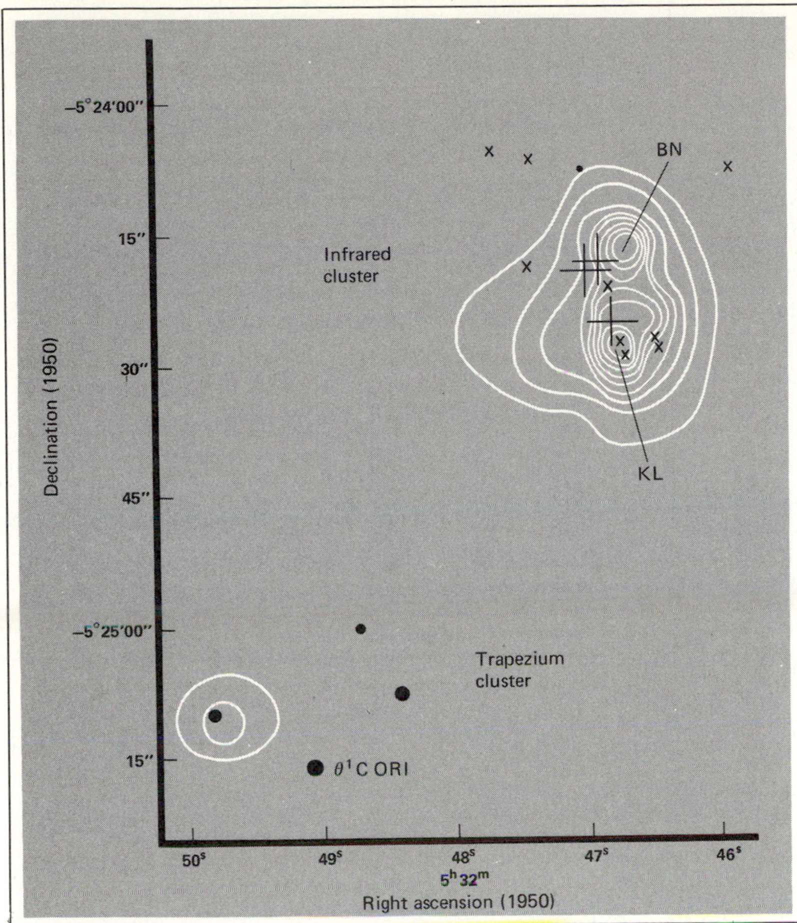

15.14

An infrared intensity contour map of the central part of the Orion Nebula. The infrared emission is concentrated in the infrared cluster (upper right). This cluster has two parts, the Kleinmann-Low (KL) source and the Becklin-Neugebauer (BN) object. The KL emission seems to come from the core of the molecular cloud. The BN object is suspected to be a protostar less than 100,000 years old. (Adapted from a diagram by E. Becklin, G. Neugebauer, and C. Wynn-Williams)

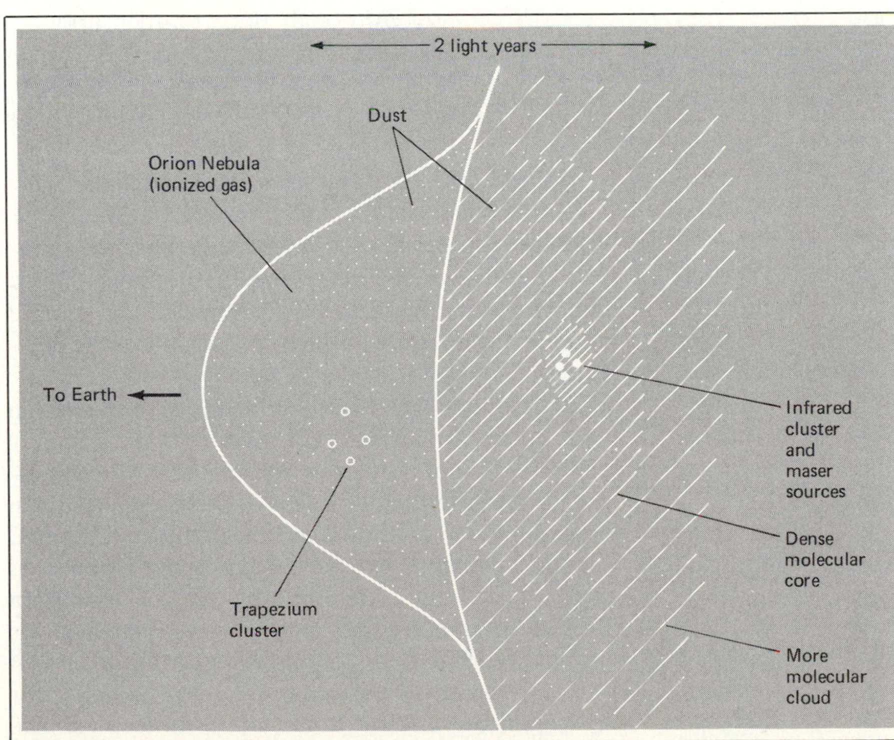

15.15

One model of the association of the Orion Nebula, its molecular cloud, and infrared sources. The Orion Nebula is a hot gas bubble, expanding outward, on the front of the molecular cloud. The infrared sources are embedded in the molecular cloud. (Based on a model by B. Zuckerman)

TABLE 15.3 Cosmic abundances of candidate elements for interstellar grains

Element	Number of atoms (per 10^6 H)
H	1,000,000
He	120,000
C	370
N	120
O	680
Ne	630
Mg	34
Si	32
S	28
Fe	26

SOURCE J. M. Greenberg, "Physics and Astrophysics of Interstellar Dust," in *Infrared Astronomy,* ed. G. Setti and G. Fazio (D. Reidel, Dordrecht, Holland, 1978).

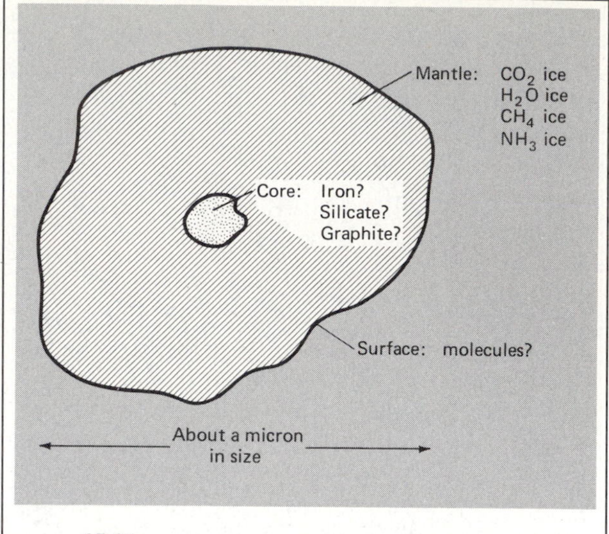

15.16
A simplified model of an interstellar dust grain. The composition is not really known; it may consist of any of the materials listed or some combination of them.

floats around one of the Trapezium stars to be visible as a weak infrared source.

The *Becklin-Neugebauer object* within the Orion Nebula is one of the strongest infrared sources there. It is less than 300 AU in diameter and emits about 1000 solar luminosities at a temperature of 600 K. The Becklin-Neugebauer object plays a key role in understanding starbirth.

The nature of interstellar dust / What is the interstellar dust made of? Astronomers have been asking this question for a number of years, yet the answers are still a bit cloudy. I'll try to present the current state of knowledge without making it appear more certain than it actually is.

One indirect but valuable clue comes from considering the cosmic abundances of candidate elements (Table 15.3); only elements that make up an appreciable fraction of the interstellar material can contribute in a large part to the dust grains. Note that abundant elements make up rather common substances: hydrogen and oxygen for water, carbon and hydrogen for methane, carbon and oxygen for carbon dioxide, nitrogen and hydrogen for ammonia, silicon and oxygen plus metals for silicates (compounds of Si and O commonly found in earth rocks). Compounds like water, methane, and carbon dioxide are loosely called *icy materials* because these materials are solids at temperatures below about 100 K (recall that they make up the bulk of the nucleus of a comet, Section 12.1.)

Now let's get down to the dust grains themselves. To account for the shape and amount of interstellar extinction, astronomers have developed *core-mantle grain models* (Fig. 15.16). The small core, about 0.05 μm in radius, can consist of silicates, iron, or graphite; silicates are most likely. The mantles, about 0.5 μm in radius, are made of icy materials, likely some mixture of them all. When grains drift into hot regions, such as an H II region, their mantles evaporate, leaving behind a bare core.

Infrared observations bolster the idea that silicates and ices (at least water ice) make up parts of interstellar grains. Some infrared sources show absorption bands near 10 and 3 μm. Silicates in terrestrial rock, meteorites, and lunar rocks have strong absorption bands at about 10 μm. Water ice absorbs well at 3 μm.

To sum up: Although uncertainties abound, we believe that interstellar dust contains elements common to the interstellar gas in the general form of ices, silicates, graphite, and metals (such as iron).

Dust and the formation of molecules / Dust and molecules are intimately associated in space. They are almost always mixed together; wherever you found a molecular cloud, you usually find a concentration of dust. This association is not chance; grains play a role in the formation of molecules.

Consider the basic problem of forming an interstellar molecule. You have to get widely separated atoms together and chemically bound—no easy task in the dilute gas of interstellar clouds. Solution? Use cold dust grains as the sticking and forming surfaces. If a hydrogen atom hits a cold grain, it will stick. Add another, and a hydrogen molecule forms, which does not stick as well as atomic hydrogen. It eventually pops off into space.

Grain surface formation seems to work for hydrogen but not for other small molecules. The main problem is how to get a molecule, once formed, off the grain's surface without destroying the molecule. Such troubles have led astrochemists to investigate chemical reactions in the gas alone. Their work indicates that chemical reactions in the gas seem to explain the formation of molecules up to those containing four atoms.

What about more complex molecules? We do not yet understand their formation. They may, like hydrogen molecules, form on grains. And the grains block out the ultraviolet light that would otherwise destroy the molecules.

To sum up what we understand so far about the formation of interstellar molecules:

1. Hydrogen molecules form on interstellar grains.
2. Molecules of up to four atoms form in the interstellar gas.
3. The formation of more complex molecules is not understood.

The formation of cosmic dust / So much for the molecules. What about the dust? Where does it come from?

Recall that interstellar grains are made basically of two materials: ices of various types, and denser solids (such as silicates and graphite). The ices solidify at a few hundred Kelvins, the denser materials at a few thousand Kelvins.

How and where are these grains made? The heavier grains are probably made in the atmospheres of cool supergiant stars. We know that such stars blow mass into space at rates of about 10^{-6} solar mass per year (Fig. 15.17). The surfaces of these *M*-stars have temperatures of only 2500 K or less. As gaseous material streams outward from them, its temperature drops, and solids can condense out of the vapor. In fact, spectra of some supergiant stars show the 10-μm silicate feature, indicating that such dust exists around them. In a rarer class of stars, in which carbon is somewhat more abundant than oxygen, graphitelike particles and particles made of silicon carbide can form in the outflowing material. Infrared spectra of these stars indicate a cloud of carbon-rich particles around them.

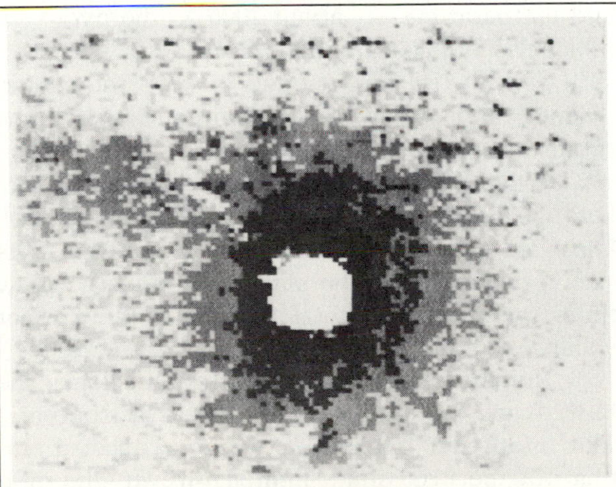

15.17
Material expanding in a shell from the red supergiant star Betelgeuse. This computer-processed infrared photo shows the star's image as a white hole in the center. The shell material appears dark; darker regions indicate denser parts of the shell. (Courtesy R. K. Honeycutt)

What about the ices that make up grain mantles (or perhaps entire grains)? It is likely that these condense on cores in the deep interiors of dense molecular clouds. Here the temperatures are low and the gas densities high, so bare grains can grow crusts of ices. A core may have to grow a mantle once every 10^8 years or so, since grains will lose their mantles in an environment where temperatures range above a few hundred Kelvins.

This discussion of dust completes your tour of the interstellar medium. I've dropped some hints about starbirth along the way. Let's deal with it now.

15.3 *Starbirth: Theoretical ideas*

Here's the big picture: Stars are born out of interstellar molecular clouds. Because these clouds contain many times the mass of a single star, they must fragment into much smaller pieces during the process of star formation. To try to cast a clear light on our present understanding of starbirth, I have divided the topic into two broad parts: theoretical ideas (this section) and observational clues (next section).

Theoretical collapse models / I'll present some highlights of models done so far. Bear in mind, however, that these theoretical calculations (done on computers) represent highly idealized models based on uncertain assumptions about how the collapse starts and the relevant physical processes. The results may depend strongly on the assumptions, so their match to reality works out only in general ways.

Newton first recognized the basic outline of star formation: the process of *gravitational collapse*. A cloud with enough mass and a low temperature will naturally contract from its own gravity. As gravitational potential energy becomes kinetic energy, the material in the cloud heats up (see Section 12.3). This eventually has two results: (1) The temperature (and density) will build up enough so that the outward pressure halts the collapse and balances gravity, and (2) the temperature eventually reaches the kindling temperature of fusion reactions; at that moment, a star is born.

Prior to the ignition of fusion reactions, while the cloud is contracting and heating, it is called a *protostar*. As a protostar evolves, its luminosity, size, and surface temperature change with time.

The evolution for protostars of different masses differs substantially. For this reason, I'll divide the discussion into two parts, one for solar mass clouds and stars and one for massive clouds and stars (which have 10 solar masses or more). Despite differences that come from different masses, the theoretical models have some common features:

1. The collapse starts out in free-fall, that is, controlled only by gravity.
2. It proceeds very unevenly; the central regions collapse more rapidly than the outer parts, and a small condensation forms at the center; this core will become a star.
3. Once the core forms, it accretes material from the in-falling envelope.
4. The star becomes visible to us, either by accreting all the surrounding material onto itself or by somehow dissipating it.

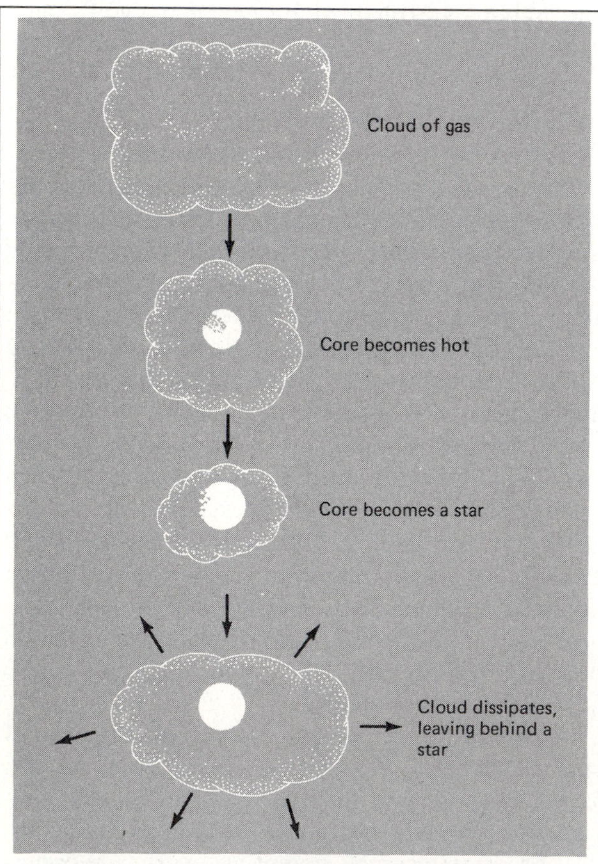

15.18
A very general scheme for the formation of a star by gravitational contraction. As an interstellar cloud contracts, the material at the core condenses faster than the envelope. When the core is hot enough, fusion reactions ignite. The cloud dissipates, revealing the new star.

With these general features in mind, let's look at one model for the formation of a sunlike star.

Solar-mass protostellar collapse / Imagine a huge interstellar cloud of gas, mostly molecular hydrogen, and dust. Assume that this cloud has sufficient mass to contract gravitationally. Its initial diameter is 5 million solar radii. The collapse begins (Fig. 15.18). Material at the cloud's center increases in density faster than at the edge. Because of the density increase, the collapse at the center speeds up (as expected from Newton's law of gravitation). It collapses faster, grows denser, and so collapses still faster. The rest of the cloud's mass is left behind in a more slowly contracting envelope.

With the rapid fall of material in the core, the hydrogen molecules gain kinetic energy. They bang into each other and also strike dust grains, transferring kinetic energy to the grains. Heated by such collisions, the dust grains radiate at infrared wavelengths. As long as this heat radiation can escape into space, the kinetic energy is dissipated, the cloud stays cool, the pressure stays low, and the collapse continues in free-fall. But at some point the density of the core reaches a critical value at which the cloud becomes opaque and traps infrared radiation. The core heats up to a few hundred Kelvins, its pressure increases, and the core's collapse slows down dramatically and stops.

Meanwhile, the envelope continues merrily falling inward, showering mass on the core. As the matter piles up, it increases the core's mass and temperature. At about 2000 K, the hydrogen molecules break up and soak up heat. The pressure doesn't rise fast enough to keep the star in balance, and gravity takes command again. The protostar again goes into free-fall.

When all the hydrogen molecules have broken up, the pressure can rise again, and the collapse stops. The protostar slowly contracts; its size is twice that of the sun, its luminosity a few times the sun's. The total time from the start of collapse to this stage is about one million years.

But the envelope, still falling in, blocks the birth from the view of the optical astronomer. This infant star hides in its womb, with dust cutting out the protostar's light. However, the light absorbed by the dust heats it so that it gives off infrared radiation. So a sign of protostars should be small, intense sources of infrared radiation in or near known clouds of gas and dust.

Eventually, like an afterbirth, the star rids itself of its cloaking cloud, which may all fall onto the young star. As the cloud dissipates, we see a *pre-main-sequence star*, one that is larger and cooler than it will be in its final state on the main sequence. The total time elapsed from the onset of collapse to reaching the main sequence is about 50 million years.

Massive protostellar collapse / The collapse of a massive protostar follows the same general scheme as that for a solar-mass one. But important differences do crop up. The main ones are as follows:

1 Fusion reactions begin *before* the accretion of the envelope stops, so a massive star begins and spends part of its main-sequence lifetime obscured from view.

2 Massive protostars do not accrete all of their original cloud's material; some is blown away; about half of the original mass ends up in the star.

3 Massive protostars are much more luminous than solar-mass ones, so they will be stronger infrared sources.

4 When the star begins fusion reactions, it emits ultraviolet photons that ionize the gas around it; this H II region can be seen by radio telescopes; as the hot H II region expands, it helps dissipate the cloak of gas and dust.

5 The protostar stage is shorter than for solar-mass stars.

Collapse with rotation / The models just described lack at least one fact of astronomical life: rotation. It is likely that interstellar clouds rotate at least a bit. Any rotating mass has *angular momentum* (see Focus 9.2). An isolated, rotating mass must conserve angular momentum. So as a spinning, spherical mass collapses gravitationally, it must spin faster. It will eventually collapse into a disk along its rotation axis, as described in Section 12.3 for the formation of the planets.

The addition of spin to theoretical models of protostar collapse makes the calculations much tougher and the results less conclusive. To give you the flavor of results to date: In some cases, a disklike core, a ring of material results. These rings turn out to be unstable in some instances; they break up into two or three blobs. Sometimes these blobs coalesce into fewer ones.

The important point to remember is this: With spin added to the models, the calculations develop rings that fragment into a few blobs. If each blob eventually becomes a star, we then have a natural explanation for the fact that most stars in the Galaxy are in binary or multiple systems—or a starting point for the formation of planets.

15.4
Starbirth: Observational clues

Enough of theoretical models. Let's now look at the real world to see how models relate to observations. It turns out that we have uncovered more information about massive-starbirth than about the birth of solar-mass stars. There's a good reason: Massive protostars have greater luminosities than solar-mass ones, and, once they reach "stardom," massive stars ionize the gas around them. Radio telescopes can then detect the ionized gas. Since all this action takes place cloaked by dust, infrared and radio observations permit us to inspect stellar wombs.

Signposts for the birth of massive stars / Let me outline what models predict should be the hallmarks, in the radio and infrared, of the birth of a massive star. With this outline as a guide, we'll then look at the observations.

First, because stars condense from molecular clouds, you need to find a molecular cloud; molecules emit at millimeter wavelengths. Second, the free-fall collapse, in its early phases, heats the dust to low temperatures, roughly 40 K. This dust emits infrared radiation that peaks at roughly 100 μm. Third, as the protostar forms, the interior dust reachs about 1000 K, and so emits with a peak at 30 μm. Fourth, as the protostar reaches the main sequence, it ionizes the hydrogen gas, and a compact H II region develops, observable at centimeter wavelengths. Fifth, as the hot, ionized gas expands, the radio and infrared intensity decreases. Finally, the H II region expands enough to blow off its dusty cloak, and the stars appear to optical view. (See Table 15.4 for a summary.)

With this scenario in mind, let's return to our old friend the Orion Nebula. The H II region around the Trapezium marks the oldest (most evolved in an evolutionary sense) part of the region. The Trapezium cluster consists, in fact, of a few hundred stars. It's the *O*- and *B*- stars of this group that ionize the gas. These massive stars are no more than a million years old. The distance between the Trapezium stars and the front edge of the molecular cloud is roughly one light year. In an evolutionary sense, the molecular cloud core that lies behind the Orion Nebula is the youngest (least evolved) part of the region.

Where is starbirth happening? Most astronomers place their bets on the Becklin-Neugebauer object. Its observed characteristics

TABLE 15.4 Possible sequence in the birth of massive stars

Evolutionary stage	*Observational signposts*	*Duration (years)*	*Important events*
1. Collapse of prestellar clouds	Cool, dense molecular clouds; OH, H_2O sources	300,000	Gravitational collapse of unstable regions set up by shock wave
2. Very young, cool stars shining from gravitational contraction	Compact far-infrared sources associated with molecular clouds	50,000 (stages 2–3)	Stars form from gravitational collapse; go through cool, luminous phase. Stars increase a lot in surface temperature, a little in luminosity.
3. Stars begin nuclear burning in cores.	Compact, near-infrared sources in or near molecular clouds		
4. Early, normal-life nuclear burning	Infrared and radio sources in or near molecular clouds	30,000	Ultraviolet photons break up molecules and ionize gas to form an H II region around star.
5. Young, expanding H II region	Weak infrared emission, diffuse centimeter-wave radio emission, optical H II region	500,000	Heated gas in H II region expanding at about 5 to 10 km/sec
6. Old, very expanded H II region	No infrared emission, very diffuse centimeter-wave emission, *OB*-stars plainly visible	2,000,000	Expansion dissipates both H II region and associated molecular cloud
7. Naked *OB*-stars	A single *OB*-star or clusters of *OB*-stars with no surrounding H II region	6,000,000	H II spreads into interstellar medium

match those expected from a massive protostar in its pre-main-sequence evolution. Observations indicate that radio and infrared emission comes from a compact H II region around the Becklin-Neugebauer object, which is a *B*-star just making it to the main sequence, surrounded by dust and gas it has just begun to ionize. It is no more than a million years old—probably less.

Messier 17—another bright H II region—presents a clearer observational view (because we happen to see it from the side) of one sequence of massive star formation. You can easily see M 17 in a telescope (Fig. 15.19 and Color Plate 10). It lies only 7000 ly away. The optical H II region marks a site of star formation some 10 million years old; contained within it is a small cluster of *O*- and *B*-stars. To the west of the nebula, nothing much appears optically. But millimeter radio observations reveal two pieces of a molecular cloud. Both fragments have temperatures above those usually found in molecular clouds, so something there is heating up the gas. In the south fragment lurks a starlike infrared source that has some of the characteristics of a protostar. To the west, connected to the two fragments next to M 17, lies a gigantic molecular cloud, some 70 by 280 ly in size and containing more than a million solar masses of material. To give you some idea of how large this cloud appears in the sky, it would take *8 full moons*, touching, to span the cloud from its east end to its west end.

To round out the observational clues, infrared photos of M 17 (Fig. 15.20) show a small cluster of *O*- and *B*-stars in the zone between the H II region and the molecular cloud. The cluster's location coincides with the peak of continuous radio emission from M 17—an indication that this dust-obscured cluster now plays an important role in the ionization of the H II region.

These observations all add up to a composite picture of a sequence of massive-star formation from giant molecular clouds (Fig. 15.21). On one side of the molecular cloud lies the oldest region of starbirth, M 17 and its immersed group of *O*-

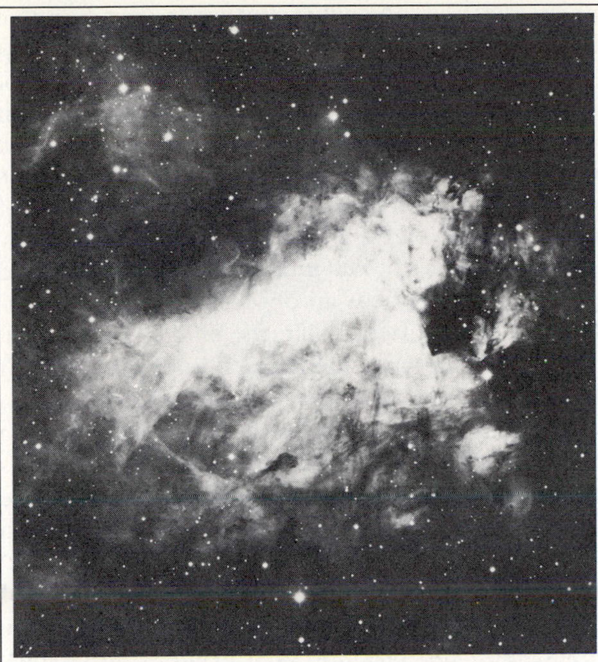

15.19
Messier 17, a nearby H II region. The area to the right where the nebula's emission suddenly cuts off indicates the presence of a molecular cloud. (Courtesy Palomar Observatory, California Institute of Technology)

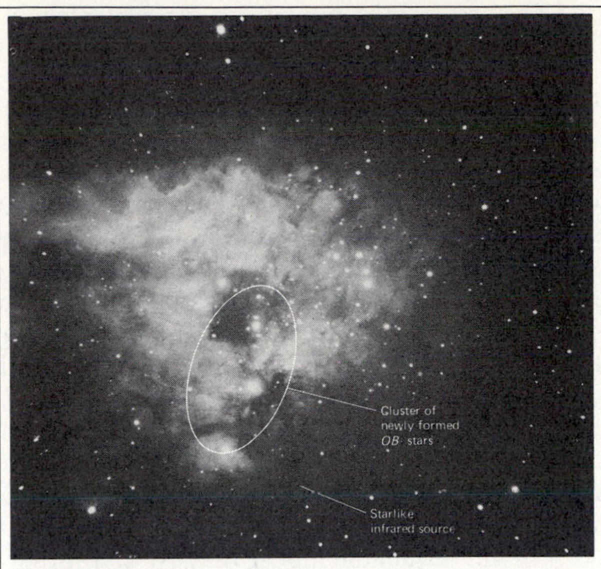

15.20
An infrared photo of the region of M 17 near the southern molecular cloud fragment. The white oval surrounds a cluster of young O- and B-stars. The line points to a starlike infrared source that may be even younger; it is located at the center of the cloud. (Courtesy M. Beetz, H. Elsasser, C. Poulakos, and R. Weinberger; photo taken with the 123-cm telescope of the Centro Astronómico Hispano-Alemán Almeria on the Calar Alto in Spain)

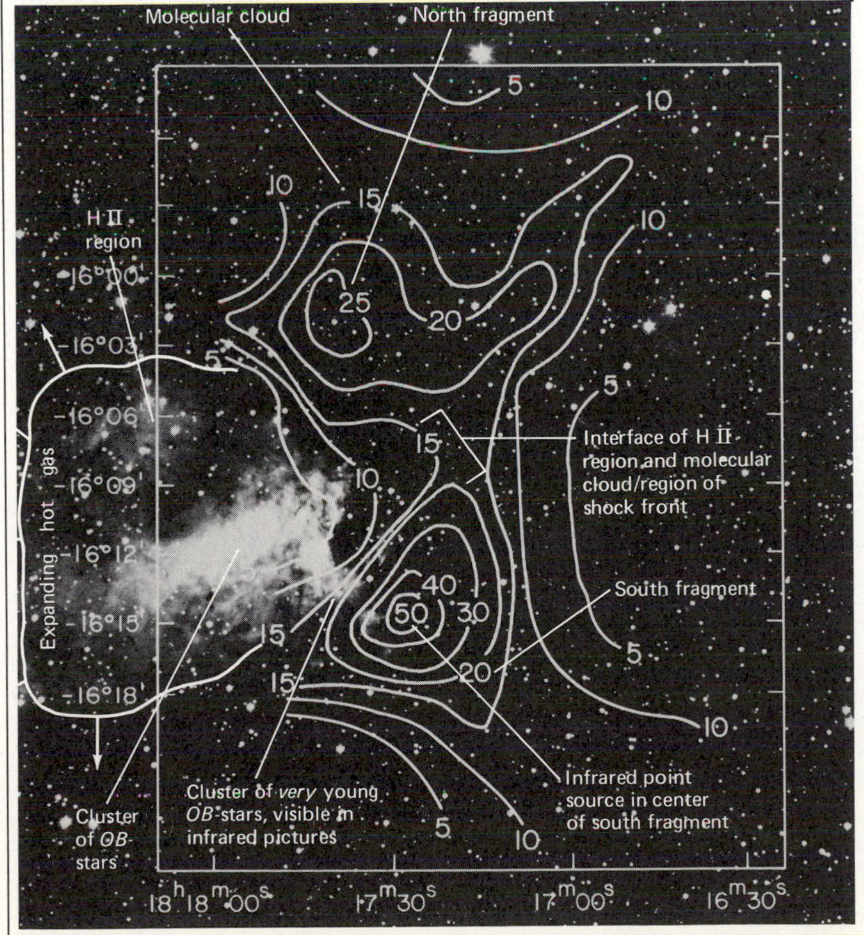

15.21
A map of the carbon monoxide emission of the M 17 region. It shows two fragments of a giant molecular cloud. An infrared source lies near the bottom peak (labeled 50). The total mass in the two pieces is about 100,000 solar masses. (Courtesy C. Lada and Harvard College Observatory; based on observations by C. Lada and colleagues)

and *B*-stars formed out of the molecular cloud. These hot stars have heated up the gas around them, ionizing it and destroying the molecules there. The hot gas of the H II region slowly expands. On the west, the expanding hot gas runs into the cold, dense molecular cloud. Here a shock wave forms. The shock wave, moving at about 10 km/sec, prompts gravitational collapse and star formation out of the molecular cloud. Moving more into the fragmented molecular cloud, the shock wave will probably trigger more star formation, and each group—of *O*- and *B*-stars—will develop an H II region and another shock wave. So the molecular cloud will finally self-destruct in an orgy of star formation. What material doesn't make it into stars is eventually dissipated by the stars that have formed.

In this sequential model, massive-star formation begins at one end of a giant molecular cloud (Fig. 15.22). (Such clouds tend to be elongated, cigar-shaped objects.) A small group of about 10 *O*- and *B*-stars forms. They evolve to the main sequence. Their ultraviolet radiation then ionizes the gas around them. The H II region, since it is hot, expands, pushing a shock wave into the molecular cloud. The gas behind the shock wave is compressed to densities sufficient to start gravitational collapse. A new group of *O*- and *B*-stars is born about a million years after the previous one. The process repeats. Small groups of massive stars are born, in this model, in a sequence of bursts across the molecular cloud.

You may have noticed in this model that once the star formation starts at an end of a molecular cloud, it propagates through it in a chain reaction. But what starts the first burst of star formation? The answer to that question is not yet clear. Perhaps it starts from the collision of molecular clouds. Or, more likely, from the blast wave of a supernova remnant crashing into an end of a molecular cloud. If so, a supernova—the death of a massive star—may ignite the birth of other massive stars.

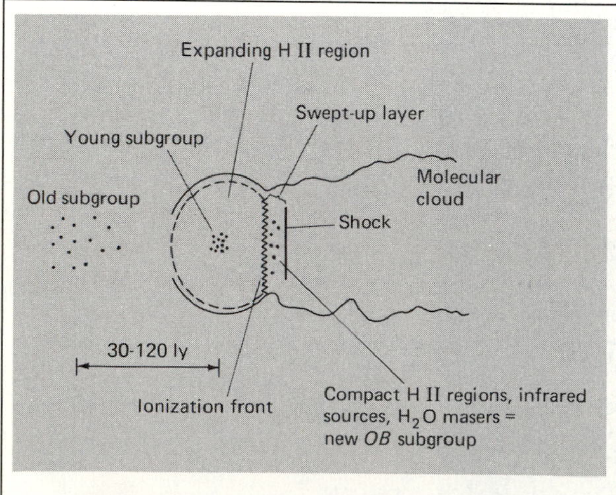

15.22
A schematic for the sequential model of massive-star formation from a giant molecular cloud. (Adapted from a figure by B. Elmegreen and C. Lada)

The birth of solar-mass stars / I wish I could paint for you as neat an observational picture for the formation of stars like the sun. But I cannot. The observational evidence is skimpy and doesn't hold together in a comprehensive way. So this discussion must be regarded as being of working models rather than finished ideas. One point is clear: Like massive stars, solar-mass stars are also born from molecular clouds. The questions are in which clouds and how?

Let's look at the H-R diagrams of clusters of stars thought to be young because of their close association with gas and dust in the form of dark clouds. (Astronomers give the name *dark cloud* to an interstellar cloud that contained so much dust that it blotted out the light of stars within it and behind it. We now know that dark clouds are one type of molecular cloud. They typically have temperatures of 10 K, densities of 10^8 atoms in a cubic meter, and masses up to a few hundred solar masses.) One such cluster, called New General Catalog (NGC) 2264, lies in front of and slightly embedded within a dark cloud. This cluster is about 2400 ly away in the constellation Monoceros, just east of Orion.

Examine the H-R diagram of NGC 2264 (Fig. 15.23). Note that the *O*- and *B*-stars of the cluster have already arrived at the main sequence. But stars from *A* to *M* in spectral type mostly fall *above* the main sequence. A natural interpretation is that all the stars in the cluster formed at approximately the same time. But the more massive ones evolved faster. These have already contracted to the main sequence, while the less massive ones (about 3 solar masses and less) are still evolving to the main sequence. They are called *pre-main sequence* stars.

Among these pre-main-sequence stars are some called *T-Tauri stars* (named after their prototype, T Tauri). Studies of T-Tauri stars indicate that most T-Tauri stars are: (1) low-mass stars (0.2 to 2.0 solar masses) now approaching the main sequence, (2) young, with ages from 100,000 to 100 million years, (3) surrounded by hot, dense envelopes, and (4) losing mass by stellar winds (analogous to the solar wind) that blow at speeds of a few hundred kilometers per second. (This last property of T-Tauri stars has been disputed

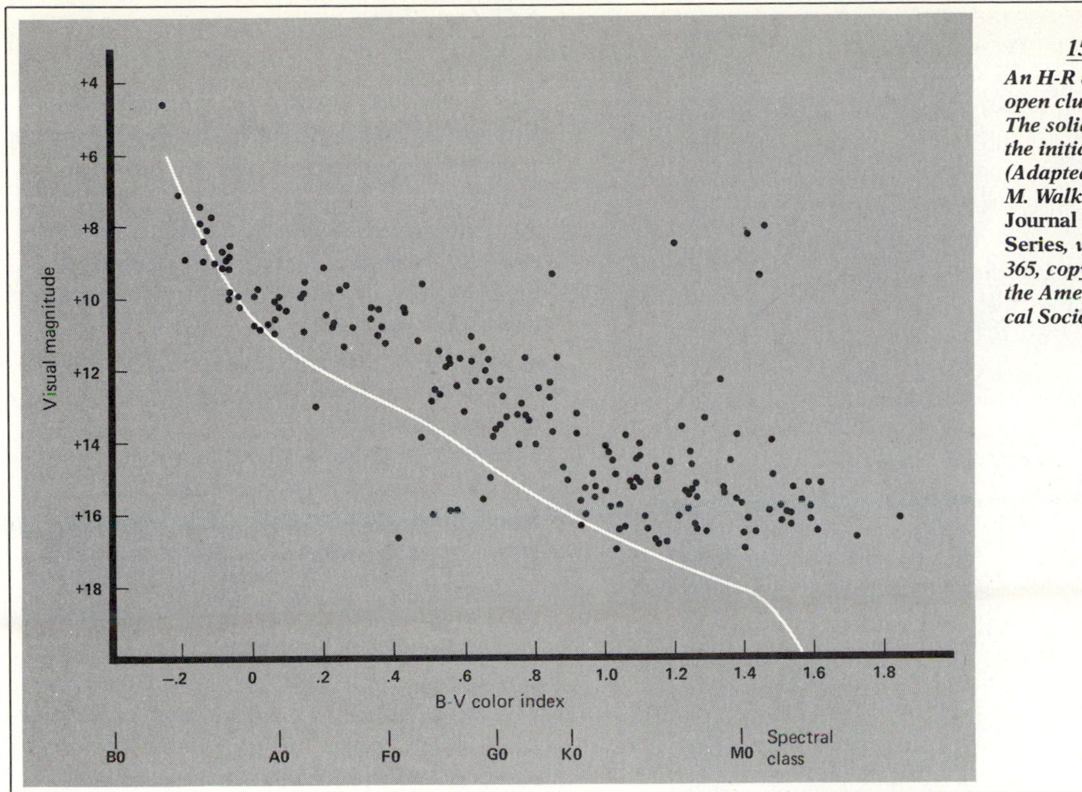

15.23

An H-R diagram for the open cluster NGC 2264. The solid line indicates the initial main sequence. (Adapted from a figure by M. Walker,* Astrophysical Journal Supplement Series, *vol. 2 (1956), p. 365, copyright © 1956 by the American Astronomical Society)

recently.) From such observations, T-Tauri stars are presumed to be solar-mass pre-main-sequence stars. T-Tauri stars can then be used as tracers of solar-mass pre-main-sequence stars. T-Tauri stars appear almost always within dark clouds.

So one view of the birth of solar-mass stars is this: They are born in fairly massive dark clouds, perhaps in giant molecular clouds along with massive stars. They form from fragments throughout the cloud, rather than at the edges as massive stars do. The birth of the massive stars—or perhaps the death of one in a supernova—sweeps away the gas and dust to reveal the stars. In this picture, most starbirth takes place in dark, massive clouds, out of which *OB* associations form. So the sun may have been born in an *OB* association, such as the one we described in Orion, perhaps blasted clean by a supernova.

There's an alternative idea that is also appealing: that solar-mass stars form out of small, isolated molecular clouds, not giant ones. These clouds are no more than a few light years in size, contain at most 1000 solar masses of gas, and are not near in space to any giant molecular clouds.

From observations so far, we are beginning to suspect that, in general, stars are born from molecular clouds; massive stars from massive clouds, and less massive stars (the majority of those in the Galaxy) from less massive molecular clouds.

Molecular outflows and starbirth / I have saved the most recent observational clue about starbirth for last because it still presents something of a puzzle but also much promise. Observations of molecules around protostars have discovered high-speed (up to 100 km/sec) flows of gas. Doppler shift measurements show that these flows tend to be bipolar: two streams moving in opposite directions from each other. The flows carry considerable mass and can span a few light years; so enormous amounts of energy push them along. The source of that energy and the origin of the flows is a mystery at the moment. Such outflows do appear associated with the birth of a massive star.

A simple model to explain the outflows envisions a young, massive star putting out a strong stellar wind. Surrounding the star is a dense disk of gas and dust. This disk would naturally channel the flow of the stellar wind so that it streams out along the thin axis of the disk, making two streams. When these two streams push enough material outward, two opposing lobes of gas should form. The early stages of this process are buried in gas and dust, but later stages might be visible to optical and radio telescopes.

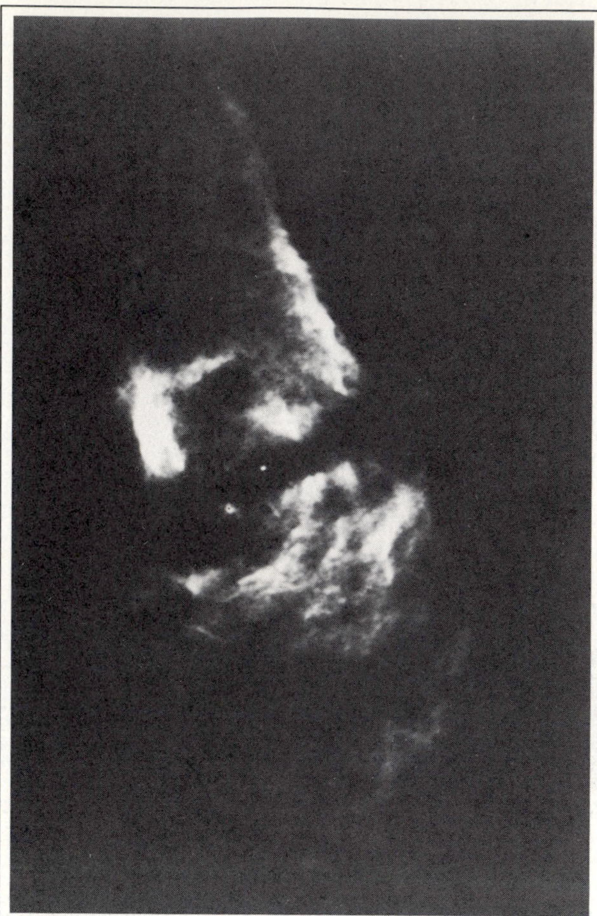

15.24
A VLA map of the bipolar region Sharpless 106 at a wavelength of 6 cm. The emission here is from ionized gas. The dot in the dark lane marks the star ionizing the gas. Note the lobes above and below the lane and the filamentary structure of the gas within them. (Courtesy J. Bally; observations by Bally, R. Snell, and R. Predmore)

The VLA has mapped the later stages of this process for the ionized gas around a young star (Fig. 15.24). The observations show lobes of material on opposite sides of the star with a dark lane cutting between them. The gas streams outward at about 75 km/sec in clumps, filaments, and shells. The star itself loses a millionth of a solar mass of material per year from a stellar wind blowing at a few hundred kilometers per second. The dark lane marks our edge-on view of the disk that surrounds the star and channels the flow.

The discovery of these two-sided flows strongly hints that disks of material typically form around massive stars during their formation. It is from such disks that planetary systems might form. So we have a clue that the nebular model for planetary formation might actually operate elsewhere in the Galaxy.

KEY CONCEPTS

1 *The interstellar medium contains both gas and dust; the gas, mostly hydrogen, comes in a variety of forms: molecules, atoms, and ions.*

2 *The gas clumps in clouds of various sizes, ranging from small clouds of atoms to the giant molecular clouds; between the clouds is a hotter intercloud gas.*

3 *A wide variety of molecules has been found in molecular clouds; some of these are organic compounds.*

4 *Supernova explosions heat up and distort the clouds of the interstellar medium.*

5 *Interstellar dust is far less abundant than the gas but does contain about 1 percent of the total mass of the interstellar medium.*

6 *Dust makes itself known by the reddening and extinction of light and also by infrared emission when it heats up.*

7 *The dust is made of grains about a micrometer in size; the core of the grains contain silicates, graphite, or iron; the mantles are made of icy materials; the grains are formed in the outflow of material from cool, supergiant stars.*

8 *Grains are associated with molecular clouds and aid the formation of molecules.*

9 *Stars are born out of molecular clouds by the process of gravitational collapse; a protostar forming in a cloud gets its energy from the conversion of gravitational potential energy to kinetic energy; the process of starbirth is hidden from our direct view, but we can infer its operation by infrared and radio observations; somehow the surrounding material must be dissipated to reveal the star.*

10 *The gravitational collapse of rotating clouds naturally results in the formation of planetary or multiple star systems.*

11 *Infrared and radio observations imply that massive stars are formed in small groups (about 10) out of giant molecular clouds in a chain-reaction sequence where one group triggers the birth of the next; the process may be started by a supernova remnant hitting the giant molecular cloud.*

12 *Solar-mass stars may be formed out of small molecular clouds or as a spin-off of the birth of massive stars from giant molecular clouds.*

STUDY EXERCISES

1 Describe one way in which astronomers observe each of the following:
(*a*) interstellar H I,
(*b*) interstellar H II,
(*c*) the coronal interstellar gas, and
(*d*) interstellar molecules. (*Objectives 1 and 2*)

2 Outline *two* ways in which astronomers "see" interstellar dust. (*Objectives 1 and 3*)

3 What evidence do we have, if any, that dense materials make up part of the interstellar dust? That icy materials do? (*Objective 4*)

4 List the observational evidence that leads to the guilt-by-association argument for the formation of massive stars in small groups from giant molecular clouds. (*Objective 5*)

5 What powers a protostar? (*Objectives 5 and 6*)

6 In a theoretical picture for the formation of a massive star, how is it that we never see a massive star directly (optically) until it reaches the main sequence? (*Objective 5*)

7 Once it reaches the main sequence, how does a massive star influence its parent cloud? (*Objective 5*)

8 What *observational* evidence do we have that solar-mass stars form from dark clouds? (*Objective 6*)

BEYOND THIS BOOK . . .

For more details on one aspect of starbirth, read "The Birth of Massive Stars" by M. Zeilik in *Scientific American*, April 1978. For another view, try "Bok Globules" by R. L. Dickman, June 1977. For a third, look at "The Newest Stars in Orion" by G. Wynn-Williams, August 1981.

For one view of the dynamic interstellar medium, read "The Structure of the Interstellar Medium" by C. Heiles in *Scientific American*, January 1978.

A good review of H II regions is in "Gaseous Nebulas" by E. Chaisson, *Scientific American*, December 1978. For a look at molecular clouds, read "Molecules in Space" by B. Turner, *Scientific American*, March 1973, which is updated in "Giant Molecular-Cloud Complexes in the Galaxy" by L. Blitz, April 1982.

For more on masers, read "Cosmic Masers" by D. Dickinson, *Scientific American*, June 1978. For another signpost of starbirth, read "Energetic Outflows from Young Star" by C. Lada in the July 1982 issue.

16 / Star lives

Not from the stars do I my judgment pluck,
And yet methinks I have astronomy . . .
WILLIAM SHAKESPEARE: Sonnet 14

Learning objectives

After studying this chapter, you should be able to:

1. **Show how the Hertzsprung-Russell diagram for many stars provides clues about the evolution of individual stars.**
2. **Describe the physical basis of a theoretical model of a star, that is, what concepts go into a star model.**
3. **Trace the evolution of a 1-solar-mass star on a Hertzsprung-Russell diagram, describing the physical changes of the star that result from changes in the star's core.**
4. **State in one sentence why stars must evolve.**
5. **Compare the evolution of a 1-solar-mass star with that of a 5-solar-mass star.**
6. **Describe the evolution, on an H-R diagram, of a cluster of stars.**
7. **Back up theoretical ideas of stellar evolution with observational evidence, with special emphasis on star clusters.**
8. **List the sequence of thermonuclear energy generation reactions in stars of different masses.**
9. **Compare and contrast galactic (open) star clusters to globular ones in terms of both their physical properties and their H-R diagrams.**
10. **Indicate how mass and chemical composition affect stellar evolution.**

Central question

How do the physical properties of stars change as they go through their normal lives?

Gravity controls the history of newborn stars. A star survives as long as it can counteract the relentless gravitational crunch. The story of this battle against gravity runs like the aging of a person from birth to death but takes millions to billions of years. Let me put this span into a human perspective. Imagine time speeded up so that one year passes in one-fifth of a second. Then the sun would live only 65 years or so. The sun's birth would be quick; only four months would pass from the start of the collapse of the sun's embryonic cloud to its establishment as an immature but full-fledged star. For about the next 60 years the sun would shine calmly as it passed through middle age. Old age would gradually fossilize the energy production of the sun. In about 5 years the elderly sun would slowly expand to almost 100 times its present size; it would become a bloated red giant. Then a sudden burst would blow off the sun's atmosphere, leaving behind a hot core that would cool quickly.

How a star lives depends mostly on how much mass it has at birth. What important stages mark its life and how long each stage lasts relate directly to a star's mass. Stars that are much more massive than the sun have short, frenetic lives.

This chapter presents, first, theoretical ideas about stellar evolution and contrasts the lives of stars like the sun to those of more massive ones. Second, it offers some observational evidence for these theoretical concoctions. You will see how stars burn, falter, and flame out. Our sun, too, will live and die. And the earth will die with it.

16.1 *Stellar evolution and the Hertzsprung-Russell diagram*

Contemporary ideas of how stars evolve focus on one general theme: As a star loses energy to space, it must change. While it lives, a star does not cool off. Most objects cool off (their temperature goes down) when they lose energy to their environment. It's natural to think a star does, too, but just the opposite is the case (for most phases of a star's life). Nuclear reactions generally supply the energy to keep a star hot. When one fuel (say hydrogen) runs out, another (say helium) can ignite, at a temperature made higher by the compression of matter by gravity. Only when a star cannot contract any more will it finally cool off. Then it's dead (see Chapter 17).

I'll approach stellar evolution from both theoretical and observational points of view. This chapter presents mostly theoretical ideas, because observations of a star's evolution do not come

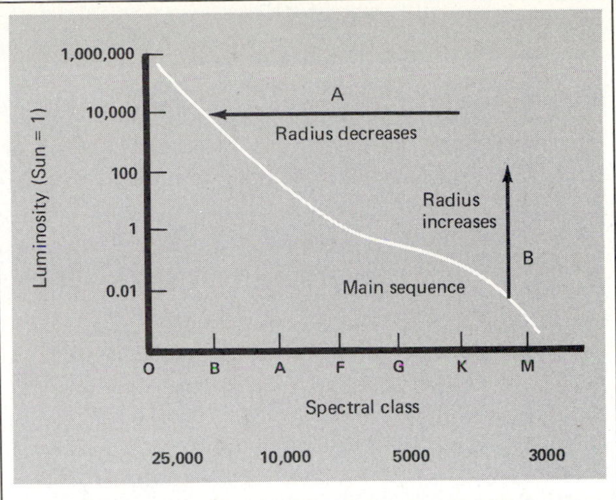

16.1
A schematic H-R diagram. A star's position on this diagram represents its surface temperature, its luminosity, and, indirectly, its size.

easily. Why? The sun's anticipated lifetime is more than *100 million* human lifetimes. So there is no way you could watch a single star, like the sun, evolve. But you can see many different stars at one time. You can organize these stars on the Hertzsprung-Russell (H-R) diagram (Fig. 16.1) if you know their luminosities and spectral types. Then you can use the H-R diagram of many stars to guess at the evolution of one star. Let's see how.

Classifying objects / Suppose you went around to ask 18- and 19-year-old males you met for their weight and height. Plot your data as a graph of weight versus height (Fig. 16.2). Note

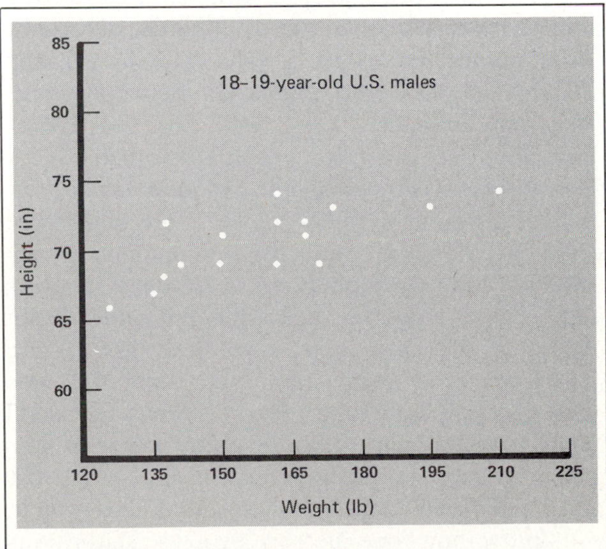

16.2
A height-weight diagram from a sample of 18- and 19-year-old U.S. males. (Data from Astronomy 10S class, University of California, Berkeley, Winter 1980)

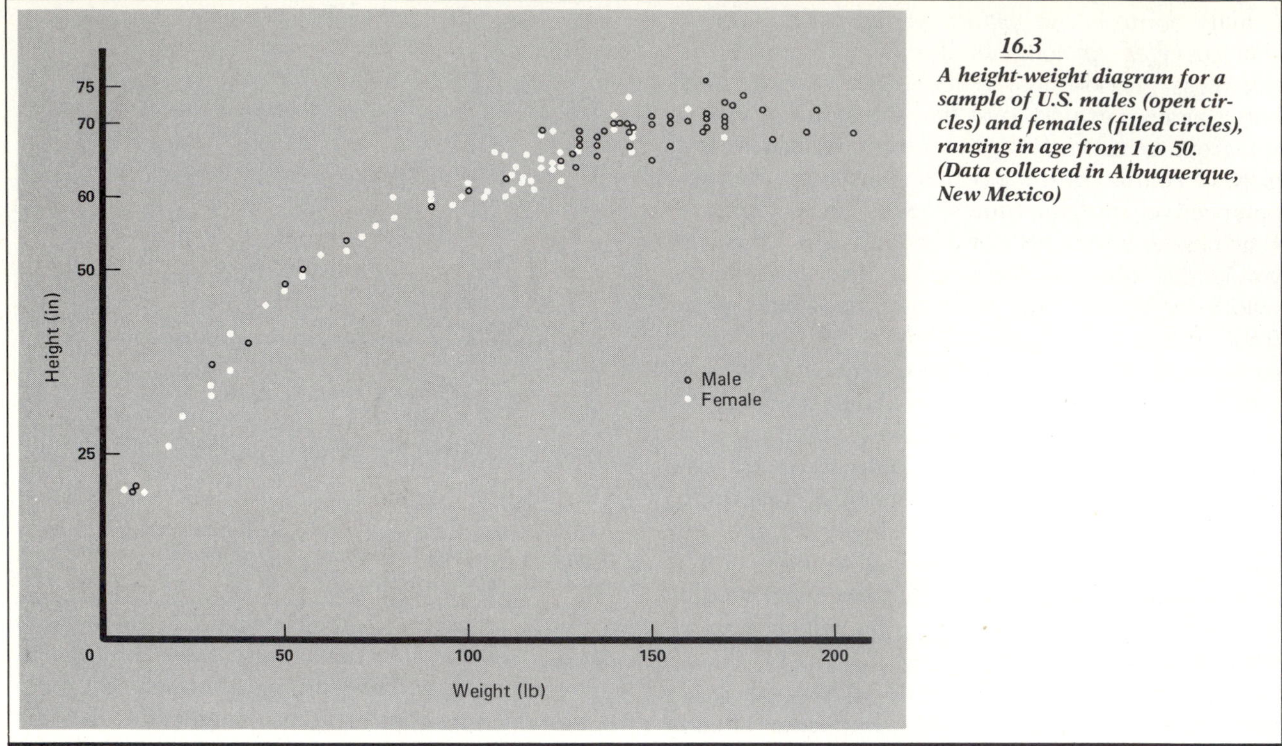

16.3
A height-weight diagram for a sample of U.S. males (open circles) and females (filled circles), ranging in age from 1 to 50. (Data collected in Albuquerque, New Mexico)

from your results that you find a trend. The points tend to fall along a line (call it a main sequence) that shows that weight generally increases with height. That result shouldn't surprise you.

Now suppose you did the same experiment, recording the height and weight of every person you encountered. Plot the data again (Fig. 16.3) and compare with the previous graph. What a difference! You still have a main sequence. But you also have other groups that don't follow the original main-sequence trend. Why the difference exists should be clear to you: The first graph includes people of the *same* age; the second, people at *different* ages.

Now here's a third graph (Fig. 16.4). It's a plot of weight versus height for an average U.S. male at different times in his life from birth to 20 years old. The line connecting the points has indicated on it the age from birth to 20 years. It shows you how a *single* person's height and weight change as he ages. You know how this growth goes from your own experience. Note that this evolution graph for one person follows the trend in the height-weight graph of many different people. You can correctly interpret the graph for many people in evolutionary terms *if*, and *only* if, you know how one person evolves. Time was implicit in that graph. In the same way, time and age implicitly play a role in an H-R diagram, in which are plotted two essential properties of stars: surface temperature and luminosity.

Time and the H-R diagram / How does the H-R diagram tell you about an evolutionary sequence of a star? Imagine that you have a large family. They have gathered together, and you take a snapshot of them all, from the newest-born to the great-grandparents. Most of the people in the picture are in their middle age (20 to 60 years old); you have a few infants, some children and

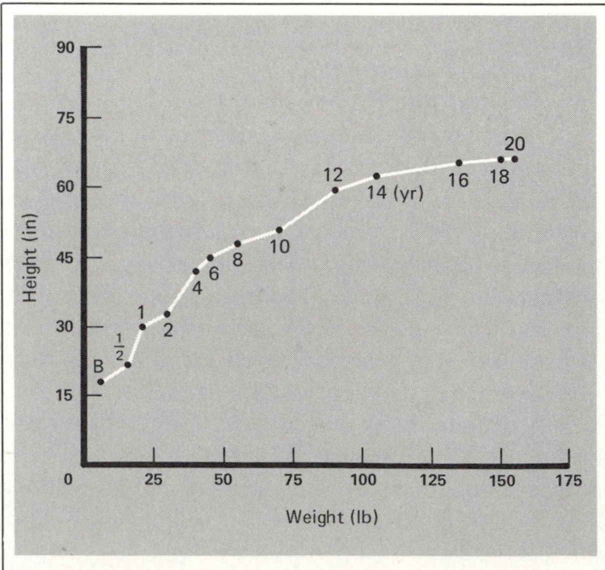

16.4
A height-weight diagram for a typical U.S. male from birth (B) to 20 years of age. (Data from the National Center for Health Statistics Growth Charts)

teenagers, and a few old people. You have so many middle-aged people because most of your life you will be middle-aged; you spend relatively less time as an infant, teenager, or old person.

Now suppose you have a collection of any objects that evolve. You believe that the collection spans an evolutionary sequence. You can estimate the relative time spent in any evolutionary stage by the relative numbers you find at that stage compared with others. (This argument holds true only if birth and death go on continuously; if no more people are born, eventually you will see only old people.)

Go back to the H-R diagram for stars (Fig. 16.1). Recall that it shows the luminosity and surface temperature of the stars. Most of the stars fall on the main sequence, so stars found here are going through the longest, most stable stage in their evolution.

The main sequence, from *O*- to *M*-stars, represents a sequence from higher to lower masses. Earlier (Section 13.6) I explained the normal process of the conversion of hydrogen to helium in the sun, which is a main-sequence star. Here's the evolutionary meaning of the main sequence: It marks stars at the stage of converting hydrogen to helium in their cores; stars remain at this stage for the greatest part of their lives. That's why we see so many main-sequence stars now. Other stages, such as becoming a red giant, must be shorter. How do we know? Because we see far fewer red giants now than main-sequence stars.

A star's mass determines how the star will evolve. So you have to examine the H-R diagram to find how stars of different masses evolve. But what is the correct interpretation of the H-R diagram for the evolution of stars? You need some hints from the physical nature of stars and theoretical calculations to make up the star models.

16.2 Stellar anatomy

What is a star? A huge, hot ball of gas, mostly hydrogen, heated by thermonuclear reactions in its core. You can imagine a star as a controlled hydrogen bomb. A star does not fly apart because gravity persistently pulls it together. All its life, a star must withstand the inward squeeze of gravity. How does it do it? By producing an outward pressure force. For most of its life, this pressure comes from the star's being hot. A star consists of gas; a hot gas has a high pressure, and this outward force balances the inward gravitational forces. This balance must hold true at every level throughout the star (Fig. 16.5); otherwise it would be unstable. So one physical requirement to model a star is that it be in balance, neither expanding nor contracting.

Second, we need to know how the material of a star behaves at different temperatures, pressures, and densities. Luckily, for most stars during most of their evolution, this behavior is the same as for an ideal gas (see Section 13.2).

Third, the star must generate energy internally. For most of a star's life, thermonuclear fusion reactions operate as the internal furnace (details to come in Section 16.4). In general, the total energy produced inside the star must equal the rate at which the energy radiates away at its surface. This balance must be true not only overall but also at each layer within the star. Otherwise the star will be unstable and expand or contract.

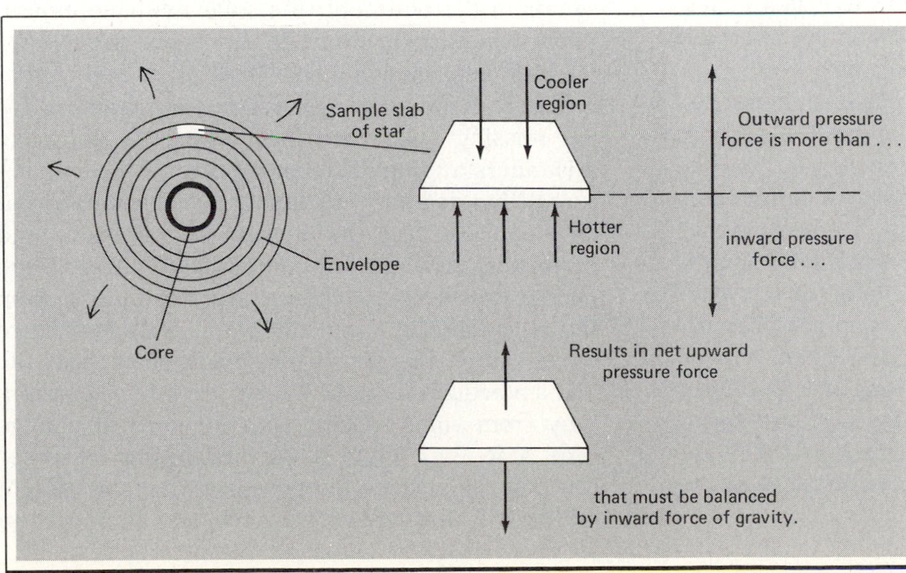

16.5

The balance of gas pressure and gravity in a star. *The outward pressure from the internal heat from fusion reactions must just balance the inward pull of gravity for the star to be stable.*

As an analogy, consider a car assembly line, on which cars pass through different assembly stages in the plant. At each stage, the rate at which the cars come in and go out must be equal. Otherwise the cars will pile up at one location. If the flow of heat through a star were uneven, the temperature of various layers would change. These temperature changes would result in pressure changes that can cause the star to expand or contract. So, generally, a star meets the condition of energy lost equals energy produced.

Finally, we need to look at how a star transports energy from its core to surface (see Focus 13.2). Basically, three methods are available: *conduction*, *convection*, and *radiation*. Which occurs depends on the star's opacity (see Section 13.4). The *opacity* of a star's material directly affects its radiative energy transport. The more opaque a star's material, the slower the flow of radiation through it. You can think of opacity as acting like insulation in a house in winter. The furnace is generating heat. The greater the house's insulation, the slower the flow of the heat to the cold outside. A slow flow keeps the exterior of the house cold and the inside hot. In a poorly-insulated house, the outside is warmer and the inside colder than in a well-insulated house. Likewise, if the opacity of a star were suddenly lowered, radiation would escape more easily, and the star would become more luminous, but its inside would become cooler. So the internal pressure would drop, and the star would contract until its balance was regained.

Whether convection occurs at any region in a star is determined by the opacity of the material there. If the material is transparent, energy will flow easily by radiation. If it is so opaque that the radiative energy flow gets bottled up, convective transport will take over and operate instead. Generally, a star's opacity depends on its chemical composition, density, and temperature.

Gravity keeps a star from expanding and just balances the outward pressure of the hot gases. At the same time, the outflow of energy is just balanced by its production. This balance cannot continue forever. A star loses energy to the cold trap of space. Thermonuclear fusion reactions supply the energy to keep it hot. Eventually the star runs out of fuel and runs out of heat, and gas pressure can no longer balance its gravity. Then it must contract. A star can survive as long as it can find a means to produce energy, to stay hot, and hence to withstand gravity. When it fails to do this, the star dies. How stars survive is the central theme of this chapter.

16.3 *Star models*

A theoretical model of a star incorporates all the physical conditions just described. A star must produce energy, be in balance between gravity and pressure, and transport energy from its core to its surface evenly.

We also need to know exactly what thermonuclear processes produce energy, as well as the mass and chemical composition of the model star. Then a few complicated equations formulate the problem, and these are solved in a consistent way to find the physical conditions (temperature, pressure, and density, for instance) in the star from the center to the surface. This catalog of values for important physical properties, such as temperature and pressure, for a specified mass and composition is called a *star model*. (This is not a real star; it's just a list of numbers!) One pair of numbers of this catalog, the luminosity and temperature at the surface, specifies a point on an H-R diagram, the *theoretical* point for a model star of given mass and chemical composition.

In practice, the construction of a star model requires much tedious calculation. High-speed electronic computers can, however, complete the calculations for one star model in a few minutes. (See Fig. 16.6 for a model for the sun.) A series of models used to describe the evolutionary sequence of one star may take hours of computer time.

Why all this fuss? Because the study of stellar evolution rests on the construction of physically reasonable models of stellar interiors. Models reveal how time enters into the H-R diagram for real stars.

Almost everything said in this chapter rides on the validity of stellar models. And one fly swims in the ointment: the solar neutrino experiment, which casts some doubt on our models for the sun (see Section 13.6). Most of us ignore that problem, hoping for some other explanation for the missing solar neutrinos besides a complete misunderstanding of stellar models.

What do star models tell us? They show that as a star evolves, its radius, temperature, and luminosity change in complicated ways. Such changes result in the change in a star's position on the H-R diagram. For example, if a star's surface temperature (but not its luminosity) increases, the star's position on the H-R diagram moves horizontally from right to left, with no vertical change (line A in Fig. 16.1). If the luminosity increases (but not the surface temperature), the star's point on the H-R diagram moves vertically from bottom

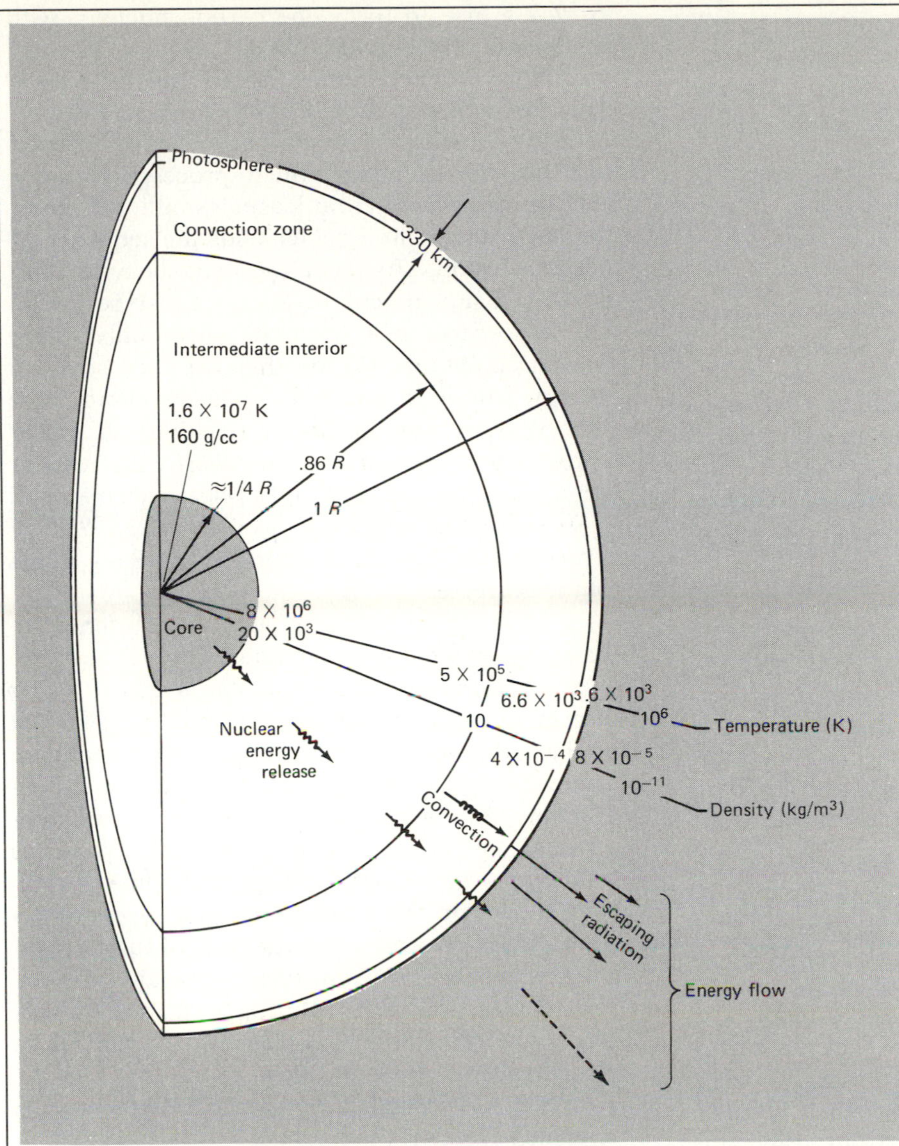

16.6

A theoretical model for the sun, for an age of 4.5 billion years. The values for the physical properties shown here have been calculated by an electronic computer. This figure shows the major regions of the sun's interior and the flow of energy through them.

to top, with no horizontal change (line B in Fig. 16.1). Both motions *on the H-R diagram* represent changes in the physical properties of the star. Note in the first case that to keep the luminosity constant at a higher temperature requires that the star's surface area, and so its radius, decrease. In the second case, with no change in temperature, the star's luminosity can go up only if its surface area, and so its radius, increases. You see that a star's surface temperature, luminosity, and radius change together. (*Warning*: What moves around on the H-R diagram is *not* the star itself but a point *representing* the luminosity and temperature of the star.)

The goal of studying stellar evolution is to understand the physical causes behind a star's evolutionary track on the H-R diagram.

16.4
Energy generation in stars

What makes a star evolve? The answer lies in the heart of the star. Here thermonuclear reactions cook lightweight elements into more complex ones by fusion. This change in chemical composition and its effects on a star's structure mark another major theme of stellar evolution.

The sun generates energy by the proton-proton (PP) reaction. Here's the essence of this reaction: Four hydrogen nuclei combine to form one helium nucleus and release a certain amount of energy. A minor part of the sun's energy comes from another reaction sequence, called the *carbon-nitrogen-oxygen cycle* (CNO cycle). In more massive stars, this cycle becomes more important than the PP cycle. If a star's central temperature is

$^{12}C + ^{1}H \rightarrow ^{13}N$ + gamma ray
$^{13}N \rightarrow ^{13}C$ + positron + neutrino
$^{13}C + ^{1}H \rightarrow ^{14}N$ + gamma ray
$^{14}N + ^{1}H \rightarrow ^{15}O$ + gamma ray
$^{15}O \rightarrow ^{15}N$ + positron + neutrino
$^{15}N + ^{1}H \rightarrow ^{12}C + ^{4}He$

16.7
The carbon-nitrogen-oxygen (CNO) cycle.

greater than about 20 million K, the CNO cycle produces more energy than the PP reaction. (Both can go on at the same time.) The net result of the CNO cycle is the same as that of the PP reaction: Four hydrogen nuclei are converted to one helium nucleus, with the release of energy.

Let's look at the CNO cycle in a little detail (Fig. 16.7). The complete cycle has six reaction steps. First, a proton (^{1}H) collides with a carbon nucleus (^{12}C) to convert it to a radioactive nitrogen isotope (^{13}N). The nitrogen nucleus emits a positron and a neutrino to become a carbon isotope (^{13}C). A proton blasts into this particle and, with the emission of a gamma-ray, turns it into stable nitrogen (^{14}N). Then a proton bangs into the nitrogen nucleus to form the radioactive oxygen isotope (^{15}O) and a gamma-ray. The oxygen isotope decays into the nitrogen isotope (^{15}N), a positron, and a neutrino. Finally, the nitrogen isotope gets knocked by a proton and splits into ordinary carbon (^{12}C) and helium (^{4}He). Note that the carbon comes back unscathed; it acts only as a catalyst, helping to glue four protons together to make helium. Total energy released to the star: 4.0×10^{-12} J. Some goes off as neutrinos.

These reactions take place only in the star's core, where temperatures are high enough to keep them going. Since a star has only a limited amount of hydrogen to burn, the core is eventually converted entirely to helium, and the CNO and PP reactions cease. What next? The core contracts and heats up. When the core's temperature gets up to roughly 100 million K, another reaction can take place: the *triple-alpha reaction,* so named because three helium nuclei (also known as *alpha particles*) fuse to form one carbon nucleus, with the release of energy (Fig. 16.8).

$3\,^{4}He \rightarrow ^{12}C$ + gamma ray

16.8
The triple-alpha process.

What happens when the helium runs out? The core contracts and heats up again. If the temperature increases enough, carbon can be fused into heavier elements. Such processes require extreme temperatures, at least 600 million K. Iron, the most stable of all nuclei, ends the sequence of nuclear fusion. To form elements heavier than iron by fusion reactions, energy must be added and absorbed; such reactions occur only under special conditions. (When they do, they soak up energy from the core.) The steady climb from hydrogen to iron in fusion reactions in stellar cores is one type of *nucleosynthesis,* the nuclear cementing process that makes heavy elements in a universe that otherwise would consist only of hydrogen and helium made in the Big-Bang origin of the universe.

You may be wondering why a star's temperature goes *up* when fusion reactions *stop*. The cause is gravitational contraction. A star is always losing energy to space by radiation. If that energy is not replaced by nuclear reactions, it must come from somewhere else—namely, from gravitational contraction. As the star contracts, some of the gravitational potential energy of its particles transforms into kinetic energy and some is radiated away. So the temperature goes up until the ignition temperature of the next set of fusion reactions is reached. Fusion turns on again, and the core contraction stops. During a star's life, short periods of gravitational contraction alternate with long spells of fusion burning. (Note that it is the *core* that is contracting, not necessarily the star as a whole.)

16.5 Theoretical evolution of a solar-mass star

Gravity instigates the birth of a star. The details of the subsequent evolution are controlled by the star's mass. If the mass is less than 0.08 solar mass, the central temperature never reaches the 10 million K needed to start the PP reaction. (This happened to the planet Jupiter.) If the mass is greater than approximately 100 solar masses, the outward pressure of radiation exceeds that of gravity, and the star blows itself apart. Between roughly 0.08 and 100 solar masses, a stable main-sequence star can form. Let's look at the evolution of a star like the sun, a 1-solar-mass star, with the sun's chemical composition.

Evolution to the main sequence / Let me briefly review a solar-mass star's pre-main-sequence evolution (see Section 15.3). The key

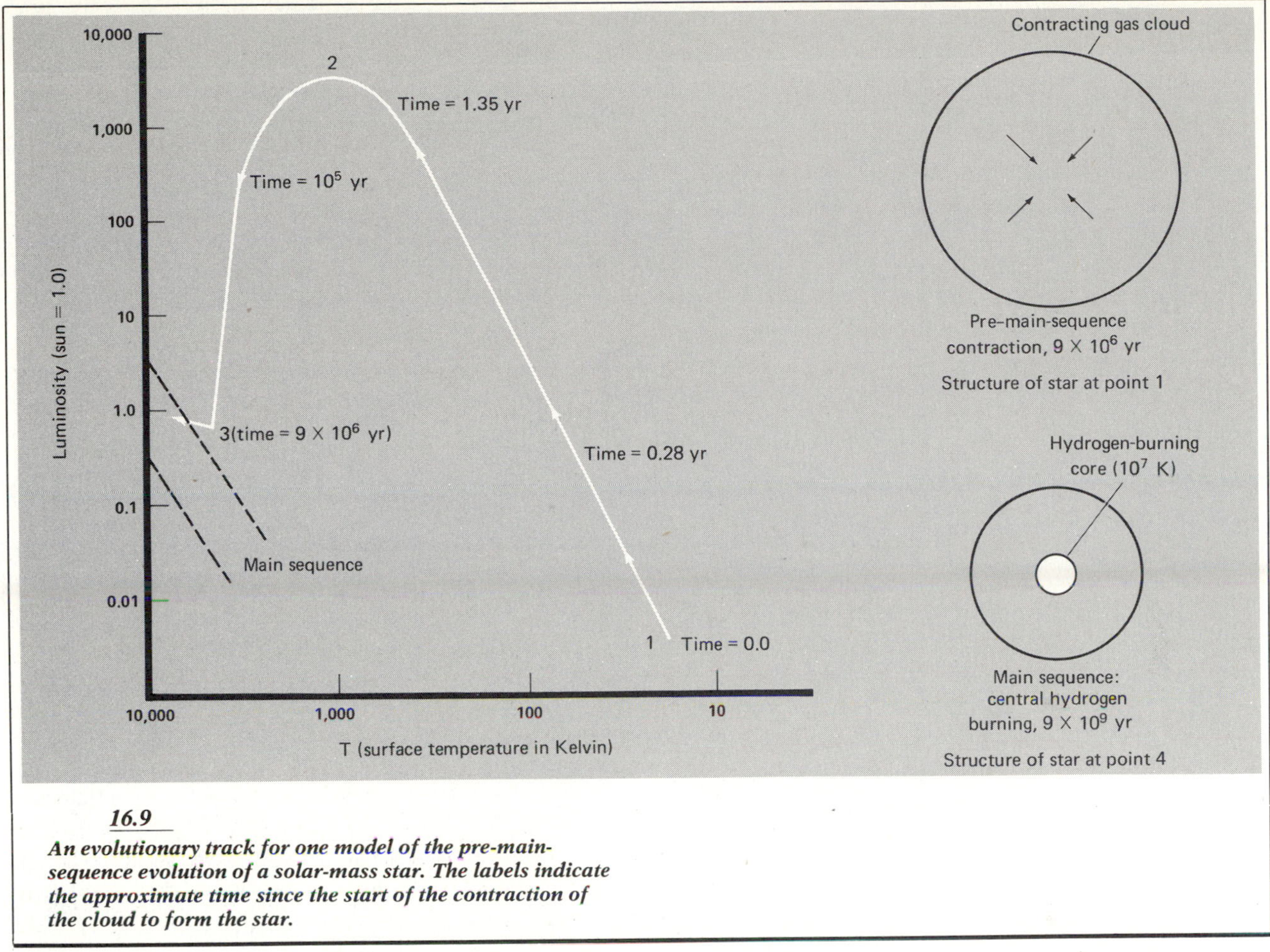

16.9

An evolutionary track for one model of the pre-main-sequence evolution of a solar-mass star. The labels indicate the approximate time since the start of the contraction of the cloud to form the star.

point to recall is that a protostar gets its energy from gravitational energy, not fusion reactions.

For a short time, a protostar has a larger radius than it will as a main-sequence star, and the surface temperature is lower (point 2 in Fig. 16.9). However, the protostar has a higher luminosity than it will when it reaches the main sequence. How can this be, if it is cooler? Because it is larger and so has more surface area to radiate energy.

At this stage, the star's temperature is so low that its opacity is relatively high (even though its density is low). Convection rather than radiation transports the energy outward. So a protostar is completely convective—a huge, bubbling ball of gas. The efficient transport of energy makes the star very luminous (points 2 to 4 in Fig. 16.9).

As the star radiates energy, it contracts to maintain its balance. As the protostar shrinks in size, its luminosity decreases (points 4 to 5 in Fig. 16.9). Its point on the H-R diagram moves downward. Meanwhile, the core continues to heat up. Eventually the core gets to a few million degrees Kelvin, high enough to start thermonuclear reactions. When the protostar gets most of its energy from thermonuclear reactions (PP reactions in the case of the sun) rather than gravitational contraction, it achieves full-fledged stardom. The star is now called a *zero-age main-sequence* (ZAMS) star. It settles down to the longest stage in its life, calmly converting hydrogen to helium in its core. The total time elapsed from initial collapse to arrival as a star on the main sequence is only 50 million years (from point 1 to point 7 in Fig. 16.9).

Evolution on the main sequence / Where the star ends up on the main sequence depends mainly on its mass. The more massive the star, the hotter and more luminous it is; the less massive the star, the cooler and less luminous it is. The main sequence consists of a series of stars of decreasing mass (but similar chemical composition), from the upper left-hand corner (*O*-stars with high mass) to the lower right-hand corner (*M*-stars with low mass). A star like the sun spends about 80 percent of its total lifetime on the main sequence as it slowly transforms its hydrogen core to helium.

How the star evolves further also depends on its mass. Because massive stars have higher luminosities, the hydrogen in them must burn faster

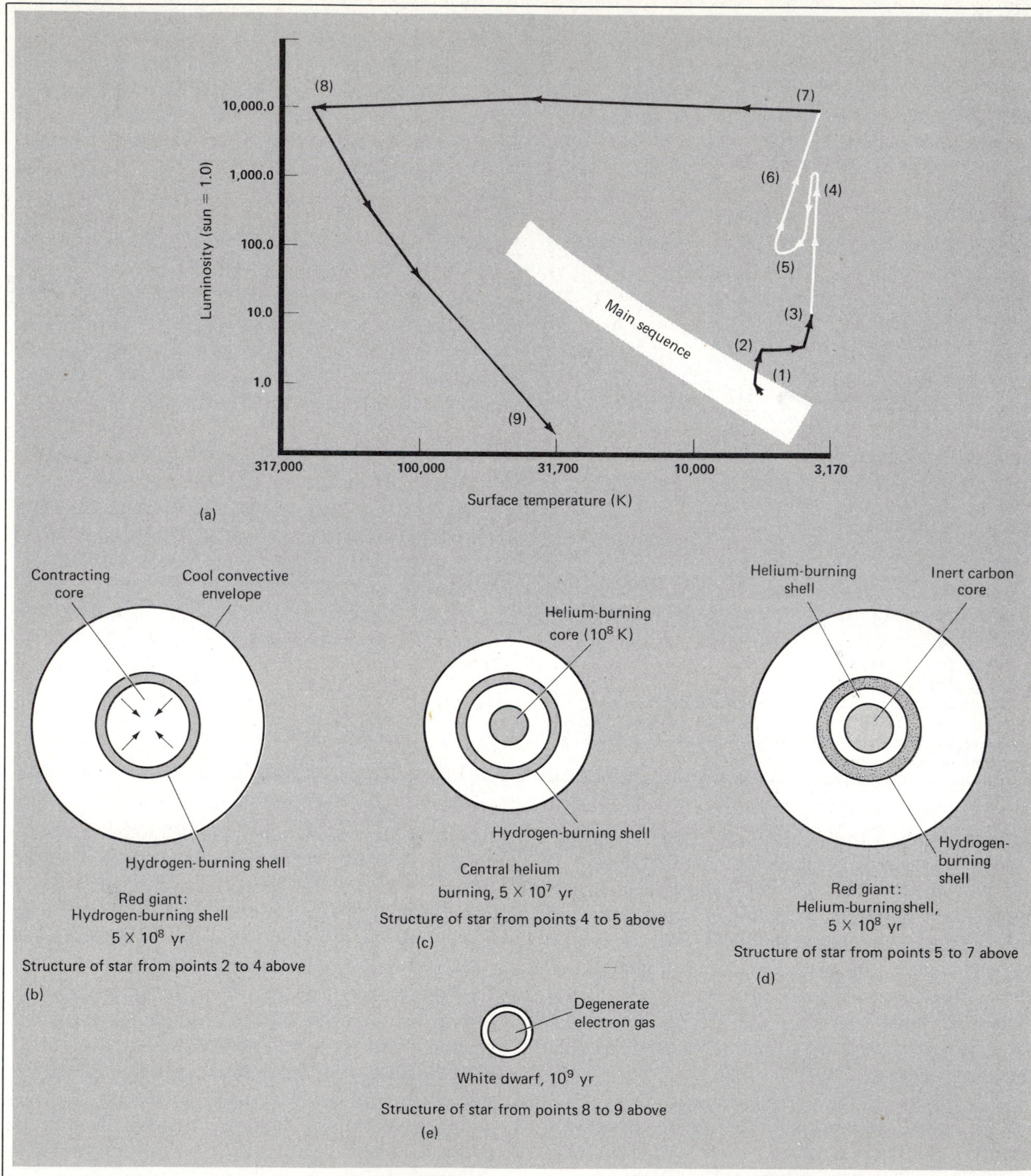

16.10

A theoretical evolutionary path of a solar-mass star off the main sequence (point 1). As the hydrogen is depleted in the core, the luminosity increases (1 to 2). When the core runs out of hydrogen, it burns hydrogen in a shell (2 to 4). Gravitational contraction heats the core until it gets hot enough to burn helium (4). When the core runs out of helium (5), it burns hydrogen and helium in a shell and becomes a red giant (5 to 7). The star throws off its outer layers (7 to 8) and becomes a white dwarf (8 to 9).

than in low-mass stars. As a result, massive stars spend less time on the main sequence, even though they have more fuel to burn, because they use it up at faster rates. Such stars are spendthrifts compared with the miserly energy generation of a star like the sun.

For example, the sun's time on the main sequence lasts about 10 *billion* years, while the same period for a 15-solar-mass star lasts only 12 *million* years. Why the great difference? Because of its greater mass, the 15-solar-mass star has a greater luminosity, about 12,000 times greater than the sun. (Recall the mass-luminosity law, Section 14.6.) So its fusion reactions must go on at a much faster rate than in the sun–also 12,000 times faster. So even though the 15-solar-mass star has more hydrogen to fuse, it does so at incredibly fast rates. Its has a shorter time on the main sequence than the sun.

The main-sequence phase ends when almost all of the hydrogen in the core has been converted to helium. During this time, the temperature in the core increases gradually, and the star contracts slightly. This results in a greater flow of energy to the surface, and the star's luminosity increases (point 1 to point 2 in Fig. 16.10; see also Table 16.1).

TABLE 16.1 ***Evolutionary phases of a solar-mass star***

H-R Position (Fig. 16.10)	*Physical processes*
1	Hydrogen core burning begins (zero-age main-sequence star).
2	Hydrogen core burning ceases; hydrogen shell burning begins.
3	Hydrogen shell burning continues; convection dominates the energy transport.
4	Helium flash occurs; helium core burning begins (red giant).
5	Helium core burning continues along with hydrogen shell burning.
6	Thermonuclear reactions in core end; helium and hydrogen shell burning continues.
7	Expansion and contraction throw off outer layers.
8	Planetary nebula.
9	All thermonuclear reactions stop; white dwarf; slow cooling.

Evolution off the main sequence / When the hydrogen in the core is used up, the thermonuclear reactions cease there. However, they keep going in a shell around the core, where fresh hydrogen still exists. At the end of fusion reactions in the core, gravity takes over and the core contracts. This heats up the layer of burning hydrogen, so the reactions go faster and produce more energy. The luminosity increases. But the layer of burning hydrogen heats up the surrounding part and causes it to expand. So the radius of the star increases, and its surface temperature decreases. The temperature fall increases the opacity, and convection carries the energy outward in the star's envelope.

The star now acquires the characteristics of a red giant; its position on the H-R diagram moves to the region of lower surface temperatures and higher luminosities (point 3 to point 4 in Fig. 16.10). In an additional 500 million years, a solar-mass star ends up with a luminosity of about 1000 solar luminosities, a surface temperature of about 3000 K, and a radius of about 100 solar radii. The sun at this stage would engulf Mercury!

Gravity has compressed the red-giant core to such a high density that it no longer behaves like an ordinary gas. The electrons in the core become a *degenerate gas* (Focus 16.1). In this state, the electrons produce a *degenerate gas pressure*, which depends only on density, not temperature, and this enables the core to attain and preserve a balance even though no fusion reactions are going on. (You will encounter degenerate gases again in Chapter 17.)

As the bloated star attains its red-giant status (point 4 in Fig. 16.10), the core temperature–which has been steadily increasing as the core contracts–hits the minimum necessary to start helium burning by the triple-alpha process. Recall that this helium core is degenerate. Once part of it ignites in the triple-alpha reaction, the heat generated by the fusion spreads rapidly throughout the core. The rest of the core quickly ignites. If the core were an ordinary gas, this explosive ignition would expand it from the rapid increase in temperature and pressure. But the core is degenerate; increased temperature does *not* increase the pressure in a degenerate gas. So the core does not expand. Instead, the increased temperature runs up the rate of the triple-alpha process, generating more energy, further increasing the temperature, and so on. This out-of-control process in the core is called the *helium flash*. When the core temperature finally reaches about 350 million K, the electrons become nondegenerate. The core expands

Focus 16.1

DEGENERATE GASES

When matter is packed to very high densities (greater than 10^8 kg/m^3), it no longer behaves in ordinary ways. In a normal gas, the particles are widely separated and rush helter-skelter into one another and rebound away (see Section 13.2). In highly compressed material, little space exists between particles. The matter is so jammed together that the electrons on the outside of the atoms are, in a sense, touching each other. The nuclei can no longer hold electrons in their usual energy levels, and the electrons move among the nuclei. But there's not much space for moving about.

Electrons abide by a quantum property called the **Pauli exclusion principle.** *It states that no two electrons can be together in exactly the same energy state. Picture a small box containing one electron in some energy state. Imagine adding another electron to our box. The electron already in it probably occupies the lowest energy state possible. So the next electron must occupy the second level available, and so on. In contrast to a low density state where many energy levels are available for occupation, a very dense gas has far fewer levels, which get filled up quickly.*

What happens if we try to cool down this dense gas by letting electrons give up kinetic energy? They can't lose very much kinetic energy, for only high energy levels are open. So no heat can be extracted from these electrons (in a sense, their temperature is zero). Yet they exert a great pressure because they move with high speeds.

A gas in this state is called a **degenerate gas** *and the pressure from the uncertainty principle in action is called the* **degenerate gas pressure.** *Unlike an ordinary gas, for which the pressure is directly proportional to temperature, degenerate gas pressure is nearly independent of the temperature.*

Electrons become degenerate at densities of about 10^8 kg/m^3. Such densities occur in stellar cores after main-sequence hydrogen burning and also in white dwarfs. When electrons are degenerate, they conduct heat very efficiently, and temperature variations are quickly smoothed out. So degenerate cores have the same temperature throughout. A high enough temperature can relieve the electrons of their degenerate condition. This requires a temperature of some 350 million K for electrons.

and cools. The whole process of helium core ignition in the helium flash takes place in a very short time—perhaps only a few minutes. It may reach a peak of some 10^{11} solar luminosities!

But no one has ever seen this helium flash in a star, for this action takes place deep in the core. The surface is hardly affected. The radius and luminosity decrease a little, and the star's point on the H-R diagram moves slightly downward and to the left. The star quietly burns helium in the core and hydrogen in a layer around the core (point 5 in Fig. 16.10). This phase is the helium-core-burning analog to the star's main-sequence phase (hydrogen core burning).

Evolution to the end / Eventually the triple-alpha process converts the core to carbon. The reaction stops in the core but continues in a layer around it. This situation—core shut down but the thermonuclear reactions going on in a layer—resembles that when the star first evolves off the main sequence. The physical processes force the same evolution; the burning layer makes the star expand. The star again becomes a red giant (point 6 in Fig. 16.10). The electrons in the core—this time carbon-rich—become degenerate again.

Because the rate of the triple-alpha reaction is very sensitive to changes in temperature, the helium-burning shell causes the star to become unstable. Here's how: Suppose the star contracts a little. The temperature and energy production in the layer increase; the pressure also increases. However, the increase in pressure more than compensates for gravity; the outer parts of the star expand. The expansion decreases the temperature, the pressure, and—most dramatically—the energy generation rate. Gravity takes command: The star contracts, the energy generation increases a lot, the star expands, and the cycle

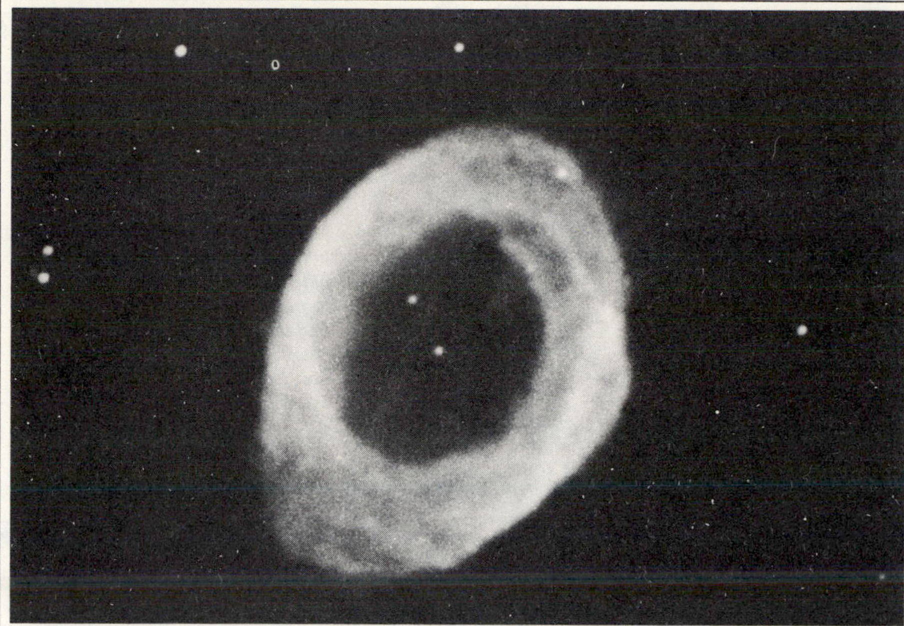

16.11
A typical planetary nebula (NGC 7293 in Aquarius). The central blue star was once the core of a red giant. The nebula, which looks like a ring but actually forms a spherical shell, was the envelope of the red giant. (Courtesy Palomar Observatory, California Institute of Technology)

repeats. The star pulsates slowly—once every tens of thousands of years.

The pulsations gradually grow larger. A final violent one ejects the cool outer layers of the star (point 7 in Fig. 16.10). A hot core is left behind. The expelled shell forms a planetary nebula, and the leftover core becomes its central star (Fig. 16.11). The nebula keeps expanding until it dissipates in the interstellar medium. What happens to the core? For a star of roughly 1 solar mass or less, the core never reaches the ignition temperature of carbon burning. Why not? Because the core has become degenerate and cannot contract and heat up to ignite carbon burning. In about 75,000 years it forms a white dwarf star, composed mostly of carbon (point 8 to point 9 in Fig. 16.10). Without energy sources, the white dwarf cools to a black dwarf, the dark culmination of a 10-billion-year biography. The degenerate electron gas pressure supports it against gravity even as it cools down.

Lower-mass stars / The evolution of stars of lower mass than the sun resembles that for the sun, with two exceptions. First, few stars less massive than the sun have had time to evolve off the main sequence. The universe simply isn't old enough. A star of 0.74 solar mass, for example, will have a luminosity about 37 percent that of the sun, according to the mass-luminosity law, and so a main-sequence lifetime of 20 billion years, longer than most estimates for the age of the universe (see Chapter 7).

Second, if the mass of a star is less than about 0.08 solar mass, it will not even reach the main sequence. Why not? Gravity is weak in such a low-mass star, and gravitational contraction does not heat it very effectively. Before it gets hot enough to start nuclear reactions, the density has risen so high that the matter becomes degenerate (Focus 16.1). Then the pressure of the degenerate electrons supports the star and keeps it from contracting any further. If gravitational contraction is prevented from heating the star, the nuclear fires can never be lit, and the star simply cools off to become an invisible black dwarf or, if the mass is *very* small, a planet.

Chemical composition and evolution / So far, I've focused on how a star's mass affects its evolution. What about its chemical composition? If a star has a much smaller percentage of heavy elements than the sun, 0.01 percent instead of 1 percent, would its evolutionary track make much different gyrations?

The answer is no. Computer models indicate that the general trends stay: The star becomes a red giant, undergoes a helium flash, sits for a while during helium core burning, returns to red giant status, then blows off its outer layers to make a planetary nebula encircling a white dwarf.

Low-metal stars do show significant differences in where they bide their time on the H-R diagram during their stage of burning helium in the core. Basically, stars with smaller percentages of heavy elements and less mass than the sun make a horizontal line on the H-R diagram during their phase of burning helium in their cores. That is, stars with the same mass but fewer heavy elements fall to higher surface temperatures (to the left on the H-R diagram; Fig. 16.12).

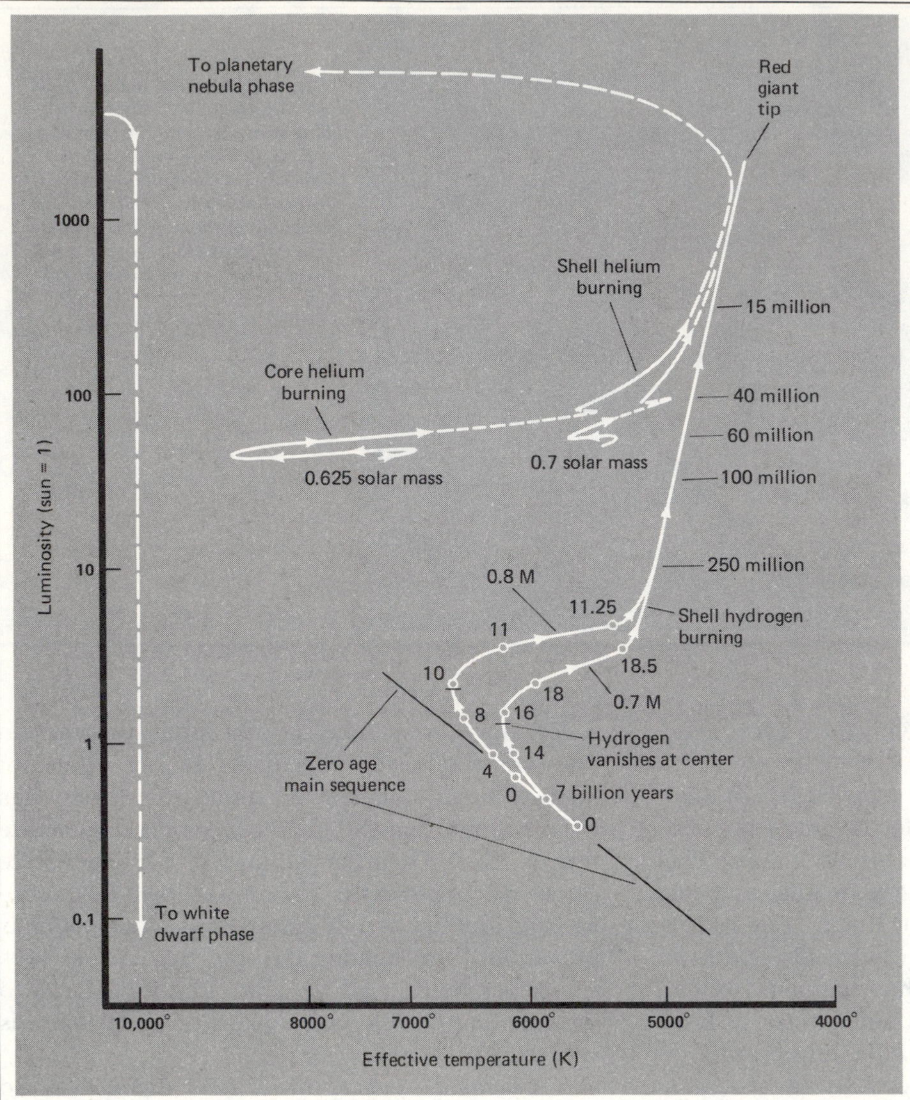

16.12
Theoretical evolutionary tracks for 0.8- and 0.7-solar-mass stars as they move off the main sequence to become red giants. These stars lose mass during their red giant phase, and so have masses of 0.7 and 0.625 solar mass during their core-helium-burning phase, just before they eject their envelopes to make planetary nebulas. (Adapted from a diagram by I. Iben, Jr., **Publications of the Astronomical Society of the Pacific,** ***vol. 83 (1971), p. 697)***

16.6 Theoretical evolution of a 5-solar-mass star

Now to examine the history of a star much more massive than the sun. I've picked a 5-solar-mass star because such are fairly common, compared to, say, 50-solar-mass stars. (Regulus, the bright star that marks the heart of Leo the Lion, is a 5-solar-mass star; see the spring constellation chart on the endpapers.) A larger mass does not change the general flow of stellar evolution, but the details (such as energy generation) do change. Massive stars differ in their evolution from less massive stars because they can reach higher temperatures in their cores. The greater temperatures have important consequences: (1) While on the main sequence, the star burns hydrogen by the CNO cycle; (2) the main-sequence lifetime is shorter; (3) the higher temperatures kindle carbon and heavier-element fusion in the core; and (4) the helium-rich core does not become degenerate.

Evolution to and on the main sequence / A 5-solar-mass star evolves to the main sequence along a track of roughly constant luminosity. The star is hot enough so that its opacity is low enough for radiative transport to be more effective than convection. The conversion of gravitational energy powers the star; it contracts and heats up. As it does, its opacity falls and energy flows out more easily. So its luminosity increases a bit as the star contracts and its core temperature climbs. In about 50 million years the star's point on the H-R diagram moves almost horizontally to the left (Fig. 16.13).

The PP reaction first ignites in the core. When the core's temperature rises to about 20 million K, the CNO cycle produces more energy each second than the PP reaction. The consumption of

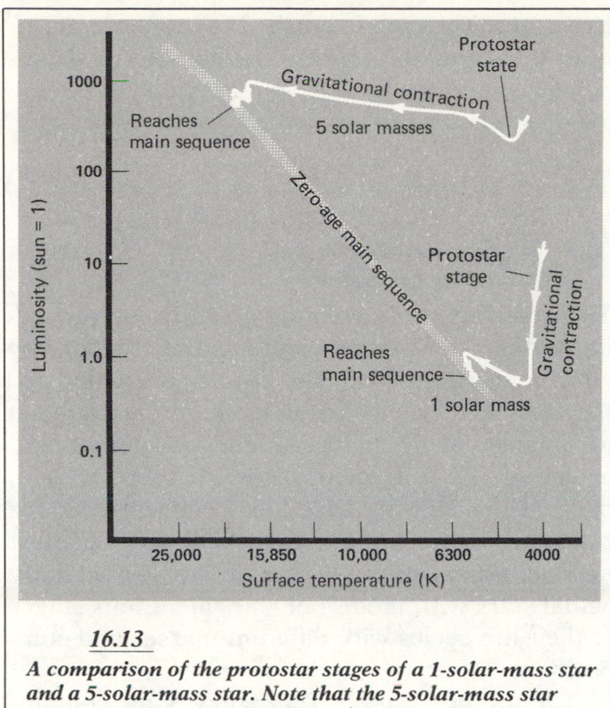

16.13
A comparison of the protostar stages of a 1-solar-mass star and a 5-solar-mass star. Note that the 5-solar-mass star approaches the main sequence at roughly constant luminosity. (Based on work by I. Iben, Jr.)

TABLE 16.2	***Masses of stars and their main-sequence type***
Mass (sun = 1)	*Main-sequence types*
20–60	*O*
4–20	*B*
0.5–4.0	*A, F, G, K*
0.1–0.5	*M*

hydrogen causes the core to shrink, its temperature to go up, and so its luminosity to increase. Stoked by the high core temperature, the CNO cycle uses up the core's hydrogen in about 60 million years, while the star's luminosity becomes twice as large (points 1 and 2 in Fig. 16.14).

Note that a 5-solar-mass star on the main sequence is more luminous and hotter than a 1-solar-mass star. Regulus, for example, is a *B*-star. Less massive stars on the main sequence are less luminous and cooler (see Table 16.2).

Evolution to the end / When the central hydrogen-fusion fires are exhausted, the star contracts. New hydrogen falls to the inner regions and ignites in a shell around the burnt-out core (point 3 in Fig. 16.14). The luminosity increases, but the radius expands, so the surface temperature drops. The star becomes a red giant. Meanwhile the core contracts until it gets hot enough to ignite the triple-alpha process (point 4 in Fig. 16.14). In contrast to a solar-mass star, the core has *not* become degenerate. So no helium flash occurs, just a relatively gentle triple-alpha ignition. (Stars with

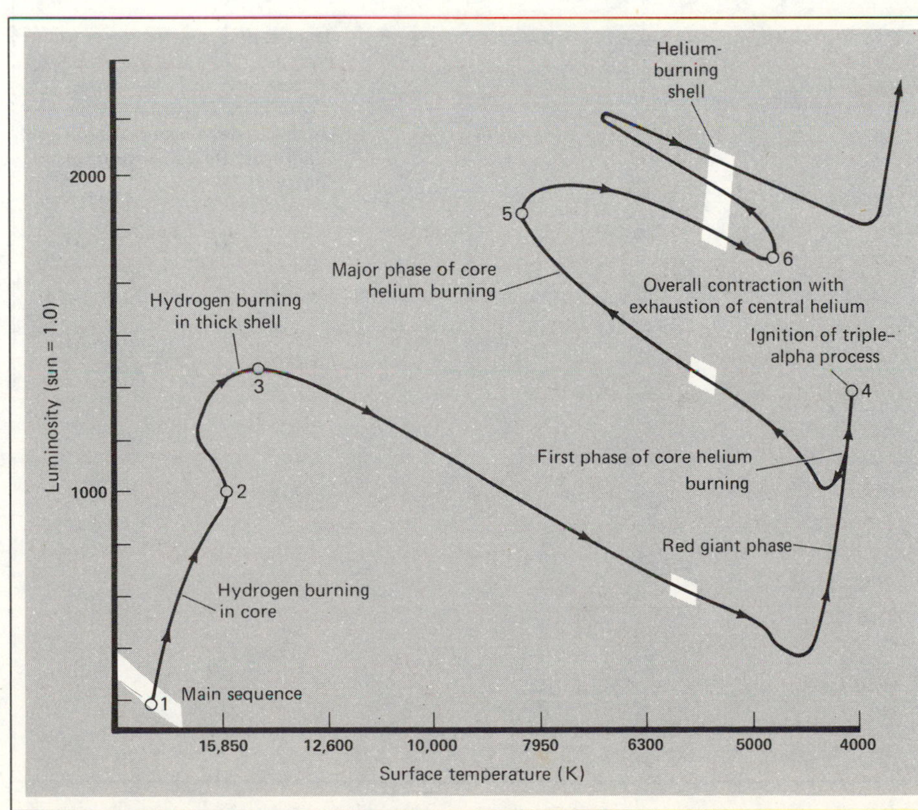

16.14
A theoretical evolutionary track for a 5-solar-mass star. The times given are those for the model star to evolve between the indicated points. Note that the star becomes a red giant every time it burns in a shell or shells around the core. Note also the alternation among core burning, shell burning, and gravitational contraction to reignite the core's fusion reactions. The details of the evolution after the last point are not clearly known. (Based on work by I. Iben, Jr.)

greater than 2.25 solar masses do not develop degenerate helium cores simply because they do not become dense enough before helium ignition takes place.)

The star burns helium in its core for about 10 million years. When the helium runs out, the star again contracts (points 5 to 6 in Fig. 16.14). Now fresh helium falls into the core to make a thin, helium-burning shell around the core. The star becomes a red giant again (beyond point 6 in Fig. 16.14). (Note that whenever fusion reactions take place in a shell rather than in the core, the star expands in size.)

This second time as a red giant may be the last. The triple-alpha process in a helium-burning shell may cause pulsations strong enough to rip off the outer layers. Or perhaps before this can happen, the core gets hot enough to burn carbon.

What next? We're really not sure of the details. We expect that a 5-solar-mass star will develop a degenerate carbon core of about 1.4 solar masses. When the core gets hot enough to ignite carbon burning, it should do so explosively in a carbon flash. This reaction may detonate the carbon in the core so swiftly that the star blows apart. Such a cataclysmic detonation may result in a *supernova*. The outer layers of the star blast into space. The core is crushed to immense densities. Whatever the physical reasons for a supernova, a supernova explosion marks the end of the life of a massive star. (More in Chapter 17.)

16.7 Observational evidence for stellar evolution

The description of evolution given in the preceding sections is based on theoretical calculations by electronic computers. The results presented for the early and late stages of evolution are subject to change. Only main-sequence evolution and evolution just off main sequence seem well in hand. How well do these ideas connect with the real astronomical world? To find out, we must look at observations to see if we can identify actual stars with models of stars at various stages of their life cycles with different masses and compositions.

Stars in groups / The Milky Way contains two contrasting types of star systems: *open clusters* (sometimes called *galactic clusters*) and *globular clusters*. How do they differ?

The Hyades and the Pleiades (see Fig. 15.12) in the constellation Taurus (see the constellation

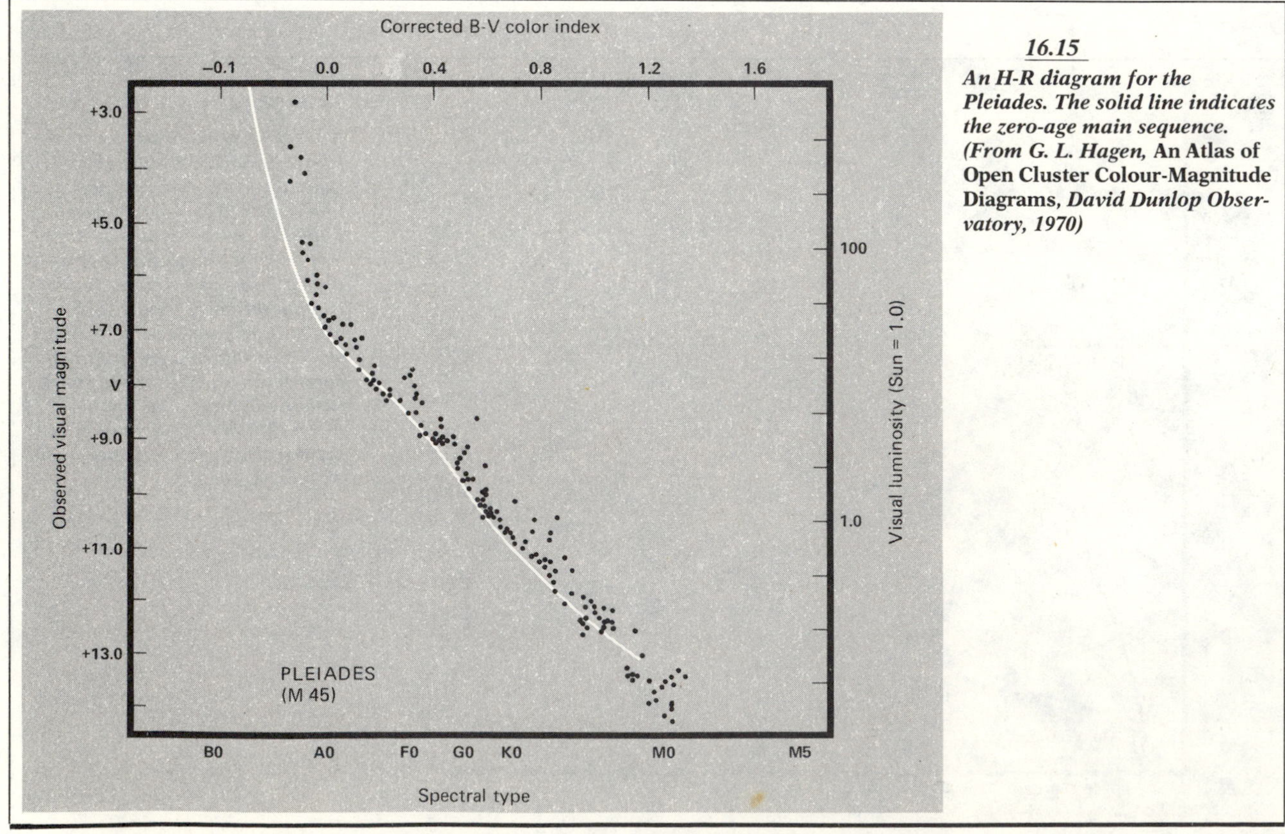

16.15
An H-R diagram for the Pleiades. The solid line indicates the zero-age main sequence. (From G. L. Hagen, An Atlas of Open Cluster Colour-Magnitude Diagrams, *David Dunlop Observatory, 1970)*

endpapers for winter) are good examples of open clusters, with loose arrays of stars. The Pleiades lie about 400 ly from the sun. The cluster contains about 100 stars within a diameter of 10 ly for an average density of about 0.1 star in a cubic light year. These statistics are pretty typical of open clusters: They contain from fewer than 100 up to a 1000 stars in a space a few tens of light years in size; so their star densities are not more than a few stars per cubic light year. Astronomers have cataloged some 1000 open clusters to date.

One key characteristic of open clusters is their H-R diagrams. The one for the Pleiades is pretty typical (Fig. 16.15). Note that the stars below spectral type *A0* fall squarely on the main sequence. Above *A0*, the stars lie above and to the right of the main sequence. Most open clusters show the same properties: The lower-mass stars fall on the main sequence, but at some point the higher-mass stars turn off it (Fig. 16.16).

Globular clusters / Globular clusters contrast dramatically with open clusters. Compare Fig. 16.17 (the globular cluster 47 Tucanae) with Fig. 15.12 (the open cluster the Pleiades). As the name implies, globular clusters have a distinct spherical shape. You can see this shape easily in a small telescope. Binoculars will show it as a miniature, fuzzy sphere, like a cotton ball; a small telescope reveals some of the brightest stars.

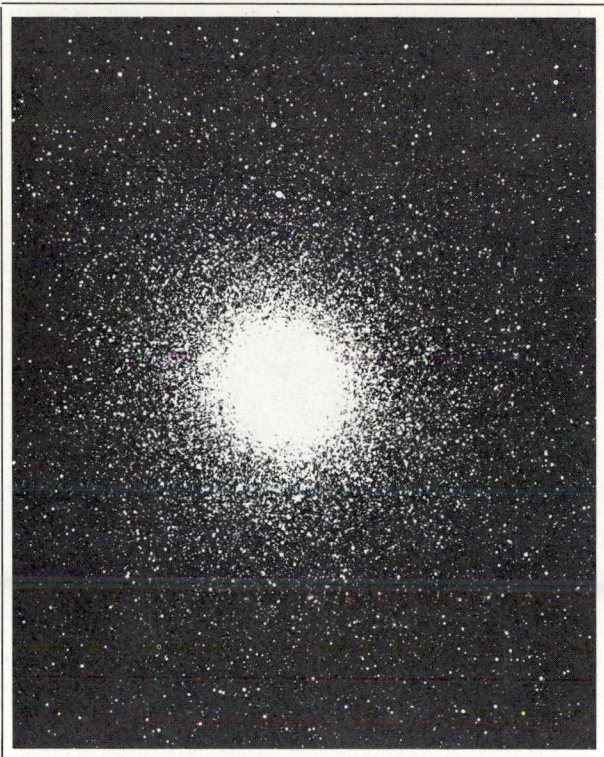

16.17
The globular cluster 47 Tucanae. (Courtesy Cerro Tololo Interamerican Observatory)

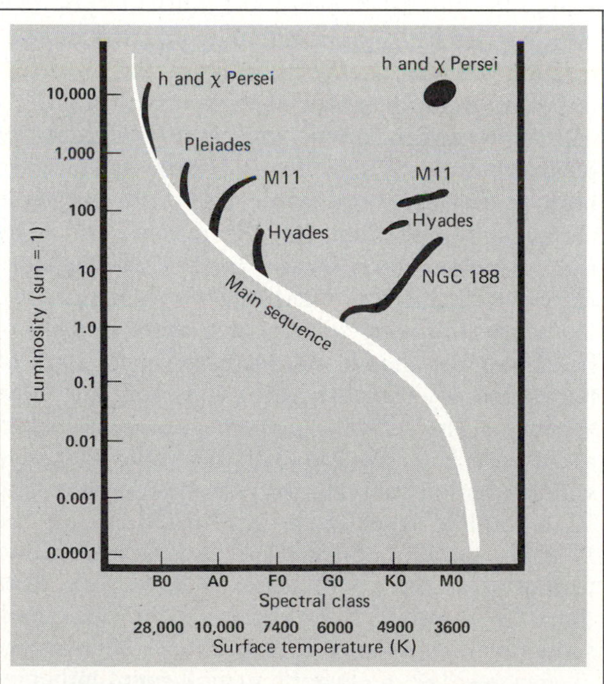

16.16
A schematic H-R diagram for some open clusters whose distances are known. Each cluster turns off the main sequence at a different point. The stars above it have evolved away from the main sequence.

Note how densely packed the stars are at the center of a globular cluster compared to its edges. In the center, the stars are packed as high as 3 to 30 per cubic light year. If the earth orbited a star in the core of a globular cluster, its nearest neighbors would be a few light *months* away. You would see thousands of stars scattered evenly over the sky. In all, a globular cluster contains 10,000 to 100,000 stars of roughly solar mass.

An H-R diagram of a globular cluster (Fig. 16.18) dramatically shows its difference from an open cluster in terms of its stellar type. The main sequence turns off to the giant branch, and the upper end of the main sequence has disappeared. A horizontal branch of stars returns from the giant region to the region of the absent upper main sequence. This slash across the H-R diagram, the *horizontal branch*, is the special signature of a globular cluster.

Table 16.3 compares and contrasts the two types of clusters.

Stellar populations / The striking difference between the H-R diagrams of open and globular clusters imply that the stellar types in the two

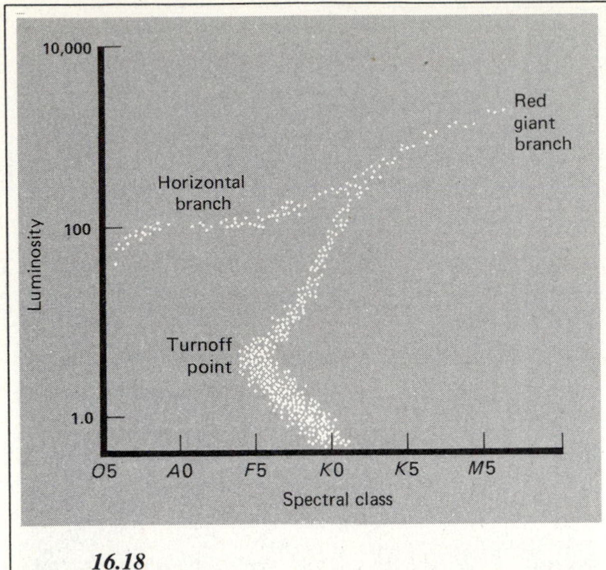

16.18
A schematic H-R diagram for a globular cluster. The stars along the lower part of the main sequence have masses less than 1 solar mass.

kinds of clusters are quite different. Astronomers call those kinds of stars found in open clusters *Population I stars* and those in globulars *Population II stars*. This distinction in stellar type was first discovered by Walter Baade in the 1940s. The brightest Population I stars are blue-white; the brightest Population II stars, red. Also, the brightest Population I stars (*O*-stars) have about 100 times the luminosity of the brightest Population II stars (red giants). The luminous Population I stars must be relatively young, since they are *O* and *B* supergiants.

A really crucial distinction cannot be seen directly in the H-R diagram: chemical composition. Spectroscopic observations show that Population I stars have essentially the same chemical composition as the sun—1 to 2 percent, by mass, heavy elements (which are all the elements more massive than hydrogen and helium). Population II stars contain about 1/100 this amount, 0.01 to 0.02 percent of the mass.

TABLE 16.3 A comparison of star groups

Characteristic	*Open clusters*	*Globular clusters*
Mass (solar masses)	10^2–10^3	10^4–10^5
Diameter (ly)	6–50	60–300
Color of brightest stars	Red to blue-white	Red
Density of stars (solar masses/ly^3)	1	100

From the description of stellar evolution so far, you should see that Population I stars must be *younger* than Population II stars, in the sense that they were born later. According to the Big Bang model for the origin of the cosmos (see Chapter 21), the original stuff of the creation was mostly hydrogen and helium—essentially no heavy elements. The heavy elements must be made in stars. When a massive star dies, it spews a lot of material back into the interstellar medium. This blown-off material has been enriched in heavy elements; from it, new stars will be born. So we know that Population II stars are *older* (formed *earlier*) than Population I stars because they have fewer heavy elements. Population I stars have been formed out of enriched, recycled material.

Warning: Not all Population I stars are luminous blue-white stars. In fact, many Population I stars are stars like the sun, and the vast majority are faint red dwarf stars (the lower right-hand end of the main sequence). Blue-white Population I stars are the most luminous and so the easiest to spot. Also, not all red giants are Population II stars, but in a group of Population II stars, the most luminous ones will be red giants. And, Population I stars have a range of ages, from a few tens of millions of years old to perhaps 8 or 10 billion years old. But all Population I stars are younger than Population II stars, and they all have a larger fraction of heavy elements, perhaps the most distinguishing characteristic of Population I stars.

Comparison with the H-R diagram of clusters / When a star cluster forms, its stars are born at essentially the same time with the same chemical composition, but the masses vary. The more massive stars, *O*- and *B*-stars, evolve more rapidly than the less massive *K*- and *M*-stars. So the more luminous stars evolve more quickly to becoming red giants. As a cluster ages, stars of lower and lower mass will evolve off the main sequence (Fig. 16.19a). At the beginning of its life, a cluster's H-R diagram will resemble that of a young open cluster (Fig. 16.19b); a little later, that of a middle-aged open cluster, such as the Pleiades. Much later, the H-R diagram is similar to that of an old open cluster (Fig. 16.19c). Note that the turnoff point away from the main sequence moves down to lower-mass stars as the cluster ages. So a cluster's turnoff point indicates its age. (To get the age in years, we need to calibrate the main sequence with stellar models. Without a calibration, we can only judge the *relative* ages of different clusters.)

Let's line up the H-R diagrams for a variety

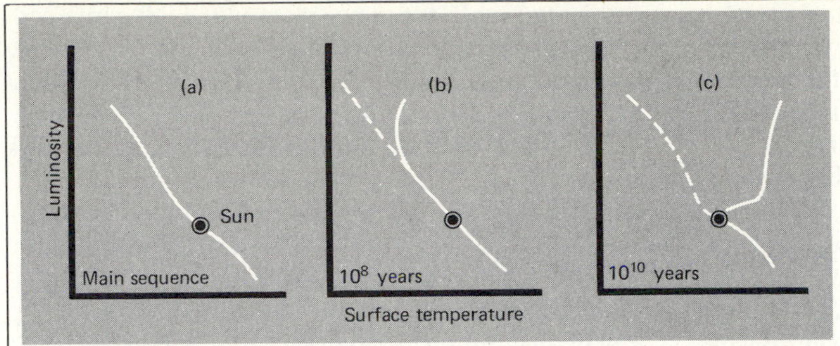

16.19

Theoretical evolution of a cluster of stars with the same chemical composition, born at the same time, but having different masses. "Sun" indicates the position of a solar-mass star.

of star clusters (Fig. 16.20). The stars in the galactic clusters do not peel off the main sequence at the same point. In some clusters (such as the Pleiades), only a few stars at the upper end of the main sequence have evolved away from it. In others, the turnoff point lies farther down the main sequence, but none is below a luminosity of about the sun's. In contrast to galactic clusters, all globular clusters (such as M 3, placed in Fig. 16.20 for comparison) have remarkably similar H-R diagrams, with roughly the same turnoff points.

What implications does this comparison have for stellar evolution? First, it says that globular clusters are older than open clusters. (M 3 *is* actually older than M 67; it appears higher up on the H-R diagram because of its Population II composition.) Second, it implies that the ages of globular clusters are roughly the same; calibrated with theoretical models, the turnoff points for globular clusters indicate that they range from 10 to 16 billion years old, give or take about 2 billion years. M 3, for instance, has an age of 12 to 14 billion years. The stars in the globular clusters are the oldest stars known. Third, open clusters have a

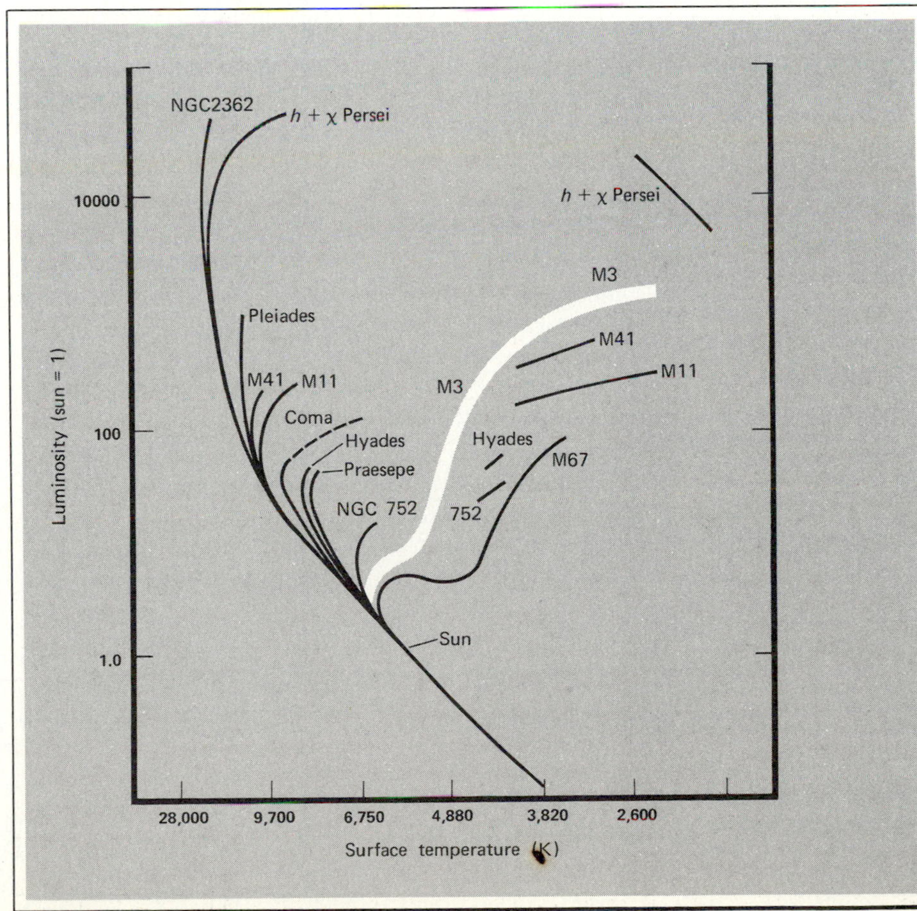

16.20

A composite H-R diagram for some galactic and globular clusters. The further down the main sequence the turnoff point occurs, the older the cluster. (Adapted from a diagram by A. Sandage)

large range of ages. The ages of open clusters can also be estimated from their turnoff points by comparing them to those found from theoretical calculations; for example, M 67 has an approximate age of 3.2 billion years, and the Hyades are much younger, not more than 700 million years old.

The connection between the evolution of an individual star and the changing appearance of the H-R diagram of a cluster of stars as it grows older merits a detailed review. The evolution of each star from the main sequence to the red giants follows similar paths, but starting from a different place on the main sequence, depending on the mass. Although the evolutionary paths are similar, the times to complete those paths are not. Stars higher up on the main sequence (that is, stars with higher mass) evolve to the red giants in a shorter time.

As a cluster ages, the most massive (and luminous) main sequence stars will evolve into red giants first, then the next most massive (next most luminous) ones will evolve, and so on. The main sequence will gradually shorten as the stars peel off, in order of mass, and evolve over into the red giant region. In fact, they last a relatively short time in this red giant phase. The more massive ones become supernovas. The less massive ones become planetary nebulas and leave white dwarf remnants behind. In either case, they stop being red giants and disappear from that part of the H-R diagram. So, in any given cluster, the red giants are stars that have just recently evolved from the main sequence; they come from the region just above the main-sequence turnoff point for that cluster and so have masses just a little larger than the stars still remaining on the main sequence. Stars that started out with still larger masses have already passed through the red giant phase.

To see this evolution clearly, let's examine

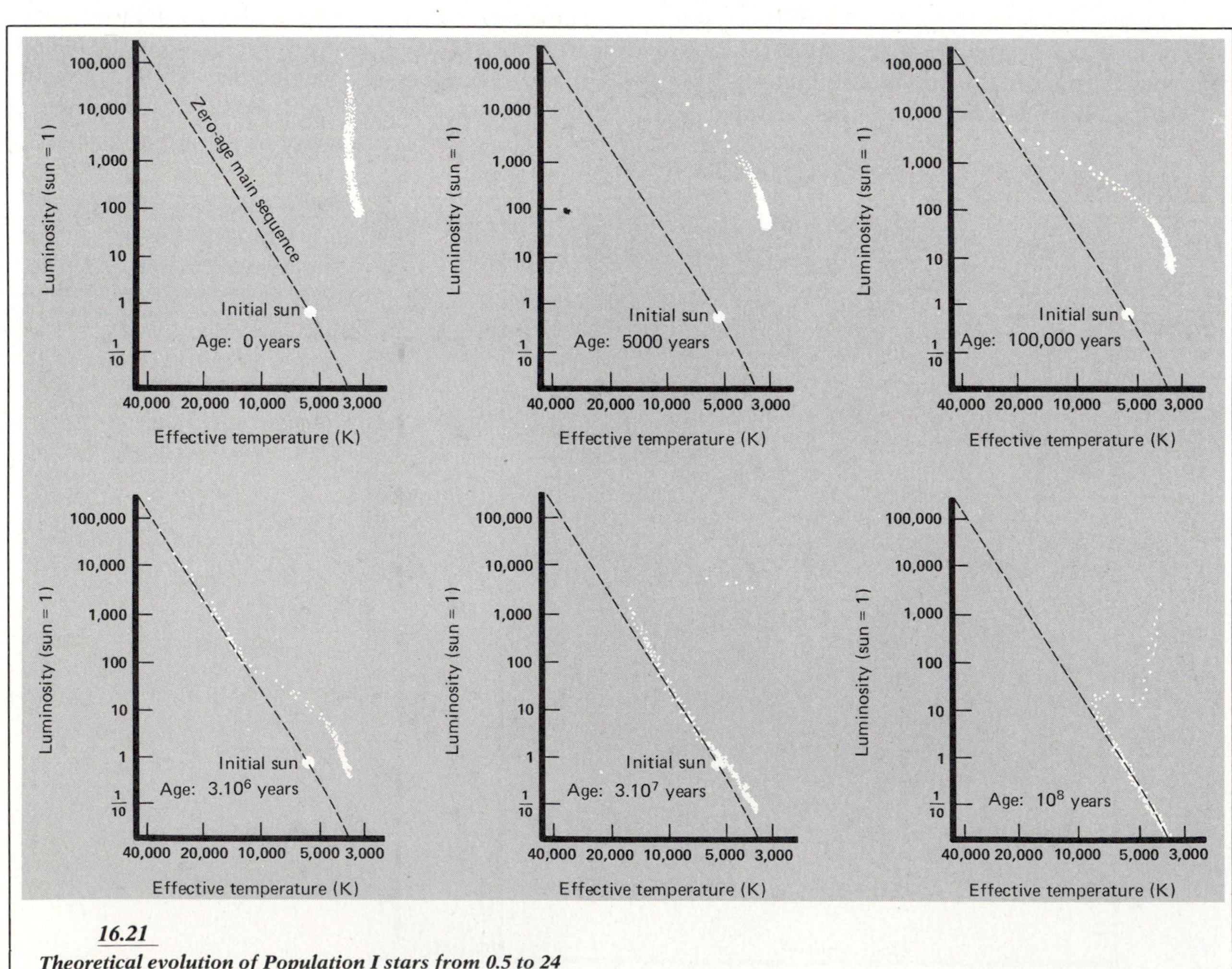

16.21

Theoretical evolution of Population I stars from 0.5 to 24 solar masses making up a hypothetical cluster with all the stars born at the same time. The age indicates time elapsed since birth. (Based on calculations by R. Kippenhahn)

the theoretical paths on the H-R diagram for a model cluster of stars (Fig. 16.21). It consists of 190 stars with masses ranging from 0.5 to 24 solar masses, with a Population I chemical composition. At birth, the stars are protostars, approaching the main sequence (Fig. 16.21a). Just 5000 years later (Fig. 16.21b), the massive stars have already moved toward the main sequence, leaving the lower-mass stars behind. After 100,000 years (Fig. 16.21c), the massive stars have hit the main sequence and commence hydrogen core burning. At 3 million years (Fig. 16.21d), the upper main sequence is established. Here the cluster's H-R diagram resembles that for the young cluster NGC 2264 (see Fig. 15.23). By 30 million years (Fig. 16.31e), the most massive stars are in the helium-core-burning stage and so are red giants. At an age of 6.6 million years, the upper main sequence is losing stars. By 100 million years (Fig. 16.21f), many stars are red giants. The cluster's H-R diagram resembles that of M 67 (Fig. 16.20). Finally, at an age of 4.2 billion years, the turnoff point has rolled down to about a solar mass.

The end result: We can infer the age of a star cluster by comparing its H-R diagram to theoretical ones. Clusters with a long main sequence and luminous red giants are young; clusters with a shorter main sequence and less luminous red giants are older. In Fig. 16.20, NGC 2362 is the youngest galactic cluster shown, M 67 the oldest. It is only by comparisons of theoretical evolutionary models with actual H-R diagram for clusters that we can date their ages. The comparison also confirms the general validity of the models.

Variable stars / A star's luminosity varies little during its main-sequence sojourn. That's about 80 percent of its lifetime. During the remaining 20 percent, a star's luminosity varies dramatically, but in times too long for us to see the change for any one star. However, astronomers have observed stars whose luminosities change over periods of a few hours to a few years. These *variable stars* (so called because their luminosities change rapidly with time) lie above the main sequence on the H-R diagram (Fig. 16.22). What is their evolutionary status? We know from model calculations that these variables are post-main-sequence stars and that those that vary regularly are in the helium-core-burning stage.

There's a staggering array of variable stars; I'll limit discussion to just three: RR Lyrae variables, cepheids, and red variables (Fig. 16.22). *RR Lyrae stars* (named after their prototype, RR Lyrae, whose period is 13.6 hours) vary in luminosity with periods of 1.5 to 24 hours (typically 12 hours). They are Population II stars, range in spectral types from *A*2 to *F*6, and have about 100 times the sun's luminosity. About 5000 RR Lyrae

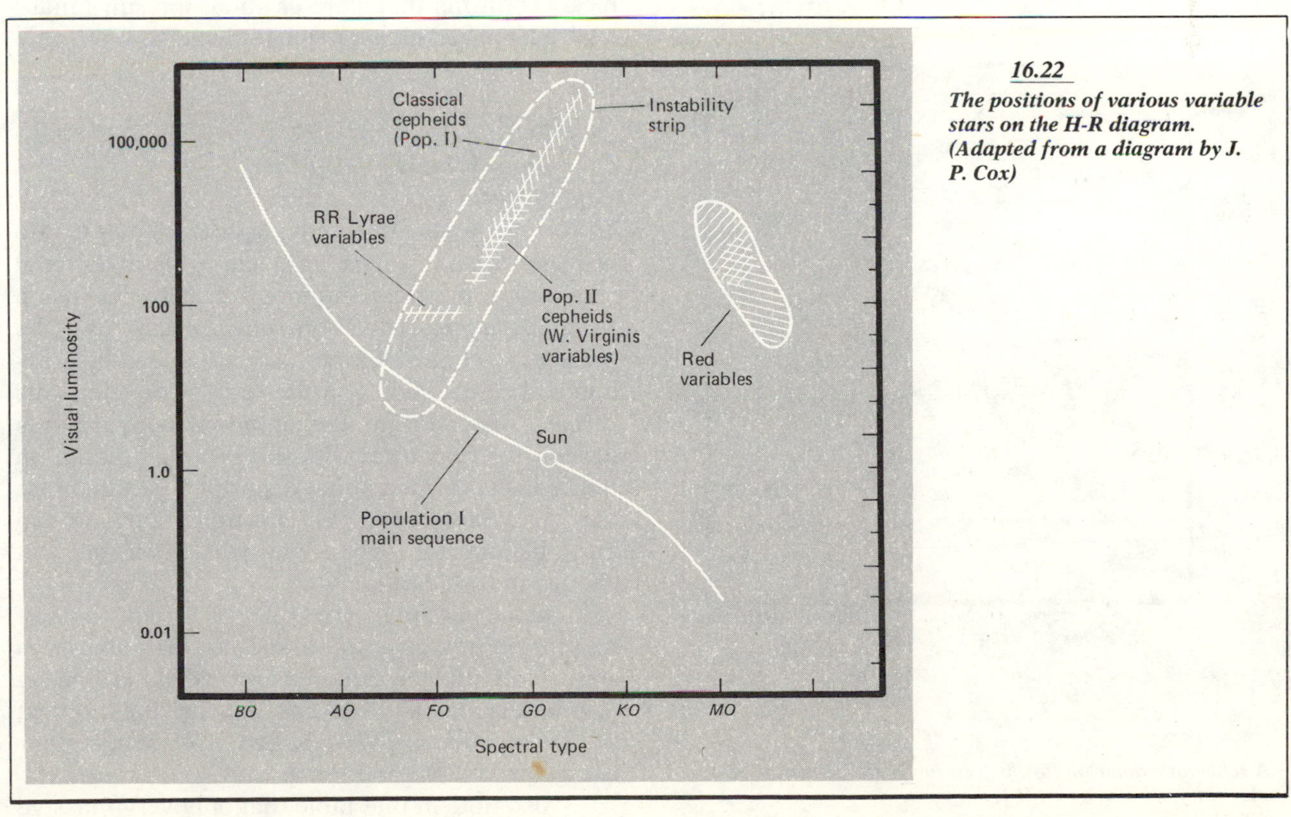

16.22

The positions of various variable stars on the H-R diagram. (Adapted from a diagram by J. P. Cox)

stars are known. *Cepheids* (named after their prototype Delta Cephei) fall into two groups: Population I, called *classical cepheids*, and Population II, sometimes called *W Virginis stars*. Classical cepheids have periods from 1 to 50 days (typically 5 to 10 days) and range from *F*6 to *K*2 in spectral type. Population II cepheids vary in periods from 2 to 45 days (typically 12 to 20 days) and range from *F*2 to *G*6 in spectral type. The RR Lyrae stars and Population I and II cepheids are *regular* or *periodic variables*; their change in luminosity with time follows a regular cycle.

In contrast, the *red variables* have irregular cycles of light variation that range from 100 to 700 days. They contain both Population I and II stars of spectral types *K* and *M* and have luminosities roughly 100 times that of the sun. They are red giant and supergiant stars.

These variables have a key physical characteristic in common: They *pulsate*. Doppler shift observations of their spectra show that as they vary, these stars expand and contract. (When the star is expanding, its surface moves toward us, producing a blue shift in the spectrum; when it contracts, the surface moves away, producing a red-shifted spectrum.) Population I cepheids, for instance, expand and contract at speeds of about 30 km/sec.

Cepheids and RR Lyrae stars lie in a region of the H-R diagram called the *instability strip*. Evolutionary tracks of low-mass (0.5 to 0.7 solar mass) Population II stars transverse this strip during their helium-core-burning phase. Theoretical tracks of Population I stars of 3 to 18 solar masses cross the upper region of this strip also during their phase of helium core burning.

The red variables, on the other hand, fall in a region of the H-R diagram where the stars undergo *shell burning* rather than core burning. Helium shell burning is unstable and results in pulsations.

So theoretical calculations indicate where on the H-R diagram we expect instability and variations; observations validate the theoretical work.

Central stars of planetary nebulas / In the scenario for the evolution of a solar-mass star, when a red giant pops off its outer layers, it leaves behind a hot, dense core. This cinder cools to form a white dwarf.

If this picture is correct, you'd expect that the central stars of planetary nebulas should fall along the evolutionary track (points 8 to 9 in Fig. 16.10) after the red giant stage. Well, they do (Fig. 16.23)! Some central stars are extremely hot and luminous; others are hot but not so luminous. Their positions on the H-R diagram fall neatly above that for white dwarfs. So the stars of planetary nebulas mark a transition between the core of a red giant and a white dwarf. This observation nicely confirms that stars of about the sun's mass evolve from red giants to white dwarfs.

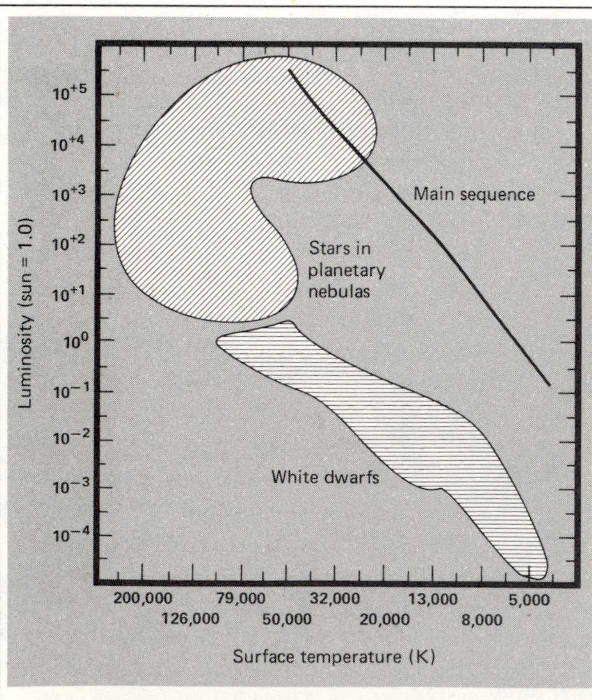

16.23
A schematic diagram for the location of the central stars of planetary nebulas on an H-R diagram.

16.8 The synthesis of elements in stars

To survive, a star must fuse lighter elements into heavier ones and so generate energy. Gravitational contraction provides the initial heat to get fusion reactions going. The more mass a star has, the greater the central temperature produced by gravitational contraction and the heavier the elements it can fuse. From the ignition temperatures needed for fusion reactions, we can set limits on the heaviest elements that a star of a certain mass can fuse (Table 16.4). For example, our sun can burn helium to carbon but will never get hot enough to fuse carbon.

Table 16.4 summarizes the principal stages of nuclear energy generation and nucleosynthesis in stars. Note that the products (or ashes) of one set of reactions usually become the fuel for the next set of reactions. What a beautiful scheme for energy production in the universe!

Also note in that table that only very massive

TABLE 16.4 Stages of thermonuclear energy generation in stars

Process	*Major products*	*Approximate temperature (K)*	*Approximate mass (solar masses)**
Hydrogen burning	Helium	$< 3 \times 10^7$	> 0.1
Helium burning	Carbon, oxygen	2×10^8	1.0–1.4
Carbon burning	Oxygen, neon, sodium, magnesium	8×10^8	1.4–5.0
Neon burning	Oxygen, magnesium	1.5×10^9	5–10
Oxygen burning	Magnesium to sulfur	2×10^9	10–20
Silicon burning	Elements near iron	3×10^9	> 20

SOURCE Adapted from a table by A. G. W. Cameron.
* Minimum mass needed for the fusion process to occur.

stars (those with masses greater than about 5 solar masses) can produce elements heavier than oxygen, neon, and sodium. Few stars have this much mass, so many stars come to the end of their nuclear evolution without having manufactured some important elements. This fact emphasizes the importance of massive stars in the scheme of cosmic evolution—they fuse heavy elements *and* throw some back into the interstellar medium.

I hope you now see where part of the periodic table of the elements (Appendix F) comes from—the part up to iron. What about the rest? You'll find out in the next chapter.

KEY CONCEPTS

1 ***A Hertzsprung-Russell diagram provides a picture of stars at different stages of their lives; time is implicit in an H-R diagram.***

2 ***Stars maintain a balance between gravity and internal pressure; to strike this balance, a star must produce energy inside; the loss of this energy to space requires that the star evolve.***

3 ***Fusion reactions normally generate a star's energy; when fusion reactions stop, gravitational contraction can produce heat and light and ignite the next stage of fusion reaction.***

4 ***As fusion reactions use up fuel, the ashes produced can become the next fuel, if the temperature gets high enough.***

5 ***A newly born star shines from energy produced by gravitational contraction; when fusion fires ignite, the star achieves the main-sequence stage of its life.***

6 ***Stars on the main sequence in the H-R diagram are fusing hydrogen to helium in their cores.***

7 ***More massive stars (upper part of the main sequence) evolve faster than less massive stars (lower part of the main sequence).***

8 ***When a star has fusion reactions occurring in a shell (or shells) around the core and not in the core, it expands in size.***

9 ***Our sun (and other solar-mass stars) will evolve to a red giant (twice) and will blow off its outer layers to make a planetary nebula; the former red giant core becomes a white dwarf, then a black dwarf.***

10 ***Medium-mass stars (5 to 10 solar masses) become red giants a number of times, then die in a supernova explosion.***

11 ***Heavy-mass stars (greater than 10 to 20 solar masses) become supergiants; they die in supernova explosions.***

12 ***A comparison of the H-R diagrams for clusters of stars confirms our basic ideas about stellar evolution; the main-sequence turnoff point indicates the age of a star cluster.***

13 ***Massive stars can fuse elements up to iron in their cores in their normal lives.***

14 ***A star's mass determines how it lives (Table 16.5); more massive stars have shorter lives.***

TABLE 16.5 A comparison of the general evolution of stars of different masses

Mass	*Evolutionary sequence*
Low ($\lesssim$ 1 solar mass)	Protostar → main sequence → red giant → planetary nebula
Middle ($\sim$ 5–10 solar masses)	Protostar → main sequence → red giant → supernova
High ($\gtrsim$ 20 solar masses)	Protostar → main sequence → supergiant → supernova

STUDY EXERCISES

1 A star like the sun consists completely of an ordinary gas. Why doesn't it suddenly collapse gravitationally? (*Objective 2*)

2 Present calculations indicate that a solar-mass protostar is much more luminous than the sun. Yet it's much cooler at the surface. How could the protosun be much cooler and yet more luminous than the present sun? (*Objective 4*)

3 How can you tell from an H-R diagram that the stars in the Pleiades cluster are younger than those in the Hyades? (*Objective 6*)

4 Why are massive stars able to fuse heavier elements than less massive stars? (*Objective 8*)

5 What evidence do we have that red giant stars become white dwarfs? (*Objective 7*)

6 What is a main difference between the evolution of a 1-solar-mass star and a 5-solar-mass star? (*Objective 5*)

7 Compare and contrast the star *types* in open and globular clusters. (*Objectives 8 and 9*)

8 In the future, our sun will likely become what kind of star and what kind of corpse? (*Objective 3*)

9 Outline the evolution of a cluster of stars containing half 5-solar-mass stars and half 1-solar-mass stars. (*Objective 6*)

BEYOND THIS BOOK . . .

"Stellar Populations" by M. and G. Burbidge in *Scientific American*, November 1958, p. 44, is a good introduction to the subject.

For more details on planetary nebulas, look at "Recent Findings About Planetary Nebulas" by Y. Terzian, *Sky and Telescope*, December 1977, p. 459.

I. Iben describes the evolution of Population II stars in "Globular Cluster Stars," *Scientific American*, July 1970.

Stars and Clusters (Harvard University Press, Cambridge, Mass., 1979) by C. Payne-Gaposchkin has good information on clusters and their relationship to stellar evolution theory.

17/Stardeath

> *I believe a leaf of grass is no less*
> *than the journey-work of the stars.*
> *WALT WHITMAN: Song of Myself*

Learning objectives

After studying this chapter, you should be able to:

1. **Compare the physical natures of white dwarfs and neutron stars and describe their place in stellar evolution.**
2. **Describe the basic physical properties of a degenerate star in contrast to an ordinary star.**
3. **Argue, with observational support, that pulsars are rapidly rotating, highly magnetic neutron stars.**
4. **Compare and contrast the observed features of a nova and a supernova.**
5. **Outline a possible model for a nova explosion that involves a binary star system.**
6. **Outline a possible model for a supernova explosion.**
7. **Cite observational evidence that the Crab Nebula is a supernova remnant and describe the effect of the pulsar on the nebula now.**
8. **Describe how synchrotron radiation is emitted and what are its observed properties.**
9. **Describe a black hole in terms of escape velocity and the speed of light.**
10. **Describe what happens to an observer falling into a black hole from the view of the in-falling observer and of an outside observer far from the black hole.**
11. **Describe how and what nucleosynthesis can occur in a supernova.**
12. **Place supernovas in the grand scheme of cosmic evolution.**

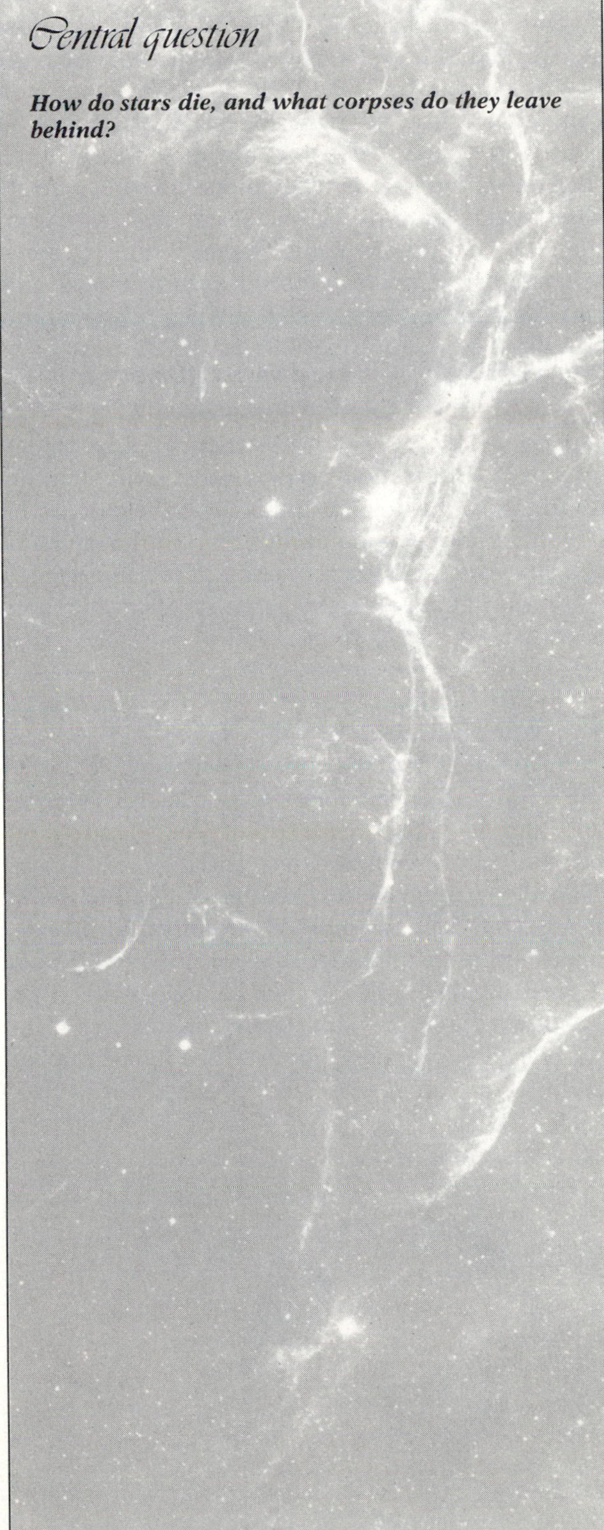

Central question

How do stars die, and what corpses do they leave behind?

How do stars die? More or less violently. Imagine, as in Chapter 16, that one year equals one-fifth of a second, so the sun would live 65 years. In this speeded-up time, you would see about 25 stars in the Milky Way wink out every second. These deaths are signaled by an ejection of a star's outer layers. These discarded shells replenish the interstellar medium, previously depleted by the formation of stars and planets. The dead star's remnant core cools. Locked tight by gravity, it forms a cinder in space. In some instances—the future sun, for example—the burned-out core becomes a white dwarf star, a solid carbon crystal. In others, the core becomes a neutron star, a smooth, spinning sphere of nuclear matter. In still others, the core may disappear through a warp in spacetime as a black hole.

Observations imply that almost all stars throw off mass before they meet their ends. A supernova is the most destructive example of mass loss. But a supernova is constructive too; in its immense explosion, many heavy elements of the universe are made and thrown to the currents of space. Supernovas spice the interstellar medium and provide the impulse to the birth of new stars.

17.1 White dwarf stars: Common corpses

The evolution of a 1-solar-mass star (see Section 16.5) illustrates the constant battle of pressure and gravitational forces. Because gravity never lapses the way thermonuclear reactions do, the final state of any star depends only on the physical properties of matter at high densities and the total mass of the star. When all the thermonuclear reactions cease, what pressure can support the star?

Physics of dense gases / The conventional model of the atom surrounds the nucleus with a cloud of electrons. The distance from the nucleus to the first electron shell is about a thousand times the diameter of the nucleus, so an atom is mostly empty space. At the high temperatures in stars, atoms are ionized, and the electrons run around free of the nuclei. As a star is crushed to higher densities in its evolution, the electrons spread to higher energies and form a *degenerate gas* (see Focus 16.1).

Recall that in an ordinary gas, the pressure depends directly on the temperature. But in a degenerate gas, the pressure is *not* tied to the temperature. It depends only on the number of particles and how tightly they are crammed together. The density at which a gas switches to a degenerate state is about 10^8 kg/m^3. A degenerate star of 1 solar mass has an average density of about 10^9 kg/m^3. If the sun were compressed to this density, it would be approximately 7000 km in radius, about the size of the earth (Fig. 17.1).

White dwarfs in theory / In 1935 Subrahmanyan Chandrasekhar applied the physics of a degenerate electron gas to the model of a star. He found that the pressure exerted by the electrons could resist the force of gravity only for stars of less than 1.4 solar masses and that such stars would have a density of about 10^9 kg/m^3. Such a star, at the end point of its thermonuclear history, is a *white dwarf*. No heavier elements are fused, and no energy produced. How does a white dwarf fend off gravity? By the outward pressure from the degenerate electron gas. What about the nuclei? They form a crystal structure embedded in the degenerate gas. If the expulsion of the outer layers of the star left a carbon core, the white dwarf would be a solid carbon crystal.

Chandrasekhar also found this result: The more massive the white dwarf, the *smaller* its radius. (This contrasts to main-sequence stars, where the more mass a star has, the larger it is.)

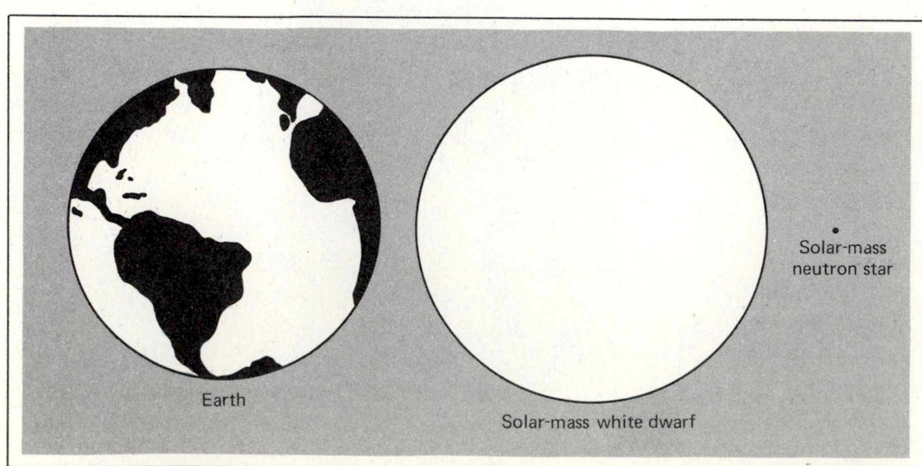

17.1
A comparison of the relative sizes of the earth, a solar-mass white dwarf, and a solar-mass neutron star.

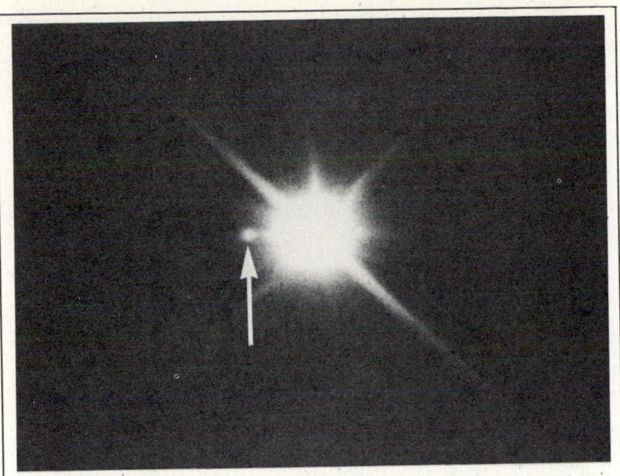

17.2
Sirius B (arrow), a white dwarf star, companion to Sirius A (brighter image). The size of the images in this photo relates only to brightness, not size. (Courtesy Lick Observatory)

How does this come about? More mass means more gravity. To balance gravity requires internal pressure. In a white dwarf, the pressure does not come from internal heating from thermonuclear reactions. Instead it arises from the nature of a degenerate gas, where greater pressures are a response to closer packing of the materials (greater density). So the more massive a white dwarf, the smaller its size.

A crucial point arrives when the mass of the white dwarf is about 1.4 solar masses; such a star has the highest density and smallest radius possible. Add a bit more mass and the gravitational forces overwhelm the degenerate electron gas pressure. The star collapses. It cannot be a stable white dwarf. This amount of mass, 1.4 solar masses, is called the *Chandrasekhar limit* and signals the point at which degenerate electron matter is crushed by gravity.

Observations of white dwarfs / In 1862 the American optician Alvan Clark observed Sirius B (Fig. 17.2), the faint companion to Sirius (see Section 14.6). Later this star was found to be a white dwarf. Sirius B has a mass of about 1 solar mass, a luminosity of about 3×10^{-3} times the sun's luminosity, and a surface temperature of about 29,500 K, and so a small radius, around 7×10^{-3} the sun's radius. Given its size and mass, Sirius B must have an average density of about *3 billion* kilograms per cubic meter!

By coincidence, the brightest star in Canis Minor (the Little Dog, near Canis Major), Procyon, also has a white dwarf companion. This companion was predicted in 1862 from the motion of Procyon and was observed in 1882. Called Procyon B, it has a mass of about 0.65 solar mass.

Most white dwarfs seen so far are actually white, but a few are yellow, and some are red. So far, about 300 stars have been identified as white dwarfs. Because of their low luminosities, white dwarfs are hard to see. This makes it hard to estimate the number of white dwarfs in the Galaxy, but they may make up 10 percent of all stars. Typical values for the physical properties of white dwarfs are: 0.8 solar mass, 0.01 solar radius (7×10^6 m), and a density of 10^9 kg/m^3.

To sum up: A white dwarf is a star with roughly the mass of the sun and the size of the earth. A degenerate electron gas supports it against the crush of gravity. No thermonuclear reactions go on in a white dwarf; it has come to the end of the line of energy production. Very slowly (over billions of years), its stored internal heat (left over from past nuclear fusion) radiates into space. Eventually, it becomes a black dwarf (not to be confused with a black hole!). Our sun will become a white dwarf, then a black dwarf—a cold corpse in space.

17.2 Neutron stars: Compact remains of massive stars

What happens to stars that have more than 1.4 solar masses of core material at the end of their evolution? Strange things happen. Degenerate electron pressure no longer supports them, and gravity crushes them to higher and higher densities. At about 10^{13} kg/m^3, the inward pressure has increased to such a value that *inverse beta decay* occurs (Fig. 17.3). This is the process by which an electron and a proton join together to form a neutron and a neutrino (an electron is sometimes called a *beta particle*). This happens both to free protons and to protons that make up the nucleus of a heavy element. At around 10^{15} kg/m^3, the neutrons begin to drip off the nuclei and to form a separate gas. At 10^{17} kg/m^3, the nuclei suddenly fall apart into a gas with proportions of 80 percent neutrons, 10 percent electrons, and 10 percent protons. At this density, the neutrons become degenerate in the same manner that electrons do at white dwarf densities. The neutrons provide a degenerate gas pressure and so balance the inward pull of gravity. This pressure allows the

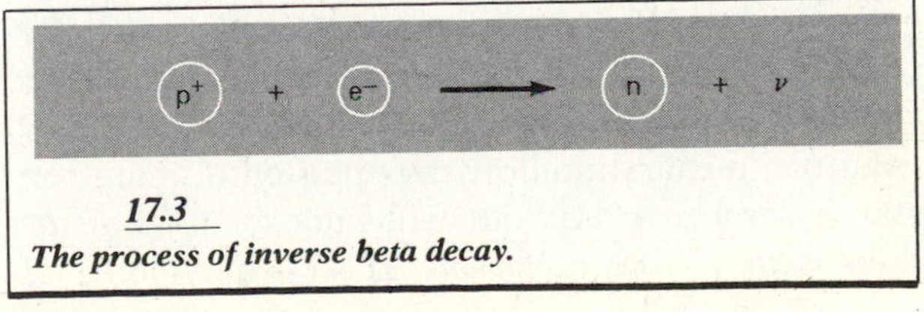

17.3
The process of inverse beta decay.

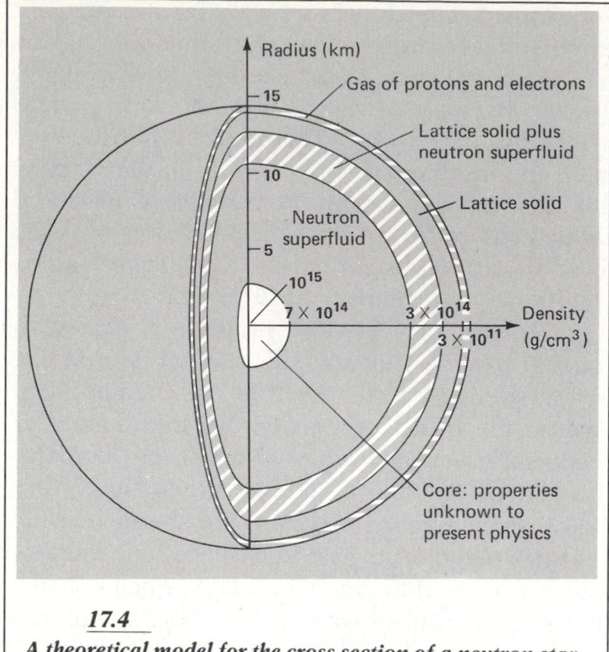

17.4
A theoretical model for the cross section of a neutron star. The solid crust is mostly iron, topped off by an atmosphere of protons and electrons only a few meters thick.

formation of a stable *neutron star*, a star composed mainly of neutrons. Its diameter will be about 10 to 20 km, depending on its mass.

A neutron star is a weird beast compared with an ordinary star. In a typical model (Fig. 17.4) with a diameter of about 15 km, the inner 12 km consists of a neutron gas at such high densities that it is a fluid. In the next 3 km out from the center is a mixture of the neutron fluid and neutron-rich nuclei, arranged in a solid lattice. The structure is a crystalline solid similar to the interior structure of a white dwarf. In the outer few meters, where the density falls quickly, the neutron star has an atmosphere of atoms, electrons, and protons. The atoms are mostly iron.

Because a neutron star is so dense, it has an enormous surface gravity. For example, a solar-mass neutron star with a radius of 12 km has a surface gravity 10^{11} times greater than that at the earth's surface! This enormous pull means that mountains on a neutron star won't be very high, a few centimeters at most. This intense gravitational field also results in a huge escape velocity, as much as about 80 percent of the speed of light. Also, objects falling onto a neutron star from a great distance have at least the escape velocity when they hit. That means that even a small mass carries a fantastic amount of kinetic energy. For example, a marshmallow dropped onto a neutron star from a few AUs out will knock into the surface with a few *megatons* (TNT equivalent) of kinetic energy!

An ordinary star with a mass at the *end* of its evolution greater than 1.4 solar masses probably ends up as a neutron star. Theoretically, a stable neutron star with a mass of less than 1.4 solar masses can also form. These low-mass neutron stars could be made in the pile-driver compression of a supernova explosion, as are most neutron stars.

Because the neutron gas is degenerate, a neutron star has the same kind of mass-radius relation as a white dwarf star: The greater the mass, the smaller the radius. In an analogy to the Chandrasekhar limit, a mass limit for neutron stars is reached when the gravitational forces overwhelm the degenerate neutron gas pressure. This limit—not known exactly, but about 5 solar masses—signals the next crushing point of matter by gravity.

Warning: Do you think neutron stars really exist? Notice that I haven't presented evidence for their existence yet. What I've sketched out are *theoretical* ideas about neutron stars, which were first worked out in the late 1930s. Evidence for the reality of neutron stars didn't crop up until almost 40 years later. That evidence involves cataclysmic explosions of stars—supernovas (Section 17.4)—and rapidly pulsing radio sources—pulsars (Section 17.6).

17.3
Novas: Mild stellar explosions

Aristotle asserted that the heavens were unchanging. This deeply ingrained principle received a hard knock in 1572 with the discovery of a new star, or *nova stella* (Latin for "new star"; the term is usually contracted to simply *nova*), which was observed extensively by Tycho Brahe. Just a few years later, Kepler kept a close watch on another nova that burst into view in the constellation Ophiuchus in 1604.

With the advent of photography and large telescopes in the nineteenth century, astronomers discovered large numbers of novas scattered throughout the sky. By the beginning of this century, a nova was no longer considered an actual new star but rather the sudden eruption of light from an existing star (Fig. 17.5). Also in this century, astronomers recognized that some of these outbursts took place with extraordinary violence. These special flare-ups are now called *supernovas* to distinguish them from ordinary novas. The novas of 1572 and 1604 are now known to have been supernovas.

Both novas and supernovas represent explosions of stars. For ordinary novas, only the outer layers of the star participate in the explosion. For supernovas, the interior regions are also involved.

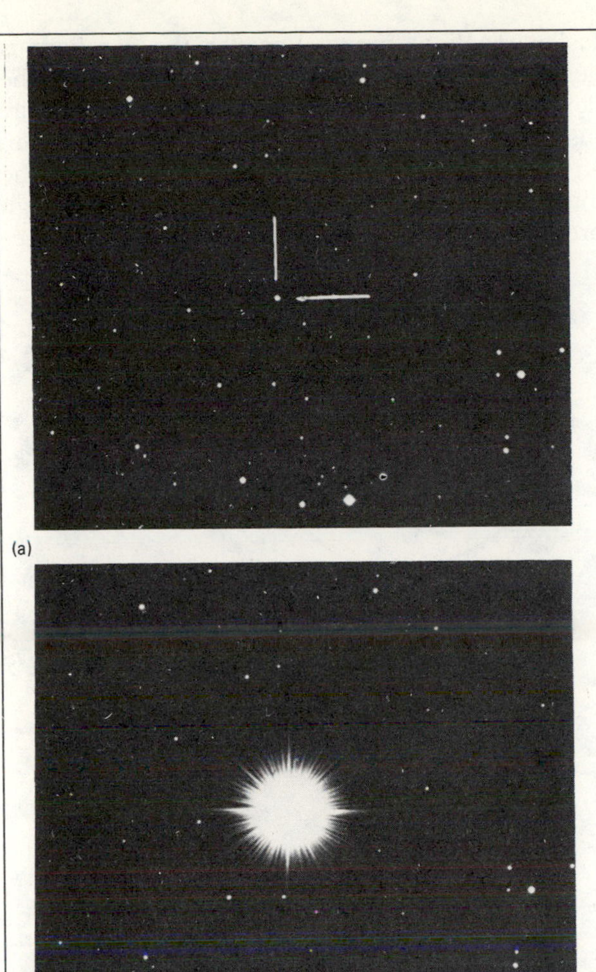

17.5
Nova Herculis 1934. These photos show the star before (a) *and during* (b) *its outburst. (Courtesy Lick Observatory)*

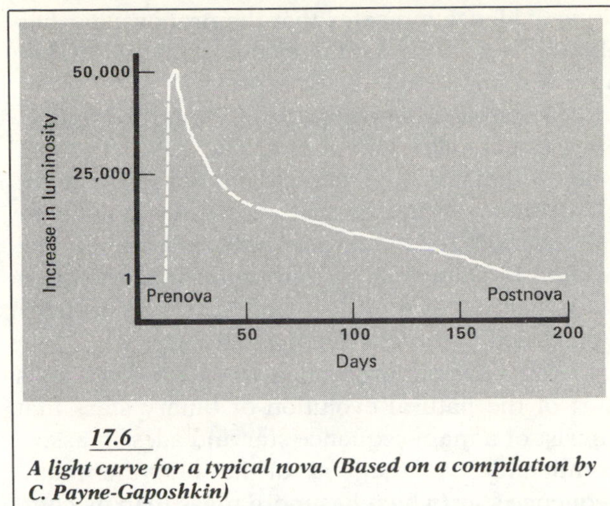

17.6
A light curve for a typical nova. (Based on a compilation by C. Payne-Gaposhkin)

Ordinary novas / In a typical nova outburst, a star in just a few days increases about 10,000 times in brightness. It stays up at peak brillance for several hours. Then the nova's light slowly declines in a few hundred days to an inconspicuous level—usually brighter, however, than the star's prenova level. A plot of a nova's rise and fall in brightness (or luminosity) is called its *light curve* (Fig. 17.6). All novas have the same general shape for their light curves: a sharp rise and a gradual decline.

A typical nova hits a peak brightness of greater than 100,000 solar luminosities. All told, a nova emits during its flare-up and demise some 10^{38} J, or about as much energy as the sun generates in about 100,000 years—emitted, instead, in a few hundred days!

A nova's spectrum undergoes pronounced and complicated changes during its outburst. I won't detail those here. Generally, the prenova star's spectrum has broad, dark absorption lines with weak or no bright emission lines. At maximum, the nova has absorption lines like an *A* or *F* supergiant star. Some time after maximum, the nova's spectrum develops emission lines similar to those from H II regions (see Section 15.1), such as the Orion Nebula. The Doppler shifts of these lines range from a few hundred kilometers per second in some novas to a few thousand kilometers per second in others.

What does the evolution of a nova's spectrum say about the outburst? First, the star's photosphere expands dramatically in size, to 100 to 300 solar radii. Second, the photosphere then collapses back onto the star. Third, a shell of material is blown off the star and rapidly expands away from it (Fig. 17.7). Overall, a nova spurts off about 10^{26} kg of material, about 10^{-4} solar mass. Since

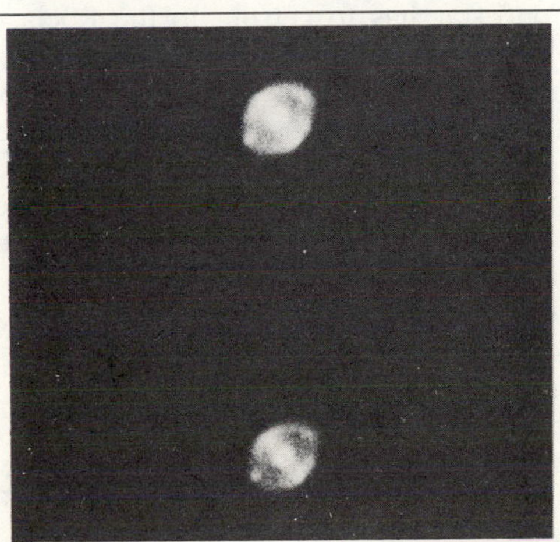

17.7
Nova Herculis 1934, photographed in red light in 1951. Note the shell of blown-off material, which is expanding into the interstellar medium. (Courtesy Palomar Observatory, California Institute of Technology)

some evidence indicates that the prenova star has about one solar mass, the ejected material makes up only a small fraction of the total.

A possible nova model / What prompts a nova to explode? One major clue: Observational studies indicate that almost all novas occur in close binary systems, that is, binary stars with short periods, such as spectroscopic binaries (see Section 14.5). In such systems, the two stars are so close that matter may flow from the more massive companion to the less massive one.

We now picture that a nova may occur as part of the natural evolution of binary stars that consist of a main-sequence star and a less massive companion. At the end of its life, the main-sequence star (which has more mass than its companion and so evolves faster) becomes a white dwarf. Remember that a white dwarf consists of a degenerate electron gas in a lattice of nuclei without hydrogen. To ignite hydrogen fusion again, it needs fresh hydrogen fuel.

Where might such material come from? In a binary star system, the material can come from the companion star. How? Around each star lies a region of space where its gravitational force dominates (Fig. 17.8). The edge of this region is called the *Roche lobe*; any matter within the Roche lobe is gravitationally bound to that star and cannot escape to the other star. But at one point the two Roche lobes touch and join, where the gravitational pull from one star just cancels that from the other. In many binary star systems, the stars orbit so close together that this common point lies very close to one or the other star. This means that a gravitational highway exists for the flow of matter between the stars.

For the mass exchange to happen, material from one star has to get out to the Roche lobe. How? When stars like the sun reach old age, they become red giants before ending up as white dwarfs. So when the less massive companion star to a white dwarf evolves, it bloats up as a red giant. Its atmosphere swells up and reaches its Roche lobe. Material then flows from the red giant to the white dwarf.

As the matter falls toward the white dwarf, it forms a disk around it. Called an *accretion disk*, the material gathers in a disklike structure from the conservation of angular momentum (see Focus 9.2). The matter spirals into the white dwarf's surface from the accretion disk. Fresh hydrogen gradually accretes on the white dwarf's surface, forming a virgin envelope. Additional material piles on, compressing and heating it. When the temperature at the bottom of the accreted layers reaches 1 million K, hydrogen fusion reactions ignite. Because the gas here is degenerate, the ignition is explosive (just like the helium flash in red giant stars). The runaway fusion reactions heat the entire layer to 10 million K; and it loses its degenerate state. Then the material expands explosively because of its high pressure, to blow the accreted material into space.

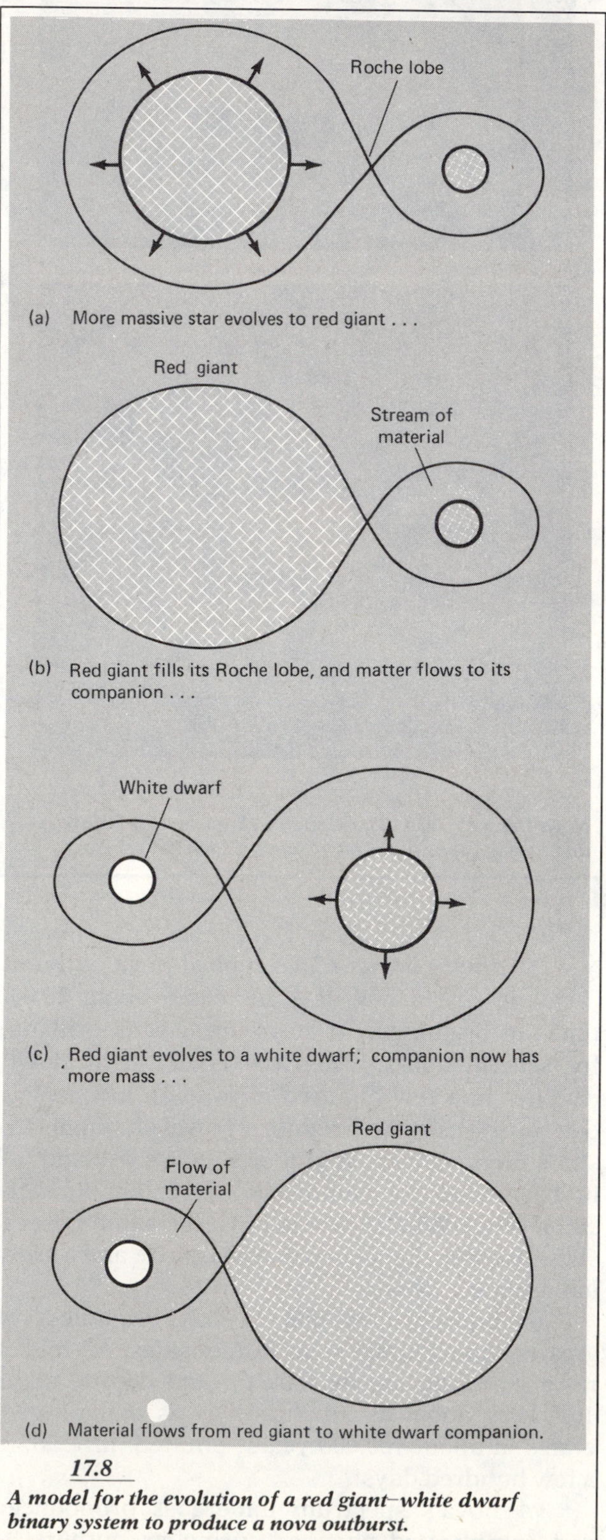
Roche lobe
(a) More massive star evolves to red giant . . .
Red giant
Stream of material
(b) Red giant fills its Roche lobe, and matter flows to its companion . . .
White dwarf
(c) Red giant evolves to a white dwarf; companion now has more mass . . .
Red giant
Flow of material
(d) Material flows from red giant to white dwarf companion.

17.8
A model for the evolution of a red giant–white dwarf binary system to produce a nova outburst.

17.9

A unique set of photos showing Nova Cygni 1975 before (arrow in 2) and during its outburst (3, 4, and 5). (Courtesy Ben Mayer)

It turns out that it doesn't take a large rate of mass falling in to set up and ignite runaway reactions in a reasonable time. For a typical nova in a binary system, just 10^{-7} solar mass a year suffices. Some studies have shown that a mere 10^{-13} solar mass a year will do the trick. This low rate can even be supplied by accretion from the general interstellar medium rather than from a companion star. That's one way a single white dwarf can become a nova.

Nova Cygni 1975 / I moved to New Mexico toward the end of August 1975. My department had a picnic in Corrales, on the west bank of the Rio Grande. It was a beauty of an August day, with the Sandia Mountains turning watermelon-pink at sunset. As I gazed up at the darkening sky, I sensed that something was wrong with the stars. Then I saw it. A new star in the constellation Cygnus near Deneb (Fig. 17.9). For a moment I was stunned. Then I recalled hearing about the discovery of this nova. But in the hustle and hassle of getting set up in a new place at a new job, I had forgotten about it. The shock of my personal discovery brought home to me the strangeness of a *nova stella* in the sky.

Nova Cygni 1975 had a light curve (Fig. 17.10) that showed a nova's typical sharp rise and decline. Prior to its nova outburst, the star was invisible on old photos. So it increased at least 16 million times in luminosity. That's unusually luminous for a nova. Yet Nova Cygni 1975 had a spectral evolution that followed the typical nova's pattern. So this outburst was a regular nova, but an extremely violent one.

So far, we have no indication that Nova Cygni 1975 is a member of a binary system. It may be one of those rare cases where a star becomes a nova from the accretion of interstellar material.

In summary: Novas may occur when material accretes onto a white dwarf. This infall heats the material, igniting runaway thermonuclear reactions that blow off the outer layers.

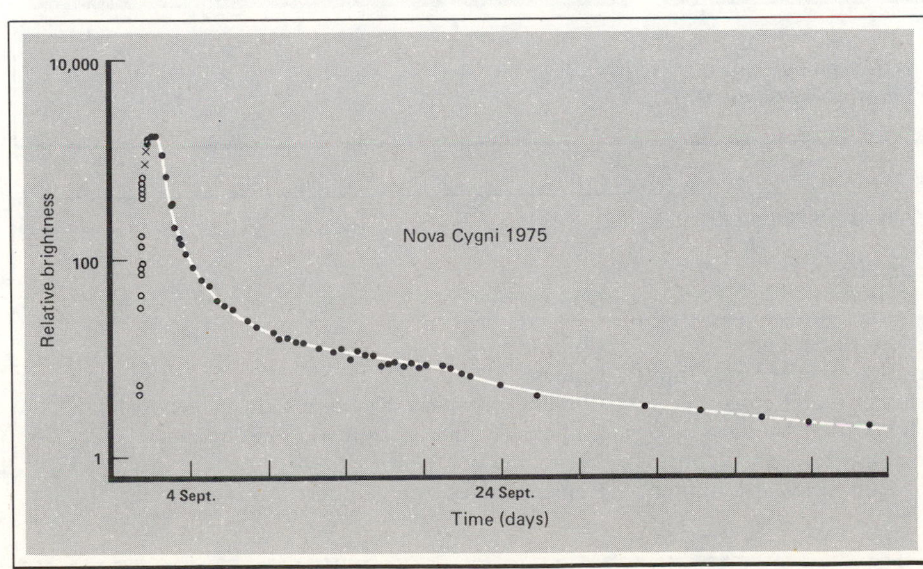

17.10

A visual light curve for Nova Cygni 1975. (From work by P. Young, H. Corwin, J. Bryan, and G. de Vaucouleurs, Astrophysical Journal, *vol. 209 (1976), p. 882, copyright © 1976 by the American Astronomical Society)*

Most novas are members of binary systems, with a red giant companion. Matter flows from the red giant to the white dwarf to set up the nova explosion. A few novas may be white dwarfs that by themselves accrete interstellar matter.

17.4 Supernovas: Cataclysmic explosions

As violent as novas may appear, they cannot match the fierce destruction of a star in a supernova. These cataclysmic explosions spew out energy in extraordinary amounts, about 10 billion times the sun's luminosity at their peak (more than 100,000 times that of a nova). According to contemporary ideas, a supernova usually signals the death of a massive star.

The name *supernova* was coined by Fritz Zwicky and Walter Baade for the extraordinary novas discovered in our own and other galaxies. Over 300 supernovas have been found in other galaxies (Fig. 17.11). Both the new stars observed by Tycho and Kepler were supernovas. Since the supernova of 1604, no such grand explosion has been seen in the Milky Way. Supernovas are such rare events that only six have been noted in the Milky Way during recorded history (Table 17.1). (Over the history of the Galaxy, about 10 billion years, hundreds of millions of supernova explosions have occurred. From this point of view, they are not rare at all.)

The supernova seen in Taurus in 1054 marks an event of continuing interest since its sighting. Chinese astronomers termed temporary celestial objects, such as novas or comets, *guest stars*. One guest star flared in the sky on July 4, 1054. Close study of Chinese and Japanese accounts of this visitor confirms that the star remained visible to the unaided eye for over 650 days in the night sky. It was visible in daylight for 23 days! The position noted by the Oriental astronomers placed the event in the constellation Taurus.

In 1731 the amateur astronomer John Brevis discovered a faint nebulosity in Taurus just above the bull's horns (Fig. 17.12). Much later, in 1928, Edwin Hubble measured the expansion rate of this nebula, which had become known as the Crab. He deduced that its expansion began about 900 years earlier. Hubble concluded that since the Crab Nebula was near the position given for the

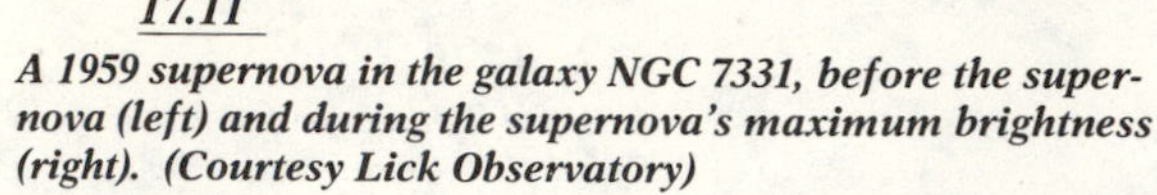

17.11
A 1959 supernova in the galaxy NGC 7331, before the supernova (left) and during the supernova's maximum brightness (right). (Courtesy Lick Observatory)

TABLE 17.1 Supernovas observed by the naked eye

Date (A.D.)	*Constellation*	*Apparent brightness*	*Observers*
185	Centaurus	Brighter than Venus	Chinese
369	Cassiopeia	Brighter than Mars or Jupiter	Chinese
1006	Lupus	Brighter than Venus	Chinese, Japanese, Koreans, Europeans, Arabians
1054	Taurus (Crab Nebula)	Brighter than Venus	Chinese, Arabians, southwestern Indians
1572	Cassiopeia	Nearly as bright as Venus	Tycho, many others
1604	Ophiuchus	Between Sirius and Jupiter	Kepler, Galileo, many others

SOURCE Adapted from a table compiled by W. C. Straka.

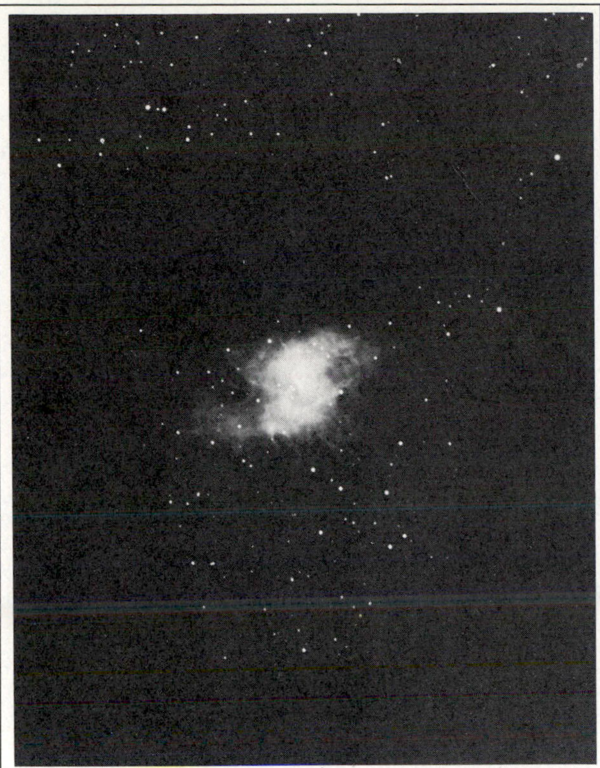

17.12
The Crab Nebula, photographed to emphasize its filamentary structure. The starless areas have been artificially blocked out. (Courtesy V. Regener, Capilla Peak Observatory, University of New Mexico)

Chinese guest star, that explosion was the source of the nebula. The Crab Nebula became the first identified supernova remnant in our Galaxy.

Classifying supernovas / Astronomers classify supernovas by the shape of their light curves into two general categories (Table 17.2): Type I, which exhibit a sharp maximum, about 10 billion solar luminosities, and die off gradually; and Type II, which have a less sharp peak at maximum, about 1 billion solar luminosities, and die away more sharply (Fig. 17.13). Studies of other galaxies have revealed that Type II supernovas occur in association with Population I stars. In confusing contrast, Type I supernovas occur in association with *both* Population I and II stars.

The total energy output from any supernova is stupendous: 10^{44} J, or approximately as much energy as the sun produces in its entire lifetime of 10 billion years. At its brightest, a supernova shines with a light of *10 billion* suns!

How often this kind of cosmic violence takes place is still debated. The rate of occurrence in any one galaxy is low, but the vast number of visible galaxies ensures that a few supernovas will be observed every year. From the rate in other galaxies, we estimate that a Type I supernova bursts forth in a galaxy, on the average, once in 60 years. Type II supernovas may be more frequent: one explosion in a galaxy roughly every 40 years. Some astronomers argue that the true frequency

TABLE 17.2 Properties of supernovas

	Type I	*Type II*
Ejected mass (solar masses)	0.5	5
Velocity of ejected mass (km/sec)	10,000	5,000
Total kinetic energy (J)	5×10^{43}	10^{44}
Visual radiated energy (J)	4×10^{42}	10^{42}
Frequency	1 in 60 yr	1 in 40 yr

SOURCE R. A. Chevalier, "The Interaction of Supernovae with the Interstellar Medium," *Annual Reviews of Astronomy and Astrophysics*, vol. 15 (1977), p. 175.

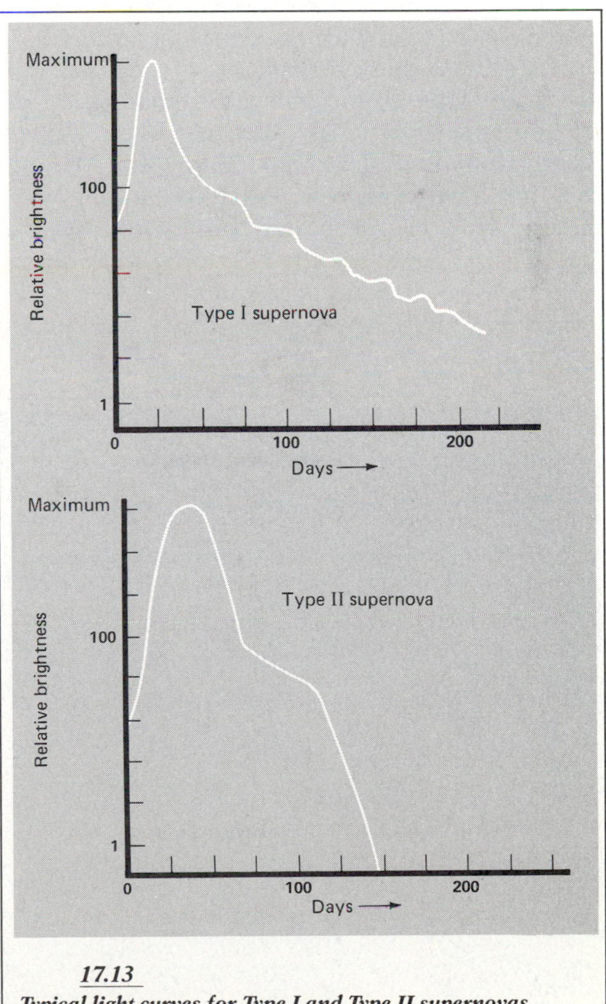

17.13
Typical light curves for Type I and Type II supernovas. (Based on diagrams by W. Straka)

must be greater because we do not observe all supernova events. In any case, one supernova every 50 years in a galaxy seems like the average supernova frequency.

If supernovas occur in a galaxy once every 50 years, why haven't we seen one in our own Milky Way Galaxy since 1604? Some may have happened, but in distant regions of the Galaxy, where their light was cut off by the interstellar dust. The next supernova in our Galaxy may not present a spectacular stellar show since its light may also be cut down by interstellar dust.

Both types of supernova violently eject a large fraction of the original star's mass at a speed of about 10,000 km/sec. At maximum brightness, a supernova can reach a size about that of the solar system—a few light hours in diameter. For about the first month after the time of maximum, the supernova's size increases dramatically. For example, one Type II supernova observed in 1970 had a radius at maximum roughly as large as the orbit of Uranus. For 30 days the star expanded at about 5000 km/sec until it attained a radius of approximately three times larger than the solar system, or one light day in diameter. After that, the star's photosphere shrank.

The origin of supernovas / What kinds of stars become supernovas? Because Type I supernovas often occur in Population II stars, these stars may have a mass about that of the sun. (Very massive stars cannot be old stars.) In contrast, Type II supernovas are thought to be stars much more massive than the sun—stars that live their normal lives as *O*- and *B*-stars. These stars probably explode once they have evolved off the main sequence and formed an iron core (the following section gives details).

Type I supernovas are really a puzzle, for it is hard to see how a solar-mass star can detonate as violently as a supernova. One idea resembles that for binary novas (Section 17.3). Imagine a binary system containing a white dwarf and a red giant star. Assume the white dwarf has a mass close to one solar mass. As material flows onto the white dwarf, it builds up and heats enough to ignite the carbon of the white dwarf. The carbon burns swiftly into nickel, cobalt, and iron—elements that are seen in the spectra of Type I supernovas.

An alternative model uses a binary white dwarf system. If one star loses mass, it expands (because it is degenerate matter) and sends more of its matter to the other star, which shrinks as it gains mass. When the temperature gets high enough, the carbon ignites in a kind of slow burn that makes the supernova explosion. This model and the one above are fairly tentative; we don't really have a good understanding of Type I supernovas yet.

Supernova remnants / A supernova bangs out a blast wave into the interstellar medium. Traveling at supersonic velocities, the shell of material creates a shock wave that plows through the interstellar gas and dust. The shock wave's collisions with the cool clouds of the interstellar medium can excite the interstellar material so that it glows. This luminous material marks a *supernova remnant.* The Loop Nebula in Cygnus (Fig. 17.14) is such a remnant. Note that it looks

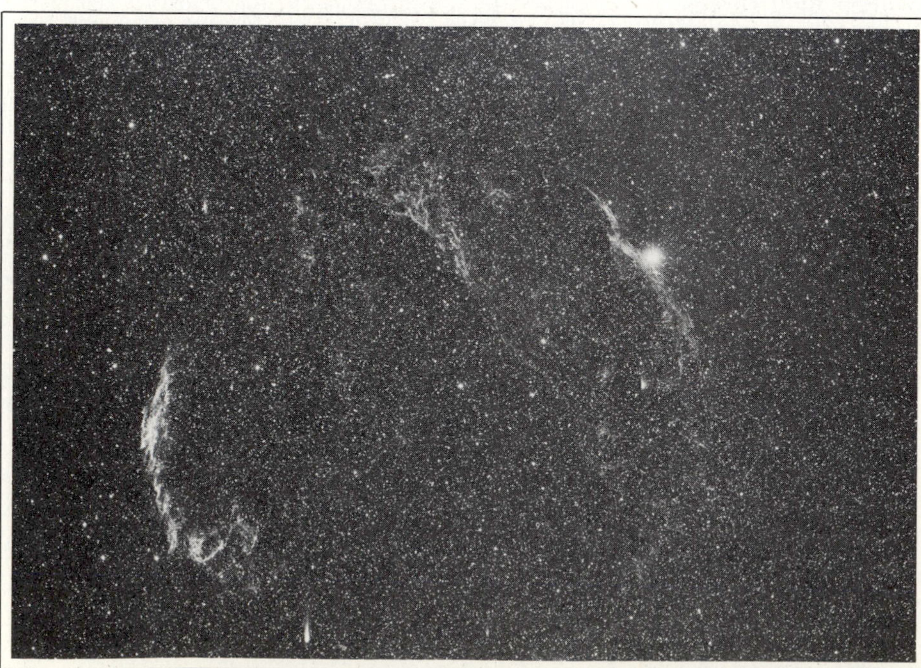

17.14

A supernova remnant in Cygnus. Note how the bright wisps almost make a circle, as if blown out from a central point. (Courtesy Palomar Observatory, California Institute of Technology)

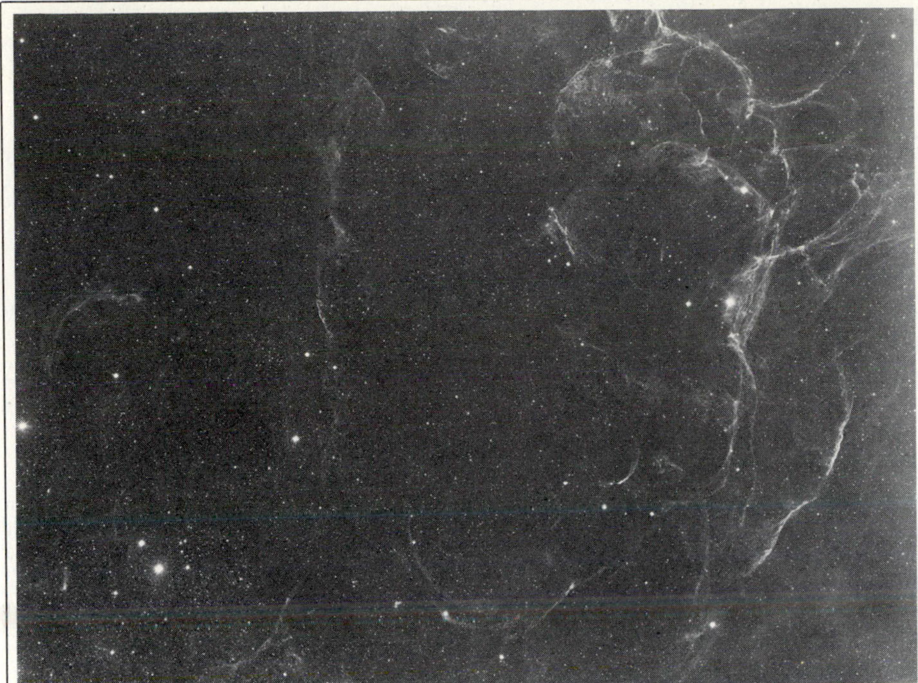

17.15
A part of the Gum Nebula, a supernova remnant. Note how the filaments form circular arcs. The entire nebula is more than 2000 light years in diameter. (Courtesy B. Bok and Steward Observatory, University of Arizona)

spherical–a shell produced by the interaction between the interstellar medium and a supernova shock wave.

The shock should also heat the interstellar gas to temperatures of about 1 million K. This hot gas will strongly emit X-rays. In fact, X-ray observations of the Cygnus Loop show X-ray emission very much like that expected from a hot gas.

A similar nebula in the southern sky, the Gum Nebula (Fig. 17.15), extends over 50° in the sky. The Gum Nebula has a diameter of about 2300 ly, its closest edge being only 300 ly from the sun. This nebula was created by the pulse of ultraviolet radiation and X-rays generated by a supernova some 11,000 to 20,000 years ago. An X-ray source, named Vela X, lies almost in the nebula's center. It is a prime suspect as the supernova site. The discovery of a pulsar (see Section 17.6) near the location of the Vela X source supports its nature as a supernova remnant. (Pulsars are believed to be the neutron stars formed as the by-product of a supernova explosion.)

Radio astronomers have one up on optical astronomers in the hunt for galactic supernova remnants: They can observe low-density excited gas that has no detectable optical emission (Fig. 17.16). For example, the radio astronomers were the first to detect Cassiopeia A, believed to be a supernova remnant. Later work with optical telescopes revealed faint patches of nebulosity at the same location–debris from the supernova.

Radio astronomers recognize supernova remnants by a special property of their radio emission. The intensity plotted versus frequency displays a *nonthermal spectrum* (see Focus 17.1 for a discussion of the difference between thermal and nonthermal spectra). If a radio source is observed at a variety of frequencies, the shape of its spectrum distinguishes between a possible supernova remnant (nonthermal) and an ordinary H II region (thermal). A nonthermal spectrum indicates that the radio emission comes from very high speed electrons accelerating in magnetic fields. This kind of emission is called *synchrotron radiation*. Its intensity and wavelength range depend on the intensity of the magnetic field and the kinetic energy of the electrons. The fact that supernova remnants display nonthermal spectra means that the synchrotron process generates their radio emission, and that means that the remnants contain magnetic fields and high-energy particles.

X-ray astronomers can observe young supernova remnants directly; over 30 have been observed so far. The huge shock waves generated by the blast plow through the interstellar medium at speeds of hundreds of kilometers per second. They compress and heat the interstellar gas to temperatures of 1 million K in the zone just behind the blast wave. This hot gas shows up in X-ray photos (Fig. 17.17), where the shells show brightness variations around their rims–an indication of the patchy nature of the interstellar medium.

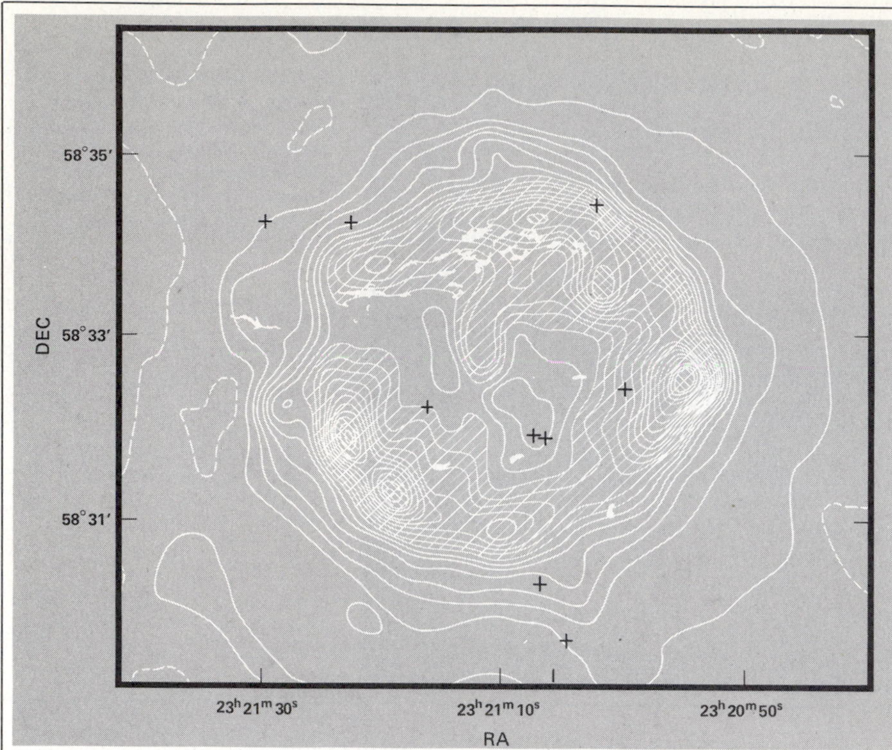

17.16

A radio intensity map of a supernova that was the "nova" observed by Tycho Brahe in 1572. It shows a shell of hot gas expanding into the interstellar medium. (Based on observations by M. Ryle, B. Elsmore, and A. Neville)

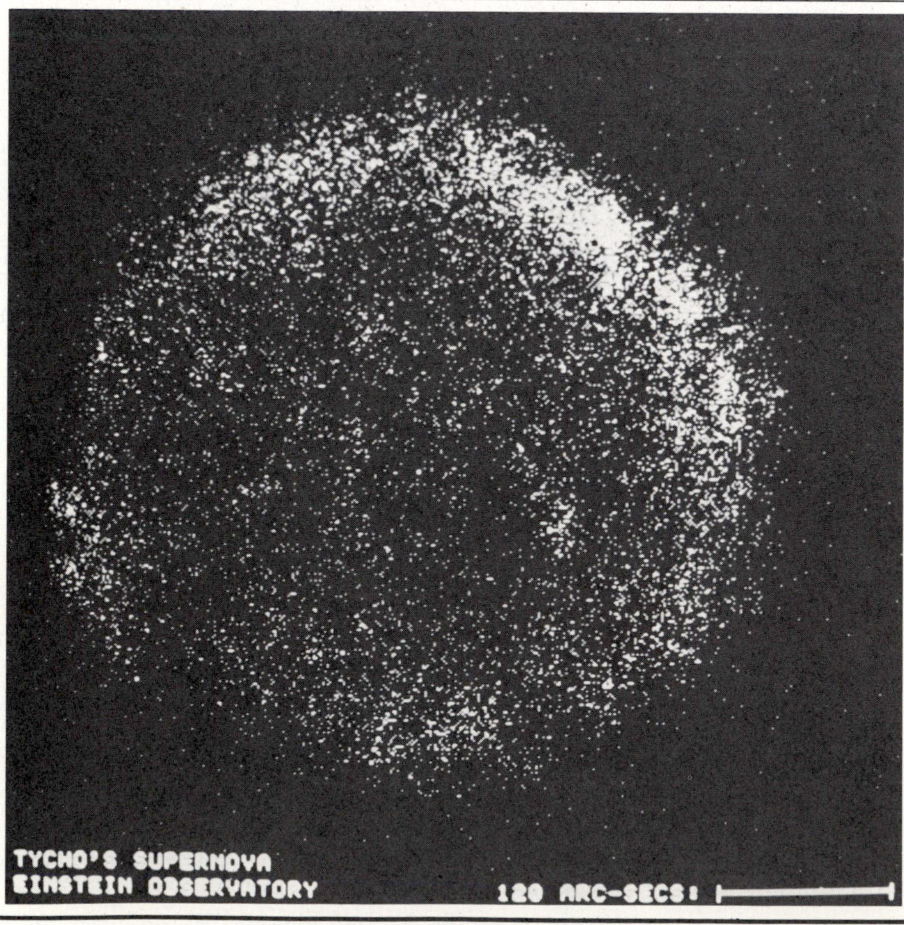

17.17

An X-ray image of Tycho's supernova remnant, taken by the Einstein X-Ray Observatory. (Courtesy P. Gorenstein and F. Seward, Center for Astrophysics)

The Crab Nebula: A supernova remnant / Chinese and Japanese astronomers carefully watched the supernova that produced the Crab Nebula. This contrasts sharply with the lack of comment by European astronomers at the time. Although ignored in Europe, the supernova may have been observed and recorded in the southwestern part of North America. One good example is a rock painting in Chaco Canyon, New Mexico, which may represent the predawn conjunction of the waning crescent moon and the supernova (Fig. 17.18). The Anasazi (see Section 1.7), who lived in Chaco then, may have made this painting to commemorate the event.

The material blown off in the explosion should be still expanding today. Indeed, photographs of the optically visible filaments in the Crab Nebula demonstrate that the gas *is* expanding (Fig. 17.19). The Doppler shift of the expanding filaments implies—if the expansion rate has been constant at 1450 km/sec—that they began expanding at around A.D. 1132. That's close to the actual date of the explosion. The fact that it is later may mean that the velocity now is higher than the average, that is, that the expansion has been accelerating. The distance to the remnant is 6500 ly.

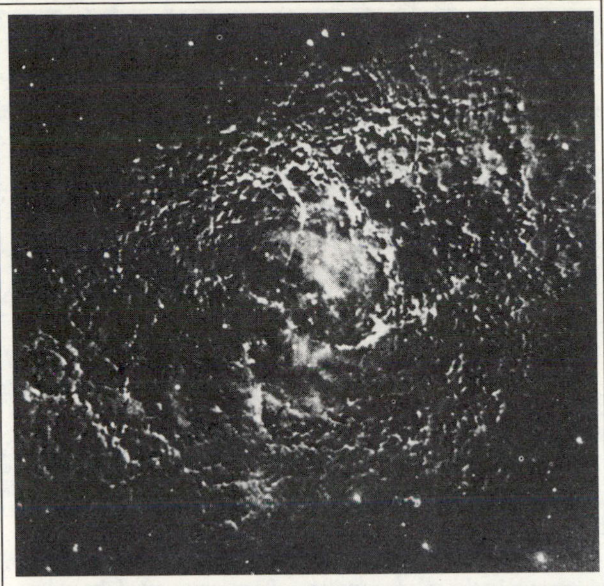

17.19

Expansion of the gas filaments in the Crab Nebula from 1959 to 1964. Here a positive print has been overlaid on a negative one, so expanding filaments look white on their outer edges, black on their inner ones. (Courtesy V. Trimble)

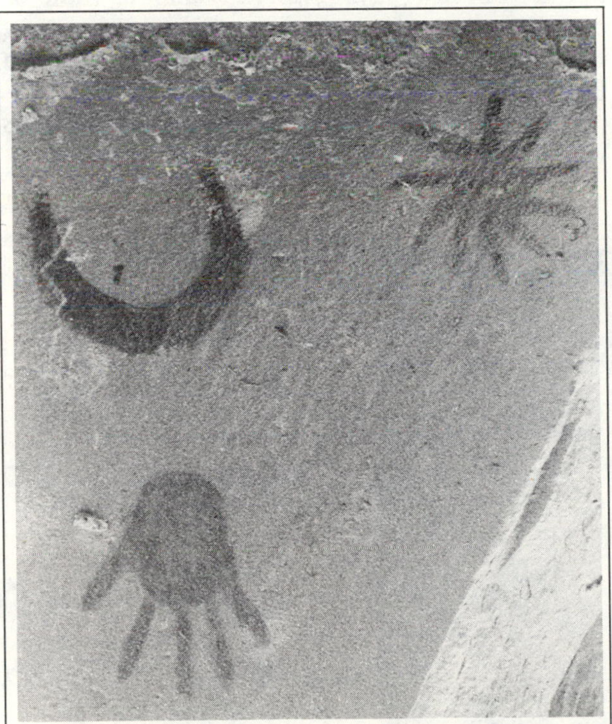

17.18

A painting in Chaco Canyon, New Mexico, that may represent the supernova of A. D. 1054. The view is looking up at the painting, which is on the underside of a rock ledge. On the morning of July 5, 1054, the supernova, brighter than Venus, rose with the waning crescent moon. (Photo by M. Zeilik)

In 1953 Josef Shklovsky resolved in part the enigma of the Crab Nebula's radio and optical emission when he suggested that the synchrotron process produced it (Focus 17.1). Synchrotron radiation requires a magnetic field and a source of energetic charged particles (such as electrons). The emission is polarized and has a nonthermal spectrum. Shklovsky's argument was clinched when Russian astronomers found that the optical emission was strongly polarized (Fig. 17.20). Later observations showed that the X-ray emission was also polarized.

What does it mean for electromagnetic radiation to be *polarized*? Light has wave properties. Waves are said to be polarized if the planes of their vibrating motion tend to be oriented in some direction—for instance, the plane of this page of paper. Light that is unpolarized has no prefered orientation: The planes of wave vibration occur in all directions in equal amounts. Synchrotron emission is usually strongly polarized.

The solution to one puzzle posed another one even more vexing: What is the source of the energetic electrons? As these electrons spiral through the magnetic field emitting synchrotron radiation, they lose energy rapidly. For the electrons pro-

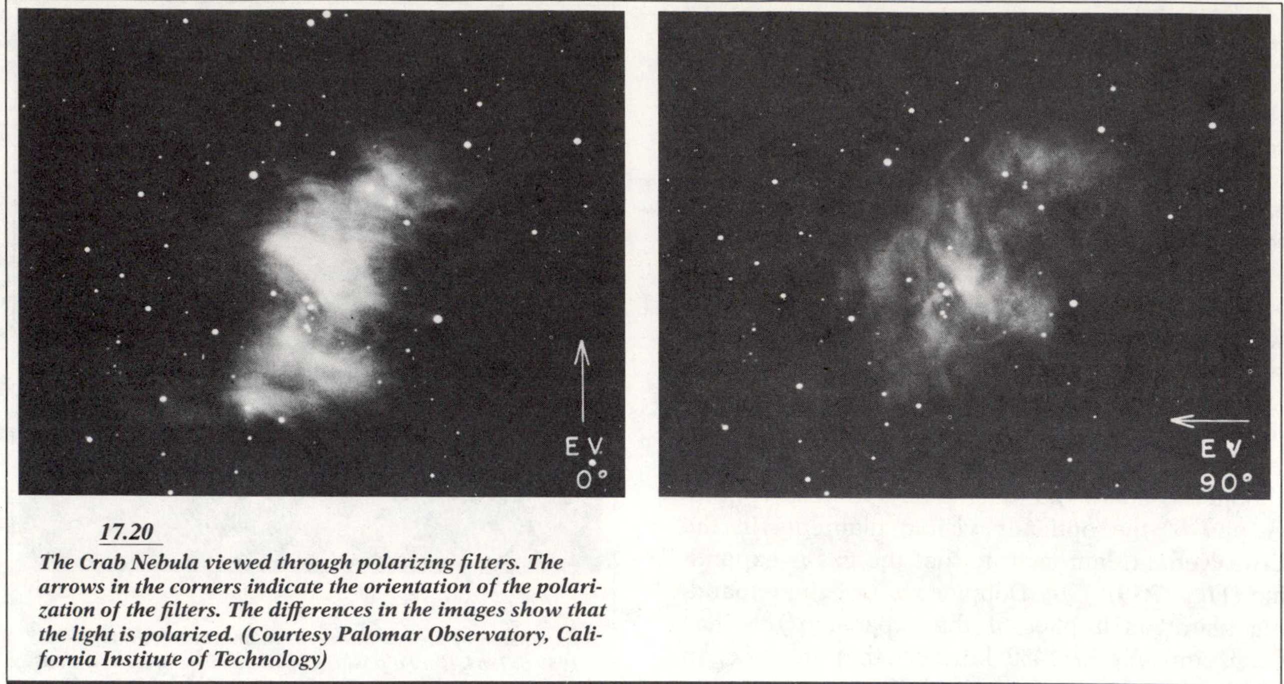

17.20
The Crab Nebula viewed through polarizing filters. The arrows in the corners indicate the orientation of the polarization of the filters. The differences in the images show that the light is polarized. (Courtesy Palomar Observatory, California Institute of Technology)

ducing the optical emission, half their energy would be drained off in only 70 years. So the supply of electrons must be continuously replenished. The problem of the electrons' source became even more acute when X-ray emission was discovered in 1963. The electrons that produce synchrotron X-ray emission have higher energies than those that produce optical emission. They also deplete their energy faster, losing half in only 7 years. The Crab Nebula emits about 100 times more energy in the form of X-rays (Fig 17.21) than as radio or optical emission, so a large amount of energy must be added to the nebula over a time of only a few years.

The energy problem disappeared in 1968 with the discovery of a pulsar in the Crab Nebula.

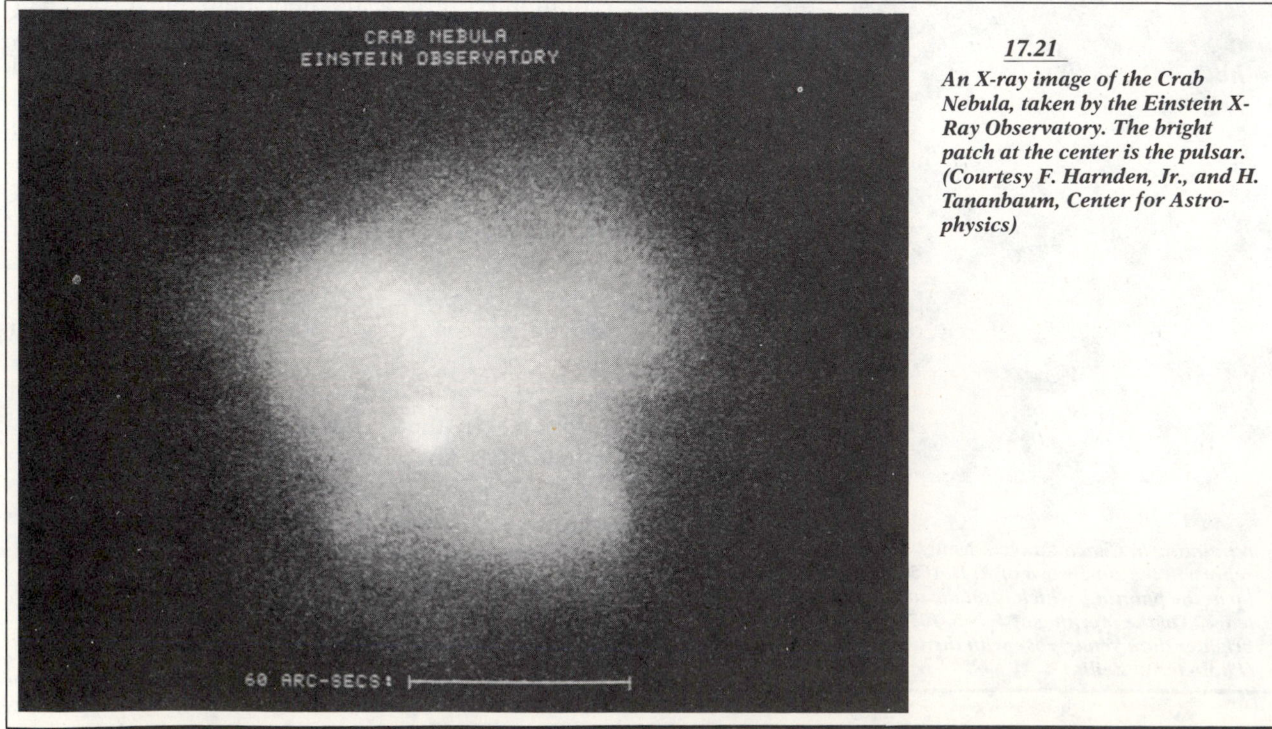

17.21
An X-ray image of the Crab Nebula, taken by the Einstein X-Ray Observatory. The bright patch at the center is the pulsar. (Courtesy F. Harnden, Jr., and H. Tananbaum, Center for Astrophysics)

Focus 17.1

THERMAL AND NONTHERMAL (SYNCHROTRON) EMISSION

The spectrum of blackbody radiation has a characteristic shape. The energy output increases from longer to shorter wavelengths, peaks at a wavelength that depends on the blackbody's temperature, and then decreases at shorter wavelengths until it hits zero.

The spectrum of blackbody emission is the archetype of thermal emission, which arises basically from the motions of the particles involved. The greater the motions, the higher the output of radiation (and the hotter the source).

Nonthermal emission does not follow the characteristic signature of blackbody radiation. In general, the nonthermal spectrum increases in intensity at longer wavelengths (Fig. F.36). Synchrotron emission is a frequently found example of nonthermal emission; it arises from the acceleration of charged particles (usually electrons) in a magnetic field. Moving charged particles interact with magnetic fields so that they spiral around the magnetic-field lines rather than traveling across them (Fig. F.37). The spiral paths are curved, so the particle is continually accelerated and thus emits electromagnetic radiation. The frequency of emission is directly related to how fast the particle spirals; the faster the spiral, the higher the frequency. Increasing the magnetic-field strength tightens the spiral and so increases the frequency.

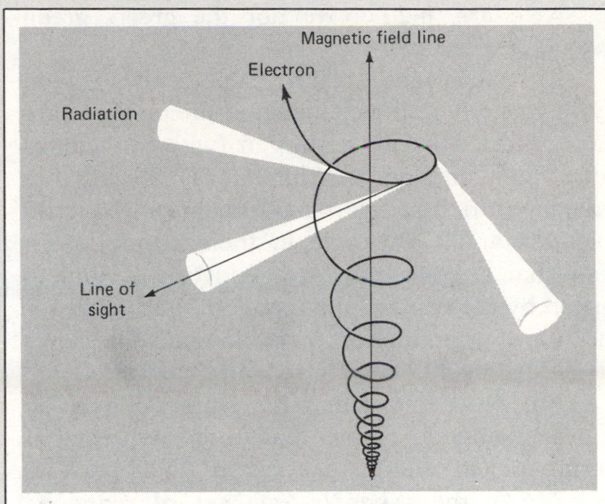

F.37
Synchrotron radiation from an electron spiraling in a magnetic field.

The velocity of the charged particle also affects the frequency directly, so more energetic particles can produce higher-frequency emission, but they also require strong fields to keep them in a tight spiral. As the particle radiates, it loses energy and generates lower-energy (longer-wavelength) radiation. So a synchrotron source needs a continually replenished supply of electrons to keep emitting at relatively short wavelengths (high-energy photons).

One important point about synchrotron emission: It is polarized. Thermal emission, such as from the sun, is not polarized. Synchrotron-emitting electrons, when viewed side on in their spiral motion, appear to be moving back and forth along almost straight lines. Their synchrotron emission has its waves more or less aligned in the same plane. So synchrotron radiation is polarized and, at visible wavelengths, can be observed as such with Polaroid filters.

The term synchrotron emission *derives from the fact that such radiation was first observed from the General Electric synchrotron, which used magnetic fields to contain electrons accelerated to high energies.*

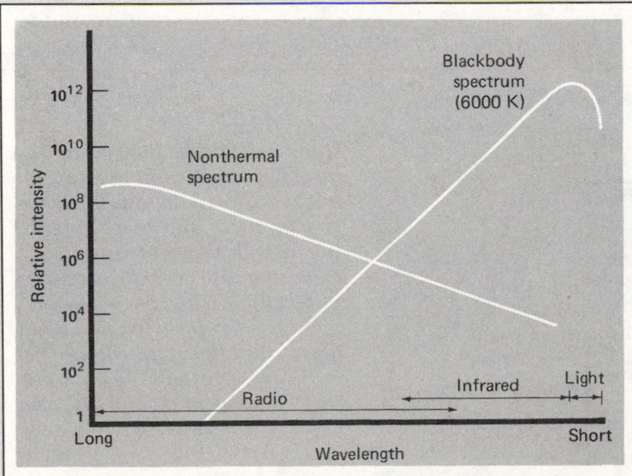

F.36
A comparison of the spectra of a thermal (blackbody) source at 6000 K and a nonthermal (synchrotron) source.

17.5 The manufacture of heavy elements

A massive star dies in a supernova. The blast can leave behind a neutron star or a black hole. So a massive star digs its own grave. Does any good come of its death? Yes. For without the deaths of massive stars, neither we nor the green, green grass would be here.

An extravagant claim? Not really, if you investigate where the elements in the universe are made. Start off with hydrogen (where it comes from you'll find out in Chapter 21). Recall that ordinary stars fuse hydrogen to helium by the PP reactions and CNO cycle in their normal lives. After that, a solar-mass star converts helium to carbon by the triple-alpha reaction. But there, for our sun, thermonuclear reactions end. The sun's core will not get hot enough to burn carbon.

But other more massive stars do burn to heavier elements. Notice that in the sequence of thermonuclear energy generation (look back at Table 16.4), the ashes of one set of reactions become fresh fuel for the next. Each step up in the fusion chain requires higher temperatures to overcome the greater repulsive electrical force of nuclei with more protons. These fusion reactions—carbon to oxygen, neon, sodium, and magnesium, and onward to silicon burning—can fire only in massive stars. The end to the fusion chain comes with iron, the most tightly bound of the normal nuclei. Iron rests at the bottom of an energy well. To split iron into lighter elements takes energy. To fuse it to heavier ones also needs energy put in. So nuclear reactions naturally stop at iron in very massive stars—greater than about 20 solar masses.

But the elements that we know do not end at iron (see Appendix F); many are heavier. Where and when are they manufactured in the course of cosmic evolution? A supernova explosion acts as nature's special workshop for forging some elements heavier than iron (Focus 17.2).

Here's one possible model of how a Type II supernova happens (Fig. 17.22). Imagine a very massive star with a core of iron. Its interior temperature decreases from the core outward. Because of this temperature decline outward, you expect the star's interior to be layered like an onion. Around the iron core is a silicon layer; here temperatures do not get high enough to fuse iron. Around that layer is one of oxygen; here temperatures are too low to fuse oxygen to silicon. These shells are still burning: In the silicon core, oxygen is being fused; in the oxygen shell, carbon; and so on.

Once the core ends up as iron, it must contract, for the fusion fires have failed. Relentlessly, gravity squeezes the core to higher temperatures and densities. When the core gets to about 5 billion K, the photons there have so much energy that they can penetrate the iron nuclei and break them down into helium. As the iron disintegrates into helium, large amounts of heat are used up. The pressure in the core no longer supports the star, and it collapses suddenly. The gravitational collapse rapidly pumps heat into the material.

Two important events occur in this collapse. First, protons and neutrons released by the disintegration of nuclei in the core pelt and penetrate remaining nuclei. These can capture neutrons and be transformed to heavier elements. Second, the layers above the core plummet inward toward the core and heat up. Suddenly, ignition temperatures of many fusion reactions are reached. They turn on explosively. A blast shock wave from this

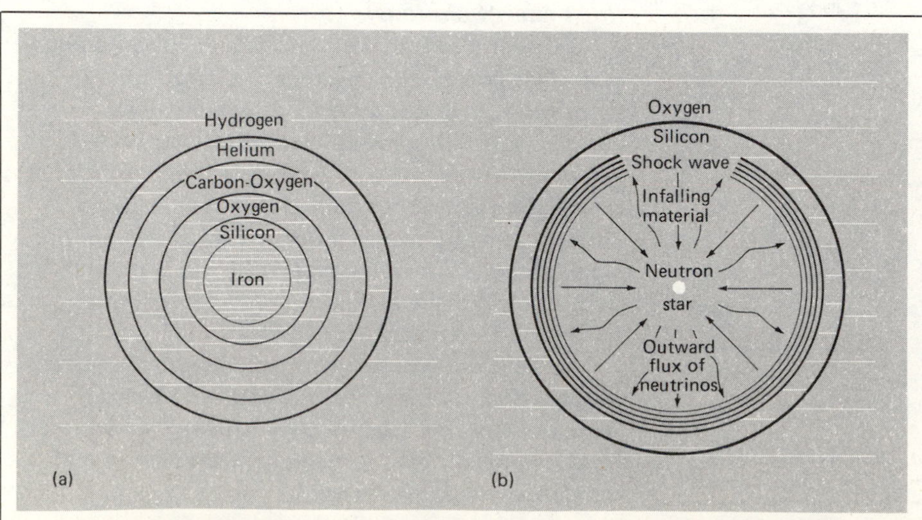

17.22

One model of the interior of a massive star becoming a supernova. Fusion reactions in shells give layers of different elements* (a). *At high enough temperatures, the iron core disintegrates into helium. This process soaks up heat from the core, and it collapses rapidly* (b), *sparking explosive ignition of fusion reactions. (Adapted from a diagram by C. Wheeler)

Focus 17.2

NUCLEOSYNTHESIS OF HEAVY ELEMENTS

Let me explain the synthesis of heavy elements in a little more detail. The most important process involves high-energy neutrons bombarding various nuclei. Because they have no charge, the neutrons have an easy time penetrating a nucleus. This nuclear capture of neutrons leads to a buildup of heavier nuclei.

Two processes take part in this buildup. Recall that a neutron under normal conditions of low density will disintegrate (in about 1000 sec) into a proton, an electron, and an antineutrino. This is called **beta decay**. *The rate of beta decay naturally divides nucleosynthesis into two processes. In one process, neutrons are captured* **faster** *than the beta-decay rate, so neutron-rich nuclei are formed. This process is called the* **rapid process (r-process** *for short). In the other, nuclei capture neutrons* **slower** *than the beta-decay rate, so proton-rich nuclei are made. This process is called the* **slow process** *(or* **s-process**). *A combination of both these processes leads to the manufacture of most of the elements and isotopes heavier than iron. In supernovas, the time scales are short, and it is mainly the r-process that is effective.*

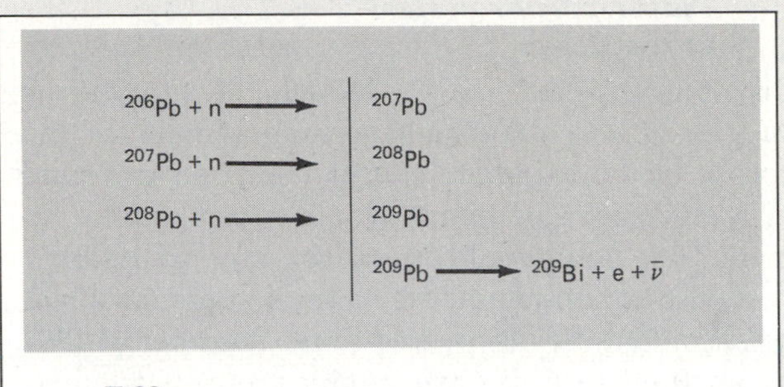

F.38
The synthesis of isotopes of lead and bismuth by neutron capture and beta decay.

A specific example: transforming lead-206 (^{206}Pb) to other isotopes of lead and finally to bismuth-209 (^{209}Bi). Sock ^{206}Pb with a neutron (Fig. F.38); it becomes ^{207}Pb. Hit it with another neutron, and then another, and you have ^{209}Pb. Each heavier isotype of lead is less stable than the previous one, and ^{209}Pb is so unstable that it decays rapidly (by beta decay) to ^{209}Bi (and an electron and an antineutrino). Then the ^{209}Bi can absorb neutrons and build up to other bismuth isotopes and finally beta-decay to the next element, and the process continues.

explosive ignition bullies its way outward from the core (Fig. 17.22). It carries material with it into the interstellar medium—material enriched with heavy elements.

Because I've emphasized nucleosynthesis in supernovas, you might be thinking that *only* supernovas make elements heavier than iron. That's not the case. Red giants also manufacture some heavy elements. While a red giant undergoes helium shell burning, small numbers of neutrons are produced and added slowly to iron to fuse heavier elements. Although this process is indeed slow, the red giant stage lasts long enough (100,000 years or so) to synthesize an appreciable amount of heavy elements. Some are also made during red giant helium flashes.

In general, elements made in red giants complement those made in supernovas to fill up the periodic table. Nucleosynthesis in red giants makes many of the elements lighter than lead but heavier than iron. Supernovas synthesize elements heavier than lead, such as uranium and thorium. (When you use electricity generated by a nuclear power plant, you're turning on your light by the fossil remains of a massive star!)

Now to support the claim about us and the green, green grass. Life as we know it (see Chapter 22) builds on carbon atoms. Stars like our sun can fuse carbon, *but* it remains locked in a corpse after death—a white dwarf. The material flung out by the sun after it becomes a red giant will come just from its outer layers. Here no fusion reactions have gone on. So this material has pretty much the same composition it started with. Only a massive star in a supernova spews into the interstellar medium newly made heavy elements. So if all stars had about the sun's mass, no carbon would get out into the interstellar medium to end up in living organisms on a planet. Your body is mostly recycled stardust. So is the grass under your feet.

Stars die to seed new life. I find that the most incredible aspect of cosmic evolution.

17.6 *Pulsars: Neutron stars in rotation*

Do neutron stars exist? Models of supernovas suggest that a neutron star may remain as the corpse of the exploded star. A neutron star found in a supernova remnant would clinch this argument. But how would a neutron star be visible? In a way not anticipated by astronomers: as a *pulsar*, accidentally discovered in the summer of 1967 by an English radio astronomy group headed by Anthony Hewish.

The detection of the pulsar ranks high on the list of those marvelous accidents of scientific discovery. The Hewish group was mapping the sky at radio wavelengths in an attempt to find quasars. Jocelyn Bell Burnell, then a graduate student in charge of the preliminary data analysis, noticed a strange signal that suddenly disappeared, only to reappear three months later. The Hewish group concentrated on this unusual signal and found radio pulses occurring at a regular rate, once every 1.33730113 sec (Fig. 17.23a). Flushed with excitement, they searched the sky for any similar signals and discovered three more objects emitting radio bursts at different rates. The Hewish group concluded that the objects must be natural phenomena and named them *pulsating stars*, or *pulsars*.

Observed pulsar characteristics / To date, a total of about 150 pulsars have been studied in detail. About an equal number have been recently discovered in a special pulsar survey, for a total of roughly 300.

For a given pulsar, the period between pulses repeats with very high accuracy, better than 1 part

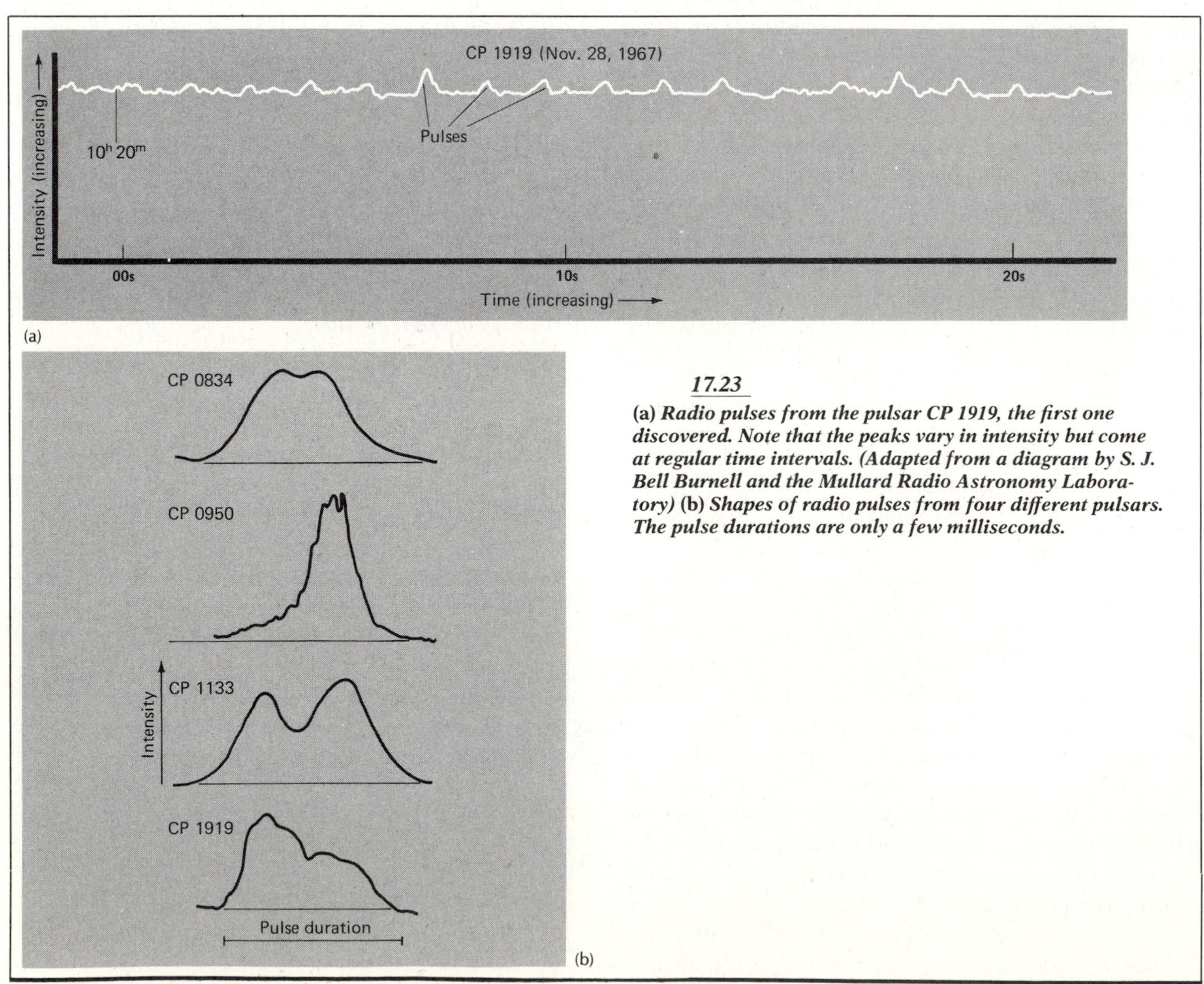

17.23

(a) ***Radio pulses from the pulsar CP 1919, the first one discovered. Note that the peaks vary in intensity but come at regular time intervals. (Adapted from a diagram by S. J. Bell Burnell and the Mullard Radio Astronomy Laboratory)*** (b) ***Shapes of radio pulses from four different pulsars. The pulse durations are only a few milliseconds.***

in 100 million. The amount of energy in a pulse, however, varies considerably; sometimes complete pulses are missing from the sequence. Although the intensity and shape vary from pulse to pulse, the average of many pulses from the same pulsar defines a unique shape (Fig. 17.23b). The average pulse typically lasts for a few tens of milliseconds.

For the well-studied pulsars, periods range from 0.0016 sec to 4.0 sec, with an average value of 0.65 sec. (Only two pulsars are known so far to have millisecond pulses: one with 1.6-msec interval, the other 6 msec.) In the cases where accurate radio observations have been made, periods have been noted to increase in a regular fashion. The rates of change have typical values of about 10^{-8} sec/yr. Such small increases can be measured only with atomic clocks, whose stability is better than 10^{-10} sec/yr.

These observed general properties of pulsars provide clues to the possible physical properties of these precise cosmic clocks. The pulse duration indicates the largest size of the bodies emitting the radiation. How? Suppose that you could switch off the sun. Because of the finite velocity of light, it would take slightly more than 2 sec for the entire solar disk to appear dark, and then another 2 sec to regain its original brightness if quickly turned on again. The part of the sun that we see when we look at the edge of the visible disk is at a greater distance (halfway around the sun) than the part at the center of the disk. So if the sun is turned off, we won't know that the edge is dark until about 2 sec after we see the center become dark. So an object the size of the sun, an average star, is too large to be a pulsar, as the pulse durations amount to a few hundred milliseconds or less. The region emitting these pulses must be less than roughly 30,000 km in size (the light-travel distance during the pulse's duration). The more typical pulse durations of a few tens of milliseconds imply sizes of roughly 3000 km or less.

What stellar objects exist at this size or smaller? And, what objects can contain the large quantities of energy that pulsars emit? A dense object in some way moving rapidly acts as a storehouse of kinetic energy. This clue points to either white dwarfs or neutron stars as the candidates for pulsars.

Physical characteristics / Here's the basic issue: Pulsar models must account for the precise clock mechanism of pulsars, that is, the extremely regular repetition of pulses. Basically, three clock mechanisms are available: revolution, pulsation, and rotation. Let's consider each in turn.

Revolution. Two very close, compact bodies can orbit with short, regular periods. But a pair of white dwarfs even in *contact* cannot orbit each other faster than once every 1.7 sec; many pulsars are known with periods shorter than this. Two neutron stars can whip around each other with shorter periods, but here a different problem crops up. According to the general theory of relativity, two massive bodies in short-period orbit should lose energy in the form of gravitational radiation. With this energy loss, the two objects come closer together and their period *decreases* (following Kepler's third law). Eventually, the two bodies would collide. But pulsar periods are known to increase, not decrease. Scratch revolution as the clock mechanism.

Pulsation. Imagine a white dwarf or neutron star expanding and contracting regularly, with one expansion and contraction equal to a pulse period. Sounds good, but it doesn't work out. How fast a spherical mass can pulsate basically depends on its density. More dense objects can pulsate more rapidly. That should seem reasonable. Imagine standing above the surface of a white dwarf star and pulling it outward. Release it; gravity pulls it in, and it pulsates about once a second. Now try the same imaginary experiment with a neutron star. Because it is denser, its surface gravity is greater, so its surface is pulled in more rapidly than a white dwarf's surface. The least massive, least dense neutron stars can pulsate no slower than once every 0.01 sec. That includes the the fastest known pulsar, which has a period of 0.0016 sec, but excludes most of the others. In general, pulsation doesn't work. White dwarfs can't pulse as fast as the fastest pulsars, and neutron stars pulsate too fast.

Rotation. Consider one rotation equal to a pulse period. Can rotation of white dwarfs explain the fastest pulsars? No. Here's the basic idea: If a spherical mass rotates too rapidly, its gravity will not be able to hold it together. Mass will fly off tangent to its equator. White dwarfs have densities of about 10^{10} kg/m^3 at most. So a white dwarf cannot rotate faster than once every 4 sec without losing mass—too slow for the fastest pulsars. In contrast, neutron stars have average densities of about 10^{17} kg/m^3. So they can rotate once every 0.001 sec without losing mass—fast enough for even the fastest pulsar. Of course, they can rotate more slowly, too.

To conclude: A rotating neutron star can provide the clock mechanism for pulsars.

Pulsars and supernovas / We need to look at supernovas or their remnants to connect neutron stars to pulsars. Have we found such links? Fortunately, we have two examples to date: a pulsar

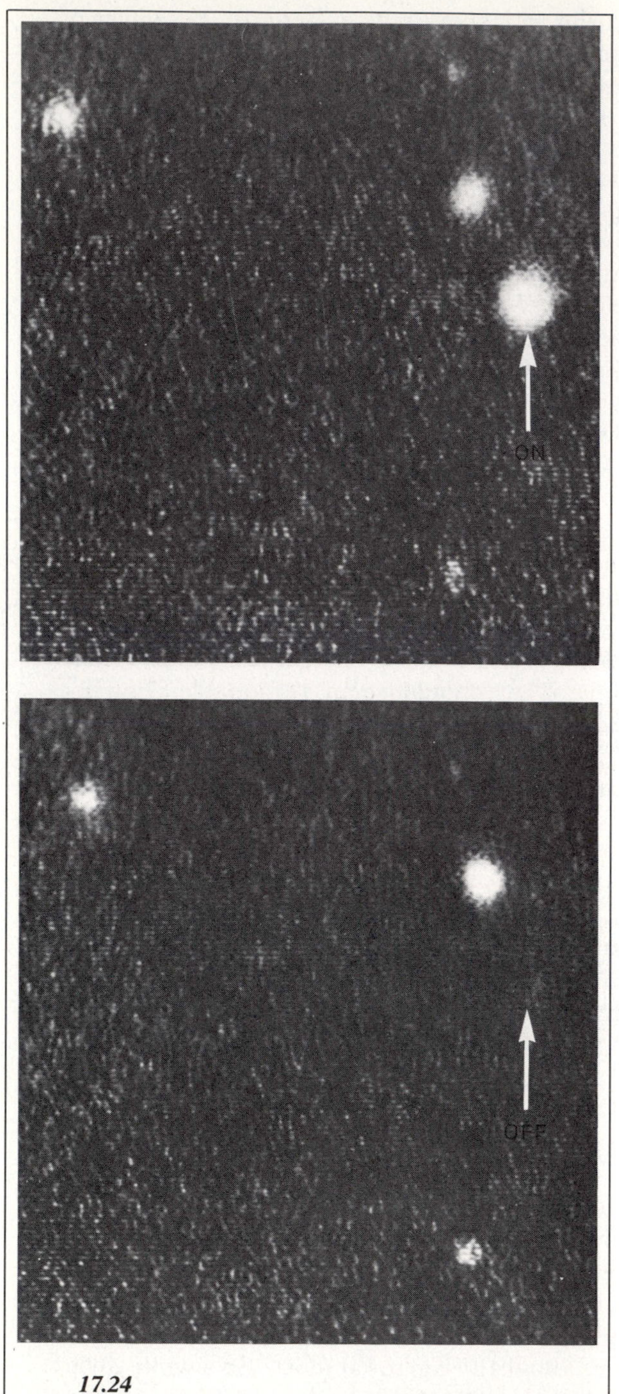

17.24
The Crab Nebula pulsar on and off at visual wavelengths. (Courtesy Lick Observatory)

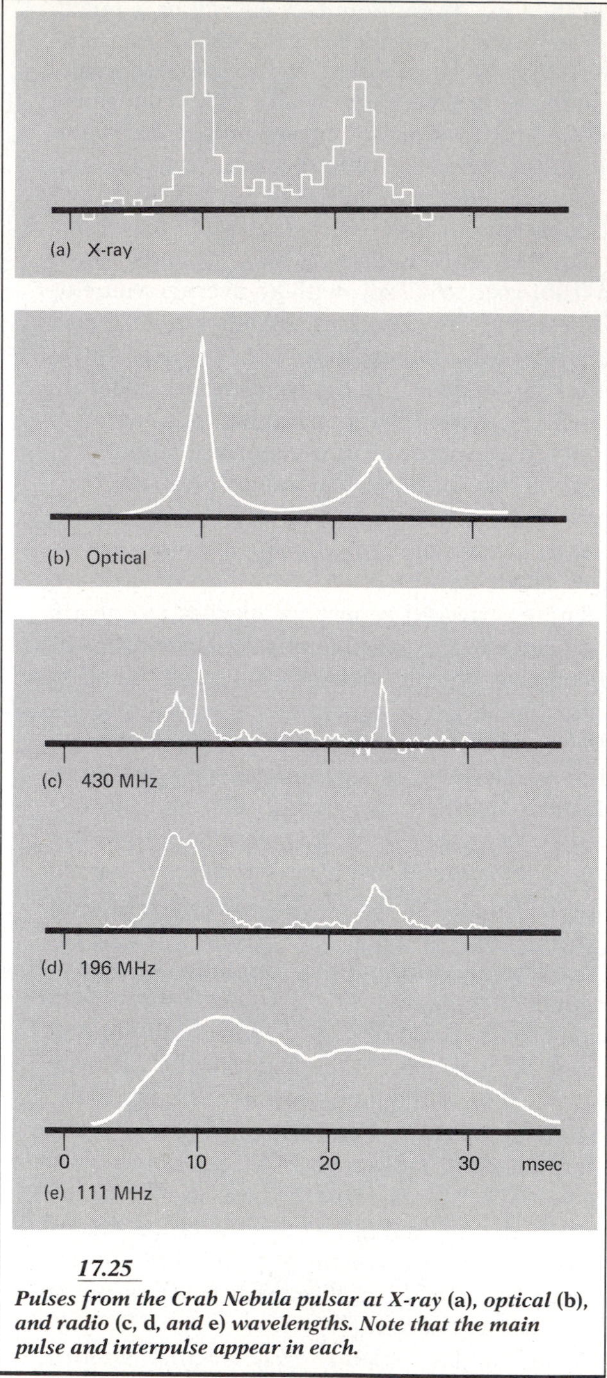

17.25
Pulses from the Crab Nebula pulsar at X-ray (a), *optical* (b), *and radio* (c, d, *and* e) *wavelengths. Note that the main pulse and interpulse appear in each.*

in the Crab Nebula and another in the Gum Nebula (see Section 17.4), both supernova remnants.

The Crab Nebula pulsar (Fig. 17.24) is called PSR 0531+21. (PSR stands for pulsar, and the numbers give the pulsar's position in the sky.) PSR 0531+21 has a fast period: 0.033 sec, or 30 pulses per second!

Two key features of the Crab pulsar: First, it is the only pulsar discovered so far to pulse not only at radio and optical wavelengths but also in the infrared, X-ray, and gamma-ray regions of the spectrum (Fig. 17.25). The total energy emitted in the pulses is about 10^{28} W. Second, the Crab pulsar was one of the first to exhibit a definite slowdown in pulse period, at a rate of about 4×10^{-13} sec/sec. That's about 10^{-5} sec/yr—fast for pulsars but less change than an electronic watch shows in a year.

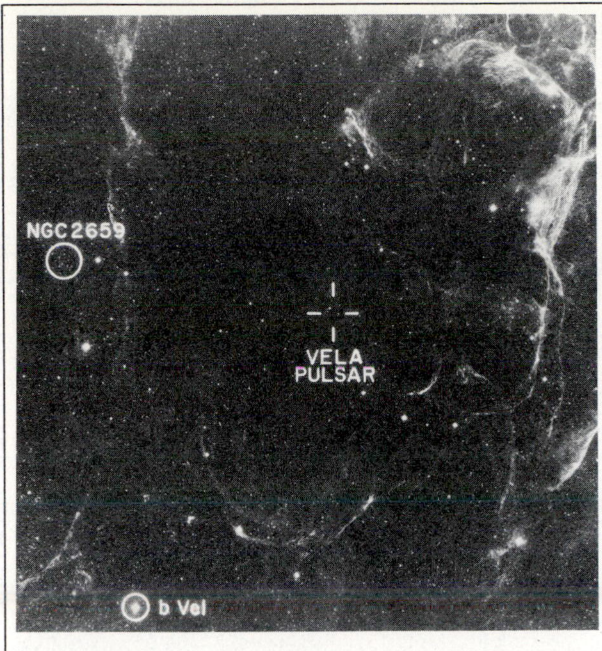

17.26
The location of the Vela pulsar in the Gum Nebula, a known supernova remnant. (Courtesy B. Bok and Steward Observatory)

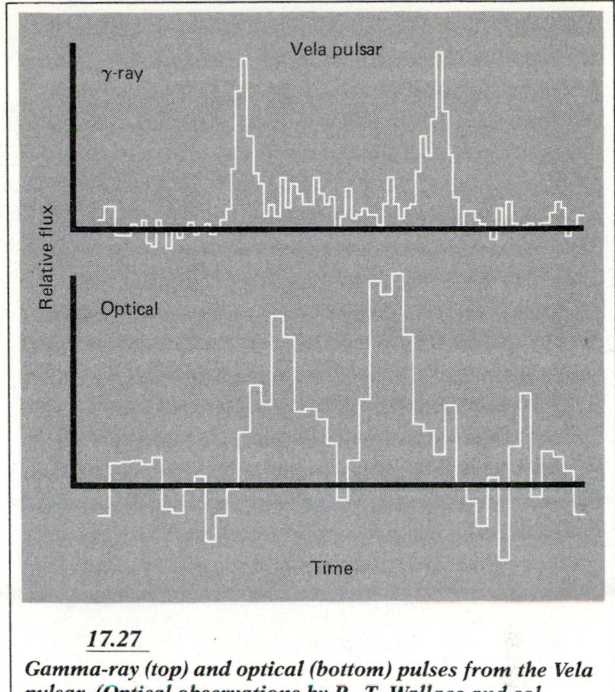

17.27
Gamma-ray (top) and optical (bottom) pulses from the Vela pulsar. (Optical observations by P. T. Wallace and colleagues; gamma-ray observations by R. Buccheri and coworkers)

The discovery of the Crab pulsar solves the energy problem of the Crab Nebula. At all wavelengths, the Crab Nebula emits about 10^{31} W. What is the source of this energy? If the pulsar is a rotating neutron star, its slowdown in period gives a change in rotational energy of about 5×10^{31} W. That's enough to power the nebula—if the rotational kinetic energy of the neutron star can somehow be converted to the kinetic and radiative energy of the nebula. In other words, the light we see now from the nebula ultimately derives from the pulsar.

If the Crab pulsar were the only one associated with a known supernova remnant, it might be written off as a chance coincidence. But astronomers know of another one: the pulsar in the constellation Vela, near the center of the Gum Nebula (Fig. 17.26). This pulsar is called PSR 0833-45.

In 1976 astronomers finally observed optical pulses from the Vela pulsar (Fig. 17.27). The pulses come every 80 msec and have two peaks, separated by about 22 msec. Also, gamma-ray telescopes have detected pulses from Vela. So this pulsar resembles the Crab pulsar in key ways: Both are rapid, both emit pulses over a wide range of the electromagnetic spectrum, and their gamma-ray pulse profiles are very similar.

The lighthouse model for pulsars / Now to tie these observations together in the accepted basic model for pulsars—a rotating, magnetic neutron star—otherwise known as the *lighthouse model*. The model has two key components: The neutron star, whose great density and fast rotation ensure a large amount of rotational energy, and a dipole magnetic field that transforms the rotational energy to electromagnetic energy. Needed is a very strong magnetic field, some 10^{12} G.

The region close to the neutron star where the magnetic field directly and strongly affects the motions of charged particles is called the pulsar's *magnetosphere*. Here all the energy-conversion action takes place. One model pictures the magnetic axis as tilted with respect to the rotational axis (Fig 17.28). As the pulsar spins, its enormous magnetic field spins with it, and this spin induces an equally enormous electric field at its surface. This electric field pulls charged particles—mostly electrons—off the solid crust of iron nuclei and electrons. The electrons flow into the magnetosphere, where they are accelerated by the rotating magnetic-field lines. The accelerated electrons emit synchrotron radiation (Focus 17.1) in a tight beam more or less along the field lines.

You can now see how a pulsar pulses without actually pulsating. (Recall the cepheids, in contrast, vary by actual expansion and contraction.) If the magnetic axis falls within our line of sight, each time a pole swings around to our view (like the spinning light of a lighthouse) we see a burst of synchrotron emission. The time between pulses

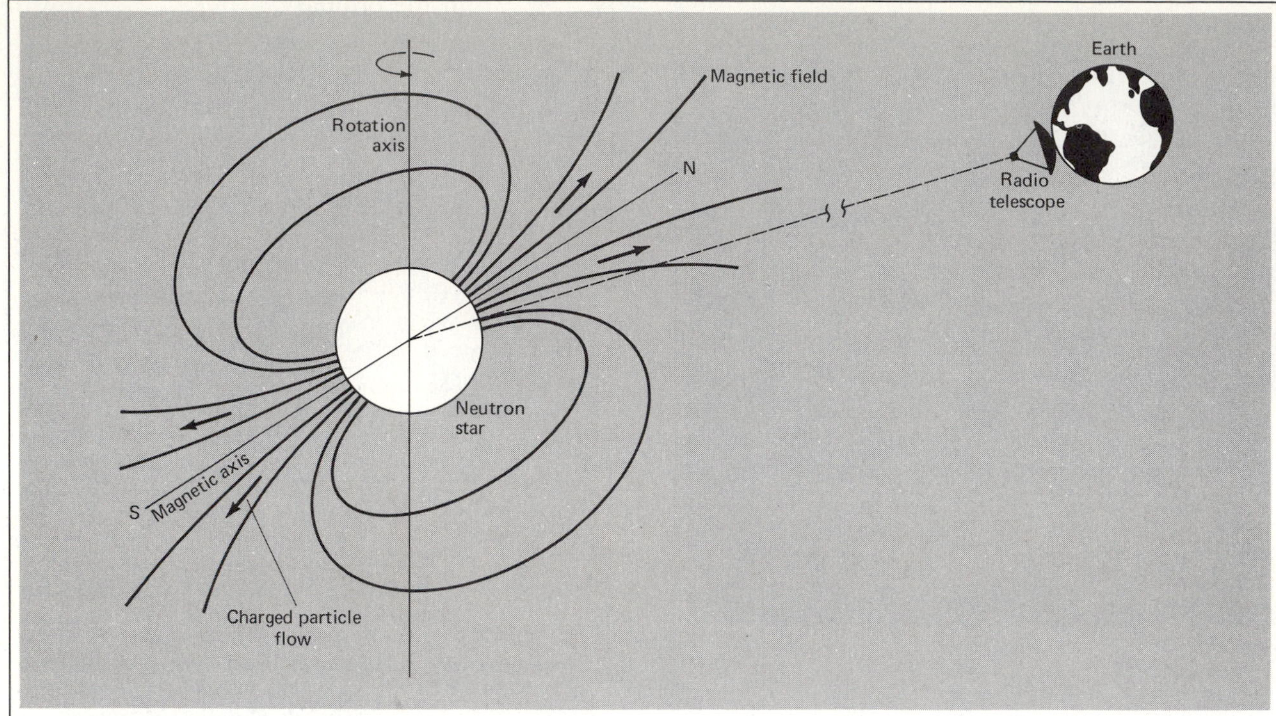

17.28
The lighthouse model of a pulsar. The magnetic axis is tilted with respect to the spin axis. Electrons from the neutron star's surface flow out along the magnetic-field lines and escape, mostly at the north and south magnetic poles. These emit synchroton radiation that we see as pulses when a pole spins across our line of sight.

is the rotation period. The duration of the pulses depends on the size of the radiating region. As the pulsar generates electromagnetic radiation, the torque from accelerating particles in its magnetic field slows down its rotation. This slowdown is observed.

Models to date have had limited success in explaining in detail the abundant observational data. The basic clock mechanism seems right, but we are not sure about the specifics of the emission mechanism. The magnetic, spinning neutron star model is a good example of a basic idea that's been generally accepted but whose details are so far somewhat vague.

Binary radio pulsars / Most stars in the Milky Way are members of binary or multiple star systems. Even after one member of a binary becomes a supernova, the system usually remains intact. So a radio pulsar could exist in a binary system. Such pulsars have been observed.

The first one found is called PSR 1913+16. This pulsar has a pulse period of only 0.059 sec. Surprisingly, its period goes through a large cyclic change in only 7.75 hr. What's going on? Such regular changes would naturally come about in a binary system of the pulsar and a companion with an orbital period of 7.75 hr. What is seen is a Doppler shift in the signal produced by the orbital motion of the system. When the pulsar is moving away from us, its pulses are spread out and come at longer intervals. When it is moving toward us, the pulses are pushed together and come at shorter intervals. (Think of the pulses per second as frequency, in analogy with electromagnetic waves.)

PSR 1913+16 lies about 15,000 ly from us. Visual and X-ray observations have so far failed to detect either the pulsar or its companion. From radio observations alone, we know that the pulsar and its companion have an orbital semimajor axis of only 700,000 km—that's only the sun's radius! Their combined masses are 2.83 solar masses, so if the pulsar has a mass of about 2 solar masses (a typical neutron star), its companion has about 0.8 solar mass. The companion might be a white dwarf. There are other pulsars in binaries, seen with X-ray telescopes, and you'll meet them later in this chapter.

SS 433: A unique neutron-star binary system / The neutron-star story takes a strange twist with a bizarre object called *SS 433*, which lies in the constellation Aquila. Radio observations show that SS 433 sits near the center of a large supernova remnant. Early radio and X-ray observations of SS 433 showed that its emission varied in a few hours.

These variations prompted optical astronomers to take spectra of SS 433. What a surprise! A very strong H-alpha line of hydrogen was visible along with other emission lines. These turned out to be highly red- and blue-shifted lines of hydrogen and helium (Fig. 17.29). Their Doppler shifts indicated radial velocities of more than 40,000 km/sec, or over 1/10 the speed of light! And the lines shifted noticeable amounts from night to night.

Later observations showed that the Doppler-shifted lines varied in a regular cycle over 164 days. In contrast, the strongest emission lines had a very small Doppler shift (only 70 km/sec) and varied over a 13-day period. A binary model explains these differences in Doppler shifts (Fig. 17.30a). An ordinary (but unseen) star orbits with a neutron star every 13 days. Their proximity funnels gas from the ordinary star onto the neutron star. The gas heats up as it falls in and produces the strongest emission lines with the small Doppler shift (from the orbital motion).

Meanwhile, the neutron star has two jets of material squirting out in opposite directions (Fig. 17.30b). The jet blasting away from us generates the highly red-shifted lines; the one pointing toward us, the blue-shifted ones. The line of the jets is tilted about 20° to the rotation axis of the neutron star; that in turn inclines about 80° with respect to the line of sight to the earth. Like the earth's axis, the neutron star's spin axis precesses in space, with a period of 164 days. As it precesses, it carries the jets around with it. This change causes the variation in the emission lines with the large Doppler shifts.

What are the jets? Radio observations with the VLA imply that they contain small blobs of ionized gas continuously blowing out from the neutron star (Fig. 17.30c). This material shoots out at some 78,000 km/sec, about 26 percent of the speed of light! The jets appear to radiate by synchrotron emission, an idea backed up by the fact

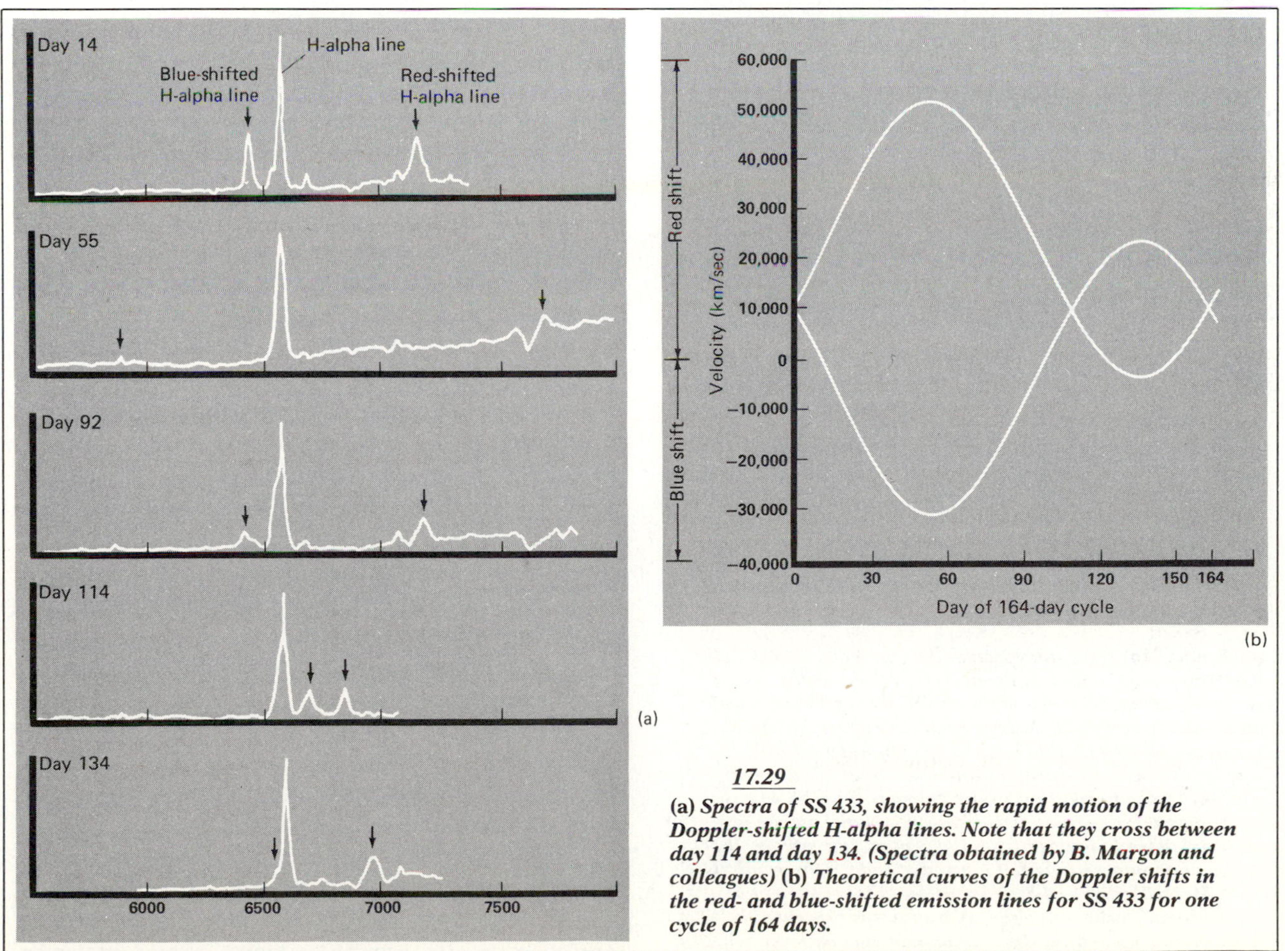

17.29

(a) ***Spectra of SS 433, showing the rapid motion of the Doppler-shifted H-alpha lines. Note that they cross between day 114 and day 134. (Spectra obtained by B. Margon and colleagues)*** **(b)** ***Theoretical curves of the Doppler shifts in the red- and blue-shifted emission lines for SS 433 for one cycle of 164 days.***

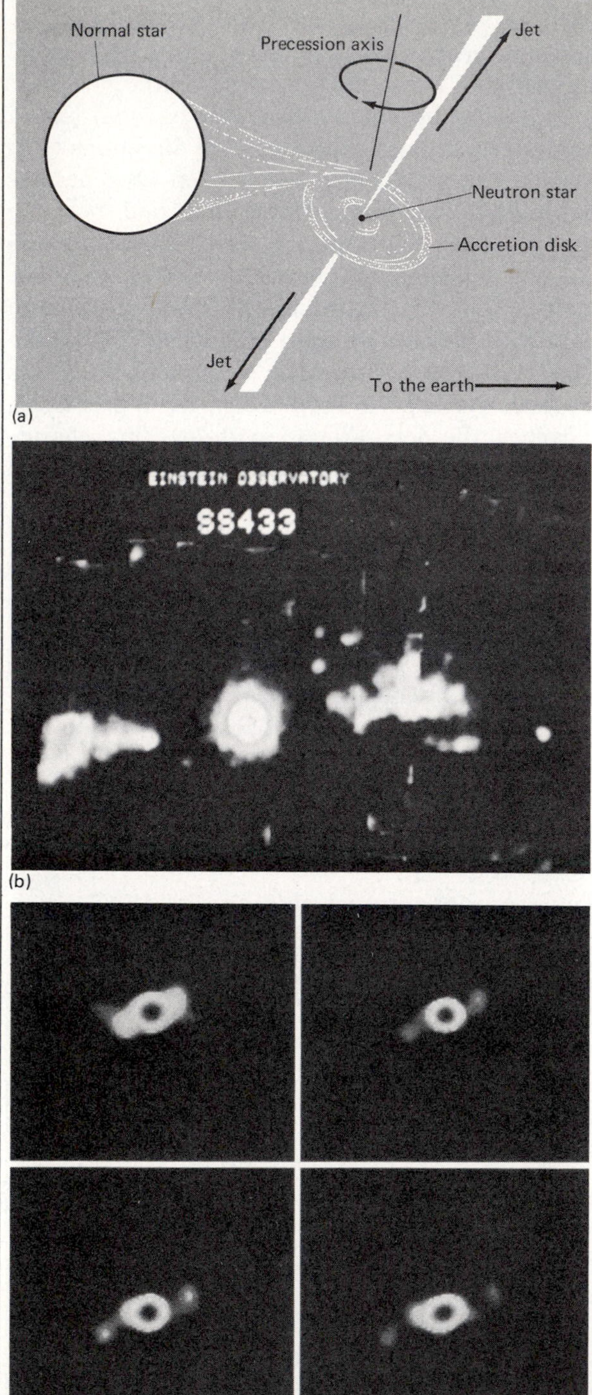

17.30

(a) *A model for the binary system that includes SS 433. Jets blast out from the face of the accretion disk, which precesses around the neutron star. (Based on a theoretical model developed by B. Margon and colleagues)* (b) *An X-ray image of SS 433, showing the jets from the central source. (Courtesy J. Grindley and the Center for Astrophysics)* (c) *VLA observations of the twin jets from SS 433 made over a four-month period in 1981. Note how the jets' material moved outward from the central star system; they did so at about one-fourth the speed of light. (Observations by R. M. Hjellming and K. J. Johnston; photo courtesy of the National Radio Astronomy Observatory, operated by Associated Universities, Inc., under contract with the National Science Foundation)*

that they are highly polarized. This relativistic jet phenomenon, which is probably produced by magnetic fields, also appears in the nuclei of galaxies (see Chapter 20). So far, SS 433 is the only *star* system producing such relativistic jets.

17.7 Black holes: The ultimate corpses

Many stars in the Milky Way have masses much greater than the neutron-star limit of about 5 solar masses. Assume that such stars do not lose enough mass during their evolution to go below this upper mass limit for a neutron star. Further assume that all thermonuclear reactions have ceased, so that the star is cold, close in temperature to absolute zero. Now ask the question: Will there be any barrier to the collapse of this material with mass greater than 5 solar masses?

No. The crush of gravity overwhelms all outward forces, including the repulsive forces between particles with the same charge. No material can withstand this final crushing point of matter. The collapse cannot be halted; the volume of the star will continue to decrease until it reaches zero. The density of the star will increase until it becomes infinite. (Neither of these events can be exactly true of a *real* object in this universe.) This theoretical collapse to a singular point of zero volume and infinite density, a *singularity*, marks a crucial limitation in our understanding of the physics of the universe.

Before a mass becomes a singularity, bizarre events occur near it. As its density increases, the paths of light rays emitted from the star are bent more and more away from straight lines going away from the star's surface. Eventually the density reaches such a high value that the light rays are wrapped around the star and do not leave. The photons are trapped by the intense gravitational field in an orbit around the star. The escape velocity from the star is then greater than the speed of light. Any additional photons emitted after the star attains this critical density can never reach an outside observer. The star becomes a *black hole* (Fig. 17.31).

The Schwarzschild radius / Let's consider the meaning of *black* in *black hole* in terms of escape velocity (see Focus 7.3). If you are familiar with rocket launches, you have probably realized that for an object to leave the earth permanently, it must leave the earth's surface with at least a minimum velocity—the escape velocity from the earth, about 11 km/sec. Imagine that you could squeeze the earth so that it would become smaller and denser. Its escape velocity would increase. Imagine the earth compressed until its escape

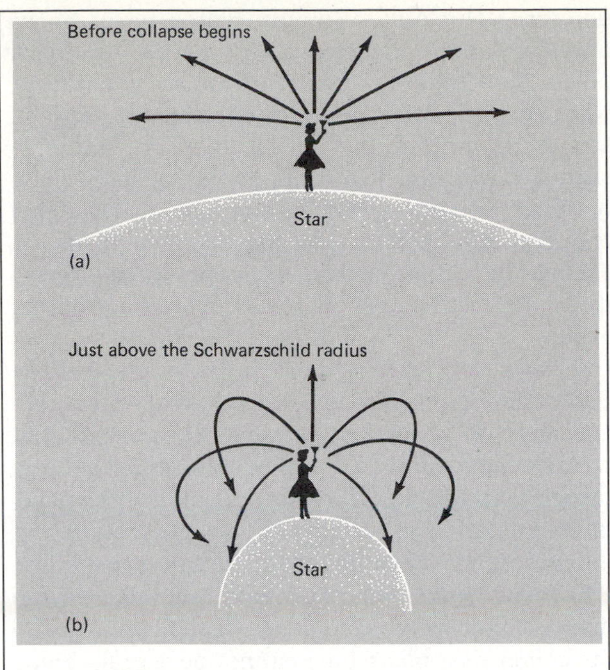

17.31
The trapping of light by the collapse of a mass into a black hole. (Adapted from a diagram by W. Kaufmann)

velocity equaled the velocity of light. Then nothing, *not even light*, emitted at its surface could escape into space. Nothing gets away, so to an outside observer, the earth would appear black.

How small must an object become to be dense enough to trap light? Einstein's general theory of relativity provides an answer. Just after the general theory's publication, the German astrophysicist Karl Schwarzschild (1873–1917) calculated this critical size, now called the *Schwarzschild radius*. For the sun, the Schwarzschild radius is about 3 km—smaller than the typical sunspot! If it were compressed to this size, the sun would have a density of about 10^{19} kg/m^3. The mass of any object gives its Schwarzschild radius directly. For 1 solar mass, it's 3 km; for 2 solar masses, 6 km; for 10 solar masses, 30 km; and so on.

How can a star get as small as its Schwarzschild radius? Two ways are possible: First, runaway gravitational collapse. If you put together more mass than about 5 solar masses, it must eventually squeeze itself into a black hole. Nothing we know about—not even the hardness of matter itself—can stop this final crushing. Second, a supernova. A star's self-destruction can slam matter into a size smaller than its Schwarzschild radius.

Note: Any size object can be made into a black hole if a force compresses it enough. But an object with a mass greater than the neutron-star limit of 5 solar masses *must* become a black hole after thermonuclear reactions have ceased, for there is no known source of pressure that can support it.

That's how black holes may form. Once a black hole forms, what happens to the matter that makes it? Einstein's general theory predicts that the matter keeps collapsing gravitationally until it has *no volume*. But it still has mass, so its density is infinite. This theoretical end to runaway gravitational collapse is called a *singularity*. The matter has literally squeezed itself so small that it occupies no space. Yet it's still there. What a paradox! How can matter *not* take up space? The general theory of relativity points to the formation of a singularity, cloaked in the center of a black hole, as the natural end of gravitational collapse. It is also where our present knowledge of physics ends.

Journey into a black hole / Put aside the puzzle of the singularity for a moment. Let me describe the theoretical properties of a black hole, both inside and outside. A person falling into a black hole meets a fate an outside observer cannot even find out about—unless the outsider drops in too. Just for fun, let's take an imaginary journey into a black hole. We'll follow the adventures of a crazy astronaut who takes the plunge and compare this trip with what an outside observer sees of it.

You and a friend start out in a spaceship orbiting far away from a 10-solar-mass black hole. Nothing peculiar here. The ship orbits the black hole in accordance with Kepler's laws, as it would any ordinary mass. In fact, Kepler's third law and the spaceship's orbit permit you to measure the hole's mass. But if you look hard for the mass, you won't see anything.

Your friend volunteers to hop in. She takes with her a laser light and an electronic watch. You and she synchronize watches. Once a second, according to her watch, she will send a laser flash back to you.

Down she goes! For a long time as she falls toward the black hole, nothing strange happens. But as she gets closer, stronger and stronger tidal gravitational forces (see Focus 9.1) stretch her out (if she falls feet first) from head to toes. Also, another tidal force squeezes her together, mostly at the shoulders. (You feel such tidal forces on the earth, but the forces is so weak that it doesn't bother you.) Near a black hole, tidal forces grow enormously. An ordinary human being would be ripped apart about 3000 km from a 10-solar-mass black hole. Let's suppose your friend is indestructible, so she can continue her trip.

Down she drops. The tidal forces get stronger fast and make her more uncomfortable. But nothing else seems strange to her. Every second on the dot she sends out a blast of laser light. Peering down, she can just make out a small black region in the sky. (A 10-solar-mass black hole has a radius of only 30 km.)

Then it happens: She crosses the Schwarzschild radius! But nothing new happens to her. No solid substance, no signs mark the edge of the black hole. However, no amount of energy can push her out of the black hole; she has crossed a one-way gate in spacetime. The trip now swiftly ends for your unfortunate friend. Quickly—in about 10^{-5} sec after she crosses the Schwarzschild radius—she crashes into a singularity (if it exists!). Crushed to zero volume, she is destroyed. Even if a singularity does not lie in the black hole's center, the mass that made the black hole probably does. So she would smash into it. Fatal end of friend's trip.

But what of your view, back in the spaceship, of your friend's adventure? You would *never* see her final destruction; in fact, you'd not even see her fall into the black hole. As she dropped closer to the black hole, you'd notice that the light from her laser was red-shifted, with the shift increasing as she fell closer to the black hole. (The light must work against gravity to get to you, so it loses energy and increases in wavelength.) Also, the time between laser flashes increases. What's happening? Compared with your watch, your friend's watch appears to slow down as she gets into regions of stronger gravity. Your watch and hers disagree about how long it takes her to travel to the black hole.

As she comes closer to the Schwarzschild radius, the watches get more and more out of sync. The times between your reception of her flashes stretches out. In fact, a laser burst sent out just as she crossed the Schwarzschild radius would take an *infinite* time to reach to you. It also would suffer an *infinite* red shift. To you, her fall would seem to grow slower and slower as she got closer to the black hole, but she would never appear to fall into it. Time slows down so much that, near a black hole, it seems to be frozen. In addition, the light gets more and more red-shifted until you can no longer detect it.

A black hole practices cosmic censorship. It prevents you from seeing your friend even fall into it. Light—our only astronomical communication medium—is cut off by the black hole. So you cannot know what happens to your friend inside of it.

Warning: Black holes aren't cosmic vacuum cleaners. Many people think that they have infinitely powerful gravitational fields that suck up everything that gets near them, scouring out the universe. Not quite. Suppose you'd never heard anything about the strange properties of black holes. Just thinking about Newtonian gravitation, what do you think would happen to the orbit of the earth around the sun if the sun suddenly shrank down to a ball 6 km across? That's right, nothing! The masses of the sun and earth haven't changed, and neither has the distance from the earth to the center of the sun. So the force of gravity on the earth hasn't changed, and neither has its orbit.

You can think of black holes as ghosts. The Schwarzschild radius is *not* a physical object. It's a region of spacetime so severely curved that weird things happen near it. Einstein's general theory actually describes the geometry of spacetime in detail; Fig. 17.32 shows a representation of this geometry. Note how badly warped spacetime is from flat (*flat* means no gravitational field, no mass). Likewise, the singularity predicted to be in the center of a black hole cannot be a real object. It's a region where space and time end in this universe. What a strange corpse from the death of a massive star!

17.8 Observing black holes

Enough for black holes as theoretical entities. How can you actually observe a black hole? With difficulty! Light emitted inside cannot get out. Light sent out close by is strongly red-shifted, so it's hard to detect. In addition, a black hole is small, only a few kilometers in size. So you'll have a hard time seeing an isolated black hole.

But a black hole surrounded by clouds of material might be observable. Any matter falling toward a black hole gains energy and heats up. (It's also squeezed by tidal forces.) Heated enough, the atoms are ionized. Gravity accelerates the ionized gas, and it emits electromagnetic radiation. If heated to a few million Kelvins or so, the material gives off X-rays. Before it is trapped in the gravitational gulf, in-falling material can send X-rays into space. So a black hole passing through an interstellar cloud or close to a star can sweep material into it and radiate. X-ray sources are good candidates for black holes.

X-rays can't penetrate the earth's atmosphere, so X-ray astronomy can only be done from space. The *Uhuru* satellite, launched in 1970 and designed to observe X-ray sources, detected about 160 strong X-ray objects. Some of these X-ray sources are prime candidates for black holes because they are binary—the X-ray source and a normal star (a potential source of in-falling material) orbit a common center of mass.

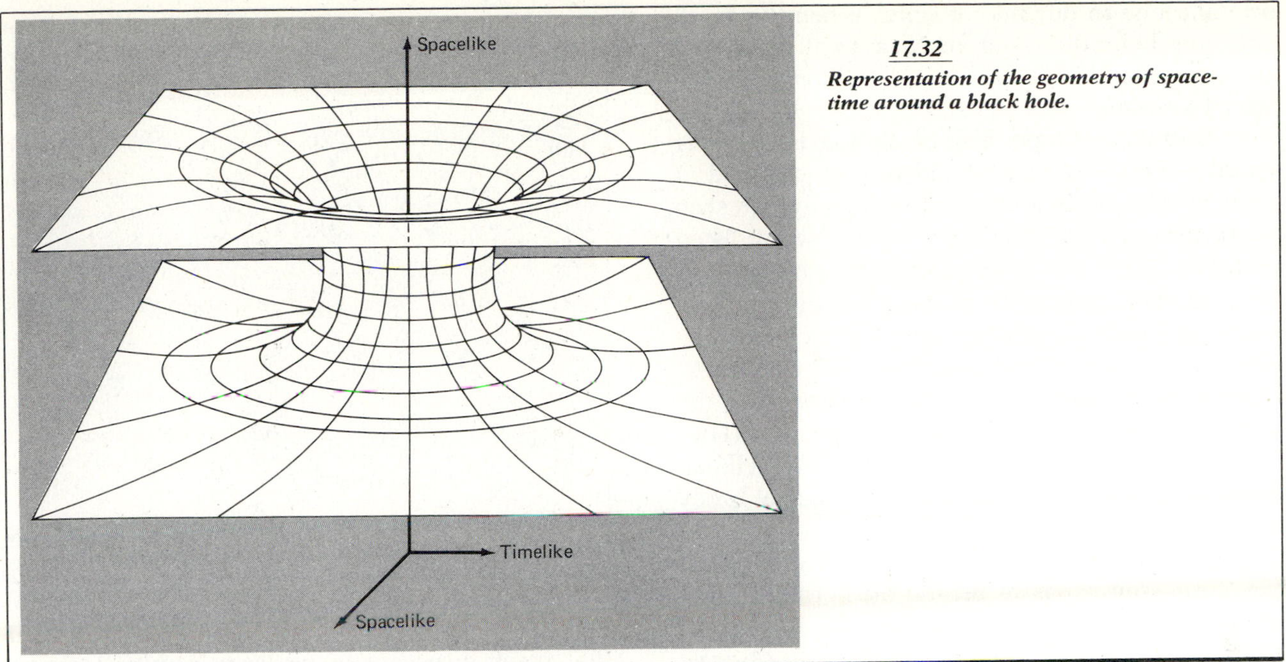

17.32
Representation of the geometry of space-time around a black hole.

Binary X-ray sources / Why are *binary* X-ray sources most suspect? Imagine a black hole orbiting a supergiant star. Suppose that they are very close together so that their orbital period is a few days or so. The star has a huge, distended atmosphere—and material blown from this atmosphere (a stellar wind) can fall into the black hole (Fig. 17.33). Falling toward the black hole, the material gains kinetic energy, heats up to 1 million K or so, and emits X-rays. Only a small region around the black hole gives off X-rays, and since the material may fall in sporadically, you might expect the intensity of the X-rays to vary quickly. Also, imagine the black hole and star with their

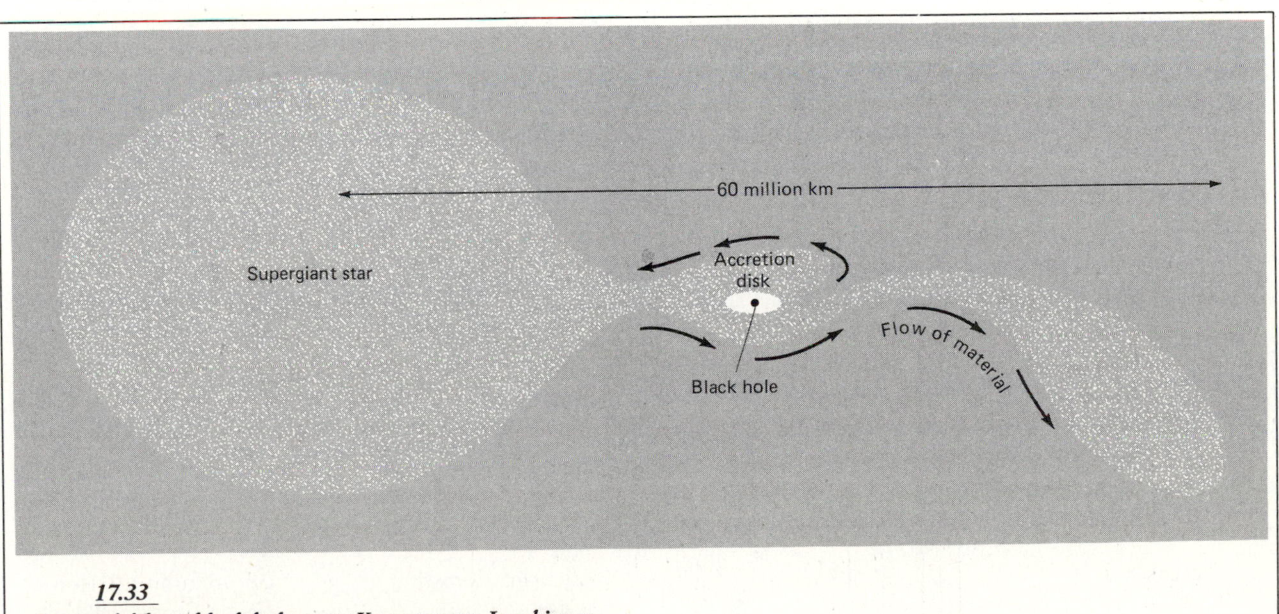

17.33
One model for a black hole as an X-ray source. In a binary system, the black hole is coupled with a giant or supergiant star. Material from the star flows to the black hole, where it falls into an accretion disk. As it falls, it heats up to about 1 million K and emits X-rays. (Adapted from a diagram by K. Thorne)

orbital plane in our line of sight. When the black hole goes behind the star, its X-rays will be cut off. In this case we would see an eclipsing X-ray binary system.

So a sign of a possible black hole is a rapidly varying X-ray source, which may be eclipsed at regular intervals in a binary system. Have we seen such variable X-ray sources? Yes—some are listed in Table 17.3. Each of these objects is believed to be a main-sequence or post-main-sequence star swinging around an X-ray source. They have X-ray luminosities in the range from 10^{29} to 10^{31} W. (That's 200 to 20,000 times the luminosity of the sun, and all in X-rays.) Three of the sources (Hercules X-1, Centaurus X-3, and Small Magellanic Cloud X-1) have short-period X-ray pulses; they are X-ray pulsars.

Five of the systems exhibit X-ray eclipses; the X-ray source passes behind the normal star as we view the system. Using spectroscopic analysis of the light from the visible star, we can observe the changes in Doppler shift, so we can find out the orbital periods—they are typically a few days. These short periods indicate that the orbits are only a few times larger than the primary stars. If we can determine the separation of the two objects, we can ascertain—from Kepler's third law—the sum of the masses (normal star plus X-ray source). If we can get an idea of the mass of the normal star from its luminosity, we can also determine the mass of the X-ray source. And if that mass turns out to be large enough (greater than 5 solar masses, the upper limit for a neutron star), the X-ray source must be a black hole.

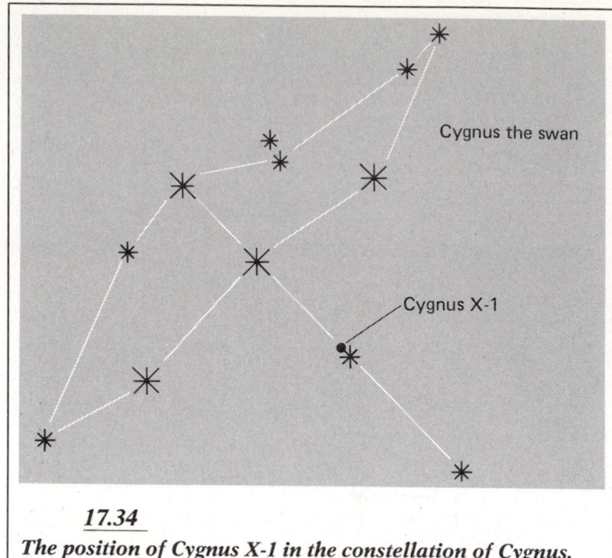

17.34
The position of Cygnus X-1 in the constellation of Cygnus.

Let's investigate one of these binary X-ray sources in detail: Cygnus X-1.

Is Cygnus X-1 a black hole? / To prove the reality of black holes, we need to observe one. So far, the most likely candidate is Cygnus X-1, a strong X-ray source in the constellation Cygnus (Fig. 17.34).

TABLE 17.3 *Some binary X-ray sources*

Name (constellation)	*Binary period (days)*	*Characteristics of X-rays*	*Characteristics of visible stars*
Cygnus X-1	5.6	Varies in duration from 0.001 to 1 sec	Blue supergiant of about 20 solar masses
Centaurus X-3	2.087	X-ray eclipses with duration of 0.488 day	Blue giant of about 16 solar masses
Small Magellanic Cloud X-1	3.89	X-ray eclipses with 0.6-day duration	Blue supergiant of about 25 solar masses
Vela X-1	8.65	X-ray eclipses with 1.7-day duration; flares lasting a few hours	Blue supergiant of about 25 solar masses
Circinus X-1	>15	X-ray eclipses lasting about a day	Not yet found
Hercules X-1	1.7	X-ray eclipses lasting 0.24 day	Companion HZ Her about 2 solar masses
Cygnus X-3	0.2	4.8-hr variations, no eclipse	None visible; infrared source with 4.8 hr variation

SOURCE Adapted from tables by H. Gursky, E. P. J. van den Heuvel, S. Rappaport, and P. C. Joss.

17.35
The blue supergiant star HDE 226868 (arrow) about which Cygnus X-1 orbits. (Courtesy J. Kristian)

Cygnus X-1 emits about 4×10^{30} W in X-rays. Observations have shown that Cygnus X-1 flickers rapidly, in less than 0.001 sec. (This variation is an intrinsic flickering of the source, not a "twinkling" like that of stars, which is caused by the earth's atmosphere.) In the same way as was done for pulsars (Section 17.6), we can interpret this observation as indicating that the X-ray-emitting region must be less than 0.001 light-second in size (less than 300 km). Astronomers have identified Cygnus X-1 as an *O*-supergiant star, that is, a hot massive star (Fig. 17.35). It is called HDE 226868 and is a supergiant star of approximately 31,000 K surface temperature.

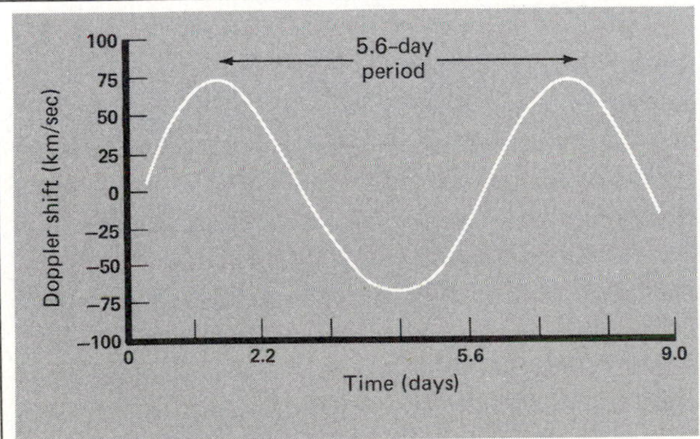

17.36
The orbital period of Cygnus X-1, inferred from the Doppler shift in the spectral lines of the blue supergiant star about which Cygnus X-1 orbits. (Based on observations by C. Bolton)

Optical observations show that the dark lines in the spectrum of the blue supergiant went through periodic Doppler shifts in 5.6 days (Fig. 17.36). So the supergiant orbits with the X-ray source about a common center of mass every 5.6 days. (If it orbited a normal star, you'd probably see the spectra of both stars, not just one.) The supergiant has a massive but optically invisible companion—Cygnus X-1.

Recall that only for binaries can we find the masses of stars directly. But we need to know the separation of the stars and their distance from the center of the mass. These distances can be worked out correctly only if we can observe *both* stars and if we know the orbital tilt with respect to our line of sight. In these two regards, the mass of Cygnus X-1 is hard to find out. We can observe the Doppler shift in the spectrum of the visible companion, but we cannot obtain the velocity of the X-ray source. And because Cygnus X-1 has not been found to eclipse, we can't pin down its orbital inclination. So we don't have enough information to determine both individual masses.

But we can make some reasonable estimates. Blue supergiant stars are typically 15 to 40 solar masses. Take 20 as typical. The orbital period and velocity of the supergiant give us a relation between the masses of the supergiant and the X-ray source, uncertain by the amount of orbital tilt. Suppose the orbital tilt were 90°; then the mass of the companion would be 4 to 5 solar masses. But the tilt can't be so high, because the star is not eclipsing. So the true velocity must be higher than observed, and the mass of the companion must be higher. A study of the brightness of the supergiant indicates that it varies a little in 5.6 days. This fact leads to an inferred orbital tilt of 30°. X-ray observations imply a tilt between 36 and 67°. If true, these values suggest that Cygnus X-1 has a mass possibly as great as 11 solar masses and as small as 6. The most likely value is about 9 solar masses. If so, and if the limit for a neutron star is 5 solar masses, Cygnus X-1 must be a black hole.

I feel that there is strong evidence here for a black hole. To sum up the chain of inference:

1. Because it emits X-rays that vary in a short time, Cygnus X-1 must be a small, dense object onto which matter is falling.
2. The blue supergiant is the star about which Cygnus X-1 revolves.
3. From the Doppler shift of its spectral lines, the blue supergiant (and so the X-ray source) has an orbital period of 5.6 days.

4 The blue supergiant has a mass of 14 to 40 solar masses.
5 If the blue supergiant has a mass of 20 solar masses, Cygnus X-1 has at least 6 solar masses, more likely 9 to 11 solar masses.
6 A small object with a mass greater than about 5 solar masses is a black hole.
7 Therefore, Cygnus X-1 is a black hole.

Not all astrophysicists accept this conclusion. For example, some argue that the upper limit on a neutron star's mass is 8 solar masses. In that case, if the mass of Cygnus X-1 is only 4 to 5 solar masses, it could be a neutron star rather than a black hole. I wish I could report the existence of black holes as certain, but in all honesty I cannot. Yet, relativity predicts black holes. To believe in general relativity pretty much requires the acceptance of black holes.

Warning: Not all binary X-ray sources contain black holes. The X-ray emission could arise from accretion disks around neutron stars. An example is Centaurus X-3 (abbreviated Cen X-3). The Uhuru satellite showed that this X-ray emission pulses every 4.84 sec. Long-term observations revealed that X-ray eclipses take place every 2.087 days and last about 0.5 day. Is Cen X-3 a black hole or a neutron star? That depends on its mass. Since Cen X-3 eclipses, we can pin down its mass better than for Cygnus X-1. It ranges between 0.6 and 1.1 solar masses—so it is not a black hole, but probably a low-mass neutron star formed in a supernova.

The fact that Cen X-3 is an X-ray pulsar also supports a neutron-star model, in analogy with the model of radio pulsars as magnetic neutron stars. The X-ray pulses might arise from accreting matter channeled into the magnetic polar regions by the intense magnetic fields (Fig. 17.37).

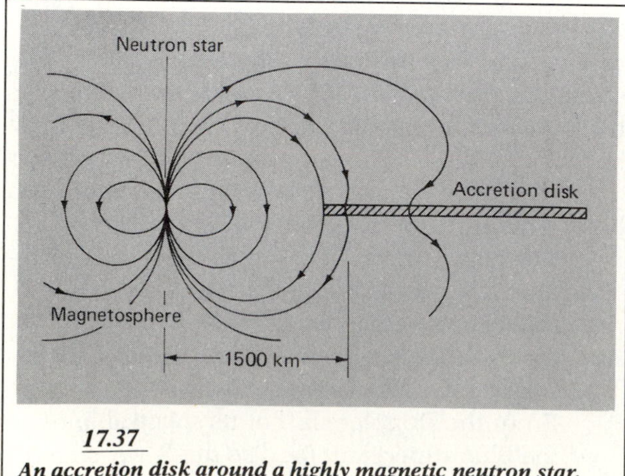

17.37
An accretion disk around a highly magnetic neutron star.

The evolution of binary X-ray systems / How does a black hole or neutron star end up in a binary system? One answer appears if you remember the basic fact of stellar evolution: The more massive a star is, the faster it evolves (and so the shorter its lifetime). With this fact in mind, let's examine some facets of one possible model for the evolution of a binary X-ray system.

Start with an ordinary binary containing a 20-solar-mass star and a 6-solar-mass star (stage 1 in Fig. 17.38). The more massive star evolves faster, becomes a red supergiant star, and fills its Roche lobe (stage 2). Matter streams from the more massive star to the less massive one. When the flow stops, the stars have switched roles (stage 3): The 6-solar-mass star is now a 17.6 one, the other a 5.4-solar-mass star—essentially the core of the former 20-solar-mass star. This core supernovas (stage 4); it leaves behind a neutron star or a black hole (stage 5). The other star now evolves rapidly (because of its increased mass). It first loses matter to a stellar wind. Any of this material falling onto the black hole generates X-rays (stage 6). Later, the star expands until it fills its Roche lobe (stage 7). Material can then flow along the gravitational bridge to the black hole and emit X-rays.

What next? The massive supergiant should also supernova. That leaves a binary black-hole system, a binary neutron-star system, or binary black hole–neutron star system. In all these cases, the matter is locked up. So the binary no longer emits X-rays. Note that the X-ray–emitting stages don't last long—at most a few million years or so.

TABLE 17.4 Death of stars

Main-sequence mass (solar masses)	*Normal life (main-sequence stars)*	*Death and final corpse*
0.1–0.5	*M*	White dwarf
0.5–4.0	*K–A*	Planetary nebula, white dwarf
4–10	*A–B*	Supernova, neutron star
10–20	*B*	Supernova, neutron star or black hole
20–60	*O*	Supernova, black hole

SOURCE Adapted from work by A. G. W. Cameron and J. C. Wheeler.

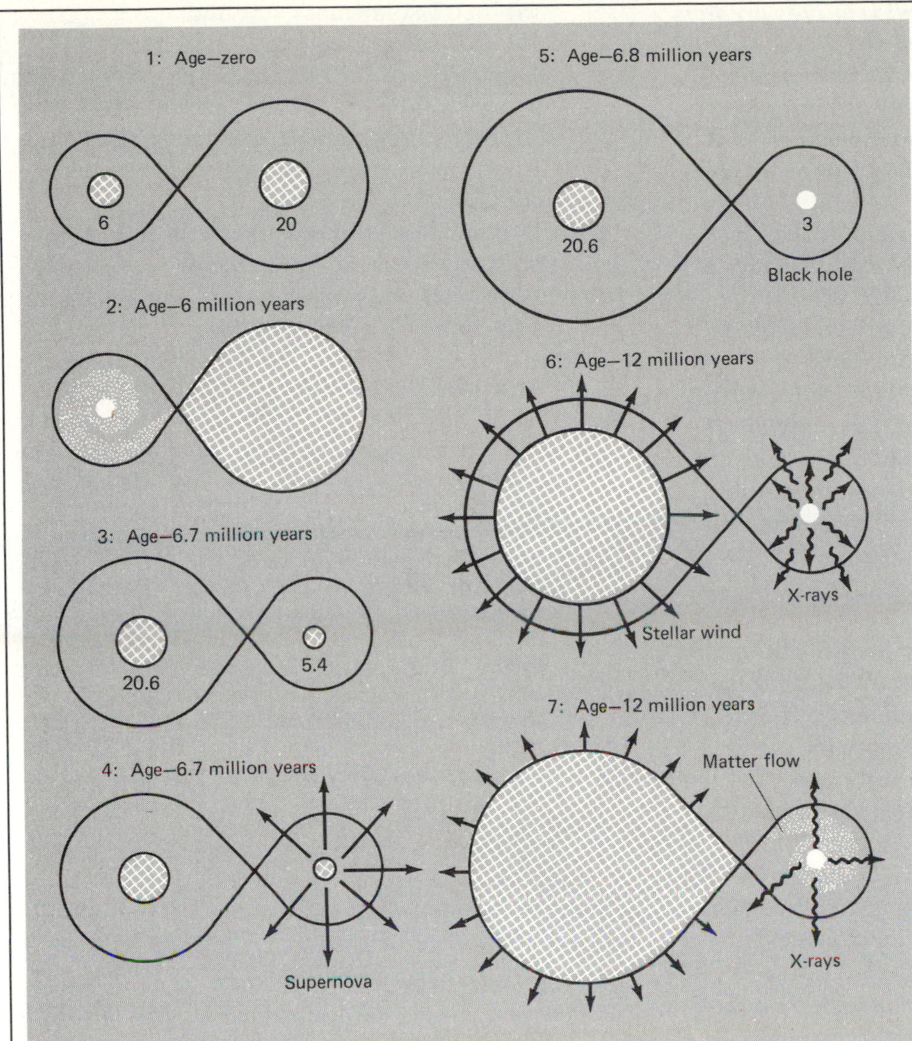

17.38

A model for the evolution of a close binary system with massive stars. Start out with 20- and 6-solar-mass stars. The more massive star evolves more quickly, and expands in size, and material flows to the companion (1 and 2), which ends up with 20.6 solar masses (3). The core of the other star supernovas (4) to leave a black hole (5). The companion evolves, expanding and developing a strong stellar wind. Matter flows into the black hole, and X-rays result (6 and 7). (Based on calculations by P. van den Heavel, C. de Loore, and J.-P. de Greve)

KEY CONCEPTS

1. ***The mass of a star at the time of its death determines the corpse it leaves behind (Table 17.4): white dwarf, up to 1.4 solar masses; neutron star, 1.4 to 5 solar masses; black hole, greater than 5 solar masses.***
2. ***White dwarfs and neutron stars are supported against gravity by the degenerate gas pressure from electrons in a white dwarf and neutrons in a neutron star.***
3. ***Novas occur in binary systems when hydrogen falls onto the surface of a white dwarf; this fresh fuel ignites explosively to produce a nova outburst.***
4. ***Supernovas (Type II) are the explosions of massive stars (greater than 5 to 10 solar masses) that have evolved iron cores; Type I supernovas are the explosions of solar-mass stars by a means not yet understood but that may involve a binary system.***
5. ***Supernova explosions make many of the elements heavier than iron and blast them into the interstellar medium (along with elements made in normal lives).***
6. ***The Crab Nebula is a supernova remnant, emitting light by the synchrotron process; its emission is powered by a pulsar in the center; that pulsar was formed in the supernova explosion of A.D. 1054.***
7. ***Pulsars are rapidly rotating neutron stars; this idea is inferred most strongly from the regular timing of the fastest pulsars; only neutron stars are small and dense enough to rotate so rapidly; pulsar emission comes from a neutron star's intense magnetic field.***
8. ***A black hole forms when a mass becomes so compacted that its escape velocity is greater than the velocity of light.***
9. ***Black holes are small: a 1-solar-mass black hole has a radius of 3 km; 2 solar masses, 6 km; 10 solar masses, 30 km; and so on.***

10 ***Einstein's theory of general relativity predicts that time appears frozen near a black hole and that a singularity resides in its center.***

11 ***Black holes can be seen only by their interaction with visible matter; an especially good circumstance would be a black hole-ordinary star binary system.***

12 ***Cygnus X-1, in a binary system, is the best candidate so far for a black hole; it emits X-rays from a hot accretion disk around the suspected black hole; most other binary X-ray sources contain neutron stars.***

13 ***Binary systems with a neutron star or black hole evolved from one very massive star (20 solar masses) that evolved to a neutron star or black hole very quickly in combination with a less massive star (5 solar masses).***

STUDY EXERCISES

1 In a short paragraph, describe the primary characteristics of a white dwarf. (*Objective 1*)

2 In a short paragraph, describe to a friend who has not studied astronomy the chief features of a neutron star. (*Objective 1*)

3 What observational evidence do we have for the actual existence of neutron stars and white dwarfs? (*Objectives 1, 2, and 3*)

4 Look around you. Of the items you see, what would not be there if supernovas didn't occur? (*Objective 8*)

5 Assuming no loss of mass, what will be the final form of

(*a*) a 0.5-solar-mass star and

(*b*) a 2-solar-mass star. (*Objective 1*)

6 Make a list of the observational evidence that supports the idea of the Crab Nebula as a supernova remnant. (*Objective 6*)

7 In what way does a black hole practice censorship? (*Objective 10*)

8 If a black hole is really black, how can it be an X-ray source? (*Objective 9*)

9 Why can't you find out what happens inside a black hole? (*Objective 10*)

BEYOND THIS BOOK . . .

We haven't had a supernova in our Galaxy recently, but we do see them in others. See "Supernovas in Other Galaxies" by R. P. Kirshner, *Scientific American*, December 1976. Also "Supernovae: Still a Challenge" by F. Reddy in *Sky and Telescope*, December 1983, p. 485.

For a personal account of the discovery of pulsars, turn to "Little Green Men, White Dwarfs, or What? " by S. J. B. Burnell in *Sky and Telescope*, March 1978, p. 218. For a different view, read A. Hewish's account in *Science*, vol. 188 (1975), p. 1079.

K. Thorne describes "The Search for Black Holes" in *Scientific American*, December 1974, p. 32.

H. Gursky and E. P. J. van den Heuvel present a complete picture of "X-Ray Emitting Double Stars" in *Scientific American*, March 1975, p. 24.

You can find more information about black holes and related beasts in *The Cosmic Frontiers of General Relativity* (Little, Brown, Boston, 1977) by W. J. Kaufmann III.

J. Shklovsky covers stardeath in *Stars: Their Birth, Life, and Death* (Freeman, San Francisco, 1978).

Epilogue part 3

This part has described our contemporary concepts in our understanding of the births, lives, and deaths of the stars. Here are the major themes:

1 A star's evolution depends mainly on its mass; more massive stars have higher core temperatures and, so higher luminosities. They evolve faster and live shorter lives. Their higher core temperatures allow them to fuse heavier elements. They are likely to die in supernova explosions.

2 Stars like the sun will evolve to become red giants, white dwarfs, and then black dwarfs. Most stars will go through this evolution because most stars contain 1 solar mass or less of material.

3 During its life, a star struggles constantly against gravity. It resists gravitational collapse by pressure from heat in its interior. Fusion reactions in the core provide this heat; they are first ignited by the heat from gravitational collapse. A star must fuse heavier and heavier elements in order to withstand gravity.

4 Stars recycle some material back into the birthplace of stars, the interstellar medium, but the medium's composition has been changed by the addition of heavier elements. Some of the material stays locked in a star's corpse, never to participate again in cosmic evolution.

5 How a star dies depends on its mass at its time of death. If its mass is less than 1.4 solar masses, its corpse takes the form of a white dwarf. If it is greater than 1.4 but less than roughly 5 solar masses, gravity crushes it to a neutron star. In either case, gravity does not defeat the star completely. But if it has greater than about 5 solar masses, gravity wins out absolutely and forms a black hole.

The stars form the crucial evolutionary links in the chain of cosmic evolution. They produce light and warmth vital to life on any planet around them. They create the elements out of which planets are made. How? By fusion reactions. What ignites these reactions? Gravitational contraction. Gravity, the driving force of the astronomical universe, squeezes matter into heavier elements in the hearts of stars.

Violence marks the death of stars. For stars of about 1 solar mass, the death rattle involves only a small fraction of the star's mass. For massive stars, almost the entire star participates in the cataclysm of a supernova. This destructive violence has constructive ends: Heavy elements are made and ejected into the interstellar medium from which stars and their planets are born. So stars go from dust to dust again. And we partake of this cosmic recycling flow.

part 4

How are the stars arranged in space? Astronomers have only recently answered this question. We have found that the sun resides in a vast pinwheel of stars—over 100 billion of them. Over 120,000 ly in diameter, this vast system is the Milky Way Galaxy. Chapter 18 deals with the Galaxy, based on our understanding of the physical nature and evolution of stars.

Beyond the Galaxy, billions of other galaxies inhabit the reaches of space—the universe is truly a universe of galaxies, which are described in Chapter 19. When you look at one galaxy through a telescope, it may appear serene. But new telescopes have recently revealed that many galaxies, especially in their cores, generate energy violently. Some appear to be blasting matter outward in narrow jets that contain particles traveling at the speed of light. This violent face of the universe (Chapter 20) is an important subtheme of cosmic evolution.

I then return to cosmology (Chapter 21) to investigate the physical evolution of the universe since the Big Bang. You will see that as the cosmos expanded and cooled, various kinds of matter froze out in forms that we can see today—or may not have seen yet! Here we find the closest connect between the smallest pieces of the cosmos—elementary particles—and the universe as a whole. That connection is made at the Big Bang. Then another look at the future of the universe and evidence that indicates that it will expand, more and more slowly, forever. The cosmos started with a bang but may end with a whimper.

The evolution of the universe, stars, and galaxies sets the stage for the chemical and biological evolution (Chapter 22) that resulted in ourselves and other living creatures on the earth. Natural processes bind simple chemicals, abundant in the cosmos, into more complex ones that serve as the foundation of life. From those complex molecules, simple life formed early on in the history of the earth. If that process crowns the course of cosmic evolution, then other planets in our Galaxy should have life on them, too.

Finally, I speculate on the future of humankind and our place in the cosmos—the place to which cosmic evolution has carried us and from which we see the future only dimly.

Galaxies and cosmic evolution

18 / The evolution of the galaxy

A broad and ample road, whose dust is gold,
And pavement stars, as stars to thee appear
Seen in the galaxy, that milky way . . .

JOHN MILTON: Paradise Lost

Learning objectives

After studying this chapter, you should be able to:

1. **Explain at least one astronomical difficulty in trying to figure out the structure of the Galaxy from our location in it.**
2. **Name the important spiral arm tracers and how they are used to map spiral structure.**
3. **Present the observational evidence for the Galaxy's having a spiral structure; that is, describe what methods astronomers use to work out the positions of spiral arms.**
4. **Sketch the rotation curve of the Galaxy, and describe how to find from it the approximate mass of the Galaxy.**
5. **Describe the sun's orbit around the galactic center and explain the techniques used to find the distance and speed of this orbit.**
6. **Explain how radio astronomers used 21-cm line observations to trace spiral arms, indicate the limitations of their method, and explain its advantage over optical observations.**
7. **Describe the contents of a typical spiral arm.**
8. **Describe the evolution of spiral arms in terms of the density-wave model for spiral structure.**
9. **Outline the evolution of the disk of the Galaxy.**
10. **Outline a model for the evolution of the halo of the Galaxy.**
11. **Make a rough sketch of the entire Galaxy, from a top and a side view, labeling the disk, spiral arms, halo, globular clusters, nucleus, and the sun's position.**
12. **Argue that a significant amount of the Galaxy's mass must exist in the halo in an unseen form.**

Central question

What evolutionary processes induce the structure of the Galaxy?

18.1
A typical spiral galaxy, Messier 101. (Courtesy Kitt Peak National Observatory)

Like a majestic cosmic pinwheel, our Milky Way Galaxy spins slowly in space. Our sun orbits the center of with over *100 billion* other stars. Since the sun is about 30,000 ly from the nucleus, it completes one revolution roughly every 250 million years.

What is the structure of this enormous system of stars? Optical astronomers can probe the structure near the sun, and radio astronomers can study regions farther away. These investigations show that the Galaxy does have a spiral structure, a pattern inferred in part from spiral designs observed in other galaxies (Fig. 18.1). The details are not yet in, but astronomers have been able to establish the broad outlines of the Galaxy's structure—a remarkable achievement, considering that we are buried within the Galaxy.

How does the Galaxy evolve? Recent theoretical work has produced new ideas about the physical cause of the Galaxy's spiral structure. The leading idea, called the *density-wave model*, links many of the galactic entities in the disk by the process of galactic evolution. The Galaxy's structure evolves—but at a rate so slow that we won't see any changes in our lifetimes.

In the past, the Galaxy looked much different than it does now. This chapter takes a brief excursion into our Galaxy's history to see how our spiral Galaxy evolved.

18.1 An overview of the Galaxy's structure

The Galaxy has three main parts: a nucleus (nuclear bulge plus core), a disk, and a halo (Fig. 18.2). The *disk*, the main body of the Galaxy, has a diameter of some 120,000 ly and a thickness of about 1000 ly. Population I stars and interstellar clouds of gas and dust inhabit the disk, the gas extending out farther than the stars. The sun resides a little above the central plane at a distance of approximately 30,000 ly from the Galaxy's center.

The *nuclear bulge* encases the central regions of the Galaxy, including the mysterious nucleus. The bulge is about 12,000 ly in diameter and 10,000 ly thick; it contains old Population I stars. The *halo* encircles the nuclear bulge and the disk. Globular clusters, containing Population II stars, make up the most obvious material in the halo, which has a diameter of at least 120,000 ly.

Note that different types of stars inhabit each part of the Galaxy: young, metal-rich Population I stars in the disk; old, metal-rich Population I stars in the bulge; and very old, metal-poor Population II stars in the halo.

We are in a bad position to observe the Galaxy's structure, for we reside in the Galaxy's disk. Imagine, for example, that you are watching

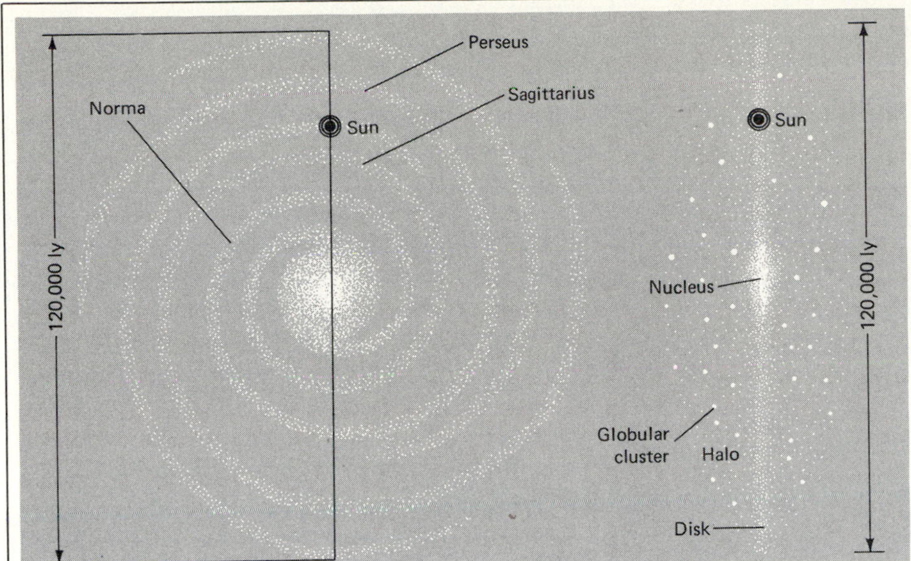

18.2

The main parts of the Milky Way Galaxy: nucleus, halo, and disk. The sun is about 33,000 ly from the center. The spiral pattern is two-armed near the center and four-armed in the outer regions. The nuclear region may have a bar across it.

a half-time show at a football game. The band has set up some elaborate formation. Up in the stands, you can easily observe what the formation looks like. But suppose you were down on the field instead, at the edge of the formation. It would at first appear to be a jumble! You could eventually figure out the shape if you could find the distances to all the band players. You could make a map of their positions by plotting the distances and directions of each person. But suppose you had to do this mapping in a fog so dense that only the closest people were visible. You would need some other method of estimating distances—perhaps by the intensity of the sound of the instruments that come through the fog.

Optical astronomers, who try to find the distances to features that mark spiral arms, find their view blocked by interstellar dust. Radio waves get through. So radio astronomers can pick up the radio emission from clouds of gas that probably mark spiral arms. But they have more difficulty in determining distances than the optical astronomers do. The results from the two techniques do not agree in all details but have uncovered the following structure.

The sun lies on the inner edge (Fig. 18.3) of a poorly defined structure called the *Cygnus arm*. (*Note:* Astronomers use the word *arm* to indicate a well-defined *segment* of a larger overall spiral-arm structure. I'll use it the same way, but you should be aware that we are discussing pieces of larger structures.) Outward, about 10,000 ly from the sun, lies an arm parallel to the Cygnus arm; since the best-observed portion is in the direction of the

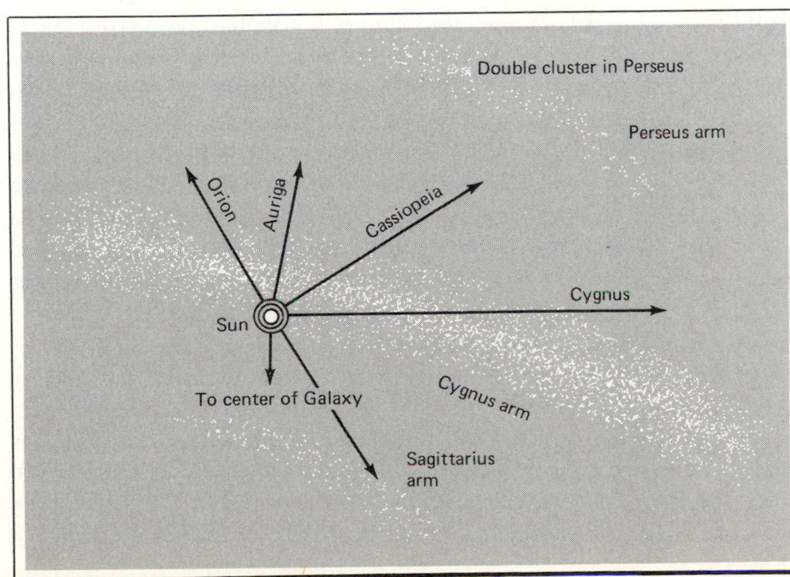

18.3

Spiral-arm segments near the sun, which lies on the inner edge of the Cygnus arm in a branch called the Orion spur.

constellation Perseus, it is called the *Perseus arm* (Fig. 18.4). At about 6000 ly interior to the sun curves the *Sagittarius arm*. Some evidence indicates that another arm may lie 13,000 ly from the sun toward the galactic center; it is called the *Centaurus arm* or the *Norma-Centaurus arm*. Finally, encircling the galactic center at a distance of about 1000 ly from the center is the *expanding arm*, so called because this innermost arm appears to be moving toward us at roughly 50 km/sec, expanding away from the center.

We cannot yet clearly see the whole scheme. But some other spiral galaxies have two arms that wind around the nucleus. We infer that our Galaxy also has two major arms (interior to the sun's orbit) and perhaps four arms (exterior to it) wound around the nucleus. The arm segments mentioned earlier are probably parts of the major arms. In particular, the four arms beyond the sun seem to be extensions of the Perseus, Cygnus, Orion, and Norma-Carina arms. The coherent, spiral pattern is clearest in these outer parts.

I've talked so far of spiral arms but haven't really defined them. A spiral arm contains many *O*- and *B*-stars. Also the overall density of material—gas, dust, and stars —inside a spiral arm is roughly 10 times that in the region between arms. The spiral arms contain most of the gas, dust, and young stars in the Galaxy. Near the sun, for example, about half of the matter is in gas and dust, the rest in stars. In contrast, for the Galaxy as a whole, only a few percent of all the material is in the form of gas and dust.

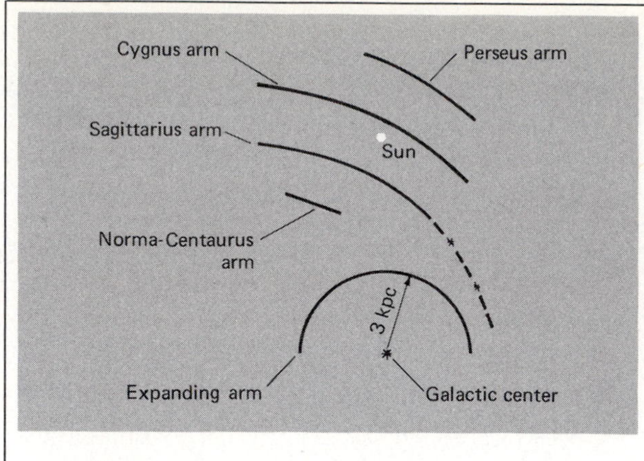

18.4
Wide-angle view of spiral arms in the sun's vicinity. The arms are named after the constellations in whose directions they lie as seen from the earth.

18.2
Galactic rotation: Matter in motion

The sun and its nearest neighboring stars are only a drop in the galactic bucket. The local stars appear to be moving in a variety of directions at a variety of speeds. In relation to the other stars, the sun moves at a speed of about 20 km/sec. But how to find out what motion the sun and nearby stars share in relation to the center of the Galaxy? And what about the orbital motions for matter in other parts of the Galaxy?

The sun's orbit and the mass of the Galaxy / How fast does the sun move around the Galaxy? We use a variety of indirect approaches to find this out. One method utilizes the motions of globular clusters. The globulars seem to orbit the Galaxy in random orbits, with a roughly spherical distribution around the nucleus. With respect to the nucleus, the average motion of all globular clusters is roughly zero. In other words, the system of globulars has no overall rotation about the galactic center (although individual clusters move rapidly, some in one direction, some in another). A study of the radial velocities of the globular clusters enables us to find the sun's motion with respect to the system of globulars. Because the system of globulars has no rotational motion with respect to the nucleus, the sun's motion found in this way is its motion with respect to the Galaxy's center. Such an analysis shows that the sun revolves around the center at about 250 km/sec.

How far is it to the sun from the Galaxy's center? That's a tough question to tackle because we cannot see the center optically. However, we can look above and below the galactic plane, where the obscuration is less, to observe objects thought to be symmetrical about the galactic center.

For example, we can use the RR Lyrae variables in globular clusters. These are Population II stars; have periods of light variation that are typically ½ day, and range in spectral type from *A*2 to *F*6. It turns out that all RR Lyrae stars have essentially the same mean luminosity no matter what their period: 50 solar luminosities. So once an RR Lyrae star is identified by the shape of its light curve, we can easily work out its distance from a measurement of its flux and the inverse-square law for light. By this technique, recent work has found the sun's distance from the galactic center to be approximately 28,000 ly, at least within a range from 24,000 to 33,000 ly. I'll use 30,000 ly, as long as you realize that the actual distance to the center is somewhat uncertain.

This technique, developed Harlow Shapley (1885–1972) around 1915, marked a crucial step in our understanding of the Galaxy and the sun's location within it (Fig. 18.5). Let's look in detail at

18.5
Harlow Shapley. (Courtesy Harvard College Observatory)

Shapley's argument, which rests on the simple observation of the distribution of globular clusters in the sky. Globular clusters concentrate in the southern sky (Fig. 18.6). From what vantage point do we, whirling around the sun, view these groups of stars? Shapley knew that our Galaxy had the shape of a flattened disk. He assumed that the globular clusters had a uniform distribution around the Galaxy's nucleus in a huge sphere that outlines the halo.

Now to infer the size of the Galaxy and the sun's location. Suppose the sun were located in the center of the Galaxy (Fig. 18.7a). Trace lines of sight in a number of directions. Because we have assumed a central vantage point in a uniform distribution of objects, every line of sight you chose should intercept the same number of globular clusters. So, the expected distribution would be uniform over the sky, but this uniformity does not in fact occur. Shapley therefore chose to give up the central location of the sun in the Galaxy (Fig. 18.7b). Now some lines of sight cut longer distances than others through the globular clusters, so more clusters are seen in these directions than along other lines of sight. The expected distribution is not symmetrical, but is most concentrated in the direction of the galactic center. So with the sun away from the center, the predicted result matched the observational one.

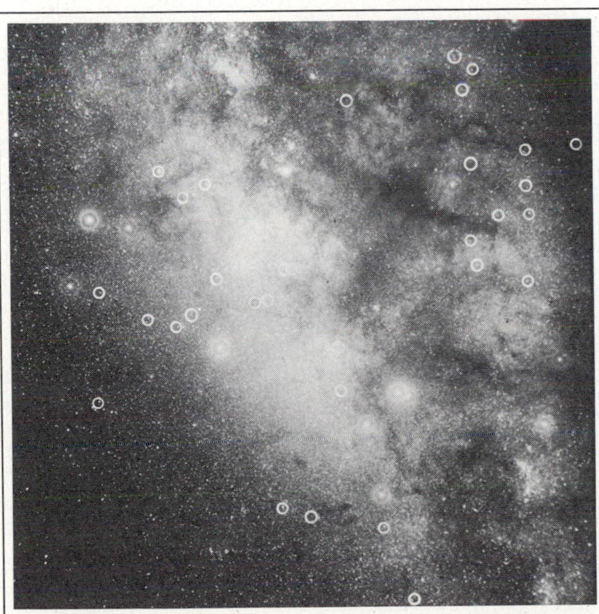

18.6
The concentration of globular clusters (within white circles) in the direction of the Galaxy's center. (Courtesy Harvard College Observatory)

The distances to the globulars can be worked out from RR Lyrae stars. Then the size of the sphere of globulars marks the extent of the halo (and so the size of the Galaxy), and the distance of the sun from the center of this sphere indicates the distance of the sun from the Galaxy's center.

Knowing the sun's velocity and distance, we apply Kepler's third law to deduce the mass of the Galaxy (Focus 18.1). The result, about 10^{11} solar masses, refers *only to the mass interior to the sun's orbit.*

Rotation curve / With the sun's orbital distance and velocity known, we find the *galactic rotation curve*—how fast an object some distance from the galactic center revolves around it. The rotation curves tells us the overall distribution of matter in the Galaxy, because gravity controls those orbital motions.

Imagine that the Galaxy's mass is mostly concentrated in the nucleus—a setup that resembles the solar system. Although the planets attract each other, each orbits about the sun. Each one responds to the overwhelming gravitational attraction of the central sun. The stars in the galactic disk revolve about the nucleus of the Galaxy much as the planets revolve around the sun. The forces are similar, so the motions should be similar. The stellar motions should follow Kepler's laws (see Section 3.5), and the velocities of the stars should decrease with increasing distance from the Galaxy's center—just as, for example, the orbital velocity of Mars is less than that of the earth.

In fact, the velocities don't follow Kepler's laws well, and that tells us an important fact about the Galaxy: The major part of the mass is *not* concentrated at the center, quite different from the case of the solar system. Of course *some* relationship exists between orbital velocity and distance

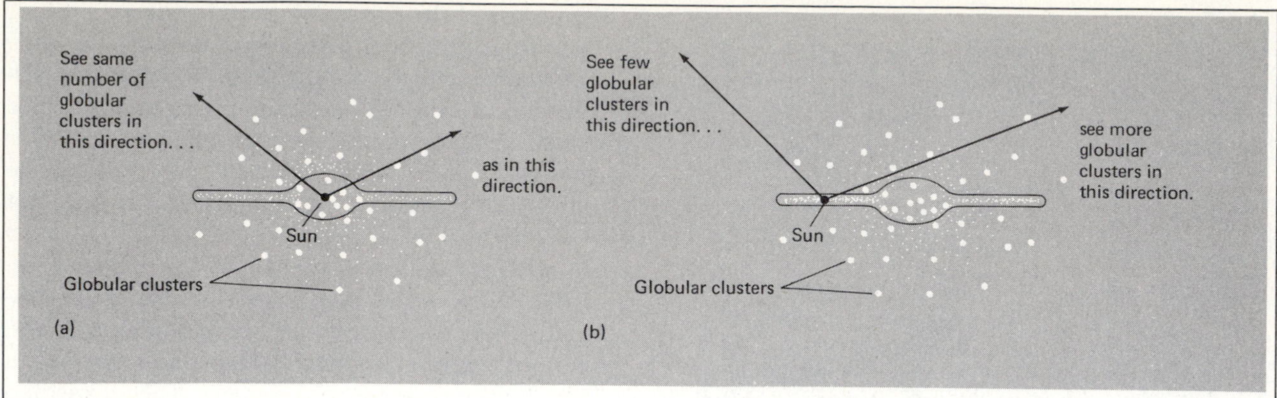

18.7

The sun's position in the Galaxy. Assume that the distribution of globular clusters is uniform around the center of the Galaxy. If the sun were at the center (a), *you would see roughly the same number of globulars in every direction in the sky. With the sun away from the center* (b), *the direction toward the center shows more globulars, as is the case.*

from the center (as expected from Newton's law of gravitation), and these differences in orbital velocities show up as systematic differences in Doppler shifts. Here we have direct evidence that the Galaxy rotates in a special way, where differing orbital speeds mark differing distances from the center.

A central concentration of mass results in Keplerian motion in the outer regions (Fig. 18.8a). In fact, the observed rotation curve (Fig. 18.8b) does not follow this simple pattern, according to 21-cm data. From close to the center out to 1000 ly, the curve rises steeply, then drops, bottoming out at about 10,000 ly. It then rises slowly and out to the position of the sun. In the outer parts of the Galaxy, from carbon monoxide observations of molecular cloud complexes, the curve rises more steeply beyond the sun, reaching almost 300 km/sec at 60,000 ly.

What does this curve say? Even the outer parts of the Galaxy do not revolve in a Keplerian fashion, so much of the Galaxy's material must lie out beyond the sun's orbit. From the rotation curve out to 60,000 ly, the Galaxy's mass is 3.4×10^{11} solar masses. So at least as much mass lies exterior to the sun as interior to it. Much of this matter is invisible. The Galaxy must have a massive halo of nonluminous matter.

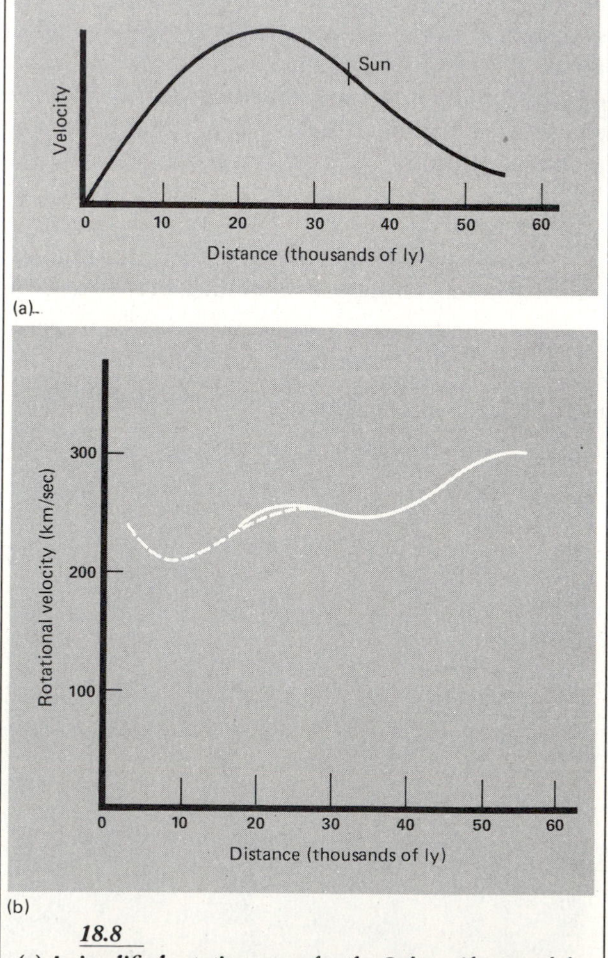

18.8

(a) *A simplified rotation curve for the Galaxy, if most of the mass were concentrated in the center.* (b) *Observed rotation curve for the Galaxy. The dashed part of the curve is from atomic hydrogen data, the solid part from carbon monoxide observations of molecular clouds. (Adapted from a diagram by L. Blitz, M. Fich, and A. Stark)*

18.3 Galactic structure from optical observations

Blue-sensitive photos taken of nearby spiral galaxies, such as M 31, the Andromeda galaxy, show the spiral arms distinctly. This appearance results

Focus 18.1

THE MASS OF THE GALAXY

The sun swings around the Galaxy at 250 km/sec at a distance of 33,000 ly from the center. Let's use this information to estimate the mass of the Galaxy. Assume that the sun moves in a circular orbit. Apply Kepler's third law in the form used for binary star systems to the Galaxy:

$$M_1 + M_2 = \frac{R^3}{P^2}$$

where R must be in units of AUs and P in years. The mass then comes out in solar masses. At 250 km/sec it takes the sun about 2.5×10^8 years to complete a circuit of the Galaxy. R is roughly 33,000 ly and 1 ly equals 6.32×10^4 AU; so 33,000 ly equals 2.1×10^9 AU. Then

$$M_1 + M_2 = \frac{(2.1 \times 10^9)^3}{(2.5 \times 10^8)^2} = \frac{9.3 \times 10^{27}}{6.3 \times 10^{16}} = 1.5 \times 10^{11} \text{ solar masses}$$

What is $M_1 + M_2$? It is the mass of the sun plus the mass of the Galaxy. The mass of the sun is so small compared with that of the Galaxy (just look at the result!) that we ignore it. So the Galaxy's mass is at least 1.5×10^{11} solar masses.

from *O*- and *B*-supergiants, H II regions, and Population I cepheids that cluster in the arms (Fig. 18.9). These objects are called *spiral tracers*.

How to apply these spiral tracers to our Galaxy to determine the location and extent of its spiral arms near the sun? It's not a simple operation! First, it requires an accurate technique for measuring the distance to each of the tracers. Second, optical observations are restricted by the blotting out of starlight by dust. Most of the interstellar dust lies concentrated in the galactic plane, so the sun sits in the thick of the interstellar smog. Third, the sun's location in the plane gives us a poor vantage point for seeing the Galaxy's spiral

18.9

Spiral arms of Messier 31, the Andromeda galaxy. Bright O- and B-stars define the spiral arms; dark, dusty regions line the inner edges of the arms. (Courtesy Palomar Observatory, California Institute of Technology)

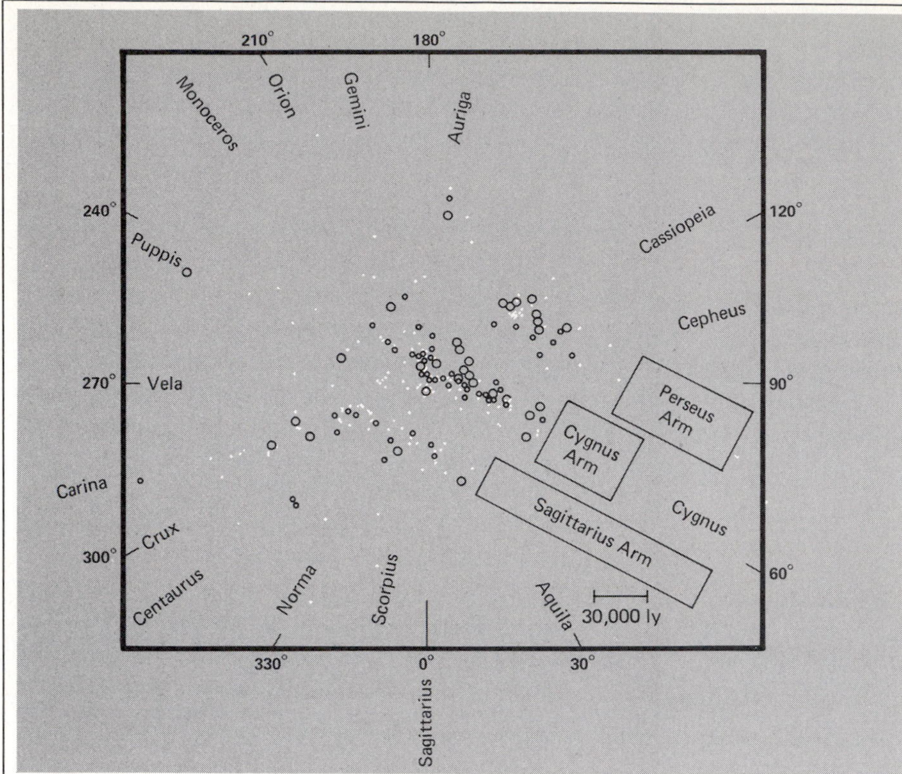

18.10
Spiral structure near the sun, as indicated by O- and B-stars, H II regions, and cepheids. Three arm segments (Perseus, Cygnus, and Sagittarius) are visible. (Based on data compiled by W. Becker and Th. Schmidt-Kaler)

structure, since we are forced to observe it edge on rather than face on.

H II regions and supergiant *O*- and *B*-stars trace the arms best because their high luminosities make them visible over large distances (Fig. 18.10). In addition, cepheids have been used effectively to delineate spiral features (Fig. 18.10). The cepheids have the advantage that their distances are easy to determine by the *period-luminosity relation*. This crucial distance-measuring technique is so important that you should know the details.

A variable star is one whose brightness changes with time. A *light curve* is a plot of the change in a star's brightness with time. A star whose light varies in a regular fashion is known as a *periodic variable*; Delta Cephei sets the standard for one such class of variables—called *cepheid variables* or *cepheids*—by the special shape of their light curves (Fig. 18.11).

For the cepheid variables, we find that a relationship exists between the luminosity of a cepheid variable and its period. This connection is called the *period-luminosity relationship* (Fig. 18.12). Basically, it shows that the longer the period of light variation of a cepheid, the more luminous it is.

We can use the relationship to find distances to cepheids. Here's how: (1) Find a cepheid (identifying it by its light curve); (2) measure its period of light variation, from peak to peak; (3) find the star's luminosity from the period-luminosity relationship; (4) measure the star's flux; and (5) calculate its distance from the inverse-square law for light.

This technique assumes that dust does *not* cut out the light from the cepheid. If it did, the flux would be less than it actually is, and you would estimate the star to be farther away than its actual distance.

Once we find the distances to cepheids, we can use that information along with their positions in the sky to plot out the locations of spiral arms.

You must take the optical maps with a bit of caution. The data extend only to about 14,000 ly

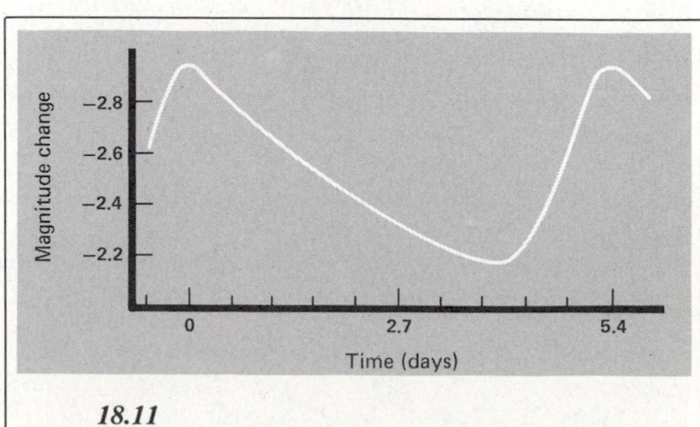

18.11
A light curve for a typical cepheid variable (Delta Cephei). It shows how the observed brightness varies with time.

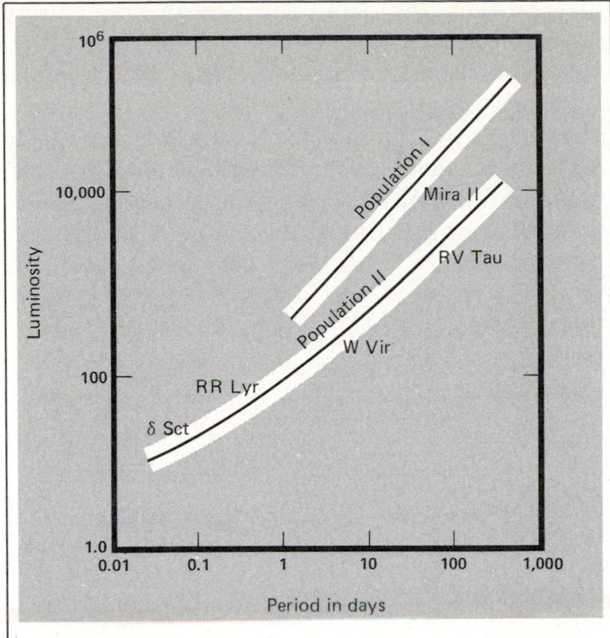

18.12
A simplified period-luminosity relationship for cepheids. Note that they fall into two groups, depending on their stellar population.

from the sun, and a larger range is needed to outline the spiral arms definitely. In addition, although the outline of spiral structure is assuredly correct, observations of other galaxies show that irregularities commonly occur. It's futile to draw a master diagram from optical data alone.

Although disagreements have arisen about the details, most optical astronomers concur that their investigations have found at least three major arm segments spaced about 7,000 ly apart. The Galaxy appears to have a spiral structure with much irregularity in the general pattern.

18.4 Exploring galactic structure by radio astronomy

The prime drawback to optical mapping of the Galaxy is obscuration from interstellar dust. Radio observations do not have this handicap because dust does not easily stop radio waves. Radio astronomers can reach far beyond the restricted range of the optical astronomers, even to the other side of the galactic nucleus. The 21-cm line from hydrogen (see Focus 15.2) is most useful to radio astronomers in diagraming the Galaxy. This radio emission comes from the concentrations of neutral hydrogen clouds, which make up the spiral arms.

We distinguish among spiral arms not only by looking in different directions (we are, after all, in the middle of the mess) but by looking at different *velocities* in the same direction. Not all 21-cm radiation arrives at the earth at exactly 21.11 cm. It is Doppler-shifted to different wavelengths because of the different velocities of the hydrogen gas clouds. These differences in velocities come mostly from the rotation of the Galaxy. So *if* we know how the Galaxy rotates (and that's the tricky part), we can translate 21-cm observations into a map of spiral structure.

What is the technique radio astronomers use? They look for 21-cm radiation from some specific positions in the Galaxy. If all the H I clouds in this direction were at rest with respect to the sun (that is, moving at the same velocity as the sun), all of their emission would pile up at exactly 21 cm. However, because of galactic rotation, the clouds along the line of sight have different radial velocities, hence different Doppler shifts, so their signals are received at slightly different wavelengths.

Here's a particular case looking outward from the sun (Fig. 18.13). Our line of sight intercepts three clouds at successively greater distances. The inner clouds travel faster than the outer ones. They all travel more slowly than the sun. So the difference between the sun's velocity and the inner cloud's velocity is the least, and its Doppler shift is the least. In contrast, the difference between the sun's velocity and velocity of the outer cloud is the most, so it has the greatest Doppler shift. Assume that each of these clouds corresponds to a piece of a spiral arm. How to arrange them at different distances from the sun?

Assume circular (or nearly circular) orbits. We know the sun's velocity around the Galaxy. When we observe 21-cm emission, we know the direction in which we are looking and can measure a radial velocity. From that radial velocity, we infer the rotational velocity of the cloud. We then look up this velocity on the galactic rotation curve to find the distance to which it corresponds.

In essence, the galactic rotation curve tells us that at a given direction in space, a certain radial velocity corresponds to a specific distance from the sun. Scanning around the galactic equator, we make a series of 21-cm observations. These can be connected to trace a spiral arm and so outline the Galaxy's structure.

Don't think this is an easy job. Conflicting radio maps have been drawn by different investigators. The heart of the problem is that the neutral gas clouds don't follow the simple scheme of circular rotation; in addition to their circular motion, they have their own random motions. Unfortunately, such noncircular motions lead to

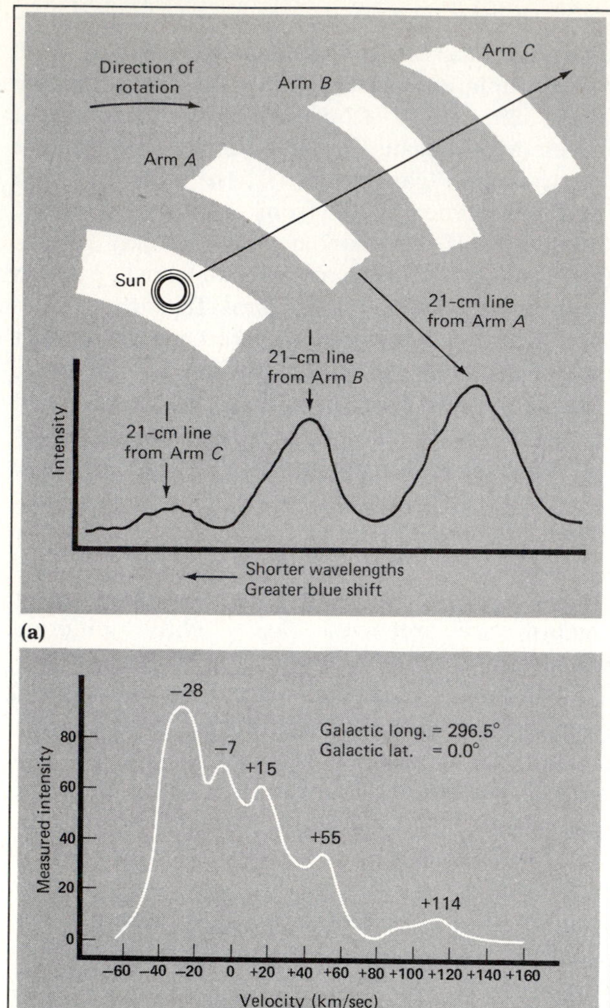

18.13

(a) ***Using 21-cm emission to trace out spiral structure. Consider looking out of the Galaxy through three spiral arms (A, B, and C). The outer arms move more slowly than the inner ones. So the velocity difference between us and A is smaller than that between us and C, and the blue shift from A is less than that from C.*** **(b)** ***Actual observations of 21-cm emission at galactic longitude 296.5°, latitude 0.0°. Five distinct clouds appear at velocities of −28, −7, 15, 55, and 114 km/sec. (Data from NRAO)***

incorrect distances and so disrupt the unity of the neutral-hydrogen spiral-arm map.

Despite such problems, the radio and optical maps coincide fairly well (Fig. 18.14). The optical results are most reliable near the sun, where the radio approach is most subject to error. Conversely, at distances greater than 13,000 ly, at least in the inner parts of the Galaxy, the radio astronomers probe regions inaccessible to optical astronomers. The two techniques complement each other. And the structure of the Galaxy derived from them shows through in its broad outline, although messy in the region interior to the sun.

Beyond the sun, the spiral-arm pattern appears more clearly, thanks to millimeter observations of giant molecular clouds. Very young stars are born from such clouds, so they make good tracers of spiral arms. Using the same technique as for the 21-cm hydrogen line, and complementing those observations, the new work shows the four arms mentioned in Section 18.1.

18.5 *The evolution of the spiral structure of the Galaxy*

For many years astronomers supposed that the spiral arms in our Galaxy and others were material arms, a coherent bunch of objects—stars, nebulas, gas, dust—somehow physically held together. Such a point of view faced two questions: (1) What holds the material in an arm together, and (2) how does an arm persist for a long time?

The persistence of an arm is hard to explain because it winds up; the outer parts of the arm rotate more slowly than the inner ones. So after a few rotations of the Galaxy, the arms should disappear. The Galaxy has turned about 20 times since the origin of the solar system; the arms are still there!

Here's an analogy to this wind-up problem. Imagine you and two friends are going to run around a track. You station yourself in the middle, one friend a few meters in from you, closer to the track's center, and the other a few meters out from you. You start running, lined up. Now insist that the friend inside run around faster than you and the one outside slower. You can guess what will happen after one or two laps: The line-up will be disrupted. The same would happen to spiral arms, if they were material arms, after a few rotations.

Astronomers have been struck by the persistence of spiral arms. We can't tell this from our Galaxy alone, for it could be that we are observing at a very special time, soon after the formation of the arms. But that would not likely be true for all galaxies. Of the brightest galaxies in the sky, over 60 percent have a spiral form (Fig. 18.1). How does this tell us that spiral arms are persistent phenomena? Assume that the spiral galaxies we see are about 10 billion years old. Consider that a galaxy's spiral structure would disappear after a few rotations. We would expect to see almost no spiral galaxies now; their rotations would have rapidly destroyed the spiral structure. Seeing so many galaxies to be spiral now implies that the structure lasts for at least some few billions of years.

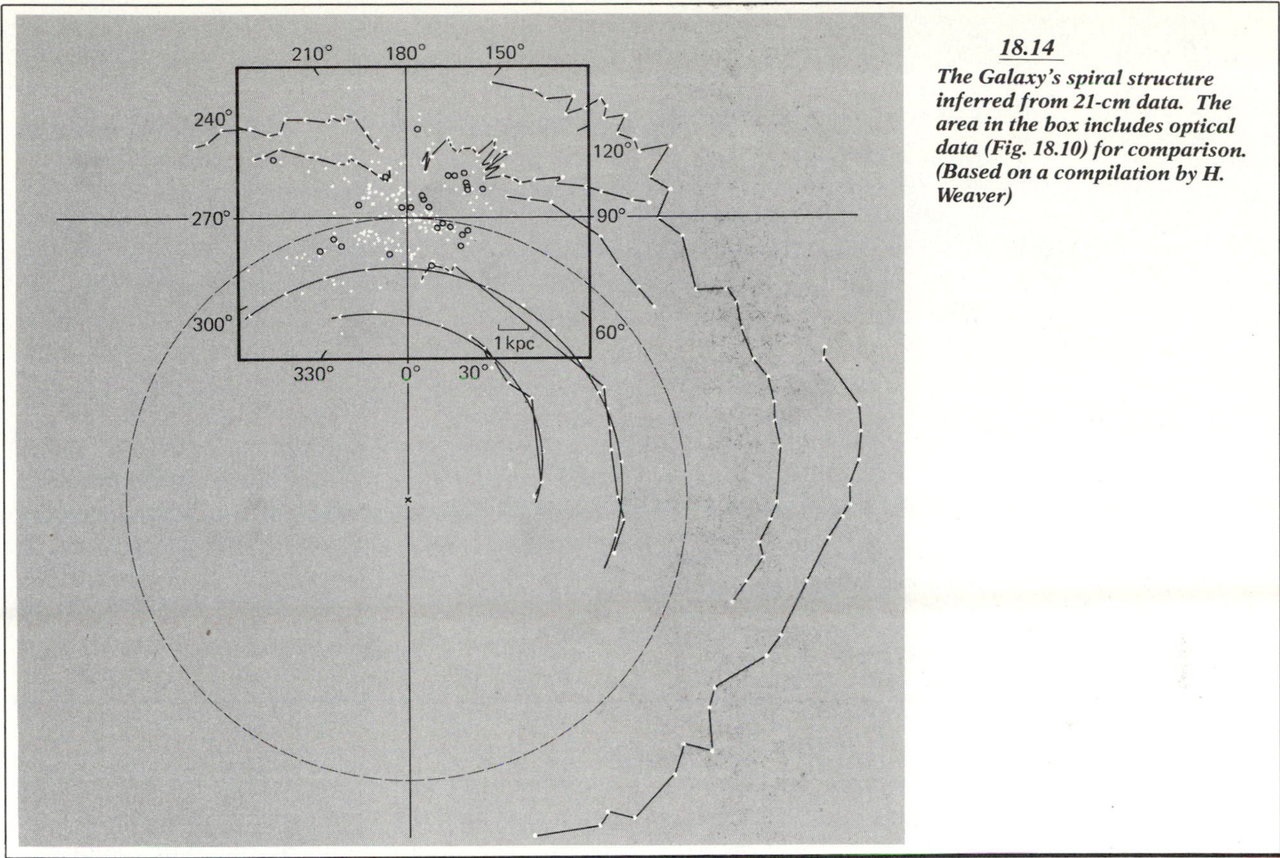

18.14
The Galaxy's spiral structure inferred from 21-cm data. The area in the box includes optical data (Fig. 18.10) for comparison. (Based on a compilation by H. Weaver)

How to explain this persistence? The contemporary attack pictures spiral arms not as material arms at all but rather as the result of a wave of higher density moving through the Galaxy's disk. This wave produces all the signposts of a spiral arm: young stars, H II regions, lanes of dust. None of these objects lasts very long. As they die and the density wave moves, new spiral-arm tracers are born from the interstellar medium. So a spiral arm always contains the same *kinds* of objects, but not the *same* objects. Any particular arm is a transient phenomenon. Individual objects rotate at the speed appropriate for their distance from the center, but the *wave* pattern rotates with a constant angular speed and does not wind up. This approach is called the *density-wave model* of spiral structure. Of the models proposed to date to explain spiral structure, it best predicts and describes the overall scheme.

What's a density wave? A sound wave is a density wave. Push against air molecules to compress them together. This first group of molecules bangs into adjacent ones in the direction of their motion, which transfers the compression to the next bunch of molecules. As this compression (density wave) travels forward, it leaves behind a trough of lower density. Two important points here: (1) A sound wave requires a source to start it, and (2) the high-density part of the wave persists even though the specific particles that make it up change at different points in the medium.

Here's an analogy: Suppose you are driving on a mountain road filled with traffic (Fig. 18.15). Everyone moves along happily at the speed limit. Ahead, an overloaded truck can go about half the maximum speed. Cars jam up just behind the trucks as the drivers wait for a clear road ahead in order to pass. When they do pass, they move along again at the speed limit, leaving the poor trucker behind. Imagine that you watched this situation from the air and concentrated on the motion of the cars. You'd see a denser region of cars just behind the truck, where they pile up for a short time; the truck moves down the road at half the average speed of the cars. You would also note that the jam-up persists, even though it does not contain the same cars. New cars get caught up in it as other cars move out.

You can think of the cars as the stars and the interstellar medium in the galactic plane, and the jam-up as the visible effect of a moving density

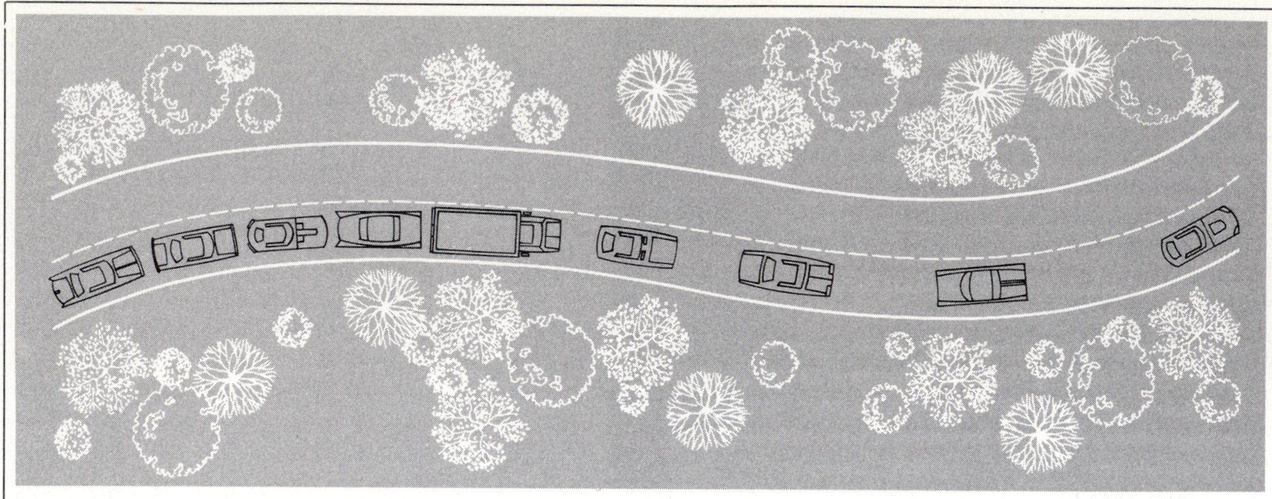

18.15
A highway bottleneck as an analogy to a density wave. The slow-moving truck has cars jammed up behind it waiting to pass. The jam is always behind the truck, but it consists of different cars at different times. From above, the jam moves more slowly than the average speed of the cars; it is a density wave.

wave. The jam-up creates a region in the disk of increased density of stars and gas—that's a spiral arm.

To apply this density-wave idea to the structure of the Galaxy *assumes* that a two-armed spiral density wave sweeps through the galactic plane. Although the wave's origin is not explained, once formed, it persists for a fairly long time. The gas in the disk piles up at the back of the wave. The buildup of pressure and density heats up the gas suddenly so that a shock wave forms along the front of a density wave. (You can make a shock wave by plowing through any medium faster than the speed of sound in that medium. For example, traveling through the air in a jet plane faster than the speed of sound creates a shock wave called a sonic boom.) The density wave travels through the gas faster than the sound speed, so a shock wave forms. The compression at the shock squeezes neutral hydrogen clouds together. This shock may initiate the collapse of the clouds to form giant molecular cloud complexes, which in turn form young stars and H II regions. Such squeezing also helps to make dust out of the gas, and a thin dust lane forms along the shock front (Fig. 18.16). The compression of the interstellar medium by the density wave forms the features associated with a spiral arm.

Warning: Just because I've been emphasizing luminous *O*- and *B*-stars (since they are good tracers of spiral arms), don't be misled into thinking that a density wave initiates the formation of

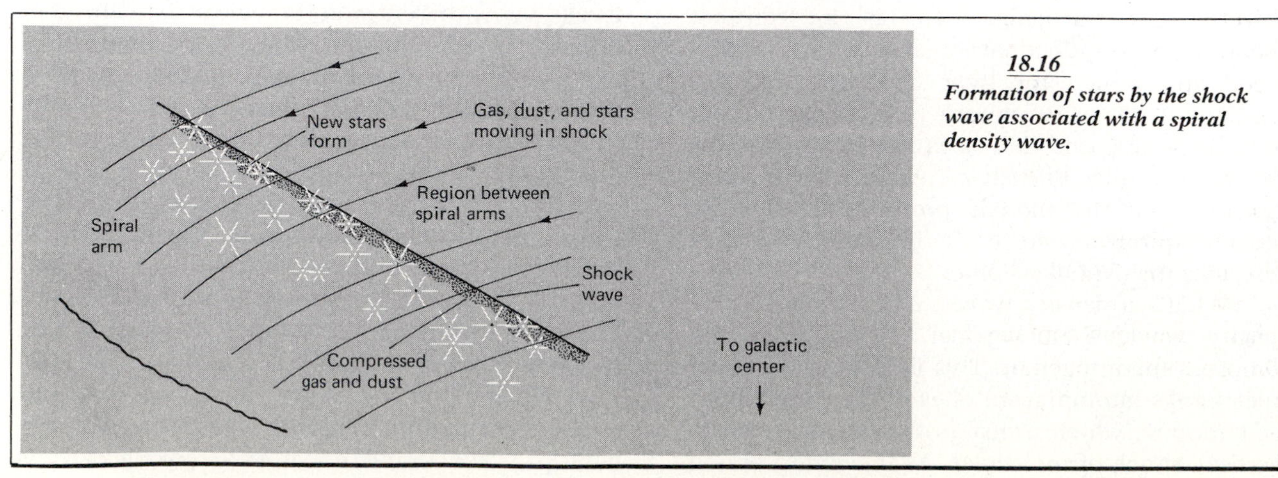

18.16
Formation of stars by the shock wave associated with a spiral density wave.

only *O*- and *B*-stars. It probably prompts the formation of stars with all possible masses. The massive ones die quickly compared with the less massive ones that are greatest in number. The less massive stars remain after the density wave moves on, but they are less conspicuous because of their low luminosities.

During the short lifetimes of the newly formed *O*- and *B*-stars, the density wave moves only a short distance. So these stars, while they last, mark the spiral arm clearly. As the density wave moves on, it provokes the formation of more stars. These take the place of the ones that have rapidly faded out. So the spiral arms persist by a continual destruction and creation maintained by the density wave.

How well does the density-wave model describe the observed spiral structure? First, it outlines the grand scheme of a overall spiral pattern that we observe in other galaxies and in our own. Second, it explains the persistence of the spiral arms in the face of galactic rotation. Third, it predicts the general features of a spiral arm. So the density-wave model succeeds fairly well in explaining the prominent features of spiral structure.

However, the model so far falls flat on a number of points. It does not explain the origin of the density waves. Nor does it clearly work out what keeps the density waves going. As the density waves ripple through the interstellar medium, they lose energy and should dissipate in about a billion years. But as evidenced by the abundance of spiral galaxies, the density waves—if they are the correct explanation—must last longer than a billion years. Some mechanism must keep supplying energy to maintain them.

To sum up the situation with the spiral density-wave model, recent observations leave little doubt that such spiral density waves exist and indeed are fairly common, but we don't yet know why. This is a puzzle to be solved in the study of our Galaxy and others.

18.6 The center of the Galaxy

Although dust largely obscures the center of the Galaxy, a few regions of low absorption open up optical glimpses of the nuclear region. In addition, we can surmise the nature of the stars in the nucleus from observations of the nuclei of other galaxies. These two kinds of observations imply that the nucleus contains mostly old Population I stars densely packed together. So jammed are these stars that if you lived in the nucleus, the nighttime sky would be as bright as twilight on the earth.

Radio, infrared, and X-ray observations can probe the nucleus. They show that the heart of the Galaxy is a bizarre place: Not only does it contain many very old Population I stars, but also some very young supergiant *M*-stars and *O*-stars. Motions of gas here suggest a high concentration of mass at the very center—perhaps a black hole. In the very center of the Galaxy lies an ultrasmall (by astronomical standards) radio source less than 140 AU (about 10 light *hours*) in size.

Radio observations / Let's look first at the continuous emission of the galactic center (Fig. 18.17). An intense radio source lies smack in the direction of the center. It is called *Sagittarius A* (*Sgr A* for short). Clustered around Sgr A—and all lying more or less along the galactic plane (or equator)—is a string of radio sources. When investigated at different radio wavelengths, these sources appear to have characteristics of H II regions: hot ionized gas around young *OB*-stars. The total extent of this region is about 300 by 850 ly, and the ultraviolet energy output from the *OB*-stars needed to keep the region ionized is at least 2×10^{33} W, or, 5 million solar luminosities.

Sgr A itself is a bit different. Some of the radio emitted here is from ionized gas. But some also comes from high-energy electrons traveling through a magnetic field—synchrotron emission (see Focus 17.1). A high-resolution radio map (Fig. 18.18a) shows that Sgr A actually consists of two separate radio sources: One, called *Sgr A East*, emits by the synchrotron process; the other, *Sgr A West*, seems more like a giant H II region. Sgr A West is associated with an agglomeration of infrared sources (to be explained shortly). Within Sgr A West lies a pointlike radio source less than 0.1″ in diameter, which appears to mark the actual core of the Galaxy. Its emission is nonthermal—probably synchrotron emissson. The ionized gas here, which amounts to a few million solar masses of material, seems to be rotating at about a few hundred kilometers a second.

The VLA has recently mapped the galactic center region (Fig. 18.18b and c). This map has been computer processed to take away the nonthermal emission from the point source within Sgr A West. What remains is the thermal emission from the inner 10 ly of the Galaxy. A curious aspect of the emission is that it has a spiral shape, as if two opposing jets of material are spurting from the core. (Chapter 20 describes the significance of jets emerging from the nuclei of galaxies.)

What about molecular radio emission? Sgr A gives off radio line emission from molecules such as CO. The observations indicate that this

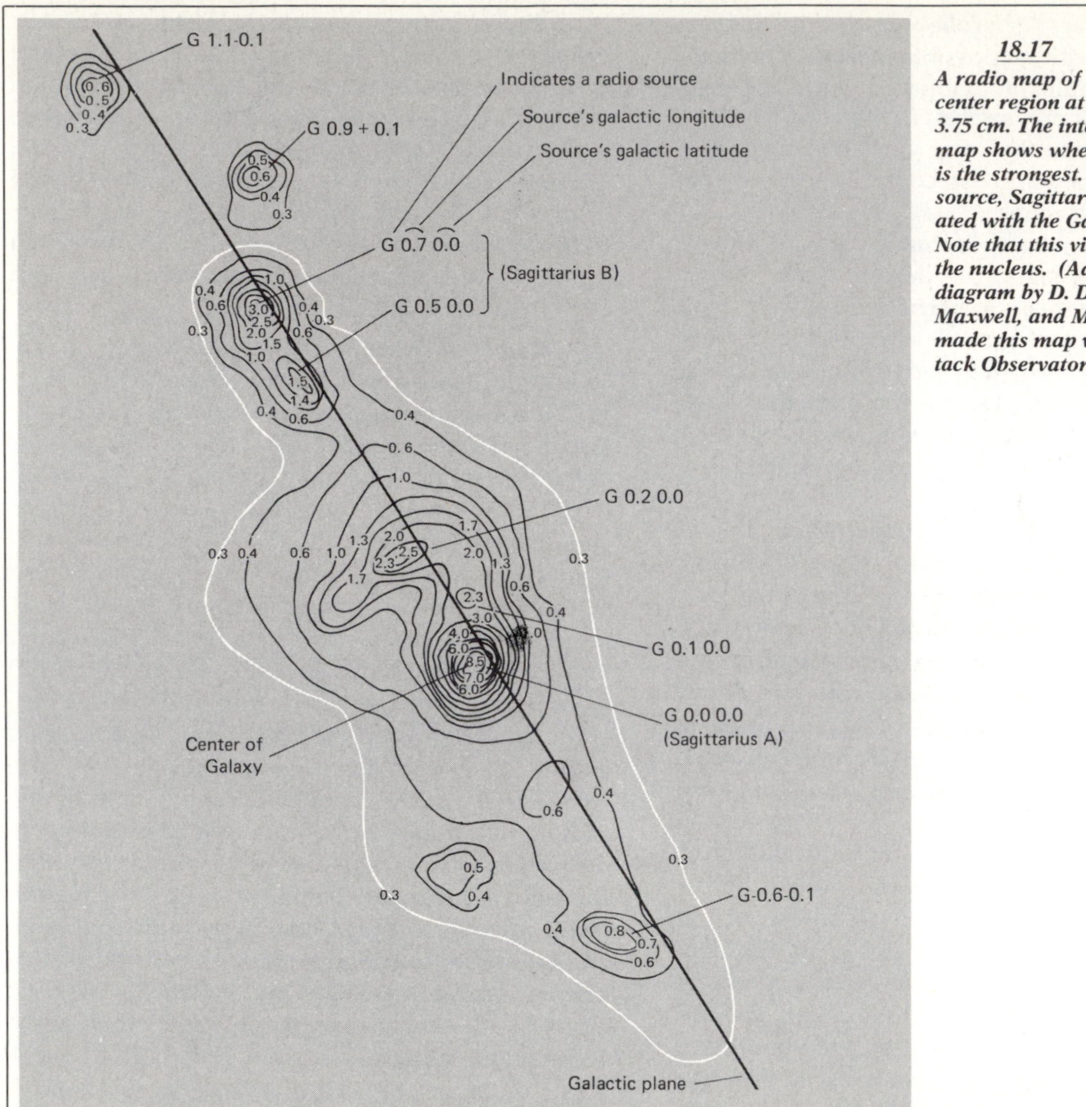

18.17
A radio map of the galactic center region at a wavelength of 3.75 cm. The intensity contour map shows where the emission is the strongest. One strong source, Sagittarius A, is associated with the Galaxy's center. Note that this view is edge on to the nucleus. (Adapted from a diagram by D. Downes, A. Maxwell, and M. L. Meeks, who made this map with the Haystack Observatory antenna)

molecular cloud, which may contain as much as a million solar masses of material, is associated more with Sgr A East than with Sgr A West, so it is not right at the center of the Galaxy.

Infrared observations / Early infrared observations showed that the galactic center region emits strongly at 2.2 μm (Fig. 18.19). The most intense part of this emission coincides with Sgr A. What is the source of this radiation? Simply the combined 2.2-μm emission from all the old Population I stars (probably mostly *K*-giants) that inhabit the galactic nucleus.

A map of the same region at 10 μm looks quite different (Fig. 18.20). Here we are seeing the infrared emission from dust that is heated by the radiation from stars. Some of the heating radiation comes from the old Population I stars. But some also derives from high-luminosity *O*-stars; the condensations in the 10-μm map are probably the locations of newly formed *O*-stars. These regions have diameters of less than a few light years—the same size as small H II regions. The combined luminosity from them, in the range from 2 to 20 μm, is roughly a million times that of the sun. In fact, over the entire infrared range, the galactic center emits about 100 million times the sun's luminosity—probably from dust heated by *O*- and *B*-stars.

Additional information comes from observations of an infrared emission line at 12.8 μm arising from singly ionized neon (Ne II). The emission comes from ionized gas in the nucleus, and dynamic information about the core can be inferred from the width and Doppler shift of the line.

What do these observations show? First, that the ionized gas concentrates in a region about 5 ly in radius. Second, the lines are very broad, an

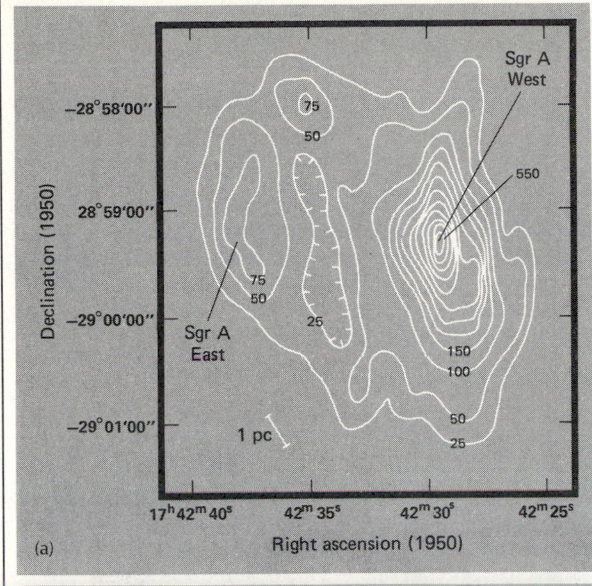

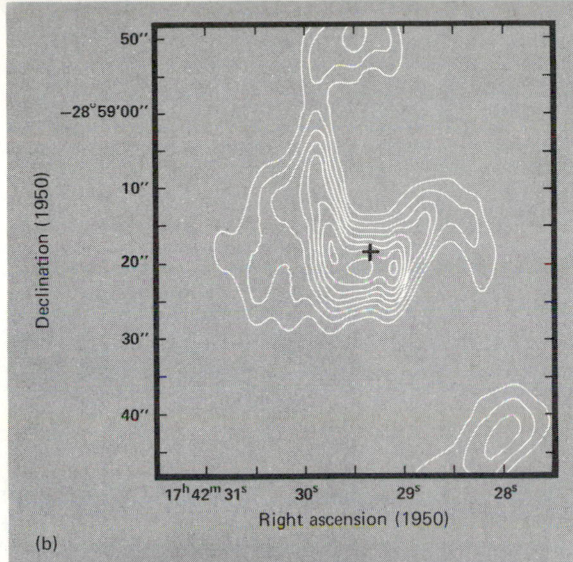

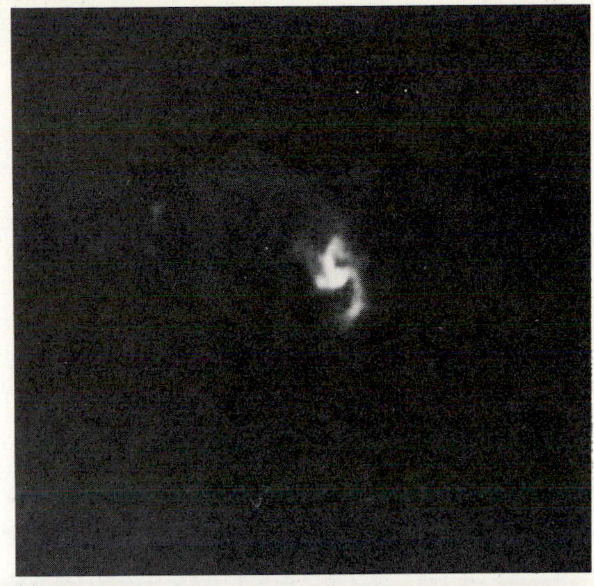

(c)

18.18

(a) *A high-resolution map of the Sagittarius A (Sgr A) source. Sgr A West appears to be a giant H II region and is at the Galaxy's core. Sgr A East is a nonthermal source. (Based on observations by R. D. Ekers, W. M. Goss, U. J. Schwarz, D. Downes, and D. Rogstad)* (b) *A map of the galactic center made with the VLA at 5 GHz. The cross indicates the position of the pointlike nonthermal source within Sgr A West. This may be the Galaxy's core. (Adapted from a diagram by R. Brown, K. Johnston, and K. Lo,* Astrophysical Journal, *vol. 250 (1981), p. 155, copyright © 1981 by the American Astronomical Society)* (c) *A computer-generated radio photo of Sgr A made from VLA observations at a wavelength of 6 cm. The area in this picture is about the same as that for* (a)*. Sgr A West shows up clearly as the white region at the right. (Courtesy R. D. Ekers; observations by R. D. Ekers, U. J. Schwarz, and W. M. Goss)*

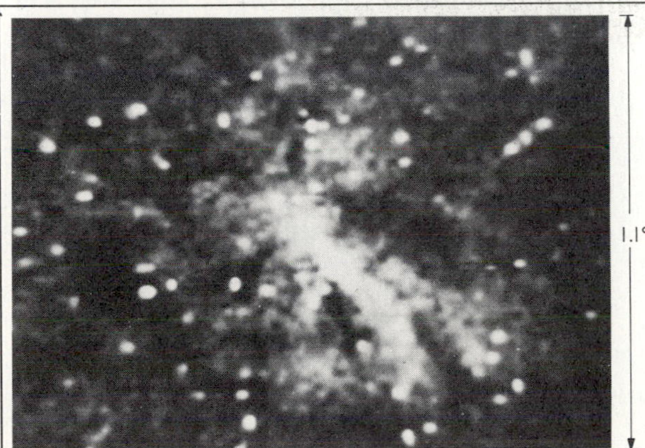

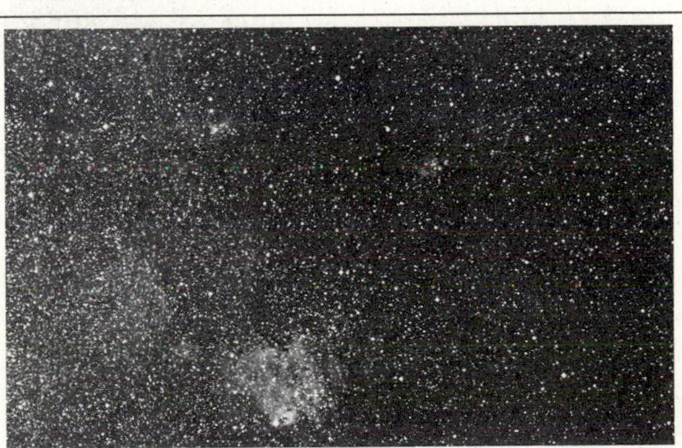

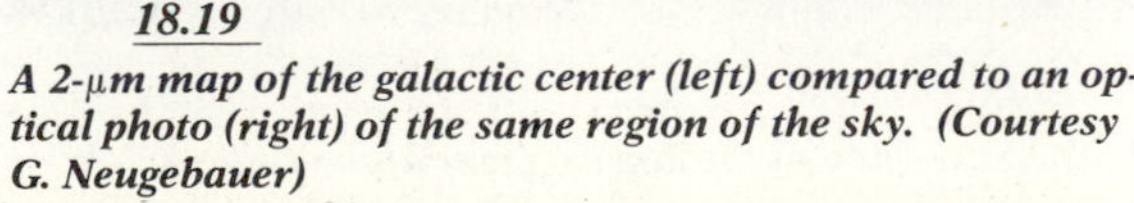

18.19

A 2-μm map of the galactic center (left) compared to an optical photo (right) of the same region of the sky. (Courtesy G. Neugebauer)

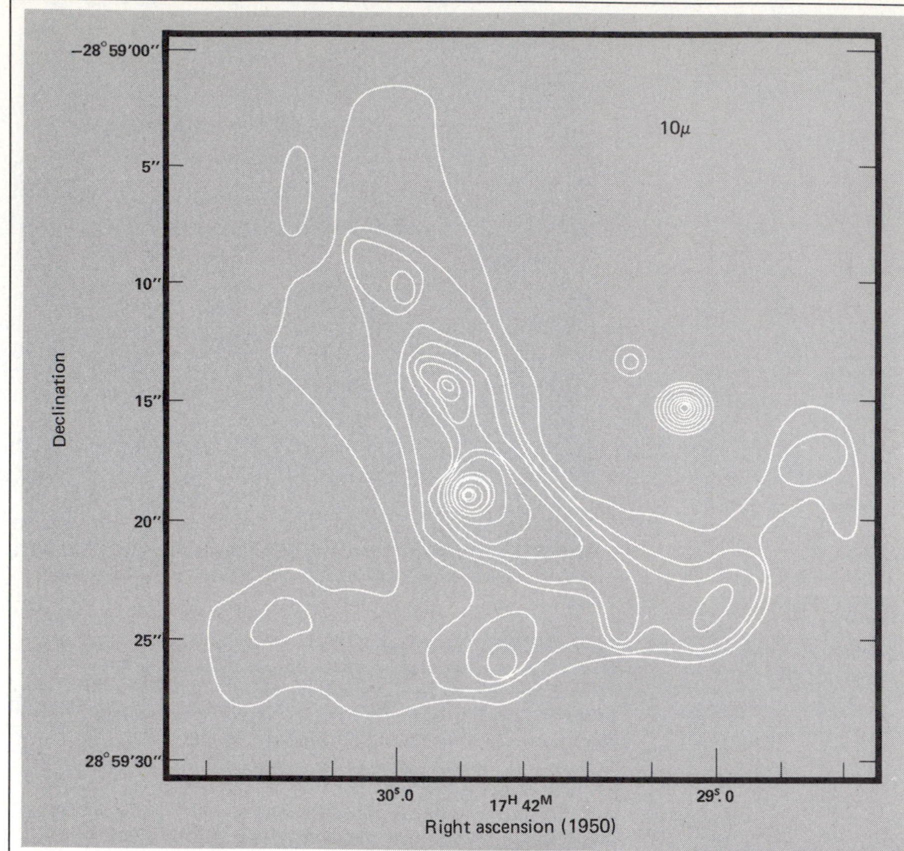

18.20
A 10-μm map of the Galaxy's center. This infrared radiation arises from dust heated by stars. (Adapted from a diagram by E. Becklin, K. Matthews, G. Neugebauer, and S. Willner, Astrophysical Journal, *vol. 219 (1978), p. 121, copyright © 1978 by the American Astronomical Society)*

indication of random motions of around 200 km/sec. Third, the lines show a systematic trend from blue to red shifts across the infrared core. This trend implies that the gas rotates at about 150 km/sec at 1.3 ly from the center. Fourth, additional observations indicate that the core contains at least seven discrete, ionized sources, many of which coincide with the peaks in the 10-μm map. These ionized zones are small—less than 1.5 ly in diameter—and contain a few solar masses of ionized material. They appear to orbit around the galactic center on an axis tilted 45° to the main rotational axis of the Galaxy. So the galactic core contains, within the Sgr A molecular cloud complex, a disk of rotating ionized gas (Fig. 18.21).

X-rays from the galactic center / Astronomers have known for a number of years that the galactic center emits X-rays. These come in the form of an extended X-ray source about 2° in size and pointlike, short-lived sources very close to the galactic center. Some of these sources give off bursts of X-rays, with the amount of enerny in each burst comparable to that from the short-lived sources. To date it is not clear if or how these sources relate to each other.

Images from the Einstein X-Ray Observatory have finally shown the galactic center in its

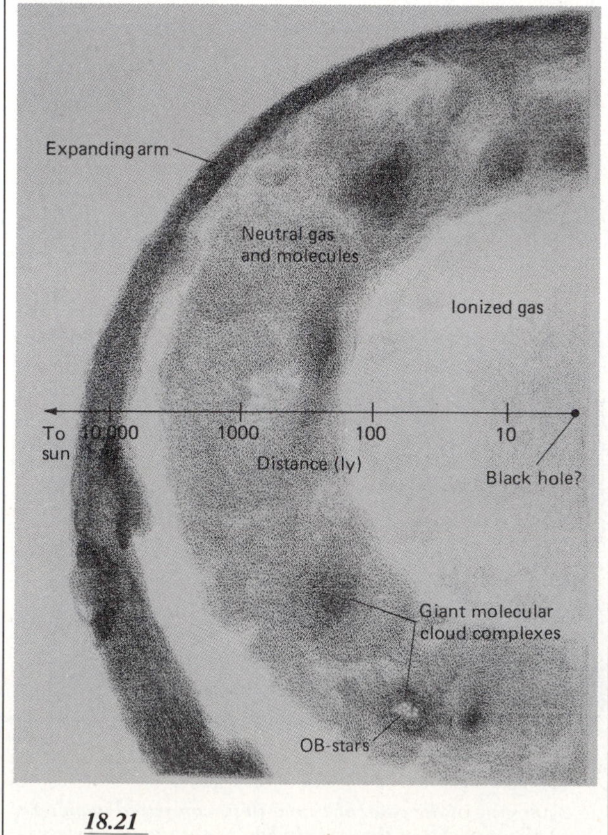

18.21
A schematic map of the inner regions of the Galaxy.

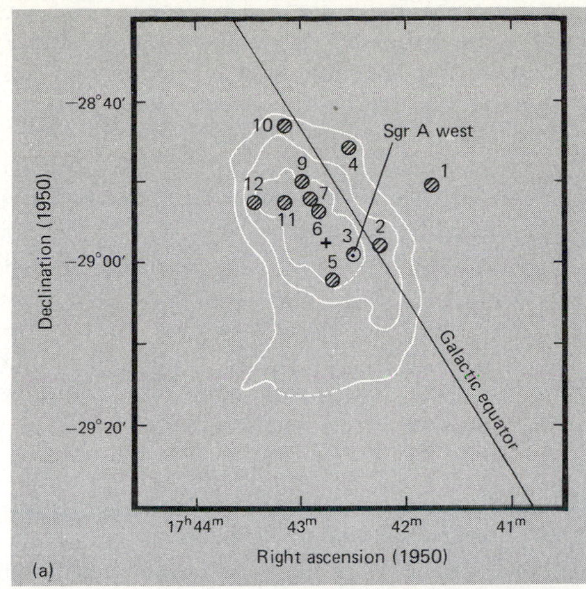

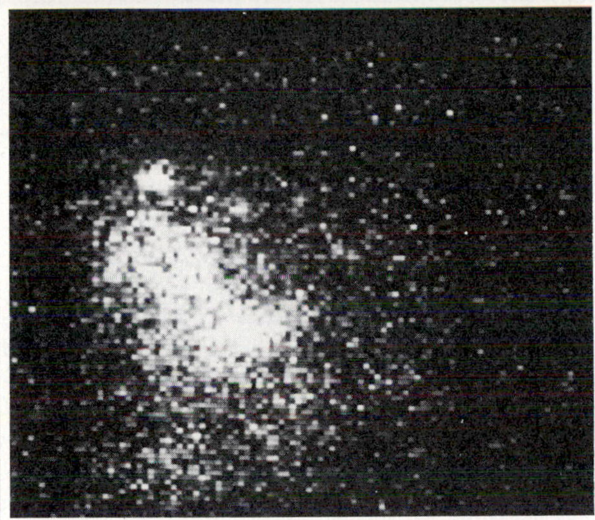

18.22

(a) ***A schematic map of the X-ray emission from the galactic center region. The shaded circles are discrete sources; the contour lines show the diffuse emission—the cross marks the center of this emission.*** **(b)** ***A computer-generated photo of the X-ray emission observed by the Einstein X-Ray Observatory. (Courtesy J. Grindley, Harvard-Smithsonian Center for Astrophysics; both figures from M. Watson, R. Willingale, J. Grindley, and P. Hertz,*** **Astrophysical Journal,** ***vol. 250 (1981), p.142, copyright © 1981 by the American Astronomical Society)***

full, high-energy glory (Fig. 18.22). Within 300 ly of the center, the X-ray emission is modest in strength. It consists of a complex of weak sources (covering some 200 by 150 ly) embedded in a halo of weaker, diffuse emission. About 12 sources are here. One coincides with Sgr A West; it has an X-ray luminosity of 10^{28} W. The other sources lie along the ridge in the same location as the cluster of infrared sources.

Does a black hole lurk in the core? / A puzzle generated by the radio and infrared-line observations arises from the rapid rotational motions near the Galaxy's core. Here it appears that the rotational velocities increase closer to the core.

Why is that a problem? Well, the rotational velocities are so high that a huge concentration of mass is needed to hold all that speedy gas together. For example, if the Galaxy's core simply contained a cluster of stars, you would expect the rotational velocities to decrease toward the core because as you get closer in, you have less and less mass to bind the moving materials gravitationally. To account for the rapid rotation requires a mass in the core of 5 million solar masses—all lumped together in a region only 0.1 ly in diameter!

What form might this mass have? One possibility—and it's very hard to come up with another—is that the mass is locked up in a black hole. If it were in the form of, say, solar-mass stars, these stars would be located, on the average, only 1 to 2 AU from each other. That seems unlikely, because stars so close, especially if many of them were red giants, would collide rather frequently.

The idea that a supermassive black hole lurks in the heart of the Galaxy has yet to be confirmed. Indirect support for the idea comes from observations that a few other galaxies may have a similar mass concentration in their nuclear region (see Chapter 19).

To recap: Infrared, radio, and X-ray telescopes can probe the galactic center directly. They have found that although the nucleus is very small—less than a few tens of light hours in diameter—it emits enormous amounts of energy, about 10 percent of the total from the Galaxy. Within it, material orbits the center at a rapid rate. At the very center lies a massive object, perhaps a black hole.

18.7 *The halo of the Galaxy*

The globular clusters outline the halo around the Galaxy. Little else is known to exist in the halo for sure. Stray stars are seen. The halo contains some gas that is hot and ionized. The halo may also contain as yet undetected objects, such as very faint low-mass stars, and extend at least 120,000 ly beyond the edge of the disk.

Globular clusters / You've already encountered the physical characteristics of globular clusters (turn back to Table 16.3). To review briefly: A globular cluster has a spherical shape (some tens to hundreds of light years in diameter) and con-

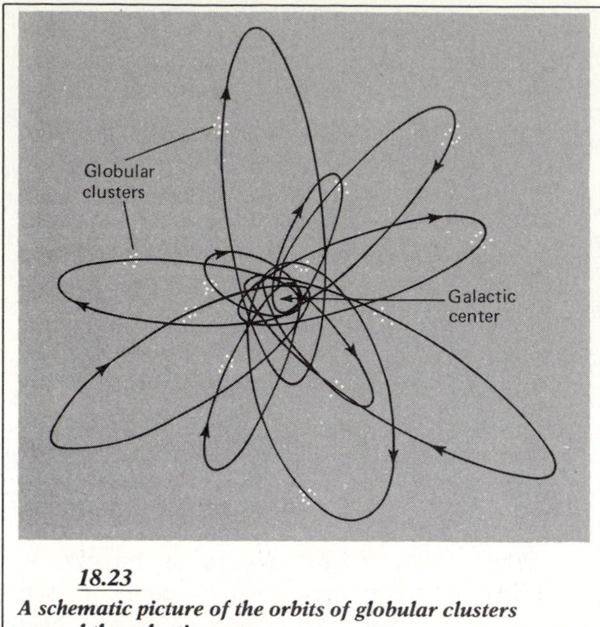

18.23
A schematic picture of the orbits of globular clusters around the galactic center.

tains up to 100,000 Population II stars, each with a little less than 1 solar mass.

The globular clusters form a sphere around the Galaxy's center. Their elliptical orbits bring them out to extreme distances of 40,000 ly from the Galaxy's nucleus. The clusters orbit at speeds about 100 km/sec, diving into and shooting out of the disk (Fig. 18.23). These passages have helped to wipe globular clusters clear of any gas and dust they once had.

Other material in the halo / The Galaxy's halo is thin and far from us, so it's hard to observe. The rotation curve indicates that the halo actually contains considerable material, perhaps much more mass than is in the visible part of the Galaxy. Using other spiral galaxies as a guide, the mass in the halo may be four or five times greater than all the mass known so far. What might it be?

Astronomers have spotted a few RR Lyrae variables above and below the galactic plane that do not belong to globular clusters. Those with periods longer than about ½ day are typically found 16,000 ly away from the plane. In addition to these stars, the halo may contain a large number of low-mass, faint red stars that are difficult to observe directly. Recent observations of a few other nearby galaxies like the Milky Way imply that they may have extensive, massive halos of faint red stars.

The halo also contains gas, but much less than the disk of the Galaxy. Observations at 21 cm show hydrogen clouds traveling with high speeds above and below the galactic plane. So the halo has some H I in it. Most of the halo's gas, however, is probably ionized hydrogen. This gas could come from the disk, blown out by supernova explosions, expanding H II regions, and stellar winds. So the halo may be fairly hot and expanding into intergalactic space.

But the halo may also contain other objects, currently unobservable. Low-mass main-sequence stars would be very hard to detect. Smaller objects, similar to planets, may also exist. Recall that hydrogen masses with less than 0.08 solar mass never get hot enough to become stars. Some astronomers have suggested numerous black holes. We just don't know what else might be there as required by the rotation curve.

18.8
A possible history of our Galaxy

We have a pretty fair idea of the architecture of the Milky Way Galaxy. What clues does this information provide about the birth and evolution of the Galaxy? The crucial clues come from two sources: (1) the chemical compositions of galactic material, and (2) its dynamics.

The process of galactic evolution links the chemistry with the dynamics. This linkage marks an important theme of cosmic evolution, because the evolution of the Galaxy results from the evolution of all the stuff that makes it up.

Populations and positions / I stated earlier that the chemical compositions of Population I and Population II stars differ considerably in their abundances of heavy elements. In general, Population II stars contain about 1 percent of the metal abundance of Population I stars.

In fact, however, we do not find a stark and simple division of metal abundances into just two groups. Rather, we find a continuous range of abundances, from about 3 percent to less than 0.1 percent for the ratio of metals to hydrogen. So, though the division into two populations is a useful tool, a continuous range of populations actually exists. And when cataloged by metal abundance, we find a striking correlation with average distance from the Galaxy's disk. *The lower the metal abundance, the farther the objects are found from the disk.*

To interpret this key observation, we need to rely on basic concepts of starbirth, stardeath, and the recycling of the interstellar medium. First, stars are born from clouds in the interstellar medium. Their atmospheric elemental abundance reflects that of the gas from which they formed. Second, stars inherit the orbital motions about the Galaxy of their parent gas and dust clouds. Third, massive stars evolve quickly and spew back into the interstellar medium heavy-element-enriched

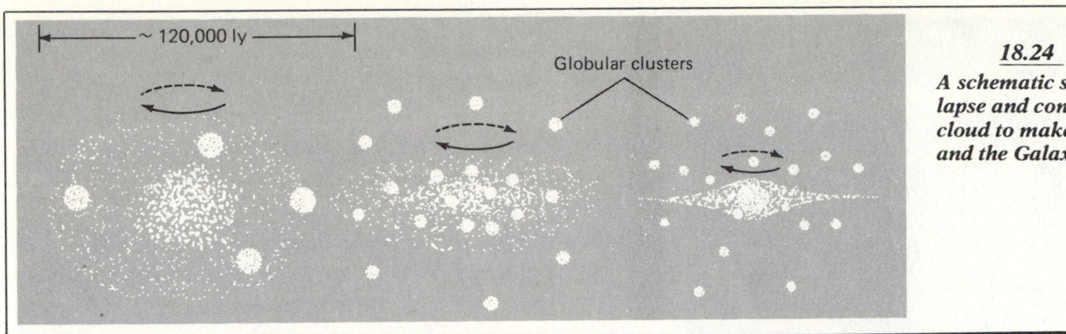

18.24
A schematic sequence of the collapse and condensation of a cloud to make globular clusters and the Galaxy's disk.

material. So as long as new stars, especially massive ones, are born, the abundance of heavy elements in the interstellar medium gradually increases as the Galaxy ages.

With these basics in mind, let's try to make sense of the observations, which show that the youngest objects (highest heavy-element abundances) hug close to the disk; the oldest objects (lowest heavy-element abundances) range far from the disk. Other objects fall in between these extremes. So the halo of the Galaxy is its oldest part, and the spiral arms its youngest.

We can estimate the Galaxy's age by finding the oldest stars in the halo. A comparison of theoretical models for globular cluster stars with H-R diagrams for them (see Chapter 16) indicates an age of 16 billion years. That's when the Galaxy formed. Let's see how it formed.

The birth of the Galaxy / Because globular clusters contain the oldest stars associated with the Galaxy, the halo marks the fossil remains of the Galaxy's birth. Within it, globulars orbit the Galaxy on extremely elongated elliptical paths. Most of the time, the globulars move slowly through the halo at the outer extremes at their orbits; only briefly do they whip in and around the nucleus. These stars exhibit the motions of the cloud from which they were formed. So the Galaxy must have been born from an initially huge gas cloud—at least 300,000 ly in radius.

Here's one model of the Galaxy's birth. Imagine a tremendous, ragged cloud of gas roughly twice as big as the Galaxy's halo today (Fig. 18.24). Its density is low. This proto-Galaxy cloud was probably turbulent, swirling around with random churning currents. Slowly at first, the cloud's self-gravity pulls it together, with its central regions getting denser faster than its outer parts. Throughout the cloud, turbulent eddies of different sizes form, break up, and die away. Eventually, the eddies become dense enough to contain sufficient mass to hold themselves together. These might be hundreds of light years in size—incipient globular clusters. Each blob then splits up to form individual stars—all born about the same time with the same chemical composition. This happened 14 to 18 billion years ago. The blobs retained some of the radial motion of contraction and ended up in somewhat eccentric orbits around the center of the sphere.

Meanwhile, not all of the gas was consumed in this burst of globular cluster formation. As this material contracted more, it fell slowly into a disk. (Sound familiar? Look back at Section 12.3 on the formation of the solar nebula.) Why a disk? Because the original cloud had a little spin, and the conservation of angular momentum (see Focus 9.2) requires that it spin faster around its rotational axis as it contracts (Fig. 18.24). The kinetic energy of the cloud slowly decreases, as gas clouds collide and heat gets radiated away. The disk thins.

As the disk forms, its density increases, and more stars form. Each burst of starbirth leaves behind representative stars at different distances from the present disk. Finally, the remaining gas and dust settle into the narrow layer we see today. Somehow density waves appear and drive the formation of spiral arms.

During this time, massive stars were manufacturing heavy elements and flinging them back into the cloud. So as stars were born in succession, each later type had more heavy elements. That enrichment continues today in the disk of the Galaxy.

One problem here. To date we have not discovered any star that contains *no* heavy elements. Where did the heavy elements contained in the oldest known stars come from? You will see (see Chapter 21) that few heavy elements were made in the Big-Bang origin of the universe. Was there then some previous generation of stars, formed before the globular clusters, which made the first heavy elements? Many astronomers think so, but then where are those objects today? They seem to have disappeared without a trace.

18.25
An artist's concept of the overall form of the Galaxy as a barred spiral with the arms delineated by H II regions. (Adapted from a diagram by G. de Vaucouleurs and W. Pence, Astronomical Journal, *vol. 83 (1976), p. 1163, copyright © 1976 by the American Astronomical Society)*

KEY CONCEPTS

1 ***The main parts of the Galaxy are the encircling halo, the flat disk, and the central nuclear bulge, which contains the nucleus.***

2 ***Different types of stars inhabit these regions: The disk contains young, metal-rich (few percent) Population I stars; the nucleus, metal-rich Population I stars; and the halo, metal-poor (few tenths of a percent) Population II stars.***

3 ***The disk contains spiral arms, at least two (interior to the sun's orbit) and possibly four (exterior to the sun's orbit), that contain concentrations of young stars, gas, and dust; the sun lies on the inner edge of one arm, and we can see pieces of other arms toward and away from the galactic center (Fig. 18.25).***

4 ***We trace out spiral arms by the use of objects found in them; optically, O- and B-stars, cepheids, and H II regions; with radio, H I clouds and molecular clouds; in all cases, the essential problem is to determine the distance to the object observed.***

5 ***The rotation curve shows how fast objects at different distances from the galactic center orbit around it; observations reveal a rapid increase in the first 1000 ly, then a gradual rise out to 60,000 ly; the fact that the curve does not follow Kepler's laws indicates that a large fraction of the Galaxy's mass lies beyond the sun; this mass has not yet been seen directly.***

6 ***The sun orbits the Galaxy at a distance of some 30,000 ly from the sun at a speed of 250 km/sec, with some uncertainty in both numbers.***

7 ***Cepheid variable stars show a period-luminosity relationship: The more luminous the cepheid, the longer the period of its light variation; this relationship is a powerful tool to infer the distances to cepheid variables.***

8 ***Radio astronomers use the Doppler shift in the 21-cm line from clouds of atomic hydrogen to infer the spiral-arm structure of the Galaxy; to do so, they assume that the clouds move along near-circular orbits (they don't) and that the rotation curve has been well observed (it hasn't).***

9 ***Any model for the evolution of the Galaxy must explain the persistence of spiral arms; the density-wave model does so by having two spiral density waves disrupt the gas of the Galaxy's disk to promote the formation of spiral arms; as the density waves plow through the disk, different material condenses into the spiral arms as the old material dissipates; so the persistence of arms is really an illusion.***

10 ***The nucleus of the Galaxy emits intense radio and infrared radiation; it contains supergiant M-stars, young massive stars, dust, and gas rotating at high speeds; to account for the motion of the gas requires a concentration of millions of solar masses of material in the inner few light years—perhaps a supermassive black hole.***

11 ***The Galaxy's halo contains globular clusters, some gas, and the invisible objects that make up the mass that shows up in the rotation curve.***

12 ***The Galaxy formed from a large, slowly spinning cloud of gas and dust; the halo formed first, then the disk; we can estimate the relative sequence of formation from the heavy-element content of stars: The more metals they contain, the younger they are.***

STUDY EXERCISES

1 What limits an optical astronomer's investigation of the Galaxy's structure? (*Objectives 1, 2, and 3*)

2 For what reason are Population I cepheids good spiral-arm tracers? (*Objectives 4 and 7*)

3 Radio astronomers need the rotation curve of the Galaxy in order to use 21-cm observations to establish its spiral structure. Why? (*Objectives 4 and 6*)

4 Argue that a spiral arm cannot be a material arm. (*Objective 7*)

5 What are the kinds of celestial objects found in spiral arms? (*Objectives 7 and 8*)

6 What characteristics of spiral arms does the density-wave model account for? In what respects is the model at present inadequate? (*Objective 8*)

7 What observational evidence do we have that a large fraction of the Galaxy's mass is not in the core, nor, in fact, within the radius of the sun's orbit? (*Objective 4*)

8 Relate the orbits of globulars and their chemical composition to the birth of the Galaxy. (*Objective 10*)

BEYOND THIS BOOK . . .

"The Arms of the Galaxy" by B. Bok in *Scientific American*, vol. 201 (1959), p. 92, is interesting to contrast to the information in *The Milky Way* by B. and P. Bok (Harvard University Press, Cambridge, Mass., 1981).

Two good books about the evolution of our ideas about the Milky Way and other galaxies are *The Discovery of the Galaxy* (Knopf, New York, 1971) by C. Whitney and *Man Discovers the Galaxies* (Science History Publications, New York, 1976) by R. Berandzen, R. Hart, and D. Seeley.

For an up-to-date look at the Galaxy, see "The Milky Way Galaxy" by B. Bok in *Scientific American*, March 1981, p. 92.

"The New Milky Way" by L. Blitz, M. Fich, and S. Kulkarni in *Science*, vol. 220 (1983), p. 1233, gives a somewhat technical picture of our current view of the Galaxy.

19/ The universe of galaxies

Look, friend at this universe
with its spiral clusters of stars
flying out all over space
like bedsprings suddenly bursting free . . .

EDWARD FIELD: "Prologue"

Learning objectives

After studying this chapter, you should be able to:

1. **Describe the general physical characteristics of spiral, elliptical, and irregular galaxies, including their differences in size, shape, mass, color, types of stars, and amount of interstellar gas and dust.**
2. **Outline the methods used to find the bulk properties of galaxies.**
3. **Describe how to use the criteria *brightness means nearness* and *smallness means farness* to estimate the relative distances to galaxies.**
4. **Indicate what observations clinched the idea that the spiral nebulas were actually other galaxies.**
5. **Describe briefly how to find distances to galaxies.**
6. **Outline a contemporary method of finding distances to distant galaxies, starting with the astronomical unit and ending with Hubble's constant.**
7. **Evaluate the weaknesses in the procedure you outlined in Objective 6 so that you can estimate the possible errors in distances to galaxies.**
8. **Show how getting distances and radial velocities for galaxies results in a value for Hubble's constant and use this value to estimate distances to galaxies.**
9. **State the range of uncertainty in the value of Hubble's constant and the implications of this uncertainty.**
10. **Define a *cluster of galaxies* and describe the appearance of clusters as photographed through a large telescope.**
11. **Evaluate the evidence for intergalactic material between and/or within clusters of galaxies.**
12. **Define a *supercluster* and describe the general layout of superclusters in space.**

Central question

What is the structure and content of galaxies, and how are they distributed throughout the universe?

19.1
A group of galaxies in Leo. (Courtesy Palomar Observatory, California Institute of Technology)

No celestial object is quite as grand as a galaxy. In a large telescope, a bright spiral galaxy is a stunning sight, a star-bright nucleus and misty swirl of spiral arms. Billions of stars caught in a whirlpool spanning hundreds of thousands of light years.

The galaxies form the basic elements of our modern cosmological vista. Their sheer numbers are almost beyond our comprehension (Fig. 19.1). The diversity of structure in these galaxies is also astounding; even more surprising is the fundamental unity found in spite of their wide variety. The fact that galaxies can be sorted into broad divisions hints at a common evolutionary process.

The galaxies form the skeleton of the universe. The crucial problem is the measurement of their distances, for many conclusions of modern cosmology hinge on them. This chapter notes that the notorious difficulties of surveying our own Galaxy are amplified when we try to appraise the vastness of the universe. In spite of present errors and problems, our vision of the universe, underpinned by the theory of general relativity, has a coherence that allows us to draw conclusions about the large-scale structure of spacetime.

19.1 *The discovery of galaxies*

For almost two centuries, astronomers hotly debated whether or not the *spiral nebulas* they saw with their large telescopes were simply clouds of gas within the Milky Way or were other galaxies like ours but far beyond it.

The controversy came to a head in April 1920, when Harlow Shapley and Heber D. Curtis (1872–1942) debated the point publicly. Shapley believed at the time that the spiral nebulas were distant parts of the Milky Way. Curtis opposed Shapley on this question and claimed that they were galaxies in their own right.

Curtis argued that the wide range of apparent angular sizes of spirals–approximately 2° for M 31, the nearest, to 10′ and less for the smallest–required a large range of distances and so they could not be part of our Galaxy. Starting from the principle of the uniformity of nature, Curtis assumed that all spirals have roughly the same physical diameter. The range in observed sizes (more or less 10 to 1) implied that the spirals must be enormous distances from the Galaxy, for if they were the same in diaMeter, the range in apparent size meant that the ones 10 times smaller must be 10 times farther off, or about 10 times the radius of the Galaxy (Fig. 19.2). So they could not be members of the Milky Way Galaxy, as Shapley argued.

In addition, Curtis noted the so-called *zone of avoidance*, a region near the plane of the Galaxy where very few spirals are visible. Curtis argued that interstellar material in the galactic plane cuts out the light from the spirals; we see fewer in the zone of avoidance because there we have to look through the disk of the Galaxy. If the spirals were actually associated with the Galaxy, they would be found concentrated in the plane–rather than avoiding it–along with the stars, galactic clusters, and H II regions. As evidence for this point of

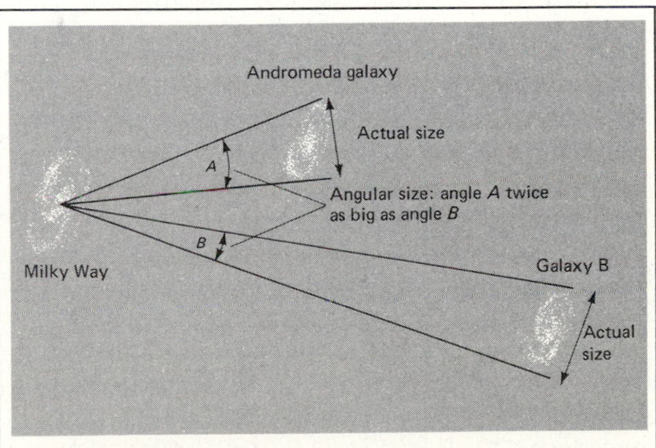

19.2
Distances to galaxies estimated by their angular sizes. If galaxies have roughly the same physical size, the more distant will have smaller angular sizes. In the case shown here, Galaxy B is about twice as far away as the Andromeda galaxy.

19.3
Dust in the plane of a spiral galaxy (Messier 104). The dark band occurs from dust cutting out starlight. (Courtesy Palomar Observatory, California Institute of Technology)

view, Curtis cited photographs of spirals showing dark lanes cutting through their planes (Fig. 19.3). If the Milky Way Galaxy and other spirals were similar in structure, Curtis reasoned, then our Galaxy must also have obscuring material collected in the plane.

Finally, Curtis pointed out that the spectra of spirals are not bright-line spectra like those of diffuse nebulas (such as the Orion Nebula) but rather resemble those from a conglomeration of stars. Instead of the bright lines expected from a thin gas if the so-called nebulas were in fact clouds, the spectra consist of faint dark lines against a bright background—the same spectrum the Galaxy would show if viewed from a great distance.

In 1924, after the Shapley-Curtis confrontation, Edwin Hubble settled the dispute conclusively by the discovery of cepheid variables in M 31. Although variables had been suspected as early as 1922, Hubble confirmed their existence in the outer arms of M 31. He derived a distance of 490,000 ly for M 31, far beyond the farthest globular clusters that marked the outer limits of the Galaxy. (Hubble's estimate was too small; recent work on cepheids in M 31 establishes a distance of 2.2 million ly.)

Warning: The principle of the *uniformity of nature* has great power, but must be used with caution. Sometimes it works; sometimes it leads you astray. For example, novas in M 31 *are* pretty much the same as those in our own Galaxy, because M 31 is a spiral galaxy very similar to the Milky Way. But the variable stars in globular clusters are *not*, it turns out, the same as the cepheids in the disk of our Galaxy, for they belong to quite different populations of stars. You can always make the assumption that things are uniform everywhere, and that can lead to new knowledge and insights, but you always have to keep looking for new ways to check that assumption and be prepared to abandon it if inconsistencies develop.

19.2 The galaxian zoo

Hubble pioneered in the field of extragalactic astronomy. He recognized that galaxies had different shapes. To catalog the differences in form, Hubble in 1926 proposed his first scheme for the classification of galaxies. Although his initial design is now considered too simple, modern classifications still use the fundamental categories of *elliptical, spiral*, and *irregular* galaxies. Let's take a look at these basic galaxy types.

Elliptical galaxies (Fig. 19.4) exhibit no spiral structure but do show an elliptical shape. Very little gas or dust appears in elliptical galaxies, and *O*- and *B*-stars are also absent. The ellipticals generally have a reddish overall color.

Hubble subdivided the ellipticals in classes from *E*0 to *E*7, according to how elliptical they appear. Imagine looking at a circular plate face on; such is the appearance of an *E*0 galaxy. Now slowly tilt the plate so that it looks more elliptical and less circular. This flattening of shape presents the same views as the sequence from *E*0 to *E*7 galaxies. Be warned that Hubble based the

19.4
A giant elliptical galaxy, Messier 87. Note the symmetry of the shape and the lack of distinct structure. The small dots are globular clusters that orbit the galaxy's nucleus. (Courtesy Lick Observatory)

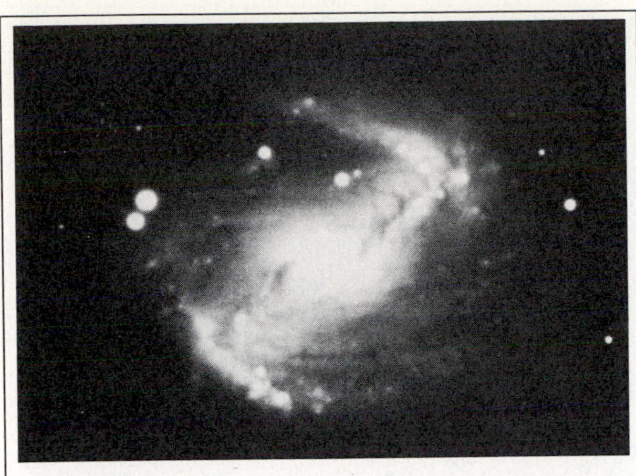

19.5
A typical barred spiral galaxy (NGC 5383). Note that the bright bar crosses the nucleus and links the two spiral arms together. (Courtesy Kitt Peak National Observatory)

classifications on the appearance of the galaxy, not its true shape. For example, an *E*7 is really a flat elliptical viewed edge on, but an *E*0 may be either a truly spherical galaxy or a flattened galaxy seen face on. One additional complication: Ellipticals come in a range of sizes, from giants to dwarfs.

Spiral galaxies display obvious spiral structure, usually with two, but sometimes more, spiral arms. One type of spiral has a prominent bar through the nucleus, the spiral arms winding out from the end of the bar (Fig. 19.5). Hubble termed the spirals without a bar *normal* and the others *barred*. (Our Galaxy may be a barred spiral.) He arranged the spiral forms in sequence according to the sizes of their nuclear region, the tightness of the spiral arms, and the degree to which the arms were resolved into patches.

Spirals come in two types: normal, denoted *S*, and barred, denoted *SB*. The normal and barred spirals are subdivided further into categories *a*, *b*, and *c* (Fig. 19.6). These types are judged by how tightly the spiral arms wind around (*a*, the tightest; *c*, the most open) and the relative size of the nucleus (*a*, the largest; *c*, the smallest). For example, the Hubble *Sa* is a normal spiral with a large nucleus and tightly coiled arms. A few galaxies appear to have the disk of a spiral but no arms. Hubble dubbed those *S*0. These are now sometimes called *lenticular galaxies* because of their shape.

Finally, as a catch-all category, Hubble designated as *irregular galaxies* those that were devoid of spiral structure or symmetry but were resolvable into distinct patches of stars (Fig. 19.7). These strange beasts fall into two groups. *Irr I* can be resolved into *O*- and *B*-stars and H II regions. Conspicuous dust clouds are usually absent. *Irr II* have the same lack of shape as Irr I. In addition, they are not resolvable into stars, don't have visible H II regions, and usually do show prominent dust lanes.

Modern classifications have expanded Hubble's format. I will stick with using Hubble's three basic categories: spirals, ellipticals, and irregulars. But don't think that this Hubble scheme includes all types of galaxies; it does not. Some galaxies stand out as peculiar in shape (Fig. 19.8) and do not fit into the three general Hubble categories. Many of these peculiar galaxies turn out to have evidence of unusual activity. Some appear to be pairs of galaxies close together, interacting gravitationally.

One more point: Fairly recently, astronomers have recognized that the same Hubble type of galaxy, say *Sb*, comes in a range of luminosities. So in analogy to stellar luminosity classes (see Section 14.5), galaxies also have luminosity classes of I, II, III, IV, and V, with the first the most luminous and the last the least luminous. An *Sc I* galaxy, for instance, is a very luminous spiral with a small nucleus and spread-out arms. It turns out that luminosity class I galaxies must be larger than class II, an dso on. So class I galaxies can be thought of as supergiant galaxies.

About 77 percent of known galaxies are spirals, 20 percent ellipticals, and 3 percent irregulars. This sample is dominated by the luminous spirals, however, which are visible at very great distances. The relative numbers in a given volume of space are quite different. A survey of a region out to 30 million ly showed that only 34 percent of the galaxies in this volume are spirals, 13 percent are ellipticals, and 54 percent are irregulars. Many of these are small galaxies of fairly low luminosity.

The schematic grouping of the galaxies marked an initial step toward delving into the far depths of the universe. But to probe the physical properties of galaxies, their masses, sizes, and luminosities, ultimately depends on a knowledge of their distances from us—distances that are as difficult to survey as they are vast to imagine.

19.3 Surveying the universe of galaxies

Only about 50 years have passed since we learned for certain that galaxies are far away islands of stars. Given this short time, you should not be surprised to discover that we know the distances to galaxies only roughly. For the nearest galaxies,

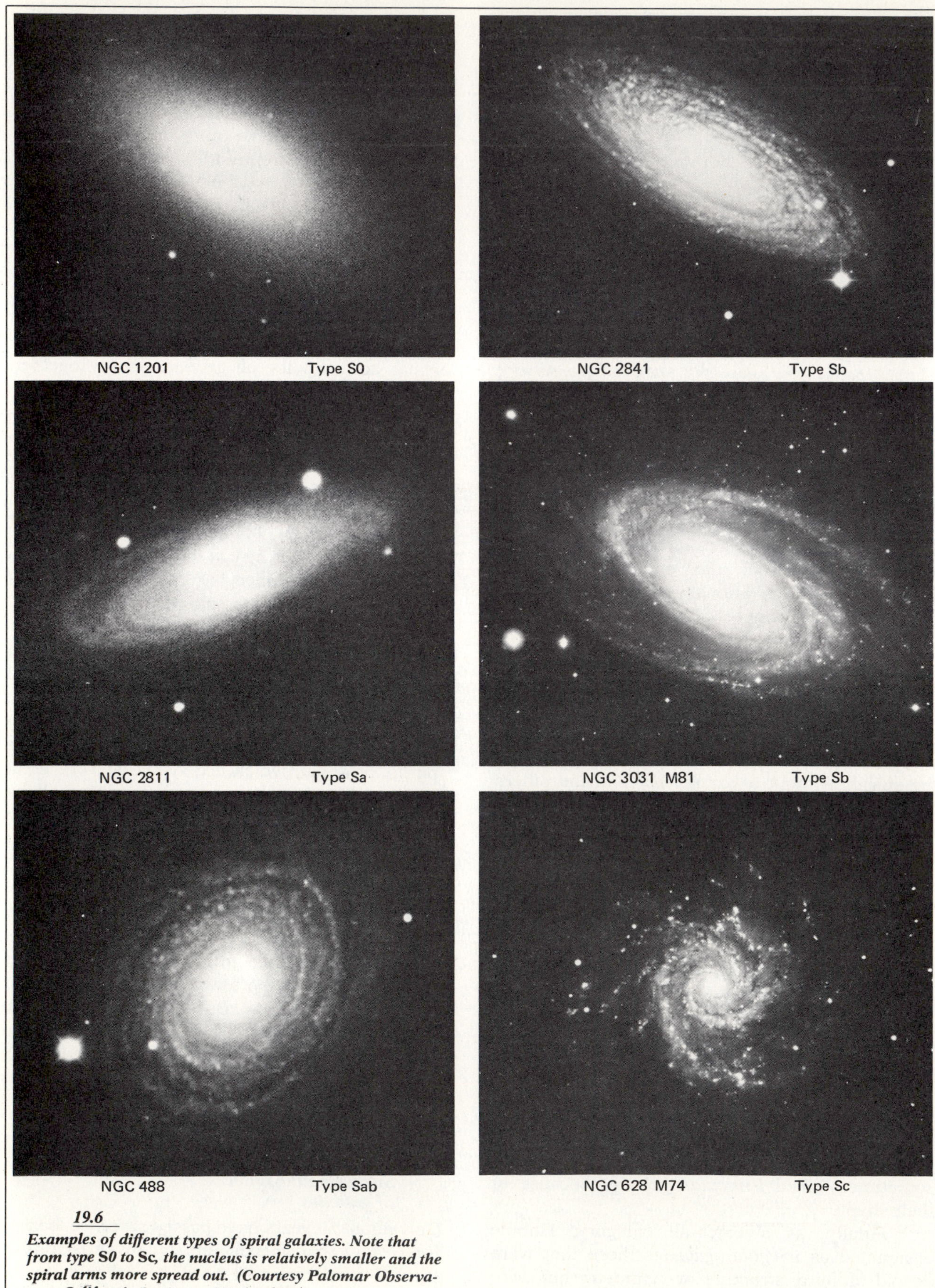

19.6
Examples of different types of spiral galaxies. Note that from type S0 to Sc, the nucleus is relatively smaller and the spiral arms more spread out. (Courtesy Palomar Observatory, California Institute of Technology)

19.7
A dwarf irregular galaxy in the constellation Sextans. (Courtesy Palomar Observatory, California Institute of Technology)

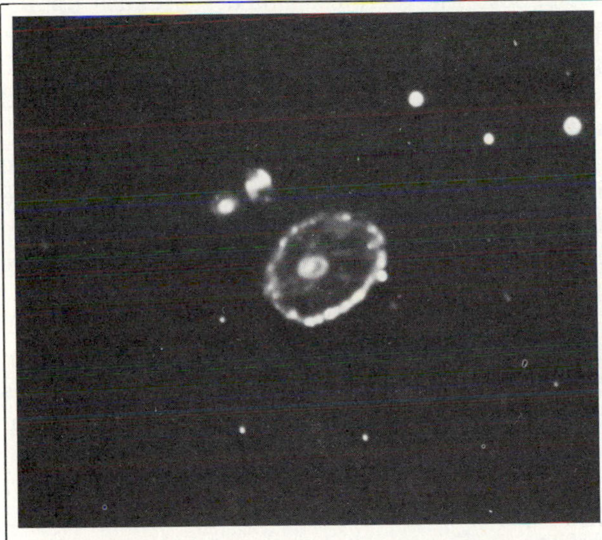

19.8
A galaxy with a peculiar shape that does not fit the simple Hubble scheme. It is called a ring galaxy. (Courtesy Kitt Peak National Observatory and Cerro Tololo Inter-american Observatory)

we have measurements that are good to about 10 percent. For the most distant visible galaxies, we are lucky if we know their distances within 50 percent of their actual value. It's a hard but essential astronomical business to survey the huge universe of galaxies.

Although the distances to galaxies have continually been revised, the essential techniques remain the same. The crudest indicators are the criteria that *brightness means nearness* and *smallness means farness*; the galaxies with the smallest angular size tend also to be the faintest and the most distant. By applying these simple criteria, you can make rough estimates of the relative distances to galaxies. For example, if you look at two galaxies, one apparently half as large as another, and if both are in reality approximately the same size, the apparently smaller galaxy must be twice as distant as the apparently larger one. It is similar for brightness: If one galaxy is 100 times fainter than another, it must be roughly 10 times farther away (recall the inverse-square law for light).

Refining this rough first approach requires the use of known physical properties of stars and galaxies inferred from theoretical models and careful observations. Each step in surveying the universe applies to certain objects and over a certain range of distances; one piles step upon step to establish a distance pyramid. This structure is very much an astronomical house of cards, for it is only as strong as its weakest lower support. Be aware that, at present, the degree of uncertainty in the distance determined to nearby galaxies is *at best* about 10 percent. As you move farther up the pyramid, the percentage of error increases.

Now to discuss the layers of the pyramid. First we need to establish the scale of the AU, say in kilometers, from solar system observations (see Chapter 13). To move out to the stars requires the use of direct parallax with the earth's orbital diameter as the baseline (see Focus 14.2). Beyond the reach of heliocentric parallaxes, astronomers rely on spectroscopic parallaxes (Section 14.5) and the cepheid period-luminosity relation (Section 18.3). So far, so good; these techniques allow a reasonable assessment of the distance to the galactic and globular clusters of the Milky Way Galaxy and so to the establishment of its size. More important, these local distances let us find out the luminosities of typical galactic occupants.

To bridge the distances to other galaxies, we must accept the assumption of the uniformity of nature: that the essential character of objects in our Galaxy (such as cepheid variable stars or supernovas) is the same for similar objects in other galaxies. That's a crucial, necessary assumption. Without it, we couldn't get anywhere. It's not a blind, unsupported assumption, however. All observations made so far are consistent with it. For example, cepheids in other galaxies have the same spectra and the same shape light curves as cepheids of the same type in our Galaxy.

Then, to find distances to galaxies, we must use identifiable objects whose *luminosities* we know. The trick here is that we want to choose objects whose luminosities are guessable. We then compare their fluxes with their luminosities to infer their distances. Unfortunately, even the largest telescopes have a limit, and some objects are too faint to be picked up. So we want to choose the *most* luminous objects in galaxies to use, as long as we know what their luminosities are.

Starting with close galaxies, we apply the period-luminosity relation to cepheids in other galaxies. The cepheids, however, are useful over a very limited range, for they are not especially luminous. And *E* and *S*0 galaxies don't have any cepheids, which are relatively massive stars. Of all visible galaxies, only about 30 are of the right kind and close enough for us to detect cepheids within them. We can use cepheids as standard candles out to approximately 13 million ly. At that distance, cepheids are fainter than the limits of photographs taken through large telescopes.

To go beyond 13 million ly requires the establishment of other standards whose visibility is greater than that of the cepheids. We can use two types of objects to make this leap: the brightest stars and the largest H II regions in galaxies (but again, elliptical and *S*0 galaxies are excepted, since they don't contain H II regions or bright *O*- and *B*-stars).

The most luminous stars work as follows: Supergiant *O*- and *B*-stars appear to have a maximum luminosity. In our own Galaxy, the brightest blue-white star (Population I) has a luminosity of 5.2×10^5 solar luminosities. In M 33, a nearby spiral galaxy whose distance we can measure by cepheids, the brightest blue-white star has a luminosity of 4×10^5 solar luminosities. For other nearby galaxies, the range is 10^5 to 10^6 solar luminosities, depending on the luminosity class of the parent galaxy. To find distances from these stars, we measure their fluxes, compare them to the inferred luminosity, and use the inverse-square law.

The designation of other distance standards follows the same strategy: Find fairly common bright objects, find their luminosities in our own or nearby galaxies, check other galaxies by methods known to be reliable, and then utilize the candle to the limits of its accuracy (Fig. 19.9).

At distances greater than 82 million ly, we can't see individual objects in galaxies (with ground-based telescopes; the space telescope will allow us to resolve stars at distances nearly 10 times farther out). What next? We use the luminosities of the galaxies themselves! Galaxies tend to lie in clusters (see Section 19.5). To ensure choosing galaxies with the same luminosity, we select one of the brightest galaxies in a cluster rather than an average one picked at random. It appears that the brightest galaxies in clusters have about the same luminosity, *if* you stick to the same *kind* of galaxy (such as large spirals). In addition, we choose the brightest galaxies because for the most distant clusters, *only* the most luminous galaxies can be seen easily.

Of course, for this method to work, we need a calibration for the luminosities. How to get it? By using cepheid variables to find the distances to nearby spiral galaxies. Then we have a sample over a range of luminosity classes that we hope represents all galaxies. (The uniformity-of-nature principle again!)

What kind of galaxies can serve as useful, far-reaching standards? In the work of Allan Sandage and Gustav Tammann, the candidates are supergiant (and so very luminous) spiral galaxies with small nuclei and spread-out spiral

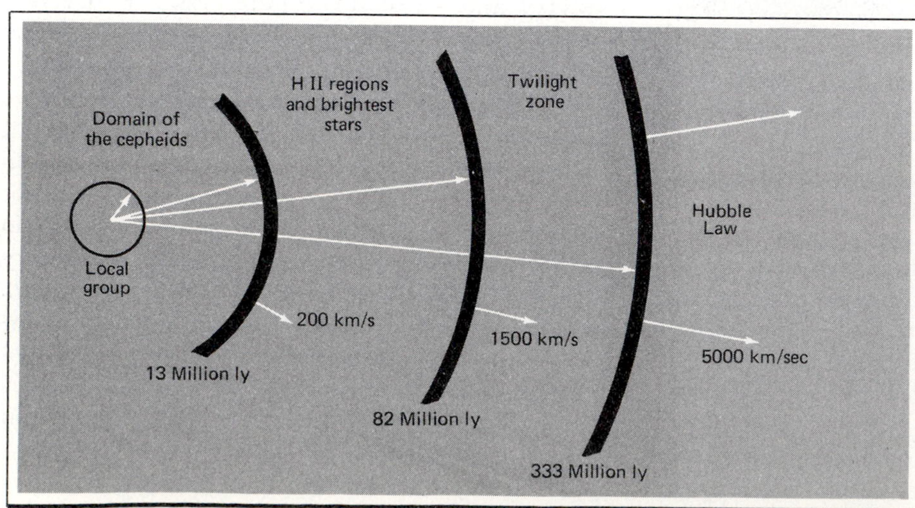

19.9
The range over which different distance indicators are usable, based on the faintest objects visible with the largest telescopes. The "twilight zone" is the range of distances beyond direct measurement but before the expansion of the cosmos can be accurately measured. (Adapted from a diagram by A. Sandage and G. Tammann,* Astrophysical Journal, *vol. 190 (1974), p. 525, copyright © 1974 by the American Astronomical Society)

arms; that is, Sc I galaxies. (M 101 is such a galaxy; see Fig. 18.1.) Sandage and Tammann calibrate the luminosity of such galaxies as about 25 billion times the sun's luminosity. Seen in distant clusters, these galaxies are relatively easy to identify because of their high luminosity and distinctive shape. With contemporary telescopes, they can be used as standards to distances of roughly 13 billion ly.

19.4
Hubble's law revisited

Thirteen billion light years is still not beyond the range of our largest telescopes. How to measure greater distances? Using the distance-velocity relationship first found by Humason and Hubble: the expansion of the universe itself (Chapter 7).

It is relatively easy to measure the red shift of a galaxy, compared with the task of finding its distance. A comparison spectrum made at the same time as the galaxy's spectrum affords a direct measurement of the shift in some prominent spectral lines (such as the H and K lines of calcium; see Fig. 19.10). As a Doppler shift (see Focus 14.3), the red shift indicates the radial velocity of a galaxy. The galaxy's distance is much tougher to get, as has been emphasized, but it is measurable out to present limits.

You can see for yourself how this method works. Look at the galaxies pictured and at their spectra with red shift indicated in Fig. 19.10. Notice that as the red shifts get larger, the galaxies appear smaller and fainter. Now plot the distances for these galaxies versus their red shifts. You can draw a straight line that represents the trend of these points and so find a value for Hubble's constant, which is the slope of the line (Fig. 19.11).

Reverse this procedure to find distances from red shifts. Suppose you were given the red shift of a galaxy. You look up this value on the red-shift axis of your plot and see to what distance it corresponds. If the red shift is larger than measured previously, you have to assume that the line you've drawn for the plotted points can be extended farther out and still be valid. Suppose a galaxy's measured red shift is 40,000 km/sec. What is its distance? If H equals 20 km/sec/Mly, then the plot gives 2.6 billion ly.

Look's simple, yes? But I must inject a note of caution. This indirect method of distance measurement rests on two crucial assumptions: (1) that the galaxies with known distances give an accurate value for Hubble's constant and (2) that the relation is a straight-line one that may be extended as far out as we like. In the face of present evidence, the first assumption seems OK but not very secure. Estimates of Hubble's constant range from 15 to 30 km/sec/Mly. The distances derived by this method can vary by a factor of 2, depending on which value of the constant is used!

The validity of the second assumption depends on the geometry of the universe (see Section 7.5). If the universe, for example, is closed (so that it will eventually stop expanding), then the Hubble plot cannot be extended very far as a straight line. It must eventually curve sharply upward, because the expansion was much faster in the past than now. (Remember that distant objects are seen now as they were in the past.)

If we knew the overall geometry (open or closed) and the present density of our universe, we could apply relativity theory to draw the expected red shift–distance relation and use it to find the distances to galaxies. But we have a circular problem here, for the observed red shift–distance relation is one of the things used to determine the geometry of the universe. At great distances, we must assume a geometry to get a distance. So our second assumption has a weak foundation.

To sum up: Finding distances to galaxies is one of the toughest jobs for astronomers. Yet it's one of the most important. Let me summarize the steps in one contemporary procedure followed by Sandage and Tammann, which yielded a value of Hubble's constant of 15 km/sec/Mly. (To determine this value took *19 years* of careful observing!) There are eight basic steps:

1 Measure the AU, using radar reflection from Venus; then the distances to nearby stars using heliocentric parallax with the AU as the baseline.

2 Determine the distance to the Hyades, a galactic cluster, using the motions of its stars, checked for consistency by comparing the brightnesses of Hyades stars with those of nearby stars with distances known by trigonometric parallax.

3 Find the distances to cepheids in our Galaxy. Search out galactic clusters that contain cepheids, compare the brightness of the stars in these clusters with that of the stars in the Hyades cluster, and find the distances to the clusters with cepheids (there are only a few). We have now calibrated the cepheid period-luminosity relation.

4 Use cepheids to determine the distances to nearby galaxies by the period-luminosity relationship.

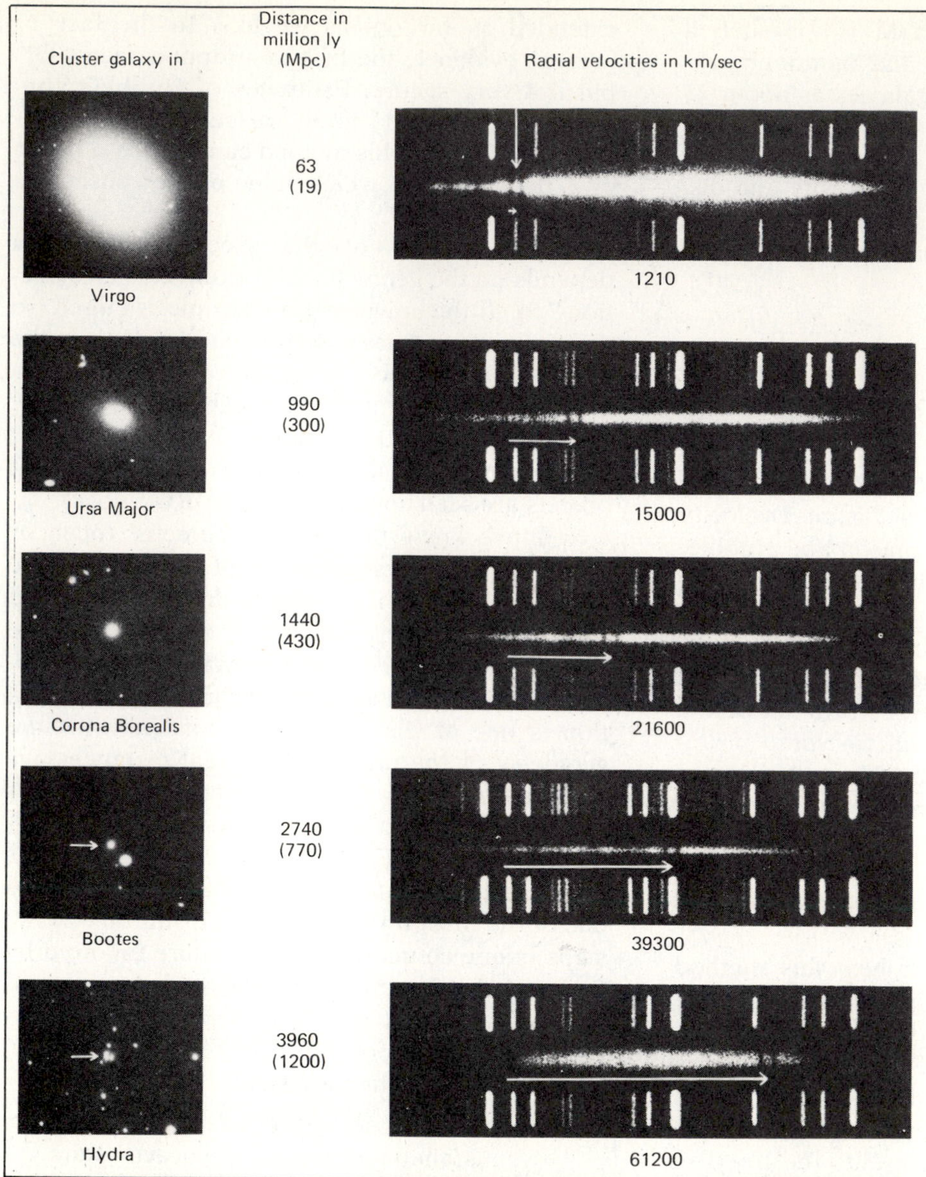

Cluster galaxy in
Distance in million ly (Mpc)
Radial velocities in km/sec
Virgo
63
(19)
1210
Ursa Major
990
(300)
15000
Corona Borealis
1440
(430)
21600
Bootes
2740
(770)
39300
Hydra
3960
(1200)
61200

19.10

Measured red shifts and distances for selected galaxies in order (top to bottom) of increasing red shift and distance. The red shifts are visible in the spectra on the right, where a white arrow indicates the size of the shift in the H and K lines of calcium; these are the two darkest lines in the spectra. (Courtesy Palomar Observatory, California Institute of Technology)

60,000
50,000
40,000
30,000
20,000
10,000
Velocity (km/sec)
0
1300
2600
3900
Distance (millions of ly)

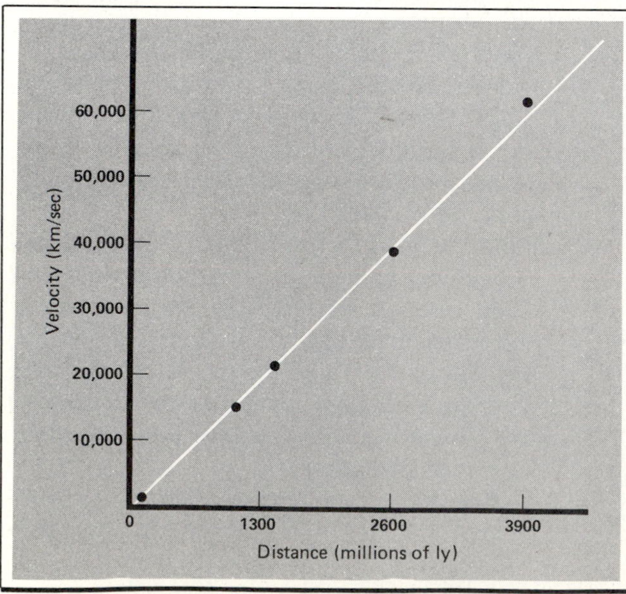

19.11

The Hubble plot for the data in Figure 19.10. The straight line corresponds to a slope for a value of Hubble's constant of 15 km/sec/Mly.

5 Measure the angular sizes of H II regions in these nearby galaxies. We find that the actual sizes of H II regions in spiral and irregular galaxies depend on the luminosity of the galaxies. Use nearby galaxies to calibrate that H II–region size-luminosity relation.

6 Extend this calibration to *supergiant*, very luminous Sc I spiral galaxies, for which the nearest, M 101, has its distance determined by different methods as a check. We now have the sizes of H II regions for Sc I galaxies. However, we must be careful because the calibration is based on only *one* galaxy, M 101.

7 Use the size-luminosity relation for H II regions to find the distances of galaxies to the limit of this method. (We can use this procedure on about 50 galaxies.) Now we know the luminosities of these galaxies and their relationships to luminosity classes.

8 Look at supergiant spiral (Sc I) galaxies—the objects we can see distinctly at the greatest distances. Use the luminosity calibration (step 7) to find their distances. Measure their red shifts. Divide their red shifts by their distances. We then have Hubble's constant.

Another way to H / Not all astronomers agree with Sandage and Tammann that *H* is about 15 km/sec/Mly. Marc Aaronson, Jeremy Mould, and John Huchra have recently developed a new technique, independent of the Sandage-Tammann steps, to find a value of *H* equal to 27 km/sec/Mly (with an uncertainty of 5 percent).

The technique rests on a relation between the luminosities of spiral galaxies and the widths of their 21-cm H I emission: The larger the line width, the greater a galaxy's luminosity. The 21-cm emission comes from the neutral gas in a spiral galaxy's disk. If we measure the 21-cm line from a spiral galaxy viewed edge on, the emission from the gas moving away from us will be red-shifted and that moving toward us will be blue-shifted, so the 21-cm line will be broader than expected if the galaxy had no rotational motion. In fact, the line width for edge-on galaxies measures the maximum rotational velocity in the disk. Most galaxies have maximum rotational velocities of 100 to 400 km/sec (depending on their type). So the relation between luminosity and 21-cm line width is really one between luminosity and rotational velocity.

Why should there be *any* relation between the maximum rotational velocity and the luminosity of a galaxy? The rotational velocity at a given distance from the center relates to the mass of a galaxy, and the luminosity of the stars in a galaxy is related to their masses. So if we stick with one type of galaxy, if the galaxies have more or less the same mix of stars of different masses, and if the degree of concentration of mass toward the center is more or less the same, we might expect the total luminosity to be related to the rotational velocity. But it is surprising that the relationship is so good, with relatively little variation from one galaxy to the next, at least among galaxies of the same type. (The relationship seems to work best for Sc galaxies.)

If properly calibrated, this relation provides a galaxy's luminosity from a measurement of its 21-cm line width. Infrared observations of nearby galaxies (such as M 31 and M 33), whose distances are known from cepheids, give the infrared luminosities. These infrared luminosities connect well with the width of the 21-cm lines from local galaxies.

Aaronson and colleagues then measured the infrared fluxes and 21-cm line widths for some nearby clusters of galaxies. Again they found a good connection between the fluxes and line widths; applying the calibration from local galaxies, they get the distances. The measured red shifts then give a Hubble's constant of 27 km/sec/Mly.

Note that a value for *H* of 27 km/sec/Mly implies a universe no older than about 10 billion years (see Focus 7.2). That result poses a problem, for the oldest stars (in globular clusters) are now thought to be some 18 billion years old.

Who's right? Both the Sandage-Tammann and Aaronson-Huchra-Mould methods probably contain unknown systematic errors. A very careful study of the calibration procedures by Gérard de Vaucouleurs results in a Hubble's constant close to 30 km/sec/Mly. But I can say pretty confidently that *H* lies in the range of 15 to 30 km/sec/Mly. The actual value may be near 20 km/sec/Mly, which wouldn't cause serious problems with the ages of stars.

19.5 General characteristics of galaxies

Let's step back a bit from our cosmological vista and study the characteristics of galaxies (Table 19.1), now that we have methods to find out their distances.

Size / Once you know the distance to a galaxy, you can find out its actual size from a measurement of its angular size. The hitch here is that the definition of the edge of a galaxy is more or less arbitrary. Different definitions of the edge

TABLE 19.1 General properties of galaxies

Property	Sprials	Irregulars	Ellipticals Giant	Ellipticals Dwarf
Size (ly)	60×10^3	23×10^3	150×10^3	10×10^3
Mass (solar masses)	100×10^9	10^9	10^{13}	10^6
Luminosity (solar luminosities)	10^{10}	10^9	10^{11}	10^8
Color	Bluish disk, Reddish halo	Bluish	Reddish	Reddish
Stars	Young disk, old halo and nucleus	Young	Old	Old
Gas and dust (% of total mass)	5	15	< 1	< 1
Fraction known (%)	77	3	20	20
Mass-luminosity ratio	5	1	10	0.01

(which is not simply where the visible stars end) result in different sizes.

Dwarf ellipticals and small irregulars tend to be the smallest galaxies—only 10,000 ly in diameter for some. Giant ellipticals can range up to 200,000 ly in size. To put this size range into perspective, imagine your height (about 2 m) to be the size of a dwarf galaxy. Then an irregular galaxy would be about twice your size, a spiral 10 times your size, and a giant elliptical some 20 times your size!

The very largest galaxies are the *supergiant* ellipticals, sometimes called *cD* galaxies. These can have radii up to 3 million ly. That's greater than the distance from our Galaxy to the Andromeda galaxy (M 31)!

Mass / To find a galaxy's total mass is no simple task. The light from a galaxy comes mostly from its stars, but much of its material may not emit visible light—or may not emit at all (black holes, for example). The most widely used methods of finding a galaxy's mass are rotation curves and binary galaxies.

The rotation curve method works as follows: A galaxy's material orbits the nucleus in a special way, so at every distance, the orbital velocity has a certain value. This rotation curve comes from the orbital motions, described by Newton's laws, that arise from the distribution of mass within the galaxy. So if we observe a galaxy's rotation curve and make up a model for that galaxy's mass distribution, we can work out the galaxy's total mass, including dark masses that we can't see directly. (We observe the rotation curve by placing a spectroscope's slit across the galaxy and measuring the Doppler shift at a number of points from the center out to the edge.)

The most massive galaxies we know of are the supergiant elliptical (cD) galaxies. They can have up to 10^{14} solar masses, 100 times more mass than our Milky Way Galaxy contains. Spiral galaxies hit as high as several 10^{12} solar masses, though their rotation curves (Fig. 19.12) become flat without dropping off out to the extent measured so far.

For binary galaxies, we again make use of the versatile Doppler shift. Imagine two galaxies orbiting about the center of mass of the binary. Assume that the orbits are stable. Then, just as with visual binary stars (see Section 14.6), we could apply Newton's form of Kepler's third law to find the masses, if we knew the distance, the angular size of the orbit, the period, and the position of the center of mass. However, galaxies revolve too slowly for us to see their actual orbits, periods, and relative centers of mass. So we cannot find the individual masses of binary galaxies. All we can measure are the present radial velocities and the separation; we don't know what part

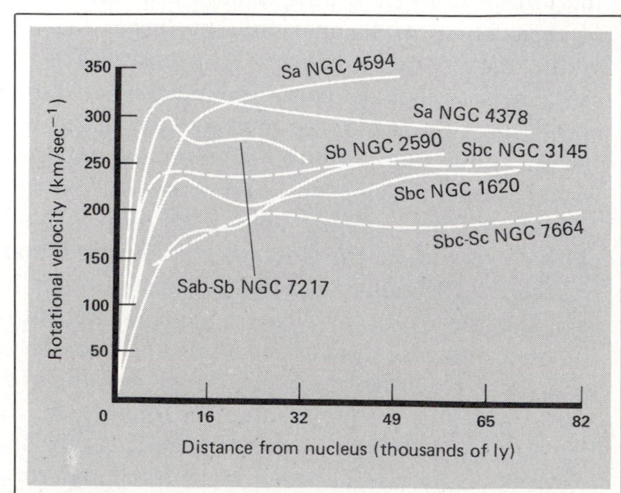

19.12
Measured rotation curves for seven spiral galaxies. (Adapted from a diagram by V. Rubin, W. K. Ford, Jr., and N. Thonnard, Astrophysical Journal (Letters), vol. 225 (1978), p. L107, copyright © 1978 by the American Astronomical Society)

of the orbit the galaxies are on, or what the inclination is, so we don't know what the true orbital velocities are. But if we examine a large sample of galaxies and assume their orbits are nearly circular and randomly oriented to our line of sight, we can estimate from these data the *average* masses of the galaxies sampled.

An investigation of 279 binary systems, mostly spirals, using the Doppler shift of the 21-cm line to get their velocities, finds an average mass of 10^{12} solar masses for these spirals (for $H = 20$ km/sec/Mly).

Warning: These two methods—rotation curves and binary galaxies—do not necessarily give the same results; masses from rotation curves are generally smaller than those from the binary galaxy method. This difference has reinforced the idea that galaxies have massive halos surrounding their bright, starry disks. New observational techniques that probe beyond the visible disk show that the rotation curves for spirals remain flat out to some 300,000 ly from their centers (Fig. 19.13). These flat curves imply directly that these galaxies have massive, invisible halos. Our own Galaxy has such a halo (see Section 18.2).

Luminosities / If we know their distances and fluxes, we can work out the luminosities. One trouble here is that it's not easy to measure a galaxy's total flux. Why not? Because a galaxy thins out gradually at its edge. It's hard to be sure that you're catching all the light from the galaxy. In addition, corrections have to be applied for light absorption: first, for that due to dust in our Galaxy, and second, for that due to dust in the galaxy itself (especially for spirals). Finally, we must correct for the fact that we view most galaxies tilted to our line of sight, so we measure some fraction of their total light output.

The luminosities of galaxies range from 2×10^5 solar luminosities for dwarf ellipticals to

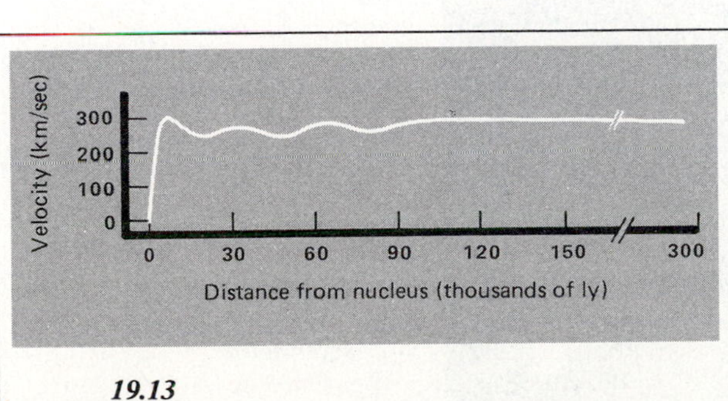

19.13
A rotation curve for a spiral galaxy out to a distance of 300,000 ly from its center. (Adapted from a diagram by V. Rubin)

10^{12} solar luminosities for supergiant (cD) ellipticals. These latter types are very rare, however. The Milky Way Galaxy, if we could see it all from space, has a luminosity somewhere near 2.5×10^{10} solar luminosities.

Mass-luminosity ratios / Divide the total mass of a galaxy by its luminosity and you have its *mass-luminosity ratio* (abbreviated M/L), an indication of the average energy output per unit solar mass from the galaxy. (It is usually expressed in units of solar masses and solar luminosities.) Modern determinations using rotation curves give 5 for the average mass-to-light ratio for spiral galaxies and about twice as much for giant ellipticals and lenticulars. (For comparison, the M/L ratio for stars in the sun's neighborhood is about 1.)

Why do ellipticals have a larger M/L? Probably because they contain a greater percentage of low-mass stars with low light output—main-sequence stars of class *M*. If true, this extra abundance of *M*-stars would mean that ellipticals should be redder in overall color than spirals (see the following paragraphs). Other possibilities are neutron stars, black holes (including perhaps a giant one in the nucleus), and dark interstellar matter, which contribute to the mass but not to luminosity.

Colors / As for stars, we can measure the colors of galaxies using various filters on a telescope. The color of a galaxy depends on the predominant stellar type in its mixture of stars; it relates directly to the kind of stars in the galaxy. For example, a galaxy with many *O*- and *B*-stars is bluer than a galaxy with many red *M*-stars.

In fact, a direct correlation exists between galaxy type and its color. Ellipticals tend to be much redder than spirals, and spirals redder than irregular galaxies. Within the spiral group, the galaxies appear redder as their nuclear bulges grow larger and their spiral arms less extensive.

The progression of color from the bluer irregulars to the redder ellipticals reflects a trend in the composition of the galaxy's population. It used to be thought that their reddish color meant that the ellipticals and nuclei of spirals contained Population II stars, like those in the globular clusters. But later work discovered that these stars, though old, nevertheless have a high metal abundance. So it is appropriate to call them old Population I. In general terms, then, an old Population I predominates in ellipticals, whereas a much younger Population I stands out in the irregulars. The mixture in the spirals is determined by the size of the nucleus (old Population I) compared with that of the spiral arms (young Population I).

(Population II is probably a minor contributor in all *large* galaxies, existing mainly in the globular clusters and galactic halo.)

Here is a way to remember the primary difference among galaxy types: Recall that our Galaxy has a halo and nucleus of red stars and spiral arms of blue stars. Imagine our Galaxy without the halo and nucleus. What remains (the arms) is like the stars, gas, and dust found in irregular galaxies. Now imagine our Galaxy stripped of its spiral arms. The remains (nucleus and halo) are typical of the composition of elliptical galaxies.

As you might suspect, a galaxy's color also relates to its overall content of gas and dust. The reddest galaxies, ellipticals, contain almost no gas and dust. The bluest galaxies, irregulars, contain the greatest percentage of gas and dust relative to their total mass.

As the population comparison implies, irregulars have both old and young stars, but ellipticals contain only old stars. Why? Because in irregulars, there is still enough gas and dust so that star formation continues; in ellipticals, it halted many years ago.

Warning: Don't think that elliptical galaxies are *completely* devoid of gas and dust. Roughly 10 percent of ellipticals have detectable emission from gas usually ionized and usually confined to the nuclear region.

To sum up: Galaxies contain stars, gas, and dust. The mixture of gas, dust, and stars, as well as the stellar types, relates closely to a galaxy's type and structure. Color and spectra measurements show that the nuclei of spirals contain an old stellar Population I, the arms much younger stars. Most elliptical galaxies have colors similar to those of spirals' nuclei, and irregular galaxies appear most akin in color and spectra to spirals' arms. Ellipticals contain very little gas and dust. In contrast, spirals and irregulars embrace extensive quantities of interstellar material. In spirals, the gas is most obvious as H I clouds or as H II regions cloaking *O*- and *B*-stars. Although the masses of galaxies are difficult to measure for distant systems, we know that in general, giant ellipticals are the most massive and irregulars the least massive.

Note that the sequence from ellipticals to irregulars marks a sequence from galaxies in which starbirth ceased long ago (ellipticals: old Population I and some Population II, no gas and dust) to those in which starbirth is carried on very actively (irregulars: young Population I, large percentage of gas and dust). This trend gives a key clue to understanding the evolution of galaxies.

19.6 Clusters of galaxies

If you ever get the chance, take a close look at wide-angle photos of the sky (such as the Palomar Sky Survey). On some photos you will see extensive swirls of gas and many stars. But on the photos of regions where the stars thin out—away from the Milky Way—you can see the tiny forms of galaxies. If you look at enough photos, you will notice that if you find one galaxy, you're likely to see others nearby. You have found that galaxies

19.14
A small part of the Virgo cluster of galaxies. (Courtesy Kitt Peak National Observatory)

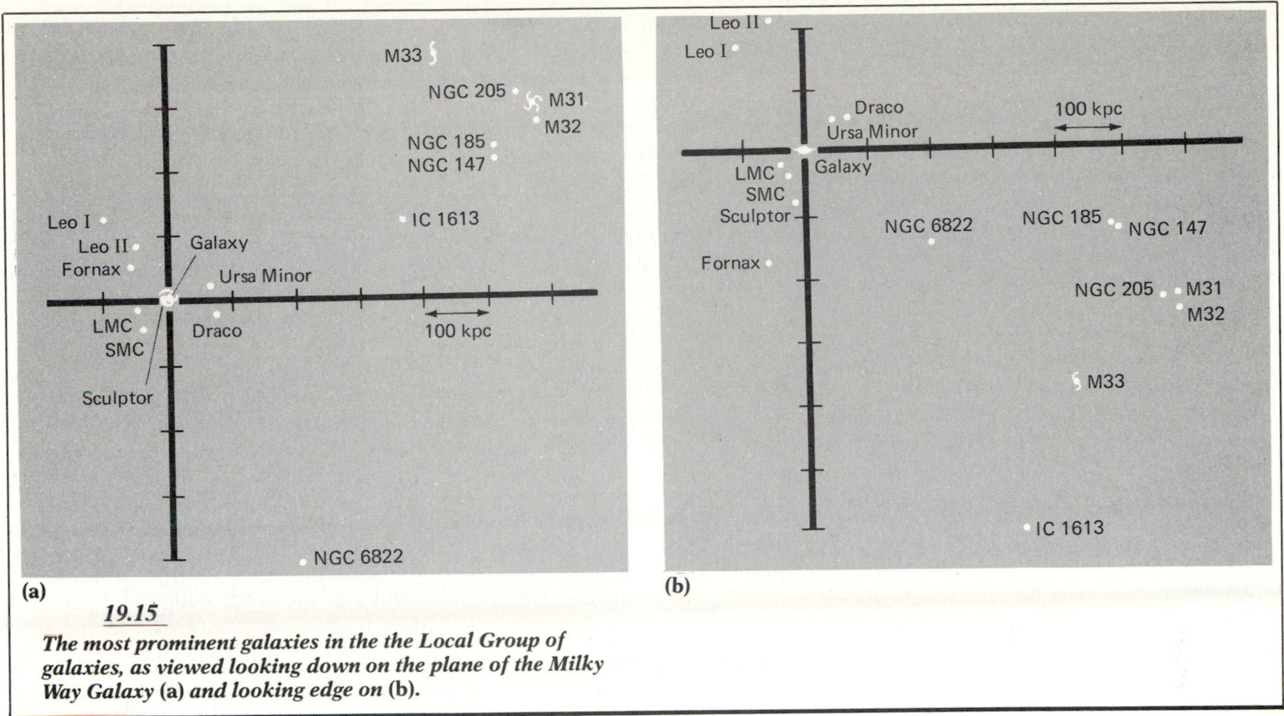

19.15
***The most prominent galaxies in the the Local Group of galaxies, as viewed looking down on the plane of the Milky Way Galaxy* (a) *and looking edge on* (b).**

tend to come in clusters (Fig. 19.14). In fact, it may be true that *all* galaxies belong to clusters—though many of these clusters may be a simple marriage of two galaxies. Our universe is one of clusters of galaxies!

The Local Group / The nearest cluster of galaxies is the one to which the Milky Way Galaxy belongs. It is called the *Local Group of galaxies* (*Local Group* for short). This aggregation takes up a volume of space nearly 3 million ly across in its long dimension (Fig. 19.15). Our Galaxy is located near one end of the Local Group, and M 31 is near the other.

As the most massive objects in the clusters, the Milky Way Galaxy and M 31 dominate its motions and secure the other members gravitationally. In fact, the Galaxy and M 31 orbit each other. The other members of the Local Group come along for the ride. The Local Group contains over 20 galaxies; they consist mostly of ellipticals (Table 19.2). Some of these ellipticals are quite faint and are the dwarf ellipticals that contrast so dramatically with the giant ellipticals found in other clusters. The obscuring matter in the Milky Way probably clouds our sight of other members, especially faint dwarf ellipticals.

Let's look briefly at some of the more important members of the Local Group.

The Large and Small Magellanic Clouds (Fig. 19.16) lie closest to the Galaxy. They are, in fact, connected to our Galaxy by a bridge of hydrogen gas. The two clouds are physically connected to each other by a large but thin envelope of neutral hydrogen. Both are distorted by tidal interactions with our Galaxy.

The Large Magellanic Cloud (abbreviated LMC) contains stars totaling about 20 billion solar masses and orbits 170,000 ly from the Galaxy. Spectra of a large number of stars in the LMC show that most stars are similar to those in the solar neighborhood, but more *O*- and *B*-stars exist. Radio studies demonstrate that the LMC has large amounts of neutral hydrogen gas, a total of approximately 3 billion solar masses. The gas is also evident from the more than 400 emission nebulas found so far. One of these nebulas, called 30 Doradus, which is illuminated by a cluster of *O*- and *B*-stars, extends over 1600 ly in diameter and contains roughly 5 million solar masses of material.

Although very small compared with the LMC, the Small Magellanic Cloud (SMC) displays a similar stellar population. It has a total mass of about 2 billion solar masses, one-tenth that of the LMC. Its distance from us is about 205,000 ly. The SMC and the LMC both orbit around the Milky Way Galaxy.

The Magellanic Clouds are not the only galaxies orbiting our Milky Way Galaxy. A swarm of six dwarf galaxies accompanies us in space. These dwarf galaxies have masses of only about a million solar masses and range from 160,000 to 980,000 ly away.

M 31 is sometimes still called the Great Nebula in Andromeda. Actually a spiral galaxy, M 31 is easily visible to the unaided eye on a dark,

TABLE 19.2 Physical properties of some galaxies in the Local Group

Name	*Type*	*Diameter (10^3 ly)*	*Distance (10^3 ly)*	*Mass (solar masses)*
Milky Way	Spiral	120.0	. . .	6.4×10^{11}
NGC 147	Dwarf elliptical	7.8	2220	?
NGC 185	Dwarf elliptical	9.5	2220	?
NGC 205	Elliptical	14.0	2220	?
M 31	Spiral	170.0	2220	4×10^{11}
M 32	Elliptical	6.8	2220	2×10^{9}
SMC	Irregular	16.0	196	?
Sculptor	Dwarf elliptical	7.5	280	3×10^{6}
IC 1613	Irregular	13.0	2220	?
M 33	Spiral	59.0	2220	2×10^{10}
Fornax	Dwarf elliptical	20.0	612	2×10^{10}
LMC	Irregular	26.0	173	?
Leo I	Dwarf elliptical	5.8	750	3×10^{6}
Leo II	Dwarf elliptical	4.2	750	10^{6}
Ursa Minor	Dwarf elliptical	7.8	222	10^{5}
Draco	Dwarf elliptical	3.3	250	10^{5}
NGC 6822	Irregular	5.5	2150	?

Note NGC stands for a listing in the *New General Catalogue,* M for the Messier Catalog.

19.16
The Large (left) and Small (right) Magellanic Clouds. These two galaxies are the closest to the Milky Way Galaxy. (By permission of Harvard College Observatory)

moonless night. With binoculars you find an elliptical, hazy patch of light. The spiral arms are too faint to see without a big telescope.

M 31 tilts 15° to the line of sight. Because of the tilt, dark lanes of obscuring material are plainly visible along with the spiral arms marked by *O*- and *B*-stars. A halo of globular clusters surrounds M 31 like bees around a hive, in a distribution like that around our galactic system. The Andromeda galaxy also has companions—a total of seven dwarf ellipticals that orbit it.

M 33 is the only other large spiral in the Local Group (Fig. 19.17). It is the closest Sc II galaxy to us, only 2.7 million ly away. In a photo, you can easily trace out wide, open spiral arms. Note that the arms are resolved into stars; most of these are blue supergiants. Because M 33 is so close, we have been able to investigate in detail its dark dust lanes, H II regions, open star clusters, novas, and cepheid variables. M 33 has an overall diameter of about 60,000 ly—only about half the size of the Milky Way Galaxy.

Other clusters of galaxies / Other clusters range from compact ones (Fig. 19.18) to rather loose arrays of galaxies. The Fornax cluster, one relatively close to us, displays a wide variety of types of galaxies, even though the total number is only 16. The huge Coma cluster (Fig. 19.19) spreads over at least 23 million ly of space and contains thousands of galaxies. Observations of even just the brightest galaxies show how common clustering is. From these observations, we find that a typical cluster contains about 100 galaxies and is separated some tens of millions of light years from its neighboring clusters.

The *Virgo cluster* stands out as one of the most stupendous in the sky. Of the 205 brightest galaxies in the Virgo cluster, the four brightest are giant ellipticals, but in all, ellipticals make up only 19 percent compared with the spirals' 68 percent. The Virgo cluster covers about 7° in the sky (14 times the diameter of the moon!), which implies that its physical diameter is some 10 million ly at a distance of 51.2 million ly.

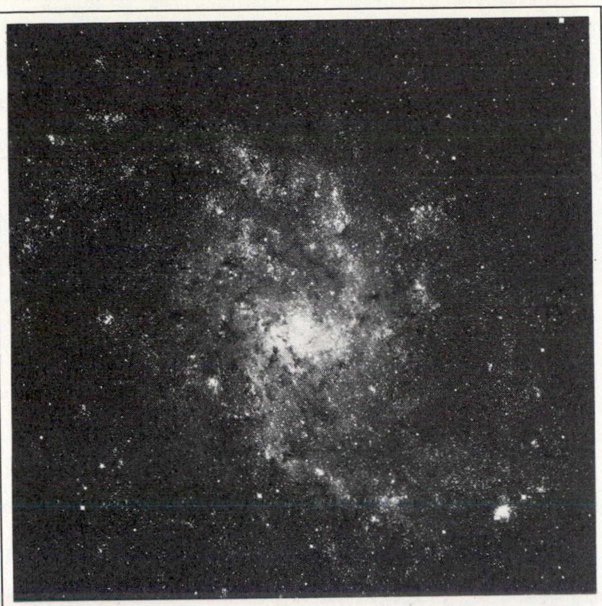

19.17
The spiral galaxy Messier 33 in Triangulum. It is Hubble type Sc. (Courtesy Lick Observatory)

19.19
A cluster of galaxies in Coma Berenices. Note how the smaller galaxies appear to cluster around the two giant ellipticals at the center and to the left. (Courtesy Kitt Peak National Observatory)

19.18
A small cluster of galaxies called Stephan's quintet. (Courtesy Lick Observatory)

Clusters and the luminosity of galaxies / From the description of the Local Group, you might suspect that very luminous galaxies are few in number compared to low-luminosity ones. Only three local galaxies are very luminous (the Milky Way Galaxy, M 31, and M 33), whereas most galaxies in the Local Group are dwarf, low-luminosity galaxies. Galaxies have a luminosity function: There are many faint galaxies and few bright ones in a cluster.

Much of a cluster's mass resides in very faint galaxies. It's difficult to estimate the masses of these clusters because not all the material in them can be seen at visual wavelengths; so adding up all the galaxies gives a lower limit to the cluster's mass. On the other hand, if the cluster is assumed to be gravitationally bound, its motions establish an upper limit on its mass. (The actual value lies between these two limits.) Masses range from 10^9 to 10^{15} solar masses. However, it is not possible to tell if all clusters are bound and stable or if they are unstable and expanding.

Superclusters / Are there clusters of clusters of galaxies? That is, do *superclusters* exist? Yes! In a catalog of 2712 clusters, 50 superclusters crop up, each containing about 11 clusters and having diameters of about 250 million ly.

Much work on superclustering has been done by Gérard de Vaucouleurs. He finds evidence for a *Local Supercluster* that has a diameter of some 100 million ly. It contains the Local Group, Virgo cluster, and Coma cluster among others, for a total mass of some 10^{15} solar masses. The Local Supercluster appears to be somewhat flattened, which may imply that it is rotating, but it may simply have been formed this way. The center of mass lies in or near the Virgo cluster; our Galaxy and M 31 lie on the outskirts.

Brent Tully and Richard Fisher have recently completed a three-dimensional map of the Local Supercluster (Fig. 19.20). It shows a rich, convoluted structure that breaks into two main clouds

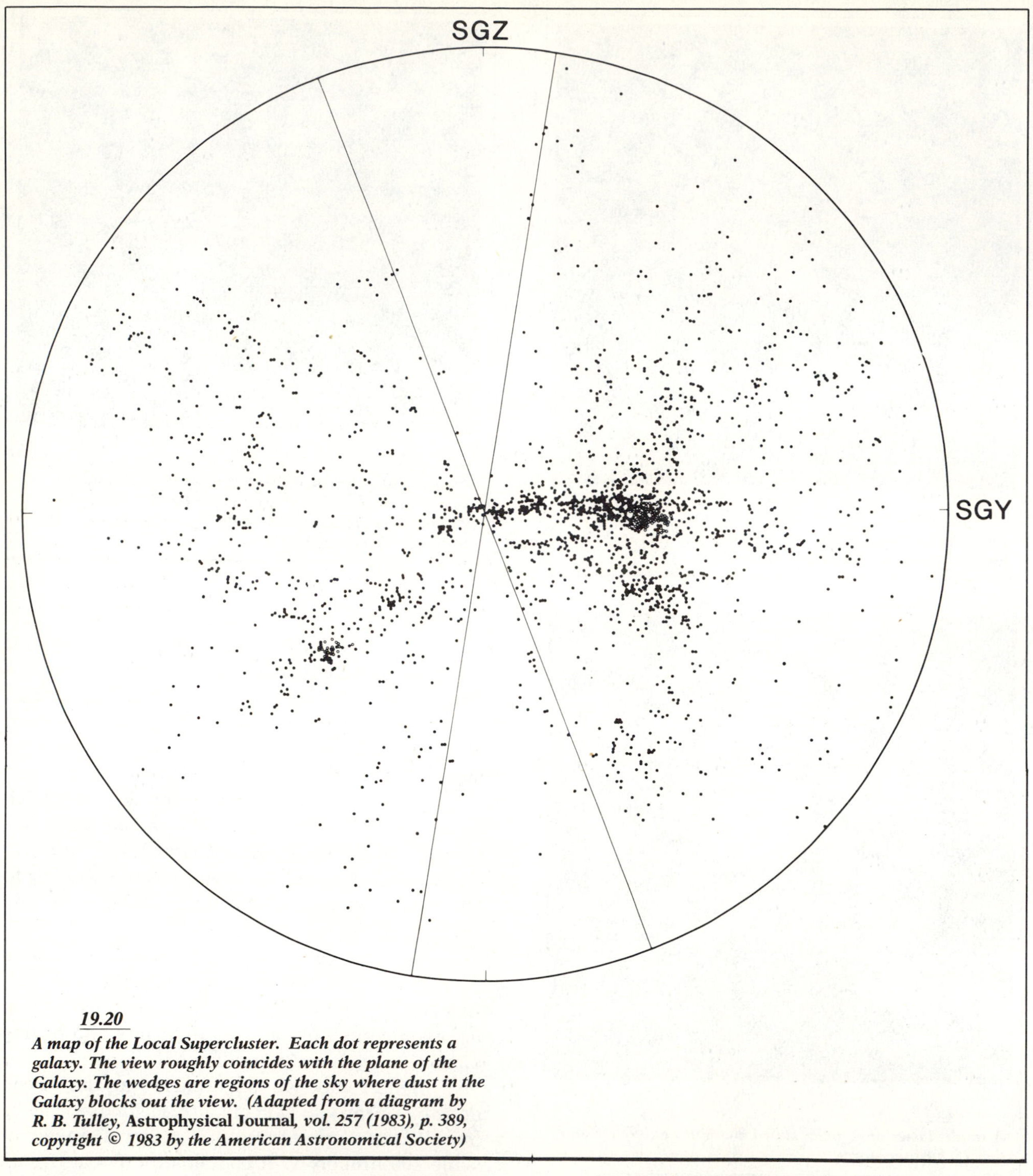

19.20

A map of the Local Supercluster. Each dot represents a galaxy. The view roughly coincides with the plane of the Galaxy. The wedges are regions of the sky where dust in the Galaxy blocks out the view. (Adapted from a diagram by R. B. Tulley, **Astrophysical Journal,** *vol. 257 (1983), p. 389, copyright © 1983 by the American Astronomical Society)*

with streamers (thin, cigar-shaped clouds of galaxies) emerging above and below the central plane. Most of the supercluster is empty space: 98 percent of the visible galaxies are restricted to 5 percent of the volume. These clouds of galaxies outline a disklike structure six times wider than it is thick—a true cosmic pancake of clusters of galaxies!

Recently, astronomers have investigated the Hercules supercluster (Fig. 19.21), which covers over 600 billion cubic light years of space. They find the supercluster lies at a distance of 720 Mly, if $H = 15$ km/sec/Mly. The supercluster itself occupies a broad band at a distance of 400 to 600 million ly. In front of the supercluster lies a void some 330 million ly deep that separates the supercluster from foreground galaxies.

The Hercules and other superclusters that have been mapped to date (Coma and Perseus) exhibit common features. First, they confirm the

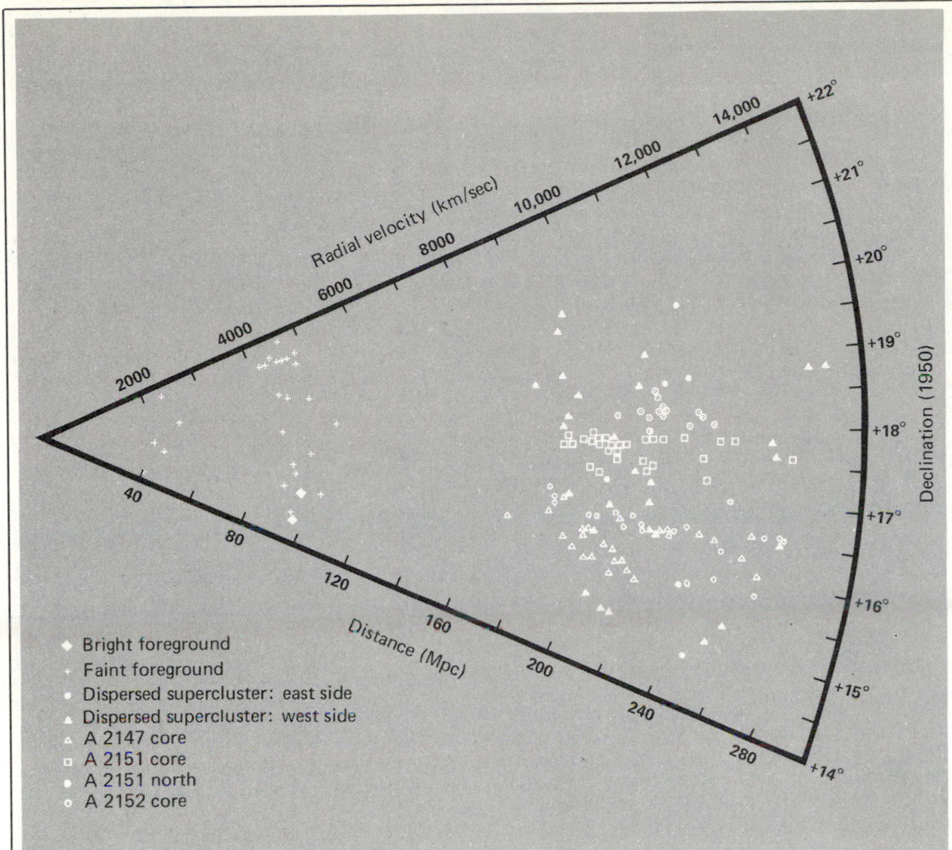

19.21
A slice of the Hercules supercluster. Shown here are the red shifts (and so distance, for H = 15 km/sec/Mly) and positions in the sky for the galaxies in the supercluster, which contains a number of clusters (labeled Abell 2147, 2151, and 2152). (Adapted from a diagram by M. Tarenghi, W. Tifft, G. Chincarini, H. Rood, and L. Thompson,* Astrophysical Journal, *vol. 234 (1979), p. 793, copyright © 1979 by the American Astronomical Society)

existence of superclusters as organized structures composed of multiple clusters of galaxies. Second, and this was a surprise, they contain large voids in which no galaxies exist. These voids must be an integral part of the process that forms superclusters—a process about which we have vague ideas right now. Third, streams of galaxies appear to connect the main concentrations in superclusters (Fig. 19.22). Although we have examined only some 2 percent of the sky to date, we are beginning to catch a glimpse of the architecture of the cosmos.

The cosmic tapestry / What does the universe of galaxies look like on the grand scale? James E. Peebles and co-workers have investigated in detail a survey of a million galaxies (Fig. 19.23) performed at Lick Observatory. (This survey took 12 years to complete!) They find that galaxies cluster in knots and filaments in a hierarchical fashion. These are chains of galaxies and clusters (looking like a chain-link fence). Nearly all galaxies are within these clusters, with huge holes between them devoid of luminous matter. Here we see the explosive imprint of the Big Bang—a filamentary texture similar to that of supernova remnants.

Galactic cannibalism / A remarkable fact about clusters of galaxies is that the spacing of galaxies is pretty close, compared to the sizes of the galaxies themselves. Consider, for instance, planets and stars in terms of sizes and spacing. In the solar system, the planets are spaced out about 100,000 times their diameters. In the Galaxy, stars are spaced out about a million times their diameters. But in a cluster of galaxies, the spacing amounts to only about 100 times a typical galaxy's diameter. Astronomically speaking, galaxies in a cluster are very crowded together. So you might wonder if they would ever pass close by one another.

Consider this additional fact: the most massive galaxies (giant ellipticals) are at least 10 million times more massive than the least massive ones (the dwarf ellipticals). Then it's not hard to imagine that the largest galaxies could disrupt the smallest ones by tidal forces (see Focus 9.1), strongly enough to destroy the structure and then pull the pieces in. Astronomers call this devouring of a smaller galaxy by a larger one *galactic cannibalism*.

A neat name, but do we have any observations to support it? Recent ones show that the supergiant elliptical galaxies (cD galaxies) do have peculiar properties; they are not simply very massive ellipticals. These properties include: (1) extensive halos, up to 3 million ly in diameter, (2)

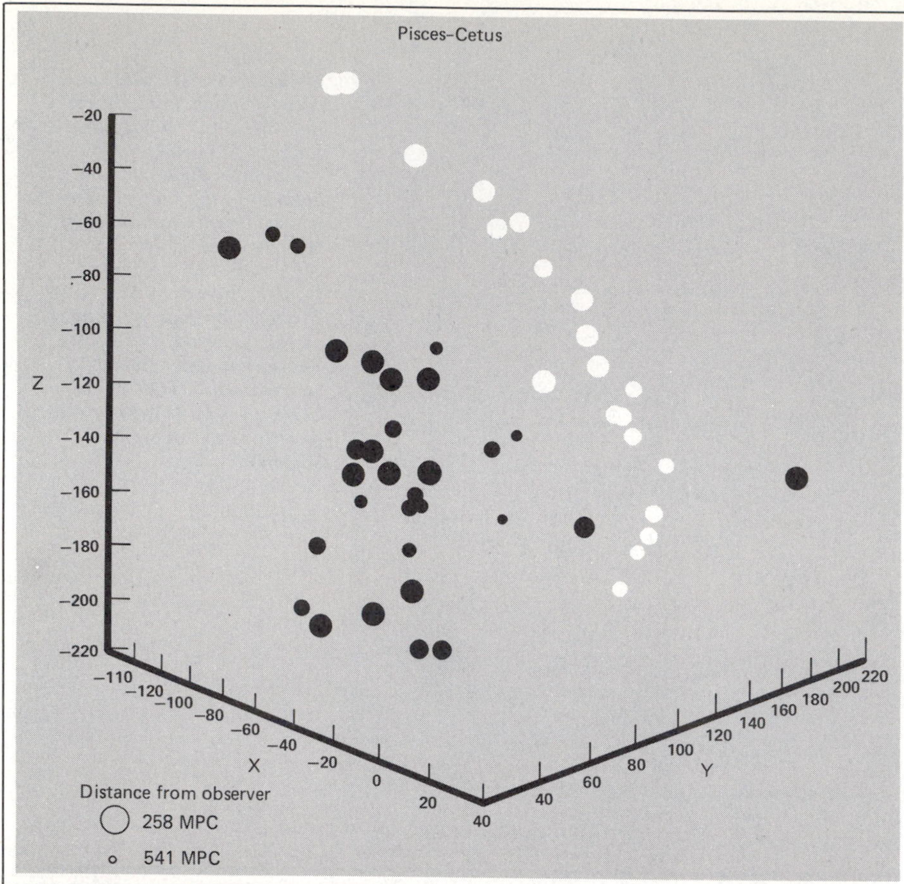

19.22

A view of a possible supercluster in the constellations Pisces and Cetus. Each circle represents a cluster of galaxies; a circle's size, its distance from an observer 1500 million ly from the center. The lighter symbols make up a chain of clusters over 800 million ly long. (Adapted from a diagram by J. Burns and D. Batuski)

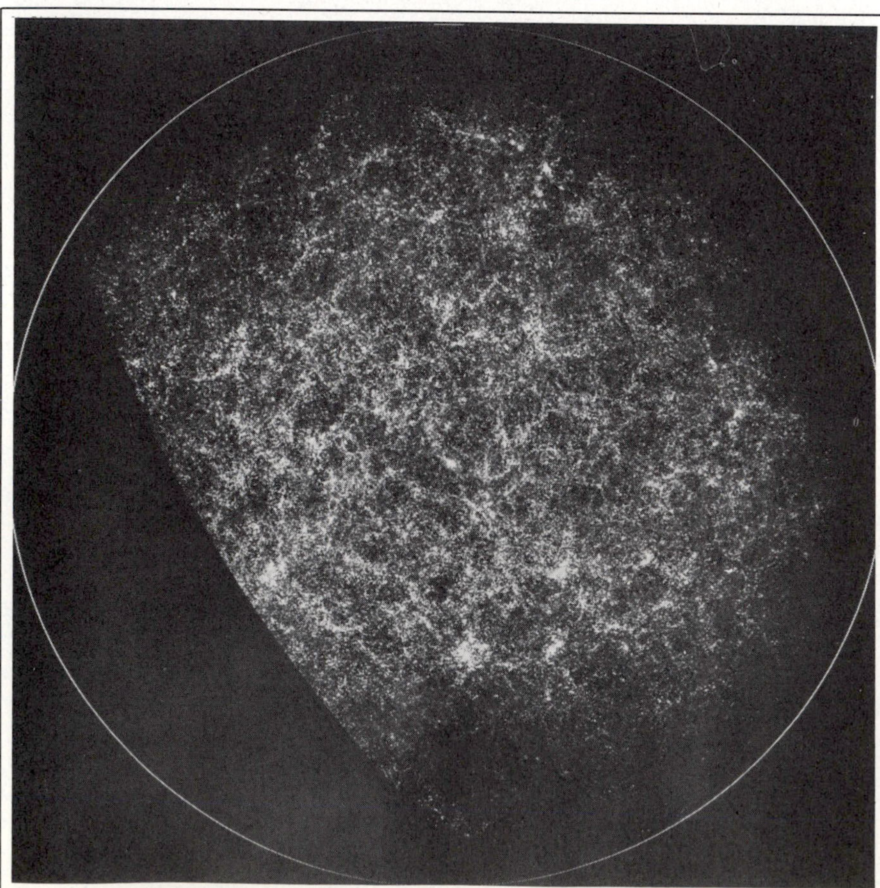

19.23

The universe of galaxies. More than a million galaxies are shown. The lighter a region, the more galaxies within it. Note the clumpy, chainlike structure. (Courtesy P. J. E. Peebles, based on the Lick Observatory Catalog by C. Shane and C. Wirtanen)

19.24
The galaxies NGC 4038 and 4039, a tidally interacting pair. (Courtesy Palomar Observatory, California Institute of Technology)

multiple nuclei near their centers, and (3) locations at the center of clusters. These observations, when combined with the motions of cD galaxies within clusters, suggest that these special galaxies form from galactic cannibalism. How? By close encounters at the centers of clusters that tidally strip material from other galaxies, which then is picked to promote the growth of a cD galaxy.

Galaxies do not have to actually merge to show the effects of tidal forces. A modestly close encounter would also have observable consequences. One, as demonstrated by the earth's tidal bulges, matter would be pulled out in tongues from both sides of each galaxy. Two, because galaxies rotate, their material after an encounter would flow off in arc-shaped streams. So we expect that bridges of material might join two tidally interacting galaxies with tails pointing away from each in opposite directions.

Some galaxies with peculiar shapes that do not fall into the standard Hubble categories show indications of tidal interactions. The pair NGC 4038 and 4039 illustrate this well (Fig. 19.24). Note a visible bridge of material between the galaxies and the tadpolelike tails extending from them. Computer simulations show that these forms result naturally from a tidal interaction.

19.7 *Intergalactic matter*

Is intergalactic space empty? Or does an intergalactic medium exist like the interstellar medium? If an intergalactic medium is present, it may contain both gas and dust. The gas (probably hydrogen) may be in neutral form, or it may be ionized. We can look for the *intergalactic medium* in two locations: (1) *between* the clusters of galaxies and (2) *within* clusters of galaxies.

Matter between clusters / To get some idea of how much material might be in an intergalactic medium, imagine the following: Take the matter from all the galaxies we can see and spread it out over the entire volume of space we can observe. This spread-out material would have a density of about 4×10^{-28} kg/m^3. (That's about 2 hydrogen atoms every 10 cubic meters.) Recall that this density is about 10 times less than that needed to close the universe (see Section 7.5). For the intergalactic medium to be significant, its density would need to be about this large. What is the evidence for such a density of dust, neutral hydrogen, or ionized hydrogen?

Consider the possibility of intergalactic dust. Such dust, if it resembled the interstellar dust in our Galaxy, would extinguish and redden the light from distant galaxies. This extinction and reddening has been searched for but not found. So intergalactic space cannot contain very much dust; the density must be less than 4×10^{-30} kg/m^3, which is 100 times less than the density of the spread-out material in galaxies.

How might we detect neutral hydrogen? Hydrogen atoms are very good absorbers of ultraviolet radiation, especially at 1216 Å, the Lyman-alpha absorption line produced when the electron is excited from its ground state to its first excited state (see Section 5.3). Such ultraviolet absorption has been sought in the spectra of distant objects at both small and large red shifts. It has not been detected. This lack of ultraviolet absorption implies that neutral hydrogen cannot have a density greater than about 10^{-9} atom/m^3 (10^{-33} kg/m^3). So if hydrogen is there, it must be ionized. Because these observations probe the universe's past, they also imply that any intergalactic gas must have remained highly ionized for most of the history of the universe.

These arguments leave ionized hydrogen (H II) as the most likely candidate for the intergalactic medium. Because intergalactic material would not have a high density, ionized hydrogen would take a very long time to find an electron and recombine. Unfortunately, detecting a low-density ionized gas is a rough job. If the gas is cool (a few thousand Kelvins), you can hunt for radio emission. If hot (a few tens of millions of Kelvins), you can search for X-ray or ultraviolet emission. X-ray observations of local superclusters show 15 sources that are probably clustered in seven superclusters. The sources consist of spots cen-

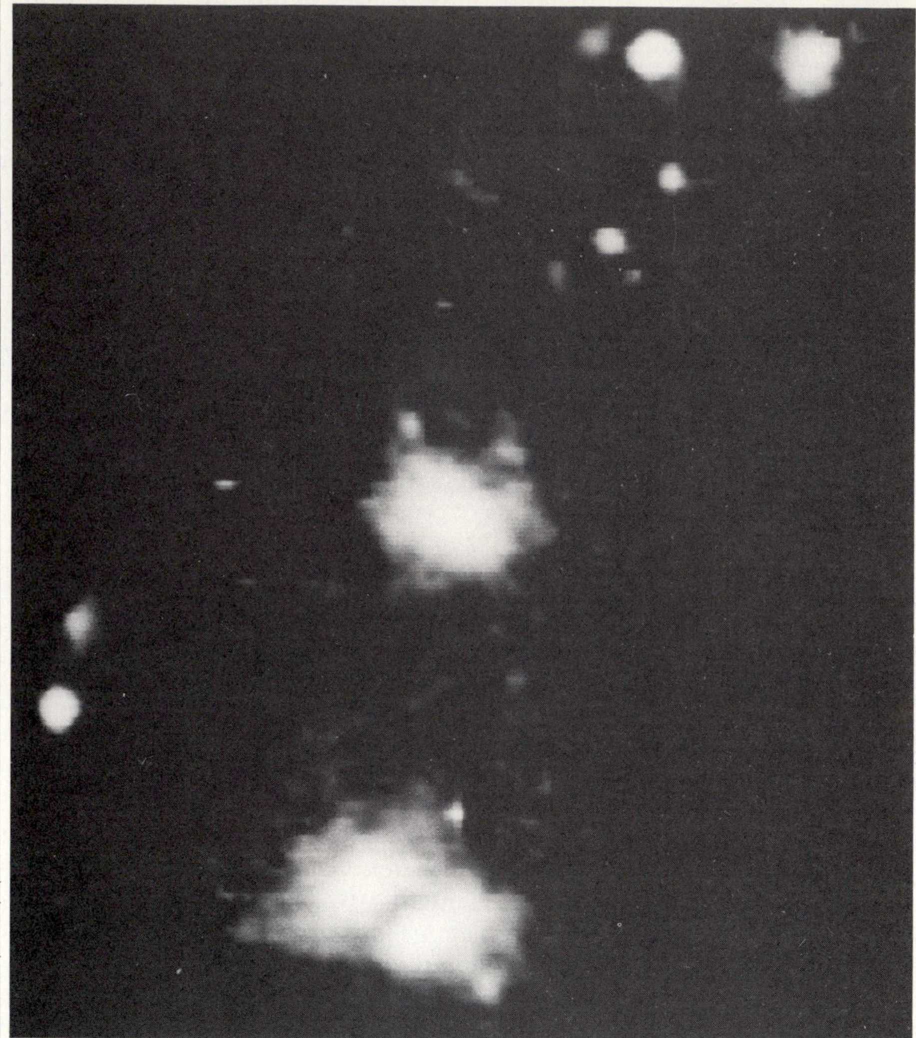

19.25

A map of the X-ray emission from a supercluster of galaxies. The brighter the region, the more intense the X-rays. The cluster at the photo's center is separated by about 18 million ly from the double cluster below it. The bright, round image to the right of center at the top is the star Canopus. (Based on observations with the Einstein X-Ray Observatory; courtesy of C. Jones and W. Forman, Smithsonian Astrophysical Observatory)

tered in rich clusters; this implies that a hot gas exists in superclusters, that it is highly clumped, and that little gas exists between clusters. The gas has a temperature of some 10^8 K and may, in total, amount to 40 percent of the density needed to close the universe.

Matter within clusters / For over 40 years astronomers have known about the puzzling characteristics of clusters of galaxies: Their masses estimated from the motions of the galaxies come out to about 10 times greater than those calculated by adding up the estimated masses of visible galaxies. In what form could this unseen material exist?

One possibility is that spiral galaxies have much larger and more massive halos than normally believed. These halos would be dark, about 1 million ly in size, and contain about 10 times the mass of the disk. The flat rotation curve of our Galaxy supports this idea.

Another good candidate for matter within clusters is ionized hydrogen gas. Recent X-ray observations back up this idea. At least 40 clusters of galaxies are known to date to emit X-rays (Fig. 19.25). The X-ray luminosities of clusters range from 10^{36} to 10^{38} W. The sizes of the X-ray–emitting cores range from 160,000 to 5 million ly. The richer clusters tend to be more luminous in X-rays.

A reasonably confirmed model for this X-ray emission is that it comes from hot, ionized gas. This model requires typical temperatures of 10 to 100 million K and densities of about 1 ion/m^3 to explain the X-ray observations. So we have reasonable evidence of intergalactic gas in clusters—about equal to the amount of mass in the galaxies themselves. But its density does not appear to be sufficient to bind the cluster gravitationally. Nor does it appear to be sufficient to close the universe.

KEY CONCEPTS

1 ***Distances to galaxies are essential to finding out their properties; they can be estimated from brightness and angular sizes.***

2 ***Galaxies come in three main types (based on their shapes): (a) ellipticals (dwarf and giant), (b) spirals (normal and barred), and (c) irregulars.***

3 ***Measuring distances to galaxies is difficult and relies on indirect schemes; the basic trick is to find very bright objects whose luminosities can be reasonably estimated, identify these in galaxies, compare their flux to their luminosity, and infer their distance from the inverse-square law for light.***

4 ***Measured red shifts and distances result in Hubble's law and Hubble's constant,*** H***; the greatest uncertainty in*** H ***arises from the uncertainties in distances;*** H ***has a value between 15 and 30 km/sec/Mly.***

5 ***Galaxies differ in terms of size, mass, luminosity, mass-luminosity ratio, and colors—all of which reflect their content of stars, gas, and dust.***

6 ***Galaxies come in clusters, containing from two to thousands, held together (at least for a while) by the gravity between the galaxies.***

7 ***The Local Group contains some 20-odd galaxies, mostly of low mass and luminosity, dominated by the Milky Way and the Andromeda galaxy; the Local Group spreads over a volume some 3 million ly in diameter.***

8 ***Clusters are grouped in superclusters; the Local Group is one small piece of the Local Supercluster (which is centered on the Virgo cluster); superclusters seem to be surrounded by voids and come in long chains; they are the largest entities in the cosmos.***

9 ***A very thin, very hot gas exists between clusters and superclusters; within clusters, hot, thin, ionized gas exists and is observed by the X-rays it emits.***

STUDY EXERCISES

1 Which galaxies appear redder, ellipticals or spirals? (*Objective 1*)

2 Which galaxies contain more gas and dust relative to their total mass, spirals or irregulars? (*Objective 1*)

3 At the same distance from us, would irregular galaxies appear larger than spiral ones? (*Objectives 1, 2, and 3*)

4 How do astronomers know that other galaxies are made of stars? (*Objective 4*)

5 In recent years the value of Hubble's constant has been revised from 30 km/sec/Mly to 15 km/sec/Mly. How does this change affect the distances to galaxies inferred from red shift and Hubble's constant? (*Objectives 6 and 8*)

6 Describe how supergiant *O*- and *B*-stars can be used to estimate distances to nearby galaxies. State the assumptions and limits of this method. (*Objective 7*)

7 Why must intergalactic gas be both ionized and hot? (*Objective 10*)

BEYOND THIS BOOK . . .

Galaxies (Harvard University Press, Cambridge, Mass., 1972) by H. Shapley, revised by P. Hodge, gives a comprehensive view of galaxies. Compare it to *The Realm of the Nebulae* (Dover, New York, 1958) by E. Hubble, first published in 1936.

Galaxies and Cosmology (McGraw-Hill, New York, 1966) by P. Hodge deals with the observed and physical properties of galaxies and their relationship to the universe.

For a gorgeous tour of the galaxies, examine *The Hubble Atlas of Galaxies* (Carnegie Institute, Washington, D.C., 1961) by A. Sandage.

"Dark Matter in Spiral Galaxies" by V. Rubin in *Scientific American*, June 1983, p. 96, presents recent knowledge from rotation curves.

S. Gregory and L. Thompson discuss "Superclusters and Voids in the Distribution of Galaxies" in *Scientific American*, March 1982, p. 106.

20 / Cosmic violence

If the radiance of a thousand suns
were to burst into the sky,
that would be
the splendor of the Mighty One . . .

The Bhagavad-Gita

Learning objectives

After studying this chapter, you should be able to:

1. **Outline the observational evidence for violent activity in our Galaxy and other galaxies, with special emphasis on synchrotron radiation.**
2. **Briefly compare and contrast active galaxies to ordinary ones (such as the Milky Way Galaxy).**
3. **List the observational characteristics of quasars.**
4. **Sketch a possible model that accounts for the observed characteristics of quasars.**
5. **Discuss the red-shift controversy for quasars, contrasting a cosmological with a noncosmological interpretation for them.**
6. **Summarize the evidence for quasars being the nuclei of distant, active galaxies.**
7. **Compare and contrast the Milky Way Galaxy, active galaxies, and quasars.**
8. **Outline the method used to estimate distances to quasars and discuss its uncertainties.**
9. **Outline a model using a supermassive black hole as the power source for active galaxies and quasars.**
10. **Describe the observational properties of jets in the nuclei of galaxies.**

Central question

What observational evidence do we have for violent activity in objects beyond the Milky Way Galaxy?

In general, the universe appears calm, caught up in the well-controlled generation of energy in stars and the strict Newtonian dance of matter. Until the middle of the twentieth century, we saw it as a gentle cosmos. Rare outbursts such as novas only occasionally shattered the stillness.

The advent of radio astronomy ripped off the veil covering a violent universe. Radio astronomers found the nucleus of our Galaxy to be a strong radio emitter. They also detected intense radio sources beyond the Galaxy, objects at enormous distances that required a tremendous outpouring of energy. By the 1970s, observational evidence demanded that violent events commonly occur in the nuclei of many types of galaxies.

Recent observations of jets spurting out from the nuclei of galaxies has reinforced this view. These jets appear as cosmic umbilical cords, channeling the power from the nucleus out to immense reservoirs of charged particles.

Quasars, starlike-looking objects with large red shifts, are the most puzzling element in the range of cosmic violence. They have the largest known red shifts of any extragalactic objects. If the red shifts of quasars result from the general expansion of the universe, the quasars must be the most energetic bodies in the universe. At the fringes of the cosmos, quasars represent some of the first-formed objects in the visible universe—and the most powerful. The mystery still remains of how quasars produce their energy.

20.1 Violence in the nucleus of the Galaxy

Galaxies appear as the most serene bodies in the universe. As we have uncovered more regions of the spectrum, some galaxies—and the nuclei of many—have acquired the aspect of a compact arena of violent events. The kernel of the nucleus is too small to investigate directly in detail in other galaxies, except M 31, and too obscured to observe optically in our Galaxy. The nature of the physical conditions and processes at the heart of a galaxy remains mostly unknown.

Evidence of violence in the Galaxy / Chapter 18 presented the structure of our home Galaxy. Here I place the results about the nucleus in the context of cosmic violence.

The nucleus lies at the center of the strong radio source Sgr A. One part of it, Sgr A West, appears to mark the actual core of the Galaxy. The central core of Sgr A West observed by radio has a size of roughly 140 AU. Spirallike structures appear to swirl out from it (look back at Fig. 18.18c).

TABLE 20.1 Emission from the galactic center

Wavelength	*Energy output (W)*
Centimeter continuous emission	10^{30}
100-μm extended source	10^{35}
10-μm sources in core	10^{32}
2-μm from stars	10^{34}
X-rays (1–10 Å)	10^{30}
Gamma-rays (10^{-4} Å)	10^{27}

Infrared observations show that the Galaxy's nucleus shines most brightly (Table 20.1) at infrared wavelengths between 2 and 350 μm. The peak of the infrared emission comes from a region only 3 ly across. High-resolution maps show that the nucleus holds a cluster of infrared sources, many less than a few light years in diameter. Here gas rotates around the nucleus at speed of a few hundreds of kilometers per second.

The galactic nucleus also puts out high-energy photons: X-rays and gamma-rays. The X-ray emission comes from an extended region around the galactic center, roughly coinciding with the source of the far-infrared emission. The X-ray output is spread over an area of a few hundred light years, with small sources embedded within it. Gamma-rays are also emitted strongly from a wide region around the galactic center itself. Both the X-ray and gamma-ray luminosities vary. Recent observations show that the gamma-rays vary over an interval of about a year.

The spectrum of the nucleus is in part nonthermal. That is, some of the emission must be by the synchrotron process. So high-energy electrons, moving at speeds close to that of light, spiraling in intense magnetic fields, must exist in the nucleus.

You'll see in what follows that most large galaxies have similar characteristics and that there exist galaxies more active than ours that display such energy output at a much higher level of violence. In other words, the nucleus of the Galaxy is a scaled-down example of the energy output of more active galaxies. The key clues are (1) emission over a wide range of wavelengths (usually nonthermal in character), (2) radio output concentrated in a small space, (3) most of the energy coming out in the infrared, and (4) variability in the emission.

Synchrotron emission revisited / The process of synchrotron emission underlies the basic physics of this chapter so firmly that you *must* have the basic concept in mind. Here it is: Whenever *charged* particles accelerate, they give off

TABLE 20.2 Properties of active galaxies

1	High luminosity, greater than 10^{37} W
2	Nonthermal emission, with excess ultraviolet, infrared, radio, and X-ray flux (compared with normal galaxies)
3	Variability over time period of a few hours to a few years
4	Peculiar photographic appearance: high brightness of nucleus and large-scale structure
5	Explosive appearance or jetlike protuberances
6	Broad emission lines and nonstellar spectrum

electromagnetic radiation. That's basically how any radio transmitter works; electronically, the transmitter forces electrons to accelerate back and forth at the frequency of transmission. These accelerated charges emit at that frequency of oscillation.

Now recall (from Focus 8.2) that magnetic fields bend the paths of charged particles—electrons more easily than protons because electrons have less mass. So as electrons move through magnetic fields, their paths are bent and the electrons accelerated. (Remember that curved motion is accelerated motion according to Newton's laws.) The electrons emit electromagnetic radiation.

In general, the faster the electrons travel, the more energetic (shorter wavelength, higher frequency) the radiation they emit. And the stronger the magnetic fields are, the more energetic the emission will be. As the electrons emit, they lose energy and slow down. So to keep a synchrotron source powered up requires a supply of electrons moving close to the speed of light.

Synchrotron emission turns out to be the source of luminosity of most of the objects presented in this chapter. We know that this is the case from the spectrum and polarization of their energy output.

20.2 Radio galaxies as active galaxies

Radio astronomers have found that many galaxies have nonthermal spectra. Such galaxies have been lumped into the category of *active galaxies*. But how active is "active"? I'll use the term here in contrast to "normal," which applies to our Galaxy. Basically, an active galaxy's spectrum does not look like that of a collection of stars. It is mostly nonthermal and has infrared, radio, ultraviolet, and X-ray outputs greater than those in the optical. (Our own Galaxy does emit at all these wavelengths, but not nearly as strongly as an active galaxy.) Table 20.2 lists some of the typical properties of active galaxies, and Fig. 20.1 compares the spectra of some active galaxies with those of a normal galaxy. Active galaxies make up only a few percent of all known galaxies.

We have many observations of active galaxies but little understanding of them right now in terms of successful models. For this reason, I'll limit myself to the discussion of the major classes of active galaxies. This section deals with the largest class: *radio galaxies*, which are those with strong radio emission.

Two principal types of radio galaxies have cropped up: *compact* and *extended*. Extended means that the radio emission is larger than a photographic image of the galaxy; compact means the radio emission is the same size or smaller. Compact radio galaxies often display very small radio sources, typically unresolved or no more than a few light years in size and always

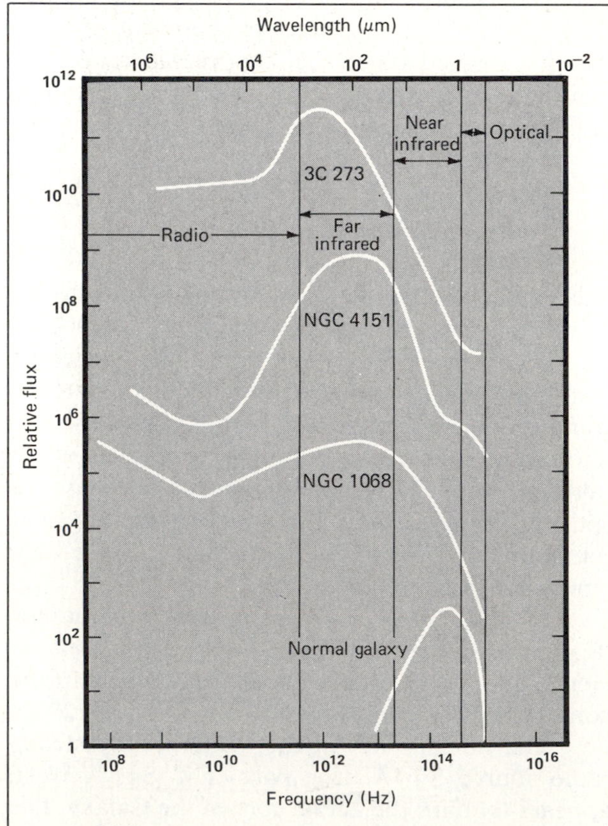

20.1
A comparison of the spectra in the radio, infrared, and optical wavelengths for a quasar, two active galaxies, and a normal galaxy. (Adapted from a diagram by R. Weymann)

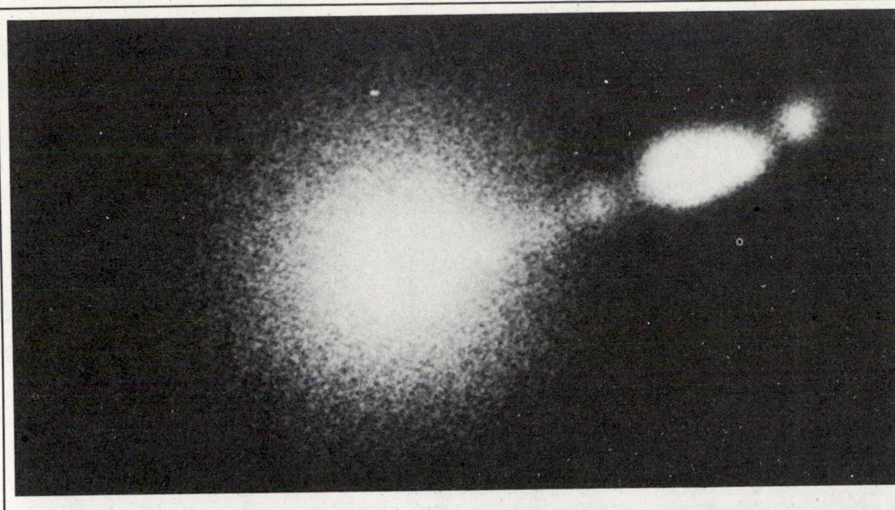

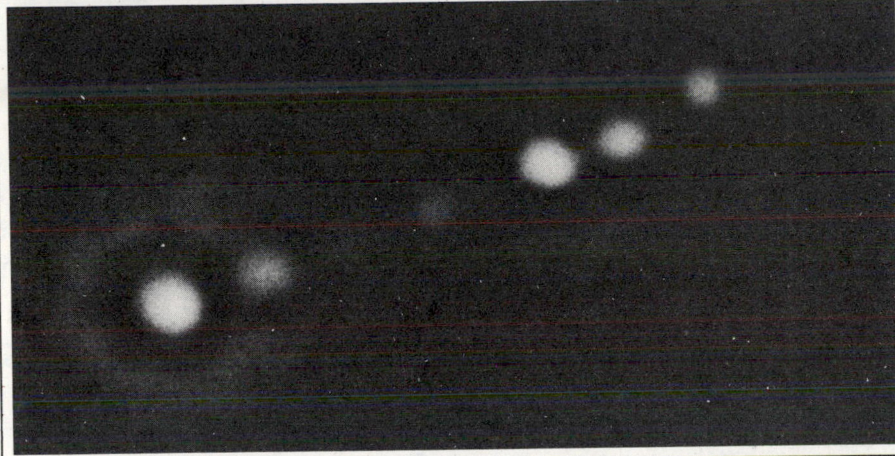

20.2
A close-up of the optical jet in M 87. The upper view is a conventional photo. The lower photo has been computer-processed to bring out the structure in the jet. Note that six blobs are clearly visible along with the nucleus. (Courtesy H. Arp)

coincident with the optical nucleus. Extended radio sources, in contrast, extend far beyond the optically visible galaxy—sometimes in two giant lobes of emission, up to millions of light years in extent, symmetrically balanced on opposite sides of the nucleus. A galaxy may show both a compact source at its nucleus and the extended lobes; it then usually also has radio jets connecting the nucleus to the extended radio emission.

Compact radio emission / Only 65 million ly away, Messier 87 (M 87) is a fine example of a compact radio source. A giant elliptical galaxy, M 87 dominates the Virgo cluster of galaxies. It presents direct evidence for violent activity. One radio source only 1.5 *months* in diameter appears in M 87's core along with a group of other compact radio sources. Poking out from the core, a remarkable, optically visible jet (Fig. 20.2) fires out into space over a length of some 6000 ly. The jet has a luminosity of roughly 10^{34} W (10^7 solar luminosities); its emission is polarized.

A beautiful photograph taken by Halton Arp shows that the jet contains at least six blobs of material, each no more than a few tens of light years in size (Fig. 20.2). Over 22 years, the blobs may have changed slightly but significantly in intensity and polarization.

M 87 also emits X-rays with about 50 times more energy than its optical emission—about 5×10^{35} W (10^9 solar luminosities) in X-rays from the whole galaxy. The jet itself also emits X-rays, in the form of a line of knots.

The VLA has mapped M 87's jet in detail and confirmed that its radio emission (Fig. 20.3) coincides with the optical and X-ray emission. The radio knots line up with the optical ones. So the jet overall emits over a wide range of frequencies, from radio to X-rays, and the knots of the jet each generate this spectrum of energies.

Other elliptical galaxies also possess nuclear jets. In fact, radio jets are common. Almost all radio galaxies of the lowest luminosities have jets (Color Plate 13).

Extended radio emission / At first glance, extended radio galaxies appear to be completely different beasts from compact radio galaxies.

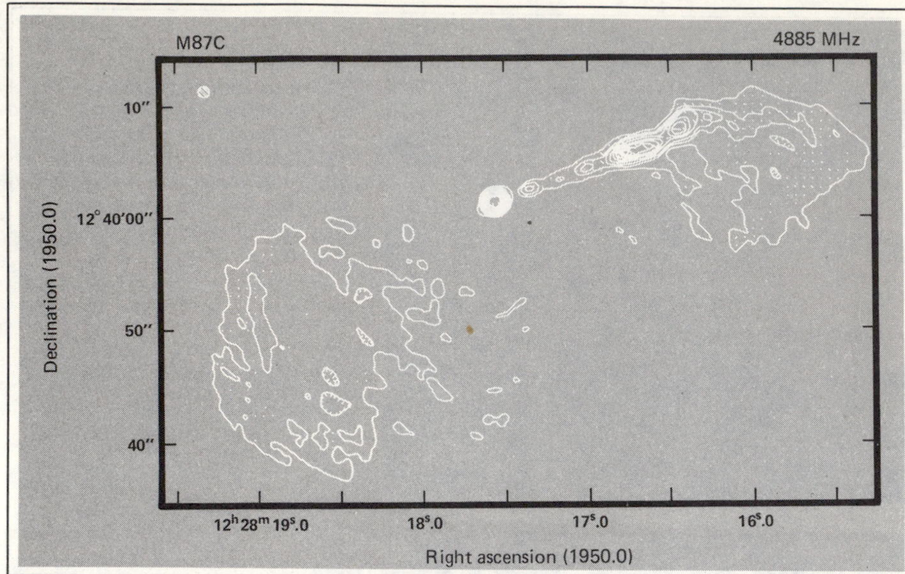

20.3
A high-resolution radio map of the central region of M 87, made with the VLA at a frequency of 5 GHz. Note the radio jet, which coincides with the optical one, extending to the right from the nucleus. (Courtesy F. Owen; observations by F. Owen and P. Hardee)

Cygnus A, one of the strongest radio sources in the sky and one of the first discovered, provides an excellent example of the typical double structure of a luminous, extended radio galaxy. Its radio output, some 1.2×10^{38} W (10^{11} solar luminosities), comes from two giant lobes set on opposite sides of the optical galaxy (Fig. 20.4). Each lobe has a diameter of 55,000 ly—about half the size of the Galaxy! They hang roughly 160,000 ly away from the central galaxy. Each lobe contains a cloud of energetic electrons and strong magnetic fields that must be storing incredible amounts of energy. Some 10^{53} J is needed to account for the radio luminosity lasting 10 million to 1 billion years. That's more energy than produced by all the stars in the Galaxy in 100 million years! And it's non-thermal!

The central galaxy of Cygnus A (Fig. 20.5) is a giant elliptical galaxy with a dust lane down its middle. It has an active nuclear region, with a

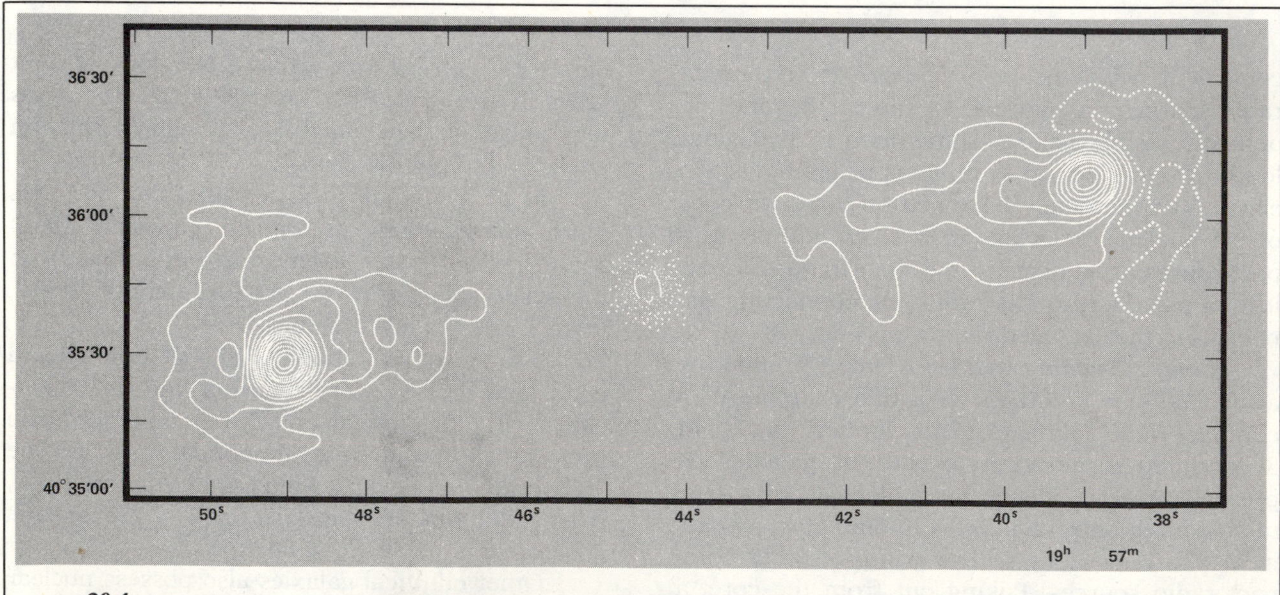

20.4
A radio map of Cygnus A, an extended, twin-lobed radio galaxy. The patch in the center represents the optically visible galaxy (Fig. 20.5). On each side of the nucleus lie immense lobes of radio emission. (Based on observations by S. Mitton and M. Ryle with the Cambridge Radio Telescope)

20.5
An optical photo of Cygnus A. Only the bright nucleus is visible, with a dust lane cutting through its middle. Compare with Centaurus A, a much closer elliptical galaxy (Fig. 20.6). (Courtesy Lick Observatory)

spectrum showing emission lines and a synchrotron continuous emission. But beyond 25,000 ly from the center, the spectrum is just that of a mix of stars. This galaxy appears to have blasted out the two clouds some 10 million to 1 billion years ago.

Centaurus A (NGC 5128) is another strong radio source somewhat similar to Cygnus A: a supergiant elliptical galaxy bisected by an irregular dust lane (Fig. 20.6). At a distance of 15 million ly, Centaurus A is the closest active galaxy. As viewed with an optical telescope, the galaxy has a diameter of a few tens of thousands of light years. Viewed with a radio telescope, Centaurus A has two huge outer lobes, 650,000 and 1,350,000 ly in size; these span over 3 million ly from tip to tip. Closer to the nucleus, another pair of radio lobes sit on the edges of the optical galaxy; these are some 33,000 ly in size.

Centaurus A also emits X-rays intensely; the source is very small and coincides with the nucleus. Remarkably, the X-ray emission varies. For instance, over a year, the X-ray luminosity has fallen one-half in value. The nucleus of Centaurus A also emits infrared and radio strongly. These observations, with those taken at X-ray energies, show that the nuclear emission is nonthermal.

Nuclear jets / The nucleus of Centaurus A has a direct connection to one inner radio lobes. X-ray observations show an X-ray jet streaming northeast from the nucleus and consisting of at least seven distinct blobs. This discovery prompted Jack Burns and colleagues to observe Centaurus A with the VLA. The VLA map (Fig. 20.7) shows radio emission along a jet that extends to one of the northern radio lobes. The jet has a bloblike structure that coincides with the X-ray blobs. More recent high-resolution radio observations reveal in the nucleus itself a very small jet, 4 ly long, that lines up with the larger one in the VLA map.

What powers the jets of M 87 and Centaurus A? Observed at a wide range of wavelengths, the spectra of both jets is nonthermal. This fact points to synchrotron-process generation of the emission from the jets. The large polarization of the jets' light in M 87 (about 25 percent) confirms

20.6
Centaurus A (NGC 5128), an active galaxy. Note the dust lane across the galaxy's middle, like that in Cygnus A. (Courtesy Kitt Peak National Observatory)

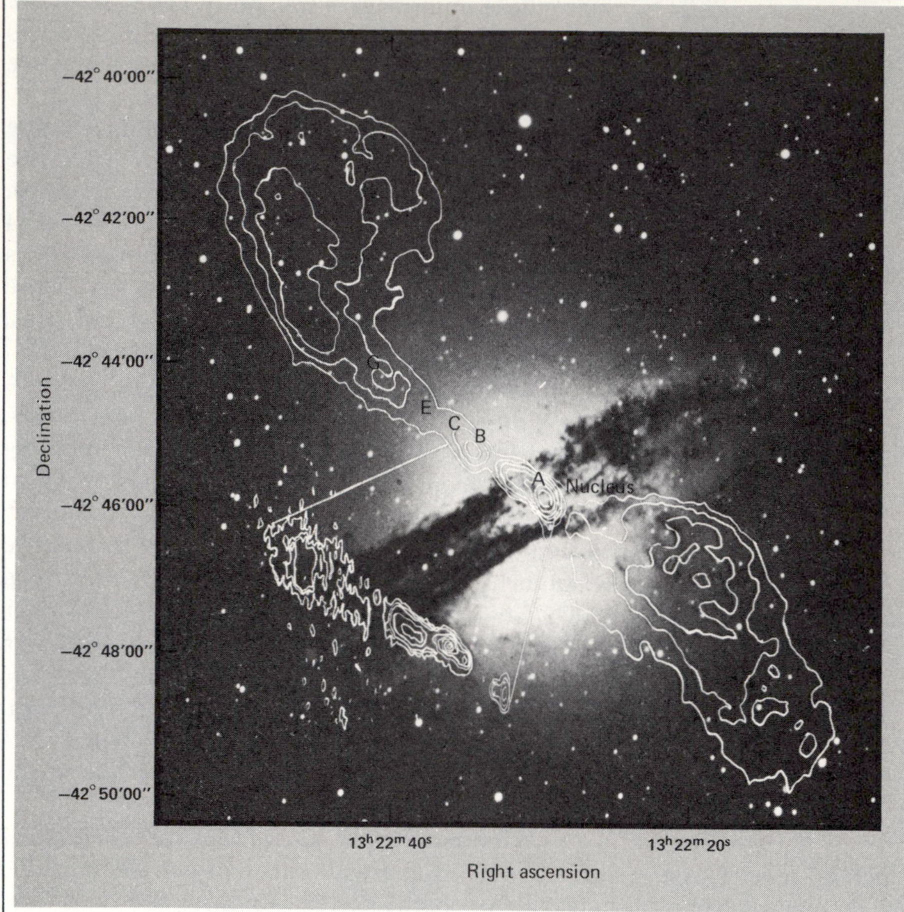

20.7
The radio contours of the VLA map of the jet and inner lobes overlaid on a photo of Centaurus A. The jet consists of a series of blobs that correspond well to the ones in X-rays. (Courtesy J. Burns)

this idea. The fact that the knots in the jets coincide at different wavelengths implies that the same electrons power all the emission from the knots. The nucleus provides the high-energy electrons, which are expelled either as a fairly constant beam of particles or a sequence of ionized blobs that are thrown out along a magnetic field. The ionized stream carries magnetic fields within it, and these help to channel the flows outward. This channel is leaky, which results in the jet's emissions being visible. That presents the major problem with this model: How do the jets keep up their emission over so long a path as the electrons lose energy?

Structures of extended radio emission / Prior to the use of aperture synthesis, array radio telescopes in the 1970s, extended radio galaxies appeared to be almost all double: 75 to 95 percent of resolved radio galaxies appear to be so, with the lobes lined up with the galaxy's center. These radio lobes are huge: Most are 150,000 to 3.3 million ly in diameter. One, called 3C 236, has lobes that extend 20 million ly end-to-end; each is about 3.3 million ly in size. One lobe could engulf the Milky Way Galaxy and the Andromeda galaxy at the same time! 3C 236 also contains a radio core that extends some 6000 ly in length. This core source has a jetlike shape and aligns closely to the axes of the giant lobes.

Radio astronomers have now found out (mostly from high-resolution observations that reveal fine details in structure) that, when classified by structure, extended radio galaxies fall into three main groups:

1. Doubles (example: Cygnus A): highest luminosities, lobes aligned through center of galaxy, bright hot spots at ends.
2. Bent doubles (example: Centaurus A): intermediate luminosities, bent through nucleus, taillike protrusions.
3. Narrow-tailed sources (example: NGC 1265; Fig. 20.8): lowest luminosities, U-shaped, rapidly moving galaxies in a cluster. (Most radio galaxies fall into this group.)

The delineation of these radio-source structures is important in two ways: (1) by providing insight into the physical processes responsible for the radio emission, and (2) by producing information about the environment in which galaxies are immersed.

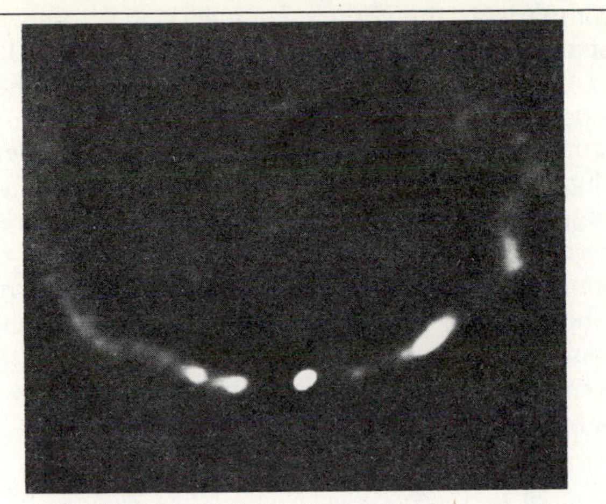

20.8
A radio picture, made with the VLA, of the nucleus of the head-tail galaxy NGC 1265. Note how the radio emission curves away and trails from the nucleus. (Courtesy J. Burns; observations by F. Owen, J. Burns, and L. Rudnick)

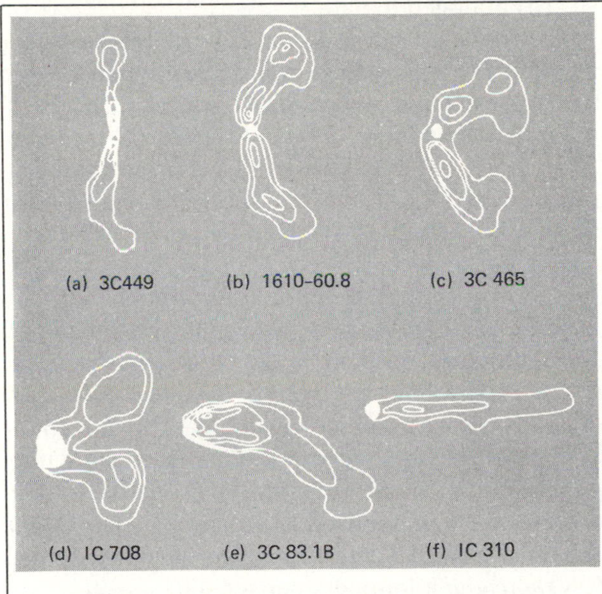

20.9
A sequence for the bending of radio tails around the nuclei of radio galaxies. (Adapted from a diagram by G. Miley)

Low-luminosity tailed or distorted sources make up the majority of extended radio sources. The *head-tail* galaxies are one subgroup. As implied by the name, head-tail radio galaxies have a radio head (around the visible galaxy) and an extended tail. NGC 1265 (Fig. 20.8), in the Perseus cluster of galaxies, is a good example. Not all galaxies with tails have them trailing. Instead, we find a sequence in the amount of bending shown by the tails (Fig. 20.9), which ranges from 180° apart (double source–Figure 20.9a) to 0° (head-tail source–Fig. 20.9f). At high resolution, the heads often contain radio jets.

How to explain this structural sequence? Recall that clusters of galaxies contain a hot, ionized intracluster medium. Imagine that a galaxy, moving rapidly through this medium, shoots out material (high-speed electrons, for instance) in a jet. As the galaxy travels along, it leaves behind a radio-visible trail–a fossil record of where it's been. Here's an analogy: Imagine driving a car slowly and blowing smoke out of an open window. The air stops the motion of the smoke, and it leaves a trail behind the car. Similarly, material flowing out of a galaxy is decelerated by the intragalactic medium, and the moving galaxy leaves it behind.

To sum up: One common type of active galaxy is a radio galaxy; many radio galaxies have emission, in the form of lobes or streams, that extends far beyond the visible galaxy (Fig. 20.10). For instance, the lobes may be a few million light years apart and thousands of light years in size. The vexing problem with these extended radio lobes is the vast amount of energy they contain: A typical lobe luminosity is 10^{37} W, while the visible elliptical galaxy with which it is associated may emit only some 10^{35} W. Another way to consider it: The lobes are energy reservoirs that are a tangle of magnetic fields and high-speed electrons, since the emission is synchrotron radiation. Then a typical lobe contains more than 10^{52} J. (For

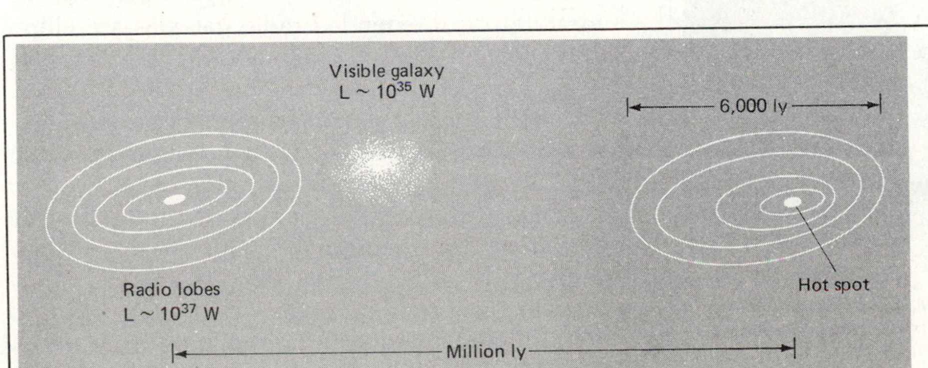

20.10
A schematic drawing of a typical twin-lobed radio galaxy. The visible galaxy is usually an elliptical one. (Adapted from a diagram by A. Bridle)

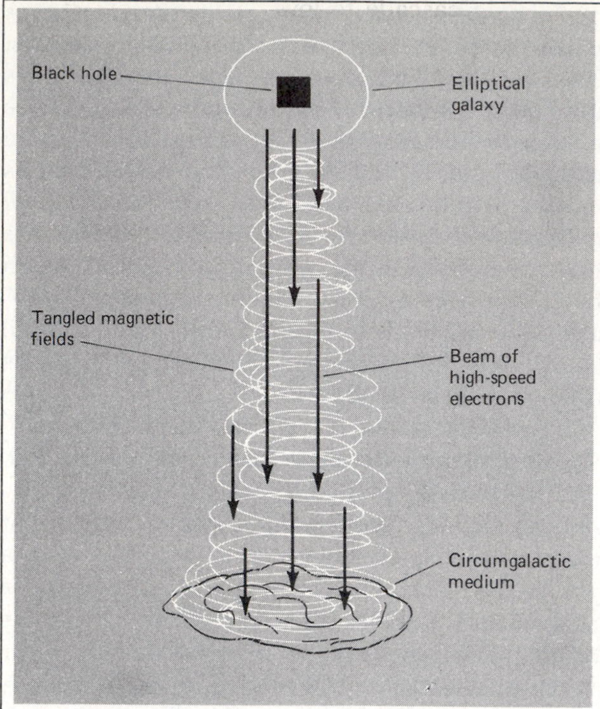

20.11
A schematic drawing of the possible connections among a radio galaxy's nuclear source, radio jet, and radio lobe. (Adapted from a diagram by A. Bridle)

20.12
The Seyfert galaxy NGC 4151. Note how bright the nucleus appears compared with the rest of the galaxy. (Courtesy Palomar Observatory, California Institute of Technology)

comparison, the complete conversion into energy of a solar mass of material releases about 10^{47} J.)

How do active radio galaxies generate these lobes? We don't know all the details yet, but a crucial clue is the radio jets from the nucleus, aligned (more or less) with the lobes. These jets suggest that high-speed electrons are channeled, perhaps in bursts, from the nucleus into the circumgalactic medium, where they pile up to form a lobe (Fig. 20.11). What's the nuclear energy source? At the moment, it's a black box that somehow converts gravitational energy into a jet of high-speed particles.

20.3 Seyfert galaxies

In 1943 Carl Seyfert noted that some spiral galaxies showed unusual, broad emission lines. These are now called *Seyfert galaxies*. Later work has added to the collection of Seyfert galaxies; some 90 are cataloged to date.

What's a Seyfert galaxy? Most galaxies have a bright nucleus, which looks fuzzy when viewed with a telescope; when photographed, a run-of-the-mill galaxy looks obviously fuzzier than a star. In contrast, a Seyfert looks like a bright star surrounded by a faint haze. In short-exposure photographs, the haze isn't seen, and a Seyfert's nucleus passes for a star!

Along with this trademark optical appearance, a Seyfert has a particular spectrum: It shows broad emission lines. If we interpret the width of the lines as due to Doppler shifts produced by motion in the emitting gas, the gas in the Seyferts has very high random velocities. Other galaxies, too, have emission lines with Doppler widths of a few hundred kilometers a second. But the lines from a Seyfert have Doppler widths of a few *thousand* kilometers per second.

Seyferts are almost always spiral galaxies (Fig. 20.12). A detailed survey of 80 Seyferts finds that only 5 to 10 percent might be ellipticals. (The small angular sizes of some Seyferts make it hard to classify them by form.) Compare this with the fact that most extended radio galaxies are ellipticals (Section 20.2). Overall, about 1 percent of all spiral galaxies (ordinary and barred) are Seyferts.

NGC 1068 is a good example of the peculiarities of Seyfert galaxies. Its continuous spectrum (Fig. 20.13) does not resemble that expected from a group of stars. In fact, much (but not all) of the spectrum is nonthermal. This observed fact implies that at least three different sources contribute to the continuous spectrum: stars, synchrotron radiation, and infrared radiation from heated dust. The optical emission from NGC 1068 is

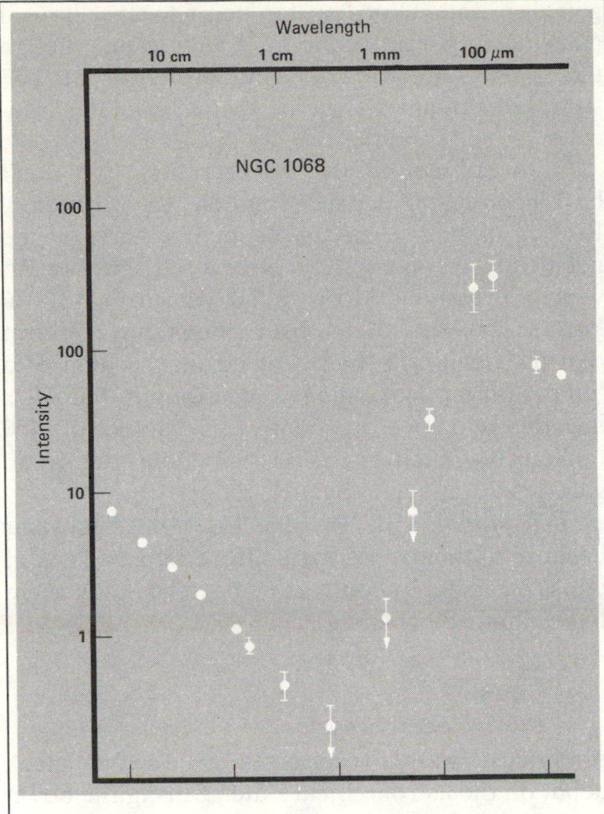

20.13
The spectrum of the Seyfert galaxy NGC 1068 in the radio and infrared wavelengths. (Adapted from a diagram by J. Elias and co-workers, Astrophysical Journal, *vol. 220 (1978), p.25, copyright © 1978 by the American Astronomical Society)*

polarized, as expected from synchrotron emission or light passing through thick dust.

Infrared astronomers have found that the bulk of NGC 1068's luminosity comes out in the infrared—at least 10^{38} W. Such infrared emission is most easily explained if the nucleus of NGC 1068 is embedded in dust. These particles can absorb ultraviolet and optical radiation from the nucleus and heat up to about 100 K. At that temperature, the dust grains give off most of their radiation in the infrared part of the spectrum.

Seyferts show strong Balmer emission lines, which are narrow in some and broad in others. The use of "broad" and "narrow" here is relative. For example, the average Balmer line width for 40 Seyfert galaxies is 3400 km/sec. But in some Seyfert galaxies, the Balmer lines have Doppler widths as great as 10,000 km/sec (3 percent of *c*)! In contrast, other Seyfert galaxies have lines only 1000 km/sec wide.

As in bright nebulas (see Focus 15.1), the Balmer lines in a Seyfert are produced by recombination of hydrogen ions (protons) with electrons, which then cascade to lower levels. What makes the lines so broad? The simplest explanation is filaments or clouds of gas moving at speeds of a few thousand kilometers a second. To explain the Balmer emission from a Seyfert requires some tens to thousands of solar masses of ionized gas in the nucleus at densities of 10^{14} ions/m^3 moving at such speeds. Additional emission then might come from a large surrounding volume of lower-density gas moving at slower speeds.

What ionizes the gas and moves it around? Probably the energy source in the nucleus that generates the synchrotron emission. That source is unknown. Also in the nucleus must reside a source of high-energy electrons and gas. Observations indicate that the nuclei of Seyferts are small, only a few light years in diameter. Gas moving at 10,000 km/sec would flow across such a tiny nucleus in only 100 years. So the gas must be replaced as it flows out. Or somehow, even at 10,000 km/sec, the nucleus must hold it in.

To sum up: Seyferts are (all?) spiral galaxies with extraordinary features. The most prominent are:

1 They have extremely small and bright nuclei.
2 Their nuclei have spectra that show emission lines not usually seen in the spectra of spiral galaxies. These bright lines do not come from stars.
3 The emission lines are very wide. Considered as Doppler shifts, the widths of the lines indicate gas motions of 500 to 4000 km/sec, an indication of violent activity.
4 Many Seyferts have compact, low-luminosity radio sources within them.

In some ways the characteristics of Seyferts resemble the Sgr A radio source in the center of the Galaxy. But Seyferts emit much more energy—at least 1000 times that of the nucleus of the Milky Way Galaxy.

20.4 *BL Lacertae objects*

I want to describe one more animal of the active galaxy zoo—one that has some similarities to radio galaxies and Seyferts. These objects are named after their prototype, BL Lacertae (BL Lac) and so are called *BL Lacertae objects*.

As a group, the BL Lac objects have the following characteristics: (1) rapid variability (as fast as one day) at radio, infrared, and visual wavelengths, (2) extremely weak or no emission lines, (3) nonthermal continuous radiation with most of the energy emitted in the infrared, and (4) strong and rapidly varying polarization. Also, BL

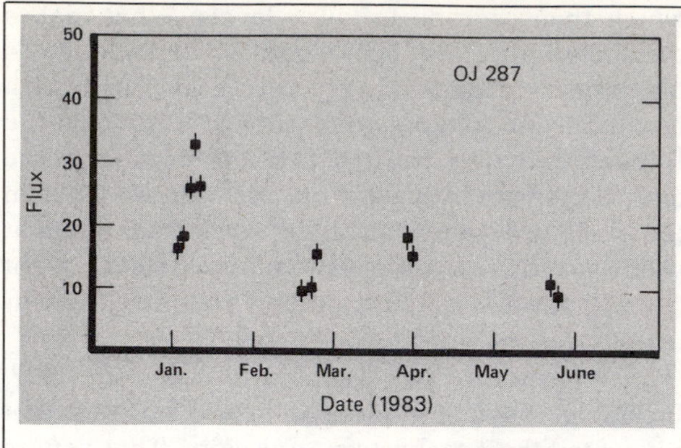

20.14
Variation in visual light output for the BL Lac object OJ 287 in 1983. Note the outburst in January. (Adapted from a diagram by P. Smith)

Lac objects generally have a starlike appearance; no structure is visible.

The BL Lac objects differ the most from other active galaxies in that their emission varies frequently, erratically, and rapidly (Fig. 20.14). For example, BL Lac itself fluctuates in luminosity by 20 times or so. Observers have noted night-to-night variations of 10 to 30 percent in luminosity. That doesn't sound like much, but imagine our Galaxy changing its light output by some 20 percent in a day. That's like 10^{10} suns turning on and off simultaneously! A few BL Lac objects have changed their brightness by as much as 100 times in luminosity. The radio emission from BL Lac objects is compact or only slightly extended. The extended radio structure is weak in contrast to intense emission from the nucleus.

What puzzles astronomers most about the BL Lac objects is that their energy variations take place in objects that show almost no emission lines in their spectra! As discussed earlier, the standard model for active galaxies pictures synchrotron emission produced by steady injection and/or bursts of high-energy electrons; that synchrotron output in the ultraviolet (and even the electrons themselves) should ionize any gas near the nucleus and produce emission lines through recombination. So where are the BL Lac's emission lines if they are powered the same way? (The same puzzle arises with radio galaxies.)

Some 40 BL Lac objects have been classified to date. They may actually not all be the same kind of beast. A few are possibly the nuclei of galaxies; some, like BL Lac itself, have a faint surrounding fuzz that might be a galaxy; others look pointlike without a hint of enveloping material. A few BL Lac objects are found in clusters of galaxies—indirect evidence that they are also galaxies. Indications so far point to an elliptical galaxy as a parent of the BL Lac nucleus, though this is not yet certain. Recent results for BL Lac show hints of a spiral galaxy rather than an elliptical one.

To compound the uncertainties, we don't have good distance determinations for very many BL Lac objects. Weak absorption features of the nebulosity around BL Lac have a red shift of 0.07, which corresponds to a radial velocity of 21,000 km/sec. In addition, observations have shown that the nebulosity has a spectrum like that of a luminous elliptical galaxy. According to Hubble's law, this velocity corresponds to a distance of 1400 million ly—much more distant than the radio galaxies discussed in Section 20.2.

To sum up: BL Lac objects are most peculiar because of their rapid variability and usual lack of emission lines. In contrast to Seyferts, they seem to be elliptical galaxies.

20.5 *Quasars:* *Unraveling the mystery*

About 20 years ago, quasars made a rather meek debut on the astronomical slate. During the boom period of radio astronomy in the late 1950s, radio astronomers—like modern Tycho Brahes—compiled catalogs replete with radio sources that were not identified with any familiar visible objects. Hunting for possible culprits in 1960, Thomas Matthews and Allan Sandage discovered a faint starlike object (hence the name *quasi-stellar object*, or *quasar*) at the position of radio object 3C 48 (object 48 from the Third Cambridge Catalogue of radio sources). This object had a spectrum of broad emission lines that could not be identified, and it emitted more ultraviolet light than an ordinary main-sequence star.

3C 48 remained a unique object until 1963, when the strong radio source 3C 273 was identified with a faint starlike object (Fig. 20.15). The emission lines of 3C 273 were just as puzzling as the emission lines from 3C 48: They coincided with no known atomic lines. Astronomers were baffled.

Quasar red shifts / Maarten Schmidt finally deciphered the spectral code of 3C 273 by recognizing the prominent emission lines as those of the hydrogen Balmer series, red-shifted by 15.8 percent compared to their normal at-rest wavelengths (Fig. 20.16). The ultraviolet spectrum of 3C 273 (Fig. 20.17) shows strong ultraviolet emission with one intense peak at 1410 Å. When shifted to shorter wavelengths by 16 percent, this line turns out to be the first line of the hydrogen Lyman series at 1216 Å. So the red shift measured in the ultraviolet confirms that found optically.

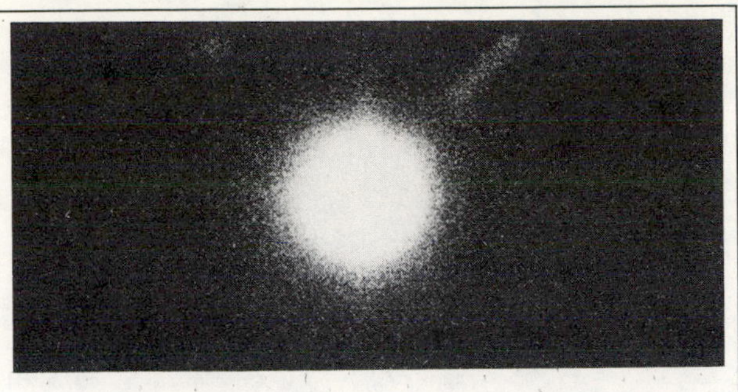

20.15

3C 273, the first quasar to have its spectrum deciphered. Note the jet sticking out of the quasar at its upper right; compare it to the one in M 87 in Fig. 20.2. (Courtesy Palomar Observatory, California Institute of Technology)

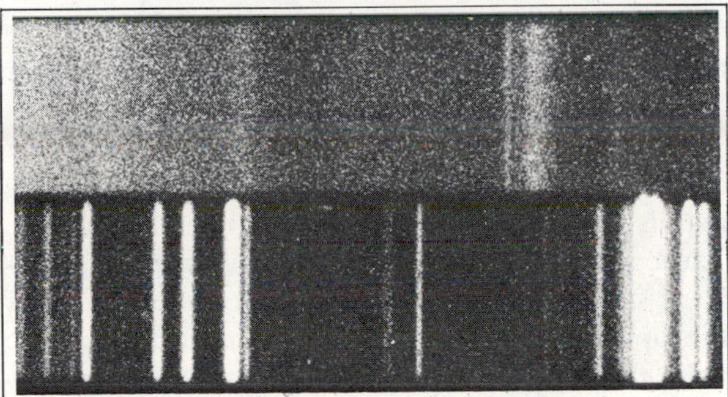

20.16

A spectrum of 3C 273, showing its large red shift. The upper spectrum is that of the quasar, the lower one a comparison spectrum that establishes the wavelength reference scale (at rest with respect to the observer). (Courtesy M. Schmidt)

After Schmidt decoded 3C 273, Jesse Greenstein applied the same analysis to 3C 48 and found that its spectrum was red-shifted by 36.7 percent. Following the initial elation of the discovery, the squabble about the nature of the red shifts began. Did the red shifts arise from the cosmic expansion, so that, according to Hubble's law, the quasars were at stupendous distances? Or were they Doppler shifts of masses expelled from the center of our Galaxy? Or from other galaxies? Or what?

Altogether, over 1500 quasars have been identified, and red shifts have been measured for most of these. On the average, one quasar appears in every 30 square degrees of sky—a patch about the size of the bowl of the Big Dipper. Some have red shifts that exceed 2.0, and a few even 3.0. For example, the quasar 4C 25.5 (4C means the Fourth Cambridge Catalogue) has emission lines shifted by 235.8 percent! If interpreted as a red shift from the expansion of the universe, the light from 4C 25.5 comes from such a distance that it must have originated about 9 billion years ago. The strongly red-shifted quasars would then be the youngest objects we can see in the universe. Note that by "youngest" I mean objects at a point in their lives far in the past, closer to the Big Bang in time.

Warning: When the measured red shift approaches 1, the simple Doppler formula (see Focus 14.3) no longer gives the correct relative velocity. For example, if applied to 4C 25.5, the formula would indicate that this quasar was

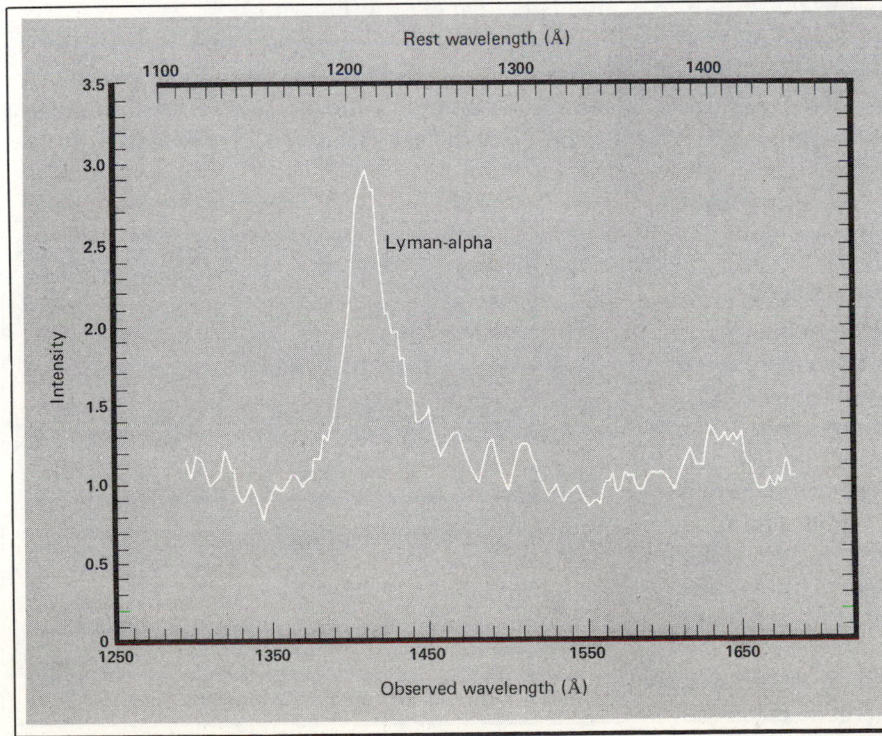

20.17

The ultraviolet spectrum of 3C 273. The red-shifted wavelengths are at the bottom and the observer's (at-rest) wavelengths at the top. The most intense line is the Lyman-alpha line from hydrogen. (Adapted from a diagram and observations by A. Davidsen, G. Hartig, and W. Fastie)

fleeing at 236 percent the speed of light! A modified formula, based on special relativity, must be used instead. Then, for example, a red shift of 2 means that the radial velocity equals 0.8 *c*.

General observed properties / Although most of the early quasars were detected because of their strong radio emission, many uncovered in later optical searches were found to be radio-quiet. The observed properties that make quasars unique and that serve as identification tags are (1) starlike appearance with a large red shift, sometimes associated with a radio source (only about 10 percent are known radio sources); (2) broad emission lines in the spectrum with absorption lines sometimes present (usually the red shift of the absorption lines is less than that of the emission lines); (3) often variable luminosity; (4) in those that are radio sources, aligned, double-lobed structures (like Cygnus A); and (5) in about 50 percent of radio-loud quasars are now found nuclear jets.

The main—and most remarkable—feature of quasars is their large red shifts. Recall that the expansion of the universe results in the red shift of light from distant galaxies as described by Hubble's law. At first glance, the most natural explanation of the quasars' red shifts is a cosmological one: Quasars participate in the universe's expansion. If so, their enormous red shifts indicate that they are very far from us, observed when the universe was very young. But if they are as far away as indicated by their red shifts, they must expend vast amounts of energy. For example, 3C 273's red shift of 16 percent, if due to the expansion of the universe, implies a distance of 3100 million ly, assuming $H = 15$ km/sec/Mly. At this distance, 3C 273 emits about 10^{14} solar luminosities, or about 40 times as much as the most luminous galaxies. (Most of this energy is in the infrared.) This light left 3C 273 about *3 billion* years ago—at the time simple life appeared on the earth!

The light from quasars / Some quasars emit radio waves intensely, and all emit visible light. What produces this emission? A key clue comes from the spectrum of quasar radiation (Fig. 20.18). The radio spectrum does not resemble that for a blackbody emitter but rather that for nonthermal, synchrotron emission. So energetic electrons (moving close to the speed of light) spiraling around magnetic lines of force produce the radio emission. Such synchrotron emission is polarized if the magnetic fields are organized rather than chaotic.

Note that synchrotron radiation requires a continuous supply of clouds of energetic electrons

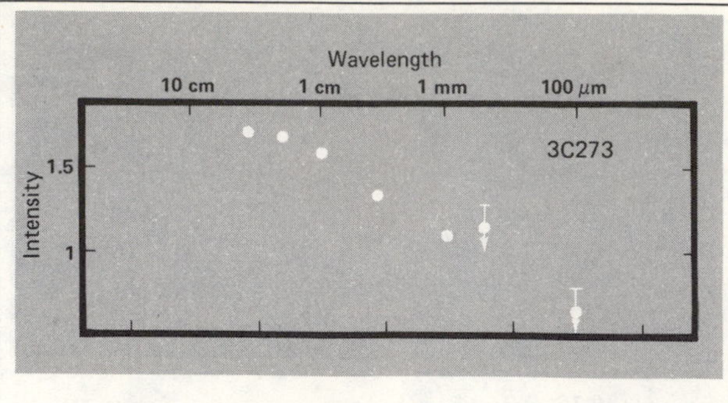

20.18
The spectrum of 3C 273 in the radio and infrared wavelengths. (Adapted from a diagram by J. Elias and coworkers, Astrophysical Journal, *vol. 220 (1978), p. 25, copyright © 1978 by the American Astronomical Society)*

and strong magnetic fields. Any model of a quasar must include these two features.

Optical observations of quasars first discovered as radio sources have added more details to the general physical picture of a quasar. The continuous optical emission is in part polarized, so synchrotron emission produces some of the optical radiation. Given a common magnetic field, the electrons that produce the optical radiation must have higher energies than those that emit the radio radiation.

Line spectra / All quasars have bright lines in their optical spectrum—the emission lines that are used to measure a quasar's red shift. Among the strongest emission lines in a quasar's spectrum are Lyman-alpha at 1216 Å and the Balmer line H-beta at 4861 Å (Fig. 20.19).

What do these emission lines tell us about the quasar? Recall that emission lines are characteristics of a hot low-density gas. For example, the Orion Nebula (see Section 15.1) is heated by the

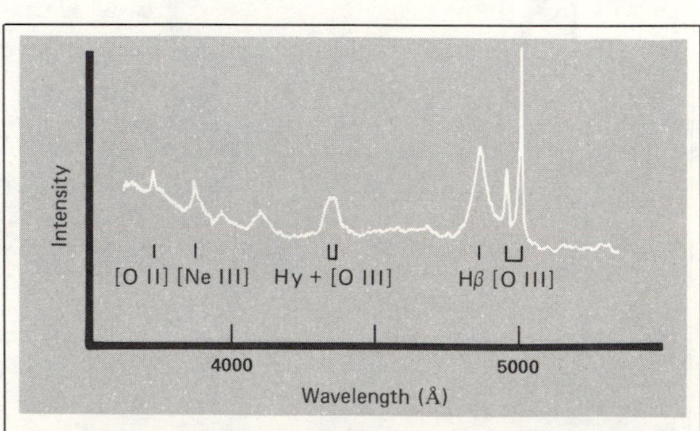

20.19
A typical quasar's spectrum, showing strong emission lines from hydrogen, oxygen, and neon. (Adapted from a diagram by J. S. Miller, Publications of the Astronomical Society of the Pacific, vol. 93 (1981), p. 681)

ultraviolet radiation from young stars and glows with a characteristic bright-line spectrum. The techniques used to analyze the physical conditions in diffuse, glowing nebulas using their emission spectra have also been applied to quasars.

The emission-line spectra of quasars indicate that a low-density cloud of gas is radiated with photons energetic enough to ionize hydrogen. This radiation probably comes from the synchrotron emission from energetic electrons, which is a plentiful source of ultraviolet photons. So to our central synchrotron source we must add clouds or filaments of gas that act to convert ultraviolet radiation (and some X-rays) from the synchrotron source into visible emission lines.

What other information comes from an analysis of a quasar's spectrum? First, the composition of the radiating gas has no surprises: Lines of hydrogen, helium, carbon, oxygen, nitrogen, and other common elements have been observed. Second, these elements are in high ionization states, a fact that reaffirms the existence of energetic photons.

Third, the emission lines are extremely broad. This fact implies that the filaments or clouds from which the emission lines originate move rapidly. Part of this broadening arises from the fact that the ions and atoms in the filaments are hot. Some of the emitting atoms (or ions) are moving toward us (relative to the average velocity of the quasar) and their emission is blue-shifted; and some are moving away from us, with consequent red shifts. As a result of blue-shifted lines blended with red-shifted lines, a given line appears broader and increases in width as the temperature (and velocity) increases. However, the emission lines from quasars are a thousand times broader than the width predicted from the broadening effect due to temperature. Most of the broadening, then, must be due to motions of the clouds or filaments themselves at relative velocities from a few thousand to 20,000 km/sec. An astronomer observing the quasar sees some blue-shifted photons, some red-shifted photons, some unshifted radiation relative to the average velocity of the quasar. Taken together, the multitude of lines blends into one broad line. (Sounds like a Seyfert galaxy, doesn't it?)

Many (but not all) quasars also have absorption lines in their spectra. Quasars with emission-line red shifts of less than 2.2 typically do not have absorption lines; those with greater red shifts have strong absorption lines. These lines are very narrow compared with the emission lines. Generally, they are identified with ionized states of common elements, such as carbon, silicon, and nitrogen.

The absorption lines of quasars present a vexing puzzle. In general, their red shifts are *less* than the shifts for the emission lines. And, even stranger, the absorption lines sometimes show more than one red shift. For example, the spectrum of the quasar PHL 938 has an emission red shift of 1.955 and absorption red shifts of 1.949, 1.945, and 0.613. The difference between red shifts of 1.955 and 0.613 amounts to a relative radial velocity difference of 0.5 the speed of light!

What produces these absorption lines? Three possibilities are: (1) gas clouds in or close to the quasar at temperatures cooler than the emission-line regions, (2) absorption by intervening (and otherwise invisible) intergalactic gas clouds, and (3) a halo of gas around an intervening galaxy. None of these possibilities has been well confirmed, but the discovery of the double quasar (see next section) supports the third model.

Variability in luminosity / Light variability was first observed for 3C 48, which showed variations of about 40 percent in 13 months. 3C 273 also has erratic variations over periods of about a year. Most radio-emitting quasars also vary in radio output over periods of years (Fig. 20.20). In contrast, about 20 percent of quasars exhibit rapid variations in light and radio output with periods on the order of days or weeks. This fact implies that the nuclear power source is very small.

20.6 The double quasar: An optical illusion

Quasars are rarely close together, so Dennis Walsh, Robert Carswell, and Ray Weymann were surprised to find two quasars only 6″ apart. They are called 0957+561A and 0957+561B. Even more surprising, the emission-line red shifts of both were essentially the same: 1.41. Later work found the absorption-line red shifts in quasar A were all the same and only slightly less than those of the emission lines. Another piece of the puzzle fell into place when Weymann and colleagues obtained high-resolution spectra, which revealed that both quasars had essentially the same absorption-line red shifts (1.391).

How mysterious! Could there be twin quasars with *exactly* the same emission and absorption red shifts, 14,000 Mly from the sun, and separated by 200,000 ly? That's not likely.

This situation forced the astronomers to propose that the quasars were not separate twins but optical images of the *same* quasar. How so? Recall (from Section 7.3) that general relativity predicts that masses deflect the paths of light rays—in essence, acting like a lens. If a very small,

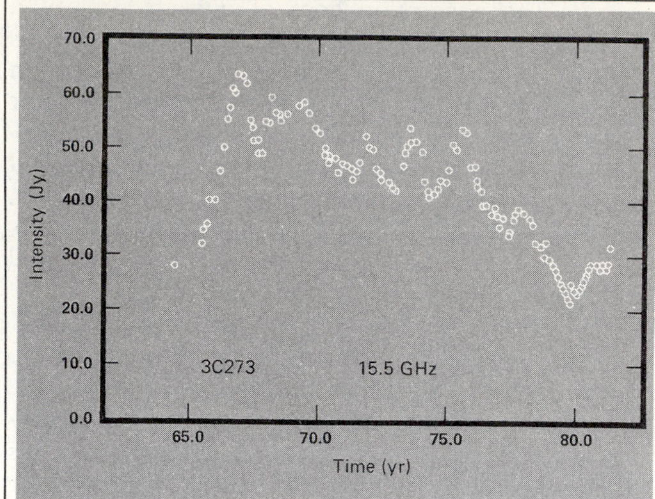

20.20
Variation of the radio emission of 3C 273 over a period from 1965 to 1982 at a frequency of 15.5 GHz. (Courtesy T. Balonek; observations by W. Dent and T. Balonek with the Haystack radio telescope)

dense mass (a black hole, for instance) lay along our line of sight to a quasar, its image would be split into two, one above and one below the quasar's actual position. This phenomenon of image making by a mass is called a *gravitational-lens* effect.

But radio astronomers found data inconsistent with this model. The VLA showed that the sources were not identical doubles, as expected from the simple gravitational-lens model. Rather, there were radio blobs to the east of the northern quasar (A) not visible west of the southern quasar (B), and no indication of any object between A and B (Fig. 20.21).

The final piece of the puzzle came from Alan Stockton. Observing with the University of Hawaii 2.2-m telescope on Mauna Kea, he obtained photographs of the quasars on a night of exceptionally good seeing. The photos show (Fig. 20.22) quasar B with a little bit of fuzz sticking out of it. This fuzz turns out to be the poorly resolved image of a faint elliptical galaxy— the gravitational lens! But since the galaxy is an extended mass, it acts like an imperfect lens and produces a complex pattern of images. (The view is complicated by the fact that the galaxy lies in a group of galaxies, each member of which helps to bend the light. This cluster is closer to us than the quasar, for the

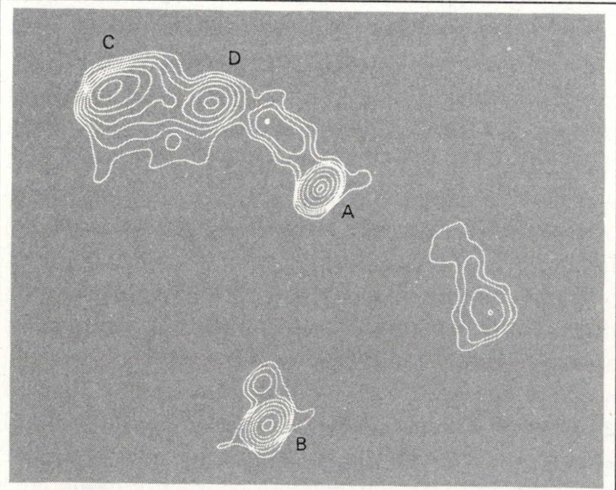

20.21
A VLA map of the radio emission from 0957+561 A and B. The double quasar is the two elliptical regions near the center (marked A and B). Note that there are blobs of emission (marked C and D) to the left of the upper quasar (A). (Based on observations by P. Greenfield, B. Burke, and D. Roberts)

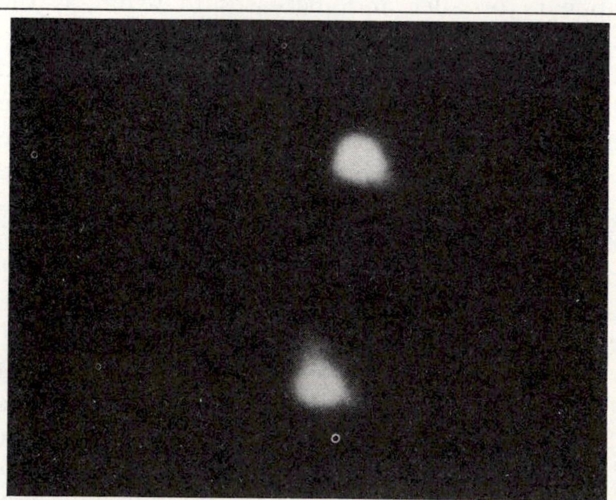

20.22
A high-resolution photo of quasars A and B. Note the little fuzz sticking out of the top of B (the lower image). This turns out to be the poorly resolved image of the galaxy making the gravitational lens. The image of B is superimposed on it. (Courtesy A. Stockton, Institute for Astronomy, University of Hawaii)

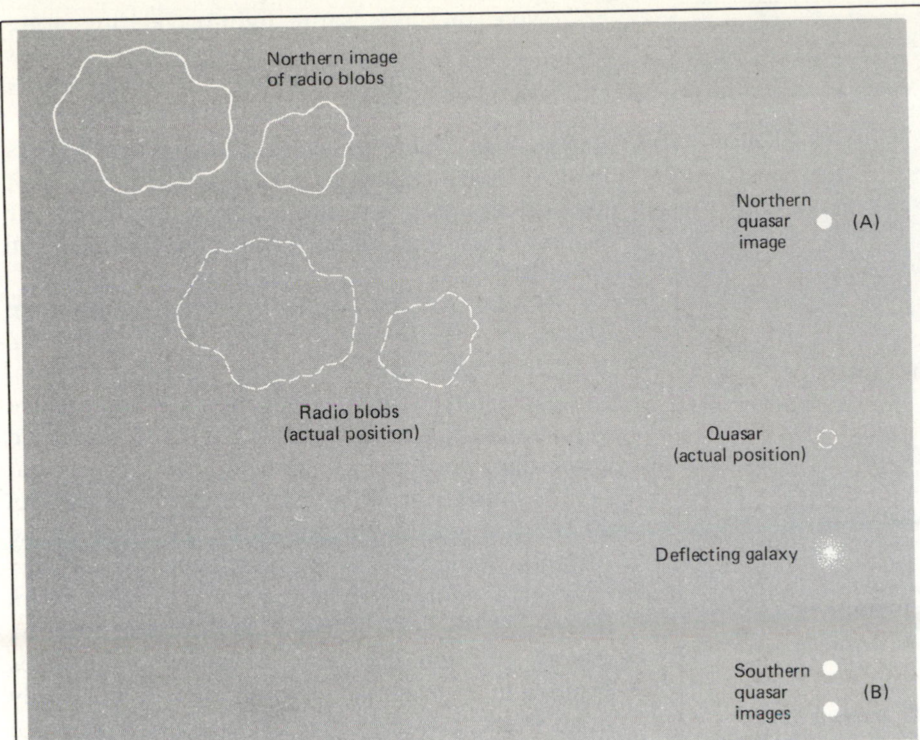

20.23

A schematic drawing of the optical illusion of the double quasar. The actual quasar and its radio emission (broken outlines) are invisible. An elliptical galaxy just below the actual quasar acts as the gravitational lens. It forms three images of the quasar, one above and two below the quasar's actual position. Only one image of the radio blobs is made, above and to the left of the actual position of the radio emission. The two lower images of the quasar lie almost on top of each other and are not visible separately.

cluster's red shift is only 0.36.) By a quirk of placement, we see only a part of the complete picture (Fig. 20.23). To add the the picture's confusion, we also see jets emerging from image A.

So the twin-quasars puzzle seems solved. We are seeing two of the three images formed by a gravitational lens, an intervening, probably elliptical, galaxy about halfway between us and the quasar (Fig. 20.24). The galaxy needs a mass of some 10^{13} solar masses to bend the light enough to form the images.

This discovery has three important implications: (1) It provides another confirmation of general relativity. (2) It proves in this case that the quasar is more distant than the galaxy, so the quasar's red shift is cosmological. (3) Gas around the galaxy creates the quasar's absorption-line spectrum; this situation may be the case for other quasars as well. At least five other candidates as double quasars have been discovered. So these implications have been strengthened.

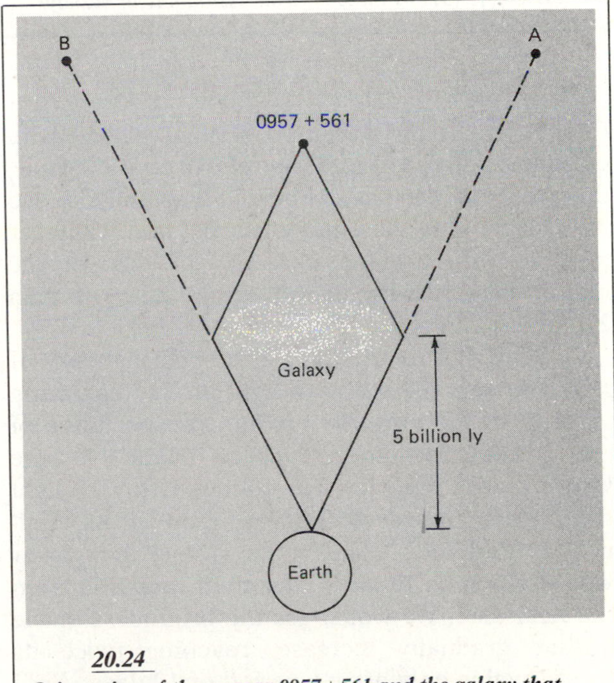

20.24

Orientation of the quasar 0957+561 and the galaxy that operates as a gravitational lens for the view from the earth.

20.7 *Troubles with quasars*

If quasars are actually billions of light years from the sun, then a quasar like 3C 273 must emit a total radiative energy of about 4×10^{40} W, an amount of energy in one second equivalent to all the energy produced by the sun in about 3 million years. A typical quasar produces up to 10,000 times as much energy as an ordinary spiral galaxy.

Not only do quasars blast out energy at enormous rates, but the energy comes from relatively small regions of space in the centers of quasars—from possibly light hours to light months to no more than light years in diameter. Two pieces of evidence point to small energy-emitting volumes.

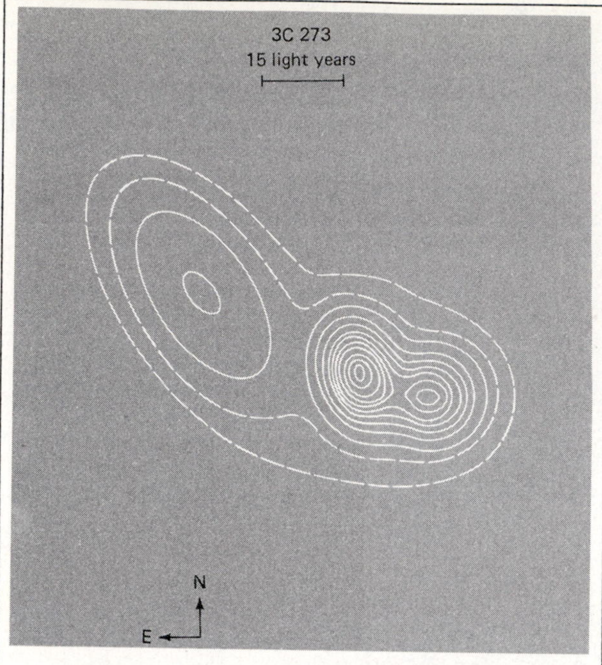

20.25
A high-resolution radio map of 3C 273, showing three distinct regions of emission. The size scale assumes a distance of 3.1 billion ly. (Observations by K. Kellerman and co-workers, Astrophysical Journal, *vol. 211 (1977), p. 658, copyright © 1977 by the American Astronomical Society)*

First, there is the variation of light output with periods of days to years. As an extreme example, the quasar 3C 446 has doubled in brightness in two days. Whatever the energy source, and whatever provoked the sudden flare-up, the size of the region that emits the energy can be no more than 2 light days across.

Why this restriction? The special theory of relativity sets the speed of light as the maximum velocity in the universe for the transportation of energy and information. Suppose that 3C 446 were larger than 2 light days in diameter, say 10 light days. Now suppose it suddenly doubled its luminosity. Radiation from the far side of the quasar would reach us 10 days later than radiation from the near side. We would see the brightness of the quasar gradually increase, reaching twice its former value only after a period of 10 days. So if 3C 446 varies in two days, it cannot be larger than 2 light days across. This conclusion arises from a fundamental property of light and does not rest on any derived property of quasars or their distances. Because quasars' light output varies on time scales of days to years, their energy-emitting regions cannot be larger than light days to light years smaller than the distance from the sun to the nearest star.

Note that this same light-travel-time argument applies to any of the active galaxies described in Section 20.2. Their variability ranges from days to months to a few years. So their emitting regions cannot be more than a few light years in size. Consider that for a moment. Quasars emit about 100 times the total energy output of our Galaxy from regions no more than light years in diameter!

Radio results obtained from international efforts of astronomers also support the conclusion that quasars must be small. Intercontinental radio interferometers (see Section 6.3) can resolve very small angular diameters. This unprecedented resolution allowed radio astronomers to see that 3C 273 consists of three separate components (Fig. 20.25). The smallest parts—if the quasar is at the distance demanded by a cosmological red shift—cannot be more than a few tens of light years across. If these observations are correct, the energy of radiation and of moving particles trapped in a quasar's small volume is fantastically huge. (Note, however, that this conclusion rests on a knowledge of the quasar's distance.)

High-resolution studies have also revealed the radio structure of quasars, especially those that are relatively near. These observations show that, for the quasars that can be resolved, the radio structures are generally symmetrical doubles, similar to Cygnus A. And some of these show nuclear jets (Fig. 20.26 and Focus 20.1).

Models of energy sources in quasars / What generates such vast energies in such small regions of space? The main model implies that all or almost all of a quasar's continuous spectrum comes from synchrotron emission—high-speed electrons gyrating in a magnetic field. As these electrons emit electromagnetic radiation, they lose energy and move more slowly. So they emit radiation with lower and lower energy. This loss of speedy electrons implies that the supply of high-energy ones must be replenished at least about every year or so.

The central energy source of a quasar must yearly blast out clouds of high-energy electrons containing a total of at least 10^{43} J. The rest of the quasar acts like a transformation machine, trapping the energy of electrons and converting it into other forms.

What energy source lies in the heart of a quasar? Remember, a model must not only explain what's seen in a general way (like a quasar's nonthermal spectrum), it must also get the numbers right (the right luminosity from the spectrum, along with the correct shape). And, finally, to be *really* successful, a model must make

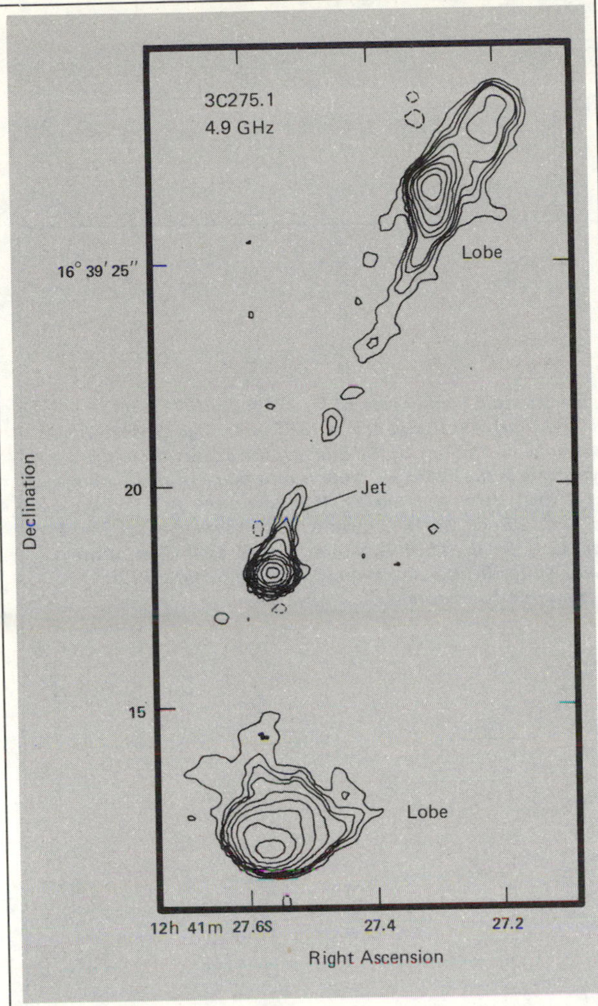

20.26

Radio emission from the quasar 3C 2751.1. Note the symmetrical, twin-lobed structure and jet. Compare Cygnus A in Fig. 20.4. (Courtesy J. Burns; observations by J. Burns and co-workers with the VLA)

specific predictions (for example, how a quasar's spectrum should evolve).

The most developed quasar models to date invoke supermassive black holes. One such model relies on a black hole of 100 million solar masses. This model takes off from that for binary X-ray sources with black holes (see Section 17.7), in which material from the normal star forms an accretion disk around the black hole before the material falls into it. A supermassive black hole in a dense galactic nucleus is fueled by the tidal disruption of passing stars (Fig. 20.27). The stellar material forms an accretion disk and radiates as it spirals into the black hole, thus powering the quasar. The model calculations show that luminosities of 10^{12} solar luminosities, about that of bright quasars, are possible. To feed the black hole requires about a solar mass of material a year.

The model must also deal with the nuclear jets now known for some quasars. One way is as follows: The accretion disk can restrict the flow of ionized gas from it, along its spin axis, either by magnetic fields or simply from the formation of a funnel of material around the black hole. This initiates a narrow beam that becomes visible as a jet once it gets beyond the main body of the quasar.

This model has a number of successful aspects. The supermassive black hole easily generates the level of quasar luminosity in a region of space only a few light years in size (the Schwarzschild radius of a 10^8-solar-mass black hole is only 3×10^8 km, or about 2 AU). And it does the energy conversion (from gravitational to radiative) with high efficiency.

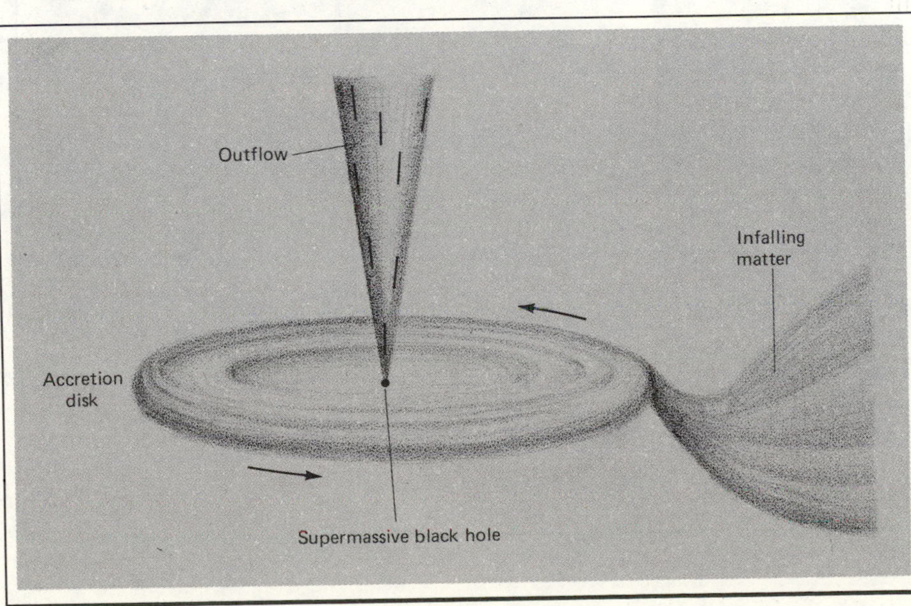

20.27

One possible model for a supermassive black hole powering a quasar or active galaxy. An accretion disk, made of infalling matter surrounds the black hole; it is thick close to the hole and thins out away from it. A sharp funnel forms where the accretion disk streams into the black hole. The disk is hot, so gas blows off it. The funnel directs the gas up along the rotation axis of the black hole, where it streams out as a jet.

Focus 20.1

FASTER THAN LIGHT?

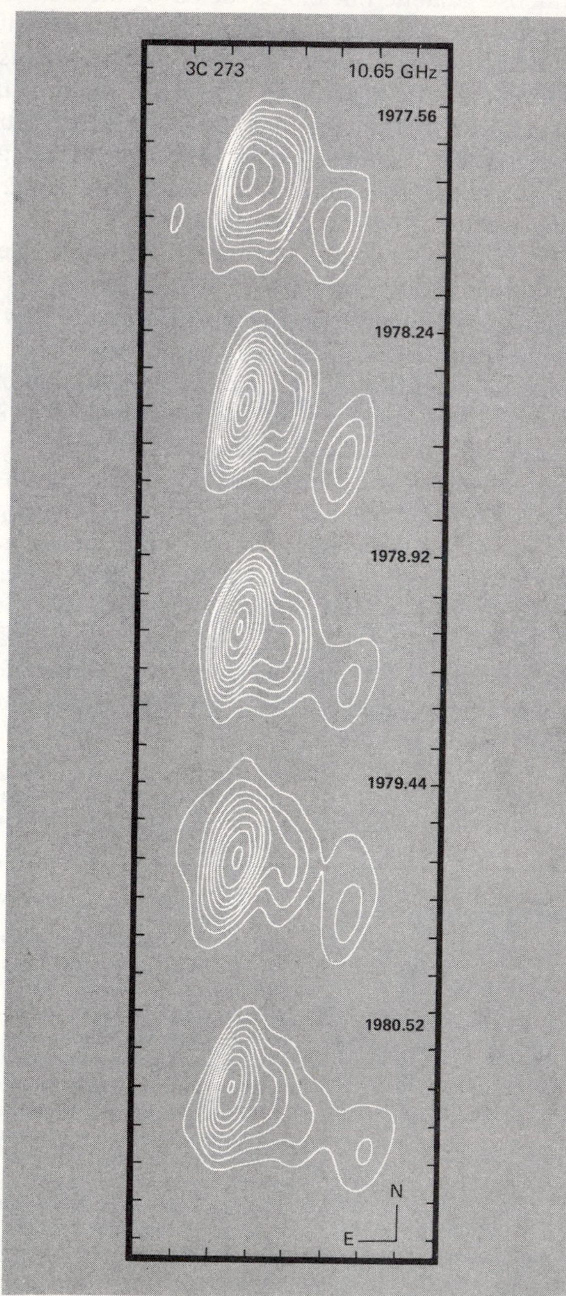

F.39

High-resolution radio maps of 3C 273 from 1977 (top) to 1980 (bottom) at a frequency of 10.65 GHz. The strongest peak is the main body of the quasar; the extension to the lower right is the radio jet. Note how a piece of this jet has moved away from the quasar. (Observations by T. J. Pearson, S. C. Unwin, M. H. Cohen, R. P. Linfield, A. C. S. Readhead, G. A. Seielstad, R. S. Simon, and R. C. Walker of the Owens Valley Radio Observatory and the National Radio Astronomy Observatory)

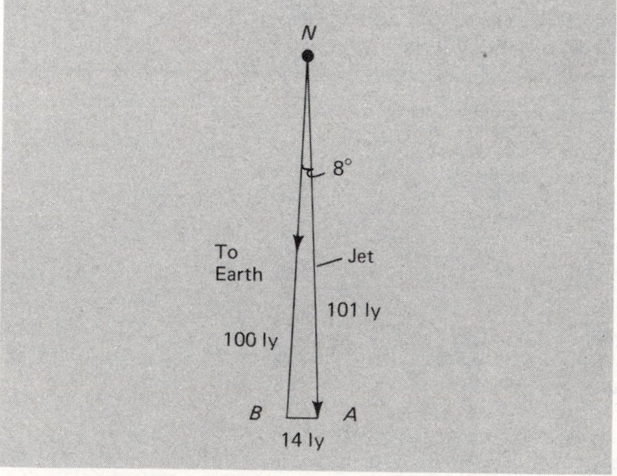

F.40

Geometry for the illusion of faster-than-light speeds. NA is the axis of the jet, tilted 8° to our line of sight (NB).

Intercontinental radio astronomy picks out the finest details of distant radio sources—even quasars. Such observations have detected changes in five quasars that, if taken face value, lead to the conclusion that parts of these objects are moving at speeds greater than that of light! Impossible! says special relativity.

Motions that seem faster than light are called **superluminal**. *The familiar quasar 3C 273 features a jet that sticks out from its core. Observations from 1977 to 1980 (Fig. F.39) show that a knot in the jet has moved away from the nucleus. If 3C 273 is at the distance implied by its red shift, the observed separation results from a motion across our line of sight of* **10 times the speed of light**! *What's happening? Like the double quasar, the superluminal effect may simply be an optical illusion.*

The sources that exhibit superluminal motions have a common feature: a radio jet from the nucleus. So they resemble nearby radio galaxies with one-sided jets (such as Centaurus A). These jets are thought to contain high-energy electrons moving close to the speed of light. If quasars have such jets, and if the jets are almost (but not quite) pointing at us, a blob of material moving out along the jet will **appear** *to travel faster than light speed.*

The superluminal effect arises from the almost head-on orientation of the jet and the finite speed of light. Here's how: Suppose a jet points at us at an angle of 8° with respect to our line of sight (Fig. F.40). Imagine that a blob of electrons ejected from the nucleus (point N) gets to point A in 101 years and the light emitted from the blob at N takes 100 years to get to point B. For an 8° angle, the separation of A and B is 14 light years. But the light from B is one year ahead of that emitted by the blob when it finally reaches A. It's taken 100 years for the light from N to reach B, 101 years to reach A.

Many years later, that light that was at B reaches us. Only one year later, that emitted at A arrives. To us, it appears that the source has moved from B to A—a distance of 14 ly—in only one year. It seems to have a speed of 14 times that of light across our line of sight. In fact, no such physical motion has occurred, only an optical illusion. And the smaller the angle of the jet to our line of sight, the greater the superluminal motion will appear. So almost any faster-than-light speed is possible because some jets will be tilted very close to our line of sight.

I would say that black-hole models have the brightest prospects for understanding a quasar's central energy engine—but they are still in the formative stages.

Quasars and active galaxies compared / You've probably noticed that quasars and active galaxies share some observed (and so some physical) characteristics. Let's make some specific comparisons.

First, radio galaxies. Most have only absorption lines in their spectra. Those that have emission lines come in two types. One type has *narrow* emission lines; the Doppler widths are roughly 500 km/sec. Narrow-line radio galaxies make up about two-thirds of the total. The other third have broad lines of hydrogen and helium, some as wide as 10,000 km/sec. Other emission lines for these galaxies tend to be narrow.

Low-red-shift quasars have optical spectra that resemble the broad-line radio galaxies in terms of the emission lines present, their widths, and the shape of the optical continuous spectrum. For instance, the hydrogen and helium lines of such quasars have widths of 4000 km/sec.

Quasars also resemble Seyferts in their emission-line spectra. Seyferts come in two types on this basis: the Seyfert 1 class that have broad hydrogen lines and relatively narrow lines from most other elements, and the Seyfert 2 class that have all narrow emission lines. The Seyfert 1 galaxies, just like the broad-line radio galaxies, look like low-red-shift quasars in terms of their emission spectra. So the physical conditions in the regions producing the spectra must be basically the same in Seyfert 1s and in low-red-shift quasars.

In addition, Seyferts look like quasars: The nuclei of both are starlike and a few nearby quasars have been shown to have galaxylike disks surrounding them. The colors of Seyferts with the largest nuclei resemble the colors of quasars. And the nuclei of Seyfert 1 galaxies vary in light over periods of months—which implies, by the light-travel-time argument, that the emitting region cannot be larger than a few light months in size.

Finally, let's compare quasars and BL Lac objects. First we have a pronounced difference: BL Lac objects do *not* have strong emission lines. So let's compare the continuous spectra of quasars and BL Lac objects. In both BL Lacs and quasars, the nonthermal nature of the continuous emission stands out most clearly compared with other active galaxies. Assuming that the nonthermal emission is synchrotron, BL Lac and the quasar 3C 279 need about the same magnetic-field intensity to account for their synchrotron spectra in the radio range.

BL Lac objects and about 15 percent of radio-bright quasars show wide variations in optical output over periods of days, weeks, or months. These swings in luminosity often occur very abruptly.

Finally, we have observations of many nuclear jets in active galaxies and quasars (over 100 are known to date). The observed properties of jets in radio galaxies and in quasars are generally similar. This implies (but does not prove) that the physical conditions producing the jets is also the same.

To sum up: Active galaxies and quasars share some of the same *observed* properties and so, by inference, some of the same *physical* properties. The most striking general aspect is the nonthermal emission from a region a few light years in size.

Are active galaxies and quasars related? / The observations suggest, but do not prove, that quasars have some connection with active galaxies. One popular idea views them as similar objects at different stages of evolution. That is, the quasar phenomenon signals very violent activity in the nucleus of a galaxy at a very early stage in its life. Other active galaxies are older quasars, and so less active. The evolutionary sequence of stages would be roughly quasar, BL Lac, Seyfert, radio galaxy stages, and finally a normal galaxy. This sequence seems likely because, in general, quasars have the greatest red shifts, Seyferts less, and radio galaxies the smallest. So if the red shifts are cosmological, then quasars are younger (less evolved) than Seyferts, which are younger than radio galaxies. (Don't get the idea that the same object starts as a quasar and then ends up as a normal galaxy. The proposed sequence is one of common evolutionary stages in which the same physical processes are at work.)

One other possible connection relates to radio structure. Quasars have high luminosities and tend to exhibit symmetrical double structures. The higher-luminosity active galaxies have similar radio shapes. So the physical processes responsible for both may be the same. The nuclear jets imply that this structure is somehow tied to violent activity in the nuclei of luminous active galaxies and quasars.

This comparison suggests that quasars are linked to active galaxies in an evolutionary fashion: Quasars may be distant active galaxies. And most (or all) galaxies may go through evolutionary stages that resemble quasars, then active galaxies, and finally normal galaxies—which have some of the properties of active galaxies but display them less violently.

What kind of galaxies have active youths? A suspicion has recently arisen that spirals and ellipticals play different roles. You have seen that strong radio sources are usually connected with elliptical galaxies. In contrast, active spirals, such as Seyferts, are usually weak radio sources. As a first guess, quasars that are weak radio sources might be spirals; those that are strong radio emitters might be ellipticals. The physical reasons for such a split are not clear.

A composite Hubble diagram (Fig. 20.28) suggests that this sequence may be the case.

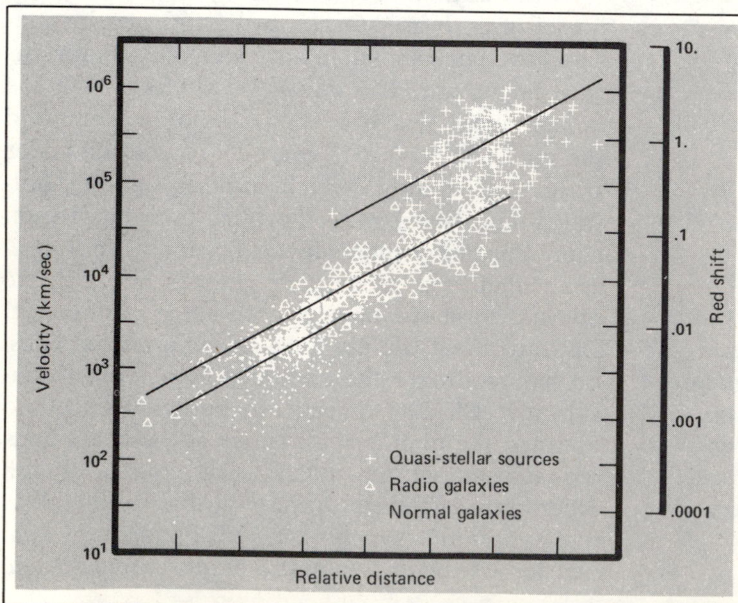

20.28
A composite Hubble diagram for quasars (crosses), radio galaxies (triangles), and normal galaxies (dots). Note how the three groups seem to merge. (Courtesy K. Lang; from K. Lang, S. Lord, J. Johanson, and P. Savage, **Astrophysical Journal,** ***vol. 202 (1975), p. 583, copyright © 1975 by the American Astronomical Society)***

When the red shifts and fluxes are plotted for normal galaxies, radio galaxies, and quasars, the three groups appear to join together. Quasar luminosities average 10^{38} W, radio galaxies 10^{37} W, and normal galaxies 10^{36} W. This decrease in luminosity and the merging of positions in the Hubble plot may indicate an evolutionary connection among the groups.

If we accept a model for the energy output of active galaxies and quasars that imagines a black hole in the nucleus eating up huge amounts of material, the nucleus should show an intense, pointlike source of light (due to a concentration of stars around the black hole), stars orbiting the center should have high velocities, and emission lines with high Doppler shifts might be visible from the in-falling matter. Observations of such effects have been reported by two research groups: One has found a high-luminosity spike in M 87's nucleus (such a spike is not found in the nucleus of normal elliptical galaxies), and the other has discovered a dramatic increase in the velocities of stars in the nucleus compared to those outside the nucleus. A model consistent with these observations is one of a 5 billion solar-mass black hole hiding in the inner 300 ly of the nucleus. The black hole would cluster the stars in the nucleus closely around it, and they would orbit at high velocities.

Given that a quasar may simply be the hyperactive nucleus of a very faraway galaxy, what about the rest of the galaxy? At cosmological distances, the disk of a quasar's parent galaxy would be too small and faint to see easily. For example, if the Andromeda galaxy were placed at a distance of 4 billion ly—equivalent to a red shift of 0.2 (small for a quasar!)—its angular diameter would be a mere 4″. Remember that the earth's atmosphere limits seeing to 1″ at best. So a distant galaxy would make an image hard to distinguish from that of a faint star.

We might, however, look for a quasar in a cluster of galaxies—a good sign that we're not looking at stars. A nice example is the quasar 3C 206, which is surrounded by some 200 faint galaxies (Fig. 20.29). The quasar's red shift is 0.206; that of a nearby pair of galaxies, 0.203 (and the same is presumed for the cluster). This result not only shows that the quasar resides in the cluster but also proves that the quasar's red shift is cosmological.

Other observations are beginning to reveal that quasars are surrounded by very faint envelopes that may well be the hard-to-detect disk of the galaxy. For example, 3C 273, the closest of the high-luminosity quasars, has a fuzzy appearance in computer-processed photos. The faint fuzz has emission lines in its spectrum with the same red shift as the quasar.

20.29
The quasar 3C 206 (arrow) surrounded by galaxies, all of which probably lie in the same cluster. (Courtesy H. Spinrad; observations by S. Wyckoff, P. Wehinger, H. Spinrad, and A. Boksenberg)

Finding black holes in nearby galaxies (such as our Galaxy and M 87) would provide evidence for the model of hyperactive nuclei. But be warned that even though the black hole models may be the best buy in terms of a conventional physics understanding of quasars and active galaxies, the models are hard to test observationally.

20.8 Alternative views about quasars and their red shifts

So far I have presented quasars' red shifts as cosmological—simply due to the expansion of the universe. But a small, hardy group of astronomers views all or part of the red shift as the result of another cause (or other causes). Don't consider this an easy step to take, for then you have abandoned the simplest explanation for quasars' red shifts. This small band has fueled the red-shift controversy about quasars.

Frankly, I believe that quasars' red shifts *are* cosmological, and so do most astronomers. So I will treat the other side's views here very briefly. They aim at discrediting at least part of quasars' red shifts as cosmological, usually on observational grounds.

Quasars might acquire large, noncosmological velocities if they were small objects shot from galaxies by gigantic explosions. Then, if the quasar is shot away from us, the red shift of the qua-

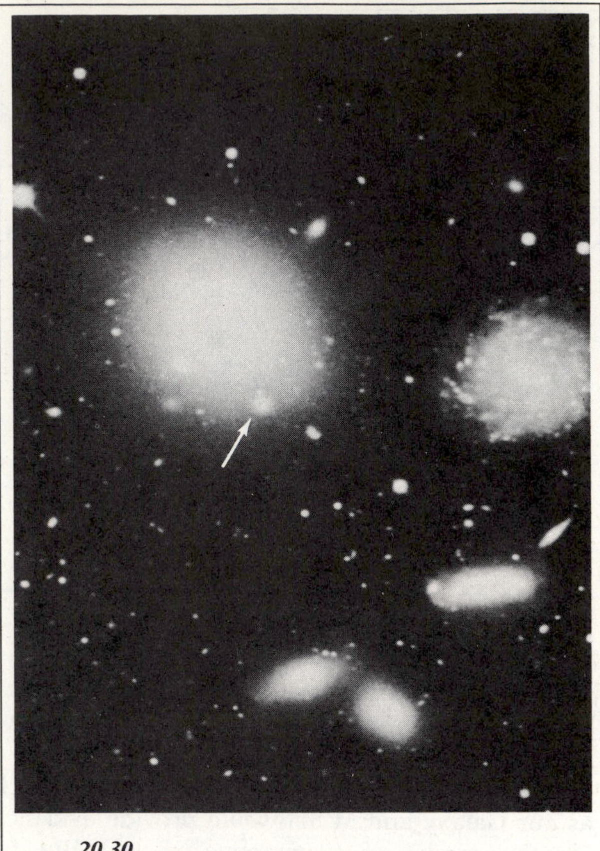

20.30
A high-red-shift object (arrow) in front of the galaxy NGC 1199. (Courtesy H. Arp)

sar would be the Doppler shift from its expulsion plus the Doppler shift of its parent galaxy. We would see a very large red shift from the combination of the two. If shot toward us, we subtract the radial velocity of the quasar from its parent galaxy; if it moves fast enough, we see a blue shift.

What do you expect observationally from a model that physically links quasars and galaxies? That you would find galaxies and quasars close together in the sky more often than by chance. But remember, when you look at the sky, you see both near and distant objects. Just because you see a galaxy and a quasar close together does not mean that they are at the same distance and physically associated.

The proponents of alternative ideas for red shifts, lead by Halton Arp, believe that observations support the association of quasars with some galaxies—and so create a suspicion of a physical connection. In one peculiar case, Arp sees a quasarlike object (red shift of 0.044) in front of an elliptical galaxy with a red shift of 0.009 (Fig. 20.30). How does he infer that the higher-red-shift object lies in front? He thinks that he sees a dark ring around it and argues that it comes from the absorption of light from the elliptical galaxy. Now, if the object were ejected by the galaxy, it could be between us and its parent galaxy, but its red shift should be less, not more. Arp concludes that the

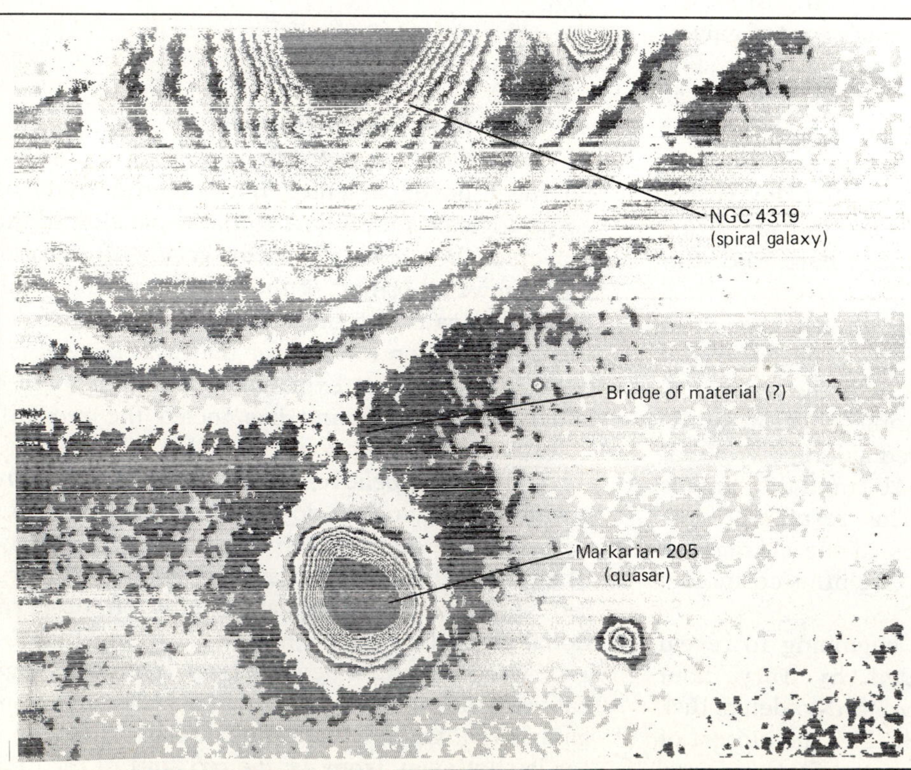

20.31
Possible evidence for a physical link between a galaxy and a quasar. This figure is a computer-generated contour map of a photo. The object at the top is the spiral galaxy NGC 4319. Below it appears the quasar Markarian 205. There seems to be a bridge of material linking the two images, but the supposed bridge may be a blending of the images as a result of the photographic process. (Courtesy R. Lynds and Kitt Peak National Observatory)

compact object was spewed out by the galaxy and that its red shift is *not* Doppler but due to some other (unknown) physical cause.

Note that proximity arguments don't *prove* physical connection. The sky contains about 300 bright galaxies and 60 bright quasars. Suppose you scattered 300 nuts and 60 grass seeds over a huge map of the sky and then asked what the chances were of a nut and seed lying closer together than some small distance. You could estimate the chances for such close pairings, and if they were actually more frequent, you could argue that a physical connection makes it so. Arp and associates argue in such a way.

Most other studies support cosmological red shifts. For example, Alan Stockton has examined a sample of 27 quasars, not intentionally selected to be near galaxies. He finds 29 galaxies within 45″ (his definition of "close") of the quasars. Among these galaxies, 13 have red shifts within 1000 km/sec of the nearby quasar. (Why an allowable spread of 1000 km/sec? Because in a cluster of galaxies, that is the typical spread in velocities. So a quasar and galaxy in the same cluster could differ that much in red shift.) Based on the known distribution of the red shifts of galaxies, the chances are only one in a *million* for the chance association of 8. The fact that 13 were found makes the cosmological nature of the red shifts in this sample virtually certain.

Observations of physical connections, such as bridges of gas and dust between galaxies and quasars, would make a compelling case for noncosmological red shifts. Such a case exists, crowned right now in controversy. In photos (Fig. 20.31) of a galaxy (NGC 4319) and a quasar (Markarian 205), Arp has argued that he finds a bridge of luminous material connecting the galaxy and the quasar. Amusingly, the galaxy has a red shift of 1800 km/sec, while the quasar has one of 21,000 km/sec!

Other astronomers contend that the bridge may be an artifact of the merged images on the photo rather than a real connection. Or that a faint star or galaxy lies in the gap and so gives the false impression of a bridge. These ideas would seem to shoot down a physical bridge. But the fight has been renewed by Jack Sulentic, who has examined by computer all the photos of this region taken with large telescopes. He claims to have confirmed that the bridge is real—it does not arise from an overlap of the images of NGC 4319 and Markarian 205 nor from a galaxy behind them that just happens to fall in the gap between them. He concludes that the bridge is an actual stream of material and represents the most conclusive evidence to date of a physical link between objects of significantly different red shifts.

So the red-shift controversy continues. But I feel it will take a tremendous amount of solid evidence to shake the cosmological view, which most astronomers find compelling.

KEY CONCEPTS

1 ***The nucleus of the Galaxy contains a small, energetic source with a partly nonthermal spectrum; it marks the core of the Galaxy.***

2 ***Continuous spectra can be divided into two types: thermal (blackbody) and nonthermal; a common form of nonthermal emission is synchrotron radiation, produced by high-speed electrons spiraling in magnetic fields.***

3 ***Active galaxies have nonthermal spectra, at least in part.***

4 ***Radio-emitting galaxies are one type of active galaxy; they fall into two classes: compact (emission from the nucleus) and extended (emission usually in two lobes widely separated from the nucleus).***

5 ***High-resolution radio observations show that in many cases, a jet of emission extends from the galaxy's nucleus to at least one of the lobes of an extended radio galaxy; the synchrotron process may be the source of the emission; it shows that high-speed electrons are somehow channeled along the jet out to the lobes.***

6 ***Seyfert galaxies, another type of active galaxy, are spiral, with bright nuclei and very broad emission lines in their spectra.***

7 ***BL Lacertae objects (also active galaxies) have nonthermal spectra, vary rapidly in luminosity, and usually have no emission lines in their spectra.***

8 ***Quasars have a starlike appearance, very large red shifts, and broad emission lines in their spectra (along with narrower dark lines); they sometimes have detectable radio emission with nuclear jets.***

9 ***If the red shifts in the emission lines from quasars arise from the expansion of the cosmos, quasars are very far away—billions of light years; if so, their luminosities are huge, up to thousands of times that of normal galaxies; yet the emitting region cannot be larger than a few tens of light years in size; we do not know yet how quasars produce so much energy in so small a space.***

10 ***Quasars and active galaxies share some of the same observational characteristics, so they might be the same types of objects***

(galaxies) at different evolutionary stages; in particular, a quasar may be a young galaxy with a hyperactive nucleus.

11 ***A few astronomers believe that the red shifts of quasars are not cosmological but arise from other causes; a little evidence to date supports this view.***

12 ***We do not yet know the energy machine in the cores of quasars and active galaxies; it may be a supermassive black hole eating up material in the nucleus; some indirect evidence supports this idea.***

STUDY EXERCISES

1 What observational evidence do we have that the synchrotron process produces some radiation from active galaxies and quasars? (*Objectives 1 and 2*)

2 Contrast the evidence for violence in the Milky Way Galaxy with that for any active galaxy. (*Objectives 2 and 7*)

3 What evidence do we have that quasars are far away? (*Objectives 3 and 5*)

4 What *observational* evidence indicates that for many quasars, their light must pass through clouds of thin, cool gases? (*Objective 4*)

5 What powers a quasar? (*Objectives 4 and 9*)

6 What connection does extended radio emission often have with the nucleus of an active galaxy? . (*Objectives 2 and 10*)

7 Make a table comparing the observed properties of radio galaxies and quasars. (*Objective 7*)

8 What physical conditions are required to produce synchrotron emission in active galaxies and quasars? (*Objective 1*)

BEYOND THIS BOOK . . .

Black Holes, Quasars, and the Universe (2d ed.) by H. Shipman (Houghton Mifflin, Boston, 1980) has an excellent presentation of the controversy over quasars.

Mercury, November-December 1974, contains "The Quasar Controversy: An Interview with Caltech Astronomer Halton Arp." Also see *The Redshift Controversy*, edited by G. Field, H. Arp, and J. Bahcall (Benjamin, New York, 1973).

R. Weymann discusses "Seyfert Galaxies" in *Scientific American*, January 1969.

For a detailed look at the double quasar, read "The Discovery of a Gravitational Lens" by F. Chaffee, *Scientific American*, November 1980. For a nice exposition on gravitational lenses, see "Charting Paths Through Gravity's Lens" by M. Gorenstein, *Sky and Telescope*, November 1983, p. 390.

For a general account about nuclear jets, read "Cosmic Jets" by R. Blandford, M. Begelman, and M. Rees, *Scientific American*, May 1982, p. 124. For a closer look at Centaurus A, try "Centaurus A: The Nearest Active Galaxy" by J. Burns and M. Price in *Scientific American*, November 1983, p. 56.

21/Cosmic history

> ***I could be bounded in a nutshell and count myself king of infinite space.***
>
> **WILLIAM SHAKESPEARE: *Hamlet***

Learning objectives

After studying this chapter, you should be able to:

1. **State the basic assumptions of cosmology.**
2. **Present, in a short paragraph, basic observations that have cosmological import.**
3. **Describe briefly the Big Bang model, focusing on its assumptions and its ability to explain fundamental cosmological observations.**
4. **Describe briefly the *observed* properties of the cosmic background radiation.**
5. **Present at least one argument for ascribing a *cosmic* origin to the background radiation and explain how it is a natural consequence of a Big Bang model.**
6. **Discuss the importance and the impact of cosmic radiation for the Big Bang model.**
7. **Outline the process of element formation in the standard Big Bang model and cite at least one observation that supports theoretical ideas.**
8. **Describe a physical basis for particle production in the young, hot universe.**
9. **Outline a history of matter and radiation from the time they both stopped interacting strongly to the present.**
10. **Describe a simple model of galaxy formation from the Big Bang and pinpoint problems with this model.**

Central question

How have the physical properties of the universe changed since its origin in the Big Bang?

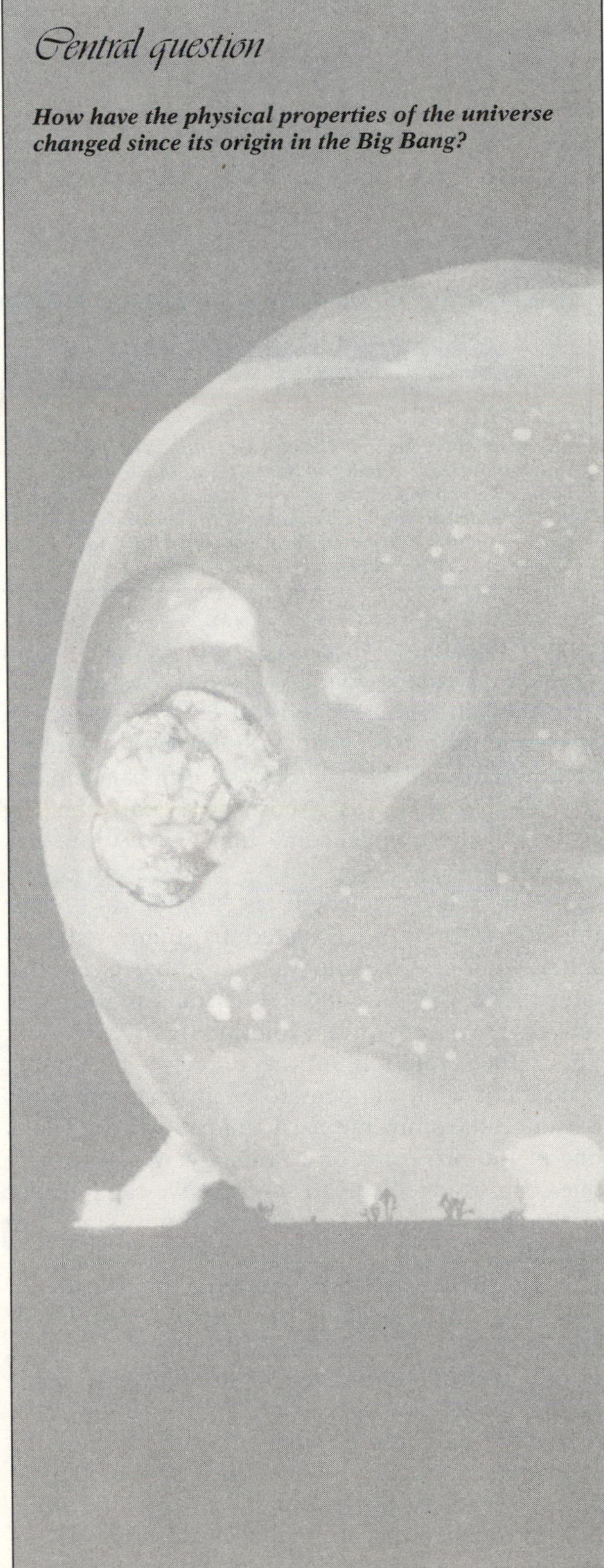

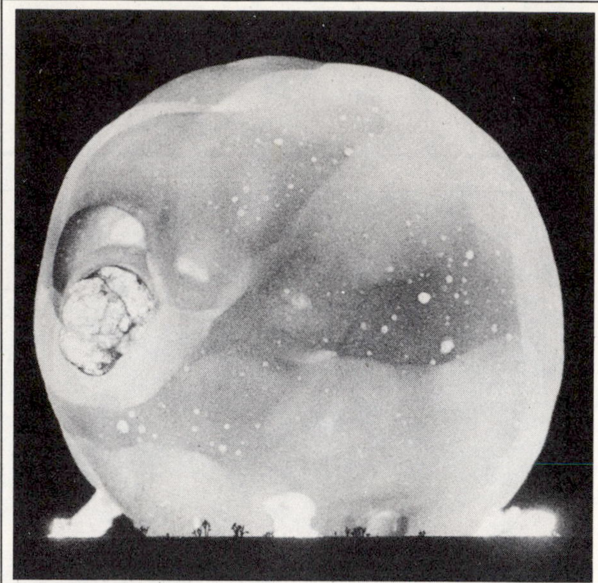

21.1
A terrestrial analog to the cosmic Big Bang. This photo shows a nuclear fireball fractions of a second after the bomb's detonation. For a very brief time interval, the temperature, density, and pressure in the fireball korrespond to those in the Big Bang. (Courtesy Harold E. Edgerton, Massachusetts Institute of Technology)

Some 20 billion years ago the cosmic bomb exploded. Perhaps you have seen movies of an H-bomb blast. Split seconds after detonation, an awesome fireball rips violently through the atmosphere. Our universe was born out of a similar fireball—but this was a cosmic fireball, the Big Bang, in whose violence all that we now see was created (Fig. 21.1).

That, in a nutshell, is a picture of our universe's creation accepted by many astronomers today. Cosmology, the subject of this chapter, is the grandiose (but human) study of universe's nature and evolution. You can't talk about the evolution of the universe by simply describing what happens to each part; you must consider the universe as a unique whole. That's one of the problems of cosmology: We have only one cosmos to look at! In contrast, we can tell a lot about stars simply because so many stars are around.

Cosmologists have been fed a meager diet of observational facts about the universe. Despite (or perhaps because of) this lack, they have been able to dream up many models of the universe. Some of the models have been quite bizarre, but only two have gathered a substantial following: the steady-state model and the Big Bang model. In part because of my personal bias, I pay little attention to the steady-state model in this chapter. I—and many other astronomers—believe that the present evidence indicates that the universe began in a Big Bang.

21.1 Cosmological assumptions

Modern astronomy has arrived at models of the cosmos in which the universe *evolves*; it is dynamic, not static. How to study the universe's evolution in detail? To do so requires a few fundamental starting assumptions, difficult-to-prove assertions about the nature of the cosmos.

First, we assume the *universality of physical laws*. This assumption covers both local (the earth and solar system) and distant regions. It means that we apply the physical laws we uncover here to all localities at all times and to the universe as a whole. A few key observations support this assumption: For example, the spectra of distant galaxies contain the same atomic spectral lines as those produced by elements found on the earth. So other galaxies are made of the same elements as here, put together in the same way. Newton's

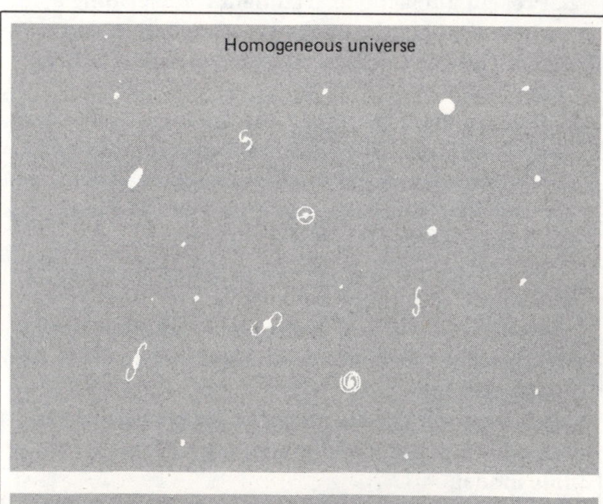

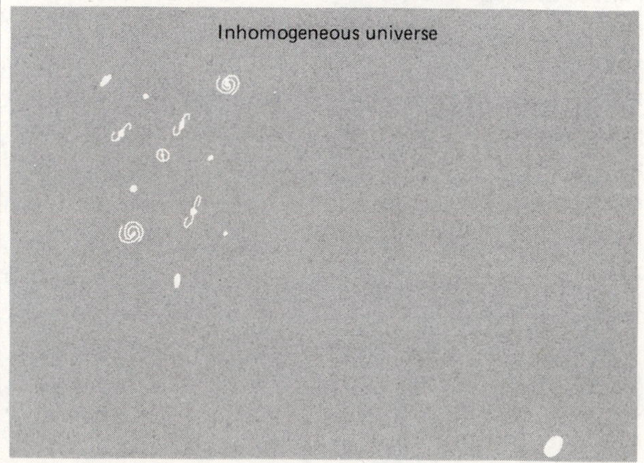

21.2
The homogeneous universe. Matter and radiation are spread out uniformly, if you look over large distances.

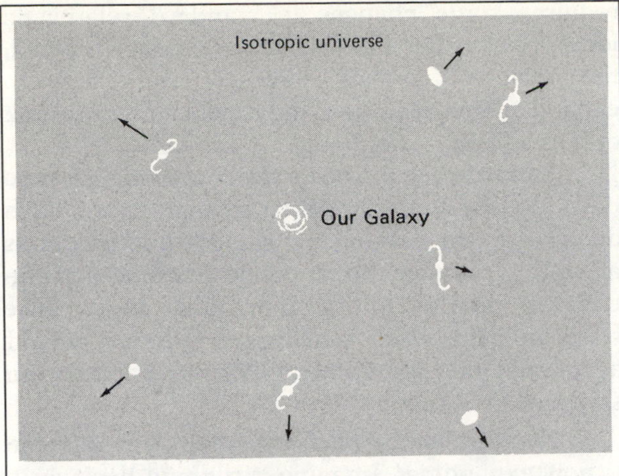

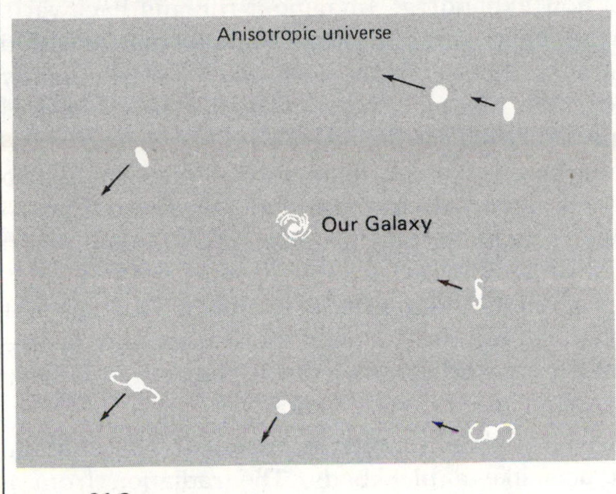

21.3
The isotropic universe. Space has the same properties in all directions. For example, the distribution of velocities of galaxies in the expanding universe is smooth, uniform, and the same in all directions.

law of gravitation correctly describes the motion of double stars and galaxies.

Second, we assume that the cosmos is *homogeneous* (Fig. 21.2). This means that matter and radiation are spread out uniformly, with no large gaps or bunches. You know that this assumption is not strictly true, for clumps of matter, such as galaxies and stars, do exist. But the cosmologist assumes that the size of the clumping is much smaller than the size of the universe. It's like looking at the earth from space: Bumps, such as mountains, are too small to be seen, so the globe looks smooth.

The third assumption is that the universe is *isotropic* (Fig. 21.3). This idea relates to a quality of space itself rather than to the matter in it. Here's one way to think of it: Space has the same properties in all directions. So no direction or place in space can be distinguished from any other by any experiment or observation. The universe has no center in space, because there is no way to tell if you are there. No direction in space provides special rewards when taken; for example, your mass does not increase as you travel in one direction or decrease when you go another way. Another example: The expansion of the universe is the same in all directions of space. Any observer would see the same Hubble's law.

These assumptions can be summed up in one sentence: The universe is *uniform*. All irregularities are ironed out. As the cosmologist Edward R. Harrison has quipped, the result is like that of the vanishing Cheshire cat in *Alice in Wonderland*: Everything is wiped out except the grin. As a result, cosmological models that rest on these assumptions ignore the structure and substance of planets, stars, and galaxies. Galaxies are considered, but only as tiny particles marking points in space, like a gas of atoms filling the universe. This gas is the cosmic grin. Such a mental simplification has real dangers, for like the Cheshire cat, what lies behind the grin may be crucial. The universe may not obey the laws we lay down for it or the models we develop for it. Observations must in the end validate our assumptions.

Finally, I've been talking rather glibly about the universe without defining it. If we consider all that can be seen with various telescopes, we are considering the *observable universe*. Yet this cannot be *all* of the universe; there are objects too faint and too far to be seen, regions of the spectrum to which we and our instruments are so far blind, and objects detectable only by their gravitational effects. So there is more to the observable universe: a *physical universe* that includes directly observable matter as well as objects we detect by effects described by the laws of physics.

The reality of the physical universe rests on the assumption that local physical laws apply to the rest of the universe at all other places and all other times. That's one of our basic assumptions; without it, we could not conceive of a physical universe at all.

21.2 A brief review of cosmology

Chapter 7 outlined the rise of relativity and its impact on cosmological ideas. It also presented some fundamental cosmological observations. These observations did not fall into a grand scheme until explained by Einstein's general theory of relativity. (This is a good place to review Chapter 7.)

What *do* we know about the universe? First, the universe *evolves*. Both the whole cosmos and its contents change with time.

Second, matter in the universe is *grouped*. Elementary particles (whatever they are) make up protons and electrons, which make up atoms. You know that atoms make up gases (molecular and atomic) and dust particles, which form stars, planets, and us. Stars come in clusters of stars, which are found in galaxies. And galaxies are grouped in clusters of galaxies, which in turn may congregate in superclusters of galaxies. And these, as far as we can tell, make up the universe itself.

Third, the universe is *expanding*. Hubble and Humason (Section 7.4) observed red shifts in the spectra of other galaxies; they made the important discovery of cosmic expansion. The observed rate of expansion now lies in the range 15 to 30 km/sec/Mly. From Einstein's theory of general relativity we know that changes in the rate of expansion relate to the average density of matter (and energy) in the universe. This average density, in turn, determines the overall geometry of space-time. Three possibilities exist for this geometry: (1) flat, an open universe with the geometry of Euclid, infinite in space and time; (2) spherical and closed, the cosmos being finite but unbounded in space and finite in time; (3) hyperbolic, again an open universe, infinite in space and time, but curved. Observations to date imply that the universe has an open geometry; the question is still unsettled. (Einstein had an aesthetic preference for a closed universe.)

Finally, we can estimate the ages of celestial objects. From Hubble's constant we find that the universe has been expanding for approximately 15 billion years (see Focus 7.2). Radioactive dating places the earth and the moon at an age of about 4.6 billion years. Astronomers estimate our Galaxy's age at some 10 to 16 billion years. The fact that we can even make these estimates, in spite of the uncertainities, is amazing: All are less than the presumed age of the universe (if H equals 15 km/sec/Mly), and they fall in a natural evolutionary sequence, from the universe to the earth. Here's a hint that the universe was born at a finite time in the past and evolved in a sequence set by known physical laws.

21.3 Contemporary cosmological models

Over the past 60 years, ingenious theoreticians have devised a bewildering array of cosmological models. Almost all fall into one of two categories: (1) the *Big Bang model*, which is the standard model based on Einstein's general theory of relativity, and (2) the *steady-state model*, which requires some changes in standard relativistic ideas. (I consider the oscillating model—in which the universe expands, then contracts, only to expand again—as a Big Bang model. You can think of it as a "bang-bang-bang . . . " model.)

I judge that the observational evidence supports the Big Bang model so strongly that I have decided not to develop the steady-state model at all in this chapter. (Some of the references at the end can provide information about the steady-state model.) Most astronomers today would agree that the Big Bang model has become the standard in use today.

The Big Bang model revisited / The universe is now expanding. Imagine it running backward. What happens? The galaxies and all matter within and without them eventually come tightly together. The extreme compression heats (just as any squeezed gas heats up) both matter and radiation to a very high temperature—high enough to break down all structure that was created previously, including all the elements fused in stars. The atoms break down into protons and electrons. In addition, the density of matter is so great that photons can travel only short distances before they are absorbed. As a result, the entire universe is opaque to its own radiation.

Now, when matter is opaque to all radiation, it acts like a blackbody. The radiation from a blackbody exhibits a characteristic shape when you study its spectrum (Fig. 21.4). In an early dense, hot state, the universe acts like a blackbody radiator. If you could have been there, you would have seen a bright fog all around, like sitting in the sun's interior (but with the radiation mainly at even shorter wavelengths, because of the higher temperature).

Imagine the universe expanding from this infernal state. It is so hot that it expands in a violent rush like an explosion; hence the name *Big Bang* for this picture of the universe. As the universe expands, its overall density and temperature decrease. (This is true for the expansion of any gas. Have you ever suddenly let a gas out of a container? If so, you noticed that the gas suddenly became cold as it expanded.) Eventually the temperature drops low enough so that protons can capture electrons to form neutral hydrogen.

This event—the formation of neutral hydrogen from an ionized gas—marks a crucial stage in the evolution of the universe. No longer ionized, the universe becomes transparent to its own radiation; the light is freed of its close interaction with matter. The radiation and matter are no longer connected. The radiation freely speeds throughout space, and the expansion dilutes it. As it is diluted,

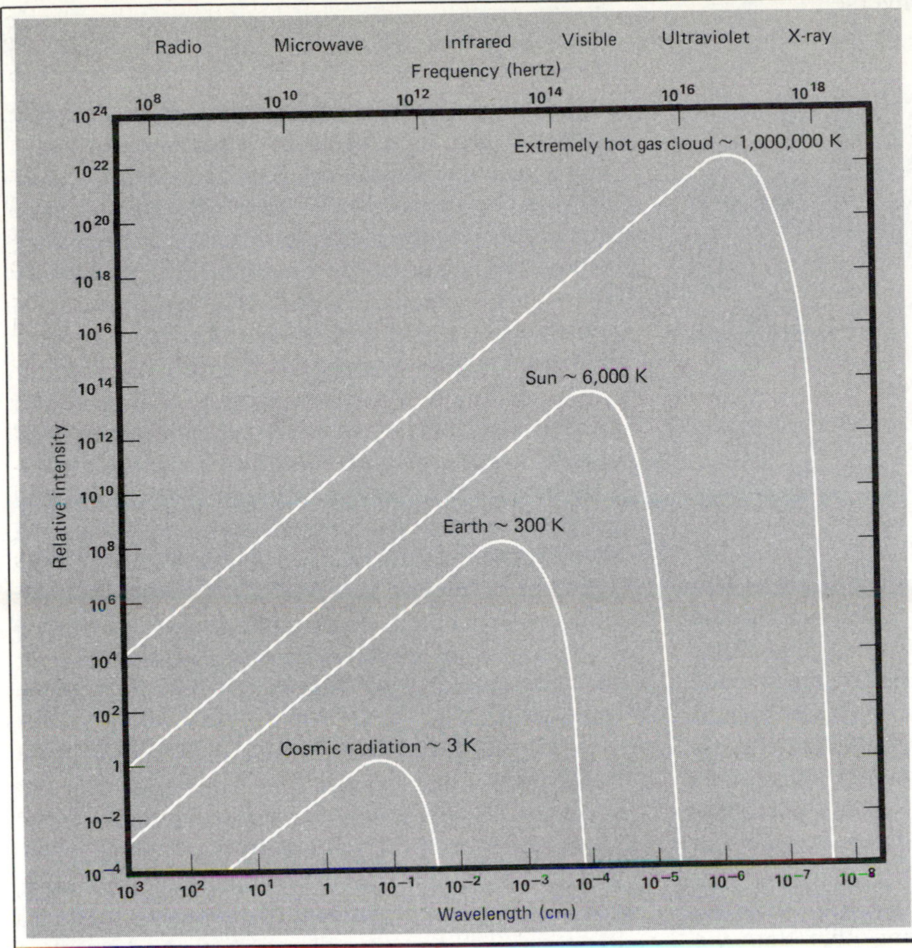

21.4
A comparison of the spectra of blackbodies at different temperatures. Note that cosmic radiation at about 3 K peaks in intensity at microwave-infrared wavelengths; hotter blackbodies peak at shorter wavelengths.

the radiation cools. It is red-shifted, just like the light of distant galaxies, for indeed, being emitted long ago, it comes from far away, and Hubble's law predicts a large red shift.

If the universe *did* begin in a hot Big Bang, debris (both matter and radiation) from the cosmic explosion of the Big Bang must now lie all around us. The matter's pretty obvious: You can see planets, stars, galaxies, and the unformed matter of interstellar gas and dust among all of these. But what about the radiation produced in the Big Bang? It's been red-shifted to a fairly low temperature. Because of its low temperature, the radiation's wavelength will be long compared with that of light. And if the universe expanded uniformly, the graph of the radiation's spectrum should show the telltale blackbody shape.

Have we seen such a cosmic radiation? Yes! Its discovery supports the hot Big Bang model.

21.4 *The cosmic background radiation*

In 1964 Robert H. Dicke, P. James E. Peebles, Peter G. Roll, and David T. Wilkinson at Princeton University were pursuing the question of the possible existence of leftover radiation from a Big Bang. The Princeton group attacked this problem: What would happen if the universe went through a hot stage, so that a high temperature decomposed any heavy nuclei into elementary particles? If a primeval fireball occurred, cosmic blackbody radiation should survive today. Peebles calculated that this fossil radiation should have a blackbody temperature of roughly 10 K.

Just as Roll and Wilkinson were building apparatus to detect the radiation, Arno Penzias and Robert Wilson, scientists with Bell Laboratories in New Jersey, detected an annoying excess radiation by means of a special low-noise radio antenna (Fig. 21.5). They intended to do a sensitive study of the radio emission from the Milky Way. The excess noise they found would affect their results, so they set about to try to eliminate it.

They tuned their radio receiver to 7.35 cm, where the radio noise from the Galaxy is very small. Still, they picked up the static. Penzias and Wilson further discovered that the noise did not change in intensity with the direction in the sky

21.5
Arno Penzias (right) and Robert Wilson (left) standing near the horn antenna with which they discovered the cosmic background radiation. (Courtesy Bell Laboratories)

they pointed, the time of day, or the season. Perplexed, they examined the antenna again and found a pair of pigeons roosting inside. These pigeons, oblivious to radio astronomy, had coated a part of the antenna with their droppings. Perhaps the coating could supply the excess noise? The birds were moved, their droppings cleaned out. But still the excess noise persisted, with an intensity at 7.35 cm equivalent to that of a blackbody at 3.5 K.

Penzias called Bernard Burke, a radio astronomer at MIT, to discuss matters other than the excess noise. Burke asked about the experiment, and Penzias explained the problem. Because Burke had heard about Peebles's work, he suggested that Penzias call the Princeton group. Penzias did. When he made contact, the Princeton group quickly concluded that the excess noise came from the cosmos—radiation left over from the Big Bang.

This intuitive, risky leap needed verification. After all, the Penzias and Wilson measurement was at a single wavelength (Fig. 21.6). But to establish the radiation as truly from a hot, dense stage in the universe's past, more observations were needed to confirm the characteristic blackbody shape of its spectrum and its uniform intensity in the sky. Soon Roll and Wilkinson added another point at another wavelength (3.2 cm) corresponding to a 2.8 K temperature, close to that observed by the Bell Labs pair, and other experimental groups later contributed additional evidence to the blackbody nature of the radiation (Fig. 21.6). The points crept up the long-wavelength side of the (presumptive) blackbody curve to the hump of the turnover point. An observed point on or over the hump would clinch the argument, but at infrared wavelengths, observations are notoriously difficult to make because of absorption by the earth's atmosphere.

Far-infrared observations made above most of the earth's atmosphere have confirmed that the spectrum of the cosmic radiation turns down at short wavelengths. The results by two different groups agree well for such difficult observations. The blackbody temperature of one result is 2.94 K; the other, 2.99 K. Other observations to date averaged together give a blackbody temperature of 2.89 K. For convenience, I'll take the radiation's blackbody temperature as simply 3 K.

To confirm its cosmic origin further, the radiation should be isotropic; that is, it should have

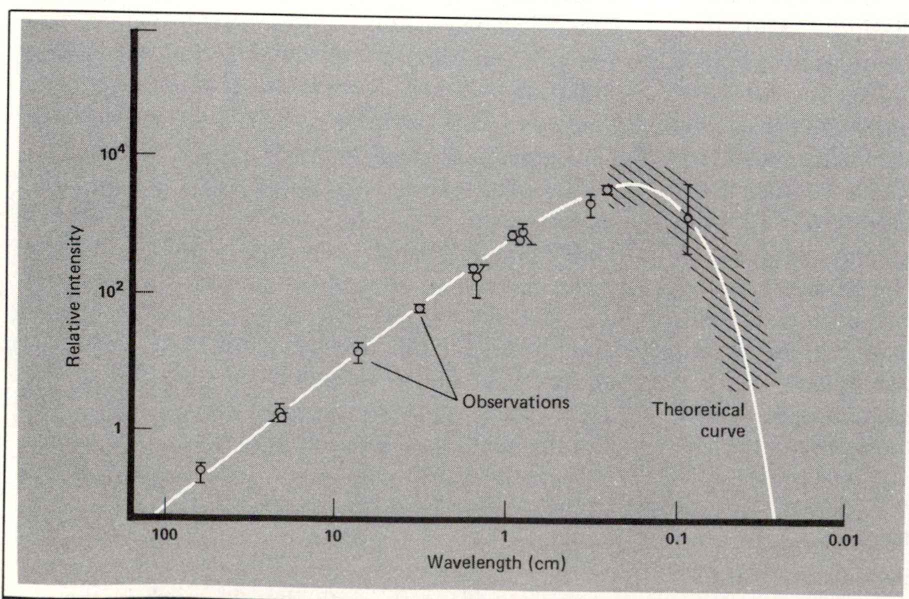

21.6
Observations of the cosmic microwave radiation. The points show measured intensities at radio and infrared wavelengths; the curve corresponds to a theoretical blackbody of 3 K. The shaded region corresponds to infrared observations by P. L. Richards and co-workers. (Adapted from a diagram by P. J. E. Peebles)

the same measured intensity from all directions in the sky. For if the radiation were from a hot primeval state, the isotropy of the universe at that time would fix the isotropy of the radiation. (Recall that Section 21.1 *assumed* isotropy—that the universe looks the same in all directions—without experimental confirmation.) The Princeton group also made some of the first observations to determine the isotropy of the radiation. R. Bruce Partridge and David Wilkinson searched for a variation in the radiation's intensity, scanning around the sky. After a year of collecting data, Partridge and Wilkinson concluded that the variation was no more than 0.5 percent. More recent observations back up the notion that the radiation is very isotropic.

This result suggested both that the radiation originated from the early universe and that, in fact, the newborn universe was very nearly isotropic. The lack of a regular variation also ruled out a local solar system or galactic source of the radiation. The evidence—equal intensity from all directions in space—implies that the radiation is cosmic, existing everywhere in space, so that it arrives at the earth uniformly from every direction. If so, it fills all space at all times. The isotropy observations clinch the interpretation of the radiation as cosmic.

This background radiation is usually given the long-winded name of *cosmic blackbody microwave radiation*: "cosmic" because it comes from all directions in space, "blackbody" because of its spectral shape, and "microwave" because its spectrum peaks at centimeter-to-millimeter wavelengths. Its discovery makes constructing models of the universe easier because it contains a wealth of information.

First, the radiation enables astronomers to glimpse the raw, young universe. We can conclude that the initial universe was indeed quite homogeneous and isotropic.

Second, the present temperature, about 3 K, and the isotropy of the radiation set severe limits on the thermal history of the universe, that is, the change of the temperature of matter and radiation with time.

Third, the radiation's presence establishes an important marker for galaxy formation. Until the radiation and matter stopped their interaction, matter could not form any clumps of large size. Only after the ionized gas recombined could matter form clumps that eventually became stars and galaxies (see Section 21.7).

To sum up: The discovery of the cosmic blackbody microwave radiation has a significance for cosmology as great as that of the discovery of the expansion of the universe (see Section 7.4). Its measured uniformity backs up our assumptions that the universe is isotropic and homogeneous. And its present measured temperature (about 3 K), combined with Einstein's equations of general relativity, allows us to work out the evolution of the universe as its temperature changes. The details of this evolution will be traced out in the next section: the standard Big Bang model that we believe has reasonable support in present observations.

21.5 *The primeval fireball*

Since the discovery of the cosmic microwave background radiation, most astronomers accept a hot big-bang model for the beginning of the universe. With the addition of experimental and theoretical knowledge on how matter behaves under hot, dense conditions, theoreticians have been able to develop step-by-step details of what can happen in a Big Bang.

Warning: Please *don't* picture the Big Bang as happening in the "center" of the universe and expanding to fill it. The Big Bang involved the *entire* universe; every place in it was *at* creation, which marked the beginning of time and space.

The hot start / Although the present temperature of the cosmic radiation is low, the amount of energy it contributes to the universe is large: Each cubic meter contains 4×10^{-14} J, equivalent ($E = mc^2$) to about 4×10^{-31} kg. For comparison, if you take the material contained in the galaxies we can see and spread it uniformly around the universe, each cubic meter would contain 4×10^{-28} kg. So for each kilogram of galactic matter, there are approximately 10^{14} J of cosmic radiation. If the energy in the radiation could be used to heat up the matter, the temperature would be greater than 10^{12} K. Here is one clue that the early universe must have been very hot.

The dominance of photons over matter requires a very hot creation, often called the *primeval fireball.* How hot was it? We don't really know. No one at present understands how matter behaves at temperatures greater than 10^{12} K. From Einstein's equations of general relativity we find that a temperature of 10^{12} K corresponds to a time of about 10^{-24} sec after creation (time "zero"). By creation I mean the actual beginning of the present expansion—the Big Bang. At that time, the universe had an average density of 5×10^{16} kg/m^3 (almost as dense as an atom's nucleus).

The primeval fireball produces such a rapid expansion that the temperature and density drop

TABLE 21.1 Particles that play a role in cosmic nucleosynthesis

Particle or antiparticle	*Symbol*	*Charge*	*Comments*
Neutrino, antineutrino	$\nu, \bar{\nu}$	0	Massless(?) particles that travel at light speed; stable(?)
Proton, antiproton	p, $\bar{\text{p}}$	+1, −1	The proton is a hydrogen nucleus; stable(?)
Electron, positron	e^-, e^+	−1, +1	Electrons surround the nucleus of an atom; stable
Neutron, antineutron	n, $\bar{\text{n}}$	0	Free neutron decays to a proton and an electron in about 1000 sec
Photon	γ	0	Packet of radiation
Deuteron	^{2}He	+1	Nucleus of deuterium, or "heavy hydrogen"; contains 1 proton, 1 neutron; stable
Helium-3	^{3}He	+2	Nucleus of an unusual type of helium; contains 2 protons, 1 neutron; stable
Helium-4	^{4}He	+2	Nucleus of ordinary helium; contains 2 protons, 2 neutrons; stable
Lithium-7	^{7}He	+3	Nucleus of the most abundant type of lithium; contains 3 protons, 4 neutrons; stable
Beryllium-7	^{7}Be	+4	Nucleus of the most common type of beryllium; contains 4 protons and 3 neutrons; unstable

rapidly. The contents of the universe (Table 21.1)–initially a flood of energetic light and heavy particles, such as protons, and their respective antiparticles–changes with temperature and time. As was mentioned earlier, the possibility that nucleosynthesis might occur on a cosmic scale motivated George Gamow (1904–1968) to investigate the situation for which he coined the term *Big Bang.* Only very limited but very important nucleosynthesis occurred in the early universe: the formation of helium and deuterium and a little bit of other light elements.

I will sketch a scenario of the young universe in which temperature plays a crucial controlling role. Each period of time can be matched with a temperature. Roughly, the universe's thermal history divides into four eras: (1) a *heavy-particle era,* when massive particles and antiparticles dominated; (2) a *light-particle era,* when particles with less mass were made; (3) a *radiation era,* when most particles had vanished and radiation was the main form of energy; and (4) a *matter era,* in which we now live, when the rest-mass energy of matter dominates (if you compare the energy in a liter in the form of photons to that as matter).

Warning: Whenever I say that the universe was "smaller" or "larger," I mean that the distance between a pair of objects is smaller or larger. The universe itself may be finite or infinite. That overall geometry does not affect what I'll say about the Big Bang.

Creation of matter from photons / Before the story unfolds, you need a little preparation about one key part: the creation of matter and antimatter from photons.

At some time in the primeval fireball, the energy of photons (essentially their temperature) must have been so high that photon collisions produced matter. The process occurs when the energy in the colliding photons equals or exceeds the mass ($E = mc^2$) of the particles produced.

Sounds bizarre? It comes directly from Einstein's relation between matter and energy, which does not restrict the *direction* of the transformation: Matter can become energy, or energy can become matter.

The creation of matter from light happens in a special way that involves both matter and antimatter. When matter and antimatter collide, they annihilate and convert to photons (Fig. 21.7). In reverse, two photons (if they have enough energy) create a matter-antimatter pair when they hit each other. Note that this process always results in *pairs* of particles.

How much energy must the photons have? At least the energy equivalent of the masses of the pair they produce. So to make protons and antiprotons takes more energy than to make elec-

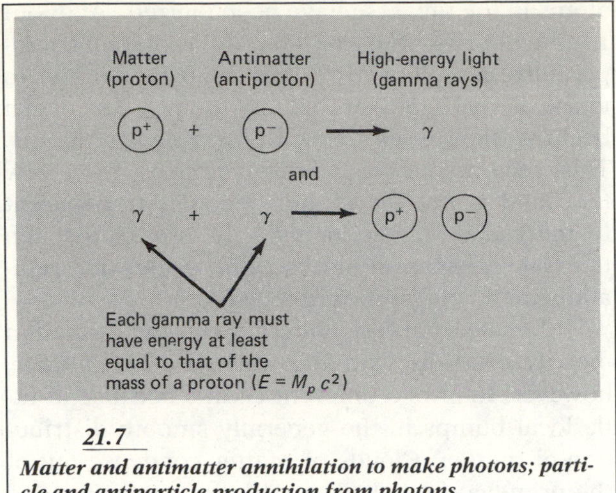

21.7
Matter and antimatter annihilation to make photons; particle and antiparticle production from photons.

trons and positrons, because protons have more mass than electrons (about 1800 times more). Now we can talk about the energy of photons in the primeval fireball in terms of their temperature. To have enough energy to make electrons requires a temperature of at least 1.2×10^{10} K; for protons, at least 2.2×10^{13} K.

Particle production from photons plays a dominant role in the very young universe. Let's see what the model tells about those times.

Temperature greater than 10^{12} K, time less than 0.0001 sec / Photons create pairs of particles and antiparticles. The photons have enough energy so that even the most massive elementary particles can be produced (Table 21.1). Annihilation also takes place, and the balance between annihilation and creation fixes the density of particles and antiparticles for the next stage. This balance between massive particle production and destruction marks the heavy-particle era. It does not last long, as the cosmos expands rapidly and the temperature declines quickly.

A critical difficulty with the model here is that the average density predicted by relativity is greater than 10^{17} kg/m³, and we really do not know how matter behaves under such extreme densities (and we certainly do not know what happens at earlier times and higher densities). But whatever happens in detail, heavy particles can be created at this stage.

So at the earliest times that we can calculate, the universe is a smooth soup of high-energy light and massive particles (Fig. 21.8).

Temperature from 10^{12} to 5×10^9 K, time from 0.0001 to 4 sec / Annihilation of protons and neutrons (and other heavy particles) with their antiparticles continues. The remaining photons, however, lack the energy to create new heavy particles. Only light particles—electrons—can be made, because these need less energy for creation. The universe enters the light-particle era.

Observations so far imply that the visible universe is mostly matter. The present imbalance of matter over antimatter requires that a few extra protons remain; that is, the early universe contained a slight excess of protons over antiprotons. (Physicists think that this happened because of an asymmetry in the physical laws obeyed by matter compared to antimatter.) Protons and electrons interact to generate equal numbers of neutrons and protons (Fig. 21.8). As the temperature decreases, the number of neutrons falls, while the

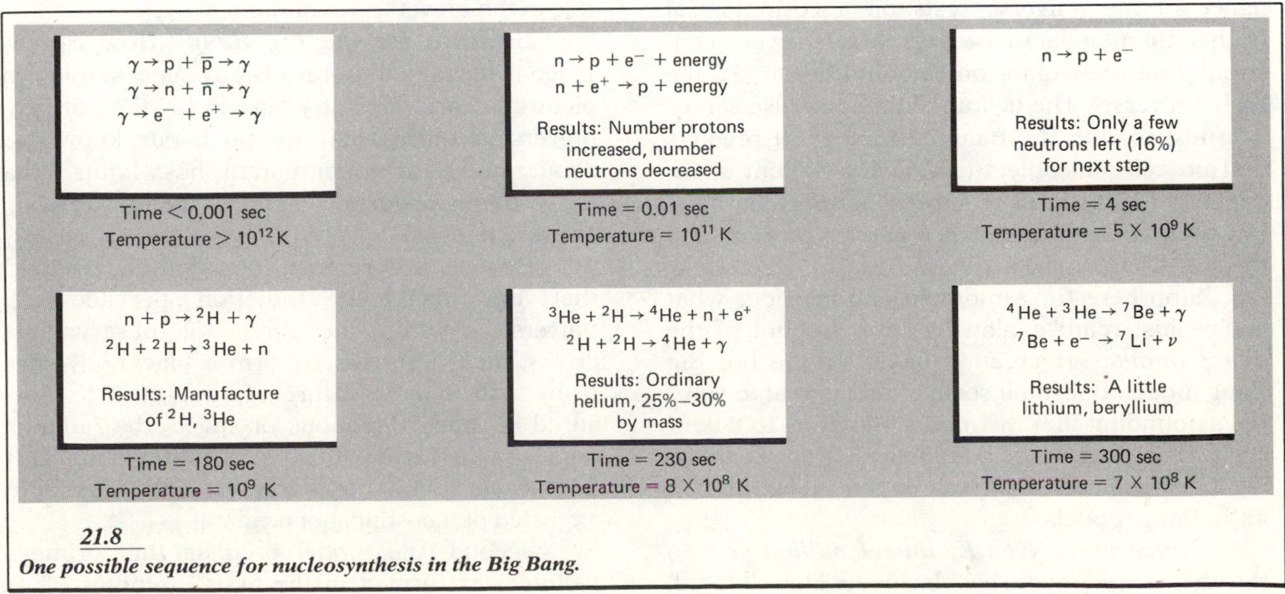

21.8
One possible sequence for nucleosynthesis in the Big Bang.

number of protons rises. (This imbalance of protons and neutrons is important later in the formation of helium.) When the temperature falls to 5×10^9 K, photons can no longer make electron-positron pairs. This temperature marks the end of the light-particle era and the beginning of the radiation era.

Neutrons, totaling 16 percent of the particles, can no longer be produced by the interaction of protons with electrons or of antiprotons with positrons. The neutrons are now free to decay into protons and electrons.

Temperature 10^9 K, time from 4 to 180 sec / In this crucial period, the remaining neutrons and protons react to form nuclei that can survive the still-high temperatures. The most important reaction involves the combination of a neutron and a proton to form deuterium. All neutrons, except those that have decayed, end up in deuterium. Once the deuterium is produced, further reactions create normal helium (Fig. 21.8) and also a little tritium (^{3}H) and ^{3}He. The net result is 25 to 30 percent helium by mass. A little bit of beryllium and lithium is created by the combination of deuterium and tritium (Fig. 21.8). Some deuterium is left over. Extremely little of the heavier elements is made.

The final helium and deuterium abundance depends on the rate of the universe's expansion. If the expansion rate were any faster, less time would have been available for neutrons to decay, and more helium would have been produced. But if the expansion rate had been very much slower, the neutrons that would have formed helium would have had a longer time to decay, so less helium would result. Helium, once formed, is tough to destroy. So the present helium abundance in the universe rests on a cosmological (rather than stellar) base, which sets severe constraints on the expansion rate and density of the early universe. The helium abundance also serves as a test for the Big Bang cosmology. It predicts that no celestial object can have a helium abundance of less than 25 to 30 percent. (Because stars form helium, the abundance can be greater than this.)

Stop here for a moment and consider what you've just read: a blow-by-blow account of the *first 3 minutes* of creation based on the hot Big Bang model. Doesn't it seem a little unreal to you? It's astounding that the model allows us to talk of such times with some confidence. What is amazing is that we can match our own universe to such a Big Bang model.

Temperature 4000 K, time 1 million yr / So far the temperature has been so high that all atoms in the universe have been ionized. At about 1 million years after creation, the radiation's temperature has plunged to only 4000 K, too low to ionize a hydrogen or helium atom. The nuclei begin to capture electrons to form neutral atoms. This recombination process happens in a few thousand years; the matter becomes transparent to the radiation. Suddenly light breaks through the now transparent matter. The matter and radiation are no longer locked together.

Freed from this interaction, the radiation merrily expands with the universe. The matter, however, follows a different course because of little local bumps in the generally smooth distribution of matter. Clouds of matter condense out of the primeval fireball. The radiation era ends, and the matter era begins. The universe has a density of 10^{-24} kg/m^3. Material condensation could not happen until the radiation and matter uncoupled as the atoms recombined. This event flags the time when galaxy formation could begin (see Section 21.7).

To sum up (Table 21.2): In a few minutes after creation, the universe expands and reaches temperatures and densities suitable for the formation of deuterium and helium. The Big Bang model predicts a helium abundance of 25 to 30 percent (by mass) and very little formation of heavier elements. (These are made later in stars.) If the Big Bang picture is correct, this abundance is the *minimum* amount of helium we should find in the universe. About 1 million years after the time of helium's formation, electrons and protons get together to form neutral atoms (almost all hydrogen and helium.) At this time, matter can clump to form the first stars, galaxies, and quasars. The radiation no longer plays a dominant role in the universe's evolution.

Evidence for the Big Bang / How can we judge if the contemporary Big Bang cosmological picture is correct? The scenario for the hot early universe requires that we push our knowledge about matter and gravitation to shaky limits. What observational evidence supports such a stretching of present ideas?

First and foremost, observations indicate that the blackbody radiation pervades the universe. Within the limits of observational errors, measurements confirm a blackbody spectrum with a temperature of roughly 3 K. Measured in many directions of space, the radiation comes to the earth with a pretty uniform intensity. The background radiation has the attributes expected of a cosmic, hot origin.

Second, the model predicts that primeval helium was formed in the first 5 minutes of the

TABLE 21.2 Sequence of events in the Big Bang model

Event	*Time*	*Density (kg/m³)*	*Temperature (K)*	*Comments*
Creation	0	?	?	Not province of present science; general relativity fails.
Heavy-particle era	10^{-44} sec	10^{97}	10^{33}	Photons make massive particles (such as protons) and antiparticles.
Light-particle era	10^{-4} sec	10^{17}	10^{12}	Photons have only enough energy to make light particles and antiparticles, such as electrons and positrons; protons and electrons combine to make neutrons.
Radiation era	10 sec	10^{7}	10^{10}	Few particles left in a sea of radiation; these partake in nucleosynthesis of deuterium, helium, lithium, and beryllium.
Matter era	10^{6} yr	10^{18}	4000	Ionized hydrogen recombines; cosmic radiation and matter decouple.
Present	10^{10} yr	10^{-31} (radiation)	3 (radiation)	Astronomers puzzle about creation.

universe's history and that the helium abundance should be 25 to 30 percent by mass. The Big Bang model sets this helium abundance as the basement level; any observation of a substantially lower amount calls the model into question. Because helium is formed in stars and once formed is difficult to destroy, the present helium abundance is probably larger than 25 to 30 percent.

Unfortunately, we can assess the present cosmic helium abundance only indirectly. We can examine material from the earth, moon, and meteorites, but any initial helium in them has escaped into space, since these bodies do not have sufficient gravity to hold down helium. Spacecraft have sampled the solar wind to find that 15 to 30 percent is helium nuclei by mass. As Chapter 13 pointed out, the solar photosphere is not hot enough to excite helium lines for direct viewing with a spectroscope, so it's not possible to measure the helium abundance in the solar photosphere. The chromosphere's higher temperatures do excite helium atoms to emission, and the observed abundance there is estimated to be 38 percent. Theoretical models of stellar evolution place the helium abundance in the sun's interior at 17 to 28 percent. Similar calculations for Jupiter yield a helium abundance of roughly 20 to 30 percent. Voyager showed that Jupiter's atmosphere contains about 20 percent helium.

Population I *O*- and *B*-stars exhibit helium absorption lines that imply a helium abundance of approximately 30 to 34 percent. H II regions surrounding hot stars have helium emission lines at optical and radio wavelengths that give abundances from 26 to 29 percent. Planetary nebulas also have strong helium emission lines that imply a helium abundance of greater than 30 percent.

But many of these objects are fairly young. They could have picked up helium made in stars. The best objects to search for primeval helium are the oldest stars now surviving, the Population II stars. Unfortunately, most Population II stars are much too cool to excite helium lines in their spectra. However, indirect evidence from star models indicates that these stars have helium abundances of about 30 percent.

Though the helium values have a considerable range, note that the helium abundance for a variety of celestial objects falls close to that predicted by the Big Bang model. The evidence for the compatibility of observations with theory is good but not conclusive, and as is usual in astrophysics, no one observation clinches the affair. But in this case, the accumulation of evidence for the hot Big Bang is impressive.

21.6 The end of time?

So much for the universe's past. What about its future? Recall (from Chapter 7) that Einstein's general theory of relativity allows the cosmos to have one of two general geometries: open or closed. If open, the universe will expand forever,

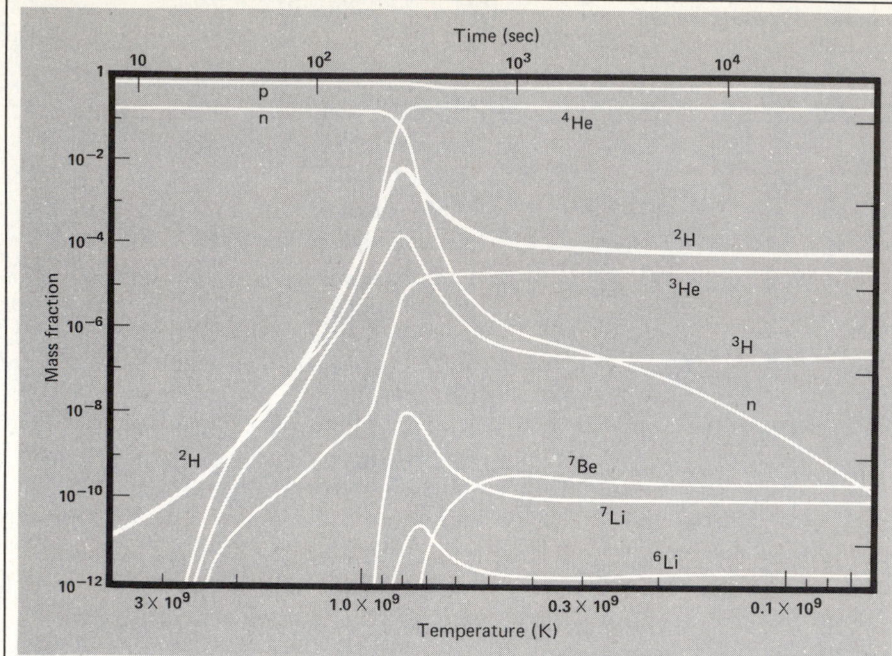

21.9
Formation of low-mass elements in the Big Bang. The top axis gives the age of the universe from time zero; the bottom axis gives the temperature; and the vertical axis gives the abundance in terms of the fraction of the total mass. Note that all the nucleosynthesis takes place in a sharp blip between 100 and 1000 sec. (Based on theoretical calculations by R. V. Wagoner)

and time will never end. If closed, however, the universe must eventually collapse, running backward through the history outlined in the preceding section.

Which fate will be ours? Chapter 7 mentioned two observational tests that indicate that the universe is open. First, the measured value of Hubble's constant (H = 20 km/sec/Mly) and Einstein's theory give a critical density for a closed universe—it's about 5×10^{-27} kg/m^3; so, on the surface, this evidence implies that the universe is open. Second, the curvature of the Hubble diagram for distant galaxies can indicate the universe's geometry. This Hubble test is tough to do, but so far, very weak evidence points to an open geometry.

The standard hot, Big Bang model provides another test, which is perhaps the strongest. The test rests on the observed present cosmic abundance of deuterium.

How can the abundance of deuterium reveal whether the universe is open, closed, or flat? To do the Big Bang model calculations, you need to put in the *present* value of Hubble's constant, the *present* temperature of the cosmic radiation, and the *present* average density of the universe. The first two items are reasonably well known, but the third is not. The amount of helium that comes out of the Big Bang calculations does not depend very much on the value used for the present density of the universe (Fig. 21.9). But the amount of deuterium that theoretical calculations predict does depend very sensitively on the value used for the present density (Fig. 21.10). So we can use the theory to turn the argument around. If we can measure the present cosmic abundance of deuterium, we can check this number against that predicted by the Big Bang model (Fig. 21.10) for various densities. The model gives the present average density of the universe, and we can compare that number with the critical density.

How can we observe deuterium? By ultraviolet observations of the interstellar medium, which show lines produced by the hydrogen molecules and HD. So we have a value for the D-to-H ratio of the interstellar medium: about 2.0×10^{-5} by mass.

Now to estimate how much original deuterium has been burned up in stars. A reasonable estimate is about half, so the original deuterium abundance was about twice that observed now, or 4×10^{-5} relative to hydrogen. Look at the Big Bang calculation (Fig. 21.10). A deuterium-to-hydrogen ratio of 4×10^{-5} implies a present cosmic density of 4×10^{-28} kg/m^3, considerably less than the critical density (5×10^{-27} kg/m^3).

Conclusion from this test: The universe is open. It will expand forever. Time will *not* end.

21.7 From Big Bang to galaxy

Up to 1 million years after creation, gravity could not clump matter into stars and galaxies. Why not? The universe was opaque to radiation, so

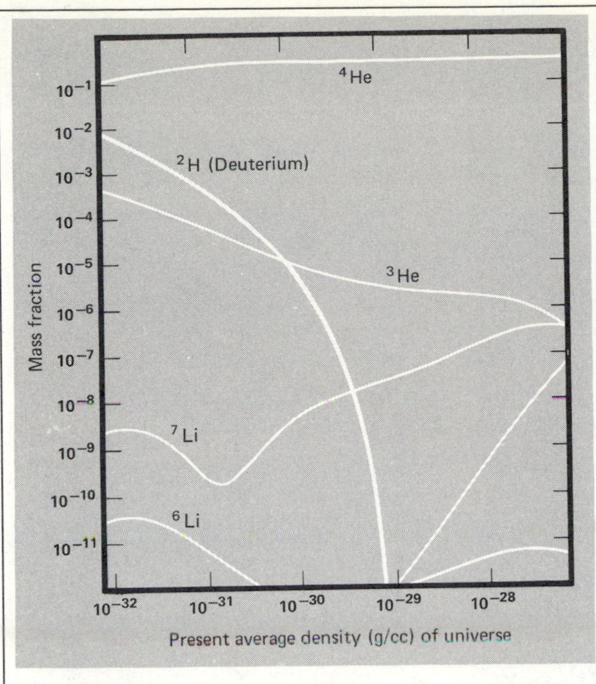

21.10

Theoretical calculations of the abundance of low-mass elements formed in the Big Bang as related to the average density of the universe now. Note how sharply the abundance of deuterium varies with different present average densities. (Based on theoretical calculations by R. V. Wagoner)

light and matter interacted strongly. Pressure from the radiation itself inhibited gravitational collapse. But once protons and electrons recombined, the universe became transparent to radiation, and radiation pressure could no longer stop gravity from doing its natural work. We refer to this time as the time of *decoupling*, when the radiation and matter were no longer coupled together.

Because we observe that the cosmic background radiation now arrives very uniformly from all directions in space, we know that at the time of decoupling, the matter and radiation in the universe must have had a very uniform distribution. But that's not the situation now. Even the most spread-out of these systems of matter—clusters of galaxies—have average densities about 100 times greater than the average density of the universe. So here's the crucial question: How did an originally very smooth universe become clumpy?

A model of galaxy formation must face a critical hurdle; it must operate effectively in at most a few billion years. When you look at very distant galaxies, you peer back in time to when the universe was billions of years younger. To date, astronomers have seen objects (the quasars) as far as some *12 billion* ly away. So the matter from the Big Bang must have formed into large clumps well before this time.

The discovery of the cosmic radiation by Penzias and Wilson forced astrophysicists to consider what happens to disturbances in a hot, dense universe filled with matter and radiation. They found that, before decoupling, the radiation played a powerful role to inhibit the growth of disturbances. A dense patch in the early universe will have a high internal pressure because the radiation adds to the pressure force that pushes the patch apart. Only very large disturbances would collapse. The radiation pressure dissipates disturbances that have masses of 10^{12} solar masses (or less) up to the time of decoupling. So before the decoupling of matter and radiation, disturbances that contained roughly the mass of a galaxy could not collapse.

Just after decoupling, it's a new show (Fig. 21.11). The radiation and the gas no longer interact, so radiation pressure no longer resists gravity. Small disturbances, amounting to only 10^5 solar masses as shown by detailed calculation, can condense out of the gas, along with disturbances of greater mass. This result gives us some hope, for the large gravitationally bound masses we see now range from 10^5 to 10^{15} solar masses (clusters of galaxies). Disturbances of this size range can condense just after decoupling. So the time of decoupling, roughly 1 million years after the Big Bang, marks the time when the galaxy formation could take place in the young universe.

There's one real weakness in these ideas: Even though disturbances can be unstable, they grow slowly, so slowly that the galaxies we see could hardly have formed by now, unless the disturbances were already fairly large in the beginning. But the smoothness of the cosmic back-

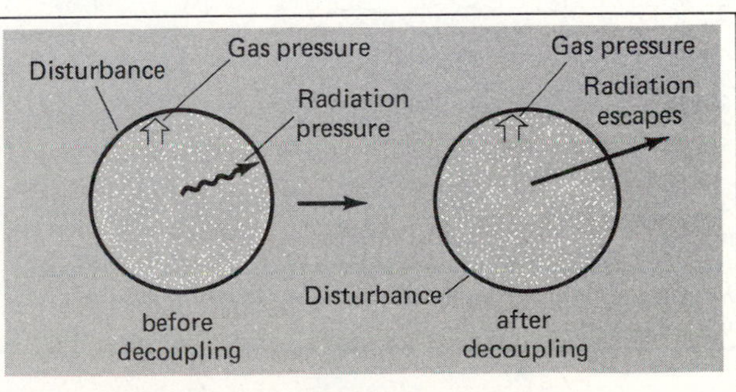

21.11

Decoupling and the growth of instabilities. Before decoupling, radiation pressure adds to the internal pressure of a disturbed patch because the gas is opaque to the radiation, and the radiation cannot escape. After decoupling, the gas is transparent, and radiation escapes, no longer contributing to the internal pressure; the region can collapse.

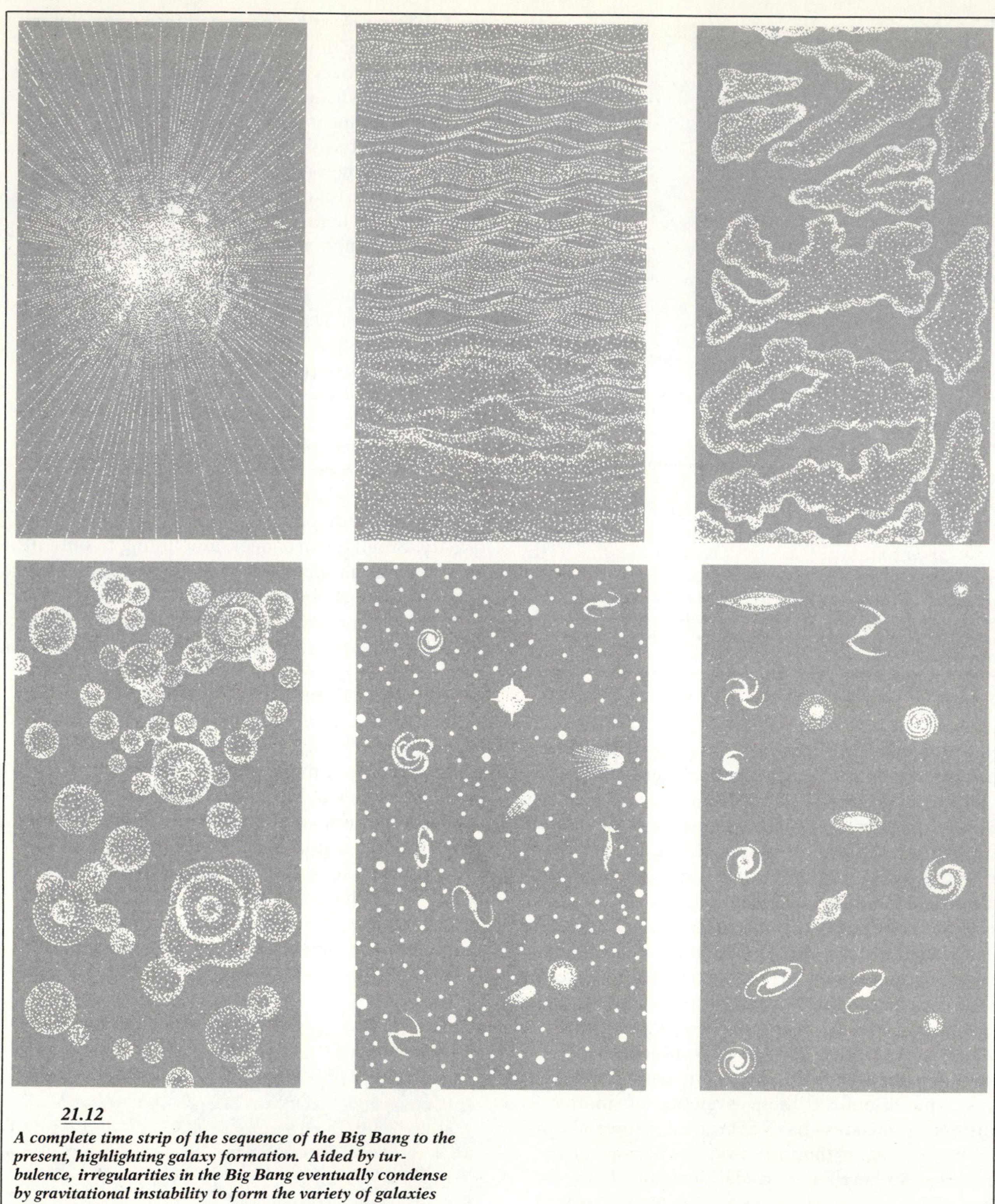

21.12
A complete time strip of the sequence of the Big Bang to the present, highlighting galaxy formation. Aided by turbulence, irregularities in the Big Bang eventually condense by gravitational instability to form the variety of galaxies we see today.

ground radiation implies that any lumps were small.

But galaxies have formed and did so early in the history of the universe. The details are obscure, however, and most ideas about galaxy formation are speculative. One conclusion appears likely: Most galaxy formation occurred in a grand burst soon after the decoupling of the radiation and matter from the primeval fireball (Fig. 21.12).

21.8
The large and the small

Although the standard cosmological model of today, the hot Big Bang model has a number of weaknesses, two of which I find especially vexing. One is the great dominance of matter over antimatter in the universe today, and the other is getting galaxies to form early and fast enough. Both problems are being tackled now in imaginative theoretical ways that unite the universe of the small—elementary particles—with the universe itself. The connection occurs in the Big Bang. To explain this connection, I'll have to digress a bit into the world of elementary particles and their interactions.

The forces of nature / Let's focus on how particles relate to each other—their interactions. We generally think of these relations in terms of forces between particles. You are familiar with two of these: gravitation and electromagnetism. These forces have one property in common: They work over infinite distances. According to Newton's law of gravitation, the most distant galaxies exert a force on you (and you on them). The same is true of objects that have a net electrical charge. These forces differ in their relative strengths. Electromagnetic forces are *much* stronger than gravity—as you know if you've lifted a nail with a magnet. The gravitational attraction of the entire earth on the nail is weaker than the force of the magnet.

Two other forces operate in nature, but these are probably not familiar to you because they work in the subatomic domain. One, called the *strong force*, holds the nuclei of atoms together. Recall that an atom's nucleus has protons tightly packed together. The electric force of each proton repells the others strongly, especially when so close—only 10^{-15} m apart. The strong force overwhelms this electric repulsion and keeps the nucleus together. The other, called the *weak force*, crops up in radioactive decay. Without it, fission would not take pmace. Like the strong force, the weak force operates over very short distances, 10^{-17} m and less.

These four forces are all that are known. As its name implies, the strong force is the strongest of them (Table 21.3), electromagnetism second, the weak force next, and gravity takes the bottom as the weakest of them all. Now, although these forces appear to operate very differently, might they have an underlying unity? That quest for a unified theory has tempted physicists for most of this century. And recently they have had some success in struggling to find a grand unified theory, fondly known as GUT. The development of GUTs (there are more than one such theory) has curiously enough affected Big Bang cosmology and helped to shore up some weaknesses.

TABLE 21.3 Properties of the fundamental forces

Force	*Strength (relative to strong)*	*Range (m)*
Strong	1	10^{-15}
Electromagnetic	1/137	Infinite
Weak	10^{-5}	10^{-17}
Gravity	6×10^{-39}	Infinite

How is that connection made? First, in the 1970s, theoreticians were able to unify the weak and the electromagnetic forces. Buoyed by their success, they next took aim at unifying the electroweak and strong forces. As part of that work, they developed the concept that a new elementary particle, called a *quark*, comprised other so-called elementary particles. For example, a proton consists of three quarks. Since the proton is now a composite particle, it can decay (as the neutron, also a composite particle, decays) with a predicted lifetime of some 10^{30} years. (Experiments are now under way to check this prediction.) GUTs also predict that the unification of strong and electroweak forces won't be apparent until energies of greater than some 10^5 J—the equivalent of converting about 10^{15} protons completely into energy. Such energies cannot be made in particle accelerators on earth, but they do occur in the Big Bang at a time of 10^{-35} sec when the temperature was 10^{26} K. The Big Bang serves as a way to test GUTs—the whole universe as a particle machine!

Note: One result of GUTs has been to simplify our view of elementary particles. We now believe that matter is composed of two classes of elementary particles: quarks and leptons. Only six particles make up the lepton group; they include electrons and neutrinos. Quarks make up all the other particles, such as protons and neutrons (and some 100 others). Six quarks are known to date. The four basic forces act on all these particles. The aim of GUTs is to reduce all particles to one kind interacting through one force—truly a grand unification!

GUTs and the cosmos / We now look at the temperature history of the Big Bang model in a new way: the temperatures at which specific aspects of particles and their forces freeze into existence. Here I use the word *freeze* in the sense that when water freezes into ice (at 0° C, or 273 K), its state changes abruptly. Water in the form of ice behaves differently than water as a liquid or gas. I have already discussed two freezings in the early cosmos (Section 21.5): the formation of simple nuclei in the nucleosynthesis era (first few minutes) and of atoms during the recombination time (1 million years). GUTs predict a very special kind of freezing at 10^{-35} sec, when a very heavy particle, called the *X-particle*, could no longer be made. These particles, because their production ceases, decay into other particles and are gone—forever. This time is called the era of GUT freezing.

These X-particles solve the antimatter problem. It results from the fact that at 10^{-3} sec, the universe contained 1 billion and 1 protons for every billion antiprotons. Why this tiny imbalance? Remember that photons create particles and antiparticles in equal pairs. Before the GUT freezing, that was true—equal numbers of X-particles and antiparticles existed. After the freezing, these particles decayed but unequally (an idea supported by some experiments). A very little more matter than antimatter was created in the decay; once this happened, nothing changed this frozen-in asymmetry. Later, the matter annihilated all the antimatter, leaving only matter in the cosmos.

Another aspect of GUT freezing relates to the galaxy-formation problem. Consider the freezing of a pond: The ice sheet does not form all at once but in patches. That is, the freezing process is not perfect but has defects. These defects have mass and survive for long times— long enough so that at the decoupling time, when matter can clump, the defects serve as the cores—the lumps—on which gravitational instability occurs.

Another way to help make galaxies relates to neutrinos. Some GUTs predict that neutrinos should have a very small mass (and a few experiments, yet unconfirmed, support this claim)—perhaps 0.001 percent that of an electron. Neutrinos freeze out at a time of 1 sec, and those relic neutrinos should be with us today. If massless, they have little effect. If they have even a slight mass, they change the universe.

First, they may contain enough mass in total to close the cosmos. Second, they can aid galaxy formation. Neutrinos with mass can start clumping by gravity well before other particles can. After recombination, atoms would gather around these neutrino clumps and so speed up the instability process. Third, massive neutrinos may congregate in clusters of galaxies and bind them gravitationally. Fourth, neutrinos can also gather in the halos of galaxies to make up the so-far invisible matter there.

Models of the clustering of dark matter, such as neutrinos, have gotten around one of the most serious objections to galaxy formation in the standard Big-Bang model (see Section 21.7). Observations rule out ordinary matter having large disturbances in it at the time of decoupling. But neutrino matter could be lumpy and would not affect the smoothness of the background radiation. Computer simulations show that neutrinos would gather into pancakelike shapes hundreds of millions of light years in diameter. Where the pancakes push into each other, knots and strings of ordinary material can collect, cool, and condense. So this pancake model for galaxy formation predicts that galaxies should occur in long chains and filaments, with pancake voids in between (Fig. 21.13). And that's what we are just beginning to see in the structure of superclusters (see Section 19.6).

So particle physics can shore up some problems with the simple primeval fireball. The key point is this: Although details of what I described will change, the connection of the very small to the Big Bang is essential to our understanding of the universe's history.

Inflation in the cosmos / The drive to integrate particle physics with cosmology reached a new fusion recently with a variation of the Big Bang theme called the *inflationary universe*. It copes with serious flaws in the Big Bang picture. I'll focus on just one, called the *flatness problem*.

Recall that the universe, in terms of general relativity, can have two basic geometries: open or closed. We can evaluate which one applies to the cosmos by examining the ratio of the measured density (of matter and energy) to the critical density predicted from Einstein's general relativity and the value of Hubble's constant. A cosmos whose actual density is exactly the critical density is flat; the ratio is 1. Now if at the Big Bang the ratio was 1, it will remain 1 forever. If, however, the value differed from 1 ever so slightly, then the ratio would be *much* different from 1 now. Surprisingly, the ratio is believed to have a value between 0.1 and 2, very close to 1. Hence, the standard Big Bang model requires special starting conditions (the geometry of the universe very, very close to flat), but does not give an explanation for it.

Enter the inflationary cosmos. It deals with the universe from a time of 10^{-45} sec to 10^{-30} sec, when massive elementary particles dominate the universe. (After 10^{-30} sec, the inflationary model melds into the standard hot Big Bang model.) During that interval, the universe undergoes a tremendous spurt of growth in size, by perhaps as much as 10^{50} times larger. (Remember, that means that the distance between two particles increased by 10^{50}. To put that increase in perspective, the dis-

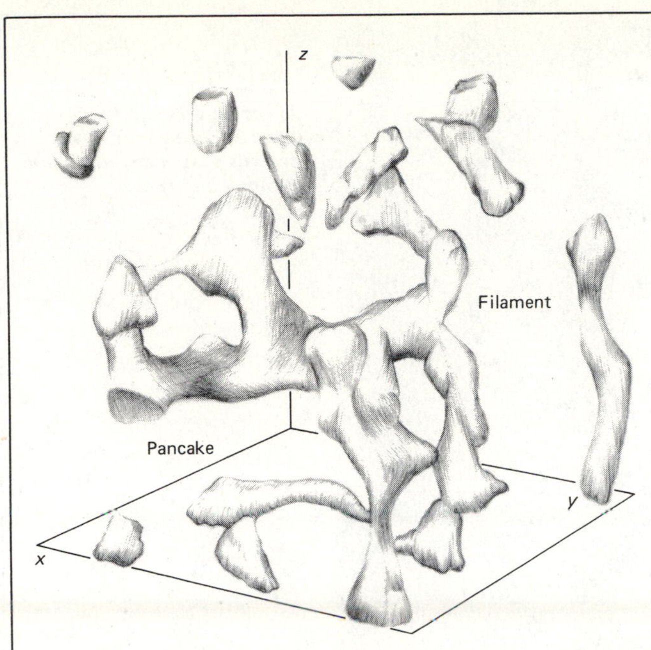

21.13
Results of a computer calculation for the clumping of dark matter in an expanding universe. Note that the matter ends up in pancakes and filaments, such as found in superclusters. (Based on calculations by J. Centrella and A. Melott)

tance from the proton to the electron in a hydrogen atom is about 10^{-10} m. If inflated by the same amount as estimated for the early universe, the distance would be 10^{40} m—or 10^{24} ly!)

This inflationary period neatly and naturally solves the flatness problem. Imagine that before inflation, spacetime contains strongly curved regions. The inflation causes these to grow to flatness automatically. As an analogy, consider the curved surface of a partially inflated balloon. The surface is clearly curved, when compared to the overall size of the balloon. Now rapidly blow up the balloon, keeping a close eye on the curvature of the surface. It becomes distinctly flatter (less curved). Similarly, when the universe inflates, curved regions become flat. So the ratio of the actual to critical density naturally reaches a value very close to 1, without the need of special assumptions.

The inflationary universe grapples with other flaws in the Big Bang model and does so modestly well, in my opinion. It still has many facets that need to be worked out. But it is the best model we have to date that unites the physics of the very large with the very small.

KEY CONCEPTS

1 ***Cosmological models assume (a) the universality of physical laws, (b) a homogeneous universe, and (c) an isotropic universe.***

2 ***Cosmological models must explain (a) the evolution of the universe, (b) the grouping of matter, and (c) the expansion of the universe.***

3 ***The main modern cosmological model is the Big Bang model (based on Einstein's theory of general relativity).***

4 ***The Big Bang model (Fig. 21.14) requires that the universe began in a hot, dense state at a finite time in the past; it predicts that now at low temperature; this radiation has been observed to be isotropic and at a temperature of about 3 K.***

5 ***The properties of the cosmic blackbody radiation, when combined with Einstein's general relativity and our present knowledge of matter, permit a detailed description of the Big Bang. In it, matter is made from photons and then interacts to form only the light elements.***

6 ***The hot Big Bang model predicts that the cosmos should have no less than 25 to 30 percent helium by mass; observations tend to support this prediction.***

7 ***How the universe ends depends on whether it is open or closed; present observations support an open universe.***

8 ***Galaxies could not form in the young universe until the temperature dropped below a few thousand Kelvins (about a million years after the expansion started); galaxies did form quickly and early on, but they could not simply grow gravitationally from small disturbances; rather, they needed the help of turbulence from the Big Bang.***

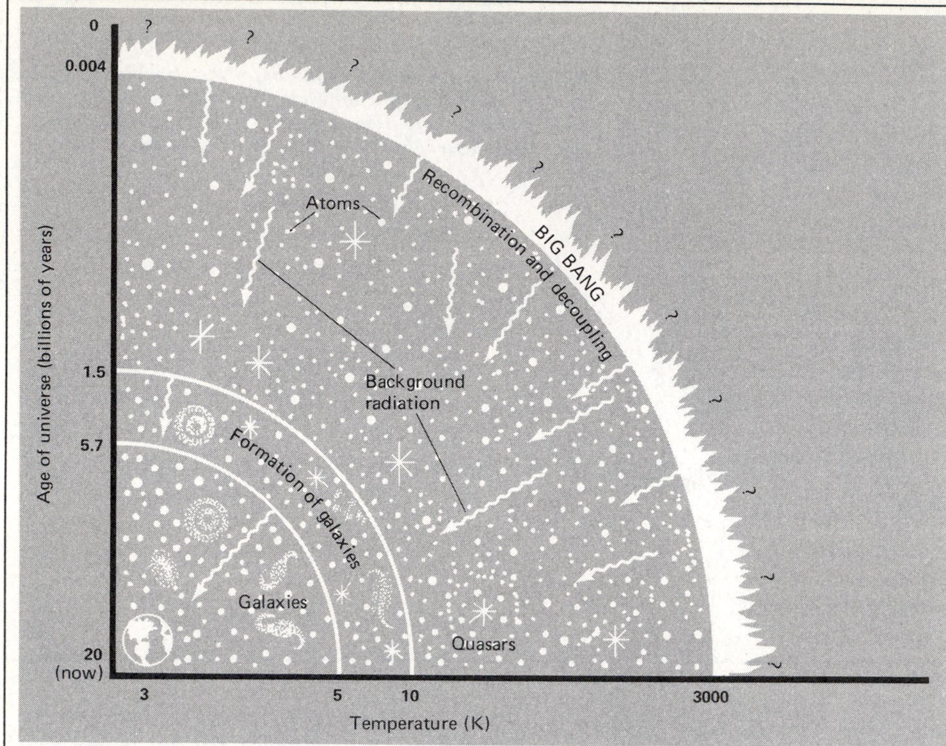

21.14
A schematic, visual history of the universe in the standard Big Bang model. We see this past in any direction we look into space.

9 ***Grand unified theories (GUTs) of elementary particles and forces provide new insights into the Big Bang and help to solve some of its major problems.***

STUDY EXERCISES

1 State in a few sentences the assumptions of the Big Bang model. (*Objectives 1 and 3*)

2 Make a short list of the fundamental cosmological observations. (*Objective 2*)

3 Interpret the observations in exercise 2 in the framework of the standard Big Bang model. (*Objective 3*)

4 Give *one* observational argument for asserting that the microwave background is cosmic in origin. (*Objectives 4 and 5*)

5 How does the discovery of the cosmic microwave background radiation confirm the Big Bang model? (*Objective 6*)

6 List the elements than can be made in a hot big bang, and give one reason why no elements heavier than lithium/beryllium are manufactured. (*Objective 9*)

7 What observational evidence do we have that backs up the standard Big Bang model? (*Objectives 3, 5, 6, 7, and 9*)

BEYOND THIS BOOK . . .

P. J. E. Peebles and D. Wilkinson discuss the discovery of the cosmic background radiation in "The Primeval Fireball" in *Scientific American*, June 1967.

J. R. Gott III, J. E. Gunn, D. N. Schramm, and B. Tinsley present an excellent discussion on "Will the Universe Expand Forever? " in *Scientific American*, March 1976. They answer yes. In a somewhat different approach, D. Dicus, J. Letaw, D. Tepliz, and V. Tepliz examine "The Future of the Universe" in *Scientific American*, March 1983.

For a comprehensive exposition of the origin of the universe in the standard Big Bang model, read *The First Three Minutes* by S. Weinberg (Basic Books, New York, 1977). A newer presentation that includes GUTs is *The Moment of Creation* (Scribner, New York, 1983) by J. Trefil.

Robert Wilson presents a personal history of the discovery of the relic radiation in "The Cosmic Microwave Background Radiation" in *Science*, August 1979, p. 866.

A good recent exposition of cosmology and the origin of galaxies is contained in *The Big Bang* by Joseph Silk (Freeman, San Francisco, 1980).

For a discussion of the imbalance of matter over antimatter in the universe and the connection to particle physics, read "The Cosmic Asymmetry Between Matter and Antimatter" by F. Wilczek, *Scientific American*, December 1980, p. 82.

22/Bios and cosmos

A zygote is a gamete's way of producing more gametes. This may be the purpose of the universe.

ROBERT HEINLEIN: Time Enough for Love

Learning objectives

After studying this chapter, you should be able to:

1. Identify at least three key characteristics of life as we know it (LAWKI).
2. State the central dogma of modern biology about life's origin.
3. Specify the chemical basis of LAWKI and pinpoint the processes that resulted in the formation of the essential elements and the deposition of those materials in the earth.
4. Outline a model for prebiological chemical evolution on the young earth and describe what experiments support this model.
5. Outline the evidence we have for the early evolution of the simplest life on the earth.
6. Describe what effects the origin of life and its expansion had on the earth's environment.
7. Sketch a time line from the creation of the universe to the present, highlighting the crucial events of physical, chemical, and biological evolution that have led up to the development of human beings on the earth.
8. Describe the results of the Viking lander experiments and discuss their implications for LAWKI.
9. Evaluate, in light of recent information from space probes and other observations, the possibility of extraterrestrial life in the solar system.
10. Assess the possibility of other planetary systems in our Galaxy, relying on both theoretical arguments and observational evidence.
11. Argue, with clearly stated assumptions, biases, and guesses, the possibility of life elsewhere in our Galaxy.
12. Argue, as in Objective 11, for one reasonable strategy to search for extraterrestrial, technologically advanced civilizations.

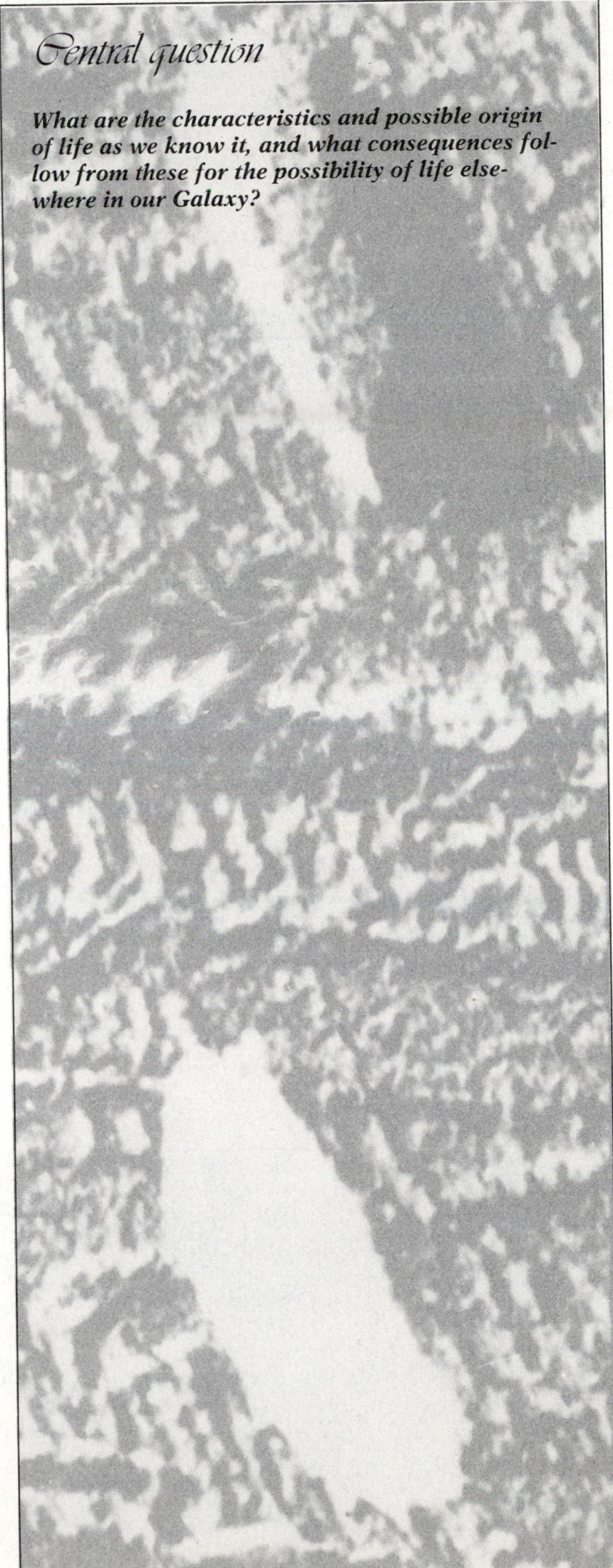

Central question

What are the characteristics and possible origin of life as we know it, and what consequences follow from these for the possibility of life elsewhere in our Galaxy?

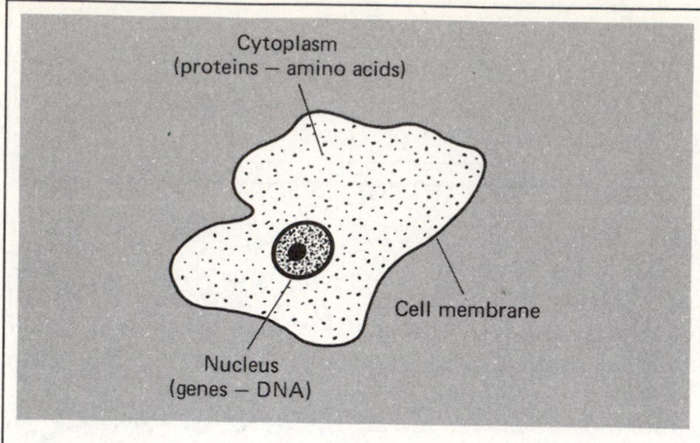

22.1
The key parts of a typical cell (simplified). The nucleus contains the genetic material, the protoplasm is the site of metabolic activity, and the cell membrane allows the passage of selected materials, such as food.

How did we get here? Is there anyone else out there?

In the span of cosmic evolution, the origin of life on the earth involved a *natural* sequence of chemical and biological evolution. The basic material was there. It needed only to be put together in a special way. The special arrangement of molecules that make a living organism on the earth results naturally from the chemical properties of matter. The origin of life may be both natural and universal.

Are we alone? How life developed here gives us some grasp on whether or not life exists elsewhere. The question is this: Can we estimate, on some reasonable basis, the chances that life exists elsewhere in our Galaxy or in other galaxies? Yes, from basic physical, chemical, and astronomical knowledge. Admittedly our estimate smacks of speculation, but it rests on a scientific basis.

This chapter will lead you seemingly far astray from astronomy into the realm of molecular biology and biological evolution. These excursions are needed to investigate the essential features of life on Spaceship Earth. They'll show you how the probable origin of life followed this sequence: physical evolution, chemical evolution, and biological evolution. Each provides clues to whether life arose elsewhere in the universe. To make the jump from here to there relies on the assumption of the uniformity of physical laws. It rests on the hope that *we are typical.* This may not be true. *We may be unique.* We hope—perhaps too romantically—that cosmic evolution has no quirks, that we are *not* alone.

22.1 The nature of life on the earth

What is life? Rather than try to define it, let's accept a useful rule of thumb: *Living things are things that reproduce, mutate, and reproduce the mutations.*

What does this mean? First, that living things have an *organization* that they pass on when they reproduce. So all living things are *organisms.* A simple example of an organism on the earth is a cell (Fig. 22.1). Second, that reproduction does not simply involve the copying of an organism. Occasionally the offspring has a genetic difference—it exhibits a *mutation*—from its parents. (The genes in a cell's nucleus carry the information code for how an organism is to be put together; if the genes change, the organism changes.)

Some mutations result in the quick death of the mutated organism. Others have no lethal effects, and the organism survives and transmits its new characteristic to its offspring. Why is this important? Because mutation provides the possibility of change. Together with the process of *natural selection,* it leads to *evolution,* the development of organisms of greater complexity, better adapted to their environment. Mutation and evolution are distinguishing features of life.

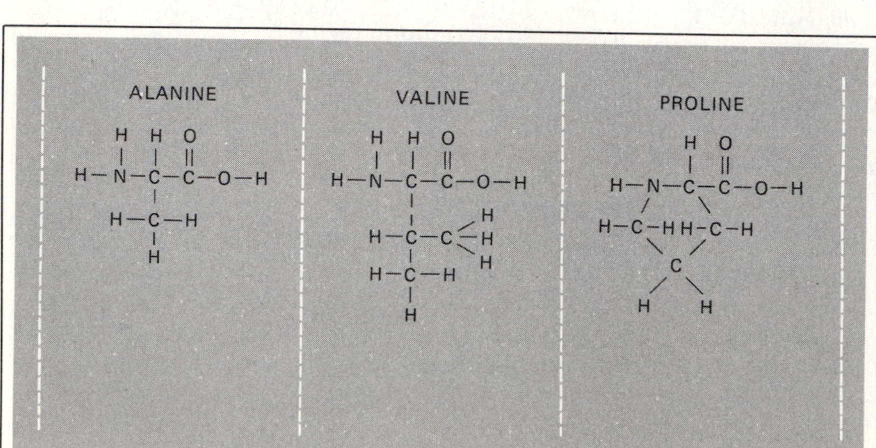

22.2
The chemical structure of three common amino acids found in life on the earth. Note that hydrogen, oxygen, carbon, and nitrogen, four of the most common elements in the cosmos, form the chemical basis.

TABLE 22.1 ***Comparative abundances of the elements in the earth's crust, the solar system (mostly the sun), and human beings****

Earth's crust		*Solar system*		*Human beings*	
Oxygen	55.1	Hydrogen	86.2	Hydrogen	63.0
Silicon	16.1	Helium	13.8	Oxygen	25.0
Hydrogen	15.5	Oxygen	7.7×10^{-4}	Carbon	10.0
Aluminum	4.9	Neon	4.3×10^{-4}	Nitrogen	1.0
Sodium	2.0	Carbon	3.5×10^{-4}	Calcium	0.5
Iron	1.6	Nitrogen	1.0×10^{-4}	Potassium	0.4
Magnesium	1.4	Silicon	3.0×10^{-4}		
Potassium	1.1				

Note The earth's crust is dominated by the silicon and oxygen that are present as silicon compounds in rocks. The solar system's composition is dominated by the sun. The large amount of hydrogen in the human body is due to the fact that animal cells are about 85 percent water by mass. Note that if neon and helium (which are chemically inert substances) are scratched from the solar system list, the elements' rank is the same as that for human beings.

* Given in percentage of the number of atoms.

Questions arise: What makes an organism? How does it acquire its organization? How does it pass this organization on? Biochemical discoveries since the 1960s have answered these questions. Two basic types of molecules operate in all terrestrial organisms; their interaction results in what we call life. The two molecular types are (1) *proteins*, which make up the organism, and (2) *nucleic acids*, which provide the information for the structure of the organism and the means to pass on this information in reproduction. Proteins are built of smaller parts called *amino acids* (Fig. 22.2). Nucleic acids also consist of smaller subunits or bases. These building blocks consist of simple combinations of the most common chemical elements in the universe: hydrogen, carbon, nitrogen, oxygen, and a few others (Table 22.1).

To understand life you must understand its chemistry, for chemical reactions provide the energy that keeps us alive. The sun is the source of this life energy. Sunlight powers photosynthesis in plants, which take in water and carbon dioxide to produce sugar (a food) and free oxygen. Sugars are carbohydrates, simple compounds of carbon, oxygen, and hydrogen. Animals eat the plants and breathe the oxygen, which burns the food slowly, to produce energy to keep them alive, growing, and reproducing. They produce water and carbon dioxide, which plants take in, and the cycle turns again. Almost all the earth's organisms live on sunshine as the ultimate energy source.

Energized by sunlight, your cells carry on chemical work using proteins as *enzymes*, which monitor and facilitate important chemical reactions in a cell. But how does a cell control the functioning of the enzymes? That's the job of the cell nucleus using the nucleic acid called *deoxyribonucleic acid*, or DNA for short. DNA is an enormous molecule; in your body, DNA strands contain *billions* of atoms. (Yet your whole body contains only one teaspoonful of DNA.) Chemically DNA consists of the bonding of four bases, sugars, and phosphates (compounds of phosphorus and oxygen) to form the DNA chain.

DNA serves as the chemical blueprint that informs the protein in cells how to function. DNA is how a cell chemically hands down its blueprint to its offspring. The offspring inherits this information.

DNA contains the instructions to make the proteins. But it needs to use proteins (enzymes) to reproduce itself. Here's the original chicken-and-egg question: Which came first, the self-replicating mechanism or the enzymes essential to make the mechanism work? Neither really—the complex interrelation of these molecules evolved from much simpler structures over a long period of time.

All terrestrial organisms have a common chemical makeup. And this makeup is closely connected to the physical evolution of the universe, in that those common elements are made in stars or were created in the Big Bang. This common chemistry suggests the central scientific idea for the origin of life: It results naturally in the evolution of the universe. In the view of modern evolutionary biology, the central dogma about life's origin is this: *Life arose from nonlife*.

22.2 The genesis of life on the earth

Modern biology sees life's origin as coming from nonliving material. Life arises as a natural consequence of the slow processes of chemical and biological evolution. What are these evolutionary processes?

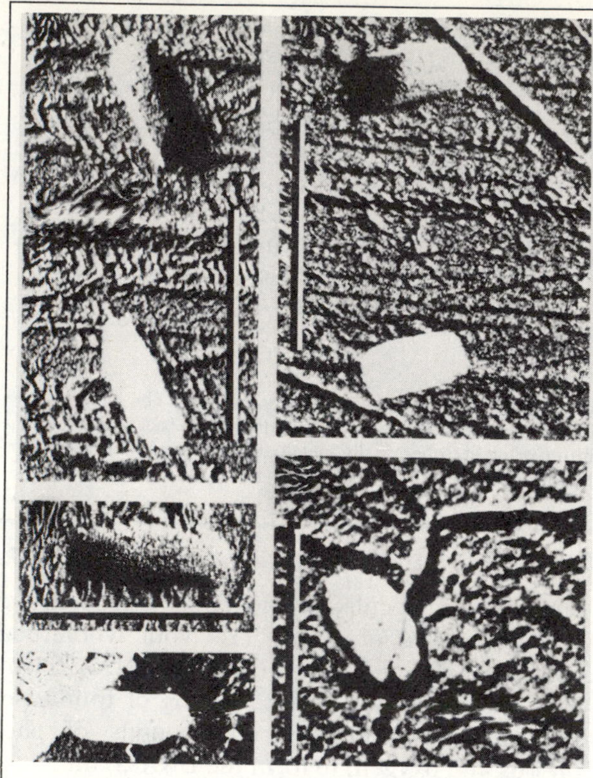

22.3
Some of the oldest fossils known. Found in the rocks of the Fig Tree formation in Africa, these relics have an age of about 3.2 billion years. Their shape is similar to that of modern bacteria. (Courtesy E. Barghoorn)

Clues from biology / In *The Origin of Species* (published in 1859), Charles Darwin (1804–1882) declared that living species are adapted for survival in their environment. No single mutation can generate the evolution of a species. Some feedback mechanism selects mutations that make a species more compatible with its environment. Although Darwin lacked modern knowledge of genetics, he patiently unearthed the general feedback mechanism: *natural selection.* (Alfred Wallace also had this idea and, after an agreement with Darwin, both published their papers at the same time.)

What is natural selection? Put too simply, modern biologists view natural selection as a consequence of successful reproduction. Individuals that are well adapted to a local environment not only survive but, more important, usually succeed in producing offspring. They have reproductive success that spreads their genetic material. Their children tend to survive and reproduce successfully. Eventually their genes dominate those available to a species, and the survival and expansion of new genetic material results in evolution. Where does this new genetic material come from? Mutation! Natural selection directs the random jumble of mutation into biological evolution. This progression assumes some primeval life form to start the process of biological evolution. At some point biological evolution began, but chemical evolution must have preceded it.

Clues from geology / As geologists pare back rock layers, they reveal images of the earth's history. These rocks sometimes trap fossils that show the life of the past. Radioactive dating techniques (see Focus 8.1) can fix the age of the rocks and the fossils in them. Careful microscopic inspection of ancient rock samples reveals the remains of bacteria and algae from 1.0 to 3.5 billion years old. These provide important clues to life's evolution on the earth.

The oldest set of rocks containing possible microfossils lies in Australia. The rocks are 3.5 billion years old. They contain remains that resemble microorganisms and layered structures that could have been built by colonies of bacteria. Another group of rocks, some 3.2 billion years old, lies in Africa. These rocks contain evidence of two distinct life forms: rod-shaped structures resembling modern bacteria (Fig. 22.3) and round cells similar to modern blue-green algae.

Many modern bacteria and all modern blue-green algae are photosynthetic. So the Fig Tree fossils appear to represent a crucial stage in biological evolution, the development of photosynthesis at about 1 billion years after the formation of the earth.

Evidence for the next evolutionary step comes from the Gunflint formation along the shores of Lake Superior in western Ontario, Canada, for which radioactive dating sets a maximum age of about 2.4 billion years. Many of the fossils here show algal structures (Fig. 22.4) similar to modern photosynthetic blue-green algae. Even more striking are structures suggestive of a cell nucleus.

The next evidence of biological evolution lies in the Bitter Springs formation in Australia. The rocks found here are approximately 1 billion years old. Three of the fossils appear to resemble modern types of green algae. Unlike the simpler blue-green algae, the green algae contain complete cell nuclei. So by this time, cells had developed the capacity to join together and mix the genetic material from different cells. Once that happened, biological evolution greatly accelerated, because the mixing of genetic material from different cells made it possible to form many new genetic combinations very quickly.

So the fossil hunters have unearthed vital clues in the biological evolution of life. The fossil discoveries imply (1) that the evolution of life as

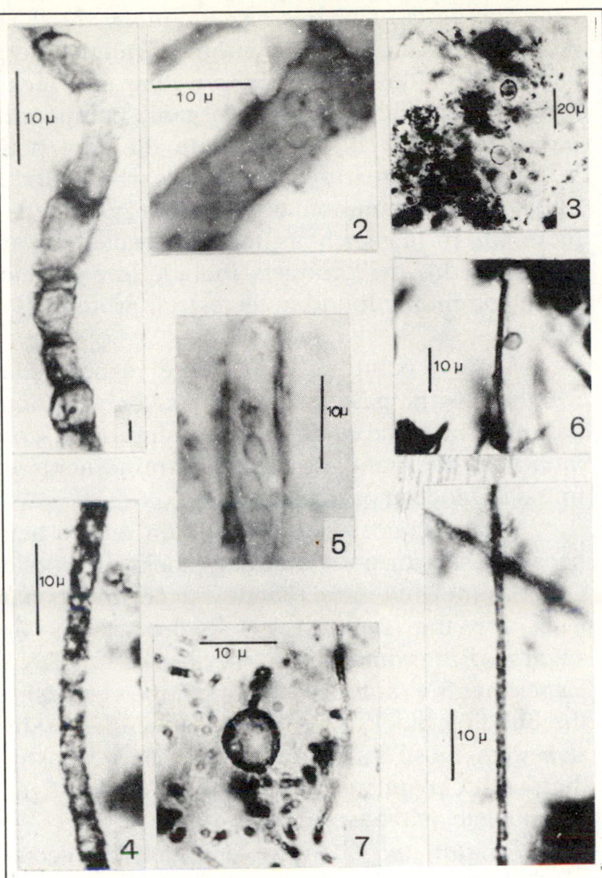

22.4
Fossils from the Gunflint rocks, about 2.4 billion years old. These resemble modern blue-green algae. (Courtesy E. Barghoorn)

we know it (LAWKI) takes a long time (*billions* of years) and (2) that chemical evolution must have been completed on the primeval earth no more than about 1 billion years after the earth formed. What must have been the conditions on the young earth for chemical evolution to take place?

Clues from astronomy / The sun reflects the average chemical composition of material in our Galaxy. The earth ended up with a composition quite different from that of the sun (Table 22.1). How did this happen? Recall (from Chapter 12) that the nebular model for the origin of the solar system pictures the protoearth forming from the accretion of planetesimals. The chemical composition of the planetesimal material depends on the temperature in the solar nebula, as given by the condensation sequence. Most of the gaseous and icy materials were left out in the agglomeration of the planetesimals to make the earth (Table 22.2). At 1 AU, it was too hot for them to condense.

Elemental depletion and the expectations from the condensation sequence give an estimate of the solar nebula's temperature at the earth's

TABLE 22.2 Depletion of elements in the earth

Element	*Depletion (abundance in earth to abundance in solar system)*
H	2.5×10^{-7}
He	10^{-14}
C	10^{-4}
N	1.3×10^{-6}
O	0.16
Ne	2.5×10^{-11}
S	0.32
Cl	0.20
Ar	5.0×10^{-7}
Kr	6.3×10^{-8}
Xe	3.2×10^{-7}

formation: a little below 700 K—probably around 600 K or so. But a 600 K condensation temperature presents a sticky problem with carbon. The condensation sequence predicts that carbon (in methane) condenses at 120 K, far cooler than 600 K. So the planetesimals forming the earth contained little methane and therefore little carbon. But the carbon *is* here; otherwise, we wouldn't be!

As yet, this carbon deposition problem is unsettled, but here's a possible way out. At high temperatures, carbon combines with oxygen to form carbon monoxide (CO). Under the proper conditions, carbon monoxide combines with hydrogen to form large hydrocarbons, such as those found in tar. If such reactions went on in the solar nebula, the tars would form on the grains that made up the protoearth. (This reaction, carbon monoxide plus hydrogen to make hydrocarbons, is used commercially to make gasoline.)

What about the early atmosphere? If the earth's surface were hot at the time—perhaps even molten—the heat would gasify volatiles within the earth, which would rise upward to the surface. (That's what happens in volcanoes today.) This process is called *outgassing* or *degassing*. Volcanoes outgas water, carbon dioxide, hydrogen sulfide, methane, and ammonia, for instance. We would expect pretty much the same materials to outgas from the primitive earth.

How fast this outgassing took place then depended on the crustal temperature. If the earth formed molten, outgassing of its secondary atmosphere happened quickly, in 100 million years or less. Although a lot of gas was produced in this way, it still did not make the present atmosphere, for this secondary atmosphere lacked oxygen. In other words, the secondary atmosphere was probably nonoxidizing and contained much carbon dioxide and water vapor.

The stage was set for the first steps in chemical evolution on the earth.

22.3 The spark of life

The earth started out with an atmosphere of simple molecules. To synthesize complex molecules from simpler ones requires free energy. Photosynthesis in plants now captures solar energy and stores it in the form of chemical bonds. On the young earth, sunlight was available, but plants were not. Where did the energy needed for synthesis of complex molecules come from? Possible sources include ultraviolet from the sun, radioactive decay, and lightning.

Solar ultraviolet at wavelengths less than 2200 Å plays a key role because it is absorbed by complex molecules and can cause the formation of still more complex ones. (The ozone layer now filters out most of the ultraviolet radiation. Because of the lack of free oxygen, our second atmosphere, that produced by outgassing, did not form an ozone layer.) As a pre-main-sequence star, the sun's surface temperature was less than now, but it may have been more active, emitting more ultraviolet as a fraction of its total energy.

Radioactive decay can also release free energy. The rate now is about 1/100 the energy input from solar ultraviolet. But 4.6 billion years ago, the fraction of radioactive materials in the earth's surface would have been greater than now—roughly three times more. So radioactive energy might have been comparable to solar energy on the early earth.

On the earth now, lightning accounts for almost as much free energy as short-wavelength ultraviolet. Before the earth cooled enough for rain to fall, probably little lightning occurred near the earth's surface. Why? The regions of thunderclouds that generate the electric charges for lightning are in the ice zone. So before the first rain, any lightning flashed at much higher altitudes than now. After a great primeval rain, which filled the oceans, lightning storms possibly raged widely over the earth's surface. Then the energy from lightning may have been more important than now.

In summary: Radioactivity from the crust, solar ultraviolet, and lightning provided sources of free energy for molecular synthesis. Some of these energy sources came in spurts and bursts rather than at constant rates. And all have the capacity to destroy as well as to help synthesize molecules. The balance between creation and destruction determines the number and kinds of molecules that could exist.

Synthesis of simple organic molecules / The critical parts of chemical synthesis are free energy and a nonoxidizing atmosphere—one that lacked free oxygen, which destroys organic compounds. Laboratory experiments validate this key point. When gaseous mixtures of water, carbon dioxide, methane, and ammonia have energy added to them (in forms such as those expected on the young earth), the products include amino acids, some commonly found in terrestrial proteins (Fig. 22.5).

A key point is that these experiments, whether with gaseous mixtures or solutions, naturally produce *most* of the amino acids common in protein and *none* of the amino acids *not* found in modern protein.

I have concentrated so far on amino acids because they form the building blocks of proteins. Other, somewhat more tentative experiments have tried for the synthesis of hydrocarbons and sugars. But what about the basics of DNA? Experiments to simulate the primitive synthesis of the building blocks of DNA have been less extensive than those for amino acids. They do show that, with phosphates added, amino acids, sugars, and nucleic-acid bases result.

To sum up: The simple organic molecules needed for LAWKI form naturally under plausible primitive-earth conditions in simulation experiments. The simple molecules can be cooked. What about the actual proteins and nucleic acids?

The synthesis of complex molecules / Here we stand on much shakier ground than with the simple molecules (and remember, we aren't yet anywhere near the complexity of living cells). Proteins and nucleic acids are not only huge molecules; they also have a special and precise architecture. How did the first of these macromolecules get together?

Let's first look at how we can synthesize such molecules in the lab. To link together amino acids, we stick together the amino group (H_2N) at one end of an amino acid and the acid group (COOH) at one end of another amino acid. The C and N then link together. A water molecule (H plus OH) must be removed. Similarly, to put together a base, sugar, and phosphate to build a nucleic acid requires the removal of a water molecule at each bonding step. So for both proteins and nucleic acids, water must be removed to build them up.

Sounds simple, but there's a hitch. In the scenario so far, the simple molecules synthesized, say in the air, fall into the primitive oceans, ponds, or lakes. The point is that molecules on a dry, solid surface cannot participate in further synthesis because they can't move around to meet

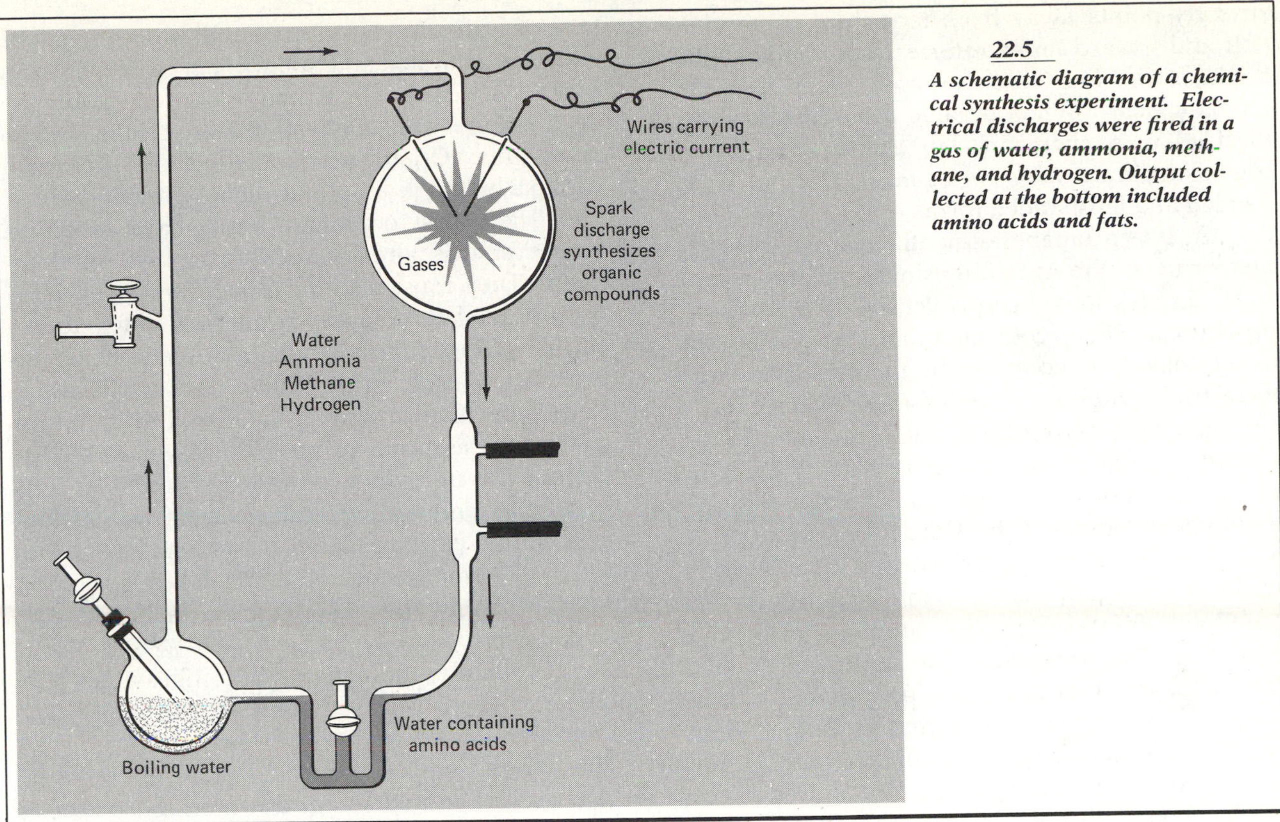

22.5
A schematic diagram of a chemical synthesis experiment. Electrical discharges were fired in a gas of water, ammonia, methane, and hydrogen. Output collected at the bottom included amino acids and fats.

other molecules. So whatever waters exist must catch the synthesized molecules and store them. But it's difficult to synthesize the more complex molecules in water, because water must be removed in the process.

Difficult, but not impossible. Tidal pools might just be a hospitable place for joining together simple molecules to make larger ones. When the tide rolls out, the water in small pockets evaporates in the sun, and molecules combine. When the tide rolls in, the water dissolves the molecules, mixing them with other molecules.

A number of attempts have been made to make proteins and nucleic acids under prebiological conditions. I won't bog you down with the chemical details; these experiments seem to be somewhat successful so far.

To sum up: The precise complexity of proteins and nucleic acids makes their synthesis difficult. We do not yet understand the specific pathways to their original production.

22.4 *Amino acids in space*

Meteorites provide some evidence to support the theories of natural synthesis of organic compounds. Of the three main classes of meteorites (see Section 12.1), carbonaceous chondrites, which contain a relatively high percentage of carbon (2 percent), make up a minority. People have regularly speculated that some of the carbon contained in these meteorites might be organic in nature.

The best chance for the discovery of extraterrestrial organic materials occurs when the sample undergoes analysis in a scrupulously clean environment soon after its fall. That chance came in 1969, when a meteorite fell in Murchison, Australia, on September 28, at about 11:00 A.M. This meteorite, a carbonaceous chondrite, was rushed to the Ames Research Laboratory of NASA and analyzed by a team of scientists headed by Cyril Ponnamperuma. The NASA group discovered five amino acids common to living protein. The quantities were small, only a few micrograms of amino acids in each gram of the meteorite.

Was this terrestrial contamination? Probably not. Organic molecules exist in two distinct forms: right-handed ones and left-handed ones, depending on the direction of the twist of the linkage of the atoms. Almost all terrestrial organic molecules are left-handed (for an as yet undiscovered reason), so earth-based contamination is expected to be left-handed. The Murchison meteorite contained just about equal quantities of right- and left-handed molecules, the left-handed forms predominating a little. This evidence strongly points away from terrestrial contamination and toward an extraterrestrial, nonbiological origin of the Murchison organic molecules. Why nonbiological? When organic molecules are synthesized in a chemistry lab (rather than by an organism), they show an equal number of right-handed and left-handed forms.

In 1969 a Japanese scientific team discovered meteorites in the Antarctic; since then, more than 1000 samples have been collected. The Antarctic provides a clean, cold environment that is relatively unlikely to contaminate the meteorites with terrestrial materials. One of these meteorites, a carbonaceous chondrite found near Allen Hills, contains amino acids free of terrestrial contamination. But it has only about 10 percent of the total amino-acid content of the Murchison meteorite.

The glut of complex molecules discovered by radio astronomers (see Section 15.1) lends further credibility to extraterrestrial, nonbiological formation of organic substances. The molecules such as formaldehyde, hydrogen cyanide, cyanoacetylene, formic acid, methyl alcohol, and methylacetylene can play a crucial role in organic chemistry. Formaldehyde and hydrogen cyanide, for instance, can be chemically combined to make amino acids.

Radio astronomers have actually searched a few molecular clouds for amino acids. No luck so far. It's a difficult observation: The amino-acid concentration in interstellar clouds is not expected to be large, and the radio frequencies of emission from amino acids are not well known. Detection efforts continue.

The interstellar medium seems a breeding ground of prebiological organic compounds. The important conclusion is this: The chemical evolution from simple compounds to complex organic substances occurs so naturally that it takes place even in the hostile environment of space, without biological aid.

22.5 From molecule to organism

I have explained how chemical evolution naturally—and perhaps inevitably—leads to the making of complex organic compounds that are the building blocks of proteins and nucleic acids. Both are needed to join together in a cell. How to make that first cell?

Quite bluntly, we don't know. Fossils cannot give information about this crucial time. And chemists have not synthesized anything as complex as a cell. What happened before this time remains something that for now we can only speculate about.

The first cells might have formed at the surface of water (such as in a small pond) or at the interface of water and solid material (such as on rocks in a tidal pool). Although we don't know the details about the origin of the first cells, we do know this: Organisms appeared on the earth within a billion years of our planet's formation.

These first organisms must have relied on food that was already present in their environment. They must also have been protected from the ultraviolet radiation streaming down to the earth's surface (ultraviolet light destroys cells; it is sometimes used in hospitals for sterilization). Remember that little oxygen existed in the atmosphere at this time, so no ozone layer existed to cut out the ultraviolet radiation. Ultraviolet light penetrates only a few centimeters of water, however, so the early organisms could survive if they were below the surface of ponds, lakes, or oceans. They probably formed at the surface and then sank down.

Gradually, the action of the ultraviolet light on water vapor released some free oxygen to the atmosphere. The ozone layer began to build up. This process cut off more and more of the ultraviolet radiation and so cut down on the synthesis of organic compounds. The first organisms had their food supply cut off at this point. Millions probably died. But a few survived—those that were able to use photosynthesis to manufacture their food. More oxygen was added to the atmosphere, some by the organisms themselves, and finally the ozone layer was thick enough to shut out almost all the ultraviolet radiation.

Life exploded on the earth by multiplying rapidly after the ozone shield was up. Why? Partly it was because the ozone shield allowed plants to spread to land. But also the new availability of oxygen allowed more efficient metabolism of compounds.

Note the opposite roles played by ultraviolet radiation at different stages in the development of life. In the very early stages of molecular evolution, ultraviolet light is invoked as a possible source of energy for the synthesis of more complex molecules from simple substances. But ultraviolet light is fatal to complex biological systems, for it can also break apart the larger and more fragile organic molecules and destroy cells. Organisms developing in water would have some protection, but before life could develop into complex cellular forms and spread over the earth, the ozone layer had to form to shield out the ultraviolet radiation. A current controversy is raging in scientific and government circles on the possible long range effects on the ozone layer of certain chemical pollutants being introduced into the atmosphere by people. If the ozone layer is des-

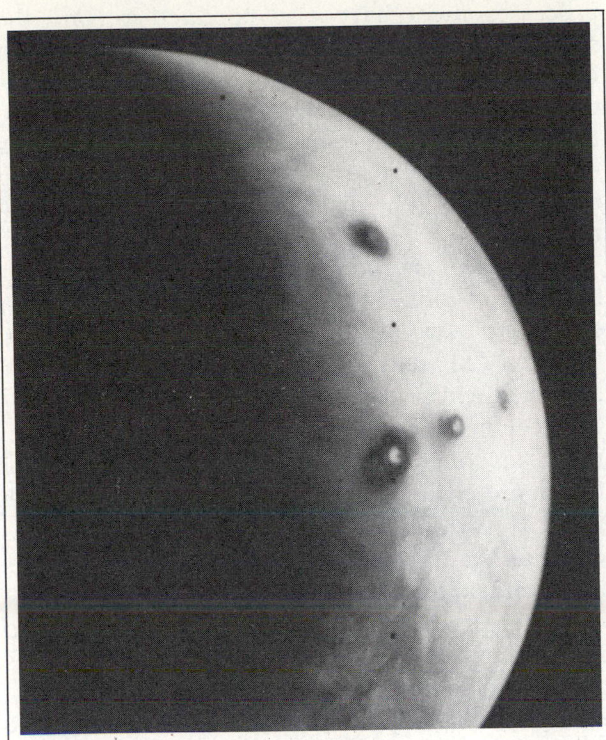

22.6
A panoramic view of the Tharsis Ridge region near the equatorial region of Mars. (Courtesy NASA)

troyed, life might have to start all over again from the simplest molecules.

Biological evolution continues today. The "somehow" that ignited biological evolution is not clearly known. But we do know that it *did* happen. With this scheme in mind, let's turn to the solar system and the Galaxy to investigate the possibility of cosmic neighbors.

22.6 The solar system as an abode of life

So far we know of only one planet in the solar system that has carbon-based life: the earth. The environments of the other planets pretty much exclude LAWKI—with the slim exception of Mars. I'll discuss Mars first and then turn to speculations about the Jovian planets as prebiological worlds. Keep in mind the essential conditions for LAWKI: temperature range from 200 to 373 K, a solvent (commonly water), a carbon chemistry, free energy (from a star), protection from ultraviolet light, and a long time to evolve.

Mars: The best chance / The fate of Martian LAWKI hinges on the abundance of surface water. The Viking missions found the surface pressure to be about 0.007 earth atmosphere—much less than that on the highest mountains on the earth's surface (in fact, you have to go about 40 km up into the earth's atmosphere to find pressure this low). At this low pressure, liquid water cannot exist on the surface. As evidenced by the polar caps, both water and carbon dioxide form solid ice. Even in these regions, abundant liquid water needed for LAWKI probably does not exist. Mars is a very, very dry planet: Even in the polar regions, the water vapor in the atmosphere, if all condensed, would form a layer only 0.1 mm thick on the surface.

Perhaps water did flow in the past. East of the volcanic ridge (Fig. 22.6) dominated by Olympus Mons stretches the series of canyons that may be due to the erosive action of liquid water in the Martian past. Some astronomers imagine that at an earlier epoch, Mars had a denser atmosphere capable of holding water vapor sufficient to generate rainfall.

This idea ties in with past volcanic activity. On the earth, volcanoes spew out large volumes of carbon dioxide and water vapor. Possibly the violent geologic episode that spawned the volcanoes at the Tharsis Ridge also injected a significant amount of volcanic gases into the Martian atmosphere. An increased atmospheric pressure would have allowed water to flow on certain regions of the surface and to cut the meandering channels. That water may now be frozen below the surface, locked up in surface minerals, or at the polar caps.

Recurring deluges or meltings may explain the origin of the laminated terrain found in the polar regions (Fig. 22.7). There, stacks of thin plates of crustal material stand about 10 km tall and up to 200 km across. Because they exist only in the polar regions, where carbon dioxide and

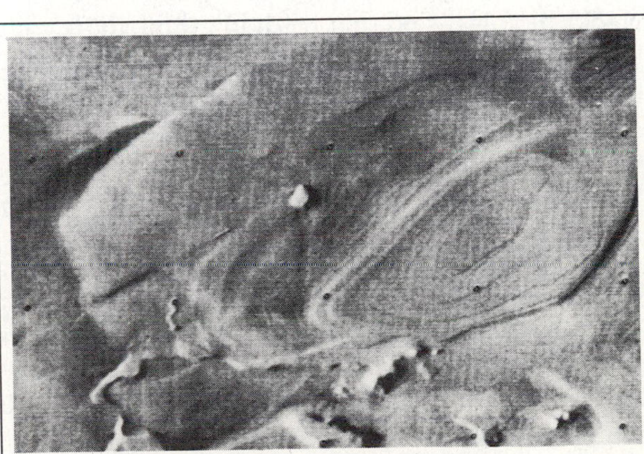

22.7
Laminated terrain near the south polar cap of Mars. These formations of surface soil probably contain water ice. The photo covers an area 47 by 60 km; the pits are about 500 m deep. (Courtesy NASA)

water ice form annually, the plates may be related to the influx and outgo of these substances. A time of denser atmosphere may have produced the laminated terrain along with the eroded channels.

An eruption of interior gases may also have been combined with astronomical effects to change the Martian environment in the past. Owing to the gravitational attraction of the other planets, Mars's orbit varies in its average distance from the sun. So the average amount of sunlight and its peak amount vary over 2-million-year periods. The variation in solar energy input affects the size of the polar caps. It is possible to imagine large polar caps accumulating during the colder periods and melting during the warmer ones, adding carbon dioxide and water to the atmosphere.

The direct test for Martian LAWKI came from the Viking landers' biology experiments. What were these results? I judge that they proved negative: In the soil that was sampled, LAWKI does *not* exist on Mars now.

Why such a negative view? Most important, both landers contained an instrument called a mass spectrometer, designed to detect and measure organic molecules, the complex chains, secured by carbon, that characterize LAWKI. At both landing sites, *no large organic molecules were found*. The instruments had the sensitivity to detect organic compounds in a concentration of just a few parts in a billion. That's about 1 million bacteria (dead or alive) in a sample—far below the concentration found in desert soils on the earth.

In light of the lack of complex molecules in the soil, the landers' three biology experiments, when they give apparently positive results, can all be explained by *chemical* reactions rather than *biological* ones. At the International Astronomical Union Meeting in Montreal in August 1979, a NASA biologist said that he didn't think there's life on Mars. Our extraterrestrial search for terrestrial life has failed so far.

22.7 The Milky Way as an abode of life

Where might life exist elsewhere in the Galaxy? The huge number of stars in the Galaxy implies planets elsewhere, if the nebular model of planetary formation is correct (see Chapter 12). Even if the probability of the genesis of life is slim, the number of possible habitats is so large that I suspect that some extraterrestrial creature has viewed the dawn of its day. If life developed from the natural evolution of the inorganic, these processes must have also operated beyond the solar system. This is the central dogma of modern biology: that life arose from nonlife. The elements of life are the most abundant in the cosmos, so there is no lack of proper ingredients. All that is required is the proper construction. This forming takes physics, chemistry, and—most important—time.

Cosmic prospecting / How to prospect for life in the Galaxy? You want to know *how much* exists and *where* is the best place to hunt for it. Here I try to answer the same two questions concerning life in the Galaxy.

Civilizations of living creatures must evolve; that's part of cosmic evolution. So the numbers of intelligent civilizations in the Galaxy changes with time. At any one time, the number of civilizations depends on the rate at which these civilizations are born and how long they last.

Here's an analogy: Suppose you are locked in a dark room filled with candles; a friend gropes about and lights one candle every 15 minutes—four per hour. Suppose each candle burns for one hour. How many candles are lit at any given time? During the first hour, the number increases from one to two to three to four. But just as the fifth one is lit, the first one goes out. As the sixth is lit, the second goes out. One goes out as each new one is lit, leaving four candles burning at any one time. If you think about it, you see that the number of observed candles is equal to the rate of candle lighting, R_c, times the lifetime of one candle, L_c, or

$$N_c = R_c \times L_c$$

So, if you know the average lifetime of a single candle and the rate at which the candles begin their life, you can anticipate the number lit at any time.

The same reasoning applies to the number of civilizations in the Galaxy at any one time: If R_{ic} is the rate of formation of intelligent civilizations and L_{ic} is their lifetime, then

$$N_{ic} = R_{ic} \times L_{ic}$$

This relation may be broken down into more specific factors, loosely independent of one another:

$$N_{ic} = R_* \, P_p \, P_e \, N_e \, P_l \, P_i \, L_{ic}$$

This equation was first put together, in somewhat different form, by radio astronomer Frank Drake, so it's called the *Drake equation*.

The meaning of each of these pieces relates directly to important facets of cosmic evolution. R_* is the rate of star formation averaged over the age of the Galaxy; P_p is the probability that once a star has formed, it will possess planets. The next factor, P_e, is the probability that the star's ecosphere will exist long enough for life to form, and N_e is the number of planets in the ecosphere. P_l is the probability that a planet in a star's ecosphere will develop life, and P_i is the probability that biological evolution will ultimately lead to intelligent life. The final term, L_{ic}, is the lifetime of this intelligent civilization. Note that these factors group into three categories: R_*, P_p, P_e, and N_e relate to astronomy and physical evolution; P_l and P_i relate to biology and chemicobiological evolution; and L_{ic} derives from what I would call speculative sociology.

Astronomical factors / The Galaxy contains a few times 10^{11} stars. These stars have formed over at least 5 billion years. So the average birth rate of stars from these figures is about 20 per year. However, the initial burst of star formation took place much more rapidly than later bursts (remember, our sun is probably a third generation star) as more material was locked up in stellar corpses. The slowdown of the birth rate pushes the initial estimate down to about 10 stars per year. I adopt 10 for R_*.

What is the chance that one of these stars will develop a planetary system? Nebular models (see Chapter 12) imply that many planets exist in the Galaxy. A collapsing gas and dust cloud must form either a star with a planetary system or a multiple-star system, perhaps also with planets. More than 50 percent of the stars in the Galaxy are in binary or other multiple-star systems. A planet in a multiple-star system may not have a stable orbit, so I exclude these from consideration. If I take a planetary system versus a multiple-star system as an either-or proposition, P_p equals 0.5.

A star's *ecosphere*—the zone in which planets must lie to have conditions suitable for life as we know it—depends primarily on the temperature of the star. The more luminous the star, the farther out the habitable zone starts (Fig. 22.8). The width of the ecosphere is also greater for luminous stars and thinner for less luminous ones. The ecosphere must persist long enough to allow the genesis and evolution of life.

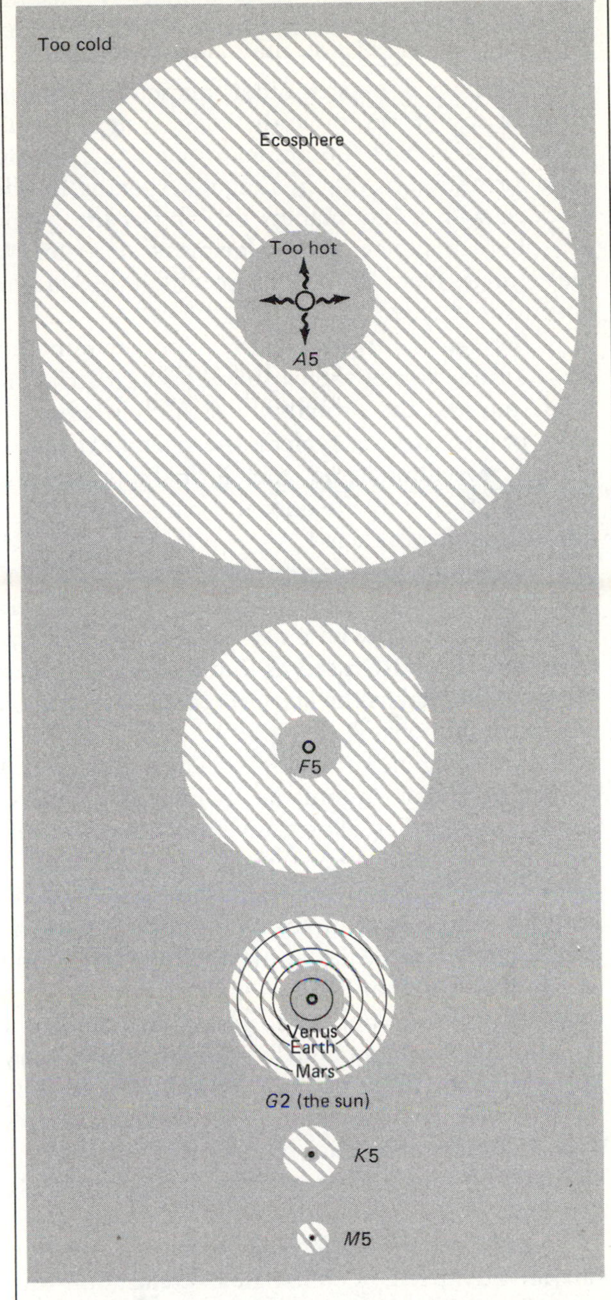

22.8
The sizes of stellar ecospheres for main-sequence stars of different spectral types. Note that the cooler the star, the thinner the ecosphere and the closer it lies to the star.

Luminous *O*- and *B*-stars live out their normal lives in about 100 million years, a time much shorter than the 3 to 4 billion years that elapsed while life evolved on the earth. By our standards, these energy spendthrifts are improper parents. It seems unlikely that attendant planetary systems to such short-lived stars would have the time to develop life. So we consider only stars whose life spans are at least equal to the sun's—spectral class *G* or cooler, a choice that fortunately includes 98 percent of all the normal stars in the Galaxy. Unfortunately, for stars cooler than spectral class *K*, the ecosphere is too small. If we throw out these cool stars, only about 8 percent of the total remains, so P_e equals 0.08. These stars are the good suns for life.

How about the number of planets in the ecosphere, N_e? Here we have only the unique example of our solar system: Three planets—Venus, the earth, and Mars—lie in that zone. If the planetary formation processes in the nebular model are universal, we'd expect other planetary systems more or less to resemble the solar system. Is this belief reasonable? Computer models of nebular-style formation results in a regular spacing for the planets, with a few planets orbiting at the magic ecosphere distance of a solar-mass star. So N_e may range from 1 to 4 or so; I'll use 3 as an optimistic choice.

Biological factors / Here's another either-or proposition. Either the existence of terrestrial life is unique and the probability of life elsewhere is zero, or the earth is typical, the normal result of cosmic evolution, and the probability of any planet's developing life once the astronomical conditions are favorable is 1.

Appealing to the uniformity of physical laws, assume that we are typical. Lab experiments (Section 22.3) have shown the natural start of chemical evolution. Because it seems that that the nature of the universe makes the start of chemical evolution inevitable, I choose P_e equal to 1.

Is intelligence inevitable? The development of multicellular organisms (which is a good feature because it allows diversification of cells for greater efficiency) requires coordination among cells, that is, some kind of nervous system, for both sensory and motor coordination. Also, the ability to learn appears at even simple levels and so aids in survival. The adaptive powers of a thinking organism appear so great that I think, if it is at all possible genetically, intelligence is very likely to be the ultimate result of natural selection. So I choose P_i equal to 1. This choice assumes that a comfortable environment persists for the billions of years needed for intelligence to develop.

Speculative sociological factors / How long can an advanced, technological civilization survive? Is intelligence flexible and complex enough to cope with adverse aspects of its technology? By our own example to date, the lifetime of an intelligent civilization may be only a few thousand years. But if it is possible that every advanced civilization steers clear of its problems, it should survive as long as the parent star. The lifetimes of civilizations encircling a *G*-type star may be about 10^{10} years. But their lifetimes may also be much, *much* shorter.

The numbers game / As I have progressed through the astronomical, biological, and sociological factors needed for a rough estimate, the footing has become shakier. I have also ignored some important factors in the analysis, such as the possible stable planetary orbits in a binary star system. I did not intend to give precise results, but rough estimates, where exact answers are not yet possible. The point is to get a feel for reasonable exclusions in the enormous range of values each element might take. My personal biases also affect the discussion; I hope I made them clear.

Not evaluating L_{ic}, I come up with

$$N_{ic} = 1 \times 0.5 \times 0.08 \times 3 \times 1 \times 1 \times L_{ic}$$
$$N_{ic} \simeq 0.1 \times L_{ic}$$

The result depends critically on L_{ic}—how long our candle remains lit. If we assume we are at the brink of destruction, then $L_{ic} \simeq 10^3$ and $N_{ic} \simeq 10^2$. Intelligent civilizations are few and far between. If we survive as long as the sun shines, then $L_{ic} \simeq 10^{10}$ and $N_{ic} \simeq 10^9$. In this case, many stars in the Galaxy have fostered an intelligent civilization!

How seriously can you take these results? Not very—basically they are speculation and shoulj be viewed very skeptically. I have delineated each of the values of the life factors in a reasonable (to me) and yet ultimately arbitrary manner. Our own example, life on the earth, may be more special than we have been willing to admit (Focus 22.1).

22.8 Neighboring solar systems?

What evidence do we have of other planetary systems? Because a planet shines by reflected light from its parent star, because it is small in size, and because a planet lies very close to its local sun, as seen from the earth, a planet's gleam would be lost in the stellar glare. So we cannot *directly* observe other planets outside the solar system with earth-based telescopes. (The space telescope will give us more power for this task.)

Instead of searching for the light from very large planets, we can hunt for the motion around the center of mass of the planet-star system. As a result of this seesaw effect, the visible star wobbles from side to side about the center of mass if a massive planet orbits it. From the observed stellar wobble and an estimate of the stellar mass, we can estimate the mass of the invisible planetary companion by the same method used to measure binary star masses.

Planetary companions? / Barnard's star, the second nearest to the sun, appears to exhibit such a corkscrew motion in its proper motion across the sky. Peter van de Kamp of the Sproul Observatory claimed that two planets, one with about 80 percent the mass of Jupiter and the other 10 percent larger than Jupiter, circle the star. The

Focus 22.1

MASS EXTINCTIONS ON THE EARTH?

Roughly 65 million years ago, a sudden trauma swept through life on this planet. In a short time—less than a million years, perhaps as swift as a thousand years—mass extinctions hit certain plants and animals. The fossil record shows an abrupt loss of ocean plankton, swimming mollusks and dinosaurs, and land animals with masses greater than 25 kg—most especially, the large walking dinosaurs. This demise was good for us, for the mammals flourished afterward. What happened then to promote these mass extinctions?

Many ideas have been proposed. One seems to be gaining the weight of reasonable evidence: that of the impact of an asteroid-sized body that caused environmental stress, which resulted in selective, world-wide extinctions. An object some 10 km in diameter with a mass of some 10^{14} kg could easily penetrate the earth's atmosphere. Its impact would release some 10^{23} J, equivalent to 10^{14} tons of TNT. Similar impacts shaped the large basins on the moon (see Sections 9.3 and 9.6). Astronomical evidence suggests that an object this large crosses the earth's orbit once every 100 million years or so.

How might such an impact influence the earth's environment? The blast could deposit small dust particles (less than 1 μm in size) high in the earth's atmosphere, where they would remain for months. Winds can circulate the dust globally. These particles would cut out a significant fraction of the sunlight reaching the earth's surface, sharply reducing photosynthesis (especially by plants in the oceans) and the overall temperature. Animals especially sensitive to temperature changes would not be able to adapt, and so would disappear.

This idea sounds very plausible from an astronomical view. Does it have any solid evidence? The main clue comes from the composition of a clay layer deposited about this time. Below it (before it in time), we find the usual range of fossils from the age of the dinosaurs. Above it (afterward), certain fossils no longer appear. The layer itself has a composition that is enriched (relative to the earth's crust) in noble metals such as iridium and gold. The overabundance of iridium in this layer appears to be a world-wide phenomenon. One source of this enrichment could be material from a large asteroid. Its impact would mix asteroidal material with terrestrial and could result in the abundances found in the clay layers.

I want to emphasize that although this scenario is reasonable, it is far from firmly proven, in my opinion. If it is supported substantially, it would carry a critical lesson for humankind. A nuclear war will not only destroy from explosions and their fall-out radiation. It will also fill the atmosphere with dust—dust that can reduce the sunlight, which would lower photosynthesis and the average temperature. The result might be a sudden, artificial winter during which many plants and animals may die—destruction far more extensive than from the bomb explosions themselves. Mass extinctions in the past should warn us to avoid such in the future. I feel that the threat of nuclear war, more than any other factor, limits the lifetime of technological civilizations. If we cannot cope with nuclear weapons, our existence on this planet may be relatively short.

lesser mass would be about 2.8 AU from the star, and the greater mass approximately 4.7 AU away. Persisting with his planet hunt, van de Kamp announced that a possible jumbo planet, six times the mass of Jupiter, may orbit the star Epsilon Eridani. This star is not only one of the closest stars to the sun, but it is also quite similar to it in spectral class. Although the supposed dark companion is much larger than any planet in our solar system, it is certainly too low in mass to be a bona fide star. Some nearby stars of solar mass also show some evidence of possible (but not yet confirmed) motions that may result from dark companions with a mass roughly similar to that of Jupiter.

Warning: The observations required to detect planets around nearby stars are *extremely* difficult to make. The wiggles sought for are only about 0.001″, or about 1/100 the size of a star's image on an astronomical photographic plate. Such minuscule changes are dramatically affected by changes in the telescopes themselves, whether produced by self-aging or conscious effort (such as cleaning). One analysis of the errors in such observations concluded that *no* good evidence supports the existence of Jovian-mass planets. The possible exception: Barnard's star.

A curious fact emerges from such investigations: The sun is the only star that we know *for certain* does have a planetary system and does not have a companion star. This brings up a key question: What fraction of stars like our sun in the Galaxy do have stellar companions?

A project at Kitt Peak National Observatory has searched 123 sunlike stars within 85 ly of the sun in an effort to find out whether or not they have companions. The technique aimed at detecting regular Doppler shifts in the spectra of these stars. Results: Over half (57 percent) of their stars had at least one stellar companion. What about the others? Most of the companions are of small mass, the number increasing as the mass gets smaller. The inference: roughly one-fifth to one-sixth of sunlike stars could have planetary-sized companions (but none was directly observed). This argument, if true, implies that the galaxy contains 15 to 20 billion stars with planets—and many of these stars resemble the sun.

Note that the foregoing discussion refers to Jovian-sized planets. We have little hope of detecting terrestrial-sized planets. Not only are the gravitational effects smaller, because of their smaller mass, but the effects would also have to be disentangled from the effects of larger bodies in the same planetary system. There is also a problem of distinguishing small effects, such as variations in radial velocity, from phenomena associated with the stars themselves. Although detection of Jovian-sized planets is conceivable, but difficult, it is even harder to directly detect terrestrial planets around other stars.

Nearby good suns / Of the few hundred stars fairly close to the sun, three stand out as good candidates for possessing planets suitable for life. The number of these nearby systems is consistent with the range of numbers developed in the preceding section for the probability of extraterrestrial life. The first is the star system closest to the earth, Alpha Centauri, a multiple system with a life-bearing-planet probability of 0.1. So if Alpha Centauri is encircled by the same number of planets as our solar system, one is likely to shelter life. Such a planet would be a true interstellar neighbor, because Alpha Centauri is only 4.3 ly from the sun.

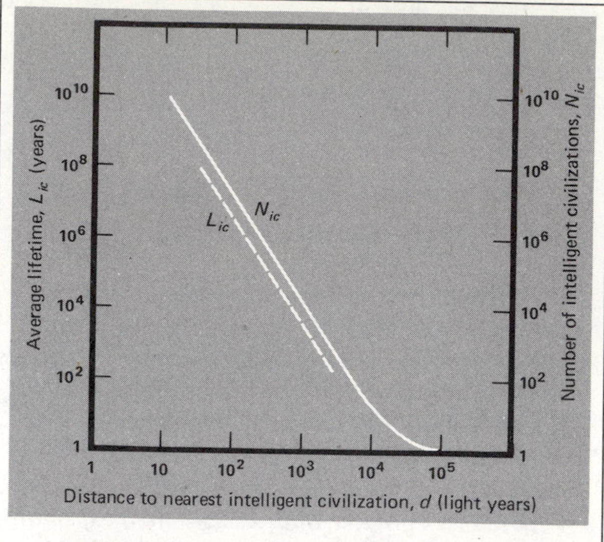

22.9
The relationship of the average lifetime of intelligent civilizations (L_{ic}), the total number of intelligent civilizations in the Galaxy now (N_{ic}), and the distance to the nearest such civilization now. (Adapted from a diagram by R. N. Bracewell)

Next in line, at a distance of 10.8 ly, is the *K*-star Epsilon Eridani. Its long main-sequence lifetime, greater than 10^{12} years, enhances the opportunity for life to be created. At almost the same distance from the solar system, 11.8 ly, is the *G*-star Tau Ceti, which has physical characteristics much like the sun's. These two stars are the most likely of those nearest to us to have planets with LAWKI.

22.9
Where are they?

Whether or not we should search for others depends on how many technologically advanced civilizations exist in the Galaxy *now*. (For this discussion, I take "technologically advanced" to mean creatures who can manipulate their environment at least to the extent that we can, so they have electricity, radios, telescopes, and so on.) If the number is large, then on the average such civilizations must be closer together than if the number is small.

As I pointed out, the key element in estimating this number, N_{ic}, is the *lifetime*, L_{ic}, of technological civilizations. For example, if the lifetime is about 100 years (which is about how long we've had a technologically advanced human culture on earth), then the average distance between galactic civilizations is roughly 10,000 ly. That makes communication practically impossible. Why? If we tried to signal by radio, for example, by sending out a message just at the moment our technology permitted, our civilization would have died while our words were still in transit. Communication is only possible (under known physical laws) if the number of civilizations is large and their lifetimes are long.

Note that N_{ic} is *not* a fixed number. It changes with time as the Galaxy evolves. For instance, for the first billion years of the Galaxy's existence, N_{ic} was probably zero, because life had not yet had time to evolve. So our estimates for now need not apply to the past or future.

How far to our galactic neighbors? / If we are it, we don't have any neighbors within 10^5 ly—the size of the Galaxy. That's for a very small N_{ic}. If N_{ic} is very large, say 10^9 to 10^{10}, then our neighbors are only a few tens of light years away (Fig. 22.9). If N_{ic} is 10^6—the compromise guess—then our neighbors live a few hundred light years away (Fig. 22.9). They are then just within reach. Note that each of these choices implies a value for L_{ic} (Fig. 22.9). If N_{ic} is very small, L_{ic} is at most a few hundred years. We are then probably on the verge of extinction. If N_{ic} is very large, L_{ic} is 10^9 to 10^{10}. Civilizations then last as long as their suns. If N_{ic} is 10^6, then L_{ic} is roughly 10^5 to 10^6, and we still may have some time on the planet earth. (But recall, L_{ic} is only an average, not necessarily the most probable lifetime. Maybe civilizations last *either* a very short time *or* an extremely long time, depending on whether or not they survive the crisis of atomic technology. In such an either-or situation, none might actually last the average time.)

To sum up: We have no direct evidence for other advanced civilizations in the Galaxy. But from cosmic evolution, we expect them. So where are they?

You could take a pessimistic stance and argue that there are none. Michael Hart is one astronomer who holds to the "we are alone" view. His opinion is based partly on computer calculations of the evolution of the earth's atmosphere (see Chapter 8). He finds that a most delicate balance must be maintained to keep temperatures in a moderate range. Hart notes that if the earth had an orbit of 0.95 AU, the greenhouse effect would run away and turn us into a Venus. On the other hand, if the earth orbited at 1.01 AU from the sun, glaciation would have iced up the earth 1.7 billion years ago. Our planet would never have gotten warm enough to foster the evolution of life. The key point is this: Early conditions on the earth may have been so special, balanced between freezing and steaming, that the chances of a similar balance elsewhere in the Galaxy is very small. If so, we may well be alone.

22.10 Searching for extraterrestrial intelligence

Let's assume that our neighbors are either very close—tens of light years—or close but not too far—hundreds of light years. How to reach or find civilizations tens to hundreds of light years away? One method stands out: radio communication.

Physically searching for life elsewhere is unrealistic, at least for now. We can't easily and cheaply make spacecraft travel at speeds close to *c*. But we do have a messenger that does go precisely at the speed of light, not slower: light itself. Photons are cheap, fast, and available. Round-trip communication to the stars within a few hundred light years takes a few hundred years.

Light encompasses a broad energy spectrum, from radio to gamma-rays. What range is best? If we require that the ideal range must have a low cost (so we can send many photons), not be absorbed readily by the interstellar medium or planetary atmospheres, be easy to collect from a large area, be easy to transmit and detect, and be expected from advanced civilizations, radio fits the bill the best.

Radio astronomers have constructed telescopes specially designed to detect very weak radio signals. These instruments can also be used to transmit, as is done in radar astronomy. So not only is radio the best range of wavelengths, but also the basic technology is available now both to send and to receive. With other civilizations too, radio will probably be the first form of technology to be developed, because it's the easiest.

The microwave band / Radio covers a wide band of the electromagnetic spectrum. What range of frequencies is best? The choice hinges on what part of the radio spectrum has the least background of natural noise, because we will be trying to detect weak signals. (Noise in this context means, for example, the incessant static you hear when you tune your AM radio to a spot between stations.)

Astronomy and physics, it turns out, naturally define a low-noise band (Fig. 22.10). At the low-frequency end (0.1 to 1 GHz), noise from the Galaxy (mostly synchrotron emission from high-speed electrons) dominates. At the high-frequency end (100 to 1000 GHz), noise in radio receivers, which comes from the quantum nature of matter and so cannot be eliminated, picks up. Between these two noise hills lies a valley of relative quiet

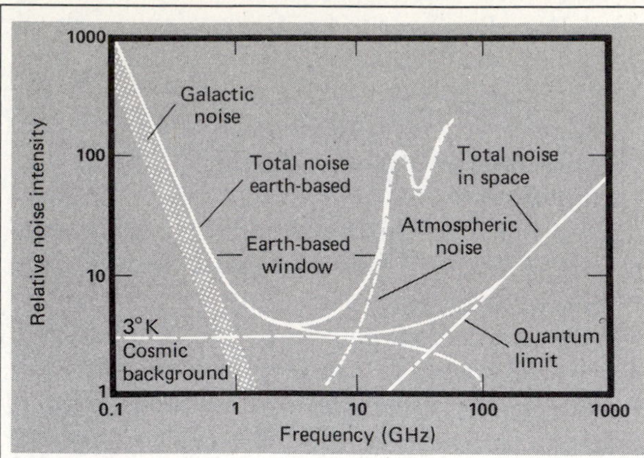

22.10
Natural background noise in the microwave region of the spectrum. At frequencies less than 1 GHz, most of the noise comes from synchrotron emission in the Galaxy (labeled "galactic noise"). At frequencies greater than 100 GHz, noise arises from the atomic nature of the materials that make up radio receivers ("quantum limit"). (Adapted from the Project Cyclops report)

(filled in a bit by the 3 K cosmic radiation; see Section 21.4) from 1 to 100 GHz, part of the *microwave* region of the radio spectrum. The earth's atmosphere fills in a bit more of this microwave noise valley at frequencies greater than 10 GHz.

Any receiver on a planet like the earth would have the same low-noise window available—1 to 10 GHz. A space receiver could have a wider window—1 to 100 GHz.

Search strategies / How to begin with the instruments at hand? First, we must decide if we are to spend our limited time listening or sending. The answer seems to be *listen*. Here's the argument: If technological civilizations live at least a thousand years, we are mere infants. Most of them will be technologically superior to us, with better facilities for sending and receiving. Since sending is technically more difficult than receiving, we use our resources most efficiently if we listen. And if we listen, we might learn something immediately, whereas if we send, it may be a long time before a reply is received (and we would have to listen for that in any case). So it is better at first to receive and listen for evidence of others. (We hope that not everybody out there is just listening.) Later, when technology improves, we can start sending.

Next, we need to decide on the search mode: Do we make a target search for specific objects or scan the whole sky hunting for sources? Remember that a radio telescope only sees a small fraction of the sky at a time—roughly 10^{-8} of the entire sky. Changing a telescope's sky position once a second, it would require only a few years to cover the sky completely, but that gives little time to listen at each position. That's an all-sky search mode.

In contrast, a target search mode could pinpoint nearby sunlike stars. Roughly, we have 10^3 such stars out to 100 ly and 10^6 to 1000 ly. These are the stars we expect may have radio emissions from technical civilizations on planets around them and so increase our chance for success.

How to effectively search the microwave band? Remember, we do not know the exact frequencies of our hypothetical transmitters. And even if they did send at, say 21.11 cm, the signals would probably be Doppler-shifted by planetary rotation and revolution—of their planet and ours. So when we search the radio band, we need to do so in a small interval of wavelength or frequency at a time. But we are stuck with a wide band to scan, a little piece at a time. Most radio telescopes can now operate a few channels at a time; a few can search 1000 channels simultaneously. What is needed is a receiver that can operate a few million channels simultaneously. Recent advances in electronics make such a receiver seem possible, but none has yet been constructed.

Of course, if we have only one antenna operating, even full time, the search will take quite a while. Recognizing this essential limitation, NASA helped to sponsor Project Cyclops, a study of the optimal feasible design for a system to detect extraterrestial signals, using state-of-the-art techniques. The result was truly a Cyclops: a circular array, 16 km in diameter, containing some thousands of 100-m radio antennas 300 m apart (Fig. 22.11). The antennas would be coordinated by and feed information to a central computer. Its receiver could handle about a million channels at a time. To cover the entire sky, two Cyclops systems would have to be built, one for each hemisphere.

What are the capabilities of such a monster system? Suppose we aimed it at the two nearest stars that might have a planetary system, Tau Ceti and Epsilon Eridani. Cyclops could detect a radio transmitter emitting at only 500 W! (A typical radio station broadcasts at more than 10,000 W.) For a search of *F–K*-stars within 100 ly, Cyclops would take about a month; for those within 1000 ly, about 50 years.

Suppose the Cyclops search fails to detect extraterrestrial intelligence? What would we have learned? First, we would have a better idea of the limits on N_{ic} and our place in cosmic evolution. Second, the search would generate a tremendous amount of radio information of astronomical

22.11
An artist's conception of the Cyclops array, 16 km in diameter, viewed from the air. (Courtesy NASA)

impact. Should we do it? I think so. To answer—to at least *try* to answer—the question, Are we alone? is a worthy effort for humankind. And so far, we have no firm results for any extraterrestrial signals, despite searches carried out since 1960.

22.11 The future of humankind

Before the sun dies, if we are still alive, we probably will have left our home planet—perhaps for others, perhaps not for planets at all. Predicting the future is always a dubious enterprise, but I think that we will have to leave the earth—long before the sun dies.

Why? Because Spaceship Earth is a finite resource. If we do not achieve a stable population for the human race—for moral, ethical, or political reasons—we have no choice but to leave.

Growth / The problem here is one of growth, constant growth that looks small but adds up quickly. Here's an example to show how constant growth rapidly gets out of hand. In 1975 the world contained about 4 billion people. Suppose we start out with one person. In 10 sec, imagine that that person doubles into 2; 10 sec later, these two split into four; and so on, with a constant doubling time of 10 sec. How long would it take for one person to be doubled to 4×10^9 people? Roughly 320 sec, or *5½ min*!

Steady increases lead to rapid increases in numbers, even for small fractional growth rates. In 1975 the growth rate of the human population was 1.9 percent a year. So the doubling time is roughly 37 years. That doesn't sound like much, but *if* the rate remained constant, the mass of people would equal the mass of the earth in only 1600 years!

Constant increases result in the rapid use of finite resources. Consider bacteria that double every minute. Suppose we place a bacterium in a bottle and note that the bottle is full at 12 noon. When was the bottle half full? One doubling time (one minute) earlier: 11:59. When was it one-quarter full? Two doubling times earlier: 11:58. Suppose at this time some farsighted bacteria leaders got together and intensively searched for more living space. At 11:59, they find an empty bottle on the shelf, which doubles their total living room. When will that new bottle be filled? *12:01.* When consumption grows steadily, even enormous increases in resources are consumed in short times.

The earth is a small planet. Even if zero population growth were achieved tomorrow, our resources would be consumed at present rates in only a few human lifetimes. We will need to leave our home planet: for space, for energy, and for natural resources.

Space colonization / When we leave the earth, where will we go? The traditional science fiction view had us journeying to and colonizing other planets in our solar system. With the possible exception of Mars, we now realize that the other worlds in the solar system are not habitable planets for us. Even if they were, don't forget the lesson of the bacteria in the bottle. Human population growth now doubles roughly every 37 years. Suppose we fill the earth to the limit. How long would it take us to fill another planet, say Mars? Right—*one* doubling time, 37 years or so, if our present growth rate continues. In only 1500 years, we would have enough people to populate 10^{11} planets, one for every star in the Galaxy!

But there's no good reason to restrict ourselves to living on planets at all. This fact has encouraged Gerard K. O'Neill to revive and develop an older idea (some aspects were foreseen by the Russian physicist Konstantin Tsiolkowsky almost 100 years ago): human habitation in space, often known as space colonies. O'Neill has aimed at making the dream of space colonization a reality with available technology.

I won't detail his plans here, but let me sketch the broad outlines. Stripped of luxuries, people need energy, air (oxygen), water, land, and (probably) gravity to live a comfortable life. With space colonies in orbit around the earth, somewhere between the earth and the moon, all these are available: energy from the sun, oxygen and raw materials from the moon, and water from the earth. (Water might also be collected from the asteroids or moons of Jupiter and Saturn.) What about gravity? It can be simulated by rotating the space colony.

The first space colonies and solar satellite stations would be built with resources from the earth. Expansion to many colonies requires cutting Mother Earth's umbilical cord. For raw resources for development, we can turn to the moon and mine it. The moon's surface, as we know from the Apollo missions, contains abundant aluminum, titanium, oxygen, and silicon. (It lacks water, so we'd have to bring hydrogen from the earth to combine with lunar oxygen to make water.) A solar power station can support mining activities. Then a reliable, inexpensive way is needed to transport the lunar materials to the colonies' orbit. O'Neill and others envision the use of a magnetic linear accelerator that would boost payloads to lunar escape velocity with accurate aim to the space colonies' vicinity. Here they would be caught in a huge net and then used for manufacturing. It is critical to this concept that the space colonies break economically from the earth. They could bankroll themselves by selling power (beamed by microwaves) to the earth.

What might a space colony look like (Fig. 22.12)? The simplest design is a cylinder. One some 3 km in diameter and 32 km long could support upward of 200,000 persons on its inner surface. Spinning once every 2 min to simulate gravity, the inside would alternate strips of land and windows. The windows would have shutters to simulate the seasons and day and night by controlling the influx of sunlight. The colony craft would be constructed of aluminum and titanium from the moon.

How long would this high frontier accommodate our population growth? At present growth rates, the local space could handle 400 to 500 years of population doubling. That may appear long, but it's only about seven human lifetimes. And continued population growth will eventually fill up the local space. For example, imagine that we fill all the space between the sun and the earth with people jammed together. How long will it be until this region is saturated? About 3000 years, at present growth rates.

The moral: The human race will leave the solar system a few thousand years from now. Unless we change our ways.

Beyond the solar system / Space colonies—"cities in space," as science fiction writer James Blish called them—could convey people between stars at speeds of 0.01 c. Imagine that we send colonies out across the Galaxy. When each arrives at a suitable star, it could build a new colony ship from local resources and send it out to the next star. A few centuries might elapse from arrival to the next departure, and another few centuries to glide to the next star. How long does it take to travel across the Galaxy? Take 100,000 ly as the diameter of the Galaxy. At 1/100 the speed of light, the time in years is 100 times the distance in light years, or 10 million years. This star-hopping process could carry people across the Galaxy and colonize it in roughly 10 million years.

Ten million years is short compared with the age of the Galaxy and the time needed for life on the earth to evolve intelligence. So we could fill the Galaxy quickly, in cosmic terms. And so could any civilization that has achieved a technological level similar to ours.

This view brings up again the question of the number of technological civilizations in the Galaxy. If colonization can happen so swiftly, and if L_{ic} is long, then it probably has happened, and "they" are everywhere. But if we take seriously the evidence of "their" absence, L_{ic} must be very short, at least less than 10^7 years. But suppose we ignore that argument. If L_{ic} *is* 10^7 or so, some 10^3 waves of colonization have rippled through the Galaxy since its formation. So, once again, where are they?

The future of technological civilizations / The Soviet radio astronomer N. S. Kardashev has classified technological civilizations by the amount of energy they control. Kardashev notes that in terrestrial history, technological advances have been made possible by an increased energy budget per person. That may also apply to the evolution of other civilizations.

In Kardashev's classification, Type I civilizations have harnessed the energy resources of their planet. Our civilization now uses some 10^{13} W. The sun provides us with about 10^{17} W. When we have mastered that amount, we will become a Type I civilization. That day will come, perhaps, with the advent of space colonies.

Type II civilizations can utilize a substantial amount of the total energy from their parent sun—about 10^{26} W. It's likely that this stage must be achieved before the civilization can leave their home solar system.

Type III civilizations are interstellar communities with the capability of controlling the stellar output of an entire galaxy—some 10^{37} W.

Within 100 years, we may achieve Type I status. Within a few thousand years, we can reach Type II. How? If we mine the moon and the asteroids, we could build a huge shell of space colonies—1 AU in radius—to catch most sunlight. A small part of this energy—1 percent or less—might be used for interstellar exploration. That could take the form of space colonies sent out on long interstellar trips. Not tied down to live on a planetary surface, they could orbit at a suitable distance from almost any good sun. The sun provides the

22.12

(a) *One possible model of a space colony, as suggested by G. K. O'Neill. The colony consists of twin cylinders, each 32 km long and 6.4 km in diameter and holding some 200,000 people. Windows and mirrors let in and control the sunlight to simulate night and day and the seasons.* (b) *Interior view of the colony. (Both courtesy NASA)*

energy, debris around it the material to produce more space colonies. Once these have achieved Type II status, they send off explorers to other stars. So we could slowly colonize the Galaxy and gain Type III status.

What next? By then a human lifetime may be much longer than now—perhaps 10 times as long. And we will have joined up with other galactic civilizations—if we find any—to make a galactic community with a common wisdom greater than we command now. Our populations will grow. We might decide to move out to other galaxies. This step is a huge one: The nearest spiral galaxy, Andromeda, is at a distance some 20 times the diameter of the Milky Way Galaxy. At near-light speeds, intergalactic ships could shelter slowly aging beings and make intergalactic trips in millions of years. Once we arrive, we might find others who, like us, have expanded to fill their galaxy. We would join up with them and others to create an intergalactic community.

This community would be a Type IV civilization—one that could manipulate the energy of clusters of galaxies, a sizable fraction of the total free radiative energy in the universe. It would take billions of years to develop.

Does the evolution of life, intelligence, and civilization end there? Perhaps not. Cosmic evolution shows us continual change. A Type IV civilization cannot remain static. I leave it to you to imagine what it might do. But I will predict that the real future will likely be wilder than you or I can imagine.

KEY CONCEPTS

1 ***Life as we know it consists of organisms that reproduce, may mutate (changes in their genetic structure), and reproduce those mutations.***
2 ***Terrestrial organisms contain proteins (the material of their construction) and nucleic acids (the blueprint of their construction).***
3 ***Life on earth has a chemical composition that (except for helium) reflects the chemistry of the cosmos.***
4 ***Natural selection shapes mutations and so drives biological evolution.***
5 ***Mutations (changes in the structure of DNA) arise from radioactivity, chemicals, and high temperatures.***
6 ***Fossils show that simple organisms existed on the earth at least 3 billion years ago and that significant evolution took billions of years more.***
7 ***Life arose (slowly) from nonlife; physical and chemical evolution must have preceded biological evolution.***
8 ***Lab experiments adding energy (sparks, ultraviolet light) to simple compounds in gaseous form (carbon dioxide, water, methane, ammonia, and so on) result in the natural synthesis of amino acids, most of which are common in life.***
9 ***Some meteorites (those rich in carbon and water) contain amino acids made by nonbiological, chemical processes.***
10 ***Interstellar molecular clouds contain simple organic molecules.***
11 ***No life has been found on Mars; it is doubtful that life exists anywhere else in the solar system.***
12 ***We can roughly estimate the number of technologically advanced civilizations in the Galaxy now by the Drake equation; that number could range from 1 to 100 billion (most likely value is 1 million); the greatest uncertainty arises from our lack of knowledge of the lifetimes of such civilizations.***
13 ***Some nearby stars may have dark companions; some of these may be planetary systems.***
14 ***The number of technologically advanced civilizations now sets the average distance between them; if that number is 1 million, the average distance is a few hundred light years.***
15 ***Light, in the form of radio microwaves (especially in the range from 1 to 10 GHz, where the background noise is least), is the most practical interstellar messenger.***
16 ***The best strategy for communication is to listen; all searches carried out to date have failed to detect extraterrestrial signals.***
17 ***Unrestrained population growth will force the human race to leave Spaceship Earth and journey to the stars.***

STUDY EXERCISES

1 Life on earth centers on carbon. Where did the carbon come from, and how did it get to the earth? (*Objectives 1, 2, and 3*)
2 How do supernovas play a crucial role in the origin of LAWKI? (*Objectives 2 and 3*)
3 What are the chances for LAWKI in the solar system now? (*Objectives 6 and 7*)
4 *O*-stars have the largest ecospheres around them, yet they are not good suns for fostering life. Why? (*Objective 8*)
5 Criticize the book's estimate of the number of intelligent civilizations in the Galaxy and come up with your own. (*Objective 9*)
6 Suppose radio astronomers announced tomorrow the discovery of amino acids in interstellar molecular clouds. How would that affect the general ideas of this chapter? (*Objectives 1 through 12*)

BEYOND THIS BOOK . . .

The Cosmic Connection by Carl Sagan (Dell, New York, 1973) is an expansive view of life in the universe.

The classic is *Intelligent Life in the Universe* by I. S. Shklovsky and C. Sagan (Holden-Day, San Francisco, 1966).

For a detailed account of the Viking lander biology results, see "The Search for Life on Mars" by N. Horowitz in *Scientific American*, October 1977, p. 52.

The September 1978 issue of *Scientific American* deals with biological evolution. Articles pertinent to this chapter are "Chemical Evolution and the Origin of Life" by R. Dickerson and "The Evolution of the Earliest Cells" by W. Schopf. See also "The Biosphere" by P. Cloud in the September 1983 issue.

The High Frontier (Bantam, New York, 1977) by G. K. O'Neill and *Colonies in Space* (Warner, New York, 1977) by T. A. Heppenheimer discuss many aspects of space colonies.

Carl Sagan presents speculations about the past and future of the human race in *The Garden of Eden* (Random House, New York, 1977).

Epilogue part 4

Assume that life as we know it is typical in the sense that it has arisen naturally in the course of cosmic evolution. Then we can hope to find the trail of that evolution, which falls into three interconnected stages: physical, chemical, and biological evolution.

We can hunt down the traces of biological evolution on the earth. Fossil evidence implies that with time, life has grown more complex on the earth and that the origin of life took a relatively long time—about a billion years to get to the first cell.

To search for evidence of the chemical evolution that preceded the biological evolution is hopeless. So we rely on general theoretical ideas and some crucial laboratory experiments. These show that if the primitive atmosphere of the earth lacked free oxygen compounds, *the addition of energy (solar ultraviolet, lightning) naturally resulted in the formation of basic organic materials, such as amino acids. The discovery of complex molecules in space and amino acids in meteorites supports the idea that such molecular formation is not a freak accident.*

Astronomical ideas underpin our understanding of physical evolution. For life as we know it, we need a planet (the earth), a star (the sun), and the proper elements (hydrogen, carbon, nitrogen, oxygen, and some others). Where did these come from? The earth, from the dust of the interstellar medium; the sun, from the gases—both dust and gas are the material lost by earlier stars, mostly in their violent ends. These explosions and normal fusion reactions in stars manufactured the chemical elements, except for hydrogen and helium, which were made in the first few minutes of the Big Bang.

This sequence in general seems appropriate for all the observable universe. So if LAWKI is typical, it must be common. We believe that statement is true. If so, our Galaxy and the entire universe teem with life.

Astronomy teaches that we are creatures of the universe, children of the stars. We are products of cosmic evolution. We are also part of its process. Perhaps we serve to make the universe aware of itself. When you and I look into space, we see the source of ourselves. And to those wide-open spaces we add hope, fear, imagination, and love.

I have tried in this last part to show you the cosmic connections that touch us all. Reading this book, you have touched me. Years from now, I hope that you'll remember that when you touch your sister, brother, parent, lover, friend—you touch the stars.

appendix A / Units

Powers of ten

Astronomers deal with quantities ranging from the truly microcosmic to the macrocosmic. It's very inconvenient to always have to write out the age of the universe as 15,000,000,000 years or the distance to the sun as 149,600,000,000 meters. To save writing time, powers-of-ten notation is used instead. For example, $10 = 10^1$; the exponent tells you how many times to multiply by 10. As another example, $10^{-2} = 1/100$; in this case the exponent is negative, so it tells you how many times to *divide* by 10. The only trick is to remember that $10^0 = 1$. Using powers-of-ten notation, the age of the universe is 15×10^{10} years, and the distance to the sun is 1.496×10^{11} meters.

The English and metric systems

You probably are familiar with the fundamental units of length, mass, and time in the English system: the yard, the pound, and the second. The other common units of the English system are often strange multiples of these fundamental units, such as the ton (2000 lb), the mile (1760 yd), the inch ($\frac{1}{36}$ yd), and the ounce ($\frac{1}{16}$ lb). Most of these units arose from accidental conventions and so have few logical relationships. Most of the world uses a much more rational system known as the metric system, with the following fundamental units:

Length:
1 meter (m)

Mass:
1 kilogram (kg)

Time:
1 second (sec *or* s)

This is the meter-kilogram-second, or mks, system. A slightly older system often used by astronomers is the centimeter-gram-second, or cgs, system, with the following fundamental units:

Length:
1 centimeter (cm)

Mass:
1 gram (g)

Time:
1 second (sec *or* s)

All of the unit relationships in the metric system are based on multiples of 10, so it is very easy to multiply, divide, and use powers-of-ten notation.

The contemporary standard for the meter uses the wavelength of orange light from krypton-86. The meter is defined as 1.65076373×10^6 times this standard wavelength. Any efficient laboratory can set up such a standard and use it accurately.

The multiples of the metric system and their associated prefixes are:

$$\frac{1}{1{,}000{,}000} = 10^{-6} = \text{micro-} \quad (\mu)$$
$$\frac{1}{1{,}000} = 10^{-3} = \text{milli-} \quad (\text{m})$$
$$\frac{1}{100} = 10^{-2} = \text{centi-} \quad (\text{c})$$
$$\frac{1}{10} = 10^{-1} = \text{deci-} \quad (\text{d})$$
$$10 = 10^{1} = \text{deca-} \quad (\text{da})$$
$$100 = 10^{2} = \text{hecto-} \quad (\text{h})$$
$$1{,}000 = 10^{3} = \text{kilo-} \quad (\text{k})$$
$$1{,}000{,}000 = 10^{6} = \text{mega-} \quad (\text{M})$$
$$100{,}000{,}000 = 10^{9} = \text{giga-} \quad (\text{G})$$

Some relationships between the metric and English system are:

Length:
1 kilometer (km) = 1000 m = 0.6214 mile
1 meter (m) = 1.094 y = 39.37 inches
1 centimeter (cm) = 0.01 m = 0.3937 inch
1 millimeter (mm) = 0.001 m = 0.03937 inch
1 mile = 1.6093 km
1 inch = 2.5400 cm

Mass:
1 metric ton = 10^6 g = 1000 kg = 2.2046×10^3 lb
1 kilogram (kg) = 10^3 g = 2.2046 lb
1 gram (g) = 0.0353 oz = 2.2046×10^{-3} lb
1 milligram (mg) = 0.001 g = 2.2046×10^{-6} lb
1 lb = 453.6 g
1 oz = 28.3495 g

Temperature scales / Scales of temperature measurement are tagged by the freezing point and boiling point of water. In the United States, the Fahrenheit (F) system is the one commonly used; water freezes at 32° F and boils at 212° F. In Europe, the Celsius (formerly centigrade) system is the common temperature system; water freezes at 0° C and boils at 100° C. The Kelvin system is

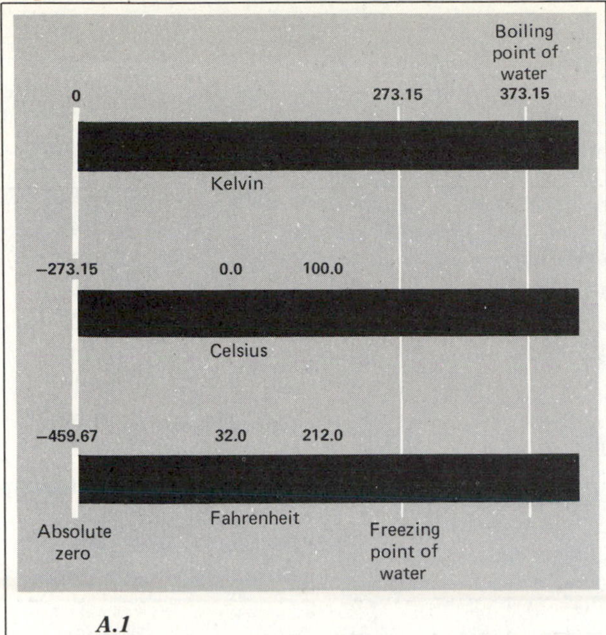

A.1
A comparison of the Kelvin, Celsius, and Fahrenheit temperature scales.

based on the idea of absolute zero, the temperature at which all random molecular motion ceases. Since 0 K is absolute zero, water freezes at 273 K and boils at 373 K. Note that the degree mark is never used with Kelvin temperatures and that the size of the degree is the same in both the Kelvin and Celsius systems (100 between the freezing and boiling points of water). To convert between the systems, recognize that 0 K $= -273°$ C $= -459°$ F and that the Celsius and Kelvin degrees are larger than Fahrenheit degrees by the factor $180/100 = 9/5$. The relationships between systems are:

$$K = C + 273$$

$$C = \frac{5}{9}(°F - 32)$$

A comparison of the three temperature scales is seen in Fig. A.1.

Astronomical distances

Although astronomers do use the metric system, they encounter distances so large that other measures are often used. In the solar system, the natural distance is the astronomical unit (AU), the average distance of the earth from the sun. The AU equals 1.496×10^8 km.

Beyond the solar system, even the astronomical unit is too small to be convenient. Astronomers then use the light year or the parsec. The light year (ly) is the distance that light travels in one year. It equals 9.46×10^{12} km. A parsec (pc) equals 206,265 AU, or 3.09×10^{13} km, or 3.26 ly.

Beyond the Galaxy, astronomers often talk in multiples of parsecs or light years. A thousand parsecs is called a kiloparsec (kpc), and a million parsecs is termed a megaparsec (Mpc). A unit not commonly used by astronomers but sometimes found in this book is the *megalight year* (Mly). Based on the discussion of metric units at the beginning of this appendix, what would you guess the Mly to be equivalent to? Right: A Mly is a million light years.

Other physical units

Another important unit you will encounter is the speed of light (c), which equals 2.9979×10^5 km/sec.

The unit of energy in the cgs system is called the erg. A mass of 2 g traveling at 1 cm/sec has an energy of 1 erg. To illustrate how small an erg is: If you place one foot up one stair, you have expended about a billion ergs. In the mks system, the energy unit is a joule (J), which equals 10^7 ergs. It takes about a joule of energy to lift an apple from the floor to a table. Table A.1 compares the energy outputs of some astronomical and familiar objects.

Power is the amount of energy coming from an object per second, so it is measured in ergs per second. (Astronomers use the word *luminosity* to describe what most physicists would call power.) A convenient and familiar unit for power (or luminosity) is the watt (W), defined as 10^7 ergs/sec, or 1 J/sec.

This book uses the gauss (G) as the unit of magnetic-field strength (rather than the SI unit tesla, which equals 10^4 G). To give you a feel for a gauss, keep in mind that the earth's magnetic field is about 0.5 G.

TABLE A.1 Comparative energy outputs

Energy source	*Energy (J)*
Big Bang	10^{68}
Radio galaxy	10^{55}
Supernova	10^{43}
Sun's radiation (1yr)	10^{34}
Volcanic explosion	10^{19}
H-bomb	10^{17}
Thunderstorm	10^{15}
Lightning flash	10^{10}
Baseball pitch	10^{2}
Hitting typewriter key	10^{-2}
Hopping flea	10^{-7}

appendix B / Planetary data

TABLE B.1 Planetary rotation rates and inclinations of rotation axes

Planet	*Rotation period (equatorial)*	*Inclination of equator to orbital plane*	*Method of measurement*
Mercury	58.65 days	7°	Radar Doppler shift
Venus	243.01 days (retrograde)	178°	Radar Doppler shift
Earth	23 hr 56 min 4.1 sec	23° 27′	Star transits
Mars	24 hr 37 min 22.6 sec	23° 59′	Optical features
Jupiter	9 hr 50.5 min	3° 05′	Optical features
Saturn	10 hr 14 min	26° 44′	Optical Doppler shift
Uranus	12–23 hr	97° 55′	Optical Doppler shift
Neptune	18–22 hr	28° 48′	Optical Doppler shift
Pluto	6.39 days	65°	Optical light variations

TABLE B.2 Distances, periods, and orbital velocities of the planets

	*Semimajor axis of orbit**		*Sidereal period*		
	(AU)	*(10^6 km)*	*(years**)*	*(days)*	*Orbital eccentricity*
Mercury	0.387	57.9	0.240	87.97	0.206
Venus	0.723	108.2	0.615	224.7	0.007
Earth	1.000	149.6	1.000	365.26	0.017
Mars	0.523	227.9	1.881	687.0	0.093
Jupiter	5.203	778.3	11.86	4,333	0.049
Saturn	9.540	1427	29.46	10,759	0.056
Uranus	19.18	2869	84.10	30,685	0.047
Neptune	30.07	4498	164.8	60,188	0.009
Pluto	39.44	5900	248.4	90,700	0.250

* Same as average distance from sun.
** Tropical years, that is, the year of seasons.

TABLE B.3 Dimensions of the planets

	Diameter		
Planet	*(km)*	*(earth diameters)*	*Method of measurement*
Mercury	4,878	0.38	Radar and optical
Venus	12,104	0.95	Optical radar
Earth	12,576	1.00	Satellite
Mars	6,787	0.53	Optical
Jupiter	143,800	10.8	Optical
Saturn	120,660	8.9	Optical
Uranus	52,290	4.0	Optical
Neptune	49,500	3.9	Optical
Pluto	3,000(?)	0.24(?)	Speckle interferometry

TABLE B.4 Masses of the planets

	Mass		
Planet	*(earth masses)*	*(kg)*	*Bulk density (kg/m^3)*
Mercury	0.05533	3.32×10^{23}	5420
Venus	0.815	4.87×10^{24}	5250
Earth	1.000	5.98×10^{24}	5520
(Moon)	0.012	7.35×10^{22}	3340
Mars	0.1074	6.42×10^{23}	3940
Jupiter	317.9	1.9×10^{27}	1314
Saturn	95.1	5.69×10^{26}	690
Uranus	14.56	8.69×10^{25}	1190
Neptune	17.24	1.03×10^{26}	1660
Pluto	0.0018	1.1×10^{22}	$\sim$ 800

TABLE B.5 Atmospheric gases

Planet	Gas
Mercury	Helium, hydrogen
Venus	Carbon dioxide, carbon monoxide, hydrogen chloride, hydrogen fluoride, water, argon, nitrogen, oxygen, hydrogen sulfide, sulfur dioxide, helium
Earth	Nitrogen, oxygen, water, argon, carbon dioxide, neon, helium, methane, krypton, nitrous oxide, ozone, xenon, hydrogen, radon
Mars	Carbon dioxide, carbon monoxide, water, oxygen, ozone, argon, nitrogen
Jupiter	Hydrogen, helium, methane, ammonia, water, carbon monoxide, acetylene, ethane, phosphine, germane
Saturn	Hydrogen, helium, methane, ammonia, acetylene, ethane, phosphine, propane
Titan	Nitrogen, methane, ethane, acetylene, ethylene, hydrogen cyanide
Uranus	Hydrogen, methane
Neptune	Hydrogen, methane, ethane
Pluto	Methane

TABLE B.6 Satellites of Mars

Satellite	Distance from center of planet (10^3 km)	Sidereal period of revolution (days)	Radius of satellite (km)	Mass (planet = 1)	Bulk density (kg/m^3)
Phobos	9.38	0.3189	$14 \times 11 \times 9$	1.5×10^{-8}	1900
Deimos	23.50	1.262	$8 \times 6 \times 5$	3.1×10^{-9}	2100

TABLE B.7 Satellites of Jupiter

Satellite		Distance from Jupiter					
Name	Number	(10^3 km)	(Jupiter radii)	Orbital period (days)	Radius of satellite (km)	Mass (planet = 1)	Bulk density (kg/m^3)
1979 J3	J16	128	1.79	0.29	20	—	—
1979 J1	J14	129	1.80	0.30	20	—	—
Almathea	J5	181	2.55	0.49	130×80	2×10^{-9}	3000
1979 J2	J15	222	3.11	0.68	40	—	—
Io	J1	422	5.95	1.77	1820	4.7×10^{-5}	3530
Europa	J2	671	9.47	3.55	1500	2.6×10^{-5}	3030
Ganymede	J3	1,070	15.10	7.15	2640	7.8×10^{-5}	1930
Callisto	J4	1,880	26.60	16.70	2500	5.6×10^{-5}	1790
Leda	J13	11,110	156	240	~ 4	5×10^{-13}	—
Himalia	J6	11,470	161	251	85	8.5×10^{-10}	1000
Lysithea	J10	11,710	164	260	~ 10	1×10^{-12}	—
Elara	J7	11,740	165	260	~ 30	4×10^{-11}	—
Ananke	J12	20,700	291	617	8(?)	7×10^{-13}	—
Carme	J11	22,350	314	692	12(?)	2×10^{-12}	—
Pasiphae	J8	23,300	327	735	14(?)	8×10^{-12}	—
Sinope	J9	23,700	333	758	10(?)	2×10^{-12}	—

TABLE B.8 Satellites of Saturn

Satellite	Distance from Saturn (10^3 km)	Distance from Saturn (Saturn radii)	Orbital period (days)	Radius of satellite (km)	Mass (planet = 1)	Bulk density (kg/m^3)
Atlas	137.67	2.28	0.602	10×20	—	—
1980 S27	139.35	2.31	0.613	$70 \times 50 \times 40$	—	—
1980 S26	141.70	2.35	0.629	$55 \times 45 \times 35$	—	—
Epimetheus	151.42	2.51	0.694	$70 \times 60 \times 50$	—	—
Janus	151.47	2.51	0.695	$110 \times 100 \times 80$	—	—
Mimas	185.54	3.08	0.942	195	6.6×10^{-8}	1200
Enceladus	238.04	3.95	1.370	250	1×10^{-7}	1200
Tethys	294.67	4.88	1.888	530	1.3×10^{-6}	1200
Telesto	294.87	4.88	1.888	$17 \times 14 \times 13$	—	—
Calypso	294.87	4.88	1.888	$17 \times 11 \times 11$	—	—
Dione	377.42	6.26	2.737	560	1.85×10^{-6}	1400
1980 S6	377.42	6.26	2.737	$18 \times 16 \times 15$	—	—
Rhea	527.07	8.74	4.518	765	4.4×10^{-6}	1300
Titan	1,221.86	20.25	15.945	2560	2.36×10^{-4}	1880
Hyperion	1,481.00	24.55	21.277	$205 \times 130 \times 110$	—	—
Iapetus	3,560.80	59.02	79.33	730	3.3×10^{-6}	1200
Phoebe	12,954	214.7	550.45	~ 110	—	—

TABLE B.9 Satellites of Uranus

Satellite	Distance from center of planet (10^3 km)	Orbital period (days)	Radius (km)
Ariel	191.8	2.52038	665
Umbriel	267.3	4.14418	555
Titania	438.7	8.70588	800
Oberon	586.6	13.46326	815
Miranda	130.1	1.414	160

TABLE B.10 Satellites of Neptune

Satellite	Distance from center of planet (10^3 km)	Orbital period (days)	Radius (km)
Triton	653.6	5.87683	1700
Nereid	5570	365	470

appendix C / Physical constants, astronomical data

Physical constants

Gravitational constant

$G = 6.673 \times 10^{-11}$ newton-m^2/kg^2

Speed of light in a vacuum

$c = 2.9979 \times 10^8$ m/sec

Planck's constant

$h = 6.62618 \times 10^{-34}$ J-sec

Wein's constant

$\sigma_w = 0.0029$ m-K

Boltzmann's constant

$k = 1.3806 \times 10^{-23}$ J/K

Stefan-Boltzmann constant

$\sigma = 5.6697 \times 10^{-8}$ W/m^2-K

Electron mass

$m_e = 9.10956 \times 10^{-31}$ kg

Proton mass

$m_p = 1.6726 \times 10^{-27}$ kg $= 1836.1\, m_e$

Neutron mass

$m_n = 1.6749 \times 10^{-27}$ kg

Mass of hydrogen atom

$m_H = 1.6735 \times 10^{-27}$ kg

Astronomical data

Astronomical unit

AU $= 1.4959789 \times 10^{11}$ m

Parsec

pc $= 206264.806$ AU
$= 3.2616$ ly
$= 3.0856 \times 10^{16}$ m

Light year

ly $= 9.46053 \times 10^{15}$ m
$= 6.324 \times 10^4$ AU

Sidereal year

y $= 3.155815 \times 10^7$ sec

Mass of sun

$M_\odot = 1.989 \times 10^{30}$ kg

Luminosity of sun

$L_\odot = 3.827 \times 10^{26}$ W

Solar constant

$S = 1370$ W/m^2

Radius of sun

$R_\odot = 6.96 \times 10^5$ km

Mass of earth

$M_\oplus = 5.9742 \times 10^{24}$ kg

Equatorial radius of earth

$R_\oplus = 6.378145 \times 10^3$ km

Mass of moon

$M_M = 7.34 \times 10^{22}$ kg

Radius of moon

$R_M = 1.738 \times 10^3$ km

appendix D / Stars within 12 light years

Name	Parallax (arcsec)	Distance (ly)	Spectral type	Apparent visual magnitude	Luminosity (sun = 1.0)
Sun	—	—	*G*2 V	−26.7	1.0
α Cen A	0.750	4.3	*G*2 V	−0.01	1.6
B			*K*1 V	1.3	0.45
C	0.772	4.2	*M*5.5	11.0	0.00006
Barnard's star	0.552	5.9	*M*3.8 V	9.5	0.00045
Wolf 359	0.431	7.6	*M*5.8	13.5	0.00002
Lalande 21185	0.402	8.1	*M*2.1 V	7.5	0.0055
Luyten 726-8AB	0.1387	8.4	*M*5.6	12.5	0.00006
Sirius A	0.377	8.6	*A*1 V	−1.5	23.5
B			wd*	8.7	0.08
Ross 154	0.345	9.4	*M*3.6	10.6	0.00048
Ross 248	0.314	10.3	*M*4.9	12.3	0.00011
ϵ Eri	0.303	10.7	*K*2 V	3.7	0.30
Luyten 789-6	0.302	10.8	*M*5.5	12.2	0.00014
Ross 128	0.301	10.8	*M*4.1	11.1	0.00036
61 Cyg A	0.292	11.2	*K*3.5 V	5.2	0.083
B			*K*4.7 V	6.0	0.040
ϵ Ind	0.291	11.2	*K*3 V	4.7	0.13
Procyon A	0.287	11.4	*F*5 IV–V	0.4	7.65
B			wd	10.7	0.00055
HD 173739	0.284	11.5	*M*3.0 V	8.9	0.0028
HD 173740			*M*3.5 V	9.7	0.0013
Groombridge 34 A	0.282	11.6	*M*1 V	8.1	0.0058
B			*M*6 V	11.0	0.00040
Lacaille 9352	0.279	11.7	*M*2 V	7.4	0.013
τ Ceti	0.273	11.9	*G*8 V	3.5	0.45

* White dwarf.

appendix E / The 20 brightest stars

Star	Name	Apparent visual magnitude	Spectral type	Absolute magnitude	Distance (ly)
α CMa A	Sirius	−1.46	*A*1 V	+1.42	8.7
α Car	Canopus	−0.72	*F*0 I–II	−3.1	98
α Boo	Arcturus	−0.06	*K*2 III	−0.3	36
α Cen A	Rigil Kentaurus	0.01	*G*2 V	+4.39	4.3
α Lyr	Vega	0.04	*A*0 V	+0.5	26.5
α Aur	Capella	0.05	*G*8 III (?)	−0.6	45
β Ori A	Rigel	0.14	*B*8 Ia	−7.1	900
α CMi A	Procyon	0.37	*F*5 IV–V	+2.7	11.3
α Ori	Betelgeuse	0.41	*M*2 Iab	−5.6	520
α Eri	Achernar	0.51	*B*3 V	−2.3	118
β Cen AB	Hadar	0.63	*B*1 III	−5.2	490
α Aql	Altair	0.77	*A*7 IV–V	+2.2	16.5
α Tau A	Aldebaran	0.86	*K*5 III	−0.7	68
α Vir	Spica	0.91	*B*1 V	−3.3	220
α Sco A	Antares	0.92	*M*1 Ib	−5.1	520
α PsA	Fomalhaut	1.15	*A*3 V	+2.0	22.6
β Gem	Pollux	1.16	*K*0 III	+1.0	35
α Cyg	Deneb	1.26	*A*2 Ia	−7.1	1600
β Cru	Beta Crucis	1.28	*B*0.5 III	−4.6	490
α Leo A	Regulus	1.36	*B*7 V	−0.7	87

appendix F/ Periodic table of the elements

Key: 11 (Atomic number) / Sodium (Name) / Na (Symbol) / 23 (Approximate atomic weight (to nearest whole number))

Period	Group IA	IIA	IIIB	IVB	VB	VIB	VIIB	VIII	VIII	VIII	IB	IIB	IIIA	IVA	VA	VIA	VIIA	VIII
1	1 Hydrogen **H** 1																	2 Helium **He** 4
2	3 Lithium **Li** 7	4 Beryllium **Be** 9											5 Boron **B** 11	6 Carbon **C** 12	7 Nitrogen **N** 14	8 Oxygen **O** 16	9 Fluorine **F** 19	10 Neon **Ne** 20
3	11 Sodium **Na** 23	12 Magnesium **Mg** 24											13 Aluminum **Al** 27	14 Silicon **Si** 28	15 Phosphorus **P** 31	16 Sulfur **S** 32	17 Chlorine **Cl** 35	18 Argon **Ar** 40
4	19 Potassium **K** 39	20 Calcium **Ca** 40	21 Scandium **Sc** 45	22 Titanium **Ti** 48	23 Vanadium **V** 51	24 Chromium **Cr** 52	25 Manganese **Mn** 55	26 Iron **Fe** 56	27 Cobalt **Co** 59	28 Nickel **Ni** 59	29 Copper **Cu** 64	30 Zinc **Zn** 65	31 Gallium **Ga** 70	32 Germanium **Ge** 73	33 Arsenic **As** 75	34 Selenium **Se** 79	35 Bromine **Br** 80	36 Krypton **Kr** 84
5	37 Rubidium **Rb** 85	38 Strontium **Sr** 88	39 Yttrium **Y** 89	40 Zirconium **Zr** 91	41 Niobium **Nb** 93	42 Molybdenum **Mo** 96	43 Technetium **Tc** 99	44 Ruthenium **Ru** 101	45 Rhodium **Rh** 103	46 Palladium **Pd** 106	47 Silver **Ag** 108	48 Cadmium **Cd** 112	49 Indium **In** 115	50 Tin **Sn** 119	51 Antimony **Sb** 122	52 Tellurium **Te** 128	53 Iodine **I** 127	54 Xenon **Xe** 131
6	55 Cesium **Cs** 133	56 Barium **Ba** 137	57 *Lanthanum **La** 139	72 Hafnium **Hf** 178	73 Tantalum **Ta** 181	74 Wolfran (tungsten) **W** 184	75 Rhenium **Re** 186	76 Osmium **Os** 190	77 Iridium **Ir** 192	78 Platinum **Pt** 195	79 Gold **Au** 197	80 Mercury **Hg** 201	81 Thallium **Tl** 204	82 Lead **Pb** 207	83 Bismuth **Bi** 209	84 Polonium **Po** 210	85 Astatine **At** 210	86 Radon **Rn** 222
7	87 Francium **Fr** 223	88 Radium **Ra** 226	89 **Actinium **Ac** 227	104 Kurcha-tovium **Ku**	105 Hahnium **Ha**													

	Period														
*Lanthanide series	6	58 Curium **Ce** 140	59 Prase-odymium **Pr** 141	60 Neodymium **Nd** 144	61 Promethium **Pm** 147	62 Samarium **Sm** 150	63 Europium **Eu** 152	64 Gadolinium **Gd** 157	65 Terbium **Tb**	66 Dysprosium **Dy** 163	67 Holmium **Ho** 165	68 Erbium **Er** 167	69 Thulium **Tm** 169	70 Ytterbium **Yb** 173	71 Lutetium **Lu** 175
**Actinide series	7	90 Thorium **Th** 232	91 Protactinium **Pa** 231	92 Uranium **U** 238	93 Neptunium **Np** 237	94 Plutonium **Pu** 242	95 Americium **Am** 243	96 Cerium **Cm** 247	97 Berkellium **Bk** 247	98 Californium **Cf** 251	99 Einsteinium **Es** 254	100 Fermium **Fm** 253	101 Mendelevium **Md** 256	102 Nobelium **No** 254	103 Lawrencium **Lr** 257

Glossary

absolute magnitude a measure of the brightness a star would have if it were to be placed at a standard distance of 32.6 light years (10 parsecs) from the earth
absorption (dark) lines colors missing in a continuous spectrum because of the absorption of those colors by atoms
absorption-line spectrum dark lines superimposed on a continuous spectrum
acceleration the rate of change of velocity with time
accretion the colliding and sticking together of small particles to make larger masses
active galaxy a galaxy characterized by a nonthermal spectrum and a large energy output
aerobe an organism that is dependent on free oxygen for its metabolism
albedo a measure of an object's reflecting power; the ratio of reflected light to incoming light, where complete reflection gives an albedo of 1
Alpha Centauri the closest star to the sun; it happens to have almost the same luminosity and surface temperature as the sun
amino acids the building blocks of proteins, consisting mostly of carbon, nitrogen, oxygen, and hydrogen atoms
anaerobe an organism that does not depend on free oxygen for its metabolism
angular momentum the tendency for bodies, because of their inertia, to keep spinning or orbiting
angular separation the observed angular distance between two celestial objects, measured in degrees, minutes, and seconds of angular measure
angular diameter the apparent diameter of an object in angular measure; the angular separation of two points on opposite sides of the object
anorthosite a basaltic mineral composed of calcium and sodium with aluminum silicate; the predominant mineral of the lunar highlands
aphelion for a body orbiting the sun, the point on its orbit that is farthest from the sun
apogee the point in its orbit where an earth satellite is farthest from the earth
apparent magnitude the brightness of a star (or any other celestial object) as seen from the earth
asteroid belt the region lying between the orbits of Mars and Jupiter, containing the majority of asteroids
asteroids (minor planets) the several thousand very small members of the solar system that revolve around the sun, generally between the orbits of Mars and Jupiter
astronomical unit (AU) the average distance between the earth and the sun; 149.6 million km or 8.3 light minutes
atmosphere a gaseous envelope surrounding a planet, or the visible layers of a star; also a unit of pressure (abbreviated ***atm***) equal to the pressure of air at sea level on the earth's surface
atmospheric extinction the decrease in light caused by passage through the atmosphere
atom the smallest particle of an element that exhibits the chemical properties of the element
AU abbreviation for *astronomical unit*
axis one of two or more reference lines in a coordinate system; also, the straight line, through the poles, about which a body rotates
Balmer series the set of transitions of electrons in a hydrogen atom between the second energy level and higher levels; also, the set of absorption or emission lines corresponding to these transitions that lie in the visible part of the spectrum
basalt an igneous rock, composed of olivine and feldspar, that makes up much of the earth's lower crust
beta decay a process of radioactive decay in which a neutron disintegrates into a proton, an electron, and a neutrino
Betelgeuse a red supergiant star with a luminosity 8000 times that of the sun
Big Bang model a picture of the evolution of the universe that postulates its origin, in an event called the Big Bang, from a hot, dense state that rapidly expanded to cooler, less dense states
binary stars two stars bound together by gravity that revolve around a common center of mass
blackbody a (hypothetical) perfect radiator of light that absorbs and re-emits all radiation incident upon it; its light output depends only on its temperature
blackbody spectrum the continuous spectrum emitted by a blackbody; the brightness at each wavelength is given by the mathematical formula known as Planck's law
black dwarf the cold remains of a white dwarf after all its thermal energy is exhausted
black hole a mass that has collapsed to such a degree that the escape velocity from its surface is greater than the speed of light, so that light is trapped by the intense gravitational field
blue shift a decrease in the wavelength of the radiation emitted by an approaching celestial body as a consequence of the Doppler effect; a shift toward the short-wavelength (blue) end of the spectrum
bolometric magnitude the magnitude of an object measured over all wavelengths
breccias rock and mineral fragments cemented together; a common part of the lunar surface
bright-line spectrum see *emission-line spectrum*
bright nebula see *diffuse nebula*
canali (Italian, "channels") term used by Giovanni

Schiaparelli to describe dark linear features seen on the surface of Mars
capture theory a model of the moon's origin, proposed about 1955, that pictures the moon as captured by the earth's gravity, after which it spiraled in toward the earth, reversed orbital direction, and spiraled outward
carbon-nitrogen-oxygen (CNO) cycle a series of thermonuclear reactions taking place in a star's core, in which carbon, nitrogen, and oxygen aid the fusion of hydrogen into helium; it is a secondary energy-production process in the sun but the major process in high-mass main-sequence stars
Cassini's division a gap about 2000 km wide in Saturn's rings, discovered in 1675 by Giovanni Cassini; now known to contain many small ringlets
catastrophic theories models for the origin of the solar system in which an improbable event involving a large mass (usually collision with another star) led to the collection of gaseous materials that became the planets
celestial pole an imaginary projection of the earth's pole onto the celestial sphere; a point about which the apparent daily rotation of the stars takes place
celestial sphere an imaginary sphere of very large radius centered on the earth on which celestial bodies appear fastened and against which their motions are charted
center of mass the balance point of a set of interacting or connected bodies
central force a force directed along a line connecting the centers of two objects
centripetal force a force required to divert a body from a straight path into a curved one, directed toward the center of the curve
cepheid variables (cepheids) stars that vary in brightness as a result of a regular variation in size; a class of variable stars for which the star Delta Cephei is the prototype
Ceres the first observed asteroid, discovered by Father Giuseppe Piazzi in 1801
Chandrasekhar limit the maximum amount of mass for a white dwarf star, about 1.4 solar masses; this amount leads to the highest density and smallest radius for a star made of a degenerate electron gas; more than this and the star collapses gravitationally
chondrite a stony meteorite characterized by the presence of small, round silicate granules (chondrules)
chondrules round silicate granules lacking volatile elements; found in chondritic meteorites, or chondrites, they are believed to be primitive solar-system materials
chromosphere the part of the sun's atmosphere just above the photosphere, hotter and less dense than the photosphere; it creates the flash spectrum seen during eclipses
circumpolar stars for an observer north of the equator, those stars that are continually above the northern horizon and never set; for a southern observer, those stars that never set below the southern horizon
closed geometry see *spherical geometry*
CNO cycle see *carbon-nitrogen-oxygen cycle*
color index the difference in the magnitudes of an object measured at two different wavelengths; a measure of the color and hence the temperature of a star
coma the bright, visible head of a comet
comets bodies of small mass that revolve around the sun, usually in highly elliptical orbits, and consist, in the dirty-snowball model, of small, solid particles (probably of rocky material) embedded in frozen gases
compound a substance composed of two or more atoms bound together by chemical forces
condensation the growth of small particles by the sticking together of atoms and molecules
condensation sequence the order in which chemical compounds can form out of a gas cloud at specific densities and temperatures
conjunction the time at which two celestial objects appear closest together in the sky
conservation of energy a fundamental principle in physics that states that the total energy of an isolated system remains constant regardless of whatever internal changes may occur
constellation an apparent arrangement of stars on the celestial sphere, usually named after ancient gods, heroes, animals, or mythological beings; now an agreed-upon region of the sky containing a group of stars
continental drift the theory that the present continents were at one time a joined landmass that fragmented and drifted apart
continuous spectrum a spectrum showing emission at all wavelengths, unbroken by either absorption lines or emission lines
convection the transfer of energy by the moving currents of a fluid
coplanar lying in the same plane
core (of the earth) the central region of the earth; it has a high density, is probably liquid, and is believed to be composed of iron and iron alloys
core (of the sun) the inner 25 percent of the sun's volume, where the temperature is great enough for thermonuclear reactions to take place
corona the outermost region of the sun's atmosphere, consisting of thin, ionized gases at a temperature of about 1 million K
cosmic blackbody microwave radiation radiation with a blackbody spectrum at a temperature of about 3 K permeating the universe; believed to be the remains of the primeval fireball in which the universe was created
cosmic rays charged atomic particles moving in space with very high energies (the particles travel close to the speed of light); most originate beyond the solar system, but some are produced in solar flares
cosmological principle the statement that the universe, averaged over a large enough volume, appears the same from any location
cosmology the study of the nature and evolution of the physical universe
cosmos the universe considered as an orderly and harmonious system
crater a circular depression of any size, usually caused by the impact by a solid body or by a surface eruption

crust the thin, outermost surface layer of a planet; on the earth, it is composed of basaltic and granitic rocks
Cygnus arm a segment of one of the spiral arms of our Galaxy; it lies about 34,000 light years out from the nucleus and contains a small branch, the Orion spur, in which the sun is located
dark-line spectrum see *absorption-line spectrum*
deferent an ancient geometric device used to account for the apparent eastward motion of the planets; a large circle, usually centered on the earth, that carries around a planet's epicycle
degenerate electron gas matter in which nuclei and electrons are packed together as much as possible and in which the electrons are no longer bound to individual atoms
degenerate gas pressure a force exerted by very dense, compacted matter that depends mostly on how dense the matter is and very little on its temperature
degenerate neutron gas matter made up of neutrons packed together as tightly as possible
Deimos the smaller of the two moons of Mars
density the amount of mass per volume in an object or region of space
density-wave theory a model for the generation of spiral structure in galaxies, which pictures density waves (similar to sound waves) plowing through the interstellar matter and sparking star formation
deoxyribonucleic acid (DNA) the basic genetic material of life as we know it; a very large molecule consisting of subunits called nucleotides
diffuse (bright) nebula a cloud of ionized gas, mostly hydrogen, with an emission-line spectrum
dirty-snowball model a model for comets that pictures the nucleus as a compact solid body of frozen materials, mixed with pieces of rocky matter, that turn into gases as a comet nears the sun, creating the head and tail
disk (of a galaxy) the flattened wheel of stars, gas, and dust outside the nucleus of a galaxy
distance (standard) candle an astronomical object of standard luminosity used to estimate distances
D-lines a pair of dark lines in the yellow region of the spectrum, produced by sodium
DNA see *deoxyribonucleic acid*
Doppler effect a change in the wavelength of waves from a source reaching an observer when the source and the observer are moving with respect to each other along the line of sight; the wavelength increases (red shift) or decreases (blue shift) according to whether the motion is away from or toward the observer
dwarf a star of relatively low light output and relatively small size; a main-sequence star of luminosity class V
eccentric an ancient geometric device used to account for nonuniform planetary motion; a point offset from the center of circular motion
eclipse the phenomenon of one body passing in front of another, cutting off its light
eclipsing binary two stars that revolve around a common center of mass, the orbits lying edge on to the line of sight, so that each star periodically passes in front of the other
ecliptic from the earth, the apparent yearly path on the celestial sphere of the sun with respect to the stars; also, the plane of the earth's orbit
ecosphere the region around a star where an orbiting planet's surface temperature is within the range for life as we know it
electromagnetic spectrum the range of all wavelengths of electromagnetic radiation
electron a negatively charged subatomic particle, usually found in orbit around the nucleus of an atom
element a substance that is made of atoms with the same chemical properties and cannot be decomposed chemically into simpler substances
ellipse a plane curve drawn so that the sum of the distances from a point on the curve to two fixed points is constant
elliptical galaxy a gravitationally bound system of stars that has rotational symmetry but no spiral structure and that contains mainly old stars and little gas or dust
elongation the angular separation of an object from the sun
emission (bright) lines light of specific wavelengths or colors emitted by atoms; sharp energy peaks in a spectrum
emission nebula see *diffuse nebula*
empirical derived from experiment or observation
energy level one of the possible states of an atom, with a specific value of energy
enzyme a protein that brings about or accelerates reactions at body temperatures without itself undergoing destruction in the process
ephemeris (pl., ephemerides) a table that gives the positions of celestial objects at various times
epicycle a small circle moving around a larger one (the deferent) used by ancient astronomers, such as Ptolemy, to account for the westward retrograde motion and other irregular motions of the planets
equant an ancient geometric device invented by Ptolemy to account for variations in planetary motion; essentially an eccentric in which the center of the circle is not the center of uniform motion
equation of state a relationship that describes the conditions in a physical system, such as an equation relating the pressure, temperature, and density of a gas
equatorial bulge the excess diameter, about 43 km, of the earth through its equator compared with the diameter through its poles
equilibrium a state of a physical system in which there is no overall change
equinox time of year of equal length of days and nights; the two times of the year when the sun crosses the celestial equator; spring (vernal) equinox occurs about March 21, and fall (autumnal) equinox about September 23
escape velocity the speed a body must achieve to break away from the gravity of another body and never return to it
Euclidean (flat) geometry geometry in which only one parallel line can be drawn through a point near another line; the sum of the angles in a triangle drawn on a flat surface is always 180°

excitation the process of raising an atom to a higher energy level
exosphere the topmost region of a planet's atmosphere, from which particles in the atmosphere can escape into space
expanding arm a segment of spiral-arm structure encircling the center of our Galaxy at a distance of about 1000 ly; it appears to be moving toward us and away from the Galaxy's center
extinction the dimming of light when it passes through some medium, such as the earth's atmosphere or interstellar material
eyepiece a magnifying lens used to view the image produced by the main light-gathering lens or telescope
fission see *nuclear fission*
fission theory the earliest of the major models for the origin of the moon, suggesting that a young, rapidly spinning, molten earth lost a piece that spiraled out into orbit and cooled down to form the moon
flash spectrum the spectrum that appears immediately before the totality of a solar eclipse as the normal absorption spectrum is replaced briefly by the chromosphere's own emission spectrum
fluorescence the process by which a high-energy photon is absorbed by an atom and re-emitted as two or more photons of lower energy
flux the amount of energy flowing through a given area in a given time (usually one second)
focus (pl., foci) the point at which light is gathered in a telescope
frame of reference a set of axes with respect to which the position or motion of something can be described or physical laws can be formulated
Fraunhofer lines the name given to absorption lines in the spectrum of a star, especially the sun
frequency the number of waves that pass a particular point in some time interval (usually a second)
fusion see *nuclear fusion*
galactic (open) cluster a small group, about ten to a few hundred, of gravitationally bound stars of Population I, found in or near the plane of the Galaxy
galactic equator the great circle along the line of the Milky Way, marking the central plane of the Galaxy
galactic latitude the angular distance north or south of the galactic equator
galactic longitude the angular distance along the galactic equator from a zero point in the direction of the galactic center
galactic rotation curve a description of how fast an object some distance from the center of a galaxy revolves around it
galaxy a huge assembly (between a million and hundreds of millions) of stars, gas, and dust that is held together by gravity; ***the Galaxy*** our own galaxy, containing the sun
Galilean moons the four largest satellites of Jupiter (Io, Europa, Ganymede, Callisto), discovered by Galileo with his telescope
gamma-ray a very high energy photon with a wavelength shorter than that of X-rays
general theory of relativity the idea developed by Albert Einstein that mass and energy determine the geometry of spacetime and that any curvature of this spacetime shows itself by what we commonly call gravitational forces
geocentric centered on the earth
geomagnetic axis the axis that connects the earth's magnetic poles; it is inclined about 20° from the geographic spin axis and does not pass through the earth's center
globular cluster a gravitationally bound group of about 10,000 to 100,000 Population II stars (of roughly solar mass), symmetrically shaped, found in the halo of the Galaxy
gnomon an ancient instrument for measuring time; most simply, a stick stuck vertically into the ground whose shadow is used to indicate the sun's position with respect to the horizon
grand unification theory (GUT) a physical theory that attempts to unite the elementary particles and the four forces in nature as the actions of one particle and one force
granule brief-lived (3 to 10 minutes) bright spots that appear as a rough texture on the solar photosphere
gravitation in Newtonian terms, a force between masses that is characterized by their acceleration toward each other; the size of the force depends directly on the product of the masses and inversely on the square of the distance between them; in Einstein's terms, the curvature of spacetime
gravitational collapse the unhindered contraction of any mass due to its own gravity
gravitational field the property of space having the potential for producing gravitational force on objects within it; characterized by the acceleration of free masses
gravitational instability the tendency for a disturbed region in a gas to undergo gravitational collapse
gravitational mass the amount of mass in an object determined by placing it in a gravitational field and finding out how much it accelerates
gravitational red shift the change to longer wavelengths that marks the loss of energy by a photon that leaves any mass
ground state the lowest energy level of an atom
H I region a region of neutral hydrogen in interstellar space
H II region a zone of ionized hydrogen in interstellar space; it usually forms a bright nebula around a hot, young star or cluster of hot stars
habitable zone same as *ecosphere*
half-life the time required for half of the radioactive atoms in a sample to disintegrate
halo (of a galaxy) the spherical region around a galaxy, not including the disk or the nucleus, containing globular clusters, some gas, and probably a few stray stars
H-alpha line the first line of the Balmer series, the set of transitions in a hydrogen atom between the second energy level and levels with higher energy; it lies in the red part of the visible spectrum
heliocentric centered on the sun
helium flash the rapid burst of energy generation with

which a star initiates helium burning by the triple-alpha process in the degenerate core of a low-mass red giant star
Hertzsprung-Russell (H-R) diagram a graphic representation of the classification of stars according to their spectral class (or color or surface temperature) and luminosity (or absolute magnitude); the physical properties of a star are correlated with its position on the diagram, so a star's evolution can be described by its change of position on the diagram with time
homogeneous having a consistent and even distribution of matter, the same in all parts
Homo sapiens humankind
horizon the intersection with the sky of a plane tangent to the earth at the location of the observer
horizontal branch a portion of the Hertzsprung-Russell diagram reached by stars of low mass after the red giant stage and typically found in a globular cluster; it ranges from yellow to red stars all having the same luminosity (about 100 times the sun's)
H-R diagram a *Hertzsprung-Russell diagram*
hydrostatic equilibrium an equilibrium characterized by the absence of mass motions, when pressure balances gravity
hyperbolic geometry an alternative to Euclidean geometry, constructed by N. I. Lobachevski on the premise that more than one parallel line can be drawn through a point near a straight line; the sum of the angles of a triangle drawn on a hyperbolic surface is always less than 180°
igneous rock rock formed by the cooling of molten lava
impact theory the idea that craters are formed by the impact of solid objects onto a surface
inertia the resistance of an object to a force acting on it because of its mass
inertial mass mass determined by subjecting an object to a known force (not gravity) and measuring the acceleration that results
interstellar medium all the gas and dust found between stars
inverse beta decay the process in which electrons and protons are forced together to form neutrons and neutrinos; the reverse process of neutron decay
ion an atom that has become electrically charged by the gain or loss of one or more electrons
ionization the process by which an atoms loses or gains electrons
ionized gas a gas that has been ionized so that it contains free electrons and charged ions; a plasma
ionosphere a layer of the earth's atmosphere ranging from about 100 to 700 kilometers above the surface where oxygen and nitrogen are ionized by sunlight, producing free electrons
irons one of the three main types of meteorites, typically made of about 90 percent iron and 9 percent nickel, with a trace of other elements
irregular galaxy a galaxy without spiral structure or rotational symmetry, containing mostly Population I stars and abundant gas and dust
isotropic having no preferred direction in space
Jeans length the minimum size a disturbance in a gas must have to result in gravitational contraction; it depends on the pressure, temperature, and density of the medium
kiloparsec one thousand parsecs
kinetic energy the ability to do work because of motion
KREEP a lunar material composed of potassium (K), rare-earth elements (REE), and phosphorus (P)
LAWKI an acronym for *life as we know it*
light curve a graph of a star's changing brightness with time
light year the distance light travels in a year, about 6 trillion miles, or 3.09×10^{13} kilometers
Local Group a gravitationally bound group of about 20 galaxies to which our Milky Way Galaxy belongs
longitudinal wave a sound wave that moves in a push-pull motion through solids, liquids, and gases with a velocity that depends on the density of the medium
luminosity the total rate at which radiative energy is given off by a celestial body, over all wavelengths; the sun's luminosity is about 4×10^{26} watts
luminosity class the categorization of stars that have the same surface temperatures but different sizes, resulting in different luminosities; based on the widths of dark lines in a star's spectrum, giant stars having narrower lines than dwarf stars
lunar soil the fine particles created by the bombardment of the lunar surface by meteorites that, with larger rock fragments, compose the lunar soil
Lyman series all transitions in a hydrogen atom to and from the lowest energy level; they involve large energy changes, corresponding to wavelengths in the ultraviolet part of the spectrum; also, the set of absorption or emission lines corresponding to these transitions
Magellanic Clouds two neighboring galaxies, the Large Magellanic Cloud (LMC) and the Small Magellanic Cloud (SMC), visible in the Southern Hemisphere to the unaided eye; companions to our Galaxy
magnetic field the property of space having the potential of exerting magnetic forces on bodies within it
magnetic lines of force a graphic representation of a magnetic field showing its direction and, by the degree of packing of the lines, its intensity
magnetosphere the region around a planet where particles from the solar wind are trapped by the planet's magnetic field
magnitude a measurement of an object's brightness; larger magnitudes represent fainter objects
main sequence the principle series of stars in the Hertzsprung-Russell diagram; such stars convert hydrogen to helium in their cores by the proton-proton process or by the carbon-nitrogen-oxygen cycle; this is the longest stage of a star's active life
mantle the major portion of the earth's interior below the crust, made of a plastic rock probably composed of olivine
mare (pl., maria) (Latin, for "sea") a lowland area on the moon that appears darker and smoother than the highland regions, probably formed by lava that solidified into basaltic rock about 3.0 to 3.5 billion years ago
mascons abnormal concentrations of mass beneath the

lunar maria; they have been detected by their effect on the orbits of moon-orbiting satellites
mass the measurement of an object's resistance to change in its motion (inertial mass); a measure of the strength of gravitational force an object can produce (gravitational mass)
mass-luminosity relation an empirical relation, for main-sequence stars, between a star's mass and its luminosity, roughly proportional to the third power of the mass
maximum elongation the greatest angular distance of an object from the sun
mechanics a branch of physics that deals with forces and their effects on bodies
megaparsec 1 million parsecs, or about 3.26 million light years
megaton an explosive force equal to that of 1 million tons of TNT (about 4×10^{15} joules)
meteor the bright streak of light that occurs when a solid particle (a meteoroid) from space enters the earth's atmosphere and is heated by friction with atmospheric particles; sometimes called a *falling star*
meteorite a solid body from space that survives a passage through the earth's atmosphere and falls to the ground
meteoroid a very small solid body moving through space in orbit around the sun
midoceanic ridge an almost continuous submarine mountain chain that extends some 64,000 kilometers through the earth's ocean basins
Milky Way the band of light that encircles the sky, caused by the blending of light from the many stars lying near the plane of the Galaxy; also sometimes used to refer to the Galaxy to which the sun belongs
minute of arc 1/60 of a degree
molecule a combination of two or more atoms bound together electrically; the smallest part of a compound that has the properties of that substance
mutation a basic change in hereditary material
natal astronomy the belief that treats the supposed influence of the stars and planets on human affairs by the use of their positions and relationships at the time of an individual's birth
natural selection the process by which individuals with genes that produce characteristics that are best adapted to their environment have greater genetic representation in future generations
nebula (Latin, for "cloud") a cloud of interstellar gas and dust
nebular theory a model for the origin of the solar system, in which an interstellar cloud of gas and dust collapsed gravitationally to form a flattened disk out of which the planets formed by accretion
neutrino an elementary particle with no mass or electric charge that travels at the speed of light and carries energy away during certain types of nuclear reactions
neutron a subatomic particle with about the mass of a proton and no electric charge; one of the main constituents of an atomic nucleus; the union of a proton and an electron
neutron star a star of extremely high density and small size that is composed mainly of very tightly packed neutrons; cannot have a mass greater than 5 solar masses
nonthermal radiation emitted energy that is not characterized by a blackbody spectrum; usually used to refer to *synchrotron radiation*
noon midday; the time halfway between sunrise and sunset when the sun reaches its highest point in the sky with respect to the horizon
Norma arm a segment of a spiral arm of our Galaxy about 12,000 light years from the sun toward the center of the Galaxy in the direction of the constellation Norma
north magnetic pole one of the two points on a star or planet from which magnetic lines of force emanate and to which the north pole of a compass points
nova (Latin, "new") a star that has a sudden outburst of energy, temporarily increasing its brightness by hundreds to thousands of times; also used in the past to refer to some stellar outbursts that modern astronomers now call *supernovas*
nuclear fission a process that releases energy from matter; in it, a heavy nucleus hit by a high-energy particle splits into two or more lighter nuclei whose combined mass is less than the original, the missing mass being converted into energy
nuclear fusion a process that releases energy from matter by the joining of nuclei of lighter elements to make heavier ones; the combined mass is less than that of the constituents, the difference appearing as energy
nucleic acid a huge spiral-shaped molecule, commonly found in the nucleus of cells, that is the chemical foundation of genetic material
nucleosynthesis the chain of thermonuclear fusion processes by which hydrogen is converted to helium, helium to carbon, and so on through all the elements of the periodic table
nucleus (of an atom) the massive central part of an atom, containing neutrons and protons, about which the electrons orbit
nucleus (of a comet) small, bright, starlike point in the head of a comet; believed to be a solid, compact (a few tens of kilometers) mass of frozen gases with some rocky material embedded in it
nucleus (of a galaxy) the central portion of a galaxy, composed of old Population I stars, some gas and dust, and, for many galaxies, a concentrated source of nonthermal radiation
objective the main light-gathering lens or mirror of a telescope
occultation the eclipse of a star or planet by the moon or another planet
Olbers's paradox the statement that if there were an infinite number of stars distributed uniformly in an infinite space, the night sky would be as bright as the surface of a star, in obvious contrast to what is observed
opacity the property of a substance that hinders (by absorption or scattering) light passing through it; opposite of transparency
open cluster same as *galactic cluster*

open geometry see *hyperbolic geometry*
opposition the time at which a celestial body lies exactly opposite the sun in the sky as seen from the earth; the time at which it has an elongation of 180°
orbital angular momentum the angular momentum of a revolving body; the product of a body's mass, orbital velocity, and the distance from the system's center of mass
orbital inclination the angle between the orbital plane of a body and some reference plane; in the case of a planet in the solar system, the reference plane is that of the earth's orbit; in the case of a satellite, the reference is usually the equatorial plane of the planet; for a double star, it is the plane perpendicular to the line of sight
organic relating to the branch of chemistry concerned with the carbon compounds of living creatures
Orion spur a small branch of the Cygnus arm in which the sun is located
ozone layer a layer of the earth's atmosphere about 40 to 60 kilometers above the surface, characterized by a high content of ozone, O_3
parallax the change in an object's apparent position when viewed from two different locations
parent meteor bodies small solid bodies, a few hundreds or thousands of kilometers in size, believed to be the source of nickel-iron meteorites; formed early in the history of the solar system and then broke up through collisions
parsec (pc) the distance an object would have to be from the earth so that its heliocentric parallax would be 1 second of arc; equal to 3.26 light years
perfect cosmological principle the statement that the universe appears the same to an observer at all locations and at all times
perigee the point in its orbit at which an earth satellite is closest to the earth
perihelion the point at which a body orbiting the sun is nearest to it
period the time interval for some regular event to take place; for example, the time required for one complete revolution of a body around another
periodic comets comets that have relatively small elliptical orbits around the sun, with periods of less than 200 years
periodic variable a star whose light varies with time in a regular fashion
period-luminosity relationship for cepheid variables, a relation between the average luminosity and the time period over which the luminosity varies; the greater the luminosity, the longer the period
Perseus arm a segment of a spiral arm that lies about 10,000 light years from the sun in the direction of the constellation Perseus
phases of the moon the monthly cycle of the changes in the moon's appearance as seen from the earth
Phobos the larger of the two moons of Mars
photodissociation the breakup of a molecule by the absorption of light with enough energy to break the molecular bonds
photon a discrete amount of light energy; the energy of a photon is related to the frequency, f, of the light by the relation $E = hf$, where h is Planck's constant
photosphere the visible surface of the sun; the region of the solar atmosphere from which visible light escapes into space
pitch angle the angle between a spiral arm's direction and the direction of circular motion about the Galaxy
planet from the Greek word for "wanderer"; any of the nine (so far known) large bodies that revolve around the sun; traditionally, any heavenly object that moved with respect to the stars (in this sense, the sun and the moon were also considered planets)
planetary nebula a thick shell of gas ejected from and moving out from an extremely hot star; thought to be the outer layers of a red giant star thrown out into space, the core of which eventually becomes a white dwarf
planetesimals asteroid-sized bodies that, in the formation of the solar system, combined with each other to form the protoplanets
plasma a gas consisting of ionized atoms and electrons
Polaris the present north pole star; the outermost star in the handle of the Little Dipper
polarization a lining up of the planes of vibration of light waves
Population I stars stars found in the disk of a spiral galaxy, especially in the spiral arms, including the most luminous, hot, and young stars, with a heavy element abundance similar to that of the sun (about 2 percent of the total); an old Population I is found in the nucleus of spiral galaxies and in elliptical galaxies
Population II stars stars found in globular clusters and the halo of a galaxy; somewhat older than any Population I stars, and contain a smaller abundance of heavy elements
positron an antimatter electron; essentially, an electron with a positive charge
potential energy the ability to do work because of position; it is storable and can later be converted into other forms of energy
PP chain see *proton-proton chain*
precession of the equinoxes the slow westward motion of the equinox points on the sky relative to the stars of the zodiac because of the wobbling of the earth's spin axis
primary the brighter of the two stars in a binary system
primeval fireball the hot, dense beginning of the universe in the Big Bang model, when most of the energy was in the form of high-energy light
principle of equivalence the fundamental idea in Einstein's general theory of relativity; the statement than one cannot distinguish between gravitational accelerations and other kinds of acceleration, or, equivalently, a statement about the equality of inertial mass and gravitational mass; a consequence is that gravitational forces can be made to vanish in a small region of spacetime by choosing an appropriate accelerated frame of reference
protein a long chain of amino acids linked by hydrogen bonds
protogalaxies clouds with enough mass that they are destined to collapse gravitationally into galaxies

proton a massive, positively charged elementary particle; one of the main constituents of the nucleus of an atom
proton-proton (PP) chain a series of thermonuclear reactions that occur in the interiors of stars, by which four hydrogen nuclei are fused into helium; this process is believed to be the primary mode of energy production in the sun
protoplanets large masses formed by the sticking together of planetesimals; the final stage in the formation of the planets out of the primeval nebula
protostar a collapsing mass of gas and dust out of which a star will be born (when thermonuclear reactions turn on) whose energy comes from gravitational contraction
pulsar a radio source that emits signals in very short, regular bursts; thought to be a highly magnetic, rotating neutron star
quanta discrete packets of energy
quasar an intense, pointlike source of light and radio waves that is characterized by large red shifts of the emission lines in its visible spectrum
radar mapping the surveying of the geographic features of a planet's surface by the reflection of radio waves from the surface
radial velocity the component of relative velocity that lies along the line of sight
radiative energy the capacity to do work that is carried by electromagnetic waves
radiation usually refers to electromagnetic waves, such as light, radio, infrared, X-rays, ultraviolet; also sometimes used to refer to atomic particles of high energy, such as electrons (beta-radiation), helium nuclei (alpha-radiation), and so on
radioactive dating a process that determines the age of an object by the rate of decay of radioactive elements within the object
radioactive decay the process by which an element fissions into lighter elements
radio-emitting galaxy a galaxy that emits large amounts of radio energy, generally characterized by two giant lobes of emission situated on opposite ends of a line drawn through the nucleus
radio-line emission sharp energy peaks at radio wavelengths, usually caused by low-energy transitions in atoms
recombination the joining of an electron to an ion; the reverse of ionization
red giant a large, cool star with a high luminosity and a low surface temperature of about 2500 K
reddening the preferential scattering or absorption of blue light by small particles, allowing more red light to pass directly through
red shift an increase in the wavelength of the radiation received from a receding celestial body as a consequence of the Doppler effect; a shift toward the long-wavelength (red) end of the spectrum
reference frame a set of coordinates by which position and motion may be specified
reflecting telescope a telescope that has a uniformly curved mirror as a primary light gatherer
reflection nebula a bright cloud of gas and dust that is visible because of the reflection of starlight by the dust
refracting telescope a telescope that uses glass lenses to gather light
resolving power the ability of a telescope to separate close stars or to pick out fine details of celestial objects
retrograde motion the apparent anomalous *westward* motion of a planet with respect to the stars, which occurs near the time of opposition (for an outer planet) or inferior conjunction (for an inner planet)
rotation the turning of a body, such as a planet, on its axis
RR Lyrae stars a class of giant, pulsating variable stars with periods of less than one day
Sagittarius arm a portion of spiral-arm structure of the Galaxy that lies about 6000 light years from the center of the Galaxy in the direction of the constellation Sagittarius
scarp a long, vertical wall running across a flat plain
Schwarzschild radius the critical size that a mass must reach to be dense enough to trap light by its gravity, that is, to become a black hole
second of arc 1/3600 of a degree, or 1/60 of a minute of arc
secondary the fainter of the two stars in a binary system
seeing the unsteadiness of the earth's atmosphere that blurs telescopic images
seismic waves sound waves traveling through and across the earth that are produced by earthquakes
seismometer an instrument used to detect earthquakes and moonquakes
Seyfert galaxy a galaxy with a bright nucleus showing broad emission lines in its spectrum
sexagesimal system a counting system based on the number 60, such as 60 minutes in an hour, or 60 minutes of arc in one degree
sidereal month the period of the moon's revolution around the earth with respect to a fixed direction in space or a fixed star; about 27.3 days
sidereal period the time interval needed by a celestial body to complete one revolution around another with respect to the background stars
signs of the zodiac the twelve equal angular divisions of 30° each into which the ecliptic is divided; each corresponds to a zodiacal constellation
silicate a compound of silicon and oxygen with other elements, very common in rocks at the earth's surface
singularity a theoretical point of zero volume and infinite density to which any mass that becomes a black hole must collapse, according to the general theory of relativity
solar day the interval of time from noon to noon
solar eclipse an eclipse of the sun by the moon, caused by the passage of the moon in front of the sun
solar mass the amount of mass in the sun, about 2×10^{28} kilograms
solar wind a stream of charged particles, mostly protons and electrons, that escapes into the sun's outer atmosphere at high speeds and streams out into the solar system

solstice the time at which the day or the night is the longest; in the Northern Hemisphere, the summer solstice (around June 21) the time of the longest day and the winter solstice (around December 21) the time of the shortest day; the dates are opposite seasons in the Southern Hemisphere
south magnetic pole a point on a star or planet from which the magnetic lines of force emanate and to which the south pole of a compass points
space a three-dimensional region in which object and events occur and have relative direction and position
spacetime looking at the universe with space and time unified; a continuous system of one time coordinate and three space coordinates by which events can be located and described
special theory of relativity Einstein's theory describing the relations between measurements of physical phenomena as viewed by observers who are in relative motion at constant velocities
spectral line a particular wavelength of light corresponding to some energy transition in an atom
spectroscope an instrument for examining spectra
spectroscopic binary two stars revolving around a common center of mass that can be identified by periodic variations in the Doppler shift of the lines of their spectra
spectroscopic parallax a technique for measuring distance by comparing the brightnesses of stars with their actual luminosities, as determined by their spectra
spectroscopy the analysis of light by separating it by wavelengths (colors)
spectrum (pl., spectra) the array of colors or wavelengths obtained when light is dispersed, as by a prism; the amount of energy given off by an object at every different wavelength
spherical (closed) geometry an alternative to Euclidean geometry, constructed by G. F. B. Riemann on the premise that no parallel lines can be drawn through a point near a straight line; the sum of the angles of a triangle drawn on a spherical surface is always greater than 180°
spin angular momentum the angular momentum of a rotating body; the product of a body's mass, rotational velocity, and radius
spinar a highly condensed, spinning object that may be the energy source for a quasar
spiral arm a structure, part of a spiral pattern in a galaxy, composed of gas, dust, and young stars, that winds out from near the galaxy's center
spiral galaxy a galaxy with spiral arms; the presumed shape of our Milky Way Galaxy
spiral tracers objects that are commonly found in spiral arms and so are used to trace spiral structure; for example, Population I cepheids, H II regions, and *OB*-stars
spontaneous generation the natural origination of living things from lifeless matter
stadium (pl., stadia) an ancient Greek unit of length, probably about 0.2 kilometer
standard candle see *distance candle*
star model a table of values of the physical characteristics (such as temperature, density, and pressure) as a function of position within a star for a specified mass, chemical composition, and age, calculated from theoretical ideas of the basic physics of stars
steady-state model a theory of the universe based on the perfect cosmological principle, in which the universe looks basically the same to all observers at all times
stellar spectral sequence a classification scheme for stars based on the strength of various lines in their spectra; the sequence runs *O-B-A-F-G-K-M*, from hottest to coolest
stones a type of meteorite made of light silicate materials
stony-irons meteorites that are a blend of nickel-iron and silicate materials
stratosphere a layer in the earth's atmosphere in which temperature changes with altitude are small and clouds are rare
summer solstice see *solstice*
sunspot a temporary cool region in the sun's photosphere with a strong magnetic field
supercluster a group of clusters of galaxies
supergiant a massive star of large size and high luminosity
supernova a stupendous explosion of a massive star, which increases its brightness hundreds of millions of times in a few days
synchrotron radiation radiation from an accelerating charged particle (usually an electron) in a magnetic field; the wavelength of the emitted radiation depends on the strength of the magnetic field and the energy of the charged particles
synodic period the interval between successive similar lineups of a celestial body with the sun, for example, between oppositions
temperature a measure of the average random speeds of the microscopic particles in a substance
thermal equilibrium steady-state situation characterized by no large-scale temperature changes
thermal radiation electromagnetic radiation due to the fact that a body is hot; often characterized by a blackbody spectrum
tidal force the difference in gravitational force affected by one body's gravitational attraction on various points of a second body, which causes the deformation of the second body
time a measure of the flow of events
Titan Saturn's largest satellite, the first satellite detected to have an atmosphere
transition (in an atom) a change in the electron arrangements in an atom, which involves a change in energy
transverse wave a wave in which the oscillatory motion is perpendicular to the direction of propagation; such sound waves cannot travel through liquids
trigonometric parallax a method of determining distances by measuring the angular position of an object as seen from the ends of a baseline having a known length
triple-alpha reaction a thermonuclear process in which

three helium nuclei (alpha particles) are fused into one carbon nucleus
troposphere the lowest level of the earth's atmosphere, reaching 10 kilometers from the surface; the area in which most of the weather takes place
turnoff point the point on the H-R diagram of a cluster at which the main sequence appears to terminate at the high-luminosity end
turbulence irregular, and sometimes violent, convective motion
two-sphere universe the basic premise of the celestial coordinate systems that the universe is composed of two concentric spheres, the earth and the celestial sphere
universality of physical laws the assumption, borne out by some evidence, that the physical laws understood locally apply throughout the universe and perhaps to the universe as a whole
universe the totality of all space and time; all that is, has been, and will be
vector a quantity that expresses magnitude and direction; for example, forces and accelerations are vector quantities
velocity the rate and direction in which distance is covered over some interval of time
visual binary two stars that revolve around a common center of mass, both of which can be seen through a telescope so their orbits can be plotted
volatiles materials, such as helium or methane, that vaporize at low temperatures
volcanic theory the theory of the formation of craters from cones left over from lava eruptions
wavelength the distance between two successive peaks or troughs of a wave
weight the total force on some mass produced by gravity
white dwarf a small, dense star that has exhausted its nuclear fuel and shines from residual heat; such stars have an upper mass limit of 1.4 solar masses, and their interior is a degenerate electron gas
Widmanstätten figures large crystal patterns that appear on the surfaces of iron meteorites when they are polished and etched
winter solstice see *solstice*
zenith the point on the celestial sphere that is located directly above the observer at 90° angular distance from the horizon
zero-age main sequence the position on the H-R diagram reached by a protostar once it derives most of its energy from thermonuclear reactions rather than from gravitational contraction
zodiac the twelve constellations through which the sun travels in its yearly motion, as seen from the earth
zone of avoidance a region near the plane of the Galaxy where very few other galaxies are visible because of obscuration by dust

Index

The night sky in JUNE

THE LAW OF BUSINESS ORGANIZATIONS

Second Edition

THE LAW OF BUSINESS ORGANIZATIONS

Second Edition

John E. Moye

WEST PUBLISHING COMPANY

St. Paul New York Los Angeles San Francisco

50 West Kellogg Boulevard
P.O. Box 3526
St. Paul, Minnesota 55165

Printed in the United States of America

Library of Congress Cataloging in Publication Data

Moye, John E.
The law of business organizations.
Includes index.
1. Corporation Law—United States. 2. Partnership—United States. I. Title. II. Series.
KF1366.M68 1982 346.73'065 81-19790
ISBN 0-314-63396-0 347.30665 AACR2

3rd Reprint—1984

With love to Fabulous Fern and our girls,

Crazy Kelly

Magic Mary

Mad-dog Megan

Rowdy Rachael

CONTENTS

3 LIMITED PARTNERSHIPS

4 THE BUSINESS CORPORATION

5 SPECIAL CORPORATE FORMS

6 FORMATION OF A CORPORATION

7 CORPORATE FINANCIAL STRUCTURE

11

AGREEMENTS REGARDING SHARE OWNERSHIP

12

CORPORATIONS IN FOREIGN JURISDICTIONS

13

CHANGES IN CORPORATE STRUCTURE AND DISSOLUTION

PREFACE

I have had the pleasure of teaching various law school classes and practice seminars for lawyers on the subject of business organizations and I also taught one of the pioneer paralegal courses in business organizations under the auspices of the Continuing Legal Education Program of the University of Denver College of Law. This book originally began as a paralegal textbook and was used in paralegal training courses. It was also used in several law schools and by practitioners who taught other lawyers in continuing legal education courses. This second edition is a greatly expanded version of the materials developed for all of those courses, improved through the helpful comments and suggestions of the now experienced paralegals and lawyers who attended or taught those classes. These materials are directed toward the training and practice of paralegals and lawyers in organizing, drafting, operating, and providing service to business organizations.

The book is designed to be used in two environments. In the first place it is intended to be used as a classroom teaching source for the training of paralegals and lawyers in the practical aspects of the law of business organizations. It also qualifies as a reference work to be used by paraprofessionals and attorneys in the practice of this field. It is probably best described as a how-to-do-it approach to the law, with an explanation of the legal principles which must be observed in counseling an enterprise. Numerous sample clauses and forms are sprinkled throughout the text to illustrate the legal rules described; procedural checklists are included for study and client contact; and sample forms from various states and reference form books are used. The textual treatment of the law is designed to acquaint or refresh the reader on the important legal topics to be considered in the drafting techniques, organization, development, and operation of an enterprise. A student should be expected to study the explanation of the law, to compare the wording in the sample clauses and forms, to suggest drafting variations based upon the law or upon different facts, and to observe the application of the law in its practical setting. A practicing paralegal or lawyer should apply the explanations of the law and the drafting examples to client variations in practice. To facilitate the use of sample forms and clauses in the preparation of assignments and documents, separate indices are included for these materials.

With the exception of the sole proprietorship, business organizations are founded upon statutory authority. In order to insure national application of this book, the textual explanation of the law is based upon the Uniform Partnership Act, the Uniform Limited Partnership Act, and the Model Business Corporation Act. The new Model Professional Corporation Act is now discussed in this second edition. Both the Uniform Limited Partnership Act and the Model Business Corporation Act have been extensively revised during the past eight years (since the first edition of this book) and the text discusses both the revisions and the former approaches of the statutes, since most states' laws continue to be based upon the prerevision statutes. Important jurisdictional variations from these statutes are noted in footnotes in many cases, but the reader is continually cautioned to analyze the appropriate state law in practice.

The uniform and model acts are set forth in Appendixes A to E, and the student should refer to the acts to become acquainted with statutory authority underlying the law as explained in the text. A practicing paraprofessional or lawyer may compare these acts with local statutes to highlight variations and to recognize additional local requirements.

The breadth of the law of business organizations necessarily results in cursory treatment of some complex principles in a book of this size. I have attempted to refer the reader to more detailed texts, particularly in matters of taxation and certain tangential matters such as estate planning for the sole proprietor. The laws of registration and regulation of securities under the federal securities acts and state blue-sky laws are certainly intimately connected with these materials, but they constitute a separate subject which deserves its own book. Accordingly, I have merely mentioned them in the appropriate places, with references to other treatises for further explanation.

The book begins with the most simple form of business enterprise, the sole proprietorship; progresses through general partnership and limited partnership; and dwells upon the most complex business organization, the corporation. Chapter 5 separately considers problems of special corporate variations, the close corporation and professional corporations, but other hybrid forms of business enterprise, such as joint ventures and joint stock associations, are not covered, partly because of space limitations and partly because of their infrequent appearance in practice. I further feel that these business forms may be explained in classroom discussion and are so identifiable with the described organizations that they may be easily comprehended.

Chapters 1, 2, and 3 contain discussions of the frequently encountered problems in organization and operation of sole proprietorships and partnerships, including procedures for formation and operation of the business, the characteristics of the enterprise, liabilities of the associated parties, taxation, dissolution, and termination. A discussion of the novel approach taken by the Revised Uniform Limited Partnership Act appears in the last section of Chapter 3 and may be used for comparison with current partnership law in most states.

The corporate materials are arranged somewhat chronologically as they usually will be met in practice. Following the identification of the intracorporate parties and a general discussion of the characteristics, powers, continuity, and taxation of the corporation (including the latest tax revision from the Economic Recovery Act of 1981), the remaining chapters consider formation procedures, development of the corporate financial structure, corporate meetings, dividends and corporate distributions, agreements affecting employment and share ownership, qualification and operation as a foreign corporation, modifications to the corporate structure, and dissolution. Of course, matters involving shareholder agreements may arise in the formation stages, and may assert that a discussion of liquidation distributions belongs with the discussion of dissolution. The material is all cross-referenced by footnotes for easy organization and understanding among the chapters.

The materials covering formation of the corporation, corporate meetings, and shareholder agreements have been considerably expanded in this second edition to illustrate variations in corporate practice and to provide the reader with more forms and checklists to use in preparing corporate documents for the client.

The placement of Chapter 10 on Employment and Compensation deserves special comment. It appears in the corporate section of the book because it involves agreements which will be executed with key employees in corporate practice, and the discussions in sections 10.09 (Pension and Profit sharing Plans), 10.10 (Stock Options), and 10.11 (Insurance Programs) are almost exclusively corporate in nature. However, the remaining sections are equally applicable to sole proprietorships and partnerships. In class discussion it may be appropriate to study some or all of the interchangeable sections when discussing proprietorships and partnerships, and to remind the students of their applicability to corporations when the unique corporate compensation schemes are considered.

Having incorporated suggestions from the various teachers, practitioners, and students who have used the first edition of this book, I am confident that this second edition covers all important areas of enterprise practice. I commend it to you with the hope that I have created a usable manual for the practice of the law of business organizations.

John E. Moye
Denver, Colorado

ACKNOWLEDGMENTS

I wish to express gratitude to some very special people who contributed significantly to the preparation of this work.

First I thank my partners at Head, Moye, Carver & Ray, John Head, Craig Carver, Pamela Ray, Ned Giles, Gary Nakarado and Ed O'Keefe, for giving me the professional support and encouragement I needed to take the time to work on this book. I also thank Darcy Rezabeck, our corporate paralegal at the firm, for her excellence and proficiency in the practice of business organizations, much of which she will recognize in some of the forms and checklists used in the book.

I gratefully acknowledge the contributions of other authors, government officials, and corporate personnel for the forms and checklists used as examples. I thank West Publishing Company for the use of their Modern Legal Forms volumes to illustrate sample clauses and agreements throughout the book. The Secretaries of State for the states of California, Colorado, Delaware, Iowa, Massachusetts, Mississippi, New Jersey, New York, North Dakota, Oregon, Pennsylvania, South Carolina, South Dakota, and Texas graciously permitted the use of official forms for their respective states. Professor Harry G. Henn of the Cornell Law School allowed the reprint of his Preincorporation Checklist from his unexcelled works on corporations, which I studied at Cornell to learn the law of business organizations. Messrs. William A. Sutherland and Paul Carrington and the Committee on Economics of Law Practice of the American Bar Association permitted the reproduction of an excellent sample partnership agreement from a previous publication. I thank Prentice Hall, Inc., Bradford-Robinson Printing Company, the University of Florida Law Review, and author Sidney A. Ward for granting permission to use their corporate materials as examples. Special thanks are due to Louis S. Kelley of the Black Hills Power and Light Company; Mr. Richard F. Atwood, General Counsel for the Coca–Cola Company; Mr. William Lee Phyfe of American Telephone and Telegraph Company; and Mr. Thomas F. Macan of General Motors Corporation for their assistance in obtaining sample corporate agreements, share certificates, and bonds for publication.

I am indebted and grateful to Nancy Stober and Mary Jo Bergman, my secretaries at Head, Moye, Carver & Ray, for their dedication in the preparation and typing of the manuscript.

Last, but certainly not least, I thank my many friends and associates who appear as characters in the sample clauses throughout the book. We decided early that blanks in these clauses make them harder to read and their assent to join the cast of business personalities in the various transactions adds the flavor of realism and is sincerely appreciated.

1

SOLE PROPRIETORSHIP

§ 1.01 CHARACTERISTICS OF THE SOLE PROPRIETORSHIP

The sole proprietorship or individual proprietorship is the simplest and most common form of business enterprise. In the sole proprietorship organization, the individual proprietor owns all business properties and carries on business for himself. The typical individual proprietor is a merchant in a small retail store or corner grocery, but even a youngster who mows lawns during summer vacation is an individual proprietor. The distinguishing characteristic of the sole proprietorship is that it is owned and managed by one person, and thereby exists as an extension of the personal life of that person. The single owner operating the business as an individual activity is a key element. If the business is conducted by co-owners, it is most likely a partnership. Unlike the corporation, a sole proprietorship requires no grant or charter from the state to exist as a going concern.

The greatest advantages to the sole proprietorship are the ease by which such an organization may be formed and the degree of flexibility in managing the business. As sole owner, the individual proprietor may operate the business as she chooses. While she may hire employees and agents to assist her, the proprietor is vested with ultimate responsibility for all decisions affecting the business. Consequently, management is usually flexible and informal.

The disadvantages of the sole proprietorship all flow from the fact that the business has complete identity with the proprietor. As a practical matter, his personal strengths and weaknesses are, necessarily, superimposed upon the business operations. Since management functions are vested in him, his management ability will have a direct effect on the success or failure of his business. That fact alone may explain why sole proprietorships are most frequently used for small limited businesses. The larger the scope of the business, the more it requires specialized business talent that few individuals could supply alone. Similarly, the identity of the individual with the business limits available business capital, and may thereby limit the size of the business. Unlike a corporation, the sole proprietorship has no shares which can be sold to outside investors. The only available methods of obtaining funds for this form of business are personal contributions of the individual proprietor and loans from financial institutions or other private sources. Further, the proprietor's ability to borrow money is limited by the potential of the business and his own personal assets, which he may have to pledge as collateral to secure a loan.

§ 1.02 LIABILITY OF THE SOLE PROPRIETOR

The law also imposes certain disadvantages on the sole proprietorship, again equating the identity of the proprietor with her business. She is personally responsible for all business losses and must bear them to the full extent of her personal resources. She is personally liable for all business liabilities. In contrast with the positions of a corporate shareholder or a limited partner, her financial risk is not limited to her investment in the enterprise but extends to all of her personal assets, including her home, car, furniture and similar property. This risk of unlimited liability may be diminished to some extent by insurance, but it may be impossible, and is at least impracticable, to insure against every conceivable business hazard. In matters involving contracts with the sole proprietorship it is possible to provide by agreement that any liability on the contract shall be limited to the business assets and shall not extend to the personal assets of the proprietor. Such agreement will provide little advantage, however, if the proprietor has contributed her personal assets for use in the business or as collateral to secure business loans.

The unlimited liability of a sole proprietorship may be a severe disadvantage to an entrepreneur with extensive personal wealth which she would prefer not to invest in the business, since absent insurance or agreement to the contrary, all personal assets must be made available to satisfy business liabilities. The problem is further compounded if the business is unusually speculative or hazardous.

On the other hand, it should be noted that the individual proprietor has full control over the extent of the business liability by virtue of her individual right to manage the business. While the law permits all partners to obligate the partnership and the officers to obligate the corporation, only the proprietor, or individuals she personally selects, may obligate the sole proprietorship.

§ 1.03 TERMINATION UPON DEATH OF PROPRIETOR

With very few exceptions specifically authorized by state statutes,[1] the sole proprietorship terminates by law upon the death of the proprietor. The owner is entitled to will his business to relatives or to an employee, but there is no assurance of continuity of the business after death. If the owner managed the business, and there are no relatives or associates willing to continue, the business will probably be liquidated. Liquidation must be accomplished by a legal representative of the deceased owner, such as a trustee or an executor, and cannot be done by agents appointed by the owner during his lifetime since, with the exception of a few ministerial acts authorized by specific statutes,[2] agents are powerless to act after the death of their principal. Because the business will be included in the personal estate of the deceased owner, a number of estate-planning considerations for the sole proprietor become essential. Generally, some authority must be granted to his personal representatives to permit them to continue the business as necessary until it may be conveniently and profitably liquidated, to employ persons to assist in liquidation and to execute all necessary documents incident to liquidation.[3] If the beneficiary of the deceased owner is willing to continue the business, a new sole proprietorship is created and it will be governed by these same rules.

§ 1.04 TAXATION OF THE SOLE PROPRIETORSHIP

The federal and state laws regarding taxation of a sole proprietorship may constitute an advantage in some cases. The law provides that all business income or loss will be treated as individual income or loss and taxed accordingly. The sole proprietor declares the business income on a separate schedule of his individual tax return. Once total income, including business income or loss, is computed, the individual income tax rates are applied. If the business is small and the owner has little income from other sources, the individual tax rates as applied to income from a sole proprietorship may be significantly lower than corporate tax rates. Similarly, if the business operates at a loss, the loss will be applied directly to offset other personal income of the sole proprietor, and will thereby result in direct tax savings. For example, if an individual received $50,000 in income from sources other than his business, qualified for $10,000 in deductions, and operated a sole proprietorship business at a net *loss* of $15,000 for the year, his taxable income would be $25,000, which, if he is married and filing a joint return, is taxed in approximately the 19% tax bracket. In contrast, if his business were incorporated, and thereby taxed under separate corporate tax rates, his personal taxable income would be $40,000, taxed in approximately the 26% tax bracket. His $15,000 business loss could not be used to offset personal income, and although it may offset future earnings of his corporation, he personally gains no advantage from the loss during this taxable year. Depending upon the amount of outside income, individual rates may also be advantageous for a small-profit business. If our sole proprietor had no other source of income but his business, and he earned $20,000 from the business during the year, his individual taxable income (after deductions) would be $5,000, which is taxed in the approximately 16% individual tax bracket. Contrast the resulting tax if his business were incorporated: the $5,000 corporate earnings will be taxed at approximately 17%.

Conversely, when personal income from other sources is high, or when the business operates at significant profit, a sole proprietorship suffers more tax than a corporation. If net taxable business income totalled $50,000 during one year, the sole proprietor, taxed as an individual would pay approximately 29.5% if married filing jointly or approximately 36% if single, of this income in tax. As a corporation the income would be taxed at 17% of the first $25,000 and 20% of the next $25,000, or an average of 18.5%. Of course, another important tax consideration for the sole proprietor is that the sole proprietorship income is taxed only once, while corporate income may be subject to "double taxation," once as corporate income and, if distributed as dividends, also taxed as income to the shareholder.[4]

If the business is the only source of income for the sole proprietor, it is possible to roughly estimate the point at which he will begin to suffer a tax disadvantage by continuing to operate as a sole proprietorship. The estimate requires a comparison of the respective percentages applied to corporate income and personal income to determine the point where each additional dollar of income is taxed less if received by a corporation than by an individual. Presently, for a single individual the corporate business organization form will provide a tax advantage if total *taxable income* exceeds approximately $14,500 during the year. Thereafter, each dollar of income is taxed at a greater percentage if earned as an individual sole proprietor than if earned by a corporation. The approximate figure for a married person filing a joint return is $21,500. Other pertinent tax considerations include the fact that the sale of sole proprietorship results in capital gains (or losses) to the proprietor based upon the value of each asset and an individual proprietor may qualify for certain retirement program deductions under a "Keogh" or "H.R. 10" Plan.[5]

In summary, the sole proprietorship offers tax advantages when the business produces relatively small profit and is the principal source of income of the proprietor; and also when a proprietor has an outside source of income and is operating the business at a loss. In other cases, when business profits exceed the very general reference points for profit discussed above, the sole proprietorship produces tax disadvantages. By way of planning, it is important to remember that a business may be commenced as a sole proprietorship to enjoy the tax benefits in the early stages of development, and subsequently be incorporated for more favorable tax rates as profits increase. However, while tax benefits are a major consideration, there are many other factors to consider in selecting the form of business enterprise.

§ 1.05 FORMATION AND OPERATION OF A SOLE PROPRIETORSHIP

Virtually no formalities are required in the formation and operation of a sole proprietorship, and this is a distinct advantage of this form of business over other forms. The sole proprietor may simply commence business by exercising his initiative.

If he intends to sell goods, he will be required to apply for a sales tax license in most jurisdictions, and, of course, any other license peculiar to his particular business must be obtained. For example, doctors must be licensed to practice medicine, a liquor license is required to sell alcoholic beverages, and so forth.[6]

If employees are contemplated, the sole proprietor must apply for a tax identification number from the Internal Revenue Service office and make arrangements to contribute to Social Security and unemloyment compensation on behalf of the employees.

A sole proprietor may conduct business under a name other than her own, and state statutes usually require registration of the assumed name by filing an affidavit or certificate for that purpose with a public official. These statutes usually also provide that the name used cannot be the same as, or deceptively similar to another registered or well-known name. The circumstances under which a particular name must be filed are subject to some fine distinctions. Generally, a firm name which contains the proprietor's surname and does not imply that other owners are associated with her need not be registered. For example, "Smith Auto Parts" or "Lyons Retail Goods" should not require filing. On the other hand, the use of the word "Company" or "Associates" implies other owners and should be registered. In any questionable case, it is better practice to register the name and avoid the problems associated with failure to file. Various penalties are prescribed for failure to register an assumed or trade name, but the usual sanction is refusal to allow the proprietor to pursue any litigation in state courts until filing has been accomplished. The filing procedure may vary by jurisdiction. Some states require a single filing with a county or state official. Others, such as California, require filing plus publication once a week for four weeks in a local newspaper.[7] Each statute should be consulted for guidance on local procedure.[8]

Sole proprietorships are permitted to do business in more than one state without additional formalities for qualification to do business. Of course, local licensing and assumed name statutes must be observed.

The operation of a sole proprietorship is extremely flexible and personal to the individual. Governmental regulation of business is found only in licensing requirements and periodic reports which may be required for certain types of business. The individual proprietor will personally determine the complexity of his business records, the need for expansion and capital improvements, salaries, and other matters affecting the policy and daily operations of the business. Compared with other business forms, the sole proprietor has considerable freedom in these matters.

NOTES

1. Some states provide statutory authority for continuity of a sole proprietorship by a proper testamentary distribution. E.g. McKinney Consol. Laws of N.Y.SCPA § 2108.
2. E.g., a bank is authorized to continue to pay checks of a deceased sole proprietor after death under U.C.C. § 4–405 until the bank learns about the death and has a reasonable opportunity to act on it.
3. See C. Rohrlich, Organizing Corporate and Other Business Enterprises, § 13.02 (1967).
4. See Section 4.06(a).
5. Contributions to retirement plans for sole proprietors are fully deductible up to a maximum amount of $2,500 or 10% of the earned income of the sole proprietor, whichever is less. See Int.Rev.Code of 1954, 26 U.S.C.A. §§ 401–405.

2

GENERAL PARTNERSHIP

§ 2.01 CHARACTERISTICS OF PARTNERSHIPS

A partnership is a special relationship between two or more persons whereby they *agree* to enter business together on their negotiated terms. The law of partnership has deep roots in ancient law and is closely related to the law of agency,[1] since each partner is an agent for the other partners and for the partnership business. The agreement is the most important element of the partnership, since it will govern all rights and responsibilities between the partners with respect to the business affairs of the firm. Partnerships are found in businesses as small as a local newstand, and in businesses as large as some multi-state enterprises.

Traditionally, the partnership has many of the legal characteristics of the sole proprietorship, and is distinguished from the proprietorship primarily by the fact that two or more persons are owners, rather than a single individual. Thus, in the pure form of partnership, each partner is personally responsible for the liabilities of the firm; management of the business affairs is vested in all partners equally; the partnership is dissolved upon death of a partner; and partnership profits and losses are taxed as though received individually by the partners. Each of these concepts, which will be explored in greater detail later, presupposes that the partnership is nothing more than an aggregation of persons who join together to own and conduct a business. In this respect the general partnership is clearly distinguished from the corporation, which has always been recognized as an entity, separate and distinct from its composite members. However, the "aggregate" theory of partnership has eroded under modern commercial law in recognition of the commercial reality of the partnership enterprise. Most partnerships operate as a business entity, the activities of which are separate from the activities of the individual partners. For example, the firm's delivery trucks are treated as belonging to the firm, not to the individual partners; the partners frequently are employed by the firm and are paid wages or salaries for their services; and the partnership conducts business and titles business property in the firm name. The Uniform Partnership Act, which governs general partnerships and has been adopted in all but a few states,[2] includes both "entity" and "aggregate" characteristics, and authorities dispute which theory prevails in most partnership problems. For purposes of this work, the question of theory must be left to be resolved by those who consider the problem a serious one. These materials will concentrate on the elements, advantages, and disadvantages of partnership, and problems in forming and conducting business under the partnership organization.

This chapter is concerned primarily with general partnerships. Another commonly recognized form of partnership is the limited partnership, where some partners enjoy limited liability in exchange for limited control over the business. Many elements of a general partnership are similar to a limited partnership, but they are discussed separately because of variations in formation and operation. Let us plunge into general partnerships, and save the limited partnerships for the next chapter.

Under the Uniform Partnership Act, a partnership is defined as "an association of two or more persons to carry on as co-owners a business for profit." [3] A few words on each element are essential to an understanding of this business form.

The "association of persons" is generally recognized to be a question of intent, and the persons must voluntarily intend to associate together in a business relationship. Their intent is usually expressed in an agreement, either written or inferred from their conduct. While the partnership may be very informal—perhaps even an oral understanding—it should be obvious that the better practice requires a comprehensive written agreement between the parties which clearly specifies their rights and responsibilities to each other and to the business.

The "persons" involved in a partnership must number two or more. A one-person business is a sole proprietorship. The persons need not be natural persons (human beings) under the Uniform Partnership Act. Partners may also include corporations, other partnerships and other associations. The Model Business Corporation Act complements this provision by permitting a corporation to be a partner.[4]

Partners must be co-owners of the business, which includes not only ownership of specific tangible assets, but also the joint rights to profit and control. The licenses and property of the partnership are held in a firm name or the partners' names jointly. The partners share profits and losses and exercise joint management privileges, the details of which are usually described in the agreement. Absent provisions to the contrary in an agreement, profits, losses and control are divided equally among the partners by law.[5]

A partnership, by definition, must be engaged in carrying on "a business for profit." The business part of this definition is further broadly defined as "every trade, occupation or profession." [6] The profit element of the partnership definition is intended to exclude charitable, religious and fraternal groups from the definition. The drafters of the Uniform Partnership Act perhaps could have more realistically stated that the business could also be operated at a loss and still be a partnership. Many partnerships do lose money. In any case, if the other elements are satisfied and the business is organized in *expectation of profit,* it will be considered a partnership.

§ 2.02 PARTNERSHIP PROPERTY

In a sole proprietorship, although the division may be informal, it is usually possible to distinguish business assets from the personal assets of the proprietor. It really does not make any difference there, however, since the business is only an extension of the proprietor's personal life, and segregation only becomes important in case of a contractural provision specifically limiting liability to the business assets of the firm. In the case of partnerships, however, a clear distinction between firm property and the personal assets of the partners is necessary. Although the personal assets of the individual partners may be vulnerable to partnership obligations, the property should be clearly divided for operating purposes.

Partnership property is first acquired from contributions by the individual partners. Partners may contribute specific assets to the firm, such as land, buildings, furniture or patents, and upon contribution the assets become partnership property. The partner may also contribute cash which is used to purchase specific assets. The cash and the assets so purchased also become firm property. The Uniform Partnership Act provides that "unless the contrary intention appears, property acquired with partnership funds is partnership property."[7] It should also follow that property purchased on credit by the firm will be partnership property. For the most part, it is not difficult to ascertain which property has been purchased with firm funds, but it may be difficult to identify property which has been contributed to the partnership unless the partnership agreement accurately reflects the intention of the parties. If a court were asked to decide whether certain property belongs to the firm or to the individual partner, it would test the intention of the parties, as found in their agreement and other overt acts.

The best guide to the intention of the partners is their written agreement. Thus, a complete description and agreed value of property which the partners contribute is essential to good draftsmanship. It is normal practice to describe the property contributed in a separate schedule which is attached to the partnership agreement and incorporated by reference.

EXAMPLE 2.02A: Contribution of Property

Nancy Stober shall contribute certain property valued at Five thousand dollars ($5,000.00). Such property is described in Schedule A attached hereto.[8]

The contribution of property to the partnership may have certain tax consequences which will be explored in detail later.[9]

Just as it is important to accurately and thoroughly describe the property contributed to the partnership, it is also important to specify which property is merely "loaned" to the partnership for its use, with the intention of retaining title in the individual partner's name. A clause covering this point should also include the period of time the firm shall be permitted to use the property, unless indefinite; any restrictions on the owner which are desirable or necessary to insure the use of the asset by the firm; and any compensation to be paid to the partner for the use of the asset.

EXAMPLE 2.02B: Property Loaned to Partnership

Craig Carver, as the owner of one 1981 Chevrolet pick-up truck, agrees to contribute to the partnership the use of such truck, with the understanding that it shall remain his separate property, and not in any event become an asset of the partnership. It is agreed that until the termination of the partnership, or until the death or retirement of Craig Carver, he will not, without the consent of all other partners, sell, assign or pledge or mortgage such property. Craig Carver further agrees that any money or rights occurring from the sale or assignment of the truck shall belong to the partnership during the term of the operation of the partnership. For the purpose of computation of profits, and not for participation in the distribution of the assets, the sum of ten thousand dollars ($10,000.00) shall be included in Craig Carver's capital account to represent the value of the truck.[10]

Thus, for property contributed at the commencement of the partnership, the intent of the parties with respect to ownership may be clearly expressed in the agreement.

The Uniform Partnership Act assists in determining the partners' intentions with respect to property subsequently acquired by the firm. It was previously noted that the Act creates a presumption that property purchased with partnership funds is firm property. Moreover, the firm may hold property in its own name. Nevertheless, it is common to title partnership property in the names of the individual partners. In such a case, the agreement should provide for this arrangement and indicate that the assets so titled are held as partnership property.

EXAMPLE 2.02C: Title to Property

Partnership property (including real estate) may, by unanimous consent of the partners, be acquired and conveyed in the name of any partner or other person as nominee for the partnership. Such property shall be recorded as partnership property in the partnership accounts.[11]

In addition to the specific provisions with respect to partnership property in the agreement, legal counsel should be sensitive to the need to provide other indicia of intent to determine ownership of the property. Partnership property should be identified in the firm's books; and all expenses, including repairs, insurance, taxes, interest on a mortgage, and so forth, should be paid by the firm. To be consistent, the firm should deduct these payments as expenses on its income tax return. Careful drafting and planning will avoid confusion regarding ownership of partnership property.

Assuming that a particular asset is partnership property and not individual property, the Uniform Partnership Act has adopted a peculiar method of legal title for such assets.

Historically, partners joined together as co-owners of a business, and a majority of jurisdictions considered all assets to be owned jointly in an ownership classification known as "tenancy in common." In the pure sense, tenancy-in-common, as its name indicates, stands for common ownership, with each owner entitled to a fraction of full title, and each owner entitled to "partition," or to sever his fractional ownership and assign or sell it to another. As the law of partnership developed, it modified this right of partition and imposed a limitation to the extent that partnership assets should first be used for partnership purposes (including satisfying obligations of the firm) and no partner would be able to otherwise alienate his fractional ownership until partnership purposes were satisfied.

The Uniform Partnership Act continues the theory that partners are co-owners of partnership property, but it creates a new ownership classification called "tenancy in partnership" [12] which better conforms to the reality that a partnership is a commercial entity using its own property for business purposes. All partnership assets are held under this form of title, and while partners are said to be co-owners, they have very limited ownership rights. In general, a partner may not possess firm property for other than partnership purposes without the consent of the other partners; a partner may not sell firm property (or her fractional interest in such property) without the consent of the other partners; the partner's individual creditors cannot apply their claims against the partner to firm assets; and the partner's heirs have no interest in the partnership assets when she dies. On the latter point, it should be noted that the *surviving partners* are vested with the deceased partner's fractional ownership at her death. To be more specific, suppose a three-person partnership owns a delivery truck which is used in their appliance business. No partner may use the truck to transport her family to a picnic without the other partner's consent. Similarly, all partners must agree to sell the truck; no single partner could sell her one-third interst in it. If one partner is being sued for a personal bill and refuses to pay, the creditor cannot use the truck to satisfy the judgment. Finally, if one partner dies, the truck belongs to the other two partners, and the deceased partner's heirs do not acquire her fractional ownership rights.

§ 2.03 PARTNER'S INTEREST IN THE PARTNERSHIP

It should be obvious that specific assets really belong to the firm, and not to the individual partners. However, each individual partner is entitled to his "interest in the partnership" which is best described as an intangible interest that includes a partner's proportionate share of the assets and his proportionate share of the liabilities together with his interest in profits and rights to management. With respect to assets and liabilities, for example, if our hypothetical appliance partnership owned $100,000 in assets, and owed $40,000 in liabilities, the total partnership interests would equal $60,000. If the partnership agreement provided that each partner shared equally, each partner's interest in the firm would be valued at $20,000. Thus, upon death of one partner, his heirs would be entitled to his interest in the firm, meaning they would have the right to be paid $20,000 from the two surviving partners. Similarly, any partner could assign his right to the $20,000 "equity" to any person outside the partnership who would be willing to purchase it.

The partner's interest in the partnership is initially determined by the partner's capital contribution to the firm. This is one reason why a value must be assigned to the contributions in the agreement. Thereafter, additional capital contributions will increase the contributing partner's interest, and subsequent profits will also increase the partner's interests as they are distributed in the proportions specified in the agreement. Conversely, if a partner withdraws funds from the partnership, the interest will be reduced. To illustrate, suppose the partners in the appliance business began as follows: Smith contributed his delivery truck, valued at $5,000; Jones contributed $10,000 in cash to buy inventory and to lease a store; and Williams did not contribute tangible property but agreed to manage the store. The agreement provides that the partners will share equally in profits and losses. At this point each partner's interest is equal to his or her contribution. Williams has no interest since she has yet to contribute anything. During the year, Smith contributed a cash register valued at $400, and Jones withdrew $500 in cash for personal reasons. At the end of the year the business showed a profit of $3,000. The partners' respective interests in the partnership will be as follows: Smith's interest is $6,400, including the value of the truck and cash register, plus his share of profit; Jones' interest is $10,500, including his initial cash contribution, plus his profit, less his withdrawal; and Williams' interest is $1,000, all of which came from profit.

This intangible "interest in the partnership" is the partner's personal property right in the firm. It is considered to be a personal asset which he owns just as he owns his home or other personal possessions. Consequently, this property interest will pass to his heirs upon death, and may be reached by his individual creditors for unpaid obligations. Similarly, a partner may assign this interest to an outsider and thereby confer his proportionate rights to profits and the value of the assets on the assignee.[13] It should be noted that an assignment of a partner's interest in the firm does not make the assignee a partner, since no person may become a partner without the consent of all other partners.[14] Nor does the assignee acquire any right to interfere with the management of the business. The assignee's sole right is to receive the profits and assets to which the assigning partner is entitled.[15]

§ 2.04 MANAGEMENT OF THE PARTNERSHIP

Right to Manage

The right to manage and control the affairs of the partnership is governed by the Uniform Partnership Act and by the agreement of the partners. In the absence of an agreement to the contrary, all general partners have equal rights in the management and conduct of partnership business.[16] Of course, it is possible (and perhaps desirable) to specify by agreement the specific management responsibilities and limitations for each partner.

Since each partner is the agent for the partnership, every act she performs on behalf of the firm must be an authorized act. Authority may come from specific provisions in the agreement itself, or from the vote of the partners in the manner specified in the agreement or by law. The Uniform Partnership Act provides that decisions regarding the ordinary matters of partnership business are to be made by majority vote of the partners.[17] Each partner, regardless of his or her contribution or share of profits, will have one vote on such matters. Thus, under the statute, even though one partner has contributed 95% of the capital and is entitled to 95% of the profits, that partner will have an equal voice with the other partners in management matters.

If the statutory management scheme is deemed desirable by the partners,—that is, equal rights in management and rule by majority vote—a clause reciting this scheme should be included in the agreement.

EXAMPLE 2.04A: Management

All partners shall have equal rights in the management of the partnership business. Decisions shall be by majority vote, each partner having one vote, except as otherwise provided in this Agreement.[18]

On the other hand, if not all partners will be actively engaged in the management of the business, it may be appropriate to appoint a managing partner or managing partners to control business affairs. In drafting such an appointment, it is good practice to specify the authority of the managing partners with reasonable detail and to provide a method for resolution of disagreement between multiple managing partners.

EXAMPLE 2.04B: Managing Partners

The management and control of the partnership business shall be vested in Fern Portnoy, James Lyons and Scott Charlton. Such managing partners shall have and are hereby given the sole power and authority:

(a) To contract and incur liabilities for and on behalf of the partnership.

(b) To borrow for and on behalf of the partnership from time to time such sum or sums of money which in their sole discretion is necessary to the conduct of the business of the partnership, and to mortgage, pledge or otherwise encumber its assets to secure the repayment of such monies so borrowed.

(c) To make all contracts for and on behalf of the partnership generally in the conduct of its business.

(d) To employ and discharge all employees, including any of the other partners who may be so employed in respect to the transaction of the partnership business.

(e) To otherwise carry on and transact or cause to be carried on and transacted, under their sole supervision and control, all of the other business of the partnership.

(f) To determine whether or not at any accounting period the profits, if any, of the partnership shall be apportioned and distributed, in whole or in part, to the partners, or retained and continued in use in the business of the partnership.

In the event of disagreement among the managing partners a decision by the majority of them shall be binding upon the partnership. If any one or two of the managing partners shall die or retire from the partnership business or become unwilling or unable to act as a managing partner, the management and control of the partnership business shall be vested in the remaining partner or partners.

It is understood and agreed that the managing partners shall consult and confer with the other partners before taking any steps resulting in any substantial change in the operation or policies of the partnership affairs, or the sale of any portion of the partnership assets other than in the usual course of business, or in any manner affecting the partnership business unusually as judged by the ordinary operation of the partnership business.[19]

With reference to the last paragraph of the example, it should be noted that the Uniform Partnership Act prohibits certain acts outside of the ordinary course of business unless all partners consent. These acts include the assignment of partnership property in trust for creditors, sale of good will of the business, confession of judgment against the firm, submission of a partnership claim to arbitration, and any other act which would make it impossible to carry on the partnership business.[20] Managing partners could not accomplish these acts unless all partners approved. In any case it is advisable to require the managing partners to refer unusual matters affecting the business to a committee of the whole for resolution rather than granting the managers unfettered discretion in these matters.

In a negative sense, it is possible to provide by agreement what the partners may *not* do without the unanimous consent of the other partners. Such a provision will usually include those acts specifically prohibited by the Uniform Partnership Act, in addition to other specific acts deemed appropriate by the partners.

EXAMPLE 2.04C: Limitations on Authority

Unless authorized by the other partners one or more but less than all the partners have no authority to:

(a) Assign the partnership property in trust for creditors or on the assignee's promise to pay the debts of the partnership.

(b) Dispose of the good will of the business.

(c) Do any other act which would make it impossible to carry on the ordinary business of a partnership.

(d) Confess a judgment.

(e) Submit a partnership claim or liability to arbitration or reference.

(f) Make, execute or deliver for the partnership any bond, mortgage, deed of trust, guarantee, indemnity bond, surety bond or accommodation paper or accommodation endorsement.

(g) Borrow money in the partnership name or use partnership property as collateral.

(h) Assign, transfer, pledge, compromise or release any claim of or debt owning to the partnership except upon payment in full.

(i) Convey any partnership real property.

(j) Pledge or transfer in any manner his interest in the partnership except to another partner.

(k) Do any of the acts for which unanimity is required by other paragraphs of this Agreement.[21]

Duties and Compensation

Partners are expected to devote their full time and attention to the activities of the partnership. This duty flows from a traditional reality that partners customarily participated in the conduct of the business, and it may have limited justification in modern practice. Moreover, the law denies partners the right to remuneration for services performed on behalf of the firm,[22] since their compensation is supposed to come from profits generated by their services and shared in a manner provided by the agreement.

If the parties desire a modification of the legal rule, matters such as salary and devotion to duty should be treated in the partnership agreement.

EXAMPLE 2.04D: Salaries and Duties

There shall be paid to each partner the following monthly salaries: To Anne Berardini, Two thousand dollars; to Michael Corrigan, One thousand five hundred dollars; etc. No increase in salaries shall be made without unanimous agreement. The payment of salaries to partners shall be an obligation of the partnership only to the extent that partnership assets are available therefor, and shall not be an obligation of the partners individually. Salaries shall, to this extent, be treated as an expense of the partnership in determining profits or losses.[23]

EXAMPLE 2.04E: Expense Allowance

An expense account, not to exceed Two hundred dollars ($200.00) per month, shall be provided for each partner for his actual, reasonable and necessary expenses, in engaging in the business and pursuits of the partnership. Each partner shall be required to keep an itemized record of such expenses and shall be paid once each month upon the submission of such statements of records.[24]

Devotion to duty may be covered by specifying the responsibilities of each partner.

EXAMPLE 2.04F: Devotion to Duty

Each partner shall devote his or her entire time and attention to the partnership business, except that each may devote reasonable time to civic, family and personal affairs; and except that Peter McLaughlin shall be permitted to pursue the business of selling magazine subscriptions during his own time and at hours other than the business hours of the partnership business.

§ 2.05 PROFITS AND LOSSES

Sharing of profit is one of the important elements of co-ownership, the crux of the partnership. We have already considered the fact that a partner is not entitled to expect compensation for his services. Hopefully, his reward is a rich share of profits from the successful business he has fostered. Unfortunately, the right to enjoy profits carries with it the obligation to bear losses—perhaps the only distressing characteristic of co-ownership.

The agreement between partners usually states the proportion in which profits and losses will be shared, and these provisions may be as simple or as complex as the parties desire. If the partners have not specified any provision for profit sharing, the Uniform Partnership Act provides that they shall be divided equally, regardless of the contributions of the parties. The Act further provides that losses shall be shared in the same proportion as profits.[25] Stated another way, if the agreement is silent, profits and losses are shared equally. If the agreement provides that profits shall be shared in a 75%/25% proportion between two partners, losses will be shared in the same proportion. It should be noted that there is no requirement that profit-loss sharing have any relationship to the respective capital contributions of the partners. For example, one partner may contribute all of the cash and property to the firm while the other only contributes services, but they may share profits and losses in any agreed proportion.

Unusual profit distribution formulas may be used in special cases. For example, the parties may agree that one partner should have a preference to profits for some compelling reason.

EXAMPLE 2.05A: Preferential Distribution of Profit

> As part of the consideration for Keith Burn joining the firm, it is understood that for a period of two years, in the annual distribution of profits, he shall receive a cumulative preference of ten per cent on his share; that is, out of the annual profits there shall first be a distribution to him up to ten per cent on his share of the capital, and also to cover any deficiency from said ten per cent in any previous years, and then a distribution pro rata to the other partners up to ten per cent on their shares of the capital, and any surplus profits shall then be distributed among the partners according to their respective shares of the capital.[26]

This example not only establishes a profit preference, but it also requires that profit be distributed in the same proportions as capital contributions. Lest the point be lost for being too obvious, although the partners are not required to distribute profits in the same proportion as they have contributed capital, they are certainly permitted to do so.

It is also good practice to include provisions for computing "net distributable profit," so as to avoid any later dispute among the parties. These provisions are particularly appropriate when the partnership is likely to incur unusual obligations which otherwise may not be included as expense items for profit computation.

EXAMPLE 2.05B: Division of Profits and Losses and Computation

> The net profits of the partnership shall be divided and the net losses of the partnership shall be borne in the following proportions: Fern Portnoy, 25%; Michael Corrigan, 10%; etc. Net profits of the partnership for any period shall be made by deducting from the gross profits disbursements made by or on behalf of the partnership for the usual and customary expenses of conducting the business, taxes chargeable to and paid by the partnership, reserves for taxes accrued but not payable, interest on all interest-bearing loans of the partnership, salaries paid to employees and partners, reserves for depreciation of partnership property and contingencies, including bad debts, allowance for accruing liabilities, and any and all other disbursements made by the partnership during such period incidental to the conduct of the business, excepting, however, payments to the partners on account of partnership profits.[27]

In drafting the partnership agreement, a distinction may be made between profits, losses, deductions, credits and cash. All of these items may be allocated the same way among partners, but a disproportionate allocation may be appropriate among partners in some cases. First, it should be understood that although these allocable items are related to each other, they are also quite separate and distinct and will have varying importance in the eyes of the various partners. A business may operate "profitably" but because of certain deductions (such as depreciation on buildings owned by the business) may incur losses for tax purposes which are shared equally unless allocated differently in the agreement. Nevertheless, because revenues have exceeded expenses for which expenditures were required (depreciation not being such an expense) there may be a surplus of cash in the business which is available for distribution to partners. Also, various tax credits might be available to the business which are to be allocated and passed through to the partners.[28]

Add to these variations the preferences of the partners. Partner A may have contributed all or most of the capital of the partnership and may prefer to receive his cash back as soon as possible with a disproportionately high allocation of distributable cash. Partner B may have a very low income and a high cash need while Partner C may have a very high income and a low need for cash. As between them, it may be appropriate to give B more cash and profits and to give C more losses and credits to maximize B's enjoyment of life and minimize the tax which C suffers because of her high income. You may recognize correctly that these disproportionate allocations are a method by which tax can be avoided. Do not think for a minute that the Internal Revenue Service is in the dark about them. The Service will not allow a disproportionate allocation without significant economic justification.[29]

For example, if Partner A contributed all of the cash to start the business it may be justifiable economically to give him a preference to cash. If Partner B does most of the work managing the business, it may be justifiable economically to give him profits and cash in higher proportion than the others. But what can you say Partner C did to justify her receipt of losses? There's the rub. You must be able to justify *all* allocations or the Internal Revenue Service may re-allocate items of profit, loss, deductions and credits in a way which will produce additional tax to the partners.

An example of a disproportionate allocation of some of the items would look like this:

EXAMPLE 2.05C: Allocations of Profits, Losses, Deductions and Credits

Except as otherwise provided herein (dealing with the allocation of the proceeds upon sale or other disposition of the assets of the Partnership), ninety-five percent (95%) of the net profits, and losses of the Partnership and each item of income, gain, loss, deduction or credit entering into the computation thereof, shall be allocated to the partners other than the managing General Partner in accordance with their respective capital contributions. Five percent (5%) of such net profits, and losses and items of income, gain, loss, deduction or credit entering into the computation thereof, shall be allocated to the Managing General Partner.

EXAMPLE 2.05D: Distribution of Cash

One hundred percent (100%) of net cash, including amounts required to be retained by the Partnership pursuant to this Agreement, shall be allocated to the partners other than the Managing General Partner and no cash shall be allocated to the Managing General Partner until the partners other than the Managing General Partner have received cash distributions totaling $1,925,000. Thereafter, ninety-five per cent (95%) of net cash shall be allocated to the partners other than the Managing General Partner and five per cent (5%) of net cash shall be allocated to the Managing General Partner throughout the term of the Partnership. The Managing General Partner shall determine the net cash available for distribution after establishing a reasonable reserve for replacements, contingencies and operating capital, and after satisfying other obligations of the Partnership then due and payable. The Managing General Partner shall distribute the net cash available for distribution no less frequently than quarterly.

§ 2.06 LIABILITY OF PARTNERS

Like sole proprietorships, general partnerships suffer the disadvantage of unlimited liability for each partner. If the assets of the partnership are inadequate to pay partnership creditors, the personal assets of the individual partners may be reached to satisfy these obligations.[30] In one sense there is an advantage to a partnership over a sole proprietorship since liabilities will be apportioned to the partners pro rata and no one person is required to bear the full responsibility. On the other hand since each partner has the capacity to bind the partnership, the potential risk of liability is proportionately increased.

The element of unlimited liability is a substantial disadvantage to hazardous and speculative enterprises, and it further imposes an unwelcome burden on a partner who enjoys substantially greater personal wealth than his fellow partners. It is possible, as with a sole proprietorship, to insure against potential liability whenever it may be anticipated. Moreover, the partnership may negotiate agreements with outsiders which provide that liability on the contract shall be limited to the partnership assets and will not extend to the individual assets of the partners. Additional protection for the partner's individual assets is provided by a rule called "marshaling of assets" which requires that firm creditors must first look to firm property for satisfaction of their obligations, and only if partnership assets are inadequate may they pursue the individual assets of the partners. Nevertheless, a partnership's potential unlimited liability, which can never be completely circumscribed by insurance, agreement, or rule for priority of assets, limits the desirability of the partnership form of business enterprise.

The Uniform Partnership Act creates an obligation of the partnership to indemnify a partner who has paid expenses or incurred liability in the ordinary course of partnership business.[31] Thus, if a particular partner used his individual assets to pay firm creditors he is entitled to be reimbursed for the appropriate share of the other partners. Although the law grants this right of indemnification, the specific authority for indemnification should be specified in the agreement.

EXAMPLE 2.06A: Indemnification

The partnership shall promptly indemnify each partner in respect of payments reasonably made and personal liabilities reasonably incurred by him in the ordinary conduct of its business, or for the preservation of its business or property.[32]

A partner who leaves the firm, by retirement, withdrawal, or for some other reason, remains individually liable for debts incurred while he was a partner.[33] As will be discussed in detail later, whenever a partner leaves the firm a technical dissolution of the partnership occurs.[34] If the remaining partners are to continue the business, certain notice is necessary to persons who transacted business with the former firm in order to relieve the withdrawn partner from future liability. Personal notice must be given to persons or companies who extended credit to the firm while the retiring partner was a member. Notice by publication will suffice for other persons who dealt with the firm.[35] Usually the remaining partners will agree to indemnify the withdrawing partner from any further liability for firm obligations. An indemnification clause as provided in Example 2.06A, above, will create the obligation if included in the original partnership agreement.

New partners to the firm are individually liable for obligations existing when they join only if they specifically agree to be. Absent an assumption of these obligations, their individual assets may not be reached to satisfy existing obligations. They are, of course, individually liable for any obligations incurred after they become partners.[36]

§ 2.07 DISSOLUTION AND TERMINATION OF THE PARTNERSHIP

The law provides that the partnership enterprise is dissolved whenever any partner ceases to be associated with the partnership business. In this respect, a partnership is very much like a sole proprietorship, where the business entity expires when the proprietor retires, dies, or otherwise disassociates herself with the business. However, the fact that the partnership is an association of two or more persons adds an element of continuity to the partnership business, since dissolution does not necessarily require termination of the business. In fact, the partnership may be technically "dissolved", yet the remaining partners (absent the partner who has disassociated herself with the business) may continue the business without her. By a strict interpretation of partnership law, a new partnership is created immediately and is governed essentially by the terms of the original agreement. In many cases, however, dissolution of the partnership requires termination and "winding-up" of the business. It is important to explore the circumstances which cause dissolution and the specific statutory authority to continue the partnership business without winding-up. The partnership agreement plays an important role in these matters, and specific clauses will be discussed.

Causes of Dissolution

Dissolution of partnership may result from a variety of causes. It should be obvious that since a partnership is created by agreement of the partners, it may also be dissolved by agreement. Their agreement may state a specific date for termination of the business, or may provide that the business should be dissolved upon the happening of a contingent event.[37] For example, the parties may agree that the partnership will be dissolved on July 1, 1999, or if the business sustains operating losses for five consecutive months, whichever occurs first. The latter contingency may be a realistic agreement insofar as the parties usually seek to avoid operating a business at a loss. The former provision, specifying a date for dissolution, is unrealistic and uncommon, since most partnerships hope to generate profits from a continuing business and the abrupt termination because of a specific provision in the original agreement could result in considerable loss of "going concern" values.

EXAMPLE 2.07A: Dissolution Contingent Upon Results of Operations

If the operation of the business over a period of five consecutive months or more discloses an average net monthly profit of less than $500.00, the managing partner is hereby authorized and empowered to negotiate a sale, exchange or other disposition of the entire partnership business upon the best possible terms available at such time, and in the event of such disposition of the partnership business the proceeds derived therefrom after the payment of the necessary costs and expenses of such disposition of the business shall be applied first to the payment of the debts of the business according to their respective legal priority, and if any balance of such proceeds shall remain after the payment and satisfaction of the debts, obligations and liabilities of the business, the same shall be divided equally between the partners. Either partner may become a purchaser of the business at any such sale.[38]

Whether or not the original agreement contains any provision for dissolution, all partners may unanimously agree to dissolve the firm at any time.[39] Better practice requires that a subsequent agreement of the partners to dissolve the firm should be reduced to writing.

EXAMPLE 2.07B: Agreement to Dissolve

The partnership heretofore subsisting between us, the undersigned Michael Corrigan, Karen Burn, and Anne Berardini, under and pursuant to the within articles of partnership, is hereby dissolved, except so far as may be necessary to continue the same for the liquidation and settlement of the business thereof. The said Michael Corrigan [or each of the undersigned] is authorized to sign in liquidation.

Dated ________, 19 ___. *Signatures]*[40]

Since one of the primary elements of partnership is a "voluntary association" of persons, any partner who no longer desires to be associated with the firm may withdraw at will, and thereby cause a dissolution.[41] If the original agreement provided that the partnership would continue for a specified term, the withdrawal of a partner at will may result in liability to the other partners for breach of the agreement, but the partnership is nevertheless dissolved. However, if the original agreement between the partners was indefinite regarding the duration of the partnership, this right or power to withdraw at will may be fully exercised, without regard to the harm it may cause to the business or the other partners. Of course, the law requires that the withdrawing partner must act in good faith in order to escape a surprise dissolution with impunity.

A partner may not have the choice of leaving the firm if the agreement provides that he may be expelled by the other partners. Such provisions may be drafted in the original agreement and usually will be conditioned upon some misconduct by a party, such as neglecting the business, refusing to pay an assessment and so on. The expulsion provision may be as general and broad as the parties desire, but provisions permitting expulsion "without cause" or "in the best interests" of the partnership are rare. If an expulsion provision is included in the agreement, the "innocent" partners may cause a dissolution of the firm by exercising their right to expel a partner.[42]

Dissolution may be required by operation of law. The partnership business may be declared unlawful, as was the case for liquor stores operated as partnerships when prohibition was imposed. More frequently, the partnership is dissolved whenever a partner dies, or becomes individually bankrupt.[43] Both of these events cause dissolution because the partner's interest in the firm, which was once his personal property, becomes the property of his heirs or his trustee in bankruptcy, none of whom are partners. In both cases, the partner has withdrawn from the firm in a sense.

Finally, dissolutions may be decreed by a court whenever any partner becomes insane or incapable of furthering the partnership business, or conducts himself in a manner so that it is impracticable to carry on the business with him, or in any other case which renders a dissolution equitable under the circumstances.[44] Thus, when partners simply cannot agree on the proper operation of the business and their constant disagreement is detrimental to the success of the business, a court may, on application by a partner, dissolve the partnership. In some cases of disagreement between partners dissolution by a court may be the only available remedy since, although a partner may terminate the partnership at will, he may be risking liability for breach of the partnership agreement by causing dissolution.

Continuation of the Partnership Despite Dissolution

As a general rule, the dissolution of the partnership requires winding-up and termination of the business and liquidation of the firm assets. There are two major exceptions to this rule: the business may be continued by the remaining partners if the dissolution was "wrongful" or if their original agreement so provided.[45]

A "wrongful" dissolution results whenever a partner has caused a dissolution without having the right to do so. If a partner has been expelled from the firm for misconduct or a violation of the partnership agreement the resulting dissolution is wrongful and the remaining partners are permitted to continue the business without liquidation. Similarly, a partner who causes a dissolution by withdrawing from the firm in violation of the agreement has caused a wrongful dissolution. Whether or not withdrawal is wrongful, you will remember, depends upon whether the agreement specifies the duration of the partnership. Whenever the business is continued by the remaining partners following a wrongful dissolution, they are required to compensate or assure compensation to the disassociated partner for his interest in the firm and to indemnify him from partnership liabilities.[46]

Even when the dissolution is innocent, it may be desirable to continue the partnership business. Particularly in the case of the untimely death of a partner, liquidation following dissolution could result in unnecessary economic loss to the remaining partners. Continuation following an innocent dissolution requires an agreement to that effect between the parties, and while the agreement may be concluded legally after dissolution has actually occurred, it is far better to provide for continuation in the detached negotiations of the original agreement of the parties.

First, the agreement should contain continuation provision authorizing the remaining partners to proceed with the business.

EXAMPLE 2.07C: Right to Continue the Business

In the event of a dissolution caused by the retirement, death, withdrawal, permanent disability, or bankruptcy of a partner, the remaining partners shall have the right to continue the partnership business under the same name by themselves or with any other person or persons they may select, but they shall pay to the other partner or his legal representatives the value of his interest in the partnership as of the date of dissolution.

Second, a method for computing the outgoing partner's interest and a method of payment should be specified.

The various methods of providing for payment of outgoing interests, be they the result of death, withdrawal, or other act dissolving the partnership, have different tax consequences which are sufficiently complex to be avoided happily here. An enterprising student may explore these various methods in detail by referring to other sources.[47] Frequently used methods of settling the outgoing interest include 1) the purchase of the interest of the outgoing partner by the continuing partners and 2) liquidating distributions from the partnership to the outgoing partner, which, in effect, result in the sale of the partner's interest to the firm itself. The latter may provide for guaranteed periodic payments or payments out of partnership profits or partnership property.

EXAMPLE 2.07D: Payments for Partnership Interest

For the interest of a retiring or deceased partner in the partnership property, including good will, the partnership shall pay to the retiring partner, or to the successor in interest of the deceased partner, $20,000.00 in each of the five years following the retirement or death of a partner, which amount may not be prepaid except with the consent of the payee. It is the intention of the partners that the payments provided under this paragraph shall qualify under 26 U.S.C.A. (1954 I.R.C.) § 736(a), and shall constitute ordinary income to the recipient and reduce the taxable income of the continuing partners.[48]

The agreement will specify the method of disposing the partner's interest, and the method of computing the compensation to him. An example of a provision permitting the purchase of the outgoing partner's interest by the remaining partners follows:

EXAMPLE 2.07E: Purchase of Partner's Interest

Upon the death, withdrawal, or insolvency of either of the partners during the existence of the partnership, the surviving or remaining partner shall purchase all the right, share and interest of the deceased, withdrawn, or insolvent partner in all the partnership business and property, and shall assume all the then existing liabilities of the partnership. The price to be paid for such purchase is hereby fixed and agreed upon as follows: It shall be the amount stated as the net value of the share in the partnership of the deceased, withdrawn or insolvent partner in the balance sheet of the first of January next preceding his death or withdrawal or insolvency, together with interest thereon from the date of said balance sheet at the rate of ten per cent per annum until paid. Such purchase price shall be paid as follows: fifty per cent thereof within six months from the date of such death, withdrawal or insolvency and the remainder at such times and in such amounts as may suit the convenience of the surviving or remaining partner; provided, that the whole thereof shall be paid within two years. The estate of the deceased partner, or the withdrawing or insolvent partner shall not be entitled to share in any increase or profits gained, nor be liable for any losses incurred, in the business after the first of January next preceding his death, withdrawal or insolvency but all such profits shall belong to, and such losses be borne by, the purchasing partner.[49]

One other important factor must be considered for purchase of a deceased partner's interest—life insurance. If the partnership agreement provides for the purchase of a deceased partner's interest, life insurance "funding" will insure that money will be available to consummate the intention of the parties when a partner dies. The partnership may apply for and purchase life insurance on the lives of the partners, or the partners may individually apply for and purchase insurance on each other. The former is commonly called an "entity purchase" plan, and the latter is referred to as a "cross-purchase" plan. Such insurance agreements usually are executed separately and are not a part of the partnership agreement. However, since they are intended to fund the purchase of the deceased partner's interest, the provisions of the insurance agreement and the insurance acquired thereby should be consistent with the buy-out arrangements specified in the partnership agreement.[50]

In addition to the clauses permitting continuation of the business and disposition of the withdrawing or deceased partner's interest, a properly drafted agreement should also require the continuing partners to assume and pay all existing partnership obligations. Moreover, since creditors will have extended credit to the firm before withdrawal, and the withdrawing partner may remain liable for future obligations unless proper notice is given, the agreement should specifically require such notice.

EXAMPLE 2.07F: Notice of Dissolution

Actual notice of dissolution shall be given to all persons who have had dealings with the partnership during the two years prior to dissolution.[51]

Finally, it may be desirable to include an anti-competition clause, to prevent, as much as possible, the withdrawing partner from competing against the firm or divulging its trade secrets.

EXAMPLE 2.07G: Non-Compete Agreement

The retiring partner shall not for a period of two years from the date of his retirement, either alone, or jointly with, or as agent for, any person, directly or indirectly, set up, exercise or carry on the trade or business of metal processing and plating within 500 miles from Denver, Colorado, and shall not set up, make or encourage any opposition to the said trade or business hereafter to be carried on by the other party or his representatives or assigns, nor do anything to the prejudice thereof, and shall not divulge to any person any of the secrets, accounts or transactions of or relating to the partnership. For any violation of this stipulation the parties bind themselves to each other in the sum of Five thousand dollars, to be deemed liquidated damages, and total extinction of this covenant, and not in the nature of a penalty.[52]

Termination and Winding Up

In the event the remaining partners do not desire to continue the business, a dissolution requires that the partnership be terminated and wound up. For these purposes, the remaining partners have the right to complete all pending partnership business and thereafter collect and dispose of the assets, pay firm creditors and enjoy whatever is left. Depending upon the size of the partnership, it may be desirable to appoint one partner to liquidate the business. Any remaining partner is eligible. If a liquidating partner is named in the original agreement there should be a successor named in the event of his death. In any case, following a dissolution, the partner's authority to act for the partnership is limited to acts necessary to wind up the firm's affairs.[53]

The Uniform Partnership Act provides that in the event of winding-up, the assets of the firm must be distributed as follows:

1. Claims of creditors other than partners;
2. Claims of partners other than for capital and profits;
3. Claims of partners in respect of capital;
4. Claims of partners in respect of profits.[54]

The partners may agree to vary this distribution priority in any manner, except that partnership creditors, who are not bound by the terms of the partnership agreement, must always be completely satisfied before the partners are entitled to share the assets among themselves.

§ 2.08 TAX CONSIDERATIONS OF GENERAL PARTNERSHIPS

A major consideration in the selection of the partnership as the appropriate form for a business enterprise is income taxation. The partnership itself pays no federal income tax. Instead, each partner is required to declare his share of the partnership income on his individual tax return. The partnership is thus treated very much like the sole proprietorship for tax purposes. The only difference is the fact that the business income will be spread over the partners in their respective proportions, rather than being applied completely to one person.

The partnership files an information tax return with the federal government, and the return is used to ascertain whether the partners have declared and paid tax on their proportionate shares of income. It does not matter whether the profits have actually been distributed to the partner during the year. Even if profits are retained in the business, the partners must declare as though all profit had been distributed. When losses are considered, this rule operates as an advantage of partnership taxation: all losses from the business are attributed personally to the partners and may be offset against other personal income.

Because of the similarity of a partnership to a sole proprietorship for tax purposes, the latter business form is not a viable alternative for tax advantage. A corporation, however, is taxed as an entity at special corporate rates, and a choice between a partnership and a corporation may depend on the tax differences. The comparison of corporations and sole proprietorships in Section 1.05 is complicated here by the fact that two or more persons are involved in the partnership. The business form which is most desirable for one person may work a disadvantage to the others. For example, if one partner has relatively insignificant outside earnings, the partnership form may be advantageous if his share of profits from the firm are taxed at a lower rate under the individual tax rate schedule than at corporate rates. It was suggested in Section 1.05 that any taxable income figure less than $21,500 for a married person filing a joint return and $14,500 for a single person would work this result. However, if his partner has significant outside earnings, the application of individual rates to his partnership income may result in payment of considerably more tax than at corporate rates. Consequently, the only general statement which may be safely advanced regarding partnership tax considerations is that each case must be separately evaluated, based upon the potential profit or loss, the expected distributions from the business and the individual financial status of the parties.

The sale of a partner's interest in the partnership results in a capital gain or loss much the same as a shareholder's sale of corporate stock.

When partnership property is sold, the capital gain or loss is attributable to the individual partner in the same proportion as profits, unless altered by agreement. Of course, since contributed property is partnership property, a special problem may arise with respect to its valuation for determining gain or loss. To illustrate the problem, consider the following case. Murlin and Short formed a partnership with equal rights to profits. Murlin contributed $5,000 in cash and Short contributed a machine which had a fair market value of $5,000, but an adjusted basis (the cost to Short plus improvements by him) of $4,000. The basis to the partnership, for computing gain or loss, is the same as the basis to the individual—$4,000. If, during the year, the machine is sold at a fair market value of $5,000, the partnership will receive a capital gain of $1,000, which is attributable to the partners in their respective proportions. Thus, both partners must claim a $500 capital gain that year. This should concern Murlin, who has now paid half the tax on Short's contributed asset. If Short sold the machine when he owned it as an individual, and contributed cash to the partnership, Short would have been individually responsible for the $1,000 gain. Similarly, the rule may work a hardship on Short. Suppose the adjusted basis of the asset is $7,000, and he contributes it when fair market value is $5,000. When the machine is sold, the capital loss of $2,000 will be shared between the partners. Murlin thereby acquires a tax benefit which Short really deserves. More complicated matters arise in determining the fair allocation of depreciation on the asset between the parties. The Internal Revenue Code offers a solution to the problem by permitting the parties to agree to the manner in which depreciation, depletion, gain or loss of contributed property will be allocated between the partners to account for variations of this sort.[55] The agreement should clearly specify the intent of the partners in this regard.

EXAMPLE 2.08A: Allocation of Tax Related Items

The partners understand that for income tax purposes the partnership's adjusted basis of property contributed by Maynard Short differs from the fair market value at which such property was accepted by the partnership at the time of its contribution. The partners agree that in determining the taxable income or loss of the partnership and the distributive share of each partner's depreciation or gain or loss with respect to such contributed property shall be shared among the partners, so as to take account, to the extent permitted by the Internal Revenue Code, of the variation between the basis of such property to the partnership and its fair market value at the time of contribution. It is the intention of the partners by this provision to take advantage of the election authorized by Section 704(c)(2) of the Internal Revenue Code.[56]

Many other tax considerations apply to the partnership form, most of which are complicated by the fact that the partnership operates as an entity, acquiring property and earning money as an apparently separate legal unit, but is treated merely as an aggregation of individuals for tax purposes. Thus, the distributions of current assets or in liquidation, determination of partnership taxable year, and transfer of partnership interests all pose unique tax problems which are ably covered by other authors.[57]

§ 2.09 FORMATION AND OPERATION OF THE GENERAL PARTNERSHIP

In many respects, the formation of a general partnership parallels the formation of the sole proprietorship. The obvious difference is the most important formality for the partnership—the agreement. In the true spirit of procrastination, the drafting of the agreement will be saved until last.

Selection of Name

A partnership may operate under any name it chooses, provided the name is not deceptively similar to another well-known company so as to constitute a deceptive trade practice. As with sole proprietorships, any partnership fictitious name must be registered in most states under the assumed name statutes.[58]

If all of the partners' surnames are used in the firm name, the name is not considered to be fictitious; but if less than all partners names are used in the firm title, it is fictitious. Thus, a partnership formed by Levine, Conviser and Chess may use all names without registration; but if the firm name is "Levine and Chess" registration is required. Similarly, the name "Levine & Co." would require registration.

The sanctions imposed on a partnership for failure to file the assumed name information are the same as those imposed upon sole proprietors.[59]

Governmental Formalities

Partnerships may conduct any legal business just as a natural person may do, and where state law imposes particular licensing requirements, the partnership must conform. Consequently, sales tax licenses must be obtained where appropriate, and the partnership must obtain any necessary licenses peculiar to the particular business conducted.

Tax identification numbers are necessary for partnerships since informational returns are filed annually with federal and state authorities. If employees are to be hired, social security and unemployment compensation laws must be considered.

Interstate Business

Partnerships are not usually subjected to any peculiar formalities for doing business in states other than the state where the firm is formed. Of course, they must comply with local licensing requirements and the name of the firm must be registered appropriately with the foreign state. At least one state, New Hampshire, has required "qualification" of a foreign partnership by insisting on a registered office and agent within the state and the payment of a fee for the privilege of doing business.[60] This qualification procedure is not popular, and is clearly a minority approach to interstate general partnership business.

The Agreement

Without the partnership agreement, the simple organizational formalities for the partnership would be extremely attractive. So far, organizing the partnership has been essentially the same as organizing the sole proprietorship. To be honest, it is possible to form a partnership without anything more than a "gentleman's agreement" to do business together. However, it should be apparent from the previous discussions of the nature of the partnership that informality is usually a fatal oversight. There is an old axiom that "one should never be a partner with a friend", recognizing the common tendency of human nature toward disagreement in business transactions. There eventually will be some discord (or at least a friendly disagreement) between the parties, and an informal agreement offers no guidance whatever in the resolution of such disputes. Thus, the only proper practice requires the drafting of a comprehensive agreement between the parties, carefully specifying their purposes, contributions, management authority, voting powers, duties, rights and responsibilities. The partnership agreement, which is also frequently called the "articles of partnership", must be tailored to the specific desires of the future partners. The following checklist with examples and references to the detailed discussions in this Chapter may be used as a guide:[61]

1. The names and addresses of the partners

EXAMPLE 2.09A: Agreement

Agreement made this ___day of _________, 19 ___, between James A. Murlin, whose address is 526 Park Avenue, New York, New York, and Maynard P. Short, whose address is 1901 K Street, N.W., Washington, D. C., (hereinafter referred to as individuals or collectively as "Partners").

2. Recitals—The background of the partners' business relationship may be stated in order to explain the agreement. Recitals of such information are inserted at the beginning of the agreement. The recitals further serve to state the intent to form a partnership and may explain the business objectives of the enterprise.

EXAMPLE 2.09B: Recitals

WHEREAS, Murlin has acquired certain business expertise in the manufacturing and marketing of rubber bicycle tires; and

WHEREAS, Short has the financial ability to contribute certain sums of money for the manufacturing and marketing of rubber bicycle tires; and

WHEREAS the parties intend to operate a business for the manufacture and marketing of rubber bicycle tires and desire to do so under the form of a partnership;

NOW, THEREFORE, it is agreed:

3. The Name of the Partnership—If the firm name does not contain the surnames of the individual partners, a tradename affidavit is required (see Section 209(a)).

EXAMPLE 2.09C: Name

The name of the Partnership shall be Shoylin Associates.

4. Place of business—The proposed offices of the firm should be stated, with permission for the partners to establish other offices as appropriate. If the firm will operate branch offices, and their location is known, they should be specified. If a multi-state business is contemplated, the partners should have authority to establish offices in other states.

EXAMPLE 2.09D: Place of Business

The principal place of business of the firm shall be located at 536 Park Avenue, New York, New York, or such other place as shall be designated by the partners from time to time. Branch offices may be located at 156 Cayuga Street, Ithaca, New York, and at a street address to be determined by agreement of the partners in Albany, New York. The partnership shall be authorized to conduct business and to establish offices in locations to be selected by agreement of the partners in the states of Connecticut, Rhode Island, and Pennsylvania.

5. Purposes—The description of the partnership business should be included in general terms, unless restrictive languange is dictated by the partners' objectives. Any intended restrictions upon the scope of partnership business should be detailed, or, following a specific description of the contemplated purposes, the agreement may provide that the part-

nership will operate "no other business." Be certain to include a provision which allows the partners to *agree* to enter into other ventures so the agreement will not be unduly restrictive.

EXAMPLE 2.09E: Nature of Business

The partnership shall engage in the business of manufacturing and marketing rubber bicycle tires, and in such other lawful business as is permitted in the jurisdiction of formation and as may be agreed upon from time to time by the partners.[62]

[or]

The partnership shall engage in the business of manufacturing and marketing rubber bicycle tires, and shall not engage in any other business or activity except as shall be directly related and incident to such business or except as shall be agreed upon from time to time by the partners.

6. Duration
   (a) The partnership may be formed for a definite term.
   (b) The partnership may be subject to termination by mutual agreement.
   (c) The partnership may be terminable at will when one partner gives the specified notice to the other partners.
   (d) The partnership may be terminated upon the completion of its purposes (e.g., sale of a parcel of real estate).
   (e) The partnership may be terminated upon the happening of a contingent event (e.g., continuous losses for a specified period).
   (f) All of the above.

EXAMPLE 2.09F: Duration

The partnership shall begin on May 1, 1981, and shall continue for the term of ten years thereafter.[63]

EXAMPLE 2.09G: Duration

The partnership shall continue for the full term of five years from the date of this agreement, and thereafter until thirty days' written notice is given by any of the partners to the others.[64]

7. Capital—The capital contributions of the partners should be described in detail. There are many possible variations of contributions (See Sections 2.02 and 2.08) but the following situations are most typical:
   (a) The partners' contributions will be in cash. The agreement should specify the amount of the contribution and the time of payment.

EXAMPLE 2.09H: Capital

The capital of the partnership shall be contributed in cash by the partners as follows:

James A. Murlin	$5,000.00
Maynard P. Short	$3,000.00

Such contribution shall be paid in full on or before May 10, 1981.[65]

   (b) One or more of the partners will contribute services. The value of the services and the treatment of the respective capital accounts should be discussed.

EXAMPLE 2.09I: Contribution of Services

James A. Murlin shall not be required to make a cash or property contribution to the partnership but shall devote his entire time to the partnership, and for such he shall be entitled to twenty per cent (20%) of the profits to be divided [*term for division*] among the partners. In the event that his monthly share of the profits shall exceed Three thousand dollars ($3,000.00), he shall contribute the excess to his capital account of business until the total amount of such contributions shall equal the capital contributions made by each of the other partners.[66]

(c) One or more of the partners will contribute tangible property. The agreed value of the property should be specified. Further, if the cost of the property to the contributing partner and the agreed value of the contribution are different, it is appropriate to consider and discuss the tax consequences of the precontribution gain or loss in the agreement (See Section 2.08).

EXAMPLE 2.09J: Contribution of Property

Maynard P. Short shall contribute property which the partners agree will be valued at Ten thousand dollars ($10,000.00). Such property is described in Schedule A attached hereto.[67]

(d) The partners may be required to furnish additional capital. The agreement should specify the circumstances under which additional contributions may be assessed (e.g., in the event of continuous losses for a specified period, or upon the vote of the majority of the partners). The partners' respective proportions for additional contributions and the procedure by which the partners will be notified of the contributions should be specified.

EXAMPLE 2.09K: Additional Capital Contributions

In the event that the cash funds of the partnership are insufficient to meet its operating expenses, the partners shall make additional capital contributions, in the same proportions in which they share the net profits of the firm.[68]

The managing partner, after determining a cash deficit, shall notify the other partners in writing at least ten (10) days prior to the date upon which such cash funds are needed and each partner shall be required to make such additional contributions on the date specified in the notice, or, if none, on the tenth day after the date of the notice.

(e) Excess contributions may be construed as advances, and be treated as loans to the firm. The authority to make such loans comes from the agreement, which should specify the need for consent of the other partners, the amount of the loan, and any desired restriction on the frequency of such advances.

EXAMPLE 2.09L: Loans and Advances

Any of the partners may, from time to time, with the consent of all of the other partners, advance sums of money to the partnership by way of loan, and each such advance shall bear interest at the rate of twelve per cent (12%) per annum.[69]

(f) Profits may be accumulated as capital.

EXAMPLE 2.09M: Accumulated Profits

Each of the partners shall be required to allow to remain in the business each year as a contribution to net worth of the partnership capital, an amount equal to thirty per cent (30%) of the partnership profits which would otherwise be distributed to him or her. Such contributions shall be allocated to or reserved in accounts for each of the partners, and shall remain in the business and be employed as capital for the business subject to further direction and order of the partners.[70]

(g) Capital may accumulate interest if the agreement so provides. If the partners agree that capital contributions should not accumulate interest, a statement to that effect should be included.

EXAMPLE 2.09N: Interest on Capital

No interest shall be paid to the partners on any contributions to capital.

or

Each of the partners shall be entitled to interest at the rate of ten per cent (10%) per annum on the amount of his respective contributions, payable semi-annually, on June 1 and December 1, of each calendar year.[71]

(h) If withdrawal of capital contributions is to be permitted, the agreement should detail the circumstances of withdrawal, limitations upon the amount of the withdrawal, if any, and any requirements for replenishing the capital account at specified times.

EXAMPLE 2.09O: Withdrawal of Capital

Each of the partners may withdraw from the partnership, for his own use, the sum of not exceeding Seven hundred dollars ($700.00) per month. If, at the close of each fiscal year, it is found that any of the partner's share withdrawn by him is in excess of his distributive share for that fiscal year, he shall forthwith refund the difference within a period not exceeding five days from the time of such determination.[72]

(i) If one partner allows profits to accumulate in a greater proportion than the others, the excees may be described as a debt to that partner, and the agreement may provide an interest rate to be applied to the excess amount.

EXAMPLE 2.09P: Individual Accumulation of Profit

All profits of the partnership during the year shall be allocated to the partners in their respective proportions in an income account which shall be subject to withdrawal by any partner from time to time. If a partner does not withdraw all of his or her income account during the year, the excess amount, not withdrawn, shall be treated as a loan to the partnership by the partner, and shall accumulate interest at the rate of ten percent (10%) per annum on the amount of the income account not withdrawn at the close of each calendar year, so long as such amount shall remain in the income account and is not withdrawn by the partner.

8. Salaries and expenses
   (a) Since partners are not ordinarily entitled to remuneration for their services (See Section 2.04), the authority to pay salaries must be

established by the agreement. Salaries may be contingent upon profits, or may be fixed by the agreement.

EXAMPLE 2.09Q: Salaries

There shall be paid to each partner the following monthly salaries: To James A. Murlin, three thousand dollars; to Maynard P. Short two thousand dollars; *etc.* No increase in salaries shall be made without unanimous agreement. The payment of salaries to partners shall be an obligation of the partnership only to the extent that partnership assets are available therfor, and shall not be an obligation of the partners individually. Salaries shall, to this extent, be treated as an expense of the partnership in determining profits or losses.[73]

(b) Expense accounts are common in business partnerships. The agreement should establish a maximum periodic amount; a procedure for submitting expenses; and a procedure for reimbursement. The agreement also should specify that only expenses incurred in furtherance of the partnership business will be reimbursed.

EXAMPLE 2.09R: Expenses

An expense account, not to exceed One hundred dollars ($100.00) per month, shall be provided for each partner for his actual, reasonable and necessary expenses, in engaging in the business and pursuits of the partnership. Each partner shall be required to keep an itemized record of such expenses and shall be paid once each month upon the submission of such statements of records.[74]

9. Profits and losses, deductions and credits—The agreement should establish a method for determining profit and loss. Simply providing for a determination of profit or loss by the partnership's accountant (or bookkeeper) using generally accepted accounting principles will create an objective standard for the determination which should avoid most disputes. It may be desirable to allow for any partner to question this determination by permitting another accountant (at the challenging partner's expense) to determine independently profits or losses, and then permitting arbitration (or some other objective determination) if the results vary more than a stated percentage from the original amount. In some cases, where extraordinary expenses are involved, such as legal fees for litigation or unusual travel and entertainment in the start-up period of the business, these expenses may be detailed in the agreement and excluded from the normal profit and loss computation so that they may be shared in some other agreed proportion.

Profits, losses, deductions and credits are usually shared in the same proportions, but not necessarily.[75] This clause is especially tailored to the desires of the partners. Some of the common schemes are:

(a) Equal sharing of profits, losses, deductions and credits.
(b) Sharing of profits, losses, deductions and credits according to the proportion of capital contributions.
(c) Deductions for depreciation and investment tax credits for certain contributed assets given to the partner who contributes the asset.
(d) All items allocated primarily to the partner who provides financial

backing until he or she receives profits (or tax benefits) equal to the capital contributed; then a primary allocation to the managing partner who is producing those profits.

(e) Losses caused by the willful neglect or default of a partner should be borne by him.

(f) A guaranty of profits may be set for certain partners. This clause usually requires the other partners to contribute any deficiency if the annual profit distribution does not exceed a certain amount.

EXAMPLE 2.09S: Allocations of Profits, Losses, Deductions and Credits

Seventy per cent (70%) of the net profits and losses of the partnership and each item of income, gain, loss, deduction or credit entering into the computation thereof, shall be allocated to the partners other than the Managing General Partner in accordance with their respective capital contributions. Thirty per cent (30%) of such net profits and losses, and items of income, gain, loss, deduction or credit entering into the computation thereof, shall be allocated to the Managing General Partner.

10. Cash Distributions—A partnership is a unique entity which allows for accumulations of cash when the business is actually producing a "paper" loss. For example, if the partnership operates an apartment building depreciation and interest may reduce the profits of the partnership (and may actually produce a loss) which rental receipts may exceed operating expenses. The agreement should provide for distributions of cash to the partners in certain proportions, and this determination should be made separately from the distribution of profits, losses, deductions and credits.

Again, the specific desires of the partners should be observed. Some of the common methods of cash distribution include:

(a) Cash is distributed in the same proportion as profits, losses, deductions and credits are shared.

(b) Cash is distributed to the partners who contributed cash or property as capital contributions, to the exclusion of any partner who is contributing only services, until a certain proportion of the contributed amount has been recovered.

(c) Cash is distributed to the partner who needs it the most, at least for a period of time.

(d) Cash is distributed to the partner who will be required to pay the most taxes as a result of partnership operations, thereby permitting that partner to pay the taxes from the cash distributed.

EXAMPLE 2.09T: Cash Distributions

One hundred per cent (100%) of the net cash of the partnership shall be allocated to the partners other than the Managing General Partner in accordance with their respective capital contributions for the first three years of the operation of the business, and no cash shall be allocated to the Managing General Partner during that time. Thereafter, twenty per cent (20%) of the net cash of the partnership shall be allocated to the partners other than the Managing General Partner and eighty per cent (80%) of the net cash shall be allocated to the Managing General Partner.

11. Books and records.
   (a) The fiscal year of the partnership must be established.

EXAMPLE 2.09U: Fiscal Year

The fiscal year of the partnership shall be from November 1 until October 31.

or

The fiscal year of the partnership shall be the calendar year.[76]

   (b) The method of accounting for the firm (accrual or cash) should be established. The agreement should further provide that generally accepted accounting principles will govern any matters not specifically covered by its terms.

EXAMPLE 2.09V: Method of Accounting

The partnership shall keep accounts on the accrual [or cash] basis. The accounts shall readily disclose items which the partners take into account separately for income tax purposes. As to matters of accounting not provided for in this agreement, generally accepted accounting principles shall govern.[77]

   (c) The location of the firm books and records must be established, and the partners' access to the books should be considered. Any restrictions upon a partner's right to inspect or copy books and records will be included here.

EXAMPLE 2.09W: Location of Books

The partnership books shall be kept at the principal place of business of the partnership, and every partner shall at all times have access to and may inspect and copy any of them.[78] [Or, all partners shall have access to such books and records only upon 72 hours prior written notice to the managing partner and during normal business hours.]

   (d) The bank accounts and other banking arrangements are stated in the agreement, including the persons authorized to sign checks, to borrow funds and otherwise to conduct banking transactions on behalf of the firm.

EXAMPLE 2.09X: Banking Arrangements

The partnership shall maintain such bank accounts as the partners shall determine. Checks shall be drawn for partnership purposes only and may be signed by any person or persons designated by the partners. All moneys received by the partnership shall be deposited in such account or accounts.[79]

   (e) Provisions should be included for the rendering of periodic reports to partners (e.g., monthly, quarterly, semi-annually or annually). The partners may be required to sign and verify the reports, subject to objection for manifest errors within a specified period of time.

EXAMPLE 2.09Y: Reports of Operations

The managing partner shall provide reports of cash activity, profit or loss, and the current balance sheet of the partnership to each partner at least quarterly, within 15 days following the close of the calendar quarter of the partnership.

Each partner shall be required to signify his or her receipt of such reports by signing a duplicate copy of the reports and returning the same to the managing partner within ten days following receipt of such reports. Any objections or questions concerning such reports must be addressed to the managing partner within 30 days following the receipt of the reports by each partner or the reports shall be deemed to be correct as to the matters presented therein.

(f) The person responsible for keeping partnership books should be named.

EXAMPLE 2.09Z: Responsibility for Records

The managing partner shall be responsible for the partnership books and records and shall maintain the same at the principal place of business of the partnership. The managing partner may, upon notice to the other partners, delegate such persons who may assume the obligations of keeping the partnership records.

(g) An audit by independent certified public accountants may be appropriate for some businesses, and the agreement will authorize the audit.

EXAMPLE 2.09AA: Audit of Books and Records

As soon as practical after the close of the partnership's calendar year, the managing partner shall engage an independent certified public accountant to audit the partnership books and records, and shall provide copies of the audit report furnished by such accountant to each of the partners.

[or]

Each partner shall be entitled to engage the services of an independent certified public accountant to audit the books and records of the partnership following the close of the partnership's calendar year. The partner so desiring such audit shall pay all expenses and fees of the accountant, and no such expenses shall be assessed against the partnership, unless the results of the audit indicate a variance of over or under ten percent of the profit and loss reported to the partners by the managing partner in any calendar year, in which event the expenses of the audit shall be borne by the partnership as a partnership expense.

12. Meetings—Partners' meeting may be established on a regular basis by the agreement, specifying the time and place for such meetings. Special meetings may be called in accordance with the agreement. A clause authorizing special meetings should consider the parties who are entitled to call special meetings, the notice required, and whether the notice must specify the purpose of the meeting.

EXAMPLE 2.09BB: Meetings

Partners meetings will be held on the second Tuesday of each month at 5:00 p.m. at the principal place of business of the partnership. No notice shall be required for the regular meetings of the partnership. A special meeting may be called by any partner upon giving three days written notice to the other partners, specifying a time and place for the meeting and the purpose of the meeting.

13. Management

(a) Method of management (See Section 2.04).

The management of the business affairs of the firm must be decided in accordance with the agreement. One or more of the partners may have a specialized business skill and this should not be overlooked. Many variations are possible for management activities, including:

(1) Each partner has an equal voice in management;
(2) Some partners (usually those with excess capital contributions) will have a greater vote than others;
(3) A committee of partners will be established to make certain decisions;
(4) A managing partner (or partners) will be appointed to control the daily business affairs of the firm;
(5) Some partners may have no management activities under the agreement. These persons are usually called "dormant" or "silent" partners.

EXAMPLE 2.09CC: Management (equal)

Management and the conduct of the business of the partnership shall be vested in all partners, and no partner shall be solely responsible for management functions. The partners shall have an equal vote on all partnership matters and all issues to be resolved in the partnership shall be determined by a majority vote.

EXAMPLE 2.09DD: Management (other than equal)

Management and the conduct of the business shall be vested in all Partners, and no Partner shall be solely responsible for management functions. The Partners shall have the following votes on Partnership matters:

Peter J. McLaughlin	11
James T. Johnston	33
Steve Forness	11
David A. French	11
Thomas Stubbs	33
Michael Theisen	1

A quorum shall be the presence at a meeting of 50 votes. No Partnership matter may be approved except at a meeting in which a quorum of votes is represented in the manner provided in this Paragraph, or as otherwise provided in this agreement.

(b) Management duties.

(1) If a managing partner is used, his or her duties should be specified in the agreement. Moreover, any management decisions which are to be referred to all of the partners should be described.

EXAMPLE 2.09EE: Duties of Managing Partner

(a) The affairs of the Partnership shall be managed and conducted by the Managing General Partner in accordance with the provisions of the Colorado Uniform Partnership Act, as amended, and subject to the terms and provisions of this Agreement.

(b) The Managing General Partner shall devote such of his time as may be necessary to select, at his sole discretion, and to acquire master recordings for the Partnership; to retain a distributor for the recordings owned by the Partnership; to supervise the activities of the distributor and to hire replacement or additional distributors if deemed necessary by the Managing General Partner; to make inspections of any physical assets owned by the Partnership and to see to it that such assets are being properly maintained; to prepare or cause to be prepared all reports of operations which are to be furnished to the Partners or which are required by any government agencies; and to do all other things which may be necessary to supervise the affairs and businesses of the Partnership in a prudent and businesslike manner in the best interest of the Partners.

(c) The Managing General Partner is hereby authorized, on behalf of the Partnership, to execute any contracts, notes, or other documents which may be required in connection with the acquisition, financing and operation of the assets and business described in this Agreement.

(d) The Managing General Partner, in addition to the other powers and rights granted to him and subject to the specific limitations imposed by this Agreement, shall have the right, upon such terms and conditions as he may deem proper, to (1) borrow money on the general credit of the Partnership for use in the Partnership business, including the right to borrow money from himself and to charge the Partnership interest on funds so borrowed, provided the interest rate to be charged by the Managing General Partner for such borrowed funds shall not exceed the rate available from commercial lenders, and provided the Managing General Partner shall not further encumber any master recordings acquired by the Partnership after their initial acquisition, other than in the ordinary course of business, without the approval of the Partners; (2) purchase personal property for use in connection with the business of the Partnership and finance such purchases, in whole or in part, by giving the seller or any other person a security interest in the property purchased; (3) make reasonable and necessary capital expenditures and improvements with respect to the assets of the Partnership and take all action reasonably necessary in connection with the management thereof; (4) establish a reasonable reserve for contingencies and operating capital from available cash flow of the Partnership; (5) contract with himself and affiliated persons on terms competitive with those which may be obtained in the open market for property or services required by the Partnership, provided, however, that the Managing General Partner shall not receive from himself or affiliated persons, or grant to himself or affiliated persons any rebates, kickbacks, or give-ups, directly or indirectly, in such transactions or agreements; (6) make reasonable and necessary expenditures for the maintenance and operation of the assets of the Partnership; and (7) enter into agreements for the management of the assets of the Partnership.

(e) The Managing General Partner shall assume a fiduciary responsibility for the safekeeping and use of all Partnership funds and assets, whether or not in his immediate possession or control, and shall not employ, or permit another to employ, such funds or assets in any manner except for the exclusive benefit of the Partnership. Partnership funds shall not be commingled with the funds of any other person or entity.

(f) The Managing General Partner shall not cause the Partnership to purchase interests in other business organizations, underwrite the securities of any other businesses, offer Partnership interests in exchange for anything other than cash or notes, or make loans to other persons or entities.

(g) Except where power or duties are reserved to the Managing General Partner, other Partnership matters shall be determined by the unanimous vote of the Partners.

EXAMPLE 2.09FF: Limitations on Powers of the Managing General Partner

The Managing General Partner shall have full, exclusive and complete discretion in the management of and control over the affairs of the Partnership; provided however, the Managing General Partner shall not take any of the following actions without the consent of all Partners:

(a) Sale, exchange, or other disposition of all or substantially all of the Partnership's assets other than in the ordinary course of business;

(b) Refinancing, recasting, increasing modifying or extending any loans secured in whole or in part by master recordings owned by the Partnership, other than in the ordinary course of business;

(c) Sale, assignment or encumbrance of the Managing General Partner's interest in the Partnership;

(d) Admission of a Successor Managing General Partner to the Partnership;

(e) Admission of additional Partners to the Partnership;

(f) Engagement of the Partnership in a business other than that specified in this Agreement; and

(g) Amendment or modification of this Agreement unless that amendment or modification is otherwise permitted under this Agreement without action of all Partners.

(2) All partners are expected to devote their time and energies to the partnership business.[80] Any deviation from this rule must be detailed in the agreement.

EXAMPLE 2.09GG: Outside Activities

No partner shall engage in, or invest or deal in the securities of, any business that in any wise competes with that of this firm, nor shall he give any time or attention to any outside business, except that of bank director, without the written consent of his copartners.[81]

(3) Each partner should be required, upon request, to account to the other partners regarding all transactions relating to the partnership business of which he has knowledge.

EXAMPLE 2.09HH: Reports of Activities

The managing partner shall, at least ten days prior to the regular meeting of the partnership, specify which of the partners are to report on areas of the partnership business within their control and responsibility. Such reports may be furnished orally at the meeting, unless the managing partner requires a written report in the notice, in which case a copy of the report shall be distributed to the partners at the meeting and appended to the minutes of the partnership meeting.

(4) The agreement may require certain partners to provide a bond for faithful performance of their management duties.

EXAMPLE 2.09II: Fidelity Bond

The managing partner shall, at the expense of the partnership, acquire and maintain a fidelity bond in the amount of $100,000.00 with an insurance company acceptable to a majority of the partners. The bond shall provide for the payment upon such bond for any willful failure or neglect of the managing partner to perform his or her duties hereunder, upon defalcation or embezzlement by the managing partner, upon the loss to the partnership of any asset as a result of the negligence of the managing partner, and upon such other terms and conditions as may required by the majority of the partners.

(c) Management formula—Depending upon the management method selected (e.g., equal voice, managing partner, etc.) the agreement will state a formula for determining partnership action. For example, if all partners have an equal voice, the formula may require a majority, two-thirds, or unanimous vote to carry action on behalf of the firm. The decision of a managing partner is usually final on matters within her control. In certain cases, unanimity of partners will be required by law.[82]

(d) Disputes—In case of a deadlocked dispute on management matters, the agreement usually requires submission of the dispute to an independent third party or to arbitration.

EXAMPLE 2.09JJ: Arbitration

All disputes and questions whatsoever which shall, either during the partnership or afterwards, arise between the partners or their respective representatives, or between any partners or between a partner and the representative of any other or others, touching these articles, or the construction or application thereof, or any clause or thing herein contained, or any account, valuation or division of assets, debts, or liabilities to be made hereunder, or as to any act, deed or omission of any partner, or as to any other matter in any way relating to the partnership business or the affairs thereof, or the rights, duties or liabilities of any person under these articles, shall be referred to a single arbitrator in case the parties agree upon one; otherwise to two arbitrators, one to be appointed by each party to the difference, or, in case of their disagreement to an umpire, to be appointed by said arbitrators.[83]

(e) Prohibited activities[84]

(1) Certain matters affecting the partnership must be decided by all partners, including assignment of the partnership property in trust for creditors, sale of the good will of the business, confession of judgment against the firm, submission of a partnership claim to arbitration, and other acts which would make it impossible to carry out the partnership business.

EXAMPLE 2.09KK: Limitation on Partner's Authority

No partner shall, without the unanimous consent of the other partners, do any of the following acts:

a) Assignment of the partnership property in trust for creditors;

b) Sale of the good will of the business or substantially all of the assets of the business;

c) Confess a judgment against the partnership or its assets;

d) Submit a partnership claim to arbitration;

e) Commit any other act which would make it impossible to carry out the partnership business.

(2) Partners may be further restricted in their power to bind the firm. An individual partner is usually not permitted to extend credit, pledge the partnership property, hire and fire employees, cause an attachment of firm property or release debts without the appropriate consensus of the other partners.

EXAMPLE 2.09LL: Limitations on Partner's Authority

No partner shall, without the consent of the others, borrow or lend money on behalf of the partnership; sell, assign, or pledge his interest in the partnership; or execute any lease, mortgage, security agreement, or endorsement on behalf of the partnership.[85] No purchase or other contract, involving a liability of more than Five thousand dollars, nor any importation from abroad, shall be made, nor any transaction out of the usual course of the retail business shall be undertaken, by either of the partners without the previous consent and approval of the other partner.[86]

(3) It may be desirable to govern the private lives of the partners in certain respects. For example, the agreement may forbid any partner from going into debt except for living necessaries; may restrict his ability to deal in securities on margin; may demand that he discharge any liens filed against him within a thirty day period, etc.

EXAMPLE 2.09MM: Limitation on Extraordinary Debts

In order to protect the property and assets of the Partnership from any claim against any Partner for personal debts owed by such Partner, each Partner shall promptly pay all debts owing by him or her and shall indemnify the Partnership from any claim that might be made to the detriment of the Partnership by any personal creditor of such Partner.

(4) The partners may be restricted in their sale or assignment of any or all of their interest in the partnership.[87] Common restrictions include the requirement of consent by the other partners, or a right of first refusal in the other partners, permitting them to purchase the interest at the offered price.

EXAMPLE 2.09NN: Sale of a Partner's Interest

In the event that a Partner desires to sell, assign, or otherwise transfer his or her share of interest in the Partnership hereby created and has obtained a bona fide offer for the sale thereof made by some person not a member of this Partnership, he or she shall first offer to sell, assign or otherwise transfer the said interest to the other Partners at the price and on the same terms as previously offered him or her, and each other Partners shall have the right to purchase his or her proportionate share of the selling Partner's interest. If any Partner does not desire to purchase the said interest on such terms or at such price, no other Partner may purchase any part of the interest, and the selling Partner may then sell, assign, or otherwise transfer his or her entire interest in the Partnership to the person making the said offer at the price offered. The intent of this provision is to require that the entire interest of a Partner be sold intact, without fractionalization. A purchaser of an interest of the Partnership shall not become a Partner without the unanimous consent of the non-selling Partners.

14. Partnership property

(a) The name in which partnership property will be titled is established by the agreement.

EXAMPLE 2.09OO: Title to Property

All assets of the partnership shall be titled in the name of the partnership, Shoylin Associates.

(b) If any property is loaned to the firm by a partner, it should be separately described and the agreement should detail the duration of the loan, any restrictions upon the disposition of the property by the partner, and any compensation to be paid to the partner for the use of his asset.[88]

EXAMPLE 2.09PP: Loans of Property by a Partner

Maynard P. Short has loaned to the partnership, and by this agreement, agrees to the exclusive use of the partnership, all of the property listed and described on Exhibit B, attached to this agreement and incorporated herein by reference. The partnership shall have exclusive use and enjoyment of the property for a period of one year from the date of this agreement, and for successive annual periods, unless, prior to the expiration of an annual term, Mr. Short gives written notice of at least 30 days to the partnership of his intention to reacquire the possession and use of the property. So long as the partnership shall be in possession of the property, Mr. Short shall not assign, sell, encumber, or otherwise deal with such property, and any such action by Mr. Short shall be deemed to be a breach of this agreement. The partnership shall pay Mr. Short the sum of $10,000.00 per year in equal quarterly installments for the use of the property described on Exhibit B.

(c) Accounting procedures for partnership property, including treatment of depreciation, repairs, insurance, taxes, interest, and other expenses should be considered.

EXAMPLE 2.09QQ: Method of Accounting for Assets

All accounting for partnership assets shall be done according to generally accepted accounting procedures, using the most conservative methods of accounting for depreciation, investment tax credits, etc. The managing partner shall be directed to provide for depreciation, repairs, insurance, taxes, interest and other reserves as necessary for partnership operations in order to meet such operating and capital expenses when they are incurred.

15. Causes of dissolution[89]
   (a) Retirement or withdrawal.
      (1) The agreement may describe the circumstances under which a partner may retire. Most agreements permit the partner to retire and withdraw at any time after a certain date or after he has reached a certain age.
      (2) Notice of retirement or withdrawal is usually required to be given to the other partners.
      (3) A non-competition clause may be appropriate to restrict the business activities of a retiring or withdrawing partner.[90]
      (4) The agreement should provide for indemnification of the retiring or withdrawing partner for all existing liabilities if the remaining partners elect to continue the business.
      (5) If the withdrawal of a partner is wrongful, any penalties to be imposed as a result of wrongful withdrawal should be specified in the agreement.

EXAMPLE 2.09RR: Retirement or Withdrawal

A Partner shall have the right, at any time during the continuance of this agreement and of the Partnership created hereby, to withdraw or retire from the said Partnership by giving three (3) months notice to the other Partners at the Partnership's place of business.

Upon giving notice, the withdrawing or retiring Partner shall be entitled to payment of his or her interest in the Partnership, the amount of which and method of payment is determined by this agreement with reference to purchase of an expelled Partner's interest. Upon the receipt of such payment, the interest of the withdrawing or retiring Partner in the Partnership shall cease and terminate.

Notwithstanding the provisions above, if the remaining Partners shall decide not to continue the business upon withdrawal or retirement of a Partner, the remaining Partners may elect to terminate and dissolve the Partnership, in which case the withdrawing or retiring Partner shall only be entitled to his or her interest in liquidation, as stated in this agreement with reference to voluntary termination.

(b) Expulsion of a partner.
   (1) The circumstances justifying expulsion of a partner must be specifically detailed.
   (2) A method for deciding upon expulsion and a procedure for notifying the expelled partner should be established. If a hearing will be permitted, the procedure for conducting the hearing should be described.
   (3) The agreement should provide for indemnification of the expelled partner for all existing liabilities if the remaining partners elect to continue the business.

EXAMPLE 2.09SS: Expulsion

A partner of this partnership may, upon the affirmative vote of the other partners, be expelled for the following acts:

a) Committing a felony under the laws of the state in which this partnership is organized;

b) Failing to cure any default or breach of this agreement after receipt of a notice of such default or breach from the other partners in writing;

c) Committing an act which is deemed to be detrimental to the business or reputation of the partnership;

d) Adjudication of insanity of the partner;

e) Competing with the business of the partnership for his or her personal account.

The partners so voting for expulsion shall give notice of expulsion, specifying the reasons therefor, to the expelled partner, who, upon receipt of such notice, shall have ten (10) days to request in writing a hearing on the matters specified in the notice. If a hearing is requested, the partner subject to expulsion shall appoint an impartial third party, and the other partners shall appoint an impartial third party, and each such third party shall appoint another impartial third party to hear such evidence or other matters as the expelled partner wishes to present on his or her behalf. Following such hearing, this panel shall determine whether the partner shall be expelled, and their decision shall be final.

The remaining partners may continue the business without the expelled partner and without liquidation of the partnership by paying the expelled partner his or her capital account, and by furnishing such indemnification and hold harmless documents as may be reasonably requested by the expelled partner for obligations of the partnership which come due following an expulsion.

(c) Bankruptcy of a partner.
  (1) Provisions should be included for the continuance of the business in case of the bankruptcy of an individual partner.
  (2) The purchase of the individual partner's interest in the partnership must be authorized.

EXAMPLE 2.09TT: Bankruptcy of a Partner

Upon the adjudication of bankruptcy of a partner, or the assignment by a partner for the benefit of his or her creditors, or the appointment of a receiver or conservator for the disposition of a partner's debts, the other partners shall have the right either to purchase the bankrupt partner's interest in the partnership or to terminate and liquidate the partnership business. If the remaining partners elect to purchase the bankrupt partner's interest, they shall serve notice of such election upon the trustee in bankruptcy, receiver, conservator or assignees of the bankrupt partner within twenty (20) days following such event, and shall pay to such person or persons the value of the bankrupt partner's interest in the partnership determined as of the day before such event occurred. If the remaining partners do not elect to purchase the bankrupt partner's interest, and instead elect to terminate the business, they shall appoint a managing partner who shall proceed with reasonable promptness to sell the property of the partnership and to liquidate the business of the partnership. The bankrupt partner's estate shall thereafter share in the proceeds of liquidation in accordance with his or her pro rata share of the proceeds thereof.

(d) Death of a partner.
  (1) Provisions should be included for the continuance of the business when one partner dies.
  (2) If the deceased partner's estate is to participate in profits of the business, the agreement should describe the extent of participation. This clause discusses the amount of profits to be distributed to the estate, the period during which such distributions are to be made, and whether profit distributions are guaranteed.
  (3) The agreement will establish the authority for the purchase of life insurance on the partners. The partners may purchase life insurance on each other (cross-purchase plan) or the firm may purchase the insurance (entity-purchase plan). The amount of the insurance to be maintained and the type of plan is described in the agreement.
  (4) If the life insurance plan is to be administered by a trustee, the trustee is named and his power and duties are defined in the agreement.

EXAMPLE 2.09UU: Death of a Partner

Upon the death of any Partner, the surviving Partners shall have the right either to purchase the decedent's interest in the Partnership or to terminate and liquidate the Partnership business. If the surviving Partners elect to purchase the decedent's interest, they shall serve notice in writing of such election, within three (3) months after the death of the decedent, upon the executor or administrator of the decedent, or, if at the time of such election no legal representative has been appointed, upon any one of the known legal heirs of the decedent at the last known address of such heir. The closing of such purchase shall be within thirty (30) days of the notice of such election.

If the surviving Partners elect to purchase the decedent's interest, the purchase price and method of payment shall be as stated in this agreement with reference to purchase of an expelled Partner's interest, except in the event insurance is in effect with respect to decedent the method of payment is provided in this section. The period from the beginning of the fiscal year in which decedent's death occurred until the end of the calendar month in which his or her death occurred shall be the period used for purposes of calculating his or her share of Partnership profits and losses in the year of death. The decedent's share of profits and losses shall also include his or her share of profits and losses of the Partnership during the period between the end of the calendar month in which death occurred and the end of the calendar month preceding the closing of purchase.

If the surviving Partners do not elect to purchase the decedent's interest, and instead elect to terminate the business, they shall appoint a managing partner who shall proceed with reasonable promptness to sell the real and personal property owned by the Partnership and to liquidate the business of the Partnership. The surviving Partners and the estate of the deceased Partner shall share in their respective proportions stated during the period of liquidation, except that the decedent's estate shall not be liable for losses in excess of the decedent's interest in the Partnership at the time of his or her death. The managing Partner shall be entitled to reasonable compensation for services performed in liquidation. Except as otherwise stated herein, the procedure as to liquidation and distribution of the Partnership assets shall be the same as stated in this agreement with reference to voluntary termination.

The Partnership may contract for life insurance protection on the lives of each of the Partners, in any amount not disproportionate to the value of each Partner's interest. Each Partner may designate the beneficiary for such life insurance. In the event of death of a Partner, insurance proceeds paid to the Partnership will be used to purchase the decedent's interest, at the purchase price determined in above, except that the payment of such price to the decedent's representatives or heirs shall be made within thirty (30) days following receipt of the insurance proceeds. Any surplus in insurance proceeds not required to purchase the decedent's interest shall be retained in the Partnership and proportionately added to the capital account of the surviving Partners. If the surviving Partners elect to liquidate the business in lieu of purchasing the decedent's interest, the proceeds of any life insurance shall be treated as an asset of the Partnership for liquidation.

(e) Other disabilities—The partnership may be dissolved when an individual partner becomes disabled, insane, or otherwise incapable of continuing in the business relationship. These special incidents of dissolution should be considered in the agreement, with provisions for continuation of the business, and purchase of the former partner's interest.

EXAMPLE 2.09VV: Disability of a Partner

In the event a partner becomes disabled, is adjudicated insane, or is otherwise unable to perform the duties required by this agreement, the remaining partners shall have the right either to purchase the disabled partner's interest in the partnership or to terminate and liquidate the disabled partner's interest in the partnership. If the surviving partners elect to purchase the disabled partner's interest, then the procedure described in this agreement for the purchase of a deceased partner's interest in the partnership shall apply. If the remaining partners elect to terminate and liquidate the business, they shall appoint a managing partner who shall proceed with reasonable promptness to terminate the business, sell the property of the partnership, and distribute the proceeds of such liquidation, after payment to creditors, in the manner provided for dissolution and liquidation in this agreement.

16. Continuation of the business—Following a dissolution, the remaining partners will have the authority to continue the business if the dissolution was wrongful. However, in other cases of dissolution, the business may be continued only if the agreement so provides. A specific clause to that effect, granting the remaining partners the right to continue the business, is appropriate. Notice of the intent to continue the business may be required to be given to the former partner or to her estate. In all cases, the withdrawing, retiring or disabled partner or the estate of the deceased partner should be indemnified from business liabilities.

17. Purchase of the partner's interest—If a partner has caused a dissolution, and the remaining partners intend to continue the business, the interest of the withdrawing, disabled or deceased partner will be purchased by the firm or by the other partners.[91]
    (a) The agreement may provide that the former partner's interest will be purchased by the other partners jointly or individually in an established proportion.
    (b) The agreement may provide for liquidating distributions to the former partner, resulting in a purchase of his interest by the firm.
    (c) The value of the partner's interest should be ascertained in accordance with a formula specified in the agreement. Independent public accountants may be necessary to make the computations under the formula. The following typical alternatives are available:
        (1) A return of the capital contribution plus interest.
        (2) A stipulated value as described in the agreement.
        (3) A formula based upon historical earnings of the partnership (earnings multiple formula).
        (4) A formula based upon the value of the assets of the partnership (book-value formula).
        (5) Appraisal by an independent third party.[92]
    (d) The extent to which good will is to be used to compute the value of the former partner's interest is established by the agreement. Good will may be ignored, or it may be considered an asset and appraised in determining the value of the interest.

(e) Payment terms should be established in the agreement. The period of time for installment payments, whether a promissory note is to be executed, and whether the obligation will be secured by assets of the partnership are appropriate topics for this provision.
(f) A fund may be withheld for a period of time for contingent claims arising before the dissolution.
(g) The treatment of the purchased interest should be discussed. It may be divided up among partners in their remaining proportion of capital contributions, equally, or by some other formula.

EXAMPLE 2.09WW: Continuation of Business and Purchase of Partner's Interest

The Partners may elect to continue the business despite a dissolution of the partnership, by purchasing the deceased, disabled, expelled or bankrupt Partner's interest, and in such case the purchase price shall be equal to the deceased, disabled, expelled or bankrupt Partner's capital account as of the date of the notice required by this agreement, plus his or her income account as of the end of the prior fiscal year, increased by his or her share of Partnership profits, or decreased by his or her share of Partnership losses, computed to the date of the notice, and decreased by withdrawals such as would have been charged to his or her income account during the present year to the date of the purchase of his or her interest. The purchase price is subject to set-off for any damages incurred as the result of the expelled or bankrupt Partner's actions, and nothing in this paragraph is intended to impair the Partnership's right to recover damages for such reasons.

The date of the notice, referred to above, shall be the date personal notice is received, or the date the certified mail is postmarked, in the case of a breach of this agreement.

18. Liquidation and winding-up (See Section 2.07)—If the business will not be continued following dissolution, liquidation and winding-up must follow in accordance with the agreement.
(a) A full and general account of the firm assets, liabilities and transactions should be authorized. An independent certified public accountant may be necessary for this purpose.
(b) A liquidating partner or committee of partners should be named. Since partners may receive remuneration for services in liquidation, the value of such services should be fixed by agreement.
(c) If the assets are capable of distribution to the partners, this may be authorized in the agreement, following the payment of business debts. Otherwise, the agreement should authorize the sale of the assets (usually in the discretion and good judgment of the liquidating partner) and the distribution of the cash received.
(d) The order of distribution of the assets is set by law.[93] If the order of distribution is to be altered, the agreement must specifically describe the new order of distribution.
(e) If the partnership has sustained a loss, so that one or more of the partners will be required to make additional capital contributions to facilitate distribution of assets, a period of time in which such

payments are to be made and the manner of payment (cash, promissory note, etc.) should be specified.

EXAMPLE 2.09XX: Distribution on termination

Upon termination of the Partnership, its affairs shall be concluded in the following manner:

a) The Managing General Partner shall proceed to the liquidation of the Partnership and the proceeds of such liquidation shall be applied and distributed in the following order of priority:

(1) To the payment of all debts and liabilities of the Partnership;

(2) To the setting up of any reserve which the Managing General Partner shall deem reasonably necessary to provide for any contingent or unforeseen liabilities or obligations of the Partnership; provided, however, that at the expiration of such period of time as the Managing General Partner shall deem advisable, the balance of such reserve remaining after the payment of such contingency shall be distributed in the manner set forth in this section;

(3) To the payment to the partners other than the Managing General Partners of an amount which, when added to any amount previously distributed to the partners other than the Managing General Partner pursuant to this agreement hereof, will equal the their aggregate capital contributions to the Partnership;

(4) Any balance then remaining shall be distributed as follows:

(i) Ninety percent (90%) of such balance to the partners other than the Managing General Partner;

(ii) Ten percent (10%) of such balance to the Managing General Partner;

b) A reasonable time shall be allowed for the orderly liquidation of the assets of the Partnership and the discharge of liabilities to creditors;

c) Each Partner shall be furnished with a statement certified by the Partnership's independent accountants which shall set forth the assets and liabilities of the Partnership as of the date of the complete liquidation.

NOTES

1. For general discussions of the law of Agency and Partnership see J. Crane and A. Bromberg, Law of Partnership §§ 2, 49–56, 68–72 (1968); and W. Seavey, Agency §§ 10A, 14A, 59 (1964).
2. 6 U.L.A. 1 (1969) and Supp. (1980). The Uniform Partnership Act (hereafter cited as U.P.A) has not been adopted in Georgia and Louisiana.
3. U.P.A. § 6(1).
4. Model Business Corporation Act (hereafter cited as M.B.C.A.) § 4(p).
5. U.P.A. § 18(a), (e).
6. U.P.A. § 2.
7. U.P.A. § 8(2).
8. West's Modern Legal Forms § 6332.
9. See Section 2.08.
10. West's Modern Legal Forms § 6342.
11. West's Modern Legal Forms § 6343.
12. U.P.A. § 25.
13. U.P.A. § 27.
14. U.P.A. § 18(g).
15. A sample assignment by a partner of his interest in the partnership appears as Form 2A in the Appendix.
16. U.P.A. § 18(e).

17. U.P.A. § 18(h).
18. West's Modern Legal Forms § 6366.
19. West's Modern Legal Forms § 6368.
20. U.P.A. § 9(3).
21. West's Modern Legal Forms § 6360.
22. U.P.A. 18(f).
23. West's Modern Legal Forms § 6347.
24. West's Modern Legal Forms § 6348.
25. U.P.A. § 18(a).
26. West's Modern Legal Forms § 6354.
27. West's Modern Legal Forms § 6353.
28. Partnership taxation is "pass-through" taxation, which means each individual partner is taxed on the partner's pro rata share of each item of profit, loss, deductions and credits. See Section 2.08.
29. Internal Revenue Code of 1954, 26 U.S.C.A. § 704(b). The Regulations under section 704(b) of the Code (adopted prior to the Tax Reform Act of 1976) outline several relevant factors that are considered in making a determination as to whether a special allocation will be recognized for federal income tax purposes. Among these factors are (i) the presence of a business purpose for the allocation, (ii) whether related items of income, gain, loss, deduction or credit from the same source are subject to the same allocation, (iii) whether the allocation was made without recognition of normal business factors, (iv) whether it was made only after the amount of the specially allocated items could reasonably be estimated, (v) the duration of the allocation, and (vi) the overall tax consequences of the allocation.
30. General partners are jointly and severally liable for damages caused by any tort or breach of trust committed by a partner within the scope of the partnership business. They are jointly liable for all other partnership obligations. U.P.A. § 15.
31. U.P.A. § 18(b).
32. West's Modern Legal Forms § 6350.
33. U.P.A. § 36(1).
34. See Section 2.07.
35. U.P.A. § 35(1)(b). Examples of the necessary notice appear as Forms 2B and 2C in the Appendix.
36. U.P.A. § 17.
37. U.P.A. § 31(1)(a).
38. West's Modern Legal Forms § 6381.
39. U.P.A. § 31(1)(c).
40. West's Modern Legal Forms § 6421.
41. U.P.A. § 31(1)(b), (2).
42. Examples of preliminary notice and final notice of expulsion appear as forms 2D and 2E in the Appendix.
43. U.P.A. § 31(3), (4), (5).
44. U.P.A. §§ 31(6), 32.
45. U.P.A. § 38(1).
46. U.P.A. § 38(2)(b).
47. See J. Crane and A. Bromberg, Law of Partnership §§ 86, 90A (1968).
48. West's Modern Legal Forms § 6303. Other examples of liquidating distributions may be found in §§ 6302, 6304, and in Article V, Sections C and D of the Complex Partnership Agreement, Form 2F in the Appendix.
49. See West's Modern Legal Forms § 6307.
50. Examples of cross-purchase and entity-purchase insurance agreements may be found in West's Modern Legal Forms §§ 6402–04.
51. West's Modern Legal Forms § 6379.
52. West's Modern Legal Forms § 6375. See also Sections 10.05 and 10.06.
53. U.P.A. § 33.
54. U.P.A. § 40.
55. Internal Revenue Code of 1954, 26 U.S.C.A. § 704(c)(2).
56. West's Modern Legal Forms § 6335.
57. See, e.g., 1 Z. Cavitch, Business Organizations §§ 7.01–8.03 (1973).
58. Examples of the trade-name affidavits and certificates appear as Forms 1A–1E in the Appendix.

59. See Section 1.05.
60. The qualification of a "foreign-partnership" is patterned after the requirements for qualification of foreign corporations, discussed in Chapter 12.
61. Other helpful references include J. Mulder and M. Volz, The Drafting of Partnership Agreements, American Law Institute (1967); R. Rowley and D. Sive, Rowley on Partnership, Vol. II, Chapter 59 (1960); and J. M. Barrett and E. Seago, Partners and Partnerships Law and Taxation, Vol. II, Appendices 1 and 10 (1956).
62. West's Modern Legal Forms § 6325.
63. West's Modern Legal Forms § 6328.
64. West's Modern Legal Forms § 6329.
65. West's Modern Legal Forms § 6331.
66. West's Modern Legal Forms § 6336.
67. West's Modern Legal Forms § 6332.
68. West's Modern Legal Forms § 6337.
69. West's Modern Legal Forms § 6340.
70. West's Modern Legal Forms § 6338.
71. West's Modern Legal Forms § 6339.
72. West's Modern Legal Forms § 6344.
73. West's Modern Legal Forms § 6348.
74. West's Modern Legal Forms § 6347.
75. See Section 2.05.
76. West's Modern Legal Forms § 6357.
77. West's Modern Legal Forms § 6355.
78. West's Modern Legal Forms § 6358.
79. West's Modern Legal Forms § 6359.
80. See Section 2.04.
81. West's Modern Legal Forms § 6365.
82. See Section 2.04.
83. West's Modern Legal Forms § 6376.
84. See Section 2.04.
85. West's Modern Legal Forms § 6361.
86. West's Modern Legal Forms § 6364.
87. See Section 2.03.
88. See Section 2.02.
89. See Section 2.07.
90. See Section 10.06.
91. See Section 2.07.
92. See, also, Shareholder Buy-Out Agreements.
93. See Section 2.07.

3

LIMITED PARTNERSHIPS

§ 3.01 CHARACTERISTICS OF THE LIMITED PARTNERSHIP

In many of the important particulars of the partnership law, the limited partnership is the same as a general partnership. It is an association of two or more persons carrying on business as co-owners for profit with one or more general partners and one or more limited partners.[1] The limited partnership enjoys certain characteristics of a corporation insofar as the limited partners are concerned, since their investment and limited liability resembles the position of a shareholder of a corporation. The general partners in a limited partnership are governed by all the rules of a general partnership as discussed in Chapter 2. It may be said that a limited partnership is a bifurcated business form, and the rights and responsibilities of the limited and general partners must be distinguished.

Most states have adopted the Uniform Limited Partnership Act [2] to regulate the formation and operation of limited partnerships. This Uniform Act was approved by the Commissioners on Uniform State Laws in 1916, and has been adopted and used in its original form by most states. In 1976 the Commissioners approved a Revised Uniform Limited Partnership Act which significantly modifies the original act and has been accepted in only a few states.[3] Since the original act remains the current basis of limited partnership law, this Chapter will refer primarily to the rules from the original act, but the last few pages refer to the novel rules of the new Act so that those among you who are prone to anticipation can learn what the law of limited partnerships may someday be.

The most important statutory requirement under both the original and revised Acts, the filing of a Limited Partnership Certificate, will be discussed in detail at a later point in this Chapter. However, it is important to note at the outset that the limited partnership may only be formed with the formality prescribed in the statute, and may not be born of a simple private agreement between the parties. Like a corporation, this form of business is imbued with quasi-public characteristics, which pose additional problems for the person drafting the documents and may operate as a trap for the unwary.

§ 3.02 THE GENERAL PARTNERS OF A LIMITED PARTNERSHIP

Each limited partnership must have at least one general partner, who faces all of the same risks and responsibilities as a partner in a general partnership. The liability exposure of the limited partners is confined to their contributions, but the general partner suffers unlimited liability, meaning that his individual assets are vulnerable to firm creditors. Similarly, the general partner has full responsibility for management and control of the partnership affairs, since limited partners are forbidden to participate in the control of the business if they are to maintain their limited liability status.[4]

One person may be a general partner and a limited partner at the same time,[5] and the authority to do this produces some benefits. In a person's status as general partner, he is fully liable for firm obligations and enjoys no limited liability. However, his contribution as a limited partner will rank with other limited partners' priorities for dissolution purposes,[6] and his limited partnership interest is freely transferable without causing a dissolution of the partnership.[7]

If the limited partnership has two or more general partners, the rights and responsibilities between them are the same as in any general partnership.

The remaining sections in this Chapter deal with the limited partnership's unique variations from a general partnership. In all respects except those specifically set forth in the following sections, a limited partnership is governed by the same rules as a general partnership.[8]

§ 3.03 LIMITED LIABILITY AND CONTRIBUTIONS

The most significant characteristic of the limited partnership is that limited partners are protected from full individual liability. The liability of the limited partner is limited to the amount of her investment as stated in the certificate,[9] and in this respect, the limited partner is almost exactly like a shareholder of a corporation. This feature makes the limited partnership particularly attractive for persons with substantial private resources which they prefer not to risk in the business enterprise. The only potential loss is the investment.

The limited partner may contribute only cash or other property. She may not contribute services.[10] This rule is based in part on the prohibition against a limited partner's participation in management. Further, since partnership creditors are confined to those contributions for the enforcement of their obligations, there should be some tangible asset for their satisfaction.

Limited liability will be observed provided the limited partner does not "hold herself out" to be a general partner, such as representing to others that she is a general partner. The Uniform Limited Partnership Act denies limited liability if the limited partner's name is used in the firm name (with some exceptions to be discussed later), or if the limited partner exercises any control over partnership affairs.[11]

§ 3.04 MANAGEMENT AND CONTROL

To preserve the limited partner's limited liability status, all management and control over partnership affairs must be vested in the general partner. This prohibition against management participation causes some uncomfortable uncertainty in the limited partnership organization under current law because it is difficult to predict the extent of participation which will defeat a partner's limited status.

There are certain activities which a general partner may never do without the consent of the limited partners, including acts in contravention of the Certificate or those interfering with the ordinary business of the partnership; possession of partnership property for other than business purposes; admission of another general partner; and confession of a judgment against the firm.[12] Certainly a limited partner's consultation in these decisions would not constitute participation in management so as to disturb his limited status. However, his partaking in any other advisory or management decisions may remove the special immunity from liability. A small minority of states, including California, Oregon and Washington, permit the limited partner to have about the same voice in management as a shareholder of a corporation without losing his limited status. Similarly, the Revised Uniform Act better defines the circumstances under which a limited partner participates in the control of the business.[13]

Limited partners are always entitled to inspect the books and to have an accounting of partnership affairs. They also have the right to be informed on all matters respecting the business of the firm.[14]

§ 3.05 ADMISSION, SUBSTITUTION AND WITHDRAWAL OF A LIMITED PARTNER

Unlike general partners, limited partners may freely come and go, with very few restrictions. If provisions are made in the partnership agreement and the certificate of limited partnership, additional limited partners may be admitted without the consent of the existing limited partners by filing an amendment to the certificate.[15] Similarly, a limited partner may withdraw from the partnership and receive a return of his contribution without causing a dissolution of the firm.[16] Of course, if the limited partner's contribution is essential to the continued operation of business, this right to withdraw may be restricted or denied by the agreement.

Both the original and revised Uniform Limited Partnership Acts permit a limited partner to withdraw and demand return of his contribution on the date specified for return of contribution in the certificate or upon giving six months notice in writing.[17] The contribution may also be returned at any time if all partners, general and limited, consent to its return. However, the investment will be returned only if the firm creditors have been paid or sufficient assets remain to pay them.[18] Unless the certificate provides otherwise, or all members consent, the limited partner has only the right to demand cash in withdrawal, even if he contributed other property to the partnership.[19] An amendment of the certificate must be filed to reflect the withdrawal.

The partnership agreement and certificate may grant authority to a limited partner to substitute a new limited partner in his place without the consent of the other partners. If the certificate does not contain such express authority, the assignment of a limited partner's interest has a similar effect as the assignment of a general partnership interest. The assignment grants the assignee the right to receive all profits and other distributions to which the limited partner is entitled, but it does not make the assignee a new partner, unless all the partners consent.[20] Any substitution of limited partners, by the power of agreement or by consent, requires an amendment to the certificate to reflect the change.[21]

§ 3.06 DISSOLUTION OF A LIMITED PARTNERSHIP

Causes of Dissolution

Dissolution of limited partnerships is very similar to dissolution of general partnerships, discussed in section 2.07. The major distinctions stem from the fact that the limited partner is not one of the key personnel of the business. He is a passive investor, like a shareholder of a corporation, and although his demise, insanity, bankruptcy, or withdrawal may be a sad event, none of those things should affect the continuation of the business. Consequently, the incapacity of a limited partner does not cause dissolution. Similarly, he may withdraw his contribution (his investment) and the partnership may continue without him. Most authorities agree, however, that misconduct by a limited partner, including any act which would adversely affect the business of the firm, would be grounds for dissolution by the other partners.

Limited partners have the right to request dissolution by decree of court when a *general* partner is insane or otherwise incapable of furthering the partnership business; or when any partner conducts himself in a manner so that it is impracticable to carry on business with him; or in any other case where dissolution may be equitable and proper.[22] An important option of the limited partner in this regard is to exercise his right to dissolve the firm if he has rightfully demanded return on his contribution but the demand has been ignored.[23]

The general partner is an integral member of the firm, and his incapacity or withdrawal results in a technical dissolution of a limited partnership of which he is a member.[24] However, the certificate may grant the remaining general partners the right to continue the business when a general partner withdraws, dies or becomes insane. Alternatively, the consent of all members may constitute authority for continuation if no statement of authority appears in the certificate.[25] Acts of misconduct by the general partner which interfere in the affairs of the partnership are grounds for dissolution by either general or limited partners.[26]

Of course, a limited partnership is the product of an agreement between the members, and it may specify a term of the partnership, upon the expiration of which it will be dissolved. Moreover, all members may consent to dissolution if the certificate is silent.[27]

Continuation of the Limited Partnership Following Dissolution

As discussed above, the right to continue the limited partnership in spite of a dissolution requires a statement to that effect in the certificate or the consent of the members.[28]

Termination and Winding Up

If a cause for dissolution occurs and the business is not continued, the limited partnership must be terminated and liquidated. The general partners, or their representatives, are responsible for winding-up. A limited partner may not participate in winding-up unless a court so orders.[29]

The Uniform Limited Partnership Act prescribes a scheme of priorities for distribution of assets of a limited partnership as follows:

1. To outside creditors of the partnership;
2. To limited partners in respect to their share of profits and income such as interest on their contributions;
3. To limited partners in respect to their contributions;
4. To general partners for claims against the partnership other than capital or profits;
5. To general partners in respect to profits;
6. To general partners in respect to capital.[30]

As with general partnerships, this scheme may be altered by the agreement of the members as long as business creditors are fully paid. Notice that the limited partners' claims are subordinated only to outside creditors, and are paid even before a general partner's loan to the firm is repaid. Limited partners are obviously preferred in distribution matters, and this constitutes a further incentive for capital investment in this form of business enterprise.[31]

§ 3.07 TAXATION OF LIMITED PARTNERSHIPS

For the most part, the limited partnership will be treated like a general partnership for tax purposes. You will recall that the general partnership acts only as a conduit through which income is deemed to be distributed to each partner in the proportions specified in the agreement. The normal limited partnership has the same treatment, and this may be an advantage to the limited partner seeking to declare losses to offset income from other sources. This tax advantage from offsetting losses is one reason why limited partnerships are favored forms of organization to conduct real estate development and operation or rental property. The accelerated depreciation allowances available for their enterprises will usually produce "paper losses" which are passes directly to the partners and "shelter" other income from tax.

Under certain circumstances, however, a limited partnership will be considered to be a corporation for tax purposes. This means that the tax advantages gained through depreciation, depletion or other losses will not be passed directly to the investor. Instead, the partnership will be considered a separate tax-paying entity and will be taxed at corporate rates. Such "restructuring" by taxing authorities occurs when the operation of the limited partnership closely resembles the operation of the corporation. For example, the certificate may authorize a limited partner to assign his interest and substitute the assignee as a limited partner in his place. This, of course, is a characteristic of corporate share ownership. Similarly, the certificate may provide for continuity of the business despite the death, incapacity, or withdrawal of the general partner. Continuity of existence is a primary corporate element. Further, if management is centralized in a few general partners, the operation of the partnership resembles the acts of a board of directors of a corporation. As previously discussed, limited partners enjoy limited liability which is another corporate attribute for shareholders. When the structure of the limited partnership includes more than two of these similar elements, it may be taxed as a corporation.[32]

§ 3.08 FORMATION AND OPERATION OF THE LIMITED PARTNERSHIP

With the singular, but extremely important exception of the limited partnership certificate, the formation of a limited partnership is the same as the formation of a general partnership. Thus, licensing requirements would have identical application to this form of business, and other state formalities must be observed. The agreement plays an even more important role in a limited partnership and the idiosyncracies of the limited partnership should receive special attention in the agreement.

Name

While the limited partnership certificate states the trade name of the firm and is a matter of public record, it may be necessary, or prudent, to file a Trade Name Affidavit or Certificate with the appropriate state official, especially if the local statute requires the limited partnership certificate to be filed in a different office than where a Trade Name Affidavit is filed. The usual restrictions on the use of a deceptively similar name which may constitute an unfair trade practice also apply to limited partnerships.

In a vast majority of jurisdictions, there is no requirement that a limited partnership name contain a "warning signal" for limited liability, such as "Limited" or "Ltd." However, there are certain statutory restrictions on the name used for limited partnerships. The use of the limited partners' surnames is prohibited if their limited liability is to be maintained, unless 1) the partnership has a general partner with the same name, or 2) the business had been carried on under a name including the limited partner's surname prior to the time he became a limited partner. Thus, a limited partnership composed of Ron Williams and Charlie Langhoff as general partners and Mary Williams, Scott Charlton and Bob Thompson as limited partners could use the name "Williams and Langhoff" even though Mary Williams is a limited partner. Similarly, if Charlie Langhoff subsequently becomes a limited partner, the firm may continue under the name "Williams and Langhoff" under the second exception.

The Partnership Agreement

Preparation of the limited partnership agreement usually will be based upon the expressed desires of the proposed general partners, since limited partners play a passive role in the formation and operation of the business. The basic form of the agreement resembles a general partnership agreement, since the limited partnership includes at least one general partner. Especially when more than one general partner will manage the business, all considerations specified in the checklist proposed in section 2.09 for general partnerships should be considered in the drafting of the limited partnership agreement.

Several special matters, raised by the specific statutory rules which govern limited partnerships, should be addressed in the agreement. The following checklist is designed to be used in addition to the matters covered in section 2.09 to draft a complete limited partnership agreement.

1. Provide for the recording of a Certificate of Limited Partnership and other necessary documents in the appropriate filing places.

EXAMPLE 3.08A: Certificate of Limited Partnership and Trade Name Affidavit, Etc.

A Certificate of Limited Partnership created hereby shall be recorded in accordance with the Limited Partnership Act in each county of Colorado in which the Partnership may establish a place of business. In addition, the General Partner shall file and publish a Trade Name Affidavit and any other notices, certificates, statements or other instruments required by any provision of any law of the State of Colorado.

2. State such provisions as are agreed for the admission of additional limited partners.

EXAMPLE 3.08B: Admission of Additional Limited Partners

Subject to any other provision of this Agreement, after the formation of the Partnership, a person may be admitted as an additional Limited Partner with the written consent of the General Partner and the recording of an amendment to the original Certificate of Limited Partnership in accordance with the requirements of the Colorado Limited Partnership Act.

3. State such provisions as are agreed for admission of substituted limited partners.

EXAMPLE 3.08C: Admission of Substituted Limited Partners

Subject to any other provision of this Agreement, after the formation of the Partnership, a person may not be admitted as a substituted Limited Partner without the written consent of the General Partner plus the recording of an amendment to the original Certificate of Limited Partnership in accordance with the requirements of the Colorado Limited Partnership Act.

4. Provide that any new partners must agree to be bound by the terms of the partnership agreement.

EXAMPLE 3.08D: Additional Partners Bound by Agreement.

Notwithstanding any other provisions of this Agreement, before any person is admitted or substituted as a Limited Partner, he shall agree in writing to be bound by all of the provisions of this Agreement.

5. Provide for additional capital contributions by limited partners if desired, and describe any restrictions or limitations on additional capital contributions.

EXAMPLE 3.08E: Limitation on Additional Capital Contributions.

After the initial capital contributions have been paid, Limited Partners may be required to contribute their proportionate share of the capital of this Partnership or such additional sums of money or property as shall be determined to be necessary by the General Partner to meet operating expenses of the Partnership when funds generated from Partnership operations are insufficient to meet such expenses. However, Limited Partners shall not be required to contribute more than twenty percent (20%) of their initial capital contributions as additional capital.

6. Describe the rights of limited partners to withdraw or reduce their capital contributions to the partnership. In addition, if limited partners will have the right to demand or receive property other than cash in return for a contribution, the circumstances under which such property would be distributed should be described.

EXAMPLE 3.08F: Withdrawal and Return of Capital.

No Limited Partner shall have the right to withdraw or reduce his contribution to the capital of the Partnership without the consent of the General Partner. No Limited Partner shall have the right to bring an action for partition against the Partnership. No Limited Partner shall have the right to demand or receive property other than cash in return for his contribution. No Limited Partner shall have priority over any other Limited Partner, either as to the return of his contribution of capital or as to profits, losses or distributions.

7. The agreement should prevent the limited partner from participating in the control of the business.

EXAMPLE 3.08G: Role of Limited Partner.

Except as otherwise provided in the Agreement, a Limited Partner shall have no part in or interfere in any manner with the conduct or control of the business of the Partnership, and shall have no right or authority to act for or by the Partnership.

8. Describe in some detail the rights, powers and obligations of the general partner, and the extent to which management may be delegated.

EXAMPLE 3.08H: Rights, Powers and Obligations of the General Partner.

The management and control of the Partnership and its business and affairs shall rest exclusively with the General Partner who shall have all the rights and powers which may be possessed by a general partner by law, and such rights and powers as are otherwise conferred by law or are necessary, advisable or convenient to the discharge of their duties under this Agreement and to the management of the business and affairs of the Partnership. Without limiting the generality of the foregoing, the General Partner shall have the following rights and powers which it may exercise.

a) To spend the capital and net income of the Partnership in the exercise of any rights or powers possessed by the General Partner hereunder.

b) To acquire, purchase, hold and sell real estate and lease the same to third parties and to enter into agreements with others with respect to such activities, which agreements may contain such terms, provisions and conditions as the General Partner in its sole and absolute discretion shall approve.

c) To borrow money to discharge the Partnership's obligations, or to protect and preserve the assets of the Partnership, or to incur any other indebtedness in the ordinary course of business and to pledge all or any of the Partnership's assets or income to secure such loans.

d) To employ a business manager or managers to manage the Partnership's affairs.

e) To execute leases, licenses, rental agreements, and use agreements, on behalf of the Partnership, of and with respect to all or any portion of the real property.

f) To delegate all or any of its duties hereunder, and in furtherance of any such delegation to appoint, employ or contract with any person it may in its sole discretion deem necessary or desirable for the transaction of the business of the Partnership, which persons may, under the supervision of the General Partner: administer the day-to-day operations of the Partnership serve as the Partnership's advisers and consultants in connection with policy decisions made by the General Partner; act as consultants, accountants, correspondents, attorneys, brokers, escrow agents, or in any other capacity deemed by the General Partner necessary or desirable; investigate, select, and on behalf of the Partnership, conduct relations with persons acting in such capacities, and enter into appropriate contracts with, or employ, or retain services performed or to be performed by, all or any of them in connection with the real estate; perform or assist in the performance of such administrative or managerial functions necessary in the management of the Partnership and its business as may be agreed upon with the General Partner; and perform such other acts or services for the Partnership as the General Partner, in its sole and absolute discretion, may approve.

9. Any limitations or restrictions on the general partners' powers should be described.

EXAMPLE 3.08I: Limitations on General Partner's Powers.

The General Partner shall not, without the written consent or ratification of the specific act by the Limited Partners:

a) Make, execute, or deliver any assignment for the benefit of creditors, or sign any confession of judgment on behalf of the Partnership.

b) Possess partnership property or assign its rights in specific partnership property for other than a Partnership purpose.

c) Act in contravention of the Agreement.

d) Conduct any act which would make it impossible to carry on the ordinary business of the Partnership.

e) Admit a person as a general partner.

f) Permit a creditor who makes a nonrecourse loan to the Partnership to acquire any interest in profits, capital or property of the Partnership other than as a secured creditor.

10. Describe the agreed rights of limited partners, consistent with their passive role in the partnership.

EXAMPLE 3.08J: Rights of the Limited Partners

Limited Partners shall have the right to:

a) Have the Partnership books kept at the principal place of business of the Partnership or such other place as designated by the General Partner, and to inspect and copy any of them in accordance with this Agreement.

b) Have dissolution of the Partnership and the liquidation and distribution of its assets upon decree of a court.

c) Have on demand true and full information of all things affecting the Partnership and a formal accounting of the Partnership affairs whenever circumstances render it just and reasonable.

11. Describe any rights which will be granted to the limited partners to remove and replace the general partner. Since the limited partners cannot take active part in management without losing limited liability, their failure to designate a new general partner should cause a dissolution of the partnership.

EXAMPLE 3.08K: Removal of General Partner.

Limited Partners shall have the right to remove the General Partner, by written vote or written consent signed and acknowledged by at least ninety percent (90%) of the then outstanding limited partnership interests, and given to the General Partner within thirty (30) days prior to the effective date of removal.

a) Removal of the General Partner shall be effective upon the substitution of the new General Partner;

b) Concurrently with such notice of removal or within thirty (30) days thereafter by notice similarly given, the Limited Partners shall, in addition, designate a new General Partner.

c) Substitution of a new General Partner shall be effective upon written acceptance of the duties and the responsibilities of General Partner hereunder. Upon effective substitution of a new General Partner, this Agreement shall be and remain in full force and effect except for the change in General Partner and the business of the Partnership shall be continued by the new General Partner. The new General Partner shall thereupon execute, acknowledge, file and publish, as appropriate, amendments to the Certificate of Limited Partnership and Trade Name Affidavit.

d) Failure of the Limited Partners to designate a General Partner within the time specified herein or failure of a new General Partner so designated to execute written acceptance of the duties and responsibilities of General Partner hereunder within ten (10) days after such designation shall dissolve the Partnership.

12. For ease of management of the partnership, it is good practice to provide that each of the limited partners grants a power of attorney to the general partner to execute documents to maintain limited partnership status in his or her name. This avoids the nuisance of attempting to locate all limited partners to obtain their signatures for documents which need to be filed to properly maintain the partnership.

EXAMPLE 3.08L: Power of Attorney

Each of the Limited Partners hereby irrevocably constitutes and appoints the General Partner as true and lawful attorney-in-fact for such Limited Partner with power and authority to act in his name and on his behalf in the execution, acknowledgment, filing and recording of documents, which shall include the following:

a) A Certificate of Limited Partnership and any amendment therto, under the laws of the State of Colorado or the laws of any other state or other jurisdiction in which such certificate or any other amendment is required to be filed.

b) Any other instrument which may be required to be filed or recorded by the Partnership under the laws of any state or by any governmental agency, or which, in the General Partner's discretion, it is advisable to file or record.

c) Any document which may be required to effect the continuation of the Partnership, the admission of an additional or substituted Limited Partner to the Partnership or the dissolution and termination of the Partnership, provided that such documents are in accordance with the terms of the Partnership Agreement.

Such Power of Attorney (i) shall be a special power of attorney coupled with an interest, shall be irrevocable and shall survive the death of such Limited Partner; (ii) may be exercised by the General Partner for each Limited Partner by a facsimile signature of the General Partner or by listing all of the Limited Partners executing any instrument with a single signature of the General Partner acting as attorney-in-fact for all of them; and (iii) shall survive the delivery of any assignment by such Limited Partner of the whole or any portions of his interest except that where the assignee of the whole thereof has been approved by the General Partner for admission to the Partnership as a substituted Limited Partner, the Power of Attorney shall survive the delivery of such assignment for the sole purpose of enabling the General Partner to execute, acknowledge and file any instrument necessary to effect such substitution.

13. Describe any limitations to be placed on transfer of limited partnership interests.

EXAMPLE 3.08M: Transfer of Limited Partnership Interests.

No heir, successor, donee, assignee, or other transferee (including a partner's spouse) of the whole or any interest in a Limited Partner's interest in the Partnership shall have the right to become a substituted Limited Partner in place of his assignor unless all of the following conditions are satisfied:

a) Upon receipt of a bona fide offer to purchase a limited partnership interest in an amount at least equal to or greater than the minimum subscription amount required by the securities laws in the respective states where the transferor and transferee reside, the holder of such interest shall communicate such offer to the General Partner. The General Partner shall have a right of first refusal to purchase such interest according to the price and terms of the bona fide offer which option must be exercised within thirty (30) days from the date of first receipt of the notice of said bona fide offer. In the event that the General Partner fail to exercise their option hereunder, the Limited Partner may transfer his interest upon the same terms as the offer and upon satisfaction of all other requirements of this Article.

b) The written instrument of assignment which has been filed with the Partnership is fully executed and acknowledged and sets forth the intention of the assignor that the assignee become a substituted Limited Partner in his place.

c) The assignor and assignee execute and acknowledge such other instruments as the General Partner may deem necessary or desirable to effect such substitution, including the written acceptance and adoption by the assignee of the provisions of this Agreement.

d) Recordation of an amendment to the Certificate of Limited Partnership in accordance with the Colorado Limited Partnership Act.

e) Payment by the transferor of all reasonable expenses of the Partnership connected with such transfer, including, but not limited to, legal fees and costs (which costs may include, for example, the cost of obtaining opinion of counsel as to the transferability of such interest or of filing any amendment to the Certificate of Limited Partnership).

f) The consent to such transfer in writing by the General Partner.

The Limited Partnership Certificate

The most troublesome formality associated with the limited partnership is the certificate which must be properly filed and maintained to assure limited liability for the limited partners. Failure to properly file and amend the certificate when necessary will prevent recognition of the limited partnership, and all members will be treated as though they belong to a general partnership.

The activities of the limited partnership resolve around this certificate and most of the original Uniform Limited Partnership Act and much of the Revised Uniform Act are concerned with its contents and maintenance.

Content. The certificate of limited partnership roughly resembles a corporation's articles of incorporation. In some respects, it is more specific and revealing. The requirements for the content of the certificate are specified in the Uniform Limited Partnership Acts, and the limited partnership agreement may be filed as the certificate if all the appropriate information is contained in the agreement. Specifically the certificate must contain:

1. The name of the partnership;
2. The character of the business;
3. The location of the principal place of business;
4. The name and place of residence of each member, general and limited partners being respectively designated;
5. The term for which the partnership is to exist;
6. The amount of cash and a description of and the agreed value of the other property contributed by each limited partner;

7. The additional contributions, if any, agreed to be made by each limited partner and the times at which or events on the happening of which they shall be made;

8. The time, if agreed upon, when the contribution of each limited partner is to be returned;

9. The share of the profits or the other compensation by way of income which each limited partner shall receive by reason of his contribution;

10. The right, if given, of a limited partner to substitute an assignee as contributor in his place, and the terms and conditions of the substitutions;

11. The right, if given, of the partners to admit additional limited partners;

12. The right, if given, of one or more of the limited partners to priority over other limited partners, as to contributions or as compensation by way of income, and the nature of such priority;

13. The right, if given, of the remaining general partner or partners to continue the business on the death, retirement or insanity of a general partner; and

14. The right, if given, of a limited partner to demand and receive property other than cash in return for his contribution.[33]

Filing. Most states require a single filing of the certificate in the office of the county clerk in the county where the principal place of business of the partnership is situated. A handful of states (and the Revised Uniform Act) designate the office of the Secretary of State as the repository for the certificate. Even fewer require filing in both places. New York requires publication once a week for six weeks in two newspapers of general circulation in the county, one of which should be in the city where the partnership is located. Of course, the appropriate state statute should be carefully reviewed in any case. Moreover, if the partnership intends to do business in more than one county or more than one state, appropriate multiple filings should be made to avoid any question of compliance with this important provision.

Amendments. During the course of operating a limited partnership there are several situations which require an amendment to the certificate. The specific matters requiring an amendment are stated in Section 24 of the Uniform Limited Partnership Act and they are, generally: changes in the name, amount or character of the limited partners' contributions, character of business or time for dissolution or return of contribution; admission of any partner; substitution of a limited partner; continuation of the business after withdrawal, death or insanity of a general partner; or any other case where an amendment is required to correct an erroneous statement in the certificate or to accurately represent the agreement between members.[34]

The amending statement must conform on the requirements specified for the certificate itself, and must be signed and sworn to by all members, including a substituted or added partner.[35] To avoid the nuisance of locating and obtaining the signature of each limited partner, a power of attorney to permit the general partner to sign for this purpose on behalf of each limited partner is appropriate in the partnership agreement.

It should be apparent that this amendment procedure is cumbersome and may be even annoying. The details for which an amendment is required are even more specific than those required for corporate amendments. For example, a corporation does not have to amend its articles of incorporation every time it acquires a new shareholder or loses a director, but the limited partnership must amend for analogous changes in personnel.

Cancellation of the Certificate. When the limited partnership is dissolved, or when all limited partners cease to be limited, the certificate of limited partnership must be cancelled. Since the limited partnership was formed in a public manner, by filing the certificate, it should be dissolved with the same formality. A certificate of cancellation is provided for this purpose, and it also must be signed by all members.[36]

§ 3.09 REFORMS PROMISED BY THE REVISED LIMITED PARTNERSHIP ACT

The Revised Uniform Limited Partnership Act was adopted by the National Conference of Commissioners on Uniform State Laws in 1976 and was intended to modernize the prior uniform law concerning limited partnerships. Nevertheless, the law was intended to retain the special characteristics of limited partnerships as they have been distinguished historically from corporations. The original act, proposed in 1916, contained many areas which are considered to be vague, misleading, and uncertain. The practical application of the original uniform act has resulted in a clearly hybrid business form, which uses some characteristics of the general partnership and some characteristics of the business corporation.

The Revised Act provides many provisions which appear to be corporate provisions, such as the right of limited partners to commence derivative litigation and the need to qualify foreign limited partnerships in a manner similar to the qualification of foreign corporations. Nevertheless, the Revised Act emphasizes that the relationship among partners is consensual and requires a degree of privity which forces the general partner to seek the approval of the partners, and in some cases unanimous approval, under circumstances that corporate management would find unthinkable.

Definitions

The Revised Act contains several definitions, which have not previously been defined by any statute, assisting in the elimination of any ambiguities in the law of limited partnership.

A "partner" may mean either a limited or general partner, and the terms "general partner" and "limited partner" are defined as persons who have been admitted to the limited partnership in those respective capacities and who are *named* in the certificate of limited partnership as a general partner or limited partner, as the case may be.[37] The definitions of a partner emphasizes that a pre-requisite of becoming a limited partner under the Revised Act is to be named in the certificate of limited partnership as a limited partner. Of course, the certificate will also have to be filed, or an amendment to the certificate properly accomplished and filed in order to protect limited partnership status.[38]

"Contribution" has been generously defined in the Revised Act. In addition to cash and property (which have always been proper contributions by a limited partner), the Revised Act permits the contribution of services rendered or a promissory note or other binding obligation to contribute cash or property or to perform services which a partner may contribute to the partnership in his or her capacity as a partner.[39] By this new definition of contribution, the limited partnership law permits investors more latitude in investing in limited partnerships than they would have in becoming shareholders for corporations. Proper corporate contributions include cash, property and services rendered, but would not permit the promise of future services or the use of a promissory note.[40]

The "partnership agreement" is defined as any valid agreement, written or oral, of the partners as to the affairs of the limited partnership and the conduct of its business. The agreement, of course, in good practice should always be written, but the definition permits the use of an oral agreement as to matters affecting the affairs and conduct of the partnership's business. The one required written document is limited partnership certificate. The prior act did not refer to the partnership agreement at all, and appeared to assume that all important matters effecting the limited partners would be set forth in the certificate of limited parthership. Under modern practice, however, it has been common for partners to enter into a comprehensive partnership agreement, only part of which was included in the certificate of limited partnership which was filed as a matter of public notice. Under the Revised Act, the certificate of limited partnership is confined principally to matters respecting the addition and withdrawal of partners and of capital and other important issues are left to the partnership agreement.[41]

Name

The revised statute governing the name of the limited partnership is very similar to statutes affecting corporate names. At last, the name of the limited partnership must contain, without abbreviation, the words "Limited Partnership" which should give notice to the world of the limited liability of certain of its members. The name may not contain any word or phrase indicating or implying that it is organized other than for a purpose stated in its certificate of limited partnership and may not be the same as or deceptively similar to the name of any corporation or other limited partnership organized or qualified under the laws of the local jurisdiction. The original rule that a limited partner's name may not be included in the name of the partnership, except under certain circumstances, is retained.[42] The name of the limited partnership may be reserved by anyone attempting to organize the limited partnership or intending to qualify a foreign limited partnership in a state. The name of the limited partnership may be reserved for a period of 120 days, and extended for additional 60 days. Moreover, similar to corporate law, the reserved and limited partnership name may be transferred by an appropriate notice of transfer.[43]

Registered Office and Agent, Books and Records

The benefit to the public and to local government agencies of a designated office for business records and an agent for service of process has long been recognized for corporations. The Revised Act requires that specified records of the limited partnership must be maintained at an office within the state, which may but need not be the principal place of the business of the limited partnership. In addition, an agent for service of process on the limited partnership who must be an individual resident of the state must also be continuously maintained by the limited partnership.[44] The partnership is required to keep in its office a current list of each partner in alphabetical order, a copy of the certificate of limited partnership and all amendments thereto (with executed powers of attorney pursuant to which any certificate has been executed), copies of the limited partnership's federal state and local income tax returns and reports for three years, and copies of any effective written partnership agreement and of any financial statements of the limited partnership for at least three years. These records are subject to inspection and copying by any partner upon reasonable request during normal business hours.[45]

Formation, Amendment and Cancellation

The Revised Act does not change significantly the substantive law concerning the formation of limited partnerships, but does organize all sections affecting the formation process in one location, namely Article 2. All provisions respecting the execution and filing of certificates of limited partnership and certificates of amendment and cancellation are now collected here. In addition, the provisions clarify sections of the original act which suffered from ambiguity.

The certificate of limited partnership is essentially the same and has been previously discussed.[46] The place of filing the certificate has been denominated to be the "Office of the Secretary of State", providing a central depository for certificates of limited partnership like articles of incorporation.

An amendment to the certificate is required in four instances:

1. A change in the amount or character of a contribution of any partner or in the partners' obligations to make a contribution;
2. The admission of a new partner;
3. The withdrawal of a partner;
4. The continuation of the business after the withdrawal of a general partner.[47]

An amendment to the certificate must occur within thirty days after the happening of any of the foregoing events. Of course, if a general partner becomes aware that any statement in the certificate is false when made or that other facts have changed making the certificate inaccurate in any respect, the certificate must be amended promptly.[48]

The general partner may consider the revised statute less strict than the original act. The general partner has no responsibility for filing an amendment to the certificate of limited partnership unless he or she *is aware* that a statement currently in the certificate is false or has become inaccurate. In addition, there is a "safe harbor" filing period of 30 days after the occurence of the event, to protect the general partner from claims of creditors or others who assert that they may have been misled by the failure to amend the certificate of limited partnership to reflect changes in any of the important facts contained in the certificate.

The certificate of limited partnership and certificates of amendment and cancellation must be executed properly in accordance with the statute. The original certificate of limited partnership must be signed by all partners named therein. However, an amendment may be signed by only one general partner and by each other partner designated in the certificate as a new partner or whose contribution is described as having been increased. Further the Revised Act now specifically permits any person to sign a certificate by an attorney-in-fact.[49]

Nevertheless, if a certificate change respecting the admission or increased contribution of a partner is signed on behalf of that partner by an attorney-in-fact, the power of attorney authorizing such execution must include special reference to the increase or to the admission to the partnership.[50]

Cancellation of the certificate of limited partnership is required upon the dissolution and commencement of winding up of the partnership or at any other time there are no limited partners.[51] Under the original act, cancellation of the certificate was required upon dissolution, but it was not certain whether a certificate should be recorded if a mere technical dissolution may have occurred but the partnership business was to continue. Under the Revised Act, only the commencement of winding up of the partnership would require a certificate of cancellation, so that if the business is to be continued following a dissolution, a certificate of cancellation need not be filed.

Status of Limited Partners

One of the most troublesome aspects of limited partnership law which has been partially resolved by the Revised Act concerns the limited partners' participation in the control of the business so as to remove his or her limited liability. The Revised Act provides that a limited partner does not participate in the control of the business simply by (1) being a contractor for or an agent or employee of the partnership or of a general partner; (2) consulting with and advising a general partner with respect to the business of the partnership; (3) acting as a surety for the partnership; (4) approving or disapproving of amendment to the partnership agreement; or (5) voting on dissolution, sale of assets, the incurrence of indebtedness, a change in the nature of the business or removal of a general partner.[52] These enumerated items are very similar to transactions normally engaged in by stockholders under corporate law. Since the acts of stockholders do not remove their corporate shield of limited liability, it is consistent with this policy that the limited partners should not lose limited liability for similar activities.

To avoid confusion that could arise because of negative inferences drawn from the activities enumerated by the statute, a separate section has been added to state that the enumeration of these activities does not mean that the possession or exercise of any other powers by a limited partner constitutes participation in the business of a limited partnership. Rather, the participation or exercise of such non-enumerated activities must in fact be shown to constitute an act of partnership control.[53] Moreover, if the limited partners' participation in the control of the business is not substantially the same as the exercise of the powers of a general partner, he is liable only to persons who transact business with the limited partnership *with actual knowledge of* his participation in control.[54]

General Partners

The Revised Act does little to modify the status or rights of a general partner of a limited partnership. The general partner of a limited partnership still has the rights and powers and is subject to the restrictions and liabilities of a partner in a partnership without limited partners.[55]

The Revised Act calls everything that a general partner may do to get out of the partnership (whether voluntary or involuntary) an "event of withdrawal". The consequences of a withdrawal by the general partner is a dissolution of the limited partnership unless there is at least one other general partner who can carry on the business or, within 90 days after withdrawal, all partners agree in writing to continue the business and to appoint a new general partner.[56]

The general partner will have "withdrawn" if he or she gives written notice of resignation to the other partners, assigns the partnership interest (thereby ceasing to be a member of the partnership) or is removed, perhaps forceably, by the other partners in accordance with the partnership agreement. It is likely that each of the foregoing events of "withdrawal" will involve a violation of the partnership agreement for which the general partner may be liable to the other partners. In addition, other events, mostly dealing with financial or personal weakness, may cause a general partner to withdraw. Specifically, if a general partner is bankrupt or takes certain action similar to bankruptcy or has been involuntarily forced into bankruptcy-type activities, the general partner will have "withdrawn". Moreover, in the case of a general partner who is a human being, death or an adjudication of incompetence will cause the general partner to have withdrawn. Finally, if the general partner is another entity, e.g. a corporation, trust or another partnership, if certain actions occur which cause the entity to terminate, the general partner will have withdrawn.[57]

Voting Rights

The original Uniform Limited Partnership Act was silent concerning the right to vote among the partners and whether limited partners must be accorded the right to vote as a class, exclusive of any vote of the general partner. The Revised Act specifically provides for voting for both limited and general partners.

The partnership agreement may grant to all or a specified group of the limited partners the right to vote (on a per capita or some other basis) upon any matter. It is possible, therefore, to permit the limited partners to vote on certain matters affecting the partnership business (subject to a question of participating in the control of the business) and to permit the vote by specifying classes of limited partners who are entitled to vote on certain matters, or by permitting all limited partners to vote with a per capita vote, or by permitting limited partners to vote on the basis of capital contributions or some other objective criteria.[58]

The partnership agreement may also grant to all or to certain identified general partners the right to vote (on a per capita or any other basis), separately or with all or any class of the limited partners on any matter.[59] This provision permits the general partner, or a specifically identified general partner, the right to vote on certain matters, either separately as a class of general partners, or together with all or a portion of the limited partners.

In sum, the Revised Act provides for complete flexability in the drafting of the partnership agreement to provide for varied voting among general and limited partners depending upon the matter under consideration at the time. The only caveat to the drafter of the partnership agreement will be to carefully observe the provisions of Section 303 of the Revised Act which considers the limited partners' participation in the control of the business. Any voting rights accorded to the limited partners must not constitute participation in control of the business or the limited partners will lose limited liability.

Dissolution and Winding Up

In case of a dissolution and winding up of the limited partnership, the Revised Act has altered distribution priorities to provide for payment to creditors, and then payment to partners for accrued but unpaid distributions due on an interim basis and upon withdrawal, and then the partners for return of their contributions, and lastly in payment of partnership interests.[60] There also is no further distinction between general and limited partners for purposes of distribution upon liquidation. A general partner is no longer subordinated by virtue of his or her control status, and is accorded equal status, similar to the treatment of a stockholder who may also be a corporate director or officer.

The distribution provisions may be varied by agreement.[61] The drafter of the agreement may, therefore, acknowledge the desires of the partners as to dissolution distributions and the rule stated in the partnership agreement will be the rule to follow for the partnership.

In the event of a surprise dissolution by the withdrawal of a general partner, it may not be necessary to wind up the affairs of the limited partnership if the limited partners would prefer to continue the business. The Revised Act permits 90 days after the withdrawal for all partners to agree in writing to continue the business of the partnership and to appoint one or more additional general partners if necessary or desired. The partnership agreement should address this point and provide for an immediate meeting of all partners in the event of a withdrawal of the general partner in order to vote on the continuation of the business and to the appointment of an additional general partner if necessary.[62]

Foreign Limited Partnerships

The revised act borrowed a number of corporate rules providing for the qualification and registration of foreign limited partnerships in other states. Any foreign limited partnership (defined as a partnership formed under the laws of some other state) must register with the Secretary of State before transacting business in a new state.[63]

An application for registration as a foreign limited partnership must contain the following items:

1. Name of the foreign limited partnership and, if different, the name under which it proposes to register and transact business in the new state (including the words "Limited Partnership" as part of the name);
2. The state and date of its formation;
3. The general character of business that it proposes to transact in the new state;
4. The name and address of any agent for service of process who is either a resident of the new state or an entity formed under the laws of and or qualified to do business in the new state;
5. An appointment of the Secretary of State of the new state as the agent of the foreign limited partnership if the otherwise appointed agent can no longer be found;
6. The address of its office; and
7. Names and addresses of all of the partners if the certificate of limited partnership filed in the state of organization does not include such a list.[64]

There are requirements for amendments to the registration certificate and for cancellation of registration when the partnership ceases to do business in the foreign state.[65]

A foreign limited partnership which transacts business without registration in a state operating under the revised act is prohibited from maintaining any action, suit or other proceeding in a court of that state until registration has occurred. Merely failing to register does not effect any contract or act that the foreign limited partnership may have conducted in the new state, nor will it effect the limited liability of limited partners in the new state.[66]

Derivative Actions

Stockholders of a corporation may, under certain circumstances, maintain a derivative action in order to enforce the rights of their corporation. A limited partner has never expressly had such a right under the original act, and many cases have considered whether limited partners were entitled to bring derivative actions, with diverse results.

The Revised Act expressly permits a limited partner to bring an action in the right of the limited partnership to recover a judgment in its favor if the general partners with authority to do so have refused to bring the action or if an effort to cause the general partners to bring such an action is not likely to succeed.[67]

NOTES

1. Uniform Limited Partnership Act (hereafter cited as U.L.P.A.) § 1, and Revised Uniform Limited Partnership Act (hereafter cited as R.U.L.P.A.) § 101(7).
2. 6 U.L.A. 561 (1969). The Uniform Limited Partnership Act has not been adopted in Louisiana.
3. Connecticut and Wyoming adopted the Revised Act in 1979.
4. U.L.P.A. § 7 and R.U.L.P.A. § 303(a).
5. U.L.P.A. § 12 and R.U.L.P.A. § 404.
6. See Section 3.06.
7. U.L.P.A. § 19(1) and R.U.L.P.A. § 702.
8. U.P.A. § 6(2). Remember that the Uniform Act most used to govern limited partnerships is the original Act written in 1916. The Revised Uniform Limited Partnership Act is a much more comprehensive statute and leaves little to be covered by the general partnership Statute. See Section 3.09.
9. U.L.P.A. § 7 and R.U.L.P.A. § 303(a).
10. U.L.P.A. § 4. Under the Revised Act, a limited partner may contribute services rendered as capital, and may also contribute a promise to contribute cash (e.g. a promissory note) or property or to perform services. R.U.L.P.A. § 501.
11. U.L.P.A. § 7. Under the Revised Act, a limited partner who knowingly permits his name to be used in the name of the limited partnership (unless his name is also the name of a general partner or the corporate name of a corporate general partner) will be liable to creditors who extend credit to the limited partnership without actual knowledge that the limited partner is not the general partner. R.U.L.P.A. §§ 102(2)(i) and 303(d). Similarly, the Revised Act better defines the circumstances under which a limited partner will have participated in the control of the business. R.U.L.P.A. §§ 303(b) and (c).
12. U.L.P.A. § 9 and R.U.L.P.A. § 403.
13. Under R.U.L.P.A. § 303(b), a limited partner is not participating in the control of the business solely by doing one or more of the following:
    a) being a contractor for or an agent or employee for of the limited partnership or of a general partner;
    b) consulting with and advising a general partner with respect to the business of the limited partnership;
    c) acting as surety for the limited partnership;
    d) approving or disapproving an amendment to the partnership agreement; or
    e) voting on one or more of the following matters:
    (i) the dissolution and winding up of the limited partnership;
    (ii) the sale, exchange, lease, mortgage, pledge, or other transfer of all or substantially all of the assets of the limited partnership other than in the ordinary course of its business;
    (iii) the incurrence of indebtedness by the limited partnerhsip other than in the ordinary course of its business;
    (iv) a change in the nature of the business;
    (v) the removal of a general partner.
    The statute further provides that the foregoing enumeration of powers does *not* mean that possession or exercise of any other powers by a limited partner is participation by him in the business of the partnership. R.U.L.P.A. § 303(c).
14. U.L.P.A. § 10 and R.U.L.P.A. § 305.
15. U.L.P.A. § 8 and R.U.L.P.A. § 301(a).

16. U.L.P.A. § 16 and R.U.L.P.A. § 603.
17. *Ibid.* Under the original Act, notice must be given to "all members," while under the Revised Act, notice is given to "each general partner."
18. U.L.P.A. § 16 and R.U.L.P.A. § 607.
19. U.L.P.A. § 16(3) and R.U.L.P.A. § 605.
20. U.L.P.A. § 19(4) and R.U.L.P.A. §§ 301(2) and 704.
21. U.L.P.A. § 19 and R.U.L.P.A. § 301(b). Forms for the Assignment and Consent to Substitution by a Limited Partner appear as Forms 3A and 3B in the Appendix. The Revised Act abandons the terminology of a "substituted" limited partner. Instead, the Revised Act refers to an assignee who has been granted the right to become a limited partner. Amendment of the certificate is discussed in Section 3.-08.
22. U.L.P.A. § 10(c). The Revised Act permits a decree of dissolution whenever "it is not reasonably practicable to carry on the business in conformity with the partnership agreement." R.U.L.P.A. § 802.
23. U.L.P.A. § 16(4)(a). The Revised Act does not permit such a drastic remedy if the contribution is not returned to a demanding limited partner. Instead, the limited partner is treated as an ordinary creditor of the partnership and may obtain a judgment for the amount of the contribution. R.U.L.P.A. § 606.
24. U.L.P.A. § 20. The Revised Act provides that a general partner will have "withdrawn" from the partnership by resigning (through written notice), assigning his interest, being removed in accordance with the agreement, becoming bankrupt (or taking action similar to bankruptcy), dying, becoming incompetent, or, (in the case of entity general partners) ceasing to be a valid entity under law. R.U.L.P.A. § 402. Then the Revised Act makes "withdrawal" an event of dissolution unless there is at least one other general partner and the certificate of the limited partnership permits the business to be carried on by the remaining general partner or, if within 90 days after the withdrawal, all partners agree in writing to continue the business and to the appointment of a new or additional general partner if necessary or desired. R.U.L.P.A. § 801(3).
25. U.L.P.A. §§ 9(1)(g) and 20. R.U.L.P.A. § 801(3) (which adds a 90 day grace period for all partners to so agree.)
26. The misconduct of a general partner would cause dissolution to be "equitable and proper" under U.L.P.A. § 10(c) and presumably would be an event which makes it reasonably impracticable "to carry on the business in conformity with the partnership agreement" under R.U.L.P.A. § 802.
27. U.P.A. § 31(c) and R.U.L.P.A. § 801(2).
28. U.L.P.A. § 20 and R.U.L.P.A. § 801(3). An example of such a continuation statement appears in Clause 26 of the Sample Limited Partnership Agreement, Form 3C, in the Appendix.
29. Not so under the Revised Act, which permits a winding up by the general partners who have not wrongfully dissolved the limited partnership (unless provided otherwise in the partnership agreement) or, if none, the limited partners to wind up the limited partnership's affairs. Further, upon the application of any partner, his legal representative or assignee, a court may wind up the limited partnership's affairs. R.U.L.P.A. § 803.
30. U.L.P.A. § 23. Under the Revised Act, the scheme for distribution of assets is considerably modified by providing that (1) to the extent the partners (general or limited) are also creditors, other than in respect to their partnership interests, they will share with other creditors, (2) once the partnership's obligation to make a distribution accrues, it must be paid before any other distributions of an "equity" nature are made, and (3) general and limited partners rank on the same level except as otherwise provided in the partnership agreement. See R.U.L.P.A. § 804.
31. Notice that for this result to occur under the Revised Act, it will be necessary to provide for preference for the limited partner distributions in the partnership agreement. See R.U.L.P.A. § 804.
32. See Treas.Reg. § 301.7702–2 (1960) for an enumeration of corporate characteristics which may affect the tax status of limited partnership.
33. U.L.P.A. § 2. The Revised Act contains essentially the same information, with the following differences:
   a) An office and an agent for service of process are required to be specified;
   b) The business address of each partner is requested, rather than the place of residence;

c) An agreed value of services to be contributed must be stated if services are to be contributed by any partners;

d) If agreed upon, the time at which or the events upon the happening of which a partner may terminate his membership in the partnership and the amount of or the method of determining the distribution to which he may be entitled respecting his partnership interest (including any terms and conditions of the termination and distribution);

e) The right of a partner to receive distributions of property or return of all or any part of the partner's contribution. See R.U.L.P.A. § 201(a). Sample limited partnership certificates under the original act and the Revised Act appear as Forms 3D and 3E in the Appendix.

34. Similar circumstances require an amendment to the limited partnership certificate under the Revised Act. Within 30 days after the happening of such events, the amendment must be filed. Further, the general partner who becomes aware that a statement in the certificate of limited partnership was false when made or that facts have changed making the certificate inaccurate at this time, shall promptly amend the certificate, but an amendment to show a change of address of a limited partner need be filed only once every 12 months. R.U.L.P.A. § 202(b)–(c).
35. U.L.P.A. § 25. The Revised Act provides only for the signature of at least one general partner and of each other partner who is a new partner or whose contribution is described as having increased. R.U.L.P.A. § 204(a)(2). A sample amendment for admission of a substituted limited partner appears as Form 3F in the Appendix.
36. U.L.P.A. § 24 and R.U.L.P.A. § 203 (requiring a cancellation not jsut upon a dissolution, but when winding-up has commenced. A sample cancellation certificate appears as Form 3G in the Appendix.
37. R.U.L.P.A. § 101(5)(6) and (8).
38. R.U.L.P.A. §§ 201, 202(b) and 202(e).
39. R.U.L.P.A. § 101(2).
40. M.B.C.A. § 19 and see Section ____.
41. R.U.L.P.A. §§ 101(9) and 201.
42. R.U.L.P.A. § 102.
43. The application for reservation for a corporate name and for notice of transfer of a corporate name, which appear as forms ____and ____in the Appendix, may be adapted for the same purposes of limited partnerships under the revised act. R.U.L.P.A. § 103.
44. R.U.L.P.A. § 104.
45. R.U.L.P.A. § 105.
46. See Section 3.08.
47. R.U.L.P.A. § 202(b).
48. R.U.L.P.A. § 202(c).
49. R.U.L.P.A. § 204.
50. R.U.L.P.A. § 204(b).
51. R.U.L.P.A. § 203.
52. R.U.L.P.A. § 303(b).
53. R.U.L.P.A. § 303(a).
54. R.U.L.P.A. § 303(a).
55. R.U.L.P.A. § 403 and U.L.P.A. § 9(1).
56. R.U.L.P.A. §§ 402 and 801.
57. R.U.L.P.A. § 402.
58. R.U.L.P.A. § 302.
59. R.U.L.P.A. § 405.
60. R.U.L.P.A. § 804.
61. R.U.L.P.A. § 804(2), (3).
62. R.U.L.P.A. § 801(3).
63. R.U.L.P.A. §§ 101(4) and 902.
64. R.U.L.P.A. § 902.
65. R.U.L.P.A. § 905 and 906.
66. R.U.L.P.A. § 907.
67. R.U.L.P.A. § 1001.

4

THE BUSINESS CORPORATION

The business corporation is the most interesting and most complex form of business enterprise, and the remainder of these materials will be concerned primarily with the joys and sorrows of doing business as a corporation. To begin, this Chapter defines the legal characteristics of the corporation, the interaction of its members in the management of its business, and some recognized advantages and disadvantages of the corporate business form. Later chapters will discuss problems of formation and organization, corporate finance, internal agreements, distributions of cash and property, qualification of foreign corporations, corporate structural changes and dissolution. The term "business corporation" excludes the many other types of corporations that may be formed under federal or state law. For example, most states authorize the formation and operation of nonprofit corporations, religious and charitable corporations, and municipal corporations, all of which have peculiar characteristics which will not be discussed in this work. The "professional corporation", formed for the purpose of practicing the learned professions, such as law, medicine, accounting, and so forth, will be considered in the next chapter.

§ 4.01 ENTITY CHARACTERISTICS OF A CORPORATION

The most significant characteristic of a corporation which distinguishes it from other forms of business enterprise is that the corporation is considered by the law to be a separate legal entity, a separate "person." The business therefore exists quite apart from its aggregate membership. It was clear that a sole proprietorship is no more than an extension of the personal life of the proprietor, its owner. Moreover, a partnership is treated as an association of individuals, and, for the most part, is also an extension of their personalities, as evidenced by the rules which prohibit the addition of a partner without unanimous consent of the other partners and require dissolution whenever a partner leaves the firm. A corporation, however, exists alone and detached, so that shareholders (its owners) may come and go without affecting its legal status. Continuing this theme, the corporation is liable for its own obligations, and the individual assets of its owners usually may not be reached for satisfaction of these obligations. It will be obvious throughout the remaining discussion of corporate characteristics that this concept of separateness creates special advantages (and occasionally causes special disadvantages) as compared with other business organizations.

Having established that the corporation is treated as a legal entity, it should be recognized that the corporation is a creature of statute. It obtains life from the applicable state law, which authorizes certain corporate powers, prescribes certain rules and requirements for the regulation of its business affairs, and controls the internal relationships between shareholders and management. Since state statutes vary considerably on their approach to corporations, the corporate structure in one state is often quite different from the corporate structure in another. For example, a Delaware corporation may have a one-person board of directors, while most states still require at least three members on the boards.[1] Similarly, in some states the initial by-laws of the corporation are adopted by the board of directors while a few permit the shareholders to do it.[2] These details are dictated by local corporate statutes, which authorize the formation and operation of a corporation as a business form. Consequently, the analysis of these statutory requirements and strict compliance with them is the touchstone of a successful corporate practice. A few words about the statutory variations and their history are appropriate.

The law of corporations was developed by each state to regulate the internal affairs of the corporations it had chartered to do business within its boundaries. As American businesses expanded, interstate operations became commonplace, and it was possible for the organizers of the corporation to shop around for a state whose corporation laws were the most permissive, so that the formation and operation of the corporation would be an easy exercise. The more strict and complex the state regulation, the less attractive that state became for establishing a corporation within its boundaries. Since it is possible to do business in one state and be incorporated in another, and since a state acquires certain benefits by having businesses incorporated under its laws (not the least of which is the authority to levy taxes) state legislatures began to recognize that they could attract corporate businesses by adopting flexible and permissive statutory provisions. New Jersey was the first state to liberalize its laws for this purpose, and Delaware followed closely. Delaware has remained the consistent leader in "mothering" corporations and its statute is considered by many to be the most modern, most permissive, and most sympathetic to the problems of corporate organization and operation.[3] In 1950 the American Bar Association Committee on Corporate Laws prepared a Model Business Corporation Act, which was initially patterned after Illinois law. The Act has since been revised extensively, also with a view toward permissiveness and flexibility, and has been used as a model by many states in their own revisions of corporate statutes. The discussion of corporate law in this book will concentrate upon the provisions of the Model Act, but unusual variations from important states will be separately noted and discussed. No state has adopted the Model Act verbatim, and consequently, there is no substitute for full and complete analysis and understanding of the particular requirements of the state statute under which incorporation is contemplated.

In addition to the statutory regulation of corporate affairs, corporate operations are also governed by certain rules and regulations adopted by the persons forming the business. The articles of incorporation and the by-laws, both of which will be discussed in detail later, are adopted by the corporate membership and will govern the activities of the corporation throughout its operation.[4] It should be noted that most state statutes are very broad in their description of corporate powers, because the statute is designed to cover every conceivable corporate form and every type of business. The articles of incorporation may contain only the essential information required by statute, or they may elaborate on specific matters to govern internal corporate affairs. If the articles are general, then the by-laws should provide specific rules for regulation of corporate activities. Of course, a properly formed corporation will have no conflict between the by-laws, the articles, and the appropriate state law. Rather, the by-laws and articles will refine and elaborate upon the concepts embodied in the state statute, and thereby provide a comprehensive and workable scheme for the regulation of the corporation. It is also important to recognize that the by-laws are adopted and modified by internal action of the corporation, and consequently, the rules contained therein are easily changed. The articles of incorporation, which are filed with the Secretary of State as public notice of the existence and structure of the corporation, may only be amended by a cumbersome amendment procedure. The most flexible regulation of internal affairs will result from drafting the rules for these corporate activities in the easily-amendable by-laws.

In sum, a properly-formed corporation exists as a legal entity, and is treated for all practical purposes as an individual person, separate and distinct from the persons who own and manage it. Its formation and operation are governed by specific state statutes and by its own articles of incorporation and by-laws, as adopted to suit the particular needs of its business.

§ 4.02 STATUTORY POWERS OF A CORPORATION

Each state law grants the corporation the necessary powers to conduct business, and to conduct any other activities necessary to the business in which it is engaged. Most statutes granting corporate powers permit the corporation to do almost everything that a private individual could do. Section 4 of the Model Business Corporation Act enumerates corporate powers as follows: "Each corporation shall have power:

(a) To have perpetual succession by its corporate name unless a limited period of duration is stated in its articles of incorporation.

(b) To sue and be sued, complain and defend, in its corporate name.

(c) To have a corporate seal which may be altered at pleasure, and to use the same by causing it, or a facsimile thereof, to be impressed or affixed or in any other manner reproduced.

(d) To purchase, take, receive, lease, or otherwise acquire, own, hold, improve, use and otherwise deal in and with, real or personal property, or any interest therein, wherever situated.

(e) To sell, convey, mortgage, pledge, lease, exchange, transfer and otherwise dispose of all or any part of its property and assets.

(f) To lend money and use its credit to assist its employees.

(g) To purchase, take, receive, subscribe for, or otherwise acquire, own, hold, vote, use, employ, sell, mortgage, lend, pledge, or otherwise dispose of, and otherwise use and deal in and with, shares or other interests in, or obligations of, other domestic or foreign corporations, associations, partnerships or individuals, or direct or indirect obligations of the United States or of any other government, state, territory, governmental district or municipality or of any instrumentality thereof.

(h) To make contracts and guarantees and incur liabilities, borrow money at such rates of interest as the corporation may determine, issue its notes, bonds, and other obligations, and secure any of its obligations by mortgage or pledge of all or any of its property, franchises and income.

(i) To lend money for its corporate purposes, invest and reinvest its funds, and take and hold real and personal property as security for the payment of funds so loaned or invested.

(j) To conduct its business, carry on its operations and have offices and exercise the powers granted by this Act, within or without this State.

(k) To elect or appoint officers and agents of the corporation, and define their duties and fix their compensation.

(l) To make and alter by-laws, not inconsistent with its articles of incorporation or with the laws of this State, for the administration and regulation of the affairs of the corporation.

(m) To make donations for the public welfare or for charitable, scientific or educational purposes.

(n) To transact any lawful business which the board of directors shall find will be in aid of governmental policy.

(o) To pay pensions and establish pension plans, pension trusts, profit sharing plans, stock bonus plans, stock option plans and other incentive plans for any or all of its directors, officers and employees.

(p) To be a promoter, partner, member, associate, or manager of any partnership, joint venture, trust or other enterprise.

(q) To have and exercise all powers necessary or convenient to effect its purposes."

Remember that the foregoing powers are conferred by statute and a corporation is permitted to do all things authorized therein. The attorney may, in his or her discretion, deem it appropriate to grant broad powers in the articles of incorporation.

EXAMPLE 4.02A: Powers

To do everything necessary and proper for the accomplishment of any of the purposes, or the attainment of any of the objects, or the furtherance of any of the powers hereinbefore set forth, either alone or in association with other corporations, firms, or individuals, and to do every other act or acts, thing or things, incidental to or growing out of or connected with the aforesaid business or powers, or any part or parts thereof; provided, the same is not inconsistent with the laws under which this corporation is organized.[5]

In most cases, however, the articles of incorporation and the by-laws of the corporation will refine the statutory powers to tailor the corporate structure to the incorporators' needs. Notice that some of the statutory powers actually suggest that elaboration is necessary in the articles of incorporation or by-laws. For example, the articles of incorporation or by-laws must define the duties of officers,[6] and they may predetermine the maximum interest rate at which the corporation may borrow funds.[7]

EXAMPLE 4.02B: Power to Borrow

To borrow money, and to make and issue notes, bonds, debentures, obligations, and evidences of indebtedness of all kinds, whether secured by mortgage, pledge, or otherwise, without limit as to amount, but with interest not to exceed 12 per cent per annum, and to secure the same by mortgage, pledge, or otherwise, and generally to make and perform agreements and contracts of every kind and description.[8]

It is also good practice to elaborate upon the corporation's power to conduct business in other states and countries.

EXAMPLE 4.02C: Power to Qualify in Foreign Jurisdictions

The company shall have power to conduct and carry on its business, or any part thereof, and to have one or more offices, and to exercise all or any of its corporate powers and rights, in the State of New York, and in the various other states, territories, colonies, and dependencies of the United States, in the District of Columbia, and in all or any foreign countries.[9]

Thus, the general grant of power under the state statute represents the maximum limits of corporate power. If incorporators or organizers intend to restrict this power, the modifications are drafted into the articles of incorporation and by-laws.

It should also be noted that subsection (q) of Section 4 grants *power* to do anything necessary to accomplish the *purposes* for which the corporation is organized. The corporate purposes are the particular business objectives which the incorporators direct their corporation to pursue, such as operating a restaurant, owning and leasing real estate and so forth. These purposes are specified in the articles of incorporation, drafted in accordance with the objectives of the incorporators, and they guide corporate management in the type of business to be conducted.[10] The permitted purposes are regulated by statute, also, but this is one place where permissiveness is rampant. The Model Act and most states permit business corporations to be organized for "any lawful purpose or purposes", except banking and insurance since these industries are usually more closely regulated by other state statutes. Consequently, if the incorporators adopt very broad corporate purposes, and authorize the corporation to transact any lawful business, the statutory corporate powers, permitting all power necessary to carry out the purposes, will grant the corporation as much power as any individual would have in conducting a business. The corporate powers enumerated and described in the Model Business Corporation Act are typical of the powers contained in most state statutes. There are, however, some important variations and details pertaining to certain powers which should be discussed.

Perpetual Duration

A vast majority of states allow a corporation to exist indefinitely, and also permit a corporation to be limited to a specific period of time, if such a restriction is deemed important by the incorporators. Most statutes require that the articles of incorporation recite the period of corporate existence, and if none is stated, the corporation will be deemed to exist perpetually. A very few states do not permit perpetual existence and specifically limit the duration of a corporation. For example, Mississippi limits the duration of a corporation to 99 years.

Power to Own and Deal with Real Property

Every state permits a corporation to acquire and hold real property in the corporate name. In several jurisdictions, however, this power is limited to such property necessary to further corporate purposes.[11] Thus, if a restaurant corporation were to acquire a larger building than it actually needed for its restaurant business, there would be a question about its power to do so. However, if it could show that the larger building was purchased with a view to future expansion of the restaurant business or is otherwise convenient and appropriate to its specified corporate purposes, its ownership would be authorized. A specific power clause in the articles of incorporation on this point may help.

EXAMPLE 4.02D: Power to Deal in Property

To the same extent as natural persons might or could do, to purchase or otherwise acquire, and to hold, own, maintain, work, develop, sell, lease, exchange, hire, convey, mortgage, or otherwise dispose of and deal in lands and leaseholds, and any interest, estate, and rights in real property, and any personal or mixed property, and any franchises, rights, licenses, or privileges necessary, convenient, or appropriate for any of the purposes herein expressed.[12]

Lend Money to Assist Employees

These provisions vary considerably among the states. Most states have no statutory power for corporate loans to employees, directors, or officers. Some completely prohibit loans to officers and directors. Those that do grant such power usually impose certain restrictions on it. Shareholder approval of the loan is frequently required, and in some cases the directors must be able to show that the transaction will be of some benefit to the corporation.[13]

Donations

Statutes authorizing corporate power to make donations specify various purposes for which donations may be made, different procedures for internal authorization of donations, and certain limitations on the amount. Usually donations may be made for charitable, educational, religious, public welfare and scientific purposes.[14] In most states the decision to donate would be made by the board of directors, but a few require shareholder approval.[15] Several states impose limitations on the amount which may be somewhat flexible, such as a "reasonable sum" in New Jersey, or may be a firm limitation, such as 5% of net income before taxes in Virginia.

Power to be a Partner or Member of Another Enterprise

Early law prohibited a corporation from becoming a partner in a separate enterprise. Over half of the states have yet to adopt statutory authority granting this power, but the judicial attitude toward the power has become increasingly favorable even without statutory support. In most cases and in some statutes this power is limited to permit a corporation to become a partner or member of another enterprise conducting a business which would be authorized in the corporate purposes—that is, a business which the corporation could lawfully conduct on its own.[16]

§ 4.03 OWNERSHIP AND MANAGEMENT OF THE CORPORATION

The corporation departs significantly from the sole proprietorship and the partnership in the areas of ownership and management of the business enterprise. As you will recall, the sole proprietor is the owner and manager of his own business. In the general partnership each partner is an owner and each partner is vested with the responsibilities of management. More analogous to a corporation, a limited partnership has investors without management control who merely contribute cash or property to the capital of the business while the general partners are responsible for management. Corporate business is managed by a board of directors and officers appointed by them. The owners of the business are the shareholders, who contribute cash, property, or services in exchange for their ownership rights, evidenced by share certificates. It is certainly possible for a shareholder to also be a director and an officer, but the rights and responsibilities of each intracorporate group are clearly segregated in corporate law and each capacity must be separately considered.

Incorporators

The incorporators are responsible for filing the articles of incorporation and securing preincorporation agreements and share subscriptions. The incorporators are usually the "promoters" of the corporation who work closely with counsel in drafting the appropriate documents to comply with the statutory requirements. The main tasks of the incorporators are to prepare and execute the articles of incorporation and to file them with the Secretary of State. Attorneys or their staff forming the corporation may act as the incorporators (sometimes referred to as "dummy" incorporators) since the act of incorporation is primarily a technical legal function.

The necessary number and qualifications of the incorporators are specified by statute. The original provision of the Model Act required that three or more incorporators were required to properly incorporate and that such persons must be over the age of twenty-one. In keeping with the trend toward "permissive" corporate statutes, the Model Act has recently been amended to require *one* or more incorporators without mention of age or other qualifications.[17] Most states require adult natural persons to incorporate and only a small minority of states have state residency requirements for the incorporators.[18] Some states, such as Pennsylvania, require that the incorporators subscribe for shares. The modern trend is to permit any one person to act as an incorporator, and over half of the states have adopted statutes permitting a single incorporator.

Directors

Section 35 of the Model Act states that the business and affairs of a corporation shall be managed by a board of directors who shall exercise all powers of the corporation unless otherwise provided in the statute or in the articles of incorporation. Thus, the board of directors is the autocratic governing body of the corporation and the directors are responsible for management of the shareholder's enterprise. The directors are usually charged with the responsibility of determining corporate policies, managing the affairs of the business and selecting and supervising the officers who handle the detailed business matters. It should be noted that the Model Act vests all corporate power in the directors "unless otherwise provided in the articles of incorporation." Therefore, the incorporators (or shareholders at a later date) may specify in the articles of incorporation that the shareholders will have the management power. Such a provision is found most frequently in "close corporations", which are discussed in detail in the next chapter.

Since directors act as the primary governing body for the shareholder-owned business, it is only fair that the shareholders are entitled to elect the directors. The first directors of the corporation must be named in the articles of incorporation in most states, and this initial board serves until the shareholders meet to elect their successors.[19] The election should occur at each annual shareholder meeting, and the directors so elected usually serve until the next directors are elected.

It is possible to "stagger" or "classify" the board of directors to insure continuity of corporate management. This procedure avoids the election of a complete new board every year by varying the term of office for each director. The Model Act authorizes as many as three classes if the board is greater than 9 members.[20] Thus, if the board has twelve members and is classified to three classes, four directors would serve until the first annual meeting, four would serve until the second annual meeting, and four would serve until the third annual meeting. When the four new directors are elected at the first annual meeting they serve for three years, until the fourth annual meeting, and the process repeats itself. Thus, shareholders will elect four new directors every year and they will join a board of eight old directors, who are presumably familiar with existing corporate policy and will insure continuity in management principles. This classification procedure is treated differently in the state statutes, but most states permit it. The number of classes and the necessary size of the board before classification is permitted are the major variants.

Any person may be a director of a corporation, and only a few states require that a director be "full" or "legal age".[21] The Model Act specifically provides that directors do not have to be shareholders of the corporation or residents of the state unless the articles of incorporation of the by-laws so require.[22] No state requires share ownership by a director, but a few impose residency requirements on at least a fraction of the board.[23] The articles of incorporation or the by-laws may impose residency or share-ownership requirements as necessary qualifications to hold the office of director. Moreover, these documents may also prescribe any other reasonable qualifications for directors, such as a minimum or maximum age or United States citizenship.

In most states, the board of directors must consist of at least three members, and the exact number is fixed in the articles of incorporation or the by-laws. The Model Act was amended in 1969 to require only one director, if the incorporators or shareholders felt that was appropriate.[24] The "one director" provision is also found in a few other states, including Delaware. A corporation may usually have an unlimited number of directors, but, of course, the greater the number of persons on the board, the more difficult it becomes to make corporate decisions. The number of directors, as fixed in the articles of incorporation or by-laws, may be increased or decreased by an appropriate amendment thereto, but the amendment may never authorize less than the minimum number of persons required by the state statute. If any vacancy occurs in the board, either by death, removal or retirement of a director or by an amendment increasing the number of directors, the vacancy may be filled under the Model Act by the affirmative vote of a majority of the remaining directors.[25] This is the only time a director is not elected by the shareholders and some states expressly reserve the power to fill the vacancy to the shareholders, especially if the vacancy has been created by the removal of a director by the shareholder.[26] A director selected to fill a vacancy serves for the remaining term of his predecessor, and a new director will be elected at the next meeting of shareholders.

Directors serve at the pleasure of the shareholders. As owners of the corporate business, the shareholders' control over the corporate director positions is probably their most important power, and the Model Act amplifies this power by permitting the shareholders to remove a director with or without cause.[27] Whether or not a director is guilty of misconduct, the shareholders may remove her at will and for whatever reason. Further, the shareholders' purge is not limited to one director at a time. They may vote to remove the entire board if that is deemed appropriate. The required vote for removal is usually a majority of the shares entitled to vote for the election of the directors.[28]

The board of directors takes action on behalf of the corporation at regular or special meetings where they consider and adopt resolutions of corporate policy. These meetings are called in accordance with the corporate by-laws and are discussed in greater detail in Chapter 8.

Generally, the board of directors is empowered to make all corporate decisions, but realistically their actions are concerned with certain special important matters, and the day-to-day activities of the corporation are left to the officers. Directors are considered by the law to be "fiduciaries," which means that all of their actions should be directed to further and protect the interests of those they serve. Their fiduciary capacity requires that they act independently, however, and they are not bound by the will of the shareholders who elect them. Of course, a director who ignores the desires of his constituents, the shareholders, always runs the risk of being removed from office or losing re-election at the next shareholder meeting. Apart from this realistic possibility of losing his job if he acts too independently, a director is only required to act in the best interests of the corporation by using his independent discretion. More specifically, a director is required to use his best judgment in determining corporate policy and in authorizing corporate action, and to avoid any act which is in conflict with his director position, or which will cause him to profit personally to the detriment of the corporation.

Since the law taints corporate action by a director with a conflict of interest, it may be necessary, if the other intra-corporate parties do not object to the director's personal interests in the negotiation of corporate business, to specifically so provide in the articles of incorporation. These exculpatory clauses are quite commonplace and should be considered if it is likely that any director may have a potential conflict of interest.

EXAMPLE 4.03A: Transactions With an Interested Director or Officer

No contract or other transaction between this corporation and one or more of its directors, officers or stockholders or between this corporation and any other corporation, firm or association in which one or more of its officers, directors or stockholders are officers, directors or stockholders shall be either void or voidable (1) if at a meeting of the board of directors or committee authorizing or ratifying the contract or transaction there is a quorum of persons not so interested and the contract or other transaction is approved by a majority of such quorum, or (2) if the contract or other transaction is ratified at an annual or special meeting of stockholders, or (3) if the contract or other transaction is just and reasonable to the corporation at the time it is made, authorized or ratified.[29]

While directors are generally vested with primary responsibility for management decisions, their powers may be delegated to officers or to an executive committee if authorized by the articles of incorporation or by-laws.

EXAMPLE 4.03B: Executive Committee

The Board of Directors may, by resolution or resolutions passed by a majority of the whole Board, designate one or more committees, each committee to consist of two or more of the directors of the Corporation, which, to the extent provided in said resolution or resolutions, shall have and may exercise the powers of the Board of Directors in the management of the business and affairs of the Corporation, and may have power to authorize the seal of the Corporation to be affixed to all papers which may require it. Such committee or committees shall have such name or names as may be determined from time to time by resolution adopted by the Board of Directors.[30]

There are, however, some specific matters upon which the directors are required to act.

Selection of Officers. Section 50 of the Model Act provides that the officers shall be elected by the board of directors at the time and the manner prescribed by the by-laws. Many important jurisdictions, including Delaware, have adopted statutes permitting the articles of incorporation or by-laws to allow for the election of officers by the shareholders. Since the officers are selected by the directors in most cases, and are required to be selected by the directors in jurisdictions following the Model Act, the directors are under a duty to supervise the officers. The directors may be liable for failure to use due care in the selection or supervision of the officer.

Determining Management Compensation. The board of directors fixes executive compensation, including that of the officers and their own, but the articles of incorporation may require shareholder approval.

EXAMPLE 4.03C: Management Compensation

No salary or other compensation for services shall be paid to any director or officer of the corporation unless and until the same shall have been approved in writing or at a duly held stockholders' meeting by stockholders owning at least seventy-five per cent in amount of the capital stock of the corporation then outstanding.[31]

By-Laws. In most states and under the Model Act, the initial by-laws of the corporation are adopted by the board of directors.[32] They also retain the power to alter, amend or repeal the by-laws or adopt new by-laws, but the articles of incorporation may reserve these rights to the shareholders. In either case, the articles should be specific on the authority desired.

EXAMPLE 4.03D: Adoption, Amendment or Repeal of Bylaws

The directors shall also have power, without the assent or vote of the stockholders, to adopt, amend or repeal bylaws relating to the business of the corporation, the conduct of its affairs and the rights or powers of its shareholders, directors or officers.[33]

Initiation of Extraordinary Corporate Matters. Extraordinary corporate matters, such as amendments of the articles of incorporation, sale or lease of all the corporate assets not in the regular course of business, merger, consolidation, and so forth, are usually initiated by the board of directors, and approved by the shareholders. These matters are obviously beyond the scope of day-to-day management, and may have considerable ramifications on the ownership rights of the shareholders. Consequently, the shareholders must approve such action by an appropriate vote after the action has been initiated by the board of directors.[34] The articles of incorporation should contain provisions respecting the directors powers in such cases.

EXAMPLE 4.03E: Disposition of Assets

> The directors shall also have power, with the consent in writing of a majority of the holders of the voting stock issued and outstanding, or upon the affirmative vote of the holders of a majority of the stock issued and outstanding having voting power, to sell, lease, or exchange all of its property and assets, including its good will and its corporate franchises, upon such terms and conditions as the Board of Directors deem expedient and for the best interests of the corporation.[35]

Declaration of Dividends. Dividends are paid to shareholders from time to time as a return on their investment. The determination of whether dividends are to be paid is a decision for the board of directors, and they are reserved broad discretion in this area.[36]

Issuance of Stock and Determination of Value. The articles of incorporation must state the number of shares of stock the corporation is authorized to issue. It is most unusual for a corporation to issue all of the authorized stock at the beginning of its corporate existence. Consequently, the subsequent determination to issue stock is a decision for the board of directors. The corporation's stock may be either par or no par value, as indicated in the articles of incorporation, and, in the case of no par stock, the value of the stock is determined from time to time by the board of directors in their good judgment.[37]

Officers

As indicated above, officers are usually selected by and receive their power from the board of directors. Section 50 of the Model Act provides that a corporation must have a president, a secretary, and a treasurer, and may have one or more vice-presidents as prescribed in the by-laws. In addition, the board of directors may appoint such other officers and assistant officers as may be necessary. Every state requires a corporation to have at least a president and a secretary and statutes frequently impose limitations on one person holding two offices. Under the Model Act, any two or more offices may be held by the same person, except that the offices of president and secretary must be held by separate persons.[38] Every state permits the offices of secretary and treasurer to be held by the same person.

The authority and responsibility of the officers is a very broad topic. Generally, officers perform whatever duties have been delegated to them by the board of directors or the by-laws,[39] and are responsible for management of the day-to-day affairs of the corporation. However, state statutes frequently prescribe certain things that officers must do. These typically include execution of articles of merger, articles of consolidation, articles of amendment and articles of dissolution[40] by the president or a vice-president and the secretary or an assistant secretary of the corporation. Similarly, the same persons usually must subscribe the certificates representing the shares of the corporation.

Officers are subject to removal at the pleasure of the board of directors, but the directors may have to establish that the best interests of the corporation will be served by removing the officer. However, the officers may have negotiated an employment contract with the corporation, and in such a case, his removal prior to the expiration of the term of the contract may subject the corporation to a lawsuit for breach of contract.[41] Considering potential liability of the corporation the removal of an officer under an employment contract certainly must be supported by a very good reason in order to establish that his removal is in the best interests of the corporation.

Shareholders

The shareholders are the owners of the corporation. They contribute capital for investment in the business, and receive in exchange a stock certificate representing their ownership interest. For purposes of the Model Act, shareholders are defined as "holders of record shares in a corporation." [42] The words "holder of record" deserve some explanation. A corporation maintains a stock transfer ledger, in which the names of the owners of shares of the corporation are registered. Those persons listed in the ledger are the "holders of record." Whenever shares are transferred, the new owner's name is entered on the stock transfer ledger and that person becomes the "holder of record". The holder of record is entitled to vote the shares, to receive dividends, and to receive a proportionate share of assets in dissolution depending, of course, on the voting, dissolution, and dividend characteristics of the stock.[43]

As owners of the corporation the shareholders enjoy certain ownership rights, but not in the same sense as a sole proprietor owns her business, or even as a general partner has ownership rights in his partnership. Rather, the shareholder's rights as an owner are strictly limited by the state corporation statute. Generally, it may be said that the shareholders' ownership rights only include their right to vote, their right to a return on their investment by way of dividends if the directors declare such distributions, and their right to share in the assets if the business is liquidated. From the previous discussion about directors and officers it should be apparent that the shareholders have little or no voice in the day-to-day management of the corporation. However, they do have the power to elect the directors, who are responsible for the selection and supervision of the officers, who are in turn responsible for the daily corporate activities. Thus, shareholders indirectly control corporate policy and activity by electing those directors who are most sympathetic to their desires. Moreover, the law requires that shareholders be consulted whenever the governing body, the board of directors, intends to modify or transform the character of the business in any manner which will affect the shareholders' ownership interests. These "fundamental" corporate changes are described in this book as extraordinary changes in corporate structure, and they include such matters as amendments to the articles, merger, consolidation, exchanges of stock, sale or exchange of assets not in the ordinary course of business, and dissolution. Shareholder control is limited, therefore, to the rights to vote in the selection of the corporate management and to be consulted in matters which may modify the character of their investment in the business. These indirect ownership rights will be explored in some detail.

Election and Removal of Directors. The initial directors of the corporation are named in the articles of incorporation and these directors usually serve until the first annual shareholders' meeting. Shareholders then elect new directors at the first meeting, and these directors serve for the prescribed term, usually one year. As previously discussed, directors are also subject to removal by the shareholders with or without cause by an appropriate vote, as prescribed by the statute. If the statute does not specifically provide for removal without cause, a clause to that effect in the articles of incorporation or by-laws is necessary if that right is deemed important.

EXAMPLE 4.03F: Removal of Directors

The stockholders of the Corporation may, at any meeting called for the purpose, remove any director from office, with or without cause, by a vote of a majority of the outstanding shares of the class of stock which elected the director or directors to be removed; provided, however, that no director shall be removed in case the votes of a sufficient number of shares are cast against his or her removal, which if cumulatively voted at an election of the entire board of directors would be sufficient to elect him or her.[44]

There is a special procedure in corporate law for the election of directors called cumulative voting, which may or may not be in effect in a particular corporation, depending upon the appropriate state law and the articles of incorporation. Some states guarantee cumulative voting by constitutional provision.[45] Others have statutes which require cumulative voting to be used unless the procedure is specifically denied in the articles of incorporation.[46] Other states, including Delaware, do not grant cumulative voting unless the articles of incorporation specifically authorize it.[47] The Model Act offers both of the latter alternatives, and allows each state to select the preferred provision.[48] Consequently, as with most of these specific points of corporate law, it is very important to review the appropriate state statute to determine the manner by which cumulative voting is authorized, and to state the desired procedure in the articles of incorporation.

EXAMPLE 4.03G: Cumulative Voting

At all elections for directors each stockholder shall be entitled to as many votes as shall equal the number of his shares of stock multiplied by the number of directors to be elected, and he may cast all of such votes for a single director, or may distribute them among the number to be voted for, or any two or more of them, as he may see fit.[49]

Amending the Articles of Incorporation. The articles of incorporation may be amended upon the suggestion of the board of directors and, in some states, upon the suggestion of a certain percentage of shareholders. In any case, the proposed amendment must be submitted to a vote of the shareholders, either at an annual meeting or a special meeting called for that purpose. Shareholder approval of such amendments to the basic "charter" of their corporation is consistent with their rights as owners.[50]

Other Extraordinary Corporation Action. Shareholder approval is also required for certain other extraordinary corporate matters, such as merger, consolidation, sale or exchange of assets out of the ordinary course of business, and dissolution of the corporation. Since these actions may significantly alter the character of the investment, a shareholder objecting to such action, may have his stock appraised and purchased. As a simple rule of thumb, any matter which may have a substantial impact on the operation of the business or the ownership rights of the shareholders requires shareholder approval. Even without a statutory mandate for shareholder approval, a sensible board of directors will request shareholder approval of major corporate decisions, perhaps for no other reason than to gauge shareholder sentiment to their activities as directors.

Right to Inspect. A corollary to the shareholder's voting right is the right to periodically "check-up" on management and inspect corporate records. Most states have statutes which permit the shareholder to inspect and copy books and records, including minutes of shareholders' meetings and other shareholder records. In order to avoid recalcitrant persons who simply buy shares to harass corporate management, these statutes also establish certain criteria as a condition to the shareholder's right to inspect. The Model Act, for example, requires share ownership for at least six months preceding a demand of inspection *or* that the demanding shareholder be a holder of record of at least 5 percent of all the outstanding shares of the corporation.[51] Thus, the demanding shareholder must be established as a shareholder or purchase a significant block of stock in order to have the right to inspect. Moreover, the demand for inspection must state the purpose of the inspection and the stated purpose must be "proper", and not for a reason conflicting with the best interests of the corporation. It may also be a good idea to grant the directors power to control inspection times and procedures in the by-laws:

EXAMPLE 4.03H: Inspection By Shareholders

The directors from time to time may determine at what times and places, and under what conditions and regulations, the accounts and books of the Corporation shall be open to the inspection of the stockholders.[52]

Preemptive Rights. The shareholder's preemptive right is the right to purchase newly issued shares of the corporation in the same proportion as his present share ownership before outsiders may purchase them. For example, suppose XYZ Corporation has three shareholders: Judi Wagner, who owns 100 shares; Gail Schoettler, who owns 200 shares; and Fern Portnoy, who owns 300 shares; and the corporation now determines that it will issue 1,500 new shares of stock. If the shareholders have preemptive rights, Wagner has the right to buy 250 shares (or ⅙th) of the new issue; Schoettler has the right to buy 500 shares of the new issue and Portnoy has the right to buy

750 shares of the new issue. If the shareholders fail to buy their allocated number of shares, the shares may then be sold to outsiders. Preemptive rights began as a common law theory designed to protect the proportionate interests of shareholders and to preserve their proportionate control over the corporation. Obviously, Wagner Schoettler and Portnoy would suffer complete loss of control if the 1,500 shares were sold to a single outsider. Of course, if control is not an important factor, they do not have to purchase their proportionate amount of the new issue, in which case the shares will be sold to other investors. Most states now specifically treat preemptive rights in their corporate statutes. Some states provide that preemptive rights are granted automatically by law unless the articles of incorporation specifically deny them.[53] Others, including Delaware, provide that the articles of incorporation must specifically grant preemptive rights or they do not exist.[54] The Model Business Corporation Act has alternative provisions taking both approaches which the states may adopt as they choose.[55]

It is generally recognized that preemptive rights are important to shareholders of close corporations and a nuisance to large, publicly held corporations. If American Telephone & Telegraph Co. had to offer a proportionate right to purchase to each of its millions of shareholders, the procedural problems and expense would be overwhelming. To assist in the flexibility of incentive compensation programs, it is also a good idea to exclude employee stock option plans from preemptive rights. In any case, the articles of incorporation should specify the corporate policy with respect to preemptive rights.

EXAMPLE 4.031: Preemptive Rights

No holder of any stock of the Corporation shall be entitled, as a matter of right, to purchase, subscribe for or otherwise acquire any new or additional shares of stock of the Corporation of any class, or any options or warrants to purchase, subscribe for or otherwise acquire any such new or additional shares, or any shares, bonds, notes, debentures or other securities convertible into or carrying options or warrants to purchase, subscribe for or otherwise acquire any such new or additional shares.[56]

Distributions. In addition to voting, inspection and preemptive rights, shareholders are entitled to a return on their investment by way of dividend distributions if the corporation makes a profit, and if the directors in their discretion and good judgment deem such a distribution desirable. The objectives of shareholder investment are to receive dividend distributions and to realize capital appreciation when the value of their stock increases.

As owners of the corporation, shareholders may share in the assets of the corporation when the business is dissolved and the corporate creditors have been paid. The remaining assets will be divided among them proportionately according to the terms of their stock.[57]

§ 4.04 LIMITED LIABILITY

One of the most attractive characteristics of the corporation is that the investors risk only the amount of their investment and are not individually responsible for corporate obligations. This limited liability advantage flows from the recognition by the law that a corporation is a separate legal person, and its debts and liabilities are personal to it.

Limited liability for the corporation should be contrasted with the full individual liability of a partner in a partnership or an owner in a sole proprietorship. The limited partnership borrows this characteristic of limited liability from the corporation in protecting the limited partners. The shareholder, who is the owner of the corporation, is liable only to the extent that she has invested in the corporation. She may lose the amount of money that she paid for the shares, but she does not expose her personal assets if the corporation should incur excessive liability. Similarly, the persons who run the corporation, the directors, officers and other corporate executives, are not personally liable for corporate obligations unless they have exceeded their authority or breached their fiduciary duties of good judgment and due care in incurring such obligations.

The protection of limited liability offered by corporations is one of the major reasons for choosing the corporate form over others, and the theory of limited liability is well established in judicial decisions. There are, however, two limitations on the principle, one practical and the other legal.

When a new corporation has been formed, it usually has not matured to an established business, and, while it may own certain assets and have good prospects for future profit, its ability to generate profits is untested. Consequently, potential creditors are understandably wary of extending credit to a new corporation. If the business is not as good as predicted, or if the managers are not as capable as they think they are, the corporation may not prosper, and a creditor may be forced to look only to the corporate assets for satisfaction of the obligation. Anticipating this problem, sophisticated creditors of the corporation will attempt to bind all available parties for the repayment of the obligation—just in case something goes wrong. In such a case, the shareholders and directors may personally obligate themselves for corporate obligations in order to persuade outsiders to advance credit to their corporation, and, of course, they then become individually responsible for the obligation. Practical realities may result in the limited liability protection being diminished by agreement.

The legal problem associated with limited liability is a theory called "piercing the corporate veil". The courts which have imposed personal liability on shareholders under the theory have recognized the corporate organization as offering a "shield" of limited liability for shareholder protection. Having found an abuse of this protection, courts are perfectly willing to pierce the shield, disregard the corporate entity, and hold the shareholder responsible for the acts of his corporation. Implicit in the finding of abuse of this protection is a finding that the shareholders have neglected to comply with the statutory requirements for proper operation of a corporation. It is possible, therefore, to advise a client in advance of ways to avoid this problem.

The typical abuses of the corporate form which reappear consistently in piercing cases are a failure by the shareholders to supply the corporation with adequate financial resources to support its operations; and a failure to observe corporate formalities, such as holding meetings for shareholders and directors, keeping separate books of the corporation, distinguishing personal assets from corporate assets, and issuing stock. Notwithstanding these typical unoffensive abuses, if the corporation is used to perpetrate fraud or for other illegal purpose, a court will have no trouble looking behind the corporate veil to hold the individual shareholders responsible, whether or not the corporation was properly funded or the formalities were observed.

It should be apparent that these problems are most likely to arise in a small, closely-held corporation rather than one of our industrial giants. The small businessman, who has formed a corporation for its limited liability benefits, is most vulnerable to piercing the veil. If he uses his family typewriter for business letters, uses excess family furniture in his office, commingles his personal funds with corporate funds, and ignores formal meetings in making corporate decisions, his corporate protection is weak. Recognize, however, that if his corporate assets are substantial, this piercing problem may never arise, even with the transgressions suggested above. The piercing doctrine is a judicial theory used to resolve litigation if necessary. If the corporation is sued by a business creditor or a victim who slipped on its snow-covered sidewalk, and if the corporate assets are adequate to pay the claim, there is no need to pierce the corporate veil to reach the shareholders' personal resources. On the other hand, if the creditor or victim will suffer because of an inadequately-financed corporation, a court will tend to reach behind the corporate "shell" to require the shareholders to pay.

As a precautionary measure, it has been suggested that all corporate clients should be advised to observe four principal objectives:

"(1) The formalities of corporate procedure, including the holding of shareholders' and directors' meetings and the keeping of minute books, should be observed.

"(2) The corporation should be operated as a really separate business and financial unit, with separate books and accounts, and without any intermingling or confusing of its funds, affairs, and transactions with those of the shareholders (whether individuals or corporations), officers, directors, or affiliated corporations in disregard of the corporate entity.

"(3) No representation or other holding out should be made which would lead outsiders to believe that the business is being conducted as a sole proprietorship or as a partnership (with the assurance of the personal liability of those forms).

"(4) The corporation should have adequate capital to meet its obligations and such contingencies as are reasonably to be expected in its business."[58]

Permit one final observation on the theory of piercing the corporate veil. This is also a frequent problem with parent-subsidiary corporations. If a large, profitable corporation is seeking to enter a risky enterprise, it may be imprudent to risk all of the corporation's profit and other assets for one questionable venture. The solution is to form a separate corporation (called a "subsidiary"), whose stock is primarily or wholly-owned by the large corporation (called a "parent corporation"), and the only risk, if the subsidiary corporation's shield of liability is observed, is the subsidiary's assets. The questions here are substantially the same as those detailed above. If the subsidiary is undercapitalized, and there is a loose separation between the parent and subsidiary, the parent may be required to respond to all liabilities and obligations of the subsidiary.

§ 4.05 CONTINUITY OF EXISTENCE AND DISSOLUTION

A corporation has the power to exist perpetually under most state statutes, and will therefore be unaffected by the death of an owner or manager or by the transfer of ownership interests. The definitive term of a sole proprietorship, which ends when the proprietor dies, and of a partnership, which is technically dissolved upon the death, withdrawal, or other incapacity of a partner, was deemed to be a disadvantage to those forms of business. The corporation does not suffer from this infirmity. It is assured indefinite life by statute, and its ownership interests (shares) are freely transferable without impairing its continuity.

Continuity of existence is an extremely important characteristic for a large corporation, since any abrupt termination of existence could result in financial tragedy. On the other hand, the continuity of a corporation may work a hardship on a minority shareholder who is dissatisfied with her investment and can find no market for her shares. This shareholder will be unable to terminate the corporate entity in order to withdraw her investment, and may simply be forced to continue in shareholder status at the mercy of the majority shareholders and management.

Dissolution of a corporation may be accomplished by agreement of the appropriate intra-corporate group (incorporators or shareholders) as provided by statute. A court may also dissolve a corporation upon request of the Attorney General, whenever the corporation has failed to comply with statutory requirements, has procured its articles by fraud, or has otherwise abused its authority. Hopefully, the court-ordered dissolutions will remain rare. The voluntary dissolutions are cumbersome and require the consensus of at least a majority of the appropriate intra-corporate group to carry a resolution for dissolution.[59]

§ 4.06 TAXATION OF CORPORATION

Corporations, like natural persons, are subject to taxation by the federal, state and local governments based upon the amount of income they earn each year. Consistent with the separate corporate personality, the corporation is regarded as a separate taxable entity for most federal and state tax purposes and is taxed on separate corporate tax rates. This separate entity taxation is a significant distinguishing characteristic of the corporation from the sole proprietorship and partnership, where income is merely funneled to the individuals who comprise the business organization and is declared as individual income for tax purposes.[60] Taxation of corporations has some advantages and some disadvantages, all of which must be carefully considered by the attorney in advising the client that the corporate form is the proper organization for a proposed business.

Double Taxation

The greatest disadvantage of corporate taxation is the concept of "double taxation." Income received by the corporation is taxed at the corporate level according to the corporate rates then in effect. The profit remaining after taxes is available to be distributed to shareholders as dividends. However, if dividends are distributed, distribution is taxed again as personal income to the shareholder. You will recall that federal individual tax rates vary from 14 per cent to 50 per cent. Corporate tax rates on income are according to the following schedules:

Tax Years Beginning in 1981

Over	—But not over	Tax rate
$ 0	$ 25,000	17
25,000	50,000	20
50,000	75,000	30
75,000	100,000	40
100,000		46

Tax Years Beginning in 1982

Over	—But not over	Tax rate
$ 0	$ 25,000	16
25,000	50,000	19
50,000	75,000	30
75,000	100,000	40
100,000		46

Tax Years Beginning in 1983 and later

Over	—But not over	Tax rate
$ 0	$ 25,000	15
25,000	50,000	18
50,000	75,000	30
75,000	100,000	40
100,000		46

Suppose a corporation distributes a dividend to a shareholder who is taxed in the lowest (14 per cent) individual tax bracket. Each dollar of the corporations profit is taxed at least 17 per cent, so the dollar is reduced at least to 83 cents at the first level. If the 83 cents is distributed as a dividend to this shareholder, it is further taxed at least 14 per cent, or 11.6 cents at the shareholder level. Each corporate dollar which is distributed as a dividend, therefore, will be reduced at least 28.6 cents by federal taxes. The disadvantage of double taxation is obvious if you consider the same business operated as a sole proprietorship, with the proprietor in an individual tax bracket of 30 per cent. While corporate tax rates appear more favorable (17 per cent versus 30 per cent) the actual effect of taxes on each distributed dollar is that the proprietorship pays less than the minimum double tax rate (28.6 per cent). If our 30% individual proprietor were a shareholder, the corporate tax bite is even worse, since the 83 cents distributed is now taxed at 30 per cent, or 24.9 cents. Thus, as an individual sole proprietor he would be paying 30 cents in taxes on each dollar he received. As a shareholder receiving distributions from his corporation, each dollar received is taxed 41.9 cents.

The double taxation problem can be considerably worse. For example, if the dividend dollar were a portion of corporate profit over $100,000, the dollar would be reduced to 54 cents at the corporate taxation level. If this 54 cents were distributed to a shareholder who is taxed in the 50 percent bracket, it would be taxed 27 cents more. Thus, the corporate dollar would be reduced a total of 81 cents in taxes, leaving the shareholder with 19 cents as cash returned!

Notice that with parent-subsidiary corporations there may be triple taxation: when the subsidiary pays its corporate tax, and distributes remaining profits to the parent as dividends where it is taxed as corporate income, and it is then distributed to the shareholder and the shareholder is taxed at individual tax rates.

Double taxation is recognized as a distinct disadvantage for the corporation as compared with other business forms especially if a significant portion of the corporate income will be paid to the shareholders as dividend distributions. Larger corporations with many stockholders simply accept the disadvantage, since the corporate form offers many other advantages which are essential to the operation of a large business. In smaller, closely-held corporations, double taxation may be minimized by several options. Whenever shareholders are officers or employees of the corporation, as is frequently the case in smaller organizations, they may be paid salaries, which are deductible as a corporate expense. The shareholder-employee is thereby compensated, but the corporate tax is not imposed on the salary and double taxation is avoided. Further, anticipating this problem, the small corporation may be structured so that much of its capital comes from loans to the business, rather than shareholder investment. Having established sufficient equity capital (money paid through shareholder investment), the remaining funds needed for the business may be raised through interest-bearing loans, and the interest is deductible to the corporation as an expense. The interest paid to the creditor (investor) is income, which substitutes for dividends but is not subject to double taxation. Similarly, shareholders of closely held corporations may purchase property and equipment and lease it to their corporation, receiving rental payments which are treated as expense to the corporation, and are taxed only as rental income to the shareholder-lessors.

Another practical approach to double taxation is to leave the corporate profits in the business and not distribute dividends. Then only corporate tax rates are applied, and while the retained profits will increase the value of the stock, resulting in a capital gains tax when it's sold, no individual income tax is applied to the profits themselves. This solution is too simple to be effective, however, since the taxing authorities have devised a penalty which encourages corporations to distribute earnings to the shareholders rather than accumulate them. The accumulated earnings tax is applied to income unreasonably retained by the corporation. The company is entitled to accumulate $250,000 ($150,000 in the case of certain service corporations), but then a penalty of tax of 27½% on the next $100,000 and 38½% thereafter is applied to income which has been retained by the corporation beyond the reasonable needs of the business.[62] The corporation has the burden of proving that income has been accumulated for reasonable business needs in order to avoid the tax.

Subchapter S Election

The Internal Revenue Code of 1954 provides that a "small business corporation" may elect not to be taxed at the corporate level but to have its income (whether distributed or not) passed through and taxed prorata to its shareholders as ordinary income.[63] This election effectively treats the corporation like a sole proprietorship or a partnership for tax purposes, and all profits are attributed proportionately to those persons who own the business. Similarly, corporate losses may be offset against other personal income of the shareholders. The election is particularly beneficial to the shareholders whose tax rates are significantly lower than the corporate tax rate, or when it is expected that most of the corporate profits will be distributed to the shareholders. The election avoids two disadvantages of corporate taxation: the profits are not "double-taxed", and corporate losses may be taken as ordinary losses thereby reducing the personal income of the shareholders.

In order to be classified as a "small business corporation" for the Subchapter S election, a corporation must meet the following requirements:

1. There may be no more than 25 shareholders (husbands and wives are treated as one stockholder, regardless of how the stock is held);
2. Shareholders must be natural persons, and cannot be another corporation or partnership, but may be an estate of a natural person or certain trusts;
3. The corporation may have only one class of stock;
4. The corporation cannot have a non-resident alien as a shareholder.

The election of a "small business corporation" refers to the number of shareholders, as indicated in the listed requirements for qualification, and has nothing to do with the size of the corporation in terms of its assets, revenue or earnings. These requirements effectively limit the Subchapter S election to "close" corporations.[64]

Under the Subchapter S election, corporate profits, whether distributed or not, must be claimed by each shareholder in the proportion of ownership interest held, and, consequently, all shareholders must consent to the election. Each shareholder should sign a separate statement of consent acknowledging the effect of the election. The statement is submitted with the form electing taxation under Subchapter S.[65]

The election to be taxed under Subchapter S is made on Form 2553 of the Internal Revenue Service,[66] and it may be made for any taxable year any time during the previous taxable year or any time during the first seventy-five days of the taxable year.[67] Thus, if the corporation determined to elect Subchapter S taxation, and had adopted September 1 to August 31 as its taxable year, the election may be made anytime during the preceding fiscal year for the following year or before November 14th for the present year.

The election may be terminated in one of several ways. The most common termination results from the corporation ceasing to qualify as a "small business corporation" under the requirements listed above, as, for example, when it acquires a twenty–sixth shareholder. It may also be terminated the same way it was made, by all shareholders consenting to revocation of the election. If any new shareholder files an affirmative refusal to consent to tax treatment under this provision, the election is also terminated.[68]

As a practical matter, Subchapter S is particularly desirable when the corporation is expected to incur losses during the first few years of operation, when shareholder's individual tax rates are lower than the corporate rates, or when corporate profits are expected to be distributed to shareholders. It should be noted that election and maintenance of Subchapter S treatment increases the burdensome formalities required for corporate existence, and will increase the legal and accounting costs incident to corporate operation.

State Income Tax

Corporations operating within any given state are also subject to the state income tax. The general rule is that a state may tax a corporation operating within its borders in a reasonable relation to the business activity conducted within the state. Thus, a corporation incorporated in Colorado will be automatically subject to the Colorado state tax if it does business there or has Colorado source income because it was originally formed in that state. If it then does business in Wyoming and Nebraska, those states may also tax its income in relation to the business conducted within their boundaries. The domestic or domicile state, in this case Colorado, usually allows a tax credit for taxes paid to other states. Various formulas are employed by the states to determine a proper allocation of tax on local business activity. As a practical matter, states attempt to devise formulas which will maximize tax revenue from business activity. For example, a state with very little localized industry usually has a formula based upon sales made within the state rather than corporate assets located within the state. On the other hand, a state with a heavy industrial population will probably tax on a percentage of total assets located within the state.

In addition to state and federal income taxes, corporations are frequently subject to other special taxes which may result in the corporation bearing a greater tax burden than other forms of business enterprise. Franchise taxes, organization and capital taxes, original issue taxes for issuance of shares of stock, and taxes on transfers of shares and other corporate securities are the most common. Proper planning in the selection of a business organization requires an analysis of the myriad charges which various states impose on the corporations operating within their boundaries.

Section 1244 Stock

The foregoing discussion has been primarily concerned with income taxes assessed against a corporation and its shareholders. The other tax ramifications of a corporation include capital gains and losses associated with the purchase and sale of stock. Each share of stock is a capital asset, and if it is sold after appreciating in value, a taxable gain is realized. Similarly, when stock is sold after depreciating in value, a capital loss is claimed. Capital losses for individuals, however, may be used only to offset capital gains, if there are any, or deducted from ordinary income up to a maximum limitation. For example, suppose a shareholder has invested $5,000 in stock of a corporation which is now bankrupt, thus making the stock worthless and resulting in a capital loss of $5,000. If the stockholder had no other capital gains that year, a maximum of $3,000 of the loss could be used in that year. Further, a long term capital loss provides only $1 of deduction for each $2 of loss. While there are certain carry-forward provisions for individual losses, it would be perferable to be able to claim the full $5,000 against ordinary income for that taxable year. Section 1244 of the Internal Revenue Code provides this effect for stock which qualifies as a "small business stock."[69]

The definition of a "small business" is different for Section 1244 stock than for a Subchapter S election. In this case, the qualification of a corporation as a "small business" depends upon the amount of money to be raised by a plan to sell Section 1244 stock and the existing equity capital of the corporation. The amount of stock offered under the plan which is intended to qualify under Section 1244 and the amounts received by the corporation as a contribution to capital or paid in surplus cannot exceed $1,000,000. Any property contributed for stock (other than money) is valued at the adjusted basis of the property to the corporation, less any liability against the property which the corporation assumed.

For example, suppose the Lyons Corporation adopts a Section 1244 plan to offer stock for an amount not in excess of $500,000. If its equity capital had been $600,000 at the time the plan was adopted, $100,000 of the stock would not qualify under Section 1244 because the equity capital plus the aggregate amount offered would exceed $1,000,000. In such a case, the maximum amount which would qualify under the plan would be $400,000. The corporation has a right to designate which stock shall qualify. However, suppose the Lyons Corporation was newly formed when it adopted the plan. In its first year, it has sold $400,000 in stock, and business successes have deposited $800,000 into equity capital. This does not destroy the qualification of the stock because the equity capital test includes only amounts paid for contributions to capital, not revenues from operations.

To qualify for Section 1244 stock, the corporation must acquire most of its income (more than 50 per cent) from sources *other than* royalties, rents, dividends, interest, annuities and transactions in stock for five years preceding the loss.[70] The effect of the plan will be lost if the business does not comply with this source-of-income provision after the stock is issued and for five years before the investor sustains a loss.

For taxable years beginning after 1979, the former requirement that the Section 1244 stock be issued pursuant to a written plan has been repealed. Nevertheless, it is good practice to prepare a plan or a written corporate resolution to indicate clearly that the shares are being sold pursuant to Section 1244.[71] If all requirements of the statute are met, all stock issued pursuant to the plan will receive ordinary loss treatment if a loss is incurred when the stock is sold. This means that the selling shareholders may use any loss on the stock to offset ordinary income during his taxable year, rather than treating the loss as a capital loss with its limited and deferred tax treatment.

Other Tax Advantages of a Corporation

Tax authorities allow a corporation to deduct certain "necessary" expenses incurred in providing fringe benefits to employees to encourage their continuous faithful performance. There are also tax advantages to the employee under "qualified" incentive plans. If these employees are also shareholders the deductibility of such expenses is unaffected.

These incentive benefits with tax advantages include share options, medical/dental reimbursement plans, qualified pension and profit-sharing plans, and life, health and accident insurance programs.[72]

Incentive compensation programs give the employee a right to participate in the success of the business, while enjoying significant tax breaks on the compensation he receives under the plan. For example, a qualified profit-sharing plan permits a corporate deduction of profits accumulated for employees under the plan, but the employee is not taxed until he receives payment, and then usually only capital gain tax. Qualified pension plans are similarly treated for tax purposes.

Insurance plans may provide a direct economic benefit to employees, who may also be shareholders, without tax on the proceeds of the insurance. The corporation may deduct the expense of paying insurance premiums for employees as an ordinary business expense. Hospital, accident, health and disability insurance plans may be maintained by the corporation with very few limitations. Group life insurance, with maximum dollar limitations per employee, may be maintained by the corporation, with the premiums treated as an expense to the corporation but not taxable to the employee.

These special insurance and incentive compensation plans, with their attendant tax advantages, are unique to the corporation, with its separate legal personality. Partnerships and sole proprietorships do not enjoy the separate entity characteristic, and therefore do not obtain tax advantages through these devices.

NOTES

1. See Section 4.03.
2. See Section 4.03.
3. Many states advertise the advantages of incorporation under their laws. Delaware sends the following synopsis of their permissive corporate laws to persons requesting information regarding incorporation:

"The Outstanding advantages of incorporating in Delaware are as follows:

"The fees payable to the State of Delaware are based upon the number of shares of authorized capital stock, with the no par shares fee one-half the par shares fee.

"The franchise tax compares favorably with that of any other State.

"Shares of stock owned by persons outside of the State are not subject to taxation.

"Shares of stock which are part of the estate of a non-resident decedent are exempt from the State Inheritance Tax Law.

"The policy of Delaware courts has always been to construe the Corporation Law liberally, to interpret any ambiguities or uncertainties in the wording of the Statutes so as to reach a reasonable and fair construction. This causes the careful investor to have confidence in the security of the investment.

"The corporation service companies throughout the nation consider the Delaware Corporation Law among the most attractive for organization purposes and the State of Delaware a valuable jurisdiction in which to organize new companies."

4. See Sections 6.05 and 6.09.
5. West's Modern Legal Forms § 2692.
6. M.B.C.A. § 4(k), and see Section 6.09. Also see Article VI, clauses 6–10, of sample By-Laws, Form 6Q in the Appendix.
7. M.B.C.A. § 4(h).
8. West's Modern Legal Forms § 2693.
9. West's Modern Legal Forms § 2698.
10. Examples of specific corporate purposes appear in the discussion of the articles of incorporation in section 6.05.
11. This restriction may appear in the state constitution, as in Oklahoma (Okla.Const.Art. XXII, § 2), or in the state corporation statutes. E.g., New Hampshire, N.H.Rev.Stat.Ann. § 294:4V.
12. West's Modern Legal Forms § 2695.
13. M.B.C.A. §§ 4(f), 47.
14. See M.B.C.A. § 4(m).
15. E.g., New Jersey, N.J.Stat.Ann. § 14A:3–4.
16. E.g., Delaware, 8 Del.Code Ann. § 122(11).
17. M.B.C.A. § 53.
18. E.g. South Dakota, S.D.Comp.Laws Ann. § 47–2–4.
19. M.B.C.A. § 36.
20. M.B.C.A. § 37.
21. E.g. Pennsylvania, 15 Pa.Stat. §§ 1401, 1402.
22. M.B.C.A. § 35.
23. E.g., Hawaii requires that at least one of the members of the board must be a resident of the state, Hawaii Rev.Stat. § 416–4.
24. M.B.C.A. § 36.
25. M.B.C.A. § 38.
26. E.g., Arkansas, Ark.Stats. §§ 64–303, 64–304; and Wisconsin, Wis.Stat.Ann. § 180.34.
27. M.B.C.A. § 39.
28. M.B.C.A. § 39. However, if cumulative voting is in effect for the election of directors, the same procedure must be followed in the removal of a director. See Section 8.08.
29. West's Modern Legal Forms § 2719.
30. West's Modern Legal Forms § 2721.
31. West's Modern Legal Forms § 2718.
32. M.B.C.A. § 27.
33. West's Modern Legal Forms § 2713.
34. See Sections 13.01–13.03.
35. West's Modern Legal Forms § 2714.
36. M.B.C.A. § 45. Corporate distributions and dividends are fully discussed in Chapter 9.
37. The concept of par or no par value stock has been eliminated in the Model Act. See M.B.C.A. § 18. Most states still maintain the distinction, however. Characteristics of corporate stock are discussed fully in Chapter 7.
38. M.B.C.A. § 50. Some states prohibit simultaneous holding of the offices of president and vice-president (e.g. Minnesota, Minn.Stat.Ann. §§ 301.30).
39. See Article VI, clauses 6–10 of the sample By-Laws, Form 6Q in the Appendix.
40. These documents are required to accomplish extraordinary corporate actions and are discussed in Chapter 13.
41. See Sections 10.01 and 10.02.
42. M.B.C.A. § 2(f).
43. These characteristics are discussed in Sections 7.10 and 7.11.
44. West's Modern Legal Forms § 2720.
45. E.g., Illinois, Ill.Const. Transition Schedule § 8.
46. E.g., Colorado, Colo.Rev.Stat. § 7–4–116(4).
47. E.g., 8 Del.Code Ann. § 214.
48. M.B.C.A. § 33. Cumulative voting is discussed in detail in Section 8.08.
49. West's Modern Legal Forms § 2712.

50. The procedure for accomplishing amendments to the articles is more fully explored in Section 13.01.
51. M.B.C.A. § 52.
52. West's Modern Legal Forms § 2731.
53. E.g. Colorado, Colo.Rev.Stat. § 7–4–110.
54. E.g. Delaware, 8 Del.Code Ann. § 102(b)(3).
55. M.B.C.A. §§ 26, 26A.
56. West's Modern Legal Forms § 2735.
57. Dividend distributions are discussed in Chapter 9; distributions in dissolution are considered in Chapter 13.
58. R. Deer, The Lawyers Basic Corporate Practice Manual § 1.02 (1971).
59. Precise elements and the procedure for Dissolution are discussed in Chapter 13.
60. See Sections 1.04 and 2.08.
61. Int.Rev.Code of 1954, 26 U.S.C.A. § 11, as amended.
62. Int.Rev.Code of 1954, 26 U.S.C.A. § 531.
63. Int.Rev.Code of 1954, 26 U.S.C.A. § 1371–79.
64. Close corporations are specifically discussed in Section 5.01.
65. A sample statement of consent appears as Form 4A in the Appendix.
66. Form 2553 is Form 4B in the Appendix.
67. Int.Rev.Code of 1954, 26 U.S.C.A. § 1372(c)(1).
68. Int.Rev.Code of 1954, 26 U.S.C.A. § 1372(e).
69. Int.Rev.Code of 1954, 26 U.S.C.A. § 1244(c)(3).
70. Ibid. at § 1244(c)(1)(C).
71. A sample Plan and Resolution to Issue Stock Under Section 1244 appear as Forms 4C and 4D in the Appendix.
72. See Chapter 10 for a full discussion of incentive benefit plans.

5

SPECIAL CORPORATE FORMS

The corporate model described in the last chapter is typical of most American business corporations, but there are some special variations which modify the general corporate characteristics in certain important particulars.

§ 5.01 CLOSE CORPORATIONS

Corporations whose shares are held by a small group of shareholders are called close corporations. The shareholders of a close corporation are frequently closely related by blood or at least by friendship. In most jurisdictions, these corporations are distinguishable from other business corporations only by the fact that share ownership is restricted to a select few persons who are intimately involved with the business and who operate the corporation with substantial shareholder participation. A significant characteristic of the close corporation is that the shareholders actively participate in the management of the business. Thus, unlike a large, publicly-held corporation, there is a mixture of management and ownership in the close corporation, and this unique relationship between the shareholders usually results in a guarded interest in maintaining ownership control through internal shareholder agreements and restrictions on the transfer of equity securities. These corporate objectives accurately suggest that the operation of a close corporation resembles the operation of a partnership. An examination of any close corporation should reveal a group of persons who might as well have been partners, but instead selected the corporate form for its limited liability and tax advantages.

The attorney's greatest challenges in the formation and operation of a close corporation are the various agreements among shareholders which are designed to perpetuate management and ownership control through voting power and share transfer restrictions. These intricate agreements are separately considered in a later chapter.[1] For now, the primary concern is the manner in which the structure and operation of the close corporation differs from other corporations. Most jurisdictions have no separate "close corporation statute" and require that close corporations be formed and operated under the normal corporation code. In these states any desired informality and owner-management must be achieved by procedures or agreements which comply with the normal statutory requirements. However, the trend toward permissiveness in modern corporate statutes has provided the close corporation with statutory authority for the desired flexibility and informality. For example, formal shareholder meetings may be avoided under Section 145 of the Model Act. Instead, action by shareholders may be taken without a meeting if all shareholders entitled to vote at the meeting sign a consent to the action in writing. To enable the shareholders of a close corporation to maintain tight personal control over corporate activities, Section 143 of the Model Act permits greater-than-normal voting requirements for shareholder action to be drafted into the articles of incorporation. If the statute normally permitted shareholder action by the vote of the majority, the articles could specify a two-thirds, three-fourths or even unanimous voting requirement to increase individual control. The Model Act further authorizes an important adjustment in management functions for the close corporation by providing that the conduct of business and affairs of a corporation will be vested in the board of directors, *unless the articles of incorporation otherwise provide*.[2] Shareholder management authority may, therefore, be specified in the articles of incorporation to the extent desired by corporate personnel. Some states, notably New York and North Carolina, allow shareholder agreements to impinge on management functions which are usually reserved to the board of directors. Thus, persons seeking close corporation status may be accommodated under several modern corporate statutes which permit sufficient flexibility in operation and control to use close corporation procedures.

A few jurisdictions have adopted more sophisticated statutory authority for close corporations by adding separate sections specifically directed to the unique operations of close corporations in the regular corporation statute.[3] The Delaware Act is illustrative of the approach to this special corporate form.

The Delaware statute defines a "close corporation" as a corporation with less than 30 shareholders of record, with all of its shares subject to a share transfer restriction, and with none of its shares offered to the "public." This latter provision is concerned with the "public offering" which requires registration under the state and federal securitites laws. If the corporation qualified as a close corporation its certificate of incorporation will state that it is a close corporation, and will restate the statutory qualifications.[4]

The significant departure from normal corporate operations authorized by the Delaware statute is that management of the close corporation may be vested in the stockholders, rather than the board of directors. A provision to this effect may be inserted in the certificate of incorporation, either at filing or by amendment if unanimously approved, or it may result from an unanimous written agreement among the shareholders.[5] If management is granted to the shareholders, the directors will be relieved from any liability for managerial acts or omissions and liability will be placed instead upon the shareholders of the corporation. Moreover, with shareholder management there is no need for a shareholder meeting to elect directors, since the shareholders are performing the directors' functions. You will recall that this owner-manager characteristic is usually found in partnerships. By adopting shareholder management in a close corporation, the parties are effectively structuring the operation of the business like a partnership. In fact, the Delaware statute goes on to provide that a written agreement among shareholders or a provision adopted in the articles of by-laws of a close corporation which arranges relations between the shareholders in a manner that is usually only appropriate among partners is permissible.[6]

The statute also covers the possibility that the close corporation is to be managed by directors, but that their activities are restricted by an agreement among shareholders. The majority of the shareholders may enter into an agreement, binding among them, which relates to the conduct of the business and affairs of the corporation even though it restricts or interferes with the discretion or powers of the board of directors.[7] Such agreements, which would be impracticable in larger corporations, are prohibited by law for regular corporations.

Another major departure from typical corporate law in the Delaware close corporation statute is the authority for the certificate of incorporation to grant any shareholder the right to have the corporation dissolved at will or upon the occurrence of a specified event or contingency.[8] This provision is similar to the right of a partner to terminate the partnership enterprise at will, and is particularly important for a close corporation, where one of the major problems may be deadlock if the shareholders cannot agree on operation of the business.

§ 5.02 PROFESSIONAL CORPORATIONS

The "learned" professions, such as law, medicine, accounting, and others traditionally have been prohibited from operating as corporations. The policy reasons behind this interdiction have never been clearly defined, but they probably grew out of the desire to limit the association of persons engaged in such professions to duly licensed practitioners and from the concern that professional persons should not be allowed to shield themselves from liability through the use of the corporate form. The obvious disadvantage to professionals who were required to practice as sole proprietors or partners is that they could not use favorable corporate tax rates and fringe-benefit plans unique to corporations. While some states enacted professional corporation statutes in 1961, it wasn't until 1969 that the Internal Revenue Service conceded that such organizations should be treated like any other corporation for income tax purposes. In 1979, the Model Business Corporation Act finally appeared with a Professional Corporation Act Supplement.[9]

It should be noted that some states permit professionals to form "professional associations" which are really partnerships with a sufficient number of corporate characteristics, such as continuity of life, centralized management, and transferability of ownership interests, so they will be taxed as a corporation.[10] Some states permit the formation of either an association or a corporation for professionals.[11] This section is primarily concerned with the professional corporation.[12]

The statutory authority for professional corporations varies widely from state to state, and several statutes include the authority to incorporate with other statutes regulating the particular profession (such as licensing and qualification statutes). These states have no single professional corporation law and the Model Professional Corporation Act has not yet been adopted in its entirety in any state. In a few jurisdictions the authority for professional legal corporations for attorneys is contained in a Supreme Court rule rather than a statute.[13] In states where the professional corporation has been added as an adjunct to the business corporation statutes, the latter statute controls except for the specific provisions of the professional corporation section. All states now allow creation of professional corporations, and all include attorneys, doctors, and dentists. Accountants, veterinarians, psychologists, engineers, and architects are also usually included, and a few permit corporate practice by registered nurses, physical therapists, pharmacists, and marriage counsellors.

The structural variations of the professional corporation from the business corporation are also treated differently in the individual state statutes, but most states have adopted certain general modifications which are the same as the provisions of the Model Professional Corporation Act.

Scope

Under the Model Professional Corporation Act (MPCA) those professions who render a service lawfully only by persons licensed under provisions of a licensing law of a state may become a "Professional Corporation" under the Act. Some of the existing state statutes under which professional persons are permitted to incorporate cover all licensed services and are not restricted to persons who are otherwise prohibited from incorporating under the business corporation law. Other existing state statutes limit those who may incorporate to specific profession describing in a single statute or a series of similar statutes each applicable to one profession. The definition in the MPCA has the effect of restricting the use of the Act to the practice of the professions. However, rather than listing designated professions, the Act follows the precedents set by many existing state statutes of defining professional services as those licensed services which may not be rendered by a corporation organized under the business corporation law.[14]

Purposes, Powers and Prohibited Activities

Most state statutes have limited the purposes of a professional corporation to the practice of a single profession because of the ethical proscriptions placed upon joint practice of various professions. The Act permits the practice of various professional services and ancillary services within a single profession, but would also permit a joint practice of various professions if this combination of professional purposes is permitted by the licensing laws of the local state.[15]

A professional corporation formed under the Act would be permitted all of the powers enumerated in the Business Corporation Act except that the professional corporation may not be a promoter, general partner or associated with a partnership, joint venture, trust or other enterprise unless it is engaged only in rendering professional services or carrying on a business permitted by the articles of incorporation of the corporation. Similarly, the professional corporation can only engage in the profession or professions and businesses permitted by its articles of incorporation. The Act permits, however, the investment of the funds in real estate, mortgages, stock and bonds or any other type of investment as part of the permitted activities of a professionnal corporation.[16]

Name

Existing state statutes vary in the selection of terms required in the corporate name as corporate designations for professional corporation. The Model Act provides that the term "Professional Corporation" or the abbreviation "P.C." is the permitted designation. As with other corporate statutes, the name of a professional corporation should not be the same as or deceptively similar to the name of any other corporation. However, the Model Act makes a special exception if the similarity results from the use in the corporate name of personal names of shareholders who are or were associated with the organization or if written consent of the other corporation who uses a similar name is filed with the Secretary of State.[17] These special provisions are intended to provide for the similarity of personal names used by professional practitioners in their practice and the likelihood that the public will not be confused significantly if professional corporations have similar names which are personal to those who practice as members of the corporation.

Share Ownership

Shares in professional corporations may only be owned by persons who are authorized to render professional services permitted by the articles of incorporation. The Model Act and only a few states permit shares to be owned by partnerships and other professional corporations who are authorized to render the professional services permitted by the articles of the corporation and by persons licensed outside of the state of incorporation.[18]

No shares of a professional corporation can be transferred or otherwise disposed of except to persons who are qualified to hold shares issued by the professional corporation. The intent of these provisions is to require that the shares of a professional corporation be held only by those persons who are licensed to practice the particular profession, so that any transfer of the shares to persons who are not so licensed would be void, against public policy and in violation of the statute.[19] To accomplish this objective, each certificate representing shares of a professional corporation should state conspicuously on its face that the shares are subject to restrictions upon transfer imposed by the statute and by the licensing authority which supervises the profession.

If a shareholder dies or becomes disqualified (for example, by losing his or her license to practice the profession) the shares should be transferred to a qualified shareholder or purchased by the corporation within a specified period of time following the shareholder's death or disqualification. The Model Act requires payment of fair value for such shares if the corporation does not establish an alternative method, and the procedure for determining fair value parallels the procedure of the Business Corporation Act with respect to the determination of rights of dissenting shareholders.[20] If shares of a deceased or disqualified shareholder have not been transferred or purchased within 10 months after the death or five months after disqualification or transfer, the shares are cancelled and the shareholder's interest becomes a creditor's claim against the corporation.[21]

Liability for Professional Activities

The principal excuse for refusing corporate status to professional service organizations was that each practitioner should be individually responsible for all professional acts, and that no professional person should be able to hide behind the corporate shield of limited liability when professional services were rendered. All existing state statutes concerning professional corporations include some provision about professional liability or professional responsibility. Most of the enabling statutes specifically provide the professional person shall be personally liable for his acts and professional acts performed under his supervision. In some cases, limited liability will be allowed when the corporation maintains a minimum amount of liability insurance.

Most states are silent as to the vicarious liability of shareholders of a professional corporation, although some statutes clearly provide that shareholder liability is limited as it would be in a business corporation. In other words, if one doctor committed malpractice in a professional corporation, he may be personally liable for his own malpractice, but his fellow shareholders of the professional corporation would not be liable individually for their colleague's malpractice. A few other states expressly state the shareholders are jointly and severally liable for obligations of the corporation. Most states simply provide that the statute does not modify any law applicable to the relationship between a person furnishing professional services and a person receiving such services including liability arising out of professional services. The Model Act affirmatively states rules for liability of the professional corporation, its employees and its shareholders resulting from negligence in the performance of professional services. A professional employee is responsible for only his personal negligence and the corporation may be liable for the conduct of professional employees within the scope of their employment or within their apparent authority.[22]

The Model Act proposes three alternative provisions as to liability of shareholders of professional corporations:

1. Limited liability as in a business corporation;
2. Vicarious personal liability as in a partnership;
3. Personal liability limited in amount and conditioned upon financial responsibility in the form of malpractice or negligence insurance or a surety bond.[23]

Most state statutes and the Model Act specifically provide that any relationship of confidence that exists between a professional person and his or her client or patient is preserved notwithstanding the use of the corporate form. In fact, any privilege applicable to communications with a professional person extends to the professional corporation and is preserved.[24]

Directors and Officers

Most states express a preference that all directors and officers be licensed to practice the particular profession involved, but where lay directors are permitted, they usually are not allowed to exercise any authority over professional matters. The Model Act requires not less than one-half the directors of a professional corporation and all the officers other than the Secretary and the Treasurer should be qualified persons (licensed to practice the particular profession) with respect to the corporation.[25]

Fundamental Changes

Professional corporations are capable of normal fundamental corporate acts, such as amending the articles of incorporation, merger, consolidation, and dissolution. Most state statutes and the Model Act provide enabling legilation to permit such activities by professional corporations, provided the professional status and purposes of the corporation and the qualifications of shareholders are observed always. For example, § 16 of the Model Act permits mergers and consolidations among professional corporations and business corporations only if every shareholder of each corporation is qualified to be a shareholder of the surviving or new corporation.

If a professional corporation ceases to render professional services, the Model Act permits the corporation to amend its articles to delete the rendering of professional services from its purposes and to conform to the requirements of the Business Corporation Act regarding its corporate name. The corporation may then continue in existence as a corporation under the Business Corporation Act.[26] This section would avoid the forced dissolution of a professional corporation whose shareholders have died or become disqualified. Rather, this corporation may continue in business under the Business Corporation Act to invest its funds or conduct any other business which may be lawfully permitted under the local law.

Foreign Professional Corporations

Many professional practices are conducted in more than one state by individuals licensed to practice in more than one state or by partnerships whose members are licensed to practice in various states. Few state statutes contain any provisions concerning foreign professional corporations, but the Model Act has specifically provided for the admission, qualification and authority to do business of professional corporations among states.

The professional corporation who seeks to practice the profession in a new state will not be entitled to avoid the professional corporation laws of the state in which it carries on its practice by incorporating in a state with more lenient professional corporation requirements. The foreign corporations must comply with the domestic state law requirements concerning corporate purposes and qualifications of shareholders, directors and officers.[27] A foreign corporation may render professional services only through persons permitted to render such services in the state.[28] Responsibility for professional services and security for professional responsibility is made applicable to foreign corporations as well as domestic corporations and foreign corporations are also subject to regulation by the local licensing authority to the same extent as domestic corporations.[29]

A professional corporation must obtain a certificate of authority if the corporation maintains an office in a state.[30] The application for a certificate of authority of a foreign professional corporation would include information required for normal business corporations, and a statement that all shareholders, not less than one-half of the directors, and all of the officers other than Secretary and Treasurer are licensed to render a professional service described in the statement of purposes of the corporation.[31]

Under the state statutes which permit a professional corporation and under the Model Act, professional persons will be entitled to the advantage of the corporate business form. Although one important advantage of corporateness is lost to the professions—limited liability, and although the statutory requirements for shareholder-director-officer qualification and operation are strict and rigidly observed, the tax advantages and operating flexability of the corporate organization make the professional corporation an attractive business form.

NOTES

1. See Chapter 11. An agreement between shareholders organizing a close corporation appears as Form 5A in the Appendix.
2. M.B.C.A. § 35.
3. E.g., Delaware, 8 Del.Code Ann. §§ 341–56.
4. 8 Del.Code Ann. §§ 342, 343.
5. Ibid. at § 351.
6. Ibid. at § 354.
7. Ibid. at § 350.
8. Ibid. at § 355(a).
9. The text of the Professional Corporation Act Supplement to the Model Business Corporation Act is reproduced in the Appendix.
10. E.g., Pennsylvania, 15 Pa.Stat. §§ 12601–12619. Forms for the formation of a professional association may be found in West's Modern Legal Forms §§ 6461–66.
11. E.g., Texas, Vernon's Ann.Tex.Civ.Stat. arts. 1528(e) and 1528(f).
12. Articles of incorporation of a Professional Medical Corporation and its application for registration appear as Forms 5B and 5C in the Appendix.
13. E.g., Colorado.

14. Professional Corporation Act supplements to the Model Business Corporation Act (hereinafter cited MPCA) § 2.
15. M.P.C.A. § 3.
16. M.P.C.A. §§ 4 and 5.
17. M.P.C.A. § 8.
18. M.P.C.A. § 9(a).
19. See M.P.C.A. § 9(c).
20. See M.P.C.A. § 10(b) and M.B.C.A. § 81.
21. M.P.C.A. § 10(g).
22. M.P.C.A. § 11(b) and (c).
23. M.P.C.A. § 11(d).
24. M.P.C.A. § 12.
25. M.P.C.A. § 14.
26. M.P.C.A. § 17.
27. M.P.C.A. § 19.
28. M.P.C.A. § 6.
29. M.P.C.A. §§ 11 and 26.
30. M.P.C.A. § 19(b).
31. M.P.C.A. § 20.

6

FORMATION OF A CORPORATION

§ 6.01 PREINCORPORATION RESPONSIBILITY

The embryo of a corporation is the business idea conceived by an individual or group of individuals. The idea may be fresh, as with entrepreneurs who simply decide to begin a business, or it may evolve from an established commercial enterprise which will continue under the corporate form. Regardless of the genesis of the idea, the attorney is consulted for the purpose of forming the appropriate structure for operation of the business. If limited liability, flexible capital structure, and tax advantages are desired, the corporate organization may be most desirable. The organizers rely upon the attorney to properly consider the advantages and disadvantages of the various business forms and to advise them of the most beneficial organization.[1]

At this point, the organizers are private individuals with a business idea, and better practice suggests that they agree among themselves in writing as to certain important matters regarding the corporation to be formed. The relationship between these organizers or promoters resembles a "joint venture", which is like a partnership. Even without a written agreement, the law imposes certain rights and responsibilities upon their relationship, including duties to disclose important information to each other and to avoid any conflict of interest which might interfere with their participation in the project. However, to avoid disputes and to facilitate the smooth incorporation of their business, a written agreement between the organizers is appropriate.[2]

The organizers or promoters are responsible for investigating the particular business opportunity and assembling the property, cash and personnel to accomplish the business objectives. Generally, the promoters will look for a suitable business establishment, negotiate a lease or purchase of it, and contract for necessary furniture, fixtures and so forth. They will search for capable employees, if needed, and may negotiate employment contracts with them. If the business opportunity is unique, patents, copyrights or trademarks must be obtained. A common denominator to each of these activities is that the promoters are acting as individuals in a joint venture relationship on behalf of a corporation yet to be formed. They cannot bind the corporation to the contracts they are negotiating because the corporation does not exist. This means that the promoters will usually be required to obligate themselves individually on these contracts. After the corporation is formed, it may adopt the contracts through appropriate action by the board of directors, but the promoters who have signed the contracts in their individual capacities will usually remain obligated for performance of the contract. Promoters considering the corporate form should always be advised of these ramifications of pre-incorporation agreements. The advice should not deter them from a corporation, however, since they would not escape individual liability by using an alternative business form.

When soliciting capital for the corporation, the promoters' activities are governed by a different set of rules. Operating capital may be obtained through loans or by sale of stock in the corporation to be formed. Loans negotiated prior to formation of the corporation are treated as ordinary preincorporation agreements, with the promoters risking individual liability for repayment. Sales of stock are accomplished by a pre-incorporation share subscription.

§ 6.02 PREINCORPORATION SHARE SUBSCRIPTIONS

Share subscriptions are offers from interested investors to purchase shares of a corporation. Preincorporation subscriptions are, as the name indicates, offers to purchase shares when the corporation is subsequently formed, and they may be necessary to proper formation of a corporation for several reasons. In a practical sense, every corporation needs capital to commence business and it is important that investors be identified and promises to purchase shares be secured prior to launching the new enterprise. In addition, from a strictly legal standpoint, some state statutes require the use of preincorporation share subscriptions in various stages of the formation procedure. One such provision, not found in the Model Act, is that the incorporators must also be subscribers.[3] In states with this provision, a prospective incorporator must tender a preincorporation share subscription in order to qualify as an incorporator. Several other states require that a corporation must have a minimum amount of paid-in capital before it may commence business. The Model Act formerly required that $1,000 must be collected as capital as a condition to doing business, but this requirement was eliminated in 1969. Several states have retained the minimum paid-in capital requirement,[4] and in these jurisdictions preincorporation share subscriptions will be used to solidify promises to contribute the amount required by statute. The corporation will not be allowed to commence business without the requisite capital. Thus, preincorporation share subscriptions will be used in most cases for practical or legal reasons to secure promises to purchase shares once the corporation is formed.

A preincorporation share subscription may be executed by anyone who has decided to invest in the company. The terms of the subscription describe an offer to purchase shares. If the corporation accepts the offer, a binding contract is created. A few states require written, signed subscriptions and, of course, it is good practice to obtain a written offer in any case. A single subscription may be executed by several subscribers, or each subscriber may execute his or her own subscription.

EXAMPLE 6.02A: Share Subscription For Several Subscribers

We, the undersigned, hereby severally subscribe for the number of shares of the capital stock of Trouble, Inc., set opposite our respective names. The Corporation is to be organized under the laws of the State of Delaware, with an authorized capital stock of $50,000.00 consisting of 5,000 shares of common stock, $10.00 par value. We further agree to pay the amount subscribed in cash on demand of the treasurer of the said Corporation as soon as it is organized [or at such times and in such amounts as may be prescribed by the Board of Directors of the Corporation].

Dated July 1, 1981.

Names	Addresses	Shares Subscribed	Amount Subscribed
______	______	______	______[5]
______	______	______	______

EXAMPLE 6.02B: Share Subscription For Single Subscriber

The Dillon Manufacturing Company to be incorporated under the laws of the State of Michigan.

Capital Stock $1,000,000.00 Shares $100.00 par value

I, the undersigned, hereby subscribe for 100 shares of the capital stock of the Dillon Manufacturing Company to be incorporated and agree to pay in cash for said stock the sum of $10,000.00 on demand of the Board of Directors of the Corporation.

This agreement is made upon the condition that eighty per cent of the capital stock of the Corporation is subscribed in good faith by solvent persons on or before the 1st day of July, 1981, and the Corporation is incorporated within 30 days thereafter.

Dated May 1, 1981.

______________[6]

These subscriptions are also assignable. Therefore, in jurisdictions where incorporators are required to be subscribers by law, the incorporators must subscribe, but if they do not intend to invest, they may assign their preincorporation share subscriptions to outsiders who have acknowledged a desire to invest.

The law presumes that the corporation is formed in reliance upon the offers of subscribers to purchase shares, especially when the statute requires minimum paid-in capital as a condition to commencing business. At common law, share subscriptions were revocable until the corporation had been formed and had accepted them by agreeing to issue shares for the amount of the subscriptions. Modern statutes now provide that the preincorporation share subscriptions are irrevocable for a period of time. Section 17 of the Model Act states that a preincorporation subscription for shares of a corporation is irrevocable for a period of six months unless otherwise provided in the terms of the subscription agreement or unless all of the subscribers consent to the revocation. The period of irrevocability varies among the state statutes from three months to one year.[7] A few states specify the period to be a stated time after the certificate of incorporation is issued.[8] If the corporation is formed during the period of irrevocability it may accept the subscription and require the subscriber to purchase the shares for the amount stated therein. In most jurisdictions acceptance occurs by action of the board of directors after the corporation is formed,[9] but Pennsylvania makes acceptance automatic upon filing the articles of incorporation,[10] and a few states make acceptance automatic upon the issuance of a certificate of incorporation.[11] When the subscription is accepted by the corporation or automatically under the statute the subscriber is usually required to pay the amount in full, but the board of directors may permit payment in installments.

EXAMPLE 6.02C: Call of Subscription

August 1, 1981

To: James Lyons
[*Address*]
Dear Sir:

At a regular meeting of the Board of Directors of Trouble, Inc., held on July 30, 1981 a resolution was duly adopted fixing the amount of calls on stock issued as partly paid and the date of payment of each call.

You are hereby notified that the first call on your subscription for partly paid stock amounts to $500.00, which sum is due and payable at the office of the Corporation on August 15, 1981.

Trouble, Inc.
By __________________ Secretary[12]

If a subscriber defaults on his contract and refuses to pay any installment when due the corporation may sell the shares to another investor, and the defaulting subscriber may be liable for breach of contract. Moreover, under the Model Act subscriber may forfeit his right to the shares if he has not paid the amount due within 20 days after a written demand has been made.[13]

EXAMPLE 6.02D: Demand For Payment and Notice of Forfeiture

September 1, 1981

To: James Lyons
[*Address*]

You are hereby notified that at a regular meeting of the Board of Directors of Trouble, Inc., the following resolution was duly adopted:

"Resolved, that the entire [or ________ per cent of the] unpaid balance on all subscriptions to the common stock of this Corporation is hereby called for payment forthwith, and the Secretary is hereby directed to demand payment from each subscriber having an unpaid balance, by mailing to him, at his last known address, a written demand requiring payment within 20 days from receipt of such notice, in default of which his shares and all previous payments thereon will be forfeited."

> Demand is hereby made upon you for payment in accordance with the provisions of the above resolution, and you are hereby notified that, in default of payment within 20 days from receipt of this notice and demand, your shares and previous payments thereon will be forfeited.
>
> Trouble, Inc.
>
> By ________________ Secretary[14]

Most statutes further permit the by-laws to prescribe other penalties for failure to pay in accordance with the subscription.

§ 6.03 SELECTION OF JURISDICTION

Preceding sections have discussed the variations in corporate statutes and the trend toward permissiveness and flexibility in the jurisdictional approach to corporate problems. Moreover, states approach corporate taxation differently, and subject corporations doing business outside the boundaries of their domestic or home state to special procedures when qualifying to do business.[15] These factors play an important role in the selection of the jurisdiction in which to incorporate.

A corporation which is formed within the state is known as a domestic corporation, while one formed in some other state is called a foreign corporation. Each state statute has provisions regulating domestic corporations and special provisions for foreign corporations. If a particular state statute contains flexible and advantageous provisions for its domestic corporations and restrictive and cumbersome procedures for foreign corporations, that state should be considered a good candidate for incorporation (domestication). Of course, the converse is also true. There are also other factors to consider, such as whether the state corporation law has been well-tested by court decisions so as to be capable of accurate interpretation; whether the state's taxation structure is acceptable; and whether the state laws will permit all desired corporate features. This last factor requires an analysis of all important points of corporate law in each jurisdiction to be considered. A 108-question checklist for jurisdiction selection appears as Form 6D in the Appendix.

Having perused the checklist, you may be reeling at the thought of the monumental task of comparing all of those points for each of 50 states, and also wondering how a corporation is ever formed if that much research is required as a preface. A couple of observations may decrease the anxiety.

First, a corporation should not consider incorporating in a state where it does not intend to do business. The exceptional case may arise when a "permissive" jurisdiction, such as Delaware, is particularly attractive for some special reason.[16] Thus, incorporators of a restaurant business in Santa Fe, New Mexico, would not consider incorporating outside of the state unless there is an extremely attractive feature of another state that is deemed particularly important for their corporate structure. To domesticate the corporation in a state where no business activity will be conducted without compelling reasons only complicates the corporate structure, increases the cost of organizational expenses and may result in double taxation. Consequently, the first predisposition in the selection of jurisdiction is to incorporate in the state where the corporation will conduct most of its business.

If business is to be conducted in only one state, that state should be the prime candidate for incorporation.

Second, when a corporation intends to conduct relatively equal amounts of business in several jurisdictions, a state by state comparison of all of the checklist items will yield a net result of advantages and disadvantages for each state, and a coin-toss decision may be appropriate. The permissive jurisdictions will stand out in relative advantages and flexibility and they should always be considered when extensive interstate business is contemplated. Remember however, that flexibility and ease of formation and operation are frequently costly. Most states with permissive corporate laws are seeking to attract corporate business—so they can impose *taxes*. Make certain that the choice of a permissive state is worth the cost.

Other than the foregoing rules, perhaps the most accurate statement which can be made regarding jurisdiction selection is that the choice depends upon the circumstances of each case. The particular needs and desires of the client on each of the enumerated points in the checklist must be considered, and the decision will depend upon the weight assigned to each element of the corporate structure.

§ 6.04 SELECTION AND RESERVATION OF CORPORATE NAME

Every corporation must have a name which indicates that it is a corporation and the selection and determination of the availability of the name should come at an early stage in the incorporation procedure. The Model Act specifically requires that all names of corporations contain the words "Corporation," "Company," "Incorporated," or "Limited," or some abbreviation of one of those words.[17] This requirement is common to most jurisdictions. A few states will not permit "Limited" or "Company," and other statutory restrictions specify certain names and titles which may not be used in a corporate name.[18] Many state statutes also prohibit a corporate name which contains any word or phrase indicating that the corporation is organized for any purpose other than the purposes stated in its articles of incorporation. Moreover, every jurisdiction forbids a corporate name which is the same as, or deceptively similar to, the name of any other domestic corporation existing under the laws of the state or any foreign corporation authorized to transact business in the state.[19] This latter requirement is really a response to unfair competition, and is designed to avoid the use of one organization's name and reputation by another in order to induce public patronage. The Great Atlantic and Pacific Tea Company, which operates A & P Food Stores, prevented a separate corporation from using the name A & P Trucking Corporation because of the possible public confusion in the names. It may not even be necessary for the name to belong to another corporation. The television entertainer, Ed Sullivan, once prevented another Edward J. Sullivan from using "Ed Sullivan Radio & TV, Inc." as his corporate name.

The prohibition against deceptive similarity has several ramifications. On the negative side, care must be taken to avoid the selection of a name which is dangerously close to another well-known company or individual. Moreover, if the new corporation intends to do business in several states, the name cannot approximate a well-known name in *any* of those states, even if the similar name is recognized only regionally, and would be unknown in all the other states. On the positive side, the selected corporate name should be one which the state courts will protect against infringement by others. For example, descriptive names, such as "Janitorial Service, Inc." or "Builders Supply Company" are very vulnerable to infringement because of their general application. Similarly, "Jones and Smith, Inc." could not expect complete protection because of the courts' reluctance to prohibit other Joneses and Smiths from using their own names. On the other hand, coined names which are selected arbitrarily, such as "Gazorninplat Corporation," "Jello, Inc." or "Sunkist Fruit Co." will receive the greatest protection from infringement by a competitor. The selection of the name is another of those individual matters for the client to decide. However, the problems of similarlity, overstating corporate purposes, and statutory requirements must be considered in making that decision.

Availability of Name

To determine whether the proposed corporate name is available for use in a particular state, counsel should consult the state agency designated as the repository of corporate names, which is usually the Secretary of State. That office will review the records to determine if the name has been used or if it is deceptively similar to the name of another corporation. The decision regarding availability is usually discretionary with the state authorities, and they will refuse to accept a reservation of name or the articles of incorporation if they feel the name is too similar to another in use. Sometimes this issue may be negotiated if the name is important to the client, and the addition of an extra word to the name may be enough to obtain permission to use the desired name. According to legend on one occasion, the Colorado Secretary of State refused the use of the name "Westwind Corporation" because another corporation was using the name, but permitted the new corporation to be formed under the name "Westwind Corporation Jr." The addition of one word—"Junior"—was enough to distinguish the names. Both corporations (apparently operating different types of businesses) lived happily ever after.

Since the name will appear on all corporate documents, the choice of the name and its reservation must come early in the incorporation procedure.

Reservation of Name

If the proposed corporate name is available, it should be reserved while the corporate papers are being prepared. Nothing can be more frustrating and embarrassing than to learn that a name is available, and to prepare all corporate documents using that name, only to learn upon filing that the name has just been taken by another organization. Most states permit reservation of a corporate name for a limited period of time (from 30 days to 12 months) for a small fee.[20] This reserves the particular name for the exclusive use of the corporation for the specified period, and a few states allow extension of the reservation for another limited period.

Under the Model Act a corporate name may be reserved by any person intending to organize a corporation, domestic or foreign; any foreign corporation intending to qualify to do business within the state; or any organized domestic or foreign corporation intending to change its name. The period of reservation provided by the Model Act is 120 days.[21]

It is usually possible to transfer the reserved name to any person by filing a notice of transfer with the same state officer. This procedure allows an attorney or a member of the attorney's staff to reserve the name for the client, and to transfer the name to the corporation when the articles of incorporation are filed.[22]

The Model Act and jurisdictions which follow it have an additional statutory provision which allows a corporation to "register" the corporate name for periods of one year at a time. Registration of the corporate name has the same effect as reservation of the name, but registration is permitted only for a corporation already organized and existing and would, therefore, be used only by foreign corporations interested in qualifying to do business within the state. An organized and existing domestic corporation would have no use for the registration procedure since the fact of incorporation reserves the name of a domestic corporation as long as the corporation remains in good standing. Of course, the registered name may not be the same as or deceptively similar to the name of a domestic or qualified foreign corporation. The initial registration is effective until the close of the calendar year in which the application for registration is filed and the fee is pro-rated for the portion of the year remaining when filed. Thereafter, a corporation may continue to renew such registration from year to year by filing an application for renewal.[23]

States without this registration procedure pose problems for multi-state corporations who want to be certain that their name is protected nationwide. Reservation of the name is not a viable alternative, since the reservation is good only for a limited period, and actual qualification to do business in every state will subject the corporation to additional regulation and taxation.[24] Consequently, many corporations use "name-saver" subsidiaries to protect their corporate name. A domestic subsidiary corporation which uses the corporate name is formed within each state, and the name is thereby permanently reserved. Since the subsidiary usually has no major assets and conducts no business, the taxation and regulation imposed on the "name-saver" corporation is minimal.

Operation Under an Assumed Name

The corporation may be formed under one name but may desire to operate under another, especially when it operates several different types of businesses and wants its various divisions to conduct business under separate names. An assumed name is also frequently used to satisfy the statutory restriction that the name may not contain any word or phrase which indicates or implies that the corporation is organized for any purpose other than those stated in its articles of incorporation.

The procedure for the use of an assumed or trade name by a corporation is very similar to that followed by a sole proprietorship or partnership.

Several state statutes allow the use of assumed names by corporations, and they usually require the filing of a statement with the Secretary of State.[25] The corporation also may have to file with county officials where business is conducted. The statutes specify whether an assumed name must contain the special corporate words ("company", "incorporated", "limited," "corporation"), and the assumed name may not be deceptively similar to any other well-known or reserved name.[26]

§ 6.05 THE ARTICLES OF INCORPORATION

The document which initiates the creation of the corporate existence and defines the corporate structure is the articles of incorporation, or corporate charter. The document is also variously called "Certificate of Incorporation", "Articles of Association" or "Articles of Agreement".[27] Recall that corporate existence is regulated first by state statute. The articles of incorporation then flesh out the statutory provisions to tailor a particular corporate structure to the needs of its incorporators.

The articles of incorporation may be thorough and detailed or they may contain only the bare essentials required by statute. The content of the articles is not arbitrarily determined. There are certain matters which are required by the state statute and these always must be included. Thereafter, *any* aspect of corporate existence may be regulated by the articles if desired, but remember that the articles, once filed, are difficult to change. The amendment procedure requires the approval of the shareholders and filing with the designated public officials.[28] Thus, if the incorporators want certain rules which will have some degree of permanence, these rules should be included in the articles of incorporation. On the other hand, if certain rules for regulation of corporate affairs are expected to change in the future, they are best reserved to the by-laws of the corporation, where the amendment procedure is considerably more convenient. These matters should be determined in an early conference with the incorporators when all statutory requirements and other drafting possibilities for the articles of incorporation are discussed. A checklist for this conference is set forth as Form 6N in the Appendix.

Statutory Requirements

The Model Act requires that the articles of incorporation set forth the following:

1. The name of the corporation.
2. The period of duration, which may be perpetual.
3. The purpose or purposes for which the corporation is organized which may be stated to be, or to include, the transaction of any or all lawful business for which corporations may be incorporated under this Act.
4. The aggregate number of shares which the corporation shall have authority to issue and if such shares are to be divided into classes, the number of shares of each class.

5. If the shares are to be divided into classes, the designation of each class and a statement of the preferences, limitations and relative rights in respect of the shares of each class.

6. If the corporation is to issue the shares of any preferred or special class in series, then the designation of each series and a statement of the variations in the relative rights and preferences as between series insofar as the same are to be fixed in the articles of incorporation, and a statement of any authority to be vested in the board of directors to establish series and fix and determine the variations in the relative rights and preferences as between series.

7. If any preemptive right is to be granted to shareholders, the provisions therefor.

8. The address of its initial registered office, and the name of its initial registered agent at such address.

9. The number of directors constituting the initial board of directors and the names and addresses of the persons who are to serve as directors until the first annual meeting of shareholders or until their successors be elected and qualify.

10. The name and address of each incorporator.

In addition to provisions required therein, the articles of incorporation may also contain provisions not inconsistent with law regarding:

1. The direction of the management of the business and the regulation of the affairs of the corporation;

2. The definition, limitation and regulation of the powers of the corporation, the directors, and the shareholders, or any class of the shareholders, including restrictions on the transfer of shares;

3. The par value of any authorized shares or class of shares;

4. Any provision which under this Act is required or permitted to be set forth in the by-laws.[29]

These requirements are typical of the statutory provisions regulating the content of the articles of incorporation in most jurisdictions, but the specific demands of each state statute should be carefully studied and scrupulously followed. It is usually unnecessary for the articles of incorporation to repeat the corporate *powers* enumerated in the statute,[30] but the articles must specify the corporate purposes. Most states print and distribute "official" forms which may be used for the articles of incorporation, and they contain the bare essentials necessary for compliance with the statute. The incorporators or counsel will usually prefer more elaboration in the articles than is permitted on the official forms.[31]

Many of the items specified as necessary ingredients to the articles of incorporation have been previously discussed or will be discussed in detail later. Nevertheless, let us wander through the requirements briefly here:

Name of the Corporation. The articles of incorporation must contain the corporate name selected by the incorporators and approved by the Secretary of State or other designated public official. If the name has not previously been reserved, the articles of incorporation will reserve the name for the use of the corporation during its existence.[32]

EXAMPLE 6.05A: Name

The name of the corporation shall be Five Points Land and Cattle Company.

Period of Duration. The articles of incorporation usually state that the corporation shall exist perpetually. It is possible to establish a specified period after which the corporate existence will automatically terminate, but this may cause an unnecessary burden in that an amendment to the articles of incorporation would be required if the owners should subsequently decide to continue the business. If perpetual existence is specified in the articles of incorporation the corporation will only terminate if dissolved according to the statutory procedure.[33]

EXAMPLE 6.05B: Period of Duration

This corporation shall exist perpetually [or] *shall terminate on* ________, 19 ___, unless dissolved according to law.

Corporate Purposes. The corporation may do only those acts which are within the scope of its stated authorized purposes. Corporate purposes should be distinguished from corporate powers, which were previously discussed in Section 4.02. The purposes are the business objectives of the corporation, and the powers are the means by which these objectives are achieved. For example, the incorporators may form a corporation to purchase and rent apartment buildings. Their corporate *purposes* would specify real estate investment, management, operation, lease, etc. The statutory *powers* provide that the corporation has power to purchase and hold property, make contracts, borrow money, etc. The powers are, therefore, the enabling authority for the corporation to pursue its purposes. The modern trend of corporate law is to permit the incorporators to adopt broad corporate purposes and thereby authorize the corporation to do any legal act. The Model Act allows the formation of a corporation for any lawful purposes, except banking and insurance.[34] Several states, including Delaware and Pennsylvania permit the articles of incorporation to authorize "any lawful activity."

In the drafting stages, the incorporators will have described the general nature of the contemplated business, such as operating a book store, manufacturing bicycles, conducting environmental services, and so forth. However, after formation the management of the corporation may decide to invest in real estate with its manufacturing profits, or to open a cafeteria next to its bookstore, and the scope of the designated purposes in the articles of incorporation becomes a critical consideration. The purpose clauses of the corporation usually specify a particular type of business, for example:

EXAMPLE 6.05C: Purposes

To buy and sell, and otherwise deal in, at both wholesale and retail, all kinds and brands of cherries; to brine and preserve maraschino cherries of every nature and character; to engage in the canning and pitting of cherries and to prepare cherries for every possible purpose and use; to engage in the buying, selling, and otherwise dealing with and in the canning, preservation and preparation of all kinds of fruits of every nature, character and description; and generally to do all acts reasonable and necessary for the furtherance of the foregoing business.[35]

EXAMPLE 6.05D: Purposes

To carry on business as jewelers, gold and silver smiths; as dealers in china, curiosities, coins, medals, bullion and precious stones; as manufacturers of and dealers in gold and silver plate, plated articles, watches, clocks, chronometers, and optical and scientific instruments and appliances of every description; and as bankers, commission agents and general merchants.[36]

These limited purposes would not allow the corporation to operate a restaurant, or to manufacture bicycles, or to invest in real estate. When the incorporators anticipate these additional activities, additional purpose clauses must be added. If the state statute is sufficiently permissive to allow the articles of incorporation to authorize "any lawful activity" without further specification and the incorporators want broad purposes, the drafting of the purpose clause is simple. However, if the state requires specificity of corporate purposes, or if the incorporators desire to restrict the corporate purposes, the drafter's job becomes more difficult. Counsel must pay close attention to detail to insure that the drafted purpose clauses in the articles of incorporation will permit the corporation to do everything necessary to operate the intended business. The purpose clauses must also anticipate expansion and give the corporation room to do everything it might be expected to do in the near future. Finally, the corporate purposes must not be overbroad, so management has no business guidance. The incorporators may restrict the corporate purposes to direct management toward specific business objectives.

Remember that the law provides implied power for the corporation to conduct any necessary act which is consistent with its stated purposes, but the law will not allow the corporation to exceed its purposes if they are restricted in the articles of incorporation. Admittedly, this is a delicate distinction. Consider our cherry fruit business described in the example purpose clause above. The corporate *powers* would permit the corporation to buy a cannery to conduct its canning and pitting operations and they may allow it to also buy an adjacent building if expansion was contemplated. However, the corporate *purposes* would not allow it to buy the adjacent building for investment purposes. Consequently, it is better practice under the modern statutes to state the corporate purposes as broadly as possible. In any case, the drafter should attempt to prepare a statement of corporate purposes which is sufficiently specific to avoid excursions into unauthorized areas of business while being sufficiently broad to allow expansion of the contemplated business without amendment of the articles of incorporation.

A word about the dangers lurking in the statement of corporate purposes may be appropriate. A corporation is not permitted to exceed its stated corporate purposes, and if it does, it is said to have committed an "ultra vires" act. The law protects the shareholders from such abuses of corporate authority by allowing their application to a court to have the unauthorized act stopped. The Attorney General may protect the interests of the state by suing to stop the act, or by dissolving the corporation for committing unauthorized acts. Moreover, directors and officers who have caused the corporation to venture forth into the unauthorized business activities may be held personally liable for any loss occasioned by such transactions.

Shares. The corporation's "equity securities," or shares, must be accurately described in the articles of incorporation. The articles of incorporation establish the number of shares that the corporation has authority to issue, and the corporation is limited to the number of shares so authorized, unless an amendment to the articles is adopted permitting the issuance of additional shares. The articles must also detail and describe the classes of equity securities, such as common stock, preferred stock, and so forth, but the articles generally do not contain any information about the corporation's "debt" securities, or bonds.

Every corporation must have at least one class of stock, and if only one class is authorized, it is usually called "common" stock. The financial structure of a corporation has infinite flexibility and may be as simple or as complex as desired, depending upon the projected financial needs of the businesss and, probably, the imagination of the drafter.[37] Presently, it is important to identify those details of the financial structure which must be described in the articles of incorporation in order to authorize the issuance of equity securities. This information typically includes:

1. The number of shares which the corporation will have authority to issue.
2. The par value of the shares to be issued or a statement that the shares are without par value (although the Model Act no longer requires such a distinction).
3. If the shares are to be divided into classes,
   (a) the designation of each class;
   (b) a statement of the preferences, limitations, and relative rights of the shares of each class;
   (c) the par value of each class, or a statement that the shares are to be without par value;
   (d) the authority, if any, of the board of directors to establish a class or series and determine variations in the rights and preferences between series.[38]

The share authorization clause in the articles of incorporation may be very simple, as when the corporation intends to issue only one class of common stock, or quite complex, when multiple classes of stock are to be issued with varying rights and preferences among the classes.

EXAMPLE 6.05E: Capital Stock

The amount of the total authorized capital stock of the Corporation shall be 50,000 shares of common stock of the par value of $1.00 per share.[39]

EXAMPLE 6.05F: Capital Stock

a) The total authorized capital stock of this Corporation shall be divided into one thousand (1000) shares of which five hundred (500) shares shall be preferred stock and shall be issued at a par value of One Hundred Dollars ($100) each; and five hundred (500) shares shall be common stock which shall be issued without par value and shall be sold at $1 per share.

b) The holders of the shares of preferred stock shall be and are entitled to receive and shall so receive dividends on the value of such stock at the rate of six per cent (6%) per annum, which shall be cumulative and which shall be set aside and paid before any dividend shall be set aside or paid upon the shares of common capital stock.

c) The voting power of the shares of capital stock in this Corporation shall be vested wholly in the holders of the shares of common capital stock. The preferred capital stock shall have no voting power whatever.

d) In the event of the liquidation or dissolution, or the winding up of the business affairs of this Corporation, the holders of the preferred shares of capital stock shall be and they are entitled to be paid first for the full and determined value of their shares, together with unpaid dividends up to the time of the payment; after the payment to the preferred stockholders, the remaining assets of the Corporation shall be distributed among the holders of the common capital stock to the extent of their respective shares.

e) This Corporation shall have the right at its option to retire the preferred stock upon ten (10) days notice, by a resolution of its Board of Directors, by paying for each share of preferred stock One Hundred Two Dollars ($102) in cash, and in addition thereto all unpaid dividends accrued thereon to the date fixed for such redemption.[40]

The capital stock structure is developed after studying many financial and practical matters, all of which are discussed in detail later.[41]

Briefly, however, the decision to issue par value or no par value shares depends upon the consideration (money, property or services) expected to be given in exchange for the shares, the organizational taxes imposed by the state, and the accounting ramifications of each value approach. Par value shares may not be sold for less than par value,[42] which means that a share of $100 par value stock can only be issued in exchange for cash, property or services valued at $100 or more. No par value shares may be sold at their "stated value," which is determined from time to time by the board of directors.[43] The no par feature adds some flexibility to the sale of shares since the board of directors may exert control over the going price of the shares. Organizational taxes are imposed in some jurisdictions on the total aggregate value of the authorized capital stock structure, and the distinction between par value and no par value shares is also important for tax computation. To compute the total aggregate value, state statutes usually place a value on no-par shares. For example, suppose the state imposes a $10 tax for each $10,000 aggregate authorized capital stock, and places a $100 value on each no-par share. A corporation could authorize 100 shares with no par value for a tax of $10; they could also set par value at $1.00 per share and authorize 10,000 shares for the same tax. A few jurisdictions with organization taxes based upon the capital stock structure actually try to discourage no-par shares by placing a high valuation on them for tax computation purposes.[44]

Finally, accounting principles require that the par value of issued shares must be placed in an account called "stated capital", and that account is restricted so that no dividends may be paid from it. However, the consideration for no-par shares may be allocated to an account called "capital surplus", and those funds may be available in special circumstances for distribution to shareholders or for repurchase of corporate shares.[45] Thus, if the corporation issued $100 par value shares for $100 cash, all of the funds must go to the restricted stated capital account; but if it issued no-par shares for the same amount, some or all of the $100 could be placed in capital surplus, a more flexible account. These accounting ramifications may be important to a corporation which requires the flexibility to be able to distribute its equity accounts before it has accumulated profits to distribute.

The decision to issue several classes of equity securities is usually based upon the attractiveness of the securities to potential investors. If shares of common stock will sell well enough to raise the needed capital, there is usually no reason to authorize other classes of stock. However, if some investors insist that their stock must have special preferences to dividends, voting or liquidation, then separate classes of securities will be necessary.

All special features of equity securities should be described in the articles of incorporation. Conversion privileges, redemption provisions and restrictions on the sale of stock should also be specifically described in the articles of incorporation as part of the capital stock structure.[46]

Preemptive Rights. The articles of incorporation usually contain a statement regarding shareholders' preemptive rights. You will recall that a shareholder's preemptive right is her common-law right to maintain her proportionate ownership interest in the corporation. If the corporation intends to issue additional shares of stock, the existing shareholders have the right to buy their proportionate share of the new stock. Some states require preemptive rights for the shareholders unless they are specifically denied in the articles of incorporation.[47] Other statutes provide that preemptive rights will not exist unless specifically granted in the articles of incorporation.[48] In any case, it is good practice to always specify the desires of the incorporators on this point.

The articles of incorporation may simply deny or grant pre-emptive rights without further elaboration.

EXAMPLE 6.05G: Preemptive Rights

No holder of any stock of the Corporation shall be entitled, as a matter of right, to purchase, subscribe for or otherwise acquire any new or additional shares of stock of the corporation of any class, or any options or warrants to purchase, subscribe for or otherwise acquire any such new or additional shares, or any shares, bonds, notes, debentures or other securities convertible into or carrying options or warrants to purchase, subscribe for or otherwise acquire any such new or additional shares.[49]

They may also distinguish preemptive rights between specified classes of equity securities.

EXAMPLE 6.05H: Preemptive Rights Among Classes

Holders of preferred stock shall have the right to subscribe for and purchase their pro rata shares of any new preferred stock which may be issued by the Corporation, but shall have no such preemptive rights with respect to new shares of common stock which may be issued. Holders of common stock shall have the right to subscribe for and purchase their pro rata shares of any new common stock which may be issued, but shall have no such preemptive rights with respect to new shares of preferred stock which may be issued.[50]

In addition, it is possible to otherwise limit, define or expand preemptive rights in the articles of incorporation. For example, preemptive rights may be limited to stock issued only for cash and may be excluded from employee stock option plans. It is also good practice to specify the scope of preemptive rights with respect to treasury shares, stock repurchased by the company which may subsequently be resold.[51]

Cumulative Voting. If shareholders are to be permitted to cumulate their shares in elections of directors, a statement to that effect in the articles of incorporation is appropriate. Cumulative voting is treated like preemptive rights in the various state statutes—that is, some grant the right unless it is specifically denied and others deny it unless specifically granted.[52] The articles always should reflect the corporate policy either way.

EXAMPLE 6.05I: Cumulative Voting

At all elections for directors each stockholder shall be entitled to as many votes as shall equal the number of his shares of stock multiplied by the number of directors to be elected, and he may cast all of such votes for a single director, or may distribute them among the number to be voted for, or any two or more of them, as he may see fit.[53]

Registered Office and Agent. The corporation must maintain a registered office and a registered agent within the state so that all legal or official matters pertaining to its corporate existence may be addressed there. The registered office does not have to be the principal place of the business of the corporation, although it frequently is. The registered agent may be any person who is located at that office.

The registered office serves many functions, and is referred to throughout state corporate laws. For example, most statutes require notices to the corporation to be addressed to the registered office, and many states require the corporation to keep the stock transfer record at the registered office.

The registered agent has the primary responsibility for receiving notices of litigation, service of process, for the corporation. If the corporation has no available registered agent, the Secretary of State receives process on behalf of the corporation, and, under the Model Act, the failure to maintain a registered agent for thirty days is grounds for the Attorney General to dissolve the corporation.[54]

Every state, except Tennessee, requires a registered office, but several, including New Hampshire, Pennsylvania and Minnesota do not require a registered agent. Again, the corporate statute of the jurisdiction where incorporation is contemplated should be carefully studied for this purpose.

EXAMPLE 6.05J: Registered Office and Agent

The registered office of the corporation shall be at 730 Seventeenth Street, Suite 600, Denver, Colorado, 80202 and the name of the initial registered agent at such address is Nancy A. Stober. Either registered office or the registered agent may be changed in a manner provided by law.

Initial Directors. The Model Act requires the articles of incorporation to name the initial board of directors and to give their addresses.[55] Regarding the structure of the board of directors the articles of incorporation may do one of three things: 1) specify the number of directors who will constitute the board; 2) specify a formula or procedure to determine the desired number; or 3) delegate this determination to the by-laws.

EXAMPLE 6.05K: Initial Board of Directors

The initial board of directors of the corporation shall consist of three directors and the names and address of the persons who shall service as directors until the first annual meeting of shareholders or until their successors are elected and shall qualify are as follows:

Name	Address
____________________	____________________
____________________	____________________
____________________	____________________

EXAMPLE 6.05L: Number of Directors

The number of directors of the corporation shall be fixed and may be altered from time to time as may be provided in the bylaws. In case of any increase in the number of directors, the additional directors may be elected by the directors or by the stockholders at an annual or special meeting, as shall be provided in the bylaws.[56]

No particular qualifications are required for directors under the Model Act, although some states require "legal age," share ownership or states citizenship.[57] Moreover, the Model Act permits a single director, but most states require three or more. The initial directors hold office until the shareholders meet to elect their successors and their successors are qualified. The written consent of the initial board of directors to serve as directors may be necessary under state law, and may be a desirable procedure in any case.

Incorporators. The incorporators are also named in the articles of incorporation, and they sign the articles. Usually the incorporators must be adults or "legal age" and they may have to meet other qualifications, such as citizenship, residency or share subscription requirements.[58]

Permissive Provisions

We have just considered certain provisions which are required to be enumerated in the articles of incorporation under most statutes. There are, however, many other provisions which *may* be included in the articles of incorporation. The Model Act allows the articles of incorporation to contain any provision for the regulation of internal affairs of the corporation which might ordinarily be set forth in the by-laws, as long as these provisions are not inconsistent with the statute.[59] By virtue of this broad statutory authority, the articles of incorporation may contain any number of various rules and regulations pertaining to the operation of the company. However, remember that provisions in the articles are more permanent than by-law provisions, since the amendment procedure for articles of incorporation is considerably more difficult.[60] The drafter should begin with this inflexibility in mind when considering miscellaneous provisions for the articles of incorporation.

Generally, the articles of incorporation may contain any regulation of internal affairs which is not inconsistent with the law. If the incorporators had devised a procedure for distributing keys to the corporate restrooms, the procedure could be posted on a bulletin board, written into the by-laws, or given special dignity (and public notice) by being drafted into the articles of incorporation. This may be a case where the inflexibility of the articles could become painfully obvious, however, if it were later discovered that the specified procedure did not cover certain corporate executives and they had to wait until an amendment could be adopted.

There are, however, several instances in the Model Act and other corporate statutes where the articles of incorporation may modify the statutory rules, but a by-law provision is ineffective for that purpose. Therefore, if the incorporators desire a modified approach to their corporate structure, certain additional provisions must be included in the articles of incorporation. The following statutory rules of the Model Act and other corporation codes may be modified or amplified only by a special provision in the articles of incorporation:

Indemnification of Officers and Directors. The corporation has the power to indemnify its management personnel from any liability or expenses incurred by reason of litigation against them in their capacities as directors, officers, or employees, of the corporation. The Model Act specifically confers this power in Section 5, and its complex provisions generally grant the right to indemnification if the individual was not negligent in the performance of his duties to the corporation and if he was acting in good faith and in a manner he reasonably believed to be in the best interests of the corporation. In jurisdictions adopting the Model Act provision, the statutory authority for indemnification obviates any need to grant such power in the articles of incorporation, but in most jurisdictions the statutory right to indemnification is considerably more limited. In many, the statute requires that a director must be successful in the litigation in order to be indemnified,[61] but the corporation may agree to indemnify further. Many persons would not agree to serve as a director, officer or employee unless they knew that the corporation would stand behind them for litigation fees, expenses, and liability incurred as a result of their employment. Consequently, the articles of incorporation should establish the scope of indemnification for corporate personnel.

EXAMPLE 6.05M: Indemnification

The Corporation shall indemnify any director, officer, or employee, or former director, officer, or employee of the Corporation, or any person who may have served at its request as a director, officer, or employee of another corporation in which it owns shares of capital stock, or of which it is a creditor, against expenses actually and necessarily incurred by him in connection with the defense of any action, suit or proceeding in which he is made a party by reason of being or having been such director, officer, or employee, except in relation to matters as to which he shall be adjudged in such action, suit, or proceeding to be liable for negligence or misconduct in the performance of duty. The Corporation may also reimburse to any director, officer, or employee the reasonable costs of settlement of any such action, suit, or proceeding, if it shall be found by a majority of a committee composed of the directors not involved in the matter in controversy (whether or not a quorum) that it was to the interests of the corporation that such settlement be made and that such director, officer, or employee was not guilty of negligence or misconduct. Such rights of indemnification and reimbursement shall not be deemed exclusive or any other rights to which such director, officer, or employee may be entitled under any bylaws, agreement, vote of shareholders, or otherwise.[62]

[or]

EXAMPLE 6.05N: Indemnification (Following Model Act Provisions)

The corporation may:

a) Indemnify any person who was or is a party or is threatened to be made a party to any threatened, pending, or completed action, suit, or proceeding, whether civil, criminal, administrative, or investigative (other than an action by or in the right of the corporation), by reason of the fact that he is or was a director, officer, employee, or agent of the corporation or is or was serving at the request of the corporation as a director, officer, employee, or agent of another corporation, partnership, joint venture, trust, or other enterprise, against expenses (including attorneys' fees), judgments, fines, and amounts paid in settlement actually and reasonably incurred by him in connection with such action, suit, or proceeding, if he acted in good faith and in a manner he reasonably believed to be in the best interest of the corporation and, with respect to any criminal action or proceeding, had no reasonable cause to believe his conduct was unlawful. The termination of any action, suit, or proceeding by judgment, order, settlement, or conviction or upon a plea of nolo contendere or its equivalent shall not of itself create a presumption that the person did not act in good faith and in a manner which he reasonably believed to be in the best interest of the corporation and, with respect to any criminal action or proceeding, had reasonable cause to believe his conduct was unlawful.

b) The corporation may indemnify any person who was or is a party or is threatened to be made a party to any threatened, pending, or completed action or suit by or in the right of the corporation to procure a judgment in its favor by reason of the fact that he is or was a director, officer, employee, or agent of the corporation or is or was serving at the request of the corporation as a director, officer, employee, or agent of another corporation, partnership, joint venture, trust or other enterprise against expenses (including attorneys' fees) actually and reasonably incurred by him in connection with the defense or settlement of such action or suit if he acted in good faith and in a manner he reasonably believed to be in the best interest of the corporation; but no indemnification shall be made in respect of any claim, issue, or matter as to which such person has been adjudged to be liable for negligence or misconduct in the performance of his duty to the corporation unless and only to the extent that the court in which such action or suit was brought determines upon application that, despite the adjudication of liability, but in view of all circumstances of the case, such person is fairly and reasonably entitled to indemnification for such expenses which such court deems proper.

c) To the extent that a director, officer, employee, or agent of a corporation has been successful on the merits in defense of any action, suit, or proceeding referred to in (A) or (B) of this Article or in defense of any claim, issue, or matter therein, he shall be indemnified against expenses (including attorneys' fees) actually and reasonably incurred by him in connection therewith.

d) Any indemnification under (A) or (B) of this Article (unless ordered by a court) and as distinguished from (C) of this Article shall be made by the corporation only as authorized in the specific case upon a determination that indemnification of the director, officer, employee, or agent is proper in the circumstances because he has met the applicable standard of conduct set forth in (A) or (B) above. Such determination shall be made by the board of directors by a majority vote of a quorum consisting of directors who were not parties to such action, suit, or proceeding, or, if such a quorum is not obtainable or, even if obtainable, if a quorum of disinterested directors so directs, by independent legal counsel in a written opinion, or by the shareholders.

e) Expenses (including attorneys' fees) incurred in defending a civil or criminal action, suit, or proceeding may be paid by the corporation in advance of the final disposition of such action, suit, or proceeding as authorized in (C) or (D) of this Article upon receipt of an undertaking by or on behalf of the director, officer, employee, or agent to repay such amount unless it is ultimately determined that he is entitled to be indemnified by the corporation as authorized in this Article.

f) The indemnification provided by this Article shall not be deemed exclusive of any other rights to which those indemnified may be entitled under any bylaw, agreement, vote of shareholders or disinterested directors, or otherwise, and any procedure provided for by any of the foregoing, both as to action in his official capacity and as to action in another capacity while holding such office, and shall continue as to a person who has ceased to be a director, officer, employee, or agent and shall inure to the benefit of heirs, executors, and administrators of such a person.

g) The corporation may purchase and maintain insurance on behalf of any person who is or was director, officer, employee or agent of the corporation or who is or was serving at the request of the corporation as a director, officer, employee, or agent of another corporation, partnership, joint venture, trust, or other enterprise against any liability asserted against him and incurred by him in any such capacity or arising out of his status as such, whether or not the corporation would have the power to indemnify him against such liability under provisions of this Article.

Purchase of Corporate Shares. Although the Model Act has been amended to eliminate statutory restrictions on purchase of corporate shares,[63] most state statutes permit the corporation to repurchase its own shares from investors, thereby creating "treasury" shares, but these statutes limit the source of funds for such purchases to unreserved and unrestricted earned surplus. This means that the corporation may purchase these shares only with accumulated profits which have not been designated for any other purpose. Under the statute, if no profits have accumulated, the corporation may not repurchase its own stock. However, the articles of incorporation may provide that capital surplus (the excess amount collected over par value, or the amount collected and designated capital surplus for no par value shares) may be used in addition to earned surplus for this purpose. There are many reasons supporting this flexibility. For example, management may desire to reduce the number of shares outstanding so as to increase the earnings-per-share figures, or they may wish to reacquire outstanding shares to hold for employee stock purchase plans. Counsel should remember that an appropriate clause in the articles of incorporation is necessary to open the capital surplus account for the repurchase of shares.

The provisions of the articles of incorporation may also have a negative impact on the corporate purchase of its own securities. The articles may restrict management by requiring that all corporate shares repurchased by the corporation must be cancelled, and cannot be resold or reissued. Management would not be bound to cancel such shares without an express provision to that effect in the articles of incorporation.

EXAMPLE 6.05O: Repurchase of Corporate Shares

The corporation shall have the power to repurchase its shares of cumulative preferred stock with any surplus then in existence which has not been otherwise reserved or restricted. [Check statutory authority for the type of surplus which may be permitted for repurchase of shares]

[and]

Upon repurchase of shares of the corporation the corporation shall cancel and retire the same, and such shares shall not be held as treasurer shares or reissued to shareholders under any circumstances.

Reserve the Right to Fix Consideration for Shares to the Shareholders. Many matters which are ordinarily determined by the directors may be reserved to the shareholders by an appropriate clause in the articles of incorporation. This is one of them. Section 18 of the Model Act vests the power to determine the price of shares to the directors, but the shareholders may exercise this power if the articles of incorporation so provide.

EXAMPLE 6.05P: Right to Fix Consideration for Shares

The shareholders of the corporation at a meeting duly called for such purpose shall fix and determine the stated value of the shares of the corporation.

Stock Rights and Options. The corporation may create stock options or stock rights which entitle the holder of the option or right to buy shares at a designated price. The articles of incorporation may restrict management in creating such options or rights and may also elaborate upon their terms, including time of exercise and price. Restrictive provision in the articles of incorporation would only be necessary if the incorporators wanted to narrow management's broad statutory authority to create such options, as is contained in Section 20 of the Model Act.

EXAMPLE 6.05Q: Restrictions on Issuance of Stock Rights and Options

The board of directors may not, without the express approval of at least the majority of the then outstanding shares of the corporation at a meeting duly called for such purpose, create or issue rights or options entitling the holders thereof to purchase from the corporation shares of any class or classes. Further, even upon such approval by the shareholders, the board of directors shall not create and issue such rights or options which shall provide for a price less than 50% of the then market value of such shares, determined by an independent certified public accountant of the corporation, or upon terms which would permit the holder of such options or rights to pay the purchase price of such shares over a period longer than six months.

Quorum and Vote of Shareholders and Directors. A majority of the shares entitled to vote is a quorum for shareholder meetings, and the affirmative vote of the majority of the quorum carries action on behalf of the shareholders under Section 32 of the Model Act. The articles of incorporation may vary these requirements in any manner except one: a quorum may never be less than one-third of the shares entitled to vote. Thus, the articles of incorporation could provide that a quorum shall be 40% of the shares entitled to vote, and shareholder action requires an affirmative vote of 75% of the shares represented; or that a quorum requires 80% of the share entitled to vote and shareholder action requires 80% of the shares represented, and so forth.

The articles of incorporation may similarly modify the quorum and vote necessary for director action under Section 40. However, a quorum or vote of directors may not be reduced below a majority, and the voting or quorum requirements may only be increased by the articles of incorporation.

EXAMPLE 6.05R: Quorum and Vote of Shareholders

The quorum of the shareholders of this corporation for each annual or special meeting of the shareholders shall be one-third of the shares then outstanding and entitled to vote. No resolution of the corporation at any meeting of the shareholders shall be adopted except by the vote of at least seventy-five percent (75%) of the shares represented in a properly called meeting at which a quorum of the shares is present.

EXAMPLE 6.05S: Vote of Directors

No resolution of the Corporation at any meeting whether regular or special, shall be adopted except by the unanimous vote of the three directors duly elected as provided herein.[64]

Directors are also permitted by statute to take action without a meeting by signing a consent to action in writing.[65] The articles of incorporation may deny this power, however, if the incorporators want their directors to act only in formal session.

Shareholder Control of By-Laws. The initial by-laws of the corporation are adopted by the board of directors at their organizational meeting, and the normal statutory rule is that the board of directors has the power to alter, amend or repeal the by-laws.[66] This power may be reserved to the shareholders in the articles of incorporation.

EXAMPLE 6.05T: Amendments to the Bylaws

The bylaws of this corporation shall not be amended, modified or altered except by the vote of the shareholders of the corporation at a meeting of the shareholders, duly called, at which a quorum is present.

Dividend Provisions. The board of directors has full discretion under the Model Act for the payment of dividends to shareholders.[67] The articles of incorporation may restrict this discretion, and may establish certain conditions which must be satisfied before dividends may be declared. Conversely, in most states the articles of incorporation may expand the corporation's ability to distribute cash or property to shareholders by expressly authorizing such distributions out of capital surplus.[68] Moreover, the articles of incorporation for a corporation whose principal business is the exploitation of natural resources, such as timber, oil wells and mines, may authorize the payment of dividends from depletion reserves, an account which reflects the reduction of the natural resources available to the corporation.[69]

EXAMPLE 6.05U: Restriction on Payment of Dividends

The board of directors of the corporation may not pay or declare a dividend during the first two years of the corporation's operation of its business. Thereafter, the board of directors may, from time to time, declare and pay dividends in accordance with the law provided that the corporation has adequate cash reserves at all times to meet six-months projected operating expenses.

EXAMPLE 6.05V: Distributions from Capital Surplus

The board of directors of the corporation may, from time to time, distribute to the shareholders out of capital surplus of the corporation a portion of the assets of the corporation, in cash or property, provided:

a) No such distribution shall be made at a time when the corporation in insolvent or when such distribution would render the corporation insolvent.

b) No such distribution shall be made to the holders of any class of shares unless all cumulative dividends accrued on all preferred classes of shares entitled to preferential dividends shall have been fully paid.

c) No such distribution shall be made to the holders of any class of shares which would reduce the remaining net assets of the corporation below the aggregate preferential amount payable in the event of an involuntary liquidation to the holders of shares having preferential rights to the assets of the corporation in the event of liquidation.

d) Each such distribution, when made, shall be identified as a distribution from capital surplus and the amount per share disclosed to the shareholders receiving the same concurrently with the distribution thereof.

EXAMPLE 6.05W: Payment of Dividends From Depletion Reserves

The board of directors may, from time to time, declare and the corporation may pay dividends in cash of the depletion reserves earned by the corporation through its business of exploiting natural resources, but such reserves and the amount per share paid from such reserves shall be disclosed to the shareholders receiving the same concurrently with the distribution thereof.

Transactions with Interested Directors. A director owes a most strict duty of loyalty to her corporation, and, in exercising her responsibilities must strive to represent the corporation without any conflict of interest. The common law looked askance at any contract formed between the corporation and a director in her personal capacity or with another corporation for which she also served as a director. When the same director appeared in the negotiations for both sides of the transaction, either personally or as a director to another corporation, the transaction was always vulnerable to a court test and would be upheld only upon a showing that it was eminently fair despite the apparent conflict. In modern corporations, common or "interlocking" directors appear very frequently, and it is good practice to include a clause in the articles of incorporation which describes the corporation's position on transactions where conflict of interest may be implied. The clause should provide that such transactions will not be considered automatically invalid, but also should not completely exculpate the directors involved. The conflict of interest protection should be preserved for the rare cases where a director has compromised the corporation for her own personal gain.

EXAMPLE 6.05X: Transactions with Interested Directors

No contract or other transaction between the corporation and any other corporation, whether or not a majority of the shares of the capital stock of such other corporation is owned by the corporation, and no act of the corporation shall in any way be affected or invalidated by the fact that any of the directors of the corporation are pecuniarily or otherwise interested in, or are directors or officers of, such other corporation; any director individually, or any firm of which such director may be a member, may be a party to, or may be pecuniarily or otherwise interested in, any contract or transaction of the corporation, provided that the fact that he or such firm is so interested shall be disclosed or shall have been known to the Board of Directors, or a majority thereof; and any director of the corporation who is also a director or officer of such other corporation, or who is so interested, may be counted in determining the existence of a quorum at any meeting of the Board of Directors of the corporation which shall authorize such contract or transaction, and may vote thereat to authorize such contract or transaction, with like force and effect as if he were not such director or officer of such other corporation or not so interested.[70]

Classification, Compensation and Qualifications of Directors. The articles of incorporation may provide for staggered terms for directors to insure continuity of management policies. A "staggered" board of directors will always have some "seasoned" members.[71] Section 37 of the Model Act permits classification of directors only if the entire board consists of nine or more members. A sample classification clause for the articles of incorporation follows:

EXAMPLE 6.05Y: Classification of Directors

At the first annual meeting of the shareholders, the members of the Board of Directors shall be divided into three classes of three members each. The members of the first class shall hold office for a term of one year; the members of the second class shall hold office for a term of two years; the members of the third class shall hold office for a term of three years. At all annual elections thereafter three directors shall be elected by the shareholders for a term of three years to succeed the three directors whose term then expires; provided that nothing herein shall be construed to prevent the election of a director to succeed himself.[72]

The compensation of the board of directors is set by the directors, unless the articles of incorporation provide otherwise.[73]

As long as the articles of incorporation are touching upon some matters relating to directors, qualifications may also be covered. Under the Model Act, directors need not have any particular qualifications to serve as such,[74] but the articles of incorporation or the by-laws may impose any reasonable qualifications for directors. It may be desirable to require that directors are shareholders, for example, or over 35 years of age, or perhaps under 35 years of age. Director qualifications should be tailored to the desires of the incorporators.

§ 6.06 FILING AND OTHER FORMALITIES

Filing Procedure

The articles of incorporation are filed with the Secretary of State or other designated public official, and the Model Act requires duplicate originals to be filed.[75] Several states also require that the articles be filed with certain designated county offices in which the corporation has it registered office, and the corporation is not properly formed unless the articles of incorporation are filed in all places required by statute.[76] After determining that the articles of incorporation are in proper form and that all fees have been paid the Secretary of State will return the duplicate copy of the articles of incorporation with the certificate of incorporation.

Miscellaneous Formalities

Each state statute treats the execution and filing of the articles of incorporation differently. All jurisdictions require that the articles of incorporation must be signed by the incorporators. The Model Act states simply that the incorporators sign the document, but acknowledgement, a procedure whereby the signatures of the incorporators must be notarized, is required in the New York statute and in several other states.[77] As previously stated, county recording of the articles of incorporation is a common formality. Some states require approval of the State Corporation Commission,[78] filing with a probate judge,[79] or publication of the articles of incorporation in a newspaper of general circulation in the county where the corporation has its registered office.[80] Finally, a state may require certain other documents to be filed with the articles of incorporation. For example, California requires the filing of an application for a permit to issue stock with the Commissioner of Corporations, and most states which require payment of a minimum amount of paid-in capital also require an affidavit of subscription or payment to accompany the articles of incorporation.

Careful analysis of the particular state statute under which the corporation is to be formed is absolutely necessary in order to insure strict compliance with its provisions.[81]

Payment of Capital

The Model Act formerly required that a certain amount of capital must be collected before a corporation may commence business, and more than half of the states have preserved the rule. In these states the payment of the preincorporation share subscriptions in the prescribed amount is a formality which must be satisfied before the corporation may commence business.

§ 6.07 CORPORATE EXISTENCE

Modern statutes have adopted simple incorporation procedures, the principal features of which are the preparation and filing of the articles of incorporation and the subsequent issuance of the certificate of incorporation. In about half the states and the Model Act corporate existence begins when the Secretary of State, after reviewing the articles of incorporation, issues a certificate of incorporation.[82] Several other jurisdictions, including Michigan, New York, Delaware, and California, provide that corporate existence begins when the articles of incorporation are *filed* with or endorsed by the appropriate state official.

The point at which the corporation is born is important to circumscribe shareholder and promoter liabilities for "corporate" obligations and to establish the beginning of corporate characteristics, such as taxation as a separate entity. When the certificate of incorporation is issued, or, in the appropriate case, when the articles are filed, the corporation is said to be a *de jure* corporation, or a corporation by law, and it acquires all power to act in accordance with the statute under which it is organized.

§ 6.08 FORMALITIES AFTER FORMATION OF A CORPORATION

Although the corporation is formed when the articles of incorporation have been filed or when a certificate of incorporation has been issued, there are several other matters which should precede commencement of the corporate business.

Organization Meetings

Organizational meetings of the incorporators and the initial directors are usually required as one of the first matters of corporate business.

Because organizational meetings are quite routine, counsel may draft the minutes in advance and use the pre-drafted minutes as an agenda for the meetings. The particular statute of each state should be consulted to determine which of the corporate groups (incorporators, directors or shareholders) are required to hold an organization meeting. Several states require only an organizational meeting of the incorporators.[83] Colorado, Pennsylvania and Texas require only an organizational meeting of the directors. In Delaware either the incorporators or the directors must have an organizational meeting, while in Illinois shareholders and directors must have an organizational meeting. Of course, there is certainly nothing wrong with holding an organizational meeting for a corporate group which is not required to meet by statute. Organizational meetings assist in establishing the air of formality which must be continually observed in corporate operations. The important point to be recognized here is the necessity to hold the statutory organizational meetings so as to be considered a properly-formed corporation. Even if corporate existence begins when the certificate of incorporation is issued or the articles are filed, a failure to observe the statutory formalities following these events may destroy the protection and special privileges of the corporation.

An organizational meeting of the incorporators may consider the acceptance of the certificate of incorporation or articles of incorporation and acknowledgment of the payment of taxes; election of initial directors (if they are not named in the articles of incorporation) and the resignation of any accommodation ("dummy") directors; authorization of the board of directors to issue shares; adoption of by-laws; transfers of any subscriptions from accommodation ("dummy") incorporators; and the transaction of any other business appropriate for incorporators to consider.

An organizational meeting of the board of directors will consider many of the same matters, and, if an organizational meeting of incorporators has been held, the board usually reviews and approves the business conducted there. In addition, the board of directors will decide other matters relating to issuance and transfer of shares, preincorporation agreements, banking arrangements, election of officers, qualification as a foreign corporation and tax plans.[84]

Corporate Supplies

The attorney's office usually orders the corporation's supplies for the newly-formed business. The corporation must maintain a minute book and a stock transfer ledger and it must have share certificates and a corporate seal. Corporation supply kits are available from many local printers and those who advertise in legal periodicals.

§ 6.09 BY-LAWS

By-laws complement the state statute and the articles of incorporation by prescribing rules to regulate the internal affairs of the corporation. Rules for the internal management which are intended to be flexible are best described in the by-laws, since they are most easily amended. On the other hand, rules which require permanence should be placed in the articles of incorporation. Interchangeability between the articles and by-laws is facilitated by the statutory rule that any provision which is required or permitted to be set forth in the by-laws may be also included in the articles of incorporation.[85] Of course, the converse is not true.

The authority to adopt by-laws is contained in the state statute, which may also suggest certain matters which should be contained in the by-laws.[86] Most states and the Model Act simply provide that the by-laws may contain any provision for the regulation and management for the affairs of the corporation which is not inconsistent with the statute or the articles of incorporation.[87] In these jurisdictions the by-laws may be as simple or as complicated as is desired. There are certain provisions which usually appear in the by-laws, such as the place of holding meetings of shareholders and the time of the annual meeting of shareholders; the number of directors, except the first board of directors; the notice to be given for directors meetings; the procedure for the election and appointment of officers; and a description of the officers' duties.

The by-laws should not be complicated with intricate procedures for corporate operation, because a complicated by-law provision may become a trap for the unwary, rather than a useful guide to corporate management. However, the by-laws should be as extensive and thorough as necessary to insure that the procedures for internal management of the corporation are fully described in writing for the officers and directors.

Initial By-Laws

The adoption of the initial by-laws is the responsibility of the incorporators, the shareholders or the board of directors depending upon the jurisdiction involved. In Delaware and New York the incorporators adopt the initial by-laws. The by-laws are then approved by the board of directors at their organizational meeting. In a few jurisdictions the shareholders adopt the initial by-laws.[88] Most jurisdictions and the Model Act provide for the adoption of the initial by-laws by the board of directors, but the articles of incorporation may reserve this power to the shareholders.[89]

The by-laws are prepared by counsel, with guidance from the incorporators and the initial directors, and they are presented for the approval of the appropriate intra-corporate group at the organizational meeting.

Content of By-Laws

Standard by-law provisions deal with the following matters:

1. Offices
   (a) Location of the principal office of the corporation;
   (b) Location of the registered office of the corporation;
   (c) Authority to change the address of the registered office by the board of directors.

EXAMPLE 6.09A: Offices

The principal office of the Corporation in the State of South Dakota shall be located in the City of Deadwood, County of Lawrence. The Corporation may have such other offices, either within or without the State of South Dakota, as the Board of Directors may designate or as the business of the Corporation may require from time to time.

The registered office of the Corporation required by The South Dakota Business Corporation Act to be maintained in the State of South Dakota may be, but need not be, identical with the principal office in the State of South Dakota, and the address of the registered office may be changed from time to time by the Board of Directors.[90]

2. Shareholders[91]
   (a) Time of the annual meeting;

EXAMPLE 6.09B: Annual Meeting

The annual meeting of the shareholders shall be held on the first Tuesday in the month of May in each year, beginning with the year 1981, at the hour of 9:00 o'clock A.M., for the purpose of electing directors and for the transaction of such other business as may come before the meeting. If the day fixed for the annual meeting shall be a legal holiday in the State of South Dakota, such meeting shall be held on the next succeeding business day. If the election of directors shall not be held on the day designated herein for any annual meeting of the shareholders, or at any adjournment thereof, the Board of Directors shall cause the election to be held at a special meeting of the shareholders as soon thereafter as conveniently may be.

(b) Procedure for calling special meetings of shareholders;

EXAMPLE 6.09C: Special Meetings

Special meetings of the shareholders, for any purpose or purposes, unless otherwise prescribed by statute, may be called by the President or by the Board of Directors, and shall be called by the President at the request of the holders of not less than one-tenth of all the outstanding shares of the corporation entitled to vote at the meeting.

(c) Place of the shareholder meetings;
(d) Authority for waiver of notice to be signed by shareholders entitled to vote at the meeting (This procedure permits a cure of defective notice or failure to give notice by obtaining written waivers from shareholders entitled to notice.);

EXAMPLE 6.09D: Place of Meeting

The Board of Directors may designate any place, either within or without the State of South Dakota, as the place of meeting for any annual meeting or for any special meeting called by the Board of Directors. A waiver of notice signed by all shareholders entitled to vote at a meeting may designate any place, either within or without the State of South Dakota, as the place for the holding of such meeting. If no designation is made, or if a special meeting be otherwise called, the place of meeting shall be the principal office of the Corporation in the State of South Dakota.

(e) Procedure for sending notice of meeting and time period for which notice is appropriate;
(f) Procedure for determining the shareholders entitled to notice or entitled to vote or entitled to receive dividends. (This procedure

states a particular time that the stock transfer books will be closed in order to determine the "holders of record.")

EXAMPLE 6.09E: Notice of Meeting

Written notice stating the place, day and hour of the meeting and, in case of a special meeting, the purpose or purposes for which the meeting is called, shall be delivered not less than ten or more than fifty days before the date of the meeting, either personally or by mail, by or at the direction of the President, or the Secretary, or the persons calling the meeting, to each shareholder or record entitled to vote at such meeting. If mailed, such notice shall be deemed to be delivered when deposited in the United States mail, addressed to the shareholder at his address as it appears on the stock transfer books of the corporation, with postage thereon prepaid.

EXAMPLE 6.09F: Closing of Transfer Books or Fixing of Record Date

For the purpose of determining shareholders entitled to notice of or to vote at any meeting of shareholders or any adjournment thereof, or shareholders entitled to receive payment of any dividend, or in order to make a determination of shareholders for any other proper purpose, the Board of Directors of the Corporation may provide that the stock transfer books shall be closed for a stated period but not to exceed, in any case, fifty days. If the stock transfer books shall be closed for the purpose of determining shareholders entitled to notice of or to vote at a meeting of shareholders, such books shall be closed for at least ten days immediately preceding such meeting. In lieu of closing the stock transfer books, the Board of Directors may fix in advance a date as the record date for any such determination of shareholders, such date in any case to be not more than fifty days and, in case of a meeting of shareholders, not less than ten days prior to the date on which the particular action, requiring such determination of shareholders, is to be taken. If the stock transfer books are not closed and no record date is fixed for the determination of shareholders entitled to notice of or to vote at a meeting of shareholders, or shareholders entitled to receive payment of a dividend, the date on which notice of the meeting is mailed or the date on which the resolution of the Board of Directors declaring such dividend is adopted, as the case may be, shall be the record date for such determination of shareholders. When a determination of shareholders entitled to vote at any meeting of shareholders has been made as provided in this section, such determination shall apply to any adjournment thereof except where the determination has been made through the closing of the stock transfer books and the stated period of closing has expired.

(g) Procedure for preparation of a voting list;
(h) Provision for examination of voting lists;

EXAMPLE 6.09G: Voting Lists

The officer or agent having charge of the stock transfer books for shares of the Corporation shall make a complete list of the shareholders entitled to vote at each meeting of shareholders or any adjournment thereof, arranged in alphabetical order, with the address of and the number of shares held by each. Such list shall be produced and kept open at the time and place of the meeting and shall be subject to the inspection of any shareholder during the whole time of the meeting for the purposes thereof.

(i) Number of shares required to constitute a quorum, and number of shares required to adjourn the meeting of shareholders;

EXAMPLE 6.09H: Quorum

A majority of the outstanding shares of the Corporation entitled to vote, represented in person or by proxy, shall constitute a quorum at a meeting of shareholders. If less than a majority of the outstanding shares are represented at a meeting, a majority of the shares so represented may adjourn the meeting from time to time without further notice. At such adjourned meeting at which a quorum shall be present or represented, any business may be transacted which might have been transacted at the meeting as originally notified. The shareholders present at a duly organized meeting may continue to transact business until adjournment, notwithstanding the withdrawal of enough shareholders to leave less than a quorum.

(j) Authorization for voting by proxy;

EXAMPLE 6.09I: Proxies

At all meetings of shareholders, a shareholder may vote in person or by proxy executed in writing by the shareholder or by his duly authorized attorney in fact. Such proxy shall be filed with the secretary of the Corporation before or at the time of the meeting. No proxy shall be valid after eleven months from the date of its execution, unless otherwise provided in the proxy.

(k) Voting entitlements of each class of stock;

EXAMPLE 6.09J: Voting of Shares.

Each outstanding share entitled to vote shall be entitled to one vote upon each matter submitted to a vote at a meeting of shareholders.

(l) Authorization to vote by representatives by holder of record (e.g., administrator, executor, agent of another corporation, etc.);

EXAMPLE 6.09K: Voting of Shares by Certain Holders.

Shares standing in the name of another corporation may be voted by such officer, agent or proxy as the bylaws of such corporation may prescribe, or, in the absence of such provision, as the board of directors of such corporation may determine.

Shares held by an administrator, executor, guardian or conservator may be voted by him, either in person or by proxy, without a transfer of such shares into his name. Shares standing in the name of a trustee may be voted by him, either in person or by proxy, but no trustee shall be entitled to vote shares held by him without a transfer of such shares into his name.

Shares standing in the name of a receiver may be voted by such receiver, and shares held by or under the control of a receiver may be voted by such receiver without the transfer thereof into his name if authority so to do be contained in an appropriate order of the court by which such receiver was appointed.

A shareholder whose shares are pledged shall be entitled to vote such shares until the shares have been transferred into the name of the pledgee, and thereafter the pledgee shall be entitled to vote the shares so transferred.

Neither shares of its own stock held by the Corporation, nor those held by another corporation of a majority of the shares entitled to vote for the election of directors of such other corporation are held by the Corporation, shall be voted at any meeting or counted in determining the total number of outstanding shares at any given time for purposes of any meeting.

(m) Informal action by the shareholders:

EXAMPLE 6.09L: Informal Action by Shareholders.

Any action required to be taken at a meeting of the shareholders, or any action which may be taken at a meeting of the shareholders, may be taken without a meeting if a consent in writing, setting forth the action so taken, shall be signed by all of the shareholders entitled to vote with respect to the subject matter thereof.

(n) Cumulative voting rights;

EXAMPLE 6.09M: Cumulative Voting.

At each election for directors every shareholder entitled to vote at such election shall have the right to vote, in person or by proxy, the number of shares owned by him for as many persons as there are directors to be elected and for whose election he has a right to vote, or to cumulate his votes by giving one candidate as many votes as the number of such directors multiplied by the number of his shares shall equal, or by distributing such votes on the same principles among any number of candidates.

3. Board of Directors
   (a) Authorization for the board of directors to manage the business;[92]

EXAMPLE 6.09N: General Powers.

The business and affairs of the Corporation shall be managed by its Board of Directors.

(b) The number, tenure and qualifications of directors;

EXAMPLE 6.09O: Number, Tenure and Qualifications.

The number of directors of the Corporation shall be nine. Each director shall hold office until the next annual meeting of shareholders and until his successor shall have been elected and qualified. Directors need not be residents of the State of South Dakota or shareholders of the Corporation.

(c) Classification of directors (if desired);

EXAMPLE 6.09P: Classification of Directors.

At the first annual meeting of the shareholders, the members of the Board of Directors shall be dividied into three classes of three members each. The members of the first class shall hold office for a term of one year; the members of the second class shall hold office for a term of two years; the members of the third class shall hold office for a term of three years. At all annual elections thereafter three directors shall be elected by the shareholders for a term of three years to succeed the three directors whose term then expires; provided that nothing herein shall be construed to prevent the election of a director to succeed himself.[93]

(d) Time and place for regular meetings; [94]

EXAMPLE 6.09Q: Regular Meetings.

A regular meeting of the Board of Directors shall be held without other notice than this By-law immediately after, and at the same place as, the annual meeting of shareholders. The Board of Directors may provide, by resolution, the time and place, either within or without the State of South Dakota, for the holding of additional regular meetings without other notice than such resolution.

(e) Procedure for calling special meetings;

EXAMPLE 6.09R: Special Meetings.

Special meetings of the Board of Directors may be called by or at the request of the President or any two directors. The person or persons authorized to call special meetings of the Board of Directors may fix any place, either within or without the State of South Dakota, as the place for holding any special meeting of the Board of Directors called by them.

(f) Procedure for giving notice of special meetings;
(g) Authorization to waive notice of any meeting;

EXAMPLE 6.09S: Notice.

Notice of any special meeting shall be given at least two days previously thereto by written notice delivered personally or mailed to each director at his business address, or by telegram. If mailed, such notice shall be deemed to be delivered when deposited in the United States mail so addressed, with postage thereon prepaid. If notice be given by telegram, such notice shall be deemed to be delivered when the telegram is delivered to the telegraph company. Any director may waive notice of any meeting. The attendance of a director at a meeting shall constitute a waiver of notice of such meeting, except where a director attends a meeting for the express purpose of objecting to the transaction of any business because the meeting is not lawfully called or convened. Neither the business to be transacted at, nor the purpose of, any regular or special meeting of the Board of Directors need be specified in the notice or waiver of notice of such meeting.

(h) The number of directors for a quorum and to adjourn the meeting;

EXAMPLE 6.09T: Quorum.

A majority of the number of directors fixed by [Example 6.09O] shall constitute a quorum for the transaction of business at any meeting of the Board of Directors, but if less than such majority is present at a meeting, a majority of the directors present may adjourn the meeting from time to time without further notice.

(i) The number of directors required to approve a certain matter;

EXAMPLE 6.09U: Manner of Acting.

The act of the majority of the directors present at a meeting at which a quorum is present shall be the act of the Board of Directors.

EXAMPLE 6.90V: Manner of Acting.

No resolution of the Corporation at any meeting whether regular or special, shall be adopted except by the unanimous vote of the directors duly elected as provided herein.[95]

(j) Informal action by the board of directors;

EXAMPLE 6.09W: Action without a Meeting.

Any action that may be taken by the Board of Directors at a meeting may be taken without a meeting if a consent in writing, setting forth the action so to be taken, shall be signed before such action by all of the directors.

(k) Procedure for removing directors and filling vacancies; [96]

EXAMPLE 6.09X: Vacancies.

Any vacancy occurring in the Board of Directors may be filled by the affirmative vote of a majority of the remaining directors though less than a quorum of the Board of Directors. A director elected to fill a vacancy shall be elected for the unexpired term of his predecessor in office. Any directorship to be filled by reason of an increase in the number of directors may be filled by election by the Board of Directors for a term of office continuing only until the next election of directors by the shareholders.

EXAMPLE 6.09Y: Removal.

The stockholders of the Corporation may, at any meeting called for the purpose, remove any director from office, with or without cause, by a vote of a majority of the outstanding shares of the class of stock which elected the director or directors to be removed; provided, however, that no director shall be removed in case the votes of a sufficient number of shares are cast against his removal, which if cumulatively voted at an election of the entire board of directors would be sufficient to elect him.[97]

(l) Compensation, and payment of expenses;

EXAMPLE 6.09Z: Compensation.

By resolution of the Board of Directors, each director may be paid his expenses, if any, of attendance at each meeting of the Board of Directors, and may be paid a stated salary as director or a fixed sum for attendance at each meeting of the Board of Directors or both. No such payment shall preclude any director from serving the Corporation in any other capacity and receiving compensation therefor.

(m) Presumption of assent when the director is present at a meeting;

EXAMPLE 6.09AA: Presumption of Assent.

A director of the Corporation who is present at a meeting of the Board of Directors at which action on any corporate matter is taken shall be presumed to have assented to the action taken unless his dissent shall be entered in the minutes of the meeting or unless he shall file his written dissent to such action with the person acting as the secretary of the meeting before the adjournment thereof or shall forward such dissent by registered mail to the Secretary of the Corporation immediately after the adjournment of the meeting. Such right to dissent shall not apply to a director who voted in favor of such action.

4. Executive Committees [98]
   (a) Authority for appointment of executive committees and delegation of authority;

EXAMPLE 6.09BB: Appointment.

The Board of Directors, by resolution adopted by a majority of the full board, may designate two or more of its members to constitute an Executive Committee. The designation of such committee and the delegation thereto of authority shall not operate to relieve the Board of Directors, or any member thereof, of any responsibility imposed by law.

EXAMPLE 6.09CC: Authority.

The Executive Committee, when the Board of Directors is not in session, shall have and may exercise all of the authority of the Board of Directors except to the extent, if any, that such authority shall be limited by the resolution appointing the Executive Committee and except also that the Executive Committee shall not have the authority of the Board of Directors in reference to amending the Articles of Incorporation, adopting a plan of merger or consolidation, recommending to the shareholders the sale, lease or other disposition of all or substantially all of the property and assets of the Corporation otherwise than in the usual and regular course of its business, recommending to the shareholders a voluntary dissolution of the Corporation or a revocation thereof, or amending the By-laws of the Corporation.

   (b) Tenure and qualifications of members of the executive committee;

EXAMPLE 6.09DD: Tenure and Qualifications.

Each member of the Executive Committee shall hold office until the next regular annual meeting of the Board of Directors following his designation and until his successor is designated as a member of the Executive Committee and is elected and qualified.

   (c) Time and place for regular meetings of the executive committee;
   (d) Procedure for calling special meetings of the executive committee;
   (e) Procedure for giving notice of a meeting to the executive committee;

EXAMPLE 6.09EE: Meetings.

Regular meetings of the Executive Committee may be held without notice at such times and places as the Executive Committee may fix from time to time by resolution. Special meetings of the Executive Committee may be called by any member thereof upon not less than one day's notice stating the place, date and hour of the meeting, which notice may be written or oral, and if mailed, shall be deemed to be delivered when deposited in the United States mail addressed to the member of the Executive Committee at his business address. Any member of the Executive Committee may waive notice of any meeting and no notice of any meeting need be given to any member thereof who attends in person. The notice of a meeting of the Executive Committee need not state the business proposed to be transacted at the meeting.

(f) Number of the members of the committee necessary to constitute a quorum, and the vote required of the committee to authorize certain acts;

EXAMPLE 6.09FF: Quorum.

A majority of the members of the Executive Committee shall constitute a quorum for the transaction of business at any meeting thereof and action of the Executive Committee must be authorized by the affirmative vote of a majority of the members present at a meeting at which a quorum is present.

(g) Informal action by the executive committee;

EXAMPLE 6.09GG: Action without a Meeting.

Any action that may be taken by the Executive Committee at a meeting may be taken without a meeting if a consent in writing, setting forth the action so to be taken, shall be signed before such action by all of the members of the Executive Committee.

(h) Procedure for filling vacancies, accepting resignations, and removal of members of the executive committee;

EXAMPLE 6.09HH: Vacancies.

Any vacancy in the Executive Committee may be filled by a resolution adopted by a majority of the full Board of Directors.

EXAMPLE 6.09II: Resignation and Removal.

Any member of the Executive Committee may be removed at any time with or without cause by resolution adopted by a majority of the full Board of Directors. Any member of the Executive Committee may resign from the Executive Committee at any time by giving written notice to the President or Secretary of the corporation, and unless otherwise specified therein, the acceptance of such resignation shall not be necesary to make it effective.

(i) Procedure for conducting executive committee meetings;

EXAMPLE 6.09JJ: Procedure.

The Executive Committee shall elect a presiding officer from its members and may fix its own rules of procedure which shall not be inconsistent with these By-laws. It shall keep regular minutes of its proceedings and report the same to the Board of Directors for its information at the meeting thereof held next after the proceedings shall have been taken.

5. Officers [99]

(a) Number of officers;

EXAMPLE 6.09KK: Number.

The officers of the Corporation shall be a President, one or more Vice Presidents (the number thereof to be determined by the Board of Directors), a Secretary, and a Treasurer, each of whom shall be elected by the Board of Directors. Such other officers and assistant officers as may be deemed necessary may be elected or appointed by the Board of Directors. Any two or more offices may be held by the same person, except the offices of President and Secretary.

(b) Procedure for election and term of office;

EXAMPLE 6.09LL: Election and Term of Office.

The officers of the Corporation to be elected by the Board of Directors shall be elected annually by the Board of Directors at the first meeting of the Board of Directors held after each annual meeting of the shareholders. If the election of officers shall not be held at such meeting, such election shall be held as soon thereafter as conveniently may be. Each officer shall hold office until his successor shall have been duly elected and shall have qualified or until his death or until he shall resign or shall have been removed in the manner hereinafter provided.

(c) Removal and filling vacancies;

EXAMPLE 6.09MM: Removal.

Any officer or agent may be removed by the Board of Directors whenever in its judgment the best interests of the Corporation will be served thereby, but such removal shall be without prejudice to the contract rights, if any, of the person so removed. Election or appointment of an officer or agent shall not of itself create contract rights.

EXAMPLE 6.09NN: Vacancies.

A vacancy in any office because of death, resignation, removal, disqualification or otherwise, may be filled by the Board of Directors for the unexpired portion of the term.

(d) Responsibilities of the officers;

EXAMPLE 6.09OO: Officers

President. The President shall be the principal executive officer of the Corporation and, subject to the control of the Board of Directors, shall in general supervise and control all of the business and affairs of the Corporation. He shall, when present, preside at all meetings of the shareholders and of the Board of Directors. He may sign, with the Secretary or any other proper officer of the corporation thereunto authorized by the Board of Directors, certificates for shares of the corporation, any deeds, mortgages, bonds, contracts, or other instruments which the Board of Directors has authorized to be executed, except in cases where the signing and execution thereof shall be expressly delegated by the Board of Directors or by these By–laws to some other officer or agent of the corporation, or shall be required by law to be otherwise signed or executed; and in general shall perform all duties incident to the office of President and such other duties as may be prescribed by the Board of Directors from time to time.

The Vice Presidents. In the absence of the President or in the event of his death, inability or refusal to act, the Vice President (or in the event there be more than one Vice President, the Vice Presidents in the order designated at the time of their election, or in the absence of any designation, then in the order of their election) shall perform the duties of the President, and when so acting, shall have all the powers of and be subject to all the restrictions upon the President. Any Vice President may sign, with the Secretary or an Assistant Secretary, certificates for shares of the Corporation; and shall perform such other duties as from time to time may be assigned to him by the President or by the Board of Directors.

The Secretary. The Secretary shall: (a) keep the minutes of the proceedings of the shareholders and of the Board of Directors in one or more books provided for that purpose; (b) see that all notices are duly given in accordance with the provisions of these By-laws or as required by law; (c) be custodian of the corporate records and of the seal of the Corporation and see that the seal of the Corporation is affixed to all documents the execution of which on behalf of the Corporation under its seal is duly authorized; (d) keep a register of the postoffice address of each shareholder which shall be furnished to the Secretary by such shareholder; (e) sign with the President, or a Vice President, certificates for shares of the Corporation, the issuance of which shall have been authorized by resolution of the Board of Directors; (f) have general charge of the stock transfer books of the Corporation; and (g) in general perform all duties incident to the office of Secretary and such other duties as from time to time may be assigned to him by the President or by the Board of Directors.

The Treasurer. The Treasurer shall: (a) have charge and custody of and be responsible for all funds and securities of the Corporation; (b) receive and give receipts for moneys due and payable to the Corporation from any source whatsoever, and deposit all such moneys in the name of the Corporation in such banks, trust companies or other depositaries as shall be selected in accordance with the provisions of these By-laws; and (c) in general perform all of the duties incident to the office of Treasurer and such other duties as from time to time may be assigned to him by the President or by the Board of Directors. If required by the Board of Directors, the Treasurer shall give a bond for the faithful discharge of his duties in such sum and with such surety or sureties as the Board of Directors shall determine.

Assistant Secretaries and Assistant Treasurers. The Assistant Secretaries, when authorized by the Board of Directors, may sign with the President or a Vice President certificates for shares of the Corporation the issuance of which shall have been authorized by a resolution of the Board of Directors. The Assistant Treasurers shall respectively, if required by the Board of Directors, give bonds for the faithful discharge of their duties in such sums and with such sureties as the Board of Directors shall determine. The Assistant Secretaries and Assistant Treasurers, in general, shall perform such duties as shall be assigned to them by the Secretary or the Treasurer, respectively, or by the President or the Board of Directors.

(e) Salaries;

EXAMPLE 6.09PP: Salaries.

The salaries of the officers shall be fixed from time to time by the Board of Directors and no officer shall be prevented from receiving such salary by reason of the fact that he is also a director of the Corporation.

[or]

EXAMPLE 6.09QQ: Salaries.

No salary or other compensation for services shall be paid to any director or officer of the corporation unless and until the same shall have been approved in writing or at a duly held stockholders' meeting by stockholders owning at least seventy-five per cent in amount of the capital stock of the corporation then outstanding.[100]

6. Authorization for Executing Contracts and Other Written Matters on Behalf of the Corporation—These provisions permit the board of directors to authorize any officer to contract on behalf of the corporation and they may further restrict the ability of management to contract loans or other indebtedness. It is also common to specify here which persons must sign checks, drafts and other evidences of indebtedness issued in the name of the corporation, and where the funds of the corporation will be deposited.

EXAMPLE 6.09RR: Authorization.

Contracts. The Board of Directors may authorize any officer or officers, agent or agents, to enter into any contract or execute and deliver any instrument in the name of and on behalf of the Corporation, and such authority may be general or confined to specific instances.

Loans. No loans shall be contracted on behalf of the Corporation and no evidences of indebtedness shall be issued in its name unless authorized by a resolution of the Board of Directors. Such authority may be general or confined to specific instances.

Checks, Drafts, etc. All checks, drafts or other orders for the payment of money, notes or other evidences of indebtedness issued in the name of the corporation, shall be signed by such officer or officers, agent or agents of the Corporation and in such manner as shall from time to time be determined by resolution of the Board of Directors.

Deposits. All funds of the Corporation not otherwise employed shall be deposited from time to time to the credit of the Corporation in such banks, trust companies or other depositaries as the Board of Directors may select.

7. Matters Involving the Certificates of Shares and Their Transfer[101]—These provisions usually permit the board of directors to determine the form of the share certificates and prescribe which of the corporate officers will be required to sign them. Section 23 of the Model Act requires that if certificates are used by the corporation, the certificates representing shares shall set forth on its face that the corporation is organized under the laws of the particular state; the name of the person to whom issued; the number and class of shares which the certificates represents; and the par value of the shares or a statement that the shares are without par value. In addition, if the corporation is authorized to issue different

classes of shares, the certificates should specify the designations, preferences, limitations and relative rights of the shares of each class. These statutory requirements need not be restated in the by-laws. If the incorporators wish to restrict the board of directors use of "uncertificated" shares, a prohibitive provisions to that effect should be stated in the by-laws. Any other special provisions respecting the transfer of shares and the method of keeping the stock transfer ledger should also be included under this by-law section.

EXAMPLE 6.09SS: Certificates for Shares.

Certificates representing shares of the Corporation shall be in such form as shall be determined by the Board of Directors. Such certificates shall be signed by the President or a Vice President and by the Secretary or an Assistant Secretary and sealed with the corporate seal or a facsimile thereof. The signatures of such officers upon a certificate may be facsimiles if the certificate is countersigned by a transfer agent, or registered by a registrar, other than the Corporation itself or one of its employees. All certificates for shares shall be consecutively numbered or otherwise identified. The name and address of the person to whom the shares represented thereby are issued, with the number of shares and date of issue, shall be entered on the stock transfer books of the Corporation. All certificates surrendered to the Corporation for transfer shall be cancelled and no new certificate shall be issued until the former certificate for a like number of shares shall have been surrendered and cancelled, except that in case of a lost, destroyed or mutilated certificate a new one may be issued therefor upon such terms and indemnity to the Corporation as the Board of Directors may prescribe.

The board of directors shall not be permitted to issue "uncertificated" shares without the express approval of at least two-thirds of the then outstanding stock entitled to vote.

EXAMPLE 6.09TT: Transfer of Shares.

Transfer of shares of the Corporation shall be made only on the stock transfer books of the Corporation by the holder of record thereof or by his legal representative, who shall furnish proper evidence of authority to transfer, or by his attorney thereunto authorized by power of attorney duly executed and filed with the Secretary of the Corporation, and on surrender for cancellation of the certificate for such shares. The person in whose name shares stand on the books of the corporation shall be deemed by the Corporation to be the owner thereof for all purposes.

EXAMPLE 6.09UU: Transfer Agent.

The Secretary of the Corporation shall act as Transfer Agent of the certificates representing the shares of common stock and preferred stock of the Corporation. He shall maintain a Stock Transfer Book, the stubs in which shall set forth, among other things, the names and addresses of the holders of all issued shares of the Corporation, the number of shares held by each, the certificate numbers representing such shares, the date of issue of the certificates representing such shares, and whether or not such shares originate from original issue or from transfer. The names and addresses of the stockholders as they appear on the stubs of the Stock Transfer Book shall be conclusive evidence as to who are the stockholders of record and as such entitled to receive notice of the meetings of stockholders; to vote at such meetings; to examine the list of the stockholders entitled to vote at meetings; to receive dividends; and to own, enjoy and exercise any other property or rights deriving from such shares against the Corporation. Each stockholder shall be responsible for notifying the Secretary in writing any change in his name or address and failure so to do will relieve the Corporation, its directors, officers and agents, from liability for failure to direct notices or other documents, or pay over or transfer dividends or other property or rights, to a name or address other than the name and address appearing on the stub of the Stock Transfer Book.[102]

8. The fiscal year of the corporation;

EXAMPLE 6.09VV: Fiscal Year.

The fiscal year of the Corporation shall begin on the first day of January and end on the thirty-first day of December in each year.

9. Authority of the board of directors to declare and pay dividends on the outstanding shares of the corporation;[103]

EXAMPLE 6.09WW: Dividends.

The Board of Directors may from time to time declare, and the Corporation may pay, dividends on its outstanding shares in the manner and upon the terms and conditions provided by law and its Articles of Incorporation.

10. Description of the corporate seal;

EXAMPLE 6.09XX: Seal.

The Board of Directors shall provide a corporate seal which shall be circular in form and shall have inscribed thereon the name of the Corporation and the state of incorporation and the words, "Corporate Seal."

11. Provisions for adopting emergency by-laws and the term for which those by-laws will be in effect;

EXAMPLE 6.09YY: Emergency By-laws.

The Emergency By-laws provided in this Article shall be operative during any emergency in the conduct of the business of the Corporation resulting from an attack on the United States or any nuclear or atomic disaster, notwithstanding any different provision in the preceding Articles of the By-laws or in the Articles of Incorporation of the Corporation or in the Business Corporation Act. To the extent not inconsistent with the provisions of this Article, the By-laws provided in the preceding Articles shall remain in effect during such emergency and upon its termination the Emergency By-laws shall cease to be operative.

During any such emergency:

(a) A meeting of the Board of Directors may be called by any officer or Director of the Corporation. Notice of the time and place of the meeting shall be given by the person calling the meeting to such of the Directors as it may be feasible to reach by any available means of communication. Such notice shall be given at such time in advance of the meeting as circumstances permit in the judgment of the person calling the meeting.

(b) At any such meeting of the Board of Directors, a quorum shall consist of ______[*here insert the particular provision desired*].

(c) The Board of Directors, either before or during any such emergency, may provide, and from time to time modify, lines of succession in the event that during such an emergency any or all officers or agents of the Corporation shall for any reason be rendered incapable of discharging their duties.

(d) The Board of Directors, either before or during any such emergency, may, effective in the emergency, change the head office or designate several alternative head offices or regional offices, or authorize the officers so to do.

No officer, Director or employee acting in accordance with these Emergency By-laws shall be liable except for willful misconduct.

These Emergency By-laws shall be subject to repeal or change by further action of the Board of Directors or by action of the shareholders, but no such repeal or change shall modify the provisions of the next preceding paragraph with regard to action taken prior to the time of such repeal or change. Any amendment of these Emergency By-laws may make any further or different provision that may be practical and necessary for the circumstances of the emergency.

12. Provisions for amending, altering, or repealing the by-laws or adopting new by-laws;

EXAMPLE 6.09ZZ: Amendment

These By-laws may be altered, amended or repealed and new By-laws may be adopted by the Board of Directors at any regular or special meeting of the Board of Directors.

Sample by-laws for a Delaware corporation appear as Form 6Q in the Appendix.

NOTES

1. The advice must consider the particular needs of the business, including ownership rights, management responsibilities, duration, need for capital, potential liability and taxation. If a corporation is selected as the appropriate business form, certain special information must be obtained from the organizers. A Preincorporation Checklist appears as Form 6A in the Appendix.
2. An example of a preincorporation agreement between promoters appears as Form 6B in the Appendix.
3. E.g., Pennsylvania, 15 Pa.Stat. §§ 1201, 1204(8).
4. About 15 states require $1,000 minimum paid-in capital (e.g., Illinois, Ill.Rev.Stat. ch. 32, § 157.47–14); a few states require $500 (e.g., Ohio, Ohio Rev.Code § 1701.04); and Arkansas requires $300 (Ark.Stat.Ann. § 64–502). A certificate of payment of capital stock is Form 6C in the Appendix.
5. West's Modern Legal Forms § 2452.
6. West's Modern Legal Forms § 2455.
7. E.g., New York, 3 months, McKinney Consol.Laws of N.Y.Bus.Corp. Law § 503; and Louisiana, one year, La.Rev.Stat.Ann. § 12:71.
8. E.g., New Jersey, six-months or 60 days after filing certificate of incorporation, N.J.Stat.Ann. § 14A:7–3.
9. See M.B.C.A. § 17.

10. 15 Pa.Stat. § 1207.
11. Oklahoma, Okla.Stat.Ann. tit. 18 § 1.31.
12. West's Modern Legal Forms § 2468.
13. M.B.C.A. § 17.
14. West's Modern Legal Forms § 2471.
15. See Chapter 12.
16. For example, it may be particularly important for management to be able to declare dividends out of current profits even though the corporation has no "earned surplus". This would be permitted in Delaware, 8 Del. Code Ann. § 170, but would not be permitted in Colorado. Colo.Rev.Stat. 1973 § 7-5-110.
17. M.B.C.A. § 8(a).
18. See 1 Prentice Hall, Corporation Reporter, Corporation Checklists ¶ 9002, subparagraph 14 under each state.
19. See, e.g., M.B.C.A. § 8(c). Some jurisdictions, and the Model Act, allow the use of a similar corporate name, provided the written consent of the holder of the name is obtained, and a distinguishing word is added to the name.
20. Many states use a 30 day period for reservation of corporate names, e. g., Maryland, Md.Ann.Code art. 23 § 6; and Massachusetts, Mass.Gen. Laws Ann. c. 156B, § 11(c). A twelve month period is rare, and is used in Minnesota (Minn.Stat.Ann. §§ 301–05(3), (4)).
21. M.B.C.A. § 9.
22. An application for Reserved Name, a Certificate of Reserved Name, a Notice of Transfer of Reserved Name and a Certificate of Transfer of Reserved Name are Forms 6E, 6F, 6G, and 6H, respectively, in the Appendix.
23. M.B.C.A. §§ 10, 11. Examples of an application for Registration, a Certificate of Registration, an Application for Renewal and a Certificate of Renewal appear as Forms 6I, 6J, 6K and 6L, respectively, in the Appendix.
24. See Chapter 12 on Qualification of Foreign Corporations.
25. E.g., Colorado, Colo.Rev.Stat. § 7-71-101.
26. A statement of assumed name is Form 6M in the Appendix.
27. The Model Business Corporation Act and most jurisdictions use the term "articles of incorporation." M.B.C.A. §§ 2(c), 54.
28. See Section 13.01.
29. M.B.C.A. § 54.
30. See Section 4.02.
31. Examples of articles of incorporation appear as Forms 6O and 6P in the Appendix.
32. See Section 6.04.
33. See Sections 13.05 and 13.06.
34. M.B.C.A. §§ 3, 54(c).
35. West's Modern Legal Forms § 2562.
36. West's Modern Legal Forms § 2613. Many other examples of corporate purpose clauses may be found in West's Modern Legal Forms §§ 2522–2680.
37. The details and flexibility of the corporate financial structure are discussed more fully in Chapter 7.
38. This summary paraphrases the requirements of M.B.C.A. § 54(d), (e) and (f), except that the Model Act does not require any statement concerning par value of shares. Most states still require this designation for shares.
39. West's Modern Legal Forms § 2752.
40. West's Modern Legal Forms § 2755.1.
41. See Chapter 7.
42. See Sections 7.05 and 7.09.
43. M.B.C.A. § 18 permits shares to be issued at a price set by the board of directors, which would be entirely in their discretion if par value were not required in the articles of incorporation. See M.B.C.A. § 54.
44. E.g. Alabama and Iowa are states which place a value of $100 on each no-par share for computation of the initial taxes. In these states, a corporation could authorize one hundred times as many $1 par value shares as no par shares for the same tax.
45. See Section 9.02.
46. See Sections 7.11 and 11.03.

47. E.g., New York, McKinney Consol. Laws of N.Y.Bus.Corp.Act § 622; and Illinois, Ill.Rev.Stat. ch. 32 §§ 157.24, 157.47-11.
48. E.g., California, West's Ann.Calif.Corp.Code § 204(a)(2); Delaware, 8 Del.Code Ann. § 102(b)(3).
49. West's Modern Legal Forms § 2735.
50. West's Modern Legal Forms § 2736.
51. Treasury shares are defined in most corporation statutes. On preemptive rights and treasury shares see clause 9 of the Delaware articles of incorporation, Form 6O in the Appendix.
52. See Sections 4.03 and 8.08.
53. West's Modern Legal Forms § 2712.
54. See M.B.C.A. § 94(d), (e), and Section 13.06.
55. M.B.C.A. §§ 36, 54(i).
56. West's Modern Legal Forms § 2711.
57. See Section 4.03.
58. See Section 4.03.
59. M.B.C.A. § 54.
60. See Section 13.01.
61. E.g., California, West's Ann.Calif.Corp.Code § 317.
62. West's Modern Legal Forms § 2723.1.
63. M.B.C.A. § 6.
64. West's Modern Legal Forms § 2803.
65. M.B.C.A. § 44.
66. M.B.C.A. § 27.
67. M.B.C.A. § 45, and see Sections 9.02 and 9.03.
68. See, e.g., Delaware, 8 Del.Code Ann. § 170.
69. See, e.g., Delaware, 8 Del.Code Ann. § 170(b).
70. West's Modern Legal Forms § 2719.1.
71. See Section 4.03.
72. West's Modern Legal Forms § 2802.
73. M.B.C.A. § 35.
74. M.B.C.A. § 35.
75. M.B.C.A. § 55.
76. E.g., Delaware, 8 Del.Code Ann. § 103.
77. McKinney Consol.Laws of N.Y.Bus.Corp. Law § 402(a).
78. E.g., Arizona, Ariz.Rev.Stat. § 10-055.
79. Alabama, 10 Ala.Code § 21(6).
80. E.g., Minnesota, Minn.Stat.Ann. § 301.06; Pennsylvania, 15 Pa.Stat. § 1205.
81. The myriad variations of filing requirements may be easily reviewed by consulting 1 Prentice Hall, Corporation Reporter, Corporation Checklists. ¶ 9002 *et seq.*
82. See M.B.C.A. § 56.
83. E.g., New York, McKinney Consol. Law of N.Y.Bus.Corp. Law § 404.
84. Checklists and a full discussion of organizational meetings are contained in Chapter 8.
85. See M.B.C.A. § 54(h).
86. Some state statutes prescribe certain specific matters which must be contained in the by-laws. E.g., California, West's Ann.Calif.Corp. Code § 1501; and New Hampshire, N.H.Rev.Stat.Ann. § 294:4.
87. See M.B.C.A. § 27.
88. E.g., Nebraska, Neb.Rev.Stat. § 21–2026.
89. See M.B.C.A. § 27.
90. This example and subsequent by-law clauses, except where provided otherwise, are copied or adopted from Form 51, Official Forms for Use under the Model Business Corporation Act, also reproduced in West's Modern Legal Forms § 2792.
91. See Shareholder Meetings, Section 8.07.
92. See Section 4.03.
93. West's Modern Legal Forms § 2802. The Model Act requires that a classification provision appear in the articles of incorporation and a by-law provision would be ineffective. See M.B.C.A. § 37 and Section 6.06. However, several states permit classification of the board

of directors to be accomplished in the by-laws. E.g., New York, McKinney Consol.Laws of N.Y.Bus.Corp.Law § 704; Pennsylvania, 15 Pa.Stat. § 1403.

94. See Directors Meetings, Section 8.06.
95. West's Modern Legal Forms § 2803.
96. See Section 4.06.
97. West's Modern Legal Forms § 2720.
98. See M.B.C.A. § 42.
99. See Section 4.03.
100. West's Modern Legal Forms § 2718.
101. See Sections 7.06 and 11.03. M.B.C.A. § 23 has been amended recently to permit "uncertificated" shares (by which share ownership is reflected by an entry in the corporate books and not by a negotiable certificate.)
102. West's Modern Legal Forms § 2806.01.
103. See Sections 9.03 and 9.04.

7

CORPORATE FINANCIAL STRUCTURE

§ 7.01 GENERALLY

Corporate capital is obtained principally from investors, creditors and shareholders, who exchange money, property, or services for "securities" issued by the corporation. The attractiveness of shares as an investment is an important advantage to the corporate form of business enterprise. In addition to corporate "equity" securities, or shares, a corporation may contract for many varied types of "debt" financing, transactions whereby the corporation borrows money from outsiders who are willing to lend funds to the corporation. Unlike shareholders, these creditors are not "owners" of the company, but their loans are generally considered to be a more conservative investment. The corporation is obligated to repay a loan from the "debt" investor but there is no obligation to repay the funds invested by the shareholders, who risk the loss of some or all of their funds. The corporate financial structure has great flexibility and corporate securities may have any number of various features which increase the quality and the attractiveness of the investment. The capitalization may be limited only to common stock, or may be some combination of equity securities, including separate classes and series, and debt securities.

§ 7.02 TYPES OF CORPORATE SECURITIES

The term "securities" has a special meaning in the law. It generally refers to a contractual-proprietary obligation which exists between a business enterprise and an investor. For purposes of the federal and state securities acts, the term may include any one of several different forms of investment obligations. In the corporate sense, securities fall into two classes:

1. securities which evidence a corporate obligation to repay money borrowed from a creditor, typically called "debt" securities or "bonds;" and
2. "equity" securities, which evidence a shareholder's ownership interest in the corporation and are usually referred to as "shares."

When a corporation borrows money, it executes a document or bond which represents the obligation of the corporation to repay the borrowed funds. Bonds may be unsecured or secured for payment with property of the corporation. An unsecured obligation, the corporate equivalent of a personal signature loan, is called a debenture. Secured bonds may be called "mortgage bonds."

Bonds always state a principal amount owed by the corporation, a date when repayment of the principal amount is due, and a provision for interest, which is usually paid periodically. Debt securities may be marketed at a higher price than the principal amount if the attractiveness of the investment creates a demand, in which case it is said that they are sold at a "premium," or if the investment is not all that attractive, at a lower price than the principal amount, in which case they are sold at a "discount." Debt securities usually do not have voting rights in the corporate affairs. Instead, they represent a loan obligation in the strict business sense and the holders are merely creditors. As a practical matter, debt securities are often issued under an agreement executed by the corporation, the outside lender, and a trustee who is usually a financial institution. The agreement is called an "indenture," and the indenture includes the terms of the obligation, the rights of the security holders and the trustee, and any conditions upon which the bonds are issued. Debt securities issued by larger corporations are freely sold on an open market, and the price of the bond depends upon the quality of the investment. Bonds frequently have several advantageous features which make them a very desirable investment, such as a high interest rate, a provision allowing conversion into common shares at a specified price, and redemption features.

Equity securities are distinguished from debt securities by the relationship between the investor and the corporation. A purchaser of equity securities, a shareholder, becomes a part owner of the corporation. The proportion that his shares bear to the total number of shares outstanding represents his fractional ownership interest. When shares are issued, the corporation, instead of creating a liability, creates a capital account which represents the equity of the corporation. Unlike a debt security, the corporation is under no obligation to repay a shareholder, and the return of the investment is usually strictly dependent upon the shareholder's ability to sell his shares to another investor. The income paid on equity securities is usually a distribution of profit of the corporation and is called a dividend. The frequency and amount of these distributions are determined within the discretion of the board of directors. An equity investment is attractive if it appears that the corporation's business will expand and be profitable, thereby increasing value of the equity security and likely resulting in dividend distributions to the shareholders. Shareholders usually have the right to vote, and they also have the right to a proportionate distribution of corporate assets upon dissolution of the corporation.

Imaginative enterpreneurs have developed all sorts of variations on these two basic types of corporate securities, and most state corporation statutes encourage inspired financial configurations by imposing very few restrictions upon the corporate financial structure.

However, since some unscrupulous enterpreneurs have duped investors with worthless securities, it should be noted at the outset that the issuance and sale of corporate securities is strictly regulated by federal and state securities laws. Any public sale of corporate shares or bonds is subject to the disclosure requirements of the Securities Act of 1933[1] or the applicable state "Blue Sky" laws. These acts and the Securities Act of 1934[2] are generally designed to fully inform a potential investor of the character and quality of the investment, and to avoid untrue statements of fact and misleading omissions about the security which may affect a decision to purchase or sell. The detailed requirements of the securities acts are beyond the scope of this work, but counsel should explore their provisions and should examine helpful research sources[3] before advising the corporate client to sell any securities.

§ 7.03 EQUITY SECURITIES

Equity securities grant the shareholder a three-pronged ownership interest in the corporation. He is entitled to a proportionate share of earnings, distributed as dividends in the discretion of the board of directors; a proportionate share of assets in corporate dissolution; and a vote on all shareholder matters, which gives indirect control over management activities. Ignoring special classes of equity securities for a moment, a common stock shareholder is entitled to share in the earnings and assets of the corporation in the proportion that the number of shares he owns bears to the total number of shares outstanding. He is also entitled to one vote for each share. Every corporate statute authorizes the issuance of a certain number of shares, and the division of those shares into classes, allowing preferences, limitations, and other special rights as specified in the articles of incorporation. The Model Act grants this general authority in Section 15.

§ 7.04 STAGES OF EQUITY SECURITIES

The articles of incorporation must state the number of shares that the corporation will have the authority to issue.[4] This number will be determined by the incorporators and counsel, considering the anticipated capital requirements of the corporation. Having established the authority, the corporation may issue up to the specified number of shares without any requirement for an amendment to the articles of incorporation. Thus, the first step in the issuance of corporate equity securities is the creation of the authority to issue them by describing the characteristics of the securities and specifying the number of shares in the articles of incorporation. The shares described in the articles of incorporation are the "authorized" shares of the corporation.

It is not necessary to issue all of the authorized shares of the corporation, and it is sometimes undesirable to do so. Of course, it will be necessary to issue the number of shares required for the minimum paid-in capital, if that is a requirement under the applicable state statute,[5] and to issue enough shares for sufficient capital to commence business even if no minimum capital requirements exist. Other authorized shares should be saved to allow for additional capital financing in future corporate operations. Shares which have been authorized and sold are issued to the holders and they are then described as "issued and outstanding" shares. Thus, a shareholder holds "authorized, issued and outstanding" shares of the company.

Shares which have been sold to investors may be reacquired by the corporation by one of several methods. The shareholder could donate or resell them to the company or the corporation may, if so authorized in the articles of incorporation, redeem the shares or convert them to other shares. When shares are reacquired by the corporation, they are called "treasury shares."[6] Treasury shares are "authorized and issued" but not "outstanding" since they are held by the corporation and not by investors.

The issuance and sale of shares is initiated by the decision of the board of directors to obtain additional capital for the corporation. The board of directors may not issue more shares than have been authorized in the articles of incorporation and they also must observe the present shareholders' preemptive rights, if they exist, by offering newly issued shares to existing shareholders in their respective proportions of share ownership before selling the shares to other investors.[7]

The distinction between the stages of equity securities is important for several reasons. In order for equity securities to be fully active (including entitlement to vote, receipt of a proportionate share of earnings, and receipt of proportionate share of assets upon dissolution) shares must be authorized, issued, and outstanding. Treasury shares are usually not counted for determining a quorum and are not entitled to a vote.[8] They also may not be entitled to any dividend distributions.[9]

Another distinguishing feature involves the consideration for shares. If shares are authorized and are being issued and sold, they usually must be sold for an amount no less than the par value, if they have par value, or the stated value for no-par value shares. Thus, if shares bear a $10 par value, the corporation may not sell them for less than $10 per share. If it should sell shares for less than that amount, the consideration is inadequate and the shares become "watered" or "discount" shares. Shareholders who purchased watered shares are assessable for the full amount of unpaid consideration. On the other hand, shares which have been authorized, issued and outstanding and have been reacquired and held by the corporation as treasury shares may be sold again for any consideration fixed by the board of directors, whether or not the consideration is equal to or less than par value.

§ 7.05 PAR VALUE OR NO PAR VALUE

Nearly all jurisdictions require a statement in the articles of incorporation indicating whether shares are to be issued for a stated par value or for no-par value. The Model Act has recently taken the revolutionary position first propounded by California that it is not necessary to state whether shares have a par value or not.[10] Most states still require such a statement and the corporation always has a choice between no-par or par value shares, except in Nebraska where all shares must have a par value.[11] The distinction between these provisions relates to the value required to be given for their purchase, and to the rates of capital franchise fees which must be paid in some states upon incorporation. In addition, shares with no-par value permit greater flexibility in allocating the amount received in exchange for the shares to certain surplus accounts in the corporate books.

Shares with a par value may be issued only for such consideration expressed in dollars, *not less than the par value*, as may be fixed from time to time by the board of directors. This provision is common in most state statutes. It means that shares with a $10 par value may be issued for $20 if someone is willing to pay that amount, but in no case may they be issued for less than $10. A handful of states make exceptions to this rule. One state has actually abolished the concept of par value completely.[12]

Shares without par value may be issued for whatever consideration may be fixed from time to time by the board of directors, although the Model Act and many states provide that the articles of incorporation may reserve the right to fix the consideration for no-par value shares to the shareholders.[13] A few jurisdictions reverse this authority—it is granted to the shareholders unless reserved to the directors in the articles of incorporation.[14] Thus, shares without par value may be issued for any amount set by the board of directors or shareholders in their good judgment. The only limitation on this authority to fix the amount of no-par shares is that they must be issued for approximately the same amount of consideration at approximately the same time. Major variations in prices of no-par shares within a short time period raise a question of breach of the director's duty of due care in dealing with shareholders. For example, suppose the corporation intended to sell no-par value shares to three investors, Burn, Bush and Bradford. In private discussions with the three investors it was determined that Burn and Bush would pay approximately $10 per share, but that Bradford could be persuaded to pay $15 per share. If 100 shares were simultaneously issued to each of the investors at those prices, Bradford would have immediate grounds for complaint. His shares were immediately "diluted" with the sales to the other two investors because the board of directors effectively reduced the stated value by $5 per share. However, if Burn and Bush had purchased their shares in January, and Bradford's purchase at $15 per share occurred in September, it may be said that the shares increased in value enough to warrant the increase in price. The "dilution" problem results when no-par value shares are sold at substantially lower prices at about the same time as other no-par shares, so as to undercut the contribution of one of the investors. Notwithstanding the foregoing, if the board of directos can establish that the varying prices were set for a good business reason, they will be permitted to rapidly adjust the prices on shares without par value.

The par value/no-par value distinctions have other ramifications in states which exact annual franchise fees based upon the aggregate authorized capital of the corporation. In some states the franchise fee is computed upon the amount stated in the articles of incorporation as the total aggregate value of authorized shares. To compute this fee, shares without par value are assigned an arbitrary value.[15] For example, if an arbitrary value of $1 per share were placed on no-par share to compute the franchise fee for a state, the franchise fee would be the same for a corporation authorized to issue 10,000 shares at $5 par value as it would be for a corporation authorized to issue 50,000 shares no-par value in that state. States which discourage the use of no-par shares impose a high statutory valuation on these shares to compute the fee. It then becomes more advantageous to set a lower par value and to authorize more shares.[16]

Certain accounting classifications also depend upon the distinction between par value and no par value shares. In some cases it may be desirable to create a capital surplus account for the corporation in the early stages of corporate existence. Assuming, for example, that the incorporators predict the desirability of repurchasing some of the shares issued by the corporation. Many statutes, and formerly the Model Act, permit the corporation to repurchase its own shares only if it has a surplus account from which it may make the purchase.[17]

The creation of a surplus account occurs in different ways. The typical surplus accounts are "earned surplus" and "capital surplus." Earned surplus is created when the corporation accumulates profits from operations. Capital surplus is an account created for surplus funds received from the sale of stock. If the corporation issues $10 par value stock for $30 per share, $10 must be placed in stated capital, and $20 may be placed in capital surplus. Any consideration in excess of par value may be placed in capital surplus, and, with no par stock, any part of the consideration may be placed in capital surplus.[18]

If there is a need for a surplus account in the early stages of the corporate operations, the par value or no par value characteristic of the stock is important. These principles may be illustrated as follows. If the corporation has issued shares with $10 par value, and those shares are sold for exactly $10 each, that amount of money must be placed in an account called "stated capital." In that event there would be no capital surplus, and if the corporation has not yet earned profits to hold as retained earnings, there would be no earned surplus. In such a case, the corporation would not be able to repurchase its own shares, since there is no surplus account. On the other hand, if the corporation had sold shares with no-par value for $10 a share, it would be possible to divert any portion of that consideration to a capital surplus account. The board of directors may, usually within sixty days after the no part shares have been issued, allocate a portion of the consideration to capital surplus. For example, capital surplus could receive $9 per share and stated capital would receive the other $1 per share. The corporation could then use the amount of capital surplus to repurchase its own shares, provided the articles of incorporation authorize the use of capital surplus for this purpose.

Dividends and other distributions to shareholders are subject to similar rules. Most states limit the payment of dividends to funds available from earned surplus with a few exceptions. However, the corporation is permitted to distribute cash or property to its shareholders out of capital surplus under certain circumstances.[19] If management intends to make a distribution to shareholders before there is sufficient earned surplus to declare a dividend, the creation of a capital surplus account is essential. No par stock will assure the ability of the board of directors to create the account immediately upon issuance of the first shares.

§ 7.06 CERTIFICATES FOR SHARES

The shares of the corporation are generally represented by certificates, examples of which appear in the Appendix. Recently, the Model Act was amended to permit uncertificated shares. This amendment, together with amendments to the Uniform Commercial Code, permit a corporation to provide by resolution that some or all of the classes and series of shares shall be "uncertificated" which means that the ownership of the shares is recorded in the corporate records, but the shares are not represented by a certificate which is generally regarded to be "negotiable". The rights and obligations of the holders of uncertificated shares and the rights and obligations of the holders of certificates representing shares of the same class and series are identical under corporate law. The only purpose of the uncertificated provision is to avoid the paper crunch anticipated as more and more corporations issue certificates for shares and these certificates are rapidly traded on over-the-counter and national stock exchanges. The new system is intended to simplify the transfer of ownership in a corporation by not requiring the transfer of a piece of negotiable paper as part of that transaction. It also is intended to guard against the loss or theft of the negotiable piece of paper (the certificate) representing ownership in the corporation.

When certificates are used, all states prescribe the content of the certificate representing shares, and the requirements of the Model Act are typical:

> "Certificates shall be signed by the chairman or vice-chairman of the board of directors or the president or a vice-president and by the treasurer or an assistant treasurer or the secretary or an assistant secretary of the corporation, and may be sealed with the seal of the corporation or a facimile thereof. Any of or all the signatures upon a certificate may be a facimile."[20]

The statute further provides that the signature of any person who was an officer and ceased to be such before the certificate is issued will have the same effect as if he were still an officer as of the date of the issuance. Each certificate representing shares must state upon its face:

1. that the corporation is organized under the laws of the particular state;
2. the name of the person to whom issued;
3. the number and class of shares and the designation of the series, if any, which such certificate represents; and
4. the par value of each share represented by such certificate or a statement that the shares are without par value.

The major variations among the state statutes with respect to requirements for share certificates are the officers who are required to sign the certificates; the circumstances under which facsimile signatures may be used; the need for a corporate seal and whether the certificate must state that the shares are fully paid.

A corporation may choose to issue shares in various classes or in series, and in such a case each certificate must describe the particular elements of each class or series. Section 23 of the Model Act requires that every share certificate of a corporation which is authorized to issue shares of more than one class shall set forth on the certificate (or state that the corporation will furnish to any shareholder upon request and without charge) a full statement of the designations, preferences, limitations, and relative rights of the shares of each class authorized to be issued. Further, if the corporation is authorized to issue preferred or special classes in series, the variations between the shares so far as they have been determined and the authority of the board of directors to fix the relative rights of the shares must be stated on the certificate. The classes and series shares will be explored in detail in the next section.

If the corporation is authorized to issue only one class of stock, therefore, its share certificates need contain only the bare statutory requirements. If the corporation has adopted a complex financial structure, including classes of common stock, preferred stock, or special classes and series of either, the certificate is more complex. The certificate may describe in detail the relative variations of each class of securities, or, as is more common, it may merely state that the corporation will furnish a full statement of these variations to any shareholder of the corporation upon request and without charge. Such a statement should also describe the authority of the board of directors to determine the rights and preferences of each class and series of shares. If the latter procedure is used, it is not necessary to amend or otherwise modify the certificates representing shares whenever the corporate financial structure is changed.

§ 7.07 CLASSIFICATIONS OF SHARES

The articles of incorporation may authorize the issuance of only one class of shares, i.e., common stock. and in that case, the shareholders of the common stock are entitled to all of the voting rights, all of the dividends, and all of the net assets in a dissolution distribution. However, the corporate financial structure may be more complicated.

Classes of Shares

The equity securities of the corporation may be divided into several classes of shares. Common stock is the basic class, and additional classes may be authorized to grant certain shareholders a preferred right to dividends or a preferred right to assets in case of corporate dissolution. Various classes of securities may also have different voting rights, such as no vote, or two votes per share, or any other formula. When more than one class is to be authorized, the articles of incorporation must set forth the designations, preferences, limitations, and relative rights of each class.

EXAMPLE 7.07A: Classes of Stock

The total number of shares of all classes of stock which the Corporation shall have authority to issue is 500,000, of which 100,000 shares shall be Class A

> common stock without par value and 400,000 shares shall be Class B common stock without par value. There shall be no distinction between the two classes, except that the holders of the Class B common stock shall have no voting power for any purpose whatsoever and the holders of the Class A common stock shall, to the exclusion of the holders of the Class B common stock, have full voting power for all purposes.[21]

As mentioned above, the share certificate must also contain this information, or contain a statement that the corporation will provide the information to any shareholder requesting the information without charge.

It may be helpful at this point to pause and consider the circumstances under which the authority to issue preferred shares will be used. The general principle to be observed is that certain investors may insist upon a superior ownership position in the corporation, and, in order to obtain necessary financing, it may be necessary to "prefer" these investors over others. For example, suppose Phil Hopkins and Terrence Conner plan to form a corporation for the operation of a restaurant. Hopkins intends to run the business and to invest his available personal capital of $25,000, and Conner is capable of investing $100,000. Assume further that no other shareholders are contemplated at this point. Conner may be willing to take a greater proportion of common stock for his $100,000 investment but that will give him 80% of the common stock to Hopkins' 20%. Conner clearly has shareholder control in such a case, and Hopkins may be hampered thereby in his efforts to run the business. At least, Hopkins will be aware of Conner's potential control, and may object to this posture. It is possible to issue each $25,000 in common stock (with voting rights) and to issue non-voting preferred shares, with dividend and liquidation preferences, to Conner for his $75,000 excess investment. Conner and Hopkins now have equal voting rights and an equal investment in the basic equity of the corporation.

Conner's excess investment is protected by his preferred status, insuring that he will receive the first distributions of profit through his dividend preference, and the first distribution of assets in dissolution through the liquidation preference. Thereafter, they share equally, just as though each had invested only $25,000.

Similar problems arise when a third investor is considered. Suppose Craig Carver has offered to invest $50,000, but has no expertise in the restaurant business, and no interest in management and control. However, Carver is concerned about two things: a high return on his investment and a right to assert a vote if the business is being managed improperly. A separate class of securities can be created for Carver's investment. He may receive first dividend preference, even over Conner's preferred, and his preference may be at a higher rate and cumulative to insure an accumulated high return on his investment every year. His securities would have no voting rights until the happening of a contingent event, such as when the gross receipts from the business drop below a certain amount for several consecutive months.

The possibilities for variations in preferred shares are endless, and each class can be structured to fit the peculiar needs of each class of investors.

The statutory authority for issuance of shares in classes usually specifies the manner in which these shares may differ from common shares. For example, the Model Act Section 15 provides that a corporation may issue shares of preferred or special classes (a) permitting the corporation to redeem the shares at a price fixed by the articles of incorporation; (b) entitling the holders to cumulative, non-cumulative or partially cumulative dividends; (c) having preference over any other class or classes of shares as to the payment of dividends; (d) having preference in the assets of the corporation upon liquidation over other class of shares; and (e) allowing the shares to be convertible into shares of any other class or series under certain specified terms provided in the statute. In addition, the articles of incorporation may limit or deny the voting rights or provide special voting rights for certain classes under the authority of Section 33. The individual variations in these rights are explored in detail in section 7.11.

Shares in Series

The articles of incorporation may authorize the division and issuance of any class of shares in series.[22] The principle behind series shares is a refinement of the theory behind preferred shares. As previously discussed, classes of preferred shares are created to meet the particular needs and demands of certain investors. Series shares do the same thing. However, you should recall that all authority to issue shares eminates from the articles of incorporation. Thus, in order to issue a special class of shares, the articles of incorporation must authorize that class, and define its designations, preferences, limitations and relative rights. If the articles of incorporation do not specifically authorize the issuance of a particular class of shares, the articles must be amended to grant this authority, and an amendment requires a board of directors' resolution, the approval of the shareholders, and the appropriate filing with the Secretary of State.[23] The corporate officials attempting to raise capital may thus be hindered by a cumbersome, time-consuming amendment procedure in tailoring the corporate securities to the requirements of potential investors. This procedure may be particularly frustrating when prompt action is essential, such as when the corporation is negotiating to acquire desirable property in exchange for its preferred shares.

Series shares are designed to avoid this problem. The board of directors may be authorized (by the articles of incorporation) to fix the terms of a series of preferred shares, without requiring a formal amendment to the articles or shareholder approval. This broad authority to vary the terms of the stock is the source of the common name, "blank stock," for series shares. With this authority, the Model Act permits directors to issue shares which vary from other shares of the same class in the following particulars: (a) the rate of dividend; (b) the price at and the terms and conditions upon which shares may be redeemed; (c) the amount payable upon shares in the event of voluntary or involuntary liquidation; (d) sinking fund provisions, if any, for the redemption for purchase of shares; (e) the terms and conditions upon which shares may be converted; and (f) voting rights.[24] In all other respects, the shares of the series must be the same as other shares of the same class.

Not all states are as permissive as the Model Act in the authority to establish series shares. Many do not allow voting rights to vary among series.[25] Others limit the authority to create a special sinking fund for the redemption of shares.[26]

The scope of the directors' authority to establish series shares is defined by the articles of incorporation. The articles could state that the board of directors has full authority to divide classes of shares into series and to determine the variations in the relative rights.

EXAMPLE 7.07B: Shares in Series

The preferred stock of the Corporation shall be issued in one or more series as may be determined from time to time by the Board of Directors. In establishing a series the Board of Directors shall give to it a distinctive designation so as to distinguish from it from the shares of all other series and classes, shall fix the number of shares in such series, and the preferences, rights and restrictions thereof. All shares of any one series shall be alike in every particular. All series shall be alike except that there may be variation as to the following: (1) the rate of dividend, (2) the price at and the terms and conditions on which shares shall be redeemed, (3) the amount payable upon shares in the event of involuntary liquidation, (4) the amount payable upon shares in the event of voluntary liquidation, (5) sinking fund provisions for the redemption of shares, and (6) the terms and conditions on which shares may be converted if the shares of any series are issued with the privilege of conversion.[27]

The articles could also limit the authority to vary series shares to certain characteristics, such as conversion privileges or dividend preferences.

The procedure for establishing a series is considerably more informal than a formal amendment to the articles of incorporation. The directors adopt a resolution stating the designation of the series and fixing the rights of the series shares. A statement is filed with the Secretary of State, quoting the resolution and acknowledging that it was adopted by the board of directors on a certain date. In some states, franchise taxes and fees will be due upon the establishment of the series shares. The resolution, as filed, is considered by the Model Act to be an amendment to the articles of incorporation.[28]

§ 7.08 FRACTIONS OF SHARES OR SCRIP

In some cases the corporation may need to fractionalize shares of stock, and the authority for the issuance of a fractional share must come from the appropriate state statute. For example, fractional shares may be required when a corporation has declared a stock dividend entitling the holders of 100 shares of stock to receive one additional share as a dividend. In that case, the holder of 150 shares of stock would be entitled to one and one-half shares. If the state statute permits fractional shares, the corporation may issue a certificate for one-half share. Alternatively, many statutes authorize the issuance of scrip, in lieu of fractional shares. Scrip is a separate certificate representing a percentage of a full share, and it entitles a shareholder to receive a certificate for one full share when he has accumulated scrip aggregating a full share.

The Model Act provides that a corporation may, but is not obligated to, issue a certificate for a fractional share, or it may issue scrip in lieu thereof. Its other alternatives are to arrange for the disposition of fractional interests, as by finding two shareholders entitled to one-half share and arranging for a sale from one to the other so a whole share will be issued, or to pay the shareholder cash equal to the fair value of the fractional interest. If a certificate for a fractional share is issued, the holder is entitled to exercise a fractional voting right, to receive a fractional share of dividends, and to participate accordingly in the corporate assets in the event of liquidation. Unless otherwise provided by the board of directors, scrip is not usually entitled to these rights. The board of directors may further provide that the scrip will become void if not exchanged for full certificates before a specified date.

State statutes take divergent positions with respect to fractional shares and scrip, and the individual statutes should be consulted. Most states authorize one or more of the options specified in the Model Act.

§ 7.09 CONSIDERATION FOR SHARES

As previously indicated, the consideration required for shares depends in part upon whether the shares have a par value or are without par value.[29] Shares with a par value usually may be issued for no less than the par value, and shares with no-par value receive a stated value as fixed from time to time by the board of directors or the shareholders, as provided by statute and the articles of incorporation. Thus, in the quantitative sense, the consideration given in exchange for shares must at least equal the par value or the stated value for no-par shares, whichever the case may be.

When the consideration for shares is cash the quantity valuation is obvious. However, when property is transferred or when services are performed in exchange for shares the law has developed at least two rules for appraising the consideration. The first, the "true value rule," requires that the property or services have an actual value at the time of the issuance of shares of no less than par value or no less than the stated value for no-par shares. This means that the property or services must be appraised and the actual value compared with the minimum requirements for the particular type of stock. The Model Act and the majority of jurisdictions follow the "absence of fraud rule," which requires that the property or services be evaluated and deemed adequate by the board of directors, who must determine in good faith that the consideration received has a value at least equal to the minimum requirements, and their determination of value will be conclusive in the absence of fraud. These rules are used to determine whether the consideration received for shares is legally adequate in quantity. It should be noted that the board of directors, in determining the value of the offered consideration, should go on record with their evaluation.[30]

Statutes also dictate which types of consideration may be given in exchange for stock. This is a question of the *quality* of the consideration. The Model Act limits the permissible consideration to money; other property, tangible or intangible; or labor or services *actually performed* for the corporation.[31] A handful of state statutes permit promissory notes to be given in exchange for shares,[32] and future services are almost uniformily excluded from proper consideration. Both promissory notes and future services are specifically prohibited by the Model Act. There is some authority to the effect that preincorporation services, the services performed by the promoters and incorporators, are not really performed for the corporation because the corporation does not yet exist. Consequently, preincorporation services should not be considered adequate consideration for stock.[33]

Section 19 of the Model Act forbids the issuance of shares until the full consideration for the shares has been received by the corporation. If shares which are not fully paid have been issued, the shareholders may be liable to the creditors of the corporation for the deficiency and in such a case, the shares are assessable. Of course, directors who vote to issue shares which are not fully paid probably have violated their fiduciary duty to the corporation and other shareholders, and may be personally liable.

§ 7.10 COMMON STOCK RIGHTS

The corporation organized with only one class of stock has only common stock, which is best described as those shares of the corporation without any special features. To authorize the issuance of common stock the articles of incorporation need only describe the number of shares authorized and their par value or a statement that the shares have no-par value. Common stock has the following rights under most state statutes.

Dividends

Dividends may be declared from time to time in the discretion of the board of directors, and may be paid in cash, property, or other shares of the corporation, provided the corporation is solvent. Common stockholders receive dividends in the same proportion that their individual shares bear to the total number of common shares outstanding. The common stockholder's right to declared dividends is limited only by the solvency of the corporation (so creditors cannot be harmed by a distribution of assets as a dividend from an insolvent corporation) and by preferred stockholders' rights to dividends, which must be paid before the common shares are entitled to dividends. The rules relating to the distribution of dividends are expanded in a later chapter.[34]

Voting Rights

The one-vote-per-share rule is codified by statute in Section 33 of the Model Act. Each outstanding share regardless of class is entitled to one vote and each fractional share is entitled to a corresponding fractional vote on each matter submitted to a vote at the meeting of the shareholders. The statute further provides, however, that the voting rights of any class may be limited or denied by the articles of incorporation. These voting rules are common to most jurisdictions. If the corporation has only one class of stock, which is not divided into series, the single class must have full voting rights. If the corporation's financial structure includes several classes or series of securities, one or more of those classes or series may have limited or greater voting rights.

Depending upon the statutory requirements and the provisions in the articles of incorporation, cumulative voting may be authorized for shares of common stock.[35]

Liquidation Rights

After a decision has been made to dissolve the corporation, corporate officials are required by law to collect and dispose of the assets, and to satisfy all liabilities and obligations. Thereafter, the remaining assets in cash or property are distributed to the shareholders according to their respective interests. If the corporation has a single class of common stock, those shareholders will receive a proportionate interest in all of the net assets following dissolution and liquidation. The liquidation rights of common stock may be subordinated by the issuance of additional classes of securities which have a preference to the assets in dissolution.

Preemptive Rights

Depending upon the state statute, preemptive rights may exist unless denied, or may not exist unless granted in the articles of incorporation. The Model Act offers both approaches in alternative provisions. The preemptive right of a shareholder is the right to purchase her prorata share of newly issued stock before they may be offered to outsiders.[36]

§ 7.11 PREFERRED STOCK RIGHTS

Preferred stock is a common term for a class of stock which has been granted a preference to one or more of the normal shareholder rights. That is, a preferred stock usually has a preference to dividends, or a preference to assets in liquidation, or both. Issuance of preferred stock must be authorized by the articles of incorporation and the terms of the articles must contain a designation of each class and a statement of the preferences, limitations, and relative rights of the shares of each class. Moreover, if management intends to issue the shares of a preferred class in series, the designation of each series and a statement of the variations in the rights and preferences between series must be contained in the articles of incorporation. The articles also must describe any authority to be vested in the board of directors to establish the series and to determine the variations in the relative rights of the series.

Preferred stock is customarily preferred in the following ways:

Dividend Preference

One of the most common attractions to preferred stock is a dividend preference, if and when the board of directors declares a dividend. Preferred shareholders may have a mandatory priority to dividends, in a predetermined amount, before the corporation may pay any dividends to the holders of any other class of stock. A dividend preference may be cumulative, noncumulative, or a compromise of the two, and may also include a "participation" provision.

Cumulative. A cumulative dividend preference means that preferred stockholders will accrue an entitlement to dividends each year, whether or not the dividends are declared and paid. If the dividends are not paid during a certain year, they accumulate in the prescribed amount, and when the board of directors finally declares a dividend, all accumulated dividends on the preferred stock must be paid before dividends may be paid on any other stock. For example, suppose that preferred stock is entitled to a cumulative dividend of $4 per share and that Alexis Levine owns 100 shares. In the first and second year the corporation pays no dividends so that that Alexis' stock now has accumulated dividends in the amount of $800. In the third year management declares a dividend on all classes of stock. The corporation must first pay Alexis $1,200 on her shares ($800 accumulated plus $400 for the present year) before it may pay any dividends to any other class of stock.

A statement of a cumulative dividend preference follows:

EXAMPLE 7:11A: Cumulative Preferred Dividends

The holders of the preferred stock shall be entitled to receive, when and as declared by the Board of Directors, out of the assets of the Corporation legally available therefor, cash dividends at the rate of $8.40 per share per year, and no more, payable quarter-annually on the first days of January, April, July and October in each year, prior to the payment of any dividends of the common stock. Such dividends shall be cumulative from the date of original issue.[37]

Noncumulative. Noncumulative preferred stock is entitled to receive a dividend preference in any given year, but if dividends are not paid during that year, the preferred shareholders lose right to those dividends (just as common shareholders have no right to undeclared dividends). Using the same example, if the preferred stock is entitled to a $4 per share noncumulative dividend, when the corporation fails to pay dividends in the first and second year, Alexis loses the right to the $800 preference she would have had during those years. In the third year if management decides to pay dividends, it must pay her $400 (that year's preferred amount) before it can pay other shareholders.

EXAMPLE 7.11B: Noncumulative Preferred Dividends

The holders of the preferred stock shall be entitled to receive out of the surplus or net profits of the Corporation, in each fiscal year, dividends at such rate or rates, not exceeding 8.4 percent per annum, as shall be determined by the Board of Directors in connection with the issue of the respective series of said stock and expressed in the stock certificates therefor, before any dividends shall be paid upon the common stock, but such dividends shall be noncumulative. No dividends shall be paid, declared, or set apart for payment on the common stock of the Corporation, in any fiscal year, unless the full dividend on the preferred stock for such year shall have been paid or provided for. However, if the Directors in the exercise of their discretion fail to declare dividends on the preferred stock in a particular fiscal year the right of such stock to dividends for that year shall be lost even though there was available surplus or net profits out of which dividends might have been lawfully declared.[38]

Other Provisions for Dividend Preferences. Perferred stock may have a "cumulative to the extent earned" preference to dividends, permitting dividends to accumulate on preferred stock if the corporation earned money and could have declared dividends in a given year, but the board of directors decided not to declare dividends that year. On the other hand, if the corporation had not earned or accumulated enough profit to declare a dividend that year, the dividend is lost and does not carry forward.

Another common provision is the right of the preferred stockholder to participate in the other dividends declared, in addition to his preference. For example, if Alexis' stock provided for a $4 per year preference, noncumulative, but participating, then in the third year when the corporation determined to pay dividends, Alexis would be entitled to $400 as a preference. She would also be entitled to share on a prorata basis with all other securities in the remaining dividends which are declared and paid.

Liquidation Preferences

Holders of preferred stock are also frequently granted a preference upon dissolution and liquidation of the corporation. The terms of the stock usually recite the right for the preferred stockholder to be paid a specified amount, plus any accrued dividends which have not been paid, before any other security holder is entitled to share in the assets upon liquidation. The preferred shareholder is thus placed in position analogous to a priority-creditor. A liquidation preference is as good a guarantee as can be made that shareholders will recoup their investment in case of liquidation. Although they are subordinate to corporate creditors, they are paid before any other investor, and are entitled to be fully satisfied before any other shareholder receives a distribution in liquidation.

Liquidation preferences may be determined at a fixed percentage of par value:

EXAMPLE 7.11C: Liquidation Preference

In the event of any liquidation, dissolution or winding up of the Corporation, either voluntary or involuntary, or in the event of any reduction of capital of the Corporation resulting in a distribution of assets to its stockholders, the holders of preferred shares shall be entitled to receive out of the assets of the Corporation, without regard to capital or the existance of a surplus of any nature, an amount equal to one hundred per cent (100%) of the par value of such preferred shares, and, in addition to such amount, a further amount equal to the dividends unpaid and accumulated thereon to the date of such distribution, whether or not earned or declared, and more, before any payment shall be made or any assets distributed to the holders of the common shares. After the making of such payments to the holders of the preferred stock, the remaining assets of the Corporation shall be distributed among the holders of the common stock alone, according to the number of shares held by each. If the assets of the Corporation distributable as aforesaid among the holders of the preferred stock shall be insufficient to permit of the payment to them of said amounts, the entire assets shall be distributed ratably among the holders of the preferred stock.[39] or the preferred shares may participate in liquidation with other securites.

They may also be based on a fixed sum, with participation rights:

EXAMPLE 7.11D: Liquidation Preference

In the event of any liquidation, dissolution or winding up of the Corporation, whether voluntary or involuntary, the holders of the preferred stock of the Corporation shall be entitled, before any assets of the Corporation shall be distributed among or paid over to the holders of the common stock, to be paid in full $100.00 per share of preferred stock, together with all accrued and unpaid dividends and with interest on said dividends at the rate of fifteen per cent (15%) per annum. After payment in full of the above preferential rights of the holders of the preferred stock, then the holders of the preferred stock and common stock shall participate equally in the division of the remaining assets of the Corporation, so that from such remaining assets the amount per share of preferred stock distributed to the holders of the preferred stock shall equal the amount per share of common stock distributed to the holders of the common stock.[40]

Voting Rights

Each corporation must have one class of shares with full voting rights, so at least one class will be entitled to vote on all matters submitted to the shareholders. A corporation with multiple classes of stock may have one class with full voting rights in shareholder matters, and another class with no voting rights at all. It is also permissible to grant greater than one-vote-per-share to one class while retaining the single vote per share for another class. These provisions are designed to establish voting control in one of the classes of securities. A typical application of these machinations may involve shareholders of a corporation who wish to retain their control, but need to issue additional shares to secure new capital. By establishing a new class of non-voting shares, the issuance of new shares will not dilute present voting control. Modifications to voting rights are described in the articles of incorporation, and they may be expanded, denied, or granted subject to a contingency. For example, preferred shares may have no voting rights unless they have not received dividends for a specified period of time, in which case they will be entitled to vote.

EXAMPLE 7.11E: Voting Rights

The holders of the preferred stock shall not have any voting power whatsoever, except upon the question of selling, conveying, transferring or otherwise disposing of the property and assets of the Corporation as an entirety, provided, however:

In the event that the Corporation shall fail to pay any dividend upon the preferred stock when it regularly becomes due, and such dividend shall remain in arrears for a period of six (6) months, the holders of the preferred stock shall have the right to vote on all matters in like manner as the holders of the common stock, during the year next ensuing, and during each year thereafter during the continuance of said default until the Corporation shall have paid all accrued dividends upon the preferred stock. The holders of the common stock shall have the right to vote on all questions to the exclusion of all other stockholders, except as herein otherwise provided.

The articles of incorporation may also reconfirm the statutory voting scheme.

EXAMPLE 7.11F: Voting Rights

The preferred stock and common stock shall have equal voting powers and the holders thereof shall be entitled to one vote in person or by proxy for each share of stock held. The common stock, however, to the exclusion of the preferred

> stock, shall have the sole voting power with respect to the determination as to whether or not the preferred stock shall be redeemed, as hereinafter provided.[41]

If no modifications of voting rights appear in the articles, the Model Act and most states grant each outstanding share (including preferred shares) one vote on each matter submitted to a vote at the meeting of the shareholders.[42]

If voting rights are denied to a particular class of shares, the holders of that class are not entitled to vote on typical shareholder business. A potential problem lurks in this rule. Consider the fact that the structure (including dividend and liquidation entitlements) of the non-voting securities is specified in the articles of incorporation and this structure may be changed by an amendment to the articles which has been approved by the shareholders. If the holders of the non-voting classes were never entitled to vote, even on such matters, they would be at the mercy of the voting shareholders, who could vote to amend away the advantageous features of their stock. Consequently, corporate statutes uniformly provide for class voting on matters which may affect the rights of the class.

Section 60 of the Model Act provides that a class will be entitled to vote on any proposed amendment which would:

1. increase or decrease the aggregate number of authorized shares of such class.
2. effect an exchange, reclassification or cancellation of all or part of the shares of such class.
3. effect an exchange, or create a right of exchange, of all or any part of the shares of another class into the shares of such class.
4. change the designations, preferences, limitations or relative rights of the shares of such class.
5. change the shares of such class into the same or a different number of shares of the same class or another class or classes.
6. create a new class of shares having rights and preferences prior and superior to the shares of such class, or increase the rights and preferences or the number of authorized shares, of any class having rights and preferences prior or superior to the shares of such class.
7. in the case of a preferred or special class of shares, divide the shares of such class into series and fix and determine the designation of such series and the variations in the relative rights and preferences between the shares of such series, or authorize the board of directors to do so.
8. limit or deny any existing preemptive rights of the shares of such class.
9. cancel or otherwise affect dividends on the shares of such class which have accrued but have not been declared.

Redemption Rights

The terms of preferred stock may include provisions for redemption of the shares. A corporation may have the right to redeem or reacquire its shares if redemption terms are included in the description of the class. The redemption feature is a greater advantage to the corporation than to the shareholders. For example, suppose the corporation issued 20% cumulative redeemable preferred stock to acquire needed capital. The dividend percentage would make this stock very attractive, but management would prefer not to continue paying a high cumulative dividend any longer than necessary. The redemption feature would permit the corporation to retire the securities when it has generated enough capital from operations to do so.

A proper redemption provision should spell out the terms of the "forced" sale from the preferred shareholders to the corporation and the procedure that must be followed to accomplish the sale. A redemption clause must appear on the articles of incorporation and on the share certificate, and it typically includes:

1. a date upon which the stock will be redeemable by the corporation;
2. a price at which the stock is redeemable, usually including a provision to the effect that an amount equal to accrued and unpaid dividends will be added to the price;
3. a period of notice preceding the date of redemption and the persons to whom notice must be given;
4. a place at which payment is to be made, and the person who will make payment;
5. a time at which payment is to be made and whether the board of directors has a right to accelerate the payment date;
6. provisions regarding the surrender of share certificates and the cancellation of all rights of the shareholder upon redemption; and
7. provisions covering the possibility that a shareholder will not surrender shares for cancellation in accordance with the redemption right.

EXAMPLE 7.11G: Redemption of Shares

The preferred stock may be redeemed in whole or in part on any quarterly dividend payment date, at the option of the Board of Directors, upon not less than sixty (60) days prior notice to the holders of record of the preferred stock, published, mailed and given in such manner and form and on such other terms and conditions as may be prescribed by the bylaws or by resolution of the Board of Directors by payment in cash for each share of the preferred stock to be redeemed of one hundred two per cent (102%) of the par amount thereof and in addition thereto all unpaid dividends accrued on such share.

From and after May 1, 1981, the Board of Directors shall retire not less than 1,000 shares of preferred stock per annum; but the Board of Directors shall first set aside a reserve to provide full dividends for the current year on all preferred stock which shall be outstanding after such purchase or retirement, and provided further that no such purchase or retirement shall be made if the capital of the Corporation would thereby be impaired.

If less than all the outstanding shares are to be redeemed, such redemption may be made by lot or pro-rata as may be prescribed by resolution of the Board of Directors; provided, however, that the Board of Directors may alternatively invite from shareholders offers to the Corporation of preferred stock at less than One hundred two dollars ($102.00), and when such offers are invited, the Board of Directors shall then be required to buy at the lowest price or prices offered, up to the amount to be purchased.

From and after the date fixed in any such notice as the date of redemption (unless default shall be made by the Corporation in the payment of the redemption price), all dividends on the preferred stock thereby called for redemption shall cease to accrue and all rights of the holders hereof as stockholders of the Corporation, except the right to receive the redemption price, shall cease and determine.

Any purchase by the Corporation of shares of its preferred stock shall not be made at prices in excess of said redemption price.[43]

EXAMPLE 7.11H: Redemption of Shares

By a unanimous vote of a full board of directors of the number fixed by the stockholders at their last annual meeting, all or any shares of common stock of the Corporation held by such holder or holders as may be designated in such vote may be called at any time for purchase, or for retirement or cancellation in connection with any reduction of capital stock, at the book value of such shares as determined by the Board of Directors as of the close of the month next preceding such vote. Such determination, including the method thereof and the matters considered therein, shall be final and conclusive.

Not less than 30 days prior to the day for which a call of common stock for purchase or for retirement or cancellation is made, notice of such call shall be mailed to each holder of shares of stock called at his address as it appears upon the books of the Corporation. The Corporation shall, not later than said day, deposit with a bank to be designated in such notice, for the account of such holder, the amount of the purchase price of the shares so called. After such notice and deposit all shares so called shall be deemed to have been transferred to the Corporation, or retired or cancelled as the case may be, and the holder shall cease to have, in respect thereof, any claim to future dividends or other rights as stockholder, and shall be entitled only to the sums so deposited for his account. Any shares so acquired by the Corporation may be held and may be disposed of at such times, in such manner and for such consideration as the Board of Directors shall determine.[44]

The redemption of corporate securities requires large disbursements of cash which may not be absorbed in normal corporate operations. In order to plan for eventual redemption, therefore, management should consider a "sinking fund" for the payment of the purchase price. The objective of the sinking fund is the same as that of a Christmas savings plan. By faithfully depositing a certain amount at periodic intervals, a lump sum will be available at the projected date of need. A clause establishing redemption of preferred stock with a sinking fund might look like this:

EXAMPLE 7.11I: Sinking Fund

a) There shall be a sinking fund for the benefit of the shares of the preferred stock. So long as there shall remain outstanding any preferred stock, the Corporation shall set aside annually, on or before October 15, 1981, and on or before October 15th in each year thereafter, as and for such sinking fund for the then current year, an amount in cash equal to the lesser of $25,000, or 2.7% of the Consolidated Net Earnings of the Corporation and its subsidiaries for the preceding calendar year (computed as hereinafter provided). So long as dividends on the preferred stock for any past quarterly dividend payment date shall not have been fully paid or declared and a sum sufficient for the payment thereof set apart, the date for the setting aside of any amounts for the sinking fund shall be postponed until all such dividends in arrears shall have been paid or declared and a sum sufficient for the payment thereof set aside, and no amounts shall be set aside for the sinking fund while such arrears shall exist. In addition to the aforesaid sinking fund payments, the Corporation shall pay out of its general funds all amounts paid in excess of $103 per share (for commissions or as, or based upon, accrued dividends) upon any purchase or redemption of preferred stock through the sinking fund, as hereinafter provided.

b) The moneys set aside for any annual installment for the sinking fund (with any amounts remaining unexpended from previous sinking fund installments), may, at the option of the Corporation, be immediately applied (but not earlier than the October 15th on or before which such installment is required to be set aside), as nearly as possible, to the redemption of shares of preferred stock at the redemption price of $103 per share plus dividends accrued thereon to the date of redemption, in the manner provided herein for the redemption of preferred stock; provided, however, that if at the time any such annual installment is set aside (but not earlier than the October 15th on or before which such installment is required to be set aside), any holder of preferred stock shall hold 5% or more of the then outstanding shares of preferred stock, then there shall promptly be redeemed from such holder the number of whole shares of preferred stock (and no more) that shall bear, as nearly as practicable, the same ratio to the total number of shares which could be redeemed pursuant to this subdivision with such moneys, as the number of shares of preferred stock then owned by such holder shall bear to the total number of shares of preferred stock then outstanding.

c) Any moneys set aside for the sinking fund, as hereinabove required, and not applied to the redemption of preferred stock as provided in the preceding clause (b) (with any amounts remaining unexpended from previous sinking fund payments) shall be applied from time to time by the Corporation to the purchase, directly or through agents, of preferred stock in the open market or at public or private sale, with or without advertisement or notice, as the Board of Directors shall in its discretion determine, at prices not exceeding $103 per share plus accrued dividends and plus the usual customary brokerage commissions payable in connection with such purchases. If at the expiration of a full period of 90 days following the date each such amount is set apart during which the Corporation shall have been entitled hereunder to purchase shares of preferred stock with such funds, there shall remain in the sinking fund amounts exceeding $5,000 in the aggregate which shall not have been expended during such periods, then the Corporation shall promptly select and call for redemption at $103 per share plus dividends accrued thereon to the date of redemption, in the manner herein provided for the redemption of preferred stock such number of shares of preferred stock as is necessary to exhaust as nearly as may be all of said moneys, except that no shares shall then be allocated for redemption from any holder if the pro rata share of the then curent sinking fund payment shall have been applied to the redemption of shares of such holder as hereinabove in paragraph (b) of this subdivision provided. Anything herein to the contrary notwithstanding, no purchase or redemption of shares of preferred stock with any moneys set aside for the sinking fund shall be made or ordered unless full cumulative dividends for all past quarterly dividend payment dates have been paid or declared and a sum sufficient for the payment thereof set aside upon all shares of preferred stock then outstanding. When no shares of preferred stock shall remain outstanding any balance remaining in the sinking fund shall be and become part of the general funds of the Corporation.[45]

The Model Act repealed its statutory restrictions on redemption of shares in 1979. Most state statutes are based on the former Model Act provisions, however, and they continue to address the redemption right in two important areas. Redemption or purchase of shares is prohibited when the corporation is insolvent or would be rendered insolvent by the redemption. Redemption is also forbidden if the transaction would reduce the net assets below the aggregate amount payable to the holders of shares having prior or equal rights to the assets of the corporation upon involuntary dissolution.

This latter provision deserves an illustration. Suppose the corporation has $1,000,000 in assets, $700,000 in liabilities and three classes of stock outstanding; 1,000 shares of 8% non-cumulative preferred with a $105 liquidation preference; 2,000 shares of 10% non-cumulative preferred redeemable at $100 per share; and 100,000 shares of common stock. The net assets of the corporation total $300,000 (assets minus liabilities). The corporation could not redeem all 2,000 shares of the 10% preferred, since that would require $200,000 and payment of that amount would reduce net assets to $100,000. The remainder is insufficient to pay the 8% preferred shareholders' liquidation preference of $105,000.

Redeemed shares are usually cancelled and restored to "authorized but unissued" status. Under the amended Model Act, the restoration of the shares to this status is automatic unless the articles of incorporation provide that the shares shall not be reissued. In such a case, the redeemed shares are eradicated by reducing the total authorized shares. A statement of cancellation must be filed upon redemption, describing the number of redeemable shares cancelled, the aggregate number of issued shares after the cancellation, the dollar amount of the stated capital after the cancellation, and the number of shares which the corporation has authority to issue after the cancellation.[46]

Conversion Privileges

Certain classes of stock may be entitled to a conversion privilege whereby a holder of those shares may convert his shares for shares of another class at a specified rate and time. A conversion feature enhances the marketability of more conservative classes of stock, because the holder has the best of both worlds. A holder of preferred shares with a conversion privilege, for example, enjoys the preferences and conservative investment protection of his preferred stock while also maintaining the option to convert to common stock if its growth rate is attractive. The shareholder who owns convertible preferred stock, therefore, receives the security of preferred shares plus the right to elect an interest in the basic equity growth of the company.

A conversion clause should appear in the articles of incorporation and on the share certificates. A conversion privilege provision should include;

1. a conversion rate specifying the number of shares of common stock (or another class) into which each share of preferred stock is convertible;
2. provisions respecting the issuance of fractional shares, since the conversion may result in a fractional share problem;[47]
3. a procedure for the method of conversion, detailing the written notice required to convert the shares and a period of time following the election that the conversion will become effective;

4. provisions for the adjustment of conversion rates, in cases of a stock dividend, stock split, or other corporate action which changes the character of the shares into which the preferred is convertible;

5. requirement of notice to preferred shareholders when the conversion rate is adjusted;

6. a reservation of an adequate number of shares of common stock, in case all conversion privileges are exercised. Remember that a corporation must have authority to issue common stock in its articles of incorporation, and the common shares must be reserved if they are to be issued to preferred shareholders.

EXAMPLE 7.11J: Conversion of Shares

The preferred stock of this Corporation of $100 par value may, at the option of the holder thereof, at any time on or before January 10, 1990, be converted into common stock of this Corporation of $100 par value upon the following terms:

a) Any holder of such preferred stock desiring to avail himself of the option for conversion of his stock shall, on or before January 10, 1990, deliver, duly endorsed in blank, the certificates representing the stock to be converted to the Secretary of the Corporation at its office, and at the same time notify the Secretary in writing over his signature that he desires to convert his stock into common stock of $100 par value pursuant to these provisions.

b) Upon receipt by the Secretary of a certificate or certificates representing such preferred stock and a notice that the holder desires to convert the same, the Corporation shall forthwith cause to be issued to the holder one share of common stock for each share of preferred stock surrendered, and shall deliver to such holder a certificate in due form for such common stock.

c) One hundred thousand shares of common stock of this Corporation shall be set aside and such shares shall be issued only in conversion of preferred stock as herein provided.

d) Shares of preferred stock which have been converted shall revert to the status of unissued shares and shall not be reissued. Such shares may be eliminated as provided by law.

e) If, at any time the convertible preferred stock of this Corporation is outstanding, the Corporation increases the number of common shares outstanding without adjusting the stated capital of the corporation, the conversion rate shall be adjusted accordingly, so as to make each share of preferred stock convertible into the same proportionate amount of common stock as it would have been convertible without such adjustment to the common stock. Each preferred shareholder shall be notified in writing of the adjusted conversion rate within thirty (30) days of such action by the Corporation.

f) These provisions for conversion of preferred stock of this Corporation shall be subject to the limitations and restrictions contained in section 57.100 of the Business Corporation Law of the State of Oregon.[48]

§ 7.12 TRANSFER AGENTS

Many corporations employ transfer agents and registrars to keep track of the record holders of the corporate securities. By statute, a stock transfer ledger must be maintained either at the place of business of the corporation or at the office of its transfer agent or registrar,[49] which is normally a bank or other financial institution which has a separate division for the explicit purpose of maintaining such records for corporations. The transfer agent is responsible for issuing new stock certificates. Blank certificates are usually kept in bulk at the agent's office and the statutory signatures are usually printed on the certificates, since the Model Act permits a facsimile signature of the officers if the certificates are countersigned by an independent transfer agent or registrar.[50] If a registrar is used separately, the transfer agent delivers newly prepared certificates to the registrar for registration and countersignature. The registrar then returns the registered and signed certificate to the transfer agent who issues it. The registrar's responsibilities include recording all certificates representing shares of stock in the corporation. If the transfer agent and registrar are separate individuals, all entries made in the stock transfer ledger are made by the registrar, and the ledger is kept at the registrar's office. Transfer agents and registrars receive corporate authority by resolution of the board of directors.

EXAMPLE 7.12A: Appointment of Transfer Agent and Registrar

Appointment of } TRANSFER AGENT
REGISTRAR

____
(Name of Corporation)

"Resolved, that The ________Bank ______________________________
is hereby appointed [sole] [Transfer Agent] [Registrar] for all of the shares

of the ________Preferred stock, and

________shares
all of the shares

of the ________Common stock
of this Company, to act in accordance with its general practice and with the regulations set forth in the pamphlet submitted to this meeting entitled 'Regulations of The ________Bank ______________________________
for the Transfer and Registration of Stock', which pamphlet the Secretary is directed to mark for identification and file with the records of the Company."

I, the undersigned, Secretary of the above named Corporation, do hereby certify that the foregoing is a true and correct copy of a resolution duly adopted by the Board of Directors of said Corporation at a meeting thereof duly called and held on ________, 19 ___, at which a quorum were present, and that said resolution has not been in any wise rescinded, annulled or revoked but the same is still in full force and effect.

And I do further certify to the following facts:

The authorized and outstanding stock of the Corporation is as follows:

Class	**Par Value**	**Authorized**	**Outstanding**
________	________	________	________
________	________	________	________

The address of the Corporation to which notices may be sent is ____.

The below named persons have been duly elected, have duly qualified, and this day are, officers of the Corporation, holding the respective offices below set opposite their names, and the signatures below set opposite their names are their genuine signatures.

______________________ **President** ______________________
______________________ **Vice-President** ______________________
______________________ **Vice-President** ______________________
______________________ **Treasurer** ______________________
______________________ **Assistant Treasurer** ______________________
______________________ **Secretary** ______________________
______________________ **Assistant Secretary** ______________________

The name and address of legal counsel for the Corporation is ______________________.

The names and addresses of all of the Transfer Agents and Registrars of the stock of the Corporation are as follows:

Class of Stock	**Transfer Agent(s)**	**Registrar(s)**
____________	____________	____________
____________	____________	____________

Witness my hand and the seal of the Corporation this ____________ day of ________, 19___.

____________________[51]
Secretary

[*Corporate Seal*]

In addition to a specification of authority to act as transfer agent or registrar such a resolution may also include a statement that the officers of the company are authorized and empowered to give instructions to the transfer agent and to the registrar and to take any other action they deem necessary to effect the issuance of the common stock.

§ 7.13 DEBT SECURITIES

Every state empowers a corporation to borrow funds for corporate purposes. Debt securities represent loans to the corporation, and a debt security holder is a creditor of the corporation. He usually enjoys no right to participate in management and he also has no right to receive profits. A debt security holder is entitled to the repayment of his loan with the prescribed interest, and if the debt remains unpaid at the time of dissolution, he will be entitled to repayment of the debt from the available assets. However, debt security holders enjoy greater security for their investment than equity security holders, since the debt holders are creditors and are entitled to be satisfied from the available assets first.

Debt securities state a principal amount, a maturity date and a periodic interest rate. The interest is the return on the investment, and usually determines the attractiveness of a debt security. State statutes do not strictly regulate debt obligations, and as a matter of law these securities are considered to be individual agreements between the debt security holder and the corporation, containing negotiated terms. Of course, the terms of debt securities ("bonds") issued by large corporations are fixed and not subject to negotiation or modification. However, in smaller corporations debt securities may contain a variety of terms which are in fact negotiated between the creditor-investor and the corporation. Nevertheless, it is possible to generalize some typical features of debt securities.

§ 7.14 TYPES OF CORPORATE DEBT SECURITIES

Unsecured Debt

A corporate debt obligation may be as simple as a promissory note, the terms of which include the amount of the debt, a promise to pay the principal at a certain time, and a promise to pay interest:

EXAMPLE 7.14A: Promissory Note

$50,000.00 November 1, 1981

One year after date, for value received, Happiness, Inc. promises to pay to Pamela A. Ray or order, payable at 1216 Charlotte Avenue, Austin, Texas, the sum of Fifty thousand DOLLARS, with interest thereon at the rate of 15 per cent. per annum from date, payable monthly until paid.

Failure to pay any installment of principal or interest when due shall cause the whole note to become due and payable at once, or the interest to be counted as principal, at the option of the holder of this note, and it shall not be necessary for the holder to declare the same due, but she may proceed to collect the same as if the whole was due and payable by its terms.

Presentment for payment, notice of dishonor, and protest are hereby waived by the maker or makers, and endorser or endorsers, and each endorser for himself guarantees the payment of this note according to its terms. No extension of payment shall release any signer or be paid by the parties liable for the payment of this note.

________________ ________________

________________ ________________

or it may be a lengthy, complex debenture obligation.

EXAMPLE 7.14B: 18% Twenty-Year Debenture
Due August 1, 2010

$1,000 No. ______

Trouble, Inc., a Colorado corporation (hereinafter called the "Company", which term includes any successor corporation under the Indenture hereinafter referred to), for value received, hereby promises to pay to the bearer, or, if this Debenture be registered as to principal, to the registered holder hereof, on August 1, 2010, the sum of One Thousand Dollars and to pay interest thereon, from the date hereof, semi-annually on June 1 and December 1 in each year, at the rate of Eighteen (18) per cent per annum. Payment of the principal of (and premium, if any) and interest on this Debenture will be made at the office or agency of the Company maintained for that purpose in Denver, Colorado, in such coin or currency of the United States of America as at the time of payment is legal tender for payment of public and private debts.

This Debenture is one of a duly authorized issue of Debentures of the Company designated as its 18% Twenty-Year Debentures Due August 1, 2010, (hereinafter called the "Debentures"), limited in aggregate principal amount to $1,500,000.00, issued and to be issued under an indebture dated August 1, 1980 (hereinafter called the "Indenture"), between the Company and Glorious Trust Company as Trustee (hereinafter called the "Trustee", which term includes any successor trustee under the Indenture), to which Indenture and all indentures supplemental thereto reference is hereby made for a statement of the respective rights thereunder of the Company, the Trustee and the holders of the Debentures and coupons, and the terms upon which the Debentures are, and are to be, authenticated and delivered.

If an Event of Default, as defined in the Indenture, shall occur, the principal of all the Debentures may be declared due and payable in the manner and with the effect provided in the Indenture.

The Indenture permits, with certain exceptions as therein provided, the amendment thereof and the modification of the rights and obligations of the Company and the rights of the holders of the Debentures under the Indenture at any time by the Company with the consent of the holders of 66⅔% in aggregate principal amount of the Debentures at the time outstanding, as defined in the Indenture. The Indenture also contains provisions permitting the holders of specified percentages in aggregate principal amount of the Debentures at the time outstanding, as defined in the Indenture, on behalf of the holders of all the Debentures, by written consent to waive compliance by the Company with certain provisions of the Indenture and certain past defaults under the Indenture and their consequences. Any such consent or waiver by the holder of this Debenture shall be conclusive and binding upon such holder and upon all future holders of this Debenture and of any Debenture issued in exchange herefor or in lieu hereof whether or not notation of such consent or waiver is made upon this Debenture.

No reference herein to the Indenture and no provision of this Debenture or of the Indenture shall alter or impair the obligation of the Company, which is absolute and unconditional, to pay the principal of (and premium, if any) and interest on this Debenture at the times, place and rate, and in the coin or currency, herein prescribed.

This Debenture is transferable by delivery, unless registered as to principal in the name of the holder in the Debenture Register of the Company. This Debenture may be so registered upon presentation hereof at the office or agency of the Company in any place where the principal hereof and interest hereon are payable, such registration being noted hereon. While registered as aforesaid, this Debenture shall be transferable on the Debenture Register of the Company by the registered holder hereof, upon like presentation of this Debenture for notation of such transfer hereon, accompanied by a written instrument of transfer in form satisfactory to the Company and the Debenture Registrar duly executed by the registered holder hereof or his attorney duly authorized in writing, all as provided in the Indenture and subject to certain limitations therein set forth; but this Debenture may be discharged from registration by being in like manner transferred to bearer, and thereupon transferability by delivery shall be restored. This Debenture shall continue to be subject to successive registrations and transfers to bearer at the option of the bearer or registered holder, as the case may be. Such registration, however, shall not affect the transferability by delivery of the coupons appertaining hereto, which shall continue to be payable to bearer and transferable by delivery. No service charge shall be made for any such registration, transfer or discharge from registration, but the Company may require payment of a sum sufficient to cover any tax or other governmental charge payable in connection therewith.

The Company, the Trustee and any agent of the Company may treat the bearer of this Debenture, or, if this Debenture is registered as herein authorized, the person in whose name the same is registered, and the bearer of any coupon appertaining hereto, as the absolute owner hereof for all purposes, whether or not this Debenture or such coupon be overdue, and neither the Company, the Trustee nor any such agent shall be affected by notice to the contrary.

The Debentures are issuable as coupon Debentures, registrable as to principal, in the denomination of $1,000 and as registered Debentures without coupons in denominations of $1,000 and any multiple thereof. As provided in the Indenture and subject to certain limitations therein set forth, Debentures are exchangeable for a like aggregate principal amount of Debentures of a different authorized kind or denomination, as requested by the holder surrendering the same, upon payment of taxes and other governmental charges.

Unless the certificate of authentication hereon has been executed by the Trustee by the manual signature of one of its authorized officers, neither this Debenture, nor any coupon appertaining hereto, shall be entitled to any benefit under the Indenture, or be valid or obligatory for any purpose.

In witness whereof, the Company has caused this Debenture to be duly executed under its corporate seal, and coupons bearing the facsimile signature of its Treasurer to be hereto annexed.

Date: ________ ____________________

By ________________ [53]

Attest:

____

In any case, these are unsecured obligations, meaning that the corporation simply borrows money on the strength of its own ability to repay. The characteristic common to the simple promissory note and the debenture bond is that there is no specific corporate property to which the creditor will be entitled if the corporation defaults.

Secured Debt

The unsecured debenture or promissory note should be contrasted with a mortgage bond or secured note in which the corporation pledges certain property as collateral to secure repayment of the obligation. If there is a default, the creditor may reach the collateral to satisfy the debt. Mortgage bonds usually have corporate land as collateral. The obligation between the creditor and the corporation is represented by a mortgage bond, or note, and a mortgage agreement, which specifies the terms under which the property will be held for the benefit of the creditor. A mortgage agreement usually requires the mortgagor (corporation) to insure the property, maintain it in good order, and to keep it free from other liens or obligations so that the creditor will have its full benefit, if necessary. Secured debt obligations may also involve personal property collateral, including equipment, inventory, accounts receivable, etc. Personal property security interests are governed by the Uniform Commercial Code which has been adopted in every state except Louisiana. The Code requires a security agreement between the creditor and the debtor and creditor.[54] Further, a financing statement, which meets the statutory requirements,[55] must be recorded in the appropriate state or county offices.[56]

§ 7.15 TRUST INDENTURE

When numerous bonds are issued at once, a trustee is usually appointed to act on behalf of the holders of the security in case of default by the corporation. The appointment of a trustee is particularly desirable for secured debt obligations. If the corporation defaults, the trustee will act on behalf of the creditors to recover the property securing the obligations. The trustee is usually a financial institution, and is appointed by the execution of a trust indenture, an agreement which specifies the rights and responsibilities of the corporation, the rights and responsibilities of the trustees, and the rights of the security holders. If the bonds are to be sold to the public, the indenture must comply with the requirements of the Trust Indenture Act of 1939, the federal Securities Act of 1933,[57] and perhaps state securities statutes. The document which evidences the obligation, the bond itself, merely refers to the trust indenture for all details of the obligation. Trust indentures are unconscionably lengthy documents, and most experienced financial institutions have standard indentures for use of their corporate customers.[58]

§ 7.16 COMMON PROVISIONS IN DEBT SECURITIES

As with stock, it is possible to introduce various privileges into debt securities which make the investment more attractive. Debt securities may also accommodate redemption or conversion provisions, and the terms of the obligation may contain a "subordination" feature which establishes a priority of one debt security over another. The investment objectives of a debt security holder are twofold: repayment of the principal and receipt of the periodic interest. The security may become more or less attractive depending upon the circumstances under which it may be subordinated (endangering repayment of principal); redeemed (terminating interest); or converted (altering the character of the investment).

Redemption

Bonds may be redeemable at the option of the board of directors at a specified time and a specified price. The redemption price for bonds is usually stated in terms of a percentage of the principal amount, and it may be determined by a declining percentage from the date of issue to maturity or a fixed percentage figure throughout. For example, a bond may be redeemable during the first year at 110% of the principal amount, during the second year at 109.5%, during the third year 109% and so forth. Of course, the closer the bond is to maturity, the redemption price will approach 100%. The declining percentage is designed to protect the bondholder by discouraging early redemption by the corporation.

The terms of bond redemption provisions are similar to redemption clauses for shares enumerated in Section 7.11, including the authority to create a sinking fund for redemption. An example of a fixed percentage redemption provision follows:

EXAMPLE 7.16A: Redemption Provisions

The Company may at its option redeem this debenture at any time hereafter upon payment of the principal amount hereof, plus a premium of five per cent (5%) of such principal amount, plus any unpaid interest payable for any fiscal year ended prior to the date of redemption, plus interest at the rate of five per cent (5%) per annum upon such principal amount for the period from the first day of the fiscal year in which redemption is so made to the date of redemption, provided that notice of such redemption, stating the time and place of redemption, shall be published at least once each week for four (4) successive weeks prior to the redemption date in a daily newspaper of general circulation published in Cook County, Illinois. Thereupon this debenture shall become due and payable at the time and place designated for redemption in such notice, and payment of the redemption price shall be paid to the bearer of this debenture upon presentation and surrender thereof and of all unpaid interest coupons annexed hereto. Unless default shall be made in the redemption of this debenture upon such presentation, interest on this debenture shall cease from and after the date of redemption so designated. If the amount necessary to redeem this debenture shall have been deposited with the Abraham Lincoln Trust Company and if the notice of redemption shall have been duly published as aforesaid, this debenture shall be conclusively deemed to have been redeemed on the date specified for redemption, and all liability of the Company hereon shall cease on such date and all rights of the holder of this debenture, except the right to receive the redemption price out of the moneys so deposited, shall cease and terminate on such date.[59]

The notice required by the redemption provision could be worded as follows:

EXAMPLE 7.16B: Notice of Redemption

Notice is hereby given that the Five Year Convertible Income Debentures due November 30, 1985 of The Nobles Corporation have been called for redemption at 110% of the principal amount thereof and will be redeemed at the office of the Abraham Lincoln Trust Company, 1198 West Adams Street, Chicago, Illinois, on May 15, 1981. From and after May 15, 1981, the holders of said Debentures will have no conversion rights or any other rights, except to receive the redemption price.[60]

Conversion

Debt securities may be convertible into equity securities, and that further enhances the value of the bond. The bond conversion feature will specify the number of shares of stock into which the bond is convertible, the procedure for conversion, a reservation of an appropriate number of common shares, and adjustments and other matters that concern the conversion privilege.

EXAMPLE 7.16C: Conversion Provisions

As provided in the indenture with Abraham Lincoln Trust Company this bond is convertible at the option of the holder thereof, at any time prior to maturity (or, if this bond is at any time called for redemption, then at any time before the date fixed for redemption), upon surrender of this bond for that purpose at the office of the Nobles Corporation, Chicago, Illinois. Conversion shall be made into common shares of the Nobles Corporation upon the basis of one common share for each $100 of principal sum of this bond, subject to the provision of the indenture as to interest on bonds converted and dividends on shares received therefor, and as to change in the conversion basis or substitution of other shares, securities, or property in the event of consolidation, merger, conveyance of assets, recapitalization, or the issuance of additional shares.

Priority and Subordination

Management of the corporation may desire to subordinate the existing debt securities in order to secure additional financing at a later time. If the bond holders agree to accept second or third place to other creditors for certain purposes, the subordination of their debt should assist in obtaining the maximum borrowing capacity for the corporation. The subordination feature would only apply if the bonds are not paid. If the corporation defaults on the obligation, the subordination clause will determine which creditors have prior rights to corporate assets to enforce their respective obligations against the corporation.

When a bond issue is already outstanding and management is attempting to obtain additional financing, subordination may be an after-thought. Corporate officials may approach existing bond holders and solicit their agreement to subordinate their debt. On the other hand, subordination may be a condition precedent to the bond obligation. For example, an investor may be willing to invest in the corporation and purchase a bond, or lend the money, provided that the corporation will subordinate any future borrowing to his debt security. Subordination provisions always include the amounts that will be subordinated (principal only, principal and interest, interest only, and so forth), and they always describe the senior obligations to which the bond is subordinated.

EXAMPLE 7.16D: Subordination Provisions

The rights of the holder hereof to the principal sum or any part thereof, and the interest due thereon, are and shall remain subject and subordinate to the claims as to principal and interest of the holders of 7¾% First Mortgage Bonds of the corporation, and upon dissolution or liquidation of the corporation no payment shall be due or payable upon this debenture until all claims of the holders of said bonds shall have been paid in full.

Voting Rights

A rare privilege accorded to bond holders in a minority of jurisdictions is the right to vote on corporate matters, if so authorized in the articles of incorporation.[61] If holders of debt securities are permitted to vote in corporate elections, they are treated like a separate class of voting shareholders.

§ 7.17 IMPORTANT CONSIDERATIONS REGARDING DEBT AND EQUITY

In planning the capitalization of the corporation, the drafter has tremendous flexibility. The necessary capital may be raised any number of ways and represented by any of the myriad debt or equity securities depending upon the expectations of the investors and the selected corporate capitalization structure. However, certain practical matters should be considered in choosing between debt and various classes of equity securities.

Anticipate Later Financing

If the capital structure of the corporation is simple in the beginning, management will have greater flexibility to raise money in the future. If the initial corporate structure has several classes of stock and various debt securities, it will be considerably more difficult to create new classes of stock later, since new classes will probably affect the rights of the existing shareholders, whose approval must be acquired for any such amendments to the articles of incorporation. It is important, therefore, to consider the future capital needs of the business at the outset, and to plan the initial capital structure with those predictions in mind.

Advantages to Common Shareholders Through the Use of "Senior" Securities

Preferred stock and bonds are commonly called "senior" securities because of their special preferential rights. Preferred shareholders are usually entitled to a dividend preference and liquidation preference, insuring a return on and a return of their investment. Similarly, holders of debt securities have a right to interest and a right to repayment of their obligation upon maturity. The common shareholders, you will recall, have no special rights to dividends, and are entitled to share in the assets in liquidation, but only after the holders of debt securities and the holders of preferred stock have been satisfied. However, the common shareholders may gain an advantage through the use of "senior" securities when the expected profit return on capital each year exceeds the payments which must be made, either in dividends or interest, to the holders of senior securities. The converse is also true. If the expected profit return on capital each year is less than the payments which must be made to the holders of senior securities, the common shareholders are at a disadvantage.

These principles may be illustrated by considering the following example. Suppose Trouble Incorporated needs $100,000 capital for the operation of its business and can reasonably predict profits *after taxes* of $20,000 or more each year. Assuming federal income tax rates of 17% on the first $25,000 profit, the profit *before taxes* must be $24,096 or better. The securities which will be issued in exchange for the $100,000 capital may be any combination of common stock, preferred stock and debt. The profit return on capital after taxes is now estimated to be 20% or more each year. If 1,000 shares of common stock are issued to raise the $100,000 capital, the earnings per share would be computed as follows, assuming the estimated profit is realized:

Profit before tax	$24,096
Federal Income Tax (17%)	4,096
Profit after tax	$20,000
Earnings per share of common stock (1,000 shares)	$20.00

However, assume investors were found who were willing to take preferred stock with $100 par value and a preferred dividend rate of 16%. The $100,000 capital may be raised by issuing 500 shares of 16% preferred stock for $50,000, and 500 shares of common stock for $50,000. The common stock earnings per share would now be computed as follows:

Profit before taxes	$24,096
Federal Income Tax (17%)	4,096
Profit after taxes	$20,000
Preferred dividends (16% of $50,000)	8,000
Profit after preferred dividends	$12,000
Earnings per share of common stock (500 shares)	$24.00

Further assume debt securities with the same interest rate (16%) were issued for the $50,000 capital instead of preferred stock. The interest paid on debt securities is deductible as an expense before taxes, and this further improves the common shareholder's earnings. The statement would look like this:

Profit before interest	$24,096
Interest (16% of $50,000)	8,000
Profit before taxes	$16,096
Federal Income Tax (17%)	2,736
Profit after taxes	$13,360
Earnings per share of common stock (500 shares)	$26.72

A combination of all three will be even better for the common shareholders. Suppose $40,000 capital is raised by the sale of 400 common shares, $30,000 by the sale of 300 shares of 16% preferred stock and $30,000 by 16% debt securities. The result would be:

Profit before interest	$24,096
Interest (16% of $30,000)	4,800
Profit before taxes	$19,296
Federal Income Tax (17%)	3,280
Profit after taxes	$16,016

Dividends for preferred stock (16% of $30,000)	$ 4,800
Profit after preferred dividends	$11,216
Earnings per share of common stock (400 shares)	$28.04

Of course, the common stock takes the full risk that profits will reach or exceed expectations. The illustrated advantage depends upon the fact that the profits after taxes will be a greater percentage of capital than the percentage return required to be paid to the senior securities. Watch what happens to the last example if profits dipped to $15.000 (before deducting interest) rather than the predicted $24,096 or better. The statement would look like this:

Profit before interest	$15,000
Interest (16% of $30,000)	4,800
Profit before taxes	$11,200
Federal Income Taxes (17%)	1,904
Profit after taxes	$ 9,296
Dividends on preferred stock (16% of $30,000)	4,800
	$ 4,496
Earnings per share of common stock (400 shares)	$11.24

Contrast this earnings figure with the statement from the original capital structure where all $100,000 had been raised by the sale of $1,000 shares of common stock.

Profit before taxes	$15,000
Federal Income Tax (17%)	2,550
Profit after taxes	$12,450
Earnings per share of common stock (1000 shares)	$12.45

In the latter case, the common shareholders benefit more from complete common stock capitalization than from combinations of common stock, preferred stock and debt.

Tax Considerations

The concept of "double taxation" and its erosion of corporate profits was discussed in section 4.06. Debt securities avoid double taxation since the interest paid on the debt is deductible as an expense to the corporation, rather than being taxed as corporate profit. In this very important respect, debt securities enjoy a tax advantage over equity securities, since dividends are taxed first as corporate profit and again as individual income.

The solution, you might propose, would be to issue as many debt securities and as few equity securities as possible. You should also expect that the tax authorities of the federal government would have thought of this. They have. A disproportionate debt to equity ratio is called "thin incorporation" and the tax authorities may characterize interest payments on debt securities as dividends on equity securities for tax purposes, disallowing the interest expense deduction and requiring tax to be paid on the resulting increased profits. This restructuring has been upheld in the more severe cases, such as where the majority shareholder had loaned considerable sums to the corporation in return for separate debt securities, and where all shareholders have loaned disproportionately large amounts to a corporation with very little investment represented in common stock. It has been suggested that a good "debt to equity" ratio should not exceed 4:1 to avoid problems of thin incorporation.[62]

NOTES

1. Securities Act of 1933, 15 U.S.C.A. §§ 77a–77aa.
2. Securities Exchange Act of 1934, 15 U.S.C.A. §§ 78a–78m.
3. See L. Loss, Securities Regulation (1961); H. Henn, Corporations, §§ 291–308 (2d Ed. 1970); R. Deer, The Lawyer's Basic Corporate Practice Manual, §§ 6.01–6.08 (1970).
4. See M.B.C.A. §§ 15, 54 and Section 6.05.
5. See Section 6.06.
6. See former M.B.C.A. §2(h).
7. See Section 4.03.
8. See, e.g. Colorado, Col.Rev.Stat. §7–4–116.
9. Compare M.B.C.A. § 33 which was amended in 1979 to remove any distinction for treasury shares, including receipt of dividends.
10. See M.B.C.A. § 54 and Section 6.05.
11. Neb.Rev.Stat. § 21–2014.
12. E.g., California, West's Ann.Calif.Corp.Code § 409.
13. M.B.C.A. § 18.
14. E.g., Hawaii, Hawaii Rev.Stat. § 416–59.
15. See, e.g., Ala.Code § 10–2–96, which assigns an arbitrary value by $100 per share on no par shares.
16. See Section 6.05.
17. See former M.B.C.A. § 6 which was amended in 1979 to remove this rule. Most states still follow the former rule.
18. See, e.g., Colorado, Colo.Rev.Stat. § 7–6–101(2).
19. Dividends and other distributions are discussed in detail in Chapter 9.
20. M.B.C.A. § 23.
21. West's Modern Legal Forms § 2755.5.
22. See M.B.C.A. § 16.
23. See Section 13.01.
24. M.B.C.A. § 16.
25. E.g., Pennsylvania, 15 Pa.Stat. § 1601.
26. E.g., New York, McKinney Consol. Laws of N.Y.Bus.Corp.Law § 502.
27. West's Modern Legal Forms § 2755.10.
28. M.B.C.A. § 16, and see Form 7B in the Appendix.
29. See Section 7.05.
30. See the director's resolution for evaluation of property or services in Section 8.05.
31. M.B.C.A. § 19.
32. E.g., Oregon, Ore.Rev.Stat. § 57.106; and Florida, Fla.Stat.Ann. § 607.054.

33. Compare New Mexico's corporation statute which specifically allows shares to be issued for preincorporation services. N.Mex.Stat.Ann. § 53–11–19.
34. See Chapter 9.
35. See Section 6.05.
36. See Sections 4.03 and 6.05.
37. West's Modern Legal Forms § 2756.
38. West's Modern Legal Forms § 2756.10.
39. West's Modern Legal Forms § 2761.
40. West's Modern Legal Forms § 2761.1.
41. West's Modern Legal Forms § 2759.
42. M.B.C.A. § 33.
43. West's Modern Legal Forms § 2757.1.
44. West's Modern Legal Forms § 2757.2.
45. West's Modern Legal Forms § 2762.
46. A sample statement of cancellation appears as Form 7C in the Appendix.
47. See Section 7.08.
48. See West's Modern Legal Forms § 2758.
49. See M.B.C.A. § 52.
50. See M.B.C.A. § 23.
51. West's Modern Legal Forms § 2957. See also § 2957.1.
53. West's Modern Legal Forms § 2979.1.
54. Uniform Commercial Code (hereinafter cited U.C.C.) § 9–203. An example of a security agreement under the U.C.C. appears as Form 7D in the Appendix.
55. U.C.C. § 9–402.
56. U.C.C. § 9–401. A sample financing statement is Form 7E in the Appendix.
57. Securities Act of 1933, 15 U.S.C.A. §§ 77a–77aa; Trust Indenture Act of 1939, 15 U.S.C.A. §§77aaa–77bbbb.
58. A skeletal trust indenture, including articles, section and subsection headings, is Form 7F in the Appendix.
59. West's Modern Legal Forms § 2972.1.
60. West's Modern Legal Forms § 2972.2.
61. E.g. Delaware, 8 Del.Code Ann. § 221; Pennsylvania, 15 Pa.Stat. § 1309.1.
62. See H. Henn, The Law of Corporations § 166 n. 15 (1970).

8

CORPORATE MEETINGS

§ 8.01 TYPES AND PURPOSES OF MEETINGS

Corporate activity is conducted through meetings of the internal corporate groups, the incorporators, the directors and shareholders. Under the common law of corporations the directors of a corporation do not act individually, but may act only as a board collectively convened. Shareholder and incorporator activities are somewhat more individual, but the democratic rule that the majority will control the minority is applied to both groups. Traditionally, corporate action may not be taken unless it has been approved by one of these groups duly convened at a meeting. In theory, then, any action taken by the corporation requires an approving resolution by the appropriate intra-corporate group, but, of course, it would be much too cumbersome to hold a meeting of one of these groups for daily business decisions. Instead, the board of directors delegates authority for the everyday business affairs of the corporation to the officers, and the board is responsible for the supervision of officer activities. A corporate decision of any magnitude should be made at a director's meeting with an appropriate resolution set forth in the minutes. Some major corporate decisions require shareholder approval.[1] There are three types of corporate meetings which may be held by the intra-corporate groups described above: organizational meetings, regular meetings, and special meetings. State statutes variously detail the need for such meetings, authority to call a meeting, notice required, and time and place for the meetings.

§ 8.02 REQUIREMENT FOR ORGANIZATIONAL MEETINGS

As previously discussed, the corporation is actually formed by the filing of articles of incorporation or the issuance of a certificate of incorporation, depending upon the law of the particular jurisdiction.[2] Thereafter, certain organizational meetings must be held, and each state statute should be consulted to determine the parties to the meeting and the business which must be conducted. The organizational meetings are, by most statutes, a required condition which should be satisfied prior to the corporation's commencing business operations.[3] Some states require only an organizational meeting of the incorporators, while others require only an organizational meeting of the directors. Still other states permit either group to hold an organizational meeting, and a few also require the shareholders to have an organizational meeting.[4] Section 57 of the Model Act provides:

> "After the issuance of the certificate of incorporation an organization meeting of the board of directors named in the articles of incorporation shall be held, either within or without this State, at the call of a majority of the directors named in the articles of incorporation, for the purposes of adopting by-laws, electing officers and transacting such other business as may come before the meeting."

§ 8.03 DIRECTORS' ORGANIZATIONAL MEETING

The Model Act formerly provided that the directors' organizational meeting was to be called by the incorporators. The Act was amended to provide that the meeting may be called by a majority of the directors named in the articles of incorporation, and several states have adopted this rule. In Delaware the call may be issued by either the incorporators or the directors.

Notice

Under the Model Act at least three days notice must be given to the directors by mail. Some states require no notice, Illinois demands for one day's notice, and Pennsylvania prescribes five days' notice.[5] It is common practice, however, to secure a waiver of notice from the initial directors to avoid the observance of the notice period.

EXAMPLE 8.03A: Waiver of Notice of Organization Meeting

We, the undersigned, constituting all of the directors of Happiness, Inc., a corporation organzied under the laws of the State of Colorado, do hereby severally waive notice of the time, place and purpose of the first meeting of directors of said corporation, and consent that the meeting be held at the corporate offices, and on the 10th day of June, 1981, at 8:30 o'clock a. m., and we do further consent to the transaction of any and all business that may properly come before the meeting.

Dated June 10, 1981

____

____

________________ [6]

Nominal Directors

The articles of incorporation need only contain the names of the persons who will serve as directors until the first meeting of shareholdes and until their successors have been elected and qualified. If the persons named are the actual directors of the corporation, they may transact all necessary director business at their organizational meeting. On the other hand, if the directors named in the articles of incorporation are only "nominal" or "dummy" directors they should not be expected to conduct any more than the formal statutory business, such as adoption of by-laws and electing officers. Other corporate business should be reserved for the consideration of the actual directors elected by the shareholders. If multiple meetings are undesirable, it is possible to have the nominal directors submit their resignation one by one at the organizational meeting, with the board of directors adopting by resolution the resignation of each "dummy" director and electing the actual director to fill the vacancy. The actual director will then serve for the period of his predecessor "dummy" director or until the next shareholder meeting when his successor will be elected.

EXAMPLE 8.03B: Resignation of a "Dummy" Director

To the Board of Happiness, Inc.
I regret that, owing to other business commitments, I am no longer able to act as a director, and I hereby tender my resignation as a director of Happiness, Inc. to take effect as and from the 10th day of June, 1981.

[Signature] [7]

EXAMPLE 8.03C: Acceptance of Resignation of a Director

RESOLVED, that the resignation of John Head as director of this Corporation be accepted to take effect on June 10, 1981, and that the secretary be and he hereby is directed to notify John Head of such acceptance.[8]

If the directors are not named in the articles of incorporation, they are elected at an organizational meeting of the incorporators. The elected directors then hold an organizational meeting after their election.

§ 8.04 INCORPORATORS' ORGANIZATIONAL MEETING

If the state statute requires an organizational meeting of the incorporators, it may also require a period of notice preceding the meeting and will usually specify the business to be conducted at the meeting. Of course, the statutory requirements must be strictly followed. Even if the statute does not specify the business which must be conducted, there are certain matters which are normally considered at all organizational meetings.

§ 8.05 BUSINESS CONDUCTED AT ORGANIZATIONAL MEETINGS

Most state statutes describe specific matters which must be on the agenda of an organizational meeting. In the Model Act the organizational meeting of the board of directors must consider the adoption of the by-laws and the election of officers. The statute further provides that the directors should consider "transacting such other business as may come before the meeting."[9] The following discussion considers the chronological order of business at an organization meeting and includes examples of how the minutes should reflect the actions taken:

Determining a Quorum and Electing a Chairman and Secretary

Counsel (or whoever is initially presiding) should be certain to determine the presence of a quorum according to the rules stated in the articles of incorporation or the statute. If a quorum is present, that fact should be noted in the minutes, and the persons present and absent should be named in the minutes. A chairman and a secretary should be elected as the first order of business of an organizational meeting. The secretary is responsible for the minutes of the meeting. The minutes are frequently prepared in advance by counsel and they may serve as an agenda for the meeting, but the secretary should review them for accuracy and add any necessary new material. The chairman is responsible for an orderly meeting.

EXAMPLE 8.05A: Recital of Quorum, Election of Chairman and Secretary

The following directors named in the articles of incorporation were present: Edward Giles, Gary Nakarado, and Terryl Gorrell. The following directors named in the articles of incorporation were absent: Michael Corrigan and Karen Burn.

The presence of the foregoing directors constituted a quorum. By unanimous vote of the directors, Edward Giles was elected Chairman of the meeting and Gary Nakarado was elected Secretary of the meeting.

Notice or Waivers of Notice

If the appropriate notice has been given to the members of the group, a copy of the notices should be presented to the meeting and attached to the minutes. If waivers of notice of the meeting have been obtained from those present, the waivers should be presented to the meeting and affixed to the minutes as an attachment.

EXAMPLE 8.05B: Recognition of Notice

The Secretary presented the waiver of the notice of the meeting signed by all of the directors, which was ordered filed with the minutes of the meeting.

Determining Actual Directors

If the directors named in the articles of incorporation are the actual directors of the corporation, no further action regarding their status is required at the organizational meeting. Some states do not require the naming of the initial directors and all states would permit "dummy" directors to be named in the articles of incorporation.

Election of the actual directors will occur at the incorporators' organizational meeting if no directors are named in the articles of incorporation. If the directors are named, but are only nominal directors, the incorporators may obtain the resignation of the nominal directors and replace them with actual directors at the incorporators' organizational meeting.

The directors' organizational meeting should be conducted by actual directors. Therefore, if nominal directors are named in the articles of incorporation, and there is no meeting of incorporators to elect actual directors, the nominal directors may be replaced by the procedure suggested in Section 8.03.

The resignations of nominal directors should be affixed to the minutes of the meeting wherein they were replaced.

EXAMPLE 8.05C: Determining Actual Directors

The Secretary announced that resignations had been received from Michael Corrigan and Karen Burn, who were nominal directors of the corporation named in the articles of incorporation.

Thereupon, upon motion duly made, seconded and unanimously adopted it was

> RESOLVED, that the resignations of Michael Corrigan and Karen Burn as directors of this corporation be accepted to take effect on the date of this meeting, and that the Secretary be and he hereby is directed to notify Mr. Corrigan and Ms. Burn of such acceptance, and to affix to the minutes of this meeting the original written resignations of these directors.
>
> FURTHER RESOLVED, that John Carver and James Burghardt shall be appointed to fill the vacancies created by the resignations of these directors, and shall be directors of this corporation to serve until the first annual meeting of the shareholders or until their successors are otherwise elected and qualified.

Presentation of Articles of Incorporation

The articles of incorporation, returned by the Secretary of State with the certificate of incorporation, should be presented to the meeting and affixed as an attachment to the minutes of the meeting. It is not necessary to have a resolution approving the articles of incorporation, but the minutes should reflect the fact that the articles were presented to the meeting and the secretary should be instructed to insert the certificate of incorporation and the copy of the articles in the minute book.

EXAMPLE 8.05D: Presentation of the Articles of Incorporation

The Chairman submitted to the meeting a copy of the Articles of Incorporation of the corporation and an original receipt showing payment of the statutory organization taxes and filing fees. The Chairman reported that the original of these Articles of Incorporation had been filed in the office of the Secretary of State, State of Nebraska, on November 1, 1981. Thereupon, upon motion duly made, seconded and unanimously adopted, it was:

> RESOLVED, that the Articles of Incorporation as presented be, and they hereby are, accepted and approved and that said Articles of Incorporation, together with the original receipt showing payment of the statutory orga-

nization taxes and filing fees, be placed in the minute book of the corporation.

Approval of the Action Taken at Previous Meetings

When the incorporators have held an organizational meeting and the board of directors subsequently conducts an organizational meeting, it is customary for the board to approve, ratify, and confirm all of the actions taken at the incorporators meeting. This has the effect of making the action of the incorporators the action of the board of directors. For example, if the incorporators have adopted the by-laws, this resolution grants the same approval by the board of directors. The minutes of the board of directors' meeting may be abbreviated in this fashion.

EXAMPLE 8.05E: Acceptance of Incorporator's Action

The Secretary presented to the meeting the minutes of the first meeting of incorporators of the Corporation together with a copy of the by-laws adopted by the incorporators at their meeting held November 10, 1981.

On motion duly made and seconded, it was unanimously

RESOLVED that the minutes of the first meeting of the incorporators of the Corporation held on November 10, 1981 be and they hereby are in all respects ratified, approved and confirmed.

FURTHER RESOLVED that the by-laws adopted by the incorporators at such first meeting hereby are adopted by this Board as and for the by-laws of this Corporation.

Approval of By-Laws

Counsel should have drafted the by-laws pursuant to the instructions of the incorporators prior to the organizational meeting. The by-laws are presented to the organizational meeting of the appropriate group for approval. The minutes must contain a resolution that the by-laws have been approved, and the secretary should be instructed to insert a copy of the by-laws in the minutes.

EXAMPLE 8.05F: Acceptance of By-laws

RESOLVED, that the By-laws submitted to the meeting be and are adopted as the By-laws of the Corporation, and that the Secretary is instructed to insert a copy of such By-laws, certified by him, in the minute book immediately following the Certificate of Incorporation with affixed duplicate original of the Articles of Incorporation.

Approval of Corporate Seal

As a part of the corporation supplies counsel should have obtained a corporate seal designed to the specifications of the incorporators. It is customary to adopt a resolution to the effect that the seal is accepted as the corporate seal, and to affix the seal to the margin in the minute book page. If any regulation of the use of the seal is comtemplated, the regulation should be specified in the resolution.

EXAMPLE 8.05G: Acceptance of the Seal

RESOLVED, that the seal now produced by the secretary, an impression whereof is now made in the minute book of the Company, be adopted as the seal of the Company, and that such seal shall not be affixed to any deed or instrument of any description, except in the presence of an officer, or director, and the secretary of the Company, who shall respectively sign said deed or instrument.[10]

Approval of Share Certificates

Share certificates are also obtained as part of the corporate supplies and a specimen certificate should be presented to the meeting. The share certificate should contain all appropriate legends if share transfer restrictions are contemplated, and it will also contain other matters unique to the particular corporation or the particular class of stock.[11] The share certificate is accepted by resolution at the organization meeting, and the secretary should insert the specimen in the minute book.

EXAMPLE 8.05H: Acceptance of Share Certificate

RESOLVED, that the form of share certificate presented at this meeting is adopted as the form of share certificate for this Corporation; and the Secretary of the meeting is instructed to append a sample of such certificate to the minutes of this meeting.

Authorization to Issue Shares

At their organizational meeting the incorporators authorize the board of directors to issue the shares of the company. The incorporators' authorizing resolution should be contained in the minutes of their meeting. The directors may adopt such a resolution at their organizational meeting authorizing the appropriate officers (as specified by statute) to issue the certificates.

EXAMPLE 8.05I: Authority to Issue Shares

RESOLVED, that the President and Secretary of this Corporation be, and they are hereby authorized to issue certificates for shares in the form as submitted to this meeting and ordered attached to these minutes.

Acceptance of Transfers of Share Subscriptions from Dummy Incorporators

As previously indicated, some states require that incorporators subscribe for shares as a condition to qualification as an incorporator.[12]

To satisfy this rule, incorporators frequently subscribe for shares they do not intend to purchase. In such cases the incorporators may assign their pre-incorporation share subscriptions to actual investors in the company, and these subscription transfers are presented to the organizational meeting of the board of directors for their approval.

A resolution reflecting the directors' approval should be included in the minutes.

EXAMPLE 8.05J: Transfer of Subscription

Dated, Kearney, Nebraska, October 31, 1981
FOR VALUE RECEIVED, I, Peter McLaughlin, hereby sell, assign and transfer unto James Johnston all my right, title and interest as subscriber to the shares of common stock of Happiness, Inc., which subscription was executed by me on the 3rd day of October, 1981, and, when accepted, entitles me to receive 500 shares of the common stock of Happiness, Inc., and I hereby direct said corporation to issue certificates for said shares of stock to and in the name of the aforesaid assignee, or his nominees or assigns.

[*Signature*]

EXAMPLE 8.05K: Adoption of Assignment

RESOLVED, that the assignment and transfer of a stock subscription from Peter McLaughlin to James Johnston, dated October 31, 1981, is hereby approved and accepted on behalf of the Corporation.

Acceptance of Share Subscriptions

The preincorporation share subscriptions are offers to the corporation to buy shares when the corporation is formed.[13] Now that formation has been accomplished, the board of directors should accept the offers on behalf of the corporation and thereby obligate the subscribers to pay for the shares they have offered to purchase. The acceptance of the share subscriptions is accomplished by a resolution and each subscription should be listed therein, specifying the number of shares the subscriber has offered to purchase, the class of the security, the par value, and the price at which the offer was tendered.

If cash has been offered for share purchases, the resolution accepting the offer need only state the amount offered when describing the consideration.

EXAMPLE 8.05L: Acceptance of Cash Subscriptions

RESOLVED, that the written offers dated March 23, 1981, pertaining to the issuance of shares of the Corporation, to wit:

Name	Number of Shares	Consideration
Fred Thomas	100	$100
Bill Jones	1000	$1000
Tom Myers	750	$750

be, and the same hereby are, in all respects accepted for and on behalf of the Corporation.

However, if property or services are offered to the corporation in exchange for shares, the board of directors must evaluate the property and services consistently with the valuation rule of the particular jurisdiction. Some states require that an actual market value be determined and used by the board in appraising property or services, but most states permit the board of directors to determine the value of the property or services in good faith considering the best interests of the corporation. The directors' determination of value is then conclusive in the absence of fraud. The Model Act takes this latter approach in Section 19. It should be noted that this determination of value is critical to the issuance of par value securities, since they cannot be sold for less than par value. Thus, if a corporation received an offer to transfer certain land in exchange for 1,000 shares which had a $10.00 par value, the board of directors must appraise the land, by the appropriate valuation rule, at an amount equal to or greater than $10,000. The resolution in the minutes should state the valuation determination.

EXAMPLE 8.05M: Acceptance of Property Subscription

RESOLVED that this Corporation hereby accepts the offer of Steven Petterson and Mary Petterson, as joint owners, to sell and convey to it good and marketable title to the fee of the premises known as 3590 E. Nobles Road, Littleton, Colorado, 80122, together with the buildings thereon, and all personal property belonging to them used in connection with the premises free and clear of all liens and encumbrances, in consideration of this Corporation's issuing and delivering to Steven Petterson, or his nominee, certificates for 150 of its fully paid and non-assessable shares of its 5½% perferred shares, $100 par value, and of its issuing and delivering to Mary Petterson, or her nominee, certificates for 150 of its fully paid and nonassessable 5½% preferred shares, $100 par value. The Board of Directors does hereby adjudge and declare that said property is of the fair value of $30,000, and that the same is necessary for the business of the corporation.

If the cash or property is not immediately tendered with the offer, it is appropriate for the directors to adopt a resolution to assess or "call" the consideration due. Unless otherwise stated in the subscription agreement, the offer is payable in full upon acceptance, but the board of directors may permit payment by installments.[14]

EXAMPLE 8.05N: Partial Call of Subscriptions

RESOLVED, that a call of fifty per cent is hereby made upon each and every share of the capital stock of the Company subscribed for, and same is to be paid by each subscriber to the treasurer of the Company, on or before the 30th day of November, 1981, (and that the president and secretary issue certificates of full-paid stock therefor).[15]

[or]

EXAMPLE 8.05O: Full Call of Subscriptions

RESOLVED, that a full call is hereby made upon each and every share of the capital stock of the Company subscribed for, and the same is to be paid by each subscriber to the treasurer of the Company, on or before the 30th day of November, 1981, (and that the president and secretary issue certificates of full-aid stock therefor).

Authorization to Issue Shares

The board of directors should authorize the officers of the company to issue the shares represented by the accepted share subscriptions. The resolution generally states that the company will issue and deliver the prescribed number of shares to the subscriber when full consideration has been received for the shares. The resolution should further state that the officers are authorized to execute common stock certificates and to register the shares in the names of the subscribers.

EXAMPLE 8.05P: Authorization to Issue Shares

RESOLVED, that the Corporation issue and deliver to those persons upon receipt of the consideration therefor, pursuant to the terms of the aforesaid offer, certificates representing the subscribed shares of the Corporation each, no par value, such shares to include the shares originally subscribed for by the subscribers to the capital stock of the Corporation, and subsequently assigned to the officers; and

FURTHER RESOLVED, that the officers of the Corporation be, and they hereby are, authorized, empowered and directed to take any and all steps, and to execute and deliver any and all instruments in connection with consummating the transaction contemplated by the aforesaid offer and in connection with carrying the foregoing resolutions into effect; and

FURTHER RESOLVED that upon the delivery to this Corporation of proper instruments of conveyance, assignment, and transfer, in such form as counsel for this Corporation may approve, the proper officers of this Corporation may approve, the proper officers of this Corporation be and they hereby are authorized and directed to issue to Mr. Petterson an appropriate certificate for 150 of its 5½% preferred shares, $100 par value, when issued as provided in the foregoing resolutions shall be fully paid and nonassessable.

If the shares are being issued in excess of par value, or if no par value shares are being sold in excess of stated value, the board of directors should resolve to allocate the excess consideration to the capital surplus account.

EXAMPLE 8.05Q: Allocation to Capital Surplus

RESOLVED, that since the corporation's common stock has a par value of $.50 per share, and the same is being sold for $1.00 per share, the excess amount over par value shall be allocated to the capital surplus account of the corporation;

AND FURTHER RESOLVED, that seventy-five percent (75%) of the consideration received for the company's no par value stock shall be allocated and applied to the capital surplus account of the corporation.

Reimbursement of Fees

The authority to pay the expenses in connection with the formation of the corporation eminates from the board of directors in its organizational meeting. The treasurer is authorized to pay all taxes, fees, and other expenses incurred and to be incurred in connection with the organization of the company and to reimburse any persons who have made expenditures on behalf of the company during the formation procedure. Legal fees are included herein, and that usually makes this a very important resolution.

EXAMPLE 8.05R: Authorization to Pay Expenses

FURTHER RESOLVED that the President of this Corporation be and she hereby is authorized to pay all charges and expenses incident to or arising out of the organization of this Corporation, including the bill of John Jacobs, Esq., for legal services in connection therewith in the sum of $500, and to reimburse any person who has made any disbursement therefor.

Adoption of Preincorporation Agreements

Prior to formation of the corporation, the incorporators may have entered contracts on behalf of the corporation to insure that the necessary resources for conducting business are available when the corporation is formed. For example, they may have leased office space or may have purchased equipment on credit. We have already discussed the general resolution by which the board of directors adopts the acts of the incorporators,[16] but any prior agreements which were intended to benefit the corporation should be adopted by a separate, specific resolution during the board of directors' organizational meeting. The resolution should summarize the terms of the agreement and clearly express the directors' approval of transaction.

EXAMPLE 8.05S: Adoption of Pre-Incorporation Agreements

RESOLVED, the board of directors has reviewed and considered an agreement on behalf of the corporation, a copy of which is attached to these minutes as Exhibit A and incorporated herein by reference. This agreement was entered into prior to the existence and formation of the corporation, and the board of directors, having considered the agreement on behalf of the corporation, does hereby adopt and accept the agreement according to its terms, and agrees that the corporation shall perform all of its obligations and be entitled to all of its rights as specified therein.

Election of Corporate Officers

The officers of the corporation are elected by the board of directors, and this is one of the prescribed matters to be considered at the directors' organizational meeting. The resolution for the election usually states that the officers will serve for a stated period of time or at the discretion of the board of directors and fixes their compensation.

EXAMPLE 8.05T: Election and Compensation of Officers

The meeting then proceeded to the election of officers to serve until the next annual meeting of stockholders or until their successors are elected and qualified. The following nominations were made and seconded:

President	Karen Gehlhausen
Vice President	Scott Charlton
Secretary	John Baker
Treasurer	Keith Burn

There being no further nominations the foregoing persons were unanimously elected to the office set opposite their respective names.

RESOLVED, that until further action by the Board of Directors, the annual salaries of the officers are fixed in the following amounts, effective as of January 1, 1982, and payable in twelve equal monthly installments:

President $50,000.00
Vice President $45,000.00
Secretary $25,000.00
Treasurer $20,000.00

Bank Resolution

The bank resolution prescribes the authority for the maintenance of a bank account and names those persons who have authority to obligate the corporation in banking matters. Every bank supplies a form for a banking resolution, which should be completed and attached to the minutes of the meeting. The minutes contain a director resolution which authorizes the opening of the bank account, and adopts by reference the provisions in the attached bank form. The director's resolution looks like this:

EXAMPLE 8.05U: Acceptance of Bank Resolutions

RESOLVED that the funds of the Corporation be deposited in the Central City Bank of Kearney, 6th Avenue Branch, and that the printed resolutions supplied by that bank, as filed in at this meeting, be attached to the minutes of this meeting and be deemed resolutions of this Corporation duly adopted by the Board of Directors.

The bank resolution looks like this:

EXAMPLE 8.05V: Bank Resolution

....

(Name of Corporation)

I HEREBY CERTIFY TO .... , that at a meeting of the Board of Directors of .... , a corporation organized under the laws of the State of .... duly called (a quorum being present) and held at the office of said corporation, No .... in the city of ....State of ....on the .... day of .... , 19 ., the following resolutions were duly adopted and are now in full force and effect:

Depositary and Signing Resolution

RESOLVED, that the above bank be designed as a depositary of this corporation and that funds of this corporation deposited in said Bank be subject to withdrawal upon checks, notes, drafts, bills of exchange, acceptances, undertakings of other orders for the payment of money when signed on behalf of this corporation by any .... of its following officers to wit:

(Number)

RESOLVED, that the above bank, is hereby authorized to pay any such orders and also to receive the same for credit of or in payment from the payee or any other holder without inquiry as to the officer or tendered in payment of his individual obligation.

Borrowing Resolution

RESOLVED, that ....

....

be and they hereby are authorized to borrow from time to time on behalf of this corporation from the above bank sums of money for such period or periods of time, and upon such terms, rates of interest and amounts as may to them in their discretion seem advisable, and to execute notes or agreements in the forms required by said Bank in the name of the corporation for the payment of any sums so borrowed.

That said officers are hereby authorized to pledge or mortgage any of the bonds, stocks or other securities, bills receivable, warehouse receipts or other property real or personal of the corporation, for the purpose of securing the payment of any moneys so borrowed; to endorse said securities and/or to issue the necessary powers of attorney and to execute loan, pledge or liability agreements in the forms required by the said bank in connection with the same.

That said officers are hereby authorized to discount with the above bank any bills receivable held by this corporation upon such terms as they may deem proper.

That the foregoing powers and authority will continue until written notice of revocation has been delivered to the above bank.

RESOLVED, that the secretary of this corporation be and he hereby is authorized to certify to the above bank, the foregoing resolutions and that the provisions thereof are in conformity with the charter and by-laws of this corporation.

I FURTHER CERTIFY that there is no provision in the charter or by-laws of said corporation limiting the power of the board of directors to pass the foregoing resolutions and that the same are in conformity with the provisions of said charter and by-laws.

I further certify that the following are the genuine signatures of the persons now holding office in said company as indicated opposite their respective signatures.

.... (Title)

.... (Title)

.... (Title)

.... (Title)

.... (Title)

IN WITNESS WHEREOF, I have hereunto set my hand as secretary of said corporation and affixed the corporate seal this ..... day of ...., 19 ....

....

(CORPORATE SEAL) *Secretary of the Corporation*

NOTE: *In case the secretary or other recording officer is authorized to sign checks, notes, etc., by the above resolutions, this certificate must also be signed by a second officer of the corporation.*

DeLano Service, Allegan, Mich. Form R-10 [A8995]

Application for Qualification as a Foreign Corporation

If management contemplates doing business in another jurisdiction, the board of directors must adopt an appropriate resolution authorizing the officers of the corporation to apply for admission and qualification of the corporation as a foreign corporation in any other jurisdiction where it plans to do business.

EXAMPLE 8.05W: Authorization of Foreign Qualification

RESOLVED that the officers of the Corporation be authorized and directed to qualify the Corporation as a foreign corporation authorized to conduct business in the State of Kansas, and in connection therewith to appoint all necessary agents or attorneys for service of process and to take all other action which may be deemed necessary or advisable.

Appointment of Resident Agents and Office

The articles of incorporation name the resident agents but their appointment should be "ratified" by a resolution of the board of directors. Similarly, the establishment of a principal office of the corporation may be resolved at the organizational meeting.

EXAMPLE 8.05X: Principal Office and Agent

RESOLVED, that the Articles of Incorporation correctly state the principal office of the corporation and that the person named therein as registered agent shall remain registered agent until subsequently changed by a resolution of the board of directors.

Designation of Counsel and Accountant

The board of directors may designate a certain attorney to act as the general counsel of the company, if appropriate, and also may specify the persons to be retained as the corporation's accountants.

EXAMPLE 8.05Y: Appointment of Counsel and Auditors

RESOLVED, that Charles Counsel be hereby appointed to act as attorney for the Company, and that he be paid the ordinary professional charges for his services as attorney.

RESOLVED, that Albert Audit & Co. be hereby appointed auditors of the Company for the ensuing year, and that the remuneration for their services as such auditors be the sum of $2,000.00.[18]

Authority to Use Assumed Name

Some states permit the corporation to conduct business under an assumed name as long as it is not deceptively similar to another reserved or registered name. Usually the corporation must file a statement of assumed name with the appropriate state official.[19] The authority to use such a name and to file the statement comes from a resolution of the directors.

EXAMPLE 8.05Z: Adoption of An Assumed Name

RESOLVED, that the corporation may use the name "Black, Inc." as an assumed business name to carry out its purposes and objects in the state of Oregon, and that the officers of the corporation as required by the statute are authorized to execute such documents as are necessary to accomplish the registration of the corporation's assumed business name.

Adoption of Section 1244 Plan

The Small Business Tax Revision Act added Section 1244 of the Internal Revenue Code to offer special loss protection for shareholders of a small corporation. Losses on "Section 1244 stock" are fully deductible as business losses up to certain dollar limits per year, instead of being treated as capital losses. The substantive rules of Section 1244 stock are explained in detail in section 4.06.

In order to qualify to issue "Section 1244 stock," a corporation must be a "small business corporation," as defined by the statute.[20]

We are here concerned with the actual adoption of a Section 1244 plan. If a new corporation qualifies as a "small business corporation" and issues "Section 1244 stock," shareholder losses from the sale of the shares will be treated as ordinary losses and not capital losses, so the shareholder may offset the lost value against his "ordinary" income, such as his wages, interest, dividends, etc. Unless the shares qualify under Section 1244, any such losses only offset capital gains and, to a very limited extent, ordinary income.

In 1978, Section 1244 was amended to eliminate the requirement of a *written* plan to issue stock under Section 1244. Nevertheless, it is good practice to indicate clearly the intention to qualify for this special protection by adopting a resolution and plan in the minutes of the organizational meeting. A proper resolution adopting a Section 1244 plan should restate the statutory requirements. Thus, it should recite that the payment of the shares will be in cash or other property, but not securities or services and should provide that the stock will be offered for sale at a price not less than par value of the shares and not in excess of an aggregate of $1,000,000.

EXAMPLE 8.05AA: Section 1244 Plan

A plan was read and (unanimously) adopted for the issuance of common stock of the corporation to qualify the same as "small business corporation" stock under the provisions of Section 1244 of the Internal Revenue Code of 1954. The Secretary was directed to place a copy of the Plan immediately following these minutes.

Plan for Issuance of Stock

1. The corporation shall offer and issue under this Plan, a maximum of 50,000 shares of its common stock at a maximum price of ten dollars ($10.00) per share.
2. This offer shall terminate by:
   (a) Complete issuance of all shares offered hereunder, or
   (b) Appropriate action terminating the same by the Board of Directors and the Stockholders, or
   (c) By the adoption of a new Plan by the Stockholders for the issuance of additional stock under Section 1244, Internal Revenue Code.
3. No increase in the basis of outstanding stock shall result from a contribution to capital hereunder.
4. No stock offered hereunder shall be issued on the exercise of a stock right, stock warrant, or stock option, unless such right, warrant, or option is applicable solely to unissued stock offered under the Plan and is exercised during the period of the Plan.
5. Stock subscribed for prior to the adoption of the Plan, including stock subscribed for prior to the date the corporation comes into existence, may be issued hereunder, provided, however, that the said stock is not in fact issued prior to the adoption of such Plan.
6. No stock shall be issued hereunder for a payment which, along or together with prior payments, exceeds the maximum amount that may be received under the plan.

7. Any offering or portion of an offer outstanding which is unissued at the time of the adoption of this Plan is herewith withdrawn. Stock rights, stock warrants, stock options or securities convertible into stock, which are outstanding at the time this Plan is adopted, are likewise herewith withdrawn.
8. Stock is issued hereunder shall be in exchange for money or other property except for stock or securities. Stock issued hereunder shall not be in return for services rendered or to be rendered to, or for the benefit of, the corporation. Stock may be issued hereunder however, in consideration for cancellation of indebtedness of the corporation unless such indebtedness is evidenced by a security, or arises out of the performance of personal services.
9. Any matters pertaining to this issue not covered under the provisions of this Plan shall be resolved in favor of the applicable law and regulations in order to qualify such issue under Section 1244 of the Internal Revenue Code. If any shares issued hereunder are finally determined not to be so qualified, such shares, and only such shares shall be deemed not to be in the Plan, and such other shares issued hereunder shall not be affected thereby.
10. The sum of the aggregate amount offered hereunder plus the equity capital of the corporation amounts to $500,000.00.
11. The date of adoption of this Plan is November 15, 1981.[21]

This plan should be copied directly into the minutes.

Subchapter S Election

In order to elect taxation under Subchapter S the corporation must again qualify as a "small business corporation," but remember that the Subchapter S definition of a "small business corporation" is different than the Section 1244 definition.[22] Under a Subchapter S election the income of the corporation is treated as ordinary income of the shareholders and thus the problem of "double taxation" of corporation income is avoided.[23] If the corporation (and the shareholders) desire to be taxed under Subchapter S, the board of directors should adopt a resolution which provides that the corporation has elected to be taxed as a small business corporation. The resolution should state that the corporation meets the statutory requirements and has elected to be taxed under that provision.

EXAMPLE 8.05BB: Subchapter S Election

WHEREAS, the corporation qualifies as a small business corporation under Section 1371(a) of the Internal Revenue Code of 1954, as amended; and

WHEREAS, the board of directors deems it to be in the best interests of the corporation and the shareholders to elect to be taxed as a small business corporation under the Internal Revenue Code of 1954, as amended, it is

RESOLVED, that the election to be so taxed be submitted to the shareholders for their consent, and that, upon obtaining said consent, the officers of the corporation shall prepare and submit the necessary documents and forms to accomplish said election under Section 1372 of the Internal Revenue Code of 1954, as amended.

The shareholder consent, duly executed, should be attached to the minutes.

EXAMPLE 8.05CC: Shareholder Consent to Subchapter S

________________, 19__

We, the undersigned, being all of the stockholders in Happiness, Inc., a Nebraska corporation, hereby consent to the election under Section 1372(a) of the Internal Revenue Code of 1954 as amended, to be treated as a small business corporation for income tax purposes, and submit the following information:

Name and Address of Corporation: Happiness, Inc., 200 West 14th Avenue, Kearney, Nebraska

Name and Address of Stockholders:	No. of Shares	Date Acquired
		[24]

Dates of Meetings

The by-laws usually permit the directors to establish their regular date for their board and a resolution establishing these dates and times is appropriate at the organizational meeting. The resolution usually also identifies the place for the meeting.

EXAMPLE 8.05DD: Place of Regular Meetings

RESOLVED that regular meetings of the Board of Directors be held at the office of the Corporation at 200 West 14th Avenue, Kearney, Nebraska, on the third Wednesday of the months of February, May, September, and December, and that a regular meeting also be held in April immediately following the annual meeting of shareholders. No notice shall be required to be given of any of these regular meetings.

Delegation of Authority to the Officers

The board of directors may adopt a resolution defining the authority of the officers. These resolutions are usually drafted broadly, and they really are not necessary. They may be useful, however, as a written record that the directors have delegated the authority. The typical rubric of the resolution includes grants of authority to the president and vice president to conduct all business on behalf of the corporation, to sign all documents necessary in the ordinary course of business in the corporate name, and to perform other necessary managerial acts on behalf of the corporation. The secretary is authorized to procure and maintain necessary corporate books and records, and to open and maintain a stock transfer ledger in accordance with the statute and by-laws. The treasurer is always authorized to pay and discharge any obligations of the corporation, and to perform all other acts necessary and proper within the financial structure of the corporation. The authority of the officers is also specified in the by-laws,[25] and by approving the by-laws, the directors accomplish the same delegation of authority.

EXAMPLE 8.05EE: Delegation of Authority to Officers

The authority of the officers of the corporation was discussed, and upon motion made and unanimously approved, the following authority is granted to the officers of the corporation, until subsequently modified by appropriate resolution of the Board of Directors:

President. The President shall be the principal executive officer of the Corporation and, subject to the control of the Board of Directors, shall in general supervise and control all of the business and affairs of the Corporation. He shall, when present, preside at all meetings of the shareholders and of the Board of Directors. He may sign, with the Secretary or any other proper officer of the corporation thereunto authorized by the Board of Directors, certificates for shares of the corporation, any deeds, mortgages, bonds, contracts, or other instruments which the Board of Directors has authorized to be executed, except in cases where the signing and execution thereof shall be expressly delegated by the Board of Directors or by the Bylaws to some other officer or agent of the corporation, or shall be required by law to be otherwise signed or executed; and in general shall perform all duties incident to the office of President and such other duties as may be prescribed by the Board of Directors from time to time.

The Vice Presidents. In the absence of the President or in the event of his death, inability or refusal to act, the Vice President (or in the event there be more than one Vice President, the Vice Presidents in the order designated at the time of their election, or in the absence of any designation, then in the order of their election) shall perform the duties of the President, and when so acting, shall have all the powers of and be subject to all the restrictions upon the President. Any Vice President may sign, with the Secretary or an Assistant Secretary, certificates for shares of the Corporation; and shall perform such other duties as from time to time may be assigned to him by the President or by the Board of Directors.

The Secretary. The Secretary shall; (a) keep the minutes of the proceedings of the shareholders and of the Board of Directors in one or more books provided for that purpose; (b) see that all notices are duly given in accordance with the provisions of the By–laws or as required by law; (c) be custodian of the corporate records and of the seal of the Corporation and see that the seal of the Corporation is affixed to all documents the execution of which on behalf of the Corporation under its seal is duly authorized; (d) keep a register of the post office address of each shareholder which shall be furnished to the Secretary by such shareholder; (e) sign with the President, or a Vice President, certificates for shares of the Corporation, the issuance of which shall have been authorized by resolution of the Board of Directors; (f) have general charge of the stock transfer books of the Corporation; and (g) in general perform all duties incident to the office of Secretary and such other duties as from time to time may be assigned to him by the President or by the Board of Directors.

The Treasurer. The Treasurer shall; (a) have charge and custody of and be responsible for all funds and securities of the Corporation; (b) receive and give receipts for moneys due and payable to the Corporation from any source whatsoever, and deposit all such moneys in the name of the Corporation in such banks, trust companies, or other depositaries as shall be selected in accordance with the provisions of the By–laws; and (c) in general perform all of the duties incident to the office of Treasurer and such other duties as from time to time may be assigned to him by the President or by the Board of Directors. If required by the Board of Directors, the Treasurer shall give a bond for the faithful discharge of his duties in such sum and with surety or sureties as the Board of Directors shall determine.

Assistant Secretaries and Assistant Treasurers. The Assistant Secretaries, when authorized by the Board of Directors, may sign with the President or a

Vice President certificates for shares of the Corporation the issuance of which shall have been authorized by a resolution of the Board of Directors. The Assistant Treasurers shall respectively, if required by the Board of Directors, give bonds for the faithful discharge of their duties in such sums and with such sureties as the Board of Directors shall determine. The Assistant Secretaries and Assistant Treasurers, in general, shall perform such duties as shall be assigned to them by the Secretary or the Treasurer, respectively, or by the President or the Board of Directors.

Adjournment, Signatures and Attachment

After all of the business has been conducted, the minutes should close with the statement that "There being no further business the meeting is adjourned."

Normally, the Secretary, who is in charge of complete and accurate minutes, will sign the minutes of the organizational meeting. It is also permissible to have all the directors, after their review, sign the minutes of the meeting signifying their approval.

Do not forget the attachments to the minutes of the meeting. A typical organizational meeting will have at least the following attachments:

1. Notice or waiver of notice of the meeting.
2. Articles of Incorporation and the Certificates of Incorporation.
3. Minutes and attachments of incorporators meeting, if appropriate.
4. The bylaws.
5. All promoter or incorporator contracts approved by the board of directors.
6. Specimen share certificates.
7. Written stock subscriptions.
8. Bills for organizational expenses.
9. A banking resolution.
10. (If Subchapter S has been elected) Internal Revenue Service form 2553 (the Election of a Small Business Corporation By The Shareholders).

§ 8.06 DIRECTORS' REGULAR AND SPECIAL MEETINGS

Directors' meetings are not strictly regulated by statute. The Model Act merely states that the directors may meet at regular or special meetings, either within or without the state, and defers most details such as notice and frequency of the meetings to the by-laws.[26] Since the statutes contain little guidance for directors' meetings, the by-law provisions should be carefully drafted to specify any desired procedures or notice for these meetings.[27] Even in states where the statutes specify certain rules regulating directors' meetings, the rule may usually be changed in the by-laws.

Matters Provided by Statute

The quorum of directors required for action by the board is specified in Section 40 of the Model Act to be the majority of the number of directors fixed by the by-laws or stated in the articles of incorporation, but either of these documents may provide that a greater number than a majority is required for a quorum. Section 40 furthur provides that the board of director action will be approved by a majority vote of those directors present, unless the articles of incorporation or by-laws state that a greater than majority vote is required. These director quorum and voting provisions are common to most jurisdictions. Observe how they work; suppose the corporation has a nine-member board of directors; if five members are present, they constitute a quorum and may conduct business. The affirmative vote of three members will carry action for the board, since they are the majority of those present, even though they represent only one-third of the total board.

Most state statutes and the Model Act provide that the attendance of a director at a meeting shall constitute a waiver of notice of the meeting unless the director attends for the express purpose of objecting to the transaction of any business because the meeting is not lawfully convened.[28] The procedures for sending notice and the content of the notice are left for determination by the by-laws.

Matters Contained in By-Laws or Resolutions

Place of Regular Meetings. The place for the regular director's meeting may be specified in the by-laws or may be left to the determination of the board of directors from time to time. If the by-laws leave the decision to the board members, a resolution should be adopted at each meeting of the board of directors specifying where the next regular meeting will be held.[29]

Call and Procedure for Special Meetings. Certain rules for special meetings of the board of directors should be detailed in the by-laws, such as the persons authorized to call the meeting and the notice which must be given. The place for special meetings of the board may also be established in the by-laws, but it is preferable to defer the selection of a meeting place to the person calling the meeting.

EXAMPLE 8.06A: Notice of a Directors' Special Meeting

To Keith Burn, Patrick Conroy, and Randall Wilson, Directors:

Pursuant to the power given me by the By-laws of Happiness, Inc., I hereby call a special meeting of its Board of Directors to be held at the corporate offices, on the 20th day of July, 1982, at 4:00 o'clock p.m., for the purpose of considering the advisability of authorizing the officers of the Company to renew the lease on the offices now occupied by them, and for such other action in regard thereto as the Board may deem advisable.

____________________, President

The undersigned hereby admit receipt of a copy of the foregoing notice and consent that the meeting may be held as called.

____
____
____

Directors[30]

Notice. Whether notice should be required for regular directors' meetings depends upon the size of the board, the directors' involvement in other corporate affairs, their proximity to the corporate offices, and their personal preferences. For example, formal notice is probably not required for a small group of directors who are also key employees of the corporation. However, notice may be necessary for a large board composed of advisory directors whose only corporate function is attendance at board meetings. Courtesy reminders should be given in any case. If formal notice is deemed desirable, the by-laws should specify the manner of giving notice and the period of time within which notice must be given.

EXAMPLE 8.06B: Notice of Directors' Regular Meeting

To: [*Name and address of director*]

You are hereby notified that the regular quarterly meeting of the Board of Directors of Happiness, Inc. will be held at the principal office of said Company at 200 West 14th Avenue, Kearney, Nebraska, on the 1st day of August, 1982, at 2:00 o'clock a. m.

[*Date*]

________________, Secretary[31]

EXAMPLE 8.06C: Notice of a Directors' Special Meeting

To: [*Name and address of director*]

You are hereby notified that a special meeting of the Board of Directors of Happiness, Inc. has been called by the president of the Company, to be held at the principal office of the Company at 200 West 14th Avenue, Kearney, Nebraska, on Monday, the 9th day of September, 1982, at 10:00 o'clock a. m., to consider the question of selling the corporate stock of Trouble Corporation, and of authorizing the officers of this Company to make the transfer.

[*Date*]

________________, Secretary[32]

Unless the by-laws so require, the notice need not specify the purpose of the meeting. The Model Act requires notice for special directors' meetings, but permits the by-laws to specify the content and time of notice.[33] It is common to provide a short period of notice for special meetings, since they are usually called to consider urgent matters and a cumbersome notice procedure is likely detrimental to the best interests of the corporation.

The notice requirements are nullified somewhat by the statutory provisions that the attendance at a meeting by a director constitutes waiver of notice, unless he or she attends only for the purpose of objecting to the call of the meeting. In addition, most state statutes provide that whenever any notice is required to be given, waiver of notice in writing signed by the person entitled to the notice, whether executed before or after the time stated therein, shall be equivalent to the giving of such notice. The Model Act contains this rule in Section 144.

EXAMPLE 8.06D: Waiver of Notice

We, the undersigned, directors of Happiness, Inc., a corporation of Colorado, do hereby waive any and all notice required by the statutes of Nebraska, or by the Articles of Incorporation or By-laws of said Corporation, of a meeting to be held on the 10th day of August, 1982, at 4:00 o'clock p. m. for the purpose of authorizing the officers of said Corporation to execute a trust deed for the benefit of creditors.

Dated __________, 19___.

[Signatures of all directors][34]

Method of Voting. Voting by directors is usually conducted in an informal manner, but formal records should be kept, particularly on matters where there is disagreement among directors. There is no particular statutory regulation of director voting, but directors must vote in person and are not permitted to vote by proxy. The by-laws may prescribe any desirable voting procedure. It is good practice to specify a voting procedure for larger boards. Directors express their vote by voice or written ballot on each resolution presented to the meeting, and the secretary of the meeting is responsible for recording the votes. If the vote is not unanimous on any particular issue, each director's position should be stated in the minutes of the meeting in case there may be a possibility of liability for their action.

§ 8.07 SHAREHOLDER MEETINGS

Frequency

Shareholder meetings are more strictly regulated by statute to protect the shareholder voice in corporate matters. Moreover, the shareholders have the responsibility of electing the directors, and this usually is done on an annual basis. Consequently, the Model Act provides that an annual meeting of the shareholders *shall* be held at such time and place as may be fixed in the by-laws.[35] This statutory provision clearly indicates that a shareholder meeting must be held every year. Moreover, the Act further provides that if the annual meeting is not held within any thirteen-month period, any shareholder may apply to a court to summarily order a meeting to be held. The various state statutes approach this issue differently. Nearly all states require annual meetings of shareholders, but failure to call the meeting triggers various alternatives. A few jurisdictions, like the Model Act, allow shareholders to apply to a court to order the meeting.[36] Most states permit a certain number of the holders of the voting shares to call a meeting.[37] All states would agree that the failure to hold the shareholders' meeting does not invalidate the acts of the corporation, constitute grounds for dissolution or otherwise impair its business operations.

In addition to the regular annual meeting, the Model Act states that special meetings of the shareholders may be called by the board of directors or the holders of not less than one-tenth of all the shares entitled to vote at the meeting. Further, the articles of incorporation or by-laws may authorize any other person to call a special shareholders' meeting.[38] The call is addressed to the Secretary of the corporation who is responsible for giving notice of the meeting.

EXAMPLE 8.07A: Call Of A Special Shareholders Meeting

To John Barker, Secretary:

We, the undersigned, Stockholders of Happiness, Inc., owning the number of shares of stock set opposite our names, pursuant to provisions of the By-laws, do hereby call a special meeting of the Stockholders of said Company to be held at the corporate offices, on the 15th day of July, 1982, at 3:00 o'clock p. m. for the purpose of removing Charles Miser as a director and for the transaction of any or all business that may be brought before the meeting and we hereby authorize and direct that you notify the Stockholders of such meeting in accordance with the provisions of the By-laws.

Dated ________, 19___.

Signature of Stockholders	Number of Shares
____________________	________________
____________________	________________

[39]

State statutes specify different persons who are entitled to call a special meeting and they particularly differ on the number of shareholders who must join in the call of their own special meeting. For example, the Ohio statute requires holders of twenty-five per cent of the voting shares to join, and the articles of incorporation may require the concurrence of up to fifty per cent of the voting shares to call a special meeting in that state.

Location of Meeting

The by-laws may fix a particular place for the shareholder meeting or they may authorize the board of directors to determine the meeting place from time to time.

The latter authority facilitates a decision by the board of directors to hold the annual shareholders' meeting near the beaches of Florida, if a winter meeting, or in the cool mountains of Colorful Colorado if the meeting is scheduled in the summertime. State statutes usually provide that if the bylaws are silent on the matter, the meetings will be held at the registered office of the corporation.

Notice

Shareholders must receive notice of all meetings, and the statutory notice procedure may be burdensome, especially if the shareholder population is large. State statutes protect the shareholders by prescribing rules for determining the persons who are entitled to receive notice and setting the periods within which notice must be sent.

Persons Entitled to Receive Notice. In order to determine the shareholders entitled to receive notice of the meeting, the board of directors may set a date at which the corporation's stock transfer books will be closed. All persons listed in the stock record at that time are identified as "holders of record," and these holders will be entitled to notice of the meeting. Instead of closing the stock transfer books, the board of directors may set a record date in advance of the meeting, and a list of shareholders entitled to receive notice will be prepared that day. Alternatively, the directors may simply direct that the notices will be mailed on a specified date, and all shareholders as of that date will receive notice.[40] Since these determinative dates are all prior to the meeting, a person holding shares at the time the notice lists are prepared could sell the shares and no longer be a shareholder at the time of the annual meeting. Nevertheless, that person will receive the notice of the meeting and will be entitled to vote at the meeting. The voting determination procedure is founded on the proposition that the corporation must draw the line somewhere, and in the interests of orderly procedure the statute merely suggests a cut-off date.

The various state statutes have a few principal differences in determining the persons who are entitled to notice and to vote at the meeting. Section 30 of the Model Act limits the period for closing the stock transfer books to fifty days before the meeting. The books must be closed for at least ten days immediately preceding the meeting. Thus, there is a period from ten to fifty days preceding the meeting within which the stock transfer books may be closed. Instead of closing the stock transfer books, the by-laws or the board of directors may fix a "record date" for determination of the shareholders entitled to notice, and, again, that date may not be more than fifty days or less than ten days prior to the date of the shareholder meeting. Under the "record date" procedure, the transfer books are not closed, but an arbitrary date is fixed—say, thirty days before the scheduled meeting—and all persons who own shares as of that date will receive notice and be entitled to vote. A final alternative is allowed: if the stock transfer books are not closed and no record date is set then the date that the notice is mailed to the shareholders will be considered to be the "record date" for determination of shareholders. Since the notice must be delivered no less than ten and no more than fifty days before the meeting,[41] the same time periods apply to this alternative. Thus, the board of directors or by-laws may simply direct that notices will be sent on the thirtieth day preceding the meeting, and that day is the "record date." This latter procedure is only feasible if the notices are prepared and sent the same day.

The major variant in these provisions among the states is the period within which the books may be closed or the record date set. The Delaware statute allows the board of directors to fix a record date not more than sixty or less than ten days before the meeting and, if no record date is fixed, the record date will be determined to be the close of business on the day preceding the day on which the notice is given. Most states have a minimum period of ten days, and the longest maximum period for identifying holders of record is seventy days preceding the meeting in Connnecticut. If these notice provisions appear complicated, you might consider incorporating in Vermont where the legislature obviated the problem by not adopting any statutory rule for the determination of shareholders who are entitled to vote at a meeting.

Advance Notice. Notice of the shareholder meeting must be written under most statutes, and should be written anyway. It must state the place, day, and hour of the meeting, and, for a special meeting, the notice usually must also indicate the purposes for which the meeting is called.[42] A few states require that notices for every meeting must state the purpose of the meeting.[43]

Notice for the meeting may be delivered to the shareholder either personally or by mail, and, if mailed, the notice is deemed to be delivered when deposited in the United States mail with postage prepaid and addressed to the shareholder at his address as it appears on the stock transfer books of the corporation.

EXAMPLE 8.07B: Notice of Annual Shareholders' Meeting

To the Stockholders of Happiness, Inc.:

The Annual Meeting of the Stockholders of Happiness, Inc., will be held in the office of the Company, at 200 West 14th Avenue, Kearney, Nebraska, on Monday, September 9, 1982, at twelve o'clock noon, for the election of three directors and for the transaction of such other business as may properly come before the meeting.

The stock transfer books of the Company will not be closed, but only stockholders of record at the close of business on August 20, 1982, will be entitled to vote.

__________________, Secretary[44]

Dated ________, 19___.

EXAMPLE 8.07C: Notice of Special Shareholders' Meeting

To the Stockholders of Happiness, Inc.

Pursuant to vote of the Board of Directors a special meeting of the Stockholders of Happiness, Inc. is hereby called to be held on Wednesday, November 8, 1982, at 10 o'clock a. m., at the principal office of the Corporation, at 200 West 14th Avenue, Kearney, Nebraska, for the following purposes:

1. To consider and act upon the question of increasing the authorized capital stock of the Corporation and of amending the Certificate of Incorporation of the Corporation accordingly, as set forth in the following resolutions of the Board of Directors passed at a meeting of said Board held on the 3rd day of October, 1982, viz.:

"Resolved, that it is advisable that the amount of the authorized capital stock of this Corporation be increased, by amendment of the Certificate of Incorporation, so as to authorize 100,000 additional shares of the common stock, of the par value of $1.00 each, and that for this purpose Article V of said Certificate of Incorporation should be amended by striking out the first two sentences thereof and substituting in lieu of said first two sentences the following, viz.: 'The total amount of the authorized capital stock of the Corporation is $500,000, divided into 500,000 shares, of the par value of $1.00 each. Of such authorized capital stock, 100,000 shares, amounting at par to $100,000, shall be preferred stock, and 400,000 shares, amounting at par to $400,000, shall be common stock.' Said Article V in all other respects to remain unchanged.

"Further resolved, that a special meeting of the stockholders be called to be held at the principal and registered office of the Corporation, to wit, at the office of 200 West 14th Avenue, Kearney, Nebraska, on November 8, 1982, at 10 o'clock a. m. to take action on the foregoing resolution.

2. To transact any other business which may properly come before the meeting.

The transfer books of the Corporation will be closed at the close of business on October 20, 1982, and reopened at 10 a. m., on November 8, 1982.

By order of the Board of Directors.

Dated ________, 19___. ________________, Secretary

If you are unable to be present at the above meeting please sign and return the enclosed proxy.[45]

Under the Model Act, the notice must be *delivered* not less than ten nor more than fifty days before the meeting.[46] Although this time period looks familiar, you should recognize that this is a different rule from the one stated above for the determination of shareholders entitled to receive notice and to vote. The shareholders entitled to notice must first be determined and then their notice must be properly delivered. For example, the board of directors are permitted to set a record date for determining shareholders entitled to receive notice during the period from ten to fifty days before the meeting. They could, therefore, legally establish the record date on the eleventh day before the meeting. However, if some of the notices were not mailed to the shareholders until the ninth day before the meeting, the delivery rule would be violated.

The time period within which the notice must be delivered varies by jurisdiction. The shortest statutory period for delivery of notice is five-days personal notice in Tennessee, and the earliest period prescribed is ninety days in Maryland. However, a few states say nothing about the time for notice and Delaware allows the period to be changed in the by-laws.

Several jurisdictions have special notice rules if certain unusual matters are to be considered at the meeting. These statutes usually specify a longer minimum time within which the notice must be given, apparently so the shareholders will have a longer period of time to consider their vote. The Model Act has two such provisions in Sections 73 and 79 which require a minimum of twenty-days notice when the shareholder meeting is being called for the purpose of considering a plan of merger or consolidation or approval of the sale of assets not in the ordinary course of business.[47]

Having waded through the notice provisions and determined the precise procedure and timing of the notice, the giving of proper notice in practice can be a fulfilling event. But if you've become mired in these rules, the question may be asked, and not necessarily rhetorically, "What happens if we just ignore this requirement and hold the meeting anyway?" Failure to give proper notice renders the meeting invalid and vulnerable to the challenge of any shareholder who did not receive proper notice. However, there are some saving provisions. Section 144 of the Model Act permits written waiver of notice by any shareholder entitled to receive such notice, and the waiver may be signed before or after the event. There is also a procedure for obtaining consent to action in writing, discussed in section 8.09, which may be a solution to the inadequate notice problem for smaller corporations.

However, unlike the rule for directors' meetings,[48] the attendance of a shareholder at a meeting for which no notice was given does not constitute waiver of notice whether or not he attends for the purpose of objecting to the lack of notice.

EXAMPLE 8.07D: Shareholders' Waiver of Notice

The undersigned, a shareholder of Happiness, Inc., hereby waives any and all notice required by the statutes of Nebraska, or the articles of incorporation or bylaws of said corporation, for a meeting to be held on the 24th day of November, 1982 at 3:00 p. m. at the corporate offices for the purposes of increasing the authorized capital stock of the corporation and amending the certificate of incorporation of the corporation accordingly.

Dated November 24, 1982.

[Signature of the Shareholder]

Proxies

Shareholders who are unable to be present at the shareholders' meeting may vote by proxy under most state laws. A proxy is a written authorization by a shareholder directing the proxy-holder to vote his shares on his behalf. Proxies, like any other agreement, may contain any limiting or expanding provisions that the shareholder desires. The proxy form is usually furnished by the management and conforms to a standard form for general authorization to vote. The Model Act and other statutes regulating proxies require that they be written and signed by the shareholder. The Model Act also permits the proxy to be signed by the shareholder's attorney-in-fact.[49]

A general proxy authorizes the proxy-holder to vote on all matters properly presented to the shareholder meeting.

EXAMPLE 8.07E: General Proxy for a Specified Meeting

I hereby constitute and appoint Ezra Brooks or Jack Daniels, or either one of them, and in place of either, in case of substitution, his substitute, attorneys and agent for me and in my name, place and stead, to vote as my proxy at the next Annual Meeting, and at any adjournment or adjournments thereof, of the Stockholders of Happiness, Inc., upon any question which may be brought before such meeting, including the election of directors, according to the number of votes I should be entitled to vote if then personally present, with full power to each of my said attorneys to appoint a substitute in his place.

Dated ________, 19___. __________________[50]

The proxy may have a stated duration, in which case it is valid for the period of time stated, but if no period is stated it automtically expires after eleven months from the date of execution under the Model Act.

EXAMPLE 8.07F: Continuing General Proxy

The undersigned hereby constitutes and appoints Jack Daniels, Bud Weiser, and John Walker, or any two of them acting jointly, as his, her, or their proxy to cast the votes of the undersigned at all general, special, and adjourned meetings of the Stockholders of Happiness, Inc. from time to time and from year to year, when the undersigned is not present at any such meeting and if present does not elect to vote in person. This proxy shall be effective for two years from the date hereof unless sooner revoked by written notice to the Secretary of the Corporation.

Dated ________, 19___. ____________________[51]

The statutory period of duration varies among the jurisdictions but proxies are revocable at will, unless they are "coupled with an interest", such as when stock is pledged to a creditor to secure repayment of a loan, and a proxy to vote the shares is given to the creditor for the duration of the security interest.

The proxy-holder is bound to vote the shares in the manner directed by the shareholder in the proxy.

Proxies are most frequently used for larger, publicly-held corporations, where many of the shareholders will not be able to attend the meeting. It is important to note that the Federal Securities Exchange Act of 1934 strictly regulates the solicitation of proxies for shareholders of a publicly-held corporation.[52] A proxy in compliance with that Act must satisfy special requirements as to wording and form.

EXAMPLE 8.07G: Public Corporation Proxy

Happiness, Inc.
Proxy Solicited by Management for Special Meeting
of Stockholders, October 10, 1981

The undersigned hereby appoints Jack Daniels, Bud Weiser and John Walker and
P each or any of them, attorneys, with powers the undersigned would possess if personally present, to vote all shares of Common Stock of the undersigned in
R Happiness, Inc. at the Special Meeting of its Stockholders to be held October 10, 1981 at 2:00 P.M., Central Daylight Saving Time, at Kearney, Nebraska, and at
O any adjournment thereof, upon the proposed amendment to the Certificate of Incorporation of the Company, which amendment is set forth in the Proxy State-
X ment and has been declared advisable by the Board of Directors, and upon a split of each outstanding share of Common Stock of the par valued of $5 into two
Y shares of Common Stock of the par value of $5 each, and upon other matters properly coming before the meeting.

The directors favor voting FOR the proposed amendment to the Certificate of Incorporation.

(Continued, and to be signed on the other side.)

(Continued from the other side.) Proxy No.

The vote of the undersigned is to be cast (please indicate)

FOR ☐ AGAINST ☐

the proposed amendment to the Certificate of Incorporation and the split of each outstanding share of Common Stock of the par value of $5 into two shares of Common Stock of the par value of $5 each.

UNLESS OTHERWISE DIRECTED THE VOTE OF THE UNDERSIGNED IS TO BE CAST "FOR" THE PROPOSED AMENDMENT TO THE CERTIFICATE OF INCORPORATION AND THE SPLIT OF EACH OUTSTANDING SHARE OF COMMON STOCK.

Receipt of Notice of Special Meeting of Stockholders and the accompanying Proxy Statement is acknowledged.

Date ____________________ 19___

[*Name and address of stockholder*]

____

____

____

Please sign above as name(s) appear(s) hereon.

(When signing as attorney, executor, administrator, trustee, guardian, etc., give title as such. If joint account, each joint owner should sign.)[53]

Quorum

The Model Act states that the majority of shares entitled to vote (represented in person or by proxy) will constitute a quorum at a shareholder meeting unless otherwise provided in the articles of incorporation. However, in no event may the articles of incorporation provide that a quorum shall be less than one-third of the shareholders entitled to vote at the meeting.[54] Most states impose a limit below which the quorum requirements cannot be reduced, and one-third is the most common limit. Louisiana permits reduction as low as one-quarter of the voting shares. The reduction in quorum requirements may be contained in the by-laws in some states.[55]

Voting of Shares

Unless the articles of incorporation provide otherwise, each outstanding share, regardless of class, is entitled to one vote. This is the provision in Section 33 the Model Act, and most states take this approach to shareholder voting. A few jurisdictions extend voting to fractional shares permitting a corresponding fractional vote on each matter submitted to the shareholders.[56]

Shareholder voting may be altered and concentrated through several devices. The articles of incorporation may provide that shares of different classes may have more or less than one vote per share—a principle called "weighted voting."[57]

To concentrate voting power, shareholders may predetermine how their shares will be voted by using a pooling agreement or voting trust.[58]

In most cases, the affirmative vote of the majority of shares present at the meeting and entitled to vote constitutes the act of shareholders. Again, this provision is subject to modification by the articles of incorporation or by-laws. Thus, if the corporation has 100,000 shares of voting stock outstanding, 50,001 shares constitute a quorum for a meeting and shareholder action would be taken if 25,001 shares were voted in favor of the proposition. The articles of incorporation could provide that one-third of the shares entitled to vote would constitute a quorum, in which case 33,334 shares could hold a meeting, and 16,668 shares could decide any issue. In both of these cases, the shares which carry the action are less than the majority of all shares entitled to vote. Conversely, if the articles of incorporation required that 80% of the voting shares must be represented to constitute a quorum, and 80% of the represented shares must vote affirmative to carry an issue, the affirmative vote by a minimum of 64,000 shares is required to constitute shareholder action. Once a quorum is present, the rule is to apply the appropriate percentage to the shares represented.

As a word of caution, the statute may require greater than a majority vote on certain matters. The necessary shareholder vote on individual issues is specified in the following section dealing with shareholder business.

The method of voting at a shareholder meeting is not generally prescribed by statute. Although any method is acceptable, including voice vote or written ballot, the latter is the preferable procedure and a ballot may be required for the election of directors. Moreover, remember that shareholder action is taken by a specified percentage of shares represented at the meeting. A voice vote or show of hands indicates the shareholders' vote *per capita*, and if there is any disagreement a ballot will be necessary to determine the number of *shares* (not shareholders) voting in favor of the proposition. A ballot should specifically describe the matter being submitted to shareholder vote, and the secretary of the meeting is responsible for tallying the votes. Some states permit the appointment of impartial judges of election if requested by the shareholders or required in the by-laws.[59] These judges are responsible for determining whether the required shareholder vote has been received.

EXAMPLE 8.07H: Ballot

For voting at a meeting of the Shareholders of Happiness, Inc., on November 24, 1982.

ISSUE NO. _______

RESOLVED, [*here state resolution to be acted upon*]

Please record my vote:

_______ FOR

_______ AGAINST

_______ ABSTAIN

____

Signature of Shareholder [optional]

____

Number of Shares

§ 8.08 SHAREHOLDER BUSINESS AND VOTE REQUIRED

Shareholders have an indirect voice in management, and, except in a close corporation, have very little direct control over the daily business affairs of the corporation. Their meetings, therefore, focus generally on receiving information about corporate business and taking action on those matters which are within the ambit of shareholder control as specified in the statute, the articles of incorporation and the by-laws.

The most important shareholder business is the election of directors, and through their right to elect the directors, shareholders indirectly control mangement policies and direction of their corporation. Moreover, shareholders are expected to periodically review management activities and they may ratify and approve management acts at their annual shareholder meeting. Most statutes also grant shareholders the right to vote directly on major decisions involving modifications in the structure of their corporation. These major "fundamental" changes usually significantly affect the ownership rights of the shareholder. It is also possible for the incorporators to grant greater control to the shareholders by special provisions in the articles of incorporation or by-laws. Shareholder involvement in each area of shareholder control depends on the particular structure of each corporation and the voting requirements imposed by the statute.

The agenda of a shareholder meeting, therefore, will vary considerably from corporation to corporation. Certain shareholder business, however, may be expected to be conducted in every case.

Special Matters Over Which Shareholders Have Control

The articles of incorporation or by-laws may reserve to the shareholders certain items of business which would otherwise be determined by the directors. The reservation of control may be as extensive as complete control over all management activities, as is permitted under the close corporation statutes enacted in Delaware and elsewhere,[60] or as limited as allowing the shareholders to amend and repeal by-laws. Depending upon the local statutory authority to place control with the shareholders, the articles of incorporation or by—laws may grant them the right to select officers; fix compensation; determine the stated value of no par stock; adopt, amend and repeal by-laws; and so forth. However, there are obvious practical limitations on shareholder control in these areas. If the shareholders are cohesive and few, they may confortably be vested with these responsibilities. On the other hand, the larger the group of shareholders, the more cumbersome it becomes to take action in these important management areas.

The shareholder vote necessary to carry action on matters reserved to their control is governed by statute and may be altered by the articles of incorporation or by-laws. The Model Act provides that such matters will be decided by a majority vote of the shares represented at the meeting and entitled to vote on the matter.[61]

Election of Directors

The election of directors is usually an item of business at each annual shareholder meeting, since the term of office for directors is generally until the next succeeding annual shareholder meeting and until the successor directors have been elected and qualified.[62] Even if the board of directors is "classified" or "staggered", a certain percentage of the board will be elected each year.

In previously discussing shareholder rights, it was noted that the shareholders may be able to cumulate their votes in the election of directors, depending upon the local statute and articles of incorporation. Some jurisdictions require the use of the cumulative voting procedure in the election of directors as a constitutional right of the shareholder. Others have statutes requiring cumulative voting unless the articles of incorporation specifically deny it. Delaware and other states deny cumulative voting under the corporation statute, but permit the articles of incorporation to grant it. The Model Act offers alternative provisions so the states following it may take either position.[63] In any case the time has come to discuss the mechanics of cumulative voting.

Cumulative voting is a procedure for voting shares in the election of directors which is designed to secure representation of the minority shareholders on the board of directors. With "straight" voting, the holders of a majority of the stock should be able to elect the directors who will represent their interests, and if the interests of the majority are inconsistent with the interests of the minority, the latter group may suffer without representation on the board. With cumulative voting, each share carries as many votes as there are vacancies to be filled on the board of directors, and the shareholder is permitted to distribute the votes for all of his shares among any candidates he chooses. For example, suppose Bilko Building Company has three directors to be elected every year and has 500 shares of stock outstanding of which Anderson owns 300 shares, Bonner owns 100 shares, and Carlyle owns 100 shares. With straight voting each person votes his shares for the candidates one at a time. Suppose three directors are to be elected at the meeting; if Anderson nominates Davis, Everett and Ford as directors; and the minority shareholders nominate Girtler to represent their interests. The votes will be probably tallied as follows with straight voting:

Anderson's 300 shares	Bonner's 100 shares	Carlyle's 100 shares
X Davis	Davis	Davis
X Everett	X Everett	X Everett
X Ford	X Ford	X Ford
Girtler	X Girtler	X Girtler

The total votes for each candidate are:

Davis	- 300
Everett	- 500
Ford	- 500
Girtler	- 200

Thus, Davis, Everett and Ford are elected to fill the three offices, and the minority shareholders have lost in their bid to elect Girtler.

Contrast the result of this election with the cumulative voting procedure. Each shareholder is entitled to as many votes as he has shares, multiplied times the number of vacancies to be filled. Thus, Anderson has 900 votes (300 shares × 3 vacancies); Bonner has 300 votes (100 × 3); and Carlyle has 300 (100 × 3). Each shareholder may parcel his available votes any way he pleases, including applying all of them to one candidate. Therefore, if Bonner and Carlyle want to be certain to elect Girtler as their director, they may apply all of their votes for that purpose. Anderson cannot prevent Girtler's election no matter how he distributes his votes. A cumulative voting ballot would look like this:

Anderson's 300 shares (900 votes)		Bonner's 100 shares (300 votes)		Carlyle's 100 shares (300 votes)	
300	Davis	0	Davis	0	Davis
300	Everett	0	Everett	0	Everett
300	Ford	0	Ford	0	Ford
0	Girtler	300	Girtler	300	Girtler

The total votes for the candidates are:

Girtler	- 600
Davis	- 300
Everett	- 300
Ford	- 300

Thus, Girtler would be elected, and a run-off election would be necessary to determine the two remaining positions between Anderson's three nominees. Anderson could have applied 601 votes for one of his candidates, thereby insuring that one of his candidates will beat Girtler, but that leaves Anderson with only 299 shares for another candidate, and Girtler still will be elected.

Notice that cumulative voting assures minority representation on the board only if the minority shareholders are cohesive and determined in electing a representative. If Bonner and Carlyle could not agree on a suitable candidate to represent the minority, they may lose the success of their combined vote.

At this point, recall that the directors may be removed with or without cause by vote of the shareholders. If a director were elected to represent the minority interests under a cumulative voting procedure, the protection of cumulative voting could be nullified if the majority shareholders could remove him after the election. Consequently, state statutes usually specify that if cumulative voting is in effect for election of directors, no director may be removed unless the same cumulative voting procedure is used in the removal action. He cannot be removed if the votes cast *against* his removal would be sufficient to elect him if cumulatively voted at an election of the entire board of directors.[64] Thus Bonner and Carlyle could vote cumulatively against Girtler's removal and prevent it the same way they elected him.

One last point regarding cumulative voting deserves attention. Recall that many states permit the board of directors to be "classified" or "staggered", so that not all directors are elected each year.[65] When this classification procedure is combined with cumulative voting, it may neutralize the protective effect of cumulative voting. Suppose in our previous example, that the corporation's three directors were staggered over a three year period. This means only one director would be elected each year. Thus, cumulative voting, which authorizes the number of votes equal to number of shares times vacancies to be filled, gives Anderson 300 votes (300 shares × 1 vacancy), Bonner 100 votes (100 × 1) and Carlyle 100 votes (100 × 1). Anderson, therefore, could always defeat Bonner and Carlyle with his 300 votes to their 200, just as with straight voting. For this reason, the Model Act permits staggering the board only if the board consists of nine or more members.[66] That way, at least three directors will be elected each year, and the effect of cumulative voting is preserved.

The ballot for the election of directors should present all nominees, and, if cumulative voting is used, should explain how to use it.

EXAMPLE 8.08A: Ballot for the Election of Directors

[STRAIGHT VOTING]

Annual Meeting of the Shareholders of Happiness, Inc. November 24, 1981.

The following persons have been nominated for the Board of Directors of Happiness, Inc. to serve until the next annual meeting of the Shareholders or until their successors have been elected and qualified.

Slate of Directors

[Nominee]
[Nominee]
[Nominee]
[Nominee]

Other nominations

____

____

[write in]

FOUR DIRECTORS ARE TO BE ELECTED. PLEASE CHECK ONLY FOUR SELECTIONS. IF MORE THAN FOUR SELECTIONS ARE MADE, THIS BALLOT WILL BE VOIDED [or] [ONLY THE FIRST FOUR SELECTIONS WILL BE COUNTED.]

Please enter the number of shares you own: __________ Shares

Voted for	Name of Nominee
________	[Nominee]
________	[Nominee]
________	[Nominee]
________	[Nominee]
________	[write in]
________	[write in]

____

Signature of Shareholder
[optional]

EXAMPLE 8.08B: Ballot for the Election of Directors

[CUMULATIVE VOTING]

Annual meeting of the Shareholders of Happiness, Inc., November 24, 1981.

The following person have been nominated for the Board of Directors of Happiness, Inc., to serve until the next annual meeting of the Shareholders or until their successors have been elected.

Slate of Directors

[Nominee]
[Nominee]
[Nominee]
[Nominee]

Other nominations

____

____
[write in]

FOUR DIRECTORS ARE TO BE ELECTED. YOU ARE ENTITLED TO CUMULATE YOUR VOTES IN THIS ELECTION. PLEASE COMPUTE THE NUMBER OF VOTES TO WHICH YOU ARE ENTITLED AS FOLLOWS:

Number of shares owned ________ × 4 = ____________Number of Votes

You may cast these votes for any or all of the nominess, by writing the number of votes you wish to cast for each nominee next to the name of the nominee. THE TOTAL VOTES CAST MAY NOT EXCEED THE NUMBER OF VOTES COMPUTED ABOVE.

IF YOU CAST MORE VOTES THAN YOU ARE ALLOWED, OF IF THE NUMBER OF VOTES IS NOT ENTERED BELOW, THIS BALLOT WILL BE VOIDED.

	Number of votes cast [write number]	Name of Nominee
	________	[Nominee]
	________	[Nominee]
	________	[Nominee]
	________	[Nominee]
	________	[write in]
Total votes cast	________	________________ Signature of Shareholder [optional]

Approval of Extraordinary Matters

Certain structural changes of a corporation require the approval of the shareholders because their ownership rights as investors may be materially affected by the action. These changes may involve major modifications to the organization or financial structure of the corporation, disposition of its assets, adjustments in the ownership characteristics of the shares, or termination of business.

The most frequent structural change in the corporation is the amendment of its articles of incorporation. The organization of the corporation is described in the articles of incorporation, and any amendment to those provisions modifies the structure and probably affects the ownership characteristics of the shareholders in some manner. An amendment as minor as changing the corporate name must be approved by the shareholders through the amendment procedure, although the actual effect on shareholder rights may be imperceptable. Other changes may have a more obvious effect on the character of the investment, such as amending the period of duration of the corporation or diminishing the scope of the business it will conduct. Other typical changes, enumerated in Section 58 of the Model Act, directly concern the shares themselves. These include amendments to change the aggregate number of shares which the corporation has authority to issue; to increase or decrease par value, or change par value shares into no par value shares and vice versa; to exchange, divide, reclassify, redesignate or cancel shares; to create new classes, with special preferences; to modify preferences or change the authority of the board of directors to establish series of shares; and to limit, deny or grant preemptive rights of shareholders. Any amendment to the articles of incorporation, therefore, must be submitted to a vote of the shareholders.[67]

Merger and consolidation of the corporation and disposition of corporate assets other than in the ordinary course of business also require special shareholder approval. In a merger, one corporation joins another, the merging corporation ceases to exist and the surviving corporation continues business with the assets, liabilities and shareholders of both corporations. In a consolidation, two corporations join to form a new corporation, and both of the original corporations cease to exist. The ownership rights of the shareholders of these constituent corporations will be modified in these transactions. If their corporation is the survivor to the merger, their corporation will probably issue new shares of stock to the shareholders of the merged corporation and that may dilute their ownership interests. If their corporation has been merged into another corporation, the survivor corporation will probably exchange its shares for their shares in the old corporation. In a consolidation procedure, the new corporation resulting from the combination will issue new securities to shareholders of both old corporations. The shareholders obviously should have some say in these matters. Similarly, if the directors of the corporation intend to sell or otherwise dispose of substantially all of the assets of the corporation it may be necessary to consult the shareholders. Certainly this is not necessary when the corporation merely sells goods from inventory, such as a department store selling television sets to its customers; but when the sale of assets is outside the scope of the ordinary course of business, as when the department store sells its display counters, cash registers and substantially all of its inventory in one transaction with another department store, the shareholders should be asked for their approval.

Finally, the ownership rights of the shareholders will certainly be affected by a dissolution of their corporation. The directors are not vested with the authority to dissolve the corporation at will if shares have been issued; voluntary dissolution is regarded as a fundamental change requiring shareholder approval. Similarly, if the shareholders have approved dissolution of their corporation. The directors may not revoke the dissolution proceedings without affirmative shareholder approval of the revocation.

The specific procedures for the approval of these corporate structural modifications are discussed in a later chapter,[68] but here we are concerned with the vote necessary for shareholder approval. The Model Act originally provided that shareholder action on these matters would be carried by the affirmative vote of the holders of two-thirds of the shares entitled to vote on the issue. If a particular class was entitled to vote on the issue as a class,[69] the two-thirds affirmative vote of that class was also required.

Many states still require this percentage of shareholder vote to approve structural modifications to the corporation. The Model Act now merely requires a majority vote by the appropriate shares for approval, in order to comply with "contemporary practice in similar institutional matters." The Model Act, like the other progressive jurisdictions which permit approval by the majority, allows the articles of incorporation to require a greater proportion of shareholder votes, if the extra shareholder protection is desired in these special areas. When a two-thirds vote is required for approval of the fundamental corporate changes, the minority shareholders holding more than one-third of the voting stock can successfully block such actions by the majority. Since the majority shareholders may be at odds with the minority, and there are cases where the majority stands to profit from such transactions, the extra protection for the minority shareholders may be important.

§ 8.09 ACTION WITHOUT A MEETING

Notwithstanding the foregoing discussion which intimates that shareholder and director meetings are necessary for effective corporate action, state law usually prescribes a written consent procedure for these intra-corporate groups to take action without a formal meeting. The Model Act permits written consent by directors in Section 44 and by shareholders in Section 145. The Act further provides that the articles of incorporation or by-laws to deny this procedure for directors, but it does not so condition action by shareholders. All but a few states have comparable provisions. Curiously, more states deny this procedure to directors than to shareholders.[70]

The statutory requirements for taking action without a meeting are that the proposed action would have been proper to submit to regular meeting and consent in writing setting forth the proposed action must be obtained for all shareholders or all directors, as the case may be. The effect of the consent is unanimous approval of the particular action involved.

EXAMPLE 8.09A: Unanimous Consent to Action of the Board of Directors

Pursuant to the provisions of the Nebraska Corporation Code, the following action is taken by the board of directors of Happiness, Inc., by unanimous written consent as if a meeting of the board of directors had been properly called pursuant to notice and all directors were present and voting in favor of such action.

> RESOLVED, that the salary of the vice president be increased from the sum of $15,000.00 per year to the sum of $20,000.00 per year.

IN WITNESS WHEREOF, we have executed this unanimous Consent of Action on the dates set forth after our respective names, effective ______________, 19_____.

______________________________	______________________________
[Director]	Date
______________________________	______________________________
[Director]	Date
______________________________	______________________________
[Director]	Date

EXAMPLE 8.09B: Unanimous Consent to Action of the Shareholders

Pursuant to the provisions of the Nebraska Corporation Code, the following action is taken by the shareholders of Happiness, Inc., by unanimous written consent, as if a meeting of the shareholders had been properly called pursuant to notice and all shareholders entitled to vote on the matters presented herein had been present and voting in favor of such action.

There are 10,000 shares entitled to vote on the matters presented herein, and the undersigned shareholders are the holders of record of all such shares on the date of this Unanimous Consent of the Shareholders.

> RESOLVED, that the action of the officers and directors of this corporation is making an investment in securities of Trouble, Inc., as set forth in the report which the corporation mailed to all stockholders of record on May 1, 1981, be and the same hereby is ratified.

IN WITNESS WHEREOF, the undersigned, constituting all of the shareholders of the corporation entitled to vote on the matters presented herein, have executed this unanimous Consent to Action of the Shareholders on the dates set forth after our respective names, effective November 10, 1981.

______________________________		______________________________
[Shareholder]	[Number of shares]	[Date]

[Shareholder]	[Number of shares]	[Date]
[Shareholder]	[Number of shares]	[Date]
[Shareholder]	[Number of shares]	[Date]

[etc]

The only major variations among the statutes with consent to action provisions are whether the articles of incorporation or by-laws may deny this procedure for directors' meetings, and the percentage of shareholders required to file written consent in order to carry the shareholder action. Most states do not authorize a limitation on the directors' rights to file written consent and act without a meeting. The Model Act, Delaware, and New Jersey allow the certificate or articles of incorporation or by-laws to alter this provision. Pennsylvania permits modification only in the by-laws.

Delaware allows for less-than-unanimous approval for shareholder consent. The statute provides that the holders of outstanding stock having at least the minimum number of votes which would be necessary to approve such action at a meeting may consent in writing and thereby bind the shareholders. The remaining shareholders are then entitled to notice of the action so taken.[71] Other states authorize the articles of incorporation to prescribe a less-than-unanimous number for shareholder written consent.[72] Otherwise, most states provide that all of the shareholders must file written consent to act without a meeting. The use of shareholder consent without a meeting should obviously be limited to small, close corporations. The requirements of unanimity and of obtaining the signatures of all shareholders makes the procedure completely impracticable for large, publicly-held corporations.

§ 8.10 MINUTES

There is no mandatory procedure for conducting meetings of shareholders and directors and, in many cases, these meetings are conducted in an informal manner. The meetings usually become more formal as the group becomes larger. The science of conducting a corporate meeting for a large intra-corporate group has fortunately been reduced to writing in an easy-to-follow systematic procedure for corporate secretaries,[73] so formal meeting procedure escapes further elaboration here.

It is always necessary to follow the statutory requirements, such as notice, voting, and solicitation of proxies for every meeting. Consequently, the minutes of the meeting should always reflect that the statutory requirements have been observed. The minutes of the meeting constitute the permanent written record of the corporate history, and all matters considered at a meeting must be carefully recorded in order to trace the origin of all corporate actions.

The secretary of the corporation is usually bestowed the privilege of keeping the minutes. The only guidelines for recording minutes are that the secretary must comply with instructions given by the board of directors and by the chairman of the meeting, and he or she must compose the minutes in such a way that they constitute an accurate and complete transcription of the action taken by the intra-corporate group at the particular meeting recorded. Otherwise, the secretary has broad discretion in the manner in which the minutes will be kept. Counsel may assist the secretary in establishing a corporate minute policy after formation of the corporation. As a very general guide it may be suggested that all minutes should include at least the following:

1. the name of the corporation;
2. the date;
3. the place where the meeting is held;
4. special statutory, articles of incorporation, or bylaw authority under which the meeting is called;
5. the persons present, the person absent, or the shares represented in person or by proxy;
6. a statement that the meeting is held pursuant to a notice or waiver which is attached to the minutes;
7. the nature of the meeting (regular or special);
8. approval of minutes of previous meetings;
9. the substance of the issues presented at the meeting, how they were submitted, and by whom;
10. the decision and vote of the intra-corporate group on each issue in resolution form;
11. the presentation of all reports, with copies attached if the report is written and a summary of the report if it is oral; and
12. a summary of the other business before the meeting.[74]

If the meeting is a directors' meeting, the directors present and absent should be named. It is also important to report correctly the directors' vote on each resolution. If the vote is unanimous, that should be stated. If any director abstained or dissented, his name should be listed, particularly if there is any possibility of the directors' personal liability for action taken. For example, if the director were voting on the issuance of a dividend but funds were not legally available for payment of the dividend, the recording of a dissent may be necessary to relieve the director from liability for illegally declared dividends. If a director is "interested" in the particular action, as she would be if the contract being considered by the board of directors is another corporation in which she is financially interested, it should be noted that she did not vote, or that she left the room during the discussion.[75]

The names of the shareholders present at a shareholder meeting and the number of shares they hold may be noted if the group is small. Otherwise, the number of shares represented in person and by proxy should be listed.

Informal activity which occurs during a meeting should not be described in unnecessary detail. It would not be appropriate, for example, to list the persons who availed themselves of the corporate punchbowl at the meeting, or to record a discussion regarding a director's new house in the mountains. However, informal information about tentative corporate plans and current business conditions may be reported if they were discussed at the meeting. In this regard, if the chairman obtains the informal consensus of the group on any particular matter, the minutes should express the affirmative reaction of those present.

NOTES

1. See Chapter 13.
2. See Section 6.07.
3. See M.B.C.A. § 57 (organization meeting of directors).
4. See Section 6.08.
5. E.g., Illinois, Ill.Rev.Stat. ch. 32 § 157.51; Pennsylvania, 15 Pa.Stat. § 1210.
6. West's Modern Legal Forms § 2852.
7. West's Modern Legal Forms § 2860.
8. West's Modern Legal Forms § 2861.
9. M.B.C.A. § 57.
10. West's Modern Legal Forms § 2886.
11. See Sections 7.06.
12. See Sections 4.03 and 6.02.
13. See Section 6.02.
14. M.B.C.A. § 17.
15. West's Modern Legal Forms § 2896.
16. See Section 8.05.
18. West's Modern Legal Forms §§ 2892–93.
19. See Section 6.04.
20. Int.Rev.Code of 1954, 26 U.S.C.A. § 1244(c)(3).
21. West's Modern Legal Forms § 2909.6.
22. See Section 4.06.
23. Substantive elements of the Subchapter S election are discussed at Section 4.06.
24. West's Modern Legal Forms § 2838.
25. See Section 6.09.
26. M.B.C.A. § 43.
27. See the sample by law provisions regulating director's meetings in Section 6.09 and Form 6Q in the Appendix.
28. See M.B.C.A. § 43.
29. See the by-law provisions in section 6.09 and Form 6Q in the Appendix, and the director resolution in Section 8.05.
30. West's Modern Legal Forms § 2855.
31. West's Modern Legal Forms § 2854.
32. West's Modern Legal Forms § 2857.
33. M.B.C.A. § 43.
34. West's Modern Legal Forms § 2858.
35. M.B.C.A. § 28.
36. E.g., Nebraska, Neb.Rev.Stat. § 21–2027 (application to a court after thirteen months without a meeting.); Iowa, Iowa Code Ann. § 496A.27 (application to a court after eighteen months without a meeting.)
37. The holders of one-tenth of the voting shares may call a special meeting if the annual meeting has not been held in Massachusetts, Mass.Gen.Laws Ann. ch. 156B §§ 33, 34.
38. M.B.C.A. § 28.
39. West's Modern Legal Forms § 2831.
40. See M.B.C.A. § 30.

41. M.B.C.A. § 29.
42. M.B.C.A. § 29.
43. E.g., New Jersey, N.J.Stat.Ann. § 14A:5–4.
44. West's Modern Legal Forms § 2823.
45. West's Modern Legal Forms § 2825.
46. M.B.C.A. § 29.
47. For further elaboration on voting procedures for these transactions see Sections 13.02 and 13.03.
48. See Section 8.06.
49. M.B.C.A. § 33.
50. West's Modern Legal Forms § 2828.
51. West's Modern Legal Forms § 2828.1.
52. Securities Exch. Act of 1934, 15 U.S.C.A. § 78n, and the Rules promulgated thereunder.
53. West's Modern Legal Forms § 2827.6. This proxy must be accompanied by a special notice and a proxy solicitation statement, prepared in accordance with the Securities Exchange Act of 1934. See West's Modern Legal Forms §§ 2827, 2827.1 and 2827.5.
54. M.B.C.A. § 32.
55. E.g., New York, McKinney Consol. Laws of N.Y.Bus.Corp. Laws § 608.
56. M.B.C.A. § 24.
57. Variations in voting rights between classes of shares is discussed in Sections 7.10 and 7.11.
58. These shareholder agreements are most frequently found in close corporations. See Section 11.02.
59. E.g., California, West's Ann.Calif.Corp. Code § 707; New York, McKinney Consol.Laws of N.Y.Bus.Corp. Law § 610.
60. See Section 5.01.
61. M.B.C.A. § 32.
62. M.B.C.A. § 36.
63. M.B.C.A. § 33 and see Section 6.05.
64. M.B.C.A. § 39.
65. See Sections 4.03 and 6.05.
66. M.B.C.A. § 37.
67. Further elaboration on the procedure to amend the articles of incorporation is contained in Section 13.01.
68. See Chapter 13.
69. See Section 7.11.
70. About eleven states have no statutory authority for written consent to action by directors. All but three states have shareholders consent to action provisions.
71. 8 Del.Code Ann. § 228. See, also, New Jersey, N.J.Stat.Ann. § 14A.5–6.
72. E.g., New York, McKinney Consol. Laws of N.Y.Bus.Corp. Law § 615; Pennsylvania, 15 Pa.Stat. § 1513.
73. B. M. Miller, 1 Manual and Guide for the Corporate Secretary, 3–35, 53–69, 107–137 (1969).
74. See the general forms for minutes of shareholder and director meetings, Forms 8A and 8B in the Appendix.
75. See M.B.C.A. § 41.

9

CORPORATE DIVIDENDS AND OTHER DISTRIBUTIONS

§ 9.01 TYPES OF CORPORATE DISTRIBUTIONS

One of the investment objectives of share ownership is the receipt of profit distributions or "dividends". The attractiveness of a stock investment is measured by this "yield" in addition to projected value appreciation of the shares as the corporate business prospers. Shares are also entitled to receive a proportionate share of the assets of the corporation when the business is dissolved and liquidated. These dividends and other distributions to shareholders are the subject matter of this chapter.

The board of directors is vested with the authority and the discretion to declare dividends from time to time. Dividends may be paid in cash, property or shares of the corporation, but payment usually is restricted to certain available funds or prohibited unless certain financial tests are met.[1] Generally, however, if the appropriate funds are available or if the financial condition of the corporation is satisfactory, the directors may distribute all sorts of desirable things to its shareholders.

Cash is always welcome and the cash dividend is the most common corporate distribution. Cash dividends are declared by director resolution which usually specifies a dollar amount per share and directs payment of the dividend to all shareholders entitled to receive it.

Property dividends are less common, but are equally available for use by the board of directors. If the corporation markets a desirable product, the product itself may be a dividend distribution. R. J. Reynolds Tobacco Company once distributed a specially prepared package of its tobacco products to its shareholders as a dividend, and one of the famous corporate cases involved a scramble to purchase shares of a distillery corporation which was reported to be preparing to issue a property dividend of its liquor products during World War II when liquor was scarce.[2] Property dividends occasionally include shares of stock of a subsidiary corporation owned by the parent corporation. There is no restriction on the type of property which may be distributed. If the corporation had a surplus of adding machine tape it could distribute that as a dividend, subject, of course, to the approval of the shareholder public relations department.

Dividends may also consist of more shares of the same corporation. Stock dividends increase the number of shares owned by the receiving shareholder, but, of course, do not affect his proportionate ownership interest. For example, the directors may declare a dividend of one share of common stock for every 100 shares, outstanding. Each shareholder will receive an extra share of stock per hundred as his corporate distribution. You should notice that the previous discussion regarding fractional shares[3] becomes relevant here. If Debbie McCarty owns 125 shares and a 1/100 stock dividend is declared, she will be entitled to receive 1¼ shares in the dividend. The corporation may issue a stock certificate or scrip representing the fraction, it may pay the fair value of the quarter share in cash, or it may arrange for the sale of the fractional interest to another shareholder similarly situated.

Dividend distributions to shareholders do not necessarily come at regularly scheduled intervals. The board of directors may declare as many or as few dividends as they deem appropriate. It should be obvious, however, that the directors who declare frequent dividends are usually more popular with the shareholders. The practice of larger corporations with established dividend policies is to declare a regular quarterly dividend on the corporate shares, as profits permit.

A dissolution distribution is usually a one-time distribution to shareholders when the corporate assets are liquidated and business is terminated. Partial liquidations are also possible, however, if the corporation ceases to operate one phase of its business and distributes assets from the discontinued operations to shareholders, continuing other active business operations thereafter.

The most important legal considerations regarding corporate distributions are the authority to declare or demand the distributions; the legally available funds out of which the distribution is paid; and the tax ramifications.

§ 9.02 SOURCES OF FUNDS FOR DISTRIBUTION

All state statutes regulate the manner in which a corporation may distribute its assets to its shareholders, and each statute identifies particular accounts from which a distribution may be made. Common sources of dividend funds are "earned surplus" or "retained earnings" or "profits". The Model Act Section 45 was amended in 1979 to permit dividends from any source so long as the corporation remains solvent and its total assets remain greater than its total liabilities (plus any amounts payable as liquidation preferences.[4] Most state statutes use "unreserved and unrestricted earned surplus" as the main source for dividend distributions, and also allow dividends to be paid from other more specialized accounts. However, in order to understand the law of corporate distributions, the distinctions between sources of funds for distributions and the accounting principles used in creating these accounts should be explored. Consider the following accounts (which are defined this way in many state statutes):

1. "Net assets" means the amount by which the total assets of a corporation exceed the total debts of the corporation.

2. "Stated capital" means, at any particular time, the sum of (a) the par value of all shares of the corporation having a par value that have been issued, (b) the amount of the consideration received by the corporation for all shares of the corporation without par value that have been issued, except such part of the consideration therefor as may have been allocated to capital surplus in a manner permitted by law, and (c) such amounts not included in clauses (a) and (b) of this paragraph as have been transferred to stated capital of the corporation, whether upon the issue of shares as a share dividend or otherwise, minus all reductions from such sum as have been effected in a manner permitted by law.

3. "Surplus" means the excess of the net assets of a corporation over its stated capital.

4. "Earned surplus" means the portion of the surplus of a corporation equal to the balance of its net profits, income, gains and losses from the date of incorporation, or from the latest date when a deficit was eliminated by an application of its capital surplus or stated capital or otherwise, after deducting subsequent distributions to shareholders and transfers to stated capital and capital surplus to the extent such distributions and transfers are made out of earned surplus. Earned surplus shall include also any portion of surplus allocated to earned surplus in mergers, consolidations or acquisitions of all or substantially all of the outstanding shares or of the property and assets of another corporation, domestic or foreign.

5. "Capital surplus" means the entire surplus of a corporation other than its earned surplus.

Now review the following corporate balance sheet with these definitions in mind:

Following the statutory definitions, "net assets" is the amount by which the total assets, $78,000, exceed the total debts, $26,000. Thus, the net assets of the corporation in this case would be $52,000.

"Stated capital" in our example equals the par value of the 5,000 common shares plus the par value of the 50 preferred shares. The "capital surplus" results from the excess consideration for which the par value shares were sold. For example, the balance sheet shows capital surplus of $20,000, which could have resulted from selling the 5,000 common shares at $4 a share, allocating the $3 per share in excess of par to capital surplus, and selling the preferred shares at $200 per share, allocating the $100 per share excess to capital surplus.

"Earned surplus" represents accumulated corporate profits which have been earned during preceding accounting periods and retained in the corporation. The term "surplus" refers to the total amount of earned surplus

THE NOBLES COMPANY

BALANCE SHEET

December 31, 1981

ASSETS			
Current Assets:			
Cash	$ 5,000		
Marketable Securities			
Apex Telephone Company, at cost	8,000		
Accounts Receivable	2,000		
Inventory	5,000		
	$20,000		
Total Current Assets			20,000
Property Plant and Equipment			58,000
Total Assets			78,000
LIABILITIES AND EQUITY			
Current Liabilities:			
Accounts Payable	3,000		
Accrued Expenses Payable	3,000		
Total Current Liabilities		6,000	
Bonds, First Mortgage 13½% Interest		20,000	
Total Liabilities			26,000
Equity			
Stated Capital			
Common Shares, $1 par value; 50,000 authorized, 5,000 issued and outstanding	5,000		
Preferred shares, 5% cumulative $100 par value; 100% of par liquidation preference; 10,000 authorized, 50 issued and outstanding	5,000		
Total Stated Capital		10,000	
Capital Surplus		20,000	
Earned Surplus		22,000	
Total Equity			52,000
Total Liabilities and Equity			78,000

and capital surplus. Thus, the total "surplus" of this corporation would be $42,000. The entire equity section of the balance sheet is also called "net worth".

There is one other important definition for analyzing the legality of corporate distributions. A corporation is "insolvent" when it is unable to pay its debts as they become due in the usual course of business.[5]

The application of these terms will become clear as we consider the statutory provisions regulating corporate distributions.

§ 9.03 CASH AND PROPERTY DIVIDENDS

As previously discussed, a dividend is a distribution of cash, other property or shares to the stockholders of the corporation in the proportion of their share ownership. Recall that various classes of stock may be treated differently with respect to dividends. Dividend preferences are common in complex corporate financial structures, and, of course, the preferences must always be observed. There is also a rule that dividends must be uniform within each class or series, meaning that each share of a given class or series will receive the same distribution as the other shares of that class or series. By way of illustration, consider the Nobles Company capital structure previously described, with 5,000 shares of common stock and 50 shares of 5% cumulative preferred with $100 par value. Suppose the directors decided to issue a dividend and to distribute $2,000 in cash to the shareholders. The preferred shareholders are entitled to be paid their preferential dividends (5% of $100 per share) so they would receive $250 in dividends first. This leaves $1,750 for distribution to the common shareholders, who should receive $.35 per share. Each shareholder of the particular class must be treated equally. The directors could not order a distribution of $.50 per share to some common shareholders and $.20 per share to others.

Shareholders' Rights to Dividends

The decision to distribute dividends is within the sole discretion of the board of directors, and no shareholder has a right to dividends until the board of directors declares it.[6] Cumulative preferred stock will accumulate dividends annually until they are finally paid, but even preferred shareholders have no right to dividends until they are declared by the board of directors. In our Nobles Company example, the 5% cumulative preferred stock will accumulate a dividend entitlement of $5 per share per year, and if the board of directors ever declares a dividend, the arrearage must be paid before the common shareholders may receive dividends.

Once the board of directors has declared a dividend through an appropriate resolution, the shareholders have a right to the dividend and it becomes a debt from the corporation to the shareholders. The procedure for declaring a dividend will be discussed after we explore the restrictions on sources of funds for dividends.

Restrictions on Payment of Dividends

Corporate statutes usually restrict the payment of dividends to funds contained in specified corporate accounts. There are also certain restrictions on payment even if funds appear to be available in the prescribed accounts. Section 45 of the Model Act prohibits the payment of dividends if the corporation is insolvent; or if the declaration or payment would be contrary to any restriction contained in the articles of incorporation.[7] If the organizers intend to restrict the directors' ability to declare and pay dividends, the law honors their restraints.

The Model Act further provides that dividends may be declared only if the corporation's total assets exceed total liabilities and (unless provided otherwise in the articles of incorporation) the amounts payable to preferred shares having a preference to assets in liquidation. In the case of the Nobles Company, this excess is $47,000: total assets ($78,000) less total liabilities ($26,000) and amounts payable in liquidation to the preferred shares (100% of par, or $5,000). The asset and liabilities values are determined from the balance sheet or based on some other reasonable and fair valuation. Thus, if the Nobles Company's stock in Apex Telephone Company were valued in the market at $20,000, an additional $12,000 would be available for dividends.

Most states continue to permit dividends to be paid in cash or property only out of the unreserved and unrestricted earned surplus of the corporation with a few exceptions, which will be discussed in a moment. In these states, therefore, a corporation may pay dividends from available funds in its earned surplus account, and in the case of our hypothetical company described above, $22,000 would be available for payment of dividends. However, there may be practical restrictions on the payment of such a dividend. A glance at the balance sheet should reveal that payment of a $22,000 divident is unrealistic considering the relatively low cash position of the corporation and the fact that its total liquid assets, including cash, securities, inventory and accounts receivable, total only $20,000. The corporation would be unable to pay a full $22,000 dividend without liquidating part of its property, plant and equipment, even though the funds are legally available for dividends up to $22,000. Thus, in every case, the maximum legal limits on dividends will be tempered by business advisability. The directors determine the latter within the bounds of the former.

Earned surplus generally represents the accumulated profits from corporate operations as determined by accepted accounting principles.

Some states, including Delaware and Nevada, provide that in addition to the funds from earned surplus, dividends may be declared and paid out of the unreserved and unrestricted "net earnings for the current fiscal year and the next preceding fiscal year taken as a single period." This means that if the corporation has no historical earned surplus, but has earnings (profits) for the past two fiscal years, dividends may legally be paid from the profit funds, without regard to the balance in the earned surplus account.

In California, the availability of funds for dividends depends on a ratio of "current assets" to "current liabilities" and management is allowed to *anticipate* receipts and expenses for the next year to make the computation.

The words "unrestricted and unreserved" deserve some comment. Earned surplus or earnings may be restricted in several ways. Suppose the corporation has borrowed money from a bank or other financial institution, and the terms of the loan agreement require that earned surplus will be maintained at a minimum level of $10,000 for the duration of the loan. In accordance with the loan agreement, that amount of earned surplus should be restricted, and may not be used for the payment of dividends. A restriction would also be imposed when the corporation purchases its own shares from outside investors to hold as treasury shares. Most statutes require that the amount of earned surplus used to purchase treasury shares be restricted for as long as the shares remain treasury shares.[8] Many statutes ignore the distinction between ordinary surplus and reserved or restricted surplus, and permit dividends to be declared and paid notwithstanding such restrictions.

Sometimes dividends may be paid out of the depletion reserves of a corporation engaged in exploiting natural resources, such as minerals, oil and gas and timber. The articles of incorporation also must expressly authorize such dividends.[9]

Cash and property dividend provisions vary extensively among the states. Most couch their restrictions on legally available funds in terms of earned surplus or earnings during a prescribed period. Several add further restrictions to prohibit dividends from unrealized appreciation and depreciation.[10]

Procedure for Declaration, Payment and Accounting

The board of directors determines the amount to be paid as dividends by the corporation in their sole discretion. Notice, however, that in the Nobles Company financial structure they must pay $5 per share for 50 shares of preferred stock as a cumulative dividend before any dividends may be paid to the common shares. If the directors declared a dividend of $3,000, they would first resolve to distribute $250 to the preferred shares, or $5 per share, and then distribute the remaining $2,750 in a dividend of $.55 per share for the common stock.

The decision to declare the dividend is made at a directors' meeting and is recorded as a resolution in the minutes:

EXAMPLE 9.03A: Resolution to Declare a Cash Dividend

RESOLVED, that a dividend of $5.00 per share be declared and paid on the preferred stock and a dividend of $.55 per share be declared and paid on the common stock of this corporation out of the unreserved and unrestricted earned surplus to the holders of stock as shown by the records of the corporation on the 15th day of June, 1982, distributable on the 1st day of July, 1982, and that the treasurer is directed forthwith to mail checks for the same to the stockholders of record.[11]

An example of the resolution for a property dividend follows:

EXAMPLE 9.03B: Resolution to Declare a Property Dividend

From the report furnished the meeting on the financial condition of the Company for the fiscal year ended December 31, 1981, and for each of the subsequent

> months, it appeared that the Company was in a position to declare and pay a dividend in property upon its outstanding shares of common stock.
>
> Thereupon, after discussion it was on motion duly made, seconded and unanimously adopted by the affirmative vote of the directors present:
>
> RESOLVED, that a dividend on the outstanding shares of common stock of this Corporation be and it hereby is declared and ordered to be paid in property of this Corporation, to-wit, shares of common stock of Apex Telephone Company owned by this Corporation at the rate of ten shares of common stock of said Apex Telephone Company for each share of common stock of this Corporation issued and outstanding. Said dividend to be payable on September 1, 1982 to the holders of record of said common stock of this Corporation at the close of business on August 10, 1982 from the net surplus of this Corporation as at the close of business on December 31, 1981, or from the net profits of this Corporation for its current fiscal year; and
>
> FURTHER RESOLVED, that the proper officers of the Corporation be and they hereby are authorized and directed in the name and on behalf of the Corporation to do or cause to be done all acts or things necessary or proper to carry out the foregoing resolution.[12]

Notice that the resolution sets a record date for a determination of stockholders who will be entitled to receive dividends. Nearly every state statute has certain rules to follow in fixing the record date, and in most states the rules are exactly the same as those regarding shareholder voting described earlier.[13] Model Act Section 30 combines the determination of shareholders for voting and dividend purposes, and permits the stock transfer books to be closed as early as fifty days before the dividend is paid. Alternatively, the directors may set a record date within the fifty day period to determine which stockholders are entitled to receive dividends. If the books are not closed and if no date is set, the date for determining "shareholders of record" is the date that the board of directors adopted the resolution declaring the dividend.

When the dividend is paid in cash, the cash amount will be reduced by $3,000, and the earned surplus account will be reduced by $3,000. This maintains the balance on the balance sheet. The dividend may also be paid in other property, such as the securities described in the foregoing example, which appear as current assets on the balance sheet. If the securities distributed were valued at $3,000, the earned surplus and the marketable securities accounts would be reduced by that amount. The value of the property distributed and charged to earned surplus is determined by the "book value", which is the value shown on the balance sheet.

Tax Ramifications of Cash or Property Dividends

In the early discussion of the corporate form it was observed that one of the disadvantages of corporate existence is the problem of double taxation, and dividends are at the heart of that problem. The corporation is taxed on its profits and if the after-tax-profits are distributed to shareholders as dividends, the dividends are income to the shareholders and are also taxed at the individual shareholder's rate. A corporation may avoid the problem of double taxation by paying salaries or consulting expenses to its shareholders wherever possible, since these items are deductible as expenses, and the money thus expended is not taxed at the corporate level. Salaries are viable alternatives to dividends when the shareholders are employees of the corporation, as is frequently the case in small, close corporations. In larger corporations, however, dividends are a necessary evil to provide the shareholders a return on their investment.

Cash dividends are declared on a shareholder's income tax return as income in the actual amount distributed. Property dividends are subject to a special rule. When property is distributed as a dividend, the shareholders must report as income the fair market value of the property received even though the "book value" of the property may be less:[44] For example, if the distributed marketable securities of the Nobles Company had a fair market value of $4,000, even though their book value is $3,000, the shareholder must report $4,000 as ordinary income.

Corporations which pay dividends aggregating more than $10 to any one person during the calendar year must file certain reports with the Internal Revenue Service, which uses these "informational" reports to determine whether the shareholders have reported the dividends on their individual income tax returns.

§ 9.04 SHARE DIVIDENDS

A share dividend is a corporate distribution of its own shares. The number of shares owned by the shareholders is increased, but the proportionate stock ownership of each shareholder does not change. A share dividend requires no modification in any of the characteristics of the shares, and it adds nothing to the shareholders ownership interest; instead, it dilutes his ownership interest by dividing his investment into more shares.

Legal Restrictions

Most states permit share dividends from two sources: the corporation's treasury shares, if shares have been purchased and held in the treasury; and authorized but unissued shares. In either case, the earned surplus account must contain available funds and it will be adjusted to reflect the dividend.

When authorized but unissued shares are used for share dividends, a transfer from surplus to stated capital occurs. Section 45 of the Model Act formerly required such a transfer and many states still follow the rule.[45] If the shares have a par value, they must be issued at an amount equal to or greater than the par value. An amount of surplus equal to the aggregate par value of the dividend shares must then be transferred to stated capital at the time the dividend is paid. If the dividend shares have no par value, the shares are issued at a stated value fixed by a resolution of the board of directors and stated capital receives an amount of surplus equal to the aggregate stated value. Further, the amount of surplus so transferred to stated capital must be disclosed to the shareholders when the dividend is paid.

In order to avoid dilution of one class in favor of another class, there is usually an additional restriction. No dividend payable in shares of any class may be paid to the holders of shares of any other class unless authorized by the articles of incorporation *or* by the affirmative vote or the written consent of the holders of at least a majority of the outstanding shares of the class in which the payment is to be made.

Using the Nobles Company as an example, note that the balance sheet reflects 10,000 preferred shares with $100 par value *authorized* but only 50 of these shares are *issued and outstanding*. There are, therefore, 9,950 preferred shares authorized but unissued. The directors may consider issuing one share of preferred stock for each share of common stock as a share dividend, but they have two hurdles to jump: the authority to issue a share dividend to holders of one class in shares of another class, and the legally available funds for the dividend. If 5,000 shares of preferred were distributed to the common shareholders, the present holders of the 50 preferred shares would have their ownership interests severely diluted. Accordingly, either the articles of incorporation must authorize such a share dividend, or the preferred shareholders must approve the dividend. The second hurdle is insurmountable in this case, since the preferred shares have a par value of $100, and that amount must be transferred from surplus to stated capital for each distributed share. The surplus of the Nobles Company, the excess of net assets and stated capital, is only $42,000. In order to issue 5,000 shares of $100 par value stock, $50,000 in surplus must be transferred to stated capital. Consequently, this share dividend could not be distributed. It should be noted in passing that a share dividend of one class of shares to the holders of another class is quite rare. Usually a share dividend will be paid to holders of each class in shares of the same class.

If the corporation has no treasury shares, and if all of the authorized shares have been issued, the articles of incorporation must be amended to supply additional authorized shares before a share dividend may be declared.

Procedure for Declaration, Payment and Accounting

The decision to declare a share dividend is made by the board of directors, who will adopt a resolution specifying the number of shares to be distributed, the proportion of distribution, a record and payment date for the persons to receive dividends, and authority for the transfer of surplus to stated capital.

EXAMPLE 9.04A: Resolution to Declare a Stock Dividend

WHEREAS, there has been accumulated from undistributed profits of the company a surplus of $22,000.00, which in the opinion of the board of directors can be advantageously used for the benefit of stockholders.

RESOLVED, that a stock dividend be and the same hereby is declared payable to stockholders of record as of the 10th day of July, 1982, one share of common stock of the par value of $1 per share to be distributed as such stock dividend to the holder of each outstanding share of common stock, and a like amount to the holder of each share of preferred stock of the par value of $100 each, said stock dividend to be extra and additional to any cash dividend now or hereinafter declared; and

FURTHER RESOLVED, that $5,050.00 of said surplus be transferred to capital to accomplish said stock dividend, and 5,050 shares of common stock of the par value of $1 per share be issued and disbursed as hereinbefore provided.[16]

On the record date, the shareholders entitled to receive the shares are identified and on the payment date, certificates are executed and distributed to these persons.

All of the accounting entries reflecting a stock dividend occur in the shareholders' equity section of the balance sheet, because the "payment" (the value of the shares) is transferred from the surplus accounts to the stated capital account. Stated capital receives the amount of the par value of the shares if the shares have a par value, or an amount determined by the board of directors for no par shares. For example, if the Nobles Company declared a one-share-per-ten common stock dividend, 500 new shares of $1 par common stock would be issued as a dividend to the holders of the 5,000 outstanding shares. The stated capital would be increased by $500, the amount equal to the par value per share times the number of shares distributed as a dividend. The earned surplus account would be reduced by $500 since "payment" is theoretically made from there.

The New York Stock Exchange imposes a special accounting rule for stock dividends from companies listed on the exchange, and generally accepted accounting principles require the use of the rule for other corporations as well. Stated simply, the accounting for a stock dividend must recognize the fair market value of the shares being distributed. In the distribution described above, suppose that the 500 common shares, with $1 par value, had a fair market value of $10 per share. If the same shares were sold, $500 would be entered in stated capital and $4,500 would be transferred to capital surplus. When the shares are distributed as a stock dividend, the proper accounting entries would be $500 to stated capital, $4,500 to capital surplus, and $5,000 from earned surplus.

If you study these accounting entries, you will see why the issuance of a stock dividend really adds nothing to the true economic interests of the shareholder. The shareholder's ownership interest in the corporation is represented by the shareholder's equity section of the balance sheet, including stated capital, capital surplus, and earned surplus. In issuing a stock dividend the corporation is merely transferring funds within the shareholder's equity accounts, and is not distributing any assets of the corporation. The advantage of a stock dividend lies in the future; if the business continues to prosper, the value of each share of stock will increase. The eternal hope of every shareholder is that all of her stock, including the shares distributed in a stock dividend, will appreciate in value, and she will realize capital gains when the stock is sold.

Tax Ramifications of Share Dividends

The tax ramifications of a stock dividend for shareholders is simple on its face, but complicated in its application. The dividend itself is a non-taxable transfer, since the shareholder receives no distribution of value from the stock dividend. Rather, the ownership interest is simply divided into more shares. Thus, no tax need be paid when a stock dividend is received. The tax problem comes later. When the shares are sold, the shareholder must compute the "basis" for the shares—that is, their cost to him—in order to compute the capital gains. Stated very simply, if shares were purchased for $5 per share, and subsequently sold for $8 per share, the basis is the cost of the shares and the capital gain is $3 per share. Stock dividends complicate this computation, because they require an adjustment to the basis of the original shares. For example, suppose a shareholder purchased 100 shares of the Nobles Company common stock at $5 per share, and subsequently received a one-for-ten stock dividend, resulting in a total of 110 shares for his original investment of $500. His basis per share is now $4.545 per share and that figure will be used to compute capital gains when the shares are sold. Further complications are apparent if the shareholder had purchased 10 shares in 1978 for $5 per share; 50 shares in 1979 for $6 per share; 25 shares in 1980 for $6.50 per share and 15 shares in 1981 for $6.75 per share. When he later receives a 10 share stock dividend, all of these figures must be adjusted. A stock dividend may be unpopular with the shareholders because of these confusing tax computations.

§ 9.05 STOCK SPLITS

A stock split is very similar to a stock dividend inasmuch as it results in a greater division of the same ownership interest for each shareholder. However, a typical stock split usually involves some modification of the capital stock structure itself. A split is normally accomplished by reducing the par value of the shares so that the aggregate stated capital of the corporation will be unaffected by the distribution. If the shares have no par value, the stock split effects a reduction of the stated value of each share. The generally accepted distinction between stock splits and stock dividends stems from the mechanics of effecting the distribution. A stock *dividend* involves a transfer of surplus to stated capital within the shareholder equity section of the balance sheet. By taking from surplus and adding to capital, the stock dividend may be considered to be a type of earnings distribution, although no value is actually distributed to the shareholders. A stock *split*, on the other hand, usually changes the par value of the shares and the stated capital remains the same. There are no shifting of funds and no distributions of earnings. This distinction between stock dividends and stock splits has very little practical effect, with the singular exception that a true stock split will require an amendment to the articles of incorporation.

Like a share dividend, a stock split involves the issuance of a certain number of shares for each share currently held. Splits may be two-for-one, three-for-one, or even 100-for-one for that matter. There are two separate procedures for accomplishing a split.

A modification of the par value of the shares is the common method of effecting a split. In a two-for-one split the par value is halved, and twice as many shares are issued for the same amount of stated capital. In the case of the common shareholders of the Nobles Company, a two-for-one stock split would reduce the par value of the common shares to $.50 per share, and result in a distribution of 5,000 additional shares for the 5,000 shares presently outstanding. There will then be 10,000 shares of common stock issued and outstanding represented by the same $5,000 in stated capital. The shareholders' equity section of the balance sheet would be changed only to indicate that there were 10,000 shares issued and outstanding, and that the par value was now $.50 per share. An amendment to the articles of incorporation is necessary, however, to reflect the new par value in the capital structure. The amendment may also require an increase in the authorized stock of the particular class. If the Nobles Company sought to issue an eleven-for-one stock split, the presently authorized shares are insufficient for the distribution, since 55,000 shares are needed. The extra shares may be authorized in the same amendment which changes par value. Shares without par value may also be involved in a split, but the amendment to the articles may not be necessary since they remain shares without par value after the split. Of course, an amendment is required if additional authorized shares are needed to accomplish the split. The initiative for the stock split begins with a resolution by the board of directors for an amendment to the articles of incorporation:

EXAMPLE 9.05A: Resolution to Amend the Articles for a Stock Split

RESOLVED, that Article Fourth of the Articles of Incorporation be and the same is hereby amended as follows:

"FOURTH: * * * ten thousand (10,000) shares shall be common stock with a par value of $.50 per share."

At the time this amendment becomes effective, and without any further action on the part of the corporation or its stockholders, each share of common stock of a par value of $1.00 per share then issued and outstanding shall be changed and reclassified into two fully paid and nonassessable shares of common stock of a par value of $.50. The capital account of the corporation shall not be increased or decreased by such change and reclassification. To reflect the said change and reclassification, each certificate representing shares of common stock of a par value of $1.00 theretofore issued and outstanding shall be cancelled, and the holder of record of each such certificate shall be entitled to receive a new certificate representing two shares of common stock of a par value of $.50, so that upon this amendment becoming effective each holder of record of a certificate representing theretofore issued and outstanding common stock of the corporation will be entitled to certificates representing in the aggregate two shares of common stock of a par value of $.50 authorized by this amendment for each share of common stock of a par value of $1.00 per share of which he or she was the holder prior to the effectiveness of this amendment.[17]

The second method of declaring a stock split is the same procedure used to issue a stock dividend. Instead of changing par value of the shares by an amendment to the articles of incorporation, an amount may be transferred from surplus to stated capital, and the new shares may be issued for the increased stated capital. The Nobles Company could accomplish a two-for-one split by transferring $5,000 from surplus to capital, thereby increasing stated capital to $10,000 and covering the issuance of 5,000 additional shares at the same par value. Although this is technically a distribution of earnings and is more appropriately described as a stock dividend, it is generally referred to as a "split" because of the doubling of shares held by each shareholder. An amendment to the articles of incorporation may still be necessary, however, if the remaining authorized but unissued shares are fewer than the additional shares required in the split.

Although the amendment procedure and the transfer of surplus to capital are alternative methods of accomplishing stock splits, there may be other rules which require "capitalization of earnings" in a stock distribution. That phrase refers to the second procedure of transferring surplus to stated capital, and the rules of the New York Stock Exchange and the Securities Exchange Act of 1934 should be consulted for the circumstances requiring capitalization of earnings in a stock split.[18]

In all other respects, stock splits are treated the same as stock dividends. Upon appropriate resolution of the board of directors, and amendment of the articles of incorporation if necessary, new certificates are issued to shareholders reflecting the additional shares of the split. The tax ramifications for the shareholders are identical to the tax treatment of stock dividends.

§ 9.06 CORPORATION'S PURCHASE OF ITS OWN SHARES

The corporation may purchase its own shares from its shareholders if the board of directors authorizes the purchase and the applicable state statute so permits. The Model Act authorizes the repurchase of shares in Section 6, and Section 45 treats a repurchase as a distribution to shareholders, so that the same restrictions that are placed upon dividends also apply here. Most states regulate the repurchase of shares as dividends are regulated, so that the funds used to purchase the shares are limited to unrestricted and unreserved earned surplus. Usually capital surplus may be also used if authorized by the articles of incorporation or by a majority vote of the shareholders.

The corporation's purchase of its own shares is a distribution to shareholders, since the corporation exchanges cash for shares, and because of the distribution characteristics, purchases of shares are regulated much the same way as other distributions. The law refuses permission to purchase shares if the corporation is insolvent or would be rendered insolvent by the purchase. However, there are some generally recognized exceptional cases when the corporation may reacquire its own shares without making a "distribution" subject to statutory regulation. These include the purchase of corporate shares to eliminate fractional shares;[19] to compromise or collect an indebtedness to the corporation; to pay dissenting shareholders entitled to payment under the law;[20] and to retire redeemable shares.[21] In all other cases, the corporate purchase of shares must be made from legally available funds, earned surplus or capital surplus in an appropriate case.

Consider the purchasing ability of the Nobles Company. Under the revised Model Act approach, the corporation may only make a distribution if it will be able to pay its debts as they come due and if total assets exceed (and after the distribution occurs will continue to exceed) total liabilities plus the amount payable to preferred shares as a liquidation preference. The balance sheet of the Nobles Company shows total assets of $78,000, total liabilities of $26,000, and a liquidation preference for preferred shares of $5,000. Thus, the corporation has a total of $47,000 in legal purchasing power, subject to its ability to pay debts as they come due. In states which use the earned surplus and capital surplus tests, the Nobles Company would have less purchasing power. The balance sheet shows unreserved and unrestricted earned surplus of $22,000, and that amount could be used by the board of directors to purchase shares. The balance sheet also shows $20,000 in capital surplus which could be used if the articles of incorporation or the holders of the majority of voting shares authorized the action. Of course, the liquidity of the company may be a practical barrier to the purchase of shares. The corporation will be limited by the relatively low cash position shown on the balance sheet under both tests.

The stock purchased by the corporation may either be cancelled or held by the corporation as treasury shares. In the latter case, most states require that the surplus used to purchase the shares must be restricted in the amount required for the purchase, and the restriction must remain in effect as long as the shares are held as treasury shares. If the shares are subsequently sold or cancelled the restriction may be removed. To illustrate, assume that the Nobles Company repurchased 1,000 shares of its common stock to hold as treasury shares at a price of $3.00 per share, using $3,000 of the earned surplus for the purchase. The earned surplus account would reflect $19,000 in available earned surplus and $3,000 in restricted earned surplus, representing the 1,000 shares of common stock held in the treasury. Thereafter, the corporation may resell the treasury shares for any price (including a price below par value). Of course, when resold, the shares are no longer treasury shares and the restriction is lifted. If the corporation sold the shares for $2.50 per share then only $2,500 would be added back to earned surplus and the loss of $500 would be permanently subtracted from that account. These transactions reflect the fact that the corporation has suffered a loss by investing in its own shares.

The corporation is also permitted to cancel the purchased shares, and former Section 68 of the Model Act prescribes the cancellation procedure used in most states. The board of directors adopts a resolution authorizing the cancellation of shares and a statement of cancellation is filed with the Secretary of State.[22] When the cancellation is completed, the stated capital account is reduced to reflect the cancellation and the earned surplus is adjusted to show that only the par value of the shares has been realized by the company. Thus if the 1,000 shares reacquired by the Nobles Company for $3,000 were cancelled, the stated capital account would show 4,000 common shares issued and outstanding with $4,000 stated capital. The earned surplus account would be adjusted as follows:

After purchase, before cancellation:

Earned surplus	
Unrestricted	$19,000
Restricted	3,000
Total	$22,000

Cancellation entries:

Remove restriction—	
Add par value of cancelled shares	$1,000
Subtract purchase price of shares	(3,000)
Net reduction	($2,000)

After cancellation:

Earned surplus:	
Unrestricted	$20,000

Those perceptive readers may have become uneasy about the corporation's purchase of its own shares, remembering that an important rule for dividend distributions is that all shareholders of the same class must be treated equally. A relevant question at this point might be the corporation is required to purchase the entire class of stock if it intends to purchase any shares. The answer is no. The corporation may purchase all or any part of a particular class, and theoretically may select the shareholders from whom the purchase is made. However, there is a taint of unfairness if the selling shareholders happen to be the directors who authorized the purchase, and the plot thickens if the corporate business suddenly takes a turn for the worse right after the sale. There are laws which protect the other shareholders in such a case. Traditional common law principles of fraud apply here, and the shareholders are further protected through the fiduciary duties owed to them by management. Finally, federal securities laws regulate "insider trading" by management, and those laws may also undo the harm. Nevertheless, the corporate statutes do not require that all shareholders of a particular class be treated equally when the corporation purchases its own shares.

§ 9.07 PARTIAL LIQUIDATIONS

A partial liquidation is a combination of a distribution of assets, usually cash, and the purchase of the corporation's own shares. In fact, any purchase of shares by a corporation is a partial liquidation, but there are special tax advantages for certain liquidations, and the tax definition of partial liquidation is the subject of this section. A tax-favored partial liquidation occurs when the corporation exchanges assets for stock, and the assets exchanged are attributable to the termination of a business which has been actively conducted for a five-year period preceding the liquidation. This could result from the closing of a complete division of the corporation and terminating the trade conducted by that division. The liquidation may also result from a contraction of the corporation's business, such as closing certain selected manufacturing plants. In either case, the corporation must continue to be actively engaged in *some* business after the liquidation in order for the distribution to qualify as a tax-favored partial liquidation.[23] If the distribution satisfies these statutory requirements, the Internal Revenue Code specifies that the distribution to the shareholders will be taxed at capital gain rates, rather than ordinary income rates. A partial liquidation is thus considered a return of capital to the shareholders.[24]

To illustrate a partial liquidation with our hypothetical corporation, the Nobles Company could sell a portion of its plant and equipment and liquidate the cash received in a distribution to the shareholders in exchange for stock. The transaction may look something like this: the company could sell $13,000 in plant equipment and continue its reduced business with the remaining plant and equipment available; the net cash from the sale may be distributed to the stockholders in exchange for a proportionate amount of stock. Since the $13,000 in plant and equipment equals approximately one-fourth of the net assets, the corporation may reacquire approximately one-quarter of the total outstanding stock in exchange for the distribution. The corporation has accomplished a purchase of stock by distributing cash assets received in a partial liquidation, and the shareholders will be taxed at capital gain rates on the shares "sold" to the corporation.

The procedure for a partial liquidation begins with a resolution of the board of directors.

EXAMPLE 9.07A: Resolution to Partially Liquidate

Whereas, this Corporation presently is the owner of sundry producing and undeveloped oil and gas leases, drilling equipment, oil payments, and miscellaneous properties;

Whereas, the Board of Directors deem it good business and advisable to make a partial liquidation of the Corporation and to pay out certain properties of the Corporation as a dividend in kind ratably to the stockholders of record on the 10th day of December, 1982; and

Whereas, the payment of a partial liquidating dividend will impair the capital stock of the Corporation.

Now, therefore, be it resolved, that a partial liquidating dividend be declared and paid as of January 2, 1983, said dividend to be paid in properties in kind and consisting of the following described properties: [*Here describe.*]

Be it further resolved, that the Certificate of Incorporation of this Corporation be amended, reducing the capital stock of the Corporation from 5,000 shares of common stock of the par value of $1.00 per share to 3,750 shares of common stock of the par value of $1.00 per share, and from 50 shares of preferred stock of the par value of $100.00 per share to 37.5 shares of preferred stock of the par value of $100.00 per share; and that of the net book value of the assets paid out as a liquidating dividend, $1,250.00 of such amount be charged against the stated capital of the common shares and $1,250.00 of such amount be charged against the stated capital of the preferred shares, and the remainder be charged against the capital surplus account.[25]

§ 9.08 DISSOLUTION AND LIQUIDATION

Distributions to shareholders are also involved in a complete disolution of the corporation, since the shareholders are entitled to receive their proportionate ownership interest in the assets after corporate creditors have been paid. Shareholders of record are identified and notice is given to creditors permitting a reasonable period for filing claims.[26]

All shareholders participate ratably in the net assets of the corporation after payments to creditors, unless the articles of incorporation provide otherwise. The articles may authorize a capital stock structure with various classes or series of shares, one or more of which may be entitled to a preference to assets in liquidation. The preference will be honored before other classes of stock are entitled to their proportionate share of assets.

Liquidation preferences are usually fixed as a percentage of par value or a specified dollar amount, and preferred shares, while entitled to the first priority to the assets, are limited to the liquidation amount specified. For example, observe that the preferred shares of the Nobles Company are entitled to a liquidation preference of 100% of par value per share. If dissolution occurred when the financial status of the corporation was as described in the balance sheet in section 9.02, the net assets of the corporation would be $52,000 at dissolution. The preferred shares would be entitled to receive $5,000 of the assets (100% of $100 par value for 50 shares). The remaining $47,000 in assets would be distributed to the common shareholders, and the preferred shareholders would not share in this distribution. It is possible, however, to create "participating preferred shares" so the preferred shareholders participate in the distribution of assets after receiving their preference. To illustrate, suppose that in addition to the 100% par value preference in liquidation, the articles of incorporation further provided that the preferred shares would participate equally with the common shares in liquidation. The preferred shares would still receive the first $5,000, and the remaining $47,000 would be distributed equally among 5,050 shares (5,000 common and 50 preferred). Each share of each class would receive $9.31 in liquidation in the second distribution. Thus, the total liquidation distribution to the preferred shareholders is $109.31 per share.

Classes of preferred stock which are entitled to cumulative *dividend* preferences may also have a preferred claim in liquidation to the extent of the dividend arrearages. This principle has undergone some tortured construction in litigation and is not firmly settled. However, the principle is simple and any interpretation problem could be avoided by careful drafting of the articles of incorporation. The Nobles Company preferred shares are entitled to a 5% cumulative dividend preference, meaning that dividends of $5.00 per share (5% of $100 par value) accumulate annually. When a dividend is finally declared, the total arrearage must be paid to the cumulative preferred shareholders before any other shareholders may receive dividends. Suppose the Nobles Company has not declared dividends for five years and is now being dissolved and liquidated. The dividends for the preferred shares have accumulated in the amount of $25 per share. The question is whether the preferred shareholders are entitled to receive their dividend arrearages as a preference in liquidation before other classes of stock are permitted to share the assets. The articles of incorporation could provide either way, but should be specific in any case. If it is intended that the preferred shares will be entitled to a liquidation preference for unpaid cumulative dividends, the articles should so specify, and should also define a "dividend arrearage" or "accumulation" as including unpaid amounts regardless of whether the corporation has ever declared a dividend or has had funds available for a declaration of a dividend. To negate the liquidation preference for unpaid cumulative dividends, the articles should specifically state that the right to unpaid accumulated dividends is lost if the corporation is dissolved and liquidated.

NOTES

1. See M.B.C.A. § 45.
2. Park & Tilford v. Schulte, 160 F.2d 984 (2d Cir. 1947).

3. See Section 7.08.
4. M.B.C.A. § 45.
5. See M.B.C.A. § 45(a).
6. M.B.C.A. § 45.
7. Dividend restrictions of the articles of incorporation were suggested as important considerations at the drafting stage. See Section 6.05.
8. See Section 9.06.
9. E.g., Nebraska, Neb.Rev.Stat. § 21–2043(2) Mississippi, Miss.Code Ann. § 79–3–83(b).
10. E.g., California, West's Ann.Calif.Corp.Code § 500; Idaho, Idaho Code Ann. § 30–1–45; and Illinois, Ill.Rev.Stat. c32, § 157.41.
11. West's Modern Legal Forms §§ 2291 and 2292.
12. West's Modern Legal Forms § 2927.1.
13. See Section 8.07.
14. Internal Revenue Code of 1954, 26 U.S.C.A. § 301.
15. The 1979 revisions to the Model Act eliminated the concepts of treasury shares, stated capital or surplus.
16. See also West's Modern Legal Forms § 2926 for a clause which authorizes the issuance of shares of one class to holders of shares of another class.
17. See West's Modern Legal Forms § 2835 for a form of resolution for a share split. The procedure to amend the articles of incorporation is detailed in Section 13.01.
18. See Rule 10(b)–12, Securities Exchange Act of 1934, 15 U.S.C.A. § 78(j).
19. See Section 7.08.
20. See Section 13.04.
21. See Section 7.11.
22. A statement of cancellation of reacquired shares appears as Form 9A in the Appendix.
23. Int.Rev.Code of 1954, 26 U.S.C.A. § 346(b).
24. Int.Rev.Code of 1954, 26 U.S.C.A. § 341.
25. West's Modern Legal Forms § 2926.5.
26. The dissolution procedure is more fully discussed in Sections 13.05–13.07.

10

EMPLOYMENT AND COMPENSATION

A corporation offers the most flexible employment and compensation possibilities of any form of business. Managerial talent may be widely distributed in the corporate structure, from directors and officers to other management executives and supervisory personnel, and the compensation schemes are equally varied, particularly because of the many tax advantages which are available through the corporate entity. This chapter is primarily devoted to employment arrangements with corporate employees, although much of what is said here also applies to employees of proprietorships and partnerships.

Most employees are hired on an informal basis, without any written contract, and they perform duties for a stated compensation, in salary or wages, until their employment is terminated. There may be certain personal motives which induce the employee's performance and provide compensatory incentives. The company may be a family business, and the employee, as a member of the family, may have a kindred incentive to do the job well. The employee also may be a shareholder, and therefore have a pecuniary interest in his own performance. If his performance contributes to business successes, his stock will become more valuable. However, these informal employee relationships do not require much planning or counsel, and because of their simplicity, are certainly appropriate arrangements for most proprietorships and many partnerships and corporations. However, they do not take advantage of the many special methods of preserving talent and compensation which are available through more elaborate employment agreements.

As the company structure becomes larger and more employees are needed, managerial talent becomes more important. "Key employees" become an integral part of the organization and their compensation and incentives assume greater significance. The company must insure that they will remain employees in order to insure a smooth reliable business organization. Key employees are not necessarily executives. They include salesmen, research and development personnel and any other employee with a special skill. The objectives of the company in retaining these people is selfish on two counts: first, without them the business may suffer mild to serious reverses until they are replaced, and second, if they decided to work for the corporation's competitors, the company not only loses valuable talent, but also risks the loss of important trade secrets and processes.

The employees also have some stake in their employment arrangements. Theirs is a question of job security, advancement and compensation. These complementary objectives of a continuing employment relationship may be satisfied through many combinations of employee agreements, incentive compensation plans and fringe benefits.

§ 10.01 EMPLOYEE AGREEMENTS GENERALLY

The employee's job security objectives may be assured by an employment contract by which the company promises to keep her employed for a period of time and she promises to perform her duties diligently for the specified period. It is generally recognized that employment contracts provide considerably greater protection for the employee than the employer. In part, this attitude stems from the fact that courts generally refuse to force a person to perform against her will. Consequently, if the employee quits before the expiration of his employment term, the employer could not successfully petition a court to order her to continue working against her will. On the other hand, if the employer fired the employee prior to the expiration of the agreement, the employee may recover the compensation she would have received had she been allowed to perform the agreement, unless the employer can prove good cause for termination. Nevertheless, an employer may gain some benefits from the employment contract insofar as its terms prescribe incentive compensation to encourage the faithful continued performance of the employee. Further, the agreement may include noncompetition clauses which prohibit the employee's entry into a competing business upon termination, or it may reserve to the company any developments or ideas discovered by the employee during employment.

Like so many other areas of the law of business organizations, employment contracts must be tailored to the particular needs of the parties, and it would be unrealistic to attempt to cover every possibility. There are, however, certain general rules which are basic to each agreement.

§ 10.02 EMPLOYMENT AGREEMENT STRUCTURE

A typical employment agreement should contain clauses describing the employee's responsibilities and duties; provisions for compensation and reimbursement of expenses; the duration of the agreement; and, in many cases, non-competition clauses, death and disability clauses, and provisions for company rights to discovery and development during employment.

Employee Duties

The contractual language which details the duties and responsibilities of the employee always depends on the particular needs of the employer, the talent and position of the employee, and the exigencies of the business. The definition of duties is extremely important, however, considering the possibility of later disputes. Not only must the employer know what is expected of him, but the employer must have some definitive guidelines upon which to measure the employee's performance. If the employee is terminated involuntarily prior to the expiration of the term of the agreement, the employer must be prepared to show that the employee failed to perform his duties under the contract. If the duties described in the agreement are ambiguous or otherwise ill-defined, this proof may be impossible.

The description of duties for top-level management positions necessarily must be broad. It would be very difficult to attempt to detail all of the duties expected of a company executive since he is hired to run the business. A general statement of management duties is unavoidable here.

EXAMPLE 10.02A: Duties

The Manager shall well and faithfully serve the Employer in such capacity as aforesaid, and shall at all times devote his whole time, attention, and energies to the management, superintendence, and improvement of said business to the utmost of his ability, and shall do and perform all such services, acts, and things connected therewith as the Employer shall from time to time direct and are of a kind properly belonging to the duties of a Manager.[1]

Other more specific provisions may be used for positions with definable boundaries. Consider the duties of a research chemist:

EXAMPLE 10.02B: Duties and Inventions

The Corporation hereby employs the Employee as a research chemist. The Employee's duties shall include the application of his skill and knowledge as a chemist towards devising new pharmaceutical products and improving existing formulas, processes, and methods employed by the Corporation. All inventions, discoveries and improvements devised or discovered by the Employee while in the employ of the Corporation shall become and remain the sole and exclusive property of the Corporation, whether discovered during or after regular working hours.[2]

A manager of a merchandising outlet could have duties prescribed as follows:

EXAMPLE 10.02C: Duties of Manager

The duties of the Manager shall be such as are assigned to him by the Company. Initially there shall be included among his duties and authority the selection of a stock of merchandise for this venture; schedule of purchases to be submitted and subject to advance approval by the Company and copies of proposed orders

> shall be submitted to the Company to be passed upon and approved. Selections for current replacements for stock and purchases for new season requirements shall likewise be made by the Manager subject to prior approval by and submission of orders to the Company, in the same manner as is above provided for the initial stock. The Manager shall also keep a perpetual inventory of merchandise on hand and take a monthly physical inventory. The Manager likewise shall have full authority to employ and discharge employees of the business, subject to approval of the Company. The Manager will refer all disputed claims, not allowed by him for adjustments or returns on complaints, to the Company. The duties and authority hereby conferred are subject to change at the pleasure of the Company.[3]

In preparing a description of duties, the employer (or, in some cases, the employee) should prepare a statement of job description, which includes all of the specific items that the employee is expected to do as part of employment. The initial job description prepared by the employer or the employee may then be modified to include other items which are related to the types of specific duties which the employee expects to perform for the employer. The list of duties should begin with the most specific duties anticipated, and become more general as the list grows longer. It is important to attempt to identify all potential duties which the employee is expected to perform, and to highlight those technical duties which are unique to this particular employee.

Each description of duties must be tailored to the specific employee and to the position which the employee will hold.

In representing the employer, it is advisable to include a phrase which permits additional duties to be assigned to the employee from time to time. For example, a duties clause may provide that the employee will perform certain duties and "all other matters connected therewith as the employer shall, from time to time direct, and are of a kind properly belonging to the duties of an employee of this type." Similarly, the duties clause may include a statement that says that "the duties and authority hereby conferred are subject to change at the pleasure of the company". These "catch-all" provisions are desirable from the employer's viewpoint in order to be certain that the talents of the employee may be directed to changing employment opportunities, and also to permit the employer to assign specific duties, which may not have been contemplated at the time the agreement was negotiated, but which will better define a breach of the agreement should an employee fail to perform them.

Obviously, the "catch all" provisions in a statement of duties work to the disadvantage of the employee, unless there are certain limits placed upon them. On behalf of an employee, it would be important to provide that any additional duties which are assigned are "of a type which properly belong to the duties of an employee in a particular position", or "are reasonable" duties. In this manner, the employee may avoid the assignment of duties which are not consistent with the overall employment or the assignment of inappropriate or distasteful duties to force the employee to breach the agreement.

The duties clause of the employment agreement may also serve to restrict authority and responsibility of the employee. Any limitations on the scope of the employee's authority should also be clearly defined. The agreement may reserve certain decisions or transactions to employees at a higher level for organizational purposes. The duties clause may also define territories, as in the case of sales personnel, or impose any other restrictions consistent with the employment relationship.

EXAMPLE 10.02D: Limitations on Authority of District Manager

The District Manager shall possess no authority not herein expressly granted and is not authorized on behalf of the Companies to make, alter or discharge contracts or binders, except as may be directed in writing by the Companies, nor to waive forfeitures, grant permits, guarantee in dividends, if any, name extra rates, extend the time of payment of any premium, waive payment in cash, or to write receipts except for first premiums, or make any endorsements on the policies of the Companies, and shall receive no further remuneration for any service except as herein provided. It is expressly stipulated and agreed that the District Manager is not authorized to incur any indebtedness or liability in the name or in behalf of the Companies for any advertising, office rent, clerk hire, or any other purposes whatsoever or to receive any moneys due or to become due the Companies exclusive of the first premium, except as may be specifically directed by the Companies and the powers of the District Manager shall extend no farther than are herein expressly stated.

Restriction

The District Manager shall not make or permit to be made by any agent, any use of the radio or insert any advertisement respecting the Companies in any paper, or other matter, magazine, newspaper, periodical, or other publication or issue any circular or paper referring to the Companies without the written consent of the Companies.[4]

EXAMPLE 10.02E: Reservation of Right to Reject Orders

No order shall be deemed binding upon the company until accepted by the company in its Illinois office in writing, and the company reserves the right to reject any order, or to cancel the same or any part thereof after acceptance, for credit or any other reason whatsoever deemed by the company to be sufficient.[5]

Provisions for Compensation

The compensation clauses of an employment contract deserve special attention. In the first place, any ambiguity here is certain to become a matter of dispute since financial matters are of utmost concern to both parties to the agreement. Moreover, the compensation provisions of the contract offer the best opportunity for the employer to insure continued faithful performance by the employee. Recall that courts are generally unwilling to force an employee to return to the job following his breach of an employment contract. Carefully planned incentive compensation provisions will motivate the employee to remain with the company. Finally, the compensation provisions may permit certain tax advantages for all parties, and these matters should be explored and explained to the client.

At this point the discussion will be limited to basic compensation schemes applicable to all types of business organizations, and to "current" incentive provisions, which encourage immediate, rather than long-term performance. The exclusively corporate compensation schemes, such as stock options, and the long-term incentive provisions, such as deferred compensation and profit sharing, will be discussed separately in later sections.

The most basic type of compensation for an employee is a salary arrangement, and contractual terms should specify at least the amount and frequency of payment.

EXAMPLE 10.02F: Compensation

In consideration of the service so to be performed the Employer agrees to pay to the Employee the sum of $25,000.00, payable in equal monthly installments for twelve consecutive months at the end of each month, until the termination of this agreement.[6]

A current incentive may supplement the salary agreement to encourage diligent performance by the employee.

EXAMPLE 10.02G: Compensation

The Employer shall pay to the Manager a salary of Thirty thousand dollars per annum, payable by monthly installments of Two thousand five hundred dollars on the tenth day of each month and shall also pay to the manager a commission of two per cent per annum on the net profits of said business, such commission to be paid within ninety days after the year accounts have been certified by the accountants employed by the employer, whose certificate as to the amount of such net profits shall be conclusive.[7]

A percentage compensation agreement, a "commission," is probably the best current incentive provision. The rate of compensation is based directly upon the employee's own performance, and the commission technique is used most effectively when applied to individual activities. However, this scheme may also be used to compensate management or supervisory employees whose performance depends in part on the efforts of others they control. For management employees, the percentage is frequently based upon profits produced under their direction. A definitional clause must define "profits" for application of the percentage. Moreover, since a determination of profits will not usually be made until the close of the business year, the employee will usually be permitted to withdraw specified sums in advance for current living expenses until his percentage share has been determined. An example of these provisions for executive compensation follows:

EXAMPLE 10.02H: Compensation

The Company shall pay to the Manager as compensation for her services one-third of the net profits arising from the Chicago business.

Net Profits

In arriving at what shall be deemed the net profits arising from the Chicago business, the following items shall be paid out of the gross profits, viz.: The rents of the premises wherein the Chicago business shall be conducted and all repairs and alterations of the same, all taxes and payments for insurance, all salaries and wages of clerks and employees other than the Manager employed in or about the Chicago business, and all charges and expenses incurred in or about the same, all debts or other moneys which shall be payable on account of the Chicago business, the interest on the capital for the time being advanced by the Company, and all losses and damages incurred in or about the Chicago business.

Drawing Account

The Manager shall have the right to draw out for her own use the sum of $2,500.00 per month on account of her salary. The balance of her one-third share in the net profits shall not be withdrawn by her until after the annual general account hereinafter mentioned shall have been made and signed.[8]

Commission provisions for an individual employee's performance require a similar approach. However, instead of profit, the percentage is usually applied to sales obtained for the company by the employee, and that criterion creates another definitional problem. The "sale" of goods involves at least four transactional stages: the order is signed; the order is approved by the company; the goods are shipped; and the customer pays. One of these four stages, or some other reasonably ascertainable point, should be selected as the time the commission is earned. If any intermediate stage is selected, the company may further specify that the commission will be withdrawn or reduced if the customer fails to pay.

EXAMPLE 10.02I: Compensation and Basis for Computing Commissions

In consideration of your services, we agree to pay you a commission of five percent on all sales during the term of your employment made to customers located in the territory covered by you. Such commissions shall be calculated on the net amount of sales, after deducting returns, allowances, freight charges, discounts, bad debts and similar items, and shall be deemed to be earned and payable only as and when orders have been shipped and actually paid for by customers. The prepayment of commissions shall not be deemed to be a waiver of the foregoing provisions, and in all cases in which commissions have been paid in advance of payment by the customer, or where returns or allowances are subsequently made, such appropriate adjustment as may be necessary shall thereafter be made.

Drawing Account and Traveling Expenses

You shall be entitled to receive and we agree to pay you a drawing account of $600 per week, which sum shall be applied against and deducted from commissions then or thereafter due you. You personally shall defray all traveling and other expenses incurred by you in connection with your employment.[9]

To be effective, current incentive programs should be directly related to the performance of the employee or of the persons under the supervision of the employee. The employee has no control over unrelated performance and thus he has no opportunity (or incentive) to improve it. Current incentive programs pose special drafting problems to insure absolute clarity. In the foregoing examples, the percentages, time for payment, formula for determining the amount to which the percentage is applied, and application of draws against earned compensation are all well defined. Each of these items must be unambiguous if the compensation provisions are to be effective. Some additional suggestions should be helpful:

1. A date or periodic date should be specified for the payment of incentive compensation. For example, if the incentive compensation is computed on an annual basis, it should be specified that payment will be made on a specific date of the following year, or within a certain number of days from the close of the business year. The exact date permits the employee to plan the receipt of income, and allows the employer to plan cash flow.

2. If accounting terms are used, such as "net profit" or "net sales", they should be defined in the agreement. Specific expense items should be named if they are to be deducted from "gross profit" to reach "net profit," or from "gross sales" to reach "net sales." Moreover, the provisions should define "profit" as before or after income taxes are deducted, whichever represents the agreement of the parties.

3. The person or persons who determine the base amount against which the percentage is applied should be named. If the company has retained independent public accountants to audit their records, they should be specified as the persons who will make these determinations, and it is good practice to specify that their determination is to be based upon "generally accepted accounting principles."

4. If the employee is permitted to draw against the incentive compensation, the agreement should cover the contingency that the draws may exceed the earned compensation. For example, suppose the manager is to receive one-third of the profits as determined at the end of the year, and he has drawn $30,000 during the year. If the profits total $99,000 when tallied, will he be required to repay the company $3,000, or will that amount accrue to be applied against the following year's profits, or will it be forgiven and deemed to be an expense of the company? The agreement should provide an answer.

5. If the incentive compensation is based upon periodic performance such as annual profits, the employee's commencement and termination during the period should be considered. A relatively small number of employees are hired on the first day of the business year. (A few more may quit on the last day, however, considering that the arduous task of taking inventory may be looming in the immediate future.) Nevertheless, some formula must be inserted for the employee who has not worked for the entire period upon which incentive compensation is based. A formula for this purpose may be drafted on any reasonable basis. For example, the employee's percentage may be applied to profit computed for the period of the year in which the employee actually worked. If the employee began working August 1, the actual profit would be computed from August 1 to December 31, and the percentage applied against that figure. Alternatively, the formula may specify a prorata determination of profit for the entire year. Here, the entire year's profit would be reduced to $5/12$, the fraction of the year from August 1 to December 31, and the percentages would be applied to that amount. Finally, a separate special formula for years of commencement and termination may

be stated, such as 1/6 of the profit for these years, or no profit percentage if the employee works for less than half of any year, and so forth.

Closely related to the percentage compensation is a "bonus" plan based upon minimum performance of the employee, her division or department. The agreement may establish certain sums to be paid at a specified time following the close of the business year if her individual performance or her section of the organization produces results above a specified minimum. The bonus amounts may be periodically increased to reward continued performance. For example, the agreement with a manager of a retail store may provide that she shall receive a bonus of $1,000 the first year that the net profits before taxes from her store exceed $10,000; $2,000 for the second consecutive year profits exceed that amount; and $3,000 for the third and subsequent consecutive years profits continue to exceed $10,000. The drafting considerations detailed above are equally applicable here. The agreement should define "net profit" or any other selected criterion, should specify the person who will determine the base figures, should provide for termination during the year, and should specify a date the bonus will be paid.

The bonus plan may be limited to an incentive for continuous employment, without considering the specific performance of the employee. For example, the agreement may discourage voluntary termination during the year:

EXAMPLE 10.02J: Compensation and Bonus

The Employee shall receive a weekly salary of $1,000. In addition, the Corporation shall pay to the Employee at the end of each year of the term a year-end bonus of not less than $10,000 (less withholding taxes, social security and other required deductions); but no part of said bonus shall be payable if the Employee shall be in default under this agreement or shall not then be in the employ of the Corporation.[10]

The compensation section of the employment agreement is also an appropriate place to describe incidental financial benefits, such as reimbursement for expenses, vacation pay and so forth.

EXAMPLE 10.02K: Expense Reimbursement

Employer shall also pay to the Salesman his reasonable expenses of traveling, board and lodging, postage, and other expenses reasonably incurred by him as such Salesman in or about the business of the Employer.

EXAMPLE 10.02L: Vacations

The Employee shall be entitled to vacations with pay in accordance with the established practices of the Corporation now or hereafter in effect for supervisory personnel.

The foregoing compensation plans are common to most employment agreements. Continued faithful performance of the employee may also be reasonably assured by the use of several available special compensation techniques discussed in subsequent sections.

Term of the Agreement

The duration of the employment agreement should be specific and, as a general rule, should be reasonably short for the company's protection. If a continuing employment relationship is contemplated, it is far better to provide for options to renew the agreement than to leave the term indefinite.

The basic term of the agreement is always simple.

EXAMPLE 10.02M: Duties and Term

The Employee agrees to give her undivided time and service in the employ of the Employer in such capacity as the Employer may direct, for the period of one year from and after the 1st day of December, 1983.[11]

Renewal provisions may be drafted one of three ways. First, the option to renew the agreement may be granted to the employer, with appropriate advance notice to the employee of the election to renew. Second, the option to renew may be vested with the employee, with appropriate advance notice to the employer of the election to renew. Third, the agreement may be automatically renewed for specified periods unless appropriate notice is given by either party to the other of the intention *not* to renew. The latter provision is most common, and most adaptable to a continuing employment relationship.

EXAMPLE 10.02N: Option to Renew

Employee grants Employer the option to renew this contract for a period of two years upon all the terms and conditions herein contained, except for the option to renew for a further period. This option may be exercised by Employer by giving Employee notice in writing at least thirty days prior to the expiration hereof; and such notice to Employee may be given by delivery to Employee personally or by mailing to Employee at the last known address.[12]

EXAMPLE 10.02O: Initial Term and Automatic Renewal

The term of this employment shall commence February 1, 1981 and shall continue for a period of two years until January 31, 1983, and thereafter shall be deemed to be renewed automatically, upon the same terms and conditions, for successive periods of one year each, until either party, at least thirty days prior to the expiration of the original term or of any extended term, shall give written notice to the other of intention not to renew such employment.[13]

Provisions for notice should be tailored for specific circumstances. Written notice should always be required, and should be sent to a specified address of each party. For the employee, the address may be specified as "the last known address, or the "address on the records of the records of the company." The time period for notice should be longer for a more specialized employee so that he may search for other employment if necessary, or so that the company may search for his replacement.

Termination of the Agreement

The employment agreement should be terminated or terminable upon the happening of certain contingent events. In many cases, salary continuation protection for an employee may be appropriate upon the happening of one of these events, and in cases where the event is also defined by an insurance policy, the employment agreement should use the definition from the insurance policy to be certain there is no conflict between the two definitions.

Cause for termination should be considered for the following contingent events:

1. If the employee is disabled for a period of time.
2. If the employee is bankrupt.
3. If the employee has been convicted of a crime.
4. If the employee is incarcerated.
5. If the employee is mentally disabled or otherwise unable to perform duties.
6. If the employee is suffering from alcoholism or drug-related disabilities.
7. If the employee breaches the agreement.
8. If the portion of the business in which the business is employee is employed is discontinued for any reason.
9. If the business is insolvent or bankrupt.
10. If a substantial portion of the business assets are destroyed.
11. If the business is sold, merged, consolidated, or dissolved for any reason.
12. If majority ownership of the business changes.
13. Any other matter which, considering the special duties of the employee, may constitute cause for termination under the agreement.

EXAMPLE 10.02P: Termination

In the event of the illness of the Manager or other cause incapacitating him from attending to his duties as manager for six consecutive weeks, the Employer may terminate this agreement without notice upon payment to the Manager of five hundred dollars in addition to all arrears of salary and commission when ascertained up to the date of such termination. In the event of a breach of this agreement or of an act of bankruptcy on the part of the Manager, the Employer may terminate this agreement without notice or payment of salary or commission as hereinbefore provided.[14]

EXAMPLE 10.02Q: Termination

This agreement shall terminate in the event of the dissolution of the firm by death or otherwise, the appointment of a receiver or trustee in bankruptcy or the filing of a petition to reorganize under the Bankruptcy Act, or the destruction by fire of the firm's warehouse at South Bend, Indiana, notwithstanding the full term of one year may not, at the happening of any of said events, have fully expired.[15]

Finally, it is possible to agree to termination without cause with due notice.

EXAMPLE 10.02R: Termination

The employment of the Editor may be terminated at any time (during the said period of two years) by either party giving to the other two calendar months' notice in writing of its or her intention to terminate the same, or by the Company upon its paying to the Editor a sum equal to two months' salary at the rate aforesaid in lieu of such notice.[16]

§ 10.03 RESTRICTIVE AND PROPRIETARY COVENANTS

While the employer may be unsuccessful in persuading a court to order an employee to continue to work for the company against his will, the employer may prevent the employee from exploiting the company by appropriating ideas that were developed during his employment for his own benefit; leaving with trade secrets, specialized confidential knowledge, or customer lists; or working for competitors. Certain restrictive or proprietary covenants in the employment agreement may protect the employer from such abuse, and, if properly drafted, these covenants will receive court protection.

The covenants generally cover three areas of possible employee exploitation: 1) using developments, inventions and ideas which the employee produced during the course of his employment; 2) maintaining confidentiality of business secrets during and after employment; and 3) competing with the company after employment. Without an agreement on each of these points, under common law the employer risks the loss of significant market advantages through the acts of unfaithful employees. No former employee would be prohibited from competing against his former employer without a specific agreement to that effect. The law has always favored fair and free competition and there is nothing implicit in an employment relationship that requires the employee to withdraw from the marketplace after he has terminated his employ. That is not to say, however, that the employee would be allowed to duplicate his former employer's secret practices and processes in future competition. The law protects trade secrets, even without an agreement prohibiting their use, but the protection is somewhat elusive and unsatisfactory if left to common law resolution alone. First, it is difficult to determine which practices, procedures and other matters are truly "trade secrets" entitled to protection. The employer will have the burden to prove that he is the "owner" of the secret, meaning it was developed by or for him, is not used by others, and is sufficiently unique so as to deserve proprietary protection. Further, the employer must show that the secret is known only to him and his employees in whom it must be confided for business purposes. Even if the matter is shown to be a trade secret, a court would have to be convinced that the employee's exploitation should be prohibited. The court would have to find that the employee's use or publication of the secret could cause irreparable harm to the employer. An agreement restricting the employee's use of trade secrets is essential for certainty of protection in this area.

The common law is also unpredictable on a question of the employer's rights in the employee's original ideas and developments discovered during the course of his employment. Certainly if the employee is hired to do research and create inventions, the employer has ownership rights in any productive research while the employee is on the job. But what happens if the employee dreams up an invention which is unrelated to the employer's research, or if he terminates his employment before the research is productive, and subsequently produces the invention on his own? Specific provisions in the agreement may anticipate and resolve these problems.

§ 10.04 EMPLOYER'S RIGHT TO EMPLOYEE WORK PRODUCT

The employer may have hired the employee for the express purpose of developing new products or business innovations. Certain specialized skills, such as research chemistry or engineering, are widely sought for this purpose, and the employment relationship is directed to the production of "inventions." The employee also may be hired to develop certain business practices and procedures which will increase efficiency and utilization of other employees' skills. These common cases should be distinguished by the test of a legally protectible work product. In the first case, an invention from the original thought of an engineer may be patentable and thereby have certain proprietary interests attached to it. In the second case, a particular procedural system devised by a time-and-motion expert may be unique, but will not be such an original creative product that a patent or copyright could be obtained. The employer's interest in each may be indistinguishable, however. The company obviously wants to retain proprietary rights in patentable inventions, and management may be equally concerned about reserving newly devised business procedures to themselves for the associated competitive advantages. While the clauses suggested here will protect the company's rights to the patentable inventions, they may not be adaptable to protection of other unpatentable work products. Rather, in the latter case, it would be more appropriate to describe the developed business procedure as a trade secret in the agreement and thereby insure some confidentiality, or to prohibit the development of a similar system for a competitor through a non-competitive covenant with the employee.

Work product clauses are most effective for inventions and other original creations which are capable of being patented or copyrighted.

A work product clause should include at least the following general components:

1. A specific grant from the employee to the employer of the right to use such work product.

2. A statement that the clause applies to any use to which the employer chooses to make of the work product.

3. A statement that the clause applies to inventions, designs, procedures and other matters in both their unperfected and improved states.

4. A statement that the clause applies to inventions, designs and other matters developed or obtained by the employee by himself or severally or jointly with other persons.

5. A statement that the clause applies to the entire period of the employment with the employer.

6. A specific description of the employee's talents out of which inventions and designs are expected to be produced.

7. A specific description of the type of inventions, designs and other matters which the employee is expected to develop.

8. A general statement that other inventions, designs and other matters which relate to the employee's product will be covered by the clause.

9. A release by the employee of any legal or equitable right to the work product.

10. A promise by the employee that all necessary documents for perfecting title will be executed and delivered to the employer on demand.

11. A representation by the employee that such work product will not infringe upon any patents or other protected rights of others.

12. Any special compensation arrangements which had been negotiated with the employee for subsequent use of the invention, design or work product.

13. "Trade secret" protection of the work product to protect against its publication before a patent, copyright or other protected proprietary rights may be obtained.

14. The relief or remedies to which the employer may be entitled for a breach of the clause, such as injunctions, liquidated damages, a constructive trust on all profits produced and so forth.

The scope of these provisions is usually determined by the nature of the employment. For example, if an employee is hired for the broad purpose of researching and developing improvements in elevators, a broad protective clause would be appropriate.

EXAMPLE 10.04A: Grant of License.

> The Employee hereby grants to the Employer the exclusive license to manufacture, sell and deal in all inventions, designs, improvements and discoveries of the Employee, whether now perfected or whether invented, improved and discovered subsequent hereto, which pertain or relate to elevators and their appliances, or capable of use in connection therewith.[17]

However, if the employment objective is more specific, such as the research and development of a valve and starter plug to improve elevator control, the clause may also be more specific.

EXAMPLE 10.04B: Inventions Property of Employer.

> The Employee agrees that all inventions, improvements, ideas, and suggestions made by him and patents obtained by him severally or jointly with any other person or persons during the entire period of his employment, and any written renewal thereof made by him with the Employer, with relation to said valve and its appurtenances, including present starting plug, or method of elevator control, and all inventions of elevator valves, plugs, or methods of elevator control and valve appliances, and to machinery for manufacturing the same, are and shall be the sole property of the Employer, free from any legal or equitable title of the Employee, and that all necessary documents for perfecting such title shall be executed by the Employee and delivered to the Employer on demand.[18]

The specificity of the clause may be a matter for negotiation between the parties. The employee in the second example may have insisted upon the narrow description so that his subsequent development of an elevator door, for example, would not belong to the employer. However, he may be willing to consent to the broader provision for increased compensation. The employer's objective, of course, is to make the subject matter of the clause all inclusive.

Work product protection clauses should always require the execution of any necessary documents for perfecting the employer's title to the inventions, as shown in the second example. This may obviate any need for an interpretation of the contract before the employer can market the invention, and the clause will place the employee in breach of the agreement if he fails to cooperate fully with the employer. It is also desirable to contract for "trade secret" protection of the invention to protect against its publication before a patent may be obtained. A clause on these points follows:

EXAMPLE 10.04C: Execution of Further Documents

I further agree, without charge to said company, but at its expense, to execute, acknowledge, and deliver all such further papers, including applications for patents, and to perform such other acts as I lawfully may, as may be necessary in the opinion of said company to obtain or maintain patents for said inventions in any and all countries and to vest title thereto in said company, its successors and assigns; and I further agree that I will not divulge to others any information I may obtain during the course of my employment relating to the formulas, processes, methods, machines, manufacturers, compositions, or inventions of said company without first obtaining written permission from said company so to do.[19]

It may be appropriate (and necessary, if the employee has his wits about him when he negotiates the agreement) for the employer to compensate the employee based upon the profitable inventions he has created for the company. The percentage compensation scale would be tailored to the demands of the parties, and may look like this:

EXAMPLE 10.04D: Compensation for Inventions

In order to recompense the above employee (hereinafter called the Employee) for meritorious inventions, the Corporation agrees on its part to examine the inventions disclosed to it by the Employee, and where said inventions, in the sole opinion of the Corporation, warrant such action, to cause United States patent applications to be filed through its attorneys covering the same, but without assuming any responsibility for the prosecution or defense of such patent applications, and further agrees to give to the Employee in any cases where it decides to license said inventions, applications or patents to others, a percentage of any money royalties which it may receive from such licenses upon the following scale:

Of the first $1,000.00 or part thereof collected in any one calendar year—40%

Of the next $2,000.00 or part thereof collected in any one calendar year—30%

Of the next $2,000.00 or part thereof collected in any one calendar year—20%

Of the next $5,000.00 or part thereof collected in any one calendar year—15%

Of all further sums collected in any one calendar year—10%

It is understood and agreed, however, that the question of when, how and to whom licenses shall be granted shall be in the sole discretion of the Corporation, and that in cases where the Corporation shall grant licenses involving, in addition to the Employee's inventions, the inventions of others, the Corporation shall have the sole right and authority to apportion the royalties received and the Employee shall receive the above percentage on the proportion awarded to his inventions.[20]

§ 10.05 TRADE SECRET PROTECTION

Beginning with the truism that various companies have various secrets, some more important than others, trade secrets protection must always be drafted to fit the particular exigencies of the company. A discount retail merchant may consider her supplier list to be a trade secret; a manufacturing company may consider the assembly process to be a trade secret; and a computer firm would treat their programs and techniques for interpretation to be trade secrets. In each of the businesses, the "secret" is an integral part of the competitive advantage, and confidentiality is deemed to be crucial to continued success of the operation. It has been frequently recognized that one of the best-kept trade secrets is the process and ingredients used to produce "Coca-Cola." In part, this mystery is a result of well-drafted employment agreements and the fact that the secret process is never completely disclosed to anyone. The Coca Cola Company's Non-Disclosure Agreement appears as Form 10A in the Appendix.

Several important rules should be followed in drafting trade secret clauses for employment contracts. First, the promise to keep the secrets should bind the employee during his employment *and* after termination of employment. Second, the clause should not only cover specific secrets that the employee uses in his duties, but also those which he may have learned from his own observations or from other employees. Third, the agreement should broadly prohibit divulging any "trade secrets, procedures, processes or knowledge of operations," and should also specifically itemize particular matters which are to be protected. Fourth, because of the difficulty of proving actual damages from the publication of a trade secret, an agreed damage clause, or "liquidated damages," should be considered. A clause with all of these ingredients follows:

EXAMPLE 10.05A: Trade Secrets.

The Salesman further covenants not to communicate during the continuance of this agreement, or at any time subsequently, any trade secrets, processes, procedures or business operations, specifically including but not limited to information relating to the secrets of the traveling, advertising and canvassing departments, nor any knowledge or secrets which he then had or might from time to time acquire pertaining to other departments of the business of the Employer, to any person not a member of the Employer's firm, except as requested in writing by the Employer. In case of violation of this covenant, the Salesman agrees to pay the Employer or its successors the sum of five thousand dollars as liquidated damages, but such payment is not to release the Salesman from the obligations undertaken, or from liability for further breach thereof.[21]

It would also be advisable to prohibit the employee's use of the secret as an individual, or the employee's direct or indirect benefit from the use of the secret, such as a shareholder, partner, employee, consultant, creditor or other participant in another business which learns about or adopts the secret from the employee.

Customer lists commonly are protected by trade secret clauses, since courts are not likely to construe customer lists as a business secret under common law so as to shelter them without an agreement. However, in highly-competitive businesses the secrecy of customer lists is an important competitive advantage. A competitor who has obtained them will be spared considerable time and expenses in locating interested customers. Depending upon the nature of the business, customer list protection may not need to extend indefinitely after termination of employment. The list may change significantly within a year or two, and the contractual provision may be so limited.

EXAMPLE 10.05B: Customer Lists.

The Employee further agrees that during the period of one (1) year immediately after the termination of his employment with the Employer he will not, either directly or indirectly, make known or divulge the names or addresses of any of the customers or patrons of Employer at the time he entered the employ of Employer or with whom he became acquainted after entering the employ of Employer, to any person, firm or corporation, and that he will not, directly or indirectly, either for himself or for any other person, firm, company or corporation, call upon, solicit, divert, or take away, or attempt to solicit, divert or take away any of the customers, business or patrons, of the Employer upon whom he called or whom he solicited or to whom he catered or with whom he became acquainted after his employment with the Employer.

The Employee hereby consents and agrees that for any violation of any of the provisions of this agreement, a restraining order and/or an injunction may issue against him in addition to any other rights the Employer may have.

In the event that the Employer is successful in any suit or proceeding brought or instituted by the Employer to enforce any of the provisions of the within agreement or on account of any damages sustained by the Employer by reason of the violation by the Employee of any of the terms and/or provisions of this Agreement to be performed by the Employee, the Employee agrees to pay to the Employer reasonable attorneys' fees to be fixed by the Court.[22]

§ 10.06 COVENANTS NOT TO COMPETE

Unlike the foregoing restrictive provisions governing the employer's ownership rights to the employee's work product and the protection of trade secrets and other confidential information, a covenant not to compete does not endeavor to solidify the company's ownership rights. Rather, this covenant is designed to prevent the employee from using his talents against his former employer. The negative objective causes some problems. Consider the plight of a research chemist who is an expert in industrial cleaning solutions. If he were a party to an employment contract which contained a clause forbidding future employment with any other industrial cleaner manufacturer, upon termination of his employment he will have lost his livelihood. His specialized technical knowledge significantly reduces his professional flexibility, and he must either breach the agreement or develop a new expertise. Of course, the company's objective is to prevent him from using his technical abilities to benefit a competitor, and the more specialized the skill, the more important it is for the company to discourage its marketability after termination. There is a bit of a tug of war here. If the restriction is too severe, a court simply won't enforce it. If it is too loose, the company can't enforce it.

Many courts are very reluctant to enforce noncompetition agreements for several reasons. First, a noncompetition agreement is a restraint of trade which is normally illegal both in common law and under state and federal antitrust statutes. However, limited noncompetition covenants are sometimes lawful if they are ancillary to an otherwise legitimate agreement. Generally, a reasonable noncompetition agreement will be enforced if it is necessary for the protection of the employer, imposes no undue hardship on the employee, and does not injure the general public. Because courts are frequently reluctant to enforce noncompetition agreements, many states have adopted statutes which severely restrict the use of such covenants. In some, all restrictive competition covenants are void except when given in connection with the sale of a business or the dissolution if a partnership.[23] Other states permit such clauses in employment contracts, but limit the effectiveness to certain justifying characteristics, such as training or advertising expenses, a license to practice, executive or management personnel, or specific territory limitations.[24]

Almost uniformly, the enforceability of a covenant not to compete depends upon whether the covenant is "reasonable", which includes the consideration of the following factors:

1. The legitimate needs of the employer for such protection.
2. The interest of society in preventing monopolies or other excessive restrictions on competition.
3. The burden placed upon the employee.
4. Whether the employee has had frequent contacts with customers or clients of the employer.
5. Whether the employer's business relied to a substantial degree on trade secrets to which the employee had access.
6. Whether the employer provided training to the employee.
7. Whether the employer's business is highly technical or complex.
8. Whether the employer's business is highly competitive.
9. Whether the employee, while employed, was a key employee, such as a manager.
10. Whether the employee provided unique services while employed by the employer.
11. Whether the covenant exceeds boundaries of time, space and type of activity which are reasonably required to give the employer the protection to which the employer is entitled.
12. Whether there is a clear disparity of bargaining power between the employer and employee.

13. Whether the employee understood the nature of the covenant at the time it was signed.

Based upon these considerations, drafting a noncompetition clause is a delicate operation.

One rule may be followed with impunity. A non-competition clause may never prohibit the employee from engaging in competitive activity indefinitely. The protection of the clause must have a reasonable basis in fact, and no employer could convince a court that competitive activity by a certain employee will forever cause irreparable harm to his business. Consequently, the clause should be limited to a specified period during which the company will be justified in keeping the employee out of the market to preserve a competitive advantage.

From the employer's standpoint, the agreement should also provide that the employee is prohibited from competition upon termination "for any reason whatsoever." Certainly an employee who left to work for a competitor prior to the expiration of his agreement could not expect much sympathy from a court. However, an employee who was fired, and must now refrain from marketing his talent is in a different position. Without the language suggested above, a court may narrowly construe an employee's "termination," and enforce the noncompetition clause only when termination results from the employee's initiative.

In order to enforce a non-competitive agreement, the employer may have to show a court that the former employee's competitive activities are causing irreparable harm to the company. This presents obvious difficulties in proof, which may be avoided by exacting a consent to injunctive relief against the employee should he violate the agreed provisions.

Finally, the competitive activities which are to be prohibited and the geographical limits of the prohibition should be specified with accuracy and clarity. If the provision is overly broad, it is less likely to be enforced in litigation involving its breach. Moreover, any ambiguity will always be resolved against the drafter, meaning the employer. For example, a provision which prohibits the employee from working in "the retail sales industry" is useless. A description of "retail sales of men's wear" is better but questionable. Equally unenforceable is a provision which prohibits the employee's competition in the "western part of the United States." The "South Dakota area" may be enforceable but requires a great deal of interpretation and would undoubtedly be limited to the boundaries of the state. The defined activities and geographical area should be consistent with the activities and market of the employer and should be specific. Consider the strengths and weaknesses of the following examples:

EXAMPLE 10.06A: Agreement Not to Compete

I also agree that I will not work for any competitive company or for myself or sell directly or indirectly milk or milk products in the same territory covered by me either as route salesman or foreman of routes for a period of at least one year after severing relations with this Company.[25]

EXAMPLE 10.06B: Covenant Not to Compete

It is further agreed by the Employee that the sale of the Employer's petroleum products in the trading area hereinbefore referred to is a valuable asset to the Employer and in order to promote the sales of petroleum products in said trading area the Employer will make expenditures through advertising and otherwise, and in consideration of the covenants and agreements herein contained the Employee agrees that in the event of the termination of this contract for any reason, with or without cause, that in such event, the Employee will not engage in the sale of gasoline, fuel oils or petroleum products, directly or indirectly, either on her own account, or as an employee for any other person, firm or corporation, in the city of Chicago, Cook County, Illinois, for a period of five (5) years following the termination of this contract.[26]

EXAMPLE 10.06C: Agreement and Covenant Not to Compete.

The Employee further covenants and agrees that at no time during the term of this employment, or for two (2) years immediately following termination thereof (regardless of whether such termination is voluntary or involuntary) will he for himself or in behalf of any other person, partnership, corporation or company, engage in the pest control business or any business engaged in the eradication and control of rats, mice, roaches, bugs, vermin, termites, beetles and other insects within the territory known as cities of Spearfish, and Belle Fourche, South Dakota, and a radius of 25 miles of each of said cities, nor will he directly or indirectly for himself, or in behalf of or in conjunction with any other person, partnership, corporation or company, solicit or attempt to solicit the business or patronage of any person, corporation, company or partnership within the said territory for the purpose of selling a service for the eradication and/or control of rats, mice, roaches, bugs, vermin, termites, beetles and other insects, and such other incidental business and service now engaged in by the Company, nor will the Employee disclose to any person, whatsoever, any of the secrets, methods or systems used by the Company in and about its business.

The Employee hereby consents and agrees that for any violation of any of the provisions of this agreement, a restraining order and/or an injunction may issue against him in addition to any other rights the Employer may have.[27]

It is better practice to specifically define the boundaries in which competition is prohibited. For example, describing county lines or city limits is preferrable to providing a point from which a radius will be computed. The larger the geographical area, the less likely the covenant will be enforced. If the employer is actually engaged in the sale of goods within a particular city, it would be inadvisable to define noncompetition as including the entire state. This is true even if the employer plans to expand operations to the entire state, since a court will be more concerned with the actual needed protection at the time the covenant was signed rather than attempting to speculate expansion plans and potential competitive problems in the future.

The clause should provide that the employee understands the nature of the covenant, and consents to the fact that the covenant will prohibit activities of the employee which may compete with the employer. Further, the covenant should specify that the employee understands that the clause is necessary for the employer's protection, and that the employee agrees that any violation of the clause will do irreparable harm to the employer.

The employee should be prohibited from directly or indirectly competing with the employer, either as an owner, manager, operator, controlling person, or being employed by, participating in or being connected in any manner with the ownership, management or operations or control of a competitor, including a position as a creditor. One of the classic avoidance techniques for these covenants involves former employees who loan money to a new corporation which will compete with the former employer. In exchange for the loan, these "creditors" receive a convertible debenture which will be converted to common stock (and majority ownership of the corporation) the day after the former employee's covenant not to compete expires.

If the covenant will be regulated by a statute in the jurisdiction in which it is to be enforced, the specific statutory reason for the covenant (such as because the employee was trained by the employer, or because the employee is part of management or executive personnel) should be stated in the covenant.

The clause should contain "severability provisions" which would allow for the removal of any objectionable portion of the clause, but the enforcement of the remainder of the clause for the employer's protection.

Consideration should be given for penalties other than injunctive relief or damages, such as loss of accrued but unpaid commissions, loss of retirement or profit sharing incentives, and similar penalties which discourage but do not forbid one from entering into competition.

The fair objective of restrictive covenants in employment contracts is to protect the legitimate interests of the company, without unduly restricting the activities of the employee. Properly drafted restrictions will never exceed the limits of necessary company protection, and they will be firm and thorough on those points. They thereby will be affored a better opportunity to do what they are supposed to do.

§ 10.07 INCENTIVE COMPENSATION PLANS

Employee incentive is a key element to current performance and continued employment with the company. Individual incentive compensation terms for individual employment contracts have been previously discussed.[28] It may be desirable, however, to use group incentive plans in lieu of or in addition to individual incentive provisions in each agreement. Of course, as the number of beneficiaries to an incentive compensation plan grows larger, it becomes less effective as a true incentive. For example, an individual saleswoman can easily be motivated by a personal incentive based directly on her sales, but her motivation is more tenuous if her sales division, comprised of many sales persons, is rewarded for the aggregate performance of all. As with personal incentive plans, a cardinal rule for the effectiveness of a group incentive plan is that the compensation must be related directly to the performance of an ascertainable division of the company. In practice these incentive plans are usually directed to key employees, since they motivate the employees to concentrate their efforts towards the employer's growth and profits, and further provide an incentive for the employees to remain with the company.

A group plan is properly administered under the direction of a committee, which, in the case of a corporation, may be composed of the directors but would not have to be. Persons who are entitled to compensation under the plan should not be members of the committee. The committee is usually vested with some discretion to determine the amount of the incentive awards and the recipients from among those eligible to participate in the plan.

The plan should begin with a statement of purpose, which usually indicates an intent to provide incentives to certain employees by enabling them to participate in the success of the company. The eligible participants in the plan must be clearly defined, as should the base accounts from which the amounts awarded under the plan will be determined. For example, if the participants under the plan are to receive a certain percentage of "profits," that word must be defined, specifying whether taxes, allocations of overhead, contingent or unusual expenses, etc. are to be deducted. The formula for compensation under the plan may refer to multiple accounts, such as a percentage of the extent to which profits exceed capital, and all stated accounts should be clearly defined.

The membership of the committee is set forth in the plan, and its duties and term of membership should be prescribed. If the committee is to have discretion in granting the incentive awards, a procedure for determination of the recipients, with specified guidelines for merit, may be provided. Alternatively, a formula could be drafted which would make the application of the awards a mechanical task for the committee, but this minimizes the flexibility of the plan by removing the committee's discretion from the incentive characteristics. True merit may be rewarded in their discretion, but a formula may not account for that.

A method for determining the amount of the fund must be included in the plan. A percentage of the defined profits is most simple and the complex formulas would defy the imagination. The fund is usually set aside in an "Incentive Compensation Reserve" account, awaiting the directions of the committee.

The payments under a corporate incentive plan may be made in cash or a stock equivalent, based upon the current market price of stock. If the stock is reported on a national exchange, a method is prescribed to compute market price, such as the "average daily opening price on the exchange during the calendar month preceding the month of award." The committee may decide whether the award is to be in cash or stock, and the committee should also have the authority to pay the award immediately or to defer payment in whole or in part. If deferral is permitted, separate accounts must be maintained for that purpose. The deferral of the award may provide certain tax benefits to the employees.[29]

Additional incentive thrust will result from imposing conditions upon payment of the compensation. For example, the terms of the plan may refuse payment to an eligible participant if he terminates his employment during the year without the consent of the company. The plan may further deny deferred payments if a former employee is subsequently employed by a competitor of the company. It is also common to provide for forfeiture of any award if the employee is discharged for misconduct. The plan should specifically state that the employee has no claim or right to be granted an award under the plan, and the plan should not be construed as granting the participant as right to be retained in the employ of the company.

The company management should be granted authority to modify or suspend the plan in their discretion, but only prospectively. They should not be able to retroactively affect the rights of employees with respect to unpaid awards previously granted. Finally, a corporate incentive plan should be approved by the shareholders especially if it contemplates the issuance of stock as an incentive award.[30]

§ 10.08 DEFERRED COMPENSATION

Key employees and executives may be plied with an incentive to remain with the company by a deferred compensation plan. These plans are designed to meet two important objectives in the employment relationship: 1) income for the employee is deferred until he or she retires or becomes incapacitated, providing necessary security and allowing receipt when the employee's income is taxed in a lower bracket; and 2) incentive to remain with the company to receive accumulated retirement benefits. The agreement for a deferred plan may be incorporated into an employee agreement or may be executed separately after a period of satisfactory employment.

The amounts of compensation to be deferred may be determined by a number of methods. A portion of base salary may be deferred, or a percentage of salary in addition to normal base salary may be used. Of course, the deferral provisions may also be tied to bonus or incentive plans, such as those described in the preceding section.

The deferred income is retained in a fund for the employee until payment at the prescribed time, usually following retirement or disability. Payment may be in a lump sum or in installments over a period of years.

EXAMPLE 10.08A: Retirement Date

The Company agrees that Roe may retire from the active and daily service of the Company upon the first day of the month nearest his sixty-fifth birthday.

Retirement Compensation

The Company agrees that commencing with the date of such retirement it will pay to Roe the sum of $18,000.00 per annum payable in equal installments of $1,500.00 each, payable upon the first business day of each calendar month. The Company agrees that it will continue to make such payments to Roe during his lifetime, and with no liability to make payments to his legal representatives, for ten (10) years and until and Roe shall have received one hundred twenty (120) monthly payments of $1,500.00 each; subject, however, to the conditions and limitations hereinafter set forth.[31]

The incentive to continue with the company results from the continual increase in the deferred fund and from the terms of the agreement which typically concludes all rights and obligations under the agreement if the employee terminates his employment relationship without the consent of the company.

EXAMPLE 10.08B: Termination of Employment

If Roe shall voluntarily terminate his employment during his lifetime and prior to his said retirement, or if his employment shall be terminated for sufficient cause as determined by the Board of Directors of the Company, this Agreement shall automatically terminate and the Company shall have no further obligation hereunder.

These agreements also frequently prescribe a forfeiture of all benefits under the plan if the employee subsequently engages in competition with the company.

EXAMPLE 10.08C: Covenant Not to Compete

Roe agrees that during such period of receipt of monthly payments from the Company he will not directly or indirectly enter into or in any manner take part in any business, profession or other endeavor either as an employee, agent, independent contractor, owner or otherwise in the City of Fort Lauderdale, Florida, which in the opinion of the directors of the Company shall be in competition with the business of the Company, which opinion of the directors shall be final and conclusive for the purposes hereof.

Forfeiture

Roe agrees that if he shall fail to observe any of the covenants hereof and shall continue to breach any covenant for a period of thirty (30) days after the Company shall have requested him to perform the same, or if he shall have entered any business described in the preceding paragraph and shall continue therein either directly or indirectly as aforesaid for a period of fifteen (15) days after the Company shall have notified him in writing at his home address that the directors of the Company have decided that such business is in competition with the Company; then, any of the provisions hereof to the contrary notwithstanding, Roe agrees that no further payments shall be due or payable by the Company hereunder either to Roe or to his said wife, and that the Company shall have no further liability hereunder.

In addition to the incentive character of the forfeiture clauses, they serve another useful purpose. The tax benefits from a deferred compensation plan may be generally stated to be that the company may deduct the payments to the deferred fund when paid, but the employee need not declare the payments as income until he receives them. The latter rule is conditioned upon the fact that the employee does not "constructively receive" the payments earlier—that is, he does not earn a vested right to the payments before they are paid to him. The conditions described above avoid the constructive receipt problem, since the employee's continuous performance and prohibition against competition are superimposed upon his right to the payment and prevent any vesting of his interest until the conditions are satisfied.

The deferred compensation plan should divert the payment of the deferred income to the employees' heirs in the case of his death.

EXAMPLE 10.08D: Payments to Widow if Husband Dies after Retirement

The Company agrees that if Roe shall so retire but shall die before receiving the said one hundred twenty (120) monthly payments, it will continue to make such monthly payments, to Jane Roe, the said wife of Roe, if she shall survive Roe, until the total payments made to Roe and his said wife shall equal $180,000.00; provided that if Jane Roe shall survive Roe but shall herself die before the said amount shall be paid by the Company, the Company shall have no liability to continue any payments hereunder beyond the first day of the month in which Jane Roe shall die.

Death of Husband after Retirement with No Widow Surviving

If Roe shall so retire and if his said wife shall not survive him, the Company shall not be required to continue any payments hereunder beyond the first day of the month in which Roe shall die.

Payments to Widow if Husband Dies before Retirement

If Roe shall die before the aforesaid retirement date, and if his said wife shall survive him, the Company agrees to make the said monthly payments hereinbefore described to the said Jane Roe commencing with the first day of the month following the month in which Roe shall die and ending when one hundred twenty (120) said monthly payments shall be made to the said Jane Roe or until and including the first day of the month in which she shall die, whichever event shall first occur.

Death of Husband before Retirement with No Widow Surviving

If Roe shall die before the said retirement date and if Jane Roe shall not survive him, the Company shall not be required to make any payments hereunder.

§ 10.09 PENSION AND PROFIT SHARING PLANS

Pension and profit sharing plans are deferred compensation plans, but they may produce additional tax benefits for the employer and the employee if they qualify under the Internal Revenue Code.[32] Both pension and profit sharing plans accumulate and defer income until some future date. The employer's objective is to induce the employee to remain with the company to receive accumulated retirement benefits. The profit sharing plan also adds current performance incentive, since the amount contributed to the plan is based upon profits produced by the employee and his fellow-workers.

Both profit sharing and pension plans are directed toward retirement or disability income and faithful performance. However, each reaches the objective by different means. The profit sharing plan is based upon profits, and the employer's contribution to the fund is couched in terms of a percentage of annual profit. From the employer's standpoint, a profit-sharing plan is less onerous. Subject to special tax rules, since the obligation to contribute depends upon the profitable operation of the business, there is no fixed obligation to contribute annually. The contribution requirements reflect the economic cycles of the business and the plan is more flexible as a result. The size of the fund will depend entirely upon the profits contributed throughout the duration of the plan. Allocation of the fund to individual employees is usually based upon their compensation during their periods of employment. These factors all result in a minimal burden for the employer and, if incentives are taken seriously, greater benefit to the employees. Younger, aggressive employees who intend to remain with the company will particularly benefit. Their efforts in producing profit increases their compensation under the plan, and as they remain with the company, their share of the profits distributed to the fund will increase as their base compensation increases. Finally, it is possible to provide for periodic withdrawals from the profit sharing accounts in addition to distribution upon retirement, death or disability.

Pension plans, on the other hand, are specifically directed toward retirement income and the employer usually is obliged to contribute on an annual basis the necessary funds. They may be either defined benefit plans (where the exact benefits upon retirement are specified and contributions must be made to reach the benefits specified) or defined contribution plans (where a certain contribution is made by the employer each year and the benefits depend on the total funds available upon retirement). The benefits of the plan provide a specified income for the employee after retirement, and the employer contributions are fixed by an amount necessary to provide income for the specified period. The employer contributions are a fixed annual obligation and are not related to profits in any way. The contribution is, therefore, a charge on operations which must be considered in estimating costs and pricing. Moreover, the contribution is usually directly based upon length of service and age, since the contribution for an older employee must be higher to accumulate enough funds for his specified retirement income under the plan. The allocation of funds among the eligible employees under the plan will be in accordance with the prescribed pension, and younger, aggressive employees are not particularly rewarded for their enthusiastic business achievements. However, the fixed obligation to contribute to the pension plan has some employee advantages that are not available with a profit sharing plan. The pension plan permits recognition of employment prior to the establishment of the plan, and accommodates immediate pensions for employees who have already reached retirement age when the plan is adopted.

The choice between a defined benefit or defined contribution pension or a profit sharing plan depends on many factors, the most important of which include the ability of the company to make the required contributions, the desired stimulation of incentive, and the characteristics of the employee group. For indecisive company managers, it may be suggested that a combination of both plans be used to benefit all employees.

Tax Ramifications

The tax treatment of profit sharing and pension plans is complicated, and should be thoroughly studied before recommending or drafting a plan. Only a few of the most important provisions are generally covered here.

Tax-favored plans must be "qualified," meaning they satisfy the statutory requirements prescribed in the Internal Revenue Code.[33] A qualified plan results in tax benefits for both the employer and the employee. Subject to certain statutory limitations, the employer is permitted to deduct contributions to the plan in the year made, just as normal compensation would be deducted as an expense, despite the fact that the compensation is not actually paid to the employee during that year.[34] Therefore, the employer suffers no tax disadvantage by paying compensation to the plan, rather than paying directly to the employee. The funds paid into the plan may be accumulated and invested during the holding period, but the income earned on the investment is exempt from any tax.[35] The funds thereby increase must faster than if they were invested by the corporation without a qualified plan or by the individual employee himself. The employee-beneficiary of the plan is not taxed on any of the funds until they are distributed to him.[36] Thus, an executive who receives a substantial salary which elevates her tax bracket at the height of her earning power will not lose a proportionate amount of the compensation under the plan by having to pay tax on this deferred income in the years she is employed. Rather, she will pay tax on distributed amounts received in years of her retirement, when her other income is minimized and her tax bracket is much lower. Moreover, lump sum payments after retirement or disability receive other favorable tax treatment.[37] Finally, the amounts contributed by the company and paid to heirs or beneficiaries upon the death of the employee can pass free of estate tax, since the employee may specify direct payment to named beneficiaries and the funds will not be included in his or her estate.

The foregoing tax benefits increase the desirability of these plans as a part of the employee compensation scheme. Qualified plans permit the employer to minimize cost while maximizing employee compensation.

Qualification

The requirements of the Internal Revenue Code and the rulings thereunder are a maze of intricate rules with myriad exceptions, definitions, inclusions and exclusions—all designed to describe the qualification procedure and standards of a profit sharing or pension plan. The purpose of the plan, amounts of benefits, participation, vesting of benefits, operation, contributions and reporting must be structured in accordance with the statutory rules.

A qualified plan must be adopted for the purpose of offering the employees a share of the profits or income of the employer or for providing a fund which will be distributed to the employees or their beneficiaries after retirement. The plan must be established and maintained by the employer for the exclusive benefit of the employees. The funds cannot be diverted for any use other than the specified purposes.

The plan may not be qualified if it is intended to benefit only a few select employees. Section 401(a) of the Code is extremely forceful in insisting that the qualified plan may not discriminate in favor of employees who are officers, shareholders or highly compensated. Generally, the contributions to the plan should be allocated in proportion to and benefits awarded in relation to the total current compensation of the participants of the plan. It is possible to base contributions and benefits on less than all compensation of the employees so long as the plan remains nondiscriminatory.

Participation in the plan depends upon compliance with two general statutory tests: age/service requirements and classes of employees.

Concerning the minimum age and service conditions, the qualified plan must permit participation for employees who have completed at least one year of service with the employer or who have reached age 25, whichever is later. One year of service means a twelve month period during which the employee works at least 1,000 hours. A part-time employee who works for more than a year but only logs 18 hours a week can be safely excluded under a qualified plan. But a full time employee who works forty hours a week and takes a 25–week vacation must be permitted to participate (unless, of course, the vacation turns out to be permanent and the vesting rules permit a forfeiture for the terminated employee.) If the plan provides that the employees who participate will be vested immediately with the benefits of the plan, an employer may provide for a three-year eligibility period. Older employees cannot be excluded from the plan simply because of age unless the plan is a defined (or target) benefit plan and the employee begins employment within five years of the normal retirement age specified in the plan. This latter rule permits an employer to hire an employee who is near retirement without having to contribute large sums to the plan to provide the defined benefit amount when the employee does retire during the next five years.

The participation of certain classes of employees is even more interesting. The statutory test has three parts. The plan must cover:

1. 70% of all eligible employees;
2. 80% or more of all the employees who are eligible to benefit under the plan if 70% or more of all the employees are eligible;
3. a specified group of employees, provided the group does not discriminate in favor of officers, stockholders, highly-paid or supervisory employees and the classification is approved by the Internal Revenue Service.

A complex illustration may serve to explain these otherwise simple rules. Suppose a corporation has 80 regular employees who work full-time, and 5 part-time employees who work less than 1,000 hours in a twelve-month period. The 80 regular employees may be classified as follows:

Age	Served over 5 years	Served 3–5 years	Served 1–3 years	Served less than 12 months
Over 60	7	2		
25–60	42	7	5	1
Under 25	3	4	2	7
	52	13	7	8

First, determine who will be potentially excludable under the age and service rules. The plan may exclude the 5 part-time employees because they work too few hours, and may exclude the 8 full-time employees who have served less than 12 months. The 9 full-time employees who have not reached age 25 may be excluded. The plan could also exclude those 2 employees who are over 60 and who have served less than 5 years but only if they began their employment when they were older than a specified age (say 60) *and* if the normal age of retirement is not more than five year older (say 65), *and* if the plan is a defined benefit plan where benefits are set in a certain amount after retirement and the employer would have to rapidly contribute great quantities of money for this employee in order to meet those benefits upon retirement.

As for participation, the first alternative required that a qualified plan must cover 70% of "all" eligible employees. In this case the plan must cover at least 45 employees, since it may exclude the 5 part-time employees and the 8 full-time employees who do not meet the minimum service requirement, and the 9 full-time employees who are under age 25. This leaves 63 employees who meet the minimum age and service requirements and 70% of them, or 45, must be covered. If the plan provided that those employees who earned more than $30,000 per year would be included as participants, and if only 40 of the employees qualified under that test, the plan would not be qualified under this first alternative because of the participation formula rules.

It might qualify, however under the second test. The second alternative requires coverage of 80% of the eligible employees with 70% of all employees eligible. In this corporation, 70% of the eligible employees is 45, as illustrated above. If 80% or more of these 45 employees earned $30,000 per year, a plan which specified participation at that salary level will be qualified. If 40 employees of this corporation earned more than $30,000 per year, that number exceeds 80% of 45 (36) and the plan satisfies the participation test.

Suppose the corporation adopts a plan which is designed to cover only those employees who work in a specified place, and only 20 employees meet that test. All is not lost even though the plan dismally fails the first two alternatives. It would still be possible that the plan could obtain special approval of the Internal Revenue Service if it is not found not to be discriminatory in favor of employees who are officers, shareholders or highly compensated. Thus, if the Service learned that the "special place" turned out to be corporate headquarters where management offices are located, the special classification would not likely succeed.

One final illustration to test your keen awareness of these rules. Could the corporation qualify a plan which covered only those 63 employees who work full-time, are over 25 years of age, and have worked at least three full years with the corporation? Yes. It meets the 70% test. Minimum age and service requirements are also met *if* the plan provides for immediate vesting.

Speaking of vesting, the qualified plan may meet one of three alternative vesting schedules under Section 411 of the Code. Before these vesting schedules are reviewed, it should be noted that vesting is the employer's incentive hammer. So long as the benefits under the plan have not vested, the employer can reasonably expect that the employee will continue to work for the employer, hoping to enjoy vesting of the benefits one day. If the benefits have vested, the employee is absolutely entitled to them, and the incentive to remain with the company may be diminished. Consequently, most employers would prefer to defer vesting as long as possible. The vesting schedules state the *maximum* periods of time which the plans may provide and still be qualified.

The first alternative is called the "ten year cliff" vesting. The plan may provide that no benefits will vest until ten years of service, and then the benefits must be 100% vested and nonforfeitable. Secondly, the plan may provide for five to fifteen-year vesting. The employee may be required to complete at least five years before any benefits become vested. At the end of five years, 25% of the benefits vest, and each year 5% more of the benefits vest until the tenth year, when each year thereafter 10% more of the benefits vest. At the end of the fifteenth year, the benefits are 100% vested and nonforfeitable.

Finally, vesting may occur according to the Rule of 45. This rule is best applied by a computer, but generally the benefits do not vest at all until the completion of five years service. Vesting thereafter depends upon the sum of age and service with the corporation, according to the following table:

If years of service equal or exceed—	and sum of age and service equals or exceeds—	then the nonforfeitable percentage is—
5	45	55
6	47	60
7	49	70
8	51	80
9	53	90
10	55	100

Section 415 of the Code places certain limits on the benefits and contributions to qualified plans which must be strictly observed. A defined benefit plan is generally limited to the lesser of $75,000 or 100% of the participant's average compensation for his or her highest three years. This $75,000 is adjusted for cost of living increases by the Treasury annually, and the limitation is also modified depending upon the age and years of service for each employee. The defined contribution plan is limited to the lesser of $25,000 or 25% of the participant's compensation for the annual contribution to the plan. This amount is also adjusted for cost of living increases, and employee contributions are taken into account in making the computation.

The statutory formalities demand that the plan be permanent, in writing and communicated to the employees. Its terms must contemplate a continuing program of contributions by the employer and distributions to the employees. While the plan may be terminated for business exigencies not within the employer's control, termination is always subject to careful scrutiny by the Internal Revenue Service to determine whether it was truly intended to be continuous at inception. Thus, there must be a valid business reason, unforeseeable when the plan was adopted, in order to terminate the plan. The written plan may be distributed to the employees, or, at least, a pamphlet describing its salient provisions should be available to the employees.

Finally, the operation of the plan is an important element to its qualification. The plan must be "funded." The funding arrangements may be accomplished by a trust, requiring contributions from the employer to an institutional trustee, who will invest the funds and distribute them in accordance with the trust agreement. It is also possible to use the funds to purchase insurance contracts or other investment contracts for the benefit of the employee plan. In sum, there must be a formal contribution and investment arrangement, distinguishable from normal corporate activities by contractual provisions, preferably with an independent third party, to operate and administer the plan.[38]

§ 10.10 STOCK OPTIONS

Stock options which are offered as additional employee compensation may also perform an incentive function with tax advantages. However, rather than monetary remuneration, the stock option is designed to give the optionee-employee an ownership interest in her corporation-employer (the optionor).

A stock option is a written instrument in which a corporation offers the right to purchase stock in the corporation to an individual at a predetermined price at any time during a stated period. The individual is under no obligation to accept the offer to purchase. As you may expect, in order to benefit the recipient, the option offers the stock to the optionee at a price other than the market price, and this feature has considerable incentive value. Much like the carrot in front of the donkey, the price is set at an amount higher than market price at first, encouraging the employee to produce profits so market price will rise above the option price, and the employee will then have the option to purchase the stock at the discount option price. The incentive factor continues even after the employee exercises the option since the value of the stock purchased should increase as the profits of the granting corporation increase. An employee compensated by stock options should be motivated to use his best efforts on behalf of the corporation since he will eventually profit by holding the option until the market value of the stock exceeds his option price. By exercising the option, he has made a discount purchase of the stock and has become a shareholder who now has a pecuniary interest in his own performance. Moreover, by using stock options the corporation is compensating an employee without incurring any immediate expense, because the issuance of an option requires no direct payment to the employee. Of course, the ownership interest of existing shareholders is affected by stock options. The exercise of the option dilutes their proportionate ownership of the company. However, if the stock option motivates the employees to continually strive for profits, the existing shareholders will also benefit from the resulting increase in stock values.

As a part of its compensation program, a corporation may adopt a stock option plan to benefit certain employees. Options granted under the plan give the recipients the right to purchase stock of the corporation at a specified price, the "option price", for a specified period of time. The offer of stock to the employee may be accepted by him during this period, and the act of acceptance is the exercise of the option. Upon payment of the option price to the corporation the employee is entitled to receive certificates for his stock.

Tax Ramifications

A stock option plan may also be "qualified" to receive special tax treatment under the Internal Revenue Code. A qualified stock option has the tax advantage of capital gains treatment for "compensation" which migh otherwise be taxed at ordinary income rates. An individual's ordinary income, such as cash compensation, is taxed at higher rates than capital gain income, and if the corporation paid the value of the option to the employee in cash, the payment would be treated as ordinary income and taxed at normal ordinary income rates. The taxing authorities recognize that the grant of a normal stock option is simply another way of compensating the employee, and they treat the benefits from an unqualified stock option as if cash had been paid, taxing the "compensation," the difference between the option price and the market price of the stock upon exercise of the option, at ordinary income rates. Since the employee is taxed on this compensation, the corporation is entitled to a corresponding deduction as an expense.

The tax scheme is different for a qualified stock option. If the option is created and granted in accordance with the statutory rules, the individual employee's tax is deferred. Rather than being taxed upon the exercise of the option or at some other designated time prior to final disposition of the stock, the tax is imposed when the stock is finally sold at a gain.[39] Moreover, if the stock acquired upon exercise of the option has been held for at least three years, any gain realized by the employee upon sale of the stock is taxed at capital gain rates.[40] The advantage belongs to the employee. The "compensation" received from a qualified stock option, the profit received when the stock is sold, is subject to a lower tax than ordinary income and the tax is deferred until the employee has actually received the cash value of the remuneration. The corporation loses its deduction for shares transferred under a qualified stock option plan.

Under the Tax Reform Act of 1969, certain portions of the option benefits are defined as "tax preference" income, and an additional tax is imposed on these amounts. The surtax on "tax preference income" is 15% of the amount by which all tax preference income of the taxpayer for the year in question exceeds $10,000 or one-half of the taxes imposed for the taxable year.[41] For a qualified stock option the tax preference income is defined to be: the difference between the fair market value of the shares purchased and the aggregate option price for such shares in the year that an option is exercised, and net capital gain in the year that the shares purchased under the option are sold.[42] This additional tax does not constitute an onerous burden on most employees, since the tax preference surtax only affects employees who receive substantial gains on the sale of stock. To illustrate, consider Amy Coover, an employee of the Nobles Company, who has an option to buy 1,000 shares of corporate stock at $10 per share. When she exercises the option, the market price of the stock is $50 per share. In that year, she has tax preference income of $40,000, the difference between market value and the option price. Further suppose that she pays $5,000 in income tax for her ordinary income and other investments that year. Her tax preference income is taxed at 15% of the amount the preference income exceeds $10,000 or the other income tax payable whichever is greater. The tax preference amount is $30,000 ($40,000 − $10,000) and her tax preference surtax is $4,500 in the year the option is exercised. If she holds the stock for more than three years, and sells it at $60 per share, her long term capital gain is $50,000, since she purchased the stock at $10 per share. Tax preference income in the year of sale is the net capital gain, or $30,000. Her surtax that year could be as high as $3,000 ($30,000 − $10,000 × .15) or $25,000.

The tax preference surtax does not diminish the desirability of the qualified stock option as a form of compensation. Amy will have paid $7,500 in extra tax in the hypothetical transaction, but she has also profited $50,000. Who could complain about that?

Qualification

The Internal Revenue Code specifies several requirements for the qualification of tax-favored stock option plans,[43] and, like qualified pension and profit-sharing plans, the qualified stock option plan must strictly accord with the statutory rules. Generally, these requirements are concerned with adoption and content of the option plan, duration of the plan, exercise of the option, the option price, and the employees covered by the plan.

A qualified stock option plan must state the number of shares which may be issued under the plan, and must describe the employees who are eligible to receive options.

The identification of shares to be issued under the plan should include the description of the shares from the articles of incorporation. The number may be specific or a formula may be prescribed to establish the total number of shares which may be issued under the plan.

EXAMPLE 10.10A: Shares Subject to the Plan

> The Committee, from time to time, may provide for the option and sale in the aggregate of up to 100,000 shares of Common Stock of the Company, without par value. Shares shall be made available from authorized and unissued or reacquired Common Stock.[44]

In order to enjoy the tax benefits of the qualified stock option it must be granted to a person employed at the time of the grant. Individuals who are employed by a parent or subsidiary corporation of the granting corporation would qualify, but a prospective employee would not. Moreover, the optionee must remain an employee at all times during the period from the date the option is granted to three months before the date the option is exercised. The continuous employment requirement does not apply to a deceased employee, but is strictly observed in all other cases. The plan may include this limitation in its provisions:

EXAMPLE 10.10B: Rights in Event of Termination of Employment

> In the event of termination of employment for any cause other than death, a participant's option shall expire within three months after his employment terminates. Nothing contained in the Plan shall confer upon any participant any right to be continued in the employ of the Company or any subsidiary of the Company or shall prevent the Company or any subsidiary from terminating his employment at any time, with or without cause.

Of course, it is always possible to state in the plan that the stock options granted under it will expire immediately upon termination of employment. The Internal Revenue Code simply provides that the option may not extend beyond three months after termination to qualify for the advantageous tax treatment. The estate of a deceased employee is not so limited unless the option specifically so provides, and it may.

EXAMPLE 10.10C: Rights in Event of Death

In the event of the death of a participant while in the employ of the Company or a subsidiary, or within three months after termination of such employment, the option theretofore granted him shall be exercisable, in whole or in part, within the next three months succeeding his death, if and to the extent the participant could have exercised it at the date of his death, by such person as shall have acquired the right to exercise such option by will or by the laws of descent and distribution.

The identification of employees covered by the plan may be satisfied by reasonably specific descriptions, for example, "supervisory employees," "all salaried employees" or "all employees of the corporation." The board of directors, or their delegates, usually have the power to determine the number of shares to be optioned to each employee within those described in the plan.

EXAMPLE 10.10D: Participants

The Committee shall determine and designate from time to time those key employees (including employees who are also officers or directors) of the Company and its subsidiaries (as defined in section 425(f) of the Internal Revenue Code of 1954, as amended) to whom options are granted and who thereby become participants in the Plan. In selecting the individuals to whom options shall be granted, as well as in determining the number of shares subject to each option, the Committee shall weigh the positions and responsibilities of the individuals being considered, the nature of their services, their present and potential contributions to the success of the Company, and such other factors as the Committee shall deem relevant to accomplish the purpose of the Plan. No option shall be granted to an employee who, immediately after such option is granted, owns stock possessing more than 5 percent of the total combined voting power or value of all classes of stock of the Company or its subsidiaries. Each grant of an option shall be evidenced by an option agreement which shall contain such terms and conditions as may be approved by the Committee and shall be signed by an officer of the Company and the employee.

A qualified stock option plan must be approved by the stockholders of the granting corporation within twelve months before or after the date the plan is adopted. A plan is usually adopted by an appropriate resolution of the board of directors.

EXAMPLE 10.10E: Resolution to Adopt a Qualified Stock Option Plan

WHEREAS, it is the belief of the Board of Directors of this Corporation that its key employees would be interested in acquiring a part of the capital stock of this Corporation, and that ownership of stock of this Corporation by its key employees would be to the advantage of this Corporation, and

WHEREAS, by the recent increase in capital stock of this Corporation, there is available for further subscriptions a total of one hundred thousand (100,000) shares of the common stock, without par value, of this Corporation,

BE IT RESOLVED, That the Board of Directors do hereby adopt the Qualified Stock Option Plan, a copy of which is made a part of this resolution, subject to the approval of the common stockholders at their next regular meeting, and

RESOLVED FURTHER, That the Board of Directors of this Corporation do hereby recommend to the stockholders that they authorize the Board of Directors to set aside a total of ten thousand (10,000) shares of the common stock of this Corporation for sale to its key employees under said Qualified Stock Option Plan.[45]

The date of the resolution adopting the plan will be the date used to determine whether shareholder approval has been obtained within the statutory period. The required shareholder approval may come within twelve months before or after the adoption of the plan by the board of directors. Consequently, the shareholder action may be the impetus which causes the board to adopt a plan. As long as shareholder approval and board adoption occur within a twelve-month period, the plan may be qualified.

The federal tax law only requires that the plan be approved by the affirmative vote of the holders of the majority of the voting stock of the corporation. The vote is taken at a regular or special stockholders' meeting and the normal shareholder voting provisions of the state corporate statute applies.[46] Further, there may be other requirements for shareholder action in the statute, articles of incorporation, or by-laws to obtain a waiver of any preemptive rights in the stock issued under the option. Shareholder approval of the plan must comply with those requirements.

A qualified stock option must be granted within ten years from the date on which the plan is adopted or the date on which the plan is approved by stockholders, whichever is earlier. Accordingly, the plan should contain an expiration date within the ten year period. Of course, the expiration of the plan has no effect upon any options previously granted pursuant to its terms.

EXAMPLE 10.10F: Termination

The Plan shall terminate ten years after its effective date, or on such earlier date as the Board of Directors may determine.

The option may not be exercised after the expiration of five years from the date it is granted. The plan should so state and each option should recite a specified period for exercise within this limitation.

EXAMPLE 10.10G: Option Period

The term of each option shall be such period as the Committee may determine, but not more than five years from the date the option is granted.

Some quick mathematics should reveal that the period from first grant to last exercise under any qualified stock option plan may not exceed 15 years. If an option were granted just before the expiration of the ten year period following adoption and shareholder approval of the plan it would remain exercisable for five additional years. Thus, the last option may be exercised almost fifteen years after the plan took effect.

The option price may not be less than the fair market value of the stock on the date the option is granted in order to qualify for special tax treatment. Thus, the employee can benefit only from the appreciation in the value of the stock following the grant of the option. Of course, this feature increases the incentive value of the option, since the employees' performance will be directed toward increasing the market value of the stock over his option price. The tax benefits of a qualified stock option will be lost if the option price is less than the market price when the option is granted.

The determination of the option price will generally be the responsibility of the board of directors or their appointed committee, and only a general statement regarding option price need be included in the option plan. It is absolutely necessary to specify that the option price will be not less than the market value of the stock on the date the option is granted. Of course, if the stock subject to the date the option has a par value, the option price also must be at least par value.

EXAMPLE 10.10H: Option Price

Shares of Common Stock of the Company shall be offered from time to time at a price to be determined by the Stock Option Committee, which price shall be not less than the fair market value of the stock on the day the option is granted.

Fair market value shall be the mean between the highest and lowest selling prices on the New York Stock Exchange on the valuation date. If there are no sales on such date, the value shall be determined by taking the mean between the highest and lowest sales upon the last preceding date on which such sales occurred. In no event shall the option price be less than the par value.

The remaining statutory requirements for qualification limit the employee-shareholder's ability to exercise other outstanding options, the ability to transfer the option to others, and the percentage ownership interest in the corporation and its affiliated corporations. The first rule is that an option may not be exercisable while there is outstanding an employee stock option previously granted to the same employee. This rule prevents the successive issuance of options at lower amounts when the market value of the company's stock is declining. The plan should recognize this limitation in its terms.

EXAMPLE 10.10I: Prior Options

A subsequent option may not be exercised by a participant while he has outstanding any other prior stock options under this or prior stock option plans which entitle him to purchase stock of the same class in the Company at a price higher than the option price of such subsequent option.

The Code also states that a qualified stock option may not be transferable by the employee—it may be exercised only by that employee. The singular exception to this rule is that transfers are permitted upon the death of the employee by the employee's will or by the laws of descent and distribution. The terms of the option should so state.

EXAMPLE 10.10J: Nonassignability

Options are not transferable otherwise than by will or the laws of descent and distribution, and are exercisable during a participant's lifetime only by him.

Finally, there is a limitation on the employee's other stock ownership in the corporation. The employee, immediately after the option is granted, may not own more than 10% of the voting power or value of all classes of stock of the granting corporation, its parent or subsidiaries. This percentage adjusts downward to 5% as the equity capital of the corporation increases to $2,000,000. The description of the employees entitled to participate should recognize the limitation, as is illustrated in the sample clause describing the participants in Example 10.10D.

Administration of the Plan

The terms of the plan will describe the persons who are to administer it. The board of directors are naturals for the administration, but they may choose to delegate administration authority to a committee appointed by the board. The terms of the plan should also prescribe a procedure for their actions, and the authority granted or restricted in the administration of the plan.

EXAMPLE 10.10K: Administration

The Plan shall be administered by a Stock Option Committee (herein called the Committee) consisting of three or more members of the Board of Directors of the Company who are not eligible to receive options under the Plan or who have waived their rights to receive options during such time as they are members of the Committee. The Committee shall be appointed annually by the Board of Directors, which may from time to time appoint additional members of the Committee or remove members and appoint new members in substitution for those previously appointed and fill vacancies however caused. A majority of the Committee shall constitute a quorum and the acts of a majority of the members present at any meeting at which a quorum is present, or acts approved in writing by all of the members, shall be deemed the action of the Committee. Subject to the provisions of the Plan, the Committee is authorized to interpret it, to prescribe, amend and rescind rules and regulations relating to it, and to make all other determinations necessary or advisable for its administration. It is intended that all options granted under the Plan be "qualified stock options" under the Internal Revenue Code of 1954, as amended.

Allotment of Shares

The Committee shall determine the number of shares to be offered from time to time to each participant except that the maximum number of shares offered to any one participant upon the initial offering at the inception of the Plan shall not exceed 2,000 shares, and the maximum number of shares which any participant may purchase pursuant to the initial offering and all subsequent offerings under the Plan shall not exceed 5,000 shares. The Committee may also prescribe a minimum number of shares which may be purchased at any one time, and the time or times shares will be issued pursuant to the exercise of options. In any offering after the initial offering, the Committee may offer available shares to new participants or to then participants or to a greater or lesser number of participants, and may include previous participants in accordance with such determination as the Committee shall make from time to time.

The administrators should have authority to adjust the provisions of the option to account for subsequent capital adjustments in the corporation. For example, a stock split or stock dividend would accordingly reduce the market price, and may cause the option price to be unrealistic. An adjustment to the option will preserve its viability.

EXAMPLE 10.10L: Adjustment upon Changes in Capitalization

In the event there is any change in the Common Stock of the Company by reason of stock dividends, stock split-ups, recapitalization, reorganizations, mergers, consolidations, combinations or exchanges of shares, or otherwise, the number of shares available for option and the shares subject to any option shall be appropriately adjusted by the Committee.

The plan should carefully limit the authority of the administration to modify its terms. If an amendment to the plan violates any of the statutory requirements for qualification, the qualified status of the plan and the options granted under it will be lost. Consequently, only technical amendments should be permitted, and the authority to amend should prohibit the types of changes which would result in a substantive modification affecting qualification.

EXAMPLE 10.10M: Amendment

The Board of Directors may at any time suspend, rescind or terminate the Plan and may amend it from time to time in such respects as it may deem advisable, provided, however, that no such amendment shall, without further approval of the stockholders of the Company, except as provided herein, (a) increase the aggregate number of shares as to which options may be granted under the Plan either to all individuals or any one individual; (b) change the minimum option purchase price; (c) increase the maximum period during which options may be exercised; (d) extend the termination date of the plan; or (e) permit the granting of options to members of the Committee. No option may be granted during any suspension of the Plan or after the Plan has been rescinded or terminated and no amendment, rescission, suspension or termination shall, without the participant's consent, alter or impair any of the rights or obligations under any option theretofore granted to him under the Plan.

In granting a stock option, the corporation is offering to issue securities which may have to be registered under federal and state securities laws. If the securities have been registered, the corporation should have no problem issuing the stock under the option. However, if the securities are not registered, it is quite important to obtain a representation from the employees that the stock is being purchased only for investment and not for distribution and resale to the general public. The sale of securities to key employees may be exempt from the registration requirements of the securities laws, but severe penalties are prescribed for sale of unregistered stock to the general public. This representation is designed to protect against these liabilities. The plan may contain a clause acknowledging the purpose of the purchase:

EXAMPLE 10.10N: Purchase for Investment

All stock of the Company purchased pursuant to any option must be purchased for investment and not with the view to the distribution or resale thereof. Each option will be granted on the understanding that any shares purchased thereunder will be so purchased, and each employee to whom an option is granted shall be required to deliver to the Company a written representation and agreement to that effect.

Finally, the plan should recite the requirements of payment and other consideration to be given by the employee for the receipt and the exercise of the option. The plan may exact a promise from the employee that he will remain employed with the company for a period of time as a condition to the privilege of receiving the option.

EXAMPLE 10.10O: Consideration for Option

Each participant shall, as consideration for the grant of the option, agree to remain in the continuous employ of the Company or one or more of its subsidiaries for at least two years from the date of the grant of such option.

Payment for Stock

Full payment for shares purchased shall be made at the time of exercising an option in whole or in part. No shares shall be issued until full payment therefor has been made and a participant shall have none of the rights of a stockholder until shares are issued to him.

§ 10.11 INSURANCE PROGRAMS

Various programs of insurance coverage are available to a corporation for the benefit of its employees. Some of these programs have unique income tax advantages under the Internal Revenue Code, and all of them are intended to benefit the beneficiaries of the employee upon the employee's death with little or no cost to the employer.

Death benefit compensation programs are frequently used by the corporations. Typically, the employer will promise to pay a portion of the employee's salary to the heirs for a period of time after death in consideration of the employee's continuous faithful performance. The amounts paid to the employee's heirs are tax free up to a certain amount[47] and the amounts paid are deductible to the corporation as business expenses. The corporation will normally apply for life insurance on the employee to cover this contingent liability, and will use the proceeds of the insurance to pay the agreed amounts to the heirs. For the employee, a death benefit agreement is like an insurance policy for the family, and the insurance is completely without cost to the employee.

Split-dollar insurance is another common plan used to benefit key personnel. The corporation and the employee join in the purchase of a life insurance policy on the employee, and the employer-corporation pays a share of the premium, usually equal to the annual increase in cash value of the policy. The employee pays the remaining premium. The corporation is named beneficiary to receive the amount of the premiums it has paid, and the employee will designate his own beneficiary for the balance. When the employee dies, the corporation receives a return of all premiums paid, and the employees's beneficiary receives the remaining face value of the policy. The advantage to the employee is that the beneficiary will receive a considerable sum while the employee's share of the premiums has been minimal. The corporation will recover all it has paid, and will have no out-of-pocket costs to provide the insurance benefit.

§ 10.12 EXECUTIVE MEDICAL EXPENSE REIMBURSEMENT PLANS

Many companies have instituted self-insured plans for executives as a method of supplementing group hospital insurance plans, major medical insurance and other forms of reimbursement which may be provided to all employees. The board based group medical insurance plans usually do not cover all expenses incurred. For example, certain amounts may not be reimbursed as a result of deductibles, co–insurance, and maximum benefit provisions.

In addition, certain expenses such as private hospital room care, routine dental work, eye glasses or contact lenses, and annual physical examinations are not covered under many group medical insurance plans covering employees generally. Employers may institute medical reimbursement plans under which all or some of these expenses are covered by the company.

Prior to 1978, the employer could choose, without adverse tax ramifications, the employee or group of employees who would participate in such an executive medical or dental expense reimbursement plan. The amounts reimbursed were not taxable to the executive but were deductible to the employer. If on the other hand, an executive paid these expenses personally, the expenses would be tax deductible only to the extent that they exceeded three percent of the executive's adjusted gross income (the amount allowed for a medical deduction on personal income tax returns). In the typical case, therefore, most or all of the expenses would not be tax deductible.

These medical expense reimbursement plans were very popular in corporations prior to 1978 to provide a tax-free fringe benefit for special executives.

The Revenue Act of 1978 provided that all or a portion of the expenses reimbursed to an executive under a discriminatory, uninsured expense reimbursement plan will be taxed to the executive. The major concern is the discriminatory nature of a plan and the fact that it was not funded by insurance.

Employers now may structure a plan three ways:

1. Operation of an uninsured plan in the same manner as prior to the 1978 Act, accepting the fact that executives will be taxed on all or part of the reimbursements to them under the plan.
2. Broaden the coverage of an existing uninsured plan to include sufficient other employees so that tax benefits can be preserved under the new nondiscriminatory coverage.

3. Insure the plan, so that benefits may be provided on a tax-free basis for the executive group only.

In drafting a new medical or dental expense reimbursement plan, it is important to focus on what constitutes a "discriminatory" plan. A plan is discriminatory if it favors "highly compensated" employees as to *either* participation under the plan in the first instance, or benefits paid under the plan. An employee is highly compensated if he or she is:

1. One of the five highest paid officers in the company;
2. A shareholder owning directly or indirectly more than ten percent of the value of the corporation's stock;
3. One of the highest paid 25% of all employees of the company. Under this test, certain employees need not be considered, as discussed below. The determination for any year is based upon the amount of each employees' compensation for the previous year.

In order to satisfy the participation requirements under the 1978 Act, a plan must satisfy one of three tests, which by now ought to be somewhat familiar. Under the first test, participation requirements are satisfied if the plan covers either 70% or more of all employees. Secondly, the plan may cover at least 80% of all eligible employees, provided at least 70% of all employees are eligible for benefits under the plan. Under the third test, a plan can meet the requirements if an employee participation classification is established by the employer and found not to be discriminatory in favor of officers, stockholders, highly-paid or supervisory employees and the classification is approved by the Internal Revenue Service.

For the purposes of these tests, certain employees need not be considered. They include employees who have completed fewer than three years in service, who have not yet attained age 25, who are part-time or seasonal employees, who are covered by a collective bargaining agreement (provided that health benefits have been bargained in good faith with the union), who are non-resident aliens receiving no earned income from the employer from sources within the United States.

To satisfy the non-discrimination rules pertaining to benefits, a self-insured plan must provide benefits of the same type and amount to all participants. This test applies to benefits which are available under the plan and not to benefits actually paid. If a plan provided medical benefits for participants generally, but provided reimbursement for eye glasses and dental care expenses only for a selected group of highly paid employees, any amounts so reimbursed for dental care expenses incurred or expenses for eye glasses would be regarded as discriminatory. However, the plan will not be found to be discriminatory merely because plan benefits are off set by benefits paid under Medicare, Workmen's Compensation or other federal or state law. It is discriminatory for a plan to place a percentage of compensation ceiling on the total benefits payable to any employee. For example, if the plan provided that total reimbursement for any employee could not exceed ten percent of that employee's compensation, benefits would be regarded as discriminatory because the effect of the reimbursement is to limit most of the reimbursement to highly paid executive employees. It is permissable to place a flat dollar amount ceiling on the amount reimbursable to any employee.

Discriminatory Self-Insured Plans

Under discriminatory self-insured plans, highly compensated individuals will be taxed on all or part of amounts reimbursed. Of course, they should prefer this treatment rather than having to pay the full costs of such expenses out of their own funds. One advantage to the executive is that he or she will not have to pay any out-of-pocket costs of the time of a medical emergency. Instead, an expense will be incurred and the amount will be payable at the time the income tax return is filed on the reimbursement plans are not subject to withholding. Estimated tax payment requirements however must be considered.

An executive will have a financial advantage by paying less in taxes then he or she would have paid had the entire amount come from personal funds. The employer, on the other hand, will continue to be entitled to a deduction for reimbursement amounts, subject, of course, to reasonable compensation rules.

Non-Discriminatory Self-Insured Plans

The principal disadvantage of a non-discriminatory plan is the additional cost of covering non-executive personnel, plus the administrative burden of demonstrating compliance with the various non-discrimination rules. If an employer engages relatively few full time non-executive employees who are over the age of 25 with at least three years of service, these considerations could be outweighed by the value of preserving the tax benefits. It may also be valuable to preserve the tax benefits if there are a large number of non-executive employees, since many may be excluded under the collecting bargaining rule.

Insured Plans

In most instances, the only opportunity for obtaining the tax benefits of a plan whose participants are substantially or entirely the executive group is to insure the plan. A number of insurance companies offer executive medical expense reimbursement contracts. These plans reimburse a covered executive for most medical costs which qualify as deductible medical expenses under the Internal Revenue Code and are not otherwise reimbursed to the executive under the company's normal health plan. Where there are deductibles, co-insurance amounts or dollar maximums under the normal health plan, the excess plan reimburses these coverage gaps as well as expenses which may not be covered at all under the basic plan, such as dental or eye care. The excess contracts usually limit the maximum which would be reimbursed to any one executive in any year. Executive excess coverage is frequently available only when the company has a basic health plan including hospitalization and major medical in place with the same insurance company. It is difficult to find such insurance on a single-contract basis.

Drafting the Plan

The following checklist should be followed for preparation of a non-discriminatory self-insured plan:

1. An introduction to the plan should recite its purpose, effective date, and the manner in which the plan will be administered.

EXAMPLE 10.12A: Purpose, Effective Date and Administration

MEDICAL AND DENTAL EXPENSE REIMBURSEMENT PLAN (the "Plan") has been established by Sickness And Health, Inc. a corporation (the "employer"), to provide for reimbursement of certain medical and dental expenses incurred by its eligible employees and their dependents.

Effective Date

The "effective date" of the Plan is July 1, 1982. The Plan year shall be the calendar year

Administration of the Plan

The Plan will be administered by the employer. Any documents required to be filed with the employer will be given or filed properly if delivered or mailed by registered mail, postage prepaid, to the employer at 7561 South Cedar Street, Atlanta, Georgia.

2. The membership should be defined in accordance with the rules against discrimination.

EXAMPLE 10.12B: Membership

Each employee of the employer will become a member in the Plan on the effective date, or on the first day of any calendar month during which he meets all of the following eligibility requirements, if such employee is not employed or does not meet such requirements, on the effective date:

a) He is an employee who is customarily employed by the employer on a full time (customarily thirty (30) hours or more per week) and permanent (customarily nine (9) months or more per calendar year) basis;

b) He has attained the age of twenty-five (25); and

c) He has completed three (3) years of service with the employer. However, if the employer has not been in existence for such three (3) year period upon meeting the other requirements hereof, then the employee shall qualify if such employee has been employed for the lesser period in which the employer has been in existence.

3. The plan benefits should be stated, with a maximum dollar limit.

EXAMPLE 10.12C: Plan Benefits

Subject to the conditions and limitations of the Plan, each calendar year, beginning on the effective date, each member will be entitled to reimbursement from the employer of the medical care costs (as defined herein) incurred during that year with respect to his family unit (as defined herein) to the extent that such costs do not exceed an amount equal to the smallest of:

a) The total medical care costs of his family unit paid during that calendar year; or

b) $1,000.

4. A definition of "Medical Care Costs" should be included.

EXAMPLE 10.12D: Medical Care Costs

The term "medical care costs" as used in the Plan means amounts paid by a member (or any other individual included in his family unit) for:

a) Diagnosis, cure, mitigation, treatment or prevention of disease, or for the purpose of affecting any structure or function of the body;

b) Transportation primarily for and essential to medical care referred to in subparagraph (a) above;

c) Insurance covering medical care referred to in subparagraphs (a) and (b) above; and

d) Any other amounts paid which are included within the meaning of "medical care" as defined in Section 213(e) of the Internal Revenue Code of 1954, or any comparable provision of any future legislation that amends, supplements or supersedes that section.

5. The beneficiaries under the plan should be defined.

EXAMPLE 10.12D: Family Unit

The term "family unit" as applied to any member means the member, his spouse and such other persons as are dependents within the meaning of Section 152 of the Internal Revenue Code of 1954, or any comparable provision of any future legislation that amends, supplements or supersedes that section.

6. The manner of making payments under the plan should be described.

EXAMPLE 10.12E: Manner of Making Payments

By the end of each calendar year or within thirty (30) days thereafter, the employer shall reimburse each member for the portion of his family unit's medicalcare costs incurred during the year that is payable to him, provided that by the end of that year or within thirty (30) days thereafter, the employer receives evidence acceptable to it that such medical care costs have been paid by the member or any other individual included in his family unit.

7. Provisions should be recited to avoid duplication of payments for the same expenses.

EXAMPLE 10.12F: Non-Duplication of Benefits

A member shall not be reimbursed for medical care costs under this Plan to the extent that such costs are paid to or for the benefit of the member, or to or for the benefit of any other individual included in his family unit, under the provisions of any public plan of health insurance or under the provisions of any other plan or insurance policy, the costs or premiums of which are directly or indirectly paid, in whole or part, by the employer.

8. If the employer is to pay all of the benefits under the plan, that fact should be stated.

EXAMPLE 10.12G: Financing Plan Benefits

Subject to the provisions of [Amendment and Termination], the employer expects and intends to pay the entire cost of the benefits provided by this Plan. No member will be required or permitted to make contributions under the Plan.

9. Such items as the employer may require to document the charges should be described.

EXAMPLE 10.12H: Information to be Furnished by Members

Members must furnish to the employer such documents, evidence, data or information as the employer considers necessary or desirable for the purpose of administering the Plan or for the employer's protection. The benefits of the Plan for each member are on the condition that he furnish full, true and complete data, evidence or other information, and that he will promptly sign any documents related to the Plan requested by the employer.

10. Any implication that the plan provides a continuing right of employment should be eliminated.

EXAMPLE 10.12I: Employee Rights

The Plan does not constitute a contract of employment and participation in the Plan will not give any member the right to be retained in the employ of the employer, nor will participation in the Plan give any member any right or claim to any benefit under the Plan, unless such right or claim has specifically accrued under the terms of the Plan.

11. Someone (usually the employer) should be able to interpret the plan, with that person's decision being final.

EXAMPLE 10.12J: Employer's Decison Final

Any interpretation of the Plan and any decision on any matter within the discretion of the employer made by the employer in good faith is binding on all persons. A misstatement or other mistake of fact shall be corrected when it becomes known, and the employer shall make such adjustments on account thereof as it considers equitable and practicable.

12. Provisions for terminated participating employees should be included.

EXAMPLE 10.12K: Benefits of Terminated Participating Employees

If a member's employment by the employer is terminated by reason of his resignation or dismissal, then any amount payable to him under the Plan immediately prior to his termination shall be paid to him. If a member's participation in the Plan is terminated by reason of his death, then any amount that has become payable to him under the Plan as of the date of his death shall, in the discretion of the employer, be paid to either his spouse or his estate.

13. Amendment and termination provisions should describe the manner in which an amendment occurs, the causes of termination and provisions for notice of either.

EXAMPLE 10.12L: Amendment and Termination

General. While the employer expects to continue the Plan, it must necessarily reserve and does reserve the right to amend the Plan from time to time or to terminate the Plan.

Amendment. If the employer exercises its right to amend the Plan, any amount that has become payable under the Plan to any person prior to the date on which such amendment is adopted shall be paid by the employer in accordance with the terms of the Plan as in effect prior to that date.

Termination. The Plan will terminate on the first to occur of the following:

a) The date it is terminated by the employer;

b) The date the employer is judicially declared bankrupt or insolvent; or

c) The dissolution, merger, consolidation or reorganization of the employer, or the sale by the employer of all or substantially all of its assets, except that in any such event arrangements may be made whereby the Plan will be continued by any successor to the employer or by any purchaser of all or substantially all of the employer's assets, in which case the successor or purchaser will be substituted for the employer under the Plan.

If the Plan is terminated in accordance with subparagraph (a) or (c) above, any amount that has become payable under the Plan to any person prior to the date of termination shall be paid by the employer in accordance with the terms of the Plan as in effect prior to its termination.

Notice of Amendment or Termination. Members will be notified of an amendment or termination within a reasonable time.

NOTES

1. West's Modern Legal Forms § 8522.
2. West's Modern Legal Forms § 8573.
3. West's Modern Legal Forms § 8525.
4. West's Modern Legal Forms § 8524.
5. West's Modern Legal Forms § 8581.
6. West's Modern Legal Forms § 8526.
7. West's Modern Legal Forms § 8522.
8. West's Modern Legal Forms § 8523.
9. West's Modern Legal Forms § 8581.
10. West's Modern Legal Forms § 8573.
11. West's Modern Legal Forms § 8526.
12. West's Modern Legal Forms § 8537.
13. West's Modern Legal Forms § 8573.
14. West's Modern Legal Forms § 8522.
15. West's Modern Legal Forms § 8526.
16. West's Modern Legal Forms § 8527.
17. West's Modern Legal Forms § 8663.
18. West's Modern Legal Forms § 8664.
19. West's Modern Legal Forms § 8681.
20. West's Modern Legal Forms § 8682.
21. West's Modern Legal Forms § 8659.
22. West's Modern Legal Forms § 8658.
23. E.g. Cal.Bus. & Prof.Code §§ 16600–02; Mich.Comp. Laws Ann. §§ 445.761, 445.766.
24. See Colo.Rev.Stat. § 8-2-113; Fla.Stat.Ann. § 542.12 and S.D. Compiled Laws Ann. §§ 53-9-8.
25. West's Modern Legal Forms § 8652.
26. West's Modern Legal Forms § 8656.
27. West's Modern Legal Forms §§ 8657, 8658.
28. See Section 10.02(B).
29. See Section 10.08.
30. A form for an incentive compensation plan appears in West's Modern Legal Forms § 3132.

31. All of the examples for this section are from West's Modern Legal Forms § 3149.
32. Int.Rev.Code of 1954, 26 U.S.C.A. § 401.
33. Id.
34. Int.Rev.Code of 1954, 26 U.S.C.A. § 404.
35. Int.Rev.Code of 1954, 26 U.S.C.A. §§ 401, 501.
36. Int.Rev.Code of 1954, 26 U.S.C.A. § 402.
37. Id. and see Prentice Hall, Pension and Profit Sharing Plans, ¶¶ 5341, 5348.
38. Forms for pension and profit sharing plans may be found in West's Modern Legal Forms §§ 6666–6687.15.
39. Int.Rev.Code of 1954, 26 U.S.C.A. § 421(a).
40. Int.Rev.Code of 1954, 26 U.S.C.A. §§ 421, 422.
41. Int.Rev.Code of 1954, 26 U.S.C.A. § 56.
42. Int.Rev.Code of 1954, 26 U.S.C.A. §57(a)(6), (9).
43. Int.Rev.Code of 1954, 26 U.S.C.A. § 422.
44. This clause and succeeding clauses in this section are taken from West's Modern Legal Forms § 3122, unless otherwise stated.
45. Prentice Hall, Corporations, Forms ¶ 31,121.
46. The Model Act and most states require the affirmative vote of the holders of a majority of the voting stock to approve such action. See M.B.C.A. § 32 and Section 8.08.
47. Int.Rev.Code of 1954, 26 U.S.C.A. § 101(b).

11

AGREEMENTS REGARDING SHARE OWNERSHIP

§ 11.01 PURPOSES AND LEGAL SUPPORT

In a previous discussion of special corporate forms, the "close corporation" was studied.[1] Ownership of the close corporation is usually held by a clique of shareholders, who may also be the acting management of the corporation. The control exercised by the shareholders is a significant characteristic of this special corporate form and this exclusive control may be protected by various provisions and agreements which prevent the sale of stock to persons outside of the select shareholder group. Share transfer restrictions are designed to preserve present ownership interests and these restrictions may be drafted as a limitation upon the capital stock structure in the articles of corporation or may be the subject matter of a shareholder agreement to which all of the shareholders and the corporation are parties.

Restrictions on the transfer of shares are recognized as a viable and legal way to retain ownership control among a closed group of people, with one important limitation. Since corporate shares are personal property of the shareholder, and since the law will not enforce an agreement which completely nullifies property rights, an agreement restricting the transfer of shares may not completely and irrevocably prohibit the sale of the stock. However, an agreement may impose all sorts of restrictions which discourage sale to outsiders, and it may require that the shares must first be offered to the corporation or the other shareholders before they may be sold elsewhere. Although the selling shareholder must be able to sell his stock in the end, he may be required to satisfy a maze of precedent conditions to do so.

There are also shareholder agreements designed to concentrate voting power, with a group of shareholders pooling their votes under a contract which prescribes in advance how their shares will be voted on certain matters. Whatever control can be wielded by the concerted action of the total shares represented by the agreement will be applied to a vote on the specified corporate issues. The typical issues covered by shareholder voting agreements include the election of directors, amendments to the articles of incorporation, mergers, consolidation, dissolution and sales or other disposition of assets. Although these agreements are not unique to close corporations, they may be used in that setting to establish a predetermined voting position on such matters as salaries and dividends, especially where the shareholders are also employees of the corporation and are receiving salaries in lieu of dividends.

Shareholder voting agreements are commonly used to protect minority shareholders from abusive action by the majority shareholders. To digress for a moment, recall that there are other provisions, statutory and by agreement, which protect the minority shareholders from the dangers of oppressive majority control. Cumulative voting is specifically designed to assure minority shareholder representation on the board of directors (but an agreement among minority shareholders may still be necessary to effectively utilize cumulative voting).[2] The articles of incorporation may also require greater-than-majority voting requirements for shareholder action. For example, if the articles of incorporation exacted a 90% affirmative vote on all shareholder matters, an 11% minority shareholder could veto effectively any unwanted action. However, the majority shareholders may not appreciate this, and the incorporators must consider their potential dissatisfaction when the articles are drafted. After all, the majority shareholders do invest the majority of capital. Some state statutes require a greater-than-majority vote for major changes in the corporate structure, such as amendments to the articles of incorporation, merger, and so forth,[3] and the minority shareholders receive some protection against modifications to their ownership interests in the corporation. Finally, judicial decisions support a cause of action by the minority shareholders against majority shareholder for oppression of the minority interest, but all would agree that it is far more desirable to avoid that oppression through concentrated voting power rather than by subsequent litigation.

The Model Act grants statutory approval of shareholder voting agreements. Section 34 states that such agreements shall be valid and enforceable according to their terms. Common law has long recognized the ability of shareholders to predetermine their position by agreement on normal shareholder business, such as electing directors. A rule has developed, however, which frowns upon any shareholder agreement which impinges on the statutory rules or usurps the power of the board of directors. For example, shareholders could not agree that a quorum for shareholder meetings shall be one-fourth of the total voting shares, since the minimum allowable by statute is one-third in most jurisdictions.[4] Similarly, a shareholder agreement that a particular person will be continuously maintained as an officer of the corporation may be ineffective because the authority to select officers belong to the directors. Several states have adopted statutory rules which would permit the latter agreement, but further add that the shareholders must bear full responsibility for managerial acts governed by their agreement.[5]

Leaving aside agreements which encroach upon director discretion or regulate other managerial acts, there is no question that shareholders may agree regarding the manner which their shares will be voted on issues normally requiring shareholder action.

§ 11.02 CONCENTRATION OF VOTING POWER

In addition to the statutory and chartered rules which protect the minority interest (cumulative voting and greater-than-majority voting requirements), shareholders may pool their votes in an agreement executed among themselves, and thereby concentrate their aggregate voting power on each shareholder issue. There is a formal and an informal way to do this. The formal voting trust insures a more reliable concentration of power, since it prevents the possibility of a divided position from a shareholder who subsequently decides to act independently. The informal pooling agreement is much easier to operate and may exist for longer periods of time.

Voting Trust

The formal approach to the concentration of shareholder voting power is a device called the voting trust. Voting trusts are permitted under the Model Act and in most jurisdictions, and the statutory requirements must be strictly followed. The voting trust is a trust arrangement in every sense of the term. The shares represented by this agreement are placed in trust, out of the hands of the shareholders, and a designated voting trustee is directed to vote the shares represented by the trust in accordance with the terms of the agreement. The duration of a voting trust is limited to a period of 10 years under the Model Act.[6] A few states increase the term to 15 years[7] and a couple permit a 21 year period.[8] Extensions are permitted in several jurisdictions. Under the Model Act the trust becomes effective when the terms of the trust agreement are deposited with the corporation at its registered office. The shareholders who are party to a voting trust surrender their shares to the trust and the voting trustee becomes the "record owner" of those shares, thereby insuring that he is notified of every shareholder meeting and that he will have the legal right to vote the shares at such meetings. The shareholder is issued a voting trust certificate representing the shares of stock he once held. Voting trust certificates may be as transferable as the stock certificates themselves although the purchaser takes subject to the terms of the trust, which are stated on the certificate, along with the name of the trustee and other important matters respecting the agreement.

The specific statutory requirements of the voting trust under the Model Act are:

1. that a written trust agreement be prepared specifying the terms and conditions of the trust and conferring upon the trustee the right to vote the shares represented by the trust;

2. that the shares represented be transferred to the trustee in return for voting trust certificates;

3. that the term of the agreement shall not be more than the statutory period;

4. that the trustees keep a record of all beneficiaries (shareholders) with their name and address and the number of shares deposited with the trust; and

5. that a counterpart of the trust agreement and the record of beneficiaries be deposited with the corporation at its registered office.[9]

Several jurisdictions omit the requirement of the trustee's record of beneficiaries, but most of the other requirements are the same.

The voting trust agreement must strictly observe the statutory requirements, for a failure to do so may invalidate the trust. The agreement should specify its duration (within the statutory limit) and may provide for termination at any time by a prescribed vote of the beneficiaries. The designation of the trustees should state any qualifications the parties intend to impose, such as requiring the trustees to be shareholders and prohibiting any director of the corporation from acting as trustee. The description of authority should carefully detail the power of the trustees to vote the stock, naming specific issues if the trust is so limited, or granting total voting power of the stock to the trustees. The decisions of the trustees may be based upon their good judgment, or the agreement may require that they obtain the consensus of a certain percentage of the beneficiaries before casting the trust vote. The trustees should also be excused from liability for their actions, except for gross negligence, and should be indemnified for expenses and liabilities incurred in the exercise of their trust power. The procedure for transfer and issuance of voting trust certificates must also be detailed. Finally, ministerial duties of the trustee in receiving and paying dividends of the stock, filing documents with the corporation, and recording voting trust certificates may be specified. A sample Voting Trust Agreement with a Voting Trust Certificate is Form 11A in the Appendix.

Stock Pooling Agreement

Another agreement designed to concentrate shareholder voting power is a pooling agreement, which may accomplish the same purpose as the voting trust, but usually is not subject to statutory regulation. Several shareholders join together and pool their respective voting interests, predetermining the manner in which the shares will be voted by the agreement.

Stock pooling agreements may be quite informal and may vary depending upon the desires of the parties, since there is virtually no statutory regulation governing their formation and operation. Section 34 of the Model Act simply provides that agreements among shareholders regarding the voting of their shares are valid and enforceable according to their terms. In principle, then, a stock pooling agreement could last indefinitely, as contrasted with the voting trusts' limited term. The pooling agreement may also be a secret, if that is desirable, since no evidence of the agreement need be deposited with the corporation.

Stock pooling agreements suffer from one serious deficiency—their enforceability. Unlike voting trust agreements, shareholders do not deposit their shares under a stock pooling agreement. They merely agree to vote the shares, which they still control, in the same manner as the other parties to the agreement. Suppose Wagner, Naylor and Shaklee enter into a stock pooling agreement that they will all vote the same way on shareholder matters, and the manner in which all votes will be cast will be determined by a majority vote between them. Suppose further that on a given issue Wagner and Naylor want the votes cast one way, but Shaklee dissents. If Shaklee refuses to abide by the decision, he may still cast the votes he controls any way he wants. The other parties may be able to sue him for breach of the pooling agreement, but the most important objective, the concentrated voting power, was lost on that issue. This could not happen in a formal voting trust.

Another problem unique to the pooling agreement arises if one of the parties to the agreement decides to sell the stock. The concentrated voting power is lost if the purchaser is not obliged to abide by the agreement. A typical response to this problem is to impose a restriction on the transfer of stock held by the parties to the agreement requiring an offer to sell the stock to be directed to the other parties before the shares may be sold to a nonparty investor.

EXAMPLE 11.02A: Restriction on Transfer of Stock

Neither party will sell any shares of stock in the corporation to any other person whomsoever, without first making a written offer to the other party hereto of all of the shares proposed to be sold, for the same price and upon the same terms and conditions as in such proposed sale, and allowing such other party a time of not less than 180 days from the date of such written offer within which to accept same.

The stock pooling agreement requires joint action of the participants in exercising their voting rights. The joint action may be required on all matters submitted to the vote of the shareholders, or may be limited to certain issues where concentration of voting power is deemed to be important, such as amendments to the articles, election of directors and so forth. An example of a joint action clause covering all shareholders matters would look like this:

EXAMPLE 11.02B: Joint Action

In exercising any voting rights to which either party may be entitled by virtue of ownership of stock held by them in the corporation each party will consult and confer with the other and the parties will act jointly in exercising such voting rights in accordance with such agreement as they may reach with respect to any matter calling for the exercise of such voting rights.[10]

The determination of how the votes will be cast is made by agreement of the participants, and, if there are more than two parties to the agreement, a formula should be prescribed for this determination. For example, the agreement could require the unanimous vote of the shareholders represented by the agreement, but this may be an impossible criterion to satisfy. A provision allowing the determination of position to be made by a majority of the shares represented by the agreement would be more practicable. If a deadlock is possible under the agreement, as it might be if only two shareholders were parties, a provision regarding arbitration is appropriate.

EXAMPLE 11.02C: Arbitration

In the event the parties fail to agree with respect to any matter covered by the preceding paragraph the question in disagreement shall be submitted for arbitration to D. S. Charlton, of Montgomery, Alabama, as arbitrator and his decision thereon shall be binding upon the parties hereto. Such arbitration shall be exercised to the end of assuring good management for the corporation. The parties may at any time by written agreement designate any other individual to act as arbitrator in lieu of said D. S. Charlton.

Duration and termination provisions should be included, reflecting the desires of the parties as to such matters.

EXAMPLE 11.02D: Duration

This agreement shall be in effect from the date hereof and shall continue in effect for a period of ten years unless sooner terminated by mutual agreement in writing by the parties hereto.[11]

Agreements to Secure Director Representation

Since corporate management is vested in the board of directors, and the board of directors are elected by the shareholders, the board is usually elected by the majority shareholders, particularly when cumulative voting is not in effect. Even if cumulative voting is used in the election of directors, the minority shareholders must unite to secure representation of the board.[12] Consequently, the minority shareholders frequently are not represented on the board of directors. An assurance of minority representation on the board may be accomplished by agreement in two ways. Using the concentration of minority voting power, a shareholder pooling agreement or voting trust may predetermine the manner in which the parties to the agreement will vote at the election of the directors. The terms of this agreement would provide that all parties to the agreement (or the trustee) would vote for a person who, according to a majority of the persons represented by the agreement (or other formula determination), would best represent the interests of the parties. The second method would be an agreement among *all* shareholders that certain positions on the board of directors will be reserved to the nominee of the minority shareholders, and that all shareholders would vote to elect the minority nominee to the board at each election

EXAMPLE 11.02E: Agreement to Segregate the Board of Directors

That the number of the members of the Board of Directors of Trouble, Inc. be reduced from five, as it now is, to the number of four, that the number of

> members of said Board of Directors shall be maintained at four in number, of which at all times two thereof shall be such persons as shall be nominated or designated by the said parties of the first part and the other two thereof shall be such persons as shall be nominated or designated by the said party of the second part. And it is further mutually agreed between the parties that at all stockholders' meetings of the said Trouble, Inc. held for the purpose of election of directors or director (in case of vacancy of the Board of Directors), that all of the said shares of stock of parties of the first part and also of party of the second part and also any additional shares of stock of the Trouble, Inc. which may be subsequently acquired by the said parties or either of them, shall be voted in such manner and for such person or persons as will keep and maintain the Board of Directors four in number, of which two thereof shall be such persons as shall be nominated or designated by said parties of the first part and two thereof shall be such persons as shall be nominated or designated by the said party of the second part.[13]

Instead of the general description, "minority nominee," it is possible to name a certain person in the shareholder agreement and agree that he or she will be continually elected as a director to represent the specialized shareholder interests. However, these agreements have been subjected to careful scrutiny by the courts, and have been declared invalid if they fail to leave room for the defeat of an incompetent director. On the other hand, a shareholder agreement will be enforceable if it states that a named director will be maintained in office as long as he faithfully and conscientiously performs his duties. Therefore, it is good practice to include "a savings clause" making the election of a particular director obligatory only if he is competent to serve in that position.

§ 11.03 SHARE TRANSFER RESTRICTIONS AND BUY-OUT AGREEMENTS

Shareholder agreements are frequently concerned with restrictions on the transfer of shares. These restrictions are not necessarily directed to the protection of special shareholder stock pooling agreement as in Example 11.02A. Usually the restrictions on transfer are intended to protect all shareholders who desire to avoid alienation or interference with corporate control. Minority shareholders could suffer greatly if majority shareholders were allowed to sell their shares and the associated control of the corporation to an outsider who is not sympathetic to the minority interest. Similarly, majority shareholders may suffer from the sale of even one share to a recalcitrant, argumentative shareholder, particularly in a close corporation where all of the shareholders must work closely to further the enterprise.

Restrictions on transfer of shares may be imposed in the articles of incorporation, the by-laws, or in separate shareholder agreements. The corporation should also be a party to the restrictive agreement. The thrust of most restrictive agreements is to prohibit the transfer of shares to an outsider without first offering to sell them to the corporation or to the other shareholders.

Agreements affecting share ownership may also specify mandatory buy-out or sell-out arrangements, directed to the corporation and the shareholders, but the objectives here are opposite from those of the share transfer restrictions. While the latter are intended to discourage the transfer of shares, the former are designed to require it. The intra-corporate groups may prefer to prohibit share ownership by a person who is not actively engaged in the business and a mandatory sell-out agreement may satisfy this objective. On the other hand, a shareholder may wish to assure his ability to sell his shares, and this can be accomplished through a mandatory buy-out agreement. In small, closely-held corporations the shares may not be marketable, and without a mandatory buy-out agreement, a shareholder has no choice but to hold the shares indefinitely with no prospect of receiving the return of his invested capital until the corporation is dissolved.

Restrictive Agreements

It should be noted at the outset that the validity of a share-transfer restriction will depend, in part, upon the nature and structure of the business conducted. The restriction must be adopted for a lawful purpose and it may be necessary to show that there is a special need for a share transfer restriction in this particular type of business or in the particular relationship between shareholders. This should not be an onerous burden, but it should be considered when a share transfer restriction is adopted by the parties. There is ample reason to support the restriction in small, close corporations, but, in any case, the agreement should recite that its purpose is to further harmonious relations between the parties and to promote the best interests of the business.

The most effective way to insure that no outsider will be admitted to a select shareholder group is to completely and forever prohibit the sale of the stock to any person at any time. This complete prohibition on transferability, while certainly restrictive, is unenforceable and against public policy. Agreements which indefinitely prohibit the exercise of personal rights are presumed to be unfair and it is virtually impossible to muster any rational reason to support such a severe restriction.

The most common restriction used to control the transfer of shares grants the corporation or the other shareholders the first option to purchase shares before they may be sold to an outsider.

Events Triggering the Restriction. A stock transfer restriction is normally triggered when a selling-shareholder has received a valid and sincere offer to purchase his shares. If the agreement does not provide for some method of determining whether an offer is valid and sincere, the shareholder may manipulate the agreement to force the corporation or the other shareholders to either buy the stock when the shareholder gives the appropriate notice, or to run the risk that the stock will be free from any transfer restriction if the corporation or the other shareholders fail to purchase the shares.

Some definition of a good faith offer to purchase should be included in every restriction, and several possibilities exist for establishing the validity of a good faith offer.

1. The agreement should provide that the offer to purchase the shares from an outside investor should be in writing.

2. The agreement may provide that the good faith offer be supported by an earnest money deposit.

3. The agreement may provide that the good faith offer be supported by an escrow of the total purchase price, the terms of which would provide that in the event the corporation or the other shareholders fail to purchase their proportionate share of the selling-shareholder's stock, the escrow proceeds would be distributed automatically to the selling-shareholder and the stock would be deemed to have been purchased at that time.

4. The agreement may or may not provide for the disclosure of the identity of the good faith purchaser, since the corporation or the other shareholders may be able to manipulate the agreement by ascertaining the identity of the good faith purchaser and attempting to thwart the sale or sell other shares to this purchaser, thereby discouraging the purchase from the proposed selling-shareholder.

EXAMPLE 11.03A: Bona Fide Offer

> Upon receipt of a bona fide offer to purchase the shares by a person not a shareholder of the corporation, a selling shareholder must follow the restrictions contained in this article prior to selling any shares of stock to the offeror. A bona fide offer shall require that the offeror place an amount equal to the purchase price of the stock in escrow, the terms of which shall require the release of said funds for the purchase of the stock if the corporation and other shareholders do not exercise their options hereunder, and contain an agreement from the offeror to be bound by the terms of these restrictions upon purchase of such shares.

Other events may also be the subject of a stock transfer restriction which prevent the free transferability of shares under certain conditions. For example, the death of a shareholder may be an event which requires that the heirs or representatives of the deceased shareholder must offer the shares to the corporation or to the other shareholders. Similarly, retirement, disability, bankruptcy or loss of a professional or occupational license necessary to the business of the corporation may similarly trigger a stock transfer restriction. In these cases, however, it is more likely that the agreement would provide for a mandatory purchase or sale upon the happening of the event. These mandatory agreements are discussed in greater detail below.

Option to Purchase. The usual option to purchase share transfer restriction requires an offer to sell directed first to the corporation, which has a right of refusal, and then to the other shareholders, who also have the right of refusal to purchase the shares. If both the corporation and the shareholders decline to purchase the shares, then the selling shareholder may sell to the outsider. Alternatively, the restriction may run only to the corporation, or may by-pass the corporation and grant the option to purchase only to the other shareholders. When the other shareholders are granted the option to purchase, the shares will usually be offered to them in the same proportion as their present ownership interests in the corporation.

An example of a provision granting a right of refusal to the corporation and then to the remaining shareholders would be worded as follows:

EXAMPLE 11.03B: Option to Purchase Shares

Should any shareholder wish to dispose of his stock, it shall first be offered to the corporation at a price no greater than a bona fide offer by any third person, and in no event at a price greater than the par value and the proportionate share of any earned surplus, and said stock shall be available to the corporation for a period of thirty days. In the event that any of the said stock is not purchased by the corporation, it shall be offered to the remaining shareholders of the same class of stock in the same proportion as their respective stock interests in said class of stock, for a like price and for a similar period of time. In the event any of the remaining shareholders declines to purchase his proportionate share of said stock, that share shall be offered to the then remaining shareholders of the same class of stock for a like price and for a similar period of time. In the event that any of said stock is not purchased by the corporation or the shareholders, the remaining stock may then be sold by the shareholder at the price of the bona fide offer of the third person.

By way of illustration, suppose the Nobles Company has three shareholders: Dworet who owns 100 shares, Rezabeck who owns 200 shares and Weiler who owns 300 shares. If Dworet desired to sell the shares and received a bona fide offer for the purchase, she must first offer to sell them to the corporation at the price determined by the formula and the corporation shall have 30 days to reject or accept the offer. If the corporation rejects the offer, she must then offer the shares to Rezabeck and Weiler, who own the same class of stock in a 2:3 ratio, since Rezabeck owns 200 shares and Weiler owns 300 shares. They may then purchase in that ratio, meaning that Rezabeck could purchase 40 shares and Weiler could purchase 60 shares. If Rezabeck declined to purchase the shares to which she is entitled, they must then be offered to Weiler or vice versa, according to the sample provision. Only if the corporation and the other two shareholders decline to purchase the shares may they be sold to the outsider.

Any variations in this scheme are permissable.

Considerations in Designating the Option to Purchase. There are a couple of practical observations to consider in drafting the restriction. The first offer to the corporation is desirable because a profitable corporation will probably have funds legally available to purchase the shares. Moreover, it is more convenient to make the offer to the corporation through a single notice than to notify all of the other shareholders. The remaining shareholders may also prefer that the purchase funds come from internal corporate operations rather than attempting to raise funds individually. However, there may be situations when the corporation is not permitted to buy its own shares because of the statutory restrictions on the funds which must be used for that purpose. If funds are not legally available, the corporation must refuse the offer, and, to preserve the viability of the restriction, the offer should then run to the individual shareholders.

If the corporation is going to exercise an option to purchase or redeem its shares, it must do so in strict compliance with statutory restrictions on funds available for repurchase. Provided the corporation is solvent (and the purchase of shares will not render it insolvent) the corporation may purchase its own shares to the extent of its unreserved and unrestricted surplus.[14] It usually may redeem shares (which have been issued as redeemable shares) so long as the redemption will not reduce its net assets below the aggregate amount payable to shareholders with liquidation rights upon involuntary liquidation and dissolution of the corporation.[15] The surplus requirements under the applicable statute have nothing to do with the availability of cash to effect a repurchase, except insofar as the corporation must be able to pay its debts and liabilities as they fall due after distributing portion of its surplus. The corporation may have sufficient cash for the stock purchase, but insufficient surplus, or there may be enough surplus on its books, but no ready cash.

The corporation may use life insurance as a method of funding a buy-out agreement, especially when the event triggering the buy-out is death. Life insurance funding may also be used when the triggering event is retirement, disability, or termination of employment. These funding options will be discussed in greater detail later.

The remaining shareholders may find raising funds to buy-out a withdrawing shareholder a difficult task, particularly if they must come up with the cash immediately on a shareholder's death or some other triggering event. The problem may be alleviated if the agreement provides for installment payments. The agreement should permit those who elect to purchase more than their prorata share to divide equitably among themselves any shares not taken on the first division. If some shares still remain, the agreement should provide whether the original number or only the untaken shares may be sold to an outsider. Another question is whether the selling shareholder may break his or her stock holdings into small blocks and offer them to several people, thus obtaining a higher price per share. If the parties wish to eliminate these possibilities, the agreement should specifically prohibit such actions.

All or Nothing Purchase. A choice must be made at the drafting stage between a restriction permitting a partial purchase of the offered stock or one requiring a purchase of all-or-nothing. Presumably Dworet has received an offer from an outsider to buy her 100 shares at a price. The question now is whether the corporation or the other shareholders may purchase only a portion of her 100 shares in exercising the right of first refusal, or whether they must buy the entire block to exercise the option. Dworet is better protected under the latter approach, since the outsider may not be interested in a purchase of only a portion of the 100 shares. However, the corporation and remaining shareholders have a greater guarantee against stock transfers if they are permitted to exercise their option in part, since they may then buy just enough of the stock to discourage the outsider, but they are not compelled to purchase all offered shares to prevent transfer. In the spirit of fairness, it is better practice to require the corporation to exercise its option in full or not at all. When the corporation refuses purchase, and the offer is made to remaining shareholders, the restriction should further specify a procedure to prevent portion purchases for the protection of the selling shareholder. When more than one shareholder is entitled to an option to purchase, it is possible that some shareholders will exercise their option and others will not. In that case a portion of the offered shares is available for sale to the outsider, but his interest in the purchase may dwindle after the number of shares has been reduced. To avoid this result, the restriction could provide that if all of the options are not exercised none of them may be, and this will preserve the block of stock intact. Alternatively, if the restrictions permits some shareholders to refuse the option without impairing the rights of the other shareholders to exercise their option, it should further specify that those shares which have been refused will be offered to the shareholders who intend to exercise their option. The latter group has evidenced an interest in purchasing their quota of the offered stock; perhaps they will also purchase the remaining shares. This second chance for internal sale, which also furthers the objectives of the remaining shareholders by granting another opportunity to avoid alienation of the shares, appears in Example 11.03B. It also should be observed that if not all shares of the selling shareholder are purchased under the agreement, the amount paid by the corporation to the selling shareholder may be taxed as a dividend.[16]

In sum, the most fair and effective share transfer restriction will first grant the corporation the right to purchase all of the offered shares at a specified price. If the corporation refuses to buy all of the shares, then the other shareholders will have a right to purchase the offered shares according to their respective proportionate stock interests at a specified price. The restriction should then provide that if some shareholders do not exercise their option then none of them may; or that the remaining unpurchased shares must be offered to those shareholders who have exercised their option before they may be sold to an outsider.

Mandatory Buy-Out or Sell-Out Provisions

The foregoing restrictions on transfer of shares are designed to avoid alienation of shares and the potential loss of control of the corporation. This is accomplished by granting the corporation or other shareholders the right of first refusal when shares are offered for sale. However, the shareholder may not be able to sell his shares to anybody (especially if he has invested in a small, closely-held corporation) and now he deserves some protection. His shareholder agreement may *require* the corporation or the other shareholders to purchase the shares under certain circumstances. Conversely, the corporation may demand the right to purchase certain shares, such as when a shareholder-employee retires or quits. Thus, depending on the purpose to be served, the mandatory buy or sell provisions may work both ways: the contract may require the corporation or other shareholders to purchase the shares, or it may require the shareholder to sell the shares.

Mandatory buy-out agreements reflect an *obligation* on the part of the corporation and other shareholders to purchase the selling shareholder's stock, as distinguished from their *option* to buy under the share transfer restriction. These mandatory provisions are designed to guarantee a market for the stock, which may be needed particularly by a minority shareholder. He cannot sell if no willing purchasers are available, and the minority shareholder is powerless to force dissolution to recoup his capital if the majority shareholders resist. The majority shareholders may also need the guarantee. Willing purchasers may be even harder to find for a larger block of stock, and while the majority shareholders could force dissolution, that may be an unwise business decision and may constitute oppression of the minority shareholders.

On the other hand, the mandatory buy-out agreement may also impose the *obligation* on the part of the shareholder or his representatives upon the occurrence of a triggering event to sell the shares to the corporation or to the other shareholders. The corporation and the other shareholders may thereby protect against the ownership of shares by persons who are strangers to the enterprise, such as the heirs of a deceased shareholder, a trustee in bankruptcy, or the representatives of a disabled shareholder.

Events Commonly Triggering Buy-Outs. A mandatory buy-out agreement is usually conditioned upon the death of a shareholder, the retirement of a shareholder at a certain age or after specified length of service to the corporation, the disability of a shareholder, the bankruptcy of a shareholder, the loss of a shareholder's occupational license, or any attempt by a shareholder to force a dissolution of the corporation under statutory dissolution sections.[17]

The event triggering the buy-out should be clearly defined, and, in many cases where life insurance or other insurance is used to fund the buy-out agreement, the definition of the contingent event should contain the same terms as the definition of that event under the insurance policies which are expected to fund the purchase. For example, if the disability of a shareholder will trigger a buy-out, the agreement should define "disability" in the same method as the insurance policy which defines the term, or, refer to the insurance policy definition in the agreement. Further, after a specified period of time of continuous disability, again as defined in the insurance policy, a buy-out will occur with the proceeds of the insurance policy.

EXAMPLE 11.03C: Purchase on Death

The Company will have the option, for a period commencing with the death of any shareholder and ending 60 days following the qualification of his or her executor or administrator, to purchase all of the shares owned by the decedent, at the price and terms provided in this agreement. The option shall be exercised by giving notice to the decedent's estate or other successor in interest in accordance with this agreement. If the option is not exercised within such 60-day period as to all shares owned by the decedent, the surviving shareholders shall have the option, for a period of 30 days commencing with the end of that 60-day period to purchase all of the shares owned by the decedent, at the price on the terms provided in this agreement. The option shall be exercised by giving notice, in accordance with this agreement, to the executor or administrator, stating the number of shares as to which it is exercised. If notice of exercise from the surviving shareholders specify in the aggregate more shares than are available for purchase by the shareholders, each shareholder shall have priority, up to the number of shares specified in his or her notice, to such proportion of those available shares as the number of Company shares he or she holds bears to the number of the Company shares held by all shareholders electing to purchase. The shares not purchased on such a priority basis shall be allocated in one or more successive allocations to those shareholders electing to purchase more than the number of shares to which they have a priority right, up to the number of shares specified in their respective notices, in the proportion that the number of shares held by each of them bears to the number of shares held by all of them. In the event this option is not exercised as to all of the shares owned by the decedent, his or her estate will hold those shares subject to the provisions of this agreement.

EXAMPLE 11.03D: Purchase on Other Events

In the event any shareholder is adjudicated a bankrupt (voluntary or involuntary), or makes an assignment for the benefit of his or her creditors, or is physically or mentally incapacitated for more than three months, the event of incapacity as described in an insurance policy now owned by the corporation with the Equitable Life Insurance Corporation, Policy No. 40-82-123, the Company and the remaining shareholders shall have the option for a period of 90 days following notice of any such event to purchase all of the shares owned by the shareholder. Notice shall be given to the shareholder or his or her representative in accordance with this agreement. The option shall be exerciseable first by the Company and thereafter by the remaining shareholders, and the price, terms of purchase, and methods of exercise of the option shall be the same as are provided in this agreement to apply in the event of death. In the event this option is not exercised as to all of the shares owned by the shareholder, he or she or his or her successor in interest will own the shares subject to the provisions of this agreement.

The compelling reasons for these types of agreements are easy to appreciate. The shares of a small corporation are not generally marketable and the beneficiaries or legatees of a deceased shareholder or the representatives of a disabled shareholder are not usually interested in holding shares in the business which he enjoyed during his lifetime or while he was productive in the business. Moreover, as an employee, the shareholder was probably receiving a salary instead of dividends, and dividends are rare in a small corporation anyway. Consequently, if the stock is not readily marketable, the salary is terminated, and there are no dividends or other emoluments of share ownership, the beneficiaries or representatives of the shareholder will receive nothing from the share ownership, unless the corporation and the other shareholders are required to purchase the shares.

The mandatory sell-out agreements are frequently adjunct to employment contracts, the terms of which contemplate issuance of shares as an incentive to performance. If the employee was misjudged and is subsequently terminated, the corporation has the right to buy back the shares. The agreement may also be separately executed to prevent continued stock ownership by any person for whatever reason. The shareholder is required to sell the shares to the corporation or to the other shareholders if they insist on the sale, with appropriate notice. The price is usually established in the agreement, as discussed in detail later, and provisions for surrender of shares should be included.

EXAMPLE 11.03E: Common Stock of One Leaving Employ of Corporation May Be Purchased

In the event that any holder of the common stock of this corporation who may now or hereafter be an officer or employee of this corporation ceases, for any reason, to be such officer or employee, and provided further that the Board of Directors shall require it, by resolution passed at a special meeting called for that express purpose on not less than 2 days notice, the corporation or any officer or any common stockholder subscribing to this agreement shall have the option, within 30 days after such person shall cease to be an officer or employee, to purchase all of the common stock held by such person ceasing to be an officer or employee, at a price to be determined by the same method as hereinabove provided, and the tender of the amount of such purchase price shall operate to transfer and vest said shares of common stock in the corporation or officer or stockholder making such tender, and the common stockholder who has thus ceased to be an officer or employee shall, upon such payment or tender, transfer, assign and set over his common stock to the officer or common stockholder exercising such option.[18]

Considerations in Designating the Mandatory Purchase Requirement. The agreement should bind the corporation and the other shareholders to the purchase. The corporation's ability to purchase shares, even though required by the agreement, is limited by most state statutes and the corporation may not have funds legally available for the purchase. In that event, the agreement should obligate the other shareholders to purchase their proportionate share of the stock, or to vote in favor of dissolution of the corporation. The latter provision would anticipate the possibility that the individual shareholders also cannot afford the purchase.

Of course, if the other shareholders are to buy the shares under the agreement, a procedure should be specified to proportion the shares they are permitted to purchase among them in the same ratio as their existing share ownership. See Example 11.03C above.

Mechanics of the Agreement

Notice Procedure. The agreement should always establish a notice procedure to advise the corporation and the other shareholders of the intended sale (in the case of the stock transfer restriction) or to advise the persons holding the shares of a shareholder subject to a mandatory buy-out agreement that the option to purchase is being exercised. This notice procedure should further state a time period for a decision to purchase or refuse and for surrender of the shares.

EXAMPLE 11.03F: Notice of Sale

The shareholder shall notify the directors of his desire to sell or transfer by notice in writing, which notice shall contain the price at which he is willing to sell. The directors shall within thirty days thereafter either accept or reject the offer by notice to him in writing. After the acceptance of the offer, the directors shall have thirty days within which to purchase the same at such valuation, but if at the expiration of thirty days, the corporation shall not have exercised the right to so purchase, the owner of the stock shall be at liberty to dispose of the same in any manner he may see fit.[19]

EXAMPLE 11.03G: Notice to Purchase

The Company shall have the option, for a period commencing with the death of any shareholder and ending 30 days following the death of the shareholder,

> to purchase any part of the shares owned by the decedent, at the price and on the terms provided in this agreement. The option shall be exercised by giving notice of it to the decedent's estate or other successor in interest in writing. Such notice shall be deemed to have duly given on the date of service if served personally on the party to whom notice is to be given, or within 72 hours after mailing if mailed to the party to whom notice is to be given by first class mail, registered or certified, postage prepaid and properly addressed to the party of his address set forth on the signature page of this agreement, or any other address that that party may designate by written notice to the other parties of this agreement.

Price Provisions. Any shareholder agreement involving the purchase and sale of stock must specify the price and payment terms applicable to the transaction. Restrictions on share transfer which require the offer of shares to the corporation and/or the remaining shareholders must establish a price for the offer. Similarly, mandatory buy-out or sell-out agreements must specify the price to be paid. There are competing interests here which frequently arise in the negotiation of the price term. The shareholder would usually prefer to receive the highest price in cash as soon as possible, while the purchaser would usually prefer to pay the lowest price over the longest period of time. Many practical considerations also arise. For one thing, extended payment provisions always involve some risk for the selling-shareholder, since the purchaser may become insolvent, may be unable to pay for another reason, or may simply refuse to pay. However, immediate payment in cash may be unrealistic depending upon the number of shares involved and the cash position of the purchaser.

It is important to recognize that the price provisions of the buy-out or restrictive agreement may accomplish several objectives. Depending upon the clients' desires, the following objectives should be considered:

1. What price will estimate most accurately the value of the stock in case of a buy-out or sell-out?
2. What price term will reflect most permanently the formula necessary to value the stock accurately?
3. What assets will be used to fund the stock purchase, and, if insurance proceeds are anticipated, will the price vary from the availability of the proceeds, and if so, in what manner?
4. Is there a desire to use the price provision as a further restriction upon the transfer of the shares?
5. Will the price provision be so unrealistic that the entire agreement will be unenforceable?

To be fair, the price provision of the stock purchase agreement should attempt to reflect accurately the true value of the stock at the time of purchase. Even if the parties intend to use the price provision as an additional restriction on the stock, so that no shareholder will be motivated to attempt to sell the stock because the price at which it must be sold under the stock transfer restriction would be prohibitive, it should be noted that courts are reluctant to enforce a stock transfer restriction which contains an unrealistically low price for the transfer of the stock. The effect of such price provisions are to prohibit the sale of the stock, since no shareholder would attempt to locate a buyer if it were necessary to sell the stock to other shareholders or to the corporation at an unrealistically low price.

There are several ways to prescribe the price and method of payment for stock purchase agreements. These terms are always subject to negotiation by the parties. It should also be noted that counsel who represents the corporation may not be able to represent ethically the interests of the individual shareholders who are intended to be parties to the agreement.

Firm Price. The agreement may establish a firm price to be paid for shares, such as $50.00 a share, which will be applied to any purchase of stock for the duration of the agreement. This practice should be discouraged except for extremely short-term agreements or at the early stages of the corporation (when the price is difficult to establish from other sources). It is probable that the stated price will become unrealistic one way or the other over an extended period of time.

EXAMPLE 11.03H: Price

The price at which the shares are to be offered to the corporation or to the remaining shareholders shall be $2.00 per share for the first year of this agreement.

Adjusted Stated Value. The agreement may provide for a stated value with a procedure for periodic adjustment. This method allows modifications in price to account for changed circumstances over an extended period of time. Usually a stated price will be coupled with a further agreement that the shareholders of the corporation will evaluate and reset the stated value on a periodic basis. Of course, as in any situation where people are negotiating for price, it is possible that the shareholders will not be able to agree on the adjustment. Therefore, the provision should include a certain formula to compute the adjustment to stated value in case the parties do not agree upon an adjustment to the price. For example, in the absence of shareholder agreement, the stated value may be increased or decreased by a percentage of the net income or loss, or a reevaluation of the assets. Alternatively, arbitration may be used to resolve issues which prevent an agreement on price of the shares.

EXAMPLE 11.03I: Agreed Price With Arbitration

The purchase price to be paid for each of the shares subject to this agreement shall be equal to the agreed value of the Company divided by the total number of shares outstanding as of the date of the price to be determined. The initial agreed value of the Company is $185,000.00, and on January 20 of each year hereafter, the parties to this agreement shall review the Company's financial condition as of the end of the preceding fiscal year and shall determine by mutual agreement the Company's fair market value, which, if agreed upon, shall be the Company's value until a different value is agreed on or otherwise established under the provisions of this agreement. If the parties are able to reach mutual agreement, they shall evidence it by placing their written and executed agreement in the minute book of the Company.

> If no valuation has been agreed upon within two years before the date of the event requiring determination of value, the value of a selling shareholder's interest shall be agreed upon by the selling shareholder or his successor in interest and the remaining shareholders. If they do not mutually agree on a value within 60 days after the date of the event requiring the determination, the value of the selling shareholder's interest shall be determined by arbitration as follows: The remaining shareholders and the selling shareholder or his successor in interest shall each name an arbitrator. If the two arbitrators cannot agree on a value, they shall appoint a third, and the decision of the majority shall be binding on all parties. Arbitration shall be in accordance with the rules of the American Arbitration Association as such rules shall be in effect at the time of arbitration.

Earnings Multiple Formula. A preferred method of determining the price of the stock is to specify a formula which will account for the success of the business, the value of the assets, and the desirability of the stock if a market existed for its sale. A common formula for evaluation of stock is an "multiple of earnings" formula.

An earnings-multiple formula establishes the price of corporate stock by multiplying the earnings of the corporation by a stated figure, which is set when the agreement is negotiated. The multiplier may fluctuate in the agreement, depending upon the number of years the shares have been held, or upon the number of shares held. For example, a multiplier of 3 times earnings may apply to blocks of 100 shares or less, held less than 2 years; 4 times earnings for blocks of 100–200 shares held less than 2 years, and so forth. The definition of "earnings" deserves attention in the agreement, which should specify whether "earnings" refers to net earnings before or after tax, and whether the "earnings" will be determined by an average of several years earnings or a current income figure. The earnings multiple formula probably has no relation to the actual market value of the stock, if one exists, or to the book value of the stock. It simply insures that if the corporation has increased its earnings during the agreed period the shareholder who desires to sell his stock will realize some benefit from that increase. On the other hand, if earnings have decreased the shareholder will have suffered by waiting to sell his shares. In the event the corporation loses money and the earnings multiple produces a negative figure, the agreement may contain a "savings" provision that in no event will the stock be valued at any less than a stated amount. This insures that the stock will always have some minimum value, and is obviously a desirable provision from the shareholders' standpoint.

The provisions which establish the earnings-multiple formula should specify a person who will determine earnings conclusively from a specified source. The company's accountant, who has prepared the financial statements under "generally accepted accounting principles" perhaps should be specified as the person who will determine and compute earnings as of the date of the buy-out agreement. Moreover, it may be necessary in small, closely held corporations, to specify certain adjustments to earnings which will more accurately reflect the true earnings of the corporation for the period averaged. In closely-held corporations, it is common to pay salaries to shareholder-employees, which may be higher than salaries normally paid for similar employees. It is also common to provide for greater than normal lease payments for shareholder-owned equipment, and greater than normal interest payments for shareholder loans. These expense figures should be addressed in the earnings formula computation, to more accurately reflect the true value of the stock based upon earnings.

The determination of an earnings multiple is frequently a negotiated matter, and ultimately depends upon a reasonable rate of return in the industry. Rates of return for certain industries are published from time to time by economic marketing sources and business brokers, and they may be used as a guide in determining the appropriate rate of return to set the multiple for earnings. The shareholders themselves are frequently capable of estimating a reasonable rate of return which, if agreed to, may be used as the multiple in the earnings formula.

EXAMPLE 11.03J: Capitalized Earnings Formula

The purchase price to be paid for each of the shares subject to this agreement shall be determined as follows:

The net profits of the Company for each of the three complete fiscal years preceding the date of determination of price for purposes of this agreement shall be adjusted by deducting from the Company's profits state and federal income taxes, lease payments to shareholders, salary payments to shareholders, and interest payments on loans from shareholders. The net profit figures for the three years, thus adjusted, shall be added, and the total shall be divided by three. The average adjusted net profit figure so obtained shall be multiplied by 10, and the result shall be divided by the number of shares of the Company's capital stock then outstanding.

Book Value Formula. A book value formula may be a more accurate estimate of the actual value of the shares, depending upon the definition of book value and the nature of the business. Usually book value is determined by dividing the net assets (total assets less total liabilities) by the number of the outstanding shares of the corporation. This means that each is entitled to a proportionate share of the assets and the purchase price of the shares will be an amount equal to this proportionate interest. In this case, of course, the purchase price will increase as net assets increase and vice versa. The accuracy of the formula is affected by the nature of the business. A highly profitable organization may operate with few assets, in which case the book value will be considerably lower than the fair "market" value of the shares. Conversely, it may be possible for a heavy asset business to have a high book value for shares that are virtually worthless.

The book value should be determined by a specified person, and the clause containing the book value formula should provide some guidance for the computation of book value. Again, if an independent accountant is used to prepare the corporate balance sheets in accordance with "generally accepted accounting principles" that person may make the sole determination of the book value at a specified time. In addition, if certain assets are undervalued on the balance sheet, such as real estate which may have appreciated considerably from the time it was purchased, or if liabilities are overstated, such as contingent liabilities which are not likely to be realized by the corporation, the book value formula should direct the accountant to adjust those figures to reflect more accurately the true value or liability.

To minimize the cost of a determination of book value at any point in time, it is advisable to provide that book value will be determined as of the date of the last financial statement prepared prior to the occurance of a contingent event. In this regard, the financial statements will be prepared on a regular basis and it is much less expensive for the corporation to be able to use a regular financial statement rather than attempting to prepare a new balance sheet only for the purpose of estimating value of the stock for the contingent event which has triggered a buy-out. It may, of course, be provided that the values contained on the last financial statement should be adjusted, upward or downward, to reflect material changes in current operations.

Since book value includes all of the assets of the corporation, care must be taken to be certain not to include the proceeds of insurance which may be payable to the corporation and which were intended to be used to purchase the shares under the shareholder agreement. For example, if the corporation is expecting to use proceeds of life insurance to fund a buy-out in case of a shareholder's death, the proceeds should be specifically excluded from a determination of book value, since the corporation will be entitled to them upon the death of the shareholder, and they will appropriately increase the corporation's net assets.

EXAMPLE 11.03K: Book Value

The purchase price to be paid for the shares subject to this agreement shall be their book value determined as of the most recent financial statement prepared by the Company's accountants with additions or subtractions for current operations up to the end of the month preceding the month in which the event requiring determination of the purchase price occurs. Book value shall be determined from the books of the Company according to generally accepted principles of cash accounting applied in a consistent manner by the accountants of the Company who customarily prepare the Company's financial statements. The Company's book value shall be equal to its assets, excluding any proceeds of insurance policies, less its assets, excluding any proceeds of insurance policies, less its liabilities, and the amount thus determined shall be divided by all shares of the Company's capital stock then outstanding.

Combination of Formulas. Since "earnings multiple" and "book value" formulas rarely accurately reflect the true "market value" of the stock, it may be appropriate to combine these formulas together with others, to attempt to accurately estimate the true value of the stock at the time of the purchase.

EXAMPLE 11.03L: Formula to Determine Value (Book Value with Earnings)

If any holder of any shares of the common stock of this Corporation desires to dispose of the same or any part thereof, he shall not transfer or otherwise dispose of the same to any person unless and until he has first complied with the provisions hereof and given the other common stockholders of the Corporation who are entitled to the benefits of this contract an opportunity to purchase the same, as herein provided. The common stockholder desiring to dispose of all or any of his stock shall give written notice of such desire to each of the officers of this Corporation within the State of Montana, stating the number of shares he desires to sell. Any officer or any other common stockholder of the Corporation entitled to the benefits of this contract may, within thirty days after the service of such notice upon the last officer to be served, elect to purchase any part or all the common stock so offered, and, in the event of the exercise of such option, the common stockholder so giving such notice of his desire to sell shall forthwith sell, assign, transfer and set over his said shares of common stock to the officer or common stockholder electing to purchase the same, and the officer or common stockholder to whom the shares are so transferred shall, at the same time, pay to the seller, as and for the purchase price thereof, the amount of the book value of said common stock as shown upon the last annual statement of the Corporation, and in addition thereto an amount equal to the stock's pro rata proportion of the net profits of the business of the Corporation for such fractional part of the fiscal year as he elapsed since the date as of which the last annual statement was made, less any dividends declared during said fractional period.

For the purpose of determining said profits, the amount of the average annual net profits of the Corporation for the two fiscal years preceding the last annual statement shall be assumed to be the amount of the net profits which the Corporation shall earn during the current fiscal year, and the amount of the net profits of the Corporation for the fractional period of the year since the last annual statement shall be considered as that proportion of the average annual net profits of said two preceding years as the length of time which has lapsed since the last annual statement bears to the period of a full year. For the purposes of this contract, until the first annual statement of the Corporation is made, the book value shall be determined on the figures at which this Corporation has purchased the business and property of Everready Associates, a copartnership, and until this Corporation has completed two fiscal years which may be used as a basis for determining the average annual net profits, as aforesaid, the net earnings of the Corporation, for the purpose of this contract, shall be determined from the average net earnings during the preceding two fiscal years of the operation of said business either by this Corporation or by the copartnership from which its business was acquired, and, for that purpose, reference shall be had to the books of said copartnership for a sufficient period prior to the organization of this Corporation to produce a two year average. If it shall be necessary to use the net profits of the copartnership as a basis, proper adjustment and allowance shall be made for the fact that no salaries were paid by said corpartnership, and that part of the capital of the Corporation is preferred stock. For the purposes of this contract, the annual statements of the Corporation shall be made up on the same plan and method as has heretofore been followed by said copartnership.[20]

Several other formulas may be used for determining the value of the shares, but they are mostly a product of the imagination of the drafter. Any formula which will establish a value for the shares and will serve the purposes of the agreement may be used. It should be noted here that the choice of the formula may significantly assist the effect of share transfer restrictions. If a share transfer restriction requires that all shares must be offered to the corporation at book value, and the book value considerably understates market price, the shareholder will be less likely to attempt to sell his shares because he risks having to accept the book value price in any case. This does not run afoul of the rule that a complete restriction on sale is prohibited since the shareholder does have the right to sell the stock. However, as a practical matter he is not likely to do so.

Matching a Bona Fide Offer. If a shareholder intends to sell his stock and is subject to a share transfer restriction, the agreement may specify, in lieu of an agreed value, adjustable agreed value, or formula evaluation of price, that the corporation will be obliged to pay a price equal to that offered to the shareholder by the outsider investor. The selling-shareholder benefits from this price determination since he will receive exactly the same price from the corporation as he would have received from the outsider. Of course, a matching price provision should only be used in a case where the agreement also requires, as is usually the case, that the shareholder must have received a good faith offer from an outsider and the offer is definite and provable to the corporation's satisfaction.

If one of the objectives of the agreement is to discourage the transfer of shares to outsiders, the matching price provision is not the best alternative, because it insures that the shareholder will receive the same consideration no matter who purchases the shares. Transfer is best discouraged by a clause giving the corporation or other shareholders the option to match the price, or to purchase at some other price, stated or determined by formula, *whichever is lower.*

EXAMPLE 11.03M: Matching Offer

The price at which the shares are to be offered to the corporation or to the remaining shareholders shall be equal to the bona fide offer received from the offeror, or equal to book value as determined by the provisions of this agreement, whichever is lower.

Appraisal. An appraisal at the time of purchase may provide the most accurate but also the most expensive determination of value. The person who is to make the appraisal should be named in the agreement, or a procedure for naming appraisers should be described. For example, the agreement could name a mutually agreeable appraiser, or could provide for the selection of a panel of appraisers who will determine the value of the stock.

The clause providing for an appraiser should also specify the appraiser's qualifications, which should indicate some familiarity with the particular industry in which the corporation conducts its business. The clause should provide for the method of payment of the appraiser's expenses, and it is fair to provide that these expenses will be paid by the corporation (thereby absorbing the cost of the appraisal among all shareholders). In the latter case, care should be taken to consider that when the corporation is paying the appraiser it may impair his or her independence in determining the actual market value of the stock.

The method of appraisal should be specified, since business appraisers may use various methods to determine the value of the business. A liquidation value may be unrealistic, since it would only include the value of the assets less the payment of the liabilities, if the assets were immediately sold for a price. A preferable method would be an appraisal based upon "going concern" value which should include consideration of good will, business reputation, the expected useful life of the assets and the liquidity of the company (its cash and current asset position projected over a period of time considering potential expenses and liabilities). Most business appraisers will apply a discount to the value of minority shares, since minority shareholders are rarely able to affect corporate policies. This potential discount for minority interests should be considered in the agreement, and if it is not desirable to discount shares simply because they represent a minority position, the discount should be excluded in the instructions to the appraiser.

EXAMPLE 11.03N: Appraisal

The purchase price to be paid for each of the shares subject to this agreement shall be determined by appraisal. Within ten days after the occurrence of the event requiring the determination of the purchase price under this agreement, the Company shall cause Levine & Company, independent appraisers, to appraise the Company and determine its value. The appraisal fee shall be paid by the Company. In making the appraisal, the appraisers shall value real estate and improvements at fair market value; machinery and equipment shall be valued at replacement costs or fair market value, whichever is lower; finished inventory shall be valued at cost or market, whichever is lower; goods in process shall be valued at cost, using cost accounting procedures customarily employed by the Company in preparing financial statements; receivables shall be valued at their face amount, less an allowance for uncollectable receivables that is reasonable in view of the past experience of the Company and the recent review of their collectability; all liabilities shall be deducted at their face value, and a reserve for contingent liabilities shall be established, if appropriate in the sole discretion of the appraiser. The value of other comparable companies, if known, shall also be considered. The value determined by appraisal shall be divided by the total number of shares of the Company's capital stock then outstanding. No discount shall be applied for the fact that the shares to be purchased under this agreement shall constitute less than 50% of the total shares then outstanding.

The appraisal provisions may be combined with other evaluation methods. The following clause uses a shareholder determination of stated value, but appraisal is used if the determination of the shareholders is not current.

EXAMPLE 11.03O: Agreed Value or Appraisal to Determine Price

For the purposes of this agreement, each share of said stock shall be regarded as having a value of One hundred dollars ($100.00). The value of said stock as above determined may be changed from time to time by an endorsement over the signatures of the stockholders in the appendix to this agreement. A determination of value whether made in this clause or in the appendix shall remain vital and controlling for the period of one year from its effective date unless within such period it is superseded by a new determination. Should the death of a stockholder occur after one year from the effective date of the last determination of value, the value at the date of death shall be determined by three appraisers, one to be appointed by the surviving stockholder(s), one by the decedent's estate, and one by the two appraisers appointed as first provided. In their process of appraisement, the appraisers shall assume that the last valuation made by the stockholders, whether in this clause or in the appendix, was true and correct as of the date it was made, and with that assumption as a point of beginning, they shall proceed to redetermine such value with reference to the relevant facts and circumstances existing at the time of the decedent's death. Notwithstanding this provision for appraisement, the surviving stockholder(s) and the decedent's estate may elect to accept as controlling the last valuation made by the stockholders, even though such valuation was not made within the year preceding the date of the decedent's death.

The value of the stock as above stated or as same may be determined from time to time hereafter is or shall be inclusive of any value referable to the good will of the corporation as a going concern.[21]

Arbitration. Rather than appraisal, the agreement may provide for an arbitration among independent arbitrators, who will conduct whatever investigation may be necessary to ascertain the value of the stock. This objective determination by independent third parties may be desirable, but probably is also expensive and will only serve to resolve a dispute rather than establish a true reflection of the value of the Company's stock.

EXAMPLE 11.03P: Value Determined by Arbitration

The shareholder shall notify the corporation of the price at which he is willing to sell the stock, which notification shall contain the name of one arbitrator. The corporation shall, within thirty days thereafter, accept the offer, or by notice to the shareholder in writing, name a second arbitrator, and these two shall name a third. It shall then be the duty of the arbitrators to ascertain the value of the stock, and if any arbitrator shall neglect or refuse to appear at any meeting appointed by the arbitrators, a majority may act in the absence of such arbitrator.[22]

Terms of Payment. The agreement should specify the procedure and terms of payment when the transfer of shares is accomplished. As previously indicated, the purchasers may prefer to extend payment over a period of time, while the seller may prefer immediate cash. Full payment in cash rarely happens, and the most common terms of payment provide for a cash down payment and installment payments for the balance, which may be represented by an interest-bearing promissory note. Installment payments may provide tax benefits to the seller, especially if a large block of valuable stock is the subject of the transfer. Whenever stock is sold, the seller is required to report any capital gain received in the year of the sale. A shareholder who sells a large block of stock at one time may incur considerable tax liability by receiving the payments in cash during the year of sale. However, if payments are to be made in installments, the shareholder need only report the amount of the proportionate capital gain represented by the installments received during the year. This will spread his capital gains over the period of installments, which may be several years. The installment-sale tax treatment now applies no matter how much of the purchase price is received during the year in which the sale is consummated and whether or not the payments extend our two or more installments.[23]

By agreeing to extend payments over a period of time, the selling shareholder risks the subsequent insolvency of the purchasers, or their unwillingness to pay. This problem can be mitigated by providing the selling shareholder with some security to protect his interest in the payments. The security may be any property pledged as collateral to secure the note, but usually consists of the stock being sold. This means that the selling shareholder may "repossess" his stock upon default of the obligation. The agreement may specify this right of repossession or may establish an escrow arrangement by placing the shares being transferred in the hands of a third party pending payment of the full purchase price. Escrow terms require the return of the shares to the seller if the obligation is defaulted. If the obligation is paid in full, the shares will be delivered to the purchaser. A clause reciting the installment sale requirements and permitting a security interest in the stock follows:

EXAMPLE 11.03Q: Payment of Purchase Price

Not less than one-half the consideration required under the preceding clauses shall be paid in cash and, for the balance, a promissory note(s) of the kind hereinafter described may be given. On failure of the purchaser to settle in the manner required within the period of sixty (60) days from the election to purchase, the seller may rescind this agreement and re-establish the situation that would have existed had it never been made.

A note given for part of the consideration shall provide for annual payments on the principal over a period not to exceed five (5) years from the date of the purchase, at the end of which time the unpaid portion of the principal shall be due and payable, and shall provide for interest at the rate of ten (10%) per centum per annum and for optional acceleration of maturity in event of a default in payment of principal or interest. The seller may require the purchaser to secure the payment of a note given for the purchase price by a pledge of all or a portion of the stock.[24]

Whether the selling shareholder will have the right to vote the shares which are security for the installment purchase is a subject of negotiation, but it is preferable to permit the selling shareholder to vote the shares unless some serious objection is raised to the contrary. It should also be recognized that the shares of the corporation, as security for the installment payment, may represent worthless collateral, since when the corporation ceases to pay on the installments, it is likely that the corporation's financial position will have deteriorated so much that the shares may be valueless. In representing a shareholder whose only security are the shares being sold, it is advisable to consider the following additional terms in the security agreement representing the shares.

1. Restrictions upon the payment of dividends, distributions in liquidation, or salaries during the time that the shares are held as security.
2. The imposition of an asset–to–liability ratio during the period that the shares are security, so that the installment payments may be accelerated if the ratio is not maintained.
3. Restrictions on the corporation's ability to borrow money, sell substantially all of its assets outside of the ordinary course of business, merge, consolidate or dissolve during the period that the shares are subject to the restriction.

4. Anti-dilution provisions which will adjust the shares held as security to reflect any stock splits, stock dividends or other capital reorganization.
5. Terms which facilitate the perfection of the security interest in the shares, such as a promise by the company to deliver necessary stock certificates or other documents which may be necessary to perfection of the security interest in the particular jurisdiction.

EXAMPLE 11.03R: Purchase With Security

The deferred portion of the purchase price for any shares purchased under this agreement shall be represented by a promissory note executed by all the purchasing shareholders providing for joint and several liability. Each maker agrees that he or she will pay his or her pro rata portion of each installment of principal and interest as it falls due. The note shall provide for payment of principal in 24 equal quarterly installments with interest on the unpaid balance at the rate of 18% per annum, with full privilege of prepayment of all or any part of the principal at any time without penalty or bonus. Any prepaid sums shall be applied against the installments thereafter falling due in inverse order of their maturity, or against all the remaining installments equally, at the option of the payers. The note shall provide that, in case of default, at the election of the holder the entire sum of principal and interest will immediately be due and payable, and that the makers shall pay reasonable attorneys' fees to the holder in the event that such suit is commenced because of default. The note shall be secured by a pledge of all the shares being purchased in the transaction to which the note relates, and of all other shares owned by the purchasing shareholders. The note shall further be secured by a deed of trust on the real property of the corporation, and a security interest in all personal property owned by the corporation. The pledge agreements and other agreements required to effect and execute such pledges shall contain such other terms and provisions as may be customary and reasonable. As long as no default occurs in payments on the note, the purchaser shall be entitled to vote the shares; however, dividends shall be paid to the holder of the note as a prepayment of principal. The purchaser shall expressly waive demand, notice of default, and notice of sale, and shall consent to public or private sell of the shares in the event of default, in whole or in lots at the option of the pledge holder, and the seller shall have the right to purchase at the sale.

It is preferable for the shareholder to obtain security other than the shares being transferred as collateral for installment payments under a share transfer agreement. The example above illustrates a security interest in personal property of the corporation and a mortgage on the corporation's real estate. If the corporation is not the purchaser, other personal or real property of the purchasing shareholders should be considered. Of course, the terms of the security must be negotiated, and the corporation's counsel should be sensitive particularly to the impairment on the corporation's borrowing power by the grant of a security to the shareholder whose shares are being purchased. Appropriate subordination provisions may be included in the security documents to permit the corporation to borrow for normal operating reasons.

Funding of the Agreement Through Insurance. The corporation may use life insurance, disability insurance or other insurance contracts as a method of funding a buy-out agreement. These fundings techniques are especially effective if the event triggering the buy-out is death or disability. Other insurance contracts are also available for retirement or termination of employment.

A principal determination is whether all or some of the shareholders should or can be insured. A problem may arise when there are differences in ages, such as one shareholder who is over 65 and others who are under 30, or when one or more of the shareholders is not insurable because of physical infirmities. When the corporation purchases the life insurance policies funding the buy-out agreement, the shareholders automatically bear the cost in proportion to their ownership interests in the corporation. If the majority shareholder is the oldest, this works to his or her disadvantage. If all of the shareholders are roughly in the same age category for insurance purposes, the cost of insurance is spread more equitably. Advantages of corporation ownership of the policies include fewer policies and more easily ensuring that the premiums are timely paid and that the policies remain in effect. The shareholders have statutory rights to access to the corporation's books and can verify the information given to them about the status of the policies.[16]

A buy-out agreement funded with life insurance usually causes the last survivor or survivors to come out ahead. The problem may be illustrated by the case of the corporation valued at $100,000.00 with four shareholders. The interest of each shareholder is worth $25,000.00 and the corporation purchases insurance policies for $25,000.00 on the life of each shareholder. The estate of the first shareholder to die recieves $25,000.00 and each remaining shareholder has a one-third interest with a value now of $33,333.33, in a corporation still worth $100,000.00. The last survivor, of course, gets the entire corporation. One way to avoid this problem is to increase the insurance on the survivor's lives, but this may be too expensive and does not entirely eliminate the windfall to the longer surviving shareholders.

The corporation may not deduct life insurance premiums paid on policies on shareholders' lives if it is directly or indirectly a beneficiary under the policy.[25] This rule ordinarily prevents the corporation from deducting premiums for life insurance used to fund the corporation's purchase of its stock whether it is the designated beneficiary or the indirect recepient of the proceeds through a trustee or a member of the decedent's family. If the shareholder is the beneficiary, some portion of the insurance proceeds may be included in the shareholder's gross estate for estate tax purposes.[26]

As alternative funding methods, the corporation may build up cash or liquid investments as a reserve with which to purchase the shareholder's interest. This creates an evaluation problem, especially when book value is considered the appropriate formula for price of the purchased shares, in that the existence of the reserve enhances the value of the corporation and therefore may increase the amount to be paid at the buy-out. Also, the corporation's inability to use the money in the reserve may handicap its day-to-day operations, and the fund may be reachable by its creditors. There is further risk that the accumulated earnings tax will be imposed on such a reserve under the Internal Revenue Code.[27]

Life insurance may also be used to fund a cross-purchase agreement of the shareholders but the tax and non-tax factors should be considered carefully. If the shareholders are employees, they may pay the premiums out of their corporate salaries, which are deductible by the corporation if the compensation is reasonable.[28] If the shareholders are using dividend income from the corporation to pay the premiums, the corporate deduction is not available. Direct payments of the insurance premiums by the corporation may be held to constitute dividends to a shareholder. To avoid estate tax problems, each shareholder should purchase insurance on the life of each other shareholder, but not on his or her own life.[29] A factor weighing against life insurance funding is the potential windfall for the surviving shareholders.

Especially when the cross-purchase agreement among shareholders is funded by life insurance, a trustee should be appointed to perform certain functions. Stock certificates may be deposited with the trustee and he or she may receive payments and handle the paperwork attendant to such transfers. The trustee also may send notices and perform calculations as to the numbers of shares the offeree may purchase. The shareholders may prefer that a disinterested person perform these functions and the agreement should provide the manner in which the trustee will be selected. The agreement should also provide the manner in which the cost for the services of the trustee will be paid.

EXAMPLE 11.03S: Cross-Purchase Insurance Agreement

In order to fund the payment of the purchase price for the shares to be purchased under this agreement on the death of any shareholder, each shareholder shall maintain in full force and effect a policy of life insurance on the life of each other shareholder in the face amount shown on Exhibit A to this agreement. Each such policy is listed and described in the Exhibit, and any additional policies hereafter acquired for the same purpose shall also be listed in the Exhibit. Each policy belongs solely to the shareholder who applied for it and, subject to the provisions of this agreement, the owner of each policy reserves all the powers and rights of ownership of it. Each such owner shall be named as the primary beneficiary of his respective policies, and shall pay all premiums on them as they become due. No shareholder shall exercise any of the powers of ownership of any of the policies by changing the named beneficiary, cancelling the policy, electing optional methods of payment, converting the policy, borrowing against it, or in any other way changing its nature, value, or the rights under the policy. Any dividends paid on any of the policies before maturity or the insured's death shall be paid to the policy owner and shall not be subject to this agreement. Receipts showing payment of premiums shall be delivered to the Secretary of the company no less than 20 days before each date upon which the respective premiums are due, and the receipts shall be held by the Secretary for inspection by all shareholders.

If one shareholder dies, that shareholder will have owned policies of insurance on the lives of fellow shareholders. Accordingly, it is desirable to provide for the disposition of any unneeded premiums policies in the agreement.

EXAMPLE 11.03T: Unneeded Insurance Policies

On the death of any shareholder, each of the surviving shareholders shall have the option for 90 days to purchase the policy of life insurance on the shareholder's life owned by the decedent. Each shareholder shall also have the right to purchase the policies on his life within 90 days after the sale or transfer of all his shares, or after termination of this agreement. This option shall be exercised by delivery of written notice of exercise to the decedent's personal representative or to the owner of the policy and paying the purchase price in cash. The purchase price shall be equal to the cash surrender value of the policy, reduced by any unpaid loans made against the policy. If the option is not exercised within that period, the policy owner may surrender the policy for its cash value or dispose of it in any other way he or she sees fit. The parties agree to execute such releases and assignments that may be necessary to effectuate the provisions of this paragraph.

Legend on Certificates. In order to insure that shareholders will not violate the agreement and provide a purchase of the shares with stock certificates free from the transfer restrictions, it is necessary to place a conspicuous legend on each certificate for the shares, the terms of which should be specified by the agreement.

Section 8–204 of Uniform Commercial Code states that a purchaser of stock which is subject to a stock transfer restriction will purchase the shares free from the transfer restriction unless the certificate contains a conspicuous notation of the restriction or unless the purchaser has actual knowledge of the restriction.

The agreement should require that each shareholder shall surrender his or her certificate representing the shares to permit the inscription of an appropriate legend. The legend may provide the actual terms of the restriction, or simply say that the shares shall not be transferred, encumbered or in any way alienated except under the terms of the agreement, referring to the agreement by date, and indicating a place at which the agreement may be inspected.

EXAMPLE 11.03U: Legend

Each share certificate, when issued, shall have a conspicuously endorsed legend on its face with the following words: "Sale, transfer, or hypothecation of the shares represented by this certificate is restricted by the provisions of a buy-out agreement among the shareholders and the Company dated November 28, 1982, a copy of which may be inspected at the principal office of the Company and all the provisions of which are incorporated by reference in this certificate." A copy of this agreement shall be delivered to the Secretary of the company, and shall be shown by the Secretary to any person making any inquiry about it.

Miscellaneous Provisions. Several additional considerations must be reviewed in preparing an agreement regarding share ownership. Since the transfer of shares under the agreement may have an effect on other corporate activities, and may be regulated with respect to securities law aspects by state and federal agencies, special issues concerning transfer of shares must be reviewed with the client and considered in drafting the agreement.

If the corporation has previously elected Subchapter S status for taxation, it may be desirable to continue the Subchapter S election even though shares are being transferred under a stock transfer agreement. Of course, each shareholder's consent is necessary to taxation of the corporation under Subchapter S, and the agreement should provide that any transferree of the shares under the share transfer agreement will execute required documents and consent to the election.

EXAMPLE 11.03V: Subchapter S Election

The Company and each of the shareholders agree to execute such documents and consents and to cause them to be delivered in a timely manner to the Internal Revenue Service in order to cause the Company to elect to be taxed as a small business corporation under sections 1371–1379 of the Internal Revenue Code. Each shareholder shall cause any transferree of any of his or her shares to file in a timely manner the required consent to the election. Notwithstanding any provision of this agreement to the contrary, no transfer of any of the Company's shares shall be made by any shareholder to any corporation, partnership, or trust, or to any other transferee, if the effect of the transfer would cause the election to be lost or revoked.

In the case of death or disability of a shareholder, his or her spouse will have certain rights to assets of the shareholder. These assets would include the shares of stock owned by the shareholder. The agreement should contemplate the potential claims to be made by spouses and hiers of the shareholder which may be inconsistent with the terms of the agreement. The shareholders should be required to take such steps that may be necessary to reconcile personal estate documents, such as wills and trusts, with the shareholder agreement.

EXAMPLE 11.03W: Spouses' Consent

I acknowledge that I have read the foregoing agreement and that I know its contents. I am aware that by its provisions my spouse agrees to sell all of his or her shares to the Company, including my community interest in them, if any, on the occurrence of certain events. I hereby consent to the sale, approve of the provisions of the agreement, and agree that those shares and my interest in them are subject to the provisions of the agreement and that I will take no action at any time to hinder operation of the agreement on those shares or my interest in them.

EXAMPLE 11.03X: Wills

Each shareholder agrees to include in his or her will a direction and authorization to his or her executor to comply with the provisions of this agreement and to sell his or her shares in accordance with this agreement; however, the failure of any shareholder to do so shall not effect the validity or enforceability of this agreement.

NOTES

1. See Section 5.01.
2. See Sections 6.05, 8.08.
3. See Chapter 13.
4. See M.B.C.A. § 32 and Section 8.07.
5. Delaware's close corporation statute allows written shareholder agreements which interfere with the discretion of the directors, but the shareholders are responsible for acts controlled by the agreement. 8 Del.Code Ann. § 350.
6. M.B.C.A. § 34.

7. E.g. Minnesota, Minn.Stat.Ann. § 301.27; Nevada, Nev.Rev.Stat. § 78.365.
8. New Jersey, N.J.Stat.Ann. § 14A:5–20; and Maine, 13A Me.Rev.Stat.Ann. § 619.
9. M.B.C.A. § 34.
10. West's Modern Legal Forms § 3013.1.
11. West's Modern Legal Forms § 3013.
12. See the example described in Section 8.08.
13. West's Modern Legal Forms § 3013.5.
14. See Section 9.06 and M.B.C.A. §§ 6 and 45.
15. See M.B.C.A. § 45.
16. I.R.C. § 302.
17. See Section 13.06.
18. West's Modern Legal Forms § 3013.
19. West's Modern Legal Forms § 2764.1.
20. West's Modern Legal Forms § 3013.
21. West's Modern Legal Forms § 3026.
22. West's Modern Legal Forms § 2764.1.
23. Int.Rev.Code of 1954, 26 U.S.C.A. § 453 as amended.
24. West's Modern Legal Forms § 3026.
25. See I.R.C. § 264(a); Reg. § 1.264–1(b).
26. See I.R.C. § 2042.
27. See I.R.C. § 531.
28. See I.R.C. § 162(a)(1).
29. See I.R.C. § 2042(2). Examples of agreements involving insurance funded buy-out provisions may be found in West's Modern Legal Forms §§ 3027–3033.

12

CORPORATIONS IN FOREIGN JURISDICTIONS

§ 12.01 SELECTION OF JURISDICTION

In the early formation stages of a corporation, a particular jurisdiction will be selected as the situs of incorporation. The selection process was described in some detail in an earlier section,[1] and it was there recognized that a multi-state business will have several jurisdictions to consider. A decision to domesticate the multi-state corporation in one particular state will usually be made after a comparison of the permissiveness and flexibilty of the various statutes. The state of incorporation is the domestic state, and the corporation is a foreign corporation to every other state. The rule to be considered in this Chapter is that a foreign corporation must *qualify* to do business in any state in which it intends to conduct business.

§ 12.02 CONSTITUTIONAL BASIS FOR QUALIFICATION

A little history and legal theory of the corporate entity are important here. A corporation is a fictitious person created under the authority of a state statute, and consequently, is a citizen of the state in which it has incorporated. Under the early common law, the corporation existed only in the state in which it incorporated, and it was not permitted to do business in any other jurisdiction. When corporate businesses began to outgrow the boundaries of their domestic states, a legal question arose as to whether a state could prevent a foreign corporation from doing business within its boundaries without incorporating therein. Alternatively, if incorporation could not be required, could a state place restrictions upon the foreign corporation's business and require it to satisfy certain conditions before it was entitled to do business in the foreign state? The Constitution of the United States guarantees to all "persons" the ability to move freely among the states without restriction, and the question then became whether a corporation may be considered to be a "person" under these provisions of the Constitution. The Supreme Court eventually decided that a corporation, whether or not it is a "person" in the constitutional sense, could not be prevented from entering another state, but could be reasonably regulated by the foreign state, under the guise of the state's power to prescribe regulations to protect its own residents. If the regulations imposed were reasonable and designed to protect the local citizenry, the regulations were permitted. If you keep this history in mind throughout this Chapter, the current regulations on foreign corporations are more easily digested. Most states require foreign corporations to disclose certain matters about their business structure by a public filing with a state official, so that the citizens of the state have access to necessary information about the organization with which they transact business. A separate area of regulation is concerned with litigation by and against the foreign corporation. These statutes are designed to subject the foreign organization to legal process within the state so that the citizens of a state may conveniently redress their complaints against the corporation. They further insure compliance with qualification requirements before the corporation may use the state courts.

In a constitutional sense, therefore, any state may impose regulations on foreign corporations, provided they are reasonably directed to the state's responsibility and power to protect its own citizens.

§ 12.03 AUTHORIZATION TO QUALIFY AS A FOREIGN CORPORATION

The authority to conduct business in a foreign state must come from within the corporation by official direction of the board of directors. Such management approval, which is usually accompanied by an enabling provision in the articles of incorporation, authorizes the corporation to submit to the necessary regulations in order to do business within a foreign jurisdiction.

The board of directors will usually adopt a resolution authorizing qualification at its organizational meeting if the plans for expansion are solidified at that time. The resolution states that the corporation may qualify to do business under the laws of other states, and that the officers of the corporation are empowered to execute any necessary documents and to pay all necessary taxes and fees in order to qualify the corporation as a foreign corporation.[2]

§ 12.04 STATUTORY PROHIBITION FROM DOING BUSINESS WITHOUT QUALIFICATION

Every state has a statute pertaining to qualification of foreign corporations. Several states have adopted the Model Act approach, prohibiting any foreign corporation from transacting business within the boundaries of the state without qualifying and receiving a certificate of authority from the appropriate state official.[3] This is a strict provision and nearly half of the states do not condition the right to transact business on the receipt of the certificate of authority. However, they specify a procedure for obtaining a certificate of authority, and further specify sanctions for failure to qualify.

It is not necessary that the foreign corporation be incorporated in a state whose laws are substantially similar to the laws of the state in which it seeks authority to do business. In fact, section 106 of the Model Act specifically prohibits denial of a certificate of authority simply because the laws of the state under which the foreign corporation is organized differ from the laws of the particular state in which it intends to qualify. Therefore, a corporation established under the permissive laws of Delaware, where the regulation of corporate management is very flexible, could not be denied admission to another state whose corporate statute restricts the activities of the intra-corporate parties. However, the foreign corporation could be denied admission if it is organized for a purpose which is unlawful in the host state. For example, a Nevada corporation organized to conduct a gambling business may be denied admission to conduct gambling operations in a state where those activities are illegal. Otherwise, any corporation may qualify to do any business in any foreign jurisdiction.

§ 12.05 TRANSACTING BUSINESS

The traditional statutory test for determining whether a corporation must qualify in a foreign jurisdiction is whether the corporation is "transacting business" within the foreign state. The transacting business test is not the most precise definition ever devised, and it has been responsible for considerable litigation. A sampling of the cases should illustrate the problem: 1) Would an Indiana corporation be doing business in Oklahoma by manufacturing equipment in Indiana, delivering it to Oklahoma and installing it there? 2) Would a Delaware corporation be transacting business in New York by engaging an answering service and a soliciting salesman in New York? 3) Would an Arizona corporation be conducting business in Texas by sending salesmen and mechanics into Texas to solicit orders and to install and repair the machinery? 4) Would a Georgia corporation be doing business in Mississippi by hiring a local Mississippi mechanic to service an ice cream dispenser for one year? The answer from the cases considering these facts as follows: 1) No; 2) Yes; 3) No; 4) Yes.

The judicial uncertainty surrounding this test is particularly unfortunate when the sanctions for failure to qualify are considered. A corporation may suffer severe penalties if it is found to have been conducting business without qualification. To obviate the problem, Section 106 of the Model Act enumerates certain activities which may be conducted by a foreign corporation without being considered to be transacting business. These activities are:

1. Maintaining or defending any action or suit or any administrative or arbitration proceeding, or effecting the settlement thereof or the settlement of claims or disputes.
2. Holding meetings of its directors or shareholders or carrying on other activities concerning its internal affairs.

3. Maintaining bank accounts.
4. Maintaining offices or agencies for the transfer, exchange and registration of its securities, or appointing and maintaining trustees or depositaries with relation to its securities.
5. Effecting sales through independent contractors.
6. Soliciting or procuring orders, whether by mail or through employees or agents or otherwise, where such orders require acceptance without this State before becoming binding contracts.
7. Creating as borrower or lender, or acquiring, indebtedness or mortgages or other security interests in real or personal property.
8. Securing or collecting debts or enforcing any rights in property securing the same.
9. Transacting any business in interstate commerce.
10. Conducting an isolated transaction completed within a period of thirty days and not in the course of a number of repeated transactions of like nature.

Only a few states detail such a comprehensive list, and many states have no list at all. Delaware provides a more specific list, which might be more helpful in solving the illustrative cases described above. The Delaware statute states that a foreign corporation shall not be required to qualify in the state:

1. If it is in the mail order or a similar business, merely receiving orders by mail or otherwise in pursuance of letters, circulars, catalogs, or other forms of advertising, or solicitation, accepting the orders outside this State;
2. If it employs salesmen, either resident or traveling, to solicit orders in this State, either by display of samples or otherwise (whether or not maintaining sales offices in this State), all orders being subject to approval at the offices of the corporation without this State, and all goods applicable to the orders being shipped in pursuance thereof from without this State to the vendee or to the seller or his agent for delivery to the vendee, and if any samples kept within this State are for display or advertising purposes only, and no sales, repairs, or replacements are made from stock on hand in this State;
3. If it sells, by contract consummated outside this State, and agrees, by the contract, to deliver into this State, machinery, plants or equipment, the construction, erection or installation of which within this State requires the supervision of technical engineers or skilled employees performing services not generally available, and as a part of the contract of sale agrees to furnish such services, and such services only, to the vendee at the time of construction, erection or installation;
4. If its business operations within this State, although not falling within the terms of paragraphs (1) (2) and (3) of this section or any of them, are nevertheless wholly interstate in character.
5. If it is an insurance company doing business in this State.
6. If it creates, as borrower or lender, or acquires, evidences of debt, mortgages or liens on real or personal property;

7. If it secures or collects debts or enforces any rights in property securing the same.[4]

Some statutory guidance may be available, therefore, on what does *not* constitute the transaction of business within the state. However, that is not to say that everything not enumerated in the statute *is* transacting business so as to require qualification. There is still room for judicial interpretation of the other corporate activities.

The safe approach should be obvious: if there is any question about the scope of the corporation's activities in a foreign jurisdiction, it should apply for admission as a foreign corporation and obtain a certificate of authority. Failure to do so may subject the corporation to the statutory sanctions for failing to qualify. The safe approach is not always the most practical, however. Qualification does impose certain special burdens on the corporation, and corporate management may be unwilling to accept them. Management should be fully advised on all ramifications of qualification, and the decision will be theirs, considering the costs, formalities, taxes and fees on one hand, and the penalties for failure to qualify on the other.

§ 12.06 SANCTIONS FOR NOT QUALIFYING

If a foreign corporation is transacting business within the state and has not received a certificate of authority from the host state, most state statutes impose certain interdictions and fines on the corporation or its management. Foreign corporations may be denied access to the local courts for any action, suit or proceeding until a certificate of authority has been obtained. This sanction is found in the Model Act and in the majority of jurisdictions. However, under the Model Act the failure to qualify does not impair the validity of any contract or act of the corporation, and it does not prevent any corporation from *defending* any action in a court of the host state. Consequently, the practical drawback of failing to qualify under the Model Act is the inability of the corporation to maintain a suit in its own name. However, there are also pecuniary disadvantages. The failure to obtain a certificate renders the corporation liable to the state for all fees and taxes which would have been imposed had it been qualified for the years during which it transacted business without a certificate of authority. In addition, the Model Act authorizes the imposition of any penalties normally levied for failure to pay the fees and these "fines" also will be exacted from the foreign corporation. The Attorney General is authorized to bring a suit to recover the amounts due under the statute.[5]

Although the prohibition from maintaining litigation and the collection of fees, taxes and fines are severe sanctions, the Model Act is somewhat liberal by comparison to sanctions imposed in other states. Arizona provides that all acts of an unauthorized foreign corporation are void. Some state statutes say that any contract entered into by an unauthorized foreign corporation is not enforceable.[6] Certain jurisdictions impose fines, instead of the normal penalties for fees, and the amount of the fine may be as high as $10,000.[7] Fines may also be levied on corporate directors and officers. Other states authorize an action by the Attorney General to enjoin the corporation from doing business.[8]

In the spirit of forgiveness, many states excuse the sanctions as soon as the corporation properly qualifies, although the relief may depend upon showing good cause for failure to qualify or court approval.

§ 12.07 APPLICATION FOR CERTIFICATE OF AUTHORITY

All state statutes describe a procedure for a qualification of a foreign corporation. In order to acquire a certificate of authority in most states, the corporation must apply to the appropriate state official, and the contents of the application vary among the jurisdictions. However, there is a common purpose behind the qualification procedures. The applications reveal necessary information about the corporate structure, its solvency, the location of its property and its business potential. Foreign corporations are required to furnish essentially the same initial and periodic information as domestic corporations.

Section 110 of the Model Act includes the following items in an application for a certificate of authority:

1. The name of the corporation and the state or country under the laws of which it is incorporated.
2. If the name of the corporation does not contain the word "corporation," "company," "incorporated," or "limited," or does not contain an abbreviation of one of such words, then the name of the corporation with the word or abbreviation which it elects to add thereto for use in this State.
3. The date of incorporation and the period of duration of the corporation.
4. The address of the principal office of the corporation in the state or country under the laws of which it is incorporated.
5. The address of the proposed registered office of the corporation in this State, and the name of its proposed registered agent in this State at such address.
6. The purpose or purposes of the corporation which it proposes to pursue in the transaction of business in this State.
7. The names and respective addresses of the directors and officers of the corporation.
8. A statement of the aggregate number of shares which the corporation has authority to issue, itemized by classes and series, if any, within a class.
9. A statement of the aggregate number of issued shares itemized by classes and series, if any, within a class.
10. An estimate, expressed in dollars, of the value of all property to be owned by the corporation for the following year, wherever located, and an estimate of the value of the property of the corporation to be located within this State during such year, and an estimate, expressed in dollars, of the gross amount of business which will be transacted by the corporation during such year, and an estimate of the gross amount thereof which will be transacted by the corporation at or from places of business in this State during such year.

11. Such additional information as may be necessary or appropriate in order to enable the Secretary of State to determine whether such corporation is entitled to a certificate of authority to transact business in this State and to determine and assess the fees and franchise taxes payable as in this Act prescribed.

A study of subsection (10) should suggest that it is designed to facilitate the discovery and evaluation of corporate property within the state and to gauge the local productivity of the foreign corporation. This information is used to estimate the tax potential of the foreign corporation, and may also assist local citizens in their litigation against the corporation.

Each state statute should be consulted for local application requirements. In addition to the information required by the Model Act, it may be necessary to file a list of other jurisdictions to which the corporation has been admitted,[9] or to file a certified sealed copy of the vote authorizing the corporation to do business within the state.[10] Statements of good standing may be required from the corporation's home state,[11] and nearly half of the states require filing of the articles of incorporation duly authenticated by the domestic state officials. In Pennsylvania it is necessary to file a separate "registry" statement for tax purposes with the application.[12] Other unusual formalities include the following: the corporation may be required to declare how long it expects to be in the state;[13] it may have to stipulate an agent who is a local resident and a member of the local bar;[14] and statements regarding the amount of paid-in capital or paid-in surplus may be required.[15]

Payment of fees and franchise taxes is almost always necessary, and some states require the filing of previous annual reports.

To reiterate, as with other matters respecting corporate existence, it is necessary to carefully examine the statute of the state in which the corporation intends to do business, and to strictly comply with the statutory requirements.

The application for certificate of authority is prepared in duplicate or triplicate, as the statute requires, and executed by the President and Secretary or one of their assistants. Under the Model Act it must be verified by one of the signatories.[16]

Most of the matters contained in the application for qualification are self-explanatory. However, a couple of the items deserve elaboration.

Corporate Name

The foreign corporation must comply with statutes regulating corporate names in the host state. The Model Act prescribes essentially the same name requirements for domestic and foreign corporations.[17] The name must contain the word "Corporation," "Company," "Incorporated," or "Limited," or an abbreviation of one of these words, and it may not contain any word or phrase which indicates or implies that it is organized for any purpose other than those enumerated in its articles of incorporation. Moreover, it cannot be the same as, or deceptively similar to, the name of any domestic corporation existing under the laws of the state or of any foreign corporation already authorized to transact business in the state. The Act permits a departure from this latter rule if the foreign corporation files one of three documents with the Secretary of State:

1. a resolution of the board of directors adopting a fictitious name which meets the above tests; or
2. the written consent of the established corporation or of the holder of a reserved or registered name to use the name which would otherwise be refused as being deceptively similar, provided one or more words are added to make the names distinguishable; or
3. a certified copy of a final court decree establishing the prior right of this foreign corporation to use the name in the state.

The first exception is not really an exception. Suppose American Can Company is a New York corporation, and is seeking to qualify to do business in Tennessee, where there is an established domestic corporation by the same name. The name of the foreign corporation cannot be the same as that of a domestic corporation or previously qualified foreign corporation. The names are the same in this case, so the New York corporation cannot be qualified in Tennessee under the circumstances. However, if the board of directors of the New York corporation adopts a *fictitous* name that is not deceptively similar to an existing name, then the corporation may be qualified under that fictitious name. The other two exceptions actually allow the use of a name provided written consent (with a distinguishing word) or a court order is obtained.

The foreign corporation seeking to qualify to do business will usually resist the fictitious name technique because the reputation of the corporate name, usually well established in other states, is lost by the adoption of an assumed name. One can only speculate at the business success of Xerox Corporation if it had been required to use a fictitious name of "Copying Machines, Inc." in some of the foreign jurisdictions where it qualified to do business. However, many state officials have adopted informal tests for variations in fictitious names which will be acceptable. For example, if a domestic corporation had reserved the name "Xerox Corp." and the real Xerox Corporation attempted to qualify to do business, many states would permit the fictitious name to be "Xerox Corporation of Delaware", simply adding the name of the state of incorporation.

Corporations may certainly buy the right to use a particular name from its owner, and that is usually what must be done to obtain written consent for its use under the second exception. Law students who have studied this exception have dreamed of immediate and great fortunes by anticipating expansion of a large established corporation into their state, filing a reservation of the corporate name, and waiting to be approached for the consent. The third exception is designed to give the corporation an alternative if the price of the consent is excessive and a prior right to the name can be shown.

The corporate name problem may be solved in advance with a little planning. You may recall that all jurisdictions permit reservation of a corporate name and several permit registration of the name.[18] Registration is specifically designed for use by foreign corporations. An organized and existing corporation may file an application for registration of its corporate name with the Secretary of State for a small fee. The name may be registered for calendar years, and the registration may be renewed. Corporate management, foreseeing corporate growth into foreign jurisdictions, would be well advised to pursue registration of the corporate name in states where registration is permitted.[19]

Registered Office and Agent

Each state requires that a qualified foreign corporation must maintain a registered office and appoint a registered agent to receive legal documents addressed to the corporation. The registered office may be, but need not be, the same as the corporation's place of business in the state; and the registered agent may be either an individual resident in the state or a domestic corporation or another foreign corporation authorized to transact business in the state. As is required of a domestic corporation, any changes in the office or agent of the foreign corporation must be filed in the office of Secretary of State.[20]

The purpose of the registered office and the agent is to facilitate the service of any process, notice, or demand required or permitted by law. A state may reasonably require a convenient way to notify a foreign corporation of any legal matters as a condition to its permission to do business in the state. For that reason, whenever a foreign corporation fails to appoint or maintain a registered agent in the state or whenever the registered agent cannot be found with due diligence, most statutes provide that the Secretary of State will be deemed to be an agent of the corporation to receive these legal documents. The Secretary of State must send one copy of the document received by registered mail to the corporation at its principal office in the state of incorporation.[21] This provision circumvents any potential escape from local complaints by failing to maintain a local agent and insures the amenability of the foreign corporation to suit in local courts.

Several "registered agent" companies are available at the request of attorneys to act as registered agents for foreign corporations and to comply with the requirements of local laws on their behalf. They include the CT Corporation System, Prentice Hall, Inc. and the United States Corporation Company.

§ 12.08 CERTIFICATE OF AUTHORITY

Upon receipt of the application for a certificate of authority, the Secretary of State or other appropriate official files the application and issues a certificate of authority. When the certificate is issued, the corporation is authorized to transact business in the state for the purposes set forth in its application as long as it remains in good standing.[22]

§ 12.09 EFFECT OF QUALIFICATION

Once a foreign corporation has received authority to do business within the foreign state, it is entitled to enjoy the same rights and privileges as a domestic corporation organized for the same purposes. In addition, it is subject to the same duties, restrictions, penalties, and liabilities as a domestic corporation. In the host state, therefore, the foreign corporate will be treated like a native, receiving no better or worse treatment than the domestic corporations. It should be noted, however, that the foreign corporation remains subject to restrictions imposed by its home state, and it must observe those restrictions while operating in the host state. For example, consider an Arkansas corporation which qualifies to do business in Pennsylvania. Arkansas law permits the corporation to be a general partner in another enterprise only if authorized in the articles of incorporation or by vote of the shareholders. Pennsylvania law permits a corporation to be a general partner without any such authorization. If the articles of incorporation do not authorize the corporation's entry into a partnership, and the shareholders have not approved such a transaction, the Arkansas corporation could not become a general partner, even in Pennsylvania, where it would otherwise be treated like a domestic corporation. The restrictions placed upon the corporation in its home state are therefore superimposed upon its operations in the foreign state.

The reverse is equally true. A Pennsylvania corporation which qualifies to do business in Arkansas is subject to the Arkansas restrictions on domestic corporations. The Pennsylvania corporation could not become a general partner in Arkansas without authorization in its articles of incorporation or the requisite shareholder vote, even though it would not be subject to those restrictions in its home state.

In addition to the restrictions upon internal affairs, a qualified foreign corporation also accepts other responsibilities within the host state.

Service of Process

A foreign corporation authorized to transact business in a state is subjected to the jurisdiction of the courts of the host state. Consequently, service of documents relating to litigation upon the registered agent of the corporation is as effective as if the corporation were incorporated within the state and had been served at its principal office.

Taxes

By qualifying to do business within a state, a foreign corporation agrees to pay taxes to the host state. A state is permitted to tax a foreign corporation under the Constitution if the corporation has "substantial contacts" within the state. The law is now quite clear that by qualifying to do business within the state the corporation creates those substantial contacts. The individual tax structures of the states are dissimilar, but there are several typical types of taxes that are imposed upon foreign corporations.

Some states impose an initial franchise tax upon filing of the application for a certificate of authority. This tax is generally based upon the aggregate amount of authorized capital stock of the corporation, similar to the measure for taxes imposed on a domestic corporation when its articles of incorporation are filed. A fee is also charged for filing the application of a foreign corporation.

Annual income taxes also are imposed upon foreign corporations. Generally, the tax formula used in any given state has been designed around the commercial character of the state and is intended to maximize tax revenues from foreign corporations. If the state is a recognized location for heavy industry, so that most foreign corporations have manufacturing or industrial plants there, the state will probably impose a tax on the value of the property of the foreign corporation located within the state. This tax formula maximizes revenue for states with an industrial character to their commerce. If the state is not heavily industrialized but has a high population, a tax may be imposed on the proportionate volume of business that the foreign corporation is transacting in the state. A tax on foreign corporations may also be computed by a formula based upon the total number of employees located within the state.

The tax provisions of each state play an important role in the selection of jurisdiction for a corporation anticipating a multi-state business. At the formation stage, the choice between incorporating or qualifying to do business in a given jurisdiction will be influenced by that state's tax attitude. For example, if a corporation plans to locate a manufacturing plant in State X and intends to sell its products to customers primarily located in State Y, it would be a mistake to incorporate in State X, where the plant is located, if that state bases its foreign corporation tax on the volume of sales transacted within the state. You may have to stop and ponder that for a minute, but the principle is simple. The corporation should operate as a foreign corporation in State X, because its tax base depends upon volume of sales and this corporation will be making most of its sales out of state.

Annual Reports

Like domestic corporations, qualified foreign corporations must file annual reports with the state. The Model Act requires that the annual reports of domestic and foreign corporations must contain certain standard information regarding the corporation, its registered offices and agents, directors and officers and the character of its business. The reported items which are used to levy taxes include:

1. A statement of the aggregate number of shares which the corporation has authority to issue, itemized by classes and series, if any, within a class;
2. A statement of the aggregate number of issued shares, itemized by classes and series;
3. A statement, expressed in dollars, of the value of all the property of the corporation, wherever located, and the value of the property of the corporation located within the state, and the statement of the gross amount of business transacted by the corporation for the period of the report;
4. Additional information as may be necessary or appropriate in order to enable the Secretary of State to determine and assess the proper amount of franchise taxes payable by the corporation.[23]

The requirements for annual reports in the Model Act are exactly the same for domestic corporations and foreign corporations, and the reports are intended to be used to assist the Secretary of State in enforcing the corporation statute and insuring compliance with its provisions, in fixing responsibility for any corporate transgressions on the named officers and directors, and in evaluating the appropriate tax to be assessed. Most states have the same reporting requirements for foreign and domestic corporations.

§ 12.10 STRUCTURAL CHANGES OF A FOREIGN CORPORATION

Every state requires reporting of all corporate structural changes such as mergers, consolidations, sale or exchange of assets, or amendments to the articles of incorporation for its domestic corporations. Similarly, qualified foreign corporations must follow certain procedures in the host state whenever a structural change occurs in the corporate organization.

Amendment to Articles of Incorporation

Model Act Section 116 provides that a qualified foreign corporation must file a statement of any amendment to its articles of incorporation with the Secretary of State of the host state within thirty days after such amendment becomes effective. The time period for filing copies or statements of amendments differs among the states. Alabama requires that such an amendment be filed within thirty days, several states permit as long as 60 days,[24] and a few states have no time limit.[25] Again, careful study of the appropriate state law is important.

The filing procedure for amendments to the articles of incorporation of a foreign corporation is amplified if the amendment changes the corporate name or the corporate purposes. Merely filing copies of the amended articles of incorporation will not suffice to authorize the foreign corporation to use another name or to pursue a new business direction under its amended corporate purposes in the host state. In addition to filing the amended articles of incorporation, the Model Act requires an amended certificate of authority to accomplish these changes.[26] A foreign corporation may change its corporate name or pursue additional or different purposes by filing a new application for an amended certificate of authority. The form and contents of the application for an amended certificate and the procedure for issuance of the amended certificate are the same as those described for the original application for a certificate of authority.[27] Time limits on filing the application for the amended certificate are frequently imposed.[28]

There is an obvious problem if a foreign corporation changes its name by amending its articles of incorporation and the new name is not available in the host state. The drafters of the Model Act took a fairly firm stand on this issue. Under section 109 if the new name is not available, the certificate of authority for the corporation will be suspended until it again changes its name to one that is available. A few states soften the harshness of this rule by an interim grace period of 180 days, during which the corporation may transact business under its old name, but by the end of the period it must change its name to a name that is available under the laws of the state. If it fails to change to an acceptable name but continues to transact business, its certificate of authority may be suspended.

Merger and Consolidation with the Foreign Corporation

A merger of a foreign corporation also requires certain additional filings in the host state when the foreign corporation is the surviving corporation after the merger. The legal consequences of a merger will be considered in detail later,[29] but stated simply, a merger is a combination of two or more corporations into one corporate entity, whereby one of the corporate parties survives the transaction and the others cease to exist. Under the Model Act and most state statutes, if the foreign corporation survives the merger, it must file a copy of the articles of merger with the Secretary of State within 30 days after the merger bcomes effective.[30] It is not necessary for the surviving foreign corporation to procure as amended certificate of authority unless it has changed its name under the merger or unless it intends to pursue additional purposes other than those which are authorized in its current certificate of authority. If either of those results are produced by the merger, the amendment procedure described above must be followed.

If a qualified foreign corporation merges with another foreign corporation which is not authorized to transact business within the host state, and the non-qualified corporation survives the merger, the surviving corporation must qualify in the foreign jurisdiction by filing an original application for a certificate of authority. The surviving corporation does not inherit the previously granted authority of the merged corporation through the merger.

Of course, if a foreign corporation merges with a domestic corporation, and the domestic corporation survives the merger, the Articles of Merger will be filed by the domestic corporation pursuant to the laws of the state.[31] However, if the foreign corporation survives the merger, certain other filings are required. The surviving foreign corporation may not be authorized to transact business within the state, and, if it intends to do so, it must file an original application for a certificate of authority. Even if it does not intend to do business within the state, the surviving foreign corporation must file the following items with the Secretary of State:

1. an agreement that it may be served with process within the state for any proceeding to enforce an obligation of the merged domestic corporation or to enforce the rights of a dissenting shareholder of the merged domestic corporation;
2. an irrevocable appointment of the Secretary of State as its agent to accept service of process for any such proceeding;
3. an agreement that it will promptly pay the dissenting shareholders of the merged domestic corporation the value of their shares under the local corporation law.[32]

If the surviving foreign corporation had been authorized to do business within the state, the terms of its authority theoretically include each of these items and no further filing should be required. However, Section 77 of the Model Act states that the agreements and appointments of agent must be filed "in every case" where a foreign corporation survives a merger with a domestic corporation.

Finally, two or more corporations may consolidate, a technique whereby the constituent corporations combine to form a single new corporation. If a qualified foreign corporation consolidates with another foreign corporation, whether qualified or unqualified, the new corporation must seek original authority to do business and does not inherit the qualified status enjoyed by any foreign corporation party to the consolidation.

§ 12.11 WITHDRAWAL OF AUTHORITY

The management of a qualified foreign corporation may decide to discontinue business operations within the host state. However, this does not mean that they may simply pull up their tent and steal away. State regulation of foreign corporations is designed to require payment of fees and taxes and to insure the availability of the foreign corporation for litigation commenced against it in the state. Consequently, the withdrawal of a foreign corporation is a formal procedure. The foreign corporation must file an application for withdrawal which, under the Model Act, provides current information regarding its capital stock structure and states that the corporation surrenders its authority to transact business in the state. It must further specifically revoke the authority of its registered agent to accept service of process and consent to service of process on the Secretary of State for any proceeding based upon a cause of action arising during the time the corporation was operating within the state. The withdrawal application also includes a post-office address to which the Secretary of State may mail a copy of any process received for the corporation and other additional information which may be necessary to enable the Secretary of State to assess any unpaid fees or franchise taxes.[33]

State statutes governing withdrawal of foreign corporations are as varied as those pertaining to admission. All statutes are directed toward full financial disclosure and amenability to service of process, and generally require that the foreign corporation tidy up its affairs before leaving the state. The statute may require proof that all corporate creditors within the state have been satisfied;[34] or a statement that the corporation no longer owns property within the state.[35] In all cases, taxes and fees must be paid as a condition to the approval of withdrawal.

Upon filing the application and the satisfaction of all statutory conditions, the appropriate state official will issue a certificate of withdrawal.[36]

§ 12.12 REVOCATION OF CERTIFICATE OF AUTHORITY

A foreign corporation's authority to do business within a state may also cease by the revocation of authority by the host state. Generally, the certificate may be revoked whenever the foreign corporation has failed to comply with the law. For example, the corporation may have failed to file annual reports, or may have failed to pay fees, franchise taxes or penalties. Other grounds for revocation include the corporation's failure to appoint and maintain a registered agent; failure to notify the state of a change in the agent or office; failure to file amendments to its articles of incorporation or articles of merger within the time prescribed; or a misrepresentation in the material matter in any application, report, affidavit or other document filed with the state. In addition to these grounds some states add abusing or exceeding the corporation's authority; violating of a state law; using an unauthorized name; or acting in a manner detrimental to the citizens of the state. The District of Columbia and Illinois have a revocation provision which looks like default; if the corporation does not conduct business or own tangible property within the state for a specified period, its certificate of authority may be revoked.

Most state statutes require notice to the corporation prior to the revocation of a certificate of authority. The Model Act directs the Secretary of State to give the corporation sixty days' notice by mail addressed to its registered office in the state, and, if the corporation corrects the specified problem within the notice period, the certificate of authority may not be revoked. However, following the sixty-day period if nothing is done, the Secretary of State may issue a certificate of revocation. The minimum notice period among the individual state statutes is twenty days,[37] and the maximum is ninety days.[38] Many states permit the remedy of the defect during the intermediate period.

A sample of a certificate of revocation appears as form 12H in the Appendix, and when the certificate is issued, the corporation's authority to transact business in the state ceases.

NOTES

1. Section 6.03.
2. See the sample resolution for organizational meetings of the board of directors at Section 8.05.
3. See M.B.C.A. § 106.
4. 8 Del.Code Ann. § 373(a).
5. M.B.C.A. § 124.
6. E.g., Arkansas, Ark.Stats. § 64–1202.
7. See schedule of penalties for doing business without qualifying, 1 Prentice Hall, Corporations ¶ 7103.
8. E.g., Delaware, 8 Del.Code Ann. § 384; and New York, McKinney Consol.Laws of N.Y.Bus.Corp.Law § 1303.
9. E.g., Illinois, Ill.Rev.Stat. c. 32 § 157.106(e); and Missouri, Vernon's Ann.Mo.Stat. § 351.580(1)(6).
10. New Hampshire, N.H.Rev.Stat.Ann. § 300:4(I)(c).
11. E.g., Nebraska, Nev.Rev.Stat. § 21–20.110.
12. 15 Pa.Stat. § 2004.
13. Oklahoma, 18 Okl.Stat.Ann. § 1.228(3).
14. Virginia, Va.Code Ann. § 13.1–109(b).
15. E.g., Illinios, Ill.Rev.Stat. Ch. 32 § 157.106(k); and Wisconsin, Wis.Stat.Ann. § 180.813(I)(j) (paid-in capital); Missouri, Vernon's Ann.Mo.Stat. § 351.580(11); and Oklahoma, 18 Okl.Stat.Ann. § 1.228(11) (paid-in surplus).
16. M.B.C.A. § 110. An example of an application for certificate of authority appears as Form 12A in the Appendix.
17. Compare M.B.C.A. §§ 8, 108.
18. See Section 6.04.

19. See M.B.C.A. § 9. Forms for registration and transfer of a corporate name appear as Forms 6I, 6J, 6K and 6L in the Appendix.
20. See M.B.C.A. §§ 113, 114. A statement for change of registered office or agent appears as Form 12B in the Appendix.
21. See M.B.C.A. § 115.
22. See M.B.C.A. § 112. An example of the certificate of authority is Form 12C in the Appendix.
23. M.B.C.A. § 125.
24. E.g., Arizona, Ariz.Rev.Stat. § 10–116(A); and Montana, Mont.Code Ann. § 35–1–1015.
25. E.g., Indiana, Burns' Ind.Ann.Stat. § 23–1–11–8.
26. M.B.C.A. § 118.
27. See Sections 12.07 and 12.08, and M.B.C.A. § 118. Examples of an application for amended certificate of authority and an amended certificate of authority are Forms 12D and 12E in the Appendix.
28. E.g., Missouri, 60 days Vernon's Ann.Mo.Stat. § 351.600(1); and Georgia, 30 days, Ga.Code § 22–1413.
29. See Section 13.02.
30. See M.B.C.A. § 117.
31. M.B.C.A. § 74.
32. M.B.C.A. § 77. The Model Act provides official forms to be used for any combination of merger or consolidation between foreign and domestic corporations. See Official Forms for Use under the Model Business Corporation Act, Forms 26–31.
33. M.B.C.A. § 119.
34. E.g., Hawaii, Hawaii Rev.Stat. § 418–14.
35. E.g., Minnesota, Minn.Stat.Ann. § 303.16.
36. Examples of the application for withdrawal and the certificate of withdrawal are Forms 12F and 12G in the appendix.
37. E.g., Connecticut, Conn.Gen.Stat.Ann. § 33–409; and Tennessee, Tenn.Code Ann. § 48–1107.
38. New Jersey, N.J.Stat.Ann., § 14A:13–10.

13

CHANGES IN CORPORATE STRUCTURE AND DISSOLUTION

Previous chapters have considered corporate activities which occur in the ordinary course of business. It was earlier noted that the board of directors, and the officers to whom they delegate authority, are vested with continuing discretion in the management of business affairs, and that the shareholders exercise only indirect control over corporate operations through their election of the directors. However, here we are concerned with extraordinary corporate activity outside the scope of corporate business routine. Each of these extraordinary matters involves structural changes to the corporation, and, in most cases, affects the ownership rights of the shareholders. Consequently, a common characteristic in each of these transactions is the requirement for shareholder approval. Moreover, the law governing extraordinary corporate activity grants special rights for shareholders in some cases, such as the right to have their shares appraised and purchased by the corporation if they disagree with the decision of management and their fellow shareholders. Special statutory procedures have been adopted by most states to regulate these structural changes, and this Chapter is devoted to them.

§ 13.01 AMENDMENT OF THE ARTICLES OF INCORPORATION

Any amendment of the articles of incorporation is a structural change of the corporation because the amendment changes the primary authorizing document for corporate existence. The corporation has the right to amend its articles of incorporation within the statutory guidelines established for the original articles of incorporation. Any provision may be inserted in an amendment if it would have been permitted in the original articles. Section 58 of the Model Act details fifteen specific amendments which may be adopted, although the section recognizes that other modifications are possible under the general corporate power to amend. Specifically, the corporation may change its name; period of duration; corporate purposes; number of authorized shares and their par value, including changing par value shares to no-par, and vice versa; and the designations, preferences, limitations or other rights of the shares. The amendment may reclassify or cancel shares; create new classes of shares; divide classes of shares; authorize or revoke authorization of the board of directors to establish series shares; and cancel dividends accrued but not declared. The amendment may also limit, grant or deny preemptive rights to shareholders.

The Model Act's broad statutory power to amend is typical of most state statutes on the subject of amendments to the articles of incorporation. Many state statutes do not detail particular amendments as thoroughly as the Model Act, but the power to amend on any absent issues may be safely implied from the general statutory authority.

Procedure

In the usual amendment procedure, the board of directors adopts a resolution which sets forth the proposed amendment and directs that it be submitted to a vote at an annual or special meeting of the shareholders.[1]

EXAMPLE 13.01A: Resolution to Change Corporate Name

RESOLVED, that Article I of the Articles of Incorporation of The Nobles Company be amended to read as follows:

"The name of this corporation is The Nobility Company."

FURTHER RESOLVED, that this amendment shall be submitted to the vote of the shareholders at a special meeting called for the purpose of considering the amendment.

Some states permit the shareholders to propose an amendment to the articles of incorporation.[2] The concerted action of a specified number of shareholders, i.e. the holders of one-tenth of the outstanding voting stock of the corporation, is required, and they may petition the board of directors to propose the amendment or may request that the president of the company call a meeting of shareholders to consider the proposed amendment.

Written notice of the proposed amendment must be given within the statutory period to each shareholder of record entitled to vote upon the proposal.[3] A few states have added special notice requirements for certain amendments, such as those changing the number of authorized shares.[4] In many cases the proposal will be submitted to the shareholders at their annual meeting and the written proposal may be included in the notice of the annual meeting. If a special meeting is called, the notice must state the reason for the meeting—that is, to consider a proposed amendment to the articles of incorporation. In jurisdictions where the shareholders may unanimously consent in writing in lieu of a meeting, that procedure may be used to consider and approve the amendment.[5]

Adoption of the Amendment

The number of shareholder votes required to approve a proposed amendment to the articles of incorporation may be greater than the number required for routine shareholder matters. Moreover, if the amendment affects the rights of the shareholders of a certain class, they also must approve the amendment, even if they otherwise have no voting rights.

The Model Act formerly required the affirmative vote of the holders of two-thirds of the shares entitled to vote, but a recent amendment to the Act reduced the vote to a majority. The reduced voting provision has been accepted in a majority of the jurisdictions which follow the Model Act.

If a proposed amendment affects the rights of the holders of a certain class of shares, they are entitled to vote as a class on its adoption. An amendment is deemed to affect the rights of a particular class when it increases or decreases the aggregate number of authorized shares of the class, or modifies the number of shares held by shareholders of the class. Changing any of the designations, preferences, limitations or rights of the shares of the class also qualify for special approval. If the proposed amendment creates a new class having superior rights to the class, provides for an exchange of shares of another class into shares of the class, or divides the class into series, class voting applies. Finally any amendment which limits or denies the pre-emptive rights of the shares of the class, or affects accrued but undeclared dividends of the class must be approved by the class.[6] In most states, a change in the par value of the shares of the class will also require a class vote. Some examples are appropriate. If the corporation has a class of common stock and a class of non-voting 6% cumulative preferred stock with a par value of $100, the holders of the preferred shares would be entitled to vote on all of the following amendments:

1. An amendment increasing par value to $200 per share;
2. An amendment changing dividends from cumulative to non-cumulative, but only if dividends have accrued at the time the amendment is proposed;
3. An amendment adding a new class of preferred stock with superior liquidation preferences to the existing preferred class;
4. An amendment permitting the directors to issue the remaining authorized shares of the preferred class in series;
5. An amendment adding an additional 1,000 authorized shares of the preferred class.

Each of these amendments directly affects the preferred shareholders by diluting their ownership interest or altering their preferred status, and, in order to pass the amendment, the holders of a majority (or two-thirds, depending upon the jurisdiction) of the shares of the class must vote affirmatively. The class has no voice on other amendments, however. If the corporation changes its stated purposes, the number of directors or its period of duration, the non-voting class may not vote, even though these amendments may indirectly affect the value or quality of the shares.

Since shareholder approval is required for adoption of an amendment to the articles of incorporation, it would obviously be difficult to amend the articles before any shares have been issued unless there were a separate procedure for that contingency. The Model Act has one in Section 59, and many states have comparable provisions. If shares have not been issued, an amendment to the articles of incorporation may be adopted by the resolution of the board of directors. Several states substitute the incorporators for directors in this instance.

Articles of Amendment

The adopted amendment is set forth in the articles of amendment which are filed with the appropriate state official. Additional fees and franchise taxes may be due under the state statute when the articles of amendment are filed.

In addition to the statement of the amendment, the Model Act requires that the articles of amendment contain information about the corporation, the number of shares entitled to vote on the amendment, the outcome of the vote, and other specified matters for special amendments.[7]

After examination of the articles of amendment to insure conformity with the law, the state officer issues a certificate of amendment, and, under the Model Act and most state statutes, the amendment becomes effective upon the issuance of the certificate.[8] In a few states the amendment is effective upon filing,[9] and an even smaller number allow the amendment to be effective at a later date as specified in the amendment.[10]

Finally, it should be noted that the additional formalities for amendment of the articles of incorporation parallel the formalities for the articles of incorporation in each jurisdiction.[11] Thus, a jurisdiction which requires that the articles be filed with a county clerk in addition to the Secretary of State will also require that the amendment to the articles be so filed. Similarly, if the state statute requires that the articles of incorporation must be published in a newspaper, the amendment to the articles must also be published.

Restated Articles of Incorporation

If the original articles of incorporation have been amended several times, it may be difficult to determine the current status of the articles by studying the files of the Secretary of State. Consequently, most statutes permit a "restatement" or "composite" of the articles of incorporation whereby all past amendments are consolidated with the original articles of incorporation into a new document which supersedes the original articles and the filed amendments. Under the Model Act shareholder approval is not necessary to restate the articles of incorporation, since it is only a mechanical process of putting the corporation's file in order. Many states require shareholder approval, and, if a new amendment is to be added in connection with the restatement, shareholder approval would be required under the Model Act as well. The procedure for restatement is specified in the statute, and a restated certificate of incorporation is issued.[12]

§ 13.02 MERGER, CONSOLIDATION AND EXCHANGE

Merger and consolidation are statutory devices for the combination of two or more corporations into one corporate entity. The resulting corporation takes over the assets, liabilities and businesses of the merging or consolidating corporations, and at least one of the corporations in the transaction will cease to exist. The corporate parties to a merger or consolidation are called "constituent" corporations, and that terminology will be used in the discussion of these transactions.

A merger is a device whereby one or more constituent corporations merge into and become a part of another constituent corporation. The corporations which merge into the other corporation cease to exist after the merger. The "surviving" corporation survives the merger, and takes over the assets and liabilities of the merging corporations. The survivor also takes over the stockholders, personnel, business contacts and other normal business activities of the terminated corporations. To illustrate, suppose the ABC Corporation and the XYZ Corporation agree to merge, and their agreement provides that the XYZ Corporation will survive the merger. When the merger is accomplished, the ABC Corporation will no longer exist, and all of its assets, liabilities and other business incidents will belong to XYZ Corporation, which maintains its original corporate structure throughout, unless the merger requires certain amendments to the structure.

A consolidation is different. In a consolidation transaction one or more constituent corporations join together to form a new corporation, pooling their assets, liabilities and business and transferring them to a new consolidated entity. The hypothetical corporations described above could consolidate by forming the LMN Corporation, and transferring all of their respective business to this new corporation. In a consolidation, all constituent corporations cease to exist and the consolidation results in a fresh new corporate kid on the block.

A more cautious combination is an exchange, neither corporation ceases to exist in an exchange, but some or all of the shares of one corporation will be exchanged for some or all of the shares of the other corporation. For example, the XYZ Corporation could exchange a certain number of its common shares for all of the preferred shares of ABC Corporation, or all of the common shares of ABC Corporation, or some combination of those transactions. Of course, if XYZ Corporation exchanged shares of its common stock for *all* shares of ABC Corporation, common and preferred, that begins to look like a merger, and it may be necessary to follow merger rules.

Merger, consolidation and exchange involve structural changes and affect share ownership in the constituent corporations. Consider the shareholders of the ABC Corporation in the merger with the XYZ Corporation. After the merger, their corporation no longer exists, and they would rightfully expect to be consulted for their approval of the transaction. The shareholders of the expiring constituent corporation will usually receive a specified number of shares of the surviving corporation or cash in return for their original shares. The shareholders of the XYZ Corporation should also approve the transaction, because their share ownership will be diluted when shares are issued to all of the shareholders of the late ABC Corporation. The consolidation and exchange transactions involve the same equities, since shareholders of both constituent corporations will probably be receiving shares of the new consolidated corporation or in exchange for their original shares.

This section will be devoted to the statutory procedures for effecting a merger, consolidation and exchange but before pressing on, permit a short digression in deference to the tax-lawyer's approach to these problems. In tax parlance the merger, consolidation or exchange may be referred to as a "reorganization", based upon definitions of the Internal Revenue Code.[13] The tax treatment of the transfers of assets depends upon the type of reorganization involved, and several different types of reorganizations are defined. The "statutory merger or consolidation"—that is, one accomplished under the state corporate statutes, as described in this section—is called an "A Reorganization". Other important types of reorganization may involve the acquisition of one corporation by another or a combination of two corporations but they fall short of the complete corporate law merger or consolidation. In "B Reorganization" one corporation swaps its voting shares for a controlling block (80–100%) of shares of another corporation. This is an exchange transaction. Both corporations continue to exist, however, so full merger or consolidation is not accomplished. The acquired corporation becomes a subsidiary of the acquiring corporation, which maintains at least an 80 percent controlling interest in the subsidiary's stock. A "C Reorganization" involves an exchange of voting shares of the acquiring corporation for substantially all of the assets of the acquired corporation. Since both corporations continue to exist, this also is not a true merger or consolidation. Neither is it an exchange because the stock was traded for assets, not for other stock. The tax statute identifies three other types of reorganizations, "D", "E" and "F Reorganizations", which need not be described here, but detailed sources are available to the inquisitive student to learn the definitions and tax aspects of the reorganizations.[14]

With that cursory explanation of the tax terminology out of the way, we may proceed with an analysis of the corporate practice requirements of a statutory merger or consolidation.

Procedure

The board of directors' resolution is the procedural starting point for a merger or consolidation. The board of directors of both corporations for a merger or consolidation and the board of directors of the corporation whose shares are to be acquired in an exchange, approves the transaction stating the names of the constituent corporations and the name of the corporation into which they intend to merge, or the name of the new corporation into which they will consolidate; the terms of the proposed merger, consolidation or the name of the corporation whose shares will be exchanged; and the manner and basis for converting the shares of the constituent corporations into shares of the exchanging corporation, or shares or cash from the surviving corporation in merger or the new corporation in consolidation. The plan approved by resolution must also state changes to be made in the articles of incorporation of the surviving corporation for merger, and the content of the articles of incorporation of the new corporation in consolidation. The plan may further include any other terms necessary to accomplish the transaction.[15]

While the resolution of the board of directors is the first statutory step toward approval of a merger or consolidation, it is really the tip of the iceberg. Notice that the resolution must contain the terms and conditions of the proposed transaction. This unassuming requirement represents the culmination of several months (maybe years) of planning, drafting and negotiation between the parties to establish those terms. Corporate management will have labored over a lengthy agreement containing terms of the structural changes which they believe will be acceptable to the shareholders and in the best business interests of all corporate parties. New corporate purposes must be drafted to account for the expanded business; the positions of the directors and officers of the constituent corporations must be placed or abandoned in the surviving or new corporation; the accounts of all corporate parties must be combined and reconciled; and by-laws must be harmonized. Certain restrictions will usually be placed on the constituent corporations during the pendency of the transaction regarding dividends, sales of stock, issuance of options or other activities out of the ordinary course of business. During various stages of these negotiations the corporate parties will frequently exchange "letters of intent" which express in writing their respective understandings of the terms of the proposed agreement. Further negotiations are conducted based upon these stated positions, and eventually the negotiations will result in the final agreement, or in abandonment of the transaction if the negotiations reach an impasse. After acceptable terms are drafted,[16] a proposed closing date will be set, considering the other preparatory procedures which must be accomplished prior to closing. Rulings on the tax ramifications of the transaction are usually required, and the impact of the securities laws on the transfers of stock should be examined. Current accounting opinions should be scheduled and financial reports will be supplemented with current information. Documents must be reviewed by the attorneys, accountants and other experts for all parties, and appropriate director and shareholder meetings must be held in accordance with state law. That brings us back to the statutory requirements, which begin with the director's resolution to approve the merger or consolidation plan.

The resolution should reflect that the plan of merger or consolidation has been presented to the meeting of directors and approved by them, should authorize appropriate corporate officers to call a meeting of shareholders to consider the plan, and should further authorize the officers to file the necessary documents to accomplish the plan if the shareholders of the constituent corporations approve it.

EXAMPLE 13.02A: Resolution to Approve Merger

RESOLVED, that the board of directors hereby recommends and approves the proposed Plan of Merger between this corporation and The Nobles Company, a Colorado corporation, substantially in the form presented to this meeting, and the directors and officers of this corporation are hereby authorized to enter into said plan by executing the same, under the seal of this corporation, and

FURTHER RESOLVED, that said plan as entered into by the directors and officers of this corporation be submitted to the holders of the common stock of this corporation at a special meeting to be called for the purpose of considering and adopting said plan on August 15, 1981 at 2:00 P.M., at the offices of the corporation, and

FURTHER RESOLVED, that July 15, 1981, is hereby fixed as the record date for the determination of the holders of the common stock entitled to notice of and to vote at such special meeting, and

FURTHER RESOLVED, that in the event said plan shall be approved and adopted at the special meeting of the shareholders of this corporation in accordance with the statutory requirements of the State of Colorado, and shall also be approved and adopted by the shareholders of The Nobles Company in accordance with the statutory requirements of the State of Colorado, then the Secretary of this corporation is hereby authorized to certify upon said plan that it has been adopted, and the President and Secretary of this corporation are hereby authorized to execute articles of merger in the name and on behalf of this corporation and under its seal and to cause the same to be filed in the Office of the Secretary of State of the State of Colorado.

The Model Act requires shareholder approval for any merger or consolidation by the shareholders of both corporations and approval of an exchange by the shareholders of the corporation whose shares are being exchanged. There are two important exceptions: if the acquiring corporation owns 90% of the stock of a subsidiary, it may merge that subsidiary into itself without shareholder approval; and if the merger has a "small impact" on the affairs of the surviving corporation, the shareholders of the surviving corporation do not vote.[17] Shareholder approval of a plan of merger, consolidation or exchange is very similar to that required for amendment of the articles of incorporation and other structural changes. The plan may be considered at either a special or annual meeting of shareholders. The Model Act requires notice to be given to every shareholder, whether or not entitled to vote, within twenty days of the meeting, and the notice must always state that the plan of merger, consolidation or exchange is to be considered at the meeting.[18] Most states require that notice be sent to every shareholder and that it contain a statement of the purpose of the meeting. Further, it may have to inform shareholders of their dissenting rights. The period of notice varies from sixty days[19] to ten days.[20]

Because these transactions affect all corporate shares, many jurisdictions permit all shares to vote on the plan, whether or not they have the right to vote on other corporate matters. The Model Act originally demanded these expanded voting rights, but was later amended to include only regular voting shares on the theory that shareholders with non-voting stock had waived the right to vote unless the shares of their particular class would be directly affected by the plan. Presently, Section 73 of the Model Act requires the affirmative vote of the holders of the majority of voting stock and the affirmative vote of the holders of shares of each class entitled to vote, based upon the same tests for class voting as applied to amendments to the articles of incorporation.[21] The majority vote concept is a relatively recent addition to the Model Act, and many states continue to require the affirmative vote of the holders of two-thirds of the voting shares.

Shareholders are almost uniformly granted the right to dissent to these transactions and to demand payment for their shares.[22]

Articles of Merger, Consolidation or Exchange

Following the shareholder approval, articles of merger or articles of consolidation are prepared and filed with the appropriate state official. Many state statutes have no provision for separate "articles" and instead require that the plan of merger or consolidation, duly certified as having been approved, be filed. Publication may also be required, paralleling the formalities for the original articles of incorporation.[23]

Section 74 of the Model Act establishes the contents of the articles of merger or articles of consolidation, and Section 72A provides the contents of the plan of exchange, including:

1. The terms and conditions of the proposed transaction, which is usually contained in a plan which attached to the articles and made a part thereof by reference;
2. The number of shares outstanding in each constituent corporation, and if the shares are entitled to vote as a class, the designation and number of shares for each class, or, a statement that a vote of shareholders is not required; and
3. The number of shares voted for and against the plan for each constituent corporation, and the same information for each class entitled to vote as such.

After the articles have been found to conform to law, and the necessary franchise taxes and fees have been paid, a certificate or merger, consolidation or exchange is issued. As with other matters involving a certificate from the Secretary of State, the Model Act makes the transaction effective when the certificate is issued.[24] Nearly half of the jurisdictions date the effectiveness by the act of filing the required documents,[25] and many also provide that the effectiveness of the transaction may be delayed until a date fixed in the plan.[26] The delayed effectiveness alternative is particularly desirable where filings are required in several states and the simultaneous filing is impracticable. The effective date may be set at a specified time and all filings completed prior to that date.

Statutory Effect

When the merger or consolidation becomes effective, all constituent corporate parties to the plan become a single corporation (the designated survivor in a merger or the new corporation in a consolidation), and the other corporations cease to exist. The surviving or new corporation has all of the rights and privileges, is vested with all assets, and is responsible for all liabilities and obligations of the constituent corporations. The articles of incorporation of the surviving corporation are deemed amended to the extent provided in the merger plan, filed as a part of the articles of merger. Thus, if the plan requires modifications to the structure of the surviving corporation, there is no need to comply separately with the statutory procedure for amendments to the articles of incorporation. In the case of consolidation, the articles of consolidation are deemed to be the articles of incorporation of the new consolidated corporation.[27]

§ 13.03 SALE, MORTGAGE OR OTHER DISPOSITION OF ASSETS

If the corporation disposes of substantially all of its assets, a "corporate shell" results, and while the basic corporate structure remains the same, the corporation becomes an organization without normal business assets. The sale, mortgage, lease, exchange or other disposition of substantially all corporate assets is considered by most states to be a structural change in the corporation which requires shareholders approval.

This type of transaction may be a part of a "C Reorganization" in tax language, where an acquiring corporation exchanges its voting stock for substantially all of the assets of the acquired corporation, or a "D Reorganization," where substantially all of the assets are transferred to a corporation which is controlled by the transferring corporation or its shareholders.[28] These transactions are not statutory mergers or consolidations since all corporations survive the transaction. However, instead of owning business assets, the transferring corporation will own voting stock of the acquiring corporation. The management of the corporation could also sell the entire corporate business to a purchaser for cash, and subsequently dissolve the corporation, distributing the cash to its shareholders. Statutes regulating these dispositions of assets are designed to secure shareholder approval if substantially all of the assets of the corporation are to be alienated from the business. To illustrate the equities of these statutes, suppose that the Nobles Company is engaged in the business of manufacturing and selling sporting goods. Its assets would include all the machinery used for manufacture, the manufacturing plant, its inventory of skis, bicycles and other sporting goods, accounts receivable, good will and so forth. If substantially all of these assets were sold to another company for cash or stock, the Nobles Company shareholders will have an entirely different investment. Instead of an investment in a growing, successful sporting goods company, they own a corporation holding cash, which will probably be distributed to them in exchange for their shares. Alternatively, their corporation may have received stock of the purchasing corporation, and while the business may be continued by the purchaser, it operates under different management, who probably have different policies and interests. The character of the investment is thus changed. The law recognizes the fairness of consulting shareholders for their approval of such transactions.

The Model Act makes several distinctions regarding these transactions in Sections 78 and 79. First, the mortgage or pledge of corporate property never requires shareholder approval. Second, the sale, exchange, lease or other disposition of substantially all of the property and assets *in the regular course of business* does *not* require shareholder approval. Third, shareholder approval is required if substantially all of the corporate assets are sold, leased, exchanged or disposed of in a transaction not within the ordinary course of business.

Most states permit the mortgage or pledge of corporate property without shareholder approval. In these transactions, the corporation continues to use the property, but has granted an interest in the property as collateral to secure a loan or other obligation. Business should continue as usual, and the property will be lost only if the corporation defaults on the obligation.

The character of the shareholder's investment will not be affected if all goes as planned—that is, the corporate business will generate enough income to pay the obligation, the mortgage or pledge will be removed, and the corporate assets will remain intact. Consequently, there is no pressing need for shareholder protection here.

Several jurisdictions provide, as does the Model Act, that a sale or other disposition of substantially all of the corporate assets, which is within the regular course of corporate business, may be accomplished by action of the board of directors without shareholder approval. The theory behind this rule is easy to understand. If the transaction is within the ordinary course of corporate business, the board of directors is already authorized to proceed with it, and shareholder approval is never required for normal business transactions. On the other hand, other state statutes do not attempt to distinguish between transactions in or out of the ordinary course of business, perhaps because the distinction is difficult to apply. However, where the distinction exists, the normalcy of the transaction may be determined by the statement of purposes in the articles of incorporation. Suppose a corporation is organized for the purpose of purchasing and selling a single parcel of real estate, anticipating a profit from the sale. When the property is sold, the transaction is within the ordinary course of business, since that is exactly what the corporation was organized to do. Most cases are not that easy, however. If the articles of incorporation of the Nobles Company stated that one of the purposes of the corporation would be to "sell, lease, transfer, exchange or otherwise deal in the assets of the corporation," the broad enabling authority may make a transfer of substantially all assets a normal corporate event, but that would certainly be subject to interpretation. From the standpoint of better corporate practice, any questionable transaction should be approved by the shareholders.

Procedure

Once the sale or other disposition of assets is characterized as a structural change (as it is in every case where the transaction is not within the ordinary course of business), the procedure for approval of the transaction parallels the approval of a merger or consolidation. The board of directors adopts a resolution recommending the transaction and directing the submission of its terms to the shareholders for their approval:

EXAMPLE 13.03A: Resolution for Sale of Assets Outside the Ordinary Course of Business

RESOLVED, that this Board do and hereby does declare that the consideration in the form of capital stock of The Nobles Company to be received in exchange for the hereinabove described properties and interests, is a full, fair and adequate consideration; that this Board do and hereby does ratify, confirm and approve all of the acts of its officers in making said agreement with The Nobles Company; and that this Board do and hereby does recommend to the stockholders of this Company that said agreement be approved by said stockholders; and

FURTHER RESOLVED, that the question of approval of said agreement with The Nobles Company be submitted to the stockholders of this Company in a special meeting called for that purpose; and to that end it is

> FURTHER RESOLVED, that a special meeting of the stockholders of this Company be called to be held at the principal office of this Company in the City of Des Moines, State of Iowa, at 10:00 o'clock, a.m., on the 10th day of October, 1981; that the secretary of this Company be and he hereby is authorized and directed to give all the stockholders of this Company proper, timely and adequate notice of the time, place and purpose of said meeting; and that books for the transfer of stock will close at the conclusion of business on September 15, 1981, and will reopen on the day following the adjournment of said meeting.

The shareholders may consider the transaction at either an annual or special meeting, but twenty days' notice, with a statement of the purpose of the meeting, is required by the Model Act.[29] Every shareholder, whether or not otherwise entitled to vote, must receive such notice. The period of notice is likely to be different in the several states, and some require a statement of the shareholder's rights if they dissent to the transaction.[30] The Model Act requires an affirmative vote of the majority of the voting shares to approve the transaction, and authorizes class voting if the transaction affects the rights of the particular class. Many states continue to adhere to a two-thirds affirmative vote requirement, and some permit all outstanding shares to vote on the transaction.[31]

Bulk Transfer Requirements

The Uniform Commercial Code contains provisions for the protection of business creditors whenever an enterprise whose principal business is to sell merchandise from stock (including those who manufacture what they sell) sells a major part of its assets out of the regular course of business. The Uniform Commercial Code has been adopted in every state except Louisiana, so its provisions must be observed whenever a corporate "enterprise", as defined, sells or otherwise transfers a major part of its assets outside the scope of ordinary business activities. Our hypothetical Nobles Company which is engaged in the manufacture and sale of sporting goods is one such enterprise, and a transfer of its assets, including inventory, equipment, materials, and supplies is subject to these requirements.

The Bulk Transfer requirements are concerned only with the creditors of the selling corporation who obviously have an interest in the assets of the business. If the business assets are sold in one transaction without their knowledge, their chances of being paid may be significantly reduced.

To comply with the Code the seller must prepare an affidavit containing the names and business addresses of all creditors and the amounts owed to them as of the date of the transfer. The seller must also prepare a schedule identifying the property transferred. The list of creditors and the schedule identifying the property must be preserved by the purchaser for a period of six months following the transfer, and it must be available for inspection and copying by any business creditor.[32] In addition to the list and schedule, notice must be sent to every creditor of the business at least ten days before the buyer takes possession of the property or pays for it, whichever occurs first.[33] Only those creditors who hold claims as of the date of the transfer should receive notice.

Failure to prepare and preserve the list and schedule or to give notice to creditors makes the transfer ineffective against the creditors of the business, meaning that they may disregard the transfer and use the property to satisfy their claims. Of course, if the creditors are paid, they have no claim against the property. Moreover, the right to disregard the transfer does not last indefinitely. No creditor may reach the property more than six months after the transferee took possession of the goods, unless the transfer has been concealed, in which case the six month period runs from the date the transfer is discovered.[34]

§ 13.04 RIGHTS OF DISSENTING SHAREHOLDERS

Under the Model Act and several statutes nearly half of the outstanding shares may be voted *against* a merger, consolidation, exchange or sale of assets and the transaction will still be approved. The majority rule controls the holders of the dissenting shares, who must live with the decision of the majority, despite the effect the transaction may have on their shares. The statutory solution to this problem is to grant the dissenting shareholders the right to have their shares appraised and purchased, with limited exceptions, if they do not want to continue as investors in the corporation. In some states this is called the "shareholder's right of appraisal" or the "shareholder's right to demand payment of the value of the stock." Every state includes some rights of dissent and payment in its corporate statute.

Circumstances Giving Rise to Appraisal Rights

The Model Act grants the right to dissent in cases of mergers, consolidations, exchanges and sales or exchange of substantially all of assets outside the ordinary course of business. In addition, dissenter's rights will apply if the articles of incorporation are amended to materially affect the rights of a shareholder (such as abolishing a preferential right, a preemptive right, a redemption right or voting right), and if the articles or bylaws so provide, they may apply to any transaction designated therein.[35] The dissent and appraisal rights are limited by two important exceptions. They do not apply to shareholders of a surviving corporation whose votes were not necessary to approve the merger. This rule covers a merger of a subsidiary corporation into a parent corporation when the parent owned 90% or more of the stock of the subsidiary before the merger and to the shareholders of the surviving corporation when the merger will have a "small impact" upon the surviving corporation.[36] The Model Act also excepts holders of shares registered on a national securities exchange at the time the shareholders entitled to vote were identified, unless the articles of incorporation provided otherwise. Delaware and a few other states[37] have a similar rule, excepting holders of shares listed on a national exchange and holders of a class of shares which has 2,000 or more shareholders of record. The assumption which forms the basis for these rules is that the shares are readily salable if they are registered on a national securities exchange (or if 2,000 other shareholders exist) so the cumbersome appraisal procedure is not necessary to satisfy the dissenting shareholder. He should be able to sell his shares at market if he decides to terminate his investment.

The Model Act includes another unusual provision allowing a shareholder to dissent as to less than all shares registered in his name, in which case his rights are to be determined as to the shares dissenting and the remaining shares are treated as if they belonged to different shareholders. On its face, the provision seems to be directed to an indecisive shareholder who is uncertain about the transaction, and will dissent as to some of his shares to recoup some of his investment, but will keep other shares in case the structural change turns out to be successful. Theoretically, this could happen, but the rule was designed to permit brokers, trustees and agents holding shares for their clients to split the shares into approving and disapproving groups depending upon their clients' wishes.

Some states are more generous with dissenting shareholders' appraisal rights than the Model Act. They extend the rights to certain other amendments to the articles of incorporation, such as changes in the corporate purposes, extension of corporate life, and changes in the capital stock structure.[38]

Procedure

Statutory provisions detailing the procedures for perfecting the shareholder's right to payment after dissent are quite complex. The first step is the dissent itself. The shareholder must file *written* objection to the transaction before or at the meeting of shareholders called to consider the transaction.[39] The objection must be made prior to the meeting in some states and a few require that the objection include a demand for appraisal and purchase of the shares.[40]

Of course, the shareholder must not vote in favor of the transaction at the meeting, and the Model Act now requires that the shareholder must *refrain* from voting at the meeting.

We must assume the transaction is approved by the other shareholders and will eventually be consummated by the corporation. After approval, the shareholder has a certain period of time in most states within which to demand payment of the fair value of his shares. This time period varies. The demand is addressed to the corporation, in case of a sale of assets; to the surviving corporation, in case of a merger; or to the new corporation in a consolidation. If the shareholder fails to object, to vote against the action, or to demand payment within the time provided, he loses his right to payment for shares, and is bound by the corporate action.

The revised Model Act now takes a little different approach. If the transaction is approved, the corporation must send notice of that fact to the shareholder who has objected and refrained from voting, together with a form to use to demand payment and a copy of the statute, stating when the demand must be made and the shares deposited. The time period in the notice cannot be less than 30 days from the date the notice was mailed. Failure to return the demand on time with the shares, forfeits the shareholder's rights to dissent and sell shares to the corporation.

The "fair value" of the shares is defined by the Model Act to be the value immediately before the effectuation of the corporate transaction, excluding appreciation or depreciation in anticipation of the transaction. This test requires an evaluation of the impact of the publicity surrounding the transaction in the appraisal of the shares, and is extremely difficult to apply. For that reason many states ignore appreciation or depreciation in value determination, or set the date for this appraisal at some other date further removed from shareholder approval.

The corporation and the shareholder are encouraged to agree upon the value of the shares, but if fair value is disputed, the Model Act prescribes an elaborate procedure for resolving the issue. The corporation gives notice of the consummation of the transaction to all demanding shareholders when the transaction is effected and remits what it believes to be a fair price for the shares, accompanied by a recent balance sheet and profit and loss statement, and a notice of the dissenter's right to supplemental payments.

If the shareholder disagrees with the estimate of fair value he may file his own estimate with the corporation within 30 days. If the corporation cannot settle with the shareholder within 60 days after demand for supplemental payment, the question of fair value will be referred to a court. The corporation must file the petition and, if it fails to do so, the corporation must pay the shareholders the price they demanded in the supplemental demand. If an action is filed with a court, all dissenters who dispute fair value are made parties to the action. The court may appoint appraisers to determine the fair value of the shares, and the court will enter a judgment for the value they set. The corporation is usually required to pay the expenses of the proceedings, but the court may assess the expenses against the dissenting shareholders if it finds that their rejection of the corporation's offered amount was arbitrary, vexatious or not in good faith, and was without justification.

Many states continue to base dissenter's rights on the old Model Act provisions which also contained elaborate procedures for these disputes. To illustrate the differences between the old and new Model Act sections, and to show dissenter's rights procedures as they exist in many states, the following table is offered:

Timetable #1	Days	Former Section 81	Revised Section 81
	-1	Date fair value determined	
Date of meeting for shareholder approval	0	Latest date for written objection to action; shareholder must vote against action	Latest date for filing written objection; shareholder must refrain from voting; if transaction approved, corporation immediately sends notice to dissenters
	10	Latest date for written demand for payment	
	30	Latest date for submitting certificates representing shares to corporation for notation	Earliest date that corporation may require shareholder to demand payment and deposit certificates

	60		If transaction not effected by now, the corporation must return certificates
Timetable #2 Date action is effected	0		Corporation remits its estimates of fair value with information required by statute
	10	Latest date for corporation notice to shareholders with an offer at a specified price	
	30	Latest date for agreement	
	60	Latest date for corporation filing petition on its own	Latest date for dissenter's demand for supplemental payment
	90	Latest date for corporation filing petition on shareholder demand; date of payment if agreement reached on fair value	Latest date for corporation to settle with shareholder or file a petition with the court. Failure to do one or the other requires payment to the dissenter at dissenter's price

Considering this complicated procedure, you should be able to understand why a corporate officer would equate the blessed glory of heaven with unanimous shareholder approval of the transaction. Further, in the spirit of the analogy, the dissenting shareholder, like a fallen angel, is banished from the corporate paradise, unless he repents. This remedy of the shareholder is said to be "exclusive", meaning that a demand for payment for his shares is the only remedy he has if he is dissatisfied with the transaction. When the demand for payment is made, the shareholder loses the right to vote or to exercise any other rights as a shareholder. He is usually not entitled to withdraw the demand unless the corporation consents, and he may regain status as a shareholder only upon withdrawal and consent, or if the corporation abandons the transaction, or if a court decides that he is not entitled to the right of payment for his shares.

§ 13.05 VOLUNTARY DISSOLUTION

The dissolution of the corporation is certainly a structural change which will affect the shareholders. Again they must be consulted for their approval.

Procedure

The corporation may be dissolved at any time after it is formed by the appropriate concurrence of its aggregate membership. A decision to dissolve may be made immediately after formation, before the corporation commences business and before shares are issued. The incorporators constitute the total aggregate membership at this point and an admission that the corporation was a bad idea is theirs to make. The Model Act procedure for dissolution prior to commencement of business and prior to the issuance of shares is very simple. The majority of the incorporators may execute and file articles of dissolution and a certificate of dissolution will be issued.[41]

A corporation which has commenced business and has issued shares may be dissolved through two typical procedures which may originate with either the shareholders or the directors of the corporation.

If the shareholders take the initiative, the Model Act allows voluntary dissolution by their unanimous written consent.[42] All shareholders, whether or not they are otherwise entitled to vote, normally must join the consent, but a few states allow the holders of only the voting shares to make the decision.[43] Unanimity is not always required.[44]

If the decision to dissolve eminates from the board of directors or if unanimous shareholder consent to the dissolution is not feasible, the corporation may be dissolved by the usual procedure for corporate structural changes—that is, by the directors and the shareholders acting in their respective meetings. The board of directors adopts a resolution recommending that the corporation be dissolved and submits this resolution to the vote of the shareholders at either an annual or special meeting.

EXAMPLE 13.05A: Resolution to Dissolve

The president made a statement as to the present plight of the Corporation, and after a confirmatory statement by the treasurer it was unanimously

RESOLVED, that the Board of Directors hereby recommend to the stockholders that in their interest this Corporation be dissolved and its affairs wound up; and

FURTHER RESOLVED, that a special meeting of the stockholders of the Corporation be held at the offices of the Corporation, on the 3rd day of December, 1981, at 3:00 o'clock p. m., to vote on the question as to whether this Corporation be dissolved, and that the secretary is hereby directed to give due notice of said meeting.[45]

The Model Act does not specify any unusual time period within which notice must be given, but the notice must state that the meeting will be for the purpose of considering the dissolution of the corporation.[46]

EXAMPLE 13.05B: Notice of a Meeting for Dissolution

The stockholders of The Nobles Company are hereby notified that a special meeting of the stockholders of said Company will be held at the corporation's offices, 200 West 14th Avenue, Denver, Colorado, on the 3rd day of December, 1981, at 3:00 o'clock p. m., to vote on the question as to whether said Company should be dissolved.

Dated November 18, 1981.

____________________, Secretary[47]

The shareholders must approve dissolution by a designated percentage of the vote. The Model Act originally required the affirmative vote of the holders of two-thirds of the outstanding shares entitled to vote on the issue, but the percentage was recently reduced to a majority of the voting shares. Class voting is also authorized under the Model Act. Many states continue to adhere to the two-thirds vote requirement, and several jurisdictions allow every share to vote on the dissolution question, whether or not otherwise entitled to vote. The minutes of the shareholder meeting would reflect the approval of the dissolution:

EXAMPLE 13.05C: Resolution to Dissolve

WHEREAS, the Board of Directors believe this Corporation should be dissolved, and have called this special meeting of the stockholders to consider the matter; and

WHEREAS, after considering the statements of officers and a report of a committee of the stockholders, it appears to be for the best interests of the stockholders of this Corporation that its business should be terminated, the Corporation dissolved, and its assets distributed according to law:

NOW, THEREFORE, the holders of record of two-thirds of the outstanding shares of this Corporation entitled to vote therein concurring therein,

RESOLVED, that this Corporation hereby elects to dissolve, and that pursuant to the Colorado Corporation Law, the president and secretary, or other proper officers, are hereby authorized to execute and file the proper certificate of dissolution with the Secretary of State, that they duly publish the certificate of the Secretary of State of said filing, and that they and the other officers of this Corporation are hereby authorized and directed to take the steps prescribed by law to complete the dissolution and to wind up the affairs of this Corporation.[48]

A dissenting shareholder's appraisal remedy is rare in dissolution, and even if authorized, is usually limited to special circumstances surrounding the dissolution.

Statement of Intent to Dissolve

The Model Act dissolution procedure has two sections designed to give advance public and private notice to outsiders that the corporation has initiated dissolution proceedings. These notices are intended to facilitiate orderly liquidation of the corporation.

The statement of intent to dissolve, the first notice filed for dissolution of a going concern, is not required for dissolution by the incorporators since one of the prerequisites to that dissolution procedure is that the corporation has not yet commenced business, and thus protection of the public is not deemed to be necessary.[49]

The statement of intent must be executed and filed with the Secretary of State if the dissolution has been approved by unanimous consent of the shareholders or by resolution of the board of directors with subsequent shareholder approval.[50] Many states require this first notice of the dissolution, although the statement may have to be published rather than filed.

Upon filing the statement, or beginning other specified statutory requirements for dissolution, the corporation must cease all normal business activity, and it may continue in business only for the purpose of winding-up its affairs. The filing of the statement of intent does not terminate corporate existence. Under the Model Act, the corporate existence continues until a certificate of dissolution has been issued by the Secretary of State or until a court has declared the corporation to be dissolved.

Notice to Creditors

After filing a statement of intent to dissolve, the corporation will proceed to collect its assets and to liquidate its business. As a part of the liquidation process the corporation gives notice of its intent to dissolve to each known creditor. This is the private notice which complements the filed statement. Through these notices, everyone who cares about the corporation should have learned about the dissolution before it becomes effective. Some states require that the corporation advertise its intention to dissolve in a newspaper, rather than sending notices directly to the creditors.

Revocation of Voluntary Dissolution Proceedings

Just as the shareholders and the corporation approve voluntary dissolution, so may they revoke it. This indecision is expensive and time-consuming, but every statute gives the intra-corporate parties the right to change their minds. The procedure for revocation usually duplicates the procedure for approval. Under the Model Act, the shareholders may, by written consent any time prior to the issuance of a certificate of dissolution, revoke the dissolution proceedings by submitting a statement of revocation to the Secretary of State.[51] The revocation may also be accomplished by act of the corporation. The board of directors may submit a resolution revoking voluntary dissolution proceedings to the vote of the shareholders. Shareholder approval of revocation of voluntary dissolution requires the same vote (either majority or two-thirds) as that required for approval of dissolution.[52] When the statement of revocation is filed with the Secretary of State, the corporation may again proceed to conduct business, as if nothing had ever happened.[53] The revocation of voluntary dissolution proceedings must occur before a certificate of dissolution is issued under the Model Act, but a few jurisdictions require the decision earlier[54] or permit the revocation much later.[55]

Articles of Dissolution

If the voluntary dissolution proceedings have not been revoked, then after payment of all corporate debts, liabilities and obligations and after distribution of the remaining corporate property and assets to the shareholders, articles of dissolution are executed and filed with the Secretary of State, with the same formality as the original articles of incorporation. The articles of dissolution must state that the corporation has been liquidated. If the corporation had been a going concern the statement recites that all debts, obligations and liabilities of the corporation have been paid and discharged or that adequate provisions have been made therefor, and that all the remaining property and assets of the corporation have been distributed among the shareholders in accordance with their respective rights and interests.[56] If the voluntary dissolution occurred prior to the issuance of shares and commencing business, the incorporators file articles of dissolution declaring those facts and confirming the return of amounts paid for subscriptions.[57] Upon the receipt and examination of the articles of dissolution the Secretary of State issues a certificate of dissolution, and at that point the existence of the corporation ceases, except for the purposes of suits or other proceedings.[58]

§ 13.06 INVOLUNTARY DISSOLUTION

By the State

A wayward corporation may be dragged, perhaps kicking and screaming, into dissolution by its creator, the state. The state is always entitled to enforce its laws, and if a corporation has failed to comply with the statutory requirements, the Attorney General may bring an action to force dissolution of the corporation. Typical corporate abuses which will justify involuntary dissolution include failing to file annual reports, failing to pay franchise taxes, procuring articles of incorporation through fraud, abusing or exceeding authority granted by law, and failing to appoint a registered agent or to notify regarding a change of its registered office within thirty days.[59] Some jurisdictions add insolvency, unfair competition or restraint of trade, persistent violations of state laws, or an existence which is detrimental to the public interest.

Most corporations will never run afoul of state law or commit other acts which support involuntary dissolution, but there are a few provisions in this area which are to be approached with caution. Even otherwise conscientious corporate officers may delay annual reports, overlook payment of franchise taxes or neglect to report changes of the registered office or registered agent. The procedures here should be explored, therefore, in an effort to save this creation over which we have labored nine chapters from untimely demise for a mere oversight.

The Model Act provisions appear to take the fairest approach to these problems. Under Section 95, the Secretary of State sends notice of any alleged transgressions to the Attorney General of the state *and* to the corporation at its registered office. The Attorney General is directed to file an action in an appropriate court for the dissolution of the corporation, unless the corporation cures the problem before the action is filed. The corporation may remedy the problem and avoid dissolution even after the action is filed but the corporation will have to pay the costs of the action. Not all states are so forgiving. While many states require notice to the corporation, only about half the jurisdictions allow the corporation to cure the defect after dissolution proceedings have been filed.

If the corporation fails to respond to the notices, or if it fights the dissolution proceedings and loses, the court may order liquidation of its business and thereafter enter a decree of dissolution which terminates corporate existence.[60]

By Shareholders

If corporate management refuses to consider dissolution and the unanimous consent of the shareholders cannot be obtained, voluntary dissolution is impossible, but an involuntary dissolution procedure may be invoked in special circumstances. This problem may arise in several situations. For example, under the Model Act if the majority of the directors are also shareholders and oppose dissolution, no matter how many other shareholders favor dissolution, it cannot be accomplished voluntarily. The approval of the director-shareholders is necessary for unanimous shareholder consent and for a director resolution for dissolution. Dissolution by voluntary proceedings is also impossible if the directors or shareholders are deadlocked.

The statutory escape is the shareholders' application to a court for liquidation of the business and a decree of dissolution. Section 97 of the Model Act grants liquidation power to a court upon application of a shareholder who can establish an unbreakable director deadlock threatening irreparable injury to the corporation; oppressive, illegal or fraudulent acts of those in control of the corporation; a shareholder deadlock in failing to elect new directors for a period of two years; or misapplication or waste of the corporate assets. These grounds are typical among corporate statutes granting the shareholder right to bring an action for involuntary liquidation and dissolution. Abandonment of the corporate business or persistent commission of ultra-vires acts are also frequently specified grounds.

If the shareholder proves the allegations in the action, the court proceeds to a judicially supervised liquidation ending in a decree of dissolution.[61] However, if the problem is solved during the course of the liquidation proceedings, they will be discontinued, and all corporate property will be returned to the corporation.[62]

By a Creditor

Creditors may also force involuntary dissolution if the corporation is insolvent and the creditor's claim is undisputed. The Model Act deems claims which have been reduced to judgment and claims which have been admitted by the corporation in writing to be undisputed.[63] The creditor is in the frustrating position of owning an uncontested debt, which corporate management can't pay because the corporation is insolvent. Moreover, management is resisting dissolution and liquidation whereby the creditor would receive some satisfaction from the assets. In such a case, the creditor may force judicial liquidation and involuntary dissolution.

§ 13.07 LIQUIDATION

Closely associated with dissolution, whether voluntary or involuntary, is the process of collecting all corporate assets, completing or terminating unexecuted contracts, paying creditors and expenses, and distributing the remains to the owners. These activities are collectively referred to as liquidation. Under the Model Act, corporate existence ceases when a decree of dissolution is entered by a court or when the Secretary of State issues a certificate of dissolution.[64] Consequently, liquidation and winding-up precedes final dissolution.

Non-judicial Liquidation

When dissolution is voluntary and does not involve judicial proceedings, corporate management is responsible for winding-up the corporate business and liquidating. This process is commenced after filing the statement of intent to dissolve and must be completed before the articles of dissolution are filed, since the articles recite that all debts, obligations and liabilities have been paid or provided for, and that the remaining assets have been distributed among the shareholders. There is no time limit on the liquidation process, but practical considerations encourage management to proceed as expeditiously as possible. Non-judicial liquidation may be conducted as informally as desired, as long as all creditors are paid and remaining assets are distributed. Special safeguards are inserted for creditors: the directors of the corporation may be personally liable if they distribute assets to the shareholders without providing for creditors,[65] and a forgotten creditor may enforce his claim against the corporation, directors or shareholders for a period of two years after dissolution.[66]

If the directors or officers become immersed in liquidation and discover dissatisfied shareholders, or hostile creditors, they may apply to have the liquidation supervised by a court.

Judicial Liquidation

Court-supervised liquidation is available to corporate management in voluntary dissolution proceedings and is also used when involuntary proceedings have been commenced by the state, the shareholders or creditors of the corporations. The court may enjoin any person who threatens to interfere with orderly proceedings, and may appoint a receiver who will carry on the corporate business and preserve its assets during the proceedings. Creditors are usually required to file their claims under oath within a prescribed time, and a hearing will be held to finally determine the claims of all parties. A liquidating receiver is then appointed with authority to collect and sell the assets of the corporation, to apply the proceeds to the expenses of the liquidation and to creditor's claims and then to distribute remaining funds to shareholders.[67]

Under the Model Act, the Court will issue a decree of dissolution when liquidation has been completed.

Liquidation Distributions

In any liquidation of a corporation, judicially supervised or conducted by management, the corporate assets will be collected and may be sold, and the proceeds are first used to pay expenses of liquidation and then creditors of the corporation. Whatever remains belongs to the shareholders of the corporation. The remnants of their corporation are distributed to them in accordance with their liquidation preferences.[68]

NOTES

1. M.B.C.A. § 59.
2. E.g., Pennsylvania, 15 Pa.Stat. § 1802; and Tennessee, Tenn.Code Ann. § 48.302.
3. See Section 8.07(C).
4. E.g., Montana, Rev.Code Mont. § 15–2253. (Thirty days' notice if the amendment increases the authorized number of shares).
5. See M.B.C.A. § 145 and Section 8.09.
6. See M.B.C.A. § 60.
7. M.B.C.A. § 61. An example of articles of amendment appears as Form 13A in the Appendix.
8. M.B.C.A. § 63. An example of a certificate of amendment appears as Form 13B in the Appendix.
9. E.g., Nebraska, Neb.Rev.Stat. § 21–2062; and Pennsylvania, 15 Pa.Stat. § 1809.
10. E.g., Florida, Fla.Stat.Ann. § 607.191(2) (not more than 90 days after filing); and N.J.Stat.Ann. § 14A:9–4(5) (not more than thirty days after filing).
11. See Section 6.06.
12. Examples of restated articles and a restated certificate of incorporation are Forms 13C and 13D in the Appendix.
13. Int.Rev.Code of 1954, 25 U.S.C.A. § 368(a)(1).
14. See B. Bittker and J. Eustice, Federal Income Taxation of Corporations and Shareholders, Chapter 14 (1971); H. Henn, Corporations § 351 (1970); R. Deer, Lawyer's Basic Practice Manual § 9.03 (1971).
15. See M.B.C.A. §§ 71, 72 and 72A.
16. Sample agreements for statutory mergers and statutory consolidations may be found in West's Modern Legal Forms §§ 3044–3049.
17. M.B.C.A. §§ 75 and 73(d). A "small impact" means that the name of the surviving corporation does not change, the stockholders of the surviving corporation have the same number of shares before and after the transaction, and the voting shares or participating shares (entitled to dividends or other distributions) in the surviving corporation do not increase more than 20% as a result of the transaction.
18. M.B.C.A. § 73.
19. E.g., New Jersey, N.J.Stat.Ann. § 14A:10–3.
20. E.g., Pennsylvania, 15 Pa.Stat. § 1902.
21. See Section 13.01.
22. See Section 13.04.
23. See Section 6.06.
24. M.B.C.A. § 76. Examples of articles of merger, articles of consolidation and certificates of each appear as Forms 13E–1 in the Appendix.
25. E.g., California, West's Ann.Calif.Corp. Code § 1103; New York, McKinney Consol. Laws of N.Y.Bus.Corp. Law § 906; and Pennsylvania, 15 Pa.Stat. § 1906.
26. E.g., Iowa, Iowa Code Ann. § 496A.73; and Michigan, Mich.Comp.Laws Ann. §§ 450.1131, 450.1707(2) (no later than 90 days after filing).
27. M.B.C.A. § 76.
28. Int.Rev.Code of 1954, 26 U.S.C.A. § 368(1)(a).
29. M.B.C.A. § 79.
30. E.g., Georgia, Ga.Code Ann. § 22.1102; and Pennsylvania, 15 Pa.Stat. § 1131(B). See section 13.04.
31. E.g., Connecticut, Conn.Gen.Stat.Ann. § 33–372; and North Carolina, N.C.Gen.Stat. § 55–112. (two-thirds of all outstanding shares necessary to approve the transaction).
32. Uniform Commercial Code (hereinafter cited as U.C.C.) § 6–104.
33. U.C.C. § 6–105. Examples of the affidavit and notice to creditors appear as Forms 13J and 13K in the Appendix.
34. U.C.C. § 6–111.

35. M.B.C.A. § 80. Dissenter's rights are not extended to mortgage, lease or other disposition of substantially all of the corporate assets, unless so provided in the articles of incorporation or bylaws.
36. M.B.C.A. §§ 73(d) and 75.
37. E.g., Florida, Fla.Stat.Ann. § 607.244; and Pennsylvania, 15 Pa.Stat. § 1515.
38. E.g., Minnesota, Minn.Stat.Ann. § 301.40 (change of purpose and extension of corporate life); New York, McKinney Consol. Laws of N.Y.Bus.Corp.Law §§ 806, 910, 1319.
39. M.B.C.A. § 81.
40. E.g., New Jersey, N.J.Stat.Ann. § 14A.11–2; and Texas, Vernon's Ann.Tex.Stat.Bus.Corp. Act art. 5.12.
41. M.B.C.A. § 82.
42. M.B.C.A. § 83.
43. E.g., Delaware, 8 Del.Code Ann. § 275(c).
44. The articles of incorporation in Michigan may permit an individual shareholder, or any specified number of shares, to dissolve the corporation at will by signing a written consent to that effect. Mich.Comp.Laws § 450.1805.
45. West's Modern Legal Forms § 3082.
46. M.B.C.A. § 84.
47. West's Modern Legal Forms § 3083.
48. West's Modern Legal Forms § 3084.
49. M.B.C.A. §§ 85 and 86.
50. M.B.C.A. § 85. Examples of statements of intent to dissolve are Forms 13L and 13M in the Appendix.
51. M.B.C.A. § 88.
52. M.B.C.A. § 89.
53. M.B.C.A. § 90. Examples of statements of revocation appear as Forms 13N and 13O in the Appendix.
54. E.g., California, West's Ann.Calif.Corp. Code § 1901 (prior to distribution of assets).
55. E.g., Delaware, 8 Del.Code Ann. § 311.
56. M.B.C.A. § 92.
57. M.B.C.A. § 82. Sample articles of dissolution appear as Forms 13P and 13Q in the Appendix.
58. In order to provide redress against the corporation or on behalf of the corporation for claims arising prior to dissolution, several states extend corporate life for a specified period beyond dissolution. A dissolved corporation continues as a body corporate for this purpose indefinitely in California, West's Ann.Calif.Corp. Code § 2001; New Jersey, N.J.Stat.Ann. § 14A:12–9; New York, McKinney Consol. Laws of N.Y.Bus.Corp. Law § 1006, and others; a period of three years in Delaware, 8 Del.Code Ann. § 278, Massachusetts, Mass.Gen. Laws Ann. c156B § 102 and others. Section 105 of the Model Act, states that the dissolution of the corporation does not impair any remedy available to or against the corporation for any right or claim arising prior to dissolution, provided the action is commenced within two years after dissolution. The corporation remains in existence under law for the limited purpose of pursuing litigation commenced within a two year period following dissolution. Examples of certificates of dissolution appear as Forms 13R and 13S in the Appendix.
59. M.B.C.A. § 94.
60. M.B.C.A. §§ 97(d) and 102.
61. M.B.C.A. §§ 97–100, 102.
62. M.B.C.A. § 101.
63. M.B.C.A. § 97(b).
64. M.B.C.A. §§ 93, 102.
65. M.B.C.A. § 48.
66. M.B.C.A. § 105 and see footnote 58.
67. M.B.C.A. § 98.
68. See Section 9.08.

APPENDIX A

UNIFORM PARTNERSHIP ACT

Sec. 26 Nature of Partner's Interest in the Partnership.
Sec. 27 Assignment of Partner's Interest.
Sec. 28 Partner's Interest Subject to Charging Order.

Part VI Dissolution and Winding Up
Sec. 29 Dissolution Defined.
Sec. 30 Partnership not Terminated by Dissolution.
Sec. 31 Causes of Dissolution.
Sec. 32 Dissolution by Decree of Court.
Sec. 33 General Effect of Dissolution on Authority of Partner.
Sec. 34 Right of Partner to Contribution from Co-partners after Dissolution.
Sec. 35 Power of Partner to Bind Partnership to Third Persons after Dissolution.
Sec. 36 Effect of Dissolution on Partner's Existing Liability.
Sec. 37 Right to Wind Up.
Sec. 38 Rights of Partners to Application of Partnership Property.
Sec. 39 Rights Where Partnership Is Dissolved for Fraud or Misrepresentation.
Sec. 40 Rules for Distribution.
Sec. 41 Liability of Persons Continuing the Business in Certain Cases.
Sec. 42 Rights of Retiring or Estate of Deceased Partner When the Business Is Continued.
Sec. 43 Accrual of Actions.

Part VII Miscellaneous Provisions
Sec. 44 When Act Takes Effect.
Sec. 45 Legislation Repealed.

(Adopted in 48 States, all except Georgia and Louisiana; the District of Columbia, the Virgin Islands, and Guam. The adoptions by Alabama and Nebraska do not follow the official text in every respect, but are substantially similar, with local variations.)

The Act consists of 7 Parts as follows:

I. Preliminary Provisions

II. Nature of Partnership

III. Relations of Partners to Persons Dealing with the Partnership

IV. Relations of Partners to One Another

V. Property Rights of a Partner

VI. Dissolution and Winding Up

VII. Miscellaneous Provisions

An Act to make uniform the Law of Partnerships
Be it enacted, etc.:

Part I **Preliminary Provisions**

Sec. 1. **Name of Act.**

This act may be cited as Uniform Partnership Act.

Sec. 2. **Definition of Terms.**

In this act, "Court" includes every court and judge having jurisdiction in the case.
"Business" includes every trade, occupation, or profession.
"Person" includes individuals, partnerships, corporations, and other associations.
"Bankrupt" includes bankrupt under the Federal Bankruptcy Act or insolvent under any state insolvent act.
"Conveyance" includes every assignment, lease, mortgage, or encumbrance.
"Real property" includes land and any interest or estate in land.

Sec. 3. **Interpretation of Knowledge and Notice.**

(1) A person has "knowledge" of a fact within the meaning of this act not only when he has actual knowledge thereof, but also when he has knowledge of such other facts as in the circumstances shows bad faith.

(2) A person has "notice" of a fact within the meaning of this act when the person who claims the benefit of the notice

(a) States the fact to such person, or

(b) Delivers through the mail, or by other means of communication, a written statement of the fact to such person or to a proper person at his place of business or residence.

Sec. 4. **Rules of Construction.**

(1) The rule that statutes in derogation of the common law are to be strictly construed shall have no application to this act.

(2) The law of estoppel shall apply under this act.

(3) The law of agency shall apply under this act.

(4) This act shall be so interpreted and construed as to effect its general purpose to make uniform the law of those states which enact it.

(5) This act shall not be construed so as to impair the obligations of any contract existing when the act goes into effect, nor to affect any action or proceedings begun or right accrued before this act takes effect.

Sec. 5. **Rules for Cases Not Provided for in this Act.**

In any case not provided for in this act the rules of law and equity, including the law merchant, shall govern.

Part II **Nature of Partnership**

Sec. 6. **Partnership Defined.**

(1) A partnership is an association of two or more persons to carry on as co-owners a business for profit.

(2) But any association formed under any other statute of this state, or any statute adopted by authority, other than the authority of this state, is not a partnership under this act, unless such association would have been a partnership in this state prior to the adoption of this act; but this act shall apply to limited partnerships except in so far as the statutes relating to such partnerships are inconsistent herewith.

Sec. 7. **Rules for Determining the Existence of a Partnership.**

In determining whether a partnership exists, these rules shall apply:

(1) Except as provided by Section 16 persons who are not partners as to each other are not partners as to third persons.

(2) Joint tenancy, tenancy in common, tenancy by the entireties, joint property, common property, or part

ownership does not of itself establish a partnership, whether such co-owners do or do not share any profits made by the use of the property.

(3) The sharing of gross returns does not of itself establish a partnership, whether or not the persons sharing them have a joint or common right or interest in any property from which the returns are derived.

(4) The receipt by a person of a share of the profits of a business is prima facie evidence that he is a partner in the business, but no such inference shall be drawn if such profits were received in payment:

(a) As a debt by installments or otherwise,

(b) As wages of an employee or rent to a landlord,

(c) As an annuity to a widow or representative of a deceased partner,

(d) As interest on a loan, though the amount of payment vary with the profits of the business.

(e) As the consideration for the sale of a good-will of a business or other property by installments or otherwise.

Sec. 8. **Partnership Property.**

(1) All property originally brought into the partnership stock or subsequently acquired by purchase or otherwise, on account of the partnership, is partnership property.

(2) Unless the contrary intention appears, property acquired with partnership funds is partnership property.

(3) Any estate in real property may be acquired in the partnership name. Title so acquired can be conveyed only in the partnership name.

(4) A conveyance to a partnership in the partnership name, though without words of inheritance, passes the entire estate of the grantor unless a contrary intent appears.

Part III **Relations of Partners to Persons Dealing with the Partnership**

Sec. 9. **Partner Agent of Partnership as to Partnership Business.**

(1) Every partner is an agent of the partnership for the purpose of its business, and the act of every partner, including the execution in the partnership name of any instrument, for apparently carrying on in the usual way the business of the partnership of which he is a member binds the partnership, unless the partner so acting has in fact no authority to act for the partnership in the particular matter, and the person with whom he is dealing has knowledge of the fact that he has no such authority.

(2) An act of a partner which is not apparently for the carrying on of the business of the partnership in the usual way does not bind the partnership unless authorized by the other partners.

(3) Unless authorized by the other partners or unless they have abandoned the business, one or more but less than all the partners have no authority to:

(a) Assign the partnership property in trust for creditors or on the assignee's promise to pay the debts of the partnership,

(b) Dispose of the good-will of the business,

(c) Do any other act which would make it impossible to carry on the ordinary business of a partnership,

(d) Confess a judgment,

(e) Submit a partnership claim or liability to arbitration or reference.

(4) No act of a partner in contravention of a restriction on authority shall bind the partnership to persons having knowledge of the restriction.

Sec. 10. **Conveyance of Real Property of the Partnership.**

(1) Where title to real property is in the partnership name, any partner may convey title to such property by a conveyance executed in the partnership name; but the partnership may recover such property unless the partner's act binds the partnership under the provisions of paragraph (1) of section 9 or unless such property has been conveyed by the grantee or a person claiming through such grantee to a holder for value without knowledge that the partner, in making the conveyance, has exceeded his authority.

(2) Where title to real property is in the name of the partnership, a conveyance executed by a partner, in his own name, passes the equitable interest of the partnership, provided the act is one within the authority of the partner under the provisions of paragraph (1) of section 9.

(3) Where title to real property is in the name of one or more but not all the partners, and the record does not disclose the right of the partnership, the partners

in whose name the title stands may convey title to such property, but the partnership may recover such property if the partners' act does not bind the partnership under the provisions of paragraph (1) of section 9, unless the purchaser or his assignee, is a holder for value, without knowledge.

(4) Where the title to real property is in the name of one or more or all the partners, or in a third person in trust for the partnership, a conveyance executed by a partner in the partnership name, or in his own name, passes the equitable interest of the partnership, provided the act is one within the authority of the partner under the provisions of paragraph (1) of section 9.

(5) Where the title to real property is in the names of all the partners a conveyance executed by all the partners passes all their rights in such property.

Sec. 11. **Partnership Bound by Admission of Partner.**

An admission or representation made by any partner concerning partnership affairs within the scope of his authority as conferred by this act is evidence against the partnership.

Sec. 12. **Partnership Charged with Knowledge of or Notice to Partner.**

Notice to any partner of any matter relating to partnership affairs, and the knowledge of the partner acting in the particular matter, acquired while a partner or then present to his mind, and the knowledge of any other partner who reasonably could and should have communicated it to the acting partner, operate as notice to or knowledge of the partnership, except in the case of a fraud on the partnership committed by or with the consent of that partner.

Sec. 13. **Partnership Bound by Partner's Wrongful Act.**

Where, by any wrongful act or omission of any partner acting in the ordinary course of the business of the partnership or with the authority of his copartners, loss or injury is caused to any person, not being a partner in the partnership, or any penalty is incurred, the partnership is liable therefor to the same extent as the partner so acting or omitting to act.

Sec. 14. **Partnership Bound by Partner's Breach of Trust.**

The partnership is bound to make good the loss:

(a) Where one partner acting within the scope of his apparent authority receives money or property of a third person and misapplies it; and

(b) Where the partnership in the course of its business receives money or property of a third person and the money or property so received is misapplied by any partner while it is in the custody of the partnership.

Sec. 15. **Nature of Partner's Liability.**

All partners are liable

(a) Jointly and severally for everything chargeable to the partnership under sections 13 and 14.

(b) Jointly for all other debts and obligations of the partnership; but any partner may enter into a separate obligation to perform a partnership contract.

Sec. 16. **Partner by Estoppel.**

(1) When a person, by words spoken or written or by conduct, represents himself, or consents to another representing him to any one, as a partner in an existing partnership or with one or more persons not actual partners, he is liable to any such person to whom such representation has been made, who has, on the faith of such representation, given credit to the actual or apparent partnership, and if he has made such representation or consented to its being made in a public manner he is liable to such person, whether the representation has or has not been made or communicated to such person so giving credit by or with the knowledge of the apparent partner making the representation or consenting to its being made.

(a) When a partnership liability results, he is liable as though he were an actual member of the partnership.

(b) When no partnership liability results, he is liable jointly with the other persons, if any, so consenting to the contract or representation as to incur liability, otherwise separately.

(2) When a person has been thus represented to be a partner in an existing partnership, or with one or more persons not actual partners, he is an agent of the persons consenting to such representation to bind them to the same extent and in the same manner as though he were a partner in fact, with respect to persons who rely upon the representation. Where all the members of the existing partnership consent to the representation, a partnership act or obligation

results; but in all other cases it is the joint act or obligation of the person acting and the persons consenting to the representation.

Sec. 17. **Liability of Incoming Partner.**

A person admitted as a partner into an existing partnership is liable for all the obligations of the partnership arising before his admission as though he had been a partner when such obligations were incurred, except that this liability shall be satisfied only out of partnership property.

Part IV **Relations of Partners to One Another**

Sec. 18. **Rules Determining Rights and Duties of Partners.**

The rights and duties of the partners in relation to the partnership shall be determined, subject to any agreement between them, by the following rules:

(a) Each partner shall be repaid his contributions, whether by way of capital or advances to the partnership property and share equally in the profits and surplus remaining after all liabilities, including those to partners, are satisfied; and must contribute towards the losses, whether of capital or otherwise, sustained by the partnership according to his share in the profits.

(b) The partnership must indemnify every partner in respect of payments made and personal liabilities reasonably incurred by him in the ordinary and proper conduct of its business, or for the preservation of its business or property.

(c) A partner, who in aid of the partnership makes any payment or advance beyond the amount of capital which he agreed to contribute, shall be paid interest from the date of the payment or advance.

(d) A partner shall receive interest on the capital contributed by him only from the date when repayment should be made.

(e) All partners have equal rights in the management and conduct of the partnership business.

(f) No partner is entitled to remuneration for acting in the partnership business, except that a surviving partner is entitled to reasonable compensation for his services in winding up the partnership affairs.

(g) No person can become a member of a partnership without the consent of all the partners.

(h) Any difference arising as to ordinary matters connected with the partnership business may be decided by a majority of the partners; but no act in contravention of any agreement between the partners may be done rightfully without the consent of all the partners.

Sec. 19. **Partnership Books.**

The partnership books shall be kept, subject to any agreement between the partners, at the principal place of business of the partnership, and every partner shall at all times have access to and may inspect and copy any of them.

Sec. 20. **Duty of Partners to Render Information.**

Partners shall render on demand true and full information of all things affecting the partnership to any partner or the legal representative of any deceased partner or partner under legal disability.

Sec. 21. **Partner Accountable as a Fiduciary.**

(1) Every partner must account to the partnership for any benefit, and hold as trustee for it any profits derived by him without the consent of the other partners from any transaction connected with the formation, conduct, or liquidation of the partnership or from any use by him of its property.

(2) This section applies also to the representatives of a deceased partner engaged in the liquidation of the affairs of the partnership as the personal representatives of the last surviving partner.

Sec. 22. **Right to an Account.**

Any partner shall have the right to a formal account as to partnership affairs:

(a) If he is wrongfully excluded from the partnership business or possession of its property by his co-partners,

(b) If the right exists under the terms of any agreement,

(c) As provided by section 21,

(d) Whenever other circumstances render it just and reasonable.

Sec. 23. **Continuation of Partnership Beyond Fixed Term.**

(1) When a partnership for a fixed term or particular undertaking is continued after the termination of such term or particular undertaking without any

express agreement, the rights and duties of the partners remain the same as they were at such termination, so far as is consistent with a partnership at will.

(2) A continuation of the business by the partners or such of them as habitually acted therein during the term, without any settlement or liquidation of the partnership affairs, is prima facie evidence of a continuation of the partnership.

Part V **Property Rights of a Partner**

Sec. 24. **Extent of Property Rights of a Partner.**

The property rights of a partner are (1) his rights in specific partnership property, (2) his interest in the partnership, and (3) his right to participate in the management.

Sec. 25. **Nature of a Partner's Right in Specific Partnership Property.**

(1) A partner is co-owner with his partners of specific partnership property holding as a tenant in partnership.

(2) The incidents of this tenancy are such that:

(a) A partner, subject to the provisions of this act and to any agreement between the partners, has an equal right with his partners to possess specific partnership property for partnership purposes; but he has no right to possess such property for any other purpose without the consent of his partners.

(b) A partner's right in specific partnership property is not assignable except in connection with the assignment of rights of all the partners in the same property.

(c) A partner's right in specific partnership property is not subject to attachment or execution, except on a claim against the partnership. When partnership property is attached for a partnership debt the partners, or any of them, or the representatives of a deceased partner, cannot claim any right under the homestead or exemption laws.

(d) On the death of a partner his right in specific partnership property vests in the surviving partner or partners, except where the deceased was the last surviving partner, when his right in such property vests in his legal representative. Such surviving partner or partners, or the legal representative of the last surviving partner, has no right to possess the partnership property for any but a partnership purpose.

(e) A partner's right in specific partnership property is not subject to dower, curtesy, or allowances to widows, heirs, or next of kin.

Sec. 26. **Nature of Partner's Interest in the Partnership.**

A partner's interest in the partnership is his share of the profits and surplus, and the same is personal property.

Sec. 27. **Assignment of Partner's Interest.**

(1) A conveyance by a partner of his interest in the partnership does not of itself dissolve the partnership, nor, as against the other partners in the absence of agreement, entitle the assignee, during the continuance of the partnership to interfere in the management or administration of the partnership business or affairs, or to require any information or account of partnership transactions, or to inspect the partnership books; but it merely entitles the assignee to receive in accordance with his contract the profits to which the assigning partner would otherwise be entitled.

(2) In case of a dissolution of the partnership, the assignee is entitled to receive his assignor's interest and may require an account from the date only of the last account agreed to by all the partners.

Sec. 28. **Partner's Interest Subject to Charging Order.**

(1) On due application to a competent court by any judgment creditor of a partner, the court which entered the judgment, order, or decree, or any other court, may charge the interest of the debtor partner with payment of the unsatisfied amount of such judgment debt with interest thereon; and may then or later appoint a receiver of his share of the profits, and of any other money due or to fall due to him in respect of the partnership, and make all other orders, directions, accounts and inquiries which the debtor partner might have made, or which the circumstances of the case may require.

(2) The interest charged may be redeemed at any time before foreclosure, or in case of a sale being directed by the court may be purchased without thereby causing a dissolution:

(a) With separate property, by any one or more of the partners, or

(b) With partnership property, by any one or more of the partners with the consent of all the partners whose interests are not so charged or sold.

(3) Nothing in this act shall be held to deprive a partner of his right, if any, under the exemption laws, as regards his interest in the partnership.

Part VI **Dissolution and Winding up**

Sec. 29. **Dissolution Defined.**

The dissolution of a partnership is the change in the relation of the partners caused by any partner ceasing to be associated in the carrying on as distinguished from the winding up of the business.

Sec. 30. **Partnership Not Terminated by Dissolution.**

On dissolution the partnership is not terminated, but continues until the winding up of partnership affairs is completed.

Sec. 31. **Causes of Dissolution.**

Dissolution is caused:

(1) Without violation of the agreement between the partners,

(a) By the termination of the definite term or particular undertaking specified in the agreement,

(b) By the express will of any partner when no definite term or particular undertaking is specified,

(c) By the express will of all the partners who have not assigned their interests or suffered them to be charged for their separate debts, either before or after the termination of any specified term or particular undertaking.

(d) By the explusion of any partner from the business bona fide in accordance with such a power conferred by the agreement between the partners;

(2) In contravention of the agreement between the partners, where the circumstances do not permit a dissolution under any other provision of this section, by the express will of any partner at any time;

(3) By any event which makes it unlawful for the business of the partnership to be carried on or for the members to carry it on in partnership;

(4) By the death of any partner;

(5) By the bankruptcy of any partner or the partnership;

(6) By decree of court under section 32.

Sec. 32. **Dissolution by Decree of Court.**

(1) On application by or for a partner the court shall decree a dissolution whenever:

(a) A partner has been declared a lunatic in any judicial proceeding or is shown to be of unsound mind,

(b) A partner becomes in any other way incapable of performing his part of the partnership contract,

(c) A partner has been guilty of such conduct as tends to affect prejudicially the carrying on of the business,

(d) A partner wilfully or persistently commits a breach of the partnership agreement, or otherwise so conducts himself in matters relating to the partnership business that it is not reasonably practicable to carry on the business in partnership with him,

(e) The business of the partnership can only be carried on at a loss,

(f) Other circumstances render a dissolution equitable.

(2) On the application of the purchaser of a partner's interest under sections 27 or 28:

(a) After the termination of the specified term or particular undertaking,

(b) At any time if the partnership was a partnership at will when the interest was assigned or when the charging order was issued.

Sec. 33. **General Effect of Dissolution on Authority of Partner.**

Except so far as may be necessary to wind up partnership affairs or to complete transactions begun but not then finished, dissolution terminates all authority of any partner to act for the partnership,

(1) With respect to the partners,

(a) When the dissolution is not by the act, bankruptcy or death of a partner; or

(b) When the dissolution is by such act, bankruptcy or death of a partner, in cases where section 34 so requires.

(2) With respect to persons not partners, as declared in section 35.

Sec. 34. **Right of Partner to Contribution From Copartners After Dissolution.**

Where the dissolution is caused by the act, death or bankruptcy of a partner, each partner is liable to his copartners for his share of any liability created by any partner acting for the partnership as if the partnership had not been dissolved unless

(a) The dissolution being by act of any partner, the partner acting for the partnership had knowledge of the dissolution, or

(b) The dissolution being by the death or bankruptcy of a partner, the partner acting for the partnership had knowledge or notice of the death or bankruptcy.

Sec. 35. **Power of Partner to Bind Partnership to Third Persons After Dissolution.**

(1) After dissolution a partner can bind the partnership except as provided in Paragraph (3)

(a) By any act appropriate for winding up partnership affairs or completing transactions unfinished at dissolution;

(b) By any transaction which would bind the partnership if dissolution had not taken place, provided the other party to the transaction

(I) Had extended credit to the partnership prior to dissolution and had no knowledge or notice of the dissolution; or

(II) Though he had not so extended credit, had nevertheless known of the partnership prior to dissolution, and, having no knowledge or notice of dissolution, the fact of dissolution had not been advertised in a newspaper of general circulation in the place (or in each place if more than one) at which the partnership business was regularly carried on.

(2) The liability of a partner under paragraph (1b) shall be satisfied out of partnership assets alone when such partner had been prior to dissolution

(a) Unknown as a partner to the person with whom the contract is made; and

(b) So far unknown and inactive in partnership affairs that the business reputation of the partnership could not be said to have been in any degree due to his connection with it.

(3) The partnership is in no case bound by any act of a partner after dissolution

(a) Where the partnership is dissolved because it is unlawful to carry on the business, unless the act is appropriate for winding up partnership affairs; or

(b) Where the partner has become bankrupt; or

(c) Where the partner has no authority to wind up partnership affairs; except by a transaction with one who

(I) Had extended credit to the partnership prior to dissolution and had no knowledge or notice of his want of authority; or

(II) Had not extended credit to the partnership prior to dissolution, and, having no knowledge or notice of his want of authority, the fact of his want of authority has not been advertised in the manner provided for advertising the fact of dissolution in paragraph (1bII).

(4) Nothing in this section shall affect the liability under section 16 of any person who after dissolution represents himself or consents to another representing him as a partner in a partnership engaged in carrying on business.

Sec. 36. **Effect of Dissolution on Partner's Existing Liability.**

(1) The dissolution of the partnership does not of itself discharge the existing liability of any partner.

(2) A partner is discharged from any existing liability upon dissolution of the partnership by an agreement to that effect between himself, the partnership creditor and the person or partnership continuing the business; and such agreement may be inferred from the course of dealing between the creditor having knowledge of the dissolution and the person or partnership continuing the business.

(3) Where a person agrees to assume the existing obligations of a dissolved partnership, the partners whose obligations have been assumed shall be discharged from any liability to any creditor of the partnership who, knowing of the agreement, consents

to a material alteration in the nature or time of payment of such obligations.

(4) The individual property of a deceased partner shall be liable for all obligations of the partnership incurred while he was a partner but subject to the prior payment of his separate debts.

Sec. 37. **Right to Wind Up.**

Unless otherwise agreed the partners who have not wrongfully dissolved the partnership or the legal representative of the last surviving partner, not bankrupt, has the right to wind up the partnership affairs; provided, however, that any partner, his legal representative or his assignee, upon cause shown, may obtain winding up by the court.

Sec. 38. **Rights of Partners to Application of Partnership Property.**

(1) When dissolution is caused in any way, except in contravention of the partnership agreement, each partner as against his co-partners and all persons claiming through them in respect of their interests in the partnership, unless otherwise agreed, may have the partnership property applied to discharge its liabilities, and the surplus applied to pay in cash the net amount owing to the respective partners. But if dissolution is caused by expulsion of a partner, bona fide under the partnership agreement and if the expelled partner is discharged from all partnership liabilities, either by payment or agreement under section 36(2), he shall receive in cash only the net amount due him from the partnership.

(2) When dissolution is caused in contravention of the partnership agreement the rights of the partners shall be as follows:

(a) Each partner who has not caused dissolution wrongfully shall have,

(I) All the rights specified in paragraph (1) of this section, and

(II) The right, as against each partner who has caused the dissolution wrongfully, to damages for breach of the agreement.

(b) The partners who have not caused the dissolution wrongfully, if they all desire to continue the business in the same name, either by themselves or jointly with others, may do so, during the agreed term for the partnership and for that purpose may possess the partnership property, provided they secure the payment by bond approved by the court, or pay to any partner who has caused the dissolution wrongfully, the value of his interest in the partnership at the dissolution, less any damages recoverable under clause (2aII) of the section, and in like manner indemnify him against all present or future partnership liabilities.

(c) A partner who has caused the dissolution wrongfully shall have:

(I) If the business is not continued under the provisions of paragraph (2b) all the rights of a partner under paragraph (1), subject to clause (2aII), of this section,

(II) If the business is continued under paragraph (2b) of this section the right as against his co-partners and all claiming through them in respect of their interests in the partnership, to have the value of his interest in the partnership, less any damages caused to his co-partners by the dissolution, ascertained and paid to him in cash, or the payment secured by bond approved by the court, and to be released from all existing liabilities of the partnership; but in ascertaining the value of the partner's interest the value of the good-will of the business shall not be considered.

Sec. 39. **Rights Where Partnership is Dissolved for Fraud or Misrepresentation.**

Where a partnership contract is rescinded on the ground of the fraud or misrepresentation of one of the parties thereto, the party entitled to rescind is, without prejudice to any other right, entitled,

(a) To a lien on, or right of retention of, the surplus of the partnership property after satisfying the partnership liabilities to third persons for any sum of money paid by him for the purchase of an interest in the partnership and for any capital or advances contributed by him; and

(b) To stand, after all liabilities to third persons have been satisfied, in the place of the creditors of the partnership for any payments made by him in respect of the partnership liabilities; and

(c) To be indemnified by the person guilty of the fraud or making the representation against all debts and liabilities of the partnership.

Sec. 40. **Rules for Distribution.**

In settling accounts between the partners after dissolution, the following rules shall be observed, subject to any agreement to the contrary:

(a) The assets of the partnership are:

(I) The partnership property,

(II) The contributions of the partners necessary for the payment of all the liabilities specified in clause (b) of this paragraph.

(b) The liabilities of the partnership shall rank in order of payment, as follows:

(I) Those owing to creditors other than partners,

(II) Those owing to partners other than for capital and profits,

(III) Those owing to partners in respect of capital,

(IV) Those owing to partners in respect of profits.

(c) The assets shall be applied in the order of their declaration in clause (a) of this paragraph to the satisfaction of the liabilities.

(d) The partners shall contribute, as provided by section 18(a) the amount necessary to satisfy the liabilities; but if any, but not all, of the partners are insolvent, or, not being subject to process, refuse to contribute, the other parties shall contribute their share of the liabilities, and, in the relative proportions in which they share the profits, the additional amount necessary to pay the liabilities.

(e) An assignee for the benefit of creditors or any person appointed by the court shall have the right to enforce the contributions specified in clause (d) of this paragraph.

(f) Any partner or his legal representative shall have the right to enforce the contributions specified in clause (d) of this paragraph, to the extent of the amount which he has paid in excess of his share of the liability.

(g) The individual property of a deceased partner shall be liable for the contributions specified in clause (d) of this paragraph.

(h) When partnership property and the individual properties of the partners are in possession of a court for distribution, partnership creditors shall have priority on partnership property and separate creditors on individual property, saving the rights of lien or secured creditors as heretofore.

(i) Where a partner has become bankrupt or his estate is insolvent the claims against his separate property shall rank in the following order:

(I) Those owing to separate creditors,

(II) Those owing to partnership creditors,

(III) Those owing to partners by way of contribution.

Sec. 41. **Liability of Persons Continuing the Business in Certain Cases.**

(1) When any new partner is admitted into an existing partnership, or when any partner retires and assigns (or the representative of the deceased partner assigns) his rights in partnership property to two or more of the partners, or to one or more of the partners and one or more third persons, if the business is continued without liquidation of the partnership affairs, creditors of the first or dissolved partnership are also creditors of the partnership so continuing the business.

(2) When all but one partner retire and assign (or the representative of a deceased partner assigns) their rights in partnership property to the remaining partner, who continues the business without liquidation of partnership affairs, either alone or with others, creditors of the dissolved partnership are also creditors of the person or partnership so continuing the business.

(3) When any partner retires or dies and the business of the dissolved partnership is continued as set forth in paragraphs (1) and (2) of this section, with the consent of the retired partners or the representative of the deceased partner, but without any assignment of his right in partnership property, rights of creditors of the dissolved partnership and of the creditors of the person or partnership continuing the business shall be as if such assignment had been made.

(4) When all the partners or their representatives assign their rights in partnership property to one or more third persons who promise to pay the debts and who continue the business of the dissolved partnership, creditors of the dissolved partnership are also creditors of the person or partnership continuing the business.

(5) When any partner wrongfully causes a dissolution and the remaining partners continue the business under the provisions of section 38(2b), either alone or with others, and without liquidation of the partner-

ship affairs, creditors of the dissolved partnership are also creditors of the person or partnership continuing the business.

(6) When a partner is expelled and the remaining partners continue the business either alone or with others, without liquidation of the partnership affairs, creditors of the dissolved partnership are also creditors of the person or partnership continuing the business.

(7) The liability of a third person becoming a partner in the partnership continuing the business, under this section, to the creditors of the dissolved partnership shall be satisfied out of partnership property only.

(8) When the business of a partnership after dissolution is continued under any conditions set forth in this section the creditors of the dissolved partnership, as against the separate creditors of the retiring or deceased partner or the representative of the deceased partner, have a prior right to any claim of the retired partner or the representative of the deceased partner against the person or partnership continuing the business, on account of the retired or deceased partner's interest in the dissolved partnership or on account of any consideration promised for such interest or for his right in partnership property.

(9) Nothing in this section shall be held to modify any right of creditors to set aside any assignment on the ground of fraud.

(10) The use by the person or partnership continuing the business of the partnership name, or the name of a deceased partner as part thereof, shall not of itself make the individual property of the deceased partner liable for any debts contracted by such person or partnership.

Sec. 42. **Rights of Retiring or Estate of Deceased Partner When the Business is Continued.**

When any partner retires or dies, and the business is continued under any of the conditions set forth in section 41 (1, 2, 3, 5, 6), or section 38(2b), without any settlement of accounts as between him or his estate and the person or partnership continuing the business, unless otherwise agreed, he or his legal representative as against such persons or partnership may have the value of his interest at the date of dissolution ascertained, and shall receive as an ordinary creditor an amount equal to the value of his interest in the dissolved partnership with interest, or, at his option or at the option of his legal representative, in lieu of interest, the profits attributable to the use of his right in the property of the dissolved partnership; provided that the creditors of the dissolved partnership as against the separate creditors, or the representative of the retired or deceased partner, shall have priority on any claim arising under this section, as provided by section 41(8) of this act.

Sec. 43. **Accrual of Actions.**

The right to an account of his interest shall accrue to any partner, or his legal representative, as against the winding up partners or the surviving partners or the person or partnership continuing the business, at the date of dissolution, in the absence of any agreement to the contrary.

Part VII **Miscellaneous Provisions**

Part VII **Miscellaneous Provisions**

Sec. 44. **When Act Takes Effect.**

This act shall take effect on the _____ day of _____ one thousand nine hundred and _____.

Sec. 45. **Legislation Repealed.**

All acts or parts of acts inconsistent with this act are hereby repealed.

APPENDIX **B**

UNIFORM LIMITED PARTNERSHIP ACT*

*The Uniform Limited Partnership Act was revised in 1976, but most states base their state statutes on the original version of the Act. This is the original version, and the Revised Act follows.

(Adopted in 46 states, all except Connecticut, Minnesota, Wyoming and Louisiana; also in the District of Columbia, and the Virgin Islands.)

An Act to Make Uniform the Law Relating to Limited Partnerships

Be it enacted, etc., as follows:

Sec. 1. **Limited Partnership Defined.**

A limited partnership is a partnership formed by two or more persons under the provisions of Section 2, having as members one or more general partners and one or more limited partners. The limited partners as such shall not be bound by the obligations of the partnership.

Sec. 2. **Formation.**

(1) Two or more persons desiring to form a limited partnership shall

(a) Sign and swear to a certificate, which shall state

I. The name of the partnership,

II. The character of the business,

III. The location of the principal place of business,

IV. The name and place of residence of each member; general and limited partners being respectively designated,

V. The term for which the partnership is to exist,

VI. The amount of cash and a description of and the agreed value of the other property contributed by each limited partner,

VII. The additional contributions, if any, agreed to be made by each limited partner and the times at which or events on the happening of which they shall be made,

VIII. The time, if agreed upon, when the contribution of each limited partner is to be returned,

IX. The share of the profits or the other compensation by way of income which each limited partner shall receive by reason of his contribution,

X. The right, if given, of a limited partner to substitute an assignee as contributor in his place, and the terms and conditions of the substitution,

XI. The right, if given, of the partners to admit additional limited partners,

XII. The right, if given, of one or more of the limited partners to priority over other limited partners, as to contributions or as to compensation by way of income, and the nature of such priority,

XIII. The right, if given, of the remaining general partner or partners to continue the business on the death, retirement or insanity of a general partner, and

XIV. The right, if given, of a limited partner to demand and receive property other than cash in return for his contribution.

(b) File for record the certificate in the office of [here designate the proper office].

(2) A limited partnership is formed if there has been substantial compliance in good faith with the requirements of paragraph (1).

Sec. 3. **Business Which May Be Carried On.**

A limited partnership may carry on any business which a partnership without limited partners may carry on, except [here designate the business to be prohibited].

Sec. 4. **Character of Limited Partner's Contribution.**

The contributions of a limited partner may be cash or other property, but not services.

Sec. 5. **A Name Not to Contain Surname of Limited Partner; Exceptions.**

(1) The surname of a limited partner shall not appear in the partnership name, unless

(a) It is also the surname of a general partner, or

(b) Prior to the time when the limited partner became such the business had been carried on under a name in which his surname appeared.

(2) A limited partner whose name appears in a partnership name contrary to the provisions of paragraph (1) is liable as a general partner to partnership creditors who extend credit to the partnership without actual knowledge that he is not a general partner.

Sec. 6. **Liability for False Statements in Certificate.**

If the certificate contains a false statement, one who suffers loss by reliance on such statement may hold liable any party to the certificate who knew the statement to be false.

(a) At the time he signed the certificate, or

(b) Subsequently, but within a sufficient time before the statement was relied upon to enable him to cancel or amend the certificate, or to file a petition for its cancellation or amendment as provided in Section 25(3).

Sec. 7. **Limited Partner Not Liable to Creditors.**

A limited partner shall not become liable as a general partner unless, in addition to the exercise of his rights and powers as a limited partner, he takes part in the control of the business.

Sec. 8. **Admission of Additional Limited Partners.**

After the formation of a limited partnership, additional limited partners may be admitted upon filing an amendment to the original certificate in accordance with the requirements of Section 25.

Sec. 9. **Rights, Powers and Liabilities of a General Partner.**

(1) A general partner shall have all the rights and powers and be subject to all the restrictions and liabilities of a partner in a partnership without limited partners, except that without the written consent or ratification of the specific act by all the limited partners, a general partner or all of the general partners have no authority to

(a) Do any act in contravention of the certificate,

(b) Do any act which would make it impossible to carry on the ordinary business of the partnership,

(c) Confess a judgment against the partnership,

(d) Possess partnership property, or assign their rights in specific partnership property, for other than a partnership purpose,

(e) Admit a person as a general partner,

(f) Admit a person as a limited partner, unless the right so to do is given in the certificate,

(g) Continue the business with partnership property on the death, retirement or insanity of a general partner, unless the right so to do is given in the certificate.

Sec. 10 **Rights of a Limited Partner.**

(1) A limited partner shall have the same rights as a general partner to

(a) Have the partnership books kept at the principal place of business of the partnership, and at all times to inspect and copy any of them,

(b) Have on demand true and full information of all things affecting the partnership, and a formal account of partnership affairs, whenever circumstances render it just and reasonable, and

(c) Have dissolution and winding up by decree of court.

(2) A limited partner shall have the right to receive a share of the profits or other compensation by way of income, and to the return of his contribution as provided in Sections 15 and 16.

Sec. 11. **Status of Person Erroneously Believing Himself a Limited Partner.**

A person who has contributed to the capital of a business conducted by a person or partnership erroneously believing that he has become a limited partner in a limited partnership, is not, by reason of his exercise of the rights of a limited partner, a general partner with the person or in the partnership carrying on the business, or bound by the obligations of such person or partnership; provided that on ascertaining the mistake he promptly renounces his interest in the profits of the business, or other compensation by way of income.

Sec. 12. **One Person Both General and Limited Partner.**

(1) A person may be a general partner and a limited partner in the same partnership at the same time.

(2) A person who is a general, and also at the same time a limited partner, shall have all the rights and powers and be subject to all the restrictions of a general partner; except that, in respect to his contribution, he shall have the rights against the other members which he would have had if he were not also a general partner.

Sec. 13. **Loans and Other Business Transactions with Limited Partner.**

(1) A limited partner also may loan money to and transact other business with the partnership, and, unless he is also a general partner, receive on account of resulting claims against the partnership, with general creditors, a pro rata share of the assets. No limited partner shall in respect to any such claim

(a) Receive or hold as collateral security any partnership property, or

(b) Receive from a general partner or the partnership any payment, conveyance, or release from liability, if at the time the assets of the partnership are not sufficient to discharge partnership liabilities to persons not claiming as general or limited partners.

(2) The receiving of collateral security, or a payment, conveyance, or release in violation of the provisions of paragraph (1) is a fraud on the creditors of the partnership.

Sec. 14. **Relation of Limited Partners Inter Se.**

Where there are several limited partners the members may agree that one or more of the limited partners shall have a priority over other limited partners as to the return of their contributions, as to their compensation by way of income, or as to any other matter. If such an agreement is made it shall be stated in the certificate, and in the absence of such a statement all the limited partners shall stand upon equal footing.

Sec. 15. **Compensation of Limited Partner.**

A limited partner may receive from the partnership the share of the profits or the compensation by way of income stipulated for in the certificate; provided, that after such payment is made, whether from the property of the partnership or that of a general partner, the partnership assets are in excess of all liabilities of the partnership except liabilities to limited partners on account of their contributions and to general partners.

Sec. 16. **Withdrawal or Reduction of Limited Partner's Contribution.**

(1) A limited partner shall not receive from a general partner or out of partnership property any part of his contribution until

(a) All liabilities of the partnership, except liabilities to general partners and to limited partners on account of their contributions, have been paid or there remains property of the partnership sufficient to pay them,

(b) The consent of all members is had, unless the return of the contribution may be rightfully demanded under the provisions of paragraph (2), and

(c) The certificate is cancelled or so amended as to set forth the withdrawal or reduction.

(2) Subject to the provisions of paragraph (1) a limited partner may rightfully demand the return of his contribution

(a) On the dissolution of a partnership, or

(b) When the date specified in the certificate for its return has arrived, or

(c) After he has given six months' notice in writing to all other members, if no time is specified in the certificate either for the return of the contribution or for the dissolution of the partnership.

(3) In the absence of any statement in the certificate to the contrary or the consent of all members, a limited partner, irrespective of the nature of his contribution, has only the right to demand and receive cash in return for his contribution.

(4) A limited partner may have the partnership dissolved and its affairs wound up when

(a) He rightfully but unsuccessfully demands the return of his contribution, or

(b) The other liabilities of the partnership have not been paid, or the partnership property is insufficient for their payment as required by paragraph (1a) and the limited partner would otherwise be entitled to the return of his contribution.

Sec. 17. **Liability of Limited Partner to Partnership.**

(1) A limited partner is liable to the partnership

(a) For the difference between his contribution as actually made and that stated in the certificate as having been made, and

(b) For any unpaid contribution which he agreed in the certificate to make in the future at the time and on the conditions stated in the certificate.

(2) A limited partner holds as trustee for the partnership

(a) Specific property stated in the certificate as contributed by him, but which was not contributed or which has been wrongfully returned, and

(b) Money or other property wrongfully paid or conveyed to him on account of his contribution.

(3) The liabilities of a limited partner as set forth in this section can be waived or compromised only by the consent of all members; but a waiver or com-

promise shall not affect the right of a creditor of a partnership, who extended credit or whose claim arose after the filing and before a cancellation or amendment of the certificate, to enforce such liabilities.

(4) When a contributor has rightfully received the return in whole or in part of the capital of his contribution, he is nevertheless liable to the partnership for any sum, not in excess of such return with interest, necessary to discharge its liabilities to all creditors who extended credit or whose claims arose before such return.

Sec. 18. **Nature of Limited Partner's Interest in Partnership.**

A limited partner's interest in the partnership is personal property.

Sec. 19. **Assignment of Limited Partner's Interest.**

(1) A limited partner's interest is assignable.

(2) A substituted limited partner is a person admitted to all the rights of a limited partner who has died or has assigned his interest in a partnership.

(3) An assignee, who does not become a substituted limited partner, has no right to require any information or account of the partnership transactions or to inspect the partnership books; he is only entitled to receive the share of the profits or other compensation by way of income, or the return of his contribution, to which his assignor would otherwise be entitled.

(4) An assignee shall have the right to become a substituted limited partner if all the members (except the assignor) consent thereto or if the assignor, being thereunto empowered by the certificate, gives the assignee that right.

(5) An assignee becomes a substituted limited partner when the certificate is appropriately amended in accordance with Section 25.

(6) The substituted limited partner has all the rights and powers, and is subject to all the restrictions and liabilities of his assignor, except those liabilities of which he was ignorant at the time he became a limited partner and which could not be ascertained from the certificate.

(7) The substitution of the assignee as a limited partner does not release the assignor from liability to the partnership under Sections 6 and 17.

Sec. 20. **Effect of Retirement, Death or Insanity of a General Partner.**

The retirement, death or insanity of a general partner dissolves the partnership, unless the business is continued by the remaining general partners

(a) Under a right so to do stated in the certificate, or

(b) With the consent of all members.

Sec. 21. **Death of Limited Partner.**

(1) On the death of a limited partner his executor or administrator shall have all the rights of a limited partner for the purpose of settling his estate, and such power as the deceased had to constitute his assignee a substituted limited partner.

(2) The estate of a deceased limited partner shall be liable for all his liabilities as a limited partner.

Sec. 22. **Rights of Creditors of Limited Partner.**

(1) On due application to a court of competent jurisdiction by any judgment creditor of a limited partner, the court may charge the interest of the indebted limited partner with payment of the unsatisfied amount of the judgment debt; and may appoint a receiver, and make all other orders, directions, and inquiries which the circumstances of the case may require.

In those states where a creditor on beginning an action can attach debts due the defendant before he has obtained a judgment against the defendant it is recommended that paragraph (1) of this section read as follows:

On due application to a court of competent jurisdiction by any creditor of a limited partner, the court may charge the interest of the indebted limited partner with payment of the unsatisfied amount of such claim; and may appoint a receiver, and make all other orders, directions, and inquiries which the circumstances of the case may require.

(2) The interest may be redeemed with the separate property of any general partner, but may not be redeemed with partnership property.

(3) The remedies conferred by paragraph (1) shall not be deemed exclusive of others which may exist.

(4) Nothing in this act shall be held to deprive a limited partner of his statutory exemption.

Sec. 23. **Distribution of Assets.**

(1) In settling accounts after dissolution the liabilities of the partnership shall be entitled to payment in the following order:

(a) Those to creditors, in the order of priority as provided by law, except those to limited partners on account of their contributions, and to general partners,

(b) Those to limited partners in respect to their share of the profits and other compensation by way of income on their contributions,

(c) Those to limited partners in respect to the capital of their contributions,

(d) Those to general partners other than for capital and profits,

(e) Those to general partners in respect to profits,

(f) Those to general partners in respect to capital.

(2) Subject to any statement in the certificate or to subsequent agreement, limited partners share in the partnership assets in respect to their claims for capital, and in respect to their claims for profits or for compensation by way of income on their contributions respectively, in proportion to the respective amounts of such claims.

Sec. 24. **When Certificate Shall be Cancelled or Amended.**

(1) The certificate shall be cancelled when the partnership is dissolved or all limited partners cease to be such.

(2) A certificate shall be amended when

(a) There is a change in the name of the partnership or in the amount or character of the contribution of any limited partner,

(b) A person is substituted as a limited partner,

(c) An additional limited partner is admitted,

(d) A person is admitted as a general partner,

(e) A general partner retires, dies or becomes insane, and the business is continued under section 20,

(f) There is a change in the character of the business of the partnership,

(g) There is a false or erroneous statement in the certificate,

(h) There is a change in the time as stated in the certificate for the dissolution of the partnership or for the return of a contribution,

(i) A time is fixed for the dissolution of the partnership, or the return of a contribution, no time having been specified in the certificate, or

(j) The members desire to make a change in any other statement in the certificate in order that it shall accurately represent the agreement between them.

Sec. 25. **Requirements for Amendment and for Cancellation of Certificate.**

(1) The writing to amend a certificate shall

(a) Conform to the requirements of Section 2(1a) as far as necessary to set forth clearly the change in the certificate which it is desired to make, and

(b) Be signed and sworn to by all members, and an amendment substituting a limited partner or adding a limited or general partner shall be signed also by the member to be substituted or added, and when a limited partner is to be substituted, the amendment shall also be signed by the assigning limited partner.

(2) The writing to cancel a certificate shall be signed by all members.

(3) A person desiring the cancellation or amendment of a certificate, if any person designated in paragraphs (1) and (2) as a person who must execute the writing refuses to do so, may petition the [here designate the proper court] to direct a cancellation or amendment thereof.

(4) If the court finds that the petitioner has a right to have the writing executed by a person who refuses to do so, it shall order the [here designate the responsible official in the office designated in Section 2] in the office where the certificate is recorded to record the cancellation or amendment of the certificate; and where the certificate is to be amended, the court shall also cause to be filed for record in said office a certified copy of its decree setting forth the amendment.

(5) A certificate is amended or cancelled when there is filed for record in the office [here designate the office designated in Section 2] where the certificate is recorded

(a) A writing in accordance with the provisions of paragraph (1), or (2) or

(b) A certified copy of the order of court in accordance with the provisions of paragraph (4).

(6) After the certificate is duly amended in accordance with this section, the amended certificate shall thereafter be for all purposes the certificate provided for by this act.

Sec. 26. **Parties to Actions.**

A contributor, unless he is a general partner, is not a proper party to proceedings by or against a partnership, except where the object is to enforce a limited partner's right against or liability to the partnership.

Sec. 27. **Name of Act.**

This act may be cited as The Uniform Limited Partnership Act.

Sec. 28. **Rules of Construction.**

(1) The rule that statutes in derogation of the common law are to be strictly construed shall have no application to this act.

(2) This act shall be so interpreted and construed as to effect its general purpose to make uniform the law of those states which enact it.

(3) This act shall not be so construed as to impair the obligations of any contract existing when the act goes into effect, nor to affect any action on proceedings begun or right accrued before this act takes effect.

Sec. 29. **Rules for Cases Not Provided for in this Act.**

In any case not provided for in this act the rules of law and equity, including the law merchant, shall govern.

Sec. 30.[1] **Provisions for Existing Limited Partnerships.**

(1) A limited partnership formed under any statute of this state prior to the adoption of this act, may become a limited partnership under this act by complying with the provisions of Section 2; provided the certificate sets forth

(a) The amount of the original contribution of each limited partner, and the time when the contribution was made, and

(b) That the property of the partnership exceeds the amount sufficient to discharge its liabilities to persons not claiming as general or limited partners by an amount greater than the sum of the contributions of its limited partners.

(2) A limited partnership formed under any statute of this state prior to the adoption of this act, until or unless it becomes a limited partnership under this act, shall continue to be governed by the provisions of [here insert proper reference to the existing limited partnership act or acts], except that such partnership shall not be renewed unless so provided in the original agreement.

Sec. 31.[1] **Act [Acts] Repealed.**

Except as affecting existing limited partnerships to the extent set forth in Section 30, the act (acts) of [here designate the existing limited partnership act or acts] is (are) hereby repealed.

[1]Sections 30, 31, will be omitted in any state which has not a limited partnership act.

APPENDIX C

REVISED UNIFORM LIMITED PARTNERSHIP ACT, 1976

(Adopted August 5, 1976, by the National Conference of Commissioners on Uniform State Laws, subject to style changes; it is intended that it will replace the existing Uniform Limited Partnership Act (Appendix B); as of publication, it has been adopted in Colorado, Connecticut, Minnesota, and Wyoming.)

Article 1
GENERAL PROVISIONS

Sec. 101. **Definitions.**

As used in this Act:

(1) "Certificate of limited partnership" means the certificate referred to in Section 201, as that certificate is amended from time to time.

(2) "Contribution" means any cash, property, or services rendered, or a promissory note or other binding obligation to contribute cash or property or to perform services, which a partner contributes to a limited partnership in his capacity as a partner.

(3) "Event of withdrawal of a general partner" means an event that causes a person to cease to be a general partner as provided in Section 402.

(4) "Foreign limited partnership" means a partnership formed under the laws of any state other than this State and having as partners one or more general partners and one or more limited partners.

(5) "General partner" means a person who has been admitted to a limited partnership as a general partner in accordance with the partnership agreement and who is named in the certificate of limited partnership as a general partner.

(6) "Limited partner" means a person who has been admitted to a limited partnership as a limited partner in accordance with the partnership agreement and who is named in the certificate of limited partnership as a limited partner.

(7) "Limited partnership" and "domestic limited partnership" mean a partnership formed by 2 or more persons under the laws of this State and having one or more general partners and one or more limited partners.

(8) "Partner" means any limited partner or general partner.

(9) "Partnership agreement" means the agreement, written or, to the extent not prohibited by law, oral or both, of the partners as to the affairs of a limited partnership and the conduct of its business.

(10) "Partnership interest" has the meaning specified in Section 701.

(11) "Person" means a natural person, partnership, limited partnership (domestic or foreign), trust, estate, association, or corporation.

(12) "State" means a state, territory, or possession of the United States, the District of Columbia, or the Commonwealth of Puerto Rico.

Sec. 102. **Name.**

The name of each limited partnership as set forth in its certificate of limited partnership:

(1) shall contain the words "limited partnership" in full;

(2) may not contain the name of a limited partner unless (i) it is also the name of a general partner or (ii) the business of the limited partnership had been carried on under that name before the admission of that limited partner;

(3) may not contain any word or phrase indicating or implying that it is organized other than for a purpose stated in its certificate of limited partnership;

(4) may not be the same as, or deceptively similar to, the name of any corporation or limited partnership organized under the laws of this State or licensed or registered as a foreign corporation or limited partnership in this State; and

(5) may not contain the following words [here insert prohibited words].

Sec. 103. **Reservation of Name.**

(a) The exclusive right to the use of a name may be reserved by:

(1) any person intending to organize a limited partnership under this Act and to adopt that name;

(2) any domestic limited partnership or any foreign limited partnership registered in this State which, in either case, intends to adopt that name;

(3) any foreign limited partnership intending to register in this State and to adopt that name; and

(4) any person intending to organize a foreign limited partnership and intending to have it registered in this State and to adopt that name.

(b) The reservation shall be made by filing with the Secretary of State an application, executed by the applicant, to reserve a specified name. If the Secretary of State finds that the name is available for use by a domestic or foreign limited partnership, he shall reserve the name for the exclusive use of the applicant for a period of 120 days. Once having reserved a name, the same applicant may not again reserve the same name until more than 60 days after the expiration of the last 120-day period for which that applicant had reserved that name. The right to the exclusive use of a name so reserved may be transferred to any other person by filing in the office of the Secretary of State a notice of the transfer, executed by the applicant for whom the name was reserved and specifying the name and address of the transferee.

Sec. 104. **Specified Office and Agent.**

Each limited partnership shall continuously maintain in this State:

(1) an office, which may but need not be a place of its business in this State, at which shall be kept the records required to be maintained by Section 105; and

(2) an agent for service of process on the limited partnership, which agent must be an individual resident of this State, a domestic corporation, or a foreign corporation authorized to do business in this State.

Sec. 105. **Records to be Kept.**

Each limited partnership shall keep at the office referred to in Section 104(1) the following: (1) a current list of the full name and last-known business address of each partner set forth in alphabetical order, (2) a copy of the certificate of limited partnership and all certificates of amendment thereto, together with executed copies of any powers of attorney pursuant to which any certificate has been executed, (3) copies of the limited partnership's federal, state, and local income tax returns and reports, if any, for the 3 most recent years, and (4) copies of any then effective written partnership agreements and of any financial statements of the limited partnership for the 3 most recent years. These records shall be available for inspection and copying at the reasonable request, and at the expense, of any partner during ordinary business hours.

Sec. 106. **Nature of Business.**

A limited partnership may carry on any business that a partnership without limited partners may carry on except [here designate prohibited activities].

Sec. 107. **Business Transactions of Partner with the Partnership.**

Except as otherwise provided in the partnership agreement, a partner may lend money to and transact other business with the limited partnership and, subject to other applicable provisions of law, has the same rights and obligations with respect thereto as a person who is not a partner.

Article 2
FORMATION; CERTIFICATE OF LIMITED PARTNERSHIP

Sec. 201. **Certificate of Limited Partnership.**

(a) Two or more persons desiring to form a limited partnership shall execute a certificate of limited partnership. The certificate shall be filed in the office of the Secretary of State and shall set forth:

(1) the name of the limited partnership;

(2) the general character of its business;

(3) the address of the office and the name and address of the agent for service of process required to be maintained by Section 104;

(4) the name and the business address of each partner (specifying the general partners and limited partners separately);

(5) the amount of cash and a description and statement of the agreed value of the other property or services contributed by each partner and which each partner has agreed to contribute in the future;

(6) the times at which or events on the happening of which any additional contributions agreed to be made by each partner are to be made;

(7) any power of a limited partner to grant an assignee of any part of his partnership interest the right to become a limited partner, and the terms and conditions of the power;

(8) if agreed upon, the time at which or the events on the happening of which a partner may termi-

nate his membership in the limited partnership and the amount of, or the method of determining, the distribution to which he may be entitled respecting his partnership interest, and the terms and conditions of the termination and distribution;

(9) any right of a partner to receive distributions of property including cash from the limited partnership;

(10) any right of a partner to receive, or of a general partner to make, distributions to a partner which include a return of all or any part of the partner's contribution;

(11) any time at which or events upon the happening of which the limited partnership is to be dissolved and its affairs wound up;

(12) any right of the remaining general partners to continue the business on the happening of an event of withdrawal of a general partner; and

(13) any other matters the partners, in their sole discretion, determine to include therein.

(b) A limited partnership is formed at the time of the filing of the certificate of limited partnership in the office of the Secretary of State or at any later time specified in the certificate of limited partnership if, in each case, there has been substantial compliance with the requirements of this section.

Sec. 202. **Amendments to Certificate.**

(a) A certificate of limited partnership is amended by filing a certificate of amendment thereto in the office of the Secretary of State. The certificate shall set forth:

(1) the name of the limited partnership;

(2) the date of filing of the certificate; and

(3) the amendments to the certificate.

(b) Within 30 days after the happening of any of the following events an amendment to a certificate of limited partnership reflecting the occurrence of the event or events shall be filed:

(1) a change in the amount or character of the contribution of any partner, or in any partner's obligation to make a contribution;

(2) the admission of a new partner;

(3) the withdrawal of a partner; and

(4) the continuation of the business under Section 801 after an event of withdrawal of a general partner.

(c) A certificate of limited partnership must be amended promptly by any general partner upon becoming aware that any statement therein was false when made or that any arrangements or other facts described have changed, making the certificate inaccurate in any respect, but amendments to show changes of addresses of limited partners need be filed only once every 12 months.

(d) A certificate of limited partnership may be amended at any time for any other proper purpose the general partners may determine.

(e) No person shall have any liability because an amendment to a certificate of limited partnership has not been filed to reflect the occurrence of any event referred to in subsection (b) of this section if the amendment is filed within the 30-day period specified in subsection (b).

Sec. 203. **Cancellation of Certificate.**

A certificate of limited partnership shall be cancelled upon the dissolution and the commencement of winding up of the limited partnership and at any other time there are no remaining limited partners. A certificate of cancellation shall be filed in the office of the Secretary of State and shall set forth:

(1) the name of the limited partnership;

(2) the date of filing of its certificate of limited partnership;

(3) the reason for filing the certificate of cancellation;

(4) the effective date (which shall be a date certain) of cancellation if it is not to be effective upon the filing of the certificate; and

(5) any other information the general partners filing the certificate may determine.

Sec. 204. **Execution of Certificates.**

(a) Each certificate required by this Article to be filed in the office of the Secretary of State shall be executed in the following manner:

(1) each original certificate of limited partnership must be signed by each partner named therein;

(2) each certificate of amendment must be signed

by at least one general partner and by each other partner who is designated in the certificate as a new partner or whose contribution is described as having been increased; and

(3) each certificate of cancellation must be signed by each general partner.

(b) Any person may sign a certificate by an attorney-in-fact, but any power of attorney to sign a certificate relating to the admission or increased contribution of a partner must specifically describe the admission or increase.

(c) The execution of a certificate by a general partner constitutes an affirmation under the penalties of perjury that the facts stated therein are true.

Sec. 205. **Amendment or Cancellation by Judicial Act.**

If the persons required by Section 204 to execute any certificate of amendment or cancellation fail or refuse to do so, any other partner, and any assignee of a partnership interest, who is adversely affected by the failure or refusal, may petition the [here designate the proper court] to direct the amendment or cancellation. If the court finds that the amendment or cancellation is proper and that the persons so designated have failed or refused to execute the certificate, it shall order the Secretary of State to record an appropriate certificate of amendment or cancellation.

Sec. 206. **Filing in the Office of the Secretary of State.**

(a) Two signed copies of the certificate of limited partnership and of any certificates of amendment or cancellation (or of any judicial decree of amendment or cancellation) shall be delivered to the Secretary of State. A person who executes a certificate as an agent or fiduciary need not exhibit evidence of his authority as a prerequisite to filing. Unless the Secretary of State finds that any certificate does not conform to law, upon receipt of all filing fees required by law the Secretary of State shall:

(1) endorse on each duplicate original the word "Filed" and the day, month, and year of the filing thereof;

(2) file one duplicate original in his office; and

(3) return the other duplicate original to the person who filed it or his representative.

(b) Upon the filing of a certificate of amendment (or judicial decree of amendment) in the office of the Secretary of State, the certificate of limited partnership shall be amended as set forth therein, and upon the effective date of a certificate of cancellation (or a judicial decree thereof), the certificate of limited partnership shall be cancelled.

Sec. 207. **Liability for False Statement in Certificate.**

If any certificate of limited partnership or certificate of amendment or cancellation contains a false statement, one who suffers loss by reliance on the statement may recover damages for the loss from:

(1) any person actually executing, or causing another to execute on his behalf, the certificate who knew, and any general partner who knew or should have known, the statement to be false at the time the certificate was executed; and

(2) any general partner who thereafter knew or should have known that any arrangements or other facts described in the certificate have changed, making the statement inaccurate in any respect, within a sufficient time before the statement was relied upon to have reasonably enabled that general partner to cancel or amend the certificate, or to file a petition for its cancellation or amendment under Section 205.

Sec. 208. **Constructive Notice.**

The fact that a certificate of limited partnership is on file in the office of the Secretary of State is constructive notice that the partnership is a limited partnership and that the persons designated therein as limited partners are limited partners, but is not constructive notice of any other fact.

Sec. 209. **Delivery of Certificates to Limited Partners.**

Upon the return by the Secretary of State pursuant to Section 206 of any certificate marked "Filed," the general partners shall promptly deliver or mail a copy of the certificate to each limited partner unless the partnership agreement provides otherwise.

Article 3 **LIMITED PARTNERS**

Sec. 301. **Admission of Additional Limited Partners.**

(a) After the filing of a limited partnership's original certificate of limited partnership, a person may be admitted as a new limited partner:

(1) in the case of a person acquiring a partnership interest directly from the limited partnership, upon compliance with the partnership agreement or, if the partnership agreement does not so provide, upon the written consent of all partners; and

(2) in the case of an assignee of a partnership interest of a partner who has the power, as provided in Section 704, to grant the assignee the right to become a limited partner, upon the exercise of that power and compliance with any conditions limiting the grant or exercise of the power.

(b) In each case under subsection (a), the person acquiring the partnership interest becomes a limited partner only upon amendment of the certificate of limited partnership reflecting that fact.

Sec. 302. **Voting.**

Subject to the provisions of Section 303, the partnership agreement may grant to all or a specified group of the limited partners the right to vote (on a per capita or any other basis) upon any matter.

Sec. 303. **Liability to Third Parties.**

(a) Except as provided in subsection (d), a limited partner as such is not liable for the obligations of a limited partnership unless, in addition to the exercise of his rights and powers as a limited partner, he takes part in the control of the business. But the limited partner's participation in the control of the business is not substantially the same as the exercise of the powers of a general partner, he is liable only to persons who transact business with the limited partnership with actual knowledge of his participation in control.

(b) A limited partner does not participate in the control of the business within the meaning of subsection (a) solely by doing one or more of the following:

(1) being a contractor for or an agent or employee of the limited partnership or of a general partner;

(2) consulting with and advising a general partner with respect to the business of the limited partnership;

(3) acting as surety for the limited partnership;

(4) approving or disapproving an amendment to the partnership agreement; and

(5) voting on one or more of the following matters:

(i) the dissolution and winding up of the limited partnership;

(ii) the sale, exchange, lease, mortgage, pledge, or other transfer of all or substantially all of the assets of the limited partnership other than in the ordinary course of its business;

(iii) the incurrence of indebtedness by the limited partnership other than in the ordinary course of its business;

(iv) a change in the nature of the business; or

(v) the removal of a general partner.

(c) The enumeration in subsection (b) shall not be construed to mean that the possession or exercise of any other powers by a limited partner constitutes participation by him in the business of the limited partnership.

(d) A limited partner who knowingly permits his name to be used in the name of the limited partnership, except under circumstances permitted by Section 102(2)(i), is liable to creditors who extend credit to the limited partnership without actual knowledge that the limited partner is not a general partner.

Sec. 304. **Person Erroneously Believing Himself a Limited Partner.**

(a) Except as provided in subsection (b) a person who makes a contribution to a business enterprise and erroneously and in good faith believes that he has become a limited partner in the enterprise is not a general partner in the enterprise and is not bound by its obligations by reason of making the contribution, receiving distributions from the enterprise, or exercising any rights of a limited partner, if, on ascertaining the mistake, he:

(1) causes an appropriate certificate of limited partnership or a certificate of amendment to be executed and filed; or

(2) withdraws from future equity participation in the enterprise.

(b) Any person who makes a contribution of the kind described in subsection (a) is liable as a general partner to any third party who transacts business with the enterprise (i) before the person withdraws and an appropriate certificate if any is filed to show the withdrawal, or (ii) before an appropriate certificate is filed to show his status as a limited partner and, in the case of an amendment, after expiration of the 30-day period for filing an amendment relating to the person as a limited partner under Section 202, but in each case only if the third party actually believed in good faith that the person was a general partner at the time of the transaction.

Sec. 305. **Information.**

Each limited partner has the right to:

(1) inspect and copy any of the partnership records required to be maintained by Section 105; and

(2) obtain from the general partners from time to time upon reasonable demand (i) true and full information regarding the state of the business and financial condition of the limited partnership, (ii) promptly after becoming available, a copy of the limited partnership's federal, state, and local income tax return for each year, and (iii) any other information regarding the affairs of the limited partnership as is just and reasonable.

Article 4
GENERAL PARTNERS

Sec. 401. **Admission.**

After the filing of a limited partnership's original certificate of limited partnership, new general partners may be admitted only with the specific written consent of each partner.

Sec. 402. **Events of Withdrawal.**

Except as otherwise approved by the specific written consent at the time of all partners, a person ceases to be a general partner of a limited partnership upon the happening of any of the following events:

(1) the general partner withdraws from the limited partnership as provided in Section 602;

(2) the general partner ceases to be a member of the limited partnership as provided in Section 702;

(3) the general partner is removed as a general partner in accordance with the partnership agreement;

(4) unless otherwise provided in the certificate of limited partnership, the general partner: makes an assignment for the benefit of creditors; files a voluntary petition in bankruptcy; is adjudicated a bankrupt or insolvent; files any petition or answer seeking for himself any reorganization, arrangement, composition, readjustment, liquidation, dissolution, or similar relief under any statute, law, or regulation; files any answer or other pleading admitting or failing to contest the material allegations of a petition filed against him in any proceeding of this nature; or seeks, consents to, or acquiesces in the appointment of any trustee, receiver, or liquidator of the general partner or of all or any substantial part of his properties;

(5) unless otherwise provided in the certificate of limited partnership, [120] days after the commencement of any proceeding against the general partner seeking any reorganization, arrangement, composition, readjustment, liquidation, dissolution, or similar relief under any statute, law, or regulation, the proceeding has not been dismissed, or if, within [90] days after the appointment without his consent or acquiescence of any trustee, receiver, or liquidator of the general partner or of all or any substantial part of his properties, the appointment is not vacated or stayed, or if, within [90] days after the expiration of any stay, the appointment is not vacated;

(6) in the case of a general partner who is a natural person

(i) his death; or

(ii) the entry by a court of competent jurisdiction adjudicating him incompetent to manage his person or his property;

(7) in the case of a general partner who is acting as such in the capacity of a trustee of a trust, the termination of the trust (but not merely the substitution of a new trustee);

(8) in the case of a general partner that is a partnership, the dissolution and commencement of winding up of the partnership;

(9) in the case of a general partner that is a corporation, the filing of a certificate of dissolution, or its equivalent, for the corporation or the revocation of its charter; and

(10) in the case of an estate, the distribution by the fiduciary of all the estate's interest in the partnership.

Sec. 403. **General Powers and Liabilities.**

Except as otherwise provided in this Act and in the partnership agreement, a general partner of a limited partnership has all the rights and powers and is subject to all the restrictions and liabilities of a partner in a partnership without limited partners.

Sec. 404. **Contributions by a General Partner.**

A general partner may make contributions to a limited partnership and share in the profits and losses of, and in distributions from, the limited partnership as a general partner. A general partner may also make contributions to and share in profits, losses, and distributions as a limited partner. A person who is both a general partner and a limited partner has all the rights and powers, and is subject to all the restrictions and liabilities, of a general partner and also has, except as otherwise provided in the partnership agreement, all powers, and is subject to the restrictions, of a limited partner to the extent he is participating in the partnership as a limited partner.

Sec. 405. **Voting.**

The partnership agreement may grant to all or a specified group of general partners the right to vote (on a per capita or any other basis), separately or with all or any class of the limited partners, on any matter.

Article 5
FINANCE

Sec. 501. **Form of Contributions.**

The contribution of a partner may be in cash, property, or services rendered, or a promissory note or other obligation to contribute cash or property or to perform services.

Sec. 502. **Liability for Contributions.**

(a) Except as otherwise provided in the certificate of limited partnership, a partner is obligated to the limited partnership to perform any promise to contribute cash or property or to perform services regardless of whether he is unable to perform because of death, disability or any other reason. If a partner does not make the required contribution of property or services, he is obligated at the option of the limited partnership to contribute cash equal to that portion of the value (as stated in the certificate of limited partnership) of the stated contribution that has not been made.

(b) Unless otherwise provided in the partnership agreement, the obligation of a partner to make a contribution or return money or other property paid or distributed in violation of this Act may be compromised only by consent of all of the partners. Notwithstanding a compromise so authorized, a creditor of a limited partnership who extends credit, or whose claim arises, after the filing of the certificate of limited partnership or an amendment thereto which, in either case, reflects the obligation and before the amendment or cancellation thereof to reflect the compromise may enforce the precompromise obligation.

Sec. 503. **Sharing of Profits and Losses.**

The profits and losses of a limited partnership shall be allocated among the partners, and among classes of partners, in the manner provided in the partnership agreement. If the partnership agreement does not so provide, profits and losses shall be allocated on the basis of the value (as stated in the certificate of limited partnership) of the contributions actually made by each partner to the extent they have not been returned.

Sec. 504. **Sharing of Distributions.**

Distributions of cash or other assets of a limited partnership shall be allocated among the partners, and among classes of partners, in the manner provided in the partnership agreement. If the partnership agreement does not so provide, distributions shall be made on the basis of the value (as stated in the certificate of limited partnership) of the contributions actually made by each partner to the extent they have not been returned.

Article 6
DISTRIBUTIONS AND WITHDRAWAL

Sec. 601. **Interim Distributions.**

Except as otherwise provided in this Article, a partner is entitled to receive distributions from a limited partnership before his withdrawal from the limited partnership and before the dissolution and winding up thereof:

(1) to the extent and at the times or upon the happening of the events specified in the partnership agreement; and

(2) if any distribution constitutes a return of any part of his contribution under Section 608(b), to the extent and at the times or upon the happening of the events specified in the certificate of limited partnership.

Sec. 602. **Withdrawal of General Partner.**

A general partner may withdraw from a limited partnership at any time by giving written notice to the other partners, but if the withdrawal violates the partnership agreement, the limited partnership may recover from the withdrawing general partner damages for breach of the partnership agreement and offset the damages against the amount otherwise distributable to him.

Sec. 603. **Withdrawal of Limited Partner.**

A limited partner may withdraw from a limited partnership at the time or upon the happening of the events specified in the certificate of limited partnership and in accordance with any procedures provided in the partnership agreement. If the certificate of limited partnership does not specify the time or the events upon the happening of which a limited partner may withdraw from the limited partnership or a definite time for the dissolution and winding up of the limited partnership, a limited partner may withdraw from the limited partnership upon not less than 6 months' prior written notice to each general partner at his address on the books of the limited partnership at its office in this State.

Sec. 604. **Distributions Upon Withdrawal.**

Except as provided in this Article, upon withdrawal any withdrawing partner is entitled to receive any distributions to which he is entitled under the partnership agreement and, if not provided, he is entitled to receive, within a reasonable time after withdrawal, the fair value of his interest in the limited partnership as of the date of withdrawal, based upon his right to share in distributions from the limited partnership.

Sec. 605. **Distributions in Kind.**

Except as provided in the certificate of limited partnership, a partner, regardless of the nature of his contribution, has no right to demand and receive any distribution from a limited partnership in any form other than cash. Except as provided in the partnership agreement, a partner may not be compelled to accept a distribution of any asset in kind from a limited partnership to the extent that the percentage of the asset distributed to him exceeds a percentage of that asset which is equal to the percentage in which he shares in distributions from the limited partnership.

Sec. 606. **Right to Distributions.**

At the time a partner becomes entitled to receive a distribution, he has the status of, and is entitled to all of the remedies available to, a creditor of the limited partnership with respect to the distribution.

Sec. 607. **Limitations on Distributions.**

A partner may not receive a distribution from a limited partnership to the extent that, after giving effect to the distribution, all liabilities of the limited partnership other than liabilities to partners on account of their partnership interests, exceed the fair value of the partnership's assets.

Sec. 608. **Liability Upon Return of Contributions.**

(a) If a partner has received the return of any part of his contribution without violation of the partnership agreement or this Act, for a period of one year thereafter he is liable to the limited partnership for the amount of his contribution returned, but only to the extent necessary to discharge the limited partnership's liabilities to creditors who extended credit to the limited partnership during the period the contribution was held by the partnership.

(b) If a partner has received the return of any part of his contribution in violation of the partnership agreement or this Act, for a period of 6 years thereafter he is liable to the limited partnership for the amount of the contribution wrongfully returned.

(c) A partner has received a return of his contribution to the extent that a distribution to him reduces his share of the fair value of the net assets of the limited partnership below the value (as set forth in the certificate of limited partnership) of his contributions which have not theretofore been distributed to him.

Article 7
ASSIGNMENT OF PARTNERSHIP INTERESTS

Sec. 701. **Nature of Partnership Interest.**

A partnership interest is a partner's share of the profits and losses of a limited partnership and the right to receive distributions of partnership assets. A partnership interest is personal property.

Sec. 702. **Assignment of Partnership Interest.**

Except as otherwise provided in the partnership agreement, a partnership interest is assignable in whole or in part. An assignment of a partnership interest does not dissolve a limited partnership nor entitle the assignee to become a partner or to exercise any of the rights thereof. An assignment only entitles the assignee to receive, to the extent assigned, any distributions to which the assignor would be entitled. Except as otherwise provided in the partnership agreement, a partner ceases to be a partner upon assignment of all his partnership interest.

Sec. 703. **Rights of Creditors.**

On due application to a court of competent jurisdiction by any judgment creditor of a partner, the court may charge the partnership interest of the partner with payment of the unsatisfied amount of the judgment debt with interest thereon. To the extent so charged, the judgment creditor has only the rights of an assignee of the partnership interest. This Act shall not be construed to deprive any partner of the benefit of any exemption laws applicable to his partnership interest.

Sec. 704. **Right of Assignee to Become Limited Partner.**

(a) An assignee of a partnership interest, including an assignee of a general partner, may become a limited partner if and to the extent that (1) the assignor gives the assignee that right in accordance with authority described in the certificate of limited partnership or, (2) in the absence of that authority, all other partners consent.

(b) An assignee who has become a limited partner has, to the extent assigned, all the rights and powers, and is subject to all the restrictions and liabilities, of a limited partner under the partnership agreement and this Act. An assignee who becomes a limited partner is also liable for the obligations of his assignor to make and return contributions as provided in Article 6, but the assignee is not obligated for liabilities unknown to the assignee at the time he became a limited partner and which could not be ascertained from the certificate of limited partnership.

(c) If an assignee of a partnership interest becomes a limited partner, the assignor is not released from the liability to the limited partnership under Sections 207 and 502.

Sec. 705. **Power of Estate of Deceased or Incompetent Partner.**

If a partner who is a natural person dies or a court of competent jurisdiction adjudges him to be incompetent to manage his person or his property, the partner's executor, administrator, guardian, conservator, or other legal representative may exercise all of the partner's rights for the purpose of settling his estate or administering his property, including any power the partner had to give an assignee the right to become a limited partner. If a partner that is a corporation, trust, or other entity other than a natural person is dissolved or terminated, those powers may be exercised by the legal representative or successor of the partner.

Article 8
DISSOLUTION

Sec. 801. **Nonjudicial Dissolution.**

A limited partnership is dissolved and its affairs shall be wound up upon the happening of the first to occur of the following:

(1) at the time or upon the happening of the events specified in the certificate of limited partnership;

(2) upon the unanimous written consent of all partners;

(3) upon the happening of an event of withdrawal of a general partner unless at the time there is at least one other general partner and the certificate of limited partnership permits the business of the limited partnership to be carried on by the remaining general partner and he does so, but the limited partnership shall not be dissolved or wound up by reason of any event of withdrawal if, within 90 days after the withdrawal, all partners agree in writing to continue the business of the limited partnership and to the appointment of one or more new general partners if necessary or desired; or

(4) upon entry of a decree of judicial dissolution in accordance with Section 802.

Sec. 802. **Dissolution by Decree of Court.**

On application by or for a partner the [here designate the proper court] court may decree a dissolution of a limited partnership whenever it is not reasonably practicable to carry on the business in conformity with the partnership agreement.

Sec. 803. **Winding Up.**

Unless otherwise provided in the partnership agreement, the general partners who have not wrongfully dissolved the limited partnership or, if none, the limited partners, may wind up the limited partnership's affairs; but any partner, his legal representative or his assignee, upon cause shown, may obtain winding up by the [here designate the proper court] court.

Sec. 804. **Distribution of Assets.**

Upon the winding up of a limited partnership, the assets shall be distributed as follows:

(1) to creditors, including partners who are creditors (to the extent otherwise permitted by law), in satisfaction of liabilities of the limited partnership other than liabilities for distributions to partners pursuant to Section 601 or 604;

(2) except as otherwise provided in the partnership agreement, to partners and ex-partners in satisfaction of liabilities for distributions pursuant to Section 601 or 604; and

(3) except as otherwise provided in the partnership agreement, to partners *first* for the return of their contributions and *second* respecting their partnership interests, in the proportions in which the partners share in distributions.

Article 9 FOREIGN LIMITED PARTNERSHIPS

Sec. 901. **Law Governing.**

Subject to the constitution and public policy of this State, the laws of the state under which a foreign limited partnership is organized govern its organization and internal affairs and the liability of its limited partners, and a foreign limited partnership may not be denied registration by reason of any difference between those laws and the laws of this State.

Sec. 902. **Registration.**

Before transacting business in this State, a foreign limited partnership shall register with the Secretary of State. In order to register, a foreign limited partnership shall submit to the Secretary of State in duplicate an application for registration as a foreign limited partnership, signed and sworn to by a general partner and setting forth:

(1) the name of the foreign limited partnership and, if different, the name under which it proposes to transact business and register in this State;

(2) the state and date of its formation;

(3) the general character of the business it proposes to transact in this State;

(4) the name and address of any agent for service of process on the foreign limited partnership whom the foreign limited partnership desires to appoint, which agent must be an individual resident of this State, a domestic corporation, or a foreign corporation authorized to do business in this State; and with a place of business in this State;

(5) a statement that the Secretary of State is appointed the agent of the foreign limited partnership for service of process if no agent has been appointed pursuant to paragraph (4) or, if appointed the agent's authority has been revoked or the agent cannot be found or served with the exercise of reasonable diligence;

(6) the address of the office required to be maintained in the state of its organization by the laws of that state or, if not so required, of the principal office of the foreign limited partnership; and

(7) if the certificate of limited partnership filed in the foreign limited partnership's state of organization is not required to include the names and business addresses of the partners, a list of the names and addresses.

Sec. 903. **Issuance of Registration.**

(a) If the Secretary of State finds that an application for registration conforms to law and all requisite fees have been paid, he shall:

(1) endorse on the application the word "Filed", and the month, day, and year of the filing thereof;

(2) file in his office one of the duplicate originals of the application; and

(3) issue a certificate of registration to transact business in this State.

(b) The certificate of registration, together with one duplicate original of the application, shall be returned to the person who filed the application or his representative.

Sec. 904. **Name.**

A foreign limited partnership may register with the Secretary of State under any name (whether or not it is the name under which it is registered in its state of organization) that includes the words "limited partnership" and that could be registered by a domestic limited partnership.

Sec. 905. **Changes and Amendments.**

If any statement in a foreign limited partnership's application for registration was false when made or any arrangements or other facts described have changed, making the application inaccurate in any respect, the foreign limited partnership shall promptly file in the office of the Secretary of State a certificate, signed and sworn to by a general partner, correcting the statement.

Sec. 906. **Cancellation of Registration.**

A foreign limited partnership may cancel its registration by filing with the Secretary of State a certificate of cancellation signed and sworn to by a general partner. A cancellation does not terminate the authority of the Secretary of State to accept service of process on the foreign limited partnership with respect to [claims for relief] [causes of action] arising out of the transaction of business in this State.

Sec. 907. **Transaction of Business Without Registration.**

(a) A foreign limited partnership transacting business in this State without registration may not maintain any action, suit, or proceeding in any court of this State until it has registered.

(b) The failure of a foreign limited partnership to register in this State does not impair the validity of any contract or act of the foreign limited partnership, and does not prevent the foreign limited partnership from defending any action, suit, or proceeding in any court of this State.

(c) A limited partner of a foreign limited partnership is not liable as a general partner of the foreign limited partnership solely by reason of the foreign limited partnership's transacting business in this State without registration.

(d) A foreign limited partnership, by transacting business in this State without registration, appoints the Secretary of State as its agent for service of process with respect to [claims for relief] [causes of action] arising out of the transaction of business in this State.

Sec. 908. **Action by [Appropriate Official].**

The [appropriate official] may bring an action to restrain a foreign limited partnership from transacting business in this State in violation of this Article.

Article 10
DERIVATIVE ACTIONS

Sec. 1001. **Right of Action.**

A limited partner may bring an action in the right of a limited partnership to recover a judgment in its favor if the general partners having authority to do so have refused to bring the action or an effort to cause those general partners to bring the action is not likely to succeed.

Sec. 1002. **Proper Plaintiff.**

In a derivative action, the plaintiff must be a partner at (1) the time of bringing the action, and (2) at the time of the transaction of which he complains or his status as a partner must have devolved upon him by operation of law or pursuant to the terms of the partnership agreement from a person who was a partner at the time of the transaction.

Sec. 1003. **Pleading.**

In any derivative action, the complaint shall set forth with particularity the effort of the plaintiff to secure initiation of the action by a general partner having authority to do so or the reasons for not making the effort.

Sec. 1004. **Expenses.**

If a derivative action is successful, in whole or in part, or anything is received by the plaintiff as a result of a judgment, compromise, or settlement of an action or claim, the court may award the plaintiff reasonable expenses, including reasonable attorney's fees, and shall direct him to account to the limited partnership for the remainder of the proceeds so received by him.

Article 11
MISCELLANEOUS

Sec. 1101. **Savings Clause.**

Sec. 1102. **Name of Act.**

This Act may be cited as the Uniform Limited Partnership Act.

Sec. 1103. **Construction and Application.**

This Act shall be so construed and applied to effect its general purpose to make uniform the law with respect to the subject of this Act among states enacting it.

Sec. 1104. **Rules for Cases Not Provided for in This Act.** In any case not provided for in this Act the provisions of the Uniform Partnership Act govern.

Sec. 1105. **Act Repealed.**
Except as affecting existing limited partnerships to the extent set forth in Section ____, the Act of [here designate the existing limited partnership act or acts] is hereby repealed.

APPENDIX **D**

MODEL BUSINESS CORPORATION ACT*

*[By The Editor] Several revisions to the Model Act were adopted in 1979 and several sections were repealed. Since most state statutes follow the Model Act and continue to use the sections which have been repealed, the former sections are reprinted here for reference but are shown to have been repealed at the beginning of each section.

Sec. 27 By-Laws.
Sec. 27A By-Laws and Other Powers in Emergency [Optional].
Sec. 28 Meetings of Shareholders.
Sec. 29 Notice of Shareholders' Meetings.
Sec. 30 Closing of Transfer Books and Fixing Record Date.
Sec. 31 Voting Record.
Sec. 32 Quorum of Shareholders.
Sec. 33 Voting of Shares.
Sec. 34 Voting Trusts and Agreements Among Shareholders.
Sec. 35 Board of Directors.
Sec. 36 Number and Election of Directors.
Sec. 37 Classification of Directors.
Sec. 38 Vacancies.
Sec. 39 Removal of Directors.
Sec. 40 Quorum of Directors.
Sec. 41 Director of Conflicts of Interest.
Sec. 42 Executive and Other Committees.
Sec. 43 Place and Notice of Directors' Meetings; Committee Meetings.
Sec. 44 Action by Directors Without a Meeting.
Sec. 45 Distributions to Shareholders.
Sec. 46 Distributions from Capital Surplus [Repealed in 1979].
Sec. 47 Loans to Employees and Directors.
Sec. 48 Liability of Directors in Certain Cases.
Sec. 49 Provisions Relating to Actions by Shareholders.
Sec. 50 Officers.
Sec. 51 Removal of Officers.
Sec. 52 Books and Records; Financial Reports to Shareholders; Examination of Records.

FORMATION OF CORPORATIONS

Sec. 53 Incorporators.
Sec. 54 Articles of Incorporation.
Sec. 55 Filing of Articles of Incorporation.
Sec. 56 Effect of Issuance of Certificate of Incorporation.
Sec. 57 Organization Meeting of Directors.

AMENDMENT

Sec. 58 Right to Amend Articles of Incorporation.
Sec. 59 Procedure to Amend Articles of Incorporation.
Sec. 60 Class Voting on Amendments.
Sec. 61 Articles of Amendment.
Sec. 62 Filing of Articles of Amendment.
Sec. 63 Effect of Certificate of Amendment.
Sec. 64 Restated Articles of Incorporation.
Sec. 65 Amendment of Articles of Incorporation in Reorganization Proceedings.
Sec. 66 Restriction on Redemption or Purchase of Redeemable Shares [Repealed in 1979].
Sec. 67 Cancellation of Redeemable Shares by Redemption or Purchase [Repealed in 1979].
Sec. 68 Cancellation of Other Reacquired Shares [Repealed in 1979].
Sec. 69 Reduction of Stated Capital in Certain Cases [Repealed in 1979].
Sec. 70 Special Provisions Relating to Surplus and Reserves [Repealed in 1979].

Sec. 113 Registered Office and Registered Agent of Foreign Corporation.
Sec. 114 Change of Registered Office or Registered Agent of Foreign Corporation.
Sec. 115 Service of Process on Foreign Corporation.
Sec. 116 Amendment to Articles of Incorporation of Foreign Corporation.
Sec. 117 Merger of Foreign Corporation Authorized to Transact Business in this State.
Sec. 118 Amended Certificate of Authority.
Sec. 119 Withdrawal of Foreign Corporation.
Sec. 120 Filing of Application for Withdrawal.
Sec. 121 Revocation of Certificate of Authority.
Sec. 122 Issuance of Certificate of Revocation.
Sec. 123 Application to Corporations Heretofore Authorized to Transact Business in this State.
Sec. 124 Transacting Business Without Certificate of Authority.

ANNUAL REPORTS
Sec. 125 Annual Report of Domestic and Foreign Corporations.
Sec. 126 Filing of Annual Report of Domestic and Foreign Corporations.

FEES, FRANCHISE TAXES AND CHARGES
Sec. 127 Fees, Franchise Taxes and Charges to be Collected by Secretary of State.
Sec. 128 Fees for Filing Documents and Issuing Certificates.
Sec. 129 Miscellaneous Charges.
Sec. 130 License Fees Payable by Domestic Corporations.
Sec. 131 License Fees Payable by Foreign Corporations.
Sec. 132 Franchise Taxes Payable by Domestic Corporations.
Sec. 133 Franchise Taxes Payable by Foreign Corporations.
Sec. 134 Assessment and Collection of Annual Franchise Taxes.

PENALTIES
Sec. 135 Penalties Imposed upon Corporations.
Sec. 136 Penalties Imposed upon Officers and Directors.

MISCELLANEOUS PROVISIONS
Sec. 137 Interrogatories by Secretary of State.
Sec. 138 Information Disclosed by Interrogatories.
Sec. 139 Powers of Secretary of State.
Sec. 140 Appeal from Secretary of State.
Sec. 141 Certificates and Certified Copies to be Received in Evidence.
Sec. 142 Forms to be Furnished by Secretary of State.
Sec. 143 Greater Voting Requirements.
Sec. 144 Waiver of Notice.
Sec. 145 Action by Shareholders Without a Meeting.
Sec. 146 Unauthorized Assumption of Corporate Powers.
Sec. 147 Application to Existing Corporations.
Sec. 148 Application to Foreign and Interstate Commerce.
Sec. 149 Reservation of Power.
Sec. 150 Effect of Repeal of Prior Acts.
Sec. 151 Effect of Invalidity of Part of this Act.
Sec. 152 Exclusivity of Certain Provisions [Optional].
Sec. 153 Repeal of Prior Acts.

§ 1. **Short Title***

This Act shall be known and may be cited as the ". . . . .† Business Corporation Act."

§ 2. **Definitions**

As used in this Act, unless the context otherwise requires, the term:

(a) "Corporation" or "domestic corporation" means a corporation for profit subject to the provisions of this Act, except a foreign corporation.

(b) "Foreign corporation" means a corporation for profit organized under laws other than the laws of this State for a purpose or purposes for which a corporation may be organized under this Act.

(c) "Articles of incorporation" means the original or restated articles of incorporation or articles of consolidation and all amendments thereto including articles of merger.

(d) "Shares" means the units into which the proprietary interests in a corporation are divided.

(e) "Subscriber" means one who subscribes for shares in a corporation, whether before or after incorporation.

(f) "Shareholder" means one who is a holder of record of shares in a corporation. If the articles of incorporation or the by-laws so provide, the board of directors may adopt by resolution a procedure whereby a shareholder of the corporation may certify in writing to the corporation that all or a portion of the shares registered in the name of such shareholder are held for the account of a specified person or persons. The resolution shall set forth (1) the classification of shareholder who may certify, (2) the purpose or purposes for which the certification may be made, (3) the form of certification and information to be contained therein, (4) if the certification is with respect to a record date or closing of the stock transfer books within which the certification must be received by the corporation and (5) such other provisions with respect to the procedure as are deemed necessary or desirable. Upon receipt by the corporation of a certification complying with the procedure, the persons specified in the certification shall be deemed, for the purpose or purposes set forth in the certification, to be the holders of record of the number of shares specified in place of the shareholder making the certification.

(g) "Authorized shares" means the shares of all classes which the corporation is authorized to issue.

(h) "Employee" includes officers but not directors. A director may accept duties which make him also an employee.

(i) "Distribution" means a direct or indirect transfer of money or other property (except its own shares) or incurrence of indebtedness, by a corporation to or for the benefit of any of its shareholders in respect of any of its shares, whether by dividend or by purchase, redemption or other acquisition of its shares, or otherwise.*

(h) "Treasury shares" means shares of a corporation which have been issued, have been subsequently acquired by and belong to the corporation, and have not, either by reason of the acquisition or thereafter, been cancelled or restored to the status of authorized but unissued shares. Treasury shares shall be deemed to be "issued" shares, but not "outstanding" shares.

(i) "Net assets" means the amount by which the total assets of a corporation exceed the total debts of the corporation.

(j) "Stated capital" means, at any particular time, the sum of (1) the par value of all shares of the corporation having a par value that have been issued, (2) the amount of the consideration received by the corporation for all shares of the corporation without par value that have been issued, except such part of the consideration therefor as may have been allocated to capital surplus in a manner permitted by law, and (3) such amounts not included in clauses (1) and (2) of this

*[By the Editor] The Model Business Corporation Act prepared by the Committee on Corporate Laws (Section of Corporation, Banking and Business Law) of the American Bar Association was originally patterned after the Illinois Business Corporation Act of 1933. It was first published as a complete act in 1950. In subsequent years several revisions, addenda and optional or alternative provisions were added. The Act was substantially revised and renumbered in 1969.
This Act should be distinguished from the Model Business Corporation Act promulgated in 1928 by the Commissioners on Uniform State Laws under the name "Uniform Business Corporation Act" and renamed Model Business Corporation Act in 1943. This Uniform Act was withdrawn in 1957.
The Model Business Corporation Act has been influential in the codification of corporation statutes in more than 35 states. However, there is no state that has totally adopted it in its current form. Moreover, since the Model Act itself has been substantially modified from time to time, there is considerable variation among the statutes of the states that used this Act as a model.

†Supply name of State.

*[By The Editor] § 2 formerly provided definitions for the following items, which are quoted here with the original subsection references:

paragraph as have been transferred to stated capital of the corporation, whether upon the issue of shares as a share dividend or otherwise, minus all reductions from such sum as have been effected in a manner permitted by law. Irrespective of the manner of designation thereof by the laws under which a foreign corporation is organized, the stated capital of a foreign corporation shall be determined on the same basis and in the same manner as the stated capital of a domestic corporation, for the purpose of computing fees, franchise taxes and other charges imposed by this Act.

(k) "Surplus" means the excess of the net assets of a corporation over its stated capital.

(l) "Earned surplus" means the portion of the surplus of a corporation equal to the balance of its net profits, income, gains and losses from the date of incorporation, or from the latest date when a deficit was eliminated by an application of its capital surplus or stated capital or otherwise, after deducting subsequent distributions to shareholders and transfers to stated capital and capital surplus to the extent such distributions and transfers are made out of earned surplus. Earned surplus shall include also any portion of surplus allocated to earned surplus in mergers, consolidations or acquisitions of all or substantially all of the outstanding shares or of the property and assets of another corporation, domestic or foreign.

(m) "Capital surplus" means the entire surplus of a corporation other than its earned surplus.

(n) "Insolvent" means inability of a corporation to pay its debts as they become due in the usual course of its business.

§ 3. **Purposes**

Corporations may be organized under this Act for any lawful purpose or purposes, except for the purpose of banking or insurance.

§ 4. **General Powers**

Each corporation shall have power:

(a) To have perpetual succession by its corporate name unless a limited period of duration is stated in its articles of incorporation.

(b) To sue and be sued, complain and defend, in its corporate name.

(c) To have a corporate seal which may be altered at pleasure, and to use the same by causing it, or a facsimile thereof, to be impressed or affixed or in any other manner reproduced.

(d) To purchase, take, receive, lease, or otherwise acquire, own, hold, improve, use and otherwise deal in and with, real or personal property, or any interest therein, wherever situated.

(e) To sell, convey, mortgage, pledge, lease, exchange, transfer and otherwise dispose of all or any part of its property and assets.

(f) To lend money and use its credit to assist its employees.

(g) To purchase, take, receive, subscribe for, or otherwise acquire, own, hold, vote, use, employ, sell, mortgage, lend, pledge, or otherwise dispose of, and otherwise use and deal in and with, shares or other interests in, or obligations of, other domestic or foreign corporations, associations, partnerships or individuals, or direct or indirect obligations of the United States or of any other government, state, territory, governmental district or municipality or of any instrumentality thereof.

(h) To make contracts and guarantees and incur liabilities, borrow money at such rates of interest as the corporation may determine, issue its notes, bonds, and other obligations, and secure any of its obligations by mortgage or pledge of all or any of its property, franchises and income.

(i) To lend money for its corporate purposes, invest and reinvest its funds, and take and hold real and personal property as security for the payment of funds so loaned or invested.

(j) To conduct its business, carry on its operations and have offices and exercise the powers granted by this Act, within or without this State.

(k) To elect or appoint officers and agents of the corporation, and define their duties and fix their compensation.

(l) To make and alter by-laws, not inconsistent with its articles of incorporation or with the laws of this State, for the administration and regulation of the affairs of the corporation.

(m) To make donations for the public welfare or for charitable, scientific or educational purposes.

(n) To transact any lawful business which the board of directors shall find will be in aid of governmental policy.

(o) To pay pensions and establish pension plans, pension trusts, profit sharing plans, stock bonus plans, stock option plans and other incentive plans for any or all of its directors, officers and employees.

(p) To be a promoter, partner, member, associate, or manager of any partnership, joint venture, trust or other enterprise.

(q) To have and exercise all powers necessary or convenient to effect its purposes.

§ 5. **Indemnification of Officers, Directors, Employees and Agents**

(a) A corporation shall have power to indemnify any person who was or is a party or is threatened to be made a party to any threatened, pending or completed action, suit or proceeding, whether civil, criminal, administrative or investigative (other than an action by or in the right of the corporation) by reason of the fact that he is or was a director, officer, employee or agent of the corporation, or is or was serving at the request of the corporation as a director, officer, employee or agent of another corporation, partnership, joint venture, trust or other enterprise, against expenses (including attorneys' fees), judgments, fines and amounts paid in settlement actually and reasonably incurred by him in connection with such action, suit or proceeding if he acted in good faith and in a manner he reasonably believed to be in or not opposed to the best interests of the corporation, and, with respect to any criminal action or proceeding, had no reasonable cause to believe his conduct was unlawful. The termination of any action, suit or proceeding by judgment, order, settlement, conviction, or upon a plea of nolo contendere or its equivalent, shall not, of itself, create a presumption that the person did not act in good faith and in a manner which he reasonably believed to be in or not opposed to the best interest of the corporation, and, with respect to any criminal action or proceeding, had reasonable cause to believe that his conduct was unlawful.

(b) A corporation shall have power to indemnify any person who was or is a party or is threatened to be made a party to any threatened, pending or completed action or suit by or in the right of the corporation to procure a judgment in its favor by reason of the fact that he is or was a director, officer, employee or agent of the corporation, or is or was serving at the request of the corporation as a director, officer, employee or agent of another corporation, partnership, joint venture, trust or other enterprise against expenses (including attorneys' fees) actually and reasonably incurred by him in connection with the defense or settlement of such action or suit if he acted in good faith and in a manner he reasonably believed to be in or not opposed to the best interests of the corporation and except that no indemnification shall be made in respect of any claim, issue or matter as to which such person shall have been adjudged to be liable for negligence or misconduct in the performance of his duty to the corporation unless and only to the extent that the court in which such action or suit was brought shall determine upon application that, despite the adjudication of liability but in view of all circumstances of the case, such person is fairly and reasonably entitled to indemnity for such expenses which such court shall deem proper.

(c) To the extent that a director, officer, employee or agent of a corporation has been successful on the merits or otherwise in defense of any action, suit or proceeding referred to in subsections (a) or (b), or in defense of any claim, issue or matter therein, he shall be indemnified against expenses (including attorneys' fees) actually and reasonably incurred by him in connection therewith.

(d) Any indemnification under subsections (a) or (b) (unless ordered by a court) shall be made by the corporation only as authorized in the specific case upon a determination that indemnification of the director, officer, employee or agent is proper in the circumstances because he has met the applicable standard of conduct set forth in subsections (a) or (b). Such determination shall be made (1) by the board of directors by a majority vote of a quorum consisting of directors who were not parties to such action, suit or proceeding, or (2) if such a quorum is not obtainable, or, even if obtainable a quorum of disinterested directors so directs, by independent legal counsel in a written opinion, or (3) by the shareholders.

(e) Expenses (including attorneys' fees) incurred in defending a civil or criminal action, suit or proceeding may be paid by the corporation in advance of the final disposition of such action, suit or proceeding as authorized in the manner provided in subsection (d) upon receipt of an undertaking by or on behalf of the director, officer, employee or agent to repay such amount unless it shall ultimately be determined that he is entitled to be indemnified by the corporation as authorized in this section.

(f) The indemnification provided by this section shall not be deemed exclusive of any other rights to which those indemnified may be entitled under any by-law, agreement, vote of shareholders or disinterested directors or otherwise, both as to action in his official capacity and as to action in another capacity while holding such office, and shall continue as to a person who has ceased to be a director, officer, employee or agent and shall inure to the benefit of the heirs, executors and administrators of such a person.

(g) A corporation shall have power to purchase and maintain insurance on behalf of any person who is or was a director, officer, employee or agent of the corporation, or is or was serving at the request of the corporation as a director, officer, employee or agent of another corporation, partnership, joint venture, trust or other enterprise against any liability asserted against him and incurred by him in any such capacity or arising out of his status as such, whether or not the corporation would have the power to indemnify him against such liability under the provisions of this section.

§ 6. **Power of Corporation to Acquire Its Own Shares**

A corporation shall have the power to acquire its own shares. All of its own shares acquired by a corporation shall, upon acquisition, constitute authorized but unissued shares, unless the articles of incorporation provide that they shall not be reissued, in which case the authorized shares shall be reduced by the number of shares acquired.

If the number of authorized shares is reduced by an acquisition, the corporation shall, not later than the time it files its next annual report under this Act with the Secretary of State, file a statement of cancellation showing the reduction in the authorized shares. The statement of cancellation shall be executed in duplicate by the corporation by its president or a vice president and by its secretary or an assistant secretary, and verified by one of the officers signing such statement, and shall set forth:

(a) The name of the corporation.

(b) The number of acquired shares cancelled, itemized by classes and series.

(c) The aggregate number of authorized shares, itemized by classes and series, after giving effect to such cancellation.

Duplicate originals of such statement shall be delivered to the Secretary of State. If the Secretary of State finds that such statement conforms to law, he shall, when all fees and franchise taxes have been paid as in this Act prescribed:

(1) Endorse on each of such duplicate originals the word "Filed", and the month, day and year of the filing thereof.

(2) File one of such duplicate originals in his office.

(3) Return the other duplicate original to the corporation or its representative.

§ 7. **Defense of Ultra Vires**

No act of a corporation and no conveyance or transfer of real or personal property to or by a corporation shall be invalid by reason of the fact that the corporation was without capacity or power to do such act or to make or receive such conveyance or transfer, but such lack of capacity or power may be asserted:

(a) In a proceeding by a shareholder against the corporation to enjoin the doing of any act or the transfer of real or personal property by or to the corporation. If the unauthorized act or transfer sought to be enjoined is being, or is to be, performed or made pursuant to a contract to which the corporation is a party, the court may, if all of the parties to the contract are parties to the proceeding and if it deems the same to be equitable, set aside and enjoin the performance of such contract, and in so doing may allow to the corporation or to the other parties to the contract, as the case may be, compensation for the loss or damage sustained by either of them which may result from the action of the court in setting aside and enjoining the performance of such contract, but anticipated profits to be derived from the performance of the contract shall not be awarded by the court as a loss or damage sustained.

(b) In a proceeding by the corporation, whether acting directly or through a receiver, trustee, or other legal representative, or through shareholders in a representative suit, against the incumbent or former officers or directors of the corporation.

(c) In a proceeding by the Attorney General, as provided in this Act, to dissolve the corporation, or in a proceeding by the Attorney General to enjoin the corporation from the transaction of unauthorized business.

§ 8. Corporate Name

The corporate name:

(a) Shall contain the word "corporation," "company," "incorporated" or "limited," or shall contain an abbreviation of one of such words.

(b) Shall not contain any word or phrase which indicates or implies that it is organized for any purpose other than one or more of the purposes contained in its articles of incorporation.

(c) Shall not be the same as, or deceptively similar to, the name of any domestic corporation existing under the laws of this State or any foreign corporation authorized to transact business in this State, or a name the exclusive right to which is, at the time, reserved in the manner provided in this Act, or the name of a corporation which has in effect a registration of its corporate name as provided in this Act, except that this provision shall not apply if the applicant files with the Secretary of State either of the following: (1) the written consent of such other corporation or holder of a reserved or registered name to use the same or deceptively similar name and one or more words are added to make such name distinguishable from such other name, or (2) a certified copy of a final decree of a court of competent jurisdiction establishing the prior right of the applicant to the use of such name in this State.

A corporation with which another corporation, domestic or foreign, is merged, or which is formed by the reorganization or consolidation of one or more domestic or foreign corporations or upon a sale, lease or other disposition to or exchange with, a domestic corporation of all or substantially all the assets of another corporation, domestic or foreign, including its name, may have the same name as that used in this State by any of such corporations if such other corporation was organized under the laws of, or is authorized to transact business in, this State.

§ 9. Reserved Name

The exclusive right to the use of a corporate name may be reserved by:

(a) Any person intending to organize a corporation under this Act.

(b) Any domestic corporation intending to change its name.

(c) Any foreign corporation intending to make application for a certificate of authority to transact business in this State.

(d) Any foreign corporation authorized to transact business in this State and intending to change its name.

(e) Any person intending to organize a foreign corporation and intending to have such corporation make application for a certificate of authority to transact business in this State.

The reservation shall be made by filing with the Secretary of State an application to reserve a specified corporate name, executed by the applicant. If the Secretary of State finds that the name is available for corporate use, he shall reserve the same for the exclusive use of the applicant for a period of one hundred and twenty days.

The right to the exclusive use of a specified corporate name so reserved may be transferred to any person or corporation by filing in the office of the Secretary of State a notice of such transfer, executed by the applicant for whom the name was reserved, and specifying the name and address of the transferee.

§ 10. Registered Name

Any corporation organized and existing under the laws of any state or territory of the United States may register its corporate name under this Act, provided its corporate name is not the same as, or deceptively similar to, the name of any domestic corporation existing under the laws of this State, or the name of any foreign corporation authorized to transact business in this State, or any corporate name reserved or registered under this Act.

Such registration shall be made by:

(a) Filing with the Secretary of State (1) an application for registration executed by the corporation by an officer thereof, setting forth the name of the corporation, the state or territory under the laws of which it is incorporated, the date of its incorporation, a statement that it is carrying on or doing business, and a brief statement of the business in which it is engaged, and (2) a certificate setting forth that such corporation is in good standing under the laws of the state or territory wherein it is organized, executed by the

Secretary of State of such state or territory or by such other official as may have custody of the records pertaining to corporations, and

(b) Paying to the Secretary of State a registration fee in the amount of .... for each month, or fraction thereof, between the date of filing such application and December 31st of the calendar year in which such application is filed.

Such registration shall be effective until the close of the calendar year in which the application for registration is filed.

§ 11. **Renewal of Registered Name**

A corporation which has in effect a registration of its corporate name, may renew such registration from year to year by annually filing an application for renewal setting forth the facts required to be set forth in an original application for registration and a certificate of good standing as required for the original registration and by paying a fee of .... A renewal application may be filed between the first day of October and the thirty-first day of December in each year, and shall extend the registration for the following calendar year.

§ 12. **Registered Office and Registered Agent**

Each corporation shall have and continuously maintain in this State:

(a) A registered office which may be, but need not be, the same as its place of business.

(b) A registered agent, which agent may be either an individual resident in this State whose business office is identical with such registered office, or a domestic corporation, or a foreign corporation authorized to transact business in this State, having a business office identical with such registered office.

§ 13. **Change of Registered Office or Registered Agent**

A corporation may change its registered office or change its registered agent, or both, upon filing in the office of the Secretary of State a statement setting forth:

(a) The name of the corporation.

(b) The address of its then registered office.

(c) If the address of its registered office is to be changed, the address to which the registered office is to be changed.

(d) The name of its then registered agent.

(e) If its registered agent is to be changed, the name of its successor registered agent.

(f) That the address of its registered office and the address of the business office of its registered agent, as changed, will be identical.

(g) That such change was authorized by resolution duly adopted by its board of directors.

Such statement shall be executed by the corporation by its president, or a vice president, and verified by him, and delivered to the Secretary of State. If the Secretary of State finds that such statement conforms to the provisions of this Act, he shall file such statement in his office, and upon such filing the change of address of the registered office, or the appointment of a new registered agent, or both, as the case may be, shall become effective.

Any registered agent of a corporation may resign as such agent upon filing a written notice thereof, executed in duplicate, with the Secretary of State, who shall forthwith mail a copy thereof to the corporation at its registered office. The appointment of such agent shall terminate upon the expiration of thirty days after receipt of such notice by the Secretary of State.

If a registered agent changes his or its business address to another place within the same ....,* he or it may change such address and the address of the registered office of any corporation of which he or it is registered agent by filing a statement as required above except that it need be signed only by the registered agent and need not be responsive to (e) or (g) and must recite that a copy of the statement has been mailed to the corporation.

§ 14. **Service of Process on Corporation**

The registered agent so appointed by a corporation shall be an agent of such corporation upon whom any process, notice or demand required or permitted by law to be served upon the corporation may be served.

Whenever a corporation shall fail to appoint or maintain a registered agent in this State, or whenever its registered agent cannot with reasonable diligence be found at the registered office, then the Secretary of State shall be an agent of such corporation upon whom any such process, notice, or demand may be

*Supply designation of jurisdiction, such as county, etc., in accordance with local practice.

served. Service on the Secretary of State of any such process, notice, or demand shall be made by delivering to and leaving with him, or with any clerk having charge of the corporation department of his office, duplicate copies of such process, notice or demand. In the event any such process, notice or demand is served on the Secretary of State, he shall immediately cause one of the copies thereof to be forwarded by registered mail, addressed to the corporation at its registered office. Any service so had on the Secretary of State shall be returnable in not less than thirty days.

The Secretary of State shall keep a record of all processes, notices and demands served upon him under this section, and shall record therein the time of such service and his action with reference thereto.

Nothing herein contained shall limit or affect the right to serve any process, notice or demand required or permitted by law to be served upon a corporation in any other manner now or hereafter permitted by law.

§ 15. **Authorized Shares**

Each corporation shall have power to create and issue the number of shares stated in its articles of incorporation. Such shares may be divided into one or more classes with such designations, preferences, limitations, and relative rights as shall be stated in the articles of incorporation. The articles of incorporation may limit or deny the voting rights of or provide special voting rights for the shares of any class to the extent not inconsistent with the provisions of this Act.

Without limiting the authority herein contained, a corporation, when so provided in its articles of incorporation, may issue shares of preferred or special classes:

(a) Subject to the right of the corporation to redeem any of such shares at the price fixed by the articles of incorporation for the redemption thereof.

(b) Entitling the holders thereof to cumulative, noncumulative or partially cumulative dividends.

(c) Having preference over any other class or classes of shares as to the payment of dividends.

(d) Having preference in the assets of the corporation over any other class or classes of shares upon the voluntary or involuntary liquidation of the corporation.

(e) Convertible into shares of any other class or into shares of any series of the same or any other class, except a class having prior or superior rights and preferences as to dividends or distribution of assets upon liquidation.

§ 16. **Issuance of Shares of Preferred or Special Classes in Series**

If the articles of incorporation so provide, the shares of any preferred or special class may be divided into and issued in series. If the shares of any such class are to be issued in series, then each series shall be so designated as to distinguish the shares thereof from the shares of all other series and classes. Any or all of the series of any such class and the variations in the relative rights and preferences as between different series may be fixed and determined by the articles of incorporation, but all shares of the same class shall be identical except as to the following relative rights and preferences, as to which there may be variations between different series:

(A) The rate of dividend.

(B) Whether shares may be redeemed and, if so, the redemption price and the terms and conditions of redemption.

(C) The amount payable upon shares in the event of voluntary and involuntary liquidation.

(D) Sinking fund provisions, if any, for the redemption or purchase of shares.

(E) The terms and conditions, if any, on which shares may be converted.

(F) Voting rights, if any.

If the articles of incorporation shall expressly vest authority in the board of directors, then, to the extent that the articles of incorporation shall not have established series and fixed and determined the variations in the relative rights and preferences as between series, the board of directors shall have authority to divide any or all of such classes into series and, within the limitations set forth in this section and in the articles of incorporation, fix and determine the relative rights and preferences of the shares of any series so established.

In order for the board of directors to establish a series, where authority so to do is contained in the articles of incorporation, the board of directors shall adopt a resolution setting forth the designation of the series and fixing and determining the relative rights and preferences thereof, or so much thereof as shall not be fixed and determined by the articles of incorporation.

Prior to the issue of any shares of a series established by resolution adopted by the board of directors, the corporation shall file in the office of the Secretary of State a statement setting forth:

(a) The name of the corporation.

(b) A copy of the resolution establishing and designating the series, and fixing and determining the relative rights and preferences thereof.

(c) The date of adoption of such resolution.

(d) That such resolution was duly adopted by the board of directors.

Such statement shall be executed in duplicate by the corporation by its president or a vice president and by its secretary or an assistant secretary, and verified by one of the officers signing such statement, and shall be delivered to the Secretary of State. If the Secretary of State finds that such statement conforms to law, he shall, when all franchise taxes and fees have been paid as in this Act prescribed:

(1) Endorse on each of such duplicate originals the word "Filed," and the month, day, and year of the filing thereof.

(2) File one of such duplicate originals in his office.

(3) Return the other duplicate original to the corporation or its representative.

Upon the filing of such statement by the Secretary of State, the resolution establishing and designating the series and fixing and determining the relative rights and preferences thereof shall become effective and shall constitute an amendment of the articles of incorporation.

§ 17. Subscriptions for Shares

A subscription for shares of a corporation to be organized shall be irrevocable for a period of six months, unless otherwise provided by the terms of the subscription agreement or unless all of the subscribers consent to the revocation of such subscription.

Unless otherwise provided in the subscription agreement, subscriptions for shares, whether made before or after the organization of a corporation, shall be paid in full at such time, or in such installments and at such times, as shall be determined by the board of directors. Any call made by the board of directors for payment on subscriptions shall be uniform as to all shares of the same class or as to all shares of the same series, as the case may be. In case of default in the payment of any installment or call when such payment is due, the corporation may proceed to collect the amount due in the same manner as any debt due the corporation. The by-laws may prescribe other penalties for failure to pay installments or calls that may become due, but no penalty working a forfeiture of a subscription, or of the amounts paid thereon, shall be declared as against any subscriber unless the amount due thereon shall remain unpaid for a period of twenty days after written demand has been made therefor. If mailed, such written demand shall be deemed to be made when deposited in the United States mail in a sealed envelope addressed to the subscriber at his last post-office address known to the corporation, with postage thereon prepaid. In the event of the sale of any shares by reason of any forfeiture, the excess of proceeds realized over the amount due and unpaid on such shares shall be paid to the delinquent subscriber or to his legal representative.

§ 18. Issuance for Shares

Subject to any restrictions in the articles of incorporation:

(a) Shares may be issued for such consideration as shall be authorized by the board of directors establishing a price (in money or other consideration) or a minimum price or general formula or method by which the price will be determined; and

(b) Upon authorization by the board of directors, the corporation may issue its own shares in exchange for or in conversion of its outstanding shares, or distribute its own shares, pro rata to its shareholders or the shareholders of one or more classes or series, to effectuate stock dividends or splits, and any such transaction shall not require consideration; provided, that no such issuance of shares of any class or series shall be made to the holders of shares of any other class or series unless it is either expressly provided for in the articles of incorporation, or is authorized by an affirmative vote or the written consent of the holders of at least a majority of the outstanding shares of the class or series in which the distribution is to be made.

§ 19. Payment for Shares

The consideration for the issuance of shares may be paid, in whole or in part, in cash, in other property, tangible or intangible, or in labor or services actually performed for the corporation. When payment of the consideration for which shares are to be issued shall have been received by the corporation, such shares shall be nonassessable.

Neither promissory notes nor future services shall constitute payment or part payment for the issuance of shares of a corporation.

In the absence of fraud in the transaction, the judgment of the board of directors or the shareholders, as the case may be, as to the value of the consideration received for shares shall be conclusive.

§ 20. **Stock Rights and Options**

Subject to any provisions in respect thereof set forth in its articles of incorporation, a corporation may create and issue, whether or not in connection with the issuance and sale of any of its shares or other securities, rights or options entitling the holders thereof to purchase from the corporation shares of any class or classes. Such rights or options shall be evidenced in such manner as the board of directors shall approve and, subject to the provisions of the articles of incorporation, shall set forth the terms upon which, the time or times within which and the price or prices at which such shares may be purchased from the corporation upon the exercise of any such right or option. If such rights or options are to be issued to directors, officers or employees as such of the corporation or of any subsidiary thereof, and not to the shareholders generally, their issuance shall be approved by the affirmative vote of the holders of a majority of the shares entitled to vote thereon or shall be authorized by and consistent with a plan approved or ratified by such a vote of shareholders. In the absence of fraud in the transaction, the judgment of the board of directors as to the adequacy of the consideration received for such rights or options shall be conclusive.

§ 21. **Determination of Amount of Stated Capital [Repealed]**

In case of the issuance by a corporation of shares having a par value, the consideration received therefor shall constitute stated capital to the extent of the par value of such shares, and the excess, if any, of such consideration shall constitute capital surplus.

In case of the issuance by a corporation of shares without par value, the entire consideration received therefor shall constitute stated capital unless the corporation shall determine as provided in this section that only a part thereof shall be stated capital. Within a period of sixty days after the issuance of any shares without par value, the board of directors may allocate to capital surplus any portion of the consideration received for the issuance of such shares. No such allocation shall be made of any portion of the consideration received for shares without par value having a preference in the assets of the corporation in the event of involuntary liquidation except the amount, if any, of such consideration in excess of such preference.

If shares have been or shall be issued by a corporation in merger or consolidation or in acquisition of all or substantially all of the outstanding shares or of the property and assets of another corporation, whether domestic or foreign, any amount that would otherwise constitute capital surplus under the foregoing provisions of this section may instead be allocated to earned surplus by the board of directors of the issuing corporation except that its aggregate earned surplus shall not exceed the sum of the earned surpluses as defined in this Act of the issuing corporation and of all other corporations, domestic or foreign, that were merged or consolidated or of which the shares or assets were acquired.

The stated capital of a corporation may be increased from time to time by resolution of the board of directors directing that all or a part of the surplus of the corporation be transferred to stated capital. The board of directors may direct that the amount of the surplus so transferred shall be deemed to be stated capital in respect of any designated class of shares.

§ 22. **Expenses of Organization, Reorganization and Financing**

The reasonable charges and expenses of organization or reorganization of a corporation, and the reasonable expenses of and compensation for the sale or underwriting of its shares, may be paid or allowed by such corporation out of the consideration received by it in payment for its shares without thereby rendering such shares assessable.

§ 23. **Shares Represented by Certificates and Uncertified Shares**

The shares of a corporation shall be represented by certificates or shall be uncertificated shares. Certificates shall be signed by the chairman or vice-chairman of the board of directors or the president or a vice president and by the treasurer or an assistant treasurer or the secretary or an assistant secretary of the corporation, and may be sealed with the seal of the corporation or a facsimile thereof. Any of or all the signatures [of the president or vice president and the secretary of assistant secretary] upon a certificate may be a facsimile. [s if the certificate is manually signed on behalf of a transfer agent or a registrar, other than the corporation itself or an employee of the corporation.] In case any officer, transfer agent or registrar who has signed or whose facsimile signature has been placed upon such certificate shall have ceased to be such officer, transfer agent or registrar before such certificate is issued, it may be issued by the corporation with the same effect as if he were such officer, transfer agent or registrar at the date of its issue.

Every certificate representing shares issued by a corporation which is authorized to issue shares of more than one class shall set forth upon the face or back of the certificate, or shall state that the corporation will furnish to any shareholder upon request and without charge, a full statement of the designations, preferences, limitations, and relative rights of the shares of each class authorized to be issued, and if the corporation is authorized to issue any preferred or special class in series, the variations in the relative rights and preferences between the shares of each such series so far as the same have been fixed and determined and the authority of the board of directors to fix and determine the relative rights and preferences of subsequent series.

Each certificate representing shares shall state upon the face thereof:

(a) That the corporation is organized under the laws of this State.

(b) The name of the person to whom issued.

(c) The number and class of shares, and the designation of the series, if any, which such certificate represents.

(d) The par value of each share represented by such certificate, or a statement that the shares are without par value.

No certificate shall be issued for any share until such share is fully paid.

Unless otherwise provided by the articles of incorporation or by-laws, the board of directors of a corporation may provide by resolution that some or all of any or all classes and series of its shares shall be uncertificated shares, provided that such resolution shall not apply to shares represented by a certificate until such certificate is surrendered to the corporation. Within a reasonable time after the issuance or transfer of uncertificated shares, the corporation shall send to the registered owner thereof a written notice containing the information required to be set forth or stated on certificates pursuant to the second and third paragraphs of this section. Except as otherwise expressly provided by law, the rights and obligations of the holders of uncertificated shares and the rights and obligations of the holders of certificates representing shares of the same class and series shall be identical.

§ 24. Fractional Shares

A corporation may (1) issue fractions of a share, either represented by a certificate or uncertificated, (2) arrange for the disposition of fractional interests by those entitled thereto, (3) pay in money the fair value of fractions of a share as of a time when those entitled to receive such fractions are determined, or (4) issue scrip in registered or bearer form which shall entitle the holder to receive a certificate for a full share or an uncertificated full share upon the surrender of such scrip aggregating a full share. A certificate for a fractional share or an uncertificated fractional share shall, but scrip shall not unless otherwise provided therein, entitle the holder to exercise voting rights, to receive dividends thereon, and to participate in any of the assets of the corporation in the event of liquidation. The board of directors may cause scrip to be issued subject to the condition that it shall become void if not exchanged for certificates representing full shares or uncertificated full shares before a specified date, or subject to the condition that the shares for which scrip is exchangeable may be sold by the corporation and the proceeds thereof distributed to the holders of scrip, or subject to any other conditions which the board of directors may deem advisable.

§ 25. Liability of Subscribers and Shareholders

A holder of or subscriber to shares of a corporation shall be under no obligation to the corporation or its creditors with respect to such shares other than the obligation to pay to the corporation the full consideration for which such shares were issued or to be issued.

Any person becoming an assignee or transferee of shares or of a subscription for shares in good faith and without knowledge or notice that the full consideration therefor has not been paid shall not be personally liable to the corporation or its creditors for any unpaid portion of such consideration.

An executor, administrator, conservator, guardian, trustee, assignee for the benefit of creditors, or receiver shall not be personally liable to the corporation as a holder of or subscriber to shares of a corporation but the estate and funds in his hands shall be so liable.

No pledgee or other holder of shares as collateral security shall be personally liable as a shareholder.

§ 26. **Shareholders' Preemptive Rights**

The shareholders of a corporation shall have no preemptive right to acquire unissued shares of the corporation, or securities of the corporation convertible into or carrying a right to subscribe to or acquire shares, except to the extent, if any, that such right is provided in the articles of incorporation.

§ 26A. **Shareholders' Preemptive Rights [Alternative]**

Except to the extent limited or denied by this section or by the articles of incorporation, shareholders shall have a preemptive right to acquire unissued shares or securities convertible into such shares or carrying a right to subscribe to or acquire shares.

Unless otherwise provided in the articles of incorporation,

(a) No preemptive right shall exist

(1) to acquire any shares issued to directors, officers or employees pursuant to approval by the affirmative vote of the holders of a majority of the shares entitled to vote thereon or when authorized by and consistent with a plan theretofore approved by such a vote of shareholders; or

(2) to acquire any shares sold otherwise than for money.

(b) Holders of shares of any class that is preferred or limited as to dividends or assets shall not be entitled to any preemptive right.

(c) Holders of shares of common stock shall not be entitled to any preemptive right to shares of any class that is preferred or limited as to dividends or assets or to any obligations, unless convertible into shares of common stock or carrying a right to subscribe to or acquire shares of common stock.

(d) Holders of common stock without voting power shall have no preemptive right to shares of common stock with voting power.

(e) The preemptive right shall be only an opportunity to acquire shares or other securities under such terms and conditions as the board of directors may fix for the purpose of providing a fair and reasonable opportunity for the exercise of such right.

§ 27. **By-Laws**

The initial by-laws of a corporation shall be adopted by its board of directors. The power to alter, amend or repeal the by-laws or adopt new by-laws, subject to repeal or change by action of the shareholders, shall be vested in the board of directors unless reserved to the shareholders by the articles of incorporation. The by-laws may contain any provisions for the regulation and management of the affairs of the corporation not inconsistent with law or the articles of incorporation.

§ 27A. **By-Laws and Other Powers in Emergency [Optional]**

The board of directors of any corporation may adopt emergency by-laws, subject to repeal or change by action of the shareholders, which shall, notwithstanding any different provision elsewhere in this Act or in the articles of incorporation or by-laws, be operative during any emergency in the conduct of the business of the corporation resulting from an attack on the United States or any nuclear or atomic disaster. The emergency by-laws may make any provision that may be practical and necessary for the circumstances of the emergency, including provisions that:

(a) A meeting of the board of directors may be called by any officer or director in such manner and under such conditions as shall be prescribed in the emergency by-laws;

(b) The director or directors in attendance at the meeting, or any greater number fixed by the emergency by-laws, shall constitute a quorum; and

(c) The officers or other persons designated on a list approved by the board of directors before the emergency, all in such order of priority and subject to such conditions, and for such period of time (not longer than reasonably necessary after the termination of the emergency) as may be provided in the emergency by-laws or in the resolution approving the list shall, to the extent required to provide a quorum at any meeting of the board of directors, be deemed directors for such meeting.

The board of directors, either before or during any such emergency, may provide, and from time to time modify, lines of succession in the event that during such an emergency any or all officers or agents of the corporation shall for any reason be rendered incapable of discharging their duties.

The board of directors, either before or during any such emergency, may, effective in the emergency, change the head office or designate several alternative head offices or regional offices, or authorize the officers so to do.

To the extent not inconsistent with any emergency by-laws so adopted, the by-laws of the corporation shall remain in effect during any such emergency and upon its termination the emergency by-laws shall cease to be operative.

Unless otherwise provided in emergency by-laws, notice of any meeting of the board of directors during any such emergency may be given only to such of the directors as it may be feasible to reach at the time and by such means as may be feasible at the time, including publication or radio.

To the extent required to constitute a quorum at any meeting of the board of directors during any such emergency, the officers of the corporation who are present shall, unless otherwise provided in emergency by-laws, be deemed, in order of rank and within the same rank in order of seniority, directors for such meeting.

No officer, director or employee acting in accordance with any emergency by-laws shall be liable except for willful misconduct. No officer, director or employee shall be liable for any action taken by him in good faith in such an emergency in furtherance of the ordinary business affairs of the corporation even though not authorized by the by-laws then in effect.

§ 28. **Meetings of Shareholders**

Meetings of shareholders may be held at such place within or without this State as may be stated in or fixed in accordance with the by-laws. If no other place is stated or so fixed, meetings shall be held at the registered office of the corporation.

An annual meeting of the shareholders shall be held at such time as may be stated in or fixed in accordance with the by-laws. If the annual meeting is not held within any thirteen-month period the Court of .... may, on the application of any shareholder, summarily order a meeting to be held.

A special meeting of the shareholders may be called by the board of directors, the holders of not less than one-tenth of all the shares entitled to vote at the meeting, or such other persons as may be authorized in the articles of incorporation or the by-laws.

§ 29. **Notice of Shareholders' Meetings**

Written notice stating the place, day and hour of the meeting and, in case of a special meeting, the purpose or purposes for which the meeting is called, shall be delivered not less than ten nor more than fifty days before the date of the meeting, either personally or by mail, by or at the direction of the president, the secretary, or the officer or persons calling the meeting, to each shareholder of record entitled to vote at such meeting. If mailed, such notice shall be deemed to be delivered when deposited in the United States mail addressed to the shareholder at his address as it appears on the stock transfer books of the corporation, with postage thereon prepaid.

§ 30. **Closing of Transfer Books and Fixing Record Date**

For the purpose of determining shareholders entitled to notice of or to vote at any meeting of shareholders or any adjournment thereof, or entitled to receive payment of any dividend, or in order to make a determination of shareholders for any other proper purpose, the board of directors of a corporation may provide that the stock transfer books shall be closed for a stated period but not to exceed, in any case, fifty days. If the stock transfer books shall be closed for the purpose of determining shareholders entitled to notice of or to vote at a meeting of shareholders, such books shall be closed for at least ten days immediately preceding such meeting. In lieu of closing the stock transfer books, the by-laws, or in the absence of an applicable by-law the board of directors, may fix in advance a date as the record date for any such determination of shareholders, such date in any case to be not more than fifty days and, in case of a meeting of shareholders, not less than ten days prior to the date on which the particular action, requiring such determination of shareholders, is to be taken. If the stock transfer books are not closed and no record date is fixed for the determination of shareholders entitled to notice of or to vote at a meeting of shareholders, or shareholders entitled to receive payment of a dividend, the date on which notice of the meeting is mailed or the date on which the resolution of the board of directors declaring such dividend is adopted, as the case may be, shall be the record date for such determination of shareholders. When a determination of shareholders entitled to vote at any meeting of shareholders has been made as provided in this section, such determination shall apply to any adjournment thereof.

§ 31. **Voting Record**

The officer or agent having charge of the stock transfer books for shares of a corporation shall make a complete record of the shareholders entitled to vote at such meeting or any adjournment thereof, arranged in alphabetical order, with the address of and the number of shares held by each. Such record shall be produced and kept open at the time and place of the meeting and shall be subject to the inspection of any shareholder during the whole time of the meeting for the purposes thereof.

Failure to comply with the requirements of this section shall not affect the validity of any action taken at such meeting.

An officer or agent having charge of the stock transfer books who shall fail to prepare the record of shareholders, or produce and keep it open for inspection at the meeting, as provided in this section, shall be liable to any shareholder suffering damage on account of such failure, to the extent of such damage.

§ 32. **Quorum of Shareholders**

Unless otherwise provided in the articles of incorporation, a majority of the shares entitled to vote, represented in person or by proxy, shall constitute a quorum at a meeting of shareholders, but in no event shall a quorum consist of less than one-third of the shares entitled to vote at the meeting. If a quorum is present, the affirmative vote of the majority of the shares represented at the meeting and entitled to vote on the subject matter shall be the act of the shareholders, unless the vote of a greater number or voting by classes is required by this Act or the articles of incorporation or by-laws.

§ 33. **Voting of Shares**

Each outstanding share, regardless of class, shall be entitled to one vote on each matter submitted to a vote at a meeting of shareholders, except as may be otherwise provided in the articles of incorporation. If the articles of incorporation provide for more or less than one vote for any share, on any matter, every reference in this Act to a majority or other proportion of shares shall refer to such a majority or other proportion of votes entitled to be cast.

Shares held by another corporation if a majority of the shares entitled to vote for the election of directors of such other corporation is held by the corporation, shall not be voted at any meeting or counted in determining the total number of outstanding shares at any given time.

A shareholder may vote either in person or by proxy executed in writing by the shareholder or by his duly authorized attorney-in-fact. No proxy shall be valid after eleven months from the date of its execution, unless otherwise provided in the proxy.

[Either of the following prefatory phrases may be inserted here: "The articles of incorporation may provide that" or "Unless the articles of incorporation otherwise provide"] . . . at each election for directors every shareholder entitled to vote at such election shall have the right to vote, in person or by proxy, the number of shares owned by him for as many persons as there are directors to be elected and for whose election he has a right to vote, or to cumulate his votes by giving one candidate as many votes as the number of such directors multiplied by the number of his shares shall equal, or by distributing such votes on the same principle among any number of such candidates.

Shares standing in the name of another corporation, domestic or foreign, may be voted by such officer, agent or proxy as the by-laws of such other corporation may prescribe, or, in the absence of such provision, as the board of directors of such other corporation may determine.

Shares held by an administrator, executor, guardian or conservator may be voted by him, either in person or by proxy, without a transfer of such shares into his name. Shares standing in the name of a trustee may be voted by him, either in person or by proxy, but no trustee shall be entitled to vote shares held by him without a transfer of such shares into his name.

Shares standing in the name of a receiver may be voted by such receiver, and shares held by or under the control of a receiver may be voted by such receiver without the transfer thereof into his name if authority so to do be contained in an appropriate order of the court by which such receiver was appointed.

A shareholder whose shares are pledged shall be entitled to vote such shares until the shares have been transferred into the name of the pledgee, and thereafter the pledgee shall be entitled to vote the shares so transferred.

On and after the date on which written notice of redemption of redeemable shares has been mailed to the holders thereof and a sum sufficient to redeem such shares has been deposited with a bank or trust company with irrevocable instruction and authority to pay the redemption price to the holders thereof upon surrender of certificates therefor, such shares shall not be entitled to vote on any matter and shall not be deemed to be outstanding shares.

§ 34. **Voting Trusts and Agreements Among Shareholders**

Any number of shareholders of a corporation may create a voting trust for the purpose of conferring upon a trustee or trustees the right to vote or otherwise represent their shares, for a period of not to exceed ten years, by entering into a written voting trust agreement specifying the terms and conditions of the voting trust, by depositing a counterpart of the agreement with the corporation at its registered office, and by transferring their shares to such trustee or trustees for the purposes of the agreement. Such trustee or trustees shall keep a record of the holders of voting trust certificates evidencing a beneficial interest in the voting trust, giving the names and addresses of all such holders and the number and class of the shares in respect of which the voting trust certificates held by each are issued, and shall deposit a copy of such record with the corporation at its registered office. The counterpart of the voting trust agreement and the copy of such record so deposited with the corporation shall be subject to the same right of examination by a shareholder of the corporation, in person or by agent or attorney, as are the books and records of the corporation, and such counterpart and such copy of such record shall be subject to examination by any holder of record of voting trust certificates, either in person or by agent or attorney, at any reasonable time for any proper purpose.

Agreements among shareholders regarding the voting of their shares shall be valid and enforceable in accordance with their terms. Such agreements shall not be subject to the provisions of this section regarding voting trusts.

§ 35. **Board of Directors**

All corporate powers shall be exercised by or under authority of, and the business and affairs of a corporation shall be managed under the direction of, a board of directors except as may be otherwise provided in this Act or the articles of incorporation. If any such provision is made in the articles of incorporation, the powers and duties conferred or imposed upon the board of directors by this Act shall be exercised or performed to such extent and by such person or persons as shall be provided in the articles of incorporation. Directors need not be residents of this State or shareholders of the corporation unless the articles of incorporation or by-laws so require. The articles of incorporation or by-laws may prescribe other qualifications for directors. The board of directors shall have authority to fix the compensation of directors unless otherwise provided in the articles of incorporation.

A director shall perform his duties as a director, including his duties as a member of any committee of the board upon which he may serve, in good faith, in a manner he reasonably believes to be in the best interests of the corporation, and with such care as an ordinarily prudent person in a like position would use under similar circumstances. In performing his duties, a director shall be entitled to rely on information, opinions, reports or statements, including financial statements and other financial data, in each case prepared or presented by:

(a) one or more officers or employees of the corporation whom the director reasonably believes to be reliable and competent in the matters presented,

(b) counsel, public accountants or other persons as to matters which the director reasonably believes to be within such person's professional or expert competence, or

(c) a committee of the board upon which he does not serve, duly designated in accordance with a provision of the articles of incorporation or the by-laws, as to matters within its designated authority, which committee the director reasonably believes to merit confidence,

but he shall not be considered to be acting in good faith if he has knowledge concerning the matter in question that would cause such reliance to be unwarranted. A person who so performs his duties shall have no liability by reason of being or having been a director of the corporation.

A director of a corporation who is present at a meeting of its board of directors at which action on any corporate matter is taken shall be presumed to have assented to the action taken unless his dissent shall be entered in the minutes of the meeting or unless he shall file his written dissent to such action with the secretary of the meeting before the adjournment thereof or shall forward such dissent by registered mail to the secretary of the corporation immediately after the adjournment of the meeting. Such right to dissent shall not apply to a director who voted in favor of such action.

§ 36. Number and Election of Directors

The board of directors of a corporation shall consist of one or more members. The number of directors shall be fixed by, or in the manner provided in, the articles of incorporation or the by-laws, except as to the number constituting the initial board of directors, which number shall be fixed by the articles of incorporation. The number of directors may be increased or decreased from time to time by amendment to, or in the manner provided in, the articles of incorporation or the by-laws, but no decrease shall have the effect of shortening the term of any incumbent director. In the absence of a by-law providing for the number of directors, the number shall be the same as that provided for in the articles of incorporation. The names and addresses of the members of the first board of directors shall be stated in the articles of incorporation. Such persons shall hold office until the first annual meeting of shareholders, and until their successors shall have been elected and qualified. At the first annual meeting of shareholders and at each annual meeting thereafter the shareholders shall elect directors to hold office until the next succeeding annual meeting, except in case of the classification of directors as permitted by this Act. Each director shall hold office for the term for which he is elected and until his successor shall have been elected and qualified.

§ 37. Classification of Directors

When the board of directors shall consist of nine or more members, in lieu of electing the whole number of directors annually, the articles of incorporation may provide that the directors be divided into either two or three classes, each class to be as nearly equal in number as possible, the term of office of directors of the first class to expire at the first annual meeting of shareholders after their election, that of the second class to expire at the second annual meeting after their election, and that of the third class, if any, to expire at the third annual meeting after their election. At each annual meeting after such classification the number of directors equal to the number of the class whose term expires at the time of such meeting shall be elected to hold office until the second succeeding annual meeting, if there be two classes, or until the third succeeding annual meeting, if there be three classes. No classification of directors shall be effective prior to the first annual meeting of shareholders.

§ 38. Vacancies

Any vacancy occurring in the board of directors may be filled by the affirmative vote of a majority of the remaining directors though less than a quorum of the board of directors. A director elected to fill a vacancy shall be elected for the unexpired term of his predecessor in office. Any directorship to be filled by reason of an increase in the number of directors may be filled by the board of directors for a term of office continuing only until the next election of directors by the shareholders.

§ 39. Removal of Directors

At a meeting of shareholders called expressly for that purpose, directors may be removed in the manner provided in this section. Any director or the entire board of directors may be removed, with or without cause, by a vote of the holders of a majority of the shares then entitled to vote at an election of directors.

In the case of a corporation having cumulative voting, if less than the entire board is to be removed, no one of the directors may be removed if the votes cast against his removal would be sufficient to elect him if then cumulatively voted at an election of the entire board of directors, or, if there be classes of directors, at an election of the class of directors of which he is a part.

Whenever the holders of the shares of any class are entitled to elect one or more directors by the provisions of the articles of incorporation, the provisions of this section shall apply, in respect to the removal of a director or directors so elected, to the vote of the holders of the outstanding shares of that class and not to the vote of the outstanding shares as a whole.

§ 40. **Quorum of Directors**

A majority of the number of directors fixed by or in the manner provided in the by-laws or in the absence of a by-law fixing or providing for the number of directors, then of the number stated in the articles of incorporation, shall constitute a quorum for the transaction of business unless a greater number is required by the articles of incorporation or the by-laws. The act of the majority of the directors present at a meeting at which a quorum is present shall be the act of the board of directors, unless the act of a greater number is required by the articles of incorporation or the by-laws.

§ 41. **Director Conflicts of Interest**

No contract or other transaction between a corporation and one or more of its directors or any other corporation, firm, association or entity in which one or more of its directors are directors or officers or are financially interested, shall be either void or voidable because of such relationship or interest or because such director or directors are present at the meeting of the board of directors or a committee thereof which authorizes, approves or ratifies such contract or transaction or because his or their votes are counted for such purpose, if:

(a) the fact of such relationship or interest is disclosed or known to the board of directors or committee which authorizes, approves or ratifies the contract or transaction by a vote or consent sufficient for the purpose without counting the votes or consents of such interested directors; or

(b) the fact of such relationship or interest is disclosed or known to the shareholders entitled to vote and they authorize, approve or ratify such contract or transaction by vote or written consent; or

(c) the contract or transaction is fair and reasonable to the corporation.

Common or interested directors may be counted in determining the presence of a quorum at a meeting of the board of directors or a committee thereof which authorizes, approves or ratifies such contract or transaction.

§ 42. **Executive and Other Committees**

If the articles of incorporation or the by-laws so provide, the board of directors, by resolution adopted by a majority of the full board of directors, may designate from among its members an executive committee and one or more other committees each of which, to the extent provided in such resolution or in the articles of incorporation or the by-laws of the corporation, shall have and may exercise all the authority of the board of directors, except that no such committee shall have authority to (i) authorize distributions, (ii) approve or recommend to shareholders actions or proposals required by this Act to be approved by shareholders, (iii) designate candidates for the office of director, for purposes of proxy solicitation or otherwise, or fill vacancies on the board of directors or any committee thereof, (iv) amend the by-laws, (v) approve a plan of merger not requiring shareholder approval, (vi) authorize or approve the reacquisition of shares unless pursuant to a general formula or method specified by the board of directors, or (vii) authorize or approve the issuance or sale of, or any contract to issue or sell, shares or designate the terms of a series of a class of shares, provided that the board of directors, having acted regarding general authorization for the issuance or sale of shares, or any contract, therefor, and, in the case of a series, the designation thereof, may, pursuant to a general formula or method specified by the board by resolution or by adoption of a stock option or other plan, authorize a committee to fix the terms of any contract for the sale of the shares and to fix the terms upon which such shares may be issued or sold, including, without limitation, the price, the dividend rate, provisions for redemption, sinking fund, conversion, voting or preferential rights, and provisions for other features of a class of shares, or a series of a class of shares, with full power in such committee to adopt any final resolution setting forth all the terms thereof and to authorize the statement of the terms of a series for filing with the Secretary of State under this Act.

Neither the designation of any such committee, the delegation thereto of authority, nor action by such committee pursuant to such authority shall alone constitute compliance by any member of the board of directors, not a member of the committee in question, with his responsibility to act in good faith, in a manner he reasonably believes to be in the best interests of the corporation, and with such care as an ordinarily prudent person in a like position would use under similar circumstances.

§ 43. **Place and Notice of Directors' Meetings; Committee Meetings**

Meetings of the board of directors, regular or special, may be held either within or without this State.

Regular meetings of the board of directors or any committee designated thereby may be held with or without notice as prescribed in the by-laws. Special meetings of the board of directors or any committee designated thereby shall be held upon such notice as is prescribed in the by-laws. Attendance of a director at a meeting shall constitute a waiver of notice of such meeting, except where a director attends a meeting for the express purpose of objecting to the transaction of any business because the meeting is not lawfully called or convened. Neither the business to be transacted at, nor the purpose of, any regular or special meeting of the board of directors or any committee designated thereby need be specified in the notice or waiver of notice of such meeting unless required by the by-laws.

Except as may be otherwise restricted by the articles of incorporation or by-laws, members of the board of directors or any committee designated thereby may participate in a meeting of such board or committee by means of a conference telephone or similar communications equipment by means of which all persons participating in the meeting can hear each other at the same time and participation by such means shall constitute presence in person at a meeting.

§ 44. **Action by Directors Without a Meeting**

Unless otherwise provided by the articles of incorporation or by-laws, any action required by this Act to be taken at a meeting of the directors of a corporation, or any action which may be taken at a meeting of the directors or of a committee, may be taken without a meeting if a consent in writing, setting forth the action so taken, shall be signed by all of the directors, or all of the members of the committee, as the case may be. Such consent shall have the same effect as a unanimous vote.

§ 45. **Distributions to Shareholders**

Subject to any restrictions in the articles of incorporation,the board of directors may authorize and the corporation may make distributions, except that no distribution may be made if, after giving effect thereto, either:

(a) the corporation would be unable to pay its debts as they become due in the usual course of its business; or

(b) the corporation's total assets would be less than the sum of its total liabilities and (unless the articles of incorporation otherwise permit) the maximum amount that then would be payable, in any liquidation, in respect of all outstanding shares having preferential rights in liqidation.

Determinations under subparagraph (b) may be based upon (i) financial statements prepared on the basis of accounting practices and principles that are reasonable in the circumstances, or (ii) a fair valuation or other method that is reasonable in the circumstances.

In the case of a purchase, redemption or other acquisition of a corporation's shares, the effect of a distribution shall be measured as of the date money or other property is transferred or debt is incurred by the corporation, or as of the date the shareholder ceases to be a shareholder of the corporation with respect to such shares, whichever is earlier. In all other cases, the effect of a distribution shall be measured as of the date of its authorization if payment occurs 120 days or less following the date of authorization, or as of the date of payment if payment occurs more than 120 days following the date of authorization.

Indebtedness of a corporation incurred or issued to a shareholder in a distribution in accordance with this Section shall be on a parity with the indebtedness of the corporation to its general unsecured creditors except to the extent subordinated by agreement.*

§ 45. **Dividends**

The board of directors of a corporation may, from time to time, declare and the corporation may pay dividends in cash, property, or its own shares, except when the corporation is insolvent or when the payment thereof would render the corporation insolvent or when the declaration or payment thereof would be contrary to

*[By the Editor] § 45 was revised in 1979. The original § 45, upon which a majority of state statutes were patterned, formerly provided:

any restriction contained in the articles of incorporation, subject to the following provisions:

(a) Dividends may be declared and paid in cash or property only out of the unreserved and unrestricted earned surplus of the corporation, except as otherwise provided in this section.

[Alternative] (a) Dividends may be declared and paid in cash or property only out of the unreserved and unrestricted earned surplus of the corporation, or out of the unreserved and unrestricted net earnings of the current fiscal year and the next preceding fiscal year taken as a single period, except as otherwise provided in this section.

(b) If the articles of incorporation of a corporation engaged in the business of exploiting natural resources so provide, dividends may be declared and paid in cash out of the depletion reserves, but each such dividend shall be identified as a distribution of such reserves and the amount per share paid from such reserves shall be disclosed to the shareholders receiving the same concurrently with the distribution thereof.

(c) Dividends may be declared and paid in its own treasury shares.

(d) Dividends may be declared and paid in its own authorized but unissued shares out of any unreserved and unrestricted surplus of the corporation upon the following conditions:

(1) If a dividend is payable in its own shares having a par value, such shares shall be issued at not less than the par value thereof and there shall be transferred to stated capital at the time such dividend is paid an amount of surplus equal to the aggregate par value of the shares to be issued as a dividend.

(2) If a dividend is payable in its own shares without par value, such shares shall be issued at such stated value as shall be fixed by the board of directors by resolution adopted at the time such dividend is declared, and there shall be transferred to stated capital at the time such dividend is paid an amount of surplus equal to the aggregate stated value so fixed in respect of such shares; and the amount per share so transferred to stated capital shall be disclosed to the shareholders receiving such dividend concurrently with the payment thereof.

(e) No dividend payable in shares of any class shall be paid to the holders of shares of any other class unless the articles of incorporation so provide or such payment is authorized by the affirmative vote or the written consent of the holders of at least a majority of the outstanding shares of the class in which the payment is to be made.

A split-up or division of the issued shares of any class into a greater number of shares of the same class without increasing the stated capital of the corporation shall not be construed to be a share dividend within the meaning of this section.

§ 46. **Distributions from Capital Surplus [Repealed]** The board of directors of a corporation may, from time to time, distribute to its shareholders out of capital surplus of the corporation a portion of its assets, in cash or property, subject to the following provisions:

(a) No such distribution shall be made at a time when the corporation is insolvent or when such distribution would render the corporation insolvent.

(b) No such distribution shall be made unless the articles of incorporation so provide or such distribution is authorized by the affirmative vote of the holders of a majority of the outstanding shares of each class whether or not entitled to vote thereon by the provisions of the articles of incorporation of the corporation.

(c) No such distribution shall be made to the holders of any class of shares unless all cumulative dividends accrued on all preferred or special classes of shares entitled to preferential dividends shall have been fully paid.

(d) No such distribution shall be made to the holders of any class of shares which would reduce the remaining net assets of the corporation below the aggregate preferential amount payable in event of involuntary liquidation to the holders of shares having preferential rights to the assets of the corporation in the event of liquidation.

(e) Each such distribution, when made, shall be identified as a distribution from capital surplus and the amount per share disclosed to the shareholders receiving the same concurrently with the distribution thereof.

The board of directors of a corporation may also, from time to time, distribute to the holders of its outstanding shares having a cumulative preferential right to receive dividends, in discharge of their cumulative dividend rights, dividends payable in cash out of the capital surplus of the corporation, if at the time the corporation has no earned surplus and is not insolvent and would not thereby be rendered insolvent. Each such distribution when made, shall be identified as a payment of cumulative dividends out of capital surplus.

§ 47. **Loans to Employees and Directors**

A corporation shall not lend money to or use its credit to assist its directors without authorization in the particular case by its shareholders, but may lend money to and use its credit to assist any employee of the corporation or of a subsidiary, including any such employee who is a director of the corporation, if the board of directors decides that such loan or assistance may benefit the corporation.

§ 48. **Liability of Directors in Certain Cases**

In addition to any other liabilities, a director who votes for or assents to any distribution contrary to the provisions of this Act or contrary to any restrictions contained in the articles of incorporation, shall, unless he complies with the standard provided in this Act for the performance of the duties of directors, be liable to the corporation, jointly and severally with all other directors so voting or assenting, for the amount of such dividend which is paid or the value of such distribution in excess of the amount of such distribution which could have been made without a violation of the provisions of this Act or the restrictions in the articles of incorporation.

Any director against whom a claim shall be asserted under or pursuant to this section for the making of a distribution and who shall be held liable thereon, shall be entitled to contribution from the shareholders who accepted or received any such distribution, knowing such distribution to have been made in violation of this Act, in proportion to the amounts received by them.

Any director against whom a claim shall be asserted under or pursuant to this section shall be entitled to contribution from any other director who voted for or assented to the action upon which the claim is asserted and who did not comply with the standard provided in this Act for the performance of the duties of directors.

§ 49. **Provisions Relating to Actions by Shareholders**

No action shall be brought in this State by a shareholder in the right of a domestic or foreign corporation unless the plaintiff was a holder of record of shares or of voting trust certificates therefor at the time of the transaction of which he complains, or his shares or voting trust certificates thereafter devolved upon him by operation of law from a person who was a holder of record at such time.

In any action hereafter instituted in the right of any domestic or foreign corporation by the holder or holders of record of shares of such corporation or of voting trust certificates therefor, the court having jurisdiction, upon final judgment and a finding that the action was brought without reasonable cause, may require the plaintiff or plaintiffs to pay to the parties named as defendant the reasonable expenses, including fees of attorneys, incurred by them in the defense of such action.

In any action now pending or hereafter instituted or maintained in the right of any domestic or foreign corporation by the holder or holders of record of less than five per cent of the outstanding shares of any class of such corporation or of voting trust certificates therefor, unless the shares or voting trust certificates so held have a market value in excess of twenty-five thousand dollars, the corporation in whose right such action is brought shall be entitled at any time before final judgment to require the plaintiff or plaintiffs to give security for the reasonable expenses, including fees of attorneys, that may be incurred by it in connection with such action or may be incurred by other parties named as defendant for which it may become legally liable. Market value shall be determined as of the date that the plaintiff institutes the action or, in the case of an intervenor, as of the date that he becomes a party to the action. The amount of such security may from time to time be increased or decreased, in the discretion of the court, upon showing that the security provided has or may become inadequate or is excessive. The corporation shall have recourse to such security in such amount as the court having jurisdiction shall determine upon the termination of such action, whether or not the court finds the action was brought without reasonable cause.

§ 50. **Officers**

The officers of a corporation shall consist of a president, one or more vice presidents as may be prescribed by the by-laws, a secretary, and a treasurer, each of whom shall be elected by the board of directors at such time and in such manner as may be prescribed by the by-laws. Such other officers and assistant officers and agents as may be deemed necessary may be elected or appointed by the board of directors or chosen in such other manner as may be prescribed by the by-laws. Any two or more offices may be held by the same person, except the offices of president and secretary.

All officers and agents of the corporation, as between themselves and the corporation, shall have such authority and perform such duties in the management of the corporation as may be provided in the by-laws, or as may be determined by resolution of the board of directors not inconsistent with the by-laws.

§ 51. **Removal of Officers**

Any officer or agent may be removed by the board of directors whenever in its judgment the best interests of the corporation will be served thereby, but such removal shall be without prejudice to the contract rights, if any, of the person so removed. Election or appointment of an officer or agent shall not of itself create contract rights.

§ 52. **Books and Records: Financial Reports to Shareholders; Examination of Records**

Each corporation shall keep correct and complete books and records of account and shall keep minutes of the proceedings of its shareholders and board of directors and shall keep at its registered office or principal place of business, or at the office of its transfer agent or registrar, a record of its shareholders, giving the names and addresses of all shareholders and the number and class of the shares held by each. Any books, records and minutes may be in written form or in any form capable of being converted into written form within a reasonable time.

Any person who shall have been a holder of record of shares or of voting trust certificates therefor at least six months immediately preceding his demand or shall be the holder of record of, or the holder of record of voting trust certificates for, at least five percent of all the outstanding shares of the corporation, upon written demand stating the purpose thereof, shall have the right to examine, in person, or by agent or attorney, at any reasonable time or times, for any proper purpose its relevant books and records of accounts, minutes, and record of shareholders and to make extracts therefrom.

Any officer or agent who, or a corporation which, shall refuse to allow any such shareholder or holder of voting trust certificates, or his agent or attorney, so to examine and make extracts from its books and records of account, minutes, and record of shareholders, for any proper purpose, shall be liable to such shareholder or holder of voting trust certificates in a penalty of ten per cent of the value of the shares owned by such shareholder, or in respect of which such voting trust certificates are issued, in addition to any other damages or remedy afforded him by law. It shall be a defense to any action for penalties under this section that the person suing therefor has within two years sold or offered for sale any list of shareholders or of holders of voting trust certificates for shares of such corporation or any other corporation or has aided or abetted any person in procuring any list of shareholders or of holders of voting trust certificates for any such purpose, or has improperly used any information secured through any prior examination of the books and records of account, or minutes, or record of shareholders or of holders of voting trust certificates for shares of such corporation or any other corporation, or was not acting in good faith or for a proper purpose in making his demand.

Nothing herein contained shall impair the power of any court of competent jurisdiction, upon proof by a shareholder or holder of voting trust certificates of proper purpose, irrespective of the period of time during which such shareholder or holder of voting trust certificates shall have been a shareholder of record or a holder of record of voting trust certificates, and irrespective of the number of shares held by him or represented by voting trust certificates held by him, to compel the production for examination by such shareholder or holder of voting trust certificates of the books and records of account, minutes and record of shareholders of a corporation.

Each corporation shall furnish to its shareholders annual financial statements, including at least a balance sheet as of the end of each fiscal year and a statement of income for such fiscal year, which shall be prepared on the basis of generally accepted accounting principles, if the corporation prepares financial statements for such fiscal year on that basis for any purpose, and may be consolidated statements of the corporation and one or more of its subsidiaries. The financial statements shall be mailed by the corporation to each of its shareholders within 120 days after the close of each fiscal year and, after such mailing and upon written request, shall be mailed by the corporation to any shareholder (or holder of a voting trust certificate for its shares) to whom a copy of the most recent annual financial statements has not previously been mailed. In the case of statements audited by a public accountant, each copy shall be accompanied by a report setting forth his opinion thereon; in other cases, each copy shall be accompanied by a statement of the president or the person in charge of the corporation's financial accounting records (1) stating his reasonable belief as to whether or not the financial statements were prepared in accordance with generally accepted accounting principles and, if not, describing the basis of presentation, and (2) describing any respects in which the financial statements were not prepared on a basis consistent with those prepared for the previous year.

§ 53. Incorporators

One or more persons, or a domestic or foreign corporation, may act as incorporator or incorporators of a corporation by signing and delivering in duplicate to the Secretary of State articles of incorporation for such corporation.

§ 54. Articles of Incorporation

The articles of incorporation shall set forth:

(a) The name of the corporation.

(b) The period of duration, which may be perpetual.

(c) The purpose or purposes for which the corporation is organized which may be stated to be, or to include, the transaction of any or all lawful business for which corporations may be incorporated under this Act.

(d) The aggregate number of shares which the corporation shall have authority to issue and, if such shares are to be divided into classes, the number of shares of each class.

(e) If the shares are to be divided into classes, the designation of each class and a statement of the preferences, limitations and relative rights in respect of the shares of each class.

(f) If the corporation is to issue the shares of any preferred or special class in series, then the designation of each series and a statement of the variations in the relative rights and preferences as between series insofar as the same are to be fixed in the articles of incorporation, and a statement of any authority to be vested in the board of directors to establish series and fix and determine the variations in the relative rights and preferences as between series.

(g) If any preemptive right is to be granted to shareholders, the provisions therefor.

(h) The address of its initial registered office, and the name of its initial registered agent at such address.

(i) The number of directors constituting the initial board of directors and the names and addresses of the persons who are to serve as directors until the first annual meeting of shareholders or until their successors be elected and qualify.

(j) The name and address of each incorporator.

In addition to provisions required therein, the articles of incorporation may also contain provisions not inconsistent with law regarding:

(1) the direction of the management of the business and the regulation of the affairs of the corporation;

(2) the definition, limitation and regulation of the powers of the corporation, the directors, and the shareholders, or any class of the shareholders, including restrictions on the transfer of shares;

(3) the par value of any authorized shares or class of shares;

(4) any provision which under this Act is required or permitted to be set forth in the by-laws.

It shall not be necessary to set forth in the articles of incorporation any of the corporate powers enumerated in this Act.

§ 55. Filing of Articles of Incorporation

Duplicate originals of the articles of incorporation shall be delivered to the Secretary of State. If the Secretary of State finds that the articles of incorporation conform to law, he shall, when all fees have been paid as in this Act prescribed:

(a) Endorse on each of such duplicate originals the word "Filed," and the month, day and year of the filing thereof.

(b) File one of such duplicate originals in his office.

(c) Issue a certificate of incorporation to which he shall affix the other duplicate original.

The certificate of incorporation, together with the duplicate original of the articles of incorporation affixed thereto by the Secretary of State, shall be returned to the incorporators or their representative.

§ 56. Effect of Issuance of Certificate of Incorporation

Upon the issuance of the certificate of incorporation, the corporate existence shall begin, and such certificate of incorporation shall be conclusive evidence that all conditions precedent required to be performed by the incorporators have been complied with and that the corporation has been incorporated under this Act, except as against this State in a proceeding to cancel or revoke the certificate of incorporation or for involuntary dissolution of the corporation.

§ 57. Organization Meeting of Directors

After the issuance of the certificate of incorporation an organization meeting of the board of directors named in the articles of incorporation shall be held, either within or without this State, at the call of a majority of the directors named in the articles of incorporation, for the purpose of adopting by-laws, electing officers and transacting such other business as may come before the meeting. The directors calling the meeting shall give at least three days' notice thereof by mail to each director so named, stating the time and place of the meeting.

§ 58. Right to Amend Articles of Incorporation

A corporation may amend its articles of incorporation, from time to time, in any and as many respects as may be desired, so long as its articles of incorporation as amended contain only such provisions as might be lawfully contained in original articles of incorporation at the time of making such amendment, and, if a change in shares or the rights of shareholders, or an exchange, reclassification or cancellation of shares or rights of shareholders is to be made, such provisions as may be necessary to effect such change, exchange, reclassification or cancellation.

In particular, and without limitation upon such general power of amendment, a corporation may amend its articles of incorporation, from time to time, so as:

(a) To change its corporate name.

(b) To change its period of duration.

(c) To change, enlarge or diminish its corporate purposes.

(d) To increase or decrease the aggregate number of shares, or shares of any class, which the corporation has authority to issue.

(e) To provide, change or eliminate any provision with respect to the par value of any shares or class of shares.

(f) To exchange, classify, reclassify or cancel all or any part of its shares, whether issued or unissued.

(g) To change the designation of all or any part of its shares, whether issued or unissued, and to change the preferences, limitations, and the relative rights in respect of all or any part of its shares, whether issued or unissued.

(h) To change the shares of any class, whether issued or unissued, into a different number of shares of the same class or into the same or a different number of shares of other classes.

(i) To create new classes of shares having rights and preferences either prior and superior or subordinate and inferior to the shares of any class then authorized, whether issued or unissued.

(j) To cancel or otherwise affect the right of the holders of the shares of any class to receive dividends which have accrued but have not been declared.

(k) To divide any preferred or special class of shares, whether issued or unissued, into series and fix and determine the designations of such series and the variations in the relative rights and preferences as between the shares of such series.

(l) To authorize the board of directors to establish, out of authorized but unissued shares, series of any preferred or special class of shares and fix and determine the relative rights and preferences of the shares of any series so established.

(m) To authorize the board of directors to fix and determine the relative rights and preferences of the authorized but unissued shares of series theretofore established in respect of which either the relative rights and preferences have not been fixed and determined or the relative rights and preferences theretofore fixed and determined are to be changed.

(n) To revoke, diminish, or enlarge the authority of the board of directors to establish series out of authorized but unissued shares of any preferred or special class and fix and determine the relative rights and preferences of the shares of any series so established.

(o) To limit, deny or grant to shareholders of any class the preemptive right to acquire additional shares of the corporation, whether then or thereafter authorized.

§ 59. Procedure to Amend Articles of Incorporation

Amendments to the articles of incorporation shall be made in the following manner:

(a) The board of directors shall adopt a resolution setting forth the proposed amendment and, if shares have been issued, directing that it be submitted to a vote at a meeting of shareholders, which may be either the annual or a special meeting. If no shares have been issued, the amendment shall be adopted by resolution of the board of directors and the provisions for adoption by shareholders shall not apply. If the corporation has only one class of shares outstanding, an amendment solely to change the number of authorized shares to effectuate a split of, or stock dividend in, the corporation's own shares, or solely to do so and to change the number of authorized shares in proportion thereto, may be adopted by the board of directors; and the provisions for adoption by shareholders shall not apply, unless otherwise provided by the articles of incorporation. The resolution may incorporate the proposed amendment in restated articles of incorporation which contain a statement that except for the designated amendment the restated articles of incorporation correctly set forth without change the corresponding provisions of the articles of incorporation as theretofore amended, and that the restated articles of incorporation together with the designated amendment supersede the original articles of incorporation and all amendments thereto.

(b) Written notice setting forth the proposed amendment or a summary of the changes to be effected thereby shall be given to each shareholder of record entitled to vote thereon within the time and in the manner provided in this Act for the giving of notice of meetings of shareholders. If the meeting be an annual meeting, the proposed amendment of such summary may be included in the notice of such annual meeting.

(c) At such meeting a vote of the shareholders entitled to vote thereon shall be taken on the proposed amendment. The proposed amendment shall be adopted upon receiving the affirmative vote of the holders of a majority of the shares entitled to vote thereon, unless any class of shares is entitled to vote thereon as a class, in which event the proposed amendment shall be adopted upon receiving the affirmative vote of the holders of a majority of the shares of each class of shares entitled to vote thereon as a class and of the total shares entitled to vote thereon.

Any number of amendments may be submitted to the shareholders, and voted upon by them, at one meeting.

§ 60. **Class Voting on Amendments**

The holders of the outstanding shares of a class shall be entitled to vote as a class upon a proposed amendment, whether or not entitled to vote thereon by the provisions of the articles of incorporation, if the amendment would:

(a) Increase or decrease the aggregate number of authorized shares of such class.

(b) Effect an exchange, reclassification or cancellation of all or part of the shares of such class.

(c) Effect an exchange, or create a right of exchange, of all or any part of the shares of another class into the shares of such class.

(d) Change the designations, preferences, limitations or relative rights of the shares of such class.

(e) Change the shares of such class, into the same or a different number of shares of the same class or another class or classes.

(f) Create a new class of shares having rights and preferences prior and superior to the shares of such class, or increase the rights and preferences or the number of authorized shares, of any class having rights and preferences prior or superior to the shares of such class.

(g) In the case of a preferred or special class of shares, divide the shares of such class into series and fix and determine the designation of such series and the variations in the relative rights and preferences between the shares of such series, or authorize the board of directors to do so.

(h) Limit or deny any existing preemptive rights of the shares of such class.

(i) Cancel or otherwise affect dividends on the shares of such class which have accrued but have not been declared.

§ 61. **Articles of Amendment**

The articles of amendment shall be executed in duplicate by the corporation by its president or a vice president and by its secretary or an assistant secretary, and verified by one of the officers signing such articles, and shall set forth:

(a) The name of the corporation.

(b) The amendments so adopted.

(c) The date of the adoption of the amendment by the shareholders, or by the board of directors where no shares have been issued.

(d) The number of shares outstanding, and the number of shares entitled to vote thereon, and if the shares of any class are entitled to vote thereon as a class, the designation and number of outstanding shares entitled to vote thereon of each such class.

(e) The number of shares voted for and against such amendment, respectively, and, if the shares of any class are entitled to vote thereon as a class, the number of shares of each such class voted for and against such amendment, respectively, or if no shares have been issued, a statement to that effect.

(f) If such amendment provides for an exchange, reclassification or cancellation of issued shares, and if the manner in which the same shall be effected is not set forth in the amendment, then a statement of the manner in which the same shall be effected.

§ 62. **Filing of Articles of Amendment**

Duplicate originals of the articles of amendment shall be delivered to the Secretary of State. If the Secretary of State finds that the articles of amendment conform to law, he shall, when all fees and franchise taxes have been paid as in this Act prescribed:

(a) Endorse on each of such duplicate originals the word "Filed," and the month, day and year of the filing thereof.

(b) File one of such duplicate originals in his office.

(c) Issue a certificate of amendment to which he shall affix the other duplicate original.

The certificate of amendment, together with the duplicate original of the articles of amendment affixed thereto by the Secretary of State, shall be returned to the corporation or its representative.

§ 63. **Effect of Certificate of Amendment**

Upon the issuance of the certificate of amendment by the Secretary of State, the amendment shall become effective and the articles of incorporation shall be deemed to be amended accordingly.

No amendment shall affect any existing cause of action in favor of or against such corporation, or any pending suit to which such corporation shall be a party, or the existing rights of persons other than shareholders; and, in the event the corporate name shall be changed by amendment, no suit brought by or against such corporation under its former name shall abate for that reason.

§ 64. **Restated Articles of Incorporation**

A domestic corporation may at any time restate its articles of incorporation as theretofore amended, by a resolution adopted by the board of directors.

Upon the adoption of such resolution, restated articles of incorporation shall be executed in duplicate by the corporation by its president or a vice president and by its secretary or assistant secretary and verified by one of the officers signing such articles and shall set forth all of the operative provisions of the articles of incorporation as theretofore amended together with a statement that the restated articles of incorporation correctly set forth without change the corresponding provisions of the articles of incorporation as theretofore amended and that the restated articles of incorporation supersede the original articles of incorporation and all amendments thereto.

Duplicate originals of the restated articles of incorporation shall be delivered to the Secretary of State. If the Secretary of State finds that such restated articles of incorporation conform to law, he shall, when all fees and franchise taxes have been paid as in this Act prescribed:

(1) Endorse on each of such duplicate originals the word "Filed," and the month, day and year of the filing thereof.

(2) File one of such duplicate originals in his office.

(3) Issue a restated certificate of incorporation, to which he shall affix the other duplicate original.

The restated certificate of incorporation, together with the duplicate original of the restated articles of incorporation affixed thereto by the Secretary of State, shall be returned to the corporation or its representative.

Upon the issuance of the restated certificate of incorporation by the Secretary of State, the restated articles of incorporation shall become effective and shall supersede the original articles of incorporation and all amendments thereto.

§ 65. **Amendment of Articles of Incorporation in Reorganization Proceedings**

Whenever a plan of reorganization of a corporation has been confirmed by decree or order of a court of competent jurisdiction in proceedings for the reorganization of such corporation, pursuant to the provisions of any applicable statute of the United States relating to reorganizations of corporations, the articles of incorporation of the corporation may be amended, in the manner provided in this section, in as many respects as may be necessary to carry out the plan and put it into effect, so long as the articles of incorporation as amended contain only such provisions as might be lawfully contained in original articles of incorporation at the time of making such amendment.

In particular and without limitation upon such general power of amendment, the articles of incorporation may be amended for such purpose so as to:

(A) Change the corporate name, period of duration or corporate purposes of the corporation;

(B) Repeal, alter or amend the by-laws of the corporation;

(C) Change the aggregate number of shares or shares of any class, which the corporation has authority to issue;

(D) Change the preferences, limitations and relative rights in respect of all or any part of the shares of the corporation, and classify, reclassify or cancel all or any part thereof, whether issued or unissued;

(E) Authorize the issuance of bonds, debentures or other obligations of the corporation, whether or not convertible into shares of any class or bearing warrants or other evidences of optional rights to purchase or subscribe for shares of any class, and fix the terms and conditions thereof; and

(F) Constitute or reconstitute and classify or reclassify the board of directors of the corporation, and appoint directors and officers in place of or in addition to all or any of the directors or officers then in office.

Amendments to the articles of incorporation pursuant to this section shall be made in the following manner:

(a) Articles of amendment approved by decree or order of such court shall be executed and verified in duplicate by such person or persons as the court shall designate or appoint for the purpose, and shall set forth the name of the corporation, the amendments of the articles of incorporation approved by the court, the date of the decree or order approving the articles of amendment, the title of the proceedings in which the decree or order was entered, and a statement that such decree or order was entered by a court having jurisdiction of the proceedings for the reorganization of the corporation pursuant to the provisions of an applicable statute of the United States.

(b) Duplicate originals of the articles of amendment shall be delivered to the Secretary of State. If the Secretary of State finds that the articles of amendment conform to law, he shall, when all fees and franchise taxes have been paid as in this Act prescribed:

(1) Endorse on each of such duplicate originals the word "Filed," and the month, day and year of the filing thereof.

(2) File one of such duplicate originals in his office.

(3) Issue a certificate of amendment to which he shall affix the other duplicate original.

The certificate of amendment, together with the duplicate original of the articles of amendment affixed thereto by the Secretary of State, shall be returned to the corporation or its representative.

Upon the issuance of the certificate of amendment by the Secretary of State, the amendment shall become effective and the articles of incorporation shall be deemed to be amended accordingly, without any action thereon by the directors or shareholders of the corporation and with the same effect as if the amendments had been adopted by unanimous action of the directors and shareholders of the corporation.

§ 66. Restriction on Redemption or Purchase of Redeemable Shares [Repealed]

No redemption or purchase of redeemable shares shall be made by a corporation when it is insolvent or when such redemption or purchase would render it insolvent, or which would reduce the net assets below the aggregate amount payable to the holders of shares having prior or equal rights to the assets of the corporation upon involuntary dissolution.

§ 67. Cancellation of Redeemable Shares by Redemption or Purchase [Repealed]

When redeemable shares of a corporation are redeemed or purchased by the corporation, the redemption or purchase shall effect a cancellation of such shares, and a statement of cancellation shall be filed as provided in this section. Thereupon such shares shall be restored to the status of authorized but unissued shares, unless the articles of incorporation provide that such shares when redeemed or purchased shall not be reissued, in which case the filing of the statement of cancellation shall constitute an amendment to the articles of incorporation and shall reduce the number of shares of the class so cancelled which the corporation is authorized to issue by the number of shares so cancelled.

The statement of cancellation shall be executed in duplicate by the corporation by its president or a vice president and by its secretary or an assistant secretary, and verified by one of the officers signing such statement, and shall set forth:

(a) The name of the corporation.

(b) The number of redeemable shares cancelled through redemption or purchase, itemized by classes and series.

(c) The aggregate number of issued shares, itemized by classes and series, after giving effect to such cancellation.

(d) The amount, expressed in dollars, of the stated capital of the corporation after giving effect to such cancellation.

(e) If the articles of incorporation provide that the cancelled shares shall not be reissued, the number of shares which the corporation will have authority to issue itemized by classes and series, after giving effect to such cancellation.

Duplicate originals of such statement shall be delivered to the Secretary of State. If the Secretary of State finds that such statement conforms to law, he shall, when all fees and franchise taxes have been paid as in this Act prescribed:

(1) Endorse on each of such duplicate originals the word "Filed," and the month, day and year of the filing thereof.

(2) File one of such duplicate originals in his office.

(3) Return the other duplicate original to the corporation or its representative.

Upon the filing of such statement of cancellation, the stated capital of the corporation shall be deemed to be reduced by that part of the stated capital which was, at the time of such cancellation, represented by the shares so cancelled.

Nothing contained in this section shall be construed to forbid a cancellation of shares or a reduction of stated capital in any other manner permitted by this Act.

§ 68. **Cancellation of Other Reacquired Shares [Repealed]**

A corporation may at any time, by resolution of its board of directors, cancel all or any part of the shares of the corporation of any class reacquired by it, other than redeemable shares redeemed or purchased, and in such event a statement of cancellation shall be filed as provided in this section.

The statement of cancellation shall be executed in duplicate by the corporation by its president or a vice president and by its secretary or an assistant secretary, and verified by one of the officers signing such statement, and shall set forth:

(a) The name of the corporation.

(b) The number of reacquired shares cancelled by resolution duly adopted by the board of directors, itemized by classes and series, and the date of its adoption.

(c) The aggregate number of issued shares, itemized by classes and series, after giving effect to such cancellation.

(d) The amount, expressed in dollars, of the stated capital of the corporation after giving effect to such cancellation.

Duplicate originals of such statement shall be delivered to the Secretary of State. If the Secretary of State finds that such statement conforms to law, he shall, when all fees and franchise taxes have been paid as in this Act prescribed:

(1) Endorse on each of such duplicate originals the word "Filed," and the month, day and year of the filing thereof.

(2) File one of such duplicate originals in his office.

(3) Return the other duplicate original to the corporation or its representative.

Upon the filing of such statement of cancellation, the stated capital of the corporation shall be deemed to be reduced by that part of the stated capital which was, at the time of such cancellation, represented by the shares so cancelled, and the shares so cancelled shall be restored to the status of authorized but unissued shares.

Nothing contained in this section shall be construed to forbid a cancellation of shares or a reduction of stated capital in any other manner permitted by this Act.

§ 69. **Reduction of Stated Capital in Certain Cases [Repealed]**

A reduction of the stated capital of a corporation, where such reduction is not accompanied by any action requiring an amendment of the articles of incorporation and not accompanied by a cancellation of shares, may be made in the following manner:

(A) The board of directors shall adopt a resolution setting forth the amount of the proposed reduction and the manner in which the reduction shall be effected, and directing that the question of such reduction be submitted to a vote at a meeting of shareholders, which may be either an annual or a special meeting.

(B) Written notice, stating that the purpose or one of the purposes of such meeting is to consider the question of reducing the stated capital of the corporation in the amount and manner proposed by the board of directors, shall be given to each shareholder of record entitled to vote thereon within the time and in the manner provided in this Act for the giving of notice of meetings of shareholders.

(C) At such meeting a vote of the shareholders entitled to vote thereon shall be taken on the question of approving the proposed reduction of stated capital, which shall require for its adoption the affirmative vote of the holders of a majority of the shares entitled to vote thereon.

When a reduction of the stated capital of a corporation has been approved as provided in this section, a statement shall be executed in duplicate by the corporation by its president or a vice president and by its secretary or an assistant secretary, and verified by one of the officers signing such statement, and shall set forth:

(a) The name of the corporation.

(b) A copy of the resolution of the shareholders approving such reduction, and the date of its adoption.

(c) The number of shares outstanding, and the number of shares entitled to vote thereon.

(d) The number of shares voted for and against such reduction, respectively.

(e) A statement of the manner in which such reduction is effected, and a statement, expressed in dollars, of the amount of stated capital of the corporation after giving effect to such reduction.

Duplicate originals of such statement shall be delivered to the Secretary of State. If the Secretary of State finds that such statement conforms to law, he shall, when all fees and franchise taxes have been paid as in this Act prescribed:

(1) Endorse on each of such duplicate originals the word "Filed," and the month, day and year of the filing thereof.

(2) File one of such duplicate originals in his office.

(3) Return the other duplicate original to the corporation or its representative.

Upon the filing of such statement, the stated capital of the corporation shall be reduced as therein set forth.

No reduction of stated capital shall be made under the provisions of this section which would reduce the amount of the aggregate stated capital of the corporation to an amount equal to or less than the aggregate preferential amounts payable upon all issued shares having a preferential right in the assets of the corporation in the event of involuntary liquidation, plus the aggregate par value of all issued shares having a par value but no preferential right in the assets of the corporation in the event of involuntary liquidation.

§ 70. **Special Provisions Relating to Surplus and Reserves [Repealed]**

The surplus, if any, created by or arising out of a reduction of the stated capital of a corporation shall be capital surplus.

The capital surplus of a corporation may be increased from time to time by resolution of the board of directors directing that all or a part of the earned surplus of the corporation be transferred to capital surplus.

A corporation may, by resolution of its board of directors, apply any part or all of its capital surplus to the reduction or elimination of any deficit arising from losses, however incurred, but only after first eliminating the earned surplus, if any, of the corporation by applying such losses against earned surplus and only to the extent that such losses exceed the earned surplus, if any. Each such application of capital surplus shall, to the extent thereof, effect a reduction of capital surplus.

A corporation may, by resolution of its board of directors, create a reserve or reserves out of its earned surplus for any proper purpose or purposes, and may abolish any such reserve in the same manner. Earned surplus of the corporation to the extent so reserved shall not be available for the payment of dividends or other distributions by the corporation except as expressly permitted by this Act.

§ 71. **Procedure for Merger**

Any two or more domestic corporations may merge into one of such corporations pursuant to a plan of merger approved in the manner provided in this Act.

The board of directors of each corporation shall, by resolution adopted by each such board, approve a plan of merger setting forth:

(a) The names of the corporations proposing to merge, and the name of the corporation into which they propose to merge, which is hereinafter designated as the surviving corporation.

(b) The terms and conditions of the proposed merger.

(c) The manner and basis of converting the shares of each corporation into shares, obligations or other securities of the surviving corporation or of any other corporation or, in whole or in part, into cash or other property.

(d) A statement of any changes in the articles of incorporation of the surviving corporation to be effected by such merger.

(e) Such other provisions with respect to the proposed merger as are deemed necessary or desirable.

§ 72. **Procedure for Consolidation**

Any two or more domestic corporations may consolidate into a new corporation pursuant to a plan of consolidation approved in the manner provided in this Act.

The board of directors of each corporation shall, by a resolution adopted by each such board, approve a plan of consolidation setting forth:

(a) The names of the corporations proposing to consolidate, and the name of the new corporation into which they propose to consolidate, which is hereinafter designated as the new corporation.

(b) The terms and conditions of the proposed consolidation.

(c) The manner and basis of converting the shares of each corporation into shares, obligations or other securities of the new corporation or of any other corporation or, in whole or in part, into cash or other property.

(d) With respect to the new corporation, all of the statements required to be set forth in articles of incorporation for corporations organized under this Act.

(e) Such other provisions with respect to the proposed consolidation as are deemed necessary or desirable.

§ 72A. **Procedure for Share Exchange**

All the issued or all the outstanding shares of one or more classes of any domestic corporation may be acquired through the exchange of all such shares of such class or classes by another domestic or foreign corporation pursuant to a plan of exchange approved in the manner provided in this Act.

The board of directors of each corporation shall, by resolution adopted by each such board, approve a plan of exchange setting forth:

(a) The name of the corporation the shares of which are proposed to be acquired by exchange and the name of the corporation to acquire the shares of such corporation in the exchange, which is hereinafter designated as the acquiring corporation.

(b) The terms and conditions of the proposed exchange.

(c) The manner and basis of exchanging the shares to be acquired for shares, obligations or other securities of the acquiring corporation or any other corporation, or, in whole or in part, for cash or other property.

(d) Such other provisions with respect to the proposed exchange as are deemed necessary or desirable.

The procedure authorized by this section shall not be deemed to limit the power of a corporation to acquire all or part of the shares of any class or classes of a corporation through a voluntary exchange or otherwise by agreement with the shareholders.

§ 73. **Approval by Shareholders**

(a) The board of directors of each corporation in the case of a merger or consolidation, and the board of directors of the corporation the shares of which are to be acquired in the case of an exchange, upon approving such plan of merger, consolidation or exchange, shall, by resolution, direct that the plan be submitted to a vote at a meeting of its shareholders, which may be either an annual or a special meeting. Written notice shall be given to each shareholder of record, whether or not entitled to vote at such meeting, not less than twenty days before such meeting, in the manner provided in this Act for the giving of notice of meetings of shareholders, and, whether the meeting be an annual or a special meeting, shall state that the purpose or one of the purposes is to condiser the proposed plan of merger, consolidation or exchange. A copy or a summary of the plan of merger, consolidation or exchange, as the case may be, shall be included in or enclosed with such notice.

(b) At each such meeting, a vote of the shareholders shall be taken on the proposed plan. The plan shall be approved upon receiving the affirmative vote of the holders of a majority of the shares entitled to vote thereon of each such corporation, unless any class of shares of any such corporation is entitled to vote thereon as a class, in which event, as to such corporation, the plan shall be approved upon receiving the affirmative vote of the holders of a majority of the shares of each class of shares entitled to vote thereon. Any class of shares of any such corporation shall be entitled to vote as a class if any such plan contains any provision which, if contained in a proposed amendment to articles of incorporation, would entitle such class of shares to vote as a class and, in the case of an exchange, if the class is included in the exchange.

(c) After such approval by a vote of the shareholders ofeach such corporation, and at any time prior to the filing of the articles of merger, consolidation or exchange, the merger, consolidation or exchange may be abandoned pursuant to provisions therefor, if any, set forth in the plan.

(d) (1) Notwithstanding the provisions of subsections (a) and (b), submission of a plan of merger to a vote at a meeting of shareholders of a surviving corporation shall not be required if—

(i) the articles of incorporation of the surviving corporation do not differ except in name from those of the corporation before the merger,

(ii) each holder of shares of the surviving corporation which were outstanding immediately before the effective date of the merger is to hold the same number of shares with identical rights immediately after,

(iii) the number of voting shares outstanding immediately after the merger, plus the number of voting shares issuable on conversion of other securities issued by virtue of the terms of the merger and on exercise of rights and warrants so issued, will not exceed by more than 20 percent the number of voting shares outstanding immediately before the merger, and

(iv) the number of participating shares outstanding immediately after the merger, plus the number of participating shares issuable on conversion of other securities issued by virtue of the terms of the merger and on exercise of rights and warrants so issued, will not exceed by more than 20 percent the number of participating shares outstanding immediately before the merger.

(2) As used in this subsection—

(i) "voting shares" means shares which entitle their holders to vote unconditionally in elections of directors;

(ii) "participating shares" means shares which entitle their holders to participate without limitation in distribution of earnings or surplus.

§ 74. Articles of Merger, Consolidation or Exchange

(a) Upon receiving the approvals required by Sections 71, 72 and 73, articles of merger or articles of consolidation shall be executed in duplicate by each corporation by its president or a vice president and by its secretary or an assistant secretary, and verified by one of the officers of each corporation signing such articles, and shall set forth:

(1) The plan of merger or the plan of consolidation;

(2) As to each corporation, either (i) the number of shares outstanding, and, if the shares of any class are entitled to vote as a class, the designation and number of outstanding shares of each such class; or (ii) a statement that the vote of shareholders is not required by virtue of subsection 73(d);

(3) As to each corporation the approval of whose shareholders is required, the number of shares voted for and against such plan, respectively, and, if the shares of any class are entitled to vote as a class,

the number of shares of each such class voted for and against such plan, respectively.

(b) Duplicate originals of the articles of merger, consolidation or exchange shall be delivered to the Secretary of State. If the Secretary of State finds that such articles conform to law, he shall, when all fees and franchise taxes have been paid as in this Act prescribed:

(1) Endorse on each of such duplicate originals the word "Filed," and the month, day and year of the filing thereof.

(2) File one of such duplicate originals in his office.

(3) Issue a certificate of merger, consolidation or exchange to which he shall affix the other duplicate original.

(c) The certificate of merger, consolidation or exchange together with the duplicate original of the articles affixed thereto by the Secretary of State, shall be returned to the surviving, new or acquiring corporation, as the case may be, or its representative.

§ 75. **Merger of Subsidiary Corporation**

Any corporation owning at least ninety per cent of the outstanding shares of each class of another corporation may merge such other corporation into itself without approval by a vote of the shareholders of either corporation. Its board of directors shall, by resolution, approve a plan of merger setting forth:

(A) The name of the subsidiary corporation and the name of the corporation owning at least ninety per cent of its shares, which is hereinafter designated as the surviving corporation.

(B) The manner and basis of converting the shares of the subsidiary corporation into shares, obligations or other securities of the surviving corporation or of any other corporation or, in whole or in part, into cash or other property.

A copy of such plan of merger shall be mailed to each shareholder of record of the subsidiary corporation.

Articles of merger shall be executed in duplicate by the surviving corporation by its president or a vice president and by its secretary or an assistant secretary, and verified by one of its officers signing such articles, and shall set forth:

(a) The plan of merger;

(b) The number of outstanding shares of each class of the subsidiary corporation and the number of such shares of each class owned by the surviving corporation; and

(c) The date of the mailing to shareholders of the subsidiary corporation of a copy of the plan of merger.

On and after the thirtieth day after the mailing of a copy of the plan of merger to shareholders of the subsidiary corporation or upon the waiver thereof by the holders of all outstanding shares duplicate originals of the articles of merger shall be delivered to the Secretary of State. If the Secretary of State finds that such articles conform to law, he shall, when all fees and franchise taxes have been paid as in this Act prescribed:

(1) Endorse on each of such duplicate originals the word "Filed," and the month, day and year of the filing thereof,

(2) File one of such duplicate originals in his office, and

(3) Issue a certificate of merger to which he shall affix the other duplicate original.

The certificate of merger, together with the duplicate original of the articles of merger affixed thereto by the Secretary of State, shall be returned to the surviving corporation or its representative.

§ 76. **Effect of Merger, Consolidation or Exchange**

Upon the issuance of the certificate of merger or the certificate of consolidation by the Secretary of State, the merger or consolidation shall be effected.

When such merger or consolidation has been effected:

(a) The several corporations parties to the plan of merger or consolidation shall be a single corporation, which, in the case of a merger, shall be that corporation designated in the plan of merger as the surviving corporation, and, in the case of a consolidation, shall be the new corporation provided for in the plan of consolidation.

(b) The separate existence of all corporations parties to the plan of merger or consolidation, except the surviving or new corporation, shall cease.

(c) Such surviving or new corporation shall have all the rights, privileges, immunities and powers and shall be subject to all the duties and liabilities of a corporation organized under this Act.

(d) Such surviving or new corporation shall thereupon and thereafter possess all the rights, privileges, immunities, and franchises, of a public as well as of a private nature, of each of the merging or consolidating corporations; and all property, real, personal and mixed, and all debts due on whatever account, including subscriptions to shares, and all other choses in action, and all and every other interest of or belonging to or due to each of the corporations so merged or consolidated, shall be taken and deemed to be transferred to and vested in such single corporation without further act or deed; and the title to any real estate, or any interest therein, vested in any of such corporations shall not revert or be in any way impaired by reason of such merger or consolidation.

(e) Such surviving or new corporation shall thenceforth be responsible and liable for all the liabilities and obligations of each of the corporations so merged or consolidated; and any claim existing or action or proceeding pending by or against any of such corporations may be prosecuted as if such merger or consolidation had not taken place, or such surviving or new corporation may be substituted in its place. Neither the rights of creditors nor any liens upon the property of any such corporation shall be impaired by such merger or consolidation.

(f) In the case of a merger, the articles of incorporation of the surviving corporation shall be deemed to be amended to the extent, if any, that changes in its articles of incorporation are stated in the plan of merger; and, in the case of a consolidation, the statements set forth in the articles of consolidation and which are required or permitted to be set forth in the articles of incorporation of corporations organized under this Act shall be deemed to be the original articles of incorporation of the new corporation.

§ 77. **Merger, Consolidation or Exchange of Shares Between Domestic and Foreign Corporations**

One or more foreign corporations and one or more domestic corporations may be merged or consolidated in the following manner, if such merger, consolidation or exchange is permitted by the laws of the state under which each such foreign corporation is organized:

(a) Each domestic corporation shall comply with the provisions of this Act with respect to the merger, consolidation or exchange, as the case may be, of domestic corporations and each foreign corporation shall comply with the applicable provisions of the laws of the state under which it is organized.

(b) If the surviving or new corporation in a merger or consolidation is to be governed by the laws of any state other than this State, it shall comply with the provisions of this Act with respect to foreign corporations if it is to transact business in this State, and in every case it shall file with the Secretary of State of this State:

(1) An agreement that it may be served with process in this State in any proceeding for the enforcement of any obligation of any domestic corporation which is a party to such merger or consolidation and in any proceeding for the enforcement of the rights of a dissenting shareholder of any such domestic corporation against the surviving or new corporation;

(2) An irrevocable appointment of the Secretary of State of this State as its agent to accept service of process in any such proceeding; and

(3) An agreement that it will promptly pay to the dissenting shareholders of any such domestic corporation, the amount, if any, to which they shall be entitled under provisions of this Act with respect to the rights of dissenting shareholders.

The effect of such merger or consolidation shall be the same as in the case of the merger or consolidation of domestic corporations, if the surviving or new corporation is to be governed by the laws of this State. If the surviving or new corporation is to be governed by the laws of any state other than this State, the effect of such merger or consolidation shall be the same as in the case of the merger or consolidation of domestic corporations except insofar as the laws of such other state provide otherwise.

At any time prior to the filing of the articles of merger or consolidation, the merger or consolidation may be abandoned pursuant to provisions therefor, if any, set forth in the plan of merger or consolidation.

§ 78. **Sale of Assets in Regular Course of Business and Mortgage or Pledge of Assets**

The sale, lease, exchange, or other disposition of all, or substantially all, the property and assets of a corporation in the usual and regular course of its business and the mortgage or pledge of any or all property and assets of a corporation whether or not in the usual and regular course of business may be made upon such terms and conditions and for such consideration, which may consist in whole or in part of cash or other property, including shares, obligations or other securities of any other corporation, domestic or foreign, as shall be authorized by its board of directors; and in any such case no authorization or consent of the shareholders shall be required.

§ 79. Sale of Assets Other Than in Regular Course of Business

A sale, lease, exchange, or other disposition of all, or substantially all, the property and assets, with or without the good will, of a corporation, if not in the usual and regular course of its business, may be made upon such terms and conditions and for such consideration, which may consist in whole or in part of cash or other property, including shares, obligations or other securities of any other corporation, domestic or foreign, as may be authorized in the following manner:

(a) The board of directors shall adopt a resolution recommending such sale, lease, exchange, or other disposition and directing the submission thereof to a vote at a meeting of shareholders, which may be either an annual or a special meeting.

(b) Written notice shall be given to each shareholder of record, whether or not entitled to vote at such meeting, not less than twenty days before such meeting, in the manner provided in this Act for the giving of notice of meetings of shareholders, and, whether the meeting be an annual or a special meeting, shall state that the purpose, or one of the purposes is to consider the proposed sale, lease, exchange, or other disposition.

(c) At such meeting the shareholders may authorize such sale, lease, exchange, or other disposition and may fix, or may authorize the board of directors to fix, any or all of the terms and conditions thereof and the consideration to be received by the corporation therefor. Such authorization shall require the affirmative vote of the holders of a majority of the shares of the corporation entitled to vote thereon, unless any class of shares is entitled to vote thereon as a class, in which event such authorization shall require the affirmative vote of the holders of a majority of the shares of each class of shares entitled to vote as a class thereon and of the total shares entitled to vote thereon.

(d) After such authorization by a vote of shareholders, the board of directors nevertheless, in its discretion, may abandon such sale, lease, exchange, or other disposition of assets, subject to the rights of third parties under any contracts relating thereto, without further action or approval by shareholders.

§ 80. Right of Shareholders to Dissent and Obtain Payment for Shares

(a) Any shareholder of a corporation shall have the right to dissent from, and to obtain payment for his shares in the event of, any of the following corporate actions:

(1) Any plan of merger or consolidation to which the corporation is a party, except as provided in subsection (c);

(2) Any sale or exchange of all or substantially all of the property and assets of the corporation not made in the usual or regular course of its business, including a sale in dissolution, but not including a sale pursuant to an order of a court having jurisdiction in the premises or a sale for cash on terms requiring that all or substantially all of the net proceeds of sale be distributed to the shareholders in accordance with their respective interests within one year after the date of sale;

(3) Any plan of exchange to which the corporation is a party as the corporation the shares of which are to be acquired;

(4) Any amendment of the articles of incorporation which materially and adversely affects the rights appurtenant to the shares of the dissenting shareholder in that it:

(i) alters or abolishes a preferential right of such shares;

(ii) creates, alters or abolishes a right in respect of the redemption of such shares, including a provision respecting a sinking fund for the redemption or repurchase of such shares;

(iii) alters or abolishes a preemptive right of the holder of such shares to acquire shares or other securities;

(iv) excludes or limits the right of the holder of such shares to vote on any matter, or to cumulate his votes, except as such right may be limited by dilution through the issuance of shares or other securities with similar voting rights; or

(5) Any other corporate action taken pursuant to a shareholder vote with respect to which the articles of incorporation, the bylaws, or a resolution of the board of directors directs that dissenting shareholders shall have a right to obtain payment for their shares.

(b) (1) A record holder of shares may assert dissenters' rights as to less than all of the shares registered in his name only if he dissents with respect to all the shares beneficially owned by any one person, and discloses the name and address of the person or persons on whose behalf he dissents. In that event, his rights shall be determined as if the shares as to which he has dissented and his other shares were registered in the names of different shareholders.

(2) A beneficial owner of shares who is not the record holder may assert dissenters' rights with respect to shares held on his behalf, and shall be treated as a dissenting shareholder under the terms of this section and section 81 if he submits to the corporation at the time of or before the assertion of these rights a written consent of the record holder.

(c) The right to obtain payment under this section shall not apply to the shareholders of the surviving corporation in a merger if a vote of the shareholders of such corporation is not necessary to authorize such merger.

(d) A shareholder of a corporation who has a right under this section to obtain payment for his shares shall have no right at law or in equity to attack the validity of the corporate action that gives rise to his right to obtain payment, nor to have the action set aside or rescinded, except when the corporate action is unlawful or fraudulent with regard to the complaining shareholder or to the corporation.

§ 81. **Procedures for Protection of Dissenters' Rights**

(a) As used in this section:

(1) "Dissenter" means a shareholder or beneficial owner who is entitled to and does assert dissenters' rights under Section 80, and who has performed every act required up to the time involved for the assertion of such rights.

(2) "Corporation" means the issuer of the shares held by the dissenter before the corporate action, or the successor by merger or consolidation of that issuer.

(3) "Fair value" of shares means their value immediately before the effectuation of the corporate action to which the dissenter objects, excluding any appreciation or depreciation in anticipation of such corporate action unless such exclusion would be inequitable.

(4) "Interest" means interest from the effective date of the corporate action until the date of payment, at the average rate currently paid by the corporation on its principal bank loans, or, if none, at such rate as is fair and equitable under all the circumstances.

(b) If a proposed corporate action which would give rise to dissenters' rights under Section 80(a) is submitted to a vote at a meeting of shareholders, the notice of meeting shall notify all shareholders that they have or may have a right to dissent and obtain payment for their shares by complying with the terms of this section, and shall be accompanied by a copy of sections 80 and 81 of this Act.

(c) If the proposed corporate action is submitted to a vote at a meeting of shareholders, any shareholder who wishes to dissent and obtain payment for his shares must file with the corporation, prior to the vote, a written notice of intention to demand that he be paid fair compensation for his shares if the proposed action is effectuated, and shall refrain from voting his shares in approval of such action. A shareholder who fails in either respect shall acquire no right to payment for his shares under this section or section 80.

(d) If the proposed corporate action is approved by the required vote at a meeting of shareholders, the corporation shall mail a further notice to all shareholders who gave due notice of intention to demand payment and who refrained from voting in favor of the proposed action. If the proposed corporate action is to be taken without a vote of shareholders, the corporation shall send to all shareholders who are entitled to dissent and demand payment for their shares a notice of the adoption of the plan of corporate action. The notice shall (1) state where and when a demand for payment must be sent and certificates of certificated shares must be deposited in order to obtain payment, (2) inform holders of uncertificated shares to what extent transfer of shares will be restricted from the time that demand for payment is received, (3) supply a form for demanding payment which includes a request for certification of the date on which the shareholder, or the person on whose behalf the shareholder dissents, acquired beneficial ownership of the shares, and (4) be accompanied by a copy of sections 80 and 81 of this Act. The time set for the demand and deposit shall be not less than 30 days from the mailing of the notice.

(e) A shareholder who fails to demand payment, or fails (in the case of certificated shares) to deposit certificates, as required by a notice pursuant to subsection (d) shall have no right under this section or section 80 to receive payment for his shares. If the shares are not represented by certificates, the corporation may restrict their transfer from the time of receipt of demand for payment until effectuation of the proposed corporate action, or the release of restrictions under the terms of subsection (f). The dissenter shall retain all other rights of a shareholder until these rights are modified by effectuation of the proposed corporate action.

(f) (1) Within 60 days after the date set for demanding payment and depositing certificates, if the corporation has not effectuated the proposed corporate action and remitted payment for shares pursuant to paragraph (3), it shall return any certificates that have been deposited, and release uncertificated shares from any transfer restrictions imposed by reason of the demand for payment.

(2) When uncertificated shares have been released from transfer restrictions, and deposited certificates have been returned, the corporation may at any later time send a new notice conforming to the requirements of subsection (d), with like effect.

(3) Immediately upon effectuation of the proposed corporate action, or upon receipt of demand for payment if the corporate action has already been effectuated, the corporation shall remit to dissenters who have made demand and (if their shares are certificated) have deposited their certificates the amount which the corporation estimates to be the fair value of the shares, with interest if any has accrued. The remittance shall be accompanied by:

(i) the corporation's closing balance sheet and statement of income for a fiscal year ending not more than 16 months before the date of remittance, together with the latest available interim financial statements;

(ii) a statement of the corporation's estimate of fair value of the shares; and

(iii) a notice of the dissenter's right to demand supplemental payment, accompanied by a copy of sections 80 and 81 of this Act.

(g) (1) If the corporation fails to remit as required by subsection (f), or if the dissenter believes that the amount remitted is less than the fair value of his shares, or that the interest is not correctly determined, he may send the corporation his own estimate of the value of the shares or of the interest, and demand payment of the deficiency.

(2) If the dissenter does not file such an estimate within 30 days after the corporation's mailing of its remittance, he shall be entitled to no more than the amount remitted.

(h) (1) Within 60 days after receiving a demand for payment pursuant to subsection (g), if any such demands for payment remain unsettled, the corporation shall file in an appropriate court a petition requesting that the fair value of the shares and interest thereon be determined by the court.

(2) An appropriate court shall be a court of competent jurisdiction in the county of this state where the registered office of the corporation is located. If, in the case of a merger or consolidation or exchange of shares, the corporation is a foreign corporation without a registered office in this state, the petition shall be filed in the county where the registered office of the domestic corporation was last located.

(3) All dissenters, wherever residing, whose demands have not been settled shall be made parties to the proceeding as in an action against their shares. A copy of the petition shall be served on each such dissenter; if a dissenter is a nonresident, the copy may be served on him by registered or certified mail or by publication as provided by law.

(4) The jurisdiction of the court shall be plenary and exclusive. The court may appoint one or more persons as appraisers to receive evidence and recommend a decision on the question of fair value. The appraisers shall have such power and authority as shall be specified in the order of their appointment or in any amendment thereof. The dissenters shall be entitled to discovery in the same manner as parties in other civil suits.

(5) All dissenters who are made parties shall be entitled to judgment for the amount by which the fair value of their shares is found to exceed the amount previously remitted, with interest.

(6) If the corporation fails to file a petition as provided in paragraph (1) of this subsection, each dissenter who made a demand and who has not already settled his claim against the corporation shall be paid by the corporation the amount demanded by him, with interest, and may sue therefor in an appropriate court.

(i) (1) The costs and expenses of any proceeding under subsection (h), including the reasonable compensation and expenses of appraisers appointed by the court, shall be determined by the court and assessed against the corporation, except that any part of the costs and expenses may be apportioned and assessed as the court may deem equitable against all or some of the dissenters who are parties and whose action in demanding supplemental payment the court finds to be arbitrary, vexatious, or not in good faith.

(2) Fees and expenses of counsel and of experts for the respective parties may be assessed as the court may deem equitable against the corporation and in favor of any or all dissenters if the corporation failed to comply substantially with the requirements of this section, and may be assessed against either the corporation or a dissenter, in favor of any other party, if the court finds that the party against whom the fees and expenses are assessed acted arbitrarily, vexatiously, or not in good faith in respect to the rights provided by this section and section 80.

(3) If the court finds that the services of counsel for any dissenter were of substantial benefit to other dissenters similarly situated, and should not be assessed against the corporation, it may award to these counsel reasonable fees to be paid out of the amounts awarded to the dissenters who were benefited.

(j) (1) Notwithstanding the foregoing provisions of this section, the corporation may elect to withhold the remittance required by subsection (f) from any dissenter with respect to shares of which the dissenter (or the person on whose behalf the dissenter acts) was not the beneficial owner on the date of the first announcement to news media or to shareholders of the terms of the proposed corporate action. With respect to such shares, the corporation shall, upon effectuating the corporate action, state to each dissenter its estimate of the fair value of the shares, state the rate of interest to be used (explaining the basis thereof), and offer to pay the resulting amounts on receiving the dissenter's agreement to accept them in full satisfaction.

(2) If the dissenter believes that the amount offered is less than the fair value of the shares and interest determined according to this section, he may within 30 days after the date of mailing of the corporation's offer, mail the corporation his own estimate of fair value and interest, and demand their payment. If the dissenter fails to do so, he shall be entitled to no more than the corporation's offer.

(3) If the dissenter makes a demand as provided in paragraph (2), the provisions of subsections (h) and (i) shall apply to further proceedings on the dissenter's demand.

§ 82. **Voluntary Dissolution by Incorporators**

A corporation which has not commenced business and which has not issued any shares, may be voluntarily dissolved by its incorporators at any time in the following manner:

(a) Articles of dissolution shall be executed in duplicate by a majority of the incorporators, and verified by them, and shall set forth:

(1) The name of the corporation.

(2) The date of issuance of its certificate of incorporation.

(3) That none of its shares has been issued.

(4) That the corporation has not commenced business.

(5) That the amount, if any, actually paid in on subscriptions for its shares, less any part thereof disbursed for necessary expenses, has been returned to those entitled thereto.

(6) That no debts of the corporation remain unpaid.

(7) That a majority of the incorporators elect that the corporation be dissolved.

(b) Duplicate originals of the articles of dissolution shall be delivered to the Secretary of State. If the Secretary of State finds that the articles of dissolution conform to law, he shall, when all fees and franchise taxes have been paid as in this Act prescribed:

(1) Endorse on each of such duplicate originals the word "Filed," and the month, day and year of the filing thereof.

(2) File one of such duplicate originals in his office.

(3) Issue a certificate of dissolution to which he shall affix the other duplicate original.

The certificate of dissolution, together with the duplicate original of the articles of dissolution affixed thereto by the Secretary of State, shall be returned to the incorporators or their representative. Upon the issuance of such certificate of dissolution by the Secretary of State, the existence of the corporation shall cease.

§ 83. Voluntary Dissolution by Consent of Shareholders

A corporation may be voluntarily dissolved by the written consent of all of its shareholders.

Upon the execution of such written consent, a statement of intent to dissolve shall be executed in duplicate by the corporation by its president or a vice president and by its secretary or an assistant secretary, and verified by one of the officers signing such statement, which statement shall set forth:

(a) The name of the corporation.

(b) The names and respective addresses of its officers.

(c) The names and respective addresses of its directors.

(d) A copy of the written consent signed by all shareholders of the corporation.

(e) A statement that such written consent has been signed by all shareholders of the corporation or signed in their names by their attorneys thereunto duly authorized.

§ 84. Voluntary Dissolution by Act of Corporation

A corporation may be dissolved by the act of the corporation, when authorized in the following manner:

(a) The board of directors shall adopt a resolution recommending that the corporation be dissolved, and directing that the question of such dissolution be submitted to a vote at a meeting of shareholders, which may be either an annual or a special meeting.

(b) Written notice shall be given to each shareholder of record entitled to vote at such meeting within the time and in the manner provided in this Act for the giving of notice of meetings of shareholders, and, whether the meeting be an annual or special meeting, shall state that the purpose, or one of the purposes, of such meeting is to consider the advisability of dissolving the corporation.

(c) At such meeting a vote of shareholders entitled to vote thereat shall be taken on a resolution to dissolve the corporation. Such resolution shall be adopted upon receiving the affirmative vote of the holders of a majority of the shares of the corporation entitled to vote thereon, unless any class of shares is entitled to vote thereon as a class, in which event the resolution shall be adopted upon receiving the affirmative vote of the holders of a majority of the shares of each class of shares entitled to vote thereon as a class and of the total shares entitled to vote thereon.

(d) Upon the adoption of such resolution, a statement of intent to dissolve shall be executed in duplicate by the corporation by its president or a vice president and by its secretary or an assistant secretary, and verified by one of the officers signing such statement, which statement shall set forth:

(1) The name of the corporation.

(2) The names and respective addresses of its officers.

(3) The names and respective addresses of its directors.

(4) A copy of the resolution adopted by the shareholders authorizing the dissolution of the corporation.

(5) The number of shares outstanding, and, if the shares of any class are entitled to vote as a class, the designation and number of outstanding shares of each such class.

(6) The number of shares voted for and against the resolution, respectively, and, if the shares of any class are entitled to vote as a class, the number of shares of each such class voted for and against the resolution, respectively.

§ 85. Filing of Statement of Intent to Dissolve

Duplicate originals of the statement of intent to dissolve, whether by consent of shareholders or by act of the corporation, shall be delivered to the Secretary of State. If the Secretary of State finds that such statement conforms to law, he shall, when all fees and franchise taxes have been paid as in this Act prescribed:

(a) Endorse on each of such duplicate originals the word "Filed," and the month, day and year of the filing thereof.

(b) File one of such duplicate originals in his office.

(c) Return the other duplicate original to the corporation or its representative.

§ 86. **Effect of Statement of Intent to Dissolve**

Upon the filing by the Secretary of State of a statement of intent to dissolve, whether by consent of shareholders or by act of the corporation, the corporation shall cease to carry on its business, except insofar as may be necessary for the winding up thereof, but its corporate existence shall continue until a certificate of dissolution has been issued by the Secretary of State or until a decree dissolving the corporation has been entered by a court of competent jurisdiction as in this Act provided.

§ 87. **Procedure after Filing of Statement of Intent to Dissolve**

After the filing by the Secretary of State of a statement of intent to dissolve:

(a) The corporation shall immediately cause notice thereof to be mailed to each known creditor of the corporation.

(b) The corporation shall proceed to collect its assets, convey and dispose of such of its properties as are not to be distributed in kind to its shareholders, pay, satisfy and discharge its liabilities and obligations and do all other acts required to liquidate its business and affairs, and, after paying or adequately providing for the payment of all its obligations, distribute the remainder of its assets, either in cash or in kind, among its shareholders according to their respective rights and interests.

(c) The corporation, at any time during the liquidation of its business and affairs, may make application to a court of competent jurisdiction within the state and judicial subdivision in which the registered office or principal place of business of the corporation is situated, to have the liquidation continued under the supervision of the court as provided in this Act.

§ 88. **Revocation of Voluntary Dissolution Proceedings by Consent of Shareholders**

By the written consent of all of its shareholders, a corporation may, at any time prior to the issuance of a certificate of dissolution by the Secretary of State, revoke voluntary dissolution proceedings theretofore taken, in the following manner:

Upon the execution of such written consent, a statement of revocation of voluntary dissolution proceedings shall be executed in duplicate by the corporation by its president or a vice president and by its secretary or an assistant secretary, and verified by one of the officers signing such statement, which statement shall set forth:

(a) The name of the corporation.

(b) The names and respective addresses of its officers.

(c) The names and respective addresses of its directors.

(d) A copy of the written consent signed by all shareholders of the corporation revoking such voluntary dissolution proceedings.

(e) That such written consent has been signed by all shareholders of the corporation or signed in their names by their attorneys thereunto duly authorized.

§ 89. **Revocation of Voluntary Dissolution Proceedings by Act of Corporation**

By the act of the corporation, a corporation may, at any time prior to the issuance of a certificate of dissolution by the Secretary of State, revoke voluntary dissolution proceedings theretofore taken, in the following manner:

(a) The board of directors shall adopt a resolution recommending that the voluntary dissolution proceedings be revoked, and directing that the question of such revocation be submitted to a vote at a special meeting of shareholders.

(b) Written notice, stating that the purpose or one of the purposes of such meeting is to consider the advisability of revoking the voluntary dissolution proceedings, shall be given to each shareholder of record entitled to vote at such meeting within the time and in the manner provided in this Act for the giving of notice of special meetings of shareholders.

(c) At such meeting a vote of the shareholders entitled to vote thereat shall be taken on a resolution to revoke the voluntary dissolution proceedings, which shall require for its adoption the affirmative vote of the holders of a majority of the shares entitled to vote thereon.

(d) Upon the adoption of such resolution, a statement of revocation of voluntary dissolution proceedings shall be executed in duplicate by the corporation by its president or a vice president and by its secretary or an assistant secretary, and verified by one of the officers signing such statement, which statement shall set forth:

(1) The name of the corporation.

(2) The names and respective addresses of its officers.

(3) The names and respective addresses of its directors.

(4) A copy of the resolution adopted by the shareholders revoking the voluntary dissolution proceedings.

(5) The number of shares outstanding.

(6) The number of shares voted for and against the resolution, respectively.

§ 90. **Filing of Statement of Revocation of Voluntary Dissolution Proceedings**

Duplicate originals of the statement of revocation of voluntary dissolution proceedings, whether by consent of shareholders or by act of the corporation, shall be delivered to the Secretary of State. If the Secretary of State finds that such statement conforms to law, he shall, when all fees and franchise taxes have been paid as in this Act prescribed:

(a) Endorse on each of such duplicate originals the word "Filed," and the month, day and year of the filing thereof.

(b) File one of such duplicate originals in his office.

(c) Return the other duplicate original to the corporation or its representative.

§ 91. **Effect of Statement of Revocation of Voluntary Dissolution Proceedings**

Upon the filing by the Secretary of State of a statement of revocation of voluntary dissolution proceedings, whether by consent of shareholders or by act of the corporation, the revocation of the voluntary dissolution proceedings shall become effective and the corporation may again carry on its business.

§ 92. **Articles of Dissolution**

If voluntary dissolution proceedings have not been revoked, then when all debts, liabilities and obligations of the corporation have been paid and discharged, or adequate provision has been made therefor, and all of the remaining property and assets of the corporation have been distributed to its shareholders, articles of dissolution shall be executed in duplicate by the corporation by its president or a vice president and by its secretary or an assistant secretary, and verified by one of the officers signing such statement, which statement shall set forth:

(a) The name of the corporation.

(b) That the Secretary of State has theretofore filed a statement of intent to dissolve the corporation, and the date on which such statement was filed.

(c) That all debts, obligations and liabilities of the corporation have been paid and discharged or that adequate provision has been made therefor.

(d) That all the remaining property and assets of the corporation have been distributed among its shareholders in accordance with their respective rights and interests.

(e) That there are no suits pending against the corporation in any court, or that adequate provision has been made for the satisfaction of any judgment, order or decree which may be entered against it in any pending suit.

§ 93. **Filing of Articles of Dissolution**

Duplicate originals of such articles of dissolution shall be delivered to the Secretary of State. If the Secretary of State finds that such articles of dissolution conform to law, he shall, when all fees and franchise taxes have been paid as in this Act prescribed:

(a) Endorse on each of such duplicate originals the word "Filed," and the month, day and year of the filing thereof.

(b) File one of such duplicate originals in his office.

(c) Issue a certificate of dissolution to which he shall affix the other duplicate original.

The certificate of dissolution, together with the duplicate original of the articles of dissolution affixed thereto by the Secretary of State, shall be returned to the representative of the dissolved corporation. Upon the issuance of such certificate of dissolution the existence of the corporation shall cease, except for the purpose of suits, other proceedings and appropriate corporate action by shareholders, directors and officers as provided in this Act.

§ 94. Involuntary Dissolution

A corporation may be dissolved involuntarily by a decree of the .... court in an action filed by the Attorney General when it is established that:

(a) The corporation has failed to file its annual report within the time required by this Act, or has failed to pay its franchise tax on or before the first day of August of the year in which such franchise tax becomes due and payable; or

(b) The corporation procured its articles of incorporation through fraud; or

(c) The corporation has continued to exceed or abuse the authority conferred upon it by law; or

(d) The corporation has failed for thirty days to appoint and maintain a registered agent in this State; or

(e) The corporation has failed for thirty days after change of its registered office or registered agent to file in the office of the Secretary of State a statement of such change.

§ 95. Notification to Attorney General

The Secretary of State, on or before the last day of December of each year, shall certify to the Attorney General the names of all corporations which have failed to file their annual reports or to pay franchise taxes in accordance with the provisions of this Act, together with the facts pertinent thereto. He shall also certify, from time to time, the names of all corporations which have given other cause for dissolution as provided in this Act, together with the facts pertinent thereto. Whenever the Secretary of State shall certify the name of a corporation to the Attorney General as having given any cause for dissolution, the Secretary of State shall concurrently mail to the corporation at its registered office a notice that such certification has been made. Upon the receipt of such certification, the Attorney General shall file an action in the name of the State against such corporation for its dissolution. Every such certificate from the Secretary of State to the Attorney General pertaining to the failure of a corporation to file an annual report or pay a franchise tax shall be taken and received in all courts as prima facie evidence of the facts therein stated. If, before action is filed, the corporation shall file its annual report or pay its franchise tax, together with all penalties thereon, or shall appoint or maintain a registered agent as provided in this Act, or shall file with the Secretary of State the required statement of change of registered office or registered agent, such fact shall be forthwith certified by the Secretary of State to the Attorney General and he shall not file an action against such corporation for such cause. If, after action is filed, the corporation shall file its annual report or pay its franchise tax, together with all penalties thereon, or shall appoint or maintain a registered agent as provided in this Act, or shall file with the Secretary of State the required statement of change of registered office or registered agent, and shall pay the costs of such action, the action for such cause shall abate.

§ 96. Venue and Process

Every action for the involuntary dissolution of a corporation shall be commenced by the Attorney General either in the .... court of the county in which the registered office of the corporation is situated, or in the .... court of .... county. Summons shall issue and be served as in other civil actions. If process is returned not found, the Attorney General shall cause publication to be made as in other civil cases in some newspaper published in the county where the registered office of the corporation is situated, containing a notice of the pendency of such action, the title of the court, the title of the action, and the date on or after which default may be entered. The Attorney General may include in one notice the names of any number of corporations against which actions are then pending in the same court. The Attorney General shall cause a copy of such notice to be mailed to the corporation at its registered office within ten days after the first publication thereof. The certificate of the Attorney General of the mailing of such notice shall be prima facie evidence thereof. Such notice shall be published at least once each week for two successive weeks, and the first publication thereof may begin at any time after the summons has been returned. Unless a corporation shall have been served with summons, no default shall be taken against it earlier than thirty days after the first publication of such notice.

§ 97. Jurisdiction of Court to Liquidate Assets and Business of Corporation

The .... courts shall have full power to liquidate the assets and business of a corporation:

(a) In an action by a shareholder when it is established:

(1) That the directors are deadlocked in the management of the corporate affairs and the shareholders are unable to break the deadlock, and that irreparable injury to the corporation is being suffered or is threatened by reason thereof; or

(2) That the acts of the directors or those in control of the corporation are illegal, oppressive or fraudulent; or

(3) That the shareholders are deadlocked in voting power, and have failed, for a period which includes at least two consecutive annual meeting dates, to elect successors to directors whose terms have expired or would have expired upon the election of their successors; or

(4) That the corporate assets are being misapplied or wasted.

(b) In an action by a creditor:

(1) When the claim of the creditor has been reduced to judgment and an execution thereon returned unsatisfied and it is established that the corporation is insolvent; or

(2) When the corporation has admitted in writing that the claim of the creditor is due and owing and it is established that the corporation is insolvent.

(c) Upon application by a corporation which has filed a statement of intent to dissolve, as provided in this Act, to have its liquidation continued under the supervision of the court.

(d) When an action has been filed by the Attorney General to dissolve a corporation and it is established that liquidation of its business and affairs should precede the entry of a decree of dissolution.

Proceedings under clause (a), (b) or (c) of this section shall be brought in the county in which the registered office or the principal office of the corporation is situated.

It shall not be necessary to make shareholders parties to any such action or proceeding unless relief is sought against them personally.

§ 98. Procedure in Liquidation of Corporation by Court

In proceedings to liquidate the assets and business of a corporation the court shall have power to issue injunctions, to appoint a receiver or receivers pendente lite, with such powers and duties as the court, from time to time, may direct, and to take such other proceedings as may be requisite to preserve the corporate assets wherever situated, and carry on the business of the corporation until a full hearing can be had.

After a hearing had upon such notice as the court may direct to be given to all parties to the proceedings and to any other parties in interest designated by the court, the court may appoint a liquidating receiver or receivers with authority to collect the assets of the corporation, including all amounts owing to the corporation by subscribers on account of any unpaid portion of the consideration for the issuance of shares. Such liquidating receiver or receivers shall have authority, subject to the order of the court, to sell, convey and dispose of all or any part of the assets of the corporation wherever situated, either at public or private sale. The assets of the corporation or the proceeds resulting from a sale, conveyance or other disposition thereof shall be applied to the expenses of such liquidation and to the payment of the liabilities and obligations of the corporation, and any remaining assets or proceeds shall be distributed among its shareholders according to their respective rights and interests. The order appointing such liquidating receiver or receivers shall state their powers and duties. Such powers and duties may be increased or diminished at any time during the proceedings.

The court shall have power to allow from time to time as expenses of the liquidation compensation to the receiver or receivers and to attorneys in the proceeding, and to direct the payment thereof out of the assets of the corporation or the proceeds of any sale or disposition of such assets.

A receiver of a corporation appointed under the provisions of this section shall have authority to sue and defend in all courts in his own name as receiver of such corporation. The court appointing such receiver shall have exclusive jurisdiction of the corporation and its property, wherever situated.

§ 99. Qualifications of Receivers

A receiver shall in all cases be a natural person or a corporation authorized to act as receiver, which corporation may be a domestic corporation or a foreign corporation authorized to transact business in this State, and shall in all cases give such bond as the court may direct with such sureties as the court may require.

§ 100. **Filing of Claims in Liquidation Proceedings**

In proceedings to liquidate the assets and business of a corporation the court may require all creditors of the corporation to file with the clerk of the court or with the receiver, in such form as the court may prescribe, proofs under oath of their respective claims. If the court requires the filing of claims it shall fix a date, which shall be not less than four months from the date of the order, as the last day for the filing of claims, and shall prescribe the notice that shall be given to creditors and claimants of the date so fixed. Prior to the date so fixed, the court may extend the time for the filing of claims. Creditors and claimants failing to file proofs of claim on or before the date so fixed may be barred, by order of court, from participating in the distribution of the assets of the corporation.

§ 101. **Discontinuance of Liquidation Proceedings**

The liquidation of the assets and business of a corporation may be discontinued at any time during the liquidation proceedings when it is established that cause for liquidation no longer exists. In such event the court shall dismiss the proceedings and direct the receiver to redeliver to the corporation all its remaining property and assets.

§ 102. **Decree of Involuntary Dissolution**

In proceedings to liquidate the assets and business of a corporation, when the costs and expenses of such proceedings and all debts, obligations and liabilities of the corporation shall have been paid and discharged and all of its remaining property and assets distributed to its shareholders, or in case its property and assets are not sufficient to satisfy and discharge such costs, expenses, debts and obligations, all the property and assets have been applied so far as they will go to their payment, the court shall enter a decree dissolving the corporation, whereupon the existence of the corporation shall cease.

§ 103. **Filing of Decree of Dissolution**

In case the court shall enter a decree dissolving a corporation, it shall be the duty of the clerk of such court to cause a certified copy of the decree to be filed with the Secretary of State. No fee shall be charged by the Secretary of State for the filing thereof.

§ 104. **Deposit with State Treasurer of Amount Due Certain Shareholders**

Upon the voluntary or involuntary dissolution of a corporation, the portion of the assets distributable to a creditor or shareholder who is unknown or cannot be found, or who is under disability and there is no person legally competent to receive such distributive portion, shall be reduced to cash and deposited with the State Treasurer and shall be paid over to such creditor or shareholder or to his legal representative upon proof satisfactory to the State Treasurer of his right thereto.

§ 105. **Survival of Remedy after Dissolution**

The dissolution of a corporation either (1) by the issuance of a certificate of dissolution by the Secretary of State, or (2) by a decree of court when the court has not liquidated the assets and business of the corporation as provided in this Act, or (3) by expiration of its period of duration, shall not take away or impair any remedy available to or against such corporation, its directors, officers, or shareholders, for any right or claim existing, or any liability incurred, prior to such dissolution if action or other proceeding thereon is commenced within two years after the date of such dissolution. Any such action or proceeding by or against the corporation may be prosecuted or defended by the corporation in its corporate name. The shareholders, directors and officers shall have power to take such corporate or other action as shall be appropriate to protect such remedy, right or claim. If such corporation was dissolved by the expiration of its period of duration, such corporation may amend its articles of incorporation at any time during such period of two years so as to extend its period of duration.

§ 106. **Admission of Foreign Corporation**

No foreign corporation shall have the right to transact business in this State until it shall have procured a certificate of authority so to do from the Secretary of State. No foreign corporation shall be entitled to procure a certificate of authority under this Act to transact in this State any business which a corporation organized under this Act is not permitted to transact. A foreign corporation shall not be denied a certificate of authority by reason of the fact that the laws of the state or country under which such corporation is organized governing its organization and internal affairs differ from the laws of this State, and nothing in this Act contained shall be construed to authorize this State to regulate the organization or the internal affairs of such corporation.

Without excluding other activities which may not constitute transacting business in this State, a foreign corporation shall not be considered to be transacting business in this State, for the purposes of this Act, by reason of carrying on in this State any one or more of the following activities:

(a) Maintaining or defending any action or suit or any administrative or arbitration proceeding, or effecting the settlement thereof or the settlement of claims or disputes.

(b) Holding meetings of its directors or shareholders or carrying on other activities concerning its internal affairs.

(c) Maintaining bank accounts.

(d) Maintaining offices or agencies for the transfer, exchange and registration of its securities, or appointing and maintaining trustees or depositaries with relation to its securities.

(e) Effecting sales through independent contractors.

(f) Soliciting or procuring orders, whether by mail or through employees or agents or otherwise, where such orders require acceptance without this State before becoming binding contracts.

(g) Creating as borrower or lender, or acquiring, indebtedness or mortgages or other security interests in real or personal property.

(h) Securing or collecting debts or enforcing any rights in property securing the same.

(i) Transacting any business in interstate commerce.

(j) Conducting an isolated transaction completed within a period of thirty days and not in the course of a number of repeated transactions of like nature.

§ 107. **Powers of Foreign Corporation**

A foreign corporation which shall have received a certificate of authority under this Act shall, until a certificate of revocation or of withdrawal shall have been issued as provided in this Act, enjoy the same, but no greater, rights and privileges as a domestic corporation organized for the purposes set forth in the application pursuant to which such certificate of authority is issued; and, except as in this Act otherwise provided, shall be subject to the same duties, restrictions, penalties and liabilities now or hereafter imposed upon a domestic corporation of like character.

§ 108. **Corporate Name of Foreign Corporation**

No certificate of authority shall be issued to a foreign corporation unless the corporate name of such corporation:

(a) Shall contain the word "corporation," "company," "incorporated," or "limited," or shall contain an abbreviation of one of such words, or such corporation shall, for use in this State, add at the end of its name one of such words or an abbreviation thereof.

(b) Shall not contain any word or phrase which indicates or implies that it is organized for any purpose other than one or more of the purposes contained in its articles of incorporation or that it is authorized or empowered to conduct the business of banking or insurance.

(c) Shall not be the same as, or deceptively similar to, the name of any domestic corporation existing under the laws of this State or any foreign corporation authorized to transact business in this State, or a name the exclusive right to which is, at the time, reserved in the manner provided in this Act, or the name of a corporation which has in effect a registration of its name as provided in this Act, except that this provision shall not apply if the foreign corporation applying for a certificate of authority files with the Secretary of State any one of the following:

(1) a resolution of its board of directors adopting a fictitious name for use in transacting business in this State which fictitious name is not deceptively similar to the name of any domestic corporation or of any foreign corporation authorized to transact business in this State or to any name reserved or registered as provided in this Act, or

(2) the written consent of such other corporation or holder of a reserved or registered name to use the same or deceptively similar name and one or more words are added to make such name distinguishable from such other name, or

(3) a certified copy of a final decree of a court of competent jurisdiction establishing the prior right of such foreign corporation to the use of such name in this State.

§ 109. **Change of Name by Foreign Corporation**

Whenever a foreign corporation which is authorized to transact business in this State shall change its name to one under which a certificate of authority would not be granted to it on application therefor, the certificate of authority of such corporation shall be suspended and it shall not thereafter transact any business in this State until it has changed its name to a name which is available to it under the laws of this State or has otherwise complied with the provisions of this Act.

§ 110. **Application for Certificate of Authority**

A foreign corporation, in order to procure a certificate of authority to transact business in this State, shall make application therefor to the Secretary of State, which application shall set forth:

(a) The name of the corporation and the state or county under the laws of which it is incorporated.

(b) If the name of the corporation does not contain the word "corporation," "company," "incorporated," or "limited," or does not contain an abbreviation of one of such words, then the name of the corporation with the word or abbreviation which it elects to add thereto for use in this State.

(c) The date of incorporation and the period of duration of the corporation.

(d) The address of the principal office of the corporation in the state or country under the laws of which it is incorporated.

(e) The address of the proposed registered office of the corporation in this State, and the name of its proposed registered agent in this State at such address.

(f) The purpose or purposes of the corporation which it proposes to pursue in the transaction of business in this State.

(g) The names and respective addresses of the directors and officers of the corporation.

(h) A statement of the aggregate number of shares which the corporation has authority to issue, itemized by classes and series, if any, within a class.

(i) A statement of the aggregate number of issued shares itemized by class and by series, if any, within each class.

(j) An estimate, expressed in dollars, of the value of all property to be owned by the corporation for the following year, wherever located, and an estimate of the value of the property of the corporation to be located within this State during such year, and an estimate, expressed in dollars, of the gross amount of business which will be transacted by the corporation during such year, and an estimate of the gross amount thereof which will be transacted by the corporation at or from places of business in this State during such year.

(k) Such additional information as may be necessary or appropriate in order to enable the Secretary of State to determine whether such corporation is entitled to a certificate of authority to transact business in this State and to determine and assess the fees and franchise taxes payable as in this Act prescribed.

Such application shall be made on forms prescribed and furnished by the Secretary of State and shall be executed in duplicate by the corporation by its president or a vice president and by its secretary or an assistant secretary, and verified by one of the officers signing such application.

§ 111. **Filing of Application for Certificate of Authority**

Duplicate originals of the application of the corporation for a certificate of authority shall be delivered to the Secretary of State, together with a copy of its articles of incorporation and all amendments thereto, duly authenticated by the proper officer of the state or country under the laws of which it is incorporated.

If the Secretary of State finds that such application conforms to law, he shall, when all fees and franchise taxes have been paid as in this Act prescribed:

(a) Endorse on each of such documents the word "Filed," and the month, day and year of the filing thereof.

(b) File in his office one of such duplicate originals of the application and the copy of the articles of incorporation and amendments thereto.

(c) Issue a certificate of authority to transact business in this State to which he shall affix the other duplicate original application.

The certificate of authority, together with the duplicate original of the application affixed thereto by the Secretary of State, shall be returned to the corporation or its representative.

§ 112. **Effect of Certificate of Authority**

Upon the issuance of a certificate of authority by the Secretary of State, the corporation shall be authorized to transact business in this State for those purposes set forth in its application, subject, however, to the right of this State to suspend or to revoke such authority as provided in this Act.

§ 113. **Registered Office and Registered Agent of Foreign Corporation**

Each foreign corporation authorized to transact business in this State shall have and continuously maintain in this State:

(a) A registered office which may be, but need not be, the same as its place of business in this State.

(b) A registered agent, which agent may be either an individual resident in this State whose business office is identical with such registered office, or a domestic corporation, or a foreign corporation authorized to transact business in this State, having a business office identical with such registered office.

§ 114. **Change of Registered Office or Registered Agent of Foreign Corporation**

A foreign corporation authorized to transact business in this State may change its registered office or change its registered agent, or both, upon filing in the office of the Secretary of State a statement setting forth:

(a) The name of the corporation.

(b) The address of its then registered office.

(c) If the address of its registered office be changed, the address to which the registered office is to be changed.

(d) The name of its then registered agent.

(e) If its registered agent be changed, the name of its successor registered agent.

(f) That the address of its registered office and the address of the business office of its registered agent, as changed, will be identical.

(g) That such change was authorized by resolution duly adopted by its board of directors.

Such statement shall be executed by the corporation by its president or a vice president, and verified by him, and delivered to the Secretary of State. If the Secretary of State finds that such statement conforms to the provisions of this Act, he shall file such statement in his office, and upon such filing the change of address of the registered office, or the appointment of a new registered agent, or both, as the case may be, shall become effective.

Any registered agent of a foreign corporation may resign as such agent upon filing a written notice thereof, executed in duplicate, with the Secretary of State, who shall forthwith mail a copy thereof to the corporation at its principal office in the state or country under the laws of which it is incorporated. The appointment of such agent shall terminate upon the expiration of thirty days after receipt of such notice by the Secretary of State.

If a registered agent changes his or its business address to another place within the same .... *, he or it may change such address and the address of the registered office of any corporation of which he or it is registered agent by filing a statement as required above except that it need be signed only by the registered agent and need not be responsive to (e) or (g) and must recite that a copy of the statement has been mailed to the corporation.

*Supply designation of jurisdiction, such as county, etc. in accordance with local practice.

§ 115. **Service of Process on Foreign Corporation**

The registered agent so appointed by a foreign corporation authorized to transact business in this State shall be an agent of such corporation upon whom any process, notice or demand required or permitted by law to be served upon the corporation may be served.

Whenever a foreign corporation authorized to transact business in this State shall fail to appoint or maintain a registered agent in this State, or whenever any such registered agent cannot with reasonable diligence be found at the registered office, or whenever the certificate of authority of a foreign corporation shall be suspended or revoked, then the Secretary of State shall be an agent of such corporation upon whom any such process, notice, or demand may be served. Service on the Secretary of State of any such process, notice or demand shall be made by delivering to and leaving with him, or with any clerk having charge of the corporation department of his office, duplicate copies of such process, notice or demand. In the event any such process, notice or demand is served on the Secretary of State, he shall immediately cause one of such copies thereof to be forwarded by registered mail, addressed to the corporation at its principal office in the state or country under the laws of which it is incorporated. Any service so had on the Secretary of State shall be returnable in not less than thirty days.

The Secretary of State shall keep a record of all processes, notices and demands served upon him under this section, and shall record therein the time of such service and his action with reference thereto.

Nothing herein contained shall limit or affect the right to serve any process, notice or demand, required or permitted by law to be served upon a foreign corporation in any other manner now or hereafter permitted by law.

§ 116. Amendment to Articles of Incorporation of Foreign Corporation

Whenever the articles of incorporation of a foreign corporation authorized to transact business in this State are amended, such foreign corporation shall, within thirty days after such amendment becomes effective, file in the office of the Secretary of State a copy of such amendment duly authenticated by the proper officer of the state or country under the laws of which it is incorporated; but the filing thereof shall not of itself enlarge or alter the purpose or purposes which such corporation is authorized to pursue in the transaction of business in this State, nor authorize such corporation to transact business in this State under any other name than the name set forth in its certificate of authority.

§ 117. Merger of Foreign Corporation Authorized to Transact Business in This State

Whenever a foreign corporation authorized to transact business in this State shall be a party to a statutory merger permitted by the laws of the state or country under the laws of which it is incorporated, and such corporation shall be the surviving corporation, it shall, within thirty days after such merger becomes effective, file with the Secretary of State a copy of the articles of merger duly authenticated by the proper officer of the state or country under the laws of which such statutory merger was effected; and it shall not be necessary for such corporation to procure either a new or amended certificate of authority to transact business in this State unless the name of such corporation be changed thereby or unless the corporation desires to pursue in this State other or additional purposes than those which it is then authorized to transact in this State.

§ 118. Amended Certificate of Authority

A foreign corporation authorized to transact business in this State shall procure an amended certificate of authority in the event it changes its corporate name, or desires to pursue in this State other or additional purposes than those set forth in its prior application for a certificate of authority, by making application therefor to the Secretary of State.

The requirements in respect to the form and contents of such application, the manner of its execution, the filing of duplicate originals thereof with the Secretary of State, the issuance of an amended certificate of authority and the effect thereof, shall be the same as in the case of an original application for a certificate of authority.

§ 119. Withdrawal of Foreign Corporation

A foreign corporation authorized to transact business in this State may withdraw from this State upon procuring from the Secretary of State a certificate of withdrawal. In order to procure such certificate of withdrawal, such foreign corporation shall deliver to the Secretary of State an application for withdrawal, which shall set forth:

(a) The name of the corporation and the state or country under the laws of which it is incorporated.

(b) That the corporation is not transacting business in this State.

(c) That the corporation surrenders its authority to transact business in this State.

(d) That the corporation revokes the authority of its registered agent in this State to accept service of process and consents that service of process in any action, suit or proceeding based upon any cause of action arising in this State during the time the corporation was authorized to transact business in this State may thereafter be made on such corporation by service thereof on the Secretary of State.

(e) A post-office address to which the Secretary of State may mail a copy of any process against the corporation that may be served on him.

(f) A statement of the aggregate number of shares

which the corporation has authority to issue, itemized by class and series, if any, within each class, as of the date of such application.

(g) A statement of the aggregate number of issued shares, itemized by class and series, if any, within each class, as of the date of such application.

(h) Such additional information as may be necessary or appropriate in order to enable the Secretary of State to determine and assess any unpaid fees or franchise taxes payable by such foreign corporation as in this Act prescribed.

The application for withdrawal shall be made on forms prescribed and furnished by the Secretary of State and shall be executed by the corporation by its president or a vice president and by its secretary or an assistant secretary, and verified by one of the officers signing the application, or, if the corporation is in the hands of a receiver or trustee, shall be executed on behalf of the corporation by such receiver or trustee and verified by him.

§ 120. Filing of Application for Withdrawal

Duplicate originals of such application for withdrawal shall be delivered to the Secretary of State. If the Secretary of State finds that such application conforms to the provisions of this Act, he shall, when all fees and franchise taxes have been paid as in this Act prescribed:

(a) Endorse on each of such duplicate originals the word "Filed," and the month, day and year of the filing thereof.

(b) File one of such duplicate originals in his office.

(c) Issue a certificate of withdrawal to which he shall affix the other duplicate original.

The certificate of withdrawal, together with the duplicate original of the application for withdrawal affixed thereto by the Secretary of State, shall be returned to the corporation or its representative. Upon the issuance of such certificate of withdrawal, the authority of the corporation to transact business in this State shall cease.

§ 121. Revocation of Certificate of Authority

The certificate of authority of a foreign corporation to transact business in this State may be revoked by the Secretary of State upon the conditions prescribed in this section when:

(a) The corporation has failed to file its annual report within the time required by this Act, or has failed to pay any fees, franchise taxes or penalties prescribed by this Act when they have become due and payable; or

(b) The corporation has failed to appoint and maintain a registered agent in this State as required by this Act; or

(c) The corporation has failed, after change of its registered office or registered agent, to file in the office of the Secretary of State a statement of such change as required by this Act; or

(d) The corporation has failed to file in the office of the Secretary of State any amendment to its articles of incorporation or any articles of merger within the time prescribed by this Act; or

(e) A misrepresentation has been made of any material matter in any application, report, affidavit, or other document submitted by such corporation pursuant to this Act.

No certificate of authority of a foreign corporation shall be revoked by the Secretary of State unless (1) he shall have given the corporation not less than sixty days' notice thereof by mail addressed to its registered office in this State, and (2) the corporation shall fail prior to revocation to file such annual report, or pay such fees, franchise taxes or penalties, or file the required statement of change of registered agent or registered office, or file such articles of amendment or articles of merger, or correct such misrepresentation.

§ 122. Issuance of Certificate of Revocation

Upon revoking any such certificate of authority, the Secretary of State shall:

(a) Issue a certificate of revocation in duplicate.

(b) File one of such certificates in his office.

(c) Mail to such corporation at its registered office in this State a notice of such revocation accompanied by one of such certificates.

Upon the issuance of such certificate of revocation, the authority of the corporation to transact business in this State shall cease.

§ 123. Application to Corporations Heretofore Authorized to Transact Business in this State

Foreign corporations which are duly authorized to transact business in this State at the time this Act takes effect, for a purpose or purposes for which a corporation might secure such authority under this Act, shall, subject to the limitations set forth in their respective certificates of authority, be entitled to all the rights and privileges applicable to foreign corporations procuring certificates of authority to transact business in this State under this Act, and from the time this Act takes effect such corporations shall be subject to all the limitations, restrictions, liabilities, and duties prescribed herein for foreign corporations procuring certificates of authority to transact business in this State under this Act.

§ 124. Transacting Business Without Certificate of Authority

No foreign corporation transacting business in this State without a certificate of authority shall be permitted to maintain any action, suit or proceeding in any court of this State, until such corporation shall have obtained a certificate of authority. Nor shall any action, suit or proceeding be maintained in any court of this State by any successor or assignee of such corporation on any right, claim or demand arising out of the transaction of business by such corporation in this State, until a certificate of authority shall have been obtained by such corporation or by a corporation which has acquired all or substantially all of its assets.

The failure of a foreign corporation to obtain a certificate of authority to transact business in this State shall not impair the validity of any contract or act of such corporation, and shall not prevent such corporation from defending any action, suit or proceeding in any court of this State.

A foreign corporation which transacts business in this State without a certificate of authority shall be liable to this State, for the years or parts thereof during which it transacted business in this State without a certificate of authority, in an amount equal to all fees and franchise taxes which would have been imposed by this Act upon such corporation had it duly applied for and received a certificate of authority to transact business in this State as required by this Act and thereafter filed all reports required by this Act, plus all penalties imposed by this Act for failure to pay such fees and franchise taxes. The Attorney General shall bring proceedings to recover all amounts due this State under the provisions of this Section.

§ 125. Annual Report of Domestic and Foreign Corporations

Each domestic corporation, and each foreign corporation authorized to transact business in this State, shall file, within the time prescribed by this Act, an annual report setting forth:

(a) The name of the corporation and the state or country under the laws of which it is incorporated.

(b) The address of the registered office of the corporation in this State, and the name of its registered agent in this State at such address, and, in case of a foreign corporation, the address of its principal office in the state or country under the laws of which it is incorporated.

(c) A brief statement of the character of the business in which the corporation is actually engaged in this State.

(d) The names and respective addresses of the directors and officers of the corporation.

(e) A statement of the aggregate number of shares which the corporation has authority to issue, itemized by class and series, if any, within each class.

(f) A statement of the aggregate number of issued shares, itemized by class and series, if any, within each class.

(g) A statement, expressed in dollars, of the value of all the property owned by the corporation, wherever located, and the value of the property of the corporation located within this State, and a statement, expressed in dollars, of the gross amount of business transacted by the corporation for the twelve months ended on the thirty-first day of December preceding the date herein provided for the filing of such report and the gross amount thereof transacted by the corporation at or from places of business in this State. If, on the thirty-first day of December preceding the time herein provided for the filing of such report, the corporation had not been in existence for a period of twelve months, or in the case of a foreign corporation had not been authorized to transact business in this State for a period of twelve months, the statement with respect to business transacted shall be furnished for the period between the date of incorporation or the date of its authorization to transact business in this State, as the case may be, and such thirty-first day of December. If all the property of the corporation is located in this State and all of its business is transacted at or from places of business in this State, then the information required by this subparagraph need not be set forth in such report.

(h) Such additional information as may be necessary or appropriate in order to enable the Secretary of State to determine and assess the proper amount of franchise taxes payable by such corporation.

Such annual report shall be made on forms prescribed and furnished by the Secretary of State, and the information therein contained shall be given as of the date of the execution of the report, except as to the information required by subparagraphs (g) and (h) which shall be given as of the close of business on the thirty-first day of December next preceding the date herein provided for the filing of such report. It shall be executed by the corporation by its president, a vice president, secretary, an assistant secretary, or treasurer, and verified by the officer executing the report, or, if the corporation is in the hands of a receiver or trustee, it shall be executed on behalf of the corporation and verified by such receiver or trustee.

§ 126. **Filing of Annual Report of Domestic and Foreign Corporations**

Such annual report of a domestic or foreign corporation shall be delivered to the Secretary of State between the first day of January and the first day of March of each year, except that the first annual report of a domestic or foreign corporation shall be filed between the first day of January and the first day of March of the year next succeeding the calendar year in which its certificate of incorporation or its certificate of authority, as the case may be, was issued by the Secretary of State. Proof to the satisfaction of the Secretary of State that prior to the first day of March such report was deposited in the United States mail in a sealed envelope, properly addressed, with postage prepaid, shall be deemed a compliance with this requirement. If the Secretary of State finds that such report conforms to the requirements of this Act, he shall file the same. If he finds that it does not so conform, he shall promptly return the same to the corporation for any necessary corrections, in which event the penalties hereinafter prescribed for failure to file such report within the time hereinabove provided shall not apply, if such report is corrected to conform to the requirements of this Act and returned to the Secretary of State within thirty days from the date on which it was mailed to the corporation by the Secretary of State.

§ 127. **Fees, Franchise Taxes and Charges to be Collected by Secretary of State**

The Secretary of State shall charge and collect in accordance with the provisions of this Act:

(a) Fees for filing documents and issuing certificates.

(b) Miscellaneous charges.

(c) License fees.

(d) Franchise taxes.

§ 128. **Fees for Filing Documents and Issuing Certificates**

The Secretary of State shall charge and collect for:

(a) Filing articles of incorporation and issuing a certificate of incorporation, ... dollars.

(b) Filing articles of amendment and issuing a certificate of amendment, ... dollars.

(c) Filing restated articles of incorporation, ... dollars.

(d) Filing articles of merger or consolidation and issuing a certificate of merger or consolidation, ... dollars.

(e) Filing an application to reserve a corporate name, ... dollars.

(f) Filing a notice of transfer of a reserved corporate name, ... dollars.

(g) Filing a statement of change of address of registered office or change of registered agent, or both, ... dollars.

(h) Filing a statement of the establishment of a series of shares, ... dollars.

(i) Filing a statement of intent to dissolve, ... dollars.

(j) Filing a statement of revocation of voluntary dissolution proceedings, ... dollars.

(k) Filing articles of dissolution, ... dollars.

(l) Filing an application of a foreign corporation for a certificate of authority to transact business in this State and issuing a certificate of authority, ... dollars.

(m) Filing an application of a foreign corporation for an amended certificate of authority to transact business in this State and issuing an amended certificate of authority, ... dollars.

(n) Filing a copy of an amendment to the articles of incorporation of a foreign corporation holding a certificate of authority to transact business in this State, ... dollars.

(o) Filing a copy of articles of merger of a foreign corporation holding a certificate of authority to transact business in this State, ... dollars.

(p) Filing an application for withdrawal of a foreign corporation and issuing a certificate of withdrawal, ... dollars.

(q) Filing any other statement or report, except an annual report, of a domestic or foreign corporation, ... dollars.

§ 129. **Miscellaneous Charges**

The Secretary of State shall charge and collect:

(a) For furnishing a certified copy of any document, instrument, or paper relating to a corporation, ... cents per page and .... dollars for the certificate and affixing the seal thereto.

(b) At the time of any service of process on him as agent of a corporation, ... dollars, which amount may be recovered as taxable costs by the party to the suit or action causing such service to be made if such party prevails in the suit or action.

§ 130. **License Fees Payable by Domestic Corporations**

The Secretary of State shall charge and collect from each domestic corporation license fees, based upon the number of shares which it will have authority to issue or the increase in the number of shares which it will have authority to issue, at the time of:

(a) Filing articles of incorporation;

(b) Filing articles of amendment increasing the number of authorized shares; and

(c) Filing articles of merger or consolidation increasing the number of authorized shares which the surviving or new corporation, if a domestic corporation, will have the authority to issue above the aggregate number of shares which the constituent domestic corporations and constituent foreign corporations authorized to transact business in this State had authority to issue.

The license fees shall be at the rate of .... cents per share up to and including the first 10,000 authorized shares, ... cents per share for each authorized share in excess of 10,000 shares up to and including 100,000 shares, and .... cents per share for each authorized share in excess of 100,000 shares.

The license fees payable on an increase in the number of authorized shares shall be imposed only on the increased number of shares, and the number of previously authorized shares shall be taken into account in determining the rate applicable to the increased number of authorized shares.

§ 131. **License Fees Payable by Foreign Corporations**

The Secretary of State shall charge and collect from each foreign corporation license fees, based upon the proportion represented in this State of the number of shares which it has authority to issue or the increase in the number of shares which it has authority to issue, at the time of:

(a) Filing an application for a certificate of authority to transact business in this State;

(b) Filing articles of amendment which increased the number of authorized shares; and

(c) Filing articles of merger or consolidation which increased the number of authorized shares which the surviving or new corporation, if a foreign corporation, has authority to issue above the aggregate number of shares which the constituent domestic corporations and constituent foreign corporations authorized to transact business in this State had authority to issue.

The license fees shall be at the rate of .... cents per share up to and including the first 10,000 authorized shares represented in this State, ... cents per share for each authorized share in excess of 10,000 shares up to and including 100,000 shares represented in this State, and .... cents per share for each authorized share in excess of 100,000 shares represented in this State.

The license fees payable on an increase in the number of authorized shares shall be imposed only on the increased number of such shares represented in this State, and the number of previously authorized shares represented in this State shall be taken into account in determining the rate applicable to the increased number of authorized shares.

The number of authorized shares represented in this State shall be that proportion of its total authorized shares which the sum of the value of its property located in this State and the gross amount of business transacted by it at or from places of business in this State bears to the sum of the value of all of its property, wherever located, and the gross amount of its business, wherever transacted. Such proportion shall be determined from information contained in the application for a certificate of authority to transact business in this State until the filing of an annual report and thereafter from information contained in the latest annual report filed by the corporation.

§ 132. Franchise Taxes Payable by Domestic Corporations

The Secretary of State shall charge and collect from each domestic corporation an initial franchise tax at the time of filing its articles of incorporation at the rate of one-twelfth of one-half of the license fee payable by such corporation under the provisions of this Act at the time of filing its articles of incorporation, for each calendar month, or fraction thereof, between the date of the issuance of the certificate of incorporation by the Secretary of State and the first day of July of the next succeeding calendar year.

The Secretary of State shall charge and collect from each domestic corporation an annual franchise tax, payable in advance for the period from July 1 in each year to July 1 in the succeeding year, beginning July 1 in the calendar year in which such corporation is required to file its first annual report under this Act, (Alternative 1: at the rate of ...... of ...... per cent of the amount represented in this State of the stated capital of the corporation, as determined in accordance with accounting practices and principles that are reasonable in the circumstances, as disclosed by the latest report filed by the corporation with the Secretary of State) (Alternative 2: at the rate of ...... cents per share up to and including the first 10,000 issued and outstanding shares, and ...... cents per share for each issued and outstanding share in excess of 10,000 shares up to and including 100,000 shares, and ...... cents per share for each issued and outstanding share in excess of 100,000 shares).

[*If Alternative 2 is enacted, the following paragraph should be deleted.*]

The amount represented in this State of the stated capital of the corporation shall be that proportion of its stated capital which the sum of the value of its property located in this State and the gross amount of business transacted by it at or from places of business in this State bears to the sum of the value of all of its property, wherever located, and the gross amount of its business, wherever transacted.

§ 133. Franchise Taxes Payable by Foreign Corporations

The Secretary of State shall charge and collect from each foreign corporation authorized to transact business in this State an initial franchise tax at the time of filing its application for a certificate of authority at the rate of one-twelfth of one-half of the license fee payable by such corporation under the provisions of this Act at the time of filing such application, for each month, or fraction thereof, between the date of the issuance of the certificate of authority by the Secretary of State and the first day of July of the next succeeding calendar year.

The Secretary of State shall charge and collect from each foreign corporation authorized to transact business in this State an annual franchise tax, payable in advance for the period from July 1 in each year to July 1 in the succeeding year, beginning July 1 in the calendar year in which such corporation is required to file its first annual report under this Act, (Alternative 1: at the rate of ...... per cent of the amount represented in this State of the stated capital of the corporation, as determined in accordance with accounting practices and principles that are reasonable in the circumstances, as disclosed by the latest annual report filed by the corporation with the Secretary of State) (Alternative 2: at a rate of ...... cents per share up to and including the first 10,000 issued and outstanding shares represented in this State, and ...... cents per share for each issued and outstanding share in excess of 10,000 shares up to and including 100,000 shares represented in this State, and ...... cents per share for each issued and outstanding share in excess of 100,000 shares represented in this State).

[*If Alternative 2 is enacted, the following paragraph should be deleted.*]

The amount represented in this State of the stated capital of the corporation shall be that proportion of its stated capital which the sum of the value of its property located in this State and the gross amount of business transacted by it at or from places of business in this State bears to the sum of the value of all of its property, wherever located, and the gross amount of its business, wherever transacted.

§ 134. **Assessment and Collection of Annual Franchise Taxes**

It shall be the duty of the Secretary of State to collect all annual franchise taxes and penalties imposed by, or assessed in accordance with, this Act.

Between the first day of March and the first day of June of each year, the Secretary of State shall assess against each corporation, domestic and foreign, required to file an annual report in such year, the franchise tax payable by it for the period from July 1 of such year to July 1 of the succeeding year in accordance with the provisions of this Act, and, if it has failed to file its annual report within the time prescribed by this Act, the penalty imposed by this Act upon such corporation for its failure so to do; and shall mail a written notice to each corporation against which such tax is assessed, addressed to such corporation at its registered office in this State, notifying the corporation (1) of the amount of franchise tax assessed against it for the ensuing year and the amount of penalty, if any, assessed against it for failure to file its annual report; (2) that objections, if any, to such assessment will be heard by the officer making the assessment on or before the fifteenth day of June of such year, upon receipt of a request from the corporation; and (3) that such tax and penalty shall be payable to the Secretary of State on the first day of July next succeeding the date of the notice. Failure to receive such notice shall not relieve the corporation of its obligation to pay the tax and any penalty assessed, or invalidate the assessment thereof.

The Secretary of State shall have power to hear and determine objections to any assessment of franchise tax at any time after such assessment and, after hearing, to change or modify any such assessment. In the event of any adjustment of franchise tax with respect to which a penalty has been assessed for failure to file an annual report, the penalty shall be adjusted in accordance with the provisions of this Act imposing such penalty.

All annual franchise taxes and all penalties for failure to file annual reports shall be due and payable on the first day of July of each year. If the annual franchise tax assessed against any corporation subject to the provisions of this Act, together with all penalties assessed thereon, shall not be paid to the Secretary of State on or before the thirty-first day of July of the year in which such tax is due and payable, the Secretary of State shall certify such fact to the Attorney General on or before the fifteenth day of November of such year, whereupon the Attorney General may institute an action against such corporation in the name of this State, in any court of competent jurisdiction, for the recovery of the amount of such franchise tax and penalties, together with the cost of suit, and prosecute the same to final judgment.

For the purpose of enforcing collection, all annual franchise taxes assessed in accordance with this Act, and all penalties assessed thereon and all interest and costs that shall accrue in connection with the collection thereof, shall be a prior and first lien on the real and personal property of the corporation from and including the first day of July of the year when such franchise taxes become due and payable until such taxes, penalties, interest, and costs shall have been paid.

§ 135. **Penalties Imposed upon Corporations**

Each corporation, domestic or foreign, that fails or refuses to file its annual report for any year within the time prescribed by this Act shall be subject to a penalty of ten per cent of the amount of the franchise tax assessed against it for the period beginning July 1 of the year in which such report should have been filed. Such penalty shall be assessed by the Secretary of State at the time of the assessment of the franchise tax. If the amount of the franchise tax as originally assessed against such corporation be thereafter adjusted in accordance with the provisions of this Act, the amount of the penalty shall be likewise adjusted to ten per cent of the amount of the adjusted franchise tax. The amount of the franchise tax and the amount of the penalty shall be separately stated in any notice to the corporation with respect thereto.

If the franchise tax assessed in accordance with the provisions of this Act shall not be paid on or before the thirty-first day of July, it shall be deemed to be delinquent, and there shall be added a penalty of one per cent for each month or part of month that the same is delinquent, commencing with the month of August.

Each corporation, domestic or foreign, that fails or refuses to answer truthfully and fully within the time prescribed by this Act interrogatories propounded by the Secretary of State in accordance with the provisions of this Act, shall be deemed to be guilty of a misdemeanor and upon conviction thereof may be fined in any amount not exceeding five hundred dollars.

§ 136. **Penalties Imposed upon Officers and Directors**

Each officer and director of a corporation, domestic or foreign, who fails or refuses within the time prescribed by this Act to answer truthfully and fully interrogatories propounded to him by the Secretary of State in accordance with the provisions of this Act, or who signs any articles, statement, report, application or other document filed with the Secretary of State which is known to such officer or director to be false in any material respect, shall be deemed to be guilty of a misdemeanor, and upon conviction thereof may be fined in any amount not exceeding .... dollars.

§ 137. **Interrogatories by Secretary of State**

The Secretary of State may propound to any corporation, domestic or foreign, subject to the provisions of this Act, and to any officer or director thereof, such interrogatories as may be reasonably necessary and proper to enable him to ascertain whether such corporation has complied with all the provisions of this Act applicable to such corporation. Such interrogatories shall be answered within thirty days after the mailing thereof, or within such additional time as shall be fixed by the Secretary of State, and the answers thereto shall be full and complete and shall be made in writing and under oath. If such interrogatories be directed to an individual they shall be answered by him, and if directed to a corporation they shall be answered by the president, vice president, secretary or assistant secretary thereof. The Secretary of State need not file any document to which such interrogatories relate until such interrogatories be answered as herein provided, and not then if the answers thereto disclose that such document is not in conformity with the provisions of this Act. The Secretary of State shall certify to the Attorney General, for such action as the Attorney General may deem appropriate, all interrogatories and answers thereto which disclose a violation of any of the provisions of this Act.

§ 138. **Information Disclosed by Interrogatories**

Interrogatories propounded by the Secretary of State and the answers thereto shall not be open to public inspection nor shall the Secretary of State disclose any facts or information obtained therefrom except insofar as his official duty may require the same to be made public or in the event such interrogatories or the answers thereto are required for evidence in any criminal proceedings or in any other action by this State.

§ 139. **Powers of Secretary of State**

The Secretary of State shall have the power and authority reasonably necessary to enable him to administer this Act efficiently and to perform the duties therein imposed upon him.

§ 140. **Appeal from Secretary of State**

If the Secretary of State shall fail to approve any articles of incorporation, amendment, merger, consolidation or dissolution, or any other document required by this Act to be approved by the Secretary of State before the same shall be filed in his office, he shall, within ten days after the delivery thereof to him, give written notice of his disapproval to the person or corporation, domestic or foreign, delivering the same, specifying the reasons therefor. From such disapproval such person or corporation may appeal to the .... court of the county in which the registered office of such corporation is, or is proposed to be, situated by filing with the clerk of such court a petition setting forth a copy of the articles or other document sought to be filed and a copy of the written disapproval thereof by the Secretary of State; whereupon the matter shall be tried de novo by the court, and the court shall either sustain the action of the Secretary of State or direct him to take such action as the court may deem proper.

If the Secretary of State shall revoke the certificate of authority to transact business in this State of any foreign corporation, pursuant to the provisions of this Act, such foreign corporation may likewise appeal to the .... court of the county where the registered office of such corporation in this State is situated, by filing with the clerk of such court a petition setting forth a copy of its certificate of authority to transact business in this State and a copy of the notice of revocation given by the Secretary of State; whereupon the matter shall be tried de novo by the court, and the court shall either sustain the action of the Secretary of State or direct him to take such action as the court may deem proper.

Appeals from all final orders and judgments entered by the .... court under this section in review of any ruling or decision of the Secretary of State may be taken as in other civil actions.

§ 141. Certificates and Certified Copies to be Received in Evidence

All certificates issued by the Secretary of State in accordance with the provisions of this Act, and all copies of documents filed in his office in accordance with the provisions of this Act when certified by him, shall be taken and received in all courts, public offices, and official bodies as prima facie evidence of the facts therein stated. A certificate by the Secretary of State under the great seal of this State, as to the existence or non-existence of the facts relating to corporations shall be taken and received in all courts, public offices, and official bodies as prima facie evidence of the existence or non-existence of the facts therein stated.

§ 142. Forms to be Furnished by Secretary of State

All reports required by this Act to be filed in the office of the Secretary of State shall be made on forms which shall be prescribed and furnished by the Secretary of State. Forms for all other documents to be filed in the office of the Secretary of State shall be furnished by the Secretary of State on request therefor, but the use thereof, unless otherwise specifically prescribed in this Act, shall not be mandatory.

§ 143. Greater Voting Requirements

Whenever, with respect to any action to be taken by the shareholders of a corporation, the articles of incorporation require the vote or concurrence of the holders of a greater proportion of the shares, or of any class or series thereof, than required by this Act with respect to such action, the provisions of the articles of incorporation shall control.

§ 144. Waiver of Notice

Whenever any notice is required to be given to any shareholder or director of a corporation under the provisions of this Act or under the provisions of the articles of incorporation or by-laws of the corporation, a waiver thereof in writing signed by the person or persons entitled to such notice, whether before or after the time stated therein, shall be equivalent to the giving of such notice.

§ 145. Action by Shareholders Without a Meeting

Any action required by this Act to be taken at a meeting of the shareholders of a corporation, or any action which may be taken at a meeting of the shareholders, may be taken without a meeting if a consent in writing, setting forth the action so taken, shall be signed by all of the shareholders entitled to vote with respect to the subject matter thereof.

Such consent shall have the same effect as a unanimous vote of shareholders, and may be stated as such in any articles or document filed with the Secretary of State under this Act.

§ 146. Unauthorized Assumption of Corporate Powers

All persons who assume to act as a corporation without authority so to do shall be jointly and severally liable for all debts and liabilities incurred or arising as a result thereof.

§ 147. Application to Existing Corporations

The provisions of this Act shall apply to all existing corporations organized under any general act of this State providing for the organization of corporations for a purpose or purposes for which a corporation might be organized under this Act, where the power has been reserved to amend, repeal or modify the act under which such corporation was organized and where such act is repealed by this Act.

§ 148. Application to Foreign and Interstate Commerce

The provisions of this Act shall apply to commerce with foreign nations and among the several states only insofar as the same may be permitted under the provisions of the Constitution of the United States.

§ 149. Reservation of Power

The ....* shall at all times have power to prescribe such regulations, provisions and limitations as it may deem advisable, which regulations, provisions and limitations shall be binding upon any and all corporations subject to the provisions of this Act, and the ....* shall have power to amend, repeal or modify this Act at pleasure.

*Insert name of legislative body.

§ 150. Effect of Repeal of Prior Acts

The repeal of a prior act by this Act shall not affect any right accrued or established, or any liability or penalty incurred, under the provisions of such act, prior to the repeal thereof.

§ 151. Effect of Invalidity of Part of this Act

If a court of competent jurisdiction shall adjudge to be invalid or unconstitutional any clause, sentence, paragraph, section or part of this Act, such judgment or decree shall not affect, impair, invalidate or nullify the remainder of this Act, but the effect thereof shall be confined to the clause, sentence, paragraph, section or part of this Act so adjudged to be invalid or unconstitutional.

§ 152. Exclusivity of Certain Provisions [Optional]

In circumstances to which section 45 and related sections of this Act are applicable, such provisions supersede the applicability of any other statutes of this state with respect to the legality of distributions.

§ 153. Repeal of Prior Acts

(insert appropriate provisions)..........

APPENDIX **E**

MODEL PROFESSIONAL CORPORATION ACT

Sec. 1 Short Title.
Sec. 2 Definitions.
Sec. 3 Purposes.
Sec. 4 Prohibited Activities.
Sec. 5 General Powers.
Sec. 6 Rendering Professional Services.
Sec. 7 Right of Corporation to Acquire Its Own Shares.
Sec. 8 Corporate Name.
Sec. 9 Issuance and Transfer or Shares; Share Certificates.
Sec. 10 Death or Disqualification of a Shareholder.
Sec. 11 Responsibilities for Professional Services.
Sec. 12 Professional Relationships; Privileged Communications.
Sec. 13 Voting of Shares.
Sec. 14 Directors and Officers.
Sec. 15 Amendments to Articles of Incorporation.
Sec. 16 Merger and Consolidation.
Sec. 17 Termination of Professional Activities.
Sec. 18 Involuntary Dissolution.
Sec. 19 Admission of Foreign Professional Corporations.
Sec. 20 Application for Certificate of Authority.
Sec. 21 Revocation of Certificate of Authority.
Sec. 22 Annual Report of Domestic and Foreign Professional Corporations.
Sec. 23 Annual Statement of Qualification of Domestic and Foreign Professional Corporations.
Sec. 24 Interrogatories by Licensing Authority.
Sec. 25 Penalties.
Sec. 26 Regulation of Professional Corporations.
Sec. 27 Application of Business Corporation Act.
Sec. 28 Application to Existing Corporations.
Sec. 29 Reservation of Power.
Sec. 30 Effect of Repeal of Prior Acts.
Sec. 31 Effect of Invalidity of Part of this Act.
Sec. 32 Repeal of Prior Acts.

Sec. 1. **Short Title.**

This Act shall be known and may be cited as the "_____Professional Corporation Act." [1]

Sec. 2. **Definitions.**

As used in this Act, unless the context otherwise requires, the term:

(1) "Professional service" means any service which may lawfully be rendered only by persons licensed under the provisions of a licensing law of this State and may not lawfully be rendered by a corporation organized under the _____ Business Corporation Act.

(2) "Licensing authority" means the officer, board, agency, court or other authority in this State which has the power to issue a license or other legal authorization to render a professional service.

(3) "Professional corporation" or "domestic professional corporation" means a corporation for profit subject to the provisions of this Act, except a foreign professional corporation.

(4) "Foreign professional corporation" means a corporation for profit organized for the purpose of rendering professional services under a law other than the law of this State.

(5) "Qualified person" means a natural person, general partnership, or professional corporation[2] which is eligible under this Act to own shares issued by a professional corporation.

(6) "Disqualified person" means any natural person, corporation, partnership, fiduciary, trust, association, government agency, or other entity which for any reason is or becomes ineligible under this Act to own shares issued by a professional corporation.

Sec. 3. **Purposes.**

(a) Except as hereinafter provided in this section professional corporations may be organized under this Act only for the purpose of rendering professional services and services ancillary thereto within a single profession.

(b) A professional corporation may be incorporated for the purpose of rendering professional services within two or more professions and for any purpose or purposes for which corporations may be organized under the _____ Business Corporation Act to the extent that such combination of professional purposes or of professional and business purposes is permitted by the licensing laws of this State applicable to such professions and rules or regulations thereunder.

Sec. 4. **Prohibited Activities**

A professional corporation shall not engage in any profession or business other than the profession or professions and businesses permitted by its articles of incorporation, except that a professional corporation may invest its funds in real estate, mortgages, stocks, bonds or any other type of investment.

Sec. 5. **General Powers**

A professional corporation shall have the powers enumerated in the _____ Business Corporation Act, except that a professional corporation may be a promoter, general partner, member, associate, or manager only of a partnership, joint venture, trust or other enterprise engaged only in rendering professional services or carrying on business permitted by the articles of incorporation of the corporation.

Sec. 6. **Rendering Professional Services**

A professional corporation, domestic or foreign, may render professional services in this State only through natural persons permitted to render such services in this State; but nothing in this Act shall be construed to require that any person who is employed by a professional corporation be licensed to perform services for which no license is otherwise required or to prohibit the rendering of professional services by a licensed natural person acting in his individual capacity, notwithstanding such person may be a shareholder, director, officer, employee or agent of a professional corporation, domestic or foreign.

Sec. 7. **Right of Corporation to Acquire Its Own Shares**

A professional corporation may purchase its own shares from a disqualified person without regard to the availability of capital or surplus for such

[1]Supply name of State as required throughout the act.

[2]Delete "professional corporation" if alternate 2 or 3 of Section 11(d) is adopted.

purchase; however, no purchase of or payment for its own shares shall be made at a time when the corporation is insolvent or when such purchase or payment would make it insolvent.

Sec. 8. **Corporate Name**

The name of a domestic professional corporation or of a foreign professional corporation authorized to transact business in this State:

(1) shall contain the words "professional corporation" or the abbreviation "P.C.";

(2) shall not contain any word or phrase which indicates or implies that it is organized for any purpose other than the purposes contained in its articles of incorporation;

(3) shall not be the same as, or deceptively similar to, the name of any domestic corporation existing under the laws of this State or any foreign corporation authorized to transact business in this State, or a name the exclusive right to which is, at the time, reserved in the manner provided in the _____ Business Corporation Act, or the name of a corporation which has in effect a registration of its corporate name as provided in the _____ Business Corporation Act; except that this provision shall not apply if:

(i) such similarity results from the use in the corporate name of personal names of its shareholders or former shareholders or of natural persons who were associated with a predecessor entity; or

(ii) the applicant files with the Secretary of State either of the following: (A) the written consent of such other corporation or holder of a reserved or registered name to use the same or deceptively similar name and one or more words are added to make such name distinguishable from such other name, or (B) a certified copy of a final decree of a court of competent jurisdiction establishing the prior right of the applicant to the use of such name in this State; and

(4) shall otherwise conform to any rule promulgated by a licensing authority having jurisdiction of a professional service described in the articles of incorporation of such corporation.

Sec. 9. **Issuance and Transfer of Shares; Share Certificates**

(a) A professional corporation may issue shares, fractional shares, and rights or options to purchase shares only to:

(1) natural persons who are authorized by law in this State or in any other state or territory of the United States or the District of Columbia to render a professional service permitted by the articles of incorporation of the corporation;

(2) general partnerships in which all the partners are qualified persons with respect to such professional corporation and in which at least one partner is authorized by law in this State to render a professional service permitted by the articles of incorporation of the corporation; and

(3) professional corporations, domestic or foreign, authorized by law in this State to render a professional service permitted by the articles of incorporation of the corporation.[3]

(b) Where deemed necessary by the licensing authority for any professional in order to prevent violations of the ethical standards of such profession, the licensing authority may by rule further restrict, condition, or abridge the authority of professional corporations to issue shares but no such rule shall, of itself, have the effect of causing a shareholder of a professional corporation at the time such rule becomes effective to become a disqualified person. All shares issued in violation of this section or any rule hereunder shall be void.

(c) A shareholder of a professional corporation may transfer or pledge shares, fractional shares, and rights or options to purchase shares of the corporation only to natural persons, general partnerships and professional corporations[4] qualified hereunder to hold shares issued directly to them by such professional corporation. Any transfer of shares in violation of this provision shall be void; however, nothing herein contained shall prohibit the transfer of shares of a professional corporation by operation or law or court decree.

[3]Delete paragraph (3) of subsection (a) if 2 or 3 of Section 11(d) is adopted.

[4]Delete "professional corporation" if alternate 2 or 3 of Section 11(d) is adopted.

(d) Every certificate representing shares of a professional corporation shall state conspicuously upon its face that the shares represented thereby are subject to restrictions on transfer imposed by this Act and are subject to such further restrictions on transfer as may be imposed by the licensing authority from time to time pursuant to this Act.

Sec. 10. **Death or Disqualification of a Shareholder**

(a) Upon the death of a shareholder of a professional corporation or if a shareholder of a professional corporation becomes a disqualified person or if shares of a professional corporation are transferred by operation of law or court decree to a disqualified person, the shares of such deceased shareholder or of such disqualified person may be transferred to a qualified person and, if not so transferred, shall be purchased or redeemed by the corporation to the extent of funds which may be legally made available for such purchase.

(b) If the price for such shares is not fixed by the articles of incorporation or by-laws of the corporation or by private agreement, the corporation, within six months after such death or thirty days after such disqualification or transfer, as the case may be, shall make a written offer to pay for such shares at a specified price deemed by such corporation to be the fair value thereof as of the date of such death, disqualification or transfer. Such offer shall be given to the executor or administrator of the estate of a deceased shareholder or to the disqualified shareholder or transferee and shall be accompanied by a balance sheet of the corporation, as of the latest available date and not more than twelve months prior to the making of such offer, and a profit and loss statement of such corporation for the twelve months' period ended on the date of such balance sheet.

(c) If within thirty days after the date of such written offer from the corporation the fair value of such shares is agreed upon between such disqualified person and the corporation, payment therefor shall be made within sixty days, or such other period as the parties may fix by agreement, after the date of such offer, upon surrender of the certificate or certificates representing such shares. Upon payment of the agreed value the disqualified persons shall cease to have any interest in such shares.

(d) If within such period of thirty days the disqualified person and the corporation do not so agree, then the corporation, within thirty days after receipt of written demand from the disqualified person given within sixty days after the date of the corporation's written offer shall, or at its election at any time within such period of sixty days may, file a petition in any court of competent jurisdiction in the county in this State where the registered office of the corporation is located requesting that the fair value of such shares be found and determined. If the corporation shall fail to institute the proceeding as herein provided, the disqualified person may do so within sixty days after delivery of such written demand to the corporation. The disqualified person, wherever residing, shall be made a party to the proceeding as an action against his shares quasi in rem. A copy of the petition shall be served on the disqualified person, if a resident of this State, and shall be served by registered or certified mail on the disqualified person, if a nonresident. Service on nonresidents shall also be made by publication as provided by law. The jurisdiction of the court shall be plenary and exclusive. The disqualified person shall be entitled to judgment against the corporation for the amount of the fair value of his shares as of the date of death, disqualification or transfer upon surrender to the corporation of the certificate or certificates representing such shares. The court may, at its discretion, order that the judgment be paid in such installments as the court may determine. The court may, if it so elects, appoint one or more persons as appraisers to receive evidence and recommend a decision on the question of fair value. The appraisers shall have such power and authority as shall be specified in the order of their appointment or an amendment thereof.

(e) The judgment shall include an allowance for interest at such rate as the court may find to be fair and equitable in all the circumstances, from the date of death, disqualification or transfer.

(f) The costs and expenses of any such proceeding shall be determined by the court and shall be assessed against the corporation, but all or any part of such costs and expenses may be apportioned and assessed as the court may deem equitable against the disqualified person if the court shall find that the action of such disqualified person in failing to accept such offer was arbitrary or vexatious or not in good faith. Such expenses shall include reasonable compensation for and reasonable expenses of the appraisers, but shall exclude the fees and expenses of counsel for and experts employed by any party; but if the fair value of the shares as determined materially exceeds the amount which the corporation offered to pay therefor, or if no offer was made, the court in its discretion may award to the disqualified person such sum as the court may determine to be reasonable compensation to any expert or experts employed by the disqualified person in the proceeding.

(g) If a purchase, redemption, or transfer of the shares of a deceased or disqualified shareholder or of a transferee who is a disqualified person is not completed within ten months after the death of the deceased shareholder or five months after the disqualification or transfer, as the case may be, the corporation shall forthwith cancel the shares on its books and the disqualified person shall have no further interest as a shareholder in the corporation other than his right to payment for such shares under this section.

(h) Shares acquired by a corporation pursuant to payment of the agreed value therefor or to payment of the judgment entered therefor, as in this section provided, may be held and disposed of by such corporation as in the case of other treasury shares.

(i) This section shall not be deemed to require the purchase of shares of a disqualified person where the period of such disqualification is for less than five months from the date of disqualification or transfer.

(j) Any provision regarding purchase, redemption or transfer of shares of a professional corporation contained in the articles of incorporation, by-laws or any private agreement shall be specifically enforceable in the courts of this State.

(k) Nothing herein contained shall prevent or relieve a professional corporation from paying pension benefits or other deferred compensation for services rendered to or on behalf of a former shareholder as otherwise permitted by law.

Sec. 11. **Responsibilities for Professional Services**

(a) Any reference to a corporation in this section shall include both domestic and foreign corporations.

(b) Every individual who renders professional services as an employee of a professional corporation shall be liable for any negligent or wrongful act or omission in which he personally participates to the same extent as if he rendered such services as a sole practitioner. An employee of a professional corporation shall not be liable for the conduct of other employees unless he is at fault in appointing, supervising, or cooperating with them.

(c) Every corporation whose employees perform professional services within the scope of their employment or of their apparent authority to act for the corporation shall be liable to the same extent as its employees.

(d) (Alternate 1) Except as otherwise provided by statute, the personal liability of a shareholder of a professional corporation shall be no greater in any respect than that of a shareholder of a corporation organized under the _____ Business Corporation Act.

(d) (Alternate 2) Except as otherwise provided by statute, if any corporation is liable under the provisions of subsection (c) of this section, every shareholder of the corporation shall be liable to the same extent as though he were a partner in a partnership and the services giving rise to liability had been rendered on behalf of the partnership.

(d) (Alternate 3) (1) Except as otherwise provided by statute, if any corporation is liable under the provisions of subsection (c) of this section, every shareholder of that corporation shall be liable to the same extent as though he were a partner in a partnership and the services giving rise to liability had been rendered on behalf of the partnership, unless the corporation has provided security for professional responsibility as provided in paragraph (2) of this subsection and the liability is satisfied to the extent contemplated by the insurance or bond which effectuates the security.

(2) A professional corporation, domestic or foreign, may provide security for professional responsibility by procuring insurance or a surety bond issued by an insurance company, or a combination thereof, as the corporation may elect. The minimum amount of security and requirements as to the form and coverage provided by the insurance policy or surety bond may be established for each profession by the licensing authority for the profession, and the minimum amount may be set to vary with the number of shareholders, the type of practice, or other variables deemed appropriate by the licensing authority. If no effective determination by the licensing authority is in effect, the minimum amount of professional responsibility security for the professional corporation shall be the product of ______[5] dollars multiplied by the number of shareholders of the professional corporation.

Sec. 12. **Professional Relationships; Privileged Communications**

(a) The relationship between an individual performing professional services as employee of a professional corporation, domestic or foreign, and a client or patient shall be the same as if the individual performed such services as a sole practitioner.

(b) The relationship between a professional corporation, domestic or foreign, performing professional services and the client or patient shall be the same as between the client or patient and the individual performing the services.

(c) Any privilege applicable to communications between a person rendering professional services and the person receiving such services recognized under the laws of this State, whether statutory or deriving from common law, shall remain inviolate and shall extend to a professional corporation, domestic or foreign, and its employees in all cases in which it shall be applicable to communications between a natural person rendering professional services on behalf of the corporation and the person receiving such services.

Sec. 13. **Voting of Shares**

No proxy for shares of a professional corporation shall be valid unless it shall be given to a qualified person. A voting trust with respect to shares of a professional corporation shall not be valid [unless all the trustees and beneficiaries thereof are qualified persons, except that a voting trust may be validly continued for a period of ten months after the death of a deceased beneficiary or for a period of five months after a beneficiary has become a disqualified person].[6]

Sec. 14. **Directors and Officers**

Not less than one-half the directors of a professional corporation and all the officers other than the secretary and the treasurer shall be qualified persons with respect to the corporation.

Sec. 15. **Amendments to Articles of Incorporation**

An administrator, executor, guardian, conservator, or receiver of the estate of a shareholder of a professional corporation who holds all of the outstanding shares of the corporation may amend the articles of incorporation by signing a written consent to such amendment. Articles of amendment so adopted shall be executed in duplicate by the corporation by such administrator, executor, guardian, conservator, or receiver and by the secretary or assistant secretary of the corporation, and verified by one of the persons signing such articles, and shall set forth:

(1) the name of the corporation;

(2) the amendments so adopted;

(3) the date of adoption of the amendment by the administrator, executor, guardian, conservator, or receiver;

(4) the number of shares outstanding; and

(5) the number of shares held by the administrator, executor, guardian, conservator, or receiver.

Sec. 16. **Merger and Consolidation**

(a) A professional corporation may merge or consolidate with another corporation, domestic or foreign, only if every shareholder of each corpo-

[5]A minimum amount to be determined by state legislature.

[6]Delete the bracketed clause if alternate 2 or 3 of Section 11(d) is adopted.

ration is qualified to be a shareholder of the surviving or new corporation.

(b) Upon the merger or consolidation of a professional corporation, if the surviving or new corporation, as the case may be, is to render professional services in this state, it shall comply with the provisions of this Act.

Sec. 17. **Termination of Professional Activities**

If a professional corporation shall cease to render professional services, it shall amend its articles of incorporation to delete from its stated purposes the rendering of professional services and to conform to the requirements of the _____ Business Corporation Act regarding its corporate name. The corporation may then continue in existence as a corporation under the _____ Business Corporation Act and shall no longer be subject to the provisions of this Act.

Sec. 18. **Involuntary Dissolution**

A professional corporation may be dissolved involuntarily by a decree of the _____ Court in an action filed by the Attorney General when it is established that the corporation has failed to comply with any provision of this Act applicable to it within sixty days after receipt of written notice of noncompliance. Each licensing authority in this State and the Secretary of State shall certify to the Attorney General, from time to time, the names of all corporations which have given cause for dissolution as provided in this Act, together with the facts pertinent thereto. Whenever the Secretary of State or any licensing authority shall certify the name of a corporation to the Attorney General as having given any cause for dissolution, the Secretary of State or such licensing authority, as the case may be, shall concurrently mail to the corporation at its registered office a notice that such certification has been made. Upon the receipt of such certification, the Attorney General shall file an action in the name of the State against such corporation for its dissolution.

Sec. 19. **Admission of Foreign Professional Corporations**

(a) A foreign professional corporation shall be entitled to procure a certificate of authority to transact business in this State only if:

(1) the name of the corporation meets the requirements of this Act:

(2) the corporation is organized only for purposes for which a professional corporation organized under this Act may be organized; and

(3) all the shareholders, not less than one-half the directors, and all the officers other than the secretary and treasurer of the corporation are qualified persons with respect to the corporation.

(b) No foreign professional corporation shall be required to obtain a certificate of authority to transact business in this State unless it shall maintain an office in this State for the conduct of business or professional practice.

Sec. 20. **Application for Certificate of Authority**

The application of a foreign professional corporation for a certificate of authority for the purpose of rendering professional services shall include a statement that all the shareholders, not less than one-half the directors, and all the officers other than the secretary and treasurer are licensed in one or more states or territories of the United States or the District of Columbia to render a professional service described in the statement of purposes of the corporation.

Sec. 21. **Revocation of Certificate of Authority**

The certificate of authority of a foreign professional corporation may be revoked by the Secretary of State if the corporation fails to comply with any provision of this Act applicable to it. Each licensing authority in this State shall certify to the Secretary of State, from time to time, the names of all foreign professional corporations which have given cause for revocation as provided in this Act, together with the facts pertinent thereto. Whenever a licensing authority shall certify the name of a corporation to the Secretary of State as having given cause for dissolution, the licensing authority shall concurrently mail to the corporation at its registered office in this State a notice that such certification has been made. No certificate of authority of a foreign professional corporation shall be revoked by the Secretary of State unless he shall have given the corporation not less than sixty days' notice thereof and the corporation shall fail prior to revocation to correct such noncompliance.

Sec. 22. **Annual Report of Domestic and Foreign Professional Corporations**

(a) The annual report of each domestic professional corporation, and each foreign professional corporation authorized to transact business in this State, filed with the Secretary of State pursuant to the _____ Business Corporation Act shall include a statement that all the shareholders, not less than one-half the directors, and all the officers other than the secretary and treasurer of the corporation are qualified persons with respect to the corporation.

(b) Financial information contained in the annual report of a professional corporation, other than the amount of stated capital of the corporation, shall not be open to public inspection nor shall the licensing authority disclose any facts or information obtained therefrom except insofar as its official duty may require the same to be made public or in the event such information is required for evidence in any criminal proceedings or in any other action by this State.

Sec. 23. **Annual Statement of Qualifications of Domestic and Foreign Professional Corporations**

(a) Each domestic professional corporation, and each foreign professional corporation, authorized to transact business in this State, shall file annually before March 1 with each licensing authority having jurisdiction over a professional service of a type described in its articles of incorporation a statement of qualification setting forth the names and respective addresses of the directors and officers of the corporation and such additional information as the licensing authority may by rule prescribe as appropriate in determining whether such corporation is complying with the provisions of this Act and rules promulgated hereunder.

(b) The licensing authority shall charge and collect a fee of _____ dollars for filing a statement of qualification pursuant to this Act.

Sec. 24. **Interrogatories by Licensing Authority**

(a) Each licensing authority of this State may propound to any professional corporation, domestic or foreign, organized to practice a profession within the jurisdiction of such licensing authority, and to any officer or director thereof, such interrogatories as may be reasonably necessary and proper to enable the licensing authority to ascertain whether such corporation has complied with all the provisions of this Act applicable to such corporation. Such interrogatories shall be answered within thirty days after the mailing thereof, or within such additional time as shall be fixed by the licensing authority, and the answers thereto shall be full and complete and shall be made in writing and under oath. If such interrogatories be directed to an individual they shall be answered by him, and if directed to a corporation they shall be answered by the president, vice president, secretary or assitant secretary thereof. The licensing authority shall certify to the Attorney General, for such action as the Attorney General may deem appropriate, all interrogatories and answers thereto which disclose a violation of any of the provisions of this Act.

(b) Interrogatories propounded by a licensing authority and the answers thereto shall not be open to public inspection nor shall the licensing authority disclose any facts of information obtained therefrom except insofar as its official duty may require the same to be made public or in the event such interrogatories or the answers thereto are required for evidence in any criminal proceedings or in any other action by this State.

Sec. 25. **Penalties**

(a) Each professional corporation, domestic or foreign, that fails or refuses to answer truthfully within the time prescribed by this Act interrogatories propounded in accordance with the provisions of this Act by the licensing authority having jurisdiction of a type of professional service described in the articles of incorporation of such corporation, shall be deemed to be guilty of a misdemeanor and upon conviction thereof may be fined in any amount not exceeding five hundred dollars.

(b) Each officer and director of a professional corporation, domestic or foreign, who fails or refuses within the time prescribed by this Act to answer truthfully and fully interrogatories propounded to him in accordance with the provisions of this Act by the licensing authority having jurisdiction of a type of professional service described in the articles of incorporation of such corporation, or who signs any articles, statement, report, application or other document filed with such licensing authority which is known to such officer or director to be false in any material respect, shall be deemed to be guilty of a misdemeanor, and upon conviction thereof may be fined in any amount not exceeding _____ dollars.

Sec. 26. **Regulation of Professional Corporations**

No professional corporation, domestic or foreign, shall begin to render professional services in this State until it has filed a copy of its articles of incorporation with each licensing authority having jurisdiction of a type of professional service described in its articles cf incorporation. Each licensing authority in this State is hereby authorized to promulgate rules in accordance with the provisions of this Act which specifically provide for the issuance of rules to the extent consistent with the public interest or required by the public health or welfare or by generally recognized standards of professional conduct. Nothing in this Act shall restrict or limit in any manner the authority or duty of a licensing authority with respect to natural persons rendering a professional service within the jurisdiction of the licensing authority, or any law, rule or regulation pertaining to standards of professional conduct.

Sec. 27. **Application of Business Corporation Act**

The provisions of the _____ Business Corporation Act shall apply to professional corporations, domestic and foreign, except to the extent such provisions are inconsistent with the provisions of this Act.

Sec. 28. **Application to Existing Corporations**

(a) The provisions of this Act shall apply to all existing corporations organized under any general act of this State which is repealed by this Act. Every such existing corporation which shall be required to amend its corporate name or purposes to comply with this Act shall deliver duly executed duplicate originals of articles of amendment or restated articles of incorporation containing such amendments to the Secretary of State within ninety days after the effective date of this Act.

(b) Any corporation organized under any act of this State which is not repealed hereby may become subject to the provisions of this Act by delivering to the Secretary of State duly executed duplicate originals of articles of amendment or restated articles of incorporation stating that the corporation elects to become subject to this Act and containing such amendment of its corporate name or purposes as may be required to comply with this Act.

(c) The provisions of this Act shall not apply to any corporation now in existence or hereafter organized under any act of this State which is not repealed hereby unless such corporation voluntarily becomes subject to this Act as herein provided, and nothing contained in this Act shall alter or affect any existing or future right or privilege permitting or not prohibiting performance of professional services through the use of any other form of business organization.

Sec. 29. **Reservation of Power**

The _____[7] shall at all times have power to prescribe such regulations, provisions and limitations as it may deem advisable, which regulations, provisions and limitations shall be binding upon any and all corporations subject to the provisions of this Act, and the _____[8] shall have power to amend, repeal or modify this Act at pleasure.

Sec. 30. **Effect of Repeal or Prior Acts**

The repeal of a prior act by this Act shall not affect any right accrued or established, or any liability or penalty incurred, under the provisions of such act, prior to the repeal thereof.

Sec. 31. **Effect of Invalidity of Part of this Act**

If a court of competent jurisdiction shall adjudge to be invalid or unconstitutional any clause, sen-

[7]Insert name of legislative body.

[8]Insert name of legislative body.

tence, paragraph, section or part of this Act, such judgment or decree shall not affect, impair, invalidate or nullify the remainder of this Act, but the effect thereof shall be confined to the clause, sentence, paragraph, section or part of this Act so adjudged to be invalid or unconstitutional.

Sec. 32. **Repeal of Prior Acts**

[*Insert appropriate provisions*]

APPENDIX *F*

FORMS

FORM 1A

TRADE NAME AFFIDAVIT (COLORADO)

Recorded at..............................o'clock...........M., ...

Reception No.. ....Recorder.

STATE OF COLORADO, }ss.
......................County of ....}

....of the

..............................County of..................................., in the State of

Colorado, ...

....being first duly

sworn, upon oath deposes and says that...

....

is the name under which a business or trade is being carried on at...

in the......................................County of..., and State of Colorado.

That the full Christian and surname and address of all the persons who are represented by the said name of..is as follows, to wit:

....

....

....

....

....

....

....

That the affiant..............................the person..........carrying on said business or trade under the name or style aforesaid.

....

....

Subscribed and sworn to before me, this..............day of..., 19.......

My commission expires.., 19.......

Witness my hand and official seal.

....
Notary Public.

NOTE—All co-partnerships and every person doing business otherwise than in his own full name should make this affidavit, which must be filed in the county in which the firm carries on its trade or business, and must be refiled whenever there is any change in the membership of the firm; and no suit can be prosecuted by such firm for the collection of any debts until such affidavit is filed.

No. 298. TRADE NAME AFFIDAVIT. Bradford Publishing Co., 1824-46 Stout Street, Denver, Colorado 2-73

No...

TRADE NAME AFFIDAVIT

OF

....

....

....

....

Trading or doing business under the name of

....

....

at ....

......................County of..
State of Colorado.

STATE OF COLORADO
..............................County of.................................. } ss.

Office of County Clerk and Recorder

I hereby certify that the within instrument was filed for record in my office at........o'clock........M.,

...., 19........,

and was duly recorded in book..........., page............

....
Recorder.

By ....
Deputy.

Fees, $.......................................

BRADFORD PUBLISHING CO., DENVER

FORM 1B

FICTITIOUS NAME STATEMENT (CALIFORNIA)

Original Copy for Filing with County Clerk of ______________________________ County

FICTITIOUS BUSINESS NAME STATEMENT

The following person (persons) is (are) doing business as:

(*) ______________________________
(FICTITIOUS BUSINESS NAME)

at (**) ______________________________

____

(***) 1. ______________________________
(FULL NAME · TYPE/PRINT)

____
(ADDRESS)

____
(CITY)

2. ______________________________
(FULL NAME · TYPE/PRINT)

____
(ADDRESS)

____
(CITY)

3. ______________________________
(FULL NAME · TYPE/PRINT)

____
(ADDRESS)

____
(CITY)

4. ______________________________
(FULL NAME · TYPE/PRINT)

____
(ADDRESS)

____
(CITY)

(****) This business is conducted by ______________________________
(i) "an individual," (ii) "a general partnership," (iii) "a limited partnership," (iv) "an unincorporated association other than a partnership," (v) "a corporation," (vi) "a business trust."

Signed ______________________________

Signature Must Also Be Typed or Printed ______________________________

This statement was filed with the County Clerk of ______________ County on ______________ (Date)

Attorney or Bank or Agent

Name ______________

Address ______________

City ______________

Telephone ______________

FILE NO. ______________

Statutory Filing Fee — $10.00
(Includes one Certification—See Page 3)

Statement Expires 5 years from Dec. 31 of year in which filed and must be Renewed then with a new Statement

- -

THE BELOW INSTRUCTIONS ARE NOT TO BE PUBLISHED (Sec. 17924 B&P)

INSTRUCTIONS FOR COMPLETION OF STATEMENT
Section 17913 Business & Professions Code

(*) **The Fictitious Name under which business is being conducted.**

(**) If the registrant has a place of business in this State, insert the street address of his principal place of business in this State. If the registrant has no place of business in this State, insert the street address of his principal place of business outside this State.

(***) If the registrant is an individual, insert his full name and residence address. If the registrant is a partnership or other association of persons, insert the full name and residence address of *each general* partner. If the registrant is a business trust, insert the full name and residence address of each trustee. If the registrant is a corporation, insert the name of the corporation as set forth in its articles of incorporation and the State of incorporation. (Attach additional sheet if necessary.)

(****) Insert whichever of the following best describes the nature of the business: "an individual", "a general partnership", "a limited partnership", "an unincorporated association other than a partnership", "a corporation", or a "business trust."

A FICTITIOUS BUSINESS NAME STATEMENT EXPIRES AT THE END OF FIVE YEARS FROM DECEMBER 31 OF THE YEAR IN WHICH IT WAS FILED. Except as provided in Section 17923, B&P Code, it expires ***40 days after any change in the facts set forth in the statement;*** except that a change in the residence address of an individual, general partner, or trustee does not cause the statement to expire.

The statement expires upon the filing of a statement of abandonment.

NOTICE TO REGISTRANT - Section 17924 Business & Professions Code

(1) Your fictitious business name statement must be published in a newspaper once a week for four successive weeks and an affidavit of publication filed with the county clerk within 30 days after publication has been accomplished. The statement should be published in a newspaper of general circulation in the county where the principal place of business is located. The statement should be published in such county in a newspaper that circulates in the area where the business is to be conducted (Sec. 17917 B&P Code)

(2) Any person who executes, files, or publishes any fictitious business name statement, knowing that such statement is false, in whole or in part, is guilty of a misdemeanor and upon conviction thereof shall be fined not to exceed five hundred dollars ($500) (Sec. 17930 B&P Code).

THE LAW...

(1) Where the asterisk (*) appears in the form, insert the fictitious business name.

(2) Where the two asterisks (**) appear in the form: If the registrant has a place of business in this state, insert the street address of his principal place of business in this state. If the registrant has no place of business in this state, insert the street address of his principal place of business outside this state.

(3) Where the three asterisks (***) appear in the form: If the registrant is an individual, insert his full name and residence address. If the registrant is a partnership or other association of persons, insert the full name and residence address of each general partner. If the registrant is a business trust, insert the full name and residence address of each trustee. If the registrant is a corporation, insert the name of the corporation as set out in its articles of incorporation and the state of incorporation.

(4) Where the four asterisks (****) appear in the form, insert whichever of the following best describes the nature of the business: (i) "an individual," (ii) "a general partnership," (iii) "a limited partnership," (iv) "an unincorporated association other than a partnership," (v) "a corporation," (vi) "a business trust."

17914. If the registrant is an individual, the statement shall be signed by the individual; if a partnership or other association of persons, by a general partner; if a business trust, by a trustee; if a corporation, by an officer.

17915. The fictitious business name statement shall be filed with the clerk of the county in which the registrant has his principal place of business in this state or, if he has no place of business in this state, with the Clerk of Sacramento County.

17916. Presentation for filing of a fictitious business name statement and one copy, tender of the filing fee, and acceptance of the statement by the county clerk constitute filing under this chapter. The county clerk shall note on the copy the file number and the date of filing the original and shall certify and deliver or send the copy to the registrant.

For Laws on Publication See Reverse Side of Attached Duplicate (Pink) Copy of This Form.

FORM 1C

PUBLICATION OF FICTITIOUS NAME (CALIFORNIA)

Duplicate Copy for Publication in the ______________________________

FICTITIOUS BUSINESS NAME STATEMENT

The following person (persons) is (are) doing business as:

(*) ______________________________
(FICTITIOUS BUSINESS NAME)

at (**) ______________________________

____

(***) 1. ______________________________
(FULL NAME · TYPE/PRINT)

____
(ADDRESS)

____
(CITY)

2. ______________________________
(FULL NAME · TYPE/PRINT)

____
(ADDRESS)

____
(CITY)

3. ______________________________
(FULL NAME · TYPE/PRINT)

____
(ADDRESS)

____
(CITY)

4. ______________________________
(FULL NAME · TYPE/PRINT)

____
(ADDRESS)

____
(CITY)

(****) This business is conducted by ______________________________

(i) "an individual," (ii) "a general partnership," (iii) "a limited partnership," (iv) "an unincorporated association other than a partnership," (v) "a corporation," (vi) "a business trust."

Signed ______________________________

Signature Must Also Be Typed or Printed ______________________________

This statement was filed with the County Clerk of ______________ County on ______________ (Date)

Attorney or Bank or Agent

FILE NO. ______________

Name ______________________________

Address ______________________________

City ______________________________

Telephone ______________________________

PUBLISH ALL ITEMS IN DOUBLE BRACKETS ABOVE

See Reverse of This Form for Publication Instructions and Law.

for FREE FORMS—

PHONE (213) 625-2141

CALIFORNIA
NEWSPAPER SERVICE BUREAU, Inc.
ESTABLISHED 1934
210 South Spring Street
Los Angeles, Calif. 90012
Affiliated with The Los Angeles Daily Journal and the Sacramento Daily Recorder

THE BELOW INSTRUCTIONS ARE NOT TO BE PUBLISHED (Sec. 17924 B&P)

INSTRUCTIONS FOR COMPLETION OF STATEMENT
Section 17913 Business & Professions Code

(*) **The Fictitious Name under which business is being conducted.**

(**) If the registrant has a place of business in this State, insert the street address of his principal place of business in this State. If the registrant has no place of business in this State, insert the street address of his principal place of business outside this State.

(***) If the registrant is an individual, insert his full name and residence address. If the registrant is a partnership or other association of persons, insert the full name and residence address of *each general* partner. If the registrant is a business trust, insert the full name and residence address of each trustee. If the registrant is a corporation, insert the name of the corporation as set forth in its articles of incorporation and the State of incorporation. (Attach additional sheet if necessary.)

(****) Insert whichever of the following best describes the nature of the business: "an individual", "a general partnership", "a limited partnership", "an unincorporated association other than a partnership", "a corporation", or a "business trust."

A FICTITIOUS BUSINESS NAME STATEMENT EXPIRES AT THE END OF FIVE YEARS FROM DECEMBER 31 OF THE YEAR IN WHICH IT WAS FILED. Except as provided in Section 17923, B&P Code, it expires ***40 days after any change in the facts set forth in the statement;*** except that a change in the residence address of an individual, general partner, or trustee does not cause the statement to expire.

The statement expires upon the filing of a statement of abandonment.

NOTICE TO REGISTRANT - Section 17924 Business & Professions Code

(1) Your fictitious business name statement must be published in a newspaper once a week for four successive weeks and an affidavit of publication filed with the county clerk within 30 days after publication has been accomplished. The statement should be published in a newspaper of general circulation in the county where the principal place of business is located. The statement should be published in such county in a newspaper that circulates in the area where the business is to be conducted (Sec. 17917 B&P Code)

(2) Any person who executes, files, or publishes any fictitious business name statement, knowing that such statement is false, in whole or in part, is guilty of a misdemeanor and upon conviction thereof shall be fined not to exceed five hundred dollars ($500) (Sec. 17930 B&P Code).

THE LAW...

17915. The fictitious business name statement shall be filed with the clerk of the county in which the registrant has his principal place of business in this state or, if he has no place of business in this state, with the Clerk of Sacramento County.

17916. Presentation for filing of a fictitious business name statement and one copy, tender of the filing fee, and acceptance of the statement by the county clerk constitute filing under this chapter. The county clerk shall note on the copy the file number and the date of filing the original and shall certify and deliver or send the copy to the registrant.

17917. (a) Within 30 days after a fictitious business name statement has been filed pursuant to this chapter, the registrant shall cause a statement in the form prescribed by subdivision (a) of Section 17913 to be published pursuant to Government Code Section 6064 in a newspaper of general circulation in the county in which the principal place of business of the registrant is located or, if there is no such newspaper in that county, then in a newspaper of general circulation in an adjoining county. If the registrant does not have a place of business in this state, the notice shall be published in a newspaper of general circulation in Sacramento County.

(b) Subject to the requirements of subdivision (a), the newspaper selected for the publication of the statement should be one that circulates in the area where the business is to be conducted.

(c) Where a new statement is required because the prior statement has expired under subdivision (a) of Section 17920, the new statement need not be published unless there has been a change in the information required in the expired statement.

(d) An affidavit showing the publication of the statement shall be filed with the county clerk within 30 days after the completion of the publication.

FORM 1D

BUSINESS CERTIFICATE (NEW YORK)

201 —Certificate of Conducting Business under an Assumed Name For Individual b JULIUS BLUMBERG, INC., LAW BLANK PUBLISHERS 80 EXCHANGE PLACE AT BROADWAY, NEW YORK

Business Certificate

I HEREBY CERTIFY *that I am conducting or transacting business under the name or designation of*

at

City or Town of *County of* *State of New York.*

*My full name is**
and I reside at

I FURTHER CERTIFY *that I am the successor in interest to*

the person or persons heretofore using such name or names to carry on or conduct or transact business.

IN WITNESS WHEREOF, *I have this* *day of* *19* *, made and signed this certificate.*

....

* Print or type name.

* If under 21 years of age, state "I am .... years of age".

STATE OF NEW YORK
COUNTY OF } *ss.:*

On this *day of* *19* *, before me personally appeared*

to me known and known to me to be the individual described in and who executed the foregoing certificate, and he thereupon duly acknowledged to me that he executed the same.

FORM 1E

ABANDONMENT OF FICTITIOUS NAME (CALIFORNIA)

CERTIFICATE OF DISCONTINUANCE OF USE AND/OR ABANDONMENT OF FICTITIOUS NAME

THE UNDERSIGNED__________ (Do) (Does) hereby certify that, effective__________ (Date)

__________ (They) (He) (She) ceased to do business under the fictitious firm name of__________

__________ (Exact Name of Business Only)

at __________ (Number) __________ (Street) __________ (City) California,

which business was formerly composed of the following person___, whose name___ in full and place___ of residence

__________ (Is) (Are) as follows, to-wit:

____

____

____

____

____

____

Certificate for transaction of business under the above fictitious name, and affidavit of publication thereof, are on file in the office of the County Clerk of__________County, under the provisions of Section 2466 of the Civil Code.

WITNESS__________ (Our) (My) hand___ this __________day of__________, 19___.

Signatures:__________

____

__________, Atty(s). __________

__________ __________

__________ __________

County Clerk's File No.__________ __________

For publication in the proper newspaper and for filing of the affidavit of publication with the appropriate County Clerk please mail two copies of this notice to

THE LOS ANGELES DAILY JOURNAL
ESTABLISHED 1888
220 W. FIRST STREET, LOS ANGELES, CALIF. 90012
TELEPHONE: MADISON 5-2141

LOS ANGELES NEWSPAPER SERVICE BUREAU, INC.
STATEWIDE LEGAL ADVERTISING CLEARING HOUSE SINCE 1934
224 W. FIRST ST., LOS ANGELES, CALIF. 90012
TELEPHONE: MADISON 5-2541

5M—8/67—A-73-74

AMENDED LAW RELATING TO NOTICES OF CEASING TO DO BUSINESS UNDER FICTITIOUS NAME

CHAPTER 268

EFFECTIVE SEPT. 9, 1953

An act to amend Section 2469.1 of the Civil Code . . . . . . relating to notices and records of cessation of doing business under fictitious name. . . . . .

SECTION 1. Section 2469.1 is added to the Civil Code to read:

2469.1. Every person and every partnership transacting business in this state under a fictitious name, or designation not showing the names of the persons interested as partners in such business, who has filed a certificate and caused the publication thereof according to the provisions of this chapter, may, upon ceasing to do business under that name, file a certificate to that effect, stating the name in full and the place of residence of such person, and stating the names in full of all the members of such partnership and their places of residence. Such certificate shall be signed by the person therein referred to, or by one or more of the partners as the case may be.

Such certificate must be published pursuant to Section 6064 of the Government Code * in a newspaper published in the county, if there be one, and if there be none in such county, then in a newspaper in an adjoining county. An affidavit showing the publication of such certificate shall be filed with the county clerk within thirty days after the completion of such publication.

Sec. 2. Section 2470 of said code is amended to read:

2470. Every county clerk must keep a register of the names of firms and persons mentioned in the certificates filed with him pursuant to this article, entering in alphabetical order the name of every such person who does business under a ficitious name, and the fictitious name, and the name of every such partnership, and of each partner therein.

Upon the abandonment of the use of a fictitious name, the clerk shall enter the fact of abandonment in the register.

* **AMENDED IN 1961 TO PROVIDE:** "Such Certificate must be published pursuant to Government Code Section 6064."

Section 6064 says: "Publication of notice pursuant to this section shall be once a week for four successive weeks." Otherwise there has been no substantive change in the law.

FORM 2A

ASSIGNMENT OF PARTNER'S INTEREST IN FIRM

Know All Men by These Presents, that for and in consideration of the sum of One Dollar ($1) and other good and valuable considerations to me in hand paid, receipt of which is hereby acknowledged, I, ______, of ______, do hereby assign to ______, of ______, all of my right, title and interest in and to a certain agreement of partnership bearing date the ______ day of ______, 19__, made and entered into by and between ______, ______, and myself; and I do hereby authorize and direct ______ to account to and with ______ for all profits, issues and income arising under the partnership agreement in the same manner and with the same force and effect as if such accounting were had and made with me personally.

In Witness Whereof, I have hereunto set my hand and seal this ______ day of ______, 19__.

______________ [*Seal*] [1]

1. West's Modern Legal Forms § 6411. Other examples of an assignment of a partner's interest in the firm, and a form for consent of the other partners to an assignment appear in West's Modern Legal Forms §§ 6412–6415.

FORM 2B

PERSONAL NOTICE OF DISSOLUTION OF PARTNERSHIP

To: ______________ Date: ______

Please be advised that the partnership between A.B., C.D. and E.F. was dissolved on the ______ day of ______, 19__, and that E.F. is no longer a member of the firm. Your account, in the amount of $______, according to our books, will be settled with A.B. and C.D. who will continue the business under the firm name of B & D.

[*Signatures of the partners*]

FORM 2C

NOTICE OF DISSOLUTION OF PARTNERSHIP BY PUBLICATION

Notice is hereby given that the partnership between A.B., C.D. and E.F. was dissolved on the ______ day of ______, 19__, so far as relates to E.F. All debts due to the partnership, and those due by them, will be settled with and by the remaining partners who will continue the business under the firm name of B. & D.

[*Date*] [*Signatures of the partners*] [2]

2. West's Modern Legal Forms § 6446.

FORM 2D

PRELIMINARY NOTICE OF EXPULSION TO PARTNER

To [*Name and address of partner*]

We hereby give you notice that we propose to exercise the power given to us by paragraph ______ of the agreement of partnership, dated the ______ day of ______, 19__, under which we are now carrying on business in partnership with you, of terminating the partnership so far as you are concerned on the ground that you have acted in a manner inconsistent with the good faith observable between partners [*or* that you have been guilty of conduct such as would be a ground for an application to the court for a dissolution of the partnership].

In order to afford you an opportunty of explaining and, if possible, satisfying us that no good cause of complaint exists, we hereby invite you to attend a meeting of the partners, to be held at ______, on ______ next, at ______ o'clock.

If you are unable to attend such meeting, we must ask you to arrange for another meeting with us, to be held at an early date and in any case within [*one week*] from the date of this notice.

Dated ______, 19__. [*Signatures of partners*] [3]

3. West's Modern Legal Forms § 6439.

FORM 2E

NOTICE OF EXPULSION TO PARTNER

To [*Name and address of partner*]

Referring to our notice to you, dated the ______ day of ______, 19__, and to the meeting of the partners held pursuant to such notice on the ______ day of ______, 19__, [*or* and in view of the fact that you neither attended the meeting to which we invited you in such notice nor have taken any other steps to meet us or explain matters], we regret to inform you that we are unable to accept as satisfactory the explanations offered by you at such meeting after hearing from us exactly what was our cause of complaint against you, and accordingly, we hereby give you notice that in exercise of the power for this purpose given to us by paragraph ______ of the agreement of partnership, dated the ______ day of ______, 19__, under which we have heretofore carried on business in partnership with you, we hereby terminate the partnership so far as you are concerned as of the date of this notice on the ground generally that [*repeat the ground as stated in the preliminary notice and add*] and more particularly on the ground that [*state shortly the facts relied on as constituting the general ground previously stated*].

Dated ______, 19__. [*Signatures of partners*] [4]

4. West's Modern Legal Forms § 6440.

FORM 2F

COMPLEX PARTNERSHIP AGREEMENT

ARTICLE I. GENERAL PROVISIONS

A. Recitals.
B. Parties.
C. Purpose.
D. Firm Name.
E. Term.
F. Location of Principal Place of Business.

ARTICLE II. CAPITAL

A. Original Capital Contributed by Partners.
B. Annual Additional Contributions to Capital.
C. Reserve for Capital Expenditures; Other Reserves.
D. Annual Reimbursements on Contributions to Capital.

ARTICLE III. PROFITS AND LOSSES OF THE FIRM; PARTICIPATION OF PARTNERS THEREIN; DRAWINGS; BONUSES

A. Units of Participation in Profits and Losses by the Respective Partners.

B. Drawing Accounts of the Respective Partners and the Extent to Which Any are Guaranteed.

C. Reserve for Bonuses and Payments Therefrom.

ARTICLE IV. MEETINGS AND VOTING OF PARTNERS

A. Meetings of Partners; Voting at Such Meetings.

B. Percentage of Votes Required for Certain Partnership Meetings; Requirement of Recommendation of the Management Committee in Advance of Certain Partnership Decisions.

ARTICLE V. CHANGES AS TO PARTNERS

A. No Classes of Partners.

B. Addition of Partners.

C. Death or Permanent Disability of a Partner.
   1. Death.
   2. Permanent Disability.

D. Permanent Withdrawal of a Partner.
   1. Notice of Withdrawal and Effective Date of Withdrawal.
   2. Possible Termination of the Firm Superseding Withdrawal Notice.
   3. Partition with and Payments to the Withdrawing Partner.

E. Retirement of Partners; Gradual Steps toward Retirement; Retirement Plans for Partners.
   1. Retired Partners; Plans for their Compensation.
   2. When a Partner Retires.
   3. Gradual Steps toward Retirement.

F. Expulsion of a Partner.
   1. Expulsion for Cause.
   2. Effects of Expulsion for Cause.
   3. Expulsion without Determining Any Cause Therefor.
   4. Effects of Expulsion without Determining Any Cause Therefor.

G. Temporary Incapacity; Leave of Absence; Temporary Withdrawal; Vacations.
   1. Temporary Incapacity or Illness.
   2. Leave of Absence.
   3. Temporary Withdrawal.
   4. Vacations.

ARTICLE VI. DUTIES OF PARTNERS

A. Devotion to Duty.

B. Charging for Services.

ARTICLE VII. MANAGEMENT

A. Authority and Membership of the Management Committee.

B. Functioning of the Management Committee and Its Subcommittees.

C. Membership in Subcommittees of the Management Committee.

ARTICLE VIII. INSURANCE; INVESTMENTS

A. Life Insurance.
B. Other Insurance.
C. Investments.

ARTICLE IX. PROPERTIES AND RECORDS

A. Firm Properties.
B. Accounting Records.

ARTICLE X. TERMINATION AND LIQUIDATION OF FIRM

A. Termination of the Firm by Voluntary Vote or Otherwise.
B. Pending Employments on Termination.
C. Liquidation of Assets.
D. Prior Opportunity of Partners to Bid for Purchase of Assets Being Liquidated.
E. Distribution of Proceeds from Liquidation.

ARTICLE XI. LEGAL EFFECT OF PROVISIONS; ARBITRATION

A. Governing Law.
B. Persons Bound.
C. Rights of Partners Not Assignable; Not to be Pledged.
D. Finality of Decisions within the Firm; Effect of Divergent or Adverse Interest Personally of Any Partner.
E. Arbitration.
F. Severability.

ARTICLE XII. AMENDMENTS

ARTICLES OF PARTNERSHIP FOR THE FIRM OF A, B & C

ARTICLE I. GENERAL PROVISIONS

Section A. Recitals. 1. The undersigned parties hereby agree this ______ day of ______, 19__, to organize a partnership under the firm name of A, B & C.

2. The effective date of this Agreement is the ______ day of ______, 19__.

Section B. Parties. A, B, C, D, E and F constitute the original partners of the firm.

Section C. Purpose. The purpose of this partnership is to engage in [*here set out nature of the business*], and any other business related thereto.

Section D. Firm Name. The name of the partnership "A, B & C" shall continue until changed in accordance with the provisions of this Agreement.

Section E. Term. The partnership shall continue from the effective date of this Agreement until dissolved in accordance with the terms hereof.

Section F. Location of Principal Place of Business. The principal place of business of the partnership shall be at ______, or at such other place or places as the partners shall hereafter determine.

ARTICLE II. CAPITAL

Section A. Original Capital Contributed by Partners. The original capital contributions of the respective partners hereunder are shown on Exhibit A attached hereto. It reflects cash contributed and property, the title of which is transferred to the firm at the current agreed market value of each item. The firm agrees to repay to each partner, at the time and as hereinafter provided, the aggregate amount he has thus contributed as original capital plus interest thereon at the rate of ______ per cent per annum on all unpaid balances.

Section B. Annual Additional Contributions to Capital. Five per cent of the net income of the firm for each fiscal year shall be withheld from distribution and credited, as additional contributions to capital, to partners, in the amount that each would have received had that sum been distributed. Interest shall be paid by the firm on all unreimbursed balances of all these additional contributions to capital at the rate of ______ per cent per annum until fully repaid.

Section C. Reserve for Capital Expenditures; Other Reserves. 1. Out of the sums contributed as additional contributions to capital for each fiscal year, there shall be set aside as of the beginning of the new fiscal year that amount, in addition to any unexpended balance in that reserve fund left over from the last year, estimated to be needed for capital expenditures of the firm during the new fiscal year. As such expenditures are incurred during that year they shall be paid for out of that reserve fund.

2. Out of the remainder of the sums contributed as additional contributions to capital for each fiscal year, there shall be set aside that amount for any other reserve fund, or to add to any existing reserve fund, estimated to be needed to meet any other anticipated obligations or commitments of the firm. As such expenditures are incurred they may be paid out of the appropriate reserve fund.

Section D. Annual Reimbursements on Contributions to Capital. 1. At the end of each fiscal year there shall be charged to firm expense for that year the amount of depreciation accrued for the year, which the firm for federal income tax purposes is entitled to deduct from firm income, and the amount of interest accrued for the year on the unreimbursed balances of contributions to capital.

2. At the end of each fiscal year, (i) the amount of such interest shall be paid to the partners entitled thereto; and (ii) cash sums aggregating the amount of such depreciation shall be paid ratably in reimbursement of contributions to capital.

3. Any amounts remaining out of the annual contributions to capital provided for in Section B of this Article, after the deduction of the reserves as provided in Section C of this Article, shall be paid to partners to reimburse them for their contributions to capital. Such reimbursements shall be made for the oldest contributions first, all repayments for contributions as of the same time being made ratably as to them.

ARTICLE III. PROFITS AND LOSSES OF THE FIRM; PARTICIPATION OF PARTNERS THEREIN; DRAWINGS; BONUSES

Section A. Units of Participation in Profits and Losses Held by the Respective Partners. Except as otherwise expressly provided in this Article, participation of partners in net profits and losses shall be on the basis of the units of participation held by each partner, which shall be as follows:

A: 30 units
B: 20 units
C: 20 units
D: 12 units
E: 8 units
F: 5 units

Upon termination of all interest in the partnership as to any partner, his units of participation and all rights thereunder shall expire. No amendment of this Agreement shall be required therefor. Otherwise no change in the aggregate number of units held by partners or in the number held by any partner shall be effected except by an appropriate amendment of this Agreement.

Section B. Drawing Accounts and the Extent to Which Any are Guaranteed. 1. The firm shall carry on its books a drawing account for each partner. As of the end of each calendar month he shall be paid the sum indicated below, which shall thereupon be charged to his drawing account.

A: $2,400.00 per month
B: 1,600.00 per month
C: 1,600.00 per month
D: 1,000.00 per month
E: 800.00 per month
F: 800.00 per month

2. As of close of each fiscal year there shall be credited to the drawing account of each partner his share of the net profits computed as provided in this Article III, less the amount of his annual contribution to capital of the firm; any reimbursements to him of contributions shall be so credited and all other debits and credits between the partner and the firm to date shall be included in the calculation. Any excess of credits over debits shall thereupon be paid to the partner.

3. If at the end of the fiscal year, after crediting to the drawing accounts of partners E and F the participation of each such partner in the net profits, there remains a deficit in his drawing account, he shall not be required to pay the amount of that deficit to the firm, but as an expense of the firm (to be shared ratably by the remaining partners who do not have the benefit of this guaranty) his account shall be credited in the amount of such deficit. Thus E and F each is guaranteed that he shall receive as a minimum his drawing account for each month of the year. Moreover, if the net profits of the year aggregate as much as the total of the drawing accounts of all partners plus any amounts credited in balancing the drawing accounts of E and F, all of the other partners shall retain the amounts of their respective drawing accounts. But, if the net profits aggregate less than the total paid in the drawing accounts plus the said amounts credited to the accounts of E and F, then A, B, C and D shall share ratably all such deficits for the year in the proportion of their respective drawing accounts, except however that D shall not be required to pay back to the firm any more than the amount that he has received in excess of the stated amounts of the drawing accounts of E and F.

4. If at the end of the fiscal year there are net profits for distribution over and above the aggregate of all the stipulated monthly drawings and payments made as the agreed annual additional contributions to capital, then the portion of such net profits not transferred to the reserve for bonuses, as provided for in the next section hereof, shall be applied first to payments to those partners who have received in their monthly drawings less than their ratable share of net profits; and thereafter the balance of net profits shall be distributed ratably to all partners in proportion to their respective units of participation.

Section C. Reserve for Bonuses and Payments Therefrom. The net profits of the firm remaining for each fiscal year after paying (or setting aside funds for paying) all expenses of the year and after paying fully the stipulated monthly drawings and making the annual agreed additional contributions to capital, shall be distributed as follows: seventy five per cent of such remaining net profits shall be distributed as heretofore provided (Sections A and B of this Article) and the remaining twenty five per cent shall be placed in a "bonus reserve". The management committee shall as promptly as convenient recommend to all partners the uses to which this fund of twenty five per cent shall be placed, and thereupon at a meeting of the firm it shall be determined to whom and in what amounts such reserved funds shall be paid. It is anticipated that normally, unless some anticipated need for the reserve fund seems to require other use of such funds in the new fiscal year immediately ahead, said reserve fund will be used for extra distributions to partners as achievement bonuses.

ARTICLE IV. MEETINGS AND VOTING OF PARTNERS

Section A. Meetings of Partners; Voting at Such Meetings. 1. A meeting of partners shall be held at any time on call of the management committee or at any time after written notice at least 10 days in advance jointly signed by any three partners, specifying the hour and purposes of the meeting. The call by the management committee may be written or oral and need not be made any period of time in advance of the meeting, nor need it specify the purposes of the meeting; except, however, that in those instances where written notice for at least a specified period of time is required by any provision of these Articles, every call or notice of such meeting shall comply with such requirement.

2. At each meeting of partners every partner shall have one vote for each unit of participation held by him, as specified in Section A of Article III of this document; a quorum for any issue at any meeting shall exist if partners holding a majority of such units are present in person or voting by proxy or written instruction. Any partner may vote on any matter (subject to provisions of paragraph 3, this Section) if not present, by general or specific proxy to a partner present or by specific instructions in writing.

3. A partner shall not vote, however, and the number of outstanding units shall be deemed to be reduced by the number he holds (for the purposes of determining on any such issue whether quorum exists or whether the requisite percentage of outstanding units have been voted in the affirmative), when he is the partner affected by any of the following issues:

(a) If the partner has given a notice of withdrawal from the firm and the partnership meeting is voting on a proposal to terminate the firm and liquidate its affairs (see Article V, Section D) the person whose notice of withdrawal is pending shall not vote and the percentage of votes for termination and liquidation shall be determined as though that partner's units of participation did not exist.

(b) If the issue before the partnership is whether a partner (i) is under permanent disability, or (ii) should be expelled from the firm, whether for cause or without determining that a cause exists or (iii) should be permitted to retire or to attain retirement by gradual steps, or (iv) should be granted a temporary withdrawal from the firm (see Article V, Sections C, E, F and G), then as to each such issue the partner involved shall not vote and the percentage of votes shall be determined as though his units of participation did not exist.

4. Excepting only as provided in paragraph 3 of this Section A of Article IV or in Section D of Article XI of this Agreement, no partner shall be disqualified from voting on any issue, notwithstanding any interest he may have therein which differs from the interest of the firm or the other partners.

Section B. Percentage of Votes Required for Certain Partnership Decisions; Requirement of Recommendation of the Management Committee in Advance of Certain Partnership Decisions.

1. As provided by Article V of this Agreement, it may be determined by partnership vote that one presently a partner (i) is under permanent disability, (ii) should be expelled from the firm, (iii) should be permitted to retire or to attain retirement by gradual steps, or (iv) should be granted temporary withdrawal from the firm; or that one not a partner presently be added as a partner (see Article V, Sections B, C, E, F and G). As to each such issue (subject in each instance to the provisions of paragraph 3 of Section A of this Article), it is required that for so determining that issue in the affirmative, affirmative votes shall be cast by partners holding at least two-thirds of the outstanding units of participation that can be voted on that issue. An affirmative recommendation of the management committee in advance is required for a vote of the partners on the addition of a new partner (see Article V, Section B) or for a vote on payments out of the bonus reserve (see Article III, Section C).

2. As provided by Article X of this Agreement, decision may be made that the firm be terminated and its affairs liquidated at any meeting held for the specific purpose of determining whether this shall be done, on the written call of the management committee or of any three partners stating the purpose of the meeting and giving at least three days' notice. For determining this issue in the affirmative (subject to the provisions of paragraph 3(a) of Section A of this Article) votes in the affirmative of partners holding at least two-thirds of the outstanding units of participation that can be voted on that issue, shall be required.

3. As provided in Article XII of this Agreement, these Articles of Partnership may be amended upon affirmative votes of partners holding at least two-thirds of the outstanding units of participation that can be voted on that issue, provided that the proposed amendment and the recommendation of the management committee with reference thereto are attached to the written notice of the meeting at which the proposed amendment is to be considered.

4. A majority of the votes cast, a quorum being present, may determine any other issue at a partnership meeting, provided no such determination shall be contrary to a provision of law or of this Agreement.

ARTICLE V. CHANGES AS TO PARTNERS

Section A. No Classes of Partners. Though their contractual rights differ, as provided in this instrument, all partners are of the same class and have identical and equal rights except as herein otherwise provided.

Section B. Addition of Partners. The management committee may from time to time propose that additional partners be invited to join the partnership, and may propose the units of participation and the drawing accounts for each, together with the proposed amendment to the Articles of Partnership, specifically providing for any drawing account, guaranties and other provisions. In each such instance:

1. There shall be given to each partner a notice of at least ten days of a meeting for all partners at which each partner shall be entitled to discuss the proposal fully; each partner shall be entitled to a postponement of that meeting up to a date not less than thirty days after the giving of the ten-day notice.

2. At that meeting the partners may by their affirmative votes (as provided in paragraph 1 of Section B of Article IV) determine that the invitation shall be extended as proposed by the management committee or with such revisions as are determined upon.

3. If the invitation is accepted, the new partner and prior partners holding at least two-thirds of the participating units entitled to vote at the meeting referred to in paragraphs 1 and 2 of this Section B, shall join in executing an amendment to these Articles of Partnership providing for the change in the partnership thus effected.

Section C. Death or Permanent Disability of a Partner.

1. *Death.* The death of a partner shall terminate all his interest in the partnership, its property and assets. The continuing firm shall pay in cash to his estate (or to his nominee or nominees in accordance with the provisions of any separate agreement entered into between him and the management committee acting for the firm) the following amounts to be paid in installments at the times indicated:

(a) On or before thirty days after the date of his death, the net amount of his capital in the firm as of the date of death plus interest on his capital to that date.

(b) Within ninety days from the date of death, an amount computed as follows:

(i) Start with his pro rata share of seventy five percent of the net profits (after reducing said profits by interest on the capital accounts of all partners to date of death) of the firm for that portion of its then current year ending with the date of death;

(ii) Add thereto any part of the remaining twenty five percent of the firm's net profits which the management committee in its discretion determines to be his fair share of such net profits as a bonus payment to him, based on the same considerations for that part of the year as are provided for any full year in Section C of Article III hereof;

(iii) Deduct from the total arrived at in (ii) above, all distributions the deceased partner had received from the firm on account of net earnings during the year; and

(iv) Adjust the remaining balance by debiting and crediting all sums owing to the firm by him or by the firm to him immediately prior to his death. If the result is a minus balance, it shall be deducted from the aggregate amount payable in monthly installments as provided in subparagraph (c) of Section C, paragraph 1.

(c) In a series of forty-two consecutive monthly installments, beginning on or before one hundred twenty days after the date of his death, a further amount which (except as otherwise herein provided) shall be the average of the sums paid to him as a partner of the firm during each of the last three complete fiscal years of the firm during which he was a partner.

(i) The computation of the sums so paid to him each year shall include all distributions to him out of net income of the firm, but without any deductions for contributions to its capital or any additions for reimbursements therefor or interest on unreimbursed contributions. If he became a retired partner or temporarily withdrawn partner, the years of retirement or of temporary withdrawal are not to be included in the computation. If he had not been a member of the firm for as long as three complete fiscal years, then there shall be paid the average of sums paid to him for two such years, and if not a member for as long as two such years, then the sums so paid to him for one year. If he had not been a member one full year, no sums shall be paid under this subparagraph (c).

(ii) The first six installments of the amount thus to be paid by the continuing firm shall each be as much as decedent's current agreed monthly drawing at the time of his death and may, at the option of the firm, be as much more as the firm shall elect. The remainder of the sum payable by the continuing firm, if any (after the payment of the first six installments) shall be paid in thirty-six monthly installments, approximately equal, beginning three hundred days after death, with interest added to each of these installments at the rate of five percent per annum from date of death until paid.

2. *Permanent Disability.* (a) The determination that a partner is permanently disabled shall terminate all his interests in the partnership and his units of participation as a partner. That determination shall be made only upon the affirmative vote by partners holding at least two-thirds of all units of participation, not including the partner whose disability is in issue or the units held by him, all in accordance with the provisions of Article IV of this agreement.

(b) As of the time of the determination of permanent disability of a partner, he shall no longer be a partner and shall no longer have any duties to perform with respect to any professional employment of the firm, nor shall he be privileged to perform any services in any such matter. His units of participation shall expire as of that time, and hence no votes at any partnership meeting may thereafter be cast by him, and he shall not be entitled to any share of profits or losses thereafter. Except for sums to be paid to him by the continuing firm as provided for in subparagraph (c) of this paragraph 2, he shall not be entitled to any payments from the firm and shall have no rights or interests in any of its properties or assets from the time of such determination. However, the partners in their discretion may vote to bestow upon him some purely honorary title such as "Partner Emeritus," without compensation.

(c) The amounts payable by the continuing firm to or for the account of a partner determined to be permanently disabled shall be computed in the same way and paid in the same manner as if he had died on the date of the determination of his permanent disability. His death before all such payments have been made shall not interrupt the continued payments by the continuing firm; but no further sums shall be owing by the firm because of his death.

Section D. Permanent Withdrawal of a Partner.

1. *Notice of Withdrawal and Effective Date of Withdrawal.* Any partner may voluntarily withdraw from the partnership at any time on notice of thirty days to the other partners. As of the expiration of the thirty day period, or sooner if mutually agreed upon, the withdrawal shall be effective.

2. *Possible Termination of the Firm Superseding Withdrawal Notice.* At any time during the pendency of a withdrawal notice and before the effective date of withdrawal, a termination of the firm may be voted in accordance with the provisions of Article X of this Agreement. If this is done, the dissolution proceedings, the liquidation of assets, and the distribution of proceeds shall ensue, and the notice of withdrawal shall be of no effect.

3. *Partition with and Payments to the Withdrawing Partner.* The withdrawing partner's right, title, and interest in the firm shall

be extinguished in consideration of the partition with and the payments to him by the continuing firm on the following bases:

(a) On the effective date of withdrawal he shall be paid the amount of his net capital in the firm plus interest thereon to that date. This payment shall be in cash unless the firm at its option elects to set aside for him and deliver to him in kind his pro rata share of all its capital assets. In the event the firm sets aside property for him it shall have a discretion as to what items to set aside, all items being valued for the purposes of partition, either by agreement between the firm and the withdrawing partner or by an independent appraisal, at current market prices.

(b) Within ninety days after the effective date of withdrawal an amount in cash shall be paid computed as follows:

(i) Start with his pro rata share of seventy-five per cent of the net profits (after reducing said profits by interest on the capital accounts of all partners to the effective date of withdrawal) of the firm for that portion of its then current year ending on the effective date of withdrawal;

(ii) Add thereto any part of the twenty-five per cent of the firm's net profits for said portion of its then current year, which the management committee fairly determines to be his fair share of such net profits as a bonus payment to him based on the same considerations for that portion of the year that are provided in Section C of Article III for any full year, and bearing in mind that the same part of twenty-five per cent of the profits from receipts of the portion of the current year, will be applied to future receipts from fees charged for services rendered before the withdrawal, pursuant to the provisions of subparagraph (c) (ii) of this paragraph 3;

(iii) Deduct from the total arrived at in (ii) above all distributions the withdrawing partner had received from the firm on account of net earnings during the year;

(iv) Adjust the balance thus arrived at by debiting the discounted value at that time of his ratable share of payments yet to accrue against the firm on account of the prior death or permanent disability of a partner; and

(v) Adjust the balance thus arrived at by debiting and crediting all sums owing to the firm by him or by the firm to him immediately prior to the effective date of withdrawal.

If the foregoing computations result in a minus balance it shall be debited against each quarterly payment later accruing to him under the provisions of subparagraphs (c) and (d); and if the debt is not thus discharged, it shall be owing by the withdrawing partner to the continuing firm.

(c) In quarter-annual installments following the withdrawal, a share of the fees collected by the firm during each quarter thereafter for services rendered by the firm prior to the effective date of withdrawal shall be paid, the amount of these quarter-annual payments to be computed as follows:

(i) Start with his pro rata share of seventy-five per cent of the gross amount of such fees collected during such quarter (after reducing same by the amount if any which the management committee of the continuing firm fairly determines to represent the share of all fees earned during that quarter by the firm which were prepaid by the client prior to the effective date of withdrawal);

(ii) Add thereto an amount which is that percentage of the figure computed under (i) immediately above, which the amount computed under (b) (ii) of this paragraph 3 bears to the figure computed under paragraph (b) (i) of this paragraph 3; and

(iii) The total amount thus arrived at shall be paid to the withdrawing partner with an accounting to him at that time of how the amount is arrived at, provided he then makes a like accounting and payment if any is due by him to the firm, in accordance with the provisions of subparagraph (d) immediately following.

(d) Subject to the right of each client to direct that any or all of his pending matters in which the firm is employed on the effective date of withdrawal shall be handled for him by the continuing firm rather than the withdrawing partner, the withdrawing partner, at his option, as to each of the current employments of the firm pending on that date for which he was the responsible partner in charge, shall then be entitled (provided he then pays the firm for all its expenditures on behalf of the client in connection with such matter for which the client then is or would later be indebted to the firm) to assume all further responsibilities to the client for that matter and to take with him all files and documents pertaining wholly to that employment. Thereafter, the withdrawing partner shall bill the client for and be entitled to collect for disbursements theretofore made and services theretofore rendered in connection with that matter as well as for subsequent services and disbursements. The withdrawing partner shall account to the continuing firm with respect to his gross collections of fees for services rendered on each such matter by the firm prior to the effective date of withdrawal, and shall pay the firm in cash, in quarter-annual installments from such collections, amounts calculated on the same basis, or as nearly as possible on the same basis, as the firm shall be accounting to the withdrawing partner and paying him in accordance with the provisions of sub-paragraph (c) immediately above.

Agreement as to Tax Effects

In view of the differences in tax results dependent upon distinctions which may not be readily apparent to lawyers outside the tax field, it is quite important that contracts with reference to the liquidation or sale of the interest of a withdrawing or disabled partner, and especially of a deceased partner, should clearly express the intention of the parties as to the tax effects anticipated by the parties to flow from their agreement. Naturally, they should be careful to see that the agreement does what they think it does. As an addition to the agreement, therefore, we suggest a paragraph pertinent to all provisions of Sections C and D of Article V:

It is contemplated by the parties to this Agreement that any payments hereunder for the interest in the firm of a withdrawing or permanently disabled or deceased partner are, to the extent that they represent payment for partnership properties, capital payments falling under Section 736(b) of the Internal Revenue Code. All other payments for the interests of such persons, including so-called "interest" payments on capital invested, are intended by the partners as payments of partnership income under Section 736(a) of the Internal Revenue Code. Each partner covenants for himself and his heirs and assigns that he will make no claims or representations with reference to the income tax nature of any such amounts that are inconsistent with the intent expressed in this subparagraph.

Section E. Retirement of Partners; Gradual Steps toward Retirement; Retirement Plans for Partners.

1. *Retired Partners; Plans for Their Compensation.* (a) A retired partner shall receive no current compensation from the firm in payment for current services, either by way of participation in distribution of net profits of the firm or agreed monthly drawings. He may receive bonuses or specifically agreed fees or shares of fees. He shall be offered, at the expense of the firm, so long as he is able and wishes to use same for at least twenty percent of the business time of each year, an office in the offices of the firm and a secretary to give him such secretarial assistance as he may require; in consideration of which he shall, whenever convenient to him, advise with and serve as consultant to any of the partners or associates of the firm. His name shall be carried on firm letterheads, in legal directories and otherwise not as an active partner of the firm but under the heading, "Of Counsel."

(b) The management committee in its discretion is authorized to pay during any year, to each retired partner as a "retirement bonus", up to twenty-five percent of his average annual income for the last three full years during which he was an active partner of the firm.

2. *When a Partner Retires.* Any partner may retire at any time upon approval by the partners, in accordance with provisions of

Article IV of this Agreement, of his request to retire. Any partner who has attained the age of seventy-five shall retire if and when requested to do so by partners holding at least two-thirds of the units or participation entitled to vote.

3. *Gradual Steps toward Retirement.* If the request of a partner that he be permitted to enter upon and carry out a plan for gradual retirement is approved by vote of the partners in accordance with the provisions of Article IV of this Agreement, a program of gradual steps toward his retirement shall be entered into and consummated, as agreed between him and the firm. Such a plan may be required of any partner at any time after he attains the age of seventy. The adoption of such a plan as to any partner will involve a program over a period of the following ten years (provided his interest in the firm is not meanwhile terminated by death, total disability, withdrawal, or expulsion; and provided said interest is not modified by an agreement between him and the firm approved by vote of the partners). During that ten-year period his duties shall be gradually reduced and hence, his units of participation and thus his share of net profits or losses, and his voting rights shall be reduced from what they are at the start of the period by eight per cent at the end of each of the first nine fiscal years, of the period, and his remaining interest in the firm shall be terminated by effecting his retirement at the end of the tenth year.

Section F. Expulsion of a Partner.

1. *Expulsion for Cause.* A partner shall be expelled for cause when it has been determined by vote of partners in accordance with the provisions of Article IV of this Agreement, that any of the following reasons for his expulsion exist:

(a) Disbarment, suspension or other major disciplinary action of any duly constituted authority.

(b) Professional misconduct or violation of the canons of professional ethics, if such misconduct continues after its desistance has been requested by the management committee.

(c) Action that injures the professional standing of the firm, if such action continues after its desistance is requested by the management committee.

(d) Insolvency or bankruptcy or assignment of assets for the benefit of creditors.

(e) Breach of any provision of the Articles of Partnership of the firm, which all other partners expressly agree is a major provision, if, after the breach has been specified as a prospective ground for expulsion by written notice given by the management committee, the same breach continues or occurs again.

(f) Any other reason which the other partners unanimously agree warrants expulsion.

2. *Effects of Expulsion for Cause.* Upon a determination that a partner be expelled for cause he shall thereby be so expelled and shall have no right or interest thereafter in the firm or any of its assets, clientele, files or records, or affairs. He shall have thereafter no further duties to the firm or any of its clients and shall be privileged to serve none of them thereafter. He shall immediately remove himself and his personal effects from the firm offices. Upon any such expulsion, the expelled partner shall be obligated not to accept employments for services from any who have been clients of the firm during the last five years preceding the determination of expulsion, the obligation not to accept such employments being a continuing one for a term of the next ensuing five years. From the time of the expulsion, the expelled partner shall have no participation whatever in the income or losses of the firm or any distribution or drawings from the net income. Realizing that the existence of any such cause for expulsion may bring disgrace on the firm and damage the firm in amounts and ways that cannot be calculated or become liquidated in amount, each partner agrees that the firm shall succeed to all of the rights of the expelled partner as hereinabove set forth and shall retain all sums unpaid by it to the expelled partner, whether accrued or not at that time; further, that the receipt and retention by the firm of all such rights and sums shall satisfy and discharge the damages of the firm, being retained as and thereby determined to be liquidated damages; no other indebtedness of the expelled partner to the firm being discharged.

3. *Expulsion without Determining Any Cause Therefor.* A partner shall be expelled immediately when, on recommendation of the management committee, it is determined by a vote of the partners as provided in Article IV that he shall be expelled without determination of any cause therefor. This method of expulsion may be employed notwithstanding the fact that grounds may exist for expulsion for cause.

4. *Effects of Expulsion without Determining Any Cause Therefor.* Upon such expulsion without determining a cause therefor, the partner so expelled shall have no right or interest thereafter in the firm or any of its assets, clientele, files or records, or affairs. He shall have thereafter no further duties to the firm or any of its clients and shall be privileged to serve none of them thereafter. He shall immediately remove himself and his personal effects from the firm offices. Except as otherwise provided in this paragraph, a partner so expelled shall be entitled to the same rights, the same payments by, and be subject to the same duties to the continuing firm as if he were then voluntarily withdrawing from the firm.

Section G. Temporary Incapacity; Leave of Absence; Temporary Withdrawal; Vacations.

1. *Temporary Incapacity or Illness.* In the event of any interruption of the performance of any partner's services to the firm or to

its clients on account of any temporary incapacity or illness, or any other reason not voluntary with him, the management committee may, in its complete discretion, make any arrangements it deems fair to the partner and to the firm, as to the period of his absence and his compensation during that period.

2. *Leave of Absence.* In the event any partner desires an interruption of the performance of his services to the firm or its clients, for any reason voluntary with him, his request shall be submitted to and may be approved by the management committee which, if the interruption shall not be for more than one year, may in its complete discretion make any arrangements it deems fair to the partner and to the firm, as to the period of his absence and his compensation during that period.

3. *Temporary Withdrawal.* If any partner desires an interruption of his services to the firm and its clients for a period longer than the management committee can, or feels that it should, approve under either of the last two paragraphs of this instrument, he may apply to the firm for a temporary withdrawal. The firm, by vote of the partners in accordance with provisions of Article IV of this document, shall determine whether the request shall be granted and if so, on what terms and conditions. During the period of any temporary withdrawal, there shall be a suspension and not a termination of the units of participation of the partner involved. Such a temporary withdrawal unless extended under the same procedure by which it was originally granted shall be for a specific time, at the expiration of which the temporarily withdrawing partner shall resume his services.

4. *Vacations.* All decisions of the firm with reference to vacations of partners in excess of ______ weeks a year for each partner are to be wholly within the discretion of the management committee.

ARTICLE VI. DUTIES OF PARTNERS

Section A. Devotion to Duty. Each partner shall devote his best efforts to serving professionally the firm and its clients. Subject to any exceptions provided in rules of the firm adopted in accordance with the provisions of Article VII, Section A of this Agreement, or any other exceptions consented to by the management committee, each partner shall devote substantially all his normal business time to such services.

Section B. Charging for Services. 1. Each partner shall charge reasonably for all services rendered by him, following generally the policies of the firm as to fees charged. However, each partner may serve without charge any member of his own family, and with the consent of the management committee any partner may serve without charge, or at less than regular charge, any civic, educational, religious, or charitable organization or project.

2. Each partner will follow rules and policies of the firm adopted in accordance with the provisions of Section A of Article VII of this Agreement relating to consideration by the firm, rather than one partner only, of fees on substantial services rendered by the firm.

3. No salaries, commissions, fees or gratuities of any substantial significance shall be accepted, directly or indirectly, by any partner personally from any client or prospective client of the firm, unless with the express consent in advance of the management committee, and the fair value of any such item received with such consent, though retained by the partner, shall be treated for accounting purposes as compensation to the firm and shall be charged against such partner as an advance on the next maturing installment or installments of his drawing account. The management committee may agree, however, to any exception to any provision of this paragraph.

ARTICLE VII. MANAGEMENT

Section A. Authority and Membership of the Management Committee. 1. Subject to the express terms of this Agreement, which as to certain specific matters provides that decisions of the firm shall be determined by the vote of the partners holding required units of participation, the complete and sole management of the firm is hereby vested in the management committee.

2. Any part or parts of the power, right, and authority vested in the management committee may, at any time and from time to time, be delegated by it to a subcommittee of one or more chosen by it. Such authority may be delegated with power in the subcommittee only to recommend to the management committee what action should be taken, or with power to act; in the latter event, action of the subcommittee shall be the action of the management committee. Any delegation may be terminated by the management committee at any time.

3. It may from time to time cause a set of the rules and policies of the firm to be distributed in an office manual to all partners, associated attorneys and employees of the firm.

4. The management committee shall consist of three partners. No one of them shall be retired (though he may be participating in gradual steps toward retirement) or the subject of pending action for expulsion. Partners subject to any of the stated disabilities shall be disqualified from election to or from acting on the management committee. Upon any such event that disqualifies from continued service a member of the committee, he shall automatically cease to be a member of the committee and shall not serve thereafter unless and until (when qualified) re-elected to fill a vacancy on the committee. There shall be an alternate member elected by the partners, and if there is a

vacancy on the committee because of death, resignation, or disqualification, the alternate shall become a member of the committee. In the event of any temporary absence of a member, the alternate may serve as a member of the committee during the period of the absence. As soon as convenient the partners shall meet and choose a successor to fill any vacancy (other than a vacancy resulting from a temporary absence of one of the four elected). In the event of any vacancy not yet filled by vote of the partners, the management committee may on its own account call on any qualified partner of its choice to serve temporarily with the committee. The tenure of one so chosen shall expire when the partners elect a successor.

5. The management committee, from the effective date of this instrument, shall consist of A, C, and E. The named alternate shall be B. Each of the four shall serve respectively until his tenure is terminated by death, resignation, disqualification, or a determination by vote of the partners that his term shall expire.

6. The tenure of every member of the committee and every alternate member shall be subject to termination without cause, by requisite vote of the partners in accordance with the provisions of Article IV of this document.

Section B. Functioning of the Management Committee and Its Subcommittees. 1. Members of the management committee shall make every reasonable effort to keep each other and the alternate advised of all pending problems, prospective decisions, and actions taken. Action of the committee shall be by majority vote. It shall not be necessary that any notice be given of the time or place of decision or of the matter to be decided. Any decision of the committee may be reversed prospectively by any subsequent action of the committee.

2. Though the committee has no obligation so to do, it may refer any matter on which all members of the committee are not in agreement to a meeting of the partners for decision.

Section C. Membership in Subcommittees of the Management Committee. The management committee shall decide what subcommittees there shall be from time to time, how many members (one or more) there shall be of each subcommittee, who the members shall be, and what the subcommittee's functions and authority shall be. The management committee may at any time modify or revise prospectively any authorized decision of any subcommittee. Any partner or any full time employee may be a member of any subcommittee.

ARTICLE VIII. INSURANCE; INVESTMENTS

Section A. Life Insurance. The management committee in its discretion shall determine from time to time what life insurance, if any, shall be carried on the lives of partners for benefit of the firm.

Section B. Other Insurance. The management committee in its discretion shall determine from time to time what other insurance, if any, the firm shall carry.

Section C. Investments. The management committee in its discretion shall determine from time to time what investments, if any, the firm shall make and all matters with reference to the proceeds of such investments, and with reference to reinvestments or changes in investment policies.

ARTICLE IX. PROPERTIES AND RECORDS

The management committee in its discretion shall make all decisions of the firm from time to time on the following subjects:

Section A. Firm Properties. [Some firms in their Articles limit the authority of the management committee or its equivalent with respect to properties. Examples:

(i) Require that the purchase of all properties, except supplies, be approved by partnership vote; or

(ii) Require such a vote for purchase of properties costing more than a specific amount; or

(iii) Require such a vote for purchase of an office site or office building; or any property not deemed necessary to the practice of law; or

(iv) Limit the amount to be spent in a year, without a partnership vote for replacements, repairs or upkeep.]

Section B. Accounting Records. [Many firms have express provisions in their partnership agreements covering one or more of the following points on this subject.

(i) Specifically requiring that the books of account be kept on a cash basis;

(ii) Specifically defining the fiscal year of the firm;

(iii) Specifically defining what financial statements shall be prepared with copies given to each partner;

(iv) Specifically requiring that partnership income tax returns be prepared and filed regularly and a copy of the same given to each partner a specific period, say at least one week, before each return is filed, and a specific period, say at least two weeks, before his personal return is due;

(v) Specifically requiring that all accounting records of the firm shall be open to inspection by each partner at any time during business hours;

(vi) Specifically requiring that the financial records of the firm shall be retained for an agreed period and shall be available for inspection or copying by anyone who was a partner at the time that such records were prepared, including one who at the time of the inspection is a former partner.]

ARTICLE X. TERMINATION AND LIQUIDATION OF FIRM

Section A. Termination of the Firm by Voluntary Vote or Otherwise. The partnership may be terminated at any time by affirmative vote of the partners at a partnership meeting, in accordance with the provisions of Article IV of this Agreement.

Section B. Pending Employments on Termination. In the event of termination of the partnership, no further services shall be rendered in the partnership name and no further business transacted for the partnership except action necessary for the winding up of its affairs, the distribution or liquidation of its assets, and the distribution of the proceeds of the liquidation. Maintenance of offices to effectuate or facilitate the winding up of the partnership affairs shall not be construed to involve a continuation of the partnership. In advance of the effective date of the termination of the partnership the management committee shall assign every uncompleted service to one or another of the partners on such terms as shall be agreeable to the clients involved and the partners to whom such matters are assigned; and the rendition of services from the effective date of the termination shall henceforth be by such individuals and other law firms, if any, in which they may respectively become partners.

Section C. Liquidation of Assets. The members of the management committee (but not including alternate members) on the effective date of the termination of the partnership, shall be the agents of the terminated partnership in liquidation, and of the individual partners, for winding up all its affairs and all business transactions of the partnership, other than the performance of incomplete professional services referred to in Section B above. Said members of the management committee shall continue to serve (unless death, incapacity, or resignation shall intervene) until the completion of the winding up and liquidation. The committee shall act by majority vote or votes. In the event of any temporary or permanent vacancy in the committee, the remaining members shall choose a third member of the committee. Members of the management committee shall not be paid for their services after the termination of the partnership in the winding up or liquidation operations. They may, out of the assets and proceeds of the assets on hand, employ such assistants as they determine appropriate, and the committee may so employ and pay any one of its members to take any such actions and render any such services in the winding up and liquidation.

Section D. Prior Opportunity of Partners to Bid for Purchase of Assets Being Liquidated. The partners holding units of participation immediately prior to the termination of the partnership may, in the discretion of the management committee, be given first opportunity over any other prospective bidder for the purchase of any of the assets, all such partners being given an equal opportunity, so that they respectively as individuals or jointly or in groups, may bid; and if the best bid by any of them, in the opinion of the management committee, is at least ninety-five per cent of the highest and best bid otherwise received, then such best bid by any partner or partners may be accepted.

Section E. Distribution of Proceeds from Liquidation. The business affairs of the partnership, in the event of the termination of the partnership, shall be wound up and liquidated as promptly as business circumstances and orderly business practices will permit. After payment of expenses incurred, the net assets and the proceeds of the liquidation shall be applied in the following order:

1. To the payment of the debts and liabilities of the partnership owing to the creditors other than partners, and the expenses of liquidation.

2. To the payment of the debts and liabilities owing to the partners other than for (i) capital, (ii) profits and (iii) any unmatured installments yet to be paid on account of the death, permanent disability, retirement (or death following retirement) or withdrawal of a partner. It is agreed that all sums to become due on installments referred to in (iii) shall be assumed ratably by each partner at the date of termination and that each shall thereafter pay his ratable share of each such installment as it becomes due.

3. To the repayment to each of the partners of his capital contributions to the firm.

4. To the payment to partners (computed on the basis of their respective units of participation at the date of termination of the firm) of all the remaining net of assets and proceeds, if any, first in whatever amounts are necessary to complete a ratable distribution for the current year, to each partner to the full extent of distributions previously received by each other partner; and second, to ratable distributions to all partners.

5. If the assets and proceeds of the liquidation are insufficient to pay all of the items referred to in paragraphs 1 and 2, but not including (i), (ii) and (iii) referred to in paragraph 2, then the management committee shall make an assessment against the partners to cover net losses of the firm and such assessments shall be paid and applied to the satisfaction of the items covered by paragraphs 1 and 2.

ARTICLE XI. LEGAL EFFECT OF PROVISIONS; ARBITRATION

Section A. Governing Law. All provisions of this Agreement shall be construed, shall be given effect and shall be enforced according to the laws of the State of ______.

Section B. Persons Bound. Each of the partners executes this Agreement with the understanding and agreement that each has hereby bound and obligated himself, his estate, and any and all claiming by, through, or under him.

Section C. Rights of Partners Not Assignable; Not to be Pledged. No partner and no one acting by authority of or for a partner may pledge, hypothecate, or in any manner transfer his interest in the partnership, or his interest in any of its assets, receivables, records, documents, files, or clientele, all such rights and interests of each partner being personal to him and non-transferable and non-assignable (except that other partners of the firm may succeed to such rights or some of them in accordance with the terms of this Agreement).

Section D. Finality of Decisions within the Firm; Effect of Personal Diverse or Adverse Interest of Any Partner. Every final decision of the firm on any matter affecting any party hereto or anyone claiming by, through or under any party, by vote of the partners or by decision of the management committee, when in accordance with the terms and provisions of this Agreement, shall be binding and conclusive. Except where it is expressly provided in this Agreement that one shall not be permitted to vote as to any such decision, there shall be no disqualification of anyone from voting who shall be entitled to vote according to the terms and provisions of this Agreement, notwithstanding any adverse or divergent interest that he may personally have in the decision; and the decision shall, nevertheless, be binding and final notwithstanding any such adverse or divergent interest held by anyone so voting. It is understood that individual partners and that members of the management committee will doubtless have divergent and may have adverse, or arguably adverse, personal interests from one another on some matters that are to be determined according to the provisions of this Agreement and have diverse or adverse interests personally from those of some party affected by the decision; all this is agreed to and waived as a disqualification. Nonetheless, anyone entitled to such a vote on any such matter may recuse himself from voting and thereupon the decision shall be made on the computation of votes to the same effect as if the one so recusing himself had as to that matter, no right to vote; and if the vote is by the partners, as if he held no units of participation. Each party having any vote on any such matters shall recuse himself on any vote if requested so to do by joint action of partners holding a majority of the units of participation then outstanding.

Section E. Arbitration. Any controversy or claim arising out of or relating to any provision of this Agreement or the breach thereof, shall be settled by arbitration in accordance with the rules then in effect of the American Arbitration Association, to the extent consistent with the laws of the State of ______. It is agreed that any party to any award rendered in any such arbitration proceedings may seek a judgment upon the award and that judgment may be entered thereon by any court having jurisdiction.

Section F. Severability. It is agreed that the invalidity or unenforceability of any Article, Section, paragraph or provision of this Agreement shall not affect the validity or enforceability of any one or more of the other Articles, Sections, paragraphs or provisions; and that the parties hereto will execute any further instruments or perform any acts which are or may be necessary to effectuate all and each of the terms and provisions of this agreement.

ARTICLE XII. AMENDMENTS

An amendment hereto may alter, revise, delete or add to any provision or provisions of this agreement. No amendment to this instrument shall be adopted or become effective unless and until it (i) has been voted in accordance with the provisions of paragraph 3 of Section B of Article IV of this Agreement; and (ii) has been executed and attached to this Agreement as a part of same.

In Witness Whereof, the parties have signed this Agreement.

[*Signatures*] [5]

5. West's Modern Legal Forms § 6257. This agreement was prepared by Paul Carrington and William A. Sutherland for the American Bar Association Standing Committee on Economics of Law Practice. The original draft printed in the ABA Economics of Law Practice Series Pamphlet Number 6, November 1961, contained the helpful comments of the authors, omitted here to save space. The authors emphasized the need to tailor each partnership agreement to the clients, and cautioned their readers to consider the various provisions of this agreement as merely suggestions for comparison and study.

FORM 3A

ASSIGNMENT OF A LIMITED PARTNER'S INTEREST

For value received, I, the undersigned, of ______, hereby assign to ______, of ______, the whole of my interest in the limited partnership of ______, conducting business under a partnership agreement dated ______, 19__. Effective upon the signing of this instrument the assignee shall be entitled to receive the share of the profits or other compensation by way of income to which I would otherwise be entitled, and to the return of my contribution to the capital of the partnership. In the event that all the other members of the partnership consent thereto, the assignee shall be entitled to all the rights which I, as a limited partner, had in the partnership.

Dated ______, 19__. [*Signature*]

[*Acknowledgment*] [6]

6. West's Modern Legal Forms § 6456.

FORM 3B

CONSENT TO SUBSTITUTION OF A LIMITED PARTNER

We, the undersigned, being all the members of the limited partnership of ______, except ______, who by an instrument dated ______, 19__, and duly acknowledged by her, has assigned her entire interest as a limited partner in this partnership to ______, of ______, do hereby consent that ______ be substituted as a limited partner in the place of ______, and entitled to all the rights which ______ had as a limited partner in this partnership pursuant to the terms of the partnership agreement dated ______, 19__.

Dated ______, 19__. [*Signatures*] [7]

7. West's Modern Legal Forms § 6457.

FORM 3C

LIMITED PARTNERSHIP AGREEMENT

Agreement of Limited Partnership made this ______ day of ______, 19__, between ______ and ______, both of ______ (herein referred to as general partners), and ______ of ______, and ______ of ______ (herein referred to as limited partners).

1. **Formation.** The parties hereby form a limited partnership pursuant to sections ______ of the [Revised Statutes] of the State of ______, known as the Uniform Limited Partnership Act.

2. **Certificate.** The parties shall forthwith sign and swear to a certificate prepared in accordance with the provisions of the Uniform Limited Partnership Act cited above, and cause the same to be filed for record in the office of [*here designate the proper office*].

3. **Name.** The name of the partnership is ______.

4. **Business.** The purpose of the partnership shall be to engage in the business of ______, and in any other business necessary and related to it.

5. **Place of Business.** The principal place of business of the partnership shall be at ______, but additional places of business may be established as the general partners shall determine.

6. **Term.** The partnership shall commence on ______, 19__, and shall continue until terminated as herein provided.

7. **Capital.** The initial capital of the partnership shall be $______. Each of the partners shall contribute in cash or in property the amount set opposite his name.

General Partners	Cash Contributions	Agreed Value of Property Contributions
______	$______	$______
______	______	______
Limited Partners		
______	______	______
______	______	______

The property contributed is described in a separate instrument attached hereto as Exhibit A.

8. **Additional Contributions to Capital.** The general partners shall make, and the limited partners shall each have the option of making, additional contributions to the capital of the partnership in such amount as the general partners deem necessary to carry on the business of the partnership.

9. **Withdrawal of Capital.** Neither a general nor a limited partner may withdraw all or any part of his capital contribution without the consent of all the general partners, provided that each limited partner may rightfully demand the return of all or part of his contribution after he has given six months' notice in writing to all the other partners. Upon any withdrawal by a limited partner the certificate of limited partnership shall be amended to reflect this change in his capital contribution.

10. **Profits and Losses.** The net profits of the partnership during each fiscal year shall be credited, and the net losses incurred by the partnership during any fiscal year shall be debited, as of the close thereof, to the capital accounts of the partners in the proportions set opposite their respective names.

General Partners	**Percentage**
______________________	__________%
______________________	__________%
Limited Partners	
______________________	__________%
______________________	__________%

Notwithstanding anything to the contrary herein contained, no limited partner shall be liable for any of the debts of the partnership or any of its losses in excess of his capital contributions to the partnership.

11. **Capital Accounts.** An individual capital account shall be maintained for each partner, to which shall be credited his contributions to capital and to which shall be debited his withdrawals from capital and his share of partnership losses.

12. **Salaries.** Each of the general partners shall receive such reasonable salaries as may from time to time be agreed upon by the general partners. These salaries shall be treated as an expense of the partnership in determining the net profit or loss in any fiscal year.

13. **Drawing Accounts.** An individual drawing account may be maintained for each partner in an amount fixed by the general part-

ners, but such drawing accounts shall be in the proportion to which the partners are entitled to share in the profits of the partnership.

14. **Management.** The general partners shall have equal rights in the management of the partnership business.

15. **Devotion to Business.** Each general partner shall devote all his normal business time and best efforts to the conduct of the business of the partnership.

16. **Limitations on General Partners' Powers.** No general partner shall, without the written consent or ratification of the specific act by all the other partners:

(a) Assign, transfer, or pledge any of the claims of or debts due to the partnership except upon payment in full, or arbitrate or consent to the arbitration of any disputes or controversies of the partnership;

(b) Make, execute, or deliver any assignment for the benefit of creditors, or sign any bond, confession of judgment, security agreement, deed, guarantee, indemnity bond, surety bond, or contract to sell or contract of sale of all or substantially all profit of the partnership;

(c) Lease or mortgage any part of partnership real estate or any interest therein, or enter into any contract for any such purpose;

(d) Pledge or hypothecate or in any manner transfer his interest in the partnership, except to the parties of this agreement;

(e) Become a surety, guarantor or accommodation party to any obligation except for partnership business;

(f) Do any act prohibited by law to be done by a single partner.

17. **Books of Account.** The partnership shall maintain adequate accounting records. All books, records and accounts of the partnership shall be kept at its principal place of business and shall be open at all times to inspection by all the partners.

18. **Accounting Basis.** The books of account shall be kept on a cash [*or* an accrual] basis.

19. **Fiscal Year.** The fiscal year of the partnership shall be the calendar year. The net profit or net loss of the partnership shall be determined in accordance with generally accepted accounting principles as soon as practicable after the close of each fiscal year.

20. **Annual Audit.** The books of account shall be audited as of the close of each fiscal year by a certified public accountant chosen by all the partners.

21. **Banking.** All the funds of the partnership shall be deposited in its name in such checking account or accounts as shall be designated

by the general partners. Checks shall be drawn on such accounts for partnership purposes only and shall be signed by any of the general partners.

22. **Assignment by Limited Partner.** Each limited partner may assign his interest in the partnership, and the assignee shall have the right to become a substituted limited partner and entitled to all the rights of the assignor if all the partners (except the assignor) consent thereto. Otherwise the assignee is only entitled to receive the share of the profits to which his assignor would be entitled.

23. **Retirement of a General Partner.** A general partner may retire from the partnership at the end of any fiscal year by giving at least 90 days' notice in writing to all the other partners.

24. **Effect of Retirement, Death or Insanity of a General Partner.** The retirement, death or insanity of a general partner dissolves the partnership, unless the business is continued by the remaining partners as herein provided.

25. **Distribution of Assets on Dissolution.** Upon dissolution of the partnership by mutual agreement or for any other reason its liabilities to creditors shall be paid in the order of priority provided by law, and the remaining assets, or the proceeds of their sale, shall be distributed in the following order:

(a) To the limited partners in proportion to their share of the profits;

(b) To the limited partners in proportion to their capital contributions;

(c) To the general partners other than for capital and profits;

(d) To the general partners in proportion to their share of the profits;

(e) To the general partners in proportion to their capital contributions.

26. **Election of Remaining Partners to Continue Business.** In the event of the retirement, death or insanity of a general partner, the remaining partners shall have the right to continue the business of the partnership under its present name either by themselves or in conjunction with any other person or persons they may select, but they shall pay to the retiring partner, or to the legal representatives of the deceased or insane partner, as the case may be, the value of his interest in the partnership, as provided in paragraph 28.

27. **Notice of Election to Continue Business.** If the remaining partners elect to continue the business of the partnership they shall

serve notice in writing of such election upon the retiring partner within two months after receipt of his notice of intention to retire, or upon the legal representatives of the deceased or insane partner within three months after the death of the decedent or the adjudication of insanity, as the case may be. If at the time of such election no legal representative has been appointed notice shall be sent to the last known address of the decedent or insane partner.

28. **Valuation of Partner's Interest.** The value of the interest of a retiring, deceased or insane partner shall be the sum of (a) his capital account, (b) his drawing account, and (c) his proportionate share of accrued net profits. If a net loss has been incurred to the date of dissolution, his share of such net loss shall be deducted. The assets of the partnership shall be valued at book value and no value shall be attributed to good will.

29. **Payment of Purchase Price.** The value of the partner's interest as determined in the above paragraph shall be paid without interest to the retiring partner, or to the legal representatives of the deceased or insane partner, as the case may be, in ______ monthly installments, commencing on the first day of the second month after the effective date of the purchase.

30. **Death of a Limited Partner.** In the event of the death of a limited partner, his personal representative during the period of administration of his estate shall succeed to his rights hereunder as a limited partner, and this interest as a limited partner may be assigned to any member of the family of the limited partner in distribution of his estate, or to any person in pursuance of a bequest in his last will and testament, and such member of the family [or person, if made by will] to whom such assignment or bequest is made, shall thereupon succeed to his interest as a limited partner and have all the rights of a substituted limited partner.

In Witness Whereof, the parties have signed and sealed this agreement.

[*Signatures and seals*] [8]

8. West's Modern Legal Forms § 6452.

FORM 3D

LIMITED PARTNERSHIP CERTIFICATE

We, the undersigned, for the purpose of forming a limited partnership pursuant to the Uniform Limited Partnership Act as set forth in Sections ______ of the ______ Code, hereby certify:

1. **Name.** The name of the partnership is ______.

2. **Character of Business.** The character of the business to be carried on is to engage in the business of ______.

3. **Place of Business.** The location of the principal place of business of the partnership is ______.

4. **General Partners.** The name and place of residence of each general partner are:

______ ______

______ ______

Limited Partners. The name and place of residence of each limited partner are:

______ ______

______ ______

5. **Term.** The term for which the partnership is to exist is indefinite [*or* from ______, 19__, to the close of business on ______, 19__, and thereafter from year to year].

6. **Initial Contribution of Each Limited Partner.** The amount of cash and a description of and the agreed value of the other property contributed by each limited partner are:

Name	Cash	Description of Other Property	Agreed Value of Other Property
______	______	______	______
______	______	______	______

7. **Additional Contributions of Each Limited Partner.** Each limited partner may (but shall not be obliged to) make such additional contributions to the capital of the partnership as may from time to time be agreed upon by the general partners.

8. **Return of Contribution to Each Limited Partner.** The contribution of each limited partner is to be returned to him as may from time to time be agreed upon by the general partners.

9. **Profit Shares of Each Limited Partner.** The share of the profits or other compensation by way of income which each limited partner shall receive by reason of his contribution is:

________________ ____________%

________________ ____________%

10. **Assignment of Limited Partner's Interest.** Each limited partner is given the right to substitute an assignee as contributor in his place, provided that the assignment is approved by all the general partners.

11. **Admission of Additional Limited Partners.** The general partners are given the right to admit additional limited partners, provided that the admissions are approved by all the general partners, but in no event other than upon a cash contribution to the partnership and upon the same terms as herein expressed.

12. **Death, Retirement or Insanity of General Partner.** In the event of the death, retirement or insanity of a general partner, the remaining general partners shall have the right to continue the business of the partnership under the same name by themselves or in conjunction with any other person or persons they may select.

13. **Right of Limited Partner to Receive Property Other Than Cash.** Each limited partner is given the right to demand and receive property other than cash in return for his contribution, and the value of such property shall be that shown on the books of the partnership.

Signed the ______ day of ______, 19__.

[*Signatures of general and limited partners*]

Subscribed and sworn to before me this ______ day of ______, 19__.

________________[9]

Notary Public

9. West's Modern Legal Forms § 6455.

FORM 3E

LIMITED PARTNERSHIP CERTIFICATE

(REVISED UNIFORM LIMITED PARTNERSHIP ACT)

We, the undersigned, for the purpose of forming a limited partnership pursuant to the Revised Uniform Limited Partnership Act as set forth in Sections ______ of the __________ Code, hereby certify:

1. **Name.** The name of the partnership is____________.

2. **Character of Business.** The character of the business to be carried on is to engage in the business of____________.

3. **Address and Agent.** The address of the office of the partnership is ____________, and the agent for service of process upon the partnership is____________.

4. **Members.** The name and the business address of each member of the partnership are as follows:

Name	Business Address	Type of Member
____________________	______________________	[General]
	______________________	
____________________	______________________	[Limited]
	______________________	
____________________	______________________	[Limited]
	______________________	

5. **Initial Contribution of Each Partner.** The amount of cash and a description and statement of the agreed value of other property or services contributed by each partner are as follows:

Name	Cash	Description of Property or Services	Agreed Value of Property or Services
________	____	____________________	__________________
________	____	____________________	__________________

6. **Additional Contributions.** The times or events which will require additional contributions to be made by each partner are as follows: ____

____

7. **Assignment of a Limited Partner's Interest.** Each limited partner is given the right to substitute an assignee as contributor in his or her place, provided that the assignment is approved by the general partners.

8. **Termination of Membership.** With sixty (60) days written notice to the general partners, any member of the partnership may terminate his or her membership in the partnership and receive a full distribution of his or her partnership interest in cash, provided, however, that no such distribution shall be made unless the assets of the partnership exceed the liabilities of the partnership on a ratio of at least 2:1.

9. **Distributions.** The partners may receive from the partnership from time to time such property of the partnership, including cash, as may be agreed upon by the general partners.

10. **Return of a Capital Contribution.** The general partners may, from time to time, as they agree distribute to the other partners such portions of the capital contributions of the other partners as the general partners may deem appropriate.

11. **Dissolution.** The partnership shall be dissolved and its affairs wound up upon the happening of any of the following:

a. Unanimous agreement by all members.

b. Death, insanity, disability or retirement of a general partner without a successor general partner having been elected within 90 days.

c. Sale or disposition of substantially all of the partnership property.

d. Any event which, in the opinion of the general partners, prevents the partnership from carrying on its ordinary business.

12. **Continuation of Business.** Notwithstanding any event of dissolution, the remaining members of the partnership may continue the business of the partnership without liquidation of the partnership by electing a successor or replacement general partner within 90 days from the event which causes the dissolution.

13. **Other Matters.** ______________________________

____

____

Dated this ________________ day of ____________, 19___.

[Signatures of general and limited partners]

Subscribed and sworn to before me this ___ day of ______, 19___.

____

Notary Public

FORM 3F

AMENDMENT OF LIMITED PARTNERSHIP CERTIFICATE

We, the undersigned, for the purpose of amending the certificate of limited partnership of ______, filed in the office of ______ on ______, 19__, hereby certify:

Whereas, the limited partner, ______, has assigned the whole of her interest in the limited partnership of ______ to ______, and ______ has been substituted as a limited partner in such partnership,

Paragraph 4 of the certificate of limited partnership is amended to read as follows:

"4. **General Partners.** The name and place of residence of each general partner are:

______ ______

______ ______

"**Limited Partners.** The name and place of residence of each limited partner are:

______ ______

______ ______"

Signed the ______ day of ______, 19__.

[*Signatures of general and limited partners including the substituted limited partner and the assigning limited partner*]

Subscribed and sworn to before me this ______ day of ______, 19__.

______[10]
Notary Public

10. West's Modern Legal Forms § 6458.

FORM 3G

STATEMENT OF CANCELLATION

Notice is hereby given that the limited partnership heretofore existing between ______, ______ and ______, under the firm name of ______, and doing the business of ______________________________

____

is hereby dissolved by mutual consent.

Dated this ______ day of ______, 19__, at ______, ______.

______________	______________
______________	______________
______________	______________
______________	______________
(General Partners)	(Limited Partners)

FORM 4A

SHAREHOLDER'S STATEMENT OF CONSENT AS TO TAXABLE STATUS UNDER SUBCHAPTER S

______, the undersigned, as a stockholder of ______ CORPORATION, hereby consents and agrees to the Corporation's election under Section 1372(a) to be treated as a "Small Business Corporation" for income tax purposes. It has been explained to me that the taxable income of the Corporation, to the extent that it exceeds dividends distributed in money out of earnings and profits of the taxable year, will be taxed directly to shareholders (rather than to the Corporation) to the extent that it would have constituted a dividend if it had been distributed on the last day of the Corporation's taxable year.

____

Shareholder

FORM 4B

INTERNAL REVENUE FORM 2553

Form **2553** (Rev. Jan. 1979) Department of the Treasury Internal Revenue Service

Election by a Small Business Corporation

(As to taxable status under subchapter S of the Internal Revenue Code)

Note: *This election under section 1372(a) (with the consent of all your shareholders) to be treated as an "electing small business corporation" for income tax purposes may be made only if the corporation meets all six of the requirements stated in instruction A. See section 1372(e) which describes certain conditions whereby the status of an electing small business corporation may be revoked or terminated.*

Name of corporation	Employer identification number (see Instruction L)	Principal business activity and specific product or service (see Instruction F)
Number and street		Election is to be effective for the taxable year beginning (Month, day, year)
City or town, State and ZIP code		Number of shares issued and outstanding (see Instruction E)
Is the corporation the outgrowth or continuation of any form of predecessor? . . . . . . ☐ Yes ☐ No If "Yes," state name of predecessor, type of organization, and period during which it was in existence ▶		Date and place of incorporation

If this election is effective for the first taxable year the corporation is in existence, complete A through H below, otherwise complete E through H.

A Date corporation first had shareholders	B Date corporation first had assets	C Date corporation began doing business	D Annual return will be filed for taxable year ending (month)

E Name and address (including ZIP code) of each shareholder	F Shareholders' Statement of Consent. We the undersigned shareholders consent to the election of the above corporation to be treated as an "electing small business corporation" under section 1372(a). (Signature of shareholders and date)	G Stock owned		H Social Security number
		Number of shares	Dates acquired	
1				
2				
3				
4				
5				
6				
7				
8				
9				
10				
11				
12				
13				
14				
15				

Note: *For this election to be valid, the consent of each shareholder must be shown above, or the consent of each shareholder must be attached to this form. (See instruction D.)*

Under penalties of perjury, I declare that I have examined this election, including accompanying schedules and statements, and to the best of my knowledge and belief it is true, correct, and complete.

Signature and Title of Officer ▶ Date ▶

Purpose

(References are to the Internal Revenue Code.)

The purpose of this election is to permit the undistributed taxable income of an "electing small business corporation" to be taxed to the shareholders rather than the corporation. The term "undistributed taxable income" means taxable income (as computed under section 1373(d)) minus the sum of (1) the tax imposed by sections 56 and 1378(a) and (2) the amount of money distributed as dividends out of earnings and profits of the taxable year.

Instructions

A. Corporations eligible to elect.—The corporation may make the election only if it meets all six of the following requirements:

1. it is a domestic corporation
2. it has no more than 15 shareholders

   Note: *For purposes of requirement 2 above, a husband and wife (and their estates) shall be treated as one shareholder.*
3. it has only individuals, estates, or certain trusts as shareholders
4. it has no nonresident alien shareholders
5. it has only one class of stock
6. it is not a member of an affiliated group of corporations (as defined in section 1504).

   Note: *A corporation is not considered a member of an affiliated group if it owns stock in a corporation that has not begun business before the close of the taxable year to which the election applies and does not have taxable income during that year.*

B. When to make the election.

1. For taxable years beginning after December 31, 1978, complete Form 2553 and file it either (1) any time during the preceding taxable year, or (2) any time during the first 75 days of the tax year. If an election is made after the first 75 days of the tax year and on or before the last day of such tax year, such election shall be treated as made for the following tax year.

2. For taxable years beginning before January 1, 1979, Form 2553 was required to be filed either (1) during the first month of that year, or (2) during the month before the first month. For example, a 1978 calendar year corporation must have made the election either in January 1978 or in December 1977 for the election to be effective for the 1978 tax year.

A prior year election which was not timely filed may be corrected if certain conditions are met. See section 5(d) of Public Law 95–628 for details concerning a perfecting election.

For purposes of this election, a new corporation's taxable year begins when it has shareholders, acquires assets, or begins doing business, whichever happens first. (See regulation 1.1372–2(b) for other details.)

The election will be effective for the taxable year for which it is made and for all later years unless it is terminated or revoked under section 1372(e).

C. Valid election.—The election will be valid only if all persons who are shareholders in such corporation on the day on which such election is made consent to such election.

D. Shareholder's statement of consent.—On the date of election, each shareholder must consent to the election either by signing the Shareholders' Statement of Consent (item F on Form 2553) or by signing a separate statement which must be attached to Form 2553 and must include:

1. the name and address of the corporation and of the shareholder,
2. the number of shares of stock owned by the shareholder,
3. the dates the shares were acquired, and
4. a statement that the shareholder consents to the corporation's election to be treated as a small business corporation under section 1372(a).

If you wish, you may incorporate the consents of all shareholders in one statement.

The consent must be signed by both husband and wife if they have a community interest in the stock or the income from it, and by each tenant in common, each joint tenant, and each tenant by the entirety.

The consent of a minor shall be made by the minor or the minor's legal guardian, or the minor's natural guardian if no legal guardian has been appointed (even in the case of stock held by a custodian for a minor under a statute patterned after the Uniform Gifts to Minors Act).

New Shareholder.—An election by a small business corporation shall terminate if a new shareholder (any person who was not a shareholder on the day on which the election was made) becomes a shareholder in such corporation and affirmatively refuses to consent to the election on or before the 60th day after the day on which the new shareholder acquires the stock.

The new shareholder's affirmative refusal to consent to the election must be filed with the Internal Revenue Service Center having jurisdiction for the area in which the principal business, office or agency of the corporation is located.

If the new shareholder is the estate of a decedent, the 60-day period for affirmatively refusing to consent to the election shall expire on the 60th day after (1) the day on which the executor or administrator of the estate qualifies, or (2) the last day of the taxable year of the corporation in which the decedent died, whichever is earlier.

Any termination of an election by reason of the affirmative refusal of any person to consent to such election shall be effective for the taxable year of the corporation in which such person becomes a shareholder in the corporation (or if later, the first taxable year for which such election would otherwise have been effective) and for all succeeding taxable years of the corporation. (See section 1372(e)(1).)

E. Number of shares issued and outstanding.—This block should contain only one figure for stock both issued and outstanding. This figure will be the number of shares of stock that have been issued to shareholders and have not been reacquired by the corporation. This number must equal the total number of shares owned by all shareholders as reported in item G of Form 2553.

F. Principal business activity and principal product or service.—In reporting the principal business activity give the one business activity that accounts for the largest percentage of "total receipts." "Total receipts" means gross sales and gross receipts, plus all other income. State the principal product or service as well as the principal business activity. For example, if the principal business activity is "Grain mill products," the principal product or service may be "cereal preparation." See Codes for Principal Business Activity at the back of the Instructions for Form 1120S, U.S. Small Business Corporation Income Tax Return.

G. Where to file.—File this election with the Internal Revenue Service Center where the corporation will file Form 1120S, U.S. Small Business Corporation Income Tax Return. **A copy should also be retained for the permanent files of the corporation.**

If the corporation's principal business, office or agency is located in ▼	Use the following Internal Revenue Service Center address ▼
New Jersey, New York City and counties of Nassau, Rockland, Suffolk, and Westchester	Holtsville, NY 00501
New York (all other counties), Connecticut, Maine, Massachusetts, New Hampshire, Rhode Island, Vermont	Andover, MA 05501
Alabama, Florida, Georgia, Mississippi, South Carolina	Atlanta, GA 31101
Michigan, Ohio	Cincinnati, OH 45999
Arkansas, Kansas, Louisiana, New Mexico, Oklahoma, Texas	Austin, TX 73301
Alaska, Arizona, Colorado, Idaho, Minnesota, Montana, Nebraska, Nevada, North Dakota, Oregon, South Dakota, Utah, Washington, Wyoming	Ogden, UT 84201
Illinois, Iowa, Missouri, Wisconsin	Kansas City, MO 64999
California, Hawaii	Fresno, CA 93888
Indiana, Kentucky, North Carolina, Tennessee, Virginia, West Virginia	Memphis, TN 37501
Delaware, District of Columbia, Maryland, Pennsylvania	Philadelphia, PA 19255

H. Signature.—This form must be signed by the president, vice president, treasurer, assistant treasurer, chief accounting officer, or any other corporate officer (such as tax officer) who is authorized to sign.

I. Election after termination or revocation.—If an election has been terminated or revoked under section 1372(e), see section 1372(f) and section 1.1372–5 of the regulations for the restrictions on eligibility to make a new election.

J. Investment credit property.—Section 47 and the regulations thereunder provide that investment credit property ceases to be investment credit property when a corporation makes a valid election under section 1372 to be an "electing small business corporation" and the tax recomputation provisions of section 47 will apply.

The corporation and its shareholders may, however, execute the agreement specified in section 1.47–4(b)(2) of the regulations so that the recapture provisions of section 1.47–1(a) of the regulations will not apply to the section 38 property.

K. Work incentive (WIN) program credit.—Section 1.50A–5(b)(1) of the regulations provides that certain WIN wages paid prior to January 1, 1979, for which the WIN credit was claimed will cease to qualify as WIN wages when a corporation makes a valid election under section 1372 to be an "electing small business corporation." Therefore, the recapture provisions of section 1.50A–3 of the regulations may apply. However, the corporation and its shareholders may execute the agreement specified in section 1.50A–5(b)(2) of the regulations so that the recapture provisions of section 1.50A–3 of the regulations will not apply to the WIN expenses because of the corporation's election under section 1372.

L. Employer identification number.—Corporations that have not applied for an employer identification number should enter "not applied for." If a number has been applied for but not received, enter "applied for."

Corporations which do not have an EIN should apply for one on Form SS–4, available from any IRS or Social Security Administration office. Send Form SS–4 to the same Internal Revenue Service Center to which Form 1120S is sent.

FORM 4C

RESOLUTION AUTHORIZING ISSUANCE OF SECTION 1244 STOCK

Whereas, the Corporation is authorized to offer and issue 500 shares of common stock, no par value, none of which has yet been issued, and the directors desire to hereafter issue 100 shares of said common stock for $20,000;

Whereas, A_______ B_______ and C_______ D_______ have expressed the wish that each be permitted to subscribe for 50 shares of said common stock at a price of $200 a share; and

Whereas, the Corporation is a domestic corporation meeting the definition of a "small business corporation" contained in section 1244 (c) (2) of the Internal Revenue Code of 1954;

Upon motion duly made, seconded, and unanimously carried, it was

Resolved, that the Corporation hereby adopts a plan, effective this date, to offer 50 shares of common stock, no par value, each to A_______ B_______ and C_______ D_______ in consideration of $10,000 to be paid by each only in money and property acceptable to the Corporation (other than stock or securities), provided, however, the said A_______ B_______ and C_______ D_______ shall within two weeks after the date of adoption of this plan notify the Corporation in writing of the acceptance of this offer, and during the said two week period the Corporation shall offer and issue only such common stock. The maximum amount to be received by the Corporation in consideration of the stock to be issued pursuant to this plan shall be $20,000. It is the intention of the officers and directors of the Corporation to comply in every respect with section 1244 of the Internal Revenue Code of 1954 pertaining to "Losses on Small Business Stock", and any questions concerning the interpretation or operation of this plan shall be resolved in such manner as will qualify the plan under said law. The officers of the Corporation are hereby authorized, empowered and directed to do and perform any and all acts and deeds necessary to carry out the plan.[11]

11. West's Modern Legal Forms § 2909.-5.

FORM 4D

PLAN FOR ISSUANCE OF SECTION 1244 STOCK

1. The corporation shall offer and issue under this Plan, a maximum of _______ shares of its common stock at a maximum price of _______ ($_______) per share.

2. This offer shall terminate, unless sooner terminated by: ____ ____

(a) Complete issuance of all shares offered hereunder, or

(b) Appropriate action terminating the same by the Board of Directors and the Stockholders, or

(c) By the adoption of a new Plan by the Stockholders for the issuance of additional stock under Section 1244, Internal Revenue Code.

3. No increase in the basis of outstanding stock shall result from a contribution to capital hereunder.

4. No stock offered hereunder shall be issued on the exercise of a stock right, stock warrant, or stock option, unless such right, warrant, or option is applicable solely to unissued stock offered under the Plan and is exercised during the period of the Plan.

5. Stock subscribed for prior to the adoption of the Plan, including stock subscribed for prior to the date the corporation comes into existence, may be issued hereunder, provided however, that the said stock is not in fact issued prior to the adoption of such Plan.

6. No stock shall be issued hereunder for a payment which, along or together with prior payments, exceeds the maximum amount that may be received under the Plan.

7. Any offering or portion of an offer outstanding which is unissued at the time of the adoption of this Plan is herewith withdrawn. Stock rights, stock warrants, stock options or securities convertible into stock, which are outstanding at the time this Plan is adopted, are likewise herewith withdrawn.

8. Stock issued hereunder shall be in exchange for money or other property except for stock or securities. Stock issued hereunder shall not be in return for services rendered or to be rendered to, or for the benefit of, the corporation. Stock may be issued hereunder however, in consideration for cancellation of indebtedness of the corporation unless such indebtedness is evidenced by a security, or arises out of the performance of personal services.

9. Any matters pertaining to this issue not covered under the provisions of this Plan shall be resolved in favor of the applicable

law and regulations in order to qualify such issue under Section 1244 of the Internal Revenue Code. If any shares issued hereunder are finally determined not to be so qualified, such shares, and only such shares shall be deemed not to be in this Plan, and such other shares issued hereunder shall not be affected thereby.

10. The sum of the aggregate amount offered hereunder plus the equity capital of the corporation amounts to $______.

11. The date of adoption of this Plan is ______, 19__.[12]

12. See West's Modern Legal Forms 2909.6.

FORM 5A

CONTRACT BETWEEN STOCKHOLDERS ORGANIZING A CLOSE CORPORATION

Agreement, made this ______ day of ______, 19__, between A____ B____ of ______, and C____ D____ of ______ (hereinafter referred to as the "Shareholders").

Whereas, the Shareholders have caused ______ Corporation to be organized as a corporation under the laws of the State of ______, and have agreed that it shall be financed and its business conducted subject to the provisions of this agreement.

Now, therefore, in consideration of the mutual covenants herein contained, it is agreed:

1. **Subscription to Stock.** The Shareholders each subscribe for and agree to purchase ______ shares each of the capital stock of the Corporation at $______ per share. These shares shall be issued and paid for within ______ days after the organization of the Corporation.

2. **Loan to Corporation.** The Shareholders each agree to loan to the Corporation the sum of $______, to be used for the purposes of the business of the Corporation, such loan to be repaid at the convenience of the Corporation, with interest thereon at ______ per cent per annum.

3. **Employment.** The Corporation shall employ A____ B____ and C____ D____ each at a salary of $______ per week. A____

B_____ and C_____ D_____ each agree to accept such employment, to devote their full time and best efforts to the business of the Corporation, and not to engage in any other competing business, directly or indirectly. Such salary shall be subject to increase or decrease and the term of employment of A_____ B_____ and C_____ D_____ may be terminated only by vote of the Board of Directors of the Corporation in accordance with the provisions contained in the Certificate of Incorporation.

4. **First Option on Termination of Employment.** In the event that either A_____ B_____ or C_____ D_____ shall at any time, for any reason whatsoever, leave the employ of the Corporation or cease to be actively engaged in the business of the Corporation, all of the shares owned by such Shareholder shall be offered for sale to the other Shareholder, who is hereby given an option for a period of ______ days from the date on which such employment or activity shall terminate, to purchase all of such shares at a price equal to the book value thereof. Book value of shares shall be computed from the books of the Corporation maintained by its regular accountant in accordance with generally accepted principles of accounting. The option hereby given shall relate to all of such shares of the offeror, and the offeree shall not have the right to purchase only part thereof. If the aforesaid offer is accepted, notwithstanding any of the foregoing provisions of this paragraph, the offeror shall receive from the offeree not less than the value of his investment in the Corporation plus the amount of any unpaid loan theretofore made by the offeror to the Corporation, with appropriate interest thereon to the date of purchase. Payments to be made under this paragraph shall be made as follows: ______ upon the acceptance of the offer; ______ ______ months thereafter; and the final ______ ______ months thereafter. Title to the shares shall pass to the offeree only upon the completion of all payments. After the payment of the first installment, the offeror shall hold such shares only as security for payment of the remaining installments, and the offeree shall have the sole right to vote the shares and to collect all dividends and other distributions thereon. Upon payment of the last installment, the shares shall be transferred of record to the offeree.

5. **Restriction on Transfer of Stock.** Each of the Shareholders expressly agrees not to transfer, sell, assign, pledge or otherwise in any manner dispose of or encumber any of his shares unless and until he shall have offered to sell all of his shares to the other Shareholder at a price to be computed and to be paid as specified in paragraph 4 above. Such offer shall be made in writing and shall continue for ______ days from the date thereof.

6. **Legend on Stock Certificates.** All stock certificates issued by the Corporation shall have marked on the face thereof "Subject to

provisions of Stockholders Agreement dated ______, 19__ restricting transfer." No dividend shall be paid on any shares transferred, pledged, assigned or encumbered in breach of this agreement.

7. **Death and Disability.** Upon the death of any Shareholder who is also an employee, his salary shall be paid to his widow or next of kin for ______ weeks following such death. If any Shareholder shall become physically incapacitated and unable to attend to his duties as an employee of the Corporation, he shall continue to receive his full salary (less the sum required to employ a substitute in his place) for a period of ______ months after the commencement of such incapacity. In the event of the death, or incapacity of any shareholder-employee for more than ______ months, the other Shareholder shall have the option, for ______ days after such death or expiration of said ______ month period, to purchase his shares at the price and on the terms provided for in paragraph 4 hereof. The life of the other Shareholder shall be insured for the benefit of the other Shareholder for $______, or for such other amount as the Shareholders may jointly agree upon. If the proceeds of such insurance payable to any Shareholder are equal to at least ______ per cent of the purchase price of the stock of the deceased Shareholder as computed in accordance with the provisions of paragraph 4, such Shareholder agrees that he will exercise his option to purchase all of the shares from the estate of the deceased Shareholder as herein provided. Upon the receipt of any such proceeds, any then remaining unpaid installments of such purchase price shall be prepaid by the purchaser, to at least the extent of such proceeds.

8. **Election of Directors.** Each Shareholder agrees, so long as he shall remain a Shareholder, to vote his shares for the election of the following four persons as Directors of the Corporation:

A____ B____ (or such other person as is designated by A____ B____)

C____ D____ (or such other person as is designated by C____ D____)

and generally to so vote at directors' and stockholders' meetings of the Corporation as to carry out and make effective all the terms and provisions of this agreement.

9. **Appointment of Officers.** So long as they are faithful, efficient and competent in the performance of their duties, the following persons shall be supported by the Shareholders for election to offices of the Corporation:

President and Treasurer	A____ B____
Vice President and Secretary	C____ D____

10. **Arbitration.** All disputes, differences and controversies arising under and in connection with this agreement shall be settled and finally determined by arbitration in the City of ______ according to the rules of the American Arbitration Association now in force or hereafter adopted.

11. **Duration.** This agreement shall continue in force during the entire period of the life of the Corporation.

12. **Successors.** This agreement and all provisions hereof shall enure to the benefit of and shall be binding upon the heirs, executors, legal representatives, next of kin, transferees and assigns of the parties hereto.

13. **Severability.** If for any reason any provision hereof shall be inoperative, the validity and effect of all other provisions shall not be affected thereby.

14. **Modifications.** No modification or waiver of any provision of this agreement shall be valid unless in writing signed by all of the parties.

In witness whereof, the parties have signed this agreement on the day and year first above written.

____

____

Confirmed and Agreed to:
______ Corporation

____
President

Attest:

______________________[13]
Secretary

13. West's Modern Legal Forms § 2432.-15.

FORM 5B

ARTICLES OF INCORPORATION OF A MEDICAL CORPORATION

ARTICLES OF INCORPORATION
OF

____

We, the undersigned, hereby associate ourselves together for the purpose of becoming a professional corporation for profit under the provisions of ______, Statutes, as amended by "The Professional Service Corporation Act" of the State of ______, and pursuant to the following Articles of Incorporation:

ARTICLE I. NAME

The name of this corporation shall be ______.

ARTICLE II. PURPOSE

The general nature of the business to be transacted by the corporation shall be and is to engage in every aspect of the general practice of medicine. The professional services involved in the corporation's practice of medicine may be rendered only through its officers, agent and employees who are duly authorized and licensed to practice medicine in the State of ______.

This corporation shall not engage in any business other than the practice of medicine. However, this corporation may invest its funds in real estate, mortgages, stocks, bonds, and other types of investments, and may own real and personal property necessary for the rendering of the professional services authorized hereby.

ARTICLE III. CAPITAL STOCK

The maximum number of shares of stock that the corporation is authorized to have outstanding at any time shall be ______ shares of the par value of one dollar ($1.00) per share, all of which shall be common stock of the same class. All stock issued shall be fully paid and non-assessable. The stockholders shall have no pre-emptive rights with respect to the stock of the corporation, and the corporation may issue and sell its common stock from time to time without offering such shares to the stockholders then holding shares of common stock. Shares of the corporation's stock and certificates therefore shall be issued only to doctors authorized and licensed to practice medicine in the State of ______.

ARTICLE IV. INITIAL CAPITAL

The amount of capital with which this corporation will begin business shall be and is the sum of ______ dollars.

ARTICLE V. DURATION

The corporation shall have perpetual existence.

ARTICLE VI. PRINCIPAL OFFICE

The principal office of this corporation shall be located in the City of ______, County of ______, State of ______, and the post office address of said principal office of the corporation shall be ______.

ARTICLE VII. NUMBER OF DIRECTORS

The number of directors of this corporation shall be not less than three (3) nor more than five (5).

ARTICLE VIII. INITIAL BOARD OF DIRECTORS

The names and post office addresses of the members of the first Board of Directors, who, subject to the provisions of the Bylaws and these Articles of Incorporation, shall hold office for the first year of the corporation's existence or until their successors are elected and have qualified, are as follows:

Names	Addresses
______	______
______	______
______	______
______	______
______	______

ARTICLE IX. SUBSCRIBERS

The name and post office address of each subscriber of these Articles of Incorporation are as follows:

Names	Addresses
______	______
______	______
______	______
______	______
______	______

The subscribers certify that the proceeds of the stock subscribed for will not be less than the amount of capital with which the corporation will begin business, as set forth in Article IV hereinabove.

ARTICLE X. STOCKHOLDERS

The stock of this corporation may be issued, owned and registered only in the name or names of an individual or individuals who are duly authorized and licensed to practice medicine in the State of ______ and who are employees, officers or agents of this corporation. In the event that a stockholder;

(a) becomes disqualified to practice medicine in this State, or

(b) is elected to a public office or accepts employment, that pursuant to law, places restrictions or limitations upon his continued rendering of professional services as a medical doctor, or

(c) ceases to be an employee, officer or agent of the corporation, or

(d) sells, transfers, hypothecates or pledges, or attempts to sell, transfer, hypothecate or pledge any shares of stock in this corporation to any person ineligible by law or by virtue of these Articles to be a shareholder in this corporation, or if such sale, transfer, hypothecation or pledge or attempt to sell, transfer, hypothecate or pledge is made in a manner prohibited by law, or in a manner inconconsistent with the provisions of these Articles, or the Bylaws of this corporation, or

(e) suffers an execution to be levied upon his stock, or such stock is subjected to judicial sale or other process, the effect of which is to vest any legal or equitable interest in such stock in some person other than the stockholder,

then the stock of such stockholder shall immediately stand forfeited and such stock shall be immediately cancelled by this corporation and the stockholder or other person in possession of such stock shall be entitled only to receive payment for the value of such stock, which said value shall be the book value thereof as of the last day of the month preceding the month in which any of the events above enumerated occurs. The stockholder whose stock so becomes forfeit and is cancelled by the corporation, shall forthwith cease to be an employee, officer, director or agent of the corporation and except to receive payment for his stock in accordance with the foregoing and payment of any other sums then lawfully due and owing to said stockholder by the corporation, such stockholder shall then and thereafter have no further financial interest of any kind in this corporation.

ARTICLE XI. DEATH OF STOCKHOLDER

Upon the death of a stockholder, his stock shall be subject to purchase by the corporation or by the other stockholders at such price and upon such terms and conditions and in such manner as may be provided for in the Bylaws of this corporation, in a manner consistent with law and these Articles.

ARTICLE XII. SALE OF STOCK

No stockholder of this corporation may sell or transfer any of such stockholder's shares of stock in this corporation except to another individual who is then duly authorized and licensed to practice medicine in the State of ______ and then only after the proposed sale or transfer shall have been first approved, at a stockholders' meeting specially called for such purpose, by such proportion, not less than a majority, of the outstanding stock excluding the shares of stock proposed to be sold or transferred, as may be provided from time to time in the Bylaws. In such stockholders' meeting, the shares of stock proposed to be sold or transferred may not be voted or counted for any purpose.

The corporation's shareholders are specifically authorized from time to time to adopt Bylaws not inconsistent herewith restraining the alienation of shares of stock of this corporation and providing for the purchase or redemption by the corporation of its shares of stock.

ARTICLE XIII. REGULATION OF BUSINESS

In furtherance of and not in limitation of the powers conferred by statute, the following specific provisions are made for the regulation of the business and the conduct of the affairs of the corporation:

1. **Management.** Subject to such restrictions, if any, as are herein expressed and such further restrictions, if any, as may be set forth in the Bylaws, the Board of Directors shall have the general management and control of the business and may exercise all of the powers of the corporation except such as may be by statute, or by the articles of incorporation or amendment thereto, or by the Bylaws as constituted from time to time, expressly conferred upon or reserved to the stockholders.

2. **Officers.** The corporation shall have such officers as may from time to time be provided in the Bylaws and such officers shall be designated in such manner and shall hold their offices for such terms and shall have such powers and duties as may be prescribed by the Bylaws or as may be determined from time to time by the Board of Directors subject to the Bylaws.

3. **Contracts.** No contract or other transaction between the corporation and any other firm, association or corporation shall be affected or invalidated by the fact that any one or more of the directors of the corporation is or are interested in or is a member, director or officer or are members, directors or officers of such firm or corporation and any director or directors individually or jointly

may be a party or parties to or may be interested in any contract or transaction of the corporation or in which the corporation is interested; and no contract, act or transaction of the corporation with any person, firm, association or corporation shall be affected or invalidated by the fact that any director or directors of the corporation is a party or are parties to or interested in such contract, act or transaction or in any way connected with such person, firm, association or corporation, and each and every person who may become a director of the corporation is hereby relieved from any liability that might otherwise exist from contracting with the corporation for the benefit of himself or any firm, association or corporation in which he may in any way be interested.

ARTICLE XIV. AMENDMENTS

This corporation reserves the right to amend, alter, change, or repeal any provision contained herein in the manner now or hereafter prescribed by law, and all rights conferred on stockholders herein are granted subject to this reservation.

In Witness Whereof, each subscriber has signed these Articles of Incorporation.

____

____

____

____

____

____

[*Acknowledgment*] [14]

14. West's Modern Legal Forms § 3158.-7.

FORM 5C

APPLICATION FOR REGISTRATION OF PROFESSIONAL CORPORATION (CALIFORNIA)

File No. ______
(To be filled in by Board)
Fee: $100.00

APPLICATION
for issuance of
CERTIFICATE OF REGISTRATION AS A MEDICAL CORPORATION
(Section 2501 of the Business and Professions Code)

1. ____
(Name of Applicant)

a professional corporation, hereby requests issuance to it of a Certificate of Registration as a medical corporation.

2. Applicant will do business as (fictitious name) ______________
____
(See Section 2393 of the Business and Professions Code and Section 13409 of the Corporations Code)

3. The corporation number assigned to the applicant by the California Secretary of State is ______.

4. Date of incorporation ______.

5. A. The address of applicant's principal office is:
____

B. The address of all other offices of applicant are:
____
____
____

C. Applicant's telephone number is: ______________
(Area Code) (Number)

6. The directors of the applicant are:

NAME	ADDRESS	PROFESSIONAL LICENSE NUMBER

(File supplemental sheet if more space required)

7. The officers of the applicant are:

   (If any officers are not licensed persons, so indicate. See Section 13403 of the Corporations Code.)

NAME	ADDRESS	PROFESSIONAL LICENSE NUMBER
President ______		
Vice-President ______		
Secretary ______		
Treasurer ______		
Asst. Secretary ______ (Need not be a licensed person. See Section 2501 of the Business and Professions Code.)		
Asst. Treasurer ______ (Need not be a licensed person. See Section 2501 of the Business and Professions Code.)		

8. The shareholders of the applicant are:

NAME	ADDRESS	PROFESSIONAL LICENSE NUMBER
(1) ______		
(2) ______		
(3) ______		
(4) ______		
(5) ______		

(File supplemental sheet if more space required)

9. The employees of the applicant rendering professional services are:

NAME	ADDRESS	PROFESSIONAL LICENSE NUMBER
(1) ______		
(2) ______		
(3) ______		
(4) ______		
(5) ______		

(File supplemental sheet if more space required)

10. The bylaws of the applicant adopted on ______ comply with Medical Corporation Rule 1378.6 in that:

    (a) Ownership of shares of the applicant may be owned only by a medical corporation, or by a licensed physician and surgeon or podiatrist, as the case may be.

(b) The income of the applicant attributable to medical services rendered while a shareholder is a disqualified person shall not in any manner accrue to the benefit of such share holder or his shares.

(c) The share certificates of the applicant contain a legend setting forth the restrictions of sections (a) and (b) above, and, where applicable, the restrictions of section (d) below.

and, where applicable:

(d) Where there are two or more shareholders in the corporation and one of the shareholders:

(1) Dies;

(2) Ceases to be an eligible shareholder; or

(3) Becomes a disqualified person as defined in Section 13401(d) of the Corporations Code, for a period exceeding ninety (90) days,

his shares shall be sold and transferred to the corporation, its shareholders, or other eligible persons, on such terms as are agreed upon. Such sale or transfer shall be not later than six (6) months after any such death and not later than ninety (90) days after the date he ceases to be an eligible shareholder, or ninety (90) days after the date he becomes a disqualified person. The requirements of subsections (a) and (b) of this section shall be set forth in the medical corporation's articles of incorporation or bylaws, except that the terms of the sale or transfer provided for in said subsection (b) need not be set forth in said articles or bylaws if they are set forth in a written agreement.

(e) The applicant and its shareholders may, but need not, agree that shares sold to it by a person who becomes a disqualified person may be resold to such person if and when he again becomes an eligible shareholder.

11. Security for claims against applicant.
(Check one)

☐ Applicant is insured as provided in Section 1378.5(a) of the Medical Corporation Rules as evidenced by the Cerificate of insurance attached as Exhibit C.

☐ Applicant is not insured.

(NOTE: Under Section 1378.5(b) of the Medical Corporation Rules all shareholders of the corporation shall be jointly and severally liable for all claims established against the corporation by its patients arising out of the rendering of, or failure to render, medical services up to the minimum amounts specified for insurance under subsection (a) hereof except during periods of time when the corporation shall provide and maintain insurance for claims against it by its patients arising out of the rendering of, or failure to render medical services.)

12. Applicant is an existing corporation and its organization, bylaws, articles of incorporation and general plan of operation are such that its affairs will be conducted in compliance with the Medical Practice Act, the Professional Corporations Act, and other applicable provisions of the Corporations Code, the Medical Corporation Rules of the Board of Medical Examiners and such other law, rules and regulations as may be applicable.

13. Enclosed herewith are the following exhibits:

    A. Articles of Incorporation, certified by the Secretary of State. (Section 2501 of the Business and Professions Code.)

    B. Bylaws certified by the Secretary of the applicant corporation. (Section 2501 of the Business and Professions Code.)

    C. Certificate of insurance.
    (Must be filed if applicant is insured. Section 1378.5(a) Medical Corporation Rules.)

    D. Notice of Liability of Shareholders (Section 1378.5(b) Medical Corporation Rules.)

    Executed this ______ day of ______, 19__.

____

[*Name of Corporation*]

By ______________________

[*Type name*]

____

[*Title of person executing*]

____

[*Signature*]

DECLARATION

I am an officer of ______________, and as such make this declara-
(Name of Applicant)

tion for and on behalf of said corporation. I have read the foregoing application and all attachments thereto and know the contents thereof, and the same are true of my own knowledge. I declare, under penalty of perjury, that the foregoing is true and correct.

Executed at ______, California, this ______ day of ______, 19___.

____
(Signature)

______________________[15]
(Title)

15. West's Modern Legal Forms § 3158.-4.

FORM 6A

PRE-INCORPORATION AGENDA AND INFORMATION SHEET

These forms are intended for use at an initial client interview as information gathering forms, and as external checklists to assist the attorney in following up with the client, the accountant, the insurance agent and others so that all aspects of the incorporation are accomplished in a timely fashion and without duplication of effort.

These forms should be filled out during the client interview and reviewed with the legal assistant when the assignment is made. A copy is to be mailed to the client, the accountant and insurance agent when acknowledging the engagement.

The Agenda should be kept in the file and reviewed periodically during the incorporation process; follow-up letters may be generated by these reviews.

PRE-INCORPORATION AGENDA

FOR

____

Item	Responsibility	To Be Completed	Date Completed
1. Reserve Corporate Name			
2. Draft and File Articles of Incorporation			
3. Prepare Initial Organizational Consent or Minutes			
4. Prepare Bylaws			
5. Additional Organization Documents:			
a) Employment Agreements			
b) Service Agreements			
c) Medical and Dental Expense Reimbursement Plan			
d) Shareholders Agreement			
e) Share Certificates			
f) Transfer Documents			
g) Bank Resolution			
h) Subchapter S Election			
6. Order Corporate Kit			
7. Send Explanatory Transmittal Letter			
8. Other:			

Client: ______________________
File No.: ______________________

PRE-INCORPORATION INFORMATION SHEET

1. Date Information Supplied: ______________________
   Parties Present: ______________________
   Lawyer: ______________________
   Client: ______________________
   Other Parties: ______________________
2. State of Incorporation: ______________________
3. Proposed Date of Incorporation: ______________________
4. Name of Corporation: ______________________
   Alternative Name: ____________/____________
   Trade Name: ______________________
5. Name Reserved? YES (__) NO (__) SHOULD BE (__)
6. Will be Incorporating a Going Business? ______________________
   If so, describe generally: ______________________
   ____
   Any Patents, Copyrights or Trademarks to be Registered or Transferred: ______________________
7. Principal Purpose of Corporation: ______________________
   ____
   ____
   (State broadly, but with sufficient specificity to meet statutory requirements.)
8. Principal Place of Business: ______________________
   ______________________ ☐ Own ☐ Lease
   Other places of significant business activity or presence:

Address	Description	Own/Lease

9. Qualification in other states required. YES (__) NO (__)
   If so, what states: ______________________
   ____
10. Registered Agent and Office in State: ______________________
    ____
    Registered Agent and Office in states in which qualified:
    ____
    ____
    (Do not commit lawyer to be registered agent.)

11. Common Stock (if preferred, explain details):

Class	Number of Shares	Par Value	Issue Price
______	______	______	______
______	______	______	______

Explain details: ______________________________

Will there be a formal stock subscription agreement: ______________________________

If no par value, amount of consideration to be allocated to capital surplus: ______________________________

Note: If stock is no par value, stated capital of the company will be equal to actual consideration paid for issued stock except that part of consideration that directors may allocate to capital surplus.

12. Anticipated Shareholders:

	Name and Phone	Address	Zip Code
(1)	______	______	______
(2)	______	______	______
(3)	______	______	______
(4)	______	______	______

13. Section 1244 Plan: ______________________________
Date of Commencement of Offer: ______________________________

Note: Section 1244 stock permits deduction of capital losses, if any, upon sale of such stock as ordinary losses against ordinary income of Section 1244 stock owner. See Section 1244 of the Internal Revenue Code for details.

14. Share Ownership and Consideration (corresponding to above stockholders):

	Number of Shares	Consideration	Date to be Paid
(1)	______	______	______
(2)	______	______	______
(3)	______	______	______
(4)	______	______	______

15. Initial Indebtedness: Secured $__________ Unsecured $__________

Note: (1) If capital structure of the company will involve debt, advised debt equity ratio should not be in excess of 3/1 and recommend 2/1 if workable for participants. Point out that any debt must specifically be treated as debt by the company, or the IRS is likely to characterize any such debt securities as stock.

Note: (2) If prospective contributor to corporate capital will take back note to secure corporate debt, make sure note will qualify as a "long-term security" under § 351 of the Internal Revenue Code. Otherwise, transfer of property to company may constitute a taxable event (i. e., a "sale") which will cause realization of probable capital gain to taxpayer-transferror.

16. Incorporators:

Name	Address (If not shown above.)	Zip Code
______	______	______
______	______	______
______	______	______
______	______	______
______	______	______

17. Directors:

Name and Phone	Address (If not shown above.)	Zip Code
______	______	______
______	______	______
______	______	______
______	______	______

18. Officers:

Name	Office	Address
______	President	______
______	Treasurer	______
______	Secretary	______
______	______	______

19. Preemptive Rights. YES (____) NO (____) Restrictions______

Note: See explanatory material in Corporate Law Notebook or in B. M. Miller, *Manual and Guide for the Corporate Secretary* 815–844 (1969), for background prior to client conference.

20. Cumulative Voting. YES (____) NO (____)

Note: See explanatory material in Corporate Law Notebook or in B. M. Miller, *supra* at 89–94, for background prior to client conference.

21. Date and Time of Annual Meeting: ______________________

____

(e. g.: "____ days following close of fiscal year" or "held each year prior to the ____ day of ______.")

For provision in Bylaws: "In the event the Board of Directors fails to so fix the date and time of such meeting, it shall be held on the ____ in ______ at ____a. m. (e. g., on the first Tuesday in March at 10:00 a. m.)

22. Bank: ______________________________________

Signatories: ______________________________________

Limitations: ______________________________________

Should we obtain banking resolutions: YES (____) NO (____)

23. Commencement of Employment, if any: ______________

Note: If business is already in progress, obtain Employer Identification Number: ______________________

24. Date of First Meeting of Board of Directors: ______________

25. Stock Transfer Restrictions? YES (_____) NO (_____) Special provisions:

____

____

Insurance Funded: YES (_____) NO (_____)
Insurance Company or Agent ____________________

26. Ownership of Real Property (list states and counties):

____

____

Ownership of Personal Property:

____

____

27. Custody of Corporate Minute Book, Stock Book, and Seal:
Client (_____) Us (_____)
Other Custodian: ____________________________

28. Supplemental Checklist

☐ COMPARE FEATURES OF CORPORATION LAWS OF FOLLOWING STATES: ____________________

____

☐ COMPARE ORGANIZATION FEES AND TAXES

☐ CHECK COSTS OF QUALIFICATION IN FOREIGN STATES

☐ CHECK ANNUAL FEES AND TAXES

☐ CHECK STATE TAX SAVINGS WHICH MAY BE EFFECTED BY SCHEDULING INCORPORATION (AND ANY QUALIFICATIONS) BEFORE OR AFTER CERTAIN DATES

29. Miscellaneous Advice to Include in Cover Letter to Client:

(a) Securities Laws Considerations (*i.e.*, investment letter): _____

____

(b) Retail Sales Tax, Use Tax, and Store Licenses: __________

____

(c) State License to do Business Compliance: ______________

____

(d) Qualification in Other States: ____________________

____

(e) State Workmen's Compensation Insurance Compliance: _____

____

(f) State Unemployment Insurance Compliance: ____________

____

(g) Obtain Federal Employer Identification Number: __________

____

(h) Employer's Tax Guide re Withholding Requirements (Federal and State): __________________________

____

(i) State Consumer Credit Compliance (Notification): ______
______

(j) Local Retail Sales Tax License: ______
______

(k) Local Use Tax License: ______
______

(l) Local Occupational Privilege Tax Compliance: ______
______

(m) "Tax Information on Subchapter S Corporations" (IRS Publication 589): ______
______

(n) "Corporations and the Federal Income Tax" (IRS Publication 542): ______
______

(o) Other Matters: ______
______
______

NOTES AND COMMENTS

______
______
______
______
______
______
______
______
______
______
______
______
______
______
______
______
______
______
______
______

______
Attorney handling
Incorporation

FORM 6B

AGREEMENT BETWEEN PROMOTERS

Agreement, made this ______ day of ______, 19__, between A____ B____ of ______, C____ D____ of ______, and E____ F____ of ______.

Whereas, the parties desire to form a corporation upon the terms and conditions set forth in this agreement.

Now, therefore, it is agreed:

1. **Formation of the Corporation.** The parties shall as soon as possible form a corporation under the laws of the State of ______.

2. **Certificate of Incorporation.** The certificate of incorporation shall provide substantially as follows:

(a) The name of the corporation shall be ______, or if this name is not available such other name as the parties shall select.

(b) The principal office or place of business of the corporation shall be located at ______. The name and address of its resident agent shall be ______.

(c) The purpose of the corporation shall be the manufacture and wholesale distribution of textile fabrics. The corporation shall have such powers as may be appropriate in connection with such a business.

(d) The names and places of residence of each of the incorporators are:

______________	______________
______________	______________
______________	______________

(e) The corporation shall have perpetual existence.

(f) The minimum amount of capital with which the corporation shall commence business is $1,000.

(g) The total number of shares of stock shall be 1,000, divided into two classes as follows:

Common Stock, $10 par value	5000 shares
Preferred Stock, $100 par value	500 shares

(h) The designations, the powers, preferences and rights, and the qualifications, limitations or restrictions of such stock are: [*Here describe*].

3. **Subscriptions of Parties.** The parties subscribe for shares of stock of the proposed corporation, as follows:

(a) Within one week after the certificate of incorporation has been filed and recorded the corporation shall issue to A____ B____

______ shares of common stock of the corporation, $10 par value, in consideration of the simultaneous execution and delivery to the corporation of a deed transferring marketable title to the following described real property, free and clear of liens: [*Here describe*].

(b) Within one week after the certificate of incorporation has been filed and recorded the corporation shall issue to C____ D____ ______ shares of common stock of the corporation, $10 par value, in consideration of the simultaneous execution of an agreement assigning to the corporation the inventions of C____ D____ relating to ______ as set forth in applications filed in the United States Patent Office and identified as follows: [*Here describe*].

(c) Within one week after the certificate of incorporation has been filed and recorded the corporation shall issue to E____ F____ ______ shares of preferred stock of the corporation, $100 par value, in consideration of the simultaneous payment by E____ F____ to the corporation of the sum of $______ in cash.

4. **Agreement to Purchase Additional Stock.** E____ F____ agrees to purchase additional preferred stock not to exceed $______ in par value if during the first two years of the operation of the corporation its net profits do not equal at least $______.

5. **Stock to Promoter for Services.** The corporation shall issue to ______ of ______, ______ shares of common stock of the corporation, par value $______, in consideration for his services in organizing the corporation.

6. **First Directors of Corporation.** The directors of the corporation for the first year shall be ______, ______ and ______.

7. **Employment Contracts.** The corporation shall employ A____ B____ as president and general manager and C____ D____ as secretary-treasurer, each for a term of 5 years, at a salary of $______ per year for A____ B____ and $______ per year for C____ D____. Their employment shall not be terminated without cause and their salary shall not be increased or decreased without unanimous approval of all directors. Written employment contracts shall be entered into with A____ B____ and C____ D____ wherein they agree to devote their time and best efforts exclusively to the business and interests of the corporation.

8. **Restrictions on Transfer of Stock.** Each of the parties agrees not to transfer, sell, assign, pledge or otherwise dispose of his shares of stock of the corporation without first obtaining the written consent of the other parties to the sale or other disposition, or without first offering to sell the shares to the corporation at a value to be determined by a board of 3 appraisers one of whom shall be appointed by each of the parties, A____ B____, C____ D____ and E____ F____.

The offer shall be in writing and shall remain open for 30 days. If the corporation fails to accept the offer within that period, a second offer also in writing shall then be made to sell the shares on similar terms to the other parties to this agreement pro-rata. If the offer be not accepted by either the corporation or the other parties the shares shall thereafter be freely transferable.

9. **Designation of Incorporators.** The parties appoint and designate ______ and ______ to act as the incorporators of the corporation and to take whatever steps are necessary to organize the corporation in accordance with the applicable laws of the State of ______. The authority which is hereby granted shall extend to the preparation, execution, and filing of such documents and other papers as are necessary in the incorporation process to carry out the terms and conditions of this agreement.

10. **Organization Expenses.** Each of the parties shall advance to ______ his pro-rata share of the funds which shall be necessary to pay the expenses and costs of incorporation. As soon as practicable after it commences business the corporation shall reimburse each of the parties for such advances.

11. **Arbitration.** All disputes, differences and controversies arising under or in connection with this agreement shall be settled and finally determined by arbitration in the City of ______ according to the Rules of the American Arbitration Association now in force or hereafter adopted.

12. **Non-Assignability of Agreement.** This agreement shall not be assignable by either party without the written consent of the other parties.

13. **Persons Bound.** The terms and conditions of this agreement shall be binding upon the parties and their respective legal representatives, successors and assigns. However, if one of the parties dies prior to the time the corporation comes into existence, this agreement shall automatically terminate.

Executed in triplicate on the date first above written.

____

____

______________________[17]

17. West's Modern Legal Forms § 2432.-10.

FORM 6C

CERTIFICATE OF PAID-IN CAPITAL (NEW JERSEY)

Form C-109—10-31-61 5 M

Filing $5.00
Recording $2.00
Total $7.00

Certificate of Payment of Capital Stock

of the .... Company.

The location of the principal office in this State is at No. .... Street,

in the .... of .... County of ....

The name of the agent therein and in charge thereof, upon whom process against this corporation may be served, is ....

In accordance with the provisions of Section 14:8-16 of the Revised Statutes, we ....

.... President,

and .... Secretary of the

.... Company,

a corporation of the State of New Jersey, do hereby certify that ....

.... dollars, being the ....

....

....

of capital stock of said company, as authorized by its Certificate of Incorporation filed in the Department of State on the .... day of ....

A. D. 19 ...., has been fully paid in: ....

dollars thereof by the purchase of property and ....

dollars thereof in cash. The capital stock of said company previously paid and reported is $ .... of Common Stock and .... of Preferred Stock.

Witness our hands the .... day of ....

A. D. 19 .....

.... *President.*

.... *Secretary.*

State of .... }
County of .... } ss.

.... President,

and .... Secretary of the

.... Company.

Being severally duly sworn, on their respective oaths depose and say that the foregoing certificate by them signed is true.

Subscribed and sworn to before me,

this .... *President.*

day of .... A. D. 19 .... *Secretary.*

Certificate of Payment

of

Capital Stock

of the

. . . .

. . . .

. . . .

. . . . **Company.**

Filed . . . . , **19** . . . .

. . . .

Secretary of State.

FORM 6D

CHECKLIST FOR SELECTION OF JURISDICTION

(1) Are there express provisions for preincorporation share subscriptions?

(2) May a corporation be formed for perpetual or limited duration?

(3) How restrictive are the provisions concerning corporate names?

(4) Are there express provisions permitting use of a similar corporate name with the consent of the existing corporation? In the case of affiliated corporations? Otherwise?

(5) Is reservation of a corporate name possible? By express statutory provisions? By administrative courtesy?

(6) For what period may a corporate name be reserved?

(7) What renewals of reservation of corporate name are possible?

(8) Is a single incorporator permissible?

(9) Are there any requirements that the incorporator or incorporators subscribe for shares? What are the qualifications required of an incorporator or incorporators with respect to: Residence? Citizenship? Age? Otherwise?

(10) May a corporation serve as an incorporator?

(11) Are there express provisions for informal action by the incorporator(s)?

(12) For what purposes may a corporation be incorporated?

(13) Are broad purposes permissible?

(14) Must specified purposes be set forth in the articles of incorporation?

(15) Are there any constitutional or statutory restrictions on corporate ownership of real property? Agricultural land? Personal property? Shares in other corporations? Are there any constitutional or statutory debt limitations?

(16) Are there express provisions on the ultra vires doctrine?

(17) How broad are the statutory general corporate powers? Do they include power to make charitable contributions irrespective of corporate benefit? To carry out retirement, incentive and benefit plans for directors, officers, and employees? To be a partner? To adopt emergency bylaws? Must the statutory general corporate powers be set forth in the articles of incorporation?

(18) What are the fees for filing or recording the articles of incorporation?

(19) What are the organization taxes? Other initial taxes?

(20) Do such taxes discriminate against shares without par value?

(21) Are filings subject to close administrative scrutiny and conservatism, with resulting delays?

(22) Is there a state stamp tax on the issuance of securities?

(23) Are "blue sky" law requirements burdensome?

(24) What, if any, is the minimum authorized or paid-in capital requirement? Must evidence of compliance be filed? Who are liable, and to what extent, for noncompliance?

(25) What qualitative and quantitative consideration requirements apply to par value shares? To shares without par value? With respect to the valuation of property or services, does the "true value" or "good faith" rule apply? Do preincorporation services satisfy such consideration requirements?

(26) To what extent may a portion of the consideration received for shares be allocated to capital surplus? Within what period after the issuance of the shares may this be done?

(27) May partly-paid shares be issued? May certificates for partly-paid shares be issued?

(28) Are there express provisions for fractions of shares? Scrip?

(29) What provisions may be made with respect to: Dividend preferences? Liquidation preferences?

(30) When two or more classes of shares are authorized, must the provisions concerning them be stated or summarized on the share certificates?

(31) Are express provisions made for issuing preferred or other "special" classes of shares in series? What are the limitations on permissible variations between series of the same class?

(32) To what extent may preferred shares be made redeemable? To what extent may common shares be made redeemable?

(33) To what extent may shares be made convertible?

(34) What are the record date provisions? Are bearer shares permissible? What rights attach to them?

(35) What are the express statutory provisions for, and judicial and administrative attitudes toward, close corporations?

(36) To what extent may voting rights of shareholders be denied or limited? Absolutely? Contingently? May shares carry multiple votes? Fractional votes?

(37) What are the minimum quorum requirements for shareholder action?

(38) Are there express provisions permitting greater-than-normal requirements for: Shareholder quorum? Shareholder vote?

(39) Are there express provisions for holding shareholder meetings outside the state? On dates to be set by board of directors? What are the notice requirements?

(40) Are there express provisions for informal action by shareholders? Unanimously? By required percentages?

(41) Is cumulative voting permissive or mandatory?

(42) What are the provisions for shareholder class voting for directors?

(43) Are there express provisions for shareholder voting agreements?

(44) Are there express provisions permitting shareholder control of directors?

(45) Are there express provisions for irrevocable proxies?

(46) Are there express provisions for voting trusts, permitting closed voting trusts and renewals?

(47) Are there express provisions for purchase and redemption by the corporation of its own shares, including use of stated capital if the purchase is made for specified purposes?

(48) Are there provisions concerning the validity and enforceability of agreements by the corporation to purchase its own shares?

(49) Is insolvency, in either the equity or the bankruptcy sense, a limitation on the redemption or purchase by the corporation of its own shares?

(50) Are there express provisions for rights and options to purchase shares, including the issuance of shares, and the share certificates therefor, even partly-paid, to directors, officers and employees? What are the judicial attitudes with respect thereto?

(51) Is shareholder approval required for the issuance of share options, either generally or to directors, officers and employees?

(52) Do preemptive rights exist or not exist absent provision in the articles of incorporation? Are they adequately defined? May they be denied, limited, amplified, or altered in the articles of incorporation?

(58) What is the minimum number of authorized directors?

(59) What are the qualifications required of directors with respect to: Residence? Citizenship? Shareholding? Age? Otherwise?

(60) May the board of directors be classified? Staggered?

(61) What are the minimum quorum requirements for board of directors action?

(62) Are there express provisions permitting greater-than-normal requirements for: Board of directors quorum? Board of directors vote?

(63) Are there express provisions for holding board of directors meetings outside the state?

(64) Are there express provisions for informal action by the board of directors? By means of conference telephone or some comparable communication technique:

(65) What are the provisions for removal of directors? For cause? Without cause?

(66) Are there express provisions for filling vacancies on the board of directors? By shareholder action? By board of directors action?

(67) Are there provisions for increasing the size of the board of directors? By shareholder action? By board of directors action?

(68) Are there provisions for filling newly-created directorships? By shareholder action? By board of directors action?

(69) Are there express provisions for executive committees of the board of directors? Other committees of the board of directors? Informal action by committees? What powers may be exercised by committees? What is the minimum number of committee members?

(70) Are there express interested directors/officers provisions?

(71) What corporate officers are required?

(72) May the same person hold more than one office?

(73) What are the required qualifications of the various officers with respect to: Residence? Citizenship? Shareholding? Being a director? Age? Otherwise?

(74) Are there provisions for the election of officers by shareholders? For the removal of officers by shareholders?

(75) To what standards are directors and officers held accountable? Standard of care? Fiduciary standards? Statutory duties? What are the possible liabilities of directors? Officers?

(76) To what extent may directors immunize themselves from liability by filing their written dissents? By reliance on records?

(77) What are the express provisions for deadlock, arbitration, and dissolution, and the judicial attitudes concerning the same?

(78) Are cash and property dividends payable out of surplus? Capital surplus? Earned surplus? Net profits?

(79) Is insolvency, in either the equity or the bankruptcy sense, a limitation on cash or property dividends?

(80) Are unrealized appreciation and depreciation recognized in computing surplus? Capital surplus? Earned surplus?

(81) Are there express "wasting assets" corporation dividend provisions?

(82) Are there express provisions for share dividends? Share splits? Other share distributions?

(83) To what extent do statutory requirements of notice or disclosure to shareholders apply in the event of: Cash or property dividends or other distributions from sources other than earned surplus? Share distributions? Reduction of stated capital by cancellation of reacquired shares? Reduction of stated capital made by board of directors? Elimination of deficit in earned surplus account by application of capital surplus ("quasi-reorganization")? Conversion of shares? Who are liable for noncompliance? Corporation? Directors or officers for subjecting corporation to liability?

(84) What are the provisions for shareholder class voting for extraordinary corporate matters? May filings effecting exordinary corporate matters have delayed effective dates?

(85) What shareholder approval is required for a sale, lease, exchange, or other disposition of corporate assets?

(86) What shareholder approval is required for a corporate mortgage or pledge?

(87) What shareholder approval is required for a corporate guaranty?

(88) Do the statutory provisions provide for expeditious amendment of the articles of incorporation? Including elimination of preemptive rights? Elimination of cumulative voting? Elimination of cumulative preferred dividend arrearages? Making nonredeemable shares redeemable? Are there provisions for "restated" articles of incorporation?

(89) What are the statutory provisions permitting merger or consolidation?

(90) Are there provisions for short-merger of a subsidiary into a parent corporation? Of a parent into a subsidiary corporation?

(91) What are the statutory provisions concerning nonjudicial dissolution?

(92) What are the statutory provisions concerning judicial dissolution?

(93) How extensive are the appraisal remedies afforded dissenting shareholders? To what extent are appraisal remedies exclusive?

(94) What are the express provisions relating to shareholder derivative actions?

(95) Are there express provisions for derivative actions by a director? By an officer? By a creditor? By others?

(96) Is there statutory differentiation between shareholder derivative actions and other actions brought by shareholders?

(97) What are the provisions for indemnification for litigation expenses of directors? Of officers? Of other corporate personnel? Are there provisions for insurance?

(98) Are the statutory indemnification provisions exclusive or not with respect to directors? Officers? Other corporate personnel?

(99) What books and records must be kept within the state?

(100) What are the requirements with respect to annual and other reports?

(101) What are the annual franchise tax rates?

(102) What are the state share transfer tax rates?

(103) Are nonresident security holders subject to local taxes? Personal property taxes? Inheritance taxes?

(104) Are there express provisions to accommodate small business investment companies?

(105) Are there express provisions to accommodate open-end investment companies ("mutual funds")?

(106) To what extent are foreign corporations doing business in the state subject to the corporate statute's regulatory provisions? Local "blue sky" laws? Local fees and taxes?

(107) To what extent has the corporate statute been construed by the courts? Are judicial and administrative attitudes sympathetic?

(108) Does the state have a statute or regulations similar to Subchapter S?[18]

18. Reprinted with permission from H. Henn, Handbook of the Law of Corporations and Other Business Enterprises 134–138 (2d ed. 1970), Copyright © 1970 by West Publishing Company.

FORM 6E

APPLICATION OF RESERVATION OF CORPORATE NAME (NEW JERSEY)

Form C-120
Rev. 7-1-71

APPLICATION FOR RESERVATION OF CORPORATE NAME

UNDER SECTION 14A:2-3, CORPORATIONS,

GENERAL, OF THE NEW JERSEY STATUTES

(For Use by Domestic or Foreign Corporations)

To: The Secretary of State
State of New Jersey

Pursuant to the provisions of Section 14A:2-3, Corporations, General, of the New Jersey Statutes, the undersigned applicant hereby applies for the reservation of a corporate name in New Jersey for a period of one hundred twenty (120) days, and for that purpose submits the following application:

1. The corporate name to be reserved is ______________________________

____.

2. The name and address* of the applicant is ___________________________

____.

(*Include zip code)

Dated: The ____________________ day of ________________, 19_____.

____
(Applicant)

____
(Print or Type Name and Title)

RESERVATION NO.:

FOR USE BY DOMESTIC OR FOREIGN CORPORATIONS

PLEASE READ CAREFULLY

BEFORE COMPLETING THIS FORM

14A:2-3. Reserved Name.

(1) The exclusive right to the use of a corporate name may be reserved upon compliance with the provisions of this section.

(2) The reservation shall be made by filing in the office of the Secretary of State an application to reserve a specified corporate name, executed by or on behalf of the applicant and setting forth the name and address of the applicant. If the Secretary of State finds that the name complies with the provisions of Section 14A:2-2, he shall reserve it for the exclusive use of the applicant for a period of 120 days from the date of filing of the application and shall issue a certificate of reservation.

(3) The right to the exclusive use of a specified corporate name so reserved may be transferred by filing in the office of the Secretary of State a notice of such transfer, executed by or on behalf of the applicant for whom the name was reserved, and specifying the name and address of the transferee.

Fees for filing in Office of the Secretary of State, State House, Trenton, N.J., 08625.

Filing Fee $20.00

NOTE: 1. All checks drawn on Out-of-State Banks must be certified.
2. No recording fees will be assessed.

TRANSACTION NO.: ____________

FOLDER NO.:

APPLICATION FOR
RESERVATION OF
CORPORATE NAME

FILED BY:

RECORDED AND FILED:

____
Recorder's Initials

FORM 6F

CERTIFICATE OF RESERVATION OF CORPORATE NAME (SOUTH DAKOTA)

STATE OF SOUTH DAKOTA

OFFICE OF THE SECRETARY OF STATE

Certificate of Reservation of Corporate Name

SPECIMEN

I, LORNA B. HERSETH, Secretary of State of the State of South Dakota, hereby certify that the corporate name of ....
has been reserved in this office for the exclusive use of ....
....
....
for a period of one hundred twenty days after the date hereof, which period shall not be extended, pursuant to the provisions of the South Dakota corporation acts.

IN TESTIMONY WHEREOF, I have hereunto set my hand and affixed the Great Seal of the State of South Dakota, at Pierre, the Capital, this .... day of .... A.D. 19......

....
Secretary of State

....
Assistant

FORM 6G

NOTICE OF TRANSFER OF RESERVED NAME (NEW JERSEY)

Form C-121
Rev. 7-1-71

NOTICE OF TRANSFER OF RESERVED CORPORATE NAME

(FOR USE BY DOMESTIC OR FOREIGN CORPORATIONS)

To: The Secretary of State

State of New Jersey

Pursuant to the provisions of Section 14A:2-3(3), Corporations, General, of the New Jersey Statutes, the undersigned hereby transfers to

____
(Name of transferee)

____
(Address of transferee, including zip code)

all rights in the following reserved corporate name:

____

The period of reservation will expire on ____________________ .
(Date)

Dated at ________________________ this _____ day of ____________, 19___ .

____
(Transferor)

____
(Print or Type Name and Title)

RESERVATION NO.:

FOR USE BY DOMESTIC OR FOREIGN CORPORATIONS

PLEASE READ CAREFULLY

BEFORE COMPLETING THIS FORM

14A:2-3. Reserved Name.

(1) The exclusive right to the use of a corporate name may be reserved upon compliance with the provisions of this section.

(2) The reservation shall be made by filing in the office of the Secretary of State an application to reserve a specified corporate name, executed by or on behalf of the applicant and setting forth the name and address of the applicant. If the Secretary of State finds that the name complies with the provisions of section 14A:2-2, he shall reserve it for the exclusive use of the applicant for a period of 120 days from the date of filing of the application and shall issue a certificate of reservation.

(3) The right to the exclusive use of a specified corporate name so reserved may be transferred by filing in the office of the Secretary of State a notice of such transfer, executed by or on behalf of the applicant for whom the name was reserved, and specifying the name and address of the transferee.

Fees for filing in Office of the Secretary of State, State House, Trenton, N.J. 08625.

Filing Fee $10.00

NOTE: 1. No recording fee will be assessed.
2. All checks drawn on Out-of-State Banks must be certified.

TRANSACTION NO.: __________

FOLDER NO.:

NOTICE OF TRANSFER
OF RESERVED CORPORATE NAME

FILED BY:

RECORDED AND FILED:

Recorder's Initials

FORM 6H

CERTIFICATE OF TRANSFER OF RESERVED CORPORATE NAME (MODEL ACT)

STATE OF ____________

OFFICE OF THE SECRETARY OF STATE

CERTIFICATE OF TRANSFER OF RESERVED CORPORATE NAME OF

____

The undersigned, as Secretary of State of the State of ________ ________, hereby certifies that the corporate name of __________ ______________________________, which was reserved in this office on ____________, 19___, for a period of one hundred twenty days thereafter, has been transferred to ____ ______________________________, whose address is ______________________________, pursuant to the provisions of Section 9 of the ____________ Business Corporation Act.

Dated ________________, 19___.

____

Secretary of State

FORM 6I

APPLICATION FOR REGISTRATION OF CORPORATE NAME (OREGON)

Application for Registration of Corporate Name

Of

____

To the Corporation Commissioner
of the State of Oregon:

Pursuant to the provisions of ORS 57.055 of the Oregon Business Corporation Act, the undersigned corporation hereby applies for the registration of its corporate name to and including December 31, 19____, and submits the following statement:

FIRST: The name of the corporation is ____

____

SECOND: It is incorporated under the laws of____

THIRD: The date of its incorporation is ____

FOURTH: It is carrying on or doing business, **but does not intend to and will not transact business of an intrastate character within the State of Oregon.**

FIFTH: The business in which it is engaged is ____

____

____

____

SIXTH: This Application is accompanied by (a) a certificate setting forth that the corporation is in good standing under the laws of ________________, wherein it is incorporated, executed by the official having custody of the records pertaining to corporations in that ________________________; and (b) a registration fee of $__________ as required by ORS 57.055.

I, the undersigned officer, declare under penalties of perjury that I have examined the foregoing and to the best of my knowledge and belief, it is true, correct and complete.

____

By ____

Dated ______________________________, 19______.   Its ______________________________

This application should be accompanied by a fee of $1.00 for each month, or fraction thereof, between date of filing and the following December 31, and forwarded to: Corporation Commissioner, Salem, Oregon 97310.

FORM 6J

CERTIFICATE OF REGISTRATION OF CORPORATE NAME (SOUTH DAKOTA)

STATE OF SOUTH DAKOTA

OFFICE OF
THE SECRETARY OF STATE

Certificate of
Registration of Corporate Name

SPECIMEN

I, LORNA B. HERSETH, Secretary of State of the State of South Dakota, hereby certify that . . . .
a corporation incorporated under the laws of the State of . . . .
. . . . has registered its corporate name in this office, pursuant to the provisions of Section 9 of the South Dakota Business Corporation Act, effective to and including December 31, 19 . . . .

IN TESTIMONY WHEREOF, I have hereunto set my hand and affixed the Great Seal of the State of South Dakota, at Pierre, the Capital, this . . . . day of
. . . . A.D. 19 . . . . . .

. . . .
Secretary of State

. . . .
Assistant

FORM 6K

APPLICATION FOR RENEWAL OF REGISTERED NAME (OREGON)

Application for
Renewal of Registration of Corporate Name

OF

____

To the Corporation Commissioner
of the State of Oregon:

Pursuant to the provisions of ORS 57.060 of the Oregon Business Corporation Act, the undersigned corporation hereby applies for a renewal of the registration of its corporate name to and including December 31, 19____, and submits the following statements:

FIRST: The name of the corporation is ______________________
____

SECOND: It is incorporated under the laws of _______________

THIRD: The date of its incorporation is ___________________

FOURTH: It is carrying on or doing business, but does not intend to and will not transact business of an intrastate character within the State of Oregon.

FIFTH: The business in which it is engaged is ____________
____
____
____

SIXTH: This Application is accompanied by (a) a certificate setting forth that the corporation is in good standing under the law of ______________________, wherein it is incorporated, executed by the official having custody of the records pertaining to corporations in that jurisdiction; and (b) a registration fee of $10 as required by said ORS 57.060.

Dated______________________, 19____.

____

By_____________________________

Its ___________________________

FORM 6L

CERTIFICATE OF RENEWAL OF REGISTERED NAME (SOUTH DAKOTA)

STATE OF SOUTH DAKOTA

GREAT SEAL OF THE STATE OF SOUTH DAKOTA 1889

OFFICE OF THE SECRETARY OF STATE

Certificate of Renewal of Registration of Corporate Name

I, LORNA B. HERSETH, Secretary of State of the State of South Dakota, hereby certify that .... a corporation incorporated under the laws of the State of .... has renewed the registration of its corporate name in this office, pursuant to the provisions of Section 10 of the South Dakota Business Corporation Act, effective to and including December 31, 19......

SPECIMEN

IN TESTIMONY WHEREOF, I have hereunto set my hand and affixed the Great Seal of the State of South Dakota, at Pierre, the Capital, this .... day of .... A.D. 19......

....
Secretary of State

....
Assistant

FORM 6M

STATEMENT OF ASSUMED NAME BY CORPORATION

SS: AN-TN-1
(Rev. 7/75)

STATE OF COLORADO)
) SS
COUNTY OF)

CERTIFICATE OF
ASSUMED OR TRADE NAME

...., a Colorado corporation, being desirous of transacting a portion of its business under an assumed or trade name as permitted by 7-71-101, Colorado Revised Statutes 1973, hereby certifies:

1. The corporate name and location of the principal office of said corporation is:

2. The name, other than its own corporate name, under which such business is carried on is:*

3. A brief description of the kind of business transacted and to be transacted under such assumed or trade name is:

IN WITNESS WHEREOF, The undersigned President and Secretary of said corporation, have this day executed this Certificate ...., 19..........

....

By..
President

Attest:

....
Secretary

Subscribed and sworn to before me this..............day of..., 19.......... .
My commission expires... .

....
Notary Public

*Any assumed name so used by any such corporation shall contain one of the words "corporation" "incorporated," "limited," or one of the abbreviations "Corp.", "Inc." or "Ltd."

SUBMIT THE ORIGINAL TYPED FORM ONLY.
Filing fee $10.00

FORM 6N

CHECKLIST FOR ARTICLES OF INCORPORATION

1. Corporate name
   - (a) Clear intended name with Secretary of State of state of incorporation
   - (b) Clear intended name with Secretary of State of other states where corporation intends to transact business
   - (c) Desirability of trade-mark search of intended name
   - (d) Desirability of filing name as a trade name
2. Location of principal office
   - (a) County
   - (b) City
   - (c) Street and number
3. Resident agent
   - (a) Name of individual or corporation
   - (b) Address (including street and number)
4. Purposes
   - (a) Nature of business in general
   - (b) Scope of activities in detail
5. Powers
   - (a) Is it desired to set out the statutory powers?
   - (b) Are there to be any limitations of the customary powers, such as the right to deal in real and personal property, to borrow money, to deal in securities, etc.?
6. Capital stock
   - (a) Total number of shares of all classes
   - (b) Number of shares of each class having a par value
     Common
     Preferred
     Other
   - (c) Par value of each class
     Common
     Preferred
     Other
   - (d) Number of shares without par value
     Common
     Preferred
     Other
   - (e) Price at which par value stock is to be sold
     Common
     Preferred
     Other

(f) Price at which no par value stock is to be sold
Common
Preferred
Other

(g) Are incorporators planning to exchange property for stock; if so, description and agreed value of property

(h) How is the price at which subsequently issued stock shall be sold to be determined?

7. Characteristics of preferred stock
   (a) Dividends
   Source
   Rate
   Dates of payment
   Cumulative or noncumulative
   (b) Redemption
   Price
   Notice
   Manner
   (c) Priority of shares in event of dissolution
   (d) Conversion rights of preferred stock
   (e) Sinking fund for redemption or retirement

8. Minimum amount of capital with which corporation will commence business

9. Incorporators
   (a) Names and addresses of each of the incorporators (dummies or principals)

10. Period of duration
    (a) Is corporation to have perpetual existence?
    (b) If for a fixed term, commencing ______; ceasing ______

11. Liability of stockholders
    (a) Is private property of stockholders to be subject to payment of corporate debts?

12. Is the certificate to include the provision for compromises and arrangements between the corporation and its creditors?

13. Is the preemptive right of stockholders to be denied?

14. Are there special situations, such as a sale of assets, which should require the approval of more than a majority of the outstanding stock?

15. Voting rights
    (a) Any limitations on the right of each stockholder to one vote for each share of stock
    (b) Is the preferred stock to have full, limited, or no voting power?

16. Is the transfer of stock to be restricted?

17. Is there to be a provision for a stock option plan?
18. Are the directors to have the power to amend bylaws?
19. Are the directors to have the power to set apart reserves?
20. Is the requirement that directors be elected by ballot to be eliminated? [19]

19. West's Modern Legal Forms § 2491.1.

FORM 60

DELAWARE ARTICLES OF INCORPORATION

CERTIFICATE OF INCORPORATION OF FINE FABRICS CORPORATION

1. **Name.** The name of the Corporation is Fine Fabrics Corporation.

2. **Registered Office and Registered Agent.** The address of the Corporation's registered office in Delaware is 100 ______ Street in the City of Wilmington and County of New Castle, and the name of its registered agent at such address is ______ Trust Company.

3. **Purposes.** The purpose of the Corporation is to engage in any lawful act or activity for which corporations may be now or hereafter organized under the General Corporation Law of Delaware.

4. **Capital Stock** [*providing for one class of par value stock*]. The Corporation is authorized to issue only one class of stock. The total number of such shares is ten thousand and the par value of each of such shares is ten dollars.

[Alternate] 4. **Capital Stock** [*providing for two classes of stock, one voting and one non-voting*]. The total number of shares of all classes of stock which the Corporation shall have authority to issue is ten thousand, all of which are to be without par value. Five thousand of such shares shall be Class A voting shares and five thousand of such shares shall be Class B non-voting shares. The Class A shares and the Class B shares shall have identical rights except that the Class B shares shall not entitle the holder thereof to vote on any matter unless specifically required by law.

[Alternate] 4. **Capital Stock** [*providing for two classes of stock, preferred and common*]. The total number of shares of all classes of capital stock which the Corporation shall have authority to issue is twenty-six million shares, of which one million shares shall be shares of Preferred Stock without par value (hereinafter called

"Preferred Stock"), and twenty-five million shares shall be shares of Common Stock of the par value of $5 per share (hereinafter called "Common Stock").

Any amendment to the Certificate of Incorporation which shall increase or decrease the authorized capital stock of the Corporation may be adopted by the affirmative vote of the holders of a majority of the outstanding shares of stock of the Corporation entitled to vote.

The designations and the powers, preferences and rights, and the qualifications, limitations or restrictions thereof, of the Preferred Stock shall be as follows:

(1) The Board of Directors is expressly authorized at any time, and from time to time, to provide for the issuance of shares of Preferred Stock in one or more series, with such voting powers, full or limited but not to exceed one vote per share, or without voting powers and with such designations, preferences and relative, participating, optional or other special rights, and qualifications, limitations or restrictions thereof, as shall be expressed in the resolution or resolutions providing for the issue thereof adopted by the Board of Directors and as are not expressed in this Certificate of Incorporation or any amendment thereto, inincluding (but without limiting the generality of the foregoing) the following:

(a) the designation of such series;

(b) The dividend rate of such series, the conditions and dates upon which such dividends shall be payable, the preference or relation which such dividends shall bear to the dividends payable on any other class or classes or on any other series of any class or classes of capital stock of the Corporation, and whether such dividends shall be cumulative or non-cumulative;

(c) whether the shares of such series shall be subject to redemption by the Corporation, and, if made subject to such redemption, the times, prices and other terms and conditions of such redemption;

(d) the terms and amount of any sinking fund provided for the purchase or redemption of the shares of such series;

(e) whether the shares of such series shall be convertible into or exchangeable for shares of any other class or classes or of any other series of any class or classes of capital stock of the Corporation, and, if provision be made for conversion or exchange, the times, prices, rates, adjustments, and other terms and conditions of such conversion or exchange;

(f) the extent, if any, to which the holders of the shares of such series shall be entitled to vote as a class or otherwise with respect to the election of directors or otherwise; provided, however, that in no event shall any holder of any series of Preferred Stock be entitled to more than one vote for each share of such Preferred Stock held by him;

(g) the restrictions and conditions, if any, upon the issue or reissue of any additional Preferred Stock ranking on a parity with or prior to such shares as to dividends or upon dissolution;

(h) the rights of the holders of the shares of such series upon the dissolution of, or upon the distribution of assets of, the Corporation, which rights may be different in the case of a voluntary dissolution than in the case of an involuntary dissolution.

(2) Except as otherwise required by law and except for such voting powers with respect to the election of directors or other matters as may be stated in the resolutions of the Board of Directors creating any series of Preferred Stock, the holders of any such series shall have no voting power whatsoever.

5. **Incorporators.** The names and mailing addresses of the incorporators are:

Name	*Mailing Address*
________________	________________
________________	________________

[Optional] 6. **Initial Directors** [*if the powers of the incorporator or incorporators are to terminate upon the filing of the certificate of incorporation*]. The names and mailing addresses of the persons who are to serve as directors until the first annual meeting of stockholders or until their successors are elected and qualify are:

Name	*Mailing Address*
________________	________________
________________	________________

[Optional] 7. **Regulatory Provisions.** The following additional provisions are inserted for the management of the business and for the conduct of the affairs of the Corporation, and creating, defining, limiting, and regulating the powers of the Corporation, the directors, and the stockholders, or any class of stockholders:

(a) *Power of Directors to Amend Bylaws.* The Board of Directors is authorized and empowered from time to time in its discretion to make, alter or repeal the bylaws of the Corporation, except as such power may be limited by any one or more bylaws of the Corporation adopted by the stockholders.

(b) *Books.* The books of the Corporation (subject to the provisions of the laws of the State of Delaware) may be kept outside of the State of Delaware at such places as from time to time may be designated by the Board of Directors.

(c) *Cumulative Voting.* At all elections of directors of the Corporation, each stockholder shall be entitled to as many votes as shall equal the number of votes which he would be entitled to cast for the election of directors with respect to his shares of stock multiplied by the number of directors to be elected, and that he may cast all of such votes for a single director or may distribute them among the number to be voted for, or for any two or more of them as he may see fit.

(d) *Consent of Stockholders in Lieu of Meeting.* Whenever the vote of stockholders at a meeting thereof is required or permitted to be taken for or in connection with any corporate action by any provision of the General Corporation Law of the State of Delaware the meeting and vote of stockholders may be dispensed with if such action is taken with the written consent of the holders of not less than a majority of all the stock entitled to be voted upon such action if a meeting were held; provided that in no case shall the written consent be by the holders of stock having less than the minimum percentage of the vote required by statute for such action, and provided that prompt notice is given to all stockholders of the taking of corporate action without a meeting and by less than unanimous written consent.

(e) *Elections of Directors.* Elections of directors need not be by written ballot.

(f) *Removal of Directors.* The stockholders may at any time, at a meeting expressly called for that purpose, remove any or all of the directors, with or without cause, by a vote of the holders of a majority of the shares then entitled to vote at an election of directors. No director may be removed when the votes cast against his removal would be sufficient to elect him if voted cumulatively at an election at which the same total number of votes were cast and the entire board were then being elected. [When by the provisions of the certificate of incorporation the holders of the shares of any class or series, voting as a class, are entitled to elect one or more directors, any director so elected may be removed only by the applicable vote of the holders of the shares of that class or series, voting as a class.]

[Optional] 8. **Creditor Arrangements.** Whenever a compromise or arrangement is proposed between this corporation and its creditors or any class of them and/or between this corporation and its stockholders or any class of them, any court of equitable jurisdiction within the State of Delaware may, on the application in a summary way of this corporation or of any creditor or stockholder thereof or on the application of any receiver or receivers appointed for this corporation under the provisions of section 291 of Title 8 of the Delaware Code or on the application of trustees in dissolution or of any receiver or receivers appointed for this corporation under the provisions of section 279 of Title 8 of the Delaware Code order a meeting of the creditors or class of creditors, and/or of the stockholders or class of stockholders of this corporation, as the case may be, to be summoned in such manner as the said court directs. If a majority in number representing three-fourths in value of the creditors or class of creditors, and/or of the stockholders or class of stockholders of this corporation, as the case may be, agree to any compromise or arrangement and to any reorganization of this corporation as consequence of such compromise or arrangement, the said compromise or arrangement and the said reorganization shall, if sanctioned by the court to which the said application has been made, be binding on all the creditors or class of creditors, and/or on all the stockholders or class of stockholders, of this corporation, as the case may be, and also on this corporation.

[Optional] 9. **Preemptive Rights.** The holders from time to time of the shares of the Corporation shall have the preemptive right to purchase, at such respective equitable prices, terms, and conditions as shall be fixed by the Board of Directors, such of the shares of the Corporation as may be issued, from time to time, over and above the issue of the first 5,000 shares of the Corporation which have never previously been sold. Such preemptive right shall apply to all shares issued after such first 5,000 shares, whether such additional shares constitute a part of the shares presently or subsequently authorized or constitute shares held in the treasury of the Corporation, and shall be exercised in the respective ratio which the number of shares held by each stockholder at the time of such issue bears to the total number of shares outstanding in the names of all stockholders at such time.

[Optional] 10. **Greater Voting Requirements.** The affirmative vote of a majority of the directors shall be necessary for the transaction of any business at any meeting of directors, except in the case of a proposal to borrow money on the Corporation's credit, in which case the favorable vote of all of the directors shall be necessary.

[Optional] 11. **Duration.** The duration of the Corporation's existence shall extend for the period beginning on the date the certificate of incorporation of the Corporation is filed with the Secretary of State of Delaware, and ending December 31, 1977.

[Optional] 12. **Personal Liability.** The stockholders shall be liable for the debts of the Corporation in the proportion that their stock bears to the total outstanding stock of the Corporation.

13. **Amendment.** The Corporation reserves the right to amend, alter, change or repeal any provision contained in the Certificate of Incorporation, in the manner now or hereafter prescribed by statute, and all rights conferred upon stockholders herein are granted subject to this reservation.

We, the undersigned, being all of the incorporators above named, for the purpose of forming a corporation pursuant to the General Corporation Law of Delaware, sign and acknowledge this certificate of incorporation this 1st day of September, 1968.

____

____

Acknowledgment

State of ________ } ss.
County of _______ }

On this 1st day of September, 1968, before me personally came _______, one of the persons who signed the foregoing certificate of incorporation, known to me personally to be such, and acknowledged that the said certificate is his act and deed and that the facts stated therein are true.

________________[20]
Notary Public

[*Seal*]

20. West's Modern Legal Forms § 2509.1.

FORM 6P

ARTICLES OF INCORPORATION (NORTH DAKOTA)

CORPORATION FOR PROFIT
SUBMIT DUPLICATE ORIGINALS

ARTICLES OF INCORPORATION

OF

____

We, the undersigned natural persons of the age of twenty-one years or more, acting as incorporators of a corporation under the North Dakota Business Corporation Act, adopt the following Articles of Incorporation for such corporation:

Article 1. The name of said corporation shall be: ______________________________
(Shall contain the word "corporation", "company", "incorporated", or "limited", or shall contain an abbreviation of one such words)

Article 2. The period of its duration is: ______________________________
("Perpetual unless limited")

Article 3. The purposes for which the corporation is organized are:

Article 4. The aggregate number of shares which the corporation shall have authority to issue is: ______________________________
(If shares consist of one class only, insert statement of par value of shares, or that all are without par value. If shares are divided into classes, insert number of shares of each class)

Total authorized capitalization is: ______________________________

Article 5. The corporation will not commence business until at least one thousand dollars has been received by it as consideration for the issuance of shares.

Article 6. Provisions limiting or denying to shareholders the preemptive right to acquire additional or treasury shares of the corporation are: ______________________________
(If preemptive rights are not to be limited or desired, insert the word "none")

Article 7. Provisions for the regulation of the internal affairs of the corporation are: ______________________________
(If no provisions for the regulation of the internal affairs of the corporation are set forth, insert the word "none")

____

Article 8. The address of the initial registered office of the corporation is: ______________________________
(Street Address and City)

____

and the name of its initial registered agent at such address is: ______________________________

____

Article 9. The number of directors constituting the initial board of directors of the corporation is

____
(State definite number—not less than 3 nor more than 15)

and the names and addresses of the persons who are to serve as directors until the first annual meeting of shareholders or until their successors are elected and shall qualify are:

Name	Street Address	City	State

Article 10. The name and address of each incorporator is:
(Not less than three)

Name	**Street Address**	**City**	**State**
______________________	______________________		
______________________	______________________		
______________________	______________________		
______________________	______________________		
______________________	______________________		

We, the above named incorporators, being first duly sworn, say that we each have read the foregoing Articles of Incorporation and know the contents thereof, and verily believe the statements made therein to be true.

______________________ ______________________

______________________ ______________________

______________________ ______________________

______________________ ______________________

Dated ______________________ 19____.

Subscribed and sworn to before me this __________ day of ______________________ 19____.

NOTARIAL SEAL

____
Notary Public

State of ______________________

My Commission Expires ______________________ 19____.

Certificate No. ______________________

Filing Date ______________________ **19**______ .

____
(Secretary of State)

____
(By Deputy)

Fees:

$25,000 capitalization or less	$25.00
$25,000 to $50,000 capitalization — **an additional**	50.00
For each $10,000 or fraction thereof over $50,000 capitalization	5.00
Filing fee in addition to above fees	16.00

TOTAL FEES: $______________

"Buy North Dakota Products"

Certificate No._________

UNITED STATES OF AMERICA

GREAT SEAL · STATE OF NORTH DAKOTA · OCTOBER 1 1889

DEPARTMENT OF STATE

State of North Dakota

CERTIFICATE OF INCORPORATION
OF

....

The undersigned, as Secretary of State of the State of North Dakota, hereby certifies that duplicate originals of Articles of Incorporation for the incorporation of

....

duly signed and verified pursuant to the provisions of the North Dakota .... Corporation Act, have been received in this office and are found to conform to law.

ACCORDINGLY the undersigned, as such Secretary of State, and by virtue of the authority vested in him by law, hereby issues this Certificate of Incorporation to

....

and attaches hereto a duplicate original of the Articles of Incorporation.

SPECIMEN

In Testimony Whereof, I have hereunto set my hand and affixed the Great Seal of the State at the Capitol in the City of Bismarck, this.................................day of

....A. D., 19.........

....

Secretary of State.

By..

Deputy.

"Buy North Dakota Products"

FORM 6Q

DELAWARE BY-LAWS

BYLAWS

OF

FINE FABRICS CORPORATION

A Delaware Corporation

ARTICLE I

Offices

The principal office of the Corporation shall be in Wilmington, Delaware. The Corporation may have offices at such other places within or without the State of Delaware as the Board of Directors may from time to time establish.

ARTICLE II

Meetings of Stockholders

Section 1. *Annual Meetings.* The annual meeting of the stockholders for the election of directors and for the transaction of such other business as properly may come before such meeting shall be held at two o'clock in the afternoon on the second Wednesday of March in each year, if not a legal holiday, or, if a legal holiday, then on the next succeeding day not a legal holiday.

Section 2. *Special Meetings.* A special meeting of the stockholders may be called at any time by the President or the Board of Directors, and shall be called by the President upon the written request of stockholders of record holding in the aggregate one-fifth or more of the outstanding shares of stock of the Corporation entitled to vote, such written request to state the purpose or purposes of the meeting and to be delivered to the President.

Section 3. *Place of Meetings.* All meetings of the stockholders shall be held at the office of the Corporation in Lincoln, Nebraska, or at such other place, within or without the State of Delaware, as shall be determined from time to time by the Board of Directors of the stockholders of the Corporation.

Section 4. *Change in Time or Place of Meetings.* The time and place specified in this Article II for the meetings of stockholders for the election of directors shall not be changed within sixty days next before the day on which such election is to be held. A notice of any such change shall be given to each stockholder at least twenty days before the election is held, in person or by letter mailed to his last known post office address.

Section 5. *Notice of Meetings.* Except as otherwise required by statute, written or printed notice of each meeting of the stockholders, whether annual or special, stating the place, day and hour thereof and the purposes for which the meeting is called, shall be given by or under the direction of the Secretary at least ten but not more than fifty days before the date fixed for such meeting, to each stockholder entitled to vote at such meeting, of record at the close of business on the day fixed by the Board of Directors as a record date for the determination of the stockholders entitled to vote at such meetings, or if no such date has been fixed, of record at the close of business on the day next preceding the day on which notice is given, by leaving such notice with him or at his residence or usual place of business or by mailing it, postage prepaid and addressed to him at his post office address as it appears on the books of the Corporation. A waiver of such notice in writing, signed by the person or persons entitled to said notice, whether before or after the time stated therein, shall be deemed equivalent to such notice. Except as otherwise required by statute, notice of any adjourned meeting of the stockholders shall not be required.

Section 6. *Quorum.* Except as otherwise required by statute, the presence at any meeting, in person or by proxy, of the holders of record of a majority of the shares then issued and outstanding and entitled to vote shall be necessary and sufficient to constitute a quorum for the transaction of business. In the absence of a quorum, a majority in interest of the stockholders entitled to vote, present in person or by proxy, or, if no stockholder entitled to vote is present in person or by proxy, any officer entitled to preside or act as secretary of such meeting, may adjourn the meeting from time to time for a period not exceeding twenty days in any one case. At any such adjourned meeting at which a quorum may be present, any business may be transacted which might have been transacted at the meeting as originally called.

Section 7. *List of Stockholders Entitled to Vote.* The officer who has charge of the stock ledger of the Corporation shall prepare and make, at least ten days before every election of directors, a complete list of the stockholders entitled to vote at said election, arranged in alphabetical order, and showing the address of each stockholder and the number of shares registered in the name of each stockholder. Such list shall be open to the examination of any stockholder during ordinary business hours, for a period of at least ten days prior to the election, either at a place within the city, town or village where the election is to be held and which place shall be specified in the notice of the meeting, or, if not so specified, at the place where said meeting is to be held, and the list shall be produced and kept at the time and place of election during the whole time thereof, and subject to the inspection of any stockholder who may be present.

Section 8. *Voting.* Except as otherwise provided by statute or by the Certificate of Incorporation, and subject to the provisions of Section 4 of Article VIII of these Bylaws, each stockholder shall at every meeting of the stockholders be entitled to one vote in person or by proxy for each share of the capital stock having voting power held by such stockholder, but no proxy shall be voted on after three years from its date, unless the proxy provides for a longer period.

At all meetings of the stockholders, except as otherwise required by statute, by the Certificate of Incorporation, or by these Bylaws, all matters shall be decided by the vote of a majority in interest of the stockholders entitled to vote present in person or by proxy.

Persons holding stock in a fiduciary capacity shall be entitled to vote the shares so held, and persons whose stock is pledged shall be entitled to vote, unless in the transfer by the pledgor on the books of the Corporation he shall have expressly empowered the pledgee to vote thereon, in which case only the pledgee or his proxy may represent said stock and vote thereon.

Shares of the capital stock of the Corporation belonging to the Corporation shall not be voted upon directly or indirectly.

Section 9. *Consent of Stockholders in Lieu of Meeting.* Whenever the vote of stockholders at a meeting thereof is required or permitted to be taken in connection with any corporate action, by any provisions of the statutes or of the Certificate of Incorporation, the meeting and vote of stockholders may be dispensed with, if all the stockholders who would have been entitled to vote upon the action if such meeting were held, shall consent in writing to such corporate action being taken.

ARTICLE III

Board of Directors

Section 1. *General Powers.* The business of the Corporation shall be managed by the Board of Directors, except as otherwise provided by statute or by the Certificate of Incorporation.

Section 2. *Number and Qualifications.* The Board of Directors shall consist of five members. Except as provided in the Certificate of Incorporation this number can be changed only by the vote or written consent of the holders of 90 per cent of the stock of the Corporation outstanding and entitled to vote. This number cannot be changed by amendment of the Bylaws of the Corporation. No director need be a stockholder.

[Alternative Clause: Indefinite Number of Directors]

Section 2. *Number and Qualifications.* The number of directors shall be not less than three nor more than fifteen, except that in case all the shares of the Corporation are owned beneficially and of record by either one or two stockholders, the number of directors may be less than three but not less than the number of stockholders. Within the limits specified, the number of directors for each corporate year shall be fixed by vote at the meeting at which they are elected. No director need be a stockholder.

Section 3. *Election and Term of Office.* The directors shall be elected annually by the stockholders, and shall hold office until their successors are respectively elected and qualified.

At all elections for directors each stockholder shall be entitled to as many votes as shall equal the number of his shares of stock multiplied by the number of directors to be elected, and he may cast all of such votes for a single director, or may distribute them among the number to be voted for, or any two or more of them, as he may see fit.

Elections of directors need not be by ballot.

Section 4. *Compensation.* The members of the Board of Directors shall be paid a fee of $______ for attendance at all annual, regular, special and adjourned meetings of the Board. No such fee shall be paid any director if absent. Any director of the Corporation may also serve the Corporation in any other capacity, and receive compensation therefor in any form. Members of special or standing committees may be allowed like compensation for attending committee meetings.

Section 5. *Removals and Resignations.* The stockholders may, at any meeting called for the purpose, by vote of two-thirds of the capital stock issued and outstanding, remove any director from office, with or without cause; provided, however, that no director shall be removed in case the votes of a sufficient number of shares are cast against his removal, which if cumulatively voted at an election of directors would be sufficient to elect him.

The stockholders may, at any meeting, by vote of a majority of such stock represented at such meeting, accept the resignation of any director.

Section 6. *Vacancies.* Any vacancy occurring in the office of director may be filled by a majority of the directors then in office, though less than a quorum, and the directors so chosen shall hold office until the next annual election and until their successors are duly elected and qualified, unless sooner displaced.

When one or more directors resign from the Board, effective at a future date, a majority of the directors then in office, including those who have so resigned, shall have power to fill such vacancy or vacancies, the vote thereon to take effect when such resignation or resignations shall become effective, and each director so chosen shall hold office as herein provided in the filling of other vacancies.

ARTICLE IV

Meetings of Board of Directors

Section 1. *Regular Meetings.* A regular meeting of the Board of Directors may be held without call or formal notice immediately after and at the same place as the annual meeting of the stockholders or any special meeting of the stockholders at which a Board of Directors is elected. Other regular meetings of the Board of Directors may be held without call or formal notice at such places within or without the State of Delaware and at such times as the Board may by vote from time to time determine.

Section 2. *Special Meetings.* Special meetings of the Board of Directors may be held at any place either within or without the State of Delaware at any time when called by the President, Treasurer, Secretary or two or more directors. Notice of the time and place thereof shall be given to each director at least three days before the meeting if by mail or at least twenty-four hours if in person or by telephone or telegraph. A waiver of such notice in writing, signed by the person or persons entitled to said notice, either before or after the time stated therein, shall be deemed equivalent to such notice. Notice of any adjourned meeting of the Board of Directors need not be given.

Section 3. *Quorum.* The presence, at any meeting, of one-third of the total number of directors, but in no case less than two directors, shall be necessary and sufficient to constitute a quorum for the transaction of business except that when a Board of one director is authorized, then one director shall constitute a quorum. Except as otherwise required by statute or by the Certificate of Incorporation, the act of a majority of the directors present at a meeting at which a quorum is present shall be the act of the Board of Directors. In the absence of a quorum, a majority of the directors present at the time and place of any meeting may adjourn such meeting from time to time until a quorum be present.

Section 4. *Consent of Directors in Lieu of Meeting.* Unless otherwise restricted by the Certificate of Incorporation, any action required or permitted to be taken at any meeting of the Board of Directors or any committee thereof may be taken without a meeting, if prior to such action a written consent thereto is signed by all members of the Board or committee, and such written consent is filed with the minutes of proceedings of the Board or committee.

ARTICLE V

Committees of Board of Directors

The Board of Directors may, by resolution passed by a majority of the whole Board, designate one or more committees, each committee to consist of two or more of the directors of the Corporation, which, to the extent provided in the resolution, shall have and may exercise the powers of the Board of Directors in the management of the business and affairs of the Corporation, and may authorize the seal of the Corporation to be affixed to all papers which may require it. Such committee or committees shall have such name or names as may be determined from time to time by resolution adopted by the Board of Directors.

The committees of the Board of Directors shall keep regular minutes of their proceedings and report the same to the Board of Directors when required.

ARTICLE VI

Officers

Section 1. *Number.* The corporation shall have a President, one or more Vice Presidents, a Secretary and a Treasurer, and such other officers, agents and factors as may be deemed necessary. One person may hold any two offices except the offices of President and Vice President and the offices of President and Secretary.

Section 2. *Election, Term of Office and Qualifications.* The officers specifically designated in Section 1 of this Article VI shall be chosen annually by the Board of Directors and shall hold office until their successors are chosen and qualified. No officer need be a director.

Section 3. *Subordinate Officers.* The Board of Directors from time to time may appoint other officers and agents, including one or more Assistant Secretaries and one or more Assistant Treasurers, each of whom shall hold office for such period, have such authority and perform such duties as are provided in these Bylaws or as the Board of Directors from time to time may determine. The Board of Directors may delegate to any officer the power to appoint any such subordinate officers, agents and factors and to prescribe their respective authorities and duties.

Section 4. *Removals and Resignations.* The Board of Directors may at any meeting called for the purpose, by vote of a majority of their entire number, remove from office any officer or agent of the Corporation, or any member of any committee appointed by the Board of Directors.

The Board of Directors may at any meeting, by vote of a majority of the directors present at such meeting, accept the resignation of any officer of the Corporation.

Section 5. *Vacancies.* Any vacancy occurring in the office of President, Vice President, Secretary, Treasurer or any other office by death, resignation, removal, or otherwise shall be filled for the unexpired portion of the term in the manner prescribed by these Bylaws for the regular election or appointment to such office.

Section 6. *The President.* The President shall be the chief executive officer of the Corporation and, subject to the direction and under the supervision of the Board of Directors, shall have general charge of the business, affairs and property of the Corporation, and control over its officers, agents and employees. The President shall preside at all meetings of the stockholders and of the Board of Directors at which he is present. The President shall do and perform such other duties and may exercise such other powers as from time to time may be assigned to him by these Bylaws or by the Board of Directors.

Section 7. *The Vice President.* At the request of the President or in the event of his absence or disability, the Vice President, or in case there shall be more than one Vice President, the Vice President designated by the President, or in the absence of such designation, the Vice President designated by the Board of Directors, shall perform all the duties of the President, and when so acting, shall have all the powers of, and be subject to all the restrictions upon, the President. Any Vice President shall perform such other duties and may exercise such other powers as from time to time may be assigned to him by these Bylaws or by the Board of Directors or the President.

Section 8. *The Secretary.* The Secretary shall

(a) record all the proceedings of the meetings of the Corporation and directors in a book to be kept for that purpose;

(b) have charge of the stock ledger (which may, however, be kept by any transfer agent or agents of the Corporation under the direction of the Secretary), an original or duplicate of which shall be kept at the principal office or place of business of the Corporation in the State of ______;

(c) prepare and make, at least ten days before every election of directors, a complete list of the stockholders entitled to vote at said election, arranged in alphabetical order;

(d) see that all notices are duly given in accordance with the provisions of these Bylaws or as required by statute;

(e) be custodian of the records of the Corporation and the Board of Directors, and of the seal of the Corporation, and see that the seal is affixed to all stock certificates prior to their issuance and to all documents the execution of which on behalf of the Corporation under its seal shall have been duly authorized;

(f) see that all books, reports, statements, certificates and the other documents and records required by law to be kept or filed are properly kept or filed; and

(g) in general, perform all duties and have all powers incident to the office of Secretary and perform such other duties and have such other powers as from time to time may be assigned to him by these Bylaws or by the Board of Directors or the President.

Section 9. *The Treasurer.* The Treasurer shall

(a) have supervision over the funds, securities, receipts, and disbursements of the Corporation;

(b) cause all moneys and other valuable effects of the Corporation to be deposited in its name and to its credit, in such depositaries as shall be selected by the Board of Directors or pursuant to authority conferred by the Board of Directors;

(c) cause the funds of the Corporation to be disbursed by checks or drafts upon the authorized depositaries of the Corporation, when such disbursements shall have been duly authorized;

(d) cause to be taken and preserved proper vouchers for all moneys disbursed;

(e) cause to be kept at the principal office of the Corporation correct books of account of all its business and transactions;

(f) render to the President or the Board of Directors, whenever requested, an account of the financial condition of the Corporation and of his transactions as Treasurer;

(g) be empowered to require from the officers or agents of the Corporation reports or statements giving such information as he may desire with respect to any and all financial transactions of the Corporation; and

(h) in general, perform all duties and have all powers incident to the office of Treasurer and perform such other duties and have such other powers as from time to time may be assigned to him by these Bylaws or by the Board of Directors or the President.

Section 10. *Assistant Secretaries and Assistant Treasurers.* The Assistant Secretaries and Assistant Treasurers shall have such duties as from time to time may be assigned to them by the Board of Directors or the President.

Section 11. *Salaries.* The salaries of the officers of the Corporation shall be fixed from time to time by the Board of Directors, except that the Board of Directors may delegate to any person the power to fix the salaries or other compensation of any officers or agents appointed in accordance with the provisions of Section 3 of this Article VI. No officer shall be prevented from receiving such salary by reason of the fact that he is also a director of the Corporation.

Section 12. *Surety Bond.* The Board of Directors may secure the fidelity of any or all of the officers of the Corporation by bond or otherwise.

ARTICLE VII

Execution of Instruments

Section 1. *Execution of Instruments Generally.* All documents, instruments or writings of any nature shall be signed, executed, verified, acknowledged and delivered by such officer or officers or such agent or agents of the Corporation and in such manner as the Board of Directors from time to time may determine.

Section 2. *Checks, Drafts, Etc.* All notes, drafts, acceptances, checks, endorsements, and all evidence of indebtedness of the Corporation whatsoever, shall be signed by such officer or officers or such agent or agents of the Corporation and in such manner as the Board of Directors from time to time may determine. Endorsements for deposit to the credit of the Corporation in any of its duly authorized depositaries shall be made in such manner as the Board of Directors from time to time may determine.

Section 3. *Proxies.* Proxies to vote with respect to shares of stock of other corporations owned by or standing in the name of the Corporation may be executed and delivered from time to time on behalf of the Corporation by the President or a Vice President and the Secretary or an Assistant Secretary of the Corporation or by any other person or persons duly authorized by the Board of Directors.

ARTICLE VIII

Capital Stock

Section 1. *Certificates of Stock.* Every holder of stock in the Corporation shall be entitled to have a certificate, signed in the name of the Corporation by the Chairman or Vice Chairman of the Board of Directors, the President or a Vice President and by the Treasurer or an Assistant Treasurer, or the Secretary or an Assistant Secretary of the Corporation, certifying the number of shares owned by him in the Corporation; provided, however, that where such certificate is signed by a transfer agent or an assistant transfer agent or by a transfer clerk acting on behalf of the Corporation and a registrar, the signature of any such Chairman or Vice Chairman of the Board of Directors, President, Vice President, Treasurer, Assistant Treasurer, Secretary or Assistant Secretary may be facsimile. In case any officer or officers who shall have signed, or whose facsimile signature or signatures shall have been used on, any such certificate or certificates shall cease to be such officer or officers of the Corporation, whether because of death, resignation or otherwise, before such certificate or certificates shall have been delivered by the Corporation, such certificate or certificates may nevertheless be adopted by the Corporation and be issued and delivered as though the person or persons who signed such certificate or certificates, or whose facsimile signature or signatures shall have been used thereon, had not ceased to be such officer or officers of the Corporation, and any such delivery shall be regarded as an adoption by the Corporation of such certificate or certificates.

Certificates of stock shall be in such form as shall, in conformity to law, be prescribed from time to time by the Board of Directors.

Section 2. *Transfer of Stock.* Shares of stock of the Corporation shall only be transferred on the books of the Corporation by the holder of record thereof or by his attorney duly authorized in writing, upon surrender to the Corporation of the certificates for such shares endorsed by the appropriate person or persons, with such evidence of the authenticity of such endorsement, transfer, authorization and other matters as the Corporation may reasonably require, and accompanied by all necessary stock transfer tax stamps. In that event it shall be the duty of the Corporation to issue a new certificate to the person entitled thereto, cancel the old certificate, and record the transaction on its books.

Section 3. *Rights of Corporation with Respect to Registered Owners.* Prior to the surrender to the Corporation of the certificates for shares of stock with a request to record the transfer of such shares, the Corporation may treat the registered owner as the person entitled to receive dividends, to vote, to receive notifications, and otherwise to exercise all the rights and powers of an owner.

Section 4. *Closing Stock Transfer Book.* The Board of Directors may close the Stock Transfer Book of the Corporation for a period not exceeding fifty days preceding the date of any meeting of stockholders or the date for payment of any dividend or the date for the allotment of rights or the date when any change or conversion or exchange of capital stock shall go into effect or for a period of not exceeding fifty days in connection with obtaining the consent of stockholders for any purpose. However, in lieu of closing the Stock Transfer Book, the Board of Directors may fix in advance a date, not exceeding fifty days preceding the date of any meeting of stockholders, or the date for the payment of any dividend, or the date for the allotment of rights, or the date when any change or conversion or exchange of capital stock shall go into effect, or a date in connection with obtaining such consent, as a record date for the determination of the stockholders entitled to notice of, and to vote at, any such meeting and any adjournment thereof, or entitled to receive payment of any such dividend, or to any such allotment of rights, or to exercise the rights in respect of any such change, conversion or exchange of capital stock, or to give such consent, and in such case such stockholders, and only such stockholders as shall be stockholders of record on the date so fixed shall be entitled to such notice of, and to vote at, such meeting and any adjournment thereof, or to receive payment of such dividend, or to receive such allotment of rights, or to exercise such rights, or to give such consent, as the case may be, notwithstanding any transfer of any stock on the books of the Corporation after any such record date fixed as aforesaid.

Section 5. *Lost, Destroyed and Stolen Certificates.* Where the owner of a certificate for shares claims that such certificate has been lost, destroyed or wrongfully taken, the Corporation shall issue a new certificate in place of the original certificate if the owner (a) so requests before the Corporation has notice that the shares have been acquired by a bona fide purchaser; (b) files with the Corporation a sufficient indemnity bond; and (c) satisfies such other reasonable requirements, including evidence of such loss, destruction, or wrongful taking, as may be imposed by the Corporation.

ARTICLE IX

Dividends

Section 1. *Sources of Dividends.* The directors of the Corporation, subject to any restrictions contained in the statutes and Certificate of Incorporation, may declare and pay dividends upon the shares of the capital stock of the Corporation either (a) out of its net assets in excess of its capital, or (b) in case there shall be no such excess, out of its net profits for the fiscal year then current or the current and preceding fiscal year.

Section 2. *Reserves.* Before the payment of any dividend, the directors of the Corporation may set apart out of any of the funds of the Corporation available for dividends a reserve or reserves for any proper purpose, and the directors may abolish any such reserve in the manner in which it was created.

Section 3. *Reliance on Corporate Records.* A director shall be fully protected in relying in good faith upon the books of account of the Corporation or statements prepared by any of its officials as to the value and amount of the assets, liabilities and net profits of the Corporation, or any other facts pertinent to the existence and amount of surplus or other funds from which dividends might properly be declared and paid.

Section 4. *Manner of Payment.* Dividends may be paid in cash, in property, or in shares of the capital stock of the Corporation at par.

ARTICLE X

Seal

The corporate seal, subject to alteration by the Board of Directors, shall be in the form of a circle and shall bear the name of the Corporation and the year of its incorporation and shall indicate its formation under the laws of the State of Delaware. Such seal may be used by causing it or a facsimile thereof to be impressed or affixed or reproduced or otherwise.

ARTICLE XI

Fiscal Year

Except as from time to time otherwise provided by the Board of Directors, the fiscal year of the Corporation shall be the calendar year.

ARTICLE XII

Amendments

Section 1. *By the Stockholders.* Except as otherwise provided in the Certificate of Incorporation or in these Bylaws, these Bylaws may be amended or repealed, or new Bylaws may be made and adopted, by a majority vote of all the stock of the Corporation issued and outstanding and entitled to vote at any annual or special meeting of the stockholders, provided that notice of intention to amend shall have been contained in the notice of meeting.

Section 2. *By the Directors.* Except as otherwise provided in the Certificate of Incorporation or in these Bylaws, these Bylaws, including amendments adopted by the stockholders, may be amended or repealed by a majority vote of the whole Board of Directors at any regular or special meeting of the Board, provided that the stockholders may from time to time specify particular provisions of the Bylaws which shall not be amended by the Board of Directors.[21]

21. West's Modern Legal Forms § 2793.

FORM 7A

BOND AND SHARE CERTIFICATES

LESS THAN 100 SHARES

NUMBER

SHARES

BLACK HILLS POWER AND LIGHT COMPANY

INCORPORATED UNDER THE LAWS OF THE STATE OF SOUTH DAKOTA

COMMON STOCK

SEE REVERSE FOR CERTAIN DEFINITIONS

THIS CERTIFIES that

is the owner of

CUSIP 092113 10 9

fully paid and non-assessable shares, having a par value of $1 per share, of the Common Stock of

BLACK HILLS POWER AND LIGHT COMPANY

CERTIFICATE OF STOCK

(hereinafter called the Corporation), transferable on the books of the Corporation by the holder hereof in person or by a duly authorized attorney upon surrender of this certificate properly endorsed. This certificate and the shares represented hereby are issued and shall be held subject to all the provisions of the Articles of Incorporation of the Corporation and all amendments thereof, copies of which are on file with the Transfer Agent, to all of which the holder, by acceptance hereof, assents. This certificate is not valid until countersigned by the Transfer Agent and registered by the Registrar.

WITNESS, the facsimile seal of the Corporation and the signatures of its duly authorized officers.

Dated,

SPECIMEN

SECRETARY

BLACK HILLS POWER AND LIGHT COMPANY CORPORATE SEAL 1941 SOUTH DAKOTA

SPECIMEN

PRESIDENT

COUNTERSIGNED:
NORTHWESTERN NATIONAL BANK OF MINNEAPOLIS
TRANSFER AGENT
BY
AUTHORIZED SIGNATURE

REGISTERED:
FIRST NATIONAL BANK OF MINNEAPOLIS
REGISTRAR
BY
AUTHORIZED SIGNATURE

BLACK HILLS POWER AND LIGHT COMPANY

Notice: The Corporation will furnish to any shareholder upon request and without charge, a full statement of the designations, preferences, limitations, and relative rights of the shares of each class of stock authorized to be issued, and a like full statement relative to any preferred or special class of stock in series which the Corporation is or may be authorized to issue, or has issued, as to the variations in the relative rights and preferences between the shares of each such series so far as the same have been fixed and determined and the authority of the Board of Directors to fix and determine the relative rights and preferences of subsequent series.

The following abbreviations, when used in the inscription on the face of this certificate, shall be construed as though they were written out in full according to applicable laws or regulations:

TEN COM—as tenants in common
TEN ENT —as tenants by the entirety
JT TEN —as joint tenants with right of survivorship and not as tenants in common

UNIF GIFT MIN ACT— ______ Custodian ______
(Cust) (Minor)
under Uniform Gifts to Minors
Act ______
(State)

Additional abbreviations may also be used though not in the above list.

For Value Received, ______ *hereby sell, assign and transfer unto*

PLEASE INSERT SOCIAL SECURITY OR OTHER IDENTIFYING NUMBER OF ASSIGNEE

______

(PLEASE PRINT OR TYPEWRITE NAME AND ADDRESS OF ASSIGNEE)

______

______

______ *Shares*

of the Stock represented by the within certificate and do hereby irrevocably constitute and appoint

______ *attorney,*

to transfer the same on the books of the within-named Corporation, with full power of substitution in the premises.

Dated ______

______

Notice: The signature to this assignment must correspond with the name as written upon the face of the certificate in every particular, without alteration or enlargement or any change whatever.

SIGNATURE GUARANTEED BY:

______

THIS SPACE MUST NOT BE COVERED IN ANY WAY

100 SHARES

NUMBER

SHARES 100

BLACK HILLS POWER AND LIGHT COMPANY

INCORPORATED UNDER THE LAWS OF THE STATE OF SOUTH DAKOTA

COMMON STOCK

SEE REVERSE FOR CERTAIN DEFINITIONS

THIS CERTIFIES that

is the owner of

CUSIP 092113 10 9

ONE HUNDRED

fully paid and non-assessable shares, having a par value of $1 per share, of the Common Stock of

BLACK HILLS POWER AND LIGHT COMPANY

CERTIFICATE OF STOCK

(hereinafter called the Corporation), transferable on the books of the Corporation by the holder hereof in person or by a duly authorized attorney upon surrender of this certificate properly endorsed. This certificate and the shares represented hereby are issued and shall be held subject to all the provisions of the Articles of Incorporation of the Corporation and all amendments thereof, copies of which are on file with the Transfer Agent, to all of which the holder, by acceptance hereof, assents. This certificate is not valid until countersigned by the Transfer Agent and registered by the Registrar.

WITNESS, the facsimile seal of the Corporation and the signatures of its duly authorized officers.

Dated,

COUNTERSIGNED:
NORTHWESTERN NATIONAL BANK OF MINNEAPOLIS
TRANSFER AGENT
BY
AUTHORIZED SIGNATURE

REGISTERED:
FIRST NATIONAL BANK OF MINNEAPOLIS
REGISTRAR
BY
AUTHORIZED SIGNATURE

SPECIMEN
SECRETARY

BLACK HILLS POWER AND LIGHT COMPANY CORPORATE SEAL 1941 SOUTH DAKOTA

SPECIMEN
PRESIDENT

© SECURITY-COLUMBIAN BANKNOTE COMPANY

BLACK HILLS POWER AND LIGHT COMPANY

Notice: The Corporation will furnish to any shareholder upon request and without charge, a full statement of the designations, preferences, limitations, and relative rights of the shares of each class of stock authorized to be issued, and a like full statement relative to any preferred or special class of stock in series which the Corporation is or may be authorized to issue, or has issued, as to the variations in the relative rights and preferences between the shares of each such series so far as the same have been fixed and determined and the authority of the Board of Directors to fix and determine the relative rights and preferences of subsequent series.

The following abbreviations, when used in the inscription on the face of this certificate, shall be construed as though they were written out in full according to applicable laws or regulations:

TEN COM—as tenants in common
TEN ENT —as tenants by the entirety
JT TEN —as joint tenants with right of survivorship and not as tenants in common

UNIF GIFT MIN ACT—______ Custodian ______
(Cust) (Minor)
under Uniform Gifts to Minors
Act______
(State)

Additional abbreviations may also be used though not in the above list.

For Value Received, ______ *hereby sell, assign and transfer unto*

PLEASE INSERT SOCIAL SECURITY OR OTHER IDENTIFYING NUMBER OF ASSIGNEE

____

(PLEASE PRINT OR TYPEWRITE NAME AND ADDRESS OF ASSIGNEE)

____

____

______________________________ *Shares*

of the Stock represented by the within certificate and do hereby irrevocably constitute and appoint

______________________________ *attorney,*

to transfer the same on the books of the within-named Corporation, with full power of substitution in the premises.

Dated ______________

____

Notice: The signature to this assignment must correspond with the name as written upon the face of the certificate in every particular, without alteration or enlargement or any change whatever.

SIGNATURE GUARANTEED BY:

____

THIS SPACE MUST NOT BE COVERED IN ANY WAY

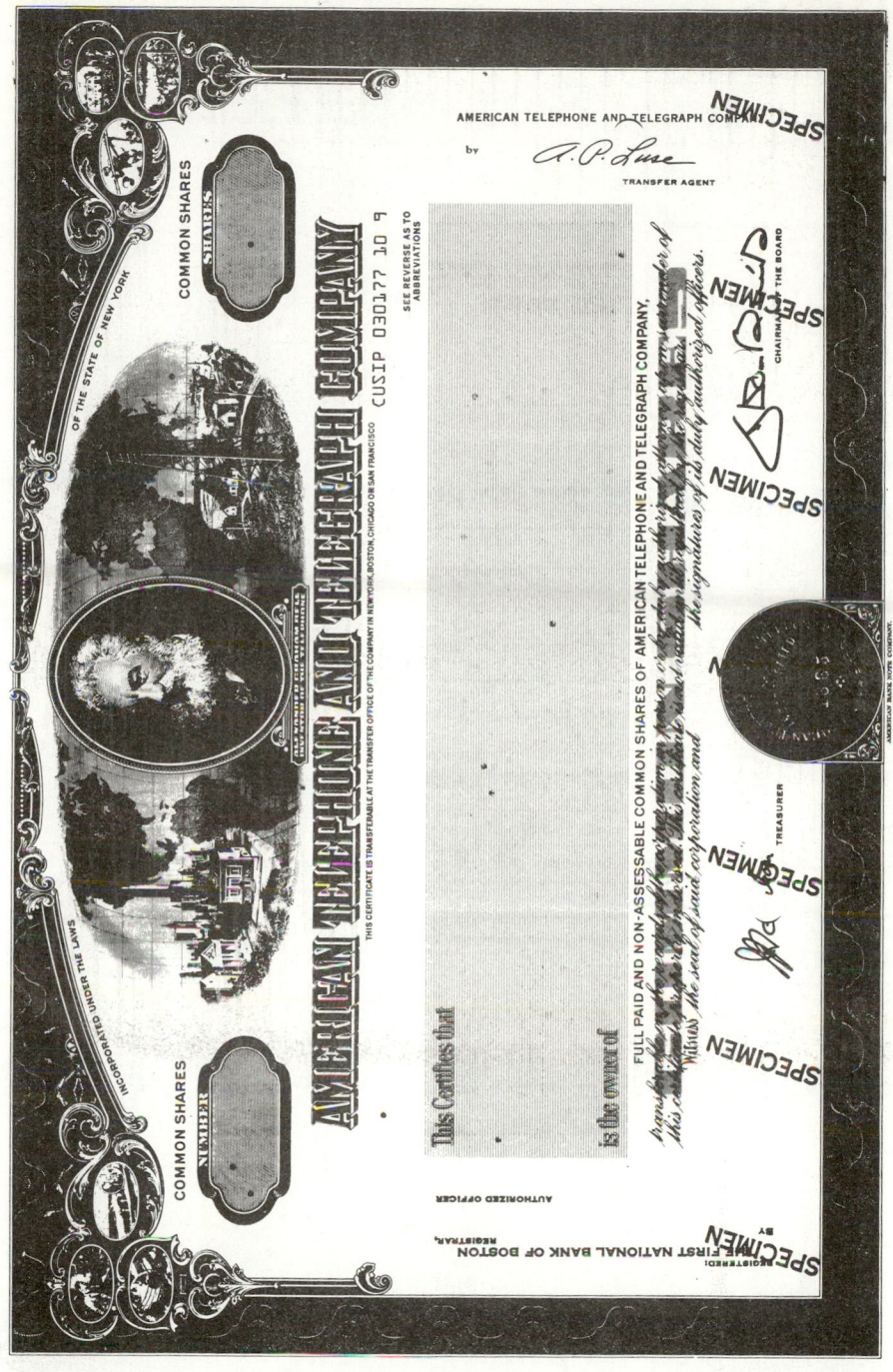

INCORPORATED UNDER THE LAWS OF THE STATE OF NEW YORK

COMMON SHARES NUMBER

COMMON SHARES SHARES

ALEXANDER GRAHAM BELL INVENTOR OF THE TELEPHONE

AMERICAN TELEPHONE AND TELEGRAPH COMPANY

THIS CERTIFICATE IS TRANSFERABLE AT THE TRANSFER OFFICE OF THE COMPANY IN NEW YORK, BOSTON, CHICAGO OR SAN FRANCISCO

CUSIP 030177 10 9

SEE REVERSE AS TO ABBREVIATIONS

This Certifies that

is the owner of

FULL PAID AND NON-ASSESSABLE COMMON SHARES OF AMERICAN TELEPHONE AND TELEGRAPH COMPANY,

transferable on the records of the corporation in person or by duly authorized attorney upon surrender of this certificate properly endorsed. This certificate is not valid until registered by the registrar.

Witness the seal of said corporation and the signatures of its duly authorized officers.

TREASURER

CHAIRMAN OF THE BOARD

AMERICAN TELEPHONE AND TELEGRAPH COMPANY

by

TRANSFER AGENT

REGISTERED:

THE FIRST NATIONAL BANK OF BOSTON

REGISTRAR,

BY

AUTHORIZED OFFICER

SPECIMEN

AMERICAN BANK NOTE COMPANY

The Company will furnish to any shareholder, upon request to its principal office or to any of its transfer offices and without charge, a full statement of the designation, relative rights, preferences and limitations of its common and preferred shares, and of each series of its preferred shares so far as the same have been fixed and the authority of the board of directors to designate and fix the relative rights, preferences and limitations of other series of its preferred shares. **As specified in such statement, the affirmative vote or consent of two-thirds of the holders of preferred shares or a series of preferred shares is required in order to take certain actions affecting preferred shares.**

For Value received, ____ *hereby sell, assign and transfer* ____ *Shares represented by the within Certificate unto*

The following abbreviations shall be construed as though the words set forth below opposite each abbreviation were written out in full where such abbreviation appears:

TEN COM — as tenants in common
TEN ENT — as tenants by the entireties
JT TEN — as joint tenants with right of survivorship and not as tenants in common

(Name) CUST (Name) UNIF - (Name) as Custodian for (Name) under the
GIFT MIN ACT (State) (State) Uniform Gifts to Minors Act

Additional abbreviations may also be used though not in the above list.

PLEASE PRINT OR TYPE TAXPAYER-IDENTIFYING NUMBER AND NAME AND ADDRESS OF ASSIGNEE	FOR AT&T USE ONLY:	
	RIN:	
SHARES	CTF. NOTE:	YR. H ☐
	CLASS ACCT:	E ☐

PLEASE PRINT OR TYPE TAXPAYER-IDENTIFYING NUMBER AND NAME AND ADDRESS OF ASSIGNEE		
	RIN:	
SHARES	CTF. NOTE:	YR. H ☐
	CLASS ACCT:	E ☐

PLEASE PRINT OR TYPE TAXPAYER-IDENTIFYING NUMBER AND NAME AND ADDRESS OF ASSIGNEE		
	RIN:	
SHARES	CTF. NOTE:	YR. H ☐
	CLASS ACCT:	E ☐

and do hereby irrevocably constitute and appoint ____ *Attorney to transfer the said shares on the records of the within named Company with full power of substitution in the premises.* ____

Dated, ____ ____

IMPORTANT { BEFORE SIGNING, READ AND COMPLY CAREFULLY WITH REQUIREMENTS PRINTED BELOW.

THE SIGNATURE(S) TO THIS ASSIGNMENT MUST CORRESPOND WITH THE NAME(S) AS WRITTEN UPON THE FACE OF THE CERTIFICATE IN EVERY PARTICULAR WITHOUT ALTERATION OR ENLARGEMENT OR ANY CHANGE WHATEVER. THE SIGNATURE(S) SHOULD BE GUARANTEED BY A COMMERCIAL BANK OR TRUST COMPANY, OR BY A NEW YORK, BOSTON, MIDWEST, PBW OR PACIFIC STOCK EXCHANGE MEMBER FIRM, WHOSE SIGNATURE IS KNOWN TO THE TRANSFER OFFICE.

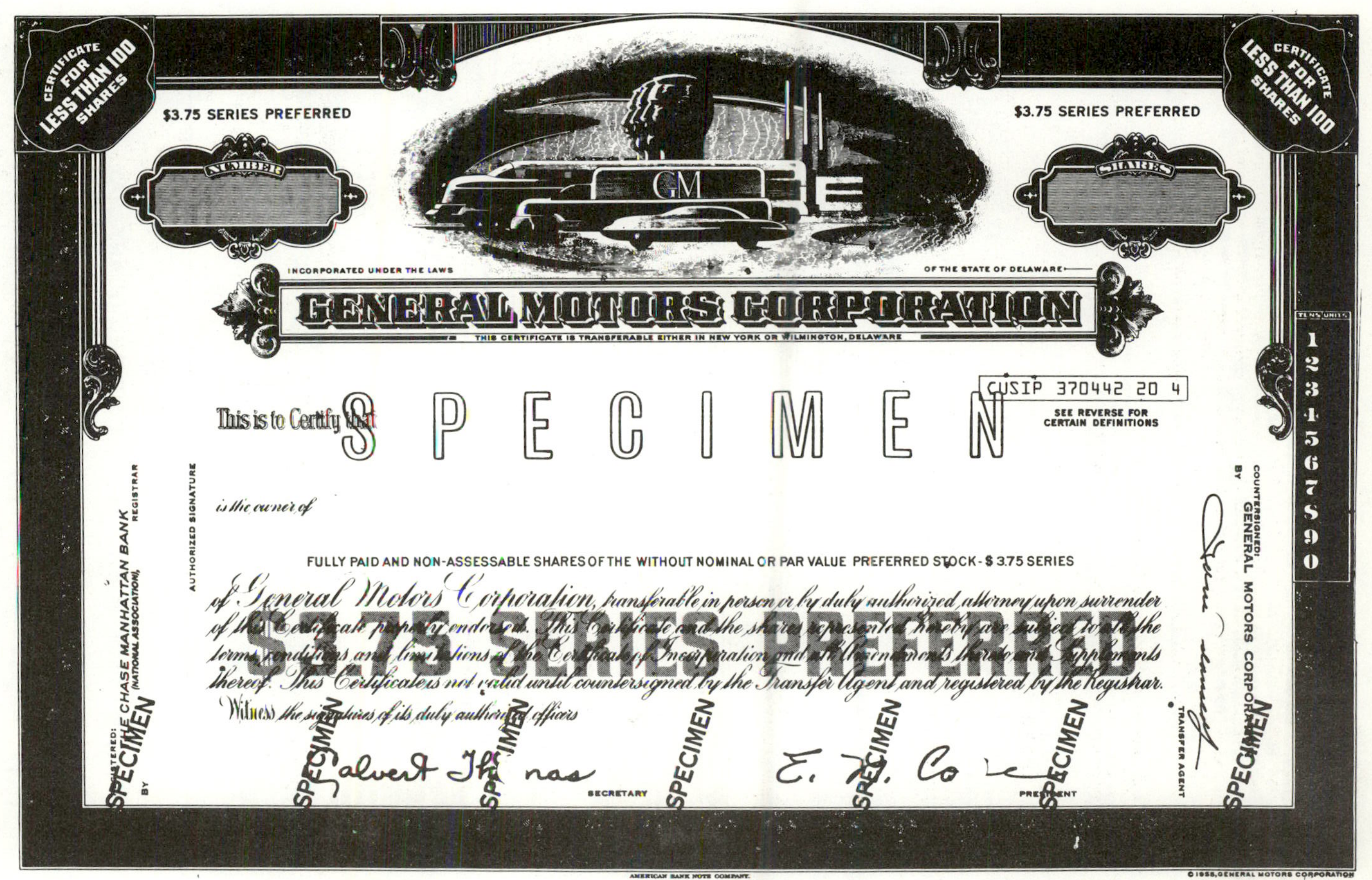

CERTIFICATE FOR LESS THAN 100 SHARES

$3.75 SERIES PREFERRED

NUMBER

SHARES

INCORPORATED UNDER THE LAWS OF THE STATE OF DELAWARE

GENERAL MOTORS CORPORATION

THIS CERTIFICATE IS TRANSFERABLE EITHER IN NEW YORK OR WILMINGTON, DELAWARE

TENS UNITS
1 2 3 4 5 6 7 8 9 0

CUSIP 370442 20 4

SEE REVERSE FOR CERTAIN DEFINITIONS

This is to Certify that SPECIMEN

is the owner of

FULLY PAID AND NON-ASSESSABLE SHARES OF THE WITHOUT NOMINAL OR PAR VALUE PREFERRED STOCK - $ 3.75 SERIES

of General Motors Corporation, transferable in person or by duly authorized attorney upon surrender of this Certificate properly endorsed. This Certificate and the shares represented hereby are subject to all the terms, conditions and limitations of the Certificate of Incorporation and all Amendments thereto and Supplements thereof. This Certificate is not valid until countersigned by the Transfer Agent and registered by the Registrar.

$3.75 SERIES PREFERRED

Witness the signatures of its duly authorized officers

SECRETARY

PRESIDENT

COUNTERSIGNED:
GENERAL MOTORS CORPORATION
BY
TRANSFER AGENT

REGISTERED:
THE CHASE MANHATTAN BANK (NATIONAL ASSOCIATION)
BY
REGISTRAR

AUTHORIZED SIGNATURE

SPECIMEN

AMERICAN BANK NOTE COMPANY

© 1955, GENERAL MOTORS CORPORATION

GENERAL MOTORS CORPORATION

The following abbreviations, when used in the inscription on the face of this certificate, shall be construed as though they were written out in full according to applicable laws or regulations:

TEN COM — as tenants in common

TEN ENT — as tenants by the entireties

JT TEN — as joint tenants with right of survivorship and not as tenants in common

UNIF GIFT MIN ACT —

.... Custodian ....
(Cust) (Minor)
under Uniform Gifts to

Minors Act....................
(State)

Additional abbreviations may also be used though not in the above list.

THE CORPORATION WILL FURNISH WITHOUT CHARGE TO EACH STOCKHOLDER WHO SO REQUESTS A STATEMENT OF THE RIGHTS, PRIVILEGES, RESTRICTIONS, VOTING POWERS, LIMITATIONS AND QUALIFICATIONS OF THE SEVERAL CLASSES OF STOCK OF THE CORPORATION. REQUESTS MAY BE DIRECTED TO THE TRANSFER AGENT, GENERAL MOTORS CORPORATION, 767 FIFTH AVENUE, NEW YORK, NEW YORK 10022.

FOR VALUE RECEIVED, THE UNDERSIGNED HEREBY SELLS, ASSIGNS AND TRANSFERS THE SHARES OF THE CAPITAL STOCK REPRESENTED BY THE WITHIN CERTIFICATE AS FOLLOWS:

PLEASE INSERT SOCIAL SECURITY OR OTHER IDENTIFYING NUMBER OF ASSIGNEE

____________ ____

________ UNTO ____
SHARES FULL NAME AND ADDRESS (INCLUDING ZIP CODE) OF ASSIGNEE SHOULD BE TYPEWRITTEN OR PRINTED LEGIBLY

________ " ____

________ " ____

________ " ____

AND HEREBY IRREVOCABLY CONSTITUTES AND APPOINTS

____ ATTORNEY

TO TRANSFER THE SAID STOCK ON THE BOOKS OF THE WITHIN-NAMED CORPORATION WITH FULL POWER OF SUBSTITUTION IN THE PREMISES.

DATED ________________

SIGN HERE ____
SIGNATURE MUST CORRESPOND WITH NAME ON FACE OF CERTIFICATE

NOTICE: THE SIGNATURE TO THIS ASSIGNMENT MUST CORRESPOND WITH THE NAME AS WRITTEN UPON THE FACE OF THE CERTIFICATE IN EVERY PARTICULAR WITHOUT ALTERATION OR ENLARGEMENT OR ANY CHANGE WHATEVER.

SIGNATURE GUARANTEED

THIS SPACE MUST NOT BE COVERED IN ANY WAY

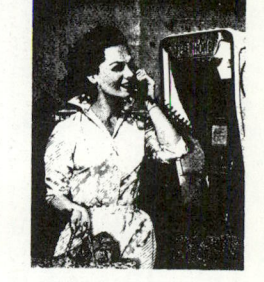

AMERICAN TELEPHONE AND TELEGRAPH COMPANY

THIRTY YEAR 8¾% DEBENTURE, DUE MAY 15, 2000

American Telephone and Telegraph Company, a New York corporation (herein referred to as the "Company"), for value received, hereby promises to pay to

8¾%

DUE MAY 15, 2000

S P E C I M E N

SEE REVERSE AS TO ABBREVIATIONS

THIS CERTIFICATE IS TRANSFERABLE AT THE OFFICE OF THE COMPANY IN NEW YORK CITY

, or registered assigns, at the office or agency of the Company in the Borough of Manhattan, The City of New York, State of New York, the principal sum of

CUSIP 030177 AX 7

on May 15, 2000, in such coin or currency of the United States of America as at the time of payment shall be legal tender for the payment of public and private debts, and to pay interest, semi-annually on May 15 and November 15, on said principal sum at the rate per annum specified in the title of this Debenture, at said office or agency, in like coin or currency, from the fifteenth day of May or November, as the case may be, to which interest on Debentures has been paid preceding the date hereof (unless the date hereof is a May 15 or a November 15 to which interest has been paid, in which case from the date hereof, or unless the date hereof is prior to the first payment of interest, in which case from May 18, 1970) until payment of said principal sum has been made or duly provided for. Notwithstanding the foregoing, unless this Debenture shall be authenticated at a time when there is an existing default in the payment of interest on the Debentures, if the date hereof is after April 15 and before the next following May 15 or is after October 15 and before the next following November 15, this Debenture shall bear interest from such May 15 or November 15, provided, however, that if the Company shall default in the payment of interest due on such May 15 or November 15, then this Debenture shall bear interest from the next preceding November 15 or May 15, as the case may be. The interest so payable on any May 15 or November 15 will, subject to certain exceptions provided in the Indenture referred to on the reverse hereof, be paid to the person in whose name this Debenture shall be registered at the close of business on the April 15 prior to such May 15 or the October 15 prior to such November 15, whether or not such April 15 or October 15 shall be a business day.

Reference is hereby made to the further provisions of this Debenture set forth on the reverse hereof and such further provisions shall for all purposes have the same effect as though fully set forth at this place.

This Debenture shall not be valid or become obligatory for any purpose until the certificate of authentication hereon shall have been executed by the Trustee under the Indenture referred to on the reverse hereof.

In Witness Whereof, American Telephone and Telegraph Company has caused this Debenture to be signed by its Chairman of the Board or by its President and by its Treasurer, each by a facsimile of his signature, and has caused a facsimile of its corporate seal to be affixed hereunto or imprinted hereon.

TRUSTEE'S CERTIFICATE OF AUTHENTICATION DATED

This is one of the Debentures described in the within-mentioned Indenture.

Chemical Bank, As Trustee

By

Authorized Officer

SPECIMEN SPECIMEN SPECIMEN SPECIMEN SPECIMEN SPECIMEN

American Telephone and Telegraph Company,

By

Treasurer

Chairman of the Board

AMERICAN BANK NOTE COMPANY.

AMERICAN TELEPHONE AND TELEGRAPH COMPANY

This Debenture is one of a duly authorized issue of Debentures of the Company, designated as set forth on the face hereof (herein referred to as the "Debentures"), limited to the aggregate principal amount of $1,569,327,000, all issued or to be issued under and pursuant to an Indenture dated May 18, 1970 (herein referred to as the "Indenture"), duly executed and delivered by the Company to Chemical Bank, Trustee (herein referred to as the "Trustee"), which Indenture and all Indentures supplemental thereto are hereby incorporated by reference in and made a part of this instrument and are hereby referred to for a description of the rights, limitation of rights, obligations, duties and immunities thereunder of the Trustee, the Company and the holders (the words "holders" or "holder" meaning the registered holders or registered holder) of the Debentures.

In case an Event of Default, as defined in the Indenture, shall have occurred and be continuing, the principal hereof may be declared, and upon such declaration shall become, due and payable, in the manner, with the effect and subject to the conditions provided in the Indenture.

The Indenture contains provisions permitting the Company and the Trustee, with the consent of the holders of not less than 66-2/3% in aggregate principal amount of the Debentures at the time outstanding, evidenced as in the Indenture provided, to execute supplemental indentures adding any provisions to or changing in any manner or eliminating any of the provisions of the Indenture or of any supplemental indenture or modifying in any manner the rights of the holders of the Debentures; provided, however, that no such supplemental indenture shall (i) extend the fixed maturity of any Debentures, or reduce the principal amount thereof, or reduce the rate or extend the time of payment of interest thereon, or reduce any premium payable upon the redemption thereof, without the consent of the holder of each Debenture so affected, or (ii) reduce the aforesaid percentage of Debentures, the consent of the holders of which is required for any such supplemental indenture, without the consent of the holders of all Debentures then outstanding. It is also provided in the Indenture that, under certain circumstances, the holders of a majority in aggregate principal amount of the Debentures at the time outstanding may on behalf of the holders of all of the Debentures waive any past default under the Indenture and its consequences, except a default in the payment of the principal of (or premium, if any) or interest on any of the Debentures. Any such consent or waiver by the holder of this Debenture (unless revoked as provided in the Indenture) shall be conclusive and binding upon such holder and upon all future holders and owners of this Debenture and of any Debenture issued in exchange or substitution therefor, irrespective of whether or not any notation of such consent or waiver is made upon this Debenture or such other Debenture.

No reference herein to the Indenture and no provision of this Debenture or of the Indenture shall alter or impair the obligation of the Company, which is absolute and unconditional, to pay the principal of (and premium, if any) and interest on this Debenture at the place, at the respective times, at the rate and in the coin or currency herein prescribed.

The Debentures are issuable as registered Debentures without coupons in denominations of $100 and any multiple of $100. At the office or agency of the Company referred to on the face hereof and in the manner and subject to the limitations provided in the Indenture, Debentures may be exchanged without a service charge for a like aggregate principal amount of Debentures of other authorized denominations.

The Debentures may be redeemed, at the option of the Company, as a whole or from time to time in part (selected by lot or otherwise in such manner as the Trustee may deem appropriate and fair), on or after May 15, 1975 and prior to maturity, upon the notice referred to below, at the following redemption prices (expressed in percentages of the principal amount): during the 12 months' periods ending May 14:

1976......	107.00%	1983......	104.55%	1990......	102.10%
1977......	106.65%	1984......	104.20%	1991......	101.75%
1978......	106.30%	1985......	103.85%	1992......	101.40%
1979......	105.95%	1986......	103.50%	1993......	101.05%
1980......	105.60%	1987......	103.15%	1994......	100.70%
1981......	105.25%	1988......	102.80%	1995......	100.35%
1982......	104.90%	1989......	102.45%		

and thereafter, 100%, together in each case with accrued interest to the date fixed for redemption. As provided in the Indenture, notice of redemption to the holders of Debentures to be redeemed as a whole or in part shall be given by mailing a notice of such redemption not less than thirty nor more than ninety days prior to the date fixed for redemption to their last addresses as they shall appear upon the register kept for that purpose.

Upon due presentment for registration of transfer of this Debenture at the above-mentioned office or agency of the Company, a new Debenture or Debentures, of authorized denominations, for a like aggregate principal amount, will be issued to the transferee as provided in the Indenture. No service charge shall be made for any such transfer, but the Company may require payment of a sum sufficient to cover any tax or other governmental charge that may be imposed in relation thereto.

The Company, the Trustee, any paying agent and any Debenture registrar may deem and treat the holder hereof as the absolute owner hereof (whether or not this Debenture shall be overdue and notwithstanding any notation of ownership or other writing hereon) for the purpose of receiving payment of or on account of the principal hereof (and premium, if any) and, subject to the provisions on the face hereof, interest hereon, and for all other purposes, and neither the Company nor the Trustee nor any paying agent nor any Debenture registrar shall be affected by any notice to the contrary.

No recourse shall be had for the payment of the principal of (or premium, if any) or the interest on this Debenture, or for any claim based hereon, or otherwise in respect hereof, or based on or in respect of the Indenture or any indenture supplemental thereto, against any incorporator, shareholder, officer or director, as such, past, present or future, of the Company or of any successor corporation, either directly or through the Company or any successor corporation, whether by virtue of any constitution, statute or rule of law or by the enforcement of any assessment or penalty or otherwise, all such liability being, by the acceptance hereof and as part of the consideration for the issue hereof, expressly waived and released.

The following abbreviations shall be construed as though the words set forth below opposite each abbreviation were written out in full where such abbreviation appears:

TEN COM – as tenants in common
TEN ENT – as tenants by the entireties
JT TEN – as joint tenants with right of survivorship and not as tenants in common

(Name) CUST (Name) UNIF GIFT MIN ACT (State) — (Name) as Custodian for (Name) under the (State) Uniform Gifts to Minors Act

Additional abbreviations may also be used though not in the above list.

FOR VALUE RECEIVED, the undersigned sells, assigns and transfers unto

PLEASE PRINT OR TYPE TAXPAYER-IDENTIFYING NUMBER	NAME AND ADDRESS INCLUDING ZIP CODE OF ASSIGNEE

____

____

RESERVED FOR A.T.&T. CO. USE			
RIN			
TRANS CODE	DEB NOTE	TOWN CODE	CLASS NON-DOM.
61			
NA 1			
CC 2			☐ 4

the within Debenture of AMERICAN TELEPHONE AND TELEGRAPH COMPANY and hereby irrevocably constitutes and appoints

____ Attorney

to transfer said Debenture on the books of said Company with full power of substitution in the premises.

____

Dated ______________________   ______________________

THE SIGNATURE(S) TO THIS ASSIGNMENT MUST CORRESPOND WITH THE NAME(S) AS WRITTEN UPON THE FACE OF THE DEBENTURE IN EVERY PARTICULAR WITHOUT ALTERATION OR ENLARGEMENT OR ANY CHANGE WHATEVER. THE SIGNATURE(S) SHOULD BE GUARANTEED BY A COMMERCIAL BANK OR TRUST COMPANY, OR BY A NEW YORK, BOSTON, MIDWEST, PHILADELPHIA-BALTIMORE-WASHINGTON OR PACIFIC COAST STOCK EXCHANGE MEMBER OR FIRM WHOSE SIGNATURE IS KNOWN TO THE TRANSFER OFFICE.

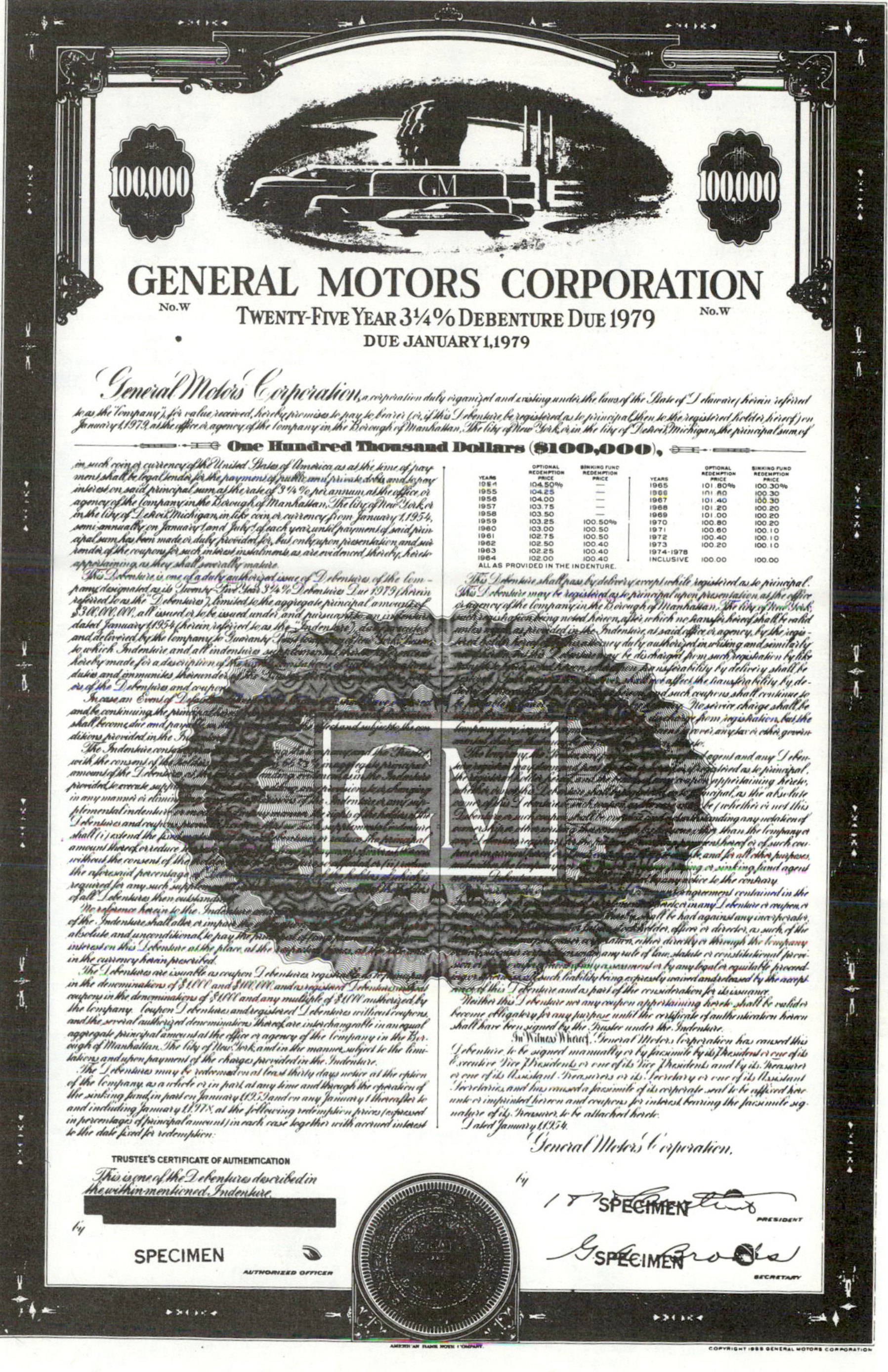

100,000 100,000

GENERAL MOTORS CORPORATION

No. W

TWENTY-FIVE YEAR 3¼% DEBENTURE DUE 1979

No. W

DUE JANUARY 1, 1979

General Motors Corporation, a corporation duly organized and existing under the laws of the State of Delaware (herein referred to as the "Company"), for value received, hereby promises to pay to bearer (or, if this Debenture be registered as to principal, then to the registered holder hereof) on January 1, 1979, at the office or agency of the Company in the Borough of Manhattan, The City of New York, or in the City of Detroit, Michigan, the principal sum of

One Hundred Thousand Dollars ($100,000),

in such coin or currency of the United States of America as at the time of payment shall be legal tender for the payment of public and private debts, and to pay interest on said principal sum at the rate of 3¼% per annum at the office or agency of the Company in the Borough of Manhattan, The City of New York, or in the City of Detroit, Michigan, in like coin or currency from January 1, 1954, semi-annually on January 1 and July 1 of each year until payment of said principal sum has been made or duly provided for, but only upon presentation and surrender of the coupons for such interest instalments as are evidenced thereby, hereto appertaining, as they shall severally mature.

This Debenture is one of a duly authorized issue of Debentures of the Company designated as its "Twenty-Five Year 3¼% Debentures Due 1979" (herein referred to as the "Debentures"), limited to the aggregate principal amount of $300,000,000, all issued or to be issued under and pursuant to an indenture dated January 1, 1954 (herein referred to as the "Indenture"), [illegible] and delivered by the Company to Guaranty [illegible] to which Indenture and all indentures [illegible] hereby made for a description of [illegible] duties and immunities thereunder [illegible] ers of the Debentures and coupon[illegible]

In case an Event of Default [illegible] and be continuing, the princip[illegible] shall become due and paya[illegible] ditions provided in the In[illegible]

The Indenture conta[illegible] with the consent of [illegible] amount of the Debentu[illegible] provided, to execute supp[illegible] in any manner or elimin[illegible] plemental indentur[illegible] Debentures and coupo[illegible] shall (i) extend the fixe[illegible] amount thereof or reduce [illegible] without the consent of the [illegible] the aforesaid percentage [illegible] required for any such supple[illegible] of all Debentures then outstandi[illegible]

No reference herein to the Inde[illegible] of the Indenture shall alter or impair [illegible] absolute and unconditional, to pay the prin[illegible] interest on this Debenture at the place, at the [illegible] in the currency herein prescribed.

The Debentures are issuable as coupon Debentures, registrable [illegible] in the denominations of $1,000 and $100,000 and as registered Debentures without coupons in the denominations of $1,000 and any multiple of $1,000 authorized by the Company. Coupon Debentures and registered Debentures without coupons, and the several authorized denominations thereof, are interchangeable in an equal aggregate principal amount at the office or agency of the Company in the Borough of Manhattan, The City of New York, and in the manner, subject to the limitations and upon payment of the charges provided in the Indenture.

The Debentures may be redeemed on at least thirty days' notice at the option of the Company, as a whole or in part, at any time and through the operation of the sinking fund, in part on January 1, 1959 and on any January 1 thereafter to and including January 1, 1978, at the following redemption prices (expressed in percentages of principal amount) in each case together with accrued interest to the date fixed for redemption:

YEARS	OPTIONAL REDEMPTION PRICE	SINKING FUND REDEMPTION PRICE	YEARS	OPTIONAL REDEMPTION PRICE	SINKING FUND REDEMPTION PRICE
1954	104.50%	—	1965	101.80%	100.30%
1955	104.25	—	1966	101.60	100.30
1956	104.00	—	1967	101.40	100.30
1957	103.75	—	1968	101.20	100.20
1958	103.50	—	1969	101.00	100.20
1959	103.25	100.50%	1970	100.80	100.20
1960	103.00	100.50	1971	100.60	100.10
1961	102.75	100.50	1972	100.40	100.10
1962	102.50	100.40	1973	100.20	100.10
1963	102.25	100.40	1974-1978 INCLUSIVE	100.00	100.00
1964	102.00	100.40			

ALL AS PROVIDED IN THE INDENTURE.

This Debenture shall pass by delivery except while registered as to principal. This Debenture may be registered as to principal upon presentation at the office or agency of the Company in the Borough of Manhattan, The City of New York, such registration being noted hereon, after which no transfer hereof shall be valid unless made as provided in the Indenture, at said office or agency, by the registered holder hereof [illegible] attorney duly authorized in writing and similarly [illegible] may be discharged from such registration by like [illegible] transferability by delivery shall be [illegible] not affect the transferability by de[illegible] and such coupons shall continue to [illegible] No service charge shall be [illegible] from registration, but the [illegible] ver, any tax or other govern[illegible]

[illegible] agent and any Deben[illegible] ered as to principal, [illegible] appertaining hereto, [illegible] ncipal, as the absolute [illegible] (whether or not this [illegible] nding any notation of [illegible] than the Company or [illegible] hereof or of such coupon [illegible] and for all other purposes, [illegible] or sinking fund agent [illegible] notice to the contrary.

[illegible] agreement contained in the [illegible] in any Debenture or coupon, or [illegible] shall be had against any incorporator, [illegible] holder, officer or director, as such, of the [illegible] ation, either directly or through the Company [illegible] any rule of law, statute or constitutional provision [illegible] any assessment or by any legal or equitable proceeding [illegible] such liability being expressly waived and released by the acceptance of this Debenture and as part of the consideration for its issuance.

Neither this Debenture nor any coupon appertaining hereto shall be valid or become obligatory for any purpose until the certificate of authentication hereon shall have been signed by the Trustee under the Indenture.

In Witness Whereof, General Motors Corporation has caused this Debenture to be signed manually or by facsimile by its President or one of its Executive Vice Presidents or one of its Vice Presidents and by its Treasurer or one of its Assistant Treasurers or its Secretary or one of its Assistant Secretaries, and has caused a facsimile of its corporate seal to be affixed hereunto or imprinted hereon and coupons for interest bearing the facsimile signature of its Treasurer to be attached hereto.

Dated January 1, 1954.

General Motors Corporation,

by

SPECIMEN

PRESIDENT

SPECIMEN

SECRETARY

TRUSTEE'S CERTIFICATE OF AUTHENTICATION

This is one of the Debentures described in the within-mentioned Indenture.

by

SPECIMEN

AUTHORIZED OFFICER

AMERICAN BANK NOTE COMPANY

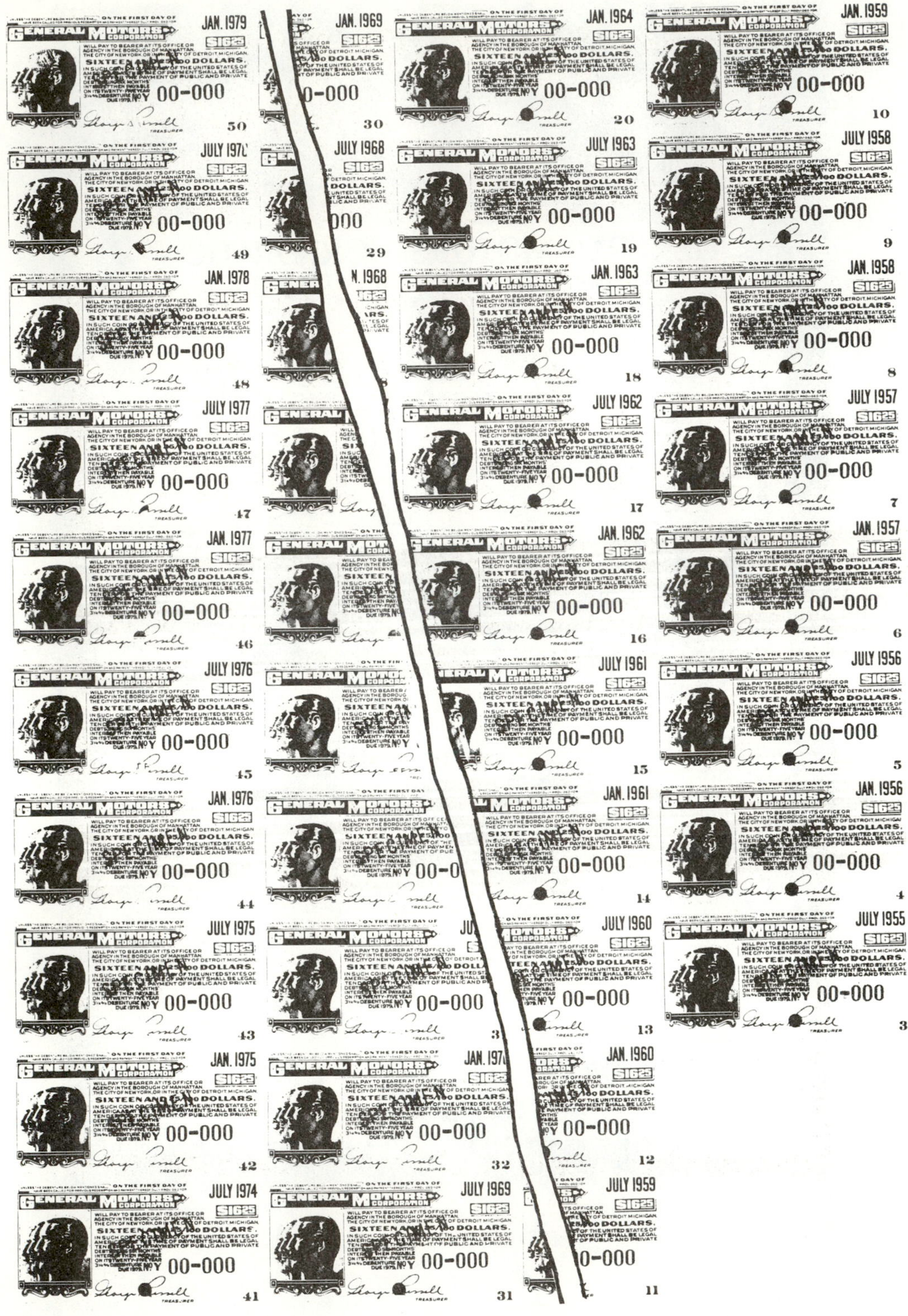
ON THE FIRST DAY OF
JAN. 1979
GENERAL MOTORS CORPORATION
$16.25
WILL PAY TO BEARER AT ITS OFFICE OR AGENCY IN THE BOROUGH OF MANHATTAN, THE CITY OF NEW YORK, OR IN THE CITY OF DETROIT MICHIGAN
SIXTEEN AND 25/100 DOLLARS,
IN SUCH COIN OR CURRENCY OF THE UNITED STATES OF AMERICA AS AT THE TIME OF PAYMENT SHALL BE LEGAL TENDER FOR THE PAYMENT OF PUBLIC AND PRIVATE DEBTS
ON ITS TWENTY-FIVE YEAR
DEBENTURE DUE 1979 NO Y 00-000
SPECIMEN
TREASURER
50
JAN. 1969 30
JAN. 1964 20
JAN. 1959 10
JULY 1970 49
JULY 1968 29
JULY 1963 19
JULY 1958 9
JAN. 1978 48
JAN. 1963 18
JAN. 1958 8
JULY 1977 47
JULY 1962 17
JULY 1957 7
JAN. 1977 46
JAN. 1962 16
JAN. 1957 6
JULY 1976 45
JULY 1961 15
JULY 1956 5
JAN. 1976 44
JAN. 1961 14
JAN. 1956 4
JULY 1975 43
JULY 1960 13
JULY 1955 3
JAN. 1975 42
JAN. 1970 32
JAN. 1960 12
JULY 1974 41
JULY 1969 31
JULY 1959 11

NO. W

GENERAL
MOTORS
CORPORATION

$1000

TWENTY-FIVE YEAR

3¼%

DEBENTURE DUE 1979

DUE JANUARY 1, 1979

INTEREST PAYABLE

JANUARY 1 AND JULY 1

PRINCIPAL AND INTEREST PAYABLE
IN THE
BOROUGH OF MANHATTAN,
THE CITY OF NEW YORK,
OR IN THE
CITY OF DETROIT,
MICHIGAN.

UNITED STATES INTERNAL REVENUE STAMPS REQUIRED BY LAW HAVE BEEN AFFIXED TO THE WITHIN-MENTIONED INDENTURE AND DULY CANCELLED.

NOTICE: NO WRITING BELOW EXCEPT BY DEBENTURE REGISTRAR.

DATE OF REGISTRATION AS TO PRINCIPAL	NAME OF REGISTERED OWNER	SIGNATURE OF REGISTRAR

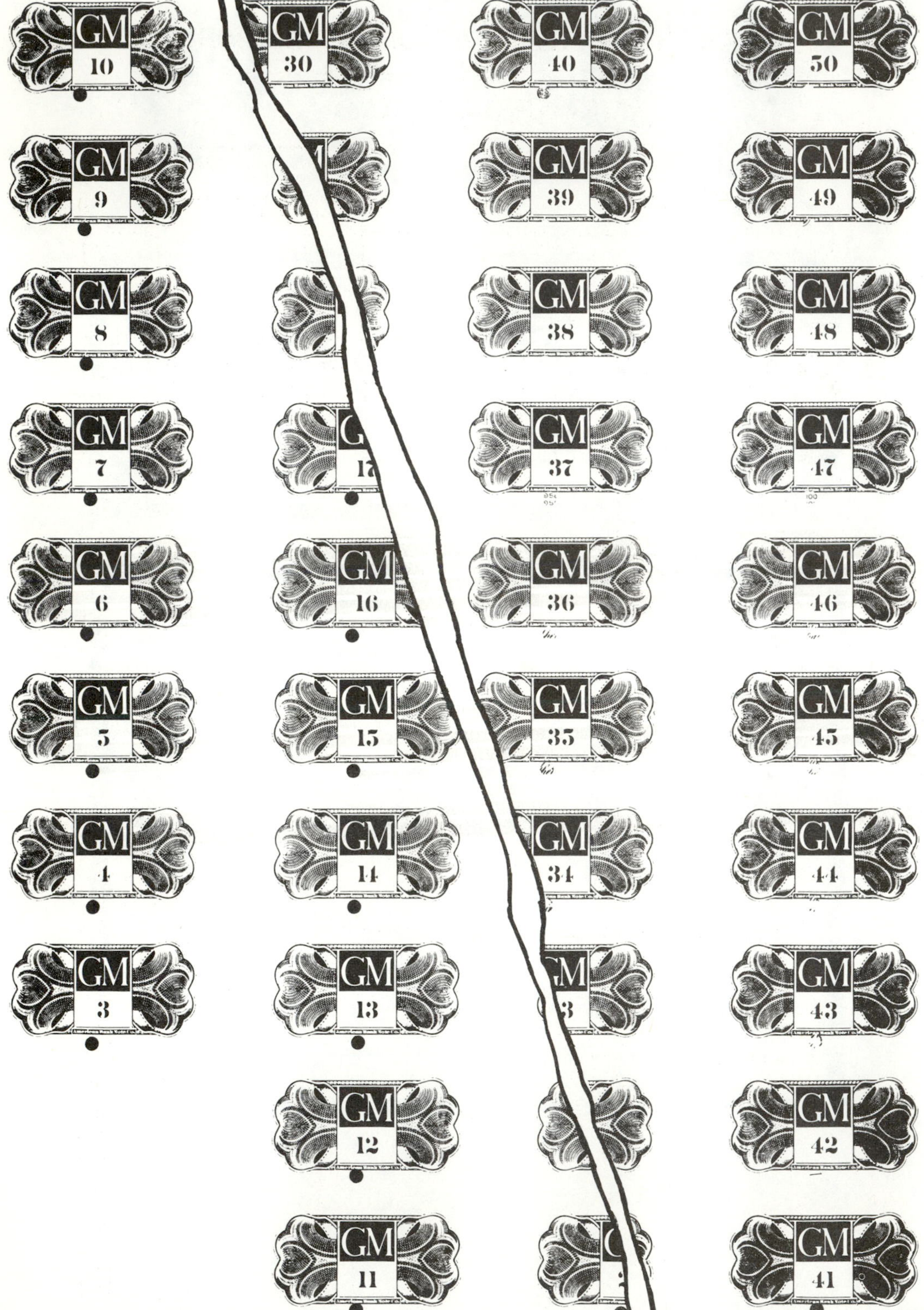

FORM 7B

STATEMENT OF RESOLUTION ESTABLISHING SERIES FOR SHARES (PENNSYLVANIA)

APPLICANT'S ACC'T NO.

DSCB:BCL—602 (Rev. 8-72)

Filing Fee: $40
AB-2
Statement Affecting Class or Series of Shares— Domestic Business Corporation

(Line for numbering)

COMMONWEALTH OF PENNSYLVANIA
DEPARTMENT OF STATE
CORPORATION BUREAU

Filed this ________ day of ________
________, 19___.
Commonwealth of Pennsylvania
Department of State

Secretary of the Commonwealth

(Box for Certification)

In compliance with the requirements of section 602 of the Business Corporation Law, act of May 5, 1933 (P. L. 364) (15 P. S. §1602), the undersigned corporation, desiring to state the voting rights, designations, preferences, qualifications, privileges, limitations, options, conversion rights, and other special rights, if any, of a class or series of a class of its shares, hereby certifies that:

1. The name of the corporation is:

____

2. (Check and complete one of the following):

☐ The resolution establishing and designating the class or series of shares and fixing and determining the relative rights and preferences thereof, set forth in full, is as follows:

☐ The resolution establishing and designating the class or series of shares and fixing and determining the relative rights and preferences thereof is set forth in full in Exhibit A attached hereto and made a part hereof.

3. The aggregate number of shares of such class or series established and designated by (a) such resolution, (b) all prior statements, if any, filed under the Business Corporation Law with respect thereto, and (c) any other provision of the Articles is ________________ shares.

4. (Check and complete one of the following):

☐ The resolution was adopted by the Board of Directors of the corporation at a duly called meeting held on the ________________ day of ________________, 19______.

DSCB:BCL—602 (Rev. 8-72)-2

☐ The resolution was adopted by a consent or consents in writing dated the ______________ day of ______________ 19____, signed by all of the Directors of the corporation and filed with the Secretary of the corporation.

IN TESTIMONY WHEREOF, the undersigned corporation has caused this statement to be signed by a duly authorized officer and its corporate seal, duly attested by another such officer, to be hereunto affixed this ______________ day of ______________, 19____.

____
(NAME OF CORPORATION)

By: ______________________________
(SIGNATURE)

____
(TITLE: PRESIDENT, VICE PRESIDENT, ETC.)

Attest:

____
(SIGNATURE)

____
(TITLE: SECRETARY, ASSISTANT SECRETARY, ETC.)

(CORPORATE SEAL)

FORM 7C

CANCELLATION OF REDEEMABLE SHARES *

Filing fee: $________.

STATEMENT OF CANCELLATION OF REDEEMABLE SHARES OF

____

To the Secretary of State
of the State of ________:

Pursuant to the provisions of Section ________ of the ________ Business Corporation Act, the undersigned corporation submits the following statement of cancellation by redemption or purchase of redeemable shares of the corporation:

FIRST: The name of the corporation is ______________________

____

SECOND: The number of redeemable shares of the corporation cancelled through redemption or purchase is ________, itemized as follows:

Class	Series	Number of Shares

THIRD: The aggregate number of issued shares of the corporation after giving effect to such cancellation is ______, itemized as follows:

Class	Series	Number of Shares

FOURTH: The amount of the stated capital of the corporation after giving effect to such cancellation is $______.

FIFTH: the number of shares which the corporation has authority to issue after giving effect to such cancellation is ______, itemized as follows:

Class	Series	Number of Shares

Dated ______, 19__.

__________________ (Note 1)

By ______________
Its ______ President
and ______________
Its ______ Secretary
(Note 2)

(Add Verification Form A)

Notes: 1. Exact corporate name of corporation making the statement.
2. Signatures and titles of officers signing for the corporation.

* The Model Act has repealed provisions relating to the cancellation of redeemable shares. (See Section 67 of the M.B.C.A.) Nevertheless, this form is used by most states whose statutes are based upon the Model Act.

FORM 7D

SECURITY AGREEMENT UNDER THE UNIFORM COMMERCIAL CODE

Form UCC 1205—Bradford Publishing Co., 1824-46 Stout Street, Denver, Colorado—5-73

STATE OF COLORADO

UNIFORM COMMERCIAL CODE — SECURITY AGREEMENT

Debtor:

Name:____

Address:
Residence:____
No. Street City State

Business:____
No. Street City State

Secured Party:

Name:____

Address:____
No. Street City State

Debtor, for consideration, hereby grants to Secured Party a security interest in the following property and any and all additions, accessions and substitutions thereto or therefor (hereinafter called the "COLLATERAL"):

To secure payment of the indebtedness evidenced by__________certain promissory note____of even date herewith, payable to the Secured Party, or order, as follows:

DEBTOR EXPRESSLY WARRANTS AND COVENANTS:

1. That except for the security interest granted hereby Debtor is, or to the extent that this agreement states that the Collateral is to be acquired after the date hereof, will be, the owner of the Collateral free from any adverse lien, security interest or encumbrances; and that Debtor will defend the Collateral against all claims and demands of all persons at anytime claiming the same or any interest therein.

2. The Collateral is used or bought primarily for:

☐ Personal, family or household purposes;
☐ Use in farming operations;
☐ Use in business.

3. That Debtor's residence is as stated above, and the Collateral will be kept at

____
No. and Street City County State

4. If any of the Collateral is crops, oil, gas, or minerals to be extracted or timber to be cut, or goods which are or are to become fixtures, said Collateral concerns the following described real estate situate in the__________

County of____________________and State of Colorado, to wit:

5. Not to sell, transfer or dispose of the Collateral, and promptly to notify Secured Party of any change in the location of the Collateral within the State of Colorado and not to remove the same from the State of Colorado without the prior written consent of the Secured Party.

6. To pay all taxes and assessments of every nature which may be levied or assessed against the Collateral.

7. Not to permit or allow any adverse lien, security interest or encumbrance whatsoever upon the Collateral and not to permit the same to be attached or replevined.

8. That the Collateral is in good condition, and that he will, at his own expense, keep the same in good condition and from time to time, forthwith, replace and repair all such parts of the Collateral as may be broken, worn out, or damaged without allowing any lien to be created upon the Collateral on account of such replacement or repairs, and that the Secured Party may examine and inspect the Collateral at any time, wherever located.

9. That he will not use the Collateral in violation of any applicable statutes, regulations or ordinances.

UNTIL DEFAULT Debtor may have possession of the Collateral and use it in any lawful manner, and upon default Secured Party shall have the immediate right to the possession of the Collateral.

DEBTOR SHALL BE IN DEFAULT under this agreement upon the happening of any of the following events or conditions:

(a) default in the payment or performancec of any obligation, covenant or liability contained or referred to herein or in any note evidencing the same;

(b) the making or furnishing of any warranty, representation or statement to Secured Party by or on behalf of Debtor which proves to have been false in any material respect when made or furnished;

(c) loss, theft, damage, destruction, sale or encumbrance to or of any of the Collateral, or the making of any levy seizure or attachment thereof or thereon;

(d) death, dissolution, termination of existence, insolvency, business failure, appointment of a receiver of any part of the property of, assignment for the benefit of creditors by, or the commencement of any proceeding under any bankruptcy or insolvency laws of, by or against Debtor or any guarantor or surety for Debtor.

UPON SUCH DEFAULT and at any time thereafter, or if it deems itself insecure, Secured Party may declare all Obligations secured hereby immediately due and payable and shall have the remedies of a secured party under Article 9 of the Colorado Uniform Commercial Code. Secured Party may require Debtor to assemble the Collateral and deliver or make it available to Secured Party at a place to be designated by Secured Party which is reasonably convenient to both parties. Expenses of retaking, holding, preparing for sale, selling or the like shall include Secured Party's reasonable attorney's fees and legal expenses.

No waiver by Secured Party of any default shall operate as a waiver of any other default or of the same default on a future occasion. The taking of this security agreement shall not waive or impair any other security said Secured Party may have or hereafter acquire for the payment of the above indebtedness, nor shall the taking of any such additional security waive or impair this security agreement; but said Secured Party may resort to any security it may have in the order it may deem proper, and notwithstanding any collateral security, Secured Party shall retain its rights of set-off against Debtor.

All rights of Secured Party hereunder shall inure to the benefit of its successors and assigns; and all promises and duties of Debtor shall bind his heirs, executors or administrators or his or its successors or assigns. If there be more than one Debtor, their liabilities hereunder shall be joint and several.

Dated this__________day of________________, 19_____.

Debtor: Secured Party:*

________________________ ________________________

________________________ ________________________

* If this Security Agreement is intended to serve as a financing statement secured party as well as the debtor must sign.

FORM 7E

FINANCING STATEMENT UNDER THE UNIFORM COMMERCIAL CODE

Bradford Publishing Co., 1824-46 Stout Street, Denver, Colorado

STATE OF COLORADO
UNIFORM COMMERCIAL CODE — FINANCING STATEMENT — COLORADO U.C.C.-1 (Rev. 1-78)
IMPORTANT — Read instructions on reverse side before filling out form

This Financing Statement is presented for filing pursuant to the Uniform Commercial Code.

1. Debtor(s) Name and Mailing Address:

2. Secured Party(ies) Name and Address:

3. For Filing Officer (Date, Time, Number, and Filing Office):

4. This Financing Statement covers the following types (or items) of property:
(WARNING: If collateral is crops, fixtures, timber, or minerals or other substances to be extracted or accounts resulting from the sale thereof, read instructions on back.)

Check only if applicable.
☐ This Financing Statement is to be filed for record in the real estate records.
☐ Products of collateral are also covered.

5. Name and address of Assignee of Secured Party:

6. This Statement is signed by the Secured Party instead of the Debtor to perfect a security interest in collateral
(Please check appropriate box)
☐ already subject to a security interest in another jurisdiction when it was brought into this state, or when the debtor's location was changed to this state.
☐ which is proceeds of the original collateral described above in which a security interest was perfected;
☐ as to which the filing has lapsed; or
☐ acquired after a change of name, identity or corporate structure of the debtor.

7. Check only if applicable: ☐ The Debtor is a transmitting utility.

____
Signature(s) of Debtor(s)

____
Signature(s) of Secured Party(ies)

Form approved by the Secretary of State and the County Clerks and Recorders Association
(1) Filing Officer Copy

COLORADO FORM U.C.C. 1 (REV. 1-78)
BRADFORD PUBLISHING CO.
DENVER, COLO.

FORM 7F

SKELETON TRUST INDENTURE

THIS INDENTURE, dated ______ ______, 19__, between ______ Corporation, a corporation organized and existing under the laws of the State of ______ (hereinafter called the Corporation), and the ______ Trust Company of ______, a corporation organized and existing under the laws of the State of ______, as Trustee (hereinafter called the Trustee), WITNESSETH:

WHEREAS, the Corporation, in the exercise of its corporate powers and for the purpose of furthering and accomplishing its corporate objects and purposes and pursuant to due corporate action, has determined to create an issue of First Mortgage Bonds, in an aggregate principal amount not exceeding $______ at any one time outstanding, and to secure the same by this Indenture; and

WHEREAS, the Corporation has determined to create an initial series of Bonds hereunder and to issue forthwith $_______ in principal amount of said initial series of Bonds to be known as "First Mortgage Bonds, _______% Series, due _______ _______, 19__", to contain such provisions as are hereinafter specified; and

WHEREAS, the text of all the First Mortgage Bonds, _______% Series, due _______ _______, 19__, of the coupons for interest to be attached thereto and of the Trustee's certificate to be endorsed thereon, is to be substantially as follows:

[*Here insert full form of bond;*] and

WHEREAS, all the requirements of law relating to the authorization of the Bonds and the execution of this Indenture and the mortgage and pledge hereby evidenced have been complied with; and all things necessary to make the Bonds, when authenticated by the Trustee and issued as in this Indenture provided, the valid and binding obligations of the Corporation, and all things necessary to constitute this Indenture a valid and binding mortgage for the security of said Bonds have been done and performed and the issue of said Bonds subject to the terms hereof and the execution of this Indenture have been in all respects duly authorized;

NOW, THEREFORE, In order to secure the payment of the principal and interest of all the Bonds at any time issued and outstanding under this Indenture, according to their tenor, purport and effect, and the performance and observance of all the covenants, agreements and conditions therein and herein contained, and to declare the terms and conditions upon which said Bonds are to be issued, authenticated, secured and held, and for and in consideration of the premises and of the purchase and acceptance of the Bonds by the holders thereof, and of the sum of _______ dollar(s) duly paid by the Trustee to the Corporation at or before the ensealing and delivery of these presents, the receipt whereof is hereby acknowledged, the Corporation has mortgaged, pledged, assigned, transferred, granted, bargained, sold, aliened, remised, released, conveyed, confirmed and set over, unto the Trustee, and its successor or successors, in the trusts hereby created, and its and their assigns, the following described properties:

[*Insert full description of properties. This usually includes all land, plants, offices and other buildings, together with all improvements and fixtures, etc., and a clause providing for after-acquired property. This is followed by the "Habendum clause":*]

TO HAVE AND TO HOLD the lands and interest in lands, estates, plants and appurtenances and other property hereby conveyed, mortgages, pledged or transferred unto the Trustee, its successors and assigns forever;

[There follows a clause excepting any specific property from the mortgage. This is followed by the "Trust" clause:]

IN TRUST NEVERTHELESS, under and subject to the conditions herein set forth, for the common and equal benefit and security of all the holders of Bonds and coupons issued and to be issued under this Indenture.

[The remainder of the indenture is divided into articles, as indicated below, each article being further divided into a number of sections and subsections:]

Article 1: Form, Execution, Delivery and Registration of the Bonds

[Authentication by Trustee; aggregate amount outstanding; recording of indenture.—Date of initial series; interest; place of payment of principal and interest; denominations.—Terms of later series.—Title of initial series; identification of later series; numbering of bonds.—Execution of supplemental indenture upon request for authentication and delivery of later series.—Registration and transfer of bonds.—Signature of bonds; use of facsimile signatures; seal; effect of Trustee's certificate.—Evidence of ownership of bonds.—Issuance of temporary bonds.—Mutilated, lost, destroyed or stolen bonds.]

Article 2: Issue of Bonds

[Authentication and delivery of initial series of bonds.—Use of deposited funds for capital expenditures.—Limitation on amount of bonds authenticated and moneys paid out for capital expenditures.—Sale of bonds reserved for authentication.—Documents required before paying out deposited moneys and authenticating and delivering bonds.—Discharge of prior lien on property.—Trustee not liable for use of bonds or deposited moneys.—Delivery of bonds in exchange for bonds cancelled.—Delivery of bonds upon surrender of bonds about to mature or called for redemption.—Cancellation of bonds converted into stock or retired through sinking fund.]

Article 3: Redemption of Bonds

[Premium paid on redemption.—Notice of redemption.—Cancellation of indenture on redemption of all outstanding bonds.—Cancellation of redeemed or reacquired bonds.]

Article 4: Sinking fund for First Mortgage Bonds, ——% Series, due ———— ————, 19——.

[Amounts to be paid into sinking fund.—Additional sinking fund equal to percentage of net profits.—Fund payments in bonds purchased by Corporation.—Application of fund to redemption of bonds.—Notice of redemption through fund.—Cancellation of bonds redeemed.]

Article 5: Particular Covenants of the Corporation

[Covenants to pay principal and interest—not to extend time for payment of interest—to subject present and after-acquired property to lien of indenture and to execute further instruments of conveyance—not to permit prior lien on property and to discharge liens—to discharge taxes and assessments—not to merge or sell assets unless purchaser assumes payment of bonds—to maintain property—to preserve corporate existence—to keep property insured—to pay expenses of Trustee—to record and file indenture—not to dispose of bonds contrary to indenture provisions—to restrict declaration of dividends, distributions and redemption of stock—to restrict purchase of stock—to maintain office for payment of principal and interest—to deliver to Trustee annual financial statements—to furnish opinion of expert as to fair value of property.]

Article 6: Release of Property Included in the Trust Estate

[Power of Corporation to sell obsolete property.—Power of Corporation to remove property.—Power of Corporation to sell limited amount of property.—Obligations in satisfaction of debt not to be subject to lien of indenture.—Power of Corporation to move, alter or remodel buildings.—Power of Corporation to amend, alter or cancel lease, license or easement.—Power of Corporation to sell or exchange for other property.—Release of trust property taken by eminent domain.—Method of release of mortgaged properties.—Application of moneys received by Trustee.—Powers of Corporation to be exercised by receiver or Trustee.]

Article 7: Events of Default—Remedies of Trust and Bondholders

[Events of default: default in payment of principal, payment of interest or sinking fund payment—involuntary bankruptcy or receivership—voluntary bankruptcy, reorganization, assignment for benefit of creditors—default in performance of covenants.—Acceleration of due date of principal; waiver of default.—Power of Trustee to take possession.—Power of Trustee to sell trust estate.—Notice of sale.—Execution of instruments and transfer to purchaser at sale.—Divesting of Corporation's title upon sale.—Suit by Trustee to enforce payment of bonds; foreclosure.—Power of bondholders to decide on remedy sought.—Payment by Corporation to Trustee for benefit of bondholders on default.—Restrictions on suits by bondholders.—Application of proceeds of sale of trust estate.—Principal of all bonds to become due on sale.—Appointment of receiver upon default.—Covenant of Corporation to waive service of process, enter appearance and consent to entry of judgment.—Waiver of Corporation of benefits of laws for stay or appraisal of trust estate.—Remedies cumulative.—Delay or omission to exercise right not waiver.—Restrictions against remedies which would surrender lien of indenture.—Power of Trustee to restrain compliance with invalid law.]

Article 8: Immunity of Incorporators, Stockholders, Officers and Directors

Article 9: Merger, Consolidation or Sale

[Covenant of Corporation not to merge, consolidate or sell assets unless new company can meet provisions of indenture.—Successor company to assume conditions of bonds and indentures.—Successor company to succeed to rights of Corporation.—Indenture to become lien on improvements by successor company and on after-acquired property.]

Article 10: Concerning the Trustee

[Power of Trustee to employ agents and attorneys.—Limitation on liability of Trustee.—Indemnification of Trustee by bondholders.—Form of request, notice or authorization to Trustee by Corporation.—Compensation of Trustee; lien for payment.—Return of moneys deposited with Trustee.—Advances by Trustee to preserve trust estate.—Conflicting interests of Trustee.—Removal of Trustee.—Appointment of successor Trustee.—Effect of merger or consolidation of Trustee.—Appointment of co-trustee.]

Article 11: Bondholders' Lists and Reports by the Corporation and the Trustee

[Corporation to furnish list of bondholders; Trustee to preserve list.—Application by bondholders for list.—Corporation to file annual and other reports.—Reports by Trustee to bondholders.—Notice by Trustee to bondholders of defaults.]

Article 12: Supplemental Indentures, Bondholders' Acts, Holdings and Apparent Authority

[Purposes for execution of supplemental indenture.—Consent to execution by proportion of bondholders.—Binding effect of supplemental indenture on non-consenting bondholders.—Revocation of consent by bondholders.—Trustee to join in supplemental indenture.—Proof of ownership of registered and bearer bonds.—Supplemental indenture to be considered part of original indenture.]

Article 13: Possession Until Default

[Corporation to retain possession until default, use income and dispose of profits.—Reversion of property to Corporation on payment of principal and interest; discharge of indenture.]

Article 14: Definitions and Miscellaneous Provisions

[Agreements binding on successors and assigns.—Definitions.—Provisions conflicting with Trust Indenture Act of 1939.—Agreements to be for exclusive benefit of parties and bondholders.—Notices to Corporation or Trustee.—Acceptance of trust by Trustee.—Appointment of attorneys for acknowledgement of indenture.]

IN WITNESS WHEREOF, ______ Corporation has caused this instrument to be signed in its corporate name and its corporate seal to be affixed by its President and its Secretary, and its corporate seal to be attested by its Secretary, by order of its Board of Directors, and the ______ Trust Company of ______, in token of its acceptance of the trusts created hereby, has caused this instrument to be signed in its corporate name and its corporate seal to be affixed by its President and its Secretary, and its corporate seal to be attested by its Secretary, by order of the Executive Committee of its Board of Directors, as of the date given at the beginning of this indenture.

_______________ Corporation

[Corporate Seal]

by _______________ President

Attest:

_______________ Secretary _______________ Secretary

The _______________ Trust Company of

_______________, Trustee

[Corporate Seal]

by _______________ President

Attest:

_______________ Secretary _______________ Secretary [22]

22. Prentice Hall, Corporations, Forms § 60351. Reprinted with the permission of Prentice Hall.

FORM 8A

SPECIAL MEETING OF THE BOARD OF DIRECTORS OF ____________, INC.

A Special Meeting of the Board of Directors of ____________, Inc., was held at ______________, ____________, ______________ on the _____ day of ____________, 19___, at ___ o'clock.

The meeting was called pursuant to section _______ of the ____________ Corporation Code [and] [or] Article ___ of the Articles of Incorporation of the corporation [and] [or] Section ___ of the Bylaws of the corporation.

The following directors were present: ______________________.
The following directors were absent: ______________________.

The presence of the foregoing directors constitutes a quorum. The following other persons were also present: ______________________.

The meeting was held pursuant to notice addressed to each director in accordance with the statute, the Articles of Incorporation, and the Bylaws of the corporation. A copy of the notice, together with the Secretary's certificate that such notice was properly mailed or delivered, is attached to the minutes of the meeting.

[or]

The meeting is held pursuant to Waiver of Notice from each director, a copy of which is attached to the minutes of the meeting.

The minutes of the meeting of the Board of Directors on ____________, 19____, were approved as read.

The President stated that the purpose of the meeting was to [here describe purpose in narrative form].

Following full discussion, upon motion duly made, seconded and unanimously adopted it was

RESOLVED, [here describe substance of resolution]

[or]

Following full discussion, upon motion duly made by ____________, seconded by ____________, the following directors voted in favor: ________________; and the following directors voted against: ________________; the following resolution:

[names]

[names]

RESOLVED, [here describe substance of resolution]

The Treasurer of the corporation reported on the financial condition of the corporation, a copy of which is attached to these minutes.

The Board of Directors informally discussed [here describe] and no action was taken at this time.

There being no further business, the meeting was adjourned.

____

Secretary

FORM 8B

SPECIAL MEETING OF THE SHAREHOLDERS OF ____________, INC.

A Special Meeting of the Shareholders of ____________, Inc., was held at ______________, ____________, ______________ on the day of ____________, 19____, at _____ o'clock.

The meeting was called pursuant to section _______ of the ____________ Corporation Code [and] [or] Article _____ of the Articles of Incorporation [and] [or] Section _____ of the Bylaws of the corporation, by the [President], [Secretary] [The holders of % of the shares entitled to vote] or [other].

[A copy of the call of the meeting dated ____________, 19____, addressed to the Secretary of the corporation and signed by the holders of ____% of the shares entitled to vote is attached to the minutes of this meeting.]

The meeting was held pursuant to notice addressed to each shareholder in accordance with the statute, the Articles of Incorporation, and the Bylaws of the corporation. A copy of the notice, together with the Secretary's certificate that such notice was properly mailed or delivered to each Shareholder, is attached to the minutes of the meeting.

The Board of Directors, by resolution dated ____________, 19____, set ____________, 19____, as the record date for the determination of the Shareholders entitled to vote at this meeting, and only Shareholders of record on that date are entitled to vote.

[or]

The meeting was held pursuant to Waiver of Notice from each Shareholder, a copy of which is attached to the minutes of this meeting.

Shareholders holding _______ Shares of record were present at the meeting. Shareholders holding _______ Shares of record were represented by proxy at the meeting, and their shares were voted by ____________, duly constituted proxy in their names.

The minutes of the Shareholder meeting on ____________, 19____, were approved as read.

The President stated that the purpose of the meeting was to [here describe purpose in narrative form].

Following full discussion, upon motion duly made, and seconded, ______ Shares voted in person in favor; ______Shares voted by proxy in favor; ______ Shares voted in person against; and ______ Shares voted by proxy against the following resolution:

RESOLVED, [here describe substance of resolution]

The Treasurer of the corporation reported on the financial condition of the corporation, a copy of which is attached to these minutes.

Several questions were raised by the Shareholders concerning [here describe informal discussion and questions]. No action was taken on these matters.

There being no further business, the meeting was adjourned.

____
Secretary

FORM 9A

STATEMENT OF CANCELLATION OF REACQUIRED SHARES (NEW JERSEY)

Form C-126 7-1-71

STATEMENT OF CANCELLATION

OF REACQUIRED SHARES OF

____

(For Use by Domestic Corporations Only)

To: The Secretary of State
State of New Jersey

Pursuant to the provisions of Section 14A:7-18, Corporations, General, of the New Jersey Statutes, the undersigned corporation hereby submits the following Statement of Cancellation of Reaquired Shares:

1. The name of the corporation is ____________________________________.

2. The number of shares cancelled is ____________ ; itemized as follows:

Class	Series	No. of Shares

(Omit the following if not applicable.)

3. If cancelled shares were not reacquired out of stated capital or by their conversion into other shares of the corporation, the date of adoption of the resolution of the board of directors cancelling such shares was on the __________ day of __________, 19______ .

4. The aggregate number of issued shares of the corporation after giving effect to such cancellation is ______________ ; itemized as follows:

Class	Series	No. of Shares

5. The amount of the stated capital of the corporation after giving effect to such cancellation is $ ______________ . (Must be set forth in dollars.)

(Use the following if the Certificate of Incorporation, or the Plan of Merger or Consolidation, in the case of shares acquired by the corporation pursuant to Section 14A: 11-1 et seq., Corporations, General, of the New Jersey Statutes (regarding rights of dissenting shareholders), provides that the cancelled shares shall not be reissued.)

6. The Certificate of Incorporation is amended pursuant to a resolution of the board of directors decreasing the aggregate number of shares which the corporation is authorized to issue by the number of shares cancelled.

The number of shares which the corporation has authority to issue, after giving effect to such cancellation is ______________ ; itemized as follows:

Class	Series	No. of Shares

(Use the following if shareholder approval is required for reduction of stated capital, pursuant to Section 14A:7-18(3), Corporations, General, of the New Jersey Statutes.)

7. The shareholders approved the reduction of stated capital on the __________

day of ______________, 19_____ .

The number of shares outstanding at the time of approving such reduction was ____________. The number of shares entitled to vote thereon was __________; itemized as follows: (If the shares of any class or series are entitled to vote as a class, set forth the number of shares of each such class and series voting for and against the reduction respectively.)

No. of Shares Voting For Reduction	No. of Shares Voting Against Reduction

Dated this __________ day of __________________, 19______.

____
(Corporate Name)

By ____________________________*
(Signature)

____
(Type or Print Name and Title)

(*May be executed by the chairman of the board, or the president, or a vice-president of the corporation.)

Fees for filing in Office of the Secretary of State, State House, Trenton, N. J. 08625.

Filing Fee $25.00

NOTE: No recording fees will be assessed.

TRANSACTION NO.: ____________

FOLDER NO.:

STATEMENT OF CANCELLATION
OF REACQUIRED SHARES OF

FILED BY:

RECORDED AND FILED:

Recorder's Initials

FORM 10A

NON-DISCLOSURE AGREEMENT

The Coca-Cola Company

NON-DISCLOSURE AGREEMENT

Covering Inventions, Discoveries, and Confidential Matter

In consideration of my employment, or my continued employment, as the case may be, by The Coca-Cola Company, a Delaware corporation (hereinafter called the Company), I agree with the Company as follows:

So long as I shall remain in the employ of the Company I will devote my whole time and ability to the service of the Company in such capacity as it shall from time to time direct, and I will perform my duties faithfully and diligently.

I will not, during my employment or thereafter, use or disclose to others without the written consent of the Company, any trade secrets, secret "know-how", confidential or secret technical information or other confidential information relative to your business, obtained by me while in the employ of the Company. Upon leaving the employ of the Company I will not take with me any confidential data, drawings, or information obtained by me as the result of my employment, or any reproductions thereof. All such Company property will be surrendered to the Company on termination or at any time on request.

I will disclose to the Company and, upon the Company's request, assign to it, without charge, all my right, title, and interest in and to any and all inventions and discoveries which I may make, solely or jointly with others, while in the employ of the Company which relate to or are useful or may be useful in connection with business of the character carried on or contemplated by the Company, and all my right, title, and interest in and to any and all domestic and foreign applications for patents covering such inventions and discoveries and any and all patents granted for such inventions and any and all reissues and extensions of such patents; and upon request of the Company whether during or subsequent to this employment I will do any and all acts and execute and deliver such instruments as may be deemed by the Company necessary or proper to vest all my right, title, and interest in and to said inventions, applications, and patents in the Company and to secure or maintain such patents, reissues and/or extensions thereof. Any inventions and discoveries relating to the Company's business made by me within one year after termination of my employment with the Company shall be deemed to be within this provision, unless I can prove that the same were conceived and made following said termination. All necessary and proper expenses in connection with the foregoing shall be borne by the Company, and if services in connection therewith are performed at the Company's request after termination of employment, the Company will pay reasonable compensation for such.

Attached hereto is a list of patent applications and unpatented inventions made prior to my employment by the Company, which I agree is a complete list and which I desire to remove from the operation of this agreement.

This agreement shall enure to the benefit of the Company, its subsidiaries, allied companies, successors and assigns or nominees of the Company, and I specifically agree to execute any and all documents considered convenient or necessary to assign, transfer, sustain and maintain inventions, discoveries, applications and patents, both in this and foreign countries.

IN WITNESS WHEREOF, I have hereunto signed my name and affixed my seal, this ______________

day of ________________________, 19____.

Witness: ________________________________(SEAL)

________________________________ ________________________________(DEPT.)

Distribution: Execute in triplicate — White copy for Department; yellow copy for Employee; and pink copy for Personnel Relations Department.

FORM 761 (1-68)

FORM 11A

VOTING TRUST AGREEMENT

Agreement, made this ______ day of ______, 19__, between ______, ______, ______, ______, and ______, hereinafter designated as Trustees, and the undersigned shareholders of ______ Company, hereinafter designated as the Beneficiaries.

Whereas, the parties do hereby agree and declare that the intent and purpose of this Agreement is to provide a means whereby the parties hereto may initiate or maintain in effect any general policy, plan, or program affecting ______ Company which the parties should determine to be to their joint benefit, interest, and advantage, and to the best interests of all stockholders of ______ Company, and to that end to elect or retain or replace any officer, executive, or employee of said corporation;

Now, therefore, the parties do hereby agree with each other as follows:

1. **Delivery of Shares to Trustees, Term of Trust.** Upon the signing of this agreement the Beneficiaries shall deliver to the Trustees the certificate or certificates representing all the shares of ______ Company now owned or controlled by them, said certificates to be endorsed in blank or accompanied by proper instruments of assignment and transfer thereof in blank. Said shares will be held by the Trustees for a period of ten years from ______, 19__ (unless this trust is sooner terminated, as hereinafter provided) in trust, however, for the Beneficiaries, their heirs, executors, administrators, successors and assigns, and at all times subject to the terms and conditions herein set forth.

2. **Additional Shares.** Any and all certificates for additional shares of ______ Company that shall hereafter during said ten year period be issued to any of the Beneficiaries shall be in like manner endorsed and delivered to the Trustees, to be held by them under the terms hereof.

3. **Voting.** During the term of this Agreement the Trustees or their successors in trust shall have the sole and exclusive voting power of the stock standing in their names as such. They shall have the power to vote the stock at all regular and special meetings of the stockholders and may vote for, do, or assent or consent to any act or proceeding which the shareholders of said corporation might or could vote for, do or assent or consent to and shall have all the powers, rights and privileges of a shareholder of said corporation. The Trustees shall consult and confer with each other, and shall make every effort to agree on how their votes are cast. The Trustees, as soon as this Agreement becomes effective, shall appoint a chairman. In any

case where shareholder action is required, the chairman may, or upon the request of any two Trustees, shall, call a meeting of the Trustees, on reasonable notice, for the purpose of reaching an agreement on the manner of voting the stock held by the Trustees, or for any other purpose deemed to be in the best interests of ______ Company. The vote of the Trustees shall always be exercised as a unit, as any four of said Trustees shall direct and determine. If any four Trustees fail to agree on any matter on which a vote of the stockholders is called for, then the question in disagreement shall be submitted for arbitration to some disinterested person (i. e., one having no financial interest in ______ Company) chosen by the affirmative vote of four of the Trustees, as sole arbitrator. If four of the Trustees are unable to agree on an arbitrator, then each of the Trustees shall nominate a similarly disinterested person as a candidate and the arbitrator shall be selected by the affirmative vote of four of the Trustees from the panel of such candidates. If any candidate receives the affirmative vote of four of the Trustees he shall be elected sole arbitrator. If no candidate receives the affirmative vote of four of the Trustees, then the candidate receiving the lowest number of votes shall be eliminated from the panel (or if there should be a tie among the low candidates, or among all the candidates, if more than two, one of such candidates shall be eliminated by lot) and the Trustees shall continue the process of voting among those remaining on the panel until one has been selected by the affirmative vote of four of the Trustees. If the voting continues to the point where no candidate receives the vote of four of the Trustees, then those two candidates receiving the highest number of votes respectively from those who voted with the majority and those who voted with the minority on the issue to be submitted to arbitration (ties among the majority and minority candidates to be decided by lot) shall be appointed arbitrators and these two shall appoint a third disinterested person as arbitrator. The decision of the arbitrator or, if more than one, a majority thereof, shall be binding upon the parties hereto and the vote of all the stock in trust shall be cast in accordance with such decision. The Beneficiaries may by unanimous written agreement designate any person as sole arbitrator who shall act during the life of this agreement.

4. **Proxies.** Any Trustee may vote in person or by proxy and a proxy in writing signed by any four of the Trustees shall be sufficient authority to the person named therein to vote all the stock held by the Trustees hereunder at any meeting, regular or special, of the stockholders of ______ Company. If at any such meeting less than four Trustees shall be present either in person or by proxy, then all of the stock held by the Trustees may be voted in accordance with the unanimous decision of those Trustees present in person or by proxy.

5. **Appointment of Successor Trustees.** In the event of the death, resignation, removal or incapacity of any of the Trustees his successor shall be named by an instrument in writing signed by a majority of the remaining Trustees. All Successor Trustees shall be clothed with all the rights, privileges, duties and powers herein conferred upon the Trustees herein named.

6. **Voting Trust Certificates.** Upon the delivery to the Trustees of said certificates representing the shares of ______ Company, the Trustees will cause the same to be transferred on the books of the corporation to themselves as Trustees and will deliver to each of the Beneficiaries a Trustees' Certificate for the number of shares delivered to said Trustees, substantially in the form hereinafter set out. Upon receipt of certificates for additional shares of ______ Company issued to any of the Beneficiaries, and upon receipt of certificates for such shares issued to other persons and which may be issued to future subscribers for shares of ______ Company, and upon compliance with the terms of this agreement by the owners of such shares, the Trustees will cause said shares to be transferred on the books of said corporation to their names as trustees, and shall deliver to each of the persons so depositing said certificates a Trustees' Certificate for the number of shares so deposited by said person.

The Trustees' Certificate shall be substantially in the following form:

Trustees' Certificate

This is to certify that the undersigned Trustees have received a certificate or certificates issued in the name of ______, evidencing the ownership of ______ shares of ______ Company, a ______ corporation, and that said shares are held subject to all the terms and conditions of that certain agreement, dated ______, 19__, by and between __________, __________, __________, __________, and __________, as Trustees, and certain shareholders of ______ Company. During the period of ten years from and after ______, 19__, the said Trustees, or their successors, shall, as provided in said agreement, possess and be entitled to exercise the right to vote and otherwise represent all of said shares for all purposes, it being agreed that no voting right shall pass to the holder hereof by virtue of the ownership of this certificate.

This certificate is assignable with the right of issuance of a new certificate of like tenor only upon the surrender to the undersigned or their successors of this certificate properly endorsed. Upon the termination of said Trust this certificate shall be surrendered to the Trustees by the holder hereof upon delivery to such holder of a stock certificate representing a like number of said shares.

In witness whereof, the undersigned Trustees have executed this Certificate this ______ day of ______, 19__.

____
____
____
____
____
Trustees

Said Trustees' Certificate, subject to the conditions hereof, may be transferred by endorsement by the person to whom issued, or by his attorney in fact, or by the administrator, executor or guardian of his estate, and delivery of the same to said Trustees; but said transfer shall not be evidence to or be binding upon said Trustees until the certificate is surrendered to them and the transfer is so entered upon their "Trustees' Certificate Book", which shall be kept by them to show the names of the parties by whom and to whom transferred, the numbers of the certificates, the number of shares and the date of transfer. No new Trustees' Certificate shall be issued until the former Trustees' Certificate for the shares represented thereby shall have been surrendered to and cancelled by said Trustees, and they shall preserve the certificates so cancelled as vouchers. In case any Trustees' Certificate shall be claimed to be lost or destroyed, a new Trustees' Certificate may be issued in lieu thereof, upon such proof of loss and such security as may be required by said Trustees.

7. **Restrictions on Transfer of Voting Trust Certificates.** Each of the Beneficiaries agrees that during the term of this agreement said Trustees' Certificates will not be sold or transferred except in accordance with Paragraph ____ of the Organization Agreement of ______ Company, dated ______, 19__, relating to the sale of shares of ______ Company, so long as said Organization Agreement remains in effect. Said Trustees' Certificates shall be regarded as stock of the ______ Company within the meaning of any provision of the Bylaws of said corporation imposing conditions or restrictions upon the sale of stock of said corporation.

8. **Dividends.** Before declaring any dividend the Board of Directors of ______ Company shall request the Trustees to certify to the Board the names of all persons who are the owners and holders of Trustees' Certificates, and the number of shares to which each of such persons is or may then be entitled as shown by the books of the Trustees and no dividend shall be declared and paid by said corporation until reasonable opportunity has been given the Trustees to submit such certificate. Said corporation is hereby irrevocably authorized and directed (a) to accept such certificate of the Trustees as true; and (b) to pay any and all dividends upon the shares enumerated in such certificate directly to the holders of the Trustees' Certificates.

In the event that any dividend paid in capital stock of the Company shall be received by the Trustees, the respective holders of Trustees' Certificates issued hereunder shall be entitled to the delivery of new or additional Trustees' Certificates to the amount of the stock received by the Trustees as such dividend upon the number of such shares of the Company represented by their respective Trustees' Certificates theretofore outstanding.

9. **Termination.** Except as herein otherwise provided the trust hereby created shall not be revoked and the powers herein delegated to the Trustees shall be irrevocable during said period of ten years from and after ______, 19__. This trust, however, shall terminate upon the vote of any four of the Trustees and their declaration in writing that said trust is terminated. Unless the Trustees by unanimous vote otherwise determine, this trust shall also terminate if and when less than 50% of the outstanding shares of ______ Company remain subject to this Trust Agreement. Upon the termination of said trust the certificates representing all of the shares so held under this agreement and then remaining in the hands of the Trustees or their successors shall be assigned to the parties then entitled thereto as shown by Trustees' Certificates then outstanding, upon surrender to the Trustees of the Trustees' Certificates representing said shares.

10. **Compensation of Trustees.** The Beneficiaries may pay a reasonable compensation to the Trustees for their service hereunder and all expenses and costs incurred by them in executing said trusts, and the Beneficiaries do agree to save and hold harmless said Trustees from any and all liability arising out of the holding by them of any of the shares of said ______ Company hereunder.

11. **Exculpatory Clause.** The Trustees shall not be liable or incur any responsibility by reason of their acts of omission or commission in the premises except for wilful misconduct or gross negligence in the execution of the trusts hereby created.

12. **Extension of Term.** At any time within one year prior to the time of expiration of this agreement, one or more Beneficiaries hereunder may, by agreement in writing and with the written consent of all of the Trustees, extend the duration of this agreement for an additional period not exceeding ten years; provided, however, that no such extension agreement shall affect the rights or obligations of persons not parties thereto.

13. **Counterparts.** This agreement may be executed in several counterparts, each of which so executed shall be deemed to be the original, and such counterparts shall together constitute one and the same instrument.

In witness whereof, the parties have hereunto set their hands or have caused their corporate names to be hereunto affixed by their officers thereunto duly authorized, the day and year first above written.

____
____
____
____
____
Trustees

______________ holding ______________ shares
______________ holding ______________ shares
______________ holding ______________ shares
______________ holding ______________ shares
______________ holding ______________ shares
______________ holding ______________ shares

Stockholders of ______ Company
Beneficiaries [25]

25. West's Modern Legal Forms § 3012.-1.

FORM 12A

APPLICATION FOR CERTIFICATE OF AUTHORITY (MISSISSIPPI)

File in Duplicate Originals

APPLICATION FOR CERTIFICATE OF AUTHORITY OF

____
(EXACT CORPORATE NAME)

To the Secretary of State
of the State of Mississippi

Pursuant to the provisions of Section 110 of the Mississippi Business Corporation Act, the undersigned corporation hereby applies for a Certificate of Authority to transact business in your State, and for that purpose submits the following statement:

FIRST: The name of the corporation is____

____

SECOND: The name which it elects to use in Mississippi is________________________

____________________________________,____________________________(Note I)

THIRD: It is incorporated under the laws of____________________________________

FOURTH: The date of its incorporation is_________________and the period of its duration is

____

FIFTH: The address of its principal office in the state or country under the laws of which it is incorporated is____

SIXTH: The address of its proposed registered office in Mississippi is__________________

____________________and the name of its proposed registered agent in Mississippi at that address is____

SEVENTH: The purpose or purposes which it proposes to pursue in the transaction of business in Mississippi are____

EIGHTH: The names and respective addresses of its directors and officers are:

Name	Office	ADDRESS
	Director	
	Director	
	Director	
	President	
	Vice President	
	Secretary	
	Treasurer	

NINTH: The aggregate number of shares which it has authority to issue, itemized by classes, par value of shares, shares without par value, and series, if any within a class, is:

Number of Shares	Class	Series	Par Value per Share or Statement that Shares are without Par Value

TENTH: The aggregate number of its issued shares, itemized by classes, par value of shares, shares without par value, and series, if any, within a class, is:

Number of Shares	Class	Series	Par Value per Share or Statement that Shares are without Par Value

ELEVENTH: The amount of its stated capital is $____________________. (Note 2)

TWELFTH: An estimate of the value of all property to be owned by it for the following year, wherever located, is $________________.

THIRTEENTH: An estimate of the value of its property to be located within Mississippi during such year is $________________.

FOURTEENTH: An estimate of the gross amount of business to be transacted by it during such year is $____________________.

FIFTEENTH: An estimate of the gross amount of business to be transacted by it at or from places of business in Mississippi during such year is $____________________.

SIXTEENTH: This Application is accompanied by a copy of its articles of incorporation and all amendments thereto, duly authenticated by the proper officer of the state or country under the laws of which it in incorporated.

Dated____________________, 19_____.

____
EXACT CORPORATE NAME

By ________________________________

Its____________________President

By ________________________________

Its____________________Secretary

STATE OF__________________________
} SS.
COUNTY OF________________________

I, ____________________________________, a notary public, do hereby certify that on this______day of________________, 19____, personally appeared before me____, who, being by me first duly sworn, declared that he is the ____ of _____________________________________, that he executed the foregoing document as __________________________________of the corporation, and that the statements therein contained are true.

____
Notary Public

My commission expires______________________.
(NOTARIAL SEAL)

Notes: 1. If the name of the corporation does not contain the word "corporation", "company", "incorporated", or "limited" or an abbreviation of one of such words, insert the name of the corporation with the word or abbreviation which it elects to add thereto for use in this State.

2. "Stated capital" means, at any particular time, the sum of (1) the par value of all shares of the corporation having a par value that have been issued, (2) the amount of the consideration received by the corporation for all shares of the corporation without par value that have been issued, except such part of the consideration therefor as may have been allocated to capital surplus in a manner permitted by law, and (3) such amounts not included in clauses (1) and (2) of this paragraph as have been transferred to stated capital of the corporation, whether upon the issue of shares as a share dividend or otherwise, minus all reductions from such sum as have been effected in a manner permitted by law.

STATE OF__________________________
} SS.
COUNTY OF________________________

I, ____________________________________, a notary public, do hereby certify that on this______day of________________, 19____, personally appeared before me____, who, being by me first duly sworn, declared that he is the ____ of _____________________________________, that he executed the foregoing document as __________________________________of the corporation, and that the statements therein contained are true.

____
Notary Public

My commission expires______________________.
(NOTARIAL SEAL)

FORM 12B

STATEMENT OF CHANGE OF REGISTERED OFFICE OR AGENT (SOUTH DAKOTA)

STATEMENT OF CHANGE OF REGISTERED OFFICE OR REGISTERED AGENT, OR BOTH, OF

....

To the Secretary of State
of the State of South Dakota:

Pursuant to the provisions of the South Dakota Corporation Acts, the undersigned corporation, organized under the laws of the State of submits the following statement for the purpose of changing its registered office or its registered agent, or both, in the State of South Dakota:

FIRST: The name of the corporation is ....

....

SECOND: The address of its previous registered office was ....

....

THIRD: The address to which its registered office is to be changed is ....

....

FOURTH: The name of its previous registered agent is ....

....

FIFTH: The name of its successor registered agent is ....

....

SIXTH: The address of its registered office and the address of the business office of its registered agent, as changed, will be identical. The address of its place of business in South Dakota is ....

....

SEVENTH: This change has been authorized by resolution duly adopted by the board of directors.

Dated ...., 19......... .

.... (Note 1)

By .... (Note 2)

Its .... President

STATE OF ....)
) ss.
COUNTY OF ....)

Before me, ... , a Notary Public in and for the said County and State, personally appeared .... who acknowledged before me that ...he is the (President) (Vice-President) of.... ...., that ...he signed the foregoing, and that the statements contained therein are true.

In witness whereof I have hereunto set my hand and seal this .... day of ...., A.D., 19................ .

....
Notary Public

My commission expires ....
(Notarial Seal)

Notes: 1. Exact corporate name of corporation making the statement.
2. Signature and title of officer signing for the corporation – must be a President or a Vice-President.

Filing fee $5.00
Submit one copy.

FORM 12C

CERTIFICATE OF AUTHORITY (MODEL ACT)

STATE OF ———
OFFICE OF THE SECRETARY OF STATE

CERTIFICATE OF AUTHORITY OF

____

The undersigned, as Secretary of State of the State of ______, hereby certifies that duplicate originals of an Application of ______ ____________________________ for a Certificate of Authority to transact business in this State, duly signed and verified pursuant to the provisions of the ______ Business Corporation Act, have been received in this office and are found to conform to law.

ACCORDINGLY the undersigned, as such Secretary of State, and by virtue of the authority vested in him by law, hereby issues this Certificate of Authority to ____________________________ to transact business in this State under the name of ______________ __________________ and attaches hereto a duplicate original of the Application for such Certificate.

Dated ______, 19__.

____
Secretary of State

FORM 12D

APPLICATION FOR AMENDED CERTIFICATE OF AUTHORITY (SOUTH CAROLINA)

APPLICATION FOR AMENDED CERTIFICATE OF AUTHORITY

STATE OF SOUTH CAROLINA
SECRETARY OF STATE

For Use by The Secretary of State	file this form in duplicate	This Space For Use by The Secretary of State
File No. ______ Fee Paid $______ Date ______ C. B. ______		

Pursuant to Section 13.8 of the South Carolina Business Corporation Act of 1962, the undersigned corporation hereby applies for an amended certificate of authority to transact business in the State of South Carolina, and for that purpose submits the following statement: (Section 12-23.8 of the 1962 supplement)

First: The name of the corporation is ______________________________

______________________________.

Second: The registered office of the Corporation in the State of South Carolina is ______________

in the City of ______________.

Third: It is incorporated under the laws of the State of ______________.

Fourth: The corporation was domesticated in the State of South Carolina on the ________ day of

______________, 19____.

Fifth: The proposed amendment to its application of authority is:

Sixth: Attached to this application is a duly authenticated copy of the amendment authorizing the change.

Date ______________ ______________
Name of Corporation

By ______________ By ______________
(Secretary or Assistant) (President or Vice President)

STATE OF ______________ }
COUNTY OF ______________ } SS:

The undersigned ______________ and ______________ do hereby certify that they are the duly elected and acting ______________ and ______________, respectively, of ______________ corporation and are authorized to execute this verification; that each of the undersigned for himself does hereby further certify that he has read the foregoing document, understands the meaning and purport of the statements therein contained and the same are true to the best of his information and belief.

Dated at ______________, this ______ day of ______________, 19____.

____
(President or Vice President)

____
(Secretary or Assistant Secretary)

NOTE: This certificate has been prepared for execution by the president (or vice president) and secretary (or assistant secretary). It may be executed by any of the persons enumerated in section 1.4 (Section 12-11.4 Supplement 1962 Code) of the South Carolina Business Corporation Act under the circumstances indicated. If anyone other than the president (or vice president) and secretary (or assistant secretary) executes the form, the wording of this verification should be changed accordingly.

Filing fees:

For amendment of Certificate of Authority	$40.00
For recording application	5.00
Total fee	$45.00

FORM 12E

AMENDED CERTIFICATE OF AUTHORITY (MODEL ACT)

STATE OF ——
OFFICE OF THE SECRETARY OF STATE

AMENDED CERTIFICATE OF AUTHORITY OF

____

The undersigned, as Secretary of State of the State of ______, hereby certifies that duplicate originals of an Application of ______ ______________________ for an Amended Certificate of Authority to transact business in this State, duly signed and verified pursuant to the provisions of the ______ Business Corporation Act, have been received in this office and are found to conform to law.

ACCORDINGLY the undersigned, as such Secretary of State, and by virtue of the authority vested in him by law, hereby issues this Amended Certificate of Authority to ______________ __________ to transact business in this State under the name of ______________________ and attaches hereto a duplicate original of the Application for such Amended Certificate.

Dated ______, 19__.

____
Secretary of State

FORM 12F

APPLICATION FOR CERTIFICATE OF WITHDRAWAL (TEXAS)

APPLICATION FOR
CERTIFICATE OF WITHDRAWAL
OF

____

To the Secretary of State
of the State of Texas

Pursuant to the provisions of Article 8.14 of the Texas Business Corporation Act, the undersigned corporation hereby applies for a Certificate of withdrawal from the State of Texas, and for that purpose submits the following statement:

1. The name of the Corporation is ____________________

2. It is incorporated under the laws of ____________________

3. It is not transacting business in the State of Texas.

4. It hereby surrenders its authority to transact business in said state.

5. It revokes the authority of its registered agent in the State of Texas to accept service of process and consents that service of process in any action, suit or proceeding based upon any cause of action arising in the State of Texas during the time it was authorized to transact business therein may thereafter be made on it by service thereof on the Secretary of State of State of Texas.

6. The post office address to which the Secretary of State may mail a copy of any process against the corporation that may be served on him is ____________________

7. All sums due or accrued by this corporation to the State of Texas have been paid.

8. All known creditors or claimants have been paid or provided for and the corporation is not involved in or threatened with litigation in any court in the State of Texas.

____

By ____________________

Its ____________ President

and ____________________

Its ____________ Secretary

STATE OF ____________ }
COUNTY OF ____________ }

I, ____________________, a notary public, do hereby certify that on this __________ day of ____________, 19____, personally appeared before me ____________________, who being by me first duly sworn, declared that he is the ____________________ of ____________________ that he signed the foregoing document as ____________________ of the corporation, and that the statements therein contained are true.

____
Notary Public

FORM 12G

CERTIFICATE OF WITHDRAWAL (MODEL ACT)

STATE OF ______
OFFICE OF THE SECRETARY OF STATE

CERTIFICATE OF WITHDRAWAL OF

____

The undersigned, as Secretary of State of the State of ______, hereby certifies that duplicate originals of an Application of ______________________________ for a Certificate of Withdrawal from this State, duly signed and verified pursuant to the provisions of the ______ Business Corporation Act, have been received in this office and are found to conform to law.

ACCORDINGLY the undersigned, as such Secretary of State, and by virtue of the authority vested in him by law, hereby issues this Certificate of Withdrawal to ______________________ ______, and attaches hereto a duplicate original of the Application for such Certificate.

Dated ______, 19__.

____
Secretary of State

FORM 12H

CERTIFICATE OF REVOCATION OF AUTHORITY (MODEL ACT)

STATE OF ______
OFFICE OF THE SECRETARY OF STATE

CERTIFICATE OF REVOCATION OF CERTIFICATE OF AUTHORITY OF

____

The undersigned, as Secretary of State of the State of ______, and by virtue of the authority vested in him by Section 122 of the ______ Business Corporation Act, hereby revokes the Certificate of

Authority of ______________________ to transact business in this State, for the following reasons: ______________

____

____

Dated ________, 19___.

____

Secretary of State

FORM 13A

ARTICLES OF AMENDMENT (PENNSYLVANIA)

APPLICANT'S ACC'T NO.

DSCB: BCL-806 (Rev. 8-72)

Filing Fee: $40
AB-2

Articles of Amendment— Domestic Business Corporation

____
(Line for numbering)

COMMONWEALTH OF PENNSYLVANIA
DEPARTMENT OF STATE
CORPORATION BUREAU

Filed this ________ day of ________
______________________, 19___.
Commonwealth of Pennsylvania
Department of State

Secretary of the Commonwealth

(Box for Certification)

In compliance with the requirements of section 806 of the Business Corporation Law, act of May 5, 1933 (P. L. 364) (15 P. S. §1806), the undersigned corporation, desiring to amend its Articles, does hereby certify that:

1. The name of the corporation is:

____

2. The location of its registered office in this Commonwealth is (the Department of State is hereby authorized to correct the following statement to conform to the records of the Department):

____
(NUMBER) (STREET)

______________________ Pennsylvania ______________
(CITY) (ZIP CODE)

3. The statute by or under which it was incorporated is:

____

4. The date of its incorporation is: ______________________

5. (Check, and if appropriate, complete one of the following):

☐ The meeting of the shareholders of the corporation at which the amendment was adopted was held at the time and place and pursuant to the kind and period of notice herein stated.

Time: The ______________ day of ______________, 19______.

Place: ______________________________________

Kind and period of notice ______________________________

____

☐ The amendment was adopted by a consent in writing, setting forth the action so taken, signed by all of the shareholders entitled to vote thereon and filed with the Secretary of the corporation.

6. At the time of the action of shareholders:

(a) The total number of shares outstanding was:

____

(b) The number of shares entitled to vote was:

____

DSCB:BCL—806 (Rev. 8-72)-2

7. In the action taken by the shareholders:

(a) The number of shares voted in favor of the amendment was:

____

(b) The number of shares voted against the amendment was:

____

8. The amendment adopted by the shareholders, set forth in full, is as follows:

IN TESTIMONY WHEREOF, the undersigned corporation has caused these Articles of Amendment to be signed by a duly authorized officer and its corporate seal, duly attested by another such officer, to be hereunto affixed this ______________ day of ______________, 19____.

Attest: ______________________________ (NAME OF CORPORATION)

______________________________ (SIGNATURE) By: ______________________________ (SIGNATURE)

______________________________ (TITLE: SECRETARY, ASSISTANT SECRETARY, ETC.) ______________________________ (TITLE: PRESIDENT, VICE PRESIDENT, ETC.)

(CORPORATE SEAL)

INSTRUCTIONS FOR COMPLETION OF FORM

A. Any necessary copies of Form DSCB:17.2 (Consent to Appropriation of Name) or Form DSCB:17.3 (Consent to Use of Similar Name) shall accompany Articles of Amendment effecting a change of name.

B. Any necessary governmental approvals shall accompany this form.

C. Where action is taken by partial written consent pursuant to the Articles, the second alternate of Paragraph 5 should be modified accordingly.

D. If the shares of any class were entitled to vote as a class, the number of shares of each class so entitled and the number of shares of all other classes entitled to vote should be set forth in Paragraph 6(b).

E. If the shares of any class were entitled to vote as a class, the number of shares of such class and the number of shares of all other classes voted for and against such amendment respectively should be set forth in Paragraphs 7(a) and 7(b).

F. BCL §807 (15 P. S. §1807) requires that the corporation shall advertise its intention to file or the filing of Articles of Amendment. Proofs of publication of such advertising should not be delivered to the Department, but should be filed with the minutes of the corporation.

FORM 13B

CERTIFICATE OF AMENDMENT (NORTH DAKOTA)

Certificate No.__________

UNITED STATES OF AMERICA

DEPARTMENT OF STATE

GREAT SEAL STATE OF NORTH DAKOTA

State of North Dakota

CERTIFICATE OF AMENDMENT

OF

____

The undersigned, as Secretary of State of the State of North Dakota, hereby certifies that duplicate originals of Articles of Amendment to the Articles of Incorporation of ____________________

____

duly signed and verified pursuant to the provisions of the North Dakota ____________________ Corporation Act, have been received in this office and are found to conform to law.

ACCORDINGLY the undersigned, as such Secretary of State, and by virtue of the authority vested in him by law, hereby issues this Certificate of Amendment to the Articles of Incorporation of ____________________

____

____

and attaches hereto a duplicate original of the Articles of Amendment.

SPECIMEN

IN TESTIMONY WHEREOF, I have hereunto set my hand and affixed the Great Seal of the State at the Capitol in the City of Bismarck, this ________ day of ________________ A.D., 19___.

____
Secretary of State.

File No.______________

ORIGINAL

By________________________, Deputy.

"Buy North Dakota Products"

FORM 13C

RESTATED ARTICLES OF INCORPORATION (NEW JERSEY)

Form C-100a 1-1-69

RESTATED CERTIFICATE OF INCORPORATION

OF

____

To: The Secretary of State

State of New Jersey

Pursuant to the provisions of Section 14A:9-5, Corporations, General, of the New Jersey Statutes, the undersigned corporation hereby executes the following Restated Certificate of Incorporation:

FIRST: The name of the corporation is ....

SECOND: The purpose or purposes for which the corporation is organized are:

(Use the following if the shares are to consist of one class only.)

THIRD: The aggregate number of shares which the corporation shall have authority to issue is .... of the par value of .... Dollars ($...........) each (*or* without par value.)

(Use the following if the shares are divided into classes, or into classes and series.)

FOURTH: The aggregate number of shares which the corporation shall have authority to issue is ...., itemized by classes, par value of shares, shares without par value, and series, if any, within a class, is:

Class	Series (if any)	Number of Shares	Par value per share or statement that shares are without par value

The relative rights, preferences and limitations of the shares of each class and series (if any), are as follows:

(If, the shares are, or are to be divided into classes, or into classes and series, insert a statement of any authority vested in the board of directors to divide the shares into classes or series, or both, and to determine or change for any class or series its designation, number or shares, relative rights, preferences and limitations.)

FIFTH: The address* of the corporation's current registered office is: ....
(*Include zip code)

.................................., and the name of its current registered agent at such address is:

....

SIXTH: The number of directors constituting the current board of directors is ....

The names and addresses of the directors are as follows:

Names	Addresses (including zip code)
....	....
....	....
....	....
....	....

SEVENTH: The duration of the corporation, if other than perpetual, is ....

(Use the following only if an effective date, not later than 30 days subsequent to the date of filing is desired.)

EIGHTH: The effective date of this Certificate shall be ....

Dated this .... day of ...., 19......

....
(Corporate Name)

By ....*

....
(Type or Print Name and Title)

(*May be executed by the chairman of the board, *or* the president, *or* a vice-president.)

CERTIFICATE REQUIRED TO BE FILED WITH THE

RESTATED CERTIFICATE OF INCORPORATION

OF

____

Pursuant to the provisions of Section 14A:9-5 (5), Corporations, General, of the New Jersey Statutes, the undersigned corporation hereby executes the following certificate:

FIRST: The name of the corporation is ....

SECOND: The Restated Certificate of Incorporation was adopted on the .... day of ...., 19......

(Use the following clause if the Restated Certificate was adopted by the shareholders.)

THIRD: At the time of the adoption of the Restated Certificate of Incorporation, the number of shares outstanding was .... The total of such shares entitled to vote thereon, and the vote of such shares was:

Total Number of Shares Entitled to Vote	Number of Shares Voted For	Against

At the time of the adoption of the Restated Certificate of Incorporation, the number of outstanding shares of each class or series entitled to vote thereon as a class and the vote of such shares, was: (if inapplicable, insert "none".)

Class or Series	Number of Shares Entitled to Vote	Number of Shares Voted For	Against

(Use the following if the Restated Certificate does not amend the Certificate of Incorporation.)

FOURTH: This Restated Certificate of Incorporation only restates and integrates and does not further amend the provisions of the Certificate of Incorporation of this corporation as heretofore amended or supplemented and there is no discrepancy between those provisions and the provisions of this Restated Certificate of Incorporation.

(Use the following if the Restated Certificate further amends the Certificate of Incorporation.)

FIFTH: This Restated Certificate of Incorporation restates and integrates and further amends the Certificate of Incorporation of this corporation by:*

(*Insert amendment or amendments adopted. If such amendment is intended to provide for an exchange, reclassification or cancellation of issued shares, insert a statement of the manner in which the same shall be effected.)

(Use the following only if an effective date, not later than 30 days subsequent to the date of filing is is desired.)

SIXTH: The effective date of this amendment shall be ....

Dated this .... day of ...., 19......

....
(Corporate Name)

By ....*

....
(Type or Print Name and Title)

(*May be executed by the chairman of the board, <u>or</u> the president, <u>or</u> a vice-president.)

FORM 13D

CERTIFICATE OF RESTATED ARTICLES OF INCORPORATION (NORTH DAKOTA)

№_____

Certificate No:______________

UNITED STATES OF AMERICA

DEPARTMENT OF STATE

GREAT SEAL · STATE OF NORTH DAKOTA

State of North Dakota

RESTATED CERTIFICATE OF INCORPORATION

OF

____

The undersigned, as Secretary of State of the State of North Dakota, hereby certifies that duplicate originals of Restated Articles of Incorporation of

____

duly signed and verified pursuant to the provisions of the North Dakota__________________________________ Act, have been received in this office and are found to conform to law.

ACCORDINGLY the undersigned, as such Secretary of State and by virtue of the authority vested in him by law, hereby issues this Restated Certificate of Incorporation to

____

and attaches hereto a duplicate original of the Restated Articles of Incorporation.

IN TESTIMONY WHEREOF, I have hereunto set my hand and affixed the Great Seal of the State at the Capitol in the City of Bismarck, this ________________ day of ________________ A.D., 19____

____
Secretary of State

By ________________________
Deputy

File No.________________

"Buy North Dakota Products"

FORM 13E

ARTICLES OF MERGER (PENNSYLVANIA)

Filed this ________ day of ________
________________, 19__.
Commonwealth of Pennsylvania
Department of State

Secretary of the Commonwealth

(Box for Certification)

APPLICANT'S ACC'T NO.

DSCB:BCL—903 (Rev. 8-72)

Filing Fee: $80 plus $20 for each party corporation in excess of two
AMB-9

Articles of Merger—Business Corporation

____
(Line for numbering)

COMMONWEALTH OF PENNSYLVANIA
DEPARTMENT OF STATE
CORPORATION BUREAU

In compliance with the requirements of section 903 of the Business Corporation Law, act of May 5, 1933 (P. L. 364) (15 P. S. §1903), the undersigned corporations, desiring to effect a merger, hereby certify that:

1. The name of the corporation surviving the merger is:

____

2. (Check and complete one of the following):

☐ The surviving corporation is a domestic corporation and the location of its registered office in this Commonwealth is (the Department of State is hereby authorized to correct the following statement to conform to the records of the Department):

____
(NUMBER) (STREET)

________________________ Pennsylvania ________
(CITY) (ZIP CODE)

☐ The surviving corporation is a foreign corporation incorporated under the laws of ________ (NAME OF JURISDICTION) ________ and the location of its office registered with such domiciliary jurisdiction is:

____
(NUMBER) (STREET)

____
(CITY) (STATE) (ZIP CODE)

3. The name and the location of the registered office of each other domestic business corporation and qualified foreign business corporation which is a party to the plan of merger are as follows:

M. BURR KEIM COMPANY, PHILADELPHIA

Page 1 of 4 pages

DSCB:BCL—903 (Rev. 8-72)-2

4. (Check, and if appropriate, complete one of the following):

☐ The plan of merger shall be effective upon filing these Articles of Merger in the Department of State.

☐ The plan of merger shall be effective on ______________ (DATE) at ______________ (HOUR).

5. The manner in which the plan of merger was adopted by each domestic corporation is as follows:

NAME OF CORPORATION	MANNER OF ADOPTION

6. (Strike out this paragraph if no foreign corporation is party to the merger.) The plan was authorized, adopted or approved, as the case may be, by the foreign corporation (or each of the foreign corporations) in accordance with the laws of the jurisdiction in which it was formed.

7. The plan of merger is set forth in Exhibit A, attached hereto and made a part hereof.

8. (Strike out this paragraph if the surviving corporation is a domestic corporation.) The Secretary of the Commonwealth and his successor in office is hereby designated as the true and lawful attorney of the surviving corporation upon whom may be served all lawful process in any action or proceeding against it for enforcement against it of any obligation of any constituent domestic corporation or any obligation arising from the merger proceedings or any action or proceeding to determine and enforce the rights of any shareholder under the provisions of section 908 of the Business Corporation Law. The surviving corporation hereby agrees that the service of process upon the Secretary of the Commonwealth shall be of the same legal force and validity as if served on the corporation and that the authority for such service of process shall continue in force as long as any of the aforesaid obligations and rights remain outstanding in this Commonwealth.

DSCB:BCL—903 (Rev. 8-72)-3

IN TESTIMONY WHEREOF, each undersigned corporation has caused these Articles of Merger to be signed by a duly authorized officer and its corporate seal, duly attested by another such officer, to be hereunto affixed this ________________ day of ________________, 19____.

____
(NAME OF CORPORATION)

By: ______________________________
(SIGNATURE)

____
(TITLE: PRESIDENT, VICE PRESIDENT, ETC.)

Attest:

____
(SIGNATURE)

____
(TITLE: SECRETARY, ASSISTANT SECRETARY, ETC.)

(CORPORATE SEAL)

____
(NAME OF CORPORATION)

By: ______________________________
(SIGNATURE)

____
(TITLE: PRESIDENT, VICE PRESIDENT, ETC.)

Attest:

____
(SIGNATURE)

____
(TITLE: SECRETARY, ASSISTANT SECRETARY, ETC.)

(CORPORATE SEAL)

DSCB:BCL—903 (Rev. 8-72)-4

INSTRUCTIONS FOR COMPLETION OF FORM:

A. If a new corporation results from the transaction the form should be rewritten as Articles of Consolidation and modified accordingly.

B. A foreign business corporation may be a party to a merger notwithstanding the fact that it has not received a certificate of authority to do business in Pennsylvania. However, if the surviving corporation is a foreign corporation which is not the holder of a Certificate of Authority under the Business Corporation Law on the effective date of the merger, there must be submitted with this form tax clearance certificates from the Department of Revenue and the Bureau of Employment Security of the Department of Labor and Industry with respect to each domestic corporation and qualified foreign corporation evidencing payment of all taxes and charges payable to the Commonwealth.

C. Any necessary copies of Form DSCB: 17.2 (Consent to Appropriation of Name) or Form DSCB: 17.3 (Consent to Use of Similar Name) shall accompany Articles of Merger effecting a change of name.

D. Any necessary governmental approvals shall accompany this form.

E. One of the following statements or the equivalent should be used in the second column of Paragraph 5 to set forth the manner of adoption:

"Adopted by action of the board of directors pursuant to section 902.1 of the Business Corporation Law."

"Approved by the affirmative vote of the shareholders entitled to vote thereon at a meeting called after at least ten days written notice to all shareholders of record, whether or not entitled to vote thereon, setting forth such purpose."

"Approved by a consent or consents in writing, setting forth the action so taken, signed by all of the shareholders entitled to vote thereon, and filed with the secretary of the corporation" (where action is taken by partial written consent pursuant to the Articles, this paragraph should be modified accordingly).

F. Where more than two corporations are parties to the merger appropriate additional corporate signatures should be added. All parties to the merger shall execute the Articles of Merger, including a nonqualified corporation which is not a surviving corporation and which is not otherwise mentioned in the body of the Articles of Merger.

FORM 13F

CERTIFICATE OF MERGER (PENNSYLVANIA)

DSCB-56 B

Department of State

TO ALL TO WHOM THESE PRESENTS SHALL COME, GREETING:

WHEREAS, Under the terms of the Business Corporation Law, approved May 5, 1933, P. L. 364, as amended, the Department of State is authorized and required to issue a

CERTIFICATE OF MERGER

evidencing the merger of one or more corporations into one of such corporations under the provisions of that law:

AND WHEREAS, The stipulations and conditions of that law relating to the merger of such corporations have been fully complied with by

THEREFORE, KNOW YE, That subject to the Constitution of this Commonwealth, and under the authority of the Business Corporation Law, approved May 5, 1933, P. L. 364, as amended, I DO BY THESE PRESENTS, which I have caused to be sealed with the Great Seal of the Commonwealth, merge the above named

which shall continue to be invested with and have and enjoy all the powers, privileges and franchises incident to a domestic business corporation, and be subject to all the duties, requirements and restrictions specified and enjoined in and by the Business Corporation Law and all other applicable laws of this Commonwealth.

GIVEN under my Hand and the Great Seal of the Commonwealth, at the City of Harrisburg, this day of in the year of our Lord one thousand nine hundred and and of the Commonwealth the one hundred and

____

Secretary of the Commonwealth

FORM 13G

ARTICLES OF CONSOLIDATION (MODEL ACT)

Filing fee: $_______

ARTICLES OF CONSOLIDATION OF DOMESTIC CORPORATIONS INTO

____

Pursuant to the provisions of Section 74 of the ____________ Business Corporation Act, the undersigned corporations adopt the following Articles of Consolidation for the purpose of consolidating them into a new corporation:

FIRST: The following Plan of Consolidation was approved by the shareholders of each of the undersigned corporations in the manner prescribed by the ____________ Business Corporation Act:

(Insert Plan of Consolidation)

SECOND: As to each of the undersigned corporations, the number of shares outstanding, and the designation and number of outstanding shares of each class entitled to vote as a class on such Plan, are as follows:

Name of Corporation	Number of Shares Outstanding	Entitled to Vote as a Class	
		Designation of Class	Number of Shares

THIRD: As to each of the undersigned corporations, the total number of shares voted for and against such Plan, respectively, and, as to each class entitled to vote thereon as a class, the number of shares of such class voted for and against such Plan, respectively, are as follows:

Name of Corporation	Number of Shares				
	Total Voted For	Total Voted Against	Entitled to Vote as a Class		
			Class	Voted For	Voted Against

Dated ________________, 19__

______________________ (Note 1)

By ____________________
Its ________ President
and ____________________ (Note 2)
Its ________ Secretary

____________________ (Note 1)

By ____________________
Its ________ President
and ____________________ (Note 2)
Its ________ Secretary

(Add Verification Form A for each corporation)

Notes: 1. Exact corporate names of respective corporations executing the Articles.
2. Signatures and titles of officers signing for the respective corporations.

FORM 13H

CERTIFICATE OF CONSOLIDATION (MODEL ACT)

STATE OF ______
OFFICE OF THE SECRETARY OF STATE

CERTIFICATE OF CONSOLIDATION OF DOMESTIC CORPORATIONS INTO

____

The undersigned, as Secretary of State of the State of ________, hereby certifies that duplicate originals of Articles of Consolidation of ____________________ ____________________ and ____________________ ____________________, domestic corporations, into ____________________, duly signed and verified pursuant to the provisions of the ________ Business Corporation Act, have been received in this office and are found to conform to law.

ACCORDINGLY the undersigned, as such Secretary of State, and by virtue of the authority vested in him by law, hereby issues

this Certificate of Consolidation of ______________________________
______________________________ and ______________________________
into ____,
and attaches hereto a duplicate original of the Articles of Consolidation.

Dated ______, 19__.

____
Secretary of State

FORM 13I

ARTICLES OF MERGER FOR A SHORT MERGER (MODEL ACT)

Filing Fee: $______

ARTICLES OF MERGER
OF DOMESTIC SUBSIDIARY CORPORATION
INTO
DOMESTIC PARENT CORPORATION

Pursuant to the provisions of Section 75 of the ______ Business Corporation Act, the undersigned corporation adopts the following Articles of Merger for the purpose of merging a subsidiary corporation into the undersigned as the surviving corporation:

FIRST: The following Plan of Merger was approved by the Board of Directors of the undersigned, as the surviving corporation, in the manner prescribed by the ______ Business Corporation Act:

(Insert Plan of Merger)

SECOND: The number of outstanding shares of each class of the subsidiary corporation and the number of such shares of each class owned by the surviving corporation are as follows:

Name of Subsidiary	Number of Shares Outstanding	Designation of Class	Number of Shares Owned by Surviving Corporation

THIRD: A copy of the Plan of Merger set forth in Article First was mailed on ______ to each shareholder of the subsidiary corporation of record on ________________________________ (Note 1).

Dated: ______, 19__.

________________________ (Note 2)

By ____________________
Its ______ President
and ____________________
Its ______ Secretary
(Note 3)

(Add Verification Form A)

Notes:
1. Insert date plan mailed to each shareholder of subsidiary and record date for mailing. If all shareholders waived such mailing, insert statement to this effect and date of waiver.
2. Exact name of parent corporation executing Articles.
3. Signatures and titles of officers signing for the corporation.

FORM 13J

NOTICE TO CREDITORS—BULK TRANSFER ACT (COLORADO)

No. 577A. Rev. '67.—Bradford Publishing Company, 1824-46 Stout Street, Denver, Colorado—7-73

NOTICE OF BULK TRANSFER

(Section 155-6-107 Colorado Revised Statutes 1963)

(UCC—Bulk Transfers)

Notice is given that a Bulk Transfer is about to be made from the transferor to the transferee named below.

The name of the transferee is:

The business address of the transferee is:

The name of the transferor is:

The business address of the transferor is:

All other business names and addresses used by the transferor within three (3) years last past so far as known to the transferee are:

*All debts of the transferor are to be paid in full as they fall due as a result of the transaction. The address to which creditors should send their bills is:

*The debts of the transferor are not to be paid in full as they fall due or the transferee is in doubt on that point.

The property to be transferred is located at and consists of

The estimated total of the transferor's debts is $

The address where the schedule of property and list of creditors (Section 155-6-104 C.R.S. 1963) may be inspected is:

**The transfer is to pay existing debts and the amount of such debts and to whom owing are as follows:

Name	Amount
....	....
....	....
....	....

**The transfer is not to pay existing debts.

‡The transfer is for new consideration in the amount of $

The time of payment is:

The place of payment is:

Signed by:

....
Transferee

....

* Strike one or the other according to fact.
**Strike one or the other according to fact.
‡ If there is no new consideration, state 'none'.

FORM 13K

AFFIDAVIT OF SELLER—BULK TRANSFER ACT (COLORADO)

No. 577. Rev. '66. Bradford Publishing Company, 1824-46 Stout Street, Denver, Colorado—6-73

STATE OF COLORADO
_______________County of_______________ } ss.

AFFIDAVIT UNDER
155-6-104 C.R.S. 1963
(UCC—Bulk Transfers)

____, **Transferor,**

makes this affidavit pursuant to Section 155-6-104 Colorado Revised Statutes 1963 in connection with proposed transfer in bulk and being first duly sworn according to law on oath deposes and says:

Exhibit A, annexed to and by reference made a part of this affidavit is a full, accurate and complete list of the names and addresses of all existing creditors of transferor with the amounts when known and also the names of all persons known to the transferor to assert claims against the transferor which are disputed by transferor.

____
Transferor

Subscribed and sworn to before me this_______________day of_______________,

19____.

My commission expires

Witness my hand and official seal.

____
Notary Public

FORM 13L

STATEMENT OF INTENT TO DISSOLVE (SHAREHOLDERS) (MODEL ACT)

Filing Fee $______

STATEMENT OF INTENT TO DISSOLVE

BY WRITTEN CONSENT OF SHAREHOLDERS

To the Secretary of State
of the State of ______:

Pursuant to the provisions of Section 83 of the ______ Business Corporation Act, the undersigned corporation submits the following statement of intent to dissolve the corporation upon written consent of all of its shareholders:

FIRST: The name of the corporation is ______________________
____

SECOND: The names and respective addresses of its officers are:

Name	Office	Address
____________	____________	____________
____________	____________	____________
____________	____________	____________
____________	____________	____________

THIRD: The names and respective addresses of its directors are:

Name	Address
________________	____________________
________________	____________________
________________	____________________

FOURTH: The following written consent to dissolution of the corporation has been signed by all of the shareholders of the corporation, or signed in their names by their respective attorneys thereunto duly authorized:

(Insert copy of Consent)

Dated ________, 19___.

______________________ (Note 1)

By ______________________
Its ________ President
and ______________________
Its ________ Secretary
(Note 2)

(Add Verification Form A)

Notes: 1. Exact corporate name of corporation making the statement.
2. Signatures and titles of officers signing for the corporation.

FORM 13M

STATEMENT OF INTENT TO DISSOLVE (CORPORATION) (MODEL ACT)

Filing fee: $________

STATEMENT OF INTENT TO DISSOLVE

____

BY ACT OF THE CORPORATION

To the Secretary of State
of the State of __________:

Pursuant to the provisions of Section 84 of the ____________ Business Corporation Act, the undersigned corporation submits the following statement of intent to dissolve the corporation by act of the corporation.

FIRST: The name of the corporation is ____________________
____

SECOND: The names and respective addresses of its officers are:

Name	Office	Address
____________	________	____________
____________	________	____________
____________	________	____________
____________	________	____________

THIRD: The names and respective addresses of its directors are:

Name	Address
______________	______________
______________	______________
______________	______________

FOURTH: The following resolution to dissolve the corporation was adopted by the shareholders of the corporation on ____________, 19__:

(Insert copy of Resolution)

FIFTH: The number of shares of the corporation outstanding at the time of such adoption was ______; and the number of shares entitled to vote thereon was:

Class	Number of Shares

(Note 1)

SIXTH: The number of shares voted for such resolution was ______; and the number of shares voted against such resolution was ______.

SEVENTH: The number of shares of each class entitled to vote thereon as a class voted for and against such resolution, respectively, was:

Class	Number of Shares Voted	
	For	Against

(Note 1)

Dated ______, 19__.

______________ (Note 2)

By ______________
Its ______ President
and ______________
Its ______ Secretary
(Note 3)

(Add Verification Form A)

Notes:
1. If inapplicable, insert "None."
2. Exact corporate name of corporation making the statement.
3. Signatures and titles of officers signing for the corporation.

FORM 13N

STATEMENT OF REVOCATION OF VOLUNTARY DISSOLUTION PROCEEDINGS (SHAREHOLDERS) (MODEL ACT)

Filing fee: $________

STATEMENT OF REVOCATION
OF
VOLUNTARY DISSOLUTION PROCEEDINGS
OF

____

BY WRITTEN CONSENT OF THE SHAREHOLDERS

To the Secretary of State
of the State of __________:

Pursuant to the provisions of Section 88 of the ____________ Business Corporation Act, the undersigned corporation submits the following statement of revocation of voluntary dissolution proceedings heretofore taken upon the written consent of all of its shareholders:

FIRST: The name of the corporation is ____________________
____

SECOND: The names and respective addresses of its officers are:

Name	Office	Address

THIRD: The names and respective addresses of its directors are:

Name	Address

FOURTH: The following written consent signed by all the shareholders of the corporation revoking its voluntary dissolution proceedings has been signed by all of the shareholders of the corporation, or signed in their names by their respective attorneys thereunto duly authorized:

(Insert copy of Consent)

Dated ____________________, 19___

____________________ (Note 1)

By ____________________
Its ______ President
and ____________________
Its ______ Secretary
(Note 2)

(Add Verification Form A)

Notes: 1. Exact corporate name of corporation making the statement.
2. Signatures and titles of officers signing for the corporation.

FORM 130

STATEMENT OF REVOCATION OF VOLUNTARY DISSOLUTION PROCEEDINGS (CORPORATION) (MODEL ACT)

Filing fee: $________

STATEMENT OF REVOCATION
OF
VOLUNTARY DISSOLUTION PROCEEDINGS
OF

____

BY ACT OF THE CORPORATION

To the Secretary of State
of the State of __________:

Pursuant to the provisions of Section 89 of the ____________ Business Corporation Act, the undersigned corporation submits the following statement of revocation of voluntary dissolution proceedings heretofore taken by act of the corporation:

FIRST: The name of the corporation is ____________________
____

SECOND: The names and respective addresses of its officers are:

Name	Office	Address
______	______	______
______	______	______
______	______	______
______	______	______

THIRD: The names and respective addresses of its directors are:

Name	Address
______	______
______	______
______	______

FOURTH: The resolution adopted by the shareholders of the corporation revoking its voluntary dissolution proceedings is as follows:

(Insert copy of Resolution)

FIFTH: The number of shares of the corporation outstanding at the time of such adoption was ______

SIXTH: The number of shares voted for such resolution was ______; and the number of shares voted against such resolution was ______.

Dated ______, 19__.

______ (Note 1)

By ______
Its ______ President
and ______
Its ______ Secretary
(Note 2)

(Add Verification Form A)

Notes: 1. Exact corporate name of corporation making the statement.
2. Signatures and titles of officers signing for the corporation.

FORM 13P

ARTICLES OF DISSOLUTION BY INCORPORATORS (MODEL ACT)

Filing fee: $______

ARTICLES OF DISSOLUTION
BY INCORPORATOR(S)
OF

____

Pursuant to the provisions of Section 82 of the ______ Business Corporation Act, the undersigned of the corporation hereinafter named, adopt the following Articles of Dissolution:

FIRST: The name of the corporation is ____

SECOND: The date of issuance of its certificate of incorporation was ______

THIRD: None of its shares has been issued.

FOURTH: The corporation has not commenced business.

FIFTH: The amount, if any, actually paid in on subscriptions for its shares, less any part thereof disposed of for necessary expenses, has been returned to those entitled thereto.

SIXTH: No debts of the corporation remain unpaid.

SEVENTH: The sole incorporator or a majority of the incorporators elects that the corporation be dissolved.

Dated ______, 19__.

____

____

____

Incorporator(s) (Note 1)

(Add Verification Form B)

Note: 1. The sole incorporator or, if more than one, a majority, must execute and verify these Articles.

FORM 13Q

ARTICLES OF DISSOLUTION (MASSACHUSETTS)

Form CD-100. 10M-4-71-049198

The Commonwealth of Massachusetts

JOHN F. X. DAVOREN

Secretary of the Commonwealth

STATE HOUSE, BOSTON, MASS.

ARTICLES OF DISSOLUTION

General Laws, Chapter 156B, Section 100

The fee for filing articles of dissolution is $25.00. Make checks payable to the Commonwealth of Massachusetts

We, , President/Vice President, and

, Clerk/Assistant Clerk of

....

(Name of Corporation)

located at ....

do hereby certify as follows:-

1. The name of the corporation and the post office address of its principal office in the Commonwealth are as set forth above.

2. The names and post office addresses of each of the directors and officers of the corporation are as follows:

Name	Post Office Address	Title

3. On , 19 , the dissolution of the corporation was duly authorized in the manner required by Section 100 of Chapter 156B of the General Laws, and notice of the proposed dissolution was duly given in the manner required by said Section.

4. The effective date of the dissolution is (1) the date of filing these articles; or (2) , 19 . [Strike out subparagraph (1) if a specific date is desired.]

*5. Other provisions deemed necessary by the corporation for its dissolution.

IN WITNESS WHEREOF AND UNDER THE PENALTIES OF PERJURY, we have hereto signed our names this day of , 19

.... President/Vice President

.... Clerk/Assistant Clerk

*If there are no such provisions, state "None". Provisions for which the space provided above is not sufficient should be set out on continuation sheets to be numbered 2A, 2B, etc. Continuation sheets shall be on 8½" wide x 11" high paper and must have a left-hand margin of 1 inch for binding. Only one side should be used.

NOTE: These articles of dissolution must be accompanied by a certificate of the Commissioner of Corporations and Taxation that all taxes due and payable by the corporation to the Commonwealth have been paid or provided for. COPIES OF NEWSPAPER PUBLICATIONS MUST ALSO ACCOMPANY THIS CERTIFICATE.

THE COMMONWEALTH OF MASSACHUSETTS

ARTICLES OF DISSOLUTION

(General Laws, Chapter 156B, Section 100)

I hereby aprove the within articles of dissolution and, the filing fee in the amount of $

having been paid, said articles are deemed to have been filed with me this

day of , 19

John F. X. Davoren

JOHN F. X. DAVOREN

Secretary of the Commonwealth
State House, Boston, Mass.

TO BE FILLED IN BY CORPORATION

PHOTO COPY OF ARTICLES OF DISSOLUTION TO BE SENT

TO:

....

....

....

Copy Mailed

FORM 13R

CERTIFICATE OF DISSOLUTION BY INCORPORATORS (MODEL ACT)

STATE OF ______________
OFFICE OF THE SECRETARY OF STATE

CERTIFICATE OF DISSOLUTION BY INCORPORATOR(S) OF

____

The undersigned, as Secretary of State of the State of ______ ________, hereby certifies that duplicate originals of Articles of Dissolution by the Incorporator(s) of ______________________ ________________________________, duly signed and verified pursuant to the provisions of the .... Business Corporation Act, have been received in this office and are found to conform to law.

ACCORDINGLY, the undersigned, as such Secretary of State, and by virtue of the authority vested in him by law hereby issues this Certificate of Dissolution of ______________________ ____________________________, and attaches hereto a duplicate original of the Articles of Dissolution.

Dated ____________________, 19__.

____
Secretary of State

FORM 13S

CERTIFICATE OF DISSOLUTION (TEXAS)

In the name and by the authority of
The State of Texas

OFFICE OF THE SECRETARY OF STATE

CERTIFICATE OF DISSOLUTION

OF

....

SPECIMEN

The undersigned, as Secretary of State of the State of Texas, hereby certifies that duplicate originals of Articles of Dissolution of the above Corporation, duly signed and verified pursuant to the provisions of the Texas Business Corporation Act, have been received in this office and are found to conform to law.

ACCORDINGLY the undersigned, as such Secretary of State, and by virtue of the authority vested in him by law, hereby issues this Certificate of Dissolution and attaches hereto a duplicate original of the Articles of Dissolution.

Dated.., 19..........

....
Secretary of State

INDEX

Share Subscriptions

Shareholders

†